Approximate Conversions
from Metric Measures

Symbol	When You Know	Multiply by	To Find	Symbol
LENGTH				
mm	millimeters	0.04	inches	in
cm	centimeters	0.4	inches	in
m	meters	3.3	feet	ft
m	meters	1.1	yards	yd
km	kilometers	0.6	miles	mi
AREA				
cm^2	square centimeters	0.16	square inches	in^2
m^2	square meters	1.2	square yards	yd^2
km^2	square kilometers	0.4	square miles	mi^2
ha	hectares (10,000 m^2)	2.5	acres	
MASS (weight)				
g	grams	0.035	ounces	oz
kg	kilograms	2.2	pounds	lb
t	tonnes (1000 kg)	1.1	short tons	
VOLUME				
ml	milliliters	0.03	fluid ounces	fl oz
l	liters	2.1	pints	pt
l	liters	1.06	quarts	qt
l	liters	0.26	gallons	gal
m^3	cubic meters	35	cubic feet	ft^3
m^3	cubic meters	1.3	cubic yards	yd^3
TEMPERATURE (exact)				
°C	Celsius temperature	9/5 (then add 32)	Fahrenheit temperature	°F

For sale by the Superintendent of Documents, U. S. Government Printing Office, Washington, D. C. 20402 (Order by SD Catalog No. C13.10: 365). Price 10¢, $6.25 per 100.

*1 in = 2.54 cm (exactly). For other exact conversions and more detailed tables, see NBS Misc. Publ. 286, Units of Weights and Measures, Price $1.50, SD Catalog No. C13.10:286.

THE PRACTICING SCIENTIST'S HANDBOOK

THE PRACTICING SCIENTIST'S HANDBOOK

A Guide for Physical and Terrestrial Scientists and Engineers

Alfred J. Moses

California Department of Justice
Bureau of Forsenic Services
Riverside Laboratory
Riverside, California

VNR VAN NOSTRAND REINHOLD COMPANY

NEW YORK   CINCINNATI   ATLANTA   DALLAS   SAN FRANCISCO
LONDON   TORONTO   MELBOURNE

Van Nostrand Reinhold Company Regional Offices:
New York Cincinnati Atlanta Dallas San Francisco

Van Nostrand Reinhold Company International Offices:
London Toronto Melbourne

Library of Congress Catalog Card Number: 77-5866
ISBN: 0-442-25584-5

Manufactured in the United States of America

Published by Van Nostrand Reinhold Company
135 West 50th Street, New York, N. Y. 10020

Published simultaneously in Canada by Van Nostrand Reinhold Ltd.

15 14 13 12 11 10 9 8 7 6 5 4 3 2

Library of Congress Cataloging in Publication Data
Moses, Alfred James, 1921–
    The practicing scientist's handbook.

    Includes index.
    1. Matter—Properties—Handbooks, manuals, etc.
2. Materials—Handbooks, manuals, etc. I. Title.
QC173.397.M67   530.4   77-5866
ISBN 0-442-25584-5

PREFACE

One of the definitions of *physics* in Webster's New International Dictionary is: "The science dealing with the property changes, interactions, etc., of matter and energy." In fact, this definition is broad enough to accommodate the chemist as well. The properties of materials (matter in its many forms), represent the response to external constraints and stimuli.

The purpose of this applications-oriented handbook is to provide a comprehensive source of materials' property data to practicing physicists, chemists, engineers and designers. In seeking to accommodate the new interest in the universe, ranging from the interior of the earth to outer space, a chapter on the environment has been included; this chapter will prove to be particularly helpful to geophysicists and geologists.

The need to keep the dimensions and cost of this handbook within reasonable bounds necessitated a careful selection of material, limitation of theory to that essential for defining terms, and adoption of a hybrid classification of materials. The brevity of the theory should be of no great handicap to readers as a number of excellent texts ably fill this gap. Similarly, minimizing the number of mathematics tables should offer no great obstacle as most scientists and engineers generally have access to these tables.

The materials classification is designed to enable the applications-oriented reader to locate information expeditiously. For example, the reader seeking information on semiconductor properties will find it in a single chapter devoted to semiconductors. In the more conventionally categorized literature he would have to search under Chemical Elements for germanium, or under Inorganic Compounds for zinc sulfide.

The handbook provides an introduction to atomic and nuclear structure, electromagnetic radiation, and materials selection criteria. Materials properties are defined and data summaries presented for each property for a number of materials to give readers a "feel" for the magnitude of the desired property and facilitate materials selection.

Properties are divided into the following groups: physical, mechanical, chemical, thermal and thermodynamic, electrical and electronic, magnetic, acoustic, optical, and nuclear. The section on nuclear radiation and radiation effects contains most of the handbook's information in these areas.

Materials have been divided into the following broad categories: chemical elements, organic compounds, inorganic compounds, alloys, refractories and supercooled liquids, composites, polymers and adhesives, semiconductors, superconductors, the environment, and miscellaneous materials. As indicated above, the chapter on the environment describes the properties of the earth, the sea, the atmosphere, meteorites, planets, stars, and space.

Special attention has been given to "space-age" materials. Various forms of such materials include fibers, films and composites. The NASA (National Aeronautics and Space Administration) and DOD (Department of Defense) report literature must be credited for much of the recent data on modern materials.

In keeping with its function of serving the practicing scientist and engineer who will not use the S.I. system of units for some time, this handbook uses the S.I. system sparingly, with the c.g.s. and English units being employed in a manner to provide maximum service to the user of the book.

ALFRED J. MOSES

ACKNOWLEDGMENT

The author is grateful to his family for their understanding of the time-consuming nature of the task of compiling this volume. Thanks are also due to the many authors for furnishing copies of their work and granting permission to use the material. The library of the California State University at Fullerton was most helpful in permitting the loan of needed material and assisting in locating same.

CONTENTS

THE PRACTICING SCIENTIST'S HANDBOOK

CHAPTER 1
INTRODUCTION

This book is designed to assist the scientist and engineer in his work, beginning with an appraisal of the safety of a proposed project.

The safety aspect of work has gained overwhelming significance, with the passage of safety legislation at all levels of government. Therefore, in contrast with earlier handbooks, this book devotes a relatively large amount of space to the subject of laboratory and industrial safety.

The booklet "Threshold Limit Values for Chemical Substances and Physical Agents in the Workroom Environment with Intended Changes for 1977" by the American Conference of Governmental Industrial Hygienists* has been reproduced in its entirety with the permission of the Conference. This booklet is revised annually, and documentation for it is available from the Conference. A list of official publications of the American Conference of Governmental Industrial Hygienists is available from the Conference at the address given below.

The information contained herein includes not only chemical hazards, but also dusts, carcinogens, heat stress, lasers, microwaves, ultraviolet radiation and noise. Sample calculations for evaluating the magnitude of selected hazards are included.

As engineers and scientists often bear some responsibility for assuring compliance by employers with the Federal or State Occupational Safety and Health Acts, it is believed that this excellent summary by the American Conference of Governmental Industrial Hygienists should be brought to their attention.

*P.O. Box 1937, Cincinnati, Ohio 45201.

TLVs®

Threshold Limit Values

for

Chemical Substances

and

Physical Agents

in the

Workroom Environment

with

Intended Changes

for

1977

Copyright 1977 by American Conference of Governmental Industrial Hygienists.

The American Conference of Governmental Industrial Hygienists will welcome requests for permission to republish or reprint these Threshold Limit Values. Requests for such permission should be directed to the Secretary-Treasurer, P.O. Box 1937, Cincinnati, Ohio 45201.

PRICE EACH

1–49	$1.50
50–199	1.25
200–999	1.10
1000–4999	.75
5000 or more	.65

Documentation of the Threshold Limit Values for Substances in Workroom Air.© A separate companion piece to the Chemical TLVs is issued by ACGIH under this title. This publication gives the pertinent scientific information and data with reference to literature sources that were used to base each limit. Each documentation also contains a statement defining the type of response against which the limit is safeguarding the worker. For a better understanding of the TLVs it is essential that the Documentation be consulted when the TLVs are being used.

Information concerning the availability of copies of the Documentation of the Threshold Limit Values for Substances in Workroom Air should be directed to the Secretary-Treasurer, ACGIH.

TLVs®

Threshold Limit Values

for

Chemical

Substances in

Workroom Air

Adopted by

ACGIH

for 1977

Any comments or questions regarding these limits should be addressed to:

Secretary-Treasurer
American Conference of Governmental
   Industrial Hygienists
P.O. Box 1937
Cincinnati OH 45201: (513) 825-0312

TABLE OF CONTENTS

PREFACE
CHEMICAL CONTAMINANTS

Threshold limit values refer to airborne concentrations of substances and represent conditions under which it is believed that nearly all workers may be repeatedly exposed day after day without adverse effect. Because of wide variation in individual susceptibility, however, a small percentage of workers may experience discomfort from some substances at concentrations at or below the threshold limit; a smaller percentage may be affected more seriously by aggravation of a pre-existing condition or by development of an occupational illness.

Tests are available (J. Occup. Med. 15: 564, 1973; Ann. N.Y. Acad. Sci., *151, Art. 2:* 968, 1968) that may be used to detect those individuals hypersusceptible to a variety of industrial chemicals (respiratory irritants, hemolytic chemicals, organic isocyanates, carbon disulfide).

Three categories of Threshold Limit Values (TLVs) are specified herein, as follows:

a) Threshold Limit Value-Time Weighted Average (TLV-TWA) — the time-weighted average concentration for a normal 8-hour workday or 40-hour workweek, to which nearly all workers may be repeatedly exposed, day after day, without adverse effect.

b) Threshold Limit Value-Short Term Exposure Limit (TLV-STEL) — the maximal concentration to which workers can be exposed for a period up to 15 minutes continuously without suffering from 1) irritation, 2) chronic or irreversible tissue change, or 3) narcosis of sufficient degree to increase accident proneness, impair self-rescue, or materially reduce work efficiency, provided that no more than four excursions per day are permitted, with at least 60 minutes between exposure periods, and provided that the daily TLV-TWA also is not exceeded. The STEL should be considered a maximal allowable concentration, or absolute ceiling, not to be exceeded at any time during the 15-minute excursion period. STELs are based on one or more of the following criteria: (1) Adopted TLVs including those with a "C" or "ceiling" limit. (2) TWA-TLV Excursion Factors listed in Appendix D. (3) Pennsylvania Short-Term Limits for Exposure to Airborne Contaminants (Penna. Dept. of Hlth., Chapter 4, Art. 432, Rev. Jan. 25, 1968). (4) OSHA Occupational Safety and Health Standards, Fed. Reg. Vol. 36, No. 105, May 29, 1971. The TWA-STEL should not be used as engineering design criterion or considered as an emergency exposure level (EEL).

c) Threshold Limit Value-Ceiling (TLV-C) — the concentration that should not be exceeded even instantaneously.

For some substances, e.g., irritant gases, only one category, the TLV-Ceiling, may be relevant. For other substances, either two or three categories may be relevant, depending upon their physiologic action. It is important to observe that if any one of these three TLVs is exceeded, a potential hazard from that substance is presumed to exist.

The TLV-TWA should be used as guides in the control of health hazards and should not be used as fine lines between safe and dangerous concentrations.

Time-weighted averages permit excursions above the limit provided they are compensated by equivalent excursions below the limit during the workday. In some instances it may be permissible to calculate the average concentration for a workweek rather than for a workday. The degree of permissible excursion is related to the magnitude of the threshold limit value of a particular substance as given in Appendix D. The relationship between threshold limit and permissible excursion is a rule of thumb and in certain cases may not apply. The amount by which threshold limits may be exceeded for short periods without injury to health depends upon a number of factors such as the nature of the contaminant, whether very high concentrations — even for short periods — produce acute poisoning, whether the effects are cumulative, the frequency with which high concentrations occur, and the duration of such periods. All factors must be taken into consideration in arriving at a decision as to whether a hazardous condition exists.

Threshold limits are based on the best available information from industrial experience, from experimental human and animal studies, and, when possible, from a combination of the three. The basis on which the values are established may differ from substance to substance; protection against impairment of health may be a guiding factor for some, whereas reasonable freedom from irritation, narcosis, nuisance or other forms of stress may form the basis for others.

The amount and nature of the information available for establishing a TLV varies from substance to substance; consequently, the precision of the estimated TLV is also subject to variation and the latest *Documentation* should be consulted in order to assess the extent of the data available for a given substance.

The committee holds to the opinion that limits based on physical irritation should be considered no less binding than those based on physical impairment. There is increasing evidence that physical irritation may initiate, promote or accelerate physical impairment through interaction with other chemical or biologic agents.

In spite of the fact that serious injury is not believed likely as a result of exposure to the threshold limit concentrations, the best practice is to maintain concentrations of all atmospheric contaminants as low as is practical.

These limits are intended for use in the practice of industrial hygiene and should be interpreted and applied only by a person trained in this discipline. They are not intended for use, or for modification for use, (1) as a relative index of hazard or toxicity, (2) in the evaluation or control of community air pollution nuisances, (3) in estimating the toxic potential of continuous, uninterrupted exposures or other extended work periods, (4) as proof or disproof of an existing disease or physical condition, or (5) for adoption by countries whose working conditions differ from those in the United States of America and where substances and processes differ.

Ceiling vs Time-Weighted Average Limits. Although the time-weighted average concentration provides the most satisfactory, practical way of monitoring airborne agents for compliance with the limits, there are certain substances for which it is inappropriate. In the latter group are substances which are predominantly fast acting and whose threshold limit is more appropriately based on this particular response. Substances with this type of response are best controlled by a ceiling "C" limit that should not be exceeded. It is implicit in these definitions that the manner of sampling to determine noncompliance with the limits for each group must differ; a single brief sample, that is applicable to a "C" limit, is not appropriate to the time-weighted limit; here, a sufficient number of samples are needed to permit a time-weighted average concentration throughout a complete cycle of operations or throughout the work shift.

Whereas the ceiling limit places a definite boundary which concentrations should not be permitted to exceed, the time-weighted average limit requires an explicit limit to the excursions that are permissible above the listed values. The magnitude of these excursions may be pegged to the magnitude of the threshold limit by an appropriate factor shown in Appendix D. It should be noted that the same factors are used by the Committee

in determining the magnitude of the value of the STELs, or whether to include or exclude a substance for a "C" listing.

"Skin" Notation. Listed substances followed by the designation "Skin" refer to the potential contribution to the overall exposure by the cutaneous route including mucous membranes and eye, either by airborne, or more particularly, by direct contact with the substance. Vehicles can alter skin absorption. This attention-calling designation is intended to suggest appropriate measures for the prevention of cutaneous absorption so that the threshold limit is not invalidated.

Mixtures. Special consideration should be given also to the application of the TLVs in assessing the health hazards which may be associated with exposure to mixtures of two or more substances. A brief discussion of basic considerations involved in developing threshold limit values for mixtures, and methods for their development, amplified by specific examples are given in Appendix C.

Nuisance Particulates. In contrast to fibrogenic dusts which cause scar tissue to be formed in lungs when inhaled in excessive amounts, so-called "nuisance" dusts have a long history of little adverse effect on lungs and do not produce significant organic disease or toxic effect when exposures are kept under reasonable control. The nuisance dusts have also been called (biologically) "inert" dusts, but the latter term is inappropriate to the extent that there is no dust which does not evoke some cellular response in the lung when inhaled in sufficient amount. However, the lung-tissue reaction caused by inhalation of nuisance dusts has the following characteristics: (1) The architecture of the air spaces remains intact. (2) Collagen (scar tissue) is not formed to a significant extent. (3) The tissue reaction is potentially reversible.

Excessive concentrations of nuisance dusts in the workroom air may seriously reduce visibility, may cause unpleasant deposits in the eyes, ears and nasal passages (Portland Cement dust), or cause injury to the skin or mucous membranes by chemical or mechanical action per se or by the rigorous skin cleansing procedures necessary for their removal.

A threshold limit of 10 mg/m^3, or 30 mppcf, of total dust < 1% quartz is recommended for substances in these categories and for which no specific threshold limits have been assigned. This limit, for a normal workday, does not apply to brief exposures at higher concentrations. Neither does it apply to those substances which may cause physiologic impairment at lower concentrations but for which a threshold limit has not yet been adopted. Some nuisance particulates are given in Appendix E.

Simple Asphyxiants — "Inert" Gases or Vapors. A number of gases and vapors, when present in high concentrations in air, act primarily as simple asphyxiants without other significant physiologic effects. A TLV may not be recommended for each simple asphyxiant because the limiting factor is the available oxygen. The minimal oxygen content should be 18 percent by volume under normal atmospheric pressure (equivalent to a partial pressure, pO_2 of 135 mm Hg). Atmospheres deficient in O_2 do not provide adequate warning and most simple asphyxiants are odorless. Several simple asphyxiants present an explosion hazard. Account should be taken of this factor in limiting the concentration of the asphyxiant. Specific examples are listed in Appendix F.

Physical Factors. It is recognized that such physical factors as heat, ultraviolet and ionizing radiation, humidity, abnormal pressure (altitude) and the like may place added stress on the body so that the effects from exposure at a threshold limit may be altered. Most of these stresses act adversely to increase the toxic response of a substance. Although most threshold limits have built-in safety factors to guard against adverse effects to moderate deviations from normal environments, the safety factors of most substances are not of such a magnitude as to take care of gross deviations. For example, continuous work at temperatures above 90°F, or overtime extending the workweek more than 25%, might be considered gross deviations. In such instances judgment must be exercised in the proper adjustments of the Threshold Limit Values.

Biologic Limit Values (BLVs). Other means exist and may be necessary for monitoring worker exposure other than reliance on the Threshold Limit Values for industrial air, namely, the Biologic Limit Values. These values represent limiting amounts of substances (or their effects) to which the worker may be exposed without hazard to health or well-being as determined in his tissues and fluids or in his exhaled breath. The biologic measurements on which the BLVs are based can furnish two kinds of information useful in the control of worker exposure: (1) measure of the individual worker's over-all exposure; (2) measure of the worker's individual and characteristic response. Measurements of response furnish a superior estimate of the physiologic status of the worker, and may be made of (a) changes in amount of some critical biochemical constituent, (b) changes in activity of a critical enzyme, (c) changes in some physiologic function. Measurement of exposure may be made by (1) determining in blood, urine, hair, nails, in body tissues and fluids, the amount of substance to which the worker was exposed; (2) determination of the amount of the metabolite(s) of the substance in tissues and fluids; (3) determination of the amount of the substance in the exhaled breath. The biologic limits may be used as an adjunct to the TLVs for air, or in place of them. The BLVs, and their associated procedures for determining compliance with them, should thus be regarded as an effective means of providing health surveillance of the worker.

Unlisted Substances. Many substances present or handled in industrial processes do not appear on the TLV list. In a number of instances the material is rarely present as a particulate, vapor or other airborne contaminant, and a TLV is not necessary. In other cases sufficient information to warrant development of a TLV, even on a tentative basis, is not available to the Committee. Other substances, of low toxicity, could be included in Appendix E pertaining to nuisance particulates. This list (as well as Appendix F) is not meant to be all inclusive; the substances serve only as examples.

In addition there are some substances of not inconsiderable toxicity, which have been omitted primarily because only a limited number of workers (e.g., employ-

ees of a single plant) are known to have potential exposure to possibly harmful concentrations.

"Notice of Intent." At the beginning of each year, proposed actions of the Committee for the forthcoming year are issued in the form of a "Notice of Intended Changes." This Notice provides not only an opportunity for comment, but solicits suggestions of substances to be added to the list. The suggestions should be accompanied by substantiating evidence. The list of Intended Changes follows the Adopted Values in the TLV booklet.

Legal Status. By publication in the Federal Register (Vol. 36, No. 105, May 29, 1971) the Threshold Limit Values for 1968 are now official federal standards for industrial air.

Reprint Permission. This publication may be reprinted provided that written permission is obtained from the Secretary-Treasurer of the Conference and that it be published in its entirety.

Substance	ADOPTED VALUES TWA		TENTATIVE VALUES STEL	
	ppm[a]	mg/m³[b]	ppm[a]	mg/m³[b]
Abate	—	10	—	20
Acetaldehyde	100	180	150	270
Acetic acid	10	25	15	37
C Acetic anhydride	5	20	—	—
Acetone	1,000	2,400	1,250	3,000
Acetonitrile	40	70	60	105
Acetylene	F	—	—	—
Acetylene dichloride, see 1. 2-Dichloroethylene	200	790	250	1,000
Acetylene tetrabromide	1	14	1.25	17.5
Acrolein	0.1	0.25	0.3	0.75
Acrylamide — Skin	—	0.3	—	0.6
Acrylonitrile — Skin	20	45	30	68
Aldrin — Skin	—	0.25	—	0.75
Allyl alcohol — Skin	2	5	4	10
Allyl chloride	1	3	2	6
Allyl glycidyl ether (AGE) — Skin	5	22	10	44
Allyl propyl disulfide	2	12	3	18
Alundum (Al₂O₃)	—	E	—	20
4-Aminodiphenyl — Skin	—	A1b	—	A1b
2-Aminoethanol, see Ethanolamine	3	6	6	12
2-Aminopyridine	0.5	2	1.5	6
Ammonia	25	18	35	27
Ammonium chloride-fume	—	10	—	20
Ammonium sulfamate (Ammate)	—	10	—	20
n-Amyl acetate	100	525	150	790
sec-Amyl acetate	125	650	150	810
Aniline — Skin	5	19	—	—
Anisidine (o-, p-isomers) — Skin	0.1	0.5	—	—
** Antimony & Compounds (as Sb)	—	(0.5)	—	—
ANTU (α-Naphthyl thiourea)	—	0.3	—	0.9
Argon	F	—	F	F
** Arsenic & compounds (as As)	—	(0.5)	—	—
Arsine	0.05	0.2	—	—

Capital letters refer to Appendices.
Footnotes (a thru h) see Page 11.
** See Notice of Intended Changes

Substance	ADOPTED VALUES TWA ppm[a]	mg/m³[b]	TENTATIVE VALUES STEL ppm[a]	mg/m³[b]
Asbestos (all forms)	—	A1a	—	A1a
Asphalt (petroleum) fumes	—	5	—	10
Azinphos methyl — Skin	—	0.2	—	0.6
Baygon (propoxur)	—	0.5	—	1.5
Barium (soluble compounds)	—	0.5	—	—
Benzene — Skin	10, A2	30, A2	—	—
Benzidine production — Skin	—	A1b	—	A1b
p-Benzoquinone, see Quinone	0.1	0.4	0.3	1.2
Benzoyl peroxide	—	5	—	—
Benz(a)pyrene	—	A2	—	A2
Benzyl chloride	1	5	—	—
Beryllium	—	0.002	—	0.025
Biphenyl	0.2	1	—	—
Bismuth telluride	—	10	—	20
Bismuth telluride, Se-doped	—	5	—	10
* Borates, tetra, sodium salts,				
Anhydrous	—	1	—	—
Decahydrate	—	5	—	—
Pentahydrate	—	1	—	—
Boron oxide	—	10	—	20
Boron tribromide	1	10	3	30
C Boron trifluoride	1	3	—	—
Bromine	0.1	0.7	0.3	2
Bromine pentafluoride	0.1	0.7	0.3	2
Bromochloromethane	200	1,050	250	1,300
Bromoform — Skin	0.5	5	—	—
Butadiene (1, 3-butadiene)	1,000	2,200	1,250	2,750
Butane	600	1,400	750	1,610
Butanethiol, see Butyl mercaptan	0.5	1.5	—	—
2-Butanone	200	590	300	885
2-Butoxy ethanol (Butyl Cellosolve) — Skin	50	240	150	720
n-Butyl acetate	150	710	200	950
sec-Butyl acetate	200	950	250	1,190
tert-Butyl acetate	200	950	250	1,190
C n-Butyl alcohol — Skin	50	150	—	—
sec-Butyl alcohol	150	450	—	—
tert-Butyl alcohol	100	300	150	450
C Butylamine — Skin	5	15	—	—
C tert-Butyl chromate (as CrO_3) — Skin	—	0.1	—	—
n-Butyl glycidyl ether (BGE)	50	270	—	—
n-Butyl lactate	5	25	—	—
Butyl mercaptan	0.5	1.5	—	—
p-tert-Butyltoluene	10	60	20	120
Cadmium, dust & salts (as Cd)	—	0.05	—	0.15
C Cadmium oxide fume (as Cd)	—	0.05	—	—
Calcium carbonate	—	E	—	20
Calcium arsenate (as As)	—	1	—	—
Calcium cyanamide	—	0.5	—	1
** Calcium hydroxide	—	5	—	—
** Calcium oxide	—	(5)	—	—
Camphor, synthetic	2	12	3	18
Caprolactam				
Dust	—	1	—	3
Vapor	5	20	10	40
* Captafol (Difolatan®) — Skin	—	0.1	—	—
Captan	—	5	—	15
Carbaryl (Sevin®)	—	5	—	10
Carbofuran (Furadan®)	—	0.1	—	—
Carbon black	—	3.5	—	7
Carbon dioxide	5,000	9,000	15,000	18,000
Carbon disulfide — Skin	20	60	30	90
Carbon monoxide	50	55	400	440
Carbon tetrabromide	0.1	1.4	0.3	4.2
Carbon tetrachloride — Skin	10	65	25	160
* Catechol (Pyrocatechol)	5	20	—	—
Cellulose (paper fiber)	—	E	—	20
Cesium hydroxide	—	2	—	—
Chlordane — Skin	—	0.5	—	2
Chlorinated camphene — Skin	—	0.5	—	1
Chlorinated diphenyl oxide	—	0.5	—	1.5
Chlorine	1	3	3	9
Chlorine dioxide	0.1	0.3	0.3	0.9
C Chlorine trifluoride	0.1	0.4	—	—
C Chloroacetaldehyde	1	3	—	—
α-Chloroacetophenone (Phenacyl chloride)	0.05	0.3	—	—
Chlorobenzene (Monochlorobenzene)	75	350	—	—
o-Chlorobenzylidene malonoitrile — Skin	0.05	0.4	—	—
Chlorobromomethane	200	1,050	250	1,300
2-Chloro-1, 3-butadiene, see β Chloroprene	25	90	35	135
Chlorodifluoromethane	1,000	3,500	1,250	4,375
Chlorodiphenyl (42% Chlorine) — Skin	—	1	—	—
Chlorodiphenyl (54% Chlorine) — Skin	—	0.5	—	1
1-Chloro, 2, 3-epoxy-propane (Epichlorhydrin)	5	20	10	40
2-Chloroethanol (Ethylene chlorohydrin)	1	3	—	—
Chloroethylene (Vinyl chloride)	A1c	—	A1c	—
** Chloroform (Trichloromethane)	(25)	(120)	—	—
bis-Chloromethyl ether	0.001	A1a	—	A1a
1-Chloro-1-nitro-propane	20	100	—	—
Chloropicrin	0.1	0.7	—	—
β-Chloroprene — Skin	25	90	35	135
Chlorpyrifos (Dursban®) — Skin	—	0.2	—	0.6
o-Chlorostyrene	50	285	75	430
o-Chlorotoluene — Skin	50	250	75	375
2-Chloro-6-(trichloromethyl pyridine (N-Serve®)	—	10	—	20
Chromates, certain insoluble forms	—	0.05A1a	—	A1a
Chromic acid and Chromates, (as Cr.)	—	0.05	—	—
Chromium, Sol. chromic, chromous salts (as Cr.)	—	0.5	—	—
Clopidol (Coyden®)	—	10	—	20
Coal tar pitch volatiles (See Particulate polycyclic aromatic hydrocarbons)	—	A1a	—	A1a
* Cobalt metal, dust and fume	—	(0.1)	—	—
Copper fume	—	0.2	—	—
Dusts & Mists	—	1	—	2
Corundum (Al_2O_3)	—	E	—	E
Cotton dust, raw	—	0.2^m	—	0.6
Crag® herbicide	—	10	—	20

Capital letters refer to Appendices.
Footnotes (a thru h) see Page 11.
*1977 Addition.
**See Notice of Intended Changes.

Substance	ADOPTED VALUES TWA ppm[a]	mg/m³[b]	TENTATIVE VALUES STEL ppm[a]	mg/m³[b]
Cresol, all isomers — Skin	5	22	—	—
Crotonaldehyde	2	6	6	18
Crufomate®	—	5	—	20
Cumene — Skin	50	245	75	365
* Cyanamide	—	2	—	—
Cyanide, as CN — Skin	—	5	—	—
Cyanogen	10	20	—	—
Cyclohexane	300	1,050	375	1,300
Cyclohexanol	50	200	—	—
Cyclohexanone	50	200	—	—
Cyclohexene	300	1,015	—	—
Cyclohexylamine — Skin	10	40	—	—
Cyclopentadiene	75	200	150	400
2, 4-D (2, 4-Diphenoxy-acetic acid)	—	10	—	20
DDT (Dichlorodiphenyl-trichloroethane)	—	1	—	3
DDVP, See Dichlorvos	0.1	1	0.3	3
Decaborane — Skin	0.05	0.3	0.15	0.9
Demeton® — Skin	0.01	0.1	0.03	0.3
Diacetone alcohol (4-hydroxy-4-methyl-2-pentanone)	50	240	75	360
1, 2-Diaminoethane, See Ethylenediamine	10	25	—	—
Diazinon — Skin	—	0.1	—	0.3
Diazomethane	0.2	0.4	—	—
Diborane	0.1	0.1	—	—
1, 2-Dibromoethane (Ethylene dibromide) — Skin	20	145	30	220
Dibrom®	—	3	—	6
2-N-Dibutylaminoethanol — Skin	2	14	4	28
Dibutyl phosphate	1	5	2	10
Dibutyl phthalate	—	5	—	10
C Dichloroacetylene	0.1	0.4	—	—
C o-Dichlorobenzene	50	300	—	—
p-Dichlorobenzene	75	450	110	675
Dichlorobenzidine — Skin	—	A2	—	A2
Dichlorodifluoromethane.	1,000	4,950	1,250	6,200
1, 3-Dichloro-5, 5-dimethyl hydantoin	—	0.2	—	0.4
1, 1-Dichloroethane	200	820	250	1,025
1, 2-Dichloroethane	50	200	75	300
1, 2-Dichloroethylene	200	790	250	1,000
Dichloroethyl ether — Skin	5	30	10	60
Dichloromethane, see Methylene chloride	200	720	250	900
** Dichloromonofluoro-methane	(1,000)	(4,200)	—	—
C 1, 1-Dichloro-1-nitroethane	10	60	—	—
1, 2-Dichloropropane, see Propylene dichloride	75	350	110	525
Dichlorotetrafluoro-ethane	1,000	7,000	1,250	8,750
Dichlorvos (DDVP) — Skin	0.1	1	0.3	3
* Dicrotophos (Bidrin®) — Skin	—	0.25	—	—
Dicyclopentadiene	5	30	—	—
Dicyclopentadienyl iron	—	10	—	20
Dieldrin — Skin	—	0.25	—	0.75
Diethylamine	25	75	—	—
Diethylaminoethanol — Skin	10	50	—	—
Diethylene triamine — Skin	1	4	—	—
Diethyl ether, see Ethyl ether	400	1,200	500	1,500
Diethyl phthalate	—	5	—	10
Difluorodibromomethane.	100	860	150	1,290
C Diglycidyl ether (DGE)	0.5	2.8	—	—
Dihydroxybenzene, see Hydroquinone	—	2	—	3
Diisobutyl ketone	25	150	—	—
Diisopropylamine — Skin	5	20	—	—
Dimethoxymethane, see Methylal	1,000	3,100	1,250	3,875
Dimethyl acetamide — Skin	10	35	15	50
Dimethylamine	10	18	—	—
Dimethylaminobenzene, see Xylidene	5	25	10	50
Dimethylaniline (N-Dimethylaniline) — Skin	5	25	10	50
Dimethylbenzene, see Xylene	100	435	150	650
Dimethyl-1, 2-dibromo-2-dichloroethyl phosphate, see Dibrom	—	3	—	6
Dimethylformamide — Skin	10	30	20	60
2, 6-Dimethylheptanone, see Diisobutyl ketone	25	150	—	—
1, 1-Dimethylhydrazine — Skin	0.5	1	1	2
Dimethylphthalate	—	5	—	10
* C Dimethyl sulfate — Skin	0.1, A2	0.5, A2	—	—
Dinitrobenzene (all isomers) — Skin	0.15	1	0.5	3
Dinitro-o-cresol — Skin	—	0.2	—	0.6
3, 5-Dinitro-o-toluamide (Zoalene®)	—	5	—	10
Dinitrotoluene — Skin	—	1.5	—	5
Dioxane, tech. grade — Skin	50	180	—	—
* Dioxathion (Delnav®)	—	0.2	—	—
Diphenyl, see Biphenyl	0.2	1	0.6	3
Diphenylamine	—	10	—	20
Diphenylmethane diisocyanate, see Methylene bisphenyl isocyanate (MDI)	0.02	0.2	0.02	0.2
Dipropylene glycol methyl ether — Skin	100	600	150	900
Diquat	—	0.5	—	1
Di-sec, octyl phthalate (Di-2-ethylhexyl-phthalate)	—	5	—	10
Disulfuram	—	2	—	5
Disyston — Skin	—	0.1	—	0.3
2, 6-Ditert. butyl-p-cresol	—	10	—	20
* Diuron	—	10	—	—
Dyfonate	—	0.1	—	—
Emery	—	E	—	20
Endosulfan (Thiodan®) — Skin	—	0.1	—	0.3
Endrin — Skin	—	0.1	—	0.3
Epichlorhydrin — Skin	5	20	10	40
EPN — Skin	—	0.5	—	1.5
1, 2-Epoxypropane, see Propylene oxide	100	240	150	360

m) See p. 12.
Capital letters refer to Appendices.
Footnotes (a thru h) see Page 11.
*1977 Addition.
**See Notice of Intended Changes.

Substance	ADOPTED VALUES TWA ppm[a]	mg/m³[b]	TENTATIVE VALUES STEL ppm[a]	mg/m³[b]
2, 3-Epoxy-1-propanol, see Glycidol	50	150	65	190
Ethane	F	—	F	—
Ethanethiol, see Ethyl mercaptan	0.5	1	1.5	3
Ethanolamine	3	6	6	12
Ethion (Nialate®) — Skin	—	0.4	—	—
2-Ethoxyethanol — Skin	100	370	150	560
2-Ethoxyethyl acetate (Cellosolve acetate) — Skin	100	540	150	810
Ethyl acetate	400	1,400	—	—
Ethyl acrylate — Skin	25	100	—	—
Ethyl alcohol (Ethanol)	1,000	1,900	—	—
Ethylamine	10	18	—	—
Ethyl sec-amyl ketone (4-Methyl-3-heptanone)	25	130	—	—
Ethyl benzene	100	435	125	545
Ethyl bromide	200	890	250	1,110
Ethylbutyl ketone (3-Heptanone)	50	230	75	345
Ethyl chloride	1,000	2,600	1,250	3,250
Ethyl ether	400	1,200	500	1,500
Ethyl formate	100	300	150	450
Ethyl mercaptan	0.5	1	—	—
Ethyl silicate	(100)	(850)	—	—
Ethylene	F	—	F	—
C Ethylene chlorohydrin — Skin	1	3	—	—
Ethylene diamine	10	25	—	—
Ethylene dibromide, see 1, 2-Dibromoethane	20	145	30	220
Ethylene dichloride, see 1, 2-Dichloroethane	50	200	75	300
Ethylene glycol, Particulate	—	10	—	20
Vapor	100	260	125	325
C Ethylene glycol dinitrate and/or Nitroglycerin — Skin	0.2[d]	—	—	—
Ethylene glycol monomethyl ether acetate (Methyl cellosolve acetate) — Skin	25	120	40	180
Ethylene oxide	50	90	75	135
Ethylenimine — Skin	0.5	1	—	—
Ethylidene chloride, see 1, 1-Dichloroethane	200	320	250	400
C Ethylidene norbornene	5	25	—	—
N-Ethylmorpholine — Skin	20	94	20	94
Fensulfothion (Dasanit)	—	0.1	—	—
Ferbam	—	10	—	20
Ferrovanadium dust	—	1	—	0.3
Fluoride (as F)	—	2.5	—	—
Fluorine	1	2	2	4
Fluorotrichloromethane	1,000	5,600	1,250	7,000
C Formaldehyde	2	3	—	—
Formamide	20	30	30	45
Formic acid	5	9	5	9
Furfural — Skin	5	20	15	60
Furfuryl alcohol — Skin	5	20	10	40
Gasoline	—	B2	—	B2
Germanium tetrahydride	0.2	0.6	0.6	1.8
Glass, fibrous[e] or dust	—	E	—	E
**C Glutaraldehyde, activated or unactivated	—	(0.25)	—	—
Glycerin mist	—	E	—	E
Glycidol (2, 3-Epoxy-1-propanol)	50	150	75	225
Glycol monoethyl ether, see 2-Ethoxyethanol	100	370	150	560
Graphite (Synthetic)	—	E	—	—
Guthion*, see Azinphos-methyl	—	0.2	—	0.6
Gypsum	—	E	—	20
Hafnium	—	0.5	—	1.5
Helium	F	—	F	—
Heptachlor — Skin	—	0.5	—	1.5
Heptane (n-Heptane)	400	1,600	500	2,000
Hexachlorocyclopentadiene	0.01	0.11	0.03	0.33
Hexachloroethane — Skin	1	10	3	30
Hexachloronaphthalene — Skin	—	0.2	—	0.6
Hexafluoroacetone	0.1	0.7	0.3	2.1
Hexane (n-hexane)	100	360	125	450
2-Hexanone, see Methyl butyl ketone — Skin	25	100	40	150
Hexone (Methyl isobutyl ketone) — Skin	100	410	125	510
sec-Hexyl acetate	50	300	—	—
C Hexylene glycol	25	125	—	—
* Hydrazine — Skin	0.1	0.1	—	—
Hydrogen	F	—	F	—
Hydrogenated terphenyls	0.5	5	—	—
Hydrogen bromide	3	10	—	—
C Hydrogen chloride	5	7	—	—
Hydrogen cyanide — Skin	10	11	15	16
Hydrogen fluoride	3	2	—	—
Hydrogen peroxide	1	1.4	2	2.8
Hydrogen selenide	0.05	0.2	—	—
Hydrogen sulfide	10	15	15	27
Hydroquinone	—	2	—	4
Indene	10	45	15	27
Indium & Compounds (as In)	—	0.1	—	0.3
C Iodine	0.1	1	—	—
Iodoform	0.2	3	0.4	0.6
Iron oxide fume	B3	5	—	10
Iron pentacarbonyl	0.01	0.08	—	—
Iron salts, soluble (as Fe)	—	1	—	2
Isoamyl acetate	100	525	125	655
Isoamyl alcohol	100	360	125	450
Isobutyl acetate	150	700	187	875
Isobutyl alcohol	50	150	75	225
C Isophorone	5	25	—	—
* Isophorone diisocyanate — Skin	0.01	0.06	—	—
Isopropyl acetate	250	950	310	1,185
Isopropyl alcohol — Skin	400	980	500	1,225
Isopropylamine	5	12	10	24
Isopropyl ether	250	1,050	310	1,320
Isopropyl glycidyl ether (IGE)	50	240	75	360
Kaolin	—	E	—	20
Ketene	0.5	0.9	1.5	2.7
Lead, inorg., fumes & dusts (as Pb)	—	0.15	—	0.45
Lead arsenate (as Pb)	—	0.15	—	0.45
Lead Chromate (as Cr)	—	0.05, A2	—	—
Limestone	—	E	—	20
Lindane — Skin	—	0.5	—	1.5
Lithium hydride	—	0.025	—	—
L.P.G. (Liquified petroleum gas)	1,000	1,800	1,250	2,250
Magnesite	—	E	—	20
Magnesium oxide fume	—	10	—	—
Malathion — Skin	—	10	—	—

Capital letters refer to Appendices.
Footnotes (a thru h) see Page 11.
*1977 Addition.
**See Notice of Intended Changes.

Substance	ADOPTED VALUES TWA ppm[a]	TWA mg/m³[b]	TENTATIVE VALUES STEL ppm[a]	STEL mg/m³[b]
Maleic anhydride	0.25	1	—	—
C Manganese & Compounds (as Mn)	—	5	—	—
Manganese cyclopentadienyl tricarbonyl (as Mn) — Skin	—	0.1	—	0.3
Marble	—	E	—	20
Mercury (Alkyl compounds) — Skin, As Hg	0.001	0.01	0.003	0.03
Mercury (All forms except alkyl) as Hg	—	0.05	—	0.15
Mesityl oxide	25	100	—	—
Methane	F	—	F	—
Methanethiol, see Methyl mercaptan	0.5	1	—	—
* Methomyl (Lannate®) — Skin	—	2.5	—	—
Methoxychlor	—	10	—	—
2-Methoxyethanol — Skin (Methyl cellosolve)	25	80	35	120
Methyl acetate	200	610	250	760
Methyl acetylene (propyne)	1,000	1,650	1,250	2,060
Methyl acetylene-propadiene mixture (MAPP)	1,000	1,800	1,250	2,250
Methyl acrylate — Skin	10	35	—	—
Methyl acrylonitrile — Skin	1	3	2	6
Methylal (dimethoxymethane)	1,000	3,100	1,250	3,875
Methyl alcohol (methanol) — Skin	200	260	250	325
Methylamine	10	12	—	—
Methyl amyl alcohol, see Methyl isobutyl carbinol	25	100	40	150
Methyl 2-cyanoacrylate	2	8	4	16
Methyl isoamyl ketone	100	475	150	710
Methyl n-amyl ketone (2-Heptanone)	100	465	150	710
Methyl bromide — Skin	15	60	—	—
Methyl butyl ketone, see 2-Hexanone	25	100	40	150
Methyl cellosolve — Skin see 2-Methoxyethanol	25	80	35	120
Methyl cellosolve acetate — Skin, see Ethylene glycol monomethyl ether acetate	25	120	35	150
Methyl chloride	100	210	125	260
Methyl chloroform	350	1,900	450	2,375
Methylcyclohexane	400	1,600	500	2,000
Methylcyclohexanol	50	235	75	350
o-Methycyclohexanone — Skin	50	230	75	345
Methylcyclopentadienyl manganese tricarbonyl (as Mn) — Skin	0.1	0.2	0.3	0.6
Methyl demeton — Skin	—	0.5	—	1.5
C Methylene bisphenyl isocyanate (MDI)	0.02	0.2	—	—
* Methylene chloride (dichloromethane)	200	720	250	900

Substance	ADOPTED VALUES TWA ppm[a]	TWA mg/m³[b]	TENTATIVE VALUES STEL ppm[a]	STEL mg/m³[b]
4, 4'-Methylene bis (2-chloraniline) — Skin	0.02, A2	—	A2	—
C Methylene bis (4-cyclohexylisocyanate)	0.01	0.11	—	—
Methyl ethyl ketone (MEK), see 2-Butanone	200	590	250	740
C Methyl ethyl ketone peroxide	0.2	1.5	—	—
Methyl formate	100	250	150	375
Methyl iodide — Skin	5	28	10	56
Methyl isobutyl carbinol — Skin	25	100	40	150
Methyl isobutyl ketone, see Hexone	100	410	125	510
Methyl isocyanate — Skin	0.02	0.05	—	—
Methyl mercaptan	0.5	1	—	—
Methyl methacrylate	100	410	125	510
Methyl parathion — Skin	—	0.2	—	0.6
Methyl propyl ketone, see 2-Pentanone	200	700	250	875
C Methyl silicate	5	30	—	—
C α-Methyl styrene	100	480	—	—
Molybdenum (as Mo) Soluble compounds	—	5	—	10
Insoluble compounds	—	10	—	20
* Monocrotophos (Azodrin®)	—	0.25	—	—
Monomethyl aniline — Skin	2	9	4	18
C Monomethyl hydrazine — Skin	0.2	0.35	—	—
Morpholine — Skin	20	70	30	105
Naphthalene	10	50	15	75
β-Naphthylamine	—	A1b	—	A1b
Neon	F	—	F	—
* Nickel carbonyl	0.05	0.35	—	—
Nickel metal	—	1	—	—
Nickel, soluble compounds (as Ni)	—	0.1	—	0.3
Nicotine — Skin	—	0.5	—	1.5
Nitric acid	2	5	4	10
Nitric oxide	25	30	35	45
p-Nitroaniline — Skin	1	6	2	12
Nitrobenzene — Skin	1	5	2	10
p-Nitrochlorobenzene — Skin	—	1	—	2
4-Nitrodiphenyl	—	A1b	—	A1b
Nitroethane	100	310	150	465
C Nitrogen dioxide	5	9	—	—
Nitrogen trifluoride	10	29	15	45
Nitroglycerin[d] — Skin	0.2	2	—	—
Nitromethane	100	250	150	375
1-Nitropropane	25	90	35	135
2-Nitropropane	25	90	—	—
N-Nitrosodimethylamine (dimethylnitrosoamine) — Skin	—	A2	—	A2
Nitrotoluene — Skin	5	30	10	60
Nitrotrichloromethane, see Chloropicrin	0.1	0.7	0.3	2
Nonane	200	1,050	250	1,300
Octachloronaphthalene — Skin	—	0.1	—	0.3
Octane	300	1,450	375	1,800
Oil mist	—	5	—	10
Osmium tetroxide (as Os)	0.0002	0.002	0.0006	0.006
Oxalic acid	—	1	—	2
Oxygen difluoride	0.05	0.1	0.15	0.3
Ozone	0.1	0.2	0.3	0.6
Paraffin wax fume	—	2	—	6

Capital letters refer to Appendices.
Footnotes (a thru h) see Page 11.
*1977 Addition.
**See Notice of Intended Changes.

Substance	ADOPTED VALUES TWA ppm[a]	mg/m³[b]	TENTATIVE VALUES STEL ppm[a]	mg/m³[b]
** Paraquat — Skin	—	(0.5)	—	—
Parathion — Skin	—	0.1	—	0.3
Particulate polycyclic aromatic hydrocarbons (PPAH) as benzene solubles	—	0.2, A1a	—	A1a
Pentaborane	0.005	0.01	0.015	0.03
Pentachloronaphthalene	—	0.5	—	1.5
Pentachlorophenol — Skin	—	0.5	—	1.5
Pentaerythritol	—	E	—	20
Pentane	600	1,800	750	2,250
2-Pentanone	200	700	250	875
Perchloroethylene — Skin	100	670	150	1,000
Perchloromethyl mercaptan	0.1	0.8	—	—
Perchloryl fluoride	3	14	6	28
Petroleum distillates (naphtha)	(e)B3	—	B3	—
Phenol — Skin	5	19	10	38
Phenothiazine — Skin	—	5	—	10
p-Phenylene diamine — Skin	—	0.1	—	—
Phenyl ether (vapor)	1	7	2	14
Phenyl ether-Diphenyl mixture (vapor)	1	7	2	14
Phenylethylene, see Styrene	100	420	125	525
Phenyl glycidyl ether (PGE)	10	60	15	90
Phenylhydrazine — Skin	5	22	10	44
C Phenylphosphine	0.05	0.25	—	—
Phorate (Thimet®) — Skin	—	0.05	—	0.15
Phosdrin (Mevinphos®) — Skin	0.01	0.1	0.03	0.3
Phosgene (carbonyl chloride)	0.10	0.4	—	—
Phosphine	0.3	0.4	1	1
Phosphoric acid	—	1	—	3
Phosphorus (yellow)	—	0.1	—	0.3
Phosphorus pentachloride	—	1	—	3
Phosphorus pentasulfide	—	1	—	3
Phosphorus trichloride	0.5	3	—	—
* Phthalic anhydride	1	6	4	24
* m-Phthalodinitrile	—	5	—	—
Picloram (Tordon®)	—	10	—	20
Picric acid — Skin	—	0.1	—	0.3
Pival® (2-Pivalyl-1, 3-indandione)	—	0.1	—	0.3
Plaster of Paris	—	E	—	20
Platinum (Soluble salts) as Pt	—	0.002	—	—
Polychlorobiphenyls, see Chlorodiphenyls	—	—	—	—
Polytetrafluoroethylene decomposition products	—	B1	—	B1
C Potassium hydroxide	—	2	—	—
Propane	F	—	F	—
β-Propiolactone	—	A2	—	A2
Propargyl alcohol — Skin	1	2	3	6
n-Propyl acetate	200	840	250	1,050
Propyl alcohol — Skin	200	500	250	625
n-Propyl nitrate	25	110	40	140
Propylene	F	—	F	—
Propylene dichloride (1, 2-Dichloropropane)	75	350	115	525
Propylene glycol monomethyl ether	100	360	150	450
Propylene imine — Skin	2	5	—	—
Propylene oxide	100	240	150	360
Propyne, see Methyl-acetylene	1,000	1,650	1,250	2,050
Pyrethrum	—	5	—	10
Pyridine	5	15	10	30
Quinone	0.1	0.4	0.3	1
RDX — Skin	—	1.5	—	3
Resorcinol	10	45	20	90
Rhodium, Metal fume and dusts (as Rh)	—	0.1	—	0.3
Soluble salts	—	0.001	—	0.003
Ronnel	—	10	—	—
Rosin core solder pyrolysis products (as formaldehyde)	—	0.1	—	0.3
Rotenone (commercial)	—	5	—	10
Rouge	—	E	—	20
* Rubber solvent	400	1,600	—	—
Selenium compounds (as Se)	—	0.2	—	—
Selenium hexafluoride, as Se	0.05	0.4	0.05	0.4
Sevin® (see Carbaryl)	—	5	—	10
Silane (see Silicon tetrahydride)	0.5	7	—	—
Silicon	—	E	—	20
Silicon carbide	—	E	—	20
Silicon tetrahydride (Silane)	0.5	0.7	1	1.5
Silver, metal and soluble compounds, as Ag	—	0.01	—	0.03
C Sodium azide	0.1	0.3	—	—
Sodium fluoroacetate (1080) — Skin	—	0.05	—	0.15
C Sodium hydroxide	—	2	—	—
Starch	—	E	—	20
Stibine	0.1	0.5	0.3	1.5
Stoddard solvent	100	575	150	720
Strychnine	—	0.15	—	0.45
Styrene, monomer (Phenylethylene)	100	420	125	525
** Succinaldehyde (see Glutaraldehyde)	—	(0.25)	—	—
C Subtilisins (Proteolytic enzymes as 100% pure crystalline enzyme)	—	0.00006(c)	—	—
Sucrose	—	E	—	20
Sulfur dioxide	5	13	—	—
Sulfur hexafluoride	1,000	6,000	1,250	7,500
Sulfuric acid	—	1	—	—
Sulfur monochloride	1	6	3	18
Sulfur pentafluoride	0.025	0.25	0.075	0.75
Sulfur tetrafluoride	0.1	0.4	0.3	1
Sulfuryl fluoride	5	20	10	40
Systox, see Demeton®	0.01	0.1	0.03	0.3
2, 4, 5-T	—	10	—	20
Tantalum	—	5	—	10
TEDP — Skin	—	0.2	—	0.6
Teflon® decomposition products	—	B1	—	B1
Tellurium	—	0.1	—	—
Tellurium hexafluoride, as Te	0.02	0.2	—	—
TEPP — Skin	0.004	0.05	0.012	0.15
C Terphenyls	1	9	—	—
1, 1, 1, 2-Tetrachloro-2, 2-difluoroethane	500	4,170	625	5,210
1, 1, 2, 2-Tetrachloro-1, 2-difluoroethane	500	4,170	625	5,210

Capital letters refer to Appendices.
Footnotes (a thru h) see Page 11.
*1977 Addition.
**See Notice of Intended Changes.

Substance	ADOPTED VALUES TWA ppm[a]	mg/m³[b]	TENTATIVE VALUES STEL ppm[a]	mg/m³[b]
1, 1, 2, 2-Tetrachloroethane — Skin	5	35	10	70
Tetrachloroethylene, see Perchloroethylene	100	670	150	1,000
Tetrachloromethane, see Carbon tetrachloride	10	65	20	130
Tetrachloronaphthalene	—	2	—	4
Tetraethyl lead (as Pb) — Skin	—	0.100[h]	—	0.3
Tetrahydrofuran	200	590	250	700
Tetramethyl lead (as Pb) — Skin	—	0.150[h]	—	0.45
Tetramethyl succinonitrile — Skin	0.5	3	1.5	9
Tetranitromethane	1	8	—	—
Tetryl (2, 4, 6-trinitrophenyl-methylnitramine) — Skin	—	1.5	—	3.0
Thallium, soluble compounds (as Tl) — Skin	—	0.1	—	—
4, 4'-Thiobis (6-tert. butyl-m-cresol)	—	10	—	20
Thiram®	—	5	—	10
Tin, inorganic compounds, except SnH₄ and SnO₂, (as Sn)	—	2	—	4
Tin, organic compounds (as Sn) — Skin	—	0.1	—	0.2
Tin oxide	—	E	—	20
Titanium dioxide	—	E	—	20
Toluene (toluol) — Skin	100	375	150	560
C Toluene-2, 4-diisocyanate (TDI)	0.02	0.14	—	—
o-Toluidine	5	22	10	44
Toxaphene, see Chlorinated camphene	—	0.5	—	1.5
Tributyl phosphate	—	5	—	5
1, 1, 1-Trichloroethane, see Methyl chloroform	350	1,900	440	2,380
1, 1, 2-Trichloroethane — Skin	10	45	20	90
Trichloroethylene	100	535	150	800
** Trichloromethane, see Chloroform	(25)	(120)	—	—
Trichloronaphthalene	—	5	—	10
1, 2, 3-Trichloropropane	50	300	150	450
1, 1, 2-Trichloro 1, 2, 2-trifluoroethane	1,000	7,600	1,250	9,500
Triethylamine	25	100	40	150
Tricyclohexyltin hydroxide (Plictran®)	—	5	—	10
Trifluoromonobromomethane	1,000	6,100	1,200	7,625
Trimethyl benzene	25	120	35	180
2, 4, 6-Trinitrophenol, see Picric acid	—	0.1	—	0.3
2, 4, 6-Trinitrophenyl-methylnitramine, see Tetryl	—	1.5	—	3.0
** Trinitrotoluene (TNT) — Skin	—	(1.5)	—	—
Triorthocresyl phosphate	—	0.1	—	0.3
Triphenyl phosphate	—	3	—	6
Tungsten & compounds, as W				
Soluble	—	1	—	3
Insoluble	—	5	—	10

Substance	ADOPTED VALUES TWA ppm[a]	mg/m³[b]	TENTATIVE VALUES STEL ppm[a]	mg/m³[b]
Turpentine	100	560	150	840
Uranium (natural) soluble & insoluble compounds, as U	—	0.2	—	0.6
Vanadium (V₂O₅), as V				
C Dust	—	0.5	—	1.5
Fume	—	0.05	—	—
Vinyl acetate	10	30	20	60
Vinyl benzene, see Styrene	100	420	150	630
Vinyl bromide	(250)	(1,100)	—	—
** Vinyl chloride	(200)	(510)	—	—
Vinyl cyanide, see Acrylonitrile	20	45	30	70
* Vinyl cyclohexene dioxide	10	60	—	—
Vinylidene chloride	10	40	20	80
Vinyl toluene	100	480	150	720
Warfarin	—	0.1	—	0.3
* Welding fumes (Total particulate)	—	5, B3	—	B3
Wood dust (nonallergenic)	—	5	—	10
Xylene (o-, m-, p-isomers) — Skin	100	435	150	655
*C m-Xylene α, α'-diamine	—	0.1	—	—
Xylidene — Skin	5	25	10	50
Yttrium	—	1	—	3
Zinc chloride fume	—	1	—	2
* Zinc chromate (as Cr)	—	0.05, A2	—	—
Zinc oxide fume	—	5	—	10
Zinc stearate	—	E	—	20
Zirconium compounds (as Zr)	—	5	—	10

Capital letters refer to Appendices.
Footnotes (a thru h) see Page 11.
o) See Page 34.
* 1977 Addition.
** See Notice of Intended Changes.

a) Parts of vapor or gas per million parts of contaminated air by volume at 25°C and 760 mm. Hg. pressure.

b) Approximate milligrams of substance per cubic meter of air.

d) An atmospheric concentration of not more than 0.02 ppm, or personal protection may be necessary to avoid headache for intermittent exposure.

e) < 7 μm in diameter.

f) As sampled by method that does not collect vapor.

g) According to analytically determined composition.

h) For control of general room air, biologic monitoring is essential for personnel control.

Radioactivity: For permissible concentrations of radio-isotopes in air, see U.S. Department of Commerce, National Bureau of Standards Handbook 69, "Maximum Permissible Body Burdens and Maximum Permissible Concentrations of Radionuclides in Air and in Water for Occupational Exposure," June 5, 1969. Also, see U.S. Department of Commerce National Bureau of Standards, Handbook 59, "Permissible Dose from External Sources of Ionizing Radiation," September 24, 1954, and addendum of April 15, 1958. A report, Basic Radiation Protection Criteria, published by the National Committee on Radiation Protection, revises and modernizes the concept of the NCRP standards of 1954, 1957 and 1958; obtainable as NCRP Rept. No. 39, P. O. Box 30175, Washington, D.C. 20014.

MINERAL DUSTS

Substance

SILICA, SiO$_2$
Crystalline
  Quartz   TLV in mppcf[i]:
$$\frac{300^{j)}}{\% \text{ quartz} + 10}$$
                  TLV for respirable dust in mg/m^3:
$$\frac{10 \text{ mg/m}^{3k)}}{\% \text{ Respirable quartz} + 2}$$
                  TLV for "total dust," respirable and nonrespirable:
$$\frac{30 \text{ mg/m}^3}{\% \text{ quartz} + 3}$$

  Cristobalite .... Use one-half the value calculated from the count or mass formulae for quartz.
  Tridymite   Use one-half the value calculated from formulae for quartz.
  Silica, fused   Use quartz formulae.
  Tripoli   Use respirable[p] mass quartz formula

Amorphous .... 20 mppcf[i]

SILICATES (< 1% quartz)
  Asbestos, all forms†   5 fibers/cc > 5μm in length[n]; A1a
  Graphite (natural)   15 mppcf
  Mica   20 mppcf
  Mineral wool fiber   10 mg/m^3
  Perlite   30 mppcf
  Portland Cement   30 mppcf
  Soapstone   20 mppcf
  Talc (nonasbestiform)   20 mppcf
  Talc (fibrous), use Asbestos limit.
  Tremolite, see Asbestos.

COAL DUST
  2 mg/m^3 (respirable dust fraction < 5% quartz).
  If > 5% quartz, use respirable mass formula.

NUISANCE PARTICULATES
(see Appendix E)

30 mppcf or 10 mg/m$^{3l)}$
of total dust < 1% quartz, or, 5 mg/m^3 respirable dust.

Conversion factors:
  mppcf × 35.3 = Million particles per cubic meter = particles per c.c.

i) Millions of particles per cubic foot of air, based on impinger samples counted by light-field technics.
j) The percentage of quartz in the formula is the amount

determined from airborne samples, except in those instances in which other methods have been shown to be applicable.

k) Both concentration and percent quartz for the application of this limit are to be determined from the fraction passing a size-selector with the following characteristics:

Aerodynamic Diameter (μm) (unit density sphere)	% passing selector
≲ 2	90
2.5	75
3.5	50
5.0	25
10	0

l) containing <1% quartz; if quartz content > 1%, use formulae for quartz.
m) Lint-free dust as measured by the vertical-elutriator, cotton-dust sampler described in the Transactions of the National Conference on Cotton Dust, J. R. Lynch, pg. 33, May 2, 1970.
n) As determined by the membrane filter method at 400–450X magnification (4 mm objective) phase contrast illumination.
o) Based on "high volume" sampling.
p) "Respirable" dust as defined by the British Medical Research Council Criteria (1) and as sampled by a device producing equivalent results (2).
  (1) Hatch, T. E. and Gross, P., Pulmonary Deposition and Retention of Inhaled Aerosols, p. 149. Academic Press, New York, New York, 1964.
  (2) Interim Guide for Respirable Mass Sampling, AIHA Aerosol Technology Committee, AHIA J. *31:* 2, 1970, p. 133.

NOTICE OF INTENDED CHANGES
(for 1977)

These substances, with their corresponding values, comprise those for which either a limit has been proposed for the first time, or for which a change in the "Adopted" listing has been proposed. In both cases, the proposed limits should be considered trial limits that will remain in the listing for a period of at least two years. If, after two years no evidence comes to light that questions the appropriateness of the values herein, the values will be reconsidered for the "Adopted" list. Documentation is available for each of these substances.

Substance	ppm[a]	mg/m$^{3b)}$
Aliphatic solvent "140 Flash"	25	150
† Aluminum metal and oxide....	—	10
† Aluminum pyro powders	—	5
† Aluminum welding fumes	—	5

*See Notice of Intended Changes.
†A more stringent TLV for crocidolite may be required.
l), n) See p. 12.

Capital letters refer to Appendices.
†1977 Addition.

Substance	ppm[a]	mg/m³[b]
† Aluminum, soluble salts.......	—	2
† Aluminum alkyls (NOC)*.......	—	2
† 3-Amino 1, 2, 4-triazole.......	A2	—
† Antimony, soluble salts (as Sb)...........................	—	2
Antimony trioxide, handling & use (as Sb)..................	—	0.5
Antimony trioxide production (as Sb).....................	—	0.05, A2
Arsenic trioxide production (as As)......................	—	0.05, A1a
Atrazine.........................	—	10
† Benomyl.........................	—	10
† Bromacil........................	—	10
Butyl acrylate...................	10	55
C Cadmium oxide production (as Cd).....................	—	0.05, A2
Calcium hydroxide...............	—	5
Calcium oxide...................	—	2
Carbonyl fluoride...............	5	15
Chloroform......................	10, A2	50
† Chloromethyl methyl ether....	A1b	A1b
Chromite ore processing (chromate), as Cr...........	—	0.05, A1a
† Cobalt metal, dust & fume (as Co).........................	—	0.05
† Cyclopentane...................	300	850
Dichloromonofluoromethane..	500	2,100
Dimethyl carbamyl chloride...	A2	A2
† Ethyl silicate...................	10	85
†C Glutaraldehyde.................	0.2	0.8
† Hexachlorobutadiene...........	A2	A2
Hexamethyl phosphoramide — Skin....	A2	A2
† Lead chromate (as Cr)........	—	0.05, A2
† Manganese fume (as Mn).....	—	1
Manganese tetroxide..........	—	1
4, 4'-Methylene dianiline......	—	A2
N, Methyl-2-pyrrolidone.......	100	400
Nickel sulfide roasting (as Ni)	—	1, A1a
Paraquat, respirable sizes.....	—	0.1
† Phenyl-beta-naphthylamine...	A2	A2
Phenyl mercaptan..............	0.5	2
Phosgene.......................	0.1	0.4
m-Phthalodinitrile..............	—	5
C Propylene glycol dinitrate-Skin................	0.2	2
Thioglycolic acid................	1	5
C 1, 2, 4-Trichlorobenzene......	5	40
Trimethyl phosphite...........	0.5	2.6
C 2, 4, 6-Trinitrotoluene (TNT).	—	0.5
Valeraldehyde..................	50	175
† Vinyl bromide..................	5	22
Vinyl chloride..................	Pending, A1c	—
† VM & P Naphtha...............	300	1,350

Capital letters refer to Appendices.
†1977 Revision or Addition.
*Not otherwise classified.

NOTICE OF INTENDED CHANGES
MINERAL DUSTS

Substance	TLV
Silica, amorphous..........	5 mg/m³ Total dust (all sampled sizes) 2 mg/m³ Respirable dust ($< 5 \mu m$)
Diatomaceous earth, natural....................	1.5 mg/m³, Respirable dust

APPENDIX A
CARCINOGENS

The Committee lists below those substances in industrial use that have proven carcinogenic in man, or have induced cancer in animals under appropriate experimental conditions. Present listing of those substances carcinogenic for man takes three forms: Those for which a TLV has been assigned (1a), those for which environmental conditions have not been sufficiently defined to assign a TLV (1b), and (1c), those whose reassignment of a TLV is awaiting more definitive data, and hence should be treated as a 1b carcinogen.

A1a. *Human Carcinogens.* Substances, or substances associated with industrial processes, recognized to have carcinogenic or cocarcinogenic potential, with an assigned TLV:

	TLV
Arsenic trioxide production	As_2O_3, 0.05 mg/m³ as As SO_2, C 5.0 ppm Sb_2O_3, 0.5 mg/m³ (as Sb)
Asbestos, all forms*	5 fibers/cc, $> 5 \mu m$ in length
bis (Chloromethyl) ether	0.001 ppm
Chromite ore processing (chromate),	0.05 mg/m³ (as Cr)
Nickel sulfide roasting, fume & dust	1.0 mg/m³(as Ni)
Particulate Polycyclic Aromatic Hydrocarbons (PPAH)	0.2 mg/m³, as benzene solubles

1b. *Human Carcinogens.* Substances, or substances associated with industrial processes, recognized to have carcinogenic potential without an assigned TLV:

4-Aminodiphenyl (p-Xenylamine)
Benzidine production
beta-Naphthylamine
4-Nitrodiphenyl

*Cigarette smoking can enhance the incidence of bronchogenic carcinoma from this and others of these substances or processes.

1c. *Human Carcinogens*. Substances with recognized carcinogenic potential awaiting reassignment of TLV pending further data acquisition:

Vinyl chloride

For the substances in 1b, no exposure or contact by any route — respiratory, skin or oral, as detected by the most sensitive methods — shall be permitted.

"No exposure or contact" means hermitizing the process or operation by the best practicable engineering methods. The worker should be properly equipped to insure virtually no contact with the carcinogen.

A2. *Industrial Substances Suspect of Carcinogenic Potential for MAN*. Chemical substances or substances associated with industrial processes, which are suspect of inducing cancer, based on either (1) limited epidemiologic evidence, exclusive of clinical reports of single cases, or (2) demonstration of carcinogenesis in one or more animal species by appropriate methods.

Antimony trioxide production*	0.05 mg/m³
Benzene — Skin	10 ppm
Benz(a)pyrene	——
Beryllium	2.0 μg/m³
Cadmium oxide production	0.05 mg/m³
Chloroform	10 ppm
Chromates of lead and zinc (as Cr)	0.05 mg/m³
3, 3′-Dichlorobenzidine	——
Dimethylcarbamyl chloride	——
1, 1-Dimethyl hydrazine	0.5 ppm
Dimethyl sulfate	1 ppm
Epichlorhydrin	5 ppm
Hexamethyl phosphoramide — Skin	——
Hydrazine	0.1 ppm
4, 4′-Methylene bis (2-chloroaniline) — Skin	0.02 ppm
4, 4′-Methylene dianiline	——
Monomethyl hydrazine	0.2 ppm
Nitrosamines	——
Propane sultone	——
beta-Propiolactone	——
Vinyl cyclohexene dioxide	10 ppm

For the above, worker exposure by all routes should be carefully controlled to levels consistent with the animal and human experience data (see Documentation), including those substances with a listed TLV.

A3. *Guidelines for the Classification of Experimental ANIMAL Carcinogens*. The following guidelines are offered in the present state of knowledge as an aid in classifying substances in the occupational environment found to be carcinogenic in experimental animals. A need was felt by the Threshold Limits Committee for such a classification in order to take the first step in developing an appropriate TLV for occupational exposure.

Determination of Approximate Threshold of Response Requirement. In order to determine in which category to classify an experimental carcinogen for the purpose of assigning an industrial air limit (TLV), an approximate threshold of neoplastic response must be determined. Because of practical experimental difficulties, a precisely defined threshold cannot be attained. For the purposes of standard-setting, this is of little moment, as an appropriate risk, or safety, factor can be applied to the approximate threshold, the magnitude of which is dependent on the degree of potency of the carcinogenic response.

To obtain the best 'practical' threshold of neoplastic response, dosage decrements should be less than logarithmic. This becomes particularly important at levels greater than 10 ppm (or corresponding mg/m³). Accordingly, after a range-finding determination has been made by logarithmic decreases, two additional dosage levels are required within the levels of "effect" and "no effect" to approximate the true threshold of neoplastic response.

The second step should attempt to establish a metabolic relationship between animal and man for the particular substance found carcinogenic in animals. If the metabolic pathways are found comparable, the substance should be classed highly suspect as a carcinogen for man. If no such relation is found, the substance should remain listed as an experimental animal carcinogen until evidence to the contrary is found.

Proposed Classification of Experimental Animal Carcinogens. Substances occurring in the occupational environment found carcinogenic for animals may be grouped into three classes, those of high, intermediate and low potency. In evaluating the incidence of animal cancers, significant incidence of cancer is defined as a neoplastic response which represents, in the judgment of the Committee, a significant excess of cancers above that occurring in negative controls.

————

EXCEPTIONS: No substance is to be considered an occupational carcinogen of any practical significance which reacts by the respiratory route at or above 1000 mg/m³ for the mouse, 2000 mg/m³ for the rat; by the dermal route, at or above 1500 mg/kg for the mouse, 3000 mg/kg for the rat; by the gastrointestinal route at or above 500 mg/kg/d for a lifetime, equivalent to about 100 g T.D. for the rat, 10g T.D. for the mouse.

————

*Cigarette smoking can enhance the incidence of respiratory cancers from this or others of these substances or processes.

These dosage limitations exclude such substances as dioxane and trichlorethylene from consideration as carcinogens.

Examples: Dioxane — rats, hepatocellular and nasal tumors from 1015 mg/kg/d, oral

Trichloroethylene — female mice, tumors (30/98 @ 900 mg/kg/d), oral

A3a. *INDUSTRIAL SUBSTANCES OF HIGH CARCINOGENIC POTENCY IN EXPERIMENTAL ANIMALS*

1. A substance to qualify as a carcinogen of high potency must fulfill one of the three following conditions in two animal species:

1a. *Respiratory*. Elicit cancer from (1) dosages below 1 mg/m³ (or equivalent ppm) via the respiratory tract in 6- 7-hour daily repeated inhalation exposures throughout lifetime; or (2) from a single intratracheally administered dose not exceeding 1 mg of particulate, or liquid, per 100 ml or less of animal minute respiratory volume;

Examples: bis-Chloromethyl ether, malignant tumors, rats, @ 0.47 mg/m³ (0.1 ppm) in 2 years;

Hexamethyl phosphoramide, nasal squamous cell carcinoma, rats, @ 0.05 ppm, in 13 months

OR

1b. *Dermal*. Elicit cancer within 20 weeks by skin-painting, twice weekly at 2 mg/kg body weight or less per application for a total dose equal to or less than 1.5 mg, in a biologically inert vehicle;

Examples: 7, 12-Dimethylbenz(a)anthracene — skin tumors @ 0.12-0.8 mg T.D. in four weeks

Benz(a)pyrene, mice 12 μg, 3X/wk for 18 mos. T.D. 2.6 mg, 90.9% skin tumors

OR

1c. *Gastrointestinal*. Elicit cancer by daily intake via the gastrointestinal tract, within six months, with a six-month holding period, at a dosage below 1 mg/kg body weight per day; total dose, rat, ≤ 50 mg; mouse, ≤ 3.5 mg;

Examples: 7, 12-Dimethylbenz(a)anthracene — mammary tumors from 10 mg 1X

3-Methylcholanthrene — Tumors @ 3 sites from 8 mg in 89 weeks

Benz(a)pyrene, mice, 3.9% leukemias, from 30 mg T.D. 198 days

2. Elicit cancer by all three routes in at least two animal species at dose levels prescribed for high or intermediate potency.

A3b. *INDUSTRIAL SUBSTANCES OF INTERMEDIATE CARCINOGENIC POTENCY IN EXPERIMENTAL ANIMALS*

To qualify as a carcinogen of intermediate potency, a substance should elicit cancer in two animal species at dosages intermediate between those described in A3a and A3c by two routes of administration.

Example: Carbamic acid Ethyl Ester
Dermal, mammary tumors, mice, 100%, 63 weeks, 500–1400 mg T.D. Gastrointestinal, various type tumors, mice 42 weeks, 320 mg T.D.

Gastrointestinal, various type tumors, rats, 60 weeks, 110–930 mg T.D.

A3c. *INDUSTRIAL SUBSTANCES OF LOW CARCINOGENIC POTENCY IN EXPERIMENTAL ANIMALS*

To qualify as a carcinogen of low potency, a substance should elicit cancer in one animal species by any *one* of three routes of administration at the following prescribed dosages and conditions:

1a. *Respiratory*. Elicit cancer from (1) dosages greater than 10 mg/m³ (or equivalent ppm) via the respiratory tract in 6- 7-hour, daily repeated inhalation exposures, for 12 months' exposure and 12 months' observation period; or (2) from intratracheally administered dosages totaling more than 10 mg of particulate or liquid per 100 ml or more of animal minute respiratory volume;

Examples: Beryl (beryllium aluminum silicate) malig. lung tumors, rats, @ 15 mg/m³ @ 17 months

Benzidine, var. tumors, rats, 10–20 mg/m³ @ > 13 mos.

OR

1b. *Dermal*. Elicit cancer by skin-painting of mice in twice weekly dosages of > 10 mg/kg body weight in a biologically inert vehicle for at least 75 weeks, i.e., ≥ 1.5g T.D.

Examples: Shale tar, mouse, 0.1 ml × 50 — g T.D. 59/60 skin tumors

Arsenic trioxide, man, dose unknown, but estimated to be high

1c. *Gastrointestinal*. Elicit cancer from daily oral dosages of 50 mg/kg/day or greater during the lifetime of the animal.

APPENDIX B
SUBSTANCES OF VARIABLE COMPOSITION

B1 *Polytetrafluoroethylene* decomposition products.* Thermal decomposition of the fluorocarbon chain in air leads to the formation of oxidized products containing carbon, fluorine and oxygen. Because these products decompose in part by hydrolysis in alkaline solution, they can be quantitatively determined in air as fluoride to provide an index of exposure. No TLV is recommended pending determination of the toxicity of the products, but air concentrations should be minimal.

B2 *Gasoline.* The composition of gasoline varies greatly and thus a single TLV for all types of these materials is no longer applicable. In general, the aromatic hydrocarbon content will determine what TLV applies. Consequently the content of benzene, other aromatics and additives should be determined to arrive at the appropriate TLV (Elkins, et al. A.I.H.A.J. *24*:99, 1963); Runion, ibid. *36*, 338, 1975).

B3 *Welding Fumes — Total Particulate*
(NOC)*
TLV, 5 mg/m³

Welding fumes cannot be classified simply. The composition and quantity of both are dependent on the alloy being welded and the process and electrodes used. Reliable analysis of fumes cannot be made without considering the nature of the welding process and system being examined; reactive metals and alloys such as aluminum and titanium are arc-welded in a protective, inert atmosphere such as argon. These arcs create relatively little fume, but an intense radiation which can produce ozone. Similar processes are used to arc-weld steels, also creating a relatively low level of fumes. Ferrous alloys also are arc-welded in oxidizing environments which generate considerable fume, and can produce carbon monoxide instead of ozone. Such fumes generally are composed of discreet particles of amorphous slags containing iron, manganese, silicon and other metallic constituents depending on the alloy system involved. Chromium and nickel compounds are found in fumes when stainless steels are arc-welded. Some coated and flux-cored electrodes are formulated with fluorides and the fumes associated with them can contain significantly more fluorides than oxides. Because of the above factors, arc-welding fumes frequently must be tested for individual constituents which are likely to be present to determine whether specific TLV's are exceeded. Conclusions based on total fume concentration are generally adequate if no toxic elements are present in welding rod, metal, or metal coating and conditions are not conducive to the formation of toxic gases.

Most welding, even with primitive ventilation, does not produce exposures inside the welding helmet above 5 mg/m³. That which does, should be controlled.

---

*Trade Names: Algoflon, Fluon, Halon, Teflon, Tetran.
*Not otherwise classified.

APPENDIX C
MIXTURES

C.1 THRESHOLD LIMIT VALUES FOR MIXTURES

When two or more hazardous substances are present, their combined effect, rather than that of either individually, should be given primary consideration. In the absence of information to the contrary, the effects of the different hazards should be considered as additive. That is, if the sum of the following fractions,

$$\frac{C_1}{T_1} + \frac{C_2}{T_2} + \ldots \frac{C_n}{T_n}$$

exceeds unity, then the threshold limit of the mixture should be considered as being exceeded. C_1 indicates the observed atmospheric concentration, and T_1 the corresponding threshold limit (See Example 1A.a. and 1A.c.).

Exceptions to the above rule may be made when there is a good reason to believe that the chief effects of the different harmful substances are not in fact additive, but *independent* as when purely local effects on different organs of the body are produced by the various components of the mixture. In such cases the threshold limit ordinarily is exceeded only when at least one member of the series $\left(\frac{C_1}{T_1} + \text{ or } + \frac{C_2}{T_2} \text{ etc.} \right)$ itself has a value exceeding unity (See Example 1A.c.).

Antagonistic action or potentiation may occur with some combinations of atmospheric contaminants. Such cases at present must be determined individually. Potentiating or antagonistic agents are not necessarily harmful by themselves. Potentiating effects of exposure to such agents by routes other than that of inhalation is also possible, e.g. imbibed alcohol and inhaled narcotic (trichloroethylene). Potentiation is characteristically exhibited at high concentrations, less probably at low.

When a given operation or process characteristically emits a number of harmful dusts, fumes, vapors or gases, it will frequently be only feasible to attempt to evaluate the hazard by measurement of a single substance. In such cases, the threshold limit used for this substance should be reduced by a suitable factor, the magnitude of which will depend on the number, toxicity and relative quantity of the other contaminants ordinarily present.

Examples of processes which are typically associated with two or more harmful atmospheric contaminants are welding, automobile repair, blasting, painting, lacquering, certain foundry operations, diesel exhausts, etc.

C.1A Examples of THRESHOLD LIMIT VALUES FOR MIXTURES

The following formulae apply only when the components in a mixture have similar toxicologic effects; they should not be used for mixtures with widely differing reactivities, e.g. hydrogen cyanide & sulfur dioxide. In such case the formula for Independent Effects (1A.c.) should be used.

1A.a. General case, where air is analyzed for each component:

a. *Additive effects. (Note: It is essential that the atmosphere be analyzed both qualitatively and quantitatively for each component present, in order to evaluate compliance or noncompliance with this calculated TLV.)*

$$\frac{C_1}{T_1} + \frac{C_2}{T_2} + \frac{C_3}{T_3} + \ldots = 1$$

Example No. 1A.a.: Air contains 5 ppm of carbonetrachloride (TLV = 10 ppm) 20 ppm of 1, 2-dichloroethane (TLV = 50 ppm) and 10 ppm of 1, 2-dibromoethane (TLV = 20 ppm)

Atmospheric concentration of mixture = 5 + 20 + 10 = 35 ppm of mixture

$$\frac{5}{10} + \frac{20}{50} + \frac{10}{20} = \frac{25 + 20 + 25}{50} = 1.4$$

Threshold Limit is exceeded. Furthurmore, the TLV of this mixture may be calculated by reducing the total fraction to 1.0; i.e.

$$\text{TLV of mixture} = \frac{35}{1.4} = 25 \text{ ppm}$$

1A.b. Special case when the source of contaminant is a liquid mixture and the atmospheric composition is *assumed* to be similar to that of the original material; e.g. on a time-weighted average exposure basis, all of the liquid (solvent) mixture eventually evaporates.

Additive effects (approximate solution)

1. The percent composition (by weight) of the liquid mixture is known, the TLVs of the constituents must be listed in mg/m³.

(Note: In order to evaluate compliance with this TLV, field sampling instruments should be calibrated, in the laboratory, for response to this specific quantitative and qualitative air-vapor mixture, and also to fractional concentrations of this mixture; e.g., 1/2 the TLV; 1/10 the TLV; 2 × the TLV; 10 × the TLV; etc.)

TLV of mixture =

$$\frac{1}{\dfrac{f_a}{TLV_a} + \dfrac{f_b}{TLV_b} + \dfrac{f_c}{TLV_c} + \ldots \dfrac{f_n}{TLV_n}}$$

Example No. 1: Liquid contains (by weight)
50% heptane: TLV = 400 ppm or 1600 mg/m³
1 mg/m³ ≡ 0.25 ppm

30% methylene chloride: TLV = 200 ppm or 720 mg/m³
1 mg/m³ ≡ 0.28 ppm

20% perchloroethylene: TLV = 100 ppm or 670 mg/m³
1 mg/m³ ≡ 0.15 ppm

$$\text{TLV of Mixture} = \frac{1}{\dfrac{0.5}{1600} + \dfrac{0.3}{720} + \dfrac{0.2}{670}}$$

$$= \frac{1}{0.00031 + 0.00042 + 0.00030}$$

$$= \frac{1}{0.00103} = 970 \text{ mg/m}^3$$

of this mixture
50% or (970) (0.5) = 485 mg/m³ is heptane
30% or (970) (0.3) = 291 mg/m³ is methylene chloride
20% or (970) (0.2) = 194 mg/m³ is perchloroethylene

These values can be converted to ppm as follows:

heptane: 485 mg/m³ × 0.25 = 121 ppm
methylene chloride: 291 mg/m³ × 0.28 = 81 ppm
perchloroethylene: 194 mg/m³ × 0.15 = 29 ppm

TLV of mixture = 121 + 81 + 29 = 231 ppm, or 970 mg/m³

1A.c. *Independent effects.*
Air contains 0.15 mg/m³ of lead (TLV, 0.15) and 0.7 mg/m³ of sulfuric acid (TLV, 1).

$$\frac{0.15}{0.15} = 1; \qquad \frac{0.7}{1} = 0.7$$

Threshold limit is not exceeded.

1B. TLV for Mixtures of Mineral Dusts.

For mixtures of biologically active mineral dusts the general formula for mixtures may be used.

For mixture containing 80% nonasbestiform talc and 20% quartz, the TLV for 100% of the mixture is given by:

$$\text{TLV} = \frac{1}{\dfrac{0.8}{20} + \dfrac{0.2}{2.7}} = 9 \text{ mppcf}$$

TLV of nonasbestiform talc (pure) = 20 mppcf
TLV of quartz (pure) =
$$\frac{300}{100 + 10} = \frac{300}{110} = 2.7 \text{ mppcf}$$

Essentially the same result will be obtained if the limit of the more (most) toxic component is used provided the effects are additive. In the above example the limit for 20% quartz is 10 mppcf.

For another mixture of 25% quartz, 25% amorphous silica and 50% talc:

25% quartz — TLV (pure) = 2.7 mppcf
25% amorphous silica — TLV (pure) = mppcf
50% talc TLV (pure) = 20 mppcf

$$TLV = \frac{1}{\dfrac{0.25}{2.7} + \dfrac{0.25}{20} + \dfrac{0.5}{20}} = 8 \text{ mppcf}$$

The limit for 25% quartz approximates 9 mppcf.

APPENDIX D
PERMISSIBLE EXCURSIONS FOR TIME-WEIGHTED AVERAGE (TWA) LIMITS

The Excursion TLV Factor in the Table automatically defines the magnitude of the permissible excursion above the limit for those substances not given a "C" designation; i.e., the TWA limits. Examples in the Table show that nitrobenzene, the TLV for which is 1 ppm, should never be allowed to exceed 3 ppm. Similarly, carbon tetrachloride, TLV = 10 ppm, should never be allowed to exceed 20 ppm. By contrast, those substances with a "C" designation are not subject to the excursion factor and must be kept at or below the TLV ceiling.

These limiting excursions are to be considered to provide a "rule-of-thumb" guidance for listed substances generally, and may not provide the most appropriate excursion for a particular substance e.g., the permissible excursion for CO is 400 ppm for 15 minutes.

For appropriate excursions for 142 substances consult Pa. Rules & Regs., Chap. 4, Art. 432, and "Acceptable Concentrations," ANSI.

Substance	TLV	Excursion Factor	Max. Conc. Permitted for short time
Nitrobenzene	1	3	3
Carbon tetrachloride	10	2	20
Trimethyl benzene	25	1.5	40
Acetone	1000	1.25	1250
Boron trifluoride	C 1	—	1
Butylamine	C 5	—	5

EXCURSION FACTORS

For all substances not bearing C notation

		Excursion Factor	
TLV>0–1	(ppm or mg/m³),	= 3
TLV>1–10	"	"	= 2
TLV>10–100	"	"	= 1.5
TLV>100–1000	"	"	= 1.25

The number of times the excursion above the TLV is permitted is governed by conformity with the Time-Weighted Average TLV.

INTERPRETATION OF MEASURED PEAK CONCENTRATIONS

With increasing use of rapid, direct-reading analytical instruments for airborne contaminants in the work area, the question of interpretation of essentially "instantaneous" peaks arises. Although no general statement can be made covering all occupational substances, the following guidelines should prove helpful, assuming peak excursions conform to time-weighted average TLV as stated above.

The toxicologic importance of momentary peak concentrations depends on whether the substance is fast or slow acting. If slow acting, as for quartz, lead, or carbon monoxide, momentary peaks are of no toxicologic concern provided, of course, they are not astronomic. On the other hand, fast-acting substances that rapidly produce disabling narcosis, e.g., H_2S, or intolerable irritation or asphyxiation, NH_3, SO_2, CO_2, or initiate sensitization — the organic isocyanates, even "instantaneous" peaks appreciably above the permissible excursion, should not be permitted, unless information exists to the contrary. Other more specific excursions will be developed in the future.

APPENDIX E
Some Nuisance Particulates[q]
TLV, 30 mppcf or 10mg/m³

Alundum (Al_2O_3)	Kaolin
Calcium carbonate	Limestone
Calcium silicate	Magnesite
Cellulose (paper fiber)	Marble
Portland Cement	Mineral Wool Fiber
Corundum (Al_2O_3)	Pentaerythritol
Emery	Plaster of Paris
Glass, fibrous[r] or dust	Rouge
Glycerin Mist	Silicon
Graphite (synthetic)	Silicon Carbide
Gypsum	Starch
Vegetable oil mists	Sucrose
(except castor, cashew	Tin Oxide
nut, or similar irritant	Titanium Dioxide
oils)	Zinc Stearate
	Zinc oxide dust

q) When toxic impurities are not present, e.g. quartz < 1%.

r) <7μm in diameter

APPENDIX F
Some Simple Asphyxiants[s]

Acetylene	Hydrogen
Argon	Methane
Butane	Neon
Ethane	Propane
Ethylene	Propylene
Helium	

s) As defined on pg. 5.

APPENDIX G

Calculations for Conversion of Particle Count Concentration (by Standard Light Field — Midget Impinger Techniques), in mppcf, to Respirable Mass Concentration (by Respirable Sampler) in mg/m^3.[†]

1. In 1967, Jacobsen and Tomb,[†] derived an empirical relationship of 5.6 mppcf to 1 milligram of respirable dust per cubic meter of air, based on 23 sets of samples, mostly coal dust. The following calculation results in an equivalence of 6.37 mppcf to 1 mg/m^3 of respirable dust. Thus, an approximate ratio of 6 mppcf to 1 mg/m^3 of respirable dust is suggested for conversion of TLVs from a count to a mass basis when the density and mass median diameter have not been determined.

2. Basic assumptions:

   a) Average density of silica containing dusts $\approx$ 2.5 gms/cm^3 (2500 mg/cm^3). Pulmonary significant dust densities may vary from 1.2 gm/cm^3 for coal dust to 3.1 gms/cm^3 for Portland Cement. Silica densities vary from 2.2 (amorphous) to 2.3 (cristobalite and tridymite) to 2.5 (alpha-quartz.) gms per cm^3.

   b) The mass median diameter (mmd) of particles collected in midget impinger samplers and counted by the standard light field technique, *and* collected in a respirable sampler is approximately 1.5 μm or 1.5 $\times$ 10^{-4} cm. This assumption is, of course, quite arbitrary since the mmd of all dust clouds is quite variable, depending on many independent parameters, such as source of dust, age of dust cloud, meteorological conditions, etc.

3. Calculation:

   a) vol. per particle: $4/3\ \pi\ r^3$; $r = 0.75 \times 10^{-4}$ cm

   $= 4/3 \cdot \pi \cdot (0.75 \times 10^{-4})^3$
   $= 1.77 \times 10^{-12}\ cm^3$

   b) wt. per particle $=$ vol. $\times$ density
   $= 1.77 \times 10^{-12}\ cm^3 \times 2.5 \times 10^3\ mg/cm^3$
   $= 4.425 \times 10^{-9}\ mg/particle$

   c) 1 particle/$ft.^3 = 35.5$ part./m^3
   (since 35.5 cu ft = 1 cu m.)
   10^6 part./ft^3 = mppcf = 35.5×10^6 part./m^3

   wt. of 1 mppcf $\equiv 35.5 \times 10^6$ part./$m^3 \times 4.425 \times 10^{-9}$ mg/part.

   1 mppcf $\equiv 0.157\ mg/m^3$
   or
   6.37 mppcf $\equiv 1\ mg/m^3$
   or approximately 6 mpccf $\equiv 1\ mg/m^3$.

4. Equivalent TLVs in mppcf and mg/m^3 (respirable mass) for Mineral Dusts.

Substance	Threshold Limit Value		
	Count mppcf	Resp. Mass mg/m^3	Total Mass* mg/m^3
Silica (SiO_2)			
Amorphous	20	(3)**	(6)
Cristobalite	1.5	0.05	0.15
Fused silica	3	0.1	0.3
Quartz	3	0.1	0.3
Tridymite	1.5	0.05	0.15
Coal Dust	(12)	2	(4)
Diatomaceous earth, natural	—	1.5	—
Graphite	15	(2.5)	(5)
Mica	20	(3)	(6)
Mineral wool fiber	—	(5)	10
Nuisance particulates	30	(5)	10
Perlite	30	(5)	(10)
Portland Cement	30	(5)	(10)
Soapstone	20	(3)	(6)
Talc (nonasbestiform)	20	(3)	(6)
Tripoli	(3)	0.1	(0.3)

*Unless otherwise specified, respirable mass is presumed to equal approximately 50% of total mass.
**All values in parentheses ( ) represent newly calculated values based on equivalence of 6 mppcf $\equiv$ 1 mg/m^3 respirable mass and respirable mass $\equiv$ 50% total mass.

[†]"Relationship Between Gravimetric Respirable Dust Concentration and Midget Impinger Number Concentration," by Murray Jacobson and T. F. Tomb, AIHAJ, 28: Nov.–Dec. 1967.

TLVs®
Threshold Limit Values
for
Physical Agents
Adopted by
ACGIH
for 1977

1977 TLV PHYSICAL AGENTS COMMITTEE

Herbert H. Jones, Central Missouri State University, Chairman
Peter A. Breysse, University of Washington
Gerald V. Coles, United Kingdom
Thomas Cummings, Ontario Dept. of Health
Irving H. Davis, Michigan Dept. of Health
Ronald D. Dobbin, NIOSH
LCDR Joseph J. Drozd, USN
Maj. George S. Kush, USAF
Edward J. Largent, OSHA
William E. Murray, NIOSH
Dr. Wordie H. Parr, NIOSH
David H. Sliney, USAEHA
Lt. Col. Robert T. Wangemann, USAEHA
Thomas K. Wilkinson, USPHS-NIH

Any comments or questions regarding these limits should be addressed to:

Secretary-Treasurer
P.O. Box 1937
Cincinnati, Ohio 45201

PREFACE

PHYSICAL AGENTS

These threshold limit values refer to levels of physical agents and represent conditions under which it is believed that nearly all workers may be repeatedly exposed day after day without adverse effect. Because of wide variations in individual susceptibility, exposure of an occasional individual at, or even below, the threshold limit may not prevent annoyance, aggravation of a pre-existing condition, or physiological damage.

These threshold limits are based on the best available information from industrial experience, from experimental human and animal studies, and when possible, from a combination of the three.

These limits are intended for use in the practice of industrial hygiene and should be interpreted and applied only by a person trained in this discipline. They are not intended for use, or for modification for use, (1) in the evaluation or control of the levels of physical agents in the community, (2) as proof or disproof of an existing physical disability, or (3) for adoption by countries whose working conditions differ from those in the United States of America.

These values are reviewed annually by the Committee on Threshold Limits for Physical Agents for revisions or additions, as further information becomes available.

Notice of Intent — At the beginning of each year, proposed actions of the Committee for the forthcoming year are issued in the form of a "Notice of Intent." This notice provides not only an opportunity for comment, but solicits suggestions of physical agents to be added to the list. The suggestions should be accompanied by substantiating evidence.

As Legislative Code — The Conference recognizes that the Threshold Limit Values may be adopted in legislative codes and regulations. If so used, the intent of the concepts contained in the Preface should be maintained and provisions should be made to keep the list current.

Reprint Permission — This publication may be reprinted provided that written permission is obtained from the Secretary-Treasurer of the Conference and that this Preface be published in its entirety along with the Threshold Limit Values.

THRESHOLD LIMIT VALUES

HEAT STRESS

These Threshold Limit Values refer to heat stress conditions under which it is believed that nearly all workers may be repeatedly exposed without adverse health effects. The TLVs shown in Table 1 are based on the assumption that nearly all acclimatized, fully clothed workers with adequate water and salt intake should be able to function effectively under the given working conditions without exceeding a deep body temperature of 38℃ (WHO technical report series #412, 1969 *Health Factors Involved in Working Under Conditions of Heat Stress*; F. N. Dukes-Dobos and A. Henschel: *"Development of Permissible Heat Exposure Limits for Occupational Work."* ASHRAE Journal, Vol. 15: No. 9, September 1973, pp. 57–62.)

Since measurement of deep body temperature is impractical for monitoring the workers' heat load, the measurement of environmental factors is required which most nearly correlate with deep body temperature and other physiological responses to heat. At the present time Wet Bulb-Globe Temperature Index (WBGT) is the simplest and most suitable technique to measure the environmental factors. WBGT values are calculated by the following equations:

1. Outdoors with solar load:
   $$WBGT = 0.7WB + 0.2GT + 0.1DB$$

2. Indoors or Outdoors with no solar load:
   $$WBGT = 0.7WB + 0.3GT$$
where:

WBGT = Wet Bulb-Globe Temperature Index
WB = Natural Wet-Bulb Temperature
DB = Dry-Bulb Temperature
GT = Globe Thermometer Temperature

The determination of WBGT requires the use of a black globe thermometer, a natural (static) wet-bulb thermometer, and a dry-bulb thermometer.

TABLE 1

Permissible Heat Exposure Threshold Limit Values
(Values are given in ℃. WBGT)

Work — Rest Regimen	Work Load		
	Light	Moderate	Heavy
Continuous work	30.0	26.7	25.0
75% Work — 25% Rest, Each hour	30.6	28.0	25.9
50% Work — 50% Rest, Each hour	31.4	29.4	27.9
25% Work — 75% Rest, Each hour	32.2	31.1	30.0

Higher heat exposures than shown in Table 1 are permissible if the workers have been undergoing medical surveillance and it has been established that they are more tolerant to work in heat than the average worker. Workers should not be permitted to continue their work when their deep body temperature exceeds 38.0℃.

APPENDIX G

HEAT STRESS

I. Measurement of the Environment

The instruments required are a dry-bulb, a natural wet-bulb, a globe thermometer, and a stand. The measurement of the environmental factors shall be performed as follows:

A. The range of the dry and the natural wet bulb thermometer shall be −5℃ to 50℃ with an accuracy of ±0.5℃. The dry bulb thermometer must be shielded from the sun and the other radiant surfaces of the environment without restricting the airflow around the bulb. The wick of the natural wet-bulb thermometer shall be kept wet with distilled water for at least 1/2 hour before the temperature reading is made. It is not enough to immerse the other end of the wick into a reservoir of distilled water and wait until the whole wick becomes wet by capillarity. The wick shall be wetted by direct application of water from a syringe 1/2 hour before each reading. The wick shall extend over the bulb of the thermometer, covering the stem about one additional bulb length. The wick should always be clean and new wicks should be washed before using.

B. A globe thermometer, consisting of a 15 cm. (6-inch) diameter hollow copper sphere, painted on the outside with a matte black finish or equivalent shall be used. The bulb or sensor of a thermometer (range −5℃ to 100℃ with an accuracy of ±0.5℃) must be fixed in the center of the sphere. The globe thermometer shall be exposed at least 25 minutes before it is read.

C. A stand shall be used to suspend the three thermometers so that they do not restrict free air flow around the bulbs, and the wet-bulb and globe thermometers are not shaded.

D. It is permissible to use any other type of temperature sensor that gives identical reading to a mercury thermometer under the same conditions.

E. The thermometers must be so placed that the readings are representative of the condition where the men work or rest, respectively.

The methodology outlined above is more fully explained in the following publications:

1. "Prevention of Heat Casualties in Marine Corps Recruits, 1955–1960, with Comparative Incidence Rates and Climatic Heat Stresses in other Training Categories," by Captain David Minard, MC, USN, Research Report No. 4, Contract No. MR005.01–0001.01, Naval Medical Research Institute, Bethesda, Maryland, 21 February 1961.

2. "Heat Casualties in the Navy and Marine Corps, 1959–1962, with Appendices on the Field Use of the Wet Bulb-Globe Temperature Index," by Captain David Minard, MC, USN, and R. L. O'Brien, HMC, USN. Research Report No. 7, Contract No. MR 005.01–0001.01, Naval Medical Research Institute, Bethesda, Maryland, 12 March 1964.

3. Minard, D.: Prevention of Heat Casualties in Marine Corps Recruits. Military Medicine *126(4)*: 261–272, 1961.

II. *Work Load Categories*

Heat produced by the body and the environmental heat together determine the total heat load. Therefore, if work is to be performed under hot environmental conditions, the workload category of each job shall be established and the heat exposure limit pertinent to the work load evaluated against the applicable standard in order to protect the worker from exposure beyond the permissible limit.

A. The work load category may be established by ranking each job into light, medium, and heavy categories on the basis of type of operation. Where the work load is ranked into one of said three categories, i.e.

(1) light work (up to 200 Kcal/hr or 800 Btu/hr): e.g., sitting or standing to control machines, performing light hand or arm work,

(2) moderate work (200–350 Kcal/hr or 800–1400 Btu/hr): e.g., walking about with moderate lifting and pushing,

(3) heavy work (350–500 Kcal/hr or 1400–2000 Btu/hr): e.g., pick and shovel work,

the permissible heat exposure limit for that work load shall be determined from Table 1.

B. The ranking of the job may be performed either by measuring the worker's metabolic rate while performing his job or by estimating his metabolic rate by the use of the scheme shown in Table 2. Tables available in the literature listed below and in other publications as well may also be utilized. When this method is used the permissible heat exposure limit can be determined by Figure 1.

1. Per-Olaf Astrand and Kaare Rodahl: "Textbook of Work Physiology" McGraw-Hill Book Company, New York, San Francisco, 1970.

2. "Ergonomies Guide to Assessment of Metabolic and Cardiac Costs of Physical Work." Amer. Ind. Hyg. Assoc. J. *32:* 560, 1971.

3. Energy Requirements for Physical Work. Purdue Farm Cardiac Project. Agricultural Experiment Station. Research Progress Report No. 30, 1961.

4. J. V. G. A. Durnin and R. Passmore: "Energy, Work and Leisure." Heinemann Educational Books, Ltd., London, 1967.

Light hand work: writing, hand knitting

Heavy hand work: typewriting

Heavy work with one arm: hammering in nails (shoemaker, upholsterer)

Light work with two arms: filing metal, planing wood, raking of a garden

Moderate work with the body: cleaning a floor, beating a carpet

TABLE 2

Assessment of Work Load

Average values of metabolic rate during different activities.

A. Body position and movement	Kcal./min.
Sitting	0.3
Standing	0.6
Walking	2.0–3.0
Walking up hill	add 0.8
	per meter (yard) rise

B. Type of Work		Average Kcal./min.	Range Kcal./min.
Hand work			
	light	0.4	0.2–1.2
	heavy	0.9	
Work with one arm			
	light	1.0	0.7–2.5
	heavy	1.8	
Work with both arms			
	light	1.5	1.0–3.5
	heavy	2.5	
Work with body			
	light	3.5	2.5–15.0
	moderate	5.0	
	heavy	7.0	
	very heavy	9.0	

Heavy work with the body: railroad track laying, digging, barking trees

Sample Calculation: Using a heavy hand tool on an assembly line

A. Walking along		2.0 Kcal./min.
B. Intermediate value between heavy work with two arms and light work with the body		3.0 Kcal./min.
		5.0 Kcal./min.
C. Add for basal metabolism		1.0 Kcal./min.
	Total	6.0 Kcal./min.

Adapted from Lehmann, G. E., A. Muller and H. Spitzer: Der Kalorienbedarf bei gewerblicher Arbeit. Arbeitsphysiol. *14:* 166, 1950.

III. *Work-Rest Regimen*

The permissible exposure limits specified in Table 1 and Figure 1 are based on the assumption that the WBGT value of the resting place is the same or very close to that of the work place. Where the WBGT of the work area is different from that of the rest area a time-weighted average value should be used for both environmental and metabolic heat. When time-weighted average values are used the appropriate curve on Figure 1 is the solid line labeled "continuous."

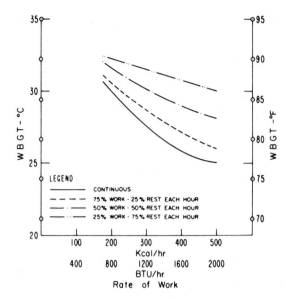

Figure 1 — Permissible Heat Exposure
Threshold Limit Value

The time-weighted average metabolic rate (M) shall be determined by the equation:

Av. M =
$$\frac{(M_1) \times (t_1) + (M_2) \times (t_2) + \ldots + (M_n) \times (t_n)}{(t_1) + (t_2) + \ldots + (t_n)}$$

Where M_1 M_2, M_n are estimated or measured metabolic rates for the various activities of the worker during the total time period. t_1, t_2, t_n are the elapsed times in minutes spent at the corresponding metabolic rate as determined by a time study.

The time-weighted average WBGT shall be determined by the equation:

Av. WBGT =
$$\frac{(WBGT_1) \times (t_1) + (WBGT_2) \times t_2 + \ldots + (WBGT_n) \times (t_n)}{(t_1) + (t_2) + \ldots + (t_n)}$$

where $WBGT_1$, $WBGT_2$, $WBGT_n$ are calculated values of WBGT for the various work and rest areas occupied during total time periods. t_1, t_2, t_n are the elapsed times in minutes spent in the corresponding areas which are determined by a time study. Where exposure to hot environmental conditions is continuous for several hours or the entire work day, the time-weighted averages shall be calculated as hourly time-weighted average i.e., $t_1 + t_2 + \ldots t_n = 60$ minutes. Where the exposure is intermittent, the time-weighted averages shall be calculated as two-hour time-weighted averages, i.e., $t_1 + t_2 + \ldots t_n = 120$ minutes.

The permissible exposure limits for continuous work are applicable where there is a work-rest regimen of a 5-day work week and an 8-hour work day with a short morning and afternoon break (approximately 15 min-

utes) and a longer lunch break (approximately 30 minutes). Higher exposure limits are permitted if additional resting time is allowed. All breaks, including unscheduled pauses and administrative or operational waiting periods during work may be counted as rest time when additional rest allowance must be given because of high environmental temperatures.

It is a common experience that when the work on a job is self-paced, the workers will spontaneously limit their hourly work load to 30–50% of their maximum physical performance capacity. They do this either by setting an appropriate work speed or by interspersing unscheduled breaks. Thus the daily average of the workers' metabolic rate seldom exceeds 330 kcal/hr. However, within an 8-hour work shift there may be periods where the workers' hourly average metabolic rate will be higher.

IV. *Water and Salt Supplementation*

During the hot season or when the worker is exposed to artificially generated heat, drinking water shall be made available to the workers in such a way that they are stimulated to frequently drink small amounts, i.e., one cup every 15–20 minutes (about 150 ml or 1/4 pint).

The water shall be kept reasonably cool (10°–15°C or 50.0°–60.0°F) and shall be placed close to the workplace so that the worker can reach it without abandoning the work area.

The workers should be encouraged to salt their food abundantly during the hot season and particularly during hot spells. If the workers are unacclimatized, salted drinking water shall be made available in a concentration of 0.1% (1g NaCl to 1.0 liter or 1 level tablespoon of salt to 15 quarts of water). The added salt shall be completely dissolved before the water is distributed, and the water shall be kept reasonably cool.

V. *Other Considerations*

A. Clothing: The permissible heat exposure TLVs are valid for light summer clothing as customarily worn by workers when working under hot environmental conditions. If special cothing is required for performing a particular job and this clothing is heavier or it impedes sweat evaporation or has higher insulation value, the worker's heat tolerance is reduced, and the permissible heat exposure limits indicated in Table 1 and Figure 1 are not applicable. For each job category where special clothing is required, the permissible heat exposure limit shall be established by an expert.

B. Acclimatization and Fitness: The recommended heat stress TLVs are valid for acclimated workers who are physically fit.

IONIZING RADIATION

See U.S. Department of Commerce National Bureau of Standards, Handbook 59, "Permissible Dose from External Sources of Ionizing Radition," September 24,

1954, and addendum of April 15, 1958. A report, Basic Radiation Protection Criteria, published by the National Committee on Radiation Protection, revises and modernizes the concept of the NCRP standards of 1954, 1957 and 1958; obtainable as NCRP Rept. No. 39, P. O. Box 4867, Washington, D.C. 20008.

LASERS

The threshold limit values are for exposure to laser radiation under conditions to which nearly all workers may be exposed without adverse effects. The values should be used as guides in the control of exposures and should not be regarded as fine lines between safe and dangerous levels. They are based on the best available information from experimental studies.

Limiting Apertures

The TLVs expressed as radiant exposure or irradiance in this section may be averaged over an aperture of 1 mm except for TLVs for the eye in the spectral range of 400–1400 nm, which should be averaged over a 7 mm limiting aperture (pupil); and except for all TLVs for wavelengths between 0.1–1 mm where the limiting aperture is 10 mm. No modification of the TLVs is permitted for pupil sizes less than 7 mm.

The TLVs for "extended sources" apply to sources

TABLE 3

Threshold Limit Value for Direct Ocular Exposures (Intrabeam Viewing) from a Laser Beam

Spectral Region	Wave Length	Exposure Time, (t) Seconds	TLV
UVC	200 nm to 280 nm	10^{-3} to 3×10^4	3 mJ $\bullet$ cm^{-2}
UVB	280 nm to 302 nm	"	3 "
	303 nm	"	4 "
	304 nm	"	6 "
	305 nm	"	10 "
	306 nm	"	16 "
	307 nm	"	25 "
	308 nm	"	40 "
	309 nm	"	63 "
	310 nm	"	100 "
	311 nm	"	160 "
	312 nm	"	250 "
	313 nm	"	400 "
	314 nm	"	630 "
	315 nm	"	1.0 J $\bullet$ cm^{-2}
UVA	315 nm to 400 nm	10 to 10^3	1.0 J $\bullet$ cm^{-2}
	" "	10^3 to 3×10^4	1.0 mW $\bullet$ cm^{-2}
Light	400 nm to 700 nm	10^{-9} to 1.8×10^{-5}	5×10^{-7} J $\bullet$ cm^{-2}
	400 nm to 700 nm	1.8×10^{-5} to 10	1.8 (t/$\sqrt[4]{t}$) mJ $\bullet$ cm^{-2}
	400 nm to 549 nm	10 to 10^4	10 mJ $\bullet$ cm^{-2}
	550 nm to 700 nm	10 to T_1	1.8 (t/$\sqrt[4]{t}$) mJ $\bullet$ cm^{-2}
	550 nm to 700 nm	T_1 to 10^4	10 C_B mJ $\bullet$ cm^{-2}
	400 nm to 700 nm	10^4 to 3×10^4	C_B μW $\bullet$ cm^{-2}
IR-A	700 nm to 1059 nm	10^{-9} to 1.8×10^{-5}	$5 C_A \times 10^{-7}$ J $\bullet$ cm^{-2}
	700 nm to 1059 nm	1.8×10^{-5} to 10^3	1.8 C_A (t/$\sqrt[4]{t}$) mJ $\bullet$ cm^{-2}
	1060 nm to 1400 nm	10^{-9} to 10^{-4}	5×10^{-6} J $\bullet$ cm^{-2}
	1060 nm to 1400 nm	10^{-4} to 10^3	9(t/$\sqrt[4]{t}$) mJ $\bullet$ cm^{-2}
	700 nm to 1400 nm	10^3 to 3×10^4	320 C_A μW $\bullet$ cm^{-2}
IR-B & C	1.4 μm to 10^3 μm	10^{-9} to 10^{-7}	10^{-2} J $\bullet$ cm^2
	" "	10^{-7} to 10	0.56 $\sqrt[4]{t}$ J $\bullet$ cm^{-2}
	" "	10 to 3×10^4	0.1 W $\bullet$ cm^{-2}

C_A – See Fig. 2, Laser TLV listing.
$C_B = 1$ for $\lambda = 400$ to 550 nm; $C_B = 10^{[0.015 (\lambda - 550)]}$ for $\lambda = 550$ to 700 nm.
$T_1 = 10$ s for $\lambda = 400$ to 550 nm; $T_1 = 10 \times 10^{[0.02 (\lambda - 550)]}$ for $\lambda = 550$ to 700 n.

which subtend an angle greater than α (Table 5) which varies with exposure time. This angle is *not* the beam divergence of the source.

Correction Factors A and B (C_A and C_B) for Eye Exposure

All TLVs in Tables 3 and 4 are to be used as given for wavelengths 400 nm to 700 nm. At all wavelengths greater than 1.06 μm and less than 1.4 μm the TLVs are to be increased by a factor of 5. TLV at wavelengths between 700 nm and 1.06 μm are to be increased by a uniformly extrapolated factor as shown in Figure 2. For certain exposure durations at wavelengths between 700–800 nm, correction factor C_B is applied.

Repetitively Pulsed Lasers

Since there are few experimental data for multiple pulses, caution must be used in the evaluation of such exposures. The protection standards for irradiance or radiant exposure in multiple pulse trains have the following limitations:

(1) The exposure from any single pulse in the train is limited to the protection standard for a single comparable pulse.

(2) The average irradiance for a group of pulses is limited to the protection standard as given in Tables 3, 4, or 6 of a single pulse of the same duration as the entire pulse group.

(3) When the Instantaneous Pulse Repetition Frequency (PRF) of any pulses within a train exceeds one, the protection standard applicable to each pulse is reduced as shown in Figure 6 for pulse durations less than 10^{-5} second. For pulses of greater duration, the following formula should be followed:

$$\text{Standard} \left(\begin{array}{c} \text{single pulse} \\ \text{in train} \end{array} \right) =$$

$$\frac{\text{Standard (pulse } n\tau)}{n}$$

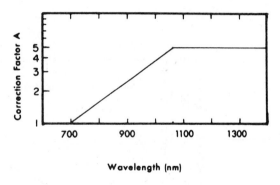

Figure 2 — TLV correction factors for laser wavelengths (eye).

where:

n = number of pulses in train
τ = duration of a single pulse in the train
Standard ($n\tau$) = protection standard of one pulse having a duration equal to $n\tau$ seconds.

MICROWAVES*

These Threshold Limit Values refer to microwave energy in the frequency range of 100 MHz to 100 GHz and represent conditions under which it is believed that nearly all workers may be repeatedly exposed without adverse effect.

These values should be used as guides in the control of exposure of microwaves and should not be regarded as a fine line between safe and dangerous levels.

---

*See Notice of Intended Changes, Page 32; Notice of Intent to Study, page 34.

TABLE 4

Threshold Limit Values for Viewing a Diffuse Reflection
of a Laser Beam or an Extended Source Laser

Spectral Region	Wave Length	Exposure Time, (t) Seconds	TLV
UV	200 nm to 400 nm	10^{-3} to 3×10^4	Same as Table 3
Light	400 nm to 700 nm	10^{-9} to 10	$10 \sqrt[3]{t}$ J • cm^{-2} • sr^{-1}
	400 nm to 549 nm	10 to 10^4	21 J • cm^{-2} • sr^{-1}
	550 nm to 700 nm	10 to T_1	$3.83 (t/ \sqrt[4]{t})$ J • cm^{-2} • sr^{-1}
	550 nm to 700 nm	T_1 to 10^4	$21/C_B$ J cm^{-2} • sr^{-1}
	400 nm to 700 nm	10^4 to 3×10^4	$2.1/C_B \times 10^{-3}$ W • cm^{-2} • sr^{-1}
IR-A	700 nm to 1400 nm	10^{-9} to 10	$10 \, C_A \sqrt[3]{t}$ J • cm^{-2} • sr^{-1}
	700 nm to 1400 nm	10 to 10^3	$3.83 \, C_A (t/ \sqrt[4]{t})$ J • cm^{-2} • sr^{-1}
	700 nm to 1400 nm	10^3 to 3×10^4	$0.64 \, C_A$ W • cm^{-2} • sr^{-1}
IR-B & C	1.4 μm to 1 mm	10^{-9} to 3×10^4	Same as Table 3

C_A, C_B, and T_1 are the same as in footnote to Table 3.

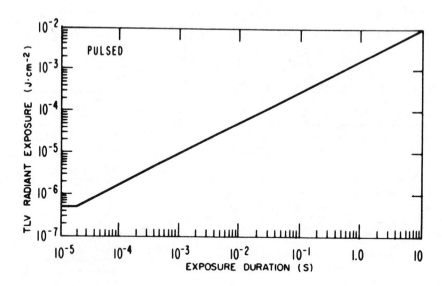

Figure 3a — TLV for intrabeam (direct) viewing of laser beam (400–700 nm).

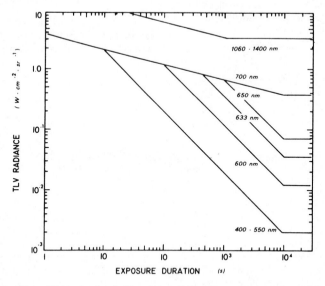

Figure 3b — TLV for intrabeam (direct) viewing of CW laser beam (400–1400 nm).

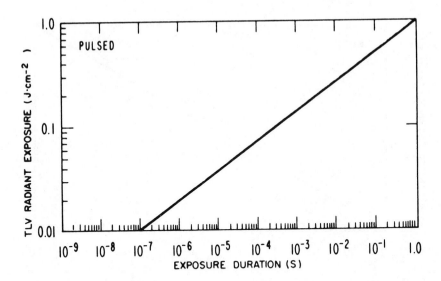

Figure 4a — TLV for laser exposure of skin and eyes for far-infrared radiation (wavelengths greater than 1.4 μm).

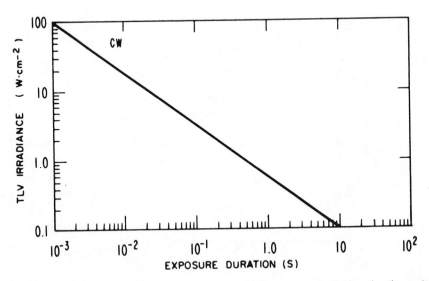

Figure 4b — TLV for CW laser exposure of skin and eyes for far-infrared radiation (wavelengths greater than 1.4 μm).

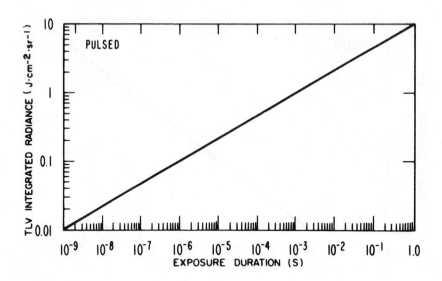

Figure 5a — TLV for extended sources or diffuse reflections of laser radiation (400–700 nm).

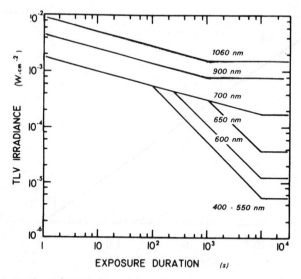

Figure 5b — TLV for extended sources or diffuse reflections of laser radiation (400–1400 nm), CW.

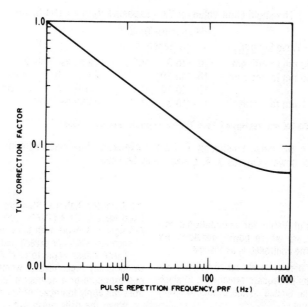

Figure 6 — Multiplicative correction factor for repetitively pulsed lasers having pulse durations less than 10^{-5} second. TLV for a single pulse of the pulse train is multiplied by the above correction factor. Correction factor for PRF greater than 1000 H_z is 0.06.

TABLE 5

Limiting Angle to Extended Source
Which May Be Used for Applying Extended Source TLVs

Exposure Duration(s)	Angle α (mrad)
10^{-9}	8.0
10^{-8}	5.4
10^{-7}	3.7
10^{-6}	2.5
10^{-5}	1.7
10^{-4}	2.2
10^{-3}	3.6
10^{-2}	5.7
10^{-1}	9.2
1.0	15
10	24
10^{2}	24
10^{3}	24
10^{4}	24

TABLE 6

Threshold Limit Value for Skin Exposure from a Laser Beam

Spectral Region	Wave Length	Exposure Time, (t) Seconds	TLV
UV	200 nm to 400 nm	10^{-3} to 3×10^4	Same as Table 3
Light &	400 nm to 1400 nm	10^{-9} to 10^{-7}	$2\,C_A \times 10^{-2}$ J • cm^{-2}
IR-A	" "	10^{-7} to 10	$1.1\,C_A \sqrt{t}$ J • cm^{-2}
IR-B & C	1.4 μm to 1 mm	10^{-9} to 3×10^4	Same as Table 3

$C_A = 1.0$ for $\lambda = 400$–700 nm; see Figure 2 Laser TLV list for greater wavelength values.

NOTE: To aid in the determination of TLV's for exposure durations requiring calculations of fractional powers Figures 3, 4, 5 and 6 may be used.

Recommended Values

The Threshold Limit Value for occupational microwave energy exposure where power densities are known and exposure time controlled is as follows:

1. For average power density levels up to but not exceeding 10 milliwatts per square centimeter, total exposure time shall be limited to the 8-hour workday (continuous exposure).
2. For average power density levels from 10 milliwatts per square centimeter up to but not exceeding 25 milliwatts per square centimeter, total exposure time shall be limited to no more than 10 minutes for any 60 minute period during an 8-hour workday (intermittent exposure).
3. For average power density levels in excess of 25 milliwatts per square centimeter, exposure is not permissible.

NOTE: For repetitively pulsed sources the average power density may be calculated by multiplying the peak power density by the duty cycle. The duty cycle is equal to the pulse duration in seconds times the pulse repetition rate in hertz.

NOISE

These threshold limit values refer to sound pressure levels and durations of exposure that represent conditions under which it is believed that nearly all workers may be repeatedly exposed without adverse effect on their ability to hear and understand normal speech. The medical profession has defined hearing impairment as an average hearing threshold level in excess of 25 decibels (ANSI-S3.6-1969) at 500, 1000, and 2000 Hz, and the limits which are given have been established to prevent a hearing loss in excess of this level. The values should be used as guides in the control of noise exposure and, due to individual susceptibility, should not be regarded as fine lines between safe and dangerous levels.

Continuous or Intermittent

The sound level shall be determined by a sound level meter, conforming as a minimum to the requirements of the American National Standard Specification for Sound Level Meters, S1.4 (1971) Type S2A, and set to use the A-weighted network with slow meter response. Duration of exposure shall not exceed that shown in Table 7.

These values apply to total duration of exposure per working day regardless of whether this is one continuous exposure or a number of short-term exposures but does not apply to impact or impulsive type of noise.

When the daily noise exposure is composed of two or more periods of noise exposure of different levels, their combined effect should be considered, rather than the individual effect of each. If the sum of the following fractions:

$$\frac{C_1}{T_1} + \frac{C_2}{T_2} + \ldots \frac{C_n}{T_n}$$

exceeds unity, then, the mixed exposure should be considered to exceed the threshold limit value, C_1 indicates the total duration of exposure at a specific noise level, and T_1 indicates the total duration of exposure permitted at that level. All on-the-job noise exposures of 80 dBA or greater shall be used in the above calculations.

Table 7
Threshold Limit Values

Duration per day Hours	Sound Level dBA[a]
16	80
8	85
4	90
2	95
1	100
1/2	105
1/4	110
1/8	115 *

*No exposure to continuous or intermittent in excess of 115 dBA

a) Sound level in decibels as measured on a sound level meter, conforming as a minimum to the requirements of the American National Standard Specification for Sound Level Meters, S1.4 (1971) Type S2A, and set to use the A-weighted network with slow meter response.

IMPULSIVE OR IMPACT NOISE

It is recommended that exposure to impulsive or impact noise shall not exceed the limits listed in Table 8 or taken from Figure 7. No exposures in excess of 140 decibels peak sound pressure level are permitted. Impulsive or impact noise is considered to be those variations in noise levels that involve maxima at intervals of greater than one per second. Where the intervals are less than one second, it should be considered continuous.

Table 8
Threshold Limit Values
Impulsive or Impact Noise

Sound Level dB*	Permitted Number of Impulses or Impacts per day
140	100
130	1000
120	10,000

*It should be recognized that the application of the TLV for noise will not protect all workers from the adverse effects of noise exposure. A hearing conservation program with audiometric testing is necessary when workers are exposed to noise at or above the TLV levels.
**Decibels peak sound pressure level.

ULTRAVIOLET RADIATION*

These threshold limit values refer to ultraviolet radiation in the spectral region between 200 and 400 nm and represent conditions under which it is believed that nearly all workers may be repeatedly exposed without adverse effect. These values for exposure of the eye or the skin apply to ultraviolet radiation from arcs, gas, and vapor discharges, fluorescent, and incandescent sources, and solar radiation, but do not apply to ultraviolet lasers.* These levels should not be used for determining exposure of photosensitive individuals to ultraviolet radiation. These values should be used as guides in the control of exposure to continuous sources where the exposure duration shall not be less than 0.1 sec.

These values should be used as guides in the control of exposure to ultraviolet sources and should not be regarded as a fine line between safe and dangerous levels.

Recommended Values:

The threshold limit value for occupational exposure to ultraviolet radiation incident upon skin or eye where irradiance values are known and exposure time is controlled are as follows:

1. For the near ultraviolet spectral region (320 to 400 nm) total irradiance incident upon the unprotected skin or eye should not exceed 1 mw/cm² for periods greater than 10^3 seconds (approximately 16 minutes) and for exposure times less than 10^3 seconds should not exceed one J/cm².

2. For the actinic ultraviolet spectral region (200 — 315 nm), radiant exposure incident upon the unprotected skin or eye should not exceed the values given in Table 9 within an 8-hour period.

3. To determine the effective irradiance of a broadband source weighted against the peak of the spectral effectiveness curve (270 nm), the following weighting formula should be used:

$$E_{eff} = \sum E_\lambda S_\lambda \Delta\lambda$$

*See Laser TLVs.

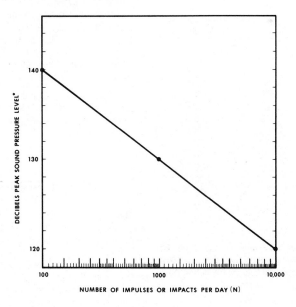

Figure 7 —

TABLE 9
Relative Spectral Effectiveness by Wavelength

Wavelength (nm)	TLV (mJ/cm²)**	Relative Spectral Effectiveness S_λ
200	100	0.03
210	40	0.075
220	25	0.12
230	16	0.19
240	10	0.30
250	7.0	0.43
254	6.0	0.5
260	4.6	0.65
270	3.0	1.0
280	3.4	0.88
290	4.7	0.64
300	10	0.30
305	50	0.06
310	200	0.015
315	1000	0.003

**1 m J/cm² = 10^{-3} J/cm²

TABLE 10
Permissible Ultraviolet Exposures

Duration of Exposure Per Day	Effective Irradiance, E_{eff} (μW/cm²)***
8 hrs.	0.1
4 hrs.	0.2
2 hrs.	0.4
1 hr.	0.8
30 min.	1.7
15 min.	3.3
10 min.	5
5 min.	10
1 min.	50
30 sec.	100
10 sec.	300
1 sec.	3,000
0.5 sec.	6,000
0.1 sec.	30,000

All the preceding TLVs for ultraviolet energy apply to sources which subtend an angle less than 80°. Sources which subtend a greater angle need to be measured only over an angle of 80°.

***1 μW/cm² = 10^{-6} W/cm²

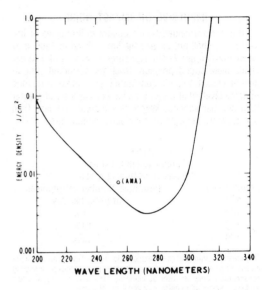

Figure 8 — Threshold Limit Values for Ultraviolet Radiation

where:

E_{eff} = effective irradiance relative to a monochromatic source at 270 nm in W/cm² (J/s/cm²)

E_λ = spectral irradiance in W/cm²/nm

S_λ = relative spectral effectiveness (unitless)

$\Delta\lambda$ = band width in nanometers

4. Permissible exposure time in seconds for exposure to actinic ultraviolet radiation incident upon the unprotected skin or eye may be computed by dividing 0.003 J/cm² by E_{eff} in W/cm². The exposure time may also be determined using Table 10 which provides exposure times corresponding to effective irradiances in μW/cm².

Conditioned (tanned) individuals can tolerate skin exposure in excess of the TLV without erythemal effects. However, such conditioning may not protect persons against skin cancer.

NOTICE OF INTENDED CHANGES
(for 1977)

These physical agents, with their corresponding values, comprise those for which either a limit has been proposed for the first time, or for which a change in the "Adopted" listing has been proposed. In both cases, the proposed limits should be considered trial limits that will remain in the listing for a period of at least one year. If after one year no evidence comes to light that questions the appropriateness of the values herein the values will be reconsidered for the "Adopted" list.

MICROWAVES

These Threshold Limit Values refer to microwave energy in the frequency range of 300 MHz to 300 GHz and represent conditions under which it is believed that nearly all workers may be repeatedly exposed without adverse effect.

Under conditions of moderate to severe heat stress, the recommended values may need to be reduced.[*] Therefore, these values should be used as guides in the control of exposure to microwave energy and should not be regarded as a fine line between safe and dangerous levels.

Recommended Values:

The Threshold Limit Value for occupational exposure to microwave energy, where power density or field intensity is known and exposure time is controlled, is as follows:

1. For exposure to continuous wave (CW) sources, the power density level shall not exceed 10 milliwatts per square centimeter (mW/cm^2) for continuous exposure, and the total exposure time shall be limited to an 8-hour workday. This power density is approximately equivalent to a free-space electric field strength of 200 volts-per-meter rms (V/m) and a free-space magnetic field strength of 0.5 ampere-per-meter rms (A/m).

2. Exposures to CW power density levels greater than 10 mW/cm^2 are permissible up to a maximum of 25 mW/cm^2 based upon an average energy density of 1 milliwatt-hour per square centimeter (mWh/cm^2) averaged over any 0.1 hour period. For example, at 25 mW/cm^2, the permissible exposure duration is approximately 2.4 minutes in any 0.1 hour period.

3. For repetitively pulsed microwave sources, the average field strength or power density is calculated by multiplying the peak-pulse value by the duty cycle. The duty cycle is equal to the pulse duration in seconds times the pulse repetition rate in Hertz. Exposure during an 8-hour workday shall not exceed the following values which are averaged over any 0.1 hour period:

Power Density	10 mW/cm^2
Energy Density	1 mWh/cm^2
Mean Squared Electric Field Strength	40,000 V^2/m^2
Mean Squared Magnetic Field Strength	0.25 A^2/m^2

4. Exposure is not permissible in CW or repetitively pulsed fields with an average power density in excess of 25 mW/cm^2 or approximate equivalent free-space field strengths of 300 V/m or 0.75 A/m.

---

*Mumford, W.W., "Heat Stress Due to R. F. Radiation," Proceedings of IEEE, Vol. 57, No. 2, Feb. 1969, pp. 171–178.

NOTICE OF INTENT TO ESTABLISH THRESHOLD LIMIT VALUES

LIGHT AND NEAR-INFRARED RADIATION

These Threshold Limit Values refer to visible and near-infrared radiation in the wavelength range of 400 nm to 1400 nm and represent conditions under which it is believed that nearly all workers may be exposed without adverse effect. These values should be used as guides in the control of exposure to light and should not be regarded as a fine line between safe and dangerous levels.

Recommended Values:

The Threshold Limit Value for occupational exposure to broad-band light and near-infrared radiation for the eye apply to exposure in any eight-hour workday and require knowledge of the spectral radiance (L_λ) and total irradiance (E) of the source as measured at the position(s) of the eye of the worker. Such detailed spectral data of a white light source is generally only required if the luminance of the source exceeds 1 cd cm^{-2}. At luminances less than this value the TLV would not be exceeded.

TABLE 11

SPECTRAL WEIGHTING FUNCTIONS FOR ASSESSING RETINAL HAZARDS FROM BROAD — BAND OPTICAL SOURCES

Wavelength (nm)	Blue-Light Hazard Function B_λ	Burn Hazard Function R_λ
400	0.10	1.0
405	0.20	2.0
410	0.40	4.0
415	0.80	8.0
420	0.90	9.0
425	0.95	9.5
430	0.98	9.8
435	1.0	10
440	1.0	10
445	0.97	9.7
450	0.94	9.4
455	0.90	9.0
460	0.80	8.0
465	0.70	7.0
470	0.62	6.2
475	0.55	5.5
480	0.45	4.5
485	0.40	4.0
490	0.22	2.2
495	0.16	1.6
500–600	$10^{[(450-\lambda)/50]}$	1.0
600–700	0.001	1.0
700–1060	0.001	$10^{[(700-\lambda)/515]}$
1060–1400	0.001	0.2

The TLV's are:

1. To protect against retinal thermal injury, the spectral radiance of the lamp weighted against the function R (Table 11) should not exceed:

$$\sum_{400}^{1400} L_\lambda R_\lambda \Delta\lambda \le \sqrt{t}/\alpha t \qquad (1)$$

where L_λ is in W cm^{-2} sr^{-1} and t is the viewing duration (or pulse duration if the lamp is pulsed) limited to 1 μs to 10 s, and α is the angular subtense of the source in radians. If the lamp is oblong, α refers to the longest dimension that can be viewed. For instance, at a viewing distance r = 100 cm from a tubular lamp of length l = 50 cm, the viewing angle is:

$$\alpha = l/r = 50/100 = 0.5 \text{ rad} \qquad (2)$$

2. To protect against retinal injury from chronic blue-light exposure the integrated spectral radiance of the lamp weighted against the blue-light hazard function B_λ (Table 11) should not exceed:

$$\sum_{400}^{1400} L_\lambda t B_\lambda \Delta\lambda \le 100 \text{ Jcm}^{-2} \text{ sr}^{-1} \ (t \le 10^4 s) \qquad (3a)$$

$$\sum_{400}^{1400} L_\lambda B_\lambda \Delta\lambda \le 10^{-2} \text{ Wcm}^{-2} \text{ sr}^{-1} \ (t > 10^4 s) \qquad (3b)$$

For a source radiance L which exceeds 2 mW cm^{-2} sr^{-1} in the blue spectral region, the permissible exposure duration t_{max} in seconds is simply:

$$t_{max} = 100 \text{ J cm}^{-2} \text{ sr}^{-1}/L \text{ (blue)} \qquad (4)$$

The latter limits are greater than the maximum permissible exposure limits for 440 nm laser radiation (see Laser TLV) because a 2–3 mm pupil is assumed rather than a 7 mm pupil for the Laser TLV.

3. *Infrared Radiation:* To avoid possible delayed effects upon the lens of the eye (cataractogenesis), the infrared radiation ($\lambda > 770$ nm) should be limited to 10 mWcm^{-2}. For an infrared heat lamp or any near-infrared source where a strong visual stimulus is absent, the near infrared (770–1400 nm) radiance as viewed by the eye should be limited to:

$$\sum_{770}^{1400} E_\lambda \Delta_\lambda = 600/\alpha \qquad (5)$$

for extended duration viewing conditions. This limit is based upon a 7 mm pupil diameter.

AIRBORNE UPPER SONIC AND ULTRASONIC ACOUSTIC RADIATION

These threshold limit values refer to sound pressure levels that represent conditions under which it is be-

lieved that nearly all workers may be repeatedly exposed without adverse effect. The values listed in Table 15 should be used as guides in the control of noise exposure and, due to individual susceptibility, should not be regarded as fine lines between safe and dangerous levels. The levels for the third octave bands centered below 20 kHz are below those which cause subjective effects. Those levels for 1/3 octaves above 20 kHz are for prevention of possible hearing losses from subharmonics of these frequencies.

TABLE 12

Permissible Ultrasound Exposure Levels

Mid-Frequency of Third-Octave Band kHz	One-Third Octave — Band Level in dB reference 0.0002 dynes/cm²
10	80
12.5	80
16	80
20	105
25	110
31.5	115
40	115
50	115

NOTICE OF INTENT TO STUDY

These agents comprise those which the Physical Agents Committee of ACGIH proposes to study during this year to determine the feasibility of establishing proposed TLVs in 1977 Comments and suggestions, accompanied by substantive evidence, are solicited.

1. *Radiofrequency Radiation.* Specifically, that portion of the spectrum from 10 MHz to 100 MHz.

2. *Microwave Radiation.* Specifically from 100 GHz to 300 GHz.

3. *Magnetic Fields.* Both pulsed and continuous.

4. *Laser Radiation.* Specifically ultraviolet radiation for pulsed exposures, and repetitively pulsed light and infrared-A laser exposures.

5. *Ultrasonic Energy.* Specifically, acoustic energy at frequencies above 10 kHz.

6. *Vibration.* Segmental and whole-body.

The American Conference of Governmental Industrial Hygienists was organized in 1938 by a group of governmental industrial hygienists who desired a medium for the free exchange of ideas, experiences and the promotion of standards and techniques in industrial health. The Conference is not an official Government Agency.

It is an organization devoted to the development of administrative and technical aspects of worker health protection. The association has contributed substantially to the development and improvement of official industrial health services to industry and labor. The committees on Industrial Ventilation and Threshold Limit Values are recognized throughout the world for their expertise and contributions to industrial hygiene.

Membership is limited to professional personnel in governmental agencies or educational institutions engaged in occupational safety and health programs. The more than 1600 members from across the United States and around the world give the organization an international scope.

Other useful sources of information about laboratory and industrial safety include:

A number of free safety booklets, published by laboratory chemical firms:

"Fisher Manual of Laboratory Safety," Fisher Scientific Co., Pittsburgh, PA, 1972.

"Safety in Handling Hazardous Chemicals," Matheson Scientific, Elk Grove Village, IL, 1971.

"A Condensed Laboratory Handbook." E.I. Du Pont de Nemours & Co., Wilmington, DE, 1971.

More completed information is provided by a number of books such as:

Sax, N. I., "Dangerous Properties of Industrial Materials," 3rd Ed., Van Nostrand Reinhold Co., New York, 1968.

Manufacturing Chemists Association, "Guide for Safety in the Chemical Laboratory," 2nd Ed., Van Nostrand Reinhold Co., New York, 1972.

MCB, "Gas Data Book," The Matheson Co., East Rutherford, NJ.

Steere, N. V., "Handbook of Laboratory Safety," 2nd Ed., Chemical Rubber Co., Cleveland, O, 1971.

Specialized data sheets and bulletins are available from the following organizations:

Manufacturing Chemists Association, 1825 Connecticut Ave., NW, Washington, DC 20009.

National Fire Protection Association, 60 Batterymarch St., Boston, MA 02110.

National Safety Council, 425 No. Michigan Ave., Chicago, IL 60611.

In closing, readers are urged to obtain information concerning hazardous chemicals and operations *first*, then plan their work with incorporation of an adequate safety margin, *before* starting laboratory work. There are still far too many accidents in laboratories and plants!

This book offers summaries of property data for a given property in Chapter I, then treating each substance in greater detail in later chapters. These later chapters are organized along physical and chemical lines as far as possible.

At the end of each chapter, and elsewhere when timely, references for data and a bibliography of additional sources of information for chapter topics are included.

Generally, the bulk of the information is taken from "References," though some information may have been obtained from literature, listed under "Bibliography." In addition, frequently consulted handbooks and encyclopedias are listed in this Section, rather than each time they are consulted.

Following is a list of general handbooks and encyclopedias touching on a broad number of areas covered in this book.

Besancon, R. M., Ed., "Encyclopedia of Physics," 2nd ed., Van Nostrand Reinhold Co., New York, 1974.

Brady, G. S., "Materials Handbook," 10th ed., McGraw-Hill, New York, 1971.

Childs, W. E., "Physical Constants," 9th ed., Halsted Press, New York, 1973.

Clark, G. L. and G. G. Hawley, Eds., "Encyclopedia of Chemistry," 2nd ed., Van Nostrand Reinhold Co., New York, 1966.

Clauser, H., Ed., "Encyclopedia of Engineering Materials and Processes," Van Nostrand Reinhold Co., New York, 1963.

Clauss, F. J., "Engineer's Guide to High-Temperature Materials," Addison-Wesley, Reading, Mass., 1969.

Cohen, E. R. et al., "Fundamental Constants of Physics," Wiley, New York, 1957.

Dean, J. A., Ed., "Lange's Handbook of Chemistry," 11th ed., McGraw-Hill, New York, 1973.

Fluegge, S. and E. Fluegge, Eds., "Encyclopedia of Physics," 54 vols., Springer-Verlag, New York, 1956-1973.

Forsythe, W. E., Ed., "Smithsonian Physical Tables," 9th rev. ed., "Smithsonian Institution, Washington, D.C., 1954.

Gardner, W. and E. I. Cooke, "Chemical Synonyms and Trade Names: a Dictionary and Commercial Handbook," 7th ed., International Publications Service, New York, 1971.

Gmelin, L., "Handbuch der anorganischen Chemie," Verlag Chemie, Weinheim, Germany, 1953-1963.

Goetzel, C. G., J. B. Rittenhouse and J. B. Singletary, "Space Materials Handbook," Addison-Wesley, Reading, Mass., 1965.

Gray, D. E., Ed., "American Institute of Physics Handbook," 3rd ed., McGraw-Hill, New York, 1972.

Hampel, C. A., Ed., "Encyclopedia of the Chemical Elements," Van Nostrand Reinhold Co., New York, 1968.

Hampel, C. A. and G. G. Hawley, "The Encyclopedia of Chemistry," 3rd ed., Van Nostrand Reinhold Co., New York, 1973.

Hellwege, K. A. and A. M. Hellwege, Eds., "Landolt-Börnstein Tables-Zahlenwerte und Funktionen aus Physik, Chemie, Astronomie, Geophysik und Technik," 24 vols., Springer-Verlag, New York, 1965-1973.

Herman, H., Ed., "Treatise on Materials Science and Technology," 3 vols., Academic Press, New York, 1972-1973.

Hittle, D. R. et al., "Sourcebook for Chemistry and Physics," Macmillan, New York, 1973.

"International Critical Tables," 7 vols., McGraw-Hill, New York, 1926-1933.

Jesson, J. P. and E. L. Mutterties, "Chemist's Guide: Basic Chemical and Physical Data," Marcel Dekker, New York, 1969.

Kaye, G. W. C. and T. H. Laby, Eds., "Tables of Physical and Chemical Constants," 13th ed., Wiley, New York, 1966.

Kent, J. A., Ed., "Riegel's Industrial Chemistry," 7th ed., Van Nostrand Reinhold Co., New York, 1973.

Kingzett, C. T., "Chemical Encyclopedia," 9th ed., Van Nostrand Reinhold Co., New York, 1967.

Kirk, R. E. and D. F. Olthmer, Eds., "Encyclopedia of Chemical Technology," 2nd ed., 24 vols., Wiley, New York, 1963-1972.

Lonung, P. Y., "Graphic Handbook of Chemistry and Metallurgy," Chemical Publishing Co., New York, 1965.

Mantell, C. E., Ed., "Engineering Materials Handbook," McGraw-Hill, New York, 1958.

Mellor, J. W., "A Comprehensive Treatise on Inorganic and Theoretical Chemistry," 16 vols. and suppl., Halsted Press, New York, 1922-1972.

Moss, J. B., "Properties of Engineering Materials," CRC Press, Cleveland, 1971.

Peters, R. L., "Materials Data Monographs," Van Nostrand Reinhold Co., New York, 1965.

Robson, J., "Basic Tables in Physics," McGraw-Hill, New York, 1967.

Shugar, G. J., "Chemical Technicians' Ready Reference Handbook," McGraw-Hill, New York, 1973.

Stecher, P. G., Ed., "The Merck Index," 8th ed., Merck & Co., Rahway, N.J., 1968.

Weast, R. C., Ed., "Handbook of Chemistry and Physics," 54th ed., Chemical Rubber Co., Cleveland, 1973.

Weissberger, A. and B. W. Rossiter, Eds., "Physical Methods of Chemistry," 10 parts, Wiley, New York, 1971-1972.

SECTION 1a
ATOMIC AND NUCLEAR STRUCTURE

This Section introduces the reader to the structure of elements and molecules and concludes with tables of particle properties. The information is presented in the following tables and figures:

Figure 1a1 Periodic Table of the Elements.

Table 1a1 Electron Configurations and Term Symbols of Atoms in their Normal States.

Table 1a2 Interatomic Distances in Elements and Molecules.

Table 1a3 Interatomic Distances in Organic Molecules.

Table 1a4 Covalent Radii of the Elements.

Table 1a5 Physical and Numerical Constants.

Table 1a6 Tables of Particle Properties.

IA	IIA										IB	IIB	IIIA	IVA	VA	VIA	VIIA	2 He 4.0026
							1 H 1.008											
3 Li 6.94	4 Be 9.0122												5 B 10.81	6 C 12.011	7 N 14.0067	8 O 15.9994	9 F 18.9984	10 Ne 20.17
11 Na 22.9898	12 Mg 24.305	IIIB	IVB	VB	VIB	VIIB	———	VIII	———		IB	IIB	13 Al 26.9815	14 Si 28.086	15 P 30.9738	16 S 32.06	17 Cl 35.453	18 A 39.948
19 K 39.102	20 Ca 40.08	21 Sc 44.956	22 Ti 47.90	23 V 50.941	24 Cr 51.996	25 Mn 54.9380	26 Fe 55.847	27 Co 58.9332	28 Ni 58.71	29 Cu 63.55	30 Zn 65.37		31 Ga 69.72	32 Ge 72.59	33 As 74.9216	34 Se 78.96	35 Br 79.904	36 Kr 83.80
37 Rb 85.47	38 Sr 87.62	39 Y 88.906	40 Zr 91.22	41 Nb 92.906	42 Mo 95.94	43 Tc 98.9062	44 Ru 101.07	45 Rh 102.905	46 Pd 106.4	47 Ag 107.868	48 Cd 112.40		49 In 114.82	50 Sn 118.69	51 Sb 121.75	52 Te 127.60	53 I 126.9045	54 Xe 131.30
55 Cs 132.905	56 Ba 137.34	57–71 Rare Earths	72 Hf 178.49	73 Ta 180.947	74 W 183.85	75 Re 186.2	76 Os 190.2	77 Ir 192.2	78 Pt 195.09	79 Au 196.967	80 Hg 200.59		81 Tl 204.37	82 Pb 207.2	83 Bi 208.981	84 Po (209)	85 At (210)	86 Rn (222)
87 Fr (223)	88 Ra (226)	89– Acti- nides	104	105														

57 La 138.91	58 Ce 140.12	59 Pr 140.908	60 Nd 144.24	61 Pm (145)	62 Sm 150.4	63 Eu 151.96	64 Gd 157.25	65 Tb 158.925	66 Dy 162.50	67 Ho 164.930	68 Er 167.26	69 Tm 168.934	70 Yb 173.04	71 Lu 174.97	Rare earths (Lanthanide series)
89 Ac (227)	90 Th 232.038	91 Pa 231.036	92 U 238.03	93 Np 237.048	94 Pu (244)	95 Am (243)	96 Cm (247)	97 Bk (247)	98 Cf (251)	99 Es (254)	100 Fm (257)	101 Md (256)	102 No (254)	103 Lr (260)	Actinide series

Numbers in parentheses are mass numbers of most stable isotope of that element. Adapted from the *Handbook of Chemistry and Physics, 1973-74.* (Particle Data Group update, April 1974.)

Figure 1a1. Periodic Table of the Elements. (Reprinted by permission of the Particle Data Group and the Editor of *Physics Letters*, from *Physics Letters*, Vol. **50B**, No. 1, April 1974.)

TABLE 1a1. ELECTRON CONFIGURATIONS AND TERM SYMBOLS OF ATOMS IN THEIR NORMAL STATES.*

Atomic Number	Element	Electronic Configuration	Ground State	Atomic Number	Element	Electronic Configuration	Ground State
1	H	$1s$	$^2S_{1/2}$	47	Ag	— $4d^{10}5s$	$^2S_{1/2}$
2	He	$1s^2$	$^1S_0$	48	Cd	— $4d^{10}5s^2$	$^1S_0$
3	Li	[He] $2s$	$^2S_{1/2}$	49	In	— $4d^{10}5s^25p$	$^2P_{1/2}$
4	Be	— $2s^2$	$^1S_0$	50	Sn	— $4d^{10}5s^25p^2$	$^3P_0$
5	B	— $2s^22p$	$^2P_{1/2}$	51	Sb	— $4d^{10}5s^25p^3$	$^4S_{3/2}$
6	C	— $2s^22p^2$	$^3P_0$	52	Te	— $4d^{10}5s^25p^4$	$^3P_2$
7	N	— $2s^22p^3$	$^4S_{3/2}$	53	I	— $4d^{10}5s^25p^5$	$^2P_{3/2}$
8	O	— $2s^22p^4$	$^3P_2$	54	Xe	— $4d^{10}5s^25p^6$	$^1S_0$
9	F	— $2s^22p^5$	$^2P_{3/2}$	55	Cs	[Xe] $6s$	$^2S_{1/2}$
10	Ne	— $2s^22p^6$	$^1S_0$	56	Ba	— $6s^2$	$^1S_0$
11	Na	[Ne] $3s$	$^2S_{1/2}$	57	La	— $5d6s^2$	$^2D_{3/2}$
12	Mg	— $3s^2$	$^1S_0$	58	Ce	— $4f^26s^2$	$^3H_4$
13	Al	— $3s^23p$	$^2P_{1/2}$	59	Pr	— $4f^36s^2$	$^4I_{9/2}$
14	Si	— $3s^23p^2$	$^3P_0$	60	Nd	— $4f^46s^2$	$^5I_4$
15	P	— $3s^23p^3$	$^4S_{3/2}$	61	Pm	— $4f^56s^2$	$^6H_{5/2}$
16	S	— $3s^23p^4$	$^3P_2$	62	Sm	— $4f^66s^2$	$^7F_0$
17	Cl	— $3s^23p^5$	$^2P_{3/2}$	63	Eu	— $4f^76s^2$	$^8S_{7/2}$
18	Ar	— $3s^23p^6$	$^1S_0$	64	Gd	— $4f^75d6s^2$	$^9D_2$
19	K	[Ar] $4s$	$^2S_{1/2}$	65	Tb	— $4f^96s^2$	$^6H_{15/2}$
20	Ca	— $4s^2$	$^1S_0$	66	Dy	— $4f^{10}6s^2$	$^5I_8$
21	Sc	— $3d4s^2$	$^2D_{3/2}$	67	Ho	— $4f^{11}6s^2$	$^4I_{15/2}$
22	Ti	— $3d^24s^2$	$^3F_2$	68	Er	— $4f^{12}6s^2$	$^3H_6$
23	V	— $3d^34s^2$	$^4F_{3/2}$	69	Tm	— $4f^{13}6s^2$	$^2F_{7/2}$
24	Cr	— $3d^54s$	$^7S_3$	70	Yb	— $4f^{14}6s^2$	$^1S_0$
25	Mn	— $3d^54s^2$	$^6S_{5/2}$	71	Lu	— $4f^{14}5d6s^2$	$^2D_{3/2}$
26	Fe	— $3d^64s^2$	$^5D_4$	72	Hf	— $4f^{14}5d^26s^2$	$^3F_2$
27	Co	— $3d^74s^2$	$^4F_{9/2}$	73	Ta	— $4f^{14}5d^36s^2$	$^4F_{3/2}$
28	Ni	— $3d^84s^2$	$^3F_4$	74	W	— $4f^{14}5d^46s^2$	$^5D_0$
29	Cu	— $3d^{10}4s$	$^2S_{1/2}$	75	Re	— $4f^{14}5d^56s^2$	$^6S_{5/2}$
30	Zn	— $3d^{10}3s^2$	$^1S_0$	76	Os	— $4f^{14}5d^66s^2$	$^5D_4$
31	Ga	— $3d^{10}4s^24p$	$^2P_{1/2}$	77	Ir	— $4f^{14}5d^76s^2$	$^4F_{9/2}$
32	Ge	— $3d^{10}4s^24p^2$	$^3P_0$	78	Pt	— $4f^{14}5d^96s$	$^3D_3$
33	As	— $3d^{10}4s^24p^3$	$^4S_{3/2}$	79	Au	[　] $6s$	$^2S_{1/2}$
34	Se	— $3d^{10}4s^24p^4$	$^3P_2$	80	Hg	— $6s^2$	$^1S_0$
35	Br	— $3d^{10}4s^24p^5$	$^2P_{3/2}$	81	Tl	— $6s^26p$	$^2P_{1/2}$
36	Kr	— $3d^{10}4s^24p^6$	$^1S_0$	82	Pb	— $6s^26p^2$	$^3P_0$
37	Rb	[Kr] $5s$	$^2S_{1/2}$	83	Bi	— $6s^26p^3$	$^4S_{3/2}$
38	Sr	— $5s^2$	$^1S_0$	84	Po	— $6s^26p^4$	$^3P_2$
39	Y	— $4d5s^2$	$^2D_{3/2}$	85	At	— $6s^26p^5$	$^2P_{3/2}$
40	Zr	— $4d^25s^2$	$^3F_2$	86	Rn	— $6s^26p^6$	$^1S_0$
41	Nb	— $4d^45s$	$^6D_{1/2}$	87	Fr	[Rn] $7s$	$^2S_{1/2}$
42	Mo	— $4d^55s$	$^7S_3$	88	Ra	— $7s^2$	$^1S_0$
43	Tc	— $4d^55s^2$	$^6S_{5/2}$	89	Ac	— $6d7s^2$	$^2D_{3/2}$
44	Ru	— $4d^75s$	$^5F_5$	90	Th	— $6d^27s^2$	$^3F_2$
45	Rh	— $4d^85s$	$^4F_{9/2}$	91	Pa	— $5f^26d7s^2$	$^4K_{11/2}$
46	Pd	— $4d^{10}$	$^1S_0$	92	U	— $5f^36d7s^2$	$^5L_6$

TABLE 1a1. (Continued)

Atomic Number	Element	Electronic Configuration	Ground State	Atomic Number	Element	Electronic Configuration	Ground State
93	Np	— $5f^46d7s^2$	$^6L_{11/2}$	98	Cf	— $5f^{10}7s^2$	$^5I_8$
94	Pu	— $5f^67s^2$	$^7F_0$	99	Es	— $5f^{11}7s^2$	$^4I_{15/2}$
95	Am	— $5f^77s^2$	$^8S_{7/2}$	100	Fm	— $5f^{12}7s^2$	$^3H_6$
96	Cm	— $5f^76d7s^2$	$^9D_2$	101	Md	— $5f^{13}7s^2$	$^2F_{7/2}$
97	Bk	— $5f^86d7s^2$	$^8H_{17/2}$	102	No	— $5f^{14}7s^2$	$^1S_0$
		or		103		— $5f^{14}6d7s^2$	$^2D_{5/2}$
		$5f^97s^2$	$^6H_{15/2}$				

*Reproduced from: Day, M. Clyde Jr. and Joe Selbin, "Theoretical Inorganic Chemistry," Van Nostrand Reinhold, New York, 1962.

TABLE 1a2. INTERATOMIC DISTANCES IN ELEMENTS AND MOLECULES.**

Molecule/Element	Bond	Interatomic Distance (Å)	Molecule/Element	Bond	Interatomic Distance (Å)
Actinium	Ac–Ac	3.756	BrF	Br–F	1.7556*
Aluminum	Al–Al	2.863	HBr	Br–H	1.408*
Antimony	Sb–Sb	2.90	DBr	Br–D	1.4144*
$SbBr_3$	Sb–Br	2.51	TBr	Br–T	1.4144*
$SbCl_3$	Sb–Cl	2.325	Cadmium	Cd–Cd	2.979
SbH_3	Sb–H	1.707	Calcium	Ca–Ca	3.947
SbD_3	Sb–D	1.702	Cerium	Ce–Ce	3.650
Arsenic	As–As	2.495	Cesium	Cs–Cs	5.309
$AsBr_3$	As–Br	2.33	Chlorine	Cl–Cl	1.988*
$AsCl_3$	As–Cl	2.161	ClF	Cl–F	1.628*
AsF_3	As–F	1.712	HCl	Cl–H	1.2745*
AsH_3	As–H	1.519	DCl	Cl–D	1.2746*
AsD_3	As–D	1.515	TCl	Cl–T	1.2746*
AsI_3	As–I	2.54	Chromium	Cr–Cr	2.498
As_4O_6	As–O	1.78	CrO_2Cl_2	Cr–Cl	2.12
As_4S_6	As–S	2.24	CrO	Cr–O	1.627*
Barium	Ba–Ba	4.347	Cobalt	Co–Co	2.506
Beryllium	Be–Be	2.226	Copper	Cu–Cu	2.556
Bismuth	Bi–Bi	3.095	$CuCl_3$	Cu–Cl	2.160
$BiBr_3$	Bi–Br	2.63	Dysprosium	Dy–Dy	3.503
$BiCl_3$	Bi–Cl	2.48	Erbium	Er–Er	3.468
BiH_3	Bi–H	1.809	Europium	Eu–Eu	3.989
BiD_3	Bi–D	1.805	Fluorine	F–F	1.418
Boron	B–B	1.589	HF	F–H	0.9170*
BBr_3	B–Br	1.87	Gadolinium	Gd–Gd	3.573
BCl_3	B–Cl	1.74	Gallium	Ga–Ga	2.442
BF_3	B–F	1.295	Germanium	Ge–Ge	2.450
B_2H_6	B–H	1.19	$GeBr_4$	Ge–Br	2.297
BN	B–N	1.281	$GeCl_4$	Ge–Cl	2.08
BO	B–O	1.2049*	GeF_4	Ge–F	1.68
BS	B–S	1.609*	GeH_4	Ge–H	1.527
Bromine	Br–Br	2.284	GeO	Ge–O	1.6507*
BrCl	Br–Cl	2.138*	Gold	Au–Au	2.884

TABLE 1a2. (Continued)

Molecule/Element	Bond	Interatomic Distance (Å)	Molecule/Element	Bond	Interatomic Distance (Å)
Hafnium (α-form)	Hf–Hf	3.127	F_2O	O–F	1.42
Helium	He–He	1.080	H_2O	O–H	0.957*
Holmium	Ho–Ho	3.486	Palladium	Pd–Pd	2.751
Hydrogen	H–H	0.74130*	Phosphorus (P_2)	P–P	1.8931*
Deuterium	D–D	0.74143*	PBr_3	P–Br	2.18
Indium	In–In	3.251	PCl_3	P–Cl	2.043
Iodine	I–I	2.6666*	PF_3	P–F	1.535
ICl	I–Cl	2.3208*	PH_3	P–H	1.415
HI	I–H	1.609*	PI_3	P–I	2.43
Iridium	Ir–Ir	2.714	PN	P–N	1.4910*
IrF_6	Ir–F	1.833	P_4O_6	P–O	1.66
Iron	Fe–Fe	2.482	Platinum	Pt–Pt	2.775
Lanthanum	La–La	3.739	Plutonium	Pu–Pu	3.026
Lead	Pb–Pb	3.500	PuF_6	Pu–F	1.969
$PbCl_2$	Pb–Cl	2.46	Polonium	Po–Po	3.359
PbH	Pb–H	1.839*	Potassium	K–K	4.544
PbI_2	Pb–I	2.79	KBr	K–Br	2.8207*
PbO	Pb–O	1.922	KCl	K–Cl	2.6666*
PbS	Pb–S	2.3948	KH	K–H	2.244*
Lithium	Li–Li	3.039	KI	K–I	3.0478*
LiH	Li–H	1.5953*	Praseodymium	Pr–Pr	3.640
LiI	Li–I	2.3919*	Protactinium	Pa–Pa	3.212
Lutetium	Lu–Lu	3.435	Rhenium	Re–Re	2.714
Magnesium	Mg–Mg	3.197	ReO_3Cl	Re–Cl	2.230
$MgCl_2$	Mg–Cl	2.18	ReO_3F	Re–F	1.86
MgF_2	Mg–F	1.77	ReO_3F	Re–O	1.692
MgO	Mg–O	1.749*	Rhodium	Rh–Rh	2.6901
Manganese			Rubidium	Rb–Rb	4.95
(1095°C)	Mn–Mn	2.731	RbBr	Rb–Br	2.9448*
MnO_3F	Mn–F	1.724	RbCl	Rb–Cl	2.7867*
MnH	Mn–H	1.7308*	RbF	Rb–F	2.2655*
MnO_3F	Mn–O	1.586	RbH	Rb–H	2.367*
Mercury	Hg–Hg	3.005	RbI	Rb–I	3.1769*
$HgBr_2$	Hg–Br	2.41	Ruthenium	Ru–Ru	2.650
$HgCl_1$	Hg–Cl	2.29	Scandium	Sc–Sc	3.212
HgH	Hg–H	1.7404*	Selenium	Se–Se	2.321
HgI_2	Hg–I	2.59	H_2Se	Se–H	1.46*
Molybdenum	Mo–Mo	2.725	Silicon	Si–Si	2.352
Neodymium	Nd–Nd	3.628	$SiBr_4$	Si–Br	2.16
Neptunium	Np–Np	2.60–2.64	$SiCl_4$	Si–Cl	2.019
Nickel	Ni–Ni	2.492	SiF_4	Si–F	1.561
Niobium	Nb–Nb	2.858	SiH_4	Si–H	1.480
Nitrogen	N–N	1.0976*	SiI_4	Si–I	2.435
NOBr	N–Br	2.14	SiN	Si–N	1.5718*
NOCl	N–Cl	1.97	SiO	Si–O	1.509*
NF_3	N–F	1.36	Silver	Ag–Ag	2.889
NH_3	N–H	1.0124*	AgCl	Ag–Cl	2.250*
ND_3	N–D	1.0108*	AgI	Ag–I	2.544*
NO_2	N–O	1.192	Sodium	Na–Na	3.716
NO	N–O	1.1502*	NaBr	Na–Br	2.5020*
Osmium	Os–Os	2.675	NaCl	Na–Cl	2.3606*
OsF_6	Os–F	1.830	NaH	Na–H	1.8873*
Oxygen	O–O	1.20741*	NaI	Na–I	2.7115*
OBr	O–Br	1.65	Strontium	Sr–Sr	4.303
OCl	O–Cl	1.546	Sulfur, (S_8)	S–S	2.05

TABLE 1a2. (Continued)

Molecule/Element	Bond	Interatomic Distance (Å)	Molecule/Element	Bond	Interatomic Distance (Å)
$SOBr_2$	S–Br	2.27	SnI_4	Sn–I	2.69
SCl_2	S–Cl	1.99	SnO_2	Sn–O	2.05
SOF_2	S–F	1.585	Titanium	Ti–Ti	2.896
SH	S–H	1.350	$TiBr_4$	Ti–Br	2.31
SO_2	S–O	1.4321	$TiCl_4$	Ti–Cl	2.185
Tantalum	Ta–Ta	2.86	TiO	Ti–O	1.620*
$TaCl_5$	Ta–Cl	2.30	Tungsten	W–W	2.741
TaO	Ta–O	1.687*	$W(CO)_6$	W–C	2.06
Technetium	Tc–Tc	2.703	WF_6	W–F	1.826
Tellurium	Te–Te	2.864	Uranium	U–U	2.77
$TeBr_2$	Te–Br	2.51	UF_6	U–F	1.994
$TeCl_4$	Te–Cl	2.33	Vanadium	V–V	2.622
TeF_6	Te–F	1.84	VCl_4	V–Cl	2.03
Terbium	Tb–Tb	3.525	Ytterbium	Yb–Yb	3.880
Thallium	Tl–Tl	3.408	Yttrium	Y–Y	3.551
$TlBr_3$	Tl–Br	2.46	YO	Y–O	1.790
$TlCl_3$	Tl–Cl	2.33	Zinc	Zn–Zn	2.665
Thorium	Th–Th	3.595	ZnH	Zn–H	1.594*
Thulium	Tm–Tm	3.447	ZnI_2	Zn–I	2.42
Tin	Sn–Sn	2.810	Zirconium	Zr–Zr	3.179
$SnBr_2$	Sn–Br	2.55	$ZrCl_4$	Zr–Cl	2.32
$SnCl_2$	Sn–Cl	2.42	ZrO	Zr–O	1.728*
SnH_4	Sn–H	1.701			

*r_e = internuclear separation with no vibration; all other values represent internuclear separations for zero-point vibration.

**Values selected from Sutton, L. E., Scientific Ed., "Tables of Interatomic Distances and Configuration in Molecules and Ions," Supplement 1956–1959, Special Publication No. 18, The Chemical Society, London, 1965.

TABLE 1a3. INTERATOMIC DISTANCES IN ORGANIC COMPOUNDS.*

Compound	Bond	Interatomic Distance (Å)	Compound	Bond	Interatomic Distance (Å)
CBr_4	C–Br	1.942	$CHCl_3$	C–H	1.073
CH_2CHBr	C–Br	1.891	CH_2CH_2	C–H	1.085
C_6H_5Br	C–Br	1.86	C_6H_6	C–H	1.084
CHCBr	C–Br	1.795	C_2H_2	C–H	1.059
CCl_4	C–Cl	1.766	CH_3I	C–I	2.139
$CHCl_3$	C–Cl	1.762	CH_3CHI	C–I	2.092
CH_2CHCl	C–Cl	1.719	C_6H_5I	C–I	2.05
C_6H_5Cl	C–Cl	1.70	CH_3CCI	C–I	1.99
CHCCl	C–Cl	1.632	CH_3NH_2	C–N	1.474
CH_3F	C–F	1.385	$C_2H_5NO_2$	C–N	1.479
CH_2CHF	C–F	1.344	HCN	C–N	1.1554
C_6H_5F	C–F	1.332	CH_3OH	C–O	1.428
$Fe(CO)_5$	C–Fe	1.84	CH_2CH_2O (Epoxide)	C–O	1.435
CH_4	C–H	1.094	HCO_2H	C–O	1.312
CD_4	C–D	1.092	CO	C–O	1.128
CH_3Br	C–H	1.095	CO_2	C–O	1.1618
CH_2Cl_2	C–H	1.068	$(CH_3)_3P$	C–P	1.841

TABLE 1a3. (Continued)

Compound	Bond	Interatomic Distance (Å)	Compound	Bond	Interatomic Distance (Å)
CH_3SH	C–S	1.819	CH_3SiH_3	C–Si	1.867
$(CH_3)_2SO$	C–S	1.80	CH_3CH_3	C–C	1.534
CH_4N_2S	C–S	1.71	CH_2CH_2	C–C	1.339
CS_2	C–S	1.5530	CHCH	C–C	1.204
$(CH_3)_2Se$	C–Se	1.98			

*Values selected from Sutton, L. E., Scientific Ed., "Tables of Interatomic Distances and Configuration in Molecules and Ions," Supplement 1956–1959, Special Publication No. 18, The Chemical Society, London, 1965.

TABLE 1a4. COVALENT RADII OF THE ELEMENTS (IN ANGSTROM UNITS)*

Element	Valence	Bond Order	Hybridization	Radius	Element	Valence	Bond Order	Hybridization	Radius
H	1	1	s	0.3754_5	P	3	1	p	1.113_5
Li	1	1	s	1.336		5	1	sp^2	1.11
Be	2	1	s	0.86			1	sp^3	1.12_5
		1	sp^3	1.07			1	pd	1.23
B	1	1	p	0.79			1	sp^3d^2	1.20
	3	1	sp^2	0.84			2	sp^3	0.872
		1	sp^3	0.92	S	2,4	1	p	1.06_8
C	4	1	sp	0.691		6	1	sp^3	1.03
		1	sp^2	0.74			1	sp^3d^2	1.10
		1	sp^3	0.772			2	$p(p\pi)$	0.914
		2	sp	0.643		4	2	$p(pd\pi)$	0.868
		2	sp^2	0.666		6	2	sp^3	0.757
		3	p	0.60	Cl	1	1	p	1.050
		3	sp	0.602		3	1	pd	1.135
N	3	1	p	0.73_7		7	2	sp^3	0.681
	5	1	sp	0.700	K	1	1	s	1.962
		1	sp^2	0.727	Ca	2	1	s	1.39
		1	sp^3	0.74_9	Sc				
	3	2	p	0.61	Ti	4	1	d^3s	1.25
	5	2	sp	0.617			1	d^5s	1.43
	3	3	p	0.60	V	4	1	d^3s	1.10
	5	3	sp	0.638		5	1	d^3s	1.17
O	2	1	p	0.745			2	d^3s	1.05
		2	p	0.654	Cr	3	1	d^2sp^3	1.45
	(6)	2	sp	0.58		6	1	d^3s	1.16
		3	p	0.599			2	d^3s	1.03
F	1	1	p	0.709	Mn	2	1	d^2sp^3	1.6
Na	1	1	s	1.539		4	1	d^2sp^3	1.20
Mg	2	1	s	1.20		7	1	d^3s	1.13
Al	1	1	p	1.22			2	d^3s	1.02
	3	1	sp^2	1.24	Fe	3	1	d^2sp^3	1.39
		1	sp^3	1.26	Co	2	1	sp^3	1.55
		1	sp^3d^2	1.44		3	1	d^2sp^3	1.35
Si	4	1	sp^3	1.176	Ni	2	1	dsp^2	1.28
		1	sp^3d^2	1.31			1	d^2sp^3	1.54
		2	sp^3	1.00	Cu	1	1	s	1.22

TABLE 1a4. (*Continued*)

Element	Valence	Bond Order	Hybridization	Radius	Element	Valence	Bond Order	Hybridization	Radius
		1	sp^3	1.38			1	sp^3d^2	1.47
	2	1	dsp^2	1.29	Sb	3	1	p	1.376
		1	d^2sp^3	1.43		5	1	sp^2	1.43
Zn	2	1	sp	1.15(?)			1	sp^3d^2	1.42(?)
		1	sp^3	1.34			1	pd	1.48
		1	sp^3d^2	1.46	Te	2	1	p	1.39
Ga	1	1	p	1.29		4	1	p^3d	1.34(?)
	3	1	sp^2	1.23			1	p^3d^3	1.55
		1	sp^3	1.27		6	1	sp^3d^2	1.36(?)
Ge	4	1	sp^3	1.225		2	2	p	1.279
		1	sp^3d^2	1.3	I	1	1	p	1.360
As	3	1	p	1.218		5	1	pd	1.42
	5	1	sp^3	1.19	Cs	1	1	s	2.18
		1	sp^3d^2	1.33(?)	Ba	2	1	s	1.52
Se	2	1	p	1.21_5	La				
	4	1	pd	1.38	Hf				
	6	1	sp^3d^2	1.22(?)	Ta	5	1	d^4s	1.37(av.)
	2	2	p	1.075			1	d^5sp	1.47(av.)
Br	1	1	p	1.193_5			1	d^5sp^2	1.48(av.)
	5	1	pd	1.28	W	6	1	d^5s	1.32
Rb	1	1	s	2.06	Re	4	1	d^2sp^3	1.44
Sr	2	1	s	1.49		6	1	d^2sp^3	1.37
Y	3					7	1	d^3s	1.25
Zr	4	1	d^3s	1.42		2		d^3s	1.22
		1	d^5s	1.53	Os	4	1	d^2sp^3	1.40
Nb	5	1	d^4s	1.37(av.)		8	2	d^3s	1.171
		1	d^5s	1.5	Ir	4	1	d^2sp^3	1.50(?)
Mo	5	1	d^4s	1.30(av.)	Pt	2	1	dsp^2	1.35
	6	1	d^3s	1.31		4	1	d^2sp^3	1.34
		1	d^5s	1.27	Au	1	1	s	1.24
		2	d^3s	1.23			1	sp	1.34
Tc						3	1	dsp^2	1.39
Ru	4	1	d^2sp^3	1.38	Hg	2	1	$s(Hg_2^{+2})$	1.27
	7	2	d^3s	1.23			1	sp	1.33
	8	2	d^3s	1.14			1	sp^3	1.54
Rh	3	1	d^2sp^3	1.48	Tl	1	1	p	1.54
Pd	2	1	dsp^2	1.30		3	1	sp^3d^2	1.58
Ag	1	1	s	1.32	Pb	2	1	p	1.50
		1	sp	1.42		4	1	sp^3	1.44
		1	sp^3	1.48		4	1	sp^3d^2	1.51
Cd	2	1	sp	1.47	Bi	3	1	p	1.53
		1	sp^3d^2	1.62			1	p^3d^3	1.58
In	1	1	p	1.47	Po	4	1	p^3d^3	1.58
	3	1	sp^2	1.47	Th	4	1	d^3s	1.69
		1	sp^3	1.43	U	6	1	d^2sp^3	1.50
		1	sp^3d^2	1.66			2	dp	1.41
Sn	2	1	p	1.45	Pu	4	1	d^2sp^3	1.72
	4	1	sp^3	1.405	Am	5	2	dp	1.42

*Reprinted from: Clifford, A. F., *et al.*, Eds., "International Encyclopedia of Chemical Science," Van Nostrand, Princeton, 1964.

TABLE 1a5. PHYSICAL AND NUMERICAL CONSTANTS.*†

PHYSICAL CONSTANTS

		Uncert. (ppm)
N	$= 6.022045(31) \times 10^{23}$ mole^{-1}	5.1
c	$= 2.99792458(1.2) \times 10^{10}$ cm sec^{-1}	0.004
e	$= 4.803242(14) \times 10^{-10}$ esu $= 1.6021892(46) \times 10^{-19}$ coulomb	2.9; 2.9
1 MeV	$= 1.6021892(46) \times 10^{-6}$ erg	2.9
$\hbar = h/2\pi$	$= 6.582173(17) \times 10^{-22}$ MeV sec $= 1.0545887(57) \times 10^{-27}$ erg sec	2.6; 5.4
$\hbar c$	$= 1.9732857(51) \times 10^{-11}$ MeV cm $= 197.32857(51)$ MeV Fermi	2.6; 2.6
	$= 0.6240078(16)$ GeV mb$^{1/2}$	2.6
α	$= e^2/\hbar c = 1/137.03604(11)$	0.82
$k_{Boltzmann}$	$= 1.380662(44) \times 10^{-16}$ erg $^\circ$K^{-1}	32
	$= 8.61735(27) \times 10^{-11}$ MeV $^\circ$K^{-1} $= 1$ eV/$11604.50(36)^\circ$K	31; 31
m_e	$= 0.5110034(14)$ MeV $= 9.109534(47) \times 10^{-31}$ kg	2.8; 5.1
m_p	$= 938.2796(27)$ MeV $= 1836.15152(70) m_e = 6.72270(31)\, m_{\pi\pm}$	2.8; 0.38; 46
	$= 1.007276470(11) m_1$ ($m_1 \equiv 1$ amu $= \frac{1}{12} m_{C12} = 931.5016(26)$MeV	0.011
m_d	$= 1875.628(5)$ MeV	3
r_e	$= e^2/m_e c^2 = 2.8179380(70)$ fermi (1 fermi $= 10^{-13}$ cm)	2.5
λ_e	$= \hbar/m_e c = r_e \alpha^{-1} = 3.8615905(64) \times 10^{-11}$ cm	1.6
$a_{\infty Bohr}$	$= \hbar^2/m_e e^2 = r_e \alpha^{-2} = 0.52917706(44)$A ($1$A $= 10^{-8}$ cm)	0.82
$\sigma_{Thompson}$	$= \frac{8}{3}\pi r_e^2 = 0.6652448(33) \times 10^{-24}$ cm^2 (10^{-24} cm^2 $= 1$ barn)	4.9
μ_{Bohr}	$= e\hbar/2m_e c = 0.57883785(95) \times 10^{-14}$ MeV gauss^{-1}	1.6
$\mu_{nucleon}$	$= e\hbar/2m_p c = 3.1524515(53) \times 10^{-18}$ MeV gauss^{-1}	1.7
$\frac{1}{2}\omega^e_{cyclotron}$	$= e/2m_e c = 8.794023(25) \times 10^6$ rad sec^{-1} gauss^{-1}	2.8
$\frac{1}{2}\omega^p_{cyclotron}$	$= e/2m_p c = 4.789378(13) \times 10^3$ rad sec^{-1} gauss^{-1}	2.8

Hydrogen-like atom (nonrelativistic, μ = reduced mass):

$$\frac{v}{c})_{rms} = \frac{ze^2}{n\hbar c}; \quad E_n = \frac{\mu}{2}v^2 = \frac{\mu z^2 e^4}{2(n\hbar)^2}; \quad a_n = \frac{n^2\hbar^2}{\mu z e^2}$$

$R_\infty = m_e e^4/2\hbar^2 = m_e c^2 \alpha^2/2 = 13.605804(36)$ eV (Rydberg)   2.6

$pc = 0.3 H\rho$ (MeV, kilogauss, cm); 0.3 (which is 10^{-11}c) enters because

there are ≈ 300 practical volts/esu volt.

1 year (sidereal)	$= 365.256$ days $= 3.1558 \times 10^7$ sec ($\approx \pi \times 10^7$ sec)
density of dry air	$= 1.205$ mg cm^{-3} (at 20°C, 760 mm)
acceleration by gravity	$= 980.62$ cm sec^{-2} (sea level, 45°)
gravitational constant	$= 6.6732(31) \times 10^{-8}$ cm^{3}g^{-1} sec^{-2}
1 calorie (thermochemical)	$= 4.184$ joules
1 atmosphere	$= 1033.2275$ g cm^{-2} $= 1.01325$ bar
1 eV per particle	$= 11604.50(36)$ $^\circ$K (from E = kT)

NUMERICAL CONSTANTS

π	$= 3.1415927$	1 rad	$= 57.2957795$ deg	$\sqrt{\pi}$	$= 1.7724539$
e	$= 2.7182818$	1/e	$= 0.3678794$	$\sqrt{2}$	$= 1.4142136$
ln2	$= 0.6931472$	ln10	$= 2.3025851$	$\sqrt{3}$	$= 1.7320508$
$\log_{10}2$	$= 0.3010300$	$\log_{10}e$	$= 0.4342945$	$\sqrt{10}$	$= 3.1622777$

*Prepared by Stanley J. Brodsky, based mainly on the adjustment of the fundamental physical constants by E. R. Cohen and B. N. Taylor, J. Phys. Chem. Ref. Data *2*, 663 (1973). The figures in parentheses correspond to the 1 standard deviation uncertainty in the last digits of the main number. (Updated April 1974.)

†Reprinted by permission of the Particle Data Group and the Editor of *Physics Letters* from: *Physics Letters*, Vol. **50B**, No. 1, April 1974.

TABLE 1a6. TABLES OF PARTICLE PROPERTIES.*
April 1974

Physics Letters, Vol. 50B, No. 1, 1974.

N. Barash-Schmidt, A. Barbaro-Galtieri, C. Bricman, V. Chaloupka, D. M. Chew,
R. L. Kelly, T. A. Lasinski, A. Rittenberg, M. Roos, A. H. Rosenfeld
P. Söding, T. G. Trippe, and F. Uchiyama

(Closing data for data: Feb. 1, 1974)

Stable Particle Table

For additional parameters, see Addendum to this table.

Quantities in italics have changed by more than one (old) standard deviation since April 1973.

Particle	$I^G(J^P)C_n$	Mass (MeV) / Mass^2 (GeV)^2	Mean Life (sec) / $c\tau$ (cm)	Partial decay mode Mode	Fraction[a]	p or p_{max}[b] (MeV/c)
γ	$0,1(1^-)^-$	$0(<2)10^{-21}$	stable	stable		
ν $\begin{smallmatrix}\nu_e\\\nu_\mu\end{smallmatrix}$	$J=\frac{1}{2}$	$0(<60\text{ eV})$ $0(<1.2)$	stable	stable		
e	$J=\frac{1}{2}$	0.5110034 $\pm.0000014$	stable $(>2\times10^{21}\text{y})$	stable		
μ	$J=\frac{1}{2}$	105.65948 $\pm.00035$ $m^2=0.01116$ $m_\mu-m_{\pi^\pm}=-33.909$ $\pm.006$	2.1994×10^{-6} $\pm.0006$ S=1.1* $c\tau=6.593\times10^4$	$e\nu\bar\nu$ $e\gamma\gamma$ $3e$ $e\gamma$	100 % $(<1.6)10^{-5}$ $(<6)10^{-9}$ $(<2.2)10^{-8}$	53 53 53 53
$\pi^\pm$	$1^-(0^-)$	139.5688 $\pm.0064$ $m^2=0.0195$	2.6030×10^{-8} $\pm.0023$ $c\tau=780.4$ $(\tau^+-\tau^-)/\bar\tau=$ $(0.05\pm0.07)\%$ (test of CPT)	$\mu\nu$ $e\nu$ $\mu\nu\gamma$ $\pi^0 e\nu$ $e\nu\gamma$ $e\nu e^+e^-$	100 % $c(1.24\pm0.03)10^{-4}$ $c(1.24\pm0.25)10^{-4}$ $c(1.02\pm0.07)10^{-8}$ $c(3.0\pm0.5)10^{-8}$ $(<3.4)10^{-8}$	30 70 30 5 70 70
π^0	$1^-(0^-)^+$	134.9645 $\pm.0074$ $m^2=0.0182$ $m_{\pi^\pm}-m_{\pi^0}=4.6043$ $\pm.0037$	0.84×10^{-16} $\pm.10$ S=2.1* $c\tau=2.5\times10^{-6}$	$\gamma\gamma$ γe^+e^- $\gamma\gamma\gamma$ $e^+e^-e^+e^-$ $\gamma\gamma\gamma\gamma$	$(98.83\pm0.05)\%$ $(1.17\pm0.05)\%$ $d(<5)10^{-6}$ $(3.47)10^{-5}$ $(<6.1)10^{-5}$	67 67 67 67 67
$K^\pm$	$\frac{1}{2}(0^-)$	493.707 ±0.037 $m^2=0.244$ $m_{K^\pm}-m_{K^0}=-3.99$ ±0.13 S=1.1*	1.2371×10^{-8} $\pm.0026$ S=1.9* $c\tau=370.8$ $(\tau^+-\tau^-)/\bar\tau=$ $(.11\pm.09)\%$ (test of CPT) S=1.2*	$\mu\nu$ $\pi\pi^0$ $\pi\pi^-\pi^+$ $\pi\pi^0\pi^0$ $\mu\pi^0\nu$ $e\pi^0\nu$ $e\pi^0\pi^0\nu$ $\pi\pi^\mp e^\pm\nu$ $\pi\pi^\pm e^\mp\nu$ $\pi\pi^\mp\mu^\pm\nu$ $\pi\pi^\pm\mu^\mp\nu$ $e\nu$ $e\nu\gamma$ $\pi\pi^0\gamma$ $\pi\pi^+\pi^-\gamma$ $\mu\pi^0\nu\gamma$ $e\pi^0\nu\gamma$ πe^+e^- $\pi^\mp e^\pm e^\pm$ $\pi\mu^+\mu^-$ $\pi\gamma\gamma$ $\pi\gamma\gamma\gamma$ $\pi\nu\bar\nu$ $\pi\gamma$ $e\pi^\mp\mu^\pm$ $e\pi^\pm\mu^\mp$ $\mu\nu\nu\bar\nu$	$(63.54\pm0.19)\%$ $(21.12\pm0.17)\%$ $(5.59\pm0.03)\%$ S=1.1* $(1.73\pm0.05)\%$ S=1.4* $(3.20\pm0.09)\%$ S=1.7* $(4.82\pm0.05)\%$ S=1.1* $(1.8\,^{+2.4}_{-0.6})10^{-5}$ $(3.7\pm0.2)10^{-5}$ $(<5)10^{-7}$ $(0.9\pm0.4)10^{-5}$ $(<3)10^{-6}$ $(1.38\pm0.20)10^{-5}$ $c(<7)10^{-5}$ $h,c(2.71\pm0.19)10^{-4}$ $c(10\pm4)10^{-5}$ $c(<6)10^{-5}$ $c(3.7\pm1.4)10^{-5}$ $(<0.26)10^{-6}$ $(<1.5)10^{-5}$ $(<2.4)10^{-6}$ $c(<3.5)10^{-5}$ $c(<3)10^{-4}$ $(<0.6)10^{-6}$ $(<4)10^{-6}$ $c(<3)10^{-8}$ $(<1.4)10^{-8}$ $c(<6)10^{-6}$	236 205 125 133 215 228 207 203 203 151 151 247 247 205 125 215 228 227 227 172 227 227 227 227 214 214 236

Stable Particle Table (*Continued*)

Particle	$I^G(J^P)C_n$	Mass (MeV) Mass^2 (GeV)^2	Mean life (sec) $c\tau$ (cm)	Partial decay mode		
				Mode	Fraction[a]	p or p_{max}[b] (MeV/c)
K^0	$\frac{1}{2}(0^-)$	497.70 ±0.13	50% K_{Short}, 50% K_{Long}			
K_S^0	$\frac{1}{2}(0^-)$	S=1.1* m^2=0.248	0.886×10^{-10} ±.007 S=2.4* $c\tau$=2.66	$\pi^+\pi^-$	(68.77±0.26)% S=1.1*	206
				$\pi^0\pi^0$	(31.23±0.26)% S=1.1*	209
				$\mu^+\mu^-$	(< 0.3) 10^{-6}	225
				e^+e^-	(<35) 10^{-5}	249
				$\pi^+\pi^-\gamma$	c(2.0 ±0.4) 10^{-3}	206
				$\gamma\gamma$	(< 0.4) 10^{-3}	249
K_L^0	$\frac{1}{2}(0^-)$		5.179×10^{-8} ±0.040 $c\tau$=1553	$\pi^0\pi^0\pi^0$	(21.3 ±0.6)% S=1.1*	139
				$\pi^+\pi^-\pi^0$	(11.9 ±0.4)% S=2.2*	133
				$\pi\mu\nu$	(27.5 ±0.5)% S=1.1*	216
				$\pi e\nu$	(39.0 ±0.6)% S=1.1*	229
		$m_{K_L}-m_{K_S}$ = $0.5403 \times 10^{10}\hbar$ sec^{-1} ±0.0035		$\pi e\nu\gamma$	c(1.3 ±0.8)%	229
				$\pi^+\pi^-$	(0.177 ±0.018)% S=4.9*	206
				$\pi^0\pi^0$	(0.093±0.019)% S=1.5*	209
				$\pi^+\pi^-\gamma$	c(< 0.4) 10^{-3}	206
				$\pi^0\gamma\gamma$	(< 2.4) 10^{-4}	231
				$\gamma\gamma$	(4.9± 0.4) 10^{-4}	249
				$e\mu$	(< 1.6) 10^{-9}	238
				$\mu^+\mu^-$	i(< 1.6) 10^{-8}	225
				e^+e^-	(< 1.6) 10^{-9}	249
				$e^+e^-\gamma$	(< 2.8) 10^{-5}	249
η	$0^+(0^-)^+$	548.8 ±0.6, S=1.4* m^2=0.301	$^e\Gamma$=(2.63±0.58)keV Neutral decays 71.1%	$\gamma\gamma$	(38.0 ±1.0)% S=1.2*	274
				$\pi^0\gamma\gamma$	e(3.1 ±1.1)% S=1.2*	258
				$3\pi^0$	(30.0 ±1.1)% S=1.1*	180
				$\pi^+\pi^-\pi^0$	(23.9 ±0.6)% S=1.1*	175
			Charged decays 28.9%	$\pi^+\pi^-\gamma$	(5.0 ±0.1)%	236
				$\pi^0e^+e^-$	(< 0.04)%	258
				$\pi^+\pi^-$	(< 0.15)%	236
				$\pi^+\pi^-e^+e^-$	(0.1 ±0.1)%	236
				$\pi^+\pi^-\pi^0\gamma$	(< 6) 10^{-4}	175
				$\pi^+\pi^-\gamma\gamma$	(< 0.2)%	236
				$\mu^+\mu^-$	(2.2 ±0.8) 10^{-5}	253
				$\mu^+\mu^-\pi^0$	(< 5) 10^{-4}	211
p	$\frac{1}{2}(\frac{1}{2}^+)$	938.2796 ±0.0027 m^2=0.8804	stable (> 2×10^{28}y)			
n	$\frac{1}{2}(\frac{1}{2}^+)$	939.5731 ±0.0027 m^2=0.8828 m_p-m_n=-1.29344 ±0.00007	918±14 $c\tau$=2.75×10^{13}	$pe\nu$	100 %	1
Λ	$0(\frac{1}{2}^+)$	1115.60 ±0.05 S=1.2* m^2=1.245	2.578×10^{-10} ±.021 S=1.6* $c\tau$=7.73	$p\pi^-$	(64.2±0.5)%	100
				$n\pi^0$	(35.8±0.5)%	104
				$pe\nu$	(8.13±0.29) 10^{-4}	163
				$p\mu\nu$	(1.57±0.35) 10^{-4}	131
				$p\pi^-\gamma$	c(0.85±0.14) 10^{-3}	100
Σ^+	$1(\frac{1}{2}^+)$	1189.37 ±0.06 S=1.8* m^2=1.415 $m_{\Sigma^+}-m_{\Sigma^-}$=-7.99 ±.08 S=1.2*	0.800×10^{-10} ±.006 $c\tau$=2.40 $\frac{\Gamma(\Sigma^+ \to \ell^+ n \nu)}{\Gamma(\Sigma^- \to \ell^- n \nu)}$<.035	$p\pi^0$	(51.6±0.7)%	189
				$n\pi^+$	(48.4±0.7)%	185
				$p\gamma$	(1.24±0.18) 10^{-3} S=1.4*	225
				$n\pi^+\gamma$	c(0.93±0.10) 10^{-3}	185
				$\Lambda e^+\nu$	(2.02±0.47) 10^{-5}	72
				$n\mu^+\nu$	(< 2.4) 10^{-5}	202
				$ne^+\nu$	(< 1.0) 10^{-5}	224
				pe^+e^-	(< 7) 10^{-6}	225
Σ^0	$1(\frac{1}{2}^+)$	1192.48 ±0.08	$<1.0 \times 10^{-14}$ $c\tau$<3×10^{-4}	$\Lambda\gamma$	100 %	74
				Λe^+e^-	d(5.45) 10^{-3}	74

Stable Particle Table (*Continued*)

Particle	$I^G(J^P)C_n$	Mass (MeV) / Mass² (GeV)²	Mean Life (sec) / cτ (cm)	Mode	Fraction[a]	p or p_{max}[b] (MeV/c)
Σ^-	$1(\tfrac{1}{2}^+)$	$m^2=1.422$ 1197.35 ±0.06 $m^2=1.434$ $m_{\Sigma^0}-m_{\Sigma^-}=-4.87$ $\pm.06$	1.482×10^{-10} $\pm.017$ S=1.5* $c\tau=4.44$	$n\pi^-$ $ne^-\nu$ $n\mu^-\nu$ $\Lambda e^-\nu$ $n\pi^-\gamma$	100 % $(\ 1.08\pm0.04)10^{-3}$ $(\ 0.45\pm0.04)10^{-3}$ $(\ 0.60\pm0.06)10^{-4}$ $c(\ 1.0\ \pm0.2\ )10^{-4}$	193 230 210 79 193
Ξ^0	$\tfrac{1}{2}(\tfrac{1}{2}^+)$f	1314.9 ±0.6 $m^2=1.729$ $m_{\Xi^0}-m_{\Xi^-}=-6.4$ $\pm.6$	2.96×10^{-10} $\pm.12$ $c\tau=8.93$	$\Lambda\pi^0$ $p\pi^-$ $pe^-\nu$ $\Sigma^+e^-\nu$ $\Sigma^-e^+\nu$ $\Sigma^+\mu^-\nu$ $\Sigma^-\mu^+\nu$ $p\mu^-\nu$	100 % $(<0.9)10^{-3}$ $(<1.3)10^{-3}$ $(<1.5)10^{-3}$ $(<1.5)10^{-3}$ $(<1.5)10^{-3}$ $(<1.5)10^{-3}$ $(<1.3)10^{-3}$	135 299 323 119 112 64 49 309
Ξ^-	$\tfrac{1}{2}(\tfrac{1}{2}^+)$f	1321.29 ±0.14 $m^2=1.746$	1.652×10^{-10} $\pm.023$ S=1.1* $c\tau=4.95$	$\Lambda\pi^-$ $\Lambda e^-\nu$ $\Sigma^0e^-\nu$ $\Lambda\mu^-\nu$ $\Sigma^0\mu^-\nu$ $n\pi^-$ $ne^-\nu$	100 % $g(\ 0.70\pm0.21)10^{-3}$ $(<0.5)10^{-3}$ $(<1.3)10^{-3}$ (<0.5)% $(<1.1)10^{-3}$ (<1.0)%	139 190 123 163 70 303 327
Ω^-	$0(\tfrac{3}{2}^+)$f	$1672.2\pm.4$ $m^2=2.797$	$1.3^{+0.3}_{-0.2}\times10^{-10}$ $c\tau=3.9$	$\left.\begin{array}{l}\Xi^0\pi^-\\ \Xi^-\pi^0\\ \Lambda K^-\end{array}\right\}$	Total of 41 events seen	293 290

Addendum To
Stable Particle Table

	Magnetic moment	
e	$1.001\ 159\ 6567$ $\pm.000\ 000\ 0035$ $\dfrac{e\hbar}{2m_e c}$	**μ Decay parameters**[j]
μ	$1.001\ 166\ 16$ $\pm.000\ 000\ 31$ $\dfrac{e\hbar}{2m_\mu c}$	$\rho = 0.752\pm0.003$ $\eta = -0.12\pm0.21$ $\xi = 0.972\pm0.013$ $\delta = 0.755\pm0.009$ $h = 1.00\pm0.13$ $\|g_A/g_V\| = 0.86^{+0.33}_{-0.11}$ $\phi = 180°\pm15°$

K±

Mode	Partial rate (sec⁻¹)	
$\mu\nu$	$(51.36\pm0.19)10^6$	S=1.2*
$\pi\pi^0$	$(17.07\pm0.15)10^6$	S=1.1*
$\pi\pi^+\pi^-$	$(\ 4.52\pm0.02)10^6$	S=1.1*
$\pi\pi^0\pi^0$	$(\ 1.40\pm0.04)10^6$	S=1.4*
$\mu\pi^0\nu$	$(\ 2.58\pm0.07)10^6$	S=1.7*
$e\pi^0\nu$	$(\ 3.90\pm0.04)10^6$	S=1.1*

$\Delta I = \tfrac{1}{2}$ rule for $K \to 3\pi$ k

$K^+\to\pi^+\pi^+\pi^-$ g=−.214±.005 S=1.7*
$K^-\to\pi^-\pi^-\pi^+$ g=−.214±.007 S=2.7*
$K^\pm\to\pi^\pm\pi^0\pi^0$ g=.522±.020 S=1.3*
$K^0_L\to\pi^0\pi^+\pi^-$ g=.610±.021 S=2.6*

K⁰_S

$\pi^+\pi^-$	$n(0.776\pm.006)10^{10}$	S=1.8*
$\pi^0\pi^0$	$n(0.352\pm.004)10^{10}$	S=1.3*

Form factors for $K_{\ell3}$ decays (see Data Card Listings for ξ, λ_+^μ and λ_0^μ)

K^+_{e3} $\lambda_+^e=.029\pm.004$ K^0_{e3} $\lambda_+^e=.026\pm.004$

K⁰_L

$\pi^0\pi^0\pi^0$	$(\ 4.11\pm0.13)10^6$	S=1.1*
$\pi^+\pi^-\pi^0$	$(\ 2.31\pm0.07)10^6$	S=2.0*
$\pi\mu\nu$	$(\ 5.31\pm0.11)10^6$	S=1.1*
$\pi e\nu$	$(\ 7.53\pm0.12)10^6$	S=1.1*
$\pi^+\pi^-$	$n(\ 3.42\pm0.36)10^4$	S=4.6*
$\pi^0\pi^0$	$n(\ 1.80\pm0.36)10^4$	S=1.5*

CP violation parameters ℓ,n

$\|\eta_{+-}\| = (2.17\pm.07)10^{-3}$ S=3.4* $\|\eta_{00}\| = (2.25\pm.09)10^{-3}$ S=1.1*
$\phi_{+-} = (46.6\pm2.5)°$ S=1.5* $\phi_{00} = (49\pm13)°$
$\|\eta_{+-0}\|^2 < 0.12$ $\|\eta_{000}\|^2 < 1.2$ $\delta=(.34\pm.01)10^{-2}$ S=1.1*

$\Delta S = -\Delta Q$
Re x=.000±.022 S=1.5*   Im x = .012±.030 S=1.2*

η

Mode	Asymmetry parameter
$\pi^+\pi^-\pi^0$	$(0.12\pm.17)$%
$\pi^+\pi^-\gamma$	$(0.88\pm.40)$%

Stable Particle Table (*Continued*)

	Magnetic moment $(e\hbar/2m_p c)$		Measured — α	Measured — ϕ(degree)	Derived — γ	Derived — Δ(degree)	g_A/g_V	g_V/g_A
p	2.7928456 ±.0000011							
n	-1.913148 ±.000066	$pe^-\nu$					$-1.250\pm.09$ $\delta=(181.1\pm1.3)°$	
Λ	-0.67 ±.06	$p\pi^-$ $n\pi^0$ $pe\nu$	0.647±0.013 0.651±0.045	(-6.5±3.5)°	0.76	$\left(7.6^{+4.0}_{-4.1}\right)°$	-0.66 ± 0.05 S=1.2*	
Σ⁺	2.62 ±.41	$p\pi^0$ $n\pi^+$ $p\gamma$	-0.979±0.016 +0.066±0.016 $-1.03^{+.52}_{-.42}$	(36±34)° (167±20)° S=1.1*	0.17 -0.97	(187±6)° $\left(-73^{+136}_{-10}\right)°$		
Σ⁻	-1.6 to 0.8	$n\pi^-$ $ne^-\nu$ $\Lambda e^-\nu$	-0.069±0.008	(10±15)°	0.98	$\left(249^{+12}_{-115}\right)°$	See Data Cds.	0.37±0.20
Ξ⁰		$\Lambda\pi^0$	-0.44±0.08 S=1.3*	(21±12)°	0.84	$\left(216^{+13}_{-19}\right)°$		
Ξ⁻	-1.93 ±.75	$\Lambda\pi^-$	-0.393±0.023 S=1.2*	(2±7)°	0.92	$\left(184\pm15\right)°$		

*S = Scale factor = $\sqrt{\chi^2/(N-1)}$, where N ≈ number of experiments. S should be ≈1. If S>1, we have enlarged the error of the mean, δx, i.e., δx→ Sδx. This convention is still inadequate, since if S≫1, the experiments are probably inconsistent, and therefore the real uncertainty is probably even greater than Sδx. See text and ideogram in Stable Particle Data Card Listings.

a. Quoted upper limits correspond to a 90% confidence level.
b. In decays with more than two bodies, p_{max} is the maximum momentum that any particle can have.
c. See Stable Particle Data Card Listings for energy limits used in this measurement.
d. Theoretical value; see also Stable Particle Data Card Listings.
e. See note in Stable Particle Data Card Listings.
f. P for Ξ and J^P for Ω not yet measured. Values reported are SU(3) predictions.
g. Assumes rate for $\Xi^-\to\Sigma^0 e^-\nu$ small compared with $\Xi^-\to\Lambda e^-\nu$.
h. The direct emission branching ratio is $(1.56\pm.35)\times10^{-5}$.
i. This upper limit is above the contradictory results of Carithers et al. $(1.0^{+.7}_{-.35})\times10^{-8}$, and Clark et al. $(<0.19)\times10^{-8}$. See note in Stable Particle Data Card Listings.
j. $|g_A/g_V|$ defined by $g_A^2 = |C_A|^2+|C'_A|^2$, $g_V^2 = |C_V|^2+|C'_V|^2$, and $\Sigma\langle\bar{e}|\Gamma_i|\mu\rangle\langle\bar{\nu}|\Gamma_i(C_i+C'_i\gamma_5)|\nu\rangle$; ϕ defined by $\cos\phi = -R_e(C^*_A C'_V+C'_A C^*_V)/g_A g_V$ [for more details, see text Section VI A].
k. The definition of the slope parameter of the Dalitz plot is as follows [see also text Section VI B.1]: $|M|^2 = 1 + g\left(\dfrac{s_3-s_0}{m_{\pi^+}^2}\right)$.
ℓ. The definition for the CP violation parameters is as follows [see also text Section VI B.3]:
$$\delta = \frac{\Gamma(K_L^0\to\ell^+) - \Gamma(K_L^0\to\ell^-)}{\Gamma(K_L^0\to\ell^+)+\Gamma(K_L^0\to\ell^-)} \qquad |\eta_{+-0}|^2 = \frac{\Gamma(K_S^0\to\pi^+\pi^-\pi^0)}{\Gamma(K_L^0\to\pi^+\pi^-\pi^0)} \qquad |\eta_{000}|^2 = \frac{\Gamma(K_S^0\to\pi^0\pi^0\pi^0)}{\Gamma(K_L^0\to\pi^0\pi^0\pi^0)}.$$
m. The definition of these quantities is as follows [for more details on sign convention, see text Section VI B]:
$$\alpha = \frac{2|s||p|\cos\Delta}{|s|^2+|p|^2}; \qquad \beta = \sqrt{1-\alpha^2}\sin\phi; \qquad g_A/g_V \text{ defined by } \langle B_f|\gamma_\lambda(g_V-g_A\gamma_5)|B_i\rangle;$$
$$\beta = \frac{-2|s||p|\sin\Delta}{|s|^2+|p|^2} \qquad \gamma = \sqrt{1-\alpha^2}\cos\phi. \qquad \delta \text{ defined by } g_A/g_V = |g_A/g_V|e^{i\delta}.$$
n. The $K_S^0\to\pi\pi$ and $K_L^0\to\pi\pi$ rates (and branching fractions) are from independent fits and do not include results of K_L^0-K_S^0 interference experiments. The $|\eta_{+-}|$ and $|\eta_{00}|$ values given in the addendum are these rates combined with the $|\eta_{+-}|$ and $|\eta_{00}|$ results from interference experiments.

Meson Table
April 1974

In addition to the entries in the Meson Table, the Meson Data Card Listings contain all substantial claims for meson resonances. See Contents of Meson Data Card Listings[1].

Quantities in italics have changed by more than one (old) standard deviation since April 1973.

Name $I^G(J^P)C_n$ estab.	Mass M (MeV)	Full Width Γ (MeV)	$\pm \Gamma M^{(a)}$ M^2 (GeV)^2	Mode	Fraction (%) [Upper limits are 1σ (%)]	p or $P_{max}^{(b)}$ (MeV/c)
$\pi^\pm(140)$ $1^-(0^-)+$	139.57	0.0	0.019483	See Stable Particle Table		
$\pi^0(135)$	134.96	7.8 eV ±.9 eV	0.018217			
$\eta(549)$ $0^+(0^-)+$	548.8 ±0.6	2.63 keV ±.58 keV	0.301 ±.000	All neutral $\pi^+\pi^-\pi^0 + \pi^+\pi^-\gamma$	71 See Stable 29 Particle Table	
ε $0^+(0^+)+$	$\lesssim 700^{(c)}$	$\gtrsim 600^{(c)}$		$\pi\pi$		
Existence of pole not established. See note on $\pi\pi$ S wave¶.						
$\rho(770)$ $1^+(1^-)-$	$770_\S$ ±10§	$150_\S$ ±10§	0.593 ±.116	$\pi\pi$ e^+e^- $\mu^+\mu^-$ For upper limits, see footnote (e)	≈100 0.0043±.0005 (d) 0.0067±.0012 (d)	359 385 370
$\omega(783)$ $0^-(1^-)-$	$782.7_\S$ ±0.6§	10.0 ±.4	0.613 ±.008	$\pi^+\pi^-\pi^0$ $\pi^+\pi^-$ $\pi^0\gamma$ e^+e^- For upper limits, see footnote (g)	90.0±0.6 S=1.2* 1.3±0.3 S=1.5* 8.7±0.5 0.0076±.0017 S=1.9*	327 366 380 391
$\eta'(958)$ or X^0 $0^+(\ ^-)+$ J=0 or 2	957.6 ±0.3	< 1 <.001	0.917	$\eta\pi\pi$ $\rho^0\gamma$ $\gamma\gamma$ For upper limits, see footnote (h)	70.6±2.5 S=1.4* 27.4±2.2 S=1.6* 1.9±0.3	234 458 479
$\delta(970)$ $1^-(0^+)\pm$	$976_\S$ ±10§	$50_\S$ ±20§	0.953 ±.049	$\eta\pi$ $\rho\pi$	seen < 25	315 139
Possibly a virtual bound state of the I = 1 $K\bar{K}$ system¶.						
$S^*(993)$ $0^+(0^+)+$	$\sim 993^{(c)}$ ±5	$40^{(c)}$ ±8	0.986 ±.040	$K\bar{K}$ $\pi\pi$		near threshold 479
See notes on $\pi\pi$ and $K\bar{K}$ S wave¶.						
$\Phi(1019)$ $0^-(1^-)-$	1019.7 ±0.3 S=1.9*	4.2 ±.2	1.040 ±.004	K^+K^- $K_L K_S$ $\pi^+\pi^-\pi^0$ (incl. $\rho\pi$) $\eta\gamma$ e^+e^- $\mu^+\mu^-$ For upper limits, see footnote (i)	46.6±2.5 S=1.6* 34.6±2.2 S=1.6* 15.8±1.5 S=1.2* 3.0±1.1 S=1.6* .032±.002 S=1.4* .025±.003	127 111 462 362 510 499
$A_1(1100)$ $1^-(1^+)+$	~ 1100	~ 300	1.21 ±.33	$\rho\pi$	~ 100	253
Broad enhancement in the $J^P=1^+$ $\rho\pi$ partial wave; not a Breit-Wigner resonance¶.						
$B(1235)$ $1^+(1^+)-$	$1237_\S$ ±10§	$120_\S$ ±20§	1.53 ±.12	$\omega\pi$ [D/S amplitude ratio = .24±.06] For upper limits, see footnote (j)	only mode seen	352

Meson Table (*Continued*)

Name	$I^G(J^P)C_n$ estab.	Mass M (MeV)	Full Width Γ (MeV)	M^2 ±ΓM[a] (GeV)²	Mode	Fraction (%) [Upper limits are 1σ (%)]	p or Pmax[b] (MeV/c)
f(1270)	$\underline{0^+(2^+)}+$	1270§ ±10§	170§ ±30§	1.61 ±.22	ππ 2π⁺2π⁻ K$\bar{\text{K}}$	83±5§ 4±1§ 4±3 S=1.5*	619 556 394
					For upper limits, see footnote (f)		
D(1285)	$\underline{0^+(A~)}+$ $J^P = 0^-,~1^+,~2^-$, with 1^+ favoured	1286§ ±10§	30§ ±20§	1.65 ±.03	K$\bar{\text{K}}$π ηππ †[δ(970)π 2π⁺2π⁻ (prob. ρ⁰π⁺π⁻)	seen seen seen] seen	305 484 245 565
A₂(1310)	$\underline{1^-(2^+)}+$	1310§ ±10§	100§ ±10§	1.72 ±.13	ρπ ηπ ωππ K$\bar{\text{K}}$ η′(958)π	71.5±1.8 S=1.2* 15.2±1.2 8.6±1.8 S=1.3* 4.7±0.6 <1	413 529 354 428 280
E(1420)	$\underline{0^+(A~)}+$	1416§ ±10§	60§ ±20§	2.01 ±.08	K$\bar{\text{K}}$π †[K*$\bar{\text{K}}$ + $\bar{\text{K}}$*K ηππ †[δ(970)π	∿ 40 ∿ 20] ∿ 60 possibly seen]	421 130 564 352
f′(1514)	$\underline{0^+(2^+)}+$	1516 ±3	40 ±10	2.29 ±.06	K$\bar{\text{K}}$ For upper limits, see footnote (k)	only mode seen	572
F₁(1540)	$\underline{1~(A~)}$	1540 ±5	40 ±15	2.37 ±.06	K*$\bar{\text{K}}$ + $\bar{\text{K}}$*K	only mode seen	321
	Evidence based on only one experiment						
ρ′(1600)	$\underline{1^+(1^-)}-$	∿ 1600	∿ 400	2.56 ±.64	4π †[ρπ⁺π⁻ ππ K$\bar{\text{K}}$	dominant seen with π⁺π⁻ in S-wave] possibly seen < 8	738 575 788 629
A₃(1640)	$\underline{1^-(2^-)}+$	∿ 1640	∿ 300	2.69 ±.49	fπ		305
	Broad enhancement in the $J^P = 2^-$ fπ partial wave; not a Breit-Wigner resonance.¶						
ω(1675)	$\underline{0^-(N~)}-$	1666§ ±10§	142§ ±20§	2.78 ±.24	ρπ 3π 5π †[ωππ	seen possibly seen possibly seen possibly seen]	647 805 778 614
g(1680)	$\underline{1^+(3^-)}-$ J^P, M and Γ from the 2π mode[ℓ].	1686§ ±20§	180§ ±30§	2.84 ±.30	2π 4π (incl. ππρ,ρρ,A₂π,ωπ) K$\bar{\text{K}}$ K$\bar{\text{K}}$π (incl. K*$\bar{\text{K}}$)	26±5§ ∿ 70 ∿ 2 ∿ 3 }(ℓ)	831 784 680 621
	See note (1) for possible heavier states.						
K⁺(494) K⁰(498)	$\underline{1/2(0^-)}$	493.71 497.70		0.244 0.248	See Stable Particle Table		
K*(892)	$\underline{1/2(1^-)}$	892.2 ±0.5	49.8 ±1.1	0.796 ±.044	Kπ Kππ Kγ	≈ 100 < 0.2 < 0.16	288 216 310
		(Charged mode; $m^0 - m^\pm = 6.1 \pm 1.5$ MeV)					

Meson Table (*Continued*)

Name $\dfrac{G}{ } \begin{array}{c|c|c} I & 0 & 1 \\ \hline - & \omega/\phi & \pi \\ \hline + & \eta & \rho \end{array}$ $I^G(J^P)C_n$ $\vdash\!\!\!\!\dashv$ estab.	Mass M (MeV)	Full Width Γ (MeV)	$\pm\Gamma M^{(a)}$ M^2 (GeV)^2	Partial decay mode		
				Mode	Fraction (%) [Upper limits are 1σ (%)]	p or $P_{max}^{(b)}$ (MeV/c)

κ   1/2(0$^+$) δ_0^1 goes slowly through 90° near 1300 MeV.

See note on Kπ S wave¶.

$\rightarrow$

| Q | $\left\{\begin{array}{l} K_A(1240)\,1/2(1^+) \\ \text{or C} \\ \hline K_A(1280\,1/2(1^+) \\ \text{to }1400 \\ \text{See note (m).} \end{array}\right.$ | 1242 ±10 seen in p̄p at rest 1280 to 1400 | 127 ±25 | 1.54 ±.16 | Kππ $\dagger[K^*\pi$ $\dagger[K\rho$ $\dagger[K(\pi\pi)_{\ell=0}$ | only mode seen large] seen] possibly seen] | |

$\rightarrow$

| K*(1420)   1/2(2$^+$) | 1421$_§$ ±5§ | 100$_§$ ±10§ | 2.02 ±.14 | Kπ $K^*\pi$ Kρ Kω Kη | 55.0±2.7 29.5±2.5 9.2±2.4 4.4±1.7 2.0±2.0 | 616 414 319 306 482 |

See note (n).

$\rightarrow$
$\rightarrow$

| L(1770)   1/2(A) | 1765$_§$ ±10§ | 140$_§$ ±50§ | 3.11 ±.25 | Kππ Kπππ $\dagger[K^*(1420)\pi$ and other subreactions¶] | dominant seen | 788 757 |

$J^P=2^-$ favoured, 1^+ and 3^+ not excluded.

$\rightarrow$

See note (1) for possible heavier states.

(1) Contents of Meson Data Card Listings

Non-strange (Y = 0)						Strange (|Y| = 1)	
entry	$I^G(J^P)C_n$		entry	$I^G(J^P)C_n$		entry	$I^G(J^P)C_n$	entry	I (J^P)

entry	$I^G(J^P)C_n$	entry	$I^G(J^P)C_n$	entry	$I^G(J^P)C_n$	entry	I (J^P)
π (140)	1$^-$(0$^-$)+	$\rightarrow$ η$_N$ (1080)	0$^+$(N)+	ρ′ (1600)	1$^+$(1$^-$)-	K (494)	1/2(0$^-$)
η (549)	0$^+$(0$^-$)$^+$	A$_1$ (1100)	1$^-$(1$^+$)+	A$_3$ (1640)	1$^-$(2$^-$)+	K* (892)	1/2(1$^-$)
ε (600)	0$^+$(0$^+$)+	$\rightarrow$ M (1150)		ω (1675)	0$^-$(N)-	κ	1/2(0$^+$)
ρ (770)	1$^+$(1$^-$)-	$\rightarrow$ A$_{1.5}$(1170)	1$^-$	g (1680)	1$^+$(3$^-$)-	Q	1/2(1$^+$)
ω (783)	0$^-$(1$^-$)-	B (1235)	1$^+$(1$^+$)-	$\rightarrow$ X (1690)		K*(1420)	1/2(2$^+$)
$\rightarrow$ M (940)		$\rightarrow$ ρ′ (1250)	1$^+$(1$^-$)-	$\rightarrow$ X (1795)	1	$\rightarrow$ K$_N$(1660)	1/2
$\rightarrow$ M (953)	$^+$	f (1270)	0$^+$(2$^+$)+	$\rightarrow$ S (1930)	1	$\rightarrow$ K$_N$(1760)	1/2
η′ (958)	0$^+$(0$^-$)+	D (1285)	0$^+$(A)+	$\rightarrow$ A$_4$ (1960)	1$^-$	L (1770)	1/2(A)
δ (970)	1$^-$(0$^+$)+	A$_2$ (1310)	1$^-$(2$^+$)+	$\rightarrow$ ρ (2100)	1$^+$	$\rightarrow$ K$_N$(1850)	
$\rightarrow$ H (990)	0$^-$(A)-	E (1420)	0$^+$(A)+	$\rightarrow$ T (2200)	1	$\rightarrow$ K*(2200)	
S* (993)	0$^+$(0$^+$)+	$\rightarrow$ X (1430)	0	$\rightarrow$ ρ (2275)	1$^+$	$\rightarrow$ K*(2800)	
φ (1019)	0$^-$(1$^-$)-	$\rightarrow$ X (1440)	1	$\rightarrow$ U (2360)	1		
$\rightarrow$ M (1033)		f′ (1514)	0$^+$(2$^+$)+	$\rightarrow$ NN̄ (2375)	0	$\rightarrow$ Exotics	
$\rightarrow$ B$_1$(1040)	1$^+$	F$_1$ (1540)	1 (A)	$\rightarrow$ X(2500-3600)			

Meson Table (*Continued*)

→   indicates an entry in Meson Data Card Listings not entered in the Meson Table. We do not regard these as established resonances.

¶   See Meson Data Card Listings.

*   Quoted error includes scale factor $S = \sqrt{\chi^2/(N-1)}$. See footnote to Stable Particle Table.

†   Square brackets indicate a subreaction of the previous (unbracketed) decay mode(s).

§   This is only an educated guess; the error given is larger than the error of the average of the published values. (See Meson Data Card Listings for the latter.)

(a) ΓM is approximately the half-width of the resonance when plotted against M^2.

(b) For decay modes into ≥ 3 particles, p_{max} is the maximum momentum that any of the particles in the final state can have. The momenta have been calculated by using the averaged central mass values, without taking into account the widths of the resonances.

(c) From pole position $(M - i\Gamma/2)$. For both ε and S* the pole is on Riemann Sheet 2.

(d) The e^+e^- branching ratio is from $e^+e^- \to \pi^+\pi^-$ experiments only. The ωρ interference is then due to ωρ mixing only, and is expected to be small. See note in Meson Data Card Listings. The $\mu^+\mu^-$ branching ratio is compiled from 3 experiments; each possibly with substantial ωρ interference. The error reflects this uncertainty; see notes in Meson Data Card Listings. If eμ universality holds, $\Gamma(\rho^0 \to \mu^+\mu^-) = \Gamma(\rho^0 \to e^+e^-) \times$ phase space correction.

(e) Empirical limits on fractions for other decay modes of ρ(765) are $\pi^\pm\gamma < 0.5\%$, $\pi^\pm\eta < 0.8\%$, $\pi^+\pi^+\pi^-\pi^- < 0.15\%$, $\pi^\pm\pi^+\pi^-\pi^0 < 0.2\%$.

(f) Empirical limits on fractions for other decay modes of f(1270) are $\eta\pi\pi < 15\%$; $K^0 K^- \pi^+ +$ c.c. $< 6\%$.

(g) Empirical limits on fractions for other decay modes of ω(783) are $\pi^+\pi^-\gamma < 5\%$. $\pi^0\pi^0\gamma < 1\%$, η + neutral(s) $< 1.5\%$, $\mu^+\mu^- < 0.02\%$, $\pi^0\mu^+\mu^- < 0.2\%$, $\eta\gamma < 0.5\%$.

(h) Empirical limits on fractions for other decay modes of η'(958): $\pi^+\pi^- < 2\%$, $\pi^+\pi^-\pi^0 < 5\%$, $\pi^+\pi^+\pi^-\pi^- < 1\%$, $\pi^+\pi^+\pi^-\pi^-\pi^0 < 1\%$, $6\pi < 1\%$, $\pi^+\pi^- e^+e^- < 0.6\%$, $\pi^0 e^+e^- < 1.3\%$, $\eta e^+e^- < 1.1\%$, $\pi^0\rho^0 < 4\%$, $\pi^0\omega + \gamma\omega < 8\%$.

(i) Empirical limits on fractions for other decay modes of φ(1019) are $\pi^+\pi^- < 0.03\%$, $\pi^+\pi^-\gamma < 0.7\%$, $\omega\gamma < 5\%$, $\rho\gamma < 2\%$, $\pi^0\gamma < 0.35\%$, $2\pi^+2\pi^-\pi^0 < 1\%$.

(j) Empirical limits on fractions for other decay modes of B(1235): $\pi\pi < 15\%$, $K\bar{K} < 2\%$, $4\pi < 50\%$, $\phi\pi < 1.5\%$, $\eta\pi < 25\%$, $(\bar{K}K)^\pm\pi^0 < 8\%$, $K_S K_S \pi^\pm < 2\%$, $K_S K_L \pi^\pm < 6\%$.

(k) Empirical limits on fractions for other decay modes of f'(1514) are $\pi^+\pi^- < 20\%$, $\eta\eta < 50\%$, $\eta\pi\pi < 30\%$, $K\bar{K}\pi + K^*\bar{K} < 35\%$, $2\pi^+2\pi^- < 32\%$.

(ℓ) We assume as a working hypothesis that peaks with $I^G = 1^+$ observed around 1.7 GeV all come from g(1680). For indications to the contrary see Meson Data Card Listings.

(m) See Q-region note in Meson Data Card Listings. Some investigators see a broad enhancement in mass (Kππ) from 1250-1400 MeV (the Q region), and others see structure. The Kη, Kω, and Kπ modes are less than a few percent.

(n) The tabulated mass of 1421 MeV comes from the Kπ mode; the Kππ mode can be contaminated with diffractively produced $Q^\pm$.

Established Nonets, and octet-singlet mixing angles from Appendix IIB, Eq. (2'). Of the two isosinglets, the "mainly octet" one is written first, followed by a semicolon.

$(J^P)C_n$	Nonet members	$\theta_{1in.}$	$\theta_{quadr.}$
$(0^-)+$ or: $(0^-)+$	π, K, η; η' π, K, η; E	24 ± 1° 16 ± 1°	10 ± 1° 6 ± 1°
$(1^-)-$	ρ, K^*, φ; ω	36 ± 1°	39 ± 1°
$(2^+)+$	A_2, $K^*(1420)$, f'; f	29 ± 2°	31 ± 2°

Baryon Table

April 1974

The following short list gives the status of all the Baryon States in the Data Card Listings. In addition to the status, the name, the nominal mass, and the quantum numbers (where known) are shown. States with three or four star status are included in the main Baryon Table, the others have been omitted because the evidence for the existence of the effect and/or for its interpretation as a resonance is open to considerable question.

N(939)	P11	****	Δ(1232)	P33	****	Λ(1116)	P01	****	Σ(1193)	P11	****	Ξ(1317)	P11	****
N(1470)	P11	****	Δ(1650)	S31	****	Λ(1330)		Dead	Σ(1385)	P13	****	Ξ(1530)	P13	****
N(1520)	D13	****	Δ(1670)	D33	***	Λ(1405)	S01	****	Σ(1440)		Dead	Ξ(1630)		**
N(1535)	S11	****	Δ(1690)	P33	*	Λ(1520)	D03	****	Σ(1480)		*	Ξ(1820)		***
N(1670)	D15	****	Δ(1890)	F35	***	Λ(1670)	S01	****	Σ(1620)	S11	**	Ξ(1940)		***
N(1688)	F15	****	Δ(1900)	S31	*	Λ(1690)	D03	****	Σ(1620)	P11	**	Ξ(2030)		**
N(1700)	S11	****	Δ(1910)	P31	***	Λ(1750)	P01	**	Σ(1670)	D13	****	Ξ(2250)		*
N(1700)	D13	**	Δ(1950)	F37	****	Λ(1815)	F05	****	Σ(1670)		**	Ξ(2500)		**
N(1780)	P11	***	Δ(1960)	D35	**	Λ(1830)	D05	***	Σ(1690)		**			
N(1810)	P13	***	Δ(2160)		**	Λ(1860)	P03	**	Σ(1750)	S11	***	Ω(1672)	P03	****
N(1990)	F17	**	Δ(2420)	H311	***	Λ(1870)	S01	***	Σ(1765)	D15	****			
N(2000)	F15	**	Δ(2850)		***	Λ(2010)	D03	**	Σ(1840)	P13	*			
N(2040)	D13	**	Δ(3230)		***	Λ(2020)	F07	**	Σ(1880)	P11	**			
N(2100)	S11	*				Λ(2100)	G07	****	Σ(1915)	F15	****			
N(2100)	D15	*	Z0(1780)	P01	*	Λ(2110)	?05	*	Σ(1940)	D13	***			
N(2190)	G17	***	Z0(1865)	D03	*	Λ(2350)		****	Σ(2000)	S11	*			
N(2220)	H19	***	Z1(1900)	P13	*	Λ(2585)		***	Σ(2030)	F17	****			
N(2650)		***	Z1(2150)		*				Σ(2070)	F15	*			
N(3030)		***	Z1(2500)		*				Σ(2080)	P13	**			
N(3245)		*							Σ(2100)	G17	**			
N(3690)		*							Σ(2250)		****			
N(3755)		*							Σ(2455)		***			
									Σ(2620)		***			
									Σ(3000)		**			

**** Good, clear, and unmistakable. *** Good, but in need of clarification or not absolutely certain.
** Needs confirmation.   * Weak.

[See notes on N's and Δ's, on possible Z^*'s, and on Y^*'s and Ξ^*'s at the beginning of those sections in the Baryon Data Card Listings; also see notes on individual resonances in the Baryon Data Card Listings.]

Particle[a]	$I \quad (J^P)$[a] ──── estab.	$\dfrac{\pi \text{ or K Beam}^b}{p_{beam}\text{(GeV/c)}}$ $\sigma = 4\pi\lambda^2$ (mb)	Mass M^c (MeV)	Full Width Γ^c (MeV)	M^2 $\pm\Gamma M^b$ (GeV^2)	Partial decay mode		
						Mode	Fraction %	p or p_{max}[d] (MeV/c)
p n	$1/2(1/2^+)$ ────		938.3 939.6		0.880 0.883	See Stable Particle Table		
N(1470)	$1/2(1/2^+)$ P'$_{11}$ ────	p = 0.66 σ = 27.8	1400 to 1470	165 to 300 (250)	2.16 ±0.37	Nπ Nη Nππ [Nε [Δπ [Nρ pγ^f nγ^f	~60 ~9 ~35 ~6]^e ~16]^e <5]^e 0.12-0.32 <0.24	420 d 368 d 177 d 435 435
N(1520)	$1/2(3/2^-)$ D'$_{13}$ ────	p = 0.74 σ = 23.5	1510 to 1540	105 to 150 (125)	2.31 ±0.19	Nπ Nππ [Nε [Nρ [Δπ Nη pγ^f nγ^f	~55 ~45 <5]^e ~18]^e ~23]^e 0.2-1.4 0.57-0.79 0.36-0.56	456 410 d d 228 d 471 471
N(1535)	$1/2(1/2^-)$ S'$_{11}$ ────	p = 0.76 σ = 22.5	1500 to 1600	50 to 160 (100)	2.36 ±0.15	Nπ Nη Nππ [Nρ [Nε pγ^f nγ	~35 ~55 ~10 ~2]^e ~2]^e .0.04-0.32 0.04-0.12	467 182 422 d d 481 481

Baryon Table (*Continued*)

Particle[a]	I (J^P)[a] estab.	π or K Beam[b] $\frac{P_{beam}(GeV/c)}{\sigma = 4\pi\lambda^2 \text{ (mb)}}$	Mass M^c (MeV)	Full Width Γ^c (MeV)	M^2 $\pm\Gamma M$[b] (GeV^2)	Mode	Fraction %	p or p_{max}[d] (MeV/c)
N(1670)[g]	1/2(5/2⁻)D'_{15}	p = 1.00 σ = 15.6	1670 to 1685	115 to 175 (140)	2.79 ±0.23	Nπ Nππ [Δπ] ΛK Nη pγ[f] nγ[f]	~40 ~60 ~50]e <1 <1h <0.04 0.01-0.13	560 525 360 200 368 572 572
N(1688)[g]	1/2(5/2⁺)F'_{15}	p = 1.03 σ = 14.9	1680 to 1690	105 to 180 (140)	2.85 ±0.24	Nπ Nππ [Nε] [Nρ] [Δπ] ΛK Nη pγ[f] nγ[f]	~60 ~40 ~13]e ~12]e ~11]e <0.1 <0.3h 0.20-0.44 <0.04	572 538 340 d 375 231 388 583 583
N(1700)[g]	1/2(1/2⁻)S''_{11}	p = 1.05 σ = 14.3	1665 to 1765	100 to 300 (200)	2.89 ±0.34	Nπ Nππ [Nε] [Nρ] [Δπ] ΛK ΣK Nη pγ[f] nγ[f]	~55 ~25 ~10]e ~10]e ~7]e ~5 <3 ~3h <0.24 <0.24	580 547 355 d 385 250 109 340 591 591
N(1780)[g]	1/2(1/2⁺)P''_{11}	p = 1.20 σ = 12.2	1650 to 1860	50 to 350 (200)	3.17 ±0.36	Nπ Nππ [Nε] [Nρ] [Δπ] ΛK ΣK Nη pγ[f] nγ[f]	~20 >40 16-40]e 20-45]e 8-20]e <7 ~7 2-20h <0.08 <0.08	633 603 440 249 448 353 267 476 643 643
N(1810)	1/2(3/2⁺)P_{13}	p = 1.26 σ = 11.5	1770 to 1860	180 to 330 (250)	3.28 ±0.45	Nπ Nππ [Nρ] ΛK Nη	~25 >50 45-55]e ~5 ~4h	652 624 297 386 503
N(2190)	1/2(7/2⁻)G_{17}	p = 2.07 σ = 6.21	2000 to 2260	150 to 325 (250)	4.80 ±0.55	Nπ	~25	888
N(2220)	1/2(9/2⁺)H_{19}	p = 2.14 σ = 5.97	2200 to 2245	260 to 330 (300)	4.93 ±0.67	Nπ	~15	905
N(2650)	1/2(?⁻)	p = 3.26 σ = 3.67	~2650	~360 (360)	7.02 ±0.95	Nπ	(J+1/2)x ~0.45j	1154
N(3030)	1/2(?)	p = 4.41 σ = 2.62	~3030	~400 (400)	9.18 ±1.21	Nπ	(J+1/2)x ~0.05j	1366
Δ(1232)[k]	3/2(3/2⁺)P'_{33}	p = 0.30 σ = 94.3	1230 to 1236	110 to 122 (110)	1.52 ±0.14	Nπ Nπ⁺π⁻ pγ[f]	~99.4 ~0 0.70-0.74	227 80 259
	Δ(++) Pole position:[k] M-iΓ/2 = (1211.6±0.7) -i(49.4±1.7)							
Δ(1650)	3/2(1/2⁻)S'_{31}	p=0.96 σ =16.4	1615 to 1695	140 to 200 (140)	2.72 ±0.23	Nπ Nππ [Nρ] [Δπ] pγ[f]	~30 ~70 10-26]e ~50]e 0.02-0.034	547 511 d 344 558

Baryon Table (*Continued*)

Particle[a]	I (J^P)[a] estab.	π or K Beam[b] p_{beam}(GeV/c) $\sigma = 4\pi\lambdabar^2$ (mb)	Mass M[c] (MeV)	Full Width Γ[c] (MeV)	M^2 $\pm\Gamma M$[b] (GeV²)	Mode	Fraction %	p or p_{max}[d] (MeV/c)
$\Delta(1670)$	$3/2(3/2^-)D_{33}$	p = 1.00 σ =15.6	1650 to 1720	190 to 270 (260)	2.79 ±0.43	Nπ N$\pi\pi$ [Nρ [$\Delta\pi$[f] pγ[f]	~15 >60 ~30]e ~40]e 0.1-0.7	560 525 d 361 572
$\Delta(1890)$	$3/2(5/2^+)F_{35}$	p = 1.42 σ = 9.88	1840 to 1920	140 to 350 (250)	3.57 ±0.47	Nπ N$\pi\pi$ [Nρ [$\Delta\pi$ ΣK pγ[f]	~17 >50 40-60]e 8-20]e <3 <0.3	704 677 403 531 400 712
$\Delta(1910)$	$3/2(1/2^+)P_{31}$	p = 1.46 σ = 9.54	1780 to 1935	200 to 340 (300)	3.65 ±0.57	Nπ N$\pi\pi$ [Nρ [$\Delta\pi$ ΣK	~25 ? small]e small]e ~6	716 691 429 545 420
$\Delta(1950)$	$3/2(7/2^+)F_{37}$	p = 1.54 σ = 8.90	1930 to 1980	170 to 270 (230)	3.80 ±0.45	Nπ N$\pi\pi$ [Nρ [$\Delta\pi$ ΣK $\Sigma(1385)$K pγ[f]	~40 >25 8-14]e 16-26]e ~2 ~1.4 0.16-0.34	741 716 471 574 460 232 749
$\Delta(2420)$	$3/2(11/2^+)$	p = 2.64 σ = 4.68	2320 to 2450	250 to 350 (300)	5.86 ±0.73	Nπ N$\pi\pi$	~11 >20	1023 1006
$\Delta(2850)$	$3/2(~?^+~)$	p = 3.85 σ = 3.05	~2850	~400 (400)	8.12 ±1.14	Nπ	(J+1/2)x ~ 0.25[j]	1266
$\Delta(3230)$	$3/2(~?~)$	p = 5.08 σ = 2.25	~3230	~440 (440)	10.43 ±1.42	Nπ	(J+1/2)x ~ 0.05[j]	1475

Z^*	Evidence for states with strangeness +1 is controversial. See the Baryon Data Card listings for discussion and display of data.							
Λ	$0(1/2^+)$		1115.6		1.245	See Stable Particle Table		
$\Lambda(1405)$	$0(1/2^-)S'_{01}$	below K⁻p threshold	1405 ±5[l]	40±10[l] (40)	1.97 ±0.06	$\Sigma\pi$	100	142
$\Lambda(1520)$	$0(3/2^-)D'_{03}$	p = 0.389 σ = 84.5	1518 ±2[l]	16 ±2[l] (16)	2.31 ±0.02	N$\overline{\text{K}}$ $\Sigma\pi$ $\Lambda\pi\pi$ $\Sigma\pi\pi$	45±1 41±1 10±.5 8±.1	234 258 250 140
$\Lambda(1670)$	$0(1/2^-)S''_{01}$	p = 0.74 σ = 28.5	1660 to 1680	23 to 40 (35)	2.79 ±0.06	N$\overline{\text{K}}$ $\Lambda\eta$ $\Sigma\pi$	15-35 15-25 30-50	410 64 393
$\Lambda(1690)$	$0(3/2^-)D''_{03}$	p = 0.78 σ = 26.1	1690 ±10[l]	30 to 70 (60)	2.86 ±0.10	N$\overline{\text{K}}$ $\Sigma\pi$ $\Lambda\pi\pi$ $\Sigma\pi\pi$	20-30 30-50 <25 <25	429 409 415 352
$\Lambda(1815)$	$0(5/2^+)F'_{05}$	p = 1.05 σ = 16.7	1820 ±5[l]	70 to 100 (85)	3.29 ±0.15	N$\overline{\text{K}}$ $\Sigma\pi$ $\Sigma(1385)\pi$	~61 ~11 15-20	542 508 362
$\Lambda(1830)$	$0(5/2^-)D'_{05}$	p = 1.09 σ = 15.8	1810 to 1840	70 to 120 (95)	3.35 ±0.17	N$\overline{\text{K}}$ $\Sigma\pi$ $\Lambda\eta$	~10 20-60 ~2	554 519 367
$\Lambda(2100)$	$0(7/2^-)G_{07}$	p = 1.68 σ = 8.68	2090 to 2120	80 to 140 (120)	4.41 ±0.25	N$\overline{\text{K}}$ $\Sigma\pi$ $\Lambda\eta$ ΞK $\Lambda\omega$	~30 ~5 <3 <3 <3	748 699 617 483 443

Baryon Table (*Continued*)

Particle[a]	I ——— estab.	$(J^P)^a$	π or K Beam[b] p_{beam}(GeV/c) $\sigma = 4\pi \lambda^2$ (mb)	Mass M^c (MeV)	Full Width Γ^c (MeV)	M^2 $\pm\Gamma M^b$ (GeV²)	Mode	Partial decay mode Fraction %	p or p_{max} [d] (MeV/c)
$\Lambda(2350)$	$\underline{0}$(?)		p = 2.29 σ = 5.85	~2350	140 to 320 (240)	5.52 ±0.56	$N\overline{K}$	(J+1/2)x ~ 0.9[j]	913
$\Lambda(2585)$	$\underline{0}$(?)		p = 2.91 σ = 4.37	~2585	~300 (300)	6.68 ±0.78	$N\overline{K}$	(J+1/2)x ~1.0[j]	1058
Σ	$\underline{1}(1/2^+)$			(+)1189.4 (0)1192.5 (-)1197.4		1.415 1.422 1.434	See Stable Particle Table		
$\Sigma(1385)$	$\underline{1}(3/2^+)P'_{13}$		below K^-p threshold	(+)1383±1 S=1.2[m] (-)1387±1 S=2.4[m]	(+)35±2 S=2.0[m] (-)42±5 S=3.5[m] (35)	1.92 ±0.05	$\Lambda\pi$ $\Sigma\pi$	88±2 12±2	208 117
$\Sigma(1670)^n$	$\underline{1}(3/2^-)D'_{13}$		p = 0.74 σ = 28.5	1670 ±10[l]	35 to 60 (50)	2.79 ±0.08	$N\overline{K}$ $\Sigma\pi$ $\Lambda\pi$	~8 30-60 ~12	410 387 447
$\Sigma(1750)$	$\underline{1}(1/2^-)S''_{11}$		p = 0.91 σ = 20.7	1700 to 1790	50 to 100 (75)	3.06 ±0.13	$N\overline{K}$ $\Lambda\pi$ $\Sigma\pi$ $\Sigma\eta$	12-45 5-18 6-19 11-44	483 507 450 54
$\Sigma(1765)$	$\underline{1}(5/2^-)D_{15}$		p = 0.94 σ = 19.6	1765 ±5[l]	~120 (120)	3.12 ±0.21	$N\overline{K}$ $\Lambda\pi$ $\Lambda(1520)\pi$ $\Sigma(1385)\pi$ $\Sigma\pi$	~41 ~13 ~15 ~10 ~1	496 518 187 315 461
$\Sigma(1915)^{g, o}$	$\underline{1}(5/2^+)F'_{15}$		p = 1.25 σ = 13.0	1900 to 1930	50 to 120 (80)	3.67 ±0.15	$N\overline{K}$ $\Lambda\pi$ $\Sigma\pi$	~14 ~6 ~6	612 619 568
$\Sigma(1940)^i$	$\underline{1}(3/2^-)D''_{13}$		p = 1.32 σ = 12.0	1865 to 1950	110 to 280 (220)	3.76 ±0.43	$N\overline{K}$ $\Lambda\pi$ $\Sigma\pi$	~21 ~4 < 7	678 680 589
$\Sigma(2030)$	$\underline{1}(7/2^+)F_{17}$		p = 1.52 σ = 9.93	2020 to 2040	120 to 170 (140)	4.12 ±0.28	$N\overline{K}$ $\Lambda\pi$ $\Sigma\pi$ ΞK	~20 ~20 ~4 < 2	700 700 652 412
$\Sigma(2250)$	$\underline{1}$(?)		p = 2.04 σ = 6.76	2245 to 2280	100 to 230 (160)	5.06 ±0.36	$N\overline{K}$	(J+1/2)x ~0.3[j]	849
$\Sigma(2455)$	$\underline{1}$(?)		p = 2.57 σ = 5.09	~2455	~ 120 (120)	6.03 ±0.29	$N\overline{K}$	(J+1/2)x ~0.2[j]	979
$\Sigma(2620)$	$\underline{1}$(?)		p = 2.95 σ = 4.30	~2620	~ 175 (175)	6.86 ±0.46	$N\overline{K}$	(J+1/2)x ~0.3[j]	1064
Ξ	$\underline{1/2}(1/2^+)$			(0)1314.9 (-)1321.3		1.729 1.746	See Stable Particle Table		
$\Xi(1530)^p$	$\underline{1/2}(3/2^+)P_{13}$			(0)1531.8±0.3 S=1.3[m] (-)1535.1±0.7	(0) 9.1±0.5 (-) 10.6±2.6 (10)	2.34 ±0.02	$\Xi\pi$	100	144
$\Xi(1820)^{p, q}$	$\underline{1/2}$(?)			1795 to 1870	12 to 100 (60)	3.31 ±0.11	$\Lambda\overline{K}$ $\Sigma\overline{K}$ $\Xi\pi$ $\Xi(1530)\pi$	seen seen seen seen	396 306 413 234
$\Xi(1940)^{p, r}$	$\underline{1/2}$(?)			1920 to 1960	40 to 140 (90)	3.76 ±0.17	$\Xi\pi$ $\Xi(1530)\pi$	seen seen	499 336
Ω^-	$\underline{0}(3/2^+)$			1672.2		2.796	See Stable Particle Table		

Baryon Table (*Continued*)

→   For convenience all Baryon States for which information exists in the Baryon Data Card Listings
are listed at the beginning of the Baryon Table. States with only a one or two star (*) rating in that
list have been omitted from the main Baryon Table; each omitted state is indicated by an arrow in
the left-hand margin of the Table. In the Listings there is an arrow under the name of each state
omitted from the Table.

a.   The names of the Baryon States in Col. 1 [such as N(1470)] contain a nominal mass which is a
rounded average of the reported values in the Data Card Listings. The convention for using primes
in the spectroscopic notation for the quantum numbers in Col. 2 [such as P'_{11}] is as follows: no
prime is attached when the Data Card Listings include only one resonance on the given Argand
diagram; when there is more than one resonance the first has been designated with a prime, the
second with a double prime, etc. The name and the quantum numbers for each state are also
given in large print at the beginning of the Data Card Listings for that state.

b.   The numbers in Col. 3 and Col. 6 are calculated using the nominal mass (see a. above) for M and
the nominal width (see c. below) for Γ.

c.   For M and Γ of most baryons we report here an interval instead of an average. Averages are
appropriate if each result is based on independent measurements, but inappropriate where the
spread in parameters arises because different models or procedures have been applied to a common set of data. A single value with an approximation sign (~) indicates that there is not enough
data to give a meaningful interval. A nominal width is included in parentheses in Col. 5; this nominal
width is used to calculate the value of ΓM given in Col. 6.

d.   For two body decay modes we give the momentum, p, of the decay products in the decaying baryon
rest frame. For decay modes into ≥ 3 particles we give the maximum momentum, p_{max}, that any
of the particles in the final state can have in this frame. The momenta are calculated using the
nominal mass (see a. above) of the decaying baryon, and of any isobars in the final state. Some
decays which would be energetically forbidden for the nominal masses actually occur because of
the finite widths of the decaying Baryon and/or isobars in the final state. In these cases, the decay momentum is omitted from Col. 9 and replaced with a reference to this footnote.

e.   Square brackets around an isobar decay mode indicate that it is a sub-reaction of the previous unbracketed decay mode. In the case of N^* and Δ decays into isobar modes we have used the isobar
model results of HERNDON 74 in addition to data from the listings (where available) to estimate
the branching fractions. The results of HERNDON 74 are shown in Table III.1 of the mini-review
preceding the N^* Data Card Listings.

f.   The tabulated radiative fractions involve a sum over two helicities (1/2, 3/2). In the case of
I = 1/2 resonances, there are two distinct isospin couplings, whence γp and γn. For conventions
and further details, see the Mini-Review preceding the Baryon Data Card Listings.

g.   Only information coming from partial-wave analyses has been used here. For the production experiments results see the Baryon Data Card Listings.

h.   Value obtained in an energy-dependent partial-wave analysis which uses a t-channel-poles-plus-
resonance parametrization.

i.   There may be more than one state in this region. The only analysis which reports an elastic
coupling (LEA 73) also finds unusually low mass and width values. Note that all branching fractions quoted here depend on the elasticity of LEA 73.

j.   This state has been seen only in total cross sections. J is not known; x is Γ_{el}/Γ.

k.   See note on $\Delta(1232)$ in the Baryon Data Card Listings. Values of mass and width are dependent
upon resonance shape used to fit the data. The pole position appear to be much less dependent
upon the parametrization used.

ℓ.   The error given here is only an educated guess; it is larger than the error of the average of the
published values (see the Baryon Data Card Listings for the latter).

m.   Quoted error includes an S (scale) factor. See first footnote to Stable Particle Table.

n.   In this energy region the situation is still confused. In addition to the effect at ~ 1670 MeV seen in
both production and formation experiments, recent formation experiments have found evidence for
fairly narrow ($\Gamma \sim 50$ MeV) S_{11} and/or P_{11} states near 1620 MeV. A narrow bump in the I = 1 $\overline{K}N$
total cross section has also been seen recently at ~ 1590 MeV. It is not clear how many states
really exist here. No one has reported a strong coupling of any of these states to $\overline{K}N$ but there is
much disagreement about branching ratios into $\pi\Lambda$ and $\pi\Sigma$. See the mini-reviews preceding the
$\Sigma(1620)$ and $\Sigma(1670)$ Data Card Listings for more information.

o.   Formation and production experiments do not agree on the $\Sigma\pi/\Lambda\pi$ ratio.

p.   Only $\Xi(1530)$ is firmly established; information on the other states comes from experiments that
have poor statistics due to the fact that the cross sections for S = - 2 states are very low. For
Ξ states, because of the meager statistics, we lower our standards and tabulate resonant effects
if they have at least a four-standard-deviation statistical significance and if they are seen by more
than one group. So $\Xi(2030)$, with main decay mode $\Sigma\overline{K}$, reported as a 3.5-standard-deviation effect,
is not tabulated. See the Baryon Data Card Listings for the other states.

q.   All four decay modes shown have been seen. Branching ratios are not quoted because there may be
more than one state here.

r.   This bump has been seen in both final states shown; it is not clear if one, or more, states are
present.

Nomenclature for the Tables of Particle Physics

Quantum Numbers:

I = isospin	Symbolic
G = G-parity	representation
J = spin	$I^G(J^P)C$
P = parity	
C = charge conjugation parity.	

PARTICLE NAME CONVENTIONS

Name	I	Y	S	G
		Mesons		
η	0	0	0	+
ω or ϕ^a	0	0	0	−
ρ	1	0	0	+
π	1	0	0	−
K^+, K^0	$\frac{1}{2}$	+1	+1	
$K^-, \overline{K}^0$	$\frac{1}{2}$	−1	−1	
		Baryons		
N	$\frac{1}{2}$	+1	0	
Δ	$\frac{3}{2}$	+1	0	
Z_0, Z_1	0, 1	+2	+1	
Λ	0	0	−1	
Σ	1	0	−1	
Ξ	$\frac{1}{2}$	−1	−2	
Ω	0	−2	−3	

[a]Since 1973, we have used the symbol ω for those $I^G = 0^-$ mesons that decay mainly into $3\pi[\omega(783)$, $\omega(1670)]$; we reserve the symbol ϕ for $\phi(1019)$ and possible future higher-mass $I^G = 0^-$ mesons that decay mainly into $K\overline{K}$.

SECTION 1b
ELECTROMAGNETIC RADIATION

This section deals very briefly with a definition of the electromagnetic spectrum, followed by appropriate conversion tables and tables listing X-ray wavelengths and X-ray atomic energy levels.

Figure 1b1. The Electromagnetic Spectrum.*

This chart presents an overview of the complete electromagnetic radiation spectrum, extending from infrasonics to cosmic rays. The wavelength, the amount of energy required to radiate one photon, a general description, the band designation, and, the normal occurrence or use are given.

*Reproduced from Graf, R. F., "Electronic Databook A Guide for designers, 2nd Ed.," Van Nostrand Reinhold Co., New York, 1974.

TABLE 1b1. CONVERSION FACTORS FROM:
WAVENUMBER (cm^{-1})—WAVELENGTH (MICRONS).

Wavenumber (cm^{-1})	Wavelength (μ)	Wavenumber (cm^{-1})	Wavelength (μ)	Wavenumber (cm^{-1})	Wavelength (μ)
5	2000	500	20.0	1×10^4	1.00
10	1000	520	19.2	1.1	0.91
15	666.7	540	18.5	1.2	0.83
20	500	560	17.9	1.3	0.77
25	400	580	17.2	1.4	0.72
30	333.3	600	16.7	1.5	0.67
35	285.7	620	16.1	1.6	0.62
40	250	640	15.6	1.7	0.59
45	222.2	660	15.2	1.8	0.56
50	200	680	14.7	1.9	0.53
55	181.8	700	14.3	2.0	0.50
60	166.7	750	13.3	2.1	0.48
65	153.8	800	12.5	2.3	0.46
70	142.9	850	11.8	2.4	0.42
75	133.3	900	11.1	2.5	0.40
80	125	950	10.5	2.6	0.38
85	117.6	1000	10.0	2.7	0.37
90	111.1	1100	9.1	2.8	0.36
95	105.3	1200	8.3	2.9	0.34
100	100.0	1300	7.6	3.0	0.33
120	83.3	1400	7.1	3.2	0.31
140	71.4	1500	6.7	3.4	0.29
160	62.5	1600	6.2	3.6	0.28
180	55.6	1700	5.9	3.8	0.26
200	50.0	1800	5.6	4.0	0.25
220	45.5	1900	5.2	4.2	0.24
240	41.7	2000	5.0	4.4	0.23
260	38.5	2200	4.5	4.6	0.22
280	35.7	2400	4.2	4.8	0.21
300	33.3	2600	3.8	5.0	0.20
320	31.2	2800	3.6	6.0	0.17
340	29.4	3000	3.3	8.0	0.12
360	27.8	3500	2.8	1×10^5	0.10
380	26.3	4000	2.5		
400	25.0	5000	2.0		
420	23.8	6000	1.7		
440	22.7	7000	1.4		
460	21.7	8000	1.2		
480	20.8	9000	1.1		

TABLE 1b2. CONVERSION FACTORS FROM: ELECTRONS VOLTS (eV)– WAVELENGTH (MICRONS)

eV	μ	eV	μ	eV	μ	eV	μ
0.1	12.4	1.8	0.69	3.5	0.35	5.2	0.24
0.2	6.2	1.9	0.65	3.6	0.34	5.3	0.23
0.3	4.13	2.0	0.62	3.7	0.34	5.4	0.23
0.4	3.1	2.1	0.59	3.8	0.33	5.5	0.23
0.5	2.48	2.2	0.56	3.9	0.32	5.6	0.22
0.6	2.07	2.3	0.54	4.0	0.31	5.7	0.22
0.7	1.77	2.4	0.52	4.1	0.30	5.8–6.0	0.21
0.8	1.55	2.5	0.50	4.2	0.30	6.1–6.3	0.20
0.9	1.38	2.6	0.48	4.3	0.29	6.4–6.7	0.19
1.0	1.24	2.7	0.46	4.4	0.28	6.8–7.0	0.18
1.1	1.13	2.8	0.44	4.5	0.28	7.1–7.5	0.17
1.2	1.03	2.9	0.43	4.6	0.27	7.6–8.0	0.16
1.3	0.95	3.0	0.41	4.7	0.26	8.1–8.5	0.15
1.4	0.89	3.1	0.40	4.8	0.26	8.6–9.1	0.14
1.5	0.83	3.2	0.39	4.9	0.25	9.2–9.9	0.13
1.6	0.78	3.3	0.38	5.0	0.25	10.0–10.8	0.12
1.7	0.73	3.4	0.36	5.1	0.24	10.9–11.8	0.11
						11.9–13.0	0.10

TABLE 1b3. X-RAY WAVELENGTHS IN NUMERICAL ORDER OF THE EMISSION LINES AND ABSORPTION EDGES. (BEARDEN, 1967)*

Wavelength Å*	p.e.	Element	Designation	keV	Wavelength Å*	p.e.	Element	Designation	keV
0.10723	1	92 U	K	Abs. Edge	115.62	0.125947	3	92 U	$K\alpha_1$	KL_{III}	98.439
0.10744	1	92 U		$KO_{II,III}$	115.39	0.12698	5	86 Rn	$K\beta_2^I$	KN_{III}	97.64
0.10780	2	92 U	$K\beta_4$	$KN_{IV,V}$	115.01	0.12719	5	86 Rn	$K\beta_2^{II}$	KN_{II}	97.47
0.10818	1	92 U	$K\beta_2^I$	KN_{III}	114.60	0.12719	5	87 Fr	$K\beta_1$	KM_{III}	97.47
0.10837	1	92 U	$K\beta_2^{II}$	KN_{II}	114.40	0.12807	5	87 Fr	$K\beta_3$	KM_{II}	96.81
0.11069	1	92 U	$K\beta_5$	$KM_{IV,V}$	112.01	0.129325	3	91 Pa	$K\alpha_1$	KL_{III}	95.868
0.11107	2	91 Pa	$K\beta_2^I$	KN_{III}	111.62	0.13052	4	85 At	$K\beta_2^I$	KN_{III}	94.99
0.11129	2	91 Pa	$K\beta_2^{II}$	KN_{II}	111.40	0.13069	5	86 Rn	$K\beta_1$	KM_{III}	94.87
0.111394	5	92 U	$K\beta_1$	KM_{III}	111.300	0.13072	4	85 At	$K\beta_2^{II}$	KN_{II}	94.84
0.112296	4	92 U	$K\beta_3$	KM_{II}	110.406	0.130968	4	92 U	$K\alpha_2$	KL_{II}	94.665
0.11307	1	90 Th	K	Abs. Edge	109.646	0.13155	5	86 Rn	$K\beta_3$	KM_{II}	94.24
0.11322	1	90 Th		$KO_{II,III}$	109.500	0.132813	2	90 Th	$K\alpha_1$	KL_{III}	93.350
0.11366	2	90 Th	$K\beta_4$	$KN_{IV,V}$	109.08	0.13418	2	84 Po	$K\beta_2^I$	KN_{III}	92.40
0.114040	9	90 Th	$K\beta_2^I$	KN_{III}	108.717	0.13432	4	85 At	$K\beta_1$	KM_{III}	92.30
0.11426	1	90 Th	$K\beta_2^{II}$	KN_{II}	108.511	0.134343	9	91 Pa	$K\alpha_2$	KL_{II}	92.287
0.114345	8	91 Pa	$K\beta_1$	KM_{III}	108.427	0.13438	2	84 Po	$K\beta_2^{II}$	KN_{II}	92.26
0.11523	2	91 Pa	$K\beta_3$	KM_{II}	107.60	0.13517	4	85 At	$K\beta_3$	KM_{II}	91.72
0.116667	9	90 Th	$K\beta_5$	$KM_{IV,V}$	106.269	0.136417	8	89 Ac	$K\alpha_1$	KL_{III}	90.884
0.11711	2	89 Ac	$K\beta_2^I$	KN_{III}	105.86	0.13694	1	83 Bi	K	Abs. Edge	90.534
0.11732	2	89 Ac	$K\beta_2^{II}$	KN_{II}	105.67	0.13709	1	83 Bi		$KO_{II,III}$	90.435
0.117396	9	90 Th	$K\beta_1$	KM_{III}	105.609	0.13759	2	83 Bi	$K\beta_4$	$KN_{IV,V}$	90.11
0.118268	3	90 Th	$K\beta_3$	KM_{II}	104.831	0.137829	2	90 Th	$K\alpha_2$	KL_{II}	89.953
0.12029	3	88 Ra	$K\beta_2^I$	KN_{III}	103.07	0.13797	1	83 Bi	$K\beta_2^I$	KN_{III}	89.864
0.12050	3	88 Ra	$K\beta_2^{II}$	KN_{II}	102.89	0.13807	2	84 Po	$K\beta_1$	KM_{III}	89.80
0.12055	2	89 Ac	$K\beta_1$	KM_{III}	102.85	0.13817	1	83 Bi	$K\beta_2^{II}$	KN_{II}	89.733
0.12143	2	89 Ac	$K\beta_3$	KM_{II}	102.10	0.13892	2	84 Po	$K\beta_3$	KM_{II}	89.25
0.12358	5	87 Fr	$K\beta_2^I$	KN_{III}	100.33	0.14014	2	88 Ra	$K\alpha_1$	KL_{III}	88.47
0.12379	5	87 Fr	$K\beta_2^{II}$	KN_{II}	100.16	0.1408	1	82 Pb		KP	88.06
0.12382	3	88 Ra	$K\beta_1$	KM_{III}	100.13	0.140880	5	82 Pb	K	Abs. Edge	88.005
0.12469	3	88 Ra	$K\beta_3$	KM_{II}	99.43	0.141012	8	82 Pb		$KO_{II,III}$	87.922

TABLE 1b3. (*Continued*)

Wavelength Å*	p.e.	Element		Designation	keV	Wavelength Å*	p.e.	Element		Designation	keV
0.14111	1	83 Bi	$K\beta_5$	$KM_{IV,V}$	87.860	0.167373	9	77 Ir	$K\beta_5{}^I$	KM_V	74.075
0.14141	2	89 Ac	$K\alpha_2$	KL_{II}	87.67	0.16759	2	77 Ir	$K\beta_5{}^{II}$	KM_{IV}	73.980
0.14155	3	82 Pb	$K\beta_4$	$KN_{IV,V}$	87.59	0.16787	1	76 Os	K	Abs. Edge	73.856
0.14191	1	82 Pb	$K\beta_2{}^I$	KN_{III}	87.364	0.16798	1	76 Os		$KO_{II,III}$	73.808
0.141948	3	83 Bi	$K\beta_1$	KM_{III}	87.343	0.16842	2	76 Os	$K\beta_4$	$KN_{IV,V}$	73.615
0.14212	2	82 Pb	$K\beta_2{}^{II}$	KN_{II}	87.23	0.168542	2	77 Ir	$K\beta_1$	KM_{III}	73.5608
0.142779	7	83 Bi	$K\beta_3$	KM_{II}	86.834	0.168906	6	76 Os	$K\beta_2{}^I$	KN_{III}	73.402
0.14399	3	87 Fr	$K\alpha_1$	KL_{III}	86.10	0.16910	1	76 Os	$K\beta_2{}^{II}$	KN_{II}	73.318
0.14495	1	81 Tl	K	Abs. Edge	85.533	0.169367	2	77 Ir	$K\beta_3$	KM_{II}	73.2027
0.14495	3	82 Pb	$K\beta_5{}^I$	KM_V	85.53	0.170136	2	81 Tl	$K\alpha_1$	KL_{III}	72.8715
0.14509	1	81 Tl		$KO_{II,III}$	85.451	0.170294	2	82 Pb	$K\alpha_2$	KL_{II}	72.8042
0.14512	2	82 Pb	$K\beta_5{}^{II}$	KM_{IV}	85.43	0.17245	1	76 Os	$K\beta_5{}^I$	KM_V	71.895
0.14512	2	88 Ra	$K\alpha_2$	KL_{II}	85.43	0.17262	1	76 Os	$K\beta_5{}^{II}$	KM_{IV}	71.824
0.14553	2	81 Tl	$K\beta_4$	$KN_{IV,V}$	85.19	0.17302	1	75 Re	K	Abs. Edge	71.658
0.14595	1	81 Tl	$K\beta_2{}^I$	KN_{III}	84.946	0.17308	1	75 Re		$KO_{II,III}$	71.633
0.145970	6	82 Pb	$K\beta_1$	KM_{III}	84.936	0.173611	3	76 Os	$K\beta_1$	KM_{III}	71.413
0.14614	1	81 Tl	$K\beta_2{}^{II}$	KN_{II}	84.836	0.17362	2	75 Re	$K\beta_4$	$KN_{IV,V}$	71.410
0.146810	4	82 Pb	$K\beta_3$	KM_{II}	84.450	0.174054	6	75 Re	$K\beta_2{}^I$	KN_{III}	71.232
0.14798	3	86 Rn	$K\alpha_1$	KL_{III}	83.78	0.17425	1	75 Re	$K\beta_2{}^{II}$	KN_{II}	71.151
0.14896	3	87 Fr	$K\alpha_2$	KL_{II}	83.23	0.174431	3	76 Os	$K\beta_3$	KM_{II}	71.077
0.14917	1	81 Tl	$K\beta_5$	$KM_{IV,V}$	83.114	0.175036	2	81 Tl	$K\alpha_2$	KL_{II}	70.8319
0.14918	1	80 Hg	K	Abs. Edge	83.109	0.175068	3	80 Hg	$K\alpha_1$	KL_{III}	70.819
0.14931	2	80 Hg		$KO_{II,III}$	83.04	0.17766	1	75 Re	$K\beta_5{}^I$	KM_V	69.786
0.14978	2	80 Hg	$K\beta_4$	$KN_{IV,V}$	82.78	0.17783	1	75 Re	$K\beta_5{}^{II}$	KM_{IV}	69.719
0.150142	5	81 Tl	$K\beta_1$	KM_{III}	82.576	0.17837	1	74 W	K	Abs. Edge	69.508
0.15020	2	80 Hg	$K\beta_2{}^I$	KN_{III}	82.54	0.178444	5	74 W		$KO_{II,III}$	69.479
0.15040	2	80 Hg	$K\beta_2{}^{II}$	KN_{II}	82.43	0.178880	3	75 Re	$K\beta_1$	KM_{III}	69.310
0.150980	6	81 Tl	$K\beta_3$	KM_{II}	82.118	0.17892	2	74 W	$K\beta_4$	$KN_{IV,V}$	69.294
0.15210	2	85 At	$K\alpha_1$	KL_{III}	81.52	0.179421	7	74 W	$K\beta_2{}^I$	KN_{III}	69.101
0.15294	3	86 Rn	$K\alpha_2$	KL_{II}	81.07	0.17960	1	74 W	$K\beta_2{}^{II}$	KN_{II}	69.031
0.15353	2	80 Hg	$K\beta_5$	$KM_{IV,V}$	80.75	0.179697	3	75 Re	$K\beta_3$	KM_{II}	68.994
0.153593	5	79 Au	K	Abs. Edge	80.720	0.179958	3	80 Hg	$K\alpha_2$	KL_{II}	68.895
0.153694	7	79 Au		$KO_{II,III}$	80.667	0.180195	2	79 Au	$K\alpha_1$	KL_{III}	68.8037
0.154224	5	79 Au	$K\beta_4$	$KN_{IV,V}$	80.391	0.183092	7	74 W	$K\beta_5{}^I$	KM_V	67.715
0.154487	3	80 Hg	$K\beta_1$	KM_{III}	80.253	0.183264	5	74 W	$K\beta_5{}^{II}$	KM_{IV}	67.652
0.154618	9	79 Au	$K\beta_2{}^I$	KN_{III}	80.185	0.18394	1	73 Ta	K	Abs. Edge	67.403
0.15483	2	79 Au	$K\beta_2{}^{II}$	KN_{II}	80.08	0.184031	7	73 Ta		$KO_{II,III}$	67.370
0.155321	3	80 Hg	$K\beta_3$	KM_{II}	79.822	0.184374	2	74 W	$K\beta_1$	KM_{III}	67.2443
0.15636	1	84 Po	$K\alpha_1$	KL_{III}	79.290	0.18451	1	73 Ta	$K\beta_4$	$KN_{IV,V}$	67.194
0.15705	2	85 At	$K\alpha_2$	KL_{II}	78.95	0.185011	8	73 Ta	$K\beta_2{}^I$	KN_{III}	67.013
0.157880	5	79 Au	$K\beta_5{}^I$	KM_V	78.529	0.185075	2	79 Au	$K\alpha_2$	KL_{II}	66.9895
0.158062	7	79 Au	$K\beta_5{}^{II}$	KM_{IV}	78.438	0.185181	2	74 W	$K\beta_3$	KM_{II}	66.9514
0.15818	1	78 Pt	K	Abs. Edge	78.381	0.185188	9	73 Ta	$K\beta_2{}^{II}$	KN_{II}	66.949
0.15826	1	78 Pt		$KO_{II,III}$	78.341	0.185511	4	78 Pt	$K\alpha_1$	KL_{III}	66.832
0.15881	2	78 Pt	$K\beta_4$	$KN_{IV,V}$	78.069	0.18672	4	79 Au		KL_I	66.40
0.158982	3	79 Au	$K\beta_1$	KM_{III}	77.984	0.188757	6	73 Ta	$K\beta_5{}^I$	KM_V	65.683
0.15920	1	78 Pt	$K\beta_2{}^I$	KN_{III}	77.878	0.188920	6	73 Ta	$K\beta_5{}^{II}$	KM_{IV}	65.626
0.15939	1	78 Pt	$K\beta_2{}^{II}$	KN_{II}	77.785	0.18982	5	72 Hf	K	Abs. Edge	65.31
0.159810	2	79 Au	$K\beta_3$	KM_{II}	77.580	0.190089	4	73 Ta	$K\beta_1$	KM_{III}	65.223
0.160789	2	83 Bi	$K\alpha_1$	KL_{III}	77.1079	0.190381	4	78 Pt	$K\alpha_2$	KL_{II}	65.122
0.16130	1	84 Po	$K\alpha_2$	KL_{II}	76.862	0.1908	2	72 Hf	$K\beta_2$	$KN_{II,III}$	64.98
0.16255	3	78 Pt	$K\beta_5{}^I$	KM_V	76.27	0.190890	2	73 Ta	$K\beta_3$	KM_{II}	64.9488
0.16271	2	78 Pt	$K\beta_5{}^{II}$	KM_{IV}	76.199	0.191047	2	77 Ir	$K\alpha_1$	KL_{III}	64.8956
0.16292	1	77 Ir	K	Abs. Edge	76.101	0.19585	5	71 Lu	K	Abs. Edge	63.31
0.163019	5	77 Ir		$KO_{II,III}$	76.053	0.19589	2	71 Lu		$KO_{II,III}$	63.293
0.16352	2	77 Ir	$K\beta_4$	$KN_{IV,V}$	75.821	0.195904	2	77 Ir	$K\alpha_2$	KL_{II}	63.2867
0.163675	3	78 Pt	$K\beta_1$	KM_{III}	75.748	0.19607	3	72 Hf	$K\beta_1$	KM_{III}	63.234
0.163956	7	77 Ir	$K\beta_2{}^I$	KN_{III}	75.619	0.196794	2	76 Os	$K\alpha_1$	KL_{III}	63.0005
0.16415	1	77 Ir	$K\beta_2{}^{II}$	KN_{II}	75.529	0.19686	4	72 Hf	$K\beta_3$	KM_{II}	62.98
0.164501	3	78 Pt	$K\beta_3$	KM_{II}	75.368	0.1969	2	71 Lu	$K\beta_2$	$KN_{II,III}$	62.97
0.165376	2	82 Pb	$K\alpha_1$	KL_{III}	74.9694	0.20084	2	71 Lu	$K\beta_5$	$KM_{IV,V}$	61.732
0.165717	2	83 Bi	$K\alpha_2$	KL_{II}	74.8148	0.201639	2	76 Os	$K\alpha_2$	KL_{II}	61.4867

TABLE 1b3. (Continued)

Wavelength Å*	p.e.	Element	Designation		keV
0.20224	5	70 Yb	K	Abs. Edge	61.30
0.20226	2	70 Yb		$KO_{II,III}$	61.298
0.20231	3	71 Lu	$K\beta_1$	KM_{III}	61.283
0.202781	2	75 Re	$K\alpha_1$	KL_{III}	61.1403
0.20309	4	71 Lu	$K\beta_2$	KM_{II}	61.05
0.2033	2	70 Yb	$K\beta_2$	$KN_{II,III}$	60.89
0.20739	2	70 Yb	$K\beta_5$	$KM_{IV,V}$	59.782
0.207611	1	75 Re	$K\alpha_2$	KL_{II}	59.7179
0.20880	5	69 Tm	K	Abs. Edge	59.38
0.20884	8	70 Yb	$K\beta_1$	KM_{III}	59.37
0.20891	2	69 Tm		$KO_{II,III}$	59.346
0.2090100	Std.	74 W	$K\alpha_1$	KL_{III}	59.31824
0.2096	1	70 Yb	$K\beta_2$	KM_{II}	59.14
0.2098	2	69 Tm	$K\beta_2$	$KN_{II,III}$	59.09
0.213828	2	74 W	$K\alpha_2$	KL_{II}	57.9817
0.21404	2	69 Tm	$K\beta_5$	$KM_{IV,V}$	57.923
0.215497	4	73 Ta	$K\alpha_1$	KL_{III}	57.532
0.21556	2	69 Tm	$K\beta_1$	KM_{III}	57.517
0.21567	1	68 Er	K	Abs. Edge	57.487
0.21581	3	68 Er		$KO_{II,III}$	57.450
0.21592	4	74 W		KL_{I}	57.42
0.21636	2	69 Tm	$K\beta_2$	KM_{II}	57.304
0.2167	2	68 Er	$K\beta_2$	$KN_{II,III}$	57.21
0.220305	8	73 Ta	$K\alpha_2$	KL_{II}	56.277
0.22124	3	68 Er	$K\beta_5$	$KM_{IV,V}$	56.040
0.222227	3	72 Hf	$K\alpha_1$	KL_{III}	55.7902
0.22266	2	68 Er	$K\beta_1$	KM_{III}	55.681
0.22291	1	67 Ho	K	Abs. Edge	55.619
0.22305	3	67 Ho		$KO_{II,III}$	55.584
0.22341	2	68 Er	$K\beta_2$	KM_{II}	55.494
0.2241	2	67 Ho	$K\beta_2$	$KN_{II,III}$	55.32
0.227024	3	72 Hf	$K\alpha_2$	KL_{II}	54.6114
0.22855	3	67 Ho	$K\beta_5$	$KM_{IV,V}$	54.246
0.229298	2	71 Lu	$K\alpha_1$	KL_{III}	54.0698
0.23012	2	67 Ho	$K\beta_1$	KM_{III}	53.877
0.23048	1	66 Dy	K	Abs. Edge	53.793
0.23056	3	66 Dy		$KO_{II,III}$	53.774
0.23083	2	67 Ho	$K\beta_2$	KM_{II}	53.711
0.2317	2	66 Dy	$K\beta_2$	$KN_{II,III}$	53.47
0.234081	2	71 Lu	$K\alpha_2$	KL_{II}	52.9650
0.23618	3	66 Dy	$K\beta_5$	$KM_{IV,V}$	52.494
0.236655	2	70 Yb	$K\alpha_1$	KL_{III}	52.3889
0.23788	2	66 Dy	$K\beta_1$	KM_{III}	52.119
0.23841	1	65 Tb	K	Abs. Edge	52.002
0.23858	3	65 Tb		$KO_{II,III}$	51.965
0.23862	2	66 Dy	$K\beta_2$	KM_{II}	51.957
0.2397	2	65 Tb	$K\beta_2$	$KN_{II,III}$	51.68
0.241424	2	70 Yb	$K\alpha_2$	KL_{II}	51.3540
0.244338	2	69 Tm	$K\alpha_1$	KL_{III}	50.7416
0.24608	2	65 Tb	$K\beta_1$	KM_{III}	50.382
0.24681	1	64 Gd	K	Abs. Edge	50.233
0.24683	2	65 Tb	$K\beta_2$	KM_{II}	50.229
0.24687	3	64 Gd		$KO_{II,III}$	50.221
0.24816	3	64 Gd	$K\beta_2$	$KN_{II,III}$	49.959
0.249095	2	69 Tm	$K\alpha_2$	KL_{II}	49.7726
0.252365	2	68 Er	$K\alpha_1$	KL_{III}	49.1277
0.25275	3	64 Gd	$K\beta_5$	$KM_{IV,V}$	49.052
0.25460	2	64 Gd	$K\beta_1$	KM_{III}	48.697
0.25534	2	64 Gd	$K\beta_2$	KM_{II}	48.555
0.25553	1	63 Eu	K	Abs. Edge	48.519
0.255645	7	63 Eu		$KO_{II,III}$	48.497
0.256923	8	63 Eu	$K\beta_2^{I}$	$KN_{II,III}$	48.256
0.257110	2	68 Er	$K\alpha_2$	KL_{II}	48.2211
0.260756	2	67 Ho	$K\alpha_1$	KL_{III}	47.5467
0.263577	5	63 Eu	$K\beta_1$	KM_{III}	47.0379
0.264332	5	63 Eu	$K\beta_2$	KM_{II}	46.9036
0.26464	5	62 Sm	K	Abs. Edge	46.849
0.26491	3	62 Sm		$KO_{II,III}$	46.801
0.265486	2	67 Ho	$K\alpha_2$	KL_{II}	46.6997
0.2662	1	62 Sm	$K\beta_2$	$KN_{II,III}$	46.57
0.269533	2	66 Dy	$K\alpha_1$	KL_{III}	45.9984
0.27111	3	62 Sm	$K\beta_5$	$KM_{IV,V}$	45.731
0.27301	2	62 Sm	$K\beta_1$	KM_{III}	45.413
0.27376	2	62 Sm	$K\beta_2$	KM_{II}	45.289
0.274247	2	66 Dy	$K\alpha_2$	KL_{II}	45.2078
0.27431	5	61 Pm	K	Abs. Edge	45.198
0.2759	1	61 Pm	$K\beta_2$	$KN_{II,III}$	44.93
0.278724	2	65 Tb	$K\alpha_1$	KL_{III}	44.4816
0.28290	3	61 Pm	$K\beta_1$	KM_{III}	43.826
0.283423	2	65 Tb	$K\alpha_2$	KL_{II}	43.7441
0.28363	4	61 Pm	$K\beta_2$	KM_{II}	43.713
0.28453	5	60 Nd	K	Abs. Edge	43.574
0.2861	1	60 Nd	$K\beta_2$	$KN_{II,III}$	43.32
0.288353	2	64 Gd	$K\alpha_1$	KL_{III}	42.9962
0.293038	2	64 Gd	$K\alpha_2$	KL_{II}	42.3089
0.293299	2	60 Nd	$K\beta_1$	KM_{III}	42.2713
0.294027	3	60 Nd	$K\beta_2$	KM_{II}	42.1665
0.29518	5	59 Pr	K	Abs. Edge	42.002
0.29679	2	59 Pr	$K\beta_2$	$KN_{II,III}$	41.773
0.298446	2	63 Eu	$K\alpha_1$	KL_{III}	41.5422
0.303118	2	63 Eu	$K\alpha_2$	KL_{II}	40.9019
0.304261	4	59 Pr	$K\beta_1$	KM_{III}	40.7482
0.304975	5	59 Pr	$K\beta_2$	KM_{II}	40.6529
0.30648	5	58 Ce	K	Abs. Edge	40.453
0.30668	2	58 Ce		$KO_{II,III}$	40.427
0.30737	2	58 Ce	$K\beta_4^{I}$	$KN_{IV,V}$	40.337
0.30816	1	58 Ce	$K\beta_2$	$KN_{II,III}$	40.233
0.309040	2	62 Sm	$K\alpha_1$	KL_{III}	40.1181
0.31342	2	58 Ce	$K\beta_5^{I}$	KM_{V}	39.558
0.31357	2	58 Ce	$K\beta_5^{II}$	KM_{IV}	39.539
0.313698	2	62 Sm	$K\alpha_2$	KL_{II}	39.5224
0.315816	2	58 Ce	$K\beta_1$	KM_{III}	39.2573
0.316520	4	58 Ce	$K\beta_2$	KM_{II}	39.1701
0.31844	5	57 La	K	Abs. Edge	38.934
0.31864	2	57 La		$KO_{II,III}$	38.909
0.31931	2	57 La	$K\beta_4^{I}$	$KN_{IV,V}$	38.828
0.320117	7	57 La	$K\beta_2$	$KN_{II,III}$	38.7299
0.320160	4	61 Pm	$K\alpha_1$	KL_{III}	38.7247
0.324803	4	61 Pm	$K\alpha_2$	KL_{II}	38.1712
0.32546	2	57 La	$K\beta_5^{I}$	KM_{V}	38.094
0.32563	2	57 La	$K\beta_5^{II}$	KM_{IV}	38.074
0.327983	3	57 La	$K\beta_1$	KM_{III}	37.8010
0.328686	4	57 La	$K\beta_2$	KM_{II}	37.7202
0.33104	1	56 Ba	K	Abs. Edge	37.452
0.33127	2	56 Ba		$KO_{II,III}$	37.426
0.331846	2	60 Nd	$K\alpha_1$	KL_{III}	37.3610
0.33229	2	56 Ba	$K\beta_4^{II}$	KN_{IV}	37.311
0.33277	1	56 Ba	$K\beta_2$	$KN_{II,III}$	37.257
0.336472	2	60 Nd	$K\alpha_2$	KL_{II}	36.8474
0.33814	2	56 Ba	$K\beta_5^{I}$	KM_{V}	36.666
0.33835	2	56 Ba	$K\beta_5^{II}$	KM_{IV}	36.643
0.340811	3	56 Ba	$K\beta_1$	KM_{III}	36.3782
0.341507	4	56 Ba	$K\beta_2$	KM_{II}	36.3040
0.344140	2	59 Pr	$K\alpha_1$	KL_{III}	36.0263

TABLE 1b3. (*Continued*)

Wavelength Å*	p.e.	Element	Designation		keV
0.34451	1	55 Cs	K	Abs. Edge	35.987
0.34611	2	55 Cs	$K\beta_2$	$KN_{II,III}$	35.822
0.348749	2	59 Pr	$K\alpha_2$	KL_{II}	35.5502
0.354364	7	55 Cs	$K\beta_1$	KM_{III}	34.9869
0.355050	4	55 Cs	$K\beta_3$	KM_{II}	34.9194
0.357092	2	58 Ce	$K\alpha_1$	KL_{III}	34.7197
0.3584	5	54 Xe	K	Abs. Edge	34.59
0.36026	3	54 Xe	$K\beta_2$	$KN_{II,III}$	34.415
0.361683	2	58 Ce	$K\alpha_2$	KL_{II}	34.2789
0.36872	2	54 Xe	$K\beta_1$	KM_{III}	33.624
0.36941	2	54 Xe	$K\beta_3$	KM_{II}	33.562
0.370737	2	57 La	$K\alpha_1$	KL_{III}	33.4418
0.37381	1	53 I	K	Abs. Edge	33.1665
0.37523	2	53 I	$K\beta_2$	$KN_{II,III}$	33.042
0.375313	2	57 La	$K\alpha_2$	KL_{II}	33.0341
0.383905	4	53 I	$K\beta_1$	KM_{III}	32.2947
0.384564	4	53 I	$K\beta_3$	KM_{II}	32.2394
0.385111	4	56 Ba	$K\alpha_1$	KL_{III}	32.1936
0.389668	5	56 Ba	$K\alpha_2$	KL_{II}	31.8171
0.38974	1	52 Te		$KO_{II,III}$	31.8114
0.38974	1	52 Te	K	Abs. Edge	31.8114
0.391102	6	52 Te	$K\beta_2$	$KN_{II,III}$	31.7004
0.399995	5	52 Te	$K\beta_1$	KM_{III}	30.9957
0.400290	4	55 Cs	$K\alpha_1$	KL_{III}	30.9728
0.400659	4	52 Te	$K\beta_3$	KM_{II}	30.9443
0.404835	4	55 Cs	$K\alpha_2$	KL_{II}	30.6251
0.40666	1	51 Sb		$KO_{II,III}$	30.4875
0.40668	1	51 Sb	K	Abs. Edge	30.4860
0.40702	1	51 Sb	$K\beta_4^{I}$	$KN_{IV,V}$	30.4604
0.407973	5	51 Sb	$K\beta_2$	$KN_{II,III}$	30.3895
0.41378	1	51 Sb	$K\beta_5^{I}$	KM_V	29.9632
0.41388	1	51 Sb	$K\beta_5^{II}$	KM_{IV}	29.9560
0.41634	2	54 Xe	$K\alpha_1$	KL_{III}	29.779
0.417085	3	51 Sb	$K\beta_1$	KM_{III}	29.7256
0.417737	4	51 Sb	$K\beta_3$	KM_{II}	29.6792
0.42087	2	54 Xe	$K\alpha_2$	KL_{II}	29.458
0.42467	3	50 Sn		$KO_{II,III}$	29.195
0.42467	1	50 Sn	K	Abs. Edge	29.1947
0.42495	3	50 Sn	$K\beta_4^{I}$	$KN_{IV,V}$	29.175
0.425915	8	50 Sn	$K\beta_2$	$KN_{II,III}$	29.1093
0.43175	3	50 Sn	$K\beta_5^{I}$	KM_V	28.716
0.43184	3	50 Sn	$K\beta_5^{II}$	KM_{IV}	28.710
0.433318	5	53 I	$K\alpha_1$	KL_{III}	28.6120
0.435236	5	50 Sn	$K\beta_1$	KM_{III}	28.4860
0.435877	5	50 Sn	$K\beta_3$	KM_{II}	28.4440
0.437829	7	53 I	$K\alpha_2$	KL_{II}	28.3172
0.44371	1	49 In	K	Abs. Edge	27.9420
0.44374	3	49 In		$KO_{II,III}$	27.940
0.44393	4	49 In	$K\beta_4^{I}$	$KN_{IV,V}$	27.928
0.44500	1	49 In	$K\beta_2$	$KN_{II,III}$	27.8608
0.45086	2	49 In	$K\beta_5^{I}$	KM_V	27.499
0.45098	2	49 In	$K\beta_5^{II}$	KM_{IV}	27.491
0.451295	3	52 Te	$K\alpha_1$	KL_{III}	27.4723
0.454545	4	49 In	$K\beta_1$	KM_{III}	27.2759
0.455181	4	49 In	$K\beta_3$	KM_{II}	27.2377
0.455784	3	52 Te	$K\alpha_2$	KL_{II}	27.2017
0.46407	1	48 Cd	K	Abs. Edge	26.7159
0.465328	7	48 Cd	$K\beta_2$	$KN_{II,III}$	26.6438
0.470354	3	51 Sb	$K\alpha_1$	KL_{III}	26.3591
0.474827	3	51 Sb	$K\alpha_2$	KL_{II}	26.1108
0.475105	6	48 Cd	$K\beta_1$	KM_{III}	26.0955
0.475730	5	48 Cd	$K\beta_3$	KM_{II}	26.0612
0.48589	1	47 Ag	K	Abs. Edge	25.5165
0.4859	9	47 Ag	$K\beta_4$	$KN_{IV,V}$	25.512
0.487032	4	47 Ag	$K\beta_2$	$KN_{II,III}$	25.4564
0.490599	3	50 Sn	$K\alpha_1$	KL_{III}	25.2713
0.49306	2	47 Ag	$K\beta_5$	$KM_{IV,V}$	25.145
0.495053	3	50 Sn	$K\alpha_2$	KL_{II}	25.0440
0.497069	4	47 Ag	$K\beta_1$	KM_{III}	24.9424
0.497685	4	47 Ag	$K\beta_3$	KM_{II}	24.9115
0.5092	1	46 Pd	K	Abs. Edge	24.348
0.5093	2	46 Pd	$K\beta_4$	$KN_{IV,V}$	24.346
0.510228	4	46 Pd	$K\beta_2$	$KN_{II,III}$	24.2991
0.512113	3	49 In	$K\alpha_1$	KL_{III}	24.2097
0.516544	3	49 In	$K\alpha_2$	KL_{II}	24.0020
0.51670	9	46 Pd	$K\beta_5$	$KM_{IV,V}$	23.995
0.520520	4	46 Pd	$K\beta_1$	KM_{III}	23.8187
0.521123	4	46 Pd	$K\beta_3$	KM_{II}	23.7911
0.53395	1	45 Rh	K	Abs. Edge	23.2198
0.53401	9	45 Rh	$K\beta_4^{I}$	$KN_{IV,V}$	23.217
0.535010	3	48 Cd	$K\alpha_1$	KL_{III}	23.1736
0.53503	2	45 Rh	$K\beta_2$	$KN_{II,III}$	23.1728
0.53513	5	45 Rh	$K\beta_2^{II}$	KN_{II}	23.168
0.5365	1	94 Pu	L_I	Abs. Edge	23.109
0.539422	3	48 Cd	$K\alpha_2$	KL_{II}	22.9841
0.54101	9	45 Rh	$K\beta_5^{I}$	KM_V	22.917
0.54118	9	45 Rh	$K\beta_5^{II}$	KM_{IV}	22.909
0.5416	1	94 Pu	$L\gamma_4$	L_IO_{III}	22.891
0.54311	2	95 Am	$L\gamma_6$	$L_{II}O_{IV}$	22.8282
0.5432	1	94 Pu	$L\gamma_4'$	L_IO_{II}	22.823
0.545605	4	45 Rh	$K\beta_1$	KM_{III}	22.7236
0.546200	4	45 Rh	$K\beta_3$	KM_{II}	22.6989
0.5544	2	95 Am	$L\gamma_2$	L_IN_{II}	22.361
0.5572	1	94 Pu	L_{II}	Abs. Edge	22.253
0.5585	5	93 Np	$L\gamma_4$	$L_IO_{II,III}$	22.20
0.5594075	6	47 Ag	$K\alpha_1$	KL_{III}	22.16292
0.55973	2	94 Pu	$L\gamma_6$	$L_{II}O_{IV}$	22.1502
0.56051	1	44 Ru	K	Abs. Edge	22.1193
0.56089	9	44 Ru	$K\beta_4$	$KN_{IV,V}$	22.104
0.56166	3	44 Ru	$K\beta_2$	$KN_{II,III}$	22.074
0.561886	9	95 Am	$L\gamma_1$	$L_{II}N_{IV}$	22.0652
0.563798	4	47 Ag	$K\alpha_2$	KL_{II}	21.9903
0.564001	9	94 Pu	$L\gamma_3$	L_IN_{III}	21.9824
0.5658	1	94 Pu	$L\gamma_8$	$L_{II}O_I$	21.914
0.56785	9	44 Ru	$K\beta_5^{I}$	KM_V	21.834
0.5680	2	44 Ru	$K\beta_5^{II}$	KM_{IV}	21.829
0.5695	1	92 U	L_I	Abs. Edge	21.771
0.5706	1	92 U	$L\gamma_{13}$	$L_IP_{II,III}$	21.729
0.57068	2	94 Pu	$L\gamma_2$	L_IN_{II}	21.7251
0.572482	4	44 Ru	$K\beta_1$	KM_{III}	21.6568
0.5725	1	92 U		$L_IO_{IV,V}$	21.657
0.573067	4	44 Ru	$K\beta_3$	KM_{II}	21.6346
0.57499	9	92 U	$L\gamma_4$	L_IO_{III}	21.562
0.576700	9	92 U	$L\gamma_4'$	L_IO_{II}	21.4984
0.57699	5	93 Np	$L\gamma_6$	$L_{II}O_{IV}$	21.488
0.578882	9	94 Pu	$L\gamma_1$	$L_{II}N_{IV}$	21.4173
0.5810	5	93 Np	$L\gamma_3$	L_IN_{III}	21.34
0.585448	3	46 Pd	$K\alpha_1$	KL_{III}	21.1771
0.5873	5	93 Np	$L\gamma_2$	L_IN_{II}	21.11
0.58906	1	43 Tc	K	Abs. Edge	21.0473
0.589821	3	46 Pd	$K\alpha_2$	KL_{II}	21.0201
0.58986	5	92 U	$L\gamma_{11}$	L_IN_V	21.019
0.59024	5	43 Tc	$K\beta_2$	$KN_{II,III}$	21.005
0.59096	5	92 U		L_IN_{IV}	20.979

TABLE 1b3. (Continued)

Wavelength Å*	p.e.	Element		Designation	keV
0.5919	1	92 U	L_{II}	Abs. Edge	20.945
0.59203	5	92 U		$L_{II}P_{IV}$	20.942
0.5930	2	92 U		$L_{II}P_{II,III}$	20.906
0.5937	1	91 Pa	$L\gamma_4$	$L_I O_{II,III}$	20.882
0.594845	9	92 U	$L\gamma_6$	$L_{II}O_{IV}$	20.8426
0.596498	9	93 Np	$L\gamma_1$	$L_{II}N_{IV}$	20.7848
0.59728	5	92 U		$L_{II}O_{III}$	20.758
0.598574	9	92 U	$L\gamma_3$	$L_I N_{III}$	20.7127
0.5988	1	94 Pu	$L\gamma_6$	$L_{II}N_I$	20.704
0.60125	5	92 U	$L\gamma_8$	$L_{II}O_I$	20.621
0.60130	4	43 Tc	$K\beta_1$	KM_{III}	20.619
0.60188	4	43 Tc	$K\beta_3$	KM_{II}	20.599
0.6031	1	92 U	Lv	$L_{II}N_{VI}$	20.556
0.605237	9	92 U	$L\gamma_2$	$L_I N_{II}$	20.4847
0.6059	1	90 Th	L_I	Abs. Edge	20.464
0.60705	8	90 Th	$L\gamma_{13}$	$L_I P_{II,III}$	20.424
0.6083	1	90 Th		$L_I O_{IV,V}$	20.383
0.61098	4	90 Th	$L\gamma_4$	$L_I O_{III}$	20.292
0.61251	4	90 Th	$L\gamma_4'$	$L_I O_{II}$	20.242
0.6133	1	91 Pa	$L\gamma_6$	$L_{II}O_{IV}$	20.216
0.613279	9	45 Rh	$K\alpha_1$	KL_{III}	20.2161
0.6146	1	90 Th		$L_I O_I$	20.174
0.614770	9	92 U	$L\gamma_1$	$L_{II}N_{IV}$	20.1671
0.6160	1	90 Th		$L_I N_{VI,VII}$	20.128
0.616	1	93 Np	$L\gamma_5$	$L_{II}N_I$	20.12
0.6169	1	91 Pa	$L\gamma_3$	$L_I N_{III}$	20.098
0.617630	4	45 Rh	$K\alpha_2$	KL_{II}	20.0737
0.61978	1	42 Mo	K	Abs. Edge	20.0039
0.62001	9	42 Mo	$K\beta_4{}^I$	$KN_{IV,V}$	19.996
0.62099	2	42 Mo	$K\beta_2$	$KN_{II,III}$	19.9652
0.62107	5	42 Mo	$K\beta_2{}^{II}$	KN_{II}	19.963
0.6228	1	92 U		$L_{II}N_{III}$	19.907
0.6239	1	91 Pa	$L\gamma_2$	$L_I N_{II}$	19.872
0.62636	9	90 Th	$L\gamma_{11}$	$L_I N_V$	19.794
0.62692	5	42 Mo	$K\beta_5{}^I$	KM_V	19.776
0.62708	5	42 Mo	$K\beta_5{}^{II}$	KM_{IV}	19.771
0.6276	1	90 Th		$L_I N_{IV}$	19.755
0.6299	1	90 Th	L_{II}	Abs. Edge	19.683
0.62991	9	90 Th		$L_{II}P_{IV}$	19.682
0.6312	1	90 Th		$L_{II}P_{II,III}$	19.642
0.6316	1	90 Th		$L_{II}P_I$	19.629
0.632288	9	42 Mo	$K\beta_1$	KM_{III}	19.6083
0.63258	4	90 Th	$L\gamma_6$	$L_{II}O_{IV}$	19.599
0.632872	2	42 Mo	$K\beta_3$	KM_{II}	19.5903
0.63358	9	91 Pa	$L\gamma_1$	$L_{II}N_{IV}$	19.568
0.63557	2	92 U	$L\gamma_5$	$L_{II}N_I$	19.5072
0.63559	4	90 Th	$L\gamma_3$	$L_I N_{III}$	19.507
0.6356	1	90 Th		$L_{II}O_{III}$	19.506
0.6369	1	90 Th		$L_{II}O_{II}$	19.466
0.63898	5	90 Th	$L\gamma_8$	$L_{II}O_I$	19.403
0.64064	9	90 Th	Lv	$L_{II}N_{VI}$	19.353
0.6416	1	94 Pu	$L\beta_9$	$L_I M_V$	19.323
0.64221	4	90 Th	$L\gamma_2$	$L_I N_{II}$	19.305
0.643083	4	44 Ru	$K\alpha_1$	KL_{III}	19.2792
0.6445	1	88 Ra	L_I	Abs. Edge	19.236
0.64513	5	88 Ra	$L\gamma_{13}$	$L_I P_{II,III}$	19.218
0.6468	1	88 Ra		$L_I O_{IV,V}$	19.167
0.647408	5	44 Ru	$K\alpha_2$	KL_{II}	19.1504
0.64755	5	90 Th		$L_I N_I$	19.146
0.6482	1	94 Pu	$L\beta_{10}$	$L_I M_{IV}$	19.126
0.64891	2	95 Am	$L\beta_3$	$L_I M_{III}$	19.1059
0.64965	5	88 Ra	$L\gamma_4$	$L_I O_{III}$	19.084

Wavelength Å*	p.e.	Element		Designation	keV
0.65131	5	88 Ra	$L\gamma_4'$	$L_I O_{II}$	19.036
0.6521	1	90 Th		$L_{II}N_V$	19.014
0.65298	1	41 Nb	K	Abs. Edge	18.9869
0.65313	3	90 Th	$L\gamma_1$	$L_{II}N_{IV}$	18.9825
0.65318	5	41 Nb	$K\beta_4$	$KN_{IV,V}$	18.981
0.65416	4	41 Nb	$K\beta_2$	$KN_{II,III}$	18.953
0.6550	1	91 Pa	$L\gamma_5$	$L_{II}N_I$	18.930
0.657655	9	95 Am	$L\beta_1$	$L_{II}M_{IV}$	18.8520
0.6620	1	90 Th		$L_{II}N_{III}$	18.729
0.6654	1	88 Ra	$L\gamma_{11}$	$L_I N_V$	18.633
0.66576	2	41 Nb	$K\beta_1$	KM_{III}	18.6225
0.66634	3	41 Nb	$K\beta_3$	KM_{II}	18.6063
0.6666	1	88 Ra		$L_I N_{IV}$	18.600
0.66871	1	94 Pu	$L\beta_3$	$L_I M_{III}$	18.5405
0.6707	1	88 Ra	L_{II}	Abs. Edge	18.486
0.6714	1	88 Ra		$L_{II}P_{II,III}$	18.466
0.6724	1	88 Ra		$L_{II}P_I$	18.439
0.67328	5	88 Ra	$L\gamma_6$	$L_{II}O_{IV}$	18.414
0.67351	9	89 Ac	$L\gamma_1$	$L_{II}N_{IV}$	18.408
0.67383	2	95 Am	$L\beta_5$	$L_{II}O_{IV,V}$	18.3996
0.67491	4	90 Th	$L\gamma_5$	$L_{II}N_I$	18.370
0.67502	3	43 Tc	$K\alpha_1$	KL_{III}	18.3671
0.67538	5	88 Ra	$L\gamma_3$	$L_I N_{III}$	18.357
0.6764	1	88 Ra		$L_{II}O_{III}$	18.330
0.67772	2	94 Pu	$L\beta_1$	$L_{II}M_{IV}$	18.2937
0.6780	1	88 Ra		$L_{II}O_{II}$	18.286
0.67932	3	43 Tc	$K\alpha_2$	KL_{II}	18.2508
0.6801	1	88 Ra	$L\gamma_8$	$L_{II}O_I$	18.230
0.681014	8	92 U	$L\beta_9$	$L_I M_V$	18.2054
0.68199	5	88 Ra	$L\gamma_2$	$L_I N_{II}$	18.179
0.68639	2	95 Am	$L\beta_4$	$L_I M_{II}$	18.0627
0.6867	1	94 Pu	L_{III}	Abs. Edge	18.054
0.6874	1	88 Ra		$L_I N_I$	18.036
0.68760	5	92 U	$L\beta_{10}$	$L_I M_{IV}$	18.031
0.68883	1	40 Zr	K	Abs. Edge	17.9989
0.68901	5	40 Zr	$K\beta_4$	$KN_{IV,V}$	17.994
0.68920	9	93 Np	$L\beta_3$	$L_I M_{III}$	17.989
0.68993	4	40 Zr	$K\beta_2$	$KN_{II,III}$	17.970
0.69068	2	94 Pu	$L\beta_5$	$L_{III}O_{IV,V}$	17.9506
0.6932	1	88 Ra		$L_{II}N_V$	17.884
0.69463	5	88 Ra	$L\gamma_1$	$L_{II}N_{IV}$	17.849
0.6959	1	40 Zr	$K\beta_5$	$KM_{IV,V}$	17.815
0.698478	9	93 Np	$L\beta_1$	$L_{II}M_{IV}$	17.7502
0.7003	1	94 Pu	L_{III}	$L_{III}O_I$	17.705
0.701390	9	95 Am	$L\beta_2$	$L_{III}N_V$	17.6765
0.70173	3	40 Zr	$K\beta_1$	KM_{III}	17.6678
0.7018	1	91 Pa	$L\beta_9$	$L_I M_V$	17.667
0.70228	4	40 Zr	$K\beta_3$	KM_{II}	17.654
0.7031	1	94 Pu	Lu	$L_{III}N_{VI,VII}$	17.635
0.70341	2	95 Am	$L\beta_{15}$	$L_{III}N_{IV}$	17.6258
0.7043	1	88 Ra		$L_{II}N_{III}$	17.604
0.70620	2	94 Pu	$L\beta_4$	$L_I M_{II}$	17.5560
0.70814	2	93 Np	$L\beta_5$	$L_{III}O_{IV,V}$	17.5081
0.7088	2	91 Pa	$L\beta_{10}$	$L_I M_{IV}$	17.492
0.709300	1	42 Mo	$K\alpha_1$	KL_{III}	17.47934
0.71029	2	92 U	$L\beta_3$	$L_I M_{III}$	17.4550
0.713590	6	42 Mo	$K\alpha_2$	KL_{II}	17.3743
0.71652	9	87 Fr	$L\gamma_1$	$L_{II}N_{IV}$	17.303
0.71774	5	88 Ra	$L\gamma_5$	$L_{II}N_I$	17.274
0.71851	2	94 Pu	$L\beta_2$	$L_{III}N_V$	17.2553
0.719984	8	92 U	$L\beta_1$	$L_{II}M_{IV}$	17.2200
0.7205	1	94 Pu	$L\beta_{15}$	$L_{III}N_{IV}$	17.208

TABLE 1b3. (*Continued*)

Wavelength Å*	p.e.	Element	Designation		keV
0.7223	1	92 U	L_{III}	Abs. Edge	17.165
0.72240	5	92 U		$L_{III}P_{IV,V}$	17.162
0.7234	1	90 Th	$L\beta_9$	$L_I M_V$	17.139
0.72426	5	92 U		$L_{III}P_{II,III}$	17.118
0.72521	5	92 U		$L_{III}P_I$	17.096
0.726305	9	92 U	$L\beta_5$	$L_{III}O_{IV,V}$	17.0701
0.72671	2	93 Np	$L\beta_4$	$L_I M_{II}$	17.0607
0.72766	5	39 Y	K	Abs. Edge	17.038
0.72776	5	39 Y	$K\beta_4$	$K N_{IV,V}$	17.036
0.72864	4	39 Y	$K\beta_2$	$K N_{II,III}$	17.0154
0.7301	1	90 Th	$L\beta_{10}$	$L_I M_{IV}$	16.981
0.7309	1	92 U		$L_{III}O_{III}$	16.962
0.73230	5	91 Pa	$L\beta_3$	$L_I M_{III}$	16.930
0.7333	1	92 U		$L_{III}O_{II}$	16.907
0.73418	2	95 Am	$L\beta_6$	$L_{III}N_I$	16.8870
0.7345	1	39 Y	$K\beta_5$	$K M_{IV,V}$	16.879
0.73602	6	92 U	$L\beta_7$	$L_{III}O_I$	16.845
0.736230	9	93 Np		$L_{III}N_V$	16.8400
0.738603	9	92 U	Lu	$L_{III}N_{VI,VII}$	16.7859
0.73928	9	86 Rn	$L\gamma_1$	$L_{II}N_{IV}$	16.770
0.74072	2	39 Y	$K\beta_1$	$K M_{III}$	16.7378
0.74126	3	39 Y	$K\beta_3$	$K M_{II}$	16.7258
0.74232	5	91 Pa	$L\beta_1$	$L_{II}M_{IV}$	16.702
0.74503	5	92 U	$L\beta_{17}$	$L_{II}M_{III}$	16.641
0.7452	2	91 Pa	$L\beta_5$	$L_{III}O_{IV,V}$	16.636
0.74620	1	41 Nb	$K\alpha_1$	$K L_{III}$	16.6151
0.747985	9	92 U	$L\beta_4$	$L_I M_{II}$	16.5753
0.75044	1	41 Nb	$K\alpha_2$	$K L_{II}$	16.5210
0.75148	2	94 Pu	$L\beta_6$	$L_{III}N_I$	16.4983
0.7546	2	91 Pa	$L\beta_7$	$L_{III}O_I$	16.431
0.754681	9	92 U	$L\beta_2$	$L_{III}N_V$	16.4283
0.75479	3	90 Th	$L\beta_3$	$L_I M_{III}$	16.4258
0.756642	9	92 U	$L\beta_{15}$	$L_{III}N_{IV}$	16.3857
0.75690	3	83 Bi	$L\gamma_{13}$	$L_I P_{II,III}$	16.3802
0.7571	1	83 Bi	L_I	Abs. Edge	16.376
0.7579	1	90 Th		$L_{II}M_V$	16.359
0.75791	5	83 Bi		$L_I O_{IV,V}$	16.358
0.7591	1	94 Pu	$L\eta$	$L_{II}M_I$	16.333
0.7607	1	90 Th	L_{III}	Abs. Edge	16.299
0.76087	9	90 Th		$L_{III}P_{IV,V}$	16.295
0.76087	3	83 Bi	$L\gamma_4$	$L_I O_{III}$	16.2947
0.76198	3	83 Bi	$L\gamma_4'$	$L_I O_{II}$	16.2709
0.7625	2	90 Th		$L_{III}P_{II,III}$	16.260
0.76289	9	85 At	$L\gamma_1$	$L_{II}N_{IV}$	16.251
0.76338	5	90 Th		$L_{III}P_I$	16.241
0.7641	5	83 Bi		$L_I N_{VI,VII}$	16.23
0.7645	2	84 Po	$L\gamma_6$	$L_{II}O_{IV}$	16.218
0.76468	5	90 Th	$L\beta_5$	$L_{III}O_{IV,V}$	16.213
0.765210	9	90 Th	$L\beta_1$	$L_{II}M_{IV}$	16.2022
0.76857	5	88 Ra	$L\beta_9$	$L_I M_V$	16.131
0.769	1	93 Np	$L\beta_6$	$L_{III}N_I$	16.13
0.7690	1	90 Th		$L_{III}O_{III}$	16.123
0.7691	1	92 U		$L_{III}N_{III}$	16.120
0.76973	5	38 Sr	K	Abs. Edge	16.107
0.7699	1	91 Pa	$L\beta_4$	$L_I M_{II}$	16.104
0.76989	5	38 Sr	$K\beta_4$	$K N_{IV,V}$	16.104
0.77081	3	38 Sr	$K\beta_2$	$K N_{II,III}$	16.0846
0.7713	1	90 Th		$L_{III}O_{II}$	16.074
0.772	1	84 Po	$L\gamma_2$	$L_I N_{II}$	16.07
0.7737	1	91 Pa	$L\beta_2$	$L_{III}N_V$	16.024
0.77437	4	90 Th	$L\beta_7$	$L_{III}O_I$	16.0105
0.77546	5	88 Ra	$L\beta_{10}$	$L_I M_{IV}$	15.988
0.7764	1	38 Sr	$K\beta_5$	$K M_{IV,V}$	15.969
0.77661	5	90 Th	Lu	$L_{III}N_{VI,VII}$	15.964
0.77728	5	83 Bi	$L\gamma_{11}$	$L_I N_V$	15.951
0.77822	9	89 Ac	$L\beta_3$	$L_I M_{III}$	15.931
0.77954	5	83 Bi		$L_I N_{IV}$	15.904
0.78017	9	92 U		$L_{III}N_{II}$	15.892
0.7809	2	93 Np	$L\eta$	$L_{II}M_I$	15.876
0.78196	5	82 Pb	L_I	Abs. Edge	15.855
0.78257	7	82 Pb		$L_I O_{IV,V}$	15.843
0.78292	2	38 Sr	$K\beta_1$	$K M_{III}$	15.8357
0.78345	3	38 Sr	$K\beta_3$	$K M_{II}$	15.8249
0.7858	1	82 Pb	$L\gamma_4$	$L_I O_{III}$	15.777
0.78593	1	40 Zr	$K\alpha_1$	$K L_{III}$	15.7751
0.78706	7	82 Pb	$L\gamma_4'$	$L_I O_{II}$	15.752
0.78748	9	84 Po	$L\gamma_1$	$L_{II}N_{IV}$	15.744
0.78838	2	92 U	$L\beta_6$	$L_{III}N_I$	15.7260
0.7884	1	82 Pb		$L_I N_{VI,VII}$	15.725
0.7887	1	83 Bi	L_{II}	Abs. Edge	15.719
0.78903	9	89 Ac	$L\beta_1$	$L_{II}M_{IV}$	15.713
0.78917	5	83 Bi	$L\gamma_2$	$L_I N_{III}$	15.7102
0.7897	1	82 Pb		$L_I O_I$	15.699
0.79015	1	40 Zr	$K\alpha_2$	$K L_{II}$	15.6909
0.79043	3	83 Bi	$L\gamma_6$	$L_{II}O_{IV}$	15.6853
0.79257	4	90 Th	$L\beta_4$	$L_I M_{II}$	15.6429
0.79257	4	90 Th	$L\beta_{17}$	$L_{II}M_{III}$	15.6429
0.79354	3	90 Th	$L\beta_2$	$L_{III}M_V$	15.6237
0.79384	5	83 Bi		$L_{II}O_{III}$	15.6178
0.79539	5	90 Th	$L\beta_{15}$	$L_{III}N_{IV}$	15.5875
0.79565	3	83 Bi	$L\gamma_2$	$L_I N_{II}$	15.5824
0.79721	9	83 Bi	Lv	$L_{II}N_{VI}$	15.552
0.7973	1	83 Bi	$L\gamma_8$	$L_{II}O_I$	15.551
0.8022	1	83 Bi		$L_I N_I$	15.456
0.80233	9	82 Pb	$L\gamma_{11}$	$L_I N_V$	15.453
0.80273	5	88 Ra	$L\beta_3$	$L_I M_{III}$	15.4449
0.8028	1	88 Ra	L_{III}	Abs. Edge	15.444
0.80364	7	82 Pb		$L_I N_{IV}$	15.427
0.8038	1	88 Ra		$L_{III}P_{II,III}$	15.425
0.8050	1	88 Ra		$L_{III}P_I$	15.402
0.80509	2	92 U	$L\eta$	$L_{II}M_I$	15.3997
0.80627	5	88 Ra	$L\beta_5$	$L_{III}O_{IV,V}$	15.3771
0.8079	1	91 Pa	$L\beta_6$	$L_{III}N_I$	15.347
0.8081	1	81 Tl	L_I	Abs. Edge	15.343
0.8082	1	90 Th		$L_{III}N_{II}$	15.341
0.80861	5	81 Tl		$L_I O_{IV,V}$	15.3327
0.81163	9	90 Th		$L_I M_I$	15.276
0.81184	5	81 Tl	$L\gamma_4$	$L_I O_{III}$	15.2716
0.81308	5	81 Tl	$L\gamma_4'$	$L_I O_{II}$	15.2482
0.81311	2	83 Bi	$L\gamma_1$	$L_{II}N_{IV}$	15.2477
0.81375	5	88 Ra	$L\beta_1$	$L_{II}M_{IV}$	15.2358
0.8147	1	82 Pb	$L\gamma_8$	$L_I N_{III}$	15.218
0.81538	5	82 Pb	L_{II}	Abs. Edge	15.2053
0.8154	2	37 Rb	$K\beta_4$	$K N_{IV,V}$	15.205
0.81554	5	37 Rb	K	Abs. Edge	15.2023
0.8158	1	81 Tl		$L_I O_I$	15.198
0.81583	5	82 Pb		$L_{II}P_I$	15.1969
0.8162	1	88 Ra	$L\beta_7$	$L_{III}O_I$	15.190
0.81645	3	37 Rb	$K\beta_2$	$K N_{II,III}$	15.1854
0.81683	5	82 Pb	$L\gamma_6$	$L_{II}O_{IV}$	15.1783
0.8186	1	88 Ra	Lu	$L_{III}N_{VI,VII}$	15.146
0.8190	2	90 Th		$L_{III}N_{II}$	15.138
0.8200	1	82 Pb		$L_{II}O_{III}$	15.120
0.8210	2	82 Pb	$L\gamma_2$	$L_I N_{II}$	15.101

TABLE 1b3. (*Continued*)

Wavelength Å*	p.e.	Element	Designation		keV	Wavelength Å*	p.e.	Element	Designation		keV
0.8219	1	37 Rb	$K\beta_5$	$KM_{IV,V}$	15.085	0.87088	5	88 Ra	$L\beta_6$	$L_{III}N_I$	14.2362
0.82327	7	82 Pb	Lv	$L_{II}N_{VI}$	15.060	0.8722	1	80 Hg	L_{II}	Abs. Edge	14.215
0.82365	5	82 Pb	$L\gamma_8$	$L_{II}O_I$	15.0527	0.87319	7	80 Hg	$L\gamma_6$	$L_{II}O_{IV}$	14.199
0.8248	1	83 Bi		$L_{II}N_{III}$	15.031	0.87526	1	38 Sr	$K\alpha_1$	$K L_{III}$	14.1650
0.82789	9	87 Fr	$L\beta_3$	$L_I M_{III}$	14.976	0.87544	7	80 Hg	$L\gamma_2$	$L_I N_{II}$	14.162
0.82790	8	90 Th	$L\beta_6$	$L_{III}N_I$	14.975	0.8758	1	80 Hg		$L_{II}O_{III}$	14.156
0.82859	7	82 Pb		$L_I N_I$	14.963	0.8784	1	80 Hg		$L_{II}O_{II}$	14.114
0.82868	2	37 Rb	$K\beta_1$	KM_{III}	14.9613	0.8785	1	36 Kr	$K\beta_1$	KM_{III}	14.112
0.82879	5	81 Tl	$L\gamma_{11}$	$L_I N_V$	14.9593	0.87885	7	80 Hg	Lv	$L_{II}N_{VI}$	14.107
0.82884	1	39 Y	$K\alpha_1$	$K L_{III}$	14.9584	0.8790	1	36 Kr	$K\beta_3$	KM_{II}	14.104
0.82921	3	37 Rb	$K\beta_3$	KM_{II}	14.9517	0.87943	1	38 Sr	$K\alpha_2$	$K L_{II}$	14.0979
0.8295	1	91 Pa	$L\eta$	$L_{II}M_I$	14.946	0.87995	7	80 Hg	$L\gamma_8$	$L_{II}O_I$	14.090
0.83001	7	81 Tl		$L_I N_{IV}$	14.937	0.87996	5	81 Tl		$L_{II}N_{III}$	14.0893
0.83305	1	39 Y	$K\alpha_2$	$K L_{II}$	14.8829	0.88028	2	94 Pu	$L\alpha_2$	$L_{III}M_{IV}$	14.0842
0.8338	1	90 Th		$L_{II}M_{II}$	14.869	0.88135	9	85 At	$L\beta_2$	$L_I M_{III}$	14.067
0.8344	9	83 Bi		$L_{II}N_{II}$	14.86	0.8827	2	80 Hg		$L_I N_I$	14.045
0.8350	2	80 Hg		$L_I O_{IV,V}$	14.847	0.88433	7	79 Au	$L\gamma_{11}$	$L_I N_V$	14.020
0.8353	1	80 Hg	L_I	Abs. Edge	14.842	0.88563	7	79 Au		$L_I N_{IV}$	13.999
0.83537	5	88 Ra	$L\beta_2$	$L_{III}N_V$	14.8414	0.8882	2	81 Tl		$L_{II}M_{II}$	13.959
0.83722	5	88 Ra	$L\beta_{15}$	$L_{III}N_{IV}$	14.8086	0.889128	9	93 Np	$L\alpha_1$	$L_{III}M_V$	13.9441
0.8382	2	82 Pb		$L_{II}N_V$	14.791	0.8931	1	78 Pt	L_I	Abs. Edge	13.883
0.83894	7	80 Hg	$L\gamma_4$	$L_I O_{III}$	14.778	0.8934	1	78 Pt		$L_I O_V$	13.878
0.83923	5	83 Bi	$L\gamma_6$	$L_{II}N_I$	14.7732	0.89349	9	85 At	$L\beta_1$	$L_{II}M_{IV}$	13.876
0.83940	9	87 Fr	$L\beta_1$	$L_{II}M_{IV}$	14.770	0.8943	1	78 Pt		$L_I O_{IV}$	13.864
0.83973	3	82 Pb	$L\gamma_1$	$L_{II}N_{IV}$	14.7644	0.89500	4	81 Tl	$L\gamma_5$	$L_{III}N_I$	13.8526
0.84013	7	80 Hg	$L\gamma_4'$	$L_I O_{II}$	14.757	0.89646	5	80 Hg	$L\gamma_1$	$L_{II}N_{IV}$	13.8301
0.84071	5	88 Ra	$L\beta_4$	$L_I M_{II}$	14.7472	0.89659	4	78 Pt	$L\gamma_4$	$L_I O_{III}$	13.8281
0.84130	4	81 Tl	$L\gamma_3$	$L_I N_{III}$	14.7368	0.89747	4	78 Pt	$L\gamma_4'$	$L_I O_{II}$	13.8145
0.8434	1	81 Tl	L_{II}	Abs. Edge	14.699	0.89783	5	79 Au	$L\gamma_8$	$L_I N_{III}$	13.8090
0.8438	1	88 Ra	$L\beta_{17}$	$L_{II}M_{III}$	14.692	0.89791	3	83 Bi	$L\beta_9$	$L_I M_V$	13.8077
0.8442	2	81 Tl	$L\gamma_6$	$L_{II}O_{IV}$	14.685	0.8995	2	78 Pt		$L_I O_I$	13.784
0.8452	2	80 Hg		$L_I O_I$	14.670	0.8996	2	84 Po	$L\beta_6$	$L_{III}O_{IV,V}$	13.782
0.84773	5	81 Tl	$L\gamma_2$	$L_I N_{II}$	14.6251	0.901045	9	93 Np	$L\alpha_2$	$L_{III}M_{IV}$	13.7597
0.848187	9	95 Am	$L\alpha_1$	$L_{III}M_V$	14.6172	0.90259	5	79 Au	L_{II}	Abs. Edge	13.7361
0.8490	1	81 Tl		$L_{II}O_{II}$	14.604	0.90297	3	79 Au	$L\gamma_6$	$L_{II}O_{IV}$	13.7304
0.85048	5	81 Tl	Lv	$L_{II}N_{VI}$	14.5777	0.90434	3	79 Au	$L\gamma_2$	$L_I N_{II}$	13.7095
0.8512	1	88 Ra		$L_{III}N_{III}$	14.566	0.90495	4	83 Bi	$L\beta_{10}$	$L_I M_{IV}$	13.7002
0.8513	2	81 Tl	$L\gamma_8$	$L_{II}O_I$	14.564	0.90638	7	79 Au		$L_{II}O_{III}$	13.69
0.85192	7	82 Pb		$L_{II}N_{III}$	14.553	0.90742	5	88 Ra	$L\eta$	$L_{II}M_I$	13.6630
0.85436	9	86 Rn	$L\beta_3$	$L_I M_{III}$	14.512	0.90746	7	79 Au		$L_{II}O_{II}$	13.662
0.85446	4	90 Th	$L\eta$	$L_{II}M_I$	14.5099	0.90837	5	79 Au	Lv	$L_{II}N_{VI}$	13.6487
0.8549	1	81 Tl		$L_I N_I$	14.503	0.90894	7	80 Hg		$L_{II}N_{III}$	13.640
0.85657	7	80 Hg	$L\gamma_{11}$	$L_I N_V$	14.474	0.9091	3	84 Po	$L\beta_3$	$L_I M_{III}$	13.638
0.858	2	87 Fr	$L\beta_2$	$L_{III}N_V$	14.45	0.90989	5	79 Au	$L\gamma_8$	$L_{II}O_I$	13.6260
0.8585	3	82 Pb		$L_{II}N_{II}$	14.442	0.910639	9	92 U	$L\alpha_1$	$L_{III}M_V$	13.6147
0.860266	9	95 Am	$L\alpha_2$	$L_{III}M_{IV}$	14.4119	0.9131	1	79 Au		$L_I N_I$	13.578
0.8618	1	88 Ra		$L_{III}N_{II}$	14.387	0.9143	2	78 Pt	$L\gamma_{11}$	$L_I N_V$	13.560
0.86376	5	79 Au	L_I	Abs. Edge	14.3537	0.9204	1	35 Br	K	Abs. Edge	13.470
0.86400	5	79 Au		$L_I O_{IV,V}$	14.3497	0.92046	2	35 Br	$K\beta_2$	$KN_{II,III}$	13.4695
0.8653	2	36 Kr	$K\beta_4$	$KN_{IV,V}$	14.328	0.9220	2	84 Po	$L\beta_1$	$L_{II}M_{IV}$	13.447
0.86552	1	36 Kr	K	Abs. Edge	14.3244	0.922558	9	92 U	$L\alpha_2$	$L_{III}M_{IV}$	13.4388
0.86605	9	86 Rn	$L\beta_1$	$L_{II}M_{IV}$	14.316	0.9234	1	83 Bi	L_{III}	Abs. Edge	13.426
0.8661	1	36 Kr	$K\beta_2$	$KN_{II,III}$	14.315	0.9236	1	77 Ir	L_I	Abs. Edge	13.423
0.86655	5	82 Pb	$L\gamma_6$	$L_{II}N_I$	14.3075	0.92413	4	83 Bi		$L_{III}P_{II,III}$	13.4159
0.86703	4	79 Au	$L\gamma_4$	$L_I O_{III}$	14.2996	0.9243	3	77 Ir		$L_I O_{IV,V}$	13.413
0.86752	3	81 Tl	$L\gamma_1$	$L_{II}N_{IV}$	14.2915	0.92453	7	80 Hg	$L\gamma_5$	$L_{III}N_I$	13.410
0.86816	4	79 Au	$L\gamma_4'$	$L_I O_{II}$	14.2809	0.9255	1	35 Br	$K\beta_5$	$KM_{IV,V}$	13.396
0.86830	2	94 Pu	$L\alpha_1$	$L_{III}M_V$	14.2786	0.925553	9	37 Rb	$K\alpha_1$	$K L_{III}$	13.3953
0.86915	7	80 Hg	$L\gamma_8$	$L_I N_{III}$	14.265	0.92556	3	83 Bi	$L\beta_5$	$L_{III}O_{IV,V}$	13.3953
0.87074	5	79 Au		$L_I O_I$	14.2385	0.92650	3	79 Au	$L\gamma_1$	$L_{II}N_{IV}$	13.3817
0.8708	2	36 Kr	$K\beta_5$	$KM_{IV,V}$	14.238	0.9268	1	82 Pb	$L\beta_9$	$L_I M_V$	13.377

TABLE 1b3. (Continued)

Wavelength Å*	p.e.	Element	Designation		keV	Wavelength Å*	p.e.	Element	Designation		keV
0.92744	3	77 Ir	$L\gamma_4$	L_IO_{III}	13.3681	0.9766	2	77 Ir		L_IN_I	12.695
0.92791	5	78 Pt	$L\gamma_3$	L_IN_{III}	13.3613	0.97690	4	83 Bi	$L\beta_4$	L_IM_{II}	12.6912
0.92831	3	77 Ir	$L\gamma_4'$	L_IO_{II}	13.3555	0.9772	3	76 Os		L_IN_{IV}	12.687
0.92937	5	84 Po	$L\beta_2$	$L_{III}N_V$	13.3404	0.9792	2	78 Pt		$L_{II}N_{II}$	12.661
0.92969	1	37 Rb	$K\alpha_2$	KL_{II}	13.3358	0.97926	5	81 Tl		$L_{III}P_{II,III}$	12.6607
0.9302	2	83 Bi		$L_{III}O_{III}$	13.328	0.9793	1	81 Tl	L_{III}	Abs. Edge	12.660
0.9312	2	84 Po	$L\beta_{15}$	$L_{III}N_{IV}$	13.314	0.97974	1	34 Se	K	Abs. Edge	12.6545
0.9323	2	83 Bi		$L_{III}O_{II}$	13.298	0.97992	5	34 Se	$K\beta_2$	$KN_{II,III}$	12.6522
0.93279	2	35 Br	$K\beta_1$	KM_{III}	13.2914	0.97993	5	89 Ac	$L\alpha_1$	$L_{III}M_V$	12.6520
0.93284	5	91 Pa	$L\alpha_1$	$L_{III}M_V$	13.2907	0.9801	1	36 Kr	$K\alpha_1$	KL_{II}	12.649
0.93327	5	35 Br	$K\beta_3$	KM_{II}	13.2845	0.98058	3	81 Tl	$L\beta_6$	$L_{III}O_{IV,V}$	12.6436
0.9339	2	82 Pb	$L\beta_{10}$	L_IM_{IV}	13.275	0.98221	7	82 Pb	$L\beta_2$	$L_{III}N_V$	12.6226
0.93414	5	78 Pt	L_{II}	Abs. Edge	13.2723	0.98280	5	83 Bi		$L_{III}N_{II}$	12.6151
0.9342	2	78 Pt	$L\gamma_6$	$L_{II}O_{IV}$	13.271	0.98291	3	82 Pb	$L\beta_1$	$L_{II}M_{IV}$	12.6137
0.93427	5	78 Pt	$L\gamma_2$	L_IN_{II}	13.2704	0.98389	7	82 Pb	$L\beta_{15}$	$L_{III}N_{IV}$	12.6011
0.93505	5	83 Bi	$L\beta_7$	$L_{III}O_I$	13.2593	0.9841	1	36 Kr	$K\alpha_2$	KL_{II}	12.598
0.93505	5	83 Bi	Lu	$L_{III}N_{VI,VII}$	13.2593	0.9843	1	34 Se	$K\beta_5$	$KM_{IV,V}$	12.595
0.93855	3	83 Bi	$L\beta_3$	L_IM_{III}	13.2098	0.98538	5	81 Tl		$L_{III}O_{III}$	12.5820
0.93931	5	78 Pt	Lv	$L_{II}N_{VI}$	13.1992	0.9871	2	80 Hg	$L\beta_9$	L_IM_V	12.560
0.9402	2	79 Au		$L_{II}N_{III}$	13.186	0.98738	5	81 Tl		$L_{III}O_{II}$	12.5566
0.9411	1	78 Pt	$L\gamma_8$	$L_{II}O_I$	13.173	0.9877	2	78 Pt	$L\gamma_5$	$L_{II}N_I$	12.552
0.94419	5	83 Bi		$L_{II}M_V$	13.1310	0.9888	1	81 Tl	Lu	$L_{III}N_{VI,VII}$	12.538
0.9446	2	77 Ir	$L\gamma_{11}$	L_IN_V	13.126	0.98913	5	83 Bi	$L\beta_{17}$	$L_{II}M_{III}$	12.5344
0.94482	5	91 Pa	$L\alpha_2$	$L_{III}M_{IV}$	13.1222	0.9894	1	75 Re	L_I	Abs. Edge	12.530
0.9455	2	78 Pt		L_IN_I	13.113	0.9900	1	75 Re		$L_IO_{IV,V}$	12.524
0.9459	2	77 Ir		L_IN_{IV}	13.108	0.99017	5	81 Tl	$L\beta_7$	$L_{III}O_I$	12.5212
0.9475	3	84 Po	$L\beta_4$	L_IM_{II}	13.086	0.99085	3	77 Ir	$L\gamma_1$	$L_{II}N_{IV}$	12.5126
0.95073	5	82 Pb	L_{III}	Abs. Edge	13.0406	0.99178	5	89 Ac	$L\alpha_2$	$L_{III}M_{IV}$	12.5008
0.95118	7	82 Pb		$L_{III}P_{II,III}$	13.0344	0.99186	5	76 Os	$L\gamma_3$	L_IN_{III}	12.4998
0.951978	9	83 Bi	$L\beta_1$	$L_{II}M_{IV}$	13.0235	0.99218	3	34 Se	$K\beta_1$	KM_{III}	12.4959
0.9526	1	82 Pb	$L\beta_6$	$L_{III}O_{IV,V}$	13.015	0.99249	5	75 Re	$L\gamma_4$	L_IO_{III}	12.4920
0.95518	4	83 Bi	$L\beta_2$	$L_{III}N_V$	12.9799	0.99268	5	34 Se	$K\beta_3$	KM_{II}	12.4896
0.95559	3	79 Au	$L\gamma_5$	$L_{II}N_I$	12.9743	0.99331	3	83 Bi	$L\beta_6$	$L_{III}N_I$	12.4816
0.9558	1	76 Os	L_I	Abs. Edge	12.972	0.99334	5	75 Re	$L\gamma_4'$	L_IO_{II}	12.4813
0.95600	3	90 Th	$L\alpha_1$	$L_{III}M_V$	12.9687	0.9962	2	80 Hg	$L\beta_{10}$	L_IM_{IV}	12.446
0.95603	5	76 Os		$L_IO_{IV,V}$	12.9683	0.9965	1	75 Re		L_IO_I	12.442
0.95675	7	81 Tl	$L\beta_9$	L_IM_V	12.9585	0.99805	5	76 Os	$L\gamma_2$	L_IN_{II}	12.4224
0.95702	5	83 Bi	$L\beta_{15}$	$L_{III}N_{IV}$	12.9549	1.0005	1	82 Pb		$L_{III}N_{III}$	12.392
0.9578	1	82 Pb		$L_{III}O_{III}$	12.945	1.0005	9	83 Bi		L_IM_I	12.39
0.95797	3	78 Pt	$L\gamma_1$	$L_{II}N_{IV}$	12.9420	1.00062	3	81 Tl	$L\beta_3$	L_IM_{III}	12.3904
0.9586	1	82 Pb		$L_{III}O_{II}$	12.934	1.00107	5	76 Os	$L\gamma_6$	$L_{II}O_{IV}$	12.3848
0.95931	5	77 Ir	$L\gamma_3$	L_IN_{III}	12.9240	1.0012	6	95 Am	Ll	$L_{III}M_I$	12.384
0.95938	8	76 Os	$L\gamma_4$	L_IO_{III}	12.923	1.0014	1	76 Os	L_{II}	Abs. Edge	12.381
0.96033	8	76 Os	$L\gamma_4'$	L_IO_{II}	12.910	1.0047	2	76 Os		$L_{II}O_{III}$	12.340
0.96133	7	82 Pb	Lu	$L_{III}N_{VI,VII}$	12.8968	1.00473	5	88 Ra	$L\alpha_1$	$L_{III}M_V$	12.3397
0.9620	1	82 Pb	$L\beta_7$	$L_{III}O_I$	12.888	1.0050	2	76 Os	Lv	$L_{II}N_{VI}$	12.337
0.96318	7	76 Os		L_IO_I	12.8721	1.0054	3	77 Ir		$L_{II}N_{III}$	12.332
0.9636	1	92 U	Ls	$L_{III}M_{III}$	12.866	1.00722	5	81 Tl		$L_{II}M_V$	12.3093
0.96389	7	81 Tl	$L\beta_{10}$	L_IM_{IV}	12.8626	1.0075	1	82 Pb	$L\beta_4$	L_IM_{II}	12.306
0.96545	3	77 Ir	$L\gamma_2$	L_IN_{II}	12.8418	1.00788	5	76 Os	$L\gamma_8$	$L_{II}O_I$	12.3012
0.96708	4	77 Ir	$L\gamma_6$	$L_{II}O_{IV}$	12.8201	1.0091	1	80 Hg	L_{III}	Abs. Edge	12.286
0.9671	1	77 Ir	L_{II}	Abs. Edge	12.820	1.00987	7	80 Hg	$L\beta_6$	$L_{III}O_{IV,V}$	12.2769
0.9672	2	84 Po	$L\beta_6$	$L_{III}N_I$	12.819	1.01031	3	81 Tl	$L\beta_2$	$L_{III}N_V$	12.2715
0.96788	2	90 Th	$L\alpha_2$	$L_{III}M_{IV}$	12.8096	1.01040	7	82 Pb		$L_{III}N_{II}$	12.2705
0.96911	7	82 Pb	$L\beta_3$	L_IM_{III}	12.7933	1.0108	1	75 Re	$L\gamma_{11}$	L_IN_V	12.266
0.96979	5	77 Ir		$L_{II}O_{III}$	12.7843	1.0112	1	90 Th	Ls	$L_{III}M_{III}$	12.261
0.97161	6	77 Ir	Lv	$L_{II}N_{VI}$	12.7603	1.0119	1	75 Re		L_IN_{IV}	12.252
0.97173	4	78 Pt		$L_{II}N_{III}$	12.7588	1.0120	2	77 Ir		L_IN_{II}	12.251
0.97321	5	83 Bi		$L_{III}N_{III}$	12.7394	1.01201	3	81 Tl	$L\beta_{15}$	$L_{III}N_{IV}$	12.2510
0.97409	3	77 Ir	$L\gamma_8$	$L_{II}O_I$	12.7279	1.01404	7	80 Hg		$L_{III}O_{III}$	12.2264
0.9747	1	82 Pb		$L_{II}M_V$	12.720	1.01513	4	81 Tl	$L\beta_1$	$L_{II}M_{IV}$	12.2133
0.9765	3	76 Os	$L\gamma_{11}$	L_IN_V	12.696	1.01558	7	80 Hg		$L_{III}O_{II}$	12.2079

TABLE 1b3. (*Continued*)

Wavelength Å*	p.e.	Element	Designation		keV
1.01656	5	88 Ra	$L\alpha_2$	$L_{III}M_{IV}$	12.1962
1.01674	7	80 Hg	Lu	$L_{III}N_{VII}$	12.1940
1.01769	7	80 Hg	Lu'	$L_{III}N_{VI}$	12.1826
1.01937	7	80 Hg	$L\beta_7$	$L_{III}O_I$	12.1625
1.02063	7	79 Au	$L\beta_9$	$L_I M_V$	12.1474
1.0210	1	82 Pb	$L\beta_6$	$L_{III}N_I$	12.143
1.02175	5	77 Ir	$L\gamma_8$	$L_{II}N_I$	12.1342
1.0223	1	82 Pb	$L\beta_{17}$	$L_{II}M_{III}$	12.127
1.0226	1	94 Pu	Ll	$L_{III}M_I$	12.124
1.02467	5	74 W	L_I	Abs. Edge	12.0996
1.0250	2	74 W		$L_I O_{IV,V}$	12.095
1.02503	5	76 Os	$L\gamma_1$	$L_{II}N_{IV}$	12.0953
1.02613	7	75 Re	$L\gamma_8$	$L_I N_{III}$	12.0824
1.02775	3	74 W	$L\gamma_4$	$L_I O_{III}$	12.0634
1.02789	7	79 Au	$L\beta_{10}$	$L_I M_{IV}$	12.0617
1.0286	1	81 Tl		$L_{III}N_{III}$	12.053
1.02863	3	74 W	$L\gamma_4'$	$L_I O_{II}$	12.0530
1.03049	5	87 Fr	$L\alpha_1$	$L_{III}M_V$	12.0313
1.0317	3	74 W		$L_I O_I$	12.017
1.03233	5	75 Re	$L\gamma_2$	$L_I N_{II}$	12.0098
1.0323	2	82 Pb		$L_I M_I$	12.010
1.03358	7	80 Hg	$L\beta_3$	$L_I M_{III}$	11.9953
1.0346	9	83 Bi		$L_I M_{II}$	11.98
1.0347	1	92 U	Lt	$L_{III}M_{II}$	11.982
1.03699	9	75 Re	$L\gamma_6$	$L_{II}O_{IV}$	11.956
1.0371	1	75 Re	L_{II}	Abs. Edge	11.954
1.03876	7	79 Au		$L_{III}P_{II,III}$	11.9355
1.03918	3	81 Tl	$L\beta_4$	$L_{II}M_{III}$	11.9306
1.0397	1	75 Re		$L_{II}O_{III}$	11.925
1.03973	5	76 Os		$L_{II}N_{III}$	11.9243
1.03974	2	35 Br	$K\alpha_1$	$K L_{III}$	11.9242
1.03975	7	80 Hg	$L\beta_2$	$L_{III}N_V$	11.9241
1.04000	5	79 Au	L_{III}	Abs. Edge	11.9212
1.0404	1	75 Re	Lv	$L_{II}N_{VI}$	11.917
1.04044	3	79 Au	$L\beta_5$	$L_{III}O_{IV,V}$	11.9163
1.04151	7	80 Hg	$L\beta_{15}$	$L_{III}N_{IV}$	11.9040
1.0420	1	75 Re		$L_I N_I$	11.899
1.04230	5	87 Fr	$L\alpha_2$	$L_{III}M_{IV}$	11.8950
1.0428	6	93 Np	Ll	$L_{III}M_I$	11.890
1.04382	2	35 Br	$K\alpha_2$	$K L_{II}$	11.8776
1.04398	5	75 Re	$L\gamma_8$	$L_{II}O_I$	11.8758
1.0450	2	79 Au		$L_{III}O_{II,III}$	11.865
1.0450	1	33 As	K	Abs. Edge	11.865
1.04500	3	33 As	$K\beta_2$	$K N_{II,III}$	11.8642
1.0458	1	74 W	$L\gamma_{11}$	$L_I N_V$	11.856
1.0468	2	74 W		$L_I N_{IV}$	11.844
1.04752	5	79 Au	Lu	$L_{III}N_{VI,VII}$	11.8357
1.04868	5	80 Hg	$L\beta_1$	$L_{II}M_{IV}$	11.8226
1.0488	1	33 As	$K\beta_5$	$K M_{IV,V}$	11.822
1.04963	5	81 Tl	$L\beta_6$	$L_{III}N_I$	11.8118
1.04974	8	79 Au	$L\beta_7$	$L_{III}O_I$	11.8106
1.05446	5	78 Pt	$L\beta_9$	$L_I M_V$	11.7577
1.05609	7	81 Tl	$L\beta_{17}$	$L_{II}M_{III}$	11.7397
1.05693	5	76 Os	$L\gamma_6$	$L_{II}N_I$	11.7303
1.05723	5	86 Rn	$L\alpha_1$	$L_{III}M_V$	11.7270
1.05730	2	33 As	$K\beta_1$	$K M_{III}$	11.7262
1.05783	5	33 As	$K\beta_3$	$K M_{II}$	11.7203
1.0585	1	80 Hg		$L_{III}N_{III}$	11.713
1.05856	3	83 Bi	$L\eta$	$L_{II}M_I$	11.7122
1.06099	5	75 Re	$L\gamma_1$	$L_{II}N_{IV}$	11.6854
1.0613	1	73 Ta	L_I	Abs. Edge	11.682
1.06183	7	78 Pt	$L\beta_{10}$	$L_I M_{IV}$	11.6762
1.06192	9	73 Ta		$L_I O_{IV,V}$	11.6752
1.06200	6	74 W	$L\gamma_8$	$L_I N_{III}$	11.6743
1.06357	9	73 Ta		$L_I N_{VI,VII}$	11.6570
1.0644	2	82 Pb		$L_{II}M_{II}$	11.648
1.0644	2	81 Tl		$L_I M_I$	11.648
1.06467	3	73 Ta	$L\gamma_4$	$L_I O_{III}$	11.6451
1.0649	2	80 Hg		$L_{III}N_{II}$	11.642
1.06544	3	73 Ta	$L\gamma_4'$	$L_I O_{II}$	11.6366
1.06712	2	92 U	Ll	$L_{III}M_I$	11.6183
1.06771	9	73 Ta		$L_I O_I$	11.6118
1.06785	9	79 Au	$L\beta_8$	$L_I M_{III}$	11.6103
1.06806	3	74 W	$L\gamma_2$	$L_I N_{II}$	11.6080
1.06899	5	86 Rn	$L\alpha_2$	$L_{III}M_{IV}$	11.5979
1.07022	3	79 Au	$L\beta_2$	$L_{III}N_V$	11.5847
1.07188	5	79 Au	$L\beta_{15}$	$L_{III}N_{IV}$	11.5667
1.07222	7	80 Hg	$L\beta_4$	$L_I M_{II}$	11.5630
1.0723	1	78 Pt	L_{III}	Abs. Edge	11.562
1.0724	2	78 Pt	$L\beta_5$	$L_{III}O_{IV,V}$	11.561
1.07448	5	74 W	$L\gamma_6$	$L_{II}O_{IV}$	11.5387
1.0745	1	74 W	L_{II}	Abs. Edge	11.538
1.0756	2	79 Au		$L_{III}M_V$	11.526
1.0761	3	78 Pt		$L_{III}O_{II,III}$	11.521
1.0767	1	75 Re		$L_{III}N_{III}$	11.515
1.0771	1	74 W	Lv	$L_{II}N_{VI}$	11.510
1.07896	5	78 Pt	Lu	$L_{III}N_{VI,VII}$	11.4908
1.0792	2	74 W		$L_{II}O_{III}$	11.488
1.07975	7	80 Hg	$L\beta_6$	$L_{III}N_I$	11.4824
1.08009	9	90 Th	Ll	$L_{III}M_{II}$	11.4788
1.08113	4	74 W	$L\gamma_8$	$L_{II}O_I$	11.4677
1.08168	3	78 Pt	$L\beta_7$	$L_{III}O_I$	11.4619
1.08205	7	73 Ta	$L\gamma_{11}$	$L_I N_V$	11.4580
1.08353	3	79 Au	$L\beta_1$	$L_{II}M_{IV}$	11.4423
1.08377	7	73 Ta		$L_I N_{IV},'$	11.4398
1.0839	1	75 Re		$L_{II}M_{II}$	11.438
1.08500	5	85 At	$L\alpha_1$	$L_{III}M_V$	11.4268
1.08975	5	77 Ir	$L\beta_9$	$L_I M_V$	11.3770
1.09026	7	79 Au		$L_{III}N_{III}$	11.3717
1.0908	1	91 Pa	Ll	$L_{III}M_I$	11.366
1.0916	5	80 Hg	$L\beta_{17}$	$L_{II}M_{III}$	11.358
1.09241	7	82 Pb	$L\eta$	$L_{II}M_I$	11.3493
1.09388	5	75 Re	$L\gamma_5$	$L_{II}N_I$	11.3341
1.09671	5	85 At	$L\alpha_2$	$L_{III}M_{IV}$	11.3048
1.09702	4	77 Ir	$L\beta_{10}$	$L_I M_{IV}$	11.3016
1.09855	3	74 W	$L\gamma_1$	$L_{II}N_{IV}$	11.2859
1.09936	4	73 Ta	$L\gamma_8$	$L_I N_{III}$	11.2776
1.0997	1	81 Tl		$L_{II}M_{II}$	11.274
1.0997	1	72 Hf	L_I	Abs. Edge	11.274
1.09968	7	79 Au		$L_{III}N_{II}$	11.2743
1.0999	2	80 Hg		$L_I M_I$	11.272
1.10086	9	72 Hf		$L_I O_{IV}$	11.2622
1.10200	3	78 Pt	$L\beta_2$	$L_{III}N_V$	11.2505
1.10303	5	72 Hf	$L\gamma_4$	$L_I O_{III}$	11.2401
1.10376	5	72 Hf	$L\gamma_4'$	$L_I O_{II}$	11.2326
1.10394	5	78 Pt	$L\beta_3$	$L_I M_{III}$	11.2308
1.10477	2	34 Se	$K\alpha_1$	$K L_{III}$	11.2224
1.1053	1	73 Ta	$L\gamma_2$	$L_I N_{II}$	11.217
1.1058	1	77 Ir	L_{III}	Abs. Edge	11.212
1.10585	3	77 Ir	$L\beta_5$	$L_{III}O_{IV,V}$	11.2114
1.10651	3	79 Au	$L\beta_4$	$L_I M_{II}$	11.2047
1.10664	9	72 Hf		$L_I O_I$	11.2034
1.10882	2	34 Se	$K\alpha_2$	$K L_{II}$	11.1814
1.10923	6	77 Ir		$L_{III}O_{II,III}$	11.1772

TABLE 1b3. (*Continued*)

Wavelength Å*	p.e.	Element	Designation	keV	Wavelength Å*	p.e.	Element	Designation	keV
1.11092	3	79 Au	$L\beta_6$	$L_{III}N_I$	11.1602	1.16487	4	75 Re	$L\beta_9$	L_IM_V	10.6433
1.11145	4	77 Ir	Lu	$L_{III}N_{VI,VII}$	11.1549	1.16545	5	77 Ir		$L_{III}N_{II}$	10.6380
1.1129	2	78 Pt		$L_{II}M_V$	11.140	1.1667	1	78 Pt	$L\beta_{17}$	$L_{II}M_{III}$	10.6265
1.1137	1	73 Ta	L_{II}	Abs. Edge	11.132	1.16719	5	88 Ra	Ll	$L_{III}M_I$	10.6222
1.11386	4	84 Po	$L\alpha_1$	$L_{III}M_V$	11.1308	1.16962	9	78 Pt		L_IM_I	10.6001
1.11388	3	73 Ta	$L\gamma_6$	$L_{II}O_{IV}$	11.1306	1.16979	8	76 Os	$L\beta_2$	$L_{III}N_V$	10.5985
1.11489	3	77 Ir	$L\beta_7$	$L_{III}O_I$	11.1205	1.1708	1	79 Au		$L_{II}M_{II}$	10.5892
1.1149	2	74 W		$L_{II}N_{III}$	11.120	1.17167	5	76 Os	$L\beta_{15}$	$L_{III}N_{IV}$	10.5816
1.11508	4	90 Th	Ll	$L_{III}M_I$	11.1186	1.17218	5	75 Re	$L\beta_{10}$	L_IM_{IV}	10.5770
1.11521	9	73 Ta		L_IN_I	11.1173	1.1729	1	73 Ta	$L\gamma_6$	$L_{II}N_I$	10.5702
1.1158	1	73 Ta	Lv	$L_{II}N_{VI}$	11.1113	1.17501	2	82 Pb	$L\alpha_1$	$L_{III}M_V$	10.5515
1.11658	5	32 Ge	K	Abs. Edge	11.1036	1.17588	1	33 As	$K\alpha_1$	KL_{III}	10.54372
1.11686	2	32 Ge	$K\beta_2$	$KN_{II,III}$	11.1008	1.17721	5	75 Re	$L\beta_5$	$L_{III}O_{IV,V}$	10.5318
1.11693	9	73 Ta		$L_{II}O_{III}$	11.1001	1.1773	1	75 Re	L_{III}	Abs. Edge	10.5306
1.11789	9	73 Ta		$L_{II}O_{II}$	11.0907	1.17788	9	72 Hf		$L_{II}N_V$	10.5258
1.1195	1	32 Ge	$K\beta_5$	$KM_{IV,V}$	11.0745	1.17796	3	77 Ir	$L\beta_6$	$L_{III}N_I$	10.5251
1.11990	2	78 Pt	$L\beta_1$	$L_{II}M_{IV}$	11.0707	1.17900	5	72 Hf	$L\gamma_1$	$L_{II}N_{IV}$	10.5158
1.1205	1	73 Ta	$L\gamma_8$	$L_{II}O_I$	11.0646	1.17953	4	71 Lu	$L\gamma_2$	L_IN_{III}	10.5110
1.12146	9	72 Hf	$L\gamma_{11}$	L_IN_V	11.0553	1.17955	7	76 Os	$L\beta_8$	L_IM_{III}	10.5108
1.1218	3	74 W		$L_{II}N_{II}$	11.052	1.17958	3	77 Ir	$L\beta_4$	L_IM_{II}	10.5106
1.12250	9	72 Hf		L_IN_{IV}	11.0451	1.17987	1	33 As	$K\alpha_2$	KL_{II}	10.50799
1.1226	2	78 Pt		$L_{III}N_{III}$	11.044	1.1815	1	75 Re	Lu	$L_{III}N_{VI,VII}$	10.4931
1.12548	5	84 Po	$L\alpha_2$	$L_{III}M_{IV}$	11.0158	1.1818	1	70 Yb	L_I	Abs. Edge	10.4904
1.12637	6	76 Os	$L\beta_9$	L_IM_V	11.0071	1.1827	1	70 Yb		$L_IO_{IV,V}$	10.4833
1.12769	3	81 Tl	$L\eta$	$L_{II}M_I$	10.9943	1.1853	1	70 Yb	$L\gamma_4$	$L_IO_{II,III}$	10.4603
1.12798	5	79 Au	$L\beta_{17}$	$L_{II}M_{III}$	10.9915	1.1853	2	71 Lu	$L\gamma_2$	L_IN_{III}	10.460
1.12894	2	32 Ge	$K\beta_1$	KM_{III}	10.9821	1.18610	5	75 Re	$L\beta_7$	$L_{III}O_I$	10.4529
1.12936	9	32 Ge	$K\beta_3$	KM_{II}	10.9780	1.18648	5	82 Pb	$L\alpha_2$	$L_{III}M_{IV}$	10.4495
1.1310	2	78 Pt		$L_{III}N_{II}$	10.962	1.1886	1	70 Yb		L_IO_I	10.4312
1.13235	3	74 W	$L\gamma_6$	$L_{II}N_I$	10.9490	1.18977	7	76 Os		$L_{II}M_V$	10.4205
1.13353	5	76 Os	$L\beta_{10}$	L_IM_{IV}	10.9376	1.1958	1	31 Ga	K	Abs. Edge	10.3682
1.13525	5	79 Au		L_IM_I	10.9210	1.19600	2	31 Ga	$K\beta_2$	$KN_{II,III}$	10.3663
1.13532	3	77 Ir	$L\beta_2$	$L_{III}N_V$	10.9203	1.19727	7	76 Os	$L\beta_1$	$L_{II}M_{IV}$	10.3553
1.13687	9	73 Ta		$L_{II}N_V$	10.9055	1.1981	2	31 Ga	$K\beta_5$	$KM_{IV,V}$	10.348
1.13707	3	77 Ir	$L\beta_{15}$	$L_{III}N_{IV}$	10.9036	1.1985	1	71 Lu	L_{II}	Abs. Edge	10.3448
1.13794	3	73 Ta	$L\gamma_1$	$L_{II}N_{IV}$	10.8952	1.1987	1	71 Lu	$L\gamma_6$	$L_{II}O_{IV}$	10.3431
1.13841	5	72 Hf	$L\gamma_3$	L_IN_{III}	10.8907	1.20086	7	76 Os		$L_{III}N_{II}$	10.3244
1.1387	5	80 Hg		$L_{II}M_{II}$	10.888	1.2014	1	71 Lu		$L_{II}O_{II,III}$	10.3198
1.1402	1	71 Lu	L_I	Abs. Edge	10.8740	1.20273	3	79 Au	$L\eta$	$L_{II}M_{II}$	10.3083
1.1405	1	76 Os	$L\beta_5$	$L_{III}O_{IV,V}$	10.8711	1.2047	1	71 Lu	$L\gamma_8$	$L_{II}O_I$	10.2915
1.1408	1	76 Os	L_{III}	Abs. Edge	10.8683	1.20479	7	74 W	$L\beta_9$	L_IM_V	10.2907
1.14085	3	77 Ir	$L\beta_3$	L_IM_{III}	10.8674	1.20660	4	75 Re	$L\beta_2$	$L_{III}N_V$	10.2752
1.14223	5	78 Pt	$L\beta_4$	L_IM_{II}	10.8543	1.2069	2	77 Ir	$L\beta_{17}$	$L_{II}M_{III}$	10.273
1.1435	1	71 Lu	$L\gamma_4$	$L_IO_{II,III}$	10.8425	1.20739	4	81 Tl	$L\alpha_1$	$L_{III}M_V$	10.2685
1.14355	5	78 Pt	$L\beta_6$	$L_{III}N_I$	10.8418	1.20789	2	31 Ga	$K\beta_1$	KM_{III}	10.2642
1.14386	2	83 Bi	$L\alpha_1$	$L_{III}M_V$	10.8388	1.20819	5	75 Re	$L\beta_{15}$	$L_{III}N_{IV}$	10.2617
1.14442	5	72 Hf	$L\gamma_2$	L_IN_{II}	10.8335	1.20835	5	31 Ga	$K\beta_3$	KM_{II}	10.2603
1.14537	7	76 Os	Lu	$L_{III}N_{VI,VII}$	10.8245	1.2102	2	77 Ir		L_IM_I	10.245
1.1489	2	77 Ir		$L_{II}M_V$	10.791	1.2105	1	83 Bi	Ls	$L_{III}M_{III}$	10.2421
1.14933	8	76 Os	$L\beta_7$	$L_{III}O_I$	10.7872	1.21218	3	74 W	$L\beta_{10}$	L_IM_{IV}	10.2279
1.1548	1	72 Hf	L_{II}	Abs. Edge	10.7362	1.213	1	78 Pt		$L_{II}M_{II}$	10.225
1.15519	5	72 Hf	$L\gamma_6$	$L_{II}O_{IV}$	10.7325	1.21349	5	76 Os	$L\beta_6$	$L_{III}N_I$	10.2169
1.1553	1	73 Ta		$L_{II}N_{III}$	10.7316	1.21537	5	72 Hf	$L\gamma_5$	$L_{II}N_I$	10.2011
1.15536	1	83 Bi	$L\alpha_2$	$L_{III}M_{IV}$	10.73091	1.21545	3	74 W	$L\beta_5$	$L_{III}O_{IV,V}$	10.2004
1.1560	3	77 Ir		$L_{III}N_{III}$	10.725	1.2155	1	74 W	L_{III}	Abs. Edge	10.1999
1.15781	3	77 Ir	$L\beta_1$	$L_{II}M_{IV}$	10.7083	1.21844	5	76 Os	$L\beta_4$	L_IM_{II}	10.1754
1.15830	9	72 Hf	Lv	$L_{II}N_{VI}$	10.7037	1.21868	5	74 W	Lu	$L_{III}N_{VI,VII}$	10.1733
1.1600	2	73 Ta		$L_{II}N_{II}$	10.688	1.21875	3	81 Tl	$L\alpha_2$	$L_{III}M_{IV}$	10.1728
1.16107	9	71 Lu	$L\gamma_{11}$	L_IN_V	10.6782	1.22031	5	75 Re	$L\beta_3$	L_IM_{III}	10.1598
1.16138	5	72 Hf	$L\gamma_8$	$L_{II}O_I$	10.6754	1.2211	2	74 W		$L_{III}O_{II,III}$	10.153
1.16227	9	71 Lu		L_IN_{IV}	10.6672	1.22228	4	71 Lu	$L\gamma_1$	$L_{II}N_{IV}$	10.1434
1.1640	1	80 Hg	$L\eta$	$L_{II}M_I$	10.6512	1.22232	5	70 Yb	$L\gamma_3$	L_IN_{III}	10.1431

TABLE 1b3. (Continued)

Wavelength Å*	p.e.	Element	Designation		keV
1.22400	4	74 W	$L\beta_7$	$L_{III}O_I$	10.1292
1.2250	1	69 Tm	L_I	Abs. Edge	10.1206
1.2263	3	69 Tm		$L_IO_{IV,V}$	10.110
1.2283	1	75 Re		$L_{III}N_{III}$	10.0933
1.22879	7	70 Yb	$L\gamma_2$	L_IN_{II}	10.0897
1.2294	2	69 Tm	$L\gamma_4$	$L_IO_{II,III}$	10.084
1.2305	1	75 Re		$L_{II}M_V$	10.0753
1.23858	2	75 Re	$L\beta_1$	$L_{II}M_{IV}$	10.0100
1.24120	5	80 Hg	$L\alpha_1$	$L_{III}M_V$	9.9888
1.24271	3	70 Yb	$L\gamma_6$	$L_{II}O_{IV}$	9.9766
1.2428	1	70 Yb	L_{II}	Abs. Edge	9.9761
1.2429	2	78 Pt	$L\eta$	$L_{II}M_I$	9.975
1.24385	7	82 Pb	Ls	$L_{III}M_{III}$	9.9675
1.24460	3	74 W	$L\beta_2$	$L_{III}N_V$	9.9615
1.2453	1	70 Yb		$L_{II}O_{II,III}$	9.9561
1.24631	3	74 W	$L\beta_{15}$	$L_{III}N_{IV}$	9.9478
1.2466	2	73 Ta	$L\beta_9$	L_IM_V	9.946
1.2480	2	76 Os	$L\beta_{17}$	$L_{II}M_{III}$	9.934
1.24923	5	70 Yb	$L\gamma_8$	$L_{II}O_I$	9.9246
1.2502	3	77 Ir		$L_{II}M_{II}$	9.917
1.25100	5	75 Re	$L\beta_6$	$L_{III}N_I$	9.9105
1.25264	7	80 Hg	$L\alpha_2$	$L_{III}M_{IV}$	9.8976
1.2537	2	73 Ta	$L\beta_{10}$	L_IM_{IV}	9.889
1.254054	9	32 Ge	$K\alpha_1$	KL_{III}	9.88642
1.2553	1	73 Ta	L_{III}	Abs. Edge	9.8766
1.2555	1	73 Ta	$L\beta_5$	$L_{III}O_{IV,V}$	9.8750
1.25778	4	73 Ta	Lu	$L_{III}N_{VI,VII}$	9.8572
1.258011	9	32 Ge	$K\alpha_2$	KL_{II}	9.85532
1.25917	5	75 Re	$L\beta_4$	L_IM_{II}	9.8463
1.2596	1	71 Lu	$L\gamma_5$	$L_{II}N_I$	9.8428
1.2601	3	73 Ta		$L_{III}O_{II,III}$	9.839
1.26269	5	74 W	$L\beta_3$	L_IM_{III}	9.8188
1.26385	5	73 Ta	$L\beta_7$	$L_{III}O_I$	9.8098
1.2672	2	74 W		$L_{III}N_{III}$	9.784
1.26769	5	70 Yb	$L\gamma_1$	$L_{II}N_{IV}$	9.7801
1.2678	2	69 Tm	$L\gamma_3$	L_IN_{III}	9.779
1.2706	1	68 Er	L_I	Abs. Edge	9.7574
1.2728	2	74 W		$L_{II}M_V$	9.741
1.2742	2	69 Tm	$L\gamma_2$	L_IN_{II}	9.730
1.2748	1	83 Bi	Lt	$L_{III}M_{II}$	9.7252
1.2752	2	68 Er	$L\gamma_4$	$L_IO_{II,III}$	9.722
1.27640	3	79 Au	$L\alpha_1$	$L_{III}M_V$	9.7133
1.2765	2	74 W		$L_{III}N_{II}$	9.712
1.27807	5	81 Tl	Ls	$L_{III}M_{III}$	9.7007
1.281809	9	74 W	$L\beta_1$	$L_{II}M_{IV}$	9.67235
1.2829	5	84 Po	Ll	$L_{III}M_I$	9.664
1.2834	1	30 Zn	K	Abs. Edge	9.6607
1.28372	2	30 Zn	$K\beta_2$	$KN_{II,III}$	9.6580
1.28448	3	77 Ir	$L\eta$	$L_{II}M_I$	9.6522
1.28454	2	73 Ta	$L\beta_2$	$L_{III}N_V$	9.6518
1.2848	1	30 Zn	$K\beta_5$	$KM_{IV,V}$	9.6501
1.28619	5	73 Ta	$L\beta_{15}$	$L_{III}N_{IV}$	9.6394
1.28772	3	79 Au	$L\alpha_2$	$L_{III}M_{IV}$	9.6280
1.2892	1	69 Tm	L_{II}	Abs. Edge	9.6171
1.28989	7	74 W	$L\beta_6$	$L_{III}N_I$	9.6117
1.29025	9	72 Hf	$L\beta_9$	L_IM_V	9.6090
1.2905	2	69 Tm	$L\gamma_6$	$L_{II}O_{IV}$	9.607
1.2927	1	75 Re	$L\beta_{17}$	$L_{II}M_{III}$	9.5910
1.2934	2	76 Os		$L_{II}M_{II}$	9.586
1.29525	2	30 Zn	$K\beta_{1,3}$	$KM_{II,III}$	9.5720
1.2972	1	72 Hf	L_{III}	Abs. Edge	9.5577
1.29761	5	72 Hf	$L\beta_5$	$L_{III}O_{IV,V}$	9.5546
1.29819	9	72 Hf	$L\beta_{10}$	L_IM_{IV}	9.5503
1.30162	5	74 W	$L\beta_4$	L_IM_{II}	9.5252
1.30165	9	72 Hf	Lu	$L_{III}N_{VI,VII}$	9.5249
1.30564	5	72 Hf	$L\beta_7$	$L_{III}O_I$	9.4958
1.3063	1	70 Yb	$L\gamma_5$	$L_{II}N_I$	9.4910
1.30678	3	73 Ta	$L\beta_3$	L_IM_{III}	9.4875
1.30767	7	82 Pb	Lt	$L_{III}M_{II}$	9.4811
1.3086	1	73 Ta		$L_{III}N_{III}$	9.4742
1.3112	2	80 Hg	Ls	$L_{III}M_{III}$	9.455
1.31304	3	78 Pt	$L\alpha_1$	$L_{III}M_V$	9.4423
1.3146	1	68 Er	$L\gamma_2$	L_IN_{III}	9.4309
1.3153	2	69 Tm	$L\gamma_1$	$L_{II}N_{IV}$	9.426
1.31610	7	83 Bi	Ll	$L_{III}M_I$	9.4204
1.3167	1	73 Ta		$L_{III}N_{II}$	9.4158
1.31897	9	73 Ta		$L_{II}M_V$	9.3998
1.3190	1	67 Ho	L_I	Abs. Edge	9.3994
1.3208	3	67 Ho		$L_IO_{IV,V}$	9.387
1.3210	2	68 Er	$L\gamma_2$	L_IN_{II}	9.385
1.3225	2	67 Ho	$L\gamma_4$	$L_IO_{II,III}$	9.374
1.32432	2	78 Pt	$L\alpha_2$	$L_{III}M_{IV}$	9.3618
1.32639	5	72 Hf	$L\beta_2$	$L_{III}N_V$	9.3473
1.32698	3	73 Ta	$L\beta_1$	$L_{II}M_{IV}$	9.3431
1.32783	5	72 Hf	$L\beta_{15}$	$L_{III}N_{IV}$	9.3371
1.32785	7	76 Os	$L\eta$	$L_{II}M_I$	9.3370
1.33094	8	73 Ta	$L\beta_6$	$L_{III}N_I$	9.3153
1.3358	1	71 Lu	$L\beta_9$	L_IM_V	9.2816
1.3365	3	74 W		L_IM_I	9.277
1.3366	1	75 Re		$L_{II}M_{II}$	9.2761
1.3386	1	68 Er	L_{II}	Abs. Edge	9.2622
1.3387	2	74 W	$L\beta_{17}$	$L_{II}M_{III}$	9.261
1.3397	3	68 Er	$L\gamma_6$	$L_{II}O_{IV}$	9.255
1.340083	9	31 Ga	$K\alpha_1$	KL_{III}	9.25174
1.3405	1	71 Lu	L_{III}	Abs. Edge	9.2490
1.34154	5	81 Tl	Lt	$L_{III}M_{II}$	9.2417
1.34183	7	71 Lu	$L\beta_5$	$L_{III}O_{IV,V}$	9.2397
1.3430	2	71 Lu	$L\beta_{10}$	L_IM_{IV}	9.232
1.34399	1	31 Ga	$K\alpha_2$	KL_{II}	9.22482
1.34524	9	71 Lu		$L_{III}O_{II,III}$	9.2163
1.34581	3	73 Ta	$L\beta_4$	L_IM_{II}	9.2124
1.34949	5	71 Lu	$L\beta_7$	$L_{III}O_I$	9.1873
1.34990	7	82 Pb	Ll	$L_{III}M_I$	9.1845
1.35053	9	72 Hf		$L_{III}N_{III}$	9.1802
1.35128	3	77 Ir	$L\alpha_1$	$L_{III}M_V$	9.1751
1.35131	7	79 Au	Ls	$L_{III}M_{III}$	9.1749
1.35300	5	72 Hf	$L\beta_3$	L_IM_{III}	9.1634
1.3558	2	69 Tm	$L\gamma_5$	$L_{II}N_I$	9.144
1.35887	9	72 Hf		$L_{III}N_{II}$	9.1239
1.36250	5	77 Ir	$L\alpha_2$	$L_{III}M_{IV}$	9.0995
1.3641	2	68 Er	$L\gamma_1$	$L_{II}N_{IV}$	9.089
1.3643	2	67 Ho	$L\gamma_3$	L_IN_{III}	9.087
1.3692	1	66 Dy	L_I	Abs. Edge	9.0548
1.3698	2	67 Ho	$L\gamma_2$	L_IN_{II}	9.051
1.37012	3	71 Lu	$L\beta_2$	$L_{III}N_V$	9.0489
1.3715	1	71 Lu	$L\beta_{15}$	$L_{III}N_{IV}$	9.0395
1.37342	5	75 Re	$L\eta$	$L_{II}M_I$	9.0272
1.37410	5	72 Hf	$L\beta_1$	$L_{II}M_{IV}$	9.0227
1.37410	5	72 Hf	$L\beta_6$	$L_{III}N_I$	9.0227
1.37459	7	66 Dy	$L\gamma_4$	$L_IO_{II,III}$	9.0195
1.3746	2	80 Hg	Lt	$L_{III}M_{II}$	9.019
1.38059	5	29 Cu	K	Abs. Edge	8.9803
1.38109	3	29 Cu	$K\beta_2$	$KM_{IV,V}$	8.9770
1.3838	1	70 Yb	$L\beta_9$	L_IM_V	8.9597

TABLE 1b3. (Continued)

Wavelength Å*	p.e.	Element	Designation		keV
1.38477	3	81 Tl	Ll	$L_{III}M_I$	8.9532
1.3862	1	70 Yb	L_{III}	Abs. Edge	8.9441
1.3864	1	73 Ta	$L\beta_{17}$	$L_{II}M_{III}$	8.9428
1.38696	7	70 Yb	$L\beta_5$	$L_{III}O_{IV,V}$	8.9390
1.3895	2	78 Pt	Ls	$L_{III}M_{III}$	8.923
1.3898	1	70 Yb		$L_{III}O_{II,III}$	8.9209
1.3905	1	67 Ho	L_{II}	Abs. Edge	8.9164
1.39121	5	76 Os	$L\alpha_1$	$L_{III}M_V$	8.9117
1.3915	1	70 Yb	$L\beta_{10}$	L_IM_{IV}	8.9100
1.39220	5	72 Hf	$L\beta_4$	L_IM_{II}	8.9054
1.392218	9	29 Cu	$K\beta_{1,3}$	$KM_{II,III}$	8.90529
1.3923	2	67 Ho	$L\gamma_6$	$L_{II}O_{IV}$	8.905
1.3926	1	29 Cu	$K\beta_3$	KM_{II}	8.9029
1.3948	1	70 Yb	$L\beta_7$	$L_{III}O_I$	8.8889
1.3983	2	67 Ho	$L\gamma_8$	$L_{II}O_I$	8.867
1.40140	5	71 Lu	$L\beta_2$	L_IM_{III}	8.8469
1.40234	5	76 Os	$L\alpha_2$	$L_{III}M_{IV}$	8.8410
1.4067	3	68 Er	$L\gamma_5$	$L_{II}N_I$	8.814
1.41366	7	79 Au	Ll	$L_{III}M_{II}$	8.7702
1.41550	5	70 Yb	$L\beta_{2,15}$	$L_{III}N_{IV,V}$	8.7588
1.41640	7	66 Dy	$L\gamma_2$	L_IN_{III}	8.7532
1.4174	2	67 Ho	$L\gamma_1$	$L_{II}N_{IV}$	8.747
1.4189	1	71 Lu	$L\beta_6$	$L_{III}N_I$	8.7376
1.42110	3	74 W	$L\eta$	$L_{II}M_I$	8.7243
1.4216	1	80 Hg	Ll	$L_{III}M_I$	8.7210
1.4223	1	65 Tb	L_I	Abs. Edge	8.7167
1.42278	7	66 Dy	$L\gamma_2$	L_IN_{II}	8.7140
1.4228	3	65 Tb		$L_IO_{IV,V}$	8.714
1.42359	3	71 Lu	$L\beta_1$	$L_{II}M_{IV}$	8.7090
1.4276	2	65 Tb	$L\gamma_4$	$L_IO_{II,III}$	8.685
1.43025	9	72 Hf		L_IM_I	8.6685
1.43048	9	73 Ta		$L_{II}M_{II}$	8.6671
1.4318	2	77 Ir	Ls	$L_{III}M_{III}$	8.659
1.43290	4	75 Re	$L\alpha_1$	$L_{III}M_V$	8.6525
1.4334	1	69 Tm	L_{III}	Abs. Edge	8.6496
1.4336	3	69 Tm	$L\beta_9$	L_IM_V	8.648
1.4349	2	69 Tm	$L\beta_5$	$L_{III}O_{IV,V}$	8.641
1.435155	7	30 Zn	$K\alpha_1$	KL_{III}	8.63886
1.43643	9	72 Hf	$L\beta_{17}$	$L_{II}M_{III}$	8.6312
1.439000	8	30 Zn	$K\alpha_2$	KL_{II}	8.61578
1.44056	5	71 Lu	$L\beta_4$	L_IM_{II}	8.6064
1.4410	3	69 Tm	$L\beta_{10}$	L_IM_{IV}	8.604
1.44396	5	75 Re	$L\alpha_2$	$L_{III}M_{IV}$	8.5862
1.4445	1	66 Dy	L_{II}	Abs. Edge	8.5830
1.44579	7	66 Dy	$L\gamma_6$	$L_{II}O_{IV}$	8.5753
1.45233	5	70 Yb	$L\beta_3$	L_IM_{III}	8.5367
1.4530	2	78 Pt	Ll	$L_{III}M_{II}$	8.533
1.45964	9	79 Au	Ll	$L_{III}M_I$	8.4939
1.4618	2	67 Ho	$L\gamma_5$	$L_{II}N_I$	8.481
1.4640	2	69 Tm	$L\beta_{2,15}$	$L_{III}N_{IV,V}$	8.468
1.4661	1	70 Yb	$L\beta_6$	$L_{III}N_I$	8.4563
1.47106	5	73 Ta	$L\eta$	$L_{II}M_I$	8.4280
1.4718	2	65 Tb	$L\gamma_2$	L_IN_{III}	8.423
1.47266	7	66 Dy	$L\gamma_1$	$L_{II}N_{IV}$	8.4188
1.4735	2	76 Os	Ls	$L_{III}M_{III}$	8.414
1.47565	5	70 Yb	$L\beta_1$	$L_{II}M_{IV}$	8.4018
1.4764	2	65 Tb	$L\gamma_2$	L_IN_{II}	8.398
1.47639	2	74 W	$L\alpha_1$	$L_{III}M_V$	8.3976
1.4784	1	64 Gd	L_I	Abs. Edge	8.3864
1.48064	9	72 Hf		$L_{II}M_{II}$	8.3735
1.4807	3	64 Gd		$L_IO_{IV,V}$	8.373
1.4835	1	68 Er	L_{III}	Abs. Edge	8.3575
1.4839	2	64 Gd	$L\gamma_4$	$L_IO_{II,III}$	8.355
1.4848	3	68 Er	$L\beta_5$	$L_{III}O_{IV,V}$	8.350
1.4855	5	68 Er	$L\beta_9$	L_IM_V	8.346
1.48743	2	74 W	$L\alpha_2$	$L_{III}M_{IV}$	8.3352
1.48807	1	28 Ni	K	Abs. Edge	8.33165
1.48862	4	28 Ni	$K\beta_5$	$KM_{IV,V}$	8.3286
1.49138	3	70 Yb	$L\beta_4$	L_IM_{II}	8.3132
1.4930	3	77 Ir	Ll	$L_{III}M_{II}$	8.304
1.4941	3	68 Er	$L\beta_7$	$L_{III}O_I$	8.298
1.4941	3	68 Er	$L\beta_{10}$	L_IM_{IV}	8.298
1.4995	2	78 Pt	Ll	$L_{III}M_I$	8.268
1.500135	8	28 Ni	$K\beta_{1,3}$	$KM_{II,III}$	8.26466
1.5023	1	65 Tb	L_{II}	Abs. Edge	8.2527
1.5035	2	65 Tb	$L\gamma_6$	$L_{II}O_{IV}$	8.246
1.5063	2	69 Tm	$L\beta_3$	L_IM_{III}	8.231
1.5097	2	65 Tb	$L\gamma_8$	$L_{II}O_I$	8.212
1.51399	9	68 Er	$L\beta_{2,15}$	$L_{III}N_{IV,V}$	8.1890
1.5162	2	69 Tm	$L\beta_6$	$L_{III}N_I$	8.177
1.5178	1	75 Re	Ls	$L_{III}M_{III}$	8.1682
1.51824	7	66 Dy	$L\gamma_5$	$L_{II}N_I$	8.1661
1.52197	2	73 Ta	$L\alpha_1$	$L_{III}M_V$	8.1461
1.52325	5	72 Hf	$L\eta$	$L_{II}M_I$	8.1393
1.5297	2	64 Gd	$L\gamma_8$	L_IN_{III}	8.105
1.5303	2	65 Tb	$L\gamma_1$	$L_{II}N_{IV}$	8.102
1.5304	2	69 Tm	$L\beta_1$	$L_{II}M_{IV}$	8.101
1.53293	2	73 Ta	$L\alpha_2$	$L_{III}M_{IV}$	8.0879
1.5331	2	64 Gd	$L\gamma_2$	L_IN_{II}	8.087
1.53333	9	71 Lu		$L_{II}M_{II}$	8.0858
1.5347	2	76 Os	Ll	$L_{III}M_{II}$	8.079
1.5368	1	67 Ho	L_{III}	Abs. Edge	8.0676
1.5378	2	67 Ho	$L\beta_5$	$L_{III}O_{IV,V}$	8.062
1.5381	1	63 Eu	L_I	Abs. Edge	8.0607
1.540562	2	29 Cu	$K\alpha_1$	KL_{III}	8.04778
1.54094	3	77 Ir	Ll	$L_{III}M_I$	8.0458
1.5439	1	63 Eu	$L\gamma_4$	$L_IO_{II,III}$	8.0304
1.544390	2	29 Cu	$K\alpha_2$	KL_{II}	8.02783
1.5448	2	69 Tm	$L\beta_4$	L_IM_{II}	8.026
1.5486	3	67 Ho	$L\beta_{10}$	L_IM_{IV}	8.006
1.5616	1	68 Er	$L\beta_3$	L_IM_{III}	7.9392
1.5632	1	64 Gd	L_{II}	Abs. Edge	7.9310
1.5642	3	74 W	Ls	$L_{III}M_{III}$	7.926
1.5644	2	64 Gd	$L\gamma_6$	$L_{II}O_{IV}$	7.925
1.5671	2	67 Ho	$L\beta_{2,15}$	$L_{III}N_{IV,V}$	7.911
1.5675	2	68 Er	$L\beta_6$	$L_{III}N_I$	7.909
1.56958	5	72 Hf	$L\alpha_1$	$L_{III}M_V$	7.8990
1.5707	2	64 Gd	$L\gamma_8$	$L_{II}O_I$	7.894
1.5779	1	71 Lu	$L\eta$	$L_{II}M_I$	7.8575
1.5787	2	65 Tb	$L\gamma_5$	$L_{II}N_I$	7.8535
1.5789	1	75 Re	Ll	$L_{III}M_{II}$	7.8525
1.58046	5	72 Hf	$L\alpha_2$	$L_{III}M_{IV}$	7.8446
1.58498	7	76 Os	Ll	$L_{III}M_I$	7.8222
1.5873	1	68 Er	$L\beta_1$	$L_{II}M_{IV}$	7.8109
1.58837	7	66 Dy	$L\beta_5$	$L_{III}O_{IV,V}$	7.8055
1.58844	9	70 Yb		$L_{II}M_{II}$	7.8052
1.5903	2	63 Eu	$L\gamma_2$	L_IN_{III}	7.7961
1.5916	1	66 Dy	L_{III}	Abs. Edge	7.7897
1.5924	2	64 Gd	$L\gamma_1$	$L_{II}N_{IV}$	7.7858
1.5961	2	63 Eu	$L\gamma_2$	L_IN_{II}	7.7677
1.59973	9	66 Dy	$L\beta_9$	L_IM_V	7.7501
1.6002	1	62 Sm	L_I	Abs. Edge	7.7478
1.6007	1	68 Er	$L\beta_4$	L_IM_{II}	7.7453
1.60447	7	66 Dy	$L\beta_7$	$L_{III}O_I$	7.7272

TABLE 1b3. (*Continued*)

Wavelength Å*	p.e.	Element		Designation	keV
1.60728	3	62 Sm	$L\gamma_4$	$L_IO_{II,III}$	7.714
1.60743	9	66 Dy	$L\beta_{10}$	L_IM_{IV}	7.7130
1.60815	1	27 Co	K	Abs. Edge	7.70954
1.60891	3	27 Co	$K\beta_5$	$KM_{IV,V}$	7.7059
1.61264	9	73 Ta	Ls	$L_{III}M_{III}$	7.6881
1.61951	3	71 Lu	$L\alpha_1$	$L_{III}M_V$	7.6555
1.6203	2	67 Ho	$L\beta_2$	L_IM_{III}	7.6519
1.62079	2	27 Co	$K\beta_{1,3}$	$KM_{II,III}$	7.64943
1.6237	2	67 Ho	$L\beta_6$	$L_{III}N_I$	7.6359
1.62369	7	66 Dy	$L\beta_{2,15}$	$L_{III}N_{IV,V}$	7.6357
1.6244	3	74 W	Ll	$L_{III}M_{II}$	7.6324
1.6271	1	63 Eu	L_{II}	Abs. Edge	7.6199
1.6282	2	63 Eu	$L\gamma_6$	$L_{II}O_{IV}$	7.6147
1.63029	5	71 Lu	$L\alpha_2$	$L_{III}M_{IV}$	7.6049
1.63056	5	75 Re	Ll	$L_{III}M_I$	7.6036
1.6346	2	63 Eu	$L\gamma_8$	$L_{II}O_I$	7.5849
1.63560	5	70 Yb	$L\eta$	$L_{II}M_I$	7.5802
1.6412	2	64 Gd	$L\gamma_6$	$L_{II}N_I$	7.5543
1.6475	2	67 Ho	$L\beta_1$	$L_{II}M_{IV}$	7.5253
1.6497	1	65 Tb	L_{III}	Abs. Edge	7.5153
1.6510	2	65 Tb	$L\beta_5$	$L_{III}O_{IV,V}$	7.5094
1.65601	3	62 Sm	$L\gamma_2$	L_IN_{III}	7.487
1.6574	2	63 Eu	$L\gamma_1$	$L_{II}N_{IV}$	7.4803
1.657910	8	28 Ni	$K\alpha_1$	KL_{III}	7.47815
1.6585	2	65 Tb	$L\beta_7$	$L_{III}O_I$	7.4753
1.6595	2	67 Ho	$L\beta_4$	L_IM_{II}	7.4708
1.66044	6	62 Sm	$L\gamma_2$	L_IN_{II}	7.467
1.661747	8	28 Ni	$K\alpha_2$	KL_{II}	7.46089
1.66346	9	72 Hf	Ls	$L_{III}M_{III}$	7.4532
1.6673	3	65 Tb	$L\beta_{10}$	L_IM_{IV}	7.436
1.6674	5	61 Pm	L_I	Abs. Edge	7.436
1.67189	4	70 Yb	$L\alpha_1$	$L_{III}M_V$	7.4156
1.67265	9	73 Ta	Lt	$L_{III}M_{II}$	7.4123
1.6782	1	74 W	Ll	$L_{III}M_I$	7.3878
1.68213	7	66 Dy	$L\beta_6$	$L_{III}N_I$	7.3705
1.6822	2	66 Dy	$L\beta_3$	L_IM_{III}	7.3702
1.68285	5	70 Yb	$L\alpha_2$	$L_{III}M_{IV}$	7.3673
1.6830	2	65 Tb	$L\beta_{2,15}$	$L_{III}N_{IV,V}$	7.3667
1.6953	1	62 Sm	L_{II}	Abs. Edge	7.3132
1.6963	2	69 Tm	$L\eta$	$L_{II}M_I$	7.3088
1.6966	9	62 Sm	$L\gamma_6$	$L_{II}O_{IV}$	7.308
1.7085	2	63 Eu	$L\gamma_5$	$L_{II}N_I$	7.2566
1.71062	7	66 Dy	$L\beta_1$	$L_{II}M_{IV}$	7.2477
1.7117	1	64 Gd	L_{III}	Abs. Edge	7.2430
1.7130	2	64 Gd	$L\beta_5$	$L_{III}O_{IV,V}$	7.2374
1.7203	2	64 Gd	$L\beta_7$	$L_{III}O_I$	7.2071
1.72103	7	66 Dy	$L\beta_4$	L_IM_{II}	7.2039
1.72305	9	72 Hf	Lt	$L_{III}M_{II}$	7.1954
1.7240	3	64 Gd	$L\beta_9$	L_IM_V	7.192
1.72724	3	62 Sm	$L\gamma_1$	$L_{II}N_{IV}$	7.178
1.7268	2	69 Tm	$L\alpha_1$	$L_{III}M_V$	7.1799
1.72841	5	73 Ta	Ll	$L_{III}M_I$	7.1731
1.7315	3	64 Gd	$L\beta_{10}$	L_IM_{IV}	7.160
1.7381	2	69 Tm	$L\alpha_2$	$L_{III}M_{IV}$	7.1331
1.7390	1	60 Nd	L_I	Abs. Edge	7.1294
1.7422	2	65 Tb	$L\beta_6$	$L_{III}N_I$	7.1163
1.74346	1	26 Fe	K	Abs. Edge	7.11120
1.7442	1	26 Fe	$K\beta_5$	$KM_{IV,V}$	7.1081
1.7445	4	60 Nd	$L\gamma_4$	$L_IO_{II,III}$	7.107
1.7455	2	64 Gd	$L\beta_{2,15}$	$L_{III}N_{IV,V}$	7.1028
1.7472	2	65 Tb	$L\beta_3$	L_IM_{III}	7.0959
1.75661	2	26 Fe	$K\beta_{1,3}$	$KM_{II,III}$	7.05798
1.7566	1	68 Er	$L\eta$	$L_{II}M_I$	7.0579
1.7676	5	61 Pm	L_{II}	Abs. Edge	7.014
1.7760	1	71 Lu	Lt	$L_{III}M_{II}$	6.9810
1.7761	1	63 Eu	L_{III}	Abs. Edge	6.9806
1.7768	3	65 Tb	$L\beta_1$	$L_{II}M_{IV}$	6.978
1.7772	2	63 Eu	$L\beta_5$	$L_{III}O_{IV,V}$	6.9763
1.77934	3	62 Sm	$L\gamma_5$	$L_{II}N_I$	6.968
1.78145	5	72 Hf	Ll	$L_{III}M_I$	6.9596
1.78425	9	68 Er	$L\alpha_1$	$L_{III}M_V$	6.9487
1.7851	2	63 Eu	$L\beta_7$	$L_{III}O_I$	6.9453
1.7864	2	65 Tb	$L\beta_4$	L_IM_{II}	6.9403
1.788965	9	27 Co	$K\alpha_1$	KL_{III}	6.93032
1.7916	3	63 Eu	$L\beta_9$	L_IM_V	6.920
1.792850	9	27 Co	$K\alpha_2$	KL_{II}	6.91530
1.7955	2	68 Er	$L\alpha_2$	$L_{III}M_{IV}$	6.9050
1.7964	4	60 Nd	$L\gamma_3$	L_IN_{III}	6.902
1.7989	9	61 Pm	$L\gamma_1$	$L_{II}N_{IV}$	6.892
1.7993	3	63 Eu	$L\beta_{10}$	L_IM_{IV}	6.890
1.8013	4	60 Nd	$L\gamma_2$	L_IN_{II}	6.883
1.8054	2	64 Gd	$L\beta_6$	$L_{III}N_I$	6.8671
1.8118	2	63 Eu	$L\beta_{2,15}$	$L_{III}N_{IV,V}$	6.8432
1.8141	5	59 Pr	L_I	Abs. Edge	6.834
1.8150	2	64 Gd	$L\beta_3$	L_IM_{III}	6.8311
1.8193	4	59 Pr	$L\gamma_4$	$L_IO_{II,III}$	6.815
1.8264	2	67 Ho	$L\eta$	$L_{II}M_I$	6.7883
1.83091	9	70 Yb	Lt	$L_{III}M_{II}$	6.7715
1.8360	1	71 Lu	Ll	$L_{III}M_I$	6.7528
1.8440	1	60 Nd	L_{II}	Abs. Edge	6.7234
1.8450	2	67 Ho	$L\alpha_1$	$L_{III}M_V$	6.7198
1.8457	1	62 Sm	L_{III}	Abs. Edge	6.7172
1.8468	2	64 Gd	$L\beta_1$	$L_{II}M_{IV}$	6.7132
1.84700	9	62 Sm	$L\beta_5$	$L_{III}O_{IV,V}$	6.7126
1.8540	2	64 Gd	$L\beta_4$	L_IM_{II}	6.6871
1.8552	5	60 Nd	$L\gamma_8$	$L_{II}O_I$	6.683
1.8561	2	67 Ho	$L\alpha_2$	$L_{III}M_{IV}$	6.6795
1.85626	3	62 Sm	$L\beta_7$	$L_{III}O_I$	6.679
1.86166	3	62 Sm	$L\beta_9$	L_IM_V	6.660
1.86990	3	62 Sm	$L\beta_{10}$	L_IM_{IV}	6.634
1.8737	2	63 Eu	$L\beta_6$	$L_{III}N_I$	6.6170
1.8740	4	59 Pr	$L\gamma_3$	L_IN_{III}	6.616
1.8779	2	60 Nd	$L\gamma_1$	$L_{II}N_{IV}$	6.6021
1.8791	4	59 Pr	$L\gamma_2$	L_IN_{II}	6.598
1.8821	3	62 Sm	$L\beta_{2,15}$	$L_{III}N_{IV,V}$	6.586
1.8867	2	63 Eu	$L\beta_3$	L_IM_{III}	6.5713
1.8934	5	58 Ce	L_I	Abs. Edge	6.548
1.89415	5	70 Yb	Ll	$L_{III}M_I$	6.5455
1.89643	5	25 Mn	K	Abs. Edge	6.5376
1.8971	1	25 Mn	$K\beta_5$	$KM_{IV,V}$	6.5352
1.89743	7	66 Dy	$L\eta$	$L_{II}M_I$	6.5342
1.8991	4	58 Ce	$L\gamma_4$	$L_IO_{II,III}$	6.528
1.90881	3	66 Dy	$L\alpha_1$	$L_{III}M_V$	6.4952
1.91021	2	25 Mn	$K\beta_{1,3}$	$KM_{II,III}$	6.49045
1.9191	1	61 Pm	L_{III}	Abs. Edge	6.4605
1.91991	3	66 Dy	$L\alpha_2$	$L_{III}M_{IV}$	6.4577
1.9203	2	63 Eu	$L\beta_1$	$L_{II}M_{IV}$	6.4564
1.9255	2	63 Eu	$L\beta_4$	L_IM_{II}	6.4389
1.9255	5	59 Pr	L_{II}	Abs. Edge	6.439
1.9355	4	60 Nd	$L\gamma_5$	$L_{II}N_I$	6.406
1.936042	9	26 Fe	$K\alpha_1$	KL_{III}	6.40384
1.9362	4	59 Pr	$L\gamma_8$	$L_{II}O_I$	6.403
1.939980	9	26 Fe	$K\alpha_2$	KL_{II}	6.39084
1.94643	3	62 Sm	$L\beta_6$	$L_{III}N_I$	6.3693

TABLE 1b3. (*Continued*)

Wavelength Å*	p.e.	Element	Designation		keV	Wavelength Å*	p.e.	Element	Designation		keV
1.9550	2	69 Tm	Ll	$L_{III}M_I$	6.3419	2.1958	5	58 Ce	$L\beta_{10}$	L_IM_{IV}	5.646
1.9553	3	58 Ce	$L\gamma_3$	L_IN_{III}	6.3409	2.1998	2	62 Sm	$L\alpha_1$	$L_{III}M_V$	5.6361
1.9559	6	61 Pm	$L\beta_{2,15}$	$L_{III}N_{IV,V}$	6.339	2.2048	1	56 Ba	L_{II}	Abs. Edge	5.6233
1.9602	3	58 Ce	$L\gamma_2$	L_IN_{II}	6.3250	2.2056	4	57 La	$L\gamma_5$	$L_{II}N_I$	5.621
1.9611	3	59 Pr	$L\gamma_1$	$L_{II}N_{IV}$	6.3221	2.2087	2	58 Ce	$L\beta_{2,15}$	$L_{III}N_{IV,V}$	5.6134
1.96241	3	62 Sm	$L\beta_3$	L_IM_{III}	6.318	2.21062	3	62 Sm	$L\alpha_2$	$L_{III}M_{IV}$	5.6090
1.9730	2	65 Tb	$L\eta$	$L_{II}M_I$	6.2839	2.2172	3	59 Pr	$L\beta_3$	L_IM_{III}	5.5918
1.9765	2	65 Tb	$L\alpha_1$	$L_{III}M_V$	6.2728	2.21824	3	62 Sm	$L\eta$	$L_{II}M_I$	5.589
1.9780	5	57 La	L_I	Abs. Edge	6.268	2.2328	2	55 Cs	$L\gamma_3$	L_IN_{III}	5.5527
1.9830	4	57 La	$L\gamma_4$	$L_IO_{II,III}$	6.252	2.2352	2	65 Tb	Ll	$L_{III}M_I$	5.5467
1.9875	2	65 Tb	$L\alpha_2$	$L_{III}M_{IV}$	6.2380	2.2371	2	55 Cs	$L\gamma_2$	L_IN_{II}	5.5420
1.9967	1	60 Nd	L_{III}	Abs. Edge	6.2092	2.2415	2	56 Ba	$L\gamma_1$	$L_{II}N_{IV}$	5.5311
1.99806	3	62 Sm	$L\beta_1$	$L_{II}M_{IV}$	6.2051	2.253	6	92 U		M_IP_{III}	5.50
2.00095	6	62 Sm	$L\beta_4$	L_IM_{II}	6.196	2.2550	4	59 Pr	$L\beta_4$	L_IM_{II}	5.4981
2.0092	3	60 Nd	$L\beta_7$	$L_{III}O_I$	6.1708	2.2588	3	59 Pr	$L\beta_1$	$L_{II}M_{IV}$	5.4889
2.0124	5	58 Ce	L_{II}	Abs. Edge	6.161	2.261	1	57 La	L_{III}	Abs. Edge	5.484
2.015	1	68 Er	Ll	$L_{III}M_I$	6.152	2.2691	1	23 V	K	Abs. Edge	5.4639
2.0165	3	60 Nd	$L\beta_9$	L_IM_V	6.1484	2.26951	6	23 V	$K\beta_5$	$KM_{IV,V}$	5.4629
2.0205	4	59 Pr	$L\gamma_5$	$L_{II}N_I$	6.136	2.2737	1	54 Xe	L_I	Abs. Edge	5.4528
2.0237	4	58 Ce	$L\gamma_8$	L_IIO_I	6.126	2.275	3	57 La	$L\beta_7$	$L_{III}O_I$	5.450
2.0237	3	60 Nd	$L\beta_{10}$	L_IM_{IV}	6.1265	2.282	3	57 La	$L\beta_9$	L_IM_V	5.434
2.0360	3	60 Nd	$L\beta_{2,15}$	$L_{III}N_{IV,V}$	6.0894	2.2818	3	58 Ce	$L\beta_6$	$L_{III}N_I$	5.4334
2.0410	4	57 La	$L\gamma_3$	L_IN_{III}	6.074	2.2822	3	61 Pm	$L\alpha_1$	$L_{III}M_V$	5.4325
2.0421	4	61 Pm	$L\beta_3$	L_IM_{III}	6.071	2.28440	2	23 V	$K\beta_{1,3}$	$KM_{II,III}$	5.42729
2.0460	4	57 La	$L\gamma_2$	L_IN_{II}	6.060	2.28970	2	24 Cr	$K\alpha_1$	KL_{III}	5.41472
2.0468	2	64 Gd	$L\alpha_1$	$L_{III}M_V$	6.0572	2.290	3	57 La	$L\beta_{10}$	L_IM_{IV}	5.415
2.0487	4	58 Ce	$L\gamma_1$	$L_{II}N_{IV}$	6.052	2.2926	4	61 Pm	$L\alpha_2$	$L_{III}M_{IV}$	5.4078
2.0494	1	64 Gd	$L\eta$	$L_{II}M_I$	6.0495	2.293606	3	24 Cr	$K\alpha_2$	KL_{II}	5.405509
2.0578	2	64 Gd	$L\alpha_2$	$L_{III}M_{IV}$	6.0250	2.3030	3	57 La	$L\beta_{2,15}$	$L_{III}N_{IV,V}$	5.3835
2.0678	5	56 Ba	L_I	Abs. Edge	5.996	2.304	7	92 U		M_IO_{III}	5.38
2.07020	5	24 Cr	K	Abs. Edge	5.9888	2.3085	3	56 Ba	$L\gamma_5$	$L_{II}N_I$	5.3707
2.07087	6	24 Cr	$K\beta_5$	$KM_{IV,V}$	5.9869	2.3109	3	58 Ce	$L\beta_3$	L_IM_{III}	5.3651
2.0756	3	56 Ba	$L\gamma_4$	$L_IO_{II,III}$	5.9733	2.3122	2	64 Gd	Ll	$L_{III}M_I$	5.3621
2.0791	5	59 Pr	L_{III}	Abs. Edge	5.963	2.3139	1	55 Cs	L_{II}	Abs. Edge	5.3581
2.0797	4	61 Pm	$L\beta_1$	$L_{II}M_{IV}$	5.961	2.3480	2	55 Cs	$L\gamma_1$	$L_{II}N_{IV}$	5.2804
2.08487	2	24 Cr	$K\beta_{1,3}$	$KM_{II,III}$	5.94671	2.3497	4	58 Ce	$L\beta_4$	L_IM_{II}	5.2765
2.0860	2	67 Ho	Ll	$L_{III}M_I$	5.9434	2.3561	3	58 Ce	$L\beta_1$	$L_{II}M_{IV}$	5.2622
2.0919	4	59 Pr	$L\beta_7$	$L_{III}O_I$	5.927	2.3629	1	56 Ba	L_{III}	Abs. Edge	5.2470
2.1004	4	59 Pr	$L\beta_9$	L_IM_V	5.903	2.3704	2	60 Nd	$L\alpha_1$	$L_{III}M_V$	5.2304
2.101820	9	25 Mn	$K\alpha_1$	KL_{III}	5.89875	2.3764	2	56 Ba	$L\beta_9$	L_IM_V	5.2171
2.1039	3	60 Nd	$L\beta_6$	$L_{III}N_I$	5.8930	2.3790	4	57 La	$L\beta_6$	$L_{III}N_I$	5.2114
2.1053	5	57 La	L_{II}	Abs. Edge	5.889	2.3806	2	56 Ba	$L\beta_7$	$L_{III}O_I$	5.2079
2.10578	2	25 Mn	$K\alpha_2$	KL_{II}	5.88765	2.3807	3	60 Nd	$L\alpha_2$	$L_{III}M_{IV}$	5.2077
2.1071	4	59 Pr	$L\beta_{10}$	L_IM_{IV}	5.884	2.3869	2	56 Ba	$L\beta_{10}$	L_IM_{IV}	5.1941
2.1103	3	58 Ce	$L\gamma_5$	$L_{II}N_I$	5.8751	2.3880	5	53 I	L_I	Abs. Edge	5.192
2.1194	4	59 Pr	$L\beta_{2,15}$	$L_{III}N_{IV,V}$	5.850	2.3913	2	53 I	$L\gamma_4$	$L_IO_{II,III}$	5.1848
2.1209	2	63 Eu	$L\alpha_1$	$L_{III}M_V$	5.8457	2.3948	2	63 Eu	Ll	$L_{III}M_I$	5.1772
2.1268	2	60 Nd	$L\beta_3$	L_IM_{III}	5.8294	2.40435	6	56 Ba	$L\beta_{2,15}$	$L_{III}N_{IV,V}$	5.1565
2.1315	2	63 Eu	$L\eta$	$L_{II}M_I$	5.8166	2.4094	4	60 Nd	$L\eta$	$L_{II}M_I$	5.1457
2.1315	2	63 Eu	$L\alpha_2$	$L_{III}M_{IV}$	5.8166	2.4105	3	57 La	$L\beta_3$	L_IM_{III}	5.1434
2.1342	2	56 Ba	$L\gamma_3$	L_IN_{III}	5.8092	2.4174	2	55 Cs	$L\gamma_5$	$L_{II}N_I$	5.1287
2.1387	2	56 Ba	$L\gamma_2$	L_IN_{II}	5.7969	2.4292	1	54 Xe	L_{II}	Abs. Edge	5.1037
2.1418	3	57 La	$L\gamma_1$	$L_{II}N_{IV}$	5.7885	2.442	9	90 Th		M_IO_{III}	5.08
2.15877	7	66 Dy	Ll	$L_{III}M_I$	5.7431	2.443	4	92 U		$M_{II}O_{IV}$	5.075
2.166	1	58 Ce	L_{III}	Abs. Edge	5.723	2.4475	2	53 I	$L\gamma_{2,3}$	$L_IN_{II,III}$	5.0657
2.1669	3	60 Nd	$L\beta_4$	L_IM_{II}	5.7216	2.4493	3	57 La	$L\beta_4$	L_IM_{II}	5.0620
2.1669	2	60 Nd	$L\beta_1$	$L_{II}M_{IV}$	5.7216	2.45891	4	57 La	$L\beta_1$	$L_{II}M_{IV}$	5.0421
2.1673	5	55 Cs	L_I	Abs. Edge	5.721	2.4630	2	59 Pr	$L\alpha_1$	$L_{III}M_V$	5.0337
2.1701	2	58 Ce	$L\beta_7$	$L_{III}O_I$	5.7132	2.4729	3	59 Pr	$L\alpha_2$	$L_{III}M_{IV}$	5.0135
2.1741	2	55 Cs	$L\gamma_4$	$L_IO_{II,III}$	5.7026	2.4740	1	55 Cs	L_{III}	Abs. Edge	5.0113
2.1885	3	58 Ce	$L\beta_9$	L_IM_V	5.6650	2.4783	2	55 Cs	$L\beta_9$	L_IM_V	5.0026
2.1906	4	59 Pr	$L\beta_6$	$L_{III}N_I$	5.660	2.4823	4	62 Sm	Ll	$L_{III}M_I$	4.9945

TABLE 1b3. *(Continued)*

Wavelength Å*	p.e.	Element	Designation		keV	Wavelength Å*	p.e.	Element	Designation		keV
2.4826	2	56 Ba	$L\beta_6$	$L_{III}N_I$	4.9939	2.8555	1	52 Te	L_{III}	Abs. Edge	4.3418
2.4849	2	55 Cs	$L\beta_7$	$L_{III}O_I$	4.9893	2.8627	3	56 Ba	$L\eta$	$L_{II}M_I$	4.3309
2.4920	2	55 Cs	$L\beta_{10}$	L_IM_{IV}	4.9752	2.8634	3	52 Te	$L\beta_7$	$L_{III}O_I$	4.3298
2.49734	5	22 Ti	K	Abs. Edge	4.96452	2.87429	9	53 I	$L\beta_3$	L_IM_{III}	4.3134
2.4985	2	22 Ti	$K\beta_5$	$KM_{IV,V}$	4.9623	2.88217	8	52 Te	$L\beta_{2,15}$	$L_{III}N_{IV,V}$	4.3017
2.50356	2	23 V	$K\alpha_1$	KL_{III}	4.95220	2.884	5	92 U	M_{III}	Abs. Edge	4.299
2.50738	2	23 V	$K\alpha_2$	KL_{II}	4.94464	2.8917	4	58 Ce	Ll	$L_{III}M_I$	4.2875
2.5099	1	52 Te	L_I	Abs. Edge	4.9397	2.8924	2	55 Cs	$L\alpha_1$	$L_{III}M_V$	4.2865
2.5113	2	52 Te	$L\gamma_4$	$L_IO_{II,III}$	4.9369	2.9020	2	55 Cs	$L\alpha_2$	$L_{III}M_{IV}$	4.2722
2.5118	2	55 Cs	$L\beta_{2,15}$	$L_{III}N_{IV,V}$	4.9359	2.910	2	91 Pa		$M_{II}N_{IV}$	4.260
2.512	3	59 Pr	$L\eta$	$L_{II}M_I$	4.935	2.91207	9	53 I	$L\beta_4$	L_IM_{II}	4.2575
2.51391	2	22 Ti	$K\beta_{1,3}$	$KM_{II,III}$	4.93181	2.92	2	92 U		M_IN_{II}	4.25
2.5164	2	56 Ba	$L\beta_3$	L_IM_{III}	4.9269	2.9260	1	49 In	L_I	Abs. Edge	4.2373
2.527	4	91 Pa		$M_{II}O_{IV}$	4.906	2.9264	2	49 In	$L\gamma_4$	$L_IO_{II,III}$	4.2367
2.5542	5	53 I	L_{II}	Abs. Edge	4.8540	2.93187	9	51 Sb	$L\gamma_5$	$L_{II}N_I$	4.2287
2.5553	2	56 Ba	$L\beta_4$	L_IM_{II}	4.8519	2.934	8	90 Th		M_IN_{III}	4.23
2.5615	2	58 Ce	$L\alpha_1$	$L_{III}M_V$	4.8402	2.93744	6	53 I	$L\beta_1$	$L_{II}M_{IV}$	4.22072
2.5674	2	52 Te	$L\gamma_{2,3}$	$L_IN_{II,III}$	4.8290	2.948	2	92 U		$M_{III}O_{IV,V}$	4.205
2.56821	5	56 Ba	$L\beta_1$	$L_{II}M_{IV}$	4.82753	2.97088	9	52 Te	$L\beta_6$	$L_{III}N_I$	4.1732
2.5706	3	58 Ce	$L\alpha_2$	$L_{III}M_{IV}$	4.8230	2.97261	9	51 Sb	$L\beta_9$	L_IM_V	4.1708
2.58244	8	53 I	$L\gamma_1$	$L_{II}N_{IV}$	4.8009	2.97917	9	51 Sb	$L\beta_{10}$	L_IM_{IV}	4.1616
2.5926	1	54 Xe	L_{III}	Abs. Edge	4.7822	2.9800	2	49 In	$L\gamma_{2,3}$	$L_IN_{II,III}$	4.1605
2.5932	2	55 Cs	$L\beta_6$	$L_{III}N_I$	4.7811	2.9823	1	50 Sn	L_{II}	Abs. Edge	4.1573
2.618	5	90 Th		$M_{II}O_{IV}$	4.735	2.9932	2	55 Cs	$L\eta$	$L_{II}M_I$	4.1421
2.6203	4	58 Ce	$L\eta$	$L_{II}M_I$	4.7315	3.0003	1	51 Sb	L_{III}	Abs. Edge	4.1323
2.6285	2	55 Cs	$L\beta_3$	L_IM_{III}	4.7167	3.00115	3	50 Sn	$L\gamma_1$	$L_{II}N_{IV}$	4.13112
2.6388	1	51 Sb	L_I	Abs. Edge	4.6984	3.0052	3	51 Sb	$L\beta_7$	$L_{III}O_I$	4.1255
2.6398	2	51 Sb	$L\gamma_4$	$L_IO_{II,III}$	4.6967	3.006	3	57 La	Ll	$L_{III}M_I$	4.124
2.65710	9	53 I	$L\gamma_5$	$L_{II}N_I$	4.6660	3.00893	9	52 Te	$L\beta_3$	L_IM_{III}	4.1204
2.66570	5	57 La	$L\alpha_1$	$L_{III}M_V$	4.65097	3.011	2	90 Th		$M_{II}N_{IV}$	4.117
2.6666	2	55 Cs	$L\beta_4$	L_IM_{II}	4.6494	3.0166	2	54 Xe	$L\alpha_1$	$L_{III}M_V$	4.1099
2.67533	5	57 La	$L\alpha_2$	$L_{III}M_{IV}$	4.63423	3.02335	3	51 Sb	$L\beta_{2,15}$	$L_{III}N_{IV,V}$	4.10078
2.6760	4	60 Nd	Ll	$L_{III}M_I$	4.6330	3.0309	1	21 Sc	$K\alpha_1$	KL_{III}	4.0906
2.6837	2	55 Cs	$L\beta_1$	$L_{II}M_{IV}$	4.6198	3.0342	1	21 Sc	$K\alpha_2$	KL_{II}	4.0861
2.6879	1	52 Te	L_{II}	Abs. Edge	4.6126	3.038	2	91 Pa		$M_{III}O_{IV,V}$	4.081
2.6953	2	51 Sb	$L\gamma_{2,3}$	$L_IN_{II,III}$	4.5999	3.04661	9	52 Te	$L\beta_4$	L_IM_{II}	4.0695
2.71241	6	52 Te	$L\gamma_1$	$L_{II}N_{IV}$	4.5709	3.068	5	90 Th	M_{III}	Abs. Edge	4.041
2.71352	9	53 I	$L\beta_9$	L_IM_V	4.5690	3.0703	1	20 Ca	K	Abs. Edge	4.0381
2.7196	5	53 I	L_{III}	Abs. Edge	4.5587	3.0746	3	20 Ca	$K\beta_5$	$KM_{IV,V}$	4.0325
2.72104	9	53 I	$L\beta_{10}$	L_IM_{IV}	4.5564	3.07677	6	52 Te	$L\beta_1$	$L_{II}M_{IV}$	4.02958
2.7288	3	53 I	$L\beta_7$	$L_{III}O_I$	4.5435	3.08475	9	50 Sn	$L\gamma_5$	$L_{II}N_I$	4.0192
2.740	3	57 La	$L\eta$	$L_{II}M_I$	4.525	3.0849	1	48 Cd	L_I	Abs. Edge	4.0190
2.74851	2	22 Ti	$K\alpha_1$	KL_{III}	4.51084	3.0897	2	20 Ca	$K\beta_{1,3}$	$KM_{II,III}$	4.0127
2.75053	8	53 I	$L\beta_{2,15}$	$L_{III}N_{IV,V}$	4.5075	3.094	5	83 Bi	M_I	Abs. Edge	4.007
2.75216	2	22 Ti	$K\alpha_2$	KL_{II}	4.50486	3.11513	9	50 Sn	$L\beta_9$	L_IM_V	3.9800
2.753	8	92 U		M_IN_{III}	4.50	3.11513	9	51 Sb	$L\beta_6$	$L_{III}N_I$	3.9800
2.762	1	21 Sc	K	Abs. Edge	4.489	3.115	7	92 U		$M_{III}O_I$	3.980
2.7634	3	21 Sc	$K\beta_5$	$KM_{IV,V}$	4.4865	3.12170	9	50 Sn	$L\beta_{10}$	L_IM_{IV}	3.9716
2.77595	5	56 Ba	$L\alpha_1$	$L_{III}M_V$	4.46626	3.131	3	90 Th		$M_{II}O_{IV,V}$	3.959
2.7769	1	50 Sn	L_I	Abs. Edge	4.4648	3.1355	2	56 Ba	Ll	$L_{III}M_I$	3.9541
2.7775	2	50 Sn	$L\gamma_4$	$L_IO_{II,III}$	4.4638	3.1377	2	48 Cd	$L\gamma_2$	L_IN_{II}	3.9513
2.7796	2	21 Sc	$K\beta_{1,3}$	$KM_{II,III}$	4.4605	3.1473	1	49 In	L_{II}	Abs. Edge	3.9393
2.7841	4	59 Pr	Ll	$L_{III}M_I$	4.4532	3.14860	6	53 I	$L\alpha_1$	$L_{III}M_V$	3.93765
2.78553	5	56 Ba	$L\alpha_2$	$L_{III}M_{IV}$	4.45090	3.15258	9	51 Sb	$L\beta_3$	L_IM_{III}	3.9327
2.79007	9	52 Te	$L\gamma_5$	$L_{II}N_I$	4.4437	3.1557	1	50 Sn	L_{III}	Abs. Edge	3.9288
2.817	2	92 U		$M_{II}N_{IV}$	4.401	3.1564	3	50 Sn	$L\beta_7$	$L_{III}O_I$	3.9279
2.8294	5	51 Sb	L_{II}	Abs. Edge	4.3819	3.15791	6	53 I	$L\alpha_2$	$L_{III}M_{IV}$	3.92604
2.8327	2	50 Sn	$L\gamma_{2,3}$	$L_IN_{II,III}$	4.3768	3.16213	4	49 In	$L\gamma_1$	$L_{II}N_{IV}$	3.92081
2.83672	9	53 I	$L\beta_6$	$L_{III}N_I$	4.3706	3.17505	3	50 Sn	$L\beta_{2,15}$	$L_{III}N_{IV,V}$	3.90486
2.83897	9	52 Te	$L\beta_9$	L_IM_V	4.3671	3.19014	9	51 Sb	$L\beta_4$	L_IM_{II}	3.8364
2.84679	9	52 Te	$L\beta_{10}$	L_IM_{IV}	4.3551	3.217	5	82 Pb	M_I	Abs. Edge	3.854
2.85159	3	51 Sb	$L\gamma_1$	$L_{II}N_{IV}$	4.34779	3.22567	4	51 Sb	$L\beta_1$	$L_{II}M_{IV}$	3.84357

TABLE 1b3. (*Continued*)

Wavelength Å*	p.e.	Element	Designation		keV	Wavelength Å*	p.e.	Element	Designation		keV
3.245	9	91 Pa		$M_{III}O_I$	3.82	3.614	2	91 Pa		$M_{III}N_{IV}$	3.430
3.24907	9	49 In	$L\gamma_5$	$L_{II}N_I$	3.8159	3.61467	9	48 Cd	$L\beta_6$	$L_{III}N_I$	3.42994
3.2564	1	47 Ag	L_I	Abs. Edge	3.8072	3.61638	9	47 Ag	$L\gamma_5$	$L_{II}N_I$	3.42832
3.2670	2	55 Cs	Ll	$L_{III}M_I$	3.7950	3.616	5	79 Au	M_I	Abs. Edge	3.428
3.26763	9	49 In	$L\beta_9$	L_IM_V	3.7942	3.629	5	45 Rh	L_I	Abs. Edge	3.417
3.26901	9	50 Sn	$L\beta_6$	$L_{III}N_I$	3.7926	3.634	5	81 Tl	M_{II}	Abs. Edge	3.412
3.27404	9	49 In	$L\beta_{10}$	L_IM_{IV}	3.7868	3.64495	9	48 Cd	$L\beta_2$	L_IM_{III}	3.40145
3.27979	9	53 I	$L\eta$	$L_{II}M_I$	3.7801	3.679	2	90 Th	$M\gamma$	$M_{III}N_V$	3.370
3.283	9	90 Th		$M_{III}O_I$	3.78	3.68203	9	48 Cd	$L\beta_4$	L_IM_{II}	3.36719
3.28920	6	52 Te	$L\alpha_1$	$L_{III}M_V$	3.76933	3.6855	2	45 Rh	$L\gamma_{2,3}$	$L_IN_{II,III}$	3.3640
3.29846	9	52 Te	$L\alpha_2$	$L_{III}M_{IV}$	3.7588	3.691	2	91 Pa		$M_{IV}O_{II}$	3.359
3.30585	3	50 Sn	$L\beta_3$	L_IM_{III}	3.7500	3.6999	1	47 Ag	L_{III}	Abs. Edge	3.35096
3.30635	9	47 Ag	$L\gamma_3$	L_IN_{III}	3.7498	3.70335	3	47 Ag	$L\beta_{2,15}$	$L_{III}N_{IV,V}$	3.34781
3.31216	9	47 Ag	$L\gamma_2$	L_IN_{II}	3.7432	3.716	1	92 U	$M\beta$	$M_{IV}N_{VI}$	3.3367
3.3237	1	49 In	L_{III}	Abs. Edge	3.7302	3.71696	9	52 Te	Ll	$L_{III}M_I$	3.33555
3.324	4	49 In	$L\beta_7$	$L_{III}O_I$	3.730	3.718	3	90 Th		$M_{III}N_{IV}$	3.335
3.3257	1	48 Cd	L_{II}	Abs. Edge	3.7280	3.7228	1	46 Pd	L_{II}	Abs. Edge	3.33031
3.329	4	92 U		$M_{II}N_I$	3.724	3.7246	2	46 Pd	$L\gamma_1$	$L_{II}N_{IV}$	3.3287
3.333	5	92 U	M_{IV}	Abs. Edge	3.720	3.729	5	90 Th	M_V	Abs. Edge	3.325
3.33564	6	48 Cd	$L\gamma_1$	$L_{II}N_{IV}$	3.71686	3.73823	4	48 Cd	$L\beta_1$	$L_{II}M_{IV}$	3.31657
3.33838	3	49 In	$L\beta_{2,15}$	$L_{III}N_{IV,V}$	3.71381	3.740	9	83 Bi		M_IN_{III}	3.315
3.34335	9	50 Sn	$L\beta_4$	L_IM_{II}	3.7083	3.7414	2	19 K	$K\alpha_1$	KL_{III}	3.3138
3.346	5	81 Tl	M_I	Abs. Edge	3.705	3.7445	2	19 K	$K\alpha_2$	KL_{II}	3.3111
3.35839	3	20 Ca	$K\alpha_1$	KL_{III}	3.69168	3.760	9	90 Th		M_VP_{III}	3.298
3.359	5	83 Bi	M_{II}	Abs. Edge	3.691	3.762	5	78 Pt	M_I	Abs. Edge	3.296
3.36166	3	20 Ca	$K\alpha_2$	KL_{II}	3.68809	3.77192	4	49 In	$L\alpha_1$	$L_{III}M_V$	3.28694
3.38487	3	50 Sn	$L\beta_1$	$L_{II}M_{IV}$	3.66280	3.78073	6	49 In	$L\alpha_2$	$L_{III}M_{IV}$	3.27929
3.42551	9	48 Cd	$L\gamma_5$	$L_{II}N_I$	3.61935	3.783	5	80 Hg	M_{II}	Abs. Edge	3.277
3.43015	9	48 Cd	$L\beta_9$	L_IM_V	3.61445	3.78876	9	50 Sn	$L\eta$	$L_{II}M_I$	3.27234
3.43606	9	49 In	$L\beta_6$	$L_{III}N_I$	3.60823	3.7920	2	46 Pd	$L\beta_9$	L_IM_V	3.2696
3.4365	1	19 K	K	Abs. Edge	3.6078	3.7988	2	46 Pd	$L\beta_{10}$	L_IM_{IV}	3.2637
3.4367	2	48 Cd	$L\beta_{10}$	L_IM_{IV}	3.6075	3.80774	9	47 Ag	$L\beta_6$	$L_{III}N_I$	3.25603
3.437	1	46 Pd	L_I	Abs. Edge	3.607	3.808	4	90 Th		$M_{IV}O_{II}$	3.256
3.43832	9	52 Te	$L\eta$	$L_{II}M_I$	3.60586	3.8222	2	46 Pd	$L\gamma_5$	$L_{II}N_I$	3.2437
3.43941	4	51 Sb	$L\alpha_1$	$L_{III}M_V$	3.60472	3.827	1	91 Pa	$M\beta$	$M_{IV}N_{VI}$	3.2397
3.441	5	91 Pa		$M_{III}N_I$	3.603	3.83313	9	47 Ag	$L\beta_2$	L_IM_{III}	3.23446
3.4413	4	19 K	$K\beta_5$	$KM_{IV,V}$	3.6027	3.834	4	83 Bi		$M_{III}N_{IV}$	3.234
3.44840	6	51 Sb	$L\alpha_2$	$L_{III}M_{IV}$	3.59532	3.835	5	44 Ru	L_I	Abs. Edge	3.233
3.4539	2	19 K	$K\beta_{1,3}$	$KM_{II,III}$	3.5896	3.87023	5	47 Ag	$L\beta_4$	L_IM_{II}	3.20346
3.46984	9	49 In	$L\beta_3$	L_IM_{III}	3.57311	3.87090	5	18 A	K	Abs. Edge	3.20290
3.478	5	80 Hg	M_I	Abs. Edge	3.565	3.872	9	82 Pb		M_IN_{III}	3.202
3.479	1	92 U	$M\gamma$	$M_{III}N_V$	3.563	3.8860	2	18 A	$K\beta_{1,3}$	$KM_{II,III}$	3.1905
3.4892	2	46 Pd	$L\gamma_{2,3}$	$L_IN_{II,III}$	3.5533	3.88826	9	51 Sb	Ll	$L_{III}M_I$	3.18860
3.492	5	82 Pb	M_{II}	Abs. Edge	3.550	3.892	9	83 Bi		M_IN_{II}	3.185
3.497	5	92 U	M_V	Abs. Edge	3.545	3.8977	2	44 Ru	$L\gamma_{2,3}$	$L_IN_{II,III}$	3.1809
3.5047	1	48 Cd	L_{III}	Abs. Edge	3.5376	3.904	5	83 Bi	M_{III}	Abs. Edge	3.176
3.50697	9	49 In	$L\beta_1$	$L_{II}M_{IV}$	3.53528	3.9074	1	46 Pd	L_{III}	Abs. Edge	3.17298
3.51408	4	48 Cd	$L\beta_{2,15}$	$L_{III}N_{IV,V}$	3.52812	3.90887	4	46 Pd	$L\beta_{2,15}$	$L_{III}N_{IV,V}$	3.17179
3.5164	1	47 Ag	L_{II}	Abs. Edge	3.5258	3.910	1	92 U	$M\alpha_1$	M_VN_{VII}	3.1708
3.521	2	92 U		$M_{III}N_{IV}$	3.521	3.915	5	77 Ir	M_I	Abs. Edge	3.167
3.52260	4	47 Ag	$L\gamma_1$	$L_{II}N_{IV}$	3.51959	3.924	1	92 U	$M\alpha_2$	M_VN_{VI}	3.1595
3.537	9	90 Th		$M_{II}N_I$	3.505	3.932	6	83 Bi		$M_{III}O_{IV,V}$	3.153
3.55531	4	49 In	$L\beta_1$	$L_{II}M_{IV}$	3.48721	3.93473	4	47 Ag	$L\beta_1$	$L_{II}M_{IV}$	3.15094
3.557	5	90 Th	M_{IV}	Abs. Edge	3.485	3.936	5	79 Au	M_{II}	Abs. Edge	3.150
3.55754	9	53 I	Ll	$L_{III}M_I$	3.48502	3.941	1	90 Th	$M\beta$	$M_{IV}N_{VI}$	3.1458
3.576	1	92 U		$M_{IV}O_{II}$	3.4666	3.9425	5	45 Rh	L_{II}	Abs. Edge	3.1448
3.577	1	91 Pa	$M\gamma$	$M_{III}N_V$	3.4657	3.9437	2	45 Rh	$L\gamma_1$	$L_{II}N_{IV}$	3.1438
3.59994	3	50 Sn	$L\alpha_1$	$L_{III}M_V$	3.44398	3.95635	4	48 Cd	$L\alpha_1$	$L_{III}M_V$	3.13373
3.60497	9	47 Ag	$L\beta_9$	L_IM_V	3.43917	3.96496	6	48 Cd	$L\alpha_2$	$L_{III}M_{IV}$	3.12691
3.60765	9	51 Sb	$L\eta$	$L_{II}M_I$	3.43661	3.968	5	82 Pb		$M_{II}N_{IV}$	3.124
3.60891	4	50 Sn	$L\alpha_2$	$L_{III}M_{IV}$	3.43542	3.98327	9	49 In	$L\eta$	$L_{II}M_I$	3.11254
3.61158	9	47 Ag	$L\beta_{10}$	L_IM_{IV}	3.43287	4.013	9	81 Tl		M_IN_{III}	3.089

TABLE 1b3. (Continued)

Wavelength Å*	p.e. Element		Designation	keV	Wavelength Å*	p.e. Element		Designation	keV
4.0162	2 46 Pd	$L\beta_4$	$L_{III}N_I$	3.0870	4.532	2 83 Bi	$M\gamma$	$M_{III}N_V$	2.735
4.022	1 91 Pa	$M\alpha_1$	M_VN_{VII}	3.0823	4.568	5 90 Th		$M_{III}N_I$	2.714
4.0346	2 46 Pd	$L\beta_3$	L_IM_{III}	3.0730	4.571	5 83 Bi		$M_{III}N_{IV}$	2.712
4.035	3 91 Pa	$M\alpha_2$	M_VN_{VI}	3.072	4.572	5 83 Bi	M_{IV}	Abs. Edge	2.711
4.0451	2 45 Rh	$L\gamma_6$	$L_{II}N_I$	3.0650	4.575	5 41 Nb	L_I	Abs. Edge	2.710
4.047	1 82 Pb	M_{III}	Abs. Edge	3.0632	4.585	5 73 Ta	M_I	Abs. Edge	2.704
4.058	5 43 Te	L_I	Abs. Edge	3.055	4.59	2 83 Bi		$M_{IV}P_{II,III}$	2.70
4.069	6 82 Pb		$M_{III}O_{IV,V}$	3.047	4.59743	9 45 Rh	$L\alpha_1$	$L_{III}M_V$	2.69674
4.0711	2 46 Pd	$L\beta_4$	L_IM_{II}	3.0454	4.601	4 78 Pt		$M_{II}N_{IV}$	2.695
4.071	5 76 Os	M_I	Abs. Edge	3.045	4.60545	9 45 Rh	$L\alpha_2$	$L_{III}M_{IV}$	2.69205
4.07165	9 50 Sn	Ll	$L_{III}M_I$	3.04499	4.620	5 75 Re	M_{II}	Abs. Edge	2.684
4.093	5 78 Pt	M_{II}	Abs. Edge	3.029	4.62058	3 44 Ru	$L\beta_1$	$L_{II}M_{IV}$	2.68323
4.105	9 83 Bi		$M_{III}O_I$	3.021	4.625	5 92 U		$M_{IV}N_{III}$	2.681
4.116	4 81 Tl		$M_{II}N_{IV}$	3.013	4.630	1 43 Tc	L_{III}	Abs. Edge	2.6780
4.1299	5 45 Rh	L_{III}	Abs. Edge	3.0021	4.631	9 77 Ir		M_IN_{III}	2.677
4.1310	2 45 Rh	$L\beta_{2,15}$	$L_{III}N_{IV,V}$	3.0013	4.6542	2 41 Nb	$L\gamma_{2,3}$	$L_IN_{II,III}$	2.6638
4.1381	9 90 Th	$M\alpha_1$	M_VN_{VII}	2.9961	4.655	8 82 Pb		$M_{II}N_I$	2.664
4.14622	5 46 Pd	$L\beta_1$	$L_{II}M_{IV}$	2.99022	4.6605	2 46 Pd	$L\eta$	$L_{II}M_I$	2.6603
4.151	2 90 Th	$M\alpha_2$	M_VN_{VI}	2.987	4.674	1 82 Pb	$M\gamma$	$M_{III}N_V$	2.6527
4.15443	3 47 Ag	$L\alpha_1$	$L_{III}M_V$	2.98431	4.686	1 78 Pt	M_{III}	Abs. Edge	2.6459
4.16294	5 47 Ag	$L\alpha_2$	$L_{III}M_{IV}$	2.97821	4.694	8 78 Pt		$M_{III}O_{IV,V}$	2.641
4.180	1 44 Ru	L_{II}	Abs. Edge	2.9663	4.703	9 79 Au		$M_{III}O_I$	2.636
4.1822	2 44 Ru	$L\gamma_1$	$L_{II}N_{IV}$	2.9645	4.7076	2 47 Ag	Ll	$L_{III}M_I$	2.6337
4.19180	5 18 A	$K\alpha_1$	$K L_{III}$	2.95770	4.715	3 82 Pb		$M_{III}N_{IV}$	2.630
4.19315	9 48 Cd	$L\eta$	$L_{II}M_I$	2.95675	4.719	1 42 Mo	L_{II}	Abs. Edge	2.6274
4.19474	5 18 A	$K\alpha_2$	$K L_{II}$	2.95563	4.7258	2 42 Mo	$L\gamma_1$	$L_{II}N_{IV}$	2.6235
4.198	1 81 Tl	M_{III}	Abs. Edge	2.9535	4.7278	1 17 Cl	$K\alpha_1$	$K L_{III}$	2.62239
4.216	6 81 Tl		$M_{III}O_{IV,V}$	2.941	4.7307	1 17 Cl	$K\alpha_2$	$K L_{II}$	2.62078
4.236	5 75 Re	M_I	Abs. Edge	2.927	4.757	5 82 Pb	M_{IV}	Abs. Edge	2.606
4.2417	2 45 Rh	$L\beta_6$	$L_{III}N_I$	2.9229	4.764	5 83 Bi	M_V	Abs. Edge	2.603
4.244	9 82 Pb		$M_{III}O_I$	2.921	4.780	4 77 Ir		$M_{II}N_{IV}$	2.594
4.2522	2 45 Rh	$L\beta_3$	L_IM_{III}	2.9157	4.79	2 76 Os		M_IN_{III}	2.59
4.260	5 77 Ir	M_{II}	Abs. Edge	2.910	4.815	5 74 W	M_{II}	Abs. Edge	2.575
4.26873	9 49 In	Ll	$L_{III}M_I$	2.90440	4.823	3 83 Bi		$M_{IV}O_{II}$	2.571
4.2873	2 44 Ru	$L\gamma_6$	$L_{II}N_I$	2.8918	4.823	4 81 Tl	$M\gamma$	$M_{III}N_V$	2.571
4.2888	2 45 Rh	$L\beta_4$	L_IM_{II}	2.8908	4.8369	2 42 Mo	$L\gamma_6$	$L_{II}N_I$	2.5632
4.300	9 79 Au		M_IN_{III}	2.883	4.84575	5 44 Ru	$L\alpha_1$	$L_{III}M_V$	2.55855
4.304	5 42 Mo	L_I	Abs. Edge	2.881	4.85381	7 44 Ru	$L\alpha_2$	$L_{III}M_{IV}$	2.55431
4.330	2 92 U		$M_{III}N_I$	2.863	4.861	1 77 Ir	M_{III}	Abs. Edge	2.5505
4.355	1 80 Hg	M_{III}	Abs. Edge	2.8469	4.865	5 81 Tl		$M_{III}N_{IV}$	2.548
4.36767	5 46 Pd	$L\alpha_1$	$L_{III}M_V$	2.83861	4.869	9 77 Ir		$M_{III}O_{IV,V}$	2.546
4.369	1 44 Ru	L_{III}	Abs. Edge	2.8377	4.876	9 78 Pt		$M_{III}O_I$	2.543
4.3718	2 44 Ru	$L\beta_{2,15}$	$L_{III}N_{IV,V}$	2.8360	4.879	5 40 Zr	L_I	Abs. Edge	2.541
4.37414	4 45 Rh	$L\beta_1$	$L_{II}M_{IV}$	2.83441	4.8873	8 43 Tc	$L\beta_1$	$L_{II}M_{IV}$	2.5368
4.37588	7 46 Pd	$L\alpha_2$	$L_{III}M_{IV}$	2.83329	4.909	1 83 Bi	$M\beta$	$M_{IV}N_{VI}$	2.5255
4.3800	2 42 Mo	$L\gamma_{2,3}$	$L_IN_{II,III}$	2.8306	4.911	5 90 Th		$M_{IV}N_{III}$	2.524
4.3971	1 17 Cl	K	Abs. Edge	2.81960	4.913	1 42 Mo	L_{III}	Abs. Edge	2.5234
4.4034	3 17 Cl	$K\beta$	KM	2.8156	4.9217	2 45 Rh	$L\eta$	$L_{II}M_I$	2.5191
4.407	5 74 W	M_I	Abs. Edge	2.813	4.9232	2 42 Mo	$L\beta_{2,15}$	$L_{III}N_{IV,V}$	2.5183
4.4183	2 47 Ag	$L\eta$	$L_{II}M_I$	2.8061	4.946	2 92 U	$M\zeta_1$	M_VN_{III}	2.507
4.432	4 79 Au		$M_{II}N_{IV}$	2.797	4.952	5 81 Tl	M_{IV}	Abs. Edge	2.504
4.433	5 76 Os	M_{II}	Abs. Edge	2.797	4.9525	3 46 Pd	Ll	$L_{III}M_I$	2.5034
4.436	1 43 Te	L_{II}	Abs. Edge	2.7948	4.9536	3 40 Zr	$L\gamma_{2,3}$	$L_IN_{II,III}$	2.5029
4.44	2 74 W		$M_IO_{II,III}$	2.79	4.955	4 76 Os		$M_{II}N_{IV}$	2.502
4.450	4 91 Pa		$M_{III}N_I$	2.786	4.955	5 82 Pb	M_V	Abs. Edge	2.502
4.460	9 78 Pt		M_IN_{III}	2.780	4.984	2 80 Hg	$M\gamma$	$M_{III}N_V$	2.4875
4.48014	9 48 Cd	Ll	$L_{III}M_I$	2.76735	5.004	9 82 Pb		$M_{IV}O_{II}$	2.477
4.4866	3 44 Ru	$L\beta_3$	L_IM_{III}	2.7634	5.0133	3 42 Mo	$L\beta_3$	L_IM_{III}	2.4730
4.4866	3 44 Ru	$L\beta_6$	$L_{III}N_I$	2.7634	5.0185	1 16 S	K	Abs. Edge	2.47048
4.518	1 79 Au	M_{III}	Abs. Edge	2.7439	5.020	5 73 Ta	M_{II}	Abs. Edge	2.470
4.522	6 79 Au		$M_{III}O_{IV,V}$	2.742	5.0233	3 16 S	$K\beta_x$	KM	2.4681
4.5230	2 44 Ru	$L\beta_4$	L_IM_{II}	2.7411	5.031	1 41 Nb	L_{II}	Abs. Edge	2.4641

TABLE 1b3. (*Continued*)

Wavelength Å*	p.e.	Element	Designation	keV	Wavelength Å*	p.e.	Element	Designation	keV
5.0316	2	16 S	$K\beta_1$	KM	2.46404	5.59	1	78 Pt	M_{IV}	Abs. Edge	2.217
5.0361	3	41 Nb	$L\gamma_1$	$L_{II}N_{IV}$	2.4618	5.592	5	38 Sr	L_I	Abs. Edge	2.217
5.043	5	76 Os	M_{III}	Abs. Edge	2.458	5.624	1	79 Au	$M\beta$	$M_{IV}N_{VI}$	2.2046
5.0488	3	42 Mo	$L\beta_4$	L_IM_{II}	2.4557	5.628	8	74 W		$M_{III}O_I$	2.203
5.0488	5	42 Mo	$L\beta_6$	$L_{III}N_I$	2.4557	5.6330	3	40 Zr	$L\beta_3$	L_IM_{III}	2.2010
5.050	2	92 U	$M\zeta_2$	$M_{IV}N_{II}$	2.4548	5.6445	3	38 Sr	$L\gamma_{2,3}$	$L_IN_{II,III}$	2.1965
5.076	1	82 Pb	$M\beta$	$M_{IV}N_{VI}$	2.4427	5.6476	9	80 Hg	$M\alpha_1$	M_VN_{VII}	2.1953
5.092	2	91 Pa	$M\zeta_1$	M_VN_{III}	2.4350	5.650	5	73 Ta	M_{III}	Abs. Edge	2.194
5.1148	3	43 Tc	$L\alpha_1$	$L_{III}M_V$	2.4240	5.6681	3	40 Zr	$L\beta_4$	L_IM_{II}	2.1873
5.118	1	83 Bi	$M\alpha_1$	M_VN_{VII}	2.4226	5.67	3	73 Ta		$M_{III}O_{IV,V}$	2.19
5.130	2	83 Bi	$M\alpha_2$	M_VN_{VI}	2.4170	5.682	4	76 Os	$M\gamma$	$M_{III}N_V$	2.182
5.145	4	79 Au	$M\gamma$	$M_{III}N_V$	2.410	5.704	8	82 Pb		$M_{III}N_I$	2.174
5.1517	3	41 Nb	$L\gamma_5$	$L_{II}N_I$	2.4066	5.7101	3	40 Zr	$L\beta_6$	$L_{III}N_I$	2.1712
5.153	5	81 Tl	M_V	Abs. Edge	2.406	5.724	5	76 Os		$M_{III}N_{IV}$	2.166
5.157	5	80 Hg	M_{IV}	Abs. Edge	2.404	5.7243	2	41 Nb	$L\alpha_1$	$L_{III}M_V$	2.16589
5.168	9	82 Pb		M_VO_{III}	2.399	5.7319	3	41 Nb	$L\alpha_2$	$L_{III}M_{IV}$	2.1630
5.172	9	74 W		M_IN_{III}	2.397	5.756	1	39 Y	L_{II}	Abs. Edge	2.1540
5.17708	8	42 Mo	$L\beta_1$	$L_{II}M_{IV}$	2.39481	5.767	9	79 Au		M_VO_{III}	2.150
5.186	5	79 Au		$M_{III}N_{IV}$	2.391	5.784	1	15 P	K	Abs. Edge	2.1435
5.193	2	91 Pa	$M\zeta_2$	$M_{IV}N_{II}$	2.3876	5.796	2	15 P	$K\beta$	KM	2.1391
5.196	9	81 Tl		$M_{IV}O_{II}$	2.386	5.81	2	76 Os		$M_{II}N_I$	2.133
5.2050	2	44 Ru	$L\eta$	$L_{II}M_I$	2.38197	5.81	1	78 Pt	M_V	Abs. Edge	2.133
5.217	5	39 Y	L_I	Abs. Edge	2.377	5.828	1	78 Pt	$M\beta$	$M_{IV}N_{VI}$	2.1273
5.2169	3	45 Rh	Ll	$L_{III}M_I$	2.3765	5.83	2	73 Ta		$M_{III}O_I$	2.126
5.230	1	41 Nb	L_{III}	Abs. Edge	2.3706	5.83	1	77 Ir	M_{IV}	Abs. Edge	2.126
5.234	5	75 Re	M_{III}	Abs. Edge	2.369	5.8360	3	40 Zr	$L\beta_1$	$L_{II}M_{IV}$	2.1244
5.2379	3	41 Nb	$L\beta_{2,15}$	$L_{III}N_{IV,V}$	2.3670	5.840	1	79 Au	$M\alpha_1$	M_VN_{VII}	2.1229
5.245	5	90 Th	$M\zeta_1$	M_VN_{III}	2.364	5.8475	3	42 Mo	$L\eta$	$L_{II}M_I$	2.1202
5.249	1	81 Tl	$M\beta$	$M_{IV}N_{VI}$	2.3621	5.854	3	79 Au	$M\alpha_2$	M_VN_{VI}	2.118
5.2830	3	39 Y	$L\gamma_{2,3}$	$L_IN_{II,III}$	2.3468	5.8754	3	39 Y	$L\gamma_5$	$L_{II}N_I$	2.1102
5.286	1	82 Pb	$M\alpha_1$	M_VN_{VII}	2.3455	5.884	8	81 Tl		$M_{III}N_I$	2.107
5.299	2	82 Pb	$M\alpha_2$	M_VN_{VI}	2.3397	5.885	2	75 Re	$M\gamma$	$M_{III}N_V$	2.1067
5.3102	3	41 Nb	$L\beta_3$	L_IM_{III}	2.3348	5.931	5	75 Re		$M_{III}N_{IV}$	2.090
5.319	4	78 Pt	$M\gamma$	$M_{III}N_V$	2.331	5.962	1	39 Y	L_{III}	Abs. Edge	2.0794
5.340	5	90 Th	$M\zeta_2$	$M_{IV}N_{II}$	2.322	5.9832	3	39 Y	$L\beta_3$	L_IM_{III}	2.0722
5.3455	3	41 Nb	$L\beta_4$	L_IM_{II}	2.3194	5.987	9	78 Pt		M_VO_{III}	2.071
5.357	4	74 W		$M_{II}N_{IV}$	2.314	6.008	5	37 Rb	L_I	Abs. Edge	2.063
5.357	5	78 Pt		$M_{III}N_{IV}$	2.314	6.0186	3	39 Y	$L\beta_4$	L_IM_{II}	2.0600
5.36	1	80 Hg	M_V	Abs. Edge	2.313	6.038	1	77 Ir	$M\beta$	$M_{IV}N_{VI}$	2.0535
5.3613	3	41 Nb	$L\beta_6$	$L_{III}N_I$	2.3125	6.0458	3	37 Rb	$L\gamma_{2,3}$	$L_IN_{II,III}$	2.0507
5.37216	7	16 S	$K\alpha_1$	KL_{III}	2.30784	6.047	1	78 Pt	$M\alpha_1$	M_VN_{VII}	2.0505
5.374	5	79 Au	M_{IV}	Abs. Edge	2.307	6.05	1	77 Ir	M_V	Abs. Edge	2.048
5.37496	8	16 S	$K\alpha_2$	KL_{II}	2.30664	6.058	3	78 Pt	$M\alpha_2$	M_VN_{VI}	2.047
5.378	1	40 Zr	L_{II}	Abs. Edge	2.3053	6.0705	2	40 Zr	$L\alpha_1$	$L_{III}M_V$	2.04236
5.3843	3	40 Zr	$L\gamma_1$	$L_{II}N_{IV}$	2.3027	6.073	5	76 Os	M_{IV}	Abs. Edge	2.042
5.40	2	73 Ta		M_IN_{III}	2.295	6.0778	3	40 Zr	$L\alpha_2$	$L_{III}M_{IV}$	2.0399
5.40655	8	42 Mo	$L\alpha_1$	$L_{III}M_V$	2.29316	6.09	2	80 Hg		$M_{III}N_I$	2.036
5.41437	8	42 Mo	$L\alpha_2$	$L_{III}M_{IV}$	2.28985	6.092	3	74 W	$M\gamma$	$M_{III}N_V$	2.035
5.4318	9	80 Hg	$M\beta$	$M_{IV}N_{VI}$	2.2825	6.0942	3	39 Y	$L\beta_6$	$L_{III}N_I$	2.0344
5.435	1	74 W	M_{III}	Abs. Edge	2.2811	6.134	4	74 W		$M_{III}N_{IV}$	2.021
5.460	1	81 Tl	$M\alpha_1$	M_VN_{VII}	2.2706	6.1508	3	42 Mo	Ll	$L_{III}M_I$	2.01568
5.472	2	81 Tl	$M\alpha_2$	M_VN_{VI}	2.2656	6.157	1	15 P	$K\alpha_1$	KL_{III}	2.0137
5.4923	3	41 Nb	$L\beta_1$	$L_{II}M_{IV}$	2.2574	6.160	1	15 P	$K\alpha_2$	KL_{II}	2.0127
5.4977	3	40 Zr	$L\gamma_5$	$L_{II}N_I$	2.2551	6.162	8	83 Bi		$M_{IV}N_{III}$	2.012
5.500	4	77 Ir	$M\gamma$	$M_{III}N_V$	2.254	6.173	1	38 Sr	L_{II}	Abs. Edge	2.0085
5.5035	3	44 Ru	Ll	$L_{III}M_I$	2.2528	6.2109	3	41 Nb	$L\eta$	$L_{II}M_I$	1.99620
5.537	8	83 Bi		$M_{III}N_I$	2.239	6.2120	3	39 Y	$L\beta_1$	$L_{II}M_{IV}$	1.99584
5.540	5	77 Ir		$M_{III}N_{IV}$	2.238	6.259	9	79 Au		$M_{III}N_I$	1.981
5.570	4	73 Ta		$M_{II}N_{IV}$	2.226	6.262	1	77 Ir	$M\alpha_1$	M_VN_{VII}	1.9799
5.579	1	40 Zr	L_{III}	Abs. Edge	2.2225	6.267	1	76 Os	$M\beta$	$M_{IV}N_{VI}$	1.9783
5.584	5	79 Au	M_V	Abs. Edge	2.220	6.275	3	77 Ir	$M\alpha_2$	M_VN_{VI}	1.9758
5.5863	3	40 Zr	$L\beta_{2,15}$	$L_{III}N_{IV,V}$	2.2194	6.28	2	74 W		$M_{II}N_I$	1.973

TABLE 1b3. (*Continued*)

Wavelength Å*	p.e.	Element	Designation		keV
6.2961	3	38 Sr	$L\gamma_5$	$L_{II}N_I$	1.96916
6.30	1	76 Os	M_V	Abs. Edge	1.967
6.312	4	73 Ta	$M\gamma$	$M_{III}N_V$	1.964
6.33	1	75 Re	M_{IV}	Abs. Edge	1.958
6.353	5	73 Ta		$M_{III}N_{IV}$	1.951
6.3672	3	38 Sr	$L\beta_3$	L_IM_{III}	1.94719
6.384	7	82 Pb		$M_{IV}N_{III}$	1.942
6.387	1	38 Sr	L_{III}	Abs. Edge	1.9411
6.4026	3	38 Sr	$L\beta_4$	L_IM_{II}	1.93643
6.4488	2	39 Y	$L\alpha_1$	$L_{III}M_V$	1.92256
6.455	9	78 Pt		$M_{III}N_I$	1.921
6.4558	3	39 Y	$L\alpha_2$	$L_{III}M_{IV}$	1.92047
6.47	1	36 Kr	L_I	Abs. Edge	1.915
6.490	1	76 Os	$M\alpha$	$M_VN_{VI,VII}$	1.9102
6.504	1	75 Re	$M\beta$	$M_{IV}N_{VI}$	1.9061
6.5176	3	41 Nb	Ll	$L_{III}M_I$	1.90225
6.5191	3	38 Sr	$L\beta_6$	$L_{III}N_I$	1.90181
6.521	4	83 Bi	$M\zeta_1$	M_VN_{III}	1.901
6.544	4	72 Hf	$M\gamma$	$M_{III}N_V$	1.895
6.560	5	75 Re	M_V	Abs. Edge	1.890
6.585	5	83 Bi	$M\zeta_2$	$M_{IV}N_{II}$	1.883
6.59	1	74 W	M_{IV}	Abs. Edge	1.880
6.6069	3	40 Zr	$L\eta$	$L_{II}M_I$	1.87654
6.6239	3	38 Sr	$L\beta_1$	$L_{II}M_{IV}$	1.87172
6.644	1	37 Rb	L_{II}	Abs. Edge	1.8661
6.669	9	77 Ir		$M_{III}N_I$	1.859
6.729	1	75 Re	$M\alpha$	$M_VN_{VI,VII}$	1.8425
6.738	1	14 Si	K	Abs. Edge	1.8400
6.740	3	82 Pb	$M\zeta_1$	M_VN_{III}	1.8395
6.7530	1	14 Si	$K\beta$	KM	1.83594
6.755	3	37 Rb	$L\gamma_5$	$L_{II}N_{IV}$	1.83532
6.757	1	74 W	$M\beta$	$M_{IV}N_{VI}$	1.8349
6.768	6	71 Lu	$M\gamma$	$M_{III}N_V$	1.832
6.7876	3	37 Rb	$L\beta_3$	L_IM_{III}	1.82659
6.802	5	82 Pb	$M\zeta_2$	$M_{IV}N_{II}$	1.823
6.806	9	74 W		$M_{IV}O_{II}$	1.822
6.8207	3	37 Rb	$L\beta_4$	L_IM_{II}	1.81771
6.83	1	74 W	M_V	Abs. Edge	1.814
6.862	1	37 Rb	L_{III}	Abs. Edge	1.8067
6.8628	2	38 Sr	$L\alpha_1$	$L_{III}M_V$	1.80656
6.8697	3	38 Sr	$L\alpha_2$	$L_{III}M_{IV}$	1.80474
6.87	1	73 Ta	M_{IV}	Abs. Edge	1.804
6.87	2	80 Hg	δ	$M_{IV}N_{III}$	1.805
6.89	2	76 Os		$M_{III}N_I$	1.798
6.9185	3	40 Zr	Ll	$L_{III}M_I$	1.79201
6.959	5	35 Br	L_I	Abs. Edge	1.781
6.974	4	81 Tl	$M\zeta_1$	M_VN_{III}	1.778
6.983	1	74 W	$M\alpha_1$	M_VN_{VII}	1.7754
6.9842	3	37 Rb	$L\beta_6$	$L_{III}N_I$	1.77517
6.992	2	74 W	$M\alpha_2$	M_VN_{VI}	1.7731
7.005	9	74 W		M_VO_{III}	1.770
7.023	1	73 Ta	$M\beta$	$M_{IV}N_{VI}$	1.7655
7.024	8	70 Yb	$M\gamma$	$M_{III}N_V$	1.765
7.032	5	81 Tl	$M\zeta_2$	$M_{IV}N_{II}$	1.763
7.0406	3	39 Y	$L\eta$	$L_{II}M_I$	1.76095
7.0759	3	37 Rb	$L\beta_1$	$L_{II}M_{IV}$	1.75217
7.09	2	73 Ta		$M_{IV}O_{II,III}$	1.748
7.101	8	79 Au		$M_{IV}N_{III}$	1.746
7.11	1	73 Ta	M_V	Abs. Edge	1.743
7.12542	9	14 Si	$K\alpha_1$	KL_{III}	1.73998
7.12791	9	14 Si	$K\alpha_2$	KL_{II}	1.73938
7.168	1	36 Kr	L_{II}	Abs. Edge	1.7297
7.250	5	36 Kr		$L_{II}N_{III}$	1.710
7.252	1	73 Ta	$M\alpha$	$M_VN_{VI,VII}$	1.7096
7.264	5	36 Kr	$L\beta_2$	L_IM_{III}	1.707
7.279	5	36 Kr	$L\gamma_5$	$L_{II}N_I$	1.703
7.30	2	73 Ta		M_VO_{III}	1.700
7.303	1	72 Hf	$M\beta$	$M_{IV}N_{VI}$	1.6976
7.304	5	36 Kr	$L\beta_4$	L_IM_{II}	1.697
7.3183	2	37 Rb	$L\alpha_1$	$L_{III}M_V$	1.69413
7.3251	3	37 Rb	$L\alpha_2$	$L_{III}M_{IV}$	1.69256
7.3563	3	39 Y	Ll	$L_{III}M_I$	1.68536
7.360	8	74 W		$M_{III}N_I$	1.684
7.371	8	78 Pt		$M_{IV}N_{III}$	1.682
7.392	1	36 Kr	L_{III}	Abs. Edge	1.6772
7.466	4	79 Au	$M\zeta_1$	M_VN_{III}	1.6605
7.503	1	34 Se	L_I	Abs. Edge	1.6525
7.510	4	36 Kr	$L\beta_6$	$L_{III}N_I$	1.6510
7.5171	3	38 Sr	$L\eta$	$L_{II}M_I$	1.64933
7.523	5	79 Au	$M\zeta_2$	$M_{IV}N_{II}$	1.648
7.539	1	72 Hf	$M\alpha$	$M_VN_{VI,VII}$	1.6446
7.546	8	68 Er	$M\gamma$	$M_{III}N_V$	1.643
7.576	3	36 Kr	$L\beta_1$	$L_{II}M_{IV}$	1.6366
7.60	1	68 Er		$M_{III}N_{IV}$	1.632
7.601	2	71 Lu	$M\beta$	$M_{IV}N_{VI}$	1.6312
7.612	9	73 Ta		$M_{III}N_I$	1.629
7.645	8	77 Ir		$M_{IV}N_{III}$	1.622
7.738	4	78 Pt	$M\zeta_1$	M_VN_{III}	1.6022
7.753	5	35 Br	L_{II}	Abs. Edge	1.599
7.767	9	35 Br	$L\beta_{3,4}$	$L_IM_{II,III}$	1.596
7.790	5	78 Pt	$M\zeta_2$	$M_{IV}N_{II}$	1.592
7.817	3	36 Kr	$L\alpha_{1,2}$	$L_{III}M_{IV,V}$	1.5860
7.8362	3	38 Sr	Ll	$L_{III}M_I$	1.58215
7.840	2	71 Lu	$M\alpha$	$M_VN_{VI,VII}$	1.5813
7.865	9	67 Ho	$M\gamma$	$M_{III}N_{IV,V}$	1.576
7.887	9	72 Hf		$M_{III}N_I$	1.572
7.909	2	70 Yb	$M\beta$	$M_{IV}N_{VI}$	1.5675
7.94813	5	13 Al	K	Abs. Edge	1.55988
7.960	2	13 Al	$K\beta$	KM	1.55745
7.984	5	35 Br	L_{III}	Abs. Edge	1.5530
8.021	4	77 Ir	$M\zeta_1$	M_VN_{III}	1.5458
8.0415	4	37 Rb	$L\eta$	$L_{II}M_I$	1.54177
8.065	5	77 Ir	$M\zeta_2$	$M_{IV}N_{II}$	1.5373
8.107	1	33 As	L_I	Abs. Edge	1.5293
8.1251	5	35 Br	$L\beta_1$	$L_{II}M_{IV}$	1.52590
8.144	9	66 Dy	$M\gamma$	$M_{III}N_{IV,V}$	1.522
8.149	5	70 Yb	$M\alpha$	$M_VN_{VI,VII}$	1.5214
8.239	8	75 Re		$M_{IV}N_{III}$	1.505
8.249	7	69 Tm	$M\beta$	$M_{IV}N_{VI}$	1.503
8.310	4	76 Os	$M\zeta_1$	M_VN_{III}	1.4919
8.321	9	34 Se	$L\beta_{3,4}$	$L_IM_{II,III}$	1.490
8.33934	9	13 Al	$K\alpha_1$	KL_{III}	1.48670
8.34173	9	13 Al	$K\alpha_2$	KL_{II}	1.48627
8.359	5	76 Os	$M\zeta_2$	$M_{IV}N_{II}$	1.4831
8.3636	4	37 Rb	Ll	$L_{III}M_I$	1.48238
8.3746	5	35 Br	$L\alpha_{1,2}$	$L_{III}M_{IV,V}$	1.48043
8.407	1	34 Se	L_{II}	Abs. Edge	1.4747
8.470	9	70 Yb		$M_{III}N_I$	1.464
8.48	1	69 Tm	$M\alpha$	$M_VN_{VI,VII}$	1.462
8.486	9	65 Tb	$M\gamma$	$M_{III}N_{IV,V}$	1.461
8.487	5	69 Tm	M_V	Abs. Edge	1.4609
8.573	8	74 W		$M_{IV}N_{III}$	1.446
8.592	3	68 Er	$M\beta$	$M_{IV}N_{VI}$	1.4430
8.60	7	92 U		$N_IP_{IV,V}$	1.44

TABLE 1b3. (*Continued*)

Wavelength Å*	p.e.	Element	Designation		keV
8.601	5	68 Er	M_{IV}	Abs. Edge	1.4415
8.629	4	75 Re	$M\zeta_1$	M_VN_{III}	1.4368
8.646	1	34 Se	L_{III}	Abs. Edge	1.4340
8.664	5	75 Re	$M\zeta_2$	$M_{IV}N_{II}$	1.4310
8.7358	5	34 Se	$L\beta_1$	$L_{II}M_{IV}$	1.41923
8.76	7	92 U		N_IP_{III}	1.42
8.773	1	32 Ge	L_I	Abs. Edge	1.4132
8.81	7	92 U		N_IP_{II}	1.41
8.82	1	68 Er	$M\alpha$	$M_VN_{VI,VII}$	1.406
8.844	9	64 Gd	$M\gamma$	$M_{III}N_{IV,V}$	1.402
8.847	5	68 Er	M_V	Abs. Edge	1.4013
8.90	2	73 Ta		$M_{IV}N_{III}$	1.393
8.929	1	33 As	$L\beta_{3,4}$	$L_IM_{II,III}$	1.3884
8.962	4	74 W	$M\zeta_1$	M_VN_{III}	1.3835
8.965	4	67 Ho	$M\beta$	$M_{IV}N_{VI}$	1.3830
8.9900	5	34 Se	$L\alpha_{1,2}$	$L_{III}M_{IV,V}$	1.37910
8.993	5	74 W	$M\zeta_2$	$M_{IV}N_{II}$	1.3787
9.125	1	33 As	L_{II}	Abs. Edge	1.3587
9.20	2	67 Ho	$M\alpha$	$M_VN_{VI,VII}$	1.348
9.211	9	63 Eu	$M\gamma$	$M_{III}N_{IV,V}$	1.346
9.255	1	35 Br	$L\eta$	$L_{II}M_I$	1.3396
9.316	4	73 Ta	$M\zeta_1$	M_VN_{III}	1.3308
9.330	5	73 Ta	$M\zeta_2$	$M_{IV}N_{II}$	1.3288
9.357	6	66 Dy	$M\beta$	$M_{IV}N_{VI}$	1.3250
9.367	1	33 As	L_{III}	Abs. Edge	1.3235
9.40	7	90 Th		N_IP_{III}	1.319
9.4141	8	33 As	$L\beta_1$	$L_{II}M_{IV}$	1.3170
9.44	7	90 Th		N_IP_{II}	1.313
9.5122	1	12 Mg	K	Abs. Edge	1.30339
9.517	5	31 Ga	L_I	Abs. Edge	1.3028
9.521	2	12 Mg	$K\beta$	KM	1.3022
9.581	2	32 Ge	$L\beta_2$	L_IM_{III}	1.2941
9.585	1	35 Br	Ll	$L_{III}M_I$	1.2935
9.59	2	66 Dy	$M\alpha$	$M_VN_{VI,VII}$	1.293
9.600	9	62 Sm	$M\gamma$	$M_{III}N_{IV,V}$	1.291
9.640	2	32 Ge	$L\beta_4$	L_IM_{II}	1.2861
9.6709	8	33 As	$L\alpha_{1,2}$	$L_{III}M_{IV,V}$	1.2820
9.686	7	72 Hf	$M\zeta_2$	$M_{IV}N_{II}$	1.2800
9.686	7	72 Hf	$M\zeta_1$	M_VN_{III}	1.2800
9.792	6	65 Tb	$M\beta$	$M_{IV}N_{VI}$	1.2661
9.8900	2	12 Mg	$K\alpha_{1,2}$	$KL_{II,III}$	1.25360
9.924	1	32 Ge	L_{II}	Abs. Edge	1.2494
9.962	1	34 Se	$L\eta$	$L_{II}M_I$	1.2446
10.00	2	65 Tb	$M\alpha$	$M_VN_{VI,VII}$	1.240
10.09	7	92 U		N_IO_{III}	1.229
10.175	1	32 Ge	$L\beta_1$	$L_{II}M_{IV}$	1.2185
10.187	1	32 Ge	L_{III}	Abs. Edge	1.2170
10.254	6	64 Gd	$M\beta$	$M_{IV}N_{VI}$	1.2091
10.294	1	34 Se	Ll	$L_{III}M_I$	1.2044
10.359	9	31 Ga	$L\beta_{3,4}$	$L_IM_{II,III}$	1.197
10.40	7	92 U		$N_{II}P_I$	1.192
10.4361	8	32 Ge	$L\alpha_{1,2}$	$L_{III}M_{IV,V}$	1.18800
10.46	3	64 Gd	$M\alpha$	$M_VN_{VI,VII}$	1.185
10.48	1	70 Yb	$M\zeta$	M_VN_{III}	1.183
10.505	9	60 Nd	$M\gamma$	$M_{III}N_{IV,V}$	1.180
10.711	5	63 Eu	M_{IV}	Abs. Edge	1.1575
10.734	1	33 As	$L\eta$	$L_{II}M_I$	1.1550
10.750	7	63 Eu	$M\beta$	$M_{IV}N_{VI}$	1.1533
10.828	5	31 Ga	L_{II}	Abs. Edge	1.1450
10.96	3	63 Eu	$M\alpha$	$M_VN_{VI,VII}$	1.131
10.998	9	59 Pr	$M\gamma$	$M_{III}N_{IV,V}$	1.1273
11.013	5	63 Eu	M_V	Abs. Edge	1.1258
11.023	2	31 Ga	$L\beta_1$	$L_{II}M_{IV}$	1.1248
11.072	1	33 As	Ll	$L_{III}M_I$	1.1198
11.07	7	90 Th		$N_{II}P_I$	1.120
11.100	1	31 Ga	L_{III}	Abs. Edge	1.1169
11.200	7	30 Zn	$L\beta_{3,4}$	$L_IM_{II,III}$	1.1070
11.27	1	62 Sm	$M\beta$	$M_{IV}N_{VI}$	1.0998
11.288	5	62 Sm	M_{IV}	Abs. Edge	1.0983
11.292	1	31 Ga	$L\alpha_{1,2}$	$L_{III}M_{IV,V}$	1.09792
11.37	1	68 Er	$M\zeta$	M_VN_{III}	1.0901
11.47	3	62 Sm	$M\alpha$	$M_VN_{VI,VII}$	1.081
11.53	1	58 Ce	$M\gamma$	$M_{III}N_{IV,V}$	1.0749
11.552	5	62 Sm	M_V	Abs. Edge	1.0732
11.56	5	90 Th		$N_{II}O_{IV}$	1.072
11.569	1	11 Na	K	Abs. Edge	1.07167
11.575	2	11 Na	$K\beta$	KM	1.0711
11.609	2	32 Ge	$L\eta$	$L_{II}M_I$	1.0680
11.862	1	30 Zn	L_{II}	Abs. Edge	1.04523
11.86	1	67 Ho	$M\zeta$	M_VN_{III}	1.0450
11.9101	9	11 Na	$K\alpha_{1,2}$	$KL_{II,III}$	1.04098
11.965	2	32 Ge	Ll	$L_{III}M_I$	1.0362
11.983	3	30 Zn	$L\beta_1$	$L_{II}M_{IV}$	1.0347
12.08	4	57 La	$M\gamma$	$M_{III}N_{IV,V}$	1.027
12.122	3	29 Cu	$L\beta_{3,4}$	$L_IM_{II,III}$	1.0228
12.131	1	30 Zn	L_{III}	Abs. Edge	1.02201
12.254	3	30 Zn	$L\alpha_{1,2}$	$L_{III}M_{IV,V}$	1.0117
12.43	2	66 Dy	$M\zeta$	M_VN_{III}	0.998
12.44	2	60 Nd	$M\beta$	$M_{IV}N_{VI}$	0.997
12.459	5	60 Nd	M_{IV}	Abs. Edge	0.9951
12.597	2	31 Ga	$L\eta$	$L_{II}M_I$	0.9842
12.68	2	60 Nd	$M\alpha$	$M_VN_{VI,VII}$	0.978
12.737	5	60 Nd	M_V	Abs. Edge	0.9734
12.75	3	56 Ba	$M\gamma$	$M_{III}N_{IV,V}$	0.973
12.90	9	92 U		$N_{III}O_V$	0.961
12.953	2	31 Ga	Ll	$L_{III}M_I$	0.9572
12.98	2	65 Tb	$M\zeta$	M_VN_{III}	0.955
13.014	1	29 Cu	L_{II}	Abs. Edge	0.95268
13.053	3	29 Cu	$L\beta_1$	$L_{II}M_{IV}$	0.9498
13.06	2	59 Pr	$M\beta$	$M_{IV}N_{VI}$	0.950
10.31	1	30 Zn	L_I	Abs. Edge	1.197
13.122	5	59 Pr	M_{IV}	Abs. Edge	0.9448
13.18	2	28 Ni	$L\beta_{3,4}$	$L_IM_{II,III}$	0.941
13.288	1	29 Cu	L_{III}	Abs. Edge	0.93306
13.30	6	83 Bi		$N_IP_{II,III}$	0.932
13.336	3	29 Cu	$L\alpha_{1,2}$	$L_{III}M_{IV,V}$	0.9297
13.343	5	59 Pr	$M\alpha$	$M_VN_{VI,VII}$	0.9292
13.394	5	59 Pr	M_V	Abs. Edge	0.9257
13.57	2	64 Gd	$M\zeta$	M_VN_{III}	0.914
13.68	2	30 Zn	$L\eta$	$L_{II}M_I$	0.906
13.75	4	58 Ce	$M\beta$	$M_{IV}N_{VI}$	0.902
13.8	1	90 Th		$N_{III}O_V$	0.897
14.02	2	30 Zn	Ll	$L_{III}M_I$	0.884
14.04	2	58 Ce	$M\alpha$	$M_VN_{VI,VII}$	0.883
14.22	2	63 Eu	$M\zeta$	M_VN_{III}	0.872
14.242	5	28 Ni	L_{II}	Abs. Edge	0.8706
14.271	6	28 Ni	$L\beta_1$	$L_{II}M_{IV}$	0.8688
14.3018	1	10 Ne	K	Abs. Edge	0.866889
14.31	3	27 Co	$L\beta_{3,4}$	$L_IM_{II,III}$	0.870
14.39	5	58 Ce		$M_VO_{II,III}$	0.862
14.452	5	10 Ne	$K\beta$	KM	0.8579
14.51	5	57 La	$M\beta$	$M_{IV}N_{VI}$	0.854
14.525	5	28 Ni	L_{III}	Abs. Edge	0.8536
14.561	3	28 Ni	$L\alpha_{1,2}$	$L_{III}M_{IV,V}$	0.8515

TABLE 1b3. (*Continued*)

Wavelength Å*	p.e.	Element	Designation		keV
14.610	3	10 Ne	$K\alpha_{1,2}$	$KL_{II,III}$	0.8486
14.88	5	57 La	$M\alpha$	$M_V N_{VI,VII}$	0.833
14.90	2	29 Cu	$L\eta$	$L_{II}M_I$	0.832
14.91	4	62 Sm	$M\zeta$	$M_V N_{III}$	0.831
15.286	9	29 Cu	Ll	$l_{III}M_I$	0.8111
15.56	1	56 Ba	M_{IV}	Abs. Edge	0.7967
15.618	5	27 Co	L_{II}	Abs. Edge	0.7938
15.65	4	26 Fe	$L\beta_{2,4}$	$L_I M_{II,III}$	0.792
15.666	8	27 Co	$L\beta_1$	$L_{II}M_{IV}$	0.7914
15.72	9	56 Ba		$M_{IV}O_{III}$	0.789
15.89	1	56 Ba	M_V	Abs. Edge	0.7801
15.91	5	56 Ba		$M_{IV}O_{II}$	0.779
15.915	5	27 Co	L_{III}	Abs. Edge	0.7790
15.93	4	52 Te	$M\gamma$	$M_{III}N_{IV,V}$	0.778
15.972	6	27 Co	$L\alpha_{1,2}$	$L_{III}M_{IV,V}$	0.7762
15.98	5	51 Sb		$M_{II}N_{IV}$	0.776
16.20	5	56 Ba		$M_V O_{III}$	0.765
16.27	3	28 Ni	$L\eta$	$L_{II}M_I$	0.762
16.46	4	60 Nd	$M\zeta$	$M_V N_{III}$	0.753
16.693	9	28 Ni	Ll	$L_{III}M_I$	0.7427
16.7	1	24 Cr	L_I	Abs. Edge	0.741
16.92	4	51 Sb	$M\gamma$	$M_{III}N_{IV,V}$	0.733
16.93	5	50 Sn		$M_{II}N_{IV}$	0.733
17.19	4	25 Mn	$L\beta_{2,4}$	$L_I M_{II,III}$	0.721
17.202	5	26 Fe	L_{II}	Abs. Edge	0.7208
17.26	1	26 Fe	$L\beta_1$	$L_{II}M_{IV}$	0.7185
17.38	4	59 Pr	$M\zeta$	$M_V N_{III}$	0.714
17.525	5	26 Fe	L_{III}	Abs. Edge	0.7074
17.59	2	26 Fe	$L\alpha_{1,2}$	$L_{III}M_{IV,V}$	0.7050
17.6	1	52 Te		$M_{II}N_I$	0.703
17.87	3	27 Co	$L\eta$	$L_{II}M_I$	0.694
17.94	5	50 Sn	$M\gamma$	$M_{III}N_{IV,V}$	0.691
17.9	1	24 Cr	L_{II}	Abs. Edge	0.691
18.292	8	27 Co	Ll	$L_{III}M_I$	0.6778
18.32	2	9 F	$K\alpha$	KL	0.6768
18.35	4	58 Ce	$M\zeta$	$M_V N_{III}$	0.676
18.8	1	51 Sb		$M_{II}N_I$	0.658
18.8	2	47 Ag		$M_I N_{II,III}$	0.658
18.96	4	24 Cr	$L\beta_{2,4}$	$L_I M_{II,III}$	0.654
19.11	2	25 Mn	$L\beta_1$	$L_{II}M_{IV}$	0.6488
19.1	1	52 Te		$M_{III}N_I$	0.648
19.40	7	48 Cd		$M_{II}N_{IV}$	0.639
19.44	5	57 La	$M\zeta$	$M_V N_{III}$	0.638
19.45	1	25 Mn	$L\alpha_{1,2}$	$L_{III}M_{IV,V}$	0.6374
19.66	5	53 I	$M_{IV,V}$	Abs. Edge	0.631
19.75	4	26 Fe	$L\eta$	$L_{II}M_I$	0.628
20.0	1	50 Sn		$M_{II}N_I$	0.619
20.1	2	46 Pd		$M_I N_{II,III}$	0.616
20.15	1	26 Fe	Ll	$L_{III}M_I$	0.6152
20.2	1	51 Sb		$M_{III}N_I$	0.612
20.47	7	48 Cd	$M\gamma$	$M_{III}N_{IV,V}$	0.606
20.64	4	56 Ba	$M\zeta$	$M_V N_{III}$	0.601
20.66	7	47 Ag		$M_{II}N_{IV}$	0.600
20.7	1	24 Cr	L_{III}	Abs. Edge	0.598
21.19	5	23 Va	$L\beta_{2,4}$	$L_I M_{II,III}$	0.585
21.27	1	24 Cr	$L\beta_1$	$L_{II}M_{IV}$	0.5828
21.34	5	52 Te		$M_{IV}O_{II,III}$	0.581
21.5	1	50 Sn		$M_{III}N_I$	0.575
21.64	3	24 Cr	$L\alpha_{1,2}$	$L_{III}M_{IV,V}$	0.5728
21.78	5	52 Te		$M_V O_{III}$	0.569
21.82	7	47 Ag	$M\gamma$	$M_{III}N_{IV,V}$	0.568
21.85	2	25 Mn	$L\eta$	$L_{II}M_I$	0.5675
22.1	1	46 Pd		$M_{II}N_{IV}$	0.560
22.29	1	25 Mn	Ll	$L_{III}M_I$	0.5563
22.9	2	48 Cd		$M_{II}N_I$	0.540
23.32	1	8 O	K	Abs. Edge	0.5317
23.3	1	46 Pd	$M\gamma$	$M_{III}N_{IV,V}$	0.531
23.62	3	8 O	$K\alpha$	KL	0.5249
23.88	4	23 Va	$L\beta_1$	$L_{II}M_{IV}$	0.5192
24.25	3	23 Va	$L\alpha_{1,2}$	$L_{III}M_{IV,V}$	0.5113
24.28	5	50 Sn	$M_{IV,V}$	Abs. Edge	0.511
24.30	3	24 Cr	$L\eta$	$L_{II}M_I$	0.5102
24.4	2	47 Ag		$M_V N_I$	0.509
24.5	1	48 Cd		$M_{III}N_I$	0.507
24.78	1	24 Cr	Ll	$L_{III}M_I$	0.5003
25.01	9	45 Rh	$M\gamma$	$M_{III}N_{IV,V}$	0.496
25.3	1	50 Sn		$M_{IV}O_{II,III}$	0.491
25.50	9	44 Ru		$M_{II}N_{IV}$	0.486
25.7	1	50 Sn		$M_V O_{III}$	0.483
26.0	1	47 Ag		$M_{III}N_I$	0.478
26.2	2	46 Pd		$M_{II}N_I$	0.474
26.72	9	52 Te	$M\zeta$	$M_{IV,V}N_{II,III}$	0.464
26.9	1	44 Ru	$M\gamma$	$M_{III}N_{IV,V}$	0.462
27.05	2	22 Ti	$L\beta_1$	$L_{II}M_{IV}$	0.4584
27.29	1	22 Ti	$L_{II,III}$	Abs. Edge	0.4544
27.34	3	23 Va	$L\eta$	$L_{II}M_I$	0.4535
27.42	2	22 Ti	$L\alpha_{1,2}$	$L_{III}M_{IV,V}$	0.4522
27.77	1	23 Va	Ll	$L_{III}M_I$	0.4465
27.9	1	46 Pd		$M_{III}N_I$	0.445
28.1	2	45 Rh		$M_{II}N_I$	0.442
28.13	5	48 Cd	$M_{IV,V}$	Abs. Edge	0.4408
28.88	8	51 Sb	$M\zeta$	$M_{IV,V}N_{II,III}$	0.429
29.8	1	45 Rh		$M_{III}N_I$	0.417
30.4	1	48 Cd		$M_{IV}O_{II,III}$	0.408
30.8	1	48 Cd		$M_V O_{III}$	0.403
30.82	5	47 Ag	M_{IV}	Abs. Edge	0.4022
30.89	3	22 Ti	$L\eta$	$L_{II}M_I$	0.4013
30.99	1	7 N	K	Abs. Edge	0.4000
31.02	2	21 Sc	$L\beta_1$	$L_{II}M_{IV}$	0.3996
31.14	5	47 Ag	M_V	Abs. Edge	0.3981
31.24	9	50 Sn	$M\zeta$	$M_{IV,V}N_{II,III}$	0.397
31.35	3	21 Sc	$L\alpha_{1,2}$	$L_{III}M_{IV,V}$	0.3954
31.36	2	22 Ti	Ll	$L_{III}M_I$	0.3953
31.60	4	7 N	$K\alpha$	KL	0.3924
31.8	1	92 U		$N_{IV}N_{VI}$	0.390
32.3	2	44 Ru		$M_{II}N_I$	0.384
33.1	2	41 Nb		$M_{II}N_{IV}$	0.375
33.5	3	47 Ag		$M_{IV,V}O_{II,III}$	0.370
33.57	9	90 Th		$N_{IV}N_{VI}$	0.3693
34.8	1	92 U		$N_V N_{VI,VII}$	0.357
34.9	2	41 Nb	$M\gamma$	$M_{III}N_{IV,V}$	0.356
35.13	2	21 Sc	$L\eta$	$L_{II}M_I$	0.3529
35.13	1	20 Ca	L_{II}	Abs. Edge	0.3529
35.3	3	42 Mo		$M_{II}N_I$	0.351
35.49	1	20 Ca	L_{III}	Abs. Edge	0.34931
35.59	3	21 Sc	Ll	$L_{III}M_I$	0.3483
35.63	1	20 Ca	$L_{II,III}$	Abs. Edge	0.34793
35.94	2	20 Ca	$L\beta_1$	$L_{II}M_{IV}$	0.3449
36.32	9	90 Th		$N_V N_{VI,VII}$	0.3414
36.33	2	20 Ca	$L\alpha_{1,2}$	$L_{III}M_{IV,V}$	0.3413
36.8	1	48 Cd	$M\zeta$	$M_{IV,V}N_{II,III}$	0.3371
37.4	2	46 Pd		$M_{IV,V}O_{II,III}$	0.332
37.5	2	42 Mo		$M_{III}N_I$	0.331
38.4	3	41 Nb		$M_{II}N_I$	0.323

TABLE 1b3. (*Continued*)

Wavelength Å*	p.e. Element	Designation		keV	Wavelength Å*	p.e. Element	Designation		keV
39.77	7 47 Ag	$M\zeta$	$M_{IV,V}N_{II,III}$	0.3117	64.38	7 42 Mo	$M\zeta$	$M_{IV,V}N_{II,III}$	0.1926
40.46	2 20 Ca	$L\eta$	$L_{II}M_I$	0.3064	65.1	7 70 Yb		$N_{IV}N_{VI}$	0.190
40.7	2 41 Nb		$M_{III}N_I$	0.305	65.5	1 45 Rh		$M_{III}M_V$	0.1892
40.9	2 45 Rh		$M_{IV,V}O_{II,III}$	0.303	65.7	2 71 Lu		$N_VN_{VI,VII}$	0.1886
40.96	2 20 Ca	Ll	$L_{III}M_I$	0.3027	67.33	9 17 Cl	$L\eta$	$L_{II}M_I$	0.1841
42.1	2 92 U		$N_{VI}O_V$	0.295	67.6	3 5 B	$K\alpha$	KL	0.1833
42.1	1 19 K	$L_{II,III}$	Abs. Edge	0.2946	67.90	9 17 Cl	Ll	$L_{III}M_I$	0.1826
42.3	2 82 Pb		$N_{IV}N_{VI}$	0.293	68.2	3 90 Th		$O_{III}P_{IV,V}$	0.1817
43.3	2 92 U		$N_{VI}O_{IV}$	0.286	68.3	1 44 Ru		$M_{III}M_V$	0.1814
43.6	1 46 Pd	$M\zeta$	$M_{IV,V}N_{II,III}$	0.2844	68.9	2 42 Mo		$M_{II}M_{IV}$	0.1798
43.68	1 6 C	K	Abs. Edge	0.28384	69.3	5 70 Yb		$N_VN_{VI,VII}$	0.179
44.7	3 6 C	$K\alpha$	KL	0.277	70.0	4 40 Zr		$M_{IV,V}O_{II,III}$	0.177
44.8	1 44 Ru		$M_{IV,V}O_{II,III}$	0.2768	72.1	3 41 Nb		$M_{II}M_{IV}$	0.1718
45.0	1 82 Pb		$N_VN_{VI,VII}$	0.2756	72.19	9 41 Nb	$M\zeta$	$M_{IV,V}N_{II,III}$	0.1717
45.2	3 80 Hg		$N_{IV}N_{VI}$	0.274	72.7	9 68 Er		$N_{IV}N_{VI}$	0.171
45.2	1 51 Sb		$M_{II}M_{IV}$	0.2743	74.9	1 42 Mo		$M_{III}M_V$	0.1656
46.48	9 39 Y		$M_{III}N_I$	0.267	76.3	7 68 Er		$N_VN_{VI,VII}$	0.163
46.5	2 81 Tl		$N_VN_{VI,VII}$	0.267	76.7	2 40 Zr		$M_{II}M_{IV}$	0.1617
46.8	2 79 Au		$N_{IV}N_{VI}$	0.265	76.9	2 35 Br		$M_{II}N_I$	0.1613
47.24	2 19 K	Ll	$L_{II}M_I$	0.2625	78.4	2 41 Nb		$M_{III}M_V$	0.1582
47.3	1 50 Sn		$M_{II}M_{IV}$	0.2621	79.8	3 35 Br		$M_{III}N_I$	0.1554
47.67	9 45 Rh	$M\zeta$	$M_{IV,V}N_{II,III}$	0.2601	80.9	3 40 Zr		$M_{III}M_V$	0.1533
47.74	1 19 K	Ll	$L_{III}M_I$	0.25971	81.5	2 39 Y		$M_{II}M_{IV}$	0.1522
47.9	3 80 Hg		$N_VN_{VI,VII}$	0.259	82.1	2 40 Zr	$M\zeta$	$M_{IV,V}N_{II,III}$	0.1511
48.1	2 78 Pt		$N_{IV}N_{VI}$	0.258	83.	1 66 Dy		$N_{IV,V}N_{VI,VII}$	0.149
48.2	1 90 Th		$N_{VI}O_V$	0.2572	83.4	3 16 S	Ll, η	$L_{II,III}M_I$	0.1487
48.5	2 39 Y		$M_{III}N_I{}'$	0.256	85.7	2 38 Sr		$M_{II}M_{IV}$	0.1447
49.4	1 79 Au		$N_VN_{VI,VII}$	0.2510	86.	1 65 Tb		$N_{IV,V}N_{VI,VII}$	0.144
49.5	1 90 Th		$N_{VI}O_{IV}$	0.2505	86.5	2 39 Y		$M_{III}M_{IV,V}$	0.1434
50.0	1 90 Th		$N_{VII}O_V$	0.2479	91.4	2 38 Sr		$M_{III}M_{IV,V}$	0.1357
50.2	1 77 Ir		$N_{IV}N_{VI}$	0.2470	91.5	2 37 Rb		$M_{II}M_{IV}$	0.1355
50.3	1 52 Te		$M_{III}M_V$	0.2465	91.6	1 83 Bi		$N_{VI}O_{IV}$	0.1354
50.9	1 78 Pt		$N_VN_{VI,VII}$	0.2436	93.2	1 83 Bi		$N_{VII}O_V$	0.1330
51.3	1 38 Sr		$M_{II}N_I$	0.2416	93.4	2 39 Y	$M\zeta$	$M_{IV,V}N_{II,III}$	0.1328
51.9	1 76 Os		$N_{IV}N_{VI}$	0.2388	94.	1 15 P	$L_{II,III}$	Abs. Edge	0.132
52.0	2 48 Cd		$M_{II}M_{IV}$	0.2384	96.7	2 37 Rb		$M_{III}M_{IV,V}$	0.1282
52.2	1 51 Sb		$M_{III}M_V$	0.2375	97.2	8 66 Dy		$N_{IV,V}O_{II,III}$	0.128
52.34	7 44 Ru	$M\zeta$	$M_{IV,V}N_{II,III}$	0.2369	98.	1 62 Sm		$N_{IV,V}N_{VI,VII}$	0.126
52.8	1 77 Ir		$N_VN_{VI,VII}$	0.2348	100.2	2 82 Pb		$N_{VI}O_V$	0.1237
53.6	1 38 Sr		$M_{III}N_I$	0.2313	102.2	4 65 Tb		$N_{IV,V}O_{II,III}$	0.1213
54.0	2 74 W		$N_{II}N_{IV}$	0.2295	102.4	1 82 Pb		$N_{VII}O_{IV}$	0.1211
54.0	1 47 Ag		$M_{II}M_{IV}$	0.2295	103.8	4 15 P		$L_{II,III}M$	0.1194
54.2	1 50 Sn		$M_{III}M_V$	0.2287	104.3	1 82 Pb		$N_{VII}O_V$	0.1189
54.7	2 76 Os		$N_VN_{VI,VII}$	0.2266	107.	1 60 Nd		$N_{IV,V}N_{VI,VII}$	0.116
54.8	2 42 Mo		$M_{IV,V}O_{II,III}$	0.2262	108.0	2 38 Sr	$M\zeta_2$	$M_{IV}N_{II,III}$	0.1148
55.8	1 74 W		$N_{IV}N_{VI}$	0.2221	108.7	1 38 Sr	$M\zeta_1$	M_VN_{III}	0.1140
55.9	1 18 A	$L\eta$	$L_{II}M_I$	0.2217	109.4	3 35 Br		$M_{II}M_{IV}$	0.1133
56.3	1 18 A	Ll	$L_{III}M_I$	0.2201	110.6	5 29 Cu	M_I	Abs. Edge	0.1121
56.5	1 46 Pd		$M_{II}M_{IV}$	0.2194	111.	1 4 Be	K	Abs. Edge	0.111
57.0	2 37 Rb		$M_{II}N_I$	0.2174	112.0	6 63 Eu		$N_{IV,V}O_{II,III}$	0.1107
58.2	1 73 Ta		$N_{IV}N_{VI}$	0.2130	113.0	1 81 Tl		$N_{VI}O_V$	0.10968
58.4	1 74 W		N_VN_{VII}	0.2122	113.	1 59 Pr		$N_{IV,V}N_{VI,VII}$	0.1095
58.7	2 48 Cd		$M_{III}M_V$	0.2111	113.8	3 35 Br		$M_{III}M_{IV,V}$	0.1089
59.3	1 45 Rh		$M_{II}M_{IV}$	0.2090	114.	1 4 Be	$K\alpha$	KL	0.1085
59.5	3 74 W		N_VN_{VI}	0.208	115.3	2 81 Tl		$N_{VI}O_{IV}$	0.1075
59.5	2 37 Rb		$M_{III}N_I$	0.2083	117.4	4 62 Sm		$N_{IV,V}O_{II,III}$	0.1056
60.5	1 47 Ag		$M_{III}M_V$	0.2048	117.7	1 81 Tl		$N_{VII}O_V$	0.10530
61.1	2 73 Ta		$N_VN_{VI,VII}$	0.2028	123.	1 14 Si	$L_{II,III}$	Abs. Edge	0.1006
61.9	2 41 Nb		$M_{IV,V}O_{II,III}$	0.2002	126.8	2 37 Rb		$M_{IV}N_{III}$	0.0978
62.2	1 44 Ru		$M_{II}M_{IV}$	0.1992	127.8	2 37 Rb	$M\zeta_2$	$M_{IV}N_{II}$	0.0970
62.9	1 46 Pd		$M_{III}M_V$	0.1970	128.7	2 37 Rb	$M\zeta_1$	M_VN_{III}	0.0964
63.0	5 71 Lu		$N_{IV}N_{VI}$	0.197	128.9	7 60 Nd		$N_{IV,V}O_{II,III}$	0.0962

TABLE 1b3. (*Continued*)

Wavelength Å*	p.e.	Element		Designation	keV
135.5	4	14 Si		$L_{II,III}M$	0.0915
136.5	4	59 Pr		$N_{IV,V}O_{II,III}$	0.0908
137.0	5	30 Zn	M_{II}	Abs. Edge	0.0905
142.5	1	13 Al	L_I	Abs. Edge	0.08701
143.9	5	30 Zn	M_{III}	Abs. Edge	0.0862
144.4	6	58 Ce		$N_{IV,V}O_{II,III}$	0.0859
144.4	3	37 Rb		M_IM_{III}	0.0859
152.6	6	57 La		$N_{IV,V}O_{II,III}$	0.0812
157.	3	30 Zn		$M_{II,III}M_{IV,V}$	0.079
159.0	2	56 Ba		$N_{IV}O_{III}$	0.07796
159.5	5	29 Cu	M_{II}	Abs. Edge	0.0777
163.3	2	56 Ba		$N_{IV}O_{II}$	0.07590
164.6	2	56 Ba		N_VO_{III}	0.07530
164.7	3	35 Br		M_IM_{III}	0.0753
166.0	5	29 Cu	M_{III}	Abs. Edge	0.0747
170.4	1	13 Al	$L_{II,III}$	Abs. Edge	0.07278
171.4	5	13 Al		$L_{II,III}M$	0.0724
173.	3	29 Cu		$M_{II,III}M_{IV,V}$	0.072
181.	5	90 Th		$O_{IV,V}Q_{II,III}$	0.068
183.8	1	55 Cs		$N_{IV}O_{II}$	0.06746
184.6	3	35 Br		M_IM_{II}	0.0672
188.4	1	28 Ni	M_{III}	Abs. Edge	0.06581
188.6	1	55 Cs		$N_{IV}O_{II}$	0.06574
189.5	3	35 Br		$M_{IV}N_{III}$	0.0654
190.3	1	55 Cs		N_VO_{III}	0.06515
190.	2	28 Ni		$M_{II,III}M_{IV,V}$	0.0651
191.1	2	35 Br	$M\zeta_2$	$M_{IV}N_{II}$	0.06488
192.6	2	35 Br	$M\zeta_1$	M_VN_{III}	0.06437
197.3	1	12 Mg	L_I	Abs. Edge	0.06284
202.	5	27 Co	$M_{II,III}$	Abs. Edge	0.061
203.	1	16 S		$L_IL_{II,III}$	0.061
214.	6	27 Co		$M_{II,III}M_{IV,V}$	0.058
224.	1	53 I	$N_{IV,V}$	Abs. Edge	0.0552
226.5	1	3 Li	K	Abs. Edge	0.05475
227.8	1	34 Se	M_V	Abs. Edge	0.05443
228.	1	3 Li	$K\alpha$	KL	0.0543
230.	2	34 Se		M_VN_{III}	0.0538
230.	1	26 Fe	$M_{II,III}$	Abs. Edge	0.0538
243.	5	26 Fe		$M_{II,III}M_{IV,V}$	0.051
249.3	1	12 Mg	L_{II}	Abs. Edge	0.04973
250.7	1	12 Mg	L_{III}	Abs. Edge	0.04945
251.5	5	12 Mg		$L_{II,III}M$	0.04929
273.	6	25 Mn		$M_{II,III}M_{IV,V}$	0.045
290.	1	13 Al		$L_IL_{II,III}$	0.0428
309.	9	24 Cr		$M_{II,III}M_{IV,V}$	0.040
317.	1	12 Mg		$L_IL_{II,III}$	0.0392
337.	9	23 V		$M_{II,III}M_{IV,V}$	0.0368
376.	1	11 Na		$L_IL_{II,III}$	0.03299
399.	5	35 Br	N_I	Abs. Edge	0.0311
405.	5	11 Na	$L_{II,III}$	Abs. Edge	0.0306
407.1	5	11 Na		$L_{II,III}M$	0.03045
417.	5	17 Cl	M_I	Abs. Edge	0.0297
444.	5	53 I	O_I	Abs. Edge	0.0279
525.	9	20 Ca		$M_{II,III}N_I$	0.0236
692.	9	19 K		$M_{II,III}N_I$	0.0179

*Reprinted with permission of the author and editor from *Revs. Mod. Phys.* **39,** (1), 78–124 (1967).

TABLE 1b4. X-RAY EMISSION LINES OF ELEMENTS (BEARDEN, 1967)*

X-ray wavelengths in Å* units and in keV. The probable error (p.e.) is the error in the last digit of wavelength. Designation indicates both conventional Siegbahn notation (if applicable) and transition, e.g., β_1 $L_{II}M_{IV}$ denotes a transition between the L_{II} and M_{IV} levels, which is the $L\beta_1$ line in Siegbahn notation.

Designation	Å*	p.e.	keV	Å*	p.e.	keV
3 Lithium				**4 Beryllium**		
$\alpha\,KL$	228.	1	0.0543	114.	1	0.1085
5 Boron				**6 Carbon**		
$\alpha\,KL$	67.6	3	0.1833	44.7	3	0.277
7 Nitrogen				**8 Oxygen**		
$\alpha\,KL$	31.6	4	0.3924	23.62	3	0.5249
9 Fluorine				**10 Neon**		
$\alpha_{1,2}\,KL_{II,III}$	18.32	2	0.6768	14.610	3	0.8486
$\beta\,KM$				14.452	5	0.8579
11 Sodium				**12 Magnesium**		
$\alpha_{1,2}\,KL_{II,III}$	11.9101	9	1.0410	9.8900	2	1.25360
$\beta\,KM$	11.575	2	1.0711	9.521	2	1.3022
$L_{II,III}M$	407.1	5	0.03045	251.5	5	0.0493
$L_1L_{II,III}$	376	1	0.0330	317	1	0.0392
13 Aluminum				**14 Silicon**		
$\alpha_2\,KL_{II}$	8.34173	9	1.48627	7.12791	9	1.73938
$\alpha_1\,KL_{III}$	8.33934	9	1.48670	7.12542	9	1.73998
$\beta\,KM$	7.960	2	1.5574	6.753	1	1.8359
$L_{II,III}$	171.4	5	0.0724	135.5	4	0.0915
$L_1L_{II,III}$	290.	1	0.0428			
15 Phosphorus				**16 Sulfur**		
$\alpha_2\,KL_{II}$	6.160†	1	2.0127	5.37496	8	2.30664
$\alpha_1\,KL_{III}$	6.157†	1	2.0137	5.37216	7	2.30784
$\beta\,KM$	5.796	2	2.1390			
$\beta_1\,KM$				5.0316	2	2.4640
$\beta_x\,KM$				5.0233	3	2.4681
$L_{II,III}M$	103.8	4	0.1194			
$l,\eta\,L_{II,III}M_I$				83.4	3	0.1487
17 Chlorine				**18 Argon**		
$\alpha_2\,KL_{II}$	4.7307	1	2.62078	4.19474	5	2.95563
$\alpha_1\,KL_{III}$	4.7278	1	2.62239	4.19180	5	2.95770
$\beta\,KM$	4.4034	3	2.8156			
$\beta_{1,3}\,KM_{II,III}$				3.8860	2	3.1905
$\eta\,L_{II}M_I$	67.33	9	0.1841	55.9†	1	0.2217
$l\,L_{III}M_I$	67.90	9	0.1826	56.3†	1	0.2201
19 Potassium				**20 Calcium**		
$\alpha_2\,KL_{II}$	3.7445	2	3.3111	3.36166	3	3.68809
$\alpha_1\,KL_{III}$	3.7414	2	3.3138	3.35839	3	3.69168
$\beta_{1,3}\,KM_{II,III}$	3.4539	2	3.5896	3.0897	2	4.0127
$\beta_5\,KM_{IV,V}$	3.4413	4	3.6027	3.0746	3	4.0325

Designation	Å*	p.e.	keV	Å*	p.e.	keV
19 Potassium (*Cont.*)				**20 Calcium** (*Cont.*)		
$\eta\,L_{II}M_I$	47.24	2	0.2625	40.46	2	0.3064
β_1				35.94	2	0.3449
$l\,L_{III}M_I$	47.74	1	0.25971	40.96	2	0.3027
$\alpha_{1,2}\,L_{III}M_{IV,V}$				36.33	2	0.3413
$M_{II,III}N_I$	692	9	0.0179	525.	9	0.0236
21 Scandium				**22 Titanium**		
$\alpha_2\,KL_{II}$	3.0342	1	4.0861	2.75216	2	4.50486
$\alpha_1\,KL_{III}$	3.0309†	1	4.0906	2.74851	2	4.51084
$\beta_{1,3}\,KM_{II,III}$	2.7796	2	4.4605	2.51391	2	4.93181
$\beta_5\,KM_{IV,V}$	2.7634	3	4.4865	2.4985	2	4.9623
$\eta\,L_{II}M_I$	35.13	2	0.3529	30.89	3	0.4013
$\beta_1\,L_{II}M_{IV}$	31.02	2	0.3996	27.05	2	0.4584
$l\,L_{III}M_I$	35.59	3	0.3483	31.36	2	0.3953
$\alpha_{1,2}\,L_{III}M_{IV,V}$	31.35	3	0.3954	27.42	2	0.4522
23 Vanadium				**24 Chromium**		
$\alpha_2\,KL_{II}$	2.50738	2	4.94464	2.293606	3	5.40551
$\alpha_1\,KL_{III}$	2.50356	2	4.95220	2.28970	2	5.41472
$\beta_{1,3}\,KM_{II,III}$	2.28440	2	5.42729	2.08487	2	5.94671
$\beta_5\,KM_{IV,V}$	2.26951	6	5.4629	2.07087	6	5.9869
$\beta_{3,4}\,L_1M_{II,III}$	21.19†	9	0.585	18.96	2	0.654
$\eta\,L_{II}M_I$	27.34	3	0.4535	24.30	3	0.5102
$\beta_1\,L_{II}M_{IV}$	23.88	4	0.5192	21.27	1	0.5828
$l\,L_{III}M_I$	27.77	1	0.4465	24.78	1	0.5003
$\alpha_{1,2}\,L_{III}M_{IV,V}$	24.25	3	0.5113	21.64	3	0.5728
$M_{II,III}M_{IV,V}$	337.	9	0.037	309.	9	0.040
25 Manganese				**26 Iron**		
$\alpha_2\,KL_{II}$	2.10578	2	5.88765	1.939980	9	6.39084
$\alpha_1\,KL_{III}$	2.101820	9	5.89875	1.936042	9	6.40384
$\beta_{1,3}\,KM_{II,III}$	1.91021	2	6.49045	1.75661	2	7.05798
$\beta_5\,KM_{IV,V}$	1.8971	2	6.5352	1.7442	1	7.1081
$\beta_{3,4}\,L_1M_{II,III}$	17.19	2	0.721	15.65	2	0.792
$\eta\,L_{II}M_I$	21.85	2	0.5675	19.75	4	0.628
$\beta_1\,L_{II}M_{IV}$	19.11	2	0.6488	17.26	1	0.7185
$l\,L_{III}M_I$	22.29	1	0.5563	20.15	1	0.6152
$\alpha_{1,2}\,L_{III}M_{IV,V}$	19.45	1	0.6374	17.59	2	0.7050
$M_{II,III}M_{IV,V}$	273.	6	0.045	243.	5	0.051
27 Cobalt				**28 Nickel**		
$\alpha_2\,KL_{II}$	1.792850	9	6.91530	1.661747	8	7.46089
$\alpha_1\,KL_{III}$	1.788965	9	6.93032	1.657910	8	7.47815
$\beta_{1,3}\,KM_{II,III}$	1.62079	2	7.64943	1.500135	8	8.26466
$\beta_5\,KM_{IV,V}$	1.60891	3	7.7059	1.48862	4	8.3286
$\beta_{3,4}\,L_1M_{II,III}$	14.31	3	0.870	13.18	1	0.941
$\eta\,L_{II}M_I$	17.87	3	0.694	16.27	3	0.762
$\beta_1\,L_{II}M_{IV}$	15.666	8	0.7914	14.271	6	0.8688
$l\,L_{III}M_I$	18.292	8	0.6778	16.693	9	0.7427
$\alpha_{1,2}\,L_{III}M_{IV,V}$	15.972	6	0.7762	14.561	3	0.8515
$M_{II,III}M_{IV,V}$	214.	6	0.058	190.	2	0.0651

TABLE 1b4. *(Continued)*

29 Copper / 30 Zinc

Designation	Å*	p.e.	keV	Å*	p.e.	keV
$\alpha_2\,KL_{II}$	1.544390	2	8.02783	1.439000	8	8.61578
$\alpha_1\,KL_{III}$	1.540562	2	8.04778	1.435155	7	8.63886
$\beta_3\,KM_{II}$	1.3926	1	8.9029			
$\beta_{1,3}\,KM_{II,III}$	1.392218	9	8.90529	1.29525	2	9.5720
$\beta_2\,KN_{II,III}$				1.28372	2	9.6580
$\beta_5\,KM_{IV,V}$	1.38109	3	8.9770	1.2848	1	9.6501
$\beta_{3,4}\,L_IM_{II,III}$	12.122	8	1.0228	11.200	7	1.1070
$\eta\,L_{II}M_I$	14.90	2	0.832	13.68	2	0.906
$\beta_1\,L_{II}M_{IV}$	13.053	3	0.9498	11.983	3	1.0347
$l\,L_{III}M_I$	15.286	9	0.8111	14.02	2	0.884
$\alpha_{1,2}\,L_{III}M_{IV,V}$	13.336	3	0.9297	12.254	3	1.0117
$M_{II,III}M_{V,V}$	173.	3	0.072	157.	3	0.079

31 Gallium / 32 Germanium

Designation	Å*	p.e.	keV	Å*	p.e.	keV
$\alpha_2\,KL_{II}$	1.34399	1	9.22482	1.258011	9	9.85532
$\alpha_1\,KL_{III}$	1.340083	9	9.25174	1.254054	9	9.88642
$\beta_3\,KM_{II}$	1.20835	5	10.2603	1.12936	9	10.9780
$\beta_1\,KM_{III}$	1.20789	2	10.2642	1.12894	2	10.9821
$\beta_2\,KN_{II,III}$	1.19600	2	10.3663	1.11686	2	11.1008
$\beta_5\,KM_{IV,V}$	1.1981	2	10.348	1.1195	1	11.0745
$\beta_4\,L_IM_{II}$				9.640	2	1.2861
$\beta_3\,L_IM_{III}$				9.581	2	1.2941
$\beta_{3,4}\,L_IM_{II,III}$	10.359†	8	1.197			
$\eta\,L_{II}M_I$	12.597	2	0.9842	11.609	2	1.0680
$\beta_1\,L_{II}M_{IV}$	11.023	2	1.1248	10.175	1	1.2185
$l\,L_{III}M_I$	12.953	2	0.9572	11.965	4	1.0362
$\alpha_{1,2}\,L_{III}M_{IV,V}$	11.292	1	1.09792	10.4361	8	1.18800

33 Arsenic / 34 Selenium

Designation	Å*	p.e.	keV	Å*	p.e.	keV
$\alpha_2\,KL_{II}$	1.17987	1	10.50799	1.10882	2	11.1814
$\alpha_1\,KL_{III}$	1.17588	1	10.54372	1.10477	2	11.2224
$\beta_3\,KM_{II}$	1.05783	5	11.7203	0.99268	5	12.4896
$\beta_1\,KM_{III}$	1.05730	2	11.7262	0.99218	3	12.4959
$\beta_2\,KN_{II,III}$	1.04500	3	11.8642	0.97992	5	12.6522
$\beta_5\,KM_{IV,V}$	1.0488	1	11.822	0.9843	1	12.595
$\beta_{3,4}\,L_IM_{II,III}$	8.929	1	1.3884	8.321†	9	1.490
$\eta\,L_{II}M_I$	10.734	1	1.1550	9.962	1	1.2446
$\beta_1\,L_{II}M_{IV}$	9.4141	8	1.3170	8.7358	5	1.41923
$l\,L_{III}M_I$	11.072	1	1.1198	10.294	1	1.2044
$\alpha_{1,2}\,L_{III}M_{IV,V}$	9.6709	8	1.2820	8.9900	5	1.37910
M_VN_{III}				230.	2	0.0538

35 Bromine / 36 Krypton

Designation	Å*	p.e.	keV	Å*	p.e.	keV
$\alpha_2\,KL_{II}$	1.04382	2	11.8776	0.9841	1	12.598
$\alpha_1\,KL_{III}$	1.03974	2	11.9242	0.9801	1	12.649
$\beta_3\,KM_{II}$	0.93327	5	13.2845	0.8790	1	14.104
$\beta_1\,KM_{III}$	0.93279	2	13.2914	0.8785	1	14.112
$\beta_2\,KN_{II,III}$	0.92046	2	13.4695	0.8661	1	14.315
$\beta_5\,KM_{IV,V}$	0.9255	1	13.396	0.8708	2	14.238
$\beta_4\,KN_{IV,V}$				0.8653	2	14.328
$\beta_4\,L_IM_{II}$				7.304	5	1.697
$\beta_3\,L_IM_{III}$				7.264	5	1.707

35 Bromine *(Cont.)* / 36 Krypton *(Cont.)*

Designation	Å*	p.e.	keV	Å*	p.e.	keV
$\beta_{3,4}\,L_IM_{II,III}$	7.767†	9	1.596			
$\eta\,L_{II}M_I$	9.255	1	1.3396			
$\beta_1\,L_{II}M_{IV}$	8.1251	5	1.52590	7.576†	3	1.6366
γ_5				7.279	5	1.703
$l\,L_{III}M_I$	9.585	1	1.2935			
$\alpha_{1,2}\,L_{III}M_{IV,V}$	8.3746	5	1.48043	7.817†	3	1.5860
β_6				7.510	4	1.6510
$L_{II}N_{III}$				7.250	5	1.710
M_IM_{II}	184.6	3	0.0672			
M_IM_{III}	164.7	3	0.0753			
$M_{II}M_{IV}$	109.4	3	0.1133			
$M_{II}N_I$	76.9	2	0.1613			
$M_{III}M_{IV,V}$	113.8	3	0.1089			
	79.8	3	0.1554			
$\zeta_2\,M_{IV}N_{II}$	191.1	2	0.06488			
$M_{IV}N_{III}$	189.5	3	0.0654			
$\zeta_1\,M_VN_{III}$	192.6	2	0.06437			

37 Rubidium / 38 Strontium

Designation	Å*	p.e.	keV	Å*	p.e.	keV
$\alpha_2\,KL_{II}$	0.92969	1	13.3358	0.87943	1	14.0979
$\alpha_1\,KL_{III}$	0.925553	9	13.3953	0.87526	1	14.1650
$\beta_3\,KM_{II}$	0.82921	3	14.9517	0.78345	3	15.8249
$\beta_1\,KM_{III}$	0.82868	2	14.9613	0.78292	2	15.8357
$\beta_2\,KN_{II,III}$	0.81645	3	15.1854	0.77081	3	16.0846
$\beta_5\,KM_{IV,V}$	0.8219	1	15.085	0.7764	1	15.969
$\beta_4\,KN_{IV,V}$	0.8154	2	15.205	0.76989	5	16.104
$\beta_4\,L_IM_{II}$	6.8207	3	1.81771	6.4026	3	1.93643
$\beta_3\,L_IM_{III}$	6.7876	3	1.82659	6.3672	3	1.94719
$\gamma_{2,3}\,L_IN_{II,III}$	6.0458	3	2.0507	5.6445	3	2.1965
$\eta\,L_{II}M_I$	8.0415	4	1.54177	7.5171	3	1.64933
$\beta_1\,L_{II}M_{IV}$	7.0759	3	1.75217	6.6239	3	1.87172
$\gamma_5\,L_{II}N_{IV}$	6.7553	3	1.83532	6.2961	3	1.96916
$l\,L_{III}M_I$	8.3636	4	1.48238	7.8362	3	1.58215
$\alpha_2\,L_{III}M_{IV}$	7.3251	3	1.69256	6.8697	3	1.80474
$\alpha_1\,L_{III}M_V$	7.3183	2	1.69413	6.8628	2	1.80656
$\beta_6\,L_{III}N_I$	6.9842	3	1.77517	6.5191	3	1.90181
M_IM_{III}	144.4	3	0.0859			
$M_{II}M_{IV}$	91.5	2	0.1355	85.7	2	0.1447
$M_{II}N_I$	57.0	2	0.2174	51.3	1	0.2416
$M_{III}M_{IV,V}$	96.7	2	0.1282	91.4	2	0.1357
	59.5	2	0.2083	53.6	1	0.2313
$\zeta_2\,M_{IV}N_{II}$	127.8	2	0.0970			
$M_{IV}N_{III}$	126.8	2	0.0978			
$\zeta_2\,M_{IV}N_{II,III}$				108.0	2	0.1148
$\zeta_1\,M_VN_{III}$	128.7	2	0.0964	108.7	1	0.1140

39 Yttrium / 40 Zirconium

Designation	Å*	p.e.	keV	Å*	p.e.	keV
$\alpha_2\,KL_{II}$	0.83305	1	14.8829	0.79015	1	15.6909
$\alpha_1\,KL_{III}$	0.82884	1	14.9584	0.78593	1	15.7751
$\beta_3\,KM_{II}$	0.74126	3	16.7258	0.70228	4	17.654
$\beta_1\,KM_{III}$	0.74072	2	16.7378	0.70173	3	17.6678
$\beta_2\,KN_{II,III}$	0.72864	4	17.0154	0.68993	4	17.970
$\beta_5\,KM_{IV,V}$	0.7345	1	16.879	0.6959	1	17.815

TABLE 1b4. (Continued)

39 Yttrium (Cont.) | 40 Zirconium (Cont.)

Designation	Å*	p.e.	keV	Å*	p.e.	keV
$\beta_4 KN_{IV,V}$	0.72776	5	17.036	0.68901	5	17.994
$\beta_4 L_I M_{II}$	6.0186	3	2.0600	5.6681	3	2.1873
$\beta_3 L_I M_{III}$	5.9832	3	2.0722	5.6330	3	2.2010
$\gamma_{2,3} L_I N_{II,III}$	5.2830	3	2.3468	4.9536	3	2.5029
$\eta L_I M_I$	7.0406	3	1.76095	6.6069	3	1.87654
$\beta_1 L_{III} M_{IV}$	6.2120	3	1.99584	5.8360	3	2.1244
$\gamma_6 L_{II} N_I$	5.8754	3	2.1102	5.4977	3	2.2551
$\gamma_1 L_{II} N_{IV}$				5.3843	3	2.3027
$l L_{III} M_I$	7.3563	3	1.68536	6.9185	3	1.79201
$\alpha_2 L_{III} M_{IV}$	6.4558	3	1.92047	6.0778	3	2.0399
$\alpha_1 L_{III} M_V$	6.4488	2	1.92256	6.0705	2	2.04236
$\beta_6 L_{III} N_I$	6.0942	3	2.0344	5.7101	3	2.1712
$\beta_{2,15}$				5.5863	3	2.2194
$M_{II} M_{IV}$	81.5	2	0.1522	76.7	2	0.1617
$M_{II} N_I$	46.48	9	0.267			
$M_{III} M_V$				80.9	3	0.1533
$M_{III} N_I$	48.5	2	0.256			
$M_{III} M_{IV,V}$	86.5	2	0.1434	82.1	2	0.1511
$\zeta M_{IV,V} N_{II,III}$	93.4	2	0.1328	70.0	4	0.177
$M_{IV,V} O_{II,III}$						

43 Technetium | 44 Ruthenium

Designation	Å*	p.e.	keV	Å*	p.e.	keV
$\alpha_2 KL_{II}$	0.67932†	3	18.2508	0.647408	5	19.1504
$\alpha_1 KL_{III}$	0.67502†	3	18.3671	0.643083	4	19.2792
$\beta_3 KM_{II}$	0.60188†	4	20.599	0.573067	4	21.6346
$\beta_1 KM_{III}$	0.60130†	4	20.619	0.572482	4	21.6568
$\beta_2 KN_{II,III}$	0.59024†	5	21.005	0.56166	3	22.074
$\beta_5^{II} KM_{IV}$				0.5680	2	21.829
$\beta_5^{I} KM_V$				0.56785	9	21.834
β_4				0.56089	9	22.104
$\beta_4 L_{II} M_{IV}$				4.5230	2	2.7411
$\beta_3 L_I M_{III}$				4.4866	3	2.7634
$\gamma_{2,3} L_I N_{II,III}$				3.8977	2	3.1809
$\eta L_{II} M_I$				5.2050	2	2.38197
$\beta_1 L_{II} M_{IV}$	4.8873†	8	2.5368	4.62058	3	2.68323
$\gamma_6 L_{II} N_I$				4.2873	2	2.8918
$\gamma_1 L_{II} N_{IV}$				4.1822	2	2.9645
$l L_{III} M_I$				5.5035	3	2.2528
$\alpha_2 L_{III} M_{IV}$				4.85381	7	2.55431
$\alpha_1 L_{III} M_V$	5.1148†	3	2.4240	4.84575	5	2.55855
$\beta_6 L_{III} N_I$				4.4866	3	2.7634
$\beta_{2,15} L_{III} N_{IV,V}$				4.3718	2	2.8360
$M_{II} M_{IV}$				62.2	1	0.1992
$M_{II} N_I$				32.3	2	0.384
$M_{II} N_{IV}$				25.50	9	0.486
$M_{III} M_V$				68.3	1	0.1814
$\gamma M_{III} N_{IV,V}$				26.9	1	0.462
$\zeta M_{IV,V} N_{II,III}$				52.34	7	0.2369
$M_{IV,V} O_{II,III}$				44.8	1	0.2768

41 Niobium | 42 Molybdenum

Designation	Å*	p.e.	keV	Å*	p.e.	keV
$\alpha_2 KL_{II}$	0.75044	1	16.5210	0.713590	6	17.3743
$\alpha_1 KL_{III}$	0.74620	1	16.6151	0.709300	1	17.47934
$\beta_3 KM_{II}$	0.66634	3	18.6063	0.632872	9	19.5903
$\beta_1 KM_{III}$	0.66576	2	18.6225	0.632288	9	19.6083
β_2^{II}				0.62107	5	19.963
$\beta_2 KN_{II,III}$	0.65416	4	18.953	0.62099	2	19.9652
$\beta_4 KN_{IV,V}$	0.65318	5	18.981			
$\beta_5^{II} KM_{IV}$				0.62708	5	19.771
$\beta_5^{I} KM_V$				0.62692	5	19.776
$\beta_4 KN_{IV,V}$				0.62001	9	19.996
$\beta_4 L_I M_{II}$	5.3455	3	2.3194	5.0488	3	2.4557
$\beta_3 L_I M_{III}$	5.3102	3	2.3348	5.0133	3	2.4730
$\gamma_{2,3} L_I N_{II,III}$	4.6542	2	2.6638	4.3800	2	2.8306
$\eta L_{II} M_I$	6.2109	3	1.99620	5.8475	3	2.1202
$\beta_1 L_{II} M_{IV}$	5.4923	3	2.2574	5.17708	8	2.39481
$\gamma_6 L_{II} N_I$	5.1517	3	2.4066	4.8369	2	2.5632
$\gamma_1 L_{II} N_{IV}$	5.0361	3	2.4618	4.7258	2	2.6235
$l L_{III} M_I$	6.5176	3	1.90225	6.1508	3	2.01568
$\alpha_2 L_{III} M_{IV}$	5.7319	3	2.1630	5.41437	8	2.28985
$\alpha_1 L_{III} M_V$	5.7243	2	2.16589	5.40655	8	2.29316
$\beta_6 L_{III} N_I$	5.3613	3	2.3125	5.0488	5	2.4557
$\beta_{2,15} L_{III} N_{IV,V}$	5.2379	3	2.3670	4.9232	2	2.5183
$M_{II} M_{IV}$	72.1	3	0.1718	68.9	2	0.1798
$M_{II} N_I$	38.4	3	0.323	35.3	3	0.351
$M_{II} N_{IV}$	33.1	2	0.375			
$M_{III} M_V$	78.4	2	0.1582	74.9	1	0.1656
$M_{III} N_I$	40.7	2	0.305	37.5	2	0.331
$\gamma M_{III} N_{IV,V}$	34.9	2	0.356			
$\zeta M_{IV,V} N_{II,III}$	72.19	9	0.1717	64.38	7	0.1926
$M_{IV,V} O_{II,III}$	61.9	2	0.2002	54.8	2	0.2262

45 Rhodium | 46 Palladium

Designation	Å*	p.e.	keV	Å*	p.e.	keV
$\alpha_2 KL_{II}$	0.617630	4	20.0737	0.589821	3	21.0201
$\alpha_1 KL_{III}$	0.613279	4	20.2161	0.585448	3	21.1771
$\beta_3 KM_{II}$	0.546200	4	22.6989	0.521123	4	23.7911
$\beta_1 KM_{III}$	0.545605	4	22.7236	0.520520	4	23.8187
$\beta_2^{II} KN_{II}$	0.53513	5	23.168			
$\beta_2 KN_{II,III}$	0.53503	2	23.1728	0.510228	4	24.2991
$\beta_5^{II} KM_{IV}$	0.54118	9	22.909			
$\beta_5^{I} KM_V$	0.54101	9	22.917			
$\beta_4 KN_{IV,V}$	0.53401	9	23.217	0.5093	2	24.346
$\beta_5 KM_{IV,V}$				0.51670	9	23.995
$\beta_4 L_I M_{II}$	4.2888	2	2.8908	4.0711	2	3.0454
$\beta_3 L_I M_{III}$	4.2522	2	2.9157	4.0346	2	3.0730
$\gamma_{2,3} L_I N_{II,III}$	3.6855	2	3.3640	3.4892	2	3.5533
$\eta L_{II} M_I$	4.9217	2	2.5191	4.6605	2	2.6603
$\beta_1 L_{II} M_{IV}$	4.37414	4	2.83441	4.14622	5	2.99022
$\gamma_6 L_{II} N_I$	4.0451	2	3.0650	3.8222	2	3.2437
$\gamma_1 L_{II} N_{IV}$	3.9437	2	3.1438	3.7246	2	3.3287
$l L_{III} M_I$	5.2169	3	2.3765	4.9525	3	2.5034
$\alpha_2 L_{III} M_{IV}$	4.60545	9	2.69205	4.37588	7	2.83329
$\alpha_1 L_{III} M_V$	4.59743	9	2.69674	4.36767	5	2.83861
$\beta_6 L_{III} N_I$	4.2417	2	2.9229	4.0162	2	3.0870
$\beta_{2,15} L_{III} N_{IV,V}$	4.1310	2	3.0013	3.90887	4	3.17179
$\beta_{10} L_I M_{IV}$				3.7988	2	3.2637

TABLE 1b4. (*Continued*)

45 Rhodium (*Cont.*) | 46 Palladium (*Cont.*)

Designation	Å*	p.e.	keV	Å*	p.e.	keV
$\beta_9\,L_IM_V$				3.7920	2	3.2696
$M_IN_{II,III}$				20.1	2	0.616
$M_{II}M_{IV}$	59.3	1	0.2090	56.5	1	0.2194
$M_{II}N_I$	28.1	2	0.442	26.2	2	0.474
$M_{II}N_{IV}$				22.1	1	0.560
$M_{II}M_V$	65.5	1	0.1892	62.9	1	0.1970
$M_{II}^IN_I$	29.8	1	0.417	27.9	1	0.445
$\gamma\,M_{III}N_{IV,V}$	25.01	9	0.496	23.3†	1	0.531
$\zeta\,M_{IV,V}N_{II,III}$	47.67	9	0.2601	43.6	1	0.2844
$M_{IV,V}O_{II,III}$	40.9	2	0.303	37.4	2	0.332

47 Silver | 48 Cadmium

Designation	Å*	p.e.	keV	Å*	p.e.	keV
$\alpha_2\,KL_{II}$	0.563798	4	21.9903	0.539422	3	22.9841
$\alpha_1\,KL_{III}$	0.5594075	6	22.16292	0.535010	3	23.1736
$\beta_3\,KM_{II}$	0.497685	4	24.9115	0.475730	5	26.0612
$\beta_1\,KM_{III}$	0.497069	4	24.9424	0.475105	6	26.0955
$\beta_2\,KN_{II,III}$	0.487032	4	25.4564	0.465328	7	26.6438
$\beta_5\,KM_{IV,V}$	0.49306	2	25.145			
$\beta_4\,KN_{IV,V}$	0.48598	3	25.512			
$\beta_4\,L_IM_{II}$	3.87023	5	3.20346	3.68203	9	3.36719
$\beta_3\,L_IM_{III}$	3.83313	9	3.23446	3.64495	9	3.40145
$\gamma_2\,L_IN_{II}$	3.31216	9	3.7432	3.1377	2	3.9513
$\gamma_3\,L_IN_{III}$	3.30635	9	3.7498			
$\eta\,L_{II}M_I$	4.4183	2	2.8061	4.19315	9	2.95675
$\beta_1\,L_{II}M_{IV}$	3.93473	3	3.15094	3.73823	4	3.31657
$\gamma_5\,L_{II}N_I$	3.61638	9	3.42832	3.42551	9	3.61935
$\gamma_1\,L_{III}N_{IV}$	3.52260	4	3.51959	3.33564	6	3.71686
$l\,L_{III}M_I$	4.7076	2	2.6337	4.48014	9	2.76735
$\alpha_2\,L_{III}M_{IV}$	4.16294	5	2.97821	3.96496	6	3.12691
$\alpha_1\,L_{III}M_V$	4.15443	3	2.98431	3.95635	4	3.13373
$\beta_6\,L_{III}N_I$	3.80774	9	3.25603	3.61467	9	3.42994
$\beta_{2,15}\,L_{III}N_{IV,V}$	3.70335	3	3.34781	3.51408	4	3.52812
$\beta_{10}\,L_IM_{IV}$	3.61158	9	3.43287	3.4367	2	3.6075
$\beta_9\,L_IM_V$	3.60497	9	3.43917	3.43015	9	3.61445
$M_IN_{II,III}$	18.8	2	0.658			
$M_{II}M_{IV}$	54.0	1	0.2295	52.0	2	0.2384
$M_{II}N_I$				22.9	2	0.540
$M_{II}N_{IV}$	20.66	7	0.600	19.40	7	0.639
$M_{III}M_V$	60.5	1	0.2048	58.7	2	0.2111
$M_{III}N_I$	26.0	1	0.478	24.5	1	0.507
$\gamma\,M_{III}N_{IV,V}$	21.82	7	0.568	20.47	7	0.606
$M_{IV}O_{II,III}$				30.4	1	0.408
$\zeta\,M_{IV,V}N_{II,III}$	39.77	7	0.3117	36.8	1	0.3371
M_VN_I	24.4	2	0.509			
M_VO_{III}				30.8	1	0.403
$M_{IV,V}O_{II,III}$	33.5	3	0.370			

49 Indium | 50 Tin

Designation	Å*	p.e.	keV	Å*	p.e.	keV
$\alpha_2\,KL_{II}$	0.516544	3	24.0020	0.495053	3	25.0440
$\alpha_1\,KL_{III}$	0.512113	3	24.2097	0.490599	3	25.2713
$\beta_3\,KM_{II}$	0.455181	4	27.2377	0.435877	5	28.4440

49 Indium (*Cont.*) | 50 Tin (*Cont.*)

Designation	Å*	p.e.	keV	Å*	p.e.	keV
$\beta_1\,KM_{III}$	0.454545	4	27.2759	0.435236	5	28.4860
$\beta_2\,KN_{II,III}$	0.44500	1	27.8608	0.425915	8	29.1093
$KO_{II,III}$	0.44374	3	27.940	0.42467	3	29.195
$\beta_5^{II}\,KM_{IV}$	0.45098	2	27.491	0.43184	3	28.710
$\beta_5^{I}\,KM_V$	0.45086	2	27.499	0.43175	3	28.716
$\beta_4\,KN_{IV,V}$	0.44393	4	27.928	0.42495	3	29.175
$\beta_4\,L_IM_{II}$	3.50697	9	3.5353	3.34335	9	3.7083
$\beta_3\,L_IM_{III}$	3.46984	9	3.5731	3.30585	3	3.7500
$\gamma_{2,3}\,L_IN_{II,III}$	2.9800	2	4.1605	2.8327	2	4.3768
$\gamma_4\,L_IO_{II,III}$	2.9264	2	4.2367	2.7775	2	4.4638
$\eta\,L_{II}M_I$	3.98327	9	3.11254	3.78876	9	3.27234
$\beta_1\,L_{II}M_{IV}$	3.55531	4	3.48721	3.38487	3	3.66280
$\gamma_5\,L_{II}N_I$	3.24907	9	3.8159	3.08475	9	4.0192
$\gamma_1\,L_{II}N_{IV}$	3.16213	4	3.92081	3.00115	3	4.13112
$l\,L_{III}M_I$	4.26873	9	2.90440	4.07165	9	3.04499
$\alpha_2\,L_{III}M_{IV}$	3.78073	6	3.27929	3.60891	4	3.43542
$\alpha_1\,L_{III}M_V$	3.77192	4	3.28694	3.59994	3	3.44398
$\beta_6\,L_{III}N_I$	3.43606	9	3.60823	3.26901	9	3.7926
$\beta_{2,15}\,L_{III}N_{IV,V}$	3.33838	3	3.71381	3.17505	3	3.90486
$\beta_7\,L_{III}O_I$	3.324	4	3.730	3.1564	3	3.9279
$\beta_{10}\,L_IM_{IV}$	3.27404	9	3.7868	3.12170	9	3.9716
$\beta_9\,L_IM_V$	3.26763	9	3.7942	3.11513	9	3.9800
$M_{II}M_V$				47.3	1	0.2621
$M_{II}N_I$				20.0	1	0.619
$M_{II}N_{IV}$				16.93	5	0.733
$M_{III}M_V$				54.2	1	0.2287
$M_{III}N_I$				21.5	1	0.575
$\gamma\,M_{III}N_{IV,V}$				17.94	5	0.691
$M_{IV}O_{III}$				25.3	1	0.491
$\zeta\,M_{IV,V}N_{II,III}$				31.24	9	0.397
M_VO_{III}				25.7	1	0.483

51 Antimony | 52 Tellurium

Designation	Å*	p.e.	keV	Å*	p.e.	keV
$\alpha_2\,KL_{II}$	0.474827	3	26.1108	0.455784	3	27.2017
$\alpha_1\,KL_{III}$	0.470354	3	26.3591	0.451295	3	27.4723
$\beta_3\,KM_{II}$	0.417737	4	29.6792	0.400659	4	30.9443
$\beta_1\,KM_{III}$	0.417085	3	29.7256	0.399995	5	30.9957
$\beta_2\,KN_{II,III}$	0.407973	5	30.3895	0.391102	6	31.7004
$KO_{II,III}$	0.40666	1	30.4875	0.38974	1	31.8114
$\beta_5^{II}\,KM_{IV}$	0.41388	1	29.9560			
$\beta_5^{I}\,KM_V$	0.41378	1	29.9632			
$\beta_4\,KN_{IV,V}$	0.40702	1	30.4604			
$\beta_4\,L_IM_{II}$	3.19014	9	3.8864	3.04661	9	4.0695
$\beta_3\,L_IM_{III}$	3.15258	9	3.9327	3.00893	9	4.1204
$\gamma_{2,3}\,L_IN_{II,III}$	2.6953	2	4.5999	2.5674	2	4.8290
$\gamma_4\,L_IO_{II,III}$	2.6398	2	4.6967	2.5113	2	4.9369
$\eta\,L_{II}M_I$	3.60765	9	3.43661	3.43832	9	3.60586
$\beta_1\,L_{II}M_{IV}$	3.22567	4	3.84357	3.07677	6	4.02958
$\gamma_5\,L_{II}N_I$	2.93187	9	4.2287	2.79007	9	4.4437
$\gamma_1\,L_{II}N_{IV}$	2.85159	3	4.34779	2.71241	6	4.5709
$l\,L_{III}M_I$	3.88826	9	3.18860	3.71696	9	3.33555
$\alpha_2\,L_{III}M_{IV}$	3.44840	6	3.59532	3.29846	9	3.7588

TABLE 1b4. (Continued)

Left panel

Designation	Å*	p.e.	keV	Å*	p.e.	keV
51 Antimony (Cont.)				**52 Tellurium (Cont.)**		
$\alpha_1\ L_{III}M_V$	3.43941	4	3.60472	3.28920	6	3.76933
$\beta_6\ L_{III}N_I$	3.11513	9	3.9800	2.97088	9	4.1732
$\beta_{2,15}\ L_{III}N_{IV,V}$	3.02335	3	4.10078	2.88217	8	4.3017
$\beta_7\ L_{III}O_I$	3.0052	3	4.1255	2.8634	3	4.3298
$\beta_{10}\ L_IM_{IV}$	2.97917	9	4.1616	2.84679	9	4.3551
$\beta_9\ L_IM_V$	2.97261	9	4.1708	2.83897	9	4.3671
$M_{II}M_{IV}$	45.2	1	0.2743			
$M_{II}N_I$	18.8	1	0.658	17.6	1	0.703
$M_{II}N_{IV}$	15.98	5	0.776			
$M_{III}M_V$	52.2	1	0.2375	50.3	1	0.2465
$M_{III}N_I$	20.2	1	0.612	19.1	1	0.648
$\gamma\ M_{III}N_{IV,V}$	16.92	4	0.733	15.93	4	0.778
$M_{IV}O_{II,III}$				21.34	5	0.581
$\zeta\ M_{IV,V}N_{II,III}$	28.88	8	0.429	26.72	9	0.464
M_VO_{III}				21.78	5	0.569
53 Iodine				**54 Xenon**		
$\alpha_2\ KL_{II}$	0.437829	7	28.3172	0.42087†	2	29.458
$\alpha_1\ KL_{III}$	0.433318	5	28.6120	0.41634†	2	29.779
$\beta_3\ KM_{II}$	0.384564	4	32.2394	0.36941†	2	33.562
$\beta_1\ KM_{III}$	0.383905	4	32.2947	0.36872†	2	33.624
$\beta_2\ KN_{II,III}$	0.37523†	2	33.042	0.36026†	3	34.415
$\beta_4\ L_IM_{II}$	2.91207	9	4.2575			
$\beta_3\ L_IM_{III}$	2.87429	9	4.3134			
$\gamma_{2,3}\ L_IN_{II,III}$	2.4475	2	5.0657			
$\gamma_4\ L_IO_{II,III}$	2.3913	2	5.1848			
$\eta\ L_{II}M_I$	3.27979	9	3.7801			
$\beta_1\ L_{II}M_{IV}$	2.93744	6	4.22072			
$\gamma_5\ L_{II}N_I$	2.65710	9	4.6660			
$\gamma_1\ L_{II}N_{IV}$	2.58244	8	4.8009			
$l\ L_{III}M_I$	3.55754	9	3.48502			
$\alpha_2\ L_{III}M_{IV}$	3.15791	6	3.92604			
$\alpha_1\ L_{III}M_V$	3.14860	6	3.93765	3.0166†	2	4.1099
$\beta_6\ L_{III}N_I$	2.83672	9	4.3706			
$\beta_{2,15}\ L_{III}N_{IV,V}$	2.75053	8	4.5075			
$\beta_7\ L_{III}O_I$	2.7288	3	4.5435			
$\beta_{10}\ L_IM_{IV}$	2.72104	9	4.5564			
$\beta_9\ L_IM_V$	2.71352	9	4.5690			
55 Cesium				**56 Barium**		
$\alpha_2\ KL_{II}$	0.404835	4	30.6251	0.389668	5	31.8171
$\alpha_1\ KL_{III}$	0.400290	4	30.9728	0.385111	4	32.1936
$\beta_3\ KM_{II}$	0.355050	4	34.9194	0.341507	4	36.3040
$\beta_1\ KM_{III}$	0.354364	7	34.9869	0.340811	3	36.3782
$\beta_2\ KN_{II,III}$	0.34611	2	35.822	0.33277	1	37.257
$KO_{II,III}$				0.33127	2	37.426
$\beta_5{}^{II}\ KM_{IV}$				0.33835	2	36.643
$\beta_5{}^{I}\ KM_V$				0.33814	2	36.666
$\beta_4\ KN_{IV,V}$				0.33229	2	37.311
$\beta_4\ L_IM_{II}$	2.6666	2	4.6494	2.5553	2	4.8519
$\beta_3\ L_IM_{III}$	2.6285	2	4.7167	2.5164	2	4.9269
$\gamma_2\ L_IN_{II}$	2.2371	2	5.5420	2.1387	2	5.7969
$\gamma_3\ L_IN_{III}$	2.2328	2	5.5527	2.1342	2	5.8092

Right panel

Designation	Å*	p.e.	keV	Å*	p.e.	keV
55 Cesium (Cont.)				**56 Barium (Cont.)**		
$\gamma_4\ L_IO_{II,III}$	2.1741	2	5.7026	2.0756	3	5.9733
$\eta\ L_{II}M_I$	2.9932	2	4.1421	2.8627	3	4.3309
$\beta_1\ L_{II}M_{IV}$	2.6837	2	4.6198	2.56821	5	4.82753
$\gamma_5\ L_{II}N_I$	2.4174	2	5.1287	2.3085	3	5.3707
$\gamma_1\ L_{II}N_{IV}$	2.3480	2	5.2804	2.2415	2	5.5311
$l\ L_{III}M_I$	3.2670	2	3.7950	3.1355	2	3.9541
$\alpha_2\ L_{III}M_{IV}$	2.9020	2	4.2722	2.78553	5	4.45090
$\alpha_1\ L_{III}M_V$	2.8924	2	4.2865	2.77595	5	4.46626
$\beta_6\ L_{III}N_I$	2.5932	2	4.7811	2.4826	2	4.9939
$\beta_{2,15}\ L_{III}N_{IV,V}$	2.5118	2	4.9359	2.40435	6	5.1565
$\beta_7\ L_{III}O_I$	2.4849	2	4.9893	2.3806	2	5.2079
$\beta_{10}\ L_IM_{IV}$	2.4920	2	4.9752	2.3869	2	5.1941
$\beta_9\ L_IM_V$	2.4783	2	5.0026	2.3764	2	5.2171
$\gamma\ M_{III}N_{IV,V}$				12.75	3	0.973
$M_{IV}O_{II}$				15.91	5	0.779
$M_{IV}O_{III}$				15.72	9	0.789
$\zeta\ M_VN_{III}$				20.64	4	0.601
M_VO_{III}				16.20	5	0.765
$N_{IV}O_{II}$	188.6	1	0.06574	163.3	2	0.07590
$N_{IV}O_{III}$	183.8	1	0.06746	159.0	2	0.07796
N_VO_{III}	190.3	1	0.06515	164.6	2	0.07530
57 Lanthanum				**58 Cerium**		
$\alpha_2\ KL_{II}$	0.375313	2	33.0341	0.361683	2	34.2789
$\alpha_1\ KL_{III}$	0.370737	2	33.4418	0.357092	2	34.7197
$\beta_3\ KM_{II}$	0.328686	4	37.7202	0.316520	4	39.1701
$\beta_1\ KM_{III}$	0.327983	3	37.8010	0.315816	2	39.2573
$\beta_2\ KN_{II,III}$	0.320117	7	38.7299	0.30816	1	40.233
$KO_{II,III}$	0.31864	2	38.909	0.30668	2	40.427
$\beta_5{}^{II}\ KM_{IV}$	0.32563	2	38.074	0.31357	2	39.539
$\beta_5{}^{I}\ KM_V$	0.32546	2	38.094	0.31342	2	39.558
$\beta_4\ KN_{IV,V}$	0.31931	2	38.828	0.30737	2	40.337
$\beta_4\ L_IM_{II}$	2.4493	3	5.0620	2.3497	4	5.2765
$\beta_3\ L_IM_{III}$	2.4105	3	5.1434	2.3109	3	5.3651
$\gamma_2\ L_IN_{II}$	2.0460	4	6.060	1.9602	3	6.3250
$\gamma_3\ L_IN_{III}$	2.0410	4	6.074	1.9553	3	6.3409
$\gamma_4\ L_IO_{II,III}$	1.9830	4	6.252	1.8991	4	6.528
$\eta\ L_{II}M_I$	2.740	3	4.525	2.6203	4	4.7315
$\beta_1\ L_{II}M_{IV}$	2.45891	5	5.0421	2.3561	3	5.2622
$\gamma_5\ L_{II}N_I$	2.2056	4	5.621	2.1103	3	5.8751
$\gamma_1\ L_{II}N_{IV}$	2.1418	3	5.7885	2.0487	4	6.052
$\gamma_8\ L_{II}O_I$				2.0237	4	6.126
$l\ L_{III}M_I$	3.006	3	4.124	2.8917	4	4.2875
$\alpha_2\ L_{III}M_{IV}$	2.67533	5	4.63423	2.5706	3	4.8230
$\alpha_1\ L_{III}M_V$	2.66570	5	4.65097	2.5615	2	4.8402
$\beta_6\ L_{III}N_I$	2.3790	4	5.2114	2.2818	3	5.4334
$\beta_{2,15}\ L_{III}N_{IV,V}$	2.3030	3	5.3835	2.2087	2	5.6134
$\beta_7\ L_{III}O_I$	2.275	3	5.450	2.1701	2	5.7132
$\beta_{10}\ L_IM_{IV}$	2.290	3	5.415	2.1958	5	5.646
$\beta_9\ L_IM_V$	2.282	3	5.434	2.1885	3	5.6650
$\gamma\ M_{III}N_{IV,V}$	12.08	4	1.027	11.53	1	1.0749
$\beta\ M_{IV}N_{VI}$	14.51	5	0.854	13.75	4	0.902
$\zeta\ M_VN_{III}$	19.44	5	0.638	18.35	4	0.676
$\alpha\ M_VN_{VI,VII}$	14.88	5	0.833	14.04	2	0.883

TABLE 1b4. (*Continued*)

Desig-nation	Å*	p.e.	keV	Å*	p.e.	keV
	57 Lanthanum (*Cont.*)			**58 Cerium** (*Cont.*)		
$M_V O_{II,III}$				14.39	5	0.862
$N_{IV,V} O_{II,III}$	152.6	6	0.0812	144.4	6	0.0859
	59 Praseodymium			**60 Neodymium**		
$\alpha_2 KL_{II}$	0.348749	2	35.5502	0.336472	2	36.8474
$\alpha_1 KL_{III}$	0.344140	2	36.0263	0.331846	2	37.3610
$\beta_3 KM_{II}$	0.304975	5	40.6529	0.294027	3	42.1665
$\beta_1 KM_{III}$	0.304261	4	40.7482	0.293299	2	42.2713
$\beta_2 KN_{II,III}$	0.29679	2	41.773	0.2861†	1	43.33
$\beta_4 L_I M_{II}$	2.2550	4	5.4981	2.1669	3	5.7216
$\beta_3 L_I M_{III}$	2.2172	3	5.5918	2.1268	2	5.8294
$\gamma_2 L_I N_{II}$	1.8791	4	6.598	1.8013	4	6.883
$\gamma_3 L_I N_{III}$	1.8740	4	6.616	1.7964	4	6.902
$\gamma_4 L_I O_{II,III}$	1.8193	4	6.815	1.7445	4	7.107
$\eta L_{II} M_I$	2.512	3	4.935	2.4094	4	5.1457
$\beta_1 L_{II} M_{IV}$	2.2588	3	5.4889	2.1669	2	5.7216
$\gamma_5 L_{II} N_I$	2.0205	4	6.136	1.9355	4	6.406
$\gamma_1 L_{II} N_{IV}$	1.9611	3	6.3221	1.8779	2	6.6021
$\gamma_8 L_{II} O_I$	1.9362	4	6.403	1.8552	5	6.683
$\ell L_{III} M_I$	2.7841	4	4.4532	2.6760	4	4.6330
$\alpha_2 L_{III} M_{IV}$	2.4729	3	5.0135	2.3807	3	5.2077
$\alpha_1 L_{III} M_V$	2.4630	2	5.0337	2.3704	2	5.2304
$\beta_6 L_{III} N_I$	2.1906	4	5.660	2.1039	3	5.8930
$\beta_{2,15} L_{III} N_{IV,V}$	2.1194	4	5.850	2.0360	3	6.0894
$\beta_7 L_{III} O_I$	2.0919	4	5.927	2.0092	3	6.1708
$\beta_{10} L_I M_{IV}$	2.1071	4	5.884	2.0237	3	6.1265
$\beta_9 L_I M_V$	2.1004	4	5.903	2.0165	3	6.1484
$\gamma M_{III} N_{IV,V}$	10.998	9	1.1273	10.505	9	1.180
$\beta M_{IV} N_{VI}$	13.06	2	0.950	12.44	2	0.997
$\zeta M_V N_{III}$	17.38	4	0.714	16.46	4	0.753
$\alpha M_V N_{VI,VII}$	13.343	5	0.9292	12.68	2	0.978
$N_{IV,V} N_{VI,VII}$	113.	1	0.1095	107.	1	0.116
$N_{IV,V} O_{II,III}$	136.5	4	0.0908	128.9	7	0.0962
	61 Promethium			**62 Samarium**		
$\alpha_2 KL_{II}$	0.324803	4	38.1712	0.313698	2	39.5224
$\alpha_1 KL_{III}$	0.320160	4	38.7247	0.309040	2	40.1181
$\beta_3 KM_{II}$	0.28363†	4	43.713	0.27376	2	45.289
$\beta_1 KM_{III}$	0.28290†	3	43.826	0.27301	2	45.413
$\beta_2 KN_{II,III}$	0.2759†	1	44.94	0.2662	1	46.58
$KO_{II,III}$				0.26491	3	46.801
$\beta_5 KM_{IV,V}$				0.27111	3	45.731
$\beta_4 L_I M_{II}$				2.00095	6	6.1963
$\beta_3 L_I M_{III}$	2.0421	4	6.071	1.96241	3	6.3180
$\gamma_2 L_I N_{II}$				1.66044	6	7.4668
$\gamma_3 L_I N_{III}$				1.65601	3	7.4867
$\gamma_4 L_I O_{II,III}$				1.60728	3	7.7137
$\eta L_{II} M_I$				2.21824	3	5.5892
$\beta_1 L_{II} M_{IV}$	2.0797	4	5.961	1.99806	3	6.2051
$\gamma_5 L_{II} N_I$				1.77934	3	6.9678
$\gamma_1 L_{II} N_{IV}$	1.7989	9	6.892	1.72724	3	7.1780
$\gamma_8 L_{II} O_{IV}$				1.6966	9	7.3076
$\ell L_{III} M_I$				2.4823	4	4.9945

Desig-nation	Å*	p.e.	keV	Å*	p.e.	keV
	61 Promethium (*Cont.*)			**62 Samarium** (*Cont.*)		
$\alpha_2 L_{III} M_{IV}$	2.2926	4	5.4078	2.21062	3	5.6084
$\alpha_1 L_{III} M_V$	2.2822	3	5.4325	2.1998	2	5.6361
$\beta_6 L_{III} N_I$				1.94643	3	6.3697
$\beta_{2,15} L_{III} N_{IV,V}$	1.9559	6	6.339	1.88221	3	6.5870
$\beta_7 L_{III} O_I$				1.85626	3	6.6791
$\beta_5 L_{III} O_{IV,V}$				1.84700	9	6.7126
$\beta_{10} L_I M_{IV}$				1.86990	3	6.6304
$\beta_9 L_I M_V$				1.86166	3	6.6597
$\gamma M_{III} N_{IV,V}$				9.600	9	1.291
$\beta M_{IV} N_{VI}$				11.27	1	1.0998
$\zeta M_V N_{III}$				14.91	4	0.831
$\alpha M_V N_{VI,VII}$				11.47	3	1.081
$N_{IV,V} N_{VI,VII}$				98.	1	0.126
$N_{IV,V} O_{II,III}$				117.4	4	0.1056
	63 Europium			**64 Gadolinium**		
$\alpha_2 KL_{II}$	0.303118	2	40.9019	0.293038	2	42.3089
$\alpha_1 KL_{III}$	0.298446	2	41.5422	0.288353	2	42.9962
$\beta_3 KM_{II}$	0.264332	5	46.9036	0.25534	2	48.555
$\beta_1 KM_{III}$	0.263577	5	47.0379	0.25460	2	48.697
$\beta_2 KN_{II,III}$	0.256923	8	48.256	0.24816	3	49.959
$KO_{II,III}$	0.255645	7	48.497	0.24687	3	50.221
$\beta_5 KM_{IV,V}$				0.25275	3	49.052
$\beta_4 L_I M_{II}$	1.9255	2	6.4389	1.8540	2	6.6871
$\beta_3 L_I M_{III}$	1.8867	2	6.5713	1.8150	2	6.8311
$\gamma_2 L_I N_{II}$	1.5961	2	7.7677	1.5331	2	8.087
$\gamma_3 L_I N_{III}$	1.5903	2	7.7961	1.5297	2	8.105
$\gamma_4 L_I O_{II,III}$	1.5439	1	8.0304	1.4839	2	8.355
$\eta L_{II} M_I$	2.1315	2	5.8166	2.0494	1	6.0495
$\beta_1 L_{II} M_{IV}$	1.9203	2	6.4564	1.8468	2	6.7132
$\gamma_5 L_{II} N_I$	1.7085	2	7.2566	1.6412	2	7.5543
$\gamma_1 L_{II} N_{IV}$	1.6574	2	7.4803	1.5924	2	7.7858
$\gamma_8 L_{II} O_I$	1.6346	2	7.5849	1.5707	2	7.894
$\gamma_6 L_{II} O_{IV}$	1.6282	2	7.6147	1.5644	2	7.925
$\ell L_{III} M_I$	2.3948	2	5.1772	2.3122	2	5.3621
$\alpha_2 L_{III} M_{IV}$	2.1315	2	5.8166	2.0578	2	6.0250
$\alpha_1 L_{III} M_V$	2.1209	2	5.8457	2.0468	2	6.0572
$\beta_6 L_{III} N_I$	1.8737	2	6.6170	1.8054	2	6.8671
$\beta_{2,15} L_{III} N_{IV,V}$	1.8118	2	6.8432	1.7455	2	7.1028
$\beta_7 L_{III} O_I$	1.7851	2	6.9453	1.7203	2	7.2071
$\beta_5 L_{III} O_{IV,V}$	1.7772	2	6.9763	1.7130	2	7.2374
$\beta_{10} L_I M_{IV}$	1.7993	3	6.890	1.7315	3	7.160
$\beta_9 L_I M_V$	1.7916	3	6.920	1.7240	3	7.192
$L_I O_{IV,V}$				1.4807	3	8.373
$\gamma M_{III} N_{IV,V}$	9.211	9	1.346	8.844	9	1.402
$\beta M_{IV} N_{VI}$	10.750	7	1.1533	10.254	6	1.2091
$\zeta M_V N_{III}$	14.22	2	0.872	13.57	2	0.914
$\alpha M_V N_{VI,VII}$	10.96	3	1.131	10.46	3	1.185
$N_{IV,V} O_{II,III}$	112.0	6	0.1107			
	65 Terbium			**66 Dysprosium**		
$\alpha_2 KL_{II}$	0.283423	2	43.7441	0.274247	2	45.2078
$\alpha_1 KL_{III}$	0.278724	2	44.4816	0.269533	2	45.9984
βKM_{II}	0.24683	2	50.229	0.23862	2	51.957

TABLE 1b4. (*Continued*)

Designation	Å*	p.e.	keV	Å*	p.e.	keV
65 Terbium (*Cont.*)				**66 Dysprosium** (*Cont.*)		
$\beta_1\,KM_{III}$	0.24608	2	50.382	0.23788	2	52.119
$\beta_2\,KN_{II,III}$	0.2397†	2	51.72	0.2317†	2	53.51
$KO_{II,III}$	0.23858	3	51.965	0.23056	3	53.774
$\beta_5\,KM_{IV,V}$				0.23618	3	52.494
$\beta_4\,L_IM_{II}$	1.7864	2	6.9403	1.72103	7	7.2039
$\beta_3\,L_IM_{III}$	1.7472	2	7.0959	1.6822	2	7.3702
$\gamma_2\,L_IN_{II}$	1.4764	2	8.398	1.42278	7	8.7140
$\gamma_3\,L_IN_{III}$	1.4718	2	8.423	1.41640	7	8.7532
$\gamma_4\,L_IO_{II,III}$	1.4276	2	8.685	1.37459	7	9.0195
$\eta\,L_{II}M_I$	1.9730	2	6.2839	1.89743	7	6.5342
$\beta_1\,L_{II}M_{IV}$	1.7768	3	6.978	1.71062	7	7.2477
$\gamma_5\,L_{II}N_I$	1.5787	2	7.8535	1.51824	7	8.1661
$\gamma_1\,L_{II}N_{IV}$	1.5303	2	8.102	1.47266	7	8.4188
$\gamma_8\,L_{II}O_I$	1.5097	2	8.212			
$\gamma_6\,L_{II}O_{IV}$	1.5035	2	8.246	1.44579	7	8.5753
$l\,L_{III}M_I$	2.2352	2	5.5467	2.15877	7	5.7431
$\alpha_2\,L_{III}M_{IV}$	1.9875	2	6.2380	1.91991	3	6.4577
$\alpha_1\,L_{III}M_V$	1.9765	2	6.2728	1.90881	4	6.4952
$\beta_6\,L_{III}N_I$	1.7422	2	7.1163	1.68213	7	7.3705
$\beta_{2,15}\,L_{III}N_{IV,V}$	1.6830	2	7.3667	1.62369	7	7.6357
$\beta_7\,L_{III}O_I$	1.6585	2	7.4753	1.60447	7	7.7272
$\beta_5\,L_{III}O_{IV,V}$	1.6510	2	7.5094	1.58837	7	7.8055
$\beta_{10}\,L_IM_{IV}$	1.6673	3	7.436	1.60743	9	7.7130
$\beta_9\,L_IM_V$				1.59973	9	7.7501
$L_IO_{IV,V}$	1.4228	3	8.714			
$\gamma\,M_{III}N_{IV,V}$	8.486	9	1.461	8.144	9	1.522
$\beta\,M_{IV}N_{VI}$	9.792	6	1.2661	9.357	6	1.3250
$\zeta\,M_VN_{III}$	12.98	2	0.955	12.43	2	0.998
$\alpha\,M_VN_{VI,VII}$	10.00	2	1.240	9.59	2	1.293
$N_{IV,V}N_{VI,VII}$	86.	1	0.144	83.	1	0.149
$N_{IV,V}O_{II,III}$	102.2	4	0.1213	97.2	8	0.128
67 Holmium				**68 Erbium**		
$\alpha_2\,KL_{II}$	0.265486	2	46.6997	0.257110	2	48.2211
$\alpha_1\,KL_{III}$	0.260756	2	47.5467	0.252365	2	49.1277
$\beta_3\,KM_{II}$	0.23083	2	53.711	0.22341	2	55.494
$\beta_1\,KM_{III}$	0.23012	2	53.877	0.22266	2	55.681
$\beta_2\,KN_{II,III}$	0.2241†	2	55.32	0.2167†	2	57.21
$KO_{II,III}$	0.22305	3	55.584	0.21581	3	57.450
$\beta_5\,KM_{IV,V}$	0.22855	3	54.246	0.22124	3	56.040
$\beta_4\,L_IM_{II}$	1.6595	2	7.4708	1.6007	1	7.7453
$\beta_3\,L_IM_{III}$	1.6203	2	7.6519	1.5616	1	7.9392
$\gamma_2\,L_IN_{II}$	1.3698	2	9.051	1.3210	2	9.385
$\gamma_3\,L_IN_{III}$	1.3643	2	9.087	1.3146	1	9.4309
$\gamma_4\,L_IO_{II,III}$	1.3225	2	9.374	1.2752	2	9.722
$\eta\,L_{II}M_I$	1.8264	2	6.7883	1.7566	1	7.0579
$\beta_1\,L_{II}M_{IV}$	1.6475	2	7.5253	1.5873	1	7.8109
$\gamma_5\,L_{II}N_I$	1.4618	2	8.481	1.4067	3	8.814
$\gamma_1\,L_{II}N_{IV}$	1.4174	2	8.747	1.3641	2	9.089
$\gamma_8\,L_{II}O_I$	1.3983	2	8.867			
$\gamma_6\,L_{II}O_{IV}$	1.3923	2	8.905	1.3397	3	9.255
$l\,L_{III}M_I$	2.0860	2	5.9434	2.015	1	6.152
$\alpha_2\,L_{III}M_{IV}$	1.8561	2	6.6795	1.7955	3	6.9050
$\alpha_1\,L_{III}M_V$	1.8450	2	6.7198	1.78425	9	6.9487
$\beta_6\,L_{III}N_I$	1.6237	2	7.6359	1.5675	2	7.909
$\beta_{2,15}\,L_{III}N_{IV,V}$	1.5671	2	7.911	1.51399	9	8.1890
$\beta_7\,L_{III}O_I$				1.4941	3	8.298
$\beta_5\,L_{III}O_{IV,V}$	1.5378	2	8.062	1.4848	3	8.350

Designation	Å*	p.e.	keV	Å*	p.e.	keV
67 Holmium (*Cont.*)				**68 Erbium** (*Cont.*)		
$\beta_{10}\,L_IM_{IV}$	1.5486	3	8.006	1.4941	3	8.298
$L_IO_{IV,V}$	1.3208	3	9.387			
$\beta_9\,L_IM_V$				1.4855	5	8.346
$M_{II}N_{IV}$				7.60	1	1.632
$\gamma\,M_{III}N_{IV,V}$	7.865	9	1.576			
$\gamma\,M_{III}$ v				7.546	8	1.643
$\beta\,M_{IV}N_{VI}$	8.965	4	1.3830	8.592	3	1.4430
$\zeta\,M_VN_{III}$	11.86	1	1.0450	11.37	1	1.0901
$\alpha\,M_VN_{VI,VII}$	9.20	2	1.348	8.82	1	1.406
$N_{IV}N_{VI}$				72.7	9	0.171
$N_VN_{VI,VII}$				76.3	7	0.163
69 Thulium				**70 Ytterbium**		
$\alpha_2\,KL_{II}$	0.249095	2	49.7726	0.241424	2	51.3540
$\alpha_1\,KL_{III}$	0.244338	2	50.7416	0.236655	2	52.3889
$\beta_3\,KM_{II}$	0.21636	2	57.304	0.2096†	1	59.14
$\beta_1\,KM_{III}$	0.21556	2	57.517	0.20884	2	59.37
$\beta_2\,KN_{II,III}$	0.2098†	2	59.09	0.2033†	2	60.98
$KO_{II,III}$	0.20891	2	59.346	0.20226	2	61.298
$\beta_5\,KM_{IV,V}$	0.21404	2	57.923	0.20739	2	59.782
$\beta_4\,L_IM_{II}$	1.5448	2	8.026	1.49138	3	8.3132
$\beta_3\,L_IM_{III}$	1.5063	2	8.231	1.45233	5	8.5367
$\gamma_2\,L_IN_{II}$	1.2742	2	9.730	1.22879	7	10.0897
$\gamma_3\,L_IN_{III}$	1.2678	2	9.779	1.22232	5	10.1431
$\gamma_4\,L_IO_{II,III}$	1.2294	2	10.084	1.1853	1	10.4603
$\eta\,L_{II}M_I$	1.6963	2	7.3088	1.63560	5	7.5802
$\beta_1\,L_{II}M_{IV}$	1.5304	2	8.101	1.47565	5	8.4018
$\gamma_5\,L_{II}N_I$	1.3558	2	9.144	1.3063	1	9.4910
$\gamma_1\,L_{II}N_{IV}$	1.3153	2	9.426	1.26769	5	9.8701
$\gamma_8\,L_{II}O_I$				1.24923	5	9.9246
$\gamma_6\,L_{II}O_{IV}$	1.2905	2	9.607	1.24271	3	9.9766
$l\,L_{III}M_I$	1.9550	2	6.3419	1.89415	5	6.5455
$\alpha_2\,L_{III}M_{IV}$	1.7381	2	7.1331	1.68285	5	7.3673
$\alpha_1\,L_{III}M_V$	1.7268†	2	7.1799	1.67189	4	7.4156
$\beta_6\,L_{III}N_I$	1.5162	2	8.177	1.4661	5	8.4563
$\beta_{2,15}\,L_{III}N_{IV,V}$	1.4640	2	8.468	1.41550	5	8.7588
$\beta_7\,L_{III}O_I$				1.3948	1	8.8889
$\beta_5\,L_{III}O_{IV,V}$	1.4349	2	8.641	1.38696	7	8.9390
$\beta_{10}\,L_IM_{IV}$	1.4410	3	8.604	1.3915	2	8.9100
$\beta_9\,L_IM_V$	1.4336	3	8.648	1.3838	1	8.9597
L_IO_I				1.1886	1	10.4312
$L_IO_{IV,V}$	1.2263	3	10.110	1.1827	1	10.4833
$L_{II}M_{II}$				1.58844	9	7.8052
$L_{II}O_{II,III}$				1.2453	1	9.9561
$t\,L_{III}M_{II}$				1.83091	9	6.7715
$L_{III}O_{II,III}$				1.3898	1	8.9209
$M_{III}N_I$				8.470	9	1.464
$\gamma\,M_{III}N_V$				7.024	8	1.765
$\beta\,M_{IV}N_{VI}$	8.249	7	1.503	7.909	2	1.5675
$\zeta\,M_VN_{III}$				10.48	1	1.183
$\alpha\,M_VN_{VI,VII}$	8.48	1	1.462	8.149	5	1.5214
$N_{IV}N_{VI}$				65.1	7	0.190
$N_VN_{VI,VII}$				69.3	5	0.179
71 Lutetium				**72 Hafnium**		
$\alpha_2\,KL_{II}$	0.234081	2	52.9650	0.227024	3	54.6114
$\alpha_1\,KL_{III}$	0.229298	2	54.0698	0.222227	3	55.7902
$\beta_3\,KM_{II}$	0.20309†	4	61.05	0.19686†	4	62.98

TABLE 1b4. (Continued)

71 Lutetium (Cont.) / 72 Hafnium (Cont.)

Designation	Å*	p.e.	keV	Å*	p.e.	keV
$\beta_1\ KM_{III}$	0.20231†	3	61.283	0.19607†	3	63.234
$\beta_2\ KN_{II,III}$	0.1969†	2	62.97	0.1908†	2	64.98
$KO_{II,III}$	0.19589	2	63.293			
$\beta_5\ KM_{IV,V}$	0.20084	2	61.732			
$\beta_4\ L_IM_{II}$	1.44056	5	8.6064	1.39220	5	8.9054
$\beta_3\ L_IM_{III}$	1.40140	5	8.8469	1.35300	5	9.1634
$\gamma_2\ L_IN_{II}$	1.1853	2	10.460	1.14442	5	10.8335
$\gamma_3\ L_IN_{III}$	1.17953	4	10.5110	1.13841	5	10.8907
$\gamma'_4\ L_IO_{II}$				1.10376	5	11.2326
$\gamma_4\ L_IO_{II,III}$	1.1435	1	10.8425	1.10303	5	11.2401
$\eta\ L_{II}M_I$	1.5779	1	7.8575	1.52325	5	8.1393
$\beta_1\ L_{II}M_{IV}$	1.42359	3	8.7090	1.37410	5	9.0227
$\gamma_5\ L_{II}N_I$	1.2596	1	9.8428	1.21537	5	10.2011
$\gamma_1\ L_{II}N_{IV}$	1.22228	4	10.1434	1.17900	5	10.5158
$\gamma_8\ L_{II}O_I$	1.2047	1	10.2915	1.16138	5	10.6754
$\gamma_6\ L_{II}O_{III}$	1.1987	1	10.3431	1.15519	5	10.7325
$l\ L_{III}M_I$	1.8360	1	6.7528	1.78145	5	6.9596
$\alpha_2\ L_{III}M_{IV}$	1.63029	5	7.6049	1.58046	5	7.8446
$\alpha_1\ L_{III}M_{IV}$	1.61951	3	7.6555	1.56958	5	7.8990
$\beta_6\ L_{III}N_I$	1.4189	1	8.7376	1.37410	5	9.0227
$\beta_{15}\ L_{III}N_{IV}$	1.3715	1	9.0395	1.32783	5	9.3371
$\beta_2\ L_{III}N_V$	1.37012	3	9.0489	1.32639	5	9.3473
$\beta_7\ L_{III}O_I$	1.34949	5	9.1873	1.30564	5	9.4958
$\beta_5\ L_{III}O_{IV,V}$	1.34183	7	9.2397	1.29761	5	9.5546
				1.43025	9	8.6685
$\beta_{10}\ L_IM_{IV}$	1.3430	2	9.232	1.29819	9	9.5503
$\beta_9\ L_IM_V$	1.3358	1	9.2816	1.29025	9	9.6090
L_IN_{IV}	1.16227	9	10.6672	1.12250	9	11.0451
$\gamma_{11}\ L_IN_V$	1.16107	9	10.6782	1.12146	9	11.0553
L_IO_I				1.10664	9	11.2034
L_IO_{IV}				1.10086	9	11.2622
$L_{II}M_{II}$	1.53333	9	8.0858	1.48064	9	8.3735
$\beta_{17}\ L_{II}M_{III}$				1.43643	9	8.6312
$L_{II}N_V$				1.17788	9	10.5258
$v\ L_{II}N_{VI}$				1.15830	9	10.7037
$L_{II}O_{II,III}$	1.2014	1	10.3198			
$t\ L_{III}M_{II}$	1.7760	1	6.9810	1.72305	9	7.1954
$s\ L_{III}M_{III}$				1.66346	9	7.4532
$L_{III}N_{II}$				1.35887	9	9.1239
$L_{III}N_{III}$				1.35053	9	9.1802
$u\ L_{III}N_{VI,VII}$				1.30165	9	9.5249
$L_{III}O_{II,III}$	1.34524	9	9.2163			
$M_{III}N_I$						
$\gamma\ M_{III}N_V$	6.768	6	1.832	6.544	4	1.895
ζ_2				9.686	7	1.2800
$\beta\ M_{IV}N_{VI}$	7.601	2	1.6312	7.303	1	1.6976
ζ_1				9.686	7	1.2800
$\alpha\ M_VN_{VI,VII}$	7.840	2	1.5813	7.539	1	1.6446
$N_{IV}N_{VI}$	63.0	5	0.197			
$N_VN_{VI,VII}$	65.7	2	0.1886			

73 Tantalum / 74 Tungsten

Designation	Å*	p.e.	keV	Å*	p.e.	keV
$\alpha_2\ KL_{II}$	0.220305	8	56.277	0.213828	2	57.9817
$\alpha_1\ KL_{III}$	0.215497	4	57.532	0.2090100	Std	59.31824
$\beta_3\ KM_{II}$	0.190890	2	64.9488	0.185181	2	66.9514
$\beta_1\ KM_{III}$	0.190089	4	65.223	0.184374	2	67.2443
$\beta_2^{II}\ KN_{II}$	0.185188	9	66.949	0.17960	1	69.031
$\beta_2^{I}\ KN_{III}$	0.185011	8	67.013	0.179421	7	69.101

73 Tantalum (Cont.) / 74 Tungsten (Cont.)

Designation	Å*	p.e.	keV	Å*	p.e.	keV
$KO_{II,III}$	0.184031	7	67.370	0.178444	5	69.479
KL_I				0.21592	4	57.42
$\beta_5^{II}\ KM_{IV}$	0.188920	6	65.626	0.183264	5	67.652
$\beta_5^{I}\ KM_V$	0.188757	6	65.683	0.183092	7	67.715
$\beta_4\ KN_{IV,V}$	0.18451	1	67.194	0.17892	2	69.294
$\beta_4\ L_IM_{II}$	1.34581	3	9.2124	1.30162	5	9.5252
$\beta_3\ L_IM_{III}$	1.30678	3	9.4875	1.26269	5	9.8188
$\gamma_2\ L_IN_{II}$	1.1053	1	11.217	1.06806	3	11.6080
$\gamma_3\ L_IN_{III}$	1.09936	4	11.2776	1.06200	3	11.6743
$\gamma'_4\ L_IO_{II}$	1.06544	3	11.6366	1.02863	3	12.0530
$\gamma_4\ L_IO_{III}$	1.06467	3	11.6451	1.02775	3	12.0634
$\eta\ L_{II}M_I$	1.47106	5	8.4280	1.42110	3	8.7243
$\beta_1\ L_{II}M_{IV}$	1.32698	3	9.3431	1.281809	9	9.67235
$\gamma_5\ L_{II}N_I$	1.1729	1	10.5702	1.13235	3	10.9490
$\gamma_1\ L_{II}N_{IV}$	1.13794	3	10.8952	1.09855	3	11.2859
$\gamma_8\ L_{II}O_I$	1.1205	1	11.0646	1.08113	4	11.4677
$\gamma_6\ L_{II}O_{IV}$	1.11388	3	11.1306	1.07448	5	11.5387
$l\ L_{III}M_I$	1.72841	5	7.1731	1.6782	1	7.3878
$\alpha_2\ L_{III}M_{IV}$	1.53293	2	8.0879	1.48743	2	8.3352
$\alpha_1\ L_{III}M_V$	1.52197	2	8.1461	1.47639	2	8.3976
$\beta_6\ L_{III}N_I$	1.33094	8	9.3153	1.28989	7	9.6117
$\beta_{15}\ L_{III}N_{IV}$	1.28619	5	9.6394	1.24631	3	9.9478
$\beta_2\ L_{III}N_V$	1.28454	2	9.6518	1.24460	3	9.9615
$\beta_7\ L_{III}O_I$	1.26385	5	9.8098	1.22400	4	10.1292
$\beta_5\ L_{III}O_{IV,V}$	1.2555	1	9.8750	1.21545	3	10.2004
L_IM_I				1.3365	3	9.277
$\beta_{10}\ L_IM_{IV}$	1.2537	2	9.889	1.21218	3	10.2279
$\beta_9\ L_IM_V$	1.2466	2	9.946	1.20479	7	10.2907
L_IN_I	1.11521	9	11.1173			
L_IN_{IV}	1.08377	7	11.4398	1.0468	2	11.844
$\gamma_{11}\ L_IN_V$	1.08205	7	11.4580	1.0458	1	11.856
$L_IN_{VI,VII}$	1.06357	9	11.6570			
L_IO_I	1.06771	9	11.6118	1.0317	3	12.017
$L_IO_{IV,V}$	1.06192	9	11.6752	1.0250	2	12.095
$L_{II}M_{II}$	1.43048	9	8.6671			
$\beta_{17}\ L_{II}M_{III}$	1.3864	1	8.9428	1.3387	2	9.261
$L_{II}M_V$	1.31897	9	9.3998	1.2728	2	9.741
$L_{II}N_{II}$	1.1600	2	10.688	1.1218	3	11.052
$L_{II}N_{III}$	1.1553	1	10.7316	1.1149	2	11.120
$L_{II}N_V$	1.13687	9	10.9055			
$v\ L_{II}N_{VI}$	1.1158	1	11.1113	1.0771	1	11.510
$L_{II}O_{II}$	1.11789	9	11.0907			
$L_{II}O_{III}$	1.11693	9	11.1001	1.0792	2	11.488
$t\ L_{III}M_{II}$	1.67265	9	7.4123	1.6244	3	7.632
$s\ L_{III}M_{III}$	1.61264	9	7.6881	1.5642	3	7.926
$L_{III}N_{II}$	1.3167	1	9.4158	1.2765	2	9.712
$L_{III}N_{III}$	1.3086	1	9.4742	1.2672	2	9.784
$u\ L_{III}N_{VI,VII}$	1.25778	4	9.8572	1.21868	5	10.1733
$L_{III}O_{II,III}$	1.2601	3	9.839	1.2211	2	10.153
M_IN_{III}	5.40	2	2.295	5.172	9	2.397
$M_IO_{II,III}$				4.44	2	2.79
$M_{II}N_I$				6.28	2	1.973
$M_{II}N_{IV}$	5.570	4	2.226	5.357	4	2.314
$M_{III}N_I$	7.612	9	1.629	7.360	8	1.684
$M_{III}N_{IV}$	6.353	5	1.951	6.134	4	2.021
$\gamma\ M_{III}N_V$	6.312	4	1.964	6.092	3	2.035
$M_{III}O_I$	5.83	2	2.126	5.628	8	2.203
$M_{III}O_{IV,V}$	5.67	3	2.19			

TABLE 1b4. (*Continued*)

Designation	Å*	p.e.	keV	Å*	p.e.	keV
	73 Tantalum (*Cont.*)			**74 Tungsten** (*Cont.*)		
$\zeta_2\ M_{IV}N_{II}$	9.330	5	1.3288	8.993	5	1.3787
$M_{IV}N_{III}$	8.90	2	1.393	8.573	8	1.446
$\beta\ M_{IV}N_{VI}$	7.023	1	1.7655	6.757	1	1.8349
$M_{IV}O_{II}$	7.09		1.748	6.806	9	1.822
$\zeta_1\ M_V N_{III}$	9.316	4	1.3308	8.962	4	1.3835
$\alpha\ M_V N_{VI,VII}$	7.252	1	1.7096			
$\alpha_2\ M_V N_{VI}$				6.992	2	1.7731
$\alpha_1\ M_V N_{VII}$				6.983	1	1.7754
$M_V O_{III}$	7.30	2	1.700	7.005	9	1.770
$N_{II}N_{IV}$				54.0	2	0.2295
$N_{IV}N_{VI}$	58.2	1	0.2130	55.8	1	0.2221
$N_V N_{VI,VII}$	61.1	2	0.2028			
$N_V N_{VI}$				59.5	3	0.208
$N_V N_{VII}$				58.4	1	0.2122
	75 Rhenium			**76 Osmium**		
$\alpha_2\ KL_{II}$	0.207611	1	59.7179	0.201639	2	61.4867
$\alpha_1\ KL_{III}$	0.202781	2	61.1403	0.196794	2	63.0005
$\beta_3\ KM_{II}$	0.179697	3	68.994	0.174431	3	71.077
$\beta_1\ KM_{III}$	0.178880	3	69.310	0.173611	3	71.413
$\beta_2^{II}\ KN_{II}$	0.17425	1	71.151	0.16910	1	73.318
$\beta_2^{I}\ KN_{III}$	0.174054	6	71.232	0.168906	6	73.402
$KO_{II,III}$	0.17308	1	71.633	0.16798	1	73.808
$\beta_5^{II}\ KM_{IV}$	0.17783	1	69.719	0.17262	1	71.824
$\beta_5^{I}\ KM_V$	0.17766	1	69.786	0.17245	1	71.895
$\beta_4\ KN_{IV,V}$	0.17362	2	71.410	0.16842	2	73.615
$\beta_4\ L_I M_{II}$	1.25917	5	9.8463	1.21844	5	10.1754
$\beta_3\ L_I M_{III}$	1.22031	5	10.1598	1.17955	7	10.5108
$\gamma_2\ L_I N_{II}$	1.03233	5	12.0098	0.99805	5	12.4224
$\gamma_3\ L_I N_{III}$	1.02613	7	12.0824	0.99186	5	12.4998
$\gamma'_4\ L_I O_{III}$	0.99334	5	12.4813	0.96033	8	12.910
$\gamma_4\ L_I O_{III}$	0.99249	5	12.4920	0.95938	8	12.923
$\eta\ L_{II}M_I$	1.37342	5	9.0272	1.32785	7	9.3370
$\beta_1\ L_{II}M_{IV}$	1.23858	2	10.0100	1.19727	7	10.3553
$\gamma_5\ L_{II}N_I$	1.09388	5	11.3341	1.05693	5	11.7303
$\gamma_1\ L_{II}N_{IV}$	1.06099	5	11.6854	1.02503	5	12.0953
$\gamma_8\ L_{II}O_I$	1.04398	5	11.8758	1.00788	5	12.3012
$\gamma_6\ L_{II}O_{IV}$	1.03699	9	11.956	1.00107	5	12.3848
$l\ L_{III}M_I$	1.63056	5	7.6036	1.58498	7	7.8222
$\alpha_2\ L_{III}M_{IV}$	1.44396	5	8.5862	1.40234	5	8.8410
$\alpha_1\ L_{III}M_V$	1.43290	4	8.6525	1.39121	5	8.9117
$\beta_6\ L_{III}N_I$	1.25100	5	9.9105	1.21349	5	10.2169
$\beta_{15}\ L_{III}N_{IV}$	1.20819	5	10.2617	1.17167	5	10.5816
$\beta_2\ L_{III}N_V$	1.20660	5	10.2752	1.16979	8	10.5985
$\beta_7\ L_{III}O_I$	1.18610	5	10.4529	1.14933	8	10.7872
$\beta_5\ L_{III}O_{IV,V}$	1.17721	5	10.5318	1.1405	1	10.8711
$\beta_{10}\ L_I M_{IV}$	1.17218	5	10.5770	1.13353	5	10.9376
$\beta_9\ L_I M_V$	1.16487	4	10.6433	1.12637	6	11.0071
$L_I N_I$	1.0420	1	11.899			
$L_I N_{IV}$	1.0119	1	12.252	0.9772	3	12.687
$\gamma_{11}\ L_I N_V$	1.0108	1	12.266	0.9765	3	12.696
$L_I O_I$	0.9965	1	12.442	0.96318	7	12.8721
$L_I O_{IV,V}$	0.9900	1	12.524	0.95603	5	12.9683
$L_{II}M_{II}$	1.3366	1	9.2761	1.2934	2	9.586
$\beta_{17}\ L_{II}M_{III}$	1.2927	1	9.5910	1.2480	2	9.934

Designation	Å*	p.e.	keV	Å*	p.e.	keV
	75 Rhenium (*Cont.*)			**76 Osmium** (*Cont.*)		
$L_{II}M_V$	1.2305	1	10.0753	1.18977	7	10.4205
$L_{II}N_{II}$	1.0839	1	11.438			
$L_{II}N_{III}$	1.0767	1	11.515	1.03973	5	11.9243
$v\ L_{II}N_{VI}$	1.0404	1	11.917	1.0050	2	12.337
$L_{II}O_{III}$	1.0397	1	11.925	1.0047	2	12.340
$t\ L_{III}M_{II}$	1.5789	1	7.8525	1.5347	2	8.079
$s\ L_{III}M_{III}$	1.5178	1	8.1682	1.4735	2	8.414
$L_{III}N_I$				1.20086	7	10.3244
$L_{III}N_{III}$	1.2283	1	10.0933			
$u\ L_{III}N_{VI,VII}$	1.1815	1	10.4931	1.14537	7	10.8245
$M_I N_{III}$				4.79	2	2.59
$M_{II}N_I$				5.81	2	2.133
$M_{II}N_{IV}$				4.955	4	2.502
$M_{III}N_I$				6.89	2	1.798
$M_{III}N_{IV}$	5.931	5	2.090	5.724	5	2.166
$\gamma\ M_{III}N_V$	5.885	2	2.1067	5.682	4	2.182
$\zeta_2\ M_{IV}N_{II}$	8.664	5	1.4310	8.359	5	1.4831
$M_{IV}N_{III}$	8.239	1	1.505			
$\beta\ M_{IV}N_{VI}$	6.504	1	1.9061	6.267	1	1.9783
$\zeta_1\ M_V N_{III}$	8.629	4	1.4368	8.310	4	1.4919
$\alpha\ M_V N_{VI,VII}$	6.729	1	1.8425	6.490	1	1.9102
$N_{IV}N_{VI}$				51.9	1	0.2388
$N_V N_{VI,VII}$				54.7	2	0.2266
	77 Iridium			**78 Platinum**		
$\alpha_2\ KL_{II}$	0.195904	2	63.2867	0.190381	4	65.122
$\alpha_1\ KL_{III}$	0.191047	2	64.8956	0.185511	4	66.832
$\beta_3\ KM_{II}$	0.169367	2	73.2027	0.164501	3	75.368
$\beta_1\ KM_{III}$	0.168542	2	73.5608	0.163675	3	75.748
$\beta_2^{II}\ KN_{II}$	0.16415	1	75.529	0.15939	1	77.785
$\beta_2^{I}\ KN_{III}$	0.163956	7	75.619	0.15920	1	77.878
$KO_{II,III}$	0.163019	1	76.053	0.15826	1	78.341
$\beta_5^{II}\ KM_{IV}$	0.16759	2	73.980	0.16271	2	76.199
$\beta_5^{I}\ KM_V$	0.167373	9	74.075	0.16255	3	76.27
$\beta_4\ KN_{IV,V}$	0.16352	2	75.821	0.15881	2	78.069
$\beta_4\ L_I M_{III}$	1.17795	3	10.5106	1.14223	5	10.8543
$\beta_1\ L_I M_{III}$	1.14085	3	10.8674	1.10394	5	11.2308
$\gamma_2\ L_I N_{II}$	0.96545	3	12.8418	0.93427	5	13.2704
$\gamma_3\ L_I N_{III}$	0.95931	5	12.9240	0.92791	5	13.3613
$\gamma'_4\ L_I O_{II}$	0.92831	3	13.3555	0.89747	4	13.8145
$\gamma_4\ L_I O_{III}$	0.92744	3	13.3681	0.89659	4	13.8281
$\eta\ L_{II}M_I$	1.28448	3	9.6522	1.2429	2	9.975
$\beta_1\ L_{II}M_{IV}$	1.15781	3	10.7083	1.11990	2	11.0707
$\gamma_5\ L_{II}N_I$	1.02175	5	12.1342	0.9877	1	12.552
$\gamma_1\ L_{II}N_{IV}$	0.99085	3	12.5126	0.95797	3	12.9420
$\gamma_8\ L_{II}O_I$	0.97409	3	12.7279	0.9411	1	13.173
$\gamma_6\ L_{II}O_{IV}$	0.96708	4	12.8201	0.9342	2	13.271
$l\ L_{III}M_I$	1.54094	3	8.0458	1.4995	2	8.268
$\alpha_2\ L_{III}M_{IV}$	1.36250	5	9.0995	1.32432	2	9.3618
$\alpha_1\ L_{III}M_V$	1.35128	3	9.1751	1.31304	2	9.4423
$\beta_6\ L_{III}N_I$	1.17796	3	10.5251	1.14355	5	10.8418
$\beta_{15}\ L_{III}N_{IV}$	1.13707	3	10.9036			
$\beta_2\ L_{III}N_V$	1.13532	3	10.9203	1.10200	3	11.2505
$\beta_7\ L_{III}O_I$	1.11489	3	11.1205	1.08168	3	11.4619

TABLE 1b4. (Continued)

Desig-nation	Å*	p.e.	keV	Å*	p.e.	keV
	77 Iridium (*Cont.*)			**78 Platinum** (*Cont.*)		
$\beta_6\,L_{III}O_{IV,V}$	1.10585	3	11.2114	1.0724	2	11.561
L_IM_I	1.2102	2	10.245	1.16962	9	10.6001
$\beta_{10}\,L_IM_{IV}$	1.09702	4	11.3016	1.06183	7	11.6762
$\beta_9\,L_IM_V$	1.08975	5	11.3770	1.05446	5	11.7877
L_IN_I	0.9766	2	12.695	0.9455	2	13.113
L_IN_{IV}	0.9459	2	13.108			
$\gamma_{11}\,L_IN_V$	0.9446	2	13.126	0.9143	2	13.560
$L_IO_{IV,V}$	0.9243	3	13.413			
L_IO_I				0.8995	2	13.784
L_IO_{IV}				0.8943	1	13.864
L_IO_V				0.8934	1	13.878
$L_{II}M_{II}$	1.2502	3	9.917	1.213	1	10.225
$\beta_{17}\,L_{II}M_{III}$	1.2069	2	10.273	1.1667	1	10.6265
$L_{II}M_V$	1.1489	2	10.791	1.1129	2	11.140
$L_{II}N_{II}$	1.0120	2	12.251	0.9792	2	12.661
$L_{II}N_{III}$	1.0054	3	12.332	0.97173	4	12.7588
$v\,L_{II}N_{VI}$	0.97161	6	12.7603	0.93931	5	13.1992
$L_{II}O_{III}$	0.96979	5	12.7843			
$t\,L_{III}M_{II}$	1.4930	3	8.304	1.4530	2	8.533
$s\,L_{III}M_{III}$	1.4318	2	8.659	1.3895	2	8.923
$L_{III}N_{II}$	1.16545	5	10.6380	1.1310	2	10.962
$L_{III}N_{III}$	1.1560	3	10.725	1.1226	2	11.044
$u\,L_{III}N_{VI,VII}$	1.11145	4	11.1549	1.07896	5	11.4908
$L_{III}O_{II,III}$	1.10923	6	11.1772	1.0761	3	11.521
M_IM_{II}	4.631†	9	2.677	4.460	9	2.780
$M_{II}N_{IV}$	4.780	4	2.594	4.601	4	2.695
$M_{III}N_I$	6.669	9	1.859	6.455	9	1.921
$M_{III}N_{IV}$	5.540	5	2.238	5.357	5	2.314
$\gamma\,M_{II}N_V$	5.500	4	2.254	5.319	4	2.331
$M_{III}O_I$				4.876	9	2.543
$M_{III}O_{IV\,V}$	4.869	9	2.546	4.694	8	2.641
$\zeta_2\,M_{IV}N_{II}$	8.065	5	1.5373	7.790	5	1.592
$M_{IV}N_{III}$	7.645	8	1.622	7.371	8	1.682
$\beta\,M_{IV}N_{VI}$	6.038	1	2.0535	5.828	1	2.1273
$\zeta_1\,M_VN_{III}$	8.021	4	1.5458	7.738	4	1.6022
$\alpha_2\,M_VN_{VI}$	6.275	3	1.9758	6.058	3	2.047
$\alpha_1\,M_VN_{VII}$	6.262	1	1.9799	6.047	1	2.0505
M_VO_{III}				5.987	9	2.071
$N_{IV}N_{VI}$	50.2	1	0.2470	48.1	2	0.258
$N_VN_{VI,VII}$	52.8	1	0.2348	50.9	1	0.2436
	79 Gold			**80 Mercury**		
$\alpha_2\,K\,L_{II}$	0.185075	2	66.9895	0.179958	3	68.895
$\alpha_1\,K\,L_{III}$	0.180195	2	68.8037	0.175068	3	70.819
$\beta_3\,K\,M_{II}$	0.159810	2	77.580	0.155321	3	79.822
$\beta_1\,K\,M_{III}$	0.158982	3	77.984	0.154487	3	80.253
$\beta_2^{II}\,K\,N_{II}$	0.15483	2	80.08	0.15040	2	82.43
$\beta_2^{I}\,K\,N_{III}$	0.154618	9	80.185	0.15020	2	82.54
$K\,O_{II,III}$	0.153694	7	80.667	0.14931	2	83.04
$K\,L_I$	0.18672	4	66.40			
$\beta_5^{II}\,K\,M_{IV}$	0.158062	7	78.438			
$\beta_5^{I}\,K\,M_V$	0.157880	5	78.529			
$\beta_5\,K\,M_{IV,V}$				0.15353	2	80.75
$\beta_4\,K\,N_{IV,V}$	0.154224	5	80.391	0.14978	2	82.78
$\beta_4\,L_IM_{II}$	1.10651	3	11.2047	1.07222	7	11.5630
$\beta_3\,L_IM_{III}$	1.06785	9	11.6103	1.03358	7	11.9953

Desig-nation	Å*	p.e.	keV	Å*	p.e.	keV
	79 Gold (*Cont.*)			**80 Mercury** (*Cont.*)		
$\gamma_2\,L_IN_{II}$	0.90434	3	13.7095	0.87544	7	14.162
$\gamma_3\,L_IN_{III}$	0.89783	5	13.8090	0.86915	7	14.265
$\gamma'\,L_IO_{II}$	0.86816	4	14.2809	0.84013	7	14.757
$\gamma_4\,L_IO_{III}$	0.86703	4	14.2996	0.83894	7	14.778
$\eta\,L_{II}M_I$	1.20273	3	10.3083	1.1640	1	10.6512
$\beta_1\,L_{II}M_{IV}$	1.08353	3	11.4423	1.04868	5	11.8226
$\gamma_5\,L_{II}N_I$	0.95559	3	12.9743	0.92453	7	13.410
$\gamma_1\,L_{II}N_{IV}$	0.92650	3	13.3817	0.89646	5	13.8301
$\gamma_8\,L_{II}O_I$	0.90989	5	13.6260	0.87995	7	14.090
$\gamma_6\,L_{II}O_{IV}$	0.90297	3	13.7304	0.87319	7	14.199
$l\,L_{III}M_I$	1.45964	9	8.4939	1.4216	1	8.7210
$\alpha_2\,L_{III}M_{IV}$	1.28772	3	9.6280	1.25264	7	9.8976
$\alpha_1\,L_{III}M_V$	1.27640	3	9.7133	1.24120	5	9.9888
$\beta_6\,L_{III}N_I$	1.11092	3	11.1602	1.07975	7	11.4824
$\beta_{15}\,L_{III}N_{IV}$	1.07188	5	11.5667	1.04151	7	11.9040
$\beta_2\,L_{III}N_V$	1.07022	3	11.5847	1.03975	7	11.9241
$\beta_7\,L_{III}O_I$	1.04974	8	11.8106	1.01937	7	12.1625
$\beta_6\,L_{III}O_{IV,V}$	1.04044	3	11.9163	1.00987	7	12.2769
L_IM_I	1.13525	5	10.9210	1.0999	2	11.272
$\beta_{10}\,L_IM_{IV}$	1.02789	7	12.0617	0.9962	2	12.446
$\beta_9\,L_IM_V$	1.02063	7	12.1474	0.9871	2	12.560
L_IN_I	0.9131	1	13.578	0.8827	2	14.045
L_IN_{IV}	0.88563	7	13.999			
$\gamma_{11}\,L_IN_V$	0.88433	7	14.020	0.85657	7	14.474
L_IO_I	0.87074	5	14.2385	0.8452	2	14.670
$L_IO_{IV,V}$	0.86400	5	14.3497	0.8350	2	14.847
$L_{II}M_{II}$	1.1708	1	10.5892	1.1387	5	10.888
$\beta_{17}\,L_{II}M_{III}$	1.12798	5	10.9915	1.0916	5	11.358
$L_{II}M_V$	1.0756	2	11.526			
$L_{II}N_{III}$	0.9402	2	13.186	0.90894	7	13.640
$v\,L_{II}N_{VI}$	0.90837	5	13.6487	0.87885	7	14.107
$L_{II}O_I$	0.90746	7	13.662	0.8784	1	14.114
$L_{II}O_{III}$	0.90638	7	13.679	0.8758	1	14.156
$t\,L_{III}M_{II}$	1.41366	7	8.7702	1.3746	2	9.019
$s\,L_{III}M_{III}$	1.35131	7	9.1749	1.3112	2	9.455
$L_{III}N_{II}$	1.09968	7	11.2743	1.0649	2	11.642
$L_{III}N_{III}$	1.09026	7	11.3717	1.0585	1	11.713
$u\,L_{III}N_{VI,VII}$	1.04752	5	11.8357			
$u'\,L_{III}N_{VI}$				1.01769	7	12.1826
$u\,L_{III}N_{VII}$				1.01674	7	12.1940
$L_{III}O_{II,III}$	1.0450	2	11.865			
$L_{III}O_{II}$				1.01558	7	12.2079
$L_{III}O_{III}$				1.01404	7	12.2264
$L_{III}P_{II,III}$	1.03876	7	11.9355			
M_IN_{III}	4.300	9	2.883			
$M_{II}N_{IV}$	4.432	4	2.797			
$M_{III}N_I$	6.259	9	1.981	6.09	2	2.036
$M_{III}N_{IV}$	5.186	5	2.391			
$\gamma\,M_{III}N_V$	5.145	4	2.410	4.984†	2	2.4875
$M_{III}O_I$	4.703	9	2.636			
$M_{III}O_{IV,V}$	4.522	6	2.742			
$\zeta_2\,M_{IV}N_{II}$	7.523	5	1.648			
$M_{IV}N_{III}$	7.101	8	1.746	6.87	2	1.805
$\beta\,M_{IV}N_{VI}$	5.624	1	2.2046	5.4318†	9	2.2825
$\zeta_1\,M_VN_{III}$	7.466	4	1.6605			
$\alpha_2\,M_VN_{VI}$	5.854	3	2.118			

TABLE 1b4. (Continued)

Designation	Å*	p.e.	keV	Å*	p.e.	keV
	79 Gold (Cont.)			**80 Mercury** (Cont.)		
$\alpha_1\ M_V N_{VII}$	5.840	1	2.1229	5.6476†	9	2.1953
$M_V O_{III}$	5.767	9	2.150			
$N_{IV} N_{VI}$	46.8	2	0.265	45.2†	3	0.274
$N_V N_{VI,VII}$	49.4	1	0.2510	47.9†	3	0.259
	81 Thallium			**82 Lead**		
$\alpha_2\ K L_{II}$	0.175036	2	70.8319	0.170294	2	72.8042
$\alpha_1\ K L_{III}$	0.170136	2	72.8715	0.165376	2	74.9694
$\beta_3\ K M_{II}$	0.150980	6	82.118	0.146810	4	84.450
$\beta_1\ K M_{III}$	0.150142	5	82.576	0.145970	6	84.936
$\beta_2{}^{II}\ K N_{II}$	0.14614	1	84.836	0.14212	2	87.23
$\beta_2{}^{I}\ K N_{III}$	0.14595	1	84.946	0.14191	1	87.364
$K O_{II,III}$	0.14509	1	85.451	0.141012	8	87.922
$K P$				0.1408	1	88.06
$\beta_5\ K M_{IV,V}$	0.14917	1	83.114			
$\beta_5{}^{II}\ K M_{IV}$				0.14512	2	85.43
$\beta_5{}^{I}\ K M_V$				0.14495	3	85.53
$\beta_4\ K N_{IV,V}$	0.14553	2	85.19	0.14155	3	87.59
$\beta_4\ L_I M_{II}$	1.03918	3	11.9306	1.0075	1	12.306
$\beta_3\ L_I M_{III}$	1.00062	3	12.3904	0.96911	7	12.7933
$\gamma_2\ L_I N_{II}$	0.84773	5	14.6251	0.8210	2	15.101
$\gamma_3\ L_I N_{III}$	0.84130	4	14.7368	0.8147	1	15.218
$\gamma'_4\ L_I O_{II}$	0.81308	5	15.2482	0.78706	7	15.752
$\gamma_4\ L_I O_{III}$	0.81184	5	15.2716	0.7858	1	15.777
$\eta\ L_{II} M_I$	1.12769	3	10.9943	1.09241	7	11.3493
$\beta_1\ L_{II} M_{IV}$	1.01513	4	12.2133	0.98291	3	12.6137
$\gamma_5\ L_{II} N_I$	0.89500	4	13.8526	0.86655	5	14.3075
$\gamma_1\ L_{II} N_{IV}$	0.86752	3	14.2915	0.83973	3	14.7644
$\gamma_8\ L_{II} O_I$	0.8513	2	14.564	0.82365	5	15.0527
$\gamma_6\ L_{II} O_{IV}$	0.8442	2	14.685	0.81683	5	15.1783
$L_{II} P_I$				0.81583	5	15.1969
$l\ L_{III} M_I$	1.38477	3	8.9532	1.34990	7	9.1845
$\alpha_2\ L_{III} M_{IV}$	1.21875	3	10.1728	1.18648	5	10.4495
$\alpha_1\ L_{III} M_V$	1.20739	4	10.2685	1.17501	2	10.5515
$\beta_6\ L_{III} N_I$	1.04963	5	11.8118	1.0210	1	12.143
$\beta_{15}\ L_{III} N_{IV}$	1.01201	3	12.2510	0.98389	7	12.6011
$\beta_2\ L_{III} N_V$	1.01031	3	12.2715	0.98221	7	12.6226
$\beta_7\ L_{III} O_I$	0.99017	5	12.5212	0.9620	1	12.888
$\beta_5\ L_{III} O_{IV,V}$	0.98058	3	12.6436	0.9526	1	13.015
$L_I M_I$	1.0644	2	11.648	1.0323	2	12.010
$\beta_{10}\ L_I M_{IV}$	0.96389	7	12.8626	0.9339	1	13.275
$\beta_9\ L_I M_V$	0.95675	7	12.9585	0.9268	1	13.377
$L_I N_I$	0.8549	1	14.503	0.82859	7	14.963
$L_I N_{IV}$	0.83001	7	14.937	0.80364	7	15.427
$\gamma_{11}\ L_I N_V$	0.82879	5	14.9593	0.80233	9	15.453
$L_I N_{VI,VII}$				0.7884	1	15.725
$L_I O_I$	0.8158	1	15.198	0.7897	1	15.699
$L_I O_{IV,V}$	0.80861	5	15.3327	0.78257	7	15.843
$L_{II} M_{II}$	1.0997	1	11.274	1.0644	2	11.648
$\beta_{17}\ L_{II} M_{III}$	1.05609	7	11.7397	1.0223	1	12.127
$L_{II} M_V$	1.00722	5	12.3093	0.9747	1	12.720
$L_{II} N_{II}$	0.882	2	14.057	0.8585	3	14.442

Designation	Å*	p.e.	keV	Å*	p.e.	keV
	81 Thallium (Cont.)			**82 Lead** (Cont.)		
$L_{II} N_{III}$	0.87996	5	14.0893	0.85192	7	14.553
$L_{II} N_V$				0.8382	2	14.791
$v\ L_{II} N_{VI}$	0.85048	5	14.5777	0.82327	7	15.060
$J_{II} O_{II}$	0.8490	1	14.604			
$L_{II} O_{II}$				0.8200	1	15.120
$t\ L_{III} M_{II}$	1.34154	5	9.2417	1.30767	7	9.4811
$s\ L_{III} M_{III}$	1.27807	5	9.7007	1.24385	7	9.9675
$L_{III} N_{II}$				1.01040	7	12.2705
$L_{III} N_{III}$	1.0286	1	12.053	1.0005	1	12.392
$u\ L_{III} N_{VI,VII}$	0.9888	1	12.538	0.96133	7	12.8968
$L_{III} O_{II}$	0.98738	5	12.5566	0.9586	1	12.934
$L_{III} O_{III}$	0.98538	5	12.5820	0.9578	1	12.945
$L_{III} P_{II,III}$	0.97926	5	12.6607	0.95118	7	13.0344
$M_I N_{III}$	4.013	9	3.089	3.872	9	3.202
$M_{II} N_I$				4.655	8	2.664
$M_{II} N_{IV}$	4.116	4	3.013	3.968	5	3.124
$M_{III} N_I$	5.884	8	2.107	5.704	8	2.174
$M_{III} N_{IV}$	4.865	5	2.548	4.715	3	2.630
$\gamma\ M_{III} N_V$	4.823	4	2.571	4.674	1	2.6527
$M_{III} O_I$				4.244	9	2.921
$M_{II} O_{IV,V}$	4.216	6	2.941	4.069	6	3.047
$\zeta_2\ M_{IV} N_{II}$	7.032	5	1.763	6.802	5	1.823
$M_{IV} N_{III}$				6.384	7	1.942
$\beta\ M_{IV} N_{VI}$	5.249	1	2.3621	5.076	1	2.4427
$M_{IV} O_{II}$	5.196	9	2.386	5.004	9	2.477
$\zeta_1\ M_V N_{III}$	6.974	4	1.778	6.740	3	1.8395
$\alpha_2\ M_V N_{VI}$	5.472	2	2.2656	5.299	2	2.3397
$\alpha_1\ M_V N_{VII}$	5.460	1	2.2706	5.286	1	2.3455
$M_V O_{III}$				5.168	9	2.399
$N_{IV} N_{VI}$				42.3	2	0.293
$N_V N_{VI,VII}$	46.5	2	0.267	45.0	1	0.2756
$N_{VI} O_{IV}$	115.3	2	0.1075	102.4	1	0.1211
$N_{VI} O_V$	113.0	1	0.10968	100.2	2	0.1237
$N_{VII} O_V$	117.7	1	0.10530	104.3	1	0.1189
	83 Bismuth			**84 Polonium**		
$\alpha_2\ K L_{II}$	0.165717	2	74.8148	0.16130†	1	76.862
$\alpha_1\ K L_{III}$	0.160789	2	77.1079	0.15636†	1	79.290
$\beta_3\ K M_{II}$	0.142779	7	86.834	0.13892†	2	89.25
$\beta_1\ K M_{III}$	0.141948	3	87.343	0.13807†	2	89.80
$\beta_2{}^{II}\ K N_{II}$	0.13817	1	89.733	0.13438†	2	92.26
$\beta_2{}^{I}\ K N_{III}$	0.13797	1	89.864	0.13418†	2	92.40
$K O_{II,III}$	0.13709	1	90.435			
$\beta_5\ K M_{IV,V}$	0.14111	1	87.860			
$\beta_4\ K N_{IV,V}$	0.13759	2	90.11			
$\beta_4\ L_I M_{II}$	0.97690	4	12.6912	0.9475	3	13.086
$\beta_3\ L_I M_{III}$	0.93855	3	13.2098	0.9091	3	13,638
$\gamma_2\ L_I N_{II}$	0.79565	3	15.5824	0.772	1	16.07
$\gamma_3\ L_I N_{III}$	0.78917	5	15.7102			
$\gamma'_4\ L_I O_{II}$	0.76198	3	16.2709			
$\gamma_4\ L_I O_{III}$	0.76087	3	16.2947			
$\gamma_{13}\ L_I P_{II,III}$	0.75690	3	16.3802			
$\eta\ L_{II} M_I$	1.05856	3	11.7122			
$\beta_1\ L_{II} M_{IV}$	0.951978	9	13.0235	0.9220	2	13.447
$\gamma_5\ L_{II} N_I$	0.83923	5	14.7732			

TABLE 1b4. (Continued)

Desig-nation	Å*	p.e.	keV	Å*	p.e.	keV
	83 Bismuth (Cont.)			**84 Pononium** (Cont.)		
$\gamma_1\ L_{II}N_{IV}$	0.81311	2	15.2477	0.78748	9	15.744
$\gamma_8\ L_{II}O_I$	0.7973	1	15.551			
$\gamma_6\ L_{II}O_{IV}$	0.79043	3	15.6853	0.7645	2	16.218
$l\ L_{III}M_I$	1.31610	7	9.4204	1.2829	5	9.664
$\alpha_2\ L_{III}M_{IV}$	1.15536	1	10.73091	1.12548†	5	11.0158
$\alpha_1\ L_{III}M_V$	1.14386	2	10.8388	1.11386	4	11.1308
$\beta_6\ L_{III}N_I$	0.99331	3	12.4816	0.9672	2	12.819
$\beta_{15}\ L_{III}N_{IV}$	0.95702	5	12.9549	0.9312	2	13.314
$\beta_2\ L_{III}N_V$	0.95518	4	12.9799	0.92937	5	13.3404
$\beta_7\ L_{III}O_I$	0.93505	5	13.2593			
$\beta_5\ L_{III}O_{IV,V}$	0.92556	3	13.3953	0.8996	2	13.782
L_IM_I	1.0005	9	12.39			
$\beta_{10}\ L_IM_{IV}$	0.90495	4	13.7002			
$\beta_9\ L_IM_V$	0.89791	3	13.8077			
L_IN_I	0.8022	1	15.456			
L_IN_{IV}	0.7795	5	15.904			
$\gamma_{11}\ L_IN_V$	0.77728	5	15.951			
$L_IN_{VI,VII}$	0.7641	5	16.23			
$L_IO_{IV,V}$	0.75791	5	16.358			
$L_{II}M_{II}$	1.0346	9	11.98			
$\beta_{17}\ L_{II}M_{III}$	0.98913	5	12.5344			
$L_{II}M_V$	0.94419	5	13.1310			
$L_{II}N_{II}$	0.8344	9	14.86			
$L_{II}N_{III}$	0.8248	1	15.031			
$v\ L_{II}N_{VI}$	0.79721	9	15.552			
$L_{II}O_{III}$	0.79384	5	.15.6178			
$t\ L_{III}M_{II}$	1.2748	1	9.7252			
$s\ L_{III}M_{III}$	1.2105	1	10.2421			
$L_{III}N_{II}$	0.98280	5	12.6151			
$L_{III}N_{III}$	0.97321	5	12.7394			
$u\ L_{III}N_{VI,VII}$	0.93505	5	13.2593			
$L_{III}O_{II}$	0.9323	2	13.298			
$L_{III}O_{III}$	0.9302	2	13.328			
$L_{III}P_{II,III}$	0.92413	4	13.4159			
M_IN_{II}	3.892	9	3.185			
M_IN_{III}	3.740	9	3.315			
$M_{II}N_{IV}$	3.834	4	3.234			
$M_{III}N_I$	5.537	8	2.239			
$M_{III}N_{IV}$	4.571	5	2.712			
$\gamma\ M_{III}N_V$	4.532	2	2.735			
$M_{III}O_I$	4.105	9	3.021			
$M_{III}O_{IV,V}$	3.932	6	3.153			
$\zeta_2\ M_{IV}N_{II}$	6.585	5	1.883			
$M_{IV}N_{III}$	6.162	8	2.012			
$\beta\ M_{IV}N_{VI}$	4.909	1	2.5255			
$M_{IV}O_{II}$	4.823	3	2.571			
$M_{IV}P_{II,III}$	4.59	2	2.70			
$\zeta_1\ M_VN_{III}$	6.521	4	1.901			
$\alpha_2\ M_VN_{VI}$	5.130	2	2.4170			
$\alpha_1\ M_VN_{VII}$	5.118	1	2.4226			
$N_IP_{II,III}$	13.30	6	0.932			
$N_{VI}O_{IV}$	91.6	1	0.1354			
$N_{VII}O_V$	93.2	1	0.1330			

Desig-nation	Å*	p.e.	keV	Å*	p.e.	keV
	85 Astatine			**86 Radon**		
$\alpha_2\ KL_{II}$	0.15705†	2	78.95	0.15294†	3	81.07
$\alpha_1\ KL_{III}$	0.15210†	2	81.52	0.14798†	3	83.78
$\beta_3\ KM_{II}$	0.13517†	4	91.72	0.13155†	5	94.24
$\beta_1\ KM_{III}$	0.13432†	4	92.30	0.13069†	5	94.87
$\beta_2^{II}\ KN_{II}$	0.13072†	4	94.84	0.12719†	5	97.47
$\beta_2^{I}\ KN_{III}$	0.13052†	4	94.99	0.12698†	5	97.64
$\beta_3\ L_IM_{III}$	0.88135†	9	14.067	0.85436†	9	14.512
$\beta_1\ L_{II}M_{IV}$	0.89349†	9	13.876	0.86605†	9	14.316
$\gamma_1\ L_{II}N_{IV}$	0.76289†	9	16.251	0.73928†	9	16.770
$\alpha_2\ L_{III}M_{IV}$	1.09671†	5	11.3048	1.06899†	5	11.5979
$\alpha_1\ L_{III}M_V$	1.08500†	5	11.4268	1.05723†	5	11.7270
	87 Francium			**88 Radium**		
$\alpha_2\ KL_{II}$	0.14896†	3	83.23	0.14512†	2	85.43
$\alpha_1\ KL_{III}$	0.14399†	3	86.10	0.14014†	2	88.47
$\beta_3\ KM_{II}$	0.12807†	5	96.81	0.12469†	3	99.43
$\beta_1\ KM_{III}$	0.12719†	5	97.47	0.12382†	3	100.13
$\beta_2^{II}\ KN_{II}$	0.12379†	5	100.16	0.12050†	3	102.89
$\beta_2^{I}\ KN_{III}$	0.12358†	5	100.33	0.12029†	3	103.07
$\beta_4\ L_IM_{II}$				0.84071	5	14.7472
$\beta_3\ L_IM_{III}$	0.82789†	9	14.976	0.80273	5	15.4449
$\gamma_2\ L_IN_{II}$				0.68199	5	18.179
$\gamma_3\ L_IN_{III}$				0.67538	5	18.357
$\gamma'_4\ L_IO_{II}$				0.65131	5	19.036
$\gamma_4\ L_IO_{III}$				0.64965	5	19.084
$\gamma_{13}\ L_IP_{II,III}$				0.64513	5	19.218
$\eta\ L_{II}M_I$				0.90742	5	13.6630
$\beta_1\ L_{II}M_{IV}$	0.83940†	9	14.770	0.81375	5	15.2358
$\gamma_5\ L_{II}N_I$				0.71774	5	17.274
$\gamma_1\ L_{II}N_{IV}$	0.71652†	9	17.303	0.69463	5	17.849
γ_8				0.6801	1	18.230
$\gamma_6\ L_{II}O_{IV}$				0.67328	5	18.414
$L_{II}P_I$				0.6724	1	18.439
$l\ L_{III}M_I$				1.16719	5	10.6222
$\alpha_2\ L_{III}M_{IV}$	1.04230	5	11.8950	1.01656	5	12.1962
$\alpha_1\ L_{III}M_V$	1.03049	5	12.0313	1.00473	5	12.3397
$\beta_6\ L_{III}N_I$				0.87088	5	14.2362
$\beta_{15}\ L_{III}N_{IV}$				0.83722	5	14.8086
$\beta_2\ L_{III}N_V$	0.858	2	14.45	0.83537	5	14.8414
$\beta_7\ L_{III}O_I$				0.8162	1	15.190
$\beta_5\ L_{III}O_{IV,V}$				0.80627	5	15.3771
$L_{III}P_I$				0.8050	1	15.402
$\beta_{10}\ L_IM_{IV}$				0.77546	5	15.988
$\beta_9\ L_IM_V$				0.76857	5	16.131
L_IN_I				0.6874	1	18.036
L_IN_{IV}				0.6666	1	18.600
$\gamma_{11}\ L_IN_V$				0.6654	1	18.633
$L_IO_{IV,V}$				0.6468	1	19.167
$\beta_{17}\ L_{II}M_{III}$				0.8438	1	14.692
$L_{II}N_{II}$				0.7043	1	17.604
$L_{II}N_V$				0.6932	1	17.884
$L_{II}O_{II}$				0.6780	1	18.286

TABLE 1b4. (*Continued*)

Designation	Å*	p.e.	keV	Å*	p.e.	keV
	87 Francium (*Cont.*)			**88 Radium** (*Cont.*)		
$L_{II}O_{III}$				0.6764	1	18.330
$L_{II}P_{II,III}$				0.6714	1	18.466
$L_{III}N_{II}$				0.8618	1	14.387
$L_{III}N_{III}$				0.8512	1	14.566
$u\ L_{III}N_{VI,VII}$				0.8186	1	15.146
$L_{III}P_{II,III}$				0.8038	1	15.425
	89 Actinium			**90 Thorium**		
$\alpha_2\ KL_{II}$	0.14141†	2	87.67	0.137829	2	89.953
$\alpha_1\ KL_{III}$	0.136417†	8	90.884	0.132813	2	93.350
$\beta_3\ KM_{II}$	0.12143†	2	102.10	0.118268	3	104.831
$\beta_1\ KM_{III}$	0.12055†	2	102.85	0.117396	9	105.609
$\beta_2^{II}\ KN_{II}$	0.11732†	2	105.67	0.11426	1	108.511
$\beta_2^{I}\ KN_{III}$	0.11711†	2	105.86	0.114040	9	108.717
$KO_{II,III}$				0.11322	1	109.500
$\beta_5\ KM_{IV,V}$				0.116667	9	106.269
$\beta_4\ KN_{IV,V}$				0.11366	2	109.08
$\beta_4\ L_1M_{II}$				0.79257	4	15.6429
$\beta_3\ L_1M_{III}$	0.77822†	9	15.931	0.75479	3	16.4258
$\gamma_2\ L_{II}N_{II}$				0.64221	4	19.305
$\gamma_3\ L_1N_{III}$				0.63559	4	19.507
$\gamma'_4\ L_1O_{II}$				0.61251	4	20.242
$\gamma_4\ L_1O_{III}$				0.61098	4	20.292
$\gamma_{13}\ L_1P_{II,III}$				0.60705	8	20.424
$\eta\ L_{II}M_1$				0.85446	1	14.5099
$\beta_1\ L_{II}M_{IV}$	0.78903†	9	15.713	0.765210	9	16.2022
$\gamma_5\ L_{II}N_1$				0.67491	4	18.370
$\gamma_1\ L_{II}N_{IV}$	0.67351†	9	18.408	0.65313	3	18.9825
$\gamma_8\ L_{II}O_1$				0.63898	5	19.403
$\gamma_6\ L_{II}O_{IV}$				0.63258	4	19.599
$L_{II}P_1$				0.6316	1	19.629
$L_{II}P_{IV}$				0.62991	9	19.682
$l\ L_{III}M_1$				1.11508	4	11.1186
$\alpha_2\ L_{III}M_{IV}$	0.99178†	5	12.5008	0.96788	2	12.8096
$\alpha_1\ L_{III}M_V$	0.97993†	5	12.6520	0.95600	3	12.9687
$\beta_6\ L_{III}N_1$				0.82790	8	14.975
$\beta_{15}\ L_{III}N_{IV}$				0.79539	5	15.5875
$\beta_2\ L_{III}N_V$				0.79354	3	15.6237
$\beta_7\ L_{III}O_1$				0.77437	4	16.0105
$\beta_5\ L_{III}O_{IV,V}$				0.76468	5	16.213
$L_{III}P_1$				0.76338	5	16.241
$L_{III}P_{IV,V}$				0.76087	9	16.295
$\beta_{10}\ L_1M_{IV}$				0.7301	1	16.981
$\beta_9\ L_1M_V$				0.7234	1	17.139
L_1N_1				0.64755	5	19.146
L_1N_{IV}				0.6276	1	19.755
$\gamma_{11}\ L_1N_V$				0.62636	9	19.794
$L_1N_{VI,VII}$				0.6160	1	20.128
L_1O_1				0.6146	1	20.174
$L_1O_{IV,V}$				0.6083	1	20.383
$L_{II}M_{II}$				0.8338	1	14.869
$\beta_{17}\ L_{II}M_{III}$				0.79257	4	15.6429
$L_{II}M_V$				0.7579	1	16.359
$L_{II}N_{III}$				0.6620	1	18.729
$L_{II}N_V$				0.6521	1	19.014

Designation	Å*	p.e.	keV	Å*	p.e.	keV
	89 Actinium (*Cont.*)			**90 Thorium** (*Cont.*)		
$v\ L_{II}N_{VI}$				0.64064	9	19.353
$L_{II}O_{II}$				0.6369	1	19.466
$L_{II}O_{III}$				0.6356	1	19.506
$L_{II}P_{II,III}$				0.6312	1	19.642
$t\ L_{III}M_1$				1.08009	9	11.4788
$s\ L_{III}M_{II}$				1.0112	1	12.261
$L_{III}N_{II}$				0.8190	2	15.138
$L_{III}N_{III}$				0.8082	1	15.341
$u\ L_{III}N_{VI,VII}$				0.77661	5	15.964
$L_{III}O_{II}$				0.7713	1	16.074
$L_{III}O_{III}$				0.7690	1	16.123
$L_{III}P_{II,III}$				0.7625	2	16.260
M_1N_{III}				2.934	8	4.23
M_1O_{III}				2.442	9	5.08
$M_{II}N_1$				3.537	9	3.505
$M_{II}N_{IV}$				3.011	2	4.117
$M_{II}O_{IV}$				2.618	5	4.735
$M_{III}N_1$				4.568	5	2.714
$M_{III}N_{IV}$				3.718	3	3.335
$\gamma\ M_{II}N_V$				3.679	2	3.370
$M_{III}O_1$				3.283	9	3.78
$M_{III}O_{IV,V}$				3.131	3	3.959
$\zeta_2\ M_V N_{II}$				5.340	5	2.322
$M_{IV}N_{III}$				4.911	5	2.524
$\beta\ M_{IV}N_{VI}$				3.941	1	3.1458
$M_{IV}O_1$				3.808	4	3.256
$\xi_1\ M_V N_{III}$				5.245	5	2.364
$\alpha_2\ M_V N_{VI}$				4.151	2	2.987
$\alpha_1\ M_V N_{VII}$				4.1381	9	2.9961
$M_V P_{III}$				3.760	9	3.298
$N\ P_{II}$				9.44	7	1.313
$N\ P_{III}$				9.40	7	1.1319
$N_{II}O_{IV}$				11.56	5	1.072
N_1P_1				11.07	7	1.120
$N_{III}O_V$				13.8	1	0.897
$N_{IV}N_{VI}$				33.57	9	0.3693
$N_V N_{VI,VII}$				36.32	9	0.3414
$N_{VI}O_{IV}$				49.5	1	0.2505
$N_{VI}O_V$				48.2	1	0.2572
$N_{VII}O_V$				50.0	1	0.2479
$O_{III}P_{IV,V}$				68.2	3	0.1817
$O_{IV,V}Q_{II,III}$				181.	5	0.068
	91 Protactinium			**92 Uranium**		
$\alpha_2\ KL_{II}$	0.134343†	9	92.287	0.130968	4	94.665
$\alpha_1\ KL_{III}$	0.129325†	3	95.868	0.125947	3	98.439
$\beta_3\ KM_{II}$	0.11523†	2	107.60	0.112296	4	110.406
$\beta_1\ KM_{III}$	0.114345†	8	108.427	0.111394	5	111.300
$\beta_2^{II}\ KN_{II}$	0.11129†	2	111.40	0.10837	1	114.40
$\beta_2^{I}\ KN_{III}$	0.11107†	2	111.62	0.10818	1	114.60
$KO_{II,III}$				0.10744	1	115.39
$\beta_5\ KM_{IV,V}$				0.11069	1	112.01
$\beta_4\ KN_{IV,V}$				0.10780	2	115.01
$\beta_4\ L_1M_1$	0.7699	1	16.104	0.747985	9	16.5753
$\beta_3\ L_1M_{III}$	0.73230	5	16.930	0.71029	2	17.4550

TABLE 1b4. (*Continued*)

91 Protactinium (*Cont.*) / 92 Uranium (*Cont.*)

Designation	Å*	p.e.	keV	Å*	p.e.	keV
γ_2 $L_I N_{II}$	0.6239	1	19.872	0.605237	9	20.4847
γ_3 $L_I N_{III}$	0.6169	1	20.098	0.598574	9	20.7127
γ'_4 $L_I O_{II}$				0.576700	9	21.4984
γ_4 $L_I O_{II,III}$	0.5937	1	20.882	0.57499	9	21.562
γ_{18}				0.5706	1	21.729
η $L_{II} M_I$	0.8295 ♪	1	14.946	0.80509	2	15.3997
β_1 $L_{II} M_{IV}$	0.74232	5	16.702	0.719984	8	17.2200
γ_5 $L_{II} N_I$	0.6550	1	18.930	0.63557	2	19.5072
γ_1 $L_{II} N_{IV}$	0.63358†	9	19.568	0.614770	9	20.1671
γ_8 $L_{II} O_I$				0.60125	5	20.621
γ_6 $L_{II} O_{IV}$	0.6133	1	20.216	0.594845	9	20.8426
$L_{II} P_{IV}$				0.59203	5	20.942
l $L_{III} M_I$	1.0908	1	11.366	1.06712	2	11.6183
α_2 $L_{III} M_{IV}$	0.94482†	5	13.1222	0.922558	9	13.4388
α_1 $L_{III} M_V$	0.93284	5	13.2907	0.910639	9	13.6147
β_6 $L_{III} N_I$	0.8079	1	15.347	0.78838	2	15.7260
β_{15} $L_{III} N_{IV}$				0.756642	9	16.3857
β_2 $L_{III} N_V$	0.7737	1	16.024	0.754681	9	16.4283
β_7 $L_{III} O_I$	0.7546	2	16.431	0.73602	6	16.845
β_5 $L_{III} O_{IV,V}$	0.7452	2	16.636	0.726305	9	17.0701
$L_{III} P_I$				0.72521	5	17.096
$L_{III} P_{IV,V}$				0.72240	5	17.162
β_{10} $L_I M_V$	0.7088	2	17.492	0.68760	5	18.031
β_9 $L_I M_V$	0.7018	1	17.667	0.681014	8	18.2054
$L_I N_{IV}$				0.59096	5	20.979
γ_{11} $L_I N_V$				0.58986	5	21.019
$L_I O_{IV,V}$				0.5725	1	21.657
β_{17} $L_{II} M_{III}$				0.74503	5	16.641
$L_{II} N_{III}$				0.6228	1	19.907
v $L_{II} N_{VI}$				0.6031	1	20.556
$L_{III} O_{III}$				0.59728	5	20.758
$L_{II} P_{II,III}$				0.5930	2	20.906
t $L_{III} M_{II}$				1.0347	1	11.982
s $L_{III} M_{III}$				0.9636	1	12.866
$L_{III} N_{II}$				0.78017	9	15.892
$L_{III} N_{III}$				0.7691	1	16.120
u $L_{III} N_{VI,VII}$				0.738603	9	16.7859
$L_{III} O_{II}$				0.7333	1	16.907
$L_{III} O_{III}$				0.7309	1	16.962
$L_{III} P_{II,III}$				0.72426	5	17.118
$M_I N_{II}$				2.92	2	4.25
$M_I N_{III}$				2.753	8	4.50
$M_I O_{III}$				2.304	7	5.38
$M_I P_{III}$				2.253	6	5.50
$M_{II} N_I$	3.441	5	3.603	3.329	4	3.724
$M_{II} N_{IV}$	2.910	2	4.260	2.817	2	4.401
$M_{II} O_{IV}$	2.527	4	4.906	2.443	4	5.075
$M_{III} N_I$	4.450	4	2.786	4.330	2	2.863
$M_{III} N_{IV}$	3.614	2	3.430	3.521	2	3.521
γ $M_{III} N_V$	3.577	1	3.4657	3.479	1	3.563
$M_{III} O_I$	3.245	9	3.82	3.115	7	3.980
$M_{III} O_{IV,V}$	3.038	2	4.081	2.948	2	4.205
ξ_2 $M_{IV} N_{II}$	5.193	2	2.3876	5.050	2	2.4548
$M_{IV} N_{III}$				4.625	5	2.681
β $M_{IV} N_{VI}$	3.827	1	3.2397	3.716	1	3.3367

91 Protactinium (*Cont.*) / 92 Uranium (*Cont.*)

Designation	Å*	p.e.	keV	Å*	p.e.	keV
$M_{IV} O_{II}$	3.691	2	3.359	3.576	1	3.4666
ξ_1 $M_V N_{III}$	5.092	2	2.4350	4.946	2	2.507
α_2 $M_V N_{VI}$	4.035	3	3.072	3.924	1	3.1595
α_1 $M_V N_{VII}$	4.022	1	3.0823	3.910	1	3.1708
$N_I O_{III}$				10.09	7	1.229
$N_I P_{II}$				8.81	7	1.41
$N_I P_{III}$				8.76	7	1.42
$N_{II} P_I$				10.40	7	1.192
$N_{III} O_V$				12.90	9	0.961
$N_{IV} N_{VI}$				31.8	1	0.390
$N_V N_{VI,VII}$				34.8	1	0.357
$N_{IV} O_V$				43.3	2	0.286
$N_{VI} O_V$				42.1	2	0.295
$N_I P_{IV,V}$				8.60	7	1.44

93 Neptunium / 94 Plutonium

Designation	Å*	p.e.	keV	Å*	p.e.	keV
β_4 $L_I M_{II}$	0.72671	2	17.0607	0.70620	2	17.5560
β_3 $L_I M_{III}$	0.68920†	9	17.989	0.66871	2	18.5405
γ_2 $L_I N_{II}$	0.5873	5	21.11	0.57068	2	21.7251
γ_3 $L_I N_{III}$	0.5810	5	21.34	0.564001	9	21.9824
γ'_4 $L_I O_{II}$				0.5432	1	22.823
γ_4 $L_I O_{II,III}$	0.5585	5	22.20	0.5416	1	22.891
η $L_{II} M_I$	0.7809	2	15.876	0.7591	1	16.333
β_1 $L_{II} M_{IV}$	0.698478	9	17.7502	0.67772	2	18.2937
γ_5 $L_{II} N_I$	0.616	1	20.12	0.5988	1	20.704
γ_1 $L_{II} N_{IV}$	0.596498	9	20.7848	0.578882	9	21.4173
γ_8				0.5658	1	21.914
γ_6 $L_{II} O_{IV}$	0.57699	5	21.488	0.55973	2	22.1502
l $L_{III} M_I$	1.0428	6	11.890	1.0226	1	12.124
α_2 $L_{III} M_{IV}$	0.901045	9	13.7597	0.88028	2	14.0842
α_1 $L_{III} M_V$	0.889128	9	13.9441	0.86830	2	14.2786
β_6 $L_{III} N_I$	0.769	1	16.13	0.75148	2	16.4983
β_{15} $L_{III} N_{IV}$				0.7205	1	17.208
β_2 $L_{III} N_V$	0.736230	9	16.8400	0.71851	2	17.2553
β_7 $L_{III} O_I$				0.7003	1	17.705
β_5 $L_{III} O_{IV,V}$	0.70814	2	17.5081	0.69068	2	17.9506
β_{10} $L_I M_{IV}$				0.6482	1	19.126
β_9 $L_I M_V$				0.6416	1	19.323
u $L_{III} N_{VI,VII}$				0.7031	1	17.635

95 Americium

Designation	Å*	p.e.	keV
β_4 $L_I M_{II}$	0.68639	2	18.0627
β_3 $L_I M_{III}$	0.64891	2	19.1059
γ_2 $L_I N_{II}$	0.5544	2	22.361
β_1 $L_{II} M_{IV}$	0.657655	9	18.8520
γ_1 $L_{II} N_{IV}$	0.561886	9	22.0652
γ_6 $L_{II} O_{IV}$	0.54311	2	22.8282
l $L_{III} M_I$	1.0012	6	12.384
α_2 $L_{III} M_{IV}$	0.860266	9	14.4119
α_1 $L_{III} M_V$	0.848187	9	14.6172
β_6 $L_{III} N_I$	0.73418	2	16.8870
β_{15} $L_{III} N_{IV}$	0.70341	2	17.6258
β_2 $L_{III} N_V$	0.701390	9	17.6765
β_4 $L_{III} O_{IV,V}$	0.67383	2	18.3996

*Reprinted with permission of the author and editor from *Revs. Mod. Phys.* Vol. 39, No. 1, 78–124, 1967.

TABLE 1b5. X-RAY ATOMIC ENERGY LEVELS (BEARDEN, 1967A)*

TABLE I. Recommended values of the atomic energy levels, and probable errors in eV. Where available, photoelectron direct measurements are listed in brackets [ ] immediately under the recommended values. The measured values of the x-ray absorption energies (from Ref. 13) are shown in parentheses ( ). Interpolated values are enclosed in angle brackets ⟨ ⟩.

Level	1 H	2 He	3 Li	4 Be	5 B	6 C	7 N	8 O
K	13.59811[a]	24.58678[b]	54.75±0.02 (54.75)	111.0±1.0 (111.0)	188.0±0.4 [188.0][c]	283.8±0.4 [283.8][c] (283.8)	401.6±0.4 [401.6][c]	532.0±0.4 [532.0][c]
L_I								23.7±0.4 [23.7][d]
$L_{II,III}$					4.7±0.9	6.4±1.9	9.2±0.6	7.1±0.8

Level	9 F	10 Ne	11 Na	12 Mg	13 Al	14 Si	15 P	16 S
K	685.4±0.4 [685.4][c]	866.9±0.3 (866.9)	1072.1±0.4 [1072.1][c] (1072.)	1305.0±0.4 [1305.0][c] (1303.)	1559.6±0.4 [1559.6][c] (1559.8)	1838.9±0.4 [1838.9][c]	2145.5±0.4 [2145.5][c]	2472.0±0.4 [2472.0][c] (2470.)
L_I	⟨31.⟩	⟨45.⟩	63.3±0.4 [63.3][d]	89.4±0.4 [89.4][d] ⟨63.⟩	117.7±0.4 [117.7][d] ⟨87.⟩	148.7±0.4 [148.7][d]	189.3±0.4 [189.3][d]	229.2±0.4 [229.2][d]
$L_{II,III}$	8.6±0.8	18.3±0.4	31.1±0.4 ⟨31.⟩	51.4±0.5 ⟨50.⟩	73.1±0.5 (72.8)	99.2±0.4 (100.6)	132.2±0.5 ⟨132.⟩	164.8±0.7

Level	17 Cl	18 Ar	19 K	20 Ca	21 Sc	22 Ti	23 V	24 Cr
K	2822.4±0.3 [2822.4][c] (2020.)	3202.9±0.3 (3202.9)	3607.4±0.4 [3607.4][c] (3607.8)	4038.1±0.4 [4038.1][c] (4038.1)	4492.8±0.4 [4492.8][c]	4966.4±0.4 [4966.4][d] (4964.5)	5465.1±0.3 [5465.1][c] (5464.)	5989.2±0.3 [5989.2][c] (5989.)
L_I	270.2±0.4 [270.2][d]	320. ⟨320.⟩[d]	377.1±0.4 [377.1][d]	437.8±0.4 [437.8][d]	500.4±0.4 [500.4][d]	563.7±0.4 [563.7][d]	628.2±0.4 [628.2][d]	694.6±0.4 [694.6][d]
L_{II}	201.6±0.3	247.3±0.3	296.3±0.4	350.0±0.4	406.7±0.4	461.5±0.4	520.5±0.3	583.7±0.3
L_{III}	200.0±0.3	245.2±0.3	293.6±0.4	346.4±0.4	402.2±0.4	455.5±0.4	512.9±0.3	574.5±0.3
M_I	17.5±0.4	25.3±0.4	33.9±0.4	43.7±0.4	53.8±0.4	60.3±0.4	66.5±0.4	74.1±0.4
$M_{II,III}$	6.8±0.4	12.4±0.3	17.8±0.4	25.4±0.4	32.3±0.5	34.6±0.4	37.8±0.3	42.5±0.3
$M_{IV,V}$					6.6±0.5	3.7	2.2±0.3	2.3±0.4

TABLE 1b5. (Continued)

	25 Mn	26 Fe	27 Co	28 Ni	29 Cu	30 Zn	31 Ga	32 Ge
K	6539.0±0.4 [6539.0]* (6538.)	7112.0±0.9 [7111.3]*,f (7111.2)	7708.9±0.3 [7708.9]* (7709.5)	8332.8±0.4 [8332.8]* (8331.6)	8978.9±0.4 [8978.9]*,** (8980.3)	9658.6±0.6 [9658.6]* (9660.7)	10367.1±0.5 [10367.1]* (10368.2)	11103.1±0.7 [11103.8]* (11103.6)
L_I	769.0±0.4 [769.0]^d	846.1±0.4 [846.1]^d	925.6±0.4 [925.6]^d	1008.1±0.4 [1008.1]^d	1096.1±0.4 [1096.0]^d	1193.6±0.9	1297.7±1.1	1414.3±0.7 [1413.6]*
L_{II}	651.4±0.4	721.1±0.9 (720.8)	793.6±0.3 (793.8)	871.9±0.4 (870.6)	951.0±0.4 [950.0]^h (953.)	1042.8±0.6 (1045.)	1142.3±0.5	1247.8±0.7 (1249.)
L_{III}	640.3±0.4	708.1±0.9 (707.4)	778.6±0.3 (779.0)	854.7±0.4 (853.6)	931.1±0.4 [931.4]^b (933.)	1019.7±0.6 (1022.)	1115.4±0.5 (1117.)	1216.7±0.7 (1217.0)
M_I	83.9±0.5	92.9±0.9	100.7±0.4	111.8±0.6	119.8±0.6	135.9±1.1	158.1±0.5	180.0±0.8
M_{II}	48.6±0.4 (54.)	54.0±0.9 (54.)	59.5±0.3 (61.)	68.1±0.4 (66.)	73.6±0.4 (75.)	86.6±0.6 (86.)	106.8±0.7	127.9±0.9
M_{III}							102.9±0.5	120.8±0.7
$M_{IV,V}$	3.3±0.5	3.6±0.9	2.9±0.3	3.6±0.4	1.6±0.4	8.1±0.6	17.4±0.5	28.7±0.7

	33 As	34 Se	35 Br	36 Kr	37 Rb	38 Sr	39 Y	40 Zr
K	11866.7±0.7 [11866.7]^i (11865.)	12657.8±0.7 [12657.8]* (12654.5)	13473.7±0.4 (13470.)	14325.6±0.8 (14324.4)	15199.7±0.3 (15202.)	16104.6±0.3 (16107.)	17038.4±0.3 (17038.)	17997.6±0.4 (17999.)
L_I	1526.5±0.8 (1529.)	1653.9±3.5 (1652.5)	1782.0±0.4 [1782.0]^j	1921.0±0.6 [1921.2]^j	2065.1±0.3 [2065.4]^j	2216.3±0.3 [2216.2]^j	2372.5±0.3 [2372.7]^j	2531.6±0.3 [2531.6]^j
L_{II}	1358.6±0.7 (1358.7)	1476.2±0.7 (1474.7)	1596.0±0.4 [1596.2]^j	1727.2±0.5 [1727.2]^j (1730.)	1863.9±0.3 [1863.4]^j	2006.8±0.3 [2006.6]^j (2008.5)	2155.5±0.3 [2155.0]^j (2154.0)	2306.7±0.3 [2306.5]^j (2305.3)
L_{III}	1323.1±0.7 (1323.5)	1435.8±0.7 (1434.0)	1549.9±0.4 [1549.7]^j	1674.9±0.5 [1674.8]^j (1677.)	1804.4±0.3 [1804.6]^j	1939.6±0.3 [1939.9]^j (1941.)	2080.0±0.3 [2080.2]^j (2079.4)	2222.3±0.3 [2222.5]^j (2222.5)
M_I	203.5±0.7	231.5±0.7	256.5±0.4		322.1±0.3	357.5±0.3	393.6±0.3	430.3±0.3
M_{II}	146.4±1.2	168.2±1.3	189.3±0.4	222.7±1.1	247.4±0.3	279.8±0.3	312.4±0.4	344.2±0.4

TABLE 1b5. (*Continued*)

	33 As	34 Se	35 Br	36 Kr	37 Rb	38 Sr	39 Y	40 Zr
M_{III}	140.5±0.8	161.9±1.0	181.5±0.4	213.8±1.1	238.5±0.3	269.1±0.3	300.3±0.4	330.5±0.4
M_{IV}	41.2±0.7	56.7±0.8	70.1±0.4	88.9±0.8	111.8±0.3	135.0±0.3	159.6±0.3	182.4±0.3
M_V			69.0±0.4		110.3±0.3	133.1±0.3	157.4±0.3	180.0±0.3
N_I			27.3±0.5	24.0±0.8	29.3±0.3	37.7±0.3	45.4±0.3	51.3±0.3
N_{II}	2.5±1.0	5.6±1.3	5.2±0.4	10.6±1.9	14.8±0.4	19.9±0.3	25.6±0.4	28.7±0.4
N_{III}			4.6±0.4		14.0±0.3			

	41 Nb	42 Mo	43 Tc	44 Ru	45 Rh	46 Pd	47 Ag	48 Cd
K	18985.6±0.4 (18987.)	19999.5±0.3 (20004.)	21044.0±0.7	22117.2±0.3 (22119.)	23219.9±0.3 (23219.8)	24350.3±0.3 (24348.)	25514.0±0.3 (25516.)	26711.2±0.3 (26716.)
L_I	2697.7±0.3 [2697.7][l]	2865.5±0.3 [2866.0][l]	3042.5±0.4 [3042.5][l]	3224.0±0.3 [3224.3][l]	3411.9±0.3 [3412.0][l] (3417.)	3604.3±0.3 [3604.6][l] (3607.)	3805.8±0.3 [3806.2][m] (3807.)	4018.0±0.3 [4018.1][m] (4019.)
L_{II}	2464.7±0.3 [2464.7][l]	2625.1±0.3 [2624.5][l] (2627.)	2793.2±0.4 [2973.2][l]	2966.9±0.3 [2966.8][l] (2966.3)	3146.1±0.3 [3146.3][l] (3145.)	3330.3±0.3 [3330.3][l] (3330.3)	3523.7±0.3 [3523.6][a,m] (3526.)	3727.0±0.3 [3727.1][m] (3728.)
L_{III}	2370.5±0.3 [2370.6][l]	2520.2±0.3 [2520.2][l] (2523.2)	2676.9±0.4 [2676.9][l]	2837.9±0.3 [2837.7][l] (2837.7)	3003.8±0.3 [3003.5][a,l] (3002.)	3173.3±0.3 [3173.0][a,l] (3173.0)	3351.1±0.3 [3350.8][a] (3351.0)	3537.5±0.3 [3537.3][a] (3537.6)
M_I	468.4±0.3	504.6±0.3		585.0±0.3	627.1±0.3	669.9±0.3	717.5±0.3	770.2±0.3
M_{II}	378.4±0.4	409.7±0.4	444.9±1.5	482.8±0.3	521.0±0.3	559.1±0.3	602.4±0.3	650.7±0.3
M_{III}	363.0±0.4	392.3±0.3	425.0±1.5	460.6±0.3	496.2±0.3	531.5±0.3	571.4±0.3	616.5±0.3
M_{IV}	207.4±0.3	230.3±0.3	256.4±0.5	283.6±0.3	311.7±0.3	340.0±0.3	372.8±0.3	410.5±0.3
M_V	204.6±0.3	227.0±0.3	252.9±0.4	279.4±0.3	307.0±0.3	334.7±0.3	366.7±0.3	403.7±0.3
N_I	58.1±0.3	61.8±0.3		74.9±0.3	81.0±0.3	86.4±0.3	95.2±0.3	107.6±0.3
N_{II}	33.9±0.4	34.8±0.4	38.9±1.9	43.1±0.4	47.9±0.4	51.1±0.4	62.6±0.3	66.9±0.4
N_{III}							55.9±0.3	
$N_{IV,V}$	3.2±0.3	1.8±0.3		2.0±0.3	2.5±0.4	1.5±0.3	3.3±0.3	9.3±0.3

TABLE 1b5. *(Continued)*

	49 In	50 Sn	51 Sb	52 Te	53 I	54 Xe	55 Cs	56 Ba
K	27939.9±0.3	29200.1±0.4 (29195.)	30491.2±0.3 (30486.)	31813.8±0.3 (31811.)	33169.4±0.4 (33167.)	34561.4±1.1 (34590.)	35984.6±0.4 (35987.)	37440.6±0.4 (37452.)
L_I	4237.5±0.3 [4237.7]^m (4237.3)	4464.7±0.3 [4464.5]^m (4464.8)	4698.3±0.3 [4698.3]^m (4698.4)	4939.2±0.3 [4939.3]^m (4939.7)	5188.1±0.3 [5188.1]^j	5452.8±0.4 (5452.8)	5714.3±0.4 [5712.7]^j (5721.)	5988.8±0.4 [5986.8]^j (5996.)
L_{II}	3938.0±0.3 [3937.8]^m (3939.3)	4156.1±0.3 [4156.2]^m (4157.)	4380.4±0.3 [4380.6]^m (4382.)	4612.0±0.3 [4612.0]^m (4612.6)	4852.1±0.3 [4852.0]^j	5103.7±0.4 (5103.7)	5359.4±0.3 [5359.5]^j (5358.)	5623.6±0.3 [5623.6]^j (5623.3)
L_{III}	3730.1±0.3 [3730.0]^m (3730.2)	3928.8±0.3 [3928.8]^m (3928.8)	4132.2±0.3 [4132.2]^m (4132.3)	4341.4±0.3 [4341.2]^m (4341.8)	4557.1±0.3 [4557.1]^j	4782.2±0.4 (4782.2)	5011.9±0.3 [5012.0]^j (5011.3)	5247.0±0.3 [5247.3]^j (5247.0)
M_I	825.6±0.3	883.8±0.3	943.7±0.3	1006.0±0.3	1072.1±0.3		1217.1±0.4	1292.8±0.4
M_{II}	702.2±0.3	756.4±0.4	811.9±0.3	869.7±0.3	930.5±0.3	999.0±2.1	1065.0±0.5	1136.7±0.5
M_{III}	664.3±0.3	714.4±0.3	765.6±0.3	818.7±0.3	874.6±0.3	937.0±2.1	997.6±0.5	1062.2±0.5
M_{IV}	450.8±0.3	493.3±0.3	536.9±0.3	582.5±0.3	631.3±0.3		739.5±0.4	796.1±0.3
M_V	443.1±0.3	484.8±0.3	527.5±0.3	572.1±0.3	619.4±0.3	672.3±0.5	725.5±0.5	780.7±0.3
N_I	121.9±0.3	136.5±0.4	152.0±0.3	168.3±0.3	186.4±0.3		230.8±0.4	253.0±0.5
N_{II}	77.4±0.4	88.6±0.4	98.4±0.5	110.2±0.5	122.7±0.5	146.7±3.1	172.3±0.6	191.8±0.7
N_{III}							161.6±0.6	179.7±0.6
N_{IV}	16.2±0.3	23.9±0.3	31.4±0.3	39.8±0.3	49.6±0.3		78.8±0.5	92.5±0.5
N_V							76.5±0.5	89.9±0.5
O_I	0.1±4.5	0.9±0.5	6.7±0.5	11.6±0.6	13.6±0.6		22.7±0.5	39.1±0.6
O_{II}							13.1±0.5	16.6±0.5
O_{III}	0.8±0.4	1.1±0.5	2.1±0.4	2.3±0.5	3.3±0.5		11.4±0.5	14.6±0.5

	57 La	58 Ce	59 Pr	60 Nd	61 Pm	62 Sm	63 Eu	64 Gd
K	38924.6±0.4 (38934.)	40443.0±0.4 (40453.)	41990.6±0.5 (42002.)	43568.9±0.4 (43574.)	45184.0±0.7 (45198.)	46834.2±0.5 (46849.)	48519.0±0.4 (48519.)	50239.1±0.5 (50233.)

TABLE 1b5. (*Continued*)

	57 La	58 Ce	59 Pr	60 Nd	61 Pm	62 Sm	63 Eu	64 Gd
L_I	6266.3±0.5 [6266.3][a]	6548.8±0.5 [6548.5][a]	6834.8±0.5 [6834.9][a]	7126.0±0.4 [7125.8][a] (7129.)	7427.9±0.8 [7427.9][a]	7736.8±0.5 [7736.2][a] (7748.)	8052.0±0.4 [8051.7][a] (8061.)	8375.6±0.5 [8375.4][a] (8386.)
L_{II}	5890.6±0.4 [5890.7][a]	6164.2±0.4 [6164.3][a]	6440.4±0.5 [6440.2][a]	6721.5±0.4 [6721.8][a] (6723.)	7012.8±0.6 [7012.8][b]·	7311.8±0.4 [7312.0][a] (7313.)	7617.1±0.4 [7617.6][b] (7620.)	7930.3±0.4 [7930.5][b] (7931.)
L_{III}	5482.7±0.4 [5482.6][a]	5723.4±0.4 [5723.6][a]	5964.3±0.4 [5964.3][a]	6207.9±0.4 [6208.0][a] (6209.)	6459.3±0.6 [6459.4][b]	6716.2±0.5 [6716.8][a] (6717.)	6976.9±0.4 [6976.7][a] (6981.)	7242.8±0.4 [7242.8][a] (7243.)
M_I	1361.3±0.3	1434.6±0.6	1511.0±0.8	1575.3±0.7		1722.8±0.8	1800.0±0.5	1880.8±0.5
M_{II}	1204.4±0.6	1272.8±0.6	1337.4±0.7	1402.8±0.6	1471.4±6.2	1540.7±1.2	1613.9±0.7	1688.3±0.7
M_{III}	1123.4±0.5	1185.4±0.5	1242.2±0.6	1297.4±0.5	1356.9±1.4	1419.8±1.1	1480.6±0.6	1544.0±0.8
M_{IV}	848.5±0.4	901.3±0.6	951.1±0.6	999.9±0.6	1051.5±0.9	1106.0±0.8	1160.6±0.6	1217.2±0.6
M_V	831.7±0.4	883.3±0.5	931.0±0.6	977.7±0.6	1026.9±1.0	1080.2±0.6	1130.9±0.6	1185.2±0.6
N_I	270.4±0.8	289.6±0.7	304.5±0.9	315.2±0.8		345.7±0.9	360.2±0.7	375.8±0.7
N_{II}	205.8±1.2	223.3±1.1	236.3±1.5	243.3±1.6	242.±16.	265.6±1.9	283.9±1.0	288.5±1.2
N_{III}	191.4±0.9	207.2±0.9	217.6±1.1	224.6±1.3		247.4±1.5	256.6±0.8	270.9±0.9
$N_{IV,V}$	98.9±0.8	110.0±0.6	113.2±0.7	117.5±0.7	120.4±2.0	129.0±1.2	133.2±0.6	140.5±0.8
$N_{VI,VII}$		0.1±1.2	2.0±0.6	1.5±0.6		5.5±1.1	0.0±3.2	0.1±3.5
O_I	32.3±7.2	37.8±1.3	37.4±1.0	37.5±0.9		37.4±1.5	31.8±0.7	36.1±0.8
$O_{II,III}$	14.4±1.2	19.8±1.2	22.3±0.7	21.1±0.8		21.3±1.5	22.0±0.6	20.3±1.2

	65 Tb	66 Dy	67 Ho	68 Er	69 Tm	70 Yb	-71 Lu	72 Hf
K	51995.7±0.5 (52002.)	53788.5±0.5 (53793.)	55617.7±0.5 (55619.)	57485.5±0.5 (57487.)	59389.6±0.5	61332.3±0.5 (61300.)	63313.8±0.5 (63310.)	65350.8±0.6 (65310.)
L_I	8708.0±0.5 [8707.6][a] (8717.)	9045.8±0.5 [9046.5][a]	9394.2±0.4 [9394.3][a] (9399.)	9751.3±0.4 [9751.5][b] (9757.)	10115.7±0.4 [10115.6][b] (10121.)	10486.4±0.4 [10487.3][a] (10490.)	10870.4±0.4 [10870.1][a] (10874.)	11270.7±0.4 [11271.6][b] (11274.)

TABLE 1b5. (Continued)

	65 Tb	66 Dy	67 Ho	68 Er	69 Tm	70 Yb	71 Lu	72 Hf
L_{II}	8251.6±0.4 [8251.8][a] (8253.)	8580.6±0.4 [8580.4][a] (8583.)	8917.8±0.4 [8918.2][a] (8916.)	9264.3±0.4 [9264.3][a] (9262.)	9616.9±0.4 [9617.1][a] (9617.1)	9978.2±0.4 [9977.9][a] (9976.)	10348.6±0.4 [10349.0][a] (10345.)	10739.4±0.4 [10738.9][a] (10736.)
L_{III}	7514.0±0.4 [7514.2][a] (7515.)	7790.1±0.4 [7789.6][a] (7789.7)	8071.1±0.4 [8070.6][a] (8068.)	8357.9±0.4 [8357.6][a] (8357.5)	8648.0±0.4 [8647.8][a] (8649.6)	8943.6±0.4 [8942.6][a] (8944.1)	9244.1±0.4 [9243.8][a]	9560.7±0.4 [9560.4][a] (9558.)
M_I	1967.5±0.6	2046.8±0.4	2128.3±0.6	2206.5±0.6	2306.8±0.7	2398.1±0.4	2491.2±0.5	2600.9±0.4
M_{II}	1767.7±0.9	1841.8±0.5	1922.8±1.0	2005.8±0.6	2089.8±1.1	2173.0±0.4	2263.5±0.4	2365.4±0.4
M_{III}	1611.3±0.8	1675.6±0.9	1741.2±0.9	1811.8±0.6	1884.5±1.1	1949.8±0.5	2023.6±0.5	2107.6±0.4
M_{IV}	1275.0±0.6	1332.5±0.4	1391.5±0.7	1453.3±0.5	1514.6±0.7	1576.3±0.4	1639.4±0.4	1716.4±0.4
M_V	1241.2±0.7	1294.9±0.4	1351.4±0.8	1409.3±0.5	1467.7±0.9	1527.8±0.4	1588.5±0.4	1661.7±0.4
N_I	397.9±0.8	416.3±0.5	435.7±0.8	449.1±1.0	471.7±0.9	487.2±0.6	506.2±0.6	538.1±0.4
N_{II}	310.2±1.2	331.8±0.6	343.5±1.4	366.2±1.5	385.9±1.6	396.7±0.7	410.1±1.8	437.0±0.5
N_{III}	285.0±1.0	292.9±0.6	306.6±0.9	320.0±0.7	336.6±1.6	343.5±0.5	359.3±0.5	380.4±0.5
N_{IV}	147.0±0.8	154.2±0.5	161.0±1.0	176.7±1.2	179.6±1.2	198.1±0.5	204.8±0.5	223.8±0.4
N_V				167.6±1.5		184.9±1.3	195.0±0.4	213.7±0.5
$N_{VI,VII}$	2.6±1.5	4.2±1.6	3.7±3.0	4.3±1.4	5.3±1.9	6.3±1.0	6.9±0.5	17.1±0.5
O_I	39.0±0.8	62.9±0.5	51.2±1.3	59.8±1.7	53.2±3.0	54.1±0.5	56.8±0.5	64.9±0.4
O_{II}	25.4±0.8	26.3±0.6	20.3±1.5	29.4±1.6	32.3±1.6	23.4±0.6	28.0±0.6	38.1±0.6
O_{III}								30.6±0.6

	73 Ta	74 W	75 Re	76 Os	77 Ir	78 Pt	79 Au	80 Hg
K	67416.4±0.6 (67403.)	69525.0±0.3 (69508.)	71676.4±0.4 (71658.)	73870.8±0.5	76111.0±0.5	78394.8±0.7 (78381.)	80724.9±0.5 (80720.)	83102.3±0.8
L_I	11681.5±0.3 [11680.2][b] (11682.)	12099.8±0.3 [12098.2][b] (12099.6)	12526.7±0.4 (12530.)	12968.0±0.4 (12972.)	13418.5±0.3 (13423.)	13879.9±0.4 (13883.)	14352.8±0.4 (14353.7)	14839.3±1.0 (14842.)

TABLE 1b5. (Continued)

	73 Ta	74 W	75 Re	76 Os	77 Ir	78 Pt	79 Au	80 Hg
L_{II}	11136.1±0.3 [11136.1]^p (11132.)	11544.0±0.3 [11541.4]^p (11538.)	11958.7±0.3 [11956.9]^p (11954.)	12385.0±0.4	12824.1±0.3 [12824.0]^{a,p} (12820.)	13272.6±0.3 [13272.6]^{a,p} (13272.3)	13733.6±0.3 [13733.5]^{a,p} (13736.)	14208.7±0.7 (14215.)
L_{III}	9881.1±0.3 [9880.3]^p (9877.7)	10206.8±0.3 [10204.2]^p (10200.)	10535.3±0.3 [10534.2]^p (10531.)	10870.9±0.3 [10870.7]^p (10868.)	11215.2±0.3 [11215.1]^{a,p} (11212.)	11563.7±0.3 [11563.7]^{a,p} (11562.)	11918.7±0.3 [11918.2]^{a,p} (11921.)	12283.9±0.4 [12284.0]^{a,p} (12286.)
M_I	2708.0±0.4	2819.6±0.4	2931.7±0.4	3048.5±0.4	3173.7±1.7	3296.0±0.9	3424.9±0.3 [3424.8]^p	3561.6±1.1
M_{II}	2468.7±0.3 [2468.6]^p	2574.9±0.3 [2575.0]^p	2681.6±0.4	2792.2±0.3 [2791.9]^p	2908.7±0.3 [2909.1]^p	3026.5±0.4 [3026.5]^p (3029.)	3147.8±0.4 [3149.5]^p	3278.5±1.3
M_{III}	2194.0±0.3 [2194.1]^p	2281.0±0.3 [2281.0]^p	2367.3±0.3 [2367.3]^p	2457.2±0.4 [2457.4]^p	2550.7±0.3 [2550.5]^p (2550.5)	2645.4±0.4 [2645.5]^p (2645.9)	2743.0±0.3 [2743.1]^p (2744.0)	2847.1±0.4 [2847.1]^p
M_{IV}	1793.2±0.3 [1793.1]^p	1871.6±0.3 [1871.4]^p	1948.9±0.3 [1948.9]^p	2030.8±0.3 [2031.0]^p	2116.1±0.3 [2116.1]^p	2201.9±0.3 [2201.9]^p	2291.1±0.3 [2291.2]^p (2307.)	2384.9±0.3 [2384.9]^p
M_V	1735.1±0.3 [1735.2]^p	1809.2±0.3 [1809.3]^p	1882.9±0.3 [1882.9]^p	1960.1±0.3 [1960.2]^p	2040.4±0.3 [2040.5]^p	2121.6±0.3 [2121.6]^p	2205.7±0.3 [2206.1]^p (2220.)	2294.9±0.3 [2294.9]^p
N_I	565.5±0.5	595.0±0.4	625.0±0.4	654.3±0.5	690.1±0.4	722.0±0.6	758.8±0.4	800.3±1.0
N_{II}	464.8±0.5	491.6±0.4	517.9±0.5	546.5±0.5	577.1±0.4	609.2±0.6	643.7±0.5	676.9±2.4
N_{III}	404.5±0.4	425.3±0.5	444.4±0.5	468.2±0.6	494.3±0.6	519.0±0.6	545.4±0.5	571.0±1.4
N_{IV}	241.3±0.4	258.8±0.4	273.7±0.5	289.4±0.5	311.4±0.4	330.8±0.5	352.0±0.4	378.3±1.0
N_V	229.3±0.3	245.4±0.4	260.2±0.4	272.8±0.6	294.9±0.4	313.3±0.4	333.9±0.4	359.8±1.2
N_{VI}	25.0±0.4	36.5±0.4	40.6±0.4	46.3±0.6	63.4±0.4	74.3±0.4	86.4±0.4	102.2±0.5
N_{VII}		33.6±0.4			60.5±0.4	71.1±0.5	82.8±0.5	98.5±0.5
O_I	71.1±0.5	77.1±0.4	82.8±0.5	83.7±0.6	95.2±0.4	101.7±0.4	107.8±0.7	120.3±1.3
O_{II}	44.9±0.4	46.8±0.5	45.6±0.7	58.0±1.1	63.0±0.6	65.3±0.7	71.7±0.7	80.5±1.3
O_{III}	36.4±0.4	35.6±0.5	34.6±0.6	45.4±1.0	50.5±0.6	51.7±0.7	53.7±0.7	57.6±1.3
$O_{IV,V}$	5.7±0.4	6.1±0.4	3.5±0.5		3.8±0.4	2.2±1.3	2.5±0.5	6.4±1.4

TABLE 1b5. *(Continued)*

	81 Tl	82 Pb	83 Bi	84 Po	85 At	86 Rn	87 Fr	88 Ra
K	85530.4±0.6 (85005.)	88004.5±0.7 (88005.)	90525.9±0.7 (90534.)	93105.0±3.8	95729.9±7.7	98404.±12.	101137.±13.	103921.9±7.2
L_I	15346.7±0.4 (15343.)	15860.8±0.5 (15855.)	16387.5±0.4 (16376.)	16939.3±9.8	17493.±29.	18049.±38.	18639.±40.	19236.7±1.5 (19236.0)
L_{II}	14697.9±0.3 [14697.3]^p (14699.)	15200.0±0.4 (15205.)	15711.1±0.3 [15708.4]^p (15719.)	16244.3±2.4	16784.7±2.5	17337.1±3.4	17906.5±3.5	18484.3±1.5 (18486.0)
L_{III}	12657.5±0.3 [12656.3]^{e,p} (12660.)	13035.2±0.3 [13034.9]^{e,p} (13041.)	13418.6±0.3 [13418.3]^{e,p} (13426.)	13813.8±1.0 ⟨13813.8⟩	14213.5±2.0 ⟨14213.5⟩	14619.4±3.0 ⟨14619.4⟩	15031.2±3.0 ⟨15031.2⟩	15444.4±1.5 (15444.0)
M_I	3704.1±0.4	3850.7±0.5	3999.1±0.3 [3999.1]^p	4149.4±3.9	⟨4317.⟩	⟨4482.⟩	⟨4652.⟩	4822.0±1.5
M_{II}	3415.7±0.3 [3415.7]^p	3554.2±0.3 [3554.2]^p	3696.3±0.3 [3696.4]^p	3854.1±9.8	4008.±28.	4159.±38.	4327.±40.	4489.5±1.8
M_{III}	2956.6±0.3 [2956.5]^p	3066.4±0.4 [3066.3]^p	3176.9±0.3 [3176.8]^p	3301.9±9.9	3426.±29.	3538.±38.	3663.±40.	3791.8±1.7
M_{IV}	2485.1±0.3 [2485.2]^p	2585.6±0.3 [2585.5]^p	2687.6±0.3 [2687.4]^p	2798.0±1.2	2908.7±2.1	3021.5±3.1	3136.2±3.1	3248.4±1.6
M_V	2389.3±0.3 [2389.4]^p	2484.0±0.3 [2484.2]^p (2502.)	2579.6±0.3 [2579.5]^p	2683.0±1.1	2786.7±2.1	2892.4±3.1	2999.9±3.1	3104.9±1.6
N_I	845.5±0.5	893.6±0.7	938.2±0.3 [938.7]^p	995.3±2.9	⟨1042.⟩	⟨1097.⟩	⟨1153.⟩	1208.4±1.6
N_{II}	721.3±0.8	763.9±0.8	805.3±0.3 [805.3]^p	851.±12.	886.±30.	929.±40.	980.±42.	1057.6±1.8
N_{III}	609.0±0.5	644.5±0.6	678.9±0.3 [678.9]^p	705.±14.	740.±30.	768.±40.	810.±43.	879.1±1.8
N_{IV}	406.6±0.4	435.2±0.5	463.6±0.3 [463.6]^p	500.2±2.4	533.2±3.2	566.6±4.0	603.3±4.1	635.9±1.6
N_V	386.2±0.5	412.9±0.6	440.0±0.3 [440.1]^p	473.4±1.3			577.±34.	602.7±1.7

TABLE 1b5. (Continued)

	81 Tl	82 Pb	83 Bi	84 Po	85 At	86 Rn	87 Fr	88 Ra
N_{VI}	122.8±0.4	142.9±0.4	161.9±0.5					298.9±2.4
N_{VII}	118.5±0.4	138.1±0.4	157.4±0.6					254.4±2.1
O_I	136.3±0.7	147.3±0.8	159.3±0.7					200.4±2.0
O_{II}	99.6±0.6	104.8±1.0	116.8±0.7					152.8±2.0
O_{III}	75.4±0.6	86.0±1.0	92.8±0.6					
O_{IV}	15.3±0.4	21.8±0.4	26.5±0.5	31.4±3.2				67.2±1.7
O_V	13.1±0.4	19.2±0.4	24.4±0.6					
P_I		3.1±1.0						43.5±2.2
$P_{II,III}$		0.7±1.0	2.7±0.7					18.8±1.8

	89 Ac	90 Th	91 Pa	92 U	93 Np	94 Pu	95 Am	96 Cm
K	106755.3±5.3	109650.9±0.9	112601.4±2.4	115606.1±1.6	118678.±33.	121818.±44.	125027.±55.	128220
L_I	19840.±18.	20472.1±0.5	21104.6±1.8	21757.4±0.3	22426.8±0.9	23097.2±1.6	23772.9±2.0	24460
		(20464.)	(21128.)	(21771.)		(23109.)	(23772.9)	
L_{II}	19083.2±2.8	19693.2±0.4	20313.7±1.5	20947.6±0.3	21600.5±0.4	22266.2±0.7	22944.0±1.0	23779
		(19683.)	(20319.)	(20945.)		(22253.)		
L_{III}	15871.0±2.0	16300.3±0.3	16733.1±1.4	17166.3±0.3	17610.0±0.4	18056.8±0.6	18504.1±0.9	18930
	(15871.0)	[16299.6][a]	(16733.)	[17168.5][r]	(17606.2)	(18053.1)	(18504.1)	
		(16299.)		(17165.)				
M_I	⟨5002.⟩	5182.3±0.3	5366.9±1.6	5548.0±0.4	5723.2±3.6	5932.9±1.4	6120.5±7.5	6288
		[5182.3][a]						
M_{II}	4656.±18.	4830.4±0.4	5000.9±2.3	5182.2±0.4	5366.2±0.7	5541.2±1.7	5710.2±2.1	5895
		[4830.6][a]		[5180.9][r]	[5366.4][a]			
M_{III}	3909.±18.	4046.1±0.4	4173.8±1.8	4303.4±0.3	4434.7±0.5	4556.6±1.5	4667.0±2.1	4797
		[4046.1][a]		[4303.6][r]	[4434.6][a]			
		(4041.)		(4299.)				
M_{IV}	3370.2±2.1	3490.8±0.3	3611.2±1.4	3727.6±0.3	3850.3±0.4	3972.6±0.6	4092.1±1.0	4227
		[3490.7][a]	(3608.)	[3728.1][r]	[3849.8][a]	[3972.7][a]		
		(3485.)		(3720.)				

TABLE 1b5. (Continued)

	89 Ac	90 Th	91 Pa	92 U	93 Np	94 Pu	95 Am	96 Cm
M_V	3219.0±2.1	3332.0±0.3 [3332.1]^a (3325.)	3441.8±1.4 (3436.)	3551.7±0.3 [3551.7]^r (3545.)	3665.8±0.4 [3664.2]^r	3778.1±0.6 [3778.0]^r	3886.9±1.0	3971
N_I	(1269.)	1329.5±0.4 [1329.8]^a	1387.1±1.9	1440.8±0.4 [1441.3]^r	1500.7±0.8 [1500.7]^r	1558.6±0.8	1617.1±1.1	1643
N_{II}	1080.±19.	1168.2±0.4 [1168.3]^a	1224.3±1.6	1272.6±0.3 [1272.5]^r	1327.7±0.8 [1327.7]^r	1372.1±1.8	1411.8±8.3	1440
N_{III}	890.±19.	967.3±0.4 [967.6]^a	1006.7±1.7	1044.9±0.3 [1044.9]^r	1086.8±0.7 [1086.8]^r	1114.8±1.6	⟨1135.7⟩	1154
N_{IV}	674.9±3.7	714.1±0.4 [714.4]^a	743.4±2.1	780.4±0.3 [779.7]^r	815.9±0.5 [817.1]^r	848.9±0.6 [848.9]^r	878.7±1.0	
N_V		676.4±0.4 [676.4]^a	708.2±1.8	737.7±0.3 [737.6]^r	770.3±0.4 [773.2]^r	801.4±0.6 [801.4]^r	827.6±1.0	
N_{VI}		344.4±0.3 [344.2]^a	371.2±1.6	391.3±0.6	415.0±0.8 [415.0]^r	445.8±1.7		
N_{VII}		335.2±0.4 [335.0]^a	359.5±1.6	380.9±0.9	404.4±0.5 [404.4]^r	432.4±2.1		
O_I		290.2±0.8	309.6±4.3	323.7±1.1		351.9±2.4		385
O_{II}		229.4±1.1	222.9±3.9	259.3±0.5	283.4±0.8 [283.4]^r	274.1±4.7		
O_{III}		181.8±0.4 [181.8]^a		195.1±1.3	206.1±0.7 [206.1]^r	206.5±4.7		
O_{IV}		94.3±0.4 [94.4]^a	94.1±2.8	105.0±0.5	109.3±0.7 [108.8]^r	116.0±1.2	115.8±1.3	
O_V		87.9±0.3 [88.1]^a		96.3±1.4	101.3±0.5 [101.4]^r	105.4±1.0	103.3±1.1	
P_I		59.5±1.1		70.7±1.2				
P_{II}		49.0±2.5		42.3±9.0				
P_{III}		43.0±2.5		32.3±9.0				

TABLE 1b5. *(Continued)*

	97 Bk	98 Cf	99 Es	100 Fm	101 Md	102 No	103 Lw
K	[131590±40][u]	135960	139490	143090	146780	150540	154380
L_I	[25275±17][u]	26110	26900	27700	28530	29380	30240
L_{II}	[24385±17][u]	25250	26020	26810	27610	28440	29280
L_{III}	[19452±20][u]	19930	20410	20900	21390	21880	22360
M_I	[6556±21][u]	6754	6977	7205	7441	7675	7900
M_{II}	[6147±31][u]	6359	6574	6793	7019	7245	7460
M_{III}	[4977±31][u]	5109	5252	5397	5546	5688	5710
M_{IV}	4366	4497	4630	4766	4903	5037	5150
M_V	4132	4253	4374	4498	4622	4741	4860
N_I	[1755±22][u]	1799	1868	1937	2010	2078	2140
N_{II}	1554	1616	1680	1747	1814	1876	1930
N_{III}	1235	1279	1321	1366	1410	1448	1480
O_I	[398±22][u]	419	435	454	472	484	490

[a] J. E. Mack, 1949, as given in C. E. Moore, *Atomic Energy Levels* (U. S. National Bureau of Standards, Washington, D. C., 1949), Vol. 1, p. 1.

[b] G. Herzberg, 1957, as given in C. E. Moore, *Atomic Energy Levels* (U. S. National Bureau of Standards, Washington, D. C., 1958), Vol. 3, p. 238.

[c] See Ref. 18.

[d] A. Fahlman, D. Hamrin, R. Nordberg, C. Nordling, and K. Siegbahn, Phys. Rev. Letters **14**, 127 (1965). See also Ref. 26.

[e] See Ref. 15.

[f] See Ref. 11.

[g] C. Nordling, Arkiv Fysik **15**, 397 (1959).

[h] E. Sokolowski, C. Nordling, and K. Siegbahn, Arkiv Fysik **12**, 301 (1957).

[i] C. Nordling and S. Hagström, Arkiv Fysik **16**, 515 (1960).

[j] I. Andersson and S. Hagström, Arkiv Fysik **27**, 161 (1964).

[k] M. O. Krause, Phys. Rev. **140**, A1845 (1965).

[l] A. Fahlman, O. Hörnfeldt, and C. Nordling, Arkiv Fysik **23**, 75 (1962).

[m] P. Bergvall, O. Hörnfeldt, and C. Nordling, Arkiv Fysik **17**, 113 (1960).

[n] P. Bergvall and S. Hagström, Arkiv Fysik **17**, 61 (1960).

[o] S. Hagström, Z. Physik **178**, 82 (1964).

[p] A. Fahlman and S. Hagström, Arkiv Fysik **27**, 69 (1964).

[q] C. Nordling and S. Hagström, Z. Physik **178**, 418 (1964).

[r] C. Nordling and S. Hagström, Arkiv Fysik **15**, 431 (1959).

[s] S. Hagström, Bull. Am. Phys. Soc. **11**, 389 (1966).

[t] A. Fahlman, K. Hamrin, R. Nordberg, C. Nordling, K. Siegbahn, and L. W. Holm, Phys. Letters **19**, 643 (1966).

[u] J. M. Hollander, M. D. Holtz, T. Novakov, and R. L. Graham, Arkiv Fysik **28**, 375 (1965).

*Reprinted with permission of the author and editor from *Revs. Mod. Phys.* Vol. 39, No. 1, 125–142, 1967.

SECTION 1c
MATERIALS CLASSES

There are many ways to classify materials, and the following examples will illustrate this contention.

"Solids," "liquids'" and "gases" represent a classification by the substances' physical state. The metallurgist looks at "ferrous" and "non-ferrous" alloys. The organic chemist may see organic and inorganic substances, where he divides the former into many categories such as "hydrocarbons," "aldehydes," "ketones," "esters," "carbohydrates," etc.

The applications-oriented materials engineer tends to view materials as substances needed to build a device.

It should be apparent to the reader by this time that each "class" of scientist and engineer designates materials in a manner most suitable to him.

The engineering materials classification, presented in Figure 1c1, represents an applications-oriented scheme that is relatively broad as it incorporates materials designations as well as applications.

FIGURE 1c1. ENGINEERING MATERIALS TECHNOLOGY.*

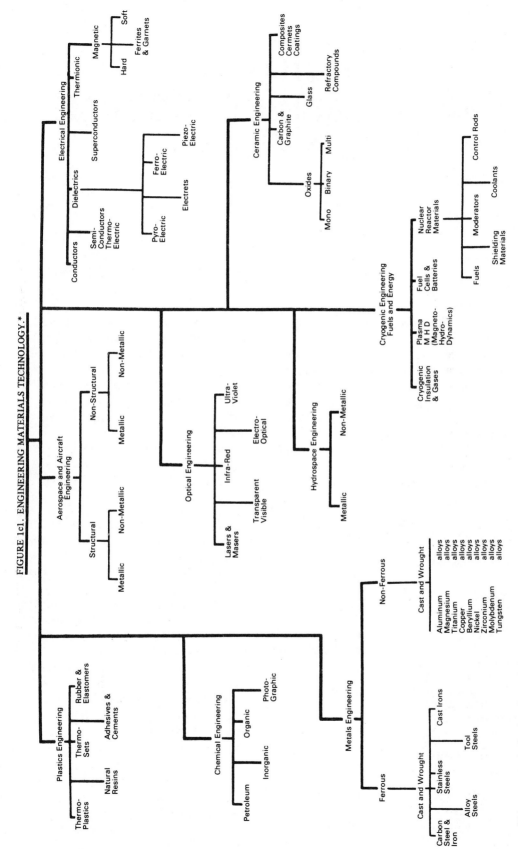

*Private communication from J. T. Milek.

111

SECTION 1d
MATERIALS SELECTION CRITERIA

Selection of a material for a given application must be based on the following logical steps: (a) establishing the performance requirements that the material must meet; (b) selecting the appropriate materials for evaluation; and (c) selecting the material that best meets the performance requirements and economic constraints.

Establishing performance requirements involves knowledge of a number of conditions which may include: (1) corrosion resistance; (2) ductility; (3) electrical resistivity; (4) environment; (5) mechanical strength; (6) specific gravity; (7) stability; (8) thermal conductivity, and (9) other properties.

Economic constraints may be imposed by: (1) materials cost; (2) availability, and (3) fabricability.

Many causes of material failure are attributable to changes that took place in the material, and Table 1d1 summarizes such changes.

TABLE 1d1. SOME CAUSES OF MATERIAL FAILURE. (Bushey, 1960)

A Material can fail because it . . .

• Bends	• Softens
• Breaks	• Sticks
• Freezes	• Burns
• Melts	• Chars
• Vaporizes	• Dissolves
• Corrodes	• Changes color

- Changes transparency
- Gains, or loses, conductivity
- Conducts, or fails to conduct, heat
- Develops friction
- Gains, or loses, porosity
- Ionizes, or fails to ionize
- Hydrates, or dehydrates
- Loses magnetization
- Changes viscosity

The relationship between materials properties, the internal structure of a material, and service conditions is evident from Figure 1d1.

Impurities in materials can have an enormous effect on the properties of such materials. For example, phosphorus in iron leads to the formation of a brittle, hard compound that lowers the melting point and affects the properties of the iron.

The effect of impurities is most apparent in semiconductors, where one part of arsenic in ten billion parts of germanium has produced a change in electrical resistivity of 15%.

Therefore, the data presented here are generally guides. The actual selection of a product should be based on specific data obtained from the vendor, or on performance tests made by the potential user when impurity-sensitive materials applications are planned.

REFERENCES

Bushey, A. H., "The Chemistry of Materials Failure," *Mat. Design Eng.* 7, 99–105 (1960).

Smith, C. O., "The Science of Engineering Materials," Prentice-Hall, Inc., Englewood Cliffs, N.J., 1969.

PROPERTIES OF MATERIALS DEPEND UPON:

Fig. 1d1. Properties of materials are related to internal structure and to service conditions (Smith 1969). Reproduced from: Charles O. Smith, "The Science of Engineering Materials," © 1969. By permission of Prentice-Hall, Inc., Englewood Cliffs, New Jersey.

Section 1e
PHYSICAL PROPERTIES

This section contains information on units of weight and measure, fundamental constants, temperature calibration points, and selected other data.

The information is presented in the following tables and figure:

SPELLING AND SYMBOLS FOR UNITS

The spelling of the names of units as adopted by the National Bureau of Standards is that given in the list below. The spelling of the metric units is in accordance with that given in the law of July 28, 1866, legalizing the Metric System in the United States.

Following the name of each unit in the list below is given the symbol that the Bureau has adopted. Attention is particularly called to the following principles:

1. No period is used with symbols for units. Whenever "in" for inch might be confused with the preposition "in", "inch" should be spelled out.

2. The exponents "2" and "3" are used to signify "square" and "cubic," respectively, instead of the symbols "sq" or "cu," which are, however, frequently used in technical literature for the U. S. Customary units.

3. The same symbol is used for both singular and plural.

TABLE 1e1. SOME UNITS AND THEIR SYMBOLS.##

Unit	Symbol	Unit	Symbol	Unit	Symbol
acre	acre	fathom	fath	millimeter	mm
are	a	foot	ft	minim	minim
barrel	bbl	furlong	furlong	ounce	oz
board foot	fbm	gallon	gal	ounce, avoirdupois	oz avdp
bushel	bu	grain	grain	ounce, liquid	liq oz
carat	c	gram	g	ounce, troy	oz tr
Celsius, degree	°C	hectare	ha	peck	peck
centare	ca	hectogram	hg	pennyweight	dwt
centigram	cg	hectoliter	hl	pint, liquid	liq pt
centiliter	cl	hectometer	hm	pound	lb
centimeter	cm	hogshead	hhd	pound, avoirdupois	lb avdp
chain	ch	hundredweight	cwt	pound, troy	lb tr
cubic centimeter	cm³	inch	in	quart, liquid	liq qt
cubic decimeter	dm³	International		rod	rod
cubic dekameter	dam³	Nautical Mile	INM	second	s
cubic foot	ft³	Kelvin, degree	°K	square centimeter	cm²
cubic hectometer	hm³	kilogram	kg	square decimeter	dm²
cubic inch	in³	kiloliter	kl	square dekameter	dam²
cubic kilometer	km³	kilometer	km	square foot	ft²
cubic meter	m³	link	link	square hectometer	hm²
cubic mile	mi³	liquid	liq	square inch	in²
cubic millimeter	mm³	liter	liter	square kilometer	km²
cubic yard	yd³	meter	m	square meter	m²
decigram	dg	microgram	μg	square mile	mi²
deciliter	dl	microinch	μin	square millimeter	mm²
decimeter	dm	microliter	μl	square yard	yd²
dekagram	dag	micron	μm	stere	stere
dekaliter	dal	mile	mi	ton, long	long ton
dekameter	dam	milligram	mg	ton, metric	t
dram, avoirdupois	dr avdp	milliliter	ml	ton, short	short ton
				yard	yd

Tables 1e1 through 1e9 are reproduced from: Chisholm, L. J., "Units of Weight and Measure, International (Metric) and U.S. Customary," NBS Mics. Publ. 286, U.S. Government Printing Office, Washington, D.C., 1967.

TABLE 1e2. PREFIXES.##

The following prefixes, in combination with the basic unit names, provide the multiples and submultiples in the International System. For example, the unit name "meter," with the prefix "kilo" added, produces "kilometer," meaning "1000 meters."

Multiples and Submultiples	Prefixes	Symbols	Pronunciations
10^{12}	tera	T	tĕr′à
10^9	giga	G	ji′gà
10^6	mega	M	mĕg′à
10^3	kilo	k	kĭl′ŏ
10^2	hecto	h	hĕk′tŏ
10	deka	da	dĕk′à
10^{-1}	deci	d	dĕs′ĭ
10^{-2}	centi	c	sĕn′tĭ
10^{-3}	milli	m	mĭl ĭ
10^{-6}	micro	μ	mī′krŏ
10^{-9}	nano	n	năn′ŏ
10^{-12}	pico	p	pē′cŏ
10^{-15}	femto	f	fĕm′tŏ
10^{-18}	atto	a	ăt′tŏ

TABLE 1e3. DEFINITIONS OF UNITS.[##]
Length

A *meter* is a unit of length equal to 1 650 763.73 wavelengths in a vacuum of the orange-red radiation of krypton 86.

A *yard* is a unit of length equal to 0.914 4 meter.

Mass

A *kilogram* is a unit of mass equal to the mass of the International Prototype Kilogram.

An *avoirdupois pound* is a unit of mass equal to 0.453 592 37 kilogram.

Capacity, or Volume

A *cubic meter* is a unit of volume equal to a cube the edges of which are 1 meter.

A *liter* is a unit of volume equal to a cubic decimeter.

A *cubic yard* is a unit of volume equal to a cube the edges of which are 1 yard.

A *gallon* is a unit of volume equal to 231 cubic inches. It is used for measuring liquids only.

A *bushel* is a unit of volume equal to 2 150.42 cubic inches. It is used for measuring dry commodities only.

Area

A *square meter* is a unit of area equal to the area of a square the sides of which are 1 meter.

A *square yard* is a unit of area equal to the area of a square the sides of which are 1 yard.

TABLE 1e3 (*Continued*)

Six units have been adopted to serve as the base for the International System:

Length _____meter
Mass_____kilogram
Time_____second
Electric current_____ampere
Thermodynamic temperature_____degree Kelvin
Light intensity_____candela

Some of the other more frequently used units of the SI and their symbols and, where applicable, their derivations are listed below.

SUPPLEMENTARY UNITS

Quantity	*Unit*	*Symbol*	*Derivation*
Plane angle	radian	rad	
Solid angle	steradian	sr	

DERIVED UNITS

Quantity	Unit	Symbol	Derivation
Area	square meter	m^2	
Volume	cubic meter	m^3	
Frequency	hertz	Hz	(s^{-1})
Density	kilogram per cubic meter	kg/m^3	
Velocity	meter per second	m/s	
Angular velocity	radian per second	rad/s	
Acceleration	meter per second squared	m/s^2	
Angular acceleration	radian per second squared	rad/s^2	
Force	newton	N	$(kg \cdot m/s^2)$
Pressure	newton per square meter	N/m^2	
Kinematic viscosity	square meter per second	m^2/s	
Dynamic viscosity	newton-second per square meter	$N \cdot s/m^2$	
Work, energy, quantity of heat	joule	J	$(N \cdot m)$
Power	watt	W	(J/s)
Electric charge	coulomb	C	$(A \cdot s)$
Voltage, potential difference, electromotive force	volt	V	(W/A)
Electric field strength	volt per meter	V/m	
Electric resistance	ohm	Ω	(V/A)
Electric capacitance	farad	F	$(A \cdot s/V)$
Magnetic flux	weber	Wb	$(V \cdot s)$
Inductance	henry	H	$(V \cdot s/A)$
Magnetic flux density	tesla	T	(Wb/m^2)
Magnetic field strength	ampere per meter	A/m	
Magnetomotive force	ampere	A	
Flux of light	lumen	lm	$(cd \cdot sr)$
Luminance	candela per square meter	cd/m^2	
Illumination	lux	lx	(lm/m^2)

TABLE 1e4. CONVERSION FACTORS*—UNITS OF LENGTH.##

To Convert from **Centimeters**	
To	Multiply by
Inches------------------------	0.393 700 8
Feet--------------------------	0.032 808 40
Yards-------------------------	0.010 936 13
Meters------------------------	**0.01**

To Convert from **Meters**	
To	Multiply by
Inches--------------------	39.370 08
Feet----------------------	3.280 840
Yards---------------------	1.093 613
Miles---------------------	0.000 621 37
Millimeters--------------	**1 000**
Centimeters--------------	**100**
Kilometers---------------	**0.001**

To Convert from **Inches**	
To	Multiply by
Feet-----------------------	0.083 333 33
Yards----------------------	0.027 777 78
Centimeters----------------	**2.54**
Meters---------------------	**0.025 4**

To Convert from **Feet**	
To	Multiply by
Inches--------------------	**12**
Yards---------------------	0.333 333 3
Miles---------------------	0.000 189 39
Centimeters--------------	**30.48**
Meters--------------------	**0.304 8**
Kilometers---------------	**0.000 304 8**

* All boldface figures are exact; the others generally are given to seven significant figures.

In using conversion factors, it is possible to perform division as well as the multiplication process shown here. Division may be particularly advantageous where more than the significant figures published here are required. Division may be performed in lieu of multiplication by using the reciprocal of any indicated multiplier as divisor. For example, to convert from centimeters to inches by division, refer to the table headed "To Convert from *Inches*" and use the factor listed at "centimeters" (*2.54*) as divisor.

To Convert from **Yards**	
To	Multiply by
Inches-------------------	**36**
Feet---------------------	**3**
Miles--------------------	0.000 568 18
Centimeters--------------	**91.44**
Meters-------------------	**0.914 4**

To Convert from **Miles**	
To	Multiply by
Inches--------------------	**63 360**
Feet----------------------	**5 280**
Yards---------------------	**1 760**
Centimeters--------------	**160 934.4**
Meters--------------------	**1 609.344**
Kilometers---------------	**1.609 344**

TABLE 1e5. CONVERSION FACTORS—UNITS OF AREA.##

To Convert from **Square Centimeters**

To	Multiply by
Square Inches	0.155 000 3
Square Feet	0.001 076 39
Square Yards	0.000 119 599
Square Meters	**0.000 1**

To Convert from **Square Meters**

To	Multiply by
Square Inches	1 550.003
Square Feet	10.763 91
Square Yards	1.195 990
Acres	0.000 247 105
Square Centimeters	**10 000**
Hectares	**0.000 1**

To Convert from **Hectares**

To	Multiply by
Square Feet	107 639.1
Square Yards	11 959.90
Acres	2.471 054
Square Miles	0.003 861 02
Square Meters	**10 000**

To Convert from **Square Inches**

To	Multiply by
Square Feet	0.006 944 44
Square Yards	0.000 771 605
Square Centimeters	**6.451 6**
Square Meters	**0.000 645 16**

To Convert from **Square Feet**

To	Multiply by
Square Inches	**144**
Square Yards	0.111 111 1
Acres	0.000 022 957
Square Centimeters	**929.030 4**
Square Meters	**0.092 903 04**

To Convert from **Square Yards**

To	Multiply by
Square Inches	**1 296**
Square Feet	**9**
Acres	0.000 206 611 6
Square Miles	0.000 000 322 830 6
Square Centimeters	**8 361.273 6**
Square Meters	**0.836 127 36**
Hectares	**0.000 083 612 736**

To Convert from **Acres**

To	Multiply by
Square Feet	**43 560**
Square Yards	**4 840**
Square Miles	0.001 562 5
Square Meters	4 046.856 422 4
Hectares	0.404 685 642 24

To Convert from **Square Miles**

To	Multiply by
Square Feet	**27 878 400**
Square Yards	**3 097 600**
Acres	**640**
Square Meters	**2 589 988.110 336**
Hectares	**258.998 811 033 6**

TABLE 1e6. CONVERSION FACTORS—UNITS OF CAPACITY, OR VOLUME, LIQUID MEASURE.##

To	To Convert from **Milliliters** Multiply by
Minims	16.230 73
Liquid Ounces	0.033 814 02
Gills	0.008 453 5
Liquid Pints	0.002 113 4
Liquid Quarts	0.001 056 7
Gallons	0.000 264 17
Cubic Inches	0.061 023 74
Liters	**0.001**

To	To Convert from **Liters** Multiply by
Liquid Ounces	33.814 02
Gills	8.453 506
Liquid Pints	2.113 376
Liquid Quarts	1.056 688
Gallons	0.264 172 05
Cubic Inches	61.023 74
Cubic Feet	0.035 314 67
Milliliters	**1 000**
Cubic Meters	**0.001**
Cubic Yards	0.001 307 95

To	To Convert from **Cubic Meters** Multiply by
Gallons	264.172 05
Cubic Inches	61 023.74
Cubic Feet	35.314 67
Liters	**1 000**
Cubic Yards	1.307 950 6

To	To Convert from **Minims** Multiply by
Liquid Ounces	0.002 083 33
Gills	0.000 520 83
Cubic Inches	0.003 759 77
Milliliters	0.061 611 52

To	To Convert from **Liquid Quarts** Multiply by
Minims	**15 360**
Liquid Ounces	**32**
Gills	**8**
Liquid Pints	**2**
Gallons	**0.25**
Cubic Inches	57.75
Cubic Feet	0.033 420 14
Milliliters	**946.352 946**
Liters	**0.946 352 946**

To	To Convert from **Gallons** Multiply by
Minims	**61 440**
Liquid Ounces	**128**
Gills	**32**
Liquid Pints	**8**
Liquid Quarts	**4**
Cubic Inches	**231**
Cubic Feet	0.133 680 6
Milliliters	**3 785.411 784**
Liters	**3.785 411 784**
Cubic Meters	**0.003 785 411 784**
Cubic Yards	0.004 951 13

TABLE 1e6 (*Continued*)

To Convert from **Gills**	
To	Multiply by
Minims	1 920
Liquid Ounces	4
Liquid Pints	0.25
Liquid Quarts	0.125
Gallons	0.031 25
Cubic Inches	7.218 75
Cubic Feet	0.004 177 517
Milliliters	118.294 118 25
Liters	0.118 294 118 25

To Convert from **Liquid Pints**	
To	Multiply by
Minims	7 680
Liquid Ounces	16
Gills	4
Liquid Quarts	0.5
Gallons	0.125
Cubic Inches	28.875
Cubic Feet	0.016 710 07
Milliliters	473.176 473
Liters	0.473 176 473

To Convert from **Liquid Ounces**	
To	Multiply by
Minims	480
Gills	0.25
Liquid Pints	0.062 5
Liquid Quarts	0.031 25
Gallons	0.007 812 5
Cubic Inches	1.804 687 5
Cubic Feet	0.001 044 38
Milliliters	29.573 53
Liters	0.029 573 53

To Convert from **Cubic Feet**	
To	Multiply by
Liquid Ounces	957.506 5
Gills	239.376 6
Liquid Pints	59.844 16
Liquid Quarts	29.922 08
Gallons	7.480 519
Cubic Inches	1 728
Liters	28.316 846 592
Cubic Meters	0.028 316 846 592
Cubic Yards	0.037 037 04

To Convert from **Cubic Inches**	
To	Multiply by
Minims	265.974 0
Liquid Ounces	0.554 112 6
Gills	0.138 528 1
Liquid Pints	0.034 632 03
Liquid Quarts	0.017 316 02
Gallons	0.004 329 0
Cubic Feet	0.000 578 7
Milliliters	16.387 064
Liters	0.016 387 064
Cubic Meters	0.000 016 387 064
Cubic Yards	0.000 021 43

To Convert from **Cubic Yards**	
To	Multiply by
Gallons	201.974 0
Cubic Inches	46 656
Cubic Feet	27
Liters	764.554 857 984
Cubic Meters	0.764 554 857 984

TABLE 1e7. CONVERSION FACTORS—UNITS OF CAPACITY, OR VOLUME, DRY MEASURE.##

To Convert from **Liters**	
To	Multiply by
Dry Pints	1.816 166
Dry Quarts	0.908 082 98
Pecks	0.113 510 4
Bushels	0.028 377 59
Dekaliters	**0.1**

To Convert from **Cubic Meters**	
To	Multiply by
Pecks	113.510 4
Bushels	28.377 59

To Convert from **Dry Quarts**	
To	Multiply by
Dry Pints	**2**
Pecks	**0.125**
Bushels	**0.031 25**
Cubic Inches	**67.200 625**
Cubic Feet	0.038 889 25
Liters	1.101 221
Dekaliters	0.110 122 1

To Convert from **Bushels**	
To	Multiply by
Dry Pints	**64**
Dry Quarts	**32**
Pecks	**4**
Cubic Inches	**2 150.42**
Cubic Feet	1.244 456
Liters	35.239 07
Dekaliters	3.523 907
Cubic Meters	0.035 239 07
Cubic Yards	0.046 090 96

To Convert from **Cubic Feet**	
To	Multiply by
Dry Pints	51.428 09
Dry Quarts	25.714 05
Pecks	3.214 256
Bushels	0.803 563 95

To Convert from **Dekaliters**	
To	Multiply by
Dry Pints	18.161 66
Dry Quarts	9.080 829 8
Pecks	1.135 104
Bushels	0.283 775 9
Cubic Inches	610.237 4
Cubic Feet	0.353 146 7
Liters	**10**

To Convert from **Dry Pints**	
To	Multiply by
Dry Quarts	**0.5**
Pecks	**0.062 5**
Bushels	**0.015 625**
Cubic Inches	**33.600 312 5**
Cubic Feet	0.019 444 63
Liters	0.550 610 47
Dekaliters	0.055 061 05

To Convert from **Pecks**	
To	Multiply by
Dry Pints	**16**
Dry Quarts	**8**
Bushels	**0.25**
Cubic Inches	**537.605**
Cubic Feet	0.311 114
Liters	8.809 767 5
Dekaliters	0.880 976 75
Cubic Meters	0.008 809 77
Cubic Yards	0.011 522 74

To Convert from **Cubic Inches**	
To	Multiply by
Dry Pints	0.029 761 6
Dry Quarts	0.014 880 8
Pecks	0.001 860 10
Bushels	0.000 465 025

To Convert from **Cubic Yards**	
To	Multiply by
Pecks	86.784 91
Bushels	21.696 227

TABLE 1e8. CONVERSION FACTORS—UNITS OF MASS.##

To Convert from Grams

To	Multiply by
Grains	15.432 36
Avoirdupois Drams	0.564 383 4
Avoirdupois Ounces	0.035 273 96
Troy Ounces	0.032 150 75
Troy Pounds	0.002 679 23
Avoirdupois Pounds	0.002 204 62
Milligrams	1 000
Kilograms	0.001

To Convert from Kilograms

To	Multiply by
Grains	15 432.36
Avoirdupois Drams	564.383 4
Avoirdupois Ounces	35.273 96
Troy Ounces	32.150 75
Troy Pounds	2.679 229
Avoirdupois Pounds	2.204 623
Grams	1 000
Short Hundredweights	0.022 046 23
Short Tons	0.001 102 31
Long Tons	0.000 984 2
Metric Tons	0.001

To Convert from Metric Tons

To	Multiply by
Avoirdupois Pounds	2 204.623
Short Hundredweights	22.046 23
Short Tons	1.102 311 3
Long Tons	0.984 206 5
Kilograms	1 000

To Convert from Grains

To	Multiply by
Avoirdupois Drams	0.036 571 43
Avoirdupois Ounces	0.002 285 71
Troy Ounces	0.002 083 33
Troy Pounds	0.000 173 61
Avoirdupois Pounds	0.000 142 86
Milligrams	64.798 91
Grams	0.064 798 91
Kilograms	0.000 064 798 91

To Convert from Avoirdupois Ounces

To	Multiply by
Grains	437.5
Avoirdupois Drams	16
Troy Ounces	0.911 458 3
Troy Pounds	0.075 954 86
Avoirdupois Pounds	0.062 5
Grams	28.349 523 125
Kilograms	0.028 349 523 125

To Convert from Avoirdupois Pounds

To	Multiply by
Grains	7 000
Avoirdupois Drams	256
Avoirdupois Ounces	16
Troy Ounces	14.583 33
Troy Pounds	1.215 278
Grams	453.592 37
Kilograms	0.453 592 37
Short Hundredweights	0.01
Short Tons	0.000 5
Long Tons	0.000 446 428 6
Metric Tons	0.000 453 592 37

To Convert from Short Hundredweights

To	Multiply by
Avoirdupois Pounds	100
Short Tons	0.05
Long Tons	0.044 642 86
Kilograms	45.359 237
Metric Tons	0.045 359 237

TABLE 1e8 (*Continued*)

To Convert from **Short Tons**	
To	Multiply by
Avoirdupois Pounds	**2 000**
Short Hundredweights	**20**
Long Tons	0.892 857 1
Kilograms	**907.184 74**
Metric Tons	**0.907 184 74**

To Convert from **Long Tons**	
To	Multiply by
Avoirdupois Ounces	**35 840**
Avoirdupois Pounds	**2 240**
Short Hundredweights	**22.4**
Short Tons	**1.12**
Kilograms	**1 016.046 908 8**
Metric Tons	**1.016 046 908 8**

To Convert from **Troy Ounces**	
To	Multiply by
Grains	**480**
Avoirdupois Drams	17.554 29
Avoirdupois Ounces	1.097 143
Troy Pounds	0.083 333 3
Avoirdupois Pounds	0.068 571 43
Grams	**31.103 476 8**

To Convert from **Troy Pounds**	
To	Multiply by
Grains	**5 760**
Avoirdupois Drams	210.651 4
Avoirdupois Ounces	13.165 71
Troy Ounces	**12**
Avoirdupois Pounds	0.822 857 1
Grams	**373.241 721 6**

TABLE 1e9. FRACTIONAL AND DECIMAL LENGTH CONVERSION.##

Length—Inches and Millimeters—Equivalents of Decimal and Binary Fractions of an Inch in Millimeters

From 1/64 to 1 Inch

½'s	¼'s	8ths	16ths	32ds	64ths	Milli-meters	Decimals of an inch	Inch	½'s	¼'s	8ths	16ths	32ds	64ths	Milli-meters	Decimals of an inch
					1	= 0.397	0.015625							33	=13.097	0.515625
				1	2	= .794	.03125						17	34	=13.494	.53125
					3	= 1.191	.046875							35	=13.891	.546875
			1	2	4	= 1.588	.0625					9	18	36	=14.288	.5625
					5	= 1.984	.078125							37	=14.684	.578125
				3	6	= 2.381	.09375						19	38	=15.081	.59375
					7	= 2.778	.109375							39	=15.478	.609375
		1	2	4	8	= 3.175	.1250				5	10	20	40	=15.875	.625
					9	= 3.572	.140625							41	=16.272	.640625
				5	10	= 3.969	.15625						21	42	=16.669	.65625
					11	= 4.366	.171875							43	=17.066	.671875
			3	6	12	= 4.762	.1875					11	22	44	=17.462	.6875
					13	= 5.159	.203125							45	=17.859	.703125
				7	14	= 5.556	.21875						23	46	=18.256	.71875
					15	= 5.953	.234375							47	=18.653	.734375
	1	2	4	8	16	= 6.350	.2500			3	6	12	24	48	=19.050	.75
					17	= 6.747	.265625							49	=19.447	.765625
				9	18	= 7.144	.28125						25	50	=19.844	.78125
					19	= 7.541	.296875							51	=20.241	.796875
			5	10	20	= 7.938	.3125					13	26	52	=20.638	.8125
					21	= 8.334	.328125							53	=21.034	.828125
				11	22	= 8.731	.34375						27	54	=21.431	.84375
					23	= 9.128	.359375							55	=21.828	.859375
		3	6	12	24	= 9.525	.3750				7	14	28	56	=22.225	.875
					25	= 9.922	.390625							57	=22.622	.890625
				13	26	=10.319	.40625						29	58	=23.019	.90625
					27	=10.716	.421875							59	=23.416	.921875
			7	14	28	=11.112	.4375					15	30	60	=23.812	.9375
					29	=11.509	.453125							61	=24.209	.953125
				15	30	=11.906	.46875						31	62	=24.606	.96875
					31	=12.303	.484375							63	=25.003	.984375
1	2	4	8	16	32	=12.700	.5	1	2	4	8	16	32	64	=25.400	1.000

TABLE 1e10.   RECOMMENDED CONSISTENT VALUES OF THE FUNDAMENTAL CONSTANTS#

Quantity	Symbol	Value	Uncertainty (ppm)
1. Permeability of Vacuum	μ_0	$4\pi \times 10^{-7}$ H m^{-1} = 12.5663706144 $\times 10^{-7}$ H m^{-1}	0.004
2. Speed of Light in Vacuum	c	2.99792458(1.2) m s^{-1}	0.008
3. Permittivity of Vacuum	$\epsilon_0 = (\mu_0 c^2)^{-1}$	8.85418782(7) $\times 10^{-12}$ F m^{-1}	0.008
4. Fine Structure Constant, $\mu_0 c e^2/2h$	α	0.0072973506(60)	0.82
	α^{-1}	137.03604(11)	0.82
5. Elementary Charge	e	1.6021892(46) $\times 10^{-19}$ C	2.9
6. Planck Constant	h	6.626176(36) $\times 10^{-34}$ J Hz^{-1}	5.4
	$h = h/2\pi$	1.0545887(57) $\times 10^{-34}$ J s	5.4
7. Avogadro Constant	N_A	6.022045(31) $\times 10^{23}$ mol^{-1}	5.1
8. Atomic Mass Unit	1 u = (10^{-3} kg mol^{-1})/N_A	1.6605655(86) $\times 10^{-27}$ kg	5.1
9. Electron Rest Mass	m_e	0.9109534(47) $\times 10^{-30}$ kg	5.1
		5.4858026(21) $\times 10^{-4}$ u	0.38
10. Muon Rest Mass	m_μ	1.883566(11) $\times 10^{-28}$ kg	5.6
		0.11342920(26) u	2.3
11. Proton Rest Mass	m_p	1.6726485(86) $\times 10^{-27}$ kg	5.1
		1.007276470(11) u	0.011
12. Neutron Rest Mass	m_n	1.6749543(86) $\times 10^{-27}$ kg	5.1
		1.008665012(37) u	0.037
13. Ratio, Proton Mass to Electron Mass	m_p/m_e	1836.15152(70)	0.38
14. Ratio, Muon Mass to Electron Mass	m_μ/m_e	206.76865(47)	2.3
15. Specific Electron Charge	e/m_e	1.7588047(49) $\times 10^{11}$ C kg^{-1}	2.8
16. Faraday Constant	$F = N_A e$	9.648456(27) $\times 10^4$ C mol^{-1}	2.8
17. Magnetic Flux Quantum	$\Phi_0 = h/2e$	2.0678506(54) $\times 10^{-15}$ Wb	2.6
	h/e	4.135701(11) $\times 10^{-15}$ J Hz^{-1} C^{-1}	2.6
18. Josephson Frequency-Voltage Ratio	$2e/h$	483.5939(13)THz V^{-1}	2.6
19. Quantum of Circulation	$h/2m_e$	3.6369455(60) $\times 10^{-4}$ J Hz^{-1} kg^{-1}	1.6
	h/m_e	7.273891(12) $\times 10^{-4}$ J Hz^{-1} kg^{-1}	1.6
20. Rydberg Constant	R_∞	1.09737317(83) $\times 10^7$ m^{-1}	0.075
21. Bohr Radius	$a_0 = \alpha/4\pi R_\infty$	0.5291706(44) $\times 10^{-10}$ m	0.82
22. Electron Compton Wavelength	$\lambda_C = \alpha^2/2R_\infty = \alpha a_0$	2.4263089(40) $\times 10^{-12}$ m	1.6
	$\lambda_C = \lambda_C/2\pi = \alpha a_0$	3.8615905(64) $\times 10^{-13}$ m	1.6
23. Classical Electron Radius	$r_e = \mu_0 e^2/4\pi m_e = \alpha \lambda_C$	2.8179380(70) $\times 10^{-15}$ m	2.5
24. Electron g-Factor	$\frac{1}{2}g_e = \mu_e/\mu_B$	1.0011596567(35)	0.0035
25. Muon g-Factor	$\frac{1}{2}g_\mu$	1.00116616(31)	0.31

TABLE 1e10 (Continued)

Quantity	Symbol	Value	Uncertainty (ppm)
26. Proton Moment in Nuclear Magnetons	μ_p/μ_N	2.7928456(11)	0.38
27. Bohr Magneton	$\mu_B = e\hbar/2m_e$	9.274078(36) × 10^{-24} J T^{-1}	3.9
28. Nuclear Magneton	$\mu_N = e\hbar/2m_p$	5.050824(20) × 10^{-27} J T^{-1}	3.9
29. Electron Magnetic Moment	μ_e	9.284832(36) × 10^{-24} J T^{-1}	3.9
30. Proton Magnetic Moment	μ_p	1.4106171(55) × 10^{-26} J T^{-1}	3.9
31. Proton Magnetic Moment in Bohr Magnetons	μ_p/μ_B	1.521032209(16) × 10^{-3}	0.011
32. Ratio, Electron to Proton Magnetic Moments	μ_e/μ_p	658.2106880(66)	0.010
33. Ratio, Muon Moment to Proton Moment	μ_μ/μ_p	3.183340(72)	2.3
34. Muon Magnetic Moment	μ_μ	4.490474(18) × 10^{-26} J T^{-1}	3.9
35. Proton Gyromagnetic Ratio	γ_p	2.6751987(75) × 10^8 s^{-1} T^{-1}	2.8
36. Diamagnetic Shielding Factor, Spherical H$_2$O Sample	$1 + \sigma(H_2O)$	1.00002563(67)	0.067
37. Proton Gyromagnetic Ratio, (Uncorrected)	γ_p'	2.6751301(75) × 10^8 s^{-1} T^{-1}	2.8
	$\gamma_p'/2\pi$	42.57602(12) MHzT^{-1}	2.8
38. Proton Moment in Nuclear Magnetons (Uncorrected)	μ_p'/μ_N	2.7927740(11)	0.38
39. Proton Compton Wavelength	$\lambda_{C,p} = h/m_p c$	1.3214099(22) × 10^{-15} m	1.7
	$\lambda_{C,p} = \lambda_{C,p}/2\pi$	2.103089(36) × 10^{-16} m	1.7
40. Neutron Compton Wavelength	$\lambda_{C,n} = h/m_n c$	1.3195909(22) × 10^{-15} m	1.7
	$\lambda_{C,n} = \lambda_{C,n}/2\pi$	2.100941(35) × 10^{-16} m	1.7
41. Molar Gas Constant	R	8.31441(26) J mol^{-1} K^{-1}	31
42. Molar Volume, Ideal Gas (T_0 = 273.15 K, p_0 = 1 atm)	$V_m = RT_0/p_0$	0.02241383(70) m^3 mol^{-1}	31
43. Boltzmann Constant	$k = R/N_A$	1.380662(44) × 10^{-23} J K^{-1}	32
44. Stefan-Boltzmann Constant	$\sigma = (\pi^2/60)k^4/\hbar^3 c^2$	5.67032(71) × 10^{-8} W m^{-2} K^{-4}	125
45. First Radiation Constant	$c_1 = 2\pi hc^2$	3.741832(20) × 10^{-16} W m^2	5.4
46. Second Radiation Constant	$c_2 = hc/k$	0.01438786(45) m K	31
47. Gravitational Constant	G	6.6720(41) × 10^{-11} N m^2 kg^{-2}	615

#Reproduced from "Recommended Consistent Values of the Fundamental Physical Constants, 1973," CODATA Bull. No. 11, Committee on Data for Science and Technol., Frankfurt, 1973.

TABLE 1e10 (Continued)

List of French names of the above quantities#

1. Perméabilité du vide
2. Vitesse de la lumière dans le vide
3. Permittivité du vide
4. Constante de structure fine
5. Charge élémentaire
6. Constante de Planck
7. Constante d'Avogadro
8. Unité de masse atomique
9. Masse au repos de l'électron
10. Masse au repos du muon
11. Masse au repos du proton
12. Masse au repos du neutron
13. Rapport de la masse du proton à celle de l'électron
14. Rapport de la masse du muon à celle de l'électron
15. Charge massique de l'électron
16. Constante de Faraday
17. Quantum de flux magnétique
18. Relation fréquence-tension dans l'effet Josephson
19. Quantum de circulation
20. Constante de Rydberg
21. Rayon de Bohr
22. Longueur d'onde de Compton pour l'électron
23. Rayon de l'électron
24. Facteur g pour l'électron

25. Facteur g pour le muon
26. Rapport du moment du proton au magnéton nucléaire
27. Magnéton de Bohr
28. Magnéton nucléaire
29. Moment magnétique de l'électron
30. Moment magnétique du proton
31. Rapport du moment magnétique du proton au magnéton de Bohr
32. Rapport du moment magnétique de l'électron à celui du proton
33. Rapport du moment magnétique du muon à celui du proton
34. Moment magnétique du muon
35. Coefficient gyromagnétique du proton
36. Facteur d'écran diamagnétique, pour une sphère d'eau
37. Coefficient gyromagnétique du proton (dans l'eau, non corrigé)
38. Rapport du moment du proton au magnéton nucléaire (non corrigé)
39. Longueur d'onde de Compton du proton
40. Longueur d'onde de Compton du neutron
41. Constante molaire des gaz
42. Volume molaire normal, gaz parfait
43. Constante de Boltzmann
44. Constante de Stefan-Boltzmann
45. Première constante de la loi du rayonnement
46. Deuxième constante de la loi du rayonnement
47. Constante de gravitation

TABLE 1e11. CONVERSION FACTORS AND MISCELLANEOUS CONSTANTS[#]

Quantity	Symbol	Value	Uncertainty (ppm)
Ratio, BIPM ampere to SI ampere	$K \equiv A_{B169}/A$	1.0000007(26)	2.6
Ratio, BIPM ohm to SI ohm	$\bar{R} \equiv \Omega_{B169}/\Omega$	0.99999947(19)	0.19
Ratio, BIPM volt to SI volt	V_{B169}/V	1.0000002(26)	2.6
Ratio, kxu to ångström, $\lambda(CuK\alpha_1) \equiv 1.537400$ kxu	Λ	1.0020772(54)	5.3
Ratio, Å* to ångström, $\lambda(WK\alpha_1) \equiv 0.2090100$ Å*	Λ^*	1.0000205(56)	5.6
Energy Equivalents			
1 μ		931.5016(26) MeV	2.8
1 proton mass		938.2796(27) MeV	2.8
1 neutron mass		939.5731(27) MeV	2.8
1 muon mass		105.65948(35) MeV	3.3
1 electron mass		0.5110034(14) MeV	2.8
1 electron volt	1 eV/k	11604.50(36) K	31
	1 eV/hc	8065.479(21) cm^{-1}	2.6
	1 eV/h	2.417696(63) $\times 10^{14}$ Hz	2.6
Voltage-wavelength product	$V\lambda$	12398.520(32) eV Å	2.6
Rydberg constant	$R_\infty hc$	13.605804(36) eV	2.6
Gas Constant	R	82.0568(26) cm^3 atm mol^{-1} K^{-1}	31
		1.98719(6) cal mol^{-1} K^{-1}	31

[#]Reproduced from "Recommended Consistent Values of the Fundamental Physical Constants, 1973," CODATA Bull. No. 11, Committee on Data for Science and Technology, Frankfurt/Main, Germany, 1973.

TABLE 1e12.   DEFINING FIXED POINTS OF THE IPTS-68[a]

Equilibrium state	Assigned value of International Practical Temperature	
	T_{68}(K)	t_{68}(°C)
Equilibrium between the solid, liquid and vapour phases of equilibrium hydrogen (triple point of equilibrium hydrogen)	13.81	−259.34
Equilibrium between the liquid and vapour phases of equilibrium hydrogen at a pressure of 33 330.6 N/m² (25/76 standard atmosphere)	17.042	−256.108
Equilibrium between the liquid and vapour phases of equilibrium hydrogen (boiling point of equilibrium hydrogen)	20.28	−252.87
Equilibrium between the liquid and vapour phases of neon (boiling point of neon)	27.102	−246.048
Equilibrium between the solid, liquid and vapour phases of oxygen (triple point of oxygen)	54.361	−218.789
Equilibrium between the liquid and vapour phases of oxygen (boiling point of oxygen)	90.188	−182.962
Equilibrium between the solid, liquid and vapour phases of water (triple point of water)[c]	273.16	0.01
Equilibrium between the liquid and vapour phases of water (boiling point of water)[b,c]	373.15	100
Equilibrium between the solid and liquid phases of zinc (freezing point of zinc)	692.73	419.58
Equilibrium between the solid and liquid phases of silver (freezing point of silver)	1235.08	961.93
Equilibrium between the solid and liquid phases of gold (freezing point of gold)	1337.58	1064.43

*Reproduced from "The International Practical Temperature Scale of 1968," *Metrologia* 5, No. 2, April 1969, by permission of Springer-Verlag New York, Inc., New York.

[a] Except for the triple points and one equilibrium hydrogen point (17.042 K) the assigned values of temperature are for equilibrium states at a pressure p_0 = 1 standard atmosphere (101 325 N/m²). In the realization of the fixed points small departures from the assigned temperatures will occur as a result of the differing immersion depths of thermometers or the failure to realize the required pressure exactly. If due allowance is made for these small temperature differences, they will not affect the accuracy of realization of the Scale.

[b] The equilibrium state between the solid and liquid phases of tin (freezing point of tin) has the assigned value of t_{68} = 231.9681 °C and may be used as an alternative to the boiling point of water.

[c] The water used should have the isotopic composition of ocean water.

TABLE 1e13.   SECONDARY REFERENCE POINTS OF IPTS-68*

Equilibrium state	International Practical Temperature	
	$T_{68}(K)$	$t_{68}(°C)$
Equilibrium between the solid, liquid and vapour phases of normal hydrogen (triple point of normal hydrogen)	13.956	259.194
Equilibrium between the liquid and vapour phases of normal hydrogen (boiling point of normal hydrogen)	20.397	−252.753

$$\lg \frac{p}{p_0} = A + \frac{B}{T_{68}} + CT_{68} + DT_{68}^2 \tag{23}$$

$A = 1.734\ 791$, $B = -44.623\ 68$ K, $C = 0.023\ 186\ 9$ K^{-1}, $D = -0.000\ 048\ 017$ K^{-2}

for the temperature range from 13.956 K to 30 K.

Equilibrium between the solid, liquid and vapour phases of neon (triple point of neon)	24.555	−248.595
Equilibrium between the liquid and vapour phases of neon		

$$\lg \frac{p}{p_0} = A + \frac{B}{T_{68}} + CT_{68} + DT_{68}^2 \tag{24}$$

$A = 4.611\ 52$, $B = -106.3851$ K, $C = -0.036\ 833\ 1$ K^{-1}, $D = 4.248\ 92 \times 10^{-4}$ K^{-2}

for the temperature range from 24.555 K to 40 K.

Equilibrium between the solid, liquid and vapour phases of nitrogen (triple point of nitrogen)	63.148	−210.002
Equilibrium between the liquid and vapour phases of nitrogen (boiling point of nitrogen)	77.348	−195.802

$$\lg \frac{p}{p_0} = A + \frac{B}{T_{68}} + C \lg \frac{T_{68}}{T_0} + DT_{68} + ET_{68}^2 \tag{25}$$

$A = 5.893\ 139$, $B = -404.131\ 05$ K, $C = -2.3749$, $D = -0.014\ 250\ 5$ K^{-1}, $E = 72.5342 \times 10^{-6}$ K^{-2}

for the temperature range from 63.148 K to 84 K.

Equilibrium between the liquid and vapour phases of oxygen

$$\lg \frac{p}{p_0} = A + \frac{B}{T_{68}} + C \lg \frac{T_{68}}{T_0} + DT_{68} + ET_{68}^2 \tag{26}$$

$A = 5.961\ 546$, $B = -467.455\ 76$ K, $C = -1.664\ 512$, $D = -0.013\ 213\ 01$ K^{-1}, $E = 50.8041 \times 10^{-6}$ K^{-2}

for the temperature range from 54.361 K to 94 K.

Equilibrium between the solid and vapour phases of carbon dioxide (sublimation point of carbon dioxide)	194.674	−78.476

$$T_{68} = \left[194.674 + 12.264\left(\frac{p}{p_0} - 1\right) - 9.15\left(\frac{p}{p_0} - 1\right)^2\right] K \tag{27}$$

for the temperature range from 194 K to 195 K.

Equilibrium between the solid and liquid phases of mercury (freezing point of mercury)[a]	234.288	−38.862
Equilibrium between ice and air-saturated water (ice point)	273.15	0
Equilibrium between the solid, liquid and vapour phases of phenoxybenzene (diphenyl ether) (triple point of phenoxybenzene)	300.02	26.87
Equilibrium between the solid, liquid and vapour phases of benzoic acid (triple point of benzoic acid)	395.52	122.37
Equilibrium between the solid and liquid phases of indium (freezing point of indium)[a]	429.784	156.634
Equilibrium between the solid and liquid phases of bismuth (freezing point of bismuth)[a]	544.592	271.442
Equilibrium between the solid and liquid phases of cadmium (freezing point of cadmium)[a]	594.258	321.108
Equilibrium between the solid and liquid phases of lead (freezing point of lead)[a]	600.652	327.502
Equilibrium between the liquid and vapour phases of mercury (boiling point of mercury)	629.81	356.66

$$t_{68} = \left[356.66 + 55.552\left(\frac{p}{p_0} - 1\right) - 23.03\left(\frac{p}{p_0} - 1\right)^2 + 14.0\left(\frac{p}{p_0} - 1\right)^3\right] °C \tag{28}$$

for $p = 90 \times 10^3$ N/m^2 to 104×10^3 N/m^2.

TABLE 1e13. *(Continued)*

Equilibrium state	International Practical Temperature	
	T_{68}(K)	t_{68}(°C)
Equilibrium between the liquid and vapour phases of sulphur (boling point of sulfur)	717.824	444.674

$$t_{68} = \left[444.674 + 69.010 \left(\frac{p}{p_0} - 1 \right) - 27.48 \left(\frac{p}{p_0} - 1 \right)^2 + 19.14 \left(\frac{p}{p_0} - 1 \right)^3 \right] °C \quad (29)$$

for $p = 90 \times 10^3$ N/m² to 104×10^3 N/m².

Equilibrium state	T_{68}(K)	t_{68}(°C)
Equilibrium between the solid and liquid phases of the copper-aluminium eutectic	821.38	548.23
Equilibrium between the solid and liquid phases of antimony (freezing point of antimony)[a]	903.89	630.74
Equilibrium between the solid and liquid phases of aluminium (freezing point of aluminium)	933.52	660.37
Equilibrium between the solid and liquid phases of copper (freezing point of copper)	1357.6	1084.5
Equilibrium between the solid and liquid phases of nickel (freezing point of nickel)	1728	1455
Equilibrium between the solid and liquid phases of cobalt (freezing point of cobalt)	1767	1494
Equilibrium between the solid and liquid phases of palladium (freezing point of palladium)	1827	1554
Equilibrium between the solid and liquid phases of platinum (freezing point of platinum)	2045	1772
Equilibrium between the solid and liquid phases of rhodium (freezing point of rhodium)	2236	1963
Equilibrium between the solid and liquid phases of iridium (freezing point of iridium)	2720	2447
Equilibrium between the solid and liquid phases of tungsten (temperature of melting tungsten)	3660	3387

*Reproduced from "The International Practical Temperature Scale of 1968," *Metrologia* 5, No. 2, April 1969, by permission of Springer-Verlag New York, Inc., New York.

TABLE 1e14. CONVERSION TABLE FOR BAROMETRIC PRESSURE UNITS*

	Atm	N/M²	bars	mb	kg/cm²	gm/cm² (cm H_2O)	mm Hg	in. Hg (" Hg)	lb/in² (psi)
1 Atmosphere =	1	1.013×10^5	1.013	1013	1.033	1033	760	29.92	14.70
1 Newton/M² (N/M²) =	$.9869 \times 10^{-5}$	1	10^{-5}	.01	1.02×10^{-5}	.0102	.0075	$.2953 \times 10^{-3}$	$.1451 \times 10^{-3}$
1 bar =	.9869	10^5	1	1000	1.02	1020	750.1	29.53	14.51
1 millibar (mb) =	$.9869 \times 10^{-3}$	100	.001	1	.00102	1.02	.7501	.02953	.01451
1 kg/cm² =	.9681	$.9807 \times 10^5$	.9807	980.7	1	1000	735	28.94	14.22
1 gm/cm² (1 cm H_2O) =	968.1	98.07	$.9807 \times 10^{-3}$	.9807	.001	1	.735	.02894	.01422
1 mm Hg =	.001316	133.3	.001333	1.333	.00136	1.36	1	.03937	.01934
1 in. Hg (" Hg) =	.0334	3386	.03386	33.86	.03453	34.53	25.4	1	.4910
1 lb/in² (psi) =	.06804	6895	.06895	68.95	.0703	70.3	51.70	2.035	1

*Reproduced from: Parker, J. F., Jr., and V. R. West, Eds., "Bioastronautics Data Book," 2nd ed., Report No. NASA SP-3006, National Aeronautics and Space Administration, Washington, D.C., 1973.

TABLE 1e15. CONVERSION FROM DEGREES BAUMÉ TO DENSITY*

DENSITY OF LIQUIDS LIGHTER THAN WATER

Conversions from Degrees Baumé to Density and Vice Versa for liquids at 60°/60°F. (15.56°C)

Formula: $\text{Density} = \dfrac{140}{130 + \text{Deg. Bé}}$

$\text{Degrees Baumé} = \dfrac{140}{\text{Density}} - 130$

DENSITY OF LIQUIDS HEAVIER THAN WATER

Conversions from Degrees Baumé to Density and Vice Versa for liquids at 60°/60°F. (15.56°C)

Formula: $\text{Density} = \dfrac{145}{145 - \text{Deg. Bé}}$

$\text{Degrees Baumé} = 145 - \dfrac{145}{\text{Density}}$

DEGREES BAUMÉ	DENSITY	DEGREES BAUMÉ	DENSITY	DEGREES BAUMÉ	DENSITY	DEGREES BAUMÉ	DENSITY
10	1.0000	46	0.7955	0	1.0000	36	1.3303
11	0.9929	47	0.7910	1	1.0069	37	1.3426
12	0.9859	48	0.7865	2	1.0140	38	1.3551
13	0.9790	49	0.7821	3	1.0211	39	1.3679
14	0.9722	50	0.7778	4	1.0284	40	1.3810
15	0.9655	51	0.7735	5	1.0357	41	1.3942
16	0.9589	52	0.7692	6	1.0432	42	1.4078
17	0.9524	53	0.7650	7	1.0507	43	1.4216
18	0.9459	54	0.7609	8	1.0584	44	1.4356
19	0.9396	55	0.7568	9	1.0662	45	1.4500
20	0.9333	56	0.7527	10	1.0741	46	1.4646
21	0.9272	57	0.7487	11	1.0821	47	1.4796
22	0.9211	58	0.7447	12	1.0902	48	1.4948
23	0.9150	59	0.7407	13	1.0985	49	1.5104
24	0.9091	60	0.7368	14	1.1069	50	1.5263
25	0.9032	61	0.7330	15	1.1154	51	1.5426
26	0.8974	62	0.7292	16	1.1240	52	1.5591
27	0.8917	63	0.7254	17	1.1328	53	1.5761
28	0.8861	64	0.7216	18	1.1417	54	1.5934
29	0.8805	65	0.7179	19	1.1508	55	1.6111
30	0.8750	66	0.7143	20	1.1600	56	1.6292
31	0.8696	67	0.7107	21	1.1694	57	1.6477
32	0.8642	68	0.7071	22	1.1789	58	1.6667
33	0.8589	69	0.7035	23	1.1885	59	1.6860
34	0.8537	70	0.7000	24	1.1983	60	1.7059
35	0.8485	71	0.6965	25	1.2083	61	1.7262
36	0.8434	72	0.6931	26	1.2185	62	1.7470
37	0.8383	73	0.6897	27	1.2288	63	1.7683
38	0.8333	74	0.6863	28	1.2393	64	1.7901
39	0.8284	75	0.6829	29	1.2500	65	1.8125
40	0.8235	76	0.6796	30	1.2609	66	1.8354
41	0.8187	77	0.6763	31	1.2719	67	1.8590
42	0.8140	78	0.6731	32	1.2832	68	1.8831
43	0.8092	79	0.6699	33	1.2946	69	1.9079
44	0.8046	80	0.6667	34	1.3063	70	1.9333
45	0.8000	81	0.6635	35	1.3182	71	1.9595

*Reproduced with permission from: "A Condensed Laboratory Handbook," Du Pont Company, Industrial Chemicals Department, Wilmington, DE, 1971.

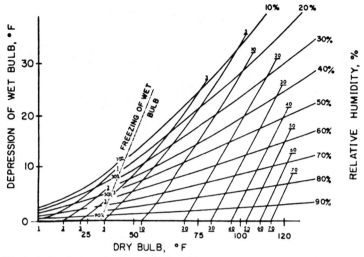

Fig. 1e1. Psychrometric data. The humidity mixing ratio in gm/m³ is shown by underlined numbers. Reproduced from Platt, R. B. and J. Griffiths, "Environmental Measurement," Van Nostrand Reinhold Co., New York, 1964.

BIBLIOGRAPHY

Brauer, G., "Handbook of Preparative Chemistry," Academic Press, New York, 1963.

Gmelin, L., "Handbuch der anorganischen Chemie," Verlag Chemie, Weinheim, Germany, 1953–1963.

Hannay, N. B., Ed., "Treatise on Solid State Chemistry," 6 vols, Plenum, New York, 1973–1974.

Jorgensen, C. K., "Oxidation Numbers and Oxidation States," Springer-Verlag, New York, 1969.

Kane, P. F. and G. B. Larrabee, Eds., "Characterization of Solid Surfaces," Plenum, New York, 1974.

Levin, E. M., C. R. Robbins and H. F. McCurdie, "Phase Diagrams for Ceramicists," American Ceramic Society, Columbus, Ohio, 1961.

Linke, W. F., Ed., "Solubilities," Vol. 1, American Chemical Society, Washington, D.C., 1965.

Linke, W. F., "Solubilities of Inorganic and Metal Organic Compounds," (Seidell), Vol. 2, American Chemical Society, Washington, D.C., 1965.

Murov, S. L., "Handbook of Photochemistry," Marcel Dekker, New York, 1973.

Pascal, P., "Noveau Traite de Chimie Minerale, "Masson, Paris, France, 1961.

Pauling, L., "The Nature of the Chemical Bond," 3rd ed., Cornell University Press, Ithaca, New York, 1960.

Riddick, J. A. and W. B. Bunger," Organic Solvents: Physical Properties and Methods of Purification," 3rd ed., Wiley, New York, 1971.

Sanderson, R. T., "Chemical Periodicity," Reinhold, New York, 1964.

Stephen, H. and T. Stephen, "Solubilities of Inorganic and Organic Compounds," Pergamon Press, New York, 1963.

Wells, A. F., "Structural Inorganic Chemistry," 3rd ed., Oxford University Press, Oxford, 1962.

Wyckoff, R. W. G., "Crystal Structures," Wiley, New York, 1963–1968.

SECTION 1f
MECHANICAL PROPERTIES OF MATERIALS

A brief survey of selected mechanical properties of a wide variety of materials is given in this Section. The information is presented in the following tables and figures:

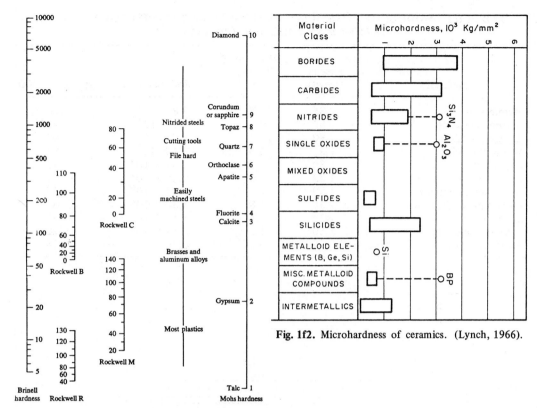

Fig. 1f1. Comparison of a number of hardness scales in current use. (From *Engineering Properties and Applications of Plastics*, by G. F. Kinney, copyright 1957, John Wiley and Sons, Inc. Used by permission.)

Fig. 1f2. Microhardness of ceramics. (Lynch, 1966).

TABLE 1f1.   MODULUS OF ELASTICITY OF MATERIALS.   (PISANI, 1964)

$$E = \frac{f}{\varepsilon} \ (\text{psi})$$

Material	Modulus of Elasticity
Steel	29–30,000,000
Wrought iron	25–28,000,000
Cast iron	12–15,000,000
Timber (with grain)	1,500,000 (average)
Aluminum	10,500,000
Concrete	2,000,000
Copper	18,000,000
Brass	13,000,000

TABLE 1f2.   MODULI OF ELASTICITY IN TENSION AND SPECIFIC GRAVITIES OF PLASTICS AND METALS.   (PISANI, 1964).

Material	Specific Gravity	Modulus of Elasticity (psi)	Modulus/S.G.	Ratio (stainless = 100)
Stainless steel	7.85	30×10^6	3.8×10^6	100
Chrome-moly steel	7.85	29	3.7×10^6	98
Aluminum alloy	2.80	10.4	3.7×10^6	98
Magnesium alloy	1.81	6.5	3.6×10^6	94
Pregwood	1.30	3.7	2.84×10^6	75
Paper laminate	1.33	3.0	2.25×10^6	59
Glass fabric laminate	1.50	2	1.34×10^6	35
Impact phenolic, molded	1.38	1.8	1.30×10^6	34
Canvas laminate	1.33	1.5	1.12×10^6	29
Asbestos laminate	1.80	1.5	0.84×10^6	22
Wood flour, phenolic, molded	1.36	0.96	0.70×10^6	18

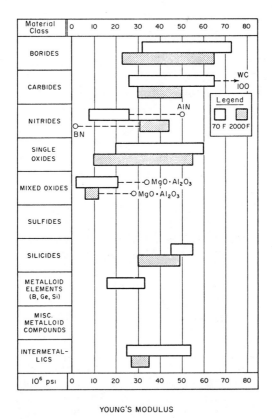

YOUNG'S MODULUS

Fig. 1f3. Modulus of elasticity of ceramics. (Lynch, 1966).

TABLE 1f3.   COMPARISON OF STRENGTH/WEIGHT PROPERTIES OF PLASTICS AND METALS (PISANI, 1964)

Material	Specific Gravity	Tensile Strength psi	Tensile/S.G.	Ratio (magnesium alloy = 100)
Magnesium alloy	1.81	46,000	25,400	100
Stainless steel	7.85	185,000	23,600	93
Pregwood	1.30	30,000	23,000	90
Chrome-moly steel	7.85	180,000	22,900	90
Aluminum alloy	2.80	62,000	22,100	87
Paper laminate	1.33	12,500	9,400	36
Glass fabric laminate	1.50	14,000	9,300	36
Canvas fabric laminate	1.33	9,500	7,100	28
Wood flour, phenolic, molded	1.36	8,500	6,200	24
Asbestos paper laminate	1.80	10,000	5,500	21
Impact phenolic molded	1.38	7,500	5,400	21

TABLE 1f4.   COMPRESSIVE STRENGTH/WEIGHT VALUES OF PLASTICS AND METALS. (PISANI, 1964)

Material	Specific Gravity	Compressive Strength psi	Comp. Str./Spec. Gr.	Ratio (canvas laminate = 100)
Canvas laminate	1.33	38,000	28,000	100
Paper laminate	1.33	35,000	26,000	93
Glass fabric laminate	1.5	40,000	26,000	93
Impact phenolic, molded	1.38	35,000	25,000	89
Wood flour, phenolic, molded	1.36	30,000	22,000	78
Asbestos laminate	1.80	38,000	21,000	75
Magnesium alloy	1.81	35,000	19,300	69
Chrome-moly steel	7.85	150,000	19,100	68
Stainless steel	7.85	150,000	19,100	68
Aluminum alloy	2.80	40,000	14,300	51
Pregwood	1.3	15,000	11,000	39

TABLE 1f5.   ULTIMATE STRENGTH OF MATERIALS (PSI). (PISANI, 1964)

Material	Tension	Compression	Shear
Structural Steel	60–80,000	60–80,000	50,000
Wrought Iron	50,000	50,000	38,000
Cast iron (gray)	18–24,000	90,000	20,000
Malleable cast iron	27–35,000	46,000	40,000
Timber (with grain)	10,000	8,000	8,000
Stone, granite	1,200	12,000	–
Nickel (3.25% Ni.) Steel	85–100,000	100,000	75,000

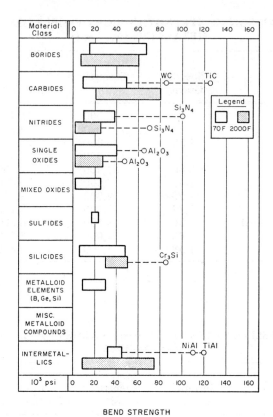

BEND STRENGTH

Fig. 1f4. Bend strength of ceramics. (Lynch, 1966).

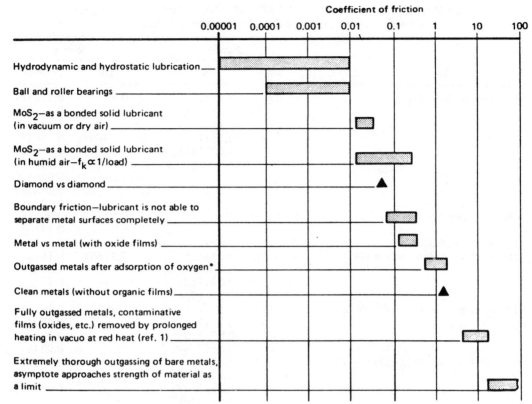

Fig. 1f5. Typical range of friction values.   (Anon., 1971).

TABLE 1f6. COEFFICIENTS OF STATIC AND SLIDING FRICTION.*
(Reference letters indicate the lubricant used; see footnote)

Materials	Static		Sliding	
	Dry	Greasy	Dry	Greasy
Hard steel on hard steel	0.78 (1)	0.11 (1, a)	0.42 (2)	0.029 (h)
	—	0.23 (1, b)	—	0.081 (c)
	—	0.15 (1, c)	—	0.080 (i)
	—	0.11 (1, d)	—	0.058 (j)
	—	0.0075 (18, p)	—	0.084 (d)
	—	0.0052 (18, h)	—	0.105 (k)
	—	—	—	0.096 (l)
	—	—	—	0.108 (m)
	—	—	—	0.12 (a)
Mild steel on mild steel	0.74 (19)	—	0.57 (3)	0.09 (a)
	—	—	—	0.19 (u)
Hard steel on graphite	0.21 (1)	0.09 (1, a)		
Hard steel on babbitt (ASTM No. 1)	0.70 (11)	0.23 (1, b)	0.33 (6)	0.16 (b)
	—	0.15 (1, c)	—	0.06 (c)
	—	0.08 (1, d)	—	0.11 (d)
	—	0.085 (1, e)		
Hard steel on babbitt (ASTM No. 8)	0.42 (11)	0.17 (1, b)	0.35 (11)	0.14 (b)
	—	0.11 (1, c)	—	0.065 (c)
	—	0.09 (1, d)	—	0.07 (d)
	—	0.08 (1, e)	—	0.08 (h)
Hard steel on babbitt (ASTM No. 10)	—	0.25 (1, b)	—	0.13 (b)
	—	0.12 (1, c)	—	0.06 (c)
	—	0.10 (1, d)	—	0.055 (d)
	—	0.11 (1, e)		
Mild steel on cadmium silver	—	—	—	0.097 (f)
Mild steel on phosphor bronze	—	—	0.34 (3)	0.173 (f)
Mild steel on copper lead	—	—	—	0.145 (f)
Mild steel on cast iron	—	0.183 (15, c)	0.23 (6)	0.133 (f)
Mild steel on lead	0.95 (11)	0.5 (1, f)	0.95 (11)	0.3 (f)
Nickel on mild steel	—	—	0.64 (3)	0.178 (x)
Aluminum on mild steel	0.61 (8)	—	0.47 (3)	
Magnesium on mild steel	—	—	0.42 (3)	
Magnesium on magnesium	0.6 (22)	0.08 (22, y)		
Teflon on Teflon	0.04 (22)	—	—	0.04 (f)
Teflon on steel	0.04 (22)	—	—	0.04 (f)
Tungsten carbide on tungsten carbide	0.2 (22)	0.12 (22, a)		
Tungsten carbide on steel	0.5 (22)	0.08 (22, a)		
Tungsten carbide on copper	0.35 (23)			
Tungsten carbide on iron	0.8 (23)			
Bonded carbide on copper	0.35 (23)			
Bonded carbide on iron	0.8 (23)			
Cadmium on mild steel	—	—	0.46 (3)	
Copper on mild steel	0.53 (8)	—	0.36 (3)	0.18 (a)
Nickel on nickel	1.10 (16)	—	0.53 (3)	0.12 (w)
Brass on mild steel	0.51 (8)	—	0.44 (6)	
Brass on cast iron	—	—	0.30 (6)	
Zinc on cast iron	0.85 (16)	—	0.21 (7)	
Magnesium on cast iron	—	—	0.25 (7)	
Copper on cast iron	1.05 (16)	—	0.29 (7)	
Tin on cast iron	—	—	0.32 (7)	
Lead on cast iron	—	—	0.43 (7)	

TABLE 1f6. (*Continued*)

Materials	Static		Sliding	
	Dry	Greasy	Dry	Greasy
Aluminum on aluminum	1.05 (16)	–	1.4 (3)	
Glass on glass	0.94 (8)	0.01 (10, p)	0.40 (3)	0.09 (a)
	–	0.005 (10, q)	–	0.116 (v)
Carbon on glass	–	–	0.18 (3)	
Garnet on mild steel	–	–	0.39 (3)	
Glass on nickel	0.78 (8)	–	0.56 (3)	
Copper on glass	0.68 (8)	–	0.53 (3)	
Cast iron on cast iron	1.10 (16)	–	0.15 (9)	0.070 (d)
	–	–	–	0.064 (n)
Bronze on cast iron	–	–	0.22 (9)	0.077 (n)
Oak on oak (parallel to grain)	0.62(9)	–	0.48 (9)	0.164 (r)
	–	–	–	0.067 (s)
Oak on oak (perpendicular)	0.54 (9)	–	0.32 (9)	0.072 (s)
Leather on oak (parallel)	0.61 (9)	–	0.52 (9)	
Cast iron on oak	–	–	0.49 (9)	0.075 (n)
Leather on cast iron	–	–	0.56 (9)	0.36 (t)
	–	–	–	0.13 (n)
Laminated plastic on steel	–	–	0.35 (12)	0.05 (t)
Fluted rubber bearing on steel	–	–	–	0.05 (t)

(*a*) Oleic acid: (*b*) Atlantic spindle oil (light mineral); (*c*) castor oil; (*d*) lard oil; (*e*) Atlantic spindle oil plus 2 percent oleic acid; (*f*) medium mineral oil; (*g*) medium mineral oil plus ½ percent oleic acid; (*h*) stearic acid; (*i*) grease (zinc oxide base); (*j*) graphite; (*k*) turbine oil plus 1 percent graphite; (*l*) turbine oil plus 1 percent stearic acid; (*m*) turbine oil (medium mineral); (*n*) olive oil; (*p*) palmitic acid; (*q*) ricinoleic acid; (*r*) dry soap; (*s*) lard; (*t*) water; (*u*) rape oil; (*v*) 3-in-1 oil; (*w*) octyl alcohol; (*x*) triolein; (*y*) 1 percent lauric acid in paraffin oil.

TABLE 1f7. STATIC FRICTION DATA FOR SOLID LUBRICANTS*

Material	Load (grams)	Coefficient of Friction 75–80°F	450–500°F	1000°F
Ag_2S	128	0.36		
Ag_2Se	326	0.30–0.35		0.8–0.9
Ag_2Te	128	0.44		
AgI	128	1.0	1.0	1.0
$AlPO_4$	128	1.0–1.33	1.0–1.31	0.31–0.80
$AlPO_4$	326	0.60		0.51
BN	326	0.3	0.15	
Bi_2S_3	128	0.23–0.60	0.21–0.77	0.18–0.38
Bi_2S_3	326	0.56–0.58	0.24–0.62	0.20–0.32
C (graphite)	326	0.14–0.30	0.06–0.12	0.20–0.27
$CdCl_2$	326	0.6	0.17	
CdS	128	0.58–1.0	0.84	0.55
CdSe	128	0.23–0.33	0.58	0.27–0.38
CdTe	128	0.46	0.40	0.33–0.60
CoS	128	0.50–1.15	0.72	0.58
$CrCl_3$	326	0.2–0.3		
Cu_2S	128	1.0	1.2	1.2
GeO_2	128	0.48–0.58		
InSe	128	0.46–0.60	0.41–0.60	0.60 (750°F)
LiF	326	0.3–0.4	0.9	0.65–0.75
$MnCl_2$	326	0.35	0.17	
MoS_2	326	0.34	0.10	
$MoSe_2$	128	0.20–0.33	0.31–0.40	
$NiCl_2$	326	0.45	0.19	
NiS	128	0.29–1.0		
PbF_2	328	0.6	0.6	
PbS	326	0.47	0.27–0.47	0.15–0.19
PbSe	128	0.40–0.67	0.25	0.25
Sb_2O_5	128	0.21–0.96		
Sb_2S_3	128	0.38	0.21–0.49	0.49
Sb_2S_5	128	0.31–0.80	0.35–1.0	
Si_3N_4	326	0.3–0.9		0.9
SnS_2	326	>0.95		
SnO	326	>0.95		
SnS	326	>0.95	0.63	
SrS	326	0.7–0.9		0.7–0.9
Ta_2S_4	326	1.15		
TiC	326	0.55		
TiS_2	326	0.7	0.6	
$TiTe_2$	128	0.36–0.58	0.86–0.95	0.77 (670°F)
Tl_2S	326	0.25–0.5		
WS_2	326	0.7–1.6	0.2	
ZnSe	326	0.4–0.6		0.55–0.70
ZnTe	128	0.60–0.68	0.30–0.40	0.58
$ZrCl_4$	128	0.45	0.48–0.96	0.27
ZrN	326	0.2–0.3		0.55–0.75
Micas				
Pyrophyllite	90	0.48–0.89	0.10–0.27 (80°F in vacuo)	
Muscovite	90	>1.0	>1.0 (80°F in vacuo)	
Talc	90	0.13–0.89	0.21–0.72 (80°F in vacuo)	
Phlogopite	90	0.38–0.64	(80°F in vacuo)	

*Data selected from Campbell, M. E., J. B. Loser and E. Sneegas, "Solid Lubricants," Report NASA SP-5059, U.S. Government Printing Office, Washington, 1966.

REFERENCES

Anon, "Lubrication, Friction and Wear," NASA Space Vehicle Design Criteria (Structures) Report No. NASA SP-8063, June 1971.

Lynch, J. F., *et al.*, "Engineering Properties of Selected Ceramic Materials," The American Ceramic Society, Inc., Columbus, Ohio, 1966.

Pisani, T. J., "Essentials of Strength of Materials," 3rd ed., Van Nostrand Reinhold, New York, 1964.

BIBLIOGRAPHY

Campbell, I. E., "High Temperature Technology," Wiley, New York, 1957.

Clauss, F. J., "Solid Lubricants and Self-lubricating Solids," Academic Press, New York, 1972.

Fontana, M. G. and R. W. Staehle, Eds., "Advances in Corrosion Science and Technology," 4 vols., Plenum Press, New York, 1970–1974.

McLean, D., "Mechanical Properties of Metals," Wiley, New York, 1962.

Sims, C. T. and W. C. Hagel, Eds., "The Superalloys," Wiley, New York, 1972.

Wigley, D. A., "Mechanical Properties of Materials at Low Temperatures," Plenum Press, New York, 1971.

SECTION 1g
CHEMICAL PROPERTIES OF MATERIALS

This Section contains a number of tables of particular interest to the chemist. The tables are entitled:

TABLE 1g1. LIST OF NAMES FOR IONS AND RADICALS.*

In inorganic chemistry substitutive names seldom are used, but the organic-chemical names are shown to draw attention to certain differences between organic and inorganic nomenclature

Atom or Group	As Neutral Molecule	As Cation or Cationic Radical[1]	As Anion	As Ligand	As Prefix for Substituent in Organic Compounds
H	monohydrogen	hydrogen	hydride	hydrido	
F	monofluorine		fluoride	fluoro	fluoro
Cl	monochlorine	chlorine	chloride	chloro	chloro
Br	monobromine	bromine	bromide	bromo	bromo
I	monoiodine	iodine	iodide	iodo	iodo
I_3			triiodide		
ClO		chlorosyl	hypochlorite	hypochlorito	
ClO_2	chlorine dioxide	chloryl	chlorite	chlorito	
ClO_3		perchloryl	chlorate	chlorato	
ClO_4			perchlorate		
IO		iodosyl	hypoiodite		iodoso
IO_2		iodyl			iodyl or iodoxy
O	monoöxygen		oxide	oxo	oxo or keto
O_2	dioxygen		O_2^{2-}: peroxide	peroxo	peroxy
			O_2^-: hyperoxide		
HO	hydroxyl		hydroxide	hydroxo	hydroxy
HO_2	(perhydroxyl)		hydrogen peroxide	hydrogen peroxo	hydroperoxy
S	monosulfur		sulfide	thio (sulfido)	thio
HS	(sulfhydryl)		hydrogen sulfide	thiolo	thiol or mercapto
S_2	disulfur		disulfide	disulfido	
SO	sulfur monoxide	sulfinyl (thionyl)			sulfinyl
SO_2	sulfur dioxide	sulfonyl (sulfuryl)	sulfoxylate		sulfonyl
SO_3	sulfur trioxide		sulfite	sulfito	
HSO_3			hydrogen sulfite	hydrogen sulfito	
S_2O_3			thiosulfate	thiosulfato	
SO_4			sulfate	sulfato	
Se	selenium		selenide	seleno	seleno
SeO		seleninyl			seleninyl
SeO_2		selenonyl			selenonyl
SeO_3	selenium trioxide		selenite	selenito	
SeO_4			selenate	selenato	
Te	tellurium		telluride	telluro	telluro
CrO_2		chromyl			
UO_2		uranyl			
NpO_2		neptunyl			
PuO_2		plutonyl			
AmO_2		americyl			
N	mononitrogen		nitride	nitrido	
N_3			azide	azido	
NH			imide	imido	imino
NH_2			amide	amido	amino
NHOH			hydroxylamide	hydroxylamido	hydroxylamino
N_2H_3			hydrazide	hydrazido	hydrazino
NO	nitrogen oxide	nitrosyl		nitrosyl	nitroso
NO_2	nitrogen dioxide	nitryl		nitro	nitro
ONO			nitrite	nitrito	
NS		thionitrosyl			
$(NS)_n$		thiazyl (e.g., trithiazyl)			
NO_3			nitrate	nitrato	
N_2O_3			hyponitrite	hyponitrito	
P	phosphorus		phosphide	phosphido	
PO		phosphoryl			
PS		thiophosphoryl			phosphoroso
PH_2O_3			hypophosphite	hypophosphito	
PHO_3			phosphite	phosphito	
PO_4			phosphate	phosphato	
AsO_4			arsenate	arsenato	
VO		vanadyl			
CO	carbon monoxide	carbonyl		carbonyl	carbonyl
CS		thiocarbonyl			
CH_3O	methoxyl		methanolate	methoxo	methoxy
C_2H_5O	ethoxyl		ethanolate	ethoxo	ethoxy
CH_3S			methanethiolate	methanethiolato	methylthio
C_2H_5S			ethanethiolate	ethanethiolato	ethylthio
CN		cyanogen	cyanide	cyano.	cyano
OCN			cyanate	cyanato	cyanato
SCN			thiocyanate	thiocyanato and isothiocyanato	thiocyanato and isothiocyanato
SeCN			selenocyanate	selenocyanato	selenocyanato
TeCN			tellurocyanate	tellurocyanato	
CO_2			carbonate	carbonato	
HCO_3			hydrogen carbonate	hydrogen carbonato	
CH_3CO_2			acetate	acetato	acetoxy
CH_3CO	acetyl	acetyl			acetyl
C_2O_4			oxalate	oxalato	

[1] If necessary, oxidation state is to be given by Stock notation.

NOTE: Although some additions might be made to this list, especially in the last column, and a few changes suggested, no attempt has been made to do so at this time. Cf. comments at such rules as **3.31**, **3.32**, **5.35**, **6.2**, and **7.312**.

*Reproduced from: Potter, J. H., Ed., "Handbook of the Engineering Sciences," Vol. 1, Van Nostrand Reinhold, New York, 1967.

TABLE 1g2. CONCENTRATION OF ACIDS AND BASES*

Common Commercial Strengths

	Molecular weight	Moles per liter	Grams per liter	Percent by weight	Specific gravity
acetic acid, glacial	60.05	17.4	1045	99.5	1.05
acetic acid	60.05	6.27	376	36	1.045
butyric acid	88.1	10.3	912	95	0.96
formic acid	46.02	23.4	1080	90	1.20
		5.75	264	25	1.06
hydriodic acid	127.9	7.57	969	57	1.70
		5.51	705	47	1.50
		0.86	110	10	1.1
hydrobromic acid	80.92	8.89	720	48	1.50
		6.82	552	40	1.38
hydrochloric acid	36.5	11.6	424	36	1.18
		2.9	105	10	1.05
hydrocyanic acid	27.03	25	676	97	0.697
		0.74	19.9	2	0.996
hydrofluoric acid	20.01	32.1	642	55	1.167
		28.8	578	50	1.155
hydrofluosilicic acid	144.1	2.65	382	30	1.27
hypophosphorous acid	66.0	9.47	625	50	1.25
		5.14	339	30	1.13
		1.57	104	10	1.04
lactic acid	90.1	11.3	1020	85	1.2
nitric acid	63.02	15.99	1008	71	1.42
		14.9	938	67	1.40
		13.3	837	61	1.37
perchloric acid	100.5	11.65	1172	70	1.67
		9.2	923	60	1.54
phosphoric acid	98	14.7	1445	85	1.70
sulfuric acid	98.1	18.0	1766	96	1.84
sulfurous acid	82.1	0.74	61.2	6	1.02
ammonia water	17.0	14.8	252	28	0.898
potassium hydroxide	56.1	13.5	757	50	1.52
		1.94	109	10	1.09
sodium carbonate	106.0	1.04	110	10	1.10
sodium hydroxide	40.0	19.1	763	50	1.53
		2.75	111	10	1.11

*Reproduced from page 1309 of "The Merck Index," Eighth Edition, copyright 1968 by Merck & Co., Inc., Rahway, New Jersey, U.S.A. with the permission of the copyright holder.

TABLE 1g3. SATURATED SOLUTIONS*

The following table provides the data for making saturated solutions of the substances listed at the temperature designated. Data are provided for making saturated solutions by weight (g of substance per 100 g of saturated solution) and by volume (g of substance per 100 ml of saturated solution and the ml of water required to make such a solution).

To make one *fluid ounce* of a saturated solution: multiply the grams of substance per 100 ml of saturated solution by 4.55 to obtain the number of grains required, by 0.01039 to obtain the number of avoirdupois ounces, by 0.00947 to obtain the number of apothecaries (Troy) ounces; also multiply the ml of water by 16.23 to obtain the number of minims, or divide by 100 to obtain the number of fluid ounces.

To make one *fluid dram:* multiply the grams of substance per 100 ml of saturated solution by 0.5682 to obtain the number of grains required; also multiply the ml of water by 0.60 to obtain the number of minims required.

Substance	Formula	Temp, °C	g/100 g satd soln	g/100 ml satd soln	ml water/ 100 ml satd soln	Specific gravity
acetanilide	$C_6H_5NHCOCH_3$	25	0.54	0.54	99.2	0.997
p-acetophenetidin	$C_6H_4(OC_2H_5)NHCH_3CO$	25	0.0766	0.0766	99.92	1.00
p-acetotoluide	$CH_3CONHC_6H_4CH_3$	25	0.12	0.12	99.7	0.9979
alanine	$CH_3CH(NH_2)COOH$	25	14.1	14.7	89.5	1.042
aluminum ammonium sulfate	$Al_2(SO_4)_3(NH_4)_2SO_4.24H_2O$	25	12.4	13	92	1.05
aluminum chloride hydrated	$AlCl_3.6H_2O$	25	55.5	75	60	1.35
aluminum fluoride	$Al_2F_6.5H_2O$	20	0.499	0.5015	100.0	1.0051
aluminum potassium sulfate	$AlK(SO_4)_2$	25	6.62	7.02	99.1	1.061
aluminum sulfate	$Al_2(SO_4)_3.18H_2O$	25	48.8	63	66	1.29
o-aminobenzoic acid	$C_6H_4NH_2COOH$	25	0.52	0.519	99.4	0.999
DL-α-amino-n-butyric acid	$CH_3CH_2CH(NH_2)COOH$	25	17.8	18.6	86.2	1.046
DL-α-aminoisobutyric acid	$(CH_3)_2C(NH_2)COOH$	25	13.3	13.7	89.5	1.031
ammonium arsenate	$NH_4H_2AsO_4$	20	32.7	40.2	83.0	1.228
ammonium benzoate	$NH_4C_7H_5O_2$	25	18.6	19.4	84.7	1.040
ammonium bromide	NH_4Br	15	41.7	53.8	75.2	1.290
ammonium carbonate		25	20	22	88	1.10
ammonium chloride	NH_4Cl	15	26.3	28.3	79.3	1.075
ammonium citrate, dibasic	$(NH_4)_2HC_6H_5O_7$	25	48.7	60.5	61.5	1.22
ammonium dichromate	$(NH_4)_2Cr_2O_7$	25	27.9	33	85	1.18
ammonium iodide	NH_4I	25	64.5	106.2	58.3	1.646
ammonium molybdate	$(NH_4)_6Mo_7O_{24}.4H_2O$	25	30.6	39	88	1.27
ammonium nitrate	NH_4NO_3	25	68.3	90.2	41.8	1.320
ammonium oxalate	$(NH_4)_2C_2O_4.H_2O$	25	4.95	5.06	97.0	1.019
ammonium perchlorate	NH_4ClO_4	25	21.1	23.7	88.7	1.123
ammonium periodate	NH_4IO_4	16	2.63	2.68	99.2	1.018
ammonium persulfate	$(NH_4)_2S_2O_8$	25	42.7	53	71	1.24
ammonium phosphate, dibasic	$(NH_4)_2HPO_4$	14.5	56.2	75.5	58.8	1.343
ammonium phosphate, monobasic	$NH_4H_2PO_4$	25	28.4	33	83	1.16
ammonium salicylate	$NH_4C_7H_5O_3$	25	50.8	58.2	56.4	1.145
ammonium silicofluoride	$(NH_4)_2SiF_6$	17.5	15.7	17.2	92.3	1.095
ammonium sulfate	$(NH_4)_2SO_4$	20	42.6	53.1	71.7	1.248
ammonium sulfite	$(NH_4)_2SO_3.H_2O$	25	39.3	47.3	73.2	1.204
ammonium thiocyanate	NH_4CNS	25	62.2	71	43	1.14
amyl alcohol	$C_5H_{11}OH$	25	2.61	2.60	96.9	0.995
aniline	$C_6H_5NH_2$	22	3.61	3.61	96.2	0.998
aniline hydrochloride	$C_6H_5NH_2.HCl$	25	49	54	56	1.10
aniline sulfate	$(C_6H_5NH_2)_2.H_2SO_4$	25	5.88	6	96	1.02
L-asparagine	$NH_2COCH_2CH(NH_2)COOH$	25	2.44	2.46	98.2	1.007
barium bromide	$BaBr_2$	20	51	87.2	83.8	1.710
barium chlorate	$Ba(ClO_3)_2$	25	28.5	36.8	92.6	1.294
barium chloride	$BaCl_2$	20	26.3	33.4	93.8	1.27
barium iodide	$BaI_2.7\frac{1}{2}H_2O$	25	68.8	157.0	71.1	2.277
barium nitrate	$Ba(NO_3)_2$	25	9.4	10.2	97.9	1.080
barium nitrite	$Ba(NO_2)_2$	17	40	59.6	89.4	1.490
barium perchlorate	$Ba(ClO_4)_2$	25	75.3	145.8	47.8	1.936
benzamide	$C_6H_5CONH_2$	25	1.33	1.33	98.6	0.999
benzoic acid	$C_7H_6O_2$	25	0.367	0.367	99.63	1.00
beryllium sulfate	$BeSO_4.4H_2O$	25	28.7	37.3	93.0	1.301
boric acid	H_3BO_3	25	4.99	5.1	97	1.02
n-butyl alcohol	$CH_3(CH_2)_2CH_2OH$	25	79.7	67.3	17.1	0.845
cadmium bromide	$CdBr_2.4H_2O$	25	52.9	94.0	83.9	1.775
cadmium chlorate	$Cd(ClO_3)_2.12H_2O$	18	76.4	174.5	54.0	2.284

Saturated Solutions (*Continued*)

Substance	Formula	Temp, °C	g/100 g satd soln	g/100 ml satd soln	ml water/ 100 ml satd soln	Specific gravity
cadmium chloride	$CdCl_2.2\frac{1}{2}H_2O$	25	54.7	97.2	80.8	1.778
cadmium iodide	CdI_2	20	45.9	73.0	86.3	1.590
cadmium sulfate	$3(CdSO_4).8H_2O$	25	43.4	70.3	91.8	1.619
calcium bromide	$CaBr_2$	20	58.8	107.2	75.0	1.82
calcium chlorate	$Ca(ClO_3)_2.2H_2O$	18	64.0	110.7	62.3	1.729
calcium chloride	$CaCl_2.6H_2O$	25	46.1	67.8	79.2	1.47
calcium chromate	$CaCrO_4.2H_2O$	18	14.3	16.4	98.7	1.149
calcium ferrocyanide	$Ca_2Fe(CN)_6$	25	36.5	49.6	86.2	1.357
calcium iodide	CaI_2	20	67.6	143.8	69.0	2.125
calcium lactate	$Ca(C_3H_5O_3)_2.5H_2O$	25	4.95	5	96	1.01
calcium nitrite	$Ca(NO_2)_2.4H_2O$	18	45.8	65.7	77.8	1.427
calcium sulfate	$CaSO_4.2H_2O$	25	0.208	0.208	99.70	0.999
camphoric acid	$C_8H_{14}(COOH)_2$	25	0.754	0.754	99.246	1.00
carbon disulfide	CS_2	22	0.173	0.173	99.63	0.998
cerium nitrate	$Ce(NO_3)_3.6H_2O$	25	63.7	119.9	68.2	1.880
cesium bromide	$CsBr$	21.4	53.1	89.8	79.5	1.693
cesium chloride	$CsCl$	25	65.7	126.3	65.9	1.923
cesium iodide	CsI	22.8	48.0	74.1	80.5	1.545
cesium nitrate	$CsNO_3$	25	21.9	26.1	92.9	1.187
cesium perchlorate	$CsClO_4$	25	2.01	2.03	99.0	1.010
cesium periodate	$CsIO_4$	15	2.10	2.13	99.5	1.017
cesium sulfate	Cs_2SO_4	25	64.5	129.8	71.7	2.013
chloral hydrate	$CCl_3CHO.H_2O$	25	79.4	120	31	1.51
chloroform	$CHCl_3$	29.4	0.703	0.705	99.57	1.0028
chromic oxide	CrO_3	18	62.5	106.3	64.0	1.703
chromium potassium sulfate	$Cr_2K_2(SO_4)_4.24H_2O$	25	19.6	22	90	1.12
citric acid	$(CH_2)_2COH(COOH)_3.H_2O$	25	67.5	88.6	42.7	1.311
cobalt chlorate	$Co(ClO_3)_2$	18	64.2	119.3	66.5	1.857
cobalt nitrate	$Co(NO_3)_2$	18	49.7	78.2	79.1	1.572
cobalt perchlorate	$Co(ClO_4)_2$	26	71.8	113.5	44.7	1.581
cupric ammonium chloride	$CuCl_2.2NH_4Cl.2H_2O$	25	30.3	35.5	82	1.17
cupric ammonium sulfate	$CuSO_4.(NH_4)_2SO_4$	19	15.3	17.3	96.0	1.131
cupric bromide	$CuBr_2$	25	55.8	102.5	81.2	1.84
cupric chlorate	$Cu(ClO_3)_2$	18	62.2	105.2	64.1	1.692
cupric chloride	$CuCl_2.2H_2O$	25	53.3	80	70	1.50
cupric nitrate	$Cu(NO_3)_2.6H_2O$	20	56.0	94.5	74.3	1.688
cupric selenate	$CuSeO_4$	21.2	14.7	17.2	99.4	1.165
cupric sulfate	$CuSO_4.5H_2O$	25	18.5	22.3	98.7	1.211
dextrose	$C_6H_{12}O_6.H_2O$	25	49.5	59	60	1.19
ether	$(C_2H_5)_2O$	22	5.45	5.34	93.0	0.985
ethyl acetate	$CH_3COOC_2H_5$	25	7.47	7.44	92.1	0.996
ferric ammonium citrate		25	67.7	97	46	1.43
ferric ammonium oxalate	$Fe(NH_4)_3(C_2O_4)_3.3H_2O$	25	51.5	65	61	1.26
ferric ammonium sulfate	$FeSO_4.(NH_4)_2SO_4$	16.5	19.1	22.4	94.3	1.165
ferric chloride	$FeCl_3$	25	73.1	131.1	48.3	1.793
ferric nitrate	$Fe(NO_3)_3$	25	46.8	70.2	79.8	1.50
ferric perchlorate	$Fe(ClO_4)_3.10H_2O$	25	79.9	132.1	33.2	1.656
ferrous sulfate	$FeSO_4.7H_2O$	25	42.1	52.8	72.7	1.255
gallic acid	$C_6H_2(OH)_3COOH.H_2O$	25	1.15	1.15	99.05	1.002
D-glutamic acid	$C_5H_9O_4N$	25	0.86	0.86	99.15	1.0002
glycine	NH_2CH_2COOH	25	20.0	21.7	86.8	1.083
hydroquinone	$C_6H_4(OH)_2$	20	6.7	6.78	94.4	1.012
m-hydroxybenzoic acid	$C_6H_4OHCOOH$	25	0.975	0.975	99.03	1.000
lactose	$C_{12}H_{22}O_{11}.H_2O$	25	15.9	17	90	1.07
lead acetate	$Pb(C_2H_3O_2)_2$	25	36.5	49.0	85.1	1.340
lead bromide	$PbBr_2$	25	0.97	0.98	99.6	1.006
lead chlorate	$Pb(ClO_3)_2$	18	60.2	117.0	77.3	1.944
lead chloride	$PbCl_2$	25	1.07	1.08	99.6	1.007
lead iodide	PbI_2	25	0.08	0.08	99.7	0.998
lead nitrate	$Pb(NO_3)_2$	25	37.1	53.6	91.0	1.445
DL-leucine	$C_6H_{13}O_2N$	25	0.976	0.975	98.9	0.999
L-leucine	$C_6H_{13}O_2N$	25	2.24	2.24	97.85	1.0012
lithium benzoate	$LiC_7H_5O_2$	25	27.7	30.4	79.6	1.100
lithium bromate	$LiBrO_3$	18	60.4	110.5	72.5	1.830

Saturated Solutions (*Continued*)

Substance	Formula	Temp, °C	g/100 g satd soln	g/100 ml satd soln	ml water/ 100 ml satd soln	Specific gravity
lithium carbonate	Li_2CO_3	15	1.36	1.38	100.0	1.014
lithium chloride	$LiCl.H_2O$	25	45.9	59.5	70.2	1.296
lithium citrate	$Li_3C_6H_5O_7$	25	31.8	38.6	82.8	1.213
lithium dichromate	$Li_2Cr_2O_7.H_2O$	18	52.6	82.9	74.8	1.574
lithium fluoride	LiF	18	0.27	0.27	99.9	1.002
lithium formate	$LiCHO_2$	18	27.9	31.8	80.4	1.140
lithium iodate	$LiIO_3$	18	44.6	69.9	86.8	1.566
lithium nitrate	$LiNO_3$	19	48.9	64.5	67.5	1.318
lithium perchlorate	$LiClO_4.3H_2O$	25	37.5	47.6	79.5	1.269
lithium salicylate	$LiC_7H_5O_3$	25	52.7	63.6	57.1	1.206
lithium sulfate	$Li_2SO_4.H_2O$	25	27.2	33	88.5	1.21
magnesium bromide	$MgBr_2.6H_2O$	18	50.1	83.1	82.8	1.655
magnesium chlorate	$Mg(ClO_3)_2$	18	56.3	90.0	69.7	1.594
magnesium chloride	$MgCl_2.6H_2O$	25	62.5	79	47.5	1.26
magnesium chromate	$MgCr_2O_4.7H_2O$	18	42.0	59.7	82.5	1.422
magnesium dichromate	$MgCrO_7.5H_2O$	25	81.0	138.8	32.6	1.712
magnesium iodate	$Mg(IO_3)_2.4H_2O$	18	6.44	6.95	100.8	1.078
magnesium iodide	$MgI_2.8H_2O$	18	59.7	114.0	77.1	1.909
magnesium molybdate	$MgMoO_4$	25	15.9	18.4	97.4	1.159
magnesium nitrate	$Mg(NO_3)_2.6H_2O$	25	42.1	58.6	80.5	1.388
magnesium perchlorate	$Mg(ClO_4)_2.6H_2O$	25	49.9	73.6	73.9	1.472
magnesium selenate	$MgSeO_4$	20	35.3	50.8	93.0	1.440
magnesium sulfate	$MgSO_4.7H_2O$	25	55.3	72	58.5	1.30
manganese chloride	$MnCl_2$	25	43.6	63.2	82.0	1.449
manganese nitrate	$Mn(NO_3)_2.6H_2O$	18	57.3	93.2	69.2	1.624
manganese silicofluoride	$MnSiF_6$	17.5	37.7	54.5	90.1	1.446
manganese sulfate	$MnSO_4$	25	39.4	59.1	90.8	1.499
mercuric acetate	$Hg(C_2H_3O_2)_2$	25	30.2	38	88	1.26
mercuric bromide	$HgBr_2$	25	0.609	0.610	99.6	1.0023
mercury bichloride	$HgCl_2$	25	6.6	6.96	98.5	1.054
methylene blue	$C_{16}H_{18}N_3ClS.3H_2O$	25	4.25	4.3	97	1.01
methyl salicylate	$C_6H_4OHCOOCH_3$	25	0.12	0.12	99.88	1.00
monochloracetic acid	$CH_2ClCOOH$	25	78.8	105	28	1.33
β-naphthalenesulfonic acid	$C_{10}H_7SO_3H$	30	56.9	67.9	51.4	1.193
nickel ammonium sulfate	$NiSO_4(NH_4)_2SO_4.6H_2O$	25	9.0	9.5	96	1.05
nickel chlorate	$Ni(ClO_3)_2$	18	56.7	94.2	72.0	1.658
nickel chlorate	$Ni(ClO_3)_2.6H_2O$	18	64.5	107.2	59.1	1.661
nickel nitrate	$Ni(NO_3)_2.6H_2O$	25	77	122	36	1.58
nickel perchlorate	$Ni(ClO_4)_2$	26	70.8	112.2	46.4	1.584
nickel perchlorate	$Ni(ClO_4)_2.9H_2O$	18	52.4	82.7	75.1	1.576
nickel sulfate	$NiSO_4.6H_2O$	25	47.3	64	71	1.35
DL-norleucine	$C_6H_{13}NO_2$	25	1.13	1.13	98.97	0.999
oxalic acid	$H_2C_2O_4.2H_2O$	25	9.81	10.3	94.2	1.044
phenol	C_6H_5OH	20	6.1	6.14	94.5	1.0057
β-phenylalanine	$C_6H_5CH_2CH(NH_2)COOH$	25	2.88	2.89	97.5	1.0035
m-phenylenediamine	$C_6H_8N_2$	20	23.1	23.8	79.3	1.032
p-phenylenediamine	$C_6H_8N_2$	20	3.69	3.70	96.67	1.0038
phenyl salicylate	$C_6H_4OHCOOC_6H_5$	25	0.015	0.015	99.84	0.999
phenyl thiourea	$CS(NH_2)NHC_6H_5$	25	0.24	0.24	99.6	0.998
phosphomolybdic acid	$20MoO_3.2H_3PO_4.48H_2O$	25	74.3	135	46	1.81
phosphotungstic acid	Approx. $20WO_3.2H_3PO_4.25H_2O$	25	71.4	160	64	2.24
potassium acetate	$KC_2H_3O_2$	25	68.7	97.1	44.3	1.413
potassium antimony tartrate	$KSbOC_4H_4O_6$	25	7.64	8.02	96.9	1.049
potassium bicarbonate	$KHCO_3$	25	26.6	31.6	87.5	1.188
potassium bitartrate	$KC_4H_5O_6$	25	0.65	0.65	99.3	0.999
potassium bromate	$KBrO_3$	25	7.53	7.89	97.5	1.054
potassium bromide	KBr	25	40.6	56.0	82.0	1.380
potassium carbonate	$K_2CO_3.1\frac{1}{2}H_2O$	25	52.9	82.2	73.5	1.559
potassium chlorate	$KClO_3$	25	8.0	8.41	96.6	1.051
potassium chloride	KCl	25	26.5	31.2	86.8	1.178
potassium chromate	K_2CrO_4	25	39.4	54.1	83.7	1.381
potassium citrate	$K_3C_6H_5O_7$	25	60.91	92.1	59.2	1.514
potassium dichromate	$K_2Cr_2O_7$	25	13.0	14.2	95.0	1.092
potassium ferricyanide	$K_3Fe(CN)_6$	22	32.1	38.1	80.8	1.187

Saturated Solutions (*Continued*)

Substance	Formula	Temp, °C	g/100 g satd soln	g/100 ml satd soln	ml water/ 100 ml satd soln	Specific gravity
potassium ferrocyanide	$K_4Fe(CN)_6$	25	24.0	28.2	89.2	1.173
potassium fluoride	$KF.2H_2O$	18	48.0	72.0	78.0	1.500
potassium formate	$KCHO_2$	18	76.8	120.6	36.4	1.571
potassium hydroxide	KOH	15	51.7	79.2	74.2	1.536
potassium iodate	KIO_3	25	8.40	8.99	98.0	1.071
potassium iodide	KI	25	59.8	103.2	69.1	1.721
potassium meta-antimonate	$KSbO_3$	18	2.73	2.81	99.7	1.025
potassium nitrate	KNO_3	25	28.0	33.4	86.0	1.193
potassium nitrite	KNO_2	20	74.3	121.5	42.3	1.649
potassium oxalate	$K_2C_2O_4.H_2O$	25	28.3	34	86	1.20
potassium perchlorate	$KClO_4$	25	2.68	2.72	99.0	1.014
potassium periodate	KIO_4	13	0.658	0.661	99.83	1.005
potassium permanganate	$KMnO_4$	25	7.10	7.43	97.3	1.046
potassium sodium tartrate	$KNaC_4H_4O_6.4H_2O$	25	39.71	51.9	78.8	1.308
potassium stannate	K_2SnO_3	15.5	42.7	69.2	92.9	1.620
potassium sulfate	K_2SO_4	25	10.83	11.8	96.9	1.086
quinine salicylate	$C_{20}H_{24}N_2O_2.C_6H_4(OH)COOH.2H_2O$	25	0.065	0.065	99.84	0.999
resorcinol	$C_6H_4(OH)_2$	25	58.8	67.2	47.2	1.142
rubidium bromate	$RbBrO_3$	16	2.15	2.18	99.4	1.016
rubidium bromide	$RbBr$	25	52.7	85.6	76.9	1.625
rubidium chloride	$RbCl$	25	48.6	72.8	77.1	1.50
rubidium iodate	$RbIO_3$	15.6	2.72	2.78	99.5	1.022
rubidium iodide	RbI	24.3	63.6	117.7	67.3	1.850
rubidium nitrate	$RbNO_3$	25	40.1	55.0	82.4	1.375
rubidium perchlorate	$RbClO_4$	25	1.88	1.90	99.3	1.012
rubidium periodate	$RbIO_4$	16	0.645	0.648	99.85	1.0052
rubidium sulfate	Rb_2SO_4	25	33.8	45.6	89.7	1.354
silicotungstic acid	$H_4SiW_{12}O_{40}$	18	90.6	258	26.8	2.843
silver acetate	$Ag(C_2H_3O_2)$	25	1.10	1.11	99.40	1.0047
silver bromate	$AgBrO_3$	25	0.204	0.2037	99.65	0.9985
silver fluoride	$AgF.2H_2O$	15.8	64.5	168.4	92.7	2.61
silver nitrate	$AgNO_3$	25	71.5	164	65.5	2.29
silver perchlorate	$AgClO_4.H_2O$	25	84.5	237.1	43.5	2.806
sodium acetate	$NaC_2H_3O_2$	25	33.6	40.5	80.0	1.205
sodium ammonium sulfate	$NaNH_4SO_4$	15	25.2	29.6	87.9	1.174
sodium arsenate	$Na_3AsO_4.12H_2O$	17	21.1	23.5	88.0	1.119
sodium benzenesulfonate	$NaC_6H_5SO_3$	25	16.4	17.6	90.1	1.076
sodium benzoate	$NaC_7H_5O_2$	25	36.0	41.5	73.9	1.152
sodium bicarbonate	$NaHCO_3$	15	8.28	8.80	97.6	1.061
sodium bisulfate	$NaHSO_4.H_2O$	25	59	87	60	1.47
sodium bromide	$NaBr.2H_2O$	25	48.6	75.0	79.4	1.542
sodium carbonate	$Na_2CO_3.10H_2O$	25	22.6	28.1	96.5	1.242
sodium chlorate	$NaClO_3$	25	51.7	74.3	69.6	1.440
sodium chloride	$NaCl$	25	26.5	31.7	88.1	1.198
sodium chromate	Na_2CrO_4	18	40.1	57.4	85.7	1.430
sodium citrate	$Na_3C_6H_5O_7.5H_2O$	25	48.1	61.2	66.0	1.272
sodium dichromate	$Na_2Cr_2O_7$	18	63.9	111.4	63.0	1.743
sodium ferrocyanide	$Na_4Fe(CN)_6$	25	17.1	19.4	93.9	1.131
sodium fluoride	NaF	25	3.98	4.14	99.7	1.038
sodium formate	$NaCHO_2$	18	44.7	58.9	73.0	1.316
sodium hydroxide	$NaOH$	25	50.8	77	74	1.51
sodium hypophosphite	NaH_2PO_2	16	52.1	72.4	66.6	1.386
sodium iodate	$NaIO_3.H_2O$	25	8.57	9.21	98.5	1.075
sodium iodide	NaI	25	64.8	124.3	67.7	1.919
sodium molybdate	Na_2MoO_4	18	39.4	56.6	87.0	1.435
sodium nitrate	$NaNO_3$	25	47.9	66.7	72.5	1.391
sodium nitrite	$NaNO_2$	20	45.8	62.3	73.8	1.359
sodium oxalate	$Na_2(CO_2)_2$	25	3.48	3.58	99.1	1.025
sodium paratungstate	$(Na_2O)_3(WO_3)_7.16H_2O$	0	26.7	35.2	96.5	1.316
sodium perchlorate	$NaClO_4$	25	67.8	114.1	54.1	1.683
sodium periodate	$NaIO_4.3H_2O$	25	12.6	13.9	96.2	1.103
sodium phenolsulfonate	$C_6H_4(OH)SO_3Na$	25	16.1	17.4	90.5	1.079
sodium phosphate dibasic	Na_2HPO_4	17	4.2	4.4	99.9	1.043
sodium phosphate tribasic	Na_3PO_4	14	9.5	10.5	99.8	1.103

Saturated Solutions (*Continued*)

Substance	Formula	Temp, °C	g/100 g satd soln	g/100 ml satd soln	ml water/ 100 ml satd soln	Specific gravity
sodium pyrophosphate	$Na_2H_2P_2O_7.6H_2O$	25	13.0	14.4	95.8	1.104
sodium salicylate	$NaC_7H_5O_3$	25	53.6	67.0	58.0	1.248
sodium selenate	Na_2SeO_4	18	29.0	38.1	93.4	1.313
sodium silicofluoride	$NaSiF_6$	20	0.773	0.737	99.76	1.0054
sodium sulfate	Na_2SO_4	25	21.8	26.4	94.5	1.208
sodium sulfate	$Na_2SO_4.10H_2O$	25	27.7	33.3	87.0	1.207
sodium sulfide	$Na_2S.9H_2O$	25	52.3	63	57	1.20
sodium sulfite, anhydrous	Na_2SO_3	25	23	28.5	95.5	1.24
sodium thiocyanate	$NaCNS$	25	62.9	87	51	1.38
sodium thiosulfate	$Na_2S_2O_3.5H_2O$	25	66.8	93	46	1.39
sodium tungstate	$Na_2WO_4.10H_2O$	18	42.0	66.1	91.3	1.573
stannous chloride	$SnCl_2$	15	72.9	133.1	49.5	1.827
strontium chlorate	$Sr(ClO_3)_2$	18	63.6	117.0	67.0	1.839
strontium chloride	$SrCl_2.6H_2O$	15	33.4	45.5	90.7	1.36
strontium iodide	$SrI_2.6H_2O$	20	64.0	137.8	77.5	2.15
strontium nitrate	$Sr(NO_3)_2$	25	44.2	65.3	82.5	1.477
strontium nitrite	$Sr(NO_2)_2$	19	39.3	56.8	87.8	1.445
strontium perchlorate	$Sr(ClO_4)_2$	25	75.6	158.5	50.8	2.084
strontium salicylate	$Sr(C_7H_5O_3)_2$	25	4.58	4.68	97.5	1.019
succinic acid	$(CH_2)_2(COOH)_2$	25	7.67	7.82	94.5	1.021
succinimide	$(CH_2CO)_2NH.H_2O$	25	30.6	32.7	74.2	1.067
sucrose	$C_{12}H_{22}O_{11}$	25	67.89	90.9	43.0	1.340
tartaric acid	$C_2H_2(OH)_2(COOH)_2$	15	58.5	76.9	54.7	1.31
tetraethyl ammonium iodide	$N(C_2H_5)_4I$	25	32.9	36.2	74.0	1.102
tetramethyl ammonium iodide	$N(CH_3)_4I$	25	5.51	5.60	96.1	1.016
thallium chloride	$TlCl$	25	0.40	0.40	99.6	1.0005
thallium nitrate	$TlNO_3$	25	10.4	11.4	98.0	1.093
thallium nitrite	$TlNO_2$	25	32.1	43.7	92.5	1.360
thallium perchlorate	$TlClO_4$	25	13.5	15.2	97.1	1.122
thallium sulfate	Tl_2SO_4	25	5.48	5.74	99.0	1.047
trichloroacetic acid	CCl_3COOH	25	92.3	149.6	12.41	1.615
uranyl chloride	UO_2Cl_2	18	76.2	208.5	65.2	2.736
uranyl nitrate	$UO_2(NO_3)_2.6H_2O$	25	68.9	120	54.5	1.74
urea	$(NH_2)_2CO$	25	53.8	62	53.5	1.15
urea phosphate	$CO(NH_2)_2.H_3PO_4$	24.5	52.4	66.1	60.1	1.26
urethan	$NH_2CO_2C_2H_5$	25	82.8	88.8	18.5	1.073
D-valine	$(CH_3)_2CHCH(NH_2)COOH$	25	8.14	8.26	93.3	1.015
DL-valine	$(CH_3)_2CHCH(NH_2)COOH$	25	6.61	6.68	94.5	1.012
zinc acetate	$Zn(C_2H_3O_2)_2$	25	25.7	30.0	86.5	1.165
zinc benzenesulfonate	$Zn(C_6H_5SO_3)_2$	25	29.5	34.9	83.4	1.182
zinc chlorate	$Zn(ClO_3)_2$	18	65.0	124.4	67.0	1.914
zinc chloride	$ZnCl_2$	25	67.5	128	61	1.89
zinc iodide	ZnI_2	18	81.2	221.3	51.2	2.725
zinc phenolsulfonate	$(C_6H_5OSO_3)_2Zn.8H_2O$	25	39.8	47.3	71.5	1.185
zinc selenate	$ZnSeO_4$	22	37.8	58.9	97.0	1.559
zinc silicofluoride	$ZnSiF_6.6H_2O$	20	32.9	47.2	96.3	1.434
zinc sulfate	$ZnSO_4.7H_2O$	25	36.7	54.6	94.7	1.492
zinc valerate	$Zn(C_5H_9O_2)_2$	25	1.27	1.27	98.8	1.001

TABLE 1g4. STANDARD SOLUTIONS*

The group of compounds in this table are available in grades of such high purity that solutions of known titer may be prepared by accurately weighing and dissolving the solid in water and diluting the solution to a fixed total volume. If stored in tightly stoppered pyrex glass, in a cool place away from strong light the titers will remain unchanged for about a year.

COMPOUND	FORMULA	MOL. WT.	WT. IN GRAMS TO PREPARE 1 L. 0.1000 N SOLU
Borax	$Na_2B_4O_7 \cdot 10H_2O$*	381.42	19.071
Potassium Acid Phthalate	$C_6H_4(COOK)(COOH)$	204.23	20.423
Potassium Bromate	$KBrO_3$	167.02	2.7836
Potassium Bromide	KBr	119.02	11.902
Potassium Dichromate	$K_2Cr_2O_7$	294.22	4.9037
Potassium Iodate	KIO_3	214.01	3.5667
Sodium Carbonate	Na_2CO_3*	105.99	5.2997
Sodium Chloride	NaCl	58.448	5.8448
Sulfamic Acid	NH_2SO_2OH	97.098	9.7098

*These materials in solution are best stored in polyethylene bottles.

*Reproduced with permission from: "A Condensed Laboratory Handbook," Du Pont Company, Industrial Chemicals Department, Wilmington, DE, 1971.

TABLE 1g5. CONSTANT HUMIDITY WITH SULFURIC ACID SOLUTIONS AT 20°C.*

DENSITY OF ACID SOLUTION	RELATIVE HUMIDITY	VAPOR PRESSURE
1.00	100.0	17.4
1.05	97.5	17.0
1.10	93.9	16.3
1.15	88.8	15.4
1.20	80.5	14.0
1.25	70.4	12.2
1.30	58.3	10.1
1.35	47.2	8.3
1.40	37.1	6.5
1.50	18.8	3.3
1.60	8.5	1.5
1.70	3.2	0.6

The relative humidities and pressures of aqueous vapor in air are given for equilibrium conditions above the indicated aqueous solutions of sulfuric acid. A large glass dessicator is a convenient constant humidity chamber. Place about 400 ml. of the appropriate acid solution in the bottom of a clean, dry dessicator. Replace the porcelain shelf, grease the ground surfaces, close lid and allow system to equilibrate.

*Reproduced with permission from: "A Condensed Laboratory Handbook," Du Pont Company, Industrial Chemicals Department, Wilmington, DE, 1971.

TABLE 1g6. CONSTANT HUMIDITY SOLUTIONS*

A saturated aqueous solution in contact with an excess of the solute when kept in an enclosed space will maintain a constant humidity at a given temperature.

Substance dissolved and solid phase	Temp, °C	% Humidity
lead nitrate, $Pb(NO_3)_2$	20	98
dibasic sodium phosphate, $Na_2HPO_4.12H_2O$	20	95
monobasic ammonium phosphate, $NH_4H_2PO_4$	20–25	93
zinc sulfate, $ZnSO_4.7H_2O$	20	90
potassium chromate, K_2CrO_4	20	88
potassium bisulfate, $KHSO_4$	20	86
potassium bromide, KBr	20	84
ammonium sulfate, $(NH_4)_2SO_4$	20	81
ammonium chloride, NH_4Cl	20–25	79
sodium acetate, $NaC_2H_3O_2.3H_2O$	20	76
sodium chlorate, $NaClO_3$	20	75
sodium nitrate, $NaNO_2$	20	66
sodium bromide, $NaBr.2H_2O$	20	58
magnesium nitrite, $Mg(NO_3)_2.6H_2O$	18.5	56
sodium dichromate, $Na_2Cr_2O_7.2H_2O$	20	52
potassium thiocyanate, $KSCN$	20	47
zinc nitrate, $Zn(NO_3)_2.6H_2O$	20	42
chromium trioxide, CrO_3	20	35
calcium chloride, $CaCl_2.6H_2O$	24.5	31
potassium acetate, $KC_2H_3O_2$	20	20
lithium chloride, $LiCl.H_2O$	20	15

*Reproduced from page 1310 of "The Merck Index," Eighth Edition, copyright 1968 by Merck & Co., Inc., Rahway, New Jersey, U.S.A. with the permission of the copyright holder.

TABLE 1g7. COOLING MIXTURES*

Substance	Quantity of substance	Quantity of water	Temp of resulting mixtures in °C	Approx temp, °F
ammonium nitrate	100	94	−4.0	25
sodium acetate	85	100	−4.7	23
sodium nitrate	75	100	−5.3	22.5
sodium thiosulfate cryst	110	100	−8.0	18
calcium chloride, $6H_2O$	100	246 (ice)	−9.0	16
sodium chloride	36	100	−10.0	14
ammonium nitrate	45	100 (ice)	−16.8	1.5
sodium nitrate	50	100 (ice)	−17.8	0
ammonium thiocyanate	133	100	−18.0	0
sodium chloride	33	100 (ice)	−21.3	−6
calcium chloride, $6H_2O$	100	123 (ice)	−21.5	−6.5
sodium bromide	66	100 (ice)	−28	−18
magnesium chloride cryst	85	100 (ice)	−34	−29
sulfuric acid (66.1% H_2SO_4)	100	109.7 (snow)	−37.0	−34.6
calcium chloride, $6H_2O$	100	81 (ice)	−40.3	−40
calcium chloride, $6H_2O$	100	70 (ice)	−55	−67
alcohol at 4°C with solid carbon dioxide			−72	−98
chloroform with solid carbon dioxide			−77	−106
acetone with solid carbon dioxide			−86	−123
ether with solid carbon dioxide			−100	−148
solid carbon dioxide (dry ice) sublimes at			−78.5	−109

*Reproduced from page 1310 of "The Merck Index," Eighth Edition, copyright 1968 by Merck & Co., Inc., Rahway, New Jersey, U.S.A. with permission of the copyright holder.

TABLE 1g8. STANDARD BUFFERS FOR CALIBRATION OF pH METERS*

Temperature °C	pH values				
	$0.1M$ Hydrochloric acid	Saturated potassium hydrogen tartrate	$0.05M$ Potassium hydrogen phthalate	$0.05M$ Phosphate	$0.01M$ Borax
0	1.10		4.01	6.98	9.46
5	1.10		4.01	6.95	9.38
10	1.10		4.00	6.92	9.33
15	1.10		4.00	6.90	9.27
20	1.10	3.56	4.00	6.88	9.22
25	1.10	3.57	4.01	6.86	9.18
30	1.10	3.58	4.01	6.85	9.14
35	1.10		4.02	6.84	9.10
40	1.10		4.03	6.84	9.07
45	1.10		4.04	6.83	9.04
50	1.11		4.06	6.83	9.01
55	1.11		4.08	6.83	8.98
60	1.11		4.10	6.84	8.96

Preparation of Solutions

0.1M Hydrochloric Acid. Prepare 0.1N hydrochloric acid by diluting reagent-grade hydrochloric acid. Determine the molarity by titration with standard alkali and adjust to $0.1000 \pm 0.005M$.

Saturated Potassium Hydrogen Tartrate. Shake an excess of reagent-grade potassium hydrogen tartrate with water at $25 \pm 5°C$ for about 3 minutes.

0.05M Potassium Hydrogen Phthalate. Dissolve 10.21 ± 0.05 g of reagent-grade potassium hydrogen phthalate in sufficient water to make 1000 ml of solution.

0.05M Phosphate. Dissolve 3.40 ± 0.01 g of reagent-grade potassium dihydrogen phosphate and 3.55 ± 0.01 g of reagent-grade anhydrous disodium hydrogen phosphate in sufficient water to make 1000 ml of solution.

0.01M Borax. Dissolve 3.81 ± 0.01 g of reagent-grade sodium tetraborate decahydrate in sufficient water to make 1000 ml of solution.

TABLE 1g9. pH VALUES OF STANDARD SOLUTIONS*

Normality	pH values			
	HCl	CH_3COOH	NaOH	NH_3
1	0.10	2.37	14.05	11.77
0.1	1.07	2.87	13.07	11.27
0.01	2.02	3.37	12.12	10.77
0.001	3.01	3.87	11.13	10.27
0.0001	4.01			

TABLE 1g10. INDICATORS FOR VOLUMETRIC WORK AND pH DETERMINATIONS*

Indicator	Chemical name	Acid color	pH range	Basic color	Preparation
methyl violet 6B	tetra- and pentamethylated p-rosaniline hydrochloride	Y	0.1–1.5	B	pH: 0.25% water
metacresol purple (acid range)	m-cresolsulfonphthalein	R	0.5–2.5	Y	pH: 0.10 g in 13.6 ml 0.02N NaOH, diluted to 250 ml with water
metanil yellow	4-phenylamino-azobenzene-3′-sulfonic acid	R	1.2–2.3	Y	pH: 0.25% in ethanol
p-xylenol blue (acid range)	1,4-dimethyl-5-hydroxybenzene-sulfonphthalein	R	1.2–2.8	Y	pH: 0.04% in ethanol
thymol blue (acid range)	thymolsulfonphthalein	R	1.2–2.8	Y	pH: 0.1 g in 10.75 ml 0.02N NaOH, diluted to 250 ml with water
tropaeolin OO	sodium p-diphenylaminoazo-benzenesulfonate	R	1.4–2.6	Y	pH: 0.1% in water vol: 1% in water
quinaldine red	2-(p-dimethylaminostyryl)-quinoline ethiodide	C	1.4–3.2	R	vol: 0.1% in ethanol
benzopurpurine 4B	ditolyl-diazo-bis-α-naphthyl-amine-4-sulfonic acid	B-V	1.3–4.0	R	pH, vol: 0.1% in water
methyl violet 6B	tetra- and pentamethylated p-rosaniline hydrochloride	B	1.5–3.2	V	pH, vol: 0.25% in water
2,4-dinitrophenol		C	2.6–4.0	Y	pH, vol: 0.1 g in 5 ml ethanol, diluted to 100 ml with water
methyl yellow	p-dimethylaminoazobenzene	R	2.9–4.0	Y	pH, vol: 0.05% in ethanol
bromphenol blue	tetrabromophenolsulfon-phthalein	Y	3.0–4.6	B	pH: 0.1 g in 7.45 ml 0.02N NaOH, diluted to 250 ml with water
tetrabromophenol blue	tetrabromophenol–tetrabro-mosulfonphthalein	Y	3.0–4.6	B	pH: 0.1 g in 5.00 ml 0.02N NaOH, diluted to 250 ml with water
direct purple	disodium 4,4′-bis(2-amino-1-naphthylazo)-2,2′-stil-benedisulfonate	B-P	3.0–4.6	R	vol: 0.1 g in 7.35 ml 0.02N NaOH, diluted to 100 ml with water
Congo red	diphenyl-diazo-bis-1-naphthyl-amine-4-sodium sulfonate	B	3.0–5.2	R	pH: 0.1% in water
methyl orange	4′-dimethylaminoazobenzene-4-sodium sulfonate	R	3.1–4.4	Y	vol: 0.1% in water
brom-chlorphenol blue	dibromodichlorophenolsulfon-phthalein	Y	3.2–4.8	B	pH: 0.1 g in 8.6 ml 0.02N NaOH, diluted to 250 ml with water vol: 0.04% in ethanol
p-ethoxychrysoidine	4′-ethoxy-2,4-diaminoazobenzene	R	3.5–5.5	Y	vol: 0.1% in ethanol
α-naphthyl red		R	3.7–5.0	Y	vol: 0.1% in ethanol
sodium alizarinsulfon-ate	dihydroxyanthraquinone sodium sulfonate	Y	3.7–5.2	V	pH, vol: 1% in water
bromcresol green	tetrabromo-m-cresolsulfon-phthalein	Y	3.8–5.4	B	pH: 0.10 g in 7.15 ml 0.02N NaOH, diluted to 250 ml with water
2,5-dinitrophenol		C	4.0–5.8	Y	pH, vol: 0.10 g in 20 ml eth-anol, then dilute to 100 ml with water
methyl red	4′-dimethylaminoazobenzene-2-carboxylic acid	R	4.2–6.2	Y	pH: 0.10 g in 18.6 ml 0.02N NaOH, diluted to 250 ml with water vol: 0.1% in ethanol
lacmoid		R	4.4–6.2	B	vol: 0.5% in ethanol
azolitmin		R	4.5–8.3	B	vol: 0.5% in water
litmus		R	4.5–8.3	B	vol: 0.5% in water
cochineal	complex hydroxyanthraquinone derivative	R	4.8–6.2	V	vol: triturate 1 g with 20 ml ethanol and 60 ml water, let stand 4 days, and filter
hematoxylin		Y	5.0–6.0	V	vol: 0.5% in ethanol
chlorphenol red	dichlorophenolsulfonphthalein	Y	5.0–6.6	R	pH: 0.1 g in 11.8 ml 0.02N NaOH, diluted to 250 ml with water vol: 0.04% in ethanol
bromcresol purple	dibromo-o-cresolsulfonphthalein	Y	5.2–6.8	Pu	pH: 0.1 g in 9.25 ml 0.02N NaOH, diluted to 250 ml with water vol: 0.02% in ethanol

Indicator	Chemical name	Acid color	pH range	Basic color	Preparation
bromphenol red	dibromophenolsulfonphthalein	Y	5.2–7.0	R	pH: 0.1 g in 9.75 ml 0.02N NaOH, diluted to 250 ml with water vol: 0.04% in ethanol
alizarin	1,2-dihydroxyanthraquinone	Y	5.5–6.8	R	vol: 0.1% in ethanol
dibromophenoltetra-bromophenolsulfon-phthalein		Y	5.6–7.2	Pu	pH: 0.1 g in 1.21 ml 0.1N NaOH, diluted to 250 ml with water
p-nitrophenol		C	5.6–7.6	Y	pH, vol: 0.25% in water
bromothymol blue	dibromothymolsulfonphthalein	Y	6.0–7.6	B	pH: 0.1 g in 8 ml 0.02N NaOH, diluted to 250 ml with water vol: 0.1% in 50% ethanol
indo-oxine	5,8-quinolinequinone-8-hydroxy-5-quinolyl-5-imide	R	6.0–8.0	B	vol: 0.05% in ethanol
curcumin		Y	6.0–8.0	Br-R	vol: saturated aq soln
quinoline blue	cyanine	C	6.6–8.6	B	vol: 1% in ethanol
phenol red	phenolsulfonphthalein	Y	6.8–8.4	R	pH: 0.1 g in 14.20 ml 0.02N NaOH, diluted to 250 ml with water vol: 0.1% in ethanol
neutral red	2-methyl-3-amino-6-dimethyl-aminophenazine	R	6.8–8.0	Y	pH, vol: 0.1 g in 70 ml ethanol, diluted to 100 ml with water
rosolic acid aurin; corallin		Y	6.8–8.2	R	pH, vol: 1% in 50% ethanol
cresol red	o-cresolsulfonphthalein	Y	7.2–8.8	R	pH: 0.1 g in 13.1 ml 0.02N NaOH, diluted to 250 ml with water vol: 0.1% in ethanol
α-naphtholphthalein		P	7.3–8.7	G	pH, vol: 0.1% in 50% ethanol
metacresol purple (alkaline range)	m-cresolsulfonphthalein	Y	7.4–9.0	P	pH: 0.1 g in 13.1 ml 0.02N NaOH, diluted to 250 ml with water vol: 0.1% in ethanol
ethylbis-2,4-dinitro-phenylacetate		C	7.5–9.1	B	vol: saturated soln in equal volumes of acetone and ethanol
tropaeolin OOO no. 1	sodium α-naphtholazobenzene-sulfonate	Y	7.6–8.9	R	vol: 0.1% in water
thymol blue (alkaline range)	thymolsulfonphthalein	Y	8.0–9.6	B	pH: 0.1 g in 10.75 ml 0.02N NaOH, diluted to 250 ml with water vol: 0.1% in ethanol
p-xylenol blue	1,4-dimethyl-5-hydroxybenzene-sulfonphthalein	Y	8.0–9.6	B	pH, vol: 0.04% in ethanol
o-cresolphthalein		C	8.2–9.8	R	pH, vol: 0.04% in ethanol
α-naphtholbenzein		Y	8.5–9.8	G	pH, vol: 1% in ethanol
phenolphthalein	3,3-bis(p-hydroxyphenyl)phtha-lide	C	8.2–10	R	vol: 1% in ethanol
thymolphthalein		C	9.3–10.5	B	pH, vol: 0.1% in ethanol
Nile blue A	aminonaphthodiethylamino-phenoxazine sulfate	B	10–11	P	vol: 0.1% in water
alizarin yellow GG	3-carboxy-4-hydroxy-3'-nitro-azobenzene	Y	10–12	L	pH, vol: 0.1% io 50% ethanol
alizarin yellow R	3-carboxy-4-hydroxy-4'-nitro-azobenzene sodium salt	Y	10.2–12.0	R	pH, vol: 0.1% in water
Poirrier's blue C4B		B	11–13	R	pH: 0.2% in water
tropaeolin O	p-benzenesulfonic acid-azoresor-cinol	Y	11–13	O	pH: 0.1% in water
nitramine	picrylnitromethylamine	C	10.8–13	Br	pH: 0.1% in 70% ethanol
1,3,5-trinitrobenzene		C	11.5–14	O	pH: 0.1% in ethanol
indigo carmine	sodium indigodisulfonate	B	11.6–14	Y	pH: 0.25% in 50% ethanol

The indicator colors are abbreviated as follows: B, blue; Br, brown; C, colorless; G, green; L, lilac; O, orange; P, pink; Pu, purple; R, red; V, violet; and Y, yellow.

References
1. Direct Purple: Taras, *Anal. Chem.* **19,** 339 (1947).
2. p-Ethoxychrysoidine: Schulek and Rozsa, *Z. Anal. Chem.* **115,** 185 (1939).
3. Indo-oxine: Berg and Becker, *Z. Anal. Chem.* **119,** 81 (1940).
4. Ethylbis-2,4-dinitrophenylacetate: Fehnel and Amstutz, *Ind. Eng. Chem.*, Anal. Ed. **16,** 53 (1944).

TABLE 1g11. MIXED INDICATORS*

Composition	Solvent	Transition pH	Acid color	Transition color	Basic color
dimethyl yellow, 0.05% + methylene blue, 0.05%	alc	3.2	blue-violet	—	green
methyl orange, 0.02% + xylene cyanole FF, 0.28%	50% alc	3.9	red	gray	green
methyl yellow, 0.08% + methylene blue, 0.004%	alc	3.9	pink	straw pink	yellow-green
methyl orange, 0.1% + indigocarmine, 0.25%	aq	4.1	violet	gray	yellow-green
bromcresol green, 0.1% + methyl orange, 0.02%	aq	4.3	orange	light green	dark green
bromcresol green, 0.075% + methyl red, 0.05%	alc	5.1	wine-red	—	green
methyl red, 0.1% + methylene blue, 0.05%	alc	5.4	red-violet	dirty blue	green
bromcresol green, 0.05% + chlorphenol red, 0.05%	aq	6.1	yellow-green	—	blue-violet
bromcresol purple, 0.05% + bromthymol blue, 0.05%	aq	6.7	yellow	violet	violet-blue
neutral red, 0.05% + methylene blue, 0.05%	alc	7.0	violet-blue	violet-blue	green
bromthymol blue, 0.05% + phenol red, 0.05%	aq	7.5	yellow	violet	dark violet
cresol red, 0.025% + thymol blue, 0.15%	aq	8.3	yellow	rose	violet
phenolphthalein, 0.033% + methyl green, 0.067%	alc	8.9	green	gray-blue	violet
phenolphthalein, 0.075% + thymol blue, 0.025%	50% alc	9.0	yellow	green	violet
phenolphthalein, 0.067% + naphtholphthalein, 0.033%	50% alc	9.6	pale rose	—	violet
phenolphthalein, 0.033% + Nile blue, 0.133%	alc	10.0	blue	violet	red
alizarin yellow, 0.033% + Nile blue, 0.133%	alc	10.8	green	—	red-brown

TABLE 1g12. UNIVERSAL INDICATORS* FOR APPROXIMATE pH DETERMINATIONS

No. 1. Dissolve 60 mg methyl yellow, 40 mg methyl red, 80 mg bromthymol blue, 100 mg thymol blue and 20 mg phenolphthalein in 100 ml of ethanol and add enough $0.1N$ NaOH to produce a yellow color.

No. 2. Dissolve 18.5 mg methyl red, 60 mg bromthymol blue and 64 mg phenolphthalein in 100 ml of 50% ethanol and add enough $0.1N$ NaOH to produce a green color.

pH	Color		pH	Color	
	No. 1	No. 2		No. 1	No. 2
1	cherry-red	red	7	yellowish-green	greenish-yellow
2	rose	red	8	green	green
3	red-orange	red	9	bluish-green	greenish-blue
4	orange-red	deeper red	10	blue	violet
5	orange	orange-red	11	—	reddish-violet
6	yellow	orange-yellow			

TABLE 1g13. OXIDATION–REDUCTION INDICATORS*

Common name	Reference	Transition potential, volts (N hydrogen electrode = 0.000)	Color Reduced form	Color Oxidized form
p-ethoxychrysoidine	1	0.76	red	yellow
diphenylamine	2	0.776	colorless	purple
diphenylbenzidine	3	0.776	colorless	purple
diphenylamine-sulfonic acid or barium salt	4	0.84	colorless	purple
naphthidine	5	—	colorless	red
dimethylferroin	6	0.97	red	yellowish-green
eriogreen B	7	0.99	yellow	orange
erioglaucin A	7	1.0	yellowish-green	red
xylene cyanole FF	11	1.0		
2,2'-dipyridyl ferrous ion	6	1.03	red	colorless
N-phenylanthranilic acid	8	1.08	colorless	pink
methylferroin	6	1.08	red	pale-blue
ferroin (o-phenanthrolineferrous ion)	9	1.12	red	pale-blue
chloroferroin	6	1.17	red	pale-blue
nitroferroin	6	1.31	red	pale greenish-blue
α-naphtholflavone	10	—	pale straw	brownish-orange

References

1. Schulek and Rozsa, *Z. Anal. Chem.* **115,** 185 (1939). This indicator is particularly suitable for titrations with bromine.
2. Knop, *J. Am. Chem. Soc.* **46,** 263 (1924).
3. Kolthoff and Sarver, *J. Am. Chem. Soc.* **52,** 4179 (1930).
4. Kolthoff and Sarver, *J. Am. Chem. Soc.* **53,** 2902 (1931).
5. Stroka and Oesper, *Ind. Eng. Chem.*, Anal. Ed. **6,** 465 (1934).
6. Smith and Richter, *Ind. Eng. Chem.*, Anal. Ed. **16,** 580 (1944).
7. Knop, *Z. Anal. Chem.* **77,** 111 (1929).
8. Syrokomsky and Stiepin, *J. Am. Chem. Soc.* **58,** 928 (1936).
9. Walden, Hammett, and Chapman, *J. Am. Chem. Soc.* **55,** 2649 (1933).
10. Belcher, *Anal. Chim. Acta* **3,** 578 (1949). This indicator is suitable for titrations with bromine.
11. Tomlinson, Alpli, Ebert, *Anal. Chem.* **23,** 286 (1951).

TABLE 1g14. IONIZATION CONSTANTS AT 25°C*

ACIDS

	FORMULA		K_{ion}	pK_a
Acetic	$HC_2H_3O_2$		1.753×10^{-5}	4.76
Adipic	$H_2C_6H_8O_4$		3.72×10^{-5}	4.43
Benzoic	$HC_7H_5O_2$		6.30×10^{-5}	4.20
Boric	H_3BO_3		5.8×10^{-10}	9.24
Carbonic	H_2CO_3	K_1	4.31×10^{-7}	6.37
		K_2	5.6×10^{-11}	10.25
Chloroacetic	$HC_2H_2O_2Cl$		1.396×10^{-3}	2.86
Citric	$H_3C_6H_5O_7$	K_1	8.7×10^{-4}	3.06
		K_2	1.8×10^{-5}	4.74
		K_3	4.0×10^{-6}	5.40
Dichloroacetic	$HC_2HO_2Cl_2$		5×10^{-2}	1.30
Formic	$HCHO_2$		1.77×10^{-4}	3.75
Hydrofluoric	HF		7.2×10^{-4}	3.14
Hydrosulfuric	H_2S	K_1	5.7×10^{-8}	7.24
		K_2	1.2×10^{-15}	14.92
Hydroxyacetic (Glycolic)	$HC_2H_3O_3$		1.52×10^{-4}	3.82
Lactic	$HC_3H_5O_3$		1.387×10^{-4}	3.86
Oxalic	$H_2C_2O_4$	K_1	6.5×10^{-2}	1.19
		K_2	6.1×10^{-5}	4.21
Phenol	HC_6H_5O		1.3×10^{-10}	9.89
Phosphoric	H_3PO_4	K_1	7.5×10^{-3}	2.12
		K_2	6.2×10^{-8}	7.21
		K_3	4.8×10^{-13}	12.32
o-Phthalic	$C_6H_4(COOH)_2$	K_1	1.3×10^{-3}	2.89
		K_2	3.9×10^{-6}	5.41
Salicylic	$H_2C_7H_4O_3$	K_1	1.06×10^{-3}	2.97
		K_2	3.6×10^{-14}	13.44
Sulfuric	H_2SO_4	K_1	$\sim 4 \times 10^{-1}$	0.40
		K_2	1.2×10^{-2}	1.92

BASES

	FORMULA	K_{ion}	pK_a
Ammonium Hydroxide	NH_4OH	1.8×10^{-5}	4.74
Diethylamine	$NH(C_2H_5)_2$	1.2×10^{-3}	2.90
Dimethylamine	$HN(CH_3)_2$	5.12×10^{-4}	3.29
Dimethylaniline	$(CH_3)_2NH \cdot C_6H_5$	1.15×10^{-9}	8.94
Ethylamine	$C_2H_5 \cdot NH_2$	5.6×10^{-4}	3.25
Hydrazine	$(NH_2)_2$	3.0×10^{-6}	5.52
Hydroxylamine	NH_2OH	1.07×10^{-8}	7.97
Methylamine	CH_3NH_2	4.38×10^{-4}	3.36
Pyridine	C_5H_5N	1.4×10^{-9}	8.85
Silver Hydroxide	$AgOH$	1.1×10^{-4}	3.96
Thiourea	CH_4N_2S	1.1×10^{-15}	14.96
Triethylamine	$(C_2H_5)_3N$	5.65×10^{-4}	3.25
Trimethylamine	$(CH_3)_3N$	5.27×10^{-5}	4.28
Urea	$CO(NH_2)_2$	1.5×10^{-14}	13.82

*Reproduced with permission from: "A Condensed Laboratory Handbook," Du Pont Company, Industrial Chemicals Department, Wilmington, DE, 1971.

**TABLE 1g15. MEAN ACTIVITY COEFFICIENTS OF 1-1 ELECTROLYTES
AT 25° (HARNED AND OWEN, 1950)**

m	LiCl	NaCl	KCl	RbCl	CsCl
0.1	0.792	0.778	0.769	0.764	0.755
.2	.761	.734	.717	.709	.693
.3	.748	.710	.687	.675	.653
.5	.742	.682	.650	.634	.604
.7	.754	.668	.626	.607	.573
1.	.781	.658	.605	.583	.543
1.5	.841	.659	.585	.559	.514
2.	.931	.671	.575	.547	.495
2.5	1.043	.692	.572	.540	.485
3.	1.174	.720	.573	.538	.480
3.5	−	.753	.576	.539	.476
4.	−	.792	.582	.541	.474
4.5	−	−	.590	.544	.474
5.	−	−	−	.547	.476

m	LiBr	NaBr	KBr	RbBr	CsBr
0.1	0.794	0.781	0.771	0.763	0.754
.2	.764	.739	.721	.706	.692
.3	.757	.717	.692	.674	.652
.5	.755	.695	.657	.634	.603
.7	.770	.687	.637	.606	.570
1.	.811	.687	.617	.579	.537
1.5	.899	.704	.601	.552	.504
2.	1.016	.732	.596	.537	.486
2.5	1.166	.770	.596	.527	.474
3.	1.352	.817	.600	.521	.468
3.5	−	.871	.606	.518	.462
4.	−	.938	.615	.517	.460
4.5	−	−	−	.517	.459
5.	−	−	−	.518	.460

m	LiI	NaI	KI	RbI	CsI
0.1	0.811	0.788	0.776	0.762	0.753
.2	.800	.752	.731	.705	.691
.3	.799	.737	.704	.673	.651
.5	.819	.726	.675	.631	.599
.7	.848	.729	.659	.602	.566
1.	.907	.739	.646	.575	.532
1.5	1.029	.772	.639	.548	.495
2.	1.196	.824	.641	.533	.470
2.5	1.423	.889	.649	.525	.450
3.	1.739	.967	.657	.519	.434
3.5	−	1.060	.667	.518	−
4.	−	−	.678	.517	−
4.5	−	−	.692	.519	−
5.	−	−	−	.520	−

TABLE 1g15. (*Continued*)

m	$LiNO_3$	$NaNO_3$	KNO_3	$RbNO_3$	$CsNO_3$
0.1	0.788	0.758	0.733	0.730	0.729
.2	.751	.702	.659	.656	.651
.3	.737	.664	.607	.603	.598
.5	.728	.615	.542	.534	.526
.7	.731	.583	.494	.484	.475
1.	.746	.548	.441	.429	.419
1.5	.783	.509	.378	.365	.354
2.	.840	.481	.327	.319	—
2.5	.903	.457	.293	.284	—
3.	.973	.438	.266	.256	—
3.5	1.052	.423	.244	.235	—
4.	—	.410	—	.216	—
4.5	—	.398	—	.200	—
5.	—	.388	—	—	—
5.5	—	.380	—	—	—
6.	—	.373	—	—	—

m	$LiAc^a$	NaAc	KAc	RbAc	CsAc
0.1	0.782	0.791	0.796	0.797	0.798
.2	.740	.755	.767	.771	.773
.3	.718	.741	.752	.759	.763
.5	.698	.740	.751	.760	.765
.7	.691	.741	.755	.769	.777
1.	.690	.757	.779	.795	.802
1.5	.709	.799	.839	.859	.868
2.	.734	.854	.910	.940	.952
2.5	.769	.920	.993	1.034	1.046
3.	.807	.993	1.086	1.139	1.153
3.5	.847	1.070	1.187	1.255	1.277
4.	.893	—	—	—	—

m	NaF	KF	NaCNS	KCNS	HI
0.1	0.764	0.774	0.787	0.769	0.818
.2	.708	.727	.750	.716	.807
.3	.675	.701	.731	.685	.811
.5	.631	.672	.715	.646	.839
.7	.602	.657	.710	.623	.883
1.	.572	.649	.712	.600	.965
1.5	—	.649	.725	.574	1.139
2.	—	.663	.751	.558	1.367
2.5	—	.684	.784	.548	1.656
3.	—	.713	.820	.542	2.025
3.5	—	.748	.860	.537	—
4.	—	.790	.911	.533	—
4.5	—	—	—	.531	—
5.	—	—	—	.529	—

^{a}Ac represents acetate radical.

TABLE 1g16. MEAN ACTIVITY COEFFICIENTS OF SOME 1-1 ELECTROLYTES, SULPHURIC ACID, CALCIUM CHLORIDE AND NITRATE AT 25° AND HIGH CONCENTRATIONS (HARNED AND OWEN, 1950)

m	NaCl	NaOH	LiCl	LiBr	LiNO$_3$	H$_2$SO$_4$	CaCl$_2$	HClO$_4$	Ca(NO$_3$)$_2$
0.1	0.778	0.766	0.790	0.796	0.788	0.2655	0.518	0.803	0.485
.2	.735	.727	.757	.766	.752	.2090	.472	.778	.426
.5	.681	.693	.739	.753	.726	.1557	.448	.769	.363
1.	.657	.679	.774	.803	.743	.1316	.500	.823	.336
1.5	.656	.683	.838	.896	.783	–	–	.923	.336
2.	.668	(0.700)	.921	1.015	.835	.1276	.792	1.055	.345
3.	.714	.774	1.156	1.341	.966	.1422	1.483	1.448	.380
4.	.783	.890	1.510	1.897	1.125	.1700	2.934	2.08	.435
5.	.874	1.060	2.02	2.74	1.310	.2081	5.89	3.11	.507
6.	.986	1.280	2.72	3.92	1.515	.2567	11.11	4.76	.592
7.	–	1.578	3.71	5.76	1.723	.3166	18.28	7.44	.690
8.	–	1.979	5.10	8.61	1.952	.386	26.02	11.83	.801
9.	–	2.51	6.96	12.92	2.19	.467	34.20	19.11	.935
10.	–	3.18	9.40	19.92	2.44	.559	43.0	30.9	1.065
11.	–	4.04	12.55	31.0	2.69	.661	–	50.1	1.184
12.	–	5.11	16.41	46.3	2.95	.770	–	80.8	1.311
14.	–	7.91	26.2	104.7	–	1.017	–	205.	1.538
16.	–	11.38	37.9	198.0	–	1.300	–	500.	1.724
18.	–	15.15	49.9	331.	–	1.608	–	–	1.917
20.	–	19.0	62.4	485.	–	1.940	–	–	2.008
22.	–	22.7	–	–	–	2.300	–	–	–
24.	–	26.1	–	–	–	–	–	–	–
26.	–	29.0	–	–	–	–	–	–	–
29.	–	33.2	–	–	–	–	–	–	–

TABLE 1g17. DISSOCIATION CONSTANTS OF SOME MODERATELY
STRONG ELECTROLYTES (HARNED AND OWEN, 1950)

Substance	K	Substance	K
HIO_3	0.17	$PbBr^+$	0.07
CCl_3COOH	.22–0.23	PbI^+	.035
Picric Acid	.16	H_2SO_3	.012–0.017
$CHCl_2COOH$	.14	$HSeO_4^-$	.01
$H_4P_2O_7$	.14	o-Nitrobenzoic acid	.006
$PbCl^+$	.06–0.1		

TABLE 1g18. DISSOCIATION CONSTANTS OF SALTS AND
COMPLEX IONS IN WATER ESTIMATED FROM
CONDUCTANCE DATA AT 25°* and 18°.

	K		$100K$
$KClO_2$	1.4	(Ca Acetate)$^+$	100*
KNO_3	1.4	$(CaOH)^+$	3.1*
$AgNO_3$	1.2	$(BaOH)^+$	23*
$TlNO_3$	.52	$MgSO_4$	0.78
$TlCl$	.300	$CaSO_4$	.53
$(CdNO_3)^+$	.394	$CuSO_4$	.50
$(CdCl)^+$	.0101	$ZnSO_4$	.53
$(PbNO_3)^+$	.0647	$CdSO_4$	.49
$(PbCl)^+$	.0304	$MgSO_4$	.63*
$(CaNO_3)^+$	.521	$CoSO_4$	.34*
$(SrNO_3)^+$	.150	$NiSO_4$	.40*
$(BaNO_3)^+$	.121	$CuSO_4$	.43*
$(LiSO_4)^-$	.229	$ZnSO_4$	.49*
$(NaSO_4)^-$	.198	Cu Malonate	.00025*
$(KSO_4)^-$	.151	Zn Malonate	.021*
$(AgSO_4)^-$	.05	Cd Malonate	.051*
$(TiSO_4)^-$	.0472	$(KFe(CN)_6)^{\equiv}$	.55*

*Values followed by an asterisk were estimated at 25°.

TABLE 1g19. SINGLE BOND ENERGIES AT 25°C (IN KCAL/MOLE).*

H—H	104.18	O—Sb	71		
H—B	ca 93	O—I	ca 48		
H—C	98.7	F—F	36.6		
H—N	93.4	F—Si	135		
H—O	110.6	F—P	117		
H—F	135	F—S	68		
H—Si	76	F—Cl	ca 61		
H—P	ca 77	F—As	111		
H—S	83	F—Se	68		
H—Cl	103.1	F—Br	61		
H—As	ca 59	F—Te	80		
H—Se	ca 66	F—I	63		
H—Br	87.4	Na—Na	18.4		
H—Te	ca 57	Si—Si	53		
H—I	71.4	Si—S	60.9		
Li—Li	27.2	Si—Cl	91		
B—C	89	Si—Br	74		
B—N	106.5	Si—I	56		
B—O	128	P—P	48		
B—F	154	P—Cl	78		
B—Cl	109	P—Br	63		
B—Br	90	P—I	44		
C—C	82.6	S—S	58.1		
C—N	72.8	S—Cl	61		
C—O	85.5	S—Br	ca 52		
C—F	116	Cl—Cl	57.87		
C—Al	61	Cl—Ge	81		
C—Si	72	Cl—As	70		
C—P	63	Cl—Se	58		
C—S	65.6	Cl—Br	52.7		
C—Cl	81	Cl—Sn	76		
C—Zn	40	Cl—Sb	74		
C—Ge	ca 44	Cl—I	51		
C—As	48	Cl—Hg	54		
C—Se	58	Cl—Bi	67		
C—Br	68	K—K	12.6		
C—Cd	32	Ge—Ge	45		
C—Sn	54	Ge—Br	66		
C—Sb	47	Ge—I	51		
C—I	51	As—As	35		
C—Hg	23	As—Br	58		
C—Pb	31	As—I	43		
C—Bi	31	Se—Se	41		
N—N	39	Br—Br	46.08		
N—O	48	Br—Sn	65		
N—F	65	Br—I	43		
N—Cl	46	Br—Hg	44		
O—O	35	Rb—Rb	11.5		
O—F	45.3	Sn—Sn	39		
O—Si	108	Sn—I	65		
O—P	ca 80	Sb—Sb	ca 29		
O—Cl	52	I—I	36.06		
O—As	72	I—Hg	35		
O—Br	ca 48	Cs—Cs	10.4		

Diatomic molecules with multiple bonds

O=O	119.1
N≡N	225.8
C=O	255.8
(carbon monoxide)	

Multiple bonds

C=C	145.8
C≡C	199.6
N=N	100
N=O	145
C=O	192.0
(CO₂)	
C=O	166
(formaldehyde)	
C=O	176
(other aldehydes)	
C=O	179
(ketones)	
C=S	128
C=N	147
C≡N	212.6

*Reproduced from: Potter, J. H., Ed., "Handbook of the Engineering Sciences," Vol. 1, Van Nostrand Reinhold, New York, 1967.

TABLE 1g20. BIMOLECULAR RATE CONSTANTS OF HYDRATED ELECTRONS (ANBAR AND NETA, 1965.)*

Reactant	pH	Rate constant $M^{-1}sec^{-1}$
e_{aq}^-	11	$4 \cdot 5 \pm 0 \cdot 07 \times 10^9$
e_{aq}^-	13	$5 \cdot 5 \pm 0 \cdot 07 \times 10^9$
H	10–11	$2 \cdot 5 \pm 0 \cdot 5 \times 10^{10}$
OH	10–11	$3 \cdot 0 \pm 0 \cdot 6 \times 10^{10}$
O^-	13	$2 \cdot 1 \pm 0 \cdot 6 \times 10^{10}$
Ag^+	7	$3 \cdot 6 \pm 0 \cdot 4 \times 10^{10}$
Ag^+		$3 \cdot 2 \times 10^{10}$
$Ag(NH_3)_2^+$	10·0	$8 \pm 1 \cdot 5 \times 10^{10}$
$Ag(CN)_2^-$	10·3	$1 \cdot 5 \pm 0 \cdot 2 \times 10^9$
Al^{3+}	6·8	$2 \cdot 0 \pm 0 \cdot 3 \times 10^9$
Al^{3+}	11·2	$4 \cdot 0 \pm 1 \cdot 0 \times 10^8$
$Al(OH)_4^-$	14	$5 \cdot 5 \pm 1 \cdot 2 \times 10^6$
As^{3+}	11	$2 \cdot 7 \times 10^8$
As^{5+}	7	$2 \cdot 1 \pm 0 \cdot 3 \times 10^8$
As^{5+}	11	$2 \cdot 0 \pm 0 \cdot 2 \times 10^8$
AsF_6^-	7	$9 \pm 1 \cdot 9 \times 10^9$
$Au(CN)_2^-$	10·5	$6 \cdot 8 \times 10^9$
BrO_3^-	7	$2 \cdot 1 \pm 0 \cdot 3 \times 10^9$
BrO_3^-	11	$3 \cdot 8 \pm 0 \cdot 5 \times 10^9$
BrO_3^-	14	$5 \cdot 8 \pm 0 \cdot 7 \times 10^9$
BrO_3^-	3 M OH^-	$5 \cdot 3 \pm 0 \cdot 6 \times 10^9$
CCl_4	7	3×10^{10}
CO	7	$\sim 1 \times 10^9$
CO_2	7	$7 \cdot 67 \pm 1 \cdot 1 \times 10^9$
HCO_3^-		$< 10^6$
$CO_3^=$		$< 10^6$
CN^-	11·0	$< 10^6$
CN^-		$< 10^8$
CNO^-	11	$1 \cdot 3 \times 10^6$
CNS^-	~ 7	$< 10^6$
Cd^{2+}		$5 \cdot 2 \times 10^{10}$
$Cd(H_2O)_4^{2+}$	6·7	$4 \cdot 8 \pm 0 \cdot 6 \times 10^{10}$
$Cd(NH_3)_4^{2+}$	10·0	$3 \cdot 1 \pm 0 \cdot 3 \times 10^{10}$
$CdCl(H_2O)_3^+$	7·0	$1 \cdot 1 \pm 0 \cdot 1 \times 10^{10}$
$CdI(H_2O)_3^+$	7·0	$1 \cdot 6 \pm 0 \cdot 2 \times 10^{10}$
$Cd(CN)_4^=$	10·3	$1 \cdot 4 \pm 0 \cdot 2 \times 10^8$
Ce^{3+}		$< 10^9$
Cl^-	~ 7	$< 10^6$
Cl^-	10	$< 10^5$
ClO^-	10	$7 \cdot 2 \times 10^9$
ClO_3^-	10	$3 \cdot 5 \times 10^8$
ClO_3^-	~ 10	4×10^6
ClO_4^-	~ 10	$< 10^6$
ClO_4^-		$< 10^5$
Co^{2+}		$1 \cdot 22 \times 10^{10}$
$Co(NH_3)_6^{3+}$	10·1	$9 \cdot 0 \times 10^{10}$
$Co(en)_3^{3+}$		$8 \cdot 2 \times 10^{10}$
$Co(NH_3)_5H_2O^{3+}$		$6 \cdot 2 \times 10^{10}$
$Co(NH_3)_5Cl^{2+}$		$5 \cdot 4 \times 10^{10}$
$Co(EDTA)$		$6 \cdot 0 \pm 0 \cdot 6 \times 10^{10}$
$Co(en)_2Cl_2^+$		$3 \cdot 2 \times 10^{10}$
$Co(CN)_6^{3-}$		$2 \cdot 7 \times 10^9$
$Co(CN)_6^{3-}$	10·3	$3 \cdot 6 \pm 0 \cdot 4 \times 10^9$
$Co(ox)_3^{3-}$		$1 \cdot 2 \times 10^{10}$

TABLE 1g20. (Continued)

Reactant	pH	Rate constant $M^{-1}sec^{-1}$
$Co(NO_2)_6^{3-}$		5.8×10^{10}
Cr^{2+}	6.9	$4.2 \pm 0.8 \times 10^{10}$
Cr^{2+}	11.2	$1.9 \pm 0.5 \times 10^{10}$
Cr^{3+}	7.1	$6.0 \pm 0.5 \times 10^{10}$
Cr^{3+}	10.9	$4.6 \pm 0.5 \times 10^{10}$
Cr^{9+}	14	$2.0 \pm 0.2 \times 10^{8}$
Cr^{3+}	3 M OH$^-$	$2.2 \pm 0.2 \times 10^{8}$
$Cr(H_2O)_6^{3+}$	6.7	$6.0 \pm 0.5 \times 10^{10}$
$Cr(en)_3^{3+}$		$5.3 \pm 0.5 \times 10^{10}$
$Cr(en)_3^{3+}$		1.3×10^{11}
$Cr(en)_2Cl_2^{+}$		$7.1 \pm 0.7 \times 10^{10}$
$Cr(EDTA)$		$2.6 \pm 0.4 \times 10^{10}$
CrF_6^{3-}	10.1	$1.4 \pm 0.2 \times 10^{10}$
$Cr(CN)_6^{3-}$	10.6	$1.5 \pm 0.2 \times 10^{10}$
$Cr_2O_7^{=}$	~7	3.3×10^{10}
$Cr_2O_7^{=}$	~13	5.4×10^{10}
Cu^{2+}	6.8	2.9×10^{10}
Cu^{2+}	7	$3.3 \pm 0.3 \times 10^{10}$
$Cu(OH)_4^{=}$	14	$5.8 \pm 0.6 \times 10^{9}$
$Cu(OH)_4^{=}$	0.3 M OH$^-$	$4.5 \pm 0.5 \times 10^{9}$
$Cu(OH)_4^{=}$	5 M OH$^-$	$3.4 \pm 0.5 \times 10^{9}$
$Cu(NH_3)_4^{2+}$	10.1	$1.8 \pm 0.3 \times 10^{10}$
$Cu(CN)_4^{=}$	10.4	$3.0 \pm 0.3 \times 10^{8}$
Dy^{3+}	5.90	4.6×10^{8}
Eb^{2+}		$< 7 \times 10^{7}$
Eu^{3+}	5.55	6.1×10^{10}
Fe^{2+}		$\sim 3.5 \times 10^{8}$
$Fe(CN)_6^{3-}$	7 and 10.3	$3.0 \pm 0.4 \times 10^{9}$
$Fe(CN)_6^{3-}$	2.5	$6.4 \pm 0.8 \times 10^{9}$
Gd^{3+}	6.05	5.5×10^{8}
H^+		2.16×10^{10}
H^+	4.0–4.6	$2.36 \pm 0.24 \times 10^{10}$
H^+	4.1–4.7	$2.26 \pm 0.21 \times 10^{10}$
H^+ in MeOH		$3.9 \pm 0.9 \times 10^{10}$
H^+ in EtOH		$2.0 \pm 0.4 \times 10^{10}$
H^+	2.1–4.3	$2.06 \pm 0.08 \times 10^{10}$
H_2	neutral	$< 10^{7}$
H_2O	neutral	$4.9 \pm 0.3 \times 10^{2}$
H_2O	10 and 13	4.1×10^{2}
H_2O	13	1.4×10^{2}
H_2O_2	7	$1.23 \pm 0.14 \times 10^{10}$
H_2O_2	7	$1.36 \pm 0.2 \times 10^{10}$
HF		$\sim 3.7 \times 10^{7}$*
Ho^{3+}	5.88	2.4×10^{9}
Ho^{3+}		$\sim 6.6 \times 10^{7}$
I_2	7	5.1×10^{10}
IO_3^{-}	7	$7.7 \pm 0.9 \times 10^{9}$
IO_3^{-}	11	$8.3 \pm 1.0 \times 10^{9}$
IO_3^{-}	14	$9.6 \pm 1.2 \times 10^{9}$
IO_3^{-}	3 M OH$^-$	$8.1 \pm 0.8 \times 10^{9}$
IO_4^{-}	7	$1.10 \pm 0.15 \times 10^{10}$
IO_4^{-}	11	$1.90 \pm 0.21 \times 10^{10}$
IO_4^{-}	14	$2.15 \pm 0.25 \times 10^{10}$
IO_4^{-}	3 M OH$^-$	$1.63 \pm 0.20 \times 10^{10}$
K^+		$< 5 \times 10^{5}$
K^+		$< 3 \times 10^{4}$

TABLE 1g20. (*Continued*)

Reactant	pH	Rate constant $M^{-1}sec^{-1}$
La^{3+}	6·98	$3·4 \times 10^8$
La^{3+}		$\sim 6·9 \times 10^8$
Lu^{3+}	6·20	$2·5 \times 10^8$
Mn^{2+}		$7·7 \times 10^7$
MnO_4^-	7	$2·2 \times 10^{10}$
MnO_4^-	13	$3·7 \times 10^{10}$
N_3^-	11	$< 5·6 \times 10^6$
NH_4^+		$\sim 2·5 \times 10^{6*}$
N_2H_4		$< 10^8$
$N_2H_5^+$		$< 3·5 \times 10^8$
NH_2OH		$< 2 \times 10^7$
N_2O	7	$8·67 \pm 0·6 \times 10^9$
N_2O	7	$5·6 \pm 2·0 \times 10^9$
NO	7	$3·14 \pm 0·2 \times 10^{10}$
NO_2^-	7·0	$4·58 \times 10^9$
NO_2^-		$3·45 \times 10^9$
NO_3^-	7·0	$1·1 \pm 0·1 \times 10^{10}$
NO_3^-		$8·15 \times 10^9$
Na^+		$< 10^6$
Na^+		$< 10^5$
Nd^{3+}	4·66	$5·9 \times 10^8$
Ni^{2+}		$2·2 \times 10^{10}$
O_2	7	$1·88 \pm 0·2 \times 10^{10}$
O_2	7	$2·16 \pm 0·25 \times 10^{10}$
O_2 in MeOH	—	$1·9 \pm 0·4 \times 10^{10}$
$H_2PO_4^-$		$< 5 \times 10^{6*}$
$H_2PO_4^-$	7	$\sim 2 \times 10^{7*}$
$H_2PO_4^-$		6×10^9
Pb^{2+}	7	$3·9 \times 10^{10}$
Pb^{2+}	11·2	$1·3 \pm 0·1 \times 10^{10}$
Pb^{2+}	14	$1·0 \pm 0·1 \times 10^{10}$
Pb^{2+}	3 M OH^-	$9·2 \pm 1·0 \times 10^9$
$PdCl_4^=$	5·0	$1·2 \pm 0·15 \times 10^{10}$
$Pd(CN)_4^=$		$2·0 \pm 0·3 \times 10^9$
Pr^{3+}	6·10	$2·9 \times 10^8$
$PtCl_4^=$		9×10^9
$PtCl_4^=$	5	$1·2 \pm 0·15 \times 10^{10}$
$Pt(CN)_4^=$	10	$3·2 \pm 0·4 \times 10^9$
$SO_4^=$	7	$< 10^6$
$S_2O_3^=$	11·9	$< 10^8$
$S_2O_8^=$	7	$1·06 \times 10^{10}$
$S_2O_8^=$	neutral	$7·6 \times 10^9$
Sb^{5+}	11	$1·2 \pm 0·2 \times 10^{10}$
Sm^{3+}	5·96	$2·5 \times 10^{10}$
Sn^{2+}	11	$3·4 \pm 0·3 \times 10^9$
Sn^{2+}	14	$6·2 \pm 0·5 \times 10^9$
Tb^{3+}	6·15	$3·7 \times 10^8$
Tl^+		$\sim 1·1 \times 10^{10}$
Tl^+	7	7×10^{10}
Tm^{3+}	6·05	3×10^9
UO_2^{2+}		$7·3 \times 10^{10}$
Y^{3+}		$\sim 2·0 \times 10^8$
Yb^{3+}	6·03	$4·3 \times 10^{10}$
Yb^{3+}		$\sim 3·7 \times 10^{10}$
$Zn(H_2O)_4^{2+}$	6·8	$1·0 \pm 0·3 \times 10^9$
Zn^{2+}		$1·5 \times 10^9$

TABLE 1g20. (*Continued*)

Reactant	pH	Rate constant $M^{-1}sec^{-1}$
Zn^{2+}	9·7	$5·6 \pm 0·7 \times 10^8$
Zn^{2+}	12	$2·0 \pm 0·3 \times 10^8$
$Zn(OH)_4^=$	14	$1·6 \pm 0·3 \times 10^7$
$Zn(OH)_4^=$	3 M OH$^-$	$7·5 \pm 1·5 \times 10^6$
$Zn(NH_3)_4^{2+}$	9·8	$6·5 \pm 0·6 \times 10^8$
$Zn(CN)_4^=$	10·5	$1·8 \pm 0·2 \times 10^8$
acetaldehyde	6·55	$3·5 \times 10^9$
acetaldehyde	11	$3·5 \times 10^9$
acetate ion	~10	$< 10^6$
acetate ion		$< 10^8$
acetic acid	5·4	$1·76 \pm 0·3 \times 10^8$
acetone	7	$5·9 \pm 0·2 \times 10^9$
acetone	11	$5·6 \pm 0·6 \times 10^9$
acetone	14	$5·2 \pm 0·6 \times 10^9$
acetone	3 M OH$^-$	$4·2 \pm 0·5 \times 10^9$
acrylamide	7	$1·8 \times 10^{10}$
adenine	7	$1·1 \times 10^{10}*$
adenosine	12·0	$1·0 \times 10^{10}$
adensoine	7	$1·3 \times 10^{10}*$
adenosine-5-phosphate	7	$4·5 \times 10^9*$
alanine		8×10^7
p-aminobenzoate ion	11	$2·1 \times 10^9$
aniline	11·94	$< 2 \times 10^7$
arabinose		$< 10^7$
arginine		$< 10^7$
aspartic acid		$< 10^7$
benzamide	11	$1·7 \times 10^{10}$
benzene	7	$< 7 \times 10^6$
benzene	11	$1·4 \times 10^7$
benzenesulphonamide	11	$1·6 \times 10^{10}$
benzenesulphonate ion	11	$4·0 \times 10^9$
benzoate ion	11	$3·1 \times 10^9$
benzoate ion	14	$2·9 \pm 0·3 \times 10^9$
benzoate ion	3 M OH$^-$	$2·4 \pm 0·3 \times 10^9$
benzonitrile	11	$1·6 \times 10^{10}$
benzoquinone	6·6	$1·25 \times 10^9$
benzyl alcohol	11	$1·3 \times 10^8$
benzyl chloride	11	$5·1 - 5·5 \times 10^9$
benzyl chloride in MeOH	—	$5·0 \pm 1·2 \times 10^9$
benzyl chloride in EtOH	—	$5·1 \pm 1·2 \times 10^9$
bromoacetate ion	11	$6·2 \pm 0·7 \times 10^9$
bromobenzene	11	$4·3 \times 10^9$
p-bromobenzoate ion	11	$7·7 \times 10^9$
2-bromoethanol	11	$1·6 \pm 0·2 \times 10^9$
o-bromophenol	11	$1·9 \times 10^9$
m-bromophenol	11	$2·7 \times 10^9$
p-bromophenol	11	$2·9 \times 10^9$
2-bromopropionate ion	11	$5·3 \times 10^9$
3-bromopropionate ion	11	$2·7 \times 10^9$
butadiene	7	$2·4 \times 10^8$
carbon disulphide	7·7	$3·1 \pm 0·3 \times 10^{10}$
carbon tetrachloride	7	$3·0 \times 10^{10}$
carbon tetrachloride	7	$3·1 \times 10^{10}$
chloroacetate ion	11	$1·2 \times 10^9$

TABLE 1g20. (*Continued*)

Reactant	pH	Rate constant $M^{-1}sec^{-1}$
chlorobenzene	11	5.0×10^8
o-chlorobenzoate ion	11	1.2×10^9
m-chlorobenzoate ion	11	5.5×10^9
p-chlorobenzoate ion	11	6.0×10^9
1-chlorobutane	11	$3.2 \pm 0.4 \times 10^8$
2-chlorobutane	11	$5.1 \pm 0.8 \times 10^8$
2-chloroethanol	11	$4.1 \pm 0.6 \times 10^8$
chloroform	7	$2.0 \pm 0.05 \times 10^{10}$
chloroform	7	3.0×10^{10}
o-chlorophenol	11	2.0×10^8
m-chlorophenol	11	5.0×10^8
p-chlorophenol	11	6.4×10^8
2-chloropropionate ion	11	$1.4 \pm 0.2 \times 10^9$
3-chloropropionate ion	11	$4.0 \pm 0.4 \times 10^8$
p-chlorotoluene	11	4.5×10^8
citric acid		$< 10^7$
cyanoacetate ion	11	4×10^7
p-cyanobenzoate ion	11	1.0×10^{10}
1-cyanonaphthalene	11	2.13×10^{10}
2-cyanonaphthalene	11	2.07×10^{10}
o-cyanophenol	11	8.2×10^9
m-cyanophenol	11	4.8×10^9
p-cyanophenol	11	2.0×10^9
p-cyanotoluene	11	1.4×10^{10}
cysteine		$< 10^7$
l-cystine	12.0	3.4×10^9
cytidine	12.0	1.2×10^{10}
cytosine	6	$\sim 7 \times 10^9$
cytosine	7	$1.1 \times 10^{10}*$
diethyl ether		$< 10^7$
diphenyl in EtOH	—	4.3×10^9
o-dichlorobenzene	11	4.7×10^9
m-dichlorobenzene	11	5.2×10^9
p-dichlorobenzene	11	5.0×10^9
ethanol		$< 10^5$
ethyl acetate		$< 10^7$
ethylene		7.6×10^6
fluoroacetate ion	11	$< 2 \times 10^6$
fluorobenzene	11	6.10×10^7
o-fluorobenzoate ion	11	3.1×10^9
m-fluorobenzoate ion	11	6.7×10^9
p-fluorobenzoate ion	11	3.8×10^9
o-fluorophenol	11	3.4×10^8
m-fluorophenol	11	2.0×10^8
p-fluorophenol	11	1.2×10^8
formaldehyde	7	$< 10^7$
formate ion	~ 10	$< 10^6$
formic acid	5.0	$1.43 \pm 0.1 \times 10^8$
fumarate ion	12–13	7.5×10^9
glutamic acid		$< 10^7$
glycine		$< 10^7$
glycine		$\sim 1.2 \times 10^8$
hydroquinone	13	$< 10^7$
o-hydroxybenzoate ion	11	3.2×10^9
m-hydroxybenzoate ion	11	1.1×10^9
p-hydroxybenzoate ion	11	4.0×10^8

TABLE 1g20. (*Continued*)

Reactant	pH	Rate constant $M^{-1}sec^{-1}$
hypoxanthine	6·6	$1·7 \times 10^{10}$
hystidine		$\sim 4·5 \times 10^7$
indole		$\sim 7 \times 10^8$
iodoacetate ion	11	$1·15 \pm 0·10 \times 10^{10}$
iodobenzene	11	$1·2 \times 10^{10}$
o-iodobenzoate ion	11	$4·6 \times 10^9$
m-iodobenzoate ion	11	$1·3 \times 10^{10}$
p-iodobenzoate ion	11	$9·1 \times 10^9$
3-iodopropionate ion	11	$6·6 \pm 0·9 \times 10^9$
p-iodotoluene	11	$1·3 \times 10^{10}$
lactate ion	11	$< 2 \times 10^6$
lactic acid		$< 10^7$
maleate ion	8·45	$1·7 \times 10^9$
maleate ion	12·7	$2·2 \times 10^9$
maleic acid	6·5	$1·2 \times 10^{10}$
methacrylate ion	10·1	$8·4 \times 10^9$
methacrylate ion	10·1	$7·2 \times 10^9$
methane		$< 10^7$
methanol	11	$< 10^7$
methanol	11	$< 10^4$
5-methylcytosine	neutral	$1·0 \times 10^{10}$
methylene blue	neutral	$3·4 \times 10^{11}$
naphthacene in EtOH	—	$1·02 \pm 0·08 \times 10^{10}$
naphthalene	7	$3·1 \times 10^8$
naphthalene	11	$5·4 \pm 0·5 \times 10^9$
naphthalene in EtOH	—	$5·4 \times 10^9$
1-naphthoate ion	11	$6·1 \times 10^9$
2-naphthoate ion	11	$9·5 \times 10^9$
1-naphthol	11	$9·6 \times 10^8$
2-naphthol	11	$1·2 \times 10^9$
2-naphthol	11	$1·8 \times 10^9$
nitrobenzene	7	$3·0 \times 10^{10}$
o-nitrophenol	11	$2·0 \times 10^{10}$
m-nitrophenol	11	$2·5 \times 10^{10}$
p-nitrophenol	11	$2·5 \times 10^{10}$
p-nitrotoluene	11	$1·9 \times 10^{10}$
orotic acid	6·56	$1·5 \times 10^{10}$
oxalate ion		$< 10^8$
oxalic acid		$< 10^7$
phenol	11	$< 4·0 \times 10^6$
phenylalanine	11	$< 10^7$
phthalate ion	12·8	$2·0 \times 10^9$
picric acid	13	$3·5 \times 10^{10}$
proline		$< 10^7$
purine	7·2	$1·7 \times 10^{10}$
pyridine	7	$1·0 \times 10^9$
pyruvate ion	12·7	$6·8 \times 10^9$
ribose		$< 10^7$
serine		6×10^7
styrene	7	$1·5 \times 10^{10}$
styrene	12·7	$1·1 \times 10^{10}$
succinic acid		$< 10^7$
sulfanilate ion	11	$4·6 \times 10^8$
terephthalate ion	11	$7·3 \times 10^9$
p-terphenyl in EtOH	—	$7·2 \pm 0·6 \times 10^9$
tetracyanoethylene	7	$1·5 \times 10^{10}$

TABLE 1g20. (*Continued*)

Reactant	pH	Rate constant $M^{-1}sec^{-1}$
tetranitromethane	6	$4.6 \pm 0.5 \times 10^{10}$
thiobarbituric acid		$\sim 6 \times 10^7$
thiophenol	11	4.7×10^7
thiourea	6.41	2.9×10^9
threonine		8×10^7
thymine	6.0	1.7×10^{10}
thymine	12.0	2.7×10^9
o-phthalate ion	13	1.8×10^9
m-phthalate ion	13	3.0×10^9
p-phthalate ion	13	7.3×10^9
o-toluate ion	11	2.7×10^8
m-toluate ion	11	2.6×10^9
p-toluate ion	11	3.0×10^9
toluene	11	1.2×10^7
p-toluenesulphonate ion	11	1.65×10^9
trichloro-acetate ion	11	$8.5 \pm 1.0 \times 10^9$
α-trichlorotoluene	11	8.3×10^9
trifluoroacetate ion	11	$< 2.6 \times 10^6$
α-trifluorotoluene	11	1.8×10^9
trinitromethyl ion	7	$3.0 \pm 0.5 \times 10^{10}$
triphenylcarbinol in EtOH	—	$2 \pm 0.4 \times 10^8$
tryptophane		$\sim 8 \times 10^8$
tryptophane		$\sim 2 \times 10^8$
uracil	6.4	7.7×10^9
uracil	12.2	2.3×10^9
uracil	7	$1.3 \times 10^{10}*$

*Reprinted with permission from M. Anbar and P. Neta, "Tables of Bimolecular Rate Constants," *Int. J. Appl. Radn. Isotopes*, Vol. 16, pp. 227–242, 1965, Pergamon Press.

TABLE 1g21. BIMOLECULAR RATE CONSTANTS OF HYDROGEN ATOMS
(ANBAR AND NETA 1965)*

Reactant	pH	Rate constant $M^{-1} sec^{-1}$	Method*
uric acid		$\sim 6 \times 10^9$	
e_{aq}^-	10–11	$\sim 3 \times 10^{10}$	p.r.
H	acid	6×10^9	p.r.
H	acid	7.5×10^9	p.r.
H	0.4–3.0	1.3×10^{10}	p.r.
H	0.1–1	1.6×10^{10}	p.r.
H	2	1×10^{10}	p.r.
OH	acid	1.2×10^{10}	p.r.
OH	0.4–3.0	3.2×10^{10}	p.r.
As^{+3}	2.7	$\sim 10^9*$	c. H_2O_2
HCO_3^-	neutral	$\sim 3 \times 10^4*$	c. methanol
$Co(NH_3)_6^{3+}$	5	$1.5 \times 10^6*$	c. D_3-methanol
$Co(NH_3)_5OH^{2+}$	7	$2 \times 10^7*$	c. 2-D-isopropanol
$Co(NH_3)_5Cl^{2+}$	5	$2 \times 10^8*$	c. 2-D-isopropanol
Cu^{2+}	1	$5.8 \times 10^7*$	c. formic acid
Cu^{2+}	1	$\sim 1.5 \times 10^8*$	c. acetone
Cu^{2+}	neutral	$6 \pm 1 \times 10^8*$	c. isopropanol
D_2	gas	4×10^4	
D_2	2	$7 \pm 3 \times 10^5*$	c. Fe^{3+}
Fe^{2+}	0.4	$1.6 \pm 0.4 \times 10^7*$	c. O_2
Fe^{2+}	2.1	$1.3 \pm 0.3 \times 10^7*$	c. O_2
Fe^{2+}	2.1	$1.3 \times 10^7*$	c. H_2O_2, Fe^{3+}
Fe^{2+}	0.1–1	2×10^7	p.r.
Fe^{3+}	0.1	$8 \times 10^7*$	c. O_2
Fe^{3+}	0.4	$1.2 \times 10^6*$	c. O_2
Fe^{3+}	1.57	$2.2 \times 10^7*$	c. O_2
Fe^{3+}	2.1	$9.5 \pm 1 \times 10^7*$	c. O_2
Fe^{3+}	2.1	$9.6 \times 10^7*$	c. D_2
Fe^{3+}	0.1	$1.6 \times 10^6*$	c. Fe^{2+}
Fe^{3+}	2	9.0×10^7	p.r.
$FeCl^{2+}$	0.1	$4.5 \times 10^9*$	c. O_2
$FeCl_2^+$	0.1	$9 \times 10^9*$	c. O_2
$Fe(CN)_6^{3-}$	1–3	$4 \pm 1 \times 10^9*$	c. formate
$Fe(CN)_6^{3-}$	neutral	$4 \pm 1 \times 10^9*$	c. isopropanol
H^+	acid	$0.5 - 5 \times 10^2$	c.H
H^+	3.5–11	$2.6 \pm 1.1 \times 10^3$	photochem.
I_2^-	3.5–11	$1.8 \pm 0.8 \times 10^7$	photochem.
N_2O	0–2	$2.2 \pm 0.6 \times 10^5*$	c.Fe^{2+}
N_2O	3.5–11	$1.25 \pm 0.5 \times 10^4$	photochem.
NO_2^-	neutral	$6 \pm 2 \times 10^8*$	c. methanol
NO_2^-	neutral	$1.3 \pm 0.2 \times 10^8*$	c. isopropanol
NO_2^-	6	$6 \times 10^8*$	c. 2-D-isopropanol
NO_3^-	11–13	$\sim 7 \times 10^6*$	c. acetate
NO_3^-	neutral	$2.4 \pm 1 \times 10^7*$	c. isopropanol
NO_3^-	6	$1.7 \times 10^7*$	c. 2-D-isopropanol
$(NH_4)_2SO_4$	neutral	$< 2 \times 10^4*$	c. D-formate
O_2	acid	1.2×10^{10}	p.r.
O_2	0.4–3.0	2.6×10^{10}	p.r.
O_2	2	1.9×10^{10}	p.r.
OH^-	7.5–12.5	$1.2 \times 10^7*$	c. chloroacetate
OH^-	11–13	$2.3 \times 10^7*$	c. formate, acetone
OH^-	11.5	1.5–2.3×10^7	p.r.
OH^-		2×10^7	p.r.

TABLE 1g21. (*Continued*)

Reactant	pH	Rate constant $M^{-1} sec^{-1}$	Method*
HO_2	acid	2×10^{10}	p.r.
H_2O_2	acid	$4 \times 10^{7}*$	c. O_2, photochem.
H_2O_2	acid	4×10^{7}	p.r.
H_2O_2	0·4–3·0	$1·6 \times 10^{8}$	p.r.
H_2O_2	2	$9 \pm 1 \times 10^{7}$	p.r.
Sn^{2+}	0·1	$\sim 8 \times 10^{10}*$	c. Fe^{3+}
Sn^{4+}	0·1	$2·5 \times 10^{6}*$	c. Fe^{3+}
acetate ion	neutral	$2·7 \times 10^{5}*$	c. NO_2^-
acetate ion	11–13	$\sim 3 \times 10^{5}*$	c. OH^-
acetate ion	neutral	$2·7 \pm 0·3 \times 10^{5}*$	c. D-formate
D_3-acetate ion	neutral	$1·2 \times 10^{4}*$	c. acetate
acetic acid	1	$1 \times 10^{5}*$	c. Fe^{3+}
acetic acid (addn)	in hexane	$6·1 \times 10^{8}*$	c. n-hexane
acetic acid (abstn)	in hexane	$1·5 \times 10^{8}*$	c. n-hexane
acetone	1	$3·5 \times 10^{5}*$	c. Fe^{+3}
acetone (addn)	in hexane	$3·2 \times 10^{8}*$	c. n-hexane
acetone (abstn)	in hexane	$1·6 \times 10^{8}*$	c. n-hexane
acetone (H abstn)	11–13	$\sim 2 \times 10^{6}*$	c. OH^-
acetone (H addn)	11–13	$\sim 4 \times 10^{5}*$	
acetone	neutral	$6 \pm 0·7 \times 10^{5}*$	c. D-formate
acetophenone (addn)	in hexane	$1·2 \times 10^{9}*$	c. n-hexane
adenine	neutral	$6 \pm 2 \times 10^{8}*$	c. isopropanol
allyl alcohol	neutral	$2·3 \pm 0·5 \times 10^{9}*$	c. isopropanol
t-amyl benzene (addn)	in hexane	$3·5 \times 10^{8}*$	c. n-hexane
aniline (addn)	in hexane	$4 \times 10^{8}*$	c. n-hexane
anthracene	in hexane	$3·1 \times 10^{9}*$	c. n-hexane
benzene	7 and 0	$\sim 7 \times 10^{7}*$	c. O_2
benzene (addn)	in hexane	$1·8 \times 10^{8}*$	c. n-hexane
benzoic ac. (addn)	in hexane	$2·9 \times 10^{9}*$	c. n-hexane
benzonitrile (addn)	in hexane	$1 \times 10^{9}*$	c. n-hexane
benzoquinone	1	$3·1 \times 10^{9}*$	c. Fe^{3+}
benzoquinone	1	$1·2 \times 10^{9}*$	c. methanol
benzylacetate	in hexane	$2·7 \times 10^{9}*$	c. n-hexane
benzyl alcohol	neutral	$6·5 \pm 1·5 \times 10^{8}*$	c. isopropanol
benzyl alcohol	in hexane	$6 \times 10^{8}*$	c. n-hexane
bibenzyl (addn)	in hexane	$4·4 \times 10^{8}*$	c. n-hexane
biphenyl (addn)	in hexane	$1·4 \times 10^{9}*$	c. n-hexane
bromoform	in hexane	$5·1 \times 10^{9}*$	c. n-hexane
n-butane	in hexane	$2·7 \times 10^{6}*$	c. various
iso-butane	in hexane	$8·5 \times 10^{6}*$	c. various
t-butanol	neutral	$1 \pm 0·2 \times 10^{5}*$	c. D-formate
butyl acetate	in hexane	$8·4 \times 10^{7}*$	c. n-hexane
t-butyl bromide (Br abstn)	in hexane	$1·5 \times 10^{9}*$	c. n-hexane
n-butyl iodide (I abstn)	in hexane	$7·6 \times 10^{9}*$	c. n-hexane
n-butyl iodide (H abstn)	in hexane	$9 \times 10^{8}*$	c. n-hexane
t-butyl iodide (I abstn)	in hexane	$6·7 \times 10^{9}*$	c. n-hexane
carbon tetrabromide	in hexane	$1 \times 10^{10}*$	c. n-hexane
carbon tetrachloride	in hexane	$1·75 \times 10^{9}*$	c. n-hexane
carbon tetraiodide	in hexane	$2·1 \times 10^{10}*$	c. n-hexane
chloroacetate ion (H abstn)	neutral	$2·6 \pm 5 \times 10^{6}*$	c. H
chloroacetate ion (Cl abstn)	neutral	$2·6 \pm 0·5 \times 10^{5}*$	c. H

TABLE 1g21. (*Continued*)

Reactant	pH	Rate constant $M^{-1} sec^{-1}$	Method*
chloroacetate ion	neutral	$1.2 \pm 0.3 \times 10^{6}*$	c. D-formate
chloroacetic acid (H abstn)	acid	$1.8 \pm 0.4 \times 10^{5}*$	c. H
chloroacetic acid (Cl abstn)	acid	$8 \times 10^{4}*$	c. H
chloroform (Cl abstn)	in hexane	$1.6 \times 10^{9}*$	c. n-hexane
chloroform (H abstn)	in hexane	$2.2 \times 10^{8}*$	c. n-hexane
cyclobutane	in hexane	$4.4 \times 10^{6}*$	c. various
cyclohexane	in hexane	$6.6 \times 10^{6}*$	c. various
cyclohexanone (addn)	in hexane	$6.8 \times 10^{8}*$	c. n-hexane
cyclohexanone (abstn)	in hexane	$2.4 \times 10^{8}*$	c. n-hexane
cyclohexene (addn)	in hexane	$4.9 \times 10^{8}*$	c. n-hexane
cyclopropane	in hexane	$3.3 \times 10^{6}*$	c. various
2-deoxy-D-ribose	neutral	$2.2 \pm 0.2 \times 10^{7}*$	c. D-formate
diethyl ketone	in hexane	$3.8 \times 10^{8}*$	c. n-hexane
ethane	in hexane	$5.0 \times 10^{4}*$	c. various
ethanol	1	$1.5 \times 10^{7}*$	c. Fe^{3+}
ethanol	1 and 3	$1.65 \times 10^{7}*$	c. $Fe(CN)_6^{3-}$
ethanol	neutral	$1.6 \pm 0.2 \times 10^{7}*$	c. D-formate
ethyl acetate	1	$6 \times 10^{5}*$	c. Fe^{3+}
ethyl acetate	in hexane	$8.3 \times 10^{7}*$	c. n-hexane
ethylene glycol	neutral	$9.5 \pm 1 \times 10^{6}*$	c. D-formate
formaldehyde	1	$5 \times 10^{6}*$	c. Fe^{+3}
formate ion	1–3	$2.5 \pm 0.5 \times 10^{8}*$	c. $Fe(CN)_6^{3-}$
formate ion	neutral	$2.2 \times 10^{8}*$	c. $Fe(CN)_6^{3-}$
formate ion	11–13	$2.5 \times 10^{8}*$	c. OH^{-}
formate ion	neutral	$1.5 \pm 0.2 \times 10^{8}*$	c. D-formate
D-formate ion	neutral	$2.2 \times 10^{7}*$	c. ethanol
D-formate ion	neutral	$3.5 \times 10^{7}*$	c. formate
formic acid	1	$1.1 \times 10^{6}*$	c. Fe^{3+}
formic acid	1	$1.1 \times 10^{6}*$	c. $Fe(CN)_6^{3-}$
D-formic acid	1	$1.5 \times 10^{5}*$	c. Fe^{3+}
glucose	1	$4 \pm 1 \times 10^{7}*$	c. $Fe(CN)_6^{3-}$
glycerol	1	$2 \times 10^{7}*$	c. $Fe(CN)_6^{3-}$
glyoxalate ion	neutral	$3.5 \times 10^{7}*$	c. D-formate
n-hexane	in hexane	$4.9 \times 10^{6}*$	c. various
hexene-1 (addn)	in hexane	$7.9 \times 10^{8}*$	c. n-hexane
hexene-1 (abstn)	in hexane	$2.4 \times 10^{8}*$	c. n-hexane
hexene-2 (addn)	in hexane	$5.2 \times 10^{8}*$	c. n-hexane
hexene-2 (abstn)	in hexane	$1.27 \times 10^{8}*$	c. n-hexane
iodoform	in hexane	$1.9 \times 10^{10}*$	c. n-hexane
methanol	1	$1.6 \times 10^{6}*$	c. Fe^{3+}
methanol	in hexane	$8.7 \times 10^{5}*$	c. n-hexane
methanol	neutral	$1.7 \times 10^{6}*$	c. $Fe(CN)_6^{3-}$
methanol	neutral	$1.6 \times 10^{6}*$	c. NO_2^{-}
methanol	neutral	$1.65 \pm 0.2 \times 10^{6}*$	c. D-formate
D_3-methanol	neutral	$8 \times 10^{4}*$	c. methanol
2-methylbutene-1 (addn)	in hexane	$1.08 \times 10^{9}*$	c. n-hexane
2-methylbutene-1 (abstn)	in hexane	$3.4 \times 10^{8}*$	c. n-hexane
methyl chloroacetate	in hexane	$2.2 \times 10^{9}*$	c. n-hexane
methyl iodoacetate	in hexane	$1 \times 10^{10}*$	c. n-hexane
naphthalene (addn)	in hexane	$1.5 \times 10^{9}*$	c. n-hexane
nitrobenzene (addn)	in hexane	$1.25 \times 10^{9}*$	c. n-hexane
n-pentane	in hexane	$3.8 \times 10^{6}*$	c. various
phenol	0	$1 - 2 \times 10^{8}*$	c. Fe^{2+}

TABLE 1g21. (*Continued*)

Reactant	pH	Rate constant $M^{-1}sec^{-1}$	Method*
phenol	in hexane	$4 \cdot 2 \times 10^8*$	c. n-hexane
propane	in hexane	$1 \cdot 6 \times 10^5*$	c. various
2-propanol	1	$5 \times 10^7*$	c. $Fe(CN)_6^{3-}$
2-propanol	neutral	$5 \pm 0 \cdot 5 \times 10^7*$	c. D-formate
2-D-2-propanol	neutral	$8 \cdot 5 \times 10^6*$	c. isopropanol
propionic acid (addn)	in hexane	$6 \cdot 9 \times 10^8*$	c. n-hexane
propionic acid (abstn)	in hexane	$1 \cdot 6 \times 10^8*$	c. n-hexane
n-propyl benzene (addn)	in hexane	$3 \cdot 3 \times 10^8*$	c. n-hexane
n-propyl bromide (Br abstn)	in hexane	$1 \cdot 24 \times 10^9*$	c. n-hexane
n-propyl bromide (H abstn)	in hexane	$3 \cdot 3 \times 10^8*$	c. n-hexane
n-propyl chloride (Cl abstn)	in hexane	$1 \cdot 4 \times 10^8*$	c. n-hexane
n-propyl chloride (H abstn)	in hexane	$9 \cdot 6 \times 10^7*$	c. n-hexane
pyridine	neutral	$6 \cdot 4 \pm 2 \times 10^8*$	c. isopropanol
sucrose	1	$3 \cdot 7 \times 10^7*$	c. Fe^{3+}
trifluoroacetic acid (addn)	in hexane	$1 \cdot 6 \times 10^9*$	c. n-hexane
trifluoroacetic acid (abstn)	in hexane	$4 \cdot 4 \times 10^8*$	c. n-hexane
o-terphenyl	in hexane	$2 \times 10^9*$	c. n-hexane
uracil	neutral	$3 \pm 0 \cdot 6 \times 10^9*$	c. isopropanol
o-xylene (addn)	in hexane	$2 \cdot 4 \times 10^8*$	c. n-hexane
m-xylene (addn)	in hexane	$2 \cdot 9 \times 10^8*$	c. n-hexane
p-xylene (addn)	in hexane	$4 \cdot 9 \times 10^8*$	c. n-hexane

*p.r. = pulse radiolysis,
    c = competing compound.

*Reprinted with permission from: M. Anbar and P. Neta, "Tables of Bimolecular Rate Constants," *Int. J. Appl. Radn. Isotopes*, **16**, pp. 227–242, 1965, Pergamon Press.

TABLE 1g22. BIMOLECULAR RATE CONSTANTS OF HYDROXYL RADICALS (ANBAR AND NETA 1965)*

Reactant	pH	Rate constant $M^{-1}sec^{-1}$	Method*
e_{aq}^-	10–11	$\sim 3 \times 10^{10}$	p. r.
H	acid	$1\cdot2 \times 10^{10}$	p. r.
H	0·4–3·0	$3\cdot2 \times 10^{10}$	p. r.
OH	neutral	$4 \pm 0\cdot9 \times 10^9$	c. H_2O_2
OH	0·4–3·0	6×10^9	p. r.
O^-	13	1×10^9	p. r.
Br^-	0	$1\cdot6 \times 10^{10}$	c. OH
Br^-	0–2	$3\cdot6 \times 10^{10}$	c. H_2
Br^-	neutral	$1\cdot3 \times 10^{8}*$	c. ethanol
Br^-	neutral	$1\cdot3 \times 10^{9}*$	c. H_2O_2
Br^-	neutral	$1\cdot3 \times 10^{8}*$	c. $CO_3^=$, p. r.
CO	0	$6\cdot8 \times 10^{8}*$	Fenton's
CO	1	$8\cdot3 \times 10^{8}*$	Fenton's
CO	7	$5\cdot9 \times 10^{8}*$	c. H_2O_2, photochem.
CO		$4\cdot5 \times 10^{8}$	
$CO_3^=$	neutral	$8 \times 10^{7}*$	c. NO_2^-, p. r.
CNS^-	neutral	$1\cdot3 \times 10^{9}*$	c. $CO_3^=$, p. r.
Ce^{+3}	acid	$2\cdot2 \times 10^{8}*$	Fenton's
Ce^{3+}	0·1	$2\cdot2 \times 10^{8}*$	c. formic acid
Cl^-	0	4×10^9	c. OH
Cl^-	0	4×10^9	p. r.
Cl^-		3×10^8	
Cl^-	3	2×10^7	p. r.
D_2	1	$1\cdot6 \times 10^{7}*$	Fenton's
Fe^{2+}	1	$3\cdot0 \times 10^{8}*$	c. H_2, Fenton's
Fe^{2+}	0	$> 10^{8}*$	p. r.
Fe^{2+}	1·57	$3\cdot2 \times 10^{8}*$	c. H_2
Fe^{2+}	2·0	$2\cdot6 \times 10^{8}*$	p. r.
Fe^{2+}	2·1	$2\cdot5 \pm 0\cdot5 \times 10^{8}*$	c. H_2
Fe^{2+}	1	$3\cdot2 \pm 0\cdot3 \times 10^{8}*$	c. H_2. Fenton's
$Fe(CN)_6^{4+}$	2·5–10·5	$2\cdot1 \times 10^{9}*$	c. formate
H_2	neutral	$4\cdot2 \times 10^{7}*$	c. H_2O_2
H_2	neutral	$4\cdot5 \pm 0\cdot4 \times 10^{7}*$	c. H_2O_2
H_2	13	$1\cdot6 \pm 0\cdot6 \times 10^{8}*$	p. r. c. $Fe(CN)_6^{4-}$
HO_2	0·4–3·0	$1\cdot5 \times 10^{10}$	p. r.
HO_2		$1\cdot1 \times 10^{10}$	p. r.
H_2O_2	2·5–7	$4\cdot5 \pm 0\cdot4 \times 10^{7}*$	c. H_2
H_2O_2	7	$4\cdot5 \pm 0\cdot4 \times 10^7$	c. OH
H_2O_2	0·4–3·0	$1\cdot2 \times 10^7$	p. r.
OH^-		$3\cdot6 \pm 1 \times 10^{8}*$	photochem.
I^-	neutral	$1\cdot6 \times 10^{9}*$	photochem.
I^-	neutral	$\sim 1\cdot2 \times 10^{9}*$	c. ethylenediamine
NO_2^-	neutral	$2\cdot5 \times 10^{9}*$	c. H_2O_2
NO_2^-	neutral	$2\cdot5 \times 10^{9}*$	c. $CO_3^=$, p. r.
$SO_3^=$	neutral	$1\cdot2 \times 10^{9}*$	c. $CO_3^=$, p. r.
HSO_3^-	neutral	$2\cdot1 \times 10^{9}*$	c. $CO_3^=$, p. r.
HSO_4^-	0·1	$3\cdot3 \times 10^{7}*$	c. formic acid
HSO_4^-	acid	$2 \pm 0\cdot4 \times 10^{7}*$	photochem.
H_2SO_4	0·1	$3\cdot5 \times 10^{5}*$	c. Ce^{3+}, formic acid
H_2SO_4	0–1	$4 \times 10^{5}*$	c. formic acid
Sn^{2+}	0·1	$2 \times 10^{9}*$	c. Fe^{2+}
Tl^+	0·1	$8\cdot5 \times 10^{9}*$	c. Ce^{3+}
Tl^+	acid	$9 \times 10^{9}*$	c. Ce^{3+}, photochem.

TABLE 1g22. (*Continued*)

Reactant	pH	Rate constant $M^{-1}sec^{-1}$	Method*
acetaldehyde	1	7×10^{8}*	Fenton's
acrylamide		1.4×10^{10}*	c. Fe^{2+}
adipic acid	1	4×10^{8}*	Fenton's
alanine	1	3.8×10^{7}*	Fenton's
alanine	6	2.25×10^{7}*	c. $Fe(CN)_6^{3-}$
alanine	9.8	6.3×10^{8}*	c. $Fe(CN)_6^{3-}$
benzamide	1	2.1×10^{9}*	Fenton's
benzene	0 and 7	$4.3 \pm 0.9 \times 10^{9}$	p. r.
benzene		$6.6 \pm 1.6 \times 10^{9}$	p. r.
benzene	1	1×10^{9}*	Fenton's
D_6-benzene	0 and 7	$4.7 \pm 0.9 \times 10^{9}$	p. r.
benzenesulfonic acid	1	1.5×10^{9}*	Fenton's
benzoic acid	1	2.2×10^{9}*	Fenton's
benzoic acid	3	$4.8 \pm 0.7 \times 10^{9}$	p. r.
t-butanol	1	2.1×10^{8}*	Fenton's
1,2-butylene glycol	1	6×10^{6}*	Fenton's
2,3-butylene glycol	1	3.2×10^{6}*	Fenton's
n-butyric acid	1	2.3×10^{8}*	Fenton's
chloral hydrate	1	1.5×10^{9}*	Fenton's
diethylamine sulfate	1	1.3×10^{8}*	Fenton's
diethyl ether	1	2×10^{9}*	Fenton's
dimethylaniline sulfate	1	2.1×10^{9}*	Fenton's
dioxane	4	4.5×10^{8}*	c. ethylenediamine
ethanol	1	1.2×10^{9}*	Fenton's
ethanol	6–10.5	2.25×10^{8}*	c. $Fe(CN)_6^{4-}$
ethanol	neutral	2.9×10^{8}*	c. $CO_3^{=}$, p. r.
ethanol		3×10^{9}	
ethylenediamine	5	$\sim 1.5 \times 10^{7}$*	c. isopropanol
ethylenediamine	4	$\sim 7 \times 10^{6}$*	c. ethanol
ethylene glycol	1	1.15×10^{8}*	Fenton's
ethyl acetate	1	3.5×10^{8}*	Fenton's
formaldehyde	1	9.6×10^{8}*	Fenton's
formaldehyde	1	2×10^{9}	
formate ion	2.5–10.5	1×10^{9}*	c. $Fe(CN)^{4-}$
formate ion	4	$\sim 2 \times 10^{8}$*	c. ethylenediamine
formic acid	3	2.5×10^{8}*	c. H_2O_2
formic acid	0	5×10^{7}	
formic acid	3	2.5×10^{8}*	photochem.
formic acid	0.1	1.2×10^{8}*	c. Ce^{3+}
formic acid	2.5–4	2.5×10^{8}*	c. $Fe(CN)_6^{4-}$
formic acid	3	3×10^{8}	
formic acid	0.4	4.2×10^{7}	
glycerol	neutral	4.2×10^{8}*	c. $CO_3^{=}$, p. r.
glycine	1	1.1×10^{7}*	Fenton's
glycine	6	2.7×10^{6}*	c. $Fe(CN)_6^{4-}$
glycine	9.5–10.5	1.3×10^{9}*	c. $Fe(CN)_6^{4-}$
glycollic acid	1	6.4×10^{8}*	Fenton's
α-hydroxybutyric acid	1	9.6×10^{8}*	Fenton's
lactic acid	1	5.5×10^{8}*	Fenton's
methanol	1	7.4×10^{8}*	Fenton's
methanol		1.8×10^{9}	
methanol	4	$\sim 3 \times 10^{8}$*	c. ethylenediamine
methyl acetate	1	2.7×10^{8}*	Fenton's
nitrobenzene	1	9.3×10^{8}*	Fenton's

TABLE 1g22. (*Continued*)

Reactant	pH	Rate constant $M^{-1}sec^{-1}$	Method*
phenol	1	$1\cdot7 \times 10^6$*	Fenton's
phenylacetic acid	1	$1\cdot5 \times 10^9$*	Fenton's
pinacol	1	4×10^8*	Fenton's
1-propanol	1	$8\cdot3 \times 10^8$*	Fenton's
2-propanol	1	$9\cdot6 \times 10^8$*	Fenton's
2-propanol	4	$\sim 3 \times 10^8$*	c. ethylenediamine
propionic acid	1	$2\cdot7 \times 10^8$*	Fenton's
iso-propyl acetate	1	4×10^8*	Fenton's
pyridine	1	$5\cdot7 \times 10^7$*	Fenton's
succinic acid	1	$9\,6 \times 10^6$*	Fenton's
tetrahydrofuran	1	2×10^9*	Fenton's
tetrahydropyran	1	$1\cdot4 \times 10^9$*	Fenton's
thioglycollic acid	1	$8\cdot6 \times 10^8$*	Fenton's
toluene	3	$7\cdot0 \pm 1\cdot5 \times 10^9$	p. r.
p-toluenesulfonate ion	4	$\sim 1\cdot5 \times 10^9$	c. ethylenediamine
triethylamine	1	$2\cdot5 \times 10^8$*	Fenton's
iso-valeric acid	1	1×10^9*	Fenton's

*p.r. = pulse radiolysis,
 c = competing compound.
*Reprinted with permission from: M. Anbar and P. Neta, "Tables of Bimolecular Rate Constants," *Int. J. Appl. Radn. Isotopes*, **16**, pp. 227–242, 1965, Pergamon Press.

REFERENCES

Anbar, M. and P. Neta, "Tables of Bimolecular Rate Constants of Hydrated Electrons, Hydrogen Atoms and Hydroxyl Radicals with Inorganic and Organic Compounds," *Int. J. Appl. Radn. and Isotopes* **16**, 227–242, 1965.

Harned, H. S. and B. B. Owen, "The Physical Chemistry of Electrolytic Solutions," Van Nostrand Reinhold, New York, 1950.

SECTION 1h
THERMAL AND THERMODYNAMIC PROPERTIES OF MATERIALS

Section 1h offers information on selected thermal and thermodynamic properties of a number of materials. It is organized as follows:

TABLE 1h1. TEMPERATURE CONVERSION TABLES AND FORMULAS.*

To convert from Fahrenheit to Celsius*—locate temperature (°F) in center column and read °C in left column.

To convert from Celsius* to Fahrenheit—locate temperature (°C) in center column and read °F in right column.

-459.4 To -70			-69 To 0			1 To 69			70 To 139			140 To 290			300 To 1000		
C		F	C		F	C		F	C		F	C		F	C		F
-273	-459.4		-56.1	-69	-92.2	-17.2	1	33.8	21.1	70	158.0	60.0	140	284.0	149	300	572
-268	-450		-55.5	-68	-90.4	-16.7	2	35.6	21.7	71	159.8	60.6	141	285.8	154	310	590
-262	-440		-55.0	-67	-88.6	-16.1	3	37.4	22.2	72	161.6	61.1	142	287.6	160	320	608
-257	-430		-54.4	-66	-86.8	-15.6	4	39.2	22.8	73	163.4	61.7	143	289.4	166	330	626
-251	-420		-53.9	-65	-85.0	-15.0	5	41.0	23.3	74	165.2	62.2	144	291.2	171	340	644
-246	-410		-53.3	-64	-83.2	-14.4	6	42.8	23.9	75	167.0	62.8	145	293.0	177	350	662
-240	-400		-52.8	-63	-81.4	-13.9	7	44.6	24.4	76	168.8	63.3	146	294.8	182	360	680
			-52.2	-62	-79.6	-13.3	8	46.4	25.0	77	170.6	63.9	147	296.6	188	370	698
			-51.7	-61	-77.8	-12.8	9	48.2	25.6	78	172.4	64.4	148	298.4	193	380	716
			-51.1	-60	-76.0				26.1	79	174.2	65.0	149	300.2	199	390	734
-234	-390		-50.6	-59	-74.2	-12.2	10	50.0	26.7	80	176.0	65.6	150	302.0	204	400	752
-229	-380		-50.0	-58	-72.4	-11.7	11	51.8	27.2	81	177.8	66.1	151	303.8	210	410	770
-223	-370		-49.5	-57	-70.6	-11.1	12	53.6	27.8	82	179.6	66.7	152	305.6	216	420	788
-218	-360		-48.9	-56	-68.8	-10.6	13	55.4	28.3	83	181.4	67.2	153	307.4	221	430	806
-212	-350		-48.4	-55	-67.0	-10.0	14	57.2	28.9	84	183.2	67.8	154	309.2	227	440	824
-207	-340		-47.8	-54	-65.2	-9.44	15	59.0	29.4	85	185.0	68.3	155	311.0	232	450	842
-201	-330		-47.3	-53	-63.4	-8.69	16	60.8	30.0	86	186.8	68.9	156	312.9	238	460	860
-196	-320		-46.7	-52	-61.6	-8.33	17	62.6	30.6	87	188.6	69.4	157	314.6	243	470	878
-190	-310		-46.2	-51	-59.8	-7.78	18	64.4	31.1	88	190.4	70.0	158	316.4	249	480	896
-184	-300		-45.6	-50	-58.0	-7.22	19	66.2	31.7	89	192.2	70.6	159	318.2	254	490	914
-179	-290		-45.0	-49	-56.2	-6.67	20	68.0	32.2	90	194.0	71.1	160	320.0	260	500	932
-173	-280		-44.4	-48	-54.4	-6.11	21	69.8	32.8	91	195.0	71.7	161	321.0	266	510	950
-169	-273	-459.4	-43.9	-47	-52.6	-5.56	22	71.6	33.3	92	197.0	72.2	162	323.0	271	520	968
-168	-270	-454	-43.3	-46	-50.8	-5.00	23	73.4	33.9	93	199.4	72.8	163	325.4	277	530	986
-162	-260	-435	-42.8	-45	-49.0	-4.44	24	75.2	34.4	94	201.2	73.3	164	327.2	282	540	1004
-157	-250	-418	-42.2	-44	-47.2	-3.89	25	77.0	35.0	95	203.0	73.9	165	329.0	288	550	1022
-151	-240	-400	-41.7	-43	-45.4	-3.33	26	78.8	35.6	96	204.8	74.4	166	330.8	293	560	1040
-146	-230	-382	-41.1	-42	-43.6	-2.78	27	80.6	36.1	97	206.6	75.0	167	332.6	299	570	1058
-140	-220	-364	-40.6	-41	-41.8	-2.22	28	82.4	36.7	98	208.4	75.6	168	334.4	304	580	1076
-134	-210	-346	-40.0	-40	-40.0	-1.67	29	84.2	37.2	99	210.2	76.1	169	336.2	310	590	1094
-129	-200	-328															
-123	-190	-310	-39.4	-39	-38.2	-1.11	30	86.0	37.8	100	212.0	76.7	170	338.0	316	600	1112
-118	-180	-292	-38.8	-38	-36.4	-0.56	31	87.8	38.3	101	213.8	77.2	171	339.8	321	610	1130
-112	-170	-274	-38.3	-37	-34.6	0	32	89.6	38.9	102	215.6	77.8	172	341.6	327	620	1148
-107	-160	-256	-37.8	-36	-32.8	0.56	33	91.4	39.4	103	217.4	78.3	173	343.4	332	630	1166
-101	-150	-238	-37.2	-35	-31.0	1.11	34	93.2	40.0	104	219.2	78.9	174	345.2	338	640	1184
-95.6	-140	-220	-36.6	-34	-29.2	1.67	35	95.0	40.6	105	221.0	79.4	175	347.0	343	650	1202
-90.0	-130	-202	-36.1	-33	-27.4	2.22	36	96.8	41.1	106	222.8	80.0	176	348.8	349	660	1220
-84.4	-120	-184	-35.5	-32	-25.6	2.78	37	98.6	41.7	107	224.6	80.6	177	350.6	354	670	1238
-78.9	-110	-166	-35.0	-31	-23.8	3.33	38	100.4	42.2	108	226.4	81.1	178	352.4	360	680	1256
-73.3	-100	-148	-34.4	-30	-22.0	3.89	39	102.2	42.8	109	228.2	81.7	179	354.2	366	690	1274
-72.6	-99	-146.2	-33.9	-29	-20.2	4.44	40	104.0	43.3	110	230.0	82.2	180	356.0	371	700	1292
-72.2	-98	-144.4	-33.3	-28	-18.4	5.00	41	105.8	43.9	111	231.8	82.8	181	357.8	377	710	1310
-71.7	-97	-142.6	-32.8	-27	-16.6	5.56	42	107.6	44.4	112	233.6	83.3	182	359.6	382	720	1328
-71.1	-96	-140.8	-32.2	-26	-14.8	6.11	43	109.4	45.0	113	235.4	83.9	183	361.4	388	730	1346
-70.6	-95	-139.0	-31.7	-25	-13.0	6.67	44	111.2	45.6	114	237.2	84.4	184	363.2	393	740	1364
-70.0	-94	-137.2	-31.1	-24	-11.2	7.22	45	113.0	46.1	115	239.0	85.0	185	365.0	399	750	1382
-69.5	-93	-135.4	-30.6	-23	-9.4	7.78	46	114.8	46.7	116	240.8	85.6	186	366.8	404	760	1400
-68.9	-92	-133.6	-30.0	-22	-7.6	8.33	47	116.6	47.2	117	242.6	86.1	187	368.6	410	770	1418
-68.4	-91	-131.8	-29.5	-21	-5.8	8.89	48	118.4	47.8	118	244.4	86.7	188	370.4	416	780	1436
-67.8	-90	-130.0	-28.9	-20	-4.0	9.44	49	120.2	48.3	119	246.2	87.2	189	372.2	421	790	1454
-67.2	-89	-128.2	-28.3	-19	-2.2	10.0	50	122.0	48.9	120	248.0	87.8	190	374.0	427	800	1472
-66.6	-88	-126.4	-27.7	-18	-0.4	10.6	51	123.8	49.4	121	249.8	88.3	191	375.8	432	810	1490
-66.1	-87	-124.6	-27.2	-17	1.4	11.1	52	125.6	50.0	122	251.6	88.9	192	377.6	438	820	1508
-65.5	-86	-122.8	-26.6	-16	3.2	11.7	53	127.4	50.6	123	253.4	89.4	193	379.4	443	830	1526
-65.0	-85	-121.0	-26.1	-15	5.0	12.2	54	129.2	51.1	124	255.2	90.0	194	381.2	449	840	1544
-64.4	-84	-119.2	-25.5	-14	6.8	12.8	55	131.0	51.7	125	257.0	90.6	195	383.0	454	850	1562
-63.9	-83	-117.4	-25.0	-13	8.6	13.3	56	132.8	52.2	126	258.8	91.1	196	384.8	460	860	1580
-63.3	-82	-115.6	-24.4	-12	10.4	13.9	57	134.6	52.8	127	260.6	91.7	197	386.6	466	870	1598
-62.8	-81	-113.8	-23.9	-11	12.2	14.4	58	136.4	53.3	128	262.4	92.2	198	388.4	471	880	1616
-62.2	-80	-112.0	-23.3	-10	14.0	15.0	59	138.2	53.9	129	264.2	92.8	199	390.2	477	890	1634

TABLE 1h1. (*Continued*)

-459.4 To -70		-69 To 0		1 To 69		70 To 139		140 To 290		300 To 1000	
C	F	C	F	C	F	C	F	C	F	C	F
- 61.7 - 79	-110.2	-22.8 - 9	15.8	15.6 60	140.0	54.4 130	266.0	93.3 200	392	482 900	1652
- 61.1 - 78	-108.4	-22.2 - 8	17.6	16.1 61	141.8	55.0 131	267.8	98.9 210	410	488 910	1670
- 60.6 - 77	-106.6	-21.7 - 7	19.4	16.7 62	143.6	55.6 132	269.6	100 212	413	493 920	1688
- 60.0 - 76	-104.8	-21.1 - 6	21.2	17.2 63	145.4	56.1 133	271.4	104 220	428	499 930	1706
- 59.5 - 75	-103.0	-20.6 - 5	23.0	17.8 64	147.2	56.7 134	273.2	110 230	446	504 940	1724
- 58.9 - 74	-101.2	-20.0 - 4	24.8	18.3 65	149.0	57.2 135	275.0	116 240	464	510 950	1742
- 58.4 - 73	- 99.4	-19.5 - 3	26.6	18.9 66	150.8	57.8 136	276.8	121 250	482	516 960	1760
- 57.8 - 72	- .97.6	-18.9 - 2	28.4	19.4 67	152.6	58.3 137	278.6	127 260	500	521 970	1778
- 57.3 - 71	- 95.8	-18.4 - 1	30.2	20.0 68	154.4	58.9 138	280.4	132 270	518	527 980	1796
- 56.7 - 70	- 94.0	-17.8 0	32.0	20.6 69	156.2	59.4 139	282.2	138 280	536	532 990	1814
								143 290	554		

1000 to 1490		1500 to 1990		2000 to 2490		2500 to 3000	
C	F	C	F	C	F	C	F
538 1000	1832	816 1500	2732	1093 2000	3632	1371 2500	4532
543 1010	1850	821 1510	2750	1099 2010	3650	1377 2510	4550
549 1020	1868	827 1520	2768	1104 2020	3668	1382 2520	4568
554 1030	1886	832 1530	2786	1110 2030	3686	1388 2530	4586
560 1040	1904	838 1540	2804	1116 2040	3704	1393 2540	4604
566 1050	1922	843 1550	2822	1121 2050	3722	1399 2550	4622
571 1060	1940	849 1560	2840	1127 2060	3740	1404 2560	4640
577 1070	1958	854 1570	2858	1132 2070	3758	1410 2570	4658
582 1080	1976	860 1580	2876	1138 2080	3776	1416 2580	4676
588 1090	1994	866 1590	2894	1143 2090	3794	1421 2590	4694
593 1100	2012	871 1600	2912	1149 2100	3812	1427 2600	4712
599 1110	2030	877 1610	2930	1154 2110	3830	1432 2610	4730
604 1120	2048	882 1620	2948	1160 2120	3848	1438 2620	4748
610 1130	2066	888 1630	2966	1166 2130	3866	1443 2630	4765
616 1140	2084	893 1640	2984	1171 2140	3884	1449 2640	4784
621 1150	2102	899 1650	3002	1177 2150	3902	1454 2650	4802
627 1160	2120	904 1660	3020	1182 2160	3920	1460 2660	4820
632 1170	2138	910 1670	3038	1188 2170	3938	1466 2670	4838
638 1180	2156	916 1680	3056	1193 2180	3956	1471 2680	4856
643 1190	2174	921 1690	3074	1199 2190	3974	1477 2690	4874
649 1200	2192	927 1700	3092	1204 2200	3992	1482 2700	4892
654 1210	2210	932 1710	3110	1210 2210	4010	1488 2710	4910
660 1220	2228	938 1720	3128	1216 2220	4028	1493 2720	4928
666 1230	2246	943 1730	3146	1221 2230	4046	1499 2730	4945
671 1240	2264	949 1740	3164	1227 2240	4064	1504 2740	4964
677 1250	2282	954 1750	3182	1232 2250	4082	1510 2750	4982
682 1260	2300	960 1760	3200	1238 2260	4100	1516 2760	5000
688 1270	2318	966 1770	3218	1243 2270	4118	1521 2770	5018
693 1280	2336	971 1780	3236	1249 2280	4136	1527 2780	5036
699 1290	2354	977 1790	3254	1254 2290	4154	1532 2790	5054
704 1300	2372	982 1800	3272	1260 2300	4172	1538 2800	5072
710 1310	2390	988 1810	3290	1266 2310	4190	1543 2810	5090
716 1320	2408	993 1820	3308	1271 2320	4208	1549 2820	5108
721 1330	2426	999 1830 3326		1277 2330 4226		1554 2830 5126	
727 1340	2444	1004 1840	3344	1282 2340 4244		1560 2840 5144	
732 1350	2462	1010 1850	3362	1288 2350	4262	1566 2850	5162
738 1360	2480	1016 1860	3380	1293 2360	4280	1571 2860	5180
743 1370	2498	1021 1870	3398	1299 2370	4298	1577 2870	5198
749 1380	2516	1027 1880	3416	1304 2380	4316	1582 2880	5216
754 1390	2534	1032 1890	3434	1310 2390	4334	1588 2890	5234
760 1400	2552	1038 1900	3452	1316 2400	4352	1593 2900	5252
766 1410	2570	1043 1910	3470	1321 2410	4370	1599 2910	5270
771 1420	2588	1049 1920	3488	1327 2420	4388	1604 2920	5288
777 1430	2606	1054 1930	3506	1332 2430	4406	1610 2930	5306
782 1440	2624	1060 1940	3524	1338 2440	4424	1616 2940	5324
788 1450	2642	1066 1950	3542	1343 2450	4442	1621 2950	5342
793 1460	2660	1071 1960	3560	1349 2460	4460	1627 2960	5360
799 1470	2678	1077 1970	3578	1354 2470	4478	1632 2970	5378
804 1480	2696	1082 1980	3596	1360 2480	4496	1638 2980	5396
810 1490	2714	1088 1990	3614	1366 2490	4514	1643 2990	5414
						1649 3000	5432

TABLE 1h1. (*Continued*)

Interpolation Factors

C		F
0.56	1	1.8
1.11	2	3.6
1.67	3	5.4
2.22	4	7.2
2.78	5	9.0
3.33	6	10.8
3.89	7	12.6
4.44	8	14.4
5.00	9	16.2
5.56	10	18.0

*The term Centigrade was officially changed to Celsius by international agreement in 1948. The Celsius scale uses the triple phase point of water, at $0.01°$ Centigrade, in place of the ice point as a reference, but for all practical purposes the two terms are interchangeable.

Given	Temperature Conversion				
	Celsius	*Fahrenheit*	*Kelvin*	*Reaumur*	*Rankine*
Cels.	—	$\left(\dfrac{9}{5}C\right) + 32$	$C + 273.16$	$\dfrac{4}{5}C$	$1.8\,(C + 273.16)$
Fahr.	$\dfrac{5}{9}(F - 32)$	—	$\left[\dfrac{5}{9}(F - 32)\right] + 273.16$	$\dfrac{4}{9}(F - 32)$	$F + 459.7$
Kelvin	$K - 273.16$	$\left[\dfrac{9}{5}(K - 273.16)\right] + 32$	—	$\dfrac{4}{5}(K - 273.16)$	$K \times 1.8$
Reau.	$Re \times \dfrac{5}{4}$	$\left(\dfrac{9}{4}Re\right) + 32$	$\left(\dfrac{5}{4}Re\right) + 273.16$	—	$\left(\dfrac{9}{4}Re\right) + 491.7$
Rank.	$\dfrac{Ra}{1.8} - 273.16$	$Ra - 459.7$	$\dfrac{Ra}{1.8}$	$\dfrac{4}{9}(Ra - 491.7)$	—

Five major temperature scales are in use at present. They are: Fahrenheit, Celsius, Kelvin (Absolute), Rankine, and Reaumur. The interrelationship among the scales is shown here.

*Reprinted from: Graf, R. F., "Electronic Design Data Book," Van Nostrand Reinhold Company, New York, 1971.

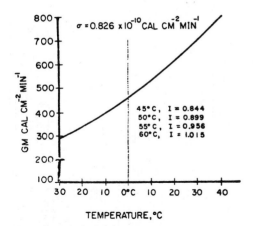

Figure 1h1. Black body radiation, I, as a function of temperature.*

*Reproduced from: Plate, R. B. and J. Griffiths, "Environmental Measurement," Van Nostrand Reinhold Co., New York, 1964.

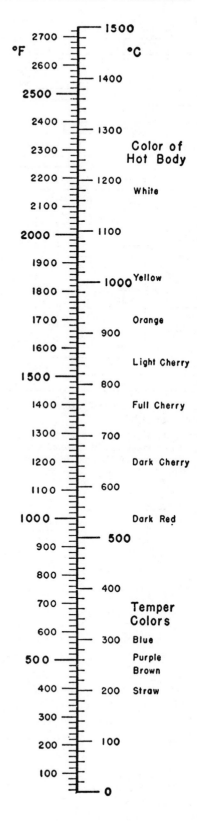

Figure 1h2. Temperature-Color   Relationship   for Steel.*

*From Digges, T. G., S. J. Rosenberg and G. W. Geil, "Heat Treatment and Properties of Iron and Steel," NBS Monograph 88, U.S. Government Printing Office, Washington, 1966.

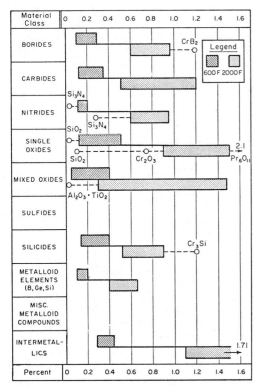

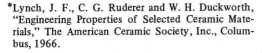

Figure 1h3. Linear thermal expansion of ceramic materials. (BMI 1966).*

*Lynch, J. F., C. G. Ruderer and W. H. Duckworth, "Engineering Properties of Selected Ceramic Materials," The American Ceramic Society, Inc., Columbus, 1966.

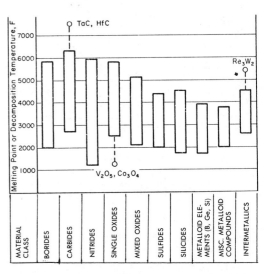

Figure 1h4. Melting or decomposition temperature of ceramic materials. (BMI 1966).

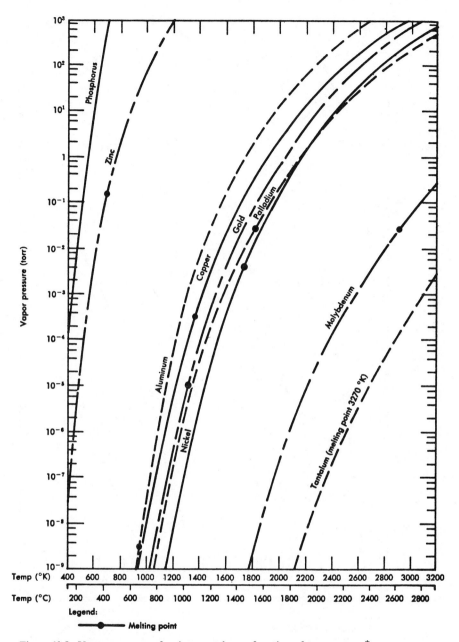

Figure 1h5. Vapor pressures of various metals as a function of temperature.*

*(After R. E. Honig and D. A. Kramer, "Vapor Pressure Data for the Solid and Liquid Elements," *RCA Rev.*, Vol. **30**, p. 285 June 1969); reprinted with permission.

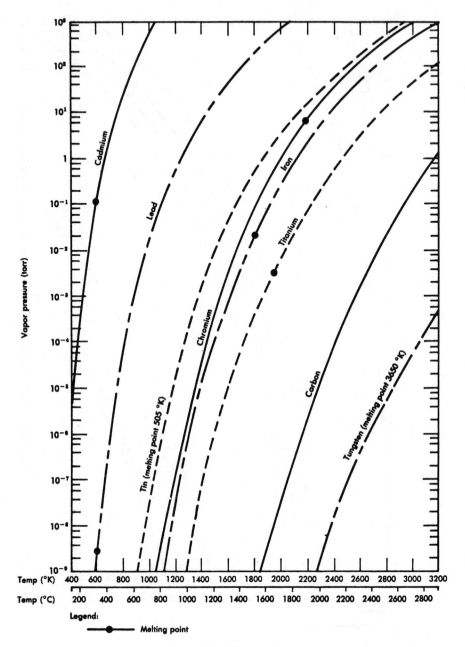

Figure 1h5. (*Continued*)

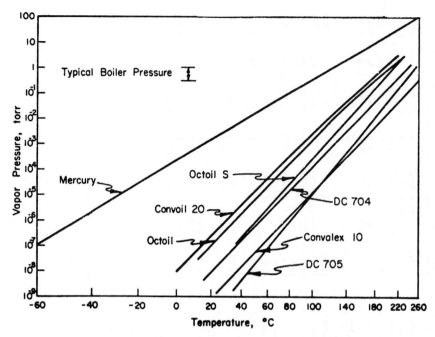

Figure 1h6. Vapor pressure of diffusion pump fluids.*

TABLE 1h2. CONVERSION FACTORS BETWEEN THE VARIOUS UNITS OF SPEED AND THROUGHPUT*

Speed	Throughput
1 liter/sec:	1 torr liter/sec:
2.12 ft^3/min	1.316 atm cm^3/sec
1000 cm^3/sec	1.27 × 10^5 micron ft^3/hr
60 liter/min	10^3 micron liter/sec
3.6 m^3/hr	

*From Santeler, D. J., D. W. Jones, D. H. Holkeboer and F. Pagano, "Vacuum Technology and Space Simulation," Report No. NASA SP-105, U.S. Government Printing Office, Washington, D.C., 1966.

TABLE 1h3. SUMMARY OF SELECTED CRITICAL PROPERTIES[a] (MATHEWS 1972)

Formula	Substance	Mol wt	Critical temperature				Critical pressure		Critical density		Critical volume		Z_c
			°K	°C	°R	°F	Atm	$lb_f/in.^2$	g/cm^3	lb_m/ft^3	$l./g\text{-}mol$	$ft^3/lb\text{-}mol$	
						A. Noble Gases							
^{4}He	Helium-4	4.0026	5.1889	−267.9611	9.3400	−450.3300	2.24	32.9	0.0698	4.358	0.0573	0.9185	0.301
^{3}He	Helium-3	3.0009	3.309	−269.841	5.956	−453.714	1.13	16.6	0.0414	2.58	0.0725	1.161	0.302
Ne	Neon	20.183	44.40	−228.75	79.92	−379.75	27.2	400	0.484	30.2	0.0417	0.668	0.311
Ar	Argon	39.948	150.8	−122.4	271.4	−188.3	48.1	707	0.533	33.3	0.0749	1.20	0.291
Kr	Krypton	83.80	209.4	−63.8	376.8	−82.9	54.3	798	0.919	57.4	0.0912	1.46	0.287
Xe	Xenon	131.30	289.733	16.583	521.519	61.849	57.64	847.1	1.11	69.3	0.118	1.89	0.286
Rn	Radon	222	377	104	679	219	62	911					
						B. Elementary Gases							
H_2	Hydrogen (normal)	2.0159	33.2	−239.9	59.8	−399.9	12.80	188.1	0.0310	1.935	0.0650	1.042	0.305
H_2	Hydrogen (equilibrium)	2.0159	32.98	−240.17	59.36	−400.31	12.76	187.5	0.0314	1.960	0.0642	1.028	0.303
HD	Hydrogen deuteride	3.024	36.0	−237.2	64.7	−395.0	14.6	215	0.0482	3.009	0.0628	1.005	0.310
D_2	Deuterium (normal)	4.032	38.4	−234.8	69.0	−390.7	16.4	241					
D_2	Deuterium (equilibrium)	4.032	38.2	−234.9	68.8	−390.9	16.3	240	0.0669	4.177	0.0603	0.965	0.314
Rb	Rubidium	85.47	2105	1832	3789	3329			0.34	21	0.25	4.0	
Ce	Cesium	132.905	2079	1806	3742	3282			0.44	27	0.30	4.8	
N_2	Nitrogen	28.013	126.2	−147.0	227.1	−232.6	33.5	429	0.313	19.5	0.0895	1.43	0.290
P	Phosphorus	30.974	994	721	1789	1329							
O_2	Oxygen	31.999	154.58	−118.57	278.24	−181.43	49.77	731.4	0.436	27.2	0.0734	1.18	0.288
O_3	Ozone	47.998	261.0	−12.1	469.9	10.2	55	808	0.54	34	0.089	1.4	0.23
S	Sulfur	32.064	1314	1041	2365	1905	116	1705					
F_2	Fluorine	37.997	144.30	−128.85	259.74	−199.93	51.47	756.4	0.574	35.8	0.0662	1.06	0.288
Cl_2	Chlorine	70.906	417	144	751	291	76	1117	0.573	35.8	0.124	1.98	0.275
Br_2	Bromine	159.818	584	311	1051	591	102	1499	1.26	78.7	0.127	2.03	0.270
I_2	Iodine	253.809	819	546	1474	1014			1.64	102	0.155	2.48	
Hg	Mercury	200.59	1735	1462	3123	2663	1587	23322					
						C. Deuterides							
ND_3	Nitrogen trideuteride	20.052	405.6	132.4	730.0	270.3							
PD_3	Phosphorus trideuteride	37.022	323.6	50.4	582.4	122.7							
AsD_3	Arsenic trideuteride	80.970	372.0	98.9	669.7	210.0							
						D. Nitrogen Compounds							
NH_3	Ammonia	17.03	405.6	132.4	730.0	270.3	111.3	1636	0.235	14.7	0.0725	1.16	0.242
N_2H_4	Hydrazine	32.05	653	380	1176	716	145	2131					
HCN	Hydrogen cyanide	27.03	456.8	183.6	822.2	362.5	53.2	782	0.195	12.2	0.139	2.22	0.197
C_2N_2	Cyanogen	52.04	400	127	720	260	59	867					
NOF_3	Trifluoramine oxide	87.00	302.6	29.5	544.8	85.1			0.593	37.0	0.169	2.70	
						E. Chalcogenides							
						1. Oxides							
H_2O	Water	18.015	647.14	373.99	1164.85	705.18	217.6	3198	0.32	20	0.056	0.90	0.23
D_2O	Heavy water	20.031	644.0	370.8	1159.1	699.4	213.8	3142	0.36	22	0.056	0.89	0.23
CO	Carbon monoxide	28.011	132.91	−140.24	239.2	−220.47	34.53	507.4	0.301	18.8	0.0931	1.49	0.295
CO_2	Carbon dioxide	44.010	304.2	31.0	547.5	87.8	72.8	1070	0.468	29.2	0.0940	1.51	0.274

N_2O	Nitrous oxide	44.013	309.56	36.41	557.21	97.54	71.5	1051	0.452	28.2	0.0974	1.56	0.274
NO	Nitric oxide	30.006	180	−93	324	−136	64	940	0.52	32	0.058	0.92	0.25
N_2O_4	Dinitrogen tetroxide	92.011	431	158	776	316	100	1470	0.55	34	0.17	2.7	0.47
SO_2	Sulfur dioxide	64.06	430.8	157.6	775.4	315.4	77.81	1144	0.525	32.8	0.122	1.95	0.268
SO_3	Sulfur trioxide	80.06	491.0	217.8	883.7	424.0	81	1190	0.63	39	0.13	2.0	0.26
Re_2O_7	Rhenium oxide	484.4	942	669	1696	1236			1.45	90.5	0.334	5.35	0.26
	2. Sulfides												
H_2S	Hydrogen sulfide	34.08	373.2	100.0	671.7	212.0	88.2	1296	0.346	21.6	0.0985	1.58	0.284
D_2S	Deuterium sulfide	36.09	372.2	99.1	670.0	210.3	78						
CS_2	Carbon disulfide	71.64	552	279	994	534		1146	0.44	27	0.16	2.6	0.28
	3. Selenides												
H_2Se	Hydrogen selenide	81.01	411	138	740	280	88	1293	0.44				
D_2Se	Deuterium selenide	83.02	412.4	139.2	742.2	282.5							
	4. Mixed												
COS	Carbonyl sulfide	60.07	375	102	675	215	58	852	0.44	27	0.14	2.2	0.26
	F. Halides												
	1. Fluorides												
HF	Hydrogen fluoride	20.01	461	188	830	370	64	940		18	0.069	1.1	0.12
BF_3	Boron trifluoride	67.81	260.8	−12.3	469.5	9.8	49.2	723					
SiF_4	Silicon tetrafluoride	104.08	259.0	−14.1	466.3	6.6	36.7	539					
N_2F_4	Tetrafluorohydrazine	104.01	309.4	36.2	556.8	97.1	37	544					
N_2F_2	cis-Difluorodiazine	66.01	272	−1	490	30							
N_2F_2	trans-Difluorodiazine	66.01	260	−13	468	8							
NF_3	Nitrogen trifluoride	71.00	234.0	−39.2	421.1	−38.7	44.7	657					
HNF_2	Difluoramine	53.01	403	130	725	265							
PF_3	Phosphorus trifluoride	87.97	271.2	−2.0	488.1	28.4	42.7	628					
OF_2	Oxygen fluoride	54.00	215.2	−58.0	387.3	−72.4							
SF_4	Sulfur tetrafluoride	108.06	364.0	90.9	655.3	195.6							
SF_6	Sulfur hexafluoride	146.05	318.69	45.54	573.64	113.97	37.10	545.2	0.736	45.9	0.198	3.18	0.267
NbF_5	Niobium pentafluoride	187.90	737	464	1327	867	62	911	1.21	75.5	0.155	2.49	0.15
UF_6	Uranium hexafluoride	352.02	505.8	232.6	910.4	450.7	46	676	1.41	88.0	0.250	4.00	0.28
	2. Chlorides												
HCl	Hydrogen chloride	36.46	324.6	51.5	584.4	124.7	82.0	1205	0.45	28	0.081	1.3	0.25
DCl	Deuterium chloride	37.47	323.4	50.3	582.2	122.5							
BCl_3	Boron trichloride	127.17	452.0	178.8	813.5	353.8	38.2	561	0.510	31.8	0.261	4.19	0.13
$AlCl_3$	Aluminum trichloride	133.34	629	356	1132	672	26	382					
$SiHCl_3$	Trichlorosilane	135.45	479	206	862	402							
$SiCl_4$	Silicon tetrachloride	169.90	507	234	913	453	37	544	0.521	32.5	0.326	5.23	0.29
$GeCl_4$	Germanium tetrachloride	214.41	552	279	994	534	38	558	0.65	41	0.33	5.3	0.28
$SnCl_4$	Stannic chloride	260.50	592.0	318.8	1065.5	605.8	37.0	544	0.742	46.3	0.351	5.62	0.267
PCl_3	Phosphorus trichloride	137.33	563	290	1014	554			0.52	32	0.26	4.2	
$AsCl_3$	Arsenic trichloride	181.28	591	318	1064	604			0.720	44.9	0.252	4.03	
$SbCl_3$	Antimony trichloride	228.11	794	521	1429	969			0.84	52	0.27	4.3	
$BiCl_3$	**Bismuth chloride**	315.34	1179	906	2122	1662	118	1734	1.21	75.5	0.261	4.17	0.32
$TiCl_4$	Titanium tetrachloride	189.71	638	365	1149	689	46	676	0.56	35	0.34	5.4	0.30
$ZrCl_4$	Zirconium tetrachloride	233.03	778	505	1401	941	56.9	836	0.730	45.6	0.319	5.11	0.284

TABLE 1h3. (Continued)

Formula	Substance	Mol wt	Critical temperature				Critical pressure		Critical density		Critical volume		Zc
			°K	°C	°R	°F	Atm	lbf/in.²	g/cm³	lbm/ft³	l./g-mol	ft³/lb-mol	
					2. Chlorides (Continued)								
HfCl₄	Hafnium tetrachloride	320.30	723	450	1302	842	57.0	838	1.05	65.6	0.304	4.88	0.292
NbCl₅	Niobium pentachloride	270.17	807	534	1453	993			0.68	42	0.40	6.4	
TaCl₅	Tantalum pentachloride	358.21	767	494	1381	921			0.89	56	0.40	6.4	
MoCl₅	Molybdenum pentachloride	273.21	850	577	1530	1070			0.74	46	0.37	5.9	
WCl₆	Tungsten hexachloride	396.57	923	650	1662	1202			0.94	59	0.42	6.8	
HgCl₂	Mercuric chloride	271.50	973	700	1752	1292			1.56	97.4	0.174	2.79	
					3. Bromides								
HBr	Hydrogen bromide	80.92	363.2	90.0	653.7	194.0	84.4	1240					
DBr	Deuterium bromide	81.93	362.0	88.8	651.5	191.8							
BBr₃	Boron tribromide	250.54	573	300	1032	572							
AlBr₃	Aluminum tribromide	266.71	763	490	1374	914	28.5	419	0.90	56	0.28	4.5	0.141
SiHBr₃	Tribromosilane	268.82	609	336	1096	636							
SiBr₄	Silicon tetrabromide	347.72	663	390	1194	734			0.860	53.7	0.310	4.97	
SbBr₃	Antimony tribromide	361.48	904	631	1627	1167							
BiBr₃	Bismuth bromide	448.71	1219	946	2194	1734			1.49	93.0	0.301	4.82	
ZrBr₄	Zirconium tetrabromide	410.86	805	532	1449	989			0.97	61	0.42	6.8	
HfBr₄	Hafnium tetrabromide	498.13	746	473	1343	883			1.20	74.9	0.415	6.65	
NbBr₅	Niobium pentabromide	492.46	1010	737	1818	1358			1.05	65.6	0.469	7.51	
TaBr₅	Tantalum pentabromide	580.49	974	701	1753	1293			1.26	78.7	0.461	7.38	
HgBr₂	Mercuric bromide	360.41	1012	739	1822	1362							
					4. Iodides								
HI	Hydrogen iodide	127.91	424.0	150.8	763.1	303.4	82	1205					
DI	Deuterium iodide	128.92	421.8	148.6	759.2	299.5							
AlI₃	Aluminum triiodide	407.69	955	682	1719	1259							
SbI₃	Antimony triiodide	502.46	1102	829	1984	1524							
ZrI₄	Zirconium tetraiodide	598.84	960	687	1728	1268			1.13	70.5	0.530	8.49	
HfI₄	Hafnium tetraiodide	686.11	916	643	1649	1189			1.30	81.1	0.528	8.45	
HgI₂	Mercuric iodide	454.90	1072	799	1930	1470							
					5. Mixed								
ClF₅	Chlorine pentafluoride	130.44	416	143	749	289	52	764	0.56	35	0.23	3.7	0.35
SiF₃Cl	Chlorotrifluorosilane	120.53	307.6	34.5	553.8	94.1	34.2	503					
SiF₂Cl₂	Dichlorodifluorosilane	136.99	369.0	95.8	664.1	204.4	34.5	507					
SiFCl₃	Trichlorofluorosilane	153.44	438.6	165.4	789.4	329.7	35.3	519					
NClF₂	Nitrogen chloride difluoride	87.46	337.4	64.3	607.4	147.7	50.8	747					
PF₂Cl	Phosphorus(III) chloride difluoride	104.42	362.4	89.2	652.2	192.5	44.6	655					
PFCl₂	Phosphorus(III) dichloride fluoride	120.88	463.0	189.8	833.3	373.6	49	720					

					G. Oxyhalides							
NOF	Nitryl fluoride	65.00	349.4	76.3	169.3	629.0	56	823	0.52	32	0.19	3.0	0.28
$COCl_2$	Phosgene	98.92	455	182	359	819							
NOCl	Nitrosyl chloride	65.46	440	167	332	792			0.60	37	0.29	4.6	
VOCl	Vanadium oxychloride	173.30	636	363	685	1145							
$WOCl_4$	Tungsten oxide tetrachloride	341.66	782	509	948	1408			1.01	63.0	0.338	5.42	
$ReOCl_4$	Rhenium oxide tetrachloride	344.0	781	508	946	1406			0.95	59	0.36	5.8	
ClO_2F	Perchloryl fluoride	102.45	368.4	95.2	203.3	663.0	53.0	779	0.637	39.8	0.161	2.58	0.282
						H. Miscellaneous							
B_2H_6	Diborane	26.67	289.8	16.6	61.9	521.6	40	588					
PH_3	Phosphine	34.00	324.8	51.6	124.9	584.6	64.5	948					
PH_4Cl	Phosphonium chloride	70.46	322.2	49.1	120.3	580.0	72.7	1068					
	Phosphonitrilic fluoride												
$(PNF_2)_3$	trimer	248.93	461.0	187.8	370.0	829.7							
$(PNF_2)_4$	tetramer	331.91	496.4	223.2	433.7	893.4							
$(PNF_2)_5$	pentamer	414.88	524.0	250.8	483.4	943.1							
AsH_3	Arsine	77.95	373.0	99.9	211.8	671.5							

[a] $°K = °C + 273.15$, $°F = (1.8)°C + 32$, $°R = °F + 459.67$, 1 atm = 14.696 lb/in.^{2}, 1 g/cm^3 = 62.48 lb/ft^3, $R = 82.056$ cm^2 atm/(deg mol).

TABLE 1h4. AZEOTROPIC MIXTURES–BINARY AZEOTROPES.*

Component A: Water, (H$_2$O); Boiling Point of 'A' at 760 torr (°C) 100.

Component B:		B.P. of 'B' at 760 torr (°C)	Azeotropic Mixture	
			B.P. at 760 torr (°C)	% A by Weight
Hydrazine	(H$_4$N$_2$)	113.8	120	32.3
Carbon tetrachloride	(CCl$_4$)	76.75	66	4.1
Carbon disulfide	(CS$_2$)	46.25	42.6	2.8
Chloroform	(CHCl$_3$)	61	56.1	2.8
Dichloromethane	(CH$_2$Cl$_2$)	43.5	38.1	1.5
Formic acid	(CH$_2$O$_2$)	100.75	107.2	22.6
Nitromethane	(CH$_3$NO$_2$)	101.2	83.59	23.6
1,1,2-Trichlorotrifluoroethane	(C$_2$Cl$_3$F$_3$)	47.5	44.5	1.0
Tetrachloroethylene	(C$_2$Cl$_4$)	121	88.5	17.2
Trichloroethylene	(C$_2$HCl$_3$)	86.2	73.4	7.02
1,1,2-Trichloroethane	(C$_2$H$_3$Cl$_3$)	113.8	86.0	16.4
Acetonitrile	(C$_2$H$_3$N)	81.6	86.0	16.5
1,2-Dichloroethane	(C$_2$H$_4$Cl$_2$)	83.5	72.28	9.2
Nitroethane	(C$_2$H$_5$NO$_2$)	114.07	87.22	28.5
Ethyl nitrate	(C$_2$H$_5$NO$_3$)	87.68	74.35	22
Ethanol	(C$_2$H$_6$O)	78.3	78.174	4.0
1,2-Ethylenediamine	(C$_2$H$_8$N$_2$)	116	118.5	20
Acrylonitrile	(C$_3$H$_3$N)	77.2	70.6	14.3
Acrolein	(C$_3$H$_4$O)	52.8	52.4	2.6
Epichlorohydrin	(C$_3$H$_5$ClO)	115.2	88.5	26
Allyl alcohol	(C$_3$H$_6$O)	96.90	88.89	27.7
Propionaldehyde	(C$_3$H$_6$O)	47.9	47.5	2
Propionic acid	(C$_3$H$_6$O$_2$)	141.4	99.1	82.2
Isopropyl alcohol	(C$_3$H$_8$O)	82.3	80.3	12.6
Propyl alcohol	(C$_3$H$_8$O)	97.3	87.76	29.1
2-Methoxyethonal	(C$_3$H$_8$O$_2$)	124.5	99.9	84.7
Methacrylic acid	(C$_4$H$_6$O$_2$)	160.5	99.3	76.9
Butyronitrile	(C$_4$H$_7$N)	117.6	88.7	32.5
Isobutyronitrile	(C$_4$H$_7$N)	103	82.5	23
2-Butanone	(C$_4$H$_8$O)	79.6	73.41	11.3
Butyraldehyde	(C$_4$H$_8$O)	74	68	6
Isobutyraldehyde	(C$_4$H$_8$O)	63.5	64.3	6.7
Butyric acid	(C$_4$H$_8$O$_2$)	162.45	99.4	81.5
Dioxane	(C$_4$H$_8$O$_2$)	101.3	87.8	17.6
Ethyl acetate	(C$_4$H$_8$O$_2$)	77.15	70.38	8.47
Isobutyric acid	(C$_4$H$_8$O$_2$)	154.5	98.8	71.8
Methyl propionate	(C$_4$H$_8$O$_2$)	79.7	71.0	8.2
Butyl alcohol	(C$_4$H$_{10}$O)	117.4	92.7	42.5
sec-Butyl alcohol	(C$_4$H$_{10}$O)	99.5	87.0	26.8
tert-Butyl alcohol	(C$_4$H$_{10}$O)	82.5	79.9	11.76
Ethyl ether	(C$_4$H$_{10}$O)	34.5	34.15	1.26
Isobutyl alcohol	(C$_4$H$_{10}$O)	107.0	89.8	33.0
2-Ethoxyethanol	(C$_4$H$_{10}$O$_2$)	135.1	99.4	71.2
Pyridine	(C$_5$H$_5$N)	115.3	93.6	41.3
2-Methylfuran	(C$_5$H$_6$O)	63.7	58.2	–
Furfuryl alcohol	(C$_5$H$_6$O$_2$)	169.35	98.5	80
Isoprene	(C$_5$H$_8$)	34.1	32.4	0.14
Methyl methacrylate	(C$_5$H$_8$O$_2$)	100.8	83	14
Cyclopentanol	(C$_5$H$_{10}$O)	140.85	96.25	58
2-Pentanone	(C$_5$H$_{10}$O)	102.3	83.3	19.5
3-Pentanone	(C$_5$H$_{10}$O)	102.05	82.9	14
Butyl formate	(C$_5$H$_{10}$O$_2$)	106.6	83.8	16.5

TABLE 1h4. (*Continued*)

Component A: Water, (H_2O); Boiling Point of 'A' at 760 torr (°C) 100.

Component B:		B.P. of 'B' at 760 torr (°C)	Azeotropic Mixture	
			B.P. at 760 torr (°C)	% A by Weight
Ethyl propionate	($C_5H_{10}O_2$)	99.15	81.2	10
Isopropyl acetate	($C_5H_{10}O_2$)	88.6	76.6	10.6
Isovaleric acid	($C_5H_{10}O_2$)	176.5	99.5	81.6
Methyl butyrate	($C_5H_{10}O_2$)	102.65	82.7	11.5
Methyl isobutyrate	($C_5H_{10}O_2$)	92.3	77.7	6.8
Propyl acetate	($C_5H_{10}O_2$)	101.6	82.4	14
Valeric acid	($C_5H_{10}O_2$)	185.5	99.8	89
Ethyl carbonate	($C_5H_{10}O_3$)	126.5	91	30
Piperdine	($C_5H_{11}N$)	105.8	92.8	35
Isoamyl nitrate	($C_5H_{11}NO_3$)	149.75	95.0	40
Petane	(C_5H_{12})	36.1	34.6	1.4
n-Amyl alcohol	($C_5H_{12}O$)	137.8	95.8	54.4
tert-Amyl alcohol	($C_5H_{12}O$)	102.25	87.35	27.5
Ethyl propyl ether	($C_5H_{12}O$)	63.6	59.5	4
Isoamyl alcohol	($C_5H_{12}O$)	132.05	95.15	49.60
3-Methyl-2-butanol	($C_5H_{12}O$)	112.9	91.0	33
2-Pentanol	($C_5H_{12}O$)	119.3	91.7	36.5
3-Pentanol	($C_5H_{12}O$)	115.4	91.7	36.0
Chlorobenzene	(C_6H_5Cl)	131.8	90.2	28.4
Benzene	(C_6H_6)	80.2	69.25	8.83
Aniline	(C_6H_7N)	184.32	41	86.6
Cyclohexanone	($C_6H_{10}O$)	155.4	95	61.6
Mesityl oxide	($C_6H_{10}O$)	129.5	91.8	34.8
Vinyl butyrate	($C_6H_{10}O_2$)	116.7	87.2	20.4
Vinyl isobutyrate	($C_6H_{10}O_2$)	105.4	83.5	17
Diallylamine	($C_6H_{11}N$)	110.5	87.2	24
Cyclohexane	(C_6H_{12})	80.8	69.5	8.4
1-Hexene	(C_6H_{12})	63.84	57.7	5.7
Butyl acetate	($C_6H_{12}O_2$)	126.2	90.2	28.7
Ethyl butyrate	($C_6H_{12}O_2$)	120.1	87.9	21.5
Ethyl isobutyrate	($C_6H_{12}O_2$)	110.1	85.2	15.2
Hexanoic acid	($C_6H_{12}O_2$)	205.7	99.8	92.1
Isobutyl acetate	($C_6H_{12}O_2$)	117.2	87.4	16.5
Paraldehyde	($C_6H_{12}O_3$)	124	90	28.5
1-Chlorohexane	($C_6H_{13}Cl$)	134.5	91.8	29.7
Hexane .	(C_6H_{14})	68.7	61.6	5.6
Butyl ethyl ether	($C_6H_{14}O$)	92.2	76.6	11.9
Hexyl alcohol	($C_6H_{14}O$)	157.1	97.8	67.2
Isopropyl ether	($C_6H_{14}O$)	69	62.2	4.5
Propyl ether	($C_6H_{14}O$)	90.7	75.4	—
Acetal	($C_6H_{14}O_2$)	103.6	82.6	14.5
Diisopropylamine	($C_6H_{15}N$)	83.86	74.1	9.2
Dipropylamine	($C_6H_{15}N$)	109	86.7	—
Toluene	(C_7H_8)	110.7	84.1	13.5
Anisole	(C_7H_8O)	153.85	95.5	40.5
Benzyl alcohol	(C_7H_8O)	205.2	99.9	91
1-Heptene	(C_7H_{14})	93.64	77.0	14.8
Amyl acetate	($C_7H_{14}O_2$)	148.8	95.2	41
sec-Amyl acetate	($C_7H_{14}O_2$)	133.5	92.0	33.2
Butyl propionate	($C_7H_{14}O_2$)	146.8	94.8	41
Heptane	(C_7H_{16})	98.4	79.2	12.9
Styrene	(C_8H_8)	145.1	93.9	40.9
Acetophenone	(C_8H_8O)	202.0	98	81.7

TABLE 1h4. (*Continued*)

Component A: Water, (H_2O); Boiling Point of 'A' at 760 torr (°C) 100.

Component B:		B.P. of 'B' at 760 torr (°C)	Azeotropic Mixture	
			B.P. at 760 torr (°C)	% A by Weight
Phenyl acetate	$(C_8H_8O_2)$	195.7	98.9	75.1
Ethylbenzene	(C_8H_{10})	136.2	92	33.0
m-Xylene	(C_8H_{10})	139.1	94.5	40
Diethyl fumarate	$(C_8H_{12}O_4)$	218.1	99.5	87.5
Ethyl maleate	$(C_8H_{12}O_4)$	223.3	99.65	88.2
Diisobutylene	(C_8H_{14})	102.3	82	12
Diethyl succinate	$(C_8H_{14}O_4)$	216.2	99.9	91
1-Octene	(C_8H_{16})	121.28	88.0	28.7
Octane	(C_8H_{18})	125.7	89.6	25.5
Isooctane	(C_8H_{18})	118	78.8	11.1
Butyl ether	(C_8H_{18})	142.6	92.9	33
Isobutyl ether	$(C_8H_{18}O)$	122.2	88.6	23
Octyl alcohol	$(C_8H_{18}O)$	195.15	99.4	90
Dibutylamine	$(C_8H_{19}N)$	159.6	97	50.5
Benzyl acetate	$(C_9H_{10}O_2)$	214.9	99.60	87.5
Ethyl benzoate	$(C_9H_{10}O_2)$	212.4	99.40	84.0
Cumene	(C_9H_{12})	152.4	95	43.8
Naphthalene	$(C_{10}H_8)$	218	98.8	84
Phenyl ether	$(C_{12}H_{10}O)$	259.3	99.33	96.75
Tributylamine	$(C_{12}H_{27}N)$	213.9	99.65	79.7
Tributyl phosphate	$(C_{12}H_{27}O_4P)$	–	100	99.4

Component A: Carbon Tetrachloride, (CCl_4); Boiling Point of 'A' at 760 torr, (°C) 76.75.

Methanol	(CH_4O)	64.7	55.7	79.44
Acetic acid	$(C_2H_4O_2)$	118.1	76	98.46
Ethanol	(C_2H_6O)	78.3	65.04	84.2
Acetone	(C_3H_6O)	56.15	56.08	11.5
Isopropyl alcohol	(C_3H_8O)	82.5	68.65	81.9
Propyl alcohol	(C_3H_8O)	97.25	73.4	92.1
Ethyl acetate	$(C_4H_8O_2)$	76.7	74.8	57
Butyl alcohol	$(C_4H_{10}O)$	117.75	76.55	97.5
sec-Butyl alcohol	$(C_4H_{10}O)$	99.5	74.6	92.4
tert-Butyl alcohol	$(C_4H_{10}O)$	82.55	71.1	83

Component A: Carbon disulfide, (CS_2); Boiling Point of 'A' at 760 torr (°C) 46.25.

Methanol	(CH_4O)	64.7	39.8	71
Ethanol	(C_2H_6O)	78.3	42.6	91
Acetone	(C_3H_6O)	56.15	39.25	67
Isopropyl alcohol	(C_3H_8O)	82.45	44.22	92.4
Propyl alcohol	(C_3H_8O)	97.1	45.65	94.5
Ethyl acetate	$(C_4H_8O_2)$	77.1	46.02	92.7
Ethyl ether	$(C_4H_{10}O)$	34.6	34.4	1
Pentane	(C_5H_{12})	36.15	35.7	11

Component A: Chloroform, $(CHCl_3)$; Boiling Point of 'A' at 760 torr (°C) 61.2.

Methanol	(CH_4O)	64.7	53.43	87.4
Ethanol	(C_2H_6O)	78.3	59.35	93
Acetone	(C_3H_6O)	56.10	64.43	78.5
Hexane	(C_6H_{14})	68.7	60.4	83.5

TABLE 1h4. (*Continued*)

Component A: Methanol, $(CH_4 O)$; Boiling Point of 'A' at 760 torr (°C) 64.7.

Component B:		B.P. of 'B' at 760 torr (°C)	Azeotropic Mixture	
			B.P. at 760 torr (°C)	% A by Weight
Tetrachloroethylene	$(C_2 Cl_4)$	121.1	63.75	63.5
Trichloroethylene	$(C_2 HCl_3)$	87	59.3	38
Bromoethane	$(C_2 H_5 Br)$	38	34.9	5.3
Acetone	$(C_3 H_6 O)$	56.15	55.5	12
Methyl acetate	$(C_3 H_6 O_2)$	57.1	53.5	19
Ethyl acetate	$(C_4 H_8 O_2)$	77.1	62.25	44
Pentane	$(C_5 H_{12})$	36	30.6	7.6
Benzene	$(C_6 H_6)$	80.1	57.50	39.1
Cyclohexane	$(C_6 H_{12})$	80	54	38
Toluene	$(C_7 H_8)$	110.7	63.8	69

Component A: Ethanol, $(C_2 H_6 O)$; Boiling Point of 'A' at 760 torr (°C) 78.3.

Component B:		B.P. of 'B' at 760 torr (°C)	B.P. at 760 torr (°C)	% A by Weight
Benzene	$(C_6 H_6)$	80.1	68.24	32.4
Cyclohexane	$(C_6 H_{12})$	80.7	64.8	29.2
1-Hexene	$(C_6 H_{12})$	63.6	54.8	17.6
n-Hexane	$(C_6 H_{14})$	68.95	58.68	21.0
Toluene	$(C_7 H_8)$	110.7	76.7	68

Component A: Acetone, $(C_3 H_6 O)$; Boiling Point of 'A' at 760 torr, (°C) 56.15.

Component B:		B.P. of 'B' at 760 torr (°C)	B.P. at 760 torr (°C)	% A by Weight
Pentane	$(C_5 H_{12})$	36.15	31.9	21
Cyclohexane	$(C_6 H_{12})$	80.75	53.0	67
Hexane	$(C_6 H_{14})$	68.95	49.8	59

Component A: Isopropyl Alcohol, $(C_3 H_8 O)$; Boiling Point of 'A' at 760 torr (°C) 82.45.

Component B:		B.P. of 'B' at 760 torr (°C)	B.P. at 760 torr (°C)	% A by Weight
Benzene	$(C_6 H_6)$	80.2	71.92	33.3
Cyclohexane	$(C_6 H_{12})$	80.7	68.80	33.0
Hexane	$(C_6 H_{14})$	68.85	62.7	23
Isopropyl ether	$(C_6 H_{14} O)$	69.0	66.2	16.3
Propyl ether	$(C_6 H_{14} O)$	90.55	78.2	52
Toluene	$(C_7 H_8)$	110.7	80.6	69

Component A: Propyl Alcohol, $(C_3 H_8 O)$; Boiling Point of 'A' at 760 torr (°C) 97.2.

Component B:		B.P. of 'B' at 760 torr (°C)	B.P. at 760 torr (°C)	% A by Weight
Dioxane	$(C_4 H_8 O_2)$	101.35	95.3	55
Benzene	$(C_6 H_6)$	80.2	77.12	16.9
Cyclohexane	$(C_6 H_{12})$	80.75	74.3	20
Hexane	$(C_6 H_{14})$	68.95	65.65	4
Propyl ether	$(C_6 H_{14} O)$	90.7	85.8	32.2
Acetal	$(C_6 H_{14} O_2)$	103.55	92.4	37
Toluene	$(C_7 H_8)$	110.7	92.6	51.5
m-Xylene	$(C_8 H_{10})$	139.2	97.08	94
p-Xylene	$(C_8 H_{10})$	138.4	96.88	92.2
Octane	$(C_8 H_{18})$	125.6	93.9	70

Component A: Benzene, $(C_6 H_6)$; Boiling Point of 'A' at 760 torr (°C) 80.15.

Component B:		B.P. of 'B' at 760 torr (°C)	B.P. at 760 torr (°C)	% A by Weight
Cyclohexane	$(C_6 H_{12})$	80.6	77.7	51.8
Heptane	$(C_7 H_{16})$	98.4	80.1	99.3

*From: Horsley, L. H., "Azeotropic Data-III," American Chemical Society, Washington, D.C., 1973.

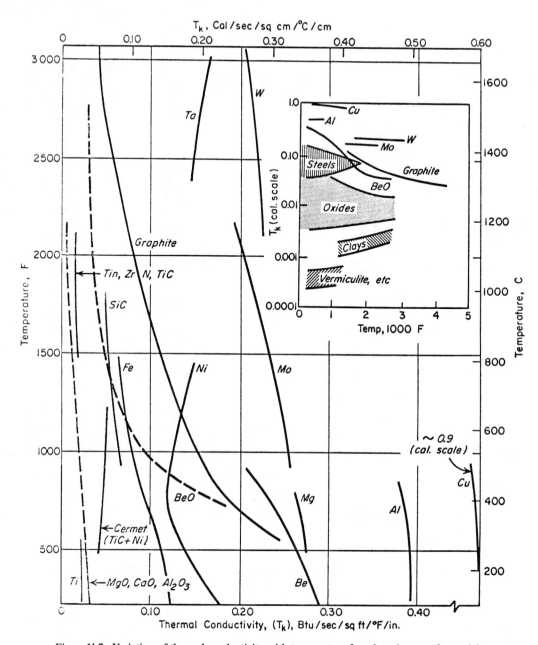

Figure 1h7. Variation of thermal conductivity with temperature for a broad range of materials.

Courtesy, W. S. Pellini, Superintendant, Metallurgy Div., U.S. Naval Research Laboratory. Reproduced from: *Materials in Design Engineering*, **52**, 177 (September 1960).

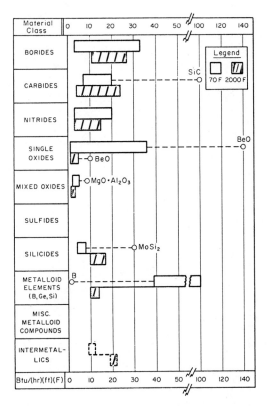

Figure 1h8. Thermal conductivity of ceramic materials. (BMI 1966).

TABLE 1h5. CONVERSION FACTORS FOR THERMAL CONDUCTIVITY.[a]*

		Watt / cm K	Watt / m K	Watt in. / in.² °R	Cal cm / cm² sec K	Kcal m / n² hr K	Cal in. / in.² sec °R	Btu in. / in.² sec °R	Btu in. / in.² hr °R	Btu ft. / ft² hr °R	Btu in. / ft² hr °R
Watt / cm K	=	1.000	100.0	1.411	0.2390	86.04	0.3373	1.338×10^{-3}	4.818	57.82	693.8
Watt / m K	=	1.000×10^{-2}	1.000	1.411×10^{-2}	2.390×10^{-3}	0.8604	3.373×10^{-3}	1.338×10^{-5}	4.818×10^{-2}	0.5782	6.938
Watt in. / in.² °R	=	0.7087	70.87	1.000	0.1694	60.97	0.2390	9.485×10^{-4}	3.414	40.97	491.7
Cal cm / cm² sec K	=	4.184	418.4	5.904	1.000	360.0	1.411	5.600×10^{-3}	20.16	241.9	2.903
Kcal m / m² hr K	=	1.162×10^{-2}	1.162	1.640×10^{-2}	2.778×10^{-3}	1.000	3.920×10^{-3}	1.555×10^{-5}	5.600×10^{-2}	0.6720	8.064
Cal in. / in.² sec °R	=	2.965	296.5	4.184	0.7087	255.1	1.000	3.968×10^{-3}	14.29	171.4	2.057
Btu in. / in.² sec °R	=	747.2	7.472×10^{4}	1054	178.6	6.429×10^{4}	252.0	1.000	3600	4.320×10^{4}	5.184×10^{5}
Btu in. / in.² hr °R	=	0.2075	20.75	0.2929	4.961×10^{-2}	67.86	7.000×10^{-2}	2.778×10^{-4}	1.000	12.00	144.0
Btu ft. / ft² hr °R	=	1.730×10^{-2}	1.730	2.441×10^{-2}	4.134×10^{-3}	0.488	5.833×10^{-3}	2.315×10^{-5}	8.333×10^{-2}	1.000	12.00
Btu in. / ft² hr °R	=	1.441×10^{-3}	0.1441	2.034×10^{-3}	3.445×10^{-4}	0.1240	4.861×10^{-4}	1.929×10^{-6}	6.944×10^{-3}	8.333×10^{-2}	1.000

[a] Units are given in terms of (1) the absolute joule per second or watt, (2) the defined thermochemical calorie = 4.184 joules, or (3) the defined British thermal unit (Btu) where 1.8 Btu/lb = 1 cal/g and therefore 1 Btu = 1,054.35 joules.

*Reproduced from Childs, G. E., L. J. Ericks and R. L. Powell, "Thermal Conductivity of Solids at Room Temperature and Below—A Review and Compilation of the Literature," NBS Monograph 131, U.S. Government Printing Office, Washington, D.C., 1973.

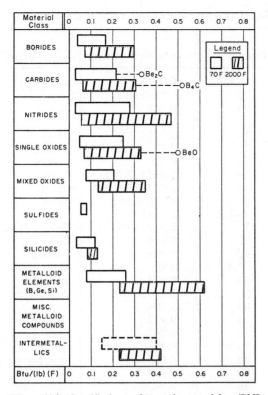

Figure 1h9. Specific heat of ceramic materials. (BMI 1966).

TABLE 1h6. TOTAL INFRARED REFLECTANCE OF METALS AND ALLOYS.*

Metal/Alloy	Temp. (°K)	Total Reflectance	Material[a]	Source Temp. (°K)
Aluminum	2.0–4.2	0.9889	foil	~300
	76	0.93–0.982	sheet, foil, film	300
	90	0.945	foil	293
Brass, yellow				
(65% Cu–35% Zn)	76	0.971	sheet	300
Cadmium	76	0.97	plating	300
Chromium	76	0.92	plating	300
Copper	2.0–4.2	0.9853, 0.9938	sheet	~300
	76	0.912–0.985	sheet	300
	90	0.965–0.981	sheet	293
Gold	2.0–4.2	0.9885	sheet	~300
	76	0.972–0.990	foil, plating, film	300
	90	0.974	foil	293
Iron	90	0.903	plating	(2–18 μ)
Monel	77.4	0.89	sheet	273
Nickel	20	0.973	plating	300
	76	0.967, 0.978	foil, plating	300
Platinum	85	0.904	–	290
Rhodium	76	0.922	plating	300
Silver	20	0.987	plating	300
	75.8	0.9916–0.9924	plating	268–367
	90	0.964, 0.977	foil, plating	293
Steel	90	0.904	foil	293
-type 302	76	0.93, 0.952	sheet	300
Tin	76	0.9876	sheet	~300
Tungsten	85	0.901	sheet	300
Zinc	76	0.98	foil	300

*Dickson, P. F. and M. C. Jones, "Infrared Reflectances of Metals at Cryogenic Temperatures—a Compilation from the Literature," NBS Tech. Note No. 348, U.S. Government Printing Office, Washington, D.C., 1966.
[a] "Sheet" denotes thick mass. Reflectances depend significantly on surface finish.

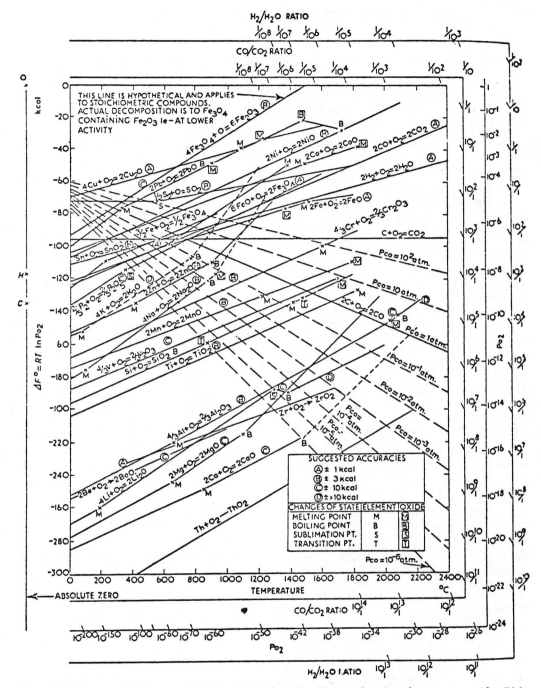

Figure 1h10. The standard free energy of formation of metal oxides as a function of temperature. After Richardson and Jeffes, 1948.*

*Reprinted from Kohl, W. H., "Handbook of Materials and Techniques for Vacuum Devices," Van Nostrand Reinhold, New York, 1967.

TABLE 1h7. HIGH-TEMPERATURE HEAT-CONTENT, HEAT-CAPACITY, AND ENTROPY DATA*

In this section, there are assembled tables of high-temperature heat contents and entropies, together with equations representing high-temperature heat contents and heat capacities. Both the tabular data and the equations are in terms of calorie-gram mole-degree Kelvin units. In the case of a substance for which the data are too limited or too inaccurate to warrant preparation of a table, only a heat-capacity equation (sometimes estimated), an average heat-capacity value for a short temperature range, or a single true heat-capacity value is given.

The tabular results all have the base temperature 298.15° K. (25° C.). The heat unit is the defined calorie (490), 1 cal.=4.1840 abs. joules or 4.1833 int. joules. Molecular weights were calculated from the Report on Atomic Weights for 1956–57 (771). Values for the molal heat capacities at 298.15° K., used in deriving heat-content equations, are from a compilation of the author (342), except in some instances in which more recent measurements are available.

The average deviation of each heat-content equation from the corresponding tabular values has been computed and is shown in parentheses, together with the temperature range of validity of the equation. Thus, the symbol (0.1 percent; 298°–2,000° K.) means that the equation deviates by an average of 0.1 percent (or less) from the tabulated values in the temperature range 298° to 2,000° K.

Heat-of-transition and heat-of-fusion values are, with few exceptions, differences in heat contents obtained by smoothing measured data above and below the specified transition and fusion temperatures. The errors in these differences sometimes are large in an absolute sense because of impurities that produce pretransition and premelting phenomena and because of other factors that produce lack of definition of the heat content curves near the points of change.

High-temperature heat content, as used in this section, always means the total heat evolved in bringing the substance from another temperature to 298.15° K. This quantity of heat, of course, depends upon the state of the substance at 298.15° K. To avoid any possible confusion, this state of reference is indicated specifically in the table heading. Results for gases always are for the hypothetical ideal state at 1 atmosphere pressure. The symbols, c, l, g, and gl, refer, respectively, to crystals, liquid, gas, and glass. The Greek letters α, β, γ, and δ are used generally to denote different crystalline states; in a few instances these symbols denote substances above and below a region of abnormal heat capacity which does not involve a change in macroscopic structure.

The references cited in this section constitute a substantially complete bibliography of high

temperature heat-content and heat-capacity data for the elements and inorganic compounds. A few trivial, old references have been omitted intentionally as they no longer are significant in the light of more recent and extensive data. Measurements given greatest weight in compilation of the tables are indicated by italicizing the names of the appropriate investigators.

ACTINIUM

ELEMENT

TABLE 1.—*Heat content and entropy of Ac(c, l)*

[Base, crystals at 298.15° K; atomic wt., 227]

T, ° K.	$H_T-H_{298.15}$ cal./mole	$S_T-S_{298.15}$ cal./deg. mole	T, ° K.	$H_T-H_{298.15}$ cal./mole	$S_T-S_{298.15}$ cal./deg. mole
400	670	1.93	1,470(l)	12,410	14.08
500	1,350	3.45	1,500	12,650	14.24
600	2,050	4.72	1,600	13,450	14.76
700	2,770	5.83	1,700	14,250	15.25
800	3,510	6.82	1,800	15,050	15.70
900	4,270	7.72	1,900	15,850	16.14
1,000	5,050	8.54	2,000	16,650	16.55
1,100	5,850	9.30	2,200	18,250	17.31
1,200	6,670	10.01	2,400	19,850	18.01
1,300	7,510	10.69	2,600	21,450	18.64
1,400	8,370	11.32	2,800	23,050	19.24
1,470 (c)	8,990	11.76	3,000	24,650	19.79

Ac(c):

$$H_T - H_{298.15} = 5.91T + 0.99 \times 10^{-3}T^2 - 1,850 \quad (0.1 \text{ percent}; 298°-1,470° \text{ K.});$$

$$C_p = 5.91 + 1.98 \times 10^{-3}T;$$

$$\Delta H_{1470} \ (fusion) = 3,420.$$

Ac(l):

$$H_T - H_{298.15} = 8.00T + 650 \quad (0.1 \text{ percent}; 1,470°-3,000° \text{ K.});$$

$$C_p = 8.00.$$

ALUMINUM AND ITS COMPOUNDS

ELEMENT

TABLE 2.—*Heat content and entropy of Al(c, l)*

[Base, crystals at 298.15° K.; atomic wt., 26.98]

T, ° K.	$H_T-H_{298.15}$ cal./mole	$S_T-S_{298.15}$ cal./deg. mole	T, ° K.	$H_T-H_{298.15}$ cal./mole	$S_T-S_{298.15}$ cal./deg. mole
400	600	1.72	1,500	10,830	13.59
500	1,230	3.14	1,600	11,530	14.04
600	1,890	4.34	1,700	12,230	14.46
700	2,580	5.40	1,800	12,930	14.86
800	3,310	6.38	1,900	13,630	15.24
900	4,060	7.26	2,000	14,330	15.60
932 (c)	4,280	7.50	2,100	15,030	15.94
932 (l)	6,850	10.26	2,200	15,730	16.27
1,000	7,330	10.75	2,300	16,430	16.58
1,100	8,030	11.42	2,400	17,130	16.88
1,200	8,730	12.03	2,500	17,830	17.16
1,300	9,430	12.59	2,600	18,530	17.44
1,400	10,130	13.11	2,700	19,230	17.70

*Reproduced from Kelley, K. K., "Contributions to the Data on Theoretical Metallurgy: XII. High-Temperature Heat-Content, Heat-Capacity, and Entropy Data for the Elements and Inorganic Compounds," Bureau of Mines Bulletin 584, U.S. Government Printing Office, Washington, 1960.

TABLE 1h7. (*Continued*)

Al(*c*):

$$H_T - H_{298.15} = 4.94T + 1.48 \times 10^{-3}T^2 - 1,604 \quad (0.6 \text{ percent;}$$
$$298°\text{-}932° \text{ K.});$$
$$C_p = 4.94 + 2.96 \times 10^{-3}T;$$
$$\Delta H_{932} \text{ (fusion)} = 2,570.$$

Al(*l*):

$$H_T - H_{298.15} = 7.00 + 330 \quad (0.1 \text{ percent; } 932°\text{-}2,700° \text{ K.});$$
$$C_p = 7.00.$$

TABLE 3.—*Heat content and entropy of Al(g)*

[Base, ideal gas at 298.15° K.; atomic wt., 26.98]

T, ° K.	$H_T - H_{298.15}$, cal./mole	$S_T - S_{298.15}$, cal./deg. mole	T, ° K.	$H_T - H_{298.15}$, cal./mole	$S_T - S_{298.15}$, cal./deg. mole
400	515	1.49	1,900	7,995	9.27
500	1,020	2.62	2,000	8,490	9.53
600	1,520	3.53	2,200	9,485	10.00
700	2,020	4.30	2,400	10,480	10.43
800	2,520	4.97	2,600	11,470	10.83
900	3,020	5.55	2,800	12,470	11.20
1,000	3,515	6.08	3,000	13,460	11.54
1,100	4,015	6.55	3,500	15,950	12.31
1,200	4,515	6.99	4,000	18,440	12.98
1,300	5,010	7.38	4,500	20,950	13.57
1,400	5,510	7.75	5,000	23,490	14.10
1,500	6,005	8.10	6,000	28,790	15.07
1,600	6,505	8.42	7,000	34,640	15.97
1,700	7,000	8.72	8,000	41,440	16.87
1,800	7,500	9.00			

Al(*g*):

$$H_T - H_{298.15} = 4.97T - 0.12 \times 10^5 T^{-1} - 1,442 \quad (0.1 \text{ percent;}$$
$$298°\text{-}5,000° \text{ K.});$$
$$C_p = 4.97 + 0.12 \times 10^5 T^{-2}.$$

OXIDES

TABLE 4.—*Heat content and entropy of AlO(g)*

[Base, ideal gas at 298.15° K.; mol. wt., 42.98]

T, ° K.	$H_T - H_{298.15}$, cal./mole	$S_T - S_{298.15}$, cal./deg. mole	T, ° K.	$H_T - H_{298.15}$, cal./mole	$S_T - S_{298.15}$, cal./deg. mole
400	770	2.22	1,000	5,775	9.80
500	1,560	3.98	1,200	7,510	11.38
600	2,375	5.46	1,400	9,260	12.73
700	3,205	6.74	1,600	11,020	13.90
800	4,050	7.87	1,800	12,790	14.94
900	4,910	8.88	2,000	14,565	15.88

AlO(*g*):

$$H_T - H_{298.15} = 8.22T + 0.22 \times 10^{-3}T^2 + 0.87 \times 10^5 T^{-1}$$
$$- 2,762 \quad (0.5 \text{ percent; } 298°\text{-}2,000° \text{ K.});$$
$$C_p = 8.22 + 0.44 \times 10^{-3}T - 0.87 \times 10^5 T^{-2}.$$

TABLE 5.—*Heat content and entropy of $Al_2O_3(c)$*

[Base, corundum at 298.15° K.; mol. wt., 101.96]

T, ° K.	$H_T - H_{298.15}$, cal./mole	$S_T - S_{298.15}$, cal./deg. mole	T, ° K.	$H_T - H_{298.15}$, cal./mole	$S_T - S_{298.15}$, cal./deg. mole
400	2,150	6.17	1,300	27,730	38.87
500	4,580	11.59	1,400	30,800	41.14
600	7,200	16.36	1,500	33,890	43.27
700	9,960	20.61	1,600	36,990	45.27
800	12,810	24.42	1,700	40,100	47.16
900	15,720	27.84	1,800	43,220	48.94
1,000	18,670	30.95	1,900	46,350	50.63
1,100	21,660	33.80	2,000	49,490	52.24
1,200	24,680	36.43			

$Al_2O_3(c)$:

$$H_T - H_{298.15} = 27.49T + 1.41 \times 10^{-3}T^2 + 8.38 \times 10^5 T^{-1}$$
$$- 11,132 \quad (0.5 \text{ percent; } 298°\text{-}1,800° \text{ K.});$$
$$C_p = 27.49 + 2.82 \times 10^{-3}T - 8.38 \times 10^5 T^{-2}.$$

CRYSTALLINE HYDRATED OXIDES

TABLE 6.—*Heat content and entropy of $Al_2O_3 \cdot H_2O(c)$*

[Base, crystals at 298.15° K.; mol. wt., 119.98]

T, ° K.	$H_T - H_{298.15}$, cal./mole	$S_T - S_{298.15}$, cal./deg. mole	T, ° K.	$H_T - H_{298.15}$, cal./mole	$S_T - S_{298.15}$, cal./deg. mole
325	770	2.47	425	3,970	11.03
350	1,515	4.69	450	4,860	13.06
375	2,295	6.83	475	5,810	15.07
400	3,115	8.96	500	6,850	17.25

$Al_2O_3 \cdot H_2O(c)$:

$$H_T - H_{298.15} = 28.87T + 4.20 \times 10^{-3}T^2 - 8,981$$
$$(4.7 \text{ percent; } 298°\text{-}500° \text{ K.});$$
$$C_p = 28.87 + 8.40 \times 10^{-3}T.$$

TABLE 7.—*Heat content and entropy of $Al_2O_3 \cdot 3H_2O(c)$*

[Base, gibbsite at 298.15° K.; mol. wt., 156.01]

T, ° K.	$H_T - H_{298.15}$, cal./mole	$S_T - S_{298.15}$, cal./deg. mole	T, ° K.	$H_T - H_{298.15}$, cal./mole	$S_T - S_{298.15}$, cal./deg. mole
325	1,180	3.79	400	5,010	14.37
350	2,390	7.38	425	6,390	17.72
375	3,680	10.93			

$Al_2O_3 \cdot 3H_2O(c)$:

$$H_T - H_{298.15} = 17.30T + 45.6 \times 10^{-3}T^2 - 9,212$$
$$(1.3 \text{ percent; } 298°\text{-}425° \text{ K.});$$
$$C_p = 17.30 + 91.2 \times 10^{-3}T.$$

CARBIDE

TABLE 8.—*Heat content and entropy of $Al_4C_3(c)$*

[Base, crystals at 298.15° K.; mol. wt., 143.95]

T, ° K.	$H_T - H_{298.15}$, cal./mole	$S_T - S_{298.15}$, cal./deg. mole	T, ° K.	$H_T - H_{298.15}$, cal./mole	$S_T - S_{298.15}$, cal./deg. mole
400	3,570	10.28	600	11,530	26.35
500	7,430	18.88			

$Al_4C_3(c)$:

$$H_T - H_{298.15} = 24.08T + 15.8 \times 10^{-3}T^2 - 8,584$$
$$(0.3 \text{ percent; } 298°\text{-}600° \text{ K.});$$
$$C_p = 24.08 + 31.6 \times 10^{-3}T.$$

NITRIDE

TABLE 9.—*Heat content and entropy of AlN(c)*

[Base, crystals at 298.15° K.; mol. wt., 40.99]

T, ° K.	$H_T - H_{298.15}$, cal./mole	$S_T - S_{298.15}$, cal./deg. mole	T, ° K.	$H_T - H_{298.15}$, cal./mole	$S_T - S_{298.15}$, cal./deg. mole
400	830	2.39	700	3,780	7.84
500	1,740	4.44	800	4,870	9.30
600	2,740	6.24	900	6,020	10.66

TABLE 1h7. (*Continued*)

AlN(*c*):

$$H_T - H_{298.15} = 5.47T + 3.90 \times 10^{-3}T^2 - 1,987 \ (0.9 \text{ percent,}$$
$$298°-900° \text{ K.});$$
$$C_p = 5.47 + 7.80 \times 10^{-3}T.$$

HYDRIDES

TABLE 10.—*Heat content and entropy of AlH(g)*

[Base, ideal gas at 298.15° K.; mol. wt., 27.99]

T, ° K.	$H_T-H_{298.15},$ cal./mole	$S_T-S_{298.15},$ cal./deg. mole	T, ° K.	$H_T-H_{298.15},$ cal./mole	$S_T-S_{298.15},$ cal./deg. mole
400	720	2.08	1,000	5,375	9.11
500	1,445	3.69	1,200	7,045	10.63
600	2,195	5.06	1,400	8,740	11.93
700	2,965	6.24	1,600	10,460	13.08
800	3,755	7.30	1,800	12,190	14.10
900	4,555	8.24	2,000	13,930	15.02

AlH(*g*):

$$H_T - H_{298.15} = 7.04T + 0.54 \times 10^{-3}T^2 + 0.33 \times 10^5 T^{-1}$$
$$- 2,258 \ (0.6 \text{ percent}; \ 298°-2,000° \text{ K.});$$
$$C_p = 7.04 + 1.08 \times 10^{-3}T - 0.33 \times 10^5 T^{-2}.$$

TABLE 11.—*Heat content and entropy of AlD(g)*

[Base, ideal gas at 298.15° K.; mol. wt., 29.00]

T, ° K.	$H_T-H_{298.15},$ cal./mole	$S_T-S_{298.15},$ cal./deg. mole	T, ° K.	$H_T-H_{298.15},$ cal./mole	$S_T-S_{298.15},$ cal./deg. mole
400	745	2.14	1,000	5,625	9.52
500	1,510	3.84	1,200	7,345	11.08
600	2,300	5.28	1,400	9,080	12.42
700	3,110	6.53	1,600	10,825	13.58
800	3,940	7.64	1,800	12,580	14.62
900	4,775	8.62	2,000	14,345	15.55

AlD(*g*):

$$H_T - H_{298.15} = 7.80T + 0.34 \times 10^{-3}T^2 + 0.74 \times 10^5 T^{-1}$$
$$- 2,604 \ (0.6 \text{ percent}; \ 298°-2,000° \text{ K.});$$
$$C_p = 7.80 + 0.68 \times 10^{-3}T - 0.74 \times 10^5 T^{-2}.$$

BROMIDES

TABLE 12.—*Heat content and entropy of AlBr(g)*

[Base, ideal gas at 298.15° K.; mol. wt., 106.90]

T, ° K.	$H_T-H_{298.15},$ cal./mole	$S_T-S_{298.15},$ cal./deg. mole	T, ° K.	$H_T-H_{298.15},$ cal./mole	$S_T-S_{298.15},$ cal./deg. mole
400	875	2.52	1,000	6,170	10.59
500	1,745	4.46	1,200	7,950	12.22
600	2,625	6.07	1,400	9,730	13.59
700	3,505	7.43	1,600	11,515	14.78
800	4,390	8.61	1,800	13,300	15.83
900	5,280	9.66	2,000	15,085	16.77

AlBr(*g*):

$$H_T - H_{298.15} = 8.88T + 0.02 \times 10^{-3}T^2 + 0.36 \times 10^5 T^{-1}$$
$$- 2,770 \ (0.1 \text{ percent}; \ 298°-2,000° \text{ K.});$$
$$C_p = 8.88 + 0.04 \times 10^{-3}T - 0.36 \times 10^5 T^{-2}.$$

TABLE 13.—*Heat content and entropy of AlBr₃(c, l)*

[Base, crystals at 298.15° K.; mol. wt., 266.73]

T, ° K.	$H_T-H_{298.15},$ cal./mole	$S_T-S_{298.15},$ cal./deg. mole	T, ° K.	$H_T-H_{298.15},$ cal./mole	$S_T-S_{298.15},$ cal./deg. mole
370.6(*c*)	1,810	5.43	400	5,390	15.00
370.6(*l*)	4,520	12.74	500	8,340	21.58

AlBr₃(*c*):

$$H_T - H_{298.15} = 18.74T + 9.33 \times 10^{-3}T^2 - 6,417 \ (0.1 \text{ percent};$$
$$298°-370.6° \text{ K.});$$
$$C_p = 18.74 + 18.66 \times 10^{-3}T;$$
$$\Delta H_{370.6}(\textit{fusion}) = 2,710.$$

AlBr₃(*l*):

$$H_T - H_{298.15} = 29.50T - 6,410 \ (0.1 \text{ percent};$$
$$370.6°-500° \text{ K.});$$
$$C_p = 29.50.$$

CHLORIDES

TABLE 14.—*Heat content and entropy of AlCl(g)*

[Base, ideal gas at 298.15° K.; mol. wt., 62.44]

T, ° K.	$H_T-H_{298.15},$ cal./mole	$S_T-S_{298.15},$ cal./deg. mole	T, ° K.	$H_T-H_{298.15},$ cal./mole	$S_T-S_{298.15},$ cal./deg. mole
400	855	2.47	1,000	6,110	10.47
500	1,715	4.38	1,200	7,890	12.09
600	2,585	5.97	1,400	9,670	13.46
700	3,460	7.32	1,600	11,450	14.65
800	4,340	8.49	1,800	13,230	15.70
900	5,225	9.54	2,000	15,015	16.64

AlCl(*g*):

$$H_T - H_{298.15} = 8.82T + 0.04 \times 10^{-3}T^2 + 0.52 \times 10^5 T^{-1}$$
$$- 2,808 \ (0.1 \text{ percent}; \ 298°-2,000° \text{ K.});$$
$$C_p = 8.82 + 0.08 \times 10^{-3}T - 0.52 \times 10^5 T^{-2}.$$

TABLE 15.—*Heat content and entropy of AlCl₃ (c, l)*

[Base, crystals at 298.15° K.; mol. wt., 133.35]

T, ° K.	$H_T-H_{298.15},$ cal./mole	$S_T-S_{298.15},$ cal./deg. mole	T, ° K.	$H_T-H_{298.15},$ cal./mole	$S_T-S_{298.15},$ cal./deg. mole
400	2,340	6.73	465.6(*l*)	12,510	28.85
450	3,610	9.72	500	13,580	31.08
465.6(*c*)	4,010	10.59			

AlCl₃(*c*):

$$H_T - H_{298.15} = 13.25T + 14.00 \times 10^{-3}T^2$$
$$- 5,195 \ (0.1 \text{ percent}; \ 298°-465.6° \text{ K.});$$
$$C_p = 13.25 + 28.00 \times 10^{-3}T;$$
$$\Delta H_{465.6} \ (\textit{fusion}) = 8,500.$$

AlCl₃(*l*):

$$H_T - H_{298.15} = 31.20T$$
$$- 2,018 \ (0.1 \text{ percent}; \ 465.6°-500° \text{ K.});$$
$$C_p = 31.20.$$

TABLE 1h7. *(Continued)*

HYDRATED CHLORIDE

$AlCl_3 \cdot 6H_2O(c)$:

$\overline{C}_p = 76.4$ (288°–327° K.).

FLUORIDES

TABLE 16.—*Heat content and entropy of AlF(g)*

[Base, ideal gas at 298.15° K.; mol. wt., 45.98]

T, ° K.	$H_T - H_{298.15}$, cal./mole	$S_T - S_{298.15}$, cal./deg. mole	T, ° K.	$H_T - H_{298.15}$, cal./mole	$S_T - S_{298.15}$, cal./deg. mole
400	795	2.29	1,000	5,890	10.02
500	1,610	4.11	1,200	7,645	11.62
600	2,445	5.63	1,400	9,410	12.98
700	3,295	6.94	1,600	11,180	14.16
800	4,155	8.09	1,800	12,950	15.21
900	5,020	9.10	2,000	14,730	16.14

$AlF(g)$:

$H_T - H_{298.15} = 8.53T + 0.12 \times 10^{-3}T^2 + 0.88 \times 10^5 T^{-1}$
$-2,849$ (0.3 percent; 298°–2,000° K.);
$C_p = 8.53 + 0.24 \times 10^{-3}T - 0.88 \times 10^5 T^{-2}$.

TABLE 17.—*Heat content and entropy of AlF$_3$(c)*

[Base, α-crystals at 298.15° K.; mol. wt., 83.98]

T, ° K.	$H_T - H_{298.15}$, cal./mole	$S_T - S_{298.15}$, cal./deg. mole	T, ° K.	$H_T - H_{298.15}$, cal./mole	$S_T - S_{298.15}$, cal./deg. mole
400	1,950	5.61	900	13,550	24.20
500	4,070	10.33	1,000	15,930	26.71
600	6,310	14.41	1,100	18,340	29.01
700	8,680	18.06	1,200	20,780	31.13
727(α)	9,360	19.02	1,300	23,250	33.11
727(β)	9,510	19.22	1,400	25,740	34.95
800	11,200	21.43			

$AlF_3(\alpha)$:

$H_T - H_{298.15} = 17.27T + 5.48 \times 10^{-3}T^2 + 2.30 \times 10^5 T^{-1}$
$-6,408$ (0.2 percent; 298°–727° K.);
$C_p = 17.27 + 10.96 \times 10^{-3}T - 2.30 \times 10^5 T^{-2}$;
ΔH_{727} *(transition)* $= 150$.

$AlF_3(\beta)$:

$H_T - H_{298.15} = 20.93T + 1.50 \times 10^{-3}T^2$
$-6,500$ (0.1 percent; 727°–1,400° K.);
$C_p = 20.93 + 3.00 \times 10^{-3}T$.

HYDRATED FLUORIDE

$AlF_3 \cdot 3.5H_2O(c)$:

$\overline{C}_p = 50.3$ (288°–326° K.).

IODIDES

TABLE 18.—*Heat content and entropy of AlI(g)*

[Base, ideal gas at 298.15° K.; mol. wt., 153.89]

T, ° K.	$H_T - H_{298.15}$, cal./mole	$S_T - S_{298.15}$, cal./deg. mole	T, ° K.	$H_T - H_{298.15}$, cal./mole	$S_T - S_{298.15}$, cal./deg. mole
400	885	2.55	1,000	6,200	10.66
500	1,760	4.50	1,200	7,980	12.28
600	2,645	6.12	1,400	9,765	13.66
700	3,530	7.48	1,600	11,550	14.85
800	4,420	8.67	1,800	13,335	15.90
900	5,310	9.72	2,000	15,125	16.84

$AlI(g)$:

$H_T - H_{298.15} = 8.94T + 0.32 \times 10^5 T^{-1}$
$-2,773$ (0.1 percent; 298°–2,000° K.);
$C_p = 8.94 - 0.32 \times 10^5 T^{-2}$.

TABLE 19.—*Heat content and entropy of AlI$_3$(c, l)*

[Base, crystals at 298.15° K.; mol. wt., 407.71]

T, ° K.	$H_T - H_{298.15}$, cal./mole	$S_T - S_{298.15}$, cal./deg. mole	T, ° K.	$H_T - H_{298.15}$, cal./mole	$S_T - S_{298.15}$, cal./deg. mole
400	2,525	7.27	464(l)	8,210	19.80
464(c)	4,230	11.22	500	9,250	21.97

$AlI_3(c)$:

$H_T - H_{298.15} = 16.88T + 11.33 \times 10^{-3}T^2$
$-6,040$ (0.1 percent; 298°–464° K.);
$C_p = 16.88 + 22.66 \times 10^{-3}T$;
ΔH_{464}(*fusion*) $= 3,980$.

$AlI_3(l)$:

$H_T - H_{298.15} = 29.00T - 5,246$ (0.1 percent; 464°–500° K.);
$C_p = 29.00$.

HYDRATED NITRATE

$Al(NO_3)_3 \cdot 6H_2O(c)$:

$C_p = 103.5$ (298° K.).

SILICATES

TABLE 20.—*Heat content and entropy of Al$_2$SiO$_5$*
(sillimanite)

[Base, crystals at 298.15° K.; mol. wt., 162.05]

T, ° K.	$H_T - H_{298.15}$, cal./mole	$S_T - S_{298.15}$, cal./deg. mole	T, ° K.	$H_T - H_{298.15}$, cal./mole	$S_T - S_{298.15}$, cal./deg. mole
400	3,300	9.49	1,100	33,400	52.06
500	6,940	17.60	1,200	37,900	55.98
600	10,900	24.82	1,300	42,500	59.66
700	15,300	31.60	1,400	47,000	62.99
800	19,900	37.74	1,500	51,600	66.17
900	24,400	43.03	1,600	56,300	69.20
1,000	28,900	47.77			

Al_2SiO_5 *(sillimanite)*:

$H_T - H_{298.15} = 40.09T + 2.93 \times 10^{-3}T^2 + 10.13 \times 10^5 T^{-1}$
$-15,611$ (1.5 percent; 298°–1,600° K.);
$C_p = 40.09 + 5.86 \times 10^{-3}T - 10.13 \times 10^5 T^{-2}$.

TABLE 21.—*Heat content and entropy of Al$_2$SiO$_5$*
(andalusite)

[Base, crystals at 298.15° K.; mol. wt., 162.05]

T, ° K.	$H_T - H_{298.15}$, cal./mole	$S_T - S_{298.15}$, cal./deg. mole	T, ° K.	$H_T - H_{298.15}$, cal./mole	$S_T - S_{298.15}$, cal./deg. mole
400	3,720	10.70	1,100	34,000	53.92
500	7,620	19.39	1,200	38,500	57.84
600	11,800	27.01	1,300	43,000	61.43
700	16,200	33.78	1,400	47,600	64.85
800	20,700	39.79	1,500	52,200	68.03
900	25,200	45.08	1,600	56,800	70.99
1,000	29,600	49.72			

TABLE 1h7. (*Continued*)

Al$_2$SiO$_5$ (*andalusite*):

$$H_T - H_{298.15} = 46.24T + 12.53 \times 10^5 T^{-1}$$
$$-17,989 \ (0.4 \ percent; \ 298°-1,600°\ K.);$$
$$C_v = 46.24 - 12.53 \times 10^5 T^{-2}.$$

TABLE 22.—*Heat content and entropy of Al$_2$SiO$_5$*
(*kyanite*)

[Base, crystals at 298.15° K.; mol. wt., 162.05]

T, ° K.	$H_T-H_{298.15}$, cal./mole	$S_T-S_{298.15}$, cal./deg. mole	T, ° K.	$H_T-H_{298.15}$, cal./mole	$S_T-S_{298.15}$, cal./deg. mole
400	3,600	10.33	1,100	33,600	52.82
500	7,400	18.80	1,200	38,300	56.91
600	11,300	25.90	1,300	43,000	60.67
700	15,600	32.52	1,400	47,800	64.23
800	20,000	38.40	1,500	52,700	67.61
900	24,500	43.69	1,600	57,700	70.83
1,000	29,000	48.44	1,700	62,800	73.92

Al$_2$SiO$_5$ (*kyanite*):

$$H_T - H_{298.15} = 45.32T + 1.17 \times 10^{-3}T^2 + 16.00$$
$$\times 10^5 T^{-1} - 18,983 \ (1.0 \ percent; \ 298°-1,700° \ K.);$$
$$C_p = 45.32 + 2.34 \times 10^{-3}T - 16.00 \times 10^5 T^{-2}.$$

TABLE 23.—*Heat content and entropy of Al$_6$Si$_2$O$_{13}$*
(*mullite*)

[Base, crystals at 298.15° K.; mol. wt., 426.06]

T, ° K.	$H_T-H_{298.15}$, cal./mole	$S_T-S_{298.15}$, cal./deg. mole	T, ° K.	$H_T-H_{298.15}$, cal./mole	$S_T-S_{298.15}$, cal./deg. mole
400	8,430	24.28	600	27,040	61.85
500	17,450	44.38			

Al$_6$Si$_2$O$_{13}$ (*mullite*):

$$H_T - H_{298.15} = 59.65T + 33.5 \times 10^{-3}T^2$$
$$-20,763 \ (0.3 \ percent; \ 298°-600° \ K.);$$
$$C_p = 59.65 + 67.0 \times 10^{-3}T.$$

SULFATE

TABLE 24.—*Heat content and entropy of*
Al$_2$(SO$_4$)$_3$(c)

[Base, crystals at 298.15° K.; mol. wt., 342.16]

T, ° K.	$H_T-H_{298.15}$, cal./mole	$S_T-S_{298.15}$, cal./deg. mole	T, ° K.	$H_T-H_{298.15}$, cal./mole	$S_T-S_{298.15}$, cal./deg. mole
400	7,180	20.63	800	42,600	81.21
500	15,260	38.63	900	52,240	92.55
600	23,860	54.30	1,000	61,980	102.82
700	33,100	68.52	1,100	71,800	112.18

Al$_2$(SO$_4$)$_3$(c):

$$H_T - H_{298.15} = 87.55T + 7.48 \times 10^{-3}T^2 + 26.88 \times 10^5 T^{-1}$$
$$-35,716 \ (0.3 \ percent; \ 298°-1,100° \ K.);$$
$$C_p = 87.55 + 14.96 \times 10^{-3}T - 26.88 \times 10^5 T^{-2}.$$

HYDRATED SULFATES

Al$_2$(SO$_4$)$_3$·6H$_2$O(c):

$$C_p = 117.8 \ (298° \ K.).$$

Al$_2$(SO$_4$)$_3$·18H$_2$O(c):

$$\overline{C_p} = 235 \ (288-325° \ K.).$$

TITANATE

TABLE 25.—*Heat content and entropy of*
Al$_2$TiO$_5$(c)

[Base, crystals at 298.15° K.; mol. wt., 181.86]

T, ° K.	$H_T-H_{298.15}$, cal./mole	$S_T-S_{298.15}$, cal./deg. mole	T, ° K.	$H_T-H_{298.15}$, cal./mole	$S_T-S_{298.15}$, cal./deg. mole
400	3,600	10.35	1,200	40,180	59.59
500	7,620	19.30	1,300	45,150	63.57
600	11,930	27.16	1,400	50,180	67.30
700	16,420	34.07	1,500	55,260	70.80
800	21,020	40.22	1,600	60,370	74.10
900	25,700	45.73	1,700	65,490	77.21
1,000	30,450	50.73	1,800	70,620	80.14
1,100	35,280	55.33			

Al$_2$Ti$_2$O$_5$(c):

$$H_T - H_{298.15} = 43.63T + 2.65 \times 10^{-3}T^2 + 11.21 \times 10^5 T^{-1}$$
$$-17,004 \ (0.4 \ percent; \ 298°-1,800°K.);$$
$$C_p = 43.63T + 5.30 \times 10^{-3}T - 11.21 \times 10^5 T^{-2}.$$

ANTIMONY AND ITS COMPOUNDS

ELEMENT

TABLE 26.—*Heat content and entropy of Sb(c, l)*

[Base, crystals at 298.15° K.; atomic wt., 121.76]

T, ° K.	$H_T-H_{298.15}$, cal./mole	$S_T-S_{298.15}$, cal./deg. mole	T, ° K.	$H_T-H_{298.15}$, cal./mole	$S_T-S_{298.15}$, cal./deg. mole
400	625	1.80	1,100	10,190	13.89
500	1,250	3.19	1,200	10,940	14.54
600	1,890	4.36	1,300	11,690	15.14
700	2,550	5.38	1,400	12,440	15.70
800	3,240	6.30	1,500	13,190	16.22
900	3,950	7.14	1,600	13,940	16.70
903(c)	3,970	7.16	1,700	14,690	17.15
903(l)	8,710	12.41	1,800	15,440	17.58
1,000	9,440	13.18	1,900	16,190	17.99

Sb(c):

$$H_T - H_{298.15} = 5.51T + 0.87 \times 10^{-3}T^2 - 1,720$$
$$(0.4 \ percent; \ 298°-903° \ K.);$$
$$C_p = 5.51 + 1.74 \times 10^{-3}T;$$
$$\Delta H_{903}(fusion) = 4,740.$$

Sb(l):

$$H_T - H_{298.15} = 7.50T + 1,940 \ (0.1 \ percent;$$
$$903°-1,900° \ K.);$$
$$C_p = 7.50.$$

TABLE 1h7. (Continued)

TABLE 27.—*Heat content and entropy of Sb(g)*

[Base, ideal gas at 298.15° K.; atomic wt., 121.76]

T, ° K.	$H_T - H_{298.15}$, cal./mole	$S_T - S_{298.15}$, cal./deg. mole	T, ° K.	$H_T - H_{298.15}$, cal./mole	$S_T - S_{298.15}$, cal./deg. mole
400	505	1.46	1,900	8,025	9.24
500	1,005	2.57	2,000	8,545	9.51
600	1,500	3.48	2,200	9,610	10.02
700	1,995	4.24	2,400	10,705	10.49
800	2,495	4.90	2,600	11,835	10.95
900	2,990	5.49	2,800	13,005	11.38
1,000	3,490	6.01	3,000	14,205	11.79
1,100	3,985	6.49	3,500	17,355	12.76
1,200	4,485	6.92	4,000	20,675	13.65
1,300	4,980	7.32	4,500	24,120	14.46
1,400	5,480	7.69	5,000	27,635	15.20
1,500	5,985	8.04	6,000	34,715	16.49
1,600	6,485	8.36	7,000	41,740	17.58
1,700	6,995	8.67	8,000	48,700	18.50
1,800	7,505	8.96			

$$\text{Sb}(g):$$
$$H_T - H_{298.15} = 4.97T - 1,482 \ (0.3 \text{ percent};$$
$$298°\text{–}2,000° \text{ K.});$$
$$C_p = 4.97.$$

TABLE 28.—*Heat content and entropy of Sb$_2$(g)*

[Base, ideal gas at 298.15° K.; mol. wt. 243.52]

T, ° K.	$H_T - H_{298.15}$, cal./mole	$S_T - S_{298.15}$, cal./deg. mole	T, ° K.	$H_T - H_{298.15}$, cal./mole	$S_T - S_{298.15}$, cal./deg. mole
400	890	2.57	1,600	11,575	14.89
500	1,775	4.54	1,800	13,360	15.94
600	2,660	6.15	2,000	15,150	16.89
700	3,545	7.52	2,200	16,935	17.74
800	4,440	8.70	2,400	18,720	18.51
900	5,330	9.75	2,600	20,510	19.23
1,000	6,220	10.70	2,800	22,300	19.89
1,200	8,005	12.33	3,000	24,085	20.51
1,400	9,790	13.70			

$$\text{Sb}_2(g):$$
$$H_T - H_{298.15} = 8.94T + 0.22 \times 10^5 T^{-1} - 2,739 \ (0.1 \text{ percent};$$
$$298°\text{–}3,000° \text{ K.});$$
$$C_p = 8.94 - 0.22 \times 10^5 T^{-2}.$$

TABLE 29.—*Heat content and entropy of Sb$_4$(g)*

[Base, ideal gas at 298.15° K.; mol. wt., 487.04]

T, ° K.	$H_T - H_{298.15}$, cal./mole	$S_T - S_{298.15}$, cal./deg. mole	T, ° K.	$H_T - H_{298.15}$, cal./mole	$S_T - S_{298.15}$, cal./deg. mole
400	1,980	5.71	1,300	19,770	28.98
500	3,940	10.09	1,400	21,760	30.46
600	5,910	13.68	1,500	23,745	31.83
700	7,890	16.73	1,600	25,730	33.11
800	9,860	19.36	1,700	27,715	34.31
900	11,840	21.69	1,800	29,700	35.45
1,000	13,830	23.79	1,900	31,685	36.52
1,100	15,810	25.68	2,000	33,670	37.54
1,200	17,790	27.40			

$$\text{Sb}_4(g):$$
$$H_T - H_{298.15} = 19.85T + 0.44 \times 10^5 T^{-1} - 6,066 \ (0.1 \text{ percent};$$
$$298°\text{–}2,000° \text{ K.});$$
$$C_p = 19.85 - 0.44 \times 10^5 T^{-2}.$$

OXIDES

TABLE 30.—*Heat content and entropy of SbO(g)*

[Base, ideal gas at 298.15° K.; mol. wt., 137.76]

T, ° K.	$H_T - H_{298.15}$, cal./mole	$S_T - S_{298.15}$, cal./deg. mole	T, ° K.	$H_T - H_{298.15}$, cal./mole	$S_T - S_{298.15}$, cal./deg. mole
400	795	2.29	1,000	6,125	10.33
500	1,615	4.12	1,200	8,040	12.07
600	2,470	5.68	1,400	9,975	13.56
700	3,350	7.04	1,600	11,915	14.86
800	4,255	8.24	1,800	13,850	16.00
900	5,180	9.33	2,000	15,780	17.02

$$\text{SbO}(g):$$
$$H_T - H_{298.15} = 8.47T + 0.42 \times 10^{-3}T^2 + 0.99 \times 10^5 T^{-1}$$
$$- 2,895 \ (0.9 \text{ percent}; 298°\text{–}2,000° \text{ K.});$$
$$C_p = 8.47 + 0.84 \times 10^{-3}T - 0.99 \times 10^5 T^{-2}.$$

$$\text{Sb}_2\text{O}_3(c):$$
$$C_p = 19.1 + 17.1 \times 10^{-3}T \ (estimated) \ (298°\text{–}929° \text{ K.}).$$

$$\text{Sb}_2\text{O}_4(c):$$
$$C_p = 22.6 + 16.2 \times 10^{-3}T \ (estimated) \ (298°\text{–}1,198° \text{ K.}).$$

$$\text{Sb}_2\text{O}_5(c):$$
$$C_p = 28.11 \ (298° \text{ K.}).$$

SULFIDE

$$\text{Sb}_2\text{S}_3(c):$$
$$C_p = 24.2 + 13.2 \times 10^{-3}T \ (estimated) \ (298°\text{–}821° \text{ K.}).$$

NITRIDE

TABLE 31.—*Heat content and entropy of SbN(g)*

[Base, ideal gas at 298.15° K.; mol. wt., 135.77]

T, ° K.	$H_T - H_{298.15}$, cal./mole	$S_T - S_{298.15}$, cal./deg. mole	T, ° K.	$H_T - H_{298.15}$, cal./mole	$S_T - S_{298.15}$, cal./deg. mole
400	775	2.23	1,000	5,795	9.84
500	1,570	4.01	1,200	7,540	11.43
600	2,390	5.50	1,400	9,295	12.78
700	3,225	6.79	1,600	10,055	13.96
800	4,075	7.92	1,800	12,825	15.00
900	4,930	8.93	2,000	14,595	15.93

$$\text{SbN}(g):$$
$$H_T - H_{298.15} = 8.26T + 0.21 \times 10^{-3}T^2 + 0.86 \times 10^5 T^{-1}$$
$$- 2,770 \ (0.4 \text{ percent}; 298°\text{–}2,000° \text{ K.});$$
$$C_p = 8.26 + 0.42 \times 10^{-3}T - 0.86 \times 10^5 T^{-2}.$$

HYDRIDES

TABLE 32.—*Heat content and entropy of SbH$_3$(g)*

[Base, ideal gas at 298.15° K.; mol. wt., 124.78]

T, ° K.	$H_T - H_{298.15}$, cal./mole	$S_T - S_{298.15}$, cal./deg. mole	T, ° K.	$H_T - H_{298.15}$, cal./mole	$S_T - S_{298.15}$, cal./deg. mole
400	1,070	3.07	1,000	9,585	15.71
500	2,250	5.70	1,200	12,965	18.79
600	3,545	8.06	1,400	16,465	21.48
700	4,935	10.20	1,600	20,070	23.89
800	6,415	12.18	1,800	23,755	26.06
900	7,970	14.01	2,000	27,485	28.02

TABLE 1h7. *(Continued)*

SbH$_3$(g):

$H_T - H_{298.15} = 11.52T + 2.29 \times 10^{-3}T^2 + 2.74 \times 10^5 T^{-1}$
$-4,557$ (1.4 percent; 298°–2,000° K.);
$C_p = 11.52 + 4.58 \times 10^{-3}T - 2.74 \times 10^5 T^{-2}$.

TABLE 33.—*Heat content and entropy of* SbD$_3$(g)

[Base, ideal gas at 298.15° K.; mol. wt., 127.80]

T, ° K.	$H_T-H_{298.15}$, cal./mole	$S_T-S_{298.15}$, cal./deg. mole	T, ° K.	$H_T-H_{298.15}$, cal./mole	$S_T-S_{298.15}$, cal./deg. mole
400	1,250	3.59	1,000	10,875	17.96
500	2,630	6.67	1,200	14,515	21.28
600	4,125	9.39	1,400	18,220	24.13
700	5,720	11.85	1,600	21,985	26.64
800	7,385	14.07	1,800	25,800	28.89
900	9,110	16.10	2,000	29,660	30.92

SbD$_3$(g):

$H_T - H_{298.15} = 14.72T + 1.50 \times 10^{-3}T^2 + 3.72 \times 10^5 T^{-1}$
$-5,770$ (1.4 percent; 298°–2,000° K.);
$C_p = 14.72 + 3.00 \times 10^{-3}T - 3.72 \times 10^5 T^{-2}$.

BROMIDE

SbBr$_3$(c):

$C_p = 17.2 + 29.3 \times 10^{-3}T$ *(estimated)* (298°–370° K.)

CHLORIDES

TABLE 34.—*Heat content and entropy of* SbCl(g)

[Base, ideal gas at 298.15° K.; mol. wt., 157.22]

T, ° K.	$H_T-H_{298.15}$, cal./mole	$S_T-S_{298.15}$, cal./deg. mole	T, ° K.	$H_T-H_{298.15}$, cal./mole	$S_T-S_{298.15}$, cal./deg. mole
400	875	2.52	1,000	6,175	10.60
500	1,750	4.47	1,200	7,955	12.22
600	2,625	6.07	1,400	9,740	13.60
700	3,510	7.43	1,600	11,525	14.79
800	4,395	8.61	1,800	13,310	15.84
900	5,285	9.66	2,000	15,095	16.78

SbCl(g):

$H_T - H_{298.15} = 8.93T + 0.39 \times 10^5 T^{-1} - 2,793$ (0.1 percent; 298°–2,000° K.);
$C_p = 8.93 - 0.39 \times 10^5 T^{-2}$.

SbCl$_3$(c):

$C_p = 10.3 + 51.1 \times 10^{-3}T$ *(estimated)* (298°–346° K.).

TABLE 35.—*Heat content and entropy of* SbCl$_3$(g)

[Base, ideal gas at 298.15° K.; mol. wt., 228.13]

T, ° K.	$H_T-H_{298.15}$, cal./mole	$S_T-S_{298.15}$, cal./deg. mole	T, ° K.	$H_T-H_{298.15}$, cal./mole	$S_T-S_{298.15}$, cal./deg. mole
400	1,915	5.52	800	9,700	18.99
500	3,840	9.82	900	11,670	21.31
600	5,780	13.36	1,000	13,640	23.3?
700	7,735	16.37			

SbCl$_3$(g):

$H_T - H_{298.15} = 19.83T + 1.20 \times 10^5 T^{-1} - 6,315$
(0.1 percent; 298°–1,000° K.);
$C_p = 19.83 - 1.20 \times 10^5 T^{-2}$.

FLUORIDE

TABLE 36.—*Heat content and entropy of* SbF(g)

[Base, ideal gas at 298.15° K.; mol. wt., 140.76]

T, ° K.	$H_T-H_{298.15}$, cal./mole	$S_T-S_{298.15}$, cal./deg. mole	T, ° K.	$H_T-H_{298.15}$, cal./mole	$S_T-S_{298.15}$, cal./deg. mole
400	830	2.39	1,000	6,025	10.29
500	1,670	4.26	1,200	7,790	11.90
600	2,525	5.82	1,400	9,565	13.27
700	3,395	7.16	1,600	11,345	14.46
800	4,265	8.33	1,800	13,125	15.50
900	5,145	9.36	2,000	14,905	16.44

SbF(g):

$H_T - H_{298.15} = 8.75T + 0.06 \times 10^{-3}T^2 + 0.73 \times 10^5 T^{-1}$
$-2,859$ (0.2 percent; 298°–2,000° K.);
$C_p = 8.75 + 0.12 \times 10^{-3}T - 0.73 \times 10^5 T^{-2}$.

ANTIMONY-BISMUTH

TABLE 37.—*Heat content and entropy of* SbBi(g)

[Base, ideal gas at 298.15° K.; mol. wt., 330.76]

T, ° K.	$H_T-H_{298.15}$, cal./mole	$S_T-S_{298.15}$, cal./deg. mole	T, ° K.	$H_T-H_{298.15}$, cal./mole	$S_T-S_{298.15}$, cal./deg. mole
400	895	2.58	1,000	6,240	10.74
500	1,785	4.57	1,200	8,025	12.37
600	2,675	6.19	1,400	9,810	13.75
700	3,565	7.56	1,600	11,600	14.94
800	4,455	8.75	1,800	13,385	15.99
900	5,345	9.80	2,000	15,170	16.93

SbBi(g):

$H_T - H_{298.15} = 8.94T + 0.14 \times 10^5 T^{-1} - 2,713$
(0.1 percent; 298°–2,000° K.);
$C_p = 8.94 - 0.14 \times 10^5 T^{-2}$.

ANTIMONY-PALLADIUM

TABLE 38.—*Heat content and entropy of* SbPd(c)

[Base, crystals at 298.15° K.; mol. wt., 228.16]

T, ° K.	$H_T-H_{298.15}$, cal./mole	$S_T-S_{298.15}$, cal./deg. mole	T, ° K.	$H_T-H_{298.15}$, cal./mole	$S_T-S_{298.15}$, cal./deg. mole
400	1,230	3.55	700	5,110	10.75
500	2,480	6.34	800	6,480	12.58
600	3,770	8.69	900	7,880	14.23

SbPd(c):

$H_T - H_{298.15} = 10.70T + 2.00 \times 10^{-3}T^2 - 3,368$
(0.2 percent; 298°–900° K.);
$C_p = 10.70 + 4.00 \times 10^{-3}T$.

TABLE 39.—*Heat content and entropy of* Sb$_2$Pd(c)

[Base, crystals at 298.15° K., mol. wt., 349.92]

T, ° K.	$H_T-H_{298.15}$, cal./mole	$S_T-S_{298.15}$, cal./deg. mole	T, ° K.	$H_T-H_{298.15}$, cal./mole	$S_T-S_{298.15}$, cal./deg. mole
400	1,800	5.19	700	7,590	15.90
500	3,620	9.24	800	9,730	18.76
600	5,550	12.76	900	11,920	21.34

TABLE 1h7. (*Continued*)

Sb$_2$Pd(c):

$$H_T - H_{298.15} = 14.05T + 4.87 \times 10^{-3}T^2 - 4,622$$
(0.4 percent; 298°–900° K.);
$$C_p = 14.05 + 9.74 \times 10^{-3}T.$$

TABLE 40.—*Heat content and entropy of* $SbPd_3(c)$

[Base, crystals at 298.15° K.; mol. wt., 440.96]

T, ° K.	$H_T - H_{298.15}$, cal./mole	$S_T - S_{298.15}$, cal./deg. mole	T, ° K.	$H_T - H_{298.15}$, cal./mole	$S_T - S_{298.15}$, cal./deg. mole
400	2,430	7.01	1,000	18,250	30.88
500	4,900	12.52	1,100	21,130	33.63
600	7,430	17.13	1,200	24,070	36.19
700	10,030	21.13	1,223(α)	24,760	36.76
800	12,700	24.70	1,223(β)	27,220	38.77
900	15,440	27.92	1,300	29,700	40.74

SbPd$_3$(α):

$$H_T - H_{298.15} = 21.54T + 3.44 \times 10^{-3}T^2 - 6,728$$
(0.1 percent; 298°–1,223° K.);
$$C_p = 21.54 + 6.88 \times 10^{-3}T;$$
$$\Delta H_{1,223}(transition) = 2,460.$$

SbPd$_3$(β):

$$H_T - H_{298.15} = 32.20T$$
$$-12,160 \ (0.1 \text{ percent}; 1,223°–1,300° \text{ K.});$$
$$C_p = 32.20.$$

ANTIMONY-PLATINUM

TABLE 41.—*Heat content and entropy of* $Sb_2Pt(c)$

[Base, crystals at 298.15° K.; mol. wt., 438.61]

T, ° K.	$H_T - H_{298.15}$, cal./mole	$S_T - S_{298.15}$, cal./deg. mole	T, ° K.	$H_T - H_{298.15}$, cal./mole	$S_T - S_{298.15}$, cal./deg. mole
400	1,700	4.90	700	7,170	15.08
500	3,490	8.90	800	9,060	17.60
600	5,320	12.23	900	10,980	19.86

Sb$_2$Pt(c):

$$H_T - H_{298.15} = 15.27T + 2.53 \times 10^{-3}T^2$$
$$-4,778 \ (0.6 \text{ percent}; 298°–900° \text{ K.});$$
$$C_p = 15.27 + 5.06 \times 10^{-3}T.$$

ARGON

ELEMENT

TABLE 42.—*Heat content and entropy of* $A(g)$

[Base, ideal gas at 298.15° K.; atomic wt., 39.94]

T, ° K.	$H_T - H_{298.15}$, cal./mole	$S_T - S_{298.15}$, cal./deg. mole	T, ° K.	$H_T - H_{298.15}$, cal./mole	$S_T - S_{298.15}$, cal./deg. mole
400	506	1.46	1,900	7,958	9.20
500	1,003	2.57	2,000	8,455	9.46
600	1,500	3.48	2,200	9,448	9.93
700	1,996	4.24	2,400	10,442	10.36
800	2,493	4.90	2,600	11,436	10.76
900	2,990	5.49	2,800	12,429	11.13
1,000	3,487	6.01	3,000	13,423	11.47
1,100	3,984	6.48	3,500	15,907	12.24
1,200	4,480	6.92	4,000	18,391	12.90
1,300	4,977	7.32	4,500	20,875	13.48
1,400	5,474	7.68	5,000	23,359	14.01
1,500	5,971	8.03	6,000	28,327	14.91
1,600	6,468	8.35	7,000	33,295	15.68
1,700	6,964	8.65	8,000	38,263	16.34
1,800	7,461	8.93			

A(g):

$$H_T - H_{298.15} = 4.968T$$
$$-1,481.2 \ (0.1 \text{ percent}; 298°–8,000° \text{ K.});$$
$$C_p = 4.968.$$

ARSENIC AND ITS COMPOUNDS

ELEMENT

TABLE 43.—*Heat content and entropy of* $As(c)$

[Base, crystals at 298.15° K.; atomic wt., 74.91]

T, ° K.	$H_T - H_{298.15}$, cal./mole	$S_T - S_{298.15}$, cal./deg. mole	T, ° K.	$H_T - H_{298.15}$, cal./mole	$S_T - S_{298.15}$, cal./deg. mole
400	610	1.76	800	3,230	6.28
500	1,240	3.16	900	3,940	7.10
600	1,880	4.33	1,000	4,660	7.90
700	2,540	5.35	1,100	5,470	8.64

As(c):

$$H_T - H_{298.15} = 5.23T + 1.11 \times 10^{-3}T^2$$
$$-1,658 \ (0.3 \text{ percent}; 298°–1,100° \text{ K.});$$
$$C_p = 5.23 + 2.22 \times 10^{-3}T.$$

TABLE 44.—*Heat content and entropy of* $As(g)$

[Base, ideal gas at 298.15° K.; atomic wt., 74.91]

T, ° K.	$H_T - H_{298.15}$, cal./mole	$S_T - S_{298.15}$, cal./deg. mole	T, ° K.	$H_T - H_{298.15}$, cal./mole	$S_T - S_{298.15}$, cal./deg. mole
400	505	1.46	1,900	7,985	9.22
500	1,005	2.57	2,000	8,490	9.48
600	1,500	3.48	2,200	9,520	9.97
700	1,995	4.24	2,400	10,565	10.42
800	2,495	4.90	2,600	11,640	10.85
900	2,990	5.49	2,800	12,740	11.26
1,000	3,490	6.01	3,000	13,870	11.65
1,100	3,985	6.49	3,500	16,840	12.56
1,200	4,480	6.92	4,000	20,005	13.41
1,300	4,980	7.32	4,500	23,340	14.20
1,400	5,475	7.68	5,000	26,805	14.92
1,500	5,975	8.03	6,000	33,935	16.22
1,600	6,475	8.35	7,000	41,120	17.33
1,700	6,975	8.66	8,000	48,240	18.28
1,800	7,480	8.94			

As(g):

$$H_T - H_{298.15} = 4.92T + 0.03 \times 10^{-3}T^2 - 0.02 \times 10^5 T^{-1}$$
$$-1,463 \ (0.2 \text{ percent}; 298°–2,400° \text{ K.});$$
$$C_p = 4.92 + 0.06 \times 10^{-3}T + 0.02 \times 10^5 T^{-2}.$$

TABLE 45.—*Heat content and entropy of* $As_2(g)$

[Base, ideal gas at 298.15° K; mol. wt., 149.82]

T, ° K.	$H_T - H_{298.15}$, cal./mole	$S_T - S_{298.15}$, cal./deg. mole	T, ° K.	$H_T - H_{298.15}$, cal./mole	$S_T - S_{298.15}$, cal./deg. mole
400	865	2.50	1,500	10,590	14.14
500	1,730	4.43	1,600	11,485	14.72
600	2,605	6.02	1,700	12,375	15.26
700	3,485	7.38	1,800	13,270	15.77
800	4,370	8.56	1,900	14,160	16.25
900	5,255	9.60	2,000	15,050	16.71
1,000	6,140	10.53	2,200	16,830	17.56
1,100	7,030	11.38	2,400	18,620	18.34
1,200	7,920	12.16	2,600	20,410	19.05
1,300	8,810	12.87	2,800	22,195	19.71
1,400	9,700	13.53	3,000	23,985	20.33

As$_2$(g):

$$H_T - H_{298.15} = 8.93T + 0.52 \times 10^5 T^{-1}$$
$$-2,837 \ (0.1 \text{ percent}; 298°–3,000° \text{ K.});$$
$$C_p = 8.93 - 0.52 \times 10^5 T^{-2}.$$

TABLE 1h7. (Continued)

TABLE 46.—Heat content and entropy of $As_4(g)$

[Base, ideal gas at 298.15° K; mol. wt., 299.64]

T, ° K.	$H_T - H_{298.15}$, cal./mole	$S_T - S_{298.15}$, cal./deg. mole	T, ° K.	$H_T - H_{298.15}$, cal./mole	$S_T - S_{298.15}$, cal./deg. mole
400	1,920	5.54	1,300	19,560	28.57
500	3,840	9.82	1,400	21,540	30.04
600	5,790	13.38	1,500	23,520	31.40
700	7,740	16.38	1,600	25,500	32.68
800	9,700	19.00	1,700	27,490	33.89
900	11,670	21.32	1,800	29,470	35.02
1,000	13,640	23.39	1,900	31,450	36.09
1,100	15,610	25.27	2,000	33,430	37.11
1,200	17,590	26.99			

$As_4(g)$:

$$H_T - H_{298.15} = 19.84T + 1.20 \times 10^5 T^{-1}$$
$$-6,318 \ (0.1 \text{ percent } 298°-2,000° \text{ K.});$$
$$C_p = 19.84 - 1.20 \times 10^5 T^{-2}.$$

OXIDES

TABLE 47.—Heat content and entropy of $AsO(g)$

[Base, ideal gas at 298.15° K; mol. wt., 90.91]

T, ° K.	$H_T - H_{298.15}$, cal./mole	$S_T - S_{298.15}$, cal./deg. mole	T, ° K.	$H_T - H_{298.15}$, cal./mole	$S_T - S_{298.15}$, cal./deg. mole
400	825	2.38	1,000	6,305	10.68
500	1,690	4.31	1,200	8,160	12.38
600	2,585	5.94	1,400	10,010	13.80
700	3,510	7.36	1,600	11,845	15.03
800	4,440	8.60	1,800	13,675	16.10
900	5,370	9.70	2,000	15,500	17.06

$AsO(g)$:

$$H_T - H_{298.15} = 9.74T - 0.15 \times 10^{-3}T^2 + 1.72 \times 10^5 T^{-1}$$
$$-3,468 \ (0.4 \text{ percent; } 298°-2,000° \text{ K.});$$
$$C_p = 9.74 - 0.30 \times 10^{-3}T - 1.72 \times 10^5 T^{-2}.$$

$As_2O_3(c)$:

$$C_p = 8.37 + 48.6 \times 10^{-3}T \ (estimated) \ (298°-548° \text{ K.}).$$

$As_2O_5(c)$:

$$C_p = 27.85 \ (298° \text{ K.}).$$

SULFIDE

$As_2S_3(c)$:

$$\overline{C_p} = 25.8 \ (293°-373° \text{ K.}).$$

NITRIDE

TABLE 48.—Heat content and entropy of $AsN(g)$

[Base, ideal gas at 298.15° K; mol. wt., 88.92]

T, ° K.	$H_T - H_{298.15}$, cal./mole	$S_T - S_{298.15}$, cal./deg. mole	T, ° K.	$H_T - H_{298.15}$, cal./mole	$S_T - S_{298.15}$, cal./deg. mole
400	760	2.19	1,000	5,710	9.68
500	1,535	3.92	1,200	7,440	11.26
600	2,340	5.39	1,400	9,185	12.60
700	3,165	6.66	1,600	10,940	13.77
800	4,005	7.78	1,800	12,700	14.81
900	4,850	8.77	2,000	14,470	15.74

$AsN(g)$:

$$H_T - H_{298.15} = 8.04T + 0.27 \times 10^{-3}T^2 + 0.83 \times 10^5 T^{-1}$$
$$-2,700 \ (0.5 \text{ percent; } 298°-2,000° \text{ K.});$$
$$C_p = 8.04 + 0.54 \times 10^{-3}T - 0.83 \times 10^5 T^{-2}.$$

HYDRIDES

TABLE 49.—Heat content and entropy of $AsH_3(g)$

[Base, ideal gas at 298.15° K.; mol. wt., 77.93]

T, ° K.	$H_T - H_{298.15}$, cal./mole	$S_T - S_{298.15}$, cal./deg. mole	T, ° K.	$H_T - H_{298.15}$, cal./mole	$S_T - S_{298.15}$, cal./deg. mole
400	1,005	2.89	1,000	9,140	14.93
500	2,110	5.35	1,200	12,425	17.92
600	3,335	7.58	1,400	15,870	20.57
700	4,665	9.63	1,600	19,415	22.94
800	6,085	11.53	1,800	23,055	25.08
900	7,580	13.29	2,000	26,745	27.02

$AsH_3(g)$:

$$H_T - H_{298.15} = 10.07T + 2.71 \times 10^{-3}T^2 + 2.20 \times 10^5 T^{-1}$$
$$-3,981 \ (1.5 \text{ percent; } 298°-2,000° \text{ K.});$$
$$C_p = 10.07 + 5.42 \times 10^{-3}T - 2.20 \times 10^5 T^{-2}.$$

TABLE 50.—Heat content and entropy of $AsD_3(g)$

[Base, ideal gas at 298.15° K.; mol. wt., 80.95]

T, ° K.	$H_T - H_{298.15}$, cal./mole	$S_T - S_{298.15}$, cal./deg. mole	T, ° K.	$H_T - H_{298.15}$, cal./mole	$S_T - S_{298.15}$, cal./deg. mole
400	1,170	3.36	1,000	10,460	17.19
500	2,480	6.28	1,200	14,020	20.44
600	3,910	8.89	1,400	17,690	23.26
700	5,440	11.24	1,600	21,430	25.76
800	7,055	13.40	1,800	25,220	27.99
900	8,735	15.38	2,000	29,035	30.00

$AsD_3(g)$:

$$H_T - H_{298.15} = 13.36T + 1.94 \times 10^{-3}T^2 + 3.42 \times 10^5 T^{-1}$$
$$-5,303 \ (1.3 \text{ percent; } 298°-2,000° \text{ K.});$$
$$C_p = 13.36 + 3.88 \times 10^{-3}T - 3.42 \times 10^5 T^{-2}.$$

CHLORIDE

$AsCl_3(l)$:

$$\overline{C_p} = 31.9 \ (287°-371° \text{ K.}).$$

TABLE 51.—Heat content and entropy of $AsCl_3(g)$

[Base, ideal gas at 298.15° K.; mol. wt., 181.28]

T, ° K.	$H_T - H_{298.15}$, cal./mole	$S_T - S_{298.15}$, cal./deg. mole	T, ° K.	$H_T - H_{298.15}$, cal./mole	$S_T - S_{298.15}$, cal./deg. mole
400	1,885	5.43	1,000	13,545	23.18
500	3,785	9.67	1,200	17,490	26.78
600	5,715	13.19	1,400	21,445	29.83
700	7,660	16.19	1,600	25,405	32.47
800	9,615	18.80	1,800	29,365	34.80
900	11,575	21.11	2,000	33,325	36.89

$AsCl_3(g)$:

$$H_T - H_{298.15} = 19.72T + 0.05 \times 10^{-3}T^2 + 1.46 \times 10^5 T^{-1}$$
$$-6,374 \ (0.1 \text{ percent; } 298°-2,000° \text{ K.});$$
$$C_p = 19.72 + 0.10 \times 10^{-3}T - 1.46 \times 10^5 T^{-2}.$$

TABLE 1h7. (*Continued*)

FLUORIDE

TABLE 52.—*Heat content and entropy of* $AsF_3(g)$

[Base, ideal gas at 298.15° K.; mol. wt. 131.91]

T, ° K.	$H_T-H_{298.15}$, cal./mole	$S_T-S_{298.15}$, cal./deg. mole	T, ° K.	$H_T-H_{298.15}$, cal./mole	$S_T-S_{298.15}$, cal./deg. mole
400	1,680	4.84	1,000	12,875	21.79
500	3,440	8.76	1,200	16,765	25.33
600	5,265	12.09	1,400	20,675	28.35
700	7,130	14.96	1,600	24,595	30.96
800	9,030	17.50	1,800	28,535	33.28
900	10,945	19.75	2,000	32,480	35.36

$AsF_3(g)$:

$$H_T-H_{298.15}=19.04T+0.26\times10^{-3}T^2+3.12\times10^5T^{-1}$$
$$-6,746 \text{ (0.3 percent; } 298°-2,000° \text{ K.)};$$
$$C_p=19.04+0.52\times10^{-3}T-3.12\times10^5T^{-2}.$$

IODIDE

TABLE 53.—*Heat content and entropy of* $AsI_3(g)$

[Base, ideal gas at 298.15° K.; mol. wt., 455.64]

T, ° K.	$H_T-H_{298.15}$, cal./mole	$S_T-S_{298.15}$, cal./deg. mole	T, ° K.	$H_T-H_{298.15}$, cal./mole	$S_T-S_{298.15}$, cal./deg. mole
400	1,980	5.71	1,000	13,820	23.78
500	3,935	10.08	1,200	17,785	27.39
600	5,905	13.67	1,400	21,755	30.45
700	7,885	16.72	1,600	25,725	33.10
800	9,865	19.36	1,800	29,695	35.44
900	11,840	21.69	2,000	33,670	37.53

$AsI_3(g)$:

$$H_T-H_{298.15}=19.87T+0.52\times10^5T^{-1}-6,099$$
$$(0.1 \text{ percent; } 298°-2,000° \text{ K.)};$$
$$C_p=19.87-0.52\times10^5T^{-2}.$$

ASTATINE

ELEMENT

TABLE 54.—*Heat content and entropy of* $At_2(c, l)$

[Base, crystals at 298.15° K.; mol. wt., 420]

T, ° K.	$H_T-H_{298.15}$, cal./mole	$S_T-S_{298.15}$, cal./deg. mole	T, ° K.	$H_T-H_{298.15}$, cal./mole	$S_T-S_{298.15}$, cal./deg. mole
400	1,425	4.11	575(l)	9,575	19.10
500	2,825	7.24	600	10,075	19.96
575(c)	3,875	9.19	650	11,075	21.56

$At_2(c)$:

$$H_T-H_{298.15}=14.00T-4,174$$
$$(0.1 \text{ percent; } 298°-575° \text{ K.)};$$
$$C_p=14.00;$$
$$\Delta H_{575} \text{ (fusion)}=5,700.$$

$At_2(l)$:

$$H_T-H_{298.15}=20.00T-1,925$$
$$(0.1 \text{ percent; } 575°-650° \text{ K.)};$$
$$C_p=20.00.$$

TABLE 55.—*Heat content and entropy of* $At(g)$

[Base, ideal gas at 298.15° K.; atomic wt., 210]

T, ° K.	$H_T-H_{298.15}$, cal./mole	$S_T-S_{298.15}$, cal./deg. mole	T, ° K.	$H_T-H_{298.15}$, cal./mole	$S_T-S_{298.15}$, cal./deg. mole
400	500	1.46	1,500	5,970	8.03
500	1,005	2.57	1,600	6,470	8.35
600	1,500	3.48	1,700	6,965	8.65
700	1,995	4.24	1,800	7,460	8.93
800	2,495	4.91	1,900	7,960	9.20
900	2,990	5.49	2,000	8,455	9.46
1,000	3,485	6.01	2,200	9,455	9.93
1,100	3,985	6.49	2,400	10,460	10.37
1,200	4,480	6.92	2,600	11,470	10.78
1,300	4,975	7.32	2,800	12,485	11.15
1,400	5,475	7.69	3,000	13,510	11.51

$At(g)$:

$$H_T-H_{298.15}=4.97T-1,482$$
$$(0.2 \text{ percent; } 298°-3,000° \text{ K.)};$$
$$C_p=4.97$$

TABLE 56.—*Heat content and entropy of* $At_2(g)$

[Base, ideal gas at 298.15° K.; mol. wt., 420]

T, ° K.	$H_T-H_{298.15}$, cal./mole	$S_T-S_{298.15}$, cal./deg. mole	T, ° K.	$H_T-H_{298.15}$, cal./mole	$S_T-S_{298.15}$, cal./deg. mole
400	910	2.63	1,500	10,880	14.58
500	1,805	4.62	1,600	11,795	15.17
600	2,710	6.27	1,700	12,715	15.73
700	3,610	7.66	1,800	13,630	16.25
800	4,515	8.87	1,900	14,550	16.75
900	5,420	9.93	2,000	15,470	17.22
1,000	6,325	10.89	2,200	17,320	18.10
1,100	7,235	11.75	2,400	19,170	18.91
1,200	8,145	12.55	2,600	21,025	19.65
1,300	9,055	13.27	2,800	22,890	20.34
1,400	9,970	13.95	3,000	24,755	20.98

$At_2(g)$:

$$H_T-H_{298.15}=8.94T+0.07\times10^{-3}T^2+0.06\times10^5T^{-1}$$
$$-2,692 \text{ (0.1 percent; } 298°-3,000° \text{ K.)};$$
$$C_p=8.94+0.14\times10^{-3}T-0.06\times10^5T^{-2}.$$

BARIUM AND ITS COMPOUNDS

ELEMENT

TABLE 57.—*Heat content and entropy of* $Ba(c, l)$

[Base, α-crystals at 298.15° K.; atomic wt., 137.36]

T, ° K.	$H_T-H_{298.15}$, cal./mole	$S_T-S_{298.15}$, cal./deg. mole	T, ° K.	$H_T-H_{298.15}$, cal./mole	$S_T-S_{298.15}$, cal./deg. mole
400	655	1.89	1,000	7,250	10.87
500	1,335	3.40	1,100	8,000	11.58
600	2,045	4.70	1,200	8,750	12.23
643(α)	2,360	5.21	1,300	9,500	12.83
643(β)	2,510	5.44	1,400	10,250	13.39
700	2,920	6.05	1,500	11,000	13.91
800	3,695	7.08	1,600	11,750	14.39
900	4,540	8.08	1,700	12,500	14.85
983(β)	5,290	8.87	1,800	13,250	15.28
983(l)	7,120	10.74	1,900	14,000	15.68

Ba(α):

$$H_T-H_{298.15}=5.36T+1.58\times10^{-3}T^2-1,739$$
$$(0.2 \text{ percent; } 298°-643° \text{ K.)};$$
$$C_p=5.36+3.16\times10^{-3}T;$$
$$\Delta H_{643} \text{ (transition)}=150.$$

TABLE 1h7. (*Continued*)

Ba(β):

$H_T - H_{298.15} = 2.60\,T + 3.43 \times 10^{-3}\,T^2 - 580$
(0.1 percent; 643°–983° K.);
$C_p = 2.60 + 6.86 \times 10^{-3}\,T$;
ΔH_{983} (*fusion*) = 1,830.

Ba(l):

$H_T - H_{298.15} = 7.50\,T - 250$ (0.1 percent; 983°–1,900° K.);
$C_p = 7.50$.

TABLE 58.—*Heat content and entropy of Ba* (*g*)

[Base, ideal gas at 298.15° K.; atomic wt., 137.36]

T, ° K.	$H_T - H_{298.15}$, cal./mole	$S_T - S_{298.15}$, cal./deg. mole	T, ° K.	$H_T - H_{298.15}$, cal./mole	$S_T - S_{298.15}$, cal./deg. mole
400	505	1.46	1,000	8,430	9.43
500	1,005	2.57	2,000	9,005	9.77
600	1,500	3.48	2,200	10,465	10.47
700	1,995	4.24	2,400	12,135	11.19
800	2,495	4.90	2,600	14,015	11.94
900	2,990	5.49	2,800	16,095	12.72
1,000	3,490	6.01	3,000	18,345	13.49
1,100	3,985	6.49	3,500	24,425	15.36
1,200	3,485	6.92	4,000	30,680	17.03
1,300	4,995	7.33	4,500	36,785	18.47
1,400	5,505	7.71	5,000	42,690	19.72
1,500	6,030	8.07	6,000	54,285	21.83
1,600	6,570	8.42	7,000	65,980	23.63
1,700	7,130	8.76	8,000	77,920	24.23
1,800	7,720	9.10			

Ba(g):

$H_T - H_{298.15} = 4.84\,T + 0.29 \times 10^{-3}\,T^2 - 0.28 \times 10^5\,T^{-1}$
$- 1,268$ (0.9 percent; 298°–2,000° K.);
$C_p = 4.84 + 0.58 \times 10^{-3}\,T + 0.28 \times 10^5\,T^{-2}$.

OXIDE

TABLE 59.—*Heat content and entropy of BaO*(*c*)

[Base, crystals at 298.15° K.; mol. wt., 153.36]

T, ° K.	$H_T - H_{298.15}$, cal./mole	$S_T - S_{298.15}$, cal./deg. mole	T, ° K.	$H_T - H_{298.15}$, cal./mole	$S_T - S_{298.15}$, cal./deg. mole
400	1,170	3.36	1,300	13,090	18.77
500	2,380	6.07	1,400	14,520	19.83
600	3,660	8.40	1,500	15,970	20.83
700	4,980	10.43	1,600	17,440	21.78
800	6,300	12.20	1,700	18,920	22.67
900	7,620	13.75	1,800	20,420	23.53
1,000	8,950	15.15	1,900	21,930	24.35
1,100	10,300	16.44	2,000	23,450	25.13
1,200	11,680	17.64			

BaO(c):

$H_T - H_{298.15} = 11.79\,T + 0.94 \times 10^{-3}\,T^2 + 0.88 \times 10^5\,T^{-1}$
$- 3,894$ (0.5 percent; 298°–2,000° K.);
$C_p = 11.79 + 1.88 \times 10^{-3}\,T - 0.88 \times 10^5\,T^{-2}$.

TABLE 60.—*Heat content and entropy of BaO*(*g*)

[Base, ideal gas at 298.15° K.; mol. wt., 153.36]

T, ° K.	$H_T - H_{298.15}$, cal./mole	$S_T - S_{298.15}$, cal./deg. mole	T, ° K.	$H_T - H_{298.15}$, cal./mole	$S_T - S_{298.15}$ cal./deg. mole
400	820	2.36	1,000	5,985	10.21
500	1,650	4.21	1,200	7,750	11.82
600	2,500	5.76	1,400	9,520	13.18
700	3,365	7.10	1,600	11,295	14.37
800	4,235	8.26	1,800	13,075	15.42
900	5,110	9.29	2,000	14,855	16.35

BaO(g):

$H_T - H_{298.15} = 8.62\,T + 0.10 \times 10^{-3}\,T^2 + 0.67 \times 10^5\,T^{-1}$
$- 2,804$ (0.3 percent; 298°–2,000° K.);
$C_p = 8.62 + 0.20 \times 10^{-3}\,T - 0.67 \times 10^5\,T^{-2}$.

HYDROXIDE

TABLE 61.—*Heat content and entropy of Ba*(*OH*)$_2$(*c, l*)

[Base, crystals at 298.15° K.; mol. wt., 171.38]

T, ° K.	$H_T - H_{287.17}$, cal./mole	$S_T - S_{298.15}$, cal./deg. mole	T, ° K.	$H_T - H_{298.15}$, cal./mole	$S_T - S_{298.15}$, cal./deg. mole
400	2,500	7.20	800	19,070	34.28
500	5,180	13.17	900	22,360	38.16
600	8,070	18.43	1,000	25,650	41.62
690(*c*)	10,860	22.76	1,100	28,940	44.76
690(*l*)	15,450	29.41	1,200	32,230	47.62
700	15,780	29.89			

Ba(OH)$_2$(*c*):

$H_T - H_{298.15} = 16.90\,T + 10.95 \times 10^{-3}\,T^2$
$- 6,012$ (0.1 percent; 298°–690° K.);
$C_p = 16.90 + 21.90 \times 10^{-3}\,T$;
ΔH_{690}(*fusion*) = 4,590.

Ba(OH)$_2$(*l*):

$H_T - H_{298.15} = 32.90\,T - 7,250$
(0.1 percent; 690°–1,200°K.);
$C_p = 32.90$.

HYDRIDE

TABLE 62.—*Heat content and entropy of BaH*(*g*)

[Base, ideal gas at 298.15° K.; mol. wt., 138.37]

T, ° K.	$H_T - H_{298.15}$, cal./mole	$S_T - S_{298.15}$, cal./deg. mole	T, ° K.	$H_T - H_{298.15}$, cal./mole	$S_T - S_{298.15}$, cal./deg. mole
400	750	2.16	1,000	5,655	9.58
500	1,520	3.88	1,200	7,375	11.15
600	2,315	5.33	1,400	9,115	12.49
700	3,130	6.58	1,600	10,865	13.66
800	3,960	7.69	1,800	12,625	14.70
900	4,800	8.68	2,000	14,390	15.63

BaH(g):

$H_T - H_{298.15} = 7.88\,T + 0.32 \times 10^{-3}\,T^2 + 0.77$
$\times 10^5\,T^{-1} - 2,636$ (0.5 percent; 298°–2,000° K.);
$C_p = 7.88 + 0.64 \times 10^{-3}\,T - 0.77 \times 10^5\,T^{-2}$.

BROMIDE

TABLE 63.—*Heat content and entropy of BaBr*(*g*)

[Base, ideal gas at 298.15° K.; mol. wt., 217.28]

T, ° K.	$H_T - H_{298.15}$, cal./mole	$S_T - S_{298.15}$, cal./deg. mole	T, ° K.	$H_T - H_{298.15}$, cal./mole	$S_T HS_{298.15}$, cal./deg. mole
400	900	2.60	1,000	6,245	10.75
500	1,785	4.57	1,200	8,030	12.38
600	2,675	6.19	1,400	9,820	13.76
700	3,570	7.57	1,600	11,605	14.95
800	4,460	8.76	1,800	13,395	16.01
900	5,355	9.82	2,000	15,180	16.95

TABLE 1h7. (*Continued*)

BaBr(g):

$$H_T - H_{298.15} = 8.94T + 0.13 \times 10^5 T^{-1}$$
$$-2,709 \ (0.1 \text{ percent; } 298°–2,000° \text{ K.});$$
$$C_p = 8.94 - 0.13 \times 10^5 T^{-2}.$$

CHLORIDES

TABLE 64.—*Heat content and entropy of BaCl(g)*

[Base, ideal gas at 298.15° K.; mol. wt., 172.82]

T, ° K.	$H_T-H_{298.15}$, cal./mole	$S_T-S_{298.15}$, cal./deg. mole	T, ° K.	$H_T-H_{298.15}$, cal./mole	$S_T-S_{298.15}$, cal./deg. mole
400	890	2.57	1,000	6,215	10.69
500	1,770	4.53	1,200	8,000	12.32
600	2,665	6.14	1,400	9,785	13.69
700	3,545	7.51	1,600	11,570	14.88
800	4,435	8.70	1,800	13,355	15.93
900	5,325	9.75	2,000	15,145	16.88

BaCl(g):

$$H_T - H_{298.15} = 8.94T + 0.25 \times 10^5 T^{-1}$$
$$-2,749 \ (0.1 \text{ percent; } 298°–2,000° \text{ K.});$$
$$C_p = 8.94 - 0.25 \times 10^5 T^{-2}.$$

BaCl$_2$(c):

$$C_p = 17.00 + 3.34 \times 10^{-3} T (\textit{estimated}) \ (298°–1,198° \text{ K.}).$$

BaCl$_2 \cdot$H$_2$O(c):

$$\overline{C_p} = 28.2 \ (273°–307° \text{ K.}).$$

BaCl$_2 \cdot$2H$_2$O(c):

$$\overline{C_p} = 37.3 \ (273°–307° \text{ K.}).$$

FLUORIDES

TABLE 65.—*Heat content and entropy of BaF(g)*

[Base, ideal gas at 298.15° K.; mol. wt., 156.36]

T, ° K.	$H_T-H_{298.15}$, cal./mole	$S_T-S_{298.15}$, cal./deg. mole	T, ° K.	$H_T-H_{298.15}$, cal./mole	$S_T-S_{298.15}$, cal./deg. mole
400	860	2.48	1,000	6,120	10.49
500	1,720	4.40	1,200	7,895	12.11
600	2,590	5.98	1,400	9,675	13.48
700	3,470	7.34	1,600	11,455	14.67
800	4,350	8.52	1,800	13,240	15.72
900	5,235	9.56	2,000	15,025	16.66

BaF(g):

$$H_T - H_{298.15} = 8.87T + 0.02 \times 10^{-3} T^2 + 0.52 \times 10^5 T^{-1}$$
$$-2,821 \ (0.1 \text{ percent; } 298°–2,000° \text{ K.});$$
$$C_p = 8.87 + 0.04 \times 10^{-3} T - 0.52 \times 10^5 T^{-2}.$$

TABLE 66.—*Heat content and entropy of BaF$_2$(c)*

[Base, crystals at 298.15° K.; mol. wt., 175.36]

T, ° K.	$H_T-H_{298.15}$, cal./mole	$S_T-S_{298.15}$, cal./deg. mole	T, ° K.	$H_T-H_{298.15}$, cal./mole	$S_T-S_{298.15}$, cal./deg. mole
400	1,850	5.33	900	11,900	21.41
500	3,700	9.45	1,000	14,200	23.83
600	5,650	13.01	1,100	16,700	26.22
700	7,650	16.09	1,200	19,400	28.56
800	9,700	18.82	1,300	22,300	30.88

BaF$_2$(c):

$$H_T - H_{298.15} = 13.98T + 5.10 \times 10^{-3} T^2$$
$$-4,621 \ (1.4 \text{ percent; } 298°–1,300° \text{ K.});$$
$$C_p = 13.98 + 10.20 \times 10^{-3} T.$$

BROMATE

Ba(BrO$_3$)$_2 \cdot$H$_2$O(c):

$$C_p = 52.90 \ (298° \text{ K.}).$$

CHLORATE

Ba(ClO$_3$)$_2 \cdot$H$_2$O:

$$\overline{C_p} = 50.6 \ (289°–320° \text{ K.}).$$

CARBONATE

TABLE 67.—*Heat content and entropy of BaCO$_3$(c)*

[Base, α–crystals at 298.15° K.; mol. wt., 197.37]

T, ° K.	$H_T-H_{298.15}$, cal./mole	$S_T-S_{298.15}$, cal./deg. mole	T, ° K.	$H_T-H_{298.15}$, cal./mole	$S_T-S_{298.15}$, cal./deg. mole
400	2,300	6.61	1,100	27,060	39.17
500	4,730	12.03	1,200	30,760	42.39
600	7,330	16.76	1,241(β)	32,270	43.62
700	10,080	21.00	1,241(γ)	33,000	44.21
800	12,980	24.87	1,300	35,240	45.98
900	16,020	28.45	1,400	39,040	48.79
1,000	19,190	31.79	1,500	42,840	51.41
1,079(α)	21,790	34.29	1,600	46,640	53.86
1,079(β)	26,280	38.45			

BaCO$_3$(α):

$$H_T - H_{298.15} = 21.50T + 5.53 \times 10^{-3} T^2 + 3.91 \times 10^5 T^{-1}$$
$$-8,213 \ (0.5 \text{ percent; } 298°–1,079° \text{ K.});$$
$$C_p = 21.50 + 11.06 \times 10^{-3} T - 3.91 \times 10^5 T^{-2};$$
$$\Delta H_{1079}(\textit{transition}) = 4,490.$$

BaCO$_3$(β):

$$H_T - H_{298.15} = 37.00T$$
$$-13,644 \ (0.1 \text{ percent; } 1,079°–1,241° \text{ K.});$$
$$C_p = 37.00;$$
$$\Delta H_{1241}(\textit{transition}) = 730.$$

BaCO$_3$(γ):

$$H_T - H_{298.15} = 38.00T$$
$$-14,160 \ (0.1 \text{ percent; } 1,241°–1,600° \text{ K.});$$
$$C_p = 38.00.$$

MOLYBDATE

BaMoO$_4$(c):

$$\overline{C_p} = 33.6 \ (273°–297° \text{ K.}).$$

NITRATE

TABLE 68.—*Heat content and entropy of Ba(NO$_3$)$_2$(c)*

[Base, crystals at 298.15° K.; mol. wt., 261.38]

T, ° K.	$H_T-H_{298.15}$, cal./mole	$S_T-S_{298.15}$, cal./deg. mole	T, ° K.	$H_T-H_{298.15}$, cal./mole	$S_T-S_{298.15}$, cal./deg. mole
400	3,960	11.39	700	18,460	38.14
500	8,410	21.30	800	24,070	45.62
600	13,240	30.10	850	27,020	49.20

Ba(NO$_3$)$_2$(c):

$$H_T - H_{298.15} = 30.05T + 17.85 \times 10^{-3} T^2 + 4.01 \times 10^5 T^{-1}$$
$$-11,891 \ (0.2 \text{ percent; } 298°–850° \text{ K.});$$
$$C_p = 30.05 + 35.70 \times 10^{-3} T - 4.01 \times 10^5 T^{-2}.$$

TABLE 1h7. (Continued)

SULFATE

TABLE 69.—*Heat content and entropy of* $BaSO_4(c)$

[Base, crystals at 298.15° K.; mol. wt., 233.43]

T, ° K.	$H_T-H_{298.15}$, cal./mole	$S_T-S_{298.15}$, cal./deg. mole	T, ° K.	$H_T-H_{298.15}$, cal./mole	$S_T-S_{298.15}$, cal./deg. mole
400	2,700	7.77	900	18,400	33.07
500	5,700	14.46	1,000	21,600	36.45
600	8,800	20.11	1,100	24,900	39.59
700	12,000	25.03	1,200	28,300	42.55
800	15,200	29.31	1,300	31,800	45.35

$BaSO_4(c)$:

$$H_T-H_{298.15}=33.80T+8.43\times10^5T^{-1}$$
$$-12,905 \ (0.5 \text{ percent}; 298°\text{–}1,300° \text{ K.});$$
$$C_p=33.80-8.43\times10^5T^{-2}.$$

THIOSULFATE

$BaS_2O_3(c)$:

$$\overline{C_p}=40.7 \ (293°\text{–}373° \text{ K.})$$

TITANATES

TABLE 70.—*Heat content and entropy of* $BaTiO_3(c)$

[Base, crystals at 298.15° K.; mol. wt., 233.26]

T, ° K.	$H_T-H_{298.15}$, cal./mole	$S_T-S_{298.15}$, cal./deg. mole	T, ° K.	$H_T-H_{298.15}$, cal./mole	$S_T-S_{298.15}$, cal./deg. mole
400	2,695	7.76	1,300	29,510	42.39
500	5,450	13.90	1,400	32,660	44.72
600	8,290	19.08	1,500	35,840	46.92
700	11,200	23.56	1,600	39,040	48.98
800	14,160	27.51	1,700	42,270	50.94
900	17,170	31.06	1,800	45,540	52.81
1,000	20,210	34.26	1,900	48,840	54.59
1,100	23,280	37.18	2,000	52,160	56.29
1,200	26,380	39.88			

$BaTiO_3(c)$:

$$H_T-H_{298.15}=29.03T+1.02\times10^{-3}T^2+4.58\times10^5T^{-1}$$
$$-10,282 \ (0.3 \text{ percent}; 298°\text{–}2,000° \text{ K.});$$
$$C_p=29.03+2.04\times10^{-3}T-4.58\times10^5T^{-2}.$$

TABLE 71.—*Heat content and entropy of* $Ba_2TiO_4(c)$

[Base, crystals at 298.15° K.; mol. wt., 386.62]

T, ° K.	$H_T-H_{298.15}$, cal./mole	$S_T-S_{298.15}$, cal./deg. mole	T, ° K.	$H_T-H_{298.15}$, cal./mole	$S_T-S_{298.15}$, cal./deg. mole
400	3,780	10.88	1,300	42,670	61.17
500	7,730	19.69	1,400	47,110	64.46
600	11,860	27.22	1,500	51,570	67.52
700	16,140	33.81	1,600	56,060	70.42
800	20,520	39.66	1,700	60,590	73.17
900	24,950	44.88	1,800	65,170	75.79
1,000	29,380	49.55	1,900	69,790	78.29
1,100	33,810	53.77	2,000	74,450	80.86
1,200	38,240	57.62			

$Ba_2TiO_4(c)$:

$$H_T-H_{298.15}=43.00T+0.80\times10^{-3}T^2+6.96\times10^5T^{-1}$$
$$-15,226 \ (0.5 \text{ percent}; 298°\text{–}2,000° \text{ K.});$$
$$C_p=43.00+1.60\times10^{-3}T-6.96\times10^5T^{-2}.$$

BERYLLIUM AND ITS COMPOUNDS

ELEMENT

TABLE 72.—*Heat content and entropy of* $Be(c, l)$

[Base, crystals at 298.15° K.; atomic wt., 9.013]

T, ° K.	$H_T-H_{298.15}$, cal./mole	$S_T-S_{298.15}$, cal./deg. mole	T, ° K.	$H_T-H_{298.15}$, cal./mole	$S_T-S_{298.15}$, cal./deg. mole
400	440	1.26	1,556(c)	7,920	9.61
500	940	2.38	1,556(l)	10,720	11.41
600	1,485	3.37	1,600	11,050	11.62
700	2,060	4.26	1,700	11,800	12.08
800	2,655	5.05	1,800	12,550	12.51
900	3,270	5.77	1,900	13,300	12.91
1,000	3,910	6.45	2,000	14,050	13.30
1,100	4,580	7.08	2,200	15,550	14.01
1,200	5,270	7.68	2,400	17,050	14.66
1,300	5,980	8.25	2,600	18,550	15.26
1,400	6,720	8.80	2,700	19,300	15.55
1,500	7,480	9.32			

$Be(c)$:

$$H_T-H_{298.15}=4.58T+1.06\times10^{-3}T^2+1.14\times10^5T^{-1}$$
$$-1,842 \ (0.4 \text{ percent}; 298°\text{–}1,556° \text{ K.});$$
$$C_p=4.58+2.12\times10^{-3}T-1.14\times10^5T^{-2};$$
$$\Delta H_{1556}(fusion)=2,800.$$

$Be(l)$:

$$H_T-H_{298.15}=7.50T$$
$$-950 \ (0.1 \text{ percent}; 1,556°\text{–}2,700° \text{ K.});$$
$$C_p=7.50.$$

TABLE 73.—*Heat content and entropy of* $Be(g)$

[Base, ideal gas at 298.15° K.; atomic wt., 9.013]

T, ° K.	$H_T-H_{298.15}$, cal./mole	$S_T-S_{298.15}$, cal./deg. mole	T, ° K.	$H_T-H_{298.15}$, cal./mole	$S_T-S_{298.15}$, cal./deg. mole
400	505	1.46	1,900	7,960	9.20
500	1,005	2.57	2,000	8,455	9.46
600	1,500	3.48	2,200	9,450	9.93
700	1,995	4.24	2,400	10,445	10.36
800	2,495	4.90	2,600	11,440	10.76
900	2,990	5.49	2,800	12,440	11.13
1,000	3,490	6.01	3,000	13,440	11.48
1,100	3,985	6.49	3,500	15,980	12.26
1,200	4,480	6.92	4,000	18,605	12.96
1,300	4,980	7.32	4,500	21,380	13.62
1,400	5,475	6.69	5,000	24,365	14.24
1,500	5,970	8.03	6,000	31,135	15.47
1,600	6,470	8.35	7,000	39,035	16.69
1,700	6,965	8.65	8,000	47,940	17.88
1,800	7,465	8.93			

$Be(g)$:

$$H_T-H_{298.15}=4.97T-1,482$$
$$(0.1 \text{ percent}; 298°\text{–}3,000° \text{ K.});$$
$$C_p=4.97.$$

OXIDE

TABLE 74.—*Heat content and entropy of* $BeO(c)$

[Base, crystals at 298.15° K.; mol. wt., 25.01]

T, ° K.	$H_T-H_{298.15}$, cal./mole	$S_T-S_{298.15}$, cal./deg. mole	T, ° K.	$H_T-H_{298.15}$, cal./mole	$S_T-S_{298.15}$, cal./deg. mole
400	730	2.09	900	5,830	10.16
500	1,580	3.98	1,000	7,010	11.40
600	2,540	5.73	1,100	8,240	12.58
700	3,600	7.36	1,200	9,510	13.68
800	4,700	8.83			

TABLE 1h7. (*Continued*)

BeO(c):

$$H_T - H_{298.15} = 8.45T + 2.00 \times 10^{-3}T^2 + 3.17 \times 10^5 T^{-1}$$
$$-3,760 \ (0.5 \ \text{percent}; \ 298°-1,200° \ \text{K.});$$
$$C_p = 8.45 + 4.00 \times 10^{-3}T - 3.17 \times 10^5 T^{-2}.$$

NITRIDE

TABLE 75.—*Heat content and entropy of* $Be_3N_2(c)$

[Base, crystals at 298.15° K.; mol. wt., 55.06]

T, ° K.	$H_T-H_{298.15}$, cal./mole	$S_T-S_{298.15}$, cal./deg. mole	T. K.	$H_T-H_{298.15}$, cal./mole	$S_T-S_{298.15}$, cal./deg. mole
400	1,840	5.29	700	9,200	18.75
500	3,930	9.94	800	12,130	22.66
600	6,420	14.47			

$Be_3N_2(c)$:

$$H_T - H_{298.15} = 7.32T + 15.40 \times 10^{-3}T^2 - 3,551$$
$$(0.6 \ \text{percent}; \ 298°-800° \ \text{K.});$$
$$C_p = 7.32 + 30.80 \times 10^{-3}T.$$

HYDRIDE

TABLE 76.—*Heat content and entropy of* $BeH(g)$

[Base, ideal gas at 298.15° K.; mol. wt., 10.02]

T, ° K.	$H_T-H_{298.15}$, cal./mole	$S_T-S_{298.15}$, cal./deg. mole	T. ° K.	$H_T-H_{298.15}$, cal./mole	$S_T-S_{298.15}$, cal./deg. mole
400	710	2.05	1,000	5,225	8.88
500	1,420	3.63	1,200	6,850	10.35
600	2,150	4.96	1,400	8,510	11.63
700	2,890	6.10	1,600	10,195	12.75
800	3,655	7.13	1,800	11,900	13.76
900	4,430	8.04	2,000	13,625	14.66

BeH(g):

$$H_T - H_{298.15} = 6.63T + 0.62 \times 10^{-3}T^2 + 0.02 \times 10^5 T^{-1}$$
$$-2,039 \ (0.6 \ \text{percent}; \ 298°-2,000° \ \text{K.});$$
$$C_p = 6.63 + 1.24 \times 10^{-3}T - 0.02 \times 10^5 T^{-2}.$$

CHLORIDE

TABLE 77.—*Heat content and entropy of* $BeCl(g)$

[Base, ideal gas at 298.15° K.; mol. wt., 44.47]

T, ° K.	$H_T-H_{298.15}$, cal./mole	$S_T-S_{298.15}$, cal./deg. mole	T, ° K.	$H_T-H_{298.15}$, cal./mole	$S_T-S_{298.15}$, cal./deg. mole
400	790	2.28	1,000	5,865	9.97
500	1,595	4.07	1,200	7,615	11.57
600	2,430	5.60	1,400	9,375	12.92
700	3,275	6.90	1,600	11,145	14.11
800	4,130	8.04	1,800	12,915	15.15
900	4,995	9.06	2,000	14,690	16.08

BeCl(g):

$$H_T - H_{298.15} = 8.39T + 0.17 \times 10^{-3}T^2 + 0.80 \times 10^5 T^{-1}$$
$$-2,785 \ (0.4 \ \text{percent}; \ 298°-2,000° \ \text{K.});$$
$$C_p = 8.39 + 0.34 \times 10^{-3}T - 0.80 \times 10^5 T^{-2}.$$

FLUORIDE

TABLE 78.—*Heat content and entropy of* $BeF(g)$

[Base, ideal gas at 298.15° K.,; mol. wt., 28.01]

T, ° K.	$H_T-H_{298.15}$, cal./mole	$S_T-S_{298.15}$, cal./deg. mole	T, ° K.	$H_T-H_{298.15}$, cal./mole	$S_T-S_{298.15}$, cal./deg. mole
400	740	2.13	1,000	5,585	9.45
500	1,495	3.82	1,200	7,295	11.01
600	2,280	5.25	1,400	9,025	12.35
700	3,085	6.49	1,600	10,770	13.51
800	3,905	7.58	1,800	12,525	14.54
900	4,740	8.56	2,000	14,280	15.47

BeF(g):

$$H_T - H_{298.15} = 7.69T + 0.37 \times 10^{-3}T^2 + 0.70 \times 10^5 T^{-1}$$
$$-2,560 \ (0.6 \ \text{percent}; \ 298°-2,000° \ \text{K.});$$
$$C_p = 7.69 + 0.74 \times 10^{-3}T - 0.70 \times 10^5 T^{-2}.$$

ALUMINATE

$BeAl_2O_4(c)$:

$$\overline{C_p} = 25.4 \ (273°-373° \ \text{K.}).$$

SILICATE

$Be_2SiO_4(c)$:

$$C_p = 22.84 \ (298° \ \text{K.}).$$

SULFATE

$BeSO_4(c)$:

$$\overline{C_p} = 20.8 \ (273°-373° \ \text{K.}).$$

BISMUTH AND ITS COMPOUNDS

ELEMENT

TABLE 79.—*Heat content and entropy of* $Bi(c, l)$

[Base, crystals at 298.15° K.; atomic wt., 209.00]

T, ° K.	$H_T-H_{298.15}$, cal./mole	$S_T-S_{298.15}$, cal./deg. mole	T, ° K.	$H_T-H_{298.15}$, cal./mole	$S_T-S_{298.15}$, cal./deg. mole
400	650	1.87	1,100	8,430	14.09
500	1,340	3.41	1,200	9,180	14.74
544.5(c)	1,660	4.02	1,300	9,930	15.34
544.5(l)	4,260	8.80	1,400	10,680	15.89
600	4,680	9.54	1,500	11,430	16.41
700	5,430	10.70	1,600	12,180	16.90
800	6,180	11.70	1,700	12,930	17.35
900	6,930	12.58	1,800	13,680	17.78
1,000	7,680	13.37			

Bi(c):

$$H_T - H_{298.15} = 4.49T + 2.70 \times 10^{-3}T^2 - 1,579 \ (0.2 \ \text{percent};$$
$$298°-544.5° \ \text{K.});$$
$$C_p = 4.49 + 5.40 \times 10^{-3}T;$$
$$\Delta H_{544.5}(fusion) = 2,600.$$

Bi(l):

$$H_T - H_{298.15} = 7.50T + 180 \ (0.1 \ \text{percent};$$
$$544.5°-1,800° \ \text{K.});$$
$$C_p = 7.50.$$

TABLE 1h7. (*Continued*)

TABLE 80.—*Heat content and entropy of Bi(g)*

[Base, ideal gas at 298.15° K.; atomic wt., 209.00]

T, ° K.	$H_T-H_{298.15}$, cal./mole	$S_T-S_{298.15}$, cal./deg. mole	T, ° K.	$H_T-H_{298.15}$, cal./mole	$S_T-S_{298.15}$, cal./deg. mole
400	505	1.46	1,500	5,970	8.03
500	1,005	2.57	1,600	6,470	8.35
600	1,500	3.48	1,700	6,965	8.65
700	1,995	4.24	1,800	7,465	8.93
800	2,495	4.90	1,900	7,965	9.20
900	2,990	5.49	2,000	8,465	9.46
1,000	3,485	6.01	2,200	9,470	9.94
1,100	3,985	6.48	2,400	10,485	10.38
1,200	4,480	6.92	2,600	11,510	10.79
1,300	4,975	7.32	2,800	12,545	11.18
1,400	5,475	7.68	3,000	13,600	11.54

Bi(*g*):

$$H_T-H_{298.15}=4.97T-1,482 \text{ (0.1 percent;}$$
$$298°-2,500° \text{ K.)};$$
$$C_p=4.97.$$

TABLE 81.—*Heat content and entropy of Bi$_2$(g)*

[Base, ideal gas at 298.15° K.; mol. wt., 418.00]

T, ° K.	$H_T-H_{298.15}$, cal./mole	$S_T-S_{298.15}$, cal./deg. mole	T, ° K.	$H_T-H_{298.15}$, cal./mole	$S_T-S_{298.15}$, cal./deg. mole
400	900	2.60	1,000	6,250	10.77
500	1,790	4.58	1,200	8,040	12.40
600	2,685	6.22	1,400	9,830	13.78
700	3,575	7.59	1,600	11,615	14.97
800	4,465	8.78	1,800	13,400	16.02
900	5,360	9.83	2,000	15,190	16.96

Bi$_2$(*g*):

$$H_T-H_{298.15}=8.94+0.10\times10^5 T^{-1}-2,699$$
$$\text{(0.1 percent; } 298°-2,000° \text{ K.)};$$
$$C_p=8.94-0.10\times10^5 T^{-2}.$$

OXIDES

TABLE 82.—*Heat content and entropy of BiO(g)*

[Base, ideal gas at 298.15° K.; mol. wt., 225.00]

T, ° K.	$H_T-H_{298.15}$, cal./mole	$S_T-S_{298.15}$, cal./deg. mole	T, ° K.	$H_T-H_{298.15}$, cal./mole	$S_T-S_{298.15}$, cal./deg. mole
400	815	2.35	1,000	5,870	10.18
500	1,645	4.20	1,200	7,730	11.78
600	2,495	5.75	1,400	9,500	13.15
700	3,355	7.07	1,600	11,275	14.33
800	4,220	8.22	1,800	13,055	15.38
9 00	5,090	9.25	2,000	14,840	16.32

BiO(*g*):

$$H_T-H_{298.15}=8.63T+0.10\times10^{-3}T^2+0.79\times10^5 T^{-1}$$
$$-2,847 \text{ (0.2 percent; } 298°-2,000° \text{ K.)};$$
$$C_p=8.63+0.20\times10^{-3}T-0.79\times10^5 T^{-2}.$$

TABLE 83.—*Heat content and entropy of Bi$_2$O$_3$(c)*

[Base, crystals at 298.15° K.; mol. wt., 466.00]

T, ° K.	$H_T-H_{298.15}$, cal./mole	$S_T-S_{298.15}$, cal./deg. mole	T. ° K.	$H_T-H_{298.15}$, cal./mole	$S_T-S_{298.15}$, cal./deg. mole
400	2,770	7.99	700	11,550	24.31
500	5,630	14.36	800	14,620	28.41
600	8,550	19.69			

Bi$_2$O$_3$(*c*):

$$H_T-H_{298.15}=24.74T+4.00\times10^{-3}T^2-7,732$$
$$\text{(0.3 percent; } 298°-800° \text{ K.)};$$
$$C_p=24.74+8.00\times10^{-3}T.$$

SULFIDE

Bi$_2$S$_3$(*c*):

$$C_p=28.90+6.10\times10^{-3}T(estimated) \text{ (298°-1,000° K.).}$$

BROMIDE

TABLE 84.—*Heat content and entropy of BiBr(g)*

[Base, ideal gas at 298.15° K.; mol. wt., 288.92]

T, ° K.	$H_T-H_{298.15}$, cal./mole	$S_T-S_{298.15}$, cal./deg. mole	T. ° K.	$H_T-H_{298.15}$, cal./mole	$S_T-S_{298.15}$, cal./deg. mole
400	900	2.60	1,000	6,240	10.75
500	1,785	4.57	1,200	8,025	12.37
600	2,675	6.19	1,400	9,815	13.75
700	3,565	7.56	1,600	11,600	14.94
800	4,455	8.75	1,800	13,390	16.00
900	5,350	9.81	2,000	15,175	16.94

BiBr(*g*):

$$H_T-H_{298.15}=8.94T+0.14\times10^5 T^{-1}-2,712$$
$$\text{(0.1 percent; } 298°-2,000° \text{ K.)};$$
$$C_p=8.94-0.14\times10^5 T^{-2}.$$

CHLORIDES

TABLE 85.—*Heat content and entropy of BiCl(g)*

[Base, ideal gas at 298.15° K.; mol. wt., 244.46]

T, ° K.	$H_T-H_{298.15}$, cal./mole	$S_T-S_{298.15}$, cal./deg. mole	T. ° K.	$H_T-H_{298.15}$, cal./mole	$S_T-S_{298.15}$, cal./deg. mole
400	885	2.55	1,000	6,205	10.67
500	1,765	4.52	1,200	7,985	12.29
600	2,650	6.13	1,400	9,770	13.67
700	3,535	7.49	1,600	11,555	14.86
800	4,425	8.68	1,800	13,340	15.91
900	5,315	9.73	2,000	15,130	16.85

BiCl(*g*):

$$H_T-H_{298.15}=8.93T+0.26\times10^5 T^{-1}-2,750 \text{ (0.1 percent;}$$
$$298°-2,000° \text{ K.)};$$
$$C_p=8.93-0.26\times10^5 T^{-2}.$$

TABLE 86.—*Heat content and entropy of BiCl$_3$(g)*

[Base, ideal gas at 298.15° K.; mol. wt., 315.37]

T, ° K.	$H_T-H_{298.15}$, cal./mole	$S_T-S_{298.15}$, cal./deg. mole	T. ° K.	$H_T-H_{298.15}$, cal./mole	$S_T-S_{298.15}$, cal./deg. mole
400	1,960	5.66	800	9,805	19.24
500	3,905	10.00	900	11,780	21.56
600	5,865	13.57	1,000	13,760	23.65
700	7,835	16.61			

BiCl$_3$(*g*):

$$H_T-H_{298.15}=19.85T+0.74\times10^5 T^{-1}-6,166 \text{ (0.1 percent;}$$
$$298°-1,000° \text{ K.)};$$
$$C_p=19.85-0.74\times10^5 T^{-2}.$$

TABLE 1h7. (*Continued*)

FLUORIDE

TABLE 87.—*Heat content and entropy of BiF(g)*

[Base, ideal gas at 298.15° K.; mol. wt., 228.00]

T, ° K.	$H_T - H_{298.15}$, cal./mole	$S_T - S_{298.15}$, cal./deg. mole	T, ° K.	$H_T - H_{298.15}$, cal./mole	$S_T - S_{298.15}$, cal./deg. mole
400	850	2.45	1,000	6,095	10.44
500	1,705	4.36	1,200	7,870	12.06
600	2,575	5.94	1,400	9,645	13.42
700	3,450	7.29	1,600	11,425	14.61
800	4,325	8.46	1,800	13,210	15.66
900	5,210	9.51	2,000	14,995	16.60

BiF(g):

$$H_T - H_{298.15} = 8.84\,T + 0.03 \times 10^{-3}\,T^2 + 0.58 \times 10^5\,T^{-1}$$
$$-2,833 \; (0.2 \text{ percent}; \; 298°\text{–}2,000° \text{ K.});$$
$$C_p = 8.84 + 0.06 \times 10^{-3}\,T - 0.58 \times 10^5\,T^{-2}.$$

IODIDE

TABLE 88.—*Heat content and entropy of BiI(g)*

[Base, ideal gas at 298.15° K.; mol. wt., 335.91]

T, ° K.	$H_T - H_{298.15}$, cal./mole	$S_T - S_{298.15}$, cal./deg. mole	T, ° K.	$H_T - H_{298.15}$, cal./mole	$S_T - S_{298.15}$, cal./deg. mole
400	905	2.61	1,000	6,255	10.78
500	1,795	4.60	1,200	8,040	12.40
600	2,685	6.22	1,400	9,830	13.78
700	3,575	7.59	1,600	11,615	14.97
800	4,470	8.78	1,800	13,405	16.03
900	5,360	9.83	2,000	15,195	16.97

BiI(g):

$$H_T - H_{298.15} = 8.94\,T + 0.08 \times 10^5\,T^{-1} - 2,692 \; (0.1 \text{ percent};$$
$$298°\text{–}2,000° \text{ K.});$$
$$C_p = 8.94 - 0.08 \times 10^5\,T^{-2}.$$

HYDRIDES

TABLE 89.—*Heat content and entropy of BiH(g)*

[Base, ideal gas at 298.15° K.; mol. wt., 210.01]

T, ° K.	$H_T - H_{298.15}$, cal./mole	$S_T - S_{298.15}$, cal./deg. mole	T, ° K.	$H_T - H_{298.15}$, cal./mole	$S_T - S_{298.15}$, cal./deg. mole
400	720	2.08	1,000	5,375	9.21
500	1,445	3.69	1,200	7,035	10.72
600	2,195	5.06	1,400	8,735	12.03
700	2,965	6.24	1,600	10,450	13.17
800	3,750	7.30	1,800	12,180	14.19
900	4,555	8.24	2,000	13,920	15.11

BiH(g):

$$H_T - H_{298.15} = 7.06\,T + 0.53 \times 10^{-3}\,T^2 + 0.33 \times 10^5\,T^{-1}$$
$$-2,263 \; (0.6 \text{ percent}; \; 298°\text{–}2,000° \text{ K.});$$
$$C_p = 7.06 + 1.06 \times 10^{-3}\,T - 0.33 \times 10^5\,T^{-2}.$$

TABLE 90.—*Heat content and entropy of BiD(g)*

[Base, ideal gas at 298.15° K.; mol. wt., 211.01]

T, ° K.	$H_T - H_{298.15}$, cal./mole	$S_T - S_{298.15}$, cal./deg. mole	T, ° K.	$H_T - H_{298.15}$, cal./mole	$S_T - S_{298.15}$, cal./deg. mole
400	745	2.15	1,000	5,630	9.54
500	1,510	3.85	1,200	7,350	11.10
600	2,305	5.30	1,400	9,085	12.44
700	3,115	6.55	1,600	10,835	13.61
800	3,940	7.65	1,800	12,590	14.64
900	4,780	8.64	2,000	14,355	15.57

BiD(g):

$$H_T - H_{298.15} = 7.81\,T + 0.34 \times 10^{-3}\,T^2 + 0.74 \times 10^5\,T^{-1}$$
$$-2,607 \; (0.6 \text{ percent}; \; 298°\text{–}2,000° \text{ K.});$$
$$C_p = 7.81 + 0.68 \times 10^{-3}\,T - 0.74 \times 10^5\,T^{-2}.$$

BORON AND ITS COMPOUNDS

ELEMENT

TABLE 91.—*Heat content and entropy of B(c, l)*

[Base, crystals at 298.15° K.; atomic wt., 10.82]

T, ° K.	$H_T - H_{298.15}$, cal./mole	$S_T - S_{298.15}$, cal./deg. mole	T, ° K.	$H_T - H_{298.15}$, cal./mole	$S_T - S_{298.15}$, cal./deg. mole
400	310	0.89	1,700	7,765	8.61
500	690	1.73	1,800	8,460	9.00
600	1,120	2.52	1,900	9,165	9.38
700	1,600	3.26	2,000	9,880	9.75
800	2,120	3.95	2,100	10,605	10.10
900	2,670	4.60	2,200	11,340	10.45
1,000	3,245	5.20	2,300(c)	12,080	10.78
1,100	3,845	5.78	2,300(l)	17,380	13.08
1,200	4,465	6.31	2,400	18,130	13.40
1,300	5,100	6.82	2,600	19,630	14.00
1,400	5,750	7.30	2,800	21,130	14.55
1,500	6,410	7.76	3,000	22,630	15.07
1,600	7,080	8.19			

B(c):

$$H_T - H_{298.15} = 4.13\,T + 0.83 \times 10^{-3}\,T^2 + 1.76 \times 10^5\,T^{-1}$$
$$-1,895 \; (2.1 \text{ percent}; \; 298°\text{–}2,300° \text{ K.});$$
$$C_p = 4.13 + 1.66 \times 10^{-3}\,T - 1.76 \times 10^5\,T^{-2};$$
$$\Delta H_{2300}(fusion) = 5,300.$$

B(l):

$$H_T - H_{298.15} = 7.50\,T + 130 \; (0.1 \text{ percent};$$
$$2,300°\text{–}3,000° \text{ K.});$$
$$C_p = 7.50.$$

TABLE 92.—*Heat content and entropy of B(amorphous)*

[Base, amorphous substance at 298.15° K.; atomic wt., 10.82]

T, ° K.	$H_T - H_{298.15}$, cal./mole	$S_T - S_{298.15}$, cal./deg. mole	T, ° K.	$H_T - H_{298.15}$, cal./mole	$S_T - S_{298.15}$, cal./deg. mole
400	350	1.00	900	3,160	5.40
500	785	1.97	1,000	3,840	6.12
600	1,295	2.90	1,100	4,535	6.78
700	1,870	3.78	1,200	5,255	7.40
800	2,500	4.62			

TABLE 1h7. (*Continued*)

B(*amorphous*):

$H_T - H_{298.15} = 3.34T + 1.98 \times 10^{-3}T^2 + 1.48 \times 10^5 T^{-1}$
$- 1,668$ (1.1 percent; 298°–1,200° K.);
$C_p = 3.34 + 3.96 \times 10^{-3}T - 1.48 \times 10^5 T^{-2}.$

BO(*g*):

$H_T - H_{298.15} = 6.89T + 0.54 \times 10^{-3}T^2 + 0.21 \times 10^5 T^{-1}$
$- 2,173$ (0.7 percent; 298°–2,500° K.);
$C_p = 6.89 + 1.08 \times 10^{-3}T - 0.21 \times 10^5 T^{-2}.$

TABLE 93.—*Heat content and entropy of B(g)*

[Base, ideal gas at 298.15° K.; atomic wt., 10.82]

T, ° K.	$H_T - H_{298.15}$, cal./mole	$S_T - S_{298.15}$, cal./deg. mole	T, ° K.	$H_T - H_{298.15}$, cal./mole	$S_T - S_{298.15}$, cal./deg. mole
400	505	1.46	1,900	7,960	9.21
500	1,005	2.57	2,000	8,460	9.46
600	1,500	3.48	2,200	9,450	9.93
700	1,995	4.24	2,400	10,445	10.37
800	2,495	4.91	2,600	11,440	10.76
900	2,990	5.49	2,800	12,435	11.13
1,000	3,490	6.02	3,000	13,430	11.48
1,100	3,985	6.49	3,500	15,915	12.24
1,200	4,480	6.92	4,000	18,400	12.91
1,300	4,980	7.32	4,500	20,890	13.49
1,400	5,475	7.69	5,000	23,410	14.02
1,500	5,975	8.03	6,000	28,505	14.95
1,600	6,470	8.35	7,000	33,775	15.76
1,700	6,965	8.65	8,000	39,300	16.50
1,800	7,465	8.94			

B(*g*):

$H_T - H_{298.15} = 4.97T - 1,482$ (0.1 percent;
298°–5,000° K.);
$C_p = 4.97.$

TABLE 94.—*Heat content and entropy of $B_2(g)$*

[Base, ideal gas at 298.15° K.; mol. wt., 21.64]

T, ° K.	$H_T - H_{298.15}$, cal./mole	$S_T - S_{298.15}$, cal./deg. mole	T, ° K.	$H_T - H_{298.15}$, cal./mole	$S_T - S_{298.15}$, cal./deg. mole
400	765	2.20	1,600	11,065	13.91
500	1,545	3.94	1,700	11,965	14.45
600	2,355	5.42	1,800	12,865	14.96
700	3,185	6.70	1,900	13,770	15.45
800	4,030	7.83	2,000	14,675	15.92
900	4,890	8.84	2,100	15,580	16.36
1,000	5,755	9.75	2,200	16,485	16.78
1,100	6,630	10.58	2,300	17,395	17.19
1,200	7,510	11.35	2,400	18,305	17.57
1,300	8,390	12.05	2,500	19,220	17.95
1,400	9,280	12.71	2,750	21,510	18.82
1,500	10,170	13.33	3,000	23,805	19.62

$B_2(g)$:

$H_T - H_{298.15} = 8.10T + 0.29 \times 10^{-3}T^2 + 0.86 \times 10^5 T^{-1}$
$- 2,729$ (0.5 percent; 298°–2,500° K.);
$C_p = 8.10 + 0.58 \times 10^{-3}T - 0.86 \times 10^5 T^{-2}.$

OXIDES

TABLE 95.—*Heat content and entropy of BO(g)*

[Base, ideal gas at 298.15° K.; mol. wt., 26.82]

T, ° K.	$H_T - H_{298.15}$, cal./mole	$S_T - S_{298.15}$, cal./deg. mole	T, ° K.	$H_T - H_{298.15}$, cal./mole	$S_T - S_{298.15}$, cal./deg. mole
400	715	2.07	1,600	10,315	12.89
500	1,430	3.66	1,700	11,175	13.42
600	2,160	4.99	1,800	12,040	13.91
700	2,915	6.15	1,900	12,910	14.38
800	3,685	7.18	2,000	13,785	14.83
900	4,475	8.11	2,100	14,660	15.26
1,000	5,280	8.96	2,200	15,540	15.66
1,100	6,095	9.74	2,300	16,420	16.06
1,200	6,925	10.46	2,400	17,305	16.43
1,300	7,760	11.13	2,500	18,195	16.80
1,400	8,605	11.75	2,750	20,415	17.64
1,500	9,455	12.34	3,000	22,645	18.42

TABLE 96.—*Heat content and entropy of $B_2O_3(c, l)$*

[Base, crystals at 298.15° K.; mol. wt., 69.64]

T, ° K.	$H_T - H_{298.15}$, cal./mole	$S_T - S_{298.15}$, cal./deg. mole	T, ° K.	$H_T - H_{298.15}$, cal./mole	$S_T - S_{298.15}$, cal./deg. mole
400	1,640	4.71	1,200	29,010	40.98
500	3,700	9.29	1,300	32,060	43.42
600	5,860	13.23	1,400	35,110	45.68
700	8,350	17.06	1,500	38,160	47.78
723(c)	8,960	17.91	1,600	41,210	49.75
723(l)	14,460	25.52	1,700	44,260	51.60
800	16,810	28.61	1,800	47,310	53.34
900	19,860	32.20	1,900	50,360	54.99
1,000	22,910	35.42	2,000	53,410	56.56
1,100	25,960	38.32			

$B_2O_3(c)$:

$H_T - H_{298.15} = 8.73T + 12.70 \times 10^{-3}T^2 + 1.31 \times 10^5 T^{-1}$
$- 4,171$ (1.0 percent; 298°–723° K.);
$C_p = 8.73 + 25.40 \times 10^{-3}T - 1.31 \times 10^5 T^{-2};$
$\Delta H_{723}(fusion) = 5,500.$

$B_2O_3(l)$:

$H_T - H_{298.15} = 30.50T - 7,740$ (0.1 percent;
723°–2,000° K.);
$C_p = 30.50.$

TABLE 97.—*Heat content and entropy of $B_2O_3(gl, l)$*

[Base, glass at 298.15° K.; mol. wt., 69.64]

T, ° K.	$H_T - H_{298.15}$, cal./mole	$S_T - S_{298.15}$, cal./deg. mole	T, ° K.	$H_T - H_{298.15}$, cal./mole	$S_T - S_{298.15}$, cal./deg. mole
400	1,620	4.64	1,200	24,650	35.27
500	3,770	9.43	1,300	27,700	37.71
600	6,400	14.21	1,400	30,750	39.97
700	9,400	18.82	1,500	33,800	42.07
723(gl)	10,100	19.81	1,600	36,850	44.04
723(l)	10,100	19.81	1,700	39,900	45.89
800	12,450	22.90	1,800	42,950	47.63
900	15,500	26.49	1,900	46,000	49.28
1,000	18,550	29.71	2,000	49,050	50.85
1,100	21,600	32.61			

$B_2O_3(gl)$:

$H_T - H_{298.15} = 2.28T + 21.05 \times 10^{-3}T^2 - 2,551$
(2.0 percent; 298°–723° K.);
$C_p = 2.28 + 42.10 \times 10^{-3}T.$

$B_2O_3(l)$:

$H_T - H_{298.15} = 30.50T - 11,950$ (0.1 percent;
723°–2,000° K.);
$C_p = 30.50.$

TABLE 1h7. (Continued)

SULFIDE

TABLE 98.—*Heat content and entropy of BS(g)*

[Base, ideal gas at 298.15° K.; mol. wt., 42.89]

T, ° K.	$H_T-H_{298.15}$, cal./mole	$S_T-S_{298.15}$, cal./deg. mole	T, ° K.	$H_T-H_{298.15}$, cal./mole	$S_T-S_{298.15}$, cal./deg. mole
400	750	2.16	1,600	10,890	13.67
500	1,515	3.87	1,700	11,780	14.20
600	2,305	5.31	1,800	12,670	14.71
700	3,120	6.56	1,900	13,560	15.19
800	3,950	7.67	2,000	14,455	15.65
900	4,795	8.67	2,100	15,350	16.09
1,000	5,650	9.57	2,200	16,245	16.51
1,100	6,510	10.39	2,300	17,140	16.91
1,200	7,380	11.14	2,400	18,040	17.29
1,300	8,250	11.84	2,500	18,940	17.66
1,400	9,130	12.49	2,750	21,195	18.52
1,500	10,010	13.10	3,000	23,460	19.30

$BS(g)$:

$$H_T-H_{298.15}=7.94T+0.30\times10^{-3}T^2+0.83\times10^5T^{-1}$$
$$-2,672\ (0.8\ \text{percent};\ 298°\text{-}3,000°\ \text{K.});$$
$$C_p=7.94+0.60\times10^{-3}T-0.83\times10^5T^{-2}.$$

CARBIDE

TABLE 99.—*Heat content and entropy of $B_4C(c)$*

[Base, crystals at 298.15° K.; mol. wt., 55.29]

T, ° K.	$H_T-H_{298.15}$, cal./mole	$S_T-S_{298.15}$, cal./deg. mole	T, ° K.	$H_T-H_{298.15}$, cal./mole	$S_T-S_{298.15}$, cal./deg. mole
400	1,640	4.69	1,300	24,700	33.62
500	3,570	8.99	1,400	27,620	35.78
600	5,780	13.01	1,500	30,580	37.82
700	8,180	16.71	1,600	33,570	39.75
800	10,720	20.10	1,700	36,600	41.59
900	13,380	23.23	1,800	39,660	43.34
1,000	16,150	26.15	1,900	42,750	45.01
1,100	18,970	28.84	2,000	45,870	46.61
1,200	21,820	31.32			

$B_4C(c)$:

$$H_T-H_{298.15}=22.99T+2.70\times10^{-3}T^2+10.72\times10^5T^{-1}$$
$$-10,690\ (0.7\ \text{percent};\ 298°\text{-}1,700°\ \text{K.});$$
$$C_p=22.99+5.40\times10^{-3}T-10.72\times10^5T^{-2}.$$

NITRIDE

TABLE 100.—*Heat content and entropy of BN(c)*

[Base, crystals at 298.15° K.; mol. wt., 24.83]

T, ° K.	$H_T-H_{298.15}$, cal./mole	$S_T-S_{298.15}$, cal./deg. mole	T, ° K.	$H_T-H_{298.15}$, cal./mole	$S_T-S_{298.15}$, cal./deg. mole
400	315	0.91	900	2,420	4.22
500	670	1.70	1,000	2,950	4.78
600	1,035	2.36	1,100	3,490	5.29
700	1,450	3.00	1,200	4,045	5.77
800	1,920	3.63			

$BN(c)$:

$$H_T-H_{298.15}=1.82T+1.81\times10^{-3}T^2-704\ (0.8\ \text{percent};$$
$$298°\text{-}1,200°\ \text{K.});$$
$$C_p=1.82+3.62\times10^{-3}T.$$

TABLE 101.—*Heat content and entropy of BN(g)*

[Base, ideal gas at 298.15° K.; mol. wt., 24.83]

T, ° K.	$H_T-H_{298.15}$, cal./mole	$S_T-S_{298.15}$, cal./deg. mole	T, ° K.	$H_T-H_{298.15}$, cal./mole	$S_T-S_{298.15}$, cal./deg. mole
400	725	2.09	1,600	10,650	13.31
500	1,465	3.74	1,700	11,530	13.84
600	2,225	5.13	1,800	12,420	14.35
700	3,010	6.34	1,900	13,310	14.83
800	3,815	7.41	2,000	14,200	15.29
900	4,635	8.38	2,100	15,095	15.73
1,000	5,470	9.26	2,200	15,990	16.14
1,100	6,315	10.06	2,300	16,890	16.54
1,200	7,170	10.81	2,400	17,795	16.93
1,300	8,030	11.50	2,500	18,700	17.30
1,400	8,900	12.14	2,750	20,965	18.16
1,500	9,770	12.74	3,000	23,240	18.95

$BN(g)$:

$$H_T-H_{298.15}=7.20T+0.53\times10^{-3}T^2+0.42\times10^5T^{-1}$$
$$-2,335\ (0.8\ \text{percent};\ 298°\text{-}2,500°\ \text{K.});$$
$$C_p=7.20+1.06\times10^{-3}T-0.42\times10^5T^{-2}.$$

HYDRIDES

TABLE 102.—*Heat content and entropy of BH(g)*

[Base, ideal gas at 298.15° K.; mol. wt., 11.83]

T, ° K.	$H_T-H_{298.15}$, cal./mole	$S_T-S_{298.15}$, cal./deg. mole	T, °K.	$H_T-H_{298.15}$, cal./mole	$S_T-S_{298.15}$, cal./deg. mole
400	710	2.05	2,000	13,625	14.62
500	1,415	3.62	2,100	14,505	15.05
600	2,135	4.93	2,200	15,395	15.46
700	2,870	6.07	2,300	16,285	15.86
800	3,620	7.07	2,400	17,185	16.24
900	4,390	7.97	2,500	18,085	16.61
1,000	5,170	8.80	2,750	20,350	17.48
1,100	5,970	9.56	3,000	22,630	18.27
1,200	6,785	10.27	3,250	24,925	19.00
1,300	7,610	10.93	3,500	27,235	19.69
1,400	8,445	11.55	3,750	29,565	20.33
1,500	9,290	12.13	4,000	31,900	20.94
1,600	10,145	12.68	4,250	34,245	21.50
1,700	11,005	13.20	4,500	36,605	22.04
1,800	11,870	13.70	4,750	38,980	22.56
1,900	12,745	14.17	5,000	41,360	23.04

$BH(g)$:

$$H_T-H_{298.15}=6.62T+0.59\times10^{-3}T^2-2,026\ (0.6\ \text{percent};$$
$$298°\text{-}2,500°\ \text{K.});$$
$$C_p=6.62+1.18\times10^{-3}T.$$

TABLE 103.—*Heat content and entropy of BD(g)*

[Base, ideal gas at 298.15° K.; mol. wt., 12.83]

T, ° K.	$H_T-H_{298.15}$, cal./mole	$S_T-S_{298.15}$, cal./deg. mole	T, ° K.	$H_T-H_{298.15}$, cal./mole	$S_T-S_{298.15}$, cal./deg. mole
400	715	2.06	1,000	5,330	9.03
500	1,435	3.66	1,200	6,985	10.54
600	2,180	5.02	1,400	8,670	11.84
700	2,940	6.20	1,600	10,380	12.98
800	3,720	7.24	1,800	12,105	14.00
900	4,520	8.18	2,000	13,840	14.91

$BD(g)$:

$$H_T-H_{298.15}=6.93T+0.56\times10^{-3}T^2+0.24\times10^5T^{-1}$$
$$-2,196\ (0.7\ \text{percent};\ 298°\text{-}2,000°\ \text{K.});$$
$$C_p=6.93+1.12\times10^{-3}T-0.24\times10^5T^{-2}.$$

TABLE 1h7. (*Continued*)

TABLE 104.—*Heat content and entropy of* $B_2H_6(g)$

[Base, ideal gas at 298.15° K.; mol. wt., 27.69]

T, ° K.	$H_T-H_{298.15}$, cal./mole	$S_T-S_{298.15}$, cal./deg. mole	T, ° K.	$H_T-H_{298.15}$, cal./mole	$S_T-S_{298.15}$, cal./deg. mole
400	1,555	4.45	1,600	38,595	43.88
500	3,460	8.69	1,700	42,440	46.21
600	5,690	12.74	1,800	46,340	48.44
700	8,215	16.63	1,900	50,290	50.58
800	10,985	20.33	2,000	54,270	52.62
900	13,970	23.84	2,100	58,280	54.58
1,000	17,130	27.17	2,200	62,330	56.46
1,100	20,445	30.33	2,300	66,400	58.27
1,200	23,890	33.32	2,400	70,490	60.01
1,300	27,445	36.17	2,500	74,595	61.68
1,400	31,090	38.87	2,750	84,950	65.63
1,500	34,810	41.43	3,000	95,390	69.26

$B_2H_6(g)$:

$$H_T - H_{298.15} = 13.68T + 9.30 \times 10^{-3}T^2 + 5.27 \times 10^5 T^{-1}$$
$$-6,673 \ (1.4 \ \text{percent}; \ 298°–1,500° \ \text{K.});$$
$$C_p = 13.68 + 18.60 \times 10^{-3}T - 5.27 \times 10^5 T^{-2}.$$

TABLE 105.—*Heat content and entropy of* $B_5H_9(g)$

[Base, ideal gas at 298.15° K.; mol. wt., 63.17]

T, ° K.	$H_T-H_{298.15}$, cal./mole	$S_T-S_{298.15}$, cal./deg. mole	T, ° K.	$H_T-H_{298.15}$, cal./mole	$S_T-S_{298.15}$, cal./deg. mole
400	2,900	8.29	1,000	33,440	52.90
500	6,625	16.58	1,100	39,800	58.96
600	10,055	24.64	1,200	46,370	64.67
700	16,050	32.33	1,300	53,100	70.06
800	21,500	39.60	1,400	59,975	75.15
900	27,315	46.45	1,500	66,980	79.98

$B_5H_9(g)$:

$$H_T - H_{298.15} = 31.26T + 16.09 \times 10^{-3}T^2 + 15.41$$
$$\times 10^5 T^{-1} - 15,919 \ (1.4 \ \text{percent}; \ 298°–1,500° \ \text{K.});$$
$$C_p = 31.26 + 32.18 \times 10^{-3}T - 15.41 \times 10^5 T^{-2}.$$

TABLE 106.—*Heat content and entropy of* $B_{10}H_{14}(g)$

[Base, ideal gas at 298.15° K.; mol. wt., 122.31]

T, ° K.	$H_T-H_{298.15}$, cal./mole	$S_T-S_{298.15}$, cal./deg. mole	T, ° K.	$H_T-H_{298.15}$, cal./mole	$S_T-S_{298.15}$, cal./deg. mole
400	4,940	14.13	800	36,880	67.85
500	11,300	28.27	900	46,925	79.67
600	18,905	42.11	1,000	57,385	90.69
700	27,490	55.32			

$B_{10}H_{14}(g)$:

$$H_T - H_{298.15} = 41.56T + 36.30 \times 10^{-3}T^2 + 20.63$$
$$\times 10^5 T^{-1} - 22,537 \ (1.0 \ \text{percent}; \ 298°–1,000° \ \text{K.});$$
$$C_p = 41.56 + 72.60 \times 10^{-3}T - 20.63 \times 10^5 T^{-2}.$$

BROMIDES

TABLE 107.—*Heat content and entropy of* $BBr(g)$

[Base, ideal gas at 298.15° K.; mol. wt., 90.74]

T, ° K.	$H_T-H_{298.15}$, cal./mole	$S_T-S_{298.15}$, cal./deg. mole	T, ° K.	$H_T-H_{298.15}$, cal./mole	$S_T-S_{298.15}$, cal./deg. mole
400	820	2.36	1,600	11,395	14.46
500	1,655	4.23	1,700	12,305	15.01
600	2,505	5.78	1,800	13,205	15.53
700	3,370	7.11	1,900	14,110	16.02
800	4,245	8.28	2,000	15,020	16.48
900	5,130	9.32	2,100	15,930	16.93
1,000	6,015	10.25	2,200	16,840	17.35
1,100	6,905	11.10	2,300	17,750	17.76
1,200	7,800	11.88	2,400	18,665	18.15
1,300	8,695	12.59	2,500	19,580	18.52
1,400	9,590	13.26	2,750	21,875	19.39
1,500	10,490	13.88	3,000	24,170	20.19

$BBr(g)$:

$$H_T - H_{298.15} = 8.72T + 0.10 \times 10^{-3}T^2 + 0.83 \times 10^5 T^{-1}$$
$$-2,887 \ (0.2 \ \text{percent}; \ 298°–3,000° \ \text{K.});$$
$$C_p = 8.72 + 0.20 \times 10^{-3}T - 0.83 \times 10^5 T^{-2}.$$

TABLE 108.—*Heat content and entropy of* $BBr_3(g)$

[Base, ideal gas at 298.15° K.; mol. wt., 250.57]

T, ° K.	$H_T-H_{298.15}$, cal./mole	$S_T-S_{298.15}$, cal./deg. mole	T, ° K.	$H_T-H_{298.15}$, cal./mole	$S_T-S_{298.15}$, cal./deg. mole
400	1,725	4.97	1,500	22,750	29.94
500	3,505	8.94	1,600	24,715	31.21
600	5,345	12.29	1,700	26,680	32.40
700	7,225	15.19	1,800	28,650	33.53
800	9,130	17.73	1,900	30,620	34.59
1,000	11,045	19.99	2,000	32,590	35.60
1,100	12,975	22.02	2,100	34,560	36.57
1,100	14,920	23.88	2,200	36,535	37.49
1,200	16,865	25.57	2,300	38,515	38.36
1,300	18,825	27.14	2,400	40,495	39.21
1,400	20,790	28.59	2,500	42,480	40.02

$BBr_3(g)$:

$$H_T - H_{298.15} = 18.92T + 0.28 \times 10^{-3}T^2 + 2.48 \times 10^5 T^{-1}$$
$$-6,498 \ (0.3 \ \text{percent}; \ 298°–2,500° \ \text{K.});$$
$$C_p = 18.92 + 0.56 \times 10^{-3}T - 2.48 \times 10^5 T^{-2}.$$

CHLORIDES

TABLE 109.—*Heat content and entropy of* $BCl(g)$

[Base, ideal gas at 298.15° K.; mol. wt., 46.28]

T, ° K.	$H_T-H_{298.15}$, cal./mole	$S_T-S_{298.15}$, cal./deg. mole	T, ° K.	$H_T-H_{298.15}$, cal./mole	$S_T-S_{298.15}$, cal./deg. mole
400	790	2.28	1,600	11,255	14.22
500	1,605	4.09	1,700	12,155	14.76
600	2,440	5.62	1,800	13,055	15.28
700	3,290	6.93	1,900	13,960	15.77
800	4,155	8.08	2,000	14,865	16.23
900	5,025	9.11	2,100	15,775	16.68
1,000	5,905	10.03	2,200	16,685	17.10
1,100	6,785	10.87	2,300	17,595	17.50
1,200	7,670	11.64	2,400	18,505	17.89
1,300	8,565	12.36	2,500	19,420	18.26
1,400	9,460	13.02	2,750	21,710	19.14
1,500	10,355	13.64	3,000	24,005	19.94

$BCl(g)$:

$$H_T - H_{298.15} = 8.55T + 0.14 \times 10^{-3}T^2 + 0.95 \times 10^5 T^{-1}$$
$$-2,880 \ (0.3 \ \text{percent}; \ 298°–3,000° \ \text{K.});$$
$$C_p = 8.55 + 0.28 \times 10^{-3}T - 0.95 \times 10^5 T^{-2}.$$

TABLE 1h7. (*Continued*)

TABLE 110.—*Heat content and entropy of $BCl_3(g)$*

[Base, ideal gas at 298.15° K.; mol. wt., 117.19]

T, ° K.	$H_T - H_{298.15}$, cal./mole	$S_T - S_{298.15}$, cal./deg. mole	T, ° K.	$H_T - H_{298.15}$, cal./mole	$S_T - S_{298.15}$, cal./deg. mole
400	1,605	4.62	1,500	22,215	28.98
500	3,300	8.40	1,600	24,175	30.25
600	5,070	11.62	1,700	26,135	31.44
700	6,895	14.44	1,800	28,085	32.55
800	8,750	16.91	1,900	30,060	33.62
900	10,635	19.13	2,000	32,020	34.62
1,000	12,540	21.14	2,100	33,975	35.58
1,100	14,460	22.97	2,200	35,955	36.50
1,200	16,390	24.65	2,300	37,925	37.37
1,300	18,325	26.20	2,400	39,885	38.21
1,400	20,270	27.64	2,500	41,865	39.02

$BCl_3(g)$:

$$H_T - H_{298.15} = 18.45T + 0.40 \times 10^{-3} T^2 + 3.28 \times 10^5 T^{-1}$$
$$-6,637 \ (0.5 \text{ percent}; 298°-2,500° \text{ K.});$$
$$C_p = 18.45 + 0.80 \times 10^{-3} T - 3.28 \times 10^5 T^{-2}.$$

FLUORIDES

TABLE 111.—*Heat content and entropy of $BF(g)$*

[Base, ideal gas at 298.15° K.; mol. wt., 29.82]

T, ° K.	$H_T - H_{298.15}$, cal./mole	$S_T - S_{298.15}$, cal./deg. mole	T, ° K.	$H_T - H_{298.15}$, cal./mole	$S_T - S_{298.15}$, cal./deg. mole
400	730	2.11	1,600	10,725	13.41
500	1,475	3.77	1,700	11,605	13.94
600	2,250	5.18	1,800	12,490	14.45
700	3,045	6.40	1,900	13,380	14.94
800	3,855	7.48	2,000	14,270	15.39
900	4,680	8.45	2,100	15,165	15.83
1,000	5,525	9.34	2,200	16,060	16.25
1,100	6,375	10.15	2,300	16,955	16.65
1,200	7,230	10.90	2,400	17,855	17.03
1,300	8,095	11.59	2,500	18,760	17.40
1,400	8,965	12.24	2,750	21,015	18.26
1,500	9,845	12.84	3,000	23,285	19.05

$BF(g)$:

$$H_T - H_{298.15} = 7.37T + 0.48 \times 10^{-3} T^2 + 0.52 \times 10^5 T^{-1}$$
$$-2,414 \ (0.7 \text{ percent}; 298°-2,500° \text{ K.});$$
$$C_p = 7.37 + 0.96 \times 10^{-3} T - 0.52 \times 10^5 T^{-2}.$$

TABLE 112.—*Heat content and entropy of $BF_3(g)$*

[Base, ideal gas at 298.15° K.; mol. wt., 67.82]

T, ° K.	$H_T - H_{298.15}$, cal./mole	$S_T - S_{298.15}$, cal./deg. mole	T, ° K.	$H_T - H_{298.15}$, cal./mole	$S_T - S_{298.15}$, cal./deg. mole
400	1,320	3.80	1,500	20,555	26.17
500	2,760	7.01	1,600	22,465	27.41
600	4,320	9.85	1,700	24,380	28.57
700	5,965	12.38	1,800	26,305	29.67
800	7,670	14.66	1,900	28,240	30.71
900	9,430	16.73	2,000	30,175	31.71
1,000	11,225	18.62	2,100	32,110	32.65
1,100	13,055	20.36	2,200	34,050	33.55
1,200	14,905	21.97	2,300	35,995	34.42
1,300	16,770	23.47	2,400	37,945	35.25
1,400	18,655	24.86	2,500	39,900	36.05

$BF_3(g)$:

$$H_T - H_{298.15} = 15.24T + 1.40 \times 10^{-3} T^2 + 3.57 \times 10^5 T^{-1}$$
$$-5,866 \ (0.9 \text{ percent}; 298°-2,200° \text{ K.});$$
$$C_p = 15.24 + 2.80 \times 10^{-3} T - 3.57 \times 10^5 T^{-2}.$$

IODIDE

TABLE 113.—*Heat content and entropy of $BI_3(g)$*

[Base, ideal gas at 298.15° K.; mol. wt., 391.55]

T, ° K.	$H_T - H_{298.15}$, cal./mole	$S_T - S_{298.15}$, cal./deg. mole	T, ° K.	$H_T - H_{298.15}$, cal./mole	$S_T - S_{298.15}$, cal./deg. mole
400	1,785	5.14	800	9,300	18.11
500	3,605	9.20	900	11,235	20.39
600	5,475	12.61	1,000	13,185	22.44
700	7,380	15.55			

$BI_3(g)$:

$$H_T - H_{298.15} = 18.79T + 0.49 \times 10^{-3} T^2 + 1.90 \times 10^5 T^{-1}$$
$$-6,283 \ (0.1 \text{ percent}; 298°-1,000° \text{ K.});$$
$$C_p = 18.79 + 0.98 \times 10^{-3} T - 1.90 \times 10^5 T^{-2}.$$

BORAZINE

TABLE 114.—*Heat content and entropy of $B_3N_3H_6(g)$*

[Base, ideal gas at 298.15° K.; mol. wt., 80.53]

T, ° K.	$H_T - H_{298.15}$, cal./mole	$S_T - S_{298.15}$, cal./deg. mole	T, ° K.	$H_T - H_{298.15}$, cal./mole	$S_T - S_{298.15}$, cal./deg. mole
400	2,735	7.83	1,000	28,620	45.89
500	6,060	15.23	1,100	33,890	50.91
600	9,900	22.22	1,200	39,330	55.64
700	14,140	28.75	1,300	44,910	60.11
800	18,710	34.85	1,400	50,610	64.33
900	23,550	40.56	1,500	56,410	68.33

$B_3N_3H_6(g)$:

$$H_T - H_{298.15} = 30.71T + 10.71 \times 10^{-3} T^2 + 12.38 \times 10^5 T^{-1}$$
$$-14,261 \ (1.1 \text{ percent}; 298°-1,500° \text{ K.});$$
$$C_p = 30.71 + 21.42 \times 10^{-3} T - 12.38 \times 10^5 T^{-2}.$$

BROMINE AND ITS COMPOUNDS

ELEMENT

TABLE 115.—*Heat content and entropy of $Br(g)$*

[Base, ideal gas at 298.15° K.; atomic wt., 79.92]

T, ° K.	$H_T - H_{298.15}$, cal./mole	$S_T - S_{298.15}$, cal./deg. mole	T, ° K.	$H_T - H_{298.15}$, cal./mole	$S_T - S_{298.15}$, cal./deg. mole
400	505	1.46	1,900	8,275	9.43
500	1,005	2.57	2,000	8,815	9.71
600	1,500	3.48	2,200	9,905	10.23
700	2,000	4.25	2,400	10,995	10.70
800	2,500	4.92	2,600	12,085	11.14
900	3,000	5.51	2,800	13,170	11.54
1,000	3,515	6.04	3,000	14,255	11.91
1,100	4,025	6.53	3,500	16,955	12.75
1,200	4,545	6.98	4,000	19,630	13.46
1,300	5,065	7.40	4,500	22,285	14.09
1,400	5,595	7.79	5,000	24,920	14.64
1,500	6,125	8.16	6,000	30,140	15.59
1,600	6,655	8.50	7,000	35,305	16.39
1,700	7,195	8.83	8,000	40,445	17.08
1,800	7,735	9.14			

$Br(g)$:

$$H_T - H_{298.15} = 4.90T + 0.11 \times 10^{-3} T^2$$
$$-1,471 \ (0.5 \text{ percent}; 298°-3,500° \text{ K.});$$
$$C_p = 4.90 + 0.22 \times 10^{-3} T.$$

TABLE 1h7. (*Continued*)

TABLE 116.—*Heat content and entropy of* $Br_2(g)$

[Base, ideal gas at 298.15° K.; mol. wt., 159.83]

T, ° K.	$H_T-H_{298.15}$, cal./mole	$S_T-S_{298.15}$, cal./deg. mole	T, ° K.	$H_T-H_{298.15}$, cal./mole	$S_T-S_{298.15}$, cal./deg. mole
400	885	2.56	1,600	11,675	14.98
500	1,770	4.52	1,700	12,585	15.53
600	2,655	6.14	1,800	13,495	16.05
700	3,550	7.52	1,900	14,410	16.54
800	4,445	8.72	2,000	15,325	17.01
900	5,345	9.77	2,100	16,235	17.46
1,000	6,245	10.72	2,200	17,150	17.88
1,100	7,145	11.58	2,300	18,070	18.29
1,200	8,050	12.37	2,400	18,990	18.68
1,300	8,955	13.09	2,500	19,905	19.05
1,400	9,860	13.76	2,750	22,210	19.94
1,500	10,770	14.39	3,000	24,515	20.74

$Br_2(g)$:

$$H_T-H_{298.15}=8.92T+0.06\times10^{-3}T^2+0.30\times10^5T^{-1}$$
$$-2,765 \ (0.1 \ \text{percent}; \ 298°\text{–}3,000° \ \text{K.});$$
$$C_p=8.92+0.12\times10^{-3}T-0.30\times10^5T^{-2}.$$

OXIDE

TABLE 117.—*Heat content and entropy of* $BrO(g)$

[Base, ideal gas at 298.15° K.; mol. wt., 95.92]

T, ° K.	$H_T-H_{298.15}$, cal./mole	$S_T-S_{298.15}$, cal./deg. mole	T, ° K.	$H_T-H_{298.15}$, cal./mole	$S_T-S_{298.15}$, cal./deg. mole
400	815	2.35	1,000	5,960	10.16
500	1,640	4.19	1,200	7,725	11.77
600	2,485	5.73	1,400	9,495	13.13
700	3,345	7.06	1,600	11,270	14.32
800	4,215	8.22	1,800	13,045	15.36
900	5,085	9.24	2,000	14,825	16.30

$BrO(g)$:

$$H_T-H_{298.15}=8.62T+0.10\times10^{-3}T^2+0.79\times10^5T^{-1}$$
$$-2,844 \ (0.2 \ \text{percent}; \ 298°\text{–}2,000° \ \text{K.});$$
$$C_p=8.62+0.20\times10^{-3}T-0.79\times10^5T^{-2}.$$

CHLORIDE

TABLE 118.—*Heat content and entropy of* $BrCl(g)$

[Base, ideal gas at 298.15° K.; mol. wt., 115.37]

T, ° K.	$H_T-H_{298.15}$, cal./mole	$S_T-S_{298.15}$, cal./deg. mole	T, ° K.	$H_T-H_{298.15}$, cal./mole	$S_T-S_{298.15}$, cal./deg. mole
400	865	2.49	1,000	6,185	10.59
500	1,735	4.43	1,200	7,985	12.24
600	2,615	6.04	1,400	9,790	13.63
700	3,500	7.40	1,600	11,605	14.84
800	4,395	8.60	1,800	13,420	15.91
900	5,285	9.65	2,000	15,245	16.87

$BrCl(g)$:

$$H_T-H_{298.15}=8.88T+0.07\times10^{-3}T^2+0.48\times10^5T^{-1}$$
$$-2,815 \ (0.1 \ \text{percent}; \ 298°\text{–}2,000° \ \text{K.});$$
$$C_p=8.88+0.14\times10^{-3}T-0.48\times10^5T^{-2}.$$

FLUORIDES

TABLE 119.—*Heat content and entropy of* $BrF(g)$

[Base, ideal gas at 298.15° K.; mol. wt., 98.92]

T, ° K.	$H_T-H_{298.15}$, cal./mole	$S_T-S_{298.15}$, cal./deg. mole	T, ° K.	$H_T-H_{298.15}$, cal./mole	$S_T-S_{298.15}$, cal./deg. mole
400	820	2.36	1,000	6,020	10.27
500	1,660	4.23	1,200	7,805	11.90
600	2,515	5.79	1,400	9,595	13.28
700	3,380	7.13	1,600	11,395	14.48
800	4,255	8.30	1,800	13,200	15.54
900	5,135	9.34	2,000	15,010	16.49

$BrF(g)$:

$$H_T-H_{298.15}=8.68T+0.12\times10^{-3}T^2+0.78\times10^5T^{-1}$$
$$-2,860 \ (0.2 \ \text{percent}; \ 298°\text{–}2,000° \ \text{K.});$$
$$C_p=8.68+0.24\times10^{-3}T-0.78\times10^5T^{-2}.$$

TABLE 120.—*Heat content and entropy of* $BrF_3(g)$

[Base, ideal gas at 298.15° K.; mol. wt., 136.92]

T, ° K.	$H_T-H_{298.15}$, cal./mole	$S_T-S_{298.15}$, cal./deg. mole	T, ° K.	$H_T-H_{298.15}$, cal./mole	$S_T-S_{298.15}$, cal./deg. mole
400	1,705	4.91	800	9,120	17.69
500	3,485	8.88	900	11,040	19.95
600	5,330	12.24	1,000	12,980	21.99
700	7,210	15.14			

$BrF_3(g)$:

$$H_T-H_{298.15}=18.73T+0.53\times10^{-3}T^2+2.76\times10^5T^{-1}$$
$$-6,557 \ (0.1 \ \text{percent}; \ 298°\text{–}1,000° \ \text{K.});$$
$$C_p=18.73+1.06\times10^{-3}T-2.76\times10^5T^{-2}.$$

TABLE 121.—*Heat content and entropy of* $BrF_5(g)$

[Base, ideal gas at 298.15° K.; mol. wt., 174.92]

T, ° K.	$H_T-H_{298.15}$, cal./mole	$S_T-S_{298.15}$, cal./deg. mole	T, ° K.	$H_T-H_{298.15}$, cal./mole	$S_T-S_{298.15}$, cal./deg. mole
400	2,595	7.47	1,000	20,375	34.36
500	5,365	13.65	1,100	23,475	37.32
600	8,255	18.92	1,200	26,580	40.02
700	11,225	23.49	1,300	29,700	42.51
800	14,245	27.52	1,400	32,830	44.82
900	17,295	31.12	1,500	35,965	46.99

$BrF_5(g)$:

$$H_T-H_{298.15}=30.03T+0.68\times10^{-3}T^2+5.78\times10^5T^{-1}$$
$$-10,952 \ (0.3 \ \text{percent}; \ 298°\text{–}1,500° \ \text{K.});$$
$$C_p=30.03+1.36\times10^{-3}T-5.78\times10^5T^{-2}.$$

CADMIUM AND ITS COMPOUNDS

ELEMENT

TABLE 122.—*Heat content and entropy of* $Cd(c, l)$

[Base, crystals at 298.15° K.; atomic wt., 112.41]

T, ° K.	$H_T-H_{298.15}$, cal./mole	$S_T-S_{298.15}$, cal./deg. mole	T, ° K.	$H_T-H_{298.15}$, cal./mole	$S_T-S_{298.15}$, cal./deg. mole
400	645	1.86	700	4,160	8.13
500	1,310	3.34	800	4,870	9.08
594(c)	1,960	4.53	900	5,580	9.92
594(l)	3,410	6.97	1,000	6,290	10.67
600	3,450	7.04	1,100	7,000	11.35

$Cd(c)$:

$$H_T-H_{298.15}=5.31T+1.47\times10^{-3}T^2$$
$$-1,714 \ (0.1 \ \text{percent}; \ 298°\text{–}594° \ \text{K.});$$
$$C_p=5.31+2.94\times10^{-3}T;$$
$$\Delta H_{594} \ (fusion)=1,450.$$

$Cd(l)$:

$$H_T-H_{298.15}=7.10T-810 \ (0.1 \ \text{percent}; \ 594°\text{–}1,100° \ \text{K.});$$
$$C_p=7.10.$$

TABLE 1h7. (*Continued*)

TABLE 123.—*Heat content and entropy of Cd(g)*

[Base, ideal gas at 298.15° K.; atomic wt., 112.41]

T, ° K.	$H_T - H_{298.15}$, cal./mole	$S_T - S_{298.15}$, cal./deg. mole	T, ° K.	$H_T - H_{298.15}$, cal./mole	$S_T - S_{298.15}$, cal./deg. mole
400	505	1.46	1,900	7,960	9.20
500	1,005	2.57	2,000	8,455	9.46
600	1,500	3.48	2,200	9,450	9.93
700	1,995	4.24	2,400	10,445	10.36
800	2,495	4.90	2,600	11,440	10.76
900	2,990	5.49	2,800	12,435	11.13
1,000	3,490	6.01	3,000	13,430	11.47
1,100	3,985	6.49	3,500	15,915	12.24
1,200	4,480	6.92	4,000	18,405	12.91
1,300	4,980	7.32	4,500	20,920	13.50
1,400	5,475	7.69	5,000	23,470	14.04
1,500	5,975	8.03	6,000	28,805	15.01
1,600	6,470	8.35	7,000	34,740	15.90
1,700	6,965	8.65	8,000	41,705	16.85
1,800	7,465	8.93			

$$\text{Cd}(g):$$
$$H_T - H_{298.15} = 4.97T - 1,482 \ (0.1 \text{ percent}; 298\text{–}5,000° \text{ K.});$$
$$C_p = 4.97.$$

OXIDE

$$\text{CdO}(c):$$
$$C_p = 9.78 + 2.02 \times 10^{-3} T \ (estimated) \ (298°\text{–}2,086° \text{ K.}).$$

SULFIDE

$$\text{CdS}(c):$$
$$C_p = 12.90T + 0.90 \times 10^{-3} T \ (estimated) \ (273°\text{–}1,273° \text{ K.}).$$

HYDRIDE

TABLE 124.—*Heat content and entropy of CdH(g)*

[Base, ideal gas at 298.15° K.; mol. wt., 113.42]

T, ° K.	$H_T - H_{298.15}$, cal./mole	$S_T - S_{298.15}$, cal./deg. mole	T, ° K.	$H_T - H_{298.15}$, cal./mole	$S_T - S_{298.15}$, cal./deg. mole
400	735	2.12	1,000	5,530	9.36
500	1,480	3.78	1,200	7,230	10.91
600	2,255	5.19	1,400	8,950	12.24
700	3,050	6.42	1,600	10,690	13.40
800	3,860	7.50	1,800	12,440	14.43
900	4,690	8.48	2,000	14,195	15.36

$$\text{CdH}(g):$$
$$H_T - H_{298.15} = 7.51T + 0.42 \times 10^{-3} T + 0.60 \times 10^5 T^{-1}$$
$$- 2,478 \ (0.6 \text{ percent}; 298°\text{–}2,000° \text{ K.});$$
$$C_p = 7.51 + 0.84 \times 10^{-3} T - 0.60 \times 10^5 T^{-2}.$$

BROMIDE

TABLE 125.—*Heat content and entropy of CdBr(g)*

[Base, ideal gas at 298.15° K.; mol. wt., 192.33]

T, ° K.	$H_T - H_{298.15}$, cal./mole	$S_T - S_{298.15}$, cal./deg. mole	T, ° K.	$H_T - H_{298.15}$, cal./mole	$S_T - S_{298.15}$, cal./deg. mole
400	895	2.58	1,000	6,235	10.73
500	1,780	4.56	1,200	8,020	12.36
600	2,670	6.18	1,400	9,805	13.74
700	3,560	7.55	1,600	11,590	14.93
800	4,450	8.74	1,800	13,380	15.98
900	5,345	9.79	2,000	15,170	16.93

$$\text{CdBr}(g):$$
$$H_T - H_{298.15} = 8.94T + 0.17 \times 10^5 T^{-1} - 2,722$$
$$(0.1 \text{ percent}; 298°\text{–}2,000° \text{ K.});$$
$$C_p = 8.94 - 0.17 \times 10^5 T^{-2}.$$

CHLORIDES

TABLE 126.—*Heat content and entropy of CdCl(g)*

[Base, ideal gas at 298.15° K.; mol. wt., 147.87]

T, ° K.	$H_T - H_{298.15}$, cal./mole	$S_T - S_{298.15}$, cal./deg. mole	T, ° K.	$H_T - H_{298.15}$, cal./mole	$S_T - S_{298.15}$, cal./deg. mole
400	880	2.54	1,000	6,195	10.65
500	1,760	4.50	1,200	7,975	12.27
600	2,640	6.11	1,400	9,760	13.64
700	3,525	7.47	1,600	11,545	14.84
800	4,415	8.66	1,800	13,330	15.89
900	5,305	9.71	2,000	15,115	16.83

$$\text{CdCl}(g):$$
$$H_T - H_{298.15} = 8.94T + 0.34 \times 10^5 T^{-1} - 2,779$$
$$(0.1 \text{ percent}; 298°\text{–}2,000° \text{ K.});$$
$$C_p = 8.94 - 0.34 \times 10^5 T^{-2}.$$

TABLE 127.—*Heat content and entropy of CdCl$_2$(c)*

[Base, crystals at 298.15° K.; mol. wt., 183.32]

T, ° K.	$H_T - H_{298.15}$, cal./mole	$S_T - S_{298.15}$, cal./deg. mole	T, ° K.	$H_T - H_{298.15}$, cal./mole	$S_T - S_{298.15}$, cal./deg. mole
400	1,780	5.13	700	7,840	16.38
500	3,720	9.46	800	9,990	19.25
600	5,750	13.16	841	10,920	20.38

$$\text{CdCl}_2(c):$$
$$H_T - H_{298.15} = 14.64T + 4.80 \times 10^{-3} T^2 - 4,792$$
$$(0.9 \text{ percent}; 298°\text{–}841° \text{ K.});$$
$$C_p = 14.64 + 9.60 \times 10^{-3} T.$$

BROMIDE

TABLE 128.—*Heat content and entropy of CdF(g)*

[Base, ideal gas at 298.15° K.; mol. wt., 131.41]

T, ° K.	$H_T - H_{298.15}$, cal./mole	$S_T - S_{298.15}$, cal./deg. mole	T, ° R.	$H_T - H_{298.15}$, cal./mole	$S_T - S_{298.15}$, cal./deg. mole
400	845	2.44	1,000	6,075	10.40
500	1,695	4.33	1,200	7,850	12.01
600	2,560	5.91	1,400	9,625	13.38
700	3,435	7.26	1,600	11,405	14.57
800	4,310	8.43	1,800	13,185	15.62
900	5,190	9.46	2,000	14,970	16.56

$$\text{CdF}(g):$$
$$H_T - H_{298.15} = 8.81T + 0.04 \times 10^{-3} T^2 + 0.63 \times 10^5 T^{-1}$$
$$- 2,842 \ (0.1 \text{ percent}; 298°\text{–}2,000° \text{ K.});$$
$$C_p = 8.81 + 0.08 \times 10^{-3} T - 0.63 \times 10^5 T^{-2}.$$

TABLE 1h7. (*Continued*)

IODIDE

TABLE 129.—*Heat content and entropy of CdI(g)*

[Base, ideal gas at 298.15° K.; mol. wt., 239.32]

T, ° K.	$H_T-H_{298.15}$, cal./mole	$S_T-S_{298.15}$, cal./deg. mole	T, ° K.	$H_T-H_{298.15}$, cal./mole	$S_T-S_{298.15}$, cal./deg. mole
400	900	2.60	1,000	6,250	10.76
500	1,790	4.58	1,200	8,040	12.40
600	2,680	6.20	1,400	9,825	13.77
700	3,575	7.58	1,600	11,615	14.97
800	4,470	8.78	1,800	13,400	16.02
900	5,360	9.83	2,000	15,190	16.96

CdI(g):

$$H_T-H_{298.15}=8.94T+0.10\times10^5T^{-1}-2,699$$
$$(0.1 \text{ percent}; 298°-2,000° \text{ K.});$$
$$C_p=8.94-0.10\times10^5T^{-2}.$$

SULFATE

CdSO$_4$(c):

$$C_p=23.81 \text{ (298° K.)}.$$

CdSO$_4\cdot$H$_2$O(c):

$$C_p=32.16 \text{ (298° K.)}.$$

CdSO$_4\cdot$8/3H$_2$O(c):

$$C_p=50.97 \text{ (298° K.)}.$$

TUNGSTATE

TABLE 130.—*Heat content and entropy of CdWO$_4$(c)*

[Base, crystals at 298.15° K.; mol. wt., 360.27]

T, ° K.	$H_T-H_{298.15}$, cal./mole	$S_T-S_{298.15}$, cal./deg. mole	T, ° K.	$H_T-H_{298.15}$, cal./mole	$S_T-S_{298.15}$, cal./deg. mole
400	3,130	9.02	800	16,790	32.47
500	6,340	16.18	900	20,550	36.90
600	9,680	22.26	1,000	24,440	41.00
700	13,170	27.64	1,100	28,480	44.85

CdWO$_4$(c):

$$H_T-H_{298.15}=25.92T+6.86\times10^{-3}T^2-8,338$$
$$(0.1 \text{ percent}; 298°-1,100° \text{ K.});$$
$$C_p=25.92+13.72\times10^{-3}T.$$

CADMIUM-ANTIMONY

TABLE 131.—*Heat content and entropy of CdSb(c, l)*

[Base, crystals at 298.15° K., mol. wt., 234.17]

T, ° K.	$H_T-H_{298.15}$, cal./mole	$S_T-S_{298.15}$, cal./deg. mole	T, ° K.	$H_T-H_{298.15}$, cal./mole	$S_T-S_{298.15}$, cal./deg. mole
400	1,190	3.42	729 (c)	6,100	12.23
500	2,500	6.34	729 (l)	13,600	22.52
600	3,970	9.02	800	14,620	23.85
700	5,600	11.53	900	16,050	25.53

CdSb(c):

$$H_T-H_{298.15}=6.45T+7.50\times10^{-3}T^2-2,590$$
$$(0.2 \text{ percent}; 298°-729° \text{ K.});$$
$$C_p=6.45+15.00\times10^{-3}T;$$
$$\Delta H_{729} \text{ (fusion)}=7,500.$$

CdSb(l):

$$H_T-H_{298.15}=14.30T+3,180 \text{ (0.1 percent; 729°-900° K.);}$$
$$C_p=14.30.$$

CALCIUM AND ITS COMPOUNDS

ELEMENT

TABLE 132.—*Heat content and entropy of Ca(c, l)*

[Base, α-crystals at 298.15° K.; atomic wt., 40.08]

T, ° K.	$H_T-H_{298.15}$, cal./mole	$S_T-S_{298.15}$, cal./deg. mole	T, ° K.	$H_T-H_{298.15}$, cal./mole	$S_T-S_{298.15}$, cal./deg. mole
400	660	1.90	1,123(β)	6,825	10.39
500	1,340	3.42	1,123(l)	8,895	12.23
600	2,055	4.72	1,200	9,465	12.72
700	2,800	5.87	1,300	10,205	13.31
713(α)	2,900	6.01	1,400	10,945	13.86
713(β)	3,170	6.39	1,500	11,685	14.37
800	3,850	7.29	1,600	12,425	14.85
900	4,690	8.28	1,700	13,165	15.30
1,000	5,605	9.24	1,800	13,905	15.72
1,100	6,590	10.18			

Ca(α):

$$H_T-H_{298.15}=5.25T+1.72\times10^{-3}T^2-1,718$$
$$(0.3 \text{ percent}; 298°-713° \text{ K.});$$
$$C_p=5.25+3.44\times10^{-3}T;$$
$$\Delta H_{713} \text{ (transition)}=270.$$

Ca(β):

$$H_T-H_{298.15}=2.68T+3.40\times10^{-3}T^2-472$$
$$(0.1 \text{ percent}; 713°-1,123° \text{ K.});$$
$$C_p=2.68+6.80\times10^{-3}T;$$
$$\Delta H_{1123} \text{ (fusion)}=2,070.$$

Ca(l):

$$H_T-H_{298.15}=7.40T+585 \text{ (0.1 percent;1,123°-1,800° K.)};$$
$$C_p=7.40.$$

TABLE 133.—*Heat content and entropy of Ca(g)*

[Base, ideal gas at 298.15° K.; atomic wt., 40.08]

T, ° K.	$H_T-H_{298.15}$, cal./mole	$S_T-S_{298.15}$, cal./deg. mole	T, ° K.	$H_T-H_{298.15}$, cal./mole	$S_T-S_{298.15}$, cal./deg. mole
400	505	1.46	1,900	7,965	9.21
500	1,005	2.57	2,000	8,465	9.46
600	1,500	3.48	2,200	9,470	9.94
700	1,995	4.24	2,400	10,490	10.39
800	2,495	4.90	2,600	11,535	10.80
900	2,990	5.49	2,800	12,615	11.20
1,000	3,490	6.01	3,000	13,745	11.59
1,100	3,985	6.49	3,500	16,880	12.56
1,200	4,480	6.92	4,000	20,670	13.57
1,300	4,980	7.32	4,500	25,250	14.64
1,400	5,475	7.69	5,000	30,685	15.79
1,500	5,975	8.03	6,000	43,870	18.18
1,600	6,470	8.35	7,000	59,365	20.57
1,700	6,970	8.65	8,000	76,140	22.81
1,800	7,465	8.94			

TABLE 1h7. (*Continued*)

Ca(g):

$$H_T - H_{298.15} = 4.97\,T - 1,482\,(0.1\,\text{percent}; 298°\text{-}2,500°\,\text{K.});$$
$$C_p = 4.97.$$

OXIDE

TABLE 134.—*Heat content and entropy of CaO(c)*

[Base, crystals at 298.15° K.; mol. wt., 56.08]

T, ° K.	$H_T - H_{298.15}$, cal./mole	$S_T - S_{298.15}$, cal./deg. mole	T, ° K.	$H_T - H_{298.15}$, cal./mole	$S_T - S_{298.15}$, cal./deg. mole
400	1,100	3.17	1,300	12,110	17.38
500	2,230	5.69	1,400	13,430	18.36
600	3,400	7.82	1,500	14,760	19.28
700	4,600	9.67	1,600	16,100	20.14
800	5,820	11.30	1,700	17,440	20.96
900	7,040	12.73	1,800	18,780	21.72
1,000	8,270	14.03	1,900	20,130	22.45
1,100	9,520	15.22	2,000	21,480	23.15
1,200	10,800	16.34			

CaO(c):

$$H_T - H_{298.15} = 11.67\,T + 0.54 \times 10^{-3}\,T^2 + 1.56 \times 10^5\,T^{-1}$$
$$-4,051\,(0.3\,\text{percent}; 298°\text{-}2,000°\,\text{K.});$$
$$C_p = 11.67 + 1.08 \times 10^{-3}\,T - 1.56 \times 10^5\,T^{-2}.$$

TABLE 135.—*Heat content and entropy of CaO(g)*

[Base, ideal gas at 298.15° K.; mol. wt., 56.08]

T, ° K.	$H_T - H_{298.15}$, cal./mole	$S_T - S_{298.15}$, cal./deg. mole	T, ° K.	$H_T - H_{298.15}$, cal./mole	$S_T - S_{298.15}$, cal./deg. mole
400	825	2.38	1,000	6,005	10.25
500	1,665	4.25	1,200	7,775	11.87
600	2,515	5.80	1,400	9,545	13.23
700	3,380	7.13	1,600	11,325	14.42
800	4,250	8.30	1,800	13,115	15.47
900	5,125	9.33	2,000	14,910	16.42

CaO(g):

$$H_T - H_{298.15} = 8.70\,T + 0.08 \times 10^{-3}\,T^2 + 0.74 \times 10^5\,T^{-1}$$
$$-2,849\,(0.2\,\text{percent}; 298°\text{-}2,000°\,\text{K.});$$
$$C_p = 8.70 + 0.16 \times 10^{-3}\,T - 0.74 \times 10^5\,T^{-2}.$$

HYDROXIDE

TABLE 136.—*Heat content and entropy of Ca(OH)$_2$(c)*

[Base, crystals at 298.15° K.; mol. wt., 74.10]

T, ° K.	$H_T - H_{298.15}$, cal./mole	$S_T - S_{298.15}$, cal./deg. mole	T, ° K.	$H_T - H_{298.15}$, cal./mole	$S_T - S_{298.15}$, cal./deg. mole
400	2,300	6.63	600	7,240	16.65
500	4,720	12.02	700	9,830	20.61

Ca(OH)$_2$(c):

$$H_T - H_{298.15} = 19.07\,T + 5.40 \times 10^{-3}\,T^2 - 6,166$$
$$(0.5\,\text{percent}; 298°\text{-}700°\,\text{K.});$$
$$C_p = 19.07 + 10.80 \times 10^{-3}\,T.$$

SULFIDE

CaS(c):

$$C_p = 10.20 + 3.80 \times 10^{-3}\,T\,\,(\textit{estimated})\,\,(298°\text{-}1,000°\,\text{K.}).$$

CARBIDE

TABLE 137.—*Heat content and entropy of CaC$_2$(c)*

[Base, α-crystals at 298.15° K,; mol. wt., 64.10]

T, ° K.	$H_T - H_{298.15}$, cal./mole	$S_T - S_{298.15}$, cal./deg. mole	T, ° K.	$H_T - H_{298.15}$, cal./mole	$S_T - S_{298.15}$, cal./deg. mole
400	1,600	4.61	1,200	16,780	25.39
500	3,260	8.31	1,300	18,560	26.82
600	4,995	11.47	1,400	20,350	28.15
700	6,760	14.19	1,500	22,150	29.39
720(α)	7,120	14.70	1,600	23,960	30.56
720(β)	8,450	16.54	1,700	25,780	31.66
800	9,790	18.31	1,800	27,610	32.71
900	11,510	20.33	1,900	29,450	33.70
1,000	13,250	22.17	2,000	31,300	34.65
1,100	15,010	23.85			

CaC$_2$(α):

$$H_T - H_{298.15} = 16.40\,T + 1.42 \times 10^{-3}\,T^2 + 2.07 \times 10^5\,T^{-1}$$
$$-5,710\,(0.1\,\text{percent}; 298°\text{-}720°);$$
$$C_p = 16.40 + 2.84 \times 10^{-3}\,T - 2.07 \times 10^5\,T^{-2};$$
$$\Delta H_{720}\,(\textit{transition}) = 1,330.$$

CaC$_2$(β):

$$H_T - H_{298.15} = 15.40\,T + 1.00 \times 10^{-3}\,T^2 - 3,156\,(0.1\,\text{percent};$$
$$720°\text{-}1,500°\,\text{K.});$$
$$C_p = 15.40 + 2.00 \times 10^{-3}\,T.$$

NITRIDE

TABLE 138.—*Heat content and entropy of Ca$_3$N$_2$(c)*

[Base, crystals at 298.15° K.; mol. wt., 148.26]

T, ° K.	$H_T - H_{298.15}$, cal./mole	$S_T - S_{298.15}$, cal./deg. mole	T, ° K.	$H_T - H_{298.15}$, cal./mole	$S_T - S_{298.15}$, cal./deg. mole
400	2,850	8.21	700	12,650	26.32
500	5,900	15.01	800	16,300	31.19
600	9,150	20.93			

Ca$_3$N$_2$(c):

$$H_T - H_{298.15} = 20.44\,T + 11.00 \times 10^{-3}\,T^2 - 7,072$$
$$(0.2\,\text{percent}; 298°\text{-}800°\,\text{K.});$$
$$C_p = 20.44 + 22.00 \times 10^{-3}\,T.$$

HYDRIDE

TABLE 139.—*Heat content and entropy of CaH(g)*

[Base, ideal gas at 298.15° K.; mol. wt., 41.09]

T, ° K.	$H_T - H_{298.15}$, cal./mole	$S_T - S_{298.15}$, cal./deg. mole	T, ° K.	$H_T - H_{298.15}$, cal./mole	$S_T - S_{298.15}$, cal./deg. mole
400	740	2.13	1,000	5,580	9.45
500	1,495	3.82	1,200	7,285	11.00
600	2,275	5.24	1,400	9,015	12.34
700	3,080	6.48	1,600	10,755	13.50
800	3,900	7.57	1,800	12,510	14.53
900	4,735	8.56	2,000	14,270	15.46

CaH (g):

$$H_T - H_{298.15} = 7.68\,T + 0.37 \times 10^{-3}\,T^2 + 0.70 \times 10^5\,T^{-1}$$
$$-2,557\,(0.6\,\text{percent}; 298°\text{-}2,000°\,\text{K.});$$
$$C_p = 7.68 + 0.74 \times 10^{-3}\,T - 0.70 \times 10^5\,T^{-2}.$$

TABLE 1h7. (Continued)

BROMIDE

TABLE 140.—*Heat content and entropy of CaBr(g)*

[Base, ideal gas at 298.15° K.; mol. wt., 120.00]

T, ° K.	$H_T-H_{298.15}$, cal./mole	$S_T-S_{298.15}$, cal./deg. mole	T, ° K.	$H_T-H_{298.15}$, cal./mole	$S_T-S_{298.15}$, cal./deg. mole
400	890	2. 57	1,000	6, 215	10. 69
500	1, 770	4. 53	1,200	8, 000	12. 32
600	2, 655	6. 14	1,400	9, 780	13. 69
700	3, 540	7. 51	1,600	11, 565	14. 88
800	4, 430	8. 70	1,800	13, 355	15. 93
900	5, 320	9. 75	2,000	15, 140	16. 87

CaBr(g):

$$H_T-H_{298.15}=8.91T+0.01\times10^{-3}T^2+0.21\times10^5T^{-1}$$
$$-2,728 \ (0.1 \text{ percent}; \ 298°-2,000° \text{ K.});$$
$$C_p=8.91+0.02\times10^{-3}T-0.21\times10^5T^{-2}.$$

CHLORIDES

TABLE 141.—*Heat content and entropy of CaCl(g)*

[Base, ideal gas at 298.15° K.; mol. wt., 75.54]

T, ° K.	$H_T-H_{298.15}$, cal./mole	$S_T-S_{298.15}$, cal./deg. mole	T, ° K.	$H_T-H_{298.15}$, cal./mole	$S_T-S_{298.15}$, cal./deg. mole
400	875	2. 52	1,000	6, 175	10. 60
500	1, 750	4. 47	1,200	7, 955	12. 22
600	2, 625	6. 07	1,400	9, 740	13. 60
700	3, 510	7. 44	1,600	11, 525	14. 79
800	4, 395	8. 62	1,800	13, 310	15. 84
900	5, 285	9. 66	2,000	15, 095	16. 78

CaCl(g):

$$H_T-H_{298.15}=8.89T+0.02\times10^{-3}T^2+0.36\times10^5T^{-1}$$
$$-2,773 \ (0.1 \text{ percent}; \ 298°-2,000° \text{ K.});$$
$$C_p=8.89+0.04\times10^{-3}T-0.36\times10^5T^{-2}.$$

TABLE 142.—*Heat content and entropy of CaCl$_2$(c, l)*

[Base, crystals at 298.15° K.; mol. wt., 110.99]

T, ° K.	$H_T-H_{298.15}$, cal./mole	$S_T-S_{298.15}$, cal./deg. mole	T, ° K.	$H_T-H_{298.15}$, cal./mole	$S_T-S_{298.15}$, cal./deg. mole
400	1, 850	5. 33	1,055(l)	21, 200	30. 18
500	3, 700	9. 46	1,100	22, 340	31. 24
600	5, 540	12. 82	1,200	24, 840	33. 42
700	7, 400	15. 68	1,300	27, 320	35. 40
800	9, 290	18. 20	1,400	29, 780	37. 22
900	11, 230	20. 49	1,500	32, 210	38. 90
1,000	13, 270	22. 64	1,600	34, 580	40. 43
1,055(c)	14, 420	23. 76	1,700	36, 860	41. 81

CaCl$_2$(c):

$$H_T-H_{298.15}=17.18T+1.52\times10^{-3}T^2+0.60\times10^5T^{-1}$$
$$-5,459 \ (0.8 \text{ percent}; \ 298°-1,055° \text{ K.});$$
$$C_p=17.18+3.04\times10^{-3}T-0.60\times10^5T^{-2};$$
$$\Delta H_{1055} \ (fusion)=6,780.$$

CaCl$_2$(l):

$$H_T-H_{298.15}=24.70T-4,858 \ (0.3 \text{ percent};$$
$$1,055°-1,700° \text{ K.});$$
$$C_p=24.70.$$

FLUORIDES

TABLE 143.—*Heat content and entropy of CaF(g)*

[Base, ideal gas at 298.15° K.; mol. wt., 59.08]

T, ° K.	$H_T-H_{298.15}$, cal./mole	$S_T-S_{298.15}$, cal./deg. mole	T, ° K.	$H_T-H_{298.15}$, cal./mole	$S_T-S_{298.15}$, cal./deg. mole
400	835	2. 41	1,000	6, 045	10. 33
500	1, 680	4. 29	1,200	7, 815	11. 95
600	2. 540	5. 86	1,400	9, 590	13. 32
700	3. 410	7. 20	1,600	11, 370	14. 50
800	4, 285	8. 37	1,800	13, 150	15. 55
900	5, 165	9. 41	2,000	14. 930	16. 49

CaF(g):

$$H_T-H_{298.15}=8.78T+0.05\times10^{-3}T^2+0.69\times10^5T^{-1}$$
$$-2,854 \ (0.2 \text{ percent}; \ 298°-2,000° \text{ K.});$$
$$C_p=8.78+0.10\times10^{-3}T-0.69\times10^5T^{-2}.$$

TABLE 144.—*Heat content and entropy of CaF$_2$(c, l)*

[Base, α-crystals at 298.15° K.; mol. wt., 78.08]

T, ° K.	$H_T-H_{298.15}$, cal./mole	$S_T-S_{298.15}$, cal./deg. mole	T, ° K.	$H_T-H_{298.15}$, cal./mole	$S_T-S_{298.15}$, cal./deg. mole
400	1, 755	5. 06	1,424(α)	23, 280	30. 79
500	3, 540	9. 04	1,424(β)	24, 420	31. 59
600	5, 400	12. 43	1,500	26, 660	33. 12
700	7, 320	15. 39	1,600	29, 620	35. 03
800	9, 280	18. 01	1,691(β)	32, 350	36. 69
900	11, 300	20. 38	1,691(l)	39, 450	40. 89
1,000	13, 380	22. 58	1,700	39, 670	41. 02
1,100	15, 550	24. 64	1,800	42, 050	42. 38
1,200	17, 850	26. 64	1,900	44, 440	43. 67
1,300	20, 230	28. 55	2,000	46, 830	44. 90
1,400	22, 680	30. 36			

CaF$_2$(α):

$$H_T-H_{298.15}=14.30T+3.64\times10^{-3}T^2-0.47\times10^5T^{-1}$$
$$-4,429 \ (0.3 \text{ percent}; \ 298°-1,424° \text{ K.});$$
$$C_p=14.30+7.28\times10^{-3}T+0.47\times10^5T^{-2};$$
$$\Delta H_{1424} \ (transition)=1,140.$$

CaF$_2$(β):

$$H_T-H_{298.15}=25.81T+1.25\times10^{-3}T^2-14,868 \ \ (0.1 \text{ per-}$$
$$\text{cent}; \ 1,424°-1,691° \text{ K.});$$
$$C_p=25.81+2.50\times10^{-3}T;$$
$$\Delta H_{1691} \ (fusion)=7,100.$$

CaF$_2$(l):

$$H_T-H_{298.15}=23.90T-965 \ (0.1 \text{ percent};$$
$$1,691°-2,000° \text{ K.});$$
$$C_p=23.90.$$

IODIDE

TABLE 145.—*Heat content and entropy of CaI(g)*

[Base, ideal gas at 298.15° K.; mol. wt., 166.99]

T, ° K.	$H_T-H_{298.15}$, cal./mole	$S_T-S_{298.15}$, cal./deg. mole	T, ° K.	$H_T-H_{298.15}$, cal./mole	$S_T-S_{298.15}$, cal./deg. mole
400	895	2. 58	1,000	6. 230	10. 72
500	1, 780	4. 55	1,200	8, 015	12. 35
600	2, 665	6. 17	1,400	9, 800	13. 73
700	3, 555	7. 54	1,600	11, 585	14. 92
800	4, 445	8. 73	1,800	13, 375	15. 97
900	5, 340	9. 78	2,000	15, 165	16. 91

TABLE 1h7. (*Continued*)

CaI(*g*):

$$H_T - H_{298.15} = 8.94T + 0.19 \times 10^5 T^{-1} - 2,729 \ (0.1 \text{ percent};$$
$$298°-2,000° \text{ K.});$$
$$C_p = 8.94 - 0.19 \times 10^5 T^{-2}.$$

CaAl₂O₄(*c*):

$$H_T - H_{298.15} = 36.00T + 2.98 \times 10^{-3} T^2 + 7.96 \times 10^5 T^{-1}$$
$$- 13,668 \ (0.3 \text{ percent}; 298°-1,800° \text{ K.});$$
$$C_p = 36.00 + 5.96 \times 10^{-3} T - 7.96 \times 10^5 T^{-2}.$$

ALUMINATES

TABLE 146.—*Heat content and entropy of 3CaO·Al₂O₃(c)*

[Base, crystals at 298.15° K.; mol. wt., 270.20]

T, ° K.	$H_T - H_{298.15}$, cal./mole	$S_T - S_{298.15}$, cal./deg. mole	T, ° K.	$H_T - H_{298.15}$, cal./mole	$S_T - S_{298.15}$, cal./deg mole
400	5,380	15.48	1,200	56,460	84.59
500	11,180	28.41	1,300	63,170	89.96
600	17,280	39.52	1,400	69,900	94.95
700	23,570	49.22	1,500	76,660	99.61
800	29,990	57.79	1,600	83,460	104.00
900	36,510	65.47	1,700	90,310	108.15
1,000	43,110	72.42	1,800	97,220	112.10
1,100	49,770	78.77			

3CaO·Al₂O₃(*c*):

$$H_T - H_{298.15} = 62.28T + 2.29 \times 10^{-3} T^2 + 12.01 \times 10^5 T^{-1}$$
$$- 22,801 \ (0.5 \text{ percent}; 298°-1,800° \text{ K.});$$
$$C_p = 62.28 + 4.58 \times 10^{-3} T - 12.01 \times 10^5 T^{-2}.$$

TABLE 147.—*Heat content and entropy of 12CaO·7Al₂O₃(c)*

[Base, α-crystals at 298.15° K.; mol. wt., 1,386.68]

T, ° K.	$H_T - H_{298.15}$, cal./mole	$S_T - S_{298.15}$, cal./deg. mole	T, ° K.	$H_T - H_{298.15}$, cal./mole	$S_T - S_{298.15}$, cal./deg. mole
400	27,800	80.0	1,300	340,250	480.2
500	58,250	147.8	1,310(α)	344,500	483.4
600	90,850	207.2	1,310(β)	344,500	483.4
700	124,650	259.3	1,400	376,700	507.2
800	158,900	305.0	1,500	414,400	533.2
900	193,600	345.9	1,600	452,600	557.9
1,000	228,750	382.9	1,700	491,400	581.4
1,100	264,500	417.0	1,800	530,900	604.0
1,200	301,150	448.9			

12CaO·7Al₂O₃(α):

$$H_T - H_{298.15} = 301.96T + 32.75 \times 10^{-3} T^2 + 55.30$$
$$\times 10^5 T^{-1} - 111,488 \ (0.6 \text{ percent}; 298°-1,310° \text{ K.});$$
$$C_p = 301.96 + 65.50 \times 10^{-3} T - 55.30 \times 10^5 T^{-2}.$$

12CaO·7Al₂O₃(β):

$$H_T - H_{298.15} = 228.52T + 49.22 \times 10^{-3} T^2 - 39,327$$
$$(0.1 \text{ percent}; 1,310°-1,800° \text{ K.});$$
$$C_p = 228.52 + 98.44 \times 10^{-3} T.$$

TABLE 148.—*Heat content and entropy of CaAl₂O₄(c)*

[Base, crystals at 298.15° K.; mol. wt., 158.04]

T, ° K.	$H_T - H_{298.15}$, cal./mole	$S_T - S_{298.15}$, cal./deg. mole	T, ° K.	$H_T - H_{298.15}$, cal./mole	$S_T - S_{298.15}$, cal./deg. mole
400	3,140	9.03	1,200	34,500	51.25
500	6,610	16.76	1,300	38,810	54.70
600	10,320	23.52	1,400	43,180	57.93
700	14,130	29.39	1,500	47,610	60.99
800	18,030	34.60	1,600	52,090	63.88
900	22,020	39.30	1,700	56,610	66.62
1,000	26,100	43.59	1,800	61,160	69.22
1,100	30,260	47.56			

TABLE 149.—*Heat content and entropy of CaO·2Al₂O₃(c)*

[Base, crystals at 298.15° K.; mol. wt., 260,00]

T, ° K.	$H_T - H_{298.15}$, cal./mole	$S_T - S_{298.15}$, cal./deg. mole	T, ° K.	$H_T - H_{298.15}$, cal./mole	$S_T - S_{298.15}$, cal./deg. mole
400	5,340	15.34	1,200	59,100	87.63
500	11,200	28.40	1,300	66,320	93.41
600	17,480	39.84	1,400	73,570	98.78
700	24,050	49.97	1,500	80,830	103.79
800	30,830	59.02	1,600	88,100	108.48
900	37,760	67.18	1,700	95,400	112.91
1,000	44,800	74.60	1,800	102,750	117.11
1,100	51,920	81.38			

CaO·2Al₂O₃(*c*):

$$H_T - H_{298.15} = 66.09T + 2.74 \times 10^{-3} T^2 + 17.80 \times 10^5 T^{-1}$$
$$- 25,918 \ (0.6 \text{ percent}; 298°-1,800° \text{ K.});$$
$$C_p = 66.09 + 5.48 \times 10^{-3} T - 17.80 \times 10^5 T^{-2}.$$

BORATES

TABLE 150.—*Heat content and entropy of 3CaO·B₂O₃(c, l)*

[Base, crystals at 298.15° K.; mol. wt., 237.88]

T, ° K.	$H_T - H_{298.15}$, cal./mole	$S_T - S_{298.15}$, cal./deg. mole	T, ° K.	$H_T - H_{298.15}$, cal./mole	$S_T - S_{298.15}$, cal./deg. mole
400	5,180	14.88	1,400	68,540	91.93
500	10,520	26.79	1,500	75,600	96.80
600	16,250	37.22	1,600	82,790	101.44
700	22,220	46.42	1,700	90,110	105.88
800	28,410	54.69	1,760(*c*)	94,550	108.44
900	34,780	62.19	1,760(*l*)	130,040	128.60
1,000	41,300	69.06	1,800	133,800	130.71
1,100	47,950	75.40	1,900	143,200	135.79
1,200	54,720	81.28	2,000	152,600	140.61
1,300	61,590	86.78			

3CaO·B₂O₃(*c*):

$$H_T - H_{298.15} = 56.44T + 5.21 \times 10^{-3} T^2 + 13.02 \times 10^5 T^{-1}$$
$$- 21,658 \ (0.4 \text{ percent}; 298°-1,760° \text{ K.});$$
$$C_p = 56.44 + 10.42 \times 10^{-3} T - 13.02 \times 10^5 T^{-2};$$
$$\Delta H_{1760}(fusion) = 35,490.$$

3CaO·B₂O₃(*l*):

$$H_T - H_{298.15} = 94.00T - 35,400 \ (0.1 \text{ percent};$$
$$1,760°-2,000° \text{ K.});$$
$$C_p = 94.00.$$

TABLE 1h7. (Continued)

TABLE 151.—*Heat content and entropy of* $2CaO \cdot B_2O_3(c, l)$

[Base, α-crystals at 298.15° K.; mol. wt., 181.80]

T, ° K.	$H_T - H_{298.15}$, cal./mole	$S_T - S_{298.15}$, cal./deg. mole	T, ° K.	$H_T - H_{298.15}$, cal./mole	$S_T - S_{298.15}$, cal./deg. mole
400	4,010	11.53	1,300	51,380	71.76
500	8,310	21.11	1,400	56,930	75.87
600	12,940	29.55	1,500	62,490	79.71
700	17,830	37.08	1,585 (β)	67,260	82.80
800	22,880	43.82	1,585 (l)	91,350	98.00
804 (α)	23,080	44.07	1,600	92,370	98.64
804 (β)	24,180	45.44	1,700	99,190	102.77
900	29,410	51.57	1,800	106,010	106.68
1,000	34,860	57.32	1,900	112,830	110.37
1,100	40,330	62.54	2,000	119,650	113.87
1,200	45,840	67.33			

$2CaO \cdot B_2O_3(\alpha)$:

$$H_T - H_{298.15} = 43.75T + 5.75 \times 10^{-3}T^2 + 10.69 \times 10^5 T^{-1}$$
$$-17,141 \ (0.4 \text{ percent}; \ 298° - 804° \text{ K.});$$
$$C_p = 43.75 + 11.50 \times 10^{-3}T - 10.69 \times 10^5 T^{-2};$$
$$\Delta H_{804}(transition) = 1,100.$$

$2CaO \cdot B_2O_3(\beta)$:

$$H_T - H_{298.15} = 52.29T + 1.20 \times 10^{-3}T^2$$
$$-18,633 \ (0.1 \text{ percent}; \ 804° - 1,585° \text{ K.});$$
$$C_p = 52.29 + 2.40 \times 10^{-3}T;$$
$$\Delta H_{1585}(fusion) = 24,090.$$

$2CaO \cdot B_2O_3(l)$:

$$H_T - H_{298.15} = 68.20T$$
$$-16,750 \ (0.1 \text{ percent}; \ 1,585° - 2,000° \text{ K.});$$
$$C_p = 68.20.$$

TABLE 152.—*Heat content and entropy of* $CaB_2O_4(c, l)$

[Base, crystals at 298.15° K.; mol. wt., 125.72]

T, ° K.	$H_T - H_{298.15}$, cal./mole	$S_T - S_{298.15}$, cal./deg. mole	T, ° K.	$H_T - H_{298.15}$, cal./mole	$S_T - S_{298.15}$, cal./deg. mole
400	2,840	8.16	1,400	41,210	54.44
500	5,950	15.09	1,435 (c)	42,740	55.52
600	9,280	21.16	1,435 (l)	60,410	67.83
700	12,820	26.61	1,500	64,420	70.58
800	16,550	31.59	1,600	70,590	74.55
900	20,420	36.14	1,700	76,760	78.29
1,000	24,410	40.35	1,800	82,930	81.82
1,100	28,480	44.22	1,900	89,100	85.15
1,200	32,630	47.84	2,000	95,270	88.32
1,300	36,870	51.23			

$CaB_2O_4(c)$:

$$H_T - H_{298.15} = 31.02T + 4.88 \times 10^{-3}T^2 + 8.07 \times 10^5 T^{-1}$$
$$-12,389 \ (0.4 \text{ percent}; \ 298° - 1,435° \text{ K.});$$
$$C_p = 31.02 + 9.76 \times 10^{-3}T - 8.07 \times 10^5 T^{-2};$$
$$\Delta H_{1435}(fusion) = 17,670.$$

$CaB_2O_4(l)$:

$$H_T - H_{298.15} = 61.70T$$
$$-28,130 \ (0.1 \text{ percent}; \ 1,435° - 2,000° \text{ K.});$$
$$C_p = 61.70.$$

TABLE 153.—*Heat content and entropy of* $CaO \cdot 2B_2O_3(c, l)$

[Base, crystals at 298.15° K.; mol. wt., 195.36]

T, ° K.	$H_T - H_{298.15}$, cal./mole	$S_T - S_{298.15}$, cal./deg. mole	T, ° K.	$H_T - H_{298.15}$, cal./mole	$S_T - S_{298.15}$, cal./deg. mole
400	4,470	12.83	1,260 (l)	86,400	104.81
500	9,500	24.03	1,300	90,650	108.13
600	15,130	34.29	1,400	101,280	116.01
700	21,170	43.59	1,500	111,910	123.34
800	27,510	52.05	1,600	122,540	130.20
900	34,090	59.79	1,700	133,170	136.65
1,000	40,870	66.94	1,800	143,800	142.72
1,100	47,820	73.56	1,900	154,430	148.47
1,200	54,920	79.74	2,000	165,060	153.92
1,260 (c)	59,340	83.33			

$CaO \cdot 2B_2O_3(c)$:

$$H_T - H_{298.15} = 51.34T + 9.58 \times 10^{-3}T^2 + 17.16 \times 10^5 T^{-1}$$
$$-21,914 \ (0.4 \text{ percent}; \ 298° - 1,260° \text{ K.});$$
$$C_p = 51.34 + 19.16 \times 10^{-3}T - 17.16 \times 10^5 T^{-2};$$
$$\Delta H_{1260}(fusion) = 27,060.$$

$CaO \cdot 2B_2O_3(l)$:

$$H_T - H_{298.15} = 106.30T$$
$$-47,540 \ (0.1 \text{ percent}; \ 1,260° - 2,000° \text{ K.});$$
$$C_p = 106.30.$$

TABLE 154.—*Heat content and entropy of* $CaO \cdot 2B_2O_3$ (gl, l)

[Base, glass at 298.15° K.; mol. wt., 195.36]

T, K° .	$H_T - H_{298.15}$, cal./mole	$S_T - S_{298.15}$, cal./deg. mole	T, K° .	$H_T - H_{298.15}$, cal./mole	$S_T - S_{298.15}$, cal./deg. mole
400	4,470	12.85	1,260 (l)	73,760	96.71
500	9,500	24.05	1,300	78,010	100.03
600	15,130	34.28	1,400	88,640	107.91
700	21,180	43.59	1,500	99,270	115.24
800	27,710	52.30	1,600	109,900	122.10
900	35,210	61.13	1,700	120,530	128.54
1,000	44,100	70.49	1,800	131,160	134.62
1,100	54,390	80.29	1,900	141,790	140.37
1,200	66,080	90.46	2,000	152,420	145.82
1,260 (gl)	73,760	96.71			

$CaO \cdot 2B_2O_3(gl)$:

$$H_T - H_{298.15} = 19.04T + 33.73 \times 10^{-3}T^2$$
$$-8,675 \ (1.5 \text{ percent}; \ 298° - 1,000° \text{ K.});$$
$$C_p = 19.04 + 67.46 \times 10^{-3}T.$$

$CaO \cdot 2B_2O_3(l)$:

$$H_T - H_{298.15} = 106.30T$$
$$-60,180 \ (0.1 \text{ percent}; \ 1,260° - 2,000° \text{ K.});$$
$$C_p = 106.30.$$

CARBONATES

TABLE 155.—*Heat content and entropy of* $CaCO_3$ *(calcite)*

[Base, crystals at 298.15° K.; mol. wt., 100.09]

T, ° K.	$H_T - H_{298.15}$, cal./mole	$S_T - S_{298.15}$, cal./deg. mole	T, ° K.	$H_T - H_{298.15}$, cal./mole	$S_T - S_{298.15}$, cal./deg. mole
400	2,220	6.38	900	15,500	27.61
500	4,610	11.71	1,000	18,430	30.70
600	7,200	16.43	1,100	21,450	33.58
700	9,890	20.57	1,200	24,550	36.27
800	12,660	24.27			

TABLE 1h7. (*Continued*)

$CaCO_3(calcite)$:

$H_T - H_{298.15} = 24.98T + 2.62 \times 10^{-3}T^2 + 6.20 \times 10^5 T^{-1}$
$-9,760$ (0.3 percent; 298°–1,200° K.);
$C_p = 24.98 + 5.24 \times 10^{-3}T - 6.20 \times 10^5 T^{-2}$.

TABLE 156.—*Heat content and entropy of $CaCO_3$ (aragonite)*

[Base, crystals at 298.15° K.; mol. wt., 100.09]

T, ° K.	$H_T - H_{298.15}$, cal./mole	$S_T - S_{298.15}$, cal./deg. mole	T, ° K.	$H_T - H_{298.15}$, cal./mole	$S_T - S_{298.15}$, cal./deg. mole
350	1,050	3.24	500	4,440	11.27
400	2,130	6.13	550	5,650	13.58
450	3,260	8.79	600	6,900	15.75

$CaCO_3(aragonite)$:

$H_T - H_{298.15} = 20.13T + 5.12 \times 10^{-3}T^2 + 3.34 \times 10^5 T^{-1}$
$-7,577$ (0.1 percent; 298°–600° K.);
$C_p = 20.13 + 10.24 \times 10^{-3}T - 3.34 \times 10^5 T^{-2}$.

DOLOMITE

$CaMg(CO_3)_2(c)$:
$\overline{C_p} = 40.09$ (299°–376° K.).

FERRITES

TABLE 157.—*Heat content and entropy of $2CaO \cdot Fe_2O_3$ (c, l)*

[Base, crystals at 298.15° K.; mol. wt., 271.86]

T, ° K.	$H_T - H_{298.15}$, cal./mole	$S_T - S_{298.15}$, cal./deg. mole	T, ° K.	$H_T - H_{298.15}$, cal./mole	$S_T - S_{298.15}$, cal./deg. mole
400	4,870	14.01	1,400	62,280	85.22
500	10,210	25.92	1,500	68,130	89.26
600	15,800	36.10	1,600	73,980	93.04
700	21,490	44.87	1,700	79,830	96.58
800	27,250	52.56	1,750(c)	82,760	98.28
900	33,060	59.41	1,750(l)	118,870	118.91
1,000	38,890	65.55	1,800	122,580	121.00
1,100	44,730	71.12	1,900	130,000	125.02
1,200	50,580	76.21	2,000	137,420	128.82
1,300	56,430	80.89			

$2CaO \cdot Fe_2O_3(c)$:

$H_T - H_{298.15} = 59.24T + 11.68 \times 10^5 T^{-1} - 21,580$
(0.6 percent; 298°–1,750° K.);
$C_p = 59.24 - 11.68 \times 10^5 T^{-2}$;
$\Delta H_{1750}(fusion) = 36,110$.

$2CaO \cdot Fe_2O_3(l)$:

$H_T - H_{298.15} = 74.20T$
$-10,980$ (0.1 percent; 1,750°–2,000° K.);
$C_p = 74.20$.

TABLE 158.—*Heat content and entropy of $CaFe_2O_4$ (c, l)*

[Base, crystals at 298.15° K.; mol. wt., 215.78]

T, ° K.	$H_T - H_{298.15}$, cal./mole	$S_T - S_{298.15}$, cal./deg. mole	T, ° K.	$H_T - H_{298.15}$, cal./mole	$S_T - S_{298.15}$, cal./deg. mole
400	3,940	11.35	1,400	46,920	64.31
500	7,900	20.18	1,500	51,540	67.50
600	11,950	27.57	1,510(c)	52,000	67.80
700	16,090	33.95	1,510(l)	77,870	84.93
800	20,310	39.58	1,600	82,810	88.11
900	24,600	44.63	1,700	88,300	91.44
1,000	28,950	49.22	1,800	93,790	94.58
1,100	33,360	53.42	1,900	99,280	97.55
1,200	37,820	57.30	2,000	104,770	100.36
1,300	42,340	60.92			

$CaFe_2O_4(c)$:

$H_T - H_{298.15} = 39.42T + 2.38 \times 10^{-3}T^2 + 3.66 \times 10^5 T^{-1}$
-13.192 (0.3 percent; 298°–1,510° K.);
$C_p = 39.42 + 4.76 \times 10^{-3}T - 3.66 \times 10^5 T^{-2}$;
$\Delta H_{1510}(fusion) = 25,870$.

$CaFe_2O_4(l)$:

$H_T - H_{298.15} = 54.90T - 5,030$ (0.1 percent;
1,510°–2,000° K.),;
$C_p = 54.90$

MOLYBDATE

$CaMoO_4(c)$

$\overline{C_p} = 33.2$ (273°–297° K.).

NITRATE

TABLE 159.—*Heat content and entropy of $CA(NO_3)_2(c)$*

[Base, crystals at 298.15° K.; mol. wt., 164.10]

T, ° K.	$H_T - H_{298.15}$, cal./mole	$S_T - S_{298.15}$, cal./deg. mole	T, ° K.	$H_T - H_{298.15}$, cal./mole	$S_T - S_{298.15}$, cal./deg. mole
400	3,960	11.42	700	18,410	38.01
500	8,340	21.15	800	23,970	45.43
600	13,150	29.91			

$Ca(NO_3)_2(c)$

$H_T - H_{298.15} = 29.37T + 18.40 \times 10^{-3}T^2 + 4.13 \times 10^5 T^{-1}$
$-11,778$ (0.3 percent; 298°–800° K.);
$C_p = 29.37 + 36.80 \times 10^{-3}T - 4.13 \times 10^5 T^{-2}$.

PHOSPHATES

TABLE 160.—*Heat content and entropy of $Ca_3(PO_4)_2(c)$*

[Base, α-crystals at 298.15° K.; mol. wt., 310.19]

T, ° K.	$H_T - H_{298.15}$, cal./mole	$S_T - S_{298.15}$, cal./deg. mole	T, ° K.	$H_T - H_{298.15}$, cal./mole	$S_T - S_{298.15}$, cal./deg. mole
400	6,000	17.27	1,200	69,000	100.32
500	12,300	31.31	1,300	78,800	108.16
600	19,100	43.70	1,373(α)	86,200	113.70
700	26,400	54.94	1,373(β)	89,900	116.39
800	34,050	65.15	1,400	92,000	117.90
900	42,050	74.56	1,500	99,900	123.35
1,000	50,600	83.57	1,600	107,800	128.45
1,100	59,600	92.14			

$Ca_3(PO_4)_2(\alpha)$:

$H_T - H_{298.15} = 48.24T + 19.84 \times 10^{-3}T^2 + 5.00 \times 10^5 T^{-1}$
$-17,823$ (0.4 percent; 298°–1,373° K.);
$C_p = 48.24 + 39.68 \times 10^{-3}T - 5.00 \times 10^5 T^{-2}$;
ΔH_{1373} (transition) $= 3,700$

$Ca_3(PO_4)_2(\beta)$:

$H_T - H_{298.15} = 79.00T - 18,600$ (0.1 percent;
1,373°–1,600° K.);
$C_p = 79.00$.

TABLE 1h7. (Continued)

TABLE 161.—*Heat content and entropy of* $Ca_2P_2O_7(c, l)$

[Base, α-crystals at 298.15° K.; mol. wt., 254.11]

T, ° K.	$H_T-H_{298.15}$, cal./mole	$S_T-S_{298.15}$, cal./deg. mole	T, ° K.	$H_T-H_{298.15}$, cal./mole	$S_T-S_{298.15}$, cal./deg. mole
400	5,010	14.40	1,300	62,040	86.93
500	10,430	26.48	1,400	69,290	92.30
600	16,150	36.90	1,413(α)	70,240	92.98
700	22,120	46.08	1,413(β)	71,840	94.11
800	28,320	54.36	1,500	78,460	98.66
900	34,720	61.89	1,600	86,080	103.58
1,000	41,300	68.83	1,626(β)	88,060	104.81
1,100	48,040	75.25	1,626(l)	112,160	119.63
1,200	54,920	81.23	1,700	119,320	123.93

$Ca_2P_2O_7(\alpha)$:

$$H_T-H_{298.15}=53.03T+7.38\times10^{-3}T^2+11.16\times10^5T^{-1}$$
$$-20,210 \text{ (0.2 percent; } 298°-1,413° \text{ K.)};$$
$$C_p=53.03+14.76\times10^{-3}T-11.16\times10^5T^{-2};$$
$$\Delta H_{1413}(transition)=1,600.$$

$Ca_2P_2O_7(\beta)$:

$$H_T-H_{298.15}=76.15T-35,760 \text{ (0.1 percent;}$$
$$1,413-1,626° \text{ K.)};$$
$$C_p=76.15;$$
$$\Delta H_{1626}(fusion)=24,100.$$

$Ca_2P_2O_7(l)$:

$$H_T-H_{298.15}=96.80T-45,240 \text{ (0.1 percent; } 1,626°-1,700°}$$
$$\text{K.)};$$
$$C_p=96.80.$$

TABLE 162.—*Heat content and entropy of* $CaP_2O_6(c)$

[Base, crystals st 298.15° K.; mol. wt., 198.03]

T, ° K.	$H_T-H_{298.15}$, cal./mole	$S_T-S_{298.15}$, cal./deg. mole	T, ° K.	$H_T-H_{298.15}$, cal./mole	$S_T-S_{298.15}$, cal./deg. mole
400	3,930	11.29	900	27,800	49.38
500	8,190	20.78	1,000	33,100	54.96
600	12,760	29.15	1,100	38,520	60.12
700	17,620	36.60	1,200	44,040	64.93
800	22,640	43.30	1,250	46,820	67.20

$CaP_2O_6(c)$:

$$H_T-H_{298.15}=43.63T+5.50\times10^{-3}T^2+10.86\times10^5T^{-1}$$
$$-17,140 \text{ (0.3 percent; } 298°-1,250° \text{ K.)};$$
$$C_p=43.63+11.00\times10^{-3}T-10.86\times10^5T^{-2}.$$

TABLE 163.—*Heat content and entropy of* CaP_2O_6 (gl, l)

[Base, glass at 298.15° K.; mol. wt., 198.03]

T, ° K.	$H_T-H_{298.15}$, cal./mole	$S_T-S_{298.15}$, cal./deg. mole	T, ° K.	$H_T-H_{298.15}$, cal./mole	$S_T-S_{298.15}$, cal./deg. mole
400	3,990	11.47	1,100	38,480	60.16
500	8,260	20.99	1,200	44,050	65.01
600	12,840	29.33	1,250(gl)	46,880	67.32
700	17,630	36.71	1,250(l)	46,880	67.32
800	22,600	43.34	1,300	49,690	69.52
900	27,730	49.38	1,400	55,310	73.69
1,090	33,030	54.97	1,500	60,940	77.57

$CaP_2O_6(gl)$:

$$H_T-H_{298.15}=41.82T+6.36\times10^{-3}T^2+8.98\times10^5T^{-1}$$
$$-16,046 \text{ (0.2 percent; } 298°-1,250° \text{ K.)};$$
$$C_p=41.82T+12.72\times10^{-3}T-8.98\times10^5T^{-2}.$$

$CaP_2O_6(l)$:

$$H_T-H_{298.15}=56.25T-23,435 \text{ (0.1 percent; } 1,250°-$$
$$1,500° \text{ K.)};$$
$$C_p=56.25.$$

TABLE 164.—*Heat content and entropy of* $Ca_{10}(PO_4)_6(OH)_2(c)$

[Base, crystals at 298.15° K.; mol. wt., 1,004.67]

T, ° K.	$H_T-H_{298.15}$, cal./mole	$S_T-S_{298.15}$, cal./deg. mole	T, ° K.	$H_T-H_{298.15}$, cal./mole	$S_T-S_{298.15}$, cal./deg. mole
400	20,350	58.5	1,000	166,100	278.0
500	42,700	108.3	1,100	192,800	303.5
600	66,000	150.9	1,200	220,100	327.2
700	90,000	187.8	1,300	247,900	349.5
800	114,700	220.8	1,400	276,300	370.5
900	140,100	250.6	1,500	305,200	390.5

$Ca_{10}(PO_4)_6(OH)_2(c)$:

$$H_T-H_{298.15}=228.52T+19.81\times10^{-3}T^2+50.00\times10^5T^{-1}$$
$$-86,664 \text{ (0.6 percent; } 298°-1,500° \text{ K.)};$$
$$C_p=228.52+39.62\times10^{-3}T-50.00\times10^5T^{-2}.$$

TABLE 165.—*Heat content and entropy of* $Ca_{10}(PO_4)_6F_2(c)$

[Base, crystals at 298.15° K.; mol. wt., 1,008.65]

T, ° K.	$H_T-H_{298.15}$, cal./mole	$S_T-S_{298.15}$, cal./deg. mole	T, ° K.	$H_T-H_{298.15}$, cal./mole	$S_T-S_{298.15}$, cal./deg. mole
400	19,800	56.9	1,100	185,600	292.6
500	41,300	104.9	1,200	211,100	314.8
600	63,700	145.7	1,300	236,800	335.4
700	86,800	181.4	1,400	262,700	354.6
800	110,800	213.3	1,500	288,800	372.6
900	135,300	242.2	1,600	315,100	389.6
1,000	160,300	268.5			

$Ca_{10}(PO_4)_6F_2(c)$:

$$H_T-H_{298.15}=226.78T+13.62\times10^{-3}T^2+49.06\times10^5T^{-1}$$
$$-85,280 \text{ (0.2 percent; } 298°-1,600° \text{ K.)};$$
$$C_p=226.78+27.24\times10^{-3}T-49.06\times10^5T^{-2}.$$

SILICATES

TABLE 166.—*Heat content and entropy of* $Ca_3SiO_5(c)$

[Base, crystals at 298.15° K.; mol. wt., 228.33]

T, ° K.	$H_T-H_{298.15}$, cal./mole	$S_T-S_{298.15}$, cal./deg. mole	T, ° K.	$H_T-H_{298.15}$, cal./mole	$S_T-S_{298.15}$, cal./deg. mole
400	4,500	12.95	1,200	48,340	71.89
500	9,340	23.74	1,300	54,340	76.69
600	14,400	32.96	1,400	60,410	81.19
700	19,720	41.15	1,500	66,570	85.44
800	25,260	48.55	1,600	72,800	89.46
900	30,900	55.19	1,700	79,150	93.31
1,000	36,620	61.22	1,800	85,620	97.00
1,100	44,420	66.74			

$Ca_3SiO_5(c)$:

$$H_T-H_{298.15}=49.85T+4.31\times10^{-3}T^2+10.15\times10^5T^{-1}$$
$$-18,651 \text{ (0.3 percent; } 298°-1,800° \text{ K)};$$
$$C_p=49.85+8.62\times10^{-3}T-10.15\times10^5T^{-2}.$$

TABLE 1h7. (Continued)

TABLE 167.—Heat content and entropy of $Ca_2SiO_4(\beta, \alpha', \alpha)$

[Base, β-crystals at 298.15° K.; mol. wt., 172.25]

$T, °$ K.	$H_T-H_{298.15}$, cal./mole	$S_T-S_{298.15}$, cal./deg. mole	$T, °$ K.	$H_T-H_{298.15}$, cal./mole	$S_T-S_{298.15}$, cal./deg. mole
400	3,335	9.59	1,300	41,290	57.99
500	6,940	17.63	1,400	45,970	61.45
600	10,790	24.64	1,500	50,780	64.77
700	14,810	30.84	1,600	55,710	67.95
800	18,940	36.35	1,700	60,780	71.03
900	23,140	41.29	1,710(α')	61,290	71.33
970(β)	26,120	44.48	1,710(α)	64,680	73.31
970(α')	26,560	44.94	1,800	69,090	75.82
1,000	27,860	46.26	1,900	73,990	78.47
1,100	32,250	50.44	2,000	78,890	80.98
1,200	36,720	54.33			

$Ca_2SiO_4(\beta)$:

$$H_T-H_{298.15}=34.87T+4.87\times10^{-3}T^2+6.26\times10^5T^{-1}$$
$$-12,929 \ (0.4 \ percent; \ 298°-970° \ K.);$$
$$C_p=34.87+9.74\times10^{-3}T-6.26\times10^5T^{-2};$$
$$\Delta H_{970}(transition)=440.$$

$Ca_2SiO_4(\alpha')$:

$$H_T-H_{298.15}=32.16T+5.51\times10^{-3}T^2-9,814$$
$$(0.1 \ percent; \ 970°-1,710° \ K.);$$
$$C_p=32.16+11.02\times10^{-3}T;$$
$$\Delta H_{1710}(transition)=3,390.$$

$Ca_2SiO_4(\alpha)$:

$$H_T-H_{298.15}=49.00T-19,110 \ (0.1 \ percent;$$
$$1,710°-2,000° \ K.);$$
$$C_p=49.00.$$

TABLE 168.—Heat content and entropy of $Ca_2SiO_4(\gamma)$

[Base, γ-crystals at 298.15° K.; mol. wt., 172.25]

$T, °$ K.	$H_T-H_{298.15}$, cal./mole	$S_T-S_{298.15}$, cal./deg. mole	$T, °$ K.	$H_T-H_{298.15}$, cal./mole	$S_T-S_{298.15}$, cal./deg. mole
400	3,270	9.41	900	22,590	40.28
500	6,760	17.19	1,000	26,890	44.81
600	10,480	23.96	1,100	31,320	49.03
700	14,380	29.97	1,120	32,220	49.84
800	18,420	35.37			

$Ca_2SiO_4(\gamma)$:

$$H_T-H_{298.15}=31.86T+6.16\times10^{-3}T^2+4.64\times10^5T^{-1}$$
$$-11,603 \ (0.2 \ percent; \ 298°-1,120° \ K.);$$
$$C_p=31.86+12.32\times10^{-3}T-4.64\times10^5T^{-2}.$$

TABLE 169.—Heat content and entropy of $CaSiO_3$ (wollastonite)

[Base, crystals at 298.15° K.; mol. wt., 116.17]

$T, °$ K.	$H_T-H_{298.15}$, cal./mole	$S_T-S_{298.15}$, cal./deg. mole	$T, °$ K.	$H_T-H_{298.15}$, cal./mole	$S_T-S_{298.15}$, cal./deg. mole
400	2,300	6.61	1,000	18,810	31.42
500	4,780	12.14	1,100	21,770	34.24
600	7,390	16.89	1,200	24,800	36.88
700	10,140	21.12	1,300	27,880	39.34
800	13,000	24.94	1,400	31,000	41.65
900	15,890	28.34	1,450	32,570	42.76

$CaSiO_3(wollastonite)$:

$$H_T-H_{298.15}=26.64T+1.80\times10^{-3}T^2+6.52\times10^5T^{-1}$$
$$-10,290 \ (0.2 \ percent; \ 298°-1,450° \ K.);$$
$$C_p=26.64+3.60\times10^{-3}T-6.52\times10^5T^{-2}.$$

TABLE 170.—Heat content and entropy of $CaSiO_3$ (pseudowollastonite)

[Base, crystals at 298.15° K.; mol. wt., 116.17]

$T, °$ K.	$H_T-H_{298.15}$, cal./mole	$S_T-S_{298.15}$, cal./deg. mole	$T, °$ K.	$H_T-H_{298.15}$, cal./mole	$S_T-S_{298.15}$, cal./deg. mole
400	2,290	6.59	1,100	21,590	33.94
500	4,710	11.98	1,200	24,580	36.55
600	7,330	16.76	1,300	27,610	38.97
700	10,040	20.93	1,400	30,690	41.25
800	12,840	24.67	1,500	33,810	43.41
900	15,710	28.04	1,600	36,970	45.44
1,000	18,630	31.12	1,700	40,170	47.38

$CaSiO_3(pseudowollastonite)$:

$$H_T-H_{298.15}=25.85T+1.97\times10^{-3}T^2+5.65\times10^5T^{-1}$$
$$-9,777 \ (0.3 \ percent; \ 298°-1,700° \ K.);$$
$$C_p=25.85+3.94\times10^{-3}T-5.65\times10^5T^{-2}.$$

TABLE 171.—Heat content and entropy of $CaSiO_3(gl)$

[Base, glass at 298.15° K.; mol. wt., 116.17]

$T, °$ K.	$H_T-H_{298.15}$, cal./mole	$S_T-S_{298.15}$, cal./deg. mole	$T, °$ K.	$H_T-H_{298.15}$, cal./mole	$S_T-S_{298.15}$, cal./deg. mole
400	2,270	6.53	800	13,130	25.12
500	4,780	12.12	900	16,260	28.81
600	7,430	16.95	1,000	19,540	32.26
700	10,200	21.21			

$CaSiO_3(gl)$:

$$H_T-H_{298.15}=21.59T+5.87\times10^{-3}T^2+4.08\times10^5T^{-1}$$
$$-8,327 \ (0.3 \ percent; \ 298°-1,000° \ K.);$$
$$C_p=21.59+11.74\times10^{-3}T-4.08\times10^5T^{-2}.$$

CALCIUM-ALUMINUM SILICATES

TABLE 172.—Heat content and entropy of $Ca_2Al_2SiO_7$ (gehlenite)

[Base, crystals at 298.15° K.; mol. wt., 274.21]

$T, °$ K.	$H_T-H_{298.15}$, cal./mole	$S_T-S_{298.15}$, cal./deg. mole	$T, °$ K.	$H_T-H_{298.15}$, cal./mole	$S_T-S_{298.15}$, cal./deg. mole
400	6,050	17.45	1,100	53,000	84.06
500	12,100	30.93	1,200	60,350	90.45
600	18,350	42.33	1,300	67,850	96.45
700	24,900	52.42	1,400	75,550	102.15
800	31,650	61.43	1,500	83,450	107.61
900	38,600	69.61	1,600	91,550	112.83
1,000	45,750	77.14			

$Ca_2Al_2SiO_7$ (gehlenite):

$$H_T-H_{298.15}=53.73T+8.84\times10^{-3}T^2+0.89\times10^5T^{-1}$$
$$-17,104 \ (0.3 \ percent; \ 298°-1,600° \ K.);$$
$$C_p=53.73+17.68\times10^{-3}T-0.89\times10^5T^{-2}.$$

TABLE 1h7. (*Continued*)

TABLE 173.—*Heat content and entropy of CaAl₂Si₂O₈ (anorthite)*

[Base, crystals at 298.15° K.; mol. wt., 278.22]

T, ° K.	$H_T-H_{298.15}$, cal./mole	$S_T-S_{298.15}$, cal./deg. mole	T, ° K.	$H_T-H_{298.15}$, cal./mole	$S_T-S_{298.15}$, cal./deg. mole
400	5,570	16.02	1,100	55,130	86.17
500	11,750	29.78	1,200	62,970	92.99
600	18,450	41.99	1,300	70,930	99.35
700	25,410	52.71	1,400	79,050	105.37
800	32,570	62.27	1,500	87,450	111.17
900	39,910	70.90	1,600	96,170	116.79
1,000	47,430	78.83	1,700	105,230	122.28

$CaAl_2Si_2O_8$ *(anorthite):*

$$H_T - H_{298.15} = 64.42T + 6.85 \times 10^{-3}T^2 + 16.89 \times 10^5 T^{-1}$$
$$-25,481 \ (0.3 \text{ percent}; \ 298°-1,700° \text{ K.});$$
$$C_p = 64.42 + 13.70 \times 10^{-3}T - 16.89 \times 10^5 T^{-2}.$$

TABLE 174.—*Heat content and entropy of CaAl₂Si₂O₈(gl).*

[Base, glass at 298.15° K.; mol. wt., 278.22]

T, ° K.	$H_T-H_{298.15}$, cal./mole	$S_T-S_{298.15}$, cal./deg. mole	T, ° K.	$H_T-H_{298.15}$, cal./mole	$S_T-S_{298.15}$, cal./deg. mole
400	5,590	16.07	800	32,870	62.77
500	11,780	29.86	900	40,250	71.46
600	18,590	42.27	1,000	47,750	79.36
700	25,630	53.11			

$CaAl_2Si_2O_8(gl)$:

$$H_T - H_{298.15} = 66.46T + 5.92 \times 10^{-3}T^2 + 18.22 \times 10^5 T^{-1}$$
$$-26,452 \ (0.4 \text{ percent}; \ 298°-1,000° \text{ K.});$$
$$C_p = 66.46 + 11.84 \times 10^{-3}T - 18.22 \times 10^5 T^{-2}.$$

CALCIUM-MAGNESIUM SILICATE

TABLE 175.—*Heat content and entropy of CaMg(SiO₃)₂(diopside)*

[Base, crystals at 298.15° K.; mol. wt., 216.58]

T, ° K.	$H_T-H_{298.15}$, cal./mole	$S_T-S_{298.15}$, cal./deg. mole	T, ° K.	$H_T-H_{298.15}$, cal./mole	$S_T-S_{298.15}$, cal./deg. mole
400	4,320	12.41	1,100	43,250	67.23
500	8,940	22.70	1,200	49,250	72.45
600	14,060	32.04	1,300	55,300	77.29
700	19,540	40.46	1,400	61,440	81.84
800	25,420	48.32	1,500	67,660	86.14
900	31,340	55.28	1,600	73,980	90.21
1,000	37,280	61.54			

$CaMg(SiO_3)_2$ *(diopside):*

$$H_T - H_{298.15} = 52.87T + 3.92 \times 10^{-3}T^2 + 15.74 \times 10^5 T^{-1}$$
$$-21,391 \ (0.9 \text{ percent}; \ 298°-1,600° \text{ K.});$$
$$C_p = 52.87 + 7.84 \times 10^{-3}T - 15.74 \times 10^5 T^{-2}.$$

TABLE 176.—*Heat content and entropy of CaMg(SiO₃)₂(gl)*

[Base, glass at 298.15° K.; mol. wt., 216.58]

T, ° K.	$H_T-H_{298.15}$, cal./mole	$S_T-S_{298.15}$, cal./deg. mole	T, ° K.	$H_T-H_{298.15}$, cal./mole	$S_T-S_{298.15}$, cal./deg. mole
400	4,470	12.85	800	25,870	49.51
500	9,370	23.77	900	31,650	56.31
600	14,650	33.39	1,000	37,590	62.57
700	20,230	41.98			

$CaMg(SiO_3)_2(gl)$:

$$H_T - H_{298.15} = 51.32T + 5.15 \times 10^{-3}T^2 + 13.24 \times 10^5 T^{-1}$$
$$-20,200 \ (0.2 \text{ percent}; \ 298°-1,000° \text{ K.});$$
$$C_p = 51.32 + 10.30 \times 10^{-3}T - 13.24 \times 10^5 T^{-2}.$$

CALCIUM-TITANIUM SILICATE

TABLE 177.—*Heat content and entropy of CaTiSiO₅ (sphene)*

[Base, crystals at 298.15° K.; mol. wt., 196.07]

T, ° K.	$H_T-H_{298.15}$, cal./mole	$S_T-S_{298.15}$, cal./deg. mole	T, ° K.	$H_T-H_{298.15}$, cal./mole	$S_T-S_{298.15}$, cal./deg. mole
400	3,750	10.78	1,400	49,350	66.51
500	7,690	19.56	1,500	54,340	69.96
600	11,860	27.16	1,600	59,400	73.22
700	16,230	33.89	1,670(c)	62,980	75.41
800	20,750	39.93	1,670(l)	92,570	93.13
900	25,380	45.38	1,700	94,570	94.32
1,000	30,070	50.32	1,800	101,250	98.14
1,100	34,800	54.83	1,900	107,930	101.75
1,200	39,580	58.99	2,000	114,610	105.18
1,300	44,430	62.87			

$CaTiSiO_5(c)$:

$$H_T - H_{298.15} = 42.39T + 2.77 \times 10^{-3}T^2 + 9.63 \times 10^5 T^{-1}$$
$$-16,115 \ (0.3 \text{ percent}; \ 298°-1,670° \text{ K.});$$
$$C_p = 42.39 + 5.54 \times 10^{-3}T - 9.63 \times 10^5 T^{-2};$$
$$\Delta H_{1670}(fusion) = 29,590.$$

$CaTiSiO_5(l)$:

$$H_T - H_{298.15} = 66.80T$$
$$-18,990 \ (0.1 \text{ percent}; \ 1,670°-2,000° \text{ K.});$$
$$C_p = 66.80.$$

SULFATE

TABLE 178.—*Heat content and entropy of CaSO₄ (anhydrite)*

[Base, crystals at 298.15° K.; mol. wt., 136.15]

T, ° K.	$H_T-H_{298.15}$, cal./mole	$S_T-S_{298.15}$, cal./deg. mole	T, ° K.	$H_T-H_{298.15}$, cal./mole	$S_T-S_{298.15}$, cal./deg. mole
400	2,600	7.49	1,000	22,850	37.11
500	5,200	13.28	1,100	27,000	41.07
600	8,050	18.47	1,200	31,300	44.81
700	11,250	23.40	1,300	35,800	48.41
800	14,850	28.20	1,400	40,500	51.89
900	18,800	32.85			

$CaSO_4$ *(anhydrite):*

$$H_T - H_{298.15} = 16.78T + 11.80 \times 10^{-3}T^2 - 6,052$$
$$(1.3 \text{ percent}; \ 298°-1,400° \text{ K.});$$
$$C_p = 16.78 + 23.60 \times 10^{-3}T.$$

$CaSO_4 \cdot \tfrac{1}{2}H_2O \ (\alpha)$:

$$C_p = 16.95 + 39.00 \times 10^{-3}T(estimated) \ (298°-450° \text{ K.}).$$

$CaSO_4 \cdot \tfrac{1}{2}H_2O(\beta)$:

$$C_p = 11.48 + 61.00 \times 10^{-3}T(estimated) \ (298°-450° \text{ K.}).$$

$CaSO_4 \cdot 2H_2O(c)$:

$$C_p = 21.84 + 76.00 \times 10^{-3}T(estimated) \ (298°-400° \text{ K.}).$$

TABLE 1h7. (*Continued*)

SULFITE

$CaSO_3(c)$:

$C_p = 21.92$ (298° K.).

TITANATES

$Ca_3Ti_2O_5(c)$:

$C_p = 57.20$ (298° K.).

TABLE 179.—*Heat content and entropy of $CaTiO_3$ (perovskite)*

[Base, α-crystals at 298.15° K.; mol. wt., 135.98]

T, °K.	$H_T - H_{298.15}$, cal./mole	$S_T - S_{298.15}$, cal./deg. mole	T, °K.	$H_T - H_{298.15}$, cal./mole	$S_T - S_{298.15}$, cal./deg. mole
400	2,680	7.71	1,400	33,030	45.10
500	5,430	13.84	1,500	36,270	47.34
600	8,300	19.07	1,530(α)	37,260	47.99
700	11,260	23.63	1,530(β)	37,810	48.35
800	14,270	27.65	1,600	40,050	49.78
900	17,310	31.23	1,700	43,250	51.72
1,000	20,380	34.47	1,800	46,460	53.56
1,100	23,490	37.44	1,900	49,680	55.29
1,200	26,640	40.18	2,000	52,910	56.94
1,300	29,820	42.72			

$CaTiO_3(\alpha)$:

$$H_T - H_{298.15} = 30.47T + 0.68 \times 10^{-3}T^2 + 6.69 \times 10^5 T^{-1}$$
$$-11,389 \ (0.7 \text{ percent}; 298°-1,530° \text{ K.});$$
$$C_p = 30.47 + 1.36 \times 10^{-3}T - 6.69 \times 10^5 T^{-2}.$$

$$\Delta H_{1530}(transition) = 550.$$

$CaTiO_3(\beta)$:

$$H_T - H_{298.15} = 32.03T - 11,197 \ (0.1 \text{ percent};$$
$$1,530°-2,000° \text{ K.});$$
$$C_p = 32.03.$$

TUNGSTATE

$CaWO_4(c)$:

$C_p = 27.8$ (292°-322° K.).

CARBON AND ITS COMPOUNDS

ELEMENT

TABLE 180.—*Heat content and entropy of $C(graphite)$*

[Base, crystals at 298.15° K.; atomic wt., 12.011]

T, °K.	$H_T - H_{298.15}$, cal./mole	$S_T - S_{298.15}$, cal./deg. mole	T, °K.	$H_T - H_{298.15}$, cal./mole	$S_T - S_{298.15}$, cal./deg. mole
400	250	0.72	1,800	7,330	7.78
500	570	1.43	1,900	7,920	8.10
600	950	2.12	2,000	8,530	8.41
700	1,370	2.77	2,100	9,130	8.70
800	1,830	3.38	2,200	9,740	8.99
900	2,320	3.95	2,300	10,350	9.26
1,000	2,820	4.48	2,400	10,970	9.52
1,100	3,340	4.98	2,500	11,600	9.78
1,200	3,880	5.45	2,750	13,170	10.38
1,300	4,430	5.89	3,000	14,770	10.93
1,400	4,990	6.30	3,250	16,390	11.45
1,500	5,560	6.70	3,500	18,020	11.93
1,600	6,150	7.08	3,750	19,660	12.39
1,700	6,740	7.44	4,000	21,310	12.81

$C(graphite)$:

$$H_T - H_{298.15} = 4.03T + 0.57 \times 10^{-3}T^2 + 2.04 \times 10^5 T^{-1}$$
$$-1,936 \ (3.0 \text{ percent}; 298°-2,500° \text{ K.});$$
$$C_p = 4.03 + 1.14 \times 10^{-3}T - 2.04 \times 10^5 T^{-2}.$$

TABLE 181.—*Heat content and entropy of $C(diamond)$*

[Base, crystals at 298.15° K.; atomic wt., 12.011]

T, °K.	$H_T - H_{298.15}$, cal./mole	$S_T - S_{298.15}$, cal./deg. mole	T, °K.	$H_T - H_{298.15}$, cal./mole	$S_T - S_{298.15}$, cal./deg. mole
400	195	0.56	900	2,160	3.60
500	470	1.17	1,000	2,650	4.12
600	820	1.80	1,100	3,160	4.60
700	1,230	2.43	1,200	3,675	5.05
800	1,680	3.04			

$C(diamond)$:

$$H_T - H_{298.15} = 2.27T + 1.53 \times 10^{-3}T^2 + 1.54 \times 10^5 T^{-1}$$
$$-1,329 \ (2.2 \text{ percent}; 298°-1,200° \text{ K.});$$
$$C_p = 2.27 + 3.06 \times 10^{-3}T - 1.54 \times 10^5 T^{-2}.$$

TABLE 182.—*Heat content and entropy of $C(g)$*

[Base, ideal gas at 298.15° K.; atomic wt., 12.011]

T, °K.	$H_T - H_{298.15}$, cal./mole	$S_T - S_{298.15}$, cal./deg. mole	T, °K.	$H_T - H_{298.15}$, cal./mole	$S_T - S_{298.15}$, cal./deg. mole
400	505	1.46	1,900	7,970	9.21
500	1,005	2.57	2,000	8,470	9.47
600	1,500	3.48	2,200	9,475	9.95
700	2,000	4.25	2,400	10,485	10.39
800	2,495	4.91	2,600	11,500	10.80
900	2,990	5.50	2,800	12,520	11.17
1,000	3,490	6.02	3,000	13,550	11.53
1,100	3,985	6.49	3,500	16,160	12.33
1,200	4,485	6.93	4,000	18,815	13.04
1,300	4,980	7.32	4,500	21,505	13.68
1,400	5,480	7.69	5,000	24,230	14.25
1,500	5,975	8.04	6,000	29,735	15.25
1,600	6,475	8.36	7,000	35,300	16.11
1,700	6,970	8.66	8,000	40,905	16.86
1,800	7,470	8.94			

$C(g)$:

$$H_T - H_{298.15} = 4.98T - 1,485 \ (0.2 \text{ percent}; 298°-3,000° \text{ K.});$$
$$C_p = 4.98.$$

TABLE 183.—*Heat content and entropy of $C_2(g)$*

[Base, ideal gas at 298.15° K.; mol. wt., 24.02]

T, °K.	$H_T - H_{298.15}$, cal./mole	$S_T - S_{298.15}$, cal./deg. mole	T, °K.	$H_T - H_{298.15}$, cal./mole	$S_T - S_{298.15}$, cal./deg. mole
400	720	2.08	1,500	9,620	12.54
500	1,445	3.69	1,600	10,490	13.10
600	2,190	5.05	1,700	11,360	13.63
700	2,960	6.24	1,800	12,230	14.13
800	3,750	7.29	1,900	13,110	14.60
900	4,560	8.25	2,000	13,990	15.05
1,000	5,380	9.11	2,200	15,750	15.89
1,100	6,210	9.90	2,400	17,530	16.67
1,200	7,050	10.63	2,600	19,310	17.38
1,300	7,900	11.31	2,800	21,090	18.04
1,400	8,760	11.95	3,000	22,890	18.66

$C_2(g)$:

$$H_T - H_{298.15} = 7.18T + 0.45 \times 10^{-3}T^2 + 0.40 \times 10^5 T^{-1}$$
$$-2,315 \ (1.0 \text{ percent}; 298°-3,000° \text{ K.});$$
$$C_p = 7.18 + 0.90 \times 10^{-3}T - 0.40 \times 10^5 T^{-2}.$$

TABLE 1h7. (*Continued*)

TABLE 184.—*Heat content and entropy of* $C_3(g)$

[Base, ideal gas at 298.15° K.; mol. wt., 36.03]

T, ° K.	$H_T-H_{298.15}$, cal./mole	$S_T-S_{298.15}$, cal./deg. mole	T, ° K.	$H_T-H_{298.15}$, cal./mole	$S_T-S_{298.15}$, cal./deg. mole
400	1,100	3.17	1,500	15,610	20.17
500	2,250	5.73	1,600	17,030	21.08
600	3,450	7.92	1,700	18,460	21.95
700	4,700	9.84	1,800	19,900	22.77
800	5,980	11.55	1,900	21,340	23.55
900	7,310	13.12	2,000	22,780	24.29
1,000	8,650	14.53	2,200	25,690	25.68
1,100	10,010	15.83	2,400	28,600	26.94
1,200	11,390	17.03	2,600	31,530	28.11
1,300	12,790	18.15	2,800	34,460	29.20
1,400	14,190	19.19	3,000	37,400	30.21

$$C_3(g):$$
$$H_T-H_{298.15}=12.40\,T+0.54\times10^{-3}T^2+2.06\times10^5T^{-1}$$
$$-4,436 \text{ (1.0 percent; } 298°-3,000° \text{ K.)};$$
$$C_p=12.40+1.08\times10^{-3}T-2.06\times10^5T^{-2}.$$

OXIDES

TABLE 185.—*Heat content and entropy of* $CO(g)$

[Base, ideal gas at 298.15° K.; mol. wt., 28.01]

T, ° K.	$H_T-H_{298.15}$, cal./mole	$S_T-S_{298.15}$, cal./deg. mole	T, ° K.	$H_T-H_{298.15}$, cal./mole	$S_T-S_{298.15}$, cal./deg. mole
400	710	2.04	2,000	13,565	14.60
500	1,415	3.63	2,100	14,430	15.02
600	2,135	4.94	2,200	15,305	15.47
700	2,875	6.07	2,300	16,175	15.81
800	3,625	7.08	2,400	17,055	16.18
900	4,400	7.99	2,500	17,935	16.54
1,000	5,185	8.82	2,750	20,140	17.38
1,100	5,985	9.58	3,000	22,360	18.16
1,200	6,795	10.28	3,250	24,590	18.87
1,300	7,620	10.94	3,500	26,825	19.53
1,400	8,450	11.56	3,750	29,070	20.15
1,500	9,285	12.14	4,000	31,325	20.74
1,600	10,130	12.68	4,250	34,580	21.28
1,700	10,980	13.20	4,500	35,840	21.80
1,800	11,835	13.69	4,750	38,105	22.29
1,900	12,700	14.15	5,000	40,375	22.76

$$CO(g):$$
$$H_T-H_{298.15}=6.79\,T+0.49\times10^{-3}T^2+0.11\times10^5T^{-1}$$
$$-2,105 \text{ (0.8 percent; } 298°-2,500° \text{ K.)};$$
$$C_p=6.79+0.98\times10^{-3}T-0.11\times10^5T^{-2}.$$

TABLE 186.—*Heat content and entropy of* $CO_2(g)$

[Base, ideal gas at 298.15° K.; mol. wt., 44.01]

T, ° K.	$H_T-H_{298.15}$, cal./mole	$S_T-S_{298.15}$, cal./deg. mole	T, ° K.	$H_T-H_{298.15}$, cal./mole	$S_T-S_{298.15}$, cal./deg. mole
400	955	2.75	2,000	21,855	22.82
500	1,985	5.05	2,100	23,300	23.53
600	3,085	7.05	2,200	24,755	24.21
700	4,255	8.83	2,300	26,210	24.85
800	5,450	10.44	2,400	27,670	25.47
900	6,700	11.92	2,500	29,140	26.08
1,000	7,985	13.27	2,750	32,825	27.48
1,100	9,295	14.52	3,000	36,530	28.77
1,200	10,630	15.68	3,250	40,260	29.96
1,300	11,985	16.76	3,500	44,005	31.07
1,400	13,360	17.78	3,750	47,760	32.11
1,500	14,750	18.74	4,000	51,530	33.08
1,600	16,150	19.64	4,250	55,325	34.00
1,700	17,560	20.50	4,500	59,115	34.87
1,800	18,985	21.31	4,750	62,930	35.70
1,900	20,415	22.09	5,000	66,750	36.48

$$CO_2(g):$$
$$H_T-H_{298.15}=10.57\,T+1.05\times10^{-3}T^2+2.06\times10^5T^{-1}$$
$$-3,936 \text{ (1.5 percent; } 298°-2,500° \text{ K.)};$$
$$C_p=10.57+2.10\times10^{-3}T-2.06\times10^5T^{-2}.$$

TABLE 187.—*Heat content and entropy of* $C_3O_2(g)$

[Base, ideal gas at 298.15° K.; mol. wt. 68.03]

T, ° K.	$H_T-H_{298.15}$, cal./mole	$S_T-S_{298.15}$, cal./deg. mole	T, ° K.	$H_T-H_{298.15}$, cal./mole	$S_T-S_{298.15}$, cal./deg. mole
400	1,710	4.92	800	9,790	18.76
500	3,550	9.02	900	12,040	21.41
600	5,650	12.68	1,000	14,330	23.82
700	7,660	15.92			

$$C_3O_2(g):$$
$$H_T-H_{298.15}=15.63\,T+4.34\times10^{-3}T^2+2.22\times10^5T^{-1}$$
$$-5,790 \text{ (0.4 percent; } 298°-1,000° \text{ K.)};$$
$$C_p=15.63+8.68\times10^{-3}T-2.22\times10^5T^{-2}.$$

SULFIDES

TABLE 188.—*Heat content and entropy of* $CS(g)$

[Base, ideal gas at 298.15° K.; mol. wt., 44.08]

T, ° K.	$H_T-H_{298.15}$, cal./mole	$S_T-S_{298.15}$, cal./deg. mole	T, ° K.	$H_T-H_{298.15}$, cal./mole	$S_T-S_{298.15}$, cal./deg. mole
400	740	2.13	1,000	5,570	9.43
500	1,495	3.81	1,200	7,275	10.99
600	2,275	5.23	1,400	9,005	12.31
700	3,075	6.47	1,600	10,745	13.48
800	3,895	7.56	1,800	12,500	14.51
900	4,725	8.54	2,000	14,260	15.44

$$CS(g):$$
$$H_T-H_{298.15}=7.39\,T+0.51\times10^{-3}T^2+0.51\times10^5T^{-1}$$
$$-2,420 \text{ (0.6 percent; } 298°-2,000° \text{ K.)};$$
$$C_p=7.39+1.02\times10^{-3}T-0.51\times10^5T^{-2}.$$

TABLE 189.—*Heat content and entropy of* $CS_2(g)$

[Base, ideal gas at 298.15° K.; mol. wt., 76.14]

T, ° K.	$H_T-H_{298.15}$, cal./mole	$S_T-S_{298.15}$, cal./deg. mole	T, ° K.	$H_T-H_{298.15}$, cal./mole	$S_T-S_{298.15}$, cal./deg. mole
400	1,165	3.36	1,200	11,930	17.90
500	2,380	6.07	1,300	13,360	19.04
600	3,655	8.39	1,400	14,790	20.10
700	4,980	10.41	1,500	16,240	21.11
800	6,335	12.24	1,600	17,700	22.05
900	7,705	13.85	1,700	19,150	22.93
1,000	9,095	15.32	1,800	20,590	23.75
1,100	10,500	16.66			

$$CS_2(g):$$
$$H_T-H_{298.15}=12.45\,T+0.80\times10^{-3}T^2+1.80\times10^5T^{-1}$$
$$-4,387 \text{ (0.5 percent; } 298°-1,800° \text{ K.)};$$
$$C_p=12.45+1.60\times10^{-3}T-1.80\times10^5T^{-2}.$$

CARBONYL SULFIDE

TABLE 190.—*Heat content and entropy of* $COS(g)$

[Base, ideal gas at 298.15° K.; mol. wt., 60.08]

T, ° K.	$H_T-H_{298.15}$, cal./mole	$S_T-S_{298.15}$, cal./deg. mole	T, ° K.	$H_T-H_{298.15}$, cal./mole	$S_T-S_{298.15}$, cal./deg. mole
400	1,070	3.08	1,200	11,300	16.85
500	2,200	5.60	1,300	12,700	17.95
600	3,395	7.78	1,400	14,090	19.00
700	4,640	9.70	1,500	15,490	19.97
800	5,925	11.41	1,600	16,910	20.88
900	7,230	12.95	1,700	18,350	21.75
1,000	8,570	14.36	1,800	19,790	22.58
1,100	9,915	15.65			

TABLE 1h7. (*Continued*)

COS(*g*):

$$H_T - H_{298.15} = 11.33T + 1.09 \times 10^{-3}T^2 + 1.83 \times 10^5 T^{-1}$$
$$-4,089 \ (0.5 \ \text{percent}; \ 298°-1,800° \ \text{K.});$$
$$C_p = 11.33 + 2.18 \times 10^{-3}T - 1.83 \times 10^5 T^{-2}.$$

CYANOGEN

TABLE 191.—*Heat content and entropy of* CN(*g*)

[Base, ideal gas at 298.15° K.; mol. wt., 26.02]

T, ° K.	$H_T-H_{298.15}$, cal./mole	$S_T-S_{298.15}$, cal./deg. mole	T, ° K.	$H_T-H_{298.15}$, cal./mole	$S_T-S_{298.15}$, cal./deg. mole
400	710	2.05	1,000	5,205	8.85
500	1,420	3.63	1,200	6,820	10.32
600	2,140	4.95	1,400	8,475	11.59
700	2,885	6.09	1,600	10,155	12.72
800	3,640	7.10	1,800	11,855	13.72
900	4,415	8.01	2,000	13,570	14.62

CN(*g*):

$$H_T - H_{298.15} = 6.60T + 0.62 \times 10^{-3}T^2 - 2,023$$
$$(0.5 \ \text{percent}; \ 298°-2,000° \ \text{K.});$$
$$C_p = 6.60 + 1.24 \times 10^{-3}T$$

TABLE 192.—*Heat content and entropy of* $C_2N_2(g)$

[Base, ideal gas at 298.15° K.; mol. wt., 52.04]

T, ° K.	$H_T-H_{298.15}$, cal./mole	$S_T-S_{298.15}$, cal./deg. mole	T, ° K.	$H_T-H_{298.15}$, cal./mole	$S_T-S_{298.15}$, cal./deg. mole
400	1,445	4.16	1,000	11,500	19.28
500	2,965	7.55	1,200	15,205	22.65
600	4,560	10.46	1,400	19,015	25.60
700	6,220	13.02	1,600	22,900	28.18
800	7,930	15.30	1,800	26,840	30.50
900	9,695	17.38	2,000	30,810	32.59

$C_2N_2(g)$:

$$H_T - H_{298.15} = 14.90T + 1.60 \times 10^{-3}T^2 + 2.04 \times 10^5 T^{-1}$$
$$-5,269 (0.7 \ \text{percent}; \ 298°-2,000° \ \text{K.});$$
$$C_p = 14.90 + 3.20 \times 10^{-3}T - 2.04 \times 10^5 T^{-2}.$$

PHOSPHIDE

TABLE 193.—*Heat content and entropy of* CP(*g*)

[Base, ideal gas at 298.15° K.; mol. wt., 42.99]

T, ° K.	$H_T-H_{298.15}$, cal./mole	$S_T-S_{298.15}$, cal./deg. mole	T, ° K.	$H_T-H_{298.15}$, cal./mole	$S_T-S_{298.15}$, cal./deg. mole
400	740	2.13	1,000	5,600	9.48
500	1,500	3.83	1,200	7,320	11.05
600	2,285	5.26	1,400	9,075	12.40
700	3,090	6.50	1,600	10,865	13.59
800	3,915	7.60	1,800	12,695	14.67
900	4,750	8.58	2,000	14,565	15.66

CP(*g*):

$$H_T - H_{.98.15} = 7.48T + 0.51 \times 10^{-3}T^2 + 0.57 \times 10^5 T^{-1}$$
$$-2,467 \ (0.5 \ \text{percent}; \ 298°-2,000° \ \text{K.});$$
$$C_p = 7.48 + 1.02 \times 10^{-3}T - 0.57 \times 10^5 T^{-2}.$$

SELENIDES

TABLE 194.—*Heat content and entropy of* CSe(*g*)

[Base, ideal gas at 298.15° K.; mol. wt., 90.97]

T, ° K.	$H_T-H_{298.15}$, cal./mole	$S_T-S_{298.15}$, cal./deg. mole	T, ° K.	$H_T-H_{298.15}$, cal./mole	$S_T-S_{298.15}$, cal./deg. mole
400	720	2.08	1,000	5,375	9.11
500	1,445	3.69	1,200	7,040	10.63
600	2,195	5.06	1,400	8,740	11.94
700	2,965	6.25	1,600	10,455	13.08
800	3,755	7.30	1,800	12,185	14.10
900	4,560	8.25	2,000	13,925	15.02

CSe(*g*):

$$H_T - H_{298.15} = 7.09T + 0.52 \times 10^{-3}T^2 + 0.35 \times 10^5 T^{-1}$$
$$-2,277 \ (0.7 \ \text{percent}; \ 298°-2,000° \ \text{K.});$$
$$C_p = 7.09 + 1.04 \times 10^{-3}T - 0.35 \times 10^5 T^{-2}.$$

TABLE 195.—*Heat content and entropy of* $CSe_2(g)$

[Base, ideal gas at 298.15° K.; mol. wt., 169.93]

T, ° K.	$H_T-H_{298.15}$, cal./mole	$S_T-S_{298.15}$, cal./deg. mole	T, ° K.	$H_T-H_{298.15}$, cal./mole	$S_T-S_{298.15}$, cal./deg. mole
400	1,255	3.62	1,000	9,485	16.07
500	2,550	6.50	1,200	12,365	18.70
600	3,890	8.94	1,400	15,270	20.94
700	5,260	11.05	1,600	18,190	22.89
800	6,650	12.91	1,800	21,125	24.62
900	8,060	14.57	2,000	24,075	26.17

$CSe_2(g)$:

$$H_T - H_{298.15} = 13.62T + 0.35 \times 10^{-3}T^2 + 1.66 \times 10^5 T^{-1}$$
$$-4,649 \ (0.5 \ \text{percent}; \ 298°-2,000° \ \text{K.});$$
$$C_p = 13.62 + 0.70 \times 10^{-3}T - 1.66 \times 10^5 T^{-2}.$$

SELENOSULFIDE

TABLE 196.—*Heat content and entropy of* CSeS(*g*)

[Base, ideal gas at 298.15° K.; mol. wt., 123.04]

T, ° K.	$H_T-H_{298.15}$, cal./mole	$S_T-S_{298.15}$, cal./deg. mole	T, ° K.	$H_T-H_{298.15}$, cal./mole	$S_T-S_{298.15}$, cal./deg. mole
400	1,210	3.48	1,000	9,285	15.69
500	2,465	6.28	1,200	12,135	18.28
600	3,770	8.66	1,400	15,025	20.51
700	5,115	10.73	1,600	17,935	22.45
800	6,485	12.56	1,800	20,855	24.17
900	7,875	14.20	2,000	23,790	25.72

CSeS(*g*):

$$H_T - H_{298.15} = 13.31T + 0.43 \times 10^{-3}T^2 + 1.90 \times 10^5 T^{-1}$$
$$-4,644 \ (0.6 \ \text{percent}; \ 298°-2,000° \ \text{K.});$$
$$C_p = 13.31 + 0.86 \times 10^{-3}T - 1.90 \times 10^5 T^{-2}.$$

TABLE 1h7. (Continued)

TELLUROSULFIDE

Table 197.—*Heat content and entropy of*
CTeS(g)

[Base, ideal gas at 298.15 K.; mol. wt., 171.69]

T, ° K.	$H_T-H_{298.15}$, cal./mole	$S_T-S_{298.15}$, cal./deg. mole	T, ° K.	$H_T-H_{298.15}$, cal./mole	$S_T-S_{298.15}$, cal./deg. mole
400	1,235	3.56	1,000	9,405	15.91
500	2,510	6.40	1,200	12,270	18.52
600	3,835	8.81	1,400	15,170	20.76
700	5,195	10.91	1,600	18,085	22.70
800	6,580	12.76	1,800	21,015	24.43
900	7,985	14.41	2,000	23,960	25.98

CTeS(*g*):

$$H_T-H_{298.15}=13.53T+0.37\times10^{-3}T^2+1.82\times10^5T^{-1}$$
$$-4,677 \text{ (0.6 percent; } 298°-2,000° \text{ K.);}$$
$$C_p=13.53+0.74\times10^{-3}T-1.82\times10^5T^{-2}.$$

CARBON-HYDROGEN COMPOUNDS

Table 198.—*Heat content and entropy of*
CH₄(g)

[Base, ideal gas at 298.15° K.; mol. wt., 16.04]

T, ° K.	$H_T-H_{298.15}$, cal./mole	$S_T-S_{298.15}$, cal./deg. mole	T, ° K.	$H_T-H_{298.15}$, cal./mole	$S_T-S_{298.15}$, cal./deg. mole
400	925	2.66	1,000	9,160	14.70
500	1,970	4.98	1,100	10,920	16.38
600	3,150	7.13	1,200	12,770	17.99
700	4,475	9.17	1,300	14,700	19.53
800	5,925	11.10	1,400	16,690	21.00
900	7,490	12.94	1,500	18,730	22.41

CH₄(*g*):

$$H_T-H_{298.15}=5.65T+5.72\times10^{-3}T^2+0.46\times10^5T^{-1}$$
$$-2,347 \text{ (1.0 percent; } 298°-1,500° \text{ K.);}$$
$$C_p=5.65+11.44\times10^{-3}T-0.46\times10^5T^{-2}.$$

Table 199.—*Heat content and entropy of*
C₂H₂(g)

[Base, ideal gas at 298.15° K.; mol. wt., 26.04]

T, ° K.	$H_T-H_{298.15}$, cal./mole	$S_T-S_{298.15}$, cal./deg. mole	T, ° K.	$H_T-H_{298.15}$, cal./mole	$S_T-S_{298.15}$, cal./deg. mole
400	1,150	3.31	1,000	9,700	16.10
500	2,400	6.09	1,200	12,970	19.07
600	3,735	8.53	1,400	16,390	21.71
700	5,140	10.69	1,600	19,940	24.08
800	6,610	12.65	1,800	23,580	26.22
900	8,130	14.44	2,000	27,300	28.18

C₂H₂(*g*):

$$H_T-H_{298.15}=12.13T+1.92\times10^{-3}T^2+2.46\times10^5T^{-1}$$
$$-4,612 \text{ (0.5 percent; } 298°-2,000° \text{ K.);}$$
$$C_p=12.13+3.84\times10^{-3}T-2.46\times10^5T^{-2}.$$

TETRABROMIDE

Table 200.—*Heat content and entropy of*
CBr₄(c, l)

[Base, α-crystals at 298.15° K; mol. wt., 331.68]

T, ° K.	$H_T-H_{298.15}$, cal./mole	$S_T-S_{298.15}$, cal./deg. mole	T, ° K.	$H_T-H_{298.15}$, cal./mole	$S_T-S_{298.15}$, cal./deg. mole
320(α)	750	2.44	363.2(l)	4,990	14.97
320(β)	2,180	6.91	400	6,340	18.51
363.2(β)	4,040	12.35	450	8,180	22.83

CBr₄(α):

$$H_T-H_{298.15}=34.50T$$
$$-10,287 \text{ (0.4 percent; } 298°-320° \text{ K.);}$$
$$C_p=34.50;$$
$$\Delta H_{320}(transition)=1,430.$$

CBr₄(β):

$$H_T-H_{298.15}=43.00T$$
$$-11,580 \text{ (0.1 percent; } 320°-363.2° \text{ K.);}$$
$$C_p=43.00;$$
$$\Delta H_{363.2}(fusion)=950.$$

CBr₄(l):

$$H_T-H_{298.15}=36.70T$$
$$-8,340 \text{ (0.1 percent; } 363.2°-450° \text{ K.);}$$
$$C_p=36.70.$$

Table 201.—*Heat content and entropy of CBr₄(g)*

[Base, ideal gas at 298.15° K; mol. wt., 331.68]

T, ° K.	$H_T-H_{298.15}$, cal./mole	$S_T-S_{298.15}$, cal./deg. mole	T, ° K.	$H_T-H_{298.15}$, cal./mole	$S_T-S_{298.15}$, cal./deg. mole
400	2,300	6.62	1,000	17,100	29.08
500	4,660	11.89	1,100	19,640	31.50
600	7,090	16.32	1,200	22,180	33.71
700	9,560	20.12	1,300	24,730	35.75
800	12,060	23.46	1,400	27,290	37.65
900	14,570	26.42	1,500	29,840	39.41

CBr₄(*g*):

$$H_T-H_{298.15}=24.86T+0.35\times10^{-3}T^2+2.91\times10^5T^{-1}$$
$$-8,419 \text{ (0.2 percent; } 298°-1,500° \text{ K.);}$$
$$C_p=24.86+0.70\times10^{-3}T-2.91\times10^5T^{-2}.$$

TETRACHLORIDE

CCl₄(l):

$$C_p=31.47 \text{ (298° K.).}$$

Table 202.—*Heat content and entropy of CCl₄(g)*

[Base, ideal gas at 298.15° K; mol. wt., 153.84]

T, ° K.	$H_T-H_{298.15}$, cal./mole	$S_T-S_{298.15}$, cal./deg. mole	T, ° K.	$H_T-H_{298.15}$, cal./mole	$S_T-S_{298.15}$, cal./deg. mole
400	2,140	6.16	1,000	16,570	27.99
500	4,395	11.19	1,100	19,080	30.38
600	6,740	15.46	1,200	21,610	32.58
700	9,150	19.17	1,300	24,140	34.61
800	11,600	22.45	1,400	26,680	36.49
900	14,080	25.37	1,500	29,220	38.24

TABLE 1h7. (*Continued*)

$CCl_4(g)$:

$$H_T - H_{298.15} = 24.17T + 0.60 \times 10^{-3}T^2 + 4.10 \times 10^5 T^{-1}$$
$$-8,635 \ (0.3 \ \text{percent}; \ 298°-1,500° \ \text{K.});$$
$$C_p = 24.17 + 1.20 \times 10^{-3}T - 4.10 \times 10^5 T^{-2}.$$

PHOSGENE

TABLE 203.—*Heat content and entropy of $COCl_2(g)$*

[Base, ideal gas at 298.15° K; mol. wt., 98.92]

T, ° K.	$H_T - H_{298.15}$, cal./mole	$S_T - S_{298.15}$, cal./deg. mole	T, ° K.	$H_T - H_{298.15}$, cal./mole	$S_T - S_{298.15}$, cal./deg. mole
400	1,545	4.45	1,300	17,700	25.24
500	3,165	8.06	1,400	19,610	26.66
600	4,855	11.14	1,500	21,530	27.98
700	6,600	13.83	1,600	23,460	29.23
800	8,400	16.23	1,700	25,390	30.40
900	10,210	18.36	1,800	27,330	31.51
1,000	12,060	20.31	1,900	29,270	32.56
1,100	13,920	22.08	2,000	31,220	33.56
1,200	15,810	23.73			

$COCl_2(g)$:

$$H_T - H_{298.15} = 16.97T + 0.82 \times 10^{-3}T^2 + 2.64 \times 10^5 T^{-1}$$
$$-6,018 \ (0.5 \ \text{percent}; \ 298°-2,000° \ \text{K.});$$
$$C_p = 16.97 + 1.64 \times 10^{-3}T - 2.64 \times 10^5 T^{-2}.$$

THIOPHOSGENE

TABLE 204.—*Heat content and entropy of $CSCl_2(g)$*

[Base, ideal gas at 298.15° K.; mol. wt., 114.99]

T, ° K.	$H_T - H_{298.15}$, cal./mole	$S_T - S_{298.15}$, cal./deg. mole	T. ° K.	$H_T - H_{298.15}$, cal./mole	$S_T - S_{298.15}$, cal./deg. mole
400	1,645	4.74	800	8,860	17.16
500	3,365	8.58	900	10,750	19.39
600	5,155	11.84	1,000	12,660	21.40
700	6,990	14.66			

$CSCl_2(g)$:

$$H_T - H_{298.15} = 17.27T + 1.20 \times 10^{-3}T^2 + 2.28 \times 10^5 T^{-1}$$
$$-6,020 \ (0.2 \ \text{percent}; \ 298°-1,000° \ \text{K.});$$
$$C_p = 17.27 + 2.40 \times 10^{-3}T - 2.28 \times 10^5 T^{-2}.$$

TETRAFLUORIDE

TABLE 205.—*Heat content and entropy of $CF_4(g)$*

[Base, ideal gas at 298.15° K.; mol. wt., 88.01]

T, ° K.	$H_T - H_{298.15}$, cal./mole	$S_T - S_{298.15}$, cal./deg. mole	T, ° K.	$H_T - H_{298.15}$, cal./mole	$S_T - S_{298.15}$, cal./deg. mole
400	1,635	4.70	1,000	14,470	23.86
500	3,470	8.79	1,100	16,850	26.12
600	5,480	12.45	1,200	19,260	28.23
700	7,610	15.73	1,300	21,700	30.18
800	9,840	18.70	1,400	24,150	32.00
900	12,130	21.40	1,500	26,620	33.70

$CF_4(g)$:

$$H_T - H_{298.15} = 19.51T + 2.20 \times 10^{-3}T^2 + 5.52 \times 10^5 T^{-1}$$
$$-7,864 \ (0.8 \ \text{percent}; \ 298°-1,500° \ \text{K.});$$
$$C_p = 19.51 + 4.40 \times 10^{-3}T - 5.52 \times 10^5 T^{-2}.$$

CARBONYL FLUORIDE

TABLE 206.—*Heat content and entropy of $COF_2(g)$*

[Base, ideal gas at 298.15° K.; mol. wt., 66.01]

T, ° K.	$H_T - H_{298.15}$, cal./mole	$S_T - S_{298.15}$, cal./deg. mole	T, ° K.	$H_T - H_{298.15}$, cal./mole	$S_T - S_{298.15}$, cal./deg. mole
400	1,240	3.56	1,000	10,850	17.93
500	2,620	6.64	1,100	12,640	19.63
600	4,130	9.39	1,200	14,460	21.22
700	5,730	11.86	1,300	16,310	22.70
800	7,390	14.07	1,400	18,170	24.08
900	9,100	16.08	1,500	20,050	25.37

$COF_2(g)$:

$$H_T - H_{298.15} = 13.81T + 2.05 \times 10^{-3}T^2 + 3.32 \times 10^5 T^{-1}$$
$$-5,413 \ (0.8 \ \text{percent}; \ 298°-1,500° \ \text{K.});$$
$$C_p = 13.81 + 4.10 \times 10^{-3}T - 3.32 \times 10^5 T^{-2}.$$

CARBONYL FLUOROCHLORIDE

TABLE 207.—*Heat content and entropy of $COFCl(g)$*

[Base, ideal gas at 298.15° K.; mol. wt., 82.47]

T, ° K.	$H_T - H_{298.15}$, cal./mole	$S_T - S_{298.15}$, cal./deg. mole	T, ° K.	$H_T - H_{298.15}$, cal./mole	$S_T - S_{298.15}$, cal./deg. mole
400	1,370	3.94	1,000	11,380	18.95
500	2,850	7.24	1,100	13,200	20.69
600	4,430	10.12	1,200	15,050	22.30
700	6,090	12.68	1,300	16,920	23.79
800	7,810	14.97	1,400	18,810	25.19
900	9,580	17.06	1,500	20,720	26.51

$COFCl(g)$:

$$H_T - H_{298.15} = 15.28T + 1.49 \times 10^{-3}T^2 + 3.24 \times 10^5 T^{-1}$$
$$-5,775 \ (0.6 \ \text{percent}; \ 298°-1,500° \ \text{K.});$$
$$C_p = 15.28 + 2.98 \times 10^{-3}T - 3.24 \times 10^5 T^{-2}.$$

CYANOGEN HALIDES

TABLE 208.—*Heat content and entropy of $CNBr(g)$*

[Base, ideal gas at 298.15° K.; mol. wt., 105.94]

T, ° K.	$H_T - H_{298.15}$, cal./mole	$S_T - S_{298.15}$, cal./deg. mole	T, ° K.	$H_T - H_{298.15}$, cal./mole	$S_T - S_{298.15}$, cal./deg. mole
400	1,175	3.39	1,000	8,910	15.07
500	2,380	6.07	1,200	11,665	17.58
600	3,630	8.35	1,400	14,475	19.74
700	4,910	10.32	1,600	17,310	21.64
800	6,220	12.07	1,800	20,185	23.33
900	7,550	13.64	2,000	23,095	24.86

$CNBr(g)$:

$$H_T - H_{298.15} = 12.20T + 0.71 \times 10^{-3}T^2 + 1.34 \times 10^5 T^{-1}$$
$$-4,150 \ (0.4 \ \text{percent}; \ 298°-2,000° \ \text{K.});$$
$$C_p = 12.20 + 1.42 \times 10^{-3}T - 1.34 \times 10^5 T^{-2}.$$

TABLE 1h7. (*Continued*)

TABLE 209.—*Heat content and entropy of CNCl(g)*

[Base, ideal gas at 298.15° K.; mol. wt., 61.48]

T, ° K.	$H_T-H_{298.15}$, cal./mole	$S_T-S_{298.15}$, cal./deg. mole	T, ° K.	$H_T-H_{298.15}$, cal./mole	$S_T-S_{298.15}$, cal./deg. mole
400	1,135	3.27	1,000	8,770	14.79
500	2,315	5.90	1,200	11,515	17.29
600	3,540	8.14	1,400	14,310	19.45
700	4,805	10.08	1,600	17,145	21.34
800	6,100	11.81	1,800	20,010	23.03
900	7,425	13.37	2,000	22,890	24.54

CNCl(g):

$$H_T-H_{298.15}=11.88T+0.82\times10^{-3}T^2+1.49\times10^5T^{-1}$$
$$-4,115 \text{ (0.5 percent; } 298°\text{--}2,000° \text{ K.);}$$
$$C_p=11.88+1.64\times10^{-3}T-1.49\times10^5T^{-2}.$$

TABLE 210.—*Heat content and entropy of CNI(g)*

[Base, ideal gas at 298.15° K; mol. wt., 152.93]

T, ° K.	$H_T-H_{298.15}$, cal./mole	$S_T-S_{298.15}$, cal./deg. mole	T, ° K.	$H_T-H_{298.15}$, cal./mole	$S_T-S_{298.15}$, cal./deg. mole
400	1,210	3.49	1,000	9,035	15.33
500	2,440	6.23	1,200	11,805	17.85
600	3,710	8.55	1,400	14,625	20.02
700	5,005	10.54	1,600	17,475	21.93
800	6,330	12.31	1,800	20,345	23.62
900	7,675	13.89	2,000	23,240	25.14

CNI(g):

$$H_T-H_{298.15}=12.30T+0.69\times10^{-3}T^2+1.04\times10^5T^{-1}$$
$$-4,077 \text{ (0.3 percent; } 298°\text{--}2,000° \text{ K.);}$$
$$C_p=12.30+1.38\times10^{-3}T-1.04\times10^5T^{-2}.$$

CERIUM AND ITS COMPOUNDS

ELEMENT

TABLE 211.—*Heat content and entropy of Ce(c, l)*

[Base, α-crystals at 298.15° K.; atomic wt., 140.13]

T, ° K.	$H_T-H_{298.15}$, cal./mole	$S_T-S_{298.15}$, cal./deg. mole	T, ° K.	$H_T-H_{298.15}$, cal./mole	$S_T-S_{298.15}$, cal./deg. mole
400	720	2.07	1,100	9,170	12.84
500	1,470	3.74	1,200	9,970	13.54
600	2,260	5.18	1,400	11,570	14.77
700	3,090	6.46	1,600	13,170	15.84
800	3,960	7.62	1,800	14,770	16.78
900	4,870	8.70	2,000	16,370	17.62
1,000	5,820	9.70	2,200	17,970	18.38
1,027(α)	6,080	9.95	2,400	19,570	19.08
1,027(β)	6,380	10.24	2,600	21,170	19.72
1,077(β)	6,790	10.63	2,800	22,770	20.32
1,077(l)	8,990	12.68	3,000	24,370	20.87

Ce(α):

$$H_T-H_{298.15}=5.70T+1.99\times10^{-3}T^2-1,876$$
$$(0.1 \text{ percent; } 298°\text{--}1,027° \text{ K.);}$$
$$C_p=5.70+3.98\times10^{-3}T;$$
$$\Delta H_{1027}(transition)=300.$$

Ce(β):

$$H_T-H_{298.15}=8.20T-2,041 \text{ (0.1 percent;}$$
$$1,027°\text{--}1,077° \text{ K.);}$$
$$C_p=8.20;$$
$$\Delta H_{1077}(fusion)=2,200.$$

Ce(l):

$$H_T-H_{298.15}=8.00T+370 \text{ (0.1 percent;}$$
$$1,077°\text{--}3,000° \text{ K.);}$$
$$C_p=8.00.$$

OXIDES

TABLE 212.—*Heat content and entropy of CeO(g)*

[Base, ideal gas at 298.15° K.; mol. wt., 156.13]

T, ° K.	$H_T-H_{298.15}$, cal./mole	$S_T-S_{298.15}$, cal./deg. mole	T, ° K.	$H_T-H_{298.15}$, cal./mole	$S_T-S_{298.15}$, cal./deg. mole
400	785	2.26	1,000	5,850	9.94
500	1,590	4.06	1,200	7,595	11.53
600	2,420	5.57	1,400	9,355	12.89
700	3,265	6.87	1,600	11,125	14.07
800	4,120	8.02	1,800	12,895	15.11
900	4,980	9.03	2,000	14,670	16.05

CeO(g):

$$H_T-H_{298.15}=8.42T+0.16\times10^{-3}T^2+0.90\times10^5T^{-1}$$
$$-2,827 \text{ (0.3 percent; } 298°\text{--}2,000° \text{ K.);}$$
$$C_p=8.42+0.32\times10^{-3}T-0.90\times10^5T^{-2}.$$

CeO$_2$(c):

$$\overline{C_p}=15.1 \text{ (273°--373° K.).}$$

FLUORIDE

TABLE 213.—*Heat content and entropy of CeF$_3$(c, l)*

[Base, crystals at 298.15° K; mol. wt., 197.13]

T, ° K.	$H_T-H_{298.15}$, cal./mole	$S_T-S_{298.15}$, cal./deg. mole	T, ° K.	$H_T-H_{298.15}$, cal./mole	$S_T-S_{298.15}$, cal./deg. mole
400	2,300	6.63	1,300	26,100	37.02
500	4,650	11.87	1,400	29,230	39.34
600	7,070	16.28	1,500	32,510	41.60
700	9,560	20.12	1,600	35,960	43.83
800	12,110	23.52	1,700	39,580	46.02
900	14,740	26.62	1,732(c)	40,760	46.71
1,000	17,440	29.46	1,732(l)	53,960	54.33
1,100	20,220	32.11	1,800	56,140	55.57
1,200	23,100	34.62			

CeF$_3$(c):

$$H_T-H_{298.15}=17.90T+5.07\times10^{-3}T^2-1.10\times10^5T^{-1}$$
$$-5,419 \text{ (0.9 percent; } 298°\text{--}1,732° \text{ K.);}$$
$$C_p=17.90+10.14\times10^{-3}T+1.10\times10^5T^{-2};$$
$$\Delta H_{1732}(fusion)=13,200.$$

CeF$_3$(l):

$$H_T-H_{298.15}=32.00T-1,460 \text{ (0.1 percent;}$$
$$1,732°\text{--}1,800° \text{ K.);}$$
$$C_p=32.00.$$

TABLE 1h7. (*Continued*)

MOLYBDATE

$Ce_2(MoO_4)_3(c)$:

$\overline{C}_p = 96.0$ (273°–297° K.).

SULFATE

$Ce_2(SO_4)_3(c)$:

$\overline{C}_p = 66.4$ (273°–373° K.).

$Ce_2(SO_4)_3 \cdot 5H_2O(c)$:

$\overline{C}_p = 131.6$ (273°–319° K.).

CESIUM AND ITS COMPOUNDS

ELEMENT

TABLE 214.—*Heat content and entropy of* $Cs(c, l)$

[Base, crystals at 298.15° K; atomic wt., 132.91]

T, ° K.	$H_T - H_{298.15}$, cal./mole	$S_T - S_{298.15}$, cal./deg. mole	T, ° K.	$H_T - H_{298.15}$, cal./mole	$S_T - S_{298.15}$, cal./deg. mole
301.8(c)	30	0.10	700	3,565	8.18
301.8(l)	540	1.79	800	4,325	9.20
400	1,285	3.93	900	5,085	10.09
500	2,045	5.62	950	5,465	10.50
600	2,805	7.01			

$Cs(c)$:

$H_T - H_{298.15} = 7.55T - 2,251$ (3 percent;

298°–301.8° K.);

$C_p = 7.55$;

$\Delta H_{301.8}(fusion) = 510$.

$Cs(l)$:

$H_T - H_{298.15} = 7.60T - 1,755$ (0.1 percent;

301.8°–950° K.);

$C_p = 7.60$.

TABLE 215.—*Heat content and entropy of* $Cs(g)$

[Base, ideal gas at 298.15° K.; atomic wt., 132.91]

T, ° K.	$H_T - H_{298.15}$, cal./mole	$S_T - S_{298.15}$, cal./deg. mole	T, ° K.	$H_T - H_{298.15}$, cal./mole	$S_T - S_{298.15}$, cal./deg. mole
400	505	1.46	1,900	7,970	9.21
500	1,005	2.57	2,000	8,475	9.47
600	1,500	3.48	2,200	9,490	9.95
700	1,995	4.24	2,400	10,520	10.40
800	2,495	4.90	2,600	11,580	10.82
900	2,990	5.49	2,800	12,670	11.23
1,000	3,490	6.01	3,000	13,810	11.62
1,100	3,985	6.49	3,500	16,975	12.60
1,200	4,480	6.92	4,000	20,895	13.64
1,300	4,980	7.32	4,500	26,050	14.85
1,400	4,475	7.69	5,000	32,865	16.28
1,500	5,975	8.03	6,000	51,305	19.63
1,600	6,470	8.35	7,000	71,945	22.81
1,700	6,970	8.65	8,000	89,840	25.21
1,800	7,470	8.94			

$Cs(g)$:

$H_T - H_{298.15} = 4.97T - 1,482$ (0.1 percent;

298°–2,000° K.);

$C_p = 4.97$

TABLE 216.—*Heat content and entropy of* $Cs_2(g)$

[Base, ideal gas at 298.15° K.; mol. wt., 265.82]

T, ° K.	$H_T - H_{298.15}$, cal./mole	$S_T - S_{298.15}$, cal./deg. mole	T, ° K.	$H_T - H_{298.15}$, cal./mole	$S_T - S_{298.15}$, cal./deg. mole
400	930	2.68	1,000	6,545	11.23
500	1,850	4.73	1,100	7,495	12.14
600	2,780	6.43	1,200	8,465	12.98
700	3,710	7.86	1,300	9,440	13.76
800	4,645	9.11	1,400	10,415	14.48
900	5,590	10.22	1,500	11,390	15.15

$Cs_2(g)$:

$H_T - H_{298.15} = 8.92T + 0.31 \times 10^{-3}T^2 - 2,687$

(0.1 percent; 298°–1,500° K.);

$C_p = 8.92 + 0.62 \times 10^{-3}T$.

HYDRIDE

TABLE 217.—*Heat content and entropy of* $CsH(g)$

[Base, ideal gas at 298.15° K.; mol. wt., 133.92]

T, ° K.	$H_T - H_{298.15}$, cal./mole	$S_T - S_{298.15}$, cal./deg. mole	T, ° K.	$H_T - H_{298.15}$, cal./mole	$S_T - S_{298.15}$, cal./deg. mole
400	785	2.26	1,000	5,485	9.94
500	1,590	4.06	1,200	7,590	11.53
600	2,415	5.56	1,400	9,350	12.88
700	3,260	6.87	1,600	11,115	14.06
800	4,115	8.01	1,800	12,885	15.10
900	4,975	9.02	2,000	14,660	16.04

$CsH(g)$:

$H_T - H_{298.15} = 8.37T + 0.17 \times 10^{-3}T^2 + 0.84 \times 10^5 T^{-1}$

$-2,792$ (0.4 percent; 298°–2,000° K.);

$C_p = 8.37 + 0.34 \times 10^{-3}T - 0.84 \times 10^5 T^{-2}$.

BROMIDE

$CsBr(c)$:

$C_p = 11.60 + 2.59 \times 10^{-3}T(estimated)$ (298°–909° K.).

CHLORIDE

$CsCl(c)$:

$C_p = 11.90 + 2.28 \times 10^{-3}T(estimated)$ (298°–915° K.).

TABLE 218.—*Heat content and entropy of* $CsCl(g)$

[Base, ideal gas at 298.15° K.; mol. wt., 168.37]

T, ° K.	$H_T - H_{298.15}$, cal./mole	$S_T - S_{298.15}$, cal./deg. mole	T, ° K.	$H_T - H_{298.15}$, cal./mole	$S_T - S_{298.15}$, cal./deg. mole
400	905	2.61	1,000	6,330	10.88
500	1,800	4.61	1,200	8,160	12.55
600	2,700	6.25	1,400	10,000	13.97
700	3,600	7.63	1,600	11,845	15.20
800	4,505	8.84	1,800	13,700	16.29
900	5,415	9.91	2,000	15,560	17.27

$CsCl(g)$:

$H_T - H_{298.15} = 8.97T + 0.09 \times 10^{-3}T^2 + 0.20 \times 10^5 T^{-1}$

$-2,749$ (0.1 percent; 298°–2,000° K.);

$C_p = 8.97 + 0.18 \times 10^{-3}T - 0.20 \times 10^5 T^{-2}$.

TABLE 1h7. (Continued)

PERCHLORATE

$CsClO_4$:

$C_p = 25.71$ (298°).

FLUORIDE

$CsF(c)$:

$C_p = 11.30 + 2.71 \times 10^{-3} T (estimated)$ (298°–988° K.).

TABLE 219.—Heat content and entropy of $CsF(g)$

[Base, ideal gas at 298.15° K.; mol. wt., 151.91]

T, °K.	$H_T - H_{298.15}$, cal./mole	$S_T - S_{298.15}$, cal./deg. mole	T, °K.	$H_T - H_{298.15}$, cal./mole	$S_T - S_{298.15}$, cal./deg. mole
400	890	2.57	1,000	6,220	10.70
500	1,775	4.54	1,200	8,005	12.33
600	2,660	6.16	1,400	9,790	13.70
700	3,545	7.52	1,600	11,575	14.89
800	4,435	8.71	1,800	13,360	15.95
900	5,325	9.76	2,000	15,150	16.89

$CsF(g)$:

$H_T - H_{298.15} = 8.94 T + 0.23 \times 10^5 T^{-1} - 2,743$

(0.1 percent; 298°–2,000° K.);

$C_p = 8.94 - 0.23 \times 10^5 T^{-2}$.

IODIDE

$CsI(c)$:

$C_p = 11.60 + 2.68 \times 10^{-3} T (estimated)$ (298°–894° K.).

TABLE 220.—Heat content and entropy of $CsI(g)$

[Base, ideal gas at 298.15° K; mol. wt., 259.82]

T, °K.	$H_T - H_{298.15}$, cal./mole	$S_T - S_{298.15}$, cal./deg. mole	T, °K.	$H_T - H_{298.15}$, cal./mole	$S_T - S_{298.15}$, cal./deg. mole
400	905	2.61	1,000	6,260	10.79
500	1,795	4.60	1,200	8,050	12.42
600	2,690	6.23	1,400	9,835	13.79
700	3,580	7.60	1,600	11,620	14.98
800	4,475	8.80	1,800	13,410	16.04
900	5,365	9.84	2,000	15,200	16.98

$CsI(g)$:

$H_T - H_{298.15} = 8.94 T + 0.06 \times 10^5 T^{-1} - 2,686$

(0.1 percent; 298°–2,000° K.);

$C_p = 8.94 - 0.06 \times 10^5 T^{-2}$.

CESIUM-ALUMINUM SULFATE

$CsAl(SO_4)_2 . 12H_2O(c)$:

$C_p = 148.1$ (298° K.).

NITRATE

$CsNO_3(c)$:

$C_p = 7.33 + 53.20 \times 10^{-3} T$ (298°–425° K.).

CHLORINE AND ITS COMPOUNDS

ELEMENT

TABLE 221.—Heat content and entropy of $Cl(g)$

[Base, ideal gas at 298.15° K.; atomic wt., 35.46]

T, °K.	$H_T - H_{298.15}$, cal./mole	$S_T - S_{298.15}$, cal./deg. mole	T, °K.	$H_T - H_{298.15}$, cal./mole	$S_T - S_{298.15}$, cal./deg. mole
400	540	1.56	1,900	8,455	9.89
500	1,080	2.76	2,000	8,965	10.12
600	1,625	3.76	2,200	9,985	10.60
700	2,170	4.60	2,400	11,000	11.04
800	2,710	5.32	2,600	12,010	11.45
900	3,245	5.95	2,800	13,020	11.82
1,000	3,780	6.51	3,000	14,030	12.17
1,100	4,310	7.02	3,500	16,545	12.94
1,200	4,835	7.48	4,000	19,050	13.62
1,300	5,360	7.90	4,500	21,550	14.20
1,400	5,880	8.28	5,000	24,050	14.73
1,500	6,400	8.64	6,000	29,045	15.64
1,600	6,915	8.97	7,000	34,030	16.41
1,700	7,430	9.28	8,000	39,020	17.08
1,800	7,945	9.58			

$Cl(g)$:

$H_T - H_{298.15} = 5.53 T - 0.08 \times 10^{-3} T^2 + 0.23 \times 10^5 T^{-1}$

$- 1,719$ (0.7 percent; 298°–5,000° K.);

$C_p = 5.53 - 0.16 \times 10^{-3} T - 0.23 \times 10^5 T^{-2}$.

TABLE 222.—Heat content and entropy of $Cl_2(g)$

[Base, ideal gas at 298.15° K.; mol. wt., 70.91]

T, °K.	$H_T - H_{298.15}$, cal./mole	$S_T - S_{298.15}$, cal./deg. mole	T, °K.	$H_T - H_{298.15}$, cal./mole	$S_T - S_{298.15}$, cal./deg. mole
400	845	2.43	1,500	10,630	14.11
500	1,700	4.34	1,600	11,535	14.69
600	2,565	5.92	1,700	12,445	15.24
700	3,445	7.28	1,800	13,360	15.77
800	4,330	8.46	1,900	14,270	16.26
900	5,220	9.51	2,000	15,185	16.73
1,000	6,115	10.45	2,200	17,020	17.60
1,100	7,015	11.30	2,400	18,860	18.40
1,200	7,915	12.09	2,600	20,710	19.14
1,300	8,815	12.81	2,800	22,560	19.83
1,400	9,720	13.48	3,000	24,415	20.47

$Cl_2(g)$:

$H_T - H_{298.15} = 8.85 T + 0.08 \times 10^{-3} T^2 + 0.68 \times 10^5 T^{-1}$

$- 2,874$ (0.2 percent; 298°–3,000° K.);

$C_p = 8.85 + 0.16 \times 10^{-3} T - 0.68 \times 10^5 T^{-2}$.

OXIDES

TABLE 223.—Heat content and entropy of $Cl_2O(g)$

[Base, ideal gas at 298.15° K.; mol. wt., 86.91]

T, °K.	$H_T - H_{298.15}$, cal./mole	$S_T - S_{298.15}$, cal./deg. mole	T, °K.	$H_T - H_{298.15}$, cal./mole	$S_T - S_{298.15}$, cal./deg. mole
400	1,155	3.33	1,000	8,875	15.00
500	2,360	6.01	1,200	11,575	17.46
600	3,615	8.30	1,400	14,295	19.56
700	4,900	10.28	1,600	17,035	21.39
800	6,210	12.03	1,800	19,775	23.00
900	7,535	13.59	2,000	22,520	24.45

TABLE 1h7. (*Continued*)

$Cl_2O(g)$:

$$H_T - H_{298.15} = 12.71T + 0.40 \times 10^{-3}T^2 + 1.86 \times 10^5 T^{-1}$$
$$-4,449 \ (0.5 \text{ percent}; 298°-2,000° \text{ K.});$$
$$C_p = 12.71 + 0.80 \times 10^{-3}T^2 - 1.86 \times 10^5 T^{-2}.$$

TABLE 224.—*Heat content and entropy of* $ClO_2(g)$

[Base, ideal gas at 298.15° K.; mol. wt., 67.46]

T, ° K.	$H_T-H_{298.15}$, cal./mole	$S_T-S_{298.15}$, cal./deg. mole	T, K.	$H_T-H_{298.15}$, cal./mole	$S_T-S_{298.15}$, cal./deg. mole
400	1,070	3.08	1,000	8,545	14.33
500	2,205	5.61	1,100	9,880	15.61
600	3,400	7.79	1,200	11,225	16.78
700	4,645	9.70	1,300	12,585	17.86
800	5,920	11.41	1,400	13,950	18.88
900	7,220	12.94	1,500	15,320	19.82

$ClO_2(g)$:

$$H_T - H_{298.15} = 11.54T + 0.90 \times 10^{-3}T^2 + 1.85 \times 10^5 T^{-1}$$
$$-4,141 \ (0.6 \text{ percent}; 298°-1,500° \text{ K.});$$
$$C_p = 11.54 + 1.80 \times 10^{-3}T - 1.85 \times 10^5 T^{-2}.$$

FLUORIDES

TABLE 225.—*Heat content and entropy of* $ClF(g)$

[Base, ideal gas at 298.15° K.; mol. wt., 54.46]

T, ° K.	$H_T-H_{298.15}$, cal./mole	$S_T-S_{298.15}$, cal./deg. mole	T, ° K.	$H_T-H_{298.15}$, cal./mole	$S_T-S_{298.15}$, cal./deg. mole
400	800	2.31	1,300	8,630	12.47
500	1,625	4.14	1,400	9,525	13.13
600	2,465	5.68	1,500	10,425	13.75
700	3,325	7.00	1,600	11,330	14.33
800	4,195	8.16	1,700	12,235	14.88
900	5,070	9.20	1,800	13,140	15.40
1,000	5,955	10.13	1,900	14,050	15.89
1,100	6,840	10.97	2,000	14,960	16.36
1,200	7,735	11.75			

$ClF(g)$:

$$H_T - H_{298.15} = 8.52T + 0.18 \times 10^{-3}T^2 + 0.85 \times 10^5 T^{-1}$$
$$-2,841 \ (0.3 \text{ percent}; 298°-2,000° \text{ K.});$$
$$C_p = 8.52 + 0.36 \times 10^{-3}T - 0.85 \times 10^5 T^{-2}.$$

TABLE 226.—*Heat content and entropy of* $ClF_3(g)$

[Base, ideal gas at 298.15° K.; mol. wt., 92.46]

T, ° K.	$H_T-H_{298.15}$, cal./mole	$S_T-S_{298.15}$, cal./deg. mole	T, ° K.	$H_T-H_{298.15}$, cal./mole	$S_T-S_{298.15}$, cal./deg. mole
400	1,665	4.79	1,000	12,830	21.68
500	3,420	8.70	1,100	14,765	23.53
600	5,240	12.02	1,200	16,715	25.24
700	7,100	14.89	1,300	18,665	26.81
800	8,990	17.41	1,400	20,610	28.25
900	10,900	19.66	1,500	22,580	29.60

$ClF_3(g)$:

$$H_T - H_{298.15} = 18.67T + 0.44 \times 10^{-3}T^2 + 3.00 \times 10^5 T^{-1}$$
$$-6,612 \ (0.3 \text{ percent}; 298°-1,500° \text{ K.});$$
$$C_p = 18.67 + 0.88 \times 10^{-3}T - 3.00 \times 10^5 T^{-2}.$$

CHROMIUM AND ITS COMPOUNDS

ELEMENT

TABLE 227.—*Heat content and entropy of* $Cr(c, l)$

[Base, crystals at 298.15° K.; atomic wt., 52.01]

T, ° K.	$H_T-H_{298.15}$, cal./mole	$S_T-S_{298.15}$, cal./deg. mole	T, ° K.	$H_T-H_{298.15}$, cal./mole	$S_T-S_{298.15}$, cal./deg. mole
400	595	1.72	1,700	10,930	12.51
500	1,220	3.11	1,800	11,980	13.11
600	1,870	4.29	1,900	13,080	13.71
700	2,530	5.31	2,000	14,220	14.29
800	3,210	6.22	2,100	15,410	14.87
900	3,910	7.03	2,176(c)	16,340	15.31
1,000	4,640	7.81	2,176(l)	21,340	17.61
1,100	5,410	8.54	2,200	21,570	17.71
1,200	6,230	9.26	2,400	23,450	18.53
1,300	7,100	9.95	2,600	25,330	19.28
1,400	8,010	10.63	2,800	27,210	19.98
1,500	8,950	11.28	3,000	29,090	20.63
1,600	9,920	11.90			

$Cr(c)$:

$$H_T - H_{298.15} = 4.16T + 1.81 \times 10^{-3}T^2 - 0.30 \times 10^5 T^{-1}$$
$$-1,301 \ (1.2 \text{ percent}; 298°-2,176° \text{ K.});$$
$$C_p = 4.16 + 3.62 \times 10^{-3}T + 0.30 \times 10^5 T^{-2};$$
$$\Delta H_{2176}(fusion) = 5,000.$$

$Cr(l)$:

$$H_T - H_{298.15} = 9.40T + 890 \ (0.1 \text{ percent};$$
$$2,176°-3,000° \text{ K.});$$
$$C_p = 9.40.$$

TABLE 228.—*Heat content and entropy of* $Cr(g)$

[Base, ideal gas at 298.15° K.; atomic wt., 52.01]

T, ° K.	$H_T-H_{298.15}$, cal./mole	$S_T-S_{298.15}$, cal./deg. mole	T, ° K.	$H_T-H_{298.15}$, cal./mole	$S_T-S_{298.15}$, cal./deg. mole
400	505	1.46	1,900	8,190	9.34
500	1,005	2.57	2,000	8,765	9.64
600	1,500	3.48	2,200	9,965	10.21
700	1,995	4.24	2,400	11,230	10.76
800	2,495	4.90	2,600	12,565	11.30
900	2,990	5.49	2,800	13,955	11.81
1,000	3,490	6.02	3,000	15,400	12.31
1,100	3,990	6.49	3,500	19,225	13.49
1,200	4,490	6.93	4,000	23,335	14.58
1,300	4,995	7.33	4,500	27,755	15.63
1,400	5,500	7.71	5,000	32,545	16.64
1,500	6,020	8.06	6,000	43,435	18.61
1,600	6,545	8.40	7,000	55,985	20.54
1,700	7,080	8.73	8,000	69,600	22.36
1,800	7,630	9.04			

$Cr(g)$:

$$H_T - H_{298.15} = 4.16T + 0.44 \times 10^{-3}T^2 - 0.48 \times 10^5 T^{-1}$$
$$-1,118 \ (1.6 \text{ percent}; 298°-3,000° \text{ K.});$$
$$C_p = 4.16 + 0.88 \times 10^{-3}T + 0.48 \times 10^5 T^{-2}.$$

OXIDE

TABLE 229.—*Heat content and entropy* $Cr_2O_3(c)$

[Base, α-crystals at 298.15° K.; mol. wt., 152.02]

T, ° K.	$H_T-H_{298.15}$, cal./mole	$S_T-S_{298.15}$, cal./deg. mole	T, ° K.	$H_T-H_{298.15}$, cal./mole	$S_T-S_{298.15}$, cal./deg. mole
298.16(β)	100	0.34	1,200	26,430	40.11
400	2,740	7.94	1,300	29,550	42.60
500	5,540	14.19	1,400	32,670	44.91
600	8,380	19.36	1,500	35,790	47.07
700	11,280	23.82	1,600	38,920	49.08
800	14,230	27.76	1,700	42,050	50.98
900	17,210	31.27	1,800	45,180	52.77
1,000	20,240	34.46	1,900	48,320	54.47
1,100	23,320	37.40	2,000	51,460	56.08

TABLE 1h7. (Continued)

$Cr_2O_3(\beta)$:

$H_T - H_{298.15} = 28.53T + 1.10 \times 10^{-3}T^2 + 3.74 \times 10^5 T^{-1}$
$- 9,758$ (0.2 percent; $298°$–$1,800°$ K.);
$C_p = 28.53 + 2.20 \times 10^{-3}T - 3.74 \times 10^5 T^{-2}$.

CARBIDES

TABLE 230.—*Heat content and entropy of* $Cr_{23}C_6(c)$

[Base, crystals at 298.15° K.; mol. wt., 1,268.30]

T, ° K.	$H_T - H_{298.15}$, cal./mole	$S_T - S_{298.15}$, cal./deg. mole	T, ° K.	$H_T - H_{298.15}$, cal./mole	$S_T - S_{298.15}$, cal./deg. mole
400	16,150	46.5	1,200	173,850	258.4
500	33,700	85.7	1,300	195,950	276.1
600	51,950	118.9	1,400	218,750	293.0
700	70,850	148.0	1,500	242,150	309.2
800	90,500	174.2	1,600	266,150	324.7
900	110,650	197.9	1,700	290,850	339.6
1,000	131,250	219.6	1,800	316,150	354.1
1,100	152,300	239.7			

$Cr_{23}C_6(c)$:

$H_T - H_{298.15} = 169.16T + 21.33 \times 10^{-3}T^2 + 28.93$
$\times 10^5 T^{-1} - 62,034$ (0.3 percent; $298°$–$1,700°$ K.);
$C_p = 169.16 + 42.66 \times 10^{-3}T - 28.93 \times 10^5 T^{-2}$.

TABLE 231.—*Heat content and entropy of* $Cr_7C_3(c)$

[Base, crystals at 298.15° K.; mol. wt., 400.10]

T, ° K.	$H_T - H_{298.15}$, cal./mole	$S_T - S_{298.15}$, cal./deg. mole	T, ° K.	$H_T - H_{298.15}$, cal./mole	$S_T - S_{298.15}$, cal./deg. mole
400	5,440	15.66	1,000	44,230	73.97
500	11,320	28.77	1,100	51,360	80.76
600	17,450	39.94	1,200	58,600	87.06
700	23,860	49.81	1,300	66,000	92.98
800	30,480	58.65	1,400	73,700	98.69
900	37,240	66.60	1,500	81,750	104.24

$Cr_7C_3(c)$:

$H_T - H_{298.15} = 56.96T + 7.27 \times 10^{-3}T^2 + 10.12 \times 10^5 T^{-1}$
$- 21,023$ (0.2 percent; $298°$–$1,500°$ K.);
$C_p = 56.96 + 14.54 \times 10^{-3}T - 10.12 \times 10^5 T^{-2}$.

TABLE 232.—*Heat content and entropy of* $Cr_3C_2(c)$

[Base, crystals at 298.15° K.; mol. wt., 180.05]

T, ° K.	$H_T - H_{298.15}$, cal./mole	$S_T - S_{298.15}$, cal./deg. mole	T, ° K.	$H_T - H_{298.15}$, cal./mole	$S_T - S_{298.15}$, cal./deg. mole
400	2,620	7.54	1,300	32,640	45.88
500	5,520	14.00	1,400	36,330	48.62
600	8,600	19.62	1,500	40,070	51.20
700	11,780	24.51	1,600	43,850	53.64
800	15,060	28.89	1,700	47,670	55.95
900	18,420	32.84	1,800	50,530	58.16
1,000	21,860	36.47	1,900	54,420	60.26
1,100	25,390	39.83	2,000	58,340	62.27
1,200	28,990	42.96			

$Cr_3C_2(c)$:

$H_T - H_{298.15} = 30.03T + 2.79 \times 10^{-3}T^2 + 7.40 \times 10^5 T^{-1}$
$- 11,683$ (0.2 percent; $298°$–$1,600°$ K.);
$C_p = 30.03 + 5.58 \times 10^{-3}T - 7.40 \times 10^5 T^{-2}$.

NITRIDES

$Cr_2N(c)$:

$C_p = 11.01 + 16.40 \times 10^{-3}T$ ($298°$–$800°$ K.).

$CrN(c)$:

$\overline{C_p} = 12.20$ ($298°$–$800°$ K.).

CHLORIDES

$CrCl_2(c)$:

$C_p = 15.23 + 5.30 \times 10^{-3}T(estimated)$ ($298°$–$1,088°$ K.).

$C_2Cl_3(c)$:

$C_p = 19.44 + 7.03 \times 10^{-3}T(estimated)$ ($298°$–$1,220°$ K.).

CARBONYL

$Cr(CO)_6(c)$:

$\overline{C_p} = 60.15$ ($293°$–$363°$ K.).

NITRATE

$Cr(NO_3)_2.9H_2O(c)$:

$C_p = 92.3$ ($300°$).

SULFATE

$Cr_2(SO_4)_3(c)$:

$C_p = 67.5$ ($300°$ K.).
$Cr(SO_4)_3 \cdot 18H_2O(c)$:
$C_p = 223.6$ ($300°$ K.).

ANTIMONIDES

$CrSb(c)$:

$C_p = 12.30 + 1.20 \times 10^{-3}T(estimated)$ ($298°$–$1,383°$ K.).

$CrSb_2(c)$:

$C_p = 19.20 + 1.84 \times 10^{-3}T(estimated)$ ($298°$–$949°$ K.).

COBALT AND ITS COMPOUNDS

ELEMENT

TABLE 233.—*Heat content and entropy of* $Co(c, l)$

[Base, α-crystals at 298.15° K.; atomic wt., 58.94]

T, ° K.	$H_T - H_{298.15}$, cal./mole	$S_T - S_{298.15}$, cal./deg. mole	T, ° K.	$H_T - H_{298.15}$, cal./mole	$S_T - S_{298.15}$, cal./deg. mole
400	625	1.80	1,500	10,510	12.93
500	1,280	3.26	1,600	11,470	13.55
600	1,975	4.52	1,700	12,420	14.12
700(α)	2,705	5.65	1,768(γ)	13,040	14.48
700(β)	2,810	5.80	1,768(l)	17,140	16.80
800	3,580	6.83	1,800	17,430	16.96
900	4,390	7.78	1,900	18,330	17.45
1,000	5,250	8.69	2,000	19,230	17.91
1,100	6,170	9.57	2,200	21,030	18.77
1,200	7,180	10.45	2,400	22,830	19.56
1,300	8,270	11.32	2,600	24,630	20.28
1,393(β)	9,400	12.16	2,800	26,430	20.95
1,393(γ)	9,400	12.16	3,000	28,230	21.57
1,400	9,480	12.22	3,200	30,030	22.15

TABLE 1h7. (Continued)

Co(α):

$H_T - H_{298.15} = 4.74T + 2.00 \times 10^{-3}T^2 - 1,591$

(0.1 percent; 298°–700° K.);

$C_p = 4.74 + 4.00 \times 10^{-3}T$;

$\Delta H_{700}(transformation) = 105$.

Co(β):

$H_T - H_{298.15} = 2.16T + 3.51 \times 10^{-3}T^2 - 422$

(0.5 percent; 700°–1,393° K.);

$C_p = 2.16 + 7.02 \times 10^{-3}T$;

$\Delta H_{1393}(magnetic\ Curie\ point) = 0$.

Co(γ):

$H_T - H_{298.15} = 17.49T - 2.46 \times 10^{-3}T^2 - 10,190$

(0.1 percent; 1,393°–1,768° K.);

$C_p = 17.49 - 4.92 \times 10^{-3}T$;

$\Delta H_{1768}(fusion) = 4,100$.

Co(l):

$H_T - H_{298.15} = 9.00T + 1,230$ (0.1 percent;

1,768°–3,200° K.);

$C_p = 9.00$.

TABLE 234.—Heat content and entropy of Co(g)

[Base, ideal gas at 298.15° K.; atomic wt., 58.94]

T, ° K.	$H_T - H_{298.15}$, cal./mole	$S_T - S_{298.15}$, cal./deg. mole	T, ° K.	$H_T - H_{298.15}$, cal./mole	$S_T - S_{298.15}$, cal./deg. mole
400	580	1.67	1,900	10,015	11.40
500	1,180	3.00	2,000	10,655	11.72
600	1,790	4.12	2,200	11,930	12.33
700	2,415	5.08	2,400	13,195	12.88
800	3,040	5.92	2,600	14,460	13.39
900	3,665	6.65	2,800	15,720	13.85
1,000	4,295	7.32	3,000	16,975	14.29
1,100	4,925	7.92	3,500	20,115	15.25
1,200	5,555	8.47	4,000	23,275	16.10
1,300	6,190	8.98	4,500	26,505	16.86
1,400	6,825	9.45	5,000	29,830	17.56
1,500	7,465	9.88	6,000	36,895	18.84
1,600	8,100	10.30	7,000	44,620	20.04
1,700	8,740	10.68	8,000	53,060	21.16
1,800	9,380	11.05			

Co(g):

$H_T - H_{298.15} = 6.38T + 0.78 \times 10^5 T^{-1} - 2,164$

(0.2 percent; 298°–5,000° K.);

$C_p = 6.38 - 0.78 \times 10^5 T^{-2}$.

OXIDES

TABLE 235.—Heat content and entropy of CoO(c)

[Base, crystals at 298.15° K.; mol. wt., 74.94]

T, ° K.	$H_T - H_{298.15}$, cal./mole	$S_T - S_{298.15}$, cal./deg. mole	T, ° K.	$H_T - H_{298.15}$, cal./mole	$S_T - S_{298.15}$, cal./deg. mole
400	1,290	3.72	1,300	13,210	19.21
500	2,570	6.58	1,400	14,640	20.27
600	3,860	8.93	1,500	16,100	21.28
700	5,160	10.93	1,600	17,600	22.25
800	6,470	12.68	1,700	19,140	23.18
900	7,790	14.24	1,800	20,730	24.09
1,000	9,120	15.64	1,900	22,360	24.97
1,100	10,460	16.92	2,000	24,020	25.82
1,200	11,820	18.10			

CoO(c):

$H_T - H_{298.15} = 11.54T + 1.02 \times 10^{-3}T^2 - 0.40 \times 10^5 T^{-1}$
$- 3,397$ (0.6 percent; 298°–2,000° K.);

$C_p = 11.54 + 2.04 \times 10^{-3}T + 0.40 \times 10^5 T^{-2}$.

TABLE 236.—Heat content and entropy of Co₃O₄(c)

[Base, crystals at 298.15° K.; mol. wt., 240.82]

T, ° K.	$H_T - H_{298.15}$, cal./mole	$S_T - S_{298.15}$, cal./deg. mole	T, ° K.	$H_T - H_{298.15}$, cal./mole	$S_T - S_{298.15}$, cal./deg. mole
400	3,270	9.40	800	18,820	36.03
500	6,850	17.38	900	23,300	41.30
600	10,660	24.32	1,000	28,250	46.52
700	14,640	30.45			

Co₃O₄(c):

$H_T - H_{298.15} = 30.84T + 8.54 \times 10^{-3}T^2 + 5.72 \times 10^5 T^{-1}$
$- 11,873$ (0.5 percent; 298°–1,000° K.):

$C_p = 30.84 + 17.08 \times 10^{-3}T - 5.72 \times 10^5 T^{-2}$.

SULFIDE

CoS(c):

$C_p = 10.60 + 2.51 \times 10^{-3}T(estimated)$ (298°–1,373° K.).

HYDRIDE

TABLE 237.—Heat content and entropy of CoH(g)

[Base, ideal gas at 298.15° K.; mol. wt., 59.95]

T, ° K.	$H_T - H_{298.15}$, cal./mole	$S_T - S_{298.15}$, cal./deg. mole	T, ° K.	$H_T - H_{298.15}$, cal./mole	$S_T - S_{298.15}$, cal./deg. mole
400	715	2.06	1,000	5,260	8.93
500	1,425	3.65	1,200	6,895	10.42
600	2,160	4.98	1,400	8,565	11.71
700	2,910	6.14	1,600	10,265	12.84
800	3,675	7.16	1,800	11,980	13.85
900	4,460	8.09	2,000	13,705	14.76

CoH(g):

$H_T - H_{298.15} = 6.71T + 0.61 \times 10^{-3}T^2 + 0.10 \times 10^5 T^{-1}$
$- 2,088$ (0.5 percent; 298°–2,000° K.);

$C_p = 6.71 + 1.22 \times 10^{-3}T - 0.10 \times 10^5 T^{-2}$.

CHLORIDE

TABLE 238.—Heat content and entropy of CoCl₂(c)

[Base, ideal gas at 298.15° K.; mol. wt., 129.85]

T, ° K.	$H_T - H_{298.15}$, cal./mole	$S_T - S_{298.15}$, cal./deg. mole	T, ° K.	$H_T - H_{298.15}$, cal./mole	$S_T - S_{298.15}$, cal./deg. mole
400	2,020	5.82	800	11,260	21.60
500	4,120	10.50	900	13,920	24.72
600	6,340	14.54	1,000	16,700	27.65
700	8,720	18.21			

CoCl₂(c):

$H_T - H_{298.15} = 14.41T + 7.30 \times 10^{-3}T^2 - 4,945$
(0.5 percent; 298°–1,000° K.);

$C_p = 14.41 + 14.60 \times 10^{-3}T$.

FLUORIDE

CoF₂(c):

$C_p = 16.44$ (298° K.).

FERRITE

FeCo₂O₄(c):

$C_p = 36.53$ (298° K.).

TABLE 1h7. (*Continued*)

NITRATE

$Co(NO_3)_2 \cdot 6H_2O(c)$:

$C_p = 108.0$ (300° K.).

SULFATE

$CoSO_4(c)$:

$C_p = 30.10 + 9.92 \times 10^{-3} T$ (*estimated*) (298°–1,200° K.).

$CoSO_4 \cdot 7H_2O(c)$:

$\overline{C_p} = 96.4$ (288°–303° K.).

HEXAMMINE HALIDES

$Co(NH_3)_6Cl_3(c)$:

$C_p = 76.6$ (298° K.).

$Co(NH_3)_6I_2(c)$:

$C_p = 69.2$ (298° K.).

$Co(NH_3)_6I_3(c)$:

$C_p = 74.3$ (298° K.).

ARSENICAL SULFIDE

$CoAsS(c)$:

$\overline{C_p} = 16.4$ (283°–373° K.).

ANTIMONIDE

$CoSb(c)$:

$C_p = 11.70 + 1.56 \times 10^{-3} T$ (*estimated*) (290°–1,464° K.).

COBALT-TIN

TABLE 239.—*Heat content and entropy of* $Co_2Sn(c)$

[Base, crystals at 298.15° K.; mol. wt., 236.58]

T, ° K.	$H_T - H_{298.15}$, cal./mole	$S_T - S_{298.15}$, cal./deg. mole	T, ° K.	$H_T - H_{298.15}$, cal./mole	$S_T - S_{298.15}$, cal./deg. mole
400	1,960	5.65	700	8,270	17.36
500	3,970	10.13	800	10,500	20.33
600	6,100	14.02	900	12,820	23.06

$Co_2Sn(c)$:

$H_T - H_{298.15} = 16.35 T + 4.20 \times 10^{-3} T^2 - 5,248$
(0.3 percent; 298°–900° K.);

$C_p = 16.35 + 8.40 \times 10^{-3} T$.

COPPER AND ITS COMPOUNDS

ELEMENT

TABLE 240.—*Heat content and entropy of* $Cu(c,l)$

[Base, crystals at 298.15° K.; atomic wt., 63.54]

T, ° K.	$H_T - H_{298.15}$, cal./mole	$S_T - S_{298.15}$, cal./deg. mole	T, ° K.	$H_T - H_{298.15}$, cal./mole	$S_T - S_{298.15}$, cal./deg. mole
400	600	1.73	1,400	10,480	12.32
500	1,215	3.10	1,500	11,230	12.84
600	1,845	4.25	1,600	11,980	13.33
700	2,480	5.23	1,700	12,730	13.78
800	3,130	6.10	1,800	13,480	14.21
900	3,800	6.89	1,900	14,230	14.62
1,000	4,490	7.61	2,000	14,980	15.00
1,100	5,190	8.28	2,200	16,480	15.71
1,200	5,895	8.90	2,400	17,980	16.37
1,300	6,615	9.47	2,600	19,480	16.97
1,357(c)	7,040	9.79	2,800	20,980	17.52
1,357(l)	10,160	12.09			

$Cu(c)$:

$H_T - H_{298.15} = 5.41 T + 0.75 \times 10^{-3} T^2 - 1,680$
(0.3 percent; 298°–1,357° K.);

$C_p = 5.41 + 1.50 \times 10^{-3} T$;

ΔH_{1357} (*fusion*) = 3,120.

$Cu(l)$:

$H_T - H_{298.15} = 7.50 T - 20$ (0.1 percent; 298°–2,800° K.);

$C_p = 7.50$.

TABLE 241.—*Heat content and entropy of* $Cu(g)$

[Base, ideal gas at 298.15° K.; atomic wt., 63.54]

T, ° K.	$H_T - H_{298.15}$, cal./mole	$S_T - S_{298.15}$, cal./deg. mole	T, ° K.	$H_T - H_{298.15}$, cal./mole	$S_T - S_{298.15}$, cal./deg. mole
400	505	1.46	1,900	7,985	9.22
500	1,005	2.57	2,000	8,495	9.48
600	1,500	3.48	2,200	9,530	9.97
700	1,995	4.24	2,400	10,590	10.43
800	2,495	4.90	2,600	11,680	10.87
900	2,990	5.49	2,800	12,815	11.29
1,000	3,490	6.01	3,000	13,995	11.70
1,100	3,985	6.49	3,500	17,155	12.67
1,200	4,480	6.92	4,000	20,600	13.59
1,300	4,980	7.32	4,500	24,290	14.46
1,400	5,475	7.69	5,000	28,150	15.27
1,500	5,975	8.03	6,000	36,160	16.73
1,600	6,475	8.35	7,000	44,450	18.01
1,700	6,975	8.66	8,000	53,175	19.17
1,800	7,480	8.94			

$Cu(g)$:

$H_T - H_{298.15} = 4.97 T - 1,482$ (0.1 percent; 298°–2,000° K.);

$C_p = 4.97$;

$H_T - H_{298.15} = 2.86 T + 0.53 \times 10^{-3} T^2 + 655$
(0.2 percent; 2,000°–5,000° K.);

$C_p = 2.86 + 1.06 \times 10^{-3} T$.

OXIDES

TABLE 242.—*Heat content and entropy of* $Cu_2O(c)$

[Base, crystals at 298.15° K.; mol. wt., 143.08]

T, ° K.	$H_T - H_{298.15}$, cal./mole	$S_T - S_{298.15}$, cal./deg. mole	T, ° K.	$H_T - H_{298.15}$, cal./mole	$S_T - S_{298.15}$, cal./deg. mole
400	1,720	4.96	900	11,000	19.98
500	3,470	8.87	1,000	13,020	22.10
600	5,280	12.26	1,100	15,120	24.10
700	7,150	15.14	1,200	17,320	26.02
800	9,050	17.68			

$Cu_2O(c)$:

$H_T - H_{298.15} = 14.90 T + 2.85 \times 10^{-3} T^2 - 4,696$
(0.2 percent; 298°–1,200° K.);

$C_p = 14.90 + 5.70 \times 10^{-3} T$.

TABLE 243.—*Heat content and entropy of* $CuO(c)$

[Base, crystals at 298.15° K.; mol. wt., 79.54]

T, ° K.	$H_T - H_{298.15}$, cal./mole	$S_T - S_{298.15}$, cal./deg. mole	T, ° K.	$H_T - H_{298.15}$, cal./mole	$S_T - S_{298.15}$, cal./deg. mole
400	1,110	3.21	900	7,320	13.15
500	2,260	5.76	1,000	8,680	14.58
600	3,460	7.95	1,100	10,120	15.96
700	4,710	9.88	1,200	11,600	17.24
800	6,000	11.60	1,250	12,360	17.86

$CuO(c)$:

$H_T - H_{298.15} = 9.27 T + 2.40 \times 10^{-3} T^2 - 2,977$
(0.3 percent; 298°–1,250° K.);

$C_p = 9.27 + 4.80 \times 10^{-3} T$.

TABLE 1h7. (*Continued*)

TABLE 244.—*Heat content and entropy of CuO(g)*

[Base, ideal gas at 298.15° K.; mol. wt., 79.54]

T, ° K.	$H_T - H_{298.15}$, cal./mole	$S_T - S_{298.15}$, cal./deg. mole	T, ° K.	$H_T - H_{298.15}$, cal./mole	$S_T - S_{298.15}$, cal./deg. mole
400	825	2.38	1,000	6,015	10.27
500	1,665	4.25	1,200	7,785	11.88
600	2,525	5.82	1,400	9,555	13.25
700	3,390	7.15	1,600	11,335	14.44
800	4,260	8.31	1,800	13,115	15.48
900	5,135	9.34	2,000	14,895	16.42

$CuO(g)$:

$$H_T - H_{298.15} = 8.72T + 0.07 \times 10^{-3}T^2 + 0.73 \times 10^5 T^{-1}$$
$$-2,851 \ (0.2 \text{ percent; } 298°-2,000° \text{ K.});$$
$$C_p = 8.72 + 0.14 \times 10^{-3}T - 0.73 \times 10^5 T^{-2}.$$

SULFIDES

TABLE 245.—*Heat content and entropy of $Cu_2S(c)$*

[Base, α-crystals at 298.15° K.; mol. wt., 159.15]

T, ° K.	$H_T - H_{298.15}$, cal./mole	$S_T - S_{298.15}$, cal./deg. mole	T. ° K.	$H_T - H_{298.15}$, cal./mole	$S_T - S_{298.15}$, cal./deg. mole
350	1,010	3.13	700	9,950	21.40
376(α)	1,520	4.53	800	11,980	24.12
376(β)	2,440	6.98	900	14,010	26.50
400	3,000	8.42	1,000	16,040	28.65
500	5,320	13.59	1,100	18,080	30.59
600	7,650	17.84	1,200	20,110	32.36
623(β)	8,180	18.71	1,300	22,140	33.98
623(γ)	8,380	19.03	1,400	24,170	35.48

$Cu_2S(\alpha)$:

$$H_T - H_{298.15} = 19.50T - 5,665 \ (0.1 \text{ percent;}$$
$$298°-376° \text{ K.});$$
$$C_p = 19.50;$$
$$\Delta H_{376}(transition) = 920.$$

$Cu_2S(\beta)$:

$$H_T - H_{298.15} = 23.25T - 6,300 \ (0.1 \text{ percent;}$$
$$370°-623° \text{ K.});$$
$$C_p = 23.25;$$
$$\Delta H_{623}(transition) = 200.$$

$Cu_2S(\gamma)$:

$$H_T - H_{2981.5} = 20.32T - 4,275 \ (0.1 \text{ percent;}$$
$$623°-1,400° \text{ K.});$$
$$C_p = 20.32.$$

$CuS(c)$:

$$C_p = 10.60 + 2.64 \times 10^{-3}T(estimated) \ (298°-1,273° \text{ K.}).$$

NITRIDE

$Cu_3N(c)$:

$$\overline{C_p} = 21.68 \ (273°-373° \text{ K.}).$$

HYDRIDES

TABLE 246.—*Heat content and entropy of CuH(g)*

[Base, ideal gas at 298.15° K.; mol. wt., 64.55]

T, ° K.	$H_T - H_{298.15}$, cal./mole	$S_T - S_{298.15}$, cal./deg. mole	T, ° K.	$H_T - H_{298.15}$, cal./mole	$S_T - S_{298.15}$, cal./deg. mole
400	715	2.06	1,000	5,275	8.95
500	1,430	3.66	1,200	6,910	10.44
600	2,160	4.99	1,400	8,580	11.73
700	2,910	6.14	1,600	10,280	12.86
800	3,680	7.17	1,800	11,995	13.87
900	4,470	8.10	2,000	13,725	14.78

$CuH(g)$:

$$H_T - H_{298.15} = 6.75T + 0.60 \times 10^{-3}T^2 + 0.11 \times 10^5 T^{-1}$$
$$-2,103 \ (0.7 \text{ percent; } 298°-2,000° \text{ K.});$$
$$C_p = 6.75 + 1.20 \times 10^{-3}T - 0.11 \times 10^5 T^{-2}.$$

TABLE 247.—*Heat content and entropy of CuD(g)*

[Base, ideal gas at 298.15° K.; mol. wt., 65.55]

T, ° K.	$H_T - H_{298.15}$, cal./mole	$S_T - S_{298.15}$, cal./deg. mole	T, ° K.	$H_T - H_{298.15}$, cal./mole	$S_T - S_{298.15}$, cal./deg. mole
400	735	2.12	1,000	5,525	9.35
500	1,480	3.78	1,200	7,225	10.90
600	2,250	5.18	1,400	8,945	12.23
700	3,045	6.41	1,600	10,685	13.39
800	3,860	7.50	1,800	12,430	14.42
900	4,685	8.47	2,000	14,185	15.34

$CuD(g)$:

$$H_T - H_{298.15} = 7.50T + 0.42 \times 10^{-3}T^2 + 0.60 \times 10^5 T^{-1}$$
$$-2,475 \ (0.6 \text{ percent; } 298°-2,000° \text{ K.});$$
$$C_p = 7.50 + 0.84 \times 10^{-3}T - 0.60 \times 10^5 T^{-2}.$$

BROMIDE

$CuBr(c)$:

$$C_p = 12.80 + 1.58 \times 10^{-3}T(estimated) \ (298°-761° \text{ K.}).$$

TABLE 248.—*Heat content and entropy of CuBr(g)*

[Base, ideal gas at 298.15° K.; mol. wt., 143.46]

T, ° K.	$H_T - H_{298.15}$, cal./mole	$S_T - S_{298.15}$, cal./deg. mole	T, ° K.	$H_T - H_{298.15}$, cal./mole	$S_T - S_{298.15}$, cal./deg. mole
400	885	2.55	1,000	6,200	10.66
500	1,765	4.51	1,200	7,985	12.28
600	2,645	6.12	1,400	9,770	13.66
700	3,530	7.48	1,600	11,550	14.85
800	4,420	8.67	1,800	13,335	15.90
900	5,310	9.72	2,000	15,125	16.84

$CuBr(g)$:

$$H_T - H_{298.15} = 8.93T + 0.28 \times 10^5 T^{-1} - 2,756$$
$$(0.1 \text{ percent; } 298°-2,000° \text{ K.});$$
$$C_p = 8.93 - 0.28 \times 10^5 T^{-2}.$$

TABLE 1h7. (*Continued*)

CHLORIDES

TABLE 249.—*Heat content and entropy of CuCl(c, l)*

[Base, crystals at 298.15° K.; mol. wt., 99.00]

T, ° K.	$H_T-H_{298.15}$, cal./mole	$S_T-S_{298.15}$, cal./deg. mole	T, ° K.	$H_T-H_{298.15}$, cal./mole	$S_T-S_{298.15}$, cal./deg. mole
400	1,280	3.68	800	10,420	18.58
500	2,720	6.89	900	12,000	20.44
600	4,385	9.92	1,000	13,580	22.10
700	6,210	12.72	1,100	15,160	23.61
703 (c)	6,270	12.81	1,200	16,740	24.99
703 (l)	8,890	16.54			

CuCl(c):

$$H_T-H_{298.15}=5.88T+9.60\times10^{-3}T^2-2,606$$

(0.3 percent; 298°–703° K.);

$$C_p=5.88+19.20\times10^{-3}T;$$

$$\Delta H_{703}(fusion)=2,620.$$

CuCl(l):

$$H_T-H_{298.15}=15.80T-2,220 \text{ (0.1 percent;}$$

703°–1,200° K.);

$$C_p=15.80.$$

TABLE 250.—*Heat content and entropy of CuCl(g)*

[Base, ideal gas at 298.15° K.; mol. wt., 99.00]

T, ° K.	$H_T-H_{298.15}$, cal./mole	$S_T-S_{298.15}$, cal./deg. mole	T, ° K.	$H_T-H_{298.15}$, cal./mole	$S_T-S_{298.15}$, cal./deg. mole
400	865	2.49	1,000	6,145	10.54
500	1,735	4.43	1,200	7,925	12.16
600	2,610	6.03	1,400	9,705	13.54
700	3,490	7.39	1,600	11,490	14.73
800	4,375	8.57	1,800	13,275	15.78
900	5,260	9.61	2,000	15,060	16.72

CuCl(g):

$$H_T-H_{298.15}=8.88T+0.02\times10^{-3}T^2+0.45\times10^5T^{-1}$$

$$-2,800 \text{ (0.1 percent; 298°–2,000° K.);}$$

$$C_p=8.88+0.04\times10^{-3}T-0.45\times10^5T^{-2}.$$

TABLE 251.—*Heat content and entropy of CuCl$_2$(c)*

[Base, crystals at 298.15° K.; mol. wt., 134.45]

T, ° K.	$H_T-H_{298.15}$, cal./mole	$S_T-S_{298.15}$, cal./deg. mole	T, ° K.	$H_T-H_{298.15}$, cal./mole	$S_T-S_{298.15}$, cal./deg. mole
400	1,990	5.74	700	8,620	18.01
500	4,080	10.40	800	11,030	21.22
600	6,290	14.42			

CuCl$_2$(c):

$$H_T-H_{298.15}=15.42T+6.00\times10^{-3}T^2-5,131$$

(0.2 percent; 298°–800° K.);

$$C_p=15.42+12.00\times10^{-3}T.$$

FLUORIDE

TABLE 252.—*Heat content and entropy of CuF(g)*

[Base, ideal gas at 298.15° K.: mol. wt., 82.54]

T, ° K.	$H_T-H_{298.15}$, cal./mole	$S_T-S_{298.15}$, cal./deg. mole	T, ° K.	$H_T-H_{298.15}$, cal./mole	$S_T-S_{298.15}$, cal./deg. mole
400	830	2.39	1,000	6,020	10.28
500	1,670	4.26	1,200	7,790	11.90
600	2,525	5.82	1,400	9,560	13.26
700	3,395	7.16	1,600	11,340	14.45
800	4,265	8.33	1,800	13,120	15.50
900	5,140	9.36	2,000	14,900	16.44

CuF(g):

$$H_T-H_{298.15}=8.73T+0.07\times10^{-3}T^2+0.74\times10^5T^{-1}$$

$$-2,857 \text{ (0.1 percent; 298°–2,000° K.);}$$

$$C_p=8.73+0.14\times10^{-3}T-0.74\times10^5T^{-2}.$$

IODIDES

CuI(c):

$$C_p=12.10+2.86\times10^{-3}T \text{ (estimated) (298°–675° K.).}$$

TABLE 253.—*Heat content and entropy of CuI(g)*

[Base, ideal gas at 298.15° K.; mol. wt., 190.45]

T, ° K.	$H_T-H_{298.15}$, cal./mole	$S_T-S_{298.15}$, cal./deg. mole	T, ° K.	$H_T-H_{298.15}$, cal./mole	$S_T-S_{298.15}$, cal./deg. mole
400	890	2.56	1,000	6,220	10.70
500	1,775	4.54	1,200	8,005	12.33
600	2,660	6.15	1,400	9,790	13.70
700	3,550	7.52	1,600	11,575	14.89
800	4,440	8.71	1,800	13,365	15.95
900	5,330	9.76	2,000	15,155	16.89

CuI(g):

$$H_T-H_{298.15}=8.94T+0.22\times10^5T^{-1}-2,739$$

(0.1 percent; 298°–2,000° K.);

$$C_p=8.94-0.22\times10^5T^{-2}.$$

CuI$_2$(c):

$$\overline{C_p}=20.1 \text{ (274°–328° K.).}$$

SELENIDE

TABLE 254.—*Heat content and entropy of Cu$_2$Se(c)*

[Base, α-crystals at 298.15° K.; mol. wt., 206.04]

T, ° K.	$H_T-H_{298.15}$, cal./mole	$S_T-S_{298.15}$, cal./deg. mole	T, ° K.	$H_T-H_{298.15}$, cal./mole	$S_T-S_{298.15}$, cal./deg. mole
350	1,100	3.40	400	3,310	9.24
383(α)	1,800	5.31	450	4,320	11.61
383(β)	2,960	8.34	500	5,530	13.74

Cu$_2$Se(α):

$$H_T-H_{298.15}=21.20T-6,321 \text{ (0.1 percent;}$$

298°–383° K.);

TABLE 1h7. (*Continued*)

$Cp = 21.20;$

$\Delta H_{383}(transition) = 1,160.$

$Cu_2Se(\beta):$

$H_T - H_{298.15} = 20.20\,T - 4,775 \ (0.2\,\text{percent}; 383°-500°\,\text{K.});$

$C_p = 20.20.$

SILICATE

$CuSiO_3 \cdot H_2O(c):$

$\overline{C_p} = 28.7 \ (292°-323°\,\text{K.}).$

SULFATE

TABLE 255.—*Heat content and entropy of* $CuSO_4(c)$

[Base, crystals at 298.15° K.; mol. wt., 159.61]

$T, °\,\text{K.}$	$H_T-H_{298.15},$ cal./mole	$S_T-S_{298.15},$ cal./deg. mole	$T, °\,\text{K.}$	$H_T-H_{298.15},$ cal./mole	$S_T-S_{298.15},$ cal./deg. mole
400	2,550	7.35	700	10,950	22.86
500	5,150	13.14	800	14,150	27.13
600	7,950	18.24	900	17,650	31.25

$CuSO_4(c):$

$H_T - H_{298.15} = 18.77\,T + 8.60\times10^{-3}\,T^2 - 6,361$

$(0.7\,\text{percent}; 298°-900°\,\text{K.});$

$C_p = 18.77 + 17.20\times10^{-3}\,T.$

$CuSO_4 \cdot H_2O(c):$

$\overline{C_p} = 32.6 \ (275°-373°\,\text{K.}).$

$CuSO_4 \cdot 3H_2O(c):$

$C_p = 49.0 \ (282°\,\text{K.}).$

$CuSO_4 \cdot 5H_2O(c):$

$C_p = 67.2 \ (282°\,\text{K.}).$

COPPER-IRON SULFIDE

$CuFeS_2(c):$

$\overline{C_p} = 24.0 \ (292°-321°\,\text{K.}).$

COPPER-ALUMINUM COMPOUNDS

TABLE 256.—*Heat content and entropy of* $Cu_3Al(c)$

[Base, crystals at 298.15° K.; mol. wt., 217.60]

$T, °\,\text{K.}$	$H_T-H_{298.15},$ cal./mole	$S_T-S_{298.15},$ cal./deg. mole	$T, °\,\text{K.}$	$H_T-H_{298.15},$ cal./mole	$S_T-S_{298.15},$ cal./deg. mole
400	2,400	6.92	700	10,010	21.01
500	4,800	12.27	800	12,760	24.68
600	7,340	16.90			

$Cu_3Al(c):$

$H_T - H_{298.15} = 20.41\,T + 4.51\times10^{-3}\,T^2 - 6,486$

$(0.4\,\text{percent}; 298°-800°\,\text{K.});$

$C_p = 20.41 + 9.02\times10^{-3}\,T.$

TABLE 257.—*Heat content and entropy of* $CuAl(c)$

[Base, crystals at 298.15° K.; mol. wt., 90.52]

$T, °\,\text{K.}$	$H_T-H_{298.15},$ cal./mole	$S_T-S_{298.15},$ cal./deg. mole	$T, °\,\text{K.}$	$H_T-H_{298.15},$ cal./mole	$S_T-S_{298.15},$ cal./deg. mole
400	1,190	3.43	700	4,960	10.43
500	2,410	6.15	750	5,630	11.36
600	3,660	8.43			

$CuAl(c):$

$H_T - H_{298.15} = 10.25\,T + 2.10\times10^{-3}\,T^2 - 3,243 \ (0.2\,\text{percent}$

$298°-750°\,\text{K.});$

$C_p = 10.25 + 4.20\times10^{-3}\,T.$

TABLE 258.—*Heat content and entropy of* $CuAl_2(c)$

[Base, crystals at 298.15° K.; mol. wt., 117.50]

$T, °\,\text{K.}$	$H_T-H_{298.15},$ cal./mole	$S_T-S_{298.15},$ cal./deg. mole	$T, °\,\text{K.}$	$H_T-H_{298.15},$ cal./mole	$S_T-S_{298.15},$ cal./deg. mole
400	1,810	5.22	700	7,500	15.81
500	3,680	9.39	800	9,430	18.39
600	5,580	12.85			

$CuAl_2(c):$

$H_T - H_{298.15} = 16.01\,T + 2.66\times10^{-3}\,T^2 - 5,010 \ (0.6\,\text{percent};$

$298°-800°\,\text{K.});$

$C_p = 16.01 + 5.32\times10^{-3}\,T.$

COPPER-ANTIMONY COMPOUNDS

TABLE 259.—*Heat content and entropy of* $Cu_3Sb(c)$

[Base, crystals at 298.15° K.; mol. wt., 312.38]

$T, °\,\text{K.}$	$H_T-H_{298.15},$ cal./mole	$S_T-S_{298.15},$ cal./deg. mole	$T, °\,\text{K.}$	$H_T-H_{298.15},$ cal./mole	$S_T-S_{298.15},$ cal./deg. mole
400	2,550	7.36	600	7,800	17.98
500	5,140	13.13	700	10,510	22.15

$Cu_3Sb(c):$

$H_T - H_{298.15} = 21.97\,T + 4.50\times10^{-3}\,T^2 - 6,897 \ (0.4\,\text{percent};$

$298°-700°\,\text{K.});$

$C_p = 21.97 + 9.00\times10^{-3}\,\text{T.}$

TABLE 260.—*Heat content and entropy of* $Cu_2Sb(c)$

[Base, crystals at 298.15° K; mol. wt., 248.84]

$T, °\,\text{K.}$	$H_T-H_{298.15},$ cal./mole	$S_T-S_{298.15},$ cal./deg. mole	$T, °\,\text{K.}$	$H_T-H_{298.15},$ cal./mole	$S_T-S_{298.15},$ cal./deg. mole
400	1,900	5.48	600	5,840	13.45
500	3,850	9.83			

$Cu_2Sb(c):$

$H_T - H_{298.15} = 16.38\,T + 3.30\times10^{-3}\,T^2 - 5,177 \ (0.2\,\text{percent};$

$298°-600°\,\text{K.});$

$C_p = 16.38 + 6.60\times10^{-3}\,T.$

COPPER-CADMIUM COMPOUNDS

TABLE 261.—*Heat content and entropy of* $Cu_2Cd_3(c,l)$

[Base, crystals at 298.15° K.; mol. wt., 464.31]

$T, °\,\text{K.}$	$H_T-H_{298.15},$ cal./mole	$S_T-S_{298.15},$ cal./deg. mole	$T, °\,\text{K.}$	$H_T-H_{298.15},$ cal./mole	$S_T-S_{298.15},$ cal./deg. mole
400	3,000	8.64	835(c)	19,360	35.69
500	6,200	15.77	835(l)	30,730	49.31
600	9,750	22.23	900	33,450	52.42
700	13,650	28.24	1,000	37,600	56.79
800	17,850	33.84			

TABLE 1h7. (*Continued*)

$Cu_2Cd_3(c)$:

$H_T - H_{298.15} = 18.84T + 15.20 \times 10^{-3}T^2 - 6,968$ (0.4 percent; 298°–835° K.);

$C_p = 18.84 + 30.40 \times 10^{-3}T$;

$\Delta H_{835}(fusion) = 11,370$.

$Cu_2Cd_3(l)$:

$H_T - H_{298.15} = 41.50T - 3,920$ (0.1 percent; 835°–1,000° K.);

$C_p = 41.50$.

TABLE 262.—*Heat content and entropy of* $Cu_5Cd_8(c, l)$

[Base, crystals at 298.15° K.; mol. wt., 1,216.98]

T, ° K.	$H_T - H_{298.15}$, cal./mole	$S_T - S_{298.15}$, cal./deg. mole	T, ° K.	$H_T - H_{298.15}$, cal./mole	$S_T - S_{298.15}$, cal./deg. mole
400	7,800	22.48	819(c)	46,440	87.72
500	16,250	41.31	819(l)	75,410	123.09
600	25,250	57.71	900	84,000	133.07
700	34,650	72.18	1,000	94,600	144.25
800	44,500	85.32			

$Cu_5Cd_8(c)$:

$H_T - H_{298.15} = 58.78T + 27.20 \times 10^{-3}T^2 - 19,943$ (0.5 percent; 298°–819° K.);

$C_p = 58.78 + 54.40 \times 10^{-3}T$;

$\Delta H_{819}(fusion) = 28,970$.

$Cu_5Cd_8(l)$:

$H_T - H_{298.15} = 106.00T - 11,400$ (0.1 percent; 819°–1,000° K.);

$C_p = 106.00$.

COPPER-PALLADIUM COMPOUNDS

TABLE 263.—*Heat content and entropy of* $Cu_3Pd(c)$

[Base, crystals at 298.15° K.; mol. wt., 297.02]

T, ° K.	$H_T - H_{298.15}$, cal./mole	$S_T - S_{298.15}$, cal./deg. mole	T, ° K.	$H_T - H_{298.15}$, cal./mole	$S_T - S_{298.15}$, cal./deg. mole
400	2,350	6.78	900	15,560	27.91
500	4,770	12.17	1,000	18,460	30.97
600	7,310	16.80	1,100	21,430	33.80
700	9,970	20.90	1,200	24,570	36.53
800	12,730	24.58			

$Cu_3Pd(c)$:

$H_T - H_{298.15} = 20.98T + 4.40 \times 10^{-3}T^2 + 1.16 \times 10^5T^{-1} - 7,035$ (0.2 percent; 298°–1,200° K.);

$C_p = 20.98 + 8.80 \times 10^{-3}T - 1.16 \times 10^5T^{-2}$.

TABLE 264.—*Heat content and entropy of* $CuPd(c)$

[Base, crystals at 298.15° K.; mol. wt., 169.94]

T, ° K.	$H_T - H_{298.15}$, cal./mole	$S_T - S_{298.15}$, cal./deg. mole	T, ° K.	$H_T - H_{298.15}$, cal./mole	$S_T - S_{298.15}$, cal./deg. mole
400	1,200	3.46	900	7,690	13.89
500	2,420	6.18	1,000	9,060	15.33
600	3,690	8.49	1,100	10,450	16.66
700	5,000	10.51	1,200	11,870	17.89
800	6,340	12.30			

$CuPd(c)$:

$H_T - H_{298.15} = 12.02T + 0.98 \times 10^{-3}T^2 + 1.16 \times 10^5T^{-1} - 4,060$ (0.2 percent; 298°–1,200° K.);

$C_p = 12.02 + 1.96 \times 10^{-3}T - 1.16 \times 10^5T^{-2}$.

SILICIDE

$Cu_3Si(c)$:

$C_p = 21.30 + 5.87 \times 10^{-3}T (estimated)$ (290°–1,135° K.).

DYSPROSIUM

ELEMENT

TABLE 265.—*Heat content and entropy of* $Dy(c, l)$

[Base, crystals at 298.15° K.; atomic wt., 162.51]

T, ° K.	$H_T - H_{298.15}$, cal./mole	$S_T - S_{298.15}$, cal./deg. mole	T, ° K.	$H_T - H_{298.15}$, cal./mole	$S_T - S_{298.15}$, cal./deg. mole
400	670	1.93	1,500	9,050	11.74
500	1,350	3.45	1,600	9,910	12.29
600	2,040	4.70	1,700	10,790	12.82
700	2,750	5.80	1,773(c)	11,440	13.20
800	3,480	6.77	1,773(l)	15,540	15.51
900	4,220	7.65	1,800	15,760	15.63
1,000	4,980	8.45	1,900	16,560	16.07
1,100	5,760	9.19	2,000	17,360	16.48
1,200	6,560	9.88	2,200	18,960	17.24
1,300	7,370	10.53	2,400	20,560	17.93
1,400	8,200	11.15	2,600	21,160	18.57

$Dy(c)$:

$H_T - H_{298.15} = 6.00T + 0.85 \times 10^{-3}T^2 - 1,864$ (0.1 percent; 298°–1,773° K.);

$C_p = 6.00 + 1.70 \times 10^{-3}T$;

$\Delta H_{1773}(fusion) = 4,100$.

$Dy(l)$:

$H_T - H_{298.15} = 8.00T + 1,360$ (0.1 percent; 1,773°–2,600° K.);

$C_p = 8.00$.

ERBIUM AND ITS COMPOUNDS

ELEMENT

TABLE 266.—*Heat content and entropy of* $Er(c, l)$

[Base, crystals at 298.15° K.; atomic wt., 167.27]

T, ° K.	$H_T - H_{298.15}$, cal./mole	$S_T - S_{298.15}$, cal./deg. mole	T, ° K.	$H_T - H_{298.15}$, cal./mole	$S_T - S_{298.15}$, cal./deg. mole
400	690	1.99	1,600	10,020	12.49
500	1,390	3.54	1,700	10,890	13.02
600	2,095	4.84	1,800(c)	11,780	13.53
700	2,820	5.95	1,800(l)	15,880	15.81
800	3,560	6.94	1,900	16,680	16.24
900	4,310	7.83	2,000	17,480	16.65
1,000	5,080	8.64	2,200	19,080	17.41
1,100	5,870	9.39	2,400	20,680	18.11
1,200	6,670	10.08	2,600	22,280	18.75
1,300	7,480	10.73	2,800	23,880	19.34
1,400	8,310	11.35	2,900	24,680	19.62
1,500	9,160	11.93			

$Er(c)$:

$H_T - H_{298.15} = 6.29T + 0.74 \times 10^{-3}T^2 - 1,941$ (0.2 percent; 298°–1,800° K.);

$C_p = 6.29 + 1.48 \times 10^{-3}T$;

$\Delta H_{1800}(fusion) = 4,100$.

TABLE 1h7. (*Continued*)

Er(l):

$H_T - H_{298.15} = 8.00T + 1,480$ (0.1 percent;
1,800°–2,900° K.);

$C_p = 8.00.$

OXIDE

Er$_2$O$_3$(c):

$\overline{C_p} = 24.86$ (273°–373° K.).

SULFATE

Er$_2$(SO$_4$)$_3$(c):

$\overline{C_p} = 64.8$ (273°–373° K.).

Er$_2$(SO$_4$)$_3$·8H$_2$O(c):

$\overline{C_p} = 138.6$ (273°–373° K.).

EUROPIUM AND ITS COMPOUNDS

ELEMENT

TABLE 267.—*Heat content and entropy of Eu(c, l)*

[Base, crystals at 298.15° K.; atomic wt., 152.0]

T, ° K.	$H_T-H_{298.15}$, cal./mole	$S_T-S_{298.15}$, cal./deg. mole	T, ° K.	$H_T-H_{298.15}$, cal./mole	$S_T-S_{298.15}$, cal./deg. mole
400	660	1.91	1,100(l)	8,270	11.45
500	1,330	3.40	1,200	9,070	12.15
600	2,020	4.66	1,300	9,870	12.79
700	2,730	5.76	1,400	10,670	13.38
800	3,460	6.73	1,500	11,470	13.93
900	4,210	7.62	1,600	12,270	14.45
1,000	4,980	8.43	1,700	13,070	14.94
1,100(c)	5,770	9.18			

Eu(c):

$H_T - H_{298.15} = 5.81T + 0.99 \times 10^{-3}T^2 - 1,820$
(0.2 percent; 298°–1,100° K.);

$C_p = 5.81 + 1.98 \times 10^{-3}T$;

$\Delta H_{1100}(fusion) = 2,500.$

Eu(l):

$H_T - H_{298.15} = 8.00T - 530$ (0.1 percent;
1,100°–1,700° K.);

$C_p = 8.00.$

TABLE 268.—*Heat content and entropy of Eu(g)*

[Base, ideal gas at 298.15° K.; atomic wt., 152.0]

T, ° K.	$H_T-H_{298.15}$, cal./mole	$S_T-S_{298.15}$, cal./deg. mole	T, ° K.	$H_T-H_{298.15}$, cal./mole	$S_T-S_{298.15}$, cal./deg. mole
400	505	1.46	1,500	5,970	8.03
500	1,005	2.57	1,600	6,470	8.35
600	1,500	3.47	1,700	6,970	8.65
700	1,995	4.24	1,800	7,470	8.94
800	2,495	4.90	1,900	7,975	9.21
900	2,990	5.49	2,000	8,480	9.47
1,000	3,485	6.01	2,200	9,510	9.96
1,100	3,985	6.48	2,400	10,580	10.42
1,200	4,480	6.92	2,600	11,700	10.87
1,300	4,975	7.31	2,800	12,900	11.32
1,400	5,475	7.68	3,000	14,195	11.76

Eu(g):

$H_T - H_{298.15} = 4.98T - 1,485$ (0.3 percent;
298°–2,200° K.);

$C_p = 4.98.$

SULFATE

Eu$_2$(SO$_4$)$_3$·8H$_2$O(c):

$C_p = 146.3$ (298° K.).

FLUORINE AND ITS COMPOUNDS

ELEMENT

TABLE 269.—*Heat content and entropy of F(g)*

[Base, ideal gas at 298.15° K.; atomic wt., 19.00]

T, ° K.	$H_T-H_{298.15}$, cal./mole	$S_T-S_{298.15}$, cal./deg. mole	T, ° K.	$H_T-H_{298.15}$, cal./mole	$S_T-S_{298.15}$, cal./deg. mole
400	550	1.59	1,900	8,195	9.60
500	1,080	2.78	2,000	8,695	9.85
600	1,605	3.73	2,200	9,695	10.33
700	2,125	4.54	2,400	10,695	10.76
800	2,640	5.22	2,600	11,690	11.16
900	3,155	5.83	2,800	12,690	11.53
1,000	3,665	6.36	3,000	13,685	11.88
1,100	4,170	6.85	3,500	16,180	12.65
1,200	4,675	7.29	4,000	18,670	13.31
1,300	5,180	7.69	4,500	21,155	13.90
1,400	5,685	8.06	5,000	23,645	14.42
1,500	6,190	8.41	6,000	28,620	15.33
1,600	6,690	8.73	7,000	33,590	16.10
1,700	7,190	9.04	8,000	38,565	16.76
1,800	7,695	9.33			

F(g):

$H_T - H_{298.15} = 5.11T - 0.02 \times 10^{-3}T^2 - 0.31 \times 10^5 T^{-1}$
$- 1,418$ (0.4 percent; 298°–5,000° K.);

$C_p = 5.11 - 0.04 \times 10^{-3}T + 0.31 \times 10^5 T^{-2}.$

TABLE 270.—*Heat content and entropy of F$_2$(g)*

[Base, ideal gas at 298.15° K.; mol. wt., 38.00]

T, ° K.	$H_T-H_{298.15}$, cal./mole	$S_T-S_{298.15}$, cal./deg. mole	T, ° K.	$H_T-H_{298.15}$, cal./mole	$S_T-S_{298.15}$, cal./deg. mole
400	785	2.26	2,000	14,990	16.29
500	1,590	4.06	2,100	15,920	16.74
600	2,420	5.57	2,200	16,850	17.17
700	3,270	6.88	2,300	17,780	17.59
800	4,135	8.03	2,400	18,715	17.99
900	5,010	9.07	2,500	19,655	18.37
1,000	5,890	10.00	2,750	22,005	19.26
1,100	6,785	10.85	3,000	24,375	20.09
1,200	7,680	11.62	3,250	26,755	20.85
1,300	8,580	12.35	3,500	29,150	21.56
1,400	9,485	13.02	3,750	31,555	22.22
1,500	10,395	13.64	4,000	33,975	22.85
1,600	11,310	14.24	4,250	36,405	23.44
1,700	12,225	14.79	4,500	38,855	24.00
1,800	13,145	15.32	4,750	41,310	24.53
1,900	14,065	15.81	5,000	43,780	25.04

F$_2$(g):

$H_T - H_{298.15} = 8.26T + 0.30 \times 10^{-3}T^2 + 0.84 \times 10^5 T^{-1}$
$- 2,771$ (0.4 percent; 298°–2,500° K.);

$C_p = 8.26 + 0.60 \times 10^{-3}T - 0.84 \times 10^5 T^{-2}.$

OXIDE

TABLE 271.—*Heat content and entropy of F$_2$O(g)*

[Base, ideal gas at 298.15° K.; mol. wt., 54.00]

T, ° K.	$H_T-H_{298.15}$, cal./mole	$S_T-S_{298.15}$, cal./deg. mole	T, ° K.	$H_T-H_{298.15}$, cal./mole	$S_T-S_{298.15}$, cal./deg. mole
400	1,110	3.19	1,000	8,725	14.69
500	2,285	5.81	1,100	10,065	15.97
600	3,515	8.05	1,200	11,410	17.13
700	4,785	10.01	1,300	12,760	18.21
800	6,080	11.74	1,400	14,125	19.23
900	7,395	13.28	1,500	15,485	20.16

TABLE 1h7. (*Continued*)

$F_2O(g)$:

$H_T - H_{298.15} = 12.48T + 0.49 \times 10^{-3}T^2 + 2.16 \times 10^5 T^{-1}$
$- 4,489$ (0.4 percent; 298°–1,500° K.);
$C_p = 12.48 + 0.98 \times 10^{-3}T - 2.16 \times 10^5 T^{-2}.$

$Fr(g)$:

$H_T - H_{298.15} = 4.98T - 1,485$ (0.3 percent;
298°–2,000° K.);
$C_p = 4.98.$

CYANIDE

TABLE 272.—*Heat content and entropy of FCN(g)*

[Base, ideal gas at 298.15° K.; mol. wt., 45.02]

T, ° K.	$H_T - H_{298.15}$, cal./mole	$S_T - S_{298.15}$, cal./deg. mole	T, ° K.	$H_T - H_{298.15}$, cal./mole	$S_T - S_{298.15}$, cal./deg. mole
400	1,085	3.12	800	5,890	11.37
500	2,210	5.63	900	7,180	12.89
600	3,395	7.79	1,000	8,500	14.28
700	4,620	9.58			

$FCN(g)$:

$H_T - H_{298.15} = 10.41T + 1.58 \times 10^{-3}T^2 + 1.04 \times 10^5 T^{-1}$
$- 3,593$ (0.2 percent; 298°–1,000° K.);
$C_p = 10.41 + 3.16 \times 10^{-3}T - 1.04 \times 10^5 T^{-2}.$

FRANCIUM

ELEMENT

TABLE 273. *Heat content and entropy of Fr (c, l)*

[Base, crystals at 298.15° K.; atomic wt., 223]

T, ° K.	$H_T - H_{298.15}$, cal./mole	$S_T - S_{298.15}$, cal./deg. mole	T, ° K.	$H_T - H_{298.15}$, cal./mole	$S_T - S_{298.15}$, cal./deg. mole
300(c)	15	0.05	600	2,790	6.98
300(l)	510	1.70	700	3,550	8.16
400	1,270	3.90	800	4,310	9.17
500	2,030	5.60	900	5,070	10.07

$Fr(c)$:

$H_T - H_{298.15} = 7.60T - 2,266$ (298°–300° K.);
$C_p = 7.60;$
$\Delta H_{300}(fusion) = 495.$

$Fr(l)$:

$H_T - H_{298.15} = 7.60T - 1,770$ (0.1 percent;
300°–900° K.);
$C_p = 7.60.$

TABLE 274.—*Heat content and entropy of Fr(g)*

[Base, ideal gas at 298.15° K.; atomic wt., 223]

T, ° K.	$H_T - H_{298.15}$, cal./mole	$S_T - S_{298.15}$, cal./deg. mole	T, ° K.	$H_T - H_{298.15}$, cal./mole	$S_T - S_{298.15}$, cal./deg. mole
400	505	1.46	1,500	5,975	8.03
500	1,005	2.57	1,600	6,475	8.35
600	1,500	3.48	1,700	6,975	8.66
700	1,995	4.24	1,800	7,480	8.95
800	2,495	4.90	1,900	7,990	9.22
900	2,990	5.49	2,000	8,505	9.49
1,000	3,485	6.01	2,200	9,555	9.99
1,100	3,985	6.49	2,400	10,650	10.46
1,200	4,480	6.92	2,600	11,805	10.93
1,300	4,980	7.32	2,800	13,035	11.38
1,400	5,475	7.69	3,000	14,345	11.83

GADOLINUM AND ITS COMPOUNDS

ELEMENT

TABLE 275.—*Heat content and entropy of Gd (c, l)*

[Base, crystals at 298.15° K.; atomic wt., 157.26]

T, ° K.	$H_T - H_{298.15}$, cal./mole	$S_T - S_{298.15}$, cal./deg. mole	T, ° K.	$H_T - H_{298.15}$, cal./mole	$S_T - S_{298.15}$, cal./deg. mole
400	780	2.29	1,500	9,480	12.46
500	1,480	3.83	1,600(c)	10,370	13.03
600	2,200	5.15	1,600(l)	14,070	15.34
700	2,940	6.30	1,800	15,670	16.29
800	3,700	7.31	2,000	17,270	17.13
900	4,480	8.22	2,200	18,870	17.89
1,000	5,270	9.06	2,400	20,470	18.59
1,100	6,080	9.83	2,600	22,070	19.23
1,200	6,900	10.54	2,800	23,670	19.82
1,300	7,740	11.22	3,000	25,270	20.37
1,400	8,600	11.85			

$Gd(c)$:

$H_T - H_{298.15} = 6.60T + 0.72 \times 10^{-3}T^2 - 2,032$
(1.0 percent; 298°–1,600° K.);
$C_p = 6.60 + 1.44 \times 10^{-3}T;$
$\Delta H_{1600}(fusion) = 3,700.$

$Gd(l)$:

$H_T - H_{298.15} = 8.00T + 1,270$ (0.1 percent;
1,600°–3,000° K.);
$C_p = 8.00.$

TABLE 276.—*Heat content and entropy of Gd(g)*

[Base, ideal gas at 298.15° K.; atomic wt., 157.26]

T, ° K.	$H_T - H_{298.15}$, cal./mole	$S_T - S_{298.15}$, cal./deg. mole	T, ° K.	$H_T - H_{298.15}$, cal./mole	$S_T - S_{298.15}$, cal./deg. mole
400	670	1.92	1,500	7,295	10.01
500	1,315	3.37	1,600	7,885	10.39
600	1,955	4.53	1,700	8,480	10.75
700	2,580	5.50	1,800	9,085	11.10
800	3,195	6.32	1,900	9,705	11.43
900	3,795	7.03	2,000	10,330	11.75
1,000	4,390	7.65	2,200	11,620	12.37
1,100	4,975	8.21	2,400	12,960	12.95
1,200	5,555	8.72	2,600	14,350	13.51
1,300	6,135	9.18	2,800	15,790	14.04
1,400	6,715	9.61	3,000	17,270	14.55

$Gd(g)$:

$H_T - H_{298.15} = 6.13T - 0.40 \times 10^5 T^{-1} - 1,694$
(2 percent; 298°–3,000° K.);
$C_p = 6.13 + 0.40 \times 10^5 T^{-2}.$

SULFATE

$Gd_2(SO_4)_3 \cdot 8H_2O(c)$:

$C_p = 140.5$ (298° K.).

TABLE 1h7. (*Continued*)

GALLIUM AND ITS COMPOUNDS

ELEMENT

TABLE 277.—*Heat content and entropy of Ga(c, l)*

[Base, crystals at 298.15° K.; atomic wt., 69.72]

T, ° K.	$H_T-H_{298.15}$, cal./mole	$S_T-S_{298.15}$, cal./deg. mole	T, ° K.	$H_T-H_{298.15}$, cal./mole	$S_T-S_{298.15}$, cal./deg. mole
303(c)____	30	0.10	1,300____	7,995	14.19
303(l)____	1,365	4.51	1,400____	8,660	14.69
400_____	2,010	6.36	1,500____	9,325	15.15
500_____	2,675	7.84	1,600____	9,990	15.57
600_____	3,340	9.05	1,700____	10,655	15.98
700_____	4,005	10.08	1,800____	11,320	16.36
800_____	4,670	10.97	1,900____	11,985	16.72
900_____	5,335	11.75	2,000____	12,650	17.06
1,000____	6,000	12.45	2,200____	13,980	17.69
1,100____	6,665	13.08	2,400____	15,310	18.27
1,200____	7,330	13.66	2,500____	15,975	18.54

Ga(c):

$$H_T-H_{298.15}=6.24T-1,860 \text{ (1.0 percent;}$$
$$298°-303° \text{ K.);}$$
$$C_p=6.24;$$
$$\Delta H_{303} \text{ (fusion)}=1,335.$$

Ga(l):

$$H_T-H_{298.15}=6.65T-650 \text{ (0.1 percent;}$$
$$303°-2,500° \text{ K.);}$$
$$C_p=6.65.$$

TABLE 278.—*Heat content and entropy of Ga(g)*

[Base, ideal gas at 298.15° K.; atomic wt., 69.72]

T, ° K.	$H_T-H_{298.15}$, cal./mole	$S_T-S_{298.15}$, cal./deg. mole	T, ° K.	$H_T-H_{298.15}$, cal./mole	$S_T-S_{298.15}$, cal./deg. mole
400_____	640	1.85	1,900____	9,095	10.94
500_____	1,290	3.29	2,000____	9,610	11.20
600_____	1,925	4.45	2,200____	10,640	11.69
700_____	2,545	5.41	2,400____	11,660	12.13
800_____	3,145	6.21	2,600____	12,675	12.54
900_____	3,730	6.90	2,800____	13,685	12.92
1,000____	4,300	7.50	3,000____	14,700	13.26
1,100____	4,855	8.03	3,500____	17,215	14.04
1,200____	5,405	8.51	4,000____	19,730	14.71
1,300____	5,945	8.94	4,500____	22,245	15.30
1,400____	6,480	9.34	5,000____	24,780	15.84
1,500____	7,010	9.70	6,000____	29,975	16.78
1,600____	7,535	10.04	7,000____	35,575	17.65
1,700____	8,060	10.36	8,000____	41,915	18.49
1,800____	8,580	10.66			

Ga(g):

$$H_T-H_{298.15}=7.08T-0.58\times10^{-3}T^2+0.60\times10^5T^{-1}$$
$$-2,261 \text{ (1.1 percent; } 298°-2,200° \text{ K.);}$$
$$C_p=7.08-1.16\times10^{-3}T-0.60\times10^5T^{-2}.$$

OXIDES

TABLE 279.—*Heat content and entropy of GaO(g)*

[Base, ideal gas at 298.15° K.; mol. wt., 85.72]

T, ° K.	$H_T-H_{298.15}$, cal./mole	$S_T-S_{298.15}$, cal./deg. mole	T, ° K.	$H_T-H_{298.15}$, cal./mole	$S_T-S_{298.15}$, cal./deg. mole
400_____	805	2.32	1,000____	5,925	10.09
500_____	1,625	4.15	1,200____	7,680	11.69
600_____	2,465	5.68	1,400____	9,445	13.05
700_____	3,320	7.00	1,600____	11,220	14.24
800_____	4,180	8.15	1,800____	12,995	15.28
900_____	5,050	9.17	2,000____	14,770	16.22

GaO(g):

$$H_T-H_{298.15}=8.56T+0.12\times10^{-3}T^2+0.84\times10^5T^{-1}$$
$$-2,845 \text{ (0.2 percent; } 298°-2,000° \text{ K.);}$$
$$C_p=8.56+0.24\times10^{-3}T-0.84\times10^5T^{-2}.$$

Ga₂O₃(c):

$$C_p=22.02 \text{ (298° K.).}$$

BROMIDE

TABLE 280.—*Heat content and entropy of GaBr(g)*

[Base, ideal gas at 298.15° K.; mol. wt., 149.64]

T, ° K.	$H_T-H_{298.15}$, cal./mole	$S_T-S_{298.15}$, cal./deg. mole	T, ° K.	$H_T-H_{298.15}$, cal./mole	$S_T-S_{298.15}$, cal./deg. mole
400_____	890	2.56	1,000____	6,225	10.70
500_____	1,775	4.54	1,200____	8,005	12.33
600_____	2,660	6.15	1,400____	9,790	13.70
700_____	3,550	7.52	1,600____	11,580	14.90
800_____	4,440	8.71	1,800____	13,365	15.95
900_____	5,330	9.76	2,000____	15,155	16.89

GaBr(g):

$$H_T-H_{298.15}=8.94T+0.22\times10^5T^{-1}-2,739 \text{ (0.1 percent;}$$
$$298°-2,000° \text{ K.);}$$
$$C_p=8.94-0.22\times10^5T^{-2}.$$

CHLORIDE

TABLE 281.—*Heat content and entropy of GaCl(g)*

[Base, ideal gas at 298.15° K.; mol. wt., 105.18]

T, ° K.	$H_T-H_{298.15}$, cal./mole	$S_T-S_{298.15}$, cal./deg. mole	T, ° K.	$H_T-H_{298.15}$, cal./mole	$S_T-S_{298.15}$, cal./deg. mole
400_____	875	2.52	1,000____	6,175	10.60
500_____	1,750	4.47	1,200____	7,955	12.23
600_____	2,630	6.08	1,400____	9,740	13.60
700_____	3,510	7.44	1,600____	11,525	14.79
800_____	4,400	8.62	1,800____	13,310	15.84
900_____	5,285	9.66	2,000____	15,095	16.79

GaCl(g):

$$H_T-H_{298.15}=8.93T+0.38\times10^5T^{-1}-2,790 \text{ (0.1 percent;}$$
$$298°-2,000° \text{ K.);}$$
$$C_p=8.93-0.38\times10^5T^{-2}.$$

IODIDE

TABLE 282.—*Heat content and entropy of GaI(g)*

[Base, ideal gas at 298.15° K.; mol. wt., 196.63]

T, ° K.	$H_T-H_{298.15}$, cal./mole	$S_T-S_{298.15}$, cal./deg. mole	T, ° K.	$H_T-H_{298.15}$, cal./mole	$S_T-S_{298.15}$, cal./deg. mole
400_____	895	2.58	1,000____	6,240	10.74
500_____	1,785	4.57	1,200____	8,025	12.37
600_____	2,675	6.19	1,400____	9,810	13.74
700_____	3,565	7.56	1,600____	11,600	14.94
800_____	4,455	8.75	1,800____	13,385	15.99
900_____	5,345	9.80	2,000____	15,175	16.94

GaI(g):

$$H_T-H_{298.15}=8.94T+0.15\times10^5T^{-1}-2,716 \text{ (0.1 percent;}$$
$$298°-2,000° \text{ K.);}$$
$$C_p=8.94-0.15\times10^5T^{-2}.$$

SULFATE

Ga₂(SO₄)₃(c):

$$\overline{C}_p=62.4 \text{ (273°-373° K.).}$$

TABLE 1h7. (*Continued*)

GERMANIUM AND ITS COMPOUNDS

ELEMENT

TABLE 283.—*Heat content and entropy of Ge(c, l)*

[Base, crystals at 298.15° K.; atomic wt., 72.60]

T, ° K.	$H_T-H_{298.15}$, cal./mole	$S_T-S_{298.15}$, cal./deg. mole	T, ° K.	$H_T-H_{298.15}$, cal./mole	$S_T-S_{298.15}$, cal./deg. mole
400_____	590	1.69	1,400_____	14,810	16.13
500_____	1,195	3.04	1,500_____	15,510	16.61
600_____	1,820	4.18	1,600_____	16,210	17.07
700_____	2,460	5.17	1,700_____	16,910	17.49
800_____	3,110	6.03	1,800_____	17,610	17.89
900_____	3,770	6.81	1,900_____	18,310	18.27
1,000_____	4,440	7.52	2,000_____	19,010	18.63
1,100_____	5,120	8.17	2,200_____	20,410	19.30
1,200_____	5,810	8.77	2,400_____	21,810	19.90
1,210.4(c)	5,880	8.83	2,600_____	23,210	20.46
1,210.4(l)	13,480	15.11	2,800_____	24,610	20.98
1,300_____	14,110	15.61	3,000_____	26,010	21.47

Ge(c):

$$H_T-H_{298.15}=5.98T+0.41\times10^{-3}T^2+0.56\times10^5T^{-1}$$
$$-2,007 \text{ (0.1 percent; } 298°-1,210.4° \text{ K.);}$$
$$C_p=5.98+0.82\times10^{-3}T-0.56\times10^5T^{-2};$$
$$\Delta H_{1,210.4}(fusion)=7,600.$$

Ge(l):

$$H_T-H_{298.15}=7.00T+5,010 \text{ (0.1 percent;}$$
$$1,210.4°-3,000° \text{ K.);}$$
$$C_p\doteq7.00.$$

TABLE 284.—*Heat content and entropy of Ge(g)*

[Base, ideal gas at 298.15° K.; atomic wt., 72.60]

T, ° K.	$H_T-H_{298.15}$, cal./mole	$S_T-S_{298.15}$, cal./deg. mole	T, ° K.	$H_T-H_{298.15}$, cal./mole	$S_T-S_{298.15}$, cal./deg. mole
400_____	755	2.18	1,900_____	9,910	12.12
500_____	1,490	3.82	2,000_____	10,460	12.40
600_____	2,205	5.12	2,200_____	11,570	12.93
700_____	2,890	6.18	2,400_____	12,675	13.41
800_____	3,550	7.06	2,600_____	13,785	13.86
900_____	4,185	7.81	2,800_____	14,900	14.27
1,000_____	4,800	8.46	3,000_____	16,015	14.65
1,100_____	5,395	9.03	3,500_____	18,815	15.52
1,200_____	5,985	9.54	4,000_____	21,615	16.26
1,300_____	6,560	10.00	4,500_____	24,410	16.92
1,400_____	7,125	10.42	5,000_____	27,200	17.51
1,500_____	7,690	10.81	6,000_____	32,755	18.52
1,600_____	8,245	11.17	7,000_____	38,330	19.38
1,700_____	8,800	11.50	8,000_____	44,045	20.15
1,800_____	9,355	11.82			

Ge(g):

$$H_T-H_{298.15}=7.77T-0.71\times10^{-3}T^2$$
$$-2,254 \text{ (1.3 percent; } 298°-2,300° \text{ K.);}$$
$$C_p=7.77-1.42\times10^{-3}T.$$

OXIDES

TABLE 285.—*Heat content and entropy of GeO(g)*

[Base, ideal gas at 298.15° K.; mol. wt., 88.60]

T, ° K.	$H_T-H_{298.15}$, cal./mole	$S_T-S_{298.15}$, cal./deg. mole	T. ° K.	$H_T-H_{298.15}$, cal./mole	$S_T-S_{298.15}$, cal./deg. mole
400_____	770	2.22	1,000_____	5,765	9.78
500_____	1,560	3.98	1,200_____	7,500	11.36
600_____	2,370	5.45	1,400_____	9,255	12.71
700_____	3,200	6.73	1,600_____	11,015	13.89
800_____	4,050	7.87	1,800_____	12,780	14.93
900_____	4,900	8.87	2,000_____	14,550	15.86

GeO(g):

$$H_T-H_{298.15}=8.18T+0.23\times10^{-3}T^2+0.85\times10^5T^{-1}$$
$$-2,743 \text{ (0.5 percent; } 298°-2,000° \text{ K.);}$$
$$C_p=8.18+0.46\times10^{-3}T-0.85\times10^5T^{-2}.$$

GeO$_2$(c, *soluble form*):

$$C_p=10.11+7.84\times10^{-3}T(estimated) \text{ (298°-1,389° K.).}$$

SULFIDE

TABLE 286.—*Heat content and entropy of GeS(g)*

[Base, ideal gas at 298.15° K.; mol. wts., 104.67]

T, ° K.	$H_T-H_{298.15}$, cal./mole	$S_T-S_{298.15}$, cal./deg. mole	T, ° K.	$H_T-H_{298.15}$, cal./mole	$S_T-S_{298.15}$, cal./deg. mole
400_____	835	2.41	1,000_____	6,050	10.34
500_____	1,685	4.30	1,200_____	7,820	11.96
600_____	2,545	5.87	1,400_____	9,595	13.32
700_____	3,415	7.21	1,600_____	11,375	14.51
800_____	4,290	8.38	1,800_____	13,155	15.56
900_____	5,170	9.42	2,000_____	14,940	16.50

GeS(g):

$$H_T-H_{298.15}=8.78T+0.05\times10^{-3}T^2+0.68\times10^5T^{-1}$$
$$-2,849 \text{ (0.2 percent; } 298°-2,000° \text{ K.);}$$
$$C_p=8.78+0.10\times10^{-3}T-0.68\times10^5T^{-2}.$$

SELENIDE

TABLE 287.—*Heat content and entropy of GeSe(g)*

[Base, ideal gas at 298.15° K.; mol. wt., 151.56]

T, ° K.	$H_T-H_{298.15}$, cal./mole	$S_T-S_{298.15}$, cal./deg. mole	T, ° K.	$H_T-H_{298.15}$, cal./mole	$S_T-S_{298.15}$, cal./deg. mole
400_____	870	2.51	1,000_____	6,155	10.56
500_____	1,740	4.45	1,200_____	7,935	12.18
600_____	2,615	6.04	1,400_____	9,715	13.56
700_____	3,495	7.40	1,600_____	11,550	14.75
800_____	4,380	8.58	1,800_____	13,285	15.80
900_____	5,265	9.62	2,000_____	15,070	16.74

GeSe(g):

$$H_T-H_{298.15}=8.88T+0.02\times10^{-3}T^2+0.42\times10^5T^{-1}$$
$$-2,790 \text{ (0.1 percent; } 298°-2,000° \text{ K.);}$$
$$C_p=8.88+0.04\times10^{-3}T-0.42\times10^5T^{-2}.$$

TELLURIDE

TABLE 288.—*Heat content and entropy of GeTe(g)*

[Base, ideal gas at 298.15° K.; mol. wt., 200.21]

T, ° K.	$H_T-H_{298.15}$, cal./mole	$S_T-S_{298.15}$, cal./deg. mole	T, ° K.	$H_T-H_{298.15}$, cal./mole	$S_T-S_{298.15}$, cal./deg. mole
400_____	885	2.55	1,000_____	6,195	10.65
500_____	1,760	4.50	1,200_____	7,980	12.28
600_____	2,645	6.12	1,400_____	9,765	13.65
700_____	3,530	7.48	1,600_____	11,550	14.84
800_____	4,415	8.66	1,800_____	13,335	15.90
900_____	5,305	9.71	2,000_____	15,120	18.84

GeTe(g):

$$H_T-H_{298.15}=8.93T+0.30\times10^5T^{-1}-2,763$$
$$\text{(0.1 percent; } 298°-2,000° \text{ K.);}$$
$$C_p=8.93-0.30\times10^5T^{-2}.$$

TABLE 1h7. (*Continued*)

HYDRIDES

TABLE 289.—*Heat content and entropy of* $GeH(g)$

[Base, ideal gas at 298.15° K.; mol. wt., 73.61]

T, ° K.	$H_T-H_{298.15}$, cal./mole	$S_T-S_{298.15}$, cal./deg. mole	T, ° K.	$H_T-H_{298.15}$, cal./mole	$S_T-S_{298.15}$, cal./deg. mole
400	715	2.06	1,000	5,265	8.94
500	1,430	3.66	1,200	6,905	10.43
600	2,160	4.99	1,400	8,575	11.72
700	2,910	6.14	1,600	10,275	12.85
800	3,680	7.17	1,800	11,990	13.86
900	4,465	8.10	2,000	13,715	14.77

$$GeH(g):$$
$$H_T-H_{298.15}=6.75T+0.60\times10^{-3}T^2+0.12\times10^5T^{-1}$$
$$-2,106 \ (0.5 \text{ percent}; 298°-2,000° \text{ K.});$$
$$C_p=6.75+1.20\times10^{-3}T-0.12\times10^5T^{-2}.$$

TABLE 290.—*Heat content and entropy of* $GeH_4(g)$

[Base, ideal gas at 298.15° K.; mol. wt. 76.63]

T, ° K.	$H_T-H_{298.15}$, cal./mole	$S_T-S_{298.15}$, cal./deg. mole	T, ° K.	$H_T-H_{298.15}$, cal./mole	$S_T-S_{298.15}$, cal./deg. mole
400	1,190	3.42	1,000	11,515	18.64
500	2,555	6.46	1,200	15,755	22.50
600	4,100	9.28	1,400	20,205	25.93
700	5,775	11.85	1,600	24,795	28.99
800	7,590	14.27	1,800	29,505	31.76
900	9,505	16.52	2,000	34,290	34.28

$$GeH_4(g):$$
$$H_T-H_{298.15}=10.17T+5.33\times10^{-3}T^2+2.45\times10^5T^{-1}$$
$$-4,328 \ (1.0 \text{ percent}; 298°-1,400° \text{ K.});$$
$$C_p=10.17+10.66\times10^{-3}T-2.45\times10^5T^{-2}.$$

BROMIDES

TABLE 291.—*Heat content and entropy of* $GeBr(g)$

[Base, ideal gas at 298.15° K.; mol. wt., 152.52]

T, ° K.	$H_T-H_{298.15}$, cal./mole	$S_T-S_{298.15}$, cal./deg. mole	T, ° K.	$H_T-H_{298.15}$, cal./mole	$S_T-S_{298.15}$, cal./deg. mole
400	925	2.66	1,000	6,725	11.50
500	1,870	4.77	1,200	8,640	13.25
600	2,835	6.53	1,400	10,535	14.71
700	3,810	8.03	1,600	12,410	15.96
800	4,785	9.34	1,800	14,270	17.05
900	5,755	10.48	2,000	16,125	18.03

$$GeBr(g):$$
$$H_T-H_{298.15}=10.28T-0.27\times10^{-3}T^2+1.11\times10^5T^{-1}$$
$$-3,413 \ (0.3 \text{ percent}; 298°-2,000° \text{ K.});$$
$$C_p=10.28-0.54\times10^{-3}T-1.11\times10^5T^{-2}.$$

TABLE 292.—*Heat content and entropy of* $GeBr_4(g)$

[Base, ideal gas at 298.15° K.; mol. wt., 392.26]

T, ° K.	$H_T-H_{298.15}$, cal./mole	$S_T-S_{298.15}$, cal./deg. mole	T, ° K.	$H_T-H_{298.15}$, cal./mole	$S_T-S_{298.15}$, cal./deg. mole
400	2,515	7.26	800	12,665	24.82
500	5,030	12.87	900	15,230	27.84
600	7,560	17.48	1,000	17,800	30.55
700	10,110	21.41			

$$GeBr_4(g):$$
$$H_T-H_{298.15}=25.66T+0.08\times10^{-3}T^2+1.22\times10^5T^{-1}$$
$$-8,067 \ (0.1 \text{ percent}; 298°-1,000° \text{ K.});$$
$$C_p=25.66+0.16\times10^{-3}T-1.22\times10^5T^{-2}.$$

CHLORIDES

TABLE 293.—*Heat content and entropy of* $GeCl(g)$

[Base, ideal gas at 298.15° K.; mol. wt., 108.06]

T, ° K.	$H_T-H_{298.15}$, cal./mole	$S_T-S_{298.15}$, cal./deg. mole	T, ° K.	$H_T-H_{298.15}$, cal./mole	$S_T-S_{298.15}$, cal./deg. mole
400	925	2.66	1,000	6,680	11.45
500	1,870	4.77	1,200	8,570	13.17
600	2,835	6.53	1,400	10,440	14.61
700	3,800	8.02	1,600	12,295	15.85
800	4,765	9.31	1,800	14,135	16.93
900	5,725	10.44	2,000	15,965	17.90

$$GeCl(g):$$
$$H_T-H_{298.15}=10.29T-0.31\times10^{-3}T^2+1.16\times10^5T^{-1}$$
$$-3,429 \ (0.2 \text{ percent}; 298°-2,000° \text{ K.});$$
$$C_p=10.29-0.62\times10^{-3}T-1.16\times10^5T^{-2}.$$

TABLE 294.—*Heat content and entropy of* $GeCl_4(g)$

[Base, ideal gas at 298.15° K.; mol. wt., 214.43]

T, ° K.	$H_T-H_{298.15}$, cal./mole	$S_T-S_{298.15}$, cal./deg. mole	T, ° K.	$H_T-H_{298.15}$, cal./mole	$S_T-S_{298.15}$, cal./deg. mole
400	2,405	6.93	800	12,385	24.17
500	4,850	12.38	900	14,925	27.16
600	7,335	16.91	1,000	17,470	29.85
700	9,850	20.79			

$$GeCl_4(g):$$
$$H_T-H_{298.15}=25.46T+0.16\times10^{-3}T^2+2.31\times10^5T^{-1}$$
$$-8,380 \ (0.1 \text{ percent}; 298°-1,000° \text{ K.});$$
$$C_p=25.46+0.32\times10^{-3}T-2.31\times10^5T^{-2}.$$

FLUORIDES

TABLE 295.—*Heat content and entropy of* $GeF(g)$

[Base, ideal gas at 298.15° K.; mol. wt., 91.60]

T, ° K.	$H_T-H_{298.15}$, cal./mole	$S_T-S_{298.15}$, cal./deg. mole	T, ° K.	$H_T-H_{298.15}$, cal./mole	$S_T-S_{298.15}$, cal./deg. mole
400	880	2.53	1,000	6,520	11.13
500	1,800	4.59	1,200	8,390	12.83
600	2,740	6.30	1,400	10,245	14.26
700	3,685	7.76	1,600	12,090	15.50
800	4,635	9.03	1,800	13,920	16.58
900	5,580	10.14	2,000	15,745	17.54

$$GeF(g):$$
$$H_T-H_{298.15}=10.17T-0.29\times10^{-3}T^2+1.50\times10^5T^{-1}$$
$$-3,510 \ (0.2 \text{ percent}; 298°-2,000° \text{ K.});$$
$$C_p=10.17-0.58\times10^{-3}T-1.50\times10^5T^{-2}.$$

TABLE 1h7. (Continued)

TABLE 296.—*Heat content and entropy of* $GeF_4(g)$

[Base, ideal gas at 298.15° K.; mol. wt., 148.60]

T, ° K.	$H_T-H_{298.15}$, cal./mole	$S_T-S_{298.15}$, cal./deg. mole	T, ° K.	$H_T-H_{298.15}$, cal./mole	$S_T-S_{298.15}$, cal./deg. mole
400	2,100	6.04	800	11,465	22.15
500	4,325	11.00	900	13,925	25.03
600	6,645	15.22	1,000	16,405	27.65
700	9,035	18.90			

$GeF_4(g)$:

$$H_T-H_{298.15}=22.72T+1.43\times10^{-3}T^2+3.56\times10^5T^{-1}$$
$$-8,095 \ (0.2 \text{ percent}; 298°-1,000° \text{ K.});$$
$$C_p=22.72+2.86\times10^{-3}T-3.56\times10^5T^{-2}.$$

IODIDE

TABLE 297.—*Heat content and entropy of* $GeI_4(g)$

[Base, ideal gas at 298.15° K.; mol. wt., 580.24]

T, ° K.	$H_T-H_{298.15}$, cal./mole	$S_T-S_{298.15}$, cal./deg. mole	T, ° K.	$H_T-H_{298.15}$, cal./mole	$S_T-S_{298.15}$, cal./deg. mole
400	2,560	7.38	800	12,785	25.09
500	5,100	13.05	900	15,355	28.11
600	7,655	17.71	1,000	17,930	30.83
700	10,215	21.65			

$GeI_4(g)$:

$$H_T-H_{298.15}=25.79T+0.76\times10^5T^{-1}-7,944$$
$$(0.1 \text{ percent}; 298°-1,000° \text{ K.});$$
$$C_p=25.79-0.76\times10^5T^{-2}.$$

GOLD AND ITS COMPOUNDS

ELEMENT

TABLE 298.—*Heat content and entropy of* $Au(c, l)$

[Base, crystals at 298.15° K.; atomic wt., 197.0]

T, ° K.	$H_T-H_{298.15}$, cal./mole	$S_T-S_{298.15}$, cal./deg. mole	T, ° K.,	$H_T-H_{298.15}$, cal./mole	$S_T-S_{298.15}$, cal./deg. mole
400	625	1.80	1,400	10,330	12.32
500	1,245	3.19	1,500	11,030	12.80
600	1,880	4.34	1,600	11,730	13.25
700	2,530	5.35	1,700	12,430	13.67
800	3,180	6.21	1,800	13,130	14.07
900	3,850	7.00	1,900	13,830	14.45
1,000	4,530	7.72	2,000	14,530	14.81
1,100	5,220	8.38	2,200	15,930	15.48
1,200	5,930	9.00	2,400	17,330	16.09
1,300	6,660	9.58	2,600	18,730	16.65
1,336(c)	6,925	9.78	2,800	20,130	17.17
1,336(l)	9,880	11.99	3,000	21,530	17.66

$Au(c)$:

$$H_T-H_{298.15}=5.66T+0.62\times10^{-3}T^2-1,743$$
$$(0.3 \text{ percent}; 298°-1,336° \text{ K.});$$
$$C_p=5.66+1.24\times10^{-3}T;$$
$$\Delta H_{1336}(fusion)=2,955.$$

$Au(l)$:

$$H_T-H_{298.15}=7.00T+530 \ (0.1\text{percent};1,336°-3,000° \text{ K.});$$
$$C_p=7.00.$$

TABLE 299.—*Heat content and entropy of* $Au(g)$

[Base, ideal gas at 298.15° K.; atomic wt., 197.0]

T, ° K.	$H_T-H_{298.15}$, cal./mole	$S_T-S_{298.15}$, cal./deg. mole	T, ° K.	$H_T-H_{298.15}$, cal./mole	$S_T-S_{298.15}$, cal./deg. mole
400	505	1.46	1,900	8,035	9.25
500	1,005	2.57	2,000	8,565	9.52
600	1,500	3.47	2,200	9,645	10.04
700	1,995	4.24	2,400	10,765	10.52
800	2,495	4.90	2,600	11,925	10.99
900	2,990	5.49	2,800	13,125	11.43
1,000	3,490	6.01	3,000	14,365	11.86
1,100	3,985	6.49	3,500	17,630	12.86
1,200	4,485	6.92	4,000	21,065	13.78
1,300	4,980	7.32	4,500	24,610	14.61
1,400	5,480	7.69	5,000	28,200	15.37
1,500	5,985	8.04	6,000	35,425	16.69
1,600	6,490	8.36	7,000	42,640	17.80
1,700	7,000	8.67	8,000	49,965	18.78
1,800	7,515	8.97			

$Au(g)$:

$$H_T-H_{298.15}=4.46T+0.24\times10^{-3}T^2-0.34\times10^5T^{-1}$$
$$-1,237 \ (1.0 \text{ percent}; 298°-3,000° \text{ K.});$$
$$C_p=4.46+0.48\times10^{-3}T+0.34\times10^5T^{-2}.$$

HYDRIDES

TABLE 300.—*Heat content and entropy of* $AuH(g)$

[Base, ideal gas at 298.15° K.; mol. wt., 198.01]

T, ° K.	$H_T-H_{298.15}$, cal./mole	$S_T-S_{298.15}$, cal./deg. mole	T, ° K.	$H_T-H_{298.15}$, cal./mole	$S_T-S_{298.15}$, cal./deg. mole
400	710	2.05	1,000	5,155	8.77
500	1,415	3.62	1,200	6,750	10.23
600	2,130	4.92	1,400	8,385	11.49
700	2,860	6.05	1,600	10,050	12.60
800	3,610	7.05	1,800	11,740	13.59
900	4,375	7.95	2,000	13,450	14.49

$AuH(g)$:

$$H_T-H_{298.15}=6.48T+0.63\times10^{-3}T^2-0.10\times10^5T^{-1}$$
$$-1,954 \ (0.4 \text{ percent}; 298°-2,000° \text{ K.});$$
$$C_p=6.48+1.26\times10^{-3}T+0.10\times10^5T^{-2}.$$

TABLE 301.—*Heat content and entropy of* $AuD(g)$

[Base, ideal gas at 298.15° K.; mol. wt., 199.01]

T, ° K.	$H_T-H_{298.15}$, cal./mole	$S_T-S_{298.15}$, cal./deg. mole	T, ° K.	$H_T-H_{298.15}$, cal./mole	$S_T-S_{298.15}$, cal./deg. mole
400	720	2.07	1,000	5,390	9.13
500	1,450	3.70	1,200	7,065	10.66
600	2,200	5.07	1,400	8,760	11.96
700	2,970	6.26	1,600	10,480	13.11
800	3,760	7.31	1,800	12,215	14.13
900	4,570	8.26	2,000	13,960	15.05

$AuD(g)$:

$$H_T-H_{298.15}=7.09T+0.53\times10^{-3}T^2+0.35\times10^5T^{-1}$$
$$-2,278 \ (0.7 \text{ percent}; 298°-2,000° \text{ K.});$$
$$C_p=7.09+1.06\times10^{-3}T-0.35\times10^5T^{-2}.$$

TABLE 1h7. (*Continued*)

CHLORIDE

TABLE 302.—*Heat content and entropy of AuCl(g)*

[Base, ideal gas at 298.15° K.; mol. wt., 232.46]

T, ° K.	$H_T-H_{298.15}$, cal./mole	$S_T-S_{298.15}$, cal./deg. mole	T. ° K.	$H_T-H_{298.15}$, cal./mole	$S_T-S_{298.15}$, cal./deg. mole
400	875	2.52	1,000	6,165	10.58
500	1,745	4.46	1,200	7,950	12.21
600	2,625	6.07	1,400	9,730	13.58
700	3,505	7.42	1,600	11,515	14.78
800	4,390	8.60	1,800	13,300	15.83
900	5,280	9.65	2,000	15,085	16.77

AuCl(*g*):

$$H_T-H_{298.15}=8.93T+0.41\times10^5T^{-1}$$
$$-2,800 \ (0.1 \text{ percent; } 298°-2,000° \text{ K.});$$
$$C_p=8.93-0.41\times10^5T^{-2}.$$

GOLD-CADMIUM

TABLE 303.—*Heat content and entropy of AuCd(c, l)*

[Base, crystals at 298.15° K.; mol. wt., 309.41]

T, ° K.	$H_T-H_{298.15}$, cal./mole	$S_T-S_{298.15}$, cal./deg. mole	T, ° K.	$H_T-H_{298.15}$, cal./mole	$S_T-S_{298.15}$, cal./deg. mole
400	1,170	3.37	900(c)	8,350	14.70
500	2,380	6.06	900(l)	12,620	19.44
600	3,670	8.41	1,000	14,060	20.96
700	5,120	10.65	1,100	15,500	22.33
800	6,700	12.75			

AuCd(*c*):

$$H_T-H_{298.15}=7.66T+5.18\times10^{-3}T^2$$
$$-2,744 \ (0.7 \text{ percent; } 298°-900° \text{ K.});$$
$$C_p=7.66+10.36\times10^{-3}T;$$
$$\Delta H_{900}(fusion)=4,270.$$

AuCd(*l*):
$$H_T-H_{298.15}=14.40T-340 \ (0.1 \text{ percent; } 900°-1,100° \text{ K.});$$
$$C_p=14.40.$$

GOLD-LEAD

TABLE 304.—*Heat content and entropy of AuPb₂(c, l)*

[Base, crystals at 298.15° K.; mol. wt., 611.42]

T, ° K.	$H_T-H_{298.15}$, cal./mole	$S_T-S_{298.15}$, cal./deg. mole	T, ° K.	$H_T-H_{298.15}$, cal./mole	$S_T-S_{298.15}$, cal./deg. mole
400	2,140	6.16	600	12,490	26.38
500	4,420	11.24	700	14,800	29.94
527(c)	5,070	12.51	800	17,110	33.02
527(l)	10,800	23.38			

AuPb₂(*c*):

$$H_T-H_{298.15}=14.80T+8.90\times10^{-3}T^2$$
$$-5,204 \ (0.1 \text{ percent; } 298°-527° \text{ K.});$$
$$C_p=14.80+17.80\times10^{-3}T;$$
$$\Delta H_{527}(fusion)=5,730.$$

AuPb₂(*l*):
$$H_T-H_{298.15}=23.10T-1,370 \ (0.1 \text{ percent; } 527°-800° \text{ K.});$$
$$C_p=23.10.$$

GOLD-ANTIMONY

TABLE 305.—*Heat content and entropy of AuSb₂(c)*

[Base, α-crystals at 298.15° K.; mol. wt., 440.52]

T, ° K.	$H_T-H_{298.15}$, cal./mole	$S_T-S_{298.15}$, cal./deg. mole	T, ° K.	$H_T-H_{298.15}$, cal./mole	$S_T-S_{298.15}$, cal./deg. mole
400	1,920	5.54	628(α)	6,360	14.30
500	3,830	9.80	628(β,γ)	6,460	14.46
600	5,790	13.37	700	8,120	16.97

AuSb₂(α):

$$H_T-H_{298.15}=17.12T+2.32\times10^{-3}T^2$$
$$-5,311 \ (0.3 \text{ percent; } 298°-628° \text{ K.});$$
$$C_p=17.12+4.64\times10^{-3}T;$$
$$\Delta H_{628}(transition)=100.$$

AuSb₂(β,γ):

$$H_T-H_{298.15}=11.47T+8.78\times10^{-3}T^2$$
$$-4,210 \ (0.1 \text{ percent; } 628°-700° \text{ K.});$$
$$C_p=11.47+17.56\times10^{-3}T.$$

GOLD-TIN

TABLE 306.—*Heat content and entropy of AuSn(c, l)*

[Base, crystals at 298.15° K.; mol. wt., 315.70]

T, ° K.	$H_T-H_{298.15}$, cal./mole	$S_T-S_{298.15}$, cal./deg. mole	T, ° K.	$H_T-H_{298.15}$, cal./mole	$S_T-S_{298.15}$, cal./deg. mole
400	1,260	3.63	691(l)	11,660	20.42
500	2,590	6.59	700	11,790	20.61
600	4,090	9.32	800	13,250	22.56
691(c)	5,490	11.49	900	14,710	24.28

AuSn(*c*):

$$H_T-H_{298.15}=8.53T+5.50\times10^{-3}T^2-3,032$$
$$(0.4 \text{ percent; } 298°-691° \text{ K.});$$
$$C_p=8.53+11.00\times10^{-3}T;$$
$$\Delta H_{691}(fusion)=6,170.$$

AuSn(*l*):
$$H_T-H_{298.15}=14.60T+1,570 \ (0.1 \text{ percent;}$$
$$691°-900° \text{ K.});$$
$$C_p=14.60.$$

GOLD-ZINC

TABLE 307.—*Heat content and entropy of AuZn(c, l)*

[Base, crystals at 298.15° K.; mol. wt., 262.38]

T, ° K.	$H_T-H_{298.15}$, cal./mole	$S_T-S_{298.15}$, cal./deg. mole	T, ° K.	$H_T-H_{298.15}$, cal./mole	$S_T-S_{298.15}$, cal./deg. mole
400	1,290	3.72	1,000	9,700	16.42
500	2,610	6.67	1,033(c)	10,200	16.91
600	3,960	9.13	1,033(l)	15,990	22.52
700	5,340	11.25	1,100	16,900	23.37
800	6,750	13.13	1,200	18,260	24.56
900	8,200	14.84			

TABLE 1h7. (*Continued*)

AuZn(*c*):

$$H_T - H_{298.15} = 11.51T + 1.78 \times 10^{-3}T^2 - 3,590$$

(0.2 percent; 298°–1,033° K.);

$$C_p = 11.51 + 3.56 \times 10^{-3}T;$$

$$\Delta H_{1033}(fusion) = 5,790.$$

AuZn(*l*):

$$H_T - H_{298.15} = 13.60T + 1,940 \text{ (0.1 percent;}$$

1,033°–1,200° K.);

$$C_p = 13.60.$$

HAFNIUM AND ITS COMPOUNDS

ELEMENT

TABLE 308.—*Heat content and entropy of Hf(c, l)*

[Base, crystals at 298.15° K.; atomic wt., 178.50]

$T,$ ° K.	$H_T - H_{298.15}$, cal./mole	$S_T - S_{298.15}$, cal./deg. mole	$T,$ ° K.	$H_T - H_{298.15}$, cal./mole	$S_T - S_{298.15}$, cal./deg. mole
400	630	1.82	1,900	11,310	12.56
500	1,260	3.22	2,000	12,120	12.97
600	1,900	4.38	2,100	12,940	13.37
700	2,550	5.38	2,200	13,770	13.76
800	3,210	6.27	2,300	14,610	14.13
900	3,890	7.07	2,400	15,460	14.50
1,000	4,580	7.79	2,495(c)	16,280	14.83
1,100	5,280	8.46	2,495(l)	22,070	17.15
1,200	5,990	9.08	2,500	22,110	17.17
1,300	6,710	9.66	2,600	22,910	17.48
1,400	7,450	10.20	2,700	23,710	17.78
1,500	8,200	10.72	2,800	24,510	18.07
1,600	8,960	11.21	2,900	25,310	18.36
1,700	9,730	11.68	3,000	26,110	18.63
1,800	10,510	12.13			

Hf(*c*):

$$H_T - H_{298.15} = 5.74T + 0.60 \times 10^{-3}T^2 - 1,765$$

(0.1 percent; 298°–2,495° K.);

$$C_p = 5.74 + 1.20 \times 10^{-3}T;$$

$$\Delta H_{2495}(fusion) = 5,790.$$

Hf(*l*):

$$H_T - H_{298.15} = 8.00T + 2,110 \text{ (0.1 percent;}$$

2,495°–3,000° K.);

$$C_p = 8.00.$$

TABLE 309.—*Heat content and entropy of Hf(g)*

[Base, ideal gas at 298.15° K.: atomic wt., 178.50]

$T,$ ° K.	$H_T - H_{298.15}$, cal./mole	$S_T - S_{298.15}$, cal./deg. mole	$T,$ ° K.	$H_T - H_{298.15}$, cal./mole	$S_T - S_{298.15}$, cal./deg. mole
400	510	1.47	1,800	9,290	10.44
500	1,010	2.60	1,900	10,020	10.83
600	1,530	3.54	2,000	10,740	11.20
700	2,070	4.37	2,100	11,460	11.55
800	2,630	5.12	2,200	12,190	11.89
900	3,220	5.81	2,300	12,910	12.21
1,000	3,830	6.45	2,400	13,630	12.52
1,100	4,460	7.05	2,500	14,340	12.81
1,200	5,110	7.62	2,600	15,050	13.09
1,300	5,780	8.15	2,700	15,760	13.35
1,400	6,460	8.66	2,800	16,460	13.61
1,500	7,160	9.14	2,900	17,160	13.86
1,600	7,860	9.60	3,000	17,860	14.09
1,700	8,570	10.03			

Hf(*g*):

$$H_T - H_{298.15} = 4.26T + 0.88 \times 10^{-3}T^2 - 0.16 \times 10^5 T^{-1}$$

$$- 1,295 \text{ (1.0 percent; 298°–2,200° K.);}$$

$$C_p = 4.26 + 1.76 \times 10^{-3}T + 0.16 \times 10^5 T^{-2}.$$

OXIDE

TABLE 310.—*Heat content and entropy of HfO₂(c)*

[Base, crystals at 298.15° K.; mol. wt., 210.50]

$T,$ ° K.	$H_T - H_{298.15}$, cal./mole	$S_T - S_{298.15}$, cal./deg. mole	$T,$ ° K.	$H_T - H_{298.15}$, cal./mole	$S_T - S_{298.15}$, cal./deg. mole
400	1,540	4.43	1,300	18,200	25.82
500	3,170	8.06	1,400	20,200	27.30
600	4,900	11.22	1,500	22,220	28.69
700	6,710	14.01	1,600	24,260	30.01
800	8,570	16.49	1,700	26,320	31.26
900	10,450	18.70	1,800	28,400	32.45
1,000	12,230	20.70	1,900	30,500	33.58
1,100	14,280	22.54	2,000	32,620	34.67
1,200	16,230	24.24			

HfO₂(*c*):

$$H_T - H_{298.15} = 17.39T + 1.04 \times 10^{-3}T^2 + 3.48 \times 10^5 T^{-1}$$

$$- 6,444 \text{ (0.3 percent; 298°–2,000° K.);}$$

$$C_p = 17.39 + 2.08 \times 10^{-3}T - 3.48 \times 10^5 T^{-2}.$$

CARBIDE

TABLE 311.—*Heat content and entropy of HfC(c)*

[Base, crystals at 298.15° K.; mol. wt., 190.51]

$T,$ ° K.	$H_T - H_{298.15}$, cal./mole	$S_T - S_{298.15}$, cal./deg. mole	$T,$ ° K.	$H_T - H_{298.15}$, cal./mole	$S_T - S_{298.15}$, cal./deg. mole
400	950	2.74	1,300	10,810	15.32
500	1,920	4.90	1,400	12,060	16.25
600	2,920	6.72	1,500	13,350	17.14
700	3,950	8.31	1,600	14,670	17.99
800	5,010	9.72	1,700	16,020	18.81
900	6,110	11.02	1,800	17,400	19.60
1,000	7,240	12.21	1,900	18,810	20.36
1,100	8,400	13.31	2,000	20,260	21.10
1,200	9,590	14.35			

HfC(*c*):

$$H_T - H_{298.15} = 8.25T + 1.59 \times 10^{-3}T^2 - 2,601 \text{ (0.1 percent;}$$

298°–2,000° K.);

$$C_p = 8.25 + 3.18 \times 10^{-3}T.$$

NITRIDE

TABLE 312.—*Heat content and entropy of HfN(c)*

[Base, crystals at 298.15° K; mol. wt., 192.51]

$T,$ ° K.	$H_T - H_{298.15}$, cal./mole	$S_T - S_{298.15}$, cal./deg. mole	$T,$ K.	$H_T - H_{298.15}$, cal./mole	$S_T - S_{298.15}$, cal./deg. mole
400	1,080	3.11	1,300	11,640	16.71
500	2,170	5.54	1,400	12,920	17.66
600	3,270	7.55	1,500	14,220	18.56
700	4,400	9.29	1,600	15,550	19.42
800	5,550	10.83	1,700	16,900	20.24
900	6,720	12.20	1,800	18,270	21.02
1,000	7,920	13.46	1,900	19,670	21.78
1,100	9,140	14.63	2,000	21,090	22.57
1,200	10,380	15.71			

HfN(*c*):

$$H_T - H_{298.15} = 9.84T + 1.11 \times 10^{-3}T^2 - 3,032 \text{ (0.1 percent;}$$

298°–2,000° K.);

$$C_p = 9.84 + 2.22 \times 10^{-3}T.$$

TABLE 1h7. (*Continued*)

TETRABROMIDE

TABLE 313.—*Heat content and entropy of HfBr$_4$(c)*

[Base, crystals at 298.15° K.; mol. wt., 498.16]

T, ° K.	$H_T-H_{298.15}$, cal./mole	$S_T-S_{298.15}$, cal./deg. mole	T, ° K.	$H_T-H_{298.15}$, cal./mole	$S_T-S_{298.15}$, cal./deg. mole
400	3,180	9.16	600	9,900	22.74
500	6,460	16.48	700	13,480	28.26

HfBr$_4$(c):
$$H_T-H_{298.15}=25.98T+7.58\times10^{-3}T^2-8,420 \text{ (0.1 percent;}$$
$$298°-700° \text{ K.);}$$
$$C_p=25.98+15.16\times10^{-3}T.$$

TABLE 314.—*Heat content and entropy of HfBr$_4$(g)*

[Base, ideal gas at 298.15° K.; mol. wt., 498.16]

T, ° K.	$H_T-H_{298.15}$, cal./mole	$S_T-S_{298.15}$, cal./deg. mole	T, ° K.	$H_T-H_{298.15}$, cal./mole	$S_T-S_{298.15}$, cal./deg. mole
400	2,590	7.47	1,000	18,000	30.98
500	5,140	13.16	1,100	20,590	33.45
600	7,700	17.83	1,200	23,180	35.70
700	10,270	21.79	1,300	25,770	37.77
800	12,840	25.22	1,400	28,370	39.70
900	15,420	28.26	1,500	30,970	41.49

HfBr$_4$(g):
$$H_T-H_{298.15}=25.29T+0.27\times10^{-3}T^2-7,564 \text{ (0.1 percent;}$$
$$298°-1,500° \text{ K.);}$$
$$C_p=25.29+0.54\times10^{-3}T.$$

TETRACHLORIDE

TABLE 315.—*Heat content and entropy of HfCl$_4$(c)*

[Base, crystals at 298.15° K.; mol. wt., 320.33]

T, ° K.	$H_T-H_{298.15}$, cal./mole	$S_T-S_{298.15}$, cal./deg. mole	T, ° K.	$H_T-H_{298.15}$, cal./mole	$S_T-S_{298.15}$, cal./deg. mole
400	3,010	8.68	600	9,070	20.95
500	6,020	15.39	700	12,160	25.71

HfCl$_4$(c):
$$H_T-H_{298.15}=31.47T+2.38\times10^5T^{-1}-10,181$$
$$\text{(0.3 percent; } 298°-700° \text{ K.);}$$
$$C_p=31.47-2.38\times10^5T^{-2}.$$

TABLE 316.—*Heat content and entropy of HfCl$_4$(g)*

[Base, ideal gas at 298.15° K.; mol. wt., 320.33]

T, ° K.	$H_T-H_{298.15}$, cal./mole	$S_T-S_{298.15}$, cal./deg. mole	T, ° K.	$H_T-H_{298.15}$, cal./mole	$S_T-S_{298.15}$, cal./deg. mole
400	2,490	7.18	1,000	17,700	30.35
500	4,980	12.74	1,100	20,270	32.80
600	7,500	17.33	1,200	22,840	35.04
700	10,030	21.23	1,300	25,420	37.10
800	12,580	24.65	1,400	28,000	39.01
900	15,140	27.65	1,500	30,590	40.80

HfCl$_4$(g):
$$H_T-H_{298.15}=24.40T+0.57\times10^{-3}T^2-7,337$$
$$\text{(0.4 percent; } 298°-1,500° \text{ K.);}$$
$$C_p=24.40+1.14\times10^{-3}T.$$

TETRAFLUORIDE

TABLE 317.—*Heat content and entropy of HfF$_4$(c)*

[Base, crystals at 298.15° K.; mol. wt., 254.50]

T, ° K.	$H_T-H_{298.15}$, cal./mole	$S_T-S_{298.15}$, cal./deg. mole	T, ° K.	$H_T-H_{298.15}$, cal./mole	$S_T-S_{298.15}$, cal./deg. mole
400	2,650	7.64	1,000	20,500	34.46
500	5,360	13.68	1,100	23,840	37.64
600	8,180	18.82	1,200	27,290	40.64
700	11,100	23.32	1,300	30,840	43.49
800	14,130	27.36	1,400	34,500	46.20
900	17,260	31.05	1,500	38,260	48.79

HfF$_4$(c):
$$H_T-H_{298.15}=22.38T+5.26\times10^{-3}T^2-7,140$$
$$\text{(0.1 percent; } 298°-1,500° \text{ K.);}$$
$$C_p=22.38+10.52\times10^{-3}T.$$

TABLE 318.—*Heat content and entropy of HfF$_4$(g)*

[Base, ideal gas at 298.15° K.; mol. wt., 254.50]

T, ° K.	$H_T-H_{298.15}$, cal./mole	$S_T-S_{298.15}$, cal./deg. mole	T, ° K.	$H_T-H_{298.15}$, cal./mole	$S_T-S_{298.15}$, cal./deg. mole
400	2,200	6.34	1,000	16,100	27.37
500	4,400	11.25	1,100	18,580	29.73
600	6,640	15.33	1,200	21,100	31.93
700	8,930	18.86	1,300	23,650	33.97
800	11,270	21.99	1,400	26,220	35.87
900	13,660	24.80	1,500	28,800	37.65

HfF$_4$(g):
$$H_T-H_{298.15}=20.06T+2.20\times10^{-3}T^2-6,176$$
$$\text{(0.2 percent; } 298°-1,500° \text{ K.);}$$
$$C_p=20.06+4.40\times10^{-3}T.$$

TETRAIODIDE

TABLE 319.—*Heat content and entropy of HfI(c)*

[Base, crystals at 298.15° K.; mol. wt., 686.14]

T, ° K.	$H_T-H_{298.15}$, cal./mole	$S_T-S_{298.15}$, cal./deg. mole	T, ° K.	$H_T-H_{298.15}$, cal./mole	$S_T-S_{298.15}$, cal./deg. mole
400	3,550	10.24	700	14,460	30.53
500	7,110	18.18	800	18,250	35.59
600	10,750	24.81			

HfI$_4$(c):
$$H_T-H_{298.15}=32.27T+3.73\times10^{-3}T^2-9,953$$
$$\text{(0.1 percent; } 298°-800° \text{ K.);}$$
$$C_p=32.27+7.46\times10^{-3}T.$$

TABLE 1h7. (*Continued*)

TABLE 320.—*Heat content and entropy of* $HfI_4(g)$

[Base, ideal gas at 298.15° K.; mol. wt., 686.14]

T, ° K.	$H_T-H_{298.15}$, cal./mole	$S_T-S_{298.15}$, cal./deg. mole	T, ° K.	$H_T-H_{298.15}$, cal./mole	$S_T-S_{298.15}$, cal./deg. mole
400	2,650	7.64	1,000	18,250	31.46
500	5,250	13.44	1,100	20,850	33.94
600	7,850	18.18	1,200	23,450	36.20
700	10,450	22.19	1,300	26,050	38.29
800	13,050	25.66	1,400	28,650	40.21
900	15,650	28.72	1,500	31,250	42.01

$HfI_4(g)$:

$$H_T-H_{298.15}=26.00T-7,752 \text{ (0.1 percent;}$$
$$298°-1,500° \text{ K.);}$$
$$C_p=26.00.$$

HELIUM
ELEMENT

TABLE 321.—*Heat content and entropy of* $He(g)$

[Base, ideal gas at 298.15° K.; atomic wt., 4.003]

T, ° K.	$H_T-H_{298.15}$, cal./mole	$S_T-S_{298.15}$, cal./deg. mole	T, ° K.	$H_T-H_{298.15}$, cal./mole	$S_T-S_{298.15}$, cal./deg. mole
400	505	1.46	1,900	7,960	9.20
500	1,005	2.57	2,000	8,455	9.46
600	1,500	3.48	2,200	9,450	9.93
700	1,995	4.24	2,400	10,445	10.36
800	2,495	4.90	2,600	11,440	10.76
900	2,990	5.49	2,800	12,430	11.13
1,000	3,490	6.01	3,000	13,425	11.47
1,100	3,895	6.49	3,500	15,910	12.24
1,200	4,480	6.92	4,000	18,395	12.90
1,300	4,980	7.32	4,500	20,880	13.49
1,400	5,475	7.69	5,000	23,365	14.01
1,500	5,970	8.03	6,000	28,335	14.92
1,600	6,470	8.35	7,000	33,305	15.68
1,700	6,965	8.65	8,000	38,280	16.35
1,800	7,465	8.94			

$He(g)$:

$$H_T-H_{298.15}=4.97T-1,482 \text{ (0.1 percent;}$$
$$298°-8,000° \text{ K.);}$$
$$C_p=4.97.$$

HOLMIUM
ELEMENT

TABLE 322.—*Heat content and entropy of* $Ho(c, l)$

[Base, crystals at 298.15° K.; atomic wt., 164.94]

T, ° K.	$H_T-H_{298.15}$, cal./mole	$S_T-S_{298.15}$, cal./deg. mole	T, ° K.	$H_T-H_{298.15}$, cal./mole	$S_T-S_{298.15}$, cal./deg. mole
400	670	1.94	1,500	9,050	11.74
500	1,350	3.44	1,600	9,910	12.29
600	2,040	4.71	1,700	10,790	12.83
700	2,750	5.80	1,773(c)	11,440	13.20
800	3,480	6.77	1,773(l)	15,540	15.52
900	4,420	7.65	1,800	15,760	15.64
1,000	4,985	8.45	2,000	17,360	16.49
1,100	5,760	9.19	2,200	18,960	17.25
1,200	6,560	9.89	2,400	20,560	17.94
1,300	7,370	10.54	2,600	22,160	18.58
1,400	8,200	11.15			

$Ho(c)$:

$$H_T-H_{298.15}=6.00T+0.85\times10^{-3}T^2-1,864 \text{ (0.1 percent;}$$
$$298°-1,773° \text{ K.);}$$
$$C_p=6.00+1.70\times10^{-3}T;$$
$$\Delta H_{1773}(fusion)=4,100.$$

$Ho(l)$:

$$H_T-H_{298.15}=8.00T+1,360 \text{ (0.1 percent;}$$
$$1,773°-2,600° \text{ K.);}$$
$$C_p=8.00.$$

HYDROGEN AND ITS COMPOUNDS
ELEMENT

TABLE 323.—*Heat content and entropy of* $H(g)$

[Base, ideal gas at 298.15° K.; atomic wt., 1.008]

T, ° K.	$H_T-H_{298.15}$, cal./mole	$S_T-S_{298.15}$, cal./deg. mole	T, ° K.	$H_T-H_{298.15}$, cal./mole	$S_T-S_{298.15}$, cal./deg. mole
400	505	1.46	1,900	7,960	9.20
500	1,005	2.57	2,000	8,455	9.46
600	1,500	3.48	2,200	9,450	9.93
700	1,995	4.24	2,400	10,445	10.36
800	2,495	4.90	2,600	11,440	10.76
900	2,990	5.49	2,800	12,430	11.13
1,000	3,490	6.02	3,000	13,425	11.47
1,100	3,985	6.49	3,500	15,910	12.24
1,200	4,480	6.92	4,000	18,395	12.90
1,300	4,980	7.32	4,500	20,880	13.49
1,400	5,475	7.69	5,000	23,365	14.01
1,500	5,970	8.03	6,000	28,335	14.92
1,600	6,470	8.35	7,000	33,305	15.68
1,700	6,965	8.65	8,000	38,285	16.35
1,800	7,465	8.93			

$H(g)$:

$$H_T-H_{298.15}=4.97T-1,482 \text{ (0.1 percent; } 298°-8,000° \text{ K);}$$
$$C_p=4.97.$$

TABLE 324.—*Heat content and entropy of* $H_2(g)$

[Base, ideal gas at 298.15° K.; mol. wt., 2.016]

T, ° K.	$H_T-H_{298.15}$, cal./mole	$S_T-S_{298.15}$, cal./deg. mole	T, ° K.	$H_T-H_{298.15}$, cal./mole	$S_T-S_{298.15}$, cal./deg. mole
400	705	2.04	2,000	12,650	13.79
500	1,405	3.60	2,100	13,470	14.19
600	2,105	4.87	2,200	14,300	14.58
700	2,810	5.96	2,300	15,135	14.95
800	3,515	6.90	2,400	15,975	15.31
900	4,225	7.74	2,500	16,825	15.66
1,000	4,940	8.49	2,750	18,980	16.48
1,100	5,670	9.18	3,000	21,160	17.24
1,200	6,405	9.82	3,250	23,375	17.94
1,300	7,150	10.42	3,500	25,610	18.61
1,400	7,905	10.98	3,750	27,870	19.23
1,500	8,670	11.51	4,000	30,150	19.82
1,600	9,445	12.01	4,250	32,445	20.38
1,700	10,235	12.49	4,500	34,755	20.90
1,800	11,030	12.94	4,750	37,085	21.41
1,900	11,835	13.38	5,000	39,425	21.89

$H_2(g)$:

$$H_T-H_{298.15}=6.52T+0.39\times10^{-3}T^2-0.12\times10^5T^{-1}$$
$$-1,938 \text{ (0.4 percent; } 298°-3,000° \text{ K.);}$$
$$C_p=6.52+0.78\times10^{-3}T+0.12\times10^5T^{-2}.$$

TABLE 325.—*Heat content and entropy of* $D_2(g)$

[Base, ideal gas at 298.15° K.; mol. wt., 4.028]

T, ° K.	$H_T-H_{298.15}$, cal./mole	$S_T-S_{298.15}$, cal./deg. mole	T, ° K.	$H_T-H_{298.15}$, cal./mole	$S_T-S_{298.15}$, cal./deg. mole
400	710	2.05	1,300	7,365	10.64
500	1,410	3.61	1,400	8,165	11.24
600	2,115	4.89	1,500	8,980	11.80
700	2,830	5.99	1,600	9,800	12.33
800	3,550	6.96	1,700	10,630	12.84
900	4,285	7.82	1,800	11,470	13.32
1,000	5,035	8.61	1,900	12,320	13.78
1,100	5,800	9.34	2,000	13,180	14.22
1,200	6,575	10.00			

TABLE 1h7. (*Continued*)

$D_2(g)$:

$H_T - H_{298.15} = 6.28\,T + 0.62 \times 10^{-3}\,T^2 - 0.29 \times 10^5\,T^{-1}$
$- 1,830$ (0.2 percent; 298°–2,000° K.);
$C_p = 6.28 + 1.24 \times 10^{-3}\,T + 0.29 \times 10^5\,T^{-2}.$

TABLE 326.—*Heat content and entropy of $T_2(g)$*

[Base, ideal gas at 298.15° K.; mol. wt., 6.034]

$T,°$ K.	$H_T - H_{298.15}$, cal./mole	$S_T - S_{298.15}$, cal./deg. mole	$T,°$ K.	$H_T - H_{298.15}$, cal./mole	$S_T - S_{298.15}$, cal./deg. mole
400	710	2.05	1,500	9,210	12.04
500	1,415	3.62	1,600	10,055	12.59
600	2,130	4.92	1,700	10,910	13.11
700	2,855	6.04	1,800	11,775	13.60
800	3,600	7.04	1,900	12,645	14.07
900	4,355	7.93	2,000	13,520	14.52
1,000	5,135	8.75	2,100	14,395	14.95
1,100	5,925	9.50	2,200	15,280	15.36
1,200	6,725	10.20	2,300	16,170	15.76
1,300	7,540	10.85	2,400	17,065	16.14
1,400	8,370	11.46	2,500	17,970	16.51

$T_2(g)$:

$H_T - H_{298.15} = 6.59\,T + 0.57 \times 10^{-3}\,T^2 - 0.04 \times 10^5\,T^{-1}$
$- 2,002$ (0.5 percent; 298°–2,500° K.);
$C_p = 6.59 + 1.14 \times 10^{-3}\,T + 0.04 \times 10^5\,T^{-2}.$

TABLE 327.—*Heat content and entropy of HD (g)*

[Base, ideal gas at 298.15° K.; mol. wt., 3.022]

$T,°$ K.	$H_T - H_{298.15}$, cal./mole	$S_T - S_{298.15}$, cal./deg. mole	$T,°$ K.	$H_T - H_{298.15}$, cal./mole	$S_T - S_{298.15}$, cal./deg. mole
400	710	2.05	1,300	7,225	10.51
500	1,410	3.61	1,400	8,000	11.09
600	2,110	4.89	1,500	8,785	11.63
700	2,815	5.97	1,600	9,580	12.14
800	3,530	6.93	1,700	10,385	12.63
900	4,250	7.78	1,800	11,205	13.10
1,000	4,975	8.55	1,900	12,030	13.55
1,100	5,710	9.25	2,000	12,865	13.98
1,200	6,460	9.90			

HD(g):

$H_T - H_{298.15} = 6.36\,T + 0.50 \times 10^{-3}\,T^2 - 0.29 \times 10^5\,T^{-1}$
$- 1,843$ (0.3 percent; 298°–2,000° K.);
$C_p = 6.36 + 1.00 \times 10^{-3}\,T + 0.29 \times 10^5\,T^{-2}.$

OXIDES

TABLE 328.—*Heat content and entropy of $H_2O(l)$*

[Base, liquid at 298.15° K.; mol. wt., 18.016]

$T,°$ K.	$H_T - H_{298.15}$, cal./mole	$S_T - S_{298.15}$, cal./deg. mole	$T,°$ K.	$H_T - H_{298.15}$, cal./mole	$S_T - S_{298.15}$, cal./deg. mole
350	934	2.89	373.15	1,353	4.05

$H_2O(l)$:

$H_T - H_{298.15} = 18.04\,T - 5,379$ (0.1 percent;
298°–373° K.);
$C_p = 18.04.$

TABLE 329.—*Heat content and entropy of $H_2O(g)$*

[Base, ideal gas at 298.15° K.; mol. wt., 18.016]

$T,°$ K.	$H_T - H_{298.15}$, cal./mole	$S_T - S_{298.15}$, cal./deg. mole	$T,°$ K.	$H_T - H_{298.15}$, cal./mole	$S_T - S_{298.15}$, cal./deg. mole
400	825	2.38	2,000	17,370	18.13
500	1,655	4.23	2,100	18,600	18.73
600	2,510	5.79	2,200	19,845	19.31
700	3,390	7.14	2,300	21,100	19.87
800	4,300	8.36	2,400	22,370	20.41
900	5,240	9.47	2,500	23,650	20.93
1,000	6,210	10.49	2,750	26,895	22.17
1,100	7,210	11.44	3,000	30,200	23.32
1,200	8,240	12.34	3,250	33,545	24.38
1,300	9,295	13.18	3,500	36,930	25.38
1,400	10,385	13.99	3,750	40,350	26.33
1,500	11,495	14.76	4,000	43,805	27.22
1,600	12,630	15.49	4,250	47,275	28.06
1,700	13,785	16.19	4,500	50,770	28.86
1,800	14,965	16.86	4,750	54,290	29.62
1,900	16,160	17.51	5,000	57,825	30.34

$H_2O(g)$:

$H_T - H_{298.15} = 7.30\,T + 1.23 \times 10^{-3}\,T^2 - 2,286$
(0.7 percent; 298°–2,750° K.);
$C_p = 7.30 + 2.46 \times 10^{-3}\,T.$

TABLE 330.—*Heat content and entropy of $D_2O(g)$*

[Base, ideal gas at 298.15° K.; mol. wt., 20.03]

$T,°$ K.	$H_T - H_{298.15}$, cal./mole	$S_T - S_{298.15}$, cal./deg. mole	$T,°$ K.	$H_T - H_{298.15}$, cal./mole	$S_T - S_{298.15}$, cal./deg. mole
400	850	2.45	2,000	18,785	19.48
500	1,720	4.39	2,100	20,095	20.12
600	2,630	6.05	2,200	21,410	20.73
700	3,575	7.50	2,300	22,735	21.32
800	4,565	8.82	2,400	24,070	21.88
900	5,595	10.04	2,500	25,410	22.43
1,000	6,665	11.16	2,750	28,795	23.72
1,100	7,765	12.21	3,000	32,225	24.92
1,200	8,900	13.20	3,250	35,675	26.02
1,300	10,065	14.13	3,500	39,160	27.05
1,400	11,255	15.01	3,750	42,665	28.02
1,500	12,465	15.85	4,000	46,195	28.93
1,600	13,695	16.64	4,250	49,740	29.79
1,700	14,945	17.40	4,500	53,300	30.61
1,800	16,215	18.12	4,750	56,875	31.38
1,900	17,495	18.82	5,000	60,465	32.12

$D_2O(g)$:

$H_T - H_{298.15} = 7.69\,T + 1.48 \times 10^{-3}\,T^2 + 0.34 \times 10^5\,T^{-1}$
$- 2,538$ (0.8 percent; 298°–2,200° K.);
$C_p = 7.69 + 2.96 \times 10^{-3}\,T - 0.34 \times 10^5\,T^{-2}.$

TABLE 331.—*Heat content and entropy of $T_2O(g)$*

[Base, ideal gas at 298.15° K.; mol. wt., 22.03]

$T,°$ K.	$H_T - H_{298.15}$, cal./mole	$S_T - S_{298.15}$, cal./deg. mole	$T,°$ K.	$H_T - H_{298.15}$, cal./mole	$S_T - S_{298.15}$, cal./deg. mole
400	870	2.51	2,000	19,545	20.28
500	1,775	4.52	2,100	20,880	20.93
600	2,720	6.25	2,200	22,220	21.56
700	3,720	7.79	2,300	23,575	22.16
800	4,760	9.18	2,400	24,930	22.74
900	5,850	10.46	2,500	26,275	23.29
1,000	6,975	11.64	2,750	29,730	24.60
1,100	8,135	12.75	3,000	33,195	25.81
1,200	9,325	13.78	3,250	36,685	26.92
1,300	10,540	14.76	3,500	40,195	27.96
1,400	11,775	15.67	3,750	43,725	28.94
1,500	13,035	16.54	4,000	47,275	29.85
1,600	14,310	17.36	4,250	50,840	30.72
1,700	15,600	18.15	4,500	54,410	31.53
1,800	16,905	18.89	4,750	58,000	32.31
1,900	18,220	19.60	5,000	61,600	33.05

TABLE 1h7. (*Continued*)

$T_2O(g)$:

$H_T - H_{298.15} = 8.02T + 1.59 \times 10^{-3}T^2 + 0.55 \times 10^5 T^{-1}$
$\quad - 2,717$ (0.8 percent; 298°–2,000° K.);
$C_p = 8.02 + 3.18 \times 10^{-3}T - 0.55 \times 10^5 T^{-2}.$

$DTO(g)$:

$H_T - H_{298.15} = 7.69T + 1.62 \times 10^{-3}T^2 + 0.36 \times 10^5 T^{-1}$
$\quad - 2,558$ (0.7 percent; 298°–2,000° K.);
$C_p = 7.69 + 3.24 \times 10^{-3}T - 0.36 \times 10^5 T^{-2}.$

TABLE 332.—*Heat content and entropy of HDO(g)*

[Base, ideal gas at 298.15° K.; mol. wt., 19.02]

T, ° K.	$H_T-H_{298.15}$, cal./mole	$S_T-S_{298.15}$, cal./deg. mole	T, ° K.	$H_T-H_{298.15}$, cal./mole	$S_T-S_{298.15}$, cal./deg. mole
400	835	2.40	2,000	18,060	18.77
500	1,680	4.29	2,100	19,325	19.39
600	2,555	5.88	2,200	20,605	19.98
700	3,470	7.29	2,300	21,895	20.55
800	4,420	8.56	2,400	23,200	21.11
900	5,400	9.72	2,500	24,510	21.64
1,000	6,420	10.79	2,750	27,825	22.91
1,100	7,470	11.79	3,000	31,190	24.08
1,200	8,550	12.73	3,250	34,585	25.16
1,300	9,660	13.62	3,500	38,020	26.18
1,400	10,800	14.46	3,750	41,485	27.14
1,500	11,960	15.26	4,000	44,975	28.04
1,600	13,145	16.03	4,250	48,480	28.89
1,700	14,345	16.76	4,500	52,010	29.70
1,800	15,570	17.46	4,750	55,555	30.46
1,900	16,805	18.12	5,000	59,120	31.19

$HDO(g)$:

$H_T - H_{298.15} = 7.74T + 1.23 \times 10^{-3}T^2 + 0.35 \times 10^5 T^{-1}$
$\quad - 2,534$ (1.0 percent; 298°–2,750° K.);
$C_p = 7.74 + 2.46 \times 10^{-3}T - 0.35 \times 10^5 T^{-2}.$

TABLE 333.—*Heat content and entropy of HTO(g)*

[Base, ideal gas at 298.15° K.; mol. wt., 20.02]

T, ° K.	$H_T-H_{298.15}$, cal./mole	$S_T-S_{298.15}$, cal./deg. mole	T, ° K.	$H_T-H_{298.15}$, cal./mole	$S_T-S_{298.15}$, cal./deg. mole
400	840	2.42	2,000	18,410	19.12
500	1,695	4.33	2,100	19,695	19.74
600	2,585	5.95	2,200	20,990	20.35
700	3,520	7.39	2,300	22,295	20.93
800	4,495	8.69	2,400	23,610	21.49
900	5,505	9.88	2,500	24,930	22.03
1,000	6,550	10.98	2,750	28,275	23.30
1,100	7,625	12.01	3,000	31,665	24.48
1,200	8,735	12.97	3,250	35,085	25.57
1,300	9,870	13.88	3,500	38,535	26.59
1,400	11,035	14.74	3,750	42,020	27.56
1,500	12,215	15.56	4,000	45,525	28.46
1,600	13,420	16.34	4,250	49,045	29.31
1,700	14,645	17.08	4,500	52,590	30.12
1,800	15,885	17.79	4,750	56,150	30.89
1,900	17,140	18.47	5,000	59,725	31.63

$HTO(g)$:

$H_T - H_{298.15} = 7.56T + 1.44 \times 10^{-3}T^2 + 0.28 \times 10^5 T^{-1}$
$\quad - 2,476$ (0.7 percent; 298°–2,200° K.);
$C_p = 7.56 + 2.88 \times 10^{-3}T - 0.28 \times 10^5 T^{-2}.$

TABLE 334.—*Heat content and entropy of DTO(g)*

[Base, ideal gas at 298.15° K.; mol. wt., 21.03]

T, ° K.	$H_T-H_{298.15}$, cal./mole	$S_T-S_{298.15}$, cal./deg. mole	T, ° K.	$H_T-H_{298.15}$, cal./mole	$S_T-S_{298.15}$, cal./deg. mole
400	860	2.48	2,000	19,155	19.87
500	1,745	4.45	2,100	20,475	20.51
600	2,670	6.14	2,200	21,805	21.13
700	3,640	7.63	2,300	23,140	21.72
800	4,660	8.99	2,400	24,485	22.29
900	5,715	10.24	2,500	25,840	22.85
1,000	6,810	11.39	2,750	29,250	24.15
1,100	7,945	12.47	3,000	32,695	25.34
1,200	9,105	13.48	3,250	36,165	26.45
1,300	10,295	14.43	3,500	39,660	27.49
1,400	11,505	15.33	3,750	43,175	28.46
1,500	12,740	16.18	4,000	46,715	29.37
1,600	13,995	16.99	4,250	50,265	30.24
1,700	15,265	17.76	4,500	53,830	31.05
1,800	16,550	18.49	4,750	57,410	31.82
1,900	17,845	19.19	5,000	61,005	32.56

TABLE 335.—*Heat content and entropy of $H_2O_2(l)$*

[Base, liquid at 298.15° K.; mol. wt., 34.02]

T, ° K.	$H_T-H_{298.15}$, cal./mole	$S_T-S_{298.15}$, cal./deg. mole	T, ° K.	$H_T-H_{298.15}$, cal./mole	$S_T-S_{298.15}$, cal./deg. mole
350	1,135	3.51	450	3,535	9.53
400	2,300	6.62			

$H_2O_2(l)$:

$H_T - H_{298.15} = 12.81T + 14.00 \times 10^{-3}T^2$
$\quad - 5,064$ (0.1 percent; 298°–450° K) ·
$C_p = 12.81 + 28.00 \times 10^{-3}T.$

TABLE 336.—*Heat content and entropy of $H_2O_2(g)$*

[Base, ideal gas at 298.15° K.; mol. wt., 34.02]

T, ° K.	$H_T-H_{298.15}$, cal./mole	$S_T-S_{298.15}$, cal./deg. mole	T, ° K.	$H_T-H_{298.15}$, cal./mole	$S_T-S_{298.15}$, cal./deg. mole
400	1,120	3.22	1,000	9,325	15.52
500	2,325	5.91	1,100	10,845	16.97
600	3,620	8.27	1,200	12,395	18.32
700	4,985	10.37	1,300	13,970	19.58
800	6,400	12.26	1,400	15,565	20.76
900	7,845	13.96	1,500	17,190	21.88

$H_2O_2(g)$:

$H_T - H_{298.15} = 11.74T + 1.71 \times 10^{-3}T^2 + 2.18 \times 10^5 T^{-1}$
$\quad - 4,383$ (0.5 percent; 298°–1,500° K.);
$C_p = 11.74 + 3.42 \times 10^{-3}T - 2.18 \times 10^5 T^{-2}.$

TABLE 337.—*Heat content and entropy of $D_2O_2(g)$*

[Base, ideal gas at 298.15° K.; mol. wt., 36.03]

T, ° K.	$H_T-H_{298.15}$, cal./mole	$S_T-S_{298.15}$, cal./deg. mole	T, ° K.	$H_T-H_{298.15}$, cal./mole	$S_T-S_{298.15}$, cal./deg. mole
400	1,210	3.48	1,000	10,050	16.74
500	2,520	6.40	1,100	11,680	18.29
600	3,920	8.95	1,200	13,340	19.74
700	5,380	11.20	1,300	15,020	21.08
800	6,900	13.23	1,400	16,730	22.35
900	8,460	15.06	1,500	18,460	23.54

$D_2O_2(g)$:

$H_T - H_{298.15} = 12.80T + 1.74 \times 10^{-3}T^2 + 2.40 \times 10^5 T^{-1}$
$\quad - 4,776$ (0.4 percent; 298°–1,500° K.);
$C_p = 12.80 + 3.48 \times 10^{-3}T - 2.40 \times 10^5 T^{-2}.$

TABLE 338.—*Heat content and entropy of $HDO_2(g)$*

[Base, ideal gas at 298.15° K.; mol. wt., 35.02]

T, ° K.	$H_T-H_{298.15}$, cal./mole	$S_T-S_{298.15}$, cal./deg. mole	T, ° K.	$H_T-H_{298.15}$, cal./mole	$S_T-S_{298.15}$, cal./deg. mole
400	1,160	3.34	1,000	9,665	16.08
500	2,410	6.12	1,100	11,235	17.58
600	3,755	8.57	1,200	12,835	18.97
700	5,160	10.74	1,300	14,460	20.27
800	6,620	12.69	1,400	16,110	21.49
900	8,125	14.46	1,500	17,770	22.64

$HDO_2(g)$:

$H_T - H_{298.15} = 12.14T + 1.78 \times 10^{-3}T^2 + 2.24 \times 10^5 T^{-1}$
$\quad - 4,529$ (0.5 percent; 298°–1,500° K.);
$C_p = 12.14 + 3.56 \times 10^{-3}T - 2.24 \times 10^5 T^{-2}.$

TABLE 1h7. (*Continued*)

TABLE 339.—*Heat content and entropy of* OH(*g*)

[Base, ideal gas at 298.15° K.; mol. wt., 17.008]

T, ° K.	$H_T-H_{298.15}$, cal./mole	$S_T-S_{298.15}$, cal./deg. mole	T, ° K.	$H_T-H_{298.15}$, cal./mole	$S_T-S_{298.15}$, cal./deg. mole
400	725	2.09	2,000	12,845	14.00
500	1,430	3.67	2,100	13,675	14.41
600	2,135	4.95	2,200	14,515	14.80
700	2,840	6.04	2,300	15,360	15.18
800	3,555	6.99	2,400	16,200	15.54
900	4,270	7.84	2,500	17,060	15.89
1,000	5,000	8.60	2,750	19,220	16.71
1,100	5,735	9.31	3,000	21,405	17.47
1,200	6,485	9.96	3,250	23,605	18.17
1,300	7,245	10.57	3,500	25,830	18.83
1,400	8,020	11.14	3,750	28,070	19.45
1,500	8,800	11.68	4,000	30,325	20.03
1,600	9,590	12.19	4,250	32,590	20.58
1,700	10,390	12.68	4,500	34,870	21.10
1,800	11,200	13.14	4,750	37,165	21.60
1,900	12,020	13.58	5,000	39,465	22.08

OH(*g*):
$$H_T-H_{298.15}=6.38T+0.47\times10^{-3}T^2-0.44\times10^5T^{-1}$$
$$-1,796 \ (0.2 \text{ percent}; \ 298°-3,000° \text{ K.});$$
$$C_p=6.38+0.94\times10^{-3}T+0.44\times10^5T^{-2}.$$

TABLE 340.—*Heat content and entropy of* OD(*g*)

[Base, ideal gas at 298.15° K.; mol. wt., 18.01]

T, ° K.	$H_T-H_{298.15}$, cal./mole	$S_T-S_{298.15}$, cal./deg. mole	T, ° K.	$H_T-H_{298.15}$, cal./mole	$S_T-S_{298.15}$, cal./deg. mole
400	725	2.10	2,000	13,355	14.44
500	1,435	3.68	2,100	14,220	14.85
600	2,150	4.98	2,200	15,090	15.26
700	2,870	6.09	2,300	15,965	15.64
800	3,605	7.08	2,400	16,840	16.02
900	4,355	7.96	2,500	17,720	16.38
1,000	5,120	8.76	2,750	19,940	17.22
1,100	5,895	9.50	3,000	22,175	18.00
1,200	6,685	10.19	3,250	24,425	18.72
1,300	7,490	10.83	3,500	26,685	19.39
1,400	8,305	11.44	3,750	28,960	20.02
1,500	9,130	12.00	4,000	31,250	20.61
1,600	9,960	12.54	4,250	33,550	21.17
1,700	10,800	13.05	4,500	35,855	21.70
1,800	11,645	13.53	4,750	38,170	22.20
1,900	12,500	14.00	5,000	40,495	22.68

OD(*g*):
$$H_T-H_{298.15}=6.62T+0.50\times10^{-3}T^2-0.22\times10^5T^{-1}$$
$$-1,944 \ (0.6 \text{ percent}; \ 298°-3,000° \text{ K.});$$
$$C_p=6.62+1.00\times10^{-3}T+0.22\times10^5T^{-2}.$$

TABLE 341.—*Heat content and entropy of* OT(*g*)

[Base, ideal gas at 298.15° K.; mol. wt., 19.02]

T, ° K.	$H_T-H_{298.15}$, cal./mole	$S_T-S_{298.15}$, cal./deg. mole	T, ° K.	$H_T-H_{298.15}$, cal./mole	$S_T-S_{298.15}$, cal./deg. mole
400	730	2.10	2,000	13,640	14.71
500	1,440	3.70	2,100	14,515	15.13
600	2,165	5.02	2,200	15,395	15.54
700	2,905	6.16	2,300	16,280	15.93
800	3,660	7.16	2,400	17,165	16.31
900	4,430	8.07	2,500	18,055	16.67
1,000	5,220	8.90	2,750	20,290	17.52
1,100	6,020	9.66	3,000	22,545	18.31
1,200	6,835	10.37	3,250	24,810	19.04
1,300	7,655	11.03	3,500	27,085	19.71
1,400	8,490	11.65	3,750	29,370	20.34
1,500	9,335	12.23	4,000	31,665	20.94
1,600	10,185	12.78	4,250	33,970	21.50
1,700	11,040	13.30	4,500	36,280	22.02
1,800	11,900	13.79	4,750	38,600	22.52
1,900	12,770	14.26	5,000	40,925	23.00

OT(*g*):
$$H_T-H_{298.15}=6.63T+0.59\times10^{-3}T^2-0.15\times10^5T^{-1}$$
$$-1,979 \ (0.5 \text{ percent}; \ 298°-2,200° \text{ K.});$$
$$C_p=6.63+1.18\times10^{-3}T+0.15\times10^5T^{-2}.$$

SULFIDES

TABLE 342.—*Heat content and entropy of* H_2S(*g*)

[Base, ideal gas at 298.15° K; mol. wt., 34.08]

T, ° K.	$H_T-H_{298.15}$, cal./mole	$S_T-S_{298.15}$, cal./deg. mole	T, ° K.	$H_T-H_{298.15}$, cal./mole	$S_T-S_{298.15}$, cal./deg. mole
400	850	2.45	2,000	18,950	19.61
500	1,720	4.39	2,100	20,265	20.25
600	2,630	6.05	2,200	21,595	20.87
700	3,585	7.52	2,300	22,935	21.47
800	4,580	8.84	2,400	24,280	22.04
900	5,620	10.07	2,500	25,635	22.59
1,000	6,695	11.20	2,750	29,055	23.89
1,100	7,810	12.27	3,000	32,510	25.10
1,200	8,960	13.26	3,250	36,000	26.21
1,300	10,135	14.19	3,500	39,515	27.25
1,400	11,335	15.09	3,750	43,060	28.23
1,500	12,560	15.94	4,000	46,625	29.16
1,600	13,805	16.74	4,250	50,205	30.03
1,700	15,070	17.51	4,500	53,805	30.85
1,800	16,345	18.24	4,750	57,420	31.63
1,900	17,640	18.94	5,000	61,055	32.38

H_2S(*g*):
$$H_T-H_{298.15}=7.81T+1.48\times10^{-3}T^2+0.46\times10^5T^{-1}$$
$$-2,614 \ (0.8 \text{ percent}; \ 298°-2,300° \text{ K.});$$
$$C_p=7.81+2.96\times10^{-3}T-0.46\times10^5T^{-2}.$$

TABLE 343.—*Heat content and entropy of* D_2S(*g*)

[Base, ideal gas at 298.15° K.; mol. wt., 36.09]

T, ° K.	$H_T-H_{298.15}$, cal./mole	$S_T-S_{298.15}$, cal./deg. mole	T, ° K.	$H_T-H_{298.15}$, cal./mole	$S_T-S_{298.15}$, cal./deg. mole
400	900	2.59	2,000	20,640	21.39
500	1,840	4.69	2,100	22,040	22.07
600	2,840	6.51	2,200	23,450	22.73
700	3,900	8.15	2,300	24,865	23.36
800	5,015	9.63	2,400	26,290	23.96
900	6,175	11.00	2,500	27,725	24.55
1,000	7,375	12.26	2,750	31,330	25.92
1,100	8,610	13.44	3,000	34,975	27.19
1,200	9,870	14.54	3,250	38,645	28.36
1,300	11,160	15.57	3,500	42,340	29.46
1,400	12,470	16.54	3,750	46,065	30.49
1,500	13,795	17.45	4,000	49,820	31.46
1,600	15,135	18.32	4,250	53,585	32.37
1,700	16,495	19.14	4,500	57,370	33.23
1,800	17,865	19.92	4,750	61,180	34.06
1,900	19,345	20.67	5,000	65,005	34.84

D_2S(*g*):
$$H_T-H_{298.15}=8.37T+1.83\times10^{-3}T^2+0.81\times10^5T^{-1}$$
$$-2,930 \ (0.8 \text{ percent}; \ 298°-1,800° \text{ K.});$$
$$C_p=8.37+3.66\times10^{-3}T-0.81\times10^5T^{-2}.$$

TABLE 344.—*Heat content and entropy of* T_2S(*g*)

[Base, ideal gas at 298.15° K.; mol. wt., 38.10]

T, ° K.	$H_T-H_{298.15}$, cal./mole	$S_T-S_{298.15}$, cal./deg. mole	T, ° K.	$H_T-H_{298.15}$, cal./mole	$S_T-S_{298.15}$, cal./deg. mole
400	940	2.70	2,000	21,040	22.03
500	1,930	4.91	2,100	22,425	22.71
600	2,990	6.84	2,200	23,815	23.35
700	4,105	8.57	2,300	25,205	23.97
800	5,275	10.13	2,400	26,605	24.57
900	6,480	11.54	2,500	28,005	25.14
1,000	7,720	12.84	2,750	31,530	26.48
1,100	8,985	14.05	3,000	35,070	27.72
1,200	10,270	15.17	3,250	38,630	28.86
1,300	11,575	16.22	3,500	42,205	29.91
1,400	12,900	17.20	3,750	45,795	30.90
1,500	14,235	18.12	4,000	49,400	31.83
1,600	15,580	18.99	4,250	53,015	32.71
1,700	16,930	19.81	4,500	56,645	33.54
1,800	18,295	20.59	4,750	60,280	34.33
1,900	19,665	21.33	5,000	63,930	35.08

TABLE 1h7. (*Continued*)

$T_2S(g)$:

$H_T - H_{298.15} = 9.31T + 1.56 \times 10^{-3}T^2 + 1.22 \times 10^5 T^{-1}$
$- 3,324$ (0.8 percent; 298°–1,800° K.);
$C_p = 9.31 + 3.12 \times 10^{-3}T - 1.22 \times 10^5 T^{-2}$.

$DTS(g)$:

$H_T - H_{298.15} = 8.80T + 1.63 \times 10^{-3}T^2 + 0.97 \times 10^5 T^{-1}$
$- 3,094$ (0.8 percent; 298°–1,800° K.);
$C_p = 8.80 + 3.26 \times 10^{-3}T - 0.97 \times 10^5 T^{-2}$.

TABLE 345.—*Heat content and entropy of HDS(g)*

[Base, ideal gas at 298.15° K.; mol. wt., 35.09]

T, ° K.	$H_T-H_{298.15}$, cal./mole	$S_T-S_{298.15}$, cal./deg. mole	T, ° K.	$H_T-H_{298.15}$, cal./mole	$S_T-S_{298.15}$, cal./deg. mole
400	865	2.50	2,000	19,450	20.20
500	1,765	4.50	2,100	20,770	20.85
600	2,715	6.23	2,200	22,100	21.47
700	3,710	7.76	2,300	23,435	22.06
800	4,755	9.16	2,400	24,775	22.63
900	5,840	10.44	2,500	26,125	23.18
1,000	6,960	11.62	2,750	29,515	24.47
1,100	8,120	12.72	3,000	32,925	25.66
1,200	9,305	13.75	3,250	36,355	26.75
1,300	10,510	14.72	3,500	39,800	27.78
1,400	11,745	15.63	3,750	43,260	28.74
1,500	12,995	16.49	4,000	46,730	29.63
1,600	14,260	17.31	4,250	50,200	30.47
1,700	15,540	18.09	4,500	53,680	31.26
1,800	16,830	18.83	4,750	56,160	32.02
1,900	18,135	19.53	5,000	60,645	32.74

$HDS(g)$:

$H_T - H_{298.15} = 8.42T + 1.37 \times 10^{-3}T^2 + 0.83 \times 10^5 T^{-1}$
$- 2,911$ (1.0 percent; 298°–2,300° K.);
$C_p = 8.42 + 2.74 \times 10^{-3}T - 0.83 \times 10^5 T^{-2}$.

TABLE 348.—*Heat content and entropy of SH(g)*

[Base, ideal gas at 298.15° K.; mol. wt., 33.07]

T, ° K.	$H_T-H_{298.15}$, cal./mole	$S_T-S_{298.15}$, cal./deg. mole	T, ° K.	$H_T-H_{298.15}$, cal./mole	$S_T-S_{298.15}$, cal./deg. mole
400	780	2.25	2,000	13,660	14.92
500	1,530	3.93	2,100	14,530	15.34
600	2,275	5.29	2,200	15,410	15.75
700	3,025	6.44	2,300	16,290	16.14
800	3,780	7.45	2,400	17,170	16.52
900	4,550	8.36	2,500	18,060	16.88
1,000	5,330	9.18	2,750	20,290	17.73
1,100	6,120	9.93	3,000	22,545	18.52
1,200	6,920	10.63	3,250	24,810	19.25
1,300	7,735	11.28	3,500	27,090	19.92
1,400	8,560	11.89	3,750	29,385	20.55
1,500	9,390	12.46	4,000	31,690	21.15
1,600	10,230	13.01	4,250	34,005	21.71
1,700	11,080	13.52	4,500	36,330	22.24
1,800	11,935	14.01	4,750	38,665	22.74
1,900	12,795	14.48	5,000	41,010	23.22

$SH(g)$:

$H_T - H_{298.15} = 6.93T + 0.43 \times 10^{-3}T^2 - 0.48 \times 10^5 T^{-1}$
$- 1,943$ (0.5 percent; 298°–3,000° K.);
$C_p = 6.93 + 0.86 \times 10^{-3}T + 0.48 \times 10^5 T^{-2}$.

TABLE 346.—*Heat content and entropy of HTS(g)*

[Base, ideal gas at 298.15° K.; mol. wt., 36.09]

T, ° K.	$H_T-H_{298.15}$, cal./mole	$S_T-S_{298.15}$, cal./deg. mole	T. ° K.	$H_T-H_{298.15}$, cal./mole	$S_T-S_{298.15}$, cal./deg. mole
400	885	2.54	2,000	19,825	20.65
500	1,805	4.60	2,100	21,155	21.30
600	2,785	6.39	2,200	22,495	21.92
700	3,815	7.97	2,300	23,840	22.52
800	4,895	9.41	2,400	25,190	23.10
900	6,010	10.73	2,500	26,545	23.65
1,000	7,160	11.94	2,750	29,950	24.95
1,100	8,345	13.07	3,000	33,380	26.14
1,200	9,555	14.12	3,250	38,825	27.24
1,300	10,785	15.10	3,500	40,280	28.27
1,400	12,035	16.03	3,750	43,750	29.23
1,500	13,300	16.90	4,000	47,230	30.13
1,600	14,580	17.73	4,250	50,710	30.97
1,700	15,875	18.51	4,500	56,195	31.76
1,800	17,185	19.26	4,750	59,685	32.52
1,900	18,500	19.97	5,000	61,175	33.24

$HTS(g)$:

$H_T - H_{298.15} = 8.37T + 1.60 \times 10^{-3}T^2 + 0.82 \times 10^5 T^{-1}$
$- 2,912$ (0.8 percent; 298°–1,900° K.);
$C_p = 8.37 + 3.20 \times 10^{-3}T - 0.82 \times 10^5 T^{-2}$.

TABLE 349.—*Heat content and entropy of SD(g)*

[Base, ideal gas at 298.15° K.; mol. wt., 34.08]

T, ° K.	$H_T-H_{298.15}$, cal./mole	$S_T-S_{298.15}$, cal./deg. mole	T, ° K.	$H_T-H_{298.15}$, cal./mole	$S_T-S_{298.15}$, cal./deg. mole
400	785	2.27	2,000	14,200	15.46
500	1,550	3.98	2,100	15,090	15.89
600	2,320	5.39	2,200	15,985	16.31
700	3,105	6.59	2,300	16,885	16.71
800	3,905	7.66	2,400	17,785	17.09
900	4,715	8.61	2,500	18,685	17.45
1,000	5,540	9.48	2,750	20,955	18.32
1,100	6,375	10.28	3,000	23,240	19.12
1,200	7,220	11.01	3,250	25,530	19.85
1,300	8,070	11.69	3,500	27,830	20.53
1,400	8,930	12.33	3,750	30,145	21.17
1,500	9,795	12.93	4,000	32,465	21.77
1,600	10,665	13.49	4,250	34,795	22.33
1,700	11,545	14.02	4,500	37,130	22.87
1,800	12,425	14.52	4,750	39,480	23.38
1,900	13,310	15.00	5,000	41,830	23.86

$SD(g)$:

$H_T - H_{298.15} = 7.22T + 0.47 \times 10^{-3}T^2 - 0.23 \times 10^5 T^{-1}$
$- 2,117$ (0.5 percent; 298°–2,500° K.);
$C_p = 7.22 + 0.94 \times 10^{-3}T + 0.23 \times 10^5 T^{-2}$.

TABLE 347.—*Heat content and entropy of DTS(g)*

[Base, ideal gas at 298.15° K.; mol. wt., 37.10]

T, ° K.	$H_T-H_{298.15}$, cal./mole	$S_T-S_{298.15}$, cal./deg. mole	T, ° K.	$H_T-H_{298.15}$, cal./mole	$S_T-S_{298.15}$, cal./deg. mole
400	915	2.64	2,000	20,515	21.44
500	1,875	4.78	2,100	21,875	22.10
600	2,900	6.64	2,200	23,235	22.73
700	3,980	8.30	2,300	24,600	23.34
800	5,110	9.81	2,400	25,970	23.93
900	6,280	11.19	2,500	27,345	24.49
1,000	7,480	12.46	2,750	30,790	25.80
1,100	8,710	13.63	3,000	34,255	27.01
1,200	9,970	14.72	3,250	37,730	28.12
1,300	11,245	15.74	3,500	41,215	29.15
1,400	12,535	16.70	3,750	44,710	30.11
1,500	13,840	17.60	4,000	48,210	31.02
1,600	15,160	18.45	4,250	51,710	31.87
1,700	16,485	19.26	4,500	55,210	32.67
1,800	17,820	20.02	4,750	58,715	33.43
1,900	19,165	20.75	5,000	62,225	34.15

TABLE 350.—*Heat content and entropy of ST(g)*

[Base, ideal gas at 298.15° K.; mol. wt., 35.08]

T, ° K.	$H_T-H_{298.15}$, cal./mole	$S_T-S_{298.15}$, cal./deg. mole	T, ° K.	$H_T-H_{298.15}$, cal./mole	$S_T-S_{298.15}$, cal./deg. mole
400	795	2.29	2,000	14,480	15.78
500	1,575	4.04	2,100	15,380	16.22
600	2,370	5.49	2,200	16,285	16.64
700	3,180	6.73	2,300	17,190	17.04
800	4,000	7.83	2,400	18,095	17.43
900	4,835	8.81	2,500	19,005	17.80
1,000	5,680	9.70	2,750	21,285	18.67
1,100	6,535	10.52	3,000	23,580	19.47
1,200	7,400	11.27	3,250	25,885	20.21
1,300	8,265	11.96	3,500	28,195	20.89
1,400	9,140	12.61	3,750	30,515	21.53
1,500	10,020	13.22	4,000	32,845	22.13
1,600	10,905	13.79	4,250	35,180	22.70
1,700	11,795	14.33	4,500	37,525	23.24
1,800	12,685	14.84	4,750	39,875	23.75
1,900	13,580	15.32	5,000	42,230	24.22

TABLE 1h7. (*Continued*)

ST(*g*):

$$H_T - H_{298.15} = 7.57T + 0.41 \times 10^{-3}T^2 - 2,293$$

(0.5 percent; 298°–2,500° K.);

$$C_p = 7.57 + 0.82 \times 10^{-3}T.$$

H$_2$S$_2$(*l*):

$$\overline{C_p} = 22.0 \ (298°–342° \text{ K.}).$$

TBr(*g*):

$$H_T - H_{298.15} = 7.23T + 0.49 \times 10^{-3}T^2 + 0.44 \times 10^5 T^{-1}$$

$$-2,347 \ (0.7 \text{ percent}; \ 298°–2,000° \text{ K.});$$

$$C_p = 7.23 + 0.98 \times 10^{-3}T - 0.44 \times 10^5 T^{-2}.$$

CHLORIDES

BROMIDES

TABLE 351.—*Heat content and entropy of HBr(g)*

[Base, ideal gas at 298.15° K.; mol. wt., 80.92]

T, ° K.	$H_T - H_{298.15}$, cal./mole	$S_T - S_{298.15}$, cal./deg. mole	T, ° K.	$H_T - H_{298.15}$, cal./mole	$S_T - S_{298.15}$, cal./deg. mole
400	710	2.05	1,300	7,465	10.76
500	1,410	3.61	1,400	8,280	11.36
600	2,120	4.90	1,500	9,105	11.93
700	2,840	6.01	1,600	9,940	12.47
800	3,575	7.00	1,700	10,785	12.98
900	4,325	7.88	1,800	11,640	13.47
1,000	5,090	8.68	1,900	12,505	13.94
1,100	5,870	9.43	2,000	13,375	14.39
1,200	6,660	10.12			

HBr(*g*):

$$H_T - H_{298.15} = 6.41T + 0.62 \times 10^{-3}T^2 - 0.15 \times 10^5 T^{-1}$$

$$-1,916 \ (0.2 \text{ percent}; \ 298°–2,000° \text{ K.});$$

$$C_p = 6.41 + 1.24 \times 10^{-3}T + 0.15 \times 10^5 T^{-2}.$$

TABLE 352.—*Heat content and entropy of DBr(g)*

[Base, ideal gas at 298.15° K.; mol. wt., 81.93]

T, ° K.	$H_T - H_{298.15}$, cal./mole	$S_T - S_{298.15}$, cal./deg. mole	T, ° K.	$H_T - H_{298.15}$, cal./mole	$S_T - S_{298.15}$, cal./deg. mole
400	715	2.06	1,000	5,280	8.96
500	1,430	3.66	1,200	6,920	10.46
600	2,165	5.00	1,400	8,590	11.75
700	2,920	6.16	1,600	10,290	12.88
800	3,690	7.19	1,800	12,015	13.89
900	4,480	8.12	2,000	13,750	14.81

DBr(*g*):

$$H_T - H_{298.15} = 6.79T + 0.59 \times 10^{-3}T^2 + 0.14 \times 10^5 T^{-1}$$

$$-2,124 \ (0.6 \text{ percent}; \ 298°–2,000° \text{ K.});$$

$$C_p = 6.79 + 1.18 \times 10^{-3}T - 0.14 \times 10^5 T^{-2}.$$

TABLE 353.—*Heat content and entropy of TBr(g)*

[Base, ideal gas at 298.15° K; mol. wt., 82.93]

T, ° K.	$H_T - H_{298.15}$, cal./mole	$S_T - S_{298.15}$, cal./deg. mole	T, ° K.	$H_T - H_{298.15}$, cal./mole	$S_T - S_{298.15}$, cal./deg. mole
400	725	2.09	1,000	5,430	9.19
500	1,455	3.72	1,200	7,105	10.72
600	2,210	5.10	1,400	8,815	12.04
700	2,990	6.30	1,600	10,540	13.19
800	3,790	7.36	1,800	12,280	14.21
900	4,605	8.32	2,000	14,025	15.13

TABLE 354.—*Heat content and entropy of HCl(g)*

[Base, ideal gas at 298.15° K; mol. wt.; 36.46]

T, ° K.	$H_T - H_{298.15}$, cal./mole	$S_T - S_{298.15}$, cal./deg. mole	T, ° K.	$H_T - H_{298.15}$, cal./mole	$S_T - S_{298.15}$, cal./deg. mole
400	710	2.04	1,300	7,355	10.64
500	1,410	3.61	1,400	8,155	11.23
600	2,110	4.89	1,500	8,965	11.79
700	2,825	5.99	1,600	9,785	12.32
800	3,545	6.95	1,700	10,610	12.82
900	4,280	7.82	1,800	11,440	13.29
1,000	5,030	8.60	1,900	12,280	13.75
1,100	5,790	9.33	2,000	13,125	14.18
1,200	6,570	10.01			

HCl(*g*):

$$H_T - H_{298.15} = 6.27T + 0.62 \times 10^{-3}T^2 - 0.30 \times 10^5 T^{-1}$$

$$-1,824 \ (0.2 \text{ percent}; \ 298°–2,000° \text{ K.});$$

$$C_p = 6.27 + 1.24 \times 10^{-3}T + 0.30 \times 10^5 T^{-2}.$$

TABLE 355.—*Heat content and entropy of DCl(g)*

[Base, ideal gas at 298.15° K.; mol. wt., 37.47]

T, ° K.	$H_T - H_{298.15}$, cal./mole	$S_T - S_{298.15}$, cal./deg. mole	T, ° K.	$H_T - H_{298.15}$, cal./mole	$S_T - S_{298.15}$, cal./deg. mole
400	710	2.05	1,000	5,190	8.82
500	1,420	3.63	1,200	6,800	10.29
600	2,140	4.95	1,400	8,450	11.56
700	2,875	6.08	1,600	10,130	12.68
800	3,630	7.08	1,800	11,830	13.68
900	4,405	8.00	2,000	13,545	14.59

DCl(*g*):

$$H_T - H_{298.15} = 6.59T + 0.62 \times 10^{-3}T^2$$

$$-2,020 \ (0.5 \text{ percent}; \ 298°–2,000° \text{ K.});$$

$$C_p = 6.59T + 1.24 \times 10^{-3}T.$$

TABLE 356.—*Heat content and entropy of TCl(g)*

[Base, idel gas at 298.15° K.; mol. wt., 38.47]

T, ° K.	$H_T - H_{298.15}$, cal./mole	$S_T - S_{298.15}$, cal./deg. mole	T, ° K.	$H_T - H_{298.15}$, cal./mole	$S_T - S_{298.15}$, cal./deg. mole
400	715	2.06	1,000	5,330	9.03
500	1,435	3.67	1,200	6,980	10.54
600	2,175	5.02	1,400	8,665	10.83
700	2,935	6.19	1,600	10,370	12.97
800	3,715	7.23	1,800	12,095	13.99
900	4,515	8.17	2,000	13,830	14.90

TCl(*g*):

$$H_T - H_{298.15} = 6.90T + 0.57 \times 10^{-3}T^2 + 0.22 \times 10^5 T^{-1}$$

$$-2,182 \ (0.7 \text{ percent}; \ 298–2,000° \text{ K.});$$

$$C_p = 6.90 + 1.14 \times 10^{-3}T - 0.22 \times 10^5 T^{-2}.$$

TABLE 1h7. (Continued)

FLUORIDES

IODIDE

TABLE 357.—Heat content and entropy of HF(g)

[Base, ideal gas at 298.15° K.; mol. wt., 20.01]

T, ° K.	$H_T-H_{298.15}$, cal./mole	$S_T-S_{298.15}$, cal./deg. mole	T, ° K.	$H_T-H_{298.15}$, cal./mole	$S_T-S_{298.15}$, cal./deg. mole
400	710	2.05	2,000	12,625	13.78
500	1,405	3.60	2,100	13,440	14.17
600	2,105	4.87	2,200	14,260	14.55
700	2,805	5.95	2,300	15,090	14.92
800	3,510	6.90	2,400	15,925	15.28
900	4,220	7.73	2,500	16,770	15.62
1,000	4,935	8.48	2,750	18,890	16.43
1,100	5,660	9.17	3,000	21,045	17.18
1,200	6,395	9.81	3,250	23,220	17.88
1,300	7,140	10.41	3,500	25,415	18.53
1,400	7,895	10.97	3,750	27,620	19.14
1,500	8,660	11.50	4,000	29,845	19.71
1,600	9,435	12.00	4,250	32,090	20.26
1,700	10,220	12.47	4,500	34,360	20.78
1,800	11,015	12.93	4,750	36,630	21.27
1,900	11,815	13.36	5,000	38,905	21.74

HF(g):

$$H_T - H_{298.15} = 6.55T + 0.36 \times 10^{-3}T^2 - 0.17 \times 10^5 T^{-1}$$
$$-1,928 \ (0.4 \ \text{percent}; \ 298°-4,000° \ \text{K.});$$
$$C_p = 6.55 + 0.72 \times 10^{-3}T + 0.17 \times 10^5 T^{-2}.$$

TABLE 358.—Heat content and entropy of DF(g)

[Base, ideal gas at 298.15° K.; mol. wt., 21.01]

T, ° K.	$H_T-H_{298.15}$, cal./mole	$S_T-S_{298.15}$, cal./deg. mole	T, ° K.	$H_T-H_{298.15}$, cal./mole	$S_T-S_{298.15}$, cal./deg. mole
400	710	2.05	2,000	13,120	14.17
500	1,405	3.61	2,100	13,970	14.59
600	2,110	4.89	2,200	14,830	14.99
700	2,820	5.98	2,300	15,690	15.37
800	3,545	6.95	2,400	16,555	15.74
900	4,280	7.81	2,500	17,425	16.09
1,000	5,025	8.60	2,750	19,620	16.93
1,100	5,790	9.33	3,000	21,835	17.70
1,200	6,565	10.00	3,250	24,060	18.41
1,300	7,350	10.63	3,500	26,305	19.08
1,400	8,145	11.22	3,750	28,565	19.70
1,500	8,955	11.78	4,000	30,830	20.29
1,600	9,770	12.30	4,250	33,110	20.84
1,700	10,595	12.81	4,500	35,395	21.36
1,800	11,430	13.28	4,750	37,695	21.86
1,900	12,270	13.74	5,000	40,000	22.33

DF(g):

$$H_T - H_{298.15} = 6.59T + 0.47 \times 10^{-3}T^2 - 0.08 \times 10^5 T^{-1}$$
$$-1,980 \ (0.5 \ \text{percent}; \ 298°-3,000° \ \text{K.});$$
$$C_p = 6.59 + 0.94 \times 10^{-3}T + 0.08 \times 10^5 T^{-2}.$$

TABLE 359.—Heat content and entropy of TF(g)

[Base, ideal gas at 298.15° K.; mol. wt., 22.02]

T, ° K.	$H_T-H_{298.15}$, cal./mole	$S_T-S_{298.15}$, cal./deg. mole	T, ° K.	$H_T-H_{298.15}$, cal./mole	$S_T-S_{298.15}$, cal./deg. mole
400	710	2.05	1,000	5,095	8.69
500	1,410	3.61	1,200	6,670	10.13
600	2,120	4.90	1,400	8,285	11.37
700	2,840	6.01	1,600	9,930	12.47
800	3,580	7.00	1,800	11,605	13.45
900	4,330	7.88	2,000	13,300	14.35

TF(g):

$$H_T - H_{298.15} = 6.37T + 0.63 \times 10^{-3}T^2 - 0.20 \times 10^5 T^{-1}$$
$$-1,888 \ (0.3 \ \text{percent}; \ 298°-2,000° \ \text{K.});$$
$$C_p = 6.37 + 1.26 \times 10^{-3}T + 0.20 \times 10^5 T^{-2}.$$

TABLE 360.—Heat content and entropy of HI(g)

[Base, ideal gas at 298.15° K.; mol. wt., 127.92]

T, ° K.	$H_T-H_{298.15}$, cal./mole	$S_T-S_{298.15}$, cal./deg. mole	T, ° K.	$H_T-H_{298.15}$, cal./mole	$S_T-S_{298.15}$, cal./deg. mole
400	710	2.05	1,300	7,605	10.93
500	1,415	3.62	1,400	8,440	11.54
600	2,135	4.93	1,500	9,285	12.12
700	2,870	6.06	1,600	10,140	12.67
800	3,620	7.07	1,700	11,005	13.20
900	4,390	7.97	1,800	11,875	13.69
1,000	5,175	8.80	1,900	12,755	14.17
1,100	5,970	9.56	2,000	13,640	14.62
1,200	6,785	10.27			

HI(g):

$$H_T - H_{298.15} = 6.39T + 0.71 \times 10^{-3}T^2 - 0.14 \times 10^5 T^{-1}$$
$$-1,921 \ (0.4 \ \text{percent}; \ 298°-2,000° \ \text{K.});$$
$$C_p = 6.39 + 1.42 \times 10^{-3}T + 0.14 \times 10^5 T^{-2}.$$

TABLE 361.—Heat content and entropy of DI(g)

[Base, ideal gas at 298.15° K.; mol. wt., 128.92]

T, ° K.	$H_T-H_{298.15}$, cal./mole	$S_T-S_{298.15}$, cal./deg. mole	T, ° K.	$H_T-H_{298.15}$, cal./mole	$S_T-S_{298.15}$, cal./deg. mole
400	725	2.09	1,000	5,390	9.13
500	1,450	3.71	1,200	7,055	10.65
600	2,200	5.07	1,400	8,755	11.96
700	2,970	6.26	1,600	10,475	13.11
800	3,760	7.32	1,800	12,205	14.13
900	4,565	8.26	2,000	13,950	15.05

DI(g):

$$H_T - H_{298.15} = 7.08T + 0.53 \times 10^{-3}T^2 + 0.34 \times 10^5 T^{-1}$$
$$-2,272 \ (0.6 \ \text{percent}; \ 298°-2,000° \ \text{K.});$$
$$C_p = 7.08 + 1.06 \times 10^{-3}T - 0.34 \times 10^5 T^{-2}.$$

SELENIDES

TABLE 362.—Heat content and entropy of $H_2Se(g)$

[Base, ideal gas at 298.15° K; mol. wt., 80.98]

T, ° K.	$H_T-H_{298.15}$, cal./mole	$S_T-S_{298.15}$, cal./deg. mole	T, ° K.	$H_T-H_{298.15}$, cal./mole	$S_T-S_{298.15}$, cal./deg. mole
400	860	2.48	1,000	6,890	11.50
500	1,750	4.46	1,200	9,210	13.62
600	2,690	6.18	1,400	11,630	15.48
700	3,675	7.69	1,600	14,120	17.14
800	4,705	9.07	1,800	16,665	18.64
900	5,780	10.33	2,000	19,245	20.00

$H_2Se(g)$:

$$H_T - H_{298.15} = 7.59T + 1.75 \times 10^{-3}T^2 + 0.31 \times 10^5 T^{-1}$$
$$-2,523 \ (0.9 \ \text{percent}; \ 298°-2,000° \ \text{K.});$$
$$C_p = 7.59 + 3.50 \times 10^{-3}T - 0.31 \times 10^5 T^{-2}.$$

TABLE 1b7. (*Continued*)

TABLE 363.—*Heat content and entropy of* $D_2Se(g)$

[Base, ideal gas at 298.15° K.; mol. wt., 82.99]

T, ° K.	$H_T-H_{298.15}$, cal./mole	$S_T-S_{298.15}$, cal./deg. mole	T, ° K.	$H_T-H_{298.15}$, cal./mole	$S_T-S_{298.15}$, cal./deg. mole
400	930	2.68	1,000	7,585	12.64
500	1,905	4.85	1,200	10,085	14.92
600	2,945	6.74	1,400	12,650	16.89
700	4,040	8.43	1,600	15,265	18.64
800	5,185	9.96	1,800	17,915	20.20
900	6,370	11.36	2,000	20,590	21.61

$D_2Se(g)$:

$$H_T-H_{298.15}=9.68T+1.20\times10^{-3}T^2+1.44\times10^5T^{-1}$$
$$-3,476 \ (1.0 \text{ percent}; 298°-2,000° \text{ K.});$$
$$C_p=9.68+2.40\times10^{-3}T-1.44\times10^5T^{-2}.$$

TABLE 364.—*Heat content and entropy of* $HDSe(g)$

[Base, ideal gas at 298.15° K.; mol. wt., 81.98]

T, ° K.	$H_T-H_{298.15}$, cal./mole	$S_T-S_{298.15}$, cal./deg. mole	T, ° K.	$H_T-H_{298.15}$, cal./mole	$S_T-S_{298.15}$, cal./deg. mole
400	890	2.56	1,000	7,215	12.03
500	1,820	4.64	1,200	9,610	14.21
600	2,810	6.44	1,400	12,100	16.13
700	3,850	8.04	1,600	14,650	17.83
800	4,930	9.48	1,800	17,240	19.36
900	6,055	10.81	2,000	19,870	20.74

$HDSe(g)$:

$$H_T-H_{298.15}=8.79T+1.37\times10^{-3}T^2+1.00\times10^5T^{-1}$$
$$-3,078 \ (1.0 \text{ percent}; 298°-2,000° \text{ K.});$$
$$C_p=8.79+2.74\times10^{-3}T-1.00\times10^5T^{-2}.$$

TELLURIDE

TABLE 365.—*Heat content and entropy of* $H_2Te(g)$

[Base, ideal gas at 298.15° K.; mole. wt., 129.63]

T, ° K.	$H_T-H_{298.15}$, cal./mole	$S_T-S_{298.15}$, cal./deg. mole	T. K.	$H_T-H_{298.15}$, cal./mole	$S_T-S_{298.15}$, cal./deg. mole
400	890	2.57	1,000	7,125	11.90
500	1,815	4.63	1,200	9,495	14.06
600	2,790	6.41	1,400	11,955	15.96
700	3,810	7.98	1,600	14,485	17.65
800	4,870	9.39	1,800	17,065	19.16
900	5,975	10.69	2,000	19,680	20.54

$H_2Te(g)$:

$$H_T-H_{298.15}=8.48T+1.44\times10^{-3}T^2+0.74\times10^5T^{-1}$$
$$-2,905 \ (0.8 \text{ percent}; 298°-2,000° \text{ K.});$$
$$C_p=8.48+2.88\times10^{-3}T-0.74\times10^5T^{-2}.$$

AZIDE

TABLE 366.—*Heat content and entropy of* $HN_3(g)$

[Base, ideal gas at 298.15° K.; mol. wt., 43.03]

T, ° K.	$H_T-H_{298.15}$, cal./mole	$S_T-S_{298.15}$, cal./deg. mole	T. ° K.	$H_T-H_{298.15}$, cal./mole	$S_T-S_{298.15}$, cal./deg. mole
400	1,095	3.15	1,000	9,565	15.76
500	2,295	5.82	1,200	12,870	18.78
600	3,600	8.20	1,400	16,280	21.40
700	4,990	10.34	1,600	19,810	23.76
800	6,460	12.30	1,800	23,420	25.88
900	7,985	14.10	2,000	27,090	27.81

$HN_3(g)$:

$$H_T-H_{298.15}=11.33T+2.31\times10^{-3}T^2+2.38\times10^5T^{-1}$$
$$-4,382 \ (0.9 \text{ percent}; 298°-1,800° \text{ K.});$$
$$C_p=11.33+4.62\times10^{-3}T-2.38\times10^5T^{-2}.$$

CYANIDES

TABLE 367.—*Heat content and entropy of* $HCN(g)$

[Base, ideal gas at 298.15° K.; mol. wt., 27.03]

T, ° K.	$H_T-H_{298.15}$, cal./mole	$S_T-S_{298.15}$, cal./deg. mole	T, ° K.	$H_T-H_{298.15}$, cal./mole	$S_T-S_{298.15}$, cal./deg. mole
400	920	2.65	2,000	20,885	21.70
500	1,895	4.82	2,100	22,315	22.39
600	2,925	6.70	2,200	23,755	23.06
700	4,005	8.36	2,300	25,210	23.71
800	5,130	9.86	2,400	26,670	24.33
900	6,295	11.23	2,500	28,135	24.93
1,000	7,495	12.50	2,750	31,840	26.34
1,100	8,730	13.67	3,000	35,590	27.65
1,200	9,995	14.77	3,250	39,375	28.86
1,300	11,285	15.81	3,500	43,190	29.99
1,400	12,600	16.78	3,750	47,035	31.05
1,500	13,935	17.70	4,000	50,910	32.05
1,600	15,295	18.58	4,250	54,800	32.99
1,700	16,670	19.41	4,500	58,715	33.89
1,800	18,060	20.21	4,750	62,655	34.74
1,900	19,465	20.97	5,000	66,615	35.55

$HCN(g)$:

$$H_T-H_{298.15}=9.41T+1.35\times10^{-3}T^2+1.44\times10^5T^{-1}$$
$$-3,409 \ (0.7 \text{ percent}; 298°-2,500° \text{ K.});$$
$$C_p=9.41+2.70\times10^{-3}T-1.44\times10^5T^{-2}.$$

TABLE 368.—*Heat content and entropy of* $DCN(g)$

[Base, ideal gas at 298.15° K.; mol. wt., 28.03]

T, ° K.	$H_T-H_{298.15}$, cal./mole	$S_T-S_{298.15}$, cal./deg. mole	T, ° K.	$H_T-H_{298.15}$, cal./mole	$S_T-S_{298.15}$, cal./deg. mole
400	975	2.81	2,000	21,620	22.59
500	2,000	5.10	2,100	23,075	23.30
600	3,080	7.06	2,200	24,540	23.98
700	4,205	8.80	2,300	26,015	24.63
800	5,380	10.36	2,400	27,495	25.26
900	6,595	11.79	2,500	28,980	25.87
1,000	7,845	13.11	2,750	32,725	27.30
1,100	9,125	14.33	3,000	36,505	28.61
1,200	10,435	15.47	3,250	40,310	29.82
1,300	11,775	16.54	3,500	44,140	30.96
1,400	13,135	17.55	3,750	48,000	32.03
1,500	14,515	18.50	4,000	51,880	33.03
1,600	15,910	19.40	4,250	55,775	33.97
1,700	17,320	20.26	4,500	59,690	34.87
1,800	18,740	21.07	4,750	63,625	35.72
1,900	20,175	21.85	5,000	67,575	36.53

TABLE 1h7. (Continued)

DCN(g):

$$H_T - H_{298.15} = 9.94T + 1.30 \times 10^{-3}T^2 + 1.35 \times 10^5 T^{-1}$$
$$-3,532 \ (0.8 \ \text{percent}; \ 298°-2,500° \ \text{K.});$$
$$C_p = 9.94 + 2.60 \times 10^{-3}T - 1.35 \times 10^5 T^{-2}.$$

TABLE 369.—*Heat content and entropy of TCN(g)*

[Base, ideal gas at 298.15° K.; mol. wt., 29.04]

T, ° K.	$H_T-H_{298.15}$, cal./mole	$S_T-S_{298.15}$, cal./deg. mole	T, ° K.	$H_T-H_{298.15}$, cal./mole	$S_T-S_{298.15}$, cal./deg. mole
400	995	2.87	2,000	22,010	23.04
500	2,040	5.20	2,100	23,480	23.75
600	3,145	7.21	2,200	24,955	24.44
700	4,300	8.99	2,300	26,440	25.10
800	5,500	10.60	2,400	27,935	25.73
900	6,745	12.06	2,500	29,430	26.35
1,000	8,025	13.41	2,750	33,200	27.78
1,100	9,335	14.66	3,000	37,000	29.10
1,200	10,670	15.82	3,250	40,830	30.33
1,300	12,030	16.91	3,500	44,685	31.47
1,400	13,415	17.93	3,750	48,560	32.54
1,500	14,815	18.90	4,000	52,460	33.55
1,600	16,230	19.81	4,250	56,375	34.50
1,700	17,655	20.68	4,500	60,310	35.40
1,800	19,095	21.50	4,750	64,260	36.25
1,900	20,550	22.29	5,000	68,230	37.07

TCN(g):

$$H_T - H_{298.15} = 10.31T + 1.25 \times 10^{-3}T^2 + 1.48 \times 10^5 T^{-1}$$
$$-3,681 \ (0.9 \ \text{percent}; \ 298°-2,500° \ \text{K.});$$
$$C_p = 10.31 + 2.50 \times 10^{-3}T - 1.48 \times 10^5 T^{-2}.$$

BORIC ACID

$H_3BO_3(c)$:

$$C_p = 19.44 \ (298.15° \ \text{K.}).$$

HYPOCHLORITES

TABLE 370.—*Heat content and entropy of HClO(g)*

[Base, ideal gas at 298.15° K.; mol. wt., 52.46]

T, ° K.	$H_T-H_{298.15}$, cal./mole	$S_T-S_{298.15}$, cal./deg. mole	T, ° K.	$H_T-H_{298.15}$, cal./mole	$S_T-S_{298.15}$, cal./deg. mole
400	940	2.71	1,000	7,395	12.43
500	1,925	4.91	1,200	9,750	14.57
600	2,955	6.78	1,400	12,175	16.44
700	4,025	8.43	1,600	14,660	18.10
800	5,120	9.89	1,800	17,190	19.59
900	6,245	11.22	2,000	19,770	20.95

HClO(g):

$$H_T - H_{298.15} = 9.72T + 0.93 \times 10^{-3}T^2 + 1.24 \times 10^5 T^{-1}$$
$$-3,397 \ (0.5 \ \text{percent}; \ 298°-2,000° \ \text{K.});$$
$$C_p = 9.72 + 1.86 \times 10^{-3}T - 1.24 \times 10^5 T^{-2}.$$

TABLE 371.—*Heat content and entropy of DClO(g)*

[Base, ideal gas at 298.15° K.; mol. wt., 53.47]

T, ° K.	$H_T-H_{298.15}$, cal./mole	$S_T-S_{298.15}$, cal./deg. mole	T, ° K.	$H_T-H_{298.15}$, cal./mole	$S_T-S_{298.15}$, cal./deg. mole
400	980	2.82	1,000	7,725	12.97
500	2,005	5.11	1,200	10,195	15.22
600	3,080	7.07	1,400	12,725	17.17
700	4,195	8.78	1,600	15,310	18.90
800	5,345	10.32	1,800	17,930	20.44
900	6,525	11.71	2,000	20,570	21.83

DClO(g):

$$H_T - H_{298.15} = 10.28T + 0.92 \times 10^{-3}T^2 + 1.46 \times 10^5 T^{-1}$$
$$-3,636 \ (0.5 \ \text{percent}; \ 298°-2,000° \ \text{K.});$$
$$C_p = 10.28 + 1.84 \times 10^{-3}T - 1.46 \times 10^5 T^{-2}.$$

ISOCYANATE

TABLE 372.—*Heat content and entropy of HNCO(g)*

[Base, ideal gas at 298.15° K.; mol. wt., 43.03]

T, ° K.	$H_T-H_{298.15}$, cal./mole	$S_T-S_{298.15}$, cal./deg. mole	T, ° K.	$H_T-H_{298.15}$, cal./mole	$S_T-S_{298.15}$, cal./deg. mole
400	1,165	3.35	1,000	9,860	16.35
500	2,430	6.17	1,200	13,170	19.36
600	3,785	8.64	1,400	16,610	22.01
700	5,215	10.84	1,600	20,155	24.38
800	6,710	12.84	1,800	23,760	26.50
900	8,255	14.66	2,000	27,435	28.44

HNCO(g):

$$H_T - H_{298.15} = 12.47T + 1.82 \times 10^{-3}T^2 + 2.52 \times 10^5 T^{-1}$$
$$-4,725 \ (0.8 \ \text{percent}; \ 298°-2,000° \ \text{K.});$$
$$C_p = 12.47 + 3.64 \times 10^{-3}T - 2.52 \times 10^5 T^{-2}.$$

NITRITE

TABLE 373.—*Heat content and entropy of $HNO_2(g)$ (cis-form)*

[Base, ideal gas at 298.15° K.; mol. wt., 47.02]

T, ° K.	$H_T-H_{298.15}$, cal./mole	$S_T-S_{298.15}$, cal./deg. mole	T, ° K.	$H_T-H_{298.15}$, cal./mole	$S_T-S_{298.15}$, cal./deg. mole
400	1,170	3.36	1,000	10,065	16.65
500	2,450	6.22	1,200	13,450	19.73
600	3,835	8.72	1,400	16,945	22.42
700	5,310	11.01	1,600	20,530	24.81
800	6,840	13.05	1,800	24,180	26.96
900	8,425	14.92	2,000	27,890	28.92

$HNO_2(g)$ (cis-form):

$$H_T - H_{298.15} = 13.07T + 1.71 \times 10^{-3}T^2 + 3.00 \times 10^5 T^{-1}$$
$$-5,055 \ (1.0 \ \text{percent}; \ 298°-2,000° \ \text{K.});$$
$$C_p = 13.07 + 3.42 \times 10^{-3}T - 3.00 \times 10^5 T^{-2}.$$

TABLE 1h7. (*Continued*)

TABLE 374.—*Heat content and entropy of* $HNO_2(g)$ *(trans-form)*

[Base, ideal gas at 298.15° K.; mol. wt., 47.02]

T, ° K.	$H_T-H_{298.15}$, cal./mole	$S_T-S_{298.15}$, cal./deg. mole	T, ° K.	$H_T-H_{298.15}$, cal./mole	$S_T-S_{298.15}$, cal./deg. mole
400	1,200	3.45	1,000	10,150	16.83
500	2,500	6.35	1,200	13,530	19.91
600	3,900	8.90	1,400	17,020	22.60
700	5,380	11.18	1,600	20,595	24.99
800	6,920	13.23	1,800	24,240	27.13
900	8,510	15.11	2,000	27,935	29.08

$HNO_2(g)$ *(trans-form)*;

$$H_T - H_{298.15} = 13.25T + 1.63 \times 10^{-3}T^2 + 2.86 \times 10^5 T^{-1}$$
$$-5,055 \ (0.8 \text{ percent}; 298°-2,000° \text{ K.});$$
$$C_p = 13.25 + 3.26 \times 10^{-3}T - 2.86 \times 10^5 T^{-2}.$$

NITRATE

$HNO_3(l)$:
$$C_p = 26.26 \ (298.15° \text{ K.}).$$
$HNO_3 \cdot H_2O(l)$:
$$C_p = 43.60 \ (298.15° \text{ K.}).$$
$HNO_3 \cdot 3H_2O(l)$:
$$C_p = 77.70 \ (298.15° \text{ K.}).$$

TABLE 375.—*Heat content and entropy of* $HNO_3(g)$

[Base, ideal gas at 298.15° K.; mol. wt., 63.02]

T, ° K.	$H_T-H_{298.15}$, cal./mole	$S_T-S_{298.15}$, cal./deg. mole	T, ° K.	$H_T-H_{298.15}$, cal./mole	$S_T-S_{298.15}$, cal./deg. mole
350	695	2.15	450	2,220	5.97
400	1,430	4.10	500	3,050	7.71

$HNO_3(g)$:

$$H_T - H_{298.15} = 8.01T + 9.59 \times 10^{-3}T^2 + 0.82 \times 10^5 T^{-1}$$
$$-3,516 \ (0.2 \text{ percent}; 298°-500° \text{ K.});$$
$$C_p = 8.01 + 19.18 \times 10^{-3}T - 0.82 \times 10^5 T^{-2}.$$

TABLE 376.—*Heat content and entropy of* $DNO_3(g)$

[Base, ideal gas at 298.15° K.; mol. wt., 64.02]

T, ° K.	$H_T-H_{298.15}$, cal./mole	$S_T-S_{298.15}$, cal./deg. mole	T, ° K.	$H_T-H_{298.15}$, cal./mole	$S_T-S_{298.15}$, cal./deg. mole
350	730	2.25	450	2,310	6.21
400	1,495	4.29	500	3,160	8.00

$DNO_3(g)$:

$$H_T - H_{298.15} = 12.90T + 5.56 \times 10^{-3}T^2 + 2.51 \times 10^5 T^{-1}$$
$$-5,182 \ (0.1 \text{ percent}; 298°-500° \text{ K.});$$
$$C_p = 12.90 + 11.12 \times 10^{-3}T - 2.51 \times 10^5 T^{-2}.$$

PHOSPHATE

$H_3PO_4(c)$:
$$C_p = 25.35 \ (298° \text{ K.}).$$
$2H_3PO_4 \cdot H_2O(c)$:
$$C_p = 60.24 \ (298° \text{ K.});$$

SULFATES

$H_2SO_4(l)$:
$$C_p = 33.2 \ (298° \text{ K.}).$$
$H_2SO_4 \cdot H_2O(l)$:
$$C_p = 51.3 \ (298° \text{ K.}).$$
$H_2SO_4 \cdot 2H_2O(l)$:
$$C_p = 62.4 \ (298° \text{ K.}).$$
$H_2SO_4 \cdot 3H_2O(l)$:
$$C_p = 76.6 \ (298° \text{ K.}).$$
$H_2SO_4 \cdot 4H_2O(l)$:
$$C_p = 92.3 \ (298° \text{ K.}).$$
$H_2SO_4 \cdot 6H_2O(l)$:
$$C_p = 127.0 \ (298° \text{ K.}).$$
$H_2SO_4 \cdot 6.5H_2O(l)$:
$$C_p = 136.3 \ (298° \text{ K.}).$$
$H_2SO_4 \cdot 8H_2O(l)$:
$$C_p = 164.3 \ (298° \text{ K.}).$$
$H_2S_2O_7(c)$:
$$C_p = 27.4 \ (291° \text{ K.}).$$
$$\Delta H_{308} \ (fusion) = 3,190.$$
$H_2S_2O_7(l)$:
$$C_p = 57.5 \ (308° \text{ K.}).$$

INDIUM AND ITS COMPOUNDS

ELEMENT

TABLE 377.—*Heat content and entropy of* $In(c, l)$

[Base, crystals at 298.15° K.; atomic wt., 114.82]

T, ° K.	$H_T-H_{298.15}$, cal./mole	$S_T-S_{298.15}$, cal./deg. mole	T, ° K.	$H_T-H_{298.15}$, cal./mole	$S_T-S_{298.15}$, cal./deg. mole
350	340	1.05	1,000	5,720	10.28
400	680	1.97	1,200	7,140	11.58
429.3(c)	890	2.47	1,400	8,560	12.67
429.3(l)	1,670	4.28	1,600	9,980	13.62
500	2,170	5.36	1,800	11,400	14.45
600	2,880	6.65	2,000	12,820	15.20
700	3,590	7.75	2,200	14,240	15.88
800	4,300	8.70	2,300	14,950	16.20
900	5,010	9.53			

$In(c)$:
$$H_T - H_{298.15} = 4.59T + 3.02 \times 10^{-3}T^2 - 1,637$$
$$(0.3 \text{ percent}; 298°-429.3° \text{ K.});$$
$$C_p = 4.59 + 6.04 \times 10^{-3}T;$$
$$\Delta H_{429.3} \ (fusion) = 780.$$
$In(l)$:
$$H_T - H_{298.15} = 7.10T - 1,380 \ (0.1 \text{ percent};$$
$$429.3-2.300° \text{ K.});$$
$$C_p = 7.10.$$

TABLE 1h7. (Continued)

TABLE 378.—Heat content and entropy of In(g)

[Base, ideal gas at 298.15° K.; atomic wt., 114.82]

T, ° K.	$H_T-H_{298.15}$, cal./mole	$S_T-S_{298.15}$, cal./deg. mole	T, ° K.	$H_T-H_{298.15}$, cal./mole	$S_T-S_{298.15}$, cal./deg. mole
400	510	1.47	1,900	9,685	10.74
500	1,025	2.62	2,000	10,285	11.05
600	1,560	3.60	2,200	11,475	11.62
700	2,130	4.47	2,400	12,640	12.12
800	2,720	5.26	2,600	13,780	12.58
900	3,340	5.99	2,800	14,905	13.00
1,000	3,970	6.65	3,000	16,015	13.38
1,100	4,615	7.27	3,500	18,735	14.22
1,200	5,265	7.83	4,000	21,410	14.93
1,300	5,910	8.35	4,500	24,060	15.56
1,400	6,555	8.83	5,000	26,730	16.12
1,500	7,195	9.27	6,000	32,345	17.14
1,600	7,830	9.68	7,000	38,925	18.15
1,700	8,455	10.06	8,000	47,335	19.27
1,800	9,075	10.41			

In(g):

$$H_T-H_{298.15}=4.15T+1.10\times10^{-3}T^2-0.14\times10^5T^{-1}$$
$$-1,288 \ (0.7 \text{ percent}; \ 298°-1,400° \text{ K.});$$
$$C_p=4.15+2.20\times10^{-3}T+0.14\times10^5T^{-2};$$
$$H_T-H_{298.15}=6.71-0.19\times10^{-3}T^2+1.40\times10^5T^{-1}$$
$$-2,453 \ (0.4 \text{ percent}; \ 1,400°-5,000° \text{ K.});$$
$$C_p=6.71-0.38\times10^{-3}T-1.40\times10^5T^{-2};$$

OXIDES

TABLE 379.—Heat content and entropy of InO(g)

[Base, ideal gas at 298.15° K.; mol. wt., 130.82]

T, ° K.	$H_T-H_{298.15}$, cal./mole	$S_T-S_{298.15}$, cal./deg. mole	T, ° K.	$H_T-H_{298.15}$, cal./mole	$S_T-S_{298.15}$, cal./deg. mole
400	815	2.35	1,000	5,970	10.18
500	1,645	4.20	1,200	7,730	11.78
600	2,490	5.74	1,400	9,495	13.14
700	3,350	7.07	1,600	12,270	14.33
800	4,215	8.22	1,800	13,045	15.37
900	5,090	9.25	2,000	14,830	16.31

InO(g):

$$H_T-H_{298.15}=8.67T+0.08\times10^{-3}T^2+0.82\times10^5T^{-1}$$
$$-2,867 \ (0.2 \text{ percent}; \ 298°-2,000° \text{ K.});$$
$$C_p=8.67T+0.16\times10^{-3}T-0.82\times10^5T^{-2}.$$

In$_2$O$_3$(c):

$$\overline{C_p}=22.4 \ (273°-373° \text{ K.}).$$

BROMIDE

TABLE 380.—Heat content and entropy of InBr(g)

[Base, ideal gas at 298.15° K.; mol. wt., 194.74]

T, ° K.	$H_T-H_{298.15}$, cal./mole	$S_T-S_{298.15}$, cal./deg. mole	T, ° K.	$H_T-H_{298.15}$, cal./mole	$S_T-S_{298.15}$, cal./deg. mole
400	900	2.60	1,000	6,240	10.74
500	1,785	4.57	1,200	8,025	12.37
600	2,670	6.19	1,400	9,810	13.75
700	3,560	7.56	1,600	11,595	14.94
800	4,445	8.75	1,800	13,385	15.99
900	5,345	9.80	2,000	15,170	16.93

InBr(g):

$$H_T-H_{298.15}=8.94T+0.15\times10^5T^{-1}-2,716 \ (0.1 \text{ percent};$$
$$298°-2,000° \text{ K.});$$
$$C_p=8.94-0.15\times10^5T^{-2}.$$

CHLORIDE

TABLE 381.—Heat content and entropy of InCl(g)

[Base, ideal gas at 298.15° K.; mol. wt., 150.28]

T, ° K.	$H_T-H_{298.15}$, cal./mole	$S_T-S_{298.15}$, cal./deg. mole	T, ° K.	$H_T-H_{298.15}$, cal./mole	$S_T-S_{298.15}$, cal./deg. mole
400	885	2.56	1,000	6,200	10.66
500	1,760	4.51	1,200	7,980	12.28
600	2,645	6.12	1,400	9,765	13.66
700	3,530	7.48	1,600	11,550	14.85
800	4,420	8.67	1,800	13,340	15.90
900	5,310	9.72	2,000	15,125	16.84

InCl(g):

$$H_T-H_{298.15}=8.93T+0.29\times10^5T^{-1}-2,760 \ (0.1 \text{ percent};$$
$$298°-2,000° \text{ K.});$$
$$C_p=8.93-0.29\times10^5T^{-2}.$$

IODIDE

TABLE 382.—Heat content and entropy of InI(g)

Base, ideal gas at 298.15° K.; mol. wt., 241.73]

T, ° K.	$H_T-H_{298.15}$, cal./mole	$S_T-S_{298.15}$, cal./deg. mole	T, ° K.	$H_T-H_{298.15}$, cal./mole	$S_T-S_{298.15}$, cal./deg. mole
400	900	2.60	1,000	6,250	10.76
500	1,790	4.58	1,200	8,040	12.40
600	2,680	6.20	1,400	9,825	13.77
700	3,575	7.58	1,600	11,615	14.97
800	4,465	8.77	1,800	13,400	16.02
900	5,360	9.83	2,000	15,185	16.96

InI(g):

$$H_T-H_{298.15}=8.94T+0.10\times10^5T^{-1}-2,699 \ (0.1 \text{ percent};$$
$$298°-2,000° \text{ K.});$$
$$C_p=8.94-0.10\times10^5T^{-2}.$$

SULFATE

In$_2$(SO$_4$)$_3$:

$$\overline{C_p}=66.8 \ (273°-373° \text{ K.}).$$

IODINE AND ITS COMPOUNDS

ELEMENT

TABLE 383.—Heat content and entropy of I$_2$(c, l, g)

[Base, crystals at 298.15° K.; mol. wt., 253.82]

T, ° K.	$H_T-H_{298.15}$, cal./mole	$S_T-S_{298.15}$, cal./deg. mole	T, ° K.	$H_T-H_{298.15}$, cal./mole	$S_T-S_{298.15}$, cal./deg. mole
350	695	2.16	1,200	22,990	47.04
386.8(c)	1,210	3.55	1,300	23,900	47.77
386.8(l)	4,980	13.30	1,400	24,810	48.45
400	5,235	13.95	1,500	25,730	49.07
450	6,195	16.21	1,600	26,640	49.66
456(l)	6,310	16.47	1,700	27,560	50.22
456(g)	16,280	38.33	1,800	28,480	50.74
500	16,670	39.15	1,900	29,400	51.24
600	17,570	40.78	2,000	30,320	51.71
700	18,470	42.17	2,200	32,160	52.59
800	19,370	43.37	2,400	34,010	53.39
900	20,270	44.43	2,600	35,860	54.13
1,000	21,180	45.39	2,800	37,720	54.82
1,100	22,080	46.25	3,000	39,580	55.46

TABLE 1h7. (*Continued*)

$I_2(c)$:

$H_T - H_{298.15} = 9.59T + 5.95 \times 10^{-3}T^2 - 3,388$

(0.2 percent; 298°–368.6° K.);

$C_p = 9.59 + 11.90 \times 10^{-3}T$;

$\Delta H_{368.6}(fusion) = 3,770$.

$I_2(l)$:

$H_T - H_{298.15} = 19.20T - 2,445$ (0.1 percent;

386.6°–456° K.);

$C_p = 19.20$;

$\Delta H_{456}(vaporization) = 9,970$.

$I_2(g)$:

$H_T - H_{298.15} = 8.94T + 0.07 \times 10^{-3}T^2 + 0.17 \times 10^5 T^{-1}$

$+ 12,151$ (0.1 percent; 456°–3,000° K.);

$C_p = 8.94 + 0.14 \times 10^{-3}T - 0.17 \times 10^5 T^{-2}$.

TABLE 384.—*Heat content and entropy of $I_2(g)$*

[Base, ideal gas at 298.15° K.; mol. wt., 253.82]

T, ° K.	$H_T - H_{298.15}$, cal./mole	$S_T - S_{298.15}$, cal./deg. mole	T, ° K.	$H_T - H_{298.15}$, cal./mole	$S_T - S_{298.15}$, cal./deg. mole
400	905	2.60	1,600	11,765	15.12
500	1,795	4.60	1,700	12,680	15.67
600	2,690	6.23	1,800	13,595	16.19
700	3,590	7.62	1,900	14,515	16.69
800	4,495	8.82	2,000	15,435	17.16
900	5,395	9.88	2,100	16,355	17.61
1,000	6,300	10.84	2,200	17,280	18.04
1,100	7,205	11.70	2,300	18,205	18.45
1,200	8,115	12.49	2,400	19,130	18.85
1,300	9,025	13.22	2,500	20,055	19.22
1,400	9,935	13.90	2,750	22,380	20.11
1,500	10,850	14.52	3,000	24,705	20.92

$I_2(g)$:

$H_T - H_{298.15} = 8.94T + 0.07 \times 10^{-3}T^2 + 0.17 \times 10^5 T^{-1}$

$- 2,729$ (0.1 percent; 298°–3,000° K.);

$C_p = 8.94 + 0.14 \times 10^{-3}T - 0.17 \times 10^5 T^{-2}$.

Table 385.—*Heat content and entropy of $I(g)$*

[Base, ideal gas at 298.15° K.; atomic wt., 126.91]

T, ° K.	$H_T - H_{298.15}$, cal./mole	$S_T - S_{298.15}$, cal./deg. mole	T, ° K.	$H_T - H_{298.15}$, cal./mole	$S_T - S_{298.15}$, cal./deg. mole
400	505	1.46	2,000	8,500	9.48
500	1,005	2.57	2,100	9,010	9.73
600	1,500	3.47	2,200	9,525	9.97
700	1,995	4.24	2,300	10,040	10.20
800	2,495	4.90	2,400	10,555	10.42
900	2,990	5.49	2,500	11,075	10.63
1,000	3,485	6.01	2,750	12,380	11.15
1,100	3,985	6.49	3,000	13,705	11.59
1,200	4,480	6.92	3,250	15,035	12.01
1,300	4,980	7.32	3,500	16,375	12.41
1,400	5,480	7.69	3,750	17,720	12.78
1,500	5,980	8.03	4,000	19,075	13.13
1,600	6,480	8.36	4,250	20,430	13.46
1,700	6,980	8.66	4,500	21,790	13.77
1,800	7,485	8.95	4,750	23,150	14.07
1,900	7,990	9.22	5,000	24,515	14.35

$I(g)$:

$H_T - H_{298.15} = 4.80T + 0.08 \times 10^{-3}T^2 - 0.11 \times 10^5 T^{-1}$

$- 1,401$ (0.3 percent; 298°–5,000° K.);

$C_p = 4.80 + 0.16 \times 10^{-3}T + 0.11 \times 10^5 T^{-2}$.

BROMIDE

TABLE 386.—*Heat content and entropy of $IBr(g)$*

[Base, ideal gas at 298.15° K.; mol. wt., 206.83]

T, ° K.	$H_T - H_{298.15}$, cal./mole	$S_T - S_{298.15}$, cal./deg. mole	T, ° K.	$H_T - H_{298.15}$, cal./mole	$S_T - S_{298.15}$, cal./deg. mole
400	895	2.58	1,300	8,990	13.16
500	1,785	4.56	1,400	9,895	13.83
600	2,675	6.19	1,500	10,805	14.46
700	3,570	7.57	1,600	11,720	15.05
800	4,470	8.77	1,700	12,630	15.60
900	5,370	9.83	1,800	13,545	16.12
1,000	6,275	10.78	1,900	14,460	16.61
1,100	7,175	11.64	2,000	15,375	17.08
1,200	8,080	12.43			

$IBr(g)$:

$H_T - H_{298.15} = 8.93T + 0.06 \times 10^{-3}T^2 + 0.22 \times 10^5 T^{-1}$

$- 2,742$ (0.1 percent; 298°–2,000° K.);

$C_p = 8.93 + 0.12 \times 10^{-3}T - 0.22 \times 10^5 T^{-2}$.

CHLORIDE

TABLE 387.—*Heat content and entropy of $ICl(g)$*

[Base, ideal gas at 298.15° K.; mol. wt., 162.37]

T, ° K.	$H_T - H_{298.15}$, cal./mole	$S_T - S_{298.15}$, cal./deg. mole	T, ° K.	$H_T - H_{298.15}$, cal./mole	$S_T - S_{298.15}$, cal./deg. mole
400	875	2.53	1,300	8,925	13.02
500	1,750	4.48	1,400	9,830	13.70
600	2,635	6.09	1,500	10,735	14.33
700	3,525	7.46	1,600	11,645	14.91
800	4,420	8.66	1,700	12,555	15.46
900	5,315	9.71	1,800	13,465	15.98
1,000	6,215	10.66	1,900	14,380	16.47
1,100	7,115	11.52	2,000	15,290	16.94
1,200	8,020	12.30			

$ICl(g)$:

$H_T - H_{298.15} = 8.92T + 0.06 \times 10^{-3}T^2 + 0.41 \times 10^5 T^{-1}$

$- 2,802$ (0.1 percent; 298°–2,000° K.);

$C_p = 8.92 + 0.12 \times 10^{-3}T - 0.41 \times 10^5 T^{-2}$.

FLUORIDES

TABLE 388.—*Heat content and entropy of $IF(g)$*

[Base, ideal gas at 298.15° K.; mol. wt., 145.91]

T, ° K.	$H_T - H_{298.15}$, cal./mole	$S_T - S_{298.15}$, cal./deg. mole	T, ° K.	$H_T - H_{298.15}$, cal./mole	$S_T - S_{298.15}$, cal./deg. mole
400	835	2.40	1,300	8,755	12.71
500	1,680	4.29	1,400	9,655	13.38
600	2,540	5.86	1,500	10,555	14.00
700	3,415	7.21	1,600	11,460	14.58
800	4,295	8.38	1,700	12,360	15.13
900	5,180	9.42	1,800	13,270	15.65
1,000	6,070	10.36	1,900	14,175	16.14
1,100	6,960	11.21	2,000	15,085	16.60
1,200	7,855	11.99			

$IF(g)$:

$H_T - H_{298.15} = 8.76 + 0.10 \times 10^{-3}T^2 + 0.73 \times 10^5 T^{-1}$

$- 2,866$ (0.1 percent; 298°–2,000° K.);

$C_p = 8.76 + 0.20 \times 10^{-3}T - 0.73 \times 10^5 T^{-2}$.

TABLE 1h7. (*Continued*)

TABLE 389.—*Heat content and entropy of* $IF_5(g)$

[Base, ideal gas at 298.15° K.; mol. wt., 221.91]

T, ° K.	$H_T-H_{298.15}$, cal./mole	$S_T-S_{298.15}$, cal./deg. mole	T, ° K.	$H_T-H_{298.15}$, cal./mole	$S_T-S_{298.15}$, cal./deg. mole
400	2,575	7.41	1,000	20,265	34.14
500	5,320	13.52	1,100	23,355	37.08
600	8,190	18.76	1,200	26,455	39.78
700	11,145	23.31	1,300	29,560	42.26
800	14,145	27.32	1,400	32,690	44.58
900	17,190	30.90	1,500	35,815	46.74

$IF_5(g)$:

$$H_T-H_{298.15}=29.66T+0.81\times10^{-3}T^2+5.73\times10^5T^{-1}$$
$$-10,837 \ (0.3 \text{ percent}; 298°\text{–}1,500° \text{ K.});$$
$$C_p=29.66+1.62\times10^{-3}T-5.73\times10^5T^{-2}.$$

TABLE 390.—*Heat content and entropy of* $IF_7(g)$

[Base, ideal gas at 298.15° K.; mol. wt., 259.91]

T, ° K.	$H_T-H_{298.15}$, cal./mole	$S_T-S_{298.15}$, cal./deg. mole	T, ° K.	$H_T-H_{298.15}$, cal./mole	$S_T-S_{298.15}$, cal./deg. mole
400	3,525	10.13	800	19,465	37.56
500	7,300	18.55	900	23,645	42.48
600	11,255	25.76	1,000	27,875	46.93
700	15,325	32.03			

$IF_7(g)$:

$$H_T-H_{298.15}=40.71T+1.35\times10^{-3}T^2+8.20\times10^5T^{-1}$$
$$-15,008 \ (0.2 \text{ percent}; 298°\text{–}1,000° \text{ K.});$$
$$C_p=40.71+2.70\times10^{-3}T-8.20\times10^5T^{-2}$$

IRIDIUM AND ITS COMPOUNDS

ELEMENT

TABLE 391.—*Heat content and entropy of* $Ir(c, l)$

[Base, crystals at 298.15° K.; atomic wt., 192.2]

T, ° K.	$H_T-H_{298.15}$, cal./mole	$S_T-S_{298.15}$, cal./deg. mole	T, ° K.	$H_T-H_{298.15}$, cal./mole	$S_T-S_{298.15}$, cal./deg. mole
400	620	1.79	1,900	11,410	12.57
500	1,235	3.16	2,000	12,240	13.00
600	1,870	4.32	2,100	13,090	13.41
700	2,525	5.33	2,200	13,950	13.81
800	3,190	6.22	2,300	14,820	14.20
900	3,860	7.00	2,400	15,710	14.58
1,000	4,545	7.73	2,500	16,620	14.95
1,100	5,250	8.40	2,600	17,540	15.31
1,200	5,970	9.03	2,700	18,470	15.66
1,300	6,700	9.61	2,727(c)	18,720	15.75
1,400	7,440	10.16	2,727(l)	25,020	18.06
1,500	8,200	10.68	2,800	25,710	18.31
1,600	8,980	11.19	2,900	26,660	18.64
1,700	9,780	11.67	3,000	27,610	18.96
1,800	10,590	12.13			

$Ir(c)$:

$$H_T-H_{298.15}=5.56T+0.71\times10^{-3}T^2$$
$$-1,721 \ (0.1 \text{ percent}; 298°\text{–}2,727° \text{ K.});$$
$$C_p=5.56+1.42\times10^{-3}T;$$
$$\Delta H_{2727}(fusion)=6,300.$$

$Ir(l)$:

$$H_T-H_{298.15}=9.50T-890$$
$$(0.1 \text{ percent}; 2,727°\text{–}3,000° \text{ K.});$$
$$C_p=9.50.$$

TABLE 392.—*Heat content and entropy of* $Ir(g)$

[Base, ideal gas at 298.15° K.; atomic wt., 192.2]

T, ° K.	$H_T-H_{298.15}$, cal./mole	$S_T-S_{298.15}$, cal./deg. mole	T, ° K.	$H_T-H_{298.15}$, cal./mole	$S_T-S_{298.15}$, cal./deg. mole
400	505	1.46	1,500	6,635	8.62
500	1,005	2.58	1,600	7,275	9.04
600	1,510	3.49	1,700	7,925	9.43
700	2,020	4.28	1,800	8,585	9.81
800	2,545	4.98	1,900	9,250	10.17
900	3,085	5.62	2,000	9,920	10.51
1,000	3,640	6.20	2,200	11,280	11.16
1,100	4,210	6.75	2,400	12,650	11.75
1,200	4,795	7.25	2,600	14,025	12.30
1,300	5,395	7.74	2,800	15,395	12.81
1,400	6,010	8.19	3,000	16,760	13.28

$Ir(g)$:

$$H_T-H_{298.15}=4.54T+0.53\times10^{-3}T^2-0.10\times10^5T^{-1}$$
$$-1,367 \ (1.0 \text{ percent}; 298°\text{–}3,000° \text{ K.});$$
$$C_p=4.54+1.06\times10^{-3}T+0.10\times10^5T^{-2}.$$

OXIDE

TABLE 393.—*Heat content and entropy of* $IrO_2(c)$

[Base, crystals at 298.15° K.; mol. wt., 224.2]

T, ° K.	$H_T-H_{298.15}$, cal./mole	$S_T-S_{298.15}$, cal./deg. mole	T, ° K.	$H_T-H_{298.15}$, cal./mole	$S_T-S_{298.15}$, cal./deg. mole
400	1,490	4.29	900	11,060	19.35
500	3,070	7.81	1,000	13,440	21.86
600	4,810	10.98	1,100	15,940	24.24
700	6,720	13.92	1,200	18,570	26.53
800	8,810	16.71	1,300	21,320	28.73

$IrO_2(c)$:

$$H_T-H_{298.15}=9.17T+7.60\times10^{-3}T^2$$
$$-3,410 \ (0.5 \text{ percent}; 298°\text{–}1,300° \text{ K.});$$
$$C_p=9.17+15.20\times10^{-3}T.$$

IRON AND ITS COMPOUNDS

ELEMENT

TABLE 394.—*Heat content and entropy of* $Fe(c, l)$

[Base, α-crystals at 298.15° K.; atomic wt., 55.85]

T, ° K.	$H_T-H_{298.15}$, cal./mole	$S_T-S_{298.15}$, cal./deg. mole	T, ° K.	$H_T-H_{298.15}$, cal./mole	$S_T-S_{298.15}$, cal./deg. mole
400	640	1.84	1,500	10,975	13.68
500	1,320	3.36	1,600	11,865	14.25
600	2,045	4.68	1,673(γ)	12,525	14.66
700	2,830	5.89	1,673(δ)	12,690	14.76
800	3,705	7.04	1,700	12,945	14.91
900	4,695	8.21	1,800	13,900	15.45
1,000	5,900	9.48	1,812(δ)	14,015	15.52
1,033(α)	6,410	9.98	1,812(l)	17,685	17.54
1,033(β)	6,410	9.98	1,900	18,610	18.04
1,100	7,225	10.74	2,000	19,665	18.58
1,183(β)	8,080	11.49	2,200	21,790	19.59
1,183(γ)	8,295	11.68	2,400	23,930	20.52
1,200	8,435	11.79	2,600	26,090	21.39
1,300	9,260	12.45	2,800	28,260	22.19
1,400	10,110	13.08	3,000	30,450	22.95

$Fe(\alpha)$

$$H_T-H_{298.15}=3.04T+3.79\times10^{-3}T^2-0.60\times10^5T^{-1}$$
$$-1,042 \ (1.0 \text{ percent}; 298°\text{–}1,033° \text{ K.});$$
$$C_p=3.04+7.58\times10^{-3}T+0.60\times10^5T^{-2}.$$
$$\Delta H_{1033}=326.$$

TABLE 1h7. (*Continued*)

(Note: There is no isothermal heat effect at the 1033° K. point, but merely a heat capacity maximum. To use this set of equations, it is necessary to add 326 calories per mole to balance the heat absorption in the maximum which cannot be accounted for in the equation for Fe(α).)

Fe(β):

$$H_T - H_{298.15} = 11.13T - 5,087 \ (0.3 \text{ percent;}$$
$$1,033-1,183° \text{ K.});$$
$$C_p = 11.13;$$
$$\Delta H_{1183}(transition) = 215.$$

Fe(γ):

$$H_T - H_{298.15} = 5.80T + 0.99 \times 10^{-3}T^2 + 49 \ (0.1 \text{ percent;}$$
$$1,183°-1,673° \text{ K.});$$
$$C_p = 5.80 + 1.98 \times 10^{-3}T;$$
$$\Delta H_{1673}(transition) = 165.$$

Fe(δ):

$$H_T - H_{298.15} = 6.74T + 0.80 \times 10^{-3}T^2 - 825\tfrac{1}{2}(0.1 \text{ percent;}$$
$$1,673°-1,812° \text{ K.});$$
$$C_p = 6.74 + 1.60 \times 10^{-3}T;$$
$$\Delta H_{1812}(fusion) = 3,670.$$

Fe(l):

$$H_T - H_{298.15} = 9.77T + 0.20 \times 10^{-3}T^2 - 670 \ (0.1 \text{ percent;}$$
$$1,812°-3,000° \text{ K.});$$
$$C_p = 9.77 + 0.40 \times 10^{-3}T.$$

TABLE 395.—*Heat content and entropy of* Fe(γ)

[Base, γ-crystals at 298.15° K.; atomic wt., 55.85]

T, ° K.	$H_T-H_{298.15}$, cal./mole	$S_T-S_{298.15}$, cal./deg. mole	T, ° K.	$H_T-H_{298.15}$, cal./mole	$S_T-S_{298.15}$, cal./deg. mole
400	660	1.90	1,300	7,400	10.53
500	1,335	3.41	1,400	8,245	11.16
600	2,020	4.66	1,500	9,115	11.76
700	2,730	5.75	1,600	10,000	12.33
800	3,455	6.72	1,700	10,910	12.88
900	4,205	7.60	1,800	11,835	13.41
1,000	4,975	8.41	1,900	12,785	13.92
1,100	5,760	9.16	2,000	13,750	14.42
1,200	6,570	9.86			

Fe(γ):

$$H_T - H_{298.15} = 5.80T + 0.99 \times 10^{-3}T^2 - 1,817 \ (0.1 \text{ percent;}$$
$$298°-2,000° \text{ K.});$$
$$C_p = 5.80 + 1.98 \times 10^{-3}T.$$

TABLE 396.—*Heat content and entropy of* Fe(g)

[Base, ideal gas at 298.15° K.; atomic wt., 55.85]

T, ° K.	$H_T-H_{298.15}$, cal./mole	$S_T-S_{298.15}$, cal./deg. mole	T, ° K.	$H_T-H_{298.15}$, cal./mole	$S_T-S_{298.15}$, cal./deg. mole
400	625	1.81	1,900	8,845	10.49
500	1,230	3.15	2,000	9,400	10.78
600	1,815	4.22	2,200	10,520	11.31
700	2,385	5.10	2,400	11,670	11.81
800	2,945	5.85	2,600	12,850	12.28
900	3,495	6.49	2,800	14,060	12.73
1,000	4,035	7.06	3,000	15,300	13.16
1,100	4,570	7.57	3,500	18,525	14.15
1,200	5,100	8.04	4,000	21,955	15.07
1,300	5,630	8.45	4,500	25,610	15.93
1,400	6,160	8.85	5,000	29,510	16.75
1,500	6,690	9.22	6,000	38,095	18.31
1,600	7,225	9.56	7,000	47,675	19.79
1,700	7,760	9.89	8,000	58,070	21.17
1,800	8,300	10.20			

Fe(g):

$$H_T - H_{298.15} = 4.18T + 0.35 \times 10^{-3}T^2 - 365 \ (0.3 \text{ percent;}$$
$$1,600°-5,000° \text{ K.});$$
$$C_p = 4.18 + 0.70 \times 10^{-3}T.$$

FERROUS OXIDE

TABLE 397.—*Heat content and entropy of* $Fe_{0.947}O(c, l)$

[Base, crystals at 298.15° K.; mol. wt. 68.89]

T, ° K.	$H_T-H_{298.15}$, cal./mole	$S_T-S_{298.15}$, cal./deg. mole	T, ° K.	$H_T-H_{298.15}$, cal./mole	$S_T-S_{298.15}$, cal./deg. mole
400	1,210	3.48	1,400	14,520	19.88
500	2,440	6.23	1,500	15,980	20.88
600	3,700	8.53	1,600	17,460	21.84
700	4,980	10.50	1,650(c)	18,210	22.30
800	6,280	12.23	1,650(l)	25,700	26.84
900	7,590	13.78	1,700	26,510	27.32
1,000	8,920	15.18	1,800	28,140	28.26
1,100	10,280	16.47	1,900	29,770	29.14
1,200	11,670	17.68	2,000	31,400	29.97
1,300	13,080	18.81			

$Fe_{0.947}O(c)$:

$$H_T - H_{298.15} = 11.66T + 1.00 \times 10^{-3}T^2 + 0.67 \times 10^5 T^{-1}$$
$$-3,790 \ (0.3 \text{ percent; } 298°-1,650° \text{ K.});$$
$$C_p = 11.66 + 2.00 \times 10^{-3}T - 0.67 \times 10^5 T^{-2};$$
$$\Delta H_{1650}(fusion) = 7,490.$$

$Fe_{0.947}O(l)$:

$$H_T - H_{298.15} = 16.30T - 1,200 \ (0.1 \text{ percent;}$$
$$1,650°-2,000° \text{ K.});$$
$$C_p = 16.30.$$

FERRIC OXIDE

TABLE 398.—*Heat content and entropy of* $Fe_2O_3(c)$

[Base, α-crystals at 298.15° K.; mol. wt., 159.70]

T, ° K.	$H_T-H_{298.15}$, cal./mole	$S_T-S_{298.15}$, cal./deg. mole	T, ° K.	$H_T-H_{298.15}$, cal./mole	$S_T-S_{298.15}$, cal./deg. mole
400	2,750	7.91	1,050(γ)	25,820	41.31
500	5,770	14.64	1,100	27,500	42.87
600	9,010	20.54	1,200	30,870	45.80
700	12,460	25.85	1,300	34,250	48.51
800	16,130	30.75	1,400	37,650	51.03
900	20,030	35.34	1,500	41,070	53.39
950(α)	22,060	37.54	1,600	44,540	55.63
950(β)	22,220	37.71	1,700	48,100	57.79
1,000	24,020	39.55	1,800	51,880	59.95
1,050(β)	25,820	41.31			

$Fe_2O_3(\alpha)$:

$$H_T - H_{298.15} = 23.49T + 9.30 \times 10^{-3}T^2 + 3.55 \times 10^5 T^{-1}$$
$$-9,021 \ (0.1 \text{ percent; } 298°-950° \text{ K.});$$
$$C_p = 23.49 + 18.60 \times 10^{-3}T - 3.55 \times 10^5 T^{-2};$$
$$\Delta H_{950}(transition) = 160.$$

$Fe_2O_3(\beta)$:

$$H_T - H_{298.15} = 36.00T - 11,980 \ (0.1 \text{ percent;}$$
$$950°-1,050° \text{ K.});$$
$$C_p = 36.00.$$
$$\Delta H_{1050}(transition) = 0.$$

$Fe_2O_3(\gamma)$:

$$H_T - H_{298.15} = 31.71T + 0.88 \times 10^{-3}T^2 - 8,446$$
$$(0.1 \text{ percent; } 1,050°-1,800° \text{ K.});$$
$$C_p = 31.71 + 1.76 \times 10^{-3}T.$$

TABLE 1h7. (*Continued*)

MAGNETITE

TABLE 399.—*Heat content and entropy of* $Fe_3O_4(c)$

[Base, α-crystals at 298.15° K.; mol. wt., 231.55]

T, ° K.	$H_T-H_{298.15}$, cal./mole	$S_T-S_{298.15}$, cal./deg. mole	T, ° K.	$H_T-H_{298.15}$, cal./mole	$S_T-S_{298.15}$, cal./deg. mole
400	3,990	11.48	1,100	40,150	62.81
500	8,320	21.12	1,200	44,950	66.99
600	13,060	29.75	1,300	49,750	70.83
700	18,340	37.88	1,400	54,550	74.39
800	24,260	45.77	1,500	59,350	77.70
900(α)	30,550	53.18	1,600	64,150	80.80
900(β)	30,550	53.18	1,700	68,950	83.71
1,000	35,350	58.24	1,800	73,750	86.45

$Fe_3O_4(\alpha)$:
$$H_T-H_{298.15}=21.88T+24.10\times10^{-3}T^2-8,666$$
$$(0.5 \text{ percent}; 298°-900° \text{ K.});$$
$$C_p=21.88+48.20\times10^{-3}T;$$
$$\Delta H_{900}(transition)=0.$$
$Fe_3O_4(\beta)$:
$$H_T-H_{298.15}=48.00T-12,650 \text{ (0.1 percent;}$$
$$900°-1,800° \text{ K.});$$
$$C_p=48.00.$$

HYDRATED OXIDE

$Fe_2O_3\cdot3H_2O(c)$:
$$\overline{C}_p=47.9 \text{ (286°-373° K.).}$$

SULFIDES

TABLE 400.—*Heat content and entropy of* $FeS(c, l)$

[Base, α-crystals at 298.15° K.; mol. wt., 87.92]

T, ° K.	$H_T-H_{298.15}$, cal./mole	$S_T-S_{298.15}$, cal./deg. mole	T, ° K.	$H_T-H_{298.15}$, cal./mole	$S_T-S_{298.15}$, cal./deg. mole
350	710	2.19	1,100	12,680	21.35
500	1,470	4.21	1,200	14,150	22.63
411(α)	1,640	4.63	1,300	15,680	23.86
411(β)	2,210	6.02	1,400	17,260	25.03
500	3,760	9.43	1,468(γ)	18,350	25.79
598(β)	5,460	12.53	1,468(l)	26,080	31.05
598(γ)	5,580	12.73	1,500	26,620	31.42
600	5,610	12.79	1,600	28,320	32.51
700	7,020	14.96	1,700	30,020	33.54
800	8,430	16.84	1,800	31,720	34.51
900	9,840	18.50	1,900	33,420	35.43
1,000	11,250	19.99	2,000	35,120	36.31

$FeS(\alpha)$:
$$H_T-H_{298.15}=5.19T+13.20\times10^{-3}T^2-2,721$$
$$(0.8 \text{ percent}; 298°-411° \text{ K.});$$
$$C_p=5.19+26.40\times10^{-3}T;$$
$$\Delta H_{411}(transition)=570.$$

$FeS(\beta)$:
$$H_T-H_{298.15}=17.40T-4,944 \text{ (0.1 percent;}$$
$$411°-598° \text{ K.});$$
$$C_p=17.40;$$
$$\Delta H_{598}(transition)=120.$$

$FeS(\gamma)$:
$$H_T-H_{298.15}=12.20T+1.19\times10^{-3}T^2-2,138$$
$$(0.3 \text{ percent}; 598°-1,468° \text{ K.});$$
$$C_p=12.20+2.38\times10^{-3}T;$$
$$\Delta H_{1468}(fusion)=7,730.$$
$FeS(l)$:
$$H_T-H_{298.15}=17.00T+1,120 \text{ (0.1 percent;}$$
$$1,468°-2,000° \text{ K.});$$
$$C_p=17.00.$$

TABLE 401.—*Heat content and entropy of* $FeS_2(c)$

[Base, crystals at 298.15° K.; mol. wt., 119.98]

T, ° K.	$H_T-H_{298.15}$, cal./mole	$S_T-S_{298.15}$, cal./deg. mole	T, ° K.	$H_T-H_{298.15}$, cal./mole	$S_T-S_{298.15}$, cal./deg. mole
400	1,670	4.81	800	8,650	16.82
500	3,350	8.55	900	10,550	19.06
600	5,060	11.67	1,000	12,520	21.14
700	6,820	14.38			

$FeS_2(c)$:
$$H_T-H_{298.15}=17.88T+0.66\times10^{-3}T^2+3.05\times10^5T^{-1}$$
$$-6,413 \text{ (1.0 percent}; 298°-1,000° \text{ K.});$$
$$C_p=17.88+1.32\times10^{-3}T-3.05\times10^5T^{-2}.$$

CARBIDE

TABLE 402.—*Heat content and entropy of* Fe_3C (c)

[Base, α-crystals at 298.15° K.; mol. wt., 179.56]

T, ° K.	$H_T-H_{298.15}$, cal./mole	$S_T-S_{298.15}$, cal./deg. mole	T, ° K.	$H_T-H_{298.15}$, cal./mole	$S_T-S_{298.15}$, cal./deg. mole
350	1,360	4.21	800	13,940	27.35
400	2,690	7.76	900	16,760	30.66
450	4,120	11.12	1,000	19,610	33.67
463(α)	4,490	11.94	1,100	22,490	36.41
463(β)	4,670	12.33	1,200	25,400	38.95
500	5,670	14.41	1,300	28,340	41.30
600	8,390	19.37	1,400	31,310	43.50
700	11,150	23.62	1,500	34,310	45.57

$Fe_3C(\alpha)$:
$$H_T-H_{298.15}=19.64T+10.00\times10^{-3}T^2-6,745$$
$$(0.5 \text{ percent}; 298°-463° \text{ K.});$$
$$C_p=19.64+20.00\times10^{-3}T;$$
$$\Delta H_{463}(transition)=180.$$

$Fe_3C(\beta)$:
$$H_T-H_{298.15}=25.62T+1.50\times10^{-3}T^2-7,515$$
$$(0.1 \text{ percent}; 463°-1,500° \text{ K.});$$
$$C_p=25.62+3.00\times10^{-3}T.$$

NITRIDES

Reference: *Sato (604, 606)* (273°–373°).

$Fe_2N(c)$:
$$C_p=14.91+6.09\times10^{-3}T(estimated) \text{ (298°-1,000° K.).}$$
$Fe_4N(c)$:
$$C_p=26.84+8.16\times10^{-3}T(estimated) \text{ (298°-1,000° K.).}$$

TABLE 1h7. (*Continued*)

SILICIDE

TABLE 403.—*Heat content and entropy of* $FeSi(c)$

[Base, crystals at 298.15° K.; mol. wt., 83.94]

T, ° K.	$H_T-H_{298.15}$, cal./mole	$S_T-S_{298.15}$, cal./deg. mole	T, ° K.	$H_T-H_{298.15}$, cal./mole	$S_T-S_{298.15}$, cal./deg. mole
400	1,240	3.58	700	5,190	10.90
500	2,490	6.36	800	6,570	12.74
600	3,820	8.79	900	7,990	14.41

$FeSi(c):$
$$H_T-H_{298.15}=10.72T+2.15\times10^{-3}T^2$$
$$-3,387\ (0.4\ \text{percent};\ 298°-900°\ \text{K.});$$
$$C_p=10.72+4.30\times10^{-3}T.$$

ARSENIDE

$FeAs_2(c):$
$$\overline{C}_p=17.8\ (283°-373°\ \text{K.}).$$

ARSENOSULFIDE

$FeAsS(c):$
$$\overline{C}_p=16.8\ (283°-373°\ \text{K.}).$$

BROMIDE

$FeBr_2(c):$
$$C_p=20.07\ (298°\ \text{K.}).$$

CHLORIDES

TABLE 404.—*Heat content and entropy of* $FeCl_2(c, l)$

[Base, crystals at 298.15° K.; mol. wt., 126.76]

T, ° K.	$H_T-H_{298.15}$, cal./mole	$S_T-S_{298.15}$, cal./deg. mole	T, ° K.	$H_T-H_{298.15}$, cal./mole	$S_T-S_{298.15}$, cal./deg. mole
400	1,930	5.57	950(c)	12,920	22.75
500	3,870	9.89	950(l)	23,200	33.67
600	5,820	13.45	1,000	24,420	34.82
700	7,800	16.50	1,100	26,860	37.14
800	9,830	19.21	1,200	29,300	39.26
900	11,880	21.62	1,300	31,740	41.22

$FeCl_2(c):$
$$H_T-H_{298.15}=18.94T+1.04\times10^{-3}T^2+1.17\times10^5T^{-1}$$
$$-6,132\ (0.5\ \text{percent};\ 298°-950°\ \text{K.});$$
$$C_p=18.94+2.08\times10^{-3}T-1.17\times10^5T^{-2};$$
$$\Delta H_{950}(fusion)=10,280.$$
$FeCl_2(l):$
$$H_T-H_{298.15}=24.40T+20\ (0.1\ \text{percent};\ 950°-1,300°\ \text{K.});$$
$$C_p=24.40.$$

TABLE 405.—*Heat content and entropy of* $FeCl_3(c, l)$

[Base, crystals at 298.15° K.; mol. wt., 162.22]

T, ° K.	$H_T-H_{298.15}$, cal./mole	$S_T-S_{298.15}$, cal./deg. mole	T, ° K.	$H_T-H_{298.15}$, cal./mole	$S_T-S_{298.15}$, cal./deg. mole
350	1,220	3.77	577(c)	7,250	17.00
400	2,500	7.18	577(l)	17,550	24.85
450	3,810	10.27	600	18,290	36.10
500	5,140	13.07	650	19,890	38.67
550	6,490	15.65	700	21,490	41.04

$FeCl_3(c):$
$$H_T-H_{298.15}=29.56T+6.11\times10^5T^{-1}$$
$$-10,863\ (0.4\ \text{percent};\ 298°-577°\ \text{K.});$$
$$C_p=29.56-6.11\times10^5T^{-2};$$
$$\Delta H_{577}(fusion)=10,300.$$
$FeCl_3(l):$
$$H_T-H_{298.15}=32.00T-910\ (0.1\ \text{percent};\ 577°-700°\ \text{K.});$$
$$C_p=32.00$$

TABLE 406.—*Heat content and entropy of* $Fe_2Cl_6(g)$

[Base, ideal gas at 298.15° K.; mol. wt., 324.44]

T, ° K.	$H_T-H_{298.15}$, cal./mole	$S_T-S_{298.15}$, cal./deg. mole	T, ° K.	$H_T-H_{298.15}$, cal./mole	$S_T-S_{298.15}$, cal./deg. mole
400	3,460	9.98	900	20,460	37.55
500	6,860	17.57	1,000	23,860	41.14
600	10,260	23.77	1,100	27,260	44.38
700	13,660	29.01	1,200	30,660	47.33
800	17,060	33.55			

$Fe_2Cl_6(g):$
$$H_T-H_{298.15}=34.00T-10,137\ (0.1\ \text{percent};$$
$$298°-1,200°\ \text{K.});$$
$$C_p=34.00.$$

FLUORIDE

$FeF_2(c):$
$$C_p=16.28\ (298°\ \text{K.}).$$

IODIDE

$FeI_2(c):$
$$C_p=20.12\ (298°\ \text{K.}).$$

ALUMINATE

$FeAl_2O_4(c):$
$$C_p=21.57+26.69\times10^{-3}T\ (298°-1,298°\ \text{K.}).$$

CARBONATE

$FeCO_3(c):$
$$C_p=11.63+26.80\times10^{-3}T\ (298°-885°\ \text{K.}).$$

TABLE 1h7. (Continued)

CHROMITE

TABLE 407.—*Heat content and entropy of* $FeCr_2O_4(c)$

[Base, crystals at 298.15° K., mol. wt., 223.87]

T, ° K.	$H_T-H_{298.15}$, cal./mole	$S_T-S_{298.15}$, cal./deg. mole	T, ° K.	$H_T-H_{298.15}$, cal./mole	$S_T-S_{298.15}$, cal./deg. mole
400	3,450	9.93	1,200	36,920	55.05
500	7,180	18.24	1,300	41,430	58.65
600	11,130	25.44	1,400	45,990	62.02
700	15,220	31.74	1,500	50,590	65.27
800	19,410	37.34	1,600	55,230	68.20
900	23,680	42.36	1,700	59,910	71.0
1,000	28,030	46.95	1,800	64,630	73.74
1,100	32,450	51.16	1,900	69,390	76.21

$FeCrO_4(c)$:

$$H_T-H_{298.15}=38.96T+2.67\times10^{-3}T^2+7.62\times10^5T^{-1}$$
$$-14,409\ (0.4\ \text{percent};\ 298°-1,900°\ \text{K.});$$
$$C_p=38.96+5.34\times10^{-3}T-7.62\times10^5T^{-2}.$$

IRON-COBALT SPINEL

$FeCo_2O_4(c)$:

$$C_p=34.27\ (298°\ \text{K.}).$$

SILICATE

TABLE 408.—*Heat content of* $Fe_2SiO_4(c,\ l)$

[Base, crystals at 298.15° K.; mol. wt., 203.79]

T, ° K.	$H_T-H_{298.15}$, cal./mole	$S_T-S_{298.15}$, cal./deg. mole	T, ° K.	$H_T-H_{298.15}$, cal./mole	$S_T-S_{298.15}$, cal./deg. mole
400	3,440	9.90	1,400	47,190	63.16
500	7,210	18.30	1,490(c)	51,690	66.28
600	11,190	25.55	1,490(l)	73,720	81.06
700	15,320	31.91	1,500	74,300	81.45
800	19,560	37.57	1,600	80,050	85.16
900	23,890	42.67	1,700	85,800	88.65
1,000	28,310	47.33	1,800	91,550	91.94
1,100	32,850	51.65	1,900	97,300	95.04
1,200	37,510	55.71	2,000	103,050	97.99
1,300	42,290	59.53			

$Fe_2SiO(c)$:

$$H_T-H_{298.15}=36.51T+4.68\times10^{-3}T^2+6.70\times10^5T^{-1}$$
$$-13,549\ (0.3\ \text{percent};\ 298°-1,490°\ \text{K.});$$
$$C_p=36.51+9.36\times10^{-3}T-6.70\times10^5T^{-2};$$
$$\Delta H_{1490}(fusion)=22,030.$$

$Fe_2SiO_4(l)$:

$$H_T-H_{298.15}=57.50T-11,950\ (0.1\ \text{percent};$$
$$1,490°-2,000°\ \text{K.});$$
$$C_p=57.50.$$

SULFATES

$FeSO_4(c)$:

$$C_p=24.03\ (298°\ \text{K.}).$$

$FeSO_4\cdot4H_2O(c)$:

$$C_p=63.6\ (282°\ \text{K.}).$$

$FeSO_4\cdot7H_2O(c)$:

$$\overline{C_p}=96.2\ (291°-319°\ \text{K.}).$$

$Fe_2(SO_4)_3(c)$:

$$\overline{C_p}=66.2\ (273°-373°\ \text{K.}).$$

TITANATES

TABLE 409.—*Heat content and entropy of* $FeTiO_3(c,\ l)$

[Base, crystals at 298.15° K.; mol. wt., 151.75]

T, ° K.	$H_T-H_{298.15}$, cal./mole	$S_T-S_{298.15}$, cal./deg. mole	T, ° K.	$H_T-H_{298.15}$, cal./mole	$S_T-S_{298.15}$, cal./deg. mole
400	2,595	7.47	1,400	33,540	45.36
500	5,330	13.57	1,500	36,920	47.69
600	8,200	18.80	1,600	40,360	49.91
700	11,130	23.31	1,640(c)	41,750	50.77
800	14,150	27.35	1,640(l)	63,420	63.98
900	17,250	31.00	1,700	66,280	65.69
1,000	20,430	34.35	1,800	71,040	68.42
1,100	23,650	37.42	1,900	75,800	70.99
1,200	26,900	40.24	2,000	80,560	73.43
1,300	30,200	42.88			

$FeTiO_3(c)$:

$$H_T-H_{298.15}=27.87T+2.18\times10^{-3}T^2+4.79\times10^5T^{-1}$$
$$-10,110\ (0.2\ \text{percent};\ 298°-1,640°\ \text{K.});$$
$$C_p=27.87+4.36\times10^{-3}T-4.79\times10^5T^{-2};$$
$$\Delta H_{1640}(fusion)=21,670.$$

$FeTiO_3(l)$:

$$H_T-H_{298.15}=47.60T-14,642\ (0.1\ \text{percent};\ 1,640°-$$
$$2,000°\ \text{K.});$$
$$C_p=47.60.$$

TABLE 410.—*Heat content and entropy of* $Fe_2TiO_4(c)$

[Base, crystals at 298.15 K.; mol. wt., 223.60]

T, ° K.	$H_T-H_{298.15}$, cal./mole	$S_T-S_{298.15}$, cal./deg. mole	T, ° K.	$H_T-H_{298.15}$, cal./mole	$S_T-S_{298.15}$, cal./deg. mole
400	3,750	10.80	1,100	34,210	53.87
500	7,610	19.40	1,200	39,180	58.20
600	11,640	26.74	1,300	44,350	62.34
700	15,850	33.23	1,400	49,760	66.35
800	20,200	39.06	1,500	55,450	70.57
900	24,740	44.38	1,600	61,460	74.15
1,000	29,400	49.29			

$Fe_2TiO_4(c)$:

$$H_T-H_{298.15}=33.34T+7.54\times10^{-3}T+3.40\times10^5T^{-1}$$
$$-11,751\ (0.7\ \text{percent};\ 298°-1,600°\ \text{K.});$$
$$C_p=33.34+15.08\times10^{-3}T-3.40\times10^5T^{-2}.$$

TABLE 411.—*Heat content and entropy of* $Fe_2TiO_5(c)$

[Base, crystals at 298.15° K., mol. wt., 239.60]

T, ° K.	$H_T-H_{298.15}$, cal./mole	$S_T-S_{298.15}$, cal./deg. mole	T, ° K.	$H_T-H_{298.15}$, cal./mole	$S_T-S_{298.15}$, cal./deg. mole
400	4,330	12.46	1,200	43,330	65.05
500	8,740	22.29	1,300	48,550	69.23
600	13,300	30.60	1,400	53,800	73.12
700	18,010	37.86	1,500	59,080	76.76
800	22,870	44.35	1,600	64,400	80.20
900	27,860	50.23	1,700	69,760	83.45
1,000	32,960	55.60	1,800	75,160	86.53
1,100	38,130	60.53			

TABLE 1h7. (Continued)

Fe$_2$TiO$_5$(c):

$H_T - H_{298.15} = 46.03T + 2.63 \times 10^{-3}T^2 + 7.41 \times 10^5 T^{-1}$
$- 16,443$ (0.4 percent; 298°–1,700° K.);

$C_p = 46.03 + 5.26 \times 10^{-3}T - 7.41 \times 10^5 T^{-2}.$

KRYPTON

ELEMENT

TABLE 412.—*Heat content and entropy of Kr(g)*

[Base, ideal gas at 298.15° K.; atomic wt., 83.80]

T, ° K.	$H_T - H_{298.15}$, cal./mole	$S_T - S_{298.15}$, cal./deg. mole	T, ° K.	$H_T - H_{298.15}$, cal./mole	$S_T - S_{298.15}$, cal./deg. mole
400	505	1.46	1,900	7,960	9.20
500	1,005	2.57	2,000	8,455	9.46
600	1,500	3.48	2,200	9,450	9.93
700	1,995	4.24	2,400	10,445	10.36
800	2,495	4.90	2,600	11,440	10.76
900	2,990	5.49	2,800	12,430	11.13
1,000	3,490	6.01	3,000	13,425	11.47
1,100	3,985	6.49	3,500	15,910	12.24
1,200	4,480	6.92	4,000	18,395	12.90
1,300	4,980	7.32	4,500	20,880	13.49
1,400	5,475	7.69	5,000	23,365	14.01
1,500	5,975	8.03	6,000	28,335	14.92
1,600	6,490	8.35	7,000	33,305	15.68
1,700	6,965	8.65	8,000	38,275	16.35
1,800	7,465	8.93			

Kr(g):

$H_T - H_{298.15} = 4.97T - 1,482$ (0.1 percent;
298°–8,000° K.);

$C_p = 4.97.$

LANTHANUM AND ITS COMPOUNDS

ELEMENT

TABLE 413.—*Heat content and entropy of La(c, l)*

[Base, crystals at 298.15° K.; atomic wt., 138.92]

T, ° K.	$H_T - H_{298.15}$, cal./mole	$S_T - S_{298.15}$, cal./deg. mole	T, ° K.	$H_T - H_{298.15}$, cal./mole	$S_T - S_{298.15}$, cal./deg. mole
400	685	1.98	1,400	11,000	13.57
500	1,370	3.51	1,500	11,800	14.13
600	2,080	4.80	1,600	12,600	14.64
700	2,800	5.91	1,700	13,400	15.13
800	3,540	6.90	1,800	14,200	15.58
900	4,290	7.78	1,900	15,000	16.02
1,000	5,060	8.59	2,000	15,800	16.43
1,100	5,840	9.33	2,200	17,400	17.19
1,193(c)	6,590	9.99	2,400	19,000	17.88
1,193(l)	9,340	12.29	2,600	20,600	18.52
1,200	9,400	12.34	2,800	22,200	19.12
1,300	10,200	12.98	3,000	23,800	19.67

La(c):

$H_T - H_{298.15} = 6.17T + 0.80 \times 10^{-3}T^2 - 1,911$

(0.1 percent; 298°–1,193° K.);

$C_p = 6.17 + 1.60 \times 10^{-3}T;$

$\Delta H_{1193}(fusion) = 2,750.$

La(l):

$H_T - H_{298.15} = 8.00T - 200$ (0.1 percent;
1,193°–3,000° K.) ;

$C_p = 8.00.$

TABLE 414.—*Heat content and entropy of La(g)*

[Base, ideal gas at 298.15° K.; atomic wt., 138.92]

T, ° K.	$H_T - H_{298.15}$, cal./mole	$S_T - S_{298.15}$, cal./deg. mole	T, ° K.	$H_T - H_{298.15}$, cal./mole	$S_T - S_{298.15}$, cal./deg. mole
400	580	1.66	1,900	11,385	12.51
500	1,185	3.02	2,000	12,150	12.91
600	1,820	4.17	2,200	13,680	13.64
700	2,475	5.18	2,400	15,210	14.30
800	3,155	6.09	2,600	16,740	14.91
900	3,850	6.91	2,800	18,270	15.48
1,000	4,565	7.66	3,000	19,810	16.01
1,100	5,300	8.36	3,500	23,705	17.21
1,200	6,045	9.01	4,000	27,690	18.28
1,300	6,800	9.61	4,500	31,785	19.24
1,400	7,560	10.18	5,000	35,995	20.13
1,500	8,320	10.70	6,000	44,710	21.72
1,600	9,085	11.20	7,000	53,620	23.09
1,700	9,855	11.66	8,000	62,500	24.28
1,800	10,620	12.10			

La(g):

$H_T - H_{298.15} = 6.19T + 0.48 \times 10^{-3}T^2 + 0.92 \times 10^5 T^{-1}$
$- 2,197$ (0.8 percent; 298°–2,000° K.);

$C_p = 6.19 + 0.96 \times 10^{-3}T - 0.92 \times 10^5 T^{-2}.$

$H_T - H_{298.15} = 7.20T + 0.10 \times 10^{-3}T^2 + 1.60 \times 10^5 T^{-1}$
$- 2,730$ (0.2 percent; 2,000°–5,000° K.);

$C_p = 7.20 + 0.20 \times 10^{-3}T - 1.60 \times 10^5 T^{-2}.$

OXIDES

TABLE 415.—*Heat control and entropy of LaO(g)*

[Base, ideal gas at 298.15° K.; mol. wt., 154.92]

T, ° K.	$H_T - H_{298.15}$, cal./mole	$S_T - S_{298.15}$, cal./deg. mole	T, ° K.	$H_T - H_{298.15}$, cal./mole	$S_T - S_{298.15}$, cal./deg. mole
400	795	2.29	1,000	5,995	10.01
500	1,605	4.10	1,200	7,640	11.61
600	2,440	5.62	1,400	9,400	12.97
700	3,290	6.93	1,600	11,170	14.15
800	4,150	8.08	1,800	12,945	15.19
900	5,015	9.10	2,000	14,720	16.13

LaO(g):

$H_T - H_{298.15} = 8.46T + 0.15 \times 10^{-3}T^2 + 0.84 \times 10^5 T^{-1}$
$- 2,817$ (0.3 percent; 298°–2,000° K.) ;

$C_p = 8.46 + 0.30 \times 10^{-3}T - 0.84 \times 10^5 T^{-2}.$

TABLE 416.—*Heat control and entropy of La$_2$O$_3$(c)*

[Base, crystals at 298.15° K.; mol. wt., 325.84]

T, ° K.	$H_T - H_{298.15}$, cal./mole	$S_T - S_{298.15}$, cal./deg. mole	T, ° K.	$H_T - H_{298.15}$, cal./mole	$S_T - S_{298.15}$, cal./deg. mole
400	2,770	7.98	1,300	30,540	43.85
500	5,630	14.36	1,400	33,830	46.29
600	8,580	19.73	1,500	37,150	48.58
700	11,590	24.37	1,600	40,500	50.74
800	14,650	28.46	1,700	43,890	52.79
900	17,750	32.11	1,800	47,310	54.75
1,000	20,890	35.42	1,900	50,760	56.61
1,100	24,070	38.45	2,000	54,240	58.40
1,200	27,290	41.25			

La$_2$O$_3$(c):

$H_T - H_{298.15} = 28.86T + 1.55 \times 10^{-3}T^2 + 3.28 \times 10^5 T^{-1}$
$- 9,843$ (0.1 percent; 298°–2,000° K.) ;

$C_p = 28.86 + 3.10 \times 10^{-3}T - 3.28 \times 10^5 T^{-2}.$

TABLE 1h7. (*Continued*)

MOLYBDATE

$La_2(MoO_4)_3(c)$:

$\overline{C_p}=86.4$ (273°–297° K.).

NITRATE

$La(NO_3)_3 \cdot 6H_2O(c)$:

$\overline{C_p}=100.7$ (273°–289° K.).

SULFATE

$La_2(SO_4)_3(c)$:

$\overline{C_p}=66.9$ (273°–373° K.).

$La_2(SO_4)_3 \cdot 9H_2O(c)$:

$\overline{C_p}=151.7$ (273°–319° K.).

LANTHANUM-MAGNESIUM NITRATE

$2La(NO_3)_3 \cdot 3Mg(NO_3)_2 \cdot 24H_2O(c)$:

$C_p=483.2$ (298° K.).

LEAD AND ITS COMPOUNDS

ELEMENT

TABLE 417.—*Heat content and entropy of Pb(c,l)*

[Base, crystals at 298.15° K.; atomic wt., 207.21]

T, ° K.	$H_T-H_{298.15}$, cal./mole	$S_T-S_{298.15}$, cal./deg. mole	T, ° K.	$H_T-H_{298.15}$, cal./mole	$S_T-S_{298.15}$, cal./deg. mole
400	655	1.89	1,200	7,415	11.48
500	1,325	3.38	1,300	8,100	12.03
600	2,015	4.64	1,400	8,780	12.53
600.6(c)	2,020	4.65	1,500	9,465	13.00
600.6(l)	3,160	6.55	1,600	10,150	13.44
700	3,885	7.66	1,700	10,840	13.86
800	4,635	8.63	1,800	11,530	14.26
900	5,320	9.47	1,900	12,230	14.64
1,000	6,025	10.21	2,000	12,940	15.00
1,100	6,725	10.88			

Pb(c):

$H_T-H_{298.15}=5.29T+1.40\times10^{-3}T^2-0.23\times10^5T^{-1}$
$-1,625$ (0.1 percent; 298°–600.6° K.);
$C_p=5.29+2.80\times10^{-3}T+0.23\times10^5T^{-2}$;
$\Delta H_{600.6}(fusion)=1,140.$

Pb(l):

$H_T-H_{298.15}=7.77T-0.36\times10^{-3}T^2-1,377$ (0.4 percent;
600.6°–2,000° K.);
$C_p=7.77-0.72\times10^{-3}T.$

TABLE 418.—*Heat content and entropy of Pb(g)*

[Base, ideal gas at 298.15° K.; atomic wt., 207.21]

T, ° K.	$H_T-H_{298.15}$, cal./mole	$S_T-S_{298.15}$, cal./deg. mole	T, ° K.	$H_T-H_{298.15}$, cal./mole	$S_T-S_{298.15}$, cal./deg. mole
400	505	1.46	1,500	6,015	8.06
500	1,005	2.57	1,600	6,535	8.40
600	1,500	3.47	1,700	7,070	8.72
700	1,995	4.24	1,800	7,620	9.03
800	2,495	4.90	1,900	8,185	9.34
900	2,990	5.49	2,000	8,765	9.64
1,000	3,490	6.01	2,200	9,985	10.22
1,100	3,985	6.49	2,400	11,290	10.78
1,200	4,485	6.92	2,600	12,680	11.34
1,300	4,990	7.33	2,800	14,155	11.89
1,400	5,500	7.70	3,000	15,705	12.42

Pb(g):

$H_T-H_{298.15}=4.63T+0.21\times10^{-3}T^2-0.18\times10^5T^{-1}$
$-1,339$ (0.7 percent; 298°–2,000° K.);
$C_p=4.63+0.42\times10^{-3}T+0.18\times10^5T^{-2}.$
$H_T-H_{298.15}=1.60T+1.07\times10^{-3}T^2+1,284$ (0.1 percent;
2,000°–3,000° K.);
$C_p=1.60+2.14\times10^{-3}T.$

TABLE 419.—*Heat content and entropy of $Pb_2(g)$*

[Base, ideal gas at 298.15° K.; mol. wt., 414.42]

T, ° K.	$H_T-H_{298.15}$, cal./mole	$S_T-S_{298.15}$, cal./deg. mole	T, ° K.	$H_T-H_{298.15}$, cal./mole	$S_T-S_{298.15}$, cal./deg. mole
400	895	2.58	1,000	6,225	10.71
500	1,775	4.55	1,200	8,010	12.34
600	2,665	6.17	1,400	9,795	13.72
700	3,550	7.53	1,600	11,580	14.91
800	4,440	8.72	1,800	13,370	15.96
900	5,335	9.77	2,000	15,160	16.90

$Pb_2(g)$:

$H_T-H_{298.15}=8.94T+0.20\times10^5T^{-1}-2,733$ (0.1 percent;
298°–2,000° K.);
$C_p=8.94-0.20\times10^5T^{-2}.$

OXIDES

TABLE 420.—*Heat content and entropy of PbO
(yellow, l)*

[Base, crystals at 298.15° K.; mol. wt., 223.21]

T, ° K.	$H_T-H_{298.15}$, cal./mole	$S_T-S_{298.15}$, cal./deg. mole	T, ° K.	$H_T-H_{298.15}$, cal./mole	$S_T-S_{298.15}$, cal./deg. mole
400	1,150	3.32	1,100	10,800	16.90
500	2,340	5.97	1,159(c)	11,720	17.72
600	3,600	8.26	1,159(l)	18,720	23.76
700	4,920	10.30	1,200	19,380	24.32
800	6,310	12.15	1,300	20,980	25.60
900	7,760	13.86	1,400	22,580	26.79
1,000	9,260	15.44	1,500	24,180	27.89

PbO(*yellow*):

$H_T-H_{298.15}=9.05T+3.20\times10^{-3}T^2-2,983$ (0.2 percent;
298°–1,159° K.);
$C_p=9.05+6.40\times10^{-3}T;$
$\Delta H_{1159}(fusion)=7,000.$

PbO(l):

$H_T-H_{298.15}=16.00T+180$ (0.1 percent; 1,159°–1,500°
K.);
$C_p=16.00.$

TABLE 421.—*Heat content and entropy of
PbO(red)*

[Base, crystals at 298.15° K.; mol. wt., 223.21]

T, ° K.	$H_T-H_{298.15}$, cal./mole	$S_T-S_{298.15}$, cal./deg. mole	T, ° K.	$H_T-H_{298.15}$, cal./mole	$S_T-S_{298.15}$, cal./deg. mole
400	1,220	3.52	700	5,060	10.65
500	2,460	6.28	800	6,420	12.46
600	3,740	8.62	900	7,820	14.11

PbO(*red*):

$H_T-H_{298.15}=10.60T+2.00\times10^{-3}T^2-3,338$
(0.1 percent; 298°–900° K.);
$C_p=10.60+4.00\times10^{-3}T.$

TABLE 1h7. (*Continued*)

TABLE 422.—*Heat content and entropy of PbO(g)*

[Base, ideal gas at 298.15° K.; mol. wt., 223.21]

T, ° K.	$H_T-H_{298.15}$, cal./mole	$S_T-S_{298.15}$, cal./deg. mole	T, ° K.	$H_T-H_{298.15}$, cal./mole	$S_T-S_{298.15}$, cal./deg. mole
400	810	2.34	1,000	5,950	10.14
500	1,635	4.18	1,200	7,710	11.75
600	2,480	5.72	1,400	9,480	13.11
700	3,340	7.04	1,600	11,250	14.29
800	4,205	8.20	1,800	13,030	15.34
900	5,075	9.22	2,000	14,810	16.28

PbO(g):

$$H_T-H_{298.15}=8.57T+0.12\times10^{-3}T^2+0.79\times10^5T^{-1}$$
$$-2,831\ (0.2\ \text{percent};\ 298°-2,000°\ \text{K.});$$
$$C_p=8.57+0.24\times10^{-3}T-0.79\times10^5T^{-2}.$$

PbO$_2$(c):
$$C_p=12.70+7.80\times10^{-3}T(estimated)\ (298°-1,000°\ \text{K.}).$$

Pb$_3$O$_4$(c):
$$C_p=35.14\ (298°\ \text{K.}).$$

Pb$_2$O$_3$(c):
$$C_p=25.74\ (298°\ \text{K.}).$$

SULFIDE

TABLE 423.—*Heat content and entropy of PbS(c)*

[Base, crystals at 298.15° K.; mol. wt., 239.28]

T, ° K.	$H_T-H_{298.15}$, cal./mole	$S_T-S_{298.15}$, cal./deg. mole	T, ° K.	$H_T-H_{298.15}$, cal./mole	$S_T-S_{298.15}$, cal./deg. mole
400	1,240	3.58	700	5,040	10.62
500	2,450	6.28	800	6,430	12.48
600	3,710	8.57	900	7,880	14.18

PbS(c):

$$H_T-H_{298.15}=10.66T+1.96\times10^{-3}T^2-3,353$$
$$(0.8\ \text{percent};\ 298°-900°\ \text{K.});$$
$$C_p=10.66+3.92\times10^{-3}T.$$

TABLE 424.—*Heat content and entropy of PbS(g)*

[Base, ideal gas at 298.15° K.; mol. wt., 239.28]

T, ° K.	$H_T-H_{298.15}$, cal./mole	$S_T-S_{298.15}$, cal./deg. mole	T, ° K.	$H_T-H_{298.15}$, cal./mole	$S_T-S_{298.15}$, cal./deg. mole
400	860	2.48	1,000	6,140	10.53
500	1,730	4.42	1,200	7,920	12.15
600	2,605	6.02	1,400	9,700	13.52
700	3,485	7.37	1,600	11,485	14.71
800	4,370	8.56	1,800	13,270	15.76
900	5,255	9.60	2,000	15,055	16.70

PbS(g):

$$H_T-H_{298.15}=8.92T+0.49\times10^5T^{-1}-2,824$$
$$(0.2\ \text{percent};\ 298°-2,000°\ \text{K.});$$
$$C_p=8.92-0.49\times10^5T^{-2}.$$

HYDRIDE

TABLE 425.—*Heat content and entropy of PbH(g)*

[Base, ideal gas at 298.15° K.; mol. wt., 208.22]

T, ° K.	$H_T-H_{298.15}$, cal./mole	$S_T-S_{298.15}$, cal./deg. mole	T, ° K.	$H_T-H_{298.15}$, cal./mole	$S_T-S_{298.15}$, cal./deg. mole
400	725	2.09	1,000	5,435	9.20
500	1,460	3.73	1,200	7,115	10.73
600	2,215	5.11	1,400	8,825	12.05
700	2,995	6.31	1,600	10,550	13.20
800	3,795	7.37	1,800	12,290	14.23
900	4,610	8.33	2,000	14,035	15.15

PbH(g):

$$H_T-H_{298.15}=7.24T+0.49\times10^{-3}T^2+0.45\times10^5T^{-1}$$
$$-2,353\ (0.6\ \text{percent};\ 298°-2,000°\ \text{K.});$$
$$C_p=7.24+0.98\times10^{-3}T-0.45\times10^5T^{-2}.$$

SELENIDE

TABLE 426.—*Heat content and entropy of PbSe(g)*

[Base, ideal gas at 298.15° K.; mol. wt., 286.17]

T, ° K.	$H_T-H_{298.15}$, cal./mole	$S_T-S_{298.15}$, cal./deg. mole	T, ° K.	$H_T-H_{298.15}$, cal./mole	$S_T-S_{298.15}$, cal./deg. mole
400	890	2.57	1,000	6,215	10.69
500	1,770	4.53	1,200	8,000	12.32
600	2,655	6.14	1,400	9,790	13.70
700	3,545	7.51	1,600	11,575	14.89
800	4,435	8.70	1,800	13,360	15.94
900	5,325	9.75	2,000	15,145	16.88

PbSe(g):

$$H_T-H_{298.15}=8.94T+0.25\times10^5T^{-1}-2,749$$
$$(0.1\ \text{percent};\ 298°-2,000°\ \text{K.});$$
$$C_p=8.94-0.25\times10^5T^{-2}.$$

TELLURIDE

TABLE 427.—*Heat content and entropy of PbTe(g)*

[Base, ideal gas at 298.15° K.; mol. wt., 334.82]

T, ° K.	$H_T-H_{298.15}$, cal./mole	$S_T-S_{298.15}$, cal./deg. mole	T, ° K.	$H_T-H_{298.15}$, cal./mole	$S_T-S_{298.15}$, cal./deg. mole
400	900	2.60	1,000	6,240	10.75
500	1,785	4.57	1,200	8,025	12.38
600	2,675	6.20	1,400	9,815	13.76
700	3,565	7.57	1,600	11,600	14.95
800	4,455	8.67	1,800	13,390	16.00
900	5,350	9.81	2,000	15,175	16.94

PbTe(g):

$$H_T-H_{298.15}=8.94T+0.14\times10^5T^{-1}-2,712$$
$$(0.1\ \text{percent};\ 298°-2,000°\ \text{K.});$$
$$C_n=8.94-0.14\times10^5T^{-2}.$$

BROMIDES

TABLE 428.—*Heat content and entropy of PbBr(g)*

[Base, ideal gas at 298.15° K.; mol. wt., 287.13]

T, ° K.	$H_T-H_{298.15}$, cal./mole	$S_T-S_{298.15}$, cal./deg. mole	T, ° K.	$H_T-H_{298.15}$, cal./mole	$S_T-S_{298.15}$, cal./deg. mole
400	900	2.60	1,000	6,240	10.75
500	1,785	4.57	1,200	8,025	12.38
600	2,675	6.20	1,400	9,815	13.76
700	3,565	7.57	1,600	11,600	14.95
800	4,460	8.76	1,800	13,390	16.00
900	5,350	9.81	2,000	15,175	16.94

PbBr(g):

$$H_T-H_{298.15}=8.94T+0.14\times10^5T^{-1}-2,712$$
$$(0.1\ \text{percent};\ 298°-2,000°\ \text{K.});$$
$$C_p=8.94-0.14\times10^5T^{-2}.$$

TABLE 1h7. (*Continued*)

TABLE 429.—*Heat content and entropy of* PbBr₂(c, l)

[Base, crystals at 298.15° K.; mol. wt., 367.04]

T, ° K.	$H_T-H_{298.15}$, cal./mole	$S_T-S_{298.15}$, cal./deg. mole	T, ° K.	$H_T-H_{298.15}$, cal./mole	$S_T-S_{298.15}$, cal./deg. mole
400	1,970	5.69	761(l)	13,570	24.25
500	3,930	10.06	800	14,640	25.62
600	5,900	13.65	900	17,400	28.87
700	7,910	16.75	1,000	20,160	31.78
761(c)	9,140	18.43			

PbBr₂(c):

$$H_T-H_{298.15}=18.59T+1.10\times10^{-3}T^2-5,640$$
(0.1 percent; 298°–761° K.);
$$C_p=18.59+2.20\times10^{-3}T;$$
$$\Delta H_{761}(fusion)=4,430.$$

PbBr₂(l):

$$H_T-H_{298.15}=27.60T-7,440 \text{ (0.1 percent;}$$
761°–1,000° K.);
$$C_p=27.60.$$

CHLORIDES

TABLE 430.—*Heat content and entropy of* PbCl(g)

[Base, ideal gas at 298.15° K.; mol. wt., 242.67]

T, ° K.	$H_T-H_{298.15}$, cal./mole	$S_T-S_{298.15}$, cal./deg. mole	T, ° K.	$H_T-H_{298.15}$, cal./mole	$S_T-S_{298.15}$, cal./deg. mole
400	885	2.55	1,000	6,205	10.67
500	1,765	4.52	1,200	7,990	12.30
600	2,650	6.13	1,400	9,775	13.67
700	3,535	7.50	1,600	11,560	14.86
800	4,425	8.68	1,800	13,345	15.91
900	5,315	9.73	2,000	15,130	16.86

PbCl(g):

$$H_T-H_{298.15}=8.94T+0.29\times10^5T^{-1}-2,763$$
(0.1 percent; 298°–2,000° K.);
$$C_p=8.94-0.29\times10^5T^{-2}.$$

TABLE 431.—*Heat content and entropy of* PbCl₂(c, l)

[Base, crystals at 298.15° K.; mol. wt., 278.12]

T, ° K.	$H_T-H_{298.15}$, cal./mole	$S_T-S_{298.15}$, cal./deg. mole	T, ° K.	$H_T-H_{298.15}$, cal./mole	$S_T-S_{298.15}$, cal./deg. mole
400	1,920	5.54	771(l)	15,370	26.46
500	3,830	9.80	800	16,160	27.48
600	5,890	13.55	900	18,880	30.68
700	8,040	16.86	1,000	21,600	33.54
771(c)	9,570	18.94			

PbCl₂(c):

$$H_T-H_{298.15}=15.96T+4.00\times10^{-3}T^2$$
$$-5,114 \text{ (0.4 percent; 298°–771° K.)};$$
$$C_p=15.96+8.00\times10^{-3}T.$$
$$\Delta H_{771}(fusion)=5,800.$$

PbCl₂(l):

$$H_T-H_{298.15}=27.20T-5,600$$
(0.1 percent; 771°–1,000° K.).
$$C_p=27.20$$

FLUORIDES

TABLE 432.—*Heat content and entropy of* PbF(g)

[Base, ideal gas at 298.15° K.; mol. wt., 226.21]

T, ° K.	$H_T-H_{298.15}$, cal./mole	$S_T-S_{298.15}$, cal./deg. mole	T, ° K.	$H_T-H_{298.15}$, cal./mole	$S_T-S_{298.15}$, cal./deg. mole
400	850	2.45	1,000	6,095	10.44
500	1,705	4.36	1,200	7,870	12.05
600	2,575	5.94	1,400	9,655	13.43
700	3,450	7.29	1,600	11,445	14.63
800	4,330	8.47	1,800	13,245	15.69
900	5,210	9.51	2,000	15,055	16.64

PbF(g):

$$H_T-H_{298.15}=8.78T+0.07\times10^{-3}T^2+0.56\times10^5T^{-1}$$
$$-2,812 \text{ (0.1 percent; 298°–2,000° K.)};$$
$$C_p=8.78+0.14\times10^{-3}T-0.56\times10^5T^{-2}.$$

PbF₂(c):

$$C_p=16.50+4.10\times10^{-3}T(estimated) \quad (298°–1,097° \text{ K.}).$$

IODIDES

TABLE 433.—*Heat content and entropy of* PbI(g)

[Base, ideal gas at 298.15° K.; mol. wt., 334.12]

T, ° K.	$H_T-H_{298.15}$, cal./mole	$S_T-S_{298.15}$, cal./deg. mole	T, ° K.	$H_T-H_{298.15}$, cal./mole	$S_T-S_{298.15}$, cal./deg. mole
400	905	2.61	1,000	6,255	10.78
500	1,795	4.60	1,200	8,040	12.40
600	2,685	6.22	1,400	9,830	13.78
700	3,575	7.59	1,600	11,615	14.97
800	4,470	8.78	1,800	13,405	16.03
900	5,360	9.83	2,000	15,195	16.97

PbI(g):

$$H_T-H_{298.15}=8.94T+0.08\times10^5T^{-1}$$
$$-2,692 \text{ (0.1 percent; 298°–2,000° K.)};$$
$$C_p=8.94-0.08\times10^5T^{-2}.$$

TABLE 434.—*Heat content and entropy of* PbI₂(c, l)

[Base, crystals at 298.15° K.; mol. wt., 461.03]

T, ° K.	$H_T-H_{298.15}$, cal./mole	$S_T-S_{298.15}$, cal./deg. mole	T, ° K.	$H_T-H_{298.15}$, cal./mole	$S_T-S_{298.15}$, cal./deg. mole
600	6,070	14.01	900	20,840	34.42
685(c)	7,860	16.80	1,000	24,080	37.83
685(l)	13,870	25.57			

PbI₂(c):

$$H_T-H_{298.15}=18.00T+2.35\times10^{-3}T^2$$
$$-5,576 \text{ (0.2 percent; 298–685° K.)};$$
$$C_p=18.00+4.70\times10^{-3}T;$$
$$\Delta H_{685}(fusion)=6,010.$$

PbI₂(l):

$$H_T-H_{298.15}=32.40T-8,320$$
(0.1 percent; 685°–1,000° K.);
$$C_p=32.40.$$

ARSENATE

Pb₃(AsO₄)₂(c):

$$\overline{C_p}=65.5 \text{ (286°–370° K.)}.$$

TABLE 1h7. (*Continued*)

BORATES

$PbB_2O_4(c)$:

$\overline{C_p}=26.5$ (288°–371° K.).

$PbB_4O_7(c)$:

$\overline{C_p}=41.4$ (288°–371° K.).

CARBONATE

$PbCO_3(c)$:

$C_p=12.39+28.60\times10^{-3}T(estimated)$ (298°–800° K.).

CHROMATE

$PbCrO_4(c)$:

$\overline{C_p}=29.1$ (292°–323° K.).

MOLYBDATE

$PbMoO_4(c)$:

$\overline{C_p}=30.4$ (292°–322° K.).

NITRATE

$Pb(NO_3)_2(c)$:

$\overline{C_p}=36.4$ (289°–320° K.).

PHOSPHATES

$Pb_3(PO_4)_2(c)$:

$C_p=61.25$ (298° K.).

$Pb_2P_2O_7(c)$:

$\overline{C_p}=48.3$ (284°–371° K.).

SILICATE

$PbSiO_3(c)$:

$C_p=21.52$ (298° K.).

$PbSiO_3(amorphous)$:

$C_p=22.43$ (298° K.).

$Pb_2SiO_4(c)$:

$C_p=32.78$ (298° K.).

SULFATE

TABLE 435.—*Heat content and entropy of* $PbSO_4(c)$

[Base, crystals at 298.15° K.; mol. wt., 303.28]

T, ° K.	$H_T-H_{298.15}$, cal./mole	$S_T-S_{298.15}$, cal./deg. mole	T, ° K.	$H_T-H_{298.15}$, cal./mole	$S_T-S_{298.15}$, cal./deg. mole
400	2,580	7.44	800	14,870	28.33
500	5,300	13.50	900	18,700	32.84
600	8,220	18.82	1,000	22,800	37.16
700	11,360	23.65	1,100	27,050	41.21

$PbSO_4(c)$:

$H_T-H_{298.15}=10.96T+15.50\times10^{-3}T^2-4.20\times10^5T^{-1}$

$-3,327$ (0.3 percent; 298°–1,100° K.);

$C_p=10.96+31.00\times10^{-3}T+4.20\times10^5T^{-2}$.

THIOSULFATE

$PbS_2O_3(c)$:

$\overline{C_p}=29.4$ (293°–373° K.).

TUNGSTATE

TABLE 436.—*Heat content and entropy of* $PbWO_4(c)$

[Base, crystals at 298.15° K.; mol. wt. 455.07]

T, ° K.	$H_T-H_{298.15}$, cal./mole	$S_T-S_{298.15}$, cal./deg. mole	T, ° K.	$H_T-H_{298.15}$, cal./mole	$S_T-S_{298.15}$, cal./deg. mole
400	3,240	9.34	800	16,900	32.86
500	6,510	16.63	900	20,550	37.16
600	9,880	22.77	1,000	24,290	41.10
700	13,340	28.11	1,100	28,130	44.76

$PbWO_4(c)$:

$H_T-H_{298.15}=28.50T+4.71\times10^{-3}T^2-8,916$

(0.1 percent; 298°–1,100° K.);

$C_p=28.50+9.42\times10^{-3}T$.

LEAD-AMMONIUM CHLORIDE

$2PbCl_2\cdot NH_4Cl(c)$:

$C_p=53.1$ (293° K.).

LEAD-THALLIUM COMPOUNDS

TABLE 437.—*Heat content and entropy of* $PbTl_7(c, l)$

[Base, crystals at 298.15° K.; mol. wt., 1,637.94]

T, ° K.	$H_T-H_{298.15}$, cal./mole	$S_T-S_{298.15}$, cal./deg. mole	T, ° K.	$H_T-H_{298.15}$, cal./mole	$S_T-S_{298.15}$, cal./deg. mole
400	4,440	12.78	607(l)	25,160	50.62
500	9,250	23.49	700	30,840	59.32
600	14,540	33.13	800	36,950	67.48
607(c)	14,930	33.77	900	43,060	74.67

$PbTl_7(c)$:

$H_T-H_{298.15}=27.16T+23.40\times10^{-3}T^2-10,178$ (0.1 percent; 298°–607° K.);

$C_p=27.16+46.80\times10^{-3}T$;

$\Delta H_{607}(fusion)=10,230$.

$PbTl_7(l)$:

$H_T-H_{298.15}=61.10T-11,930$ (0.1 percent; 607°–900° K.);

$C_p=61.10$.

TABLE 438.—*Heat content and entropy of* $Pb_3Tl_5(c, l)$

[Base, crystals at 298.15° K.; mol. wt., 1,643.58]

T, ° K.	$H_T-H_{298.15}$, cal./mole	$S_T-S_{298.15}$, cal./deg. mole	T, ° K.	$H_T-H_{298.15}$, cal./mole	$S_T-S_{298.15}$, cal./deg. mole
400	5,290	15.24	652(l)	30,510	59.69
500	10,840	27.61	700	33,580	64.24
600	16,800	38.47	800	39,980	72.79
652(c)	20,050	43.65	900	46,380	80.32

$Pb_3Tl_5(c)$:

$H_T-H_{298.15}=38.14T+19.50\times10^{-3}T^2-13,105$ (0.1 percent; 298°–625° K.);

$C_p=38.14+39.00\times10^{-3}T$;

$\Delta H_{652}(fusion)=10,460$.

$Pb_3Tl_5(l)$:

$H_T-H_{298.15}=64.00T-11,220$ (0.1 percent; 652°–900° K.);

$C_p=64.00$.

TABLE 1h7. (Continued)

LITHIUM AND ITS COMPOUNDS

ELEMENT

TABLE 439.—*Heat content and entropy of Li(c, l)*

[Base, crystals at 298.15° K.; atomic wt., 6.940]

T, °K.	$H_T-H_{298.15}$, cal./mole	$S_T-S_{298.15}$, cal./deg. mole	T, °K.	$H_T-H_{298.15}$, cal./mole	$S_T-S_{298.15}$, cal./deg. mole
350	315	0.97	900	4,845	9.08
400	630	1.82	1,000	5,535	9.80
453.7(c)	1,000	2.68	1,100	6,225	10.46
453.7(l)	1,715	4.26	1,200	6,910	11.06
500	2,050	4.96	1,300	7,595	11.61
600	2,765	6.26	1,400	8,280	12.12
700	3,465	7.34	1,500	8,960	12.58
800	4,155	8.27	1,600	9,645	13.03

Li(c):
$$H_T-H_{298.15}=1.64T+5.55\times10^{-3}T^2-0.84\times10^5T^{-1}-701$$
$$(0.1 \text{ percent}; 298°–453.7° \text{ K.});$$
$$C_p=1.64+11.10\times10^{-3}T+0.84\times10^5T^{-2};$$
$$\Delta H_{453.7}(fusion)=715.$$

Li(l):
$$H_T-H_{298.15}=6.78T-0.99\times10^5T^{-1}-1,143 \ (0.1 \text{ percent};$$
$$453.7°–1,600° \text{ K.});$$
$$C_p=6.78+0.99\times10^5T^{-2}.$$

TABLE 440.—*Heat content and entropy of Li(g)*

[Base, ideal gas at 298.15° K.; atomic wt., 6.940]

T, °K.	$H_T-H_{298.15}$, cal./mole	$S_T-S_{298.15}$, cal./deg. mole	T, °K.	$H_T-H_{298.15}$, cal./mole	$S_T-S_{298.15}$, cal./deg. mole
400	505	1.46	1,900	7,960	9.20
500	1,005	2.57	2,000	8,460	9.46
600	1,500	3.48	2,200	9,460	9.94
700	1,995	4.24	2,400	10,460	10.37
800	2,495	4.90	2,600	11,470	10.78
900	2,990	5.49	2,800	12,495	11.16
1,000	3,490	6.01	3,000	13,525	11.51
1,100	3,985	6.49	3,500	16,190	12.33
1,200	4,480	6.92	4,000	19,010	13.08
1,300	4,980	7.32	4,500	22,030	13.80
1,400	5,475	7.69	5,000	25,315	14.49
1,500	5,975	8.03	6,000	33,070	15.90
1,600	6,470	8.35	7,000	43,240	17.46
1,700	6,965	8.65	8,000	56,510	19.23
1,800	7,465	8.94			

Li(g):
$$H_T-H_{298.15}=4.97T-1,482 \ (0.2 \text{ percent};$$
$$298°–3,000° \text{ K.});$$
$$C_p=4.97.$$

TABLE 441.—*Heat content and entropy of Li₂(g)*

[Base, ideal gas at 298.15° K.; mol. wt., 13.88]

T, °K.	$H_T-H_{298.15}$, cal./mole	$S_T-S_{298.15}$, cal./deg. mole	T, °K.	$H_T-H_{298.15}$, cal./mole	$S_T-S_{298.15}$, cal./deg. mole
400	890	2.59	1,300	9,115	13.30
500	1,780	4.57	1,400	10,050	14.00
600	2,675	6.20	1,500	10,985	14.64
700	3,580	7.61	1,600	11,925	15.25
800	4,490	8.81	1,700	12,865	15.82
900	5,410	9.90	1,800	13,815	16.36
1,000	6,330	10.87	1,900	14,770	16.88
1,100	7,250	11.74	2,000	15,725	17.37
1,200	8,180	12.55			

Li₂(g):
$$H_T-H_{298.15}=8.93T+0.16\times10^{-3}T^2+0.36\times10^5T^{-1}$$
$$-2,797 \ (0.1 \text{ percent}; 298°–2,000° \text{ K.});$$
$$C_p=8.93+0.32\times10^{-3}T-0.36\times10^5T^{-2}.$$

OXIDE

TABLE 442.—*Heat content and entropy of Li₂O(c)*

[Base, crystals at 298.15° K.; mol. wt., 29.88]

T, °K.	$H_T-H_{298.15}$, cal./mole	$S_T-S_{298.15}$, cal./deg. mole	T, °K.	$H_T-H_{298.15}$, cal./mole	$S_T-S_{298.15}$, cal./deg. mole
400	1,445	4.15	900	10,430	18.48
500	3,045	7.72	1,000	12,460	20.61
600	4,765	10.85	1,100	14,540	22.60
700	6,580	13.64	1,200	16,660	24.44
800	8,470	16.17			

Li₂O(c):
$$H_T-H_{298.15}=14.94T+3.04\times10^{-3}T^2+3.38\times10^5T^{-1}$$
$$-5,858 \ (0.2 \text{ percent}; 298°–1,200° \text{ K.});$$
$$C_p=14.94+6.08\times10^{-3}T-3.38\times10^5T^{-2}.$$

HYDROXIDE

TABLE 443—*Heat content and entropy of LiOH(c, l)*

[Base crystals at 298.15° K.; mol. wt., 23.95]

T, °K.	$H_T-H_{298.15}$, cal./mole	$S_T-S_{298.15}$, cal./deg. mole	T, °K.	$H_T-H_{298.15}$, cal./mole	$S_T-S_{298.15}$, cal./deg. mole
400	1,320	3.79	744.3(l)	11,800	20.28
500	2,780	7.05	800	12,950	21.77
600	4,355	9.92	900	15,030	24.22
700	6,030	12.50	1,000	17,100	26.40
744.3(c)	6.800	13.56			

LiOH(c):
$$H_T-H_{298.15}=11.99T+4.12\times10^{-3}T^2+2.27\times10^5T^{-1}$$
$$-4,702 \ (0.1 \text{ percent}; 298°–744.3° \text{ K.});$$
$$C_p=11.99+8.24\times10^{-3}T-2.27\times10^5T^{-2};$$
$$\Delta H_{744.3}(fusion)=5,000.$$
LiOH(l):
$$H_T-H_{298.15}=20.74T-3,638 \ (0.1 \text{ percent};$$
$$744.3°–1,000° \text{ K.});$$
$$C_p=20.74.$$
LiOH·H₂O(c):
$$C_p=19.00 \ (298° \text{ K.}).$$

NITRIDE

TABLE 444.—*Heat content and entropy of Li₃N(c)*

[Base, crystals at 298.15° K.; mol. wt., 34.83]

T, °K.	$H_T-H_{298.15}$, cal./mole	$S_T-S_{298.15}$, cal./deg. mole	T, °K.	$H_T-H_{298.15}$, cal./mole	$S_T-S_{298.15}$, cal./deg. mole
400	2,000	5.76	700	9,360	19.30
500	4,200	10.66	800	12,190	23.07
600	6,680	15.17			

TABLE 1h7. (Continued)

Li$_3$N(c):

$$H_T - H_{298.15} = 11.73T + 11.50 \times 10^{-3}T^2 - 4,520$$

(0.4 percent; 298°–800° K.);

$$C_p = 11.73 + 23.00 \times 10^{-3}T.$$

HYDRIDES

TABLE 445.—*Heat content and entropy of LiH(g)*

[Base, ideal gas at 298.15° K.; mol. wt., 7.948]

T, ° K.	$H_T - H_{298.15}$, cal./mole	$S_T - S_{298.15}$, cal./deg. mole	T, ° K.	$H_T - H_{298.15}$, cal./mole	$S_T - S_{298.15}$, cal./deg. mole
400	735	2.12	1,000	5,515	9.34
500	1,480	3.78	1,200	7,215	10.89
600	2,250	5.18	1,400	8,935	12.22
700	3,045	6.41	1,600	10,670	13.38
800	3,855	7.49	1,800	12,420	14.41
900	4,680	8.46	2,000	14,175	15.33

LiH(g):

$$H_T - H_{298.15} = 7.48T + 0.43 \times 10^{-3}T^2 + 0.58 \times 10^5 T^{-1}$$

$$-2,463 \ (0.6 \ \text{percent}; \ 298°–2,000° \ \text{K.});$$

$$C_p = 7.48 + 0.86 \times 10^{-3}T - 0.58 \times 10^5 T^{-2}.$$

LiH(c):

$$C_p = 8.28 \ (298° \ \text{K.}).$$

TABLE 446.—*Heat content and entropy of LiD(g)*

[Base, ideal gas at 298.15° K.; mol. wt., 8.954]

T, ° K.	$H_T - H_{298.15}$, cal./mole	$S_T - S_{298.15}$, cal./deg. mole	T,° K.	$H_T - H_{298.15}$, cal./mole	$S_T - S_{298.15}$, cal./deg. mole
400	760	2.19	1,000	5,725	9.71
500	1,545	3.94	1,200	7,460	11.29
600	2,355	5.42	1,400	9,205	12.64
700	3,180	6.69	1,600	10,965	13.81
800	4,020	7.81	1,800	12,730	14.85
900	4,870	8.81	2,000	14,500	15.78

LiD(g):

$$H_T - H_{298.15} = 8.11T + 0.25 \times 10^{-3}T^2 + 0.85 \times 10^5 T^{-1}$$

$$-2,725 \ (0.5 \ \text{percent}; \ 298°–2,000° \ \text{K.});$$

$$C_p = 8.11 + 0.50 \times 10^{-3}T - 0.85 \times 10^5 T^{-2}.$$

BROMIDE

LiBr(c):

$$C_p = 11.50 + 3.02 \times 10^{-3}T(estimated) \ (298°–825° \ \text{K.}).$$

LiBr·H$_2$O(c):

$$\overline{C_p} = 22.6 \ (278°–368° \ \text{K.}).$$

TABLE 447.—*Heat content and entropy of LiBr(g)*

[Base, ideal gas at 298.15° K.; mol. wt., 86.86]

T, ° K.	$H_T - H_{298.15}$, cal./mole	$S_T - S_{298.15}$, cal./deg. mole	T, ° K.	$H_T - H_{298.15}$, cal./mole	$S_T - S_{298.15}$, cal./deg. mole
400	840	2.42	1,000	6,115	10.44
500	1,695	4.33	1,200	7,920	12.08
600	2,560	5.90	1,400	9,735	13.48
700	3,435	7.25	1,600	11,560	14.70
800	4,320	8.43	1,800	13,390	15.78
900	5,215	9.49	2,000	15,230	16.75

LiBr(g):

$$H_T - H_{298.15} = 8.79T + 0.12 \times 10^{-3}T^2 + 0.70 \times 10^5 T^{-1}$$

$$-2,866 \ (0.2 \ \text{percent}; \ 298°–2,000° \ \text{K.});$$

$$C_p = 8.79 + 0.24 \times 10^{-3}T - 0.70 \times 10^5 T^{-2}.$$

CHLORIDE

LiCl(c):

$$C_p = 11.00 + 3.40 \times 10^{-3}T(estimated) \ (298°–887° \ \text{K.}).$$

LiCl·H$_2$O(c):

$$\overline{C_p} = 23.5 \ (279°–360° \ \text{K.}).$$

TABLE 448.—*Heat content and entropy of LiCl(g)*

[Base, ideal gas at 298.15° K.; mol. wt. 42.40]

T, ° K.	$H_T - H_{298.15}$, cal./mole	$S_T - S_{298.15}$, cal./deg. mole	T, ° K.	$H_T - H_{298.15}$, cal./mole	$S_T - S_{298.15}$, cal./deg. mole
400	820	2.36	1,000	5,990	10.22
500	1,655	4.22	1,200	7,755	11.83
600	2,505	5.77	1,400	9,525	13.19
700	3,365	7.10	1,600	11,300	14.38
800	4,235	8.26	1,800	13,080	15.43
900	5,110	9.29	2,000	14,860	16.36

LiCl(g):

$$H_T - H_{298.15} = 8.69T + 0.08 \times 10^{-3}T^2 + 0.79 \times 10^5 T^{-1}$$

$$-2,863 \ (0.2 \ \text{percent}; \ 298°–2,000° \ \text{K.});$$

$$C_p = 8.69 + 0.16 \times 10^{-3}T - 0.79 \times 10^5 T^{-2}.$$

FLUORIDE

TABLE 449.—*Heat content and entropy of LiF(c, l)*

[Base, crystals at 298.15° K.; mol. wt. 25.94]

T, ° K.	$H_T - H_{298.15}$, cal./mole	$S_T - S_{298.15}$, cal./deg. mole	T, ° K.	$H_T - H_{298.15}$, cal./mole	$S_T - S_{298.15}$, cal./deg. mole
400	1,085	3.12	1,121.3 (c)	10,510	16.28
500	2,235	5.68	1,121.3 (l)	16,980	22.05
600	3,445	7.89	1,150	17,425	22.43
700	4,700	9.82	1,200	18,200	23.10
800	5,995	11.55	1,300	19,750	24.34
900	7,340	13.14	1,400	21,300	25.49
1,000	8,735	14.60	1,500	22,850	26.56
1,100	10,190	15.99			

LiF(c):

$$H_T - H_{298.15} = 10.41T + 1.95 \times 10^{-3}T^2 + 1.38 \times 10^5 T^{-1}$$

$$-3,740 \ (0.2 \ \text{percent}; \ 298°–1,121.3° \ \text{K.});$$

$$C_p = 10.41 + 3.90 \times 10^{-3}T - 1.38 \times 10^5 T^{-2};$$

$$\Delta H_{1121.3}(fusion) = 6,470.$$

LiF(l):

$$H_T - H_{298.15} = 15.50T - 400 \ (0.1 \ \text{percent}; \ 1,121.3°–1,500°$$

K.);

$$C_p = 15.50.$$

IODIDE

LiI(c):

$$C_p = 12.30 + 2.44 \times 10^{-3}T \ (estimated) \ (298°–713° \ \text{K.}).$$

LiI·H$_2$O(c):

$$\overline{C_p} = 23.6 \ (277°–359° \ \text{K.}).$$

LiI·2H$_2$O(c):

$$\overline{C_p} = 32.9 \ (277°–346° \ \text{K.}).$$

LiI·3H$_2$O(c):

$$\overline{C_p} = 43.2 \ (277°–347° \ \text{K.}).$$

TABLE 1h7. (*Continued*)

TABLE 450.—*Heat content and entropy of LiI(g)*

[Base, ideal gas at 298.15° K.; mol. wt., 133.85]

T, ° K.	$H_T-H_{298.15}$, cal./mole	$S_T-S_{298.15}$, cal./deg. mole	T, K.	$H_T-H_{298.15}$, cal./mole	$S_T-S_{298.15}$, cal./deg. mole
400	855	2.46	1,000	6,165	10.55
500	1,725	4.41	1,200	7,970	12.20
600	2,605	6.01	1,400	9,795	13.61
700	3,490	7.37	1,600	11,645	14.84
800	4,375	8.55	1,800	13,520	15.94
900	5,270	9.61	2,000	15,405	16.94

LiI(g):

$$H_T-H_{298.15}=8.76T+0.17\times10^{-3}T^2+0.58\times10^5T^{-1}$$
$$-2,821\ (0.2\ \text{percent};\ 298°-2,000°\ \text{K.});$$
$$C_p=8.76+0.34\times10^{-3}T-0.58\times10^5T^{-2}.$$

ALUMINATE

TABLE 451.—*Heat content and entropy of LiAlO₂(c)*

[Base, crystals at 298.15° K.; mol. wt., 65.92]

T, ° K.	$H_T-H_{298.15}$, cal./mole	$S_T-S_{298.15}$, cal./deg. mole	T, ° K.	$H_T-H_{298.15}$, cal./mole	$S_T-S_{298.15}$, cal./deg. mole
400	1,790	5.14	1,300	22,910	32.17
500	3,810	9.64	1,400	25,480	34.07
600	6,000	13.63	1,500	28,080	35.86
700	8,280	17.15	1,600	30,710	37.56
800	10,620	20.27	1,700	33,370	39.17
900	13,000	23.07	1,800	36,060	40.71
1,000	15,420	25.62	1,900	38,780	42.18
1,100	17,880	27.97	2,000	41,530	43.59
1,200	20,380	30.14			

LiAlO₂(c):

$$H_T-H_{298.15}=22.08+1.45\times10^{-3}T^2+6.00\times10^5T^{-1}$$
$$-8,724\ (0.5\ \text{percent};\ 298°-2,000°\ \text{K.});$$
$$C_p=22.08+2.90\times10^{-3}T-6.00\times10^5T^{-2}.$$

BOROHYDRIDE

LiBH₄(c):

$$C_p=19.73\ (298.15°\ \text{K.}).$$

CARBONATE

Li₂CO₃(c):

$$C_p=23.28\ (298.15°\ \text{K.}).$$

FERRITE

LiFeO₂(c):

$$C_p=19.81\ (298.15°\ \text{K.}).$$

NITRATE

TABLE 452.—*Heat content and entropy of LiNO₃(c, l)*

[Base, crystals at 298.15° K.; mol. wt., 68.95]

T, ° K.	$H_T-H_{298.15}$, cal./mole	$S_T-S_{298.15}$, cal./deg. mole	T, ° K.	$H_T-H_{298.15}$, cal./mole	$S_T-S_{298.15}$, cal./deg. mole
350	1,130	3.50	525(c)	5,380	13.29
400	2,280	6.57	525(l)	11,500	24.95
450	3,480	9.39	550	12,160	26.18
500	4,730	12.03	600	13,490	28.49

LiNO₃(c):

$$H_T-H_{298.15}=14.98T+10.60\times10^{-3}T^2-5,409$$
$$(0.1\ \text{percent};\ 298°-525°\ \text{K.});$$
$$C_p=14.98+21.20\times10^{-3}T;$$
$$\Delta H_{525}(fusion)=6,120.$$

LiNO₃(l):

$$H_T-H_{298.15}=26.60T-2,470\ (0.1\ \text{percent};$$
$$525°-600°\ \text{K.});$$
$$C_p=26.60.$$

TITANATE

TABLE 453.—*Heat content and entropy of Li₂TiO₃(c, l)*

[Base, crystals at 298.15° K.; mol. wt., 109.78]

T, ° K.	$H_T-H_{298.15}$, cal./mole	$S_T-S_{298.15}$, cal./deg. mole	T, ° K.	$H_T-H_{298.15}$, cal./mole	$S_T-S_{298.15}$, cal./deg. mole
400	2,900	8.34	1,400	38,600	52.11
500	6,040	15.34	1,485(α)	41,850	53.36
600	9,370	21.41	1,485(β)	44,600	56.21
700	12,810	26.71	1,500	45,230	56.63
800	16,330	31.41	1,600	49,490	59.38
900	19,920	35.63	1,700	53,830	62.01
1,000	23,570	39.48	1,800	58,250	64.54
1,100	27,270	43.00	1,820(β)	59,140	65.03
1,200	31,020	46.27	1,820(l)	85,470	79.50
1,300	34,800	49.29	1,850	86,910	80.28

LiTiO₃(α):

$$H_T-H_{298.15}=33.16T+2.06\times10^{-3}T^2+6.98\times10^5T^{-1}$$
$$-12,411\ (0.3\ \text{percent};\ 298°-1,485°\ \text{K.});$$
$$C_p=33.16+4.12\times10^{-3}T-6.98\times10^5T^{-2};$$
$$\Delta H_{1485}(transition)=2,750.$$

LiTiO₃(β):

$$H_T-H_{298.15}=30.20T+4.00\times10^{-3}T^2-9,070$$
$$(0.1\ \text{percent};\ 1,485°-1,820°\ \text{K.});$$
$$C_p=30.20+8.00\times10^{-3}T;$$
$$\Delta H_{1820}(fusion)=26,330.$$

Li₂TiO₃(l):

$$H_T-H_{298.15}=48.00T-1,890\ (0.1\ \text{percent};$$
$$1,820°-1,850°\ \text{K.});$$
$$C_p=48.00.$$

COMPOUNDS WITH OTHER ALKALI METALS

TABLE 454.—*Heat content and entropy of LiK(g)*

[Base, ideal gas at 298.15° K.; mol. wt., 46.04]

T, ° K.	$H_T-H_{298.15}$, cal./mole	$S_T-S_{298.15}$, cal./deg. mole	T, ° K.	$H_T-H_{298.15}$, cal./mole	$S_T-S_{298.15}$, cal./deg. mole
400	900	2.60	1,000	6,240	10.75
500	1,785	4.57	1,200	8,030	12.38
600	2,675	6.20	1,400	9,815	13.75
700	3,565	7.57	1,600	11,600	14.95
800	4,455	8.76	1,800	13,390	16.00
900	5,350	9.81	2,000	15,180	16.94

LiK(g):

$$H_T-H_{298.15}=8.94T+0.14\times10^5T^{-1}-2,712$$
$$(0.1\ \text{percent};\ 298°-2,000°\ \text{K.});$$
$$C_p=8.94-0.14\times10^5T^{-2}.$$

TABLE 1h7. *(Continued)*

TABLE 455.—*Heat content and entropy of* $LiRb(g)$

[Base, ideal gas at 298.15° K.; mol. wt., 92.42]

T, ° K.	$H_T - H_{298.15}$, cal./mole	$S_T - S_{298.15}$, cal./deg. mole	T, ° K.	$H_T - H_{298.15}$, cal./mole	$S_T - S_{298.15}$, cal./deg. mole
400	900	2.59	1,000	6,250	10.76
500	1,790	4.58	1,200	8,035	12.39
600	2,680	6.20	1,400	9,820	13.76
700	3,570	7.58	1,600	11,610	14.96
800	4,465	8.77	1,800	13,395	16.01
900	5,355	9.82	2,000	15,185	16.95

$LiRb(g)$:

$H_T - H_{298.15} = 8.94T + 0.11 \times 10^5 T^{-1} - 2,702$

(0.1 percent; 298°–2,000° K.);

$C_p = 8.94 - 0.11 \times 10^5 T^{-2}$.

TABLE 456.—*Heat content and entropy of* $LiCs(g)$

[Base, ideal gas at 298.15° K.; mol. wt., 139.85]

T, ° K.	$H_T - H_{298.15}$, cal./mole	$S_T - S_{298.15}$, cal./deg. mole	T, ° K.	$H_T - H_{298.15}$, cal./mole	$S_T - S_{298.15}$, cal./mole
400	900	2.60	1,000	6,255	10.77
500	1,790	4.58	1,200	8,040	12.40
600	2,685	6.22	1,400	9,830	13.78
700	3,575	7.59	1,600	11,615	14.97
800	4,465	8.78	1,800	13,405	16.02
900	5,360	9.83	2,000	15,190	16.96

$LiCs(g)$:

$H_T - H_{298.15} = 8.94T + 0.09 \times 10^5 T^{-1} - 2,696$

(0.1 percent; 298°–2,000° K.);

$C_p = 8.94 - 0.09 \times 10^5 T^{-2}$.

LITHIUM-CADMIUM

$LiCd(c)$:

$\overline{C_p} = 12.5$ (293°–473° K.).

LITHIUM-GALLIUM

$LiGa(c)$:

$\overline{C_p} = 12.75$ (823°–973° K.).

LITHIUM-INDIUM

$LiIn(c)$:

$C_p = 7.03 + 7.22 \times 10^{-3} T$ (298°–903° K.).

LUTETIUM

ELEMENT

TABLE 457.—*Heat content and entropy of* $Lu(c, l)$

[Base, crystals at 298.15° K.; atomic wt., 174.99]

T, ° K.	$H_T - H_{298.15}$, cal./mole	$S_T - S_{298.15}$, cal./deg. mole	T, ° K.	$H_T - H_{298.15}$, cal./mole	$S_T - S_{298.15}$, cal./mole
400	665	1.91	1,400	8,010	10.93
500	1,330	3.40	1,500	8,830	11.50
600	2,015	4.65	1,600	9,660	12.03
700	2,710	5.72	1,700	10,510	12.55
800	3,425	6.67	1,800	11,370	13.04
900	4,150	7.53	1,900	12,250	13.51
1,000	4,890	8.31	2,000(c)	13,140	13.97
1,100	5,650	9.03	2,000(l)	17,740	16.27
1,200	6,420	9.71	2,100	18,540	16.66
1,300	7,210	10.34	2,200	19,340	17.03

$Lu(c)$:

$H_T - H_{298.15} = 6.00T + 0.75 \times 10^{-3} T^2 - 1,856$

(0.1 percent; 298°–2,000° K.);

$C_p = 6.00 + 1.50 \times 10^{-3} T$;

$\Delta H_{2,000}(fusion) = 4,600$.

$Lu(l)$:

$H_T - H_{298.15} = 8.00T + 1,740$ (0.1 percent;

2,000°–2,200° K.);

$C_p = 8.00$.

TABLE 458.—*Heat content and entropy of* $Lu(g)$

[Base, ideal gas at 298.15° K.; atomic wt., 174.99]

T, ° K.	$H_T - H_{298.15}$, cal./mole	$S_T - S_{298.15}$, cal./deg. mole	T, ° K.	$H_T - H_{298.15}$, cal./mole	$S_T - S_{298.15}$, cal./deg. mole
400	510	1.48	1,500	7,100	9.19
500	1,030	2.63	1,600	7,730	9.60
600	1,570	3.62	1,700	8,355	9.98
700	2,135	4.49	1,800	8,975	10.33
800	2,725	5.27	1,900	9,595	10.67
900	3,330	5.99	2,000	10,210	10.99
1,000	3,950	6.64	2,200	11,435	11.57
1,100	4,575	7.24	2,400	12,650	12.10
1,200	5,205	7.79	2,600	13,860	12.58
1,300	5,835	8.29	2,800	15,060	13.03
1,400	6,470	8.76	3,000	16,260	13.44

$Lu(g)$:

$H_T - H_{298.15} = 6.24T - 0.03 \times 10^{-3} T^2 + 1.08 \times 10^5 T^{-1}$

$- 2,220$ (0.2 percent; 1,600°–3,000° K.);

$C_p = 6.24 - 0.06 \times 10^{-3} T - 1.08 \times 10^5 T^{-2}$.

MAGNESIUM AND ITS COMPOUNDS

ELEMENT

TABLE 459.—*Heat content and entropy of* $Mg(c, l)$

[Base, ideal gas at 298.15° K.; atomic wt., 24.32]

T, ° K.	$H_T - H_{298.15}$, cal./mole	$S_T - S_{298.15}$, cal./deg. mole	T, ° K.	$H_T - H_{298.15}$, cal./mole	$S_T - S_{298.15}$, cal./deg. mole
400	615	1.78	923(l)	6,415	9.86
500	1,255	3.20	1,000	7,020	10.48
600	1,920	4.41	1,100	7,800	11.23
700	2,610	5.48	1,200	8,580	11.91
800	3,330	6.44	1,300	9,360	12.53
900	4,095	7.34	1,400	10,140	13.11
923(c)	4,275	7.54			

$Mg(c)$:

$H_T - H_{298.15} = 4.97T + 1.52 \times 10^{-3} T^2 - 0.04 \times 10^5 T^{-1}$

$- 1,604$ (0.2 percent; 298°–923° K.);

$C_p = 4.97 + 3.04 \times 10^{-3} T + 0.04 \times 10^5 T^{-2}$;

$\Delta H_{923}(fusion) = 2,140$.

$Mg(l)$:

$H_T - H_{298.15} = 7.80T - 780$ (0.1 percent;

923°–1,400° K.);

$C_p = 7.80$.

TABLE 1h7. (Continued)

TABLE 460.—*Heat content and entropy of Mg(g)*

[Base, ideal gas at 298.15° K.; atomic wt., 24.32]

T, ° K.	$H_T - H_{298.15}$, cal./mole	$S_T - S_{298.15}$, cal./deg. mole	T, ° K.	$H_T - H_{298.15}$, cal./mole	$S_T - S_{298.15}$, cal./deg. mole
400	505	1.46	1,900	7,960	9.20
500	1,005	2.57	2,000	8,455	9.46
600	1,500	3.46	2,200	9,450	9.93
700	1,995	4.24	2,400	10,445	10.36
800	2,495	4.90	2,600	11,440	10.76
900	2,990	5.49	2,800	12,440	11.13
1,000	3,490	6.01	3,000	13,440	11.48
1,100	3,985	6.49	3,500	15,980	12.26
1,200	4,480	6.92	4,000	18,610	12.96
1,300	4,980	7.32	4,500	21,400	13.62
1,400	5,475	7.69	5,000	24,410	14.25
1,500	5,970	8.03	6,000	31,355	15.52
1,600	6,470	8.35	7,000	39,955	16.84
1,700	6,965	8.65	8,000	50,805	18.28
1,800	7,465	8.93			

$Mg(g)$:

$$H_T - H_{298.15} = 4.97T - 1,482 \quad (0.2 \text{ percent};$$
$$298° - 3,500° \text{ K.});$$
$$C_p = 4.97.$$

OXIDE

TABLE 461.—*Heat content and entropy of MgO(c)*

[Base, crystals at 298.15° K.; mol. wt., 40.32]

T, ° K.	$H_T - H_{298.15}$, cal./mole	$S_T - S_{298.15}$, cal./deg. mole	T, ° K.	$H_T - H_{298.15}$, cal./mole	$S_T - S_{298.15}$, cal./deg. mole
400	965	2.78	1,300	11,310	15.98
500	1,975	5.03	1,400	12,570	16.92
600	3,020	6.94	1,500	13,830	17.79
700	4,100	8.60	1,600	15,090	18.60
800	5,225	10.10	1,700	16,350	19.36
900	6,390	11.47	1,800	17,610	20.08
1,000	7,580	12.73	1,900	18,870	20.76
1,100	8,800	13.89	2,000	20,130	21.41
1,200	10,050	14.98	2,100	21,390	22.02

$MgO(c)$:

$$H_T - H_{298.15} = 10.18T + 0.87 \times 10^{-3}T^2 + 1.48 \times 10^5 T^{-1}$$
$$-3,609 \quad (0.8 \text{ percent}; 298° - 2,100° \text{ K.});$$
$$C_p = 10.18 + 1.74 \times 10^{-3}T - 1.48 \times 10^5 T^{-2}.$$

HYDROXIDE

TABLE 462.—*Heat content and entropy of Mg(OH)₂(c)*

[Base, crystals at 298.15° K.; mol. wt., 58.34]

T, ° K.	$H_T - H_{298.15}$, cal./mole	$S_T - S_{298.15}$, cal./deg. mole	T, ° K.	$H_T - H_{298.15}$, cal./mole	$S_T - S_{298.15}$, cal./deg. mole
400	1,890	5.44	600	6,080	13.89
500	3,890	9.90			

$Mg(OH)_2(c)$:

$$H_T - H_{298.15} = 13.04T + 7.90 \times 10^{-3}T^2$$
$$-4,590 \quad (0.2 \text{ percent}; 298° - 600° \text{ K.});$$
$$C_p = 13.04 + 15.80 \times 10^{-3}T.$$

SULFIDE

TABLE 463.—*Heat content and entropy of MgS(g)*

[Base, ideal gas at 298.15° K.; mol. wt., 56.39]

T, ° K.	$H_T - H_{298.15}$, cal./mole	$S_T - S_{298.15}$, cal./deg. mole	T, ° K.	$H_T - H_{298.15}$, cal./mole	$S_T - S_{298.15}$, cal./deg. mole
400	850	2.45	1,000	6,085	10.42
500	1,705	4.36	1,200	7,860	12.04
600	2,570	5.94	1,400	9,635	13.41
700	3,440	7.28	1,600	11,415	14.59
800	4,320	8.45	1,800	13,200	15.64
900	5,200	9.49	2,000	14,985	16.58

$MgS(g)$:

$$H_T - H_{298.15} = 8.82T + 0.04 \times 10^{-3}T^2 + 0.61 \times 10^5 T^{-1}$$
$$-2,838 \quad (0.1 \text{ percent}; 298° - 2,000° \text{ K.});$$
$$C_p = 8.82 + 0.08 \times 10^{-3}T - 0.61 \times 10^5 T^{-2}.$$

NITRIDE

TABLE 464.—*Heat content and entropy of Mg₃N₂(c)*

[Base, α-crystals at 298.15° K.; mol. wt., 100.98]

T, ° K.	$H_T - H_{298.15}$, cal./mole	$S_T - S_{298.15}$, cal./deg. mole	T, ° K.	$H_T - H_{298.15}$, cal./mole	$S_T - S_{298.15}$, cal./deg. mole
400	2,510	7.24	900	16,550	29.70
500	5,100	13.01	1,000	19,570	32.89
600	7,790	17.92	1,061(β)	21,460	34.72
700	10,590	22.23	1,061(γ)	21,680	34.93
800	13,510	26.13	1,100	22,790	35.96
823(α)	14,190	26.96	1,200	25,640	38.44
823(β)	14,300	27.09	1,300	28,490	40.72

$Mg_3N_2(\alpha)$:

$$H_T - H_{298.15} = 20.77T + 5.60 \times 10^{-3}T^2$$
$$-6,690 \quad (0.1 \text{ percent}; 298 - 823° \text{ K.});$$
$$C_p = 20.77 + 11.20 \times 10^{-3}T;$$
$$\Delta H_{823}(transition) = 110.$$

$Mg_3N_2(\beta)$:

$$H_T - H_{298.15} = 20.07T + 5.33 \times 10^{-3}T^2$$
$$-5,830 \quad (0.1 \text{ percent}; 823° - 1,061° \text{ K.});$$
$$C_p = 20.07 + 10.66 \times 10^{-3}T;$$
$$\Delta H_{1061}(transition) = 220.$$

$Mg_3N_2(\gamma)$:

$$H_T - H_{298.15} = 28.50T$$
$$-8,560 \quad (0.1 \text{ percent}; 1,061° - 1,300° \text{ K.});$$
$$C_p = 28.50.$$

SILICIDE

$Mg_2Si(c)$:

$$C_p = 15.40 + 4.15 \times 10^{-3}T(estimated) \quad (298° - 1,343° \text{ K.}).$$

TABLE 1h7. (*Continued*)

HYDRIDES

CHLORIDES

TABLE 465.—*Heat content and entropy of MgH(g)*

[Base, ideal gas at 298.15° K.; mol. wt., 25.33]

T, ° K.	$H_T - H_{298.15}$, cal./mole	$S_T - S_{298.15}$, cal./deg. mole	T, ° K.	$H_T - H_{298.15}$, cal./mole	$S_T - S_{298.15}$, cal./deg. mole
400	730	2.11	1,000	5,475	9.27
500	1,465	3.75	1,200	7,165	10.81
600	2,230	5.14	1,400	8,880	12.13
700	3,020	6.36	1,600	10,610	13.29
800	3,825	7.43	1,800	12,350	14.31
900	4,645	8.40	2,000	14,105	15.24

$$\text{MgH}(g):$$
$$H_T - H_{298.15} = 7.38T + 0.45 \times 10^{-3}T^2 + 0.53 \times 10^5 T^{-1}$$
$$-2,418 \ (0.7 \ \text{percent}; \ 298°\text{-}2,000° \ \text{K.});$$
$$C_p = 7.38 + 0.90 \times 10^{-3}T - 0.53 \times 10^5 T^{-2}.$$

TABLE 466.—*Heat content and entropy of MgD(g)*

[Base, ideal gas at 298.15° K.; mol. wt., 26.33]

T, ° K.	$H_T - H_{298.15}$, cal./mole	$S_T - S_{298.15}$, cal./deg. mole	T, ° K.	$H_T - H_{298.15}$, cal./mole	$S_T - S_{298.15}$, cal./deg. mole
400	770	2.22	1,000	5,730	9.73
500	1,550	3.96	1,200	7,460	11.30
600	2,355	5.43	1,400	9,205	12.65
700	3,180	6.70	1,600	10,965	13.82
800	4,020	7.82	1,800	12,725	14.86
900	4,870	8.82	2,000	14,495	15.79

$$\text{MgD}(g):$$
$$H_T - H_{298.15} = 8.13T + 0.24 \times 10^{-3}T^2 + 0.87 \times 10^5 T^{-1}$$
$$-2,737 \ (0.3 \ \text{percent}; \ 298°\text{-}2,000° \ \text{K.});$$
$$C_p = 8.13 + 0.48 \times 10^{-3}T - 0.87 \times 10^5 T^{-2}.$$

TABLE 468.—*Heat content and entropy of $MgCl_2(c, l)$*

[Base, crystals at 298.15° K.; mol. wt., 95.23]

T, ° K.	$H_T - H_{298.15}$, cal./mole	$S_T - S_{298.15}$, cal./deg. mole	T, ° K.	$H_T - H_{298.15}$, cal./mole	$S_T - S_{298.15}$, cal./deg. mole
400	1,800	5.19	1,000	23,750	33.27
500	3,650	9.31	1,100	25,960	35.38
600	5,555	12.79	1,200	28,170	37.30
700	7,480	15.75	1,300	30,380	39.07
800	9,420	18.34	1,400	32,590	40.70
900	11,380	20.65	1,500	34,800	42.23
987(c)	13,160	22.54	1,600	37,010	43.66
987(l)	23,460	32.98	1,700	39,220	45.00

$$\text{MgCl}_2(c):$$
$$H_T - H_{298.15} = 18.90T + 0.71 \times 10^{-3}T^2 + 2.06 \times 10^5 T^{-1}$$
$$-6,389 \ (0.1 \ \text{percent}; \ 298°\text{-}987° \ \text{K.});$$
$$C_p = 18.90 + 1.42 \times 10^{-3}T - 2.06 \times 10^5 T^{-2};$$
$$\Delta_{H_{987}}(fusion) = 10,300.$$

$$\text{MgCl}_2(l):$$
$$H_T - H_{298.15} = 22.10T - 1,650 \ (0.1 \ \text{percent};$$
$$987°\text{-}1,700° \ \text{K.});$$
$$C_p = 22.10.$$

$$\text{MgCl}_2 \cdot \text{H}_2\text{O}(c):$$
$$C_p = 21.75 + 19.45 \times 10^{-3}T(estimated) \ (298°\text{-}650° \ \text{K.}).$$

$$\text{MgCl}_2 \cdot 2\text{H}_2\text{O}(c):$$
$$C_p = 29.91 + 27.31 \times 10^{-3}T(estimated) \ (298°\text{-}500° \ \text{K.}).$$

$$\text{MgCl}_2 \cdot 4\text{H}_2\text{O}(c):$$
$$C_p = 44.83 + 43.03 \times 10^{-3}T(estimated) \ (298°\text{-}450° \ \text{K.}).$$

$$\text{MgCl}_2 \cdot 6\text{H}_2\text{O}(c):$$
$$C_p = 57.78 + 58.74 \times 10^{-3}T(estimated) \ (298°\text{-}385° \ \text{K.}).$$

BROMIDE

TABLE 467.—*Heat content and entropy of MgBr(g)*

[Base, ideal gas at 298.15° K.; mol. wt., 104.24]

T, ° K.	$H_T - H_{298.15}$, cal./mole	$S_T - S_{298.15}$, cal./deg. mole	T, ° K.	$H_T - H_{298.15}$, cal./mole	$S_T - S_{298.15}$, cal./deg. mole
400	875	2.52	1,000	6,170	10.60
500	1,745	4.46	1,200	7,950	12.22
600	2,625	6.07	1,400	9,735	13.59
700	3,510	7.43	1,600	11,520	14.78
800	4,395	8.61	1,800	13,305	15.84
900	5,285	9.66	2,000	15,090	16.78

$$\text{MgBr}(g):$$
$$H_T - H_{298.15} = 8.92T + 0.38 \times 10^5 T^{-1} - 2,787$$
$$(0.1 \ \text{percent}; \ 298°\text{-}2,000° \ \text{K.});$$
$$C_p = 8.92 - 0.38 \times 10^5 T^{-2}.$$

TABLE 469.—*Heat content and entropy of MgCl (g)*

[Base, ideal gas at 298.15° K.; mol. wt., 59.78]

T, ° K.	$H_T - H_{298.15}$, cal./mole	$S_T - S_{298.15}$, cal./deg. mole	T, ° K.	$H_T - H_{298.15}$, cal./mole	$S_T - S_{298.15}$, cal./deg. mole
400	860	2.48	1,000	6,120	10.49
500	1,720	4.40	1,200	7,900	12.11
600	2,590	5.98	1,400	9,680	13.49
700	3,470	7.34	1,600	11,460	14.67
800	4,350	8.52	1,800	13,245	15.72
900	5,235	9.56	2,000	15,030	16.66

$$\text{MgCl}(g):$$
$$H_T - H_{298.15} = 8.87T + 0.02 \times 10^{-3}T^2 + 0.52 \times 10^5 T^{-1}$$
$$-2,821 \ (0.1 \ \text{percent}; \ 298°\text{-}2,000° \ \text{K.});$$
$$C_p = 8.87 + 0.04 \times 10^{-3}T - 0.52 \times 10^5 T^{-2}.$$

TABLE 1h7. (*Continued*)

FLUORIDES

TABLE 470.—*Heat content and entropy of MgF$_2$(c, l)*

[Base, crystals at 298.15° K., mol. wt., 62.32]

T, ° K.	$H_T-H_{298.15}$, cal./mole	$S_T-S_{298.15}$, cal./deg. mole	T, ° K.	$H_T-H_{298.15}$, cal./mole	$S_T-S_{298.15}$, cal./deg. mole
400	1,645	4.74	1,400	20,460	27.82
500	3,320	8.47	1,500	22,490	29.22
600	5,080	11.68	1,536(c)	23,220	29.70
700	6,890	14.47	1,536(l)	37,120	38.75
800	8,720	16.92	1,600	38,560	39.66
900	10,590	19.12	1,700	40,820	41.03
1,000	12,510	21.14	1,800	43,080	42.33
1,100	14,450	22.99	1,900	45,340	43.55
1,200	16,430	24.71	2,000	47,600	44.71
1,300	18,440	26.32			

MgF$_2$(c):

$$H_T-H_{298.15}=16.93T+1.26\times10^{-3}T^2+2.20\times10^5T^{-1}$$
$$-5,898 \text{ (0.2 percent; } 298°-1,536° \text{ K.)};$$
$$C_p=16.93+2.52\times10^{-3}T-2.20\times10^5T^{-2};$$
$$\Delta H_{1536}(fusion)=13,900.$$

MgF$_2$(l):

$$H_T-H_{298.15}=22.60T+2,400 \text{ (0.1 percent; } 1,536°-2,000° \text{ K.)};$$
$$C_p=22.60.$$

TABLE 471.—*Heat content and entropy of MgF (g)*

[Base, ideal gas at 298.15° K.; mol. wt. 43.32]

T, ° K.	$H_T-H_{298.15}$, cal./mole	$S_T-S_{298.15}$, cal./deg. mole	T, ° K.	$H_T-H_{298.15}$, cal./mole	$S_T-S_{298.15}$, cal./deg. mole
400	810	2.33	1,000	5,955	10.15
500	1,640	4.19	1,200	7,715	11.75
600	2,485	5.73	1,400	9,485	13.12
700	3,345	7.05	1,600	11,260	14.31
800	4,210	8.21	1,800	13,035	15.35
900	5,080	9.23	2,000	14,815	16.29

MgF(g):

$$H_T-H_{298.15}=8.64T+0.09\times10^{-3}T^2+0.82\times10^5T^{-1}$$
$$-2,859 \text{ (0.2 percent; } 298°-2,000° \text{ K.)};$$
$$C_p=8.64+0.18\times10^{-3}T-0.82\times10^5T^{-2}.$$

IODIDE

TABLE 472.—*Heat content and entropy of MgI(g)*

[Base, ideal gas at 298.15° K.; mol. wt., 151.23]

T, ° K.	$H_T-H_{298.15}$, cal./mole	$S_T-S_{298.15}$, cal./deg. mole	T, ° K.	$H_T-H_{298.15}$, cal./mole	$S_T-S_{298.15}$, cal./deg. mole
400	885	2.55	1,000	6,200	10.66
500	1,765	4.51	1,200	7,985	12.29
600	2,645	6.12	1,400	9,770	13.66
700	3,535	7.49	1,600	11,555	14.86
800	4,425	8.68	1,800	13,340	15.91
900	5,315	9.73	2,000	15,130	16.85

MgI(g):

$$H_T-H_{298.15}=8.94T+0.30\times10^5T^{-1}-2,766 \text{ (0.1 percent; }$$
$$298°-2,000° \text{ K.)};$$
$$C_p=8.94-0.30\times10^5T^{-2}.$$

HYDROXYCHLORIDE

MgOHCl(c):
$$C_p=13.40+14.47\times10^{-3}T(estimated) \text{ (298°-850° K.)}$$

ALUMINATE

TABLE 473.—*Heat content and entropy of MgAl$_2$O$_4$(c)*

[Base, crystals at 298.15° K.; mol. wt., 142.28]

T, ° K.	$H_T-H_{298.15}$, cal./mole	$S_T-S_{298.15}$, cal./deg. mole	T, ° K.	$H_T-H_{298.15}$, cal./mole	$S_T-S_{298.15}$, cal./deg. mole
400	3,150	9.05	1,300	39,490	55.42
500	6,650	16.85	1,400	44,030	58.78
600	10,350	23.59	1,500	48,620	61.95
700	14,190	29.51	1,600	53,230	64.92
800	18,150	34.79	1,700	57,850	67.72
900	22,220	39.59	1,800	62,480	70.37
1,000	26,390	43.98	1,900	67,120	72.88
1,100	30,660	48.05	2,000	71,770	75.26
1,200	35,030	51.85			

MgAl$_2$O$_4$(c):

$$H_T-H_{298.15}=36.80T+3.20\times10^{-3}T^2+9.78\times10^5T^{-1}$$
$$-14,537 \text{ (0.2 percent; } 298°-1,800° \text{ K.)};$$
$$C_p=36.80+6.40\times10^{-3}T-9.78\times10^5T^{-2}.$$

CARBONATE

TABLE 474.—*Heat content and entropy of MgCO$_3$(c)*

[Base, crystals at 298.15° K.; mol. wt., 84.33]

T, ° K.	$H_T-H_{298.15}$, cal./mole	$S_T-S_{298.15}$, cal./deg. mole	T, ° K.	$H_T-H_{298.15}$, cal./mole	$S_T-S_{298.15}$, cal./deg. mole
400	2,060	5.92	700	9,450	19.54
500	4,300	10.91	750	10,820	21.43
600	6,790	15.45			

MgCO$_3$(c):

$$H_T-H_{298.15}=18.62T+6.90\times10^{-3}T^2+4.16\times10^5T^{-1}$$
$$-7,560 \text{ (0.4 percent; } 298°-750° \text{ K.)};$$
$$C_p=18.62+13.80\times10^{-3}T-4.16\times10^5T^{-2}.$$

CHROMITE

TABLE 475.—*Heat content and entropy of MgCr$_2$O$_4$(c)*

[Base, crystals at 298.15° K.; mol. wt., 192.34]

T, ° K.	$H_T-H_{298.15}$, cal./mole	$S_T-S_{298.15}$, cal./deg. mole	T, ° K.	$H_T-H_{298.15}$, cal./mole	$S_T-S_{298.15}$, cal./deg. mole
400	3,350	9.64	1,300	40,490	57.39
500	7,040	17.86	1,400	44,890	60.66
600	10,930	24.95	1,500	49,340	63.73
700	14,940	31.13	1,600	53,840	66.73
800	19,060	36.63	1,700	58,370	69.37
900	23,260	41.57	1,800	62,930	71.98
1,000	27,520	46.06	1,900	67,520	74.46
1,100	31,810	50.15	2,000	72,140	76.83
1,200	36,130	53.91			

MgCr$_2$O$_4$(c):

$$H_T-H_{298.15}=40.02T+1.78\times10^{-3}T^2+9.58\times10^5T^{-1}$$
$$-15,303 \text{ (0.2 percent; } 298°-1,800° \text{ K.)};$$
$$C_p=40.02+3.56\times10^{-3}T-9.58\times10^5T^{-2}.$$

TABLE 1h7. (Continued)

FERRITE

TABLE 476.—*Heat content and entropy of* $MgFe_2O_4(c)$

[Base, α-crystals at 298.15° K.; mol. wt., 200.02]

T, ° K.	$H_T - H_{298.15}$, cal./mole	$S_T - S_{298.15}$, cal./deg. mole	T, ° K.	$H_T - H_{298.15}$, cal./mole	$S_T - S_{298.15}$, cal./deg. mole
400	3,860	11.10	1,230(β)	41,250	61.23
500	7,870	20.03	1,230(γ)	41,600	61.51
600	12,270	28.04	1,300	44,630	63.91
665(α)	15,600	33.31	1,400	49,000	67.15
665(β)	15,600	33.31	1,500	53,520	70.27
700	17,190	35.64	1,600	58,190	73.28
800	21,730	41.70	1,700	63,000	76.20
900	26,270	47.05	1,800	67,960	79.03
1,000	30,810	51.83	1,900	73,060	81.79
1,100	35,350	56.16	2,000	78,300	84.47
1,200	39,890	60.11			

$MgFe_2O_4(\alpha)$:

$$H_T - H_{298.15} = 21.06\,T + 22.29 \times 10^{-3}\,T^2 - 8,260$$
$$(1.2 \text{ percent}; 298°\text{–}665° \text{ K.});$$
$$C_p = 21.06 + 44.58 \times 10^{-3}\,T;$$
$$\Delta H_{665}(transition) = 0.$$

$MgFe_2O_4(\beta)$:

$$H_T - H_{298.15} = 45.40\,T - 14,590 \ (0.1 \text{ percent};$$
$$665°\text{–}1,230° \text{ K.});$$
$$C_p = 45.40;$$
$$\Delta H_{1230}(transition) = 350.$$

$MgFe_2O_4(\gamma)$:

$$H_T - H_{298.15} = 25.67 + 6.79 \times 10^{-3}\,T^2 - 247$$
$$(0.1 \text{ percent}; 1,230°\text{–}2,000° \text{ K.});$$
$$C_p = 25.67 + 13.58 \times 10^{-3}\,T.$$

NITRATE

TABLE 477.—*Heat content and entropy of* $Mg(NO_3)_2(c)$

[Base, crystals at 298.15° K.; mol. wt., 148.34]

T, ° K.	$H_T - H_{298.15}$, cal./mole	$S_T - S_{298.15}$, cal./deg. mole	T, ° K.	$H_T - H_{298.15}$, cal./mole	$S_T - S_{298.15}$, cal./deg. mole
350	1,820	5.62	500	8,150	20.58
400	3,780	10.86	550	10,570	25.19
450	5,870	15.77	600	13,120	29.62

$Mg(NO_3)_2(c)$:

$$H_T - H_{298.15} = 10.68\,T + 35.60 \times 10^{-3}\,T^2 - 1.79 \times 10^5\,T^{-1}$$
$$-5,748 \ (0.4 \text{ percent}; 298°\text{–}600° \text{ K.});$$
$$C_p = 10.68 + 71.20 \times 10^{-3}\,T + 1.79 \times 10^5\,T^{-2}.$$

SILICATES

TABLE 478.—*Heat content and entropy of* $MgSiO_3(clinoenstatite)$

[Base, crystals at 298.15° K.; mol. wt., 100.41]

T, ° K.	$H_T - H_{298.15}$, cal./mole	$S_T - S_{298.15}$, cal./deg. mole	T, ° K.	$H_T - H_{298.15}$, cal./mole	$S_T - S_{298.15}$, cal./deg. mole
400	2,140	6.16	1,200	23,890	35.27
500	4,480	11.37	1,300	26,890	37.66
600	6,980	15.92	1,400	29,910	39.90
700	9,600	19.95	1,500	32,940	42.00
800	12,300	23.55	1,600	35,970	43.95
900	15,090	26.84	1,700	39,010	45.79
1,000	17,970	29.87	1,800	42,060	47.53
1,100	20,910	32.67			

$MgSiO_3(clinoenstatite)$:

$$H_T - H_{298.15} = 24.55\,T + 2.37 \times 10^{-3}\,T^2 + 6.28 \times 10^5\,T^{-1}$$
$$-9,637 \ (0.3 \text{ percent}; 298°\text{–}1,600° \text{ K.});$$
$$C_p = 24.55 + 4.74 \times 10^{-3}\,T - 6.28 \times 10^5\,T^{-2}.$$

TABLE 479.—*Heat content and entropy of* $MgSiO_3(amphibole\text{-}type)$

[Base, crystals at 298.15° K.; mol. wt., 100.41]

T, ° K.	$H_T - H_{298.15}$, cal./mole	$S_T - S_{298.15}$, cal./deg. mole	T, ° K.	$H_T - H_{298.15}$, cal./mole	$S_T - S_{298.15}$, cal./deg. mole
400	2,190	6.30	1,000	18,390	30.58
500	4,570	11.50	1,100	21,350	33.40
600	7,120	16.25	1,200	24,360	36.02
700	9,820	20.40	1,300	27,420	38.46
800	12,620	24.14	1,400	30,520	40.76
900	15,480	27.51			

$MgSiO_3(amphibole\text{-}type)$:

$$H_T - H_{298.15} = 24.54\,T + 2.72 \times 10^{-3}\,T^2 + 5.87 \times 10^5\,T^{-1}$$
$$-9,527 \ (0.3 \text{ percent}; 298°\text{–}1,400° \text{ K.});$$
$$C_p = 24.54 + 5.44 \times 10^{-3}\,T - 5.87 \times 10^5\,T^{-2}.$$

TABLE 480.—*Heat content and entropy of* $MgSiO_3(pyroxene\text{-}type)$

[Base, crystals at 298.15° K.; mol. wt., 100.41]

T, ° K.	$H_T - H_{298.15}$, cal./mole	$S_T - S_{298.15}$, cal./deg. mole	T, ° K.	$H_T - H_{298.15}$, cal./mole	$S_T - S_{298.15}$, cal./deg. mole
400	2,190	6.30	700	9,910	20.55
500	4,570	11.60	800	12,790	24.40
600	7,160	16.32			

$MgSiO_3(pyroxene\text{-}type)$:

$$H_T - H_{298.15} = 20.59\,T + 6.00 \times 10^{-3}\,T^2 + 4.06 \times 10^5\,T^{-1}$$
$$-8,034 \ (0.2 \text{ percent}; 298°\text{–}800° \text{ K.});$$
$$C_p = 20.59 + 12.00 \times 10^{-3}\,T - 4.06 \times 10^5\,T^{-2}.$$

TABLE 481.—*Heat content and entropy of* $MgSiO_3(gl)$

[Base, glass at 298.15° K.; mol. wt., 100.41]

T, ° K.	$H_T - H_{298.15}$, cal./mole	$S_T - S_{298.15}$, cal./deg. mole	T, ° K.	$H_T - H_{298.15}$, cal./mole	$S_T - S_{298.15}$, cal./deg. mole
400	2,200	6.33	800	12,700	24.27
500	4,570	11.61	900	15,620	27.71
600	7,150	16.31	1,000	18,610	30.86
700	9,870	20.50			

$MgSiO_3(gl)$:

$$H_T - H_{298.15} = 21.89\,T + 4.77 \times 10^{-3}\,T^2 + 4.43 \times 10^5\,T^{-1}$$
$$-8,436 \ (0.3 \text{ percent}; 298°\text{–}1,000° \text{ K.});$$
$$C_p = 21.89 + 9.54 \times 10^{-3}\,T - 4.43 \times 10^5\,T^{-2}.$$

TABLE 482.—*Heat content and entropy of* $Mg_2SiO_4(c)$

[Base, crystals at 298.15° K.; mol. wt., 140.73]

T, ° K.	$H_T - H_{298.15}$, cal./mole	$S_T - S_{298.15}$, cal./deg. mole	T, ° K.	$H_T - H_{298.15}$, cal./mole	$S_T - S_{298.15}$, cal./deg. mole
400	3,100	8.91	1,300	39,000	54.72
500	6,520	16.63	1,400	43,450	58.02
600	10,180	23.20	1,500	47,950	61.12
700	14,010	29.10	1,600	52,470	64.04
800	17,960	34.37	1,700	57,000	66.80
900	22,000	39.13	1,800	61,540	69.39
1,000	26,130	43.48	1,900	66,090	71.85
1,100	30,340	47.49	2,000	70,650	74.19
1,200	34,630	51.22			

TABLE 1h7. (*Continued*)

$Mg_2SiO_4(c)$:

$H_T - H_{298.15} = 35.81T + 3.27 \times 10^{-3}T^2 + 8.52 \times 10^5 T^{-1}$
$- 13,825$ (0.4 percent; $298°-1,800°$ K.);
$C_p = 35.81 + 6.54 \times 10^{-3}T - 8.52 \times 10^5 T^{-2}$.

SULFATE

$MgSO_4(c)$:

$C_p = 23.05$ ($298°$ K.).

$MgSO_4 \cdot H_2O(c)$:

$C_p = 33.2$ ($282°$ K.).

$MgSO_4 \cdot 6H_2O(c)$:

$C_p = 83.20$ ($298°$ K.).
$\overline{C}_p = 85.47$ ($298°-320°$ K.).

$MgSO_4 \cdot 7H_2O(c)$:

$\overline{C}_p = 89.1$ ($291°-319°$ K.).

TITANATES

TABLE 483.—*Heat content and entropy of*
$MgTiO_3(c)$

[Base, crystals at 298.15° K.; mol. wt., 120.22]

T, ° K.	$H_T - H_{298.15}$, cal./mole	$S_T - S_{298.15}$, cal./deg. mole	T, ° K.	$H_T - H_{298.15}$, cal./mole	$S_T - S_{298.15}$, cal./deg. mole
400	2,500	7.19	1,300	29,190	41.45
500	5,130	13.05	1,400	32,390	43.85
600	7,900	18.10	1,500	35,660	46.10
700	10,790	22.55	1,600	39,010	48.26
800	13,740	26.49	1,700	42,450	50.35
900	16,750	30.03	1,800	45,980	52.37
1,000	19,800	33.25	1,900	49,600	54.33
1,100	22,900	36.20	2,000	53,310	56.23
1,200	26,030	38.93			

$MgTiO_3(c)$:

$H_T - H_{298.15} = 28.29T + 1.64 \times 10^{-3}T^2 + 6.53 \times 10^5 T^{-1}$
$- 10,771$ (0.4 percent; $298°-1,800°$ K.);
$C_p = 28.29 + 3.28 \times 10^{-3}T - 6.53 \times 10^5 T^{-2}$.

TABLE 484.—*Heat content and entropy of*
$Mg_2TiO_4(c)$

[Base, crystals at 298.15° K.; mol. wt., 160.54]

T, ° K.	$H_T - H_{298.15}$, cal./mole	$S_T - S_{298.15}$, cal./deg. mole	T, ° K.	$H_T - H_{298.15}$, cal./mole	$S_T - S_{298.15}$, cal./deg. mole
400	3,340	9.61	1,300	41,200	57.88
500	6,990	17.74	1,400	45,960	61.41
600	10,850	24.77	1,500	50,760	64.72
700	14,840	30.92	1,600	55,600	67.84
800	18,930	36.38	1,700	60,470	70.79
900	23,120	41.31	1,800	65,370	73.59
1,000	27,430	45.86	1,900	70,300	76.26
1,100	31,910	50.13	2,000	75,250	78.80
1,200	36,510	54.13			

$Mg_2TiO_4(c)$:

$H_T - H_{298.15} = 35.96T + 4.27 \times 10^{-3}T^2 + 6.89 \times 10^5 T^{-1}$
$- 13,412$ (0.3 percent; $298°-1,800°$ K.);
$C_p = 35.96 + 8.54 \times 10^{-3}T - 6.89 \times 10^5 T^{-2}$.

TABLE 485.—*Heat content and entropy of*
$MgTi_2O_5(c)$

[Base, crystals at 298.15° K.; mol. wt., 200.12]

T, ° K.	$H_T - H_{298.15}$, cal./mole	$S_T - S_{298.15}$, cal./deg. mole	T, ° K.	$H_T - H_{298.15}$, cal./mole	$S_T - S_{298.15}$, cal./deg. mole
400	3,780	10.87	1,300	46,090	65.05
500	7,910	20.08	1,400	51,410	69.00
600	12,290	28.06	1,500	56,850	72.75
700	16,830	35.06	1,600	62,370	76.31
800	21,470	41.25	1,700	67,940	79.68
900	26,200	46.82	1,800	73,530	82.88
1,000	31,010	51.89	1,900	79,130	85.91
1,100	35,910	56.56	2,000	84,740	88.78
1,200	40,930	60.92			

$MgTi_2O_5(c)$:

$H_T - H_{298.15} = 40.68T + 4.60 \times 10^{-3}T^2 + 7.35 \times 10^5 T^{-1}$
$- 15,003$ (0.3 percent; $298°-2,000°$ K.);
$C_p = 40.68 + 9.20 \times 10^{-3}T - 7.35 \times 10^5 T^{-2}$.

BORACITE

TABLE 486.—*Heat content and entropy of*
$6\ MgO \cdot MgCl_2 \cdot 8B_2O_3(c)$

[Base, α-crystals at 298.15° K.; mol. wt., 894.27]

T, ° K.	$H_T - H_{298.15}$, cal./mole	$S_T - S_{298.15}$, cal./deg. mole	T, ° K.	$H_T - H_{298.15}$, cal./mole	$S_T - S_{298.15}$, cal./deg. mole
350	9,800	30.27	538(β)	57,730	137.74
400	20,600	59.08	550	60,990	143.75
450	32,300	86.61	600	74,590	167.42
500	45,050	113.46	650	88,190	189.15
538(α)	55,440	133.48			

$6MgO \cdot MgCl_2 \cdot 8B_2O_3(\alpha)$:

$H_T - H_{298.15} = 56.40T + 209.0 \times 10^{-3}T^2$
$- 35,394$ (0.4 percent; $298°-538°$ K.);
$C_p = 56.40 + 418.0 \times 10^{-3}T$;
$\Delta H_{538}(transition) = 2,290$.

$6MgO \cdot MgCl_2 \cdot 8B_2O_3(\beta)$:

$H_T - H_{298.15} = 272.0T$
$- 88,610$ (0.1 percent; $538°-650°$ K.);
$C_p = 272.0$.

MAGNESIUM-ALUMINUM COMPOUND

$Mg_4Al_3(c)$:

$C_p = 34.40 + 19.80 \times 10^{-3}T$ (*estimated*) ($298°-736°$ K.).

MAGNESIUM-ANTIMONY COMPOUNDS

$Mg_3Sb_2(c)$:

$C_p = 28.10 + 5.60 \times 10^{-3}T$ (*estimated*) ($298°-1,234°$ K.)

MAGNESIUM-BORON COMPOUND

$MgB_2(c)$:

$C_p = 11.43$ ($298°$ K.).

$MgB_4(c)$:

$C_p = 16.81$ ($298°$ K.).

TABLE 1h7. (*Continued*)

MAGNESIUM-CADMIUM COMPOUNDS

MgCd(c):

$C_p = 12.35$ (298° K.).

MgCd$_3$(c):

$C_p = 28.24$ (298° K.).

Mg$_3$Cd(c):

$C_p = 24.04$ (298° K.).

MAGNESIUM-COPPER COMPOUNDS

TABLE 487.—*Heat content and entropy of* MgCu$_2$(c)

[Base, crystals at 298.15° K.; mol. wt., 151.40]

T, ° K.	$H_T - H_{298.15}$, cal./mole	$S_T - S_{298.15}$, cal./deg. mole	T, ° K.	$H_T - H_{298.15}$, cal./mole	$S_T - S_{298.15}$, cal./deg. mole
400	1,810	5.22	700	7,600	15.94
500	3,630	9.28	800	9,870	18.96
600	5,550	12.78	900	12,370	21.91

MgCu$_2$(c):

$H_T - H_{298.15} = 14.67 T + 4.25 \times 10^{-3} T^2$
$- 4,752$ (1.5 percent; 298°–900° K.);
$C_p = 14.67 + 8.50 \times 10^{-3} T$.

Mg$_2$Cu(c):

$C_p = 15.50 + 6.54 \times 10^{-3} T (estimated)$ (298°–843° K.).

MAGNESIUM-GOLD COMPOUNDS

MgAu(c):

$C_p = 11.30 + 1.89 \times 10^{-3} T (estimated)$ (298°–1,433° K.).

Mg$_2$Au(c):

$C_p = 16.20 + 4.52 \times 10^{-3} T (estimated)$ (298°–1,073° K.).

Mg$_3$Au(c):

$C_p = 21.20 + 6.15 \times 10^{-3} T (estimated)$ (298°–1,103° K.).

MAGNESIUM-NICKEL COMPOUND

TABLE 488.—*Heat content and entropy of* MgNi$_2$(c)

[Base, crystals at 298.15° K.; mol. wt., 141.74]

T, ° K.	$H_T - H_{298.15}$, cal./mole	$S_T - S_{298.15}$, cal./deg. mole	T, ° K.	$H_T - H_{298.15}$, cal./mole	$S_T - S_{298.15}$, cal./deg. mole
400	1,860	5.36	700	7,780	16.34
500	3,740	9.56	800	9,890	19.16
600	5,720	13.17	900	12,030	21.67

MgNi$_2$(c):

$H_T - H_{298.15} = 15.67 T + 3.65 \times 10^{-3} T^2 - 4,996$
(0.3 percent; 298°–900° K.);
$C_p = 15.67 + 7.30 \times 10^{-3} T$.

MAGNESIUM-SILVER COMPOUND

TABLE 489.—*Heat content and entropy of* MgAg(c)

[Base, crystals at 298.15° K.; mol. wt., 132.20]

T, ° K.	$H_T - H_{298.15}$, cal./mole	$S_T - S_{298.15}$, cal./deg. mole	T, ° K.	$H_T - H_{298.15}$, cal./mole	$S_T - S_{298.15}$, cal./deg. mole
400	1,230	3.55	700	5,090	10.71
500	2,470	6.31	800	6,460	12.54
600	3,750	8.65	900	7,850	14.17

MgAg(c):

$H_T - H_{298.15} = 10.54 T + 2.12 \times 10^{-3} T^2 - 3,331$
(0.2 percent; 298°–900° K.);
$C_p = 10.54 + 4.24 \times 10^{-3} T$.

MAGNESIUM-ZINC COMPOUND

TABLE 490.—*Heat content and entropy of* MgZn$_2$(c)

[Base, crystals at 298.15° K.; mol. wt., 155.08]

T, ° K.	$H_T - H_{298.15}$, cal./mole	$S_T - S_{298.15}$, cal./deg. mole	T, ° K.	$H_T - H_{298.15}$, cal./mole	$S_T - S_{298.15}$, cal./deg. mole
400	1,830	5.28	700	7,690	16.15
500	3,720	9.49	800	9,770	18.96
600	5,670	13.04			

MgZn$_2$(c):

$H_T - H_{298.15} = 15.55 T + 3.60 \times 10^{-3} T^2 - 4,956$
(0.2 percent; 298°–800° K.);
$C_p = 15.55 + 7.20 \times 10^{-3} T$.

MANGANESE AND ITS COMPOUNDS

ELEMENT

TABLE 491.—*Heat content and entropy of* Mn (c, l)

[Base, α-crystals at 298.15° K.; atomic wt., 54.94]

T, ° K.	$H_T - H_{298.15}$, cal./mole	$S_T - S_{298.15}$, cal./deg. mole	T, ° K.	$H_T - H_{298.15}$, cal./mole	$S_T - S_{298.15}$, cal./deg. mole
400	690	1.99	1,410(γ)	10,330	13.21
500	1,385	3.54	1,410(δ)	10,760	13.51
600	2,120	4.88	1,500	11,780	14.21
700	2,895	6.07	1,517(δ)	11,970	14.34
800	3,715	7.16	1,517(l)	15,470	16.65
900	4,570	8.17	1,600	16,380	17.23
1,000(α)	5,450	9.10	1,700	17,480	17.90
1,000(β)	5,985	9.63	1,800	18,580	18.53
1,100	6,890	10.50	1,900	19,680	19.12
1,200	7,795	11.28	2,000	20,780	19.69
1,300	8,715	12.02	2,100	21,880	20.22
1,374(β)	9,395	12.53	2,200	22,980	20.74
1,374(γ)	9,940	12.93	2,300	24,080	21.22
1,400	10,220	13.13			

Mn(α):

$H_T - H_{298.15} = 5.70 T + 1.69 \times 10^{-3} T^2 + 0.37 \times 10^5 T^{-1}$
$- 1,974$ (0.7 percent; 298°–1,000° K.);
$C_p = 5.70 + 3.38 \times 10^{-3} T - 0.37 \times 10^5 T^{-2}$;
$\Delta H_{1000}(transition) = 535.$

TABLE 1h7. (*Continued*)

Mn(β):

$$H_T - H_{298.15} = 8.33\,T + 0.33 \times 10^{-3}\,T^2 - 2,675$$

(0.1 percent; 1,000°–1,374° K.);

$$C_p = 8.33 + 0.66 \times 10^{-3}\,T;$$

$$\Delta H_{1374}(transition) = 545.$$

Mn(γ):

$$H_T - H_{298.15} = 10.70\,T - 4,760 \ (0.1 \ \text{percent};$$

$$1,374°\text{–}1,410° \ \text{K.});$$

$$C_p = 10.70;$$

$$\Delta H_{1410}(transition) = 430.$$

Mn(δ):

$$H_T - H_{298.15} = 11.30\,T - 5,170 \ (0.1 \ \text{percent};$$

$$1,410°\text{–}1,517° \ \text{K.});$$

$$C_p = 11.30;$$

$$\Delta H_{1517}(fusion) = 3,500.$$

Mn(l):

$$H_T - H_{298.15} = 11.00\,T - 1,220 \ (0.1 \ \text{percent};$$

$$1,517°\text{–}2,300° \ \text{K.});$$

$$C_p = 11.00.$$

TABLE 492.—*Heat content and entropy of* Mn(γ)

[Base, γ-crystals at 298.15° K.; atomic wt., 54.94]

T, ° K.	$H_T - H_{298.15}$, cal./mole	$S_T - S_{298.15}$, cal./deg. mole	T, ° K.	$H_T - H_{298.15}$, cal./mole	$S_T - S_{298.15}$, cal./deg. mole
400	705	2.03	1,000	5,750	9.57
500	1,445	3.68	1,100	6,720	10.50
600	2,230	5.11	1,200	7,730	11.38
700	3,050	6.38	1,300	8,780	12.22
800	3,915	7.53	1,374	9,570	12.81
900	4,815	8.59			

Mn(γ):

$$H_T - H_{298.15} = 6.03\,T + 1.78 \times 10^{-3}\,T^2 + 0.44 \times 10^5\,T^{-1}$$

$$-2,104 \ (0.1 \ \text{percent}; \ 298°\text{–}1,374° \ \text{K.});$$

$$C_p = 6.03 + 3.56 \times 10^{-3}\,T - 0.44 \times 10^5\,T^{-2}.$$

TABLE 493.—*Heat content and entropy of* Mn(g)

[Base, ideal gas at 298.15° K.; atomic wt., 54.94]

T, ° K.	$H_T - H_{298.15}$, cal./mole	$S_T - S_{298.15}$, cal./deg. mole	T, ° K.	$H_T - H_{298.15}$, cal./mole	$S_T - S_{298.15}$, cal./deg. mole
400	505	1.46	1,900	7,960	9.20
500	1,005	2.57	2,000	8,460	9.46
600	1,500	3.48	2,200	9,455	9.93
700	1,995	4.24	2,400	10,455	10.37
800	2,495	4.90	2,600	11,465	10.77
900	2,990	5.49	2,800	12,485	11.15
1,000	3,490	6.01	3,000	13,525	11.51
1,100	3,985	6.49	3,500	16,255	12.35
1,200	4,480	6.92	4,000	19,290	13.16
1,300	4,980	7.32	4,500	22,775	13.98
1,400	5,475	7.68	5,000	26,840	14.84
1,500	5,970	8.03	6,000	36,925	16.67
1,600	6,470	8.35	7,000	49,415	18.59
1,700	6,965	8.65	8,000	63,560	20.48
1,800	7,465	8.93			

Mn(g):

$$H_T - H_{298.15} = 4.97\,T - 1,482 \ (0.2 \ \text{percent};$$

$$298°\text{–}3,000° \ \text{K.});$$

$$C_p = 4.97.$$

OXIDES

TABLE 494.—*Heat content and entropy of* MnO(c)

[Base, crystals at 298.15° K.; mol. wt., 70.94]

T, ° K.	$H_T - H_{298.15}$, cal./mole	$S_T - S_{298.15}$, cal./deg. mole	T, ° K.	$H_T - H_{298.15}$, cal./mole	$S_T - S_{298.15}$, cal./deg. mole
400	1,130	3.26	1,300	12,470	17.83
500	2,280	5.82	1,400	13,840	18.85
600	3,470	7.99	1,500	15,210	19.80
700	4,680	9.86	1,600	16,590	20.69
800	5,900	11.49	1,700	17,970	21.52
900	7,150	12.96	1,800	19,350	22.31
1,000	8,430	14.31	1,900	20,740	23.06
1,100	9,750	15.56	2,000	22,130	23.78
1,200	11,100	16.74			

MnO(c):

$$H_T - H_{298.15} = 11.11\,T + 0.97 \times 10^{-3}\,T^2 + 0.88 \times 10^5\,T^{-1}$$

$$-3,694 \ (0.3 \ \text{percent}; \ 298°\text{–}1,800° \ \text{K.});$$

$$C_p = 11.11 + 1.94 \times 10^{-3}\,T - 0.88 \times 10^5\,T^{-2}.$$

TABLE 495.—*Heat content and entropy of* MnO(g)

[Base, ideal gas at 298.15° K.; mol. wt., 70.94]

T, ° K.	$H_T - H_{298.15}$, cal./mole	$S_T - S_{298.15}$, cal./deg. mole	T, ° K.	$H_T - H_{298.15}$, cal./mole	$S_T - S_{298.15}$, cal./deg. mole
400	790	2.28	1,000	5,870	9.98
500	1,600	4.08	1,200	7,620	11.58
600	2,430	5.60	1,400	9,380	12.94
700	3,275	6.90	1,600	11,150	14.12
800	4,135	8.05	1,800	12,920	15.16
900	5,000	9.07	2,000	14,695	16.10

MnO(g):

$$H_T - H_{298.15} = 8.45\,T + 0.15 \times 10^{-3}\,T^2 + 0.87 \times 10^5\,T^{-1}$$

$$-2,825 \ (0.4 \ \text{percent}; \ 298°\text{–}2,000° \ \text{K.});$$

$$C_p = 8.45 + 0.30 \times 10^{-3}\,T - 0.87 \times 10^5\,T^{-2}.$$

TABLE 496.—*Heat content and entropy of* MnO$_2$(c)

[Base, crystals at 298.15° K.; mol. wt., 86.94]

T, ° K.	$H_T - H_{298.15}$, cal./mole	$S_T - S_{298.15}$, cal./deg. mole	T, ° K.	$H_T - H_{298.15}$, cal./mole	$S_T - S_{298.15}$, cal./deg. mole
400	1,445	4.16	700	6,415	13.36
500	3,020	7.67	800	8,185	15.73
600	4,685	10.70			

MnO$_2$(c):

$$H_T - H_{298.15} = 16.60\,T + 1.22 \times 10^{-3}\,T^2 + 3.88 \times 10^5\,T^{-1}$$

$$-6,359 \ (0.1 \ \text{percent}; \ 298°\text{–}800° \ \text{K.});$$

$$C_p = 16.60 + 2.44 \times 10^{-3}\,T - 3.88 \times 10^5\,T^{-2}.$$

TABLE 497.—*Heat content and entropy of* Mn$_2$O$_3$(c)

[Base, crystals at 298.15° K.; mol wt., 157.88]

T, ° K.	$H_T - H_{298.15}$, cal./mole	$S_T - S_{298.15}$, cal./deg. mole	T, ° K.	$H_T - H_{298.15}$, cal./mole	$S_T - S_{298.15}$, cal./deg. mole
400	2,550	7.33	1,000	20,420	34.15
500	5,220	13.28	1,100	23,740	37.31
600	8,040	18.42	1,200	27,150	40.28
700	10,990	22.97	1,300	30,650	43.08
800	14,040	27.04	1,350	32,430	44.43
900	17,190	30.75			

TABLE 1h7. (*Continued*)

$Mn_2O_3(c)$:

$H_T - H_{298.15} = 24.73T + 4.19 \times 10^{-3}T^2 + 3.23 \times 10^5 T^{-1}$
$-8,829$ (0.1 percent; 298°–1,350° K.);
$C_p = 24.73 + 8.38 \times 10^{-3}T - 3.23 \times 10^5 T^{-2}$.

TABLE 498.—*Heat content and entropy of* $Mn_3O_4(c)$

[Base, α-crystals at 298.15° K.; mol. wt., 228.82]

T, ° K.	$H_T - H_{298.15}$, cal./mole	$S_T - S_{298.15}$, cal./deg. mole	T, ° K.	$H_T - H_{298.15}$, cal./mole	$S_T - S_{298.15}$, cal./deg. mole
400	3,730	10.75	1,300	42,510	60.43
500	7,590	19.36	1,400	47,620	64.22
600	11,590	26.65	1,445(α)	49,960	65.87
700	15,740	33.04	1,445(β)	54,930	69.31
800	19,980	38.70	1,500	57,690	71.18
900	24,250	43.73	1,600	62,710	74.42
1,000	28,570	48.28	1,700	67,730	77.46
1,100	33,020	52.52	1,800	72,750	80.34
1,200	37,650	56.55			

$Mn_3O_4(\alpha)$:

$H_T - H_{298.15} = 34.64T + 5.41 \times 10^{-3}T^2 + 2.20 \times 10^5 T^{-1}$
$-11,547$ (0.5 percent; 298°–1,445° K.);
$C_p = 34.64 + 10.82 \times 10^{-3}T - 2.20 \times 10^5 T^{-2}$;
$\Delta H_{1445}(transition) = 4,970$.

$Mn_3O_4(\beta)$:

$H_T - H_{298.15} = 50.20T - 17,610$
(0.1 percent; 1,445°–1,800° K.);
$C_p = 50.20$.

HYDRATED SESQUIOXIDE

$Mn_2O_3 \cdot H_2O(c)$:
$\overline{C_p} = 31.0$ (290°–325° K.).

SULFIDE

TABLE 499.—*Heat content and entropy of* $MnS(c, l)$

[Base, crystals at 298.15° K.; mol. wt., 87.01]

T, ° K.	$H_T - H_{298.15}$, cal./mole	$S_T - S_{298.15}$, cal./deg. mole	T, ° K.	$H_T - H_{298.15}$, cal./mole	$S_T - S_{298.15}$, cal./deg. mole
400	1,220	3.52	1,400	14,130	19.54
500	2,440	6.24	1,500	15,530	20.50
600	3,690	8.52	1,600	16,970	21.43
700	4,970	10.49	1,700	18,450	22.33
800	6,260	12.21	1,800	19,970	23.20
900	7,560	13.74	1,803(c)	20,020	23.22
1,000	8,850	15.10	1,803(l)	26,260	26.68
1,100	10,150	16.34	1,900	27,810	27.52
1,200	11,450	17.47	2,000	29,410	28.34
1,300	12,770	18.53			

$MnS(c)$:

$H_T - H_{298.15} = 11.40T + 0.90 \times 10^{-3}T^2$
$-3,479$ (0.5 percent; 298°–1,803° K.);
$C_p = 11.40 + 1.80 \times 10^{-3}T$;
$\Delta H_{1803}(fusion) = 6,240$.

$MnS(l)$:

$H_T - H_{298.15} = 16.00T - 2,590$
(0.1 percent; 1,803°–2,000° K.);
$C_p = 16.00$.

CARBIDE

TABLE 500.—*Heat content and entropy of* $Mn_3C(c)$

[Base, α-crystals at 298.15° K.; mol. wt., 176.83]

T, ° K.	$H_T - H_{298.15}$, cal./mole	$S_T - S_{298.15}$, cal./deg. mole	T, ° K.	$H_T - H_{298.15}$, cal./mole	$S_T - S_{298.15}$, cal./deg. mole
400	2,450	7.05	1,100	22,400	35.37
500	5,020	12.78	1,200	25,540	38.10
600	7,700	17.67	1,300	28,740	40.66
700	10,490	21.96	1,310(α)	29,060	40.91
800	13,350	25.78	1,310(β)	32,630	43.64
900	16,300	29.25	1,400	36,050	46.16
1,000	19,320	32.44	1,500	39,850	48.78

$Mn_3C(\alpha)$:

$H_T - H_{298.15} = 25.26T + 2.80 \times 10^{-3}T^2 + 4.07 \times 10^5 T^{-1}$
$-9,145$ (0.2 percent; 298°–1,310° K.);
$C_p = 25.26 + 5.60 \times 10^{-3}T - 4.07 \times 10^5 T^{-2}$;
$\Delta H_{1310}(transition) = 3,570$.

$Mn_3C(\beta)$:

$H_T - H_{298.15} = 38.00T - 17,150$ (0.1 percent;
1,310°–1,500° K.);
$C_p = 38.00$.

NITRIDES

TABLE 501.—*Heat content and entropy of* $Mn_4N(c)$

[Base, crystals at 298.15° K.; mol. wt., 233.77]

T, ° K.	$H_T - H_{298.15}$, cal./mole	$S_T - S_{298.15}$, cal./deg. mole	T, ° K.	$H_T - H_{298.15}$, cal./mole	$S_T - S_{298.15}$, cal./deg. mole
400	3,250	9.36	700	14,640	30.35
500	6,720	17.10	800	19,000	36.17
600	10,520	24.02			

$Mn_4N(c)$:

$H_T - H_{298.15} = 21.15T + 15.25 \times 10^{-3}T^2 - 7,661$
(0.2 percent; 298°–800° K.);
$C_p = 21.15 + 30.50 \times 10^{-3}T$.

TABLE 502.—*Heat content and entropy of* $Mn_3N_2(c)$

[Base, crystals at 298.15° K.; mol. wt., 192.84]

T, ° K.	$H_T - H_{298.15}$, cal./mole	$S_T - S_{298.15}$, cal./deg. mole	T, ° K.	$H_T - H_{298.15}$, cal./mole	$S_T - S_{298.15}$, cal./deg. mole
400	3,070	8.85	700	13,470	28.06
500	6,300	16.05	800	17,350	33.24
600	9,750	22.33			

$Mn_3N_2(c)$:

$H_T - H_{298.15} = 22.32T + 11.20 \times 10^{-3}T^2 - 7,650$
(0.2 percent; 298°–800° K.);
$C_p = 22.32 + 22.40 \times 10^{-3}T$.

TABLE 1h7. (*Continued*)

TABLE 503.—*Heat content and entropy of* $Mn_5N_2(c)$

[Base, crystals at 298.15° K.; mol. wt., 302.72]

T, ° K.	$H_T-H_{298.15}$, cal./mole	$S_T-S_{298.15}$, cal./deg. mole	T, ° K.	$H_T-H_{298.15}$, cal./mole	$S_T-S_{298.15}$, cal./deg. mole
400	4,480	12.90	700	20,040	41.61
500	9,240	23.51	800	25,840	49.35
600	14,460	33.02			

$Mn_5N_2(c)$:

$$H_T-H_{298.15}=30.55T+19.20\times10^{-3}T^2-10,815$$
$$(0.3 \text{ percent; } 298°-800° \text{ K.});$$
$$C_p=30.55+38.40\times10^{-3}T.$$

SELENIDE

$MnSe(c)$:

$$C_p=12.20 \text{ (298° K.).}$$

TELLURIDE

$MnTe(c)$:

$$C_p=17.40 \text{ (298° K.).}$$

BROMIDE

TABLE 504.—*Heat content and entropy of* $MnBr(g)$

[Base, ideal gas at 298.15° K.; mol. wt., 134.86]

T, ° K.	$H_T-H_{298.15}$, cal./mole	$S_T-S_{298.15}$, cal./deg. mole	T, ° K.	$H_T-H_{298.15}$, cal./mole	$S_T-S_{298.15}$, cal./deg. mole
400	890	2.56	1,000	6,210	10.68
500	1,770	4.53	1,200	7,995	12.31
600	2,655	6.14	1,400	9,780	13.68
700	3,540	7.51	1,600	11,565	14.87
800	4,430	8.69	1,800	13,355	15.93
900	5,320	9.74	2,000	15,140	16.87

$MnBr(g)$:

$$H_T-H_{298.15}=8.94T+0.27\times10^5T^{-1}-2,756$$
$$(0.1 \text{ percent; } 298°-2,000° \text{ K.})$$
$$C_p=8.94-0.27\times10^5T^{-2}.$$

CHLORIDES

TABLE 505.—*Heat content and entropy of* $MnCl_2$ (c, l)

[Base, crystals at 298.15° K.; mol. wt., 125.85]

T, ° K.	$H_T-H_{298.15}$, cal./mole	$S_T-S_{298.15}$, cal./deg. mole	T, ° K.	$H_T-H_{298.15}$, cal./mole	$S_T-S_{298.15}$, cal./deg. mole
400	1,850	5.33	923(l)	21,140	31.40
500	3,730	9.52	1,000	22,880	33.21
600	5,640	13.00	1,100	25,140	35.36
700	7,590	16.01	1,200	27,400	37.33
800	9,600	18.69	1,300	29,660	39.14
900	11,680	21.14	1,400	31,920	40.81
923(c)	12,170	21.68			

$MnCl_2(c)$:

$$H_T-H_{298.15}=18.04T+1.58\times10^{-3}T^2+1.37\times10^5T^{-1}$$
$$-5,979 \text{ (0.5 percent; } 298°-923° \text{ K.});$$
$$C_p=18.04+3.16\times10^{-3}T-1.37\times10^5T^{-2};$$
$$\Delta H_{923}(fusion)=8,970.$$

$MnCl_2(l)$:

$$H_T-H_{298.15}=22.60T+280 \text{ (0.1 percent;}$$
$$923°-1,400° \text{ K.});$$
$$C_p=22.60.$$

TABLE 506.—*Heat content and entropy of* $MnCl(g)$

[Base, ideal gas at 298.15° K.; mol. wt., 90.40]

T, ° K.	$H_T-H_{298.15}$, cal./mole	$S_T-S_{298.15}$, cal./deg. mole	T, ° K.	$H_T-H_{298.15}$, cal./mole	$S_T-S_{298.15}$, cal./deg. mole
400	875	2.52	1,000	6,165	10.59
500	1,745	4.46	1,200	7,945	12.21
600	2,620	6.06	1,400	9,730	13.58
700	3,505	7.42	1,600	11,515	14.78
800	4,390	8.60	1,800	13,300	15.83
900	5,275	9.65	2,000	15,085	16.77

$MnCl(g)$:

$$H_T-H_{298.15}=8.89T+0.02\times10^{-3}T^2+0.39\times10^5T^{-1}$$
$$-2,783 \text{ (0.1 percent; } 298°-2,000° \text{ K.});$$
$$C_p=8.89+0.04\times10^{-3}T-0.39\times10^5T^{-2}.$$

FLUORIDES

$MnF_2(c)$:

$$C_p=16.24 \text{ (298° K.).}$$

TABLE 507.—*Heat content and entropy of* $MnF(g)$

[Base, ideal gas at 298.15° K.; mol. wt., 73.94]

T, ° K.	$H_T-H_{298.15}$, cal./mole	$S_T-S_{298.15}$, cal./deg. mole	T, ° K.	$H_T-H_{298.15}$, cal./mole	$S_T-S_{298.15}$, cal./deg. mole
400	830	2.39	1,000	6,025	10.29
500	1,670	4.26	1,200	7,790	11.90
600	2,530	5.83	1,400	9,565	13.27
700	3,395	7.16	1,600	11,340	14.45
800	4,265	8.33	1,800	13,120	15.50
900	5,145	9.36	2,000	14,905	16.44

$MnF(g)$:

$$_T-H_{298.15}=8.75T+0.06\times10^{-3}T^2+0.74\times10^5T^{-1}$$
$$-2,862 \text{ (0.2 percent; } 298°-2,000° \text{ K.});$$
$$C_p=8.75+0.12\times10^{-3}T-0.74\times10^5T^{-2}.$$

IODIDE

TABLE 508.—*Heat content and entropy of* $MnI(g)$

[Base, ideal gas at 298.15° K.; mol. wt., 181.85]

T, ° K.	$H_T-H_{298.15}$, cal./mole	$S_T-S_{298.15}$, cal./deg. mole	T, ° K.	$H_T-H_{298.15}$, cal./mole	$S_T-S_{298.15}$, cal./deg. mole
400	895	2.58	1,000	6,235	10.73
500	1,780	4.56	1,200	8,020	12.36
600	2,670	6.18	1,400	9,805	13.73
700	3,560	7.55	1,600	11,590	14.93
800	4,450	8.74	1,800	13,380	15.98
900	5,340	9.79	2,000	15,165	16.92

$MnI(g)$:

$$H_T-H_{298.15}=8.94T+0.18\times10^5T^{-1}-2,726$$
$$(0.1 \text{ percent; } 298°-2,000° \text{ K.});$$
$$C_p=8.94-0.18\times10^5T^{-2}.$$

TABLE 1h7. (*Continued*)

ALUMINATE

MnAl₂O₄(c):

$C_p = 26.40 + 12.90 \times 10^{-3} T$ (estimated)   (298°–1,298° K.).

CARBONATE

TABLE 509.—*Heat content and entropy of* $MnCO_3(c)$

[Base, crystals at 298.15° K.; mol. wt., 114.95]

T, ° K.	$H_T - H_{298.15}$, cal./mole	$S_T - S_{298.15}$, cal./deg. mole	T, ° K.	$H_T - H_{298.15}$, cal./mole	$S_T - S_{298.15}$, cal./deg. mole
400	2,160	6.22	600	7,095	16.17
500	4,550	11.54	700	9,800	20.34

MnCO₃(c):

$H_T - H_{298.15} = 21.99 T + 4.65 \times 10^{-3} T^2 + 4.69 \times 10^5 T^{-1} -$

8,543 (0.2 percent; 298°–700° K.);

$C_p = 21.99 + 9.30 \times 10^{-3} T - 4.69 \times 10^5 T^{-2}$.

CHROMITE

MnCr₂O₄(c):

$C_p = 25.20 + 17.80 \times 10^{-3} T$ (estimated)   (298°–1,298° K.).

FERRITE

MnFe₂O₄(c):

$C_p = 27.30 + 25.60 \times 10^{-3} T$ (estimated)   (298°–1,298° K.).

SILICATE

TABLE 510.—*Heat content and entropy of* $MnSiO_3(c)$

[Base, crystals at 298.15° K.; mol. wt., 131.03]

T, ° K.	$H_T - H_{298.15}$, cal./mole	$S_T - S_{298.15}$, cal./deg. mole	T, ° K.	$H_T - H_{298.15}$, cal./mole	$S_T - S_{298.15}$, cal./deg. mole
400	2,300	6.62	1,000	18,890	31.57
500	4,800	12.19	1,100	21,850	34.40
600	7,430	16.98	1,200	24,870	37.02
700	10,200	21.25	1,300	27,950	39.48
800	13,070	25.08	1,400	31,090	41.81
900	15,970	28.49	1,500	34,300	44.03

MnSiO₃(c):

$H_T - H_{298.15} = 26.42 T + 1.94 \times 10^{-3} T^2 + 6.16 \times 10^5 T^{-1}$

$-10,116$ (0.2 percent; 298°–1,500° K.);

$C_p = 26.42 + 3.88 \times 10^{-3} T - 6.16 \times 10^5 T^{-2}$.

SULFATE

TABLE 511.—*Heat content and entropy of* $MnSO_4(c)$

[Base, crystals at 298.15° K.; mol. wt., 151.01]

T, ° K.	$H_T - H_{298.15}$, cal./mole	$S_T - S_{298.15}$, cal./deg. mole	T, ° K.	$H_T - H_{298.15}$, cal./mole	$S_T - S_{298.15}$, cal./deg. mole
400	2,680	7.70	800	15,690	29.96
500	5,630	14.28	900	19,280	34.19
600	8,850	20.14	1,000	22,970	38.08
700	12,210	25.32	1,100	26,740	41.67

MnSO₄(c):

$H_T - H_{298.15} = 29.26 T + 4.46 \times 10^{-3} T^2 + 7.04 \times 10^5 T^{-1}$

$-11,482$ (0.3 percent; 298°–1,100° K.);

$C_p = 29.26 + 8.92 \times 10^{-3} T - 7.04 \times 10^5 T^{-2}$.

MnSO₄·5H₂O(c):

$\overline{C_p} = 77.9$ (290°–319° K.).

DITHIONATE

MnS₂O₆·2H₂O(c):

$C_p = 57.60$ (298° K.).

MERCURY AND ITS COMPOUNDS

ELEMENT

TABLE 512.—*Heat content and entropy of* $Hg(l, g)$

[Base, liquid at 298.15° K.; atomic wt., 200.61]

T, ° K.	$H_T - H_{298.15}$, cal./mole	$S_T - S_{298.15}$, cal./deg. mole	T, ° K.	$H_T - H_{298.15}$, cal./mole	$S_T - S_{298.15}$, cal./deg. mole
400	670	1.94	1,400	20,140	31.32
500	1,325	3.40	1,500	20,635	31.66
600	1,975	4.58	1,600	21,135	31.98
629.9(l)	2,170	4.90	1,700	21,630	32.28
629.9(g)	16,310	27.35	1,800	22,130	32.56
700	16,660	27.87	1,900	22,625	32.83
800	17,160	28.53	2,000	23,120	33.09
900	17,655	29.12	2,200	24,115	33.56
1,000	18,155	29.64	2,400	25,110	33.99
1,100	18,650	30.12	2,600	26,105	34.39
1,200	19,145	30.55	2,800	27,095	34.76
1,300	19,645	30.95	3,000	28,090	35.10

Hg(l):

$H_T - H_{298.15} = 6.44 T - 0.19 \times 10^5 T^{-1} - 1,856$

(0.1 percent; 298°–629.9° K.);

$C_p = 6.44 + 0.19 \times 10^5 T^{-2}$;

$\Delta H_{629.9}$(vaporization) = 14,140.

Hg(g):

$H_T - H_{298.15} = 4.97 T + 13,180$

(0.1 percent; 629.9°–3,000° K.);

$C_p = 4.97$.

TABLE 513.—*Heat content and entropy of* $Hg(g)$

[Base, ideal gas at 298.15° K.; atomic wt., 200.61]

T, ° K.	$H_T - H_{298.15}$, cal./mole	$S_T - S_{298.15}$, cal./deg. mole	T, ° K.	$H_T - H_{298.15}$, cal./mole	$S_T - S_{298.15}$, cal./deg. mole
400	505	1.46	1,900	7,960	9.20
500	1,005	2.57	2,000	8,455	9.46
600	1,500	3.48	2,200	9,450	9.93
700	1,995	4.24	2,400	10,445	10.36
800	2,495	4.90	2,600	11,440	10.76
900	2,990	5.49	2,800	12,430	11.13
1,000	3,490	6.01	3,000	13,425	11.47
1,100	3,895	6.49	3,500	15,910	12.24
1,200	4,480	6.92	4,000	18,395	12.90
1,300	4,980	7.32	4,500	20,885	13.49
1,400	5,475	7.69	5,000	23,375	14.01
1,500	5,970	8.03	6,000	28,395	14.93
1,600	6,470	8.35	7,000	33,550	15.72
1,700	6,965	8.65	8,000	40,055	16.46
1,800	7,465	8.93			

Hg(g):

$H_T - H_{298.15} = 4.97 T - 1,482$

(0.1 percent; 298°–6,000° K.);

$C_p = 4.97$.

TABLE 1h7. (*Continued*)

TABLE 514.—*Heat content and entropy of* $Hg_2(g)$

[Base, ideal gas at 298.15° K.; mol. wt., 401.22]

T, ° K.	$H_T-H_{298.15}$, cal./mole	$S_T-S_{298.15}$, cal./deg. mole	T, ° K.	$H_T-H_{298.15}$, cal./mole	$S_T-S_{298.15}$, cal./deg. mole
400	910	2.63	1,000	6,275	10.82
500	1,805	4.62	1,200	8,065	12.45
600	2,700	6.26	1,400	9,850	13.83
700	3,590	7.63	1,600	11,640	15.02
800	4,485	8.82	1,800	13,425	16.07
900	5,380	9.88	2,000	15,215	17.02

$Hg_2(g)$:

$$H_T-H_{298.15}=8.94T-2,665$$
$$(0.1 \text{ percent}; 298°-2,000° \text{ K.});$$
$$C_p=8.94.$$

OXIDE

$HgO(red)$;
$$C_p=8.33+7.37\times10^{-3}T(estimated)$$
$$(298°-769° \text{ K.}).$$

SULFIDE

$HgS(c)$:
$$C_p=10.90+3.65\times10^{-3}T(estimated)$$
$$(298°-853° \text{ K.}).$$

HYDRIDES

TABLE 515.—*Heat content and entropy of* $HgH(g)$

[Base, ideal gas at 298.15° K.; mol. wt., 201.62]

T, ° K.	$H_T-H_{298.15}$, cal./mole	$S_T-S_{298.15}$, cal./deg. mole	T, ° K.	$H_T-H_{298.15}$, cal./mole	$S_T-S_{298.15}$, cal./deg. mole
400	740	2.13	1,000	5,600	9.48
500	1,500	3.83	1,200	7,315	11.04
600	2,285	5.26	1,400	9,045	12.38
700	3,090	6.50	1,600	10,790	13.54
800	3,915	7.60	1,800	12,545	14.58
900	4,755	8.59	2,000	14,305	15.50

$HgH(g)$:

$$H_T-H_{298.15}=7.75T+0.35\times10^{-3}T^2+0.71\times10^5T^{-1}$$
$$-2,580 \text{ (0.7 percent; } 298°-2,000° \text{ K.});$$
$$C_p=7.75+0.70\times10^{-3}T-0.71\times10^5T^{-2}.$$

TABLE 516.—*Heat content and entropy of* $HgD(g)$

[Base, ideal gas at 298.15° K.; mol. wt., 202.62]

T, ° K.	$H_T-H_{298.15}$, cal./mole	$S_T-S_{298.15}$, cal./deg. mole	T, ° K.	$H_T-H_{298.15}$, cal./mole	$S_T-S_{298.15}$, cal./deg. mole
400	780	2.25	1,000	5,820	9.89
500	1,580	4.03	1,200	7,570	11.49
600	2,405	5.53	1,400	9,325	12.84
700	3,245	6.83	1,600	11,090	14.02
800	4,095	7.97	1,800	12,860	15.06
900	4,955	8.98	2,000	14,630	15.99

$HgD(g)$:

$$H_T-H_{298.15}=8.35T+0.18\times10^{-3}T^2+0.88\times10^5T^{-1}$$
$$-2,801 \text{ (0.4 percent; } 298°-2,000° \text{ K.});$$
$$C_p=8.35+0.36\times10^{-3}T-0.88\times10^5T^{-2}.$$

BROMIDES

TABLE 517.—*Heat content and entropy of* $HgBr(g)$

[Base, ideal gas at 298.15° K. mol. wt., 280.53]

T, ° K.	$H_T-H_{298.15}$, cal./mole	$S_T-S_{298.15}$, cal./deg. mole	T, ° K.	$H_T-H_{298.15}$, cal./mole	$S_T-S_{298.15}$, cal./deg. mole
400	900	2.59	1,000	6,250	10.76
500	1,790	4.58	1,200	8,035	12.39
600	2,680	6.20	1,400	9,825	13.77
700	3,570	7.57	1,600	11,610	14.96
800	4,465	8.77	1,800	13,395	16.01
900	5,355	9.82	2,000	15,185	16.95

$HgBr(g)$:

$$H_T-H_{298.15}=8.94T+0.11\times10^5T^{-1}-2,702$$
$$(0.1 \text{ percent}; 298°-2,000° \text{ K.});$$
$$C_p=8.94-0.11\times10^5T^{-2}.$$

TABLE 518.—*Heat content and entropy of* $HgBr_2(g)$

[Base, ideal gas at 298.15° K.; mol. wt. 360.44]

T, ° K.	$H_T-H_{298.15}$, cal./mole	$S_T-S_{298.15}$, cal./deg. mole	T, ° K.	$H_T-H_{298.15}$, cal./mole	$S_T-S_{298.15}$, cal./deg. mole
400	1,480	4.27	800	7,380	14.48
500	2,945	7.54	900	8,870	16.23
600	4,420	10.23	1,000	10,350	17.79
700	5,900	12.50			

$HgBr_2(g)$:

$$H_T-H_{298.15}=14.89T+0.44\times10^5T^{-1}-4,587$$
$$(0.1 \text{ percent}; 298°-1,000° \text{ K.});$$
$$C_p=14.89-0.44\times10^5T^{-2}.$$

CHLORIDES

$HgCl(c)$:
$$C_p=11.05+3.70\times10^{-3}T(estimated) \ (298°-798° \text{ K.}).$$

TABLE 519.—*Heat content and entropy of* $HgCl(g)$

[Base, ideal gas at 298.15° K.; mol. wt., 236.07]

T, ° K.	$H_T-H_{298.15}$, cal./mole	$S_T-S_{298.15}$, cal./deg. mole	T, ° K.	$H_T-H_{298.15}$, cal./mole	$S_T-S_{298.15}$, cal./deg. mole
400	890	2.57	1,000	6,210	10.68
500	1,770	4.53	1,200	7,995	12.31
600	2,655	6.14	1,400	9,780	13.69
700	3,540	7.51	1,600	11,565	14.88
800	4,430	8.70	1,800	13,350	15.93
900	5,320	9.74	2,000	15,140	16.87

$HgCl(g)$:

$$H_T-H_{298.15}=8.94T+0.26\times10^5T^{-1}-2,753$$
$$(0.1 \text{ percent}; 298°-2,000° \text{ K.});$$
$$C_p=8.94-0.26\times10^5T^{-2}.$$

$HgCl_2(c)$:
$$C_p=15.28+10.40\times10^{-3}T(estimated) \ (298°-550° \text{ K.}).$$

TABLE 1h7. (Continued)

TABLE 520.—*Heat content and entropy of* $HgCl_2(g)$

[Base, ideal gas at 298.15° K.; mol. wt., 271.52]

T, ° K.	$H_T - H_{298.15}$, cal./mole	$S_T - S_{298.15}$, cal./deg. mole	T, ° K.	$H_T - H_{298.15}$, cal./mole	$S_T - S_{298.15}$, cal./deg. mole
400	1,435	4.14	800	7,270	14.23
500	2,875	7.35	900	8,745	15.97
600	4,330	10.00	1,000	10,220	17.53
700	5,795	12.26			

$HgCl_2(g)$:

$$H_T - H_{298.15} = 14.66T + 0.12 \times 10^{-3}T^2 + 0.75 \times 10^5 T^{-1}$$
$$-4,633 \text{ (0.1 percent; } 298°\text{--}1,000° \text{ K.);}$$
$$C_p = 14.66 + 0.24 \times 10^{-3}T - 0.75 \times 10^5 T^{-2}.$$

FLUORIDE

TABLE 521.—*Heat content and entropy of* $HgF(g)$

[Base, ideal gas at 298.15° K.; mol. wt., 219.61]

T, ° K.	$H_T - H_{298.15}$, cal./mole	$S_T - S_{298.15}$, cal./deg. mole	T, ° K.	$H_T - H_{298.15}$, cal./mole	$S_T - S_{298.15}$, cal./deg. mole
400	855	2.47	1,000	6,110	10.47
500	1,715	4.38	1,200	7,885	12.09
600	2,585	5.97	1,400	9,665	13.46
700	3,460	7.32	1,600	11,445	14.65
800	4,340	8.50	1,800	13,225	15.70
900	5,225	9.54	2,000	15,010	16.64

$HgF(g)$:

$$H_T - H_{298.15} = 8.87T + 0.02 \times 10^{-3}T^2 + 0.57 \times 10^5 T^{-1}$$
$$-2,838 \text{ (0.1 percent; } 298°\text{--}2,000° \text{ K.);}$$
$$C_p = 8.87 + 0.04 \times 10^{-3}T - 0.57 \times 10^5 T^{-2}.$$

IODIDE

$HgI(c)$:

$$C_p = 11.40 + 4.61 \times 10^{-3}T(estimated) \text{ (298°--563° K.).}$$

TABLE 522.—*Heat content and entropy of* $HgI(g)$

[Base, ideal gas at 298.15° K.; mol. wt., 327.52]

T, ° K.	$H_T - H_{298.15}$, cal./mole	$S_T - S_{298.15}$, cal./deg. mole	T, ° K.	$H_T - H_{298.15}$, cal./mole	$S_T - S_{298.15}$, cal./deg. mole
400	905	2.61	1,000	6,265	10.79
500	1,800	4.61	1,200	8,050	12.42
600	2,690	6.23	1,400	9,840	13.80
700	3,585	7.61	1,600	11,625	14.99
800	4,475	8.80	1,800	13,415	16.05
900	5,370	9.85	2,000	15,205	16.99

$HgI(g)$:

$$H_T - H_{298.15} = 8.94T + 0.05 \times 10^5 T^{-1} - 2,682 \text{ (0.1 percent;}$$
$$298°\text{--}2,000° \text{ K.);}$$
$$C_p = 8.94 - 0.05 \times 10^5 T^{-2}.$$

TABLE 523.—*Heat content and entropy of* $HgI_2(c, l)$

[Base, crystals at 298.15° K.; mol. wt., 454.43]

T, ° K.	$H_T - H_{298.15}$, cal./mole	$S_T - S_{298.15}$, cal./deg. mole	T, ° K.	$H_T - H_{298.15}$, cal./mole	$S_T - S_{298.15}$, cal./deg. mole
350	960	2.97	500	4,550	11.55
400	1,885	5.44	523(β)	5,015	12.46
403(α)	1,940	5.58	523(l)	9,515	21.06
403(β)	2,590	7.19	550	10,190	22.32
450	3,540	9.42	600	11,440	24.50

$HgI_2(\alpha)$:

$$H_T - H_{298.15} = 18.50T - 5,516 \text{ (0.1 percent; } 298°\text{--}403° \text{ K.);}$$
$$C_p = 18.50;$$
$$\Delta H_{403}(transition) = 650.$$

$HgI_2(\beta)$:

$$H_T - H_{298.15} = 20.20T - 5,550 \text{ (0.1 percent; } 403°\text{--}523° \text{ K.);}$$
$$C_p = 20.20;$$
$$\Delta H_{523}(fusion) = 4,500.$$

$HgI_2(l)$:

$$H_T - H_{298.15} = 25.00T - 3,560 \text{ (0.1 percent; } 523°\text{--}600° \text{ K.);}$$
$$C_p = 25.00.$$

TABLE 524.—*Heat content and entropy of* $HgI_2(g)$

[Base, ideal gas at 298.15° K.; mol. wt., 454.43]

T, ° K.	$H_T - H_{298.15}$, cal./mole	$S_T - S_{298.15}$, cal./deg. mole	T, ° K.	$H_T - H_{298.15}$, cal./mole	$S_T - S_{298.15}$, cal./deg. mole
400	1,495	4.31	800	7,420	14.58
500	2,970	7.61	900	8,910	16.32
600	4,450	10.30	1,000	10,400	17.90
700	5,935	12.59			

$HgI_2(g)$:

$$H_T - H_{298.15} = 14.90T + 0.27 \times 10^5 T^{-1} - 4,533$$
$$\text{(0.1 percent; } 298°\text{--}1,000° \text{ K.);}$$
$$C_p = 14.90 - 0.27 \times 10^5 T^{-2}.$$

CYANIDE

$Hg(CN)_2(c)$:

$$\overline{C_p} = 25.3 \text{ (285°--319° K.).}$$

SULFATE

$Hg_2SO_4(c)$:

$$C_p = 31.55 \text{ (298° K.).}$$

MERCURY-THALLIUM COMPOUND

TABLE 525.—*Heat content and entropy of* $HgTl(g)$

[Base, ideal gas at 298.15° K.; mol. wt., 405 00]

T, ° K.	$H_T - H_{298.15}$, cal./mole	$S_T - S_{298.15}$, cal./deg. mole	T, ° K.	$H_T - H_{298.15}$, cal./mole	$S_T - S_{298.15}$, cal./deg. mole
400	910	2.63	1,000	6,275	10.82
500	1,805	4.62	1,200	8,060	12.45
600	2,700	6.26	1,400	9,850	13.83
700	3,595	7.64	1,600	11,640	15.02
800	4,490	8.83	1,800	13,430	16.08
900	5,380	9.88	2,000	15,220	17.02

$HgTl(g)$:

$$H_T - H_{298.15} = 8.94T - 2,665 \text{ (0.1 percent;}$$
$$298°\text{--}2,000° \text{ K.);}$$
$$C_p = 8.94.$$

TABLE 1h7. (Continued)

MOLYBDENUM AND ITS COMPOUNDS

ELEMENT

TABLE 526.—Heat content and entropy of Mo(c, l)

[Base, crystals at 298.15° K.; atomic wt., 95.95]

T, ° K.	$H_T - H_{298.15}$, cal./mole	$S_T - S_{298.15}$, cal./deg. mole	T, ° K.	$H_T - H_{298.15}$, cal./mole	$S_T - S_{298.15}$, cal./deg. mole
400	595	1.71	1,900	11,200	12.27
500	1,205	3.06	2,000	12,040	12.70
600	1,825	4.20	2,100	12,900	13.12
700	2,460	5.17	2,200	13,770	13.53
800	3,100	6.02	2,300	14,670	13.92
900	3,750	6.79	2,400	15,580	14.31
1,000	4,410	7.49	2,500	16,510	14.69
1,100	5,090	8.13	2,600	17,460	15.06
1,200	5,790	8.74	2,700	18,420	15.42
1,300	6,510	9.31	2,800	19,400	15.78
1,400	7,250	9.85	2,890 (c)	20,290	16.09
1,500	8,000	10.38	2,890 (l)	26,940	18.39
1,600	8,780	10.88	2,900	27,040	18.43
1,700	9,570	11.36	3,000	28,040	18.77
1,800	10,380	11.82			

Mo(c):

$$H_T - H_{298.15} = 5.18T + 0.83 \times 10^{-3}T^2 - 1,618$$
$$(0.5 \text{ percent}; 298°–2,890° \text{ K.});$$
$$C_p = 5.18 + 1.66 \times 10^{-3}T;$$
$$\Delta H_{2890}(fusion) = 6,650.$$

Mo(l):

$$H_T - H_{298.15} = 10.00T - 1,960 \ (0.1 \text{ percent};$$
$$2,890°–3,000° \text{ K.});$$
$$C_p = 10.00.$$

TABLE 527.—Heat content and entropy of Mo(g)

[Base, ideal gas at 298.15° K; atomic wt., 95.95]

T, ° K.	$H_T - H_{298.15}$, cal./mole	$S_T - S_{298.15}$, cal./deg. mole	T, ° K.	$H_T - H_{298.15}$, cal./mole	$S_T - S_{298.15}$, cal./deg. mole
400	505	1.46	1,500	5,975	8.03
500	1,005	2.57	1,600	6,470	8.35
600	1,500	3.48	1,700	6,975	8.66
700	1,995	4.24	1,800	7,475	8.94
800	2,495	4.91	1,900	7,980	9.22
900	2,990	5.49	2,000	8,490	9.48
1,000	3,485	6.01	2,200	9,530	9.97
1,100	3,985	6.49	2,400	10,600	10.44
1,200	4,480	6.92	2,600	11,710	10.88
1,300	4,980	7.32	2,800	12,880	11.31
1,400	5,475	7.69	3,000	14,115	11.74

Mo(g):

$$H_T - H_{298.15} = 4.97T - 1,482 \ (0.1 \text{ percent};$$
$$298°–1,800° \text{ K.});$$
$$C_p = 4.97.$$
$$H_T - H_{298.15} = 3.56T + 0.40 \times 10^{-3}T^2 - 230$$
$$(0.2 \text{ percent}; 1,800°–3,000° \text{ K.});$$
$$C_p = 3.56 + 0.80 \times 10^{-3}T.$$

OXIDE

TABLE 528.—Heat content and entropy of MoO₃ (c, l)

[Base, crystals at 298.15° K.; mol. wt., 143.95]

T, ° K.	$H_T - H_{298.15}$, cal./mole	$S_T - S_{298.15}$, cal./deg. mole	T, ° K.	$H_T - H_{298.15}$, cal./mole	$S_T - S_{298.15}$, cal./deg. mole
400	1,935	5.56	1,068(c)	17,670	28.26
500	4,035	10.24	1,068(l)	30,170	39.96
600	6,260	14.30	1,100	31,200	40.92
700	8,570	17.86	1,200	34,400	43.70
800	10,940	21.02	1,300	37,600	46.26
900	13,390	23.91	1,400	40,800	48.63
1,000	15,920	26.57	1,500	44,000	50.84

MoO₃(c):

$$H_T - H_{298.15} = 20.73T + 2.59 \times 10^{-3}T^2 + 4.18 \times 10^5 T^{-1}$$
$$-7,813 \ (0.1 \text{ percent}; 298°–1,068° \text{ K.});$$
$$C_p = 20.73 + 5.18 \times 10^{-3}T - 4.18 \times 10^5 T^{-2};$$
$$\Delta H_{1068}(fusion) = 12,500.$$

MoO₃(l):

$$H_T - H_{298.15} = 32.00T - 4,000 \ (0.1 \text{ percent};$$
$$1,068°–1,500° \text{ K.});$$
$$C_p = 32.00.$$

SULFIDE

MoS₂(c):

$$C_p = 11.20 + 13.50 \times 10^{-3}T(estimated) \ (298°–729° \text{ K.}).$$

NITRIDE

TABLE 529.—Heat content and entropy of Mo₂N(c)

[Base, crystals at 298.15° K.; mol. wt., 205.91]

T, ° K.	$H_T - H_{298.15}$, cal./mole	$S_T - S_{298.15}$, cal./deg. mole	T, ° K.	$H_T - H_{298.15}$, cal./mole	$S_T - S_{298.15}$, cal./deg. mole
400	1,610	4.64	700	7,290	15.13
500	3,360	8.54	800	9,370	17.90
600	5,280	12.03			

Mo₂N(c):

$$H_T - H_{298.15} = 11.19T + 6.90 \times 10^{-3}T^2 - 3,950$$
$$(0.7 \text{ percent}; 298°–800° \text{ K.});$$
$$C_p = 11.19 + 13.80 \times 10^{-3}T.$$

SILICIDES

TABLE 530.—Heat content and entropy of MoSi₂(c)

[Base, crystals at 298.15° K.; mol. wt., 152.13]

T, ° .K	$H_T - H_{298.15}$, cal./mole	$S_T - S_{298.15}$, cal./deg. mole	T, ° K.	$H_T - H_{298.15}$, cal./mole	$S_T - S_{298.15}$, cal./deg. mole
400	1,625	4.68	900	10,410	18.81
500	3,290	8.40	1,000	12,280	20.78
600	5,010	11.53	1,100	14,190	22.60
700	6,770	14.24	1,200	16,120	24.28
800	8,570	16.64			

$MoSi_2(c)$:

$H_T - H_{298.15} = 15.75T + 1.62 \times 10^{-3}T^2 + 1.08 \times 10^5 T^{-1}$
$- 5,202$ (0.1 percent; 298°–1,200° K.);
$C_p = 15.75 + 3.24 \times 10^{-3}T - 1.08 \times 10^5 T^{-2}.$

TABLE 531.—*Heat content and entropy of* $Mo_3Si(c)$

[Base, crystals at 298.15° K.; mol. wt., 315.94]

T,° K.	$H_T - H_{298.15}$, cal./mole	$S_T - S_{298.15}$, cal./deg. mole	T,° K.	$H_T - H_{298.15}$, cal./mole	$S_T - S_{298.15}$, cal./deg. mole
400	2,320	6.69	1,000	17,250	29.29
500	4,670	11.93	1,100	19,870	31.78
600	7,090	16.34	1,200	22,580	34.14
700	9,580	20.17	1,300	25,370	36.37
800	12,120	23.56	1,400	28,240	38.50
900	14,680	26.58	1,500	31,190	40.53

$Mo_3Si(c)$:

$H_T - H_{298.15} = 21.98T + 2.29 \times 10^{-3}T^2 + 1.00 \times 10^5 T^{-1}$
$- 7,092$ (0.2 percent; 298°–1,500° K.);
$C_p = 21.98T + 4.58 \times 10^{-3}T - 1.00 \times 10^5 T^{-2}.$

FLUORIDE

TABLE 532.—*Heat content and entropy of* $MoF_6(g)$

[Base, ideal gas at 298.15° K.; mol. wt., 209.95]

T,° K.	$H_T - H_{298.15}$, cal./mole	$S_T - S_{298.15}$, cal./deg. mole	T,° K.	$H_T - H_{298.15}$, cal./mole	$S_T - S_{298.15}$, cal./deg. mole
400	3,060	8.81	1,300	35,080	50.17
500	6,320	16.07	1,400	38,780	52.91
600	9,730	22.29	1,500	42,490	55.47
700	13,230	27.68	1,600	46,210	57.87
800	16,790	32.43	1,700	49,940	60.13
900	20,400	36.68	1,800	53,680	62.27
1,000	24,040	40.52	1,900	57,420	64.29
1,100	27,700	44.01	2,000	61,160	66.21
1,200	31,380	47.21			

$MoF_6(g)$:

$H_T - H_{298.15} = 35.80T + 0.59 \times 10^{-3}T^2 + 6.97 \times 10^5 T^{-1}$
$- 13,064$ (0.3 percent; 298°–2,000° K.);
$C_p = 35.80 + 1.18 \times 10^{-3}T - 6.97 \times 10^5 T^{-2}.$

CARBONYL

$Mo(CO)_6(c)$:

$C_p = 57.9$ (298° K.);
$\overline{C}_p = 61.3$ (293°–363° K.).

NEODYMIUM AND ITS COMPOUNDS

ELEMENT

TABLE 533.—*Heat content and entropy of* $Nd(c, l)$

[Base, α-crystals at 298.15° K.; atomic wt., 144.27]

T,° K.	$H_T - H_{298.15}$, cal./mole	$S_T - S_{298.15}$, cal./deg. mole	T,° K.	$H_T - H_{298.15}$, cal./mole	$S_T - S_{298.15}$, cal./deg. mole
400	755	2.18	1,297(*l*)	12,550	15.66
500	1,560	3.97	1,300	12,570	15.67
600	2,420	5.54	1,400	13,370	16.27
700	3,330	6.94	1,600	14,970	17.33
800	4,290	8.22	1,800	16,570	18.27
900	5,300	9.41	2,000	18,170	19.12
1,000	6,370	10.53	2,200	19,770	19.88
1,100	7,490	11.60	2,400	21,370	20.57
1,141(α)	7,970	12.03	2,600	22,970	21.21
1,141(β)	8,310	12.33	2,800	24,570	21.81
1,200	8,780	12.73	3,000	26,170	22.36
1,297(β)	9,560	13.35			

$Nd(\alpha)$:

$H_T - H_{298.15} = 5.61T + 2.67 \times 10^{-3}T^2 - 1,910$ (0.2 percent; 298°–1,141° K.);
$C_p = 5.61 + 5.34 \times 10^{-3}T;$
$\Delta H_{1141}(transition) = 340.$

$Nd(\beta)$:

$H_T - H_{298.15} = 8.00T - 820$ (0.1 percent; 1,141°–1,297° K.);
$C_p = 8.00;$
$\Delta H_{1297}(fusion) = 2,990.$

$Nd(l)$:

$H_T - H_{298.15} = 8.00T + 2,170$ (0.1 percent; 1,297°–3,000° K.);
$C_p = 8.00.$

TABLE 534.—*Heat content and entropy of* $Nd(g)$

[Base, ideal gas at 298.15° K.; atomic wt., 144.27]

T,° K.	$H_T - H_{298.15}$, cal./mole	$S_T - S_{298.15}$, cal./deg. mole	T,° K.	$H_T - H_{298.15}$, cal./mole	$S_T - S_{298.15}$, cal./deg. mole
400	560	1.61	1,500	7,845	10.19
500	1,145	2.91	1,600	8,530	10.63
600	1,760	4.04	1,700	9,210	11.05
700	2,395	5.02	1,800	9,885	11.43
800	3,055	5.90	1,900	10,555	11.79
900	3,725	6.69	2,000	11,220	12.14
1,000	4,400	7.40	2,200	12,535	12.76
1,100	5,085	8.05	2,400	13,820	13.32
1,200	5,775	8.65	2,600	15,085	13.83
1,300	6,465	9.20	2,800	16,330	14.29
1,400	7,155	9.72	3,000	17,560	14.71

$Nd(g)$:

$H_T - H_{298.15} = 5.98T + 0.42 \times 10^{-3}T^2 + 0.84 \times 10^5 T^{-1}$
$- 2,102$ (0.6 percent; 298°–1,400° K.);
$C_p = 5.98 + 0.84 \times 10^{-3}T - 0.84 \times 10^5 T^{-2};$
$H_T - H_{298.15} = 7.69T - 0.27 \times 10^{-3}T^2 - 3,082$ (0.1 percent; 1,400°–3,000° K.);
$C_p = 7.69 - 0.54 \times 10^{-3}T.$

OXIDE

TABLE 535.—*Heat content and entropy of* $Nd_2O_3(c)$

[Base, crystals at 298.15° K.; mol. wt., 336.54]

T,° K.	$H_T - H_{298.15}$, cal./mole	$S_T - S_{298.15}$, cal./deg. mole	T,° K.	$H_T - H_{298.15}$, cal./mole	$S_T - S_{298.15}$, cal./deg. mole
400	2,800	8.06	1,000	21,990	36.99
500	5,760	14.66	1,100	25,450	40.29
600	8,830	20.25	1,200	28,970	43.35
700	12,000	25.14	1,300	32,550	46.22
800	15,260	29.49	1,400	36,190	48.91
900	18,590	33.41	1,500	39,880	51.46

$Nd_2O_3(c)$:

$H_T - H_{298.15} = 28.99T + 2.88 \times 10^{-3}T^2 + 4.16 \times 10^5 T^{-1}$
$- 10,295$ (0.1 percent; 298°–1,500° K.);
$C_p = 28.99 + 5.76 \times 10^{-3}T - 4.16 \times 10^5 T^{-2}.$

TABLE 1h7. (*Continued*)

NEON

ELEMENT

TABLE 536.—*Heat content and entropy of* Ne(*g*)

[Base, ideal gas at 298.15° K.; atomic wt., 20.18]

T, ° K.	$H_T-H_{298.15}$, cal./mole	$S_T-S_{298.15}$, cal./deg. mole	T, ° K.	$H_T-H_{298.15}$, cal./mole	$S_T-S_{298.15}$, cal./deg. mole
400	505	1.46	1,900	7,960	9.20
500	1,005	2.57	2,000	8,455	9.46
600	1,500	3.48	2,200	9,450	9.93
700	1,995	4.24	2,400	10,445	10.36
800	2,495	4.96	2,600	11,440	10.76
900	2,990	5.49	2,800	12,430	11.13
1,000	3,490	6.01	3,000	13,425	11.47
1,100	3,985	6.49	3,500	15,910	12.24
1,200	4,480	6.92	4,000	18,395	12.90
1,300	4,980	7.32	4,500	20,880	13.49
1,400	5,475	7.69	5,000	23,365	14.01
1,500	5,970	8.03	6,000	28,335	14.92
1,600	6,470	8.35	7,000	33,305	15.68
1,700	6,965	8.65	8,000	38,275	16.35
1,800	7,465	8.93			

Ne(*g*):

$$H_T-H_{298.15}=4.97T-1,482 \quad (0.1 \text{ percent}; 298°-8,000° K.);$$

$$C_p=4.97.$$

NICKEL AND ITS COMPOUNDS

ELEMENT

TABLE 537.—*Heat content and entropy of* Ni(*c, l*)

[Base, crystals at 298.15° K.; atomic wt., 58.71]

T, ° K.	$H_T-H_{298.15}$, cal./mole	$S_T-S_{298.15}$, cal./deg. mole	T, ° K.	$H_T-H_{298.15}$, cal./mole	$S_T-S_{298.15}$, cal./deg. mole
400	665	1.91	1,500	9,320	12.16
500	1,380	3.51	1,600	10,210	12.73
600	2,180	4.96	1,700	11,110	13.28
633(α)	2,460	5.42	1,725(β)	11,330	13.40
633(β)	2,460	5.42	1,725(l)	15,540	15.84
700	2,940	6.14	1,800	16,230	16.24
800	3,690	7.14	2,000	18,070	17.21
900	4,445	8.03	2,200	19,910	18.08
1,000	5,210	8.84	2,400	21,750	18.88
1,100	5,985	9.58	2,600	23,590	19.62
1,200	6,780	10.27	2,800	25,430	20.30
1,300	7,600	10.93	3,000	27,270	20.93
1,400	8,450	11.56	3,200	29,110	21.52

Ni(α):

$$H_T-H_{298.15}=4.06T+3.52\times10^{-3}T^2-1,523$$
$$(0.3 \text{ percent}; 298°-633° K.);$$
$$C_p=4.06+7.04\times10^{-3}T;$$
$$\Delta H_{633}(transition)=0.$$

Ni(β):

$$H_T-H_{298.15}=6.00T+0.90\times10^{-3}T^2-1,701$$
$$(0.2 \text{ percent}; 633°-1,725° K.);$$
$$C_p=6.00+1.80\times10^{-3}T;$$
$$\Delta H_{1725}(fusion)=4,210.$$

Ni(l):

$$H_T-H_{298.15}=9.20T-330 \quad (0.1 \text{ percent};$$
$$1,725°-3,200° K.);$$
$$C_p=9.20.$$

TABLE 538.—*Heat content and entropy of* Ni(*g*)

[Base, ideal gas at 298.15° K.; atomic wt., 58.71]

T, ° K.	$H_T-H_{298.15}$, cal./mole	$S_T-S_{298.15}$, cal./deg. mole	T, ° K.	$H_T-H_{298.15}$, cal./mole	$S_T-S_{298.15}$, cal./deg. mole
400	575	1.66	1,900	9,310	10.77
500	1,150	2.94	2,000	9,865	11.05
600	1,740	4.01	2,200	10,970	11.58
700	2,335	4.93	2,400	12,065	12.06
800	2,930	5.73	2,600	13,155	12.49
900	3,525	6.43	2,800	14,235	12.89
1,000	4,120	7.06	3,000	15,310	13.26
1,100	4,715	7.62	3,500	17,985	14.09
1,200	5,300	8.13	4,000	20,665	14.80
1,300	5,885	8.60	4,500	23,370	15.44
1,400	6,465	9.03	5,000	26,135	16.02
1,500	7,040	9.43	6,000	31,940	17.08
1,600	7,615	9.80	7,000	38,305	18.06
1,700	8,185	10.14	8,000	45,415	19.01
1,800	8,750	10.46			

Ni(*g*):

$$H_T-H_{298.15}=5.99T+0.36\times10^5T^{-1}-1,907$$
$$(0.3 \text{ percent}; 298°-1,000° K.);$$
$$C_p=5.99-0.36\times10^5T^{-2}.$$
$$H_T-H_{298.15}=6.07T-0.11\times10^{-3}T^2-4,120$$
$$(0.4 \text{ percent}; 1,000°-4,500° K.);$$
$$C_p=6.07-0.22\times10^{-3}T.$$

OXIDE

TABLE 539.—*Heat content and entropy of* NiO(*c*)

[Base, a-crystals at 298.15° K.; mol. wt., 74.71]

T, ° K.	$H_T-H_{298.15}$, cal./mole	$S_T-S_{298.15}$, cal./deg. mole	T, ° K.	$H_T-H_{298.15}$, cal./mole	$S_T-S_{298.15}$, cal./deg. mole
400	1,165	3.35	1,100	10,370	16.76
500	2,535	6.39	1,200	11,700	17.91
525(α)	2,940	7.18	1,300	13,060	19.00
525(β)	2,940	7.18	1,400	14,450	20.03
565(β)	3,495	8.20	1,500	15,860	21.00
565(γ)	3,495	8.20	1,600	17,300	21.93
600	3,940	8.97	1,700	18,770	22.82
700	5,220	10.94	1,800	20,260	23.68
800	6,500	12.65	1,900	21,770	24.49
900	7,780	14.16	2,000	23,300	25.28
1,000	9,070	15.52			

NiO(α):

$$H_T-H_{298.15}=-4.99T+18.79\times10^{-3}T^2-3.89\times10^5T^{-1}$$
$$+1,122 \quad (0.4 \text{ percent}; 298°-525° K.);$$
$$C_p=-4.99+37.58\times10^{-3}T+3.89\times10^5T^{-2};$$
$$\Delta H_{525}(transition)=0.$$

NiO(β):

$$H_T-H_{298.15}=13.88T-4,347 \ (0.1 \text{ percent}; 525°-565° K.);$$
$$C_p=13.88;$$
$$\Delta H_{565}(transition)=0.$$

NiO(γ):

$$H_T-H_{298.15}=11.18T+1.01\times10^{-3}T^2$$
$$-3,144 \quad (0.4 \text{ percent}; 565°-2,000° K.);$$
$$C_p=11.18+2.02\times10^{-3}T.$$

TABLE 1h7. (*Continued*)

TABLE 540.—*Heat content and entropy of NiO(g)*

[Base, ideal gas at 298.15° K.; mol. wt., 74.71]

T, ° K.	$H_T - H_{298.15}$, cal./mole	$S_T - S_{298.15}$, cal./deg. mole	T, ° K.	$H_T - H_{298.15}$, cal./mole	$S_T - S_{298.15}$, cal./deg. mole
400	830	2.39	1,000	6,020	10.28
500	1,670	4.26	1,200	7,790	11.90
600	2,525	5.82	1,400	9,560	13.26
700	3,395	7.16	1,600	11,340	14.45
800	4,265	8.33	1,800	13,120	15.50
900	5,140	9.36	2,000	14,900	16.44

NiO(g):

$$H_T - H_{298.15} = 8.73T + 0.07 \times 10^{-3} T^2 + 0.74 \times 10^5 T^{-1}$$
$$-2,857 \ (0.1 \text{ percent}; 298°-2,000° \text{ K.});$$
$$C_p = 8.73 + 0.14 \times 10^{-3} T - 0.74 \times 10^5 T^{-2}.$$

SULFIDE

TABLE 541.—*Heat content and entropy of NiS(c)*

[Base, crystals at 298.15° K.; mol. wt., 90.78]

T, ° K.	$H_T - H_{298.15}$, cal./mole	$S_T - S_{298.15}$, cal./deg. mole	T, ° K.	$H_T - H_{298.15}$, cal./mole	$S_T - S_{298.15}$, cal./deg. mole
400	1,170	3.37	600	3,660	8.40
500	2,380	6.07			

NiS(c):

$$H_T - H_{298.15} = 9.25T + 3.20 \times 10^{-3} T^2$$
$$-3,042 \ (0.1 \text{ percent}; 298°-600° \text{ K.});$$
$$C_p = 9.25 + 6.40 \times 10^{-3} T.$$

SILICIDES

NiSi(c):
$$C_p = 10.00 + 3.12 \times 10^{-3} T (estimated) \quad (298°-1,273° \text{ K.}).$$

Ni₂Si(c):

$$C_p = 15.80 + 3.29 \times 10^{-3} T (estimated) \quad (298°-1,582° \text{ K.}).$$

TELLURIDE

TABLE 542.—*Heat content and entropy of NiTe(c)*

[Base, crystals at 298.15° K.; mol. wt., 186.32]

T, ° K.	$H_T - H_{298.15}$, cal./mole	$S_T - S_{298.15}$, cal./deg. mole	T, ° K.	$H_T - H_{298.15}$, cal./mole	$S_T - S_{298.15}$, cal./deg. mole
400	1,300	3.75	600	3,930	9.07
500	2,600	6.65	700	5,310	11.20

NiTe(c):

$$H_T - H_{298.15} = 11.57T + 1.65 \times 10^{-3} T^2$$
$$-3,596 \ (0.2 \text{ percent}; 298°-700° \text{ K.});$$
$$C_p = 11.57 + 3.30 \times 10^{-3} T.$$

CHLORIDE

TABLE 543.—*Heat content and entropy of NiCl₂(c, l)*

[Base, crystals at 298.15° K.; mol. wt., 129.62]

T, ° K.	$H_T - H_{298.15}$, cal./mole	$S_T - S_{298.15}$, cal./deg. mole	T, ° K.	$H_T - H_{298.15}$, cal./mole	$S_T - S_{298.15}$, cal./deg. mole
400	1,800	5.18	1,100	15,390	24.65
500	3,650	9.31	1,200	17,510	26.50
600	5,545	12.76	1,300	19,750	28.29
700	7,465	15.72	1,303(c)	19,820	28.34
800	9,400	18.30	1,303(l)	38,290	42.52
900	11,360	20.61	1,350	39,420	43.37
1,000	13,350	22.71	1,400	40,620	44.24

NiCl₂(c):

$$H_T - H_{298.15} = 17.50T + 1.58 \times 10^{-3} T^2 + 1.19 \times 10^5 T^{-1}$$
$$-5,757 \ (0.5 \text{ percent}; 298°-1,303° \text{ K.});$$
$$C_p = 17.50 + 3.16 \times 10^{-3} T - 1.19 \times 10^5 T^{-2};$$
$$\Delta H_{1303}(fusion) = 18,470.$$

NiCl₂(l):

$$H_T - H_{298.15} = 24.00T + 7,020 \ (0.1 \text{ percent};$$
$$1,303°-1,400° \text{ K.});$$
$$C_p = 24.00.$$

FLUORIDE

NiF₂(c):

$$C_p = 15.31 \ (298° \text{ K.}).$$

CARBONYL

Ni(CO)₄(l):

$$C_p = 48.88 \ (298° \text{ K.}).$$

TABLE 544.—*Heat content and entropy of Ni(CO)₄(g)*

[Base, ideal gas at 298.15° K.; mol. wt., 170.75]

T, ° K.	$H_T - H_{298.15}$, cal./mole	$S_T - S_{298.15}$, cal./deg. mole	T, ° K.	$H_T - H_{298.15}$, cal./mole	$S_T - S_{298.15}$, cal./deg. mole
350	1,850	5.72	450	5,590	15.11
400	3,700	10.66	500	7,570	19.28

Ni(CO)₄(g):

$$H_T - H_{298.15} = 26.80T + 13.40 \times 10^{-3} T^2 - 9,182$$
$$(0.3 \text{ percent}; 298°-500° \text{ K.});$$
$$C_p = 26.80 + 26.80 \times 10^{-3} T.$$

FERRITE

NiFe₂O₄(c):

$$C_p = 34.81 \ (298° \text{ K.}).$$

NITRATE

Ni(NO₃)₂·6H₂O(c):
$$C_p = 110 \ (298° \text{ K.}).$$

Ni(NO₃)₂·6NH₃(c):
$$C_p = 96.0 \ (298° \text{ K.}).$$

TABLE 1h7. (Continued)

SULFATE

$NiSO_4(c)$:

$$C_p = 30.10 + 9.92 \times 10^{-3} T (estimated) \quad (298°-1,200° \text{ K.}).$$

$NiSO_4 \cdot 6H_2O(c)$:

$$\overline{C_p} = 82 \quad (291°-325° \text{ K.}).$$

NICKEL-TIN COMPOUND

TABLE 545.—*Heat content and entropy of* $Ni_3Sn(c)$

[Base, crystals at 298.15° K.; mol. wt., 294.83]

T, ° K.	$H_T - H_{298.15}$, cal./mole	$S_T - S_{298.15}$, cal./deg. mole	T, ° K.	$H_T - H_{298.15}$, cal./mole	$S_T - S_{298.15}$, cal./deg. mole
400	2,480	7.15	700	10,440	21.91
500	5,010	12.79	800	13,280	25.70
600	7,670	17.64	900	16,160	29.09

$Ni_3Sn(c)$:

$$H_T - H_{298.15} = 20.78 T + 5.10 \times 10^{-3} T^2 - 6,649$$
$$(0.2 \text{ percent; } 298°-900° \text{ K.});$$
$$C_p = 20.78 + 10.20 \times 10^{-3} T.$$

NIOBIUM AND ITS COMPOUNDS

ELEMENT

TABLE 546.—*Heat content and entropy of* $Nb(c, l)$

[Base, crystals at 298.15° K.; atomic wt., 92.91]

T, ° K.	$H_T - H_{298.15}$, cal./mole	$S_T - S_{298.15}$, cal./deg. mole	T, ° K.	$H_T - H_{298.15}$, cal./mole	$S_T - S_{298.15}$, cal./deg. mole
400	610	1.76	1,900	10,760	12.02
500	1,215	3.11	2,000	11,510	12.40
600	1,835	4.24	2,100	12,270	12.78
700	2,470	5.22	2,200	13,050	13.14
800	3,110	6.07	2,300	13,830	13.48
900	3,750	6.83	2,400	14,620	13.82
1,000	4,400	7.51	2,500	15,420	14.15
1,100	5,070	8.15	2,600	16,230	14.47
1,200	5,760	8.75	2,700	17,050	14.78
1,300	6,450	9.30	2,770(c)	17,630	14.99
1,400	7,160	9.83	2,770(l)	24,030	17.30
1,500	7,870	10.32	2,800	24,270	17.39
1,600	8,580	10.78	2,900	25,070	17.67
1,700	9,300	11.21	3,000	25,870	17.94
1,800	10,020	11.64			

$Nb(c)$:

$$H_T - H_{298.15} = 5.66 T + 0.48 \times 10^{-3} T^2 - 1,730$$
$$(0.2 \text{ percent; } 298°-2,770° \text{ K.});$$
$$C_p = 5.66 + 0.96 \times 10^{-3} T;$$
$$\Delta H_{2770} = 6,400.$$

$Nb(l)$:

$$H_T - H_{298.15} = 8.00 T + 1,870 \quad (0.1 \text{ percent; }$$
$$2,770°-3,000° \text{ K.});$$
$$C_p = 8.00.$$

TABLE 547.—*Heat content and entropy of* $Nb(g)$

[Base, ideal gas at 298.15° K.; atomic wt., 92.91]

T, ° K.	$H_T - H_{298.15}$, cal./mole	$S_T - S_{298.15}$, cal./deg. mole	T, ° K.	$H_T - H_{298.15}$, cal./mole	$S_T - S_{298.15}$, cal./deg. mole
400	730	2.10	1,900	10,050	12.04
500	1,430	3.67	2,000	10,650	12.34
600	2,110	4.91	2,200	11,855	12.92
700	2,770	5.93	2,400	13,090	13.46
800	3,420	6.79	2,600	14,355	13.96
900	4,050	7.54	2,800	15,660	14.45
1,000	4,675	8.20	3,000	17,005	14.91
1,100	5,290	8.78	3,500	20,550	16.00
1,200	5,895	9.31	4,000	24,355	17.02
1,300	6,500	9.79	4,500	28,380	17.97
1,400	7,095	10.23	5,000	32,590	18.85
1,500	7,685	10.64	6,000	41,435	20.46
1,600	8,280	11.02	7,000	50,650	21.88
1,700	8,870	11.38	8,000	60,055	23.14
1,800	9,460	11.72			

$Nb(g)$:

$$H_T - H_{298.15} = 6.97 T - 0.34 \times 10^{-3} T^2 - 0.42 \times 10^5 T^{-1}$$
$$-1,907 \quad (0.7 \text{ percent; } 298°-2,000° \text{ K.});$$
$$C_p = 6.97 - 0.68 \times 10^{-3} T + 0.42 \times 10^5 T^{-2};$$
$$H_T - H_{298.15} = 4.02 T + 0.47 \times 10^{-3} T^2 + 730$$
$$(0.1 \text{ percent; } 2,000°-5,000° \text{ K.});$$
$$C_p = 4.02 + 0.94 \times 10^{-3} T.$$

OXIDES

$NbO_2(c)$:

$$C_p = 12.50 + 4.17 \times 10^{-3} T (estimated) \quad (298°-1,800° \text{ K.}).$$

TABLE 548.—*Heat content and entropy of* $Nb_2O_5(c, l)$

[Base, crystals at 298.15° K.; mol. wt., 265.82]

T, ° K.	$H_T - H_{298.15}$, cal./mole	$S_T - S_{298.15}$, cal./deg. mole	T, ° K.	$H_T - H_{298.15}$, cal./mole	$S_T - S_{298.15}$, cal./deg. mole
400	3,410	9.81	1,400	43,860	59.41
500	6,970	17.75	1,500	48,230	62.43
600	10,730	24.60	1,600	52,670	65.29
700	14,630	30.61	1,700	57,190	68.03
800	18,630	35.95	1,785(c)	61,090	70.27
900	22,710	40.76	1,785(l)	85,680	84.05
1,000	26,860	45.13	1,800	86,550	84.53
1,100	31,050	49.12	1,900	92,340	87.66
1,200	35,280	52.80	2,000	98,130	90.63
1,300	39,550	56.22			

$Nb_2O_5(c)$:

$$H_T - H_{298.15} = 36.90 T + 2.56 \times 10^{-3} T^2 + 6.10 \times 10^5 T^{-1}$$
$$-13,275 \quad (0.3 \text{ percent; } 298°-1,785° \text{ K.});$$
$$C_p = 36.90 + 5.12 \times 10^{-3} T - 6.10 \times 10^5 T^{-2};$$
$$\Delta H_{1785} (fusion) = 24,590.$$

$Nb_2O_5(l)$:

$$H_T - H_{298.15} = 57.90 - 17,670 \quad (0.1 \text{ percent; }$$
$$1,785°-2,000° \text{ K.});$$
$$C_p = 57.90.$$

NITRIDE

TABLE 549.—*Heat content and entropy of* $NbN(c)$

[Base, crystals at 298.15° K.; mol. wt., 106.92]

T, ° K.	$H_T - H_{298.15}$, cal./mole	$S_T - S_{298.15}$, cal./deg. mole	T, ° K.	$H_T - H_{298.15}$, cal./mole	$S_T - S_{298.15}$, cal./deg. mole
400	1,080	3.11	600	3,355	7.71
500	2,190	5.59			

TABLE 1h7. (*Continued*)

NbN(*c*):

$$H_T - H_{298.15} = 8.69T + 2.70 \times 10^{-3}T^2 - 2,831$$
(0.2 percent; 298°–600° K.);
$$C_p = 8.69 + 5.40 \times 10^{-3}T.$$

FLUORIDE

NbF$_5$(*c*):

$$C_p = 31.6 \ (298° \ \text{K.}).$$

NITROGEN AND ITS COMPOUNDS

ELEMENT

TABLE 550.—*Heat content and entropy of* $N_2(g)$

[Base, ideal gas at 298.15° K.; mol. wt., 28.016]

T, ° K.	$H_T - H_{298.15}$, cal./mole	$S_T - S_{298.15}$, cal./deg. mole	T, ° K.	$H_T - H_{298.15}$, cal./mole	$S_T - S_{298.15}$, cal./deg. mole
400	710	2.05	2,000	13,425	14.46
500	1,415	3.62	2,100	14,285	14.88
600	2,125	4.92	2,200	15,150	15.28
700	2,855	6.04	2,300	16,020	15.67
800	3,595	7.03	2,400	16,890	16.04
900	4,355	7.92	2,500	17,765	16.40
1,000	5,130	8.74	2,750	19,965	17.23
1,100	5,920	9.49	3,000	22,175	18.00
1,200	6,720	10.19	3,250	24,395	18.71
1,300	7,530	10.84	3,500	26,625	19.37
1,400	8,355	11.45	3,750	28,860	19.99
1,500	9,180	12.02	4,000	31,105	20.57
1,600	10,020	12.56	4,250	33,355	21.12
1,700	10,860	13.07	4,500	35,615	21.63
1,800	11,710	13.56	4,750	37,880	22.12
1,900	12,565	14.02	5,000	40,150	22.59

$N_2(g)$:

$$H_T - H_{298.15} = 6.83T + 0.45 \times 10^{-3}T^2 + 0.12 \times 10^5 T^{-1}$$
$$-2,117 \ (0.7 \ \text{percent}; \ 298°–3,000° \ \text{K.});$$
$$C_p = 6.83 + 0.90 \times 10^{-3}T - 0.12 \times 10^5 T^{-2}.$$

TABLE 551.—*Heat content and entropy of* $N(g)$

[Base, ideal gas at 298.15° K.; atomic wt., 14.008]

T, ° K.	$H_T - H_{298.15}$, cal./mole	$S_T - S_{298.15}$, cal./deg. mole	T, ° K.	$H_T - H_{298.15}$, cal./mole	$S_T - S_{298.15}$, cal./deg. mole
400	505	1.46	1,900	7,960	9.20
500	1,005	2.57	2,000	8,455	9.46
600	1,500	3.48	2,200	9,450	9.93
700	1,995	4.24	2,400	10,445	10.36
800	2,495	4.90	2,600	11,440	10.76
900	2,990	5.49	2,800	12,440	11.13
1,000	3,490	6.01	3,000	13,440	11.48
1,100	3,985	6.49	3,500	15,960	12.26
1,200	4,480	6.92	4,000	18,535	12.94
1,300	4,980	7.32	4,500	21,185	13.57
1,400	5,475	7.69	5,000	23,935	14.15
1,500	5,970	8.03	6,000	29,785	15.21
1,600	6,470	8.35	7,000	36,130	16.19
1,700	6,965	8.65	8,000	42,895	17.10
1,800	7,465	8.93			

$N(g)$:

$$H_T - H_{298.15} = 4.97T - 1,482 \ (0.1 \ \text{percent};$$
$$298°–3,000° \ \text{K.});$$

$$C_p = 4.97;$$

$$H_T - H_{298.15} = 3.47T + 0.22 \times 10^{-3}T^2 + 1,070 \ (0.2 \ \text{percent};$$
$$3,000°–8,000° \ \text{K.});$$
$$C_p = 3.47 + 0.44 \times 10^{-3}T.$$

NITROUS OXIDE

TABLE 552.—*Heat content and entropy of* $N_2O(g)$

[Base, ideal gas at 298.15° K.; mol. wt., 44.02]

T, ° K.	$H_T - H_{298.15}$, cal./mole	$S_T - S_{298.15}$, cal./deg. mole	T, ° K.	$H_T - H_{298.15}$, cal./mole	$S_T - S_{298.15}$, cal./deg. mole
400	990	2.85	1,000	8,145	13.58
500	2,055	5.22	1,200	10,815	16.01
600	3,175	7.27	1,400	13,555	18.12
700	4,360	9.09	1,600	16,345	19.98
800	5,585	10.73	1,800	19,170	21.65
900	6,855	12.22	2,000	22,030	23.15

$N_2O(g)$:

$$H_T - H_{298.15} = 10.92T + 1.03 \times 10^{-3}T^2 + 2.04 \times 10^5 T^{-1}$$
$$-4,032 \ (1.0 \ \text{percent}; \ 298°–2,000° \ \text{K.});$$
$$C_p = 10.92 + 2.06 \times 10^{-3}T - 2.04 \times 10^5 T^{-2}.$$

NITRIC OXIDE

TABLE 553.—*Heat content and entropy of* $NO(g)$

[Base, ideal gas at 298.15° K.; mol. wt., 30.008]

T, ° K.	$H_T - H_{298.15}$, cal./mole	$S_T - S_{298.15}$, cal./deg. mole	T, ° K.	$H_T - H_{298.15}$, cal./mole	$S_T - S_{298.15}$, cal./deg. mole
400	725	2.10	2,000	13,835	14.91
500	1,450	3.71	2,100	14,715	15.34
600	2,185	5.05	2,200	15,595	15.75
700	2,940	6.22	2,300	16,480	16.15
800	3,715	7.25	2,400	17,365	16.52
900	4,505	8.18	2,500	18,255	16.89
1,000	5,310	9.03	2,750	20,490	17.74
1,100	6,130	9.81	3,000	22,730	18.52
1,200	6,960	10.53	3,250	24,980	19.24
1,300	7,800	11.20	3,500	27,235	19.91
1,400	8,645	11.83	3,750	29,500	20.53
1,500	9,500	12.42	4,000	31,780	21.12
1,600	10,360	12.98	4,250	34,065	21.67
1,700	11,220	13.50	4,500	36,360	22.20
1,800	12,090	13.99	4,750	38,665	22.69
1,900	12,960	14.46	5,000	40,985	23.16

$NO(g)$:

$$H_T - H_{298.15} = 7.03T + 0.46 \times 10^{-3}T^2 + 0.14 \times 10^5 T^{-1}$$
$$-2,184 \ (1 \ \text{percent}; \ 298°–2,500° \ \text{K.});$$
$$C_p = 7.03 + 0.92 \times 10^{-3}T - 0.14 \times 10^5 T^{-2}.$$

OTHER OXIDES

TABLE 554.—*Heat content and entropy of* $NO_2(g)$

[Base, ideal gas at 298.15° K.; mol. wt., 46.01]

T, ° K.	$H_T - H_{298.15}$, cal./mole	$S_T - S_{298.15}$, cal./deg. mole	T, ° K.	$H_T - H_{298.15}$, cal./mole	$S_T - S_{298.15}$, cal./deg. mole
400	940	2.71	1,300	11,615	16.30
500	1,945	4.95	1,400	12,925	17.27
600	3,015	6.90	1,500	14,250	18.19
700	4,150	8.64	1,600	15,580	19.05
800	5,330	10.22	1,700	16,915	19.85
900	6,535	11.64	1,800	18,250	20.62
1,000	7,775	12.95	1,900	19,590	21.34
1,100	9,035	14.15	2,000	20,940	22.03
1,200	10,315	15.26			

$NO_2(g)$:

$$H_T - H_{298.15} = 10.07T + 1.14 \times 10^{-3}T^2 + 1.67 \times 10^5 T^{-1}$$
$$-3,664 \ (1.0 \ \text{percent}; \ 298°–2,000° \ \text{K.});$$
$$C_p = 10.07 + 2.28 \times 10^{-3}T - 1.67 \times 10^5 T^{-2}.$$

TABLE 1h7. (*Continued*)

TABLE 555.—*Heat content and entropy of* $N_2O_4(g)$

[Base, ideal gas at 298.15° K.; mol. wt., 92.02]

T, ° K.	$H_T-H_{298.15}$, cal./mole	$S_T-S_{298.15}$, cal./deg. mole	T, ° K.	$H_T-H_{298.15}$, cal./mole	$S_T-S_{298.15}$, cal./deg. mole
400	2,060	5.93	800	11,980	22.88
500	4,310	10.94	900	14,730	26.12
600	6,740	15.36	1,000	17,560	29.10
700	9,300	19.30			

$N_2O_4(g)$:

$$H_T - H_{298.15} = 20.05T + 4.75 \times 10^{-3}T^2 + 3.56 \times 10^5 T^{-1}$$
$$-7,594 \ (0.4 \text{ percent}; 298°-1,000° \text{ K.});$$
$$C_p = 20.05 + 9.50 \times 10^{-3}T - 3.56 \times 10^5 T^{-2}.$$

$N_2O_5(c)$:

$$C_p = 34.2 \ (298° \text{ K.}).$$

NITROSYL CHLORIDE

TABLE 558.—*Heat content and entropy of* $NOCl(g)$

[Base, ideal gas at 298.15° K.; mol. wt., 65.46]

T, ° K.	$H_T-H_{298.15}$, cal./mole	$S_T-S_{298.15}$, cal./deg. mole	T, ° K.	$H_T-H_{298.15}$, cal./mole	$S_T-S_{298.15}$, cal./deg. mole
400	1,115	3.21	1,000	8,460	14.31
500	2,260	5.76	1,100	9,760	15.55
600	3,445	7.93	1,200	11,070	16.69
700	4,660	9.80	1,300	12,390	17.75
800	5,905	11.46	1,400	13,720	18.73
900	7,175	12.95	1,500	15,060	19.66

$NOCl(g)$:

$$H_T - H_{298.15} = 11.45T + 0.74 \times 10^{-3}T^2 + 1.12 \times 10^5 T^{-1}$$
$$-3,855 \ (0.4 \text{ percent}; 298°-1,500° \text{ K.});$$
$$C_p = 11.45 + 1.48 \times 10^{-3}T - 1.12 \times 10^5 T^{-2}.$$

SULFIDE

TABLE 556.—*Heat content and entropy of* $NS(g)$

[Base, ideal gas at 298.15° K.; mol. wt., 46.07]

T, ° K.	$H_T-H_{298.15}$, cal./mole	$S_T-S_{298.15}$, cal./deg. mole	T, ° K.	$H_T-H_{298.15}$, cal./mole	$S_T-S_{298.15}$, cal./deg. mole
400	780	2.25	1,000	5,720	9.73
500	1,560	3.99	1,200	7,440	11.30
600	2,365	5.46	1,400	9,180	12.64
700	3,185	6.72	1,600	10,930	13.81
800	4,020	7.83	1,800	12,690	14.85
900	4,865	8.83	2,000	14,455	15.78

$NS(g)$:

$$H_T - H_{298.15} = 7.90T + 0.29 \times 10^{-3}T^2 + 0.44 \times 10^5 T^{-1}$$
$$-2,529 \ (0.6 \text{ percent}; 298°-2,000° \text{ K.});$$
$$C_p = 7.90 + 0.58 \times 10^{-3}T - 0.44 \times 10^5 T^{-2}.$$

NITROSYL FLUORIDE

TABLE 559.—*Heat content and entropy of* $NOF(g)$

[Base, ideal gas at 298.15° K.; mol. wt., 49.01]

T, ° K.	$H_T-H_{298.15}$, cal./mole	$S_T-S_{298.15}$, cal./deg. mole	T, ° K.	$H_T-H_{298.15}$, cal./mole	$S_T-S_{298.15}$, cal./deg. mole
400	1,050	3.02	1,000	8,215	13.82
500	2,145	5.46	1,100	9,500	15.04
600	3,290	7.55	1,200	10,800	16.17
700	4,480	9.39	1,300	12,115	17.23
800	5,700	11.01	1,400	13,435	18.20
900	6,945	12.48	1,500	14,765	19.12

$NOF(g)$:

$$H_T - H_{298.15} = 11.06T + 0.87 \times 10^{-3}T + 1.52 \times 10^5 T^{-1}$$
$$-3,885 \ (0.4 \text{ percent}; 298°-1,500° \text{ K.});$$
$$C_p = 11.06 + 1.74 \times 10^{-3}T - 1.52 \times 10^5 T^{-2}.$$

NITROSYL BROMIDE

TABLE 557.—*Heat content and entropy of* $NOBr(g)$

[Base, ideal gas at 298.15° K.; mol. wt., 109.92]

T, ° K.	$H_T-H_{298.15}$, cal./mole	$S_T-S_{298.15}$, cal./deg. mole	T, ° K.	$H_T-H_{298.15}$, cal./mole	$S_T-S_{298.15}$, cal./deg. mole
400	1,135	3.27	1,000	8,525	14.44
500	2,290	5.84	1,100	9,825	15.68
600	3,485	8.02	1,200	11,140	16.82
700	4,710	9.91	1,300	12,465	17.88
800	5,960	11.58	1,400	13,795	18.87
900	7,235	13.08	1,500	15,135	19.79

$NOBr(g)$:

$$H_T - H_{298.15} = 11.53T + 0.71 \times 10^{-3}T^2 + 0.98 \times 10^5 T^{-1}$$
$$-3,830 \ (0.4 \text{ percent}; 298°-1,500° \text{ K.});$$
$$C_p = 11.53 + 1.42 \times 10^{-3}T - 0.98 \times 10^5 T^{-2}.$$

NITRYL CHLORIDE

TABLE 560.—*Heat content and entropy of* $NO_2Cl(g)$

[Base, ideal gas at 298.15° K.; mol. wt., 81.46]

T, ° K.	$H_T-H_{298.15}$, cal./mole	$S_T-S_{298.15}$, cal./deg. mole	T, ° K.	$H_T-H_{298.15}$, cal./mole	$S_T-S_{298.15}$, cal./deg. mole
400	1,375	3.95	800	7,825	15.00
500	2,855	7.25	900	9,595	17.09
600	4,445	10.15	1,000	11,395	18.98
700	6,105	12.71			

$NO_2Cl(g)$:

$$H_T - H_{298.15} = 13.86 + 2.43 \times 10^{-3}T^2 + 2.32 \times 10^5 T^{-1}$$
$$-5,126 \ (0.4 \text{ percent}; 298°-1,000° \text{ K.});$$
$$C_p = 13.86 + 4.86 \times 10^{-3}T - 2.32 \times 10^5 T^{-2}.$$

TABLE 1h7. (*Continued*)

FLUORIDE

TABLE 561.—*Heat content and entropy of* $NF_3(g)$

[Base, ideal gas at 298.15° K.; mol. wt., 73.01]

T, ° K.	$H_T-H_{298.15}$, cal./mole	$S_T-S_{298.15}$, cal./deg. mole	T, ° K.	$H_T-H_{298.15}$, cal./mole	$S_T-S_{298.15}$, cal./deg. mole
400	1,410	4.05	1,000	11,850	19.73
500	2,970	7.53	1,100	13,730	21.52
600	4,620	10.53	1,200	15,630	23.17
700	6,350	13.20	1,300	17,540	24.70
800	8,150	15.60	1,400	19,460	26.12
900	9,990	17.77	1,500	21,400	27.46

$NF_3(g)$:

$H_T-H_{298.15}=16.84T+1.06\times10^{-3}T^2+4.21\times10^5T^{-1}$
$-6,527$ (0.6 percent; 298°–1,500° K.);
$C_p=16.84+2.12\times10^{-3}T-4.21\times10^5T^{-2}$.

AMMONIA

TABLE 562.—*Heat content and entropy of* $NH_3(g)$

[Base, ideal gas at 298.15° K.; mol. wt., 17.032]

T, ° K.	$H_T-H_{298.15}$, cal./mole	$S_T-S_{298.15}$, cal./deg. mole	T, ° K.	$H_T-H_{298.15}$, cal./mole	$S_T-S_{298.15}$, cal./deg. mole
400	895	2.58	1,000	7,680	12.65
500	1,845	4.70	1,200	10,450	15.18
600	2,885	6.59	1,400	13,420	17.46
700	3,975	8.27	1,600	16,565	19.56
800	5,145	9.83	1,800	19,810	21.48
900	6,380	11.28	2,000	23,195	23.25

$NH_3(g)$:

$H_T-H_{298.15}=7.11T+3.00\times10^{-3}T^2+0.37\times10^5T^{-1}$
$-2,511$ (0.7 percent; 298°–1,800 °K.);
$C_p=7.11+6.00\times10^{-3}T-0.37\times10^5T^{-2}$.

AMMONIUM ION GAS

TABLE 563.—*Heat content and entropy of* $NH_4^+(g)$

[Base, ideal gas at 298.15° K.; mol. wt., 18.040]

T, ° K.	$H_T-H_{298.15}$, cal./mole	$S_T-S_{298.15}$, cal./deg. mole	T, ° K.	$H_T-H_{298.15}$, cal./mole	$S_T-S_{298.15}$, cal./deg. mole
400	895	2.58	800	5,665	10.64
500	1,895	4.80	900	7,180	12.42
600	3,020	6.85	1,000	8,795	14.12
700	4,280	8.79			

$NH_4^+(g)$:

$H_T-H_{298.15}=3.81T+6.58\times10^{-3}T^2-0.54\times10^5T^{-1}$
$-1,540$ (0.3 percent; 298°–1,000° K.);
$C_p=3.81+13.16\times10^{-3}T+0.54\times10^5T^{-2}$.

AMMONIUM OXIDE

$(NH_4)_2O(l)$:
$C_p=56.36$ (298° K.).

AMMONIUM HYDROXIDE

$NH_4OH(l)$:
$C_p=37.02$ (298° K.).

AMMONIUM BROMIDE

$NH_4Br(c)$:
$C_p=22.7$ (273°).

AMMONIUM CHLORIDE

TABLE 564.—*Heat content and entropy of* $NH_4Cl(c)$

[Base, α-crystals at 298.15° K.; mol. wt., 53.50]

T, ° K.	$H_T-H_{298.15}$, cal./mole	$S_T-S_{298.15}$, cal./deg. mole	T, ° K.	$H_T-H_{298.15}$, cal./mole	$S_T-S_{298.15}$, cal./deg. mole
350	1,150	3.56	457.7(α)	3,810	10.16
400	2,340	6.73	457.7(β)	4,750	12.21
450	3,610	9.72	500	5,650	14.09

$NH_4Cl(\alpha)$:
$H_T-H_{298.15}=11.80T+16.00\times10^{-3}T^2$
$-4,940$ (0.1 percent; 298°–457.7° K.);
$C_p=11.80+32.00\times10^{-3}T$;
$\Delta H_{457.7}(transition)=940$.

$NH_4Cl(\beta)$:
$H_T-H_{298.15}=5.00T+17.00\times10^{-3}T^2$
$-1,100$ (0.1 percent; 457.7°–500°);
$C_p=5.00+34.00\times10^{-3}T$.

AMMONIUM FLUORIDE

$NH_4F(c)$:
$C_p=15.60$ (298° K.).

AMMONIUM IODIDE

$NH_4I(c)$:
$C_p=19.0$ (273° K.).

AMMONIUM NITRATE

TABLE 565.—*Heat content and entropy of* $NH_4NO_3(c, l)$

[Base, α-crystals at 298.15° K.; mol. wt., 80.05]

T, ° K.	$H_T-H_{298.15}$, cal./mole	$S_T-S_{298.15}$, cal./deg. mole	T, ° K.	$H_T-H_{298.15}$, cal./mole	$S_T-S_{298.15}$, cal./deg. mole
305.3(α)	240	0.80	398.4,(δ)	4,830	13.66
305.3(β)	620	2.04	400	4,900	13.83
325	1,180	3.82	425	6,040	16.60
350	1,890	5.92	442.8(δ)	6,860	18.48
357.4(β)	2,100	6.52	442.8(l)	8,160	21.42
357.4(γ)	2,420	7.41	450	8,440	22.05
375	3,020	9.05	500	10,360	26.09
398.4(γ)	3,820	11.12	550	12,290	29.77

$NH_4NO_3(\alpha)$:
$H_T-H_{298.15}=33.60T$
$-10,018$ (0.1 percent; 298°–305.3° K.);
$C_p=33.60$;
$\Delta H_{305.3}(transition)=380$.

TABLE 1h7. (Continued)

$NH_4NO_3(\beta)$:

$H_T - H_{298.15} = 28.40$

$-8,051$ (0.1 percent; 305.3°–357.4° K.);

$C_p = 28.40$;

$\Delta H_{357.4}(transition) = 320.$

$NH_4NO_3(\gamma)$:

$H_T - H_{298.15} = 34.10 T$

$-9,767$ (0.1 percent; 357.4°–398.4° K.);

$C_p = 34.10$;

$\Delta H_{398.4}(transition) = 1,010.$

$NH_4NO_3(\delta)$:

$H_T - H_{298.15} = 45.60 T$

$-13,337$ (0.1 percent; 398.4°–442.8° K.);

$C_p = 45.60$;

$\Delta H_{442.8}(fusion) = 1,300.$

$NH_4NO_3(l)$:

$H_T - H_{298.15} = 38.50 T$

$-8,887$ (0.1 percent; 442.8°–550° K.);

$C_p = 38.50$

AMMONIUM PHOSPHATES

$(NH_4)_2HPO_4(c)$:

$\overline{C}_p = 45.0$ (273°–373° K.).

$NH_4H_2PO_4(c)$:

$C_p = 34.00$ (298° K.).

$(NH_4)_3PO_4 \cdot H_3PO_4(c)$:

$\overline{C}_p = 81.8$ (273°–373° K.)

AMMONIUM SULFATE

TABLE 566.—*Heat content and entropy of*
$(NH_4)_2SO_4(c)$

[Base, crystals at 298.15° K.; mol. wt., 132.15]

T, ° K.	$H_T - H_{298.15}$, cal./mole	$S_T - S_{298.15}$, cal./deg. mole	T, ° K.	$H_T - H_{298.15}$, cal./mole	$S_T - S_{298.15}$, cal./deg. mole
350	2,510	7.76	500	10,340	26.28
400	4,970	14.32	550	13,330	31.98
450	7,570	20.44	600	16,590	37.65

$(NH_4)_2SO_4(c)$:

$H_T - H_{298.15} = 24.77 T + 33.60 \times 10^{-3} T^2$

$-10,372$ (1.2 percent; 298°–600°)

$C_p = 24.77 T + 67.20 \times 10^{-3} T.$

$(ND_4)_2SO_4(c)$:

$C_p = 51.5$ (298° K.).

AMMONIUM BISULFATE

TABLE 567.—*Heat content and entropy of*
$NH_4HSO_4(c, l)$

[Base, crystals at 298.15° K.; mol. wt., 115.11]

T, ° K.	$H_T - H_{298.15}$, cal./mole	$S_T - S_{298.15}$, cal./deg. mole	T, ° K.	$H_T - H_{298.15}$, cal./mole	$S_T - S_{298.15}$, cal./deg. mole
350	1,880	5.81	450	9,650	24.87
400	3,900	11.20	500	12,220	30.29
417(c)	4,630	12.98	550	15,000	35.58
417(l)	8,040	21.16	600	17,940	40.70

$NH_4HSO_4(c)$:

$H_T - H_{298.15} = 10.00 T + 40.50 \times 10^{-3} T^2$

$-6,582$ (0.1 percent; 298°–417° K.);

$C_p = 10.00 + 81.00 \times 10^{-3} T$;

$\Delta H_{417}(fusion) = 3,410.$

$NH_4HSO_4(l)$:

$H_T - H_{298.15} = 16.06 T + 37.40 \times 10^{-3} T^2$

$-5,160$ (0.1 percent; 417°–600° K.);

$C_p = 16.06 + 74.80 \times 10^{-3} T.$

AMMONIUM VANADATE

TABLE 568.—*Heat content and entropy of*
$NH_4VO_3(c)$

[Base, crystals at 298.15° K.; mol. wt., 116.99]

T, ° K.	$H_T - H_{298.15}$, cal./mole	$S_T - S_{298.15}$, cal./deg. mole	T, ° K.	$H_T - H_{298.15}$, cal./mole	$S_T - S_{298.15}$, cal./deg. mole
350	1,730	5.35	500	7,420	18.85
400	3,580	10.29	550	9,390	22.60
450	5,480	14.75			

$NH_4VO_3(c)$:

$H_T - H_{298.15} = 45.42 T + 12.90 \times 10^5 T^{-1}$

$-17,869$ (0.8 percent; 298°–550° K.);

$C_p = 45.42 - 12.90 \times 10^5 T^{-2}.$

AMMONIUM-ALUMINUM SULFATE

TABLE 569.—*Heat content and entropy of*
$NH_4Al(SO_4)_2(c)$

[Base, crystals at 298.15° K.; mol. wt., 237.15]

T, ° K.	$H_T - H_{298.15}$, cal./mole	$S_T - S_{298.15}$, cal./deg. mole	T, ° K.	$H_T - H_{298.15}$, cal./mole	$S_T - S_{298.15}$, cal./deg. mole
400	6,240	17.94	600	20,140	46.02
500	13,010	33.02	700	27,540	57.41

$NH_4Al(SO_4)_2(c)$:

$H_T - H_{298.15} = 79.77 T + 22.80 \times 10^5 T^{-1} - 31,431$

(0.5 percent; 298°–700° K.);

$C_p = 79.77 - 22.80 \times 10^5 T^{-2}.$

$NH_4Al(SO_4)_2 \cdot 12H_2O(c)$:

$C_p = 163.3$ (298° K.).

AMMONIUM ANALOG OF ALUNITE

TABLE 570.—*Heat content and entropy of*
$(NH_4)_2O \cdot 3Al_2O_3 \cdot 4SO_3 \cdot 6H_2O(c)$

[Base, crystals at 298.15° K.; mol. wt., 786.32]

T, ° K.	$H_T - H_{298.15}$, cal./mole	$S_T - S_{298.15}$, cal./deg. mole	T, ° K.	$H_T - H_{298.15}$, cal./mole	$S_T - S_{298.15}$, cal./deg. mole
350	10,550	32.60	500	48,100	121.29
400	22,100	63.42	550	61,900	147.59
450	34,700	93.04			

$(NH_4)_2O \cdot 3Al_2O_3 \cdot 4SO_3 \cdot 6H_2O(c)$:

$H_T - H_{298.15} = 245.70 T + 64.80 \times 10^{-3} T^2 + 88.05$

$\times 10^5 T^{-1} - 108,548$ (0.2 percent; 298°–550° K.);

$C_p = 245.70 + 129.60 \times 10^{-3} T - 88.05 \times 10^5 T^{-2}.$

TABLE 1h7. (*Continued*)

AMMONIUM BASIC ALUM

TABLE 571.—*Heat content and entropy of* $(NH_4)_2O \cdot 3Al_2O_3 \cdot 5SO_3 \cdot 9H_2O(c)$

[Base, crystals at 298.15°K.; mol. wt., 920.43]

T, °K.	$H_T-H_{298.15}$, cal./mole	$S_T-S_{298.15}$, cal./deg. mole	T, °K.	$H_T-H_{298.15}$, cal./mole	$S_T-S_{298.15}$, cal./deg. mole
350	13,400	41.42	450	42,950	115.47
400	27,700	79.79			

$(NH_4)_2O \cdot 3Al_2O_3 \cdot 5SO_3 \cdot 9H_2O(c)$:

$$H_T-H_{298.15}=147.60T+178.20\times10^{-3}T^2-59,848$$
$$(0.9 \text{ percent}; 298°-450° \text{ K.});$$
$$C_p=147.60+356.40\times10^{-3}T.$$

AMMONIUM-DIHYDROGEN ARSENATE

$NH_4H_2AsO_4(c)$:
$$C_p=36.12 \ (298° \text{ K.}).$$

AMMONIUM-CHROMIUM SULFATE

$NH_4Cr(SO_4)_2 \cdot 12H_2O(c)$:
$$C_p=168.55 \ (298° \text{ K.}).$$

OSMIUM AND ITS COMPOUNDS

ELEMENT

TABLE 572.—*Heat content and entropy of* $Os(c)$

[Base, crystals at 298.15° K.; atomic wt., 190.2]

T, °K.	$H_T-H_{298.15}$, cal./mole	$S_T-S_{298.15}$, cal./deg. mole	T, °K.	$H_T-H_{298.15}$, cal./mole	$S_T-S_{298.15}$, cal./deg. mole
400	610	1.76	1,500	7,780	10.23
500	1,210	3.10	1,600	8,490	10.69
600	1,830	4.23	1,700	9,210	11.13
700	2,460	5.20	1,800	9,930	11.54
800	3,100	6.06	1,900	10,660	11.94
900	3,740	6.81	2,000	11,400	12.31
1,000	4,390	7.49	2,200	12,910	13.03
1,100	5,050	8.12	2,400	14,450	13.70
1,200	5,710	8.70	2,600	16,030	14.34
1,300	6,380	9.23	2,800	17,650	14.94
1,400	7,070	9.74	3,000	19,300	15.50

$Os(c)$:

$$H_T-H_{298.15}=5.69T+0.44\times10^{-3}T^2-1,736$$
$$(0.2 \text{ percent}; 298°-3,000° \text{ K.});$$
$$C_p=5.69+0.88\times10^{-3}T.$$

TABLE 573.—*Heat content and entropy of* $Os(g)$

[Base, ideal gas at 298.15° K.; atomic wt. 190.2]

T, °K.	$H_T-H_{298.15}$, cal./mole	$S_T-S_{298.15}$, cal./deg. mole	T, °K.	$H_T-H_{298.15}$, cal./mole	$S_T-S_{298.15}$, cal./deg. mole
400	505	1.46	1,500	6,580	8.56
500	1,005	2.57	1,600	7,225	8.98
600	1,505	3.49	1,700	7,885	9.38
700	2,015	4.27	1,800	8,555	9.76
800	2,530	4.96	1,900	9,240	10.13
900	3,060	5.59	2,000	9,935	10.48
1,000	3,605	6.16	2,200	11,350	11.16
1,100	4,165	6.69	2,400	12,790	11.79
1,200	4,745	7.20	2,600	14,260	12.37
1,300	5,340	7.67	2,800	15,750	12.93
1,400	5,950	8.13	3,000	17,250	13.44

$Os(g)$:

$$H_T-H_{298.15}=4.02T+0.76\times10^{-3}T^2-0.43\times10^5T^{-1}$$
$$-1,122 \ (0.4 \text{ percent}; 298°-2,000° \text{ K.});$$
$$C_p=4.02+1.52\times10^{-3}T+0.43\times10^5T^{-2};$$
$$H_T-H_{298.15}=5.82T+0.30\times10^{-3}T^2-2,904$$
$$(0.1 \text{ percent}; 2,000°-3,000° \text{ K.});$$
$$C_p=5.82+0.60\times10^{-3}T.$$

OXIDE

TABLE 574.—*Heat content and entropy of* $OsO_4(g)$

[Base, ideal gas at 298.15° K.; mol. wt., 254.2]

T, °K.	$H_T-H_{298.15}$, cal./mole	$S_T-S_{298.15}$, cal./deg. mole	T, °K.	$H_T-H_{298.15}$, cal./mole	$S_T-S_{298.15}$, cal./deg. mole
400	1,925	5.54	800	10,870	20.89
500	4,005	10.17	900	13,270	23.72
600	6,220	14.21	1,000	15,705	26.28
700	8,520	17.75			

$OsO_4(g)$:

$$H_T-H_{298.15}=20.55T+2.44\times10^{-3}T^2+3.82\times10^5T^{-1}$$
$$-7,625 \ (0.3 \text{ percent}; 298°-1,000° \text{ K.});$$
$$C_p=20.55+4.88\times10^{-3}T-3.82\times10^5T^{-2}.$$

OXYGEN AND ITS COMPOUNDS

ELEMENT

TABLE 575.—*Heat content and entropy of* $O_2(g)$

[Base, ideal gas at 298.15° K.; mol. wt., 32.00]

T, °K.	$H_T-H_{298.15}$, cal./mole	$S_T-S_{298.15}$, cal./deg. mole	T, °K.	$H_T-H_{298.15}$, cal./mole	$S_T-S_{298.15}$, cal./deg. mole
400	725	2.09	2,000	14,150	15.21
500	1,455	3.72	2,100	15,055	15.65
600	2,210	5.10	2,200	15,965	16.08
700	2,990	6.29	2,300	16,880	16.48
800	3,785	7.36	2,400	17,805	16.88
900	4,600	8.32	2,500	18,730	17.25
1,000	5,425	9.19	2,750	21,070	18.15
1,100	6,265	9.99	3,000	23,445	18.97
1,200	7,115	10.73	3,250	25,845	19.74
1,300	7,970	11.42	3,500	28,270	20.46
1,400	8,835	12.06	3,750	30,725	21.14
1,500	9,705	12.66	4,000	33,195	21.78
1,600	10,585	13.22	4,250	35,685	22.35
1,700	11,465	13.76	4,500	38,200	22.95
1,800	12,355	14.26	4,750	40,725	23.50
1,900	13,250	14.75	5,000	43,250	24.02

$O_2(g)$:

$$H_T-H_{298.15}=7.16T+0.50\times10^{-3}T^2+0.40\times10^5T^{-1}$$
$$-2,313 \ (0.8 \text{ percent}; 298°-3,000° \text{ K.});$$
$$C_p=7.16+1.00\times10^{-3}T-0.40\times10^5T^{-2}.$$

TABLE 1h7. (*Continued*)

TABLE 576.—*Heat content and entropy of* $O(g)$

[Base, ideal gas at 298.15° K.; atomic wt., 16.00]

T, ° K.	$H_T-H_{298.15}$, cal./mole	$S_T-S_{298.15}$, cal./deg. mole	T, ° K.	$H_T-H_{298.15}$, cal./mole	$S_T-S_{298.15}$, cal./deg. mole
400	530	1.52	1,900	8,040	9.35
500	1,040	2.66	2,000	8,540	9.61
600	1,545	3.59	2,200	9,535	10.08
700	2,050	4.36	2,400	10,530	10.52
800	2,550	5.03	2,600	11,525	10.92
900	3,050	5.62	2,800	12,525	11.29
1,000	3,555	6.15	3,000	13,525	11.63
1,100	4,050	6.63	3,500	16,035	12.40
1,200	4,550	7.06	4,000	18,570	13.08
1,300	5,050	7.46	4,500	21,130	13.68
1,400	5,550	7.83	5,000	23,720	14.23
1,500	6,050	8.18	6,000	28,990	15.19
1,600	6,545	8.50	7,000	34,365	16.02
1,700	7,045	8.80	8,000	39,810	16.75
1,800	7,540	9.08			

$O(g)$:
$$H_T-H_{298.15}=4.98T-0.24\times10^5T^{-1}-1,404$$
$$(0.2 \text{ percent; } 298°\text{–}3,000° \text{ K.});$$
$$C_p=4.98+0.24\times10^5T^{-2}.$$

OZONE

TABLE 577.—*Heat content and entropy of* $O_3(g)$

[Base, ideal gas at 298.15° K.; mol. wt., 48.00]

T, ° K.	$H_T-H_{298.15}$, cal./mole	$S_T-S_{298.15}$, cal./deg. mole	T, ° K.	$H_T-H_{298.15}$, cal./mole	$S_T-S_{298.15}$, cal./deg. mole
400	1,010	2.90	1,000	8,285	13.84
500	2,095	5.32	1,100	9,590	15.08
600	3,255	7.44	1,200	10,915	16.23
700	4,460	9.30	1,300	12,255	17.30
800	5,710	10.96	1,400	13,595	18.30
900	6,985	12.47	1,500	14,940	19.22

$O_3(g)$:
$$H_T-H_{298.15}=11.23T+0.96\times10^{-3}T^2+2.16\times10^5T^{-1}$$
$$-4,158 \text{ (0.6 percent; } 298°\text{–}1,500° \text{ K.});$$
$$C_p=11.23+1.92\times10^{-3}T-2.16\times10^5T^{-2}.$$

PALLADIUM AND ITS COMPOUNDS

ELEMENT

TABLE 578.—*Heat content and entropy of* $Pd(c, l)$

[Base, crystals at 298.15° K.; atomic wt., 106.4]

T, ° K.	$H_T-H_{298.15}$, cal./mole	$S_T-S_{298.15}$, cal./deg. mole	T, ° K.	$H_T-H_{298.15}$, cal./mole	$S_T-S_{298.15}$, cal./deg. mole
400	640	1.85	1,600	9,250	11.54
500	1,280	3.28	1,700	10,060	12.03
600	1,940	4.48	1,800	10,980	12.50
700	2,610	5.51	1,823(c)	11,080	12.61
800	3,290	6.42	1,823(l)	15,280	14.91
900	3,980	7.23	1,900	15,920	15.26
1,000	4,690	7.98	2,000	16,750	15.68
1,100	5,420	8.68	2,200	18,410	16.47
1,200	6,170	9.33	2,400	20,070	17.20
1,300	6,930	9.94	2,600	21,730	17.86
1,400	7,690	10.50	2,800	23,390	18.47
1,500	8,460	11.03	3,000	25,050	19.05

$Pd(c)$:
$$H_T-H_{298.15}=5.80T+0.69\times10^{-3}T^2-1,791$$
$$(0.2 \text{ percent; } 298°\text{–}1,823° \text{ K.});$$
$$C_p=5.80+1.38\times10^{-3}T;$$
$$\Delta H_{1823}(fusion)=4,200.$$

$Pd(l)$:
$$H_T-H_{298.15}=8.30T+150 \text{ (0.1 percent; } 1,823°\text{–}3,000°\text{K.});$$
$$C_p=8.30.$$

TABLE 579.—*Heat content and entropy of* $Pd(g)$

[Base, ideal gas at 298.15° K.; atomic wt., 106.4]

T, ° K.	$H_T-H_{298.15}$, cal./deg.	$S_T-S_{298.15}$, cal./deg. mole	T, ° K.	$H_T-H_{298.15}$, cal./deg.	$S_T-S_{298.15}$, cal./deg. mole
400	505	1.46	1,900	9,170	9.97
500	1,005	2.57	2,000	10,010	10.40
600	1,500	3.48	2,200	11,820	11.26
700	1,995	4.24	2,400	13,775	12.11
800	2,495	4.91	2,600	15,820	12.93
900	2,995	5.50	2,800	17,910	13.70
1,000	3,500	6.03	3,000	20,005	14.42
1,100	4,015	6.52	3,500	25,070	15.99
1,200	4,540	6.98	4,000	29,750	17.24
1,300	5,090	7.42	4,500	34,040	18.25
1,400	5,670	7.84	5,000	38,030	19.09
1,500	6,280	8.27	6,000	45,475	20.45
1,600	6,935	8.69	7,000	52,715	21.57
1,700	7,635	9.11	8,000	60,185	22.56
1,800	8,375	9.54			

$Pd(g)$:
$$H_T-H_{298.15}=14.93T-0.74\times10^{-3}T^2$$
$$-18,130 \text{ (0.1 percent; } 2,800°\text{–}5,000° \text{ K.});$$
$$C_p=14.93-1.48\times10^{-3}T.$$

OXIDE

$PdO(c)$:
$$C_p=3.30+14.2\times10^{-3}T.$$

PHOSPHORUS AND ITS COMPOUNDS

ELEMENT

TABLE 580.—*Heat content and entropy of* $P(red)$

[Base, crystals at 298.15° K.; atomic wt., 30.975]

T, ° K.	$H_T-H_{298.15}$, cal./mole	$S_T-S_{298.15}$, cal./deg. mole	T, ° K.	$H_T-H_{298.15}$, cal./mole	$S_T-S_{298.15}$, cal./deg. mole
400	620	1.79	700	2,690	5.62
500	1,270	3.24	800	3,440	6.62
600	1,970	4.51			

$P(red)$:
$$H_T-H_{298.15}=4.74T+1.95\times10^{-3}T^2$$
$$-1,587 \text{ (0.3 percent; } 298°\text{–}800° \text{ K.});$$
$$C_p=4.74+3.90\times10^{-3}T.$$

TABLE 581.—*Heat content and entropy of* $P_4(white, l)$

[Base, crystals at 298.15° K.; mol. wt., 123.90]

T, ° K.	$H_T-H_{298.15}$, cal./mole	$S_T-S_{298.15}$, cal./deg. mole	T, ° K.	$H_T-H_{298.15}$, cal./mole	$S_T-S_{298.15}$, cal./deg. mole
317.4(c)	434	1.41	370	2,270	6.90
317.4(l)	1,034	3.30	400	2,975	8.74
350	1,800	5.60			

TABLE 1h7. (*Continued*)

$P_4(white)$:

$$H_T - H_{298.15} = 22.50T$$
$$-6,708 \ (0.1 \text{ percent}; \ 298°-317.4° \text{ K.});$$
$$C_p = 22.50;$$
$$\Delta H_{317.4}(fusion) = 600.$$

$P_4(l)$:

$$H_T - H_{298.15} = 23.50T$$
$$-6,425 \ (0.1 \text{ percent}; \ 314.7°-400° \text{ K.});$$
$$C_p = 23.50.$$

TABLE 582.—*Heat content and entropy of* $P(g)$

[Base, ideal gas at 298.15° K.; atomic wt., 30.975]

T, ° K.	$H_T-H_{298.15}$, cal./mole	$S_T-S_{298.15}$, cal./deg. mole	T, ° K.	$H_T-H_{298.15}$, cal./mole	$S_T-S_{298.15}$, cal./deg. mole
400	505	1.46	1,900	7,975	9.21
500	1,005	2.57	2,000	8,480	9.47
600	1,500	3.48	2,200	9,500	9.96
700	1,995	4.24	2,400	10,535	10.41
800	2,495	4.90	2,600	11,590	10.83
900	2,990	5.49	2,800	12,670	11.23
1,000	3,490	6.01	3,000	13,780	11.61
1,100	3,985	6.49	3,500	16,685	12.51
1,200	4,480	6.92	4,000	19,790	13.34
1,300	4,980	7.32	4,500	23,070	14.11
1,400	5,475	7.69	5,000	26,495	14.83
1,500	5,975	8.03	6,000	33,605	16.13
1,600	6,470	8.35	7,000	40,825	17.24
1,700	6,970	8.65	8,000	47,980	18.20
1,800	7,475	8.94			

$P(g)$:

$$H_T - H_{298.15} = 4.97T - 1,482 \ (0.1 \text{ percent}; \ 298°-2,000° \text{ K.});$$
$$C_p = 4.97.$$

TABLE 583.—*Heat content and entropy of* $P_2(g)$

[Base, ideal gas at 298.15° K.; mol. wt., 61.95]

T, ° K.	$H_T-H_{298.15}$, cal./mole	$S_T-S_{298.15}$, cal./deg. mole	T, ° K.	$H_T-H_{298.15}$, cal./mole	$S_T-S_{298.15}$, cal./deg. mole
400	800	2.31	1,000	5,910	10.06
500	1,620	4.14	1,200	7,660	11.65
600	2,450	5.65	1,400	9,430	13.02
700	3,300	6.96	1,600	11,200	14.20
800	4,160	8.11	1,800	12,970	15.24
900	5,030	9.13	2,000	14,750	16.18

$P_2(g)$:

$$H_T - H_{298.15} = 8.31T + 0.23 \times 10^{-3}T^2 + 0.72 \times 10^5 T^{-1}$$
$$-2,740 \ (0.4 \text{ percent}; \ 298°-2,000° \text{ K.});$$
$$C_p = 8.31 + 0.46 \times 10^{-3}T - 0.72 \times 10^5 T^{-2}.$$

TABLE 584.—*Heat content and entropy of* $P_4(g)$

[Base, ideal gas at 298.15° K.; mol. wt., 123.90]

T, ° K.	$H_T-H_{298.15}$, cal./mole	$S_T-S_{298.15}$, cal./deg. mole	T, ° K.	$H_T-H_{298.15}$, cal./mole	$S_T-S_{298.15}$, cal./deg. mole
400	1,710	4.93	1,300	18,900	27.21
500	3,500	8.92	1,400	20,860	28.67
600	5,360	12.31	1,500	22,830	30.03
700	7,240	15.21	1,600	24,800	31.30
800	9,150	17.76	1,700	26,770	32.50
900	11,080	20.03	1,800	28,740	33.62
1,000	13,020	22.07	1,900	30,720	34.69
1,100	14,980	23.94	2,000	32,690	35.70
1,200	16,940	25.65			

$P_4(g)$:

$$H_T - H_{298.15} = 18.93T + 0.43 \times 10^{-3}T^2 + 2.81 \times 10^5 T^{-1}$$
$$-6,625 \ (0.2 \text{ percent}; \ 298°-1,500° \text{ K.});$$
$$C_p = 18.93 + 0.86 \times 10^{-3}T - 2.81 \times 10^5 T^{-2}.$$

OXIDES

TABLE 585.—*Heat content and entropy of* $P_4O_{10}(c, g)$

[Base, crystals at 298.15° K.; mol. wt., 283.90]

T, ° K.	$H_T-H_{298.15}$, cal./mole	$S_T-S_{298.15}$, cal./deg. mole	T, ° K.	$H_T-H_{298.15}$, cal./mole	$S_T-S_{298.15}$, cal./deg. mole
400	5,550	15.97	900	59,650	102.51
500	12,080	30.47	1,000	67,050	110.31
600	19,700	44.34	1,100	74,400	117.32
631 (c)	22,270	48.51	1,200	81,750	123.72
631 (g)	39,870	76.40	1,300	89,100	129.59
700	44,950	84.05	1,400	96,450	135.04
800	52,300	93.86	1,500	103,800	140.11

$P_4O_{10}(c)$:

$$H_T - H_{298.15} = 16.75T + 54.00 \times 10^{-3}T^2 - 9,794$$
$$(0.1 \text{ percent}; \ 298°-631° \text{ K.});$$
$$C_p = 16.75 + 108.00 \times 10^{-3}T;$$
$$\Delta H_{631}(sublimation) = 17,600.$$

$P_4O_{10}(g)$:

$$H_T - H_{298.15} = 73.60T - 6,570 \ (0.1 \text{ percent};$$
$$631°-1,500° \text{ K.});$$
$$C_p = 73.60.$$

TABLE 586.—*Heat content and entropy of* $PO(g)$

[Base, ideal gas at 298.15° K.; mol. wt., 46.98]

T, ° K.	$H_T-H_{298.15}$, cal./mole	$S_T-S_{298.15}$, cal./deg. mole	T, ° K.	$H_T-H_{298.15}$, cal./mole	$S_T-S_{298.15}$, cal./deg. mole
400	780	2.25	1,000	5,710	9.72
500	1,560	3.99	1,200	7,430	11.29
600	2,360	5.45	1,400	9,170	12.63
700	3,180	6.71	1,600	10,920	13.79
800	4,015	7.83	1,800	12,680	14.83
900	4,860	8.82	2,000	14,445	15.76

$PO(g)$:

$$H_T - H_{298.15} = 7.87T + 0.30 \times 10^{-3}T^2 + 0.41 \times 10^5 T^{-1}$$
$$-2,511 \ (0.6 \text{ percent}; \ 298°-2,000° \text{ K.});$$
$$C_p = 7.87 + 0.60 \times 10^{-3}T - 0.41 \times 10^5 T^{-2}.$$

NITRIDE

TABLE 587.—*Heat content and entropy of* $PN(g)$

[Base, ideal gas at 298.15° K.; mol. wt., 44.98]

T, ° K.	$H_T-H_{298.15}$, cal./mole	$S_T-S_{298.15}$, cal./deg. mole	T, ° K.	$H_T-H_{298.15}$, cal./mole	$S_T-S_{298.15}$, cal./deg. mole
400	735	2.12	1,000	5,535	9.37
500	1,485	3.80	1,200	7,235	10.92
600	2,225	5.20	1,400	8,960	12.25
700	3,055	6.43	1,600	10,700	13.41
800	3,870	7.52	1,800	12,445	14.44
900	4,695	8.49	2,000	14,205	15.37

$PN(g)$:

$$H_T - H_{298.15} = 7.44T + 0.45 \times 10^{-3}T^2 + 0.55 \times 10^5 T^{-1}$$
$$-2,443 \ (0.6 \text{ percent}; \ 298°-2,000° \text{ K.});$$
$$C_p = 7.44T + 0.90 \times 10^{-3}T - 0.55 \times 10^5 T^{-2}.$$

TABLE 1h7. (*Continued*)

PHOSPHINE

TABLE 588.—*Heat content and entropy of* $PH_3(g)$

[Base, ideal gas at 298.15° K.; mol. wt., 34.00]

T, ° K.	$H_T-H_{298.15}$, cal./mole	$S_T-S_{298.15}$, cal./deg. mole	T, ° K.	$H_T-H_{298.15}$, cal./mole	$S_T-S_{298.15}$, cal./deg. mole
400	960	2.76	1,000	8,715	14.24
500	2,020	5.12	1,200	11,875	17.12
600	3,185	7.25	1,400	15,210	19.68
700	4,440	9.18	1,600	18,680	22.00
800	5,790	10.98	1,800	22,230	24.09
900	7,220	12.66	2,000	25,860	26.00

$PH_3(g)$:

$$H_T-H_{298.15}=9.11T+2.86\times10^{-3}T^2+1.71\times10^5T^{-1}$$
$$-3,544 \ (1.4 \text{ percent}; 298°-2,000° \text{ K.});$$
$$C_p=9.11+5.72\times10^{-3}T-1.71\times10^5T^{-2}.$$

PHOSPHONIUM ION GAS

TABLE 589.—*Heat content and entropy of* $PH_4^+(g)$

[Base, ideal gas at 298.15° K.; mol. wt., 35.01]

T, ° K.	$H_T-H_{298.15}$, cal./mole	$S_T-S_{298.15}$, cal./deg. mole	T, ° K.	$H_T-H_{298.15}$, cal./mole	$S_T-S_{298.15}$, cal./deg. mole
400	1,125	3.23	800	7,195	13.52
500	2,415	6.10	900	9,035	15.68
600	3,875	8.76	1,000	10,970	17.72
700	5,470	11.22			

$PH_4^+(g)$:

$$H_T-H_{298.15}=8.21T+6.20\times10^{-3}T^2+1.68\times10^5T^{-1}$$
$$-3,562 \ (0.3 \text{ percent}; 298°-1,000° \text{ K.});$$
$$C_p=8.21+12.40\times10^{-3}T-1.68\times10^5T^{-2}.$$

BROMIDE

TABLE 590.—*Heat content and entropy of* $PBr_3(g)$

[Base, ideal gas at 298.15° K.; mol. wt., 270.72]

T, ° K.	$H_T-H_{298.15}$, cal./mole	$S_T-S_{298.15}$, cal./deg. mole	T, ° K.	$H_T-H_{298.15}$, cal./mole	$S_T-S_{298.15}$, cal./deg. mole
400	1,910	5.51	800	9,640	18.87
500	3,810	9.75	900	11,600	21.18
600	5,730	13.25	1,000	13,565	23.25
700	7,680	16.25			

$PBr_3(g)$:

$$H_T-H_{298.15}=19.81T+1.43\times10^5T^{-1}-6,386$$
$$(0.2 \text{ percent}; 298°-1,000° \text{ K.});$$
$$C_p=19.81-1.43\times10^5T^{-2}.$$

CHLORIDES

TABLE 591.—*Heat content and entropy of* $PCl_3(g)$

[Base, ideal gas at 298.15° K.; mol. wt., 137.35]

T, ° K.	$H_T-H_{298.15}$, cal./mole	$S_T-S_{298.15}$, cal./deg. mole	T, ° K.	$H_T-H_{298.15}$, cal./mole	$S_T-S_{298.15}$, cal./deg. mole
400	1,810	5.22	800	9,420	18.35
500	3,665	9.36	900	11,365	20.65
600	5,560	12.81	1,000	13,315	22.70
700	7,480	15.77			

$PCl_3(g)$:

$$H_T-H_{298.15}=19.15T+0.37\times10^{-3}T^2+1.91\times10^5T^{-1}$$
$$-6,383 \ (0.1 \text{ percent}; 298°-1,000° \text{ K.});$$
$$C_p=19.15+0.74\times10^{-3}T-1.91\times10^5T^{-2}.$$

TABLE 592.—*Heat content and entropy of* $PCl_5(g)$

[Base, ideal gas at 298.15° K.; mol. wt., 208.26]

T, ° K.	$H_T-H_{298.15}$, cal./mole	$S_T-S_{298.15}$, cal./deg. mole	T, ° K.	$H_T-H_{298.15}$, cal./mole	$S_T-S_{298.15}$, cal./deg. mole
400	2,835	8.15	1,000	21,110	35.89
500	5,755	14.67	1,100	24,240	38.88
600	8,760	20.14	1,200	27,370	41.60
700	11,810	24.84	1,300	30,510	44.11
800	14,890	28.96	1,400	33,660	46.45
900	17,990	32.61	1,500	36,810	48.62

$PCl_5(g)$:

$$H_T-H_{298.15}=31.02T+0.28\times10^{-3}T^2+3.94\times10^5T^{-1}$$
$$-10,595 \ (0.1 \text{ percent}; 298°-1,500° \text{ K.});$$
$$C_p=31.02+0.56\times10^{-3}T-3.94\times10^5T^{-2}.$$

FLUORIDE

TABLE 593.—*Heat content and entropy of* $PF_3(g)$

[Base, ideal gas at 298.15° K.; mol. wt., 87.98]

T, ° K.	$H_T-H_{298.15}$, cal./mole	$S_T-S_{298.15}$, cal./deg. mole	T, ° K.	$H_T-H_{298.15}$, cal./mole	$S_T-S_{298.15}$, cal./deg. mole
400	1,520	4.37	1,000	12,280	20.59
500	3,160	8.03	1,100	14,180	22.40
600	4,890	11.18	1,200	16,090	24.06
700	6,680	13.94	1,300	18,010	25.60
800	8,520	16.39	1,400	19,940	27.03
900	10,390	18.60	1,500	21,870	28.36

$PF_3(g)$:

$$H_T-H_{298.15}=17.72T+0.72\times10^{-3}T^2+3.66\times10^5T^{-1}$$
$$-6,575 \ (0.5 \text{ percent}; 298°-1,500° \text{ K.});$$
$$C_p=17.72+1.44\times10^{-3}T-3.66\times10^5T^{-2}.$$

IODIDE

TABLE 594.—*Heat content and entropy of* $PI_3(g)$

[Base, ideal gas at 298.15° K.; mol. wt., 411.70]

T, ° K.	$H_T-H_{298.15}$, cal./mole	$S_T-S_{298.15}$, cal./deg. mole	T, ° K.	$H_T-H_{298.15}$, cal./mole	$S_T-S_{298.15}$, cal./deg. mole
400	1,935	5.58	800	9,750	19.10
500	3,870	9.90	900	11,720	21.42
600	5,820	13.45	1,000	13,695	23.50
700	7,780	16.47			

$PI_3(g)$:

$$H_T-H_{298.15}=19.72T+0.08\times10^{-3}T^2+0.92\times10^5T^{-1}$$
$$-6,195 \ (0.1 \text{ percent}; 298°-1,000° \text{ K.});$$
$$C_p=19.72+0.16\times10^{-3}T-0.92\times10^5T^{-2}.$$

TABLE 1h7. (*Continued*)

OXYCHLORIDE

TABLE 595.—*Heat content and entropy of POCl₃(g)*

$POCl_3(g)$

[Base, ideal gas at 298.15° K.; mol. wt., 153.35]

T, ° K.	$H_T-H_{298.15}$, cal./mole	$S_T-S_{298.15}$, cal./deg. mole	T, ° K.	$H_T-H_{298.15}$, cal./mole	$S_T-S_{298.15}$, cal./deg. mole
400	2,145	6.18	800	11,545	22.37
500	4,390	11.19	900	14,005	25.26
600	6,725	15.44	1,000	16,485	27.88
700	9,115	19.12			

$$POCl_3(g):$$
$$H_T-H_{298.15}=22.12T+1.80\times10^{-3}T^2+2.69\times10^5T^{-1}$$
$$-7,657 \ (0.2 \text{ percent; } 298°-1,000° \text{ K.});$$
$$C_p=22.12+3.60\times10^{-3}T-2.69\times10^5T^{-2}.$$

OXYFLUORIDE

TABLE 596.—*Heat content and entropy of POF₃(g)*

$POF_3(g)$

[Base, ideal gas at 298.15° K.; mol. wt., 103.98]

T, ° K.	$H_T-H_{298.15}$, cal./mole	$S_T-S_{298.15}$, cal./deg. mole	T, ° K.	$H_T-H_{298.15}$, cal./mole	$S_T-S_{298.15}$, cal./deg. mole
400	1,805	5.19	800	10,440	19.98
500	3,785	9.60	900	12,790	22.74
600	5,910	13.47	1,000	15,180	25.26
700	8,140	16.91			

$$POF_3(g):$$
$$H_T-H_{298.15}=19.00T+3.11\times10^{-3}T^2+3.96\times10^5T^{-1}$$
$$-7,269 \ (0.3 \text{ percent; } 298°-1,000° \text{ K.});$$
$$C_p=19.00+6.22\times10^{-3}T-3.96\times10^5T^{-2}.$$

SULFOCHLORIDE

TABLE 597.—*Heat content and entropy of PSCl₃(g)*

$PSCl_3(g)$

[Base, ideal gas at 298.15° K.; mol. wt., 169.41]

T, ° K.	$H_T-H_{298.15}$, cal./mole	$S_T-S_{298.15}$, cal./deg. mole	T, ° K.	$H_T-H_{298.15}$, cal./mole	$S_T-S_{298.15}$, cal./deg. mole
400	2,240	6.46	800	11,900	23.11
500	4,570	11.65	900	14,400	26.05
600	6,970	16.03	1,000	16,900	28.69
700	9,420	19.80			

$$PSCl_3(g):$$
$$H_T-H_{298.15}=24.28T+0.74\times10^{-3}T^2+3.33\times10^5T^{-1}$$
$$-8,422 \ (0.1 \text{ percent; } 298°-1,000° \text{ K.});$$
$$C_p=24.28+1.48\times10^{-3}T-3.33\times10^5T^{-2}.$$

SULFOFLUORIDE

TABLE 598.—*Heat content and entropy of PSF₃(g)*

$PSF_3(g)$

[Base, ideal gas at 298.15° K.; mol. wt., 120.04]

T, ° K.	$H_T-H_{298.15}$, cal./mole	$S_T-S_{298.15}$, cal./deg. mole	T, ° K.	$H_T-H_{298.15}$, cal./mole	$S_T-S_{298.15}$, cal./deg. mole
400	1,960	5.64	800	11,030	21.21
500	4,080	10.36	900	13,455	24.07
600	6,325	14.45	1,000	15,910	26.66
700	8,650	18.04			

$$PSF_3(g):$$
$$H_T-H_{298.15}=21.60T+1.92\times10^{-3}T^2+4.24\times10^5T^{-1}$$
$$-8,033 \ (0.3 \text{ percent; } 298°-1,000° \text{ K.});$$
$$C_p=21.60+3.84\times10^{-3}T-4.24\times10^5T^{-2}.$$

PLATINUM AND ITS COMPOUNDS

ELEMENT

TABLE 599.—*Heat content and entroy of Pt(c, l)*

[Base, crystals at 298.15° K.; atomic wt., 195.09]

T, ° K.	$H_T-H_{298.15}$, cal./mole	$S_T-S_{298.15}$, cal./deg. mole	T, ° K.	$H_T-H_{298.15}$, cal./mole	$S_T-S_{298.15}$, cal./deg. mole
400	645	1.86	1,700	9,940	11.91
500	1,280	3.28	1,800	10,740	12.37
600	1,920	4.44	1,900	11,550	12.81
700	2,580	5.46	2,000	12,370	13.23
800	3,260	6.37	2,043(c)	12,730	13.40
900	3,950	7.18	2,043(l)	17,430	15.70
1,000	4,660	7.93	2,100	17,900	15.93
1,100	5,380	8.61	2,200	18,730	16.32
1,200	6,110	9.25	2,400	20,390	17.04
1,300	6,850	9.84	2,600	22,050	17.70
1,400	7,600	10.39	2,800	23,700	18.32
1,500	8,370	10.93	3,000	25,370	18.89
1,600	9,150	11.43			

$$Pt(c):$$
$$H_T-H_{298.15}=5.81T+0.63\times10^{-3}T^2-0.06\times10^5T^{-1}$$
$$-1,768 \ (0.3 \text{ percent; } 298°-2,043° \text{ K.});$$
$$C_p=5.81+1.26\times10^{-3}T+0.06\times10^5T^{-2};$$
$$\Delta H_{2043}(fusion)=4,700.$$

$$Pt(l):$$
$$H_T-H_{298.15}=8.30T+470 \ (0.1 \text{ percent;}$$
$$2,043°-3,000° \text{ K.});$$
$$C_p=8.30.$$

TABLE 600.—*Heat content and entropy of Pt(g)*

[Base, ideal gas at 298.15° K.; atomic wt., 195.09]

T, ° K.	$H_T-H_{298.15}$, cal./mole	$S_T-S_{298.15}$, cal./deg. mole	T, ° K.	$H_T-H_{298.15}$, cal./mole	$S_T-S_{298.15}$, cal./deg. mole
400	645	1.86	1,900	9,130	10.95
500	1,290	3.30	2,000	9,665	11.22
600	1,925	4.46	2,200	10,730	11.73
700	2,540	5.41	2,400	11,810	12.20
800	3,140	6.21	2,600	12,890	12.64
900	3,720	6.89	2,800	13,985	13.04
1,000	4,285	7.49	3,000	15,090	13.42
1,100	4,840	8.02	3,500	17,900	14.29
1,200	5,390	8.49	4,000	20,770	15.06
1,300	5,935	8.93	4,500	23,705	15.75
1,400	6,470	9.33	5,000	26,690	16.38
1,500	7,005	9.70	6,000	32,825	17.49
1,600	7,540	10.04	7,000	39,210	18.48
1,700	8,070	10.36	8,000	45,915	19.37
1,800	8,600	10.66			

$$Pt(g):$$
$$H_T-H_{298.15}=4.84T+0.12\times10^{-3}T^2-510$$
$$(0.1 \text{ percent; } 2,000°-8,000° \text{ K.});$$
$$C_p=4.84+0.24\times10^{-3}T.$$

SULFIDES

$$PtS(c):$$
$$C_p=11.14+2.86\times10^{-3}T \ (298°-1,000° \text{ K.}).$$

$$PtS_2(c):$$
$$C_p=13.86+7.14\times10^{-3}T \ (298°-1,000° \text{ K.}).$$

TABLE 1h7. (*Continued*)

POLONIUM
ELEMENT

TABLE 601.—*Heat content and entropy*
of Po(c, l)

[Base, crystals at 298.15° K.; atomic wt., 210]

T, ° K.	$H_T-H_{298.15}$, cal./mole	$S_T-S_{298.15}$, cal./deg. mole	T, ° K.	$H_T-H_{298.15}$, cal./mole	$S_T-S_{298.15}$, cal./deg. mole
400	670	1.92	800	6,620	12.71
500	1,370	3.49	900	7,370	13.60
527(c)	1,570	3.88	1,000	8,120	14.39
527(l)	4,570	9.57	1,100	8,870	15.10
600	5,120	10.55	1,200	9,620	15.75
700	5,870	11.71			

Po(c):

$$H_T-H_{298.15}=4.85T+2.44\times10^{-3}T^2-1,663$$

(0.3 percent; 298°–527° K.);

$$C_p=4.85+4.88\times10^{-3}T;$$

$$\Delta H_{527}(fusion)=3,000.$$

Po(l):

$$H_T-H_{298.15}=7.50T+620 \ (0.1 \ percent; \ 527°-1,200° \ K.).$$

$$C_p=7.50.$$

TABLE 602.—*Heat content and entropy*
of Po(g)

[Base, ideal gas at 298.15° K.; atomic wt., 210]

T, ° K.	$H_T-H_{298.15}$, cal./mole	$S_T-S_{298.15}$, cal./deg. mole	T, ° K.	$H_T-H_{298.15}$, cal./mole	$S_T-S_{298.15}$, cal./deg. mole
400	505	1.46	1,500	5,975	8.03
500	1,005	2.57	1,600	6,475	8.35
600	1,500	3.48	1,700	6,975	8.66
700	1,995	4.24	1,800	7,480	8.95
800	2,495	4.90	1,900	7,985	9.22
900	2,990	5.49	2,000	8,495	9.48
1,000	3,485	6.01	2,200	9,520	9.97
1,100	3,985	6.49	2,400	10,565	10.42
1,200	4,480	6.92	2,600	11,635	10.86
1,300	4,980	7.32	2,800	12,725	11.27
1,400	5,475	7.69	3,000	13,840	11.66

Po(g):

$$H_T-H_{298.15}=4.97T-1,482 \ (0.2 \ percent;$$

298°–2,000° K.);

$$C_p=4.97.$$

TABLE 603.—*Heat content and entropy*
of Po₂(g)

[Base, ideal gas at 298.15° K.; mol. wt., 420]

T, ° K.	$H_T-H_{298.15}$, cal./mole	$S_T-S_{298.15}$, cal./deg. mole	T, ° K.	$H_T-H_{298.15}$, cal./mole	$S_T-S_{298.15}$, cal./deg. mole
400	900	2.60	1,000	6,250	10.77
500	1,790	4.58	1,100	7,150	11.63
600	2,680	6.21	1,200	8,040	12.40
700	3,580	7.59	1,300	8,940	13.12
800	4,470	8.78	1,400	9,830	13.78
900	5,360	9.83	1,500	10,720	14.40

Po₂(g):

$$H_T-H_{298.15}=8.91T-2,657 \ (0.2 \ percent;$$

298°–1,500° K.);

$$C_p=8.91.$$

POTASSIUM AND ITS COMPOUNDS
ELEMENT

TABLE 604.—*Heat content and entropy*
of K(c, l)

[Base, crystals at 298.15° K.; atomic wt., 39.10]

T, ° K.	$H_T-H_{298.15}$, cal./mole	$S_T-S_{298.15}$, cal./deg. mole	T, ° K.	$H_T-H_{298.15}$, cal./mole	$S_T-S_{298.15}$, cal./deg. mole
336.4 (c)	286	0.90	700	3,515	7.97
336.4(l)	844	2.56	800	4,225	8.92
400	1,330	3.89	900	4,940	9.76
500	2,075	5.55	1,000	5,665	10.53
600	2,805	6.88	1,100	6,400	11.23

K(c):

$$H_T-H_{298.15}=1.34T+9.70\times10^{-3}T^2-1,262$$

(0.1 percent; 298°–336.4° K.);

$$C_p=1.34+19.40\times10^{-3}T;$$

$$\Delta H_{336.4}(fusion)=558.$$

K(l):

$$H_T-H_{298.15}=7.06T-0.70\times10^5T^{-1}-1,323$$

(0.3 percent; 336.4°–1,100° K.);

$$C_p=7.06+0.70\times10^5T^{-2}.$$

TABLE 605.—*Heat content and entropy of K(g)*

[Base, ideal gas at 298.15° K.; atomic wt., 39.10]

T, ° K.	$H_T-H_{298.15}$, cal./mole	$S_T-S_{298.15}$, cal./deg. mole	T, ° K.	$H_T-H_{298.15}$, cal./mole	$S_T-S_{298.15}$, cal./deg. mole
400	505	1.46	1,900	7,965	9.21
500	1,005	2.57	2,000	8,465	9.46
600	1,500	3.48	2,200	9,475	9.94
700	1,995	4.24	2,400	10,490	10.39
800	2,495	4.90	2,600	11,525	10.80
900	2,990	5.49	2,800	12,580	11.19
1,000	3,490	6.01	3,000	13,660	11.56
1,100	3,985	6.48	3,500	16,540	12.45
1,200	4,480	6.92	4,000	19,810	13.32
1,300	4,980	7.32	4,500	23,750	14.25
1,400	5,475	7.69	5,000	28,720	15.29
1,500	5,975	8.03	6,000	42,950	17.87
1,600	6,470	8.35	7,000	62,210	20.83
1,700	6,970	8.65	8,000	82,400	23.53
1,800	7,465	8.94			

K(g):

$$H_T-H_{298.15}=4.97T-1,482$$

(0.1 percent; 298°–2,000° K.);

$$C_p=4.97.$$

TABLE 606.—*Heat content and entropy of K₂(g)*

[Base, ideal gas at 298.15° K.; mol. wt., 78.20]

T, ° K.	$H_T-H_{298.15}$, cal./mole	$S_T-S_{298.15}$, cal./deg. mole	T, ° K.	$H_T-H_{298.15}$, cal./mole	$S_T-S_{298.15}$, cal./deg. mole
400	925	2.67	1,600	12,240	15.65
500	1,840	4.71	1,700	13,215	16.24
600	2,760	6.39	1,800	14,200	16.80
700	3,685	7.82	1,900	15,180	17.33
800	4,615	9.06	2,000	16,170	17.84
900	5,550	10.16	2,100	17,165	18.33
1,000	6,490	11.15	2,200	18,165	18.79
1,100	7,440	12.05	2,300	19,170	19.24
1,200	8,390	12.88	2,400	20,180	19.67
1,300	9,345	13.65	2,500	21,195	20.08
1,400	10,305	14.36	2,750	23,760	21.06
1,500	11,270	15.02	3,000	26,350	21.96

TABLE 1h7. (*Continued*)

$K_2(g)$:

$$H_T-H_{298.15}=8.91T+0.26\times10^{-3}T^2-2,680$$
(0.1 percent; 298°–3,000° K.);
$$C_p=8.91+0.52\times10^{-3}T.$$

OXIDE

$KO_2(c)$:
$$C_p=18.53 \text{ (298° K.)}.$$

HYDRIDES

TABLE 607.—*Heat content and entropy of KH(g)*

[Base, ideal gas at 298.15° K.; mol. wt., 40.11]

T, ° K.	$H_T-H_{298.15}$, cal./mole	$S_T-S_{298.15}$, cal./deg. mole	T, ° K.	$H_T-H_{298.15}$, cal./mole	$S_T-S_{298.15}$, cal./deg. mole
400	770	2.22	1,000	5,780	9.81
500	1,565	3.99	1,200	7,525	11.40
600	2,380	5.48	1,400	9,275	12.75
700	3,215	6.76	1,600	11,035	13.92
800	4,060	7.89	1,800	12,800	14.96
900	4,915	8.90	2,000	14,575	15.90

$KH(g)$:

$$H_T-H_{298.15}=8.23T+0.22\times10^{-3}T^2+0.87\times10^5T^{-1}$$
$$-2,765 \text{ (0.4 percent; 298°–2,000° K.)};$$
$$C_p=8.23+0.44\times10^{-3}T-0.87\times10^5T^{-2}.$$

TABLE 608.—*Heat content and entropy of KD(g)*

[Base, ideal gas at 298.15° K.; mol. wt., 41.11]

T, ° K.	$H_T-H_{298.15}$, cal./mole	$S_T-S_{298.15}$, cal./deg. mole	T, ° K.	$H_T-H_{298.15}$, cal./mole	$S_T-S_{298.15}$, cal./deg. mole
400	815	2.35	1,000	5,970	10.18
500	1,645	4.20	1,200	7,730	11.78
600	2,490	5.74	1,400	9,500	13.15
700	3,350	7.07	1,600	11,275	14.33
800	4,220	8.23	1,800	13,050	15.38
900	5,090	9.25	2,000	14,830	16.32

$KD(g)$:

$$H_T-H_{298.15}=8.65T+0.09\times10^{-3}T^2+0.80\times10^5T^{-1}$$
$$-2,855 \text{ (0.2 percent; 298°–2,000° K.)};$$
$$C_p=8.65+0.18\times10^{-3}T-0.80\times10^5T^{-2}.$$

BROMIDE

TABLE 609.—*Heat content and entropy of KBr(c)*

[Base, crystals at 298.15° K.; mol. wt., 119.02]

T, ° K.	$H_T-H_{298.15}$, cal./mole	$S_T-S_{298.15}$, cal./deg. mole	T, ° K.	$H_T-H_{298.15}$, cal./mole	$S_T-S_{298.15}$, cal./deg. mole
400	1,280	3.68	800	6,700	13.02
500	2,580	6.59	900	8,150	14.73
600	3,920	9.03	1,000	9,640	16.30
700	5,290	11.14			

$KBr(c)$:

$$H_T-H_{298.15}=10.65T+2.26\times10^{-3}T^2-0.49\times10^5T^{-1}$$
$$-3,212 \text{ (0.3 percent; 298°–1,000° K.)};$$
$$C_p=10.65+4.52\times10^{-3}T+0.49\times10^5T^{-2}.$$

TABLE 610.—*Heat content and entropy of KBr(g)*

[Base, ideal gas at 298.15° K.; mol. wt., 119.02]

T, ° K.	$H_T-H_{298.15}$, cal./mole	$S_T-S_{298.15}$, cal./deg. mole	T, ° K.	$H_T-H_{298.15}$, cal./mole	$S_T-S_{298.15}$, cal./deg. mole
400	905	2.61	1,000	6,330	10.88
500	1,800	4.61	1,200	8,160	12.55
600	2,700	6.25	1,400	10,000	13.97
700	3,605	7.64	1,600	11,850	15.20
800	4,510	8.85	1,800	13,705	16.29
900	5,420	9.92	2,000	15,570	17.28

$KBr(g)$:

$$H_T-H_{298.15}=8.98T+0.09\times10^{-3}T^2+0.22\times10^5T^{-1}$$
$$-2,759 \text{ (0.1 percent; 298°–2,000° K.)};$$
$$C_p=8.98+0.18\times10^{-3}T-0.22\times10^5T^{-2}.$$

CHLORIDE

TABLE 611.—*Heat content and entropy of KCl(c,l)*

[Base, crystals at 298.15° K.; mol. wt., 74.56]

T, ° K.	$H_T-H_{298.15}$, cal./mole	$S_T-S_{298.15}$, cal./deg. mole	T, ° K.	$H_T-H_{298.15}$, cal./mole	$S_T-S_{298.15}$, cal./deg. mole
400	1,260	3.64	1,043(c)	10,150	16.66
500	2,520	6.45	1,043(l)	16,250	22.51
600	3,810	8.80	1,100	17,160	23.36
700	5,150	10.86	1,200	18,760	24.75
800	6,550	12.73	1,300	20,360	26.03
900	8,000	14.44	1,400	21,960	27.22
1,000	9,500	16.02	1,500	23,560	28.32

$KCl(c)$:

$$H_T-H_{298.15}=9.89T+2.60\times10^{-3}T^2-0.77\times10^5T^{-1}$$
$$-2,922 \text{ (0.2 percent; 298°–1,043° K.)};$$
$$C_p=9.89+5.20\times10^{-3}T+0.77\times10^5T^{-2};$$
$$\Delta H_{1043}(fusion)=6,100.$$

$KCl(l)$:

$$H_T-H_{298.15}=16.00T-440 \text{ (0.1 percent;}$$
$$1,043°–1,500° K.)};$$
$$C_p=16.00.$$

TABLE 612.—*Heat content and entropy of KCl(g)*

[Base, ideal gas at 298.15° K.; mol. wt., 75.46]

T, ° K.	$H_T-H_{298.15}$, cal./mole	$S_T-S_{298.15}$, cal./deg. mole	T, ° K.	$H_T-H_{298.15}$, cal./mole	$S_T-S_{298.15}$, cal./deg. mole
400	895	2.58	1,000	6,300	10.82
500	1,785	4.56	1,200	8,125	12.48
600	2,680	6.20	1,400	9,955	12.89
700	3,580	7.59	1,600	11,800	15.12
800	4,480	8.79	1,800	13,665	16.22
900	5,390	9.86	2,000	15,545	17.21

$KCl(g)$:

$$H_T-H_{298.15}=8.93T+0.11\times10^{-3}T^2+0.28\times10^5T^{-1}$$
$$-2,766 \text{ (0.1 percent; 298°–2,000° K.)};$$
$$C_p=8.93+0.22\times10^{-3}T-0.28\times10^5T^{-2}.$$

TABLE 1h7. (Continued)

FLUORIDE

TABLE 613.—Heat content and entropy of KF(c, l)

[Base, crystals at 298.15° K.; mol. wt., 58.10]

T, ° K.	$H_T-H_{298.15}$, cal./mole	$S_T-S_{298.15}$, cal./deg. mole	T, ° K.	$H_T-H_{298.15}$, cal./mole	$S_T-S_{298.15}$, cal./deg. mole
400	1,220	3.51	1,130(l)	17,770	23.26
500	2,490	6.34	1,200	18,890	24.22
600	3,780	8.70	1,300	20,490	25.50
700	5,080	10.70	1,400	22,090	26.69
800	6,400	12.46	1,500	23,690	27.79
900	7,760	14.06	1,600	25,290	28.82
1,000	9,150	15.53	1,700	26,890	29.79
1,100	10,580	16.89	1,800	28,490	30.71
1,130(c)	11,020	17.29			

$$KF(c):$$
$$H_T-H_{298.15}=11.88T+1.11\times10^{-3}T^2+0.72\times10^5T^{-1}$$
$$-3,882 \ (0.3 \ \text{percent}; \ 298°-1,130° \ \text{K.});$$
$$C_p=11.88+2.22\times10^{-3}T-0.72\times10^5T^{-2};$$
$$\Delta H_{1130}(fusion)=6,750.$$

$$KF(l):$$
$$H_T-H_{298.15}=16.00T-310 \ (0.1 \ \text{percent};$$
$$1,130°-1,800° \ \text{K.});$$
$$C_p=16.00.$$

TABLE 614.—Heat content and entropy of KF(g)

[Base, ideal gas at 298.15° K.; mol. wt., 58.10]

T, ° K.	$H_T-H_{298.15}$, cal./mole	$S_T-S_{298.15}$, cal./deg. mole	T, ° K.	$H_T-H_{298.15}$, cal./mole	$S_T-S_{298.15}$, cal./deg. mole
400	870	2.51	1,000	6,160	10.57
500	1,740	4.45	1,200	7,940	12.19
600	2,620	6.05	1,400	9,725	13.57
700	3,500	7.41	1,600	11,510	14.76
800	4,385	8.59	1,800	13,295	15.81
900	5,275	9.64	2,000	15,080	16.75

$$KF(g):$$
$$H_T-H_{298.15}=8.92T+0.42\times10^5T^{-1}-2,800$$
$$(0.2 \ \text{percent}; \ 298°-2,000° \ \text{K.});$$
$$C_p=8.92-0.42\times10^5T^{-2}.$$

IODIDE

TABLE 615.—Heat content and entropy of KI(c)

[Base, crystals at 298.15° K.; mol. wt., 166.01]

T, ° K.	$H_T-H_{298.15}$, cal./mole	$S_T-S_{298.15}$, cal./deg. mole	T, ° K.	$H_T-H_{298.15}$, cal./mole	$S_T-S_{298.15}$, cal./deg. mole
400	1,290	3.72	800	6,800	13.22
500	2,630	6.71	900	8,280	14.96
600	3,990	9.19	950	9,030	15.77
700	5,370	11.31			

$$KI(c):$$
$$H_T-H_{298.15}=11.36T+2.00\times10^{-3}T^2-3,565$$
$$(0.4 \ \text{percent}; \ 298°-950° \ \text{K.});$$
$$C_p=11.36+4.00\times10^{-3}T.$$

TABLE 616.—Heat content and entropy of KI(g)

[Base, ideal gas at 298.15° K.; mol. wt., 166.01]

T, ° K.	$H_T-H_{298.15}$, cal./mole	$S_T-S_{298.15}$, cal./deg. mole	T, ° K.	$H_T-H_{298.15}$, cal./mole	$S_T-S_{298.15}$, cal./deg. mole
400	900	2.60	1,000	6,240	10.74
500	1,785	4.57	1,200	8,025	12.37
600	2,675	6.19	1,400	9,810	13.75
700	3,565	7.56	1,600	11,600	14.94
800	4,455	8.75	1,800	13,390	16.00
900	5,350	9.81	2,000	15,175	16.94

$$KI(g):$$
$$H_T-H_{298.15}=8.94T+0.14\times10^5T^{-1}-2,712$$
$$(0.1 \ \text{percent}; \ 298°-2,000° \ \text{K.});$$
$$C_p=8.94-0.14\times10^5T^{-2}.$$

HYDROSULFIDE

TABLE 617.—Heat content and entropy of KSH(c)

[Base, α-crystals at 298.15° K.; mol. wt., 72.17]

T, ° K.	$H_T-H_{298.15}$, cal./mole	$S_T-S_{298.15}$, cal./deg. mole	T, ° K.	$H_T-H_{298.15}$, cal./mole	$S_T-S_{298.15}$, cal./deg. mole
350	990	3.05	455(β)	3,720	9.73
400	2,050	5.88	500	4,730	11.84
455(α)	3,250	8.69			

$$KSH(\alpha):$$
$$H_T-H_{298.15}=7.65T+17.36\times10^{-3}T^2-3,824$$
$$(0.9 \ \text{percent}; \ 298°-455° \ \text{K.});$$
$$C_p=7.65+34.72\times10^{-3}T;$$
$$\Delta H_{455}(transition)=470.$$

$$KSH(\beta):$$
$$H_T-H_{298.15}=22.50T-6,518 \ (0.1 \ \text{percent};$$
$$455°-500° \ \text{K.});$$
$$C_p=22.50.$$

HYDROSELENIDE

TABLE 618.—Heat content and entropy of KSeH(c)

[Base, α-crystals at 298.15° K.; mol. wt., 119.07]

T, ° K.	$H_T-H_{298.15}$, cal./mole	$S_T-S_{298.15}$, cal./deg. mole	T, ° K.	$H_T-H_{298.15}$, cal./mole	$S_T-S_{298.15}$, cal./deg. mole
350	900	2.78	450(β)	3,260	8.61
400	1,880	5.40	500	4,380	10.97
450(α)	2,980	7.99			

$$KSeH(\alpha):$$
$$H_T-H_{298.15}=4.19T+20.65\times10^{-3}T^2-3,085$$
$$(0.1 \ \text{percent}; \ 298°-450° \ \text{K.});$$
$$C_p=4.19+41.30\times10^{-3}T;$$
$$\Delta H_{450}(transition)=280.$$
$$KSeH(\beta):$$
$$H_T-H_{298.15}=22.50T-6,865$$
$$(0.1 \ \text{percent}; \ 450°-500° \ \text{K.});$$
$$C_p=22.50.$$

TABLE 1h7. (*Continued*)

POTASSIUM-COPPER CHLORIDE

$K_2CuCl_4 \cdot 2H_2O(c)$:

$\overline{C}_p = 63.0$ (292°–323° K.).

POTASSIUM-MAGNESIUM CHLORIDE

TABLE 619.—*Heat content and entropy of* $KMgCl_3(c, l)$

[Base, crystals at 298.15° K.; mol. wt., 169.79]

T, ° K.	$H_T - H_{298.15}$, cal./mole	$S_T - S_{298.15}$, cal./deg. mole	T, ° K.	$H_T - H_{298.15}$, cal./mole	$S_T - S_{298.15}$, cal./deg. mole
400	3,270	9.43	760(l)	28,960	49.08
500	6,650	16.96	800	30,560	51.13
600	10,200	23.43	900	34,560	55.84
700	13,950	29.20	1,000	38,560	60.05
760(c)	16,270	32.38			

$KMgCl_3(c)$:

$H_T - H_{298.15} = 25.92T + 8.80 \times 10^{-3}T^2 - 8,510$

(0.1 percent; 298°–760° K.);

$C_p = 25.92 + 17.60 \times 10^{-3}T$;

$\Delta H_{760}(fusion) = 12,690$.

$KMgCl_3(l)$:

$H_T - H_{298.15} = 40.00T - 1,440$ (0.1 percent;

760°–1,000° K.);

$C_p = 40.00$.

POTASSIUM-PLATINUM CHLORIDE

$K_2PtCl_6(c)$:

$C_p = 49.26$ (298° K.).

POTASSIUM-TIN CHLORIDE

$K_2SnCl_6(c)$:

$\overline{C}_p = 54.5$ (292°–323° K.).

POTASSIUM-ZINC CHLORIDE

$K_2ZnCl_4(c)$:

$\overline{C}_p = 43.4$ (286°–319° K.).

POTASSIUM-HYDROGEN FLUORIDE

TABLE 620.—*Heat content and entropy of* $KHF_2(c, l)$

Base, α-crystals at 298.15° K.; mol. wt., 78.11]

T, ° K.	$H_T - H_{298.15}$, cal./mole	$S_T - S_{298.15}$, cal./deg. mole	T, ° K.	$H_T - H_{298.15}$, cal./mole	$S_T - S_{298.15}$, cal./deg. mole
350	975	3.01	500	6,860	16.30
400	1,980	5.69	512(β)	7,150	16.88
450	3,040	8.19	512(l)	8,720	19.94
469(α)	3,460	9.10	550	9,670	21.73
469(β)	6,120	14.78			

$KHF_2(\alpha)$:

$H_T - H_{298.15} = 11.76T + 11.07 \times 10^{-3}T^2 - 4,490$

(0.4 percent; 298°–469° K.);

$C_p = 11.76 + 22.14 \times 10^{-3}T$;

$\Delta H_{469}(transition) = 2,660$.

$KHF_2(\beta)$:

$H_T - H_{298.15} = 23.96T - 5,120$ (0.1 percent;

469°–512° K.);

$C_p = 23.96$;

$\Delta H_{512}(fusion) = 1,570$.

$KHF_2(l)$:

$H_T - H_{298.15} = 25.00T - 4,081$ (0.1 percent;

512°–550° K.);

$C_p = 25.00$.

ARSENATES

$KAsO_3(c)$:

$\overline{C}_p = 25.3$ (290°–372°).

$KH_2AsO_4(c)$:

$C_p = 30.29$ (298° K.).

BORATES

$KBO_2(c)$:

$C_p = 12.60 + 12.60 \times 10^{-3}T(estimated)$ (273°–1,220° K.).

$K_2B_4O_7(c)$:

$\overline{C}_p = 51.3$ (290°–372° K.).

BOROHYDRIDE

TABLE 621.—*Heat content and entropy of* $KBH_4(c)$

[Base, crystals at 298.15° K.; mol. wt., 53.95]

T, ° K.	$H_T - H_{298.15}$, cal./mole	$S_T - S_{298.15}$, cal./deg. mole	T, ° K.	$H_T - H_{298.15}$, cal./mole	$S_T - S_{298.15}$, cal./deg. mole
400	2,420	6.98	600	7,325	16.90
500	4,845	12.39	700	9,955	20.96

$KBH_4(c)$:

$H_T - H_{298.15} = 20.57T + 4.21 \times 10^{-3}T^2 - 6,507$

(0.5 percent; 298°–700° K.);

$C_p = 20.57 + 8.42 \times 10^{-3}T$.

BROMATE

$KBrO_3(c)$:

$C_p = 25.07$ (298° K.).

CARBONATE

$\overline{C}_p = 29.9$ (295°–373° K.).

CHLORATE

$KClO_3(c)$:

$C_p = 23.96$ (298° K.).

PERCHLORATE

$KClO_4(c)$:

$C_p = 26.33$ (298° K.).

TABLE 1h7. (Continued)

CHROMATE

$K_2CrO_4(c)$:

$C_p = 34.90$ (298° K.).

DICHROMATE

TABLE 622.—*Heat content and entropy of* $K_2Cr_2O_7(c, l)$

[Base, crystals at 298.15° K.; mol. wt., 294.22]

T, ° K.	$H_T - H_{298.15}$, cal./mole	$S_T - S_{298.15}$, cal./deg. mole	T, ° K.	$H_T - H_{298.15}$, cal./mole	$S_T - S_{298.15}$, cal./deg. mole
400	5,700	16.41	671(l)	32,080	62.86
500	11,800	30.00	700	34,900	66.98
600	18,500	42.20	800	44,630	79.93
671(c)	23,570	50.18			

$K_2Cr_2O_7(c)$:

$H_T - H_{298.15} = 36.66T + 27.40 \times 10^{-3}T^2 - 13,366$
(0.2 percent; 298°–671° K.);
$C_p = 36.66 + 54.80 \times 10^{-3}T$;
$\Delta H_{671}(fusion) = 8,510$.

$K_2Cr_2O_7(l)$:

$H_T - H_{298.15} = 97.30T - 33,210$ (0.1 percent;
671°–800° K.);
$C_p = 97.30$.

IODATE

$KIO_3(c)$:

$C_p = 25.42$ (298° K.).

PERMANGANATE

$KMnO_4(c)$:

$C_p = 28.10$ (298° K.).

NITRATE

TABLE 623.—*Heat content and entropy of* $KNO_3(c, l)$

[Base, α-crystals at 298.15° K.; mol. wt., 101.11]

T, ° K.	$H_T - H_{298.15}$, cal./mole	$S_T - S_{298.15}$, cal./deg. mole	T, ° K.	$H_T - H_{298.15}$, cal./mole	$S_T - S_{298.15}$, cal./deg. mole
350	1,230	3.80	600	9,660	22.36
400	2,490	7.17	611(β)	9,970	22.87
401(α)	2,520	7.24	611(l)	12,770	27.45
401(β)	3,920	10.73	700	15,400	31.46
500	6,780	17.11			

$KNO_3(\alpha)$:

$H_T - H_{298.15} = 14.55T + 14.20 \times 10^{-3}T^2 - 5,600$
(0.1 percent; 298°–401° K.);
$C_p = 14.55 + 28.40 \times 10^{-3}T$;
$\Delta H_{401}(transition) = 1,400$.

$KNO_3(\beta)$:

$H_T - H_{298.15} = 28.80T - 7,625$ (0.1 percent;
401°–611° K.);
$C_p = 28.80$;
$\Delta H_{611}(fusion) = 2,800$.

$KNO_3(l)$:

$H_T - H_{298.15} = 29.50T - 5,250$ (0.1 percent;
611°–700° K.);
$C_p = 29.50$.

PHOSPHATES

$K_4P_2O_7(c)$:

$\overline{C_p} = 63.1$ (290°–371° K.).

$KH_2PO_4(c)$:

$C_p = 27.86$ (298° K.).

PERRHENATE

$KReO_4(c)$:

$C_p = 29.30$ (298° K.).

SULFATE

TABLE 624.—*Heat content and entropy of* $K_2SO_4(c, l)$

[Base, α-crystals at 298.15° K.; 174.27]

T, ° K.	$H_T - H_{298.15}$, cal./mole	$S_T - S_{298.15}$, cal./deg. mole	T, ° K.	$H_T - H_{298.15}$, cal./mole	$S_T - S_{298.15}$, cal./deg. mole
400	3,410	9.82	1,100	36,320	55.77
500	7,150	18.15	1,200	41,220	60.04
600	11,120	25.38	1,300	46,260	64.06
700	15,390	31.95	1,342(β)	48,410	65.70
800	20,100	38.24	1,342(l)	57,470	72.45
856(α)	22,780	41.48	1,400	60,240	74.46
856(β)	24,920	43.98	1,500	65,020	77.77
900	26,920	46.25	1,600	69,800	80.85
1,000	31,600	51.18	1,700	74,580	83.74

$K_2SO_4(\alpha)$:

$H_T - H_{298.15} = 28.77T + 11.90 \times 10^{-3}T^2 + 4.26 \times 10^5 T^{-1}$
$- 11,064$ (0.4 percent; 298°–856° K.);
$C_p = 28.77 + 23.80 \times 10^{-3}T - 4.26 \times 10^5 T^{-2}$;
$\Delta H_{856}(transition) = 2,140$.

$K_2SO_4(\beta)$:

$H_T - H_{298.15} = 33.60T + 6.70 \times 10^{-3}T^2 - 8,747$
(0.1 percent; 856°–1,342° K.);
$C_p = 33.60 + 13.40 \times 10^{-3}T$;
$\Delta H_{1342}(fusion) = 9,060$.

$K_2SO_4(l)$:

$H_T - H_{298.15} = 47.80T - 6,680$ (0.1 percent;
1,342°–1,700° K.);
$C_p = 47.80$.

BISULFATE

$\overline{C_p} = 33.2$ (292°–324° K.).

THIOSULFATE

$K_2S_2O_3(c)$:

$\overline{C_p} = 37.5$ (293°–373° K.).

POTASSIUM-ALUMINUM SILICATES

$KAlSiO_4(c)$ (*kaliophilite*):

$C_p = 28.63$ (298° K.).

$KAlSi_2O_6(c)$ (*leucite*):

$C_p = 39.23$ (298° K.).

TABLE 1h7. (*Continued*)

TABLE 625.—*Heat content and entropy of* $KAlSi_3O_8(c)$ *(microcline and orthoclase)*

[Base, crystals at 298.15° K.; mol. wt., 278.35]

T, ° K.	$H_T-H_{298.15}$, cal./mole	$S_T-S_{298.15}$, cal./deg. mole	T, ° K.	$H_T-H_{298.15}$, cal./mole	$S_T-S_{298.15}$, cal./deg. mole
400	5,500	15.81	1,000	46,900	77.72
500	11,550	29.29	1,100	54,500	84.97
600	17,950	40.95	1,200	62,200	91.67
700	24,800	51.50	1,300	70,000	97.90
800	32,000	61.11	1,400	77,900	103.76
900	39,400	69.82			

$KAlSi_3O_8(c)$ *(microcline and orthoclase)*:
$$H_T-H_{298.15}=63.83T+6.45\times10^{-3}T^2+17.05\times10^5T^{-1}$$
$$-25,323\ (0.4\ \text{percent};\ 298°-1,400°\ \text{K.});$$
$$C_p=63.83+12.90\times10^{-3}T-17.05\times10^5T^{-2}.$$

TABLE 626.—*Heat content and entropy of* $KAlSi_3O_8(gl)$

(Base, glass at 298.15° K.; mol. wt., 278.35]

T, ° K.	$H_T-H_{298.15}$, cal./mole	$S_T-S_{298.15}$, cal./deg. mole	T, ° K.	$H_T-H_{298.15}$, cal./mole	$S_T-S_{298.15}$, cal./deg. mole
400	5,700	16.40	1,000	48,050	79.81
500	11,950	30.32	1,100	55,950	87.34
600	18,650	42.53	1,200	63,950	94.30
700	25,550	53.16	1,300	72,150	100.86
800	32,750	62.77	1,400	80,550	107.08
900	40,250	71.59			

$KAlSi_3O_8(gl)$:
$$H_T-H_{298.15}=61.96T+8.58\times10^{-3}T^2+14.29\times10^5T^{-1}$$
$$-24,029\ (0.2\ \text{percent};\ 298°-1,400°\ \text{K.});$$
$$C_p=61.96+17.16\times10^{-3}T-14.29\times10^5T^{-2}.$$

FLUORPHLOGOPITE MICA

TABLE 627.—*Heat content and entropy of* $KMg_3AlSi_3O_{10}F_2(c, l)$

[Base, crystals at 298.15° K., mol. wt., 421.31]

T, ° K.	$H_T-H_{298.15}$, cal./mole	$S_T-S_{298.15}$, cal./deg. mole	T, ° K.	$H_T-H_{298.15}$, cal./mole	$S_T-S_{298.15}$, cal./deg. mole
400	8,930	25.68	1,300	109,240	154.13
500	18,680	47.41	1,400	121,510	163.22
600	29,030	66.27	1,500	133,980	171.82
700	39,810	82.88	1,600	146,650	180.00
800	50,870	97.64	1,670(c)	155,620	185.49
900	62,150	110.93	1,670(l)	229,420	229.68
1,000	73,630	123.02	1,700	234,350	232.60
1,100	85,300	134.14	1,800	250,740	241.97
1,200	97,170	144.47			

$KMg_3AlSi_3O_{10}F_2(c)$:
$$H_T-H_{298.15}=100.86T+8.58\times10^{-3}T^2+21.46\times10^5T^{-1}$$
$$-38,032\ (0.3\ \text{percent};\ 298°-1,670°\ \text{K.});$$
$$C_p=100.86+17.16\times10^{-3}T-21.46\times10^5T^{-2};$$
$$\Delta H_{1670}(fusion)=73,800.$$

$KMg_3AlSi_3O_{10}F_2(l)$:
$$H_T-H_{298.15}=164.00T-44,460\ (0.1\ \text{percent};$$
$$1,670°-1,800°\ \text{K.});$$
$$C_p=164.00.$$

POTASSIUM-ALUMINUM SULFATE

TABLE 628.—*Heat content and entropy of* $KAl(SO_4)_2(c)$

[Base, crystals at 298.15° K.; mol. wt., 258.21]

T, ° K.	$H_T-H_{298.15}$, cal./mole	$S_T-S_{298.15}$, cal./deg. mole	T, ° K.	$H_T-H_{298.15}$, cal./mole	$S_T-S_{298.15}$, cal./deg. mole
400	5,270	15.15	800	30,570	58.40
500	10,990	27.89	900	37,640	66.72
600	17,200	39.21	1,000	44,950	74.42
700	23,750	49.29	1,100	52,420	81.54

$KAl(SO_4)_2(c)$:
$$H_T-H_{298.15}=55.96T+9.84\times10^{-3}T^2+13.96\times10^5T^{-1}$$
$$-22,241\ (0.4\ \text{percent};\ 298°-1,100°\ \text{K.});$$
$$C_p=55.96+19.68\times10^{-3}T-13.96\times10^5T^{-2}.$$
$$KAl(SO_4)_2\cdot12H_2O(c):$$
$$C_p=155.6\ (298°\ \text{K.}).$$

ALUNITE

TABLE 629.—*Heat content and entropy of* $K_2O\cdot3Al_2O_3\cdot4SO_3\cdot6H_2O(c)$ *(natural)*

[Base, crystals at 298.15° K.; mol. wt., 828.44]

T, ° K.	$H_T-H_{298.15}$, cal./mole	$S_T-S_{298.15}$, cal./deg. mole	T, ° K.	$H_T-H_{298.15}$, cal./mole	$S_T-S_{298.15}$, cal./deg. mole
350	10,000	30.92	550	56,050	134.17
400	20,550	59.07	600	68,900	156.53
450	31,850	85.65	650	82,050	177.55
500	43,750	110.63	700	95,400	197.34

$K_2O\cdot3Al_2O_3\cdot4SO_3\cdot6H_2O(c)$ *(natural)*:
$$H_T-H_{298.15}=231.00T+39.30\times10^{-3}T^2+67.83\times10^5T^{-1}$$
$$-95,116\ (0.2\ \text{percent};\ 298°-700°\ \text{K.});$$
$$C_p=231.00+78.60\times10^{-3}T-67.83\times10^5T^{-2}.$$

TABLE 630.—*Heat content and entropy of* $K_2O\cdot3Al_2O_3\cdot4SO_3\cdot6H_2O(c)$ *(synthetic)*

[Base, crystals at 298.15° K.; mol. wt., 828.44]

T, ° K.	$H_T-H_{298.15}$, cal./mole	$S_T-S_{298.15}$, cal./deg. mole	T, ° K.	$H_T-H_{298.15}$, cal./mole	$S_T-S_{298.15}$, cal./deg. mole
350	10,300	31.84	550	60,700	145.06
400	22,050	63.19	600	74,100	168.39
450	34,600	92.70	650	87,600	189.95
500	47,500	119.90			

$K_2O\cdot3Al_2O_3\cdot4SO_3\cdot6H_2O(c)$ *(synthetic)*:
$$H_T-H_{298.15}=306.90T+109.90\times10^5T^{-1}-128,363$$
$$(0.8\ \text{percent};\ 298°-650°\ \text{K.});$$
$$C_p=306.90-109.90\times10^5T^{-2}.$$

POTASSIUM BASIC ALUM

TABLE 631.—*Heat content and entropy of* $K_2O\cdot3Al_2O_3\cdot5SO_3\cdot9H_2O(c)$

[Base, crystals at 298.15° K.; mol. wt., 962.55]

T, ° K.	$H_T-H_{298.15}$, cal./mole	$S_T-S_{298.15}$, cal./deg. mole	T, ° K.	$H_T-H_{298.15}$, cal./mole	$S_T-S_{298.15}$, cal./deg. mole
350	13,200	40.83	450	41,150	110.90
400	26,800	77.13	500	56,100	142.41

TABLE 1h7. (*Continued*)

$K_2O \cdot 3Al_2O_3 \cdot 5SO_3 \cdot 9H_2O(c)$:

$H_T - H_{298.15} = 176.90T + 125.80 \times 10^{-3}T^2 - 63,925$

(0.6 percent; 298°–500° K.);

$C_p = 176.90 + 251.60 \times 10^{-3}T$.

POTASSIUM-CHROMIUM SULFATE

$KCr(SO_4)_2 \cdot 12H_2O(c)$:

$\overline{C_p} = 161.8$ (292°–324° K.).

POTASSIUM-MAGNESIUM SULFATE

$K_2Mg(SO_4)_2 \cdot 6H_2O(c)$:

$\overline{C_p} = 106$ (292°–323° K.).

POTASSIUM-NICKEL SULFATE

$K_2Ni(SO_4)_2 \cdot 6H_2O(c)$:

$\overline{C_p} = 107$ (298°–319° K.).

POTASSIUM-ZINC SULFATE

$K_2Zn(SO_4)_2 \cdot 6H_2O(c)$:

$\overline{C_p} = 120$ (293°–317° K.).

FERROCYANIDE

$K_4Fe(CN)_6(c)$:

$\overline{C_p} = 80.1$ (273°–319°).

$K_4Fe(CN)_6 \cdot 3H_2O(c)$:

$\overline{C_p} = 115.5$ (273°–310° K.).

FERRICYANIDE

$K_3Fe(CN)_6(c)$:

$C_p = 75.65$ (298° K.).

COBALTICYANIDE

$K_3Co(CN)_6(c)$:

$C_p = 73.65$ (298° K.).

POTASSIUM-ZINC CYANIDE

$K_2Zn(CN)_4(c)$:

$\overline{C_p} = 57.5$ (287°–319° K.).

PRASEODYMIUM AND ITS COMPOUNDS

ELEMENT

TABLE 632.—*Heat content and entropy of* $Pr(c, l)$

[Base, α-crystals at 298.15° K.; atomic wt., 140.92]

T, ° K.	$H_T - H_{298.15}$, cal./mole	$S_T - S_{298.15}$, cal./deg. mole	T, ° K.	$H_T - H_{298.15}$, cal./mole	$S_T - S_{298.15}$, cal./deg. mole
400	670	1.93	1,208(l)	10,160	13.08
500	1,370	3.49	1,300	10,900	13.67
600	2,090	4.80	1,400	11,700	14.26
700	2,850	5.97	1,600	13,300	15.33
800	3,640	7.03	1,800	14,900	16.27
900	4,460	7.99	2,000	16,500	17.12
1,000	5,320	8.90	2,200	18,100	17.88
1,071(α)	5,950	9.51	2,400	19,700	18.58
1,071(β)	6,270	9.81	2,600	21,300	19.22
1,100	6,500	10.02	2,800	22,900	19.81
1,200	7,300	10.72	3,000	24,500	20.36
1,208(β)	7,360	10.77			

$Pr(\alpha)$:

$H_T - H_{298.15} = 5.50T + 1.60 \times 10^{-3}T^2 - 1,782$

(0.2 percent; 298°–1,071° K.);

$C_p = 5.50 + 3.20 \times 10^{-3}T$;

$\Delta H_{1071}(transition) = 320$.

$Pr(\beta)$:

$H_T - H_{298.15} = 8.00T - 2,300$ (0.1 percent;

1,071°–1,208° K.);

$C_p = 8.00$;

$\Delta H_{1208}(fusion) = 2,800$.

$Pr(l)$:

$H_T - H_{298.15} = 8.00T + 500$ (0.1 percent;

1,208°–3,000° K.);

$C_p = 8.00$.

OXIDE

TABLE 633.—*Heat content and entropy of* $Pr_6O_{11}(c)$

[Base, crystals at 298.15° K.; mol. wt., 1,021.52]

T, ° K.	$H_T - H_{298.15}$, cal./mole	$S_T - S_{298.15}$, cal./deg. mole	T, ° K.	$H_T - H_{298.15}$, cal./mole	$S_T - S_{298.15}$, cal./deg. mole
400	9,840	28.34	900	64,700	116.37
500	20,080	51.17	1,000	76,610	128.92
600	30,740	70.60	1,100	88,810	140.54
700	41,750	87.56	1,200	101,280	151.39
800	53,080	102.68			

$Pr_6O_{11}(c)$:

$H_T - H_{298.15} = 95.29T + 13.09 \times 10^{-3}T^2 + 9.31 \times 10^5 T^{-1}$

$- 32,697$ (0.1 percent; 298°–1,200° K.);

$C_p = 95.29 + 26.18 \times 10^{-3}T - 9.31 \times 10^5 T^{-2}$.

PROMETHIUM

ELEMENT

TABLE 634.—*Heat content and entropy of* $Pm(c, l)$

[Base, crystals at 298.15° K.; atomic wt., 145]

T, ° K.	$H_T - H_{298.15}$, cal./mole	$S_T - S_{298.15}$, cal./deg. mole	T, ° K.	$H_T - H_{298.15}$, cal./mole	$S_T - S_{298.15}$, cal./deg. mole
400	670	1.94	1,300(l)	10,760	13.28
500	1,360	3.48	1,400	11,560	13.87
600	2,070	4.77	1,600	13,160	14.94
700	2,810	5.91	1,800	14,760	15.88
800	3,570	6.93	2,000	16,360	16.73
900	4,360	7.86	2,200	17,960	17.49
1,000	5,170	8.71	2,400	19,560	18.19
1,100	6,010	9.51	2,600	21,160	18.83
1,200	6,870	10.26	2,800	22,760	19.42
1,300(c)	7,760	10.97	3,000	24,360	19.99

$Pm(c)$:

$H_T - H_{298.15} = 5.76T + 1.24 \times 10^{-3}T^2 - 1,828$

(0.2 percent; 298°–1,300° K.);

$C_p = 5.76 + 2.48 \times 10^{-3}T$;

$\Delta H_{1300}(fusion) = 3,000$.

$Pm(l)$:

$H_T - H_{298.15} = 8.00T + 360$ (0.1 percent;

1,300°–3,000° K.);

$C_p = 8.00$.

TABLE 1h7. (*Continued*)

PROTACTINIUM

ELEMENT

TABLE 635.—*Heat content and entropy of Pa(c, l)*

[Base, crystals at 298.15° K.; atomic wt., 231]

T, ° K.	$H_T - H_{298.15}$, cal./mole	$S_T - S_{298.15}$, cal./deg. mole	T, ° K.	$H_T - H_{298.15}$, cal./mole	$S_T - S_{298.15}$, cal./deg. mole
400	710	2.04	1,500(c)	10,330	13.13
500	1,430	3.65	1,500(l)	13,830	15.47
600	2,190	5.03	1,600	14,830	16.11
700	2,970	6.24	1,800	16,830	17.29
800	3,790	7.33	2,000	18,830	18.34
900	4,630	8.32	2,200	20,830	19.30
1,000	5,510	9.24	2,400	22,830	20.17
1,100	6,410	10.11	2,600	24,830	20.97
1,200	7,350	10.92	2,800	26,830	21.71
1,300	8,310	11.69	3,000	28,830	22.40
1,400	9,310	12.43			

Pa(c):

$$H_T - H_{298.15} = 5.90T + 1.50 \times 10^{-3}T^2 - 1,892$$
$$(0.1 \text{ percent}; \ 298° - 1,500° \text{ K.});$$
$$C_p = 5.90 + 3.00 \times 10^{-3}T:$$
$$\Delta H_{1500}(fusion) = 3,500.$$

Pa(l):

$$H_T - H_{298.15} = 10.00T - 1,170 \ (0.1 \text{ percent};$$
$$1,500° - 3,000° \text{ K.});$$
$$C_p = 10.00.$$

RADIUM

ELEMENT

TABLE 636.—*Heat content and entropy of Ra(c, l)*

[Base, crystals at 298.15° K.; atomic wt., 226.05]

T, ° K.	$H_T - H_{298.15}$, cal./mole	$S_T - S_{298.15}$, cal./deg. mole	T, ° K.	$H_T - H_{298.15}$, cal./mole	$S_T - S_{298.15}$, cal./deg. mole
400	690	1.98	973(l)	7,770	11.60
500	1,410	3.59	1,000	7,970	11.80
600	2,190	5.01	1,200	9,470	13.17
700	3,010	6.28	1,400	10,970	14.32
800	3,890	7.44	1,600	12,470	15.32
900	4,810	8.53	1,800	13,970	16.20
973(c)	5,520	9.29			

Ra(c):

$$H_T - H_{298.15} = 5.00T + 2.50 \times 10^{-3}T^2 - 1,713$$
$$(0.2 \text{ percent}; \ 298° - 973° \text{ K.});$$
$$C_p = 5.00 + 5.00 \times 10^{-3}T;$$
$$\Delta H_{973}(fusion) = 2,250.$$

Ra(l):

$$H_T - H_{298.15} = 7.50T + 470 \ (0.1 \text{ percent},$$
$$973° - 1,800° \text{ K.}).$$
$$C_p = 7.50.$$

TABLE 637.—*Heat content and entropy of Ra(g)*

[Base, ideal gas at 298.15° K.; atomic wt., 226.05]

T, ° K.	$H_T - H_{298.15}$, cal./mole	$S_T - S_{298.15}$, cal./deg. mole	T, ° K.	$H_T - H_{298.15}$, cal./mole	$S_T - S_{298.15}$, cal./deg. mole
400	505	1.46	1,500	5,975	8.03
500	1,005	2.57	1,600	6,475	8.36
600	1,500	3.48	1,700	6,980	8.66
700	1,995	4.24	1,800	7,480	8.95
800	2,495	4.90	1,900	7,990	9.22
900	2,990	5.49	2,000	8,500	9.49
1,000	3,485	6.01	2,200	9,550	9.99
1,100	3,985	6.49	2,400	10,650	10.46
1,200	4,480	6.92	2,600	11,820	10.93
1,300	4,980	7.32	2,800	13,070	11.40
1,400	5,475	7.69	3,000	14,445	11.87

Ra(g):

$$H_T - H_{298.15} = 4.97T - 1,482 \ (0.2 \text{ percent}; \ 298° - 2,000°$$
$$\text{K.});$$
$$C_p = 4.97.$$

RADON

ELEMENT

TABLE 638.—*Heat content and entropy of Rn(g)*

[Base, ideal gas at 298.15° K.; atomic wt., 222]

T, ° K.	$H_T - H_{298.15}$, cal./mole	$S_T - S_{298.15}$, cal./deg. mole	T, ° K.	$H_T - H_{298.15}$, cal./mole	$S_T - S_{298.15}$, cal./deg. mole
400	505	1.46	1,600	6,470	8.35
500	1,005	2.57	1,800	7,460	8.94
600	1,500	3.48	2,000	8,455	9.46
700	1,995	4.24	2,200	9,450	9.93
800	2,495	4.90	2,400	10,440	10.36
900	2,990	5.49	2,600	11,435	10.76
1,000	3,485	6.01	2,800	12,430	11.13
1,200	4,480	6.92	3,000	13,420	11.47
1,400	5,475	7.69			

Rn(g):

$$H_T - H_{298.15} = 4.97T - 1,482 \ (0.1 \text{ percent}; \ 298° - 3,000°$$
$$\text{K.});$$
$$C_p = 4.97.$$

RHENIUM AND ITS COMPOUNDS

ELEMENT

TABLE 639.—*Heat content and entropy of Re(c)*

[Base, crystals at 298.15° K.; atomic wt., 186.22]

T, ° K.	$H_T - H_{298.15}$, cal./mole	$S_T - S_{298.15}$, cal./deg. mole	T, ° K.	$H_T - H_{298.15}$, cal./mole	$S_T - S_{298.15}$, cal./deg. mole
400	620	1.79	1,500	8,220	10.72
500	1,240	3.17	1,600	8,990	11.22
600	1,890	4.36	1,700	9,770	11.69
700	2,550	5.37	1,800	10,560	12.14
800	3,210	6.25	1,900	11,370	12.58
900	3,880	7.04	2,000	12,180	12.99
1,000	4,570	7.77	2,200	13,850	13.79
1,100	5,270	8.44	2,400	15,580	14.54
1,200	5,980	9.05	2,600	17,360	15.25
1,300	6,710	9.64	2,800	19,200	15.93
1,400	7,460	10.19	3,000	21,080	16.58

TABLE 1h7. (*Continued*)

Re(*c*):

$$H_T - H_{298.15} = 5.66\,T + 0.65 \times 10^{-3}\,T^2 - 1,745 \text{ (0.2 percent;}$$
$$298°-3,000° \text{ K.);}$$
$$C_p = 5.66 + 1.30 \times 10^{-3}\,T.$$

TABLE 640.—*Heat content and entropy of Re(g)*

[Base, ideal gas at 298.15° K.; atomic wt., 186.22]

T, ° K.	$H_T - H_{298.15}$, cal./mole	$S_T - S_{298.15}$, cal./deg. mole	T, ° K.	$H_T - H_{298.15}$, cal./mole	$S_T - S_{298.15}$, cal./deg. mole
400	505	1.46	1,500	5,970	8.03
500	1,005	2.57	1,600	6,470	8.35
600	1,500	3.48	1,700	6,970	8.65
700	1,995	4.24	1,800	7,470	8.94
800	2,495	4.90	1,900	7,975	9.21
900	2,990	5.49	2,000	8,485	9.47
1,000	3,485	6.01	2,200	9,515	9.96
1,100	3,985	6.49	2,400	10,575	10.43
1,200	4,480	6.92	2,600	11,675	10.87
1,300	4,975	7.32	2,800	12,835	11.30
1,400	5,475	7.69	3,000	14,065	11.72

Re(*g*):

$$H_T - H_{298.15} = 4.97\,T - 1,482 \text{ (0.1 percent;}$$
$$298°-2,000° \text{ K.);}$$
$$C_p = 4.97.$$

$$H_T - H_{298.15} = 2.58\,T + 0.60 \times 10^{-3}\,T^2 + 925$$
$$\text{(0.1 percent; } 2,000°-3,000° \text{ K.);}$$
$$C_p = 2.58 + 1.20 \times 10^{-3}\,T.$$

OXIDE

 Re$_2$O$_7$(*c*):

$$C_p = 39.73 \text{ (298° K.).}$$

FLUORIDE

TABLE 641.—*Heat content and entropy of ReF₆(g)*

[Base, ideal gas at 298.15° K.; mol. wt., 300.22]

T, ° K.	$H_T - H_{298.15}$, cal./mole	$S_T - S_{298.15}$, cal./deg. mole	T, ° K.	$H_T - H_{298.15}$, cal./mole	$S_T - S_{298.15}$, cal./deg. mole
350	1,530	4.72	450	4,700	12.67
400	3,060	8.81	500	6,410	16.27

ReF₆(*g*):

$$H_T - H_{298.15} = 19.14\,T + 15.80 \times 10^{-3}\,T^2 - 7,110$$
$$\text{(0.3 percent; } 298°-500° \text{ K.);}$$
$$C_p = 19.14 + 31.60 \times 10^{-3}\,T.$$

RHODIUM AND ITS COMPOUNDS

ELEMENT

TABLE 642.—*Heat content and entropy of Rh(c, l)*

[Base, crystals at 298.15° K.; mol. wt., 102.91]

T, ° K.	$H_T - H_{298.15}$, cal./mole	$S_T - S_{298.15}$, cal./deg. mole	T, ° K.	$H_T - H_{298.15}$, cal./mole	$S_T - S_{298.15}$, cal./deg. mole
400	630	1.81	1,700	10,600	12.44
500	1,260	3.22	1,800	11,500	12.95
600	1,920	4.42	1,900	12,420	13.45
700	2,600	5.47	2,000	13,370	13.94
800	3,300	6.40	2,100	14,350	14.41
900	4,030	7.26	2,200	15,340	14.87
1,000	4,790	8.06	2,239 (*c*)	15,730	15.05
1,100	5,570	8.80	2,239 (*l*)	20,930	17.37
1,200	6,370	9.50	2,400	22,540	18.07
1,300	7,180	10.15	2,600	24,540	18.87
1,400	8,010	10.76	2,800	26,540	19.61
1,500	8,860	11.35	3,000	28,540	20.30
1,600	9,720	11.90			

Rh(*c*):

$$H_T - H_{298.15} = 5.49\,T + 1.03 \times 10^{-3}\,T^2 - 1,727$$
$$\text{(0.4 percent; } 298°-2,239° \text{ K.);}$$
$$C_p = 5.49 + 2.06 \times 10^{-3}\,T;$$
$$\Delta H_{2239} = 5,200.$$

Rh(*l*):

$$H_T - H_{298.15} = 10.00\,T - 1,460 \text{ (0.1 percent;}$$
$$2,239°-3,000° \text{ K.);}$$
$$C_p = 10.00.$$

TABLE 643.—*Heat content and entropy of Rh(g)*

[Base, ideal gas at 298.15° K.; mol. wt., 102.91]

T, ° K.	$H_T - H_{298.15}$, cal./mole	$S_T - S_{298.15}$, cal./deg. mole	T, ° K.	$H_T - H_{298.15}$, cal./mole	$S_T - S_{298.15}$, cal./deg. mole
400	520	1.50	1,900	9,915	10.94
500	1,045	2.67	2,000	10,580	11.28
600	1,595	3.68	2,200	11,905	11.91
700	2,170	4.56	2,400	13,230	12.49
800	2,765	5.35	2,600	14,550	13.02
900	3,375	6.07	2,800	15,875	13.51
1,000	4,005	6.73	3,000	17,205	13.96
1,100	4,640	7.34	3,500	20,540	14.99
1,200	5,290	7.91	4,000	23,900	15.89
1,300	5,940	8.43	4,500	27,305	16.69
1,400	6,600	8.91	5,000	30,745	17.42
1,500	7,260	9.37	6,000	37,760	18.70
1,600	7,925	9.80	7,000	45,000	19.81
1,700	8,585	10.20	8,000	52,530	20.82
1,800	9,250	10.58			

Rh(*g*):

$$H_T - H_{298.15} = 6.31\,T + 0.06 \times 10^{-3}\,T^2 - 2,260$$
$$\text{(0.2 percent; } 1,800°-7,000° \text{ K.);}$$
$$C_p = 6.31 + 0.12 \times 10^{-3}\,T.$$

OXIDES

Rh₂O(*c*):

$$C_p = 15.59 + 6.47 \times 10^{-3}\,T \text{ (273°-1,273° K.).}$$

RhO(*c*):

$$C_p = 9.84 + 5.53 \times 10^{-3}\,T \text{ (273°-1,273° K.).}$$

Rh₂O₃(*c*):

$$C_p = 20.73 + 13.80 \times 10^{-3}\,T \text{ (273°-1,273° K.).}$$

RUBIDIUM AND ITS COMPOUNDS

ELEMENT

TABLE 644.—*Heat content and entropy of Rb(c, l)*

[Base, crystals at 298.15° K.; atomic wt., 85.48]

T, ° K.	$H_T - H_{298.15}$, cal./mole	$S_T - S_{298.15}$, cal./deg. mole	T, ° K.	$H_T - H_{298.15}$, cal./mole	$S_T - S_{298.15}$, cal./deg. mole
312(*c*)	105	0.34	700	3,575	8.20
312(*l*)	665	2.14	800	4,325	9.20
400	1,325	4.00	900	5,075	10.08
500	2,075	5.68	1,000	5,825	10.87
600	2,825	7.04			

Rb(*c*):

$$H_T - H_{298.15} = 7.58\,T - 2,260 \text{ (0.1 percent;}$$
$$298°-312° \text{ K.);}$$
$$C_p = 7.58;$$
$$\Delta H_{312} = 560.$$

TABLE 1h7. (Continued)

Rb(l):

$$H_T - H_{298.15} = 7.50T - 1,675 \text{ (0.1 percent;}$$
$$312°-1,000° \text{ K.)};$$
$$C_p = 7.50.$$

TABLE 645.—*Heat content and entropy of* Rb(g)

[Base, ideal gas at 298.15° K.; atomic wt., 85.48]

T, ° K.	$H_T-H_{298.15}$, cal./mole	$S_T-S_{298.15}$, cal./deg. mole	T, ° K.	$H_T-H_{298.15}$, cal./mole	$S_T-S_{298.15}$, cal./deg. mole
400	505	1.46	1,900	7,970	9.21
500	1,005	2.57	2,000	8,470	9.46
600	1,500	3.48	2,200	9,480	9.95
700	1,995	4.24	2,400	10,500	10.39
800	2,495	4.90	2,600	11,540	10.81
900	2,990	5.49	2,800	12,610	11.20
1,000	3,490	6.01	3,000	13,710	11.58
1,100	3,985	6.49	3,500	16,710	12.51
1,200	4,480	6.92	4,000	20,330	13.47
1,300	4,980	7.32	4,500	25,105	14.59
1,400	5,475	7.69	5,000	31,710	15.98
1,500	5,975	8.03	6,000	51,515	19.57
1,600	6,470	8.35	7,000	75,715	23.30
1,700	6,970	8.65	8,000	96,820	26.12
1,800	7,470	8.94			

Rb(g):

$$H_T - H_{298.15} = 4.97T - 1,482 \text{ (0.1 percent;}$$
$$298°-2,000° \text{ K.)};$$
$$C_p = 4.97.$$

TABLE 646.—*Heat content and entropy of* Rb$_2$(g)

[Base, ideal gas at 298.15° K.; mol. wt., 170.96]

T, ° K.	$H_T-H_{298.15}$, cal./mole	$S_T-S_{298.15}$, cal./deg. mole	T, ° K.	$H_T-H_{298.15}$, cal./mole	$S_T-S_{298.15}$, cal./deg. mole
400	925	2.67	1,000	6,470	11.12
500	1,840	4.71	1,100	7,415	12.02
600	2,755	6.38	1,200	8,360	12.84
700	3,680	7.80	1,300	9,310	13.60
800	4,605	9.04	1,400	10,260	14.31
900	5,535	10.13	1,500	11,220	14.97

Rb$_2$(g):

$$H_T - H_{298.15} = 8.92T + 0.23 \times 10^{-3}T^2 - 2,680$$
$$\text{(0.1 percent; } 298°-1,500° \text{ K.)};$$
$$C_p = 8.92 + 0.46 \times 10^{-3}T.$$

HYDRIDE

TABLE 647.—*Heat content and entropy of* RbH(g)

[Base, ideal gas at 298.15° K.; mol. wt., 86.49]

T, ° K.	$H_T-H_{298.15}$, cal./mole	$S_T-S_{298.15}$, cal./deg. mole	T, ° K.	$H_T-H_{298.15}$, cal./mole	$S_T-S_{298.15}$, cal./deg. mole
400	780	2.25	1,000	5,810	9.87
500	1,575	4.02	1,200	7,555	11.46
600	2,395	5.51	1,400	9,315	12.82
700	3,235	6.81	1,600	11,075	13.99
800	4,085	7.94	1,800	12,845	15.04
900	4,945	8.96	2,000	14,620	15.97

RbH(g):

$$H_T - H_{298.15} = 8.29T + 0.20 \times 10^{-3}T^2 + 0.86 \times 10^5 T^{-1}$$
$$-2,778 \text{ (0.4 percent; } 298°-2,000° \text{ K.)};$$
$$C_p = 8.29 + 0.40 \times 10^{-3}T - 0.86 \times 10^5 T^{-2}.$$

BROMIDE

RbBr(c)·

$$C_p = 11.89 + 2.22 \times 10^{-3}T \text{(estimated)} \text{ (298°-950° K.).}$$

CHLORIDE

RbCl(c):

$$C_p = 11.50 + 2.49 \times 10^{-3}T \text{ (estimated)} \text{ (298°-990° K.).}$$

TABLE 648.—*Heat content and entropy of* RbCl(g)

[Base, ideal gas at 298.15° K.; mol. wt., 120.94]

T, ° K.	$H_T-H_{298.15}$, cal./mole	$S_T-S_{298.15}$, cal./deg. mole	T, ° K.	$H_T-H_{298.15}$, cal./mole	$S_T-S_{298.15}$, cal./deg. mole
400	900	2.60	1,000	6,325	10.87
500	1,795	4.59	1,200	8,155	12.53
600	2,695	6.23	1,400	9,990	13.95
700	3,600	7.63	1,600	11,835	15.18
800	4,505	8.83	1,800	13,690	16.27
900	5,415	9.91	2,000	15,555	17.26

RbCl(g):

$$H_T - H_{298.15} = 8.97T + 0.09 \times 10^{-3}T^2 + 0.22 \times 10^5 T^{-1}$$
$$-2,756 \text{ (0.1 percent; } 298°-2,000° \text{ K.)};$$
$$C_p = 8.97 + 0.18 \times 10^{-3}T - 0.22 \times 10^5 T^{-2}.$$

FLUORIDE

RbF(c):

$$C_p = 11.33 + 2.55 \times 10^{-3}T \text{ (estimated)} \text{ (298°-1,048° K.).}$$

IODIDE

RbI(c):

$$C_p = 11.93 + 2.27 \times 10^{-3}T \text{ (estimated)} \text{ (298°-911° K.).}$$

HYDROSULFIDE

TABLE 649.—*Heat content and entropy of* RbSH(c)

[Base, α-crystals at 298.15° K.; mol. wt., 118.55]

T, ° K.	$H_T-H_{298.15}$, cal./mole	$S_T-S_{298.15}$, cal./deg. mole	T, ° K.	$H_T-H_{298.15}$, cal./mole	$S_T-S_{298.15}$, cal./deg. mole
350	980	3.03	403(β)	2,380	6.68
400	1,920	5.54	450	3,340	8.93
403(α)	1,980	5.69	500	4,360	11.08

RbSH(α):

$$H_T - H_{298.15} = 18.90T - 5,635 \text{ (0.2 percent;}$$
$$298°-403° \text{ K.)};$$
$$C_p = 18.90;$$
$$\Delta H_{403}(transition) = 400.$$

RbSH(β):

$$H_T - H_{298.15} = 20.40T - 5,841 \text{ (0.1 percent;}$$
$$403°-500° \text{ K.)};$$
$$C_p = 20.40.$$

TABLE 1h7. (*Continued*)

HYDROSELENIDE

Table 650.—*Heat content and entropy of RbSeH(c)*

[Base, α-crystals at 298.15° K.; mol. wt., 165.45]

T, ° K.	$H_T-H_{298.15}$, cal./mole	$S_T-S_{298.15}$, cal./deg. mole	T, ° K.	$H_T-H_{298.15}$, cal./mole	$S_T-S_{298.15}$, cal./deg. mole
350	850	2.63	420(β)	2,210	6.12
400	1,670	4.82	450	2,760	7.39
420(α)	2,000	5.62	500	3,670	9.31

RbSeH(α):

$$H_T-H_{298.15}=16.40\,T-4,890 \text{ (0.1 percent;}$$
$$298°-420° \text{ K.)};$$
$$C_p=16.40;$$
$$\Delta H_{420}(transition)=210.$$

RbSeH(β):

$$H_T-H_{298.15}=18.30\,T-5,476 \text{ (0.1 percent;}$$
$$420°-500° \text{ K.)};$$
$$C_p=18.30.$$

CARBONATE

$Rb_2CO_3(c)$:

$$\overline{C_p}=28.4 \ (291°-320° \text{ K.}).$$

CHLORATE

$RbClO_3(c)$:

$$C_p=24.66 \ (298° \text{ K.}).$$

RUTHENIUM

ELEMENT

Table 651.—*Heat content and entropy of Ru(c, l)*

[Base, α-crystals at 298.15° K.; atomic wt., 101.1]

T, ° K.	$H_T-H_{298.15}$, cal./mole	$S_T-S_{298.15}$, cal./deg. mole	T, ° K.	$H_T-H_{298.15}$, cal./mole	$S_T-S_{298.15}$, cal./deg. mole
400	590	1.70	1,700	9,400	11.21
500	1,180	3.02	1,773(γ)	9,930	11.51
600	1,780	4.11	1,773(δ)	10,250	11.69
700	2,400	5.06	1,800	10,450	11.80
800	3,040	5.92	1,900	11,200	12.21
900	3,690	6.68	2,000	11,950	12.59
1,000	4,360	7.39	2,100	12,700	12.96
1,100	5,050	8.05	2,200	13,450	13.31
1,200	5,750	8.66	2,300	14,200	13.64
1,300	6,460	9.23	2,400	14,950	13.96
1,308(α)	6,520	9.27	2,500	15,700	14.27
1,308(β)	6,580	9.32	2,600	16,450	14.56
1,400	7,240	9.81	2,700(δ)	17,200	14.84
1,473(β)	7,770	10.18	2,700(l)	23,400	17.14
1,473(γ)	7,770	10.18	2,800	24,150	17.41
1,500	7,960	10.31	2,900	24,900	17.68
1,600	8,680	10.77	3,000	25,650	17.93

Ru(α):

$$H_T-H_{298.15}=5.25\,T+0.75\times10^{-3}T^2-1,632$$
$$(0.3 \text{ percent}; 298°-1,308° \text{ K.)};$$
$$C_p=5.25+1.50\times10^{-3}T;$$
$$\Delta H_{1308}(transition)=60.$$

Ru(β):

$$H_T-H_{298.15}=7.20\,T-2,840 \text{ (0.1 percent;}$$
$$1,308°-1,473° \text{ K.)};$$

$$C_p=7.20;$$
$$\Delta H_{1473}(transition)=0.$$

Ru(γ):

$$H_T-H_{298.15}=7.20\,T-2,840 \text{ (0.1 percent;}$$
$$1,473°-1,773° \text{ K.)};$$
$$C_p=7.20;$$
$$\Delta H_{1773}=320.$$

Ru(δ):

$$H_T-H_{298.15}=7.50\,T-3,050 \text{ (0.1 percent;}$$
$$1,773°-2,700° \text{ K.)};$$
$$C_p=7.50.$$
$$\Delta H_{2700}(fusion)=6,200.$$

Ru(l):

$$H_T-H_{298.15}=7.50\,T+3,150 \text{ (0.1 percent;}$$
$$2,700°-3,000° \text{ K.)};$$
$$C_p=7.50.$$

Table 652.—*Heat content and entropy of Ru(g)*

[Base, ideal gas at 298.15° K.; atomic wt., 101.1]

T, ° K.	$H_T-H_{298.15}$, cal./mole	$S_T-S_{298.15}$, cal./deg. mole	T, ° K.	$H_T-H_{298.15}$, cal./mole	$S_T-S_{298.15}$, cal./deg. mole
400	535	1.55	1,500	7,290	9.54
500	1,095	2.79	1,600	7,920	9.95
600	1,675	3.85	1,700	8,555	10.33
700	2,275	4.77	1,800	9,190	10.70
800	2,890	5.59	1,900	9,835	11.05
900	3,515	6.33	2,000	10,480	11.38
1,000	4,140	6.99	2,200	11,800	12.01
1,100	4,770	7.59	2,400	13,145	12.59
1,200	5,400	8.14	2,600	14,530	13.14
1,300	6,030	8.64	2,800	15,945	13.67
1,400	6,660	9.11	3,000	17,395	14.17

Ru(g):

$$H_T-H_{298.15}=5.98\,T+0.15\times10^{-3}T^2+0.86\times10^5T^{-1}$$
$$-2,085 \ (0.6 \text{ percent}; 298°-3,000° \text{ K.)};$$
$$C_p=5.98+0.30\times10^{-3}T-0.86\times10^5T^{-2}.$$

SAMARIUM AND ITS COMPOUNDS

ELEMENT

Table 653.—*Heat content and entropy of Sm(c, l)*

[Base, α-crystals at 298.15° K.; atomic wt., 150.35]

T, ° K.	$H_T-H_{298.15}$, cal./mole	$S_T-S_{298.15}$, cal./deg. mole	T, ° K.	$H_T-H_{298.15}$, cal./mole	$S_T-S_{298.15}$, cal./deg. mole
400	675	1.95	1,200	7,350	10.70
500	1,370	3.49	1,300	8,150	11.34
600	2,090	4.80	1,325(β)	8,350	11.49
700	2,835	5.95	1,325(l)	11,400	13.79
800	3,610	6.99	1,400	12,000	14.23
900	4,415	7.94	1,500	12,800	14.78
1,000	5,250	8.81	1,600	13,600	15.30
1,100	6,110	9.63	1,700	14,400	15.78
1,190(α)	6,910	10.33	1,800	15,200	16.24
1,190(β)	7,270	10.63			

Sm(α):

$$H_T-H_{298.15}=5.65\,T+1.41\times10^{-3}T^2-1,810 \text{ (0.1 percent;}$$
$$298°-1,190° \text{ K.)};$$
$$C_p=5.65+2.82\times10^{-3}T;$$
$$\Delta H_{1190}(transition)=360.$$

TABLE 1h7. (*Continued*)

Sm(β):

$H_T - H_{298.15} = 8.00T - 2,250$ (0.1 percent; 1,190°–1,325° K.);

$C_p = 8.00$;

$\Delta H_{1325}(fusion) = 3,050$.

Sm(l):

$H_T - H_{298.15} = 8.00T + 800$ (0.1 percent; 1,325°–1,800° K.);

$C_p = 8.00$.

Sc(l):

$H_T - H_{298.15} = 8.00T - 250$ (0.1 percent; 1,673°–2,700° K.);

$C_p = 8.00$.

TABLE 656.—*Heat content and entropy of Sc(g)*

[Base, ideal gas at 298.15° K.; atomic wt., 44.96]

T, ° K.	$H_T - H_{298.15}$, cal./mole	$S_T - S_{298.15}$, cal./d3g. mole	T, ° K.	$H_T - H_{298.15}$, cal./mole	$S_T - S_{298.15}$, cal./deg. mole
400	530	1.53	1,900	8,060	9.37
500	1,045	2.77	2,000	8,565	9.63
600	1,550	3.60	2,200	9,600	10.12
700	2,050	4.37	2,400	10,655	10.58
800	2,555	5.04	2,600	11,750	11.02
900	3,055	5.64	2,800	12,895	11.45
1,000	3,555	6.16	3,000	14,100	11.86
1,100	4,055	6.64	3,500	17,460	12.90
1,200	4,555	7.07	4,000	21,385	13.94
1,300	5,055	7.47	4,500	25,895	15.00
1,400	5,550	7.84	5,000	30,900	16.06
1,500	6,050	8.19	6,000	41,850	18.05
1,600	6,550	8.51	7,000	53,165	19.80
1,700	7,050	8.81	8,000	64,205	21.27
1,800	7,555	9.10			

TABLE 654.—*Heat content and entropy of Sm(g)*

[Base, ideal gas at 298.15° K.; atomic wt., 150.35]

T, ° K.	$H_T - H_{298.15}$, cal./mole	$S_T - S_{298.15}$, cal./deg. mole	T, ° K.	$H_T - H_{298.15}$, cal./mole	$S_T - S_{298.15}$, cal./deg. mole
400	740	2.14	1,500	8,700	11.76
500	1,470	3.77	1,600	9,375	12.19
600	2,205	5.11	1,700	10,035	12.59
700	2,945	6.25	1,800	10,690	12.97
800	3,685	7.23	1,900	11,335	13.32
900	4,425	8.10	2,000	11,970	13.64
1,000	5,155	8.88	2,200	13,220	14.24
1,100	5,885	9.57	2,400	14,445	14.77
1,200	6,605	10.20	2,600	15,655	15.26
1,300	7,315	10.77	2,800	16,865	15.70
1,400	8,010	11.28	3,000	18,075	16.12

Sm(g):

$H_T - H_{298.15} = 8.01T - 0.39 \times 10^{-3}T^2 + 0.48 \times 10^5 T^{-1}$

$- 2,514$ (0.4 percent; 298°–3,000° K.);

$C_p = 8.01 - 0.78 \times 10^{-3}T - 0.48 \times 10^5 T^{-2}$.

Sc(g):

$H_T - H_{298.15} = 4.60T + 0.17 \times 10^{-3}T^2 - 0.52 \times 10^5 T^{-1}$

$- 1,212$ (1.0 percent; 298°–3,000° K.);

$C_p = 4.60 + 0.34 \times 10^{-3}T + 0.52 \times 10^5 T^{-2}$.

$H_T - H_{298.15} = 2.66T + 0.71 \times 10^{-3}T^2 - 1.96 \times 10^5 T^{-1} - 199$

(1.0 percent; 3,000°–7,000° K.);

$C_p = 2.66 + 1.42 \times 10^{-3}T + 1.96 \times 10^5 T^{-2}$.

SULFATE

Sm$_2$(SO$_4$)$_3$·8H$_2$O(c):

$C_p = 144.9$ (298° K.).

OXIDES

TABLE 657.—*Heat content and entropy of ScO(g)*

[Base, ideal gas at 298.15° K.; mol. wt., 60.96]

T, ° K.	$H_T - H_{298.15}$, cal./mole	$S_T - S_{298.15}$, cal./deg. mole	T, ° K.	$H_T - H_{298.15}$, cal./mole	$S_T - S_{298.15}$, cal./deg. mole
400	770	2.22	1,000	5,775	9.82
500	1,560	3.99	1,200	7,515	11.41
600	2,375	5.48	1,400	9,265	12.76
700	3,210	6.77	1,600	11,025	13.94
800	4,055	7.90	1,800	12,790	14.98
900	4,910	8.91	2,000	14,565	15.92

ScO(g):

$H_T - H_{298.15} = 8.22T + 0.22 \times 10^{-3}T^2 + 0.87 \times 10^5 T^{-1}$

$- 2,762$ (0.5 percent; 298°–2,000° K.);

$C_p = 8.22 + 0.44 \times 10^{-3}T - 0.87 \times 10^5 T^{-2}$.

Sc$_2$O$_3$(c):

$\overline{C_p} = 21.1$ (273°–373° K.).

SCANDIUM AND ITS COMPOUNDS

ELEMENT

TABLE 655.—*Heat content and entropy of Sc(c, l)*

[Base, crystals at 298.15° K.; atomic wt., 44.96]

T, ° K.	$H_T - H_{298.15}$, cal./mole	$S_T - S_{298.15}$, cal./deg. mole	T, ° K.	$H_T - H_{298.15}$, cal./mole	$S_T - S_{298.15}$, cal./deg. mole
400	615	1.78	1,500	8,010	10.49
500	1,235	3.16	1,600	8,740	10.96
600	1,860	4.30	1,673(c)	9,280	11.29
700	2,500	5.29	1,673(l)	13,130	13.59
800	3,150	6.15	1,700	13,350	13.72
900	3,810	6.93	1,800	14,150	14.18
1,000	4,480	7.64	2,000	15,750	15.02
1,100	5,170	8.29	2,200	17,350	15.79
1,200	5,860	8.89	2,400	18,950	16.48
1,300	6,560	9.46	2,600	20,550	17.12
1,400	7,280	9.99	2,700	21,350	17.42

Sc(c):

$H_T - H_{298.15} = 5.68T + 0.54 \times 10^{-3}T^2 - 1,741$ (0.2 percent; 298°–1,673° K.);

$C_p = 5.68 + 1.08 \times 10^{-3}T$;

$\Delta H_{1673}(fusion) = 3,850$.

SULFATE

Sc$_2$(SO$_4$)$_3$(c):

$\overline{C_p} = 62.0$ (273°–373° K.).

SCANDIUM-AMMONIUM FLUORIDE

Sc(NH$_4$)$_3$F$_6$(c):

$\overline{C_p} = 90.0$ (273°–289° K.).

TABLE 1h7. (*Continued*)

SELENIUM AND ITS COMPOUNDS

ELEMENT

TABLE 658.—*Heat content and entropy of Se(c, l)*

[Base, crystals at 298.15° K.; atomic wt., 78.96]

T, ° K.	$H_T - H_{298.15}$, cal./mole	$S_T - S_{298.15}$, cal./deg. mole	T, ° K.	$H_T - H_{298.15}$, cal./mole	$S_T - S_{298.15}$, cal./deg. mole
350	320	0.99	500	2,680	6.15
400	650	1.87	600	3,520	7.68
450	1,000	2.69	700	4,360	8.97
490(c)	1,300	3.33	800	5,200	10.10
490(l)	2,600	5.98	900	6,040	11.08

Se(c):
$$H_T - H_{298.15} = 3.30T + 4.40 \times 10^{-3}T^2 - 1,375$$
$$(0.2 \text{ percent}; 298°-490° \text{ K.});$$
$$C_p = 3.30 + 8.80 \times 10^{-3}T;$$
$$\Delta H_{490}(fusion) = 1,300.$$

Se(l):
$$H_T - H_{298.15} = 8.40T - 1,520 \ (0.1 \text{ percent}; 490°-900° \text{ K.});$$
$$C_p = 8.40.$$

TABLE 659.—*Heat content and entropy of Se(g)*

[Base, ideal gas at 298.15° K.; atomic wt., 78.96]

T, ° K.	$H_T - H_{298.15}$, cal./mole	$S_T - S_{298.15}$, cal./deg. mole	T, ° K.	$H_T - H_{298.15}$, cal./mole	$S_T - S_{298.15}$, cal./deg. mole
400	510	1.47	1,900	8,810	9.95
500	1,015	2.60	2,000	9,375	10.24
600	1,530	3.54	2,200	10,495	10.77
700	2,060	4.35	2,400	11,615	11.26
800	2,600	5.08	2,600	12,735	11.71
900	3,155	5.73	2,800	13,860	12.13
1,000	3,710	6.31	3,000	14,980	12.51
1,100	4,275	6.85	3,500	17,795	13.38
1,200	4,840	7.34	4,000	20,620	14.14
1,300	5,410	7.80	4,500	23,450	14.80
1,400	5,980	8.22	5,000	26,290	15.40
1,500	6,550	8.61	6,000	31,950	16.43
1,600	7,115	8.98	7,000	37,590	17.30
1,700	7,680	9.32	8,000	43,220	18.05
1,800	8,245	9.64			

Se(g):
$$H_T - H_{298.15} = 5.13T + 0.18 \times 10^{-3}T^2 + 0.22 \times 10^5 T^{-1}$$
$$-1,619 \ (0.8 \text{ percent}; 298°-2,000° \text{ K.});$$
$$C_p = 5.13 + 0.36 \times 10^{-3}T - 0.22 \times 10^5 T^{-2}.$$
$$H_T - H_{298.15} = 5.57T + 0.01 \times 10^{-3}T^2 + 0.52 \times 10^5 T^{-1}$$
$$-1,836 \ (0.1 \text{ percent}; 2,000°-8,000° \text{ K.});$$
$$C_p = 5.57 + 0.02 \times 10^{-3}T - 0.52 \times 10^5 T^{-2}.$$

TABLE 660.—*Heat content and entropy of Se₂(g)*

[Base, ideal gas at 298.15° K.; mol. wt., 157.92]

T, ° K.	$H_T - H_{298.15}$, cal./mole	$S_T - S_{298.15}$, cal./deg. mole	T, ° K.	$H_T - H_{298.15}$, cal./mole	$S_T - S_{298.15}$, cal./deg. mole
400	875	2.52	1,500	10,720	14.30
500	1,750	4.47	1,600	11,640	14.89
600	2,630	6.08	1,700	12,560	15.45
700	3,515	7.45	1,800	13,490	15.98
800	4,405	8.64	1,900	14,420	16.48
900	5,300	9.69	2,000	15,350	16.96
1,000	6,195	10.63	2,200	17,240	17.86
1,100	7,095	11.49	2,400	19,130	18.69
1,200	7,995	12.27	2,600	21,050	19.45
1,300	8,900	13.00	2,800	22,970	20.16
1,400	9,810	13.67	3,000	24,910	20.83

Se₂(g):
$$H_T - H_{298.15} = 8.73T + 0.16 \times 10^{-3}T^2 + 0.34 \times 10^5 T^{-1}$$
$$-2,731 \ (0.2 \text{ percent}; 298°-3,000° \text{ K.});$$
$$C_p = 8.73 + 0.32 \times 10^{-3}T - 0.34 \times 10^5 T^{-2}.$$

TABLE 661.—*Heat content and entropy of Se₆(g)*

[Base, ideal gas at 298.15° K.; mol. wt., 473.76]

T, ° K.	$H_T - H_{298.15}$, cal./mole	$S_T - S_{298.15}$, cal./deg. mole	T, ° K.	$H_T - H_{298.15}$, cal./mole	$S_T - S_{298.15}$, cal./deg. mole
400	3,000	8.65	1,000	21,640	37.00
500	6,030	15.41	1,100	24,810	40.02
600	9,100	21.01	1,200	27,980	42.78
700	12,200	25.78	1,300	31,170	45.33
800	15,330	29.96	1,400	34,360	47.70
900	18,480	33.67	1,500	37,560	49.90

Se₆(g):
$$H_T - H_{298.15} = 30.74T + 0.52 \times 10^{-3}T^2 + 1.82 \times 10^5 T^{-1}$$
$$-9,822 \ (0.1 \text{ percent}; 298°-1,500° \text{ K.});$$
$$C_p = 30.74 + 1.04 \times 10^{-3}T - 1.82 \times 10^5 T^{-2}.$$

OXIDE

TABLE 662.—*Heat content and entropy of SeO(g)*

[Base, ideal gas at 298.15° K.; mol. wt., 94.96]

T, ° K.	$H_T - H_{298.15}$, cal./mole	$S_T - S_{298.15}$, cal./deg. mole	T, ° K.	$H_T - H_{298.15}$, cal./mole	$S_T - S_{298.15}$, cal./deg. mole
400	780	2.25	1,000	5,820	9.89
500	1,580	4.03	1,200	7,565	11.48
600	2,405	5.53	1,400	9,320	12.83
700	3,245	6.83	1,600	11,085	14.01
800	4,095	7.97	1,800	12,855	15.05
900	4,955	8.98	2,000	14,630	15.99

SeO(g):
$$H_T - H_{298.15} = 8.35T + 0.18 \times 10^{-3}T^2 + 0.88 \times 10^5 T^{-1}$$
$$-2,801 \ (0.4 \text{ percent}; 298°-2,000° \text{ K.});$$
$$C_p = 8.35 + 0.36 \times 10^{-3}T - 0.88 \times 10^5 T^{-2}.$$

FLUORIDE

TABLE 663.—*Heat content and entropy of SeF₆(g)*

[Base, ideal gas at 298.15° K.; mol. wt., 192.96]

T, ° K.	$H_T - H_{298.15}$, cal./mole	$S_T - S_{298.15}$, cal./deg. mole	T, ° K.	$H_T - H_{298.15}$, cal./mole	$S_T - S_{298.15}$, cal./deg. mole
400	2,895	8.33	1,000	23,500	39.40
500	6,040	15.34	1,200	30,800	46.05
600	9,360	21.39	1,400	38,160	51.72
700	12,805	26.70	1,600	45,570	56.67
800	16,320	31.39	1,800	53,010	61.05
900	19,890	35.59	2,000	60,480	64.98

SeF₆(g):
$$H_T - H_{298.15} = 34.86T + 0.98 \times 10^{-3}T^2 + 8.13 \times 10^5 T^{-1}$$
$$-13,207 \ (0.4 \text{ percent}; 298°-2,000° \text{ K.});$$
$$C_p = 34.86 + 1.96 \times 10^{-3}T - 8.13 \times 10^5 T^{-2}.$$

SILICON AND ITS COMPOUNDS

ELEMENT

TABLE 664.—*Heat content and entropy of Si(c, l)*

[Base, crystals at 298.15° K.; atomic wt., 28.09]

T, ° K.	$H_T-H_{298.15}$, cal./mole	$S_T-S_{298.15}$, cal./deg. mole	T, ° K.	$H_T-H_{298.15}$, cal./mole	$S_T-S_{298.15}$, cal./deg. mole
400	515	1.48	1,400	6,680	9.04
500	1,060	2.69	1,500	7,340	9.49
600	1,630	3.73	1,600	8,010	9.92
700	2,220	4.64	1,685(c)	8,580	10.27
800	2,830	5.46	1,685(l)	20,680	17.45
900	3,450	6.19	1,700	20,770	17.50
1,000	4,080	6.85	1,800	21,380	17.85
1,100	4,720	7.46	1,900	21,990	18.18
1,200	5,360	8.02	2,000	22,600	18.49
1,300	6,020	8.55			

$Si(c)$:

$$H_T-H_{298.15}=5.70T+0.35\times10^{-3}T^2+1.04\times10^5T^{-1}$$
$$-2,079 \text{ (0.3 percent; 298°–1,685° K.)};$$
$$C_p=5.70+0.70\times10^{-3}T-1.04\times10^5T^{-2};$$
$$\Delta H_{1685}(fusion)=12,100.$$

$Si(l)$:

$$H_T-H_{298.15}=6.10T+10,400 \text{ (0.1 percent; 1,685°–2,000°}$$
$$\text{K.)};$$
$$C_p=6.10.$$

TABLE 665.—*Heat content and entropy of Si(g)*

[Base, ideal gas at 298.15° K.; atomic wt., 28.09]

T, ° K.	$H_T-H_{298.15}$, cal./mole	$S_T-S_{298.15}$, cal./deg. mole	T, ° K.	$H_T-H_{298.15}$, cal./mole	$S_T-S_{298.15}$, cal./deg. mole
400	535	1.54	1,900	8,140	9.43
500	1,045	2.68	2,000	8,665	9.70
600	1,555	3.61	2,200	9,715	10.20
700	2,060	4.39	2,400	10,780	10.67
800	2,560	5.06	2,600	11,850	11.10
900	3,060	5.65	2,800	12,935	11.50
1,000	3,565	6.18	3,000	14,020	11.87
1,100	4,065	6.66	3,500	16,765	12.71
1,200	4,565	7.09	4,000	19,520	13.45
1,300	5,070	7.49	4,500	22,280	14.10
1,400	5,575	7.87	5,000	25,040	14.69
1,500	6,085	8.22	6,000	30,555	15.69
1,600	6,595	8.55	7,000	36,105	16.55
1,700	7,105	8.86	8,000	41,845	17.31
1,800	7,625	9.15			

$Si(g)$:

$$H_T-H_{298.15}=4.82T+0.09\times10^{-3}T^2-0.42\times10^5T^{-1}$$
$$-1,304 \text{ (0.3 percent; 298°–5,000° K.)};$$
$$C_p=4.82+0.18\times10^{-3}T+0.42\times10^5T^{-2}.$$

OXIDES

TABLE 666.—*Heat content and entropy of SiO_2 (quartz)*

[Base, α-crystals at 298.15° K.; mol. wt., 60.09]

T, ° K.	$H_T-H_{298.15}$, cal./mole	$S_T-S_{298.15}$, cal./deg. mole	T, ° K.	$H_T-H_{298.15}$, cal./mole	$S_T-S_{298.15}$, cal./deg. mole
400	1,200	3.45	1,200	14,250	20.95
500	2,560	6.48	1,300	15,940	22.30
600	4,040	9.17	1,400	17,640	23.56
700	5,630	11.62	1,500	19,360	24.75
800	7,320	13.88	1,600	21,100	25.88
848(α)	8,170	14.91	1,700	22,860	26.94
848(β)	8,460	15.25	1,800	24,630	27.96
900	9,300	16.21	1,900	26,420	28.93
1,000	10,920	17.92	2,000	28,220	29.85
1,100	12,570	19.49			

$SiO_2(\alpha\text{-}quartz)$:

$$H_T-H_{298.15}=11.22T+4.10\times10^{-3}T^2+2.70\times10^5T^{-1}$$
$$-4,615 \text{ (0.1 percent; 298°–848° K.)};$$
$$C_p=11.22+8.20\times10^{-3}T-2.70\times10^5T^{-2};$$
$$\Delta H_{848}(transition)=290.$$

$SiO_2(\beta\text{-}quartz)$:

$$H_T-H_{298.15}=14.41T+0.97\times10^{-3}T^2-4,455$$
$$\text{(0.1 percent; 848°–2,000° K.)};$$
$$C_p=14.41+1.94\times10^{-3}T.$$

TABLE 667.—*Heat content and entropy of SiO_2 (cristobalite)*

[Base, α-crystals at 298.15° K.; mol. wt., 60.09]

T, ° K.	$H_T-H_{298.15}$, cal./mole	$S_T-S_{298.15}$, cal./deg. mole	T, ° K.	$H_T-H_{298.15}$, cal./mole	$S_T-S_{298.15}$, cal./deg. mole
400	1,210	3.48	1,200	14,080	20.90
500	2,560	6.48	1,300	15,790	22.27
523(α)	2,910	7.16	1,400	17,510	23.54
523(β)	3,110	7.54	1,500	19,240	24.74
600	4,310	9.68	1,600	20,990	25.87
700	5,850	12.05	1,700	22,750	26.93
800	7,460	14.20	1,800	24,530	27.95
900	9,090	16.12	1,900	26,320	28.92
1,000	10,730	17.85	2,000	28,120	29.84
1,100	12,390	19.43			

$SiO_2(\alpha\text{-}cristobalite)$:

$$H_T-H_{298.15}=4.28T+10.53\times10^{-3}T^2-2,212$$
$$\text{(1.0 percent; 298°–523° K.)};$$
$$C_p=4.28+21.06\times10^{-3}T;$$
$$\Delta H_{523}(transition)=200.$$

$SiO_2(\beta\text{-}cristobalite)$:

$$H_T-H_{298.15}=14.40T+1.02\times10^{-3}T^2-4,696$$
$$\text{(0.2 percent; 523°–2,000° K.)};$$
$$C_p=14.40+2.04\times10^{-3}T.$$

TABLE 668.—*Heat content and entropy of SiO_2 (tridymite)*

[Base, α-crystals at 298.15° K.; mol. wt., 60.09]

T, ° K.	$H_T-H_{298.15}$, cal./mole	$S_T-S_{298.15}$, cal./deg. mole	T, ° K.	$H_T-H_{298.15}$, cal./mole	$S_T-S_{298.15}$, cal./deg. mole
350	585	1.81	1,100	12,250	19.25
390(α)	1,085	3.16	1,200	13,940	20.72
390(β)	1,125	3.26	1,300	15,650	22.09
400	1,270	3.63	1,400	17,370	23.37
500	2,710	6.84	1,500	19,100	24.56
600	4,170	9.50	1,600	20,850	25.69
700	5,710	11.87	1,700	22,610	26.75
800	7,320	14.02	1,800	24,390	27.77
900	8,950	15.94	1,900	26,180	28.74
1,000	10,590	17.67	2,000	27,980	29.66

$SiO_2(\alpha\text{-}tridymite)$:

$$H_T-H_{298.15}=3.27T+12.40\times10^{-3}T^2-2,077$$
$$\text{(0.2 percent; 298°–390° K.)};$$
$$C_p=3.27+24.80\times10^{-3}T;$$
$$\Delta H_{390}(transition)=40.$$

$SiO_2(\beta\text{-}tridymite)$:

$$H_T-H_{298.15}=13.64T+1.32\times10^{-3}T^2-4,395$$
$$\text{(0.7 percent; 390°–2,000° K.)};$$
$$C_p=13.64+2.64\times10^{-3}T.$$

TABLE 1h7. (Continued)

TABLE 669.—*Heat content and entropy of SiO$_2$(gl)*

[Base, glass at 298.15° K.; mol. wt., 60.09]

T, ° K.	$H_T-H_{298.15}$, cal./mole	$S_T-S_{298.15}$, cal./deg. mole	T, ° K.	$H_T-H_{298.15}$, cal./mole	$S_T-S_{298.15}$, cal./deg. mole
400	1,230	3.54	1,300	15,450	21.57
500	2,550	6.48	1,400	17,240	22.90
600	3,950	9.03	1,500	19,080	24.17
700	5,430	11.31	1,600	20,980	25.39
800	6,990	13.39	1,700	22,930	26.57
900	8,610	15.30	1,800	24,920	27.71
1,000	10,280	17.06	1,900	26,950	28.81
1,100	11,980	18.68	2,000	29,010	29.87
1,200	13,700	20.17			

SiO$_2$(gl):

$$H_T-H_{298.15}=13.38\,T+1.84\times10^{-3}T^2+3.45\times10^5T^{-1}$$
$$-5,310\ (0.4\ \text{percent};\ 298°-2,000°\ \text{K.});$$
$$C_p=13.38+3.68\times10^{-3}T-3.45\times10^5T^{-2}.$$

TABLE 670.—*Heat content and entropy of SiO(g)*

[Base, ideal gas at 298.15° K.; mol. wt., 44.09]

T, ° K.	$H_T-H_{298.15}$, cal./mole	$S_T-S_{298.15}$, cal./deg. mole	T, ° K.	$H_T-H_{298.15}$, cal./mole	$S_T-S_{298.15}$, cal./deg. mole
400	740	2.13	1,300	8,165	11.72
500	1,500	3.83	1,400	9,035	12.36
600	2,285	5.26	1,500	9,905	12.96
700	3,090	6.50	1,600	10,780	13.53
800	3,910	7.59	1,700	11,655	14.06
900	4,745	8.58	1,800	12,530	14.56
1,000	5,590	9.47	1,900	13,410	15.03
1,100	6,445	10.28	2,000	14,290	15.49
1,200	7,300	11.03			

SiO(g):

$$H_T-H_{298.15}=7.70\,T+0.37\times10^{-3}T^2+0.70\times10^5T^{-1}$$
$$-2,563\ (0.5\ \text{percent};\ 298°-2,000°\ \text{K.});$$
$$C_p=7.70+0.74\times10^{-3}T-0.70\times10^5T^{-2}.$$

SULFIDE

TABLE 671.—*Heat content and entropy of SiS(g)*

[Base, ideal gas at 298.15° K.; mol. wt., 60.16]

T, ° K.	$H_T-H_{298.15}$, cal./mole	$S_T-S_{298.15}$, cal./deg. mole	T, ° K.	$H_T-H_{298.15}$, cal./mole	$S_T-S_{298.15}$, cal./deg. mole
400	805	2.32	1,000	5,930	10.10
500	1,625	4.15	1,200	7,690	11.71
600	2,470	5.69	1,400	9,455	13.07
700	3,325	7.01	1,600	11,225	14.25
800	4,185	8.16	1,800	13,000	15.29
900	5,055	9.18	2,000	14,780	16.23

SiS(g):

$$H_T-H_{298.15}=8.58\,T+0.11\times10^{-3}T^2+0.84\times10^5T^{-1}$$
$$-2,850\ (0.3\ \text{percent};\ 298°-2,000°\ \text{K.});$$
$$C_p=8.58+0.22\times10^{-3}T-0.84\times10^5T^{-2}.$$

SELENIDE

TABLE 672.—*Heat content and entropy of SiSe(g)*

[Base, ideal gas at 298.15° K.; mol. wt., 107.05]

T, ° K.	$H_T-H_{298.15}$, cal./mole	$S_T-S_{298.15}$, cal./deg. mole	T, ° K.	$H_T-H_{298.15}$, cal./mole	$S_T-S_{298.15}$, cal./deg. mole
400	835	2.41	1,000	6,050	10.34
500	1,685	4.30	1,200	7,820	11.95
600	2,545	5.87	1,400	9,595	13.32
700	3,410	7.20	1,600	11,370	14.51
800	4,285	8.37	1,800	13,150	15.56
900	5,165	9.41	2,000	14,935	16.50

SiSe(g):

$$H_T-H_{298.15}=8.78\,T+0.05\times10^{-3}T^2+0.69\times10^5T^{-1}$$
$$-2,854\ (0.1\ \text{percent};\ 298°-2,000°\ \text{K.});$$
$$C_p=8.78+0.10\times10^{-3}T-0.69\times10^5T^{-2}.$$

TELLURIDE

TABLE 673.—*Heat content and entropy of SiTe(g)*

[Base, ideal gas at 298.15° K.; mol. wt., 155.70]

T, ° K.	$H_T-H_{298.15}$, cal./mole	$S_T-S_{298.15}$, cal./deg. mole	T, ° K.	$H_T-H_{298.15}$, cal./mole	$S_T-S_{298.15}$, cal./deg. mole
400	855	2.46	1,000	6,110	10.47
500	1,715	4.38	1,200	7,885	12.09
600	2,585	5.97	1,400	9,665	13.46
700	3,460	7.32	1,600	11,450	14.65
800	4,340	8.49	1,800	13,230	15.70
900	5,225	9.54	2,000	15,015	16.64

SiTe(g):

$$H_T-H_{298.15}=8.85\,T+0.03\times10^{-3}T^2+0.55\times10^5T^{-1}$$
$$-2,826\ (0.1\ \text{percent};\ 298°-2,000°\ \text{K.});$$
$$C_p=8.85+0.06\times10^{-3}T-0.55\times10^5T^{-2}.$$

CARBIDE

TABLE 674.—*Heat content and entropy of SiC (hexagonal)*

[Base, crystals at 298.15° K.; mol. wt., 40.10]

T, ° K.	$H_T-H_{298.15}$, cal./mole	$S_T-S_{298.15}$, cal./deg. mole	T, ° K.	$H_T-H_{298.15}$, cal./mole	$S_T-S_{298.15}$, cal./deg. mole
400	740	2.12	1,300	10,560	14.58
500	1,630	4.10	1,400	11,790	15.49
600	2,620	5.91	1,500	13,040	16.35
700	3,660	7.51	1,600	14,310	17.17
800	4,740	8.95	1,700	15,590	17.95
900	5,860	10.27	1,800	16,880	18.69
1,000	7,010	11.48	1,900	18,180	19.39
1,100	8,170	12.58	2,000	19,490	20.06
1,200	9,350	13.61			

SiC(hexagonal):

$$H_T-H_{298.15}=9.93\,T+0.96\times10^{-3}T^2+3.66\times10^5T^{-1}$$
$$-4,274\ (0.6\ \text{percent};\ 298°-2,000°\ \text{K.});$$
$$C_p=9.93+1.92\times10^{-3}T-3.66\times10^5T^{-2}.$$

TABLE 675.—*Heat content and entropy of SiC (cubic)*

[Base, crystals at 298.15° K.; mol. wt., 40.10]

T, ° K.	$H_T-H_{298.15}$, cal./mole	$S_T-S_{298.15}$, cal./deg. mole	T, ° K.	$H_T-H_{298.15}$, cal./mole	$S_T-S_{298.15}$, cal./deg. mole
400	750	2.15	1,300	10,530	14.56
500	1,640	4.13	1,400	11,750	15.47
600	2,620	5.91	1,500	12,990	16.32
700	3,660	7.52	1,600	14,240	17.13
800	4,740	8.96	1,700	15,520	17.90
900	5,860	10.28	1,800	16,820	18.64
1,000	7,000	11.48	1,900	18,140	19.36
1,100	8,160	12.58	2,000	19,480	20.04
1,200	9,330	13.60			

SiC(cubic):

$$H_T-H_{298.15}=9.97\,T+0.91\times10^{-3}T^2+3.64\times10^5T^{-1}$$
$$-4,274\ (0.5\ \text{percent};\ 298°-2,000°\ \text{K.});$$
$$C_p=9.97+1.82\times10^{-3}T-3.64\times10^5T^{-2}.$$

TABLE 1h7. (*Continued*)

NITRIDES

TABLE 676.—*Heat content and entropy of* $SiN(g)$

[Base, ideal gas at 298.15° K.; mol. wt., 42.10]

T, ° K.	$H_T - H_{298.15}$, cal./mole	$S_T - S_{298.15}$, cal./deg. mole	T, ° K.	$H_T - H_{298.15}$, cal./mole	$S_T - S_{298.15}$, cal./deg. mole
400	750	2.16	1,000	5,655	9.58
500	1,520	3.88	1,200	7,375	11.15
600	2,315	5.33	1,400	9,115	12.49
700	3,130	6.58	1,600	10,865	13.66
800	3,960	7.69	1,800	12,625	14.70
900	4,800	8.68	2,000	14,390	15.63

$SiN(g)$:

$$H_T - H_{298.15} = 7.88T + 0.32 \times 10^{-3} T^2 + 0.76 \times 10^5 T^{-1}$$
$$-2,633 \ (0.6 \text{ percent}; 298°-2,000° \text{ K.});$$
$$C_p = 7.88 + 0.64 \times 10^{-3} T - 0.76 \times 10^5 T^{-2}.$$

TABLE 677.—*Heat content and entropy of* $Si_3N_4(c)$

[Base, crystals at 298.15° K.; mol. wt., 140.30]

T, ° K.	$H_T - H_{298.15}$, cal./mole	$S_T - S_{298.15}$, cal./deg. mole	T, ° K.	$H_T - H_{298.15}$, cal./mole	$S_T - S_{298.15}$, cal./deg. mole
400	2,500	7.20	700	11,550	23.90
500	5,300	13.44	800	14,950	28.44
600	8,300	18.90	900	18,550	32.68

$Si_3N_4(c)$:

$$H_T - H_{298.15} = 16.83T + 11.80 \times 10^{-3} T^2 - 6,067$$
$$(0.6 \text{ percent}; 298°-900° \text{ K.});$$
$$C_p = 16.83 + 23.60 \times 10^{-3} T.$$

HYDRIDES

TABLE 678.—*Heat content and entropy of* $SiH(g)$

[Base, ideal gas at 298.15° K.; mol. wt., 29.10]

T, ° K.	$H_T - H_{298.15}$, cal./mole	$S_T - S_{298.15}$, cal./deg. mole	T, ° K.	$H_T - H_{298.15}$, cal./mole	$S_T - S_{298.15}$, cal./deg. mole
400	710	2.05	1,000	5,195	8.83
500	1,420	3.63	1,200	6,810	10.30
600	2,140	4.94	1,400	8,455	11.57
700	2,880	6.08	1,600	10,140	12.70
800	3,635	7.09	1,800	11,840	13.70
900	4,410	8.01	2,000	13,550	14.60

$SiH(g)$:

$$H_T - H_{298.15} = 6.59T + 0.62 \times 10^{-3} T^2 - 2,020$$
$$(0.5 \text{ percent}; 298°-2,000° \text{ K.});$$
$$C_p = 6.59 + 1.24 \times 10^{-3} T.$$

TABLE 679.—*Heat content and entropy of* $SiH_4(g)$

[Base, ideal gas at 298.15° K.; mol. wt., 32.12]

T, ° K.	$H_T - H_{298.15}$, cal./mole	$S_T - S_{298.15}$, cal./deg. mole	T, ° K.	$H_T - H_{298.15}$, cal./mole	$S_T - S_{298.15}$, cal./deg. mole
400	1,150	3.31	1,000	11,255	18.18
500	2,470	6.24	1,200	15,425	21.98
600	3,975	8.99	1,400	19,845	25.38
700	5,615	11.51	1,600	24,400	28.42
800	7,390	13.88	1,800	29,075	31.17
900	9,275	16.10	2,000	33,840	33.68

$SiH_4(g)$:

$$H_T - H_{298.15} = 11.05T + 4.39 \times 10^{-3} T^2 + 3.05 \times 10^5 T^{-1}$$
$$-4,708 \ (1.4 \text{ percent}; 298°-1,800° \text{ K.});$$
$$C_p = 11.05 + 8.78 \times 10^{-3} T - 3.05 \times 10^5 T^{-2}.$$

BROMIDES

TABLE 680.—*Heat content and entropy of* $SiBr(g)$

[Base, ideal gas at 298.15° K.; mol. wt., 108.01]

T, ° K.	$H_T - H_{298.15}$, cal./mole	$S_T - S_{298.15}$, cal./deg. mole	T, ° K.	$H_T - H_{298.15}$, cal./mole	$S_T - S_{298.15}$, cal./deg. mole
400	945	2.73	1,000	6,430	11.12
500	1,870	4.79	1,200	8,235	12.76
600	2,790	6.47	1,400	10,035	14.15
700	3,700	7.87	1,600	11,835	15.35
800	4,610	9.08	1,800	13,630	16.41
900	5,520	10.16	2,000	15,420	17.35

$SiBr(g)$:

$$H_T - H_{298.15} = 9.32T - 0.11 \times 10^{-3} T^2 + 0.04 \times 10^5 T^{-1}$$
$$-2,782 \ (0.3 \text{ percent}; 298°-2,000° \text{ K.});$$
$$C_p = 9.32 - 0.22 \times 10^{-3} T - 0.04 \times 10^5 T^{-2}.$$

TABLE 681.—*Heat content and entropy of* $SiBr_4(g)$

[Base, ideal gas at 298.15° K.; mol. wt., 347.75]

T, ° K.	$H_T - H_{298.15}$, cal./mole	$S_T - S_{298.15}$, cal./deg. mole	T, ° K.	$H_T - H_{298.15}$, cal./mole	$S_T - S_{298.15}$, cal./deg. mole
400	2,425	7.00	800	12,420	24.27
500	4,875	12.46	900	14,960	27.26
600	7,365	17.00	1,000	17,510	29.95
700	9,885	20.88			

$SiBr_4(g)$:

$$H_T - H_{298.15} = 25.19T + 0.32 \times 10^{-3} T^2 + 1.94 \times 10^5 T^{-1}$$
$$-8,190 \ (0.1 \text{ percent}; 298°-1,000° \text{ K.});$$
$$C_p = 25.19 + 0.64 \times 10^{-3} T - 1.94 \times 10^5 T^{-2}.$$

CHLORIDES

TABLE 682.—*Heat content and entropy data for* $SiCl(g)$

[Base, ideal gas at 298.15° K.; mol. wt., 63.55]

T, ° K.	$H_T - H_{298.15}$, cal./mole	$S_T - S_{298.15}$, cal./deg. mole	T, ° K.	$H_T - H_{298.15}$, cal./mole	$S_T - S_{298.15}$, cal./deg. mole
400	875	2.52	1,000	6,170	10.60
500	1,750	4.47	1,200	7,950	12.22
600	2,630	6.08	1,400	9,730	13.59
700	3,510	7.44	1,600	11,520	14.79
800	4,395	8.62	1,800	13,300	15.84
900	5,280	9.66	2,000	15,090	16.78

$SiCl(g)$:

$$H_T - H_{298.15} = 8.85 + 0.03 \times 10^{-3} T^2 + 0.30 \times 10^5 T^{-1}$$
$$-2,742 \ (0.1 \text{ percent}; 298°-2,000° \text{ K.});$$
$$C_p = 8.85 + 0.06 \times 10^{-3} T - 0.30 \times 10^5 T^{-2}.$$

TABLE 1h7. (*Continued*)

TABLE 683.—*Heat content and entropy of* $SiCl_4(g)$

[Base, ideal gas at 298.15° K.; mol. wt., 169.92]

T, ° K.	$H_T - H_{298.15}$, cal./mole	$S_T - S_{298.15}$, cal./deg. mole	T, ° K.	$H_T - H_{298.15}$, cal./mole	$S_T - S_{298.15}$, cal./deg. mole
400	2,295	6.62	800	12,055	23.45
500	4,655	11.88	900	14,570	26.41
600	7,085	16.31	1,000	17,100	29.08
700	9,560	20.12			

$$SiCl_4(g):$$
$$H_T - H_{298.15} = 24.25\,T + 0.82 \times 10^{-3}\,T^2 + 2.75 \times 10^5\,T^{-1}$$
$$-8,225 \;(0.1 \text{ percent; } 298°\text{-}2,000° \text{ K.});$$
$$C_p = 24.25 + 1.64 \times 10^{-3}\,T - 2.75 \times 10^5\,T^{-2}.$$

FLUORIDES

TABLE 684.—*Heat content and entropy of* $SiF(g)$

[Base, ideal gas at 298.15° K.; mol. wt., 47.09]

T, ° K.	$H_T - H_{298.15}$, cal./mole	$S_T - S_{298.15}$, cal./deg. mole	T, ° K.	$H_T - H_{298.15}$, cal./mole	$S_T - S_{298.15}$, cal./deg. mole
400	805	2.32	1,000	5,915	10.08
500	1,625	4.15	1,200	7,670	11.68
600	2,465	5.68	1,400	9,430	13.04
700	3,315	6.99	1,600	11,200	14.22
800	4,175	8.14	1,800	12,975	15.26
900	5,045	9.16	2,000	14,755	16.20

$$SiF(g):$$
$$H_T - H_{298.15} = 8.49\,T + 0.13 \times 10^{-3}\,T^2 + 0.70 \times 10^5\,T^{-1}$$
$$-2,778 \;(0.4 \text{ percent; } 298°\text{-}2,000° \text{ K.});$$
$$C_p = 8.49 + 0.26 \times 10^{-3}\,T - 0.70 \times 10^5\,T^{-2}.$$

TABLE 685.—*Heat content and entropy of* $SiF_4(g)$

[Base, ideal gas at 298.15° K.; mol. wt., 104.09]

T, ° K.	$H_T - H_{298.15}$, cal./mole	$S_T - S_{298.15}$, cal./deg. mole	T, ° K.	$H_T - H_{298.15}$, cal./mole	$S_T - S_{298.15}$, cal./deg. mole
400	1,905	5.48	1,000	15,610	26.10
500	3,965	10.07	1,200	20,545	30.60
600	6,160	14.07	1,400	25,535	34.44
700	8,445	17.59	1,600	30,570	37.80
800	10,795	20.73	1,800	35,630	40.78
900	13,185	23.54	2,000	40,720	43.46

$$SiF_4(g):$$
$$H_T - H_{298.15} = 21.95\,T + 1.33 \times 10^{-4}\,T^2 + 4.72 \times 10^5\,T^{-1}$$
$$-8,246 \;(0.7 \text{ percent; } 298°\text{-}2,000° \text{ K.});$$
$$C_p = 21.95 + 2.66 \times 10^{-3}\,T - 4.72 \times 10^5\,T^{-2}.$$

IODIDE

TABLE 686.—*Heat content and entropy of* $SiI_4(g)$

[Base, ideal gas at 298.15° K.; mol. wt., 535.73]

T, ° K.	$H_T - H_{298.15}$, cal./mole	$S_T - S_{298.15}$, cal./deg. mole	T, ° K.	$H_T - H_{298.15}$, cal./mole	$S_T - S_{298.15}$, cal./deg. mole
400	2,485	7.16	800	12,595	24.65
500	4,985	12.74	900	15,155	27.67
600	7,505	17.33	1,000	17,715	30.37
700	10,045	21.25			

$$SiI_4(g):$$
$$H_T - H_{298.15} = 25.57\,T + 0.12 \times 10^{-3}\,T^2 + 1.44 \times 10^5\,T^{-1}$$
$$-8,117 \;(0.1 \text{ percent; } 298°\text{-}1,000° \text{ K.});$$
$$C_p = 25.57 + 0.24 \times 10^{-3}\,T - 1.44 \times 10^5\,T^{-2}.$$

SILVER AND ITS COMPOUNDS
ELEMENT

TABLE 687.—*Heat content and entropy of* $Ag(c, l)$

[Base, crystals at 298.15° K.; atomic wt., 107.88]

T, ° K.	$H_T - H_{298.15}$, cal./mole	$S_T - S_{298.15}$, cal./deg. mole	T, ° K.	$H_T - H_{298.15}$, cal./mole	$S_T - S_{298.15}$, cal./deg. mole
400	615	1.78	1,300	9,650	12.03
500	1,240	3.17	1,400	10,380	12.57
600	1,885	4.35	1,500	11,110	13.07
700	2,535	5.35	1,600	11,840	13.54
800	3,195	6.23	1,700	12,570	13.98
900	3,880	7.04	1,800	13,300	14.40
1,000	4,585	7.78	1,900	14,030	14.80
1,100	5,310	8.47	2,000	14,760	15.17
1,200	6,060	9.12	2,200	16,220	15.87
1,234(c)	6,315	9.33	2,400	17,680	16.50
1,234(l)	9,170	11.64			

$$Ag(c):$$
$$H_T - H_{298.15} = 5.09\,T + 1.02 \times 10^{-3}\,T^2 - 0.36 \times 10^5\,T^{-1}$$
$$-1,488 \;(0.3 \text{ percent; } 298°\text{-}1,234° \text{ K.});$$
$$C_p = 5.09 + 2.04 \times 10^{-3}\,T + 0.36 \times 10^5\,T^{-2};$$
$$\Delta H_{1234}(fusion) = 2,855.$$
$$Ag(l):$$
$$H_T - H_{298.15} = 7.30\,T + 160 \;(0.1 \text{ percent;}$$
$$1,234°\text{-}2,400° \text{ K.});$$
$$C_p = 7.30.$$

TABLE 688.—*Heat content and entropy of* $Ag(g)$

[Base, ideal gas at 298.15° K.; atomic wt., 107.88]

T, ° K.	$H_T - H_{298.15}$, cal./mole	$S_T - S_{298.15}$, cal./deg. mole	T, ° K.	$H_T - H_{298.15}$, cal./mole	$S_T - S_{298.15}$, cal./deg. mole
400	505	1.46	1,900	7,960	9.20
500	1,005	2.57	2,000	8,455	9.46
600	1,500	3.48	2,200	9,450	9.93
700	1,995	4.24	2,400	10,445	10.36
800	2,495	4.90	2,600	11,440	10.76
900	2,990	5.49	2,800	12,430	11.13
1,000	3,490	6.01	3,000	13,425	11.47
1,100	3,985	6.49	3,500	15,915	12.24
1,200	4,480	6.92	4,000	18,405	12.91
1,300	4,980	7.32	4,500	20,915	13.50
1,400	5,475	7.69	5,000	23,465	14.03
1,500	5,970	8.03	6,000	28,785	15.00
1,600	6,470	8.35	7,000	34,700	15.91
1,700	6,965	8.65	8,000	41,665	16.84
1,800	7,465	8.93			

$$Ag(g):$$
$$H_T - H_{298.15} = 4.97\,T - 1,482 \;(0.1 \text{ percent;}$$
$$298°\text{-}4,500° \text{ K.});$$
$$C_p = 4.97.$$

OXIDES

TABLE 689.—*Heat content and entropy of* $Ag_2O(c)$

[Base, crystals at 298.15° K.; mol. wt., 231.76]

T, ° K.	$H_T - H_{298.15}$, cal./mole	$S_T - S_{298.15}$, cal./deg. mole	T, ° K.	$H_T - H_{298.15}$, cal./mole	$S_T - S_{298.15}$, cal./deg. mole
350	840	2.60	450	2,570	6.94
400	1,680	4.84	500	3,490	8.87

TABLE 1h7. (*Continued*)

$Ag_2O(c)$:

$$H_T - H_{298.15} = 11.13T + 7.74 \times 10^{-3}T^2 - 4,006$$
(0.2 percent; 298°–500° K.);
$$C_p = 11.13 + 15.48 \times 10^{-3}T.$$

TABLE 690.—*Heat content and entropy of AgO(g)*

[Base, ideal gas at 298.15° K.; mol. wt., 123.88]

T, ° K.	$H_T - H_{298.15}$, cal./mole	$S_T - S_{298.15}$, cal./deg. mole	T, ° K.	$H_T - H_{298.15}$, cal./mole	$S_T - S_{298.15}$, cal./deg. mole
400	855	2.47	1,000	6,105	10.46
500	1,710	4.37	1,200	7,880	12.08
600	2,580	5.96	1,400	9,660	13.45
700	3,460	7.32	1,600	11,440	14.64
800	4,340	8.49	1,800	13,225	15.69
900	5,220	9.53	2,000	15,010	16.63

$AgO(g)$:

$$H_T - H_{298.15} = 8.84T + 0.03 \times 10^{-3}T^2 + 0.54 \times 10^5 T^{-1}$$
$$- 2,820 \text{ (0.1 percent; 298°–2,000° K.);}$$
$$C_p = 8.84 + 0.06 \times 10^{-3}T - 0.54 \times 10^5 T^{-2}.$$

SULFIDE

TABLE 691.—*Heat content and entropy of Ag₂S(c)*

[Base, α-crystals at 298.15° K.; mol. wt., 247.83]

T, ° K.	$H_T - H_{298.15}$, cal./mole	$S_T - S_{298.15}$, cal./deg. mole	T, ° K.	$H_T - H_{298.15}$, cal./mole	$S_T - S_{298.15}$, cal./deg. mole
400	1,970	5.66	700	9,460	19.98
452(α)	3,085	8.28	800	11,630	22.88
452(β)	4,095	10.51	900	13,790	25.42
500	5,140	12.71	1,000	15,960	27.70
600	7,300	16.65			

$Ag_2S(\alpha)$:

$$H_T - H_{298.15} = 10.13T + 13.20 \times 10^{-3}T^2 - 4,194$$
(0.1 percent; 298°–452° K.);
$$C_p = 10.13 + 26.40 \times 10^{-3}T;$$
$$\Delta H_{452}(transition) = 1,010.$$

$Ag_2S(\beta)$:

$$H_T - H_{298.15} = 21.64T - 5,685 \text{ (0.1 percent;}$$
452°–1,000° K.);
$$C_p = 21.64.$$

HYDRIDE

TABLE 692.—*Heat content and entropy of AgH(g)*

[Base, ideal gas at 298.15° K.; mol. wt., 108.89]

T, ° K.	$H_T - H_{298.15}$, cal./mole	$S_T - S_{298.15}$, cal./deg. mole	T, ° K.	$H_T - H_{298.15}$, cal./mole	$S_T - S_{298.15}$, cal./deg. mole
400	720	2.08	1,000	5,345	9.06
500	1,440	3.69	1,200	7,005	10.58
600	2,185	5.04	1,400	8,695	11.88
700	2,945	6.21	1,600	10,405	13.02
800	3,730	7.26	1,800	12,130	14.03
900	4,530	8.20	2,000	13,870	14.95

$AgH(g)$:

$$H_T - H_{298.15} = 6.97T + 0.55 \times 10^{-3}T^2 + 0.28 \times 10^5 T^{-1}$$
$$- 2,221 \text{ (0.6 percent; 298°–2,000° K.);}$$
$$C_p = 6.97 + 1.10 \times 10^{-3}T - 0.28 \times 10^5 T^{-2}.$$

BROMIDE

TABLE 693.—*Heat content and entropy of AgBr(c, l)*

[Base, crystals at 298.15° K.; mol. wt., 187.80]

T, ° K.	$H_T - H_{298.15}$, cal./mole	$S_T - S_{298.15}$, cal./deg. mole	T, ° K.	$H_T - H_{298.15}$, cal./mole	$S_T - S_{298.15}$, cal./deg. mole
400	1,355	3.90	703(l)	8,520	16.16
500	2,840	7.21	800	9,970	18.08
600	4,480	10.19	900	11,460	19.83
700	6,275	12.96	1,000	12,950	21.40
703(c)	6,330	13.04			

$AgBr(c)$:

$$H_T - H_{298.15} = 7.93T + 7.70 \times 10^{-3}T^2 - 3,049$$
(0.1 percent; 298°–703° K.);
$$C_p = 7.93 + 15.40 \times 10^{-3}T;$$
$$\Delta H_{703}(fusion) = 2,190.$$

$AgBr(l)$:

$$H_T - H_{298.15} = 14.90T - 1,950 \text{ (0.1 percent;}$$
703°–1,000° K.);
$$C_p = 14.90.$$

TABLE 694.—*Heat content and entropy of AgBr(g)*

[Base, ideal gas at 298.15° K.; mol. wt., 187.80]

T, ° K.	$H_T - H_{298.15}$, cal./mole	$S_T - S_{298.15}$, cal./deg. mole	T, ° K.	$H_T - H_{298.15}$, cal./mole	$S_T - S_{298.15}$, cal./deg. mole
400	895	2.58	1,000	6,230	10.72
500	1,780	4.56	1,200	8,010	12.34
600	2,665	6.17	1,400	9,795	13.72
700	3,555	7.54	1,600	11,585	14.92
800	4,445	8.73	1,800	13,370	15.97
900	5,335	9.78	2,000	15,160	16.91

$AgBr(g)$:

$$H_T - H_{298.15} = 8.94T + 0.20 \times 10^5 T^{-1} - 2,733$$
(0.1 percent; 298°–2,000° K.);
$$C_p = 8.94 - 0.20 \times 10^5 T^{-2}.$$

CHLORIDE

TABLE 695.—*Heat content and entropy of AgCl(c, l)*

[Base, crystals at 298.15° K.; mol. wt., 143.34]

T, ° K.	$H_T - H_{298.15}$, cal./mole	$S_T - S_{298.15}$, cal./deg. mole	T, ° K.	$H_T - H_{298.15}$, cal./mole	$S_T - S_{298.15}$, cal./deg. mole
400	1,320	3.81	728(l)	9,160	16.68
500	2,720	6.93	800	10,310	18.19
600	4,150	9.54	900	11,910	20.07
700	5,660	11.86	1,000	12,510	21.76
728(c)	6,080	12.45			

$AgCl(c)$:

$$H_T - H_{298.15} = 14.88T + 0.50 \times 10^{-3}T^2 + 2.70 \times 10^5 T^{-1}$$
$$- 5,387 \text{ (0.2 percent; 298°–728° K.);}$$
$$C_p = 14.88 + 1.00 \times 10^{-3}T - 2.70 \times 10^5 T^{-2};$$
$$\Delta H_{728}(fusion) = 3,080.$$

$AgCl(l)$:

$$H_T - H_{298.15} = 16.00T - 2,490 \text{ (0.1 percent;}$$
728°–1,000° K.);
$$C_p = 16.00.$$

TABLE 1h7. (Continued)

TABLE 696.—*Heat content and entropy of AgCl(g)*

[Base, ideal gas at 298.15° K.; mol. wt., 143.34]

T, ° K.	$H_T-H_{298.15}$, cal./mole	$S_T-S_{298.15}$, cal./deg. mole	T, ° K.	$H_T-H_{298.15}$, cal./mole	$S_T-S_{298.15}$, cal./deg. mole
400	880	2.54	1,000	6,185	10.63
500	1,775	4.49	1,200	7,970	12.26
600	2,635	6.09	1,400	9,755	13.63
700	3,520	7.46	1,600	11,540	14.82
800	4,410	8.65	1,800	13,325	15.87
900	5,295	9.69	2,000	15,110	16.81

AgCl(g):

$$H_T - H_{298.15} = 8.93T + 0.34 \times 10^5 T^{-1} - 2,777 \quad (0.1 \text{ percent};$$
$$298°\text{-}2,000° \text{ K.});$$
$$C_p = 8.93 - 0.34 \times 10^5 T^{-2}.$$

FLUORIDE

AgF·2H₂O(c):
$$C_p = 31 \ (285° \text{ K.}).$$

AgF·4H₂O(c):
$$C_p = 50 \ (285° \text{ K.}).$$

IODIDE

TABLE 697.—*Heat content and entropy of AgI(c)*

[Base, α-crystals at 298.15° K.; mol. wt., 234.79]

T, ° K.	$H_T-H_{298.15}$, cal./mole	$S_T-S_{298.15}$, cal./deg. mole	T, ° K.	$H_T-H_{298.15}$, cal./mole	$S_T-S_{298.15}$, cal./deg. mole
400	1,450	4.17	600	5,670	13.24
423(α)	1,810	5.04	700	7,020	15.31
423(β)	3,280	8.52	800	8,370	17.11
500	4,320	10.77			

AgI(α):
$$H_T - H_{298.15} = 5.82T + 12.05 \times 10^{-3} T^2 - 2,806 \quad (0.1 \text{ percent};$$
$$298°\text{-}423° \text{ K.});$$
$$C_p = 5.82 + 24.10 \times 10^{-3} T;$$
$$\Delta H_{423}(transition) = 1,470.$$

AgI(β):
$$H_T - H_{298.15} = 13.50T - 2,430 \quad (0.1 \text{ percent};$$
$$423°\text{-}800° \text{ K.});$$
$$C_p = 13.50.$$

TABLE 698.—*Heat content and entropy of AgI(g)*

[Base, ideal gas at 298.15° K.; mol. wt., 234.79]

T, ° K.	$H_T-H_{298.15}$, cal./mole	$S_T-S_{298.15}$, cal./deg. mole	T, ° K.	$H_T-H_{298.15}$, cal./mole	$S_T-S_{298.15}$, cal./deg. mole
400	900	2.60	1,000	6,240	10.74
500	1,785	4.57	1,200	8,025	12.37
600	2,675	6.19	1,400	9,815	13.75
700	3,565	7.56	1,600	11,600	14.94
800	4,460	8.76	1,800	13,390	16.00
900	5,350	9.81	2,000	15,180	16.94

AgI(g):
$$H_T - H_{298.15} = 8.94T + 0.13 \times 10^5 T^{-1} - 2,709 \quad (0.1 \text{ percent};$$
$$298°\text{-}2,000° \text{ K.});$$
$$C_p = 8.94 - 0.13 \times 10^5 T^{-2}.$$

CARBONATE

TABLE 699.—*Heat content and entropy of Ag₂CO₃(c)*

[Base, crystals at 298.15° K.; mol. wt., 275.77]

T, ° K.	$H_T-H_{298.15}$, cal./mole	$S_T-S_{298.15}$, cal./deg. mole	T, ° K.	$H_T-H_{298.15}$, cal./mole	$S_T-S_{298.15}$, cal./deg. mole
350	1,420	4.39	450	4,350	11.74
400	2,860	8.23	500	5,900	15.00

AgCO₃(c):
$$H_T - H_{298.15} = 19.57T + 12.18 \times 10^{-3} T^2 - 6,918 \quad (0.2 \text{ per-}$$
$$\text{cent}; 298°\text{-}500° \text{ K.});$$
$$C_p = 19.57 + 24.36 \times 10^{-3} T.$$

CHROMATE

Ag₂CrO₄(c):
$$C_p = 34.00 \ (298° \text{ K.}).$$

CHLORITE

AgClO₂(c):
$$C_p = 20.87 \ (298° \text{ K.}).$$

CYANATE

AgCNO(c):
$$\overline{C}_p = 18.7 \ (273°\text{-}353° \text{ K.}).$$

(AgCNO)₃(c):
$$\overline{C}_p = 44.1 \ (273°\text{-}353° \text{ K.})$$

IODATE

AgIO₃(c):
$$C_p = 24.60 \ (298° \text{ K.}).$$

NITRATE

TABLE 700.—*Heat content and entropy of AgNO₃(c, l)*

[Base, α-crystals at 298.15° K.; mol. wt., 169.89]

T, ° K.	$H_T-H_{298.15}$, cal./mole	$S_T-S_{298.15}$, cal./deg. mole	T, ° K.	$H_T-H_{298.15}$, cal./mole	$S_T-S_{298.15}$, cal./deg. mole
400	2,500	7.18	484(l)	8,090	19.31
433(α)	3,410	9.37	500	8,580	20.31
433(β)	4,020	10.78	600	11,640	25.89
484(β)	5,330	13.61			

AgNO₃(α):
$$H_T - H_{298.15} = 8.76T + 22.60 \times 10^{-3} T^2 - 4,621$$
$$(0.1 \text{ percent}; 298\text{-}433° \text{ K.});$$
$$C_p = 8.76 + 45.20 \times 10^{-3} T;$$
$$\Delta H_{433}(transition) = 610.$$

AgNO₃(β):
$$H_T - H_{298.15} = 25.50T - 7,015 \quad (0.1 \text{ percent};$$
$$433°\text{-}484° \text{ K.});$$
$$C_p = 25.50;$$
$$\Delta H_{484}(fusion) = 2,760.$$

TABLE 1h7. (*Continued*)

$AgNO_3(l)$:

$$H_T - H_{298.15} = 30.60T - 6,720 \text{ (0.1 percent;}$$
$$484°-600° \text{ K.);}$$
$$C_p = 30.60.$$

NITRITE

$AgNO_2(c)$:

$$C_p = 19.17 \text{ (298° K.).}$$

PHOSPHATE

$Ag_3PO_4(c)$:

$$\overline{C_p} = 37.5 \text{ (293°-325° K.).}$$

SILICATE

Ag_2SiO_3 (*amorphous*):

$$C_p = 30.36 \text{ (298° K.).}$$

SULFATE

$Ag_2SO_4(c)$:

$$C_p = 23.10 + 27.90 \times 10^{-3}T(\text{estimated}) \text{ (298°-930° K.).}$$

SILVER-HYDROGEN PARAPERIODATE

$Ag_2H_3IO_6(c)$:

$$C_p = 47.34 \text{ (298° K.).}$$

SILVER-MERCURY IODIDE

$Ag_2HgI_4(c)$:

$$C_p = 47.4 \text{ (298° K.).}$$

SILVER-ALUMINUM COMPOUNDS

TABLE 701.—*Heat content and entropy of* $AgAl_{12}(c)$

[Base, crystals at 298.15° K.; mol. wt., 431.64]

T, ° K.	$H_T-H_{298.15}$, cal./mole	$S_T-S_{298.15}$, cal./deg. mole	T, ° K.	$H_T-H_{298.15}$, cal./mole	$S_T-S_{298.15}$, cal./deg. mole
400	8,000	23.05	700	35,250	73.23
500	16,250	41.44	800	46,120	87.74
600	25,250	57.84			

$AgAl_{12}(c)$:

$$H_T - H_{298.15} = 59.80T + 27.00 \times 10^{-3}T^2 - 20,229$$
$$(1.0 \text{ percent; } 298°-800° \text{ K.);}$$
$$C_p = 59.80 + 54.00 \times 10^{-3}T.$$

TABLE 702.—*Heat content and entropy of* $Ag_2Al(c)$

[Base, crystals at 298.15° K.; mol. wt., 242.74]

T, ° K.	$H_T-H_{298.15}$, cal./mole	$S_T-S_{298.15}$, cal./deg. mole	T, ° K.	$H_T-H_{298.15}$, cal./mole	$S_T-S_{298.15}$, cal./deg. mole
400	1,890	5.45	700	7,700	16.25
500	3,780	9.66	800	9,840	19.10
600	5,710	13.18	900	12,100	21.76

$Ag_2Al(c)$:

$$H_T - H_{298.12} = 16.07T + 3.24 \times 10^{-3}T^2 - 5,079$$
$$(0.6 \text{ percent; } 298°-900° \text{ K.);}$$
$$C_p = 16.07 + 6.48 \times 10^{-3}T.$$

TABLE 703.—*Heat content and entropy of* $Ag_3Al(c)$

[Base, α-crystals at 298.15° K.; mol. wt., 350.62]

T, ° K.	$H_T-H_{298.15}$, cal./mole	$S_T-S_{298.15}$, cal./deg. mole	T, ° K.	$H_T-H_{298.15}$, cal./mole	$S_T-S_{298.15}$, cal./deg. mole
400	2,520	7.27	800	12,900	25.13
500	5,000	12.80	883(α)	15,180	27.84
600	7,540	17.43	883(β)	16,280	29.09
700	10,200	21.53	900	16,750	29.62

$Ag_3Al(α)$:

$$H_T - H_{298.15} = 22.22T + 3.16 \times 10^{-3}T^2$$
$$-6,906 \text{ (0.4 percent; } 298°-883° \text{ K.);}$$
$$C_p = 22.22 + 6.32 \times 10^{-3}T;$$
$$\Delta H_{883}(\text{transition}) = 1,100.$$

$Ag_3Al(β)$:

$$H_T - H_{298.15} = 27.80T - 8,270 \text{ (0.1 percent; } 883°-900° \text{ K.);}$$
$$C_p = 27.80.$$

SILVER-ANTIMONY COMPOUND

TABLE 704.—*Heat content and entropy of* $Ag_3Sb(c)$

(Base, crystals at 298.15° K.; mol. wt., 445.40)

T, ° K.	$H_T-H_{298.15}$, cal./mole	$S_T-S_{298.15}$, cal./deg. mole	T, ° K.	$H_T-H_{298.15}$, cal./mole	$S_T-S_{298.15}$, cal./deg. mole
400	2,570	7.40	600	8,070	18.51
500	5,230	13.33	700	11,080	23.14

$Ag_3Sb(c)$:

$$H_T - H_{298.15} = 19.53T + 8.00 \times 10^{-3}T^2$$
$$-6,534(0.2 \text{ percent; } 298°-700° \text{ K.);}$$
$$C_p = 19.53 + 16.00 \times 10^{-3}T.$$

SELENIDE

TABLE 705.—*Heat content and entropy of* $Ag_2Se(c)$

[Base, α-crystals at 298.15° K.; mol. wt., 294.72]

T, ° K.	$H_T-H_{298.15}$, cal./mole	$S_T-S_{298.15}$, cal./deg. mole	T, ° K.	$H_T-H_{298.15}$, cal./mole	$S_T-S_{298.15}$, cal./deg. mole
400	2,120	6.10	406(β)	3,860	10.39
406 (α)	2,250	6.42	500	5,780	14.64

$Ag_2Se(α)$:

$$H_T - H_{298.15} = 15.35T + 7.79 \times 10^{-3}T^2$$
$$-5,269(0.2 \text{ percent; } 298°-406° \text{ K.);}$$
$$C_p = 15.35 + 15.58 \times 10^{-3}T;$$
$$\Delta H_{406}(\text{transition}) = 1,610.$$

$Ag_2Se(β)$:

$$H_T - H_{298.15} = 20.40T - 4,420 \text{ (0.1 percent;}$$
$$406°-500° \text{ K.);}$$
$$C_p = 20.40.$$

TABLE 1h7. (*Continued*)

TELLURIDE

TABLE 706.—*Heat content and entropy of $Ag_2Te(c)$*

[Base, α-crystals at 298.15° K.; mol. wt., 343.37]

T, ° K.	$H_T - H_{298.15}$, cal./mole	$S_T - S_{298.15}$, cal./deg. mole	T, ° K.	$H_T - H_{298.15}$, cal./mole	$S_T - S_{298.15}$, cal./deg. mole
400	2,350	6.78	500	4,750	12.17
410 (α)	2,585	7.36	600	6,970	16.22
410(β)	2,750	7.76	700	9,190	19.64

$Ag_2Te(α)$:

$$H_T - H_{298.15} = 23.10T - 6,887 \ (0.1 \text{ percent}; 298°-410° \text{ K.});$$

$$C_p = 23.10;$$

$$\Delta H_{410}(transition) = 165.$$

$Ag_2Te(β)$:

$$H_T - H_{298.15} = 22.20T - 6,350 \ (0.1 \text{ percent}; 410°-700° \text{ K.});$$

$$C_p = 22.20.$$

SODIUM AND ITS COMPOUNDS

ELEMENT

TABLE 707.—*Heat content and entropy of $Na(c, l, g)$*

[Base, crystals at 298.15° K.; atomic wt., 22.99]

T, ° K.	$H_T - H_{298.15}$, cal./mole	$S_T - S_{298.15}$, cal./deg. mole	T, ° K.	$H_T - H_{298.15}$, cal./mole	$S_T - S_{298.15}$, cal./deg. mole
350	360	1.11	1,178(*l*)	6,840	11.47
371(*c*)	514	1.54	1,178(*g*)	30,220	31.32
371(*l*)	1,136	3.21	1,200	30,330	31.41
400	1,355	3.78	1,300	30,825	31.80
500	2,095	5.44	1,400	31,325	32.18
600	2,820	6.76	1,500	31,820	32.52
700	3,520	7.84	1,600	32,315	32.84
800	4,220	8.77	1,700	32,815	33.14
900	4,910	9.58	1,800	33,310	33.42
1,000	5,595	10.31	1,900	33,805	33.69
1,100	6,295	10.98	2,000	34,305	33.94

$Na(c)$:

$$H_T - H_{298.15} = 4.02T + 4.52 \times 10^{-3} T^2$$
$$-1,599(0.3 \text{ percent}; 298°-371° \text{ K.});$$

$$C_p = 4.02 + 9.04 \times 10^{-3}T;$$

$$\Delta H_{371}(fusion) = 622.$$

$Na(l)$:

$$H_T - H_{298.15} = 6.83T - 1.08 \times 10^5 T^{-1}$$
$$-1,107(0.2 \text{ percent}; 371°-1,178° \text{ K.});$$

$$C_p = 6.83 + 1.08 \times 10^5 T^{-2}.$$

$$\Delta H_{1178}(vaporization) = 23,380$$

$Na(g)$:

$$H_T - H_{298.15} = 4.97T + 24,365 \ (0.1 \text{ percent};$$
$$1,178°-2,000° \text{ K.});$$

$$C_p = 4.97.$$

TABLE 708.—*Heat content and entropy of $Na(g)$*

[Base, ideal gas at 298.15° K.; atomic wt., 22.99]

T, ° K.	$H_T - H_{298.15}$, cal./mole	$S_T - S_{298.15}$, cal./deg. mole	T, ° K.	$H_T - H_{298.15}$, cal./mole	$S_T - S_{298.15}$, cal./deg. mole
400	505	1.46	1,900	7,960	9.20
500	1,005	2.57	2,000	8,460	9.46
600	1,500	3.48	2,200	9,455	9.93
700	1,995	4.24	2,400	10,450	10.37
800	2,495	4.90	2,600	11,450	10.77
900	2,990	5.49	2,800	12,455	11.14
1,000	3,490	6.01	3,000	13,470	11.49
1,100	3,985	6.49	3,500	16,055	12.29
1,200	4,480	6.92	4,000	18,770	13.01
1,300	4,980	7.32	4,500	21,700	13.70
1,400	5,475	7.69	5,000	25,015	14.40
1,500	5,970	8.03	6,000	33,755	15.98
1,600	6,470	8.35	7,000	46,850	18.02
1,700	6,965	8.65	8,000	65,295	20.45
1,800	7,460	8.93			

$Na(g)$:

$$H_T - H_{298.15} = 4.97T - 1,482 \ (0.1 \text{ percent};$$
$$298°-3,000° \text{ K.});$$

$$C_p = 4.97.$$

TABLE 709.—*Heat content and entropy of $Na_2(g)$*

[Base, ideal gas at 298.15° K.; mol. wt., 45.98]

T, ° K.	$H_T - H_{298.15}$, cal./mole	$S_T - S_{298.15}$, cal./deg. mole	T, ° K.	$H_T - H_{298.15}$, cal./mole	$S_T - S_{298.15}$, cal./deg. mole
400	920	2.64	1,300	9,240	13.51
500	1,825	4.67	1,400	10,185	14.21
600	2,735	6.33	1,500	11,135	14.87
700	3,655	7.75	1,600	12,090	15.48
800	4,580	8.99	1,800	14,010	16.61
900	5,500	10.07	2,000	15,945	17.63
1,000	6,435	11.06	2,200	17,895	18.56
1,100	7,360	11.94	2,400	19,860	19.41
1,200	8,305	12.76	2,600	21,845	20.21

$Na_2(g)$:

$$H_T - H_{298.15} = 8.96T + 0.18 \times 10^{-3} T^2 + 0.10 \times 10^5 T^{-1}$$
$$-2,721 \ (0.1 \text{ percent}; 298°-2,600° \text{ K.});$$

$$C_p = 8.96 + 0.36 \times 10^{-3}T - 0.10 \times 10^5 T^{-2}.$$

OXIDES

TABLE 710.—*Heat content and entropy of $Na_2O(c)$*

[Base, crystals at 298.15° K.; mol. wt., 61.98]

T, ° K.	$H_T - H_{298.15}$, cal./mole	$S_T - S_{298.15}$, cal./deg. mole	T, ° K.	$H_T - H_{298.15}$, cal./mole	$S_T - S_{298.15}$, cal./deg. mole
400	1,750	5.05	800	9,350	18.16
500	3,600	9.17	900	11,350	20.52
600	5,500	12.63	1,000	13,500	22.78
700	7,400	15.56	1,100	15,750	24.93

$Na_2O(c)$:

$$H_T - H_{298.15} = 15.70T + 2.70 \times 10^{-3} T^2 - 4,921 \ (0.7 \text{ percent};$$
$$298°-1,100° \text{ K.});$$

$$C_p = 15.70 + 5.40 \times 10^{-3}T.$$

$NaO_2(c)$:

$$C_p = 17.24 \ (298° \text{ K.}).$$

$Na_2O_2(c)$:

$$C_p = 21.35 \ (298° \text{ K.}).$$

TABLE 1h7. (*Continued*)

HYDROXIDE

TABLE 711.—*Heat content and entropy of NaOH(c, l)*

[Base, α-crystals at 298.15° K.; mol. wt., 40.00]

T, ° K.	$H_T - H_{298.15}$, cal./mole	$S_T - S_{298.15}$, cal./deg. mole	T, ° K.	$H_T - H_{298.15}$, cal./mole	$S_T - S_{298.15}$, cal./deg. mole
400	1,510	4.35	592.3(*l*)	8,010	16.60
500	3,175	8.06	600	8,170	16.86
566(α)	4,430	10.41	700	10,220	20.02
566(β)	5,950	13.10	800	12,225	22.74
575	6,135	13.42	900	14,275	25.12
592.3(β)	6,490	14.03	1,000	16,285	27.24

NaOH(α):

$H_T - H_{298.15} = 0.24T + 16.21 \times 10^{-3}T^2 - 3.87 \times 10^5 T^{-1}$
$\quad -215$ (0.2 percent; 298°–566° K.);
$C_p = 0.24 + 32.42 \times 10^{-3}T + 3.87 \times 10^5 T^{-2}$;
$\quad \Delta H_{566}(transition) = 1,520.$

NaOH(β):

$H_T - H_{298.15} = 20.53T - 5,670$ (0.1 percent; 566°–592.3° K.);
$\quad C_p = 20.53$;
$\quad \Delta H_{592.3}(fusion) = 1,520.$

NaOH(*l*):

$H_T - H_{298.15} = 21.40T - 0.69 \times 10^{-3}T^2 - 4,423$ (0.1 percent; 592.3°–1,000° K.);
$\quad C_p = 21.40 - 1.38 \times 10^{-3}T.$

SULFIDE

TABLE 712.—*Heat content and entropy of Na₂S(c)*

[Base, crystals at 298.15° K.; mol. wt., 78.05]

T, ° K.	$H_T - H_{298.15}$, cal./mole	$S_T - S_{298.15}$, cal./deg. mole	T, ° K.	$H_T - H_{298.15}$, cal./mole	$S_T - S_{298.15}$, cal./deg. mole
400	2,050	5.91	800	10,400	20.36
500	4,100	10.49	900	12,500	22.83
600	6,200	14.32	1,000	14,650	25.09
700	8,300	17.55			

Na₂S(c):

$H_T - H_{298.15} = 19.81T + 0.82 \times 10^{-3}T^2 - 5,979$ (0.4 percent; 298°–1,000° K.);
$\quad C_p = 19.81 + 1.64 \times 10^{-3}T.$

HYDRIDES

TABLE 713.—*Heat content and entropy of NaH(g)*

[Base, ideal gas at 298.15° K.; mol. wt., 24.00]

T, ° K.	$H_T - H_{298.15}$, cal./mole	$S_T - S_{298.15}$, cal./deg. mole	T, ° K.	$H_T - H_{298.15}$, cal./mole	$S_T - S_{298.15}$, cal./deg. mole
400	750	2.16	1,000	5,660	9.59
500	1,520	3.88	1,200	7,380	11.15
600	2,315	5.33	1,400	9,120	12.49
700	3,130	6.58	1,600	10,870	13.66
800	3,960	7.69	1,800	12,630	14.70
900	4,805	8.69	2,000	14,395	15.63

NaH(g):

$H_T - H_{298.15} = 7.90T + 0.31 \times 10^{-3}T^2 + 0.78 \times 10^5 T^{-1}$
$\quad -2,645$ (0.6 percent; 298°–2,000° K.);
$\quad C_p = 7.90 + 0.62 \times 10^{-3}T - 0.78 \times 10^5 T^{-2}.$

TABLE 714.—*Heat content and entropy of NaD(g)*

[Base, ideal gas at 298.15° K.; mol. wt., 25.00]

T, ° K.	$H_T - H_{298.15}$, cal./mole	$S_T - S_{298.15}$, cal./deg. mole	T, ° K.	$H_T - H_{298.15}$, cal./mole	$S_T - S_{298.15}$, cal./deg. mole
400	790	2.28	1,000	5,870	9.98
500	1,600	4.08	1,200	7,625	11.58
600	2,435	5.60	1,400	9,385	12.94
700	3,280	6.91	1,600	11,155	14.12
800	4,135	8.05	1,800	12,925	15.16
900	5,000	9.07	2,000	14,700	16.10

NaD(g):

$H_T - H_{298.15} = 8.45T + 0.15 \times 10^{-3}T^2 + 0.86 \times 10^5 T^{-1}$
$\quad -2,821$ (0.4 percent; 298°–2,000° K.);
$\quad C_p = 8.45 + 0.30 \times 10^{-3}T - 0.86 \times 10^5 T^{-2}.$

AZIDE

NaN₃(c):

$\overline{C_p} = 19.1$ (273°–373° K.).

BROMIDE

TABLE 715.—*Heat content and entropy of NaBr(c)*

[Base, crystals at 298.15° K.; mol. wt., 102.91]

T, ° K.	$H_T - H_{298.15}$, cal./mole	$S_T - S_{298.15}$, cal./deg. mole	T, ° K.	$H_T - H_{298.15}$, cal./mole	$S_T - S_{298.15}$, cal./deg. mole
350	650	2.01	500	2,565	6.56
400	1,285	3.71	550	3,215	7.80
450	1,920	5.20			

NaBr(c):

$H_T - H_{298.15} = 11.87T + 1.05 \times 10^{-3}T^2 - 3,632$
$\quad$ (0.1 percent; 298°–550° K.);
$\quad C_p = 11.87 + 2.10 \times 10^{-3}T.$

TABLE 716.—*Heat content and entropy of NaBr(g)*

[Base, ideal gas at 298.15° K.; mol. wt., 102.91]

T, ° K.	$H_T - H_{298.15}$, cal./mole	$S_T - S_{298.15}$, cal./deg. mole	T, ° K.	$H_T - H_{298.15}$, cal./mole	$S_T - S_{298.15}$, cal./deg. mole
400	890	2.57	1,000	6,290	10.80
500	1,780	4.55	1,200	8,110	12.46
600	2,675	6.19	1,400	9,940	13.87
700	3,575	7.57	1,600	11,780	15.10
800	4,480	8.78	1,800	13,630	16.19
900	5,385	9.85	2,000	15,490	17.71

NaBr(g):

$H_T - H_{298.15} = 8.95T + 0.09 \times 10^{-3}T^2 + 0.32 \times 10^5 T^{-1}$
$\quad -2,784$ (0.1 percent; 298°–2,000° K.);
$\quad C_p = 8.95 + 0.18 \times 10^{-3}T - 0.32 \times 10^5 T^{-2}.$

TABLE 1h7. (*Continued*)

CHLORIDE

TABLE 717.—*Heat content and entropy of NaCl(c, l)*

[Base, crystals at 298.15° K.; mol. wt., 58.45]

T, ° K.	$H_T-H_{298.15}$, cal./mole	$S_T-S_{298.15}$, cal./deg. mole	T, ° K.	$H_T-H_{298.15}$, cal./mole	$S_T-S_{298.15}$, cal./deg. mole
400	1,240	3.58	1,073(l)	17,430	23.44
500	2,510	6.41	1,100	17,860	23.84
600	3,830	8.82	1,200	19,460	25.24
700	5,190	10.91	1,300	21,060	26.52
800	6,590	12.78	1,400	22,660	27.70
900	8,020	14.46	1,500	24,260	28.81
1,000	9,480	16.00	1,600	25,860	29.84
1,073(c)	10,580	17.06	1,700	27,460	30.81

NaCl(c):

$$H_T-H_{298.15}=10.98T+1.95\times10^{-3}T^2-3,447$$
$$(0.5 \text{ percent}; 298°-1,073° \text{ K.});$$
$$C_p=10.98+3.90\times10^{-3}T.$$
$$\Delta H_{1073}(fusion)=6,850.$$

NaCl(l):

$$H_T-H_{298.15}=16.00T+260 \ (0.1 \text{ percent};$$
$$1,073°-1,700° \text{ K.});$$
$$C_p=16.00.$$

TABLE 718.—*Heat content and entropy of NaCl(g)*

[Base, ideal gas at 298.15° K.; mol. wt., 58.45]

T, ° K.	$H_T-H_{298.15}$, cal./mole	$S_T-S_{298.15}$, cal./deg. mole	T, ° K.	$H_T-H_{298.15}$, cal./mole	$S_T-S_{298.15}$, cal./deg. mole
400	880	2.54	1,000	6,250	10.72
500	1,760	4.50	1,200	8,070	12.38
600	2,650	6.12	1,400	9,900	13.79
700	3,545	7.50	1,600	11,740	15.02
800	4,445	8.70	1,800	13,585	16.10
900	5,345	9.77	2,000	15,440	17.08

NaCl(g):

$$H_T-H_{298.15}=8.91T+0.10\times10^{-3}T^2+0.40\times10^5T^{-1}$$
$$-2,800 \ (0.1 \text{ percent}; 298°-2,000° \text{ K.});$$
$$C_p=8.91+0.20\times10^{-3}T-0.40\times10^5T^{-2}.$$

FLUORIDE

TABLE 719.—*Heat content and entropy of NaF(c, l)*

[Base, crystals at 298.15° K.; mol. wt., 41.99]

T, ° K.	$H_T-H_{298.15}$, cal./mole	$S_T-S_{298.15}$, cal./deg. mole	T, ° K.	$H_T-H_{298.15}$, cal./mole	$S_T-S_{298.15}$, cal./deg. mole
400	1,180	3.40	1,285(c)	13,210	18.85
500	2,380	6.08	1,285(l)	21,240	25.10
600	3,620	8.34	1,300	21,490	25.30
700	4,900	10.31	1,400	23,130	26.51
800	6,220	12.07	1,500	24,770	27.64
900	7,570	13.66	1,600	26,410	28.70
1,000	8,970	15.13	1,700	28,050	29.70
1,100	10,410	16.50	1,800	29,690	30.64
1,200	11,900	17.80	1,900	31,330	31.52

NaF(c):

$$H_T-H_{298.15}=10.40T+1.94\times10^{-3}T^2+0.33\times10^5T^{-1}$$
$$-3,384 \ (0.3 \text{ percent}; 298°-1,285° \text{ K.});$$
$$C_p=10.40+3.88\times10^{-3}T-0.33\times10^5T^{-2};$$
$$\Delta H_{1285}(fusion)=8,030.$$

NaF(l):

$$H_T-H_{298.15}=16.40T+170 \ (0.1 \text{ percent};$$
$$1,285°-1,900° \text{ K.});$$
$$C_p=16.40.$$

IODIDE

NaI(c):

$$C_p=12.50+1.62\times10^{-3}T(estimated) \ (298°-936° \text{ K.}).$$

TABLE 720.—*Heat content and entropy of NaI(g)*

[Base, ideal gas at 298.15° K.; mol. wt., 149.90]

T, ° K.	$H_T-H_{298.15}$, cal./mole	$S_T-S_{298.15}$, cal./deg. mole	T, ° K.	$H_T-H_{298.15}$, cal./mole	$S_T-S_{298.15}$, cal./deg. mole
400	895	2.58	1,000	6,320	10.85
500	1,790	4.58	1,200	8,145	12.52
600	2,690	6.22	1,400	9,975	13.93
700	3,595	7.62	1,600	11,815	15.16
800	4,500	8.83	1,800	13,665	16.24
900	5,410	9.89	2,000	15,520	17.22

NaI(g):

$$H_T-H_{298.15}=9.04T+0.06\times10^{-3}T+0.34\times10^5T^{-1}$$
$$-2,815 \ (0.1 \text{ percent}; 298°-2,000° \text{ K.});$$
$$C_p=9.04+0.12\times10^{-3}T-0.34\times10^5T^{-2}.$$

ALUMINATE

TABLE 721.—*Heat content and entropy of NaAlO_2(c)*

[Base, α-crystals at 298.15° K.; mol. wt., 81.97]

T, ° K.	$H_T-H_{298.15}$, cal./mole	$S_T-S_{298.15}$, cal./deg. mole	T, ° K.	$H_T-H_{298.15}$, cal./mole	$S_T-S_{298.15}$, cal./deg. mole
400	1,910	5.49	1,000	15,960	26.61
500	3,990	10.13	1,100	18,430	28.96
600	6,200	14.16	1,200	20,940	31.15
700	8,490	17.68	1,300	23,500	33.19
740(α)	9,440	19.00	1,400	26,100	35.12
740(β)	9,750	19.42	1,500	28,730	36.94
800	11,160	21.26	1,600	31,400	38.66
900	13,540	24.06	1,700	34,110	40.30

NaAlO_2(α):

$$H_T-H_{298.15}=19.18T+3.57\times10^{-3}T^2+3.36\times10^5T^{-1}$$
$$-7,163 \ (0.2 \text{ percent}; 298°-740° \text{ K.});$$
$$C_p=19.18+7.14\times10^{-3}T-3.36\times10^5T^{-2};$$
$$\Delta H_{740}(transition)=310.$$

NaAlO_2(β):

$$H_T-H_{298.15}=20.21T+2.12\times10^{-3}T^2-6,363$$
$$(0.1 \text{ percent}; 740°-1,700° \text{ K.});$$
$$C_p=20.21+4.24\times10^{-3}T.$$

BORATES

NaBO_2(c):

$$C_p=11.88+13.01\times10^{-3}T(estimated) \ (298°-1,239° \text{ K.}).$$

Na_2B_4O_7(c):

$$C_p=44.64 \ (298° \text{ K.}).$$

Na_2B_4O_7(gl):

$$C_p=44.42 \ (298° \text{ K.}).$$

Na_2B_4O_7·10H_2O(c):

$$\overline{C_p}=147 \ (292°-323° \text{ K.}).$$

TABLE 1h7. (Continued)

BOROHYDRIDE

TABLE 722.—Heat content and entropy of $NaBH_4(c)$

[Base, crystals at 298.15° K., mol. wt., 37.84]

T, ° K.	$H_T-H_{298.15}$, cal./mole	$S_T-S_{298.15}$, cal./deg. mole	T, ° K.	$H_T-H_{298.15}$, cal./mole	$S_T-S_{298.15}$, cal./deg. mole
400	2,205	6.35	600	7,070	16.16
500	4,555	11.58	700	9,725	20.25

$NaBH_4(c)$:

$H_T-H_{298.15}=17.42T+7.25\times10^{-3}T^2+0.96\times10^5T^{-1}$

$-6,160$ (0.1 percent; 298°–700° K.);

$C_p=17.42+14.50\times10^{-3}T-0.96\times10^5T^{-2}$.

CHLORATE

TABLE 723.—Heat content and entropy of $NaClO_3(c, l)$

[Base, crystals at 298.15° K.; mol. wt., 106.45]

T, ° K.	$H_T-H_{298.15}$, cal./mole	$S_T-S_{298.15}$, cal./deg. mole	T, ° K.	$H_T-H_{298.15}$, cal./mole	$S_T-S_{298.15}$, cal./deg. mole
350	1,300	4.02	528(c)	6,520	15.98
400	2,650	7.62	528(l)	11,920	26.21
450	4,080	10.99	550	12,620	27.51
500	5,620	14.23	600	14,210	30.28

$NaClO_3(c)$:

$H_T-H_{298.15}=13.07T+18.50\times10^{-3}T^2-5,541$

(0.1 percent; 298°–528° K.);

$C_p=13.07+37.00\times10^{-3}T$;

$\Delta H_{528}(fusion)=5,400$.

$NaClO_3(l)$:

$H_T-H_{298.15}=31.80T-4,870$ (0.1 percent;

528°–600° K.);

$C_p=31.80$.

CARBONATE

TABLE 724.—Heat content and entropy of $Na_2CO_3(c, l)$

[Base, crystals at 298.15° K.; mol. wt. 105.99]

T, ° K.	$H_T-H_{298.15}$, cal./mole	$S_T-S_{298.15}$, cal./deg. mole	T, ° K.	$H_T-H_{298.15}$, cal./mole	$S_T-S_{298.15}$, cal./deg. mole
400	2,750	7.91	1,100	29,400	45.49
500	5,900	14.93	1,124(c)	30,400	46.39
600	9,500	21.48	1,124(l)	37,500	52.71
700	13,300	27.34	1,200	40,900	55.63
800	17,250	32.61	1,300	45,400	59.24
900	21,250	37.32	1,400	49,900	62.57
1,000	25,300	41.58	1,500	54,400	65.68

$Na_2CO_3(c)$:

$H_T-H_{298.15}=27.13T+7.81\times10^{-3}T^2+4.78\times10^5T^{-1}$

$-10,386$ (2.0 percent; 298°–1,124° K.);

$C_p=27.13+15.62\times10^{-3}T-4.78\times10^5T^{-2}$;

$\Delta H_{1124}(fusion)=7,100$.

$Na_2CO_3(l)$:

$H_T-H_{298.15}=45.00T-13,100$ (0.1 percent;

1,124°–1,500° K.);

$C_p=45.00$.

$Na_2CO_3\cdot10H_2O(c)$:

$C_p=128$ (284° K.).

BICARBONATE

TABLE 725.—Heat content and entropy of $NaHCO_3$ (c)

[Base, crystals at 298.15° K.; mol. wt., 84.01]

T, ° K.	$H_T-H_{298.15}$, cal./mole	$S_T-S_{298.15}$, cal./deg. mole	T, ° K.	$H_T-H_{298.15}$, cal./mole	$S_T-S_{298.15}$, cal./deg. mole
350	1,140	3.52	400	2,320	6.67

$NaHCO_3(c)$:

$H_T-H_{298.15}=10.19T+18.03\times10^{-3}T^2-4,641$

(0.3 percent; 298°–400° K.);

$C_p=10.19+36.06\times10^{-3}T$.

CYANATE

$NaCNO(c)$:

$\overline{C_p}=13.1$ (273°–353° K.).

$(NaCNO)_3(c)$:

$\overline{C_p}=32.2$ (273°–353° K.).

DICHROMATE

$Na_2Cr_2O_7(c)$:

$\overline{C_p}=65.73$ (423°–520° K.);

$\overline{C_p}=69.75$ (533°–583° K.).

FERRITE

TABLE 726.—Heat content and entropy of $NaFeO_2(c, l)$

[Base, α-crystals at 298.15° K.; mol. wt., 110.84]

T, ° K.	$H_T-H_{298.15}$, cal./mole	$S_T-S_{298.15}$, cal./deg. mole	T, ° K.	$H_T-H_{298.15}$, cal./mole	$S_T-S_{298.15}$, cal./deg. mole
400	2,180	6.27	1,270(γ)	26,300	37.38
500	4,540	11.53	1,300	27,160	38.05
600	7,050	16.10	1,400	30,040	40.18
700	9,680	20.15	1,500	32,920	42.17
800	12,430	23.82	1,600	35,800	44.03
870(α)	14,580	26.40	1,620(γ)	36,380	44.39
870(β)	14,580	26.40	1,620(l)	48,140	51.65
900	15,390	27.31	1,700	51,120	53.44
1,000	18,130	30.20	1,800	54,850	55.57
1,100	20,920	32.86	1,900	58,580	57.59
1,200	23,760	35.33	2,000	62,310	59.50
1,270(β)	25,780	36.97			

$NaFeO_2(\alpha)$:

$H_T-H_{298.15}=17.92T+7.05\times10^{-3}T^2+1.72\times10^5T^{-1}$

$-6,546$ (0.3 percent; 298°–870° K.);

$C_p=17.92+14.10\times10^{-3}T-1.72\times10^5T^{-2}$;

$\Delta H_{870}(transition)=0$.

$NaFeO_2(\beta)$:

$H_T-H_{298.15}=22.55T+2.55\times10^{-3}T^2-6,969$ (0.1 percent;

870°–1,270° K.);

TABLE 1h7. (*Continued*)

$C_p = 22.55 + 5.10 \times 10^{-3} T$;

$\Delta H_{1270}(transition) = 520$.

$NaFeO_2(\gamma)$:

$H_T - H_{298.15} = 28.80 T - 10,276$ (0.2 percent; 1,270°–1,620° K.);

$C_p = 28.80$;

$\Delta H_{1620}(fusion) = 11,760$.

$NaFeO_2(l)$:

$H_T - H_{298.15} = 37.30 T - 12,286$ (0.1 percent; 1,620°–2,000° K.);

$C_p = 37.30$.

NITRATE

TABLE 727.—*Heat content and entropy of* $NaNO_3(c, l)$

[Base, α-crystals at 298.15° K.; mol. wt., 85.00]

T, ° K.	$H_T - H_{298.15}$, cal./mole	$S_T - S_{298.15}$, cal./deg. mole	T, ° K.	$H_T - H_{298.15}$, cal./mole	$S_T - S_{298.15}$, cal./deg. mole
400	2,495	7.16	579.2(β)	9,140	20.59
500	5,575	14.01	579.2(l)	12,630	26.62
549.2(α)	7,260	17.22	600	13,400	27.93
549.2(β)	8,070	18.70	700	17,100	33.63

$NaNO_3(\alpha)$:

$H_T - H_{298.15} = 6.34 T + 26.66 \times 10^{-3} T^2 - 4,260$ (1.0 percent; 298°–549.2° K.);

$C_p = 6.34 + 53.32 \times 10^{-3} T$;

$\Delta H_{549.2}(transition) = 810$.

$NaNO_3(\beta)$:

$H_T - H_{298.15} = 35.70 T - 11,536$ (0.1 percent; 549.2°–579.2° K.);

$C_p = 35.70$;

$\Delta H_{579.2}(fusion) = 3,490$.

$NaNO_3(l)$·

$H_T - H_{298.15} = 37.00 T - 8,800$ (0.1 percent; 579.2°–700° K.);

$C_p = 37.00$.

PHOSPHATES

$NaPO_3(c)$:

$\overline{C_p} = 22.1$ (290°–319° K.).

$Na_4P_2O_7(c)$:

$\overline{C_p} = 60.7$ (290°–371° K.).

$Na_2HPO_4 \cdot 7H_2O(c)$:

$\overline{C_p} = 86.6$ (275°–308° K.).

$Na_2HPO_4 \cdot 12H_2O(c)$:

$\overline{C_p} = 133.3$ (275°–308° K.)

SILICATES

TABLE 728.—*Heat content and entropy of* $Na_2SiO_3(c, l)$

[Base, crystals at 298.15° K.; mol. wt., 122.07]

T, ° K.	$H_T - H_{298.15}$, cal./mole	$S_T - S_{298.15}$, cal./deg. mole	T, ° K.	$H_T - H_{298.15}$, cal./mole	$S_T - S_{298.15}$, cal./deg. mole
400	3,080	8.86	1,361(c)	39,870	54.16
500	6,300	16.03	1,361(l)	52,340	63.33
600	9,650	22.14	1,400	54,010	64.53
700	13,190	27.59	1,500	58,290	67.49
800	16,910	32.56	1,600	62,570	70.25
900	20,750	37.07	1,700	66,850	72.84
1,000	24,700	41.24	1,800	71,130	75.29
1,100	28,770	45.12	1,900	75,410	77.61
1,200	32,940	48.75	2,000	79,690	79.80
1,300	37,210	52.16			

$Na_2SiO_3(c)$:

$H_T - H_{298.15} = 31.14 T + 4.80 \times 10^{-3} T^2 + 6.47 \times 10^5 T^{-1}$
$- 11,881$ (0.7 percent; 298°–1,361° K.);

$C_p = 31.14 + 9.60 \times 10^{-3} T - 6.47 \times 10^5 T^{-2}$;

$\Delta H_{1361}(fusion) = 12,470$.

$Na_2SiO_3(l)$:

$H_T - H_{298.15} = 42.80 T - 5,910$ (0.1 percent; 298°–2,000° K.);

$C_p = 42.80$.

TABLE 729.—*Heat content and entropy of* $Na_2Si_2O_5(c)$

[Base, crystals at 298.15° K.; mol. wt., 182.16]

T, ° K.	$H_T - H_{298.15}$, cal./mole	$S_T - S_{298.15}$, cal./deg. mole	T, ° K.	$H_T - H_{298.15}$, cal./mole	$S_T - S_{298.15}$, cal./deg. mole
400	4,410	12.68	900	30,400	54.07
500	9,040	22.99	1,000	36,320	60.31
600	13,980	31.99	1,100	42,430	66.13
700	19,190	40.02	1,147	45,360	68.74
800	24,670	47.33			

$Na_2Si_2O_5(c)$:

$H_T - H_{298.15} = 44.38 T + 8.43 \times 10^{-3} T^2 + 10.67 \times 10^5 T^{-1}$
$- 17,560$ (1.0 percent; 298°–1,147° K.);

$C_p = 44.38 + 16.86 \times 10^{-3} T - 10.67 \times 10^5 T^{-2}$.

TABLE 730.—*Heat content and entropy of* $Na_2Si_2O_5(gl, l)$

[Base, glass at 298.15° K.; mol. wt., 182.16]

T, ° K.	$H_T - H_{298.15}$, cal./mole	$S_T - S_{298.15}$, cal./deg. mole	T, ° K.	$H_T - H_{298.15}$, cal./mole	$S_T - S_{298.15}$, cal./deg. mole
400	4,410	12.70	1,200	51,270	74.60
500	9,090	23.13	1,300	57,500	79.57
600	14,150	32.35	1,400	63,740	84.20
700	19,570	40.70	1,500	69,980	88.51
800	25,350	48.41	1,600	76,210	92.53
900	31,480	55.62	1,700	82,440	96.30
1,000	37,940	62.43	1,800	88,680	99.88
1,100	44,710	68.88	1,900	94,920	103.25
1,147(gl)	47,960	71.78	2,000	101,150	106.44
1,147(l)	47,960	71.78			

$Na_2Si_2O_5(gl)$:

$H_T - H_{298.15} = 32.91 T + 16.66 \times 10^{-3} T^2 + 1.64 \times 10^5 T^{-1}$
$- 11,843$ (0.2 percent; 298°–1,147° K.);

$C_p = 32.91 + 33.32 \times 10^{-3} T - 1.64 \times 10^5 T^{-2}$.

TABLE 1h7. (Continued)

$Na_2Si_2O_5(l)$:

$$H_T - H_{298.15} = 62.35T - 23,550 \ (0.1 \text{ percent};$$
$$1,147°-2,000° \text{ K.});$$
$$C_p = 62.35.$$

$Na_4SiO_4(c)$:

$$C_p = 43.79 \ (298° \text{ K.}).$$

SULFATE

TABLE 731.—*Heat content and entropy of* Na_2SO_4
(III, I, l)

[Base, III-crystals at 298.15° K.; mol. wt., 142.05]

T, ° K.	$H_T-H_{298.15}$, cal./mole	$S_T-S_{298.15}$, cal./deg. mole	T, ° K.	$H_T-H_{298.15}$, cal./mole	$S_T-S_{298.15}$, cal./deg. mole
400	3,370	9.69	1,157 (l)	44,400	63.94
500	7,280	18.39	1,200	46,430	65.66
514 (III)	7,870	19.55	1,300	51,140	69.43
514 (I)	9,660	23.03	1,400	55,860	72.93
600	13,080	29.18	1,500	60,580	76.18
700	17,260	35.62	1,600	65,300	79.23
800	21,600	41.42	1,700	70,020	82.09
900	26,120	46.74	1,800	74,730	84.78
1,000	30,850	51.72	1,900	79,450	87.33
1,100	35,830	56.46	2,000	84,170	89.75
1,157 (I)	38,730	59.04			

$Na_2SO_4(III)$:

$$H_T - H_{298.15} = 14.97T + 26.45 \times 10^{-3}T^2 - 6,815$$
$$(0.5 \text{ percent}; 298°-514° \text{ K.});$$
$$C_p = 14.97 + 52.90 \times 10^{-3}T;$$
$$\Delta H_{514}(transition) = 1,790.$$

$Na_2SO_4(I)$:

$$H_T - H_{298.15} = 29.06T + 9.67 \times 10^{-3}T^2 - 7,837$$
$$(0.1 \text{ percent}; 514°-1,157° \text{ K.});$$
$$C_p = 29.06 + 19.34 \times 10^{-3}T;$$
$$\Delta H_{1155}(fusion) = 5,670.$$

$Na_2SO_4(l)$:

$$H_T - H_{298.15} = 47.18T - 10,190 \ (0.1 \text{ percent};$$
$$1,157°-2,000° \text{ K.});$$
$$C_p = 47.18.$$

TABLE 732.—*Heat content and entropy of* $Na_2SO_4(V)$

[Base, V-crystals at 298.15° K.; mol. wt., 142.05]

T, ° K.	$H_T-H_{298.15}$, cal./mole	$S_T-S_{298.15}$, cal./deg. mole	T, ° K.	$H_T-H_{298.15}$, cal./mole	$S_T-S_{298.15}$, cal./deg. mole
350	1,670	5.16	450	5,150	13.88
400	3,335	9.60			

$Na_2SO_4(V)$:

$$H_T - H_{298.15} = 16.74T + 22.94 \times 10^{-3}T^2 - 7,030$$
$$(0.7 \text{ percent}; 298°-450° \text{ K.});$$
$$C_p = 16.74 + 45.88 \times 10^{-3}T.$$

$Na_2SO_4 \cdot 10H_2O(c)$:

$$C_p = 137.25 \ (298° \text{ K.}).$$

SULFITE

$Na_2SO_3(c)$:

$$C_p = 28.71 \ (298° \text{ K.}).$$

THIOSULFATE

$Na_2S_2O_3(c)$:

$$\overline{C_p} = 34.9 \ (293°-373° \text{ K.}).$$

$Na_2S_2O_3 \cdot 5H_2O(c)$:

$$\overline{C_p} = 86.2 \ (273°-307° \text{ K.}).$$

TITANATES

TABLE 733.—*Heat content and entropy of* $Na_2TiO_3(c, l)$

[Base, α-crystals at 298.15° K.; mol. wt., 141.88]

T, ° K.	$H_T-H_{298.15}$, cal./mole	$S_T-S_{298.15}$, cal./deg. mole	T, ° K.	$H_T-H_{298.15}$, cal./mole	$S_T-S_{298.15}$, cal./deg. mole
400	3,300	9.51	1,300	40,220	56.44
500	6,750	17.20	1,303 (β)	40,360	56.55
560 (α)	8,920	21.30	1,303 (l)	57,170	69.45
560 (β)	9,320	22.01	1,400	61,720	72.82
600	10,750	24.48	1,500	66,410	76.06
700	14,450	30.18	1,600	71,100	79.08
800	18,320	35.34	1,700	75,790	81.93
900	22,360	40.10	1,800	80,480	84.61
1,000	26,570	44.53	1,900	85,170	87.14
1,100	30,950	48.71	2,000	89,860	89.55
1,200	35,500	52.67			

$Na_2TiO_3(α)$:

$$H_T - H_{298.15} = 25.18T + 10.36 \times 10^{-3}T^2 - 8,429$$
$$(0.2 \text{ percent}; 298°-560° \text{ K.});$$
$$C_p = 25.18 + 20.72 \times 10^{-3}T;$$
$$\Delta H_{560}(transition) = 400.$$

$Na_2TiO_3(β)$:

$$H_T - H_{298.15} = 25.95T + 8.50 \times 10^{-3}T^2 - 7,880$$
$$(0.1 \text{ percent}; 560°-1,303° \text{ K.});$$
$$C_p = 25.95 + 17.00 \times 10^{-3}T;$$
$$\Delta H_{1303}(fusion) = 16,810.$$

$Na_2TiO_3(l)$:

$$H_T - H_{298.15} = 46.90T - 3,940 \ (0.1 \text{ percent};$$
$$1,303°-2,000° \text{ K.});$$
$$C_p = 46.90.$$

TABLE 734.—*Heat content and entropy of* $Na_2Ti_2O_5(c, l)$

[Base, crystals at 298.15° K.; mol. wt., 221.78]

T, ° K.	$H_T-H_{298.15}$, cal./mole	$S_T-S_{298.15}$, cal./deg. mole	T, ° K.	$H_T-H_{298.15}$, cal./mole	$S_T-S_{298.15}$, cal./deg. mole
400	4,880	14.07	1,258 (l)	77,670	96.20
500	9,900	25.26	1,300	80,550	98.44
600	15,070	34.69	1,400	87,400	103.52
700	20,350	42.82	1,500	94,250	108.25
800	25,730	50.00	1,600	101,100	112.67
900	31,200	56.44	1,700	107,950	116.82
1,000	36,750	62.29	1,800	114,800	120.74
1,100	42,380	67.66	1,900	121,650	124.44
1,200	48,090	72.63	2,000	128,500	127.95
1,258 (c)	51,440	75.35			

TABLE 1h7. (*Continued*)

$Na_2Ti_2O_5(c)$:

$H_T - H_{298.15} = 49.32T + 3.53 \times 10^{-3}T^2 + 4.60 \times 10^5 T^{-1}$
$- 16,561$ (0.1 percent; 298°–1,258° K.);

$C_p = 49.32 + 7.06 \times 10^{-3}T - 4.60 \times 10^5 T^{-2}$;

$\Delta H_{1258}(fusion) = 26,230$.

$Na_2Ti_2O_5(l)$:

$H_T - H_{298.15} = 68.50T - 8,500$ (0.1 percent; 1,258°–2,000° K.);

$C_p = 68.50$.

TABLE 735.—*Heat content and entropy of* $Na_2Ti_3O_7(c, l)$

[Base, crystals at 298.15° K.; mol. wt., 301.68]

T, ° K.	$H_T - H_{298.15}$, cal./mole	$S_T - S_{298.15}$, cal./deg. mole	T, ° K.	$H_T - H_{298.15}$, cal./mole	$S_T - S_{298.15}$, cal./deg. mole
400	6,360	18.34	1,400	78,390	106.85
500	12,900	32.92	1,401(c)	78,470	106.90
600	19,650	45.22	1,401(l)	115,570	133.39
700	26,550	55.85	1,500	124,890	139.82
800	33,590	65.25	1,600	134,300	145.89
900	40,760	73.69	1,700	143,720	151.59
1,000	48,060	81.38	1,800	153,140	156.97
1,100	55,470	88.44	1,900	162,550	162.06
1,200	63,000	95.00	2,000	171,960	166.89
1,300	70,640	101.10			

$Na_2Ti_3O_7(c)$:

$H_T - H_{298.15} = 63.46T + 5.32 \times 10^{-3}T^2 + 5.64 \times 10^5 T^{-1}$
$- 21,285$ (0.1 percent; 298°–1,401° K.);

$C_p = 63.46 + 10.64 \times 10^{-3}T - 5.64 \times 10^5 T^{-2}$;

$\Delta H_{1401}(fusion) = 37,100$.

$Na_2Ti_3O_7(l)$:

$H_T - H_{298.15} = 94.15T - 16,335$ (0.1 percent; 1,401°–2,000° K.);

$C_p = 94.15$.

CRYOLITE

TABLE 736.—*Heat content and entropy of* $Na_3AlF_6(c, l)$

[Base, α-crystals at 298.15° K.; mol. wt., 209.95]

T, ° K.	$H_T - H_{298.15}$, cal./mole	$S_T - S_{298.15}$, cal./deg. mole	T, ° K.	$H_T - H_{298.15}$, cal./mole	$S_T - S_{298.15}$, cal./deg. mole
400	5,510	15.86	1,000	46,240	76.41
500	11,270	28.70	1,100	53,120	82.96
600	17,380	39.83	1,200	60,160	89.09
700	23,820	49.76	1,300(β)	67,360	94.85
800	30,590	58.79	1,300(l)	95,000	116.11
845(α)	33,730	62.61	1,400	104,340	123.03
845(β)	35,890	65.17	1,500	113,680	129.47
900	39,520	69.33			

$Na_3AlF_6(\alpha)$:

$H_T - H_{298.15} = 45.95T + 14.73 \times 10^{-3}T^2 + 2.78 \times 10^5 T^{-1}$
$- 15,942$ (0.2 percent; 298°–845° K.);

$C_p = 45.95 + 29.46 \times 10^{-3}T - 2.78 \times 10^5 T^{-2}$;

$\Delta H_{845}(transition) = 2,160$.

$Na_3AlF_6(\beta)$:

$H_T - H_{298.15} = 52.15T + 7.93 \times 10^{-3}T^2 - 13,840$ (0.1 percent; 845°–1,300° K.);

$C_p = 52.15 + 15.86 \times 10^{-3}T$;

$\Delta H_{1300}(fusion) = 27,640$.

$Na_3AlF_6(l)$:

$H_T - H_{298.15} = 93.40T - 26,420$ (0.1 percent; 1,300°–1,500° K.);

$C_p = 93.40$.

SODIUM-HYDROGEN SULFIDE

TABLE 737.—*Heat content and entropy of* $NaSH(c)$

[Base, α-crystals at 298.15° K.; mol. wt., 56.06]

T, ° K.	$H_T - H_{298.15}$, cal./mole	$S_T - S_{298.15}$, cal./deg. mole	T, ° K.	$H_T - H_{298.15}$, cal./mole	$S_T - S_{298.15}$, cal./deg. mole
350	1,010	3.12	358(β)	1,870	5.53
358(α)	1,170	3.58	400	2,530	7.27

$NaSH(\alpha)$:

$H_T - H_{298.15} = 19.50T - 5,814$ (0.2 percent; 298°–358° K.);

$C_p = 19.50$;

$\Delta H_{358}(transition) = 700$.

$NaSH(\beta)$:

$H_T - H_{298.15} = 15.70T - 3,750$ (0.1 percent; 358°–400° K.);

$C_p = 15.70$.

SODIUM-HYDROGEN SELENIDE

TABLE 738.—*Heat content and entropy of* $NaSeH(c)$

[Base, α-crystals at 298.15° K.; mol. wt., 102.96]

T, ° K.	$H_T - H_{298.15}$, cal./mole	$S_T - S_{298.15}$, cal./deg. mole	T, ° K.	$H_T - H_{298.15}$, cal./mole	$S_T - S_{298.15}$, cal./deg. mole
350	990	3.06	359(β)	1,860	5.49
359(α)	1,160	3.54	400	2,520	7.23

$NaSeH(\alpha)$:

$H_T - H_{298.15} = 19.10T - 5,695$ (0.1 percent; 298°–359° K.);

$C_p = 19.10$;

$\Delta H_{359}(transition) = 700$.

$NaSeH(\beta)$:

$H_T - H_{298.15} = 16.10T - 3,920$ (0.1 percent; 359°–400° K.);

$C_p = 16.10$.

SODIUM-ALUMINUM SILICATES

TABLE 739.—*Heat content of* $NaAlSiO_4(nephelite)$

[Base, α-crystals at 298.15° K.; mol. wt., 142.06]

T, ° K.	$H_T - H_{298.15}$, cal./mole	$S_T - S_{298.15}$, cal./deg. mole	T, ° K.	$H_T - H_{298.15}$, cal./mole	$S_T - S_{298.15}$, cal./deg. mole
350	1,530	4.73	1,000	26,050	43.62
400	3,095	8.90	1,100	30,370	47.74
450	4,915	13.18	1,180(β)	34,200	51.10
467(α)	5,680	14.85	1,180(γ)	34,200	51.10
467(β)	5,680	14.85	1,200	35,050	51.81
500	6,280	17.21	1,300	39,330	55.24
600	10,420	23.77	1,400	43,620	58.42
700	14,150	29.51	1,500	47,920	61.39
800	18,000	34.65	1,525	49,000	62.10
900	21,970	39.33			

TABLE 1h7. *(Continued)*

NaAlSiO$_4$(α-nephelite):

$$H_T - H_{298.15} = 6.63T + 35.30 \times 10^{-3}T^2 - 5,115$$

(1.5 percent; 298°–467° K.);

$$C_p = 6.63 + 70.60 \times 10^{-3}T;$$

$$\Delta H_{467}(transition) = 0.$$

NaAlSiO$_4$(β-nephelite):

$$H_T - H_{298.15} = 26.79T + 8.02 \times 10^{-3}T^2 - 8,580$$

(0.3 percent; 467°–1,180° K.);

$$C_p = 26.79 + 16.04 \times 10^{-3}T;$$

$$\Delta H_{1180}(transition) = 0.$$

NaAlSiO$_4$(γ-nephelite);

$$H_T - H_{298.15} = 41.11T + 0.66 \times 10^{-3}T^2 - 15,229$$

(0.1 percent; 1,180°–1,525° K.);

$$C_p = 41.11 + 1.32 \times 10^{-3}T.$$

TABLE 740.—*Heat content and entropy of* NaAlSiO$_4$(*carnegieite*)

[Base, α-crystals at 298.15° K.; mol. wt., 142.06]

T, ° K.	$H_T-H_{298.15}$, cal./mole	$S_T-S_{298.15}$, cal./deg. mole	T, ° K.	$H_T-H_{298.15}$, cal./mole	$S_T-S_{298.15}$, cal./deg. mole
400	3,110	8.94	1,000	29,190	46.85
500	6,540	16.58	1,100	33,650	51.10
600	10,260	23.35	1,200	38,120	54.98
700	14,210	29.44	1,300	42,590	58.56
800	18,350	34.96	1,400	47,070	61.89
900	22,660	40.04	1,500	51,560	64.99
980(α)	26,260	43.87	1,600	56,070	67.90
980(β)	28,300	45.95	1,700	60,600	70.64

NaAlSiO$_4$(α-carnegieite):

$$H_T - H_{298.15} = 29.54T + 8.58 \times 10^{-3}T^2 + 5.83 \times 10^5 T^{-1}$$

$$-11,525 \text{ (0.2 percent; 298°–980° K.)};$$

$$C_p = 29.54 + 17.16 \times 10^{-3}T - 5.83 \times 10^5 T^{-2};$$

$$\Delta H_{980}(transition) = 2,040.$$

NaAlSiO$_4$(β-carnegieite):

$$H_T - H_{298.15} = 43.44T + 0.53 \times 10^{-3}T^2 - 14,780$$

(0.1 percent; 980°–1,700° K.);

$$C_p = 43.44 + 1.06 \times 10^{-3}T.$$

TABLE 741.—*Heat content and entropy of* NaAlSi$_2$O$_6$(*jadeite*)

[Base, crystals at 298.15° K.; mol. wt., 202.15]

T, ° K.	$H_T-H_{298.15}$, cal./mole	$S_T-S_{298.15}$, cal./deg. mole	T, ° K.	$H_T-H_{298.15}$, cal./mole	$S_T-S_{298.15}$, cal./deg. mole
400	4,250	12.20	900	30,490	54.12
500	8,970	22.72	1,000	36,240	60.18
600	14,040	31.96	1,100	42,120	65.78
700	19,360	40.15	1,200	48,160	71.04
800	24,860	47.49			

NaAlSi$_2$O$_6$(*jadeite*):

$$H_T - H_{298.15} = 48.16T + 5.71 \times 10^{-3}T^2 + 11.87 \times 10^5 T^{-1}$$

$$-18,848 \text{ (0.3 percent; 298°–1,200° K.)};$$

$$C_p = 48.16 + 11.42 \times 10^{-3}T - 11.87 \times 10^5 T^{-2}.$$

TABLE 742.—*Heat content and entropy* of NaAlSi$_3$O$_8$(*albite*)

[Base, crystals at 298.15° K.; mol. wt., 262.24]

T, ° K.	$H_T-H_{298.15}$, cal./mole	$S_T-S_{298.15}$, cal./deg. mole	T, ° K.	$H_T-H_{298.15}$, cal./mole	$S_T-S_{298.15}$, cal./deg. mole
400	5,410	15.55	1,000	46,220	76.74
500	11,390	28.87	1,100	53,720	83.88
600	17,900	40.73	1,200	61,340	90.51
700	24,690	51.20	1,300	69,060	96.69
800	31,690	60.54	1,400	76,860	102.47
900	38,870	68.99			

NaAlSi$_3$O$_8$ (*albite*):

$$H_T - H_{298.15} = 61.70T + 6.95 \times 10^{-3}T^2 + 15.01 \times 10^5 T^{-1}$$

$$-24,048 \text{ (0.4 percent; 298°–1,400° K.)};$$

$$C_p = 61.70 + 13.90 \times 10^{-3}T - 15.01 \times 10^5 T^{-2}.$$

TABLE 743.—*Heat content and entropy* of NaAlSi$_3$O$_8$(*gl*)

[Base, glass at 298.15° K.; mol. wt., 262.24]

T, ° K.	$H_T-H_{298.15}$, cal./mole	$S_T-S_{298.15}$, cal./deg. mole	T, ° K.	$H_T-H_{298.15}$, cal./mole	$S_T-S_{298.15}$, cal./deg. mole
400	5,540	15.93	900	39,680	70.39
500	11,580	29.38	1,000	47,220	78.34
600	18,220	41.48	1,100	55,080	85.83
700	25,200	52.22	1,200	63,380	93.05
800	32,360	61.78			

NaAlSi$_3$O$_8$(*gl*):

$$H_T - H_{298.15} = 61.31T + 9.00 \times 10^{-3}T^2 + 16.16 \times 10^5 T^{-1}$$

$$-24,500 \text{ (0.3 percent; 298°–1,200° K.)};$$

$$C_p = 61.31 + 18.00 \times 10^{-3}T - 16.16 \times 10^5 T^{-2}.$$

NaAlSi$_2$O$_6$·H$_2$O(*analcite*):

$$C_p = 50.17 \text{ (298° K.).}$$

SODIUM-POTASSIUM COMPOUND

Na$_2$K(*c*):

$$C_p = 21.16 \text{ (298° K.).}$$

Na$_2$K(*l*):

$$\overline{C_p} = 24.22 \text{ (305°–320° K.).}$$

STRONTIUM AND ITS COMPOUNDS

ELEMENT

TABLE 744.—*Heat content and entropy* of Sr(*c, l*)

[Base, α-crystals at 298.15° K.; atomic wt., 87.63]

T, ° K.	$H_T-H_{298.15}$, cal./mole	$S_T-S_{298.15}$, cal./deg. mole	T, ° K.	$H_T-H_{298.15}$, cal./mole	$S_T-S_{298.15}$, cal./deg. mole
400	660	1.90	1,043 (β)	5,930	9.48
500	1,340	3.42	1,043(l)	8,330	11.78
600	2,050	4.72	1,100	8,750	12.17
700	2,800	5.87	1,200	9,490	12.82
800	3,580	6.91	1,300	10,230	13.41
862 (α)	4,080	7.51	1,400	10,970	13.96
862 (β)	4,280	7.74	1,500	11,710	14.47
900	4,610	8.12	1,600	12,450	14.95
1,000	5,520	9.08			

TABLE 1h7. (*Continued*)

$Sr(\alpha)$:

$H_T - H_{298.15} = 5.31T + 1.66 \times 10^{-3}T^2 - 1,731$

(0.1 percent; 298°–862° K.);

$C_p = 5.31 + 3.32 \times 10^{-3}T;$

ΔH_{862} (*transition*) $= 200.$

$Sr(\beta)$:

$H_T - H_{298.15} = 9.12T - 3,582$ (0.2 percent;

862°–1,043° K.);

$C_p = 9.12;$

ΔH_{1043} (*fusion*) $= 2,400.$

$Sr(l)$:

$H_T - H_{298.15} = 7.40T + 610$ (0.1 percent;

1,043°–1,600° K.);

$C_p = 7.40.$

TABLE 745.—*Heat content and entropy of* $Sr(g)$

[Base, ideal gas at 298.15°K.; atomic wt., 87.63]

T, ° K.	$H_T - H_{298.15}$, cal./mole	$S_T - S_{298.15}$, cal./deg. mole	T, ° K.	$H_T - H_{298.15}$, cal./mole	$S_T - S_{298.15}$, cal./deg. mole
400	505	1.46	1,900	7,965	9.21
500	1,005	2.57	2,000	8,470	9.46
600	1,500	3.49	2,200	9,480	9.95
700	1,995	4.24	2,400	10,515	10.40
800	2,495	4.90	2,600	11,585	10.82
900	2,990	5.49	2,800	12,710	11.24
1,000	3,490	6.01	3,000	13,900	11.65
1,100	3,985	6.49	3,500	17,305	12.70
1,200	4,480	6.92	4,000	21,525	13.82
1,300	4,980	7.32	4,500	26,685	15.04
1,400	5,475	7.69	5,000	32,760	16.32
1,500	5,975	8.03	6,000	47,025	18.91
1,600	6,470	8.35	7,000	62,925	21.36
1,700	6,970	8.65	8,000	79,195	23.53
1,800	7,465	8.94			

$Sr(g)$:

$H_T - H_{298.15} = 4.97T - 1,482$ (0.1 percent;

298°–2,000° K.);

$C_p = 4.97.$

OXIDE

TABLE 746.—*Heat content and entropy of* $SrO(c)$

[Base, crystals at 298.15° K.; mol. wt., 103.63]

T, ° K.	$H_T - H_{298.15}$, cal./mole	$S_T - S_{298.15}$, cal./deg. mole	T, ° K.	$H_T - H_{298.15}$, cal./mole	$S_T - S_{298.15}$, cal./deg. mole
400	1,170	3.37	1,200	11,400	17.21
500	2,340	5.98	1,300	12,740	18.28
600	3,550	8.18	1,400	14,060	19.28
700	4,800	10.11	1,500	15,450	20.22
800	6,090	11.83	1,600	16,820	21.11
900	7,410	13.39	1,700	18,200	21.94
1,000	8,740	14.79	1,800	19,590	22.74
1,100	10,070	16.06			

$SrO(c)$:

$H_T - H_{298.15} = 12.13T + 0.63 \times 10^{-3}T^2 + 1.55 \times 10^5 T^{-1}$

$-4,192$ (0.5 percent; 298°–1,800° K.);

$C_p = 12.13 + 1.26 \times 10^{-3}T - 1.55 \times 10^5 T^{-2}.$

TABLE 747.—*Heat content and entropy of* $SrO(g)$

[Base, ideal gas at 298.15° K.; mol. wt., 103.63]

T, ° K.	$H_T - H_{298.15}$, cal./mole	$S_T - S_{298.15}$, cal./deg. mole	T, ° K.	$H_T - H_{298.15}$, cal./mole	$S_T - S_{298.15}$, cal./deg. mole
400	825	2.38	1,000	6,000	10.24
500	1,660	4.24	1,200	7,765	11.85
600	2,510	5.79	1,400	9,535	13.22
700	3,375	7.12	1,600	11,315	14.40
800	4,245	8.28	1,800	13,095	15.45
900	5,120	9.31	2,000	14,875	16.39

$SrO(g)$:

$H_T - H_{298.15} = 8.69T + 0.08 \times 10^{-3}T^2 + 0.76 \times 10^5 T^{-1}$

$-2,853$ (0.2 percent; 298°–2,000° K.);

$C_p = 8.69 + 0.16 \times 10^{-3}T - 0.76 \times 10^5 T^{-2}.$

HYDROXIDE

TABLE 748.—*Heat content and entropy of* $Sr(OH)_2(c, l)$

[Base, crystals at 298.15° K.; mol. wt., 121.65]

T, ° K.	$H_T - H_{298.15}$, cal./mole	$S_T - S_{298.15}$, cal./deg. mole	T, ° K.	$H_T - H_{298.15}$, cal./mole	$S_T - S_{298.15}$, cal./deg. mole
400	1,970	5.66	808(*l*)	18,800	31.44
500	4,230	10.69	900	22,160	35.37
600	6,830	15.42	1,000	25,810	39.22
700	9,770	19.94	1,100	29,460	42.70
800	13,040	24.30	1,200	33,110	45.87
808(*c*)	13,310	24.64			

$Sr(OH)_2(c)$:

$H_T - H_{298.15} = 7.64T + 16.70 \times 10^{-3}T^2 - 3,762$

(0.1 percent; 298°–808° K.);

$C_p = 7.64 + 33.40 \times 10^{-3}T;$

ΔH_{808} (*fusion*) $= 5,490.$

$Sr(OH)_2(l)$:

$H_T - H_{298.15} = 36.50T - 10,690$ (0.1 percent;

808°–1,200° K.);

$C_p = 36.50.$

HYDRIDE

TABLE 749.—*Heat content and entropy of* $SrH(g)$

[Base, ideal gas at 298.15° K.; mol. wt., 88.64]

T, ° K.	$H_T - H_{298.15}$, cal./mole	$S_T - S_{298.15}$, cal./deg. mole	T, ° K.	$H_T - H_{298.15}$, cal./mole	$S_T - S_{298.15}$, cal./deg. mole
400	745	2.15	1,000	5,630	9.54
500	1,510	3.85	1,200	7,350	11.10
600	2,300	5.29	1,400	9,085	12.44
700	3,110	6.54	1,600	10,835	13.61
800	3,940	7.65	1,800	12,595	14.64
900	4,780	8.64	2,000	14,355	15.57

$SrH(g)$:

$H_T - H_{298.15} = 7.81T + 0.34 \times 10^{-3}T^2 + 0.74 \times 10^5 T^{-1}$

$-2,607$ (0.6 percent; 298°–2,000° K.);

$C_p = 7.81 + 0.68 \times 10^{-3}T - 0.74 \times 10^5 T^{-2}.$

TABLE 1h7. (Continued)

BROMIDES

SrBr₂(c):

$C_p=18.13+3.10\times10^{-3}T(estimated)$ (298°–926° K.).

SrBr₂·H₂O(c):

$\overline{C}_p=28.9$ (276°–370° K.).

SrBr₂·6H₂O(c):

$\overline{C}_p=82.1$ (276°–327° K.).

TABLE 750.—*Heat content and entropy of SrBr(g)*

[Base, ideal gas at 298.15° K.; mol. wt., 167.55]

T, ° K.	$H_T-H_{298.15}$, cal./mole	$S_T-S_{298.15}$, cal./deg. mole	T, ° K.	$H_T-H_{298.15}$, cal./mole	$S_T-S_{298.15}$, cal./deg. mole
400	900	2.60	1,000	6,240	10.74
500	1,785	4.57	1,200	8,025	12.37
600	2,675	6.19	1,400	9,810	13.75
700	3,565	7.57	1,600	11,600	14.95
800	4,455	8.75	1,800	13,385	16.00
900	5,345	9.80	2,000	15,175	16.94

SrBr(g):

$H_T-H_{298.15}=8.94T+0.14\times10^5T^{-1}-2,712$

(0.1 percent; 298°–2,000° K.);

$C_p=8.94-0.14\times10^5T^{-2}$.

CHLORIDES

SrCl₂(c):

$C_p=18.20+2.45\times10^{-3}T(estimated)$ (298°–1,145° K.);

SrCl₂·H₂O(c):

$\overline{C}_p=28.7$ (276°–365° K.).

SrCl₂·2H₂O(c):

$\overline{C}_p=38.3$ (276°–366° K.).

TABLE 751.—*Heat content and entropy of SrCl(g)*

[Base, ideal gas at 298.15° K.; mol. wt., 123.09]

T, ° K.	$H_T-H_{298.15}$, cal./mole	$S_T-S_{298.15}$, cal./deg. mole	T, ° K.	$H_T-H_{298.15}$, cal./mole	$S_T-S_{298.15}$, cal./deg. mole
400	885	2.56	1,000	6,205	10.67
500	1,765	4.52	1,200	7,990	12.30
600	2,650	6.13	1,400	9,775	13.67
700	3,535	7.50	1,600	11,560	14.86
800	4,425	8.68	1,800	13,345	15.91
900	5,315	9.73	2,000	15,135	16.86

SrCl(g):

$H_T-H_{298.15}=8.94T+0.29\times10^5T^{-1}-2,763$

(0.1 percent; 298°–2,000° K.);

$C_p=8.94-0.29\times10^5T^{-2}$.

FLUORIDE

TABLE 752.—*Heat content and entropy of SrF(g)*

[Base, ideal gas at 298.15° K.; mol. wt., 106.63]

T, ° K.	$H_T-H_{298.15}$, cal./mole	$S_T-S_{298.15}$, cal./deg. mole	T, ° K.	$H_T-H_{298.15}$, cal./mole	$S_T-S_{298.15}$, cal./deg. mole
400	850	2.45	1,000	6,100	10.45
500	1,710	4.37	1,200	7,875	12.06
600	2,580	5.96	1,400	9,655	13.44
700	3,450	7.30	1,600	11,435	14.62
800	4,330	8.47	1,800	13,215	15.67
900	5,215	9.52	2,000	15,000	16.61

SrF(g):

$H_T-H_{298.15}=8.84T+0.03\times10^{-3}T^2+0.56\times10^5T^{-1}$
$-2,826$ (0.2 percent; 298°–2,000° K.);

$C_p=8.84+0.06\times10^{-3}T-0.56\times10^5T^{-2}$.

IODIDES

SrI₂(c):

$C_p=18.63+3.01\times10^{-3}T(estimated)$ (298°–788° K.).

SrI₂·H₂O(c):

$\overline{C}_p=28.5$ (276°–363° K.).

SrI₂·2H₂O(c):

$\overline{C}_p=39.1$ (275°–336° K.).

SrI₂·6H₂O(c):

$\overline{C}_p=84.9$ (275°–334° K.).

TABLE 753.—*Heat content and entropy of SrI(g)*

[Base, ideal gas at 298.15° K.; mol. wt., 214.54]

T, ° K.	$H_T-H_{298.15}$, cal./mole	$S_T-S_{298.15}$, cal./deg. mole	T, ° K.	$H_T-H_{298.15}$, cal./mole	$S_T-S_{298.15}$, cal./deg. mole
400	900	2.60	1,000	6,250	10.76
500	1,790	4.58	1,200	8,040	12.39
600	2,680	6.20	1,400	9,825	13.77
700	3,575	7.58	1,600	11,615	14.97
800	4,465	8.77	1,800	13,400	16.02
900	5,360	9.82	2,000	15,190	16.96

SrI(g):

$H_T-H_{298.15}=8.94T+0.10\times10^5T^{-1}-2,699$

(0.1 percent; 298°–2,000° K.);

$C_p=8.94-0.10\times10^5T^{-2}$.

CARBONATE

TABLE 754.—*Heat content and entropy of SrCO₃(c)*

[Base, α-crystals at 298.15° K.; mol. wt., 147.64]

T, ° K.	$H_T-H_{298.15}$, cal./mole	$S_T-S_{298.15}$, cal./deg. mole	T, ° K.	$H_T-H_{298.15}$, cal./mole	$S_T-S_{298.15}$, cal./deg. mole
400	2,270	6.52	1,100	21,160	33.17
500	4,620	11.76	1,197(α)	24,110	35.74
600	7,100	16.28	1,197(β)	28,810	39.67
700	9,700	20.29	1,200	28,920	39.76
800	12,420	23.92	1,300	32,380	42.53
900	15,250	27.25	1,400	35,840	45.09
1,000	18,170	30.32	1,500	39,300	47.48

SrCO₃(α):

$H_T-H_{298.15}=23.52T+3.16\times10^{-3}T^2+5.08\times10^5T^{-1}$
$-8,997$ (0.8 percent; 298°–1,197° K.);

$C_p=23.52+6.32\times10^{-3}T-5.08\times10^5T^{-2}$;

$\Delta H_{1197}(transition)=4,700.$

SrCO₃(β):

$H_T-H_{298.15}=34.60T-12,600$ (0.1 percent;
1,197°–1,500° K.);

$C_p=34.60.$

MOLYBDATE

SrMoO₄(c):

$\overline{C}_p=36.9$ (273°–297° K.).

TABLE 1h7. (*Continued*)

NITRATE

$Sr(NO_3)_2(c)$:

$\overline{C_p}=38.3$ (290°–320° K.).

SULFATE

$SrSO_4(c)$:

$C_p=21.80+13.30\times10^{-3}T(estimated)$ (298°–1,500° K.).

TITANATES

TABLE 755.—*Heat content and entropy of* $SrTiO_3(c)$

[Base, crystals at 298.15° K.; mol. wt., 183.53]

T, ° K.	$H_T-H_{298.15}$, cal./mole	$S_T-S_{298.15}$, cal./deg. mole	T, ° K.	$H_T-H_{298.15}$, cal./mole	$S_T-S_{298.15}$, cal./deg. mole
400	2,530	7.28	1,200	25,570	38.48
500	5,170	13.17	1,300	28,590	40.90
600	7,920	18.18	1,400	31,620	43.14
700	10,750	22.54	1,500	34,660	45.24
800	13,640	26.40	1,600	37,700	47.20
900	16,580	29.86	1,700	40,750	49.05
1,000	19,560	33.00	1,800	43,830	50.81
1,100	22,560	35.86			

$SrTiO_3(c)$:

$H_T-H_{298.15}=28.23T+0.88\times10^{-3}T^2+4.66\times10^5T^{-1}$

$\qquad-10,058$ (0.3 percent; 298°–1,800° K.);

$C_p=28.23+1.76\times10^{-3}T-4.66\times10^5T^{-2}$.

TABLE 756.—*Heat content and entropy of* $Sr_2TiO_4(c)$

[Base, crystals at 298.15° K.; mol. wt., 287.16]

T, ° K.	$H_T-H_{298.15}$, cal./mole	$S_T-S_{298.15}$, cal./deg. mole	T, ° K.	$H_T-H_{298.15}$, cal./mole	$S_T-S_{298.15}$, cal./deg. mole
400	3,610	10.40	1,200	36,230	54.60
500	7,370	18.78	1,300	40,530	58.04
600	11,270	25.89	1,400	44,850	61.24
700	15,280	32.07	1,500	49,180	64.23
800	19,370	37.53	1,600	53,520	67.03
900	23,520	42.42	1,700	57,870	69.67
1,000	27,720	46.84	1,800	62,230	72.16
1,100	31,960	50.88			

$Sr_2TiO_4(c)$:

$H_T-H_{298.15}=38.45T+1.92\times10^{-3}T^2+4.67\times10^5T^{-1}$

$\qquad-13,201$ (0.4 percent; 298°–1,800° K.);

$C_n=38.45+3.84\times10^{-3}T-4.67\times10^5T^{-2}$.

SULFUR AND ITS COMPOUNDS

ELEMENT

TABLE 757.—*Heat content and entropy of* $S(c, l)$

[Base, rh-crystals at 298.15° K.; atomic wt., 32.07]

T, ° K.	$H_T-H_{298.15}$, cal./mole	$S_T-S_{298.15}$, cal./deg. mole
350	290	0.90
368.6(rh)	400	1.20
368.6(mon)	485	1.43
392(mon)	630	1.82
392(l)	965	2.67
400	1,030	2.83
500	1,940	4.85
600	2,780	6.38
700	3,650	7.72
717.8	3,810	7.95

S(rh):

$H_T-H_{298.15}=3.58T+3.12\times10^{-3}T^2-1,345$ (0.2 percent;

$\qquad$298°–368.6° K.);

$\qquad C_p=3.58+6.24\times10^{-3}T$;

$\qquad \Delta H_{368.8}=85.$

S(mon):

$H_T-H_{298.15}=6.20T-1,800$ (0.1 percent;

$\qquad$368.6°–392° K.);

$\qquad C_p=6.20.$

$\qquad \Delta H_{392}(fusion)=335.$

S(l):

$H_T-H_{298.15}=8.73T-2,457$ (0.6 percent;

$\qquad$392°–717.8° K.);

$\qquad C_p=8.73.$

TABLE 758.—*Heat content and entropy of* $S(g)$

[Base, ideal gas at 298.15° K.; atomic wt., 32.07]

T, ° K.	$H_T-H_{298.15}$, cal./mole	$S_T-S_{298.15}$, cal./deg. mole	T, ° K.	$H_T-H_{298.15}$, cal./mole	$S_T-S_{298.15}$, cal./deg. mole
400	570	1.65	1,900	8,320	9.79
500	1,120	2.88	2,000	8,830	10.06
600	1,660	3.86	2,200	9,850	10.54
700	2,190	4.68	2,400	10,875	10.99
800	2,715	5.38	2,600	11,910	11.40
900	3,235	5.99	2,800	12,950	11.79
1,000	3,750	6.53	3,000	13,995	12.15
1,100	4,260	7.02	3,500	16,650	12.96
1,200	4,770	7.46	4,000	19,340	13.68
1,300	5,280	7.87	4,500	22,065	14.32
1,400	5,790	8.25	5,000	24,810	14.90
1,500	6,295	8.60	6,000	30,330	15.91
1,600	6,800	8.92	7,000	35,860	16.76
1,700	7,305	9.23	8,000	41,390	17.50
1,800	7,815	9.52			

S(g):

$H_T-H_{298.15}=5.26T-0.05\times10^{-3}T^2-0.36\times10^5T^{-1}$

$\qquad-1,443$ (0.6 percent; 298°–2,400° K.);

$\qquad C_p=5.26-0.10\times10^{-3}T+0.36\times10^5T^{-2}.$

$H_T-H_{298.15}=4.96T+0.05\times10^{-3}T^2-0.60\times10^5T^{-1}$

$\qquad-1,282$ (0.2 percent; 2,400°–8,000° K.);

$\qquad C_p=4.96+0.10\times10^{-3}T+0.60\times10^5T^{-2}.$

TABLE 759.—*Heat content and entropy of* $S_2(g)$

[Base, ideal gas at 298.15° K.; mol. wt. 64.13]

T, ° K.	$H_T-H_{298.15}$, cal./mole	$S_T-S_{298.15}$, cal./deg. mole	T, ° K.	$H_T-H_{298.15}$, cal./mole	$S_T-S_{298.15}$, cal./deg. mole
400	810	2.34	1,500	10,430	13.79
500	1,640	4.18	1,600	11,325	14.37
600	2,485	5.72	1,700	12,225	14.91
700	3,345	7.05	1,800	13,125	15.43
800	4,220	8.22	1,900	14,025	15.92
900	5,095	9.25	2,000	14,925	16.38
1,000	5,975	10.17	2,200	16,730	17.24
1,100	6,855	11.01	2,400	18,545	18.03
1,200	7,745	11.79	2,600	20,360	18.76
1,300	8,635	12.50	2,800	22,175	19.43
1,400	9,535	13.17	3,000	23,995	20.06

$S_2(g)$:

$H_T-H_{298.15}=8.72T+0.08\times10^{-3}T^2+0.90\times10^5T^{-1}$

$\qquad-2,909$ (0.3 percent; 298°–3,000° K.);

$\qquad C_p=8.72+0.16\times10^{-3}T-0.90\times10^5T^{-2}.$

TABLE 1h7. (*Continued*)

TABLE 760.—*Heat content and entropy of $S_8(g)$*

[Base, ideal gas at 298.15° K.; mol. wt. 256.53]

T, ° K.	$H_T-H_{298.15},$ cal./mole	$S_T-S_{298.15},$ cal./deg. mole	T, ° K.	$H_T-H_{298.15},$ cal./mole	$S_T-S_{298.15},$ cal./deg. mole
400	3,935	11.32	800	20,580	40.06
500	7,985	20.37	900	24,850	45.09
600	12,125	27.90	1,000	29,140	49.62
700	16,330	34.39			

$$S_8(g):$$
$$H_T - H_{298.15} = 42.54T + 0.52 \times 10^{-3}T^2 + 5.04 \times 10^5 T^{-1}$$
$$-14,420 \ (0.1 \text{ percent}; \ 298°-1,000° \text{ K.});$$
$$C_p = 42.54 + 1.04 \times 10^{-3}T - 5.04 \times 10^5 T^{-2}.$$

MONOXIDE

TABLE 761.—*Heat content and entropy $SO(g)$*

[Base, ideal gas at 298.15° K.; mol. wt. 48.07]

T, ° K.	$H_T-H_{298.15},$ cal./mole	$S_T-S_{298.15},$ cal./deg. mole	T, ° K.	$H_T-H_{298.15},$ cal./mole	$S_T-S_{298.15},$ cal./deg. mole
400	755	2.17	1,500	10,045	13.16
500	1,525	3.89	1,600	10,925	13.73
600	2,325	5.35	1,700	11,810	14.27
700	3,145	6.62	1,800	12,695	14.77
800	3,980	7.73	1,900	13,580	15.25
900	4,830	8.73	2,000	14,465	15.70
1,000	5,685	9.63	2,200	16,245	16.55
1,100	6,545	10.45	2,400	18,030	17.33
1,200	7,415	11.21	2,600	19,820	18.05
1,300	8,285	11.91	2,800	21,610	18.71
1,400	9,165	12.56	3,000	23,405	19.33

$$SO(g):$$
$$H_T - H_{298.15} = 8.26T + 0.16 \times 10^{-3}T^2 + 1.00 \times 10^5 T^{-1}$$
$$-2,812 \ (0.7 \text{ percent}; \ 298°-3,000° \text{ K.});$$
$$C_p = 8.26 + 0.32 \times 10^{-3}T - 1.00 \times 10^5 T^{-2}.$$

DIOXIDE

TABLE 762.—*Heat content and entropy of $SO_2(g)$*

[Base, ideal gas at 295.15° K.; mol. wt. 64.07]

T, ° K.	$H_T-H_{298.15},$ cal./mole	$S_T-S_{298.15},$ cal./deg. mole	T, ° K.	$H_T-H_{298.15},$ cal./mole	$S_T-S_{298.15},$ cal./deg. mole
400	1,015	2.92	1,300	12,215	17.24
500	2,090	5.32	1,400	13,565	18.24
600	3,240	7.41	1,500	14,925	19.18
700	4,440	9.26	1,600	16,290	20.06
800	5,675	10.91	1,700	17,660	20.89
900	6,940	12.40	1,800	19,035	21.68
1,000	8,230	13.76	1,900	20,415	22.42
1,100	9,545	15.01	2,000	21,800	23.13
1,200	10,875	16.17			

$$SO_2(g):$$
$$H_T - H_{298.15} = 11.04T + 0.94 \times 10^{-3}T^2 + 1.84 \times 10^5 T^{-1}$$
$$-3,992 \ (0.8 \text{ percent}; \ 298°-2,000° \text{ K.});$$
$$C_p = 11.04 + 1.88 \times 10^{-3}T - 1.84 \times 10^5 T^{-2}.$$

TRIOXIDE

TABLE 763.—*Heat content and entropy of $SO_3(g)$*

[Base, ideal gas at 298.15° K.; mol. wt. 80.07]

T, ° K.	$H_T-H_{298.15},$ cal./mole	$S_T-S_{298.15},$ cal./deg. mole	T, ° K.	$H_T-H_{298.15},$ cal./mole	$S_T-S_{298.15},$ cal./deg. mole
400	1,330	3.82	1,000	11,860	19.52
500	2,820	7.13	1,100	13,860	21.43
600	4,450	10.11	1,200	15,900	23.20
700	6,190	12.79	1,300	17,980	24.86
800	8,010	15.23	1,400	20,090	26.42
900	9,900	17.46	1,500	22,230	27.88

$$SO_3(g):$$
$$H_T - H_{298.15} = 13.90T + 3.05 \times 10^{-3}T^2 + 3.22 \times 10^5 T^{-1}$$
$$-5,495 \ (0.7 \text{ percent}; \ 298°-1,500° \text{ K.});$$
$$C_p = 13.90 + 6.10 \times 10^{-3}T - 3.22 \times 10^5 T^{-2}.$$

CHLORIDES

TABLE 764.—*Heat content and entropy of $S_2Cl_2(g)$*

[Base, ideal gas at 298.15° K.; mol. wt., 135.05]

T, ° K.	$H_T-H_{298.15},$ cal./mole	$S_T-S_{298.15},$ cal./deg. mole	T, ° K.	$H_T-H_{298.15},$ cal./mole	$S_T-S_{298.15},$ cal./deg. mole
400	1,840	5.30	800	9,590	18.68
500	3,730	9.52	900	11,580	21.03
600	5,660	13.03	1,000	13,570	23.12
700	7,610	16.04			

$$S_2Cl_2(g):$$
$$H_T - H_{298.15} = 19.43T + 0.41 \times 10^{-3}T^2 + 1.87 \times 10^5 T^{-1}$$
$$-6,457 \ (0.2 \text{ percent}; \ 298°-1,000° \text{ K.});$$
$$C_p = 19.43 + 0.82 \times 10^{-3}T - 1.87 \times 10^5 T^{-2}.$$

TABLE 765.—*Heat content and entropy of $SCl_2(g)$*

[Base, ideal gas at 298.15° K.; mol. wt., 102.98]

T, ° K.	$H_T-H_{298.15},$ cal./mole	$S_T-S_{298.15},$ cal./deg. mole	T, ° K.	$H_T-H_{298.15},$ cal./mole	$S_T-S_{298.15},$ cal./deg. mole
400	1,275	3.67	1,000	9,340	15.93
500	2,580	6.58	1,200	12,090	18.44
600	3,905	9.00	1,400	14,845	20.56
700	5,250	11.07	1,600	17,610	22.41
800	6,610	12.89	1,800	20,380	24.04
900	7,975	14.50	2,000	23,155	25.50

$$SCl_2(g):$$
$$H_T - H_{298.15} = 13.68T + 0.07 \times 10^{-3}T^2 + 1.39 \times 10^5 T^{-1}$$
$$-4,551 \ (0.1 \text{ percent}; \ 298°-2,000° \text{ K.});$$
$$C_p = 13.68 + 0.14 \times 10^{-3}T - 1.39 \times 10^5 T^{-2}.$$

TABLE 1h7. (*Continued*)

HEXAFLUORIDE

TABLE 766.—*Heat content and entropy of $SF_6(g)$*

[Base, ideal gas at 298.15° K.; mol. wt., 146.07]

T, ° K.	$H_T - H_{298.15}$, cal./mole	$S_T - S_{298.15}$, cal./deg. mole	T, ° K.	$H_T - H_{298.15}$, cal./mole	$S_T - S_{298.15}$, cal./deg. mole
400	2,595	7.46	1,000	22,420	37.20
500	5,515	13.96	1,200	29,600	43.47
600	8,675	19.72	1,400	36,890	49.36
700	11,980	24.81	1,600	44,250	54.27
800	15,390	29.36	1,800	51,640	58.63
900	18,870	33.46	2,000	59,070	62.54

$SF_6(g)$:

$H_T - H_{298.15} = 31.89T + 2.10 \times 10^{-3}T^2 + 9.01 \times 10^5 T^{-1}$
$-12,717$ (0.8 percent; 298°–2,000° K.);
$C_p = 31.89 + 4.20 \times 10^{-3}T - 9.01 \times 10^5 T^{-2}.$

OXYBROMIDE

TABLE 767.—*Heat content and entropy of $SOBr_2(g)$*

[Base, ideal gas at 298.15° K.; mol. wt., 207.90]

T, ° K.	$H_T - H_{298.15}$, cal./mole	$S_T - S_{298.15}$, cal./deg. mole	T, ° K.	$H_T - H_{298.15}$, cal./mole	$S_T - S_{298.15}$, cal./deg. mole
400	1,750	5.04	800	9,155	17.82
500	3,540	9.03	900	11,075	20.08
600	5,380	12.39	1,000	13,000	22.10
700	7,255	15.28			

$SOBr_2(g)$:

$H_T - H_{298.15} = 18.12T + 0.75 \times 10^{-3}T^2 + 1.70 \times 10^5 T^{-1}$
$-6,039$ (0.2 percent; 298°–1,000° K.);
$C_p = 18.12 + 1.50 \times 10^{-3}T - 1.70 \times 10^5 T^{-2}.$

OXYCHLORIDES

TABLE 768.—*Heat content and entropy of $SOCl_2(g)$*

[Base, ideal gas at 298.15° K.; mol. wt., 118.98]

T, ° K.	$H_T - H_{298.15}$, cal./mole	$S_T - S_{298.15}$, cal./deg. mole	T, ° K.	$H_T - H_{298.15}$, cal./mole	$S_T - S_{298.15}$, cal./deg. mole
400	1,680	4.84	800	8,960	17.39
500	3,430	8.74	900	10,860	19.62
600	5,240	12.04	1,000	12,780	21.65
700	7,080	14.88			

$SOCl_2(g)$:

$H_T - H_{298.15} = 17.53T + 1.03 \times 10^{-3}T^2 + 1.96 \times 10^5 T^{-1}$
$-5,976$ (0.2 percent; 298°–1,000° K.);
$C_p = 17.53 + 2.06 \times 10^{-3}T - 1.96 \times 10^5 T^{-2}.$

TABLE 769.—*Heat content and entropy of $SO_2Cl_2(g)$*

[Base, ideal gas at 298.15° K.; mol. wt., 134.98]

T, ° K.	$H_T - H_{298.15}$, cal./mole	$S_T - S_{298.15}$, cal./deg. mole	T, ° K.	$H_T - H_{298.15}$, cal./mole	$S_T - S_{298.15}$, cal./deg. mole
400	1,985	5.71	800	10,950	21.11
500	4,080	10.38	900	13,340	23.92
600	6,300	14.43	1,000	15,770	26.48
700	8,600	17.97			

$SO_2Cl_2(g)$:

$H_T - H_{298.15} = 21.10T + 1.93 \times 10^{-3}T^2 + 3.40 \times 10^5 T^{-1}$
$-7,603$ (0.3 percent; 298°–1,000° K.);
$C_p = 21.10 + 3.86 \times 10^{-3}T - 3.40 \times 10^5 T^{-2}.$

SULFURYL FLUORIDE

TABLE 770.—*Heat content and entropy of $SO_2F_2(g)$*

[Base, ideal gas at 298.15° H.; mol. wt., 102.07]

T, ° K.	$H_T - H_{298.15}$, cal./mole	$S_T - S_{298.15}$, cal./deg. mole	T, ° K.	$H_T - H_{298.15}$, cal./mole	$S_T - S_{298.15}$, cal./deg. mole
400	1,755	5.04	800	10,225	19.54
500	3,685	9.34	900	12,550	22.28
600	5,770	13.14	1,000	14,920	24.78
700	7,960	16.52			

$SO_2F_2(g)$:

$H_T - H_{298.15} = 19.12T + 2.76 \times 10^{-3}T^2 + 4.32 \times 10^5 T^{-1}$
$-7,395$ (0.5 percent; 298°–1,000° K.);
$C_p = 19.12 + 5.52 \times 10^{-3}T - 4.32 \times 10^5 T^{-2}.$

TANTALUM AND ITS COMPOUNDS

ELEMENT

TABLE 771.—*Heat content and entropy of $Ta(c)$*

[Base, crystals at 298.15° K.; atomic wt., 180.95]

T, ° K.	$H_T - H_{298.15}$, cal./mole	$S_T - S_{298.15}$, cal./deg. mole	T, ° K.	$H_T - H_{298.15}$, cal./mole	$S_T - S_{298.15}$, cal./deg. mole
400	630	1.81	1,500	7,930	10.50
500	1,260	3.22	1,600	8,620	10.94
600	1,900	4.39	1,700	9,310	11.36
700	2,550	5.39	1,800	10,010	11.76
800	3,210	6.27	1,900	10,720	12.14
900	3,870	7.05	2,000	11,430	12.51
1,000	4,530	7.74	2,200	12,860	13.19
1,100	5,200	8.38	2,400	14,300	13.82
1,200	5,880	8.97	2,600	15,770	14.40
1,300	6,550	9.51	2,800	17,250	14.95
1,400	7,240	10.02	3,000	18,740	15.46

$Ta(c)$:

$H_T - H_{298.15} = 6.31T + 0.20 \times 10^{-3}T^2 + 0.32 \times 10^5 T^{-1}$
$-2,006$ (0.1 percent; 298°–3,000° K.);
$C_p = 6.31 + 0.40 \times 10^{-3}T - 0.32 \times 10^5 T^{-2}.$

TABLE 772.—*Heat content and entropy of $Ta(g)$*

[Base, ideal gas at 298.15° K.; atomic wt.; 180.95]

T, ° K.	$H_T - H_{298.15}$, cal./mole	$S_T - S_{298.15}$, cal./deg. mole	T, ° K.	$H_T - H_{298.15}$, cal./mole	$S_T - S_{298.15}$, cal./deg. mole
400	510	1.48	1,500	7,565	9.58
500	1,030	2.63	1,600	8,320	10.07
600	1,570	3.62	1,700	9,090	10.53
700	2,140	4.49	1,800	9,865	10.98
800	2,735	5.29	1,900	10,655	11.40
900	3,360	6.02	2,000	11,445	11.81
1,000	4,010	6.71	2,200	13,050	12.57
1,100	4,685	7.35	2,400	14,675	13.28
1,200	5,375	7.95	2,600	16,320	13.94
1,300	6,090	8.53	2,800	17,975	14.55
1,400	6,820	9.07	3,000	19,650	15.13

$Ta(g)$:

$H_T - H_{298.15} = 4.59T + 0.94 \times 10^{-3}T^2 + 0.14 \times 10^5 T^{-1}$
$-1,499$ (1.2 percent; 298°–2,000° K.);

TABLE 1h7. (*Continued*)

$$C_p = 4.59 + 1.88 \times 10^{-3}T - 0.14 \times 10^5 T^{-2}.$$
$$H_T - H_{298.15} = 5.68T + 0.52 \times 10^{-3}T^2 + 0.88 \times 10^5 T^{-1}$$
$$-2,035 \ (0.2 \text{ percent}; 2,000°-3,000° \text{ K.});$$
$$C_p = 5.68 + 1.04 \times 10^{-3}T - 0.88 \times 10^5 T^{-2}.$$

OXIDE

TABLE 773.—*Heat content and entropy of* $Ta_2O_5(c)$

[Base, crystals at 298.15° K.; mol. wt., 441.90]

T, ° K.	$H_T - H_{298.15}$, cal./mole	$S_T - S_{298.15}$, cal./deg. mole	T, ° K.	$H_T - H_{298.15}$, cal./mole	$S_T - S_{298.15}$, cal./deg. mole
400	3,430	9.87	1,300	40,880	57.86
500	7,070	17.98	1,400	45,390	61.20
600	10,950	25.05	1,500	49,970	64.36
700	14,990	31.28	1,600	54,630	67.37
800	19,130	36.81	1,700	59,380	70.25
900	23,340	41.76	1,800	64,220	73.01
1,000	27,630	46.29	1,900	69,150	75.67
1,100	31,990	50.44	2,000	74,170	78.25
1,200	36,410	54.29			

$Ta_2O_5(c)$:
$$H_T - H_{298.15} = 37.00T + 3.28 \times 10^{-3}T^2 + 5.92 \times 10^5 T^{-1}$$
$$-13,309 \ (0.4 \text{ percent}; 298°-2,000° \text{ K.});$$
$$C_p = 37.00 + 6.56 \times 10^{-3}T - 5.92 \times 10^5 T^{-2}.$$

CARBIDE

TaC(c):
$$C_p = 8.79 \ (298° \text{ K.}).$$

NITRIDE

TABLE 774.—*Heat content and entropy of* TaN(c)

[Base, crystals at 298.15° K.; mol. wt., 194.96]

T, ° K.	$H_T - H_{298.15}$, cal./mole	$S_T - S_{298.15}$, cal./deg. mole	T, ° K.	$H_T - H_{298.15}$, cal./mole	$S_T - S_{298.15}$, cal./deg. mole
400	1,050	3.03	700	4,680	9.74
500	2,190	5.57	800	6,030	11.54
600	3,400	7.77			

TaN(c):
$$H_T - H_{298.15} = 7.73T + 3.90 \times 10^{-3}T^2 - 2,651$$
$$(0.4 \text{ percent}; 298°-800° \text{ K.});$$
$$C_p = 7.73 + 7.80 \times 10^{-3}T.$$

TECHNETIUM

ELEMENT

TABLE 775.—*Heat content and entropy of* Tc(c, l)

[Base, crystals at 298.15° K.; atomic wt., 99]

T, ° K.	$H_T - H_{298.15}$, cal./mole	$S_T - S_{298.15}$, cal./deg. mole	T, ° K.	$H_T - H_{298.15}$, cal./mole	$S_T - S_{298.15}$, cal./deg. mole
400	600	1.73	1,600	9,240	11.35
500	1,210	3.10	1,700	10,090	11.86
600	1,840	4.24	1,800	10,960	12.36
700	2,490	5.25	1,900	11,850	12.84
800	3,160	6.14	2,000	12,760	13.31
900	3,850	6.95	2,200	14,640	14.20
1,000	4,560	7.70	2,400(c)	16,600	15.06
1,100	5,290	8.40	2,400(l)	22,100	17.35
1,200	6,040	9.05	2,600	24,100	18.15
1,300	6,810	9.67	2,800	26,100	18.89
1,400	7,600	10.25	3,000	28,100	19.58
1,500	8,410	10.81			

Tc(c):
$$H_T - H_{298.15} = 5.20T + 1.00 \times 10^{-3}T^2 - 1,639$$
$$(0.1 \text{ percent}; 298°-2,400° \text{ K.});$$
$$C_p = 5.20 + 2.00 \times 10^{-3}T;$$
$$\Delta H_{2400}(fusion) = 5,500.$$

Tc(l):
$$H_T - H_{298.15} = 10.00T - 1,900 \ (0.1 \text{ percent};$$
$$2,400°-3,000° \text{ K.});$$
$$C_p = 10.00.$$

TABLE 776.—*Heat content and entropy of* Tc(g)

[Base, ideal gas at 298.15° K.; atomic wt., 99]

T, ° K.	$H_T - H_{298.15}$, cal./mole	$S_T - S_{298.15}$, cal./deg. mole	T, ° K.	$H_T - H_{298.15}$, cal./mole	$S_T - S_{298.15}$, cal./deg. mole
400	505	1.46	1,500	7,675	9.60
500	1,010	2.59	1,600	8,445	10.10
600	1,530	3.54	1,700	9,205	10.56
700	2,080	4.38	1,800	9,960	10.99
800	2,665	5.16	1,900	10,705	11.40
900	3,295	5.90	2,000	11,440	11.77
1,000	3,960	6.60	2,200	12,880	12.46
1,100	4,665	7.27	2,400	14,280	13.07
1,200	5,395	7.91	2,600	15,645	13.62
1,300	6,145	8.51	2,800	16,985	14.11
1,400	6,910	9.08	3,000	18,300	14.57

Tc(g):
$$H_T - H_{298.15} = 3.52T + 1.51 \times 10^{-3}T^2 - 0.48 \times 10^5 T^{-1}$$
$$-1,023 \ (0.8 \text{ percent}; 298°-1,600° \text{ K.});$$
$$C_p = 3.52 + 3.02 \times 10^{-3}T + 0.48 \times 10^5 T^{-2}.$$
$$H_T - H_{298.15} = 6.40T + 0.20 \times 10^{-3}T^2 + 1.38 \times 10^5 T^{-1}$$
$$-2,389 \ (0.8 \text{ percent}; 1,600°-3,000° \text{ K.});$$
$$C_p = 6.40 + 0.40 \times 10^{-3}T - 1.38 \times 10^5 T^{-2}.$$

TELLURIUM AND ITS COMPOUNDS

ELEMENT

TABLE 777.—*Heat content and entropy of* Te(c, l)

[Base, crystals at 298.15° K.; atomic wt., 127.61]

T, ° K.	$H_T - H_{298.15}$, cal./mole	$S_T - S_{298.15}$, cal./deg. mole	T, ° K.	$H_T - H_{298.15}$, cal./mole	$S_T - S_{298.15}$, cal./deg. mole
400	655	1.89	800	7,960	12.98
500	1,345	3.42	900	8,860	14.04
600	2,095	4.79	1,000	9,760	14.99
700	2,895	6.02	1,100	10,660	15.84
723(c)	3,085	6.29	1,200	11,560	16.63
723(l)	7,265	12.07			

Te(c):
$$H_T - H_{298.15} = 4.57T + 2.64 \times 10^{-3}T^2 - 1,597$$
$$(0.2 \text{ percent}; 298°-723° \text{ K.});$$
$$C_p = 4.57 + 5.28 \times 10^{-3}T;$$
$$\Delta H_{723}(fusion) = 4,180.$$

Te(l)
$$H_T - H_{298.15} = 9.00T + 760 \ (0.1 \text{ percent};$$
$$723°-1,200° \text{ K.});$$
$$C_p = 9.00.$$

TABLE 1h7. (Continued)

TABLE 778.—*Heat content and entropy of Te(g)*

[Base, ideal gas at 298.15° K.; atomic wt., 127.61]

T, ° K.	$H_T-H_{298.15}$, cal./mole	$S_T-S_{298.15}$, cal./deg. mole	T, ° K.	$H_T-H_{298.15}$, cal./mole	$S_T-S_{298.15}$, cal./deg. mole
400	505	1.46	1,500	6,085	8.12
500	1,005	2.57	1,600	6,620	8.47
600	1,500	3.48	1,700	7,160	8.80
700	1,995	4.24	1,800	7,710	9.11
800	2,495	4.91	1,900	8,260	9.41
900	2,995	5.50	2,000	8,820	9.69
1,000	3,500	6.03	2,200	9,950	10.23
1,100	4,005	6.51	2,400	11,095	10.73
1,200	4,515	6.96	2,600	12,260	11.20
1,300	5,035	7.37	2,800	13,435	11.63
1,400	5,555	7.76	3,000	14,625	12.04

Te(g):

$$H_T-H_{298.15}=4.64T+0.22\times10^{-3}T^2-0.18\times10^5T^{-1}$$
$$-1,343 \text{ (0.4 percent; } 298°-3,000° \text{ K.);}$$
$$C_p=4.64+0.44\times10^{-3}T+0.18\times10^5T^{-2}.$$

TABLE 779.—*Heat content and entropy of Te₂(g)*

[Base, ideal gas at 298.15° K.; mol. wt., 255.22]

T, ° K.	$H_T-H_{298.15}$, cal./mole	$S_T-S_{298.15}$, cal./deg. mole	T, ° K.	$H_T-H_{298.15}$, cal./mole	$S_T-S_{298.15}$, cal./deg. mole
400	895	2.58	1,500	10,690	14.34
500	1,775	4.55	1,600	11,585	14.91
600	2,665	6.17	1,700	12,475	15.46
700	3,555	7.54	1,800	13,370	15.96
800	4,445	8.73	1,900	14,265	16.45
900	5,335	9.78	2,000	15,160	16.91
1,000	6,225	10.72	2,200	16,950	17.76
1,100	7,115	11.57	2,400	18,730	18.54
1,200	8,010	12.34	2,600	20,520	19.26
1,300	8,905	13.06	2,800	22,310	19.92
1,400	9,795	13.72	3,000	24,100	20.54

Te₂(g):

$$H_T-H_{298.15}=8.94T+0.20\times10^5T^{-1}-2,733$$
$$\text{(0.1 percent; } 298°-3,000° \text{ K.);}$$
$$C_p=8.94-0.20\times10^5T^{-2}.$$

OXIDE

TABLE 780.—*Heat content and entropy of TeO(g)*

[Base, ideal gas at 298.15° K.; mol. wt., 143.61]

T, ° K.	$H_T-H_{298.15}$, cal./mole	$S_T-S_{298.15}$, cal./deg. mole	T, ° K.	$H_T-H_{298.15}$, cal./mole	$S_T-S_{298.15}$, cal./deg. mole
400	795	2.29	1,000	5,895	10.03
500	1,610	4.11	1,200	7,650	11.63
600	2,445	5.63	1,400	9,410	12.99
700	3,295	6.94	1,600	11,180	14.17
800	4,155	8.09	1,800	12,955	15.22
900	5,020	9.11	2,000	14,730	16.15

TeO(g):

$$H_T-H_{298.15}=8.44T+0.16\times10^{-3}T^2+0.83\times10^5T^{-1}$$
$$-2,809 \text{ (0.3 percent; } 298°-2,000° \text{ K.);}$$
$$C_p=8.44+0.32\times10^{-3}T-0.83\times10^5T^{-2}.$$

TETRACHLORIDE

TABLE 781.—*Heat content and entropy of TeCl₄(c, l)*

[Base, crystals at 298.15° K.; mol. wt., 269.44]

T, ° K.	$H_T-H_{298.15}$, cal./mole	$S_T-S_{298.15}$, cal./deg. mole	T, ° K.	$H_T-H_{298.15}$, cal./mole	$S_T-S_{298.15}$, cal./deg. mole
350	1,720	5.33	500	11,260	26.34
400	3,380	9.76	550	13,920	31.41
450	5,040	13.67	600	16,580	36.04
497(c)	6,600	16.97	650	19,240	40.29
497(l)	11,100	26.02			

TeCl₄(c):

$$H_T-H_{298.15}=33.20T-9,899 \text{ (0.1 percent;}$$
$$298°-497° \text{ K.);}$$
$$C_p=33.20;$$
$$\Delta H_{497}=4,500.$$

TeCl₄(l):

$$H_T-H_{298.15}=53.20T-15,340 \text{ (0.1 percent;}$$
$$497°-650° \text{ K.);}$$
$$C_p=53.20.$$

HEXAFLUORIDE

TABLE 782.—*Heat content and entropy of TeF₆(g)*

[Base, ideal gas at 298.15° K.; mol. wt., 241.61]

T, ° K.	$H_T-H_{298.15}$, cal./mole	$S_T-S_{298.15}$, cal./deg. mole	T, ° K.	$H_T-H_{298.15}$, cal./mole	$S_T-S_{298.15}$, cal./deg. mole
400	3,030	8.72	1,000	23,925	40.29
500	6,265	15.94	1,200	31,265	46.99
600	9,655	22.11	1,400	38,660	52.68
700	13,140	27.48	1,600	46,090	57.64
800	16,700	32.24	1,800	53,540	62.03
900	20,295	36.47	2,000	61,020	65.97

TeF₆(g):

$$H_T-H_{298.15}=35.33T+0.81\times10^{-3}T^2+7.00\times10^5T^{-1}$$
$$-12,953 \text{ (0.4 percent; } 298°-2,000° \text{ K.);}$$
$$C_p=35.33+1.62\times10^{-3}T-7.00\times10^5T^{-2}.$$

TERBIUM

ELEMENT

Table 783.—*Heat content and entropy of Tb(c, l)*

[Base, crystals at 298.15° K.; atomic wt., 158.93]

T, ° K.	$H_T-H_{298.15}$, cal./mole	$S_T-S_{298.15}$, cal./deg. mole	T, ° K.	$H_T-H_{298.15}$, cal./mole	$S_T-S_{298.15}$, cal./deg. mole
400	675	1.95	1,500	9,160	11.86
500	1,360	3.46	1,600	10,040	12.42
600	2,055	4.74	1,700(c)	10,930	12.97
700	2,770	5.84	1,700(l)	14,830	15.26
800	3,510	6.82	1,800	15,630	15.71
900	4,260	7.71	1,900	16,430	16.15
1,000	5,030	8.52	2,000	17,230	16.56
1,100	5,820	9.28	2,200	18,830	17.32
1,200	6,630	9.98	2,400	20,430	18.02
1,300	7,450	10.64	2,600	22,030	18.66
1,400	8,300	11.26	2,800	23,630	19.25

TABLE 1h7. (*Continued*)

Tb(*c*):

$$H_T - H_{298.15} = 6.00T + 0.90 \times 10^{-3} T^2 - 1,869 \quad (0.1 \text{ percent;}$$
$$298°-1,700° \text{ K.});$$
$$C_p = 6.00 + 1.80 \times 10^{-3} T;$$
$$\Delta H_{1700}(fusion) = 3,900.$$

Tb(*l*):

$$H_T - H_{298.15} = 8.00T + 1,230 \quad (0.1 \text{ percent;}$$
$$1,700°-2,800° \text{ K.});$$
$$C_p = 8.00.$$

THALLIUM AND ITS COMPOUNDS

ELEMENT

TABLE 784.—*Heat content and entropy of Tl(c, l)*

[Base, α-crystals at 298.15° K.; atomic wt., 204.39]

T, ° K.	$H_T-H_{298.15}$, cal./mole	$S_T-S_{298.15}$, cal./deg. mole	T, ° K.	$H_T-H_{298.15}$, cal./mole	$S_T-S_{298.15}$, cal./deg. mole
400	660	1.90	900	5,360	9.64
500	1,340	3.42	1,000	6,090	10.41
507 (α)	1,390	3.52	1,100	6,820	11.10
507 (β)	1,480	3.70	1,200	7,550	11.74
577 (β)	1,980	4.62	1,300	8,280	12.32
577 (l)	3,000	6.39	1,400	9,010	12.86
600	3,170	6.68	1,500	9,740	13.37
700	3,900	7.80	1,600	10,470	13.84
800	4,630	8.78	1,700	11,200	14.28

Tl(α):

$$H_T - H_{298.15} = 5.26T + 1.73 \times 10^{-3} T^2 - 1,722 \quad (0.1 \text{ percent;}$$
$$298°-507° \text{ K.});$$
$$C_p = 5.26 + 3.46 \times 10^{-3} T;$$
$$\Delta H_{507}(transition) = 90.$$

Tl(β):

$$H_T - H_{298.15} = 7.15T - 2,145 \quad (0.1 \text{ percent; } 507°-577° \text{ K.});$$
$$C_p = 7.15;$$
$$\Delta H_{577}(fusion) = 1,020.$$

Tl(*l*):

$$H_T - H_{298.15} = 7.30T - 1,210 \quad (0.1 \text{ percent;}$$
$$577°-1,700° \text{ K.});$$
$$C_p = 7.30.$$

TABLE 785.—*Heat content and entropy of Tl(g)*

[Base, ideal gas at 298.15° K.; atomic wt., 204.39]

T, ° K.	$H_T-H_{298.15}$, cal./mole	$S_T-S_{298.15}$, cal./deg. mole	T, ° K.	$H_T-H_{298.15}$, cal./mole	$S_T-S_{298.15}$, cal./deg. mole
400	505	1.46	1,500	5,995	8.04
500	1,005	2.57	1,600	6,510	8.37
600	1,500	3.48	1,700	7,025	8.69
700	1,995	4.24	1,800	7,550	8.99
800	2,495	4.90	1,900	8,080	9.27
900	2,990	5.49	2,000	8,615	9.55
1,000	3,485	6.01	2,200	9,720	10.07
1,100	3,985	6.49	2,400	10,850	10.57
1,200	4,485	6.92	2,600	12,015	11.03
1,300	4,985	7.32	2,800	13,215	11.47
1,400	5,490	7.69	3,000	14,435	11.90

Tl(*g*):

$$H_T - H_{298.15} = 4.59T + 0.20 \times 10^{-3} T^2 - 0.24 \times 10^5 T^{-1}$$
$$-1,306 \quad (0.7 \text{ percent; } 298°-3,000° \text{ K.});$$
$$C_p = 4.59 + 0.40 \times 10^{-3} T + 0.24 \times 10^5 T^{-2}.$$

HYDRIDE

TABLE 786.—*Heat content and entropy of TlH(g)*

[Base, ideal gas at 298.15° K.; mol. wt., 205.40]

T, ° K.	$H_T-H_{298.15}$, cal./mole	$S_T-S_{298.15}$, cal./deg. mole	T, ° K.	$H_T-H_{298.15}$, cal./mole	$S_T-S_{298.15}$, cal./deg. mole
400	735	2.12	1,000	5,525	9.36
500	1,480	3.78	1,200	7,225	10.91
600	2,255	5.19	1,400	8,945	12.23
700	3,050	6.42	1,600	10,685	13.39
800	3,860	7.50	1,800	12,435	14.42
900	4,685	8.47	2,000	14,190	15.35

TlH(*g*):

$$H_T - H_{298.15} = 7.49T + 0.43 \times 10^{-3} T^2 + 0.60 \times 10^5 T^{-1}$$
$$-2,473 \quad (0.6 \text{ percent; } 298°-2,000° \text{ K.});$$
$$C_p = 7.49 + 0.86 \times 10^{-3} T - 0.60 \times 10^5 T^{-2}.$$

BROMIDE

TABLE 787.—*Heat content and entropy of TlBr (c, l)*

[Base, crystals at 298.15° K.; mol. wt., 284.31]

T, ° K.	$H_T-H_{298.15}$, cal./mole	$S_T-S_{298.15}$, cal./deg. mole	T, ° K.	$H_T-H_{298.15}$, cal./mole	$S_T-S_{298.15}$, cal./deg. mole
400	1,290	3.72	733 (l)	9,710	17.16
500	2,570	6.58	800	10,780	18.55
600	3,860	8.93	900	12,380	20.43
700	5,160	10.93	1,000	13,980	22.12
733 (c)	5,590	11.54			

TlBr(*c*):

$$H_T - H_{298.15} = 12.24T + 0.60 \times 10^{-3} T^2 - 3,703$$
$$(0.1 \text{ percent; } 298°-733° \text{ K.});$$
$$C_p = 12.24 + 1.20 \times 10^{-3} T;$$
$$\Delta H_{733}(fusion) = 4,120.$$

TlBr(*l*):

$$H_T - H_{298.15} = 16.00T - 2,020 \quad (0.1 \text{ percent;}$$
$$733°-1,000° \text{ K.}).$$
$$C_p = 16.00.$$

TABLE 788.—*Heat content and entropy of TlBr (g)*

[Base, ideal gas at 298.15° K.; mol. wt., 284.31]

T, ° K.	$H_T-H_{298.15}$, cal./mole	$S_T-S_{298.15}$, cal./deg. mole	T, ° K.	$H_T-H_{298.15}$, cal./mole	$S_T-S_{298.15}$, cal./deg. mole
400	900	2.59	1,000	6,250	10.76
500	1,790	4.58	1,200	8,035	12.39
600	2,680	6.20	1,400	9,820	13.76
700	3,570	7.57	1,600	11,610	14.96
800	4,460	8.76	1,800	13,395	16.01
900	5,335	9.82	2,000	15,180	16.95

TlBr(*g*):

$$H_T - H_{298.15} = 8.94T + 0.11 \times 10^5 T^{-1} - 2,702$$
$$(0.1 \text{ percent; } 298°-2,000° \text{ K.});$$
$$C_p = 8.94 - 0.11 \times 10^5 T^{-2}.$$

TABLE 1h7. (*Continued*)

CHLORIDE

TABLE 789.—*Heat content and entropy of TlCl*
(*c, l*)

[Base, crystals at 298.15° K.; mol. wt., 239.85]

T, ° K.	$H_T-H_{298.15}$, cal./mole	$S_T-S_{298.15}$, cal./deg. mole	T, ° K.	$H_T-H_{298.15}$, cal./mole	$S_T-S_{298.15}$, cal./deg. mole
400	1,300	3.75	700 (*l*)	9,170	16.70
500	2,600	6.65	800	10,590	18.60
600	3,900	9.02	900	12,010	20.27
700 (*c*)	5,220	11.06	1,000	13,430	21.76

TlCl(*c*):
$$H_T-H_{298.15}=12.00\,T+1.00\times10^{-3}\,T^2-3,667$$
(0.4 percent; 298°–700° K.);
$$C_p=12.00+2.00\times10^{-3}\,T;$$
$$\Delta H_{700}(fusion)=3,950.$$

TlCl(*l*):
$$H_T-H_{298.15}=14.20\,T-770 \text{ (0.1 percent;}$$
700°–1,000° K.);
$$C_p=14.20.$$

TABLE 790.—*Heat content and entropy of TlCl*
(*g*)

[Base, ideal gas at 298.15° K.; mol. wt., 239.85]

T, ° K.	$H_T-H_{298.15}$, cal./mole	$S_T-S_{298.15}$, cal./deg. mole	T, ° K.	$H_T-H_{298.15}$, cal./mole	$S_T-S_{298.15}$, cal./deg. mole
400	890	2.57	1,000	6,215	10.68
500	1,770	4.53	1,200	8,000	12.31
600	2,655	6.14	1,400	9,780	13.69
700	3,540	7.50	1,600	11,565	14.88
800	4,430	8.69	1,800	13,355	15.94
900	5,320	9.74	2,000	15,140	16.88

TlCl(*g*):
$$H_T-H_{298.15}=8.94\,T+0.25\times10^5\,T^{-1}-2,748$$
(0.1 percent; 298°–2,000° K.);
$$C_p=8.94-0.25\times10^5\,T^{-2}.$$

FLUORIDE

TABLE 791.—*Heat content and entropy of TlF(g)*

[Base, ideal gas at 298.15° K. mol. wt., 223.39]

T, ° K.	$H_T-H_{298.15}$, cal./mole	$S_T-S_{298.15}$, cal./deg. mole	T, ° K.	$H_T-H_{298.15}$, cal./mole	$S_T-S_{298.15}$, cal./deg. mole
400	855	2.47	1,000	6,115	10.48
500	1,715	4.38	1,200	7,890	12.10
600	2,585	5.97	1,400	9,670	13.47
700	3,465	7.33	1,600	11,450	14.65
800	4,345	8.50	1,800	13,235	15.71
900	5,230	9.54	2,000	15,020	16.65

TlF(*g*):
$$H_T-H_{298.15}=8.91\,T+0.57\times10^5\,T^{-1}-2,848$$
(0.2 percent; 298°–2,000° K.);
$$C_p=8.91-0.57\times10^5\,T^{-2}.$$

IODIDE

TABLE 792.—*Heat content and entropy of TlI(g)*

[Base, ideal gas at 298.15° K.; mol. wt., 331.30]

T, ° K.	$H_T-H_{298.15}$, cal./mole	$S_T-S_{298.15}$, cal./deg. mole	T, ° K.	$H_T-H_{298.15}$, cal./mole	$S_T-S_{298.15}$, cal./deg. mole
400	905	2.61	1,000	6,260	10.78
500	1,795	4.60	1,200	8,045	12.41
600	2,685	6.22	1,400	9,835	13.79
700	3,580	7.60	1,600	11,620	14.98
800	4,470	8.78	1,800	13,410	16.04
900	5,365	9.84	2,000	15,195	16.98

TlI(*g*):
$$H_T-H_{298.15}=8.94\,T+0.07\times10^5\,T^{-1}-2,689$$
(0.1 percent; 298°–2,000° K.);
$$C_p=8.94-0.07\times10^5\,T^{-2}.$$

NITRATE

TABLE 793.—*Heat content and entropy of*
$TlNO_3(c)$

[Base, α-crystals at 298.15° K., mol. wt., 266.40]

T, ° K.	$H_T-H_{298.15}$, cal./mole	$S_T-S_{298.15}$, cal./deg. mole	T, ° K.	$H_T-H_{298.15}$, cal./mole	$S_T-S_{298.15}$, cal./deg. mole
334(α)	875	2.77	417(γ)	4,130	11.21
334(β)	945	2.98	450	5,330	13.98
400	2,690	7.73	480	6,410	16.30
417(β)	3,140	8.84			

$TlNO_3(α)$:
$$H_T-H_{298.15}=24.40\,T-7,275 \text{ (0.1 percent;}$$
298°–334° K.);
$$C_p=24.40;$$
$$\Delta H_{334}(transition)=70.$$

$TlNO_3(β)$:
$$H_T-H_{298.15}=26.50\,T-7,908 \text{ (0.1 percent;}$$
334°–417° K.);
$$C_p=26.50;$$
$$\Delta H_{417}(transition)=990.$$

$TlNO_3(γ)$:
$$H_T-H_{298.15}=36.20\,T-10,962 \text{ (0.1 percent;}$$
417°–480° K.);
$$C_p=36.20.$$

THALLIUM-BISMUTH COMPOUNDS

TABLE 794.—*Heat content and entropy of*
$TlBi_2(c, l)$

[Base, crystals at 298.15° K.; mol. wt., 622.39]

T, ° K.	$H_T-H_{298.15}$, cal./mole	$S_T-S_{298.15}$, cal./deg. mole	T, ° K.	$H_T-H_{298.15}$, cal./mole	$S_T-S_{298.15}$, cal./deg. mole
400	1,800	5.20	500	9,020	20.41
450	2,680	7.27	600	11,270	24.51
484(*c*)	3,280	8.55	700	13,520	27.98
484(*l*)	8,660	19.67	800	15,770	30.98

TABLE 1h7. (*Continued*)

$TlBi_2(c)$:

$$H_T - H_{298.15} = 17.65T - 5,262 \text{ (0.1 percent;}$$
$$298°-484° \text{ K.);}$$
$$C_p = 17.65;$$
$$\Delta H_{484}(fusion) = 5,380.$$

$TlBi_2(l)$:

$$H_T - H_{298.15} = 22.50T - 2,230 \text{ (0.1 percent;}$$
$$484°-800° \text{ K.);}$$
$$C_p = 22.50.$$

TABLE 795.—*Heat content and entropy of* $Tl_2Bi_3(c, l)$

[Base, crystals at 298.15° K.; mol. wt., 1035.78]

T, ° K.	$H_T-H_{298.15}$, cal./mole	$S_T-S_{298.15}$, cal./deg. mole	T, ° K.	$H_T-H_{298.15}$, cal./mole	$S_T-S_{298.15}$, cal./deg. mole
400	3,210	9.27	500	14,550	33.10
450	4,780	12.96	600	18,130	39.63
487(c)	5,950	15.46	700	21,710	45.15
487(l)	14,080	32.15	800	25,290	50.02

$Tl_2Bi_3(c)$:

$$H_T - H_{298.15} = 31.50T - 9,392 \text{ (0.1 percent;}$$
$$298°-487° \text{ K.);}$$
$$C_p = 31.50;$$
$$\Delta H_{487}(fusion) = 8,130.$$

$Tl_2Bi_3(l)$:

$$H_T - H_{298.15} = 35.80T - 3,350 \text{ (0.1 percent;}$$
$$487°-800° \text{ K.);}$$
$$C_p = 35.80.$$

THORIUM AND ITS COMPOUNDS

ELEMENT

TABLE 796.—*Heat content and entropy of* $Th(c, l)$

[Base, α-crystals at 298.15° K.; atomic wt., 232.05]

T, ° K.	$H_T-H_{298.15}$, cal./mole	$S_T-S_{298.15}$, cal./deg. mole	T, ° K.	$H_T-H_{298.15}$, cal./mole	$S_T-S_{298.15}$, cal./deg. mole
400	690	1.99	1,673(α)	13,280	15.18
500	1,410	3.59	1,673(β)	13,950	15.58
600	2,180	4.99	1,700	14,250	15.76
700	2,990	6.25	1,800	15,350	16.39
800	3,850	7.39	1,900	16,450	16.98
900	4,760	8.46	1,968(β)	17,200	17.37
1,000	5,710	9.46	1,968(l)	21,700	19.66
1,100	6,700	10.41	2,000	22,050	19.83
1,200	7,740	11.31	2,200	24,250	20.88
1,300	8,830	12.18	2,400	26,450	21.84
1,400	9,960	13.02	2,600	28,650	22.72
1,500	11,140	13.83	2,800	30,850	23.53
1,600	12,360	14.62	3,000	33,050	24.29

$Th(\alpha)$:

$$H_T - H_{298.15} = 5.17T + 2.28 \times 10^{-3}T^2 - 1,744 \text{ (0.1 percent;}$$
$$298°-1,673° \text{ K.);}$$
$$C_p = 5.17 + 4.56 \times 10^{-3}T;$$
$$\Delta H_{1673}(transition) = 670.$$

$Th(\beta)$:

$$H_T - H_{298.15} = 11.00T - 4,450 \text{ (0.1 percent; } 1,673°-1,968°$$

K.);
$$C_p = 11.00;$$
$$\Delta H_{1968}(fusion) = 4,500.$$

$Th(l)$·

$$H_T - H_{298.15} = 11.00T + 50 \text{ (0.1 percent; } 1,968°-3,000°$$
K.);
$$C_p = 11.00.$$

OXIDE

TABLE 797.—*Heat content and entropy of* $ThO_2(c)$

[Base, crystals at 298.15° K.; mol. wt., 264.05

T, ° K.	$H_T-H_{298.15}$, cal./mole	$S_T-S_{298.15}$, cal./deg. mole	T, ° K.	$H_T-H_{298.15}$, cal./mole	$S_T-S_{298.15}$, cal./deg. mole
400	1,600	4.61	1,300	17,800	25.39
500	3,210	8.20	1,400	19,760	26.84
600	4,890	11.26	1,500	21,740	28.21
700	6,620	13.92	1,600	23,740	29.50
800	8,390	16.29	1,700	25,750	30.72
900	10,200	18.42	1,800	27,770	31.87
1,000	12,050	20.37	1,900	29,800	32.97
1,100	13,940	22.17	2,000	31,840	34.02
1,200	15,860	23.84			

$ThO_2(c)$:

$$H_T - H_{298.15} = 15.84T + 1.44 \times 10^{-3}T^2 + 1.60 \times 10^5 T^{-1}$$
$$-5,387 \text{ (0.4 percent; } 298°-2,000° \text{ K.);}$$
$$C_p = 15.84 + 2.88 \times 10^{-3}T - 1.60 \times 10^5 T^{-2}.$$

NITRIDE

TABLE 798.—*Heat content and entropy of* $Th_3N_4(c)$

[Base, crystals at 298.15° K.; mol. wt., 752.18]

T, ° K.	$H_T-H_{298.15}$, cal./mole	$S_T-S_{298.15}$, cal./deg. mole	T, ° K.	$H_T-H_{298.15}$, cal./mole	$S_T-S_{298.15}$, cal./deg. mole
400	3,960	11.41	700	17,540	36.50
500	8,180	20.81	800	22,540	43.17
600	12,720	29.08			

$Th_3N_4(c)$:

$$H_T - H_{298.15} = 27.78T + 15.90 \times 10^{-3}T^2 - 9,696 \text{ (0.3 per-}$$
$$\text{cent; } 298°-800° \text{ K.);}$$
$$C_p = 27.78 + 31.80 \times 10^{-3}T.$$

FLUORIDE

$ThF_4(c)$:
$$C_p = 26.46 \text{ (298° K.).}$$

SULFATE

$Th(SO_4)_2(c)$:
$$\overline{C_p} = 41.2 \text{ (273°-373° K.).}$$

TABLE 1h7. (*Continued*)

THULIUM

ELEMENT

TABLE 799.—*Heat content and entropy of Tm(c, l)*

[Base, crystals at 298.15° K.; atomic wt., 168.94]

T, ° K.	$H_T - H_{298.15}$, cal./mole	$S_T - S_{298.15}$, cal./deg. mole	T, ° K.	$H_T - H_{298.15}$, cal./mole	$S_T - S_{298.15}$, cal./deg. mole
400	660	1.91	1,400	8,010	10.93
500	1,330	3.40	1,500	8,830	11.50
600	2,010	4.65	1,600	9,660	12.03
700	2,710	5.72	1,700	10,510	12.55
800	3,420	6.67	1,800	11,370	13.04
900	4,150	7.53	1,900(c)	12,250	13.52
1,000	4,890	8.31	1,900(l)	16,650	15.83
1,100	5,650	9.03	2,000	17,450	16.24
1,200	6,420	9.71	2,200	19,050	17.01
1,300	7,210	10.34	2,400	20,650	17.71

Tm(c):

$$H_T - H_{298.45} = 6.00T + 0.75 \times 10^{-3}T^2 - 1,856$$

(0.1 percent; 298°–1,900° K.);

$$C_p = 6.00 + 1.50 \times 10^{-3}T;$$

$$\Delta H_{1900}(fusion) = 4,400.$$

Tm(l):

$$H_T - H_{298.15} = 8.00T + 1,450 \text{ (0.1 percent;}$$

1,900°–2,400° K.);

$$C_p = 8.00.$$

TIN AND ITS COMPOUNDS

ELEMENT

TABLE 800.—*Heat content and entropy of Sn(c, l)*

[Base, crystals at 298.15° K.; atomic wt., 118.70]

T, ° K.	$H_T - H_{298.15}$, cal./mole	$S_T - S_{298.15}$, cal./deg. mole	T, ° K.	$H_T - H_{298.15}$, cal./mole	$S_T - S_{298.15}$, cal./deg. mole
350	340	1.05	1,300	8,960	13.94
400	680	1.96	1,400	9,690	14.48
450	1,030	2.78	1,500	10,420	14.99
500	1,400	3.56	1,600	11,150	15.46
505(c)	1,440	3.64	1,700	11,880	15.90
505(l)	3,160	7.05	1,800	12,610	16.32
600	3,850	8.30	1,900	13,340	16.71
700	4,580	9.43	2,000	14,070	17.09
800	5,310	10.40	2,200	15,530	17.78
900	6,040	11.26	2,400	16,990	18.42
1,000	6,770	12.03	2,600	18,450	19.00
1,100	7,500	12.73	2,900	19,910	19.54
1,200	8,230	13.36	3,000	21,370	20.04

Sn(c):

$$H_T - H_{298.15} = 4.42T + 3.15 \times 10^{-3}T^2 - 1,598$$

(0.8 percent; 298°–505° K.);

$$C_p = 4.42 + 6.30 \times 10^{-3}T;$$

$$\Delta H_{505}(fusion) = 1,720.$$

Sn(l):

$$H_T - H_{298.15} = 7.30 - 530 \text{ (0.1 percent;}$$

505°–3,000° K.);

$$C_p = 7.30.$$

TABLE 801.—*Heat content and entropy of Sn(g)*

[Base, ideal gas at 298.15° K.; atomic wt., 118.70]

T, ° K.	$H_T - H_{298.15}$, cal./mole	$S_T - S_{298.15}$, cal./deg. mole	T, ° K.	$H_T - H_{298.15}$, cal./mole	$S_T - S_{298.15}$, cal./deg. mole
400	535	1.54	2,000	12,200	12.99
500	1,110	2.82	2,100	12,885	13.32
600	1,750	3.98	2,200	13,555	13.64
700	2,440	5.05	2,300	14,220	13.93
800	3,180	6.03	2,400	14,880	14.21
900	3,950	6.94	2,500	15,530	14.48
1,000	4,735	7.77	2,750	17,140	15.09
1,100	5,525	8.52	3,000	18,715	15.64
1,200	6,315	9.21	3,250	20,275	16.14
1,300	7,095	9.84	3,500	21,810	16.60
1,400	7,865	10.41	3,750	23,330	17.01
1,500	8,620	10.93	4,000	24,835	17.40
1,600	9,365	11.41	4,250	26,325	17.76
1,700	10,090	11.85	4,500	27,810	18.10
1,800	10,805	12.26	4,750	29,280	18.42
1,900	11,510	12.64	5,000	30,745	18.72

Sn(g):

$$H_T - H_{298.15} = 8.31T - 0.31 \times 10^{-3}T^2 + 2.70 \times 10^5 T^{-1}$$

$$-3,356 \text{ (0.6 percent; 1,400°–5,000° K.);}$$

$$C_p = 8.31 - 0.62 \times 10^{-3}T - 2.70 \times 10^5 T^{-2}.$$

OXIDES

SnO(c):

$$C_p = 9.55 + 3.50 \times 10^{-3}T (estimated, (298°–1,273° \text{ K.).}$$

TABLE 802.—*Heat content and entropy of SnO(g)*

[Base, ideal gas at 298.15° K.; mol. wt., 134.70]

T, ° K.	$H_T - H_{298.15}$, cal./mole	$S_T - S_{298.15}$, cal./deg. mole	T, ° K.	$H_T - H_{298.15}$, cal./mole	$S_T - S_{298.15}$, cal./deg. mole
400	790	2.28	1,000	5,875	9.99
500	1,605	4.09	1,200	7,625	11.59
600	2,435	5.61	1,400	9,390	12.95
700	3,280	6.91	1,600	11,155	14.12
800	4,140	8.06	1,800	12,930	15.17
900	5,005	9.08	2,000	14,705	16.10

SnO(g):

$$H_T - H_{298.15} = 8.42T + 0.16 \times 10^{-3}T^2 + 0.84 \times 10^5 T^{-1}$$

$$-2,806 \text{ (0.4 percent; 298°–2,000° K.);}$$

$$C_p = 8.42 + 0.32 \times 10^{-3}T - 0.84 \times 10^5 T^{-2}.$$

TABLE 803.—*Heat content and entropy of SnO$_2$(c)*

[Base, crystals at 298.15° K.; mol. wt., 150.70]

T, ° K.	$H_T - H_{298.15}$, cal./mole	$S_T - S_{298.15}$, cal./deg. mole	T, ° K.	$H_T - H_{298.15}$, cal./mole	$S_T - S_{298.15}$, cal./deg. mole
400	1,510	4.34	1,000	12,210	20.38
500	3,100	7.88	1,100	14,190	22.27
600	4,780	10.95	1,200	16,210	24.03
700	6,550	13.68	1,300	18,260	25.67
800	8,390	16.13	1,400	20,340	27.21
900	10,280	18.35	1,500	22,440	28.66

SnO$_2$(c):

$$H_T - H_{298.15} = 17.66T + 1.20 \times 10^{-3}T^2 + 5.16 \times 10^5 T^{-1}$$

$$-7,103 \text{ (0.8 percent; 298°–1,500° K.);}$$

$$C_p = 17.66 + 2.40 \times 10^{-3}T - 5.16 \times 10^5 T^{-2}.$$

TABLE 1h7. (*Continued*)

SULFIDES

TABLE 804.—*Heat content and entropy of SnS (c, l)*

[Base, α-crystals at 298.15° K.; mol. wt., 150.77]

T, ° K.	$H_T - H_{298.15}$, cal./mole	$S_T - S_{298.15}$, cal./deg. mole	T, ° K.	$H_T - H_{298.15}$, cal./mole	$S_T - S_{298.15}$, cal./deg. mole
400	1,210	3.49	900	8,140	14.50
500	2,450	6.25	1,000	9,470	15.90
600	3,750	8.62	1,100	10,840	17.21
700	5,090	10.69	1,153 (β)	11,580	17.86
800	6,520	12.60	1,153 (l)	19,130	24.41
875 (α)	7,650	13.94	1,200	19,970	25.13
875 (β)	7,810	14.13	1,250	20,860	25.85

SnS(α):
$$H_T - H_{298.15} = 8.53T + 3.74 \times 10^{-3}T^2 - 0.90 \times 10^5 T^{-1}$$
$$-2,574 \ (0.2 \text{ percent}; 298°-875° \text{ K.});$$
$$C_p = 8.53 + 7.48 \times 10^{-3}T + 0.90 \times 10^5 T^{-2};$$
$$\Delta H_{875}(transition) = 160.$$

SnS(β):
$$H_T - H_{298.15} = 9.78T + 1.87 \times 10^{-3}T^2 - 2,180$$
$$(0.1 \text{ percent}; 875°-1,153° \text{ K.});$$
$$C_p = 9.78 + 3.74 \times 10^{-3}T;$$
$$\Delta H_{1153}(fusion) = 7,550.$$

SnS(l):
$$H_T - H_{298.15} = 17.90T - 1,510 \ (0.1 \text{ percent};$$
$$1,153°-1,250° \text{ K.});$$
$$C_p = 17.90.$$

TABLE 805.—*Heat content and entropy of SnS(g)*

[Base, ideal gas at 298.15° K.; mol. wt., 150.77]

T, ° K.	$H_T - H_{298.15}$, cal./mole	$S_T - S_{298.15}$, cal./deg. mole	T, ° K.	$H_T - H_{298.15}$, cal./mole	$S_T - S_{298.15}$, cal./deg. mole
400	855	2.47	1,000	6,105	10.46
500	1,710	4.37	1,200	7,880	12.08
600	2,580	5.96	1,400	9,660	13.45
700	3,455	7.31	1,600	11,440	14.64
800	4,335	8.48	1,800	13,225	15.69
900	5,220	9.52	2,000	15,010	16.63

SnS(g):
$$H_T - H_{298.15} = 8.83T + 0.04 \times 10^{-3}T^2 + 0.55 \times 10^5 T^{-1}$$
$$-2,821 \ (0.1 \text{ percent}; 298°-2,000° \text{ K.});$$
$$C_p = 8.83 + 0.08 \times 10^{-3}T - 0.55 \times 10^5 T^{-2}.$$

TABLE 806.—*Heat content and entropy of SnS₂(c)*

[Base, crystals at 298.15° K.; mol. wt., 182.83]

T, ° K.	$H_T - H_{298.15}$, cal./mole	$S_T - S_{298.15}$, cal./deg. mole	T, ° K.	$H_T - H_{298.15}$, cal./mole	$S_T - S_{298.15}$, cal./deg. mole
400	1,740	5.02	800	8,930	17.41
500	3,470	8.88	900	10,840	19.66
600	5,250	12.12	1,000	12,810	21.73
700	7,070	14.93			

SnS₂(c):
$$H_T - H_{298.15} = 15.51T + 2.10 \times 10^{-3}T^2 - 4,811$$
$$(0.2 \text{ percent}; 298°-1,000° \text{ K.});$$
$$C_p = 15.51 + 4.20 \times 10^{-3}T.$$

HYDRIDE

TABLE 807.—*Heat content and entropy of SnH(g)*

[Base, ideal gas at 298.15° K.; mol. wt., 119.71]

T, ° K.	$H_T - H_{298.15}$, cal./mole	$S_T - S_{298.15}$, cal./deg. mole	T, ° K.	$H_T - H_{298.15}$, cal./mole	$S_T - S_{298.15}$, cal./deg. mole
400	725	2.09	1,000	5,660	9.49
500	1,460	3.73	1,200	7,495	11.16
600	2,235	5.14	1,400	9,370	12.60
700	3,045	6.39	1,600	11,260	13.87
800	3,890	7.52	1,800	13,155	14.98
900	4,765	8.55	2,000	15,045	15.98

SnH(g):
$$H_T - H_{298.15} = 7.47T + 0.65 \times 10^{-3}T^2 + 0.75 \times 10^5 T^{-1}$$
$$-2,537 \ (1.5 \text{ percent}; 298°-2,000° \text{ K.});$$
$$C_p = 7.47 + 1.30 \times 10^{-3}T - 0.75 \times 10^5 T^{-2}.$$

BROMIDES

TABLE 808.—*Heat content and entropy of SnBr(g)*

[Base, ideal gas at 298.15° K.; mol. wt., 198.62]

T, ° K.	$H_T - H_{298.15}$, cal./mole	$S_T - S_{298.15}$, cal./deg. mole	T, ° K.	$H_T - H_{298.15}$, cal./mole	$S_T - S_{298.15}$, cal./deg. mole
400	895	2.58	1,000	6,425	10.98
500	1,785	4.57	1,200	8,360	12.74
600	2,685	6.21	1,400	10,315	14.25
700	3,600	7.62	1,600	12,275	15.56
800	4,530	8.86	1,800	14,235	16.71
900	5,470	9.97	2,000	16,190	17.74

SnBr(g):
$$H_T - H_{298.15} = 8.82T + 0.32 \times 10^{-3}T^2 + 0.24 \times 10^5 T^{-1}$$
$$-2,739 \ (0.5 \text{ percent}; 298°-2,000° \text{ K.});$$
$$C_p = 8.82 + 0.64 \times 10^{-3}T - 0.24 \times 10^5 T^{-2}.$$

TABLE 809.—*Heat content and entropy of SnBr₄(g)*

[Base, ideal gas at 298.15° K.; mol. wt., 438.36]

T, ° K.	$H_T - H_{298.15}$, cal./mole	$S_T - S_{298.15}$, cal./deg. mole	T, ° K.	$H_T - H_{298.15}$, cal./mole	$S_T - S_{298.15}$, cal./deg. mole
400	2,545	7.35	800	12,745	25.00
500	5,075	12.99	900	15,310	28.02
600	7,620	17.63	1,000	17,880	30.73
700	10,180	21.57			

SnBr₄(g):
$$H_T - H_{298.15} = 25.80T + 0.97 \times 10^5 T^{-1} - 8,018$$
$$(0.1 \text{ percent}; 298°-1,000° \text{ K.});$$
$$C_p = 25.80 - 0.97 \times 10^5 T^{-2}.$$

CHLORIDES

SnCl₂(c):
$$C_p = 15.82 + 10.36 \times 10^{-3}T(estimated) \ (298°-500° \text{ K.}).$$

TABLE 1h7. (*Continued*)

TABLE 810.—*Heat content and entropy of SnCl(g)*

[Base, ideal gas at 298.15° K.; mol. wt., 154.16]

T, ° K.	$H_T-H_{298.15}$, cal./mole	$S_T-S_{298.15}$, cal./deg. mole	T, ° K.	$H_T-H_{298.15}$, cal./mole	$S_T-S_{298.15}$, cal./deg. mole
400	875	2.52	1,000	6,400	10.89
500	1,755	4.49	1,200	8,340	12.66
600	2,655	6.13	1,400	10,280	14.16
700	3,565	7.53	1,600	12,240	15.47
800	4,495	8.77	1,800	14,200	16.62
900	5,440	9.88	2,000	16,150	17.65

SnCl(g):

$$H_T-H_{298.15}=8.91T+0.29\times10^{-3}T^2+0.50\times10^5T^{-1}$$
$$-2,850 \ (0.7 \text{ percent}; \ 298°-2,000° \text{ K.});$$
$$C_p=8.91+0.58\times10^{-3}T-0.50\times10^5T^{-2}.$$

TABLE 811.—*Heat content and entropy of SnCl$_4$(g)*

[Base, ideal gas at 298.15° K.; mol. wt., 260.53]

T, ° K.	$H_T-H_{298.15}$, cal./mole	$S_T-S_{298.15}$, cal./deg. mole	T, ° K.	$H_T-H_{298.15}$, cal./mole	$S_T-S_{298.15}$, cal./deg. mole
400	2,450	7.07	800	12,495	24.43
500	4,925	12.59	900	15,045	27.43
600	7,425	17.14	1,000	17,600	30.13
700	9,955	21.04			

SnCl$_4$(g):

$$H_T-H_{298.15}=25.57T+0.10\times10^{-3}T^2+1.87\times10^5T^{-1}$$
$$-8,260 \ (0.1 \text{ percent}; \ 298°-1,000° \text{ K.});$$
$$C_p=25.57+0.20\times10^{-3}T-1.87\times10^5T^{-2}.$$

FLUORIDE

TABLE 812.—*Heat content and entropy of SnF(g)*

[Base, ideal gas at 298.15° K.; mol. wt., 137.70]

T, ° K.	$H_T-H_{298.15}$, cal./mole	$S_T-S_{298.15}$, cal./deg. mole	T, ° K.	$H_T-H_{298.15}$, cal./mole	$S_T-S_{298.15}$, cal./deg. mole
400	835	2.40	1,000	6,275	10.63
500	1,690	4.31	1,200	8,205	12.39
600	2,570	5.92	1,400	10,155	13.90
700	3,470	7.30	1,600	12,105	15.20
800	4,390	8.53	1,800	14,050	16.34
900	5.325	9.63	2,000	15,985	17.37

SnF(g):

$$H_T-H_{296.15}=8.84T+0.31\times10^{-3}T^2+0.87\times10^5T^{-1}$$
$$-2,955 \ (0.8 \text{ percent}; \ 298°-2,000° \text{ K.});$$
$$C_p=8.84+0.62\times10^{-3}T-0.87\times10^5T^{-2}.$$

IODIDE

TABLE 813.—*Heat content and entropy of SnI$_4$(c, l)*

[Base, crystals at 298.15° K.; mol. wt., 626.34]

T, ° K.	$H_T-H_{298.15}$, cal./mole	$S_T-S_{298.15}$, cal./deg. mole	T, ° K.	$H_T-H_{298.15}$, cal./mole	$S_T-S_{298.15}$, cal./deg. mole
350	1,610	4.98	418(l)	8,470	21.87
400	3,255	9.37	500	11,760	29.05
418(c)	3,870	10.87	600	15,770	36.36

SnI$_4$(c):

$$H_T-H_{298.15}=19.40T+18.00\times10^{-3}T^2-7,384 \ (0.1 \text{ percent}; \ 298°-418° \text{ K.});$$
$$C_p=19.40+36.00\times10^{-3}T;$$
$$\Delta H_{418}(fusion)=4,600.$$

SnI$_4$(l):

$$H_T-H_{298.15}=40.10T-8,290 \ (0.1 \text{ percent}; \ 418°-600° \text{ K.});$$
$$C_p=40.10.$$

TABLE 814.—*Heat content and entropy of SnI$_4$(g)*

[Base, ideal gas at 298.15° K.; mol. wt., 626.34]

T, ° K.	$H_T-H_{298.15}$, cal./mole	$S_T-S_{298.15}$, cal./deg. mole	T, ° K.	$H_T-H_{298.15}$, cal./mole	$S_T-S_{298.15}$, cal./deg. mole
400	2,580	7.44	800	12,840	25.21
500	5,135	13.14	900	15,415	28.25
600	7,700	17.82	1,000	17,990	30.96
700	10,270	21.78			

SnI$_4$(g):

$$H_T-H_{298.15}=25.82T+0.56\times10^5T^{-1}-7,886 \ (0.1 \text{ percent}; \ 298°-1,000° \text{ K.});$$
$$C_p=25.82-0.56\times10^5T^{-2}.$$

SELENIDE

TABLE 815.—*Heat content and entropy of SnSe(g)*

[Base, ideal gas at 298.15° K.; mol. wt., 197.66]

T, ° K.	$H_T-H_{298.15}$, cal./mole	$S_T-S_{298.15}$, cal./deg. mole	T, ° K.	$H_T-H_{298.15}$, cal./mole	$S_T-S_{298.15}$, cal./deg. mole
400	880	2.54	1,000	6,195	10.65
500	1,760	4.50	1,200	7,975	12.27
600	2,640	6.11	1,400	9,760	13.64
700	3,525	7.47	1,600	11,545	14.84
800	4,415	8.66	1,800	13,330	15.89
900	5,305	9.71	2,000	15,115	16.83

SnSe(g):

$$H_T-H_{298.15}=8.93T+0.31\times10^5T^{-1}-2,766 \ (0.1 \text{ percent}; \ 298°-2,000° \text{ K.});$$
$$C_p=8.93-0.31\times10^5T^{-2}.$$

TELLURIDES

TABLE 816.—*Heat content and entropy of SnTe(g)*

[Base, ideal gas at 298.15° K.; mol. wt., 246.31]

T, ° K.	$H_T-H_{298.15}$, cal./mole	$S_T-S_{298.15}$, cal./deg. mole	T, ° K.	$H_T-H_{298.15}$, cal./mole	$S_T-S_{298.15}$, cal./deg. mole
400	890	2.56	1,000	6,225	10.70
500	1,775	4.54	1,200	8,010	12.33
600	2,660	6.15	1,400	9,795	13.71
700	3,550	7.52	1,600	11,580	14.90
800	4,440	8.71	1,800	13,365	15.95
900	5,330	9.76	2,000	15,155	16.89

SnTe(g):

$$H_T-H_{298.15}=8.94T+0.22\times10^5T^{-1}-2,739$$
$$(0.1 \text{ percent}; \ 298°-2,000° \text{ K.});$$
$$C_p=8.94-0.22\times10^5T^{-2}.$$

TABLE 1h7. (*Continued*)

TABLE 817.—*Heat content and entropy of* $SnTe_2(c)$

[Base, crystals at 298.15° K.; mol. wt., 373.92]

T, ° K.	$H_T-H_{298.15}$, cal./mole	$S_T-S_{298.15}$, cal./deg. mole	T, ° K.	$H_T-H_{298.15}$, cal./mole	$S_T-S_{298.15}$, cal./deg. mole
400	1,840	5.31	600	5,610	12.93
500	3,680	9.41			

$SnTe_2(c)$:

$$H_T-H_{298.15}=15.35T+3.60\times10^{-3}T^2-4,897$$

$$(0.3 \text{ percent}; 298°-600° \text{ K.});$$

$$C_p=15.35+7.20\times10^{-3}T.$$

TIN-PLATINUM COMPOUND

TABLE 818.—*Heat content and entropy of* $SnPt(c)$

[Base, crystals at 298.15° K.; mol. wt., 313.79]

T, ° K.	$H_T-H_{298.15}$, cal./mole	$S_T-S_{298.15}$, cal./deg. mole	T, ° K.	$H_T-H_{298.15}$, cal./mole	$S_T-S_{298.15}$, cal./deg. mole
400	1,230	3.55	1,000	8,910	15.18
500	2,450	6.27	1,100	10,270	16.48
600	3,700	8.55	1,200	11,650	17.68
700	4,970	10.50	1,300	13,040	18.79
800	6,270	12.24	1,400	14,430	19.82
900	7,580	13.78			

$SnPt(c)$:

$$H_T-H_{298.15}=11.26T+1.10\times10^{-3}T^2-3,455$$

$$(0.2 \text{ percent}; 298°-1,400° \text{ K.});$$

$$C_p=11.26+2.20\times10^{-3}T.$$

TITANIUM AND ITS COMPOUNDS

ELEMENT

TABLE 819.—*Heat content and entropy of* $Ti(c, l)$

[Base, α-crystals at 298.15° K.; atomic wt., 47.90]

T, ° K.	$H_T-H_{298.15}$, cal./mole	$S_T-S_{298.15}$, cal./deg. mole	T, ° K.	$H_T-H_{298.15}$, cal./mole	$S_T-S_{298.15}$, cal./deg. mole
400	625	1.80	1,500	9,600	12.05
500	1,250	3.20	1,600	10,350	12.53
600	1,920	4.42	1,700	11,100	12.99
700	2,610	5.48	1,800	11,850	13.42
800	3,330	6.44	1,900	12,600	13.82
900	4,070	7.31	1,940(β)	12,900	13.98
1,000	4,840	8.12	1,940(l)	17,360	16.28
1,100	5,630	8.87	2,000	17,840	16.52
1,155 (α)	6,070	9.26	2,200	19,440	17.29
1,155 (β)	7,020	10.09	2,400	21,040	17.98
1,200	7,350	10.38	2,600	22,640	18.62
1,300	8,100	10.98	2,800	24,240	19.21
1,400	8,850	11.53	3,000	25,840	19.77

$Ti(\alpha)$:

$$H_T-H_{298.15}=5.25T+1.26\times10^{-3}T^2-1,677$$

$$(0.3 \text{ percent}; 298°-1,155° \text{ K.});$$

$$C_p=5.25+2.52\times10^{-3}T;$$

$$\Delta H_{1155}(transition)=950.$$

$Ti(\beta)$:

$$H_T-H_{298.15}=7.50T-1,650 \ (0.1 \text{ percent};$$

$$1,155°-1,940° \text{ K.});$$

$$C_p=7.50;$$

$$\Delta H_{1940}(fusion)=4,460.$$

$Ti(l)$:

$$H_T-H_{298.15}=8.00T+1,840 \ (0.1 \text{ percent};$$

$$1,940°-3,000° \text{ K.});$$

$$C_p=8.00.$$

TABLE 820.—*Heat and content and entropy of* $Ti(g)$

[Base, ideal gas at 298.15° K.; atomic wt., 47.90]

T, ° K.	$H_T-H_{298.15}$, cal./mole	$S_T-S_{298.15}$, cal./deg. mole	T, ° K.	$H_T-H_{298.15}$, cal./mole	$S_T-S_{298.15}$, cal./deg. mole
400	575	1.67	1,900	8,495	9.87
500	1,120	2.88	2,000	9,075	10.16
600	1,650	3.84	2,200	10,275	10.74
700	2,170	4.64	2,400	11,530	11.28
800	2,685	5.33	2,600	12,840	11.79
900	3,195	5.94	2,800	14,210	12.31
1,000	3,705	6.47	3,000	15,635	12.81
1,100	4,215	6.96	3,500	19,460	13.98
1,200	4,730	7.40	4,000	23,630	15.10
1,300	5,245	7.82	4,500	28,110	16.15
1,400	5,765	8.20	5,000	32,855	17.15
1,500	6,290	8.57	6,000	42,895	18.98
1,600	6,825	8.91	7,000	53,320	20.59
1,700	7,370	9.24	8,000	63,830	21.99
1,800	7,930	9.56			

$Ti(g)$:

$$H_T-H_{298.15}=4.72T+0.20\times10^{-3}T^2-0.90\times10^5T^{-1}$$

$$-1,123 \ (0.6 \text{ percent}; 298°-2,000° \text{ K.});$$

$$C_p=4.72+0.40\times10^{-3}T+0.90\times10^5T^{-2}.$$

$$H_T-H_{298.15}=3.72T+0.59\times10^{-3}T^2-1.58\times10^5T^{-1}$$

$$-632 \ (0.6 \text{ percent}; 2,000°-6,000° \text{ K.});$$

$$C_p=3.72+1.18\times10^{-3}T+1.58\times10^5T^{-2}.$$

OXIDES

TABLE 821.—*Heat content and entropy of* $TiO(c)$

[Base, α-crystals at 298.15° K.; mol. wt., 63.90]

T, ° K.	$H_T-H_{298.15}$, cal./mole	$S_T-S_{298.15}$, cal./deg. mole	T, ° K.	$H_T-H_{298.15}$, cal./mole	$S_T-S_{298.15}$, cal./deg. mole
400	1,080	3.11	1,264(β)	13,270	18.46
500	2,220	5.65	1,300	13,840	18.90
600	3,410	7.82	1,400	15,430	20.08
700	4,640	9.71	1,500	17,050	21.20
800	5,910	11.41	1,600	18,700	22.26
900	7,230	12.96	1,700	20,380	23.28
1,000	8,600	14.40	1,800	22,090	24.26
1,100	10,020	15.75	1,900	23,830	25.20
1,200	11,490	17.03	2,000	25,600	26.11
1,264(α)	12,450	17.81			

$TiO(\alpha)$:

$$H_T-H_{298.15}=10.57T+1.80\times10^{-3}T^2+1.86\times10^5T^{-1}$$

$$-3,935 \ (0.8 \text{ percent}; 298°-1,264° \text{ K.});$$

$$C_p=10.57+3.60\times10^{-3}T-1.86\times10^5T^{-2};$$

$$\Delta H_{1264}(transition)=820.$$

$TiO(\beta)$:

$$H_T-H_{298.15}=11.85T+1.50\times10^{-3}T^2-4,100$$

$$(0.1 \text{ percent}; 1,264°-2,000° \text{ K.});$$

$$C_p=11.85+3.00\times10^{-3}T.$$

TABLE 1h7. (Continued)

TABLE 822.—Heat content and entropy of TiO(g)

[Base, ideal gas at 298.15° K.; mol. wt., 63.90]

T, ° K.	$H_T-H_{298.15}$, cal./mole	$S_T-S_{298.15}$, cal./deg. mole	T, ° K.	$H_T-H_{298.15}$, cal./mole	$S_T-S_{298.15}$, cal./deg. mole
400	780	2.24	2,000	14,665	15.98
500	1,570	4.01	2,100	15,565	16.42
600	2,390	5.50	2,200	16,465	16.84
700	3,225	6.79	2,300	17,365	17.24
800	4,070	7.92	2,400	18,265	17.62
900	4,930	8.93	2,500	19,165	17.99
1,000	5,800	9.85	2,750	20,435	18.85
1,100	6,670	10.68	3,000	23,710	19.65
1,200	7,550	11.44	3,250	26,000	20.38
1,300	8,430	12.14	3,500	28,290	21.06
1,400	9,310	12.80	3,750	30,595	21.69
1,500	10,200	13.41	4,000	32,910	22.29
1,600	11,090	13.99	4,250	35,240	22.86
1,700	11,980	14.53	4,500	37,580	23.39
1,800	12,875	15.04	4,750	39,930	23.90
1,900	13,770	15.52	5,000	42,295	24.39

TiO(g):

$$H_T-H_{298.15}=8.38T+0.13\times10^{-3}T^2+0.87\times10^5T^{-1}$$
$$-2,802\ (0.7\ \text{percent};\ 298°\text{--}5,000°\ \text{K.});$$
$$C_p=8.38+0.26\times10^{-3}T-0.87\times10^5T^{-2}.$$

TABLE 823.—Heat content and entropy of TiO₂ (rutile)

[Base, crystals at 298.15° K.; mol. wt., 79.90]

T, ° K.	$H_T-H_{298.15}$, cal./mole	$S_T-S_{298.15}$, cal./deg. mole	T, ° K.	$H_T-H_{298.15}$, cal./mole	$S_T-S_{298.15}$, cal./deg. mole
400	1,540	4.43	1,300	17,000	24.38
500	3,100	7.91	1,400	18,820	25.73
600	4,735	10.89	1,500	20,660	27.00
700	6,440	13.51	1,600	22,530	28.20
800	8,160	15.81	1,700	24,420	29.35
900	9,900	17.86	1,800	26,340	30.44
1,000	11,650	19.70	1,900	28,280	31.49
1,100	13,420	21.39	2,000	30,250	32.50
1,200	15,200	22.94			

TiO₂(rutile):

$$H_T-H_{298.15}=17.97T+0.14\times10^{-3}T^2+4.35\times10^5T^{-1}$$
$$-6,829\ (0.8\ \text{percent};\ 298°\text{--}1,800°\ \text{K.});$$
$$C_p=17.97+0.28\times10^{-3}T-4.35\times10^5T^{-2}.$$

TABLE 824.—Heat content and entropy of TiO₂ (anatase)

[Base, crystals at 298.15° K.; mol. wt., 79.90]

T, ° K.	$H_T-H_{298.15}$, cal./mole	$S_T-S_{298.15}$, cal./deg. mole	T, ° K.	$H_T-H_{298.15}$, cal./mole	$S_T-S_{298.15}$, cal./deg. mole
400	1,540	4.43	900	9,930	17.89
500	3,100	7.91	1,000	11,720	19.78
600	4,735	10.89	1,100	13,530	21.51
700	6,440	13.52	1,200	15,350	23.09
800	8,170	15.82	1,300	17,180	24.55

TiO₂(anatase):

$$H_T-H_{298.15}=17.83T+0.25\times10^{-3}T^2+4.23\times10^5T^{-1}$$
$$-6,757\ (0.7\ \text{percent};\ 298°\text{--}1,300°\ \text{K.});$$
$$C_p=17.83+0.50\times10^{-3}T-4.23\times10^5T^{-2}.$$

TABLE 825.—Heat content and entropy of Ti₂O₃(c)

[Base, α-crystals at 298.15° K.; mol. wt., 143.80]

T, ° K.	$H_T-H_{298.15}$, cal./mole	$S_T-S_{298.15}$, cal./deg. mole	T, ° K.	$H_T-H_{298.15}$, cal./mole	$S_T-S_{298.15}$, cal./deg. mole
400	2,610	7.49	1,200	29,800	44.46
473(α)	4,885	12.70	1,300	33,360	47.30
473(β)	5,100	13.15	1,400	36,950	49.96
500	5,935	14.87	1,500	40,560	52.46
600	9,140	20.71	1,600	44,180	54.79
700	12,440	25.79	1,700	47,830	57.00
800	15,830	30.31	1,800	51,490	59.10
900	19,270	34.36	1,900	55,170	61.09
1,000	22,740	38.02	2,000	58,870	62.99
1,100	26,260	41.38			

Ti₂O₃(α):

$$H_T-H_{298.15}=7.31T+26.76\times10^{-3}T^2-4,558\ (0.8\ \text{percent};$$
$$298°\text{--}473°\ \text{K.});$$
$$C_p=7.31+53.52\times10^{-3}T;$$
$$\Delta H_{473}(transition)=215.$$

Ti₂O₃(β):

$$H_T-H_{298.15}=34.68T+0.65\times10^{-3}T^2+10.20\times10^5T^{-1}$$
$$-13,605\ (0.1\ \text{percent};\ 473°\text{--}2,000°\ \text{K.});$$
$$C_p=34.68+1.30\times10^{-3}T-10.20\times10^5T^{-2}.$$

TABLE 826.—Heat content and entropy of Ti₃O₅(c)

[Base, α-crystals at 298.15° K.; mol. wt., 223.70]

T, ° K.	$H_T-H_{298.15}$, cal./mole	$S_T-S_{298.15}$, cal./deg. mole	T, ° K.	$H_T-H_{298.15}$, cal./mole	$S_T-S_{298.15}$, cal./deg. mole
400	4,660	13.43	1,200	45,510	70.89
450(α)	7,060	19.08	1,300	50,660	75.01
450(β)	9,300	24.06	1,400	55,810	78.83
500	11,570	28.85	1,500	60,970	82.39
600	16,220	37.32	1,600	66,130	85.72
700	20,880	44.50	1,700	71,300	88.85
800	25,500	50.74	1,800	76,470	91.80
900	30,290	56.31	1,900	81,650	94.60
1,000	35,230	61.52	2,000	86,830	97.26
1,100	40,370	66.42			

Ti₃O₅(α):

$$H_T-H_{298.15}=35.47T+14.75\times10^{-3}T^2-11,887\ (0.1\ \text{percent};\ 298°\text{--}450°\ \text{K.});$$
$$C_p=35.47+29.50\times10^{-3}T;$$
$$\Delta H_{450}(transition)=2,240.$$

Ti₃O₅(β):

$$H_T-H_{298.15}=41.60T+4.00\times10^{-3}T^2-10,230\ (0.3\ \text{percent};\ 450°\text{--}1,600°\ \text{K.});$$
$$C_p=41.60+8.00\times10^{-3}T.$$

CARBIDE

TABLE 827.—Heat content and entropy of TiC(c)

[Base, crystals at 298.15° K.; mol. wt., 59.91]

T, ° K.	$H_T-H_{298.15}$, cal./mole	$S_T-S_{298.15}$, cal./deg. mole	T, ° K.	$H_T-H_{298.15}$, cal./mole	$S_T-S_{298.15}$, cal./deg. mole
400	945	2.72	1,300	11,590	16.35
500	1,975	5.01	1,400	12,860	17.29
600	3,085	7.03	1,500	14,130	18.17
700	4,225	8.79	1,600	15,400	18.99
800	5,395	10.35	1,700	16,670	19.76
900	6,600	11.77	1,800	17,940	20.49
1,000	7,830	13.07	1,900	19,220	21.18
1,100	9,080	14.26	2,000	20,500	21.84
1,200	10,330	15.35			

TABLE 1h7. (Continued)

TiC(c):

$$H_T - H_{298.15} = 11.83T + 0.40 \times 10^{-3}T^2 + 3.58 \times 10^5 T^{-1}$$
$$-4,763 \ (0.4 \text{ percent}; \ 298°\text{–}2,000° \text{ K.});$$
$$C_p = 11.83 + 0.80 \times 10^{-3}T - 3.58 \times 10^5 T^{-2}.$$

TiB₂(c):

$$H_T - H_{298.15} = 14.99T + 1.87 \times 10^{-3}T^2 + 4.98 \times 10^5 T^{-1}$$
$$-6,306 \ (0.3 \text{ percent}; \ 298°\text{–}1,000° \text{ K.});$$
$$C_p = 14.99 + 3.74 \times 10^{-3}T - 4.98 \times 10^5 T^{-2}.$$

NITRIDE

TABLE 828—Heat content and entropy of TiN(c)

[Base, crystals at 298.15° K.; mol. wt., 61.91]

T, ° K.	$H_T-H_{298.15}$, cal./mole	$S_T-S_{298.15}$, cal./deg. mole	T, ° K.	$H_T-H_{298.15}$, cal./mole	$S_T-S_{298.15}$, cal./deg. mole
400	1,000	2.88	1,300	11,910	16.89
500	2,090	5.31	1,400	13,230	17.87
600	3,230	7.39	1,500	14,550	18.78
700	4,400	9.19	1,600	15,870	19.63
800	5,590	10.78	1,700	17,190	20.43
900	6,810	12.21	1,800	18,510	21.19
1,000	8,050	13.52	1,900	19,840	21.91
1,100	9,310	14.72	2,000	21,170	22.59
1,200	10,600	15.84			

TiN(c):

$$H_T - H_{298.15} = 11.91T + 0.47 \times 10^{-3}T^2 + 2.96 \times 10^5 T^{-1}$$
$$-4,586 \ (0.3 \text{ percent}; \ 298°\text{–}2,000° \text{ K.});$$
$$C_p = 11.91 + 0.94 \times 10^{-3}T - 2.96 \times 10^5 T^{-2}.$$

BROMIDES

TABLE 831.—Heat content and entropy of TiBr₂(c)

[Base, crystals at 298.15° K.; mol. wt., 207.73]

T, ° K.	$H_T-H_{298.15}$, cal./mole	$S_T-S_{298.15}$, cal./deg. mole	T, ° K.	$H_T-H_{298.15}$, cal./mole	$S_T-S_{298.15}$, cal./deg. mole
400	1,930	5.57	900	11,810	21.52
500	3,850	9.85	1,000	13,870	23.69
600	5,800	13.41	1,100	15,960	25.68
700	7,780	16.46	1,200	18,080	27.53
800	9,780	19.13			

TiBr₂(c):

$$H_T - H_{298.15} = 17.99T + 1.37 \times 10^{-3}T^2 - 5,486$$
$$(0.1 \text{ percent}; \ 298°\text{–}1,200° \text{ K.});$$
$$C_p = 17.99 + 2.74 \times 10^{-3}T.$$

SULFIDE

TABLE 829.—Heat content and entropy of TiS₂(c)

[Base, α-crystals at 298.15° K.; mol. wt., 112.03]

T, ° K.	$H_T-H_{298.15}$, cal./mole	$S_T-S_{298.15}$, cal./deg. mole	T, ° K.	$H_T-H_{298.15}$, cal./mole	$S_T-S_{298.15}$, cal./deg. mole
350	860	2.65	700	7,210	15.22
400	1,800	5.16	800	9,090	17.73
420(α)	2,180	6.09	900	11,000	19.98
420(β)	2,180	6.09	1,000	12,970	22.05
500	3,540	9.05	1,100	15,020	24.00
600	5,350	12.35	1,200	17,170	25.87

TiS₂(α):

$$H_T - H_{298.15} = 8.08T + 13.67 \times 10^{-3}T^2 - 3,624 \ (1.0 \text{ percent}$$
$$298°\text{–}420° \text{ K.});$$
$$C_p = 8.08 + 27.34 \times 10^{-3}T;$$
$$\Delta H_{420}(transition) = 0.$$

TiS₂(β):

$$H_T - H_{298.15} = 14.99T + 2.57 \times 10^{-3}T - 4,569 \ (0.3 \text{ percent};$$
$$420°\text{–}1,200° \text{ K.});$$
$$C_p = 14.99 + 5.14 \times 10^{-3}T.$$

BORIDE

TABLE 830.—Heat content and entropy of TiB₂(c)

[Base, crystals at 298.15° K.; mol. wt., 69.54]

T, ° K.	$H_T-H_{298.15}$, cal./mole	$S_T-S_{298.15}$, cal./deg. mole	T, ° K.	$H_T-H_{298.15}$, cal./mole	$S_T-S_{298.15}$, cal./deg. mole
400	1,245	3.58	800	7,485	14.24
500	2,660	6.72	900	9,240	16.31
600	4,190	9.51	1,000	11,050	18.21
700	5,800	11.99			

TABLE 832.—Heat content and entropy of TiBr₃(c)

[Base, crystals at 298.15° K.; mol. wt., 287.65]

T, ° K.	$H_T-H_{298.15}$, cal./mole	$S_T-S_{298.15}$, cal./deg. mole	T, ° K.	$H_T-H_{298.15}$, cal./mole	$S_T-S_{298.15}$, cal./deg. mole
400	2,580	7.44	900	15,800	28.79
500	5,160	13.20	1,000	18,540	31.68
600	7,760	17.94	1,100	21,330	34.33
700	10,410	22.02	1,200	24,140	36.78
800	13,080	25.59			

TiBr₃(c):

$$H_T - H_{298.15} = 24.15T + 1.75 \times 10^{-3}T^2 - 7,356$$
$$(0.1 \text{ percent}; \ 298°\text{–}1,200° \text{ K.});$$
$$C_p = 24.15 + 3.50 \times 10^{-3}T.$$

TABLE 833.—Heat content and entropy of TiBr₄(c, l)

[Base, crystals at 298.15° K.; mol. wt., 367.56]

T, ° K.	$H_T-H_{298.15}$, cal./mole	$S_T-S_{298.15}$, cal./deg. mole	T, ° K.	$H_T-H_{298.15}$, cal./mole	$S_T-S_{298.15}$, cal./deg. mole
311.4(c)	420	1.38	400	6,720	20.37
311.4(l)	3,500	11.27	450	8,530	24.63
350	4,900	15.51	500	10,340	28.44

TiBr₄(c):

$$H_T - H_{298.15} = 31.70T - 9,451 \ (0.1 \text{ percent};$$
$$298°\text{–}311.4° \text{ K.});$$
$$C_p = 31.70;$$
$$\Delta H_{311.4}(fusion) = 3,080.$$

TiBr₄(l):

$$H_T - H_{298.15} = 36.30T - 7,805 \ (0.1 \text{ percent};$$
$$311.4°\text{–}500° \text{ K.});$$
$$C_p = 36.30.$$

TABLE 1h7. (*Continued*)

TABLE 834.—*Heat content and entropy of TiBr₄(g)*

[Base, ideal gas at 298.15° K.; mol. wt., 367.56]

T, ° K.	$H_T-H_{298.15}$, cal./mole	$S_T-S_{298.15}$, cal./deg. mole	T, ° K.	$H_T-H_{298.15}$, cal./mole	$S_T-S_{298.15}$, cal./deg. mole
400	2,490	7.18	1,300	25,420	37.11
500	4,980	12.73	1,400	28,000	39.02
600	7,510	17.34	1,500	30,580	40.80
700	10,050	21.26	1,600	33,150	42.46
800	12,600	24.66	1,700	35,730	44.03
900	15,160	27.68	1,800	38,310	45.50
1,000	17,720	30.38	1,900	40,890	46.90
1,100	20,290	32.83	2,000	43,460	48.21
1,200	22,850	35.05			

TiBr₄(g):

$$H_T-H_{298.15}=25.71T+0.04\times10^{-3}T^2+1.54\times10^5T^{-1}$$
$$-8,186 \text{ (0.1 percent; } 298°-2,000° \text{ K.);}$$
$$C_p=25.71+0.08\times10^{-3}T-1.54\times10^5T^{-2}.$$

CHLORIDES

TABLE 835.—*Heat content and entropy of TiCl(g)*

[Base, ideal gas at 298.15° K.; mol. wt., 83.36]

T, ° K.	$H_T-H_{298.15}$, cal./mole	$S_T-S_{298.15}$, cal./deg. mole	T, ° K.	$H_T-H_{298.15}$, cal./mole	$S_T-S_{298.15}$, cal./deg. mole
400	860	2.48	1,000	6,130	10.51
500	1,725	4.41	1,200	7,910	12.13
600	2,600	6.00	1,400	9,690	13.50
700	3,475	7.35	1,600	11,470	14.69
800	4,360	8.53	1,800	13,255	15.74
900	5,245	9.57	2,000	15,040	16.68

TiCl(g):

$$H_T-H_{298.15}=8.88T+0.02\times10^{-3}T^2+0.51\times10^5T^{-1}$$
$$-2,820 \text{ (0.1 percent; } 298°-2,000° \text{ K.);}$$
$$C_p=8.88+0.04\times10^{-3}T-0.51\times10^5T^{-2}.$$

TABLE 836.—*Heat content and entropy of TiCl₂(c)*

[Base, crystals at 298.15° K.; mol wt., 118.81]

T, ° K.	$H_T-H_{298.15}$, cal./mole	$S_T-S_{298.15}$, cal./deg. mole	T, ° K.	$H_T-H_{298.15}$, cal./mole	$S_T-S_{298.15}$, cal./deg. mole
400	1,790	5.16	900	11,270	20.41
500	3,600	9.20	1,000	13,300	22.55
600	5,450	12.57	1,100	15,380	24.53
700	7,340	15.48	1,200	17,500	26.38
800	9,280	18.07			

TiCl₂(c):

$$H_T-H_{298.15}=16.02T+2.26\times10^{-3}T^2-4,977$$
$$\text{(0.1 percent; } 298°-1,200° \text{ K.);}$$
$$C_p=16.02+4.52\times10^{-3}T.$$

TABLE 837.—*Heat content and entropy of TiCl₃(c)*

[Base, crystals at 298.15° K.; mol. wt., 154.27]

T, ° K.	$H_T-H_{298.15}$, cal./mole	$S_T-S_{298.15}$, cal./deg. mole	T, ° K.	$H_T-H_{298.15}$, cal./mole	$S_T-S_{298.15}$, cal./deg. mole
400	2,390	6.89	900	14,660	26.71
500	4,780	12.22	1,100	17,190	29.37
600	7,200	16.63	1,100	19,740	31.80
700	9,660	20.42	1,200	22,310	34.04
800	12,150	23.75			

TiCl₃(c) ·

$$H_T-H_{298.15}=22.97T+1.29\times10^{-3}T^2+0.46\times10^5T^{-1}$$
$$-7,117 \text{ (0.1 percent; } 298°-1,200° \text{ K.);}$$
$$C_p=22.97+2.58\times10^{-3}T-0.46\times10^5T^{-2}.$$

TABLE 838.—*Heat content and entropy of TiCl₄(l)*

[Base, liquid at 298.15° K.; mol. wt., 189.73]

T, ° K.	$H_T-H_{298.15}$, cal./mole	$S_T-S_{298.15}$, cal./deg. mole	T, ° K.	$H_T-H_{298.15}$, cal./mole	$S_T-S_{298.15}$, cal./deg. mole
350	1,945	6.01	450	5,700	15.45
400	3,820	11.02	500	7,575	19.40

TiCl₄(l):

$$H_T-H_{298.15}=37.53T-11,190 \text{ (0.1 percent;}$$
$$298°-500° \text{ K.);}$$
$$C_p=37.53.$$

TABLE 839.—*Heat content and entropy of TiCl₄(g)*

[Base, ideal gas at 298.15° K.; mol. wt., 189.73]

T, ° K.	$H_T-H_{298.15}$, cal./mole	$S_T-S_{298.15}$, cal./deg. mole	T, ° K.	$H_T-H_{298.15}$, cal./mole	$S_T-S_{298.15}$, cal./deg. mole
400	2,395	6.91	1,300	25,100	36.47
500	4,830	12.34	1,400	27,665	38.37
600	7,310	16.86	1,500	30,230	40.14
700	9,815	20.72	1,600	32,800	41.80
800	12,340	24.09	1,700	35,370	43.36
900	14,880	27.08	1,800	37,945	44.83
1,000	17,425	29.76	1,900	40,520	46.22
1,100	19,975	32.19	2,000	43,100	47.54
1,200	22,535	34.42			

TiCl₄(g):

$$H_T-H_{298.15}=25.45T+0.12\times10^{-3}T^2+2.36\times10^5T^{-1}$$
$$-8,390 \text{ (0.1 percent; } 298°-2,000° \text{ K.);}$$
$$C_p=25.45+0.24\times10^{-3}T-2.36\times10^5T^{-2}.$$

IODIDES

TABLE 840.—*Heat content and entropy of TiI₂(c)*

[Base, crystals at 298.15° K.; mol. wt., 301.72]

T, ° K.	$H_T-H_{298.15}$, cal./mole	$S_T-S_{298.15}$, cal./deg. mole	T, ° K.	$H_T-H_{298.15}$, cal./mole	$S_T-S_{298.15}$, cal./deg. mole
400	2,110	6.09	900	12,720	23.24
500	4,200	10.75	1,000	14,890	25.54
600	6,300	14.58	1,100	17,080	27.63
700	8,420	17.85	1,200	19,290	29.55
800	10,560	20.70			

TiI₂(c):

$$H_T-H_{298.15}=20.09T+0.87\times10^{-3}T^2-6,067$$
$$\text{(0.1 percent; } 298°-1,200° \text{ K.);}$$
$$C_p=20.09+1.74\times10^{-3}T.$$

TABLE 841.—*Heat content and entropy of TiI₃(c)*

[Base, crystals at 298.15° K.; mol. wt., 428.63]

T, ° K.	$H_T-H_{298.15}$, cal./mole	$S_T-S_{298.15}$, cal./deg. mole	T, ° K.	$H_T-H_{298.15}$, cal./mole	$S_T-S_{298.15}$, cal./deg. mole
400	2,850	8.22	900	17,110	31.29
500	5,670	14.51	1,000	20,020	34.36
600	8,500	19.67	1,100	22,940	37.14
700	11,350	24.07	1,200	25,880	39.70
800	14,220	27.90			

TiI₃(c):

$$H_T-H_{298.15}=27.39T+0.87\times10^{-3}T^2-8,244$$
$$\text{(0.1 percent; } 298°-1,200° \text{ K.);}$$
$$C_p=27.39+1.74\times10^{-3}T.$$

TABLE 1h7. (*Continued*)

TABLE 842.—*Heat content and entropy of* $TiI_4(c, l)$

[Base, crystals at 298.15° K.; mol. wt., 507.64]

T, ° K.	$H_T-H_{298.15}$, cal./mole	$S_T-S_{298.15}$, cal./deg. mole	T, ° K.	$H_T-H_{298.15}$, cal./mole	$S_T-S_{298.15}$, cal./deg. mole
400	3,540	10.22	500	11,450	28.40
423(c)	4,340	12.16	600	15,250	35.32
423(l)	8,520	22.04	650	17,150	38.36

$TiI_4(c)$:

$$H_T - H_{298.15} = 34.76T - 10,364 \text{ (0.1 percent;}$$
$$298°–423° \text{ K.)};$$
$$C_p = 34.76;$$
$$\Delta H_{423}(fusion) = 4,180.$$

$TiI_4(l)$:

$$H_T - H_{298.15} = 38.00T - 7,550 \text{ (0.1 percent;}$$
$$423°–600° \text{ K.)};$$
$$C_p = 38.00.$$

TABLE 843.—*Heat content and entropy of* $TiI_4(g)$

[Base, ideal gas at 298.15° K.; mol. wt., 507.64]

T, ° K.	$H_T-H_{298.15}$, cal./mole	$S_T-S_{298.15}$, cal./deg. mole	T, ° K.	$H_T-H_{298.15}$, cal./mole	$S_T-S_{298.15}$, cal./deg. mole
400	2,650	7.64	1,300	26,050	38.29
500	5,250	13.44	1,400	28,650	40.21
600	7,850	18.18	1,500	31,250	42.01
700	10,450	22.19	1,600	33,850	43.68
800	13,050	25.66	1,700	36,450	45.26
900	15,650	28.72	1,800	39,050	46.75
1,000	18,250	31.46	1,900	41,650	48.15
1,100	20,850	33.94	2,000	44,250	49.49
1,200	23,450	36.20			

$TiI_4(g)$:

$$H_T - H_{298.15} = 26.00T - 7,752 \text{ (0.1 percent;}$$
$$298°–2,000° \text{ K.)};$$
$$C_p = 26.00.$$

TUNGSTEN AND ITS COMPOUNDS

ELEMENT

TABLE 844.—*Heat content and entropy of* $W(c)$

[Base, crystals at 298.15° K.; atomic wt., 183.86]

T, ° K.	$H_T-H_{298.15}$, cal./mole	$S_T-S_{298.15}$, cal./deg. mole	T, ° K.	$H_T-H_{298.15}$, cal./mole	$S_T-S_{298.15}$, cal./deg. mole
400	615	1.77	1,500	7,730	10.18
500	1,220	3.12	1,600	8,430	10.63
600	1,830	4.23	1,700	9,130	11.06
700	2,450	5.19	1,800	9,840	11.47
800	3,080	6.03	1,900	10,550	11.85
900	3,710	6.77	2,000	11,270	12.22
1,000	4,360	7.46	2,200	12,720	12.91
1,100	5,010	8.08	2,400	14,190	13.55
1,200	5,670	8.65	2,600	14,680	14.14
1,300	6,340	9.19	2,800	17,190	14.70
1,400	7,030	9.70	3,000	18,720	15.23

$W(c)$:

$$H_T - H_{298.15} = 5.74T + 0.38 \times 10^{-3}T^2 - 1,745$$
$$\text{(0.4 percent; 298°–3,000° K.)};$$
$$C_p = 5.74 + 0.76 \times 10^{-3}T.$$

TABLE 845.—*Heat content and entropy of* $W(g)$

[Base, ideal gas at 298.15° K.; atomic wt., 183.86]

T, ° K.	$H_T-H_{298.15}$, cal./mole	$S_T-S_{298.15}$, cal./deg. mole	T, ° K.	$H_T-H_{298.15}$, cal./mole	$S_T-S_{298.15}$, cal./deg. mole
400	540	1.55	1,500	10,150	12.47
500	1,130	2.86	1,600	11,040	13.04
600	1,805	4.09	1,700	11,895	13.56
700	2,580	5.28	1,800	12,730	14.04
800	3,440	6.44	1,900	13,540	14.47
900	4,375	7.53	2,000	14,325	14.88
1,000	5,350	8.56	2,200	15,855	15.61
1,100	6,340	9.50	2,400	17,330	16.25
1,200	7,325	10.36	2,600	18,780	16.83
1,300	8,290	11.14	2,800	20,210	17.36
1,400	9,235	11.84	3,000	21,635	17.85

$W(g)$:

$$H_T - H_{298.15} = 10.70T - 0.66 \times 10^{-3}T^2 + 4.64 \times 10^5 T^{-1}$$
$$-4,688 \text{ (0.3 percent; 1,600°–3,000° K.)};$$
$$C_p = 10.70 - 1.32 \times 10^{-3}T - 4.64 \times 10^5 T^{-2}.$$

OXIDE

$WO_3(c)$:

$$C_p = 17.75 + 5.87 \times 10^{-3}T(estimated) \text{ (298°–1,746° K.)}.$$

CARBONYL

$W(CO)_6(c)$:

$$\overline{C_p} = 60.50 \text{ (293°–363° K.)}.$$

FLUORIDE

TABLE 846.—*Heat content and entropy of* $WF_6(g)$

[Base, ideal gas at 298.15° K.; mol. wt., 297.86]

T, ° K.	$H_T-H_{298.15}$, cal./mole	$S_T-S_{298.15}$, cal./deg. mole	T, ° K.	$H_T-H_{298.15}$, cal./mole	$S_T-S_{298.15}$, cal./deg. mole
400	3,075	8.85	800	16,830	32.52
500	6,340	16.13	900	20,440	36.77
600	9,750	22.34	1,000	24,090	40.61
700	13,260	27.75			

$WF_6(g)$:

$$H_T - H_{298.15} = 33.69T + 2.00 \times 10^{-3}T^2 + 5.70 \times 10^5 T^{-1}$$
$$-12,134 \text{ (0.2 percent; 298°–1,000° K.)};$$
$$C_p = 33.69 + 4.00 \times 10^{-3}T - 5.70 \times 10^5 T^{-2}.$$

URANIUM AND ITS COMPOUNDS

ELEMENT

TABLE 847.—*Heat content and entropy of* $U(c, l)$

[Base, α-crystals at 298.15° K.; atomic wt., 238.07]

T, ° K.	$H_T-H_{298.15}$, cal./mole	$S_T-S_{298.15}$, cal./deg. mole	T, ° K.	$H_T-H_{298.15}$, cal./mole	$S_T-S_{298.15}$, cal./deg. mole
400	690	1.99	1,400	11,700	15.02
500	1,430	3.64	1,406(γ)	11,760	15.07
600	2,230	5.09	1,406(l)	15,010	17.38
700	3,100	6.44	1,500	15,870	17.97
800	4,050	7.71	1,600	16,790	18.56
900	5,090	8.93	1,700	17,710	19.12
935(α)	5,470	9.35	1,800	18,630	19.65
935(β)	6,170	10.10	1,900	19,550	20.15
1,000	6,830	10.78	2,000	20,470	20.62
1,045(β)	7,290	11.23	2,200	22,310	21.49
1,045(γ)	8,435	12.33	2,400	24,150	22.29
1,100	8,940	12.81	2,600	25,990	23.03
1,200	9,860	13.61	2,800	27,830	23.71
1,300	10,780	14.34	3,000	29,670	24.34

TABLE 1h7. (*Continued*)

U(α):

$H_T - H_{298.15} = 3.39T + 4.01 \times 10^{-3}T^2 - 0.70 \times 10^5 T^{-1}$
$- 1,132$ (0.2 percent; 298°–935° K.);

$C_p = 3.39 + 8.02 \times 10^{-3}T + 0.70 \times 10^5 T^{-2}$;

$\Delta H_{935}(transition) = 700$.

U(β):

$H_T - H_{298.15} = 10.18T - 3,348$ (0.1 percent;
935°–1,045° K.);

$C_p = 10.18$;

$\Delta H_{1045}(transition) = 1,145$.

U(γ):

$H_T - H_{298.15} = 9.20T - 1,180$ (0.1 percent;
1,045°–1,406° K.);

$C_p = 9.20$;

$\Delta H_{1406}(fusion) = 3,250$.

U(l):

$H_T - H_{298.15} = 9.20T + 2,070$ (0.1 percent;
1,406°–3,000° K.);

$C_p = 9.20$.

TABLE 848.—*Heat content and entropy of* U(g)

[Base, ideal gas at 298.15° K.; atomic wt., 238.07]

T, ° K.	$H_T-H_{298.15}$, cal./mole	$S_T-S_{298.15}$, cal./deg. mole	T, ° K.	$H_T-H_{298.15}$, cal./mole	$S_T-S_{298.15}$, cal./deg. mole
400	580	1.67	1,500	6,905	9.21
500	1,150	2.94	1,600	7,535	9.61
600	1,715	3.97	1,700	8,170	10.00
700	2,270	4.83	1,800	8,815	10.37
800	2,825	5.57	1,900	9,470	10.72
900	3,380	6.22	2,000	10,125	11.06
1,000	3,940	6.81	2,200	11,455	11.69
1,100	4,510	7.35	2,400	12,790	12.27
1,200	5,090	7.86	2,600	14,130	12.80
1,300	5,685	8.33	2,800	15,465	13.30
1,400	6,290	8.78	3,000	16,795	13.76

U(g):

$H_T - H_{298.15} = 5.16T + 0.32 \times 10^{-3}T^2 - 0.26 \times 10^5 T^{-1}$
$- 1,480$ (0.7 percent; 298°–3,000° K.);

$C_p = 5.16 + 0.64 \times 10^{-3}T + 0.26 \times 10^5 T^{-2}$.

OXIDES

TABLE 849.—*Heat content and entropy of* UO$_2$(c)

[Base, crystals at 298.15° K.; mol. wt., 270.07]

T, ° K.	$H_T-H_{298.15}$, cal./mole	$S_T-S_{298.15}$, cal./deg. mole	T, ° K.	$H_T-H_{298.15}$, cal./mole	$S_T-S_{298.15}$, cal./deg. mole
400	1,680	4.84	1,300	19,510	27.79
500	3,470	8.83	1,400	21,620	29.35
600	5,340	12.23	1,500	23,750	30.82
700	7,280	15.22	1,600	25,900	32.21
800	9,250	17.85	1,700	28,070	33.53
900	11,250	20.20	1,800	30,260	34.78
1,000	13,280	22.34	1,900	32,470	35.97
1,100	15,340	24.31	2,000	34,700	37.12
1,200	17,420	26.12			

UO$_2$(c):

$H_T - H_{298.15} = 19.20T + 0.81 \times 10^{-3}T^2 + 3.96 \times 10^5 T^{-1}$
$- 7,125$ (0.1 percent; 298°–2,000° K.);

$C_p = 19.20 + 1.62 \times 10^{-3}T - 3.96 \times 10^5 T^{-2}$.

TABLE 850.—*Heat content and entropy of* UO$_3$(c)

[Base, crystals at 298.15° K.; mol. wt., 286.07]

T, ° K.	$H_T-H_{298.15}$, cal./mole	$S_T-S_{298.15}$, cal./deg. mole	T, ° K.	$H_T-H_{298.15}$, cal./mole	$S_T-S_{298.15}$, cal./deg. mole
400	2,090	6.02	800	11,160	21.64
500	4,260	10.86	900	13,540	24.44
600	6,510	14.96	1,000	15,960	26.99
700	8,820	18.52			

UO$_3$(c):

$H_T - H_{298.15} = 22.09T + 1.27 \times 10^{-3}T^2 + 2.97 \times 10^5 T^{-1}$
$- 7,695$ (0.1 percent; 298°–1,000° K.);

$C_p = 22.09 + 2.54 \times 10^{-3}T - 2.97 \times 10^5 T^{-2}$.

U$_3$O$_8$(c):

$\overline{C_p} = 59.8$ (275°–315° K.).

U$_4$O$_9$(c):

$C_p = 70.11$ (298° K.).

CHLORIDES

TABLE 851.—*Heat content and entropy of* UCl$_3$(c)

[Base, crystals at 298.15° K.; mol. wt., 344.44]

T, ° K.	$H_T-H_{298.15}$, cal./mole	$S_T-S_{298.15}$, cal./deg. mole	T, ° K.	$H_T-H_{298.15}$, cal./mole	$S_T-S_{298.15}$, cal./deg. mole
400	2,500	7.22	800	12,810	24.99
500	5,000	12.79	900	15,570	28.24
600	7,540	17.42	1,000	18,430	31.25
700	10,140	21.43			

UCl$_3$(c):

$H_T - H_{298.15} = 20.98T + 3.72 \times 10^{-3}T^2 - 1.16 \times 10^5 T^{-1}$
$- 6,197$ (0.1 percent; 298°–1,000° K.);

$C_p = 20.98 + 7.44 \times 10^{-3}T + 1.16 \times 10^5 T^{-2}$.

TABLE 852.—*Heat content and entropy of* UCl$_4$(c)

[Base, crystals at 298.15° K.; mol. wt., 379.90]

T, ° K.	$H_T-H_{298.15}$, cal./mole	$S_T-S_{298.15}$, cal./deg. mole	T, ° K.	$H_T-H_{298.15}$, cal./mole	$S_T-S_{298.15}$, cal./deg. mole
400	3,030	8.74	600	9,330	21.50
500	6,150	15.70	700	12,630	26.58

UCl$_4$(c):

$H_T - H_{298.15} = 26.64T + 4.80 \times 10^{-3}T^2 - 8,369$
(0.3 percent; 298°–700° K.);

$C_p = 26.64 + 9.60 \times 10^{-3}T$.

FLUORIDES

UF$_4$(c):

$C_p = 27.92$ (298° K.).

UF$_6$(c):

$\overline{C_p} = 41.61$ (298°–337.2° K.);

$\Delta H_{337.2}(fusion) = 4,590$.

TABLE 1h7. (Continued)

$UF_6(l)$:

$\overline{C_p}=46.34$ (337.2°–370° K.).

TABLE 853.—Heat content and entropy of $UF_6(g)$

[Base, ideal gas at 298.15° K.; mol. wt., 352.07]

T, ° K.	$H_T-H_{298.15}$, cal./mole	$S_T-S_{298.15}$, cal./deg. mole	T, ° K.	$H_T-H_{298.15}$, cal./mole	$S_T-S_{298.15}$, cal./deg. mole
400	3,300	9.50	800	17,460	33.92
500	6,710	17.11	900	21,130	38.24
600	10,240	23.54	1,000	24,820	42.13
700	13,830	29.07			

$UF_6(g)$;

$H_T-H_{298.15}=35.61T+1.01\times10^{-3}T^2+4.63\times10^5T^{-1}$
$-12,260$ (0.1 percent; 298°–1,000° K.);
$C_p=35.61+2.02\times10^{-3}T-4.63\times10^5T^{-2}$.

OXYBROMIDE

$UOBr_2(c)$:

$C_p=23.42$ (298° K.).

OXYCHLORIDES

$UOCl_2(c)$:

$C_p=22.72$ (298° K.).

$UO_2Cl_2(c)$:

$Cp=25.78$ (298° K.).

OXYFLUORIDE

$UO_2F_2(c)$:

$\overline{C_p}=26.05$ (298°–425° K.).

URANYL NITRATE

$UO_2(NO_3)_2 \cdot 6H_2O(c)$:

$C_p=111.6$ (298° K.).

VANADIUM AND ITS COMPOUNDS

ELEMENT

TABLE 854.—Heat content and entropy of $V(c, l)$

[Base, crystals at 298.15° K.; atomic wt., 50.95]

T, ° K.	$H_T-H_{298.15}$, cal./mole	$S_T-S_{298.15}$, cal./deg. mole	T, ° K.	$H_T-H_{298.15}$, cal./mole	$S_T-S_{298.15}$, cal./deg. mole
400	620	1.79	1,700	10,470	12.24
500	1,250	3.19	1,800	11,430	12.79
600	1,910	4.38	1,900	12,410	13.32
700	2,570	5.41	2,000	13,410	13.83
800	3,250	6.32	2,100	14,450	14.34
900	3,940	7.13	2,190(c)	15,400	14.68
1,000	4,660	7.89	2,190(l)	20,450	16.99
1,100	5,400	8.59	2,200	20,550	17.03
1,200	6,170	9.26	2,400	22,450	17.86
1,300	6,970	9.90	2,600	24,350	18.62
1,400	7,800	10.52	2,800	26,250	19.33
1,500	8,650	11.10	3,000	28,150	19.98
1,600	9,550	11.68			

$V(c)$:

$H_T-H_{298.15}=4.90T+1.29\times10^{-3}T^2-0.20\times10^5T^{-1}$
$-1,509$ (0.9 percent; 298°–2,190° K.);
$C_p=4.90+2.58\times10^{-3}T+0.20\times10^5T^{-2}$;
$\Delta H_{2190}(fusion)=5,050$.

$V(l)$:

$H_T-H_{298.15}=9.50T-350$ (0.1 percent;
2,190°–3,000° K.);
$C_p=9.50$.

TABLE 855.—Heat content and entropy of $V(g)$

[Base, ideal gas at 298.15° K.; atomic wt., 50.95]

T, ° K.	$H_T-H_{298.15}$, cal./mole	$S_T-S_{298.15}$, cal./deg. mole	T, ° K.	$H_T-H_{298.15}$, cal./mole	$S_T-S_{298.15}$, cal./deg. mole
400	615	1.78	1,900	9,510	10.99
500	1,195	3.08	2,000	10,095	11.29
600	1,775	4.12	2,200	11,270	11.85
700	2,360	5.03	2,400	12,445	12.36
800	2,950	5.82	2,600	13,630	12.84
900	3,550	6.53	2,800	14,840	13.28
1,000	4,150	7.16	3,000	16,075	13.71
1,100	4,755	7.74	3,500	19,310	14.71
1,200	5,360	8.26	4,000	22,820	15.64
1,300	5,960	8.74	4,500	26,630	16.54
1,400	6,560	9.19	5,000	30,850	17.41
1,500	7,155	9.60	6,000	39,805	19.05
1,600	7,750	9.98	7,000	49,675	20.58
1,700	8,340	10.34	8,000	60,015	21.96
1,800	8,925	10.67			

$V(g)$:

$H_T-H_{298.15}=5.02T+0.46\times10^{-3}T^2-0.82\times10^5T^{-1}$
$-1,261$ (0.2 percent; 298°–1,200° K.);
$C_p=5.02+0.92\times10^{-3}T+0.82\times10^5T^{-2}$.
$H_T-H_{298.15}=5.78T+0.04\times10^{-3}T^2-0.38\times10^5T^{-1}$
$-1,599$ (0.3 percent; 1,200–3,000° K.);
$C_p=5.78+0.08\times10^{-3}T+0.38\times10^5T^{-2}$.

OXIDES

TABLE 856.—Heat content and entropy of $VO(c)$

[Base, crystals at 298.15° K.; mol. wt., 66.95]

T, ° K.	$H_T-H_{298.15}$, cal./mole	$S_T-S_{298.15}$, cal./deg. mole	T, ° K.	$H_T-H_{298.15}$, cal./mole	$S_T-S_{298.15}$, cal./deg. mole
400	1,160	3.34	1,300	13,610	19.22
500	2,380	6.06	1,400	15,170	20.38
600	3,640	8.35	1,500	16,760	21.48
700	4,940	10.36	1,600	18,370	22.52
800	6,280	12.15	1,700	20,000	23.50
900	7,660	13.77	1,800	21,650	24.45
1,000	9,090	15.28	1,900	23,320	25.35
1,100	10,560	16.68	2,000	25,010	26.22
1,200	12,070	17.99			

$VO(c)$:

$H_T-H_{298.15}=11.32T+1.61\times10^{-3}T^2+1.26\times10^5T^{-1}$
$-3,941$ (0.3 percent; 298°–2,000° K.);
$C_p=11.32+3.22\times10^{-3}T-1.26\times10^5T^{-2}$.

TABLE 857.—Heat content and entropy of $VO(g)$

[Base, ideal gas at 298.15° K.; mol. wt., 66.95]

T, ° K.	$H_T-H_{298.15}$, cal./mole	$S_T-S_{298.15}$, cal./deg. mole	T, ° K.	$H_T-H_{298.15}$, cal./mole	$S_T-S_{298.15}$, cal./deg. mole
400	765	2.20	1,300	8,350	12.02
500	1,550	3.96	1,400	9,225	12.67
600	2,360	5.43	1,500	10,100	13.28
700	3,190	6.71	1,600	10,980	13.85
800	4,030	7.83	1,700	11,860	14.38
900	4,880	8.83	1,800	12,750	14.89
1,000	5,740	9.74	1,900	13,630	15.36
1,100	6,605	10.56	2,000	14,520	15.82
1,200	7,475	11.32			

TABLE 1h7. (*Continued*)

VO(*g*):

$$H_T - H_{298.15} = 8.20T + 0.21 \times 10^{-3}T^2 + 0.90 \times 10^5 T^{-1}$$
$$-2,766 \ (0.4 \text{ percent}; \ 298°-2,000° \text{ K.});$$
$$C_p = 8.20 + 0.42 \times 10^{-3}T - 0.90 \times 10^5 T^{-2}.$$

TABLE 858.—*Heat content and entropy of* $V_2O_3(c)$

[Base, crystals at 298.15° K.; mol. wt., 149.90]

T, ° K.	$H_T-H_{298.15}$, cal./mole	$S_T-S_{298.15}$, cal./deg. mole	T, ° K.	$H_T-H_{298.15}$, cal./mole	$S_T-S_{298.15}$, cal./deg. mole
400	2,720	7.83	1,300	31,360	44.70
500	5,590	14.23	1,400	34,940	47.35
600	8,600	19.72	1,500	38,670	49.92
700	11,700	24.49	1,600	42,480	52.38
800	14,870	28.72	1,700	46,370	54.73
900	18,100	32.53	1,800	50,350	57.01
1,000	21,370	35.97	1,900	54,410	59.20
1,100	24,660	39.11	2,000	58,550	61.33
1,200	27,960	41.98			

$V_2O_3(c)$:

$$H_T - H_{298.15} = 29.35T + 2.38 \times 10^{-3}T^2 + 5.42 \times 10^5 T^{-1}$$
$$-10,780 \ (0.6 \text{ percent}; \ 298°-1,800° \text{ K.});$$
$$C_p = 29.35 + 4.76 \times 10^{-3}T - 5.42 \times 10^5 T^{-2}.$$

TABLE 859.—*Heat content and entropy of* $V_2O_4(c, l)$

[Base, α-crystals at 298.15° K.; mol. wt., 165.90]

T, ° K.	$H_T-H_{298.15}$, cal./mole	$S_T-S_{298.15}$, cal./deg. mole	T, ° K.	$H_T-H_{298.15}$, cal./mole	$S_T-S_{298.15}$, cal./deg. mole
345(α)	1,400	4.36	1,300	38,630	57.97
345(β)	3,450	10.30	1,400	42,600	60.92
400	5,270	15.20	1,500	46,590	63.67
500	8,600	22.62	1,600	50,620	66.27
600	12,000	28.79	1,700	54,710	68.75
700	15,560	34.29	1,800	58,850	71.12
800	19,230	39.19	1,818(β)	59,600	71.53
900	22,990	43.61	1,818(l)	86,810	86.50
1,000	26,830	47.66	1,900	90,990	88.75
1,100	30,730	51.38	2,000	96,090	91.37
1,200	34,670	54.81			

$V_2O_4(α)$:

$$H_T - H_{298.15} = 29.91T - 8,918 \ (0.1 \text{ percent};$$
$$298°-345° \text{ K.});$$
$$C_p = 29.91;$$
$$\Delta H_{345}(transition) = 2,050.$$

$V_2O_4(β)$:

$$H_T - H_{298.15} = 35.70T + 1.70 \times 10^{-3}T^2 + 7.89 \times 10^5 T^{-1}$$
$$-11,355 \ (0.4 \text{ percent}; \ 345°-1,818° \text{ K.});$$
$$C_p = 35.70 + 3.40 \times 10^{-3}T - 7.89 \times 10^5 T^{-2};$$
$$\Delta H_{1818}(fusion) = 27,210.$$

$V_2O_4(l)$:

$$H_T - H_{298.15} = 51.00T - 5,910 \ (0.1 \text{ percent};$$
$$1,818°-2,000° \text{ K.});$$
$$C_p = 51.00.$$

TABLE 860.—*Heat content and entropy of* $V_2O_5(c, l)$

[Base, crystals at 298.15° K.; mol. wt., 181.90]

T, ° K.	$H_T-H_{298.15}$, cal./mole	$S_T-S_{298.15}$, cal./deg. mole	T, ° K.	$H_T-H_{298.15}$, cal./mole	$S_T-S_{298.15}$, cal./deg. mole
400	3,650	10.49	1,200	52,700	72.00
500	7,400	18.85	1,300	57,260	75.65
600	11,290	25.94	1,400	61,820	79.03
700	15,290	32.10	1,500	66,380	82.18
800	19,390	37.58	1,600	70,940	85.12
900	23,590	42.52	1,700	75,500	87.88
943(c)	25,420	44.51	1,800	80,060	90.49
943(l)	40,980	61.01	1,900	84,620	92.95
1,000	43,580	63.69	2,000	89,180	95.29
1,100	48,140	68.03			

$V_2O_5(c)$:

$$H_T - H_{298.15} = 46.54T - 1.95 \times 10^{-3}T^2 + 13.22 \times 10^5 T^{-1}$$
$$-18,137 \ (1.2 \text{ percent}; \ 298°-943° \text{ K.});$$
$$C_p = 46.54 - 3.90 \times 10^{-3}T - 13.22 \times 10^5 T^{-2};$$
$$\Delta H_{943}(fusion) = 15,560.$$

$V_2O_5(l)$:

$$H_T - H_{298.15} = 45.60T - 2,020 \ (0.1 \text{ percent};$$
$$943°-2,000° \text{ K.});$$
$$C_p = 45.60.$$

CARBIDE

TABLE 861.—*Heat content and entropy of* $VC(c)$

[Base, crystals at 298.15° K.; mol. wt., 62.96]

T, ° K.	$H_T-H_{298.15}$, cal./mole	$S_T-S_{298.15}$, cal./deg. mole	T, ° K.	$H_T-H_{298.15}$, cal./mole	$S_T-S_{298.15}$, cal./deg. mole
400	890	2.55	1,300	11,380	15.81
500	1,850	4.70	1,400	12,720	16.80
600	2,870	6.55	1,500	14,080	17.74
700	3,950	8.22	1,600	15,450	18.63
800	5,090	9.74	1,700	16,830	19.47
900	6,280	11.14	1,800	18,220	20.26
1,000	7,510	12.43	1,900	19,620	21.02
1,100	8,770	13.64	2,000	21,030	21.74
1,200	10,060	14.76			

$VC(c)$:

$$H_T - H_{298.15} = 9.18T + 1.65 \times 10^{-3}T^2 + 1.95 \times 10^5 T^{-1}$$
$$-3,538 \ (0.4 \text{ percent}; \ 298°-1,700° \text{ K.});$$
$$C_p = 9.18 + 3.30 \times 10^{-3}T - 1.95 \times 10^5 T^{-2}.$$

NITRIDE

TABLE 862.—*Heat content and entropy of* $VN(c)$

[Base, crystals at 298.15° K.; mol. wt., 64.96]

T, ° K.	$H_T-H_{298.15}$, cal./mole	$S_T-S_{298.15}$, cal./deg. mole	T, ° K.	$H_T-H_{298.15}$, cal./mole	$S_T-S_{298.15}$, cal./deg. mole
400	1,010	2.91	1,300	12,090	17.06
500	2,080	5.30	1,400	13,450	18.07
600	3,200	7.34	1,500	14,820	19.01
700	4,370	9.14	1,600	16,200	19.90
800	5,590	10.76	1,700	17,590	20.74
900	6,850	12.25	1,800	18,990	21.54
1,000	8,130	13.60	1,900	20,400	22.31
1,100	9,430	14.84	2,000	21,820	23.04
1,200	10,750	15.99			

TABLE 1h7. (*Continued*)

VN(*c*):

$H_T - H_{298.15} = 10.94T + 1.05 \times 10^{-3}T^2 + 2.21 \times 10^5 T^{-1}$
  $-4,096$ (0.3 percent; 298°–1,800° K.);
$C_p = 10.94 + 2.10 \times 10^{-3}T - 2.21 \times 10^5 T^{-2}$.

CHLORIDES

TABLE 863.—*Heat content and entropy of* $VCl_2(c)$

[Base, crystals at 298.15° K.; mol. wt., 121.86]

T, ° K.	$H_T-H_{298.15}$, cal./mole	$S_T-S_{298.15}$, cal./deg. mole	T, ° K.	$H_T-H_{298.15}$, cal./mole	$S_T-S_{298.15}$, cal./deg. mole
400	1,840	5.30	900	11,200	20.36
500	3,620	9.27	1,000	13,180	22.45
600	5,450	12.61	1,100	15,190	24.36
700	7,330	15.50	1,200	17,220	26.13
800	9,250	18.07	1,300	19,270	27.77

$VCl_2(c)$:

$H_T - H_{298.15} = 17.25T + 1.36 \times 10^{-3}T^2 + 0.71 \times 10^5 T^{-1}$
  $-5,502$ (0.4 percent; 298°–1,300° K.);
$C_p = 17.25 + 2.72 \times 10^{-3}T - 0.71 \times 10^5 T^{-2}$.

TABLE 864.—*Heat content and entropy of* $VCl_3(c)$

[Base, crystals at 298.15° K.; mol. wt., 157.32]

T, ° K.	$H_T-H_{298.15}$, cal./mole	$S_T-S_{298.15}$, cal./deg. mole	T, ° K.	$H_T-H_{298.15}$, cal./mole	$S_T-S_{298.15}$, cal./deg. mole
400	2,360	6.80	700	9,700	20.43
500	4,730	12.09	800	12,270	23.85
600	7,180	16.55	900	14,860	26.90

$VCl_3(c)$:

$H_T - H_{298.15} = 22.99T + 1.96 \times 10^{-3}T^2 + 1.68 \times 10^5 T^{-1}$
  $-7,592$ (0.2 percent; 298°–900° K.);
$C_p = 22.99 + 3.92 \times 10^{-3}T - 1.68 \times 10^5 T^{-2}$.

OXYCHLORIDE

TABLE 865.—*Heat content and entropy of* $VOCl_3(g)$

[Base, ideal gas at 298.15° K.; mol. wt., 173.32]

T, ° K.	$H_T-H_{298.15}$, cal./mole	$S_T-S_{298.15}$, cal./deg. mole	T, ° K.	$H_T-H_{298.15}$, cal./mole	$S_T-S_{298.15}$, cal./deg. mole
400	2,270	6.54	800	11,930	23.20
500	4,605	11.75	900	14,435	26.15
600	7,010	16.13	1,000	16,950	28.80
700	9,455	19.90			

$VOCl_3(g)$:

$H_T - H_{298.15} = 24.00T + 0.80 \times 10^{-3}T^2 + 2.66 \times 10^5 T^{-1}$
  $-8,119$ (0.1 percent; 298°–1,000° K.);
$C_p = 24.00 + 1.60 \times 10^{-3}T - 2.66 \times 10^5 T^{-2}$.

XENON

ELEMENT

TABLE 866.—*Heat content and entropy of* $Xe(g)$

[Base, ideal gas at 298.15° K.; atomic wt., 131.30]

T, ° K.	$H_T-H_{298.15}$, cal./mole	$S_T-S_{298.15}$, cal./deg. mole	T, ° K.	$H_T-H_{298.15}$, cal./mole	$S_T-S_{298.15}$, cal./deg. mole
400	505	1.46	1,900	7,960	9.20
500	1,005	2.57	2,000	8,455	9.46
600	1,500	3.48	2,200	9,450	9.93
700	1,995	4.24	2,400	10,445	10.36
800	2,495	4.90	2,600	11,440	10.76
900	2,990	5.49	2,800	12,435	11.13
1,000	3,490	6.01	3,000	13,425	11.47
1,100	3,985	6.49	3,500	15,910	12.24
1,200	4,480	6.92	4,000	18,395	12.90
1,300	4,980	7.32	4,500	20,880	13.49
1,400	5,475	7.69	5,000	23,365	14.01
1,500	5,975	8.03	6,000	28,335	14.92
1,600	6,470	8.35	7,000	33,310	15.68
1,700	6,965	8.65	8,000	38,305	16.35
1,800	7,465	8.94			

$Xe(g)$:

$H_T - H_{298.15} = 4.97T - 1,482$ (0.1 percent;
  298°–8,000° K.);
$C_p = 4.97$.

YITTERBIUM AND ITS COMPOUNDS

ELEMENT

TABLE 867.—*Heat content and entropy of* $Yb(c, l)$

[Base, α-crystals at 298.15° K.; atomic wt., 173.04]

T, ° K.	$H_T-H_{298.15}$, cal./mole	$S_T-S_{298.15}$, cal./deg. mole	T, ° K.	$H_T-H_{298.15}$, cal./mole	$S_T-S_{298.15}$, cal./deg. mole
400	620	1.79	1,097(*l*)	7,930	10.92
500	1,250	3.19	1,100	7,950	10.94
600	1,900	4.38	1,200	8,700	11.59
700	2,570	5.41	1,300	9,450	12.19
800	3,260	6.33	1,400	10,200	12.75
900	3,970	7.17	1,500	10,950	13.27
1,000	4,700	7.94	1,600	11,700	13.75
1,071(*α*)	5,230	8.45	1,700	12,450	14.20
1,071(*β*)	5,530	8.73	1,800	13,200	14.64
1,097(*β*)	5,730	8.92			

$Yb(\alpha)$:

$H_T - H_{298.15} = 5.41T + 0.99 \times 10^{-3}T^2 - 1,701$
  (0.1 percent; 298°–1,071° K.);
$C_p = 5.41 + 1.98 \times 10^{-3}T$;
$\Delta H_{1071}(transition) = 300$.

$Yb(\beta)$:

$H_T - H_{298.15} = 7.70T - 2,717$ (0.1 percent;
  1,071°–1,097° K.);
$C_p = 7.70$.
$\Delta H_{1097}(fusion) = 2,200$.

$Yb(l)$:

$H_T - H_{298.15} = 7.50T - 300$ (0.1 percent;
  1,097–1,800° K.);
$C_p = 7.50$.

TABLE 1h7. (*Continued*)

TABLE 868.—*Heat content and entropy of* $Yb(g)$

[Base, ideal gas at 298.15° K.; atomic wt., 173.04]

T, ° K.	$H_T-H_{298.15}$, cal./mole	$S_T-S_{298.15}$, cal./deg. mole	T, ° K.	$H_T-H_{298.15}$, cal./mole	$S_T-S_{298.15}$, cal./deg. mole
400	505	1.46	1,500	5,970	8.03
500	1,005	2.57	1,600	6,470	8.35
600	1,500	3.48	1,700	6,965	8.65
700	1,995	4.24	1,800	7,460	8.93
800	2,495	4.91	1,900	7,960	9.20
900	2,990	5.49	2,000	8,455	9.46
1,000	3,485	6.01	2,200	9,450	9.93
1,100	3,985	6.49	2,400	10,450	10.37
1,200	4,480	6.92	2,600	11,450	10.77
1,300	4,975	7.32	2,800	12,465	11.14
1,400	5,475	7.69	3,000	13,485	11.50

$Yb(g)$:

$$H_T-H_{298.15}=4.97T-1,482 \text{ (0.2 percent;}$$
$$298°-3,000° \text{ K.);}$$
$$C_p=4.97.$$

OXIDE

$Yb_2O_3(c)$:

$$\overline{C_p}=25.5 \ (273°-373° \text{ K.}).$$

CHLORIDE

TABLE 869.—*Heat content and entropy of* $YbCl(g)$

[Base, ideal gas at 298.15° K.; mol. wt., 208.50]

T, ° K.	$H_T-H_{298.15}$, cal./mole	$S_T-S_{298.15}$, cal./deg. mole	T, ° K.	$H_T-H_{298.15}$, cal./mole	$S_T-S_{298.15}$, cal./deg. mole
400	890	2.56	1,000	6,210	10.68
500	1,770	4.53	1,200	7,995	12.31
600	2,655	6.14	1,400	9,780	13.68
700	3,540	7.51	1,600	11,565	14.88
800	4,430	8.69	1,800	13,350	15.93
900	5,320	9.74	2,000	15,140	16.87

$YbCl(g)$:

$$H_T-H_{298.15}=8.94T+0.27\times10^5T^{-1}-2,756$$
$$\text{(0.1 percent; 298°-1,000° K.);}$$
$$C_p=8.94-0.27\times10^5T^{-2}.$$

SULFATE

$Yb_2(SO_4)_3(c)$:

$$\overline{C_p}=65.9 \ (273°-373° \text{ K.}).$$

$Yb_2(SO_4)_3 \cdot 8H_2O(c)$:

$$\overline{C_p}=139.2 \ (273°-319° \text{ K.}).$$

YTTRIUM AND ITS COMPOUNDS

ELEMENT

TABLE 870.—*Heat content and entropy of* $Y(c, l)$

[Base, crystals at 298.15° K.; atomic wt., 88.92]

T, ° K.	$H_T-H_{298.15}$, cal./mole	$S_T-S_{298.15}$, cal./deg. mole	T, ° K.	$H_T-H_{298.15}$, cal./mole	$S_T-S_{298.15}$, cal./deg. mole
400	615	1.78	1,600	8,680	10.90
500	1,235	3.15	1,700	9,420	11.35
600	1,860	4.29	1,773(c)	9,960	11.66
700	2,495	5.27	1,773(l)	14,060	13.97
800	3,140	6.14	1,800	14,280	14.10
900	3,800	6.91	1,900	15,080	14.53
1,000	4,465	7.61	2,000	15,880	14.94
1,100	5,140	8.26	2,200	17,480	15.70
1,200	5,830	8.86	2,400	19,080	16.40
1,300	6,525	9.41	2,600	20,680	17.04
1,400	7,235	9.94	2,800	22,280	17.63
1,500	7,955	10.43	3,000	23,880	18.18

$Y(c)$:

$$H_T-H_{298.15}=5.72T+0.50\times10^{-3}T^2-1,750$$
$$\text{(0.2 percent; 298°-1,773° K.);}$$
$$C_p=5.72+1.00\times10^{-3}T;$$
$$\Delta H_{1773}(fusion)=4,100.$$

$Y(l)$:

$$H_T-H_{298.15}=8.00T-120 \text{ (0.1 percent;}$$
$$1,773°-3,000° \text{ K.);}$$
$$C_p=8.00.$$

TABLE 871.—*Heat content and entropy of* $Y(g)$

[Base, ideal gas at 298.15° K.; atomic wt., 88.92]

T, ° K.	$H_T-H_{298.15}$, cal./mole	$S_T-S_{298.15}$, cal./deg. mole	T, ° K.	$H_T-H_{298.15}$, cal./mole	$S_T-S_{298.15}$, cal./deg. mole
400	625	1.80	1,900	8,595	10.26
500	1,220	3.13	2,000	9,115	10.52
600	1,790	4.17	2,200	10,175	11.03
700	2,350	5.03	2,400	11,270	11.50
800	2,890	5.76	2,600	12,415	11.96
900	3,425	6.39	2,800	13,620	12.41
1,000	3,955	6.94	3,000	14,895	12.85
1,100	4,475	7.44	3,500	18,475	13.95
1,200	4,995	7.89	4,000	22,655	15.06
1,300	5,510	8.31	4,500	27,390	16.18
1,400	6,025	8.69	5,000	32,560	17.27
1,500	6,535	9.04	6,000	43,665	19.27
1,600	7,050	9.37	7,000	54,665	20.98
1,700	7,560	9.68	8,000	65,370	22.42
1,800	8,075	9.98			

$Y(g)$:

$$H_T-H_{298.15}=2.77T+0.74\times10^{-3}T^2-2.64\times10^5T^{-1}$$
$$-6 \text{ (0.7 percent; 3,000°-6,000° K.);}$$
$$C_p=2.77+1.48\times10^{-3}T+2.64\times10^5T^{-2}.$$

OXIDES

$Y_2O_3(c)$:

$$\overline{C_p}=23.2 \ (273°-373° \text{ K.}).$$

TABLE 872.—*Heat content and entropy of* $YO(g)$

[Base, ideal gas at 298.15° K.; mol. wt., 104.92]

T, ° K.	$H_T-H_{298.15}$, cal./mole	$S_T-S_{298.15}$, cal./deg. mole	T, ° K.	$H_T-H_{298.15}$, cal./mole	$S_T-S_{298.15}$, cal./deg. mole
400	790	2.27	1,000	5,855	9.96
500	1,595	4.07	1,200	7,605	11.55
600	2,425	5.58	1,400	9,365	12.91
700	3,270	6.89	1,600	11,135	14.09
800	4,125	8.03	1,800	12,905	15.13
900	4,990	9.05	2,000	14,680	16.06

$YO(g)$:

$$H_T-H_{298.15}=8.41T+0.16\times10^{-3}T^2+0.84\times10^5T^{-1}$$
$$-2,803 \text{ (0.4 percent; 298°-2,000° K.);}$$
$$C_p=8.41+0.32\times10^{-3}T-0.84\times10^5T^{-2}.$$

MOLYBDATE

$Y_2(MoO_4)_3(c)$:

$$\overline{C_p}=105 \ (273°-297° \text{ K.}).$$

NITRATE

$Y(NO_3)_3(c)$:

$$\overline{C_p}=75.7 \ (273°-289° \text{ K.}).$$

TABLE 1h7. (*Continued*)

SULFATE

$Y_2(SO_4)_3(c)$:
$\overline{C}_p = 61.5$ (273°–373° K.).

$Y_2(SO_4)_3 \cdot 8H_2O(c)$:
$\overline{C}_p = 137.7$ (273°–319° K.).

ZINC AND ITS COMPOUNDS

ELEMENT

TABLE 873.—*Heat content and entropy of* $Zn(c, l)$

[Base, crystals at 298.15° K.; atomic wt., 65.38]

T, ° K.	$H_T - H_{298.15}$, cal./mole	$S_T - S_{298.15}$, cal./deg. mole	T, ° K.	$H_T - H_{298.15}$, cal./mole	$S_T - S_{298.15}$, cal./deg. mole
400	625	1.80	800	5,150	9.08
500	1,270	3.24	900	5,900	9.96
600	1,940	4.46	1,000	6,650	10.75
692.7(c)	2,580	5.45	1,100	7,400	11.47
692.7(l)	4,345	8.00	1,200	8,150	12.12
700	4,400	8.08			

$Zn(c)$:
$$H_T - H_{298.15} = 5.35T + 1.20 \times 10^{-3}T^2 - 1,702 \text{ (0.3 percent;}$$
$$298°–692.7° \text{ K.);}$$
$$C_p = 5.35 + 2.40 \times 10^{-3}T;$$
$$\Delta H_{692.7}(fusion) = 1,765.$$

$Zn(l)$:
$$H_T - H_{298.15} = 7.50T - 850 \text{ (0.1 percent;}$$
$$692.7°–1,200° \text{ K.);}$$
$$C_p = 7.50.$$

TABLE 874.—*Heat content and entropy of* $Zn(g)$

[Base, ideal gas at 298.15° K., atomic wt., 65.38]

T, ° K.	$H_T - H_{298.15}$, cal./mole	$S_T - S_{298.15}$, cal./deg. mole	T, ° K.	$H_T - H_{298.15}$, cal./mole	$S_T - S_{298.15}$, cal./deg. mole
400	505	1.46	1,900	7,960	9.20
500	1,005	2.57	2,000	8,455	9.46
600	1,500	3.48	2,200	9,450	9.93
700	1,995	4.24	2,400	10,445	10.36
800	2,495	4.90	2,600	11,440	10.76
900	2,990	5.49	2,800	12,435	11.13
1,000	3,490	6.01	3,000	13,425	11.47
1,100	3,985	6.49	3,500	15,910	12.24
1,200	4,480	6.92	4,000	18,400	12.90
1,300	4,980	7.32	4,500	20,905	13.49
1,400	5,475	7.69	5,000	23,435	14.03
1,500	5,975	8.03	6,000	28,675	14.98
1,600	6,470	8.35	7,000	34,385	15.86
1,700	6,965	8.65	8,000	40,905	16.73
1,800	7,465	8.94			

$Zn(g)$:
$$H_T - H_{298.15} = 4.97T - 1,482 \quad (0.1 \text{ percent;}$$
$$298°–5,000° \text{ K.);}$$
$$C_p = 4.97.$$

OXIDE

TABLE 875.—*Heat content and entropy of* $ZnO(c)$

[Base, crystals at 298.15° K.; mol. wt., 81.38]

T, ° K.	$H_T - H_{298.15}$, cal./mole	$S_T - S_{298.15}$, cal./deg. mole	T, ° K.	$H_T - H_{298.15}$, cal./mole	$S_T - S_{298.15}$, cal./deg. mole
400	1,070	3.03	1,300	12,120	17.29
500	2,190	5.58	1,400	13,450	18.28
600	3,350	7.69	1,500	14,800	19.21
700	4,530	9.51	1,600	16,160	20.09
800	5,740	11.13	1,700	17,530	20.92
900	6,970	12.57	1,800	18,910	21.71
1,000	8,220	13.89	1,900	20,300	22.46
1,100	9,500	15.11	2,000	21,700	23.18
1,200	10,800	16.24			

$ZnO(c)$:
$$H_T - H_{298.15} = 11.71T + 0.61 \times 10^{-3}T^2 + 2.18 \times 10^5 T^{-1}$$
$$-4,277 \text{ (0.4 percent; } 298°–2,000° \text{ K.);}$$
$$C_p = 11.71 + 1.22 \times 10^{-3}T - 2.18 \times 10^5 T^{-2}.$$

TABLE 876.—*Heat content and entropy of* $ZnO(g)$

[Base, ideal gas at 298.15° K.; mol. wt., 81.38]

T, ° K.	$H_T - H_{298.15}$, cal./mole	$S_T - S_{298.15}$, cal./deg. mole	T, ° K.	$H_T - H_{298.15}$, cal./mole	$S_T - S_{298.15}$, cal./deg. mole
400	795	2.28	1,000	5,875	9.99
500	1,605	4.10	1,200	7,630	11.59
600	2,435	5.61	1,400	9,390	12.95
700	3,285	6.92	1,600	11,160	14.13
800	4,140	8.06	1,800	12,930	15.17
900	5,005	9.08	2,000	14,710	16.11

$ZnO(g)$:
$$H_T - H_{298.15} = 8.40T + 0.17 \times 10^{-3}T^2 + 0.82 \times 10^5 T^{-1}$$
$$-2,795 \text{ (0.3 percent; } 298°–2,000° \text{ K.);}$$
$$C_p = 8.40 + 0.34 \times 10^{-3}T - 0.82 \times 10^5 T^{-2}.$$

HYDROXIDE

$Zn(OH)_2(c)$:
$\overline{C}_p = 17.71$ (290°–323° K.).

SULFIDE

TABLE 877.—*Heat content and entropy of* $ZnS(c)$

[Base, crystals at 298.15° K.; mol. wt., 97.45]

T, ° K.	$H_T - H_{298.15}$, cal./mole	$S_T - S_{298.15}$, cal./deg. mole	T, ° K.	$H_T - H_{298.15}$, cal./mole	$S_T - S_{298.15}$, cal./deg. mole
400	1,160	3.34	900	7,530	13.62
500	2,370	6.04	1,000	8,820	14.98
600	3,660	8.39	1,100	10,110	16.21
700	4,950	10.38	1,200	11,410	17.34
800	6,240	12.10			

$ZnS(c)$:
$$H_T - H_{298.15} = 12.16T + 0.62 \times 10^{-3}T^2 + 1.36 \times 10^5 T^{-1}$$
$$-4,137 \text{ (0.8 percent; } 298°–1,200° \text{ K.);}$$
$$C_p = 12.16 + 1.24 \times 10^{-3}T - 1.36 \times 10^5 T^{-2}.$$

TABLE 1h7. (*Continued*)

NITRIDE

TABLE 878.—*Heat content and entropy of* $Zn_3N_2(c)$

[Base, crystals at 298.15° K.; mol. wt., 224.16]

T, ° K.	$H_T-H_{298.15}$, cal./mole	$S_T-S_{298.15}$, cal./deg. mole	T, ° K.	$H_T-H_{298.15}$, cal./mole	$S_T-S_{298.15}$, cal./deg. mole
400	2,770	7.98	600	8,880	20.30
500	5,700	14.51	700	12,180	25.38

$Zn_3N_2(c)$:

$$H_T-H_{298.15}=19.93\,T+10.40\times10^{-3}T^2$$
$$-6,867 \ (0.2 \text{ percent; } 298°\text{–}700° \text{ K.});$$
$$C_p=19.93+20.80\times10^{-3}T.$$

HYDRIDE

TABLE 879.—*Heat content and entropy of* $ZnH(g)$

[Base, ideal gas at 298.15° K.; mol. wt., 66.39]

T, ° K.	$H_T-H_{298.15}$, cal./mole	$S_T-S_{298.15}$, cal./deg. mole	T, ° K.	$H_T-H_{298.15}$, cal./mole	$S_T-S_{298.15}$, cal./deg. mole
400	725	2.09	1,000	5,445	9.22
500	1,460	3.73	1,200	7,120	10.74
600	2,220	5.11	1,400	8,830	12.06
700	3,000	6.32	1,600	10,560	13.22
800	3,800	7.38	1,800	12,300	14.24
900	4,615	8.34	2,000	14,045	15.16

$ZnH(g)$:

$$H_T-H_{298.15}=7.27\,T+0.48\times10^{-3}T^2+0.46\times10^5 T^{-1}$$
$$-2,365 \ (0.7 \text{ percent; } 298°\text{–}2,000° \text{ K.});$$
$$C_p=7.27+0.96\times10^{-3}T-0.46\times10^5 T^{-2}.$$

BROMIDE

TABLE 880.—*Heat content and entropy of* $ZnBr(g)$

[Base, ideal gas at 298.15° K.; mol. wt., 145.30]

T, ° K.	$H_T-H_{298.15}$, cal./mole	$S_T-S_{298.15}$, cal./deg. mole	T, ° K.	$H_T-H_{298.15}$, cal./mole	$S_I-S_{298.15}$, cal./deg. mole
400	900	2.59	1,000	6,240	10.74
500	1,785	4.57	1,200	8,025	12.37
600	2,670	6.18	1,400	9,810	13.75
700	3,560	7.56	1,600	11,595	14.94
800	4,455	8.75	1,800	13,385	15.99
900	5,345	9.80	2,000	15,170	16.93

$ZnBr(g)$:

$$H_T-H_{298.15}=8.94\,T+0.15\times10^5 T^{-1}$$
$$-2,716 \ (0.1 \text{ percent; } 298°\text{–}2,000° \text{ K.});$$
$$C_p=8.94-0.15\times10^5 T^{-2}.$$

CHLORIDES

$ZnCl_2(c)$:

$$C_p=15.00+10.85\times10^{-3}T(estimated) \ (298°\text{–}556° \text{ K.}).$$

TABLE 881.—*Heat content and entropy of* $ZnCl(g)$

[Base, ideal gas at 298.15° K.; mol. wt., 100.84]

T, ° K.	$H_T-H_{298.15}$, cal./mole	$S_T-S_{298.15}$, cal./deg. mole	T, ° K.	$H_T-H_{298.15}$, cal./mole	$S_T-S_{298.15}$, cal./deg. mole
400	875	2.52	1,000	6,165	10.59
500	1,745	4.46	1,200	7,945	12.21
600	2,620	6.06	1,400	9,725	13.58
700	3,500	7.42	1,600	11,510	14.77
800	4,385	8.60	1,800	13,295	15.82
900	5,275	9.65	2,000	15,080	16.76

$ZnCl(g)$:

$$H_T-H_{298.15}=8.93\,T+0.43\times10^5 T^{-1}$$
$$-2,807 \ (0.1 \text{ percent; } 298°\text{–}2,000° \text{ K.});$$
$$C_p=8.93-0.43\times10^5 T^{-2}.$$

FLUORIDES

$ZnF_2(c)$:

$$C_p=13.87+6.27\times10^{-3}T(estimated) \ (298°\text{–}1,145° \text{ K.}).$$

TABLE 882.—*Heat content and entropy of* $ZnF(g)$

[Base, ideal gas at 298.15° K.; mol. wt., 84.38]

T, ° K.	$H_T-H_{298.15}$, cal./mole	$S_T-S_{298.15}$, cal./deg. mole	T, ° K.	$H_T-H_{298.15}$, cal./mole	$S_T-S_{298.15}$, cal./deg. mole
400	830	2.39	1,000	6,015	10.27
500	1,670	4.26	1,200	7,785	11.89
600	2,525	5.82	1,400	9,560	12.26
700	3,390	7.16	1,600	11,335	14.44
800	4,260	8.32	1,800	13,110	15.48
900	5,135	9.35	2,000	14,890	16.42

$ZnF(g)$:

$$H_T-H_{298.15}=8.72\,T+0.07\times10^{-3}T^2+0.73\times10^5 T^{-1}-$$
$$2,850 \ (0.1 \text{ percent; } 298°\text{–}2,000° \text{ K.});$$
$$C_p=8.72+0.14\times10^{-3}T-0.73\times10^5 T^{-2}.$$

ALUMINATE

$ZnAl_2O_4(c)$:

$$C_p=24.40+20.30\times10^{-3}T(estimated) \ (298°\text{–}1,298° \text{ K.})$$

CARBONATE

$ZnCO_3(c)$:

$$C_p=9.30+33.00\times10^{-3}T(estimated) \ (298°\text{–}780° \text{ K.}).$$

CHROMITE

$ZnCr_2O_4(c)$:

$$C_p=25.50+21.70\times10^{-3}T(estimated) \ (298°\text{–}1,298° \text{ K.}).$$

FERRITE

$ZnFe_2O_4(c)$:

$$C_p=27.71+17.72\times10^{-3}T(estimated) \ (298°\text{–}1,298° \text{ K.}).$$

SILICATE

$Zn_2SiO_4(c)$:

$$C_p=29.48 \ (298° \text{ K.}).$$

SULFATE

TABLE 883.—*Heat content and entropy of* $ZnSO_4(c)$

[Base, crystals at 298.15° K.; mol. wt., 161.45]

T, ° K.	$H_T-H_{298.15}$, cal./mole	$S_T-S_{298.15}$, cal./deg. mole	T, ° K.	$H_T-H_{298.15}$, cal./mole	$S_T-S_{298.15}$, cal./deg. mole
400	2,520	7.26	800	14,220	27.16
500	5,120	13.05	900	17,770	31.34
600	7,870	18.06	1,000	21,620	35.39
700	10,920	22.76			

$ZnSO_4(c)$:

$$H_T-H_{298.15}=17.07\,T+10.40\times10^{-3}T^2-6,014$$

TABLE 1h7. (*Continued*)

(0.8 percent; 298°–1,000° K.);

$$C_p = 17.07 + 20.80 \times 10^{-3}T.$$

ZnSO$_4$·H$_2$O(c):

$$C_p = 34.7 \ (282° \text{ K.}).$$

ZnSO$_4$·6H$_2$O(c):

$$C_p = 85.02 \ (298° \text{ K.}).$$

ZnSO$_4$·7H$_2$O(c):

$$C_p = 91.17 \ (298° \text{ K.}).$$

TITANATE

TABLE 884.—*Heat content and entropy of Zn$_2$TiO$_4$(c)*

[Base, crystals at 298.15° K.; mol. wt., 242.66]

T, ° K.	$H_T - H_{298.15}$, cal./mole	$S_T - S_{298.15}$, cal./deg. mole	T, ° K.	$H_T - H_{298.15}$, cal./mole	$S_T - S_{298.15}$, cal./deg. mole
400	3,550	10.21	1,300	42,560	60.04
500	7,290	18.55	1,400	47,250	63.52
600	11,250	25.77	1,500	51,950	66.76
700	15,400	32.16	1,600	56,680	69.81
800	19,710	37.92	1,700	61,450	72.70
900	24,140	43.13	1,800	66,270	75.46
1,000	28,660	47.89	1,900	75,130	78.09
1,100	33,250	52.27	2,000	80,030	80.60
1,200	37,890	56.30			

Zn$_2$TiO$_4$(c):

$$H_T - H_{298.15} = 39.82T + 2.77 \times 10^{-3}T^2 + 7.69 \times 10^5 T^{-1}$$
$$- 14,698 \ (0.8 \text{ percent; } 298°-1,800° \text{ K.});$$
$$C_p = 39.82 + 5.54 \times 10^{-3}T - 7.69 \times 10^5 T^{-2}.$$

ZINC-ANTIMONY COMPOUND

ZnSb(c):

$$C_p = 11.50 + 3.13 \times 10^{-3}T(estimated) \ (298°-810° \text{ K.}).$$

ZIRCONIUM AND ITS COMPOUNDS

ELEMENT

TABLE 885.—*Heat content and entropy of Zr(c, l)*

[Base, crystals at 298.15° K.; atomic wt., 91.22]

T, ° K.	$H_T - H_{298.15}$, cal./mole	$S_T - S_{298.15}$, cal./deg. mole	T, ° K.	$H_T - H_{298.15}$, cal./mole	$S_T - S_{298.15}$, cal./deg. mole
400	650	1.87	1,600	10,680	12.98
500	1,330	3.39	1,700	11,470	13.46
600	2,030	4.66	1,800	12,260	13.91
700	2,740	5.76	1,900	13,050	14.33
800	3,460	6.72	2,000	13,840	14.74
900	4,200	7.59	2,100	14,630	15.12
1,000	4,980	8.41	2,130(c)	14,870	15.24
1,100	5,800	9.20	2,130(l)	19,770	17.54
1,135(α)	6,090	9.46	2,200	20,330	17.80
1,135(β)	7,005	10.26	2,400	21,930	18.49
1,200	7,520	10.70	2,600	23,530	19.13
1,300	8,310	11.33	2,800	25,130	19.72
1,400	9,100	11.92	3,000	26,730	20.28
1,500	9,890	12.47			

Zr(α):

$$H_T - H_{298.15} = 6.50T + 0.71 \times 10^{-3}T^2 + 0.82 \times 10^5 T^{-1}$$
$$- 2,276 \ (0.7 \text{ percent; } 298°-1,135° \text{ K.});$$
$$C_p = 6.50 + 1.42 \times 10^{-3}T - 0.82 \times 10^5 T^{-2};$$
$$\Delta H_{1135}(transition) = 915.$$

Zr(β):

$$H_T - H_{298.15} = 7.90T - 1,960 \ (0.1 \text{ percent;}$$
$$1,135°-2,130° \text{ K.});$$

$$C_p = 7.90.$$
$$\Delta H_{2130}(fusion) = 4,900.$$

Zr(l):

$$H_T - H_{298.15} = 8.00T + 2,730 \ (0.1 \text{ percent;}$$
$$2,130°-3,000° \text{ K.});$$
$$C_p = 8.00.$$

TABLE 886.—*Heat content and entropy of Zr(g)*

[Base, ideal gas at 298.15° K.; atomic wt., 91.22]

T, ° K.	$H_T - H_{298.15}$, cal./mole	$S_T - S_{298.15}$, cal./deg. mole	T, ° K.	$H_T - H_{298.15}$, cal./mole	$S_T - S_{298.15}$, cal./deg. mole
400	665	1.92	1,900	10,325	11.92
500	1,325	3.39	2,000	11,025	12.28
600	1,980	4.58	2,200	12,450	12.96
700	2,620	5.57	2,400	13,905	13.59
800	3,245	6.40	2,600	15,390	14.18
900	3,870	7.13	2,800	16,900	14.74
1,000	4,475	7.78	3,000	18,435	15.27
1,100	5,085	8.36	3,500	22,410	16.50
1,200	5,705	8.90	4,000	26,575	17.61
1,300	6,335	9.40	4,500	30,925	18.63
1,400	6,970	9.87	5,000	35,430	19.58
1,500	7,620	10.32	6,000	44,750	21.28
1,600	8,280	10.75	7,000	54,235	22.74
1,700	8,950	11.15	8,000	63,645	24.00
1,800	9,630	11.54			

Zr(g):

$$H_T - H_{298.15} = 7.01T - 0.35 \times 10^{-3}T^2 + 0.38 \times 10^5 T^{-1}$$
$$- 2,186 \ (0.7 \text{ percent; } 298°-1,400° \text{ K.});$$
$$C_p = 7.01 - 0.70 \times 10^{-3}T - 0.38 \times 10^5 T^{-2}.$$
$$H_T - H_{298.15} = 5.59T + 0.36 \times 10^{-3}T^2 - 0.50 \times 10^5 T^{-1}$$
$$- 1,531 \ (0.3 \text{ percent; } 1,400°-6,000° \text{ K.});$$
$$C_p = 5.59 + 0.72 \times 10^{-3}T + 0.50 \times 10^5 T^{-2}.$$

OXIDE

TABLE 887.—*Heat content and entropy of ZrO$_2$(c)*

[Base, α-crystals at 298.15° K.; mol. wt., 123.22]

T, ° K.	$H_T - H_{298.15}$, cal./mole	$S_T - S_{298.15}$, cal./deg. mole	T, ° K.	$H_T - H_{298.15}$, cal./mole	$S_T - S_{298.15}$, cal./deg. mole
400	1,475	4.24	1,400	19,150	25.94
500	3,050	7.75	1,478(α)	20,620	26.96
600	4,690	10.74	1,478(β)	22,040	27.92
700	6,380	13.34	1,500	22,430	28.18
800	8,120	15.68	1,600	24,210	29.33
900	9,910	17.79	1,700	25,990	30.41
1,000	11,730	19.70	1,800	27,770	31.43
1,100	13,570	21.46	1,900	29,550	32.39
1,200	15,420	23.07	2,000	31,330	33.30
1,300	17,280	24.55			

ZrO$_2$(α):

$$H_T - H_{298.15} = 16.64T + 0.90 \times 10^{-3}T^2 + 3.36 \times 10^5 T^{-1}$$
$$- 6,168 \ (0.2 \text{ percent; } 298°-1,478° \text{ K.});$$
$$C_p = 16.64 + 1.80 \times 10^{-3}T - 3.36 \times 10^5 T^{-2};$$
$$\Delta H_{1478}(transition) = 1,420.$$

ZrO$_2$(β):

$$H_T - H_{298.15} = 17.80T - 4,267 \ (0.1 \text{ percent; } 1,478°-2,000°$$
$$\text{ K.});$$

$$C_p = 17.80.$$

TABLE 1h7. (*Continued*)

NITRIDES

TABLE 888.—*Heat content and entropy of* $ZrN(c)$

[Base, crystals at 298.15° K.; mol. wt., 105.23]

T, ° K.	$H_T-H_{298.15}$, cal./mole	$S_T-S_{298.15}$, cal./deg. mole	T, ° K.	$H_T-H_{298.15}$, cal./mole	$S_T-S_{298.15}$, cal./deg. mole
400	1,040	3.00	1,300	12,060	17.14
500	2,120	5.40	1,400	13,370	18.11
600	3,260	7.48	1,500	14,690	19.02
700	4,450	9.31	1,600	16,020	19.88
800	5,670	10.94	1,700	17,360	20.69
900	6,920	12.41	1,800	18,710	21.46
1,000	8,190	13.75	1,900	20,070	22.19
1,100	9,470	14.97	2,000	21,440	22.90
1,200	10,760	16.10			

ZrN(c):

$$H_T-H_{298.15}=11.10T+0.84\times10^{-3}T^2+1.72\times10^5T^{-1}$$
$$-3,961 \text{ (0.4 percent; } 298°-1,700° \text{ K.)};$$
$$C_p=11.10+1.68\times10^{-3}T-1.72\times10^5T^{-2}.$$

TABLE 889.—*Heat content and entropy of* $Zr_3N_2(c)$

[Base, crystals at 298.15° K.; mol. wt., 301.68]

T, ° K.	$H_T-H_{298.15}$, cal./mole	$S_T-S_{298.15}$, cal./deg. mole	T, ° K.	$H_T-H_{298.15}$, cal./mole	$S_T-S_{298.15}$, cal./deg. mole
400	3,120	8.99	700	13,970	29.01
500	6,460	16.43	800	18,020	34.41
600	10,060	22.99			

Zr$_3$N$_2$(c):

$$H_T-H_{298.15}=21.64T+13.00\times10^{-3}T^2-7,608 \text{ (0.2 percent; } 298°-800° \text{ K.)};$$
$$C_p=21.64+26.00\times10^{-3}T.$$

HYDRIDE

TABLE 890.—*Heat content and entropy of* $ZrH(c)$

[Base, crystals at 298.15° K.; mol. wt., 92.23]

T, ° K.	$H_T-H_{298.15}$, cal./mole	$S_T-S_{298.15}$, cal./deg. mole	T, ° K.	$H_T-H_{298.15}$, cal./mole	$S_T-S_{298.15}$, cal./deg. mole
400	810	2.33	700	4,060	8.28
500	1,750	4.42	800	5,430	10.11
600	2,830	6.39			

ZrH(c):

$$H_T-H_{298.15}=2.95T+7.17\times10^{-3}T^2-1,517 \text{ (0.1 percent; } 298°-800° \text{ K.)};$$
$$C_p=2.95+14.34\times10^{-3}T.$$

CHLORIDE

TABLE 891.—*Heat content and entropy of* $ZrCl_4(c)$

[Base, crystals at 298.15° K.; mol. wt., 233.05]

T, ° K.	$H_T-H_{298.15}$, cal./mole	$S_T-S_{298.15}$, cal./deg. mole	T, ° K.	$H_T-H_{298.15}$, cal./mole	$S_T-S_{298.15}$, cal./deg. mole
350	1,495	4.62	550	7,580	18.36
400	3,005	8.65	600	9,120	21.04
450	4,525	12.23	700	12,220	25.81
500	6,050	15.44			

ZrCl$_4$(c):

$$H_T-H_{298.15}=31.92T+2.91\times10^5T^{-1}-10,493 \text{ (0.3 percent; } 298°-700° \text{ K.)};$$
$$C_p=31.92-2.91\times10^5T^{-2}.$$

TABLE 892.—*Heat content and entropy of* $ZrCl_4(g)$

[Base, ideal gas at 298.15° K.; mol. wt., 233.05]

T, ° K.	$H_T-H_{298.15}$, cal./mole	$S_T-S_{298.15}$, cal./deg. mole	T, ° K.	$H_T-H_{298.15}$, cal./mole	$S_T-S_{298.15}$, cal./deg. mole
400	2,440	7.03	1,000	17,560	30.04
500	4,900	12.52	1,200	22,680	34.71
600	7,400	17.08	1,400	27,820	38.67
700	9,920	20.96	1,600	32,960	42.10
800	12,460	24.35	1,800	38,100	45.13
900	15,010	27.36	2,000	43,250	47.84

ZrCl$_4$(g):

$$H_T-H_{298.15}=25.60T+0.08\times10^{-3}T^2+2.04\times10^5T^{-1}$$
$$-8,324 \text{ (0.1 percent; } 298°-2,000° \text{ K.)};$$
$$C_p=25.60+0.16\times10^{-3}T-2.04\times10^5T^{-2}.$$

SILICATE

TABLE 893.—*Heat content and entropy of* $ZrSiO_4(c)$

[Base, crystals at 298.15° K.; mol. wt., 183.31]

T, ° K.	$H_T-H_{298.15}$, cal./mole	$S_T-S_{298.15}$, cal./deg. mole	T, ° K.	$H_T-H_{298.15}$, cal./mole	$S_T-S_{298.15}$, cal./deg. mole
400	2,620	7.53	1,300	32,790	46.08
500	5,460	13.86	1,400	36,380	48.74
600	8,550	19.48	1,500	39,990	51.23
700	11,800	24.50	1,600	43,630	53.58
800	15,180	29.01	1,700	47,290	55.80
900	18,640	33.08	1,800	50,980	57.91
1,000	22,140	36.77	1,900	54,690	59.91
1,100	25,670	40.13	2,000	58,420	61.82
1,200	29,220	43.22			

ZrSiO$_4$(c):

$$H_T-H_{298.15}=31.48T+1.96\times10^{-3}T^2+8.08\times10^5T^{-1}$$
$$-12,270 \text{ (0.7 percent; } 298°-1,800° \text{ K.)};$$
$$C_p=31.48+3.92\times10^{-3}T-8.08\times10^5T^{-2}.$$

BIBLIOGRAPHY

Barlin, I. and O. Knacke, "Thermochemical Properties of Inorganic Sunstances," Verlag Stahleisen, Düsseldorf, Germany, 1973.

Childs, G. E., L. J. Ericks, and R. L. Powell, "Thermal Conductivity of Solids at Room Temperature and Below," NBS Monograph 131, U.S. Government Printing Office, Washington, D.C., 1973.

Cox, J. D. and Pilcher, "Thermochemistry of Organic and Organometallic Compounds," Academic Press, London, 1970.

Hamblin, F. D., "Abridged Thermodynamic and Thermochemical Tables," Pergamon Press, Elmsford, N.Y., 1968.

Hilsenrath, J., et al., "Tables of Thermodynamic and Transport Properties of Air, Argon, Carbon Dioxide, Carbon Monoxide, Hydrogen, Nitrogen, Oxygen and Steam," Pergamon Press, New York, 1960.

Hultgren, P., R. L. Orr, P. D. Anderson, and K. K. Kelley, "Selected Values of Thermodynamic Properties of Metals and Alloys," Wiley, New York, 1963.

Janz, G. J., "Thermodynamic Properties of Organic Compounds," rev. ed., Academic Press, New York, 1967.

Johnson, V. J., Ed., "A Compendium of the Properties of Materials at Low Temperatures (Phase I)," Office of Technical Services, Dept. of Commerce, Washington, D.C., 1960.

Karapet'yants, M. K. and M. L. Karapet'yants, "Handbook of Thermodynamic Constants of Inorganic and Compounds," Halsted Press, New York, 1968.

Kelley, K. K. and E. G. King, "Contribution to the Data on Theoretical Metallurgy, XIV. Entropies of the Elements and Inorganic Compounds," U.S. Bureau of Mines Bull. 592, U.S. Government Printing Office, Washington, D.C., 1961.

Kubaschewski, O., E. L. Evans, and C. B. Alcock, "Metallurgical Thermochemistry," 4th ed., Pergamon Press, New York, 1967.

Mc Bride, B. J., S. Heimel, J. G. Ehlers, and S. Gordon, "Thermodynamic Properties to 6000° K for 210 Substances Involving the First 18 Elements," Report NASA SP-3001, Office of Technical Services, U.S. Department of Commerce, Washington, D.C., 1963.

Rand, M. H. and O. Kubaschewski, "The Thermochemical Properties of Uranium Compounds," Oliver & Boyd, London, 1963.

Rossini, F. D., D. D. Wagman, W. H. Evans, S. Levin, and I. Jaffe, "Selected Values of Chemical Thermodynamic Properties," U.S. National Bureau of Standards Cir. 500, U.S. Government Printing Office, Washington, D.C., 1952.

Schramm, R. E., A. F. Clark, and R. P. Reed, "A Compilation and Evaluation of Mechanical, Thermal and Electrical Properties of Selected Polymers," NBS Monograph 132, U.S. Government Printing Office, Washington, D.C., 1973.

Stull, D. R. and H. Prophet, "JANAF Thermochemical Tables," 2nd ed., NBS Report No. NSRDS-NBS37, U.S. Government Printing Office, Washington, D.C., 1971.

Stulle, D.R. and G. C. Sinke, "Thermodynamic Properties of the Elements," American Chemical Society, Washington, D.C., 1956.

Touloukian, Y. S. and C. Y. Ho, Eds., "Thermophysical Properties of Matter," 13 vols., Plenum Press, New York, 1970-1973.

Utermark, W. and W. Schicke, "Melting Point Tables of Organic Compounds," 2nd ed., Halsted Press, New York, 1963.

Wagman, D. C., W. H. Evans, I. Halow, V. B. Parker, S. M. Bailey, and R. H. Schumm, "Selected Values of Chemical Thermodynamic Properties," NBS Technical Note 270, U.S. Government Printing Office, Washington, D.C., 1965-1974.

Wicks, C. D. and F. E. Block, "Thermodynamic Properties of 65 Elements," U.S. Bureau of Mines Bull. 605, U.S. Government Printing Office, Washington, D.C., 1963.

SECTION 1i
ELECTRICAL AND ELECTRONIC PROPERTIES

This Section describes the electrical and electronic properties of selected materials, and is divided into the following tables and figures:

TABLE 1i1. IONIZATION POTENTIALS
OF THE ELEMENTS.*

Atomic Number	Symbol	Ionization Potentials (ev)							
		I	II	III	IV	V	VI	VII	VIII
1	H	13.595							
2	He	24.580	54.403						
3	Li	5.390	75.619	122.420					
4	Be	9.320	18.206	153.850	217.657				
5	B	8.296	25.149	37.920	259.298	340.127			
6	C	11.264	24.376	47.864	64.476	391.986	489.84		
7	N	14.54	29.605	47.426	77.450	97.863	551.925	666.83	
8	O	13.614	35.146	54.934	77.394	113.873	138.080	739.114	871.12
9	F	17.418	34.98	62.646	87.23	114.214	157.117	185.139	953.60
10	Ne	21.559	41.07	64	97.16	126.4	157.91		

Atomic No.	Symbol	Ionization Potentials		Atomic No.	Symbol	Ionization Potentials	
		I	II			I	II
11	Na	5.138	47.29	52	Te	9.01	21.5
12	Mg	7.644	15.03	53	I	10.44	19.0
13	Al	5.984	18.823	54	Xe	12.127	21.21
14	Si	8.149	16.34	55	Cs	3.893	25.1
15	P	10.55	19.65	56	Ba	5.210	10.001
16	S	10.357	23.4	57	La	5.61	11.43
17	Cl	13.01	23.80	58	Ce	(6.91)	14.8
18	Ar	15.755	27.62	59	Pr	(5.76)	
19	K	4.339	31.81	60	Nd	6.3	
20	Ca	6.111	11.87	61	Pm		
21	Sc	6.56	12.80	62	Sm	5.6	11.2
22	Ti	6.83	13.57	63	Eu	5.67	11.24
23	V	6.74	14.65	64	Gd	6.16	12.0
24	Cr	6.764	16.49	65	Tb	(6.74)	
25	Mn	7.432	15.64	66	Dy	(6.82)	
26	Fe	7.90	16.18	67	Ho		
27	Co	7.86	17.05	68	Er		
28	Ni	7.633	18.15	69	Tm		
29	Cu	7.724	20.29	70	Yb	6.22	12.10
30	Zn	9.391	17.96	71	Lu	6.15	14.7
31	Ga	6.00	20.51	72	Hf	7	15
32	Ge	7.88	15.93	73	Ta	7.7	16.2
33	As	9.81	20.2	74	W	7.98	17.7
34	Se	9.75	21.5	75	Re	7.87	16.6
35	Br	11.84	21.6	76	Os	8.7	17
36	Kr	13.996	24.56	77	Ir	9.2	
37	Rb	4.176	27.5	78	Pt	9.0	18.56
38	Sr	5.692	11.027	79	Au	9.22	20.5
39	Y	6.377	12.233	80	Hg	10.434	18.751
40	Zr	6.835	12.916	81	Tl	6.106	20.42
41	Nb	6.881	13.895	82	Pb	7.415	15.028
42	Mo	7.131	15.72	83	Bi	7.287	19.3
43	Tc	7.23	14.87	84	Po	8.43	
44	Ru	7.365	16.597	85	At		
45	Rh	7.461	15.92	86	Rn	10.745	
46	Pd	8.33	19.42	87	Fr		
47	Ag	7.574	21.48	88	Ra	5.277	10.14
48	Cd	8.991	16.904	89	Ac		
49	In	5.785	18.828	90	Th		
50	Sn	7.332	14.63	91	Pa		
51	Sb	8.639	19.0	92	U	4.0	

*Reproduced from Day, M. Clyde Jr. and Joe Selbin, "Theoretical Inorganic Chemistry," Reinhold Publ. Corp., New York, 1962.

TABLE 1i2. OXIDATION-REDUCTION COUPLES IN ACID AND BASIC SOLUTIONS

ACID SOLUTIONS

Couple	E^0 (volts)	Couple	E^0 (volts)
$HN_3 = \frac{3}{2} N_2 + H^+ + e^-$	3.09	$2Ta + 5H_2O = Ta_2O_5$	
$Li = Li^+ + e^-$	3.045	$+ 10H^+ + 10\ e^-$	0.81
$K = K^+ + e^-$	2.925	$Zn = Zn^{2+} + 2e^-$	0.763
$Rb = Rb^+ + e^-$	2.925	$Tl + I^- = TlI + e^-$	0.753
$Cs = Cs^+ + e^-$	2.923	$Cr = Cr^{3+} + 3e^-$	0.74
$Ra = Ra^{2+} + 2e^-$	2.92	$H_2Te = Te + 2H^+ + 2e^-$	0.72
$Ba = Ba^{2+} + 2e^-$	2.90	$Tl + Br^- = TlBr + e^-$	0.658
$Sr = Sr^{2+} + 2e^-$	2.89	$2Nb + 5H_2O = Nb_2O_5$	
$Ca = Ca^{2+} + 2e^-$	2.87	$+ 10H^+ + 10e^-$	0.65
$Na = Na^+ + e^-$	2.714	$U^{3+} = U^{4+} + e^-$	0.61
$La = La^{3+} + 3e^-$	2.52	$AsH_3 = As + 3H^+ + 3e^-$	0.60
$Ce = Ce^{3+} + 3e^-$	2.48	$Tl + Cl^- = TlCl + e^-$	0.557
$Nd = Nd^{3+} + 3e^-$	2.44	$Ga = Ga^{3+} + 3e^-$	0.53
$Sm = Sm^{3+} + 3e^-$	2.41	$SbH_3(g) = Sb + 3H^+$	
$Gd = Gd^{3+} + 3e^-$	2.40	$+ 3e^-$	0.51
$Mg = Mg^{2+} + 2e^-$	2.37	$P + 2H_2O = H_3PO_2 + H^+$	
$Y = Y^{3+} + 3e^-$	2.37	$+ e^-$	0.51
$Am = Am^{3+} + 3e^-$	2.32	$H_3PO_2 + H_2O = H_3PO_3$	
$Lu = Lu^{3+} + 3e^-$	2.25	$+ 2H^+ + 2e^-$	0.50
$H^- = \frac{1}{2} H_2 + e^-$	2.25	$Fe = Fe^{2+} + 2e^-$	0.440
$H(g) = H^+ + e^-$	2.10	$Eu^{2+} = Eu^{3+} + e^-$	0.43
$Sc = Sc^{3+} + 3e^-$	2.08	$Cr^{2+} = Cr^{3+} + e^-$	0.41
$Pu = Pu^{3+} + 3e^-$	2.07	$Cd = Cd^{2+} + 2e^-$	0.403
$Al + 6F^- = AlF_6^{3-} + 3e^-$	2.07	$H_2Se = Se + 2H^+ + 2e^-$	0.40
$Th = Th^{4+} + 4e^-$	1.90	$Ti^{2+} = Ti^{3+} + e^-$	0.37 ca.
$Np = Np^{3+} + 3e^-$	1.86	$Pb + 2I^- = PbI_2 + 2e^-$	0.365
$Be = Be^{2+} + 2e^-$	1.85	$Pb + SO_4^{2-} = PbSO_4$	
$U = U^{3+} + 3e^-$	1.80	$+ 2e^-$	0.356
$Hf = Hf^{4+} + 4e^-$	1.70	$In = In^{3+} + 3e^-$	0.342
$Al = Al^{3+} + 3e^-$	1.66	$Tl = Tl^+ + e^-$	0.3363
$Ti = Ti^{2+} + 2e^-$	1.63	$\frac{1}{2} C_2N_2 + H_2O = HCNO$	
$Zr = Zr^{4+} + 4e^-$	1.53	$+ H^+ + e^-$	0.33
$Si + 6F^- = SiF_6^{2-} + 4e^-$	1.2	$Pt + H_2S = PtS + 2H^+$	
$Ti + 6F^- = TiF_6^{2-} + 4e^-$	1.19	$+ 2e^-$	0.30
$Mn = Mn^{2+} + 2e^-$	1.18	$Pb + 2Br^- = PbBr_2 + 2e^-$	0.280
$V = V^{2+} + 2e^-$	1.18 ca.	$Co = Co^{2+} + 2e^-$	0.277
$Nb = Nb^{3+} + 3e^-$	1.1 ca.	$H_3PO_3 + H_2O = H_3PO_4$	
$Ti + H_2O = TiO^{2+} + 2H^+$		$+ 2H^+ + 2e^-$	0.276
$+ 4e^-$	0.89	$Pb + 2Cl^- = PbCl_2 + 2e^-$	0.268
$B + 3H_2O = H_3BO_3$		$V^{2+} = V^{3+} + e^-$	0.255
$+ 3H^+ + 3e^-$	0.87	$V + 4H_2O = V(OH)_4^+$	
$Si + 2H_2O = SiO_2 + 4H^+$		$+ 4H^+ + 5e^-$	0.253
$+ 4e^-$	0.86	$Sn + 6F^- = SnF_6^{2-} + 4e^-$	0.25

TABLE 1i2. (*Continued*)

ACID SOLUTIONS — (Continued)

Couple	E^0 (volts)	Couple	E^0 (volts)
$Ni = Ni^{2+} + 2e^-$	0.250	$Hg + 4Br^- = HgBr_4^{2-}$	
$N_2H_5^+ = N_2 + 5H^+ + 4e^-$	0.23	$+ 2e^-$	−0.21
$S_2O_6^{2-} + 2H_2O = 2SO_4^{2-}$		$Ag + Cl^- = AgCl + e^-$	−0.222
$+ 4H^+ + 2e^-$	0.22	$(CH_3)_2SO + H_2O$	
$Mo = Mo^{3+} + 3e^-$	0.2 *ca.*	$= (CH_3)_2SO_2$	
$HCOOH(aq) = CO_2$		$+ 2H^+ + 2e^-$	−0.23
$+ 2H^+ + 2e^-$	0.196	$As + 2H_2O = HAsO_2(aq)$	
$Cu + I^- = CuI + e^-$	0.185	$+ 3H^+ + 3e^-$	−0.247
$Ag + I^- = AgI + e^-$	0.151	$Re + 2H_2O = ReO_2$	
$Ge + 2H_2O = GeO_2$		$+ 4H^+ + 4e^-$	−0.252
$+ 4H^+ + 4e^-$	0.15	$Bi + H_2O = BiO^+ + 2H^+$	
$Sn = Sn^{2+} + 2e^-$	0.136	$+ 3e^-$	−0.32
$HO_2 = O_2 + H^+ + e^-$	0.13	$U^{4+} + 2H_2O = UO_2^{2+}$	
$Pb = Pb^{2+} + 2e^-$	0.126	$+ 4H^+ + 2e^-$	−0.334
$W + 3H_2O = WO_3(c) + 6H^+$		$Cu = Cu^{2+} + 2e^-$	−0.337
$+ 6e^-$	0.09	$Ag + IO_3^- = AgIO_3 + e^-$	−0.35
$HS_2O_4^- + 2H_2O = 2H_2SO_3$		$Fe(CN)_6^{4-} = Fe(CN)_6^{3-}$	
$+ H^+ + 2e^-$	0.08	$+ e^-$	−0.36
$Hg + 4I^- = HgI_4^{2-} + 2e^-$	0.04	$V^{3+} + H_2O = VO^{2+}$	
$H_2 = 2H^+ + 2e^-$	0.00	$+ 2H^+ + e^-$	−0.361
$Ag + 2S_2O_3^{2-}$		$Re + 4H_2O = ReO_4^-$	
$= Ag(S_2O_3)_2^{3-} + e^-$	−0.01	$+ 8H^+ + 7e^-$	−0.363
$Cu + Br^- = CuBr + e^-$	−0.033	$HCN(aq) = \frac{1}{2} C_2N_2 + H^+$	
$UO_2^+ = UO_2^{2+} + e^-$	−0.05	$+ e^-$	−0.37
$HCHO(aq) + H_2O$		$S_2O_3^{2-} + 3H_2O = 2H_2SO_3$	
$= HCOOH(aq) + 2H^+$		$+ 2H^+ + 4e^-$	−0.40
$+ 2e^-$	−0.056	$Rh + 6Cl^- = RhCl_6^{3-}$	
$PH_3(g) = P + 3H^+ + 3e^-$	−0.06	$+ 3e^-$	−0.44
$Ag + Br^- = AgBr + e^-$	−0.095	$2Ag + CrO_4^{2-} = Ag_2CrO_4$	
$Ti^{3+} + H_2O = TiO^{2+}$		$+ 2e^-$	−0.446
$+ 2H^+ + e^-$	−0.1	$S + 3H_2O = H_2SO_3 + 4H^+$	
$SiH_4 = Si + 4H^+ + 4e^-$	−0.102	$+ 4e^-$	−0.45
$CH_4 = C + 4H^+ + 4e^-$	−0.13	$Sb_2O_4 + H_2O = Sb_2O_5$	
$Cu + Cl^- = CuCl + e^-$	−0.137	$+ 2H^+ + 2e^-$	−0.48
$H_2S = S + 2H^+ + 2e^-$	−0.141	$2Ag + MoO_4^{2-}$	
$Np^{3+} = Np^{4+} + e^-$	−0.147	$= Ag_2MoO_4 + 2e^-$	−0.49
$Sn^{2+} = Sn^{4+} + 2e^-$	−0.15	$2NH_3OH^+ = H_2N_2O_2$	
$2Sb + 3H_2O = Sb_2O_3$		$+ 6H^+ + 4e^-$	−0.496
$+ 6H^+ + 6e^-$	−0.152	$ReO_2 + 2H_2O = ReO_4^-$	
$Cu^+ = Cu^{2+} + e^-$	−0.153	$+ 4H^+ + 3e^-$	−0.51
$Bi + H_2O + Cl^- = BiOCl$		$S_4O_6^{2-} + 6H_2O = 4H_2SO_3$	
$+ 2H^+ + 3e^-$	−0.16	$+ 4H^+ + 6e^-$	−0.51
$H_2SO_3 + H_2O = SO_4^{2-}$		$C_2H_6 = C_2H_4 + 2H^+$	
$+ 4H^+ + 2e^-$	−0.17	$+ 2e^-$	−0.52
$CH_3OH(aq) = HCHO(aq)$		$Cu = Cu^+ + e^-$	−0.521
$+ 2H^+ + 2e^-$	−0.19		

TABLE 1i2. (*Continued*)

ACID SOLUTIONS — (Continued)

Couple	E^0 (volts)	Couple	E^0 (volts)
$Te + 2H_2O = TeO_2(c)$ $+ 4H^+ + 4e^-$	-0.529	$Pt + 4Cl^- = PtCl_4{}^{2-} + 2e^-$	-0.73
$2I^- = I_2 + 2e^-$	-0.5355	$C_2H_4 = C_2H_2 + 2H^+ + 2e^-$	-0.73
$3I^- = I_3{}^- + 2e^-$	-0.536	$Se + 3H_2O = H_2SeO_3$	
$CuCl = Cu^{2+} + Cl^- + e^-$	-0.538	$+ 4H^+ + 4e^-$	-0.74
$Ag + BrO_3{}^- = AgBrO_3$ $+ e^-$	-0.55	$Np^{4+} + 2H_2O = NpO_2{}^+$ $+ 4H^+ + e^-$	-0.75
$Te + 2H_2O = TeOOH^+$ $+ 3H^+ + 4e^-$	-0.559	$2CNS^- = (CNS)_2 + 2e^-$	-0.77
$HAsO_2 + 2H_2O = H_3AsO_4$		$Ir + 6Cl^- = IrCl_6{}^{3-} + 3e^-$	-0.77
$+ 2H^+ + 2e^-$	-0.559	$Fe^{2+} = Fe^{3+} + e^-$	-0.771
$Ag + NO_2{}^- = AgNO_2 + e^-$	-0.564	$2Hg = Hg_2{}^{2+} + 2e^-$	-0.789
$MnO_4{}^{2-} = MnO_4{}^- + e^-$	-0.564	$Ag = Ag^+ + e^-$	-0.7991
$2H_2SO_3 = S_2O_6{}^{2-} + 4H^+$ $+ 2e^-$	-0.57	$N_2O_4 + 2H_2O = 2NO_3{}^-$ $+ 4H^+ + 2e^-$	-0.80
$Pt + 4Br^- = PtBr_4{}^{2-}$ $+ 2e^-$	-0.58	$Rh = Rh^{3+} + 3e^-$	-0.8 *ca.*
$2SbO^+ + 3H_2O = Sb_2O_5$ $+ 6H^+ + 4e^-$	-0.581	$Os + 4H_2O = OsO_4(c) +$ $8H^+ + 8e^-$	-0.85
$CH_4 + H_2O = CH_3OH(aq)$ $+ 2H^+ + 2e^-$	-0.586	$H_2N_2O_2 + 2H_2O = 2HNO_2$ $+ 4H^+ + 4e^-$	-0.86
$Pd + 4Br^- = PdBr_4{}^{2-}$ $+ 2e^-$	-0.6	$CuI = Cu^{2+} + I^- + e^-$	-0.86
$Ru + 5Cl^- = RuCl_5{}^{2-}$ $+ 3e^-$	-0.60	$Au + 4Br^- = AuBr_4{}^-$ $+ 3e^-$	-0.87
$U^{4+} + 2H_2O = UO_2{}^+$ $+ 4H^+ + e^-$	-0.62	$Hg_2{}^{2+} = 2Hg^{2+} + 2e^-$	-0.920
$Pd + 4Cl^- = PdCl_4{}^{2-}$ $+ 2e^-$	-0.62	$PuO_2{}^+ = PuO_2{}^{2+} + e^-$	-0.93
$CuBr = Cu^{2+} + Br^- + e^-$	-0.640	$HNO_2 + H_2O = NO_3{}^-$ $+ 3H^+ + 2e^-$	-0.94
$Ag + C_2H_3O_2{}^-$ $= AgC_2H_3O_2 + e^-$	-0.643	$NO + 2H_2O = NO_3{}^-$ $+ 4H^+ + 4e^-$	-0.96
$2Ag + SO_4{}^{2-} = Ag_2SO_4$ $+ 2e^-$	-0.653	$Au + 2Br^- = AuBr_2{}^- + e^-$	-0.96
$Au + 4CNS^-$ $= Au(CNS)_4{}^- + 3e^-$	-0.66	$Pu^{3+} + Pu^{4+} + e^-$	-0.97
$PtCl_4{}^{2-} + 2Cl^- = PtCl_6{}^{2-}$ $+ 2e^-$	-0.68	$Pt + 2H_2O = Pt(OH)_2$ $+ 2H^+ + 2e^-$	-0.98
$H_2O_2 = O_2 + 2H^+ + 2e^-$	-0.682	$Pd = Pd^{2+} + 2e^-$	-0.987
$2NH_4{}^+ = HN_3 + 11H^+$ $+ 8e^-$	-0.69	$IrBr_6{}^{4-} = IrBr_6{}^{3-} + e^-$	-0.99
$H_2Te = Te + 2H^+ + 2e^-$	-0.70	$NO + H_2O = HNO_2 + H^+$ $+ e^-$	-1.00
$H_2N_2O_2 = 2NO + 2H^+$ $+ 2e^-$	-0.71	$Au + 4Cl^- = AuCl_4{}^-$ $+ 3e^-$	-1.00
$OH + H_2O = H_2O_2 + H^+$ $+ e^-$	-0.72	$VO^{2+} + 3H_2O = V(OH)_4{}^+$ $+ 2H^+ + e^-$	-1.00
		$IrCl_6{}^{3-} = IrCl_6{}^{2-} + e^-$	-1.017
		$TeO_2 + 4H_2O = H_6TeO_6 (c)$ $+ 2H^+ + 2e^-$	-1.02
		$2NO + 2H_2O = N_2O_4$ $+ 4H^+ + 4e^-$	-1.03
		$Pu^{4+} + 2H_2O = PuO_2{}^{2+}$ $+ 4H^+ + 2e^-$	-1.04

TABLE 1i2. (*Continued*)

ACID SOLUTIONS — (Continued)

Couple	E^0 (volts)	Couple	E^0 (volts)
$2Cl^- + \frac{1}{2} I_2 = ICl_2^- + e^-$	-1.06	$Au = Au^{3+} + 3e^-$	-1.50
$2Br^- = Br_2(1) + 2e^-$	-1.0652	$H_2O_2 = HO_2 + H^+ + e^-$	-1.5
$2HNO_2 = N_2O_4 + 2H^+$		$Mn^{2+} = Mn^{3+} + e^-$	-1.51
$\quad + 2e^-$	-1.07	$Mn^{2+} + 4H_2O = MnO_4^-$	
$Cu(CN)_2^- = Cu^{2+}$		$\quad + 8H^+ + 5e^-$	-1.51
$\quad + 2CN^- + e^-$	-1.12	$\frac{1}{2} Br_2 + 3H_2O = BrO_3^-$	
$Pu^{4+} + 2H_2O = PuO_2^+$		$\quad + 6H^+ + 5e^-$	-1.52
$\quad + 4H^+ + e^-$	-1.15	$\frac{1}{2} Br_2 + H_2O = HBrO$	
$H_2SeO_3 + H_2O = SeO_4^{2-}$		$\quad + H^+ + e^-$	-1.59
$\quad + 4H^+ + 2e^-$	-1.15	$2BiO^+ + 2H_2O = Bi_2O_4$	
$NpO_2^+ = NpO_2^{2+} + e^-$	-1.15	$\quad + 4H^+ + 2e^-$	-1.59
$4Cl^- + C + 4H^+ = CCl_4$		$IO_3^- + 3H_2O = H_5IO_6$	
$\quad + 4H^+ + 4e^-$	-1.18	$\quad + H^+ + 2e^-$	-1.6
$ClO_3^- + H_2O = ClO_4^-$		$Bk^{3+} = Bk^{4+} + e^-$	-1.6
$\quad + 2H^+ + 2e^-$	-1.19	$Ce^{3+} = Ce^{4+} + e^-$	-1.61
$\frac{1}{2} I_2 + 3H_2O = IO_3^-$		$\frac{1}{2} Cl_2 + H_2O = HClO$	
$\quad + 6H^+ + 5e^-$	-1.195	$\quad + H^+ + e^-$	-1.63
$HClO_2 + H_2O = ClO_3^-$		$AmO_2^+ = AmO_2^{2+} + e^-$	-1.64
$\quad + 3H^+ + 2e^-$	-1.21	$HClO + H_2O = HClO_2$	
$2H_2O = O_2 + 4H^+ + 4e^-$	-1.229	$\quad + 2H^+ + 2e^-$	-1.64
$2S + 2Cl^- = S_2Cl_2 + 2e^-$	-1.23	$Au = Au^+ + e^-$	$-1.68\ ca.$
$Mn^{2+} + 2H_2O = MnO_2$		$Ni^{2+} + 2H_2O = NiO_2$	
$\quad + 4H^+ + 2e^-$	-1.23	$\quad + 4H^+ + 2e^-$	-1.68
$Tl^+ = Tl^{3+} + 2e^-$	-1.25	$PbSO_4 + 2H_2O = PbO_2$	
$Am^{4+} + 2H_2O = AmO_2^+$		$\quad + SO_4^{2-} + 4H^+ + 2e^-$	-1.685
$\quad + 4H^+ + e^-$	-1.26	$Am^{3+} + 2H_2O = AmO_2^{2+}$	
$2NH_4^+ = N_2H_5^+ + 3H^+$		$\quad + 4H^+ + 3e^-$	-1.69
$\quad + 2e^-$	-1.275	$MnO_2 + 2H_2O = MnO_4^-$	
$HClO_2 = ClO_2 + H^+ + e^-$	-1.275	$\quad + 4H^+ + 3e^-$	-1.695
$PdCl_4^{2-} + 2Cl^- = PdCl_6^{2-}$		$Am^{3+} + 2H_2O = AmO_2^+$	
$\quad + 2e^-$	-1.288	$\quad + 4H^+ + 2e^-$	-1.725
$N_2O + 3H_2O = 2HNO_2$		$2H_2O = H_2O_2 + 2H^+$	
$\quad + 4H^+ + 4e^-$	-1.29	$\quad + 2e^-$	-1.77
$2Cr^{3+} + 7H_2O = Cr_2O_7^{2-}$		$Co^{2+} = Co^{3+} + e^-$	-1.82
$\quad + 14H^+ + 6e^-$	-1.33	$Fe^{3+} + 4H_2O = FeO_4^{2-}$	
$NH_4^+ + H_2O = NH_3OH^+$		$\quad + 8H^+ + 3e^-$	-1.9
$\quad + 2H^+ + 2e^-$	-1.35	$NH_4^+ + N_2 = HN_3 + 3H^+$	
$2Cl^- = Cl_2 + 2e^-$	-1.3595	$\quad + 2e^-$	-1.96
$N_2H_5^+ + 2H_2O$		$Ag^+ = Ag^{2+} + e^-$	-1.98
$\quad = 2NH_3OH^+ + H^+ + 2e^-$	-1.42	$2SO_4^{2-} = S_2O_8^{2-} + 2e^-$	-2.01
$Au + 3H_2O = Au(OH)_3$		$O_2 + H_2O = O_3 + 2H^+$	
$\quad + 3H^+ + 3e^-$	-1.45	$\quad + 2e^-$	-2.07
$\frac{1}{2} I_2 + H_2O = HIO + H^+$		$H_2O + 2F^- = F_2O + 2H^+$	
$\quad + e^-$	-1.45	$\quad + 4e^-$	-2.1
$Pb^{2+} + 2H_2O = PbO_2$		$Am^{3+} = Am^{4+} + e^-$	-2.18
$\quad + 4H^+ + 2e^-$	-1.455	$H_2O = O(g) + 2H^+ + 2e^-$	-2.42

TABLE 1i2. (*Continued*)

ACID SOLUTIONS — (Continued)

Couple	E^0 (volts)	Couple	E^0 (volts)
$H_2O = OH + H^+ + e^-$	-2.8	$2F^- = F_2 + 2e^-$	-2.87
$N_2 + 2H_2O = H_2N_2O_2$		$2HF(aq) = F_2 + 2H^+$	
$\quad + 2H^+ + 2e^-$	-2.85	$\quad + 2e^-$	-3.06

BASIC SOLUTIONS

Couple	E^0 (volts)	Couple	E^0 (volts)
$Ca + 2OH^- = Ca(OH)_2$		$= Na_2UO_4 + 4H_2O + 2e^-$	1.61
$\quad + 2e^-$	3.03	$H_2PO_2^- + 3OH^- = HPO_3^{2-}$	
$Sr + 2OH^- + 8H_2O$		$\quad + 2H_2O + 2e^-$	1.57
$\quad = Sr(OH)_2 \cdot 8H_2O + 2e^-$	2.99	$Mn + 2OH^- = Mn(OH)_2$	
$Ba + 2OH^- + 8H_2O$		$\quad + 2e^-$	1.55
$\quad = Ba(OH)_2 \cdot 8H_2O + 2e^-$	2.97	$Mn + CO_3^{2-} = MnCO_3$	
$H(g) + OH^- = H_2O + e^-$	2.93	$\quad + 2e^-$	1.48
$La + 3OH^- = La(OH)_3$		$Zn + S^{2-} = ZnS + 2e^-$	1.44
$\quad + 3e^-$	2.90	$Cr + 3OH^- = Cr(OH)_3$	
$Lu + 3OH^- = Lu(OH)_3$		$\quad + 3e^-$	1.3
$\quad + 3e^-$	2.72	$Zn + 4CN^- = Zn(CN)_4^{2-}$	
$Mg + 2OH^- = Mg(OH)_2$		$\quad + 2e^-$	1.26
$\quad + 2e^-$	2.69	$Zn + 2OH^- = Zn(OH)_2$	
$2Be + 6OH^- = Be_2O_3^{2-}$		$\quad + 2e^-$	1.245
$\quad + 3H_2O + 4e^-$	2.62	$Ga + 4OH^- = H_2GaO_3^-$	
$Sc + 3OH^- = Sc(OH)_3$		$\quad + H_2O + 3e^-$	1.22
$\quad + 3e^-$	2.6 *ca.*	$Zn + 4OH^- = ZnO_2^{2-}$	
$Hf + 4OH^- = HfO(OH)_2$		$\quad + 2H_2O + 2e^-$	1.216
$\quad + H_2O + 4e^-$	2.50	$Cr + 4OH^- = CrO_2^-$	
$Th + 4OH^- = Th(OH)_4$		$\quad + 2H_2O + 3e^-$	1.2
$\quad + 4e^-$	2.48	$Cd + S^{2-} = CdS + 2e^-$	1.21
$Pu + 3OH^- = Pu(OH)_3$		$6V + 33OH^- = 16H_2O$	
$\quad + 3e^-$	2.42	$\quad + HV_6O_{17}^{3-} + 30e^-$	1.15
$U + 4OH^- = UO_2 + 2H_2O$		$Te^{2-} = Te + 2e^-$	1.14
$\quad + 4e^-$	2.39	$HPO_3^{2-} + 3OH^- = PO_4^{3-}$	
$Zr + 4OH^- = H_2ZrO_3$		$\quad + 2H_2O + 2e^-$	1.12
$\quad + H_2O + 4e^-$	2.36	$S_2O_4^{2-} + 4OH^- = 2SO_3^{2-}$	
$Al + 4OH^- = H_2AlO_3^-$		$\quad + 2H_2O + 2e^-$	1.12
$\quad + H_2O + 3e^-$	2.35	$Zn + CO_3^{2-} = ZnCO_3$	
$U(OH)_3 + OH^- = U(OH)_4$		$\quad + 2e^-$	1.06
$\quad + e^-$	2.2	$W + 8OH^- = WO_4^{2-}$	
$U + 3OH^- = U(OH)_3$		$\quad + 4H_2O + 6e^-$	1.05
$\quad + 3e^-$	2.17	$Mo + 8OH^- = MoO_4^{2-}$	
$P + 2OH^- = H_2PO_2^- + e^-$	2.05	$\quad 4H_2O + 6e^-$	1.05
$B + 4OH^- = H_2BO_3^-$		$Cd + 4CN^- = Cd(CN)_4^{2-}$	
$\quad + 3e^-$	1.79	$\quad + 2e^-$	1.03
$Si + 6OH^- = SiO_3^{2-}$		$Zn + 4NH_3 = Zn(NH_3)_4^{2+}$	
$\quad + 3H_2O + 4e^-$	1.70	$\quad + 2e^-$	1.03
$U(OH)_4 = 2Na^+ + 4OH^-$		$Fe + S^{2-} = FeS_{(\alpha +)} 2e^-$	1.01

TABLE 1i2. (*Continued*)

BASIC SOLUTIONS — (Continued)

Couple	E⁰ (volts)	Couple	E⁰ (volts)
$In + 3OH^- = In(OH)_3$ $+ 3e^-$	1.0	$Cd + 4NH_3 = Cd(NH_3)_4{}^{2+}$ $+ 2e^-$	0.597
$CN^- + 2OH^- = CNO^-$ $+ H_2O + 2e^-$	0.97	$ReO_2 + 4OH^- = ReO_4{}^-$ $+ 2H_2O + 3e^-$	0.594
$2Tl + S^{2-} = Tl_2S + 2e^-$	0.96	$Re + 8OH^- = ReO_4{}^-$	
$Pb + S^{2-} = PbS + 2e^-$	0.95	$+ 4H_2O + 7e^-$	0.584
$Pu(OH)_3 + OH^-$ $= Pu(OH)_4 + e^-$	0.95	$S_2O_3{}^{2-} + 6OH^- = 2SO_3{}^{2-}$ $+ 3H_2O + 4e^-$	0.58
$Sn + S^{2-} = SnS + 2e^-$	0.94	$Re + 4OH^- = ReO_2$	
$SO_3{}^{2-} + 2OH^- = SO_4{}^{2-}$ $+ H_2O + 2e^-$	0.93	$+ 2H_2O + 4e^-$	0.576
$Se^{2-} = Se + 2e^-$	0.92	$Te + 6OH^- = TeO_3{}^{2-}$ $+ 3H_2O + 4e^-$	0.57
$Sn + 3OH^- = HSnO_2{}^-$ $+ H_2O + 2e^-$	0.91	$Fe(OH)_2 + OH^-$ $= Fe(OH)_3 + e^-$	0.56
$Ge + 5OH^- = HGeO_3{}^-$ $+ 2H_2O + 4e^-$	0.9	$O_2{}^- = O_2 + e^-$	0.56
		$2Cu + S^{2-} = Cu_2S + 2e^-$	0.54
$HSnO_2{}^- + H_2O + 3OH^-$ $= Sn(OH)_6{}^{2-} + 2e^-$	0.90	$Pb + 3OH^- = HPbO_2{}^-$ $+ H_2O + 2e^-$	0.54
$PH_3 + 3OH^- = P + 3H_2O$ $+ 3e^-$	0.89	$Pb + CO_3{}^{2-} = PbCO_3$ $+ 2e^-$	0.506
$Fe + 2OH^- = Fe(OH)_2$ $+ 2e^-$	0.877	$S^{2-} = S + 2e^-$	0.48
$Ni + S^{2-} = NiS_{(\alpha)} + 2e^-$	0.83	$Ni + 6NH_3(aq)$ $= Ni(NH_3)_6{}^{2+} + 2e^-$	0.47
$H_2 + 2OH^- = 2H_2O + 2e^-$	0.828	$Ni + CO_3{}^{2-} = NiCO_3 + 2e^-$	0.45
$Cd + 2OH^- = Cd(OH)_2$ $+ 2e^-$	0.809	$2Bi + 6OH^- = Bi_2O_3$ $+ 3H_2O + 6e^-$	0.44
$Fe + CO_3{}^{2-} = FeCO_3$ $+ 2e^-$	0.756	$Cu + 2CN^- = Cu(CN)_2{}^-$ $+ e^-$	0.43
$Cd + CO_3{}^{2-} = CdCO_3$ $+ 2e^-$	0.74	$Hg + 4CN^- = Hg(CN)_4{}^{2-}$ $+ 2e^-$	0.37
$Co + 2OH^- = Co(OH)_2$ $+ 2e^-$	0.73	$Se + 6OH^- = SeO_3{}^{2-}$ $+ 3H_2O + 4e^-$	0.366
$Hg + S^{2-} = HgS + 2e^-$	0.72	$2Cu + 2OH^- = Cu_2O$ $+ H_2O + 2e^-$	0.358
$Ni + 2OH^- = Ni(OH)_2$ $+ 2e^-$	0.72	$Tl + OH^- = Tl(OH) + e^-$	0.3445
$2Ag + S^{2-} = Ag_2S + 2e^-$	0.69	$Ag + 2CN^- = Ag(CN)_2{}^-$ $+ e^-$	0.31
$As + 4OH^- = AsO_2{}^-$ $+ 2H_2O + 3e^-$	0.68	$Cu + CNS^- = Cu(CNS)$ $+ e^-$	0.27
$AsO_2{}^- + 4OH^- = AsO_4{}^{3-}$ $+ 2H_2O + 2e^-$	0.67	$OH + 2OH^- = HO_2{}^-$ $+ H_2O + e^-$	0.24
$2FeS + S^{2-} = Fe_2S_3 + 2e^-$	0.67	$Cr(OH)_3 + 5OH^-$	
$Sb + 4OH^- = SbO_2{}^-$ $+ 2H_2O + 3e^-$	0.66	$= CrO_4{}^{2-} + 4H_2O + 3e^-$	0.13
$Co + CO_3{}^{2-} = CoCO_3$ $+ 2e^-$	0.64	$Cu + 2NH_3 = Cu(NH_3)_2{}^+$ $+ e^-$	0.12

TABLE 1i2. (*Continued*)

BASIC SOLUTIONS — (Continued)

Couple	E^0 (volts)	Couple	E^0 (volts)
$Cu_2O + 2OH^- + H_2O$ $= 2Cu(OH)_2 + 2e^-$	0.080	$2Ag + 2OH^- = Ag_2O$ $+ H_2O + 2e^-$	−0.344
$HO_2^- + OH^- = O_2 + H_2O$ $+ 2e^-$	0.076	$ClO_3^- + 2OH^- = ClO_4^-$ $+ H_2O + 2e^-$	−0.36
$TlOH + 2OH^- = Tl(OH)_3$ $+ 2e^-$	0.05	$Ag + 2NH_3 = Ag(NH_3)_2^+$ $+ e^-$	−0.373
$Ag + CN^- = AgCN + e^-$	0.017	$TeO_3^{2-} + 2OH^- = TeO_4^{2-}$ $+ H_2O + 2e^-$	−0.4
$Mn(OH)_2 + 2OH^-$ $= MnO_2 + 2H_2O + 2e^-$	0.05	$OH^- + HO_2^- = O_2^-$ $+ H_2O + e^-$	−0.4
$NO_2^- + 2OH^- = NO_3^-$ $+ H_2O + 2e^-$	−0.01	$4OH^- = O_2 + 2H_2O + 4e^-$	−0.401
$Os + 9OH^- = HOsO_5^-$ $+ 4H_2O + 8e^-$	−0.02	$2Ag + CO_3^{2-} = Ag_2CO_3$ $+ 2e^-$	−0.47
$2Rh + 6OH^- = Rh_2O_3$ $+ 3H_2O + 6e^-$	−0.04	$Ni(OH)_2 + 2OH^- = NiO_2$ $+ 2H_2O + 2e^-$	−0.49
$SeO_3^{2-} + 2OH^- = SeO_4^{2-}$ $+ H_2O + 2e^-$	−0.05	$I^- + 2OH^- = IO^- + H_2O$ $+ 2e^-$	−0.49
$Pd + 2OH^- = Pd(OH)_2$ $+ 2e^-$	−0.07	$Ag_2O + 2OH^- = 2AgO$ $+ H_2O + 2e^-$	−0.57
$2S_2O_3^{2-} = S_4O_6^{2-} + 2e^-$	−0.08	$MnO_2 + 4OH^- = MnO_4^{2-}$ $+ 2H_2O + 2e^-$	−0.60
$Hg + 2OH^- = HgO(r)$ $+ H_2O + 2e^-$	−0.098	$RuO_4^{2-} = RuO_4^- + e^-$	−0.60
$2NH_4OH + 2OH^- = N_2H_4$ $+ 4H_2O + 2e^-$	−0.1	$Br^- + 6OH^- = BrO_3^-$ $+ 3H_2O + 6e^-$	−0.61
$2Ir + 6OH^- = Ir_2O_3$ $+ 3H_2O + 6e^-$	−0.1	$ClO^- + 2OH^- = ClO_2^-$ $+ H_2O + 2e^-$	−0.66
$Co(NH_3)_6^{2+} = Co(NH_3)_6^{3+}$ $+ e^-$	−0.1	$IO_3^- + 3OH^- = H_3IO_6^{2-}$ $+ 2e^-$	−0.7
$Mn(OH)_2 + OH^-$ $= Mn(OH)_3 + e^-$	−0.1	$N_2H_4 + 2OH^- = 2NH_2OH$ $+ 2e^-$	−0.73
$Pt + 2OH^- = Pt(OH)_2$ $+ 2e^-$	−0.15	$2AgO + 2OH^- = Ag_2O_3$ $+ H_2O + 2e^-$	−0.74
$Co(OH)_2 + OH^-$ $= Co(OH)_3 + e^-$	−0.17	$Br^- + 2OH^- = BrO^-$ $+ H_2O + 2e^-$	−0.76
$PbO(r) + 2OH^- = PbO_2$ $+ H_2O + 2e^-$	−0.248	$3OH^- = HO_2^- + H_2O$ $+ 2e^-$	−0.88
$I^- + 6OH^- = IO_3^-$ $+ 3H_2O + 6e^-$	−0.26	$Cl^- + 2OH^- = ClO^-$ $+ H_2O + 2e^-$	−0.89
$PuO_2OH + OH^-$ $= PuO_2(OH)_2 + 2e^-$	−0.26	$FeO_2^- + 4OH^- = FeO_4^{2-}$ $+ 2H_2O + 3e^-$	−0.9
$Ag + 2SO_3^{2-} = Ag(SO_3)_2^{3-}$ $+ e^-$	−0.30	$ClO_2^- = ClO_2 + e^-$	−1.16
$ClO_2^- + 2OH^- = ClO_3^-$ $+ H_2O + 2e^-$	−0.33	$O_2 + 2OH^- = O_3 + H_2O$ $+ 2e^-$	−1.24
		$OH^- = OH + e^-$	−2.0

* The values reported in this table were taken directly from Latimer, "The Oxidation States of the Elements and Their Potentials in Aqueous Solutions," 2nd ed., Prentice-Hall, Inc., New York, N.Y., 1952.

TABLE 1i3. SELECTED DIPOLE MOMENTS*

(a) Selected moments (compounds not containing carbon)

Formula	Compound name	Selected moment (debyes)		Formula	Compound name	Selected moment (debyes)	
AgCl	Silver chloride	5.73	C	ClNa	Sodium chloride	9.00	A
AlCl	Aluminum chloride	x	Q				
AlF	Aluminum fluoride	1.53	D	ClNO	Nitrosyl chloride	(1.9)	Q
AsCl$_3$	Arsenic trichloride	1.59	C				
AsF$_3$	Arsenic trifluoride	2.59	B	ClNO$_2$	Nitryl chloride	0.53	A
AsH$_3$	Arsine	0.20	C	ClTl	Thallium chloride	4.44	B
As$_2$O$_3$	Diarsenic trioxide	x	Q				
BCl$_3$	Boron trichloride	0	S	Cl$_2$F$_3$P	Dichlorotrifluorophosphorus	0.68	C
BF$_3$	Boron trifluoride	0	S	Cl$_2$H$_2$Si	Dichlorosilane	1.17	B
				Cl$_2$Hg	Mercury dichloride	0	S
B$_2$H$_6$	Diborane	0	S	Cl$_2$OS	Thionyl chloride	1.45	B
B$_2$H$_6$N$_3$	Triborotriazine (Borazine)	0	S	Cl$_2$O$_2$S	Sulfuryl chloride	1.81	B
B$_5$H$_9$	Pentaborane	2.13	B	Cl$_3$F$_2$P	Trichlorodifluorophosphorus	0	S
BaO	Barium oxide	7.95	A	Cl$_3$HSi	Trichlorosilane	0.86	B
BrCl	Bromine chloride	(0.6)	Q	Cl$_3$P	Phosphorus trichloride	0.78	C
BrF	Bromine fluoride	(1.3)	Q	Cl$_4$FP	Tetrachlorofluorophosphorus	0.21	B
BrF$_3$	Bromine trifluoride	(1.1)	Q	Cl$_4$Ge	Germanium tetrachloride	0	S
				Cl$_4$Si	Silicon tetrachloride	0	S
				Cl$_4$Sn	Tin tetrachloride	0	S
BrF$_5$	Bromine pentafluoride	1.51	D	Cl$_4$Ti	Titanium tetrachloride	0	S
BrH	Hydrogen bromide	0.82	B	CsF	Cesium fluoride	7.88	A
BrH$_3$Si	Bromosilane	1.33	B	CsI	Cesium iodide	x	Q
BrK	Potassium bromide	10.41	B	FH	Hydrogen fluoride	1.82	A
BrLi	Lithium bromide	7.27	A				
BrNO	Nitrosyl bromide	(1.8)	Q				
BrRb	Rubidium bromide	x	Q				
Br$_2$Hg	Mercury dibromide	0	S	FH$_3$Si	Fluorosilane	1.27	B
Br$_4$Sn	Tin tetrabromide	0	S	FH$_3$Si$_2$	Fluorodisilane	1.26	A
ClCs	Cesium chloride	10.42	A	FK	Potassium fluoride	8.60	A
				FLi	Lithium fluoride	6.33	A
ClF	Chlorine fluoride	0.88	C				
ClFO$_3$	Perchloryl fluoride	0.023	A	FMnO$_3$	Permanganyl fluoride	(1.5)	Q
				FNO	Nitrosyl fluoride	1.81	B
				FNO$_2$	Nitryl fluoride	(0.47)	Q
ClF$_3$	Chlorine trifluoride	0.6	D	FNa	Sodium fluoride	8.16	A
ClGeH$_3$	Chlorogermane	2.13	A	FO$_3$Re	Perrhenyl fluoride	0.85	D
				FRb	Rubidium fluoride	8.55	A
ClH	Hydrogen chloride	1.08	B	FTl	Thallium fluoride	4.23	A
				F$_2$HN	Difluoramine	1.92	A
				F$_2$H$_2$Si	Difluorosilane	1.55	A
ClH$_3$Si	Chlorosilane	1.31	A	F$_2$N$_2$	cis-Difluorodiazine	0.16	A
				F$_2$O	Oxygen difluoride	0.297	A
ClI	Iodine chloride	(0.6)	Q	F$_2$OS	Thionyl fluoride	1.63	A
				F$_2$O$_2$	Dioxygen difluoride	1.44	C
				F$_2$O$_2$S	Sulfuryl fluoride	1.12	B
ClK	Potassium chloride	10.27	A	F$_2$S$_2$	Sulfur monofluoride (S=SF$_2$ isomer)	1.03	C
				F$_2$S$_2$	Sulfur monofluoride (FSSF isomer)	1.45	B
				F$_2$Si	Silicon difluoride	1.23	B
				F$_3$HSi	Trifluorosilane	1.27	B
ClLi	Lithium chloride	7.13	A	F$_3$N	Nitrogen trifluoride	0.235	A
				F$_3$NS	Nitridotrifluorosulfur	1.91	B
				F$_3$OP	Phosphoryl fluoride	1.76	B
				F$_3$P	Phosphorus trifluoride	1.03	A

*Reproduced from: Nelson, R. D., Jr., D. R. Lide, Jr. and A. A. Maryott," Selected Values of Electrical Dipole Moments for Molecules in the Gas Phase," National Bureau of Standards Report No. NSRDS-NBS 10, Superintendent of Documents, Washington, D. C. (1967).

TABLE 1i3. *(Continued)*

Formula	Compound name	Selected moment (debyes)	
F_3PS	Thiophosphoryl fluoride..................	0.64	B
F_4N_2	Tetrafluorohydrazine, *gauche* conformation.	0.26	B
F_4S	Sulfur tetrafluoride........................	0.632	A
F_4Si	Silicon tetrafluoride.......................	0	S
F_5P	Phosphorus pentafluoride.................	0	S
F_5I	Iodine pentafluoride.......................	2.18	C
F_6S	Sulfur hexafluoride........................	0	S
F_6Se	Selenium hexafluoride.....................	0	S
F_6Te	Tellurium hexafluoride....................	0	S
F_6U	Uranium hexafluoride......................	0	S
$F_{10}S_2$	Disulfur decafluoride......................	x	Q
HI	Hydrogen iodide............................	0.44	B
HLi	Lithium hydride............................	5.88	A
HN	Imidyl radical..............................		
HNO_3	Nitric acid..................................	2.17	A
HN_3	Hydrogen azide (hydrazoic acid)........	(0.8)	Q
HO	Hydroxyl radical...........................	1.66	A
$H_2N_2O_2$	Nitroamine (nitramide)...................	(3.6)	Q
H_2O	Water.......................................	1.85	A
H_2O_2	Hydrogen peroxide.........................	2.2	D
H_2S	Hydrogen sulfide...........................	0.97	A
H_2Se	Hydrogen selenide.........................	x	Q
H_3N	Ammonia....................................	1.47	A
H_3P	Phosphine...................................	0.58	A
H_3Sb	Stibine......................................	0.12	C
H_4N_2	Hydrazine...................................	1.75	C
H_4Si	Silane.......................................	0	S
H_6OSi_2	Disilyl ether (disiloxane).................	0.24	B
H_6Si_2	Disilane....................................	0	S
H_9NSi_3	Trisilylamine...............................	⩽0.1	D
HgI_2	Mercury diiodide...........................	0	S
IK	Potassium iodide...........................	x	Q
ILi	Lithium iodide.............................	7.43	A
INa	Sodium iodide..............................	x	Q
ITl	Thallium iodide............................	x	Q
I_4Sn	Tin tetraiodide.............................	0	S
NO	Nitrogen monoxide (nitric oxide)......	0.153	A

Formula	Compound name	Selected moment (debyes)	
NO_2	Nitrogen dioxide............................	0.316	A
N_2O	Dinitrogen oxide (nitrous oxide)........	0.167	A
OS	Sulfur monoxide............................	1.55	A
OS_2	Disulfur monoxide.........................	1.47	B
OSr	Strontium oxide............................	8.90	A
O_2S	Sulfur dioxide..............................	1.63	A
O_3	Ozone.......................................	0.53	B
O_3S	Sulfur trioxide.............................	0	S
O_4Os	Osmium tetroxide..........................	0	S

(*b*) Selected dipole moments (compounds containing carbon)

Formula	Compound name	Selected moment (debyes)	
$CBrF_3$	Bromotrifluoromethane....................	0.65	C
CBr_2F_2	Dibromodifluoromethane..................	0.66	C
$CClF_3$	Chlorotrifluoromethane....................	0.50	A
CClN	Cyanogen chloride..........................	2.82	B
CCl_2F_2	Dichlorodifluoromethane..................	0.51	C
CCl_2O	Carbonyl chloride (phosgene)............	1.17	A
CCl_2S	Thiocarbonyl chloride......................	0.29	C
CCl_3F	Trichlorofluoromethane...................	0.45	C
CCl_3NO_2	Trichloronitromethane....................	1.89	C
CCl_4	Carbon tetrachloride.......................	0	S
CFN	Cyanogen fluoride..........................	2.17	C
CF_2	Carbon difluoride...........................	0.46	B
CF_2O	Carbonyl fluoride..........................	0.95	A
CF_3I	Iodotrifluoromethane......................	0.92	C
CF_3NO	Trifluornitrosomethane....................	(0.3)	Q
CF_3NO_2	Trifluoronitromethane.....................	1.44	C
CF_4	Carbon tetrafluoride.......................	0	S
CN_4O_8	Tetranitromethane..........................	0	S
CO	Carbon monoxide...........................	0.112	A
COS	Carbonyl sulfide............................	0.712	A

TABLE 1i3. (*Continued*)

Formula	Compound name	Selected moment (debyes)		Formula	Compound name	Selected moment (debyes)	
COSe	Carbonyl selenide	0.73	B	CH$_3$F$_3$Ge	Methyl trifluorogermane	x	Q
CO$_2$	Carbon dioxide	0	S	CH$_3$I	Iodomethane	1.62	B
CS	Carbon monosulfide	1.98	A				
CSTe	Thiocarbonyl telluride	0.17	A				
CS$_2$	Carbon disulfide	0	S	CH$_3$NO	Hydroxyliminomethane (formaldoxime).	0.44	A
CHBrF$_2$	Bromodifluoromethane	1.50	D	CH$_3$NO	Formyl amide (formamide)	3.73	B
CHBr$_3$	Tribromomethane	0.99	B				
CHClF$_2$	Chlorodifluoromethane	1.42	B	CH$_3$NOS	Methyl sulfinylamine	1.70	B
				CH$_3$NO$_2$	Nitritomethane (methyl nitrite)		i
CHCl$_2$F	Dichlorofluoromethane	1.29	B	CH$_3$NO$_2$	Nitromethane	3.46	A
CHCl$_3$	Trichloromethane (chloroform)	1.01	B				
CHFO	Formyl fluoride	2.02	A	CH$_3$NO$_3$	Methyl nitrate	3.12	B
CHF$_3$	Trifluoromethane	1.65	A	CH$_3$N$_3$	Methyl azide	2.17	B
				CH$_4$	Methane	0	S
				CH$_4$F$_2$Si	Methyl difluorosilane	2.11	A
				CH$_4$O	Methanol	1.70	A
CHN	Hydrogen cyanide	2.98	A				
				CH$_4$S	Methanethiol (methyl mercaptan)	1.52	C
				CH$_5$FSi	Methyl monofluorosilane	1.71	A
				CH$_5$N	Methyl amine	1.31	B
CHNO	Hydrogen isocyanate	(1.6)	Q				
CHNS	Hydrogen isothiocyanate	(1.7)	Q				
CHP	Methylidyne phosphide (methinophosphide)	0.390	A	CH$_5$P	Methyl phosphine	1.10	A
CH$_2$Br$_2$	Dibromomethane	1.43	B	CH$_6$Ge	Methyl germane	0.643	A
CH$_2$ClF	Chlorofluoromethane	1.82	B	CH$_6$OSi	Methoxysilane	1.17	B
CH$_2$ClNO$_2$	Chloronitromethane	2.91	B	CH$_6$Si	Methyl silane	0.735	A
CH$_2$Cl$_2$	Dichloromethane	1.60	B				
				CH$_6$Sn	Methyl stannane	0.68	C
CH$_2$F$_2$	Difluoromethane	1.97	A	C$_2$ClF$_3$	Chlorotrifluoroethylene	0.40	D
CH$_2$N$_2$	Cyanogen amide (cyanamide)	4.27	C				
CH$_2$N$_2$	Diazomethane	1.50	A	C$_2$ClF$_5$	Chloropentafluoroethane	0.52	C
CH$_2$N$_2$	Diazirine	1.59	C	C$_2$Cl$_2$F$_4$	1,2-Dichlorotetrafluoroethane	(0.5)	Qi
CH$_2$O	Methanal (formaldehyde)	2.33	A				
				C$_2$F$_3$N	Trifluorocyanomethane	1.33	D
				C$_2$F$_6$	Hexafluoroethane	0	S
				C$_2$F$_6$O	Bis(trifluoromethyl) ether	0.54	D
				C$_2$N$_2$	Dicyanogen (cyanogen)	0	S
CH$_2$O$_2$	Methanoic acid (formic acid)	1.41	A	C$_2$N$_2$S	Dicyano sulfide	3.02	A
				C$_2$HBr	Bromoacetylene	≤0.10	D
CH$_3$BF$_2$	Methyl difluoroborane	1.66	B	C$_2$HCl	Chloroacetylene	0.44	A
				C$_2$HCl$_5$	Pentachloroethane	0.92	C
CH$_3$BO	Carbonyl borane	1.80	B	C$_2$HF	Fluoroacetylene	0.73	C
CH$_3$Br	Bromomethane	1.81	A	C$_2$HF$_3$	Trifluoroethylene	1.40	C
				C$_2$HF$_3$O$_2$	Trifluoroethanoic acid (trifluoroacetic acid).	2.28	D
				C$_2$HF$_5$	Pentafluoroethane	1.54	C
				C$_2$H$_2$	Acetylene	0	S
CH$_3$Cl	Chloromethane	1.87	A				
				C$_2$H$_2$Cl$_2$	1,1-Dichloroethylene	1.34	A
				C$_2$H$_2$Cl$_2$	*cis*-1,2-Dichloroethylene	1.90	A
				C$_2$H$_2$Cl$_2$F$_2$	1,1-Dichloro-2,2-difluoroethane		i
				C$_2$H$_2$Cl$_2$O	Chloroacetyl chloride	2.23	Ci
				C$_2$H$_2$Cl$_3$F	1,1,2-Trichloro-2-fluoroethane		i
CH$_3$F	Fluoromethane	1.85	A	C$_2$H$_2$Cl$_4$	1,1,2,2-Tetrachloroethane	1.32	Ci

TABLE 1i3. (*Continued*)

Formula	Compound name	Selected moment (debyes)		Formula	Compound name	Selected moment (debyes)
C₂H₂FN	Fluorocyanomethane........................	3.43	C	C₂H₄Si	Silyl acetylene...............................	0.316	A
C₂H₂F₂	1,1-Difluoroethylene........................	1.38	A	C₂H₅Br	Bromoethane.................................	2.03	A
C₂H₂F₂	*cis*-1,2-Difluoroethylene....................	2.42	A	C₂H₅BrO	Bromomethoxymethane....................	2.05	Ci
C₂H₂N₂O	1,2,5-Oxadiazole.............................	3.38	A	C₂H₅Cl	Chloroethane.................................	2.05	A
C₂H₂N₂O	1,3,4-Oxadiazole.............................	3.04	B				
C₂H₂N₂S	1,2,5-Thiadiazole.............................	1.56	A				
C₂H₂N₂S	1,3,4-Thiadiazole.............................	3.29	B				
C₂H₂O	Methylene carbonyl (ketene).............	1.42	B	C₂H₅ClO	2-Chloroethanol.............................	1.78	Ci
				C₂H₅ClO	Chloromethoxymethane....................		i
C₂H₃Br	Bromoethylene..............................	1.42	B				
C₂H₃Cl	Chloroethylene..............................	1.45	B	C₂H₅F	Fluoroethane.................................	1.94	B
C₂H₃ClF₂	1-Chloro-1,1-difluoroethane..............	2.14	B	C₂H₅I	Iodoethane...................................	1.91	B
C₂H₃ClO	Acetyl chloride..............................	2.72	C				
C₂H₃ClO₂	Methyl chloroformate......................		i				
C₂H₃Cl₃	1,1,1-Trichloroethane......................	1.78	B	C₂H₅N	Iminoethane (ethyleneimine).............	1.90	A
				C₂H₅N	Methyliminomethane (CH₃N=CH₂)...	1.53	A
C₂H₃Cl₃	1,1,2-Trichloroethane......................		i	C₂H₅NO	Acetyl amine (acetamide).................	3.76	Bi
C₂H₃F	Fluoroethylene..............................	1.43	A	C₂H₅NO	Methylaminomethanal (*N*-methyl-	3.83	Bi
C₂H₃FO	Acetyl fluoride..............................	2.96	A		formamide).		
C₂H₃F₃	1,1,1-Trifluoroethane......................	2.32	B	C₂H₅NO₂	Nitritoethane (ethyl nitrite)..............,	2.40	Ci
C₂H₃F₃	1,1,2-Trifluoroethane......................	1.58	B	C₂H₅NO₂	Nitroethane.................................	3.65	B
C₂H₃I	Iodoethylene...............................	1.30	D	C₂H₆	Ethane.......................................	0	S
C₂H₃N	Cyanomethane (acetonitrile).............	3.92	A				
				C₂H₆AlCl	Dimethylaluminum chloride..............	x	Q
				C₂H₆BF	Dimethyl fluoroborane....................	1.32	C
				C₂H₆O	Ethanol......................................	1.69	Bi
C₂H₃N	Isocyanomethane...........................	3.85	B				
C₂H₃NO	Methyl isocyanate..........................	(2.8)	Q	C₂H₆O	Dimethyl ether..............................	1.30	A
C₂H₃NS	Methyl thiocyanate.........................	(4.0)	Q				
C₂H₄	Ethylene.....................................	0	S	C₂H₆OS	Dimethylsulfoxide...........................	3.96	A
				C₂H₆O₂	1,2-Ethanediol (ethylene glycol)........	2.28	Ci
C₂H₄BrCl	1-Bromo-2-chloroethane...................		i	C₂H₆O₂S	Dimethyl sulfoxylate	4.49	B
C₂H₄Br₂	1,2-Dibromoethane.........................		i		(dimethyl sulfone).		
				C₂H₆S	Ethanethiol..................................	1.58	Bi
C₂H₄ClF	1-Chloro-2-fluoroethane,	2.72	C	C₂H₆S	Dimethyl sulfide............................	1.50	A
	gauche conformation.			C₂H₆Si	Silyl ethylene...............................	0.66	A
				C₂H₇B₅	2,4-Dicarbaheptaborane....................	1.32	B
				C₂H₇N	Aminoethane (ethyl amine)..............	1.22	Ci
C₂H₄ClF	1-Chloro-2-fluoroethane....................		i	C₂H₇N	Dimethyl amine.............................	1.03	B
C₂H₄ClNO₂	1-Chloro-1-nitroethane.....................	3.27	B				
C₂H₄Cl₂	1,1-Dichloroethane.........................	2.06	B				
C₂H₄Cl₂	1,2-Dichloroethane.........................		i	C₂H₇P	Ethyl phosphine............................	1.17	Bi
				C₂H₇P	Dimethyl phosphine........................	1.23	A
				C₂H₈N₂	1,2-Diaminoethane.........................	1.99	Ci
C₂H₄Cl₂O	Bis(chloromethyl) ether...................		i	C₂H₈Si	Dimethyl silane.............................	0.75	A
C₂H₄F₂	1,1-Difluoroethane.........................	2.27	B	C₂H₈Si	Ethyl silane.................................	0.81	B
				C₃Cl₃F₃	1,1,2-Trichloro-3,3,3-trifluoropropene. .	1.28	D
C₂H₄Ge	Germyl acetylene...........................	0.136	A	C₃O₂	Dicarbonyl carbon (carbon	0	S
					suboxide).		
C₂H₄O	Oxirane (ethylene oxide)..................	1.89	A	C₃HF₃	3,3,3-Trifluoropropyne....................	2.36	B
				C₃HF₇	1,1,2,2,3,3,3-Heptafluoropropane........	1.62	Di
				C₃HN	Cyanoacetylene.............................	3.72	A
C₂H₄O	Ethanal (acetaldehyde)....................	2.69	B				
				C₃H₂N₂	Dicyanomethane............................	3.73	A
				C₃H₂O	Propynal......................................	2.47	B
				C₃H₂O₃	Vinylene carbonate.........................	4.55	A
C₂H₄O₂	Ethanoic acid (acetic acid)...............	1.74	C				
C₂H₄O₂	Methyl methanoate (methyl	1.77	B				
	formate).			C₃H₃Br	3-Bromopropyne............................	1.54	C
C₂H₄S	Thiirane (ethylene sulfide)...............	1.85	A	C₃H₃Cl	3-Chloropropyne............................	1.68	C

TABLE 1i3. (*Continued*)

Formula	Compound name	Selected moment (debyes)	
C₃H₃F₃	3,3,3-Trifluoropropene	2.45	B
C₃H₃N	Cyanoethylene	3.87	B
C₃H₃NO	Acetyl cyanide	3.45	B
C₃H₃NS	Thiazole	1.62	B
C₃H₄	Cyclopropene	0.45	A
C₃H₄	Propyne	0.781	A
C₃H₄	Propadiene (allene)	0	S
C₃H₄Cl₂	1,1-Dichlorocyclopropane	1.58	B
C₃H₄Cl₂	1,3-Dichloropropene, bp 104 °C isomer.		i
C₃H₄Cl₂	1,3-Dichloropropene, bp 112 °C isomer.		i
C₃H₄Cl₂	2,3-Dichloropropene		i
C₃H₄Cl₂O	1,1-Dichloropropanone		i
C₃H₄Cl₄	1,1,2,2-Tetrachloropropane		i
C₃H₄O	Ethylidene carbonyl (methyl ketene)	1.79	B
C₃H₄O	Propenal, *trans* conformation (acrolein).	3.12	B
C₃H₄O₂	2-Oxoöxetane (β-propiolactone)	4.18	A
C₃H₄O₂	Vinyl formate	1.49	A
C₃H₅Br	3-Bromopropene	(1.9)	Q
C₃H₅Cl	2-Chloropropene	1.66	B
C₃H₅Cl	*cis*-1-Chloropropene	1.67	C
C₃H₅Cl	*trans*-1-Chloropropene	1.97	C
C₃H₅Cl	3-Chloropropene	1.94	Ci
C₃H₅ClO	Chloropropanone		i
C₃H₅ClO₂	Ethyl chloroformate		i
C₃H₅Cl₃	1,2,2-Trichloropropane		i
C₃H₅F	*cis*-1-Fluoropropene	1.46	B
C₃H₅F	*trans*-1-Fluoropropene	(1.9)	Q
C₃H₅F	2-Fluoropropene	1.61	B
C₃H₅F	3-Fluoropropene, *cis* conformation	1.76	A
C₃H₅F	3-Fluoropropene, *gauche* conformation.	1.94	A
C₃H₅N	Cyanoethane (propionitrile)	4.02	A
C₃H₅NO₂	3-Nitritopropene (allyl nitrite)		i
C₃H₆	Cyclopropane	0	S
C₃H₆	Propene	0.366	A
C₃H₆Br₂	1,2-Dibromopropane		i
C₃H₆ClNO₂	1-Chloro-1-nitropropane	3.48	Bi
C₃H₆Cl₂	1,2-Dichloropropane		i
C₃H₆Cl₂	1,3-Dichloropropane	2.08	Bi
C₃H₆Cl₂	2,2-Dichloropropane	2.27	C
C₃H₆O	Oxetane (trimethylene oxide)	1.94	A
C₃H₆O	Methyl oxirane (propylene oxide)	2.01	A
C₃H₆O	Propanone (acetone)	2.88	A
C₃H₆O	2-Propen-1-ol (allyl alcohol)	1.60	C
C₃H₆O	Propanal, *cis* conformation (propionaldehyde).	2.52	B
C₃H₆O₂	Propanoic acid	1.75	Ci
C₃H₆O₂	Methyl acetate	1.72	Ci

Formula	Compound name	Selected moment (debyes)	
C₃H₆O₂	Ethyl formate	1.93	Ci
C₃H₆O₃	Dimethyl carbonate		i
C₃H₆O₃	1,3,5-Trioxane	2.08	A
C₃H₆S	Thietane (trimethylene sulfide)	1.85	C
C₃H₆S	Methyl thiirane (propylene sulfide)	1.95	A
C₃H₇Br	1-Bromopropane	2.18	Ci
C₃H₇Br	2-Bromopropane	2.21	C
C₃H₇Cl	1-Chloropropane	2.05	Bi
C₃H₇Cl	2-Chloropropane	2.17	C
C₃H₇F	1-Fluoropropane, *gauche* conformation.	1.90	C
C₃H₇F	1-Fluoropropane, *trans* conformation.	2.05	B
C₃H₇I	1-Iodopropane	2.04	Ci
C₃H₇N	3-Aminopropene	(1.2)	Q
C₃H₇NO	N,N-Dimethylformamide	3.82	Bi
C₃H₇NO	Acetyl methylamine (N-Methylacetamide).	3.73	Bi
C₃H₇NO₂	1-Nitritopropane (n-propyl nitrite)		i
C₃H₇NO₂	2-Nitritopropane (isopropyl nitrite)		i
C₃H₇NO₂	1-Nitropropane	3.66	Bi
C₃H₇NO₂	2-Nitropropane	3.73	B
C₃H₈	Propane	0.084	A
C₃H₈O	1-Propanol	1.68	Bi
C₃H₈O	2-Propanol	1.66	Bi
C₃H₈O	Methoxyethane (methyl ethyl ether)	1.23	Ci
C₃H₈O₂	Dimethoxymethane		i
C₃H₉Al	Trimethyl aluminum	x	Q
C₃H₉As	Trimethyl arsine	0.86	B
C₃H₉BF₃N	Trimethyl amine — boron trifluoride complex.	(5)	Q
C₃H₉N	Trimethyl amine	0.612	A
C₃H₉N	1-Aminopropane (n-propylamine)	1.17	Ci
C₃H₉P	Trimethyl phosphine	1.19	A
C₃H₁₀Si	Trimethyl silane	0.525	A
C₄Cl₃F₇	2,2,3-Trichloroheptafluorobutane	(0.9)	Qi
C₄F₈	Perfluorocyclobutane	0	S
C₄F₈O	Perfluoroöxolane (perfluorotetramethylene oxide).	0.56	D
C₄F₁₀O	Perfluoro(ethoxyethane)	0.47	Di
C₄H₂N₂	*trans*-1,2-Dicyanoethylene	0	S
C₄H₄	1-Buten-3-yne	(0.4)	Q
C₄H₄Cl₂	1,4-Dichloro-2-butyne	2.10	Bi
C₄H₄F₂	1,1-Difluoro-1,3-butadiene (*trans* conformation).	1.29	A
C₄H₄N₂	1,2-Dicyanoethane		i
C₄H₄O	Furan	0.66	A
C₄H₄O	3-Butyne-2-one	(2.4)	Q
C₄H₄O₂	Diketene	3.53	B
C₄H₄S	Thiophene	0.55	C
C₄H₅Cl	4-Chloro-1,2-butadiene	2.02	Ci
C₄H₅Cl	1-Chloro-2-butyne	2.19	C
C₄H₅F	2-Fluoro-1,3-butadiene (*trans* conformation).	1.42	A
C₄H₅N	Pyrrole	1.84	C

TABLE 1i3. (Continued)

Formula	Compound name	Selected moment (debyes)		Formula	Compound name	Selected moment (debyes)	
C_4H_5N	cis-1-Cyanopropene	4.08	B	C_5F_{12}	Perfluoro (2-methylbutane)	≤ 0.10	Di
C_4H_5N	trans-1-Cyanopropene	4.50	B	C_5H_5N	Pyridine	2.19	B
C_4H_5N	2-Cyanopropene (methacrylonitrile)	3.69	C				
C_4H_6	Cyclobutene	0.132	A				
C_4H_6	1-Butyne	0.80	C	C_5H_5N	1-Cyano-1,3-butadiene	3.90	Ci
C_4H_6	1,2-Butadiene	0.403	A	C_5H_6	1,3-Cyclopentadiene	0.419	A
C_4H_6	1,3-Butadiene	0	S				
				C_5H_8	Cyclopentene	0.20	B
C_4H_6O	Cyclobutanone	2.99	B				
C_4H_6O	trans-2-Butenal (crotonaldehyde)	3.67	Bi				
C_4H_6O	2-Methylpropenal (methacrolein)	2.68	C	C_5H_8	1-Pentyne	0.81	Ci
C_4H_6O	3-Butene-2-one	3.16	B				
$C_4H_6O_2$	2,3-Butanedione	(0)	Qi				
$C_4H_6O_3$	Acetyl acetate (acetic anhydride)	(2.8)	Qi	C_5H_8	trans-1,3-Pentadiene	0.68	Di
C_4H_6S	Divinyl sulfide	1.20	D	C_5H_8	2-Methyl-1,3-butadiene (trans conformation).	0.25	A
C_4H_7Cl	1-Chloro-2-methylpropene	1.95	Bi				
				$C_5H_8Br_4$	1,3-Dibromo-2,2-bis(bromomethyl) propane.	x	Q
$C_4H_7Cl_3$	1,1,2-Trichloro-2-methylpropane	1.86	Ci	C_5H_8O	Cyclopentanone	(3.3)	Q
C_4H_7F	Fluorocyclobutane	1.94	A	$C_5H_8O_2$	Acetylacetone		Ci
C_4H_7N	1-Cyanopropane	4.07	Bi	C_5H_9N	1-Cyanobutane	4.12	Bi
C_4H_8	1-Butene	0.34	Ci	C_5H_9N	2-Cyano-2-methylpropane	3.95	A
				C_5H_{10}	Ethylcyclopropane	(0.5)	Q
				C_5H_{10}	1-Pentene	(0.5)	Q
C_4H_8	cis-2-Butene	0.3	D	$C_5H_{10}O_3$	Diethyl carbonate	1.10	Ci
C_4H_8	trans-2-Butene	0	S	$C_5H_{11}Br$	1-Bromopentane	2.20	Ci
C_4H_8	2-Methylpropene	0.50	A	$C_5H_{11}Cl$	1-Chloropentane	2.16	Ci
				C_5H_{12}	Pentane	< 0.1	Di
$C_4H_8Br_2$	1,2-Dibromo-2-methylpropane		i				
$C_4H_8Cl_2$	1,4-Dichlorobutane	2.22	Ci	C_5H_{12}	2-Methylbutane	0.13	C
$C_4H_8Cl_2$	1,2-Dichloro-2-methylpropane		i				
C_4H_8O	Tetrahydrofuran	1.63	C				
C_4H_8O	cis-2,3-Dimethyloxirane	2.03	A	C_5H_{12}	2,2-Dimethylpropane	0	S
C_4H_8O	Butanal	2.72	Bi	$C_5H_{12}O_2$	Diethoxymethane		i
$C_4H_8O_2$	1,4-Dioxane	0	S	$C_5H_{12}O_4$	Tetramethoxymethane	0	Qi
				$C_6H_2Cl_2O_2$	2,5-Dichloro-1,4-cyclohexadienedione	0	S
				C_6H_4BrF	p-Bromofluorobenzene	(0.5)	Q
$C_4H_8O_2$	Ethyl acetate	1.78	Ci	$C_6H_4ClNO_2$	o-Chloronitrobenzene	4.64	B
C_4H_9Br	1-Bromobutane	2.08	Ci	$C_6H_4ClNO_2$	m-Chloronitrobenzene	3.73	B
C_4H_9Br	2-Bromobutane	2.23	Ci	$C_6H_4ClNO_2$	p-Chloronitrobenzene	2.83	B
				$C_6H_4Cl_2$	o-Dichlorobenzene	2.50	B
C_4H_9Cl	1-Chlorobutane	2.05	Bi				
				$C_6H_4Cl_2$	m-Dichlorobenzene	1.72	C
C_4H_9Cl	2-Chlorobutane	2.04	Ci	$C_6H_4Cl_2$	p-Dichlorobenzene	0	S
C_4H_9Cl	1-Chloro-2-methylpropane	2.00	Ci	C_6H_4FI	p-Iodofluorobenzene	0.89	D
C_4H_9Cl	2-Chloro-2-methylpropane	2.13	B	$C_6H_4FNO_2$	p-Fluoronitrobenzene	2.87	B
C_4H_9F	2-Fluoro-2-methylpropane	1.96	A	$C_6H_4F_2$	m-Difluorobenzene	1.58	B
C_4H_9I	1-Iodobutane	2.12	Ci	$C_6H_4N_2O_4$	p-Dinitrobenzene	0	S
C_4H_9NO	Propanoyl methylamine (N-methylpropionamide).	3.61	Bi	$C_6H_4O_2$	1,4-Cyclohexadienedione (p-benzoquinone).	0	S
C_4H_9NO	Acetyl dimethylamine (N,N-dimethylacetamide).	3.81	Bi	C_6H_5Br	Bromobenzene	1.70	B
$C_4H_9NO_2$	2-Nitrito-2-methylpropane (t-butyl nitrite).	2.74	Ci	C_6H_5Cl	Chlorobenzene	1.69	B
$C_4H_9NO_2$	1-Nitrobutane	3.59	Bi				
$C_4H_9NO_2$	2-Nitro-2-methylpropane	3.71	B				
C_4H_{10}	Butane	≤ 0.05	Ci				
C_4H_{10}	2-Methylpropane	0.132	A	C_6H_5ClO	o-Chlorophenol		i
				C_6H_5ClO	p-Chlorophenol	2.11	C
				C_6H_5F	Fluorobenzene	1.60	C
$C_4H_{10}O$	1-Butanol	1.66	Bi				
				C_6H_5I	Iodobenzene	1.70	C
$C_4H_{10}O$	2-Methylpropan-1-ol (isobutanol)	1.64	C				
$C_4H_{10}O$	Diethyl ether	1.15	Bi	$C_6H_5NO_2$	Nitrobenzene	4.22	B
				C_6H_6	Benzene	0	S
$C_4H_{10}S$	Diethyl sulfide	1.54	Ci				
$C_4H_{11}N$	1-Aminobutane (n-butylamine)	(1.0)	Qi				
$C_4H_{11}N$	Diethyl amine	0.92	Ci				
C_5F_{12}	Perfluoropentane	≤ 0.10	Di	$C_6H_6N_2O_2$	p-Aminonitrobenzene (p-nitroaniline)	x	Q

TABLE 1i3. (*Continued*)

Formula	Compound name	Selected moment (debyes)	
C_6H_6O	Phenol..................................	1.45	C
$C_6H_6O_2$	p-Dihydroxybenzene..................	x	Q
C_6H_7N	Aminobenzene (aniline)...............	1.53	C
C_6H_8	1,3-Cyclohexadiene..................	0.44	B
C_6H_8	1,5-Hexadiyne.........................		i
$C_6H_8N_2$	o-Diaminobenzene....................	1.53	Di
$C_6H_8N_2$	m-Diaminobenzene....................	1.81	Di
$C_6H_8N_2$	p-Diaminobenzene....................	1.53	Di
$C_6H_8O_2$	1,4-Cyclohexanedione................	(1.4)	Qi
C_6H_{10}	Cyclohexene...........................	0.55	D
C_6H_{10}	1-Hexyne..............................	0.83	Ci
C_6H_{10}	2-Methyl-1,3-pentadiene.............	0.65	D
C_6H_{10}	3-Methyl-1,3-pentadiene.............	0.63	Di
C_6H_{10}	3,3-Dimethyl-1-butyne...............	0.66	A
C_6H_{10}	2-Ethyl-1,3-butadiene................	(0.45)	Qi
C_6H_{10}	2,3-Dimethyl-1,3-butadiene..........	0.52	Di
$C_6H_{10}BrCl$	trans-1-Bromo-2-chlorocyclohexane....		i
$C_6H_{10}Br_2$	trans-1,2-Dibromocyclohexane..........		i
$C_6H_{10}Cl_2$	cis-1e,2a-Dichlorocyclohexane..........	3.11	C
$C_6H_{10}Cl_2$	trans-1,2-Dichlorocyclohexane.........		i
$C_6H_{10}O_3$	Ethyl acetoacetate....................		i
$C_6H_{12}N_2$	Diisopropylidene hydrazine (dimethyl ketazine).	1.53	Bi
$C_6H_{12}O_2$	Pentyl formate (n-amyl formate)........	1.90	Ci
$C_6H_{12}O_3$	2,4,6-Trimethyl-1,3,5-trioxane (paraldehyde).	1.43	C
C_6H_{14}	Hexane.................................	≤ 0.1	Di
$C_6H_{14}O$	Dipropyl ether........................	1.21	Ci
$C_6H_{14}O$	Diisopropyl ether.....................	1.13	Di
$C_6H_{14}O_2$	1,1-Diethoxyethane....................		i
$C_6H_{15}N$	Triethyl amine........................	0.66	Ci
$C_6H_{18}OSi_2$	Bis(trimethylsilyl) ether..............	(0.66)	Qi
$C_7H_4ClF_3$	o-Chloro(trifluoromethyl)benzene......	3.46	B
$C_7H_4ClF_3$	p-Chloro(trifluoromethyl)benzene......	1.58	C
$C_7H_4N_2O_2$	p-Cyanonitrobenzene..................	(0)	Q
$C_7H_5F_3$	(Trifluoromethyl)benzene.............	2.86	B
C_7H_5N	Cyanobenzene (benzonitrile)...........	4.18	B
C_7H_7Cl	o-Chlorotoluene.......................	1.56	C
C_7H_7Cl	p-Chlorotoluene.......................	2.21	B
C_7H_7F	o-Fluorotoluene.......................	1.37	C
C_7H_7F	m-Fluorotoluene.......................	1.86	C
C_7H_7F	p-Fluorotoluene.......................	2.00	C
$C_7H_7NO_3$	o-Nitro(methoxy)benzene..............	4.83	Bi
$C_7H_7NO_3$	m-Nitro(methoxy)benzene..............	4.55	Bi
$C_7H_7NO_3$	p-Nitro(methoxy)benzene..............	5.26	B
C_7H_8	1,3,5-Cycloheptatriene................	0.25	C
C_7H_8	Toluene...............................	0.36	C
C_7H_8O	Phenylmethanol (benzyl alcohol)......	1.71	C
C_7H_8O	Methoxybenzene (anisole).............	1.38	C
C_7H_9NO	o-Amino(methoxy)benzene.............	1.61	Ci
C_7H_{10}	1,6-Heptadiyne.......................		i
C_7H_{12}	1-Heptyne.............................	0.86	i
C_7H_{14}	Methylcyclohexane....................	(0)	Qi
$C_7H_{14}O$	Methoxycyclohexane..................	1.35	Di
$C_7H_{14}O_2$	Pentyl acetate (n-amyl acetate)........	1.75	Ci
$C_7H_{15}Br$	1-Bromoheptane.......................	2.16	Ci
C_7H_{16}	Heptane...............................	≤ 0.1	Di
$C_8H_4N_2$	p-Dicyanobenzene....................	0	S
C_8H_8	Styrene...............................	(0)	Q

Formula	Compound name	Selected moment (debyes)	
C_8H_8O	Acetylbenzene (acetophenone).........	3.02	B
$C_8H_8O_2$	2,5-Dimethyl-1,4-cyclohexadienedione.	0	S
$C_8H_9NO_2$	Methyl (o-aminophenyl)formate........	x	Q
$C_8H_9NO_2$	Methyl (m-aminophenyl)formate.......	x	Q
$C_8H_9NO_2$	Methyl (p-aminophenyl)formate........	x	Q
C_8H_{10}	Ethylbenzene..........................	0.59	C
C_8H_{10}	o-Xylene..............................	0.62	C
C_8H_{10}	p-Xylene..............................	0	S
$C_8H_{10}O$	Ethoxybenzene (phenetole)............	1.45	Di
$C_8H_{11}N$	(Dimethylamino)benzene..............	1.68	D
C_8H_{12}	1,7-Octadiyne.........................		i
$C_8H_{12}O_2$	Tetramethylcyclobutane-1,3-dione......	0	S
$C_8H_{12}O_2$	Ethyl 2,4-hexadienoate (ethyl sorbate).	(2.07)	Qi
$C_8H_{14}O_4$	Diethyl butanedioate (diethyl succinate).		i
C_8H_{16}	Ethylcyclohexane......................	(0)	Q
C_8H_{18}	Octane................................	≤ 0.1	Di
$C_8H_{18}O$	Dibutyl ether.........................	1.17	Ci
C_9H_7N	Quinoline.............................	2.29	C
C_9H_7N	Isoquinoline..........................	2.73	C
$C_9H_{10}O_2$	Ethyl phenylformate (ethyl benzoate)..	2.00	Ci
C_9H_{12}	Isopropylbenzene.....................	(0.79)	Q
$C_9H_{12}O_8$	Tetrakis(methyl carboxylate)methane..	x	Q
C_9H_{14}	1,8-Nonadiyne........................		i
C_9H_{18}	Isopropylcyclohexane.................	(0)	Q
$C_{10}H_8$	Azulene...............................	0.80	B
$C_{10}H_{14}$	t-Butylbenzene........................	(0.83)	Q
$C_{10}H_{14}BeO_4$	Bis(2,4-pentanedionato) beryllium......	0	S
$C_{10}H_{18}O$	1,3,3-Trimethyl-2-oxabicyclo[2.2.2]octane.	1.42	D
$C_{10}H_{20}$	t-Butylcyclohexane....................	(0)	Q
$C_{11}H_{16}$	p-t-Butyltoluene......................	x	Q
$C_{12}H_8Br_2O$	Bis(p-bromophenyl) ether..............	1.02	D
$C_{12}H_9BrO$	p-Bromophenoxybenzene..............	1.98	C
$C_{12}H_9NO_3$	p-Nitrophenoxybenzene...............	4.54	B
$C_{12}H_{10}$	Acenaphthene.........................	(0.85)	Q
$C_{12}H_{10}$	Phenylbenzene (diphenyl).............	0	S
$C_{12}H_{10}O$	Diphenyl ether........................	(1.3)	Q
$C_{13}H_{10}O$	Diphenyl carbonyl....................	x	Q
$C_{13}H_{11}BrO$	p-Bromophenoxy-p-toluene............	2.45	C
$C_{13}H_{12}$	Diphenylmethane.....................	x	Q
$C_{13}H_{20}O_8$	Tetrakis(acetoxymethyl)methane........	x	Q
$C_{14}H_{14}O$	Bis(p-tolyl) ether.....................	1.54	C
$C_{15}H_{21}AlO_6$	Tris(2,4-pentanedionato) aluminum.....	0	S
$C_{15}H_{21}CrO_6$	Tris(2,4-pentanedionato) chromium (III).	0	S
$C_{15}H_{21}FeO_6$	Tris(2,4-pentanedionato) iron (III)......	0	S
$C_{20}H_{28}O_8Th$	Tetrakis(2,4-pentanedionato) thorium.	0	S

Abbreviations:

Code symbol	Estimated accuracy of value
A	± 1% or, for μ < 1.0 D, ± 0.01 D
B	± 2% or, for μ < 1.0 D, ± 0.02 D
C	± 5% or, for μ < 1.0 D, ± 0.05 D
D	± 10% or, for μ < 1.0 D, ± 0.10 D
Q	Questionable value
S	$\mu \equiv 0$ on grounds of molecular symmetry
i	The significance of these values may involve some ambiguity because of the possibility of different conformations or spatial isomers.

TABLE 1i4. LIMITING IONIC CONDUCTANCES IN WATER AT 25°.
(HARNED & OWEN 1958)

Cation	λ^0_+*	Cation	λ^0_+*	Anion	λ^0_-	Anion	λ^0_-
H^+	349.8			OH^-	197.6	$CNCH_2CO_2^-$	41.8
Li^+	38.69			Cl^-	76.34	$CH_3CH_2CO_2^-$	35.8
Na^+	50.11	$\frac{1}{2} Sr^{++}$	59.46	Br^-	78.3	$CH_3(CH_2)CO_2^-$	32.6
K^+	73.52	$\frac{1}{2} Ba^{++}$	63.64	I^-	76.8	$C_6H_5CO_2^-$	32.3
NH_4^+	73.4			NO_3^-	71.44	$HC_2O_4^-$	40.2
Ag^+	61.92			ClO_4^-	68.0	$\frac{1}{2} C_2O_4^{--}$	74.2
Tl^+	74.7	$\frac{1}{2} Cu^{++}$	43.	HCO_3^-	44.5	$\frac{1}{2} SO_4^{--}$	80.
		$\frac{1}{2} Zn^{++}$	53.	HCO_2^-	54.6	$\frac{1}{2} Fe(CN)_6^{--}$	101.
$\frac{1}{2} Mg^{+++}$	53.06	$\frac{1}{3} La^{+++}$	69.5	$CH_3CO_2^-$	40.9	$\frac{1}{4} Fe(CN)_6^{----}$	111.
$\frac{1}{2} Ca^{++}$	59.50	$\frac{1}{3} Co(NH_3)_6^{+++}$	102.	$ClCH_2CO_2^-$	39.8		

*$ohm^{-1} cm^2$.

TABLE 1i5. THE EQUIVALENT CONDUCTANCES OF ELECTROLYTES IN
AQUEOUS SOLUTIONS AT 25°[†] (HARNED & OWEN 1958)

	c* = 0	0.0005	0.001	0.005	0.01	0.02	0.05	0.1
HCl	426.16	422.74	421.36	415.80	412.00	407.24	399.09	391.32
LiCL	115.03	113.15	112.40	109.40	107.32	104.65	100.11	95.86
NaCl	126.45	124.50	123.74	120.65	118.51	115.76	111.06	106.74
KCl	149.86	147.81	146.95	143.55	141.27	138.34	133.37	128.96
NH_4Cl	149.7	. . . . . .	. . . . . .	. . . . . .	141.28	138.33	133.29	128.75
KBr	151.9	. . . . . .	. . . . . .	146.09	143.43	140.48	135.68	131.39
NaI	126.94	125.36	124.25	121.25	119.24	116.70	112.79	108.78
KI	150.38	. . . . . .	. . . . . .	144.37	142.18	139.45	134.97	131.11
KNO_3	144.96	142.77	141.84	138.48	132.82	132.41	126.31	120.40
$KHCO_3$	118.00	116.10	115.34	112.24	110.08	107.22	. . . . . .	. . . . . .
NaO_2CCH_3	91.0	89.2	88.5	85.72	83.76	81.24	76.92	72.80
$NaO_2C(CH_3)_2CH_3$	82.70	81.04	80.31	77.58	75.76	73.39	69.32	65.27
NaOH	247.8	245.6	244.7	240.8	238.0	. . . . . .	. . . . . .	. . . . . .
$AgNO_3$	133.36	131.36	130.51	127.20	124.76	121.41	115.24	109.14
$MgCl_2$	129.40	125.61	124.11	118.31	114.55	110.04	103.08	97.10
$CaCl_2$	135.84	131.93	130.36	124.25	120.36	115.65	108.47	102.46
$SrCl_2$	135.80	131.90	130.33	124.24	120.29	115.54	108.25	102.19
$BaCl_2$	139.98	135.96	134.34	128.02	123.94	119.09	111.48	105.19
Na_2SO_4	129.9	125.74	124.15	117.15	112.44	106.78	97.75	89.98
$CuSO_4$	133.6	121.6	115.26	94.07	83.12	72.20	59.05	50.58
$ZnSO_4$	132.8	121.4	115.53	95.49	84.91	74.24	61.20	52.64
$LaCl_3$	145.8	139.6	137.0	127.5	121.8	115.3	106.2	99.1
$K_3Fe(CN)_6$	174.5	166.4	163.1	150.7	. . . . . .	. . . . . .	. . . . . .	. . . . . .
$K_4Fe(CN)_6$	184.5	. . . . . .	167.24	146.09	134.83	122.82	107.70	97.87

*Concentrations expressed in equivalents per liter.
[†]$ohm^{-1} cm^2$.

TABLE 1i6. THERMOCOUPLE DATA*

INTRODUCTION

Revision of the International Practical Temperature Scale led to the generation of revised reference data for thermocouples by the National Bureau of Standards. The following information is excerpted from NBS Monograph 125:*

Thermocouple characteristics

THERMOCOUPLE COMBINATIONS

Type designation [a]	Temperature Range, °C	Materials
B	0 to 1820	*Platinum*-30% rhodium versus *platinum*-6% rhodium.
E	−270 to 1000	*Nickel*-chromium alloy versus a *copper*-nickel alloy.
J	−210 to 1200	Iron versus another slightly different *copper*-nickel alloy.
K	−270 to 1372	*Nickel*-chromium alloy versus *nickel*-aluminum alloy.
R	−50 to 1767	*Platinum*-13% rhodium versus platinum.
S	−50 to 1767	*Platinum*-10% rhodium versus platinum.
T	−270 to 400	Copper versus a *copper*-nickel alloy.

SINGLE-LEG MATERIALS

...N	The negative wire in a combination.
...P	The positive wire in a combination.
BN	*Platinum*-nominal 6 wt% rhodium.
BP	*Platinum*-nominal 30 wt% rhodium.
EN or TN	A *copper*-nickel alloy, constantan: Cupron[1], Advance[3], ThermoKanthal JN[2]; nominally 55% Cu, 45% Ni; often referred to as Adams constantan.
EP or KP	A *nickel*-chromium alloy: Chromel[4], Tophel[1], T–1[3], ThermoKanthal KP[2]; nominally 90% Ni, 10% Cr.
JN	A *copper*-nickel alloy similar to, but not always interchangeable with, EN and TN; SAMA specification.
JP	Iron: ThermoKanthal JP[2]; nominally 99.5% Fe.
KN	A *nickel*-aluminum alloy: Alumel[4], Nial[1], T–2[3], ThermoKanthal KN[2]; nominally 95% Ni, 2% Al, 2% Mn, 1% Si.
RN, SN	High-purity platinum.
RP	*Platinum*-13 wt% rhodium.
SP	*Platinum*-10 wt% rhodium.
TP	Copper, usually Electrolytic Tough Pitch.

[a] The letter designations used in this Monograph follow the recommendations of Committee E–20 of the American Society for Testing and Materials. The letter type, e.g., Type T, designates the thermoelectric properties, not the precise chemical composition. Thermocouples of a given type may have variations in composition as long as the resultant thermoelectric properties remain within the established limits of error.

Registered Trademarks:
[1]   Trademark—Wilbur B. Driver Co.
[2]   Trademark—Kanthal Corp.
[3]   Trademark—Driver-Harris Co.
[4]   Trademark—Hoskins Manufacturing Co.

The use of trade names does not constitute an endorsement of any manufacturer's products. All materials manufactured in compliance with the established thermoelectric voltage standards are equally acceptable.

*Powell, R. L., W. J. Hall, C. H. Hyink, Jr., L. L. Sparks, G. W. Burns, M. G. Scroger and H. H. Plumb," Thermocouple Reference Tables Based on IPTS-68," National Bureau of Standards Monograph 125, U.S. Department of Commerce, Washington, DC, 1974.

TABLE 1i6. (Continued)

TYPE B—*Platinum–30% Rhodium Alloy Versus Platinum–6% Rhodium Alloy Thermocouples*

Type B thermocouples—thermoelectric voltages, E(T), Seebeck coefficients, S(T), and first derivative of the Seebeck coefficients, dS/dT, reference junctions at 0 °C

T °C	E μV	S μV/°C	dS/dT nV/°C²	T °C	E μV	S μV/°C	dS/dT nV/°C²	T °C	E μV	S μV/°C	dS/dT nV/°C²
0	0.00	-0.247	11.82	60	6.19	0.449	11.39	120	53.40	1.122	11.07
1	-0.24	-0.235	11.81	61	6.64	0.460	11.38	121	54.53	1.133	11.06
2	-0.47	-0.223	11.80	62	7.11	0.471	11.37	122	55.66	1.144	11.06
3	-0.69	-0.211	11.79	63	7.59	0.483	11.37	123	56.81	1.155	11.05
4	-0.89	-0.200	11.79	64	8.07	0.494	11.36	124	57.97	1.166	11.05
5	-1.09	-0.188	11.78	65	8.57	0.506	11.36	125	59.15	1.177	11.04
6	-1.27	-0.176	11.77	66	9.09	0.517	11.35	126	60.33	1.188	11.04
7	-1.44	-0.164	11.76	67	9.61	0.528	11.34	127	61.52	1.199	11.03
8	-1.60	-0.152	11.75	68	10.14	0.540	11.34	128	62.73	1.210	11.03
9	-1.74	-0.141	11.75	69	10.69	0.551	11.33	129	63.94	1.221	11.02
10	-1.88	-0.129	11.74	70	11.24	0.562	11.33	130	65.17	1.232	11.02
11	-2.00	-0.117	11.73	71	11.81	0.574	11.32	131	66.41	1.243	11.02
12	-2.11	-0.106	11.72	72	12.39	0.585	11.31	132	67.66	1.254	11.01
13	-2.21	-0.094	11.71	73	12.98	0.596	11.31	133	68.92	1.265	11.01
14	-2.30	-0.082	11.71	74	13.58	0.608	11.30	134	70.19	1.276	11.00
15	-2.38	-0.070	11.70	75	14.20	0.619	11.30	135	71.47	1.287	11.00
16	-2.44	-0.059	11.69	76	14.82	0.630	11.29	136	72.76	1.298	10.99
17	-2.49	-0.047	11.68	77	15.46	0.641	11.29	137	74.07	1.309	10.99
18	-2.53	-0.035	11.67	78	16.10	0.653	11.28	138	75.38	1.320	10.98
19	-2.56	-0.024	11.67	79	16.76	0.664	11.27	139	76.71	1.331	10.98
20	-2.58	-0.012	11.66	80	17.43	0.675	11.27	140	78.04	1.342	10.98
21	-2.59	-0.000	11.65	81	18.11	0.686	11.26	141	79.39	1.353	10.97
22	-2.58	0.011	11.64	82	18.80	0.698	11.26	142	80.75	1.364	10.97
23	-2.57	0.023	11.64	83	19.51	0.709	11.25	143	82.12	1.375	10.96
24	-2.54	0.035	11.63	84	20.22	0.720	11.25	144	83.50	1.386	10.96
25	-2.50	0.046	11.62	85	20.95	0.731	11.24	145	84.89	1.397	10.95
26	-2.44	0.058	11.61	86	21.69	0.743	11.24	146	86.29	1.408	10.95
27	-2.38	0.069	11.61	87	22.43	0.754	11.23	147	87.71	1.419	10.95
28	-2.31	0.081	11.60	88	23.19	0.765	11.23	148	89.13	1.430	10.94
29	-2.22	0.093	11.59	89	23.96	0.776	11.22	149	90.57	1.441	10.94
30	-2.12	0.104	11.58	90	24.75	0.788	11.22	150	92.01	1.452	10.93
31	-2.01	0.116	11.58	91	25.54	0.799	11.21	151	93.47	1.463	10.93
32	-1.89	0.127	11.57	92	26.34	0.810	11.20	152	94.94	1.474	10.92
33	-1.76	0.139	11.56	93	27.16	0.821	11.20	153	96.42	1.485	10.92
34	-1.61	0.151	11.56	94	27.99	0.832	11.19	154	97.91	1.495	10.92
35	-1.45	0.162	11.55	95	28.82	0.844	11.19	155	99.41	1.506	10.91
36	-1.29	0.174	11.54	96	29.67	0.855	11.18	156	100.92	1.517	10.91
37	-1.11	0.185	11.54	97	30.53	0.866	11.18	157	102.44	1.528	10.90
38	-0.92	0.197	11.53	98	31.41	0.877	11.17	158	103.98	1.539	10.90
39	-0.71	0.208	11.52	99	32.29	0.888	11.17	159	105.52	1.550	10.90
40	-0.50	0.220	11.51	100	33.18	0.900	11.16	160	107.08	1.561	10.89
41	-0.27	0.231	11.51	101	34.09	0.911	11.16	161	108.64	1.572	10.89
42	-0.04	0.243	11.50	102	35.00	0.922	11.15	162	110.22	1.583	10.88
43	0.21	0.254	11.49	103	35.93	0.933	11.15	163	111.81	1.594	10.88
44	0.47	0.266	11.49	104	36.87	0.944	11.14	164	113.41	1.604	10.87
45	0.74	0.277	11.48	105	37.82	0.955	11.14	165	115.02	1.615	10.87
46	1.03	0.289	11.47	106	38.78	0.966	11.13	166	116.64	1.626	10.87
47	1.32	0.300	11.47	107	39.75	0.978	11.13	167	118.27	1.637	10.86
48	1.63	0.312	11.46	108	40.74	0.989	11.12	168	119.91	1.648	10.86
49	1.94	0.323	11.45	109	41.73	1.000	11.12	169	121.56	1.659	10.85
50	2.27	0.335	11.45	110	42.74	1.011	11.11	170	123.23	1.670	10.85
51	2.61	0.346	11.44	111	43.75	1.022	11.11	171	124.90	1.680	10.84
52	2.96	0.357	11.44	112	44.78	1.033	11.10	172	126.59	1.691	10.84
53	3.33	0.369	11.43	113	45.82	1.044	11.10	173	128.29	1.702	10.84
54	3.70	0.380	11.42	114	46.87	1.055	11.09	174	129.99	1.713	10.83
55	4.09	0.392	11.42	115	47.93	1.066	11.09	175	131.71	1.724	10.83
56	4.48	0.403	11.41	116	49.00	1.078	11.09	176	133.44	1.735	10.83
57	4.89	0.415	11.40	117	50.08	1.089	11.08	177	135.18	1.745	10.82
58	5.31	0.426	11.40	118	51.18	1.100	11.08	178	136.93	1.756	10.82
59	5.75	0.437	11.39	119	52.28	1.111	11.07	179	138.69	1.767	10.81
60	6.19	0.449	11.39	120	53.40	1.122	11.07	180	140.47	1.778	10.81

TABLE 1i6. (Continued)

Type B thermocouples—*thermoelectric voltages,* E(T), *Seebeck coefficients,* S(T), *and first derivative of the Seebeck coefficients,* dS/dT, *reference junctions at 0 °C—Continued*

T °C	E μV	S μV/°C	dS/dT nV/°C²	T °C	E μV	S μV/°C	dS/dT nV/°C²	T °C	E μV	S μV/°C	dS/dT nV/°C²
180	140.47	1.778	10.81	240	266.46	2.419	10.58	300	430.53	3.047	10.35
181	142.25	1.789	10.81	241	268.88	2.430	10.58	301	433.58	3.058	10.35
182	144.04	1.799	10.80	242	271.32	2.441	10.57	302	436.65	3.068	10.34
183	145.85	1.810	10.80	243	273.76	2.451	10.57	303	439.72	3.078	10.34
184	147.66	1.821	10.79	244	276.22	2.462	10.56	304	442.80	3.089	10.33
185	149.49	1.832	10.79	245	278.69	2.472	10.56	305	445.90	3.099	10.33
186	151.33	1.843	10.79	246	281.16	2.483	10.56	306	449.00	3.109	10.33
187	153.18	1.853	10.78	247	283.65	2.493	10.55	307	452.12	3.120	10.32
188	155.04	1.864	10.78	248	286.15	2.504	10.55	308	455.24	3.130	10.32
189	156.91	1.875	10.77	249	288.66	2.515	10.54	309	458.38	3.140	10.31
190	158.79	1.886	10.77	250	291.18	2.525	10.54	310	461.52	3.151	10.31
191	160.68	1.897	10.77	251	293.71	2.536	10.54	311	464.68	3.161	10.31
192	162.58	1.907	10.76	252	296.25	2.546	10.53	312	467.84	3.171	10.30
193	164.49	1.918	10.76	253	298.80	2.557	10.53	313	471.02	3.182	10.30
194	166.41	1.929	10.75	254	301.36	2.567	10.53	314	474.21	3.192	10.29
195	168.35	1.940	10.75	255	303.94	2.578	10.52	315	477.40	3.202	10.29
196	170.29	1.950	10.75	256	306.52	2.588	10.52	316	480.61	3.212	10.29
197	172.25	1.961	10.74	257	309.11	2.599	10.51	317	483.83	3.223	10.28
198	174.22	1.972	10.74	258	311.72	2.609	10.51	318	487.06	3.233	10.28
199	176.19	1.983	10.74	259	314.33	2.620	10.51	319	490.29	3.243	10.27
200	178.18	1.993	10.73	260	316.96	2.630	10.50	320	493.54	3.254	10.27
201	180.18	2.004	10.73	261	319.59	2.641	10.50	321	496.80	3.264	10.27
202	182.19	2.015	10.72	262	322.24	2.651	10.50	322	500.07	3.274	10.26
203	184.21	2.025	10.72	263	324.90	2.662	10.49	323	503.35	3.284	10.26
204	186.24	2.036	10.72	264	327.56	2.672	10.49	324	506.64	3.295	10.25
205	188.28	2.047	10.71	265	330.24	2.683	10.48	325	509.94	3.305	10.25
206	190.33	2.058	10.71	266	332.93	2.693	10.48	326	513.25	3.315	10.25
207	192.40	2.068	10.70	267	335.63	2.704	10.48	327	516.57	3.325	10.24
208	194.47	2.079	10.70	268	338.34	2.714	10.47	328	519.90	3.336	10.24
209	196.55	2.090	10.70	269	341.06	2.725	10.47	329	523.24	3.346	10.23
210	198.65	2.100	10.69	270	343.79	2.735	10.47	330	526.59	3.356	10.23
211	200.76	2.111	10.69	271	346.53	2.746	10.46	331	529.95	3.366	10.23
212	202.87	2.122	10.69	272	349.28	2.756	10.46	332	533.32	3.377	10.22
213	205.00	2.132	10.68	273	352.04	2.767	10.45	333	536.71	3.387	10.22
214	207.14	2.143	10.68	274	354.81	2.777	10.45	334	540.10	3.397	10.21
215	209.29	2.154	10.67	275	357.59	2.787	10.45	335	543.50	3.407	10.21
216	211.44	2.165	10.67	276	360.38	2.798	10.44	336	546.91	3.417	10.21
217	213.61	2.175	10.67	277	363.19	2.808	10.44	337	550.33	3.428	10.20
218	215.80	2.186	10.66	278	366.00	2.819	10.43	338	553.77	3.438	10.20
219	217.99	2.196	10.66	279	368.82	2.829	10.43	339	557.21	3.448	10.19
220	220.19	2.207	10.65	280	371.66	2.840	10.43	340	560.66	3.458	10.19
221	222.40	2.218	10.65	281	374.50	2.850	10.42	341	564.13	3.468	10.19
222	224.62	2.228	10.65	282	377.36	2.860	10.42	342	567.60	3.479	10.18
223	226.86	2.239	10.64	283	380.22	2.871	10.42	343	571.08	3.489	10.18
224	229.10	2.250	10.64	284	383.10	2.881	10.41	344	574.58	3.499	10.17
225	231.36	2.260	10.64	285	385.99	2.892	10.41	345	578.08	3.509	10.17
226	233.62	2.271	10.63	286	388.88	2.902	10.40	346	581.60	3.519	10.17
227	235.90	2.282	10.63	287	391.79	2.912	10.40	347	585.12	3.529	10.16
228	238.19	2.292	10.62	288	394.71	2.923	10.40	348	588.65	3.540	10.16
229	240.48	2.303	10.62	289	397.64	2.933	10.39	349	592.20	3.550	10.15
230	242.79	2.314	10.62	290	400.58	2.944	10.39	350	595.75	3.560	10.15
231	245.11	2.324	10.61	291	403.52	2.954	10.38	351	599.32	3.570	10.14
232	247.44	2.335	10.61	292	406.48	2.964	10.38	352	602.89	3.580	10.14
233	249.78	2.345	10.61	293	409.45	2.975	10.38	353	606.48	3.590	10.14
234	252.13	2.356	10.60	294	412.43	2.985	10.37	354	610.07	3.600	10.13
235	254.49	2.367	10.60	295	415.42	2.996	10.37	355	613.68	3.611	10.13
236	256.86	2.377	10.59	296	418.42	3.006	10.36	356	617.30	3.621	10.12
237	259.25	2.388	10.59	297	421.44	3.016	10.36	357	620.92	3.631	10.12
238	261.64	2.398	10.59	298	424.46	3.027	10.36	358	624.56	3.641	10.12
239	264.04	2.409	10.58	299	427.49	3.037	10.35	359	628.20	3.651	10.11
240	266.46	2.419	10.58	300	430.53	3.047	10.35	360	631.86	3.661	10.11

TABLE 1i6. (*Continued*)

Type B thermocouples—*thermoelectric voltages, E(T), Seebeck coefficients, S(T), and first derivative of the Seebeck coefficients, dS/dT, reference junctions at 0 °C—Continued*

T °C	E μV	S μV/°C	dS/dT nV/°C^2	T °C	E μV	S μV/°C	dS/dT nV/°C^2	T °C	E μV	S μV/°C	dS/dT nV/°C^2
360	631.86	3.661	10.11	420	869.57	4.260	9.85	480	1142.71	4.842	9.56
361	635.53	3.671	10.10	421	873.83	4.270	9.84	481	1147.56	4.852	9.56
362	639.20	3.681	10.10	422	878.11	4.279	9.84	482	1152.41	4.861	9.55
363	642.89	3.691	10.09	423	882.39	4.289	9.83	483	1157.28	4.871	9.55
364	646.58	3.702	10.09	424	886.68	4.299	9.83	484	1162.16	4.880	9.54
365	650.29	3.712	10.09	425	890.99	4.309	9.82	485	1167.04	4.890	9.54
366	654.01	3.722	10.08	426	895.30	4.319	9.82	486	1171.94	4.899	9.54
367	657.73	3.732	10.08	427	899.63	4.329	9.81	487	1176.84	4.909	9.53
368	661.47	3.742	10.07	428	903.96	4.338	9.81	488	1181.75	4.919	9.53
369	665.22	3.752	10.07	429	908.30	4.348	9.80	489	1186.68	4.928	9.52
370	668.98	3.762	10.06	430	912.66	4.358	9.80	490	1191.61	4.938	9.52
371	672.74	3.772	10.06	431	917.02	4.368	9.80	491	1196.55	4.947	9.51
372	676.52	3.782	10.06	432	921.39	4.378	9.79	492	1201.50	4.957	9.51
373	680.31	3.792	10.05	433	925.77	4.387	9.79	493	1206.47	4.966	9.50
374	684.10	3.802	10.05	434	930.17	4.397	9.78	494	1211.44	4.976	9.50
375	687.91	3.812	10.04	435	934.57	4.407	9.78	495	1216.42	4.985	9.49
376	691.73	3.822	10.04	436	938.98	4.417	9.77	496	1221.41	4.995	9.49
377	695.56	3.832	10.03	437	943.40	4.427	9.77	497	1226.41	5.004	9.48
378	699.39	3.842	10.03	438	947.83	4.436	9.76	498	1231.42	5.014	9.48
379	703.24	3.852	10.03	439	952.27	4.446	9.76	499	1236.43	5.023	9.47
380	707.10	3.862	10.02	440	956.73	4.456	9.75	500	1241.46	5.033	9.47
381	710.97	3.872	10.02	441	961.19	4.466	9.75	501	1246.50	5.042	9.46
382	714.84	3.882	10.01	442	965.66	4.475	9.74	502	1251.55	5.051	9.46
383	718.73	3.892	10.01	443	970.14	4.485	9.74	503	1256.60	5.061	9.45
384	722.63	3.902	10.00	444	974.63	4.495	9.74	504	1261.67	5.070	9.45
385	726.54	3.912	10.00	445	979.13	4.505	9.73	505	1266.74	5.080	9.44
386	730.45	3.922	10.00	446	983.64	4.514	9.73	506	1271.83	5.089	9.44
387	734.38	3.932	9.99	447	988.16	4.524	9.72	507	1276.92	5.099	9.43
388	738.32	3.942	9.99	448	992.68	4.534	9.72	508	1282.02	5.108	9.43
389	742.27	3.952	9.98	449	997.22	4.543	9.71	509	1287.14	5.118	9.42
390	746.22	3.962	9.98	450	1001.77	4.553	9.71	510	1292.26	5.127	9.42
391	750.19	3.972	9.97	451	1006.33	4.563	9.70	511	1297.39	5.136	9.41
392	754.17	3.982	9.97	452	1010.90	4.573	9.70	512	1302.53	5.146	9.41
393	758.15	3.992	9.97	453	1015.47	4.582	9.69	513	1307.68	5.155	9.40
394	762.15	4.002	9.96	454	1020.06	4.592	9.69	514	1312.84	5.165	9.40
395	766.16	4.012	9.96	455	1024.66	4.602	9.68	515	1318.01	5.174	9.39
396	770.18	4.022	9.95	456	1029.26	4.611	9.68	516	1323.19	5.183	9.39
397	774.20	4.032	9.95	457	1033.88	4.621	9.67	517	1328.38	5.193	9.38
398	778.24	4.042	9.94	458	1038.51	4.631	9.67	518	1333.58	5.202	9.38
399	782.29	4.052	9.94	459	1043.14	4.640	9.66	519	1338.78	5.211	9.37
400	786.35	4.062	9.93	460	1047.79	4.650	9.66	520	1344.00	5.221	9.37
401	790.41	4.072	9.93	461	1052.44	4.660	9.66	521	1349.22	5.230	9.36
402	794.49	4.082	9.93	462	1057.11	4.669	9.65	522	1354.46	5.240	9.36
403	798.58	4.092	9.92	463	1061.78	4.679	9.65	523	1359.70	5.249	9.35
404	802.67	4.102	9.92	464	1066.46	4.689	9.64	524	1364.96	5.258	9.35
405	806.78	4.112	9.91	465	1071.16	4.698	9.64	525	1370.22	5.268	9.34
406	810.90	4.122	9.91	466	1075.86	4.708	9.63	526	1375.49	5.277	9.34
407	815.02	4.131	9.90	467	1080.57	4.717	9.63	527	1380.77	5.286	9.33
408	819.16	4.141	9.90	468	1085.29	4.727	9.62	528	1386.06	5.296	9.33
409	823.31	4.151	9.89	469	1090.03	4.737	9.62	529	1391.37	5.305	9.32
410	827.46	4.161	9.89	470	1094.77	4.746	9.61	530	1396.67	5.314	9.32
411	831.63	4.171	9.89	471	1099.52	4.756	9.61	531	1401.99	5.324	9.31
412	835.80	4.181	9.88	472	1104.28	4.766	9.60	532	1407.32	5.333	9.31
413	839.99	4.191	9.88	473	1109.05	4.775	9.60	533	1412.66	5.342	9.30
414	844.18	4.201	9.87	474	1113.83	4.785	9.59	534	1418.01	5.351	9.30
415	848.39	4.211	9.87	475	1118.62	4.794	9.59	535	1423.36	5.361	9.29
416	852.61	4.220	9.86	476	1123.42	4.804	9.58	536	1428.73	5.370	9.29
417	856.83	4.230	9.86	477	1128.23	4.813	9.58	537	1434.10	5.379	9.28
418	861.07	4.240	9.85	478	1133.05	4.823	9.57	538	1439.49	5.389	9.28
419	865.31	4.250	9.85	479	1137.87	4.833	9.57	539	1444.88	5.398	9.27
420	869.57	4.260	9.85	480	1142.71	4.842	9.56	540	1450.28	5.407	9.27

TABLE 1i6. (Continued)

Type B thermocouples—*thermoelectric voltages, E(T), Seebeck coefficients, S(T), and first derivative of the Seebeck coefficients, dS/dT, reference junctions at 0 °C—Continued*

T °C	E µV	S µV/°C	dS/dT nV/°C²	T °C	E µV	S µV/°C	dS/dT nV/°C²	T °C	E µV	S µV/°C	dS/dT nV/°C²
540	1450.28	5.407	9.27	600	1791.21	5.954	8.96	660	2164.37	6.482	8.64
541	1455.69	5.416	9.26	601	1797.17	5.963	8.95	661	2170.86	6.490	8.63
542	1461.12	5.426	9.26	602	1803.13	5.972	8.95	662	2177.35	6.499	8.63
543	1466.55	5.435	9.25	603	1809.11	5.981	8.94	663	2183.85	6.508	8.62
544	1471.99	5.444	9.25	604	1815.09	5.990	8.93	664	2190.37	6.516	8.62
545	1477.43	5.453	9.24	605	1821.09	5.999	8.93	665	2196.89	6.525	8.61
546	1482.89	5.463	9.24	606	1827.09	6.008	8.92	666	2203.41	6.533	8.61
547	1488.36	5.472	9.23	607	1833.10	6.016	8.92	667	2209.95	6.542	8.60
548	1493.84	5.481	9.23	608	1839.12	6.025	8.91	668	2216.50	6.551	8.60
549	1499.32	5.490	9.22	609	1845.15	6.034	8.91	669	2223.05	6.559	8.59
550	1504.82	5.500	9.21	610	1851.19	6.043	8.90	670	2229.62	6.568	8.59
551	1510.32	5.509	9.21	611	1857.24	6.052	8.90	671	2236.19	6.576	8.58
552	1515.83	5.518	9.20	612	1863.30	6.061	8.89	672	2242.77	6.585	8.57
553	1521.36	5.527	9.20	613	1869.36	6.070	8.89	673	2249.36	6.594	8.57
554	1526.89	5.536	9.19	614	1875.44	6.079	8.88	674	2255.96	6.602	8.56
555	1532.43	5.546	9.19	615	1881.52	6.088	8.88	675	2262.56	6.611	8.56
556	1537.98	5.555	9.18	616	1887.61	6.096	8.87	676	2269.18	6.619	8.55
557	1543.54	5.564	9.18	617	1893.71	6.105	8.87	677	2275.80	6.628	8.55
558	1549.11	5.573	9.17	618	1899.82	6.114	8.86	678	2282.43	6.636	8.54
559	1554.69	5.582	9.17	619	1905.94	6.123	8.86	679	2289.07	6.645	8.54
560	1560.27	5.591	9.16	620	1912.07	6.132	8.85	680	2295.72	6.653	8.53
561	1565.87	5.601	9.16	621	1918.20	6.141	8.84	681	2302.38	6.662	8.53
562	1571.47	5.610	9.15	622	1924.35	6.150	8.84	682	2309.05	6.670	8.52
563	1577.09	5.619	9.15	623	1930.50	6.158	8.83	683	2315.72	6.679	8.52
564	1582.71	5.628	9.14	624	1936.67	6.167	8.83	684	2322.41	6.687	8.51
565	1588.34	5.637	9.14	625	1942.84	6.176	8.82	685	2329.10	6.696	8.51
566	1593.99	5.646	9.13	626	1949.02	6.185	8.82	686	2335.80	6.705	8.50
567	1599.64	5.655	9.13	627	1955.21	6.194	8.81	687	2342.51	6.713	8.49
568	1605.30	5.665	9.12	628	1961.41	6.203	8.81	688	2349.22	6.721	8.49
569	1610.97	5.674	9.12	629	1967.61	6.211	8.80	689	2355.95	6.730	8.48
570	1616.64	5.683	9.11	630	1973.83	6.220	8.80	690	2362.68	6.738	8.48
571	1622.33	5.692	9.11	631	1980.05	6.229	8.79	691	2369.43	6.747	8.47
572	1628.03	5.701	9.10	632	1986.29	6.238	8.79	692	2376.18	6.755	8.47
573	1633.73	5.710	9.10	633	1992.53	6.247	8.78	693	2382.94	6.764	8.46
574	1639.45	5.719	9.09	634	1998.78	6.255	8.78	694	2389.71	6.772	8.46
575	1645.17	5.728	9.09	635	2005.04	6.264	8.77	695	2396.48	6.781	8.45
576	1650.91	5.737	9.08	636	2011.31	6.273	8.77	696	2403.27	6.789	8.45
577	1656.65	5.747	9.08	637	2017.59	6.282	8.76	697	2410.06	6.798	8.44
578	1662.40	5.756	9.07	638	2023.87	6.290	8.76	698	2416.86	6.806	8.44
579	1668.16	5.765	9.07	639	2030.17	6.299	8.75	699	2423.67	6.815	8.43
580	1673.93	5.774	9.06	640	2036.47	6.308	8.74	700	2430.49	6.823	8.43
581	1679.71	5.783	9.05	641	2042.78	6.317	8.74	701	2437.32	6.831	8.42
582	1685.49	5.792	9.05	642	2049.10	6.325	8.73	702	2444.15	6.840	8.41
583	1691.29	5.801	9.04	643	2055.43	6.334	8.73	703	2451.00	6.848	8.41
584	1697.09	5.810	9.04	644	2061.77	6.343	8.72	704	2457.85	6.857	8.40
585	1702.91	5.819	9.03	645	2068.12	6.352	8.72	705	2464.71	6.865	8.40
586	1708.73	5.828	9.03	646	2074.47	6.360	8.71	706	2471.58	6.873	8.39
587	1714.57	5.837	9.02	647	2080.84	6.369	8.71	707	2478.46	6.882	8.39
588	1720.41	5.846	9.02	648	2087.21	6.378	8.70	708	2485.34	6.890	8.38
589	1726.26	5.855	9.01	649	2093.59	6.386	8.70	709	2492.24	6.899	8.38
590	1732.12	5.864	9.01	650	2099.99	6.395	8.69	710	2499.14	6.907	8.37
591	1737.99	5.873	9.00	651	2106.38	6.404	8.69	711	2506.05	6.915	8.37
592	1743.86	5.882	9.00	652	2112.79	6.412	8.68	712	2512.97	6.924	8.36
593	1749.75	5.891	8.99	653	2119.21	6.421	8.68	713	2519.90	6.932	8.36
594	1755.65	5.900	8.99	654	2125.63	6.430	8.67	714	2526.84	6.940	8.35
595	1761.55	5.909	8.98	655	2132.07	6.438	8.66	715	2533.78	6.949	8.35
596	1767.46	5.918	8.98	656	2138.51	6.447	8.66	716	2540.73	6.957	8.34
597	1773.39	5.927	8.97	657	2144.96	6.456	8.65	717	2547.70	6.965	8.33
598	1779.32	5.936	8.97	658	2151.42	6.464	8.65	718	2554.66	6.974	8.33
599	1785.26	5.945	8.96	659	2157.89	6.473	8.64	719	2561.64	6.982	8.32
600	1791.21	5.954	8.96	660	2164.37	6.482	8.64	720	2568.63	6.990	8.32

TABLE 1i6. (Continued)

Type B thermocouples—*thermoelectric voltages,* E(T), *Seebeck coefficients,* S(T), **and first** *derivative of the Seebeck coefficients,* dS/dT, *reference junctions at 0 °C—Continued*

T °C	E μV	S μV/°C	dS/dT nV/°C²	T °C	E μV	S μV/°C	dS/dT nV/°C²	T °C	E μV	S μV/°C	dS/dT nV/°C²
720	2568.63	6.990	8.32	780	3002.84	7.480	8.00	840	3465.84	7.950	7.68
721	2575.62	6.999	8.31	781	3010.32	7.488	7.99	841	3473.80	7.958	7.68
722	2582.63	7.007	8.31	782	3017.81	7.496	7.99	842	3481.76	7.966	7.67
723	2589.64	7.015	8.30	783	3025.31	7.504	7.98	843	3489.73	7.973	7.67
724	2596.66	7.024	8.30	784	3032.82	7.512	7.98	844	3497.71	7.981	7.66
725	2603.68	7.032	8.29	785	3040.34	7.520	7.97	845	3505.69	7.989	7.66
726	2610.72	7.040	8.29	786	3047.86	7.528	7.97	846	3513.69	7.996	7.65
727	2617.77	7.049	8.28	787	3055.39	7.536	7.96	847	3521.69	8.004	7.65
728	2624.82	7.057	8.28	788	3062.93	7.544	7.96	848	3529.69	8.012	7.64
729	2631.88	7.065	8.27	789	3070.48	7.552	7.95	849	3537.71	8.019	7.64
730	2638.95	7.073	8.27	790	3078.04	7.560	7.95	850	3545.73	8.027	7.63
731	2646.03	7.082	8.26	791	3085.60	7.568	7.94	851	3553.76	8.035	7.62
732	2653.11	7.090	8.25	792	3093.17	7.576	7.94	852	3561.80	8.042	7.62
733	2660.21	7.098	8.25	793	3100.75	7.584	7.93	853	3569.85	8.050	7.61
734	2667.31	7.106	8.24	794	3108.34	7.591	7.93	854	3577.90	8.058	7.61
735	2674.42	7.115	8.24	795	3115.93	7.599	7.92	855	3585.96	8.065	7.60
736	2681.54	7.123	8.23	796	3123.54	7.607	7.92	856	3594.03	8.073	7.60
737	2688.66	7.131	8.23	797	3131.15	7.615	7.91	857	3602.11	8.080	7.59
738	2695.80	7.139	8.22	798	3138.77	7.623	7.90	858	3610.19	8.088	7.59
739	2702.94	7.148	8.22	799	3146.39	7.631	7.90	859	3618.28	8.095	7.58
740	2710.09	7.156	8.21	800	3154.03	7.639	7.89	860	3626.38	8.103	7.58
741	2717.25	7.164	8.21	801	3161.67	7.647	7.89	861	3634.49	8.111	7.57
742	2724.42	7.172	8.20	802	3169.32	7.655	7.88	862	3642.60	8.118	7.57
743	2731.60	7.180	8.20	803	3176.98	7.663	7.88	863	3650.73	8.126	7.56
744	2738.78	7.189	8.19	804	3184.65	7.670	7.87	864	3658.86	8.133	7.56
745	2745.98	7.197	8.19	805	3192.32	7.678	7.87	865	3666.99	8.141	7.55
746	2753.18	7.205	8.18	806	3200.00	7.686	7.86	866	3675.14	8.148	7.55
747	2760.38	7.213	8.18	807	3207.69	7.694	7.86	867	3683.29	8.156	7.54
748	2767.60	7.221	8.17	808	3215.39	7.702	7.85	868	3691.45	8.164	7.54
749	2774.83	7.229	8.16	809	3223.10	7.710	7.85	869	3699.62	8.171	7.53
750	2782.06	7.238	8.16	810	3230.81	7.718	7.84	870	3707.79	8.179	7.52
751	2789.30	7.246	8.15	811	3238.53	7.725	7.84	871	3715.97	8.186	7.52
752	2796.55	7.254	8.15	812	3246.26	7.733	7.83	872	3724.16	8.194	7.51
753	2803.81	7.262	8.14	813	3254.00	7.741	7.83	873	3732.36	8.201	7.51
754	2811.08	7.270	8.14	814	3261.74	7.749	7.82	874	3740.57	8.209	7.50
755	2818.35	7.278	8.13	815	3269.50	7.757	7.82	875	3748.78	8.216	7.50
756	2825.63	7.286	8.13	816	3277.26	7.765	7.81	876	3757.00	8.224	7.49
757	2832.92	7.295	8.12	817	3285.03	7.772	7.80	877	3765.23	8.231	7.49
758	2840.22	7.303	8.12	818	3292.80	7.780	7.80	878	3773.46	8.239	7.48
759	2847.53	7.311	8.11	819	3300.59	7.788	7.79	879	3781.70	8.246	7.48
760	2854.84	7.319	8.11	820	3308.38	7.796	7.79	880	3789.95	8.254	7.47
761	2862.17	7.327	8.10	821	3316.18	7.804	7.78	881	3798.21	8.261	7.47
762	2869.50	7.335	8.10	822	3323.99	7.811	7.78	882	3806.48	8.268	7.46
763	2876.84	7.343	8.09	823	3331.80	7.819	7.77	883	3814.75	8.276	7.46
764	2884.18	7.351	8.08	824	3339.62	7.827	7.77	884	3823.03	8.283	7.45
765	2891.54	7.359	8.08	825	3347.46	7.835	7.76	885	3831.31	8.291	7.45
766	2898.90	7.367	8.07	826	3355.29	7.842	7.76	886	3839.61	8.298	7.44
767	2906.27	7.376	8.07	827	3363.14	7.850	7.75	887	3847.91	8.306	7.43
768	2913.65	7.384	8.06	828	3370.99	7.858	7.75	888	3856.22	8.313	7.43
769	2921.04	7.392	8.06	829	3378.86	7.866	7.74	889	3864.54	8.321	7.42
770	2928.44	7.400	8.05	830	3386.73	7.873	7.74	890	3872.86	8.328	7.42
771	2935.84	7.408	8.05	831	3394.60	7.881	7.73	891	3881.19	8.335	7.41
772	2943.25	7.416	8.04	832	3402.49	7.889	7.73	892	3889.53	8.343	7.41
773	2950.67	7.424	8.04	833	3410.38	7.897	7.72	893	3897.88	8.350	7.40
774	2958.10	7.432	8.03	834	3418.28	7.904	7.71	894	3906.23	8.358	7.40
775	2965.54	7.440	8.03	835	3426.19	7.912	7.71	895	3914.59	8.365	7.39
776	2972.98	7.448	8.02	836	3434.10	7.920	7.70	896	3922.96	8.372	7.39
777	2980.43	7.456	8.02	837	3442.03	7.927	7.70	897	3931.34	8.380	7.38
778	2987.89	7.464	8.01	838	3449.96	7.935	7.69	898	3939.72	8.387	7.38
779	2995.36	7.472	8.01	839	3457.90	7.943	7.69	899	3948.11	8.395	7.37
780	3002.84	7.480	8.00	840	3465.84	7.950	7.68	900	3956.51	8.402	7.37

TABLE 1i6. (*Continued*)

Type B thermocouples—*thermoelectric voltages, E(T), Seebeck coefficients, S(T), and first derivative of the Seebeck coefficients, dS/dT, reference junctions at 0 °C—Continued*

T °C	E μV	S μV/°C	dS/dT nV/°C²	T °C	E μV	S μV/°C	dS/dT nV/°C²	T °C	E μV	S μV/°C	dS/dT nV/°C²
900	3956.51	8.402	7.37	960	4473.69	8.834	7.04	1020	5016.22	9.247	6.71
901	3964.92	8.409	7.36	961	4482.53	8.841	7.04	1021	5025.47	9.253	6.70
902	3973.33	8.417	7.35	962	4491.38	8.848	7.03	1022	5034.73	9.260	6.69
903	3981.75	8.424	7.35	963	4500.23	8.855	7.03	1023	5044.00	9.267	6.69
904	3990.18	8.431	7.34	964	4509.09	8.862	7.02	1024	5053.27	9.274	6.68
905	3998.61	8.439	7.34	965	4517.95	8.869	7.02	1025	5062.54	9.280	6.68
906	4007.06	8.446	7.33	966	4526.83	8.876	7.01	1026	5071.83	9.287	6.67
907	4015.51	8.453	7.33	967	4535.71	8.883	7.00	1027	5081.12	9.294	6.67
908	4023.96	8.461	7.32	968	4544.59	8.890	7.00	1028	5090.41	9.300	6.66
909	4032.43	8.468	7.32	969	4553.49	8.897	6.99	1029	5099.72	9.307	6.65
910	4040.90	8.475	7.31	970	4562.39	8.904	6.99	1030	5109.03	9.314	6.65
911	4049.38	8.483	7.31	971	4571.30	8.911	6.98	1031	5118.34	9.320	6.64
912	4057.86	8.490	7.30	972	4580.21	8.918	6.98	1032	5127.67	9.327	6.64
913	4066.36	8.497	7.30	973	4589.13	8.925	6.97	1033	5137.00	9.333	6.63
914	4074.86	8.505	7.29	974	4598.06	8.932	6.97	1034	5146.33	9.340	6.63
915	4083.37	8.512	7.29	975	4607.00	8.939	6.96	1035	5155.68	9.347	6.62
916	4091.88	8.519	7.28	976	4615.94	8.946	6.95	1036	5165.03	9.353	6.61
917	4100.40	8.526	7.27	977	4624.89	8.953	6.95	1037	5174.38	9.360	6.61
918	4108.93	8.534	7.27	978	4633.85	8.960	6.94	1038	5183.75	9.367	6.60
919	4117.47	8.541	7.26	979	4642.81	8.967	6.94	1039	5193.12	9.373	6.60
920	4126.02	8.548	7.26	980	4651.78	8.974	6.93	1040	5202.49	9.380	6.59
921	4134.57	8.555	7.25	981	4660.76	8.981	6.93	1041	5211.88	9.386	6.58
922	4143.13	8.563	7.25	982	4669.74	8.988	6.92	1042	5221.27	9.393	6.58
923	4151.69	8.570	7.24	983	4678.73	8.995	6.92	1043	5230.66	9.399	6.57
924	4160.27	8.577	7.24	984	4687.73	9.002	6.91	1044	5240.07	9.406	6.57
925	4168.85	8.584	7.23	985	4696.74	9.009	6.90	1045	5249.47	9.413	6.56
926	4177.44	8.592	7.23	986	4705.75	9.015	6.90	1046	5258.89	9.419	6.55
927	4186.03	8.599	7.22	987	4714.77	9.022	6.89	1047	5268.31	9.426	6.55
928	4194.63	8.606	7.22	988	4723.79	9.029	6.89	1048	5277.74	9.432	6.54
929	4203.24	8.613	7.21	989	4732.83	9.036	6.88	1049	5287.18	9.439	6.54
930	4211.86	8.621	7.21	990	4741.86	9.043	6.88	1050	5296.62	9.445	6.53
931	4220.48	8.628	7.20	991	4750.91	9.050	6.87	1051	5306.07	9.452	6.52
932	4229.12	8.635	7.19	992	4759.96	9.057	6.87	1052	5315.52	9.458	6.52
933	4237.75	8.642	7.19	993	4769.02	9.064	6.86	1053	5324.99	9.465	6.51
934	4246.40	8.649	7.18	994	4778.09	9.070	6.85	1054	5334.45	9.471	6.51
935	4255.05	8.656	7.18	995	4787.17	9.077	6.85	1055	5343.93	9.478	6.50
936	4263.71	8.664	7.17	996	4796.25	9.084	6.84	1056	5353.41	9.484	6.49
937	4272.38	8.671	7.17	997	4805.33	9.091	6.84	1057	5362.90	9.491	6.49
938	4281.05	8.678	7.16	998	4814.43	9.098	6.83	1058	5372.39	9.497	6.48
939	4289.74	8.685	7.16	999	4823.53	9.105	6.83	1059	5381.89	9.504	6.48
940	4298.43	8.692	7.15	1000	4832.64	9.112	6.82	1060	5391.40	9.510	6.47
941	4307.12	8.699	7.15	1001	4841.75	9.118	6.81	1061	5400.91	9.517	6.46
942	4315.82	8.707	7.14	1002	4850.87	9.125	6.81	1062	5410.43	9.523	6.46
943	4324.53	8.714	7.14	1003	4860.00	9.132	6.80	1063	5419.96	9.530	6.45
944	4333.25	8.721	7.13	1004	4869.14	9.139	6.80	1064	5429.49	9.536	6.45
945	4341.98	8.728	7.12	1005	4878.28	9.146	6.79	1065	5439.03	9.543	6.44
946	4350.71	8.735	7.12	1006	4887.43	9.152	6.79	1066	5448.58	9.549	6.43
947	4359.45	8.742	7.11	1007	4896.59	9.159	6.78	1067	5458.13	9.555	6.43
948	4368.19	8.749	7.11	1008	4905.75	9.166	6.78	1068	5467.69	9.562	6.42
949	4376.94	8.756	7.10	1009	4914.92	9.173	6.77	1069	5477.25	9.568	6.41
950	4385.70	8.764	7.10	1010	4924.09	9.179	6.76	1070	5486.82	9.575	6.41
951	4394.47	8.771	7.09	1011	4933.28	9.186	6.76	1071	5496.40	9.581	6.40
952	4403.25	8.778	7.09	1012	4942.47	9.193	6.75	1072	5505.99	9.588	6.40
953	4412.03	8.785	7.08	1013	4951.66	9.200	6.75	1073	5515.58	9.594	6.39
954	4420.82	8.792	7.08	1014	4960.86	9.206	6.74	1074	5525.18	9.600	6.38
955	4429.61	8.799	7.07	1015	4970.07	9.213	6.74	1075	5534.78	9.607	6.38
956	4438.41	8.806	7.06	1016	4979.29	9.220	6.73	1076	5544.39	9.613	6.37
957	4447.22	8.813	7.06	1017	4988.51	9.227	6.72	1077	5554.00	9.619	6.37
958	4456.04	8.820	7.05	1018	4997.74	9.233	6.72	1078	5563.63	9.626	6.36
959	4464.86	8.827	7.05	1019	5006.98	9.240	6.71	1079	5573.26	9.632	6.35
960	4473.69	8.834	7.04	1020	5016.22	9.247	6.71	1080	5582.89	9.639	6.35

TABLE 1i6. (Continued)

Type B thermocouples—*thermoelectric voltages,* E(T), *Seebeck coefficients,* S(T), **and first** *derivative of the Seebeck coefficients,* dS/dT, *reference junctions at 0 °C—Continued*

T °C	E μV	S μV/°C	dS/dT nV/°C²	T °C	E μV	S μV/°C	dS/dT nV/°C²	T °C	E μV	S μV/°C	dS/dT nV/°C²
1080	5582.89	9.639	6.35	1140	6172.40	10.008	5.95	1200	6783.31	10.352	5.51
1081	5592.53	9.645	6.34	1141	6182.41	10.014	5.94	1201	6793.66	10.357	5.50
1082	5602.18	9.651	6.33	1142	6192.42	10.020	5.94	1202	6804.02	10.363	5.49
1083	5611.84	9.658	6.33	1143	6202.45	10.025	5.93	1203	6814.39	10.368	5.48
1084	5621.50	9.664	6.32	1144	6212.47	10.031	5.92	1204	6824.76	10.374	5.47
1085	5631.16	9.670	6.32	1145	6222.51	10.037	5.92	1205	6835.14	10.379	5.47
1086	5640.84	9.676	6.31	1146	6232.55	10.043	5.91	1206	6845.52	10.385	5.46
1087	5650.52	9.683	6.30	1147	6242.60	10.049	5.90	1207	6855.91	10.390	5.45
1088	5660.20	9.689	6.30	1148	6252.65	10.055	5.89	1208	6866.30	10.395	5.44
1089	5669.89	9.695	6.29	1149	6262.71	10.061	5.89	1209	6876.70	10.401	5.43
1090	5679.59	9.702	6.28	1150	6272.77	10.067	5.88	1210	6887.10	10.406	5.43
1091	5689.30	9.708	6.28	1151	6282.84	10.073	5.87	1211	6897.51	10.412	5.42
1092	5699.01	9.714	6.27	1152	6292.91	10.079	5.87	1212	6907.92	10.417	5.41
1093	5708.73	9.720	6.26	1153	6303.00	10.084	5.86	1213	6918.34	10.423	5.40
1094	5718.45	9.727	6.26	1154	6313.08	10.090	5.85	1214	6928.77	10.428	5.39
1095	5728.18	9.733	6.25	1155	6323.18	10.096	5.85	1215	6939.20	10.433	5.39
1096	5737.92	9.739	6.25	1156	6333.28	10.102	5.84	1216	6949.64	10.439	5.38
1097	5747.66	9.746	6.24	1157	6343.38	10.108	5.83	1217	6960.08	10.444	5.37
1098	5757.41	9.752	6.23	1158	6353.49	10.114	5.82	1218	6970.52	10.449	5.36
1099	5767.16	9.758	6.23	1159	6363.61	10.119	5.82	1219	6980.98	10.455	5.35
1100	5776.92	9.764	6.22	1160	6373.73	10.125	5.81	1220	6991.43	10.460	5.34
1101	5786.69	9.770	6.21	1161	6383.86	10.131	5.80	1221	7001.90	10.466	5.34
1102	5796.46	9.777	6.21	1162	6393.99	10.137	5.79	1222	7012.36	10.471	5.33
1103	5806.24	9.783	6.20	1163	6404.13	10.143	5.79	1223	7022.84	10.476	5.32
1104	5816.03	9.789	6.19	1164	6414.28	10.148	5.78	1224	7033.32	10.481	5.31
1105	5825.82	9.795	6.19	1165	6424.43	10.154	5.77	1225	7043.80	10.487	5.30
1106	5835.62	9.801	6.18	1166	6434.59	10.160	5.77	1226	7054.29	10.492	5.29
1107	5845.42	9.808	6.17	1167	6444.75	10.166	5.76	1227	7064.79	10.497	5.29
1108	5855.24	9.814	6.17	1168	6454.92	10.172	5.75	1228	7075.29	10.503	5.28
1109	5865.05	9.820	6.16	1169	6465.09	10.177	5.74	1229	7085.79	10.508	5.27
1110	5874.87	9.826	6.15	1170	6475.27	10.183	5.74	1230	7096.30	10.513	5.26
1111	5884.70	9.832	6.15	1171	6485.46	10.189	5.73	1231	7106.82	10.518	5.25
1112	5894.54	9.838	6.14	1172	6495.65	10.194	5.72	1232	7117.34	10.524	5.24
1113	5904.38	9.844	6.13	1173	6505.85	10.200	5.71	1233	7127.86	10.529	5.24
1114	5914.23	9.851	6.13	1174	6516.05	10.206	5.71	1234	7138.40	10.534	5.23
1115	5924.08	9.857	6.12	1175	6526.26	10.212	5.70	1235	7148.93	10.539	5.22
1116	5933.94	9.863	6.11	1176	6536.47	10.217	5.69	1236	7159.48	10.545	5.21
1117	5943.81	9.869	6.11	1177	6546.69	10.223	5.68	1237	7170.02	10.550	5.20
1118	5953.68	9.875	6.10	1178	6556.92	10.229	5.68	1238	7180.57	10.555	5.19
1119	5963.56	9.881	6.09	1179	6567.15	10.234	5.67	1239	7191.13	10.560	5.18
1120	5973.44	9.887	6.09	1180	6577.39	10.240	5.66	1240	7201.70	10.565	5.18
1121	5983.33	9.893	6.08	1181	6587.63	10.246	5.65	1241	7212.26	10.571	5.17
1122	5993.23	9.899	6.07	1182	6597.88	10.251	5.65	1242	7222.84	10.576	5.16
1123	6003.13	9.906	6.07	1183	6608.13	10.257	5.64	1243	7233.41	10.581	5.15
1124	6013.04	9.912	6.06	1184	6618.39	10.263	5.63	1244	7244.00	10.586	5.14
1125	6022.95	9.918	6.05	1185	6628.66	10.268	5.62	1245	7254.59	10.591	5.13
1126	6032.88	9.924	6.05	1186	6638.93	10.274	5.62	1246	7265.18	10.596	5.12
1127	6042.80	9.930	6.04	1187	6649.21	10.279	5.61	1247	7275.78	10.601	5.11
1128	6052.73	9.936	6.03	1188	6659.49	10.285	5.60	1248	7286.38	10.607	5.11
1129	6062.67	9.942	6.03	1189	6669.78	10.291	5.59	1249	7296.99	10.612	5.10
1130	6072.62	9.948	6.02	1190	6680.07	10.296	5.58	1250	7307.61	10.617	5.09
1131	6082.57	9.954	6.01	1191	6690.37	10.302	5.58	1251	7318.23	10.622	5.08
1132	6092.53	9.960	6.01	1192	6700.67	10.307	5.57	1252	7328.85	10.627	5.07
1133	6102.49	9.966	6.00	1193	6710.98	10.313	5.56	1253	7339.48	10.632	5.06
1134	6112.46	9.972	5.99	1194	6721.30	10.318	5.55	1254	7350.11	10.637	5.05
1135	6122.43	9.978	5.99	1195	6731.62	10.324	5.55	1255	7360.75	10.642	5.04
1136	6132.41	9.984	5.98	1196	6741.95	10.330	5.54	1256	7371.40	10.647	5.03
1137	6142.40	9.990	5.97	1197	6752.28	10.335	5.53	1257	7382.05	10.652	5.03
1138	6152.39	9.996	5.97	1198	6762.62	10.341	5.52	1258	7392.70	10.657	5.02
1139	6162.39	10.002	5.96	1199	6772.96	10.346	5.51	1259	7403.36	10.662	5.01
1140	6172.40	10.008	5.95	1200	6783.31	10.352	5.51	1260	7414.03	10.667	5.00

TABLE 1i6. (Continued)

Type B thermocouples—*thermoelectric voltages*, $E(T)$, *Seebeck coefficients*, $S(T)$, *and first derivative of the Seebeck coefficients*, dS/dT, *reference junctions at 0 °C—Continued*

T °C	E μV	S μV/°C	dS/dT nV/°C^2	T °C	E μV	S μV/°C	dS/dT nV/°C^2	T °C	E μV	S μV/°C	dS/dT nV/°C^2
1260	7414.03	10.667	5.00	1320	8062.71	10.950	4.41	1380	8727.27	11.195	3.75
1261	7424.70	10.672	4.99	1321	8073.67	10.954	4.40	1381	8738.47	11.199	3.73
1262	7435.37	10.677	4.98	1322	8084.62	10.959	4.39	1382	8749.67	11.203	3.72
1263	7446.05	10.682	4.97	1323	8095.58	10.963	4.38	1383	8760.87	11.206	3.71
1264	7456.74	10.687	4.96	1324	8106.55	10.968	4.37	1384	8772.08	11.210	3.70
1265	7467.42	10.692	4.95	1325	8117.52	10.972	4.36	1385	8783.29	11.214	3.69
1266	7478.12	10.697	4.94	1326	8128.49	10.976	4.35	1386	8794.51	11.217	3.67
1267	7488.82	10.702	4.93	1327	8139.47	10.981	4.34	1387	8805.73	11.221	3.66
1268	7499.52	10.707	4.92	1328	8150.45	10.985	4.33	1388	8816.95	11.225	3.65
1269	7510.23	10.712	4.92	1329	8161.44	10.989	4.32	1389	8828.18	11.228	3.64
1270	7520.95	10.717	4.91	1330	8172.43	10.994	4.31	1390	8839.41	11.232	3.63
1271	7531.67	10.722	4.90	1331	8183.43	10.998	4.30	1391	8850.64	11.236	3.61
1272	7542.39	10.726	4.89	1332	8194.43	11.002	4.29	1392	8861.88	11.239	3.60
1273	7553.12	10.731	4.88	1333	8205.43	11.006	4.28	1393	8873.12	11.243	3.59
1274	7563.85	10.736	4.87	1334	8216.44	11.011	4.27	1394	8884.36	11.246	3.58
1275	7574.59	10.741	4.86	1335	8227.45	11.015	4.26	1395	8895.61	11.250	3.57
1276	7585.33	10.746	4.85	1336	8238.47	11.019	4.25	1396	8906.86	11.254	3.55
1277	7596.08	10.751	4.84	1337	8249.49	11.023	4.23	1397	8918.12	11.257	3.54
1278	7606.84	10.756	4.83	1338	8260.52	11.028	4.22	1398	8929.38	11.261	3.53
1279	7617.59	10.760	4.82	1339	8271.55	11.032	4.21	1399	8940.64	11.264	3.52
1280	7628.36	10.765	4.81	1340	8282.58	11.036	4.20	1400	8951.91	11.268	3.50
1281	7639.12	10.770	4.80	1341	8293.62	11.040	4.19	1401	8963.18	11.271	3.49
1282	7649.90	10.775	4.79	1342	8304.66	11.045	4.18	1402	8974.45	11.275	3.48
1283	7660.67	10.780	4.78	1343	8315.71	11.049	4.17	1403	8985.73	11.278	3.47
1284	7671.46	10.784	4.77	1344	8326.76	11.053	4.16	1404	8997.01	11.282	3.45
1285	7682.24	10.789	4.76	1345	8337.81	11.057	4.15	1405	9008.29	11.285	3.44
1286	7693.03	10.794	4.76	1346	8348.87	11.061	4.14	1406	9019.58	11.289	3.43
1287	7703.83	10.799	4.75	1347	8359.94	11.065	4.12	1407	9030.87	11.292	3.42
1288	7714.63	10.803	4.74	1348	8371.00	11.069	4.11	1408	9042.16	11.295	3.40
1289	7725.44	10.808	4.73	1349	8382.08	11.073	4.10	1409	9053.46	11.299	3.39
1290	7736.25	10.813	4.72	1350	8393.15	11.078	4.09	1410	9064.76	11.302	3.38
1291	7747.06	10.818	4.71	1351	8404.23	11.082	4.08	1411	9076.06	11.306	3.37
1292	7757.88	10.822	4.70	1352	8415.31	11.086	4.07	1412	9087.37	11.309	3.35
1293	7768.71	10.827	4.69	1353	8426.40	11.090	4.06	1413	9098.68	11.312	3.34
1294	7779.54	10.832	4.68	1354	8437.49	11.094	4.05	1414	9109.99	11.316	3.33
1295	7790.37	10.836	4.67	1355	8448.59	11.098	4.04	1415	9121.31	11.319	3.32
1296	7801.21	10.841	4.66	1356	8459.69	11.102	4.02	1416	9132.63	11.322	3.30
1297	7812.05	10.846	4.65	1357	8470.79	11.106	4.01	1417	9143.95	11.325	3.29
1298	7822.90	10.850	4.64	1358	8481.90	11.110	4.00	1418	9155.28	11.329	3.28
1299	7833.75	10.855	4.63	1359	8493.01	11.114	3.99	1419	9166.61	11.332	3.26
1300	7844.61	10.860	4.62	1360	8504.13	11.118	3.98	1420	9177.95	11.335	3.25
1301	7855.47	10.864	4.61	1361	8515.25	11.122	3.97	1421	9189.28	11.339	3.24
1302	7866.34	10.869	4.60	1362	8526.37	11.126	3.96	1422	9200.62	11.342	3.23
1303	7877.21	10.873	4.59	1363	8537.50	11.130	3.94	1423	9211.97	11.345	3.21
1304	7888.09	10.878	4.58	1364	8548.63	11.134	3.93	1424	9223.31	11.348	3.20
1305	7898.97	10.883	4.57	1365	8559.77	11.138	3.92	1425	9234.66	11.351	3.19
1306	7909.85	10.887	4.56	1366	8570.91	11.142	3.91	1426	9246.02	11.355	3.17
1307	7920.74	10.892	4.55	1367	8582.05	11.146	3.90	1427	9257.37	11.358	3.16
1308	7931.64	10.896	4.54	1368	8593.20	11.149	3.89	1428	9268.73	11.361	3.15
1309	7942.53	10.901	4.53	1369	8604.35	11.153	3.88	1429	9280.09	11.364	3.13
1310	7953.44	10.905	4.52	1370	8615.51	11.157	3.86	1430	9291.46	11.367	3.12
1311	7964.34	10.910	4.51	1371	8626.67	11.161	3.85	1431	9302.83	11.370	3.11
1312	7975.26	10.914	4.50	1372	8637.83	11.165	3.84	1432	9314.20	11.373	3.09
1313	7986.17	10.919	4.49	1373	8648.99	11.169	3.83	1433	9325.57	11.376	3.08
1314	7997.09	10.923	4.48	1374	8660.17	11.173	3.82	1434	9336.95	11.380	3.07
1315	8008.02	10.928	4.47	1375	8671.34	11.176	3.81	1435	9348.33	11.383	3.05
1316	8018.95	10.932	4.46	1376	8682.52	11.180	3.79	1436	9359.72	11.386	3.04
1317	8029.88	10.937	4.45	1377	8693.70	11.184	3.78	1437	9371.10	11.389	3.03
1318	8040.82	10.941	4.44	1378	8704.89	11.188	3.77	1438	9382.49	11.392	3.01
1319	8051.77	10.946	4.43	1379	8716.08	11.191	3.76	1439	9393.89	11.395	3.00
1320	8062.71	10.950	4.41	1380	8727.27	11.195	3.75	1440	9405.28	11.398	2.99

TABLE 1i6. (Continued)

Type B thermocouples—*thermoelectric voltages*, E(T), *Seebeck coefficients*, S(T), *and first derivative of the Seebeck coefficients*, dS/dT, *reference junctions at 0 °C—Continued*

T °C	E μV	S μV/°C	dS/dT nV/°C^2	T °C	E μV	S μV/°C	dS/dT nV/°C^2	T °C	E μV	S μV/°C	dS/dT nV/°C^2
1440	9405.28	11.398	2.99	1500	10094.03	11.552	2.14	1560	10790.46	11.653	1.22
1441	9416.68	11.401	2.97	1501	10105.58	11.554	2.13	1561	10802.11	11.654	1.20
1442	9428.09	11.404	2.96	1502	10117.14	11.556	2.11	1562	10813.77	11.655	1.18
1443	9439.49	11.407	2.95	1503	10128.69	11.558	2.10	1563	10825.42	11.657	1.17
1444	9450.90	11.410	2.93	1504	10140.25	11.560	2.08	1564	10837.08	11.658	1.15
1445	9462.31	11.412	2.92	1505	10151.82	11.563	2.07	1565	10848.74	11.659	1.14
1446	9473.72	11.415	2.91	1506	10163.38	11.565	2.05	1566	10860.40	11.660	1.12
1447	9485.14	11.418	2.89	1507	10174.94	11.567	2.04	1567	10872.06	11.661	1.10
1448	9496.56	11.421	2.88	1508	10186.51	11.569	2.02	1568	10883.72	11.662	1.09
1449	9507.98	11.424	2.87	1509	10198.08	11.571	2.01	1569	10895.38	11.663	1.07
1450	9519.41	11.427	2.85	1510	10209.65	11.573	1.99	1570	10907.05	11.664	1.06
1451	9530.84	11.430	2.84	1511	10221.23	11.575	1.98	1571	10918.71	11.666	1.04
1452	9542.27	11.433	2.83	1512	10232.80	11.577	1.96	1572	10930.38	11.667	1.02
1453	9553.70	11.435	2.81	1513	10244.38	11.579	1.95	1573	10942.05	11.668	1.01
1454	9565.14	11.438	2.80	1514	10255.96	11.581	1.93	1574	10953.71	11.669	0.99
1455	9576.58	11.441	2.78	1515	10267.54	11.582	1.92	1575	10965.38	11.670	0.98
1456	9588.02	11.444	2.77	1516	10279.13	11.584	1.90	1576	10977.05	11.671	0.96
1457	9599.47	11.447	2.76	1517	10290.71	11.586	1.89	1577	10988.72	11.671	0.94
1458	9610.91	11.449	2.74	1518	10302.30	11.588	1.87	1578	11000.40	11.672	0.93
1459	9622.36	11.452	2.73	1519	10313.89	11.590	1.86	1579	11012.07	11.673	0.91
1460	9633.82	11.455	2.72	1520	10325.48	11.592	1.84	1580	11023.74	11.674	0.89
1461	9645.27	11.457	2.70	1521	10337.07	11.594	1.83	1581	11035.42	11.675	0.88
1462	9656.73	11.460	2.69	1522	10348.67	11.595	1.81	1582	11047.09	11.676	0.86
1463	9668.19	11.463	2.67	1523	10360.26	11.597	1.80	1583	11058.77	11.677	0.85
1464	9679.66	11.466	2.66	1524	10371.86	11.599	1.78	1584	11070.45	11.678	0.83
1465	9691.13	11.468	2.65	1525	10383.46	11.601	1.76	1585	11082.13	11.678	0.81
1466	9702.59	11.471	2.63	1526	10395.06	11.603	1.75	1586	11093.80	11.679	0.80
1467	9714.07	11.473	2.62	1527	10406.66	11.604	1.73	1587	11105.48	11.680	0.78
1468	9725.54	11.476	2.60	1528	10418.27	11.606	1.72	1588	11117.16	11.681	0.76
1469	9737.02	11.479	2.59	1529	10429.88	11.608	1.70	1589	11128.85	11.682	0.75
1470	9748.50	11.481	2.58	1530	10441.49	11.609	1.69	1590	11140.53	11.682	0.73
1471	9759.98	11.484	2.56	1531	10453.10	11.611	1.67	1591	11152.21	11.683	0.72
1472	9771.47	11.486	2.55	1532	10464.71	11.613	1.66	1592	11163.89	11.684	0.70
1473	9782.95	11.489	2.53	1533	10476.32	11.614	1.64	1593	11175.58	11.684	0.68
1474	9794.44	11.491	2.52	1534	10487.94	11.616	1.63	1594	11187.26	11.685	0.67
1475	9805.94	11.494	2.50	1535	10499.55	11.618	1.61	1595	11198.95	11.686	0.65
1476	9817.43	11.496	2.49	1536	10511.17	11.619	1.59	1596	11210.63	11.686	0.63
1477	9828.93	11.499	2.48	1537	10522.79	11.621	1.58	1597	11222.32	11.687	0.62
1478	9840.43	11.501	2.46	1538	10534.41	11.622	1.56	1598	11234.01	11.688	0.60
1479	9851.93	11.504	2.45	1539	10546.04	11.624	1.55	1599	11245.70	11.688	0.58
1480	9863.44	11.506	2.43	1540	10557.66	11.626	1.53	1600	11257.38	11.689	0.57
1481	9874.94	11.509	2.42	1541	10569.29	11.627	1.52	1601	11269.07	11.689	0.55
1482	9886.45	11.511	2.40	1542	10580.92	11.629	1.50	1602	11280.76	11.690	0.53
1483	9897.97	11.513	2.39	1543	10592.55	11.630	1.49	1603	11292.45	11.690	0.52
1484	9909.48	11.516	2.38	1544	10604.18	11.632	1.47	1604	11304.14	11.691	0.50
1485	9921.00	11.518	2.36	1545	10615.81	11.633	1.45	1605	11315.84	11.691	0.49
1486	9932.52	11.521	2.35	1546	10627.44	11.635	1.44	1606	11327.53	11.692	0.47
1487	9944.04	11.523	2.33	1547	10639.08	11.636	1.42	1607	11339.22	11.692	0.45
1488	9955.56	11.525	2.32	1548	10650.71	11.637	1.41	1608	11350.91	11.693	0.44
1489	9967.09	11.528	2.30	1549	10662.35	11.639	1.39	1609	11362.61	11.693	0.42
1490	9978.62	11.530	2.29	1550	10673.99	11.640	1.38	1610	11374.30	11.694	0.40
1491	9990.15	11.532	2.27	1551	10685.63	11.641	1.36	1611	11385.99	11.694	0.39
1492	10001.68	11.534	2.26	1552	10697.28	11.643	1.34	1612	11397.69	11.694	0.37
1493	10013.22	11.537	2.24	1553	10708.92	11.644	1.33	1613	11409.38	11.695	0.35
1494	10024.76	11.539	2.23	1554	10720.56	11.646	1.31	1614	11421.08	11.695	0.34
1495	10036.30	11.541	2.22	1555	10732.21	11.647	1.30	1615	11432.77	11.696	0.32
1496	10047.84	11.543	2.20	1556	10743.86	11.648	1.28	1616	11444.47	11.696	0.30
1497	10059.38	11.546	2.19	1557	10755.51	11.649	1.26	1617	11456.16	11.696	0.29
1498	10070.93	11.548	2.17	1558	10767.16	11.651	1.25	1618	11467.86	11.696	0.27
1499	10082.48	11.550	2.16	1559	10778.81	11.652	1.23	1619	11479.56	11.697	0.25
1500	10094.03	11.552	2.14	1560	10790.46	11.653	1.22	1620	11491.25	11.697	0.24

TABLE 1i6. (*Continued*)

Type B thermocouples—thermoelectric voltages, E(T), *Seebeck coefficients,* S(T), *and first derivative of the Seebeck coefficients,* dS/dT, *reference junctions at 0 °C—Continued*

T °C	E μV	S μV/°C	dS/dT nV/°C^2	T °C	E μV	S μV/°C	dS/dT nV/°C^2	T °C	E μV	S μV/°C	dS/dT nV/°C^2
1620	11491.25	11.697	0.24	1680	12192.89	11.681	−0.76	1740	12891.80	11.606	−1.73
1621	11502.95	11.697	0.22	1681	12204.57	11.680	−0.78	1741	12903.40	11.604	−1.74
1622	11514.65	11.697	0.20	1682	12216.25	11.680	−0.80	1742	12915.01	11.603	−1.76
1623	11526.35	11.698	0.19	1683	12227.93	11.679	−0.81	1743	12926.61	11.601	−1.77
1624	11538.04	11.698	0.17	1684	12239.61	11.678	−0.83	1744	12938.21	11.599	−1.79
1625	11549.74	11.698	0.15	1685	12251.29	11.677	−0.85	1745	12949.81	11.597	−1.80
1626	11561.44	11.698	0.14	1686	12262.97	11.676	−0.86	1746	12961.40	11.595	−1.82
1627	11573.14	11.698	0.12	1687	12274.64	11.675	−0.88	1747	12973.00	11.594	−1.83
1628	11584.83	11.698	0.10	1688	12286.32	11.674	−0.89	1748	12984.59	11.592	−1.85
1629	11596.53	11.698	0.09	1689	12297.99	11.674	−0.91	1749	12996.18	11.590	−1.86
1630	11608.23	11.698	0.07	1690	12309.66	11.673	−0.93	1750	13007.77	11.588	−1.88
1631	11619.93	11.698	0.05	1691	12321.34	11.672	−0.94	1751	13019.36	11.586	−1.89
1632	11631.63	11.699	0.04	1692	12333.01	11.671	−0.96	1752	13030.94	11.584	−1.91
1633	11643.33	11.699	0.02	1693	12344.68	11.670	−0.98	1753	13042.52	11.582	−1.92
1634	11655.03	11.699	0.00	1694	12356.35	11.669	−0.99	1754	13054.11	11.580	−1.94
1635	11666.72	11.699	−0.01	1695	12368.01	11.668	−1.01	1755	13065.68	11.578	−1.95
1636	11678.42	11.699	−0.03	1696	12379.68	11.667	−1.03	1756	13077.26	11.576	−1.97
1637	11690.12	11.699	−0.05	1697	12391.35	11.666	−1.04	1757	13088.84	11.574	−1.98
1638	11701.82	11.698	−0.06	1698	12403.01	11.665	−1.06	1758	13100.41	11.573	−2.00
1639	11713.52	11.698	−0.08	1699	12414.68	11.664	−1.07	1759	13111.98	11.570	−2.01
1640	11725.22	11.698	−0.10	1700	12426.34	11.663	−1.09	1760	13123.55	11.568	−2.03
1641	11736.91	11.698	−0.11	1701	12438.00	11.661	−1.11	1761	13135.12	11.566	−2.04
1642	11748.61	11.698	−0.13	1702	12449.66	11.660	−1.12	1762	13146.69	11.564	−2.06
1643	11760.31	11.698	−0.15	1703	12461.32	11.659	−1.14	1763	13158.25	11.562	−2.07
1644	11772.01	11.698	−0.16	1704	12472.98	11.658	−1.16	1764	13169.81	11.560	−2.09
1645	11783.71	11.698	−0.18	1705	12484.64	11.657	−1.17	1765	13181.37	11.558	−2.10
1646	11795.40	11.697	−0.20	1706	12496.30	11.656	−1.19	1766	13192.93	11.556	−2.12
1647	11807.10	11.697	−0.21	1707	12507.95	11.655	−1.20	1767	13204.48	11.554	−2.13
1648	11818.80	11.697	−0.23	1708	12519.60	11.653	−1.22	1768	13216.03	11.552	−2.15
1649	11830.50	11.697	−0.25	1709	12531.26	11.652	−1.24	1769	13227.58	11.550	−2.16
1650	11842.19	11.696	−0.26	1710	12542.91	11.651	−1.25	1770	13239.13	11.547	−2.17
1651	11853.89	11.696	−0.28	1711	12554.56	11.650	−1.27	1771	13250.68	11.545	−2.19
1652	11865.58	11.696	−0.30	1712	12566.21	11.648	−1.29	1772	13262.22	11.543	−2.20
1653	11877.28	11.696	−0.31	1713	12577.86	11.647	−1.30	1773	13273.77	11.541	−2.22
1654	11888.98	11.695	−0.33	1714	12589.50	11.646	−1.32	1774	13285.31	11.539	−2.23
1655	11900.67	11.695	−0.35	1715	12601.15	11.644	−1.33	1775	13296.84	11.536	−2.24
1656	11912.37	11.695	−0.36	1716	12612.79	11.643	−1.35	1776	13308.38	11.534	−2.26
1657	11924.06	11.694	−0.38	1717	12624.43	11.642	−1.37	1777	13319.91	11.532	−2.27
1658	11935.75	11.694	−0.40	1718	12636.07	11.640	−1.38	1778	13331.44	11.530	−2.29
1659	11947.45	11.693	−0.41	1719	12647.71	11.639	−1.40	1779	13342.97	11.527	−2.30
1660	11959.14	11.693	−0.43	1720	12659.35	11.637	−1.41	1780	13354.50	11.525	−2.31
1661	11970.83	11.693	−0.45	1721	12670.99	11.636	−1.43	1781	13366.02	11.523	−2.33
1662	11982.53	11.692	−0.46	1722	12682.62	11.635	−1.45	1782	13377.54	11.520	−2.34
1663	11994.22	11.692	−0.48	1723	12694.26	11.633	−1.46	1783	13389.06	11.518	−2.36
1664	12005.91	11.691	−0.50	1724	12705.89	11.632	−1.48	1784	13400.58	11.516	−2.37
1665	12017.60	11.691	−0.51	1725	12717.52	11.630	−1.49	1785	13412.09	11.513	−2.38
1666	12029.29	11.690	−0.53	1726	12729.15	11.629	−1.51	1786	13423.60	11.511	−2.40
1667	12040.98	11.690	−0.55	1727	12740.78	11.627	−1.52	1787	13435.11	11.508	−2.41
1668	12052.67	11.689	−0.56	1728	12752.40	11.626	−1.54	1788	13446.62	11.506	−2.42
1669	12064.36	11.688	−0.58	1729	12764.03	11.624	−1.56	1789	13458.13	11.504	−2.44
1670	12076.05	11.688	−0.60	1730	12775.65	11.623	−1.57	1790	13469.63	11.501	−2.45
1671	12087.73	11.687	−0.61	1731	12787.27	11.621	−1.59	1791	13481.13	11.499	−2.46
1672	12099.42	11.687	−0.63	1732	12798.89	11.619	−1.60	1792	13492.63	11.496	−2.48
1673	12111.11	11.686	−0.65	1733	12810.51	11.618	−1.62	1793	13504.12	11.494	−2.49
1674	12122.79	11.685	−0.66	1734	12822.13	11.616	−1.63	1794	13515.61	11.491	−2.50
1675	12134.48	11.685	−0.68	1735	12833.75	11.614	−1.65	1795	13527.10	11.489	−2.52
1676	12146.16	11.684	−0.70	1736	12845.36	11.613	−1.67	1796	13538.59	11.486	−2.53
1677	12157.85	11.683	−0.71	1737	12856.97	11.611	−1.68	1797	13550.08	11.484	−2.54
1678	12169.53	11.683	−0.73	1738	12868.58	11.609	−1.70	1798	13561.56	11.481	−2.56
1679	12181.21	11.682	−0.75	1739	12880.19	11.608	−1.71	1799	13573.04	11.479	−2.57
1680	12192.89	11.681	−0.76	1740	12891.80	11.606	−1.73	1800	13584.52	11.476	−2.58

TABLE 1i6. *(Continued)*

Type B thermocouples—thermoelectric voltages, E(T), Seebeck coefficients, S(T), and first derivative of the Seebeck coefficients, dS/dT, reference junctions at 0 °C—Continued

T °C	E μV	S μV/°C	dS/dT nV/°C²	T °C	E μV	S μV/°C	dS/dT nV/°C²	T °C	E μV	S μV/°C	dS/dT nV/°C²
1800	13584.52	11.476	-2.58	1810	13699.14	11.450	-2.71	1820	13813.50	11.422	-2.83
1801	13595.99	11.473	-2.59	1811	13710.59	11.447	-2.72				
1802	13607.46	11.471	-2.61	1812	13722.04	11.444	-2.73				
1803	13618.93	11.468	-2.62	1813	13733.48	11.441	-2.74				
1804	13630.40	11.466	-2.63	1814	13744.92	11.439	-2.76				
1805	13641.86	11.463	-2.65	1815	13756.36	11.436	-2.77				
1806	13653.33	11.460	-2.66	1816	13767.79	11.433	-2.78				
1807	13664.78	11.458	-2.67	1817	13779.22	11.430	-2.79				
1808	13676.24	11.455	-2.68	1818	13790.65	11.428	-2.80				
1809	13687.69	11.452	-2.70	1819	13802.08	11.425	-2.81				
1810	13699.14	11.450	-2.71	1820	13813.50	11.422	-2.83				

TYPE E—*Nickel*–Chromium Alloy Versus *Copper*–Nickel Alloy Thermocouples

Type E thermocouples—thermoelectric voltages, E(T), Seebeck coefficients, S(T), and first derivative of the Seebeck coefficients, dS/dT, reference junctions at 0 °C

T °C	E μV	S μV/°C	dS/dT nV/°C²	T °C	E μV	S μV/°C	dS/dT nV/°C²	T °C	E μV	S μV/°C	dS/dT nV/°C²
-270	-9835.03	1.549	514.24	-240	-9604.15	13.199	346.27	-210	-9062.86	22.507	272.79
-269	-9833.22	2.052	492.52	-239	-9590.78	13.544	344.35	-209	-9040.21	22.779	270.59
-268	-9830.93	2.535	473.56	-238	-9577.06	13.888	342.35	-208	-9017.30	23.048	268.44
-267	-9828.16	3.000	457.05	-237	-9563.00	14.229	340.26	-207	-8994.12	23.316	266.35
-266	-9824.93	3.450	442.72	-236	-9548.61	14.568	338.10	-206	-8970.67	23.581	264.30
-265	-9821.27	3.886	430.32	-235	-9533.87	14.905	335.86	-205	-8946.96	23.844	262.31
-264	-9817.17	4.311	419.62	-234	-9518.80	15.240	333.55	-204	-8922.98	24.106	260.37
-263	-9812.65	4.726	410.41	-233	-9503.39	15.572	331.18	-203	-8898.75	24.365	258.49
-262	-9807.72	5.132	402.52	-232	-9487.65	15.902	328.74	-202	-8874.25	24.623	256.65
-261	-9802.39	5.531	395.77	-231	-9471.59	16.230	326.25	-201	-8849.50	24.878	254.87
-260	-9796.66	5.924	390.00	-230	-9455.19	16.555	323.72	-200	-8824.50	25.132	253.13
-259	-9790.54	6.311	385.07	-229	-9438.48	16.877	321.14	-199	-8799.24	25.385	251.44
-258	-9784.04	6.694	380.87	-228	-9421.44	17.197	318.53	-198	-8773.73	25.635	249.80
-257	-9777.15	7.073	377.27	-227	-9404.09	17.514	315.90	-197	-8747.97	25.884	248.20
-256	-9769.89	7.449	374.19	-226	-9386.41	17.829	313.24	-196	-8721.96	26.132	246.64
-255	-9762.25	7.822	371.52	-225	-9368.43	18.141	310.58	-195	-8695.70	26.378	245.13
-254	-9754.25	8.192	369.20	-224	-9350.13	18.450	307.91	-194	-8669.20	26.622	243.65
-253	-9745.87	8.560	367.15	-223	-9331.53	18.756	305.23	-193	-8642.46	26.865	242.22
-252	-9737.13	8.927	365.31	-222	-9312.62	19.060	302.57	-192	-8615.47	27.106	240.82
-251	-9728.02	9.291	363.63	-221	-9293.41	19.362	299.92	-191	-8588.25	27.347	239.45
-250	-9718.55	9.654	362.06	-220	-9273.90	19.660	297.28	-190	-8560.78	27.585	238.12
-249	-9708.71	10.015	360.56	-219	-9254.09	19.956	294.67	-189	-8533.08	27.823	236.81
-248	-9698.52	10.375	359.10	-218	-9233.99	20.250	292.08	-188	-8505.14	28.059	235.54
-247	-9687.96	10.733	357.65	-217	-9213.59	20.540	289.53	-187	-8476.96	28.294	234.29
-246	-9677.05	11.090	356.19	-216	-9192.91	20.829	287.01	-186	-8448.55	28.528	233.06
-245	-9665.78	11.446	354.69	-215	-9171.94	21.114	284.53	-185	-8419.90	28.760	231.86
-244	-9654.16	11.800	353.15	-214	-9150.68	21.398	282.09	-184	-8391.03	28.991	230.69
-243	-9642.18	12.152	351.54	-213	-9129.14	21.679	279.69	-183	-8361.92	29.221	229.53
-242	-9629.85	12.503	349.86	-212	-9107.32	21.957	277.34	-182	-8332.59	29.450	228.39
-241	-9617.18	12.852	348.11	-211	-9085.23	22.233	275.04	-181	-8303.02	29.678	227.27
-240	-9604.15	13.199	346.27	-210	-9062.86	22.507	272.79	-180	-8273.23	29.905	226.16

TABLE 1i6. (Continued)

Type E thermocouples—thermoelectric voltages, E(T), Seebeck coefficients, S(T), and first derivative of the Seebeck coefficients, dS/dT, reference junctions at 0 °C—Continued

T °C	E μV	S μV/°C	dS/dT nV/°C²	T °C	E μV	S μV/°C	dS/dT nV/°C²	T °C	E μV	S μV/°C	dS/dT nV/°C²
-180	-8273.23	29.905	226.16	-120	-6106.80	41.797	175.08	-60	-3305.75	51.207	139.22
-179	-8243.21	30.131	225.07	-119	-6064.91	41.972	174.47	-59	-3254.48	51.346	138.70
-178	-8212.97	30.355	223.99	-118	-6022.85	42.146	173.85	-58	-3203.06	51.484	138.19
-177	-8182.50	30.579	222.92	-117	-5980.62	42.319	173.24	-57	-3151.51	51.622	137.69
-176	-8151.81	30.801	221.87	-116	-5938.21	42.492	172.64	-56	-3099.82	51.760	137.18
-175	-8120.90	31.022	220.82	-115	-5895.64	42.665	172.03	-55	-3047.99	51.897	136.68
-174	-8089.77	31.243	219.79	-114	-5852.88	42.836	171.43	-54	-2996.02	52.033	136.19
-173	-8058.42	31.462	218.76	-113	-5809.96	43.008	170.83	-53	-2943.92	52.169	135.69
-172	-8026.85	31.680	217.74	-112	-5766.87	43.178	170.23	-52	-2891.69	52.305	135.20
-171	-7995.06	31.897	216.73	-111	-5723.61	43.348	169.63	-51	-2839.31	52.440	134.71
-170	-7963.05	32.114	215.73	-110	-5680.17	43.517	169.03	-50	-2786.81	52.574	134.22
-169	-7930.83	32.329	214.73	-109	-5636.57	43.686	168.43	-49	-2734.17	52.708	133.73
-168	-7898.39	32.543	213.74	-108	-5592.80	43.854	167.83	-48	-2681.39	52.841	133.24
-167	-7865.74	32.756	212.75	-107	-5548.86	44.022	167.23	-47	-2628.48	52.974	132.76
-166	-7832.88	32.969	211.78	-106	-5504.76	44.189	166.63	-46	-2575.44	53.107	132.27
-165	-7799.81	33.180	210.80	-105	-5460.49	44.355	166.02	-45	-2522.27	53.239	131.79
-164	-7766.52	33.390	209.84	-104	-5416.05	44.521	165.42	-44	-2468.97	53.371	131.30
-163	-7733.03	33.600	208.87	-103	-5371.45	44.686	164.81	-43	-2415.53	53.502	130.82
-162	-7699.32	33.808	207.92	-102	-5326.68	44.850	164.20	-42	-2361.96	53.632	130.33
-161	-7665.41	34.015	206.97	-101	-5281.74	45.014	163.59	-41	-2308.27	53.762	129.85
-160	-7631.29	34.222	206.02	-100	-5236.65	45.178	162.98	-40	-2254.44	53.892	129.37
-159	-7596.97	34.427	205.09	-99	-5191.39	45.340	162.36	-39	-2200.48	54.021	128.89
-158	-7562.44	34.632	204.15	-98	-5145.97	45.502	161.75	-38	-2146.40	54.150	128.41
-157	-7527.70	34.836	203.23	-97	-5100.39	45.664	161.13	-37	-2092.18	54.278	127.94
-156	-7492.77	35.038	202.31	-96	-5054.64	45.825	160.51	-36	-2037.84	54.406	127.46
-155	-7457.63	35.240	201.40	-95	-5008.74	45.985	159.89	-35	-1983.37	54.533	126.99
-154	-7422.29	35.441	200.50	-94	-4962.67	46.144	159.26	-34	-1928.78	54.660	126.53
-153	-7386.75	35.641	199.60	-93	-4916.45	46.303	158.64	-33	-1874.05	54.786	126.06
-152	-7351.00	35.840	198.71	-92	-4870.07	46.462	158.02	-32	-1819.20	54.912	125.61
-151	-7315.06	36.039	197.83	-91	-4823.53	46.619	157.39	-31	-1764.23	55.037	125.16
-150	-7278.93	36.236	196.96	-90	-4776.83	46.776	156.76	-30	-1709.13	55.162	124.71
-149	-7242.59	36.433	196.10	-89	-4729.97	46.933	156.14	-29	-1653.91	55.286	124.28
-148	-7206.06	36.628	195.24	-88	-4682.96	47.089	155.51	-28	-1598.56	55.411	123.85
-147	-7169.34	36.823	194.40	-87	-4635.80	47.244	154.89	-27	-1543.09	55.534	123.43
-146	-7132.42	37.017	193.56	-86	-4588.48	47.398	154.26	-26	-1487.49	55.657	123.03
-145	-7095.30	37.210	192.74	-85	-4541.00	47.552	153.64	-25	-1431.77	55.780	122.63
-144	-7058.00	37.403	191.92	-84	-4493.37	47.706	153.01	-24	-1375.93	55.903	122.24
-143	-7020.50	37.594	191.11	-83	-4445.59	47.858	152.39	-23	-1319.97	56.025	121.86
-142	-6982.81	37.785	190.32	-82	-4397.65	48.010	151.77	-22	-1263.88	56.146	121.49
-141	-6944.93	37.975	189.53	-81	-4349.57	48.162	151.16	-21	-1207.67	56.268	121.13
-140	-6906.86	38.164	188.75	-80	-4301.33	48.313	150.54	-20	-1151.34	56.389	120.78
-139	-6868.60	38.352	187.99	-79	-4252.94	48.463	149.93	-19	-1094.90	56.509	120.43
-138	-6830.15	38.540	187.23	-78	-4204.40	48.613	149.33	-18	-1038.33	56.630	120.09
-137	-6791.52	38.727	186.48	-77	-4155.72	48.762	148.72	-17	-981.64	56.749	119.74
-136	-6752.70	38.913	185.75	-76	-4106.88	48.910	148.12	-16	-924.83	56.869	119.39
-135	-6713.70	39.098	185.02	-75	-4057.90	49.058	147.53	-15	-867.90	56.988	119.03
-134	-6674.50	39.283	184.30	-74	-4008.77	49.205	146.94	-14	-810.85	57.107	118.66
-133	-6635.13	39.467	183.59	-73	-3959.49	49.352	146.35	-13	-753.68	57.226	118.26
-132	-6595.57	39.650	182.89	-72	-3910.06	49.498	145.77	-12	-696.40	57.344	117.82
-131	-6555.83	39.833	182.20	-71	-3860.49	49.643	145.20	-11	-639.00	57.461	117.34
-130	-6515.91	40.014	181.52	-70	-3810.78	49.788	144.63	-10	-581.48	57.578	116.80
-129	-6475.80	40.196	180.85	-69	-3760.92	49.933	144.06	-9	-523.84	57.695	116.19
-128	-6435.52	40.376	180.18	-68	-3710.91	50.076	143.50	-8	-466.09	57.811	115.48
-127	-6395.05	40.556	179.52	-67	-3660.76	50.220	142.95	-7	-408.22	57.926	114.66
-126	-6354.40	40.735	178.87	-66	-3610.47	50.362	142.40	-6	-350.24	58.040	113.71
-125	-6313.58	40.914	178.22	-65	-3560.04	50.504	141.86	-5	-292.14	58.153	112.59
-124	-6272.58	41.092	177.58	-64	-3509.46	50.646	141.32	-4	-233.93	58.265	111.28
-123	-6231.40	41.269	176.95	-63	-3458.75	50.787	140.79	-3	-175.61	58.375	109.74
-122	-6190.04	41.446	176.32	-62	-3407.89	50.928	140.26	-2	-117.18	58.484	107.93
-121	-6148.51	41.622	175.70	-61	-3356.89	51.068	139.74	-1	-58.64	58.591	105.82
-120	-6106.80	41.797	175.08	-60	-3305.75	51.207	139.22	0	0.00	58.696	86.22

TABLE 1i6. (Continued)

Type E thermocouples—*thermoelectric voltages,* E(T), *Seebeck coefficients,* S(T), *and first derivative of the Seebeck coefficients,* dS/dT, *reference junctions at 0 °C*—*Continued*

T °C	E μV	S μV/°C	dS/dT nV/°C²	T °C	E μV	S μV/°C	dS/dT nV/°C²	T °C	E μV	S μV/°C	dS/dT nV/°C²
0	0.0	58.696	86.22	60	3683.4	64.110	89.26	120	7682.9	69.054	74.18
1	58.7	58.782	86.56	61	3747.6	64.199	89.09	121	7752.0	69.128	73.87
2	117.6	58.869	86.88	62	3811.8	64.288	88.93	122	7821.2	69.202	73.57
3	176.5	58.956	87.19	63	3876.1	64.377	88.76	123	7890.4	69.276	73.27
4	235.5	59.043	87.49	64	3940.5	64.465	88.58	124	7959.7	69.349	72.96
5	294.6	59.131	87.78	65	4005.1	64.554	88.40	125	8029.1	69.421	72.66
6	353.7	59.219	88.06	66	4069.7	64.642	88.22	126	8098.6	69.494	72.35
7	413.0	59.307	88.32	67	4134.3	64.730	88.03	127	8168.1	69.566	72.05
8	472.4	59.396	88.57	68	4199.1	64.818	87.84	128	8237.7	69.638	71.74
9	531.8	59.484	88.81	69	4264.0	64.906	87.64	129	8307.4	69.710	71.43
10	591.3	59.573	89.04	70	4328.9	64.993	87.44	130	8377.1	69.781	71.13
11	650.9	59.662	89.25	71	4394.0	65.081	87.24	131	8446.9	69.852	70.82
12	710.6	59.752	89.46	72	4459.1	65.168	87.03	132	8516.8	69.923	70.51
13	770.4	59.841	89.66	73	4524.3	65.255	86.82	133	8586.8	69.993	70.20
14	830.3	59.931	89.84	74	4589.6	65.341	86.60	134	8656.8	70.063	69.90
15	890.3	60.021	90.01	75	4655.0	65.428	86.39	135	8726.9	70.133	69.59
16	950.4	60.111	90.18	76	4720.5	65.514	86.16	136	8797.1	70.202	69.28
17	1010.5	60.201	90.33	77	4786.0	65.600	85.94	137	8867.3	70.271	68.97
18	1070.8	60.292	90.47	78	4851.7	65.686	85.71	138	8937.6	70.340	68.66
19	1131.1	60.382	90.61	79	4917.4	65.772	85.48	139	9008.0	70.409	68.36
20	1191.5	60.473	90.73	80	4983.2	65.857	85.25	140	9078.4	70.477	68.05
21	1252.1	60.564	90.84	81	5049.1	65.942	85.01	141	9148.9	70.545	67.74
22	1312.7	60.655	90.95	82	5115.1	66.027	84.77	142	9219.5	70.612	67.43
23	1373.4	60.746	91.04	83	5181.2	66.112	84.53	143	9290.2	70.680	67.12
24	1434.2	60.837	91.13	84	5247.3	66.196	84.28	144	9360.9	70.747	66.82
25	1495.0	60.928	91.21	85	5313.5	66.280	84.03	145	9431.7	70.813	66.51
26	1556.0	61.019	91.28	86	5379.9	66.364	83.78	146	9502.5	70.880	66.20
27	1617.1	61.110	91.33	87	5446.3	66.448	83.53	147	9573.4	70.946	65.90
28	1678.2	61.202	91.39	88	5512.8	66.531	83.28	148	9644.4	71.011	65.59
29	1739.5	61.293	91.43	89	5579.3	66.614	83.02	149	9715.4	71.077	65.29
30	1800.8	61.385	91.46	90	5646.0	66.697	82.76	150	9786.5	71.142	64.98
31	1862.2	61.476	91.49	91	5712.7	66.780	82.50	151	9857.7	71.207	64.67
32	1923.8	61.568	91.51	92	5779.6	66.862	82.23	152	9929.0	71.271	64.37
33	1985.4	61.659	91.52	93	5846.5	66.944	81.96	153	10000.3	71.335	64.06
34	2047.1	61.751	91.52	94	5913.4	67.026	81.70	154	10071.6	71.399	63.76
35	2108.9	61.842	91.51	95	5980.5	67.108	81.42	155	10143.1	71.463	63.45
36	2170.9	61.934	91.50	96	6047.7	67.189	81.15	156	10214.6	71.526	63.15
37	2232.8	62.025	91.48	97	6114.9	67.270	80.88	157	10286.1	71.589	62.85
38	2294.8	62.117	91.45	98	6182.2	67.351	80.60	158	10357.7	71.652	62.55
39	2357.0	62.208	91.42	99	6249.6	67.431	80.32	159	10429.4	71.714	62.24
40	2419.2	62.299	91.38	100	6317.1	67.511	80.04	160	10501.2	71.776	61.94
41	2481.6	62.391	91.33	101	6384.6	67.591	79.76	161	10573.0	71.838	61.64
42	2544.0	62.482	91.27	102	6452.2	67.671	79.48	162	10644.8	71.900	61.34
43	2606.5	62.573	91.21	103	6520.0	67.750	79.19	163	10716.8	71.961	61.04
44	2669.2	62.664	91.14	104	6587.7	67.829	78.91	164	10788.8	72.022	60.74
45	2731.9	62.756	91.07	105	6655.6	67.908	78.62	165	10860.8	72.082	60.44
46	2794.7	62.847	90.99	106	6723.6	67.987	78.33	166	10932.9	72.143	60.15
47	2857.6	62.938	90.90	107	6791.6	68.065	78.04	167	11005.1	72.203	59.85
48	2920.6	63.028	90.81	108	6859.7	68.143	77.75	168	11077.3	72.262	59.55
49	2983.6	63.119	90.71	109	6927.9	68.220	77.46	169	11149.6	72.322	59.26
50	3046.8	63.210	90.60	110	6996.1	68.298	77.17	170	11222.0	72.381	58.96
51	3110.0	63.300	90.49	111	7064.5	68.375	76.87	171	11294.4	72.440	58.67
52	3173.4	63.391	90.37	112	7132.9	68.451	76.58	172	11366.9	72.498	58.37
53	3236.8	63.481	90.25	113	7201.4	68.528	76.28	173	11439.4	72.557	58.08
54	3300.4	63.571	90.13	114	7269.9	68.604	75.98	174	11512.0	72.614	57.79
55	3364.0	63.661	89.99	115	7338.6	68.680	75.68	175	11584.6	72.672	57.50
56	3427.7	63.751	89.86	116	7407.3	68.755	75.38	176	11657.3	72.729	57.21
57	3491.5	63.841	89.71	117	7476.1	68.831	75.08	177	11730.1	72.787	56.92
58	3555.4	63.931	89.57	118	7545.0	68.905	74.78	178	11802.9	72.843	56.63
59	3619.3	64.020	89.41	119	7613.9	68.980	74.48	179	11875.8	72.900	56.34
60	3683.4	64.110	89.26	120	7682.9	69.054	74.18	180	11948.7	72.956	56.06

TABLE 1i6. (Continued)

Type E thermocouples—*thermoelectric voltages*, E(T), *Seebeck coefficients*, S(T), *and first deriative of the Seebeck coefficients*, dS/dT, *reference junctions at 0 °C—Continued*

T °C	E μV	S μV/°C	dS/dT nV/°C²	T °C	E μV	S μV/°C	dS/dT nV/°C²	T °C	E μV	S μV/°C	dS/dT nV/°C²
180	11948.7	72.956	56.06	240	16417.2	75.839	40.67	300	21033.1	77.910	28.91
181	12021.7	73.012	55.77	241	16493.0	75.879	40.44	301	21111.1	77.939	28.74
182	12094.7	73.068	55.48	242	16568.9	75.920	40.22	302	21189.0	77.967	28.57
183	12167.8	73.123	55.20	243	16644.9	75.960	40.00	303	21267.0	77.996	28.40
184	12241.0	73.178	54.92	244	16720.8	76.000	39.78	304	21345.0	78.024	28.23
185	12314.2	73.233	54.63	245	16796.8	76.039	39.56	305	21423.0	78.052	28.07
186	12387.4	73.287	54.35	246	16872.9	76.079	39.34	306	21501.1	78.080	27.90
187	12460.7	73.341	54.07	247	16949.0	76.118	39.12	307	21579.2	78.108	27.73
188	12534.1	73.395	53.79	248	17025.1	76.157	38.90	308	21657.3	78.136	27.57
189	12607.5	73.449	53.51	249	17101.3	76.196	38.69	309	21735.5	78.163	27.40
190	12681.0	73.502	53.24	250	17177.5	76.234	38.47	310	21813.6	78.190	27.24
191	12754.5	73.555	52.96	251	17253.8	76.273	38.26	311	21891.8	78.218	27.08
192	12828.1	73.608	52.68	252	17330.1	76.311	38.04	312	21970.1	78.245	26.91
193	12901.7	73.661	52.41	253	17406.4	76.349	37.83	313	22048.3	78.271	26.75
194	12975.4	73.713	52.13	254	17482.8	76.387	37.62	314	22126.6	78.298	26.59
195	13049.2	73.765	51.86	255	17559.2	76.424	37.41	315	22204.9	78.325	26.43
196	13123.0	73.817	51.59	256	17635.6	76.461	37.20	316	22283.3	78.351	26.27
197	13196.8	73.868	51.32	257	17712.1	76.499	36.99	317	22361.6	78.377	26.11
198	13270.7	73.919	51.05	258	17788.6	76.535	36.79	318	22440.0	78.403	25.95
199	13344.6	73.970	50.78	259	17865.2	76.572	36.58	319	22518.4	78.429	25.79
200	13418.6	74.021	50.51	260	17941.8	76.609	36.38	320	22596.9	78.455	25.64
201	13492.7	74.071	50.24	261	18018.4	76.645	36.17	321	22675.4	78.480	25.48
202	13566.8	74.121	49.98	262	18095.1	76.681	35.97	322	22753.8	78.506	25.32
203	13640.9	74.171	49.71	263	18171.8	76.717	35.77	323	22832.4	78.531	25.17
204	13715.1	74.221	49.45	264	18248.5	76.752	35.57	324	22910.9	78.556	25.01
205	13789.4	74.270	49.19	265	18325.3	76.788	35.37	325	22989.5	78.581	24.86
206	13863.7	74.319	48.93	266	18402.1	76.823	35.17	326	23068.1	78.606	24.70
207	13938.0	74.368	48.67	267	18478.9	76.858	34.97	327	23146.7	78.630	24.55
208	14012.4	74.417	48.41	268	18555.8	76.893	34.77	328	23225.3	78.655	24.39
209	14086.8	74.465	48.15	269	18632.7	76.928	34.57	329	23304.0	78.679	24.24
210	14161.3	74.513	47.89	270	18709.6	76.962	34.38	330	23382.7	78.703	24.09
211	14235.9	74.561	47.63	271	18786.6	76.997	34.18	331	23461.4	78.727	23.94
212	14310.5	74.608	47.38	272	18863.6	77.031	33.99	332	23540.1	78.751	23.79
213	14385.1	74.655	47.12	273	18940.7	77.065	33.80	333	23618.9	78.775	23.64
214	14459.8	74.702	46.87	274	19017.8	77.098	33.61	334	23697.7	78.798	23.49
215	14534.5	74.749	46.62	275	19094.9	77.132	33.41	335	23776.5	78.822	23.34
216	14609.3	74.796	46.37	276	19172.0	77.165	33.22	336	23855.3	78.845	23.19
217	14684.1	74.842	46.12	277	19249.2	77.198	33.03	337	23934.2	78.868	23.04
218	14758.9	74.888	45.87	278	19326.4	77.231	32.85	338	24013.1	78.891	22.89
219	14833.9	74.934	45.62	279	19403.7	77.264	32.66	339	24092.0	78.914	22.75
220	14908.8	74.979	45.37	280	19480.9	77.296	32.47	340	24170.9	78.937	22.60
221	14983.8	75.024	45.13	281	19558.3	77.329	32.29	341	24249.8	78.959	22.45
222	15058.9	75.069	44.88	282	19635.6	77.361	32.10	342	24328.8	78.982	22.31
223	15134.0	75.114	44.64	283	19713.0	77.393	31.92	343	24407.8	79.004	22.16
224	15209.1	75.159	44.40	284	19790.4	77.425	31.73	344	24486.8	79.026	22.02
225	15284.3	75.203	44.16	285	19867.8	77.457	31.55	345	24565.9	79.048	21.87
226	15359.5	75.247	43.92	286	19945.3	77.488	31.37	346	24644.9	79.070	21.73
227	15434.8	75.291	43.68	287	20022.8	77.519	31.19	347	24724.0	79.091	21.58
228	15510.1	75.334	43.44	288	20100.3	77.550	31.01	348	24803.1	79.113	21.44
229	15585.4	75.378	43.20	289	20177.9	77.581	30.83	349	24882.2	79.134	21.30
230	15660.8	75.421	42.97	290	20255.5	77.612	30.65	350	24961.4	79.155	21.15
231	15736.3	75.464	42.73	291	20333.1	77.643	30.47	351	25040.5	79.177	21.01
232	15811.8	75.506	42.50	292	20410.8	77.673	30.30	352	25119.7	79.197	20.87
233	15887.3	75.549	42.27	293	20488.5	77.703	30.12	353	25198.9	79.218	20.73
234	15962.9	75.591	42.03	294	20566.2	77.733	29.94	354	25278.2	79.239	20.59
235	16038.5	75.633	41.80	295	20643.9	77.763	29.77	355	25357.4	79.259	20.45
236	16114.1	75.674	41.57	296	20721.7	77.793	29.60	356	25436.7	79.280	20.31
237	16189.8	75.716	41.35	297	20799.5	77.822	29.42	357	25516.0	79.300	20.17
238	16265.6	75.757	41.12	298	20877.4	77.852	29.25	358	25595.3	79.320	20.03
239	16341.3	75.798	40.89	299	20955.2	77.881	29.08	359	25674.6	79.340	19.89
240	16417.2	75.839	40.67	300	21033.1	77.910	28.91	360	25754.0	79.360	19.75

TABLE 1i6. (Continued)

Type E thermocouples—*thermoelectric voltages,* E(T), *Seebeck coefficients,* S(T), *and first deriative of the Seebeck coefficients,* dS/dT, *reference junctions at 0 °C—Continued*

T °C	E μV	S μV/°C	dS/dT nV/°C²	T °C	E μV	S μV/°C	dS/dT nV/°C²	T °C	E μV	S μV/°C	dS/dT nV/°C²
360	25754.0	79.360	19.75	420	30546.3	80.307	11.97	480	35381.9	80.810	4.92
361	25833.3	79.380	19.61	421	30626.6	80.319	11.84	481	35462.7	80.815	4.81
362	25912.7	79.399	19.48	422	30706.9	80.331	11.72	482	35543.6	80.820	4.70
363	25992.1	79.419	19.34	423	30787.3	80.342	11.60	483	35624.4	80.825	4.59
364	26071.6	79.438	19.20	424	30867.6	80.354	11.48	484	35705.2	80.829	4.48
365	26151.0	79.457	19.06	425	30948.0	80.365	11.35	485	35786.0	80.834	4.37
366	26230.5	79.476	18.93	426	31028.4	80.377	11.23	486	35866.9	80.838	4.26
367	26310.0	79.495	18.79	427	31108.7	80.388	11.11	487	35947.7	80.842	4.15
368	26389.5	79.514	18.66	428	31189.1	80.399	10.99	488	36028.6	80.846	4.04
369	26469.0	79.532	18.52	429	31269.5	80.410	10.87	489	36109.4	80.850	3.93
370	26548.5	79.551	18.39	430	31349.9	80.420	10.75	490	36190.3	80.854	3.82
371	26628.1	79.569	18.25	431	31430.4	80.431	10.63	491	36271.1	80.858	3.71
372	26707.7	79.587	18.12	432	31510.8	80.442	10.51	492	36352.0	80.861	3.60
373	26787.3	79.605	17.98	433	31591.3	80.452	10.39	493	36432.8	80.865	3.49
374	26866.9	79.623	17.85	434	31671.7	80.463	10.26	494	36513.7	80.868	3.38
375	26946.5	79.641	17.71	435	31752.2	80.473	10.14	495	36594.6	80.872	3.27
376	27026.2	79.659	17.58	436	31832.7	80.483	10.02	496	36675.4	80.875	3.16
377	27105.8	79.676	17.45	437	31913.1	80.493	9.90	497	36756.3	80.878	3.06
378	27185.5	79.693	17.32	438	31993.6	80.503	9.78	498	36837.2	80.881	2.95
379	27265.2	79.711	17.18	439	32074.2	80.512	9.67	499	36918.1	80.884	2.84
380	27344.9	79.728	17.05	440	32154.7	80.522	9.55	500	36999.0	80.887	2.73
381	27424.7	79.745	16.92	441	32235.2	80.531	9.43	501	37079.9	80.889	2.63
382	27504.4	79.762	16.79	442	32315.7	80.541	9.31	502	37160.8	80.892	2.52
383	27584.2	79.778	16.66	443	32396.3	80.550	9.19	503	37241.6	80.894	2.41
384	27664.0	79.795	16.53	444	32476.8	80.559	9.07	504	37322.5	80.897	2.31
385	27743.8	79.811	16.39	445	32557.4	80.568	8.95	505	37403.4	80.899	2.20
386	27823.6	79.828	16.26	446	32638.0	80.577	8.83	506	37484.3	80.901	2.10
387	27903.4	79.844	16.13	447	32718.6	80.586	8.72	507	37565.2	80.903	1.99
388	27983.3	79.860	16.00	448	32799.1	80.594	8.60	508	37646.1	80.905	1.88
389	28063.2	79.876	15.87	449	32879.7	80.603	8.48	509	37727.1	80.907	1.78
390	28143.0	79.892	15.74	450	32960.3	80.611	8.36	510	37808.0	80.909	1.68
391	28222.9	79.907	15.62	451	33041.0	80.620	8.24	511	37888.9	80.910	1.57
392	28302.9	79.923	15.49	452	33121.6	80.628	8.13	512	37969.8	80.912	1.47
393	28382.8	79.938	15.36	453	33202.2	80.636	8.01	513	38050.7	80.913	1.36
394	28462.7	79.954	15.23	454	33282.9	80.644	7.89	514	38131.6	80.915	1.26
395	28542.7	79.969	15.10	455	33363.5	80.652	7.78	515	38212.5	80.916	1.16
396	28622.7	79.984	14.97	456	33444.2	80.660	7.66	516	38293.4	80.917	1.05
397	28702.7	79.999	14.85	457	33524.8	80.667	7.54	517	38374.4	80.918	0.95
398	28782.7	80.014	14.72	458	33605.5	80.675	7.43	518	38455.3	80.919	0.85
399	28862.7	80.028	14.59	459	33686.2	80.682	7.31	519	38536.2	80.920	0.75
400	28942.7	80.043	14.46	460	33766.9	80.689	7.20	520	38617.1	80.920	0.64
401	29022.8	80.057	14.34	461	33847.6	80.696	7.08	521	38698.0	80.921	0.54
402	29102.8	80.071	14.21	462	33928.3	80.703	6.97	522	38779.0	80.921	0.44
403	29182.9	80.086	14.08	463	34009.0	80.710	6.85	523	38859.9	80.922	0.34
404	29263.0	80.100	13.96	464	34089.7	80.717	6.74	524	38940.8	80.922	0.24
405	29343.1	80.113	13.83	465	34170.4	80.724	6.62	525	39021.7	80.922	0.14
406	29423.2	80.127	13.71	466	34251.1	80.730	6.51	526	39102.6	80.922	0.04
407	29503.4	80.141	13.58	467	34331.9	80.737	6.39	527	39183.6	80.922	−0.06
408	29583.5	80.154	13.46	468	34412.6	80.743	6.28	528	39264.5	80.922	−0.16
409	29663.7	80.168	13.33	469	34493.3	80.749	6.16	529	39345.4	80.922	−0.26
410	29743.9	80.181	13.21	470	34574.1	80.755	6.05	530	39426.3	80.922	−0.36
411	29824.0	80.194	13.08	471	34654.9	80.761	5.94	531	39507.3	80.921	−0.45
412	29904.2	80.207	12.96	472	34735.6	80.767	5.82	532	39588.2	80.921	−0.55
413	29984.5	80.220	12.83	473	34816.4	80.773	5.71	533	39669.1	80.920	−0.65
414	30064.7	80.233	12.71	474	34897.2	80.779	5.60	534	39750.0	80.919	−0.75
415	30144.9	80.246	12.58	475	34977.9	80.784	5.48	535	39830.9	80.919	−0.84
416	30225.2	80.258	12.46	476	35058.7	80.790	5.37	536	39911.9	80.918	−0.94
417	30305.4	80.270	12.34	477	35139.5	80.795	5.26	537	39992.8	80.917	−1.04
418	30385.7	80.283	12.21	478	35220.3	80.800	5.15	538	40073.7	80.916	−1.13
419	30466.0	80.295	12.09	479	35301.1	80.805	5.03	539	40154.6	80.915	−1.23
420	30546.3	80.307	11.97	480	35381.9	80.810	4.92	540	40235.5	80.913	−1.33

TABLE 1i6. (*Continued*)

Type E thermocouples—thermoelectric voltages, E(T), *Seebeck coefficients,* S(T), *and first deriative of the Seebeck coefficients,* dS/dT, *reference junctions at 0 °C—Continued*

T °C	E μV	S μV/°C	dS/dT nV/°C^2	T °C	E μV	S μV/°C	dS/dT nV/°C^2	T °C	E μV	S μV/°C	dS/dT nV/°C^2
540	40235.5	80.913	-1.33	600	45084.7	80.676	-6.36	660	49911.5	80.181	-9.89
541	40316.4	80.912	-1.42	601	45165.4	80.669	-6.43	661	49991.6	80.171	-9.94
542	40397.3	80.910	-1.52	602	45246.0	80.663	-6.50	662	50071.8	80.161	-9.98
543	40478.2	80.909	-1.61	603	45326.7	80.656	-6.57	663	50151.9	80.151	-10.03
544	40559.2	80.907	-1.70	604	45407.3	80.650	-6.64	664	50232.1	80.141	-10.08
545	40640.1	80.905	-1.80	605	45488.0	80.643	-6.71	665	50312.2	80.131	-10.12
546	40721.0	80.904	-1.89	606	45568.6	80.636	-6.78	666	50392.4	80.121	-10.17
547	40801.9	80.902	-1.98	607	45649.3	80.629	-6.85	667	50472.5	80.111	-10.21
548	40882.8	80.900	-2.08	608	45729.9	80.623	-6.91	668	50552.6	80.100	-10.25
549	40963.7	80.898	-2.17	609	45810.5	80.616	-6.98	669	50632.7	80.090	-10.30
550	41044.6	80.895	-2.26	610	45891.1	80.609	-7.05	670	50712.8	80.080	-10.34
551	41125.5	80.893	-2.35	611	45971.7	80.602	-7.12	671	50792.8	80.069	-10.39
552	41206.4	80.891	-2.44	612	46052.3	80.594	-7.18	672	50872.9	80.059	-10.43
553	41287.2	80.888	-2.53	613	46132.9	80.587	-7.25	673	50952.9	80.048	-10.47
554	41368.1	80.886	-2.62	614	46213.5	80.580	-7.31	674	51033.0	80.038	-10.51
555	41449.0	80.883	-2.71	615	46294.1	80.573	-7.38	675	51113.0	80.027	-10.56
556	41529.9	80.880	-2.80	616	46374.6	80.565	-7.45	676	51193.0	80.017	-10.60
557	41610.8	80.877	-2.89	617	46455.2	80.558	-7.51	677	51273.1	80.006	-10.64
558	41691.6	80.874	-2.98	618	46535.8	80.550	-7.57	678	51353.1	79.996	-10.68
559	41772.5	80.871	-3.07	619	46616.3	80.543	-7.64	679	51433.1	79.985	-10.72
560	41853.4	80.868	-3.16	620	46696.8	80.535	-7.70	680	51513.0	79.974	-10.76
561	41934.3	80.865	-3.25	621	46777.4	80.527	-7.76	681	51593.0	79.963	-10.80
562	42015.1	80.862	-3.33	622	46857.9	80.519	-7.83	682	51673.0	79.953	-10.84
563	42096.0	80.858	-3.42	623	46938.4	80.511	-7.89	683	51752.9	79.942	-10.88
564	42176.8	80.855	-3.51	624	47018.9	80.504	-7.95	684	51832.8	79.931	-10.92
565	42257.7	80.851	-3.59	625	47099.4	80.496	-8.01	685	51912.8	79.920	-10.96
566	42338.5	80.848	-3.68	626	47179.9	80.488	-8.07	686	51992.7	79.909	-11.00
567	42419.4	80.844	-3.77	627	47260.4	80.479	-8.13	687	52072.6	79.898	-11.04
568	42500.2	80.840	-3.85	628	47340.9	80.471	-8.19	688	52152.5	79.887	-11.08
569	42581.1	80.836	-3.93	629	47421.3	80.463	-8.25	689	52232.4	79.876	-11.12
570	42661.9	80.832	-4.02	630	47501.8	80.455	-8.31	690	52312.2	79.865	-11.16
571	42742.7	80.828	-4.10	631	47582.2	80.446	-8.37	691	52392.1	79.853	-11.19
572	42823.6	80.824	-4.19	632	47662.7	80.438	-8.43	692	52471.9	79.842	-11.23
573	42904.4	80.820	-4.27	633	47743.1	80.430	-8.48	693	52551.8	79.831	-11.27
574	42985.2	80.816	-4.35	634	47823.5	80.421	-8.54	694	52631.6	79.820	-11.30
575	43066.0	80.811	-4.43	635	47904.0	80.413	-8.60	695	52711.4	79.808	-11.34
576	43146.8	80.807	-4.52	636	47984.4	80.404	-8.65	696	52791.2	79.797	-11.38
577	43227.6	80.802	-4.60	637	48064.8	80.395	-8.71	697	52871.0	79.786	-11.41
578	43308.4	80.797	-4.68	638	48145.2	80.386	-8.77	698	52950.8	79.774	-11.45
579	43389.2	80.793	-4.76	639	48225.5	80.378	-8.82	699	53030.6	79.763	-11.49
580	43470.0	80.788	-4.84	640	48305.9	80.369	-8.88	700	53110.3	79.751	-11.52
581	43550.8	80.783	-4.92	641	48386.3	80.360	-8.93	701	53190.1	79.740	-11.56
582	43631.6	80.778	-5.00	642	48466.6	80.351	-8.98	702	53269.8	79.728	-11.59
583	43712.3	80.773	-5.08	643	48547.0	80.342	-9.04	703	53349.5	79.716	-11.63
584	43793.1	80.768	-5.16	644	48627.3	80.333	-9.09	704	53429.2	79.705	-11.66
585	43873.9	80.763	-5.23	645	48707.7	80.324	-9.14	705	53508.9	79.693	-11.70
586	43954.6	80.757	-5.31	646	48788.0	80.315	-9.20	706	53588.6	79.681	-11.73
587	44035.4	80.752	-5.39	647	48868.3	80.305	-9.25	707	53668.3	79.670	-11.77
588	44116.1	80.747	-5.47	648	48948.6	80.296	-9.30	708	53747.9	79.658	-11.80
589	44196.9	80.741	-5.54	649	49028.9	80.287	-9.35	709	53827.6	79.646	-11.84
590	44277.6	80.736	-5.62	650	49109.2	80.277	-9.40	710	53907.2	79.634	-11.87
591	44358.4	80.730	-5.69	651	49189.4	80.268	-9.45	711	53986.9	79.622	-11.90
592	44439.1	80.724	-5.77	652	49269.7	80.258	-9.50	712	54066.5	79.610	-11.94
593	44519.8	80.718	-5.84	653	49349.9	80.249	-9.55	713	54146.1	79.598	-11.97
594	44600.5	80.713	-5.92	654	49430.2	80.239	-9.60	714	54225.7	79.586	-12.00
595	44681.2	80.707	-5.99	655	49510.4	80.230	-9.65	715	54305.3	79.574	-12.04
596	44761.9	80.701	-6.07	656	49590.6	80.220	-9.70	716	54384.8	79.562	-12.07
597	44842.6	80.694	-6.14	657	49670.9	80.210	-9.75	717	54464.4	79.550	-12.10
598	44923.3	80.688	-6.21	658	49751.1	80.201	-9.80	718	54543.9	79.538	-12.14
599	45004.0	80.682	-6.28	659	49831.3	80.191	-9.84	719	54623.5	79.526	-12.17
600	45084.7	80.676	-6.36	660	49911.5	80.181	-9.89	720	54703.0	79.514	-12.20

TABLE 1i6. (*Continued*)

Type E thermocouples—*thermoelectric voltages*, E(T), *Seebeck coefficients*, S(T), *and first derivative of the Seebeck coefficients*, dS/dT, *reference junctions at 0 °C*—*Continued*

T °C	E μV	S μV/°C	dS/dT nV/°C^2	T °C	E μV	S μV/°C	dS/dT nV/°C^2	T °C	E μV	S μV/°C	dS/dT nV/°C^2
720	54703.0	79.514	-12.20	780	59450.7	78.723	-14.18	840	64147.1	77.799	-16.77
721	54782.5	79.502	-12.23	781	59529.4	78.709	-14.21	841	64224.9	77.782	-16.82
722	54862.0	79.489	-12.27	782	59608.1	78.695	-14.25	842	64302.7	77.766	-16.87
723	54941.5	79.477	-12.30	783	59686.8	78.681	-14.29	843	64380.5	77.749	-16.92
724	55020.9	79.465	-12.33	784	59765.5	78.666	-14.33	844	64458.2	77.732	-16.98
725	55100.4	79.452	-12.36	785	59844.1	78.652	-14.36	845	64535.9	77.715	-17.03
726	55179.8	79.440	-12.40	786	59922.8	78.638	-14.40	846	64613.6	77.698	-17.08
727	55259.3	79.428	-12.43	787	60001.4	78.623	-14.44	847	64691.3	77.681	-17.13
728	55338.7	79.415	-12.46	788	60080.0	78.609	-14.48	848	64769.0	77.663	-17.18
729	55418.1	79.403	-12.49	789	60158.6	78.594	-14.51	849	64846.7	77.646	-17.23
730	55497.5	79.390	-12.52	790	60237.2	78.580	-14.55	850	64924.3	77.629	-17.28
731	55576.9	79.378	-12.56	791	60315.8	78.565	-14.59	851	65001.9	77.612	-17.33
732	55656.3	79.365	-12.59	792	60394.3	78.551	-14.63	852	65079.5	77.594	-17.38
733	55735.6	79.352	-12.62	793	60472.9	78.536	-14.67	853	65157.1	77.577	-17.43
734	55815.0	79.340	-12.65	794	60551.4	78.521	-14.71	854	65234.7	77.559	-17.48
735	55894.3	79.327	-12.68	795	60629.9	78.507	-14.75	855	65312.2	77.542	-17.53
736	55973.6	79.314	-12.72	796	60708.4	78.492	-14.79	856	65389.8	77.524	-17.59
737	56052.9	79.302	-12.75	797	60786.9	78.477	-14.83	857	65467.3	77.507	-17.64
738	56132.2	79.289	-12.78	798	60865.4	78.462	-14.87	858	65544.8	77.489	-17.69
739	56211.5	79.276	-12.81	799	60943.8	78.447	-14.91	859	65622.2	77.471	-17.74
740	56290.8	79.263	-12.84	800	61022.3	78.432	-14.95	860	65699.7	77.454	-17.79
741	56370.0	79.250	-12.87	801	61100.7	78.417	-14.99	861	65777.2	77.436	-17.84
742	56449.3	79.238	-12.91	802	61179.1	78.402	-15.03	862	65854.6	77.418	-17.89
743	56528.5	79.225	-12.94	803	61257.5	78.387	-15.07	863	65932.0	77.400	-17.94
744	56607.7	79.212	-12.97	804	61335.9	78.372	-15.11	864	66009.4	77.382	-17.99
745	56686.9	79.199	-13.00	805	61414.2	78.357	-15.15	865	66086.8	77.364	-18.04
746	56766.1	79.186	-13.03	806	61492.6	78.342	-15.20	866	66164.1	77.346	-18.09
747	56845.3	79.173	-13.07	807	61570.9	78.327	-15.24	867	66241.4	77.328	-18.15
748	56924.5	79.160	-13.10	808	61649.2	78.311	-15.28	868	66318.8	77.310	-18.20
749	57003.6	79.146	-13.13	809	61727.6	78.296	-15.33	869	66396.1	77.291	-18.25
750	57082.8	79.133	-13.16	810	61805.8	78.281	-15.37	870	66473.3	77.273	-18.30
751	57161.9	79.120	-13.19	811	61884.1	78.265	-15.41	871	66550.6	77.255	-18.34
752	57241.0	79.107	-13.23	812	61962.4	78.250	-15.46	872	66627.9	77.236	-18.39
753	57320.1	79.094	-13.26	813	62040.6	78.235	-15.50	873	66705.1	77.218	-18.44
754	57399.2	79.080	-13.29	814	62118.8	78.219	-15.54	874	66782.3	77.200	-18.49
755	57478.3	79.067	-13.32	815	62197.1	78.203	-15.59	875	66859.5	77.181	-18.54
756	57557.3	79.054	-13.36	816	62275.2	78.188	-15.63	876	66936.7	77.162	-18.59
757	57636.4	79.040	-13.39	817	62353.4	78.172	-15.68	877	67013.8	77.144	-18.64
758	57715.4	79.027	-13.42	818	62431.6	78.157	-15.72	878	67091.0	77.125	-18.68
759	57794.4	79.014	-13.46	819	62509.7	78.141	-15.77	879	67168.1	77.106	-18.73
760	57873.4	79.000	-13.49	820	62587.9	78.125	-15.81	880	67245.2	77.088	-18.78
761	57952.4	78.987	-13.52	821	62666.0	78.109	-15.86	881	67322.2	77.069	-18.82
762	58031.4	78.973	-13.55	822	62744.1	78.093	-15.91	882	67399.3	77.050	-18.87
763	58110.4	78.959	-13.59	823	62822.2	78.077	-15.95	883	67476.3	77.031	-18.92
764	58189.3	78.946	-13.62	824	62900.2	78.061	-16.00	884	67553.4	77.012	-18.96
765	58268.3	78.932	-13.66	825	62978.3	78.045	-16.05	885	67630.4	76.993	-19.01
766	58347.2	78.919	-13.69	826	63056.3	78.029	-16.09	886	67707.3	76.974	-19.05
767	58426.1	78.905	-13.72	827	63134.4	78.013	-16.14	887	67784.3	76.955	-19.09
768	58505.0	78.891	-13.76	828	63212.4	77.997	-16.19	888	67861.3	76.936	-19.14
769	58583.9	78.877	-13.79	829	63290.4	77.981	-16.24	889	67938.2	76.917	-19.18
770	58662.7	78.863	-13.83	830	63368.3	77.964	-16.28	890	68015.1	76.898	-19.22
771	58741.6	78.850	-13.86	831	63446.3	77.948	-16.33	891	68092.0	76.878	-19.26
772	58820.4	78.836	-13.89	832	63524.2	77.932	-16.38	892	68168.8	76.859	-19.30
773	58899.3	78.822	-13.93	833	63602.1	77.915	-16.43	893	68245.7	76.840	-19.34
774	58978.1	78.808	-13.96	834	63680.1	77.899	-16.48	894	68322.5	76.820	-19.38
775	59056.9	78.794	-14.00	835	63757.9	77.882	-16.53	895	68399.3	76.801	-19.42
776	59135.7	78.780	-14.04	836	63835.8	77.866	-16.58	896	68476.1	76.782	-19.46
777	59214.5	78.766	-14.07	837	63913.7	77.849	-16.63	897	68552.9	76.762	-19.49
778	59293.2	78.752	-14.11	838	63991.5	77.833	-16.67	898	68629.6	76.743	-19.53
779	59372.0	78.738	-14.14	839	64069.3	77.816	-16.72	899	68706.4	76.723	-19.56
780	59450.7	78.723	-14.18	840	64147.1	77.799	-16.77	900	68783.1	76.704	-19.60

TABLE 1i6. (*Continued*)

Type E thermocouples—thermoelectric voltages, E(T), Seebeck coefficients, S(T), and first derivative of the Seebeck coefficients, dS/dT, reference junctions at 0 °C—Continued

T °C	E μV	S μV/°C	dS/dT nV/°C²	T °C	E μV	S μV/°C	dS/dT nV/°C²	T °C	E μV	S μV/°C	dS/dT nV/°C²
900	68783.1	76.704	-19.60	935	71455.6	76.006	-19.91	970	74103.9	75.346	-17.07
901	68859.8	76.684	-19.63	936	71531.5	75.986	-19.88	971	74179.3	75.329	-16.91
902	68936.5	76.664	-19.66	937	71607.5	75.967	-19.86	972	74254.6	75.312	-16.75
903	69013.1	76.645	-19.69	938	71683.5	75.947	-19.83	973	74329.9	75.296	-16.59
904	69089.8	76.625	-19.72	939	71759.4	75.927	-19.79	974	74405.2	75.279	-16.42
905	69166.4	76.605	-19.75	940	71835.3	75.907	-19.76	975	74480.5	75.263	-16.24
906	69243.0	76.585	-19.78	941	71911.2	75.887	-19.72	976	74555.7	75.247	-16.06
907	69319.5	76.566	-19.80	942	71987.1	75.868	-19.67	977	74631.0	75.231	-15.87
908	69396.1	76.546	-19.83	943	72063.0	75.848	-19.63	978	74706.2	75.215	-15.68
909	69472.6	76.526	-19.85	944	72138.8	75.828	-19.58	979	74781.4	75.199	-15.48
910	69549.1	76.506	-19.88	945	72214.6	75.809	-19.53	980	74856.6	75.184	-15.28
911	69625.6	76.486	-19.90	946	72290.4	75.789	-19.48	981	74931.7	75.169	-15.06
912	69702.1	76.466	-19.92	947	72366.2	75.770	-19.42	982	75006.9	75.154	-14.85
913	69778.6	76.446	-19.94	948	72442.0	75.751	-19.36	983	75082.1	75.139	-14.63
914	69855.0	76.426	-19.95	949	72517.7	75.731	-19.30	984	75157.2	75.125	-14.40
915	69931.4	76.406	-19.97	950	72593.4	75.712	-19.23	985	75232.3	75.110	-14.16
916	70007.8	76.386	-19.98	951	72669.1	75.693	-19.16	986	75307.4	75.096	-13.92
917	70084.2	76.366	-20.00	952	72744.8	75.674	-19.09	987	75382.5	75.083	-13.67
918	70160.6	76.346	-20.01	953	72820.5	75.655	-19.01	988	75457.6	75.069	-13.41
919	70236.9	76.326	-20.02	954	72896.1	75.636	-18.93	989	75532.6	75.056	-13.15
920	70313.2	76.306	-20.03	955	72971.7	75.617	-18.84	990	75607.7	75.043	-12.88
921	70389.5	76.286	-20.03	956	73047.4	75.598	-18.75	991	75682.7	75.030	-12.61
922	70465.8	76.266	-20.04	957	73122.9	75.579	-18.66	992	75757.7	75.017	-12.32
923	70542.0	76.246	-20.04	958	73198.5	75.561	-18.56	993	75832.8	75.005	-12.03
924	70618.3	76.226	-20.04	959	73274.1	75.542	-18.46	994	75907.8	74.993	-11.73
925	70694.5	76.206	-20.04	960	73349.6	75.524	-18.36	995	75982.7	74.982	-11.43
926	70770.7	76.186	-20.04	961	73425.1	75.505	-18.25	996	76057.7	74.971	-11.11
927	70846.9	76.166	-20.03	962	73500.6	75.487	-18.14	997	76132.7	74.960	-10.79
928	70923.0	76.146	-20.02	963	73576.1	75.469	-18.02	998	76207.6	74.949	-10.46
929	70999.2	76.126	-20.01	964	73651.5	75.451	-17.90	999	76282.6	74.939	-10.12
930	71075.3	76.106	-20.00	965	73727.0	75.433	-17.77	1000	76357.5	74.929	-9.78
931	71151.4	76.086	-19.99	966	73802.4	75.416	-17.64				
932	71227.4	76.066	-19.97	967	73877.8	75.398	-17.51				
933	71303.5	76.046	-19.95	968	73953.2	75.381	-17.37				
934	71379.5	76.026	-19.93	969	74028.6	75.363	-17.22				
935	71455.6	76.006	-19.91	970	74103.9	75.346	-17.07				

TYPE J—Iron Versus *Copper*–Nickel Alloy (SAMA) Thermocouples

Type J thermocouples—thermoelectric voltages, E(T), Seebeck coefficients, S(T), and first derivative of the Seebeck coefficients, dS/dT, reference junctions at 0 °C

T °C	E μV	S μV/°C	dS/dT nV/°C²	T °C	E μV	S μV/°C	dS/dT nV/°C²	T °C	E μV	S μV/°C	dS/dT nV/°C²
-210	-8095.6	19.126	283.95	-200	-7890.5	21.873	265.58	-190	-7658.7	24.441	248.29
-209	-8076.3	19.409	282.06	-199	-7868.5	22.137	263.80	-189	-7634.2	24.689	246.61
-208	-8056.8	19.690	280.19	-198	-7846.2	22.400	262.04	-188	-7609.4	24.934	244.95
-207	-8037.0	19.969	278.32	-197	-7823.7	22.662	260.28	-187	-7584.3	25.179	243.30
-206	-8016.9	20.247	276.47	-196	-7800.9	22.921	258.54	-186	-7559.0	25.421	241.66
-205	-7996.5	20.522	274.63	-195	-7777.8	23.179	256.80	-185	-7533.5	25.662	240.02
-204	-7975.8	20.796	272.79	-194	-7754.5	23.435	255.08	-184	-7507.7	25.901	238.40
-203	-7954.9	21.068	270.97	-193	-7730.9	23.689	253.37	-183	-7481.7	26.139	236.79
-202	-7933.7	21.338	269.17	-192	-7707.1	23.941	251.66	-182	-7455.4	26.375	235.19
-201	-7912.2	21.606	267.37	-191	-7683.1	24.192	249.97	-181	-7428.9	26.609	233.59
-200	-7890.5	21.873	265.58	-190	-7658.7	24.441	248.29	-180	-7402.2	26.842	232.01

TABLE 1i6. (*Continued*)

Type J thermocouples—*thermoelectric voltages, E(T), Seebeck coefficients, S(T), and first derivative of the Seebeck coefficients, dS/dT, reference junctions at 0 °C—Continued*

T °C	E μV	S μV/°C	dS/dT nV/°C^2	T °C	E μV	S μV/°C	dS/dT nV/°C^2	T °C	E μV	S μV/°C	dS/dT nV/°C^2
-180	-7402.2	26.842	232.01	-120	-5426.0	38.240	152.79	-60	-2892.5	45.669	98.27
-179	-7375.2	27.073	230.44	-119	-5387.7	38.392	151.70	-59	-2846.7	45.767	97.53
-178	-7348.0	27.303	228.87	-118	-5349.3	38.543	150.63	-58	-2800.9	45.864	96.79
-177	-7320.6	27.531	227.32	-117	-5310.6	38.693	149.55	-57	-2755.0	45.961	96.05
-176	-7293.0	27.757	225.77	-116	-5271.9	38.842	148.49	-56	-2709.0	46.056	95.32
-175	-7265.1	27.982	224.23	-115	-5233.0	38.990	147.43	-55	-2662.9	46.151	94.59
-174	-7237.0	28.206	222.71	-114	-5193.9	39.137	146.37	-54	-2616.7	46.245	93.87
-173	-7208.7	28.428	221.19	-113	-5154.7	39.283	145.33	-53	-2570.4	46.339	93.15
-172	-7180.2	28.648	219.68	-112	-5115.3	39.428	144.29	-52	-2524.0	46.432	92.43
-171	-7151.4	28.867	218.18	-111	-5075.8	39.571	143.25	-51	-2477.5	46.524	91.72
-170	-7122.4	29.085	216.69	-110	-5036.2	39.714	142.22	-50	-2431.0	46.615	91.02
-169	-7093.2	29.301	215.21	-109	-4996.4	39.856	141.20	-49	-2384.3	46.706	90.32
-168	-7063.8	29.515	213.74	-108	-4956.5	39.996	140.19	-48	-2337.6	46.796	89.62
-167	-7034.2	29.728	212.28	-107	-4916.4	40.136	139.18	-47	-2290.7	46.885	88.93
-166	-7004.4	29.940	210.83	-106	-4876.2	40.275	138.18	-46	-2243.8	46.974	88.24
-165	-6974.3	30.150	209.38	-105	-4835.9	40.413	137.18	-45	-2196.8	47.062	87.56
-164	-6944.1	30.358	207.94	-104	-4795.4	40.549	136.19	-44	-2149.7	47.149	86.88
-163	-6913.6	30.566	206.52	-103	-4754.8	40.685	135.21	-43	-2102.5	47.235	86.21
-162	-6882.9	30.771	205.10	-102	-4714.0	40.820	134.23	-42	-2055.2	47.321	85.54
-161	-6852.1	30.976	203.69	-101	-4673.1	40.953	133.26	-41	-2007.8	47.406	84.87
-160	-6821.0	31.179	202.29	-100	-4632.1	41.086	132.30	-40	-1960.4	47.491	84.21
-159	-6789.7	31.380	200.90	-99	-4590.9	41.218	131.34	-39	-1912.9	47.575	83.55
-158	-6758.2	31.581	199.51	-98	-4549.7	41.349	130.39	-38	-1865.4	47.658	82.90
-157	-6726.6	31.779	198.14	-97	-4508.2	41.479	129.44	-37	-1817.5	47.741	82.25
-156	-6694.7	31.977	196.77	-96	-4466.7	41.608	128.50	-36	-1769.8	47.823	81.60
-155	-6662.6	32.173	195.41	-95	-4425.0	41.736	127.56	-35	-1721.9	47.904	80.96
-154	-6630.3	32.368	194.06	-94	-4383.2	41.863	126.63	-34	-1673.9	47.985	80.32
-153	-6597.9	32.561	192.72	-93	-4341.3	41.989	125.71	-33	-1625.9	48.065	79.69
-152	-6565.2	32.753	191.38	-92	-4299.3	42.114	124.79	-32	-1577.8	48.144	79.06
-151	-6532.4	32.944	190.06	-91	-4257.1	42.239	123.88	-31	-1529.6	48.223	78.43
-150	-6499.3	33.133	188.74	-90	-4214.8	42.362	122.98	-30	-1481.4	48.301	77.81
-149	-6466.1	33.321	187.43	-89	-4172.4	42.485	122.08	-29	-1433.0	48.378	77.19
-148	-6432.7	33.508	186.13	-88	-4129.8	42.606	121.18	-28	-1384.6	48.455	76.58
-147	-6399.1	33.694	184.84	-87	-4087.1	42.727	120.29	-27	-1336.1	48.531	75.97
-146	-6365.3	33.878	183.55	-86	-4044.4	42.847	119.41	-26	-1287.6	48.607	75.36
-145	-6331.3	34.061	182.28	-85	-4001.4	42.966	118.53	-25	-1238.9	48.682	74.76
-144	-6297.2	34.242	181.01	-84	-3958.4	43.084	117.66	-24	-1190.2	48.757	74.16
-143	-6262.8	34.423	179.75	-83	-3915.3	43.201	116.79	-23	-1141.4	48.830	73.56
-142	-6228.3	34.602	178.49	-82	-3872.0	43.317	115.93	-22	-1092.5	48.904	72.97
-141	-6193.6	34.780	177.25	-81	-3828.6	43.433	115.07	-21	-1043.6	48.976	72.38
-140	-6158.8	34.956	176.01	-80	-3785.2	43.548	114.22	-20	-994.6	49.049	71.80
-139	-6123.7	35.132	174.78	-79	-3741.6	43.661	113.37	-19	-945.5	49.120	71.22
-138	-6088.5	35.306	173.55	-78	-3697.8	43.774	112.53	-18	-896.3	49.191	70.64
-137	-6053.1	35.479	172.34	-77	-3654.0	43.886	111.70	-17	-847.1	49.261	70.07
-136	-6017.5	35.651	171.13	-76	-3610.1	43.998	110.87	-16	-797.8	49.331	69.50
-135	-5981.8	35.821	169.93	-75	-3566.0	44.108	110.04	-15	-748.4	49.400	68.93
-134	-5945.9	35.990	168.74	-74	-3521.8	44.218	109.22	-14	-699.0	49.469	68.37
-133	-5909.8	36.159	167.55	-73	-3477.6	44.327	108.41	-13	-649.5	49.537	67.81
-132	-5873.6	36.326	166.37	-72	-3433.2	44.435	107.60	-12	-599.9	49.605	67.26
-131	-5837.2	36.491	165.20	-71	-3388.7	44.542	106.80	-11	-550.3	49.672	66.70
-130	-5800.6	36.656	164.04	-70	-3344.1	44.648	106.00	-10	-500.6	49.738	66.15
-129	-5763.9	36.819	162.88	-69	-3299.4	44.754	105.20	-9	-450.8	49.804	65.61
-128	-5727.0	36.982	161.73	-68	-3254.6	44.859	104.41	-8	-401.0	49.869	65.07
-127	-5689.9	37.143	160.59	-67	-3209.7	44.963	103.63	-7	-351.1	49.934	64.53
-126	-5652.7	37.303	159.46	-66	-3164.7	45.066	102.85	-6	-301.1	49.998	63.99
-125	-5615.3	37.462	158.33	-65	-3119.6	45.168	102.07	-5	-251.1	50.062	63.46
-124	-5577.8	37.620	157.21	-64	-3074.3	45.270	101.30	-4	-201.0	50.125	62.93
-123	-5540.1	37.776	156.09	-63	-3029.0	45.371	100.54	-3	-150.8	50.188	62.41
-122	-5502.2	37.932	154.99	-62	-2983.6	45.471	99.78	-2	-100.6	50.250	61.89
-121	-5464.2	38.086	153.88	-61	-2938.1	45.571	99.02	-1	-50.3	50.312	61.37
-120	-5426.0	38.240	152.79	-60	-2892.5	45.669	98.27	0	0.0	50.373	60.85

TABLE 1i6. (Continued)

Type J thermocouples—thermoelectric voltages, E(T), Seebeck coefficients, S(T), and first derivative of the Seebeck coefficients, dS/dT, reference junctions at 0 °C—Continued

T °C	E μV	S μV/°C	dS/dT nV/°C²	T °C	E μV	S μV/°C	dS/dT nV/°C²	T °C	E μV	S μV/°C	dS/dT nV/°C²
0	0.0	50.373	60.85	60	3115.0	53.204	35.11	120	6358.8	54.747	17.46
1	50.4	50.433	60.34	61	3168.2	53.239	34.76	121	6413.6	54.764	17.22
2	100.9	50.493	59.83	62	3221.5	53.273	34.41	122	6468.3	54.782	16.98
3	151.4	50.553	59.32	63	3274.8	53.308	34.06	123	6523.1	54.798	16.75
4	202.0	50.612	58.82	64	3328.1	53.341	33.72	124	6577.9	54.815	16.51
5	252.6	50.671	58.32	65	3381.4	53.375	33.37	125	6632.8	54.831	16.28
6	303.3	50.729	57.82	66	3434.8	53.408	33.03	126	6687.6	54.848	16.05
7	354.1	50.786	57.33	67	3488.3	53.441	32.69	127	6742.5	54.864	15.81
8	404.9	50.843	56.84	68	3541.7	53.474	32.35	128	6797.3	54.879	15.59
9	455.8	50.900	56.35	69	3595.2	53.506	32.02	129	6852.2	54.895	15.36
10	506.7	50.956	55.87	70	3648.7	53.538	31.68	130	6907.1	54.910	15.13
11	557.7	51.012	55.39	71	3702.3	53.569	31.35	131	6962.0	54.925	14.91
12	608.7	51.067	54.91	72	3755.9	53.600	31.02	132	7017.0	54.940	14.69
13	659.8	51.122	54.43	73	3809.5	53.631	30.70	133	7071.9	54.954	14.47
14	711.0	51.176	53.96	74	3863.1	53.662	30.37	134	7126.9	54.969	14.25
15	762.2	51.229	53.49	75	3916.8	53.692	30.05	135	7181.8	54.983	14.03
16	813.4	51.283	53.02	76	3970.5	53.722	29.73	136	7236.8	54.997	13.81
17	864.7	51.335	52.56	77	4024.3	53.751	29.41	137	7291.8	55.010	13.60
18	916.1	51.388	52.10	78	4078.0	53.781	29.09	138	7346.9	55.024	13.38
19	967.5	51.440	51.64	79	4131.8	53.810	28.78	139	7401.9	55.037	13.17
20	1019.0	51.491	51.19	80	4185.6	53.838	28.46	140	7456.9	55.050	12.96
21	1070.5	51.542	50.73	81	4239.5	53.866	28.15	141	7512.0	55.063	12.75
22	1122.0	51.593	50.28	82	4293.4	53.894	27.84	142	7567.1	55.076	12.55
23	1173.7	51.643	49.84	83	4347.3	53.922	27.54	143	7622.1	55.088	12.34
24	1225.3	51.692	49.39	84	4401.2	53.950	27.23	144	7677.2	55.100	12.14
25	1277.0	51.741	48.95	85	4455.2	53.977	26.93	145	7732.3	55.113	11.93
26	1328.8	51.790	48.51	86	4509.2	54.003	26.63	146	7787.5	55.124	11.73
27	1380.6	51.838	48.08	87	4563.2	54.030	26.33	147	7842.6	55.136	11.53
28	1432.5	51.886	47.64	88	4617.2	54.056	26.03	148	7897.7	55.147	11.33
29	1484.4	51.934	47.21	89	4671.3	54.082	25.73	149	7952.9	55.159	11.14
30	1536.4	51.981	46.78	90	4725.4	54.107	25.44	150	8008.1	55.170	10.94
31	1588.4	52.027	46.36	91	4779.5	54.133	25.15	151	8063.2	55.181	10.75
32	1640.4	52.073	45.94	92	4833.7	54.158	24.86	152	8118.4	55.191	10.55
33	1692.5	52.119	45.52	93	4887.8	54.183	24.57	153	8173.6	55.202	10.36
34	1744.6	52.164	45.10	94	4942.0	54.207	24.28	154	8228.8	55.212	10.17
35	1796.8	52.209	44.69	95	4996.2	54.231	24.00	155	8284.0	55.222	9.98
36	1849.1	52.254	44.27	96	5050.5	54.255	23.72	156	8339.3	55.232	9.80
37	1901.3	52.298	43.86	97	5104.8	54.279	23.44	157	8394.5	55.242	9.61
38	1953.7	52.342	43.46	98	5159.0	54.302	23.16	158	8449.7	55.251	9.43
39	2006.0	52.385	43.05	99	5213.4	54.325	22.88	159	8505.0	55.260	9.25
40	2058.4	52.428	42.65	100	5267.7	54.348	22.61	160	8560.3	55.270	9.06
41	2110.9	52.470	42.25	101	5322.1	54.370	22.33	161	8615.5	55.279	8.88
42	2163.4	52.512	41.85	102	5376.4	54.392	22.06	162	8670.8	55.287	8.71
43	2215.9	52.554	41.46	103	5430.8	54.414	21.79	163	8726.1	55.296	8.53
44	2268.5	52.595	41.07	104	5485.3	54.436	21.52	164	8781.4	55.304	8.35
45	2321.1	52.636	40.68	105	5539.7	54.457	21.25	165	8836.7	55.313	8.18
46	2373.8	52.676	40.29	106	5594.2	54.478	20.99	166	8892.0	55.321	8.00
47	2426.4	52.717	39.90	107	5648.7	54.499	20.73	167	8947.4	55.329	7.83
48	2479.2	52.756	39.52	108	5703.2	54.520	20.47	168	9002.7	55.336	7.66
49	2532.0	52.796	39.14	109	5757.7	54.540	20.21	169	9058.0	55.344	7.49
50	2584.8	52.835	38.76	110	5812.3	54.560	19.95	170	9113.4	55.351	7.33
51	2637.6	52.873	38.39	111	5866.8	54.580	19.69	171	9168.7	55.359	7.16
52	2690.5	52.911	38.02	112	5921.4	54.600	19.44	172	9224.1	55.366	6.99
53	2743.5	52.949	37.65	113	5976.0	54.619	19.18	173	9279.5	55.373	6.83
54	2796.4	52.987	37.28	114	6030.7	54.638	18.93	174	9334.8	55.379	6.67
55	2849.4	53.024	36.91	115	6085.3	54.657	18.68	175	9390.2	55.386	6.51
56	2902.5	53.060	36.55	116	6140.0	54.675	18.44	176	9445.6	55.392	6.35
57	2955.5	53.097	36.18	117	6194.6	54.694	18.19	177	9501.0	55.399	6.19
58	3008.7	53.133	35.83	113	6249.4	54.712	17.94	178	9556.4	55.405	6.03
59	3061.8	53.168	35.47	119	6304.1	54.730	17.70	179	9611.8	55.411	5.88
60	3115.0	53.204	35.11	120	6358.8	54.747	17.46	180	9667.2	55.417	5.72

TABLE 1i6. (*Continued*)

Type J thermocouples—*thermoelectric voltages*, E(T), *Seebeck coefficients*, S(T), *and first derivative of the Seebeck coefficients*, dS/dT, *reference junctions at* 0 °C—*Continued*

T °C	E μV	S μV/°C	dS/dT nV/°C²	T °C	E μV	S μV/°C	dS/dT nV/°C²	T °C	E μV	S μV/°C	dS/dT nV/°C²
180	9667.2	55.417	5.72	240	12997.7	55.529	-1.21	300	16325.1	55.359	-3.76
181	9722.7	55.422	5.57	241	13053.2	55.528	-1.29	301	16380.5	55.355	-3.77
182	9778.1	55.428	5.42	242	13108.7	55.527	-1.36	302	16435.8	55.351	-3.78
183	9833.5	55.433	5.27	243	13164.3	55.526	-1.44	303	16491.2	55.348	-3.78
184	9888.9	55.438	5.12	244	13219.8	55.524	-1.51	304	16546.5	55.344	-3.78
185	9944.4	55.443	4.97	245	13275.3	55.522	-1.59	305	16601.9	55.340	-3.79
186	9999.8	55.448	4.82	246	13330.8	55.521	-1.66	306	16657.2	55.336	-3.79
187	10055.3	55.453	4.68	247	13386.4	55.519	-1.73	307	16712.5	55.332	-3.78
188	10110.7	55.458	4.54	248	13441.9	55.517	-1.80	308	16767.9	55.329	-3.78
189	10166.2	55.462	4.39	249	13497.4	55.516	-1.87	309	16823.2	55.325	-3.78
190	10221.7	55.466	4.25	250	13552.9	55.514	-1.93	310	16878.5	55.321	-3.78
191	10277.1	55.470	4.11	251	13608.4	55.512	-2.00	311	16933.8	55.317	-3.77
192	10332.6	55.475	3.97	252	13663.9	55.510	-2.06	312	16989.1	55.314	-3.76
193	10388.1	55.478	3.84	253	13719.4	55.508	-2.13	313	17044.5	55.310	-3.76
194	10443.6	55.482	3.70	254	13775.0	55.505	-2.19	314	17099.8	55.306	-3.75
195	10499.0	55.486	3.56	255	13830.5	55.503	-2.25	315	17155.1	55.302	-3.74
196	10554.5	55.489	3.43	256	13886.0	55.501	-2.31	316	17210.4	55.299	-3.73
197	10610.0	55.493	3.30	257	13941.5	55.499	-2.37	317	17265.7	55.295	-3.71
198	10665.5	55.496	3.17	258	13997.0	55.496	-2.43	318	17321.0	55.291	-3.70
199	10721.0	55.499	3.04	259	14052.5	55.494	-2.48	319	17376.2	55.287	-3.69
200	10776.5	55.502	2.91	260	14107.9	55.491	-2.54	320	17431.5	55.284	-3.67
201	10832.0	55.505	2.78	261	14163.4	55.489	-2.59	321	17486.8	55.280	-3.65
202	10887.5	55.508	2.66	262	14218.9	55.486	-2.64	322	17542.1	55.276	-3.63
203	10943.0	55.510	2.53	263	14274.4	55.483	-2.70	323	17597.4	55.273	-3.61
204	10998.5	55.513	2.41	264	14329.9	55.481	-2.75	324	17652.6	55.269	-3.59
205	11054.1	55.515	2.29	265	14385.4	55.478	-2.79	325	17707.9	55.266	-3.57
206	11109.6	55.517	2.17	266	14440.8	55.475	-2.84	326	17763.2	55.262	-3.55
207	11165.1	55.519	2.05	267	14496.3	55.472	-2.89	327	17818.4	55.259	-3.52
208	11220.6	55.521	1.93	268	14551.8	55.469	-2.93	328	17873.7	55.255	-3.50
209	11276.1	55.523	1.81	269	14607.3	55.466	-2.98	329	17928.9	55.252	-3.47
210	11331.7	55.525	1.69	270	14662.7	55.463	-3.02	330	17984.2	55.248	-3.45
211	11387.2	55.527	1.58	271	14718.2	55.460	-3.06	331	18039.4	55.245	-3.42
212	11442.7	55.528	1.47	272	14773.6	55.457	-3.11	332	18094.7	55.241	-3.39
213	11498.2	55.530	1.35	273	14829.1	55.454	-3.14	333	18149.9	55.238	-3.36
214	11553.8	55.531	1.24	274	14884.6	55.451	-3.18	334	18205.2	55.235	-3.32
215	11609.3	55.532	1.13	275	14940.0	55.448	-3.22	335	18260.4	55.231	-3.29
216	11664.8	55.533	1.03	276	14995.4	55.444	-3.26	336	18315.6	55.228	-3.26
217	11720.4	55.534	0.92	277	15050.9	55.441	-3.29	337	18370.8	55.225	-3.22
218	11775.9	55.535	0.81	278	15106.3	55.438	-3.32	338	18426.1	55.221	-3.18
219	11831.4	55.536	0.71	279	15161.8	55.435	-3.36	339	18481.3	55.218	-3.14
220	11887.0	55.536	0.60	280	15217.2	55.431	-3.39	340	18536.5	55.215	-3.11
221	11942.5	55.537	0.50	281	15272.6	55.428	-3.42	341	18591.7	55.212	-3.06
222	11998.0	55.537	0.40	282	15328.1	55.424	-3.45	342	18646.9	55.209	-3.02
223	12053.6	55.538	0.30	283	15383.5	55.421	-3.48	343	18702.1	55.206	-2.98
224	12109.1	55.538	0.20	284	15438.9	55.417	-3.50	344	18757.3	55.203	-2.94
225	12164.7	55.538	0.11	285	15494.3	55.414	-3.53	345	18812.5	55.200	-2.89
226	12220.2	55.538	0.01	286	15549.7	55.410	-3.55	346	18867.7	55.197	-2.85
227	12275.7	55.538	-0.09	287	15605.1	55.407	-3.57	347	18922.9	55.195	-2.80
228	12331.3	55.538	-0.18	288	15660.5	55.403	-3.60	348	18978.1	55.192	-2.75
229	12386.8	55.538	-0.27	289	15715.9	55.400	-3.62	349	19033.3	55.189	-2.70
230	12442.4	55.537	-0.36	290	15771.3	55.396	-3.64	350	19088.5	55.186	-2.65
231	12497.9	55.537	-0.45	291	15826.7	55.392	-3.65	351	19143.7	55.184	-2.60
232	12553.4	55.537	-0.54	292	15882.1	55.389	-3.67	352	19198.9	55.181	-2.54
233	12609.0	55.536	-0.63	293	15937.5	55.385	-3.69	353	19254.1	55.179	-2.49
234	12664.5	55.535	-0.72	294	15992.9	55.381	-3.70	354	19309.2	55.176	-2.43
235	12720.0	55.535	-0.80	295	16048.3	55.378	-3.72	355	19364.4	55.174	-2.38
236	12775.6	55.534	-0.88	296	16103.6	55.374	-3.73	356	19419.6	55.171	-2.32
237	12831.1	55.533	-0.97	297	16159.0	55.370	-3.74	357	19474.7	55.169	-2.26
238	12886.6	55.532	-1.05	298	16214.4	55.366	-3.75	358	19529.9	55.167	-2.20
239	12942.2	55.531	-1.13	299	16269.7	55.363	-3.76	359	19585.1	55.165	-2.14
240	12997.7	55.529	-1.21	300	16325.1	55.359	-3.76	360	19640.2	55.163	-2.08

TABLE 1i6. (*Continued*)

Type J thermocouples—*thermoelectric voltages,* E(T), *Seebeck coefficients,* S(T), *and first derivative of the Seebeck coefficients,* dS/dT, *reference junctions at 0 °C—Continued*

T °C	E μV	S μV/°C	dS/dT nV/°C²	T °C	E μV	S μV/°C	dS/dT nV/°C²	T °C	E μV	S μV/°C	dS/dT nV/°C²
360	19640.2	55.163	-2.08	420	22949.1	55.190	3.64	480	26272.1	55.667	12.73
361	19695.4	55.161	-2.02	421	23004.3	55.194	3.77	481	26327.8	55.680	12.90
362	19750.6	55.159	-1.95	422	23059.5	55.198	3.90	482	26383.5	55.693	13.08
363	19805.7	55.157	-1.89	423	23114.7	55.202	4.02	483	26439.2	55.706	13.25
364	19860.9	55.155	-1.82	424	23169.9	55.206	4.15	484	26494.9	55.720	13.43
365	19916.0	55.153	-1.75	425	23225.1	55.210	4.28	485	26550.6	55.733	13.60
366	19971.2	55.151	-1.68	426	23280.3	55.215	4.42	486	26606.4	55.747	13.78
367	20026.3	55.150	-1.61	427	23335.5	55.219	4.55	487	26662.1	55.761	13.95
368	20081.5	55.148	-1.54	428	23390.8	55.224	4.68	488	26717.9	55.775	14.13
369	20136.6	55.147	-1.47	429	23446.0	55.228	4.82	489	26773.7	55.789	14.31
370	20191.8	55.145	-1.40	430	23501.2	55.233	4.95	490	26829.5	55.804	14.49
371	20246.9	55.144	-1.32	431	23556.5	55.238	5.09	491	26885.3	55.818	14.66
372	20302.1	55.142	-1.25	432	23611.7	55.243	5.22	492	26941.1	55.833	14.84
373	20357.2	55.141	-1.17	433	23666.9	55.249	5.36	493	26996.9	55.848	15.02
374	20412.3	55.140	-1.09	434	23722.2	55.254	5.50	494	27052.8	55.863	15.20
375	20467.5	55.139	-1.01	435	23777.5	55.260	5.64	495	27108.7	55.878	15.38
376	20522.6	55.138	-0.93	436	23832.7	55.265	5.78	496	27164.5	55.894	15.56
377	20577.8	55.137	-0.85	437	23888.0	55.271	5.92	497	27220.4	55.909	15.74
378	20632.9	55.136	-0.77	438	23943.3	55.277	6.06	498	27276.4	55.925	15.92
379	20688.0	55.136	-0.69	439	23998.5	55.283	6.21	499	27332.3	55.941	16.10
380	20743.2	55.135	-0.60	440	24053.8	55.290	6.35	500	27388.2	55.957	16.28
381	20798.3	55.135	-0.52	441	24109.1	55.296	6.49	501	27444.2	55.974	16.47
382	20853.4	55.134	-0.43	442	24164.4	55.303	6.64	502	27500.2	55.990	16.65
383	20908.6	55.134	-0.34	443	24219.7	55.309	6.79	503	27556.2	56.007	16.83
384	20963.7	55.133	-0.25	444	24275.0	55.316	6.93	504	27612.2	56.024	17.01
385	21018.8	55.133	-0.16	445	24330.4	55.323	7.08	505	27668.2	56.041	17.20
386	21074.0	55.133	-0.07	446	24385.7	55.330	7.23	506	27724.3	56.058	17.38
387	21129.1	55.133	0.02	447	24441.0	55.338	7.38	507	27780.4	56.076	17.56
388	21184.2	55.133	0.11	448	24496.4	55.345	7.53	508	27836.4	56.093	17.75
389	21239.4	55.133	0.21	449	24551.7	55.353	7.68	509	27892.5	56.111	17.93
390	21294.5	55.133	0.30	450	24607.1	55.361	7.83	510	27948.7	56.129	18.12
391	21349.6	55.134	0.40	451	24662.4	55.368	7.99	511	28004.8	56.148	18.30
392	21404.8	55.134	0.50	452	24717.8	55.376	8.14	512	28061.0	56.166	18.49
393	21459.9	55.135	0.60	453	24773.2	55.385	8.30	513	28117.1	56.185	18.67
394	21515.0	55.135	0.69	454	24828.6	55.393	8.45	514	28173.3	56.203	18.86
395	21570.2	55.136	0.80	455	24884.0	55.402	8.61	515	28229.5	56.222	19.04
396	21625.3	55.137	0.90	456	24939.4	55.410	8.76	516	28285.8	56.241	19.23
397	21680.5	55.138	1.00	457	24994.8	55.419	8.92	517	28342.0	56.261	19.41
398	21735.6	55.139	1.10	458	25050.2	55.428	9.08	518	28398.3	56.280	19.60
399	21790.7	55.140	1.21	459	25105.6	55.437	9.24	519	28454.6	56.300	19.79
400	21845.9	55.141	1.31	460	25161.1	55.447	9.40	520	28510.9	56.320	19.97
401	21901.0	55.143	1.42	461	25216.5	55.456	9.56	521	28567.2	56.340	20.16
402	21956.2	55.144	1.53	462	25272.0	55.466	9.72	522	28623.6	56.360	20.34
403	22011.3	55.146	1.64	463	25327.5	55.476	9.88	523	28679.9	56.381	20.53
404	22066.5	55.148	1.75	464	25383.0	55.486	10.04	524	28736.3	56.401	20.72
405	22121.6	55.149	1.86	465	25438.4	55.496	10.21	525	28792.7	56.422	20.90
406	22176.8	55.151	1.97	466	25493.9	55.506	10.37	526	28849.2	56.443	21.09
407	22231.9	55.153	2.08	467	25549.5	55.516	10.54	527	28905.6	56.464	21.28
408	22287.1	55.155	2.20	468	25605.0	55.527	10.70	528	28962.1	56.485	21.46
409	22342.2	55.158	2.31	469	25660.5	55.538	10.87	529	29018.6	56.507	21.65
410	22397.4	55.160	2.43	470	25716.1	55.549	11.03	530	29075.1	56.529	21.84
411	22452.5	55.163	2.54	471	25771.6	55.560	11.20	531	29131.7	56.551	22.02
412	22507.7	55.165	2.66	472	25827.2	55.571	11.37	532	29188.2	56.573	22.21
413	22562.9	55.168	2.78	473	25882.7	55.583	11.54	533	29244.8	56.595	22.40
414	22618.0	55.171	2.90	474	25938.3	55.594	11.71	534	29301.4	56.618	22.58
415	22673.2	55.174	3.02	475	25993.9	55.606	11.88	535	29358.0	56.640	22.77
416	22728.4	55.177	3.14	476	26049.5	55.618	12.05	536	29414.7	56.663	22.95
417	22783.6	55.180	3.27	477	26105.2	55.630	12.22	537	29471.4	56.686	23.14
418	22838.7	55.183	3.39	478	26160.8	55.642	12.39	538	29528.1	56.709	23.33
419	22893.9	55.187	3.52	479	26216.5	55.655	12.56	539	29584.8	56.733	23.51
420	22949.1	55.190	3.64	480	26272.1	55.667	12.73	540	29641.5	56.756	23.70

TABLE 1i6. (Continued)

Type J thermocouples—thermoelectric voltages, E(T), Seebeck coefficients, S(T), and first derivative of the Seebeck coefficients, dS/dT, reference junctions at 0 °C—Continued

T °C	E µV	S µV/°C	dS/dT nV/°C²	T °C	E µV	S µV/°C	dS/dT nV/°C²	T °C	E µV	S µV/°C	dS/dT nV/°C²
540	29641.5	56.756	23.70	600	33096.0	58.496	33.83	660	36670.8	60.715	38.84
541	29698.3	56.780	23.88	601	33154.5	58.529	33.97	661	36731.6	60.754	38.85
542	29755.1	56.804	24.07	602	33213.1	58.564	34.11	662	36792.3	60.793	38.85
543	29811.9	56.828	24.25	603	33271.7	58.598	34.25	663	36853.1	60.831	38.85
544	29868.8	56.853	24.44	604	33330.3	58.632	34.38	664	36914.0	60.870	38.85
545	29925.6	56.877	24.62	605	33388.9	58.666	34.52	665	36974.9	60.909	38.84
546	29982.5	56.902	24.81	606	33447.6	58.701	34.65	666	37035.8	60.948	38.83
547	30039.4	56.927	24.99	607	33506.3	58.736	34.78	667	37096.8	60.987	38.82
548	30096.4	56.952	25.17	608	33565.1	58.771	34.91	668	37157.8	61.026	38.81
549	30153.3	56.977	25.36	609	33623.9	58.806	35.04	669	37218.8	61.064	38.79
550	30210.3	57.003	25.54	610	33682.7	58.841	35.16	670	37279.9	61.103	38.77
551	30267.3	57.028	25.72	611	33741.6	58.876	35.29	671	37341.0	61.142	38.74
552	30324.4	57.054	25.90	612	33800.5	58.911	35.41	672	37402.2	61.181	38.71
553	30381.4	57.080	26.08	613	33859.4	58.947	35.53	673	37463.4	61.219	38.68
554	30438.5	57.106	26.27	614	33918.4	58.982	35.65	674	37524.6	61.258	38.64
555	30495.7	57.133	26.45	615	33977.4	59.018	35.77	675	37585.9	61.297	38.60
556	30552.8	57.159	26.63	616	34036.4	59.054	35.89	676	37647.2	61.335	38.56
557	30610.0	57.186	26.81	617	34095.5	59.090	36.00	677	37708.6	61.374	38.51
558	30667.2	57.213	26.99	618	34154.6	59.126	36.11	678	37770.0	61.412	38.46
559	30724.4	57.240	27.16	619	34213.7	59.162	36.23	679	37831.4	61.451	38.41
560	30781.6	57.267	27.34	620	34272.9	59.198	36.33	680	37892.9	61.489	38.35
561	30838.9	57.295	27.52	621	34332.1	59.235	36.44	681	37954.4	61.527	38.29
562	30896.2	57.322	27.70	622	34391.4	59.271	36.55	682	38015.9	61.566	38.22
563	30953.6	57.350	27.88	623	34450.6	59.308	36.65	683	38077.5	61.604	38.15
564	31010.9	57.378	28.05	624	34510.0	59.345	36.75	684	38139.1	61.642	38.08
565	31068.3	57.406	28.23	625	34569.3	59.381	36.85	685	38200.8	61.680	38.00
566	31125.7	57.434	28.40	626	34628.7	59.418	36.95	686	38262.5	61.718	37.92
567	31183.2	57.463	28.58	627	34688.2	59.455	37.04	687	38324.2	61.756	37.83
568	31240.7	57.492	28.75	628	34747.6	59.492	37.14	688	38386.0	61.794	37.74
569	31298.2	57.520	28.92	629	34807.2	59.529	37.23	689	38447.8	61.831	37.65
570	31355.7	57.549	29.09	630	34866.7	59.567	37.32	690	38509.7	61.869	37.55
571	31413.3	57.579	29.27	631	34926.3	59.604	37.40	691	38571.6	61.906	37.45
572	31470.9	57.608	29.44	632	34985.9	59.642	37.49	692	38633.5	61.944	37.34
573	31528.5	57.637	29.61	633	35045.6	59.679	37.57	693	38695.4	61.981	37.23
574	31586.1	57.667	29.78	634	35105.3	59.717	37.65	694	38757.4	62.018	37.11
575	31643.8	57.697	29.94	635	35165.0	59.754	37.73	695	38819.5	62.055	36.99
576	31701.5	57.727	30.11	636	35224.8	59.792	37.80	696	38881.6	62.092	36.87
577	31759.3	57.757	30.28	637	35284.6	59.830	37.88	697	38943.7	62.129	36.74
578	31817.1	57.788	30.44	638	35344.4	59.868	37.95	698	39005.8	62.166	36.61
579	31874.9	57.818	30.61	639	35404.3	59.906	38.01	699	39068.0	62.202	36.47
580	31932.7	57.849	30.77	640	35464.3	59.944	38.08	700	39130.2	62.239	36.33
581	31990.6	57.880	30.94	641	35524.2	59.982	38.14	701	39192.5	62.275	36.18
582	32048.5	57.911	31.10	642	35584.2	60.020	38.20	702	39254.8	62.311	36.03
583	32106.4	57.942	31.26	643	35644.3	60.058	38.26	703	39317.1	62.347	35.87
584	32164.3	57.973	31.42	644	35704.3	60.097	38.32	704	39379.5	62.383	35.71
585	32222.3	58.005	31.58	645	35764.5	60.135	38.37	705	39441.9	62.418	35.55
586	32280.3	58.036	31.74	646	35824.6	60.173	38.42	706	39504.3	62.454	35.38
587	32338.4	58.068	31.89	647	35884.8	60.212	38.47	707	39566.8	62.489	35.20
588	32396.5	58.100	32.05	648	35945.0	60.250	38.52	708	39629.3	62.524	35.02
589	32454.6	58.132	32.21	649	36005.3	60.289	38.56	709	39691.8	62.559	34.84
590	32512.7	58.165	32.36	650	36065.6	60.328	38.60	710	39754.4	62.594	34.65
591	32570.9	58.197	32.51	651	36126.0	60.366	38.64	711	39817.0	62.629	34.45
592	32629.1	58.230	32.66	652	36186.3	60.405	38.67	712	39879.7	62.663	34.25
593	32687.4	58.262	32.81	653	36246.8	60.443	38.70	713	39942.3	62.697	34.04
594	32745.7	58.295	32.96	654	36307.2	60.482	38.73	714	40005.0	62.731	33.83
595	32804.0	58.328	33.11	655	36367.7	60.521	38.75	715	40067.8	62.765	33.62
596	32862.3	58.361	33.26	656	36428.3	60.560	38.78	716	40130.6	62.798	33.40
597	32920.7	58.395	33.40	657	36488.8	60.598	38.80	717	40193.4	62.832	33.17
598	32979.1	58.428	33.55	658	36549.5	60.637	38.81	718	40256.2	62.865	32.94
599	33037.6	58.462	33.69	659	36610.1	60.676	38.83	719	40319.1	62.897	32.70
600	33096.0	58.496	33.83	660	36670.8	60.715	38.84	720	40382.0	62.930	32.46

TABLE 1i6. (Continued)

Type J thermocouples—thermoelectric voltages, E(T), Seebeck coefficients, S(T), and first derivative of the Seebeck coefficients, dS/dT, reference junctions at 0 °C—Continued

T °C	E μV	S μV/°C	dS/dT nV/°C²	T °C	E μV	S μV/°C	dS/dT nV/°C²	T °C	E μV	S μV/°C	dS/dT nV/°C²
720	40382.0	62.930	32.46	735	41329.5	63.386	28.13	750	42283.2	63.767	22.41
721	40445.0	62.962	32.21	736	41392.9	63.414	27.80	751	42347.0	63.789	21.98
722	40508.0	62.994	31.96	737	41456.3	63.442	27.45	752	42410.8	63.811	21.53
723	40571.0	63.026	31.70	738	41519.8	63.469	27.11	753	42474.6	63.832	21.08
724	40634.0	63.058	31.43	739	41583.3	63.496	26.75	754	42538.5	63.853	20.63
725	40697.1	63.089	31.16	740	41646.8	63.522	26.39	755	42602.3	63.873	20.16
726	40760.2	63.120	30.88	741	41710.3	63.549	26.02	756	42666.2	63.893	19.69
727	40823.3	63.151	30.60	742	41773.9	63.574	25.65	757	42730.1	63.913	19.22
728	40886.5	63.181	30.31	743	41837.4	63.600	25.27	758	42794.1	63.932	18.73
729	40949.7	63.211	30.02	744	41901.1	63.625	24.88	759	42858.0	63.950	18.24
730	41012.9	63.241	29.72	745	41964.7	63.650	24.48	760	42922.0	63.968	17.74
731	41076.2	63.271	29.42	746	42028.4	63.674	24.08				
732	41139.5	63.300	29.10	747	42092.0	63.698	23.68				
733	41202.8	63.329	28.79	748	42155.8	63.721	23.26				
734	41266.1	63.358	28.46	749	42219.5	63.744	22.84				
735	41329.5	63.386	28.13	750	42283.2	63.767	22.41				

TABLE 1i6. (*Continued*)

Type J thermocouples extended range—thermoelectric voltages, E(T), *Seebeck coefficients,* S(T), *and first derivative of the Seebeck coefficients dS/dT, reference junctions at 0 °C*

T °C	E µV	S µV/°C	dS/dT nV/°C²	T °C	E µV	S µV/°C	dS/dT nV/°C²	T °C	E µV	S µV/°C	dS/dT nV/°C²
				800	45498	64.616	−0.05	850	48716	63.904	−25.38
				801	45562	64.616	−0.74	851	48780	63.878	−25.71
				802	45627	64.615	−1.43	852	48844	63.852	−26.03
				803	45692	64.613	−2.11	853	48908	63.826	−26.34
				804	45756	64.610	−2.78	854	48971	63.800	−26.65
				805	45821	64.607	−3.44	855	49035	63.773	−26.96
				806	45885	64.603	−4.09	856	49099	63.746	−27.25
				807	45950	64.599	−4.74	857	49163	63.718	−27.54
				808	46015	64.594	−5.37	858	49226	63.691	−27.82
				809	46079	64.588	−6.00	859	49290	63.663	−28.10
				810	46144	64.582	−6.62	860	49354	63.634	−28.37
761	42985	64.001	33.86	811	46208	64.575	−7.24	861	49417	63.606	−28.64
762	43049	64.035	32.81	812	46273	64.567	−7.84	862	49481	63.577	−28.90
763	43113	64.067	31.78	813	46337	64.559	−8.44	863	49544	63.548	−29.15
764	43178	64.098	30.75	814	46402	64.551	−9.03	864	49608	63.519	−29.40
765	43242	64.129	29.73	815	46467	64.541	−9.61	865	49671	63.489	−29.64
766	43306	64.158	28.73	816	46531	64.531	−10.19	866	49735	63.459	−29.88
767	43370	64.186	27.73	817	46596	64.521	−10.75	867	49798	63.429	−30.11
768	43434	64.213	26.74	818	46660	64.510	−11.31	868	49862	63.399	−30.33
769	43498	64.239	25.76	819	46725	64.498	−11.86	869	49925	63.369	−30.55
770	43563	64.265	24.80	820	46789	64.486	−12.41	870	49989	63.338	−30.76
771	43627	64.289	23.84	821	46854	64.473	−12.94	871	50052	63.307	−30.97
772	43691	64.312	22.89	822	46918	64.460	−13.47	872	50115	63.276	−31.17
773	43756	64.335	21.95	823	46983	64.447	−13.99	873	50178	63.245	−31.37
774	43820	64.356	21.02	824	47047	64.432	−14.51	874	50242	63.213	−31.56
775	43884	64.377	20.09	825	47111	64.418	−15.01	875	50305	63.182	−31.74
776	43949	64.397	19.18	826	47176	64.402	−15.51	876	50368	63.150	−31.92
777	44013	64.415	18.28	827	47240	64.386	−16.00	877	50431	63.118	−32.09
778	44077	64.433	17.38	828	47305	64.370	−16.49	878	50494	63.086	−32.26
779	44142	64.450	16.50	829	47369	64.354	−16.97	879	50557	63.053	−32.42
780	44206	64.466	15.62	830	47433	64.336	−17.44	880	50620	63.021	−32.58
781	44271	64.481	14.76	831	47498	64.319	−17.90	881	50683	62.988	−32.73
782	44335	64.496	13.90	832	47562	64.301	−18.35	882	50746	62.956	−32.88
783	44400	64.509	13.05	833	47626	64.282	−18.80	883	50809	62.923	−33.02
784	44464	64.522	12.21	834	47691	64.263	−19.25	884	50872	62.889	−33.16
785	44529	64.533	11.38	835	47755	64.243	−19.68	885	50935	62.856	−33.29
786	44593	64.544	10.55	836	47819	64.224	−20.11	886	50998	62.823	−33.42
787	44658	64.555	9.74	837	47883	64.203	−20.53	887	51061	62.789	−33.54
788	44723	64.564	8.94	838	47947	64.183	−20.94	888	51123	62.756	−33.66
789	44787	64.572	8.14	839	48012	64.161	−21.35	889	51186	62.722	−33.77
790	44852	64.580	7.35	840	48076	64.140	−21.75	890	51249	62.688	−33.87
791	44916	64.587	6.57	841	48140	64.118	−22.14	891	51312	62.654	−33.98
792	44981	64.593	5.80	842	48204	64.096	−22.53	892	51374	62.620	−34.07
793	45045	64.599	5.04	843	48268	64.073	−22.91	893	51437	62.586	−34.16
794	45110	64.603	4.29	844	48332	64.050	−23.28	894	51499	62.552	−34.25
795	45175	64.607	3.55	845	48396	64.026	−23.65	895	51562	62.518	−34.33
796	45239	64.611	2.81	846	48460	64.002	−24.01	896	51624	62.483	−34.41
797	45304	64.613	2.08	847	48524	63.978	−24.36	897	51687	62.449	−34.48
798	45368	64.615	1.36	848	48588	63.954	−24.71	898	51749	62.414	−34.55
799	45433	64.616	0.65	849	48652	63.929	−25.05	899	51812	62.380	−34.62
800	45498	64.616	−0.05	850	48716	63.904	−25.38	900	51874	62.345	−34.67

TABLE 1i6. (*Continued*)

Type J thermocouples extended range—*thermoelectric voltages,* E(T), *Seebeck coefficients,*
S(T), *and first derivative of the Seebeck coefficients,* dS/dT,
reference junctions at 0 °C—Continued

T °C	E μV	S μV/°C	dS/dT nV/°C^2	T °C	E μV	S μV/°C	dS/dT nV/°C^2	T °C	E μV	S μV/°C	dS/dT nV/°C^2
900	51874	62.345	-34.67	960	55553	60.306	-31.25	1020	59121	58.760	-19.58
901	51936	62.310	-34.73	961	55613	60.275	-31.10	1021	59180	58.740	-19.36
902	51999	62.276	-34.78	962	55673	60.244	-30.95	1022	59238	58.721	-19.15
903	52061	62.241	-34.82	963	55733	60.213	-30.80	1023	59297	58.702	-18.93
904	52123	62.206	-34.87	964	55794	60.182	-30.64	1024	59356	58.683	-18.71
905	52185	62.171	-34.90	965	55854	60.151	-30.48	1025	59415	58.665	-18.49
906	52248	62.136	-34.93	966	55914	60.121	-30.32	1026	59473	58.646	-18.27
907	52310	62.101	-34.96	967	55974	60.091	-30.16	1027	59532	58.628	-18.06
908	52372	62.066	-34.99	968	56034	60.061	-30.00	1028	59590	58.610	-17.84
909	52434	62.031	-35.01	969	56094	60.031	-29.83	1029	59649	58.592	-17.62
910	52496	61.996	-35.02	970	56154	60.001	-29.66	1030	59708	58.575	-17.40
911	52558	61.961	-35.03	971	56214	59.971	-29.49	1031	59766	58.558	-17.19
912	52620	61.926	-35.04	972	56274	59.942	-29.32	1032	59825	58.540	-16.97
913	52682	61.891	-35.04	973	56334	59.913	-29.15	1033	59883	58.524	-16.75
914	52744	61.856	-35.04	974	56394	59.884	-28.97	1034	59942	58.507	-16.54
915	52805	61.821	-35.04	975	56454	59.855	-28.79	1035	60000	58.490	-16.32
916	52867	61.786	-35.03	976	56514	59.826	-28.61	1036	60059	58.474	-16.11
917	52929	61.751	-35.01	977	56573	59.798	-28.43	1037	60117	58.458	-15.90
918	52991	61.716	-35.00	978	56633	59.769	-28.25	1038	60176	58.442	-15.68
919	53052	61.681	-34.98	979	56693	59.741	-28.07	1039	60234	58.427	-15.47
920	53114	61.646	-34.95	980	56753	59.713	-27.88	1040	60293	58.412	-15.26
921	53176	61.611	-34.92	981	56812	59.685	-27.69	1041	60351	58.396	-15.05
922	53237	61.576	-34.89	982	56872	59.658	-27.50	1042	60409	58.381	-14.84
923	53299	61.541	-34.85	983	56932	59.630	-27.31	1043	60468	58.367	-14.63
924	53360	61.507	-34.81	984	56991	59.603	-27.12	1044	60526	58.352	-14.42
925	53422	61.472	-34.77	985	57051	59.576	-26.93	1045	60584	58.338	-14.21
926	53483	61.437	-34.72	986	57110	59.549	-26.73	1046	60643	58.324	-14.00
927	53545	61.402	-34.67	987	57170	59.523	-26.54	1047	60701	58.310	-13.80
928	53606	61.368	-34.62	988	57229	59.496	-26.34	1048	60759	58.296	-13.59
929	53667	61.333	-34.56	989	57289	59.470	-26.14	1049	60818	58.283	-13.39
930	53729	61.299	-34.50	990	57348	59.444	-25.94	1050	60876	58.269	-13.18
931	53790	61.264	-34.44	991	57408	59.418	-25.74	1051	60934	58.256	-12.98
932	53851	61.230	-34.37	992	57467	59.392	-25.54	1052	60992	58.243	-12.78
933	53912	61.195	-34.30	993	57527	59.367	-25.34	1053	61051	58.231	-12.58
934	53974	61.161	-34.22	994	57586	59.342	-25.13	1054	61109	58.218	-12.38
935	54035	61.127	-34.15	995	57645	59.317	-24.93	1055	61167	58.206	-12.18
936	54096	61.093	-34.07	996	57705	59.292	-24.72	1056	61225	58.194	-11.99
937	54157	61.059	-33.98	997	57764	59.267	-24.52	1057	61284	58.182	-11.79
938	54218	61.025	-33.89	998	57823	59.243	-24.31	1058	61342	58.170	-11.60
939	54279	60.991	-33.80	999	57882	59.219	-24.10	1059	61400	58.159	-11.41
940	54340	60.957	-33.71	1000	57942	59.195	-23.89	1060	61458	58.148	-11.22
941	54401	60.924	-33.61	1001	58001	59.171	-23.68	1061	61516	58.136	-11.03
942	54462	60.890	-33.51	1002	58060	59.147	-23.47	1062	61574	58.125	-10.84
943	54523	60.857	-33.41	1003	58119	59.124	-23.26	1063	61632	58.115	-10.65
944	54584	60.823	-33.31	1004	58178	59.101	-23.04	1064	61691	58.104	-10.47
945	54644	60.790	-33.20	1005	58237	59.078	-22.83	1065	61749	58.094	-10.29
946	54705	60.757	-33.09	1006	58296	59.055	-22.62	1066	61807	58.084	-10.10
947	54766	60.724	-32.97	1007	58355	59.033	-22.40	1067	61865	58.074	-9.92
948	54827	60.691	-32.86	1008	58414	59.010	-22.19	1068	61923	58.064	-9.75
949	54887	60.658	-32.74	1009	58473	58.988	-21.97	1069	61981	58.054	-9.57
950	54948	60.625	-32.62	1010	58532	58.966	-21.76	1070	62039	58.045	-9.40
951	55009	60.593	-32.49	1011	58591	58.945	-21.54	1071	62097	58.035	-9.22
952	55069	60.560	-32.36	1012	58650	58.923	-21.32	1072	62155	58.026	-9.05
953	55130	60.528	-32.23	1013	58709	58.902	-21.11	1073	62213	58.017	-8.88
954	55190	60.496	-32.10	1014	58768	58.881	-20.89	1074	62271	58.008	-8.72
955	55251	60.464	-31.97	1015	58827	58.860	-20.67	1075	62329	58.000	-8.55
956	55311	60.432	-31.83	1016	58886	58.840	-20.45	1076	62387	57.991	-8.39
957	55372	60.400	-31.69	1017	58945	58.819	-20.24	1077	62445	57.983	-8.23
958	55432	60.369	-31.55	1018	59003	58.799	-20.02	1078	62503	57.975	-8.07
959	55492	60.337	-31.40	1019	59062	58.779	-19.80	1079	62561	57.967	-7.91
960	55553	60.306	-31.25	1020	59121	58.760	-19.58	1080	62619	57.959	-7.76

TABLE 1i6. (Continued)

Type J thermocouples extended range—*thermoelectric voltages,* E(T), *Seebeck coefficients,* S(T), *and first derivative of the Seebeck coefficients,* dS/dT, *reference junctions at 0 °C—Continued*

T °C	E μV	S μV/°C	dS/dT nV/°C²	T °C	E μV	S μV/°C	dS/dT nV/°C²	T °C	E μV	S μV/°C	dS/dT nV/°C²
1080	62619	57.959	−7.76	1120	64932	57.744	−3.79	1160	67239	57.577	−5.75
1081	62677	57.951	−7.61	1121	64990	57.740	−3.76	1161	67297	57.571	−5.90
1082	62735	57.944	−7.46	1122	65048	57.736	−3.73	1162	67354	57.565	−6.05
1083	62793	57.936	−7.31	1123	65106	57.733	−3.70	1163	67412	57.559	−6.20
1084	62851	57.929	−7.16	1124	65163	57.729	−3.68	1164	67469	57.553	−6.37
1085	62909	57.922	−7.02	1125	65221	57.725	−3.66	1165	67527	57.546	−6.53
1086	62967	57.915	−6.88	1126	65279	57.722	−3.65	1166	67585	57.539	−6.71
1087	63024	57.908	−6.74	1127	65337	57.718	−3.64	1167	67642	57.533	−6.88
1088	63082	57.902	−6.61	1128	65394	57.714	−3.63	1168	67700	57.526	−7.07
1089	63140	57.895	−6.47	1129	65452	57.711	−3.63	1169	67757	57.519	−7.25
1090	63198	57.889	−6.34	1130	65510	57.707	−3.63	1170	67815	57.511	−7.45
1091	63256	57.882	−6.21	1131	65567	57.703	−3.64	1171	67872	57.504	−7.65
1092	63314	57.876	−6.09	1132	65625	57.700	−3.65	1172	67930	57.496	−7.85
1093	63372	57.870	−5.97	1133	65683	57.696	−3.66	1173	67987	57.488	−8.07
1094	63430	57.864	−5.84	1134	65741	57.692	−3.68	1174	68045	57.480	−8.28
1095	63488	57.859	−5.73	1135	65798	57.689	−3.70	1175	68102	57.471	−8.51
1096	63545	57.853	−5.61	1136	65856	57.685	−3.73	1176	68160	57.463	−8.73
1097	63603	57.847	−5.50	1137	65914	57.681	−3.76	1177	68217	57.454	−8.97
1098	63661	57.842	−5.39	1138	65971	57.677	−3.79	1178	68274	57.445	−9.21
1099	63719	57.837	−5.29	1139	66029	57.674	−3.83	1179	68332	57.435	−9.45
1100	63777	57.831	−5.18	1140	66087	57.670	−3.88	1180	68389	57.426	−9.71
1101	63835	57.826	−5.08	1141	66144	57.666	−3.93	1181	68447	57.416	−9.96
1102	63892	57.821	−4.99	1142	66202	57.662	−3.98	1182	68504	57.406	−10.23
1103	63950	57.816	−4.89	1143	66260	57.658	−4.04	1183	68562	57.396	−10.50
1104	64008	57.811	−4.80	1144	66317	57.654	−4.10	1184	68619	57.385	−10.77
1105	64066	57.807	−4.71	1145	66375	57.650	−4.17	1185	68676	57.374	−11.05
1106	64124	57.802	−4.63	1146	66433	57.646	−4.24	1186	68734	57.363	−11.34
1107	64181	57.797	−4.55	1147	66490	57.641	−4.32	1187	68791	57.351	−11.64
1108	64239	57.793	−4.47	1148	66548	57.637	−4.40	1188	68848	57.340	−11.94
1109	64297	57.788	−4.39	1149	66605	57.632	−4.48	1189	68906	57.327	−12.24
1110	64355	57.784	−4.32	1150	66663	57.628	−4.57	1190	68963	57.315	−12.56
1111	64413	57.780	−4.25	1151	66721	57.623	−4.67	1191	69020	57.302	−12.88
1112	64470	57.776	−4.19	1152	66778	57.619	−4.77	1192	69078	57.289	−13.20
1113	64528	57.771	−4.12	1153	66836	57.614	−4.87	1193	69135	57.276	−13.53
1114	64586	57.767	−4.07	1154	66894	57.609	−4.98	1194	69192	57.262	−13.87
1115	64644	57.763	−4.01	1155	66951	57.604	−5.10	1195	69249	57.248	−14.22
1116	64701	57.759	−3.96	1156	67009	57.599	−5.22	1196	69307	57.234	−14.57
1117	64759	57.755	−3.91	1157	67066	57.593	−5.34	1197	69364	57.219	−14.93
1118	64817	57.751	−3.87	1158	67124	57.588	−5.48	1198	69421	57.204	−15.29
1119	64875	57.748	−3.83	1159	67182	57.582	−5.61	1199	69478	57.188	−15.66
1120	64932	57.744	−3.79	1160	67239	57.577	−5.75	1200	69536	57.173	−16.04

TABLE 1i6. (*Continued*)

TYPE K—*Nickel*–Chromium Alloy Versus *Nickel*–Aluminum Alloy Thermocouples

Type K thermocouples—*thermoelectric voltages,* E(T), *Seebeck coefficients,* S(T), **and first** *derivative of the Seebeck coefficients,* dS/dT, *reference junctions at 0 °C*

T °C	E μV	S μV/°C	dS/dT nV/°C^2	T °C	E μV	S μV/°C	dS/dT nV/°C^2	T °C	E μV	S μV/°C	dS/dT nV/°C^2
−270	−6457.82	0.739	168.65	−240	−6343.92	7.096	222.90	−210	−6034.63	13.373	193.87
−269	−6456.99	0.911	174.56	−239	−6336.72	7.319	222.30	−209	−6021.16	13.566	192.89
−268	−6455.99	1.088	180.02	−238	−6329.29	7.541	221.63	−208	−6007.49	13.759	191.92
−267	−6454.81	1.270	185.07	−237	−6321.63	7.762	220.91	−207	−5993.64	13.950	190.96
−266	−6453.45	1.458	189.72	−236	−6313.76	7.983	220.14	−206	−5979.59	14.141	190.02
−265	−6451.90	1.650	193.98	−235	−6305.67	8.202	219.32	−205	−5965.36	14.330	189.08
−264	−6450.15	1.846	197.89	−234	−6297.36	8.421	218.47	−204	−5950.93	14.519	188.17
−263	−6448.20	2.045	201.46	−233	−6288.83	8.639	217.57	−203	−5936.32	14.707	187.26
−262	−6446.06	2.248	204.70	−232	−6280.08	8.856	216.65	−202	−5921.52	14.893	186.37
−261	−6443.71	2.455	207.64	−231	−6271.12	9.073	215.70	−201	−5906.53	15.079	185.49
−260	−6441.15	2.664	210.29	−230	−6261.93	9.288	214.72	−200	−5891.36	15.264	184.63
−259	−6438.38	2.875	212.67	−229	−6252.54	9.502	213.72	−199	−5876.00	15.449	183.78
−258	−6435.40	3.089	214.79	−228	−6242.93	9.715	212.70	−198	−5860.46	15.632	182.94
−257	−6432.20	3.305	216.67	−227	−6233.11	9.927	211.67	−197	−5844.74	15.814	182.12
−256	−6428.79	3.522	218.31	−226	−6223.08	10.139	210.62	−196	−5828.84	15.996	181.30
−255	−6425.15	3.741	219.74	−225	−6212.83	10.349	209.57	−195	−5812.75	16.177	180.50
−254	−6421.30	3.962	220.97	−224	−6202.38	10.558	208.51	−194	−5796.48	16.357	179.72
−253	−6417.23	4.183	222.01	−223	−6191.72	10.766	207.44	−193	−5780.03	16.537	178.94
−252	−6412.94	4.406	222.87	−222	−6180.85	10.973	206.37	−192	−5763.41	16.715	178.18
−251	−6408.42	4.629	223.57	−221	−6169.77	11.178	205.30	−191	−5746.61	16.893	177.43
−250	−6403.68	4.853	224.10	−220	−6158.49	11.383	204.23	−190	−5729.62	17.070	176.69
−249	−6398.71	5.077	224.49	−219	−6147.01	11.587	203.17	−189	−5712.47	17.246	175.96
−248	−6393.53	5.302	224.75	−218	−6135.32	11.789	202.10	−188	−5695.13	17.422	175.24
−247	−6388.11	5.526	224.87	−217	−6123.43	11.991	201.04	−187	−5677.62	17.597	174.53
−246	−6382.47	5.751	224.88	−216	−6111.34	12.192	199.99	−186	−5659.94	17.771	173.83
−245	−6376.61	5.976	224.78	−215	−6099.05	12.391	198.95	−185	−5642.08	17.944	173.14
−244	−6370.52	6.201	224.58	−214	−6086.56	12.589	197.91	−184	−5624.05	18.117	172.46
−243	−6364.21	6.425	224.28	−213	−6073.87	12.787	196.89	−183	−5605.85	18.289	171.79
−242	−6357.67	6.649	223.90	−212	−6060.98	12.983	195.87	−182	−5587.47	18.461	171.12
−241	−6350.91	6.873	223.44	−211	−6047.90	13.179	194.86	−181	−5568.92	18.632	170.47
−240	−6343.92	7.096	222.90	−210	−6034.63	13.373	193.87	−180	−5550.21	18.802	169.82

TABLE 1i6. *(Continued)*

Type K thermocouples—*thermoelectric voltages, E(T), Seebeck coefficients, S(T),* **and first** *derivative of the Seebeck coefficients, dS/dT, reference junctions at 0 °C*—*Continued*

T °C	E μV	S μV/°C	dS/dT nV/°C²	T °C	E μV	S μV/°C	dS/dT nV/°C²	T °C	E μV	S μV/°C	dS/dT nV/°C²
−180	−5550.21	18.802	169.82	−120	−4137.85	27.927	134.45	−60	−2242.51	34.872	96.56
−179	−5531.32	18.971	169.18	−119	−4109.86	28.061	133.84	−59	−2207.59	34.968	95.90
−178	−5512.27	19.140	168.54	−118	−4081.73	28.195	133.23	−58	−2172.58	35.064	95.24
−177	−5493.04	19.308	167.91	−117	−4053.47	28.328	132.62	−57	−2137.47	35.159	94.59
−176	−5473.65	19.476	167.29	−116	−4025.08	28.460	132.01	−56	−2102.26	35.253	93.93
−175	−5454.09	19.643	166.67	−115	−3996.55	28.592	131.40	−55	−2066.96	35.347	93.27
−174	−5434.36	19.809	166.06	−114	−3967.89	28.723	130.79	−54	−2031.57	35.439	92.62
−173	−5414.47	19.975	165.45	−113	−3939.10	28.853	130.17	−53	−1996.08	35.532	91.96
−172	−5394.41	20.140	164.85	−112	−3910.19	28.983	129.56	−52	−1960.50	35.623	91.30
−171	−5374.19	20.305	164.25	−111	−3881.14	29.113	128.95	−51	−1924.83	35.714	90.65
−170	−5353.80	20.469	163.66	−110	−3851.96	29.241	128.33	−50	−1889.07	35.805	89.99
−169	−5333.25	20.632	163.06	−109	−3822.66	29.369	127.71	−49	−1853.23	35.894	89.33
−168	−5312.54	20.795	162.47	−108	−3793.22	29.497	127.10	−48	−1817.29	35.983	88.68
−167	−5291.66	20.957	161.89	−107	−3763.66	29.623	126.48	−47	−1781.26	36.072	88.02
−166	−5270.63	21.119	161.31	−106	−3733.98	29.750	125.86	−46	−1745.14	36.159	87.37
−165	−5249.43	21.280	160.72	−105	−3704.16	29.875	125.24	−45	−1708.94	36.246	86.71
−164	−5228.07	21.440	160.15	−104	−3674.23	30.000	124.62	−44	−1672.65	36.333	86.06
−163	−5206.55	21.600	159.57	−103	−3644.16	30.124	124.00	−43	−1636.27	36.419	85.40
−162	−5184.87	21.759	158.99	−102	−3613.98	30.248	123.38	−42	−1599.81	36.504	84.75
−161	−5163.03	21.918	158.42	−101	−3583.67	30.371	122.76	−41	−1563.27	36.588	84.10
−160	−5141.03	22.076	157.85	−100	−3553.23	30.494	122.14	−40	−1526.64	36.672	83.45
−159	−5118.88	22.234	157.27	−99	−3522.68	30.615	121.51	−39	−1489.92	36.755	82.80
−158	−5096.57	22.391	156.70	−98	−3492.00	30.737	120.89	−38	−1453.13	36.837	82.15
−157	−5074.10	22.547	156.13	−97	−3461.21	30.857	120.26	−37	−1416.25	36.919	81.50
−156	−5051.47	22.703	155.56	−96	−3430.29	30.977	119.64	−36	−1379.29	37.000	80.85
−155	−5028.69	22.858	154.99	−95	−3399.25	31.096	119.01	−35	−1342.25	37.081	80.20
−154	−5005.76	23.013	154.42	−94	−3368.10	31.215	118.39	−34	−1305.13	37.161	79.55
−153	−4982.67	23.167	153.85	−93	−3336.82	31.333	117.76	−33	−1267.93	37.240	78.91
−152	−4959.42	23.320	153.27	−92	−3305.43	31.451	117.13	−32	−1230.65	37.319	78.26
−151	−4936.03	23.473	152.70	−91	−3273.92	31.567	116.50	−31	−1193.29	37.397	77.61
−150	−4912.48	23.626	152.13	−90	−3242.30	31.684	115.87	−30	−1155.85	37.474	76.97
−149	−4888.77	23.778	151.56	−89	−3210.55	31.799	115.24	−29	−1118.34	37.551	76.32
−148	−4864.92	23.929	150.98	−88	−3178.70	31.914	114.61	−28	−1080.75	37.627	75.67
−147	−4840.92	24.080	150.41	−87	−3146.73	32.028	113.97	−27	−1043.09	37.702	75.03
−146	−4816.76	24.230	149.83	−86	−3114.64	32.142	113.34	−26	−1005.35	37.777	74.38
−145	−4792.46	24.379	149.26	−85	−3082.44	32.255	112.71	−25	−967.54	37.851	73.73
−144	−4768.00	24.528	148.68	−84	−3050.13	32.368	112.07	−24	−929.65	37.924	73.08
−143	−4743.40	24.677	148.10	−83	−3017.71	32.479	111.43	−23	−891.69	37.997	72.43
−142	−4718.65	24.825	147.52	−82	−2985.17	32.590	110.80	−22	−853.66	38.069	71.77
−141	−4693.75	24.972	146.94	−81	−2952.53	32.701	110.16	−21	−815.55	38.140	71.11
−140	−4668.71	25.118	146.36	−80	−2919.77	32.811	109.52	−20	−777.38	38.211	70.45
−139	−4643.52	25.264	145.77	−79	−2886.91	32.920	108.88	−19	−739.13	38.281	69.79
−138	−4618.18	25.410	145.19	−78	−2853.93	33.028	108.24	−18	−700.81	38.351	69.12
−137	−4592.70	25.555	144.60	−77	−2820.85	33.136	107.60	−17	−662.43	38.419	68.44
−136	−4567.07	25.699	144.01	−76	−2787.66	33.244	106.95	−16	−623.97	38.488	67.76
−135	−4541.30	25.843	143.42	−75	−2754.36	33.350	106.31	−15	−585.45	38.555	67.07
−134	−4515.38	25.986	142.83	−74	−2720.96	33.456	105.67	−14	−546.87	38.622	66.37
−133	−4489.33	26.129	142.24	−73	−2687.45	33.562	105.02	−13	−508.21	38.688	65.66
−132	−4463.13	26.270	141.65	−72	−2653.84	33.666	104.37	−12	−469.49	38.753	64.94
−131	−4436.78	26.412	141.06	−71	−2620.12	33.770	103.73	−11	−430.70	38.818	64.21
−130	−4410.30	26.553	140.46	−70	−2586.30	33.874	103.08	−10	−391.86	38.881	63.47
−129	−4383.68	26.693	139.87	−69	−2552.37	33.977	102.43	−9	−352.94	38.945	62.71
−128	−4356.92	26.832	139.27	−68	−2518.34	34.079	101.78	−8	−313.97	39.007	61.94
−127	−4330.02	26.971	138.67	−67	−2484.21	34.180	101.13	−7	−274.93	39.068	61.15
−126	−4302.97	27.110	138.07	−66	−2449.98	34.281	100.48	−6	−235.83	39.129	60.33
−125	−4275.80	27.247	137.47	−65	−2415.65	34.381	99.83	−5	−196.67	39.189	59.50
−124	−4248.48	27.385	136.87	−64	−2381.22	34.481	99.17	−4	−157.45	39.248	58.64
−123	−4221.03	27.521	136.26	−63	−2346.69	34.579	98.52	−3	−118.17	39.306	57.76
−122	−4193.44	27.657	135.66	−62	−2312.06	34.678	97.86	−2	−78.84	39.364	56.85
−121	−4165.71	27.792	135.05	−61	−2277.34	34.775	97.21	−1	−39.45	39.420	55.91
−120	−4137.85	27.927	134.45	−60	−2242.51	34.872	96.56	0	0.00	39.475	54.93

TABLE 1i6. (Continued)

Type K thermocouples—*thermoelectric voltages, E(T), Seebeck coefficients, S(T), and first derivative of the Seebeck coefficients, dS/dT, reference junctions at 0 °C—Continued*

T °C	E μV	S μV/°C	dS/dT nV/°C²	T °C	E μV	S μV/°C	dS/dT nV/°C²	T °C	E μV	S μV/°C	dS/dT nV/°C²
0	0.0	39.475	45.65	60	2435.7	41.407	14.48	120	4919.0	40.969	-23.60
1	39.5	39.521	45.28	61	2477.1	41.421	13.73	121	4959.9	40.946	-23.80
2	79.0	39.566	44.92	62	2518.5	41.434	12.97	122	5000.8	40.922	-23.97
3	118.6	39.611	44.55	63	2560.0	41.447	12.21	123	5041.8	40.898	-24.13
4	158.3	39.655	44.19	64	2601.4	41.458	11.45	124	5082.6	40.874	-24.26
5	197.9	39.699	43.82	65	2642.9	41.470	10.67	125	5123.5	40.849	-24.38
6	237.7	39.743	43.45	66	2684.3	41.480	9.90	126	5164.3	40.825	-24.46
7	277.4	39.786	43.08	67	2725.8	41.489	9.12	127	5205.2	40.800	-24.53
8	317.2	39.829	42.71	68	2767.3	41.498	8.34	128	5245.9	40.776	-24.57
9	357.1	39.871	42.33	69	2808.8	41.506	7.55	129	5286.7	40.751	-24.60
10	397.0	39.914	41.95	70	2850.3	41.513	6.77	130	5327.4	40.727	-24.59
11	436.9	39.955	41.57	71	2891.8	41.520	5.98	131	5368.2	40.702	-24.57
12	476.9	39.997	41.18	72	2933.4	41.525	5.19	132	5408.8	40.678	-24.52
13	516.9	40.038	40.79	73	2974.9	41.530	4.40	133	5449.5	40.653	-24.46
14	557.0	40.078	40.40	74	3016.4	41.534	3.61	134	5490.2	40.629	-24.37
15	597.1	40.118	40.00	75	3058.0	41.537	2.81	135	5530.8	40.604	-24.25
16	637.2	40.158	39.60	76	3099.5	41.540	2.02	136	5571.4	40.580	-24.12
17	677.4	40.198	39.19	77	3141.0	41.541	1.23	137	5611.9	40.556	-23.96
18	717.6	40.237	38.78	78	3182.6	41.542	0.44	138	5652.5	40.532	-23.79
19	757.8	40.275	38.36	79	3224.1	41.542	-0.34	139	5693.0	40.509	-23.59
20	798.1	40.313	37.94	80	3265.7	41.541	-1.12	140	5733.5	40.485	-23.37
21	838.5	40.351	37.51	81	3307.2	41.540	-1.90	141	5774.0	40.462	-23.14
22	878.8	40.388	37.07	82	3348.8	41.538	-2.68	142	5814.4	40.439	-22.88
23	919.3	40.425	36.63	83	3390.3	41.534	-3.45	143	5854.8	40.416	-22.60
24	959.7	40.462	36.18	84	3431.8	41.531	-4.22	144	5895.2	40.394	-22.30
25	1000.2	40.498	35.72	85	3473.3	41.526	-4.98	145	5935.6	40.371	-21.99
26	1040.7	40.533	35.26	86	3514.9	41.521	-5.73	146	5976.0	40.350	-21.65
27	1081.2	40.568	34.79	87	3556.4	41.515	-6.48	147	6016.3	40.328	-21.30
28	1121.8	40.603	34.31	88	3597.9	41.508	-7.21	148	6056.6	40.307	-20.93
29	1162.4	40.637	33.82	89	3639.4	41.500	-7.95	149	6096.9	40.286	-20.55
30	1203.1	40.670	33.33	90	3680.9	41.492	-8.67	150	6137.2	40.266	-20.14
31	1243.8	40.703	32.82	91	3722.4	41.483	-9.38	151	6177.5	40.246	-19.72
32	1284.5	40.736	32.31	92	3763.9	41.473	-10.08	152	6217.7	40.226	-19.29
33	1325.3	40.768	31.79	93	3805.3	41.463	-10.78	153	6257.9	40.207	-18.84
34	1366.0	40.800	31.27	94	3846.8	41.452	-11.46	154	6298.1	40.189	-18.36
35	1406.9	40.831	30.73	95	3888.2	41.440	-12.13	155	6338.3	40.171	-17.90
36	1447.7	40.861	30.18	96	3929.7	41.427	-12.78	156	6378.5	40.153	-17.41
37	1488.6	40.891	29.63	97	3971.1	41.414	-13.43	157	6418.6	40.136	-16.90
38	1529.5	40.920	29.07	98	4012.5	41.400	-14.06	158	6458.7	40.119	-16.39
39	1570.4	40.949	28.49	99	4053.9	41.386	-14.68	159	6498.9	40.103	-15.86
40	1611.4	40.977	27.91	100	4095.3	41.371	-15.28	160	6538.9	40.087	-15.32
41	1652.4	41.005	27.32	101	4136.6	41.356	-15.87	161	6579.0	40.072	-14.77
42	1693.4	41.032	26.72	102	4178.0	41.339	-16.44	162	6619.1	40.058	-14.22
43	1734.4	41.058	26.12	103	4219.3	41.323	-17.00	163	6659.1	40.044	-13.65
44	1775.5	41.084	25.50	104	4260.6	41.305	-17.54	164	6699.2	40.031	-13.07
45	1816.6	41.109	24.87	105	4301.9	41.288	-18.06	165	6739.2	40.018	-12.49
46	1857.7	41.134	24.24	106	4343.2	41.269	-18.57	166	6779.2	40.006	-11.90
47	1898.9	41.158	23.59	107	4384.5	41.250	-19.05	167	6819.2	39.994	-11.30
48	1940.0	41.181	22.94	108	4425.7	41.231	-19.52	168	6859.2	39.983	-10.70
49	1981.2	41.204	22.28	109	4466.9	41.211	-19.97	169	6899.2	39.973	-10.09
50	2022.4	41.226	21.61	110	4508.1	41.191	-20.40	170	6939.2	39.963	-9.47
51	2063.7	41.247	20.93	111	4549.3	41.171	-20.82	171	6979.1	39.954	-8.86
52	2104.9	41.267	20.24	112	4590.5	41.150	-21.21	172	7019.1	39.945	-8.24
53	2146.2	41.287	19.55	113	4631.6	41.128	-21.58	173	7059.0	39.937	-7.61
54	2187.5	41.307	18.85	114	4672.7	41.106	-21.93	174	7098.9	39.930	-6.99
55	2228.8	41.325	18.14	115	4713.8	41.084	-22.26	175	7138.9	39.923	-6.36
56	2270.2	41.343	17.42	116	4754.9	41.062	-22.57	176	7178.8	39.917	-5.73
57	2311.5	41.360	16.69	117	4795.9	41.039	-22.86	177	7218.7	39.912	-5.11
58	2352.9	41.376	15.96	118	4837.0	41.016	-23.13	178	7258.6	39.907	-4.48
59	2394.3	41.392	15.22	119	4878.0	40.993	-23.37	179	7298.5	39.903	-3.85
60	2435.7	41.407	14.48	120	4919.0	40.969	-23.60	180	7338.4	39.899	-3.23

TABLE 1i6. (Continued)

Type K thermocouples—thermoelectric voltages, E(T), *Seebeck coefficients,* S(T), *and first derivative of the Seebeck coefficients,* dS/dT, *reference junctions at 0 °C—Continued*

T °C	E μV	S μV/°C	dS/dT nV/°C²	T °C	E μV	S μV/°C	dS/dT nV/°C²	T °C	E μV	S μV/°C	dS/dT nV/°C²
180	7338.4	39.899	-3.23	240	9745.2	40.542	18.03	300	12207.4	41.459	11.21
181	7378.3	39.896	-2.60	241	9785.7	40.560	18.05	301	12248.9	41.470	11.09
182	7418.2	39.894	-1.98	242	9826.3	40.578	18.05	302	12290.4	41.481	10.96
183	7458.1	39.892	-1.36	243	9866.9	40.596	18.05	303	12331.9	41.492	10.83
184	7498.0	39.891	-0.75	244	9907.5	40.614	18.03	304	12373.4	41.503	10.71
185	7537.9	39.891	-0.14	245	9948.1	40.632	18.01	305	12414.9	41.514	10.59
186	7577.8	39.891	0.46	246	9988.7	40.650	17.98	306	12456.4	41.524	10.47
187	7617.7	39.892	1.06	247	10029.4	40.668	17.95	307	12497.9	41.534	10.35
188	7657.6	39.893	1.66	248	10070.1	40.686	17.90	308	12539.5	41.545	10.24
189	7697.4	39.895	2.24	249	10110.8	40.704	17.85	309	12581.0	41.555	10.13
190	7737.3	39.898	2.82	250	10151.5	40.722	17.80	310	12622.6	41.565	10.02
191	7777.2	39.901	3.40	251	10192.2	40.739	17.73	311	12664.1	41.575	9.91
192	7817.1	39.904	3.96	252	10233.0	40.757	17.66	312	12705.7	41.585	9.80
193	7857.1	39.909	4.52	253	10273.7	40.775	17.58	313	12747.3	41.595	9.70
194	7897.0	39.914	5.07	254	10314.5	40.792	17.50	314	12788.9	41.604	9.60
195	7936.9	39.919	5.61	255	10355.3	40.810	17.41	315	12830.5	41.614	9.50
196	7976.8	39.925	6.14	256	10396.1	40.827	17.32	316	12872.1	41.623	9.40
197	8016.7	39.931	6.67	257	10437.0	40.844	17.22	317	12913.8	41.633	9.30
198	8056.7	39.938	7.18	258	10477.8	40.862	17.12	318	12955.4	41.642	9.21
199	8096.6	39.946	7.68	259	10518.7	40.879	17.01	319	12997.0	41.651	9.12
200	8136.6	39.953	8.17	260	10559.6	40.896	16.90	320	13038.7	41.660	9.03
201	8176.5	39.962	8.66	261	10600.5	40.912	16.78	321	13080.4	41.669	8.94
202	8216.5	39.971	9.13	262	10641.4	40.929	16.66	322	13122.0	41.678	8.86
203	8256.5	39.980	9.59	263	10682.3	40.946	16.53	323	13163.7	41.687	8.77
204	8296.4	39.990	10.03	264	10723.3	40.962	16.41	324	13205.4	41.695	8.69
205	8336.4	40.000	10.47	265	10764.3	40.979	16.28	325	13247.1	41.704	8.61
206	8376.4	40.011	10.89	266	10805.2	40.995	16.14	326	13288.8	41.713	8.53
207	8416.5	40.022	11.31	267	10846.2	41.011	16.01	327	13330.5	41.721	8.46
208	8456.5	40.033	11.71	268	10887.3	41.027	15.87	328	13372.3	41.730	8.38
209	8496.5	40.045	12.09	269	10928.3	41.043	15.73	329	13414.0	41.738	8.31
210	8536.6	40.058	12.47	270	10969.3	41.058	15.59	330	13455.7	41.746	8.24
211	8576.6	40.070	12.83	271	11010.4	41.074	15.44	331	13497.5	41.754	8.17
212	8616.7	40.083	13.18	272	11051.5	41.089	15.30	332	13539.3	41.763	8.11
213	8656.8	40.097	13.52	273	11092.6	41.104	15.15	333	13581.0	41.771	8.04
214	8696.9	40.110	13.85	274	11133.7	41.119	15.00	334	13622.8	41.779	7.98
215	8737.0	40.124	14.16	275	11174.8	41.134	14.85	335	13664.6	41.787	7.92
216	8777.2	40.139	14.46	276	11216.0	41.149	14.70	336	13706.4	41.795	7.86
217	8817.3	40.153	14.75	277	11257.1	41.164	14.55	337	13748.2	41.802	7.80
218	8857.5	40.168	15.02	278	11298.3	41.178	14.40	338	13790.0	41.810	7.74
219	8897.6	40.183	15.28	279	11339.5	41.193	14.25	339	13831.8	41.818	7.68
220	8937.8	40.199	15.53	280	11380.7	41.207	14.09	340	13873.6	41.825	7.63
221	8978.0	40.214	15.77	281	11421.9	41.221	13.94	341	13915.4	41.833	7.58
222	9018.3	40.230	15.99	282	11463.1	41.235	13.79	342	13957.3	41.841	7.52
223	9058.5	40.246	16.21	283	11504.4	41.248	13.64	343	13999.1	41.848	7.47
224	9098.8	40.263	16.40	284	11545.6	41.262	13.49	344	14041.0	41.856	7.43
225	9139.0	40.279	16.59	285	11586.9	41.275	13.34	345	14082.8	41.863	7.38
226	9179.3	40.296	16.77	286	11628.2	41.289	13.19	346	14124.7	41.870	7.33
227	9219.6	40.313	16.93	287	11669.5	41.302	13.04	347	14166.6	41.878	7.29
228	9259.9	40.330	17.08	288	11710.8	41.315	12.89	348	14208.4	41.885	7.24
229	9300.3	40.347	17.22	289	11752.1	41.327	12.74	349	14250.3	41.892	7.20
230	9340.6	40.364	17.35	290	11793.4	41.340	12.60	350	14292.2	41.899	7.16
231	9381.0	40.382	17.46	291	11834.8	41.353	12.45	351	14334.1	41.906	7.11
232	9421.4	40.399	17.57	292	11876.1	41.365	12.31	352	14376.0	41.914	7.07
233	9461.8	40.417	17.66	293	11917.5	41.377	12.17	353	14418.0	41.921	7.04
234	9502.2	40.434	17.75	294	11958.9	41.389	12.02	354	14459.9	41.928	7.00
235	9542.7	40.452	17.82	295	12000.3	41.401	11.88	355	14501.8	41.935	6.96
236	9583.1	40.470	17.88	296	12041.7	41.413	11.75	356	14543.8	41.942	6.92
237	9623.6	40.488	17.93	297	12083.1	41.425	11.61	357	14585.7	41.948	6.89
238	9664.1	40.506	17.98	298	12124.5	41.436	11.48	358	14627.7	41.955	6.85
239	9704.6	40.524	18.01	299	12166.0	41.448	11.34	359	14669.6	41.962	6.82
240	9745.2	40.542	18.03	300	12207.4	41.459	11.21	360	14711.6	41.969	6.78

TABLE 1i6. (Continued)

Type K thermocouples—*thermoelectric voltages*, E(T), *Seebeck coefficients*, S(T), *and first derivative of the Seebeck coefficients*, dS/dT, *reference junctions at 0 °C—Continued*

T °C	E μV	S μV/°C	dS/dT nV/°C²	T °C	E μV	S μV/°C	dS/dT nV/°C²	T °C	E μV	S μV/°C	dS/dT nV/°C²
360	14711.6	41.969	6.78	420	17240.9	42.325	5.11	480	19788.4	42.566	2.76
361	14753.6	41.976	6.75	421	17283.2	42.331	5.08	481	19830.9	42.569	2.72
362	14795.5	41.982	6.72	422	17325.6	42.336	5.05	482	19873.5	42.572	2.67
363	14837.5	41.989	6.69	423	17367.9	42.341	5.01	483	19916.1	42.574	2.62
364	14879.5	41.996	6.65	424	17410.2	42.346	4.98	484	19958.6	42.577	2.58
365	14921.5	42.002	6.62	425	17452.6	42.351	4.95	485	20001.2	42.580	2.53
366	14963.5	42.009	6.59	426	17494.9	42.356	4.92	486	20043.8	42.582	2.48
367	15005.5	42.016	6.56	427	17537.3	42.360	4.88	487	20086.4	42.585	2.44
368	15047.5	42.022	6.53	428	17579.7	42.365	4.85	488	20129.0	42.587	2.39
369	15089.6	42.029	6.50	429	17622.0	42.370	4.82	489	20171.6	42.589	2.34
370	15131.6	42.035	6.48	430	17664.4	42.375	4.78	490	20214.1	42.592	2.29
371	15173.6	42.042	6.45	431	17706.8	42.380	4.75	491	20256.7	42.594	2.24
372	15215.7	42.048	6.42	432	17749.2	42.384	4.72	492	20299.3	42.596	2.19
373	15257.7	42.054	6.39	433	17791.5	42.389	4.68	493	20341.9	42.598	2.15
374	15299.8	42.061	6.36	434	17833.9	42.394	4.65	494	20384.5	42.600	2.10
375	15341.9	42.067	6.34	435	17876.3	42.398	4.61	495	20427.1	42.602	2.05
376	15383.9	42.074	6.31	436	17918.7	42.403	4.58	496	20469.7	42.605	2.00
377	15426.0	42.080	6.28	437	17961.1	42.408	4.54	497	20512.3	42.606	1.95
378	15468.1	42.086	6.26	438	18003.6	42.412	4.51	498	20555.0	42.608	1.90
379	15510.2	42.092	6.23	439	18046.0	42.417	4.47	499	20597.6	42.610	1.85
380	15552.3	42.099	6.20	440	18088.4	42.421	4.43	500	20640.2	42.612	1.80
381	15594.4	42.105	6.18	441	18130.8	42.425	4.40	501	20682.8	42.614	1.75
382	15636.5	42.111	6.15	442	18173.2	42.430	4.36	502	20725.4	42.616	1.70
383	15678.6	42.117	6.13	443	18215.7	42.434	4.32	503	20768.0	42.617	1.65
384	15720.7	42.123	6.10	444	18258.1	42.439	4.28	504	20810.6	42.619	1.60
385	15762.8	42.129	6.07	445	18300.5	42.443	4.25	505	20853.3	42.620	1.55
386	15805.0	42.135	6.05	446	18343.0	42.447	4.21	506	20895.9	42.622	1.50
387	15847.1	42.141	6.02	447	18385.4	42.451	4.17	507	20938.5	42.623	1.44
388	15889.3	42.147	6.00	448	18427.9	42.455	4.13	508	20981.1	42.625	1.39
389	15931.4	42.153	5.97	449	18470.3	42.459	4.09	509	21023.7	42.626	1.34
390	15973.6	42.159	5.94	450	18512.8	42.464	4.05	510	21066.4	42.628	1.29
391	16015.7	42.165	5.92	451	18555.3	42.468	4.01	511	21109.0	42.629	1.24
392	16057.9	42.171	5.89	452	18597.7	42.472	3.97	512	21151.6	42.630	1.19
393	16100.1	42.177	5.87	453	18640.2	42.476	3.94	513	21194.3	42.631	1.14
394	16142.2	42.183	5.84	454	18682.7	42.479	3.89	514	21236.9	42.632	1.08
395	16184.4	42.189	5.82	455	18725.2	42.483	3.85	515	21279.5	42.633	1.03
396	16226.6	42.194	5.79	456	18767.7	42.487	3.81	516	21322.2	42.634	0.98
397	16268.8	42.200	5.76	457	18810.2	42.491	3.77	517	21364.8	42.635	0.93
398	16311.0	42.206	5.74	458	18852.6	42.495	3.73	518	21407.4	42.636	0.88
399	16353.2	42.212	5.71	459	18895.1	42.498	3.69	519	21450.1	42.637	0.82
400	16395.4	42.217	5.68	460	18937.6	42.502	3.65	520	21492.7	42.638	0.77
401	16437.7	42.223	5.66	461	18980.1	42.506	3.61	521	21535.3	42.639	0.72
402	16479.9	42.229	5.63	462	19022.7	42.509	3.56	522	21578.0	42.639	0.66
403	16522.1	42.234	5.60	463	19065.2	42.513	3.52	523	21620.6	42.640	0.61
404	16564.4	42.240	5.57	464	19107.7	42.516	3.48	524	21663.3	42.641	0.56
405	16606.6	42.246	5.55	465	19150.2	42.520	3.44	525	21705.9	42.641	0.51
406	16648.9	42.251	5.52	466	19192.7	42.523	3.39	526	21748.5	42.642	0.45
407	16691.1	42.257	5.49	467	19235.2	42.527	3.35	527	21791.2	42.642	0.40
408	16733.4	42.262	5.46	468	19277.8	42.530	3.31	528	21833.8	42.642	0.35
409	16775.6	42.267	5.43	469	19320.3	42.533	3.26	529	21876.5	42.643	0.29
410	16817.9	42.273	5.41	470	19362.8	42.536	3.22	530	21919.1	42.643	0.24
411	16860.2	42.278	5.38	471	19405.4	42.540	3.17	531	21961.8	42.643	0.19
412	16902.5	42.284	5.35	472	19447.9	42.543	3.13	532	22004.4	42.643	0.13
413	16944.7	42.289	5.32	473	19490.5	42.546	3.08	533	22047.0	42.643	0.08
414	16987.0	42.294	5.29	474	19533.0	42.549	3.04	534	22089.7	42.643	0.02
415	17029.3	42.300	5.26	475	19575.6	42.552	2.99	535	22132.3	42.643	-0.03
416	17071.6	42.305	5.23	476	19618.1	42.555	2.95	536	22175.0	42.643	-0.08
417	17113.9	42.310	5.20	477	19660.7	42.558	2.90	537	22217.6	42.643	-0.14
418	17156.3	42.315	5.17	478	19703.2	42.561	2.86	538	22260.3	42.643	-0.19
419	17198.6	42.320	5.14	479	19745.8	42.564	2.81	539	22302.9	42.643	-0.25
420	17240.9	42.325	5.11	480	19788.4	42.566	2.76	540	22345.5	42.643	-0.30

TABLE 1i6. (Continued)

Type K thermocouples—thermoelectric voltages, E(T), *Seebeck coefficients*, S(T), *and first derivative of the Seebeck coefficients*, dS/dT, *reference junctions at 0 °C—Continued*

T °C	E μV	S μV/°C	dS/dT nV/°C²	T °C	E μV	S μV/°C	dS/dT nV/°C²	T °C	E μV	S μV/°C	dS/dT nV/°C²
540	22345.5	42.643	−0.30	600	24901.6	42.527	−3.55	660	27445.0	42.224	−6.46
541	22388.2	42.642	−0.35	601	24944.1	42.523	−3.60	661	27487.2	42.217	−6.51
542	22430.8	42.642	−0.41	602	24986.6	42.519	−3.65	662	27529.4	42.211	−6.55
543	22473.5	42.641	−0.46	603	25029.2	42.516	−3.71	663	27571.6	42.204	−6.59
544	22516.1	42.641	−0.52	604	25071.7	42.512	−3.76	664	27613.8	42.197	−6.63
545	22558.8	42.640	−0.57	605	25114.2	42.508	−3.81	665	27656.0	42.191	−6.68
546	22601.4	42.640	−0.63	606	25156.7	42.504	−3.86	666	27698.2	42.184	−6.72
547	22644.0	42.639	−0.68	607	25199.2	42.501	−3.92	667	27740.4	42.177	−6.76
548	22686.7	42.638	−0.74	608	25241.7	42.497	−3.97	668	27782.6	42.170	−6.80
549	22729.3	42.638	−0.79	609	25284.2	42.493	−4.02	669	27824.7	42.164	−6.84
550	22771.9	42.637	−0.84	610	25326.7	42.489	−4.07	670	27866.9	42.157	−6.89
551	22814.6	42.636	−0.90	611	25369.2	42.484	−4.12	671	27909.0	42.150	−6.93
552	22857.2	42.635	−0.95	612	25411.7	42.480	−4.17	672	27951.2	42.143	−6.97
553	22899.9	42.634	−1.01	613	25454.1	42.476	−4.22	673	27993.3	42.136	−7.01
554	22942.5	42.633	−1.06	614	25496.6	42.472	−4.28	674	28035.5	42.129	−7.05
555	22985.1	42.632	−1.12	615	25539.1	42.468	−4.33	675	28077.6	42.122	−7.09
556	23027.8	42.631	−1.17	616	25581.5	42.463	−4.38	676	28119.7	42.115	−7.13
557	23070.4	42.630	−1.23	617	25624.0	42.459	−4.43	677	28161.8	42.108	−7.17
558	23113.0	42.628	−1.28	618	25666.5	42.454	−4.48	678	28203.9	42.100	−7.21
559	23155.6	42.627	−1.34	619	25708.9	42.450	−4.53	679	28246.0	42.093	−7.25
560	23198.3	42.626	−1.39	620	25751.4	42.445	−4.58	680	28288.1	42.086	−7.29
561	23240.9	42.624	−1.45	621	25793.8	42.441	−4.63	681	28330.2	42.079	−7.33
562	23283.5	42.623	−1.50	622	25836.2	42.436	−4.68	682	28372.3	42.071	−7.37
563	23326.1	42.621	−1.55	623	25878.7	42.431	−4.73	683	28414.3	42.064	−7.40
564	23368.8	42.620	−1.61	624	25921.1	42.427	−4.78	684	28456.4	42.056	−7.44
565	23411.4	42.618	−1.66	625	25963.5	42.422	−4.83	685	28498.4	42.049	−7.48
566	23454.0	42.616	−1.72	626	26005.9	42.417	−4.88	686	28540.5	42.041	−7.52
567	23496.6	42.615	−1.77	627	26048.4	42.412	−4.93	687	28582.5	42.034	−7.56
568	23539.2	42.613	−1.83	628	26090.8	42.407	−4.98	688	28624.6	42.026	−7.59
569	23581.8	42.611	−1.88	629	26133.2	42.402	−5.03	689	28666.6	42.019	−7.63
570	23624.4	42.609	−1.94	630	26175.6	42.397	−5.08	690	28708.6	42.011	−7.67
571	23667.1	42.607	−1.99	631	26218.0	42.392	−5.12	691	28750.6	42.003	−7.70
572	23709.7	42.605	−2.05	632	26260.4	42.387	−5.17	692	28792.6	41.996	−7.74
573	23752.3	42.603	−2.10	633	26302.7	42.382	−5.22	693	28834.6	41.988	−7.78
574	23794.9	42.601	−2.15	634	26345.1	42.376	−5.27	694	28876.6	41.980	−7.81
575	23837.5	42.599	−2.21	635	26387.5	42.371	−5.32	695	28918.6	41.972	−7.85
576	23880.1	42.597	−2.26	636	26429.9	42.366	−5.37	696	28960.5	41.964	−7.88
577	23922.7	42.594	−2.32	637	26472.2	42.360	−5.41	697	29002.5	41.957	−7.92
578	23965.3	42.592	−2.37	638	26514.6	42.355	−5.46	698	29044.4	41.949	−7.95
579	24007.8	42.589	−2.43	639	26556.9	42.349	−5.51	699	29086.4	41.941	−7.99
580	24050.4	42.587	−2.48	640	26599.3	42.344	−5.56	700	29128.3	41.933	−8.02
581	24093.0	42.585	−2.53	641	26641.6	42.338	−5.60	701	29170.2	41.925	−8.06
582	24135.6	42.582	−2.59	642	26684.0	42.333	−5.65	702	29212.2	41.917	−8.09
583	24178.2	42.579	−2.64	643	26726.3	42.327	−5.70	703	29254.1	41.908	−8.12
584	24220.8	42.577	−2.70	644	26768.6	42.321	−5.74	704	29296.0	41.900	−8.16
585	24263.3	42.574	−2.75	645	26810.9	42.315	−5.79	705	29337.9	41.892	−8.19
586	24305.9	42.571	−2.80	646	26853.2	42.310	−5.84	706	29379.8	41.884	−8.22
587	24348.5	42.568	−2.86	647	26895.5	42.304	−5.88	707	29421.6	41.876	−8.26
588	24391.0	42.565	−2.91	648	26937.9	42.298	−5.93	708	29463.5	41.867	−8.29
589	24433.6	42.563	−2.96	649	26980.1	42.292	−5.97	709	29505.4	41.859	−8.32
590	24476.2	42.560	−3.02	650	27022.4	42.286	−6.02	710	29547.2	41.851	−8.35
591	24518.7	42.556	−3.07	651	27064.7	42.280	−6.06	711	29589.1	41.842	−8.39
592	24561.3	42.553	−3.13	652	27107.0	42.274	−6.11	712	29630.9	41.834	−8.42
593	24603.8	42.550	−3.18	653	27149.3	42.268	−6.15	713	29672.8	41.826	−8.45
594	24646.4	42.547	−3.23	654	27191.5	42.262	−6.20	714	29714.6	41.817	−8.48
595	24688.9	42.544	−3.28	655	27233.8	42.255	−6.24	715	29756.4	41.809	−8.51
596	24731.5	42.540	−3.34	656	27276.0	42.249	−6.29	716	29798.2	41.800	−8.54
597	24774.0	42.537	−3.39	657	27318.3	42.243	−6.33	717	29840.0	41.791	−8.57
598	24816.5	42.534	−3.44	658	27360.5	42.236	−6.37	718	29881.8	41.783	−8.60
599	24859.1	42.530	−3.50	659	27402.8	42.230	−6.42	719	29923.6	41.774	−8.63
600	24901.6	42.527	−3.55	660	27445.0	42.224	−6.46	720	29965.3	41.766	−8.66

TABLE 1i6. (*Continued*)

Type K thermocouples--*thermoelectric voltages,* E(T), *Seebeck coefficients,* S(T), *and first derivative of the Seebeck coefficients,* dS/dT, *reference junctions at 0 °C—Continued*

T °C	E µV	S µV/°C	dS/dT nV/°C^2	T °C	E µV	S µV/°C	dS/dT nV/°C^2	T °C	E µV	S µV/°C	dS/dT nV/°C^2
720	29965.3	41.766	-8.66	780	32454.7	41.202	-9.98	840	34908.5	40.585	-10.47
721	30007.1	41.757	-8.69	781	32495.9	41.192	-9.99	841	34949.1	40.574	-10.48
722	30048.8	41.748	-8.72	782	32537.1	41.182	-10.01	842	34989.7	40.564	-10.48
723	30090.6	41.740	-8.75	783	32578.3	41.172	-10.02	843	35030.2	40.553	-10.48
724	30132.3	41.731	-8.78	784	32619.5	41.162	-10.03	844	35070.8	40.543	-10.48
725	30174.0	41.722	-8.81	785	32660.6	41.152	-10.05	845	35111.3	40.532	-10.49
726	30215.8	41.713	-8.83	786	32701.8	41.142	-10.06	846	35151.8	40.522	-10.49
727	30257.5	41.704	-8.86	787	32742.9	41.132	-10.07	847	35192.3	40.512	-10.49
728	30299.2	41.695	-8.89	788	32784.0	41.122	-10.09	848	35232.8	40.501	-10.49
729	30340.9	41.687	-8.92	789	32825.2	41.112	-10.10	849	35273.3	40.491	-10.49
730	30382.5	41.678	-8.94	790	32866.3	41.102	-10.11	850	35313.8	40.480	-10.49
731	30424.2	41.669	-8.97	791	32907.4	41.091	-10.12	851	35354.3	40.470	-10.49
732	30465.9	41.660	-9.00	792	32948.4	41.081	-10.14	852	35394.8	40.459	-10.50
733	30507.5	41.651	-9.02	793	32989.5	41.071	-10.15	853	35435.2	40.449	-10.50
734	30549.2	41.642	-9.05	794	33030.6	41.061	-10.16	854	35475.7	40.438	-10.50
735	30590.8	41.633	-9.07	795	33071.6	41.051	-10.17	855	35516.1	40.428	-10.50
736	30632.4	41.623	-9.10	796	33112.7	41.041	-10.18	856	35556.5	40.417	-10.50
737	30674.1	41.614	-9.13	797	33153.7	41.030	-10.19	857	35596.9	40.407	-10.50
738	30715.7	41.605	-9.15	798	33194.8	41.020	-10.20	858	35637.3	40.396	-10.50
739	30757.3	41.596	-9.17	799	33235.8	41.010	-10.21	859	35677.7	40.386	-10.50
740	30798.9	41.587	-9.20	800	33276.8	41.000	-10.22	860	35718.1	40.375	-10.50
741	30840.4	41.578	-9.22	801	33317.8	40.990	-10.23	861	35758.5	40.365	-10.50
742	30882.0	41.568	-9.25	802	33358.8	40.979	-10.24	862	35798.8	40.354	-10.50
743	30923.6	41.559	-9.27	803	33399.7	40.969	-10.25	863	35839.2	40.344	-10.50
744	30965.1	41.550	-9.30	804	33440.7	40.959	-10.26	864	35879.5	40.333	-10.50
745	31006.7	41.541	-9.32	805	33481.6	40.949	-10.27	865	35919.8	40.323	-10.50
746	31048.2	41.531	-9.34	806	33522.6	40.938	-10.28	866	35960.2	40.312	-10.50
747	31089.7	41.522	-9.36	807	33563.5	40.928	-10.29	867	36000.5	40.302	-10.50
748	31131.3	41.513	-9.39	808	33604.4	40.918	-10.30	868	36040.8	40.291	-10.49
749	31172.8	41.503	-9.41	809	33645.4	40.907	-10.31	869	36081.1	40.281	-10.49
750	31214.3	41.494	-9.43	810	33686.3	40.897	-10.32	870	36121.3	40.270	-10.49
751	31255.8	41.484	-9.45	811	33727.2	40.887	-10.32	871	36161.6	40.260	-10.49
752	31297.2	41.475	-9.47	812	33768.0	40.876	-10.33	872	36201.8	40.249	-10.49
753	31338.7	41.465	-9.50	813	33808.9	40.866	-10.34	873	36242.1	40.239	-10.49
754	31380.2	41.456	-9.52	814	33849.8	40.856	-10.35	874	36282.3	40.228	-10.49
755	31421.6	41.446	-9.54	815	33890.6	40.845	-10.35	875	36322.5	40.218	-10.49
756	31463.1	41.437	-9.56	816	33931.5	40.835	-10.36	876	36362.8	40.207	-10.48
757	31504.5	41.427	-9.58	817	33972.3	40.825	-10.37	877	36403.0	40.197	-10.48
758	31545.9	41.418	-9.60	818	34013.1	40.814	-10.37	878	36443.2	40.186	-10.48
759	31587.3	41.408	-9.62	819	34053.9	40.804	-10.38	879	36483.3	40.176	-10.48
760	31628.7	41.398	-9.64	820	34094.7	40.794	-10.39	880	36523.5	40.165	-10.48
761	31670.1	41.389	-9.66	821	34135.5	40.783	-10.39	881	36563.7	40.155	-10.47
762	31711.5	41.379	-9.68	822	34176.3	40.773	-10.40	882	36603.8	40.144	-10.47
763	31752.9	41.369	-9.69	823	34217.0	40.762	-10.40	883	36644.0	40.134	-10.47
764	31794.2	41.360	-9.71	824	34257.8	40.752	-10.41	884	36684.1	40.123	-10.47
765	31835.6	41.350	-9.73	825	34298.6	40.742	-10.41	885	36724.2	40.113	-10.47
766	31876.9	41.340	-9.75	826	34339.3	40.731	-10.42	886	36764.3	40.102	-10.46
767	31918.3	41.330	-9.77	827	34380.0	40.721	-10.42	887	36804.4	40.092	-10.46
768	31959.6	41.321	-9.79	828	34420.7	40.710	-10.43	888	36844.5	40.082	-10.46
769	32000.9	41.311	-9.80	829	34461.4	40.700	-10.43	889	36884.6	40.071	-10.46
770	32042.2	41.301	-9.82	830	34502.1	40.689	-10.44	890	36924.6	40.061	-10.45
771	32083.5	41.291	-9.84	831	34542.8	40.679	-10.44	891	36964.7	40.050	-10.45
772	32124.8	41.281	-9.85	832	34583.5	40.669	-10.45	892	37004.7	40.040	-10.45
773	32166.1	41.271	-9.87	833	34624.2	40.658	-10.45	893	37044.8	40.029	-10.44
774	32207.4	41.262	-9.89	834	34664.8	40.648	-10.45	894	37084.8	40.019	-10.44
775	32248.6	41.252	-9.90	835	34705.4	40.637	-10.46	895	37124.8	40.008	-10.44
776	32289.9	41.242	-9.92	836	34746.1	40.627	-10.46	896	37164.8	39.998	-10.44
777	32331.1	41.232	-9.93	837	34786.7	40.616	-10.46	897	37204.8	39.987	-10.43
778	32372.3	41.222	-9.95	838	34827.3	40.606	-10.47	898	37244.8	39.977	-10.43
779	32413.5	41.212	-9.96	839	34867.9	40.595	-10.47	899	37284.8	39.967	-10.43
780	32454.7	41.202	-9.98	840	34908.5	40.585	-10.47	900	37324.7	39.956	-10.42

TABLE 1i6. *(Continued)*

Type K thermocouples—*thermoelectric voltages, E(T), Seebeck coefficients, S(T), and first derivative of the Seebeck coefficients, dS/dT, reference junctions at 0 °C—Continued*

T °C	E μV	S μV/°C	dS/dT nV/°C^2	T °C	E μV	S μV/°C	dS/dT nV/°C^2	T °C	E μV	S μV/°C	dS/dT nV/°C^2
900	37324.7	39.956	-10.42	960	39703.4	39.337	-10.25	1020	42045.2	38.719	-10.43
901	37364.7	39.946	-10.42	961	39742.8	39.326	-10.25	1021	42083.9	38.709	-10.43
902	37404.6	39.935	-10.42	962	39782.1	39.316	-10.25	1022	42122.6	38.698	-10.44
903	37444.5	39.925	-10.41	963	39821.4	39.306	-10.25	1023	42161.3	38.688	-10.45
904	37484.5	39.915	-10.41	964	39860.7	39.296	-10.25	1024	42200.0	38.677	-10.46
905	37524.4	39.904	-10.41	965	39900.0	39.285	-10.25	1025	42238.6	38.667	-10.47
906	37564.3	39.894	-10.40	966	39939.3	39.275	-10.24	1026	42277.3	38.656	-10.48
907	37604.2	39.883	-10.40	967	39978.6	39.265	-10.24	1027	42315.9	38.646	-10.49
908	37644.0	39.873	-10.40	968	40017.8	39.255	-10.24	1028	42354.6	38.636	-10.50
909	37683.9	39.863	-10.39	969	40057.1	39.244	-10.24	1029	42393.2	38.625	-10.51
910	37723.8	39.852	-10.39	970	40096.3	39.234	-10.24	1030	42431.8	38.614	-10.52
911	37763.6	39.842	-10.39	971	40135.5	39.224	-10.24	1031	42470.4	38.604	-10.53
912	37803.4	39.831	-10.38	972	40174.7	39.214	-10.24	1032	42509.0	38.593	-10.54
913	37843.3	39.821	-10.38	973	40214.0	39.203	-10.24	1033	42547.6	38.583	-10.55
914	37883.1	39.811	-10.38	974	40253.2	39.193	-10.24	1034	42586.2	38.572	-10.56
915	37922.9	39.800	-10.37	975	40292.3	39.183	-10.24	1035	42624.8	38.562	-10.57
916	37962.7	39.790	-10.37	976	40331.5	39.173	-10.24	1036	42663.3	38.551	-10.58
917	38002.5	39.779	-10.37	977	40370.7	39.162	-10.24	1037	42701.9	38.541	-10.59
918	38042.2	39.769	-10.36	978	40409.8	39.152	-10.25	1038	42740.4	38.530	-10.61
919	38082.0	39.759	-10.36	979	40449.0	39.142	-10.25	1039	42778.9	38.519	-10.62
920	38121.8	39.748	-10.36	980	40488.1	39.132	-10.25	1040	42817.5	38.509	-10.63
921	38161.5	39.738	-10.35	981	40527.3	39.121	-10.25	1041	42856.0	38.498	-10.64
922	38201.2	39.728	-10.35	982	40566.4	39.111	-10.25	1042	42894.5	38.487	-10.66
923	38241.0	39.717	-10.35	983	40605.5	39.101	-10.25	1043	42932.9	38.477	-10.67
924	38280.7	39.707	-10.34	984	40644.6	39.091	-10.25	1044	42971.4	38.466	-10.68
925	38320.4	39.697	-10.34	985	40683.7	39.080	-10.26	1045	43009.9	38.455	-10.70
926	38360.1	39.686	-10.34	986	40722.7	39.070	-10.26	1046	43048.3	38.445	-10.71
927	38399.7	39.676	-10.33	987	40761.8	39.060	-10.26	1047	43086.8	38.434	-10.73
928	38439.4	39.666	-10.33	988	40800.9	39.050	-10.26	1048	43125.2	38.423	-10.74
929	38479.1	39.655	-10.33	989	40839.9	39.039	-10.26	1049	43163.6	38.413	-10.75
930	38518.7	39.645	-10.32	990	40878.9	39.029	-10.27	1050	43202.0	38.402	-10.77
931	38558.4	39.635	-10.32	991	40918.0	39.019	-10.27	1051	43240.4	38.391	-10.78
932	38598.0	39.624	-10.32	992	40957.0	39.009	-10.27	1052	43278.8	38.380	-10.80
933	38637.6	39.614	-10.31	993	40996.0	38.998	-10.28	1053	43317.2	38.369	-10.81
934	38677.2	39.604	-10.31	994	41035.0	38.988	-10.28	1054	43355.5	38.359	-10.83
935	38716.8	39.593	-10.31	995	41074.0	38.978	-10.28	1055	43393.9	38.348	-10.85
936	38756.4	39.583	-10.30	996	41112.9	38.967	-10.29	1056	43432.2	38.337	-10.86
937	38796.0	39.573	-10.30	997	41151.9	38.957	-10.29	1057	43470.6	38.326	-10.88
938	38835.6	39.563	-10.30	998	41190.8	38.947	-10.29	1058	43508.9	38.315	-10.90
939	38875.1	39.552	-10.29	999	41229.8	38.937	-10.30	1059	43547.2	38.304	-10.91
940	38914.7	39.542	-10.29	1000	41268.7	38.926	-10.30	1060	43585.5	38.293	-10.93
941	38954.2	39.532	-10.29	1001	41307.6	38.916	-10.31	1061	43623.8	38.282	-10.95
942	38993.7	39.521	-10.29	1002	41346.5	38.906	-10.31	1062	43662.1	38.271	-10.96
943	39033.2	39.511	-10.28	1003	41385.4	38.895	-10.31	1063	43700.3	38.260	-10.98
944	39072.7	39.501	-10.28	1004	41424.3	38.885	-10.32	1064	43738.6	38.249	-11.00
945	39112.2	39.491	-10.28	1005	41463.2	38.875	-10.32	1065	43776.8	38.238	-11.02
946	39151.7	39.480	-10.28	1006	41502.1	38.864	-10.33	1066	43815.1	38.227	-11.04
947	39191.2	39.470	-10.27	1007	41540.9	38.854	-10.34	1067	43853.3	38.216	-11.06
948	39230.7	39.460	-10.27	1008	41579.8	38.844	-10.34	1068	43891.5	38.205	-11.07
949	39270.1	39.449	-10.27	1009	41618.6	38.833	-10.35	1069	43929.7	38.194	-11.09
950	39309.6	39.439	-10.27	1010	41657.5	38.823	-10.35	1070	43967.9	38.183	-11.11
951	39349.0	39.429	-10.27	1011	41696.3	38.813	-10.36	1071	44006.0	38.172	-11.13
952	39388.4	39.419	-10.26	1012	41735.1	38.802	-10.37	1072	44044.2	38.161	-11.15
953	39427.8	39.408	-10.26	1013	41773.9	38.792	-10.37	1073	44082.4	38.150	-11.17
954	39467.2	39.398	-10.26	1014	41812.7	38.782	-10.38	1074	44120.5	38.139	-11.19
955	39506.6	39.388	-10.26	1015	41851.4	38.771	-10.39	1075	44158.6	38.127	-11.21
956	39546.0	39.378	-10.26	1016	41890.2	38.761	-10.39	1076	44196.8	38.116	-11.24
957	39585.4	39.367	-10.25	1017	41929.0	38.750	-10.40	1077	44234.9	38.105	-11.26
958	39624.8	39.357	-10.25	1018	41967.7	38.740	-10.41	1078	44273.0	38.094	-11.28
959	39664.1	39.347	-10.25	1019	42006.4	38.730	-10.42	1079	44311.1	38.082	-11.30
960	39703.4	39.337	-10.25	1020	42045.2	38.719	-10.43	1080	44349.1	38.071	-11.32

TABLE 1i6. (*Continued*)

Type K thermocouples—thermoelectric voltages, E(T), *Seebeck coefficients,* S(T), *and first derivative of the Seebeck coefficients,* dS/dT, *reference junctions at 0 °C—Continued*

T °C	E μV	S μV/°C	dS/dT nV/°C²	T °C	E μV	S μV/°C	dS/dT nV/°C²	T °C	E μV	S μV/°C	dS/dT nV/°C²
1080	44349.1	38.071	-11.32	1140	46612.1	37.344	-13.03	1200	48828.0	36.499	-15.12
1081	44387.2	38.060	-11.34	1141	46649.4	37.331	-13.06	1201	48864.5	36.484	-15.16
1082	44425.3	38.048	-11.37	1142	46686.8	37.318	-13.10	1202	48901.0	36.469	-15.19
1083	44463.3	38.037	-11.39	1143	46724.1	37.305	-13.13	1203	48937.5	36.454	-15.22
1084	44501.3	38.025	-11.41	1144	46761.4	37.292	-13.17	1204	48973.9	36.438	-15.25
1085	44539.4	38.014	-11.44	1145	46798.7	37.278	-13.20	1205	49010.3	36.423	-15.29
1086	44577.4	38.003	-11.46	1146	46835.9	37.265	-13.23	1206	49046.8	36.408	-15.32
1087	44615.4	37.991	-11.48	1147	46873.2	37.252	-13.27	1207	49083.2	36.392	-15.35
1088	44653.3	37.980	-11.51	1148	46910.4	37.239	-13.30	1208	49119.5	36.377	-15.38
1089	44691.3	37.968	-11.53	1149	46947.7	37.225	-13.34	1209	49155.9	36.362	-15.41
1090	44729.3	37.957	-11.55	1150	46984.9	37.212	-13.37	1210	49192.3	36.346	-15.44
1091	44767.2	37.945	-11.58	1151	47022.1	37.199	-13.41	1211	49228.6	36.331	-15.47
1092	44805.2	37.933	-11.60	1152	47059.3	37.185	-13.44	1212	49264.9	36.315	-15.50
1093	44843.1	37.922	-11.63	1153	47096.5	37.172	-13.48	1213	49301.2	36.300	-15.53
1094	44881.0	37.910	-11.65	1154	47133.6	37.158	-13.51	1214	49337.5	36.284	-15.56
1095	44918.9	37.899	-11.68	1155	47170.8	37.145	-13.55	1215	49373.8	36.269	-15.59
1096	44956.8	37.887	-11.70	1156	47207.9	37.131	-13.58	1216	49410.1	36.253	-15.62
1097	44994.7	37.875	-11.73	1157	47245.0	37.118	-13.62	1217	49446.3	36.238	-15.65
1098	45032.6	37.863	-11.76	1158	47282.2	37.104	-13.65	1218	49482.5	36.222	-15.67
1099	45070.4	37.852	-11.78	1159	47319.3	37.090	-13.69	1219	49518.8	36.206	-15.70
1100	45108.3	37.840	-11.81	1160	47356.3	37.077	-13.73	1220	49555.0	36.190	-15.73
1101	45146.1	37.828	-11.84	1161	47393.4	37.063	-13.76	1221	49591.1	36.175	-15.76
1102	45183.9	37.816	-11.86	1162	47430.5	37.049	-13.80	1222	49627.3	36.159	-15.78
1103	45221.7	37.804	-11.89	1163	47467.5	37.035	-13.83	1223	49663.5	36.143	-15.81
1104	45259.5	37.792	-11.92	1164	47504.5	37.021	-13.87	1224	49699.6	36.127	-15.83
1105	45297.3	37.780	-11.94	1165	47541.5	37.007	-13.90	1225	49735.7	36.111	-15.86
1106	45335.1	37.768	-11.97	1166	47578.5	36.994	-13.94	1226	49771.8	36.096	-15.88
1107	45372.9	37.756	-12.00	1167	47615.5	36.980	-13.97	1227	49807.9	36.080	-15.91
1108	45410.6	37.744	-12.03	1168	47652.5	36.966	-14.01	1228	49844.0	36.064	-15.93
1109	45448.3	37.732	-12.06	1169	47689.5	36.952	-14.05	1229	49880.0	36.048	-15.96
1110	45486.1	37.720	-12.09	1170	47726.4	36.937	-14.08	1230	49916.1	36.032	-15.98
1111	45523.8	37.708	-12.11	1171	47763.3	36.923	-14.12	1231	49952.1	36.016	-16.00
1112	45561.5	37.696	-12.14	1172	47800.3	36.909	-14.15	1232	49988.1	36.000	-16.02
1113	45599.2	37.684	-12.17	1173	47837.2	36.895	-14.19	1233	50024.1	35.984	-16.04
1114	45636.9	37.672	-12.20	1174	47874.0	36.881	-14.22	1234	50060.1	35.968	-16.07
1115	45674.5	37.660	-12.23	1175	47910.9	36.867	-14.26	1235	50096.0	35.952	-16.09
1116	45712.2	37.647	-12.26	1176	47947.8	36.852	-14.30	1236	50132.0	35.936	-16.11
1117	45749.8	37.635	-12.29	1177	47984.6	36.838	-14.33	1237	50167.9	35.919	-16.13
1118	45787.4	37.623	-12.32	1178	48021.5	36.824	-14.37	1238	50203.8	35.903	-16.14
1119	45825.1	37.610	-12.35	1179	48058.3	36.809	-14.40	1239	50239.7	35.887	-16.16
1120	45862.7	37.598	-12.38	1180	48095.1	36.795	-14.44	1240	50275.6	35.871	-16.18
1121	45900.3	37.586	-12.41	1181	48131.9	36.780	-14.47	1241	50311.4	35.855	-16.20
1122	45937.8	37.573	-12.44	1182	48168.6	36.766	-14.51	1242	50347.3	35.839	-16.21
1123	45975.4	37.561	-12.48	1183	48205.4	36.751	-14.54	1243	50383.1	35.822	-16.23
1124	46013.0	37.548	-12.51	1184	48242.1	36.737	-14.58	1244	50418.9	35.806	-16.25
1125	46050.5	37.536	-12.54	1185	48278.9	36.722	-14.61	1245	50454.7	35.790	-16.26
1126	46088.0	37.523	-12.57	1186	48315.6	36.708	-14.65	1246	50490.5	35.774	-16.28
1127	46125.5	37.511	-12.60	1187	48352.3	36.693	-14.68	1247	50526.3	35.757	-16.29
1128	46163.1	37.498	-12.63	1188	48389.0	36.678	-14.72	1248	50562.0	35.741	-16.30
1129	46200.5	37.485	-12.67	1189	48425.6	36.663	-14.75	1249	50597.8	35.725	-16.31
1130	46238.0	37.473	-12.70	1190	48462.3	36.649	-14.79	1250	50633.5	35.708	-16.33
1131	46275.5	37.460	-12.73	1191	48498.9	36.634	-14.82	1251	50669.2	35.692	-16.34
1132	46312.9	37.447	-12.76	1192	48535.6	36.619	-14.86	1252	50704.9	35.676	-16.35
1133	46350.4	37.434	-12.80	1193	48572.2	36.604	-14.89	1253	50740.5	35.659	-16.36
1134	46387.8	37.422	-12.83	1194	48608.8	36.589	-14.92	1254	50776.2	35.643	-16.36
1135	46425.2	37.409	-12.86	1195	48645.4	36.574	-14.96	1255	50811.8	35.627	-16.37
1136	46462.6	37.396	-12.90	1196	48681.9	36.559	-14.99	1256	50847.4	35.610	-16.38
1137	46500.0	37.383	-12.93	1197	48718.5	36.544	-15.03	1257	50883.0	35.594	-16.38
1138	46537.4	37.370	-12.96	1198	48755.0	36.529	-15.06	1258	50918.6	35.578	-16.39
1139	46574.8	37.357	-13.00	1199	48791.5	36.514	-15.09	1259	50954.2	35.561	-16.40
1140	46612.1	37.344	-13.03	1200	48828.0	36.499	-15.12	1260	50989.7	35.545	-16.40

TABLE 1i6. (Continued)

Type K thermocouples—thermoelectric voltages, E(T), Seebeck coefficients, S(T), and first derivative of the Seebeck coefficients, dS/dT, reference junctions at 0 °C—Continued

T °C	E μV	S μV/°C	dS/dT nV/°C²	T °C	E μV	S μV/°C	dS/dT nV/°C²	T °C	E μV	S μV/°C	dS/dT nV/°C²
1260	50989.7	35.545	-16.40	1300	52398.5	34.897	-15.71	1340	53782.4	34.319	-12.64
1261	51025.3	35.528	-16.40	1301	52433.4	34.881	-15.67	1341	53816.7	34.307	-12.52
1262	51060.8	35.512	-16.40	1302	52468.2	34.865	-15.62	1342	53851.0	34.294	-12.40
1263	51096.3	35.496	-16.41	1303	52503.1	34.850	-15.58	1343	53885.3	34.282	-12.28
1264	51131.8	35.479	-16.41	1304	52537.9	34.834	-15.53	1344	53919.6	34.270	-12.15
1265	51167.3	35.463	-16.41	1305	52572.8	34.819	-15.48	1345	53953.8	34.258	-12.03
1266	51202.7	35.446	-16.40	1306	52607.6	34.803	-15.43	1346	53988.1	34.246	-11.90
1267	51238.2	35.430	-16.40	1307	52642.4	34.788	-15.38	1347	54022.3	34.234	-11.76
1268	51273.6	35.414	-16.40	1308	52677.2	34.772	-15.33	1348	54056.6	34.222	-11.63
1269	51309.0	35.397	-16.39	1309	52711.9	34.757	-15.27	1349	54090.8	34.211	-11.49
1270	51344.4	35.381	-16.39	1310	52746.7	34.742	-15.21	1350	54125.0	34.199	-11.35
1271	51379.7	35.364	-16.38	1311	52781.4	34.727	-15.16	1351	54159.2	34.188	-11.21
1272	51415.1	35.348	-16.38	1312	52816.1	34.711	-15.10	1352	54193.4	34.177	-11.06
1273	51450.4	35.332	-16.37	1313	52850.8	34.696	-15.03	1353	54227.5	34.166	-10.91
1274	51485.8	35.315	-16.36	1314	52885.5	34.681	-14.97	1354	54261.7	34.155	-10.76
1275	51521.1	35.299	-16.35	1315	52920.2	34.666	-14.90	1355	54295.8	34.144	-10.61
1276	51556.4	35.283	-16.34	1316	52954.8	34.652	-14.84	1356	54330.0	34.134	-10.45
1277	51591.6	35.266	-16.32	1317	52989.5	34.637	-14.77	1357	54364.1	34.123	-10.30
1278	51626.9	35.250	-16.31	1318	53024.1	34.622	-14.70	1358	54398.2	34.113	-10.13
1279	51662.1	35.234	-16.30	1319	53058.7	34.607	-14.63	1359	54432.3	34.103	-9.97
1280	51697.4	35.217	-16.28	1320	53093.3	34.593	-14.55	1360	54466.4	34.093	-9.80
1281	51732.6	35.201	-16.26	1321	53127.9	34.578	-14.47	1361	54500.5	34.084	-9.63
1282	51767.8	35.185	-16.25	1322	53162.5	34.564	-14.40	1362	54534.6	34.074	-9.46
1283	51802.9	35.169	-16.23	1323	53197.1	34.549	-14.32	1363	54568.7	34.065	-9.28
1284	51838.1	35.152	-16.21	1324	53231.6	34.535	-14.23	1364	54602.7	34.055	-9.11
1285	51873.2	35.136	-16.19	1325	53266.1	34.521	-14.15	1365	54636.8	34.046	-8.92
1286	51908.4	35.120	-16.16	1326	53300.6	34.507	-14.06	1366	54670.8	34.038	-8.74
1287	51943.5	35.104	-16.14	1327	53335.1	34.493	-13.98	1367	54704.9	34.029	-8.55
1288	51978.6	35.088	-16.12	1328	53369.6	34.479	-13.89	1368	54738.9	34.021	-8.36
1289	52013.7	35.072	-16.09	1329	53404.1	34.465	-13.79	1369	54772.9	34.012	-8.17
1290	52048.7	35.055	-16.06	1330	53438.6	34.451	-13.70	1370	54806.9	34.004	-7.97
1291	52083.8	35.039	-16.03	1331	53473.0	34.438	-13.60	1371	54840.9	33.996	-7.77
1292	52118.8	35.023	-16.00	1332	53507.4	34.424	-13.51	1372	54874.9	33.989	-7.57
1293	52153.8	35.007	-15.97	1333	53541.8	34.411	-13.41				
1294	52188.8	34.991	-15.94	1334	53576.2	34.397	-13.30				
1295	52223.8	34.976	-15.90	1335	53610.6	34.384	-13.20				
1296	52258.8	34.960	-15.87	1336	53645.0	34.371	-13.09				
1297	52293.7	34.944	-15.83	1337	53679.4	34.358	-12.98				
1298	52328.7	34.928	-15.79	1338	53713.7	34.345	-12.87				
1299	52363.6	34.912	-15.75	1339	53748.1	34.332	-12.76				
1300	52398.5	34.897	-15.71	1340	53782.4	34.319	-12.64				

TYPE R—*Platinum*–13% Rhodium Alloy Versus Platinum Thermocouples

Type R thermocouples—thermoelectric voltages, E(T), Seebeck cofficients, S(T), and first derivative of the Seebeck coefficients, dS/dT, reference junctions at 0 °C

T °C	E μV	S μV/°C	dS/dT nV/°C²	T °C	E μV	S μV/°C	dS/dT nV/°C²	T °C	E μV	S μV/°C	dS/dT nV/°C²
				-40	-187.67	4.051	34.34	-20	-100.02	4.703	30.88
				-39	-183.61	4.085	34.16	-19	-95.30	4.733	30.72
				-38	-179.50	4.119	33.97	-18	-90.55	4.764	30.56
				-37	-175.37	4.153	33.79	-17	-85.77	4.795	30.40
				-36	-171.20	4.187	33.61	-16	-80.96	4.825	30.24
				-35	-166.99	4.221	33.44	-15	-76.12	4.855	30.08
				-34	-162.76	4.254	33.26	-14	-71.25	4.885	29.93
				-33	-158.48	4.287	33.08	-13	-66.35	4.915	29.77
				-32	-154.18	4.320	32.91	-12	-61.42	4.945	29.61
				-31	-149.84	4.353	32.73	-11	-56.46	4.974	29.46
-50	-226.44	3.698	36.23	-30	-145.48	4.386	32.56	-10	-51.48	5.004	29.31
-49	-222.72	3.735	36.03	-29	-141.07	4.418	32.39	-9	-46.46	5.033	29.15
-48	-218.97	3.770	35.84	-28	-136.64	4.450	32.22	-8	-41.41	5.062	29.00
-47	-215.18	3.806	35.65	-27	-132.17	4.482	32.05	-7	-36.33	5.091	28.85
-46	-211.36	3.842	35.46	-26	-127.67	4.514	31.88	-6	-31.23	5.120	28.70
-45	-207.50	3.877	35.27	-25	-123.14	4.546	31.71	-5	-26.09	5.148	28.55
-44	-203.60	3.912	35.08	-24	-118.58	4.578	31.54	-4	-20.93	5.177	28.41
-43	-199.67	3.947	34.89	-23	-113.99	4.609	31.38	-3	-15.74	5.205	28.26
-42	-195.71	3.982	34.71	-22	-109.36	4.641	31.21	-2	-10.52	5.233	28.11
-41	-191.71	4.017	34.52	-21	-104.71	4.672	31.05	-1	-5.28	5.261	27.97
-40	-187.67	4.051	34.34	-20	-100.02	4.703	30.88	0	0.00	5.289	27.82

TABLE 1i6. (*Continued*)

Type R thermocouples—*thermoelectric voltages,* E(T), *Seebeck coefficients,* S(T), *and first derivative of the Seebeck coefficients,* dS/dT, *reference junctions at 0 °C—Continued*

T °C	E μV	S μV/°C	dS/dT nV/°C²	T °C	E μV	S μV/°C	dS/dT nV/°C²	T °C	E μV	S μV/°C	dS/dT nV/°C²
0	0.00	5.289	27.82	60	362.68	6.728	20.57	120	800.04	7.800	15.47
1	5.30	5.317	27.68	61	369.42	6.748	20.47	121	807.85	7.816	15.40
2	10.63	5.344	27.54	62	376.18	6.769	20.37	122	815.67	7.831	15.33
3	15.99	5.372	27.39	63	382.95	6.789	20.27	123	823.51	7.846	15.26
4	21.38	5.399	27.25	64	389.75	6.809	20.17	124	831.36	7.861	15.19
5	26.79	5.426	27.11	65	396.57	6.829	20.07	125	839.23	7.877	15.12
6	32.23	5.454	26.97	66	403.41	6.849	19.97	126	847.12	7.892	15.06
7	37.70	5.480	26.83	67	410.27	6.869	19.88	127	855.02	7.907	14.99
8	43.19	5.507	26.70	68	417.15	6.889	19.78	128	862.93	7.922	14.92
9	48.71	5.534	26.56	69	424.05	6.909	19.68	129	870.86	7.937	14.85
10	54.26	5.560	26.42	70	430.97	6.929	19.59	130	878.80	7.951	14.78
11	59.83	5.587	26.29	71	437.91	6.948	19.49	131	886.76	7.966	14.72
12	65.43	5.613	26.15	72	444.87	6.967	19.40	132	894.74	7.981	14.65
13	71.06	5.639	26.02	73	451.84	6.987	19.31	133	902.72	7.995	14.58
14	76.71	5.665	25.89	74	458.84	7.006	19.21	134	910.73	8.010	14.52
15	82.39	5.691	25.76	75	465.85	7.025	19.12	135	918.74	8.024	14.45
16	88.09	5.716	25.63	76	472.89	7.044	19.03	136	926.78	8.039	14.39
17	93.82	5.742	25.49	77	479.94	7.063	18.94	137	934.82	8.053	14.32
18	99.58	5.767	25.37	78	487.02	7.082	18.85	138	942.88	8.068	14.26
19	105.36	5.793	25.24	79	494.11	7.101	18.76	139	950.96	8.082	14.19
20	111.16	5.818	25.11	80	501.22	7.120	18.67	140	959.05	8.096	14.13
21	116.99	5.843	24.98	81	508.35	7.138	18.58	141	967.15	8.110	14.07
22	122.85	5.868	24.85	82	515.49	7.157	18.49	142	975.27	8.124	14.01
23	128.73	5.893	24.73	83	522.66	7.175	18.40	143	983.40	8.138	13.94
24	134.63	5.917	24.60	84	529.85	7.194	18.31	144	991.54	8.152	13.88
25	140.56	5.942	24.48	85	537.05	7.212	18.23	145	999.70	8.166	13.82
26	146.52	5.966	24.36	86	544.27	7.230	18.14	146	1007.87	8.180	13.76
27	152.49	5.991	24.23	87	551.51	7.248	18.05	147	1016.06	8.193	13.70
28	158.50	6.015	24.11	88	558.77	7.266	17.97	148	1024.26	8.207	13.63
29	164.52	6.039	23.99	89	566.04	7.284	17.88	149	1032.47	8.221	13.57
30	170.57	6.063	23.87	90	573.33	7.302	17.80	150	1040.70	8.234	13.51
31	176.65	6.087	23.75	91	580.64	7.320	17.72	151	1048.94	8.248	13.45
32	182.75	6.110	23.63	92	587.97	7.337	17.63	152	1057.20	8.261	13.39
33	188.87	6.134	23.51	93	595.32	7.355	17.55	153	1065.47	8.274	13.34
34	195.01	6.157	23.39	94	602.68	7.373	17.47	154	1073.75	8.288	13.28
35	201.18	6.181	23.28	95	610.06	7.390	17.38	155	1082.04	8.301	13.22
36	207.38	6.204	23.16	96	617.46	7.407	17.30	156	1090.35	8.314	13.16
37	213.59	6.227	23.04	97	624.88	7.425	17.22	157	1098.67	8.327	13.10
38	219.83	6.250	22.93	98	632.31	7.442	17.14	158	1107.00	8.340	13.04
39	226.09	6.273	22.82	99	639.76	7.459	17.06	159	1115.35	8.353	12.99
40	232.38	6.296	22.70	100	647.23	7.476	16.98	160	1123.71	8.366	12.93
41	238.68	6.318	22.59	101	654.71	7.493	16.90	161	1132.08	8.379	12.87
42	245.01	6.341	22.48	102	662.22	7.510	16.82	162	1140.47	8.392	12.82
43	251.36	6.363	22.36	103	669.73	7.526	16.74	163	1148.87	8.405	12.76
44	257.74	6.385	22.25	104	677.27	7.543	16.66	164	1157.28	8.418	12.71
45	264.13	6.408	22.14	105	684.82	7.560	16.59	165	1165.70	8.430	12.65
46	270.55	6.430	22.03	106	692.39	7.576	16.51	166	1174.14	8.443	12.59
47	276.99	6.452	21.93	107	699.97	7.593	16.43	167	1182.59	8.455	12.54
48	283.46	6.474	21.82	108	707.57	7.609	16.36	168	1191.05	8.468	12.49
49	289.94	6.495	21.71	109	715.19	7.626	16.28	169	1199.52	8.480	12.43
50	296.45	6.517	21.60	110	722.83	7.642	16.21	170	1208.01	8.493	12.38
51	302.97	6.539	21.50	111	730.48	7.658	16.13	171	1216.51	8.505	12.32
52	309.52	6.560	21.39	112	738.14	7.674	16.06	172	1225.02	8.518	12.27
53	316.09	6.581	21.28	113	745.82	7.690	15.98	173	1233.54	8.530	12.22
54	322.69	6.603	21.18	114	753.52	7.706	15.91	174	1242.08	8.542	12.17
55	329.30	6.624	21.08	115	761.24	7.722	15.83	175	1250.63	8.554	12.11
56	335.93	6.645	20.97	116	768.96	7.738	15.76	176	1259.19	8.566	12.06
57	342.59	6.666	20.87	117	776.71	7.753	15.69	177	1267.76	8.578	12.01
58	349.27	6.686	20.77	118	784.47	7.769	15.62	178	1276.34	8.590	11.96
59	355.96	6.707	20.67	119	792.25	7.785	15.55	179	1284.94	8.602	11.91
60	362.68	6.728	20.57	120	800.04	7.800	15.47	180	1293.55	8.614	11.85

TABLE 1i6. *(Continued)*

Type R thermocouples—*thermoelectric voltages,* E(T), *Seebeck coefficients,* S(T), *and first derivative of the Seebeck coefficients,* dS/dT, *reference junctions at 0 °C—Continued*

T °C	E μV	S μV/°C	dS/dT nV/°C²	T °C	E μV	S μV/°C	dS/dT nV/°C²	T °C	E μV	S μV/°C	dS/dT nV/°C²
180	1293.55	8.614	11.85	240	1830.04	9.243	9.25	300	2400.05	9.739	7.39
181	1302.17	8.626	11.80	241	1839.28	9.252	9.21	301	2409.79	9.746	7.36
182	1310.80	8.638	11.75	242	1848.54	9.261	9.18	302	2419.54	9.754	7.34
183	1319.44	8.649	11.70	243	1857.81	9.271	9.14	303	2429.30	9.761	7.31
184	1328.10	8.661	11.65	244	1867.08	9.280	9.11	304	2439.06	9.768	7.29
185	1336.77	8.673	11.60	245	1876.36	9.289	9.07	305	2448.83	9.775	7.26
186	1345.44	8.684	11.55	246	1885.66	9.298	9.04	306	2458.61	9.783	7.24
187	1354.13	8.696	11.51	247	1894.96	9.307	9.00	307	2468.40	9.790	7.21
188	1362.84	8.707	11.46	248	1904.27	9.316	8.96	308	2478.19	9.797	7.19
189	1371.55	8.719	11.41	249	1913.59	9.325	8.93	309	2487.99	9.804	7.16
190	1380.27	8.730	11.36	250	1922.92	9.334	8.89	310	2497.80	9.811	7.14
191	1389.01	8.741	11.31	251	1932.26	9.343	8.86	311	2507.62	9.819	7.11
192	1397.76	8.753	11.26	252	1941.61	9.351	8.83	312	2517.44	9.826	7.09
193	1406.51	8.764	11.22	253	1950.96	9.360	8.79	313	2527.27	9.833	7.06
194	1415.28	8.775	11.17	254	1960.33	9.369	8.76	314	2537.10	9.840	7.04
195	1424.06	8.786	11.12	255	1969.70	9.378	8.72	315	2546.95	9.847	7.02
196	1432.86	8.797	11.08	256	1979.08	9.386	8.69	316	2556.80	9.854	6.99
197	1441.66	8.808	11.03	257	1988.47	9.395	8.66	317	2566.65	9.861	6.97
198	1450.47	8.819	10.98	258	1997.87	9.404	8.62	318	2576.52	9.868	6.95
199	1459.30	8.830	10.94	259	2007.28	9.412	8.59	319	2586.39	9.875	6.92
200	1468.13	8.841	10.89	260	2016.70	9.421	8.56	320	2596.27	9.882	6.90
201	1476.98	8.852	10.85	261	2026.12	9.429	8.52	321	2606.15	9.889	6.88
202	1485.84	8.863	10.80	262	2035.56	9.438	8.49	322	2616.05	9.895	6.86
203	1494.71	8.874	10.75	263	2045.00	9.446	8.46	323	2625.94	9.902	6.83
204	1503.59	8.884	10.71	264	2054.45	9.455	8.43	324	2635.85	9.909	6.81
205	1512.47	8.895	10.67	265	2063.91	9.463	8.40	325	2645.76	9.916	6.79
206	1521.38	8.906	10.62	266	2073.37	9.472	8.36	326	2655.68	9.923	6.77
207	1530.29	8.916	10.58	267	2082.85	9.480	8.33	327	2665.61	9.929	6.75
208	1539.21	8.927	10.53	268	2092.33	9.488	8.30	328	2675.54	9.936	6.72
209	1548.14	8.937	10.49	269	2101.83	9.497	8.27	329	2685.48	9.943	6.70
210	1557.08	8.948	10.45	270	2111.33	9.505	8.24	330	2695.43	9.950	6.68
211	1566.04	8.958	10.40	271	2120.84	9.513	8.21	331	2705.38	9.956	6.66
212	1575.00	8.969	10.36	272	2130.35	9.521	8.18	332	2715.34	9.963	6.64
213	1583.97	8.979	10.32	273	2139.88	9.529	8.15	333	2725.31	9.970	6.62
214	1592.96	8.989	10.28	274	2149.41	9.538	8.12	334	2735.28	9.976	6.60
215	1601.95	9.000	10.23	275	2158.95	9.546	8.09	335	2745.26	9.983	6.58
216	1610.96	9.010	10.19	276	2168.50	9.554	8.06	336	2755.24	9.989	6.56
217	1619.97	9.020	10.15	277	2178.06	9.562	8.03	337	2765.24	9.996	6.54
218	1629.00	9.030	10.11	278	2187.63	9.570	8.00	338	2775.24	10.002	6.52
219	1638.03	9.040	10.07	279	2197.20	9.578	7.97	339	2785.24	10.009	6.50
220	1647.08	9.050	10.03	280	2206.78	9.586	7.94	340	2795.25	10.015	6.48
221	1656.13	9.060	9.99	281	2216.37	9.594	7.91	341	2805.27	10.022	6.46
222	1665.20	9.070	9.94	282	2225.97	9.602	7.88	342	2815.30	10.028	6.44
223	1674.27	9.080	9.90	283	2235.58	9.609	7.85	343	2825.33	10.035	6.42
224	1683.36	9.090	9.86	284	2245.19	9.617	7.82	344	2835.37	10.041	6.40
225	1692.45	9.100	9.82	285	2254.81	9.625	7.79	345	2845.41	10.047	6.38
226	1701.56	9.110	9.78	286	2264.44	9.633	7.77	346	2855.46	10.054	6.36
227	1710.67	9.119	9.75	287	2274.08	9.641	7.74	347	2865.52	10.060	6.34
228	1719.80	9.129	9.71	288	2283.72	9.648	7.71	348	2875.58	10.067	6.32
229	1728.93	9.139	9.67	289	2293.37	9.656	7.68	349	2885.65	10.073	6.30
230	1738.08	9.149	9.63	290	2303.03	9.664	7.65	350	2895.73	10.079	6.28
231	1747.23	9.158	9.59	291	2312.70	9.671	7.63	351	2905.81	10.085	6.27
232	1756.39	9.168	9.55	292	2322.37	9.679	7.60	352	2915.90	10.092	6.25
233	1765.56	9.177	9.51	293	2332.06	9.687	7.57	353	2925.99	10.098	6.23
234	1774.75	9.187	9.47	294	2341.75	9.694	7.55	354	2936.09	10.104	6.21
235	1783.94	9.196	9.44	295	2351.45	9.702	7.52	355	2946.20	10.110	6.19
236	1793.14	9.206	9.40	296	2361.15	9.709	7.49	356	2956.31	10.117	6.18
237	1802.35	9.215	9.36	297	2370.86	9.717	7.47	357	2966.43	10.123	6.16
238	1811.57	9.224	9.33	298	2380.58	9.724	7.44	358	2976.56	10.129	6.14
239	1820.80	9.234	9.29	299	2390.31	9.731	7.41	359	2986.69	10.135	6.12
240	1830.04	9.243	9.25	300	2400.05	9.739	7.39	360	2996.83	10.141	6.11

TABLE 1i6. (*Continued*)

*Type **R** thermocouples*—*thermoelectric voltages,* E(T), *Seebeck coefficients,* S(T), *and first derivative of the Seebeck coefficients,* dS/dT, *reference junctions at 0 °C—Continued*

T °C	E μV	S μV/°C	dS/dT nV/°C²	T °C	E μV	S μV/°C	dS/dT nV/°C²	T °C	E μV	S μV/°C	dS/dT nV/°C²
360	2996.83	10.141	6.11	420	3615.74	10.481	5.31	480	4253.87	10.786	4.89
361	3006.97	10.147	6.09	421	3626.22	10.487	5.30	481	4264.66	10.790	4.88
362	3017.12	10.153	6.07	422	3636.71	10.492	5.29	482	4275.45	10.795	4.88
363	3027.28	10.159	6.06	423	3647.21	10.497	5.28	483	4286.25	10.800	4.87
364	3037.44	10.165	6.04	424	3657.71	10.502	5.27	484	4297.05	10.805	4.87
365	3047.61	10.171	6.02	425	3668.21	10.508	5.26	485	4307.86	10.810	4.86
366	3057.79	10.177	6.01	426	3678.72	10.513	5.25	486	4318.67	10.815	4.86
367	3067.97	10.183	5.99	427	3689.24	10.518	5.24	487	4329.49	10.820	4.86
368	3078.15	10.189	5.97	428	3699.76	10.523	5.23	488	4340.31	10.824	4.85
369	3088.34	10.195	5.96	429	3710.29	10.529	5.22	489	4351.14	10.829	4.85
370	3098.54	10.201	5.94	430	3720.82	10.534	5.21	490	4361.97	10.834	4.84
371	3108.75	10.207	5.93	431	3731.35	10.539	5.20	491	4372.80	10.839	4.84
372	3118.96	10.213	5.91	432	3741.89	10.544	5.20	492	4383.65	10.844	4.83
373	3129.17	10.219	5.90	433	3752.44	10.549	5.19	493	4394.49	10.849	4.83
374	3139.40	10.225	5.88	434	3762.99	10.555	5.18	494	4405.34	10.853	4.83
375	3149.62	10.231	5.86	435	3773.55	10.560	5.17	495	4416.20	10.858	4.82
376	3159.86	10.237	5.85	436	3784.11	10.565	5.16	496	4427.06	10.863	4.82
377	3170.10	10.243	5.83	437	3794.68	10.570	5.15	497	4437.93	10.868	4.82
378	3180.34	10.248	5.82	438	3805.25	10.575	5.15	498	4448.80	10.873	4.81
379	3190.59	10.254	5.81	439	3815.83	10.580	5.14	499	4459.67	10.878	4.81
380	3200.85	10.260	5.79	440	3826.41	10.586	5.13	500	4470.55	10.882	4.80
381	3211.11	10.266	5.78	441	3837.00	10.591	5.12	501	4481.44	10.887	4.80
382	3221.38	10.272	5.76	442	3847.60	10.596	5.12	502	4492.33	10.892	4.80
383	3231.66	10.277	5.75	443	3858.19	10.601	5.11	503	4503.22	10.897	4.79
384	3241.94	10.283	5.73	444	3868.80	10.606	5.10	504	4514.12	10.902	4.79
385	3252.22	10.289	5.72	445	3879.41	10.611	5.09	505	4525.02	10.906	4.79
386	3262.51	10.294	5.71	446	3890.02	10.616	5.09	506	4535.93	10.911	4.78
387	3272.81	10.300	5.69	447	3900.64	10.621	5.08	507	4546.85	10.916	4.78
388	3283.11	10.306	5.68	448	3911.26	10.626	5.07	508	4557.76	10.921	4.78
389	3293.42	10.312	5.67	449	3921.89	10.631	5.06	509	4568.69	10.925	4.77
390	3303.74	10.317	5.65	450	3932.53	10.637	5.06	510	4579.62	10.930	4.77
391	3314.06	10.323	5.64	451	3943.17	10.642	5.05	511	4590.55	10.935	4.77
392	3324.38	10.328	5.63	452	3953.81	10.647	5.04	512	4601.49	10.940	4.76
393	3334.71	10.334	5.61	453	3964.46	10.652	5.04	513	4612.43	10.945	4.76
394	3345.05	10.340	5.60	454	3975.11	10.657	5.03	514	4623.37	10.949	4.76
395	3355.39	10.345	5.59	455	3985.77	10.662	5.02	515	4634.33	10.954	4.75
396	3365.74	10.351	5.57	456	3996.44	10.667	5.02	516	4645.28	10.959	4.75
397	3376.10	10.356	5.56	457	4007.11	10.672	5.01	517	4656.24	10.964	4.75
398	3386.45	10.362	5.55	458	4017.78	10.677	5.00	518	4667.21	10.968	4.75
399	3396.82	10.368	5.54	459	4028.46	10.682	5.00	519	4678.18	10.973	4.74
400	3407.19	10.373	5.53	460	4039.14	10.687	4.99	520	4689.16	10.978	4.74
401	3417.57	10.379	5.51	461	4049.83	10.692	4.99	521	4700.14	10.983	4.74
402	3427.95	10.384	5.50	462	4060.53	10.697	4.98	522	4711.12	10.987	4.73
403	3438.33	10.390	5.49	463	4071.23	10.702	4.97	523	4722.11	10.992	4.73
404	3448.73	10.395	5.48	464	4081.93	10.707	4.97	524	4733.10	10.997	4.73
405	3459.12	10.401	5.47	465	4092.64	10.712	4.96	525	4744.10	11.001	4.72
406	3469.53	10.406	5.46	466	4103.35	10.717	4.96	526	4755.11	11.006	4.72
407	3479.94	10.411	5.44	467	4114.07	10.722	4.95	527	4766.12	11.011	4.72
408	3490.35	10.417	5.43	468	4124.80	10.727	4.95	528	4777.13	11.016	4.72
409	3500.77	10.422	5.42	469	4135.53	10.731	4.94	529	4788.15	11.020	4.71
410	3511.19	10.428	5.41	470	4146.26	10.736	4.94	530	4799.17	11.025	4.71
411	3521.62	10.433	5.40	471	4157.00	10.741	4.93	531	4810.20	11.030	4.71
412	3532.06	10.439	5.39	472	4167.74	10.746	4.93	532	4821.23	11.034	4.70
413	3542.50	10.444	5.38	473	4178.49	10.751	4.92	533	4832.27	11.039	4.70
414	3552.95	10.449	5.37	474	4189.24	10.756	4.92	534	4843.31	11.044	4.70
415	3563.40	10.455	5.36	475	4200.00	10.761	4.91	535	4854.35	11.049	4.69
416	3573.86	10.460	5.35	476	4210.77	10.766	4.91	536	4865.41	11.053	4.69
417	3584.32	10.465	5.34	477	4221.54	10.771	4.90	537	4876.46	11.058	4.69
418	3594.79	10.471	5.33	478	4232.31	10.776	4.90	538	4887.52	11.063	4.69
419	3605.26	10.476	5.32	479	4243.09	10.781	4.89	539	4898.59	11.067	4.68
420	3615.74	10.481	5.31	480	4253.87	10.786	4.89	540	4909.66	11.072	4.68

TABLE 1i6. (*Continued*)

Type R thermocouples—*thermoelectric voltages*, E(T), *Seebeck coefficients*, S(T), *and first derivative of the Seebeck coefficients*, dS/dT, *reference junctions at 0 °C—Continued*

T °C	E μV	S μV/°C	dS/dT nV/°C²	T °C	E μV	S μV/°C	dS/dT nV/°C²	T °C	E μV	S μV/°C	dS/dT nV/°C²
540	4909.66	11.072	4.68	600	5582.27	11.346	4.40	660	6271.59	11.641	4.91
541	4920.73	11.077	4.68	601	5593.62	11.350	4.39	661	6283.24	11.646	4.91
542	4931.81	11.081	4.67	602	5604.97	11.355	4.38	662	6294.89	11.651	4.91
543	4942.89	11.086	4.67	603	5616.33	11.359	4.38	663	6306.54	11.656	4.91
544	4953.98	11.091	4.67	604	5627.69	11.363	4.37	664	6318.20	11.661	4.91
545	4965.07	11.095	4.66	605	5639.05	11.368	4.36	665	6329.86	11.666	4.91
546	4976.17	11.100	4.66	606	5650.42	11.372	4.35	666	6341.53	11.671	4.90
547	4987.27	11.105	4.66	607	5661.80	11.376	4.34	667	6353.20	11.676	4.90
548	4998.38	11.109	4.66	608	5673.18	11.381	4.33	668	6364.88	11.681	4.90
549	5009.49	11.114	4.65	609	5684.56	11.385	4.33	669	6376.57	11.686	4.90
550	5020.61	11.119	4.65	610	5695.95	11.389	4.32	670	6388.25	11.690	4.90
551	5031.73	11.123	4.65	611	5707.34	11.394	4.31	671	6399.95	11.695	4.90
552	5042.86	11.128	4.64	612	5718.73	11.398	4.30	672	6411.64	11.700	4.89
553	5053.99	11.133	4.64	613	5730.13	11.402	4.29	673	6423.35	11.705	4.89
554	5065.12	11.137	4.64	614	5741.54	11.407	4.28	674	6435.05	11.710	4.89
555	5076.26	11.142	4.63	615	5752.95	11.411	4.27	675	6446.77	11.715	4.89
556	5087.40	11.146	4.63	616	5764.36	11.415	4.26	676	6458.48	11.720	4.89
557	5098.55	11.151	4.62	617	5775.78	11.419	4.25	677	6470.21	11.725	4.89
558	5109.71	11.156	4.62	618	5787.20	11.424	4.24	678	6481.93	11.730	4.89
559	5120.87	11.160	4.62	619	5798.62	11.428	4.23	679	6493.67	11.735	4.88
560	5132.03	11.165	4.61	620	5810.05	11.432	4.22	680	6505.40	11.739	4.88
561	5143.20	11.170	4.61	621	5821.49	11.436	4.20	681	6517.14	11.744	4.88
562	5154.37	11.174	4.61	622	5832.93	11.440	4.19	682	6528.89	11.749	4.88
563	5165.54	11.179	4.60	623	5844.37	11.445	4.18	683	6540.64	11.754	4.88
564	5176.72	11.183	4.60	624	5855.82	11.449	4.17	684	6552.40	11.759	4.88
565	5187.91	11.188	4.59	625	5867.27	11.453	4.16	685	6564.16	11.764	4.87
566	5199.10	11.193	4.59	626	5878.72	11.457	4.15	686	6575.93	11.769	4.87
567	5210.30	11.197	4.59	627	5890.18	11.461	4.13	687	6587.70	11.774	4.87
568	5221.49	11.202	4.58	628	5901.64	11.465	4.12	688	6599.47	11.778	4.87
569	5232.70	11.206	4.58	629	5913.11	11.470	4.11	689	6611.26	11.783	4.87
570	5243.91	11.211	4.57	630	5924.58	11.474	4.09	690	6623.04	11.788	4.87
571	5255.12	11.215	4.57	631	5936.06	11.498	4.96	691	6634.83	11.793	4.86
572	5266.34	11.220	4.56	632	5947.57	11.503	4.96	692	6646.63	11.798	4.86
573	5277.56	11.225	4.56	633	5959.07	11.508	4.96	693	6658.43	11.803	4.86
574	5288.79	11.229	4.56	634	5970.58	11.513	4.96	694	6670.23	11.808	4.86
575	5300.02	11.234	4.55	635	5982.10	11.518	4.95	695	6682.04	11.812	4.86
576	5311.26	11.238	4.55	636	5993.62	11.523	4.95	696	6693.86	11.817	4.86
577	5322.50	11.243	4.54	637	6005.14	11.528	4.95	697	6705.68	11.822	4.85
578	5333.74	11.247	4.54	638	6016.67	11.533	4.95	698	6717.50	11.827	4.85
579	5344.99	11.252	4.53	639	6028.21	11.538	4.95	699	6729.33	11.832	4.85
580	5356.24	11.256	4.53	640	6039.75	11.543	4.95	700	6741.17	11.837	4.85
581	5367.50	11.261	4.52	641	6051.30	11.548	4.94	701	6753.00	11.842	4.85
582	5378.77	11.265	4.52	642	6062.85	11.553	4.94	702	6764.85	11.846	4.85
583	5390.03	11.270	4.51	643	6074.40	11.558	4.94	703	6776.70	11.851	4.84
584	5401.31	11.274	4.50	644	6085.96	11.563	4.94	704	6788.55	11.856	4.84
585	5412.58	11.279	4.50	645	6097.53	11.568	4.94	705	6800.41	11.861	4.84
586	5423.86	11.283	4.49	646	6109.10	11.572	4.94	706	6812.27	11.866	4.84
587	5435.15	11.288	4.49	647	6120.67	11.577	4.94	707	6824.14	11.871	4.84
588	5446.44	11.292	4.48	648	6132.25	11.582	4.93	708	6836.01	11.875	4.84
589	5457.74	11.297	4.48	649	6143.84	11.587	4.93	709	6847.89	11.880	4.84
590	5469.03	11.301	4.47	650	6155.43	11.592	4.93	710	6859.77	11.885	4.83
591	5480.34	11.306	4.46	651	6167.02	11.597	4.93	711	6871.66	11.890	4.83
592	5491.65	11.310	4.46	652	6178.62	11.602	4.93	712	6883.55	11.895	4.83
593	5502.96	11.315	4.45	653	6190.22	11.607	4.93	713	6895.45	11.900	4.83
594	5514.28	11.319	4.44	654	6201.83	11.612	4.92	714	6907.35	11.904	4.83
595	5525.60	11.324	4.44	655	6213.45	11.617	4.92	715	6919.26	11.909	4.83
596	5536.92	11.328	4.43	656	6225.07	11.622	4.92	716	6931.17	11.914	4.82
597	5548.25	11.333	4.42	657	6236.69	11.627	4.92	717	6943.09	11.919	4.82
598	5559.59	11.337	4.42	658	6248.32	11.632	4.92	718	6955.01	11.924	4.82
599	5570.93	11.341	4.41	659	6259.95	11.637	4.92	719	6966.94	11.929	4.82
600	5582.27	11.346	4.40	660	6271.59	11.641	4.91	720	6978.87	11.933	4.82

TABLE 1i6. (Continued)

Type R thermocouples—thermoelectric voltages, E(T), *Seebeck coefficients,* S(T), *and first derivative of the Seebeck coefficients,* dS/dT, *reference junctions at 0 °C—Continued*

T °C	E μV	S μV/°C	dS/dT nV/°C²	T °C	E μV	S μV/°C	dS/dT nV/°C²	T °C	E μV	S μV/°C	dS/dT nV/°C²
720	6978.87	11.933	4.82	780	7703.48	12.220	4.72	840	8445.09	12.500	4.62
721	6990.80	11.938	4.82	781	7715.71	12.224	4.72	841	8457.60	12.504	4.62
722	7002.74	11.943	4.81	782	7727.93	12.229	4.72	842	8470.10	12.509	4.62
723	7014.69	11.948	4.81	783	7740.16	12.234	4.72	843	8482.61	12.514	4.62
724	7026.64	11.953	4.81	784	7752.40	12.238	4.71	844	8495.13	12.518	4.62
725	7038.59	11.957	4.81	785	7764.64	12.243	4.71	845	8507.65	12.523	4.62
726	7050.55	11.962	4.81	786	7776.89	12.248	4.71	846	8520.18	12.528	4.61
727	7062.52	11.967	4.81	787	7789.14	12.253	4.71	847	8532.71	12.532	4.61
728	7074.49	11.972	4.80	788	7801.39	12.257	4.71	848	8545.24	12.537	4.61
729	7086.46	11.977	4.80	789	7813.65	12.262	4.71	849	8557.78	12.541	4.61
730	7098.44	11.981	4.80	790	7825.91	12.267	4.70	850	8570.32	12.546	4.61
731	7110.43	11.986	4.80	791	7838.18	12.271	4.70	851	8582.87	12.551	4.61
732	7122.41	11.991	4.80	792	7850.46	12.276	4.70	852	8595.42	12.555	4.60
733	7134.41	11.996	4.80	793	7862.74	12.281	4.70	853	8607.98	12.560	4.60
734	7146.41	12.001	4.79	794	7875.02	12.285	4.70	854	8620.54	12.564	4.60
735	7158.41	12.005	4.79	795	7887.31	12.290	4.70	855	8633.11	12.569	4.60
736	7170.42	12.010	4.79	796	7899.60	12.295	4.69	856	8645.68	12.574	4.60
737	7182.43	12.015	4.79	797	7911.90	12.300	4.69	857	8658.26	12.578	4.60
738	7194.45	12.020	4.79	798	7924.20	12.304	4.69	858	8670.84	12.583	4.59
739	7206.47	12.025	4.79	799	7936.50	12.309	4.69	859	8683.42	12.587	4.59
740	7218.50	12.029	4.79	800	7948.82	12.314	4.69	860	8696.01	12.592	4.59
741	7230.53	12.034	4.78	801	7961.13	12.318	4.69	861	8708.61	12.597	4.59
742	7242.56	12.039	4.78	802	7973.45	12.323	4.69	862	8721.21	12.601	4.59
743	7254.61	12.044	4.78	803	7985.78	12.328	4.68	863	8733.81	12.606	4.59
744	7266.65	12.049	4.78	804	7998.11	12.332	4.68	864	8746.42	12.610	4.59
745	7278.70	12.053	4.78	805	8010.44	12.337	4.68	865	8759.03	12.615	4.58
746	7290.76	12.058	4.78	806	8022.78	12.342	4.68	866	8771.65	12.620	4.58
747	7302.82	12.063	4.77	807	8035.13	12.346	4.68	867	8784.27	12.624	4.58
748	7314.88	12.068	4.77	808	8047.47	12.351	4.68	868	8796.90	12.629	4.58
749	7326.95	12.072	4.77	809	8059.83	12.356	4.67	869	8809.53	12.633	4.58
750	7339.03	12.077	4.77	810	8072.19	12.360	4.67	870	8822.16	12.638	4.58
751	7351.11	12.082	4.77	811	8084.55	12.365	4.67	871	8834.80	12.642	4.57
752	7363.19	12.087	4.77	812	8096.92	12.370	4.67	872	8847.45	12.647	4.57
753	7375.28	12.091	4.76	813	8109.29	12.374	4.67	873	8860.10	12.652	4.57
754	7387.38	12.096	4.76	814	8121.67	12.379	4.67	874	8872.75	12.656	4.57
755	7399.47	12.101	4.76	815	8134.05	12.384	4.66	875	8885.41	12.661	4.57
756	7411.58	12.106	4.76	816	8146.43	12.388	4.66	876	8898.07	12.665	4.57
757	7423.69	12.111	4.76	817	8158.82	12.393	4.66	877	8910.74	12.670	4.56
758	7435.80	12.115	4.75	818	8171.22	12.398	4.66	878	8923.41	12.674	4.56
759	7447.92	12.120	4.75	819	8183.62	12.402	4.66	879	8936.09	12.679	4.56
760	7460.04	12.125	4.75	820	8196.02	12.407	4.66	880	8948.77	12.684	4.56
761	7472.17	12.130	4.75	821	8208.43	12.412	4.65	881	8961.46	12.688	4.56
762	7484.30	12.134	4.75	822	8220.85	12.416	4.65	882	8974.15	12.693	4.56
763	7496.43	12.139	4.75	823	8233.27	12.421	4.65	883	8986.84	12.697	4.55
764	7508.58	12.144	4.75	824	8245.69	12.426	4.65	884	8999.54	12.702	4.55
765	7520.72	12.149	4.74	825	8258.12	12.430	4.65	885	9012.24	12.706	4.55
766	7532.87	12.153	4.74	826	8270.55	12.435	4.65	886	9024.95	12.711	4.55
767	7545.03	12.158	4.74	827	8282.99	12.440	4.64	887	9037.67	12.715	4.55
768	7557.19	12.163	4.74	828	8295.43	12.444	4.64	888	9050.38	12.720	4.55
769	7569.35	12.167	4.74	829	8307.88	12.449	4.64	889	9063.11	12.724	4.54
770	7581.52	12.172	4.74	830	8320.33	12.454	4.64	890	9075.83	12.729	4.54
771	7593.70	12.177	4.74	831	8332.78	12.458	4.64	891	9088.56	12.734	4.54
772	7605.88	12.182	4.73	832	8345.24	12.463	4.64	892	9101.30	12.738	4.54
773	7618.06	12.186	4.73	833	8357.71	12.467	4.64	893	9114.04	12.743	4.54
774	7630.25	12.191	4.73	834	8370.18	12.472	4.63	894	9126.79	12.747	4.54
775	7642.44	12.196	4.73	835	8382.65	12.477	4.63	895	9139.54	12.752	4.54
776	7654.64	12.201	4.73	836	8395.13	12.481	4.63	896	9152.29	12.756	4.53
777	7666.85	12.205	4.73	837	8407.62	12.486	4.63	897	9165.05	12.761	4.53
778	7679.05	12.210	4.72	838	8420.10	12.491	4.63	898	9177.81	12.765	4.53
779	7691.27	12.215	4.72	839	8432.60	12.495	4.63	899	9190.58	12.770	4.53
780	7703.48	12.220	4.72	840	8445.09	12.500	4.62	900	9203.35	12.774	4.53

TABLE 1i6. (*Continued*)

Type R thermocouples—*thermoelectric voltages*, E(T), *Seebeck coefficients*, S(T), *and first derivative of the Seebeck coefficients*, dS/dT, *reference junctions at 0 °C—Continued*

T °C	E μV	S μV/°C	dS/dT nV/°C²	T °C	E μV	S μV/°C	dS/dT nV/°C²	T °C	E μV	S μV/°C	dS/dT nV/°C²
900	9203.35	12.774	4.53	960	9977.90	13.043	4.43	1020	10768.41	13.306	4.33
901	9216.13	12.779	4.53	961	9990.95	13.048	4.43	1021	10781.72	13.310	4.33
902	9228.91	12.783	4.52	962	10004.00	13.052	4.43	1022	10795.03	13.315	4.33
903	9241.69	12.788	4.52	963	10017.05	13.056	4.43	1023	10808.35	13.319	4.33
904	9254.48	12.792	4.52	964	10030.11	13.061	4.42	1024	10821.67	13.323	4.33
905	9267.28	12.797	4.52	965	10043.18	13.065	4.42	1025	10834.99	13.328	4.33
906	9280.08	12.802	4.52	966	10056.24	13.070	4.42	1026	10848.32	13.332	4.32
907	9292.88	12.806	4.52	967	10069.32	13.074	4.42	1027	10861.66	13.336	4.32
908	9305.69	12.811	4.51	968	10082.39	13.079	4.42	1028	10875.00	13.341	4.32
909	9318.50	12.815	4.51	969	10095.47	13.083	4.42	1029	10888.34	13.345	4.32
910	9331.32	12.820	4.51	970	10108.56	13.087	4.41	1030	10901.69	13.349	4.32
911	9344.14	12.824	4.51	971	10121.65	13.092	4.41	1031	10915.04	13.354	4.32
912	9356.97	12.829	4.51	972	10134.74	13.096	4.41	1032	10928.39	13.358	4.31
913	9369.80	12.833	4.51	973	10147.84	13.101	4.41	1033	10941.75	13.362	4.31
914	9382.63	12.838	4.50	974	10160.94	13.105	4.41	1034	10955.12	13.367	4.31
915	9395.47	12.842	4.50	975	10174.05	13.109	4.41	1035	10968.49	13.371	4.31
916	9408.32	12.847	4.50	976	10187.16	13.114	4.40	1036	10981.86	13.375	4.31
917	9421.17	12.851	4.50	977	10200.28	13.118	4.40	1037	10995.24	13.379	4.31
918	9434.02	12.856	4.50	978	10213.40	13.123	4.40	1038	11008.62	13.384	4.30
919	9446.88	12.860	4.50	979	10226.52	13.127	4.40	1039	11022.00	13.388	4.30
920	9459.74	12.865	4.49	980	10239.65	13.131	4.40	1040	11035.39	13.392	4.30
921	9472.61	12.869	4.49	981	10252.78	13.136	4.40	1041	11048.79	13.397	4.30
922	9485.48	12.874	4.49	982	10265.92	13.140	4.39	1042	11062.19	13.401	4.30
923	9498.36	12.878	4.49	983	10279.07	13.145	4.39	1043	11075.59	13.405	4.30
924	9511.24	12.883	4.49	984	10292.21	13.149	4.39	1044	11089.00	13.410	4.29
925	9524.12	12.887	4.49	985	10305.36	13.153	4.39	1045	11102.41	13.414	4.29
926	9537.01	12.892	4.49	986	10318.52	13.158	4.39	1046	11115.83	13.418	4.29
927	9549.90	12.896	4.48	987	10331.68	13.162	4.39	1047	11129.25	13.422	4.29
928	9562.80	12.901	4.48	988	10344.84	13.167	4.39	1048	11142.67	13.427	4.29
929	9575.71	12.905	4.48	989	10358.01	13.171	4.38	1049	11156.10	13.431	4.29
930	9588.61	12.909	4.48	990	10371.18	13.175	4.38	1050	11169.53	13.435	4.29
931	9601.52	12.914	4.48	991	10384.36	13.180	4.38	1051	11182.97	13.440	4.28
932	9614.44	12.918	4.48	992	10397.54	13.184	4.38	1052	11196.41	13.444	4.28
933	9627.36	12.923	4.47	993	10410.73	13.188	4.38	1053	11209.86	13.448	4.28
934	9640.29	12.927	4.47	994	10423.92	13.193	4.38	1054	11223.31	13.452	4.28
935	9653.22	12.932	4.47	995	10437.12	13.197	4.37	1055	11236.76	13.457	4.28
936	9666.15	12.936	4.47	996	10450.32	13.202	4.37	1056	11250.22	13.461	4.28
937	9679.09	12.941	4.47	997	10463.52	13.206	4.37	1057	11263.68	13.465	4.27
938	9692.03	12.945	4.47	998	10476.73	13.210	4.37	1058	11277.15	13.470	4.27
939	9704.98	12.950	4.46	999	10489.94	13.215	4.37	1059	11290.62	13.474	4.27
940	9717.93	12.954	4.46	1000	10503.16	13.219	4.37	1060	11304.10	13.478	4.27
941	9730.89	12.959	4.46	1001	10516.38	13.223	4.36	1061	11317.58	13.482	4.27
942	9743.85	12.963	4.46	1002	10529.60	13.228	4.36	1062	11331.06	13.487	4.27
943	9756.81	12.968	4.46	1003	10542.83	13.232	4.36	1063	11344.55	13.491	4.26
944	9769.78	12.972	4.46	1004	10556.07	13.236	4.36	1064	11358.05	13.495	4.26
945	9782.76	12.976	4.45	1005	10569.31	13.241	4.36	1065	11371.54	13.499	3.81
946	9795.74	12.981	4.45	1006	10582.55	13.245	4.36	1066	11385.04	13.503	3.80
947	9808.72	12.985	4.45	1007	10595.80	13.250	4.35	1067	11398.55	13.507	3.78
948	9821.71	12.990	4.45	1008	10609.05	13.254	4.35	1068	11412.06	13.511	3.77
949	9834.70	12.994	4.45	1009	10622.30	13.258	4.35	1069	11425.57	13.514	3.76
950	9847.70	12.999	4.45	1010	10635.57	13.263	4.35	1070	11439.09	13.518	3.75
951	9860.70	13.003	4.44	1011	10648.83	13.267	4.35	1071	11452.61	13.522	3.74
952	9873.70	13.008	4.44	1012	10662.10	13.271	4.35	1072	11466.13	13.526	3.73
953	9886.71	13.012	4.44	1013	10675.37	13.276	4.34	1073	11479.66	13.529	3.71
954	9899.73	13.017	4.44	1014	10688.65	13.280	4.34	1074	11493.19	13.533	3.70
955	9912.74	13.021	4.44	1015	10701.93	13.284	4.34	1075	11506.72	13.537	3.69
956	9925.77	13.025	4.44	1016	10715.22	13.289	4.34	1076	11520.26	13.540	3.68
957	9938.80	13.030	4.44	1017	10728.51	13.293	4.34	1077	11533.80	13.544	3.67
958	9951.83	13.034	4.43	1018	10741.81	13.297	4.34	1078	11547.35	13.548	3.66
959	9964.86	13.039	4.43	1019	10755.10	13.302	4.34	1079	11560.90	13.551	3.64
960	9977.90	13.043	4.43	1020	10768.41	13.306	4.33	1080	11574.45	13.555	3.63

TABLE 1i6. *(Continued)*

Type R thermocouples—*thermoelectric voltages*, E(T), *Seebeck coefficients*, S(T), *and first derivative of the Seebeck coefficients*, dS/dT, *reference junctions at 0 °C—Continued*

T °C	E µV	S µV/°C	dS/dT nV/°C²	T °C	E µV	S µV/°C	dS/dT nV/°C²	T °C	E µV	S µV/°C	dS/dT nV/°C²
1080	11574.45	13.555	3.63	1140	12393.87	13.752	2.94	1200	13223.87	13.907	2.24
1081	11588.01	13.559	3.62	1141	12407.63	13.755	2.93	1201	13237.78	13.910	2.23
1082	11601.57	13.562	3.61	1142	12421.38	13.758	2.91	1202	13251.69	13.912	2.22
1083	11615.13	13.566	3.60	1143	12435.14	13.761	2.90	1203	13265.60	13.914	2.21
1084	11628.70	13.569	3.59	1144	12448.91	13.764	2.89	1204	13279.52	13.916	2.19
1085	11642.27	13.573	3.58	1145	12462.67	13.767	2.88	1205	13293.43	13.918	2.18
1086	11655.85	13.577	3.56	1146	12476.44	13.770	2.87	1206	13307.35	13.921	2.17
1087	11669.43	13.580	3.55	1147	12490.21	13.772	2.86	1207	13321.27	13.923	2.16
1088	11683.01	13.584	3.54	1148	12503.98	13.775	2.84	1208	13335.20	13.925	2.15
1089	11696.59	13.587	3.53	1149	12517.76	13.778	2.83	1209	13349.12	13.927	2.14
1090	11710.18	13.591	3.52	1150	12531.54	13.781	2.82	1210	13363.05	13.929	2.12
1091	11723.78	13.594	3.51	1151	12545.32	13.784	2.81	1211	13376.98	13.931	2.11
1092	11737.37	13.598	3.49	1152	12559.11	13.787	2.80	1212	13390.92	13.933	2.10
1093	11750.97	13.601	3.48	1153	12572.89	13.789	2.79	1213	13404.85	13.936	2.09
1094	11764.57	13.605	3.47	1154	12586.69	13.792	2.77	1214	13418.79	13.938	2.08
1095	11778.18	13.608	3.46	1155	12600.48	13.795	2.76	1215	13432.73	13.940	2.07
1096	11791.79	13.612	3.45	1156	12614.28	13.798	2.75	1216	13446.67	13.942	2.06
1097	11805.40	13.615	3.44	1157	12628.07	13.800	2.74	1217	13460.61	13.944	2.04
1098	11819.02	13.618	3.42	1158	12641.88	13.803	2.73	1218	13474.55	13.946	2.03
1099	11832.64	13.622	3.41	1159	12655.68	13.806	2.72	1219	13488.50	13.948	2.02
1100	11846.26	13.625	3.40	1160	12669.49	13.809	2.71	1220	13502.45	13.950	2.01
1101	11859.89	13.629	3.39	1161	12683.30	13.811	2.69	1221	13516.40	13.952	2.00
1102	11873.52	13.632	3.38	1162	12697.11	13.814	2.68	1222	13530.35	13.954	1.99
1103	11887.15	13.635	3.37	1163	12710.93	13.817	2.67	1223	13544.31	13.956	1.97
1104	11900.79	13.639	3.35	1164	12724.74	13.819	2.66	1224	13558.27	13.958	1.96
1105	11914.43	13.642	3.34	1165	12738.56	13.822	2.65	1225	13572.22	13.960	1.95
1106	11928.08	13.646	3.33	1166	12752.39	13.825	2.64	1226	13586.19	13.962	1.94
1107	11941.72	13.649	3.32	1167	12766.21	13.827	2.62	1227	13600.15	13.964	1.93
1108	11955.37	13.652	3.31	1168	12780.04	13.830	2.61	1228	13614.11	13.966	1.92
1109	11969.03	13.655	3.30	1169	12793.87	13.832	2.60	1229	13628.08	13.968	1.90
1110	11982.69	13.659	3.29	1170	12807.71	13.835	2.59	1230	13642.05	13.969	1.89
1111	11996.35	13.662	3.27	1171	12821.54	13.838	2.58	1231	13656.02	13.971	1.88
1112	12010.01	13.665	3.26	1172	12835.38	13.840	2.57	1232	13669.99	13.973	1.87
1113	12023.68	13.669	3.25	1173	12849.22	13.843	2.55	1233	13683.96	13.975	1.86
1114	12037.35	13.672	3.24	1174	12863.07	13.845	2.54	1234	13697.94	13.977	1.85
1115	12051.02	13.675	3.23	1175	12876.91	13.848	2.53	1235	13711.92	13.979	1.83
1116	12064.70	13.678	3.22	1176	12890.76	13.850	2.52	1236	13725.90	13.981	1.82
1117	12078.38	13.681	3.20	1177	12904.61	13.853	2.51	1237	13739.88	13.982	1.81
1118	12092.06	13.685	3.19	1178	12918.47	13.855	2.50	1238	13753.86	13.984	1.80
1119	12105.75	13.688	3.18	1179	12932.32	13.858	2.48	1239	13767.85	13.986	1.79
1120	12119.43	13.691	3.17	1180	12946.18	13.860	2.47	1240	13781.83	13.988	1.78
1121	12133.13	13.694	3.16	1181	12960.04	13.863	2.46	1241	13795.82	13.990	1.77
1122	12146.82	13.697	3.15	1182	12973.91	13.865	2.45	1242	13809.81	13.991	1.75
1123	12160.52	13.700	3.13	1183	12987.78	13.868	2.44	1243	13823.81	13.993	1.74
1124	12174.22	13.704	3.12	1184	13001.64	13.870	2.43	1244	13837.80	13.995	1.73
1125	12187.93	13.707	3.11	1185	13015.52	13.873	2.42	1245	13851.80	13.997	1.72
1126	12201.64	13.710	3.10	1186	13029.39	13.875	2.40	1246	13865.79	13.998	1.71
1127	12215.35	13.713	3.09	1187	13043.27	13.877	2.39	1247	13879.79	14.000	1.70
1128	12229.06	13.716	3.08	1188	13057.14	13.880	2.38	1248	13893.79	14.002	1.68
1129	12242.78	13.719	3.06	1189	13071.02	13.882	2.37	1249	13907.80	14.003	1.67
1130	12256.50	13.722	3.05	1190	13084.91	13.884	2.36	1250	13921.80	14.005	1.66
1131	12270.23	13.725	3.04	1191	13098.79	13.887	2.35	1251	13935.81	14.007	1.65
1132	12283.95	13.728	3.03	1192	13112.68	13.889	2.33	1252	13949.81	14.008	1.64
1133	12297.68	13.731	3.02	1193	13126.57	13.891	2.32	1253	13963.82	14.010	1.63
1134	12311.41	13.734	3.01	1194	13140.46	13.894	2.31	1254	13977.83	14.012	1.61
1135	12325.15	13.737	3.00	1195	13154.36	13.896	2.30	1255	13991.85	14.013	1.60
1136	12338.89	13.740	2.98	1196	13168.26	13.898	2.29	1256	14005.86	14.015	1.59
1137	12352.63	13.743	2.97	1197	13182.16	13.901	2.28	1257	14019.87	14.016	1.58
1138	12366.38	13.746	2.96	1198	13196.06	13.903	2.26	1258	14033.89	14.018	1.57
1139	12380.12	13.749	2.95	1199	13209.96	13.905	2.25	1259	14047.91	14.019	1.56
1140	12393.87	13.752	2.94	1200	13223.87	13.907	2.24	1260	14061.93	14.021	1.54

TABLE 1i6. (*Continued*)

Type **R** *thermocouples*—*thermoelectric voltages,* E(T), *Seebeck coefficients,* S(T), *a... first derivative of the Seebeck coefficients,* dS/dT, *reference junctions at 0 °C—Continued*

T °C	E μV	S μV/°C	dS/dT nV/°C²	T °C	E μV	S μV/°C	dS/dT nV/°C²	T °C	E μV	S μV/°C	dS/dT nV/°C²
1260	14061.93	14.021	1.54	1320	14905.55	14.093	0.85	1380	15752.23	14.123	0.15
1261	14075.95	14.023	1.53	1321	14919.65	14.094	0.84	1381	15766.36	14.123	0.14
1262	14089.98	14.024	1.52	1322	14933.74	14.094	0.83	1382	15780.48	14.123	0.13
1263	14104.00	14.026	1.51	1323	14947.84	14.095	0.81	1383	15794.60	14.123	0.12
1264	14118.03	14.027	1.50	1324	14961.93	14.096	0.80	1384	15808.73	14.123	0.11
1265	14132.05	14.029	1.49	1325	14976.03	14.097	0.79	1385	15822.85	14.123	0.09
1266	14146.08	14.030	1.48	1326	14990.13	14.098	0.78	1386	15836.97	14.124	0.08
1267	14160.11	14.032	1.46	1327	15004.22	14.098	0.77	1387	15851.10	14.124	0.07
1268	14174.15	14.033	1.45	1328	15018.32	14.099	0.76	1388	15865.22	14.124	0.06
1269	14188.18	14.034	1.44	1329	15032.42	14.100	0.74	1389	15879.34	14.124	0.05
1270	14202.22	14.036	1.43	1330	15046.52	14.101	0.73	1390	15893.47	14.124	0.04
1271	14216.25	14.037	1.42	1331	15060.62	14.101	0.72	1391	15907.59	14.124	0.02
1272	14230.29	14.039	1.41	1332	15074.73	14.102	0.71	1392	15921.71	14.124	0.01
1273	14244.33	14.040	1.39	1333	15088.83	14.103	0.70	1393	15935.84	14.124	0.00
1274	14258.37	14.042	1.38	1334	15102.93	14.104	0.69	1394	15949.96	14.124	-0.01
1275	14272.41	14.043	1.37	1335	15117.04	14.104	0.67	1395	15964.09	14.124	-0.02
1276	14286.46	14.044	1.36	1336	15131.14	14.105	0.66	1396	15978.21	14.124	-0.03
1277	14300.50	14.046	1.35	1337	15145.25	14.106	0.65	1397	15992.33	14.124	-0.04
1278	14314.55	14.047	1.34	1338	15159.35	14.106	0.64	1398	16006.46	14.124	-0.06
1279	14328.60	14.048	1.32	1339	15173.46	14.107	0.63	1399	16020.58	14.124	-0.07
1280	14342.64	14.050	1.31	1340	15187.56	14.107	0.62	1400	16034.71	14.124	-0.08
1281	14356.69	14.051	1.30	1341	15201.67	14.108	0.60	1401	16048.83	14.123	-0.09
1282	14370.75	14.052	1.29	1342	15215.78	14.109	0.59	1402	16062.95	14.123	-0.10
1283	14384.80	14.053	1.28	1343	15229.89	14.109	0.58	1403	16077.08	14.123	-0.11
1284	14398.85	14.055	1.27	1344	15244.00	14.110	0.57	1404	16091.20	14.123	-0.13
1285	14412.91	14.056	1.25	1345	15258.11	14.110	0.56	1405	16105.32	14.123	-0.14
1286	14426.97	14.057	1.24	1346	15272.22	14.111	0.55	1406	16119.44	14.123	-0.15
1287	14441.02	14.058	1.23	1347	15286.33	14.112	0.54	1407	16133.57	14.123	-0.16
1288	14455.08	14.060	1.22	1348	15300.44	14.112	0.52	1408	16147.69	14.123	-0.17
1289	14469.14	14.061	1.21	1349	15314.56	14.113	0.51	1409	16161.81	14.122	-0.18
1290	14483.20	14.062	1.20	1350	15328.67	14.113	0.50	1410	16175.94	14.122	-0.20
1291	14497.27	14.063	1.19	1351	15342.78	14.114	0.49	1411	16190.06	14.122	-0.21
1292	14511.33	14.065	1.17	1352	15356.90	14.114	0.48	1412	16204.18	14.122	-0.22
1293	14525.40	14.066	1.16	1353	15371.01	14.115	0.47	1413	16218.30	14.122	-0.23
1294	14539.46	14.067	1.15	1354	15385.12	14.115	0.45	1414	16232.42	14.121	-0.24
1295	14553.53	14.068	1.14	1355	15399.24	14.115	0.44	1415	16246.54	14.121	-0.25
1296	14567.60	14.069	1.13	1356	15413.36	14.116	0.43	1416	16260.66	14.121	-0.27
1297	14581.67	14.070	1.12	1357	15427.47	14.116	0.42	1417	16274.79	14.121	-0.28
1298	14595.74	14.071	1.10	1358	15441.59	14.117	0.41	1418	16288.91	14.120	-0.29
1299	14609.81	14.072	1.09	1359	15455.70	14.117	0.40	1419	16303.03	14.120	-0.30
1300	14623.88	14.074	1.08	1360	15469.82	14.117	0.38	1420	16317.15	14.120	-0.31
1301	14637.96	14.075	1.07	1361	15483.94	14.118	0.37	1421	16331.26	14.119	-0.32
1302	14652.03	14.076	1.06	1362	15498.06	14.118	0.36	1422	16345.38	14.119	-0.33
1303	14666.11	14.077	1.05	1363	15512.18	14.119	0.35	1423	16359.50	14.119	-0.35
1304	14680.19	14.078	1.03	1364	15526.29	14.119	0.34	1424	16373.62	14.118	-0.36
1305	14694.26	14.079	1.02	1365	15540.41	14.119	0.33	1425	16387.74	14.118	-0.37
1306	14708.34	14.080	1.01	1366	15554.53	14.120	0.31	1426	16401.86	14.118	-0.38
1307	14722.42	14.081	1.00	1367	15568.65	14.120	0.30	1427	16415.97	14.117	-0.39
1308	14736.50	14.082	0.99	1368	15582.77	14.120	0.29	1428	16430.09	14.117	-0.40
1309	14750.59	14.083	0.98	1369	15596.89	14.120	0.28	1429	16444.21	14.116	-0.42
1310	14764.67	14.084	0.96	1370	15611.01	14.121	0.27	1430	16458.32	14.116	-0.43
1311	14778.75	14.085	0.95	1371	15625.14	14.121	0.26	1431	16472.44	14.116	-0.44
1312	14792.84	14.086	0.94	1372	15639.26	14.121	0.25	1432	16486.56	14.115	-0.45
1313	14806.93	14.087	0.93	1373	15653.38	14.121	0.23	1433	16500.67	14.115	-0.46
1314	14821.01	14.088	0.92	1374	15667.50	14.122	0.22	1434	16514.78	14.114	-0.47
1315	14835.10	14.088	0.91	1375	15681.62	14.122	0.21	1435	16528.90	14.114	-0.49
1316	14849.19	14.089	0.90	1376	15695.74	14.122	0.20	1436	16543.01	14.113	-0.50
1317	14863.28	14.090	0.88	1377	15709.87	14.122	0.19	1437	16557.13	14.113	-0.51
1318	14877.37	14.091	0.87	1378	15723.99	14.123	0.18	1438	16571.24	14.112	-0.52
1319	14891.46	14.092	0.86	1379	15738.11	14.123	0.16	1439	16585.35	14.112	-0.53
1320	14905.55	14.093	0.85	1380	15752.23	14.123	0.15	1440	16599.46	14.111	-0.54

TABLE 1i6. (Continued)

Type R thermocouples—thermoelectric voltages, E(T), Seebeck coefficients, S(T), and first derivative of the Seebeck coefficients, dS/dT, reference junctions at 0 °C—Continued

T °C	E μV	S μV/°C	dS/dT nV/°C²	T °C	E μV	S μV/°C	dS/dT nV/°C²	T °C	E μV	S μV/°C	dS/dT nV/°C²
1440	16599.46	14.111	-0.54	1500	17444.73	14.058	-1.24	1560	18285.54	13.962	-1.94
1441	16613.57	14.111	-0.56	1501	17458.79	14.056	-1.25	1561	18299.50	13.960	-1.95
1442	16627.68	14.110	-0.57	1502	17472.84	14.055	-1.26	1562	18313.46	13.958	-1.96
1443	16641.79	14.109	-0.58	1503	17486.90	14.054	-1.27	1563	18327.42	13.956	-1.97
1444	16655.90	14.109	-0.59	1504	17500.95	14.053	-1.29	1564	18341.37	13.954	-1.98
1445	16670.01	14.108	-0.60	1505	17515.00	14.051	-1.30	1565	18355.33	13.952	-1.99
1446	16684.12	14.108	-0.61	1506	17529.05	14.050	-1.31	1566	18369.28	13.950	-2.01
1447	16698.22	14.107	-0.62	1507	17543.10	14.049	-1.32	1567	18383.23	13.948	-2.02
1448	16712.33	14.106	-0.64	1508	17557.15	14.047	-1.33	1568	18397.17	13.946	-2.03
1449	16726.44	14.106	-0.65	1509	17571.20	14.046	-1.34	1569	18411.12	13.944	-2.04
1450	16740.54	14.105	-0.66	1510	17585.24	14.045	-1.36	1570	18425.06	13.942	-2.05
1451	16754.65	14.104	-0.67	1511	17599.29	14.043	-1.37	1571	18439.00	13.940	-2.06
1452	16768.75	14.104	-0.68	1512	17613.33	14.042	-1.38	1572	18452.94	13.938	-2.08
1453	16782.86	14.103	-0.69	1513	17627.37	14.041	-1.39	1573	18466.88	13.936	-2.09
1454	16796.96	14.102	-0.71	1514	17641.41	14.039	-1.40	1574	18480.82	13.934	-2.10
1455	16811.06	14.102	-0.72	1515	17655.45	14.038	-1.41	1575	18494.75	13.932	-2.11
1456	16825.16	14.101	-0.73	1516	17669.49	14.036	-1.43	1576	18508.68	13.930	-2.12
1457	16839.26	14.100	-0.74	1517	17683.52	14.035	-1.44	1577	18522.61	13.928	-2.13
1458	16853.36	14.099	-0.75	1518	17697.56	14.033	-1.45	1578	18536.53	13.926	-2.14
1459	16867.46	14.099	-0.76	1519	17711.59	14.032	-1.46	1579	18550.46	13.923	-2.16
1460	16881.56	14.098	-0.78	1520	17725.62	14.030	-1.47	1580	18564.38	13.921	-2.17
1461	16895.66	14.097	-0.79	1521	17739.65	14.029	-1.48	1581	18578.30	13.919	-2.18
1462	16909.75	14.096	-0.80	1522	17753.68	14.028	-1.50	1582	18592.22	13.917	-2.19
1463	16923.85	14.096	-0.81	1523	17767.70	14.026	-1.51	1583	18606.14	13.915	-2.20
1464	16937.94	14.095	-0.82	1524	17781.73	14.025	-1.52	1584	18620.05	13.913	-2.21
1465	16952.04	14.094	-0.83	1525	17795.75	14.023	-1.53	1585	18633.96	13.910	-2.23
1466	16966.13	14.093	-0.85	1526	17809.78	14.021	-1.54	1586	18647.87	13.908	-2.24
1467	16980.22	14.092	-0.86	1527	17823.80	14.020	-1.55	1587	18661.78	13.906	-2.25
1468	16994.32	14.091	-0.87	1528	17837.82	14.018	-1.56	1588	18675.68	13.904	-2.26
1469	17008.41	14.090	-0.88	1529	17851.83	14.017	-1.58	1589	18689.58	13.901	-2.27
1470	17022.50	14.090	-0.89	1530	17865.85	14.015	-1.59	1590	18703.48	13.899	-2.28
1471	17036.59	14.089	-0.90	1531	17879.86	14.014	-1.60	1591	18717.38	13.897	-2.30
1472	17050.67	14.088	-0.92	1532	17893.88	14.012	-1.61	1592	18731.28	13.894	-2.31
1473	17064.76	14.087	-0.93	1533	17907.89	14.010	-1.62	1593	18745.17	13.892	-2.32
1474	17078.85	14.086	-0.94	1534	17921.90	14.009	-1.63	1594	18759.06	13.890	-2.33
1475	17092.93	14.085	-0.95	1535	17935.90	14.007	-1.65	1595	18772.95	13.887	-2.34
1476	17107.02	14.084	-0.96	1536	17949.91	14.005	-1.66	1596	18786.84	13.885	-2.35
1477	17121.10	14.083	-0.97	1537	17963.90	14.004	-1.67	1597	18800.72	13.883	-2.37
1478	17135.18	14.082	-0.98	1538	17977.92	14.002	-1.68	1598	18814.60	13.880	-2.38
1479	17149.27	14.081	-1.00	1539	17991.92	14.000	-1.69	1599	18828.48	13.878	-2.39
1480	17163.35	14.080	-1.01	1540	18005.92	13.999	-1.70	1600	18842.36	13.876	-2.40
1481	17177.43	14.079	-1.02	1541	18019.92	13.997	-1.72	1601	18856.23	13.873	-2.41
1482	17191.50	14.078	-1.03	1542	18033.91	13.995	-1.73	1602	18870.10	13.871	-2.42
1483	17205.58	14.077	-1.04	1543	16047.91	13.994	-1.74	1603	18883.97	13.868	-2.44
1484	17219.66	14.076	-1.05	1544	18061.90	13.992	-1.75	1604	18897.84	13.866	-2.45
1485	17233.73	14.075	-1.07	1545	18075.89	13.990	-1.76	1605	18911.71	13.863	-2.46
1486	17247.81	14.074	-1.08	1546	18089.88	13.988	-1.77	1606	18925.57	13.861	-2.47
1487	17261.88	14.073	-1.09	1547	18103.87	13.987	-1.79	1607	18939.43	13.859	-2.48
1488	17275.95	14.072	-1.10	1548	18117.85	13.985	-1.80	1608	18953.29	13.856	-2.49
1489	17290.03	14.071	-1.11	1549	18131.84	13.983	-1.81	1609	18967.14	13.854	-2.50
1490	17304.10	14.069	-1.12	1550	18145.82	13.981	-1.82	1610	18980.99	13.851	-2.52
1491	17318.16	14.068	-1.14	1551	18159.80	13.979	-1.83	1611	18994.84	13.848	-2.53
1492	17332.23	14.067	-1.15	1552	18173.78	13.977	-1.84	1612	19008.69	13.846	-2.54
1493	17346.30	14.066	-1.16	1553	18187.75	13.976	-1.85	1613	19022.53	13.843	-2.55
1494	17360.36	14.065	-1.17	1554	18201.73	13.974	-1.87	1614	19036.38	13.841	-2.56
1495	17374.43	14.064	-1.18	1555	18215.70	13.972	-1.88	1615	19050.22	13.838	-2.57
1496	17388.49	14.062	-1.19	1556	18229.67	13.970	-1.89	1616	19064.05	13.836	-2.59
1497	17402.55	14.061	-1.21	1557	18243.64	13.968	-1.90	1617	19077.89	13.833	-2.60
1498	17416.61	14.060	-1.22	1558	18257.61	13.966	-1.91	1618	19091.72	13.831	-2.61
1499	17430.67	14.059	-1.23	1559	18271.57	13.964	-1.92	1619	19105.55	13.828	-2.62
1500	17444.73	14.058	-1.24	1560	18285.54	13.962	-1.94	1620	19119.37	13.825	-2.63

TABLE 1i6. (Continued)

Type R thermocouples—thermoelectric voltages, E(T), Seebeck coefficients, S(T) and first derivative of the Seebeck coefficients, dS/dT, reference junctions at 0 °C—Continued

T °C	E μV	S μV/°C	dS/dT nV/°C²	T °C	E μV	S μV/°C	dS/dT nV/°C²	T °C	E μV	S μV/°C	dS/dT nV/°C²
1620	19119.37	13.825	−2.63	1680	19943.67	13.633	−5.16	1740	20747.53	13.082	−13.18
1621	19133.20	13.823	−2.64	1681	19957.30	13.627	−5.29	1741	20760.60	13.069	−13.32
1622	19147.02	13.820	−2.66	1682	19970.92	13.622	−5.43	1742	20773.66	13.056	−13.45
1623	19160.84	13.817	−2.67	1683	19984.54	13.617	−5.56	1743	20786.71	13.042	−13.59
1624	19174.65	13.815	−2.68	1684	19998.15	13.611	−5.70	1744	20799.75	13.029	−13.72
1625	19188.47	13.812	−2.69	1685	20011.76	13.605	−5.83	1745	20812.77	13.015	−13.85
1626	19202.28	13.809	−2.70	1686	20025.37	13.599	−5.96	1746	20825.78	13.001	−13.99
1627	19216.09	13.807	−2.71	1687	20038.96	13.593	−6.10	1747	20838.77	12.987	−14.12
1628	19229.89	13.804	−2.73	1688	20052.55	13.587	−6.23	1748	20851.75	12.973	−14.25
1629	19243.69	13.801	−2.74	1689	20066.14	13.581	−6.36	1749	20864.72	12.958	−14.39
1630	19257.49	13.798	−2.75	1690	20079.71	13.574	−6.50	1750	20877.67	12.944	−14.52
1631	19271.29	13.796	−2.76	1691	20093.28	13.568	−6.63	1751	20890.60	12.929	−14.66
1632	19285.09	13.793	−2.77	1692	20106.85	13.561	−6.77	1752	20903.53	12.915	−14.79
1633	19298.88	13.790	−2.78	1693	20120.41	13.554	−6.90	1753	20916.43	12.900	−14.92
1634	19312.67	13.787	−2.79	1694	20133.96	13.547	−7.03	1754	20929.33	12.885	−15.06
1635	19326.45	13.784	−2.81	1695	20147.50	13.540	−7.17	1755	20942.20	12.870	−15.19
1636	19340.23	13.782	−2.82	1696	20161.04	13.533	−7.30	1756	20955.06	12.854	−15.32
1637	19354.01	13.779	−2.83	1697	20174.57	13.526	−7.43	1757	20967.91	12.839	−15.46
1638	19367.79	13.776	−2.84	1698	20188.09	13.518	−7.57	1758	20980.74	12.823	−15.59
1639	19381.57	13.773	−2.85	1699	20201.60	13.511	−7.70	1759	20993.56	12.808	−15.73
1640	19395.34	13.770	−2.86	1700	20215.11	13.503	−7.84	1760	21006.36	12.792	−15.86
1641	19409.11	13.767	−2.88	1701	20228.61	13.495	−7.97	1761	21019.14	12.776	−15.99
1642	19422.87	13.765	−2.89	1702	20242.10	13.487	−8.10	1762	21031.91	12.760	−16.13
1643	19436.64	13.762	−2.90	1703	20255.58	13.479	−8.24	1763	21044.66	12.744	−16.26
1644	19450.40	13.759	−2.91	1704	20269.06	13.470	−8.37	1764	21057.40	12.727	−16.39
1645	19464.15	13.756	−2.92	1705	20282.52	13.462	−8.50	1765	21070.12	12.711	−16.53
1646	19477.91	13.753	−2.93	1706	20295.98	13.453	−8.64	1766	21082.82	12.694	−16.66
1647	19491.66	13.750	−2.95	1707	20309.43	13.445	−8.77	1767	21095.51	12.678	−16.80
1648	19505.41	13.747	−2.96	1708	20322.87	13.436	−8.91	1768	21108.17	12.661	−16.93
1649	19519.15	13.744	−2.97	1709	20336.30	13.427	−9.04				
1650	19532.90	13.741	−2.98	1710	20349.72	13.418	−9.17				
1651	19546.64	13.738	−2.99	1711	20363.14	13.408	−9.31				
1652	19560.37	13.735	−3.00	1712	20376.54	13.399	−9.44				
1653	19574.11	13.732	−3.02	1713	20389.94	13.390	−9.57				
1654	19587.84	13.729	−3.03	1714	20403.32	13.380	−9.71				
1655	19601.56	13.726	−3.04	1715	20416.70	13.370	−9.84				
1656	19615.29	13.723	−3.05	1716	20430.06	13.360	−9.97				
1657	19629.01	13.720	−3.06	1717	20443.42	13.350	−10.11				
1658	19642.73	13.717	−3.07	1718	20456.76	13.340	−10.24				
1659	19656.44	13.714	−3.08	1719	20470.10	13.330	−10.38				
1660	19670.16	13.711	−3.10	1720	20483.42	13.319	−10.51				
1661	19683.87	13.708	−3.11	1721	20496.73	13.309	−10.64				
1662	19697.57	13.704	−3.12	1722	20510.04	13.298	−10.78				
1663	19711.27	13.701	−3.13	1723	20523.33	13.287	−10.91				
1664	19724.97	13.698	−3.14	1724	20536.61	13.276	−11.04				
1665	19738.67	13.695	−3.15	1725	20549.88	13.265	−11.18				
1666	19752.36	13.692	−3.29	1726	20563.14	13.254	−11.31				
1667	19766.05	13.688	−3.42	1727	20576.39	13.242	−11.45				
1668	19779.74	13.685	−3.56	1728	20589.63	13.231	−11.58				
1669	19793.42	13.681	−3.69	1729	20602.85	13.219	−11.71				
1670	19807.10	13.678	−3.82	1730	20616.07	13.208	−11.85				
1671	19820.78	13.674	−3.96	1731	20629.27	13.196	−11.98				
1672	19834.45	13.670	−4.09	1732	20642.46	13.184	−12.11				
1673	19848.12	13.666	−4.22	1733	20655.63	13.171	−12.25				
1674	19861.78	13.661	−4.36	1734	20668.80	13.159	−12.38				
1675	19875.44	13.657	−4.49	1735	20681.95	13.147	−12.52				
1676	19889.10	13.652	−4.63	1736	20695.09	13.134	−12.65				
1677	19902.75	13.648	−4.76	1737	20708.22	13.121	−12.78				
1678	19916.39	13.643	−4.89	1738	20721.34	13.108	−12.92				
1679	19930.03	13.638	−5.03	1739	20734.44	13.095	−13.05				
1680	19943.67	13.633	−5.16	1740	20747.53	13.082	−13.18				

TABLE 1i6. (*Continued*)

Type S—*Platinum*–10% Rhodium Alloy Versus Platinum Thermocouples

Type S thermocouples—*thermoelectric voltages*, E(T), *Seebeck coefficients*, S(T), *and first derivative of the Seebeck coefficients*, dS/dT, *reference junctions at 0 °C*

T °C	E μV	S μV/°C	dS/dT nV/°C²	T °C	E μV	S μV/°C	dS/dT nV/°C²	T °C	E μV	S μV/°C	dS/dT nV/°C²
				-40	-194.44	4.283	31.00	-20	-102.80	4.871	27.87
				-39	-190.14	4.314	30.84	-19	-97.91	4.899	27.73
				-38	-185.81	4.344	30.68	-18	-93.00	4.926	27.58
				-37	-181.45	4.375	30.51	-17	-88.06	4.954	27.43
				-36	-177.06	4.405	30.35	-16	-83.09	4.981	27.28
				-35	-172.64	4.436	30.19	-15	-78.10	5.008	27.14
				-34	-168.19	4.466	30.03	-14	-73.08	5.036	26.99
				-33	-163.71	4.496	29.87	-13	-68.03	5.062	26.85
				-32	-159.20	4.526	29.71	-12	-62.95	5.089	26.71
				-31	-154.66	4.555	29.56	-11	-57.85	5.116	26.56
-50	-235.69	3.964	32.69	-30	-150.09	4.585	29.40	-10	-52.72	5.142	26.42
-49	-231.71	3.997	32.51	-29	-145.49	4.614	29.24	-9	-47.57	5.169	26.28
-48	-227.69	4.029	32.34	-28	-140.86	4.643	29.09	-8	-42.38	5.195	26.14
-47	-223.65	4.062	32.17	-27	-136.20	4.672	28.93	-7	-37.18	5.221	26.00
-46	-219.57	4.094	32.00	-26	-131.52	4.701	28.78	-6	-31.94	5.247	25.86
-45	-215.46	4.126	31.83	-25	-126.80	4.730	28.63	-5	-26.68	5.273	25.72
-44	-211.32	4.157	31.67	-24	-122.06	4.758	28.48	-4	-21.40	5.298	25.58
-43	-207.15	4.189	31.50	-23	-117.29	4.787	28.32	-3	-16.09	5.324	25.45
-42	-202.94	4.220	31.33	-22	-112.49	4.815	28.17	-2	-10.75	5.349	25.31
-41	-198.71	4.252	31.17	-21	-107.66	4.843	28.02	-1	-5.39	5.374	25.17
-40	-194.44	4.283	31.00	-20	-102.80	4.871	27.87	0	0.00	5.400	25.04

TABLE 1i6. (Continued)

Type S thermocouples—thermoelectric voltages, E(T), Seebeck coefficients, S(T), and first derivative of the Seebeck coefficients, dS/dT, reference junctions at 0 °C—Continued

T °C	E μV	S μV/°C	dS/dT nV/°C²	T °C	E μV	S μV/°C	dS/dT nV/°C²	T °C	E μV	S μV/°C	dS/dT nV/°C²
0	0.00	5.400	25.04	60	364.55	6.683	18.09	120	794.81	7.609	13.07
1	5.41	5.425	24.91	61	371.24	6.701	18.00	121	802.43	7.622	13.00
2	10.85	5.449	24.77	62	377.95	6.719	17.90	122	810.05	7.635	12.93
3	16.31	5.474	24.64	63	384.68	6.737	17.80	123	817.70	7.648	12.86
4	21.80	5.499	24.51	64	391.42	6.754	17.70	124	825.35	7.661	12.79
5	27.31	5.523	24.37	65	398.19	6.772	17.61	125	833.02	7.674	12.72
6	32.84	5.547	24.24	66	404.97	6.789	17.51	126	840.70	7.686	12.66
7	38.40	5.572	24.11	67	411.77	6.807	17.42	127	848.39	7.699	12.59
8	43.99	5.596	23.98	68	418.58	6.824	17.32	128	856.10	7.712	12.52
9	49.59	5.620	23.85	69	425.41	6.842	17.23	129	863.81	7.724	12.45
10	55.23	5.643	23.73	70	432.26	6.859	17.14	130	871.54	7.736	12.39
11	60.88	5.667	23.60	71	439.13	6.876	17.04	131	879.29	7.749	12.32
12	66.56	5.691	23.47	72	446.02	6.893	16.95	132	887.04	7.761	12.25
13	72.26	5.714	23.35	73	452.92	6.910	16.86	133	894.81	7.773	12.19
14	77.99	5.737	23.22	74	459.84	6.927	16.77	134	902.59	7.786	12.12
15	83.74	5.760	23.09	75	466.77	6.943	16.68	135	910.38	7.798	12.06
16	89.51	5.783	22.97	76	473.72	6.960	16.59	136	918.18	7.810	12.00
17	95.30	5.806	22.85	77	480.69	6.976	16.50	137	926.00	7.822	11.93
18	101.12	5.829	22.72	78	487.67	6.993	16.41	138	933.83	7.833	11.87
19	106.96	5.852	22.60	79	494.68	7.009	16.32	139	941.67	7.845	11.80
20	112.82	5.874	22.48	80	501.69	7.026	16.23	140	949.52	7.857	11.74
21	118.71	5.897	22.36	81	508.73	7.042	16.14	141	957.38	7.869	11.68
22	124.62	5.919	22.24	82	515.78	7.058	16.05	142	965.26	7.880	11.62
23	130.55	5.941	22.12	83	522.84	7.074	15.97	143	973.14	7.892	11.56
24	136.50	5.963	22.00	84	529.92	7.090	15.88	144	981.04	7.904	11.49
25	142.47	5.985	21.88	85	537.02	7.106	15.80	145	988.95	7.915	11.43
26	148.47	6.007	21.76	86	544.14	7.121	15.71	146	996.87	7.926	11.37
27	154.49	6.029	21.64	87	551.27	7.137	15.62	147	1004.80	7.938	11.31
28	160.53	6.050	21.53	88	558.41	7.153	15.54	148	1012.75	7.949	11.25
29	166.59	6.072	21.41	89	565.57	7.168	15.46	149	1020.70	7.960	11.19
30	172.67	6.093	21.29	90	572.75	7.184	15.37	150	1028.67	7.971	11.13
31	178.78	6.114	21.18	91	579.94	7.199	15.29	151	1036.64	7.983	11.07
32	184.90	6.135	21.06	92	587.14	7.214	15.21	152	1044.63	7.994	11.02
33	191.05	6.157	20.95	93	594.37	7.229	15.12	153	1052.63	8.005	10.96
34	197.21	6.177	20.84	94	601.60	7.244	15.04	154	1060.64	8.015	10.90
35	203.40	6.198	20.72	95	608.85	7.259	14.96	155	1068.66	8.026	10.84
36	209.61	6.219	20.61	96	616.12	7.274	14.88	156	1076.69	8.037	10.78
37	215.84	6.239	20.50	97	623.40	7.289	14.80	157	1084.73	8.048	10.73
38	222.09	6.260	20.39	98	630.70	7.304	14.72	158	1092.79	8.059	10.67
39	228.36	6.280	20.28	99	638.01	7.319	14.64	159	1100.85	8.069	10.61
40	234.65	6.300	20.17	100	645.34	7.333	14.56	160	1108.93	8.080	10.56
41	240.96	6.321	20.06	101	652.68	7.348	14.48	161	1117.01	8.090	10.50
42	247.29	6.341	19.95	102	660.03	7.362	14.40	162	1125.11	8.101	10.45
43	253.64	6.360	19.84	103	667.40	7.377	14.33	163	1133.21	8.111	10.39
44	260.01	6.380	19.74	104	674.79	7.391	14.25	164	1141.33	8.122	10.34
45	266.40	6.400	19.63	105	682.18	7.405	14.17	165	1149.46	8.132	10.28
46	272.81	6.419	19.52	106	689.60	7.419	14.10	166	1157.59	8.142	10.23
47	279.24	6.439	19.42	107	697.02	7.433	14.02	167	1165.74	8.152	10.18
48	285.69	6.458	19.31	108	704.46	7.447	13.94	168	1173.90	8.163	10.12
49	292.16	6.478	19.21	109	711.92	7.461	13.87	169	1182.07	8.173	10.07
50	298.64	6.497	19.10	110	719.38	7.475	13.79	170	1190.24	8.183	10.02
51	305.15	6.516	19.00	111	726.87	7.489	13.72	171	1198.43	8.193	9.97
52	311.67	6.535	18.90	112	734.36	7.502	13.65	172	1206.63	8.203	9.91
53	318.22	6.554	18.80	113	741.87	7.516	13.57	173	1214.84	8.213	9.86
54	324.78	6.572	18.69	114	749.39	7.530	13.50	174	1223.05	8.222	9.81
55	331.36	6.591	18.59	115	756.93	7.543	13.43	175	1231.28	8.232	9.76
56	337.96	6.610	18.49	116	764.48	7.556	13.36	176	1239.52	8.242	9.71
57	344.58	6.628	18.39	117	772.04	7.570	13.28	177	1247.77	8.252	9.66
58	351.22	6.646	18.29	118	779.62	7.583	13.21	178	1256.02	8.261	9.61
59	357.88	6.665	18.19	119	787.21	7.596	13.14	179	1264.29	8.271	9.56
60	364.55	6.683	18.09	120	794.81	7.609	13.07	180	1272.56	8.280	9.51

TABLE 1i6. (*Continued*)

Type S thermocouples—thermoelectric voltages, E(T), Seebeck coefficients, S(T), and first derivative of the Seebeck coefficients, dS/dT, reference junctions at 0 °C—Continued

T °C	E μV	S μV/°C	dS/dT nV/°C²	T °C	E μV	S μV/°C	dS/dT nV/°C²	T °C	E μV	S μV/°C	dS/dT nV/°C²
180	1272.56	8.280	9.51	240	1784.87	8.772	7.02	300	2322.68	9.138	5.31
181	1280.85	8.290	9.46	241	1793.64	8.779	6.99	301	2331.82	9.144	5.29
182	1289.14	8.299	9.41	242	1802.42	8.786	6.95	302	2340.97	9.149	5.26
183	1297.45	8.309	9.36	243	1811.21	8.793	6.92	303	2350.12	9.154	5.24
184	1305.76	8.318	9.31	244	1820.01	8.799	6.89	304	2359.28	9.159	5.22
185	1314.08	8.327	9.26	245	1828.81	8.806	6.85	305	2368.44	9.165	5.19
186	1322.41	8.336	9.22	246	1837.62	8.813	6.82	306	2377.61	9.170	5.17
187	1330.76	8.346	9.17	247	1846.44	8.820	6.79	307	2386.78	9.175	5.15
188	1339.11	8.355	9.12	248	1855.26	8.827	6.76	308	2395.96	9.180	5.13
189	1347.47	8.364	9.07	249	1864.09	8.834	6.72	309	2405.14	9.185	5.10
190	1355.83	8.373	9.03	250	1872.93	8.840	6.69	310	2414.33	9.190	5.08
191	1364.21	8.382	8.98	251	1881.77	8.847	6.66	311	2423.52	9.195	5.06
192	1372.60	8.391	8.94	252	1890.62	8.854	6.63	312	2432.72	9.200	5.04
193	1380.99	8.400	8.89	253	1899.48	8.860	6.59	313	2441.92	9.206	5.02
194	1389.40	8.409	8.84	254	1908.34	8.867	6.56	314	2451.13	9.211	5.00
195	1397.81	8.418	8.80	255	1917.21	8.873	6.53	315	2460.34	9.216	4.97
196	1406.23	8.426	8.75	256	1926.09	8.880	6.50	316	2469.56	9.220	4.95
197	1414.66	8.435	8.71	257	1934.97	8.886	6.47	317	2478.78	9.225	4.93
198	1423.10	8.444	8.66	258	1943.86	8.893	6.44	318	2488.01	9.230	4.91
199	1431.55	8.452	8.62	259	1952.76	8.899	6.41	319	2497.25	9.235	4.89
200	1440.01	8.461	8.58	260	1961.66	8.906	6.38	320	2506.48	9.240	4.87
201	1448.47	8.470	8.53	261	1970.57	8.912	6.35	321	2515.73	9.245	4.85
202	1456.95	8.478	8.49	262	1979.49	8.918	6.32	322	2524.97	9.250	4.83
203	1465.43	8.487	8.45	263	1988.41	8.925	6.29	323	2534.23	9.255	4.81
204	1473.92	8.495	8.40	264	1997.33	8.931	6.26	324	2543.48	9.259	4.79
205	1482.42	8.503	8.36	265	2006.27	8.937	6.23	325	2552.74	9.264	4.77
206	1490.93	8.512	8.32	266	2015.21	8.943	6.20	326	2562.01	9.269	4.75
207	1499.44	8.520	8.28	267	2024.15	8.949	6.17	327	2571.28	9.274	4.73
208	1507.97	8.528	8.23	268	2033.11	8.956	6.14	328	2580.56	9.278	4.71
209	1516.50	8.536	8.19	269	2042.07	8.962	6.11	329	2589.84	9.283	4.69
210	1525.04	8.545	8.15	270	2051.03	8.968	6.09	330	2599.12	9.288	4.67
211	1533.59	8.553	8.11	271	2060.00	8.974	6.06	331	2608.41	9.292	4.65
212	1542.14	8.561	8.07	272	2068.98	8.980	6.03	332	2617.71	9.297	4.63
213	1550.71	8.569	8.03	273	2077.96	8.986	6.00	333	2627.01	9.302	4.61
214	1559.28	8.577	7.99	274	2086.95	8.992	5.97	334	2636.31	9.306	4.59
215	1567.86	8.585	7.95	275	2095.95	8.998	5.95	335	2645.62	9.311	4.58
216	1576.45	8.593	7.91	276	2104.95	9.004	5.92	336	2654.93	9.315	4.56
217	1585.05	8.601	7.87	277	2113.95	9.010	5.89	337	2664.25	9.320	4.54
218	1593.65	8.609	7.83	278	2122.97	9.016	5.86	338	2673.58	9.325	4.52
219	1602.27	8.616	7.79	279	2131.98	9.021	5.84	339	2682.90	9.329	4.50
220	1610.89	8.624	7.75	280	2141.01	9.027	5.81	340	2692.23	9.334	4.48
221	1619.51	8.632	7.71	281	2150.04	9.033	5.78	341	2701.57	9.338	4.47
222	1628.15	8.640	7.67	282	2159.08	9.039	5.76	342	2710.91	9.343	4.45
223	1636.79	8.647	7.64	283	2168.12	9.045	5.73	343	2720.25	9.347	4.43
224	1645.44	8.655	7.60	284	2177.16	9.050	5.71	344	2729.60	9.351	4.41
225	1654.10	8.662	7.56	285	2186.22	9.056	5.68	345	2738.96	9.356	4.39
226	1662.77	8.670	7.52	286	2195.28	9.062	5.65	346	2748.31	9.360	4.38
227	1671.44	8.677	7.49	287	2204.34	9.067	5.63	347	2757.68	9.365	4.36
228	1680.12	8.685	7.45	288	2213.41	9.073	5.60	348	2767.04	9.369	4.34
229	1688.81	8.692	7.41	289	2222.49	9.079	5.58	349	2776.41	9.373	4.33
230	1697.51	8.700	7.38	290	2231.57	9.084	5.55	350	2785.79	9.378	4.31
231	1706.21	8.707	7.34	291	2240.66	9.090	5.53	351	2795.17	9.382	4.29
232	1714.92	8.714	7.30	292	2249.75	9.095	5.50	352	2804.55	9.386	4.28
233	1723.64	8.722	7.27	293	2258.85	9.101	5.48	353	2813.94	9.390	4.26
234	1732.37	8.729	7.23	294	2267.95	9.106	5.45	354	2823.33	9.395	4.24
235	1741.10	8.736	7.20	295	2277.06	9.112	5.43	355	2832.73	9.399	4.23
236	1749.84	8.743	7.16	296	2286.17	9.117	5.41	356	2842.13	9.403	4.21
237	1758.58	8.750	7.13	297	2295.29	9.122	5.38	357	2851.54	9.407	4.19
238	1767.34	8.758	7.09	298	2304.42	9.128	5.36	358	2860.95	9.411	4.18
239	1776.10	8.765	7.06	299	2313.55	9.133	5.33	359	2870.36	9.416	4.16
240	1784.87	8.772	7.02	300	2322.68	9.138	5.31	360	2879.78	9.420	4.15

TABLE 1i6. (Continued)

Type S thermocouples—thermoelectric voltages, E(T), *Seebeck coefficients,* S(T), *and first derivative of the Seebeck coefficients,* dS/dT, *reference junctions at* 0 °C—*Continued*

T °C	E μV	S μV/°C	dS/dT nV/°C²	T °C	E μV	S μV/°C	dS/dT nV/°C²	T °C	E μV	S μV/°C	dS/dT nV/°C²
360	2879.78	9.420	4.15	420	3451.92	9.644	3.39	480	4036.35	9.833	2.96
361	2889.20	9.424	4.13	421	3461.56	9.647	3.38	481	4046.18	9.836	2.96
362	2898.63	9.428	4.11	422	3471.21	9.651	3.37	482	4056.02	9.839	2.95
363	2908.06	9.432	4.10	423	3480.87	9.654	3.36	483	4065.86	9.842	2.95
364	2917.49	9.436	4.08	424	3490.52	9.657	3.35	484	4075.70	9.845	2.95
365	2926.93	9.440	4.07	425	3500.18	9.661	3.34	485	4085.55	9.848	2.94
366	2936.37	9.444	4.05	426	3509.84	9.664	3.33	486	4095.40	9.851	2.94
367	2945.82	9.448	4.04	427	3519.51	9.667	3.32	487	4105.25	9.853	2.94
368	2955.27	9.452	4.02	428	3529.18	9.671	3.31	488	4115.11	9.856	2.93
369	2964.72	9.456	4.01	429	3538.85	9.674	3.30	489	4124.96	9.859	2.93
370	2974.18	9.460	3.99	430	3548.53	9.677	3.29	490	4134.83	9.862	2.93
371	2983.64	9.464	3.98	431	3558.21	9.681	3.28	491	4144.69	9.865	2.92
372	2993.11	9.468	3.97	432	3567.89	9.684	3.28	492	4154.56	9.868	2.92
373	3002.58	9.472	3.95	433	3577.57	9.687	3.27	493	4164.43	9.871	2.92
374	3012.05	9.476	3.94	434	3587.26	9.690	3.26	494	4174.30	9.874	2.91
375	3021.53	9.480	3.92	435	3596.95	9.694	3.25	495	4184.17	9.877	2.91
376	3031.02	9.484	3.91	436	3606.65	9.697	3.24	496	4194.05	9.880	2.91
377	3040.50	9.488	3.89	437	3616.35	9.700	3.23	497	4203.93	9.883	2.90
378	3049.99	9.492	3.88	438	3626.05	9.703	3.22	498	4213.82	9.886	2.90
379	3059.49	9.496	3.87	439	3635.76	9.707	3.22	499	4223.70	9.888	2.90
380	3068.98	9.500	3.85	440	3645.46	9.710	3.21	500	4233.59	9.891	2.90
381	3078.48	9.504	3.84	441	3655.18	9.713	3.20	501	4243.49	9.894	2.90
382	3087.99	9.507	3.83	442	3664.89	9.716	3.19	502	4253.38	9.897	2.89
383	3097.50	9.511	3.81	443	3674.61	9.719	3.18	503	4263.28	9.900	2.89
384	3107.01	9.515	3.80	444	3684.33	9.723	3.18	504	4273.18	9.903	2.89
385	3116.53	9.519	3.79	445	3694.05	9.726	3.17	505	4283.09	9.906	2.89
386	3126.05	9.523	3.77	446	3703.78	9.729	3.16	506	4292.99	9.909	2.89
387	3135.57	9.526	3.76	447	3713.51	9.732	3.15	507	4302.90	9.912	2.88
388	3145.10	9.530	3.75	448	3723.24	9.735	3.15	508	4312.82	9.914	2.88
389	3154.64	9.534	3.73	449	3732.98	9.738	3.14	509	4322.73	9.917	2.88
390	3164.17	9.538	3.72	450	3742.72	9.742	3.13	510	4332.65	9.920	2.88
391	3173.71	9.541	3.71	451	3752.46	9.745	3.13	511	4342.57	9.923	2.88
392	3183.25	9.545	3.70	452	3762.21	9.748	3.12	512	4352.50	9.926	2.88
393	3192.80	9.549	3.68	453	3771.96	9.751	3.11	513	4362.43	9.929	2.88
394	3202.35	9.552	3.67	454	3781.71	9.754	3.11	514	4372.36	9.932	2.88
395	3211.90	9.556	3.66	455	3791.47	9.757	3.10	515	4382.29	9.935	2.87
396	3221.46	9.560	3.65	456	3801.23	9.760	3.09	516	4392.22	9.937	2.87
397	3231.02	9.563	3.64	457	3810.99	9.763	3.09	517	4402.16	9.940	2.87
398	3240.59	9.567	3.62	458	3820.75	9.766	3.08	518	4412.11	9.943	2.87
399	3250.16	9.571	3.61	459	3830.52	9.769	3.07	519	4422.05	9.946	2.87
400	3259.73	9.574	3.60	460	3840.29	9.773	3.07	520	4432.00	9.949	2.87
401	3269.31	9.578	3.59	461	3850.07	9.776	3.06	521	4441.95	9.952	2.87
402	3278.89	9.581	3.58	462	3859.84	9.779	3.06	522	4451.90	9.955	2.87
403	3288.47	9.585	3.57	463	3869.62	9.782	3.05	523	4461.86	9.958	2.87
404	3298.06	9.588	3.55	464	3879.41	9.785	3.04	524	4471.82	9.960	2.87
405	3307.65	9.592	3.54	465	3889.19	9.788	3.04	525	4481.78	9.963	2.87
406	3317.24	9.596	3.53	466	3898.98	9.791	3.03	526	4491.74	9.966	2.87
407	3326.84	9.599	3.52	467	3908.78	9.794	3.03	527	4501.71	9.969	2.87
408	3336.44	9.603	3.51	468	3918.57	9.797	3.02	528	4511.68	9.972	2.87
409	3346.04	9.606	3.50	469	3928.37	9.800	3.02	529	4521.65	9.975	2.88
410	3355.65	9.610	3.49	470	3938.17	9.803	3.01	530	4531.63	9.978	2.88
411	3365.26	9.613	3.48	471	3947.98	9.806	3.01	531	4541.61	9.981	2.88
412	3374.88	9.617	3.47	472	3957.78	9.809	3.00	532	4551.59	9.983	2.88
413	3384.49	9.620	3.46	473	3967.59	9.812	3.00	533	4561.58	9.986	2.88
414	3394.12	9.623	3.45	474	3977.41	9.815	2.99	534	4571.57	9.989	2.88
415	3403.74	9.627	3.44	475	3987.22	9.818	2.99	535	4581.56	9.992	2.88
416	3413.37	9.630	3.43	476	3997.04	9.821	2.98	536	4591.55	9.995	2.88
417	3423.00	9.634	3.42	477	4006.86	9.824	2.98	537	4601.55	9.998	2.89
418	3432.64	9.637	3.41	478	4016.69	9.827	2.97	538	4611.55	10.001	2.89
419	3442.28	9.641	3.40	479	4026.52	9.830	2.97	539	4621.55	10.004	2.89
420	3451.92	9.644	3.39	480	4036.35	9.833	2.96	540	4631.55	10.007	2.89

TABLE 1i6. (Continued)

Type S thermocouples—*thermoelectric voltages,* E(T), *Seebeck coefficients,* S(T), *and first derivative of the Seebeck coefficients,* dS/dT, *reference junctions at 0 °C*—*Continued*

T °C	E μV	S μV/°C	dS/dT nV/°C²	T °C	E μV	S μV/°C	dS/dT nV/°C²	T °C	E μV	S μV/°C	dS/dT nV/°C²
540	4631.55	10.007	2.89	600	5237.30	10.189	3.27	660	5855.27	10.409	3.29
541	4641.56	10.009	2.89	601	5247.49	10.192	3.28	661	5865.68	10.413	3.29
542	4651.57	10.012	2.90	602	5257.68	10.195	3.29	662	5876.10	10.416	3.29
543	4661.59	10.015	2.90	603	5267.88	10.199	3.30	663	5886.52	10.419	3.29
544	4671.60	10.018	2.90	604	5278.08	10.202	3.31	664	5896.94	10.423	3.29
545	4681.62	10.021	2.90	605	5288.28	10.205	3.32	665	5907.36	10.426	3.29
546	4691.64	10.024	2.91	606	5298.49	10.208	3.33	666	5917.79	10.429	3.29
547	4701.67	10.027	2.91	607	5308.70	10.212	3.34	667	5928.22	10.433	3.29
548	4711.70	10.030	2.91	608	5318.91	10.215	3.36	668	5938.65	10.436	3.29
549	4721.73	10.033	2.92	609	5329.13	10.219	3.37	669	5949.09	10.439	3.29
550	4731.76	10.036	2.92	610	5339.35	10.222	3.38	670	5959.53	10.442	3.29
551	4741.80	10.039	2.92	611	5349.57	10.225	3.39	671	5969.98	10.446	3.29
552	4751.84	10.041	2.93	612	5359.80	10.229	3.41	672	5980.42	10.449	3.29
553	4761.88	10.044	2.93	613	5370.03	10.232	3.42	673	5990.87	10.452	3.29
554	4771.93	10.047	2.93	614	5380.27	10.236	3.43	674	6001.33	10.456	3.29
555	4781.98	10.050	2.94	615	5390.50	10.239	3.45	675	6011.78	10.459	3.29
556	4792.03	10.053	2.94	616	5400.74	10.242	3.46	676	6022.25	10.462	3.29
557	4802.08	10.056	2.95	617	5410.99	10.246	3.47	677	6032.71	10.465	3.29
558	4812.14	10.059	2.95	618	5421.23	10.249	3.49	678	6043.18	10.469	3.29
559	4822.20	10.062	2.96	619	5431.49	10.253	3.50	679	6053.65	10.472	3.29
560	4832.27	10.065	2.96	620	5441.74	10.256	3.52	680	6064.12	10.475	3.29
561	4842.33	10.068	2.96	621	5452.00	10.260	3.53	681	6074.60	10.479	3.29
562	4852.40	10.071	2.97	622	5462.26	10.263	3.54	682	6085.08	10.482	3.29
563	4862.47	10.074	2.97	623	5472.53	10.267	3.56	683	6095.56	10.485	3.29
564	4872.55	10.077	2.98	624	5482.79	10.271	3.57	684	6106.05	10.488	3.29
565	4882.63	10.080	2.99	625	5493.07	10.274	3.59	685	6116.54	10.492	3.29
566	4892.71	10.083	2.99	626	5503.34	10.278	3.61	686	6127.03	10.495	3.29
567	4902.79	10.086	3.00	627	5513.62	10.281	3.62	687	6137.53	10.498	3.29
568	4912.88	10.089	3.00	628	5523.91	10.285	3.64	688	6148.03	10.502	3.29
569	4922.97	10.092	3.01	629	5534.19	10.289	3.65	689	6158.53	10.505	3.29
570	4933.06	10.095	3.01	630	5544.48	10.292	3.67	690	6169.04	10.508	3.29
571	4943.16	10.098	3.02	631	5554.78	10.314	3.29	691	6179.55	10.511	3.29
572	4953.26	10.101	3.03	632	5565.10	10.317	3.29	692	6190.06	10.515	3.29
573	4963.36	10.104	3.03	633	5575.42	10.321	3.29	693	6200.58	10.518	3.29
574	4973.47	10.107	3.04	634	5585.74	10.324	3.29	694	6211.10	10.521	3.29
575	4983.58	10.110	3.05	635	5596.06	10.327	3.29	695	6221.62	10.525	3.29
576	4993.69	10.113	3.05	636	5606.39	10.330	3.29	696	6232.15	10.528	3.29
577	5003.80	10.116	3.06	637	5616.72	10.334	3.29	697	6242.68	10.531	3.29
578	5013.92	10.119	3.07	638	5627.06	10.337	3.29	698	6253.21	10.535	3.29
579	5024.04	10.122	3.07	639	5637.40	10.340	3.29	699	6263.74	10.538	3.29
580	5034.16	10.125	3.08	640	5647.74	10.344	3.29	700	6274.28	10.541	3.29
581	5044.29	10.128	3.09	641	5658.09	10.347	3.29	701	6284.83	10.544	3.29
582	5054.42	10.131	3.10	642	5668.44	10.350	3.29	702	6295.37	10.548	3.29
583	5064.55	10.135	3.11	643	5678.79	10.354	3.29	703	6305.92	10.551	3.29
584	5074.69	10.138	3.11	644	5689.14	10.357	3.29	704	6316.47	10.554	3.29
585	5084.83	10.141	3.12	645	5699.50	10.360	3.29	705	6327.03	10.558	3.29
586	5094.97	10.144	3.13	646	5709.86	10.363	3.29	706	6337.59	10.561	3.29
587	5105.12	10.147	3.14	647	5720.23	10.367	3.29	707	6348.15	10.564	3.29
588	5115.27	10.150	3.15	648	5730.60	10.370	3.29	708	6358.72	10.567	3.29
589	5125.42	10.153	3.16	649	5740.97	10.373	3.29	709	6369.29	10.571	3.29
590	5135.57	10.157	3.17	650	5751.34	10.377	3.29	710	6379.86	10.574	3.29
591	5145.73	10.160	3.17	651	5761.72	10.380	3.29	711	6390.43	10.577	3.29
592	5155.89	10.163	3.18	652	5772.10	10.383	3.29	712	6401.01	10.581	3.29
593	5166.06	10.166	3.19	653	5782.49	10.386	3.29	713	6411.60	10.584	3.29
594	5176.22	10.169	3.20	654	5792.87	10.390	3.29	714	6422.18	10.587	3.29
595	5186.40	10.172	3.21	655	5803.27	10.393	3.29	715	6432.77	10.590	3.29
596	5196.57	10.176	3.22	656	5813.66	10.396	3.29	716	6443.36	10.594	3.29
597	5206.75	10.179	3.23	657	5824.06	10.400	3.29	717	6453.96	10.597	3.29
598	5216.93	10.182	3.24	658	5834.46	10.403	3.29	718	6464.56	10.600	3.29
599	5227.11	10.185	3.25	659	5844.86	10.406	3.29	719	6475.16	10.604	3.29
600	5237.30	10.189	3.27	660	5855.27	10.409	3.29	720	6485.76	10.607	3.29

TABLE 1i6. (Continued)

Type S thermocouples—*thermoelectric voltages, E(T), Seebeck coefficients, S(T), and first derivative of the Seebeck coefficients, dS/dT, reference junctions at 0 °C*—*Continued*

T °C	E μV	S μV/°C	dS/dT nV/°C²	T °C	E μV	S μV/°C	dS/dT nV/°C²	T °C	E μV	S μV/°C	dS/dT nV/°C²
720	6485.76	10.607	3.29	780	7128.10	10.804	3.29	840	7782.29	11.002	3.29
721	6496.37	10.610	3.29	781	7138.91	10.808	3.29	841	7793.29	11.005	3.29
722	6506.98	10.613	3.29	782	7149.72	10.811	3.29	842	7804.30	11.008	3.29
723	6517.60	10.617	3.29	783	7160.53	10.814	3.29	843	7815.31	11.012	3.29
724	6528.22	10.620	3.29	784	7171.35	10.818	3.29	844	7826.32	11.015	3.29
725	6538.84	10.623	3.29	785	7182.17	10.821	3.29	845	7837.34	11.018	3.29
726	6549.46	10.627	3.29	786	7192.99	10.824	3.29	846	7848.36	11.022	3.29
727	6560.09	10.630	3.29	787	7203.81	10.827	3.29	847	7859.38	11.025	3.29
728	6570.72	10.633	3.29	788	7214.64	10.831	3.29	848	7870.41	11.028	3.29
729	6581.36	10.637	3.29	789	7225.47	10.834	3.29	849	7881.44	11.031	3.29
730	6592.00	10.640	3.29	790	7236.31	10.837	3.29	850	7892.47	11.035	3.29
731	6602.64	10.643	3.29	791	7247.15	10.841	3.29	851	7903.51	11.038	3.29
732	6613.28	10.646	3.29	792	7257.99	10.844	3.29	852	7914.55	11.041	3.29
733	6623.93	10.650	3.29	793	7268.84	10.847	3.29	853	7925.59	11.045	3.29
734	6634.58	10.653	3.29	794	7279.69	10.850	3.29	854	7936.64	11.048	3.29
735	6645.24	10.656	3.29	795	7290.54	10.854	3.29	855	7947.68	11.051	3.29
736	6655.90	10.660	3.29	796	7301.39	10.857	3.29	856	7958.74	11.054	3.29
737	6666.56	10.663	3.29	797	7312.25	10.860	3.29	857	7969.79	11.058	3.29
738	6677.22	10.666	3.29	798	7323.11	10.864	3.29	858	7980.85	11.061	3.29
739	6687.89	10.669	3.29	799	7333.98	10.867	3.29	859	7991.92	11.064	3.29
740	6698.56	10.673	3.29	800	7344.85	10.870	3.29	860	8002.98	11.068	3.29
741	6709.23	10.676	3.29	801	7355.72	10.873	3.29	861	8014.05	11.071	3.29
742	6719.91	10.679	3.29	802	7366.59	10.877	3.29	862	8025.12	11.074	3.29
743	6730.59	10.683	3.29	803	7377.47	10.880	3.29	863	8036.20	11.077	3.29
744	6741.28	10.686	3.29	804	7388.35	10.883	3.29	864	8047.28	11.081	3.29
745	6751.97	10.689	3.29	805	7399.24	10.887	3.29	865	8058.36	11.084	3.29
746	6762.66	10.692	3.29	806	7410.13	10.890	3.29	866	8069.45	11.087	3.29
747	6773.35	10.696	3.29	807	7421.02	10.893	3.29	867	8080.54	11.091	3.29
748	6784.05	10.699	3.29	808	7431.91	10.897	3.29	868	8091.63	11.094	3.29
749	6794.75	10.702	3.29	809	7442.81	10.900	3.29	869	8102.72	11.097	3.29
750	6805.45	10.706	3.29	810	7453.71	10.903	3.29	870	8113.82	11.101	3.29
751	6816.16	10.709	3.29	811	7464.62	10.906	3.29	871	8124.92	11.104	3.29
752	6826.87	10.712	3.29	812	7475.53	10.910	3.29	872	8136.03	11.107	3.29
753	6837.58	10.716	3.29	813	7486.44	10.913	3.29	873	8147.14	11.110	3.29
754	6848.30	10.719	3.29	814	7497.35	10.916	3.29	874	8158.25	11.114	3.29
755	6859.02	10.722	3.29	815	7508.27	10.920	3.29	875	8169.37	11.117	3.29
756	6869.75	10.725	3.29	816	7519.19	10.923	3.29	876	8180.49	11.120	3.29
757	6880.47	10.729	3.29	817	7530.12	10.926	3.29	877	8191.61	11.124	3.29
758	6891.20	10.732	3.29	818	7541.04	10.929	3.29	878	8202.73	11.127	3.29
759	6901.94	10.735	3.29	819	7551.98	10.933	3.29	879	8213.86	11.130	3.29
760	6912.67	10.739	3.29	820	7562.91	10.936	3.29	880	8224.99	11.133	3.29
761	6923.41	10.742	3.29	821	7573.85	10.939	3.29	881	8236.13	11.137	3.29
762	6934.16	10.745	3.29	822	7584.79	10.943	3.29	882	8247.27	11.140	3.29
763	6944.90	10.748	3.29	823	7595.73	10.946	3.29	883	8258.41	11.143	3.29
764	6955.65	10.752	3.29	824	7606.68	10.949	3.29	884	8269.55	11.147	3.29
765	6966.41	10.755	3.29	825	7617.63	10.952	3.29	885	8280.70	11.150	3.29
766	6977.16	10.758	3.29	826	7628.58	10.956	3.29	886	8291.85	11.153	3.29
767	6987.92	10.762	3.29	827	7639.54	10.959	3.29	887	8303.01	11.156	3.29
768	6998.69	10.765	3.29	828	7650.50	10.962	3.29	888	8314.17	11.160	3.29
769	7009.45	10.768	3.29	829	7661.47	10.966	3.29	889	8325.33	11.163	3.29
770	7020.22	10.771	3.29	830	7672.43	10.969	3.29	890	8336.49	11.166	3.29
771	7031.00	10.775	3.29	831	7683.40	10.972	3.29	891	8347.66	11.170	3.29
772	7041.77	10.778	3.29	832	7694.38	10.975	3.29	892	8358.83	11.173	3.29
773	7052.55	10.781	3.29	833	7705.36	10.979	3.29	893	8370.01	11.176	3.29
774	7063.34	10.785	3.29	834	7716.34	10.982	3.29	894	8381.18	11.180	3.29
775	7074.12	10.788	3.29	835	7727.32	10.985	3.29	895	8392.36	11.183	3.29
776	7084.91	10.791	3.29	836	7738.31	10.989	3.29	896	8403.55	11.186	3.29
777	7095.70	10.794	3.29	837	7749.30	10.992	3.29	897	8414.74	11.189	3.29
778	7106.50	10.798	3.29	838	7760.29	10.995	3.29	898	8425.93	11.193	3.29
779	7117.30	10.801	3.29	839	7771.29	10.999	3.29	899	8437.12	11.196	3.29
780	7128.10	10.804	3.29	840	7782.29	11.002	3.29	900	8448.32	11.199	3.29

TABLE 1i6. (*Continued*)

Type S thermocouples—*thermoelectric voltages*, $E(T)$, *Seebeck coefficients*, $S(T)$, *and first derivative of the Seebeck coefficients*, dS/dT, *reference junctions at* $0\ °C$—*Continued*

T °C	E μV	S μV/°C	dS/dT nV/°C²	T °C	E μV	S μV/°C	dS/dT nV/°C²	T °C	E μV	S μV/°C	dS/dT nV/°C²
900	8448.32	11.199	3.29	960	9126.20	11.397	3.29	1020	9815.92	11.594	3.29
901	8459.52	11.203	3.29	961	9137.60	11.400	3.29	1021	9827.52	11.597	3.29
902	8470.72	11.206	3.29	962	9149.00	11.403	3.29	1022	9839.12	11.601	3.29
903	8431.93	11.209	3.29	963	9160.40	11.407	3.29	1023	9850.72	11.604	3.29
904	8493.14	11.212	3.29	964	9171.81	11.410	3.29	1024	9862.33	11.607	3.29
905	8504.36	11.216	3.29	965	9183.22	11.413	3.29	1025	9873.94	11.611	3.29
906	8515.57	11.219	3.29	966	9194.64	11.416	3.29	1026	9885.55	11.614	3.29
907	8526.79	11.222	3.29	967	9206.06	11.420	3.29	1027	9897.16	11.617	3.29
908	8538.02	11.226	3.29	968	9217.48	11.423	3.29	1028	9908.78	11.620	3.29
909	8549.25	11.229	3.29	969	9228.90	11.426	3.29	1029	9920.40	11.624	3.29
910	8560.48	11.232	3.29	970	9240.33	11.430	3.29	1030	9932.03	11.627	3.29
911	8571.71	11.235	3.29	971	9251.76	11.433	3.29	1031	9943.66	11.630	3.29
912	8582.95	11.239	3.29	972	9263.20	11.436	3.29	1032	9955.29	11.634	3.29
913	8594.19	11.242	3.29	973	9274.63	11.439	3.29	1033	9966.93	11.637	3.29
914	8605.43	11.245	3.29	974	9286.07	11.443	3.29	1034	9978.56	11.640	3.29
915	8616.68	11.249	3.29	975	9297.52	11.446	3.29	1035	9990.21	11.644	3.29
916	8627.93	11.252	3.29	976	9308.97	11.449	3.29	1036	10001.85	11.647	3.29
917	8639.18	11.255	3.29	977	9320.42	11.453	3.29	1037	10013.50	11.650	3.29
918	8650.44	11.258	3.29	978	9331.87	11.456	3.29	1038	10025.15	11.653	3.29
919	8661.70	11.262	3.29	979	9343.33	11.459	3.29	1039	10036.81	11.657	3.29
920	8672.96	11.265	3.29	980	9354.79	11.463	3.29	1040	10048.47	11.660	3.29
921	8684.23	11.268	3.29	981	9366.25	11.466	3.29	1041	10060.13	11.663	3.29
922	8695.50	11.272	3.29	982	9377.72	11.469	3.29	1042	10071.79	11.667	3.29
923	8706.77	11.275	3.29	983	9389.19	11.472	3.29	1043	10083.46	11.670	3.29
924	8718.05	11.278	3.29	984	9400.67	11.476	3.29	1044	10095.13	11.673	3.29
925	8729.33	11.282	3.29	985	9412.14	11.479	3.29	1045	10106.81	11.676	3.29
926	8740.61	11.285	3.29	986	9423.62	11.482	3.29	1046	10118.48	11.680	3.29
927	8751.90	11.288	3.29	987	9435.11	11.486	3.29	1047	10130.17	11.683	3.29
928	8763.19	11.291	3.29	988	9446.60	11.489	3.29	1048	10141.85	11.686	3.29
929	8774.48	11.295	3.29	989	9458.09	11.492	3.29	1049	10153.54	11.690	3.29
930	8785.78	11.298	3.29	990	9469.58	11.495	3.29	1050	10165.23	11.693	3.29
931	8797.08	11.301	3.29	991	9481.08	11.499	3.29	1051	10176.92	11.696	3.29
932	8808.38	11.305	3.29	992	9492.58	11.502	3.29	1052	10188.62	11.699	3.29
933	8819.69	11.308	3.29	993	9504.08	11.505	3.29	1053	10200.32	11.703	3.29
934	8831.00	11.311	3.29	994	9515.59	11.509	3.29	1054	10212.03	11.706	3.29
935	8842.31	11.314	3.29	995	9527.10	11.512	3.29	1055	10223.73	11.709	3.29
936	8853.63	11.318	3.29	996	9538.61	11.515	3.29	1056	10235.45	11.713	3.29
937	8864.94	11.321	3.29	997	9550.13	11.518	3.29	1057	10247.16	11.716	3.29
938	8876.27	11.324	3.29	998	9561.65	11.522	3.29	1058	10258.88	11.719	3.29
939	8887.59	11.328	3.29	999	9573.17	11.525	3.29	1059	10270.60	11.722	3.29
940	8898.92	11.331	3.29	1000	9584.70	11.528	3.29	1060	10282.32	11.726	3.29
941	8910.25	11.334	3.29	1001	9596.23	11.532	3.29	1061	10294.05	11.729	3.29
942	8921.59	11.337	3.29	1002	9607.76	11.535	3.29	1062	10305.78	11.732	3.29
943	8932.93	11.341	3.29	1003	9619.30	11.538	3.29	1063	10317.51	11.736	3.29
944	8944.27	11.344	3.29	1004	9630.84	11.541	3.29	1064	10329.25	11.739	3.29
945	8955.62	11.347	3.29	1005	9642.38	11.545	3.29	1065	10340.99	11.742	2.72
946	8966.97	11.351	3.29	1006	9653.93	11.548	3.29	1066	10352.74	11.745	2.71
947	8978.32	11.354	3.29	1007	9665.48	11.551	3.29	1067	10364.48	11.747	2.70
948	8989.67	11.357	3.29	1008	9677.03	11.555	3.29	1068	10376.23	11.750	2.69
949	9001.03	11.361	3.29	1009	9688.59	11.558	3.29	1069	10387.98	11.753	2.68
950	9012.40	11.364	3.29	1010	9700.15	11.561	3.29	1070	10399.74	11.755	2.67
951	9023.76	11.367	3.29	1011	9711.71	11.565	3.29	1071	10411.49	11.758	2.66
952	9035.13	11.370	3.29	1012	9723.28	11.568	3.29	1072	10423.25	11.761	2.65
953	9046.50	11.374	3.29	1013	9734.85	11.571	3.29	1073	10435.01	11.763	2.64
954	9057.88	11.377	3.29	1014	9746.42	11.574	3.29	1074	10446.78	11.766	2.63
955	9069.26	11.380	3.29	1015	9757.99	11.578	3.29	1075	10458.55	11.769	2.62
956	9080.64	11.384	3.29	1016	9769.57	11.581	3.29	1076	10470.32	11.771	2.61
957	9092.02	11.387	3.29	1017	9781.16	11.584	3.29	1077	10482.09	11.774	2.61
958	9103.41	11.390	3.29	1018	9792.74	11.588	3.29	1078	10493.86	11.776	2.60
959	9114.80	11.393	3.29	1019	9804.33	11.591	3.29	1079	10505.64	11.779	2.59
960	9126.20	11.397	3.29	1020	9815.92	11.594	3.29	1080	10517.42	11.782	2.58

TABLE 1i6. (*Continued*)

Type S thermocouples—thermoelectric voltages, E(T), Seebeck coefficients, S(T), and first derivative of the Seebeck coefficients, dS/dT, reference junctions at 0 °C—Continued

T °C	E μV	S μV/°C	dS/dT nV/°C²	T °C	E μV	S μV/°C	dS/dT nV/°C²	T °C	E μV	S μV/°C	dS/dT nV/°C²
1080	10517.42	11.782	2.58	1140	11228.62	11.919	2.01	1200	11947.05	12.023	1.44
1081	10529.20	11.784	2.57	1141	11240.54	11.921	2.00	1201	11959.08	12.024	1.44
1082	10540.99	11.787	2.56	1142	11252.46	11.923	1.99	1202	11971.10	12.026	1.43
1083	10552.78	11.789	2.55	1143	11264.38	11.925	1.98	1203	11983.13	12.027	1.42
1084	10564.57	11.792	2.54	1144	11276.31	11.927	1.97	1204	11995.16	12.029	1.41
1085	10576.36	11.794	2.53	1145	11288.24	11.929	1.96	1205	12007.19	12.030	1.40
1086	10588.16	11.797	2.52	1146	11300.17	11.931	1.95	1206	12019.22	12.031	1.39
1087	10599.96	11.799	2.51	1147	11312.10	11.933	1.94	1207	12031.25	12.033	1.38
1088	10611.76	11.802	2.50	1148	11324.04	11.935	1.94	1208	12043.28	12.034	1.37
1089	10623.56	11.804	2.49	1149	11335.97	11.937	1.93	1209	12055.32	12.036	1.36
1090	10635.37	11.807	2.48	1150	11347.91	11.939	1.92	1210	12067.35	12.037	1.35
1091	10647.17	11.809	2.47	1151	11359.85	11.941	1.91	1211	12079.39	12.038	1.34
1092	10658.98	11.812	2.46	1152	11371.79	11.943	1.90	1212	12091.43	12.040	1.33
1093	10670.80	11.814	2.45	1153	11383.73	11.945	1.89	1213	12103.47	12.041	1.32
1094	10682.61	11.817	2.44	1154	11395.68	11.946	1.88	1214	12115.51	12.042	1.31
1095	10694.43	11.819	2.44	1155	11407.63	11.948	1.87	1215	12127.55	12.044	1.30
1096	10706.25	11.822	2.43	1156	11419.58	11.950	1.86	1216	12139.60	12.045	1.29
1097	10718.07	11.824	2.42	1157	11431.53	11.952	1.85	1217	12151.64	12.046	1.28
1098	10729.90	11.826	2.41	1158	11443.48	11.954	1.84	1218	12163.69	12.047	1.27
1099	10741.73	11.829	2.40	1159	11455.44	11.956	1.83	1219	12175.74	12.049	1.27
1100	10753.56	11.831	2.39	1160	11467.39	11.958	1.82	1220	12187.79	12.050	1.26
1101	10765.39	11.834	2.38	1161	11479.35	11.959	1.81	1221	12199.84	12.051	1.25
1102	10777.22	11.836	2.37	1162	11491.31	11.961	1.80	1222	12211.89	12.052	1.24
1103	10789.06	11.838	2.36	1163	11503.27	11.963	1.79	1223	12223.94	12.054	1.23
1104	10800.90	11.841	2.35	1164	11515.24	11.965	1.78	1224	12236.00	12.055	1.22
1105	10812.74	11.843	2.34	1165	11527.20	11.967	1.77	1225	12248.05	12.056	1.21
1106	10824.59	11.845	2.33	1166	11539.17	11.968	1.77	1226	12260.11	12.057	1.20
1107	10836.43	11.848	2.32	1167	11551.14	11.970	1.76	1227	12272.17	12.058	1.19
1108	10848.28	11.850	2.31	1168	11563.11	11.972	1.75	1228	12284.23	12.060	1.18
1109	10860.13	11.852	2.30	1169	11575.08	11.974	1.74	1229	12296.29	12.061	1.17
1110	10871.99	11.855	2.29	1170	11587.06	11.975	1.73	1230	12308.35	12.062	1.16
1111	10883.84	11.857	2.28	1171	11599.03	11.977	1.72	1231	12320.41	12.063	1.15
1112	10895.70	11.859	2.27	1172	11611.01	11.979	1.71	1232	12332.47	12.064	1.14
1113	10907.56	11.862	2.27	1173	11622.99	11.980	1.70	1233	12344.54	12.065	1.13
1114	10919.42	11.864	2.26	1174	11634.97	11.982	1.69	1234	12356.61	12.067	1.12
1115	10931.29	11.866	2.25	1175	11646.96	11.984	1.68	1235	12368.67	12.068	1.11
1116	10943.16	11.868	2.24	1176	11658.94	11.986	1.67	1236	12380.74	12.069	1.11
1117	10955.03	11.871	2.23	1177	11670.93	11.987	1.66	1237	12392.81	12.070	1.10
1118	10966.90	11.873	2.22	1178	11682.91	11.989	1.65	1238	12404.88	12.071	1.09
1119	10978.77	11.875	2.21	1179	11694.90	11.990	1.64	1239	12416.95	12.072	1.08
1120	10990.65	11.877	2.20	1180	11706.90	11.992	1.63	1240	12429.02	12.073	1.07
1121	11002.53	11.879	2.19	1181	11718.89	11.994	1.62	1241	12441.10	12.074	1.06
1122	11014.41	11.882	2.18	1182	11730.88	11.995	1.61	1242	12453.17	12.075	1.05
1123	11026.29	11.884	2.17	1183	11742.88	11.997	1.61	1243	12465.25	12.076	1.04
1124	11038.17	11.886	2.16	1184	11754.88	11.999	1.60	1244	12477.33	12.077	1.03
1125	11050.06	11.888	2.15	1185	11766.88	12.000	1.59	1245	12489.40	12.078	1.02
1126	11061.95	11.890	2.14	1186	11778.88	12.002	1.58	1246	12501.48	12.079	1.01
1127	11073.84	11.892	2.13	1187	11790.88	12.003	1.57	1247	12513.56	12.080	1.00
1128	11085.73	11.894	2.12	1188	11802.88	12.005	1.56	1248	12525.64	12.081	0.99
1129	11097.63	11.897	2.11	1189	11814.89	12.006	1.55	1249	12537.73	12.082	0.98
1130	11109.53	11.899	2.11	1190	11826.90	12.008	1.54	1250	12549.81	12.083	0.97
1131	11121.43	11.901	2.10	1191	11838.91	12.010	1.53	1251	12561.89	12.084	0.96
1132	11133.33	11.903	2.09	1192	11850.92	12.011	1.52	1252	12573.98	12.085	0.95
1133	11145.23	11.905	2.08	1193	11862.93	12.013	1.51	1253	12586.06	12.086	0.94
1134	11157.14	11.907	2.07	1194	11874.94	12.014	1.50	1254	12598.15	12.087	0.94
1135	11169.05	11.909	2.06	1195	11886.96	12.016	1.49	1255	12610.24	12.088	0.93
1136	11180.96	11.911	2.05	1196	11898.97	12.017	1.48	1256	12622.33	12.089	0.92
1137	11192.87	11.913	2.04	1197	11910.99	12.019	1.47	1257	12634.41	12.090	0.91
1138	11204.78	11.915	2.03	1198	11923.01	12.020	1.46	1258	12646.51	12.091	0.90
1139	11216.70	11.917	2.02	1199	11935.03	12.021	1.45	1259	12658.60	12.092	0.89
1140	11228.62	11.919	2.01	1200	11947.05	12.023	1.44	1260	12670.69	12.093	0.88

TABLE 1i6. (Continued)

Type S thermocouples—thermoelectric voltages, E(T), Seebeck coefficients, S(T), and first derivative of the Seebeck coefficients, dS/dT, reference junctions at 0 °C—Continued

T °C	E µV	S µV/°C	dS/dT nV/°C²	T °C	E µV	S µV/°C	dS/dT nV/°C²	T °C	E µV	S µV/°C	dS/dT nV/°C²
1260	12670.69	12.093	0.88	1320	13397.49	12.128	0.31	1380	14125.41	12.130	-0.25
1261	12682.78	12.093	0.87	1321	13409.62	12.129	0.30	1381	14137.54	12.130	-0.26
1262	12694.88	12.094	0.86	1322	13421.75	12.129	0.29	1382	14149.67	12.130	-0.27
1263	12706.97	12.095	0.85	1323	13433.87	12.129	0.28	1383	14161.80	12.129	-0.28
1264	12719.07	12.096	0.84	1324	13446.00	12.130	0.28	1384	14173.93	12.129	-0.29
1265	12731.16	12.097	0.83	1325	13458.13	12.130	0.27	1385	14186.06	12.129	-0.30
1266	12743.26	12.098	0.82	1326	13470.26	12.130	0.26	1386	14198.19	12.128	-0.31
1267	12755.36	12.099	0.81	1327	13482.39	12.130	0.25	1387	14210.32	12.128	-0.32
1268	12767.46	12.099	0.80	1328	13494.52	12.131	0.24	1388	14222.45	12.128	-0.33
1269	12779.56	12.100	0.79	1329	13506.66	12.131	0.23	1389	14234.57	12.127	-0.34
1270	12791.66	12.101	0.78	1330	13518.79	12.131	0.22	1390	14246.70	12.127	-0.35
1271	12803.76	12.102	0.78	1331	13530.92	12.131	0.21	1391	14258.83	12.127	-0.36
1272	12815.86	12.102	0.77	1332	13543.05	12.131	0.20	1392	14270.96	12.126	-0.37
1273	12827.96	12.103	0.76	1333	13555.18	12.132	0.19	1393	14283.08	12.126	-0.38
1274	12840.07	12.104	0.75	1334	13567.31	12.132	0.18	1394	14295.21	12.126	-0.39
1275	12852.17	12.105	0.74	1335	13579.44	12.132	0.17	1395	14307.33	12.125	-0.39
1276	12864.28	12.105	0.73	1336	13591.58	12.132	0.16	1396	14319.46	12.125	-0.40
1277	12876.38	12.106	0.72	1337	13603.71	12.132	0.15	1397	14331.58	12.124	-0.41
1278	12888.49	12.107	0.71	1338	13615.84	12.132	0.14	1398	14343.71	12.124	-0.42
1279	12900.60	12.108	0.70	1339	13627.97	12.133	0.13	1399	14355.83	12.124	-0.43
1280	12912.70	12.108	0.69	1340	13640.11	12.133	0.12	1400	14367.95	12.123	-0.44
1281	12924.81	12.109	0.68	1341	13652.24	12.133	0.11	1401	14380.08	12.123	-0.45
1282	12936.92	12.110	0.67	1342	13664.37	12.133	0.11	1402	14392.20	12.122	-0.46
1283	12949.03	12.110	0.66	1343	13676.50	12.133	0.10	1403	14404.32	12.122	-0.47
1284	12961.14	12.111	0.65	1344	13688.64	12.133	0.09	1404	14416.44	12.121	-0.48
1285	12973.25	12.112	0.64	1345	13700.77	12.133	0.08	1405	14428.56	12.121	-0.49
1286	12985.37	12.112	0.63	1346	13712.90	12.133	0.07	1406	14440.69	12.120	-0.50
1287	12997.48	12.113	0.62	1347	13725.04	12.133	0.06	1407	14452.81	12.120	-0.51
1288	13009.59	12.114	0.61	1348	13737.17	12.133	0.05	1408	14464.92	12.119	-0.52
1289	13021.71	12.114	0.61	1349	13749.30	12.133	0.04	1409	14477.04	12.119	-0.53
1290	13033.82	12.115	0.60	1350	13761.44	12.134	0.03	1410	14489.16	12.118	-0.54
1291	13045.94	12.115	0.59	1351	13773.57	12.134	0.02	1411	14501.28	12.118	-0.55
1292	13058.05	12.116	0.58	1352	13785.71	12.134	0.01	1412	14513.40	12.117	-0.56
1293	13070.17	12.116	0.57	1353	13797.84	12.134	0.00	1413	14525.52	12.117	-0.56
1294	13082.28	12.117	0.56	1354	13809.97	12.134	-0.01	1414	14537.63	12.116	-0.57
1295	13094.40	12.118	0.55	1355	13822.11	12.134	-0.02	1415	14549.75	12.116	-0.58
1296	13106.52	12.118	0.54	1356	13834.24	12.134	-0.03	1416	14561.86	12.115	-0.59
1297	13118.64	12.119	0.53	1357	13846.37	12.133	-0.04	1417	14573.98	12.114	-0.60
1298	13130.76	12.119	0.52	1358	13858.51	12.133	-0.05	1418	14586.09	12.114	-0.61
1299	13142.88	12.120	0.51	1359	13870.64	12.133	-0.06	1419	14598.20	12.113	-0.62
1300	13155.00	12.120	0.50	1360	13882.77	12.133	-0.06	1420	14610.32	12.112	-0.63
1301	13167.12	12.121	0.49	1361	13894.91	12.133	-0.07	1421	14622.43	12.112	-0.64
1302	13179.24	12.121	0.48	1362	13907.04	12.133	-0.08	1422	14634.54	12.111	-0.65
1303	13191.36	12.122	0.47	1363	13919.17	12.133	-0.09	1423	14646.65	12.111	-0.66
1304	13203.48	12.122	0.46	1364	13931.31	12.133	-0.10	1424	14658.76	12.110	-0.67
1305	13215.60	12.123	0.45	1365	13943.44	12.133	-0.11	1425	14670.87	12.109	-0.68
1306	13227.73	12.123	0.44	1366	13955.57	12.133	-0.12	1426	14682.98	12.109	-0.69
1307	13239.85	12.124	0.44	1367	13967.70	12.133	-0.13	1427	14695.09	12.108	-0.70
1308	13251.97	12.124	0.43	1368	13979.84	12.133	-0.14	1428	14707.20	12.107	-0.71
1309	13264.10	12.124	0.42	1369	13991.97	12.132	-0.15	1429	14719.30	12.106	-0.72
1310	13276.22	12.125	0.41	1370	14004.10	12.132	-0.16	1430	14731.41	12.106	-0.72
1311	13288.35	12.125	0.40	1371	14016.23	12.132	-0.17	1431	14743.51	12.105	-0.73
1312	13300.47	12.126	0.39	1372	14028.37	12.132	-0.18	1432	14755.62	12.104	-0.74
1313	13312.60	12.126	0.38	1373	14040.50	12.132	-0.19	1433	14767.72	12.103	-0.75
1314	13324.72	12.126	0.37	1374	14052.63	12.132	-0.20	1434	14779.83	12.103	-0.76
1315	13336.85	12.127	0.36	1375	14064.76	12.131	-0.21	1435	14791.93	12.102	-0.77
1316	13348.98	12.127	0.35	1376	14076.89	12.131	-0.22	1436	14804.03	12.101	-0.78
1317	13361.10	12.127	0.34	1377	14089.02	12.131	-0.22	1437	14816.13	12.100	-0.79
1318	13373.23	12.128	0.33	1378	14101.15	12.131	-0.23	1438	14828.23	12.100	-0.80
1319	13385.36	12.128	0.32	1379	14113.28	12.130	-0.24	1439	14840.33	12.099	-0.81
1320	13397.49	12.128	0.31	1380	14125.41	12.130	-0.25	1440	14852.43	12.098	-0.82

TABLE 1i6. (*Continued*)

Type S thermocouples—*thermoelectric voltages*, E(T), *Seebeck coefficients*, S(T), *and first derivative of the Seebeck coefficients*, dS/dT, *reference junctions at* 0 °C—*Continued*

T °C	E μV	S μV/°C	dS/dT nV/°C²	T °C	E μV	S μV/°C	dS/dT nV/°C²	T °C	E μV	S μV/°C	dS/dT nV/°C²
1440	14852.43	12.098	-0.82	1500	15576.49	12.032	-1.39	1560	16295.57	11.932	-1.95
1441	14864.53	12.097	-0.83	1501	15588.52	12.030	-1.39	1561	16307.50	11.930	-1.96
1442	14876.62	12.096	-0.84	1502	15600.55	12.029	-1.40	1562	16319.43	11.928	-1.97
1443	14888.72	12.095	-0.85	1503	15612.58	12.028	-1.41	1563	16331.36	11.926	-1.98
1444	14900.81	12.095	-0.86	1504	15624.61	12.026	-1.42	1564	16343.28	11.924	-1.99
1445	14912.91	12.094	-0.87	1505	15636.63	12.025	-1.43	1565	16355.21	11.922	-2.00
1446	14925.00	12.093	-0.88	1506	15648.66	12.023	-1.44	1566	16367.13	11.920	-2.01
1447	14937.09	12.092	-0.89	1507	15660.68	12.022	-1.45	1567	16379.04	11.918	-2.02
1448	14949.19	12.091	-0.89	1508	15672.70	12.020	-1.46	1568	16390.96	11.916	-2.03
1449	14961.28	12.090	-0.90	1509	15684.72	12.019	-1.47	1569	16402.88	11.914	-2.04
1450	14973.37	12.089	-0.91	1510	15696.74	12.018	-1.48	1570	16414.79	11.912	-2.05
1451	14985.45	12.088	-0.92	1511	15708.76	12.016	-1.49	1571	16426.70	11.910	-2.06
1452	14997.54	12.087	-0.93	1512	15720.77	12.015	-1.50	1572	16438.61	11.908	-2.06
1453	15009.63	12.087	-0.94	1513	15732.79	12.013	-1.51	1573	16450.52	11.906	-2.07
1454	15021.72	12.086	-0.95	1514	15744.80	12.012	-1.52	1574	16462.42	11.904	-2.08
1455	15033.80	12.085	-0.96	1515	15756.81	12.010	-1.53	1575	16474.32	11.901	-2.09
1456	15045.88	12.084	-0.97	1516	15768.82	12.008	-1.54	1576	16486.22	11.899	-2.10
1457	15057.97	12.083	-0.98	1517	15780.83	12.007	-1.55	1577	16498.12	11.897	-2.11
1458	15070.05	12.082	-0.99	1518	15792.83	12.005	-1.56	1578	16510.02	11.895	-2.12
1459	15082.13	12.081	-1.00	1519	15804.84	12.004	-1.56	1579	16521.91	11.893	-2.13
1460	15094.21	12.080	-1.01	1520	15816.84	12.002	-1.57	1580	16533.80	11.891	-2.14
1461	15106.29	12.079	-1.02	1521	15828.84	12.001	-1.58	1581	16545.69	11.889	-2.15
1462	15118.37	12.078	-1.03	1522	15840.84	11.999	-1.59	1582	16557.58	11.887	-2.16
1463	15130.45	12.077	-1.04	1523	15852.84	11.997	-1.60	1583	16569.47	11.884	-2.17
1464	15142.52	12.076	-1.05	1524	15864.84	11.996	-1.61	1584	16581.35	11.882	-2.18
1465	15154.60	12.075	-1.06	1525	15876.83	11.994	-1.62	1585	16593.23	11.880	-2.19
1466	15166.67	12.073	-1.06	1526	15888.83	11.993	-1.63	1586	16605.11	11.878	-2.20
1467	15178.74	12.072	-1.07	1527	15900.82	11.991	-1.64	1587	16616.99	11.876	-2.21
1468	15190.82	12.071	-1.08	1528	15912.81	11.989	-1.65	1588	16628.86	11.873	-2.22
1469	15202.89	12.070	-1.09	1529	15924.80	11.988	-1.66	1589	16640.73	11.871	-2.22
1470	15214.96	12.069	-1.10	1530	15936.78	11.986	-1.67	1590	16652.60	11.869	-2.23
1471	15227.03	12.068	-1.11	1531	15948.77	11.984	-1.68	1591	16664.47	11.867	-2.24
1472	15239.09	12.067	-1.12	1532	15960.75	11.983	-1.69	1592	16676.34	11.864	-2.25
1473	15251.16	12.066	-1.13	1533	15972.73	11.981	-1.70	1593	16688.20	11.862	-2.26
1474	15263.22	12.065	-1.14	1534	15984.71	11.979	-1.71	1594	16700.06	11.860	-2.27
1475	15275.29	12.064	-1.15	1535	15996.69	11.978	-1.72	1595	16711.92	11.858	-2.28
1476	15287.35	12.062	-1.16	1536	16008.67	11.976	-1.72	1596	16723.78	11.855	-2.29
1477	15299.41	12.061	-1.17	1537	16020.64	11.974	-1.73	1597	16735.63	11.853	-2.30
1478	15311.47	12.060	-1.18	1538	16032.62	11.972	-1.74	1598	16747.48	11.851	-2.31
1479	15323.53	12.059	-1.19	1539	16044.59	11.971	-1.75	1599	16759.33	11.848	-2.32
1480	15335.59	12.058	-1.20	1540	16056.56	11.969	-1.76	1600	16771.18	11.846	-2.33
1481	15347.65	12.056	-1.21	1541	16068.53	11.967	-1.77	1601	16783.02	11.844	-2.34
1482	15359.70	12.055	-1.22	1542	16080.49	11.965	-1.78	1602	16794.87	11.841	-2.35
1483	15371.76	12.054	-1.22	1543	16092.46	11.964	-1.79	1603	16806.71	11.839	-2.36
1484	15383.81	12.053	-1.23	1544	16104.42	11.962	-1.80	1604	16818.55	11.837	-2.37
1485	15395.86	12.052	-1.24	1545	16116.38	11.960	-1.81	1605	16830.38	11.834	-2.38
1486	15407.92	12.050	-1.25	1546	16128.34	11.958	-1.82	1606	16842.21	11.832	-2.39
1487	15419.97	12.049	-1.26	1547	16140.30	11.956	-1.83	1607	16854.04	11.830	-2.39
1488	15432.01	12.048	-1.27	1548	16152.25	11.954	-1.84	1608	16865.87	11.827	-2.40
1489	15444.06	12.047	-1.28	1549	16164.21	11.953	-1.85	1609	16877.70	11.825	-2.41
1490	15456.11	12.045	-1.29	1550	16176.16	11.951	-1.86	1610	16889.52	11.822	-2.42
1491	15468.15	12.044	-1.30	1551	16188.11	11.949	-1.87	1611	16901.34	11.820	-2.43
1492	15480.19	12.043	-1.31	1552	16200.06	11.947	-1.88	1612	16913.16	11.818	-2.44
1493	15492.24	12.041	-1.32	1553	16212.00	11.945	-1.89	1613	16924.98	11.815	-2.45
1494	15504.28	12.040	-1.33	1554	16223.95	11.943	-1.89	1614	16936.79	11.813	-2.46
1495	15516.32	12.039	-1.34	1555	16235.89	11.941	-1.90	1615	16948.60	11.810	-2.47
1496	15528.35	12.037	-1.35	1556	16247.83	11.939	-1.91	1616	16960.41	11.808	-2.48
1497	15540.39	12.036	-1.36	1557	16259.77	11.938	-1.92	1617	16972.22	11.805	-2.49
1498	15552.43	12.035	-1.37	1558	16271.70	11.936	-1.93	1618	16984.02	11.803	-2.50
1499	15564.46	12.033	-1.38	1559	16283.64	11.934	-1.94	1619	16995.83	11.800	-2.51
1500	15576.49	12.032	-1.39	1560	16295.57	11.932	-1.95	1620	17007.62	11.798	-2.52

TABLE 1i6. (*Continued*)

Type S thermocouples—*thermoelectric voltages*, E(T), *Seebeck coefficients*, S(T), *and first derivative of the Seebeck coefficients*, dS/dT, *reference junctions at 0 °C*—*Continued*

T °C	E μV	S μV/°C	dS/dT nV/°C^2	T °C	E μV	S μV/°C	dS/dT nV/°C^2	T °C	E μV	S μV/°C	dS/dT nV/°C^2
1620	17007.62	11.798	-2.52	1680	17710.54	11.616	-4.97	1740	18393.68	11.075	-13.06
1621	17019.42	11.795	-2.53	1681	17722.16	11.611	-5.10	1741	18404.75	11.062	-13.20
1622	17031.21	11.793	-2.54	1682	17733.77	11.605	-5.24	1742	18415.80	11.048	-13.33
1623	17043.01	11.790	-2.55	1683	17745.37	11.600	-5.37	1743	18426.85	11.035	-13.47
1624	17054.79	11.788	-2.56	1684	17756.97	11.595	-5.51	1744	18437.87	11.021	-13.60
1625	17066.58	11.785	-2.56	1685	17768.56	11.589	-5.64	1745	18448.89	11.008	-13.74
1626	17078.36	11.782	-2.57	1686	17780.14	11.583	-5.78	1746	18459.89	10.994	-13.87
1627	17090.15	11.780	-2.58	1687	17791.72	11.577	-5.91	1747	18470.88	10.980	-14.01
1628	17101.92	11.777	-2.59	1688	17803.30	11.572	-6.05	1748	18481.85	10.966	-14.14
1629	17113.70	11.775	-2.60	1689	17814.87	11.565	-6.18	1749	18492.81	10.952	-14.28
1630	17125.47	11.772	-2.61	1690	17826.43	11.559	-6.32	1750	18503.75	10.937	-14.41
1631	17137.24	11.769	-2.62	1691	17837.99	11.553	-6.45	1751	18514.68	10.923	-14.55
1632	17149.01	11.767	-2.63	1692	17849.54	11.546	-6.59	1752	18525.60	10.908	-14.68
1633	17160.78	11.764	-2.64	1693	17861.08	11.540	-6.72	1753	18536.50	10.893	-14.82
1634	17172.54	11.762	-2.65	1694	17872.61	11.533	-6.86	1754	18547.38	10.879	-14.95
1635	17184.30	11.759	-2.66	1695	17884.14	11.526	-6.99	1755	18558.25	10.864	-15.09
1636	17196.06	11.756	-2.67	1696	17895.67	11.519	-7.13	1756	18569.11	10.848	-15.22
1637	17207.81	11.754	-2.68	1697	17907.18	11.512	-7.26	1757	18579.95	10.833	-15.36
1638	17219.57	11.751	-2.69	1698	17918.69	11.504	-7.40	1758	18590.78	10.818	-15.49
1639	17231.32	11.748	-2.70	1699	17930.19	11.497	-7.53	1759	18601.59	10.802	-15.63
1640	17243.06	11.745	-2.71	1700	17941.68	11.489	-7.67	1760	18612.38	10.786	-15.76
1641	17254.81	11.743	-2.72	1701	17953.17	11.482	-7.80	1761	18623.16	10.771	-15.90
1642	17266.55	11.740	-2.72	1702	17964.65	11.474	-7.94	1762	18633.92	10.755	-16.03
1643	17278.29	11.737	-2.73	1703	17976.12	11.466	-8.07	1763	18644.67	10.738	-16.17
1644	17290.02	11.735	-2.74	1704	17987.58	11.457	-8.21	1764	18655.40	10.722	-16.30
1645	17301.76	11.732	-2.75	1705	17999.03	11.449	-8.34	1765	18666.11	10.706	-16.44
1646	17313.49	11.729	-2.76	1706	18010.48	11.441	-8.48	1766	18676.81	10.689	-16.57
1647	17325.21	11.726	-2.77	1707	18021.91	11.432	-8.61	1767	18687.49	10.673	-16.71
1648	17336.94	11.724	-2.78	1708	18033.34	11.424	-8.75	1768	18698.16	10.656	-16.84
1649	17348.66	11.721	-2.79	1709	18044.76	11.415	-8.88				
1650	17360.38	11.718	-2.80	1710	18056.17	11.406	-9.02				
1651	17372.10	11.715	-2.81	1711	18067.57	11.397	-9.15				
1652	17383.81	11.712	-2.82	1712	18078.96	11.388	-9.29				
1653	17395.52	11.709	-2.83	1713	18090.35	11.378	-9.42				
1654	17407.23	11.707	-2.84	1714	18101.72	11.369	-9.56				
1655	17418.93	11.704	-2.85	1715	18113.08	11.359	-9.69				
1656	17430.64	11.701	-2.86	1716	18124.44	11.349	-9.82				
1657	17442.34	11.698	-2.87	1717	18135.78	11.339	-9.96				
1658	17454.03	11.695	-2.88	1718	18147.12	11.329	-10.09				
1659	17465.73	11.692	-2.89	1719	18158.44	11.319	-10.23				
1660	17477.42	11.689	-2.89	1720	18169.75	11.309	-10.36				
1661	17489.11	11.687	-2.90	1721	18181.06	11.299	-10.50				
1662	17500.79	11.684	-2.91	1722	18192.35	11.288	-10.63				
1663	17512.47	11.681	-2.92	1723	18203.63	11.277	-10.77				
1664	17524.15	11.678	-2.93	1724	18214.91	11.266	-10.90				
1665	17535.83	11.675	-2.94	1725	18226.17	11.255	-11.04				
1666	17547.50	11.672	-3.08	1726	18237.42	11.244	-11.17				
1667	17559.17	11.669	-3.21	1727	18248.66	11.233	-11.31				
1668	17570.84	11.665	-3.35	1728	18259.88	11.222	-11.44				
1669	17582.50	11.662	-3.48	1729	18271.10	11.210	-11.58				
1670	17594.16	11.658	-3.62	1730	18282.30	11.199	-11.71				
1671	17605.82	11.655	-3.75	1731	18293.50	11.187	-11.85				
1672	17617.47	11.651	-3.89	1732	18304.68	11.175	-11.98				
1673	17629.12	11.647	-4.02	1733	18315.85	11.163	-12.12				
1674	17640.77	11.643	-4.16	1734	18327.00	11.151	-12.25				
1675	17652.41	11.639	-4.29	1735	18338.15	11.138	-12.39				
1676	17664.04	11.634	-4.43	1736	18349.28	11.126	-12.52				
1677	17675.68	11.630	-4.56	1737	18360.40	11.113	-12.66				
1678	17687.30	11.625	-4.70	1738	18371.51	11.101	-12.79				
1679	17698.93	11.620	-4.83	1739	18382.60	11.088	-12.93				
1680	17710.54	11.616	-4.97	1740	18393.68	11.075	-13.06				

TABLE 1i6. (*Continued*)

TYPE T—Copper Versus *Copper–Nickel Alloy·*Thermocouples

Type T thermocouples—thermoelectric voltages, E(T), *Seebeck coefficients,* S(T), *and first derivative of the Seebeck coefficients,* dS/dT, *reference junctions at 0 °C*

T °C	E μV	S μV/°C	dS/dT nV/°C²	T °C	E μV	S μV/°C	dS/dT nV/°C²	T °C	E μV	S μV/°C	dS/dT nV/°C²
-270	-6257.59	1.016	384.94	-240	-6105.09	8.726	230.04	-210	-5753.25	14.305	149.59
-269	-6256.38	1.385	354.18	-239	-6096.25	8.954	227.51	-209	-5738.87	14.454	148.23
-268	-6254.82	1.726	328.78	-238	-6087.18	9.180	224.81	-208	-5724.34	14.602	146.96
-267	-6252.93	2.044	308.01	-237	-6077.89	9.404	221.96	-207	-5709.67	14.748	145.79
-266	-6250.74	2.343	291.24	-236	-6068.38	9.624	218.98	-206	-5694.85	14.893	144.70
-265	-6248.26	2.628	277.85	-235	-6058.64	9.842	215.90	-205	-5679.88	15.038	143.71
-264	-6245.49	2.900	267.34	-234	-6048.69	10.056	212.73	-204	-5664.77	15.181	142.79
-263	-6242.45	3.163	259.25	-233	-6038.53	10.267	209.49	-203	-5649.52	15.323	141.95
-262	-6239.17	3.419	253.16	-232	-6028.16	10.475	206.21	-202	-5634.13	15.465	141.18
-261	-6235.62	3.670	248.73	-231	-6017.58	10.680	202.91	-201	-5618.59	15.605	140.47
-260	-6231.83	3.917	245.63	-230	-6006.80	10.881	199.61	-200	-5602.92	15.746	139.82
-259	-6227.79	4.162	243.58	-229	-5995.82	11.079	196.31	-199	-5587.10	15.885	139.23
-258	-6223.51	4.405	242.34	-228	-5984.64	11.274	193.05	-198	-5571.15	16.024	138.68
-257	-6218.98	4.646	241.72	-227	-5973.27	11.465	189.83	-197	-5555.05	16.163	138.18
-256	-6214.21	4.888	241.53	-226	-5961.71	11.653	186.66	-196	-5538.82	16.301	137.72
-255	-6209.20	5.130	241.63	-225	-5949.97	11.838	183.56	-195	-5522.45	16.438	137.29
-254	-6203.95	5.371	241.90	-224	-5938.04	12.020	180.54	-194	-5505.95	16.575	136.90
-253	-6198.46	5.614	242.21	-223	-5925.93	12.199	177.62	-193	-5489.30	16.712	136.53
-252	-6192.72	5.856	242.49	-222	-5913.64	12.376	174.78	-192	-5472.52	16.848	136.18
-251	-6186.75	6.098	242.68	-221	-5901.18	12.549	172.05	-191	-5455.61	16.984	135.85
-250	-6180.53	6.341	242.71	-220	-5888.54	12.720	169.43	-190	-5438.55	17.120	135.54
-249	-6174.07	6.584	242.54	-219	-5875.74	12.888	166.92	-189	-5421.37	17.255	135.25
-248	-6167.36	6.826	242.16	-218	-5862.77	13.054	164.53	-188	-5404.04	17.390	134.96
-247	-6160.41	7.068	241.52	-217	-5849.63	13.217	162.25	-187	-5386.59	17.525	134.68
-246	-6153.22	7.309	240.63	-216	-5836.33	13.378	160.09	-186	-5368.99	17.660	134.41
-245	-6145.80	7.549	239.48	-215	-5822.88	13.537	158.06	-185	-5351.27	17.794	134.14
-244	-6138.13	7.788	238.08	-214	-5809.26	13.694	156.14	-184	-5333.40	17.928	133.88
-243	-6130.22	8.025	236.41	-213	-5795.49	13.850	154.33	-183	-5315.41	18.062	133.61
-242	-6122.08	8.261	234.52	-212	-5781.56	14.003	152.64	-182	-5297.28	18.195	133.35
-241	-6113.70	8.494	232.38	-211	-5767.48	14.155	151.06	-181	-5279.02	18.328	133.09
-240	-6105.09	8.726	230.04	-210	-5753.25	14.305	149.59	-180	-5260.62	18.461	132.82

TABLE 1i6. (Continued)

Type T thermocouples—thermoelectric voltages, E(T), Seebeck coefficients, S(T), and first derivative of the Seebeck coefficients, dS/dT, reference junctions at 0 °C—Continued

T °C	E μV	S μV/°C	dS/dT nV/°C²	T °C	E μV	S μV/°C	dS/dT nV/°C²	T °C	E μV	S μV/°C	dS/dT nV/°C²
-180	-5260.62	18.461	132.82	-120	-3922.62	26.026	120.94	-60	-2152.41	32.840	106.55
-179	-5242.10	18.594	132.55	-119	-3896.54	26.147	120.72	-59	-2119.52	32.946	106.25
-178	-5223.44	18.727	132.29	-118	-3870.33	26.268	120.49	-58	-2086.52	33.052	105.94
-177	-5204.64	18.859	132.01	-117	-3844.00	26.388	120.26	-57	-2053.42	33.158	105.63
-176	-5185.72	18.991	131.74	-116	-3817.56	26.508	120.02	-56	-2020.21	33.264	105.31
-175	-5166.66	19.122	131.47	-115	-3790.99	26.628	119.77	-55	-1986.89	33.369	104.98
-174	-5147.47	19.254	131.19	-114	-3764.30	26.748	119.52	-54	-1953.47	33.474	104.66
-173	-5128.16	19.385	130.91	-113	-3737.49	26.867	119.26	-53	-1919.94	33.578	104.32
-172	-5108.71	19.515	130.63	-112	-3710.57	26.986	119.00	-52	-1886.31	33.682	103.99
-171	-5089.12	19.646	130.35	-111	-3683.52	27.105	118.73	-51	-1852.58	33.786	103.66
-170	-5069.41	19.776	130.07	-110	-3656.36	27.224	118.46	-50	-1818.74	33.889	103.33
-169	-5049.57	19.906	129.79	-109	-3629.07	27.342	118.19	-49	-1784.80	33.993	103.00
-168	-5029.60	20.036	129.51	-108	-3601.67	27.460	117.91	-48	-1750.75	34.095	102.68
-167	-5009.50	20.165	129.24	-107	-3574.15	27.578	117.64	-47	-1716.61	34.198	102.36
-166	-4989.27	20.294	128.96	-106	-3546.52	27.695	117.36	-46	-1682.36	34.300	102.05
-165	-4968.91	20.423	128.70	-105	-3518.76	27.812	117.08	-45	-1648.01	34.402	101.75
-164	-4948.43	20.551	128.43	-104	-3490.89	27.929	116.80	-44	-1613.55	34.504	101.46
-163	-4927.81	20.680	128.17	-103	-3462.90	28.046	116.52	-43	-1579.00	34.605	101.18
-162	-4907.07	20.808	127.91	-102	-3434.80	28.162	116.24	-42	-1544.34	34.706	100.92
-161	-4886.20	20.936	127.67	-101	-3406.58	28.278	115.96	-41	-1509.59	34.807	100.67
-160	-4865.20	21.063	127.42	-100	-3378.24	28.394	115.69	-40	-1474.73	34.907	100.43
-159	-4844.07	21.190	127.19	-99	-3349.79	28.510	115.42	-39	-1439.77	35.008	100.22
-158	-4822.81	21.318	126.96	-98	-3321.22	28.625	115.15	-38	-1404.72	35.108	100.01
-157	-4801.43	21.444	126.74	-97	-3292.54	28.740	114.88	-37	-1369.56	35.208	99.83
-156	-4779.93	21.571	126.52	-96	-3263.74	28.855	114.62	-36	-1334.30	35.308	99.66
-155	-4758.29	21.697	126.31	-95	-3234.83	28.969	114.37	-35	-1298.94	35.407	99.52
-154	-4736.53	21.824	126.12	-94	-3205.80	29.084	114.12	-34	-1263.49	35.507	99.38
-153	-4714.64	21.950	125.92	-93	-3176.66	29.198	113.87	-33	-1227.93	35.606	99.26
-152	-4692.63	22.075	125.74	-92	-3147.41	29.311	113.63	-32	-1192.27	35.705	99.16
-151	-4670.49	22.201	125.57	-91	-3118.04	29.425	113.40	-31	-1156.52	35.804	99.07
-150	-4648.23	22.327	125.40	-90	-3088.56	29.538	113.17	-30	-1120.67	35.903	98.99
-149	-4625.84	22.452	125.23	-89	-3058.96	29.651	112.95	-29	-1084.71	36.002	98.91
-148	-4603.33	22.577	125.08	-88	-3029.26	29.764	112.73	-28	-1048.66	36.101	98.84
-147	-4580.69	22.702	124.93	-87	-2999.44	29.877	112.51	-27	-1012.51	36.200	98.76
-146	-4557.92	22.827	124.79	-86	-2969.50	29.989	112.30	-26	-976.26	36.299	98.69
-145	-4535.03	22.952	124.65	-85	-2939.46	30.101	112.10	-25	-939.91	36.397	98.60
-144	-4512.02	23.076	124.51	-84	-2909.30	30.213	111.89	-24	-903.47	36.496	98.51
-143	-4488.88	23.201	124.38	-83	-2879.03	30.325	111.70	-23	-866.92	36.594	98.37
-142	-4465.62	23.325	124.26	-82	-2848.65	30.437	111.50	-22	-830.28	36.693	98.23
-141	-4442.23	23.449	124.13	-81	-2818.16	30.548	111.31	-21	-793.54	36.791	98.05
-140	-4418.72	23.573	124.01	-80	-2787.55	30.659	111.11	-20	-756.70	36.889	97.83
-139	-4395.08	23.697	123.89	-79	-2756.84	30.770	110.92	-19	-719.76	36.986	97.58
-138	-4371.32	23.821	123.77	-78	-2726.01	30.881	110.73	-18	-682.73	37.084	97.27
-137	-4347.44	23.945	123.65	-77	-2695.08	30.992	110.54	-17	-645.59	37.181	96.91
-136	-4323.43	24.068	123.52	-76	-2664.03	31.102	110.35	-16	-608.36	37.278	96.50
-135	-4299.30	24.192	123.40	-75	-2632.87	31.213	110.15	-15	-571.04	37.374	96.03
-134	-4275.05	24.315	123.27	-74	-2601.61	31.323	109.96	-14	-533.62	37.470	95.49
-133	-4250.67	24.438	123.14	-73	-2570.23	31.432	109.76	-13	-496.10	37.565	94.90
-132	-4226.17	24.561	123.01	-72	-2538.74	31.542	109.55	-12	-458.49	37.659	94.25
-131	-4201.55	24.684	122.87	-71	-2507.14	31.652	109.34	-11	-420.78	37.753	93.54
-130	-4176.81	24.807	122.73	-70	-2475.44	31.761	109.12	-10	-382.98	37.846	92.79
-129	-4151.94	24.930	122.58	-69	-2443.62	31.870	108.90	-9	-345.09	37.939	92.01
-128	-4126.95	25.052	122.42	-68	-2411.70	31.979	108.67	-8	-307.10	38.030	91.20
-127	-4101.83	25.175	122.26	-67	-2379.67	32.087	108.43	-7	-269.03	38.121	90.41
-126	-4076.60	25.297	122.09	-66	-2347.52	32.195	108.19	-6	-230.86	38.211	89.64
-125	-4051.24	25.419	121.92	-65	-2315.27	32.303	107.94	-5	-192.61	38.301	88.94
-124	-4025.76	25.541	121.74	-64	-2282.92	32.411	107.68	-4	-154.26	38.389	88.35
-123	-4000.16	25.662	121.55	-63	-2250.45	32.519	107.41	-3	-115.83	38.477	87.92
-122	-3974.43	25.784	121.35	-62	-2217.88	32.626	107.13	-2	-77.31	38.565	87.70
-121	-3948.59	25.905	121.15	-61	-2185.20	32.733	106.85	-1	-38.70	38.653	87.79
-120	-3922.62	26.026	120.94	-60	-2152.41	32.840	106.55	0	0.00	38.741	66.38

TABLE 1i6. (*Continued*)

Type T thermocouples—*thermoelectric voltages, E(T), Seebeck coefficients, S(T), and first derivative of the Seebeck coefficients, dS/dT, reference junctions at 0 °C—Continued*

T °C	E μV	S μV/°C	dS/dT nV/°C²	T °C	E μV	S μV/°C	dS/dT nV/°C²	T °C	E μV	S μV/°C	dS/dT nV/°C²
0	0.0	38.741	66.38	60	2467.5	43.649	83.20	120	5227.0	48.181	68.26
1	38.8	38.808	67.60	61	2511.2	43.733	82.98	121	5275.3	48.250	68.06
2	77.6	38.876	68.76	62	2555.0	43.815	82.75	122	5323.5	48.318	67.86
3	116.5	38.945	69.88	63	2598.8	43.898	82.52	123	5371.9	48.385	67.67
4	155.5	39.016	70.94	64	2642.8	43.980	82.28	124	5420.3	48.453	67.47
5	194.6	39.087	71.96	65	2686.8	44.063	82.04	125	5468.8	48.520	67.28
6	233.7	39.160	72.94	66	2730.9	44.145	81.79	126	5517.3	48.587	67.09
7	272.9	39.233	73.86	67	2775.1	44.226	81.55	127	5566.0	48.654	66.91
8	312.1	39.307	74.75	68	2819.3	44.308	81.29	128	5614.7	48.721	66.72
9	351.5	39.382	75.59	69	2863.7	44.389	81.04	129	5663.4	48.788	66.54
10	390.9	39.458	76.39	70	2908.1	44.470	80.78	130	5712.2	48.854	66.36
11	430.4	39.535	77.15	71	2952.6	44.550	80.52	131	5761.1	48.921	66.19
12	470.0	39.613	77.86	72	2997.2	44.631	80.26	132	5810.1	48.987	66.01
13	509.6	39.691	78.55	73	3041.9	44.711	79.99	133	5859.1	49.053	65.84
14	549.4	39.770	79.19	74	3086.6	44.791	79.73	134	5908.2	49.118	65.67
15	589.2	39.849	79.80	75	3131.5	44.870	79.46	135	5957.3	49.184	65.50
16	629.1	39.929	80.37	76	3176.4	44.950	79.19	136	6006.5	49.249	65.33
17	669.0	40.010	80.91	77	3221.4	45.029	78.92	137	6055.8	49.315	65.17
18	709.1	40.091	81.41	78	3266.4	45.107	78.65	138	6105.2	49.380	65.01
19	749.2	40.173	81.88	79	3311.6	45.186	78.38	139	6154.6	49.445	64.85
20	789.4	40.255	82.33	80	3356.8	45.264	78.10	140	6204.1	49.509	64.69
21	829.7	40.338	82.74	81	3402.1	45.342	77.83	141	6253.6	49.574	64.53
22	870.1	40.420	83.12	82	3447.5	45.420	77.56	142	6303.2	49.639	64.38
23	910.6	40.504	83.47	83	3492.9	45.497	77.29	143	6352.9	49.703	64.23
24	951.1	40.587	83.80	84	3538.5	45.574	77.01	144	6402.6	49.767	64.07
25	991.7	40.671	84.10	85	3584.1	45.651	76.74	145	6452.4	49.831	63.93
26	1032.5	40.756	84.37	86	3629.8	45.728	76.47	146	6502.3	49.895	63.78
27	1073.3	40.840	84.62	87	3675.5	45.804	76.20	147	6552.2	49.959	63.63
28	1114.1	40.925	84.84	88	3721.4	45.880	75.93	148	6602.2	50.022	63.49
29	1155.1	41.010	85.04	89	3767.3	45.956	75.66	149	6652.2	50.086	63.34
30	1196.2	41.095	85.22	90	3813.3	46.032	75.39	150	6702.4	50.149	63.20
31	1237.3	41.180	85.37	91	3859.4	46.107	75.13	151	6752.5	50.212	63.06
32	1278.5	41.266	85.50	92	3905.5	46.182	74.86	152	6802.8	50.275	62.92
33	1319.8	41.351	85.62	93	3951.7	46.257	74.60	153	6853.1	50.338	62.78
34	1361.2	41.437	85.71	94	3998.0	46.331	74.34	154	6903.5	50.400	62.65
35	1402.7	41.523	85.79	95	4044.4	46.405	74.08	155	6953.9	50.463	62.51
36	1444.3	41.608	85.84	96	4090.8	46.479	73.82	156	7004.4	50.526	62.37
37	1485.9	41.694	85.88	97	4137.4	46.553	73.56	157	7054.9	50.588	62.24
38	1527.7	41.780	85.90	98	4183.9	46.626	73.30	158	7105.6	50.650	62.11
39	1569.5	41.866	85.91	99	4230.6	46.700	73.05	159	7156.2	50.712	61.98
40	1611.4	41.952	85.89	100	4277.3	46.773	72.80	160	7207.0	50.774	61.84
41	1653.4	42.038	85.87	101	4324.2	46.845	72.55	161	7257.8	50.836	61.71
42	1695.5	42.124	85.83	102	4371.0	46.918	72.30	162	7308.7	50.897	61.58
43	1737.6	42.209	85.77	103	4418.0	46.990	72.06	163	7359.6	50.959	61.46
44	1779.9	42.295	85.70	104	4465.0	47.062	71.81	164	7410.6	51.020	61.33
45	1822.2	42.381	85.62	105	4512.1	47.133	71.57	165	7461.6	51.082	61.20
46	1864.6	42.466	85.53	106	4559.3	47.205	71.34	166	7512.7	51.143	61.07
47	1907.1	42.552	85.42	107	4606.5	47.276	71.10	167	7563.9	51.204	60.94
48	1949.7	42.637	85.31	108	4653.8	47.347	70.87	168	7615.2	51.265	60.82
49	1992.4	42.723	85.18	109	4701.2	47.418	70.64	169	7666.4	51.325	60.69
50	2035.2	42.808	85.04	110	4748.7	47.488	70.41	170	7717.8	51.386	60.57
51	2078.0	42.893	84.89	111	4796.2	47.559	70.18	171	7769.2	51.446	60.44
52	2121.0	42.977	84.74	112	4843.8	47.629	69.96	172	7820.7	51.507	60.32
53	2164.0	43.062	84.57	113	4891.4	47.699	69.74	173	7872.2	51.567	60.19
54	2207.1	43.147	84.40	114	4939.2	47.768	69.52	174	7923.8	51.627	60.07
55	2250.3	43.231	84.21	115	4987.0	47.838	69.30	175	7975.5	51.687	59.94
56	2293.6	43.315	84.02	116	5034.9	47.907	69.09	176	8027.2	51.747	59.82
57	2336.9	43.399	83.83	117	5082.8	47.976	68.88	177	8079.0	51.807	59.69
58	2380.4	43.483	83.62	118	5130.8	48.045	68.67	178	8130.8	51.866	59.57
59	2423.9	43.566	83.41	119	5178.9	48.113	68.46	179	8182.7	51.926	59.44
60	2467.5	43.649	83.20	120	5227.0	48.181	68.26	180	8234.7	51.985	59.32

TABLE 1i6. (*Continued*)

Type T thermocouples—*thermoelectric voltages,* E(T), *Seebeck coefficients,* S(T), *and first derivative of the Seebeck coefficients,* dS/dT, *reference junctions at 0 °C—Continued*

T °C	E μV	S μV/°C	dS/dT nV/°C²	T °C	E μV	S μV/°C	dS/dT nV/°C²	T °C	E μV	S μV/°C	dS/dT nV/°C²
180	8234.7	51.985	59.32	240	11455.8	55.300	50.80	300	14859.8	58.086	42.86
181	8286.7	52.045	59.19	241	11511.1	55.351	50.64	301	14917.9	58.129	42.78
182	8338.8	52.104	59.07	242	11566.5	55.401	50.48	302	14976.1	58.172	42.69
183	8390.9	52.163	58.94	243	11621.9	55.452	50.32	303	15034.3	58.215	42.62
184	8443.1	52.222	58.81	244	11677.4	55.502	50.17	304	15092.5	58.257	42.54
185	8495.3	52.280	58.69	245	11732.9	55.552	50.01	305	15150.8	58.300	42.46
186	8547.6	52.339	58.56	246	11788.5	55.602	49.85	306	15209.1	58.342	42.39
187	8600.0	52.397	58.43	247	11844.1	55.652	49.69	307	15267.5	58.385	42.32
188	8652.4	52.456	58.31	248	11899.8	55.701	49.54	308	15325.9	58.427	42.25
189	8704.9	52.514	58.18	249	11955.5	55.751	49.38	309	15384.3	58.469	42.19
190	8757.5	52.572	58.05	250	12011.3	55.800	49.22	310	15442.8	58.511	42.12
191	8810.1	52.630	57.92	251	12067.1	55.849	49.07	311	15501.3	58.553	42.06
192	8862.7	52.688	57.79	252	12123.0	55.898	48.91	312	15559.9	58.595	41.99
193	8915.4	52.746	57.66	253	12178.9	55.947	48.76	313	15618.5	58.637	41.93
194	8968.2	52.803	57.53	254	12234.9	55.996	48.60	314	15677.2	58.679	41.88
195	9021.1	52.861	57.40	255	12290.9	56.044	48.45	315	15735.9	58.721	41.82
196	9073.9	52.918	57.27	256	12347.0	56.093	48.30	316	15794.6	58.763	41.76
197	9126.9	52.975	57.13	257	12403.1	56.141	48.14	317	15853.4	58.805	41.71
198	9179.9	53.032	57.00	258	12459.3	56.189	47.99	318	15912.2	58.846	41.65
199	9233.0	53.089	56.87	259	12515.5	56.237	47.84	319	15971.1	58.888	41.60
200	9286.1	53.146	56.73	260	12571.7	56.285	47.69	320	16030.0	58.929	41.55
201	9339.2	53.203	56.60	261	12628.0	56.332	47.54	321	16089.0	58.971	41.50
202	9392.5	53.259	56.46	262	12684.4	56.380	47.39	322	16148.0	59.012	41.45
203	9445.8	53.316	56.32	263	12740.8	56.427	47.25	323	16207.0	59.054	41.40
204	9499.1	53.372	56.19	264	12797.3	56.474	47.10	324	16266.1	59.095	41.35
205	9552.5	53.428	56.05	265	12853.8	56.521	46.95	325	16325.2	59.137	41.31
206	9606.0	53.484	55.91	266	12910.3	56.568	46.81	326	16384.3	59.178	41.26
207	9659.5	53.540	55.77	267	12966.9	56.615	46.67	327	16443.5	59.219	41.21
208	9713.0	53.596	55.63	268	13023.5	56.661	46.52	328	16502.8	59.260	41.17
209	9766.7	53.651	55.49	269	13080.2	56.708	46.38	329	16562.1	59.301	41.12
210	9820.3	53.707	55.35	270	13136.9	56.754	46.24	330	16621.4	59.343	41.07
211	9874.1	53.762	55.20	271	13193.7	56.800	46.11	331	16680.7	59.384	41.02
212	9927.9	53.817	55.06	272	13250.5	56.846	45.97	332	16740.1	59.425	40.98
213	9981.7	53.872	54.92	273	13307.4	56.892	45.84	333	16799.6	59.466	40.93
214	10035.6	53.927	54.77	274	13364.3	56.938	45.70	334	16859.1	59.506	40.88
215	10089.6	53.982	54.63	275	13421.3	56.984	45.57	335	16918.6	59.547	40.83
216	10143.6	54.036	54.48	276	13478.3	57.029	45.44	336	16978.2	59.588	40.77
217	10197.6	54.091	54.33	277	13535.3	57.075	45.31	337	17037.8	59.629	40.72
218	10251.8	54.145	54.18	278	13592.4	57.120	45.18	338	17097.4	59.670	40.67
219	10305.9	54.199	54.04	279	13649.6	57.165	45.06	339	17157.1	59.710	40.61
220	10360.2	54.253	53.89	280	13706.8	57.210	44.93	340	17216.8	59.751	40.55
221	10414.4	54.307	53.74	281	13764.0	57.255	44.81	341	17276.6	59.791	40.49
222	10468.8	54.360	53.59	282	13821.3	57.300	44.69	342	17336.4	59.832	40.42
223	10523.2	54.414	53.44	283	13878.6	57.344	44.57	343	17396.3	59.872	40.35
224	10577.6	54.467	53.28	284	13936.0	57.389	44.46	344	17456.2	59.912	40.28
225	10632.1	54.520	53.13	285	13993.4	57.433	44.34	345	17516.1	59.953	40.21
226	10686.6	54.573	52.98	286	14050.8	57.477	44.23	346	17576.1	59.993	40.13
227	10741.2	54.626	52.82	287	14108.3	57.522	44.12	347	17636.1	60.033	40.05
228	10795.9	54.679	52.67	288	14165.9	57.566	44.01	348	17696.1	60.073	39.97
229	10850.6	54.732	52.52	289	14223.5	57.610	43.90	349	17756.2	60.113	39.87
230	10905.4	54.784	52.36	290	14281.1	57.653	43.80	350	17816.4	60.153	39.78
231	10960.2	54.836	52.21	291	14338.8	57.697	43.69	351	17876.5	60.192	39.68
232	11015.0	54.889	52.05	292	14396.5	57.741	43.59	352	17936.8	60.232	39.57
233	11069.9	54.941	51.89	293	14454.3	57.784	43.49	353	17997.0	60.272	39.46
234	11124.9	54.992	51.74	294	14512.1	57.828	43.40	354	18057.3	60.311	39.34
235	11179.9	55.044	51.58	295	14569.9	57.871	43.30	355	18117.6	60.350	39.22
236	11235.0	55.096	51.42	296	14627.8	57.914	43.21	356	18178.0	60.389	39.08
237	11290.1	55.147	51.27	297	14685.7	57.958	43.12	357	18238.4	60.428	38.94
238	11345.3	55.198	51.11	298	14743.7	58.001	43.03	358	18298.9	60.467	38.80
239	11400.5	55.249	50.95	299	14801.7	58.044	42.94	359	18359.3	60.506	38.64
240	11455.8	55.300	50.80	300	14859.8	58.086	42.86	360	18419.9	60.545	38.47

TABLE 1i6. *(Continued)*

*Type **T** thermocouples—thermoelectric voltages,* E(T), *Seebeck coefficients,* S(T), *and first derivative of the Seebeck coefficients,* dS/dT, *reference junctions at 0 °C—Continued*

T °C	E μV	S μV/°C	dS/dT nV/°C^2	T °C	E μV	S μV/°C	dS/dT nV/°C^2	T °C	E μV	S μV/°C	dS/dT nV/°C^2
360	18419.9	60.545	38.47	375	19332.2	61.096	34.64	390	20252.3	61.564	26.88
361	18480.4	60.583	38.30	376	19393.4	61.131	34.26	391	20313.9	61.591	26.16
362	18541.0	60.621	38.12	377	19454.5	61.165	33.88	392	20375.5	61.616	25.42
363	18601.7	60.659	37.92	378	19515.7	61.199	33.47	393	20437.2	61.641	24.65
364	18662.3	60.697	37.72	379	19576.9	61.232	33.04	394	20498.8	61.666	23.84
365	18723.1	60.735	37.50	380	19638.2	61.265	32.59	395	20560.5	61.689	23.01
366	18783.8	60.772	37.28	381	19699.4	61.297	32.13	396	20622.2	61.712	22.14
367	18844.6	60.809	37.04	382	19760.7	61.329	31.64	397	20683.9	61.733	21.23
368	18905.4	60.846	36.79	383	19822.1	61.360	31.13	398	20745.7	61.754	20.29
369	18966.3	60.883	36.52	384	19883.5	61.391	30.60	399	20807.4	61.774	19.32
370	19027.2	60.919	36.25	385	19944.9	61.422	30.04	400	20869.2	61.793	18.31
371	19088.1	60.955	35.96	386	20006.3	61.451	29.46				
372	19149.1	60.991	35.65	387	20067.8	61.480	28.85				
373	19210.1	61.027	35.33	388	20129.3	61.509	28.22				
374	19271.2	61.062	34.99	389	20190.8	61.537	27.56				
375	19332.2	61.096	34.64	390	20252.3	61.564	26.88				

ALUMINUM WIRE TABLES

The data on aluminum wire are reproduced
from Peterson, C., J. L. Thomas and H. Cook,
Eds., "Aluminum Wire Tables," NBS Handbook
109, U.S. Government Printing Office, Washington, D.C., 1972.

TABLE 1i7. CONVERSION TABLE FOR ELECTRICAL RESISTIVITIES OF ALUMINUM WIRE

	To obtain values in—						
Given values at 20°C in	Ohm g/m² multiply by	Ohm lb/mi² multiply by	Ohm mm²/m multiply by	Microhm-cm multiply by	Microhm-in multiply by	Ohm-cir mil/ft multiply by	Percent conductivity divide into
Ohm g/m²	–	5709.8	0.36996	36.996	14.566	222.55	4.6600
Ohm lb/mi²	0.00017514	–	.000064795	0.0064795	0.0025510	0.038976	26,609
Ohm mm²/m	2.7030	15,434	–	100	39.371	601.53	1.7241
Microhm-cm	0.02703	154.34	.01	–	.39371	6.0153	172.41
Microhm-in	0.068656	392.01	.0254	2.54	–	15.279	67.879
Ohm-cir mil/ft	0.0044935	25.657	.0016624	0.16624	.065451	–	1037.1
Percent conductivity	Divide into 4.6600	Divide into 26,609	Divide into 1.7241	Divide into 172.41	Divide into 67.879	Divide into 1037.1	

TABLE 1i8. WIRE TABLE, SOLID ALUMINUM, EC-0—DATA AT 20°C—ENGLISH UNITS.

AWG	Diameter Mils	Cross section Circular mils	Cross section Square inch	Ohms per 1000 feet	Feet per ohm	Pounds per 1000 feet	Feet per pound	Ohms per pound	Pounds per ohm
500,000	707.10	500000.	.3927	0.03356	29790.	460.2	2.173	.00007294	13710.
450,000	670.80	450000.	.3534	0.03730	26810.	414.1	2.415	.00009006	11100.
400,000	632.50	400000.	.3142	0.04195	23840.	368.2	2.716	.0001139	8777.
350,000	591.60	350000.	.2749	0.04795	20860.	322.1	3.105	.0001489	6718.
300,000	547.70	300000.	.2356	0.05594	17870.	276.1	3.622	.0002026	4935.
250,000	500.00	250000.	.1963	0.06713	14900.	230.1	4.346	.0002918	3428.
0000	460.00	211600.	.1662	0.07931	12610.	194.7	5.135	.0004073	2455.
000	409.60	167800.	.1318	0.1000	9997.	154.4	6.476	.0006478	1544.
00	364.80	133100.	.1045	0.1261	7930.	122.?	8.165	.001030	971.2
0	324.90	105600.	.08291	0.1590	6290.	97.15	10.29	.001636	611.1
1	289.30	83690.	.06573	0.2005	4987.	77.03	12.98	.002603	384.1
2	257.60	66360.	.05212	0.2529	3954.	61.07	16.37	.004141	241.5
3	229.40	52620.	.04133	0.3189	3136.	48.43	20.65	.006585	151.9
4	204.30	41740.	.03278	0.4021	2487.	38.41	26.03	.01047	95.54
5	181.90	33090.	.02599	0.5072	1972.	30.45	32.84	.01666	60.04
6	162.00	26240.	.02061	0.6395	1564.	24.15	41.40	.02648	37.77
7	144.30	20820.	.01635	0.8060	1241.	19.16	52.18	.04206	23.78
8	128.50	16510.	.01297	1.0160	983.9	15.20	65.80	.06688	14.95
9	114.40	13090.	.01028	1.282	779.8	12.04	83.02	0.1065	9.393
10	101.90	10380.	.008155	1.616	618.7	9.556	104.6	0.1691	5.913
11	90.70	8226.	.006461	2.040	490.2	7.571	132.1	0.2694	3.711
12	80.80	6529.	.005128	2.571	389.0	6.008	166.4	0.4278	2.337
13	72.00	5184.	.004072	3.237	308.9	4.771	209.6	0.6785	1.474
14	64.10	4109.	.003227	4.084	244.8	3.781	264.4	1.080	0.9258
15	57.10	3260.	.002561	5.147	194.3	3.001	333.3	1.715	0.5830
16	50.80	2581.	.002027	6.503	153.8	2.375	421.0	2.738	0.3652
17	45.30	2052.	.001612	8.178	122.3	1.889	529.5	4.330	0.2309
18	40.30	1624.	.001276	10.33	96.78	1.495	669.0	6.913	0.1447
19	35.90	1289.	.001012	13.02	76.80	1.186	843.1	10.98	0.09109
20	32.00	1024.	.0008042	16.39	61.02	0.9424	1061.	17.39	0.05750
21	28.50	812.2	.0006379	20.66	48.40	0.7475	1338.	27.64	0.03618
22	25.30	640.1	.0005027	26.22	38.14	0.5891	1698.	44.51	0.02247
23	22.60	510.8	.0004011	32.86	30.43	0.4701	2127.	69.90	0.01431
24	20.10	404.0	.0003173	41.54	24.07	0.3718	2689.	111.7	0.008951
25	17.90	320.4	.0002516	52.38	19.09	0.2949	3391.	177.6	0.005630
26	15.90	252.8	.0001986	66.38	15.06	0.2327	4298.	285.3	0.003505
27	14.20	201.6	.0001584	83.23	12.02	0.1856	5389.	448.5	0.002230
28	12.60	158.8	.0001247	105.7	9.460	0.1461	6844.	723.5	0.001382
29	11.30	127.7	.0001003	131.4	7.609	0.1175	8509.	1118.	0.0008942
30	10.00	100.00	.00007854	167.8	5.959	0.09203	10870.	1823.	0.0005484
31	8.90	79.21	.00006221	211.9	4.720	0.07290	13720.	2906.	0.0003441
32	8.00	64.00	.00005027	262.2	3.814	0.05890	16980.	4452.	0.0002246

33	7.10	50.41	.00003959	332.9	3.004	0.04639	21550.	7176.	0.0001394
34	6.30	39.69	.00003117	422.8	2.365	0.03653	27380.	11580.	0.00008639
35	5.60	31.36	.00002463	535.1	1.869	0.02886	34650.	18540.	0.00005393
36	5.00	25.00	.00001963	671.3	1.490	0.02301	43460.	29180.	0.00003428
37	4.50	20.25	.00001590	828.7	1.207	0.01864	53660.	44470.	0.00002249
38	4.00	16.00	.00001257	1049.	0.9534	0.01473	67910.	71230.	0.00001404
39	3.50	12.25	.000009621	1370.	0.7299	0.01127	88700.	121500.	.00008229
40	3.10	9.610	.000007548	1746.	0.5726	0.008844	113100.	197400.	.000005065
41	2.80	7.840	.000006158	2141.	0.4672	0.007215	138600.	296700.	.000003371
42	2.50	6.250	.000004909	2685.	0.3724	0.005752	173900.	466800.	.000002142
43	2.20	4.840	.000003801	3467.	0.2884	0.004454	224500.	778400.	.000001285
44	2.00	4.000	.000003142	4195.	0.2384	0.003681	271600.	1140000.	.0000008774
45	1.76	3.098	.000002433	5418.	0.1846	0.002851	350800.	1900000.	.0000005262
46	1.57	2.465	.000001936	6808.	0.1469	0.002269	440800.	3001000.	.0000003332
47	1.40	1.960	.000001539	8562.	0.1168	0.001804	554400.	4747000.	.0000002107
48	1.24	1.538	.000001208	10910.	0.09162	0.001415	706700.	7713000.	.0000001297
49	1.11	1.232	.0000009677	13620.	0.07342	0.001134	881900.	12010000.	.00000008325
50	0.99	0.9801	.0000007698	17120.	0.05840	0.0009020	1109000.	18980000.	.00000005268
51	0.88	0.7744	.0000006082	21670.	0.04614	0.0007127	1403000.	30410000.	.0000003289
52	0.78	0.6084	.0000004778	27580.	0.03625	0.0005599	1786000.	49260000.	.0000002030
53	0.70	0.4900	.0000003848	34250.	0.02920	0.0004510	2217000.	75950000.	.0000001317
54	0.62	0.3844	.0000003019	43660.	0.02291	0.0003538	2827000.	123400000.	.00000008103
55	0.55	0.3025	.0000002376	55480.	0.01803	0.0002784	3592000.	193300000.	.00000005018
56	0.49	0.2401	.0000001886	69900.	0.01431	0.0002210	4525000.	316300000.	.00000003161

TABLE 1i9. WIRE TABLE, SOLID ALUMINUM, EC–0—DATA AT 20°C—METRIC UNITS

Gage	Diameter (mm)	Area (sq mm)	Ohms per kilometer	Meters per ohm	Kilograms per kilometer	Meters per gram	Ohms per kilogram	Grams per ohm
500,000	17.960	253.3	0.1101	9081.	684.8	0.001460	0.0001608	6219000.
450,000	17.040	228.0	0.1224	8173.	616.3	0.001623	0.0001985	5037000.
400,000	16.070	202.7	0.1376	7266.	547.9	0.001825	0.0002512	3981000.
350,000	15.030	177.3	0.1573	6357.	479.4	0.002086	0.0003282	3047000.
300,000	13.910	152.0	0.1835	5448.	410.9	0.002434	0.0004467	2239000.
250,000	12.700	126.7	0.2202	4541.	342.4	0.002920	0.0006432	1555000.
0000	11.680	107.2	0.2602	3843.	289.8	0.003450	0.0008978	1114000.
000	10.400	85.01	0.3282	3047.	229.8	0.004352	0.001428	700200.
00	9.2660	67.43	0.4137	2417.	182.3	0.005486	0.002270	440600.
0	8.2520	53.49	0.5216	1917.	144.6	0.006917	0.003608	277200.
1	7.3480	42.41	0.6578	1520.	114.6	0.008724	0.005739	174300.
2	6.5430	33.62	0.8297	1205.	90.89	0.01100	0.009129	109500.
3	5.8270	26.67	1.046	955.8	72.08	0.01387	0.01452	68890.
4	5.1890	21.15	1.319	758.1	57.17	0.01749	0.02307	43340.
5	4.6200	16.77	1.664	601.0	45.32	0.02207	0.03672	27230.
6	4.1150	13.30	2.098	476.7	35.94	0.02782	0.05836	17130.
7	3.6650	10.55	2.644	378.2	28.52	0.03506	0.09271	10790.
8	3.2640	8.367	3.334	299.9	22.62	0.04422	0.1474	6783.
9	2.9060	6.631	4.207	237.7	17.92	0.05579	0.2347	4261.
10	2.5880	5.261	5.302	188.6	14.22	0.07031	0.3728	2682.
11	2.3040	4.168	6.693	149.4	11.27	0.08875	0.5940	1684.
12	2.0520	3.308	8.433	118.6	8.942	0.1118	0.9431	1060.
13	1.8290	2.627	10.62	94.16	7.100	0.1408	1.496	668.5
14	1.6280	2.082	13.40	74.63	5.628	0.1777	2.381	420.0
15	1.4500	1.652	16.89	59.22	4.466	0.2239	3.782	264.4
16	1.2900	1.308	21.33	46.87	3.535	0.2829	6.036	165.7
17	1.1510	1.040	26.83	37.27	2.811	0.3558	9.546	104.8
18	1.0240	0.8229	33.90	29.50	2.224	0.4496	15.24	65.62
19	0.9119	0.6531	42.72	23.41	1.765	0.5665	24.20	41.32
20	0.8128	0.5189	53.77	18.60	1.403	0.7130	38.34	26.08
21	0.7239	0.4116	67.78	14.75	1.112	0.8989	60.93	16.41
22	0.6426	0.3243	86.01	11.63	0.8767	1.141	98.11	10.19
23	0.5740	0.2588	107.8	9.277	0.6996	1.429	154.1	6.490
24	0.5105	0.2047	136.3	7.338	0.5533	1.807	246.3	4.060
25	0.4547	0.1624	171.8	5.820	0.4388	2.279	391.6	2.554
26	0.4039	0.1281	217.8	4.592	0.3463	2.888	629.0	1.590
27	0.3607	0.1022	273.0	3.662	0.2762	3.621	988.7	1.011
28	0.3200	0.08045	346.8	2.884	0.2174	4.599	1595.	0.6270
29	0.2870	0.06470	431.2	2.319	0.1749	5.718	2465.	0.4056
30	0.2540	0.05067	550.6	1.816	0.1370	7.301	4020.	0.2488
31	0.2261	0.04014	695.0	1.439	0.1085	9.218	6407.	0.1561
32	0.2032	0.03243	860.3	1.162	0.08766	11.41	9814.	0.1019

33	0.1803	0.02554	1092.	0.9156	0.06904	14.48	15820.	0.06322
34	0.1600	0.02011	1387.	0.7209	0.05436	18.40	25520.	0.03919
35	0.1422	0.01589	1756.	0.5696	0.04295	23.28	40880.	0.02446
36	0.1270	0.01267	2202.	0.4541	0.03424	29.20	64320.	0.01555
37	0.1143	0.01026	2719.	0.3678	0.02774	36.06	98030.	0.01020
38	0.1016	0.008107	3441.	0.2906	0.02191	45.63	157000.	0.006368
39	0.08890	0.006207	4494.	0.2225	0.01678	59.60	267900.	0.003733
40	0.07874	0.004869	5729.	0.1745	0.01316	75.98	435300.	0.002297
41	0.07112	0.003973	7023.	0.1424	0.01074	93.13	654000.	0.001529
42	0.06350	0.003167	8809.	0.1135	0.008560	116.8	1029000.	0.0009717
43	0.05588	0.002452	11380.	0.08791	0.006629	150.9	1716000.	0.0005827
44	0.05080	0.002027	13760.	0.07265	0.005479	182.5	2512000.	0.0003980
45	0.04470	0.001570	17770.	0.05626	0.004243	235.7	4189000.	0.0002387
46	0.03988	0.001249	22340.	0.04477	0.003376	296.2	6616000.	0.0001511
47	0.03556	0.0009931	28090.	0.03560	0.002684	372.5	10460000.	0.00009557
48	0.03150	0.0007791	35810.	0.02793	0.002106	474.8	17000000.	0.00005881
49	0.02819	0.0006243	44690.	0.02238	0.001688	592.6	26480000.	0.00003776
50	0.02515	0.0004966	56180.	0.01780	0.001342	744.9	41850000.	0.00002390
51	0.02235	0.0003924	71100.	0.01407	0.001061	942.8	67030000.	0.00001492
52	0.01981	0.0003083	90500.	0.01105	0.0008333	1200.	108600000.	0.000009208
53	0.01778	0.0002483	112400.	0.008900	0.0006711	1490.	167400000.	0.000005973
54	0.01575	0.0001948	143200.	0.006982	0.0005265	1899.	272000000.	0.000003676
55	0.01397	0.0001533	182000.	0.005494	0.0004143	2414.	439300000.	0.000002276
56	0.01245	0.0001217	229300.	0.004361	0.0003288	3041.	697300000.	0.000001434

TABLE 1i10. INTERNATIONAL ANNEALED COPPER STANDARD VALUES FOR RESISTIVITY, TEMPERATURE COEFFICIENT, AND DENSITY*

Property	Value at Temperature of			
	$0°C$	$15°C$	$20°C$	$25°C$
Resistivity, ohm-gram/meter^2	0.14133_2	0.15029_0	0.15328	0.15626_2
Temperature Coefficient of Resistance per $°C$	0.004265	0.004009	0.00393	0.003854
Density, grams/cm^3	8.90		8.89	

*Tables 1i10–1i17 inclusive are reproduced from "Copper Wire Tables," NBS Handbook 100, U.S. Government Printing Office, Washington, D.C. (1966).

TABLE 1i11. TEMPERATURE COEFFICIENTS OF COPPER FOR DIFFERENT INITIAL CELSIUS (CENTIGRADE) TEMPERATURES AND DIFFERENT CONDUCTIVITIES*

Ohm-gram/ meter$_2$ at 20 °C	Percent conduc- tivity	α_0	α_{15}	α_{20}	α_{25}	α_{30}	α_{50}	T
0.161 34	95	0.004 03	0.003 80	0.003 73	0.003 67	0.003 60	0.003 36	247.8
.159 66	96	.004 08	.003 85	.003 77	.003 70	.003 64	.003 39	245.1
.158 02	97	.004 13	.003 89	.003 81	.003 74	.003 67	.003 42	242.3
.157 21	97.5	.004 15	.003 91	.003 83	.003 76	.003 69	.003 44	241.0
.156 40	98	.004 17	.003 93	.003 85	.003 78	.003 71	.003 45	239.6
.154 82	99	.004 22	.003 97	.003 89	.003 82	.003 74	.003 48	237.0
.153 28	100	.004 27	.004 01	.003 93	.003 85	.003 78	.003 52	234.5
.151 76	101	.004 31	.004 05	.003 97	.003 89	.003 82	.003 55	231.9
.150 27	102	.004 36	.004 09	.004 01	.003 93	.003 85	.003 58	229.5

NOTE.—The fundamental relation between resistance and temperature is the following:

$$R_t = R_{t_1} (1 + \alpha_{t_1} [t - t_1]),$$

where α_{t_1} is the "temperature coefficient," and t_1 is the "initial temperature" or "temperature of reference."

The values of α in the above table exhibit the fact that the temperature coefficient of copper is proportional to the conductivity. The table was calculated by means of the following formula, which holds for any percent conductivity, n, within commercial ranges, and for Celsius temperatures. (n is considered to be expressed decimally; e. g., if percent conductivity = 99 percent, $n = 0.99$.)

$$\alpha_{t_1} = \frac{1}{\frac{1}{n(0.00393)} + (t_1 - 20)}.$$

The quantity T in the last column of the above table presents an easy way of remembering the temperature coefficient, its usefulness being evident from the following formulas:

$$t - t_1 = \frac{R_t - R_{t_1}}{R_{t_1}}(T + t^1)$$

$$\frac{R_t}{R_{t_1}} = 1 + \frac{t - t_1}{T + t_1} = \frac{T + t}{T + t_1}.$$

TABLE 1i12. REDUCTION OF OBSERVATIONS ON COPPER RESISTIVITY TO STANDARD TEMPERATURE

Temperature °C	Corrections to change resistivity to 20 °C				Factors to change resistance to 20 °C			Temperature °C
	Ohm-gram/meter³	Microhm—cm	Ohm-pound/mile²	Microhm—inch	For 96 percent conductivity	For 98 percent conductivity	For 100 percent conductivity	
0	+0.011 94	+0.1361	+68.20	+0.053 58	1.0816	1.0834	1.0853	0
5	+.008 96	+.1021	+51.15	+.040 18	1.0600	1.0613	1.0626	5
10	+.005 97	+.0681	+34.10	+.026 79	1.0392	1.0401	1.0409	10
11	+.005 37	+.0612	+30.69	+.024 11	1.0352	1.0359	1.0367	11
12	+.004 78	+.0544	+27.28	+.021 43	1.0311	1.0318	1.0325	12
13	+.004 18	+.0476	+23.87	+.018 75	1.0271	1.0277	1.0283	13
14	+.003 58	+.0408	+20.46	+.016 07	1.0232	1.0237	1.0242	14
15	+.002 99	+.0340	+17.05	+.013 40	1.0192	1.0196	1.0200	15
16	+.002 39	+.0272	+13.64	+.010 72	1.0153	1.0156	1.0160	16
17	+.001 79	+.0204	+10.23	+.008 04	1.0114	1.0117	1.0119	17
18	+.001 19	+.0136	+6.82	+.005 36	1.0076	1.0078	1.0079	18
19	+.000 60	+.0068	+3.41	+.002 68	1.0038	1.0039	1.0039	19
20	0	0	0	0	1.000	1.0000	1.0000	20
21	−.000 60	−.0068	−3.41	−.002 68	0.9962	0.9962	0.9961	21
22	−.001 19	−.0136	−6.82	−.005 36	.9925	.9924	.9922	22
23	−.001 79	−.0204	−10.23	−.008 04	.9888	.9886	.9883	23
24	−.002 39	−.0272	−13.64	−.010 72	.9851	.9848	.9845	24
25	−.002 99	−.0304	−17.05	−.013 40	.9815	.9811	.9807	25
26	−.003 58	−.0408	−20.46	−.016 07	.9779	.9774	.9770	26
27	−.004 18	−.0476	−23.87	−.018 75	.9743	.9737	.9732	27
28	−.004 78	−.0544	−27.28	−.021 43	.9707	.9701	.9695	28
29	−.005 37	−.0612	−30.69	−.024 11	.9672	.9665	.9658	29
30	−.005 97	−.0681	−34.10	−.026 79	.9636	.9629	.9622	30
35	−.008 96	−.1021	−51.15	−.040 18	.9464	.9454	.9443	35
40	−.011 94	−.1361	−68.20	−.053 58	.9298	.9285	.9271	40
45	−.014 93	−.1701	−85.25	−.066 98	.9138	.9122	.9105	45
50	−.017 92	−.2042	−102.30	−.080 37	.8983	.8964	.8945	50
55	−.020 90	−.2382	−119.35	−.093 76	.8833	.8812	.8791	55
60	−.023 89	−.2722	−136.40	−.107 16	.8689	.8665	.8642	60
65	−.026 87	−.3062	−153.45	−.120 56	.8549	.8523	.8497	65
70	−.029 86	−.3403	−170.50	−.133 95	.8413	.8385	.8358	70
75	−.032 85	−.3743	−187.55	−.147 34	.8281	.8252	.8223	75

TABLE 1i13. CONVERSION TABLE FOR ELECTRICAL RESISTIVITIES
Standard annealed copper.

Given values at at 20 °C in	To obtain values in—						
	Ohm g/m³	Ohm lb/mile²	Ohm mm²/m	Microhm-cm	Microhm-in.	Ohm-cir mil/ft	% conductivity
	multiply by	multiply by	multiply by	multiply by	multiply by	multiply by	divide into
Ohm g/m²_____	----------	5709.8	0.112 48	11.248	4.4284	67.660	15.328
Ohm lb/mi²_____	0.000 175 14	----------	.000 019 700	0.001 970 0	0.000 775 6	0.011 850	87 520
Ohm mm²/m_____	8.8900	50 763	----------	100	39.371	601.53	1.7241
Microhm-cm_____	0.088 900	507.63	0.010 000	----------	0.393 71	6.0153	172.41
Microhm-in_____	.225 81	1289.4	.025 400	2.5400	----------	15.279	67.879
Ohm-cir mil/ft___	.014 780	84.389	.001 662 4	0.166 24	.065 451	----------	1037.1
% conductivity__	divide into 15.328	divide into 87 520	divide into 1.7241	divide into 172.41	divide into 67.879	divide into 1037.1	----------

TABLE 1i14. WIRE TABLE, STANDARD ANNEALED COPPER
American Wire Gage. English units. Values at 20°C.

Gage	Diameter in mils	Cross section — Circular mils	Cross section — Square inch	Ohms per 1,000 feet	Feet per ohm	Pounds per 1,000 feet	Feet per pound	Ohms per pound	Pounds per ohm
0000	460.0	211 600	0.1662	0.049 01	20 400	640.5	1.561	0.000 076 52	13 070
000	409.6	167 800	.1318	.061 82	16 180	507.8	1.969	.000 121 7	8215
00	364.8	133 100	.1045	.077 93	12 830	402.8	2.482	.000 193 5	5169
0	324.9	105 600	.082 91	.098 25	10 180	319.5	3.130	.000 307 5	3252
1	289.3	83 690	.065 73	.1239	8070	253.3	3.947	.000 489 1	2044
2	257.6	66 360	.052 12	.1563	6398	200.9	4.978	.000 778 1	1285
3	229.4	52 620	.041 33	.1971	5074	159.3	6.278	.001 237	808.3
4	204.3	41 740	.032 78	.2485	4024	126.3	7.915	.001 967	508.5
5	181.9	33 090	.025 99	.3134	3190	100.2	9.984	.003 130	319.5
6	162.0	26 240	.020 61	.3952	2530	79.44	12.59	.004 975	201.0
7	144.3	20 820	.016 35	.4981	2008	63.03	15.87	.007 902	126.5
8	128.5	16 510	.012 97	.6281	1592	49.98	20.01	.012 57	79.58
9	114.4	13 090	.010 28	.7925	1262	39.62	25.24	.020 00	49.99
10	101.9	10 380	.008 155	.9988	1001	31.43	31.82	.031 78	31.47
11	90.7	8230	.006 46	1.26	793	24.9	40.2	.0506	19.8
12	80.8	6530	.005 13	1.59	629	19.8	50.6	.0804	12.4
13	72.0	5180	.004 07	2.00	500	15.7	63.7	.127	7.84
14	64.1	4110	.003 23	2.52	396	12.4	80.4	.203	4.93
15	57.1	3260	.002 56	3.18	314	9.87	101	.322	3.10
16	50.8	2580	.002 03	4.02	249	7.81	128	.514	1.94
17	45.3	2050	.001 61	5.05	198	6.21	161	.814	1.23
18	40.3	1620	.001 28	6.39	157	4.92	203	1.30	0.770
19	35.9	1200	.001 01	8.05	124	3.90	256	2.06	.485
20	32.0	1020	.000 804	10.1	98.7	3.10	323	3.27	.306
21	28.5	812	.000 638	12.8	78.3	2.46	407	5.19	.193
22	25.3	640	.000 503	16.2	61.7	1.94	516	8.36	.120
23	22.6	511	.000 401	20.3	49.2	1.55	647	13.1	.0761
24	20.1	404	.000 317	25.7	39.0	1.22	818	21.0	.0476
25	17.9	320	.000 252	32.4	30.9	0.970	1030	33.4	.0300
26	15.9	253	.000 199	41.0	24.4	.765	1310	53.6	.0187
27	14.2	202	.000 158	51.4	19.4	.610	1640	84.3	.0119
28	12.6	159	.000 125	65.3	15.3	.481	2080	136	.007 36
29	11.3	128	.000 100	81.2	12.3	.387	2590	210	.004 76
30	10.0	100	.000 078 5	104	9.64	.303	3300	343	.002 92
31	8.9	79.2	.000 062 2	131	7.64	.240	4170	546	.001 83
32	8.0	64.0	.000 050 3	162	6.17	.194	5160	836	.001 20
33	7.1	50.4	.000 039 6	206	4.86	.153	6550	1350	.000 742
34	6.3	39.7	.000 031 2	261	3.83	.120	8320	2170	.000 460
35	5.6	31.4	.000 024 6	331	3.02	.0949	10 500	3480	.000 287
36	5.0	25.0	.000 019 6	415	2.41	.0757	13 200	5480	.000 182
37	4.5	20.2	.000 015 9	512	1.95	.0613	16 300	8360	.000 120
38	4.0	16.0	.000 012 6	648	1.54	.0484	20 600	13 400	.000 074 7
39	3.5	12.2	.000 009 62	847	1.18	.0371	27 000	22 800	.000 043 8
40	3.1	9.61	.000 007 55	1080	0.927	.0291	34 400	37 100	.000 027 0
41	2.8	7.84	.000 006 16	1320	.756	.0237	42 100	55 700	.000 017 9
42	2.5	6.25	.000 004 91	1660	.603	.0189	52 900	87 700	.000 011 4
43	2.2	4.84	.000 003 80	2140	.467	.0147	68 300	146 000	.000 006 84
44	2.0	4.00	.000 003 14	2590	.386	.0121	82 600	214 000	.000 004 67
45	1.76	3.10	.000 002 43	3350	.299	.00938	107 000	357 000	.000 002 80
46	1.57	2.46	.000 001 94	4210	.238	.00746	134 000	564 000	.000 001 77
47	1.40	1.96	.000 001 54	5290	.189	.00593	169 000	892 000	.000 001 12
48	1.24	1.54	.000 001 21	6750	.148	.00465	215 000	1 450 000	.000 000 690
49	1.11	1.23	.000 000 968	8420	.119	.00373	268 000	2 260 000	.000 000 443
50	0.99	0.980	.000 000 770	10600	.0945	.00297	337 000	3 570 000	.000 000 280
51	0.88	0.774	.000 000 608	13400	.0747	.00234	427 000	5 710 000	.000 000 175
52	0.78	0.608	.000 000 478	17000	.0587	.00184	543 000	9 260 000	.000 000 108
53	0.70	0.490	.000 000 385	21200	.0472	.00148	674 000	14 300 000	.000 000 070 1
54	0.62	0.384	.000 000 302	27000	.0371	.00116	859 000	23 200 000	.000 000 043 1
55	0.55	0.302	.000 000 238	34300	.0292	.000916	1 090 000	37 400 000	.000 000 026 7
56	0.49	0.240	.000 000 189	43200	.0232	.000727	1 380 000	59 400 000	.000 000 016 8

NOTE 1.—The fundamental resistivity used in calculating the tables is the International Annealed Copper Standard, viz, 0.153 28 ohm-g/m² at 20 °C. The temperature coefficient, for this particular resistivity, is $\alpha_{20} = 0.003\ 93$ per °C. or $\alpha_0 = 0.004\ 27$. However, the temperature coefficient is proportional to the conductivity, and hence the change of resistivity per °C is a constant, 0.000 597 ohm-g/m². The "constant mass" temperature coefficient of any sample is

$$\alpha_t = \frac{0.000\ 597 + 0.000\ 005}{\text{resistivity in ohm-g/m² at } t\ °C}$$

The density is 8.89 g/cm³ at 20 °C.

NOTE 2.—The values given in the table are only for annealed copper of the standard resistivity. The user of the table must apply the proper correction for copper of any other resistivity. Hard-drawn copper may be taken as about 2.5 percent higher resistivity than annealed copper.

TABLE 1i15. WIRE TABLE, STANDARD ANNEALED COPPER
American Wire Gage. English units.
Ohms per 1,000 feet. 0 to 200°C.

Gage	Diameter at 20 °C mils	Cross section at 20 °C		Ohms per 1,000 feet [14] at the temperature of—						
		Circular mils	Square inch	0 °C	20 °C	25 °C	50 °C	75 °C	100 °C	200 °C
0000	460.0	211 600	0.1662	0.045 16	0.049 01	0.049 98	0.054 79	0.059 61	0.064 42	0.0836 9
000	409.6	167 800	.1318	.056 96	.061 82	.063 03	.069 11	.075 18	.081 25	.1055
00	364.8	133 100	.1045	.071 81	.077 93	.079 46	.087 12	.094 78	.1024	.1331
0	324.9	105 600	.0829 1	.090.53	.098 25	.1002	.1098	.1195	.1291	.1678
1	289.3	83 690	.065 73	.1142	.1239	.1264	.1385	.1507	.1629	.2116
2	257.6	66 360	.052 12	.1440	.1563	.1594	.1747	.1901	.2054	.2669
3	229.4	52 620	.041 33	.1816	.1971	.2010	.2203	.2397	.2590	.3365
4	204.3	41 740	.032 78	.2289	.2485	.2534	.2778	.3022	.3266	.4243
5	181.9	33 090	.025 99	.2888	.3134	.3196	.3504	.3812	.4120	.5352
6	162.0	26 240	.020 61	.3641	.3952	.4029	.4418	.4806	.5194	.6747
7	144.3	20 820	.016 35	.4589	.4981	.5079	.5568	.6057	.6547	.8504
8	128.5	16 510	.012 97	.5787	.6281	.6404	.7021	.7639	.8256	1.072
9	114.4	13 090	.010 28	.7302	.7925	.8080	.8859	.9637	1.042	1.353
10	101.9	10 380	.008 155	.9203	.9988	1.018	1.117	1.215	1.313	1.705
11	90.7	8230	.006 46	1.16	1.26	1.29	1.41	1.53	1.66	2.15
12	80.8	6530	.005 13	1.46	1.59	1.62	1.78	1.93	2.09	2.71
13	72.0	5180	.004 07	1.84	2.00	2.04	2.24	2.43	2.63	3.42
14	64.1	4110	.003 23	2.33	2.52	2.57	2.82	3.07	3.32	4.31
15	57.1	3260	.002 56	2.93	3.18	3.24	3.56	3.87	4.18	5.43
16	50.8	2580	.002 03	3.70	4.02	4.10	4.49	4.89	5.28	6.86
17	45.3	2050	.001 61	4.66	5.05	5.15	5.65	6.15	6.64	8.63
18	40.3	1620	.001 28	5.88	6.39	6.51	7.14	7.77	8.39	10.9
19	35.9	1290	.001 01	7.41	8.05	8.21	9.00	9.79	10.6	13.7
20	32.0	1020	.000 804	9.33	10.1	10.3	11.3	12.3	13.3	17.3
21	28.5	812	.000 638	11.8	12.8	13.0	14.3	15.5	16.8	21.8
22	25.3	640	.000 503	14.9	16.2	16.5	18.1	19.7	21.3	27.7
23	22.6	511	.000 401	18.7	20.3	20.7	22.7	24.7	26.7	34.7
24	20.1	404	.000 317	23.7	25.7	26.2	28.7	31.2	33.7	43.8
25	17.9	320	.000 252	29.8	32.4	33.0	36.2	39.4	42.5	55.3
26	15.9	253	.000 199	37.8	41.0	41.8	45.9	49.9	53.9	70.0
27	14.2	202	.000 158	47.4	51.4	52.4	57.5	62.6	67.6	87.8
28	12.6	159	.000 125	60.2	65.3	66.6	73.0	79.4	85.9	112
29	11.3	128	.000 100	74.8	81.2	82.8	90.8	98.8	107	139
30	10.0	100	.000 078 5	95.6	104	106	116	126	136	177
31	8.9	79.2	.000 062 2	121	131	134	146	159	172	224
32	8.0	64.0	.000 050 3	149	162	165	181	197	213	277
33	7.1	50.4	.000 039 6	190	206	210	230	250	270	351
34	6.3	39.7	.000 031 2	241	261	266	292	318	343	446
35	5.6	31.4	.000 024 6	305	331	337	370	402	435	565
36	5.0	25.0	.000 019 6	382	415	423	464	505	545	708
37	4.5	20.2	.000 015 9	472	512	522	573	623	673	874
38	4.0	16.0	.000 012 6	597	648	661	725	788	852	1110
39	3.5	12.9	.000 009 62	780	847	863	946	1030	1110	1450
40	3.1	9.61	.000 007 55	994	1080	1100	1210	1310	1420	1840
41	2.8	7.84	.000 006 16	1220	1320	1350	1480	1610	1740	2260
42	2.5	6.25	.000 004 91	1530	1660	1690	1860	2020	2180	2830
43	2.2	4.84	.000 003 80	1970	2140	2180	2400	2610	2820	3660
44	2.0	4.00	.000 003 14	2390	2590	2640	2900	3150	3410	4430
45	1.76	3.10	.000 002 43	3080	3350	3410	3740	4070	4400	5720
46	1.57	2.46	.000 001 94	3880	4210	4290	4700	5120	5530	7180
47	1.40	1.96	.000 001 54	4880	5290	5400	5920	6440	6960	9030
48	1.24	1.54	.000 001 21	6210	6750	6880	7540	8200	8870	11500
49	1.11	1.23	.000 000 968	7760	8420	8580	9410	10200	11100	14400
50	0.99	0.980	.000 000 770	9750	10600	10800	11800	12900	13900	18100
51	0.88	0.774	.000 000 608	12300	13400	13700	15000	16300	17600	22900
52	0.78	0.608	.000 000 478	15700	17000	17400	19100	20700	22400	29100
53	0.70	0.490	.000 000 385	19500	21200	21600	23700	25700	27800	36100
54	0.62	0.384	.000 000 302	24900	27000	27500	30200	32800	35500	46100
55	0.55	0.302	.000 000 238	31600	34300	35000	38300	41700	45100	58500
56	0.49	0.240	.000 000 189	39800	43200	44000	48300	52500	56800	73800

[14] Resistance at the stated temperatures of a wire whose length is 1,000 feet at 20 °C.

TABLE 1i16. BARE CONCENTRIC-LAY STRANDED CONDUCTORS OF STANDARD ANNEALED COPPER
English units.

Nominal size of conductor		Ohms per 1,000 feet		Standard concentric stranding (Class B)				Flexible concentric stranding (Class C)		
Circular mils	AWG	25 °C (=77 °F)	65 °C (=149 °F)	Pounds per 1000 feet	Number of wires	Diameter of wires	Outside diameter	Number of wires	Diameter of wires	Outside diameter
						Mils	*Mils*		*Mils*	*Mils*
5 000 000	----	0.002 22	0.002 56	15 890	217	151.8	2580	271	135.8	2580
4 500 000	----	.002 47	.002 85	14 300	217	144.0	2450	271	128.9	2450
4 000 000	----	.002 75	.003 17	12 600	217	135.8	2310	271	121.5	2310
3 500 000	----	.003 14	.003 63	11 020	169	143.9	2160	217	127.0	2160
3 000 000	----	.003 63	.004 19	9 349	169	133.2	2000	217	117.6	2000
2 500 000	----	.004 36	.005 03	7 794	127	140.3	1820	169	121.6	1820
2 000 000	----	.005 39	.006 22	6 176	127	125.5	1630	169	108.8	1630
1 900 000	----	.005 68	.006 55	5 865	127	122.3	1590	169	106.0	1590
1 800 000	----	.005 99	.006 92	5 562	127	119.1	1550	169	103.2	1450
1 700 000	....	.006 34	.007 32	5 249	127	115.7	1500	169	100.3	1500
1 600 000	....	.006 74	.007 78	4 936	127	112.2	1460	169	97.3	1460
1 500 000	....	.007 19	.008 30	4 632	91	128.4	1410	127	108.7	1410
1 400 000	....	.007 70	.008 89	4 320	91	124.0	1360	127	105.0	1360
1 300 000	....	.008 30	.009 58	4 012	91	119.5	1310	127	101.2	1320
1 200 000	....	.008 99	.010 4	3 703	91	114.8	1260	127	97.2	1260
1 100 000	....	.009 81	.011 3	3 394	91	109.9	1210	127	93.1	1210
1 000 000	....	.010 8	.012 4	3 086	61	128.0	1150	91	104.8	1150
950 000	----	.011 4	.013 1	2 933	61	124.8	1120	91	102.2	1120
900 000	----	.012 0	.013 8	2 780	61	121.5	1090	91	99.4	1090
850 000	----	.012 7	.014 6	2 622	61	118.0	1060	91	96.6	1060
800 000	----	.013 5	.015 6	2 469	61	114.5	1030	91	93.8	1030
750 000	----	.014 4	.016 6	2 316	61	110.9	998	91	90.8	1000
700 000	----	.015 4	.017 8	2 160	61	107.1	964	91	87.7	965
650 000	----	.016 6	.019 2	2 006	61	103.2	929	91	84.5	930
600 000	----	.018 0	.020 7	1 850	61	99.2	893	91	81.2	893
550 000	----	.019 6	.022 6	1 700	61	95.0	855	91	77.7	855
500 000	----	.021 6	.024 9	1 542	37	116.2	813	61	90.5	814
450 000	----	.024 0	.027 7	1 390	37	110.3	772	61	85.9	773
400 000	----	.027 0	.031 1	1 236	37	104.0	728	61	81.0	729
350 000	----	.030 8	.035 6	1 080	37	97.3	681	61	75.7	681
300 000	----	.036 0	.041 5	925	37	90.0	630	61	70.1	631
250 000	----	.043 1	.049 8	772	37	82.2	575	61	64.0	576
211 600	0000	.050 9	.058 7	653	19	105.5	528	37	75.6	529
167 800	000	.064 2	.074 1	518	19	94.0	470	37	67.3	471
133 100	00	.081 1	.093 6	411	19	83.7	418	37	60.0	420
105 600	0	.102	.117	326	19	74.5	372	37	53.4	374
83 690	1	.129	.149	259	19	66.4	332	37	47.6	333
66 360	2	.162	.187	205	7	97.4	292	19	59.1	296
52 620	3	.205	.237	162	7	86.7	260	19	52.6	263
41 740	4	.259	.299	129	7	77.2	232	19	46.9	234
33 090	5	.326	.376	102	7	68.8	206	19	41.7	208
26 240	6	.410	.473	80.9	7	61.2	184	19	37.2	186
20 820	7	.519	.599	64.2	7	54.5	164	19	33.1	166
16 510	8	.654	.755	51.0	7	48.6	146	19	29.5	148

NOTE 1.—The fundamental resistivity used in calculating the table is the International Annealed Copper Standard, viz, 0.15328 ohm-g/m² at 20 °C. The temperature coefficient is given in table 3. The density is 8.89 grams per cubic centimeter at 20 °C.

NOTE 2.—The values given for "Ohms per 1,000 feet" and "Pounds per 1,000 feet" are 2 to 5 percent greater than for a solid rod of cross section equal to the total cross section of the wires of the stranded conductor. See p. 12. The values of "pounds per 1,000 feet" are correct for Class B stranding and approximate for Class C stranding. The "ohms per 1,000 feet" are approximate for either stranding.

TABLE 1i17. COMPLETE WIRE TABLE, STANDARD ANNEALED COPPER, 20°C
American Wire Gage. Metric units.

Gage	Diameter	Cross section	Ohms per kilometer	Meters per ohm	Kilograms per kilometer	Meters per gram	Ohms per kilogram	Grams per ohm
	m m	*sq mm*						
0000	11.68	107.2	0.160 8	6 219	953.2	0.001 049	0.000 168 7	5 928 000
000	10.40	85.01	.202 8	4 931	755.8	.001 323	.000 268 4	3 726 000
00	9.266	67.43	.255 7	3 911	599.5	.001 668	.000 426 5	2 345 000
0	8.252	53.49	.322 3	3 102	475.5	.002 103	.000 677 9	1 475 000
1	7.348	42.41	.406 5	2 460	377.0	.002 652	.001 078	927 300
2	6.543	33.62	.512 8	1 950	298.9	.003 345	.001 715	582 900
3	5.827	26.67	.646 6	1 547	237.1	.004 218	.002 728	366 600
4	5.189	21.15	.815 2	1 227	188.0	.005 319	.004 336	230 600
5	4.620	16.77	1.028	972.4	149.0	.006 709	.006 900	144 900
6	4.115	13.30	1.297	771.3	118.2	.008 459	.010 97	91 180
7	3.665	10.55	1.634	612.0	93.80	.010 66	.017 42	57 400
8	3.264	8.367	2.061	485.3	74.38	.013 44	.027 70	36 100
9	2.906	6.631	2.600	384.6	58.95	.016 96	.044 10	22 680
10	2.588	5.261	3.277	305.2	46.77	.021 38	.070 06	14 270
11	2.30	4.17	4.14	242	37.1	.027 0	.112	8 960
12	2.05	3.31	5.21	192	29.4	.034 0	.177	5 640
13	1.83	2.63	6.56	152	23.4	.042 8	.281	3 560
14	1.63	2.08	8.28	121	18.5	.054 0	.447	2 230
15	1.45	1.65	10.4	95.8	14.7	.068 1	.711	1 410
16	1.29	1.31	13.2	75.8	11.6	.086 0	1.13	882
17	1.15	1.04	16.6	60.3	9.24	.108	1.79	557
18	1.02	0.823	21.0	47.7	7.32	.137	2.86	349
19	0.912	.653	26.4	37.9	5.81	.172	4.55	220
20	.813	.519	33.2	30.1	4.61	.217	7.20	139
21	.724	.412	41.9	23.9	3.66	.273	11.4	87.3
22	.643	.324	53.2	18.8	2.88	.347	18.4	54.2
23	.574	.259	66.6	15.0	2.30	.435	29.0	34.5
24	.511	.205	84.2	11.9	1.82	.549	46.3	21.6
25	.455	.162	106	9.42	1.44	.693	73.6	13.6
26	.404	.128	135	7.43	1.14	.878	118	8.46
27	.361	.102	169	5.93	0.908	1.10	186	5.38
28	.320	.080 4	214	4.67	.715	1.40	300	3.34
29	.287	.064 7	266	3.75	.575	1.74	463	2.16
30	.254	.050 7	340	2.94	.450	2.22	755	1.32
31	.226	.040 1	430	2.33	.357	2.80	1 200	0.831
32	.203	.032 4	532	1.88	.288	3.47	1 840	.542
33	.180	.025 5	675	1.48	.227	4.40	2 970	.336
34	.160	.020 1	857	1.17	.179	5.59	4 800	.209
35	.142	.015 9	1 090	0.922	.141	7.08	7 680	.130
36	.127	.012 7	1 360	.735	.113	8.88	12 100	.082 7
37	.114	.010 3	1 680	.595	.091 2	11.0	18 400	.054 3
38	.102	.008 11	2 130	.470	.072 1	13.9	29 500	.033 9
39	.089	.006 21	2 780	.360	.055 2	18.1	50 300	.019 9
40	.079	.004 87	3 540	.282	.043 3	23.1	81 800	.012 2
41	.071	.003 97	4 340	.230	.035 3	28.3	123 000	.008 14
42	.064	.003 17	5 440	.184	.028 2	35.5	193 000	.005 17
43	.056	.002 45	7 030	.142	.021 8	45.9	322 000	.003 10
44	.051	.002 03	8 510	.118	.018 0	55.5	472 000	.002 12
45	.0047	.001 57	11 000	.0910	.014 0	71.7	787 000	.001 27
46	.0399	.001 25	13 800	.0724	.011 1	90.1	1 240 000	.000 804
47	.0356	.000 993	17 400	.0576	.008 83	113	1 970 000	.000 509
48	.0315	.000 779	22 100	.0452	.006 93	144	3 200 000	.000 313
49	.0282	.000 624	27 600	.0362	.005 55	180	4 980 000	.000 201
50	.0251	.000 497	34 700	.0288	.004 41	226	7 860 000	.000 127
51	.0224	.000 392	43 900	.0228	.003 49	287	12 600 000	.000 079 4
52	.0198	.000 308	55 900	.0179	.002 74	365	20 400 000	.000 049 0
53	.0178	.000 248	69 400	.0144	.002 21	453	31 500 000	.000 031 8
54	.0157	.000 195	88 500	.0113	.001 73	578	51 100 000	.000 019 6
55	.0140	.000 153	112 000	.008 89	.001 36	734	82 500 000	.000 012 1
56	.0124	.000 122	142 000	.007 06	.001 08	925	131 000 000	.000 007 63

DIELECTRIC CONSTANTS

The following set of three summaries of dielectric constants is reproduced from Akawie, R.I. and J. T. Milek, "Dielectric Constants of Rubbers, Plastics and Ceramics: a Design Guide," Interim Report No. 67, Electronic Properties Information Center, Hughes Aircraft Co., 1969.

Dielectric constant values apply to one Megahertz, unless otherwise denoted by the following letters:

(a) Frequency unknown
(b) 60 Hertz
(c) 1 Kilohertz
(d) 100 Kilohertz
(e) 9375 Megahertz

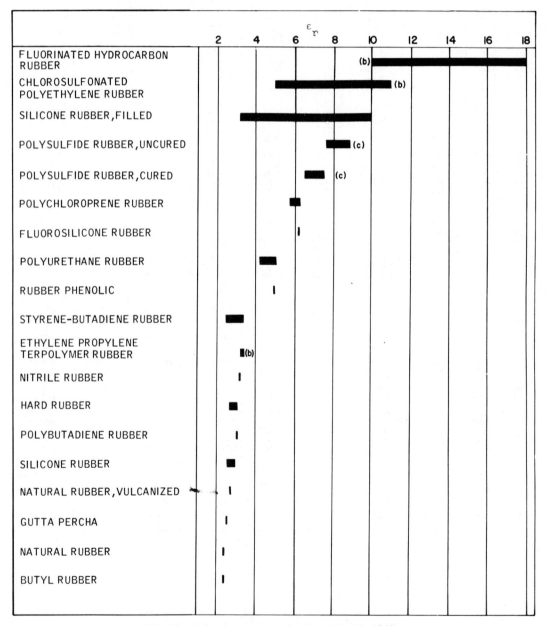

Fig. 1i1. Dielectric constants of rubbers (Akawie 1969).

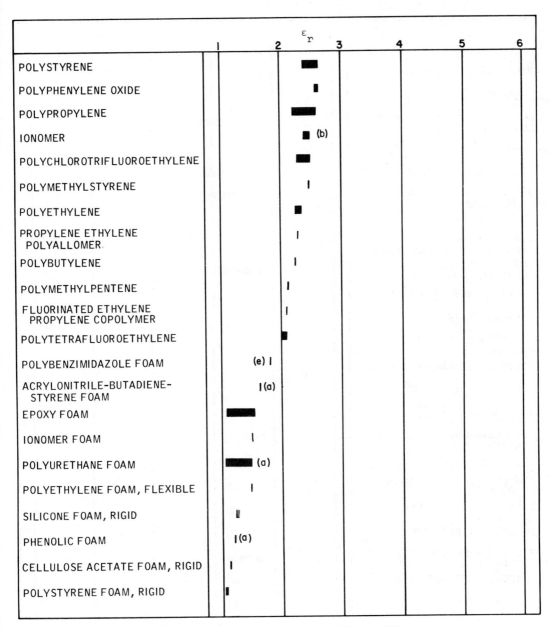

Fig. 1i2. Dielectric constants of plastics (Akawie 1969).

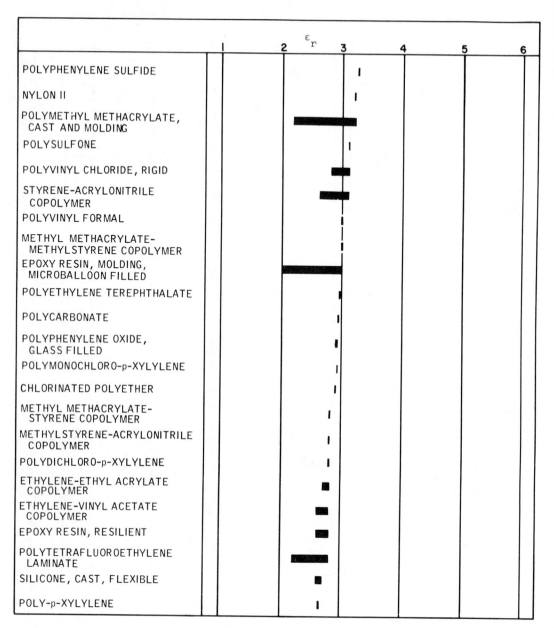

Fig. 1i2. (*Continued*)

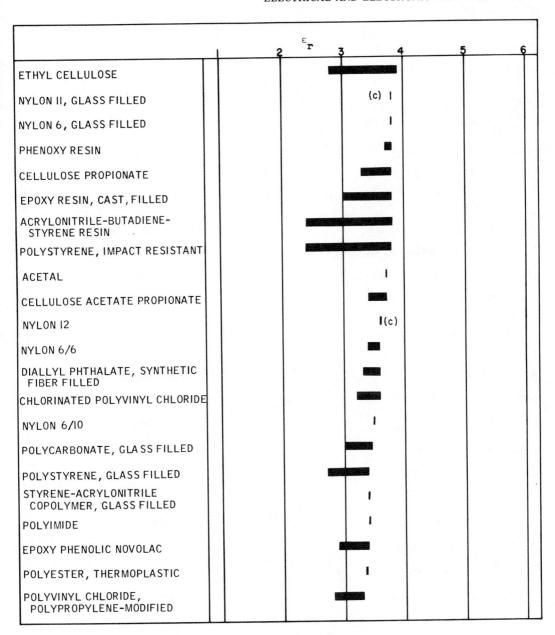

Fig. 1i2. (*Continued*)

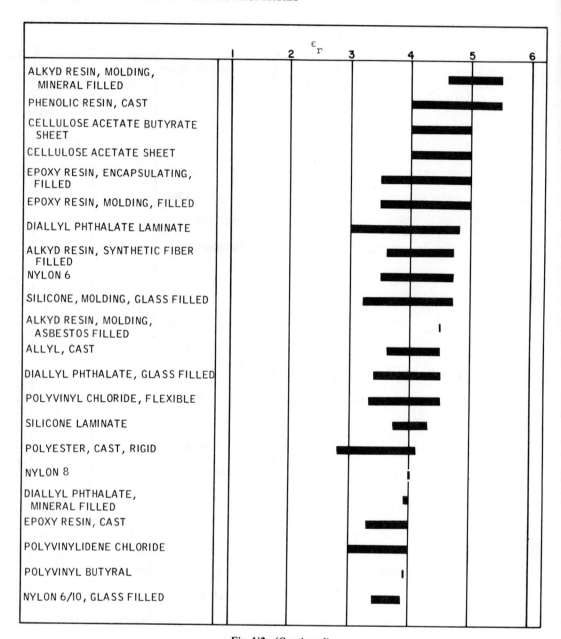

Fig. 1i2. (*Continued*)

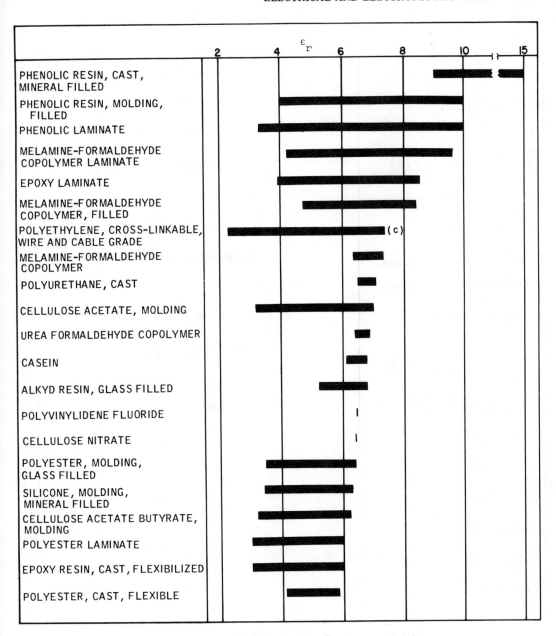

Fig. 1i2. (*Continued*)

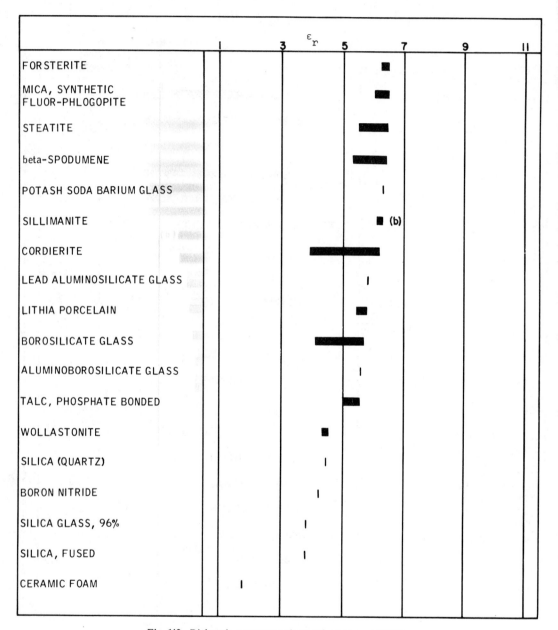

Fig. 1i3. Dielectric constants of ceramics (Akawai 1969).

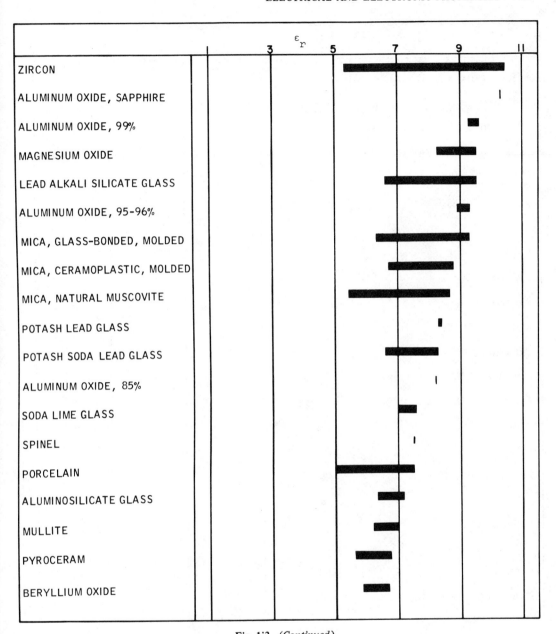

Fig. 1i3. (*Continued*)

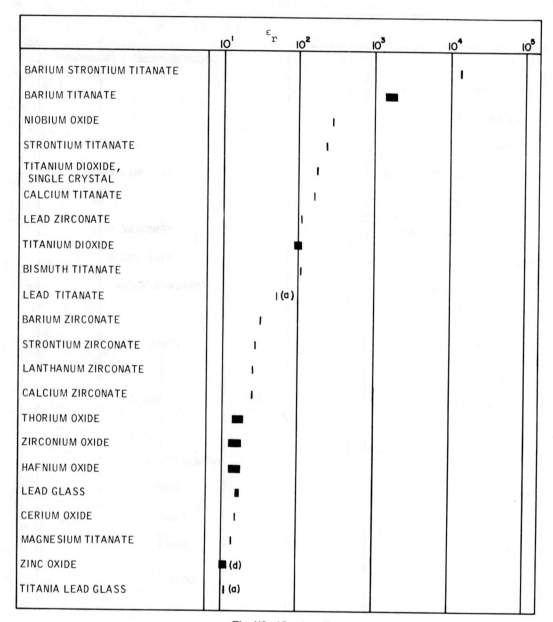

Fig. 1i3. (*Continued*)

PROPERTIES OF PIEZOELECTRIC MATERIALS

Piezoelectric materials are divided into ceramic and non-ceramic substances, and the information is drawn largely from the following sources:

Brown, C. S., R. C. Kell, R. Taylor and L. A. Thomas, "Piezoelectric Materials, a Review of Progress," *Proc. IEE*, **109B**, No. 43, 99–114 (Jan. 1962).

Mason, W. P., "Piezoelectric Crystals and their Appli-

cation to Ultrasonics," Van Nostrand Reinhold, New York, 1950.

Jaffe, H., "Piezoelectric Applications of Ferroelectrics," *IEEE Trans. Electron Devices*, **ED-16**, No. 6, 557–561 (June 1969).

Jaffe, H. and D. Berlincourt, "Piezoelectric Transducer Materials," *Proc. IEEE*, **53**, 1372–1386 (Oct. 1965).

Milek, J. T. and S. J. Welles, "Linear Electrooptic Modulator Materials," U.S. Govt. Report AD 704556, Jan. 1970.

TABLE 1i18. PROPERTIES OF CERAMIC PIEZOELECTRIC MATERIALS

Material	Density g/cm^3	Curie Temp., $°K$	Relative Permittivity, ϵ_{33}/ϵ_0	Mechanical Q-Factor (radial mode) Q_r	Elastic Compliance $10^{-12} m^2/N$ S_{11}	Electromechanical Coupling Factor			Piezoelectric Strain Constant			
									10^{-12} C/N		10^{-3} Vm/N	
						k_r	k_{31}	k_{33}	d_{31}	d_{33}	g_{31}	g_{33}
$BaTiO_3$	5.72	388	1700	400	8.55	0.35	0.21	0.49	−79	191	4.7	11.4
$(Ba_{0.80}Pb_{0.08}Ca_{0.12})\ TiO_3$	5.3	433	600	350	13	0.22	0.12*	0.30*	−35	90*	−7.3	18*
$Ba(Ti_{0.95}Zr_{0.05})O_3$	5.44	378	1400	200	11	0.28	0.15*	0.40*	−60	150*	−4.8	12.1*
$Pb\ Nb_2O_6$	6.0	843	225	11		0.07		0.42	11	80		
$(Pb_{0.70}Ba_{0.30})\ Nb_2O_6$	5.9	613	900	350	10.9	0.24	0.13*	0.33	−40	100*	−5.0	12.5*
$(Pb_{0.60}Ba_{0.40})\ Nb_2O_6$	5.9	533	1500	250	11.5	0.38	0.22*	0.55*	−90	220*	−6.8	16.5*
$Pb(Ti_{0.48}Zr_{0.52})\ O_3$		643	730		13.8				−93.5	223	−14.5	34.5
$Pb(Ti_{0.48}Zr_{0.52})\ O_3$†	7.6	593	1200	500	14.6	0.55	0.30	0.76	−130	300	−12.3	28.3
$(Na_{0.8}Cd_{0.1})\ NbO_3$	4.3	513	2000		10.5	0.30	0.17*	0.42*	−80	200*	−4.5	11.3*
$(Na_{0.75}Cd_{0.125})\ NbO_3$	4.4	473	2000		9.0	0.26	0.14*	0.35*	−60	150*	−3.4	8.5*

†Commercial grade.
*Estimates.

TABLE 1i19. PROPERTIES OF NON-CERAMIC PIEZOELECTRIC MATERIALS

Material	Class#	Density, g/cm^3	Curie Temp., °K	Relative Permittivity			Electromechanical Coupling Factor				Elastic Compliance, 10^{-12}m^2/N			
				ϵ_{11}/ϵ_0	ϵ_{22}/ϵ_0	ϵ_{33}/ϵ_0	k_{14}	k_{15}	k_{33}	k_{36}	S_{11}	S_{12}	S_{13}	S_{14}
Aluminum phosphate	32	2.57		6.05							16.1	−0.1	−8.3	8.9
Ammonium dideuterium phosphate	4̄2m		242	73		22					19.3	2	−11.5	
Ammonium dihydrogen arsenate	4̄2m	2.31	216	75		12	0.136			0.24	19.3	6.5	−14.1	
Ammonium dihydrogen phosphate	4̄2m	1.80	148	57.6		14	0.006			0.33	18.1	1.9	−11.8	
Barium sodium niobate	mm2	5.39	833	235	247	51		0.21			5.30	−1.98	−1.20	
Cadmium sulfide	6mm	4.82		9.35		10.33					20.69	9.99	−5.81	
Dipotassium tartrate	2			6.49	5.80	6.44					47.5	−17.4	−8	
Ethylenediamine tartrate	2			4.96	7.74	5.26					33.4	−3	−32.8	
Gallium arsenide	4̄3m	5.32					0.092				11.76	3.67		
Lithium niobate	3m	4.64	1483	84		30		0.446	0.327		5.20	−0.44	−1.45	0.87
Lithium sulfate monohydrate	2	1.47		5.16	10.3	4.95					22.9	−5.4	−7.5	
Lithium tantalate	3m	7.45	933	52		45		0.311	0.20		4.68	−0.16	−1.17	1.10
Potassium dideuterium phosphate	4̄2m	2.34	222	57		51				0.22				
Potassium dihydrogen arsenate	4̄2m	2.87	95.6	52		22	0.095			0.13	19	2	1	
Potassium dihydrogen phosphate	4̄2m	2.34	123	43.7		21.8	0.008			0.121	15.1	1.8	−4.0	
α-Quartz	32	2.65		4.52		4.64					12.77	−1.79	−1.22	4.50
β-Quartz**	622										9.05	−0.81	−2.76	
Rochelle salt	222	1.79	297	1100	11.1	9.2					52*	−15.3*	−17*	
Tourmaline	3m	2.9–3.2		8.2		7.5					3.85	−0.48	−0.71	0.45
Zinc oxide	6m	5.67				~8.2					7.86	−3.43	−2.21	

TABLE 1i19. (Continued)

S_{15}	S_{22}	S_{33}	S_{44}	S_{55}	S_{66}	d_{11}	d_{14}	d_{15}	d_{16}	d_{21}	d_{22}	d_{25}	d_{31}	d_{32}	d_{33}	d_{35}	d_{36}
		16.1	53		32.2	±3.3	±1.5										
		44	110		164		10										75
		48.6	146		156		41										31
			115.3		164.6		1.76										−48.3
	5.14	8.33	15.4	15.2	13.2		42						−7	−6	37		
		16.97	66.49		61.36		−13.98						−5.18		10.32		
−7.5	35.3	24.0	113.5	102	122.5		23.0		13	5.4	−4.5	6.3					−8.1
−17	36.5	100.2	191.8	122	191.4		−10.0		−12.2	10.1	2.2	−17.9					−18.4
		16.83					2.6										
		4.89	10.8		11.3			74			21		−0.86		16.2		
−2.1	22.5	22.8	71.3	64.0	36.1		−3.9		−3.1	−0.2	16.3	−7.1					1.1
		4.14	9.0		9.7			26			8.5		−3.0		9.2		
		77.7			164.7		3.3										52
	27	86	152				23.5										22
		19.5	78.1		162		1.3										−20.9
		9.60	20.04		29.12	2.31	0.73										
		10.55	27.59		19.72		1.82										
	35*	33.5*	79*	330*	102*		2300					−56					12
		6.36	15.4						−3.63		−0.33		−0.34	−1.83			
		6.94	23.57		22.58				−8.3				−5.0		12.4		

*At 30°C.
**At 625°C.
#Three-dimensional point groups, as defined in "International Tables for X-Ray Crystallography," Kynoch Press, Birmingham, England (1952).

Units: m = meter; N = Newton = 10^5 dynes; C = Coulomb = 3×10^9 statcoulombs; ϵ_0 = dielectric constant of free space = 8.854×10^{-12} Farad/meter.

TABLE 1i20. PROPERTIES OF LINEAR ELECTRO-OPTIC MODERATOR MATERIALS*

Material		Cuprous Chloride	Bismuth Germanium Oxide	Lithium Niobate		Lithium Tantalate	Proustite	Ammonium Dihydrogen Arsenate (ADA)	
Chemical Formula		CuCl	$Bi_{12}GeO_{20}$	$LiNbO_3$		$LiTaO_3$	Ag_3AsS_3	$NH_4H_2AsO_4$	
Crystal Symmetry		Cubic		Trigonal					
		$\bar{4}3m\,(T_d)$	23 (T)	$3m\,(C_{3v})$					
Curie Temperature (°C)				1195		660		-57 [c]	
Transmission Range, >50% (microns)		0.4 to 20	0.5 to 7.5	0.4 to 5		0.3 to 6	0.6 to 13	0.3 to >0.75	
λ (Å)		5460	5100	6328		6328	6328	5460	
Refractive Indices		1.99	2.55	n_o	2.291	2.177	3.019	n_o	1.580
				n_e	2.200	2.182	2.739	n_e	1.525
Electrooptic Coefficients [d] (10^{-12} m/V)	Unclamped, Constant Stress (T)	r_{41} 5.6	3	r_{13}	9.6		2.8	r_{41}	
				r_{22}	6.7		3.2		
				r_{33}	31		1.6	r_{63}	9.2
				r_{42}	32				
				r_c	15.8	21.6	1.2		
	Clamped, Constant Strain (S)	r_{41}		r_{13}	8.6	7.0		r_{41}	
				r_{22}	3.4	1.07			
				r_{33}	30.8	30		r_{63}	
				r_{42}	28				
				r_c	18.6	23.5	5.1		
Relative Dielectric Constants	Constant Stress (T)	10		ϵ_{33}/ϵ_o	30	45	20	ϵ_{33}/ϵ_o	14
				ϵ_{11}/ϵ_o	84	52	16.5	ϵ_{11}/ϵ_o	75
	Constant Strain (S)	8.3	38	ϵ_{33}/ϵ_o	29	43	18	ϵ_{33}/ϵ_o	73
				ϵ_{11}/ϵ_o	44	41	14.5	ϵ_{11}/ϵ_o	74
Loss Tangents	At 1 kHz					< 0.05		$\tan\delta_c$	7
								$\tan\delta_a$	0.9
	At 9.2 GHz	0.002		$\tan\delta_c$	0.01 [e]	0.002 [e]	0.005 [f]	$\tan\delta_c$	
				$\tan\delta_a$	0.08 [e]		0.02 [f]	$\tan\delta_a$	
Piezoelectric Constants (10^{-12} C/N or m/V)				d_{15}	68-74	26		d_{14}	41
				d_{22}	21	8			
				d_{31}	-1	-3		d_{36}	31
				d_{33}	6-16	9			
Electromechanical Coupling Coefficients			0.30	k_{15}	0.45	0.31		k_{14}	0.136
				k_{22}	0.25-0.32				
				k_{31}	0.023-0.087			k_{36}	0.24
				k_{33}	0.33-0.47	0.20-0.30			
Nonlinear Second Harmonic Generation Coefficients, at 1.06 microns (Relative to $d_{36}^{2\omega} = 1$ for KDP)				$d_{22}^{2\omega}$	6.3	4.3	50	$d_{14}^{2\omega}$	
				$d_{31}^{2\omega}$	11.9	2.6	30		
				$d_{33}^{2\omega}$	83	40		$d_{36}^{2\omega}$	

(a) Properties Depend upon Degree of Deuteration

(b) Below T_c

(c) Antiferroelectric Transition

(d) At Refractive Index wavelength

$r_c = (n_e/n_o)^3\, r_{33} - r_{13}$

$r_{c1} = r_{22} - (n_x/n_y)^3\, r_{12}$

$r_{c3} = r_{22} - (n_z/n_y)^3\, r_{32}$

(e) At 100 MHz

(f) At 20 MHz

TABLE 1i20. (Continued)

Ammonium Dihydrogen Phosphate (ADP)	Potassium Dihydrogen Arsenate (KDA)	Potassium Dideuterium Phosphate (KDDP) [a]	Potassium Dihydrogen Phosphate (KDP)	Potassium Tantalate Niobate (KTN)			Barium Sodium Niobate		Calcium Pyroniobate	
$NH_4H_2PO_4$	KH_2AsO_4	KD_2PO_4	KH_2PO_4	$KTa_xNb_{1-x}O_3$			$Ba_2NaNb_5O_{15}$		$Ca_2Nb_2O_7$	
Tetragonal							Orthorhombic		Monoclinic	
$\bar{4}2m$ (D_{2d})				$4mm$ (C_{4v}) [b]			$mm2$ (C_{2v})		2 (C_2)	
-125 [c]	-177	-53	-150	10 (x = 0.65)			560			
0.2 to 1.2	0.3 to 1.6	0.2 to 2	0.3 to 1.4	0.2 to 1.5			0.4 to 5			
6328	5460	5460	5460	5460			6328		6328	
				T_c-T	16°C	36°C	n_x	2.326	1.97	
1.522	1.571	1.508	1.512	n_o	2.281	2.275	n_y	2.324	2.16	
1.478	1.521	1.468	1.470	n_e	2.318	2.318	n_z	2.221	2.17	
24	12.5	8.8	8.7	r_{42}	3000	14,000	r_{13}	18.3	r_{12}	
									r_{22}	
							r_{23}	13.5	r_{32}	
									r_{41}	
8.5	10.9	26.4	10; 5	r_c	730	480	r_{33}	56	r_{c1}	12
									r_{c3}	14
									r_{12}	6.7
									r_{22}	25
									r_{32}	6.4
									r_{41}	2.7
4.5			9.5						r_{c1}	20
									r_{c3}	19
16	21.0	51	21.6	8000 (T_c = 271°K; x = 0.66)			ϵ_{33}/ϵ_o	51		
56	53.7	57	44				ϵ_{11}/ϵ_o	235		
14	20.6	50	19.6				ϵ_{22}/ϵ_o	247		
							ϵ_{33}/ϵ_o	32	45	
55	53	57	42				ϵ_{11}/ϵ_o	222	45	
							ϵ_{22}/ϵ_o	227		
0.04	0.25	0.42	0.005	0.01 (x = 0.80)						
0.0046	0.14		0.0015							
0.006	0.008	0.13	0.0074							
0.0072	0.007	0.02	0.0045							
1.7	25	3.3	1.3				d_{15}	42		
							d_{24}	52		
49	22	58	23				d_{31}	-7		
							d_{32}	-6		
							d_{33}	37		
0.006	0.095		0.008				k_{15}	0.21		
							k_{24}	0.25		
							k_{31}	0.14	k_{33}	0.30
0.33	0.13	0.22	0.121				k_{32}	0.13		
							k_t	0.57		
0.98	1.12	0.91	1.01				$d_{31}^{2\omega}$	29		
							$d_{32}^{2\omega}$	33.9		
0.99	1.06	0.92	1.00				$d_{33}^{2\omega}$	34.4		

*Reproduced from Milek, J. T. and S. J. Welles, "Linear Electro-optic Modulator Materials," Report No. AD 704 556, National Technical Information Service, Springfield, Va., 1970.

TABLE 1i21. FUNDAMENTAL DATA FOR FERROELECTRICS.

| Composition | Symmetry | | Structural Data | | | | | | Curie-Weiss Law | |
| | | | Lattice Parameters | | | | | | | |
Perovskite Type	Class	Group	a Å	b Å	c Å	c/a	°K	References	$x10^4$ C	$T_c - T_o$
$BaTiO_3$	cubic	m3m	4.01				473			
	tetra	4mm	3.992		4.036	1.010	293	P1	17.3	11
	ortho	mm	5.682	5.669	3.990	1.006	263	J1		
	rhomb	3m	3.998				173			
$SrTiO_3$	cubic		3.905					J1	8.3	38
$PbTiO_3$	cubic	m3m						P1		
	tetra	4mm	3.904		4.150	1.063	293	M1	11.0	
$CdTiO_3$	ortho		10.834	10.695	7.615		293	J1	4.5	
$KNbO_3$	cubic	m3m	4.021					P1		
	tetra	4mm	3.997		1.017	1.017		J1	24.0	58
	ortho	mm2	4.038		3.971	0.983		M1		
	rhomb	3m	4.016							
$NaNbO_3$	cubic		3.942					P1		
	tetra		2x3.933		4x3.942	1.0015		J1		
	tetra		2x3.924		4x3.924	1.0023		M1		
	ortho	222	5.568	5.505	15.518	1.009	293	M21		
	monoc		5.564	5.548	7.812		113			
$AgNbO_3$	cubic		3.96				923	P1		
	tetra		3.95		3.96		823	J1	18	
	ortho		3.944	3.944	3.926	0.993	393			
$KTaO_3$	cubic	m3m	3.989					P1	6.1	14
								J1		
$NaTaO_3$	ortho		5.513	5.494	2x3.875		293	P1		
	ortho		3.890			0.998		M21		
$AgTaO_3$	ortho		3.931			0.992	293	J1		
$RbTaO_3$	tetra							M21		
$LiTaO_3$	trig	3m						P1		
								J1		
$NaVO_3$								J1		
$AgVO_3$								J1		
Complex Perovskite Type										
WO_3	tetra	4/mmm	5.272		3.920		1223	M21		
	monoc	2	7.274	7.501	3.824		293	P1		
								J1		
$Pb(Sc_{0.51}Nb_{0.5})O_3$	tetra		4.074		4.082	1.002		J1		
$BaNb_{1.5}Zr_{0.25}O_{5.25}$	tetra		12.670		4.017			G10		
	ortho	222								
Misc. Oxides										
$Sr_2Ta_2O_7$	tetra		10.63		10.91			P1		
$PbNb_2O_6$	tetra		12.46		3.907			P1	30	
	ortho		17.51	17.82	2x3.86	1.017	298	J1		
								M1		
$PbTa_2O_6$	ortho							P1		
	ortho		17.68	17.72	2x3.877	1.002		J1	15	
								M1		
$LiNbO_3$	rhomb	3	5.492					P1		
								M21		

*Reproduced from Gruver, R. M., et al., "State-of-the-Art Review on Ferroelectric Materials," U.S. Government Report No. AD 801 027, 1966.

TABLE 1i21. *(Continued)*

| Thermodynamic Data at Transition Temperature | | | | Dielectric Data | | | | | | | |
| Phase Change | | Entropy Change | Enthalpy Change | Spontaneous Polarization | | Dielectric Peak | | Coercive Field | Dielectric Constant (at Room Temp.) | | |
T °K	Or-der	ΔS Cal/mol°C	ΔH Kcal/mol	Polar Axis	Maximum Ps		Temp. °K	Ec	ϵ_a/ϵ_o	ϵ_b/ϵ_o	ϵ_c/ϵ_o
393	1	0.12-0.13	0.049	[001]	26	$\epsilon_a\approx10,000$	393	0.5	$\approx5,000$	$\approx5,000$	≈160
278	1	0.06-0.09	0.021	[110]	at 296°	7.000	278	at 298°			
183	1	0.04-0.07	0.011	[111]		4,000	183				
						10K					
82					3	2,000				≈250	
43					at 4.2°						
763	1	1.51	1.51	[001]	80	$\epsilon_c\approx800$					
173		1.51		[001]	at 173°	1M at 2v				≈50	
50-60										≈250	
708	1	0.19-0.27	0.19	[001]	26	$\epsilon_a\approx4,500$		0.5			≈500
498	1	0.17	0.085	[110]	at 683°	2,000					
						900					
263	1	0.14	0.032	[111]		10K at 5v					
913	1	0.7	0.050			$\epsilon_a\approx2.700$	835	≈60			
835	1								76	76	≈670
627	1			A		10K at 5v					
73				[001]	12						
823				[001]	very	900	613				
598					small	400	333	30			
						500K					
708						$\epsilon\approx6,000$					
13						1K at 10v					
903											
853											
753											
758											
643											
523											
(723)				trig c	12	≈300		10			≈170
					at 573°			at 573°			
(653)											
(448)											
1183		0.24	0.280	NF		$\epsilon_a\approx90,000$					
1013		0.45	0.450	A							
603				A		$\epsilon_a\approx40,000$					
223				F	10 at 223°						
363											
						$\approx20,000$					
193											
843	1			[010]	Pr=0.6	$\epsilon_c\approx b$		17			
						22,000					
						500K					
533				_ to c	8 -10	$\epsilon_c\approx4,000$		25			≈300
						10K					
(723)											

TABLE 1i21. (*Continued*)

Composition	Symmetry		Lattice Parameters						Curie-Weiss Law	
								Structural Data		
Misc. Oxides	Class	Group	a Å	b Å	c Å	c/a	°K	References	$\times 10^4$ C	$T_c - T_o$
$Cd_2Nb_2O_7$	cubic	m3m	10.372				298	P1		
	tetra		10.364			1.0005	123	M21	7	
	tetra		10.378			1.0011	133	M1		
$Sr_2Ta_2O_7$	tetra		10.63		10.91			P1		
$PbNb_2O_6$	tetra		12.46		3.907			P1		
	ortho		17.51	17.82	2x3.86	1.017	298	J1	30	
								M21		
$PbTa_2O_6$	ortho							P1		
	ortho		17.68	17.72	2x3.877	1.002		J1	15	
								M1		
$LiNbO_3$	rhomb	3	5.492					P1		
								M1		
$Pb_2Ta_2O_9$	tetra							P1		
$Pb_3(MgNb_2)O_9$	tetra							P1		
$PbBi_2Ta_2O_9$	tetra				25.4			P1		
$BaBi_3Ti_2NbO_{12}$	tetra							P1		
$PbBi_3Ti_2NbO_{12}$	tetra							P1		
$BaBi_4Ti_4O_{15}$	tetra							P1		
$PbBi_4Ti_4O_{15}$	tetra				41.4			P1		
$BiFeO_3$	rhomb		3.963					R1		
$Ba(Al_{1.4}Li_{0.6})$ $(O_{2.8}F_{1.2})$	hex		10.44		8.77		298	J1		
$BaThO_3$	cubic		4.497					B1		
MnO_2	tetra							P1		
$Bi_4Ti_3O_{12}$	ortho		5.411		32.83	1.007	298	S20		
$PbBi_2Nb_2O_9$	tetra		5.492	5.503	25.53			J1		
Sulfates										
$(NH_4)_2SO_4$	ortho	mmm	7.729	10.560	5.951		298	P1		
	ortho	mm						J1		
								M1		
NH_4HSO_4	monoc	2/m	14.51	4.54	14.90			P1		
	monoc	m	14.26	4.62	14.80			J1		
	tric	1	14.24	4.56	14.81			M1		
$RbHSO_4$	monoc	2/m	14.36	4.62	14.81			P1		
	monoc	m						J1, M1		
$Li(N_2H_5)SO_4$	ortho	mm	8.97	9.91	5.18			P1		
								J1		
$(NH_4)_2Cd_2(SO_4)_3$	monoc	2(?)						P1		
	cubic	23	10.360					J1		
$(NH_4)Fe(SO_4)_2 \cdot 12H_2O$	cubic	23	12.318				298	P1		
	monoc	2						J1		
Phosphates										
KH_2PO_4	tetra	42m	7.453	7.453	6.959		273	P1	0.33	0
	tetra	42m	10.495		6.919		123	J1		
	ortho	mm	10.458	10.540	6.918		77			
KD_2PO_4	tetra	42m	7.453		6.930		273	M1, P1		
								J1, M21		
$(NH_4)_2BeF_4$	ortho	mm	7.49	10.39	5.89		298	P1		
								J1, M1		

TABLE 1i21. (*Continued*)

Thermodynamic Data at Transition Temperature			Dielectric Data							
Phase Change	Entropy Change	Enthalpy Change	Spontaneous Polarization		Dielectric Peak		Coercive Field	Dielectric Constant (at Room Temp.)		
T °K / Order	ΔS Cal/mol°C	ΔH Kcal/mol	Polar Axis	Maximum Ps		Temp. °K	Ec	ϵ_a/ϵ_o	ϵ_b/ϵ_o	ϵ_c/ϵ_o
185 1	0.09	0.018	[111]	6 at 88° 10K at 5v	$\epsilon_{111}\approx1200$		10 at 88°		≈310	
193										
843 1			[010]	Pr=0.6	$\epsilon_c\approx b$ 22,000 500K		17			
533			⊥ to c	8-10	$\epsilon_c\approx4000$ 10K		25		≈300	
(723)										
703										
543										
563										
663										
843										
1123										
(423)				0.1 at 298°			5 at 298°		10	
260										
323			tetra c		ϵ_c 20x10^5 20K at 5v					
948			ortho b	6 at 498°	ϵ_b 570 10K at 10v		30 at 498°			
823										
223 1	4.2	0.93	ortho c	0.45 <223°	$\epsilon_c\approx155$ 10K at 5v	223	4 at 213° 12 at 200°	10	9	9
270 2	0.5	0.12	monoc c	0.87 >154°	$\epsilon_c\approx1420$ 16-6 10K	270	0.15 at 260°			16
154 1	2.1	0.34				154	1.1 at 154°			
258 2			monoc c	0.65 at 107°	$\epsilon_c\approx240$ 10K at 5v		3 at 173° 22 at 123°	7	8	10
None (77 to 413)			ortho c [001]	0.3 at 298°	none 77°-413°		0.32 at 298°			14
95 1		1.00	[100]	0.5 at 93°	100≈40 10K		15 at 93°			
83				0.4			33			
123 2	0.74	0.087	tetra c	4.6 <125°	$\epsilon_c\approx47,500$ 800 at 200v		(see J1)			50
213	0.47	0.100	tetra c	4.8				88	88	90
176 1	1.90	0.31	ortho b	0.22 at 153°	$\epsilon_b\approx55$ 10K at 5v		1.6 at 153° 12 at 135°	9	10	9

TABLE 1i21. (Continued)

Composition	Symmetry		Structural Data						Curie-Weiss Law		
			Lattice Parameters								
Phosphates (Cont.)	Class	Group	a Å	b Å	c Å	c/a	°K	References	$\times 10^4$ C	T_c	T_o
$(ND_4)_2BeF_4$	ortho							P1 J1, M1			
$LiH_3(SeO_3)_2$	monoc	m	6.26	7.89	5.43			P1 J1, M1			
$NaH_3(SeO_3)_2$	monoc	2/m						P1 J1, M1			
Miscellaneous Ferroelectrics											
$K_4Fe(CN)_6 \cdot 3H_2O$	monoc	2/m	9.38	16.84	9.40			P1 J1, M1			
KNO_3	rhomb ortho	3m mmm	4.365 6.431	9.164	5.414		299	P1 J1, M1			
$RbNO_3$	cubic trig		4.36 10.48		7.45			D1			
$NaNO_2$	ortho ortho	mmm mm	5.33 5.390	5.68 5.578	3.69 3.570			P1 J1, M1			
$Ca_2B_6O_{11} \cdot 5H_2O$	monoc	2/m	8.743	11.264	6.102 $\beta=110°7'$			P1 J1, M1	0.05		
SbSI								F1			
Questionable Ferroelectrics											
$Mg_3B_6O_{12}Cl$	cubic ortho	43m mm						P1 J1			
Antiferroelectrics											
$PbZrO_3$	ortho		5.88	11.76	8.20	0.988		P1 J1, M1	12	185	
$PbHfO_3$	ortho		4.136		4.099	0.991		P1 J1, M1			
Pb_2MgWO_3								P1			
$(NH_4)_3H(SO_4)_2$	monoc	2/m m l	14.51	4.54	14.90			P1 J1 M1			
$NH_4H_2PO_4$	tetra ortho	42 222	7.479		7.516		273	P1 J1 M21			

TABLE 1i21. (*Continued*)

Thermodynamic Data at Transition Temperature				Dielectric Data						
Phase Change	Or-der	Entropy Change	Enthalpy Change	Spontaneous Polarization		Dielectric Peak	Coercive Field	Dielectric Constant (at Room Temp.)		
T °K		ΔS Cal/mol°C	ΔH Kcal/mol	Polar Axis	Maximum Ps		Temp. °K / Ec	ϵ_a/ϵ_o	ϵ_b/ϵ_o	ϵ_c/ϵ_o
179		2.27	0.38	ortho c	0.19 at 163°			9	10	9
none				$\perp$ to [001]	15 (363-77°)	none	1.4 at 293° / 27 at 73°	29	13	30
194	1	0.97	0.19	[103]	6.5 >94° / 4.3 <94°	$\epsilon_{132}\approx 250$	194 / 32 at 183° / 22 at 80°			
248	2			[101]	1.4 at 233°	$\epsilon_{101}\approx 1500$ 1M				
397 383	1		≈1.1	trig ortho b	6.3 at 394°	$\epsilon_c\approx 44$ 1M	4.5 at 394°	7		
437				trig c						
433	1			ortho b	6.4 at 416°		2.3			7.4
270	2			monoc b	0.65 at 203°	$\epsilon_{b}>700$ 1K	2 at 203°		20	
295					25	50,000	100 at 0°			
538				c	0.002	$\epsilon\approx 25$ 1M	20			≈10
503	1	0.87	0.44	tetra a		$\epsilon_a\approx 2500$ 1M at 10v				
488 436 298	1			ortho a		550				
312										
247 143				b		65 49	247 143			16
148	1	1.05	0.154	tetra a		$\epsilon_a\approx 85$ $\epsilon_c\approx 30$ 800 at 200v				

Symbols used in Table 1i21:

Lattice parameters are given in Ångstrom units. c/a is the axial ratio followed by the temperature at which the data were obtained in degrees Kelvin. The Curie constant ($C \times 10^{4°}$) and Curie-Weiss temp. difference ($T_c - T_0$) are given in °K. T indicates the transition temperature for each compound and includes the Curie temperature where reported. When known, the order of the transition (1st or 2nd order) follows the T temperature. P_s is the maximum spontaneous polarization expressed in units of 10^{-6} coulombs/centimeter squared. It is followed by the temperature in degrees Kelvin at which the maximum occurs. ϵ, the dielectric peak, is given at the temperature in degrees Kelvin where it occurs. The crystallographic axis of the measurement, the frequency in cycles per second and the external field in volts/mil are given when reported. (K equals one thousand cycles/second, M equals one million cycles/second and v equals volts/mil). E_c, the coercive field, is given in kilovolts/centimeter. The small signal dielectric constants, ϵ_a, ϵ_b and ϵ_c, are given at room temperature.

TABLE 1i22. WORK FUNCTIONS OF THE ELEMENTS (KOHL 1967)*

Group I—	Li	Na	K	Rb		Cs	Cu	Ag	Au
ϕ_0, eV	2.4	2.3	2.2	2.15		1.9	4.42†	4.5	4.9
T_e, °K	750	440	360	340		320	1040	1010	1200
$\phi_0/T_e \times 10^3$	3.2	5.0	6.1	6.3		5.9	4.3†	4.5	4.1
Group II—	Be	Mg	Ca	Sr		Ba	Zn	Cd	Hg
ϕ_0, eV	3.67†	3.6	3.2	2.6		2.5	4.3	4.1	4.5
T_e, °K	1100	550	700	620		680	500	420	250
$\phi_0/T_e \times 10^3$	3.3†	6.5	4.6	3.9		3.5	8.6	9.8	18
Group III—	Al		Ga				La	Ce	Pr
ϕ_0, eV	4.2		3.8				3.3	2.8	2.7
T_e, °K	1170		—				—	—	—
$\phi_0/T_e \times 10^3$	3.6		—				—	—	—
Group IV—	Ti	Zr	Hf	Th	C	Si	Ge	Sn	Pb
ϕ_0, eV	3.9	3.57†	3.65†	3.4	4.4	3.6	4.8	4.4	4.0
$A_0\epsilon^{-\alpha}/k$	—	120†	31.9†	70	48	8	—	—	—
T_e, °K	1600	—	2350	1910	2400	—	—	1180	760
$\phi_0/T_e \times 10^3$	2.4	—	1.6†	1.8	1.8	—	—	3.7	5.2
Group V—	V	Cb	Ta	As	Sb	Bi			
ϕ_0, eV	4.1	4.0	4.1	5.2	4.0	4.6			
$A_0\epsilon^{-\alpha}/k$	—	37	37	—	—	—			
T_e, °K	—	—	2680	—	730	750			
$\phi_0/T_e \times 10^3$	—	—	1.5	—	5.5	6.1			
Group VI—	Cr	Mo	W	U					
ϕ_0, eV	3.90†	4.2	4.5	3.3					
$A_0\epsilon^{-\alpha}/k$	48	55	70	---					
T_e, °K	1040	2230	2860	—					
$\phi_0/T_e \times 10^3$	3.8†	1.9	1.6	—					
Group VII—	Mn	Re							
ϕ_0, eV	4.0	5.1†							
$A_0\epsilon^{-\alpha}/k$	—·	52†							
T_e, °K	1000	—							
$\phi_0/T_e \times 10^3$	4.0	—							
Group VIII—	Fe	Co	Ni	Rh	Pd	Os		Ir	Pt
ϕ_0, eV	4.5	4.4	4.41†	4.80†	4.99†	4.7		5.40†	5.32†
$A_0\epsilon^{-\alpha}/k$	26	41	30	33	60	—		63	32
T_e, °K	1320	1420	1330	2640†	—	—		—	1860
$\phi_0/T_e \times 10^3$	3.4	3.1	3.3†	1.83†	—	—		—	2.8

*Data selected from: Wright, D. A., "A Survey of Present Knowledge of Thermionic Emitters," *Proc. Inst. Elec. Engrs. 100*, Pt. 3, 125–142 (May 1953); values marked by a dagger are from: Wilson, R. G., "Vacuum Thermionic Work Functions of Polycrystalline Be, Ti, Cr, Fe, Ni, Cu, Pt, and Type 304 Stainless Steel," *J. Appl. Phys.*, 37, 2261–2267 (May 1966).

In the above table, ϕ_0 = work function at T of 0°K
  A = Dushman's constant, having a theoretical value of 120.4/cm² deg²
  $\alpha = d\phi/dT$, the temperature coefficient of the true work function
  ϕ = true work function, eV
  k = Boltzmann's constant, 8.6×10^{-5} eV/°K
  T_e = vapor pressure temperature, °K, for p = 10^{-5} torr
  $\phi \cong \phi_0 + \alpha T$

TABLE 1i23. SECONDARY EMISSION DATA FOR METALS (KOHL 1967)*

Atomic Number	Chemical Symbol	δ_{max}	E_{pmax} (Volts)	E_{pI} ($\delta = 1$) (Volts)	E_{pII} ($\delta = 1$) (Volts)
3	Li	0.5	85	—	—
4	Be	0.5	200	—	—
11	Na	0.82	300	—	—
12	Mg	0.95	300	—	—
13	Al	0.95	300	—	—
19	K	0.7	200	—	—
22	Ti	0.9	280	—	—
26	Fe	1.3	(400)	120	1,400
27	Co	1.2	(500)	200	—
28	Ni	1.35	550	150	1,750
29	Cu	1.3	600	200	1,500
37	Rb	0.9	350	—	—
40	Zr	1.1	350	175	(600)
41	Cb	1.2	375	175	1,100
42	Mo	1.25	375	150	1,300
46	Pd	>1.3	>250	120	—
47	Ag	1.47	800	150	>2,000
48	Cd	1.14	450	300	700
50	Sn	1.35	500	—	—
51	Sb	1.3	600	250	2,000
55	Cs	0.72	400	—	—
56	Ba	0.82	400	—	—
73	Ta	1.3	600	250	>2,000
74	W	1.35	650	250	1,500
78	Pt	1.5	750	350	3,000
79	Au	1.45	800	150	>2,000
80	Hg	1.3	600	350	>1,200
81	Tl	1.7	650	70	>1,500
82	Pb	1.1	500	250	1,000
83	Bi	1.5	900	80	>2,000
90	Ta	1.1	800	—	—

*Data from: Kollath, R., "Secondary Emission from Solids Irradiated by Electrons," Handbuch der Physik, *21*, 232–303, Springer Verlag, Berlin (1956).

δ = secondary electron emission yield
E_p = incident electron energy
$E_{p\,max}$ = incident electron energy for maximum yield
E_{pI} = incident electron energy for first crossover
E_{pII} = incident electron energy for second crossover.

TABLE 1i24. SECONDARY EMISSION DATA FOR SEMICONDUCTORS AND INSULATORS (KOHL 1967)*

Material	δ_{max}	E_{pmax} (Volts)
Semiconductor elements		
Ge (single crystal)	1.2–1.4	400
Si (single crystal)	1.1	250
Se (amorphous)	1.3	400
Se (crystal)	1.35–1.40	400
C (diamond)	2.8	750
C (graphite)	1	250
B	1.2	150
Semiconductor compounds		
Cu_2O	1.19–1.25	400
PbS	1.2	500
MoS_2	1.10	
MoO_2	1.09–1.33	
WS_2	0.96–1.04	
Ag_2O	0.98–1.18	
ZnS	1.8	350
Intermetallic compounds		
$SbCs_3$	5–6.4	700
SbCs	1.9	550
$BiCs_3$	6–7	1,000
Bi_2Cs	1.9	1,000
GeCs	7	700
Rb_3Sb	7.1	450
Insulators		
LiF (evaporated layer)	5.6	
NaF (layer)	5.7	
NaCl (layer)	6–6.8	600
NaCl (single crystal)	14	1,200
NaBr (layer)	6.2–6.5	
NaBr (single crystal)	24	1,800
NaI (layer)	5.5	
KCl (layer)	7.5	1,200
KCl (single crystal)	12	
KI (layer)	5.5	
KI (single crystal)	10.5	1,600
RbCl (layer)	5.8	
KBr (single crystal)	12–14.7	1,800
BeO	3.4	2,000
MgO (layer)	4	400
MgO (single crystal)	23	1,200
BaO (layer)	4.8	400
BaO-SrO (layer)	5–12	1,400
Al_2O_3 (layer)	1.5–9	350–1,300
SiO_2 (quartz)	2.4	400
Mica	2.4	300–384
Glasses		
Technical glasses	2–3	300–420
Pyrex	2.3	340–400
Quartz-glass	2.9	420

*Data from: Hachenberg, O. and W. Brauer, "Secondary Electron Emission from Solids," Advances in Electronics and Electron Physics, Vol. XI, Marton, L., Ed., Academic Press, New York, pp. 413–449 (1959).

TABLE 1i25. TRIBOELECTRIC SERIES ACCORDING TO McLEAN*

Number	Positive end
1	Polyester resin
2	Polymethylmethacrylate
3	Glass ("Pyrex")
4	Muscovite mica
5	Melamine-glass laminate
6	Molded wood flour phenolic
7	Cassiterite (a tin ore)
8	Steel, copper, aluminum, and silver
9	Barium titanate
10	Hard fiber
11	Polyester resin
12	Nylon
13	Cellulose acetate
14	Butyl rubber
15	Epoxy resin ("Epon 828")
16	Glass-bonded mica
17	Phenolic-glass laminate
18	Steatite
19	Silicone rubber
20	Polystyrene
21	Polyethylene
22	Polytetrafluoroethylene ("Teflon")
23	Polyethylene terephthalate ("Mylar")
24	"Teflon"-glass laminate
25	Polychlorotrifluoroethylene
26	Unplasticized polyvinyl chloride
	Negative end

*Reproduced from Baer, E., Ed., "Engineering Design for Plastics," Van Nostrand Reinhold, New York, 1964.

REFERENCES

Harned, H. S. and B. B. Owen, "The Physical Chemistry of Electrolytic Solutions," Reinhold, New York, 1950.

Kohl, W. H., "Handbook of Materials and Techniques for Vacuum Devices," Van Nostrand Reinhold Co., New York, 1967.

BIBLIOGRAPHY

Albers, W. A., Jr., Ed., "The Physics of Opto-Electronic Materials," Plenum Press, New York, 1971.

Beam, W. R., "Electronics in Solids," McGraw-Hill, New York, 1965.

Bube, R. H., "Photoconductivity of Solids," Wiley, New York, 1960.

Callaway, J., "Electronic Energy Bands in Solids," Academic Press, New York, 1964.

Charlot, G., et al., "Selected Constants: Oxidation-Reduction Potentials of Inorganic Substances in Aqueous Solution," Butterworth, London, 1972.

Conway, B. E., "Electrochemical Data," Elsevier, Amsterdam, 1952.

Fomenko, V. S. and G. V. Samsonov, Eds., "Handbook of Thermionic Properties: Electronic Work Functions and Richardson Constants of Elements and Compounds," Plenum Press, New York, 1966.

Harper, C. A., "Handbook of Materials and Processes for Electronics," McGraw-Hill, New York, 1970.

Hurd, C. M., "The Hall Effect in Metals and Alloys," Plenum Press, New York, 1972.

Jaffe, B., W. R. Cook, Jr., and H. Jaffe, "Piezoelectric Ceramics," Academic Press, London, 1971.

Larach, S., "Photoelectronic Materials and Devices," Van Nostrand Reinhold Co., New York, 1965.

Latimer, W. M., "The Oxidation States of the Elements and their Potentials in Aqueous Solution," 2nd ed., Prentice-Hall, Englewood Cliffs, N.J., 1952.

Leeds, M. A., Ed., "Electronic Properties of Composite Materials," Plenum Press, New York, 1972.

Long, D., "Energy Bands in Semiconductors," Wiley, New York, 1968.

Korn, G. A., "Basic Tables in Electrical Engineering," McGraw-Hill, New York, 1965.

Maissel, L. and R. Glang, "Handbook for Thin Film Technology," McGraw-Hill, New York, 1970.

Maryott, A. A. and F. Buckley, "Table of Dielectric Constants and Electric Dipole Moments of Substances in the Gaseous State," NBS Cir. 537, U.S. Government Printing Office, Washington, 1953.

Mattiat, O. E., "Ultrasonic Transducer Materials," Plenum Press, New York, 1971.

McClellan, A. L., "Tables of Experimental Dipole Moments," W. H. Freeman, London, 1963.

McIntosh, R. L., "Dielectric Behavior of Physically Adsorbed Gases," Marcel Dekker, New York, 1966.

Meaden, G. T., "The Electrical Resistance of Metals," Plenum Press, New York, 1965.

Parsons, R., "Handbook of Electrochemical Constants," Butterworth, London, 1955.

Peacock, T. E., "Electronic Properties of Aromatic and Heterocyclic Molecules," Academic Press, New York, 1965.

Stanley, J. K., Ed., "Electrical and Magnetic Properties of Metals," American Society of Metals, Metals Park, Ohio, 1963.

Tallan, N. M., Ed., "Electrical Conductivity in Ceramics and Glass," Marcel Dekker, New York, 1973.

Von Aulock, W. H., "Handbook of Microwave Ferrite Materials," Academic Press, New York, 1965.

MAGNETIC PROPERTIES OF MATERIALS

This Section provides a summary of selected magnetic, magnetooptic and nuclear resonance properties of a variety of materials.

The information is arranged as follows:

*The information for these tables is drawn from most of the references and selected commercial literature.

TABLE 1j1. PROPERTIES OF PERMANENT MAGNET MATERIALS–STEELS.

Material		Material Composition	Coercive Force, H_c ($k \cdot$ oersteds)	Residual Flux Density, B_r (gauss)	Energy Product $(BH)_{max}$ (M gauss-oersteds)
Carbon,	1%	0.9 C, 1 Mn, bal. Fe	0.05	10,000	0.2
Cobalt,	3%	3.25 Co, 4 Cr, 1 C, bal. Fe	0.08	9,700	0.38
	6%	6 Co, 9 Cr, 1.5 Mo, 1.05 C, bal. Fe	0.145	7,500	0.45
	9%	9 Co, 9 Cr, 1.5 Mo, 1.05 C, bal. Fe	0.122	7,800	0.41
	17%	18.5 Co, 3.75 Cr, 5 W, 0.75 C, bal. Fe	0.150	9,500	0.65
	36%	38 Co, 3.8 Cr, 5 W, 0.75 C, bal. Fe	0.240	10,000	1.0
Chromium,	1%	1 Cr, 0.6 C, bal. Fe	0.045	9,000	0.2
	3.5%	3.5 Cr, 0.9 C, 0.3 Mn, bal. Fe	0.065	9,700	0.3
	6%	6 Cr, 1.05 C, bal. Fe	0.070	9,800	0.3
Tungsten,	5%	5 W, 0.7 C, 0.3 Mn	0.070	10,300	0.3
	6%	6 W, 0.4 C, bal. Fe	0.063	10,500	0.3
New KS		27 Co, 18 Ni, 7 Ti, 3.7 Al, bal. Fe	0.785	7,150	2.03

TABLE 1j2. PROPERTIES OF PERMANENT MAGNET MATERIALS—MISC. ALLOYS.

Material	Material Composition	Coercive Force, H_c ($k \cdot$ oersteds)	Residual Flux Density B_r (gauss)	Energy Product $(BH)_{max}$ (M gauss-oersteds)
Alnico 1	12 Al, 21 Ni, 5 Co, 3 Cu, bal. Fe	0.47	7,200	1.4
Alnico 2	10 Al, 19 Ni, 13 Co, 3 Cu, bal. Fe	0.56	7,500	1.7
Alnico 3	12 Al, 25 Ni, 3 Cu, bal. Fe	0.48	7,000	1.35
Alnico 4	12 Al, 27 Ni, 5 Co, bal. Fe	0.72	5,600	1.35
Alnico 5	8 Al, 14 Ni, 24 Co, 3 Cu, bal. Fe	0.64	12,800	5.5
Alnico 6	8 Al, 16 Ni, 24 Co, 3 Cu, 1 Ti, bal. Fe	0.78	10,500	3.9
Alnico 7	8.5 Al, 18 Ni, 24 Co, 3.25 Cu, 5 Ti, bal. Fe	1.05	7,700	2.85
Alnico 8	35 Co, 15 Ni, 4 Cu, 5 Ti, 7 Al, bal. Fe	1.65	8,200	5.3
Alnico 9	35 Co, 15 Ni, 4 Cu, 5 Ti, 7 Al, bal. Fe	1.5	10,500	9.0
Alnico 12	8 Al, 20 Ni, 19 Co, 5.5 Cu, 5.5 Ti, bal. Fe	0.95	5,500	1.6
Alnico 2(S)	10 Al, 19 Ni, 13 Co, 3 Cu, bal. Fe	0.55	7,100	1.5
Alnico 4(S)	12 Al, 27 Ni, 5 Co, bal. Fe	0.70	5,200	1.2
Alnico 5(S)	8 Al, 14 Ni, 24 Co, 3 Cu, bal. Fe	0.62	10,900	3.95
Alnico 6(S)	8 Al, 16 Ni, 24 Co, 3 Cu, 1 Ti, bal. Fe	0.79	9,400	2.95
Alnico 8(S)	35 Co, 15 Ni, 4 Cu, 5 Ti, 7 Al, bal. Fe	1.5	7,400	4.00
Cobalt-platinum	50 at. % Co, 50 at. % Pt	4.45	6,450	9.2
Cunife 1	60 Cu, 20 Ni, 20 Fe	0.50	5,400	1.3
Cunife 2	50 Cu, 20 Ni, 27.5 Fe, 2.5 Co	0.26	7,300	0.78
E.S.D. 31	20.7 Fe, 11.6 Co, 67.7 Pb	1.0	5,000	2.3
E.S.D. 32	18.3 Fe, 10.3 Co, 72.4 Pb	0.96	6,800	3.0
E.S.D. 41	20.7 Fe, 11.6 Co, 67.7 Pb	0.97	3,600	1.1
E.S.D. 42	18.3 Fe, 10.3 Co, 72.4 Pb	0.83	4,800	1.25
Indalloy	12 Co, 17 Mo, bal. Fe	0.24	9,000	0.9
Nipermag	12 Al, 32 Ni, 0.4 Ti, bal. Fe	0.66	5,600	1.34
Oerstit 120	27 Ni, 12.5 Al, bal. Fe	0.50	5,700	1.1
Oerstit 1000	19 Co, 17.5 Ni, 7.5 Al, bal. Fe	0.975	5,200	1.1
Permet	30 Co, 25 Ni, 45 Cu	0.80	2,500	0.5
Platinum-cobalt	23 Co, 77 Pt	2.6	4,500	8.0
Platinum-iron	78 Pt, 22 Fe	1.57	5,800	3.0
Remalloy (Comol)	12 Co, 17 Mo, bal. Fe	0.25	10,500	1.1
Remalloy 2	12 Co, 20 Mo, bal. Fe	0.36	9,200	1.5
Silmanal	9 Mn, 4 Al, 87 Ag	6.0	0.55	0.08
Tromolit	11 Co, 24 Ni, 11 Al, 3.5 Cu, bal. Fe	0.615	3,700	0.7
Vicalloy 1	52 Co, 10 V, bal. Fe	0.30	8,800	1.0
Vicalloy 2	52 Co, 14 V, bal. Fe	0.51	10,000	3.5

TABLE 1j3. PROPERTIES OF PERMANENT MAGNET MATERIALS—CERAMICS.

Material	Material Composition	Coercive Force, H_c ($k \cdot$ oersteds)	Residual Flux Density, B_r (gauss)	Energy Product $(BH)_{max}$ (M gauss-oersteds)[#]
Ferrites: (structure MO $\cdot$ nFe$_2$O$_3$, where M = typically an alkaline earth).				
Barium ferrite	BaO $\cdot$ 6Fe$_2$O$_3$ (ordered)	2.4	3950	3.5
	(isotropic)	1.85	2250	1.2
Strontium ferrite	SrO $\cdot$ 6Fe$_2$O$_3$ (ordered)	3.3	3425	2.9
Garnets: (structure M$_3$Fe$_2$(FeO$_4$)$_3$, where M may be a rare earth or yttrium)				
Yttrium Iron Garnet	Y$_3$Fe$_2$(FeO$_4$)$_3$	0.0006	1390	

[#]Megagauss-Oersted.

TABLE Ij4. GLOSSARY OF MAGNETIC TERMS*

Ampere Turn. A unit of magnetomotive force. It is a product of the number of turns on the coil and amperes passing through the turns.

Anisotropic, Magnetic. A material having preferred orientation so that the magnetic characteristics are better along one axis than along any other axis.

CGS System. A system of measurement in which the centimeter, gram and second are fundamental units. The cgs electromagnetic system is used in this manual.

Coercive Force—H_c. The magnetizing force that must be applied to a magnetic material in a direction opposite to the residual induction to reduce the induction to zero.

Coercive Force, Intrinsic—H_{ci}. The magnetizing force which must be applied to a magnetic material in a direction opposite to the residual induction to reduce the intrinsic induction to zero.

Demagnetization. The partial or complete reduction of induction.

Demagnetization Curve. That portion of the normal hysteresis loop in the second quadrant showing the induction in a magnetic material as related to the magnetizing force.

Dimension Ratio—L/d. The ratio of the length of a magnet in the direction of magnetization to its diameter, or the ratio of the length of the magnet to the diameter of a circle which has an area equal to the cross sectional area of the magnet.

Energy Product Curve. It is the graphical representation of the external energy produced by a magnet and is the product of the flux density and demagnetizing force as shown on the normal demagnetization curve. The maximum of this product as shown on such a curve is known as (B_dH_d) max. This value divided by 8π gives the theoretical optimum magnetic energy in ergs per cubic centimeter of material which can be set up in any external magnetic circuit associated with it.

Ferromagnetic. A material which in general exhibits hysteresis phenomena and whose permeability is dependent upon the magnetizing force.

Flux, Magnetic—$\emptyset$. The physical manifestation of a condition existing in a medium or material subjected to a magnetizing influence. The quantity is characterized by the fact that an electromotive force is induced in a conductor surrounding the flux during any time that the flux changes in magnitude. The unit in the cgs system is the maxwell.

Flux Density—B. The number of lines or maxwells per unit area in a section normal to the direction of the flux.

Gap. That portion of the magnetic circuit that does not contain ferromagnetic material, e.g. an air gap.

Gauss. Unit of flux density.

$$\text{Gauss} = \frac{\text{total flux in maxwell}}{\text{area in sq. cm.}}$$

Gilbert. A cgs unit of magnetomotive force. The magnetomotive force required to produce one maxwell magnetic flux in a magnetic circuit of unit reluctance. Magnetomotive force in gilberts $= 0.4\pi$ ampere-turns.

High Energy Materials. This term refers to magnetic materials having a comparatively high energy product. Permanent magnet materials are of this class and have been known also as hard magnet materials.

Hysteresis, Magnetic. A property of a magnetic material by virtue of which the magnetic induction for a given magnetizing force depends upon the previous conditions of magnetization.

Hysteresis Loop. A normal hysteresis loop is the graphical representation of the relationship between the magnetizing force and the resultant induced magnetization of a ferromagnetic material when the magnetizing force is carried through a complete cycle of equal and opposite values under cyclic conditions.

Incremental Permeability—μ_Δ. The ratio of cyclic change in induction to the cyclic change in magnetizing force from any position on the magnetization curve or hysteresis loop.

Induction, Intrinsic—B_i. The excess of the induction in a magnetic material over the induction in vacuum, for a given value of magnetizing force. The equation for intrinsic induction is $B_i = B - \mu_v H$.

Induction, Magnetic—B. The magnetic flux per unit area of a section normal to the direction of flux. The unit of measurement for flux density in the cgs system is the gauss.

Isotropic, Magnetic. A material having the same magnetic characteristics along any axis or direction.

Kilogauss. One Kilogauss equals 1000 gauss.

Leakage Flux. That portion of the magnetic field that is not useful.

Leakage Factor—σ. The ratio of the total flux produced in the neutral section of the magnet to the useful flux.

Line. A term commonly used interchangeably for a maxwell.

Magnetizing Force—H. The magnetomotive force per unit length at any given point in a magnetic circuit. In the cgs system the unit is the oersted and defined by equation:

$$\frac{\text{Magnetomotive force in gilberts}}{\text{Length in centimeters}}$$

Maxwell. The cgs unit of magnetic flux.

Oersted. The cgs unit of magnetizing force.

Permeability—μ. The ratio of the magnetic induction in a given medium to the induction which would be produced in a vacuum with the same magnetizing force. In the cgs system permeability is given by the equation:

$$\text{Permeability} = \frac{\text{Magnetic induction in gausses}}{\text{Magnetizing force in oersteds}}$$

Permeance P. The ratio of the flux through any cross section of a tubular portion of a magnetic circuit bounded by lines of force and by two equipotential surfaces to the magnetic potential difference between the surfaces taken within the portion under consideration. The defining equation in the cgs system is:

$$\text{Permeance} = \frac{\text{Magnetic flux in maxwells}}{\text{Magnetomotive force in gilberts}}$$

Permeance Coefficient—P. Ratio of the total external permeance to that of the permeance of the space occupied by the magnet. ($P = B_d/H_d$).

Reluctance—R. The reciprocal of permeance.

Reluctance Factor—r_f. Ratio of mmf along the magnet to the mmf along the gap.

Remanence—B_d. The magnetic induction which remains in a magnetic circuit after the removal of an applied magnetomotive force. If there is an air gap in the magnetic circuit, the remanence will be less than the residual induction.

Residual Induction—B_r. The magnetic induction corresponding to zero magnetizing force in a magnetic material which is in a symmetrically, cyclically magnetized condition.

Saturation. The condition under which all elementary moments have become oriented in one direction. A magnetic material is saturated when an increase in the applied magnetizing force produces no increase in intrinsic induction.

Stabilization. The process of subjecting magnets to various conditions such as heat, shock or demagnetizing conditions so that magnet will produce a constant magnetic field.

Tractive Force. The force which a permanent magnet exerts on a ferromagnetic object.

*Reprinted from "Design & Application of Permanent Magnets," Manual 7, with permission by Indiana General, Div. of Electronic Memories & Magnetics Corp., Valparaiso, Indiana.

TABLE 1j5. MAGNETIC SUSCEPTIBILITY OF SOME MATERIALS.*

If I is the intensity of magnetization produced in a substance by a field strength H then the magnetic susceptibility $\kappa = I/H$. This is generally referred to the unit mass; italicized figures refer to the unit volume. The susceptibility depends greatly upon the purity of the substance, especially its freedom from iron. The mass susceptibility of a solution containing p percent by weight of a water-free substance (susceptibility κ) is $\kappa_s = (p/100)\kappa + (1 - p/100)\kappa_0$. ($\kappa_0 =$ susceptibility of water.)

Substance	$\kappa \times 10^6$	Temp. °C	Remarks	Substance	$\kappa \times 10^6$	Temp. °C	Remarks
Ag	—.19	18		Li	+.38		
AgCl	—.28			Mo	+.04	18	
Air, 1 Atm	+.024	15		Mg	+.55	18	
Al	+.65	18		MgSO₄	—.40		
Al₂K₂(SO₄)₂24H₂O	—1.0		Cryst.	Mn	+11.	18	
A, 1 Atm	—.10	0		MnCl₂	+122.	18	Sol'n
As	—.3	18		MnSO₄	+100.	18	"
Au	—.15	18		N₂, 1 Atm	.001	16	
B	—.71	18		NH₃	—1.1		
BaCl₂	—.36	20		Na	+.51	18	
Be	+.79	75	Powd.	NaCl	—.50	20	
Bi	—1.4	18		Na₂CO₃	—.19	17	Powd.
Br	—.38	18		Na₂CO₃·10 H₂O	—.46	17	"
C, arc-carbon	—2.0	18		Nb	+1.3	18	
C, diamond	—.49	18		NiCl₂	+40.	18	Sol'n
CH₄, 1 Atm	+.001	16		NiSO₄	+30.	20	"
CO₂, 1 Atm	+.002	16		O₂, 1 Atm	+.120	20	
CS₂	—.77·	18		Os	+.04	20	
CaO	—.27	16	Powd.	P, white	—.90	20	
CaCl₂	—.40	19	"	P, red	—.50	20	
CaCO₃, marble	—.7			Pb	—.12	20	
Cd	—.17	18		PbCl₂	—.25	15	Powd.
CeBr₃	+6.3	18		Pd	+5.8	18	
Cl₂, 1 Atm	—.59	16		PrCl₃	+13.	18	Sol'n
CoCl₂	+90.	18	Sol'n	Pt	+1.1	18	
CoBr₂	+47.	18	"	PtCl₄	.0	22	Sol'n
CoI₂	+33.	18	"	Rh	+1.1	18	
CoSO₄	+57.	19	"	S	—.48	18	
Co(NO₃)₂	+57.	18	"	SO₂ 1 Atm	—.30	16	
Cr	+3.7	18		Sb	—.94	18	
CsCl	—.28	18	Powd.	Se	—.32	18	
Cu	—.09	18		Si	—.12	18	Cryst.
CuCl₂	+12.	20	Sol'n	SoO₂, Quartz	—.44	20	
CuSO₄	+10.	20	Sol'n	—Glass	—.5±		
CuS	+.16	17	Powd.	Sn	+.03	20	
FeCl₂	+90.	18	Sol'n	SrCl₂	—.42	20	Sol'n
FeCl₃	+90.	18	"	Ta	+.93	18	
FeSO₄	+82.	20	"	Te	—.32	20	
Fe₂(NO₃)₆	+50.	18	"	Th	+.18	18	
He, 1 Atm	—.002	0		Ti	+3.1	18	
H₂, 1 Atm	.000	16		V	+1.5	18	
H₂, 40 Atm	.000	16		W	+.33	20	
H₂O	—.79	20		Zn	—.15	18	
HCl	—.80	20		ZnSO₄	—.40		
H₂SO₄	+.78	20		Zr	—.45	18	
HNO₃	—.70	20		CH₃OH	—.73		
Hg	—.19	20		C₂H₅OH	—.80		
I	—.4	20		C₃H₇OH	—.80		
In	.1±	18		C₆H₅OC₆H₅	—.60	20	
Ir	+.15	18		CHCl₃	—.58		
K	+.40	20		C₆H₆	—.78		
KCl	—.50	20		Ebonite	+1.1		
KBr	—.40	20		Glycerine	—.64	22	
KI	—.38	20		Sugar	—.57		
KOH	—.35	22	Sol'n	Paraffin	—.58		
K₂SO₄	—.42	20		Petroleum	—.91		
KMnO₄	+2.0			Toluene	—.77		
KNO₃	—.33	20		Wood	—.2-5		
K₂CO₃	—.50	20	Sol'n	Xylene	—.81		

*Reproduced from: Forsythe, W. E., "Smithsonian Physical Tables," 9th rev. ed., The Smithsonian Institution Press, Washington, 1969.

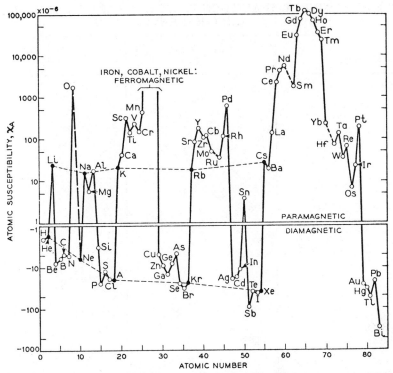

Fig. 1j1. Atomic susceptibilities of the elements at room temperature. Dotted lines connect alkali atoms and rare gases. (From Bozorth 1951.)

**TABLE Ij6. MAGNETIC TRANSITION TEMPERATURES OF MATERIALS
(CONNOLLY AND COPENHAVER, 1970)**

Material	Curie Temp. ($^\circ$K)	Neel Temp. ($^\circ$K)
Ag-5 at. % Eu	6.5 ± 0.5	
$AgCrS_2$		50
$AgCrSe_2$		50
AgDy		53–65
AgEr		10–24
AgF_2		163
$AgFeCl_4$		13.1 ± 0.2
AgGd		137–140
AgHo		32–42
$Ag_{0.5}La_{0.5}Fe_{12}O_{19}$	708	
Ag-Mn		
5.30% Mn		11.4
1.75% Mn		1.9–1.2
AgNd		24–28
AgTb		102–122
AgTm		9–20
$AlCo_2O_4$		30
$AlCr_2$		598 ± 5
$Al_{0.4}Cr_{1.6}BeO_4$		20 ± 2
AlNiCo (22Co-2Ti)	1143	
Au_4Cr		~380
$Au_{80}Cr_{20}$		
ordered		~400
disordered		~270
AuDy		24–34
AuEr		13–19
Au-5 at. % Eu	6.5 ± 0.5	
AuFe	300	
Au-20 at. % Fe	~210	
AuGd		42–50
AuHo		13–16
AuMn		493; 500; 513*
Au_2Mn		363
Au_3Mn		140
Au_4Mn	363; 371*	
Au_5Mn_2	100	348; 353*
$Au_{48.1}Mn_{51.9}$		500
Au_2MnAl	220; 252–258*	
$Au_2Mn_{2-x}Al_x$		
$x = 0.6$	314	415
$x = 1.0$	147	65
$Au_4Mn_{1-x}Cr_x$		
$x = 0.2$	325	
$x = 0.6$	225	
$x = 0.8$	150	
Au_2MnIn	140	233
$Au_2Mn_{1.25}In_{0.75}$		233
$Au_2Mn_{1.75}In_{0.25}$		~403
AuTb		48–62
AuTm		8–19
Au_4V	43; 56*	
$Au_{80}V_{19}Fe$	43	
$Au_{80}V_{15}Fe_5$	66	

TABLE Ij6 (*Continued*)

Material	Curie Temp. ($^\circ$K)	Neel Temp. ($^\circ$K)
$Au_{80}V_{11}Fe_9$	26	
Ba_2CoF_6		250
$Ba_2Co_2Fe_{28}O_{46}$	740 ± 4	
Ba_2CoWo_6		17
$BaCo_{0.5}W_{0.5}O_3$		17
$BaFeF_4$		60 ± 2
Ba_2FeMoO_6	334	351
$BaFeO_3$	180	
$BaFe_{12}O_{19}$	738; 740*	709.5 (H ⊥ c), 713.3 (H ∥ c)
$BaFe_{12}O_{19}F_2$	710	
$BaFe_{0.5}Re_{0.5}O_3$	316	
Ba_2FeSbO_6		12
$Ba_3FeU_2O_9$	123	
$BaMnF_4$		~25
$BaNiF_4$		~150
Ba_2NiF_6		200
$BaNi_{0.5}W_{0.5}O_3$		55
$Ba_{0.4}Sr_{1.6}Zn_2Fe_{12}O_{22}$	400	380
$BiCrO_3$		123
$BiFeO_3$		643; 645; 653*
$Bi_2Fe_4O_9$		256 ± 4
Bi-MnBi (0.50 wt. % Mn)	240	
$BiMnO_3$	103; 110*	
$BiMn_2O_5$		52
$CaCoSiO_4$		16
$CaCo_{0.5}W_{0.5}O_3$		26
$CaCrO_3$		90
$CaCr_2O_4$ (β)		80
Ca_2FeAlO_5		333 ± 15
$Ca_2Fe_{0.5}Ga_{1.5}O_5$		28 ± 2
Ca_2FeGaO_5		315 ± 4
$Ca_2Fe_{1.5}Ga_{0.5}O_5$		560 ± 3
$Ca_2Fe_2GaO_5$		730 ± 2
Ca_2FeMoO_6		398
$CaFeO_3$		120
$CaFe_2O_4$		180; 200*
$Ca_2Fe_2O_5$		720; 725; 730*
$CaFe_{0.5}Re_{0.5}O_3$	538	
Ca_2MnO_4	114	
Ca_2FeSbO_6		31
$CaFeSiO_4$		<8
$Ca_{0.8}La_{0.2}MnO_3$		165
$CaMnO_3$		123
$CaMn_2O_4$		225 ± 5
Ca_2MnO_4	114	114
$Ca_3Mn_2O_7$		120
$Ca_4Mn_3O_{10}$		125
$CaMnSiO_4$		9
$CaO \cdot 2Fe_2O_3$	403	
$Ca_{0.75}Sr_{0.25}MnO_{2.98}$		150
$CaRuO_3$		110 ± 10
$CdCr_2O_4$		9
$CdCr_2S_4$	80–97*	
$CdCr_2Se_4$	121–142*	55

TABLE Ij6 (*Continued*)

Material	Curie Temp. ($^\circ$K)	Neel Temp. ($^\circ$K)
$CdFe_2$	782	
$Cd_xMn_{3-x}O_4$		
$x = 1.0$	<4	
$x = 0.8$	5	
$x = 0.6$	15	
$x = 0.4$	24.5	
Ce		13
CeAg	9	
CeAl		10 (2.2 kOe), 9*
$CeAl_2$	8	4
$CeAl_4$	9	
CeAs		7.5; 8*
CeBi		25
CeC_2		33
$CeCl_3$	0.345	~0.2
$CeCo_3$	78	
$CeCo_5$	464–737*	
Ce_2Co_7	151	
Ce_2Co_{17}	1068; 1083*	
$CeCrO_3$		257
CeCu		2.7
$CeFe_2$	226–878*	
$CeFe_5$	228	
$CeFe_7$	93	
Ce_2Fe_{17}	91	
$CeFeO_3$	719	
CeGe		10
$CeH_{2.0}$		<4.3
$CeH_{2.35}$		5.3; 0.1
$CeH_{2.6}$		<4.3
$CeIn_3$		10; 11*
Ce_2Ni_7	48	
CeP		9; 10*
CeS		7
CeSb		18
CeSe		12
Ce_3Se_4	10	
CeTe		10
Ce-Y (55.0 Ce-45.0 Y)		37
CeZn		36
$CeZn_2$		7
Co	1382–1400*	
$Co_2Al_3B_6$	406	
$Co_{20}Al_3B_6$	409	
$CoAlFeO_4$	420	
$CoAlO_4$		6
$CoAlRhO_4$		24
CoB (not ferromag.)	477	
Co_2B	429; 433*	
Co_3B	747	
$Co_3B_2O_6$		30; 37*
$Co_3B_7O_{13}Br$		20
$Co_3B_7O_{13}Cl$		15
$Co_3B_7O_{13}I$		38

TABLE Ij6 (*Continued*)

Material	Curie Temp. (°K)	Neel Temp. (°K)
$CoBr_2$		19
$CoBr_2 \cdot 2H_2O$		9.5
$CoBr_2 \cdot 6H_2O$		2.91–3.2*
$CoCl_2$		24.9
$CoCl_2 \cdot 6D_2O$		2.40
$CoCl_2 \cdot 2H_2O$		17.5; 18*
$CoCl_2 \cdot 6H_2O$		2.28; 2.29*
$CoCrO_4$		14
$CoCr_2O_4$	95–100*	
$CoCr_2S_4$	227–240*	300 (NiAs type)
$CoCs_3Br_5$		0.282
$CoCs_3Cl_5$		0.523
$Co_{3.5}CuFe_{0.5}Ce$		>993
$Co_{0.46}Cu_{0.54}O$		285
CoF_2		37.7
CoF_3		460
$CoFeCoO_4$	450	
$CoFe_2O_4$	673–769	792 (quenched), 798 (slowly cooled)
$(Co_{1-x}Fe)_2P$		
$\quad x = 0.2$	~140	
$\quad x = 0.3$	~286	
$\quad x = 0.4$	~407	
$\quad x = 0.6$	~453	
$(Co_{0.38}Fe_{0.62})_2P$	459	
$CoGeO_3$		41
$Co(H_2O)_6 \cdot SiF_3$		0.15
CoI_2		12; 3*
$Co_3La_2(NO_3)_{12} \cdot 24H_2O$		0.181
$Co_{0.8}Mg_{0.2}O$		230
$Co_{0.67}Mg_{0.33}O$		155
$Co_{0.65}Mn_{0.35}$	≅140	13
Co_2MnAl	697	
Co_2MnGa	694	
Co_2MnGe	905	
$CoMnO_3$	391	
$Co_{1.8}Mn_{1.2}O_4$	191	
$CoMn_2O_4$	95–105	70
$(Co_{1-x}Mn_x)_2P$		
$\quad x = 0.2$	~200; ~218	
$\quad x = 0.3$	~218	
$\quad x = 0.5$	~583	
$\quad x = 0.6$	~445	
Co_2MnSi	985	
Co_2MnSn	811; 829*	
$CoMoO_4$		5
Co_3N		11
$Co(NC_5H_5)_2Cl_2$		3.7
$Co(NH_4)_2(SO_4)_2 \cdot 6H_2O$		0.084
$Co_{0.67}Ni_{0.33}O$		358
$Co_{0.9}Ni_{0.1}O$		311
CoO		271–328.6*
$CoO(I)$		288
$CoO(II)$		270 ± 10

TABLE Ij6 *(Continued)*

Material	Curie Temp. ($^\circ$K)	Neel Temp. ($^\circ$K)
Co_3O_4		$40; 33.0 \pm 1.0*$
$CoO_{1-x}ZnO_xFe_2O_3$		
$\quad x = 0$	~800	
$\quad x = 0.2$	~695	
$\quad x = 0.4$	~550	
$\quad x = 0.6$	~380	
$CoPt$	813	
$CoRh_{0.35}Cr_{1.65}S_4$	200	
$CoRh_{1.5}Cr_{0.5}S_4$	~50	360 ± 10
$CoRh_2O_4$		27
Co_2RuO_4		~20
CoS		358
CoS_2	110–133*	
$CoSo_4$		12
$Co(S_xSe_{1-x})_2$		
$\quad x = 0.40$		90
$\quad x = 0.70$		90
$\quad x = 0.90$	~42	
$CoSb$		40
$CoSe_2$		93
$CoSeO_4$		30
$CoSe_{0.25}S_{1.75}$	~17	
$CoSiF_6 \cdot 6H_2O$		0.20
$CoSiO_3$		50
Co_2SiO_4		49
$Co_{21}Sn_2B_6$	425	
$Co_{0.67}Te$		115
Co-Ti		
$\quad$ 21.4 at. % Ti	38	
$\quad$ 28.0 at. % Ti	42	
$\quad$ 29.0 at. % Ti	44	
$\quad$ 32.8 at. % Ti	17	
Co_2Ti		43
$CoTiO_3$		38; 42*
$CoUO_4$		12
$CoVO_3$		142 (triclinic)
Co_2VO_4	158; 160*	
CoV_2O_4	145	
$CoWO_4$		55
$Co_{0.7}Zn_{0.3}Fe_2O_4$	~591	
Cr		120 (AF_2 phase), 311 (AF_1 phase)
$CrAs$		280; 823*
Cr_2As		$393–438 \pm 15*$
Cr_3As_2	213–257*	
CrB_2		$86; 88.0 \pm 2.0*$
$CrBe_{12}$	~50	
$CrBO_3$		15
$Cr_3B_7O_{13}Br$		50
$Cr_3B_7O_{13}Cl$		25
$Cr_3B_7O_{13}I$		95
Cr_2BeO_4		28
$CrCl_2$		20–40*
$CrCl_3$		16.8

TABLE Ij6 (*Continued*)

Material	Curie Temp. (°K)	Neel Temp. (°K)
Cr-1% Co		300
CrF_2		53
CrF_3		80
CrFe		308
Cr-Fe		
20% Fe	81 ± 5	
18.2% Fe	50 ± 5	
0.49 at. % Fe		311–241
0.5 wt. % Fe		300
0.78 wt. % Fe		296
2.3 wt. % Fe		260
3.3 at. % Fe		248
4.7 % Fe		260
$Cr_{2-x}Fe_xAl$		
$x = 0$		663
$x = 0.05$		603
$x = 0.09$		413
$x = 0.15$		310
$x = 0.24$		0
Cr(0.5% Fe, 0.16% Ni)		~300
$Cr_{0.75}Fe_{0.25}Sb$		680
CrGe		62
$CrGe_2$	98; 100*	
$Cr_{11}Ge_{19}$	86	
$CrK(SO_4)_2 \cdot 12H_2O$		0.004
CrMn		
0.35% Mn		460
0.12% Mn		320
0.44% Mn		450
0.5% Mn		375
1.03% Mn		510
2.1% Mn		520
$Cr_{0.3}Mn_{0.7}As$		190
$Cr_{0.4}Mn_{0.6}As$	160	
$Cr_{0.7}Mn_{0.3}As$		263
$CrMn_2B_4$	440	
$CrMn_2O_4$	65	
CrN		273
Cr-1% Ni		200
CrO_2	378–398*	
Cr_2O_3		300–308*
Cr_2O_5		<100
Cr_2O_3-0.35% Fe_2O_3		289
Cr-Os		
0.3 at. % Os		359
2.0 at. % Os		566
$Cr_5Pd_{75}Si_{20}$		10
$Cr_9Pd_{71}Si_{20}$	60	
Cr-2 at. % Pt		563
Cr-70% Pt	687	
$Cr_{0.85}Re_{0.15}$		540
Cr-Ru		
0.9 at. % Ru		507
4.8 at. % Ru		558

TABLE Ij6 *(Continued)*

Material	Curie Temp. ($^\circ$K)	Neel Temp. ($^\circ$K)
CrS	303	460
$CrS_{1.17}$	310	152; 153*
Cr_2S_3		
rhombic	120	
trigonal	110; 833*	
Cr_3S_4		215–280*
Cr_5S_6	303	150–168*
Cr_7S_8		~125
CrSb		693–723*
CrSe		200–285*
Cr_3Se_4		80
Cr-0.9 at. % Si		241
Cr-0.4 at. % Si		277
CrTe	239–340*	150
film	343–345*	
Cr_2Te_3	172–303*	
Cr_3Te_4	329	80
Cr_5Te_6	327	
Cr_7Te_8		
disordered	361	
ordered	350	
Cr_2TeO_6		105 ± 5
$CrTe_{1-x}Sb_x$		
x = 0.25	~290	~100
x = 0.50	~220	~260
x = 0.75		500
x = 0.9		700
Cr–V		
1.0% V		210
0.45 at. % V		268
$Cr_{1-x}V_xO_2$		
x = 0.00	389.5	
x = 0.02	385.5	
x = 0.05	381.5	
x = 0.08	344	
$CrVO_4$		50
Cr_2WO_6		69
$CsCoF_3$		8
$CsCuCl_3$		10.4
$Cs_3Cu_2Cl_7 \cdot 2H_2O$		1.620 ± 0.005
$CsFeF_3$	60; 62 ± 2	
$CsMnCl_3$		69 ± 3
$CsMnCl_3 \cdot 2H_2O$		4.8; 4.88*
$Cs_2MnCl_4 \cdot 2H_2O$		1.8; 1.86*
$CsMnF_3$		53.5; 64*
$CsNiCl_3$		4.5
$CsNiF_3$	111	
$Cu_3B_7O_{13}Br$		24
$Cu_3B_7O_{13}Cl$		20
$CuBr_2$		189; 193*
$Cu(C_2H_3O_2)_2$		270
$Cu(C_{12}H_{23}O_2)_2$		230
$Cu_3(CO_3)_2 \cdot (OH)_2$		1.86
$CuCl_2$		70

TABLE Ij6 (*Continued*)

Material	Curie Temp. (°K)	Neel Temp. (°K)
$CuCl_2 \cdot 2D_2O$		4.3
$CuCl_2 \cdot 2H_2O$		4.3
$CuCrMnO_4$	45	
$CuCrO_2$		32 ± 1
$CuCr_2O_4$	133; 135*	
$CuCrS_2$		39
$CuCr_2S_4$	375–420*	
$CuCr_2Se_4$	429–460*	
$CuCr_2Se_3Br$	274; 354*	
$CuCr_2Se_3Cl$	383	
$CuCr_2Te_4$	329; 365*	
$CuCr_2Te_3Br$	282	
$CuCr_2Te_3I$	281; 294*	
CuDy		62
CuEr		10–15
CuF_2		68.7 ± 2
$CuF_2 \cdot 2H_2O$		10.9; 26*
$CuFeCl_4$		20.5; 21.7*
$CuFeO_2$		19; 25*
$CuFe_2O_4$	~723	780 ± 20
CuGd		135–145
$Cu(HCO_2)_2 \cdot 4H_2O$		17
CuHo		27
$CuK_2Cl_4 \cdot 2H_2O$	0.88	
$CuK_2(SO_4)_2 \cdot 6H_2O$		0.05; 0.0295*
$Cu_3La_2(NO_3)_{12} \cdot 24H_2O$		0.089
$CuMg_{0.5}Mn_{1.5}O_4$	57	
Cu-Mn		
4.82% Mn		11.5
1.40% Mn		4
Cu_2MnAl	600, 630*	
Cu_2MnIn	500, 520*	
$CuMn_2O_4$	25	
$Cu_{1.5}Mn_{1.5}O_4$	80	
CuMnSb		55
Cu_2MnSn	528	
$Cu(NH_3)_4SO_4 \cdot H_2O$		0.37–0.43*
$Cu(NH_4)_2Br_4 \cdot 2H_2O$	1.73, 1.74*	
normal	1.789 ± 0.001	
deuterated	1.804 ± 0.001	
$Cu(NH_4)_2Cl_4 \cdot 2H_2O$	0.70	
$Cu(NO_3)_2 \cdot 2.5H_2O$		3.2
Cu-Ni 24 at. % Ni	346	
$CuNi_{0.5}Mn_{1.5}O_4$	150	
CuO		230
$CuRb_2Cl_4 \cdot 2H_2O$	1.02	
$CuRhMnO_4$	35	
$CuRh_2O_4$	830	25
$CuSO_4$		34.5
$CuSO_4 \cdot 5H_2O$		0.029
Cu_2Sb		373
$CuSeO_4 \cdot 5H_2O$		0.046
$CuSiF_6 \cdot 6H_2O$	0.07	
$CuSiO_3 \cdot H_2O$		21

TABLE Ij6 (*Continued*)

Material	Curie Temp. (°K)	Neel Temp. (°K)
CuTb		100–118
CuTm		10–28
$CuV_xCr_{2-x}S_4$		
$x = 0.25$	267	
$x = 0.75$	46	
$x = 1.75$	8	
$CuWO_4$		90
Dy	84.7–87*	173–184*
DyAg		51–63*
$DyAg_2$		
α		15.0
β		9.5
DyAl		20
$DyAl_2$	53–70*	
Dy_3Al_2	76	
$Dy_3Al_{0.5}Fe_{4.5}O_{12}$	490	
$DyAlO_3$		3.5 ± 0.2
$Dy_3Al_5O_{12}$		2.53
DyAs		8.5
$DyAsO_4$		2.50 ± 0.10
DyAu		14
$DyAu_2$		
α		33.8
β		25.5
DyBi		12
DyC	165	
DyC_2		59
Dy_3C		170
$Dy_{1.3}Cd$	105	
$DyCo_3$	450	
$DyCo_5$	966; 1125*	
Dy_2Co_7	647	
Dy_2Co_{17}	1152; 1189*	
Dy_4Co_3	55	
Dy_2CrSbO_7	16	
DyCu		61; 64*
$DyCu_2$		24
$Dy_{0.6}Er_{0.4}Al_2$	49	
$DyFe_2$	638; 663*	
$DyFe_3$	600	
Dy_2Fe_{17}	363	
Dy_6Fe_{23}	524	
$Dy_3Fe_5O_{12}$	551–563*	2.5175
DyGe		36
$DyGe_{1.67}$		12
$DyGe_2$		28
Dy_5Ge_3		40
Dy_5Ge_4		40
DyH_2		8
Dy-50% Ho		157
Dy-100% Ho	80	
$DyIn_3$		23; 24*
$DyIr_2$	23	
$DyMn_5$	430	

TABLE Ij6 (*Continued*)

Material	Curie Temp. (°K)	Neel Temp. (°K)
$DyMn_{12}$		110
Dy_6Mn_{23}	443	
$DyMnO_3$		<30
$Dy_2Mn_4O_9$		8
DyN	17.5; −26*	
Dy_2NbO_7		1.0
DyNi	48	
$DyNi_2$	30; 32*	
$DyNi_3$	69	
$DyNi_5$	15	
Dy_2Ni_7	81	
Dy_2Ni_{17}	168; 604*	
Dy_3Ni		35
Dy_2O_3		1.20
$Dy(OH)_3$	3.50	
Dy_2O_2S		5.85
$DyOs_2$	15	
$DyPO_4$		3.39; 3.40*
$Dy_{5.07}Pd_{1.93}$	~25	41
$DyPt_2$	4.2	
$DyPt_3$		13.2 ± 0.3
DySb		9.5
$DySi_2$		17
Dy_5Si_4	140	
Dy_2TiO_5		1.55
$Dy_2Ti_2O_7$		1.3
$DyVO_4$		3.0; 3.05 ± 0.10
Dy-Y		
5% Y	91	
20.0% Y		59
50.0% Y		108
90.0% Y		163
DyZn	144	
Er	19–20*	79–88*
ErAg		15–21*
ErAl		10; 13*
$ErAl_2$	12–24*	
$ErAl_3$		5
Er_3Al_2		9
$Er_3AlFe_4O_{12}$	420	
ErAs		3.5
ErBi		3.9
ErC	50	
Er_3C		80
$Er_{1.2}Cd$	14	
$ErCo_2$	35–39*	
$ErCo_3$	401	
$ErCo_5$	986; 1050*	
$ErCo_6$	1050	
Er_2Co_7	644; 670*	
Er_2Co_{17}	1160–1193*	
Er_4Co_3	25	
$ErCrO_3$		16.8; 133.2
Er_2CrSbO_7	10	

TABLE Ij6 (*Continued*)

Material	Curie Temp. (°K)	Neel Temp. (°K)
Er-Dy		
10% Dy	31	93
50% Dy	~140	
90% Dy	65	
$ErFe_2$	473–590*	
$ErFe_3$	455–550*	
Er_2Fe_{17}	293–310*	
Er_6Fe_{23}	491–495*	
$ErFeO_3$		636–638*
$Er_3Fe_5O_{12}$	551.4–556*	
$Er_{0.95}Ho_{0.05}$	21	
$Er_{0.50}Ho_{0.50}$	33	104
$Er_{0.9}Ho_{0.1}$		88
$ErIn_3$		6
$ErIr_2$	3	
$ErMn_5$	415	
Er_6Mn_{23}	415	
$ErMnO_3$		79
ErN	3.4–6*	
$ErNi$	10	
$ErNi_2$	14.21*	
$ErNi_3$	62	
$ErNi_5$	13	
Er_2Ni_7	67	
Er_2Ni_{17}	166; 602*	
Er_3Ni		9
Er_2O_3		3.36–4*
$ErOs_2$	3	
ErP		3.1
$ErPt_2$	4.2	
$ErRu_2$	8	
$ErSb$		3.5; 3.7*
Er_5Si_4	25	
$Er_{0.20}Tb_{0.80}$	150	~205
$Er_{0.90}Tb_{0.10}$	25	
Er-Th		
5% Th	80	80
10% Th	74	
20% Th	48	
$Er_2Ti_2O_7$		1.25
Er-Y		
10% Y		78
50% Y		56
70% Y		30 ± 5
$ErZn$	50	
Eu		91–103*
$EuAl_4$	14	13
$Eu_3Al_2O_6$	10	
$Eu_5Al_2O_8$	6	
EuB_6	8.5	
$Eu_{0.45}Ca_{0.55}O$	44	
$EuCrO_3$		181
$EuCu_2$		13–15
EuF_2		2; 19.5*

TABLE Ij6 (*Continued*)

Material	Curie Temp. (°K)	Neel Temp. (°K)
$EuFeO_3$		662
$Eu_3 Fe_5 O_{12}$	563; 566*	
$EuFeS_2$		823
$EuGd_2 O_4$		4.5
$EuGd_2 S_4$	6	
EuH_2	24	
EuI_2	5	
$Eu_{0.1} La_{0.9} Al_2$	11	
$Eu_{0.4} La_{0.6} Al_2$	<4	
$EuLu_2 O_4$	7.5; 9.5*	
$EuMnO_3$		45
EuO	69.2	
$Eu_3 O_4$	7.8, 77*	5–7.5*
EuP		4
$EuPd_2$	80	
$EuPt_2$	105	
EuS	14.5–16.5*	
$EuSe$	1.6–7*	4.4–5.8*
$Eu_2 SiO_4$	7	
$Eu_3 SiO_5$	19	
$EuTb_2 Se_4$		3
$EuTe$		6–11*
$EuTiO_3$		5.3
$EuY_2 Fe_5 O_{12}$	556	
$EuZn_2$		20; 23*
Fe	1040–1044*	
$FeAl$ (Fe rich)	923	
$Fe_3 Al$	773	750
$FeAlO_3$	<300	
$FeAl_2 O_4$		8
$Fe_2 As$		323–368*
FeB	598	
$FeBO_3$	388	348.35 ± 0.2
$Fe_3 BO_6$	508	
$FeBr_2$		11
$Fe_3 C$	483	
$FeCo$	1253	
$FeCO_3$		20–38.5*
$FeC_2 O_4 \cdot 2H_2 O$		~22
$FeCl_3$		9.75–16*
$FeCr_2 O_4$	69; 88*	80
$FeCr_2 S_4$	177–193*	172–200 (NiAs type)
$FeCr_2 Se_4$		4.2
FeF_2	363	78–90*
FeF_3	362	362.4–394*
$FeGa_{1.3}$	483; 697	
$FeGe$	280	400–412*
$FeGe_2$		190–315*
$Fe_2 Ge$	478	
$Fe_3 Ge$	755 (cubic) 655 (hexag.)	
$Fe_5 Ge_3$	485	
FeI_2		10
$Fe_2 MgO_4$	653	
Fe-Mn (0.5–2.4 at % Mn)	1042.0–1017.8	

TABLE Ij6 *(Continued)*

Material	Curie Temp. (°K)	Neel Temp. (°K)
FeMnAs		463
$FeMn_2Ge$	233	
Fe_2MnGe	433	
$FeMn_2O_4$	117–390*	55
Fe-Mo (0.5–3.0 at % Mo)	1042.6–1032.4	
Fe-Ni		
95.5% Ni	683.0	
81% Ni	834.0	
80% Ni	840	
80% Ni (γ Fe)	200	
77% Ni	876.0	
60% Ni	880	
50% Ni	786.1	
40% Ni	648	
$FeNi_3$	865	
Fe-Ni-Sb spinels	590–858	
FeO		186–198*
Fe_2O_3		
(α)	948–963*	953–998*
(γ)	743	—
(δ)	—	950 (P → AF), 250 (AF → AF)
(ϵ)	484	
Fe_3O_4	848–858	
$5Fe_2O_3 \cdot 3Dy_2O_3$	551.4	
$5Fe_2O_3 \cdot 3Er_2O_3$	551.4	
$5Fe_2O_3 \cdot 3Gd_2O_3$	574.6	
$5Fe_2O_3 \cdot 3Y_2O_3$	562.7	
$5Fe_2O_3 \cdot 3Yb_2O_3$	542.6	
FeOF		315
FeOOH		
(α)		353–393
(β)		110–300
(γ)		50–73*
(δ)	450	
FeP	215	126
Fe_2P (hexagonal)	278; 306*	
Fe_3P	686; 716*	
FeP_2		250
FePd	738	
Fe-Pd (hydrogen-free)		
2.8 at. % Fe	95 ± 3	
4.0 at. % Fe	106 ± 3	
7.2 at. % Fe	168 ± 3	
10.3 at. % Fe	246 ± 3	
$FePd_3$	540	
FePt	743	
Fe-Pt		
65.5 % Fe	255 ± 2	
70.0 % Fe		145 ± 1
76.0 % Fe		212
Fe_3Pt	453	
FeRh	668	328–338*
Fe-Rh, 50% Rh	678	360
$Fe_{0.49}Rh_{0.51}$	667	310

TABLE Ij6 (*Continued*)

Material	Curie Temp. (°K)	Neel Temp. (°K)
FeS		593
$Fe_{0.902}S$	530	600
Fe_3S_4	580	
Fe_7S_8	578; 593*	
$FeSO_4$		21; 23*
$FeSb_2$		773
$FeSb_2O_4$		46 ± 1
FeSe		847
Fe_7Se_8	425–483*	475
FeSi		443
Fe-Si (0.9–7.4 at. % Si)	1043.9–1012.6	
Fe_3Si	808	
Fe_5Si_3	381; 385*	
FeSn		365–373*
$Fe_{1.3}Sn$	676	
$FeSn_2$		377–384*
Fe_3Sn	743	
Fe_3Sn_2	612	
Fe_5Sn_3	583	
FeTe		63; 70*
$Fe_{1.11}Te$	479	
$FeTe_2$		83; 85*
Fe_2TeO_6		201; 218.5*
Fe-Ti		
30 at. % Ti	~380	
38 at. % Ti		~260
$FeTiO_3$		55–68
Fe_2TiO_4	142	
$FeTi_2S_4$	60.2	
Fe-53.3% V	280	
Fe_2VO_4	440	
FeV_2O_4	109	
Fe-Zn		
0.2 at. % Zn	981.3 ± 0.5	
0.3 at. % Zn	925.0 ± 1.0	
Fe_2Zr	628	
Ga		2.3 to 1.1 (no transition)
$GaAl_2$	176	
$Ga_{0.5}Fe_{2.5}O_4$	686 ± 3	
Gd	289–295.5*	
GdAg		138–150*
GdAl		42
$GdAl_2$	151–182*	
Gd_3Al_2	275; 282*	
GdAs		19–25*
$GdAlO_3$		3.69–4.0*
GdAu		
@ 19.3 kOe		31
@ 5.5 kOe		37
GdB_6		18
GdBi		28; 32*
Gd_4Bi_3	340	
$GdBr_3$	2	2
GdC	313–373	
Gd_3C	500	

TABLE Ij6 (*Continued*)

Material	Curie Temp. (°K)	Neel Temp. (°K)
GdCd	262	
Gd-30% Ce	160	110
GdCl$_3$	2.20	
GdCl$_3 \cdot$ 6H$_2$O		0.185 ± 0.001
GdCo$_2$	404; 408*	
GdCo$_3$	612	
GdCo$_5$	1008; 1030*	
Gd$_2$Co$_7$	762; 775*	
Gd$_2$Co$_{17}$	1209; 1213*	
Gd$_4$Co$_3$	230	
Gd$_2$CoMnO$_6$	104	~115
Gd-Y		
10 at. % Y	285	
16.6 at. % Y	269	
50 at. % Y	226	
58 at. % Y	202; 203*	217
61 at. % Y	193	217
71.3 at. % Y	165	197.5
87.5 at. % Y	120	190
GdFe$_2$	490–813*	
GdFe$_3$	728	
GdFe$_5$	455	
Gd$_2$Fe$_{17}$	459–472*	
Gd$_6$Fe$_{23}$	468, 659*	
Gd$_3$Fe$_5$O$_{12}$	556–574.6*	90 (N$_1$), 328 (N$_2$)
GdGe		62
GdGe$_2$		28
Gd$_5$Ge$_3$		48
Gd$_5$Ge$_4$		15; 45
GdH$_2$		21
GdIn		28
GdIn$_3$		45
GdIr$_2$	88; 90*	
Gd-30% La	185	120
Gd-Lu		
20.0 % Lu	249	
35.7 % Lu	150	
40.0 % Lu	95	192
80.0 % Lu		100
GdMn$_2$	300; 575*	
GdMn$_5$	465; 468*	
Gd$_6$Mn$_{23}$	468; 478*	
GdN	60–72*	
GdNbO$_4$		1.67
Gd-50% Nd	145	100
GdNi	73	
GdNi$_2$	75–90*	
GdNi$_3$	116	
GdNi$_5$	27; 36*	
Gd$_2$Ni$_7$	118	
Gd$_2$Ni$_{17}$	205; 623*	
Gd$_3$Ni		100
Gd$_2$O$_3$		~1.6
Gd(OH)$_3$		2.0
GdOS$_2$	66	

TABLE Ij6 (*Continued*)

Material	Curie Temp. (°K)	Neel Temp. (°K)
Gd_2O_2S		5.7
$GdOs_2$	66	
GdP		15
$GdPO_4$		225
GdPd	39.5 ± 1	
$GdPd_2$	335	
Gd_5Pd_2	335	
Gd-35% Pr	165	110
$GdPt_2$	4.2	
$GdRh_2$	>77	
$GdRu_2$	83	
GdS		50
$Gd_2(SO_4)_3 \cdot 8H_2O$		0.182 ± 0.001
GdSb		28
Gd_4Sb_3	260	
Gd-Sc		
10 % Sc	251	
20 % Sc	212 ± 4	
28 % Sc	171	
31 % Sc	60	150
50 % Sc		103 ± 4
65 % Sc		60
75 % Sc		46 ± 4
85 % Sc		<4.2
GdSe		60
Gd_2Se_3		6
$GdSi_2$		27
Gd_5Si_4	336	
Gd-Y		
10.0 % Y	281	
20 % Y	254 ± 4	
33.3 % Y	211	
40 % Y	84 ± 4	196 ± 4
50 % Y		182
90 % Y		59 ± 4
GdZn	270	
$GdZn_2$	68	
Ho	19–23*	82.3–133*
HoAg		32; 33*
$HoAg_2$		5.7
HoAl	26	
$HoAl_2$	27–42*	
Ho_3Al_2		33
HoAs		4.8
HoAu @ 2.3 kOe		10
$HoC_{0.5}$	95	
HoC_2		26
$Ho_{1.3}Cd$	30	
$HoCl_3$		2.15
$HoCo_2$	87; 95	
$HoCo_3$	418	
$HoCo_5$	1000; 1025*	
Ho_2Co_7	644; 670*	
Ho_2Co_{17}	1173; 1183*	
Ho_4Co_3	44	

TABLE Ij6 (*Continued*)

Material	Curie Temp. (°K)	Neel Temp. (°K)
HoCu		28
HoCu$_2$		9
Ho$_{0.5}$Er$_{0.5}$	33	104
HoEr	35	
HoFe$_2$	603; 608*	
HoFe$_3$	567	
Ho$_2$Fe$_{17}$	319–325*	
Ho$_6$Fe$_{23}$	501	
HoFeO$_3$		639
Ho$_3$Fe$_5$O$_{12}$	567	
HoGe		18
HoGe$_2$		11
Ho$_5$Ge$_3$	10	
HoH$_2$		8
HoIn$_3$		11
HoIr$_2$	12	
HoMn$_5$	425; 434*	
Ho$_6$Mn$_{23}$	434	
HoMnO$_3$		76
HoMn$_2$O$_5$		46
HoN	13–19*	
HoNi	31	
HoNi$_2$	22; 23*	
HoNi$_3$	66	
HoNi$_5$	10; 22*	
Ho$_2$Ni$_7$	70	
Ho$_2$Ni$_{17}$	162; 611*	
Ho$_3$Ni		20
Ho(OH)$_3$	2.55	
Ho$_2$O$_2$S		2.5
HoOs$_2$	9	
HoP	5.5	
Ho$_{5.04}$Pd$_{1.96}$	~10	27
HoPt$_2$	4.2	
HoSb		5.5; 9*
HoSi$_2$		18
Ho$_5$Si$_4$	76	
Ho-Tb 20% Ho	171	213
Ho-Th 90% Th	44	90
Ho$_2$Ti$_2$O$_7$		1.3
Ho-Y		
0–90% Y		26–133
0–80% Y		23–132
19.6% Y	20	
50.0 % Y	24	
67.7% Y	25	57
HoZn	80	
HoZn$_2$		12
KCoF$_3$		109.5–144*
K$_2$CoF$_4$		107–125*
K$_2$CuCl$_4$ · 2H$_2$O	1	
KCuF$_3$		220–243*
K$_3$Fe(CN)$_6$		0.129
KFeF$_3$		112–115*
K$_2$FeO$_4$		3.6

TABLE Ij6 (Continued)

Material	Curie Temp. (°K)	Neel Temp. (°K)
$KFe_{11}O_{17}$		803
K_2IrCl_6	3.5 ± 0.04	3.05–3.08*
$KMnCl_3$		100 ± 3
$KMnCl_3 \cdot 2H_2O$		2.70; 2.74*
$KMnF_3$		88
K_2MnF_4		45.0 ± 1
K_3MoCl_6		4.7–6.6*
$KNiF_3$		250–280*
KO_2		7
K_2ReBr_6		15.3
K_2ReCl_6		11.9–12.3*
$LaCo_5$	840	
$LaCo_{13}$	1290	
La_2Co_7	490	
$LaCo_4Cu$	690	
La_2CoMnO_6	280	
$LaCo_4Ni$	690	
$LaCoO_3$		80
$LaCrO_3$	300	~4.2; 282–320*
$LaErO_3$	7	2.4
$LaFeO_3$		738–750*
La-Gd 69.9% Gd		155
$LaMnO_3$		100; 131*
$LaVO_3$		137
$LiCoPO_4$		23 ± 2
$LiCuCl_3 \cdot 2H_2O$		4.4–5.9*
$LiFeF_6$		105
$LiFePO_4$		50
$LiMnPO_4$		35–42*
$Li_{0.1}Mn_{0.9}Se$		70
$LiNiPO_4$		23 ± 2
(Li_2O, Al_2O_3, MnO_2) ferrites	703–863	
Lu_2Co_7	459	
Lu_2Co_{17}	1210	
$LuFe_2$	589	
$LuFe_3$	529	
Lu_2Fe_{17}	235–308*	
Lu_6Fe_{23}	485	
$LuFeO_3$		623
$Lu_3Fe_5O_{12}$	549 ± 2	
$LuMnO_3$		91
Lu_2Ni_{17}	601	
$MgCr_2O_4$		15
$MgFeAlO_4$	273	
$MgFe_2O_4$	490; 520	
$Mg_{0.5}Mn_{2.5}O_4$	20	
MgV_2O_4		45
Mn		95–100*
MnAlGe	518	
MnAs	307–318*	400
Mn_2As		573; 583*
$MnAs_{0.9}P_{0.1}$	~240	~120 (meta)
MnAu		370; 520*
$MnAu_2$		363–373*
$MnAu_3$		145; 150*

TABLE Ij6 (*Continued*)

Material	Curie Temp. ($^\circ$K)	Neel Temp. ($^\circ$K)
$MnAu_4$	360	
MnB	578	
MnB_2	140–157*	
Mn_3B_4		392; 393*
$Mn_3B_2O_6$		35
MnBi	633	
$MnBr_2$		2.16
$MnBr_2 \cdot 4H_2O$		2.124–2.136*
$Mn(C_2H_5)_2$		134
$MnCO_3$		31.5–32.5*
$MnCo_2O_4$	203	
$Mn_{0.5}Co_{0.5}O$		210
$Mn_{0.25}Co_{0.75}O$		251
$Mn_{0.75}Co_{0.25}O$		165
Mn-2% Cr		73.2
$MnCr_2O_4$	42; 43*	43
$MnCr_2S_4$	66; 103*	70 (NiAs type)
Mn_3CuN	149	
MnF_2		67.3
MnF_3		43–47*
$MnFeO_3$		58; 78*
$MnFe_2O_4$	555–638*	
Mn-Ga 34 at. % Ga	690	
Mn_2Ga	690	
Mn_3Ga	470	
Mn_3GaC	246–251*	164
Mn_3GaN		298
$MnGa_2O_4$		33
Mn_3Ge	28	350
Mn_3Ge_2	300	
Mn_5Ge_3	293–311*	
Mn_2Ge_2Cu	612	
$MnGeO_3$		10
MnHg		198–460*
MnI_2		3.40
Mn_3In	583	
$Mn_3La_2(NO_3)_{12} \cdot 24H_2O$		0.230
$MnMoO_4$		13
Mn_2N		301
$Mn(NH_4)_2(SO_4)_2 \cdot 6H_2O$		0.14–0.176*
MnNi		1140; 1073 ± 40*
Mn-30% Ni		400
$MnNi_3$		
ordered	750	
disordered	132	
MnO		116–125*
MnO_2		84–94*
Mn_2O_3		80
Mn_3O_4	30 46*	
MnOOH		40
MnP	291–298*	47–50*
Mn_2P		103; 110*
Mn_3P		115
$Mn_2P_2O_7$		14
MnPd		813–825*

TABLE Ij6 (*Continued*)

Material	Curie Temp. (°K)	Neel Temp. (°K)
Mn-Pd 25.3 at. % Pd		220 ± 10
$Mn_2 Pd_3$		653
$Mn_{30} Pd_{70}$		220–235
MnPt		973 ± 10
$Mn_{1.2} Pt_{0.80}$	540 ± 15	
MnRh		~170
$Mn_3 Rh$		855 ± 10
$MnRh_2 O_4$		15
MnS		165
MnS_2		48.2
$MnSO_4$		11.5
$MnSO_4 \cdot H_2 O$		16
MnSb	556–587*	
$Mn_2 Sb$	550	
$Mn_{53} Sb_{47}$	450	
MnSe		147–247*
$MnSe_2$		~75; ~90*
$MnSeO_4$		20; 28*
MnSi	30; 38*	
$Mn_5 Si_3$	60	68
$Mn_5 SiC$	284	
$MnSiF_6 \cdot 6H_2 O$		0.17
$MnSiO_3$		7
$Mn_2 SiO_4$		50
$Mn_{1.5} Sn$	269	
$Mn_2 Sn$	263	
$MnSn_2$		323–330*; 73
$Mn_3 Sn$		420
MnTe	260	306.6–310*
$MnTe_2$		80–87.2*
$MnTiO_3$		41–64*
$MnTiO_3$ II		24 ± 1
$Mn_2 TiO_4$	~77	
$MnUO_4$		12
$MnV_2 O_4$	56	52
$MnWO_4$		14.4; 16*
$Mn_3 ZnN$		190
MoF_3		185
$MoMn_2 B_4$	590	
Nd	29	
$NdAl_2$	65–76*	
$Nd_{1.3} Cd$	105	
$NdCl_3$	1.745	1.035
$NdCo_2$	116	
$NdCo_3$	395	
$NdCo_5$	910–925*	
$Nd_2 Co_7$	609	
$Nd_2 Co_{17}$	1150; 1166*	
$Nd_2 CrSbO_7$	8	
$NdFe_7$	327–332*	573
$Nd_2 Fe_{17}$	327	
$NdFeO_3$		687; 760*
NdGe	28	
$NdGe_2$	3.6	
$Nd_5 Ge_3$	45; 58.4*	

TABLE Ij6 (*Continued*)

Material	Curie Temp. (°K)	Neel Temp. (°K)
NdH_2	9.5	
$NdIn_3$		7
$NdIr_2$	11.8	
$NdMn_2$		86
$NdMn_{12}$		135 ± 2
$NdMnO_3$		
N_1		85
N_2		1.50
NdN	19–35*	
$NdNbO_4$		25
NdNi	35	
$NdNi_2$	10–20*	
$NdNi_3$	27	
$NdNi_5$	9	
$Nd_2 Ni_7$	87	
$Nd_3 Ni$		15
$Nd(OH)_3$		1.7
$NdOs_2$	23	
$NdPt_2$	4.2	
$NdRh_2$	8.1	
$NdRu_2$	28	
NdS		8–8.6*
$Nd_3 S_4$	50	
NdSb		16
NdSe		10.6; 14*
$NdSn_3$		4.7
$Nd_2 Sn_2 O_7$		0.91
NdTe		10.2; 13*
NdZn		148
$NdZn_2$	24	
Ni	628.3 ± 1	
$Ni_3 B_2 O_6$		49
$Ni_3 B_7 O_{13} Cl$		15
$Ni_3 B_7 O_{13} I$	60; 64*	120
$NiBr_2$		60
$NiBr_2 \cdot 6H_2 O$		6.5
$NiBr_2 \cdot 6NH_3$		0.61; 0.7*
$NiCO_3$		30
$NiCl_2$		50; 52*
$NiCl_2 \cdot 6H_2 O$		5.34–5.8
$NiCl_2 \cdot 6NH_3$		1.45
$NiCo_2 O_4$	350	
Ni-Cr 5.6 at. % Cr	324	
$NiCrO_4$		23
$NiCr_2 O_4$	60; 78*	
$NiCr_2 S_4$ (NiAs type)		200
Ni-Cu		
40 at. % Cu	173; 216*	
50 at. % Cu	110	
NiF_2		73–83*
$NiFe_2 O_4$	858	
$Ni(HCOO)_2 \cdot 2H_2 O$		15.7
NiI_2		75
$Ni(IO_3)_2 \cdot 2H_2 O$	$\cong 3.2$	4.20
$Ni_3 La_2 (NO_3)_{12} \cdot 24H_2 O$	0.393	

TABLE Ij6 (*Continued*)

Material	Curie Temp. (°K)	Neel Temp. (°K)
Ni-11.3 at. % Mn	500	
Ni_3Mn	~610	
Ni_2MnGa	379	
Ni_2MnIn	223–228; 323*	
$NiMnO_3$	437	
$NiMn_2O_4$	113–162	70
Ni-Pd		
20% Pd	583	
50% Pd	448; 467*	
80% Pd	250	
Ni-Pt		
40 at. % Pt	213	
70 at. % Pt	20	
$NiRh_2O_4$	360	18
NiS		150–265*
$NiSO_4$		37
$NiSeO_4$		27
Ni_2SiO_4		34
$NiWO_4$		67
NpC	200; 220*	310
NpN	82	
NpO_2		25
O_2		24
$PbCrO_3$		240; 250*
$PbCr_2S_4$	138	
Pd		~90
Pd-Co		
0.07 at. % Co	1.55 ± 0.1	
0.098 ± 0.005 at. % Co	0.79 ± 0.01	
0.19 at. % Co	6.5 ± 0.5	
0.20 ± 0.01 at. % Co	2.95 ± 0.05	
0.49 at. % Co	18.8 ± 3	
0.51 ± 0.02 at. % Co	16.2 ± 0.2	
1.05 ± 0.05 at. % Co	44.0 ± 0.5	
1.91 at. % Co	90 ± 1	
4.5 at. % Co	186 ± 1	
Pd-Fe		
0.10 ± 0.005 at. % Fe	0.78 ± 0.002	
0.16 at. % Fe	2.8	
0.41 at. % Fe	13.0	
0.45 at. % Fe	13.0	
0.5 at. % Fe	~23	
0.78 at. % Fe	32.6	
0.90 at. % Fe	32.6	
1 at. % Fe	37	
2.5 at. % Fe	90	
3 at. % Fe	114	
7 at. % Fe	200	
10 at. % Fe	237	
Pd_3Fe	540	
$Pd_{94}Fe_6$	173	
Pd-Mn		
1.05 at. % Mn	3.90 ± 0.02	
2.40 at. % Mn	7.35 ± 0.02	
2.91 at. % Mn	7.71 ± 0.02	

TABLE Ij6 (*Continued*)

Material	Curie Temp. (°K)	Neel Temp. (°K)
Pd_3Mn		170
Pd_3Mn_2		593
Pd_2MnAl		240
Pd_2MnIn		142
Pd_2MnSb	247–251	
Pd_2MnSn	185–189	
Pr	8.7	23
PrAg	14	
PrAl		19; 20*
$PrAl_2$	31–38.5*	
Pr_2Al		18
Pr_3Al	16	
$PrAl_2$–$PrCo_2$		
0 mol. % $PrCo_2$	35	
15 mol. % $PrCo_2$		6.5
22.5 mol. % $PrCo_2$	24	
90 mol. % $PrCo_2$	45	
PrB_6		~7.5
PrC_2		15
$PrCl_3$		0.428
$PrCo_3$	349	
$PrCo_5$	639–912*	
Pr_2Co_7	574	
Pr_2Co_{17}	1160; 1171*	
$PrCrO_3$		238.7
$PrFe_7$	280; 283*	
Pr_2Fe_{17}	282; 287*	
$PrFeO_3$		707
Pr-Gd		
63.8 at. % Gd	278	
70.5 at. % Gd		55
92.7 at. % Gd	254	
PrGe	39	
$PrGe_2$	19	
Pr_5Ge_3		12
PrI_2	16	
$PrMnO_3$		
N_1		91
N_2		1.50
$PrMn_2O_5$		46
PrNi	20	
$PrNi_2$	8	
$PrNi_3$	20	
Pr_2Ni_7	85	
Pr_3Ni		2
PrO_2		14
$PrRh_2$	8.6	
$PrRu_2$	40	
PrS		16
$PrSn_3$		8.5; 8.6*
Pt-0.8% Co	2.8	
Pt-Fe		
17.4 % Fe	305	
24.9 % Fe	425	
28 % Fe	~230	<130

TABLE Ij6 (*Continued*)

Material	Curie Temp. (°K)	Neel Temp. (°K)
30 % Fe	300	
34.3 % Fe	250	
Pt_3Fe		170
$Pt_{35}Mn_{65}$	~250	~710
Pu_2C_3		120
$PuGe_2$	34.5	
PuP	126	
PuS		4.5
PuS_2		15
Pu_2S_3, α		7
Pu_3S_4		10
RE_2Co_{17}#	1068–1213	
$RE_3Fe_5O_{12}$#	~530–555	
$RbCoCl_5$		1.14
$RbCoF_3$		32; 150*
Rb_2CoF_4		101; 130*
$RbFeF_3$	87	75–102*
Rb_2FeF_4		50–60
$RbMnCl_3$		85–92*
$Rb_2MnCl_4 \cdot 2H_2O$		2.24
$RbMnF_3$		83
Rb_2MnF_4		38.4
$RbNiCl_3$		11
$Rb_2NiCl_4 \cdot 2H_2O$		4.60 ± 0.02
$RbNiF_3$	139; 145*	
Sc-39 at. % Er		1.3
Sc-18 at. % Ho		1.3
Sc-24% In	6	
Sc-25 at. % Tb		1.3
Sm		14; 15; 106*
$SmAl_2$	122	
$SmAlO_3$		1.30
SmC	~30	
Sm_3C		30
$SmCo_5$	747–1020*	
Sm_2Co_7	713	
Sm_2Co_{17}	1190; 1195*	
$SmCrO_3$		192.5
Sm_2CrSbO_7	12	
SmCu		40
$SmFe_2$	674; 675*	
$SmFe_3$	651	
$SmFe_7$	399	
Sm_2Fe_{17}	395	
$SmFeO_3$	660; 673*	672; 674*
$Sm_3Fe_5O_{12}$	562; 578*	
$Sm_3Ga_5O_{12}$		0.967
SmGe		40
$SmIn_3$		16
$SmIr_2$	37	
$SmMn_5$	439; 440*	
Sm_6Mn_{23}	439	
$SmMnO_3, N_1$		60
SmN		13–18.2*
SmNi	45	

TABLE Ij6 *(Continued)*

Material	Curie Temp. (°K)	Neel Temp. (°K)
$SmNi_2$	21; 77*	
$SmNi_3$	85	
$SmNi_5$	25	
Sm_2Ni_{17}	186; 641*	
$SmOs_2$	34	
$SmSn_3$		12
$SmZn$	125	
$SmZn_2$		45
$SnMn_2O_4$	53	
Sr_2CoSbO_6		~35
Sr_2CoUO_6		7
Sr_2CoWO_6		22
Sr_2CrSbO_6		9
Sr_2FeMoO_6	422	
$SrFe_{12}O_{19}$	750	
Sr_2FeSbO_6		21
Sr_2FeWO_6		16
$SrGd_2O_4$		2.8
Sr_2MnWO_6		10
Sr_2NiMoO_6		71
Sr_2NiWO_6		59
$SrRuO_3$	160 ± 2; 164*	
$SrTb_2O_4$		5.9
$SrTb_2S_4$		6
$TaMn_2B_4$	780	
$TaSe_2$		130
Tb	210–135*	228–234*
$TbAg$		100–106 ± 2*
$TbAg_2$		35
$TbAl$		72
$TbAl_2$	111–121*	
Tb_3Al_2	190	
$TbAlO_3$		3; 8; 3.95 ± 0.10*
$Tb_3Al_5O_{12}$		1.35; 1.5*
$TbAs$		10.5; 12*
$TbAu$		40 (19.3 kOe)
$TbAu_2$		42.5
$TbBi$		17
TbC_2		66
Tb_2C	266 ± 2	
$Tb_{1.2}Cd$	185	
$TbCo_2$	256; 237	
$TbCo_3$	506	
$TbCo_5$	980	
Tb_2Co_7	693; 717*	
Tb_2Co_{17}	1180; 1194*	
$TbCoO_3$		3.31; 3.35*
$TbCrO_3$		117.6
N_1		158
N_2		4
N_3		3.05
Tb_2CrSbO_7	15	
$TbCu$		115; 117*
$TbCu_2$		54
$TbFe_2$	695	

TABLE Ij6 (*Continued*)

Material	Curie Temp. ($^\circ$K)	Neel Temp. ($^\circ$K)
$TbFe_3$	648	
$Tb_2 Fe_{17}$	408; 409*	
$Tb_6 Fe_{23}$	574	
$Tb_3 Fe_5 O_{12}$	553–568*	
$TbGa$	155	
$TbGe$		48
$TbGe_2$		42
$Tb_5 Ge_3$		85
$Tb_5 Ge_4$		30
$TbH_{1.98}$		40
TbH_2		40
$TbHo$	~79	~182
Tb-50% Ho	78	183
$TbIn$		190
$TbIn_3$		36
$TbIr_2$	45	
$Tb_{0.4} La_{0.6}$		68
$TbLu$	180 ± 10	
Tb-Lu		
4.7 % Lu	180	
0–77% Lu		228–80
$TbMn_2$		41
$TbMn_5$	445	
TbN	33.7–42*	
$TbNbO_4$		1.82
$TbNi$	50	
$TbNi_2$	45; 46*	
$TbNi_3$	98	
$TbNi_5$	27	
$Tb_2 Ni_7$	101	
$Tb_2 Ni_{17}$	178; 615*	
$Tb_3 Ni$		62
TbO_2		3
$Tb_2 O_3$		2.0
$Tb(OH)_3$	3.72	
$TbOs_2$	34	
$Tb_2 O_2 S$		7.7
TbP		8; 9*
$TbPO_4$		~415
$Tb_{5.10} Pd_{1.90}$	~30	62
$TbPt_2$	4.2	
$TbSb$		14–16.5*
Tb-Sc		
10 % Sc	159	200
20 % Sc	<15	174
49 % Sc		104
75 % Sc		<1.3
$TbSi$		57
$TbSi_2$		17
$Tb_5 Si_4$	225	
Tb-22% Tm		205
Tb-Y		
0–95% Y		228–25
30–90% Y		188–50
10% Y	175	

TABLE Ij6 (Continued)

Material	Curie Temp. ($^\circ$K)	Neel Temp. ($^\circ$K)
TbZn	206	
TbZn$_2$		25; 55*
ThCo$_5$	630	
ThCo$_3$Fe$_2$	690	
Th-36 at. % Er		3.5 ± 0.5
Th-80 at. % Pr		<1.3
TiCl$_3$		~100; 217 ± 2
TiCr$_2$Te$_4$	214	
TiFe$_2$	323–424	280
TiH$_2$		~300
Ti$_3$O$_5$		432–462*
TlCoF$_3$		125
Tl$_2$CoF$_4$		130
TlCuF$_3$		235
TlFeCl$_4$		11.2 ± 0.2
TlFeF$_3$	~77	78; 100*
TlMnF$_3$		76; 85*
TlNiF$_3$	129; 150*	
Tl$_2$NiF$_4$		~190
Tm	22–38*	51–60*
TmAg		9.5 ± 1; 10*
TmAl		10
Tm$_3$Al$_2$		3
TmCo$_3$	370	
TmCo$_5$	1020	
Tm$_2$Co$_{17}$	1182; 1185*	
TmCrO$_3$		~4; 123.7
Tm$_2$CrSbO$_7$	6	
TmCu		11
TmFe$_2$	613	
TmFe$_3$	539	
Tm$_2$Fe$_{17}$	248–313.5*	
Tm$_6$Fe$_{23}$	475	
TmFeO$_3$		632
Tm$_3$Fe$_5$O$_{12}$	549 ±2	
TmIr$_2$	1	
Tm$_{0.53}$Lu$_{0.47}$		38
Tm$_{0.78}$Lu$_{0.22}$	11	49
TmMn$_2$	12 ± 2	
TmNi	4	
TmNi$_2$	14	
TmNi$_3$	43	
TmNi$_5$	7	
Tm$_2$Ni$_{17}$	152; 603*	
Tm$_3$Ni	12	
Tm-Y		
45 % Y		37
85.1 % Y		20 ± 8
U		36
UAs		128
UAs$_2$		283
U$_3$As$_4$	198; 205 ± 50	
UAs-USe		
0 mole % USe		128
8 mole % Use	184	

TABLE Ij6 (*Continued*)

Material	Curie Temp. ($^\circ$K)	Neel Temp. ($^\circ$K)
35 mole % USe	~138	132
40 mole % USe	~140	134
UBi		285 ± 5; 290*
UBi_2		183
U_3Bi_4	108	
U_2C_3		50; 59*
UCl_3		~23
UD	172	
UFe_2	172	
$UFeO_4$	42; 47 ± 2	55
UGa_2	126	
UGe_2	52	260
U_3Ge_4	94	
UH_3, β	174	
UI_3		2.6
UN		50–60*
UO_2		28.7–30.8*
UOS		55
UOSe		72
UOTe		162
UP	~125	121–130*
UP_2		203; 206 ± 2*
U_3P_4	136.5–165*	
UP-US		
0.2 US	~90	~140
0.4 US	~130	~180
0.6 US	~159	~200
0.8 US	~175	~210–215
US	178 ± 2; 180*	220
USb		213; 217 ± 4
USb_2		206
U_3Sb_4	148; 149*	
USe	160.5–210*	
USe_2	13.1	
U_2Se_3	180	
U_3Se_4	160	
UTe	103; 123*	
UTe_2		<77
UTe_3		<77
U_2Te_3	122	
U_3Te_4	105	
VBO_3	32.5	
$VO_{1.147}$		4.6
$VO_{1.257}$		7.0
$VO_{1.50}$		170
$VO_{2.16}$		154
VO_2		338–345 ± 3*
V_2O_3 (monoclinic)		168; 170*
V_3O_5		235 ± 3; 253 ± 3*
V_4O_7		130 ± 3
V_5O_9		162
V_6O_{13}		154
$VOSO_4$		
α	~4.2	
β		25

TABLE Ij6 (*Continued*)

Material	Curie Temp. (°K)	Neel Temp. (°K)
$VS_{1.33}$		473
VSe		~400
$VSe_{0.90}$		~300
V-Te 64.3-66 at. % Te		410–480
WMn_2B_4	560	
WV_2O_6		370 ± 30
$Y_3AlFe_4O_{12}$	430	
$Y_2BiFe_5O_{12}$	591	
YCo_2	2, 291–296*	
YCo_3	284; 301*	
YCo_5	630–975*	
Y_2Co_7	639	
Y_2Co_{17}	1167; 1213*	
Y_4Co_3	13	
$YCrO_3$		140.6–142*
Y_2CrSbO_7	15	
YFe_2	533–552*	
Y_2Fe_{17}	301; 302.5*	
Y_6Fe_{23}	471; 484*	
$YFeO_3$	643	640; 643*
$Y_3Fe_5O_{12}$	545–551*	
polycrystal	552–562.7*	
powder	570 ± 2	
YMn_5	486 ± 3; 490*	
$YMnO_3$		46
$YNd_2Fe_5O_{12}$	557	
YNi_3	33	
Y_2Ni_7	58	
YVO_3		110
$YbCo_5$	973	
$YbCrO_3$		3.05; 158
$YbFeO_3$		625; 627*
$YbMn_2O_5$		46
$YbNi_3$	<20	
$YbNi_5$	1.4	
Yb_2O_3		2.25; 2.3*
$ZnCr_2O_4$		15; 16*
$ZnCr_2S_4$		18
$ZnCr_2Se_4$	129.5	20
$ZnFe_2O_4$	780	9; 15*
$ZnMn_2As_2$	~30	
$ZnMn_2O_4$		330
$ZrFe_2$	610–798	
$ZrZn_{1.9}$	27.5 ± 0.5	

*Indicates data from several sources.
#RE = Gd, Tb, Dy, Ho, Yb.

TABLE 1j7. NUCLEAR MAGNETIC RESONANCE TABLE.*

Isotope	NMR Frequency MHz at 14,092 G	NMR Frequency MHz at 23,487 G	Natural Abundance (%)	Relative Sensitivity for Equal Number of Nuclei at Constant Field	Spin* 1 in multiples of $h/2\pi$	Electric Quadrupole Moment Q in multiples of 10^{-24} cm^2
H^1	60.000	100.00	99.985	1.00	1/2	–
H^2	9.2104	15.351	1.5×10^{-2}	9.65×10^{-3}	1	2.73×10^{-3}
Li^7	23.317	38.862	92.58	0.293	3/2	-3×10^{-2}
Be^9	8.4321	14.054	100.	1.39×10^{-2}	(-)3/2	5.2×10^{-2}
B^{10}	6.4479	10.746	19.58	1.99×10^{-2}	3	7.4×10^{-2}
B^{11}	19.250	32.084	80.42	0.165	3/2	3.55×10^{-2}
C^{13}	15.087	25.144	1.108	1.59×10^{-2}	1/2	–
N^{14}	4.3343	7.2238	99.63	1.01×10^{-3}	1	7.1×10^{-2}
N^{15}	6.0798	10.133	0.37	1.04×10^{-3}	(-)1/2	–
O^{17}	8.134	13.56	3.7×10^{-2}	2.91×10^{-2}	(-)5/2	-2.6×10^{-2}
F^{19}	56.446	94.077	100.	0.833	1/2	–
Na^{23}	15.871	26.452	100.	9.25×10^{-2}	3/2	0.14 – 0.15
Al^{27}	15.634	26.057	100.	0.206	5/2	0.149
Si^{29}	11.919	19.865	4.70	7.84×10^{-3}	(-)1/2	–
P^{31}	24.288	40.481	100.	6.63×10^{-2}	1/2	–
S^{33}	4.6018	7.6696	0.76	2.26×10^{-3}	3/2	-6.4×10^{-2}
Cl^{35}	5.8790	9.7983	75.53	4.70×10^{-3}	3/2	-7.89×10^{-2}
Cl^{37}	4.893	8.155	24.47	2.71×10^{-3}	3/2	-6.21×10^{-2}
V^{51}	15.77	26.28	99.76	0.382	7/2	-4×10^{-2}
Mn^{55}	14.798	24.664	100.	0.175	5/2	0.55
Co^{59}	14.168	23.614	100.	0.277	7/2	0.40
Cu^{63}	15.903	26.506	69.09	9.31×10^{-2}	3/2	-0.16
Cu^{65}	17.036	28.394	30.91	0.114	3/2	-0.15
As^{75}	10.276	17.127	100.	2.51×10^{-2}	3/2	0.3
Se^{77}	11.44	19.07	7.58	6.93×10^{-3}	1/2	–
Br^{79}	15.032	25.054	50.54	7.86×10^{-2}	3/2	0.33
Br^{81}	16.204	27.006	49.46	9.85×10^{-2}	3/2	0.28
Rb^{87}	19.632	32.720	27.85	0.175	3/2	0.13
Nb^{93}	14.666	24.443	100.	0.482	9/2	-0.2
Sn^{117}	21.376	35.626	7.61	4.52×10^{-2}	(-)1/2	–
Sn^{119}	22.363	37.272	8.58	5.18×10^{-2}	(-)1/2	–
Sb^{121}	14.359	23.931	57.25	0.160	5/2	-0.5
I^{127}	12.004	20.007	100.	9.34×10^{-2}	5/2	-0.69
Cs^{133}	7.8702	13.117	100.	4.74×10^{-2}	7/2	-3×10^{-3}
Pt^{195}	12.90	21.50	33.8	9.94×10^{-3}	1/2	–
Hg^{199}	10.696	17.827	16.84	5.67×10^{-3}	1/2	–
Hg^{201}	3.9598	6.5998	13.22	1.44×10^{-3}	(-)3/2	0.50
Tl^{203}	34.290	57.150	29.50	0.187	1/2	–
Tl^{205}	34.625	57.709	70.50	0.192	1/2	–
Pb^{207}	12.553	20.922	22.6	9.16×10^{-3}	1/2	–
Bi^{209}	9.6418	16.070	100.	0.137	9/2	-0.4

*Reproduced from "NMR Table" by permission of the copyright owner, Varian Associates, Inc., Palo Alto, California.

KNIGHT SHIFT

The electronic properties of liquid metals, in particular, may be investigated by their "Knight Shift." The latter has been defined (Cusack, 1963) as:

$$K = \frac{8\pi}{3} \chi_p M P_F$$

where

χ_p = paramagnetic susceptibility per unit mass of electrons

M = the mass of one electron

P_F = the average probability density at the nucleus for electrons at the Fermi surface.

K = the fractional difference in the magnetic resonance frequency of a nucleus when it is measured in a metal and in a nonconducting salt.

Tables 1j8 and 1j9 offer Knight shift data, expressed in per cent, and are presented here through the courtesy of Dr. G. C. Carter of the NBS Alloy Data Center.

TABLE 1j8.

Knight Shifts at the Melting Point
Alloy Data Center
NOV. 1973

Legend:

Al	13
933	← melting point (°K)
0.164	← isotropic Knight shift in solid (%)
0.161	← Knight shift in liquid (%)

NUMBERS IN PARENTHESES ARE THEORETICAL

Each cell below lists: atomic number, symbol, melting point (°K), isotropic Knight shift in solid (%), Knight shift in liquid (%). Values in parentheses are theoretical.

I A	II A	III A	IV A	V A	VI A	VII A	VIII A	IX A	X A	I B	II B	III B	IV B	V B	VI B	VII B	INERT GASES
1 H																	2 He
3 Li 454 / 0.026 / 0.026	4 Be											5 B ~2500	6 C (0.09)	7 N	8 O	9 F	10 Ne
11 Na 371 / 0.113 / 0.116	12 Mg											13 Al 933 / 0.164 / 0.161	14 Si 1685 / (0.19)	15 P	16 S	17 Cl	18 Ar
19 K 336 / 0.26 / 0.27	20 Ca	21 Sc	22 Ti	23 V	24 Cr	25 Mn	26 Fe	27 Co	28 Ni	29 Cu 1356 / 0.255 / 0.265	30 Zn	31 Ga 303 / 0.155 / 0.455	32 Ge 1209 / (0.5)	33 As 1090 / 0.318 ~50 kbar	34 Se 490 / (0.48)	35 Br	36 Kr
37 Rb 312 / 0.653 / 0.662	38 Sr 1045 / (0.9) / (0.7)	39 Y	40 Nb	41 Mo		43 Tc	44 Ru	45 Rh	46 Pd	47 Ag 1234 / 0.55 / 0.58	48 Cd 594 / 0.58 / 0.79	49 In 430 / 0.82 / 0.79	50 Sn 505 / 0.75 / 0.73	51 Sb 904 / 0.71	52 Te 723 / ~-0.06 / 0.35	53 I	54 Xe
55 Cs 302 / 1.49 / 1.46	56 Ba	La–Lu	72 Hf	73 Ta	74 W	75 Re	76 Os	77 Ir	78 Pt	79 Au	80 Hg 234 / 2.71 / 2.71	81 Tl 576 / 1.6 / 1.6	82 Pb 601 / 1.54 / 1.53	83 Bi 554 / 1.41	84 Po	85 At	86 Rn
87 Fr	88 Ra	Ac–Lw															

The NBS Alloy Data Center is a part of the National Standard Reference Data System. Literature reference are available upon request

TABLE 1j9.

Knight Shifts at 4°K and 300°K

Alloy Data Center
NOV. 1973

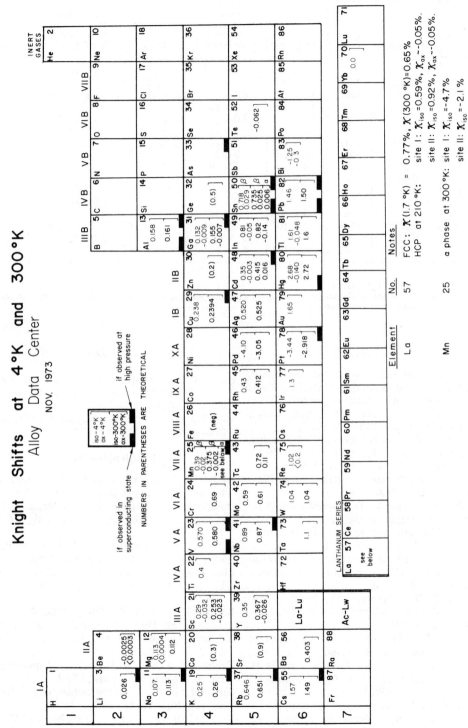

The NBS Alloy Data Center is a part of the National Standard Reference Data System
Literature reference are available upon request

TABLE 1j10. MAGNETOOPTIC PROPERTIES OF SELECTED MATERIALS (FREISER 1968, HUNT 1969).

Material	Wavelength (microns)	Temp. (°K)	Faraday Effect			Kerr Effect	
			F (deg/cm)	Absorption Coeff. α (cm^{-1})	F/α (deg)	Reflection Coeff. (r_{sp})	Incident Angle (deg. θ_i)
Fe	1.0	300	5.1×10^5	1.6×10^5	3.2	10.5×10^{-4}	60
	0.546	300	3.5×10^5	7.6×10^5	0.46		
Co	0.546	300	3.6×10^5	8.5×10^5	0.42	6.5×10^{-4}	
Ni	0.400	300	7.2×10^5	2.1×10^5	3.4		45
Permalloy	0.5	300	1.2×10^5	3.0×10^5	0.40	4×10^{-4}	45
MnBi	0.63	300	5.3×10^5	3.8×10^5	1.4	4×10^{-3}*	37
CrBr$_3$	0.493	1.5	2×10^5	2×10^3	100		
YIG**	1.20	300	2.4×10^2	0.069	3.5×10^3		
GdIG	0.52	300	4×10^3	3×10^3	1.3		
	0.60	300	2.5×10^3				
RbNiF$_3$	0.5	77	4×10^2	2×10^1	20		
EuO	0.55	12				4.6×10^{-3}	
	0.68	8	5×10^5	1×10^5	5		
	1.2	60	2×10^5	1×10^2	2.0×10^3		
EuSe	0.625	8				6.9×10^{-3}	30
	0.755	4.2	1.4×10^5	4.5×10^1	3.1×10^3		30

*Polar effect.
**Yttrium Iron Garnet (Y$_3$Fe$_5$O$_{12}$).
Note: A major part of this table has been reprinted with the permission of the IEEE from the *IEEE Trans. on Magnetics*, Vol. MAG-5, #4, Dec. 1969.

REFERENCES

Anonymous, "Metalic, Magnetic, Shielding, Conducting, and Soldering Materials, Parts and Supplies," in *Insulation/Circuits Directory-Encyclopedia*, Lake Publishing Corp., Libertyville, Ill., pp. 340–356, 1970.

Becker, J. J., F. E. Luborsky and D. L. Martin, "Permanent Magnet Materials," *IEEE Trans. Magnetics*, AMG-4, 84–99, June 1968.

Bozorth, R. M., "Ferromagnetism," Van Nostrand, Princeton, N.J., 1951.

Connolly, T. F. and E. D. Copenhaver, "Bibliography of Magnetic Materials and Tabulation of Magnetic Transition Temperatures," Report No. ORNL-RMIC-7 (Rev. 2), Clearinghouse for Federal Scientific and Technical Information, National Bureau of Standards, Springfield, Va., 1970.

Cochrane, J. H., "Surveying the Field of Permanent Magnets," *Machine Design*, pp. 194–200 (Sept. 15, 1966).

Cusack, N. E., "The Electronic Properties of Liquid Metals," in: Stickland, A. E., Ed., *Reports on Progress in Physics*, The Institute of Physics and The Physical Society, Vol. 26, pp. 361–409, 1963.

Freiser, M. J., "A Survey of Magnetooptic Effects," *IEEE Trans. Magnetics*, MAG-4, #2, pp. 152–161, 1968.

Gould, J. E., "Permanent Magnet Applica-

tions," *IEEE Trans. Magnetics*, MAG-5, pp. 812–821, 1969.

Hunt, R. L., "Magnetooptics, Lasers and Memory Systems," *IEEE Trans. Magnetics*, MAG-5, #4, pp. 700–716, 1969.

Standley, K. J., "Oxide Magnetic Materials," Clarendon Press, Oxford, 1972.

BIBLIOGRAPHY

Anderson, J. C., "Magnetism and Magnetic Materials," Chapman & Hall, London, 1968.

Cullity, B. D., "Introduction to Magnetic Materials," Addison-Wesley, Reading, Mass., 1972.

Earnshaw, A., "Introduction to Magnetochemistry," Academic Press, London, 1968.

Elliott, R. J., Ed., "Magnetic Properties of Rare Earth Metals," Plenum Press, New York, 1972.

Rado, G. T. and H. Suhl, "Magnetism: A Treatise on Modern Theory and Materials," 5 vols., Academic Press, New York, 1963–1973.

Tebble, R. S. and D. J. Craik, "Magnetic Materials," Wiley, New York, 1969.

Selwood, P. H., "Magnetochemistry," Interscience, New York, 1956.

Von Aulock, W. H., Ed., "Handbook of Microwave Ferrite Materials," Academic Press, New York, 1965.

Section 1k
ACOUSTIC PROPERTIES OF MATERIALS

The human response to sound intensity, the velocity of sound in various substances, and selected sound absorption coefficients are summarized in the following tables:

As most of these tables were assembled from many sources, specific references are not given.

TABLE 1k1. TABLE OF SOUND INTENSITY LEVELS.

Description or Effect	Sound Pressure (dyn/cm²)	Sound Intensity at Eardrum (W/cm²)	Intensity Level (dB above 10^{-16} W/cm²)	Familiar Sources of Sound (number in parentheses shows distance from source)
Impairs hearing		10^{-1}	150	
				jet engine
Pain	2040	10^{-2}	140	largest air raid siren (100 ft)
Threshold of pain		10^{-3}	130	level of painful sound
				pneumatic hammer (5 ft) airplane 1600 rpm (18 ft from propeller)
Threshold of discomfort	204	10^{-4}	120	automobile horn
				engine room of submarine (at full speed) bass drum (maximum)
Deafening		10^{-5}	110	boiler factory loud bus horn thunder clap subway (express passing a local station)
Discomfort begins	20.4	10^{-6}	100	can manufacturing plant
				very loud musical peaks noisiest spot at Niagera Falls
Very loud		10^{-7}	90	loudest orchestral music noisy factory heavy street traffic
	2.04	10^{-8}	80	loud speech police whistle very loud radio
				average factory average orchestral volume
Loud		10^{-9}	70	busy street noisy restaurant average conversation (3 ft)
				quiet typewriter
	0.204	10^{-10}	60	average (quiet) office hotel lobby quiet residential street
				soft violin solo
Moderate		10^{-11}	50	church quiet automobile
	0.0204	10^{-12}	40	average residence lowest orchestral volume
				quiet suburban garden
Faint		10^{-13}	30	average whisper
				very quiet residence
	0.00204	10^{-14}	20	faint whisper (5 ft)
				ordinary breathing (1 ft) outdoor minimum (rustle of leaves)

TABLE 1k1. (*Continued*)

Description or Effect	Sound Pressure (dyn/cm^2)	Sound Intensity at Eardrum (W/cm^2)	Intensity Level (dB above 10^{-16} W/cm^2)	Familiar Sources of Sound (number in parentheses shows distance from source)
Very faint		10^{-15}	10	anechoic room
				normal threshold of hearing
Threshold of hearing	0.000204	10^{-16}	0	reference level

*From Graf, R. F., "*Electronic Design Data Book*," Van Nostrand Reinhold Co., New York, 1971.

TABLE 1k2. VELOCITY OF SOUND IN GASES AND VAPORS AT 0°C AND ONE ATMOSPHERE.

Gas/Vapor	Formula	Velocity (km/sec)
Air	–	0.3316
Ammonia	NH_3	0.415
Argon	Ar	0.3078
Carbon Dioxide	CO_2	0.258
Carbon Monoxide	CO	0.338
Chlorine	Cl_2	0.206
Ethanol	C_2H_6O	0.231
Ether	$C_4H_{10}O$	0.179
Ethylene	C_2H_4	0.317
Helium	He	0.969
Hydrogen	H_2	1.237
	D_2	0.890
Hydrogen Bromide	HBr	0.198
Hydrogen Chloride	HCl	0.296
Hydrogen Iodide	HI	0.157
Hydrogen Sulfide	H_2S	0.289
Methane	CH_4	0.430
Neon	Ne	0.433
Nitric Oxide	NO	0.324
Nitrogen	N_2	0.334
Nitrous Oxide	N_2O	0.263
Oxygen	O_2	0.316
Silicon Tetrafluoride	SiF_4	0.167
Sulfur Dioxide	SO_2	0.213
Water	H_2O	0.401

TABLE 1k3. VELOCITY OF SOUND
IN ORGANIC LIQUIDS.

Liquid	Formula	Temp (°C)	Sound Velocity (km/sec)
Acetaldehyde	C_2H_4O	0	1.207
Acetic Acid	$C_2H_4O_2$	20	1.164
Acetic Anhydride	$C_4H_6O_3$	30	1.249
Acetone	C_3H_6O	20	1.189
Acetonitrile	C_2H_3N	20	1.300
Acetylacetone	$C_5H_8O_2$	20	1.383
Acetyl Chloride	C_2H_3ClO	20	1.060
Acrolein	C_3H_4O	20	1.190
Allyl Alcohol	C_3H_6O	30	1.126
Aniline	C_6H_7N	20	1.659
Anisole	C_7H_8O	20	1.425
Benzaldehyde	C_7H_6O	20	1.479
Benzene	C_6H_6	20	1.324
Bromal	C_2HBr_3O	20	0.966
Bromobenzene	C_6H_5Br	20	1.170
Bromobutane	C_4H_9Br	20	1.019
Bromoethane	C_2H_5Br	20	0.900
Bromoform	$CHBr_3$	20	0.931
2-Butanone	C_4H_8O	20	1.217
Butyl Alcohol	$C_4H_{10}O$	20	1.263
Isobutyl Alcohol	$C_4H_{10}O$	20	1.212
sec-Butyl Alcohol	$C_4H_{10}O$	30	1.197
tert-Butyl Alcohol	$C_4H_{10}O$	30	1.098
Butyl Acetate	$C_6H_{12}O_2$	20	1.226
Isobutyl Acetate	$C_6H_{12}O_2$	20	1.182
Butyric Acid	$C_4H_8O_2$	20	1.203
Carbon Disulfide	CS_2	20	1.157
Carbon Tetrachloride	CCl_4	20	0.938
Castor Oil	—	20	1.54
Chlorobenzene	C_6H_5Cl	20	1.289
1-Chlorobutane	C_4H_9Cl	20	1.140
Chloroform	$CHCl_3$	20	1.001
o-Cresol	C_7H_8O	20	1.541
m-Cresol	C_7H_8O	20	1.500
Cyclohexane	C_6H_{12}	20	1.277
Cyclohexanol	$C_6H_{12}O$	30	1.444
n-Decane	$C_{10}H_{22}$	20	1.255
Dibromomethane	CH_2Br_2	20	0.963
Dichloromethane	CH_2Cl_2	20	1.093
Diiodomethane	CH_2I_2	20	0.973
Dimethyl Sulfate	$C_2H_6O_4S$	20	1.255
1,4-Dioxane	$C_4H_8O_2$	20	1.366
Ethanol	C_2H_6O	20	1.159
Ethylene Glycol	$C_2H_6O_2$	20	1.667
Ethyl Acetate	$C_4H_8O_2$	20	1.177
Ethyl Ether	$C_4H_{10}O$	20	1.006
Ethyl Formate	$C_3H_6O_2$	20	1.160
Ethyl Proprionate	$C_5H_{10}O_2$	20	1.183
Fluorobenzene	C_6H_5F	20	1.189
Formamide	CH_3NO	25	1.622
Furan	C_4H_4O	30	1.105
Furfuryl Alcohol	$C_5H_6O_2$	30	1.434
Glycerol	$C_3H_8O_3$	20	1.895
n-Hexane	C_6H_{14}	20	1.116
Hexyl Alcohol	$C_6H_{14}O$	20	1.322

TABLE 1k3. (*Continued*)

Liquid	Formula	Temp (°C)	Sound Velocity (km/sec)
Iodobenzene	C_6H_5I	20	1.114
n-Iodobutane	C_4H_9I	20	0.972
Iodomethane	CH_3I	20	0.834
Isoprene	C_5H_8	15	1.095
Methanol	CH_4O	20	1.120
Methyl Acetate	$C_3H_6O_2$	20	1.182
Nitrobenzene	$C_6H_5NO_2$	20	1.475
Nitromethane	CH_3NO_2	20	1.346
Octane	C_8H_{18}	20	1.197
Iso-Octane	C_8H_{18}	20	1.111
Octene	C_8H_{16}	20	1.184
Paraldehyde	$C_6H_{12}O_3$	20	1.192
n-Pentane	C_5H_{12}	20	1.030
Pentene-1	C_5H_{10}	20	1.014
n-Pentyl Alcohol	$C_5H_{12}O$	20	1.294
Iso-Pentyl Alcohol	$C_5H_{12}O$	20	1.260
Phenol	C_6H_6O	100	1.274
n-Propyl Alcohol	C_3H_8O	20	1.222
Isopropyl Alcohol	C_3H_8O	20	1.170
Isopropylamine	C_3H_9N	20	1.089
Propyl Ether	$C_6H_{14}O$	20	1.112
Pyridine	C_5H_5N	20	1.441
Pyrrole	C_4H_5N	30	1.438
Styrene	C_8H_8	20	1.354
Tetrachloroethylene	C_2Cl_4	20	1.053
Tetrahydrofuran	C_4H_8O	30	1.255
Thioacetic Acid	C_2H_4OS	20	1.168
Toluene	C_7H_8	20	1.328
Trichloroethylene	C_2HCl_3	20	1.049
Triethylamine	$C_6H_{15}N$	20	1.143
Turpentine	—	20	1.33
Vinyl Acetate	$C_4H_6O_2$	20	1.152
o-Xylene	C_8H_{10}	20	1.360
m-Xylene	C_8H_{10}	20	1.340
p-Xylene	C_8H_{10}	20	1.330

TABLE 1k4. VELOCITY OF SOUND IN INORGANIC LIQUIDS

Liquid	Formula	Temperature (°C)	Sound Velocity (km/sec)
Mercury	Hg	20	1.451
Silicone DC 500, 0.65 cs	—	30	0.8732
5.0 cs	—	30	0.9538
50 cs	—	30	0.9816
Silicon Tetrachloride	$SiCl_4$	30	0.7662
Water, distilled	H_2O	0	1.4035
		20	1.4831
		40	1.5295
		60	1.5515
		80	1.5546
		94	1.5490
Water, Sea	H_2O	15	1.50

TABLE 1k5. VELOCITY OF SOUND IN INORGANIC SOLIDS AT ROOM TEMPERATURE

Solid	Longitudinal Velocity (km/sec)	Solid	Longitudinal Velocity (km/sec)
Aluminum	6.360	Lithium Iodide	2.85
Aluminum Antimonide	4.85	Lithium Niobate	7.20
Aluminum Oxide	10.8	Lithium Tantalate	5.82
Aluminum Oside (Sapphire)	11.21	Magnesium	5.700
Ammonium Dihydrogen Phosphate	4.65	Magnesium Aluminate	10.0
Antimony	3.140	Magnesium Oxide	9.5
Arsenic Sulfide	2.6	Magnesium Silicate	8.48
Barium	2.080	Manganese	5.560
Barium Nitrate	2.96	Mercury Selenide	1.6
Barium Sulfate	4.30	Mercury Telluride	2.76
Barium Titanate	~6.5	Molybdenum	6.650
Beryllium	12.60	Neodymium	2.720
Beryllium Oxide	12.15	Nickel	5.810
Bismuth	2.290	Niobium	5.100
Bismuth Germanium Oxide, $(Bi_{12}GeO_{20})$	3.42	Osmium	5.480
		Palladium	5.478
Bismuth Silicon Oxide, $(Bi_{12}SiO_{20})$	3.83	Platinum	4.075
		Potassium	2.23
Cadmium	2.980	Potassium Aluminum Sulfate	3.87
Cadmium Sulfide	4.50	Potassium Bromide	3.08
Calcium	4.180 (75°C)	Potassium Chloride	3.88
Calcium Barium Titanate, $(CaBaTiO_3)$	5.11	Potassium Dihydrogen Phosphate	4.75
		Potassium Fluoride	4.64
Calcium Carbonate	6.45	Potassium Iodide	2.57
Carbon (diamond)	18.1	Praseodymium	2.660
Cerium	2.30	Rhenium	5.360
Cesium	1.090	Rhodium	6.194
Chromium	6.50	Rubidium	1.430
Chromium Iron Oxide, (Cr_2FeO_4)	8.65	Rubidium Bromide	2.58
Cobalt	5.730	Rubidium Chloride	3.08
Copper	4.760	Rubidium Fluoride	3.95
Dysprosium	2.960	Rubidium Iodide	2.24
Erbium	3.080	Ruthenium	6.534
Gadolinium	2.907	Samarium	2.700
Gallium	3.030	Silicon [100]	8.396
Gallium Antimonide	4.26	[110]	9.101
Gallium Arsenide	5.09	Silicon Dioxide (quartz)	5.75
Gallium Phosphide	6.32	(fused)	5.968
Germanium [001]	4.9138	Silver	3.636
[110]	5.3996	Sodium	3.310
Germanium Oxide	6.58	Sodium Bromate	4.00
Gold	3.281	Sodium Bromide	3.33
Hafnium	3.670	Sodium Chlorate	4.29
Holmium	3.040	Sodium Chloride	4.53
Hydrogen Oxide (ice), 0°C	3.43	Sodium Fluoride	5.67
Indium	2.460	Sodium Iodide	2.72
Indium Antimonide	3.66	Sodium Nitrate	5.31
α-Iodic Acid	3.08	Strontium	2.780
Iridium	5.380	Strontium Sulfate	5.02
Iron	5.950	Tantalum	4.240
Iron Oxide, (Fe_3O_4)	7.38	Tantalum Monocarbide	4.99
Lanthanum	2.770	Tellurium Oxide	3.86
Lead	2.16	Terbium	2.920
Lead Molybdate	3.83	Thallium	1.630
Lead Nitrate	3.36	Thallium Bromide	2.13
Lead Sulfide	3.75	Thallium Chloride	2.26
Lead Telluride	2.104	Thallium Sulfate	2.47
Lithium	~5.5	Thorium	2.850
Lithium Bromide	3.62	Thorium Dioxide	6.02
Lithium Chloride	5.26	Tin (white)	3.30
Lithium Fluoride	7.20	Tin Telluride	3.23

TABLE 1k5. (*Continued*)

Solid	Longitudinal Velocity (km/sec)	Solid	Longitudinal Velocity (km/sec)
Titanium	5.99	Yttrium	4.280
Titanium Carbide	10.03	Zinc	3.890
Titanium Diboride	11.3	Zinc Oxide	6.40
Titanium Oxide	9.15	Zinc Selenide	4.27
Tungsten	5.320	Zinc Sulfide	5.22
α-Uranium	3.44	Zinc Telluride	3.80
Uranium Dioxide	5.44	Zirconium	4.65
Vandium	6.00	Zirconium Carbide	8.24
Ytterbium	1.820	Zirconium Silicate	10.1

TABLE 1k6. VELOCITY OF SOUND IN POLYMERS AT ROOM TEMPERATURE

Material	Longitudinal Velocity (km/sec)
Acetate Butyrate	3.7–6.78
Acrylic Resin	2.67
Cellulose	1.4
Cellulose Acetate, Sheet	1.0
Cellulose Acetate Butyrate	3.7
Ethyl Cellulose	1.4
Lucite	1.7
Phenolic Resin	1.42
Polyamide	2.62
Polyethylene	1.83
Polystyrene	2.34
Rubber, soft	0.07
Rubber, hard	1.45
Styrene Resin	1.7

TABLE 1k7. VELOCITY OF SOUND IN WOOD AT 10 TO 20°C TEMPERATURE

Wood	Velocity (km/sec) Along grain	Along rings	Across rings
Ash	4.670	1.260	1.390
Beach	3.340	1.415	1.840
Elm	4.120	1.013	1.420
Fir	4.640		
Mahogany	4.135		
Maple	4.110		
Oak	3.850		
Pine	3.320		
Poplar	4.280		
Sycamore	4.460		

TABLE 1k8. SOUND ABSORPTION OF SEVERAL FORMS OF $\frac{1}{4}$-INCH WOOL FELT**

Wool Felt*	Sound Absorption Coefficients at Frequency (cps) 125	250	500	1000	2000	4000
1-Ply Felt With No Coating	0.02	0.02	0.13	0.28	0.55	0.98
2-Ply Felt With Black Vinyl Septum	0.01	0.04	0.23	0.60	0.77	0.77
1-Ply Felt With Perfor. Black Vinyl Face	0.01	0.02	0.23	0.72	0.55	0.44
2-Ply Felt With Black Vinyl Septum and Black Vinyl Face	0.04	0.05	0.29	0.62	0.95	0.95

*Flame Retardant Treated.
LaFaro, Helen W., "Engineer's Guide to Felt," *Mat. Eng.*, **74, (6), 34–42 (Nov. 1971).

REFERENCES

Albers, V. M., "Underwater Acoustics Handbook," 2nd ed., Pennsylvania State University, University Park, 1965.

Degussa, "Edelmetall-Taschenbuch," Degussa, Frankfurt am Main, Germany, 1967.

Forsythe, W. E., "Smithsonian Physical Tables," 9th ed., Smithsonian Institution Press, Washington, 1969.

Hampel, C. A., Ed., "Encyclopedia of the Chemical Elements," Van Nostrand Reinhold, New York, 1968.

Mason, W. P., "Piezoelectric Crystals and their Application to Ultrasonics," Van Nostrand Reinhold, New York, 1950.

Mason, W. P., "Physical Acoustics and the Properties of Solids," Van Nostrand Reinhold, New York, 1958.

Mason, W. P., Ed., "Physical Acoustics," 10 vols., Academic Press, New York, 1966–1973.

Mattiat, Q. E., "Ultrasonic Transducer Materials," Plenum Press, New York, 1971.

Pinnow, D. A., "Guide Lines for the Selection of Acoustooptic Materials," *IEEE Trans. Quantum Electronics,* **QE-6,** (4), 223–238 (Apr. 1970).

Ritchie, P. D., Ed., "Physics of Plastics," Van Nostrand Reinhold, New York, 1965.

Schaafs, W. *et al.*, Eds., "Landolt-Börnstein New Series, Group II, 'Atomic and Molecular Physics,' Vol. 5, 'Molecular Acoustics,'" Springer-Verlag New York, New York, 1967.

Simmons, G. and H. Wang, "Single Crystal Elastic Constants and Calculated Aggregate Properties—A Handbook," 2nd ed., M.I.T. Press, Cambridge, Mass., 1971.

Stephens, R. W. B. and A. E. Bate, "Acoustics and Vibrational Physics," St. Martin's Press, New York, 1966.

SECTION 11
OPTICAL PROPERTIES
OF MATERIALS

This Section describes the optical and electro-optical properties of selected materials. The refractive indices of minerals are incorporated in a table of properties in Chapter XII.

The information in Section 11 is organized as follows:

TABLE 111. FLAME SPECTRA OF THE ELEMENTS*

Legend: AH air-hydrogen, OH oxyhydrogen, OA oxyacetylene, OC oxycyanogen; solvent is water. AHn, OHn, OAn, the corresponding flames with nonaqueous solvent is indicated, for each group of data, by the following symbols: AA acetylacetone, Ac acetone, Bu Butanol, Bz benzene, C chloroform, EA 1:2:1 ether-alcohol-water, G gasoline, H hexone, K kerosene, M methanol, N naphtha, P isopropanol; subscript number indicates percentage, e.g., Ac_{60} = 60% acetone + 40% water. Other notations: ( ) very doubtful; a arc line, b band peak, c continuum or very wide band, d double, f intercombination line, i inner cone emission, k with an S-1 photomultiplier, m multiple, p resonance line, r head of red-degraded band, s spark line, t triple, u unclassified line, v head of violet-degraded band, w wide or diffuse band, x with uncoated mirror in burner housing (mirror normally has a silicone coating).

Notes: (1) Seen in absorption against an effectively hotter OH band. (2) Data to be provided by Kniseley, Fassel and Curry, *Spectrochimica Acta.* (3) Obscured by CN bands. (4) Bands emitted only from iodate solution. (5) For 1P28 photomultiplier with red filter. (6) The arc lines 382.94, 383.23, 383.83 sometimes appear within this band. (7) The stronger bands 386–388 are interfered with by CH 387, 389.

Species	Character	Wavelength mμ	Intensity in various flames						
			AH	AHn	OH	OHn	OA	OAn	OC
Ag	p	328.07	100		100	100 EA	50	30 EA	25
Ag	p	338.29	250		170	170	50	40	20
Al	p	394.40	pc 0.3		0.8	20 H	1	100 H	25
Al	p	396.15	pc 0.35		1.0	30	2	200	30
AlO	r	467.2	rc 0.8		2	10	0.15	17	0.0
AlO	r	484.2	1.5		3	50	0.3	70	0.0
AlO	r	510.2	rc 1		2	10	0.2	17	0.0
As	a	228.81 x		2.5					0.1
As	a	234.98 x		4			(0.1) i	i	0.03
Au	p	242.80 x	0.1		1		1.1		
Au	p	267.60 x	0.3		2		1.7		
BO_2	db	494	5		30	50 Bu_4	5	8 M_{50}	0.0
	b	518.0	8		50	80	6	15	0.0
	b	547.6	11		60	90	15	17	0.0
	b	579	7		30	70	10	10	0.0
Ba	p	553.56	170		40	100 Bu_4	50	50 N	5
Ba^+	s	455.40	c 8		8	20	10	100	100
	sb	493.41	c 80		25	40	50	100	70
BaOH	b	488	100		30	40	50	5	0
	b	513	150		30	50	70	5	0
	b	830	200 k		30	25 Bz	5	(300)	0
	b	873	80 k		30	25		(300)	0
Be	p	234.86 x	0.0		0.0		0.0		10
BeO	r	470.9	rc 0.2		0.7		0.25		
	r	473.3			0.5		0.2		
Bi	dp	223.0 x	0.2	7 P			0.01 i[1]		0.0
	p	306.77	0.2	0.2	0.017	0.005 P	0.05		3
	a	472.26	0.5	0.00	0.25				0.0

Element		λ							
Ca	p	422.67	250		1000	1700 Bu$_4$	250	1000 EA	400
Ca$^+$	s	393.37			sc 25	2500 Bu$_4$	sc 30	700 EA	500
CaOH	b	554	500		1700	1000	170	300	10
	b	602	100		700	5000	250	1000	3
	b	622	500		2500		500		10
Cd	p	228.80 x	0.2	20 P	1	17 P$_{50}$	0.25		20
	f	326.11	20	6	2	5 Ac	0.25	5 Ac	0.4
CeO	bc	494	c 40		7	5 Ac	c 0.7	2	
(Ce)	c	550–600	70		10	10	0.7		
Co	a	340.51	40		20	30 Bu$_4$	5		9
	dpf	341.25	40		25	30	3		9
	dap	345.4	45		17	50	7		50
	a	350.23	30		35	30	3		7
	tap	352.8	35		20	50	7		10
	da	387.35	30			30	4		3
Cr	a	357.87	40	15 P$_{25}$	80	80 N	20	1000 H	50
	p	425.43	120	15	100	150 Ac$_{60}$	20	900	120
	p	427.48	110	8	80	110	17	550	110
	p	428.97	100		70	80	12	300	80
	ta	520.6	ac 170		70	70 Bu$_4$	10	100	80
	rc	579.4			20	120 Bu$_4$	15		0
CrO	a	455.54	250		25	30 Bu$_4$	0.3		0.03
Cs	p	852.11 k	20	40 P$_{25}$	1000	2000	1000		40
	p	894.35 k		25	300	700	300		10
Cu	p	324.75	1000		100	500 M	100	300 K	30
	p	327.40	300		100	500	100	300	20
	w	537	60		17	20 Bu$_4$	6		0
CuOH	mr	526.3	40		70	80 H			0
DyO	b	540.0	100		60	70			
	b	572.9			120	150			
	b	583.3			120	150			
ErO	b	504			30	55 H		2	
Eu	b	552			50	80			
	p	459.40			25	35 H	10	2	
	p	462.72			20	30	10	2	
(EuOH)	bc	598			100	120 H	(10)		
	wc	623			70	100	(10)		
	wc	647			25	150	(10)		
	w	702			50	100	(10)		
Fe	a	302.06	4	5 P$_{25}$	2.5	50 H	1.5	100 AA	50
	p	371.99	80	11	40	50 Bu$_4$	15	70	20
	(d)p	373.71	80	10	30	50	8	50	14
	tp	374.7	70	8	25	50	6	50	11
	(d)p	385.99	70	9	35	50	11	70	3

TABLE 111. (Continued)

Species	Character	Wavelength mμ	Intensity in various flames						
			AH	AHn	OH	OHn	OA	OAn	OC
FeO	rc	564.7	c 170		40	120 Bu$_4$	10	3 M$_{50}$	0
	rc	581.9	c 170		40	120	11	3	0
Ga	p	403.30	10		100	150 Bu$_4$	10		
	p	417.21	20		200	300	20		
GdO	r	461.6			17	30 H		2	
	mrc	580.7	(5)		50	120	(10)		
	mrc	598.7	(10)		80	250	(20)		
	bc	622	(10)		70	250	(20)		
Ge	u	259.25 x		0.7 P		0.025 P$_{50}$	0.015 i	0.025 i P$_{50}$	
	dp	265.14 x	0.00	2	0.01	0.03	0.04 i	0.05 i	
Hg	f	253.65 x	1.7	2.5 P	0.3	1.7 Ac	{0.2 / 1.2 i}	2	0.15
HoO	r	515.7			50	50 H			
	b	527			50	50			
	b	532.0			50	50			
	b	565.9			120	170			
IO	rc	484.5			(1)^4				
	rc	513.1			(1)				
	rc	530.8			(1)				
In	p	410.18	150	50 P$_{25}$	200	300 Bu$_4$	50		
	p	451.13	250	80	350	500	70		
K	da	404.5	30	1 P$_{25}$	70	100 Bu$_4$	10		
	p	766.49 k	10,000		30,000	50,000	{30,000		}300
	p	769.90 k					20,000}		
LaO	mr	438.0	5		25	25 H	17	80 H	
	mr	442.3	5		25	30	17	80	
	r	560.0	17		70	100	17	170	
	mr	743 k	30		20	1000	100	120	
	mr	792 k	25		20	1000	100	150	
Li	a	610.36	30		20	30 Bu$_4$	10		25
	p	670.78 s	10,000	1500 P$_{25}$	50,000	70,000	10,000	2	3000
LuO	mr	466.2			30	80 H			
	mr	517.0			50	110			
Mg	p	285.21	100	20 P$_{25}$	100	250 Ac	70	100 EA	1000
MgOH	b	370.2	500	60	100	100 EA	5		0
	mb^6	381–383	500	50	80	80	3.5		0

		λ							
Mn	tp	403.2	500		1000	1500 N	500		120
MnO	r	538.9	100		50	80 Bu_4	8		0.2
Mo	r	558.6	120		80	120	17		0.15
	p	379.93	c 2.5	0.0 P_{80}	c 0.8	10 iN	(0.5) i	20 $H_{50}Ac_{20}P_{20}$	3
(MoO₂)	p	386.41	c 3	0.0	c 0.8	10 i	(0.3) i	20	1[3]
NC	p	390.30	c 3	0.0	c 0.8	5 i	(0.25) i	20	11
NO	c	550–600	25		10		(3)		0
	dv	385.37				0.17 iN			
	dv	226.3 x		0.4 P					
	dv	236.3 x		0.5					
Na	da	330.3	2.5		20	10 Bu_4	10		0.5
	dp	589.2	30,000		50,000	70,000	25,000		2000
	da	819 k			5	10	20		10
(Nb)	c	550	5		1		0.3		
NdO	mb	660			10	100 H	10		
	bc	691			10	500	5	2	
	bc	702 k			10	500	10		
	tbc	712 k			10	500	10		
Ni	p	341.48	80	25 P_{25}	50	60 Bu_4	10		50
	dp	346.0	45	15	20	30	7		20
	p	349.30	25	10	17	25	7		20
	p	351.51	45	15	22	30	9		20
	p	352.45	80	25	50	80	15		30
	a	361.94	40	10	22	30	5		20
PO	vb	238.3 x	0.4	8 P					
	v	246.4 x	0.5	10				(50) C	
	c	520	30						
(P)	a	363.96	1.5	2 G	5	7 G	0.1	5 G	2.5
Pb	a	368.35	3	4	10	15	0.2	10	7
	a	405.78	3	4	10	15	0.3	10	7
Pd	a	340.46	45		70	200 H	10		80
	a	360.95	30		45	90	7		50
	a	363.47	55		80	150	10		100
(PmO)	b	(640)			(10)	(50) H	(1)		
	b	(680)			(10)	(50)	(1)		
PrO	rc	576.3			15	11 H		2	
	dr	603			10	20			
	bc	709.5 k			(10)	50			
	drc	735 k			(10)	30			
Pt	p	265.95 x	0.11		0.8		0.5		8
	p	306.47	0.25		1		0.7		7
Ra	p	482.59			(5)				
Ra^+	s	381.44			(3)				

TABLE 111. (Continued)

Species	Character	Wavelength mμ	Intensity in various flames						
			AH	AHn	OH	OHn	OA	OAn	OC
(RaOH)	b	627			(5)				
	b	665			(5)				
Rb	a	420.19	20		35	50 Bu$_4$	2		0.1
	p	780.02 k	3500		3500	5000	2000		100
	p	794.76 k	2500		2500	3000	1700		50
Re	p	488.92			(0.1)				
	p	527.55			(0.1)				
Rh	p	339.68	pc 4		8	12 P	5	8 P	
	p	343.49	2.5		15	25	9	17	
	p	350.25	pc 3.5		8	10	5	8	
	a	365.80	ac 6		11	15	6	9	
	p	369.24	7		35	40	15	22	
	a	370.09	ac 6		11	11	4.5	8	
Ru	p	349.89			5	30 M$_{50}$	3		
	p	372.80			20	150	30		
	tp	379.9			10	100	25		
Sb	p	217.59 x	0.07	17 P					0.2
	a	231.15 x	0.01	25					
	a	252.85 x	0.025	8					
	a	259.81 x		7					
ScO	r	485.8			20	60 H	5	2	
	r	581.2			40	150			
	tr	607.3			250	1700	30		
	tr	611.0			200	1200	25		
SiO	r	241.4 x		0.00 P$_{50}$		(0.1)			
	r	248.7 x		0.00		(0.1)			
	r	256.4 x		0.00		(0.1)			
SmO	mbc	614			40	80 H	2	2	
	mbc	624			30	80	2.5		
	tbc	642			20	100	2		
	mbc	652			20	110	3		
Sn	a	235.48 x		15 P		1.5 iP	1.7 i		8
	a	242.95 x		15		1.5 i	0.2 i		10
	a	270.65 x	0.08	12	0.012		{ 0.05 i		
	a	284.00	0.08	11	(0.015)		{ 0.4 i		
	p	303.41	0.4	12	0.05		0.2		
SnO	r	358.5	3	0.5 P$_{25}$	0.9	0.22 Ac$_{60}$	0.2		
Sr$^+$	p	460.73	500	150 P$_{25}$	1000	2000 Ac$_{60}$	200		150
Sr$^+$	s	407.77			30	50 Bu$_4$	17		400

Species		λ							
SrOH	b	605	5000	250 P_{25}	1000	1500 Bu_4	100		
	b	666	500		700	1000	40		
	b	680	250		1000	1000	30	2	
TbO	b	534			40	60 H	(10)		
	bc	573			50	100	(10)		
	bc	592.1			80	170			
	bc	598.0			70	150			
Te	da	238.5 x	0.01	1 P	0.3		0.000		
TeO	drc	371.4	3.5		0.2		c 0.015		
	drc	377.3	3.5		0.25		c 0.015		
	trc	382.7	3.5		0.25		c 0.017		
	trc	388.4	3.5				c 0.017		
TiO	rc	516.7			40		1.5	1.5 P	
	rc	544.9			45		1.5	1.5	
	rc	575.9			45		1.5	1.5	
	bc	715			20		10		
Tl	p	377.57	50		100	170 Bu_4	10		6
	a	535.05	30		70	100	5		8
TmO	b	491			30	50 H			
	b	538			35	60		2	
	b	541.5			35	60			
(UO_2)	c	550	15		5	5			
V	ta	318.4		0.00 P_{50}					23
VO	rc	522.9	80	c 17 P_{50}		80 Bu_4	5		0.0
	rc	547.0	100	c 17	40	100	7		0.0
	rc	573.7	110	c 30	40	110	10		0.0
YO	r	481.8			30	50 H		2	
	mr	599			300	1000	30		
	mr	615			300	800	30		
Yb	a	398.80	70		25	120 H	(10)	70 H	
	pb	555.65	250		70	60	(10)	50	
(YbOH)	b	498.1	200		50	50 H	(5)	2	
	b	532.5	250		80	80	(5)		
Zn	b	572.5	300	0.15 P_{50}	110	100	(10)		
	a	213.86 x	0.04		0.06	1 Ac	i	c 6 Ac	6
	p	481.05	c 0.3		c 0.06	0.6 iN		4	
(Zn)	c	520–600	0.5		0.12	12 Ac			0
ZrO	bc	564			1.2		(0.1)		0
	bc	574			1.2		(0.1)		0

*Reproduced from: Clark, George L., Ed., "The Encyclopedia of Spectroscopy," Van Nostrand Reinhold, New York, 1960. References to this table may be found in this Encyclopedia. The intensities are computed for the Beckman DU spectrophotometer with flame and photomultiplier attachments; the intensity $J = 100(R − B)/B$, where B is the blank reading for the pure solvent and R is the reading for a solution containing 10 mg of the element per liter for the slit width at which, for the blank, shot-effect noise equals flame flicker.

CHARACTERISTIC GROUP WAVE NUMBERS OF HYDROCARBONS—I.
(JONES, 1959)

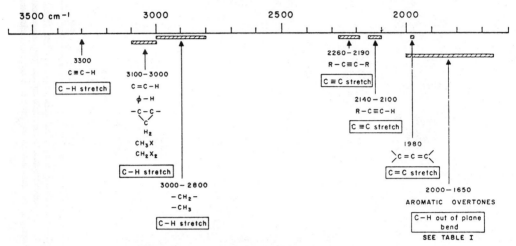

Reproduced with the permission of the National Research Council of Canada.

CHARACTERISTIC GROUP WAVE NUMBERS OF HYDROCARBONS—II.
(JONES, 1959)

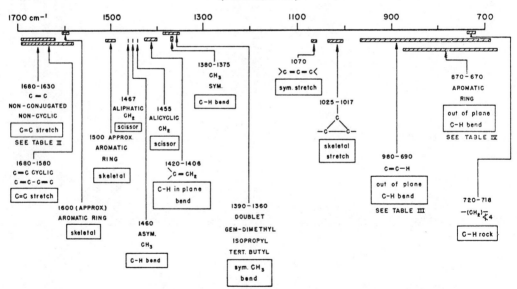

Reproduced with the permission of the National Research Council of Canada.

Fig. 111. Infrared spectra of organic compounds.

CHARACTERISTIC GROUP WAVE NUMBERS OF OXYGENATED COMPONDS—I.
(JONES, 1959)

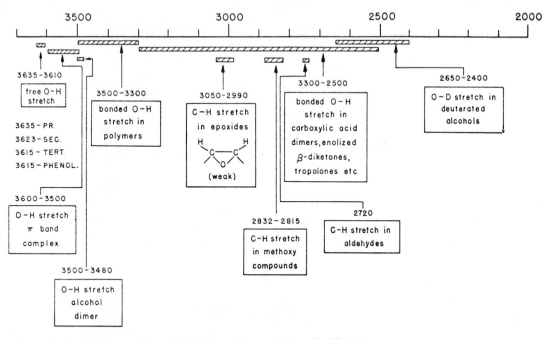

Reproduced with the permission of the National Research Council of Canada.

CHARACTERISTIC GROUP WAVE NUMBERS OF OXYGENATED COMPOUNDS—II. (JONES, 1959)

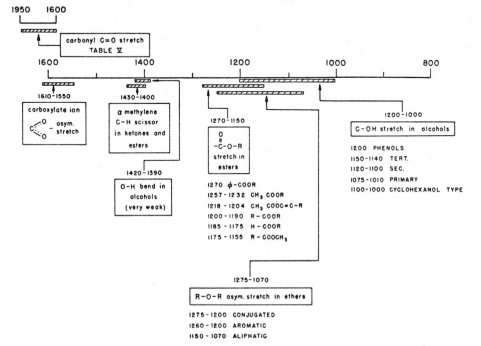

Reproduced with the permission of the National Research Council of Canada.

Fig. 111 (*Continued*).

CHARACTERISTIC GROUP WAVE NUMBERS OF NITROGEN, PHOSPHORUS, SULFUR COMPOUNDS—I. (JONES, 1959)

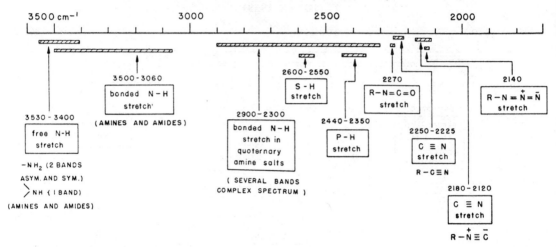

Reproduced with the permission of the National Research Council of Canada.

CHARACTERISTIC GROUP WAVE NUMBERS OF NITROGEN, PHOSPHORUS, SULFUR COMPOUNDS—II. (JONES, 1959)

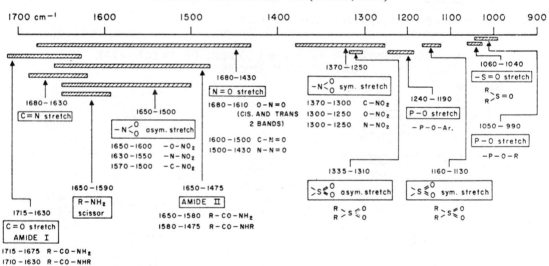

Reproduced with the permission of the National Research Council of Canada.

Fig. 111 (*Continued*).

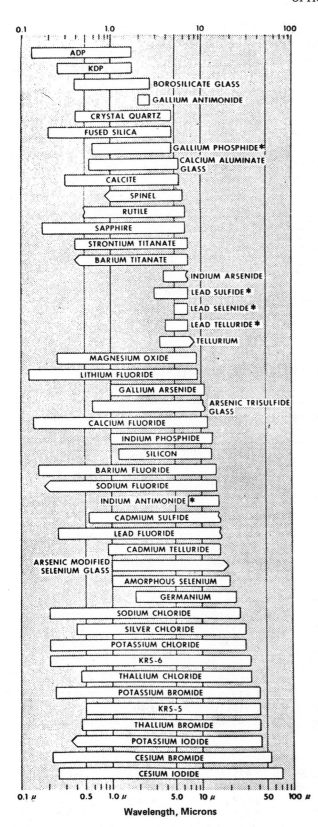

Fig. 112. The transmission regions of selected optical materials. (Regions are for 10 per cent external transmittance or better, for a 2-mm sample at room temperature.)

*Maximum external transmittance less than 10%. (BALLARD 1969)

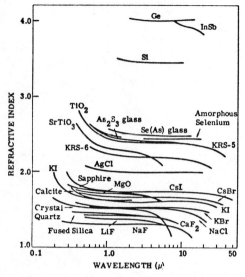

Fig. 113. Refractive index values. (Wolfe, 1965).

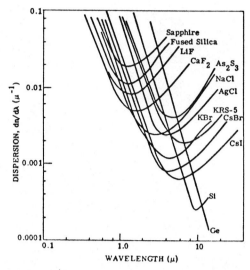

Fig. 114. $dn/d\lambda$ versus λ for selected materials. (Wolfe, (1965).

TABLE 112. REFRACTIVE INDEX OF GLASSES. (BALLARD, 1959)

Silicon Dioxide, Fused $(297°K)$

Wavelength (microns)	Refractive Index (n)	Wavelength (microns)	Refractive Index (n)	Wavelength (microns)	Refractive Index (n)
0.34	1.47877	0.56	1.459561	1.50	1.444687
0.35	1.47701	0.57	1.459168	1.60	1.443492
0.36	1.47540	0.58	1.458794	1.70	1.442250
0.37	1.47393	0.59	1.458437	1.80	1.440954
0.38	1.47258	0.60	1.458096	1.90	1.439597
0.39	1.47135	0.61	1.457769	2.00	1.438174
0.40	1.470208	0.62	1.457456	2.10	1.436680
0.41	1.469155	0.63	1.457156	2.20	1.435111
0.42	1.468179	0.64	1.456868	2.30	1.433462
0.43	1.467273	0.65	1.456591	2.40	1.431730
0.44	1.466429	0.66	1.456324	2.50	1.429911
0.45	1.465642	0.67	1.456066	2.60	1.428001
0.46	1.464908	0.68	1.455818	2.70	1.425995
0.47	1.464220	0.69	1.455579	2.80	1.423891
0.48	1.463573	0.70	1.455347	2.90	1.421684
0.49	1.462965	0.80	1.453371	3.00	1.41937
0.50	1.462394	0.90	1.451808	3.10	1.41694
0.51	1.461856	1.00	1.450473	3.20	1.41440
0.52	1.461346	1.10	1.449261	3.30	1.41173
0.53	1.460863	1.20	1.448110	3.40	1.40893
0.54	1.460406	1.30	1.446980	3.50	1.40601
0.55	1.459973	1.40	1.445845		

Arsenic Trisulfide Glass $(298°K)$

0.576960	2.66324	1.01398	2.47568	4.258	2.41013
0.653857	2.59762	1.395055	2.44379	6.238	2.40221
0.69075	2.56697	1.81308	2.42999	8.662	2.39035
0.85212	2.50611	3.4188	2.41374	11.035	2.37365

TABLE 113. REFRACTIVE INDEX OF HALIDES.

Barium Fluoride (Single Crystal, 298°K) (Ballard, 1959)

Wavelength (microns)	Refractive Index (n)	Wavelength (microns)	Refractive Index (n)	Wavelength (microns)	Refractive Index (n)
0.2536	1.5122	0.57696	1.4748	2.32542	1.4636
0.28035	1.5067	0.57907	1.4747	2.576	1.4626
0.28936	1.5039	0.58926	1.4744	2.673	1.4624
0.29673	1.5019	0.64385	1.4730	3.2434	1.4602
0.30215	1.5005	0.65628	1.4728	3.422	1.4594
0.3130	1.4978	0.70652	1.4718	3.7067	1.4588
0.32546	1.4952	0.76786	1.4708	5.138	1.4501
0.33415	1.4936	1.01398	1.4685	5.3034	1.4491
0.34036	1.4926	1.12866	1.4678	5.343	1.4488
0.34662	1.4916	1.36726	1.4668	5.549	1.4474
0.361051	1.4894	1.52952	1.4662	5.7663	1.4459
0.365015	1.4889	1.681	1.4656	6.16	1.4436
0.404656	1.4844	1.918	1.4649	9.724	1.4209
0.435834	1.4818	1.9701	1.4647	11.035	1.4142
0.54607	1.4759	2.1526	1.4641		

Calcium Fluoride (Single Crystal, 293°K) (Ballard, 1959)

Wavelength (microns)	Refractive Index (n)	Wavelength (microns)	Refractive Index (n)	Wavelength (microns)	Refractive Index (n)
0.48615	1.43704	1.900	1.42439	3.5359	1.41376
0.58758	1.43388	1.9153	1.42431	3.8306	1.41119
0.58932	1.43384	1.9644	1.42407	4.000	1.40963
0.65630	1.43249	2.0582	1.42360	4.1252	1.40847
0.68671	1.43200	2.0626	1.42357	4.2500	1.40722
0.72818	1.43143	2.1608	1.42306	4.4000	1.40568
0.76653	1.43093	2.250	1.42258	4.6000	1.40357
0.88400	1.42980	2.3573	1.42198	4.7146	1.40233
1.0140	1.42884	2.450	1.42143	4.8000	1.40130
1.08304	1.42843	2.5537	1.42080	5.000	1.39908
1.1000	1.42834	2.6519	1.42018	5.3036	1.39522
1.1786	1.42789	2.700	1.41988	5.8932	1.38712
1.250	1.42752	2.750	1.41956	6.4825	1.37824
1.3756	1.42689	2.800	1.41923	7.0718	1.36805
1.4733	1.42642	2.880	1.41890	7.6612	1.35675
1.5715	1.42596	2.9466	1.41823	8.2505	1.34440
1.650	1.42558	3.0500	1.41750	8.8398	1.33075
1.7680	1.42502	3.0980	1.41714	9.4291	1.31605
1.8400	1.42468	3.2413	1.41610		
1.8688	1.42454	3.4000	1.41487		

Cesium Bromide (Single Crystal, 300°K) (Ballard, 1959)

Wavelength (microns)	Refractive Index (n)	Wavelength (microns)	Refractive Index (n)	Wavelength (microns)	Refractive Index (n)
0.5	1.70896	15.0	1.65468	30.0	1.60947
1.0	1.67793	16.0	1.65272	31.0	1.60510
2.0	1.67061	17.0	1.65062	32.0	1.60053
3.0	1.66901	18.0	1.64838	33.0	1.59576
4.0	1.66813	19.0	1.64600	34.0	1.59078
5.0	1.66737	20.0	1.64348	35.0	1.58558
6.0	1.66659	21.0	1.64080	36.0	1.58016
7.0	1.66573	22.0	1.63798	37.0	1.57450
8.0	1.66477	23.0	1.63500	38.0	1.56860
9.0	1.66370	24.0	1.63186	39.0	1.56245
10.0	1.66251	25.0	1.62856		
11.0	1.66120	26.0	1.62509		
12.0	1.65976	27.0	1.62146		
13.0	1.65820	28.0	1.61764		
14.0	1.65651	29.0	1.61365		

TABLE 113. (*Continued*)

Cesium Iodide (Single Crystal, 297°K) (Ballard, 1959)

Wavelength (microns)	Refractive Index (n)	Wavelength (microns)	Refractive Index (n)	Wavelength (microns)	Refractive Index (n)
0.500	1.80635	17.0	1.73190	34.0	1.69717
1.00	1.75721	18.0	1.73056	35.0	1.69427
2.00	1.74616	19.0	1.72913	36.0	1.69127
3.00	1.74400	20.0	1.72762	37.0	1.68815
4.00	1.74305	21.0	1.72602	38.0	1.68493
5.00	1.74239	22.0	1.72435	39.0	1.68159
6.00	1.74181	23.0	1.72258	40.0	1.67814
7.00	1.74122	24.0	1.72073	41.0	1.67457
8.00	1.74059	25.0	1.71880	42.0	1.67088
9.00	1.73991	26.0	1.71677	43.0	1.66707
10.0	1.73916	27.0	1.71465	44.0	1.66312
11.0	1.73835	28.0	1.71244	45.0	1.65905
12.0	1.73746	29.0	1.71014	46.0	1.65485
13.0	1.73650	30.0	1.70774	47.0	1.65051
14.0	1.73547	31.0	1.70525	48.0	1.64602
15.0	1.73436	32.0	1.70266	49.0	1.64139
16.0	1.73317	33.0	1.69996	50.0	1.63662

Cuprous Chloride (Single Crystal, 298°K) (Smakula, 1967)

Wavelength (microns)	Refractive Index (n)	Absorption Coeff. (cm^{-1})	Wavelength (microns)	Refractive Index (n)	Absorption Coeff. (cm^{-1})
0.5350*	1.996		20.0	1.72	4.39
0.6710*	1.933		20.8	1.72	6.64
2.5	1.90		21.7	1.68	10.5
3.3	1.90		22.7	1.64	15.9
5.0	1.90		23.8	1.61	22.4
6.7	1.89		25.0	1.57	29.9
10.0	1.88	0.01	26.3	1.49	40.1
14.3	1.86	0.11	27.8	1.46	53.6
16.7	1.78	0.29	29.4	1.43	80.4
18.2	1.73	1.15	31.3	1.39	218

*other data.

Lithium Fluoride (Single Crystal, 298°K) (Ballard, 1959)

Wavelength (microns)	Refractive Index (n)	Wavelength (microns)	Refractive Index (n)	Wavelength (microns)	Refractive Index (n)
0.5	1.39430	2.4	1.37446	4.3	1.34319
0.6	1.39181	2.5	1.37327	4.4	1.34100
0.7	1.39017	2.6	1.37203	4.5	1.33875
0.8	1.38896	2.7	1.37075	4.6	1.33645
0.9	1.38797	2.8	1.36942	4.7	1.33408
1.0	1.38711	2.9	1.36804	4.8	1.33165
1.1	1.38631	3.0	1.36660	4.9	1.32916
1.2	1.38554	3.1	1.36512	5.0	1.32661
1.3	1.38477	3.2	1.36359	5.1	1.32399
1.4	1.38400	3.3	1.36201	5.2	1.32131
1.5	1.38320	3.4	1.36037	5.3	1.31856
1.6	1.38238	3.5	1.35868	5.4	1.31575
1.7	1.38153	3.6	1.35693	5.5	1.31287

TABLE 113. (*Continued*)

Wavelength (microns)	Refractive Index (*n*)	Wavelength (microns)	Refractive Index (*n*)	Wavelength (microns)	Refractive Index (*n*)
1.8	1.38064	3.7	1.35514	5.6	1.30993
1.9	1.37971	3.8	1.35329	5.7	1.30692
2.0	1.37875	3.9	1.35138	5.8	1.30384
2.1	1.37774	4.0	1.34942	5.9	1.30068
2.2	1.37669	4.1	1.34740	6.0	1.29745
2.3	1.37560	4.2	1.34533		

Potassium Bromide (Single Crystal, 295°K) (Ballard, 1959)

0.404656	1.589752	1.7012	1.53901	14.29	1.51505
0.435835	1.581479	2.44	1.53733	14.98	1.51280
0.496133	1.571791	2.73	1.53693	17.40	1.50390
0.508582	1.568475	3.419	1.53612	18.16	1.50076
0.546074	1.563928	4.258	1.53523	19.01	1.49703
0.587562	1.559965	6.238	1.53288	19.91	1.49288
0.653847	1.555858	6.692	1.53225	21.18	1.48655
0.706520	1.552447	8.662	1.52903	21.83	1.48311
1.01398	1.54408	9.724	1.52695	23.86	1.47140
1.12866	1.54258	11.035	1.52404	25.14	1.46324
1.36728	1.54061	11.862	1.52200		

Potassium Iodide (Single Crystal, 0.248 to 1.083 μ at 293°K, above 1.083 μ at 311°K). (Ballard, 1959).

0.302	1.82769	8.84	1.6218	21	1.5930
0.405	1.71843	10.02	1.6201	22	1.5895
0.546	1.67310	11.79	1.6172	23	1.5858
0.768	1.6494	12.97	1.6150	24	1.5819
1.014	1.6396	14.14	1.6127	25	1.5775
2.36	1.6295	15.91	1.6085	26	1.5729
3.54	1.6275	18.10	1.6030	27	1.5681
4.13	1.6268	19	1.5997	28	1.5629
5.89	1.6252	20	1.5964	29	1.5571
7.66	1.6235				

Silver Chloride (Single Crystal, 298°K) (Ballard, 1959)

0.5	2.09648	7.5	1.99021	14.5	1.95467
1.0	2.02239	8.0	1.98847	15.0	1.95113
1.5	2.01047	8.5	1.98661	15.5	1.94743
2.0	2.00615	9.0	1.98464	16.0	1.94358
2.5	2.00386	9.5	1.98255	16.5	1.93958
3.0	2.00230	10.0	1.98034	17.0	1.93542
3.5	2.00102	10.5	1.97801	17.5	1.93109
4.0	1.99983	11.0	1.97556	18.0	1.92660
4.5	1.99866	11.5	1.97297	18.5	1.92194
5.0	1.99745	12.0	1.97026	19.0	1.91710
5.5	1.99618	12.5	1.96742	19.5	1.91208
6.0	1.99483	13.0	1.96444	20.0	1.90688
6.5	1.99339	13.5	1.96133	20.5	1.90149
7.0	1.99185	14.0	1.95807		

Sodium Chloride (Single Crystal, 293°K) (Ballard, 1959)

0.589	1.54427	1.6848	1.52764	7.22	1.51020
0.6400	1.54141	1.7670	1.52736	7.59	1.50850
0.6874	1.53930	2.0736	1.52649	7.6611	1.50822

TABLE 113. (*Continued*)

Wavelength (microns)	Refractive Index (n)	Wavelength (microns)	Refractive Index (n)	Wavelength (microns)	Refractive Index (n)
0.7604	1.53682	2.1824	1.52621	7.9558	1.50665
0.7858	1.53607	2.2464	1.52606	8.04	1.5064
0.8835	1.53395	2.3560	1.52579	8.8398	1.50192
0.9033	1.53361	2.6505	1.52512	9.00	1.50100
0.9724	1.53253	2.9466	1.52466	9.50	1.49980
1.0084	1.53206	3.2736	1.52371	10.0184	1.49462
1.0540	1.53153	3.5359	1.52312	11.7864	1.48171
1.0810	1.53123	3.6288	1.52286	12.50	1.47568
1.1058	1.53098	3.8192	1.52238	12.9650	1.47160
1.1420	1.53063	4.1230	1.52156	13.50	1.4666
1.1786	1.53031	4.7120	1.51979	14.1436	1.46044
1.2016	1.53014	5.0092	1.51883	14.7330	1.45427
1.2604	1.52971	5.3009	1.51790	15.3223	1.44743
1.3126	1.52937	5.8932	1.51593	15.9116	1.44090
1.4874	1.52845	6.4825	1.51347	17.93	1.4149
1.5552	1.52815	6.80	1.51200	20.57	1.3735
1.6368	1.52781	7.0718	1.51093	22.3	1.3403

Sodium Fluoride (Single Crystal, 0.186 to 1.083 μ at 293°K, above 1.083 μ at 291°K) (Ballard, 1959)

Wavelength (microns)	Refractive Index (n)	Wavelength (microns)	Refractive Index (n)	Wavelength (microns)	Refractive Index (n)
0.186	1.3930	0.811	1.32272	8.1	1.269
0.199	1.3805	0.912	1.32198	9.1	1.262
0.203	1.3772	1.014	1.32150	10.3	1.233
0.302	1.34232	2.0	1.317	11.3	1.209
0.405	1.33194	3.1	1.313	12.5	1.180
0.486	1.32818	4.1	1.308	13.8	1.142
0.546	1.32640	5.1	1.301	15.1	1.093
0.589	1.32549	6.1	1.292	16.7	1.034
0.707	1.32372	7.1	1.281	17.3	1.000

Thallium Bromoiodide (KRS-5) (Single Crystal, 298°K) (Ballard, 1959)

Wavelength (microns)	Refractive Index (n)	Wavelength (microns)	Refractive Index (n)	Wavelength (microns)	Refractive Index (n)
0.540	2.68059	13.0	2.36371	27.0	2.30676
1.00	2.44620	14.0	2.36101	28.0	2.30098
1.50	2.40774	15.0	2.35812	29.0	2.29495
2.00	2.39498	16.0	2.35502	30.0	2.28867
3.00	2.38574	17.0	2.35173	31.0	2.28212
4.00	2.38204	18.0	2.34822	32.0	2.27531
5.00	2.37979	19.0	2.34451	33.0	2.26823
6.00	2.37797	20.0	2.34058	34.0	2.26087
7.00	2.37627	21.0	2.33643	35.0	2.25322
8.00	2.37452	22.0	2.33206	36.0	2.24528
9.00	2.37267	23.0	2.32746	37.0	2.23705
10.0	2.37069	24.0	2.32264	38.0	2.22850
11.0	2.36854	25.0	2.31758	39.0	2.21965
12.0	2.36622	26.0	2.31229	40.0	2.21047

Thallium Bromochloride (KRS-6) (Single Crystal, 293°K) (Ballard, 1959)

Wavelength (microns)	Refractive Index (n)	Wavelength (microns)	Refractive Index (n)	Wavelength (microns)	Refractive Index (n)
0.6	2.3294	2.2	2.2039	11.0	2.1723
0.7	2.2982	2.4	2.2024	12.0	2.1674
0.8	2.2660	2.6	2.2011	13.0	2.1620
0.9	2.2510	2.8	2.2001	14.0	2.1563
1.0	2.2404	3.0	2.1990	15.0	2.1504
1.1	2.2321	3.5	2.1972	16.0	2.1442
1.2	2.2255	4.0	2.1956	17.0	2.1377

TABLE 113. (*Continued*)

Wavelength (microns)	Refractive Index (*n*)	Wavelength (microns)	Refractive Index (*n*)	Wavelength (microns)	Refractive Index (*n*)
1.3	2.2212	4.5	2.1942	18.0	2.1309
1.4	2.2176	5.0	2.1928	19.0	2.1236
1.5	2.2148	6.0	2.1900	20.0	2.1154
1.6	2.2124	7.0	2.1870	21.0	2.1067
1.7	2.2103	8.0	2.1839	22.0	2.0976
1.8	2.2086	9.0	2.1805	23.0	2.0869
1.9	2.2071	10.0	2.1767	24.0	2.0752
2.0	2.2059				

TABLE 114. REFRACTIVE INDEX OF OXIDES. (BALLARD, 1959)

Aluminum Oxide (Single Crystal, 297°K) Ordinary Ray.

Wavelength (microns)	Refractive Index (*n*)	Wavelength (microns)	Refractive Index (*n*)	Wavelength (microns)	Refractive Index (*n*)
0.26520	1.8336	0.64385	1.7655	2.4374	1.7228
0.28035	1.8243	0.706519	1.7630	3.2432	1.7044
0.28936	1.8195	0.85212	1.7588	3.2656	1.7036
0.29673	1.8159	0.89440	1.7579	3.303	1.7023
0.30215	1.8135	1.01398	1.7555	3.4188	1.6982
0.3130	1.8091	1.12866	1.7534	3.5078	1.6950
0.33415	1.8018	1.36728	1.7494	3.70	1.6875
0.34662	1.7981	1.39506	1.7489	4.258	1.6637
0.361051	1.7945	1.52952	1.7466	4.954	1.6266
0.365015	1.7936	1.6932	1.7437	5.145(	1.6151
0.39064	1.7883	1.70913	1.7434	5.349	1.6020
0.404656	1.7858	1.81307	1.7414	5.419	1.5973
0.435834	1.7812	1.9701	1.7383	5.577	1.5864
0.54607	1.7708	2.1526	1.7344		
0.576960	1.7688	2.24929	1.7323		
0.579066	1.7587	2.32542	1.7306		

Magnesium Oxide (Single Crystal, 296.5°K)

0.36117	1.77318	1.6932	1.71281	3.3033	1.68526
0.360515	1.77186	1.7092	1.71258	3.5075	1.68055
1.01398	1.72259	1.81307	1.71108	4.258	1.66039
1.12866	1.72059	1.97009	1.70885	5.138	1.63138
1.36728	1.71715	2.24929	1.70470	5.35	1.62404
1.52952	1.71496	2.32542	1.70350		

Titanium Oxide (Single Crystal, 298°K)

0.4358	2.853	0.6907	2.555	3.0000	2.380
0.4916	2.723	0.7082	2.548	3.5000	2.367
0.4960	2.715	1.0140	2.483	4.0000	2.350
0.5461	2.652	1.5296	2.451	4.5000	2.322
0.5770	2.623	2.0000	2.399	5.0000	2.290
0.5791	2.621	2.5000	2.387	5.500	2.200

TABLE 114. (*Continued*)

Silicon Dioxide (Single Crystal, hexagonal; 291°K)

Wavelength (microns)	Refractive Index		Wavelength (microns)	Refractive Index	
	ordinary ray (n_o)	extraordinary ray (n_e)		ordinary ray (n_o)	extraordinary ray (n_e)
0.185	1.65751	1.68988	1.5414	1.52781	1.53630
0.198	1.65087	1.66394	1.6815	1.52583	1.53422
0.231	1.61395	1.62555	1.7614	1.52468	1.53301
0.340	1.56747	1.57737	1.9457	1.52184	1.53004
0.394	1.55846	1.56805	2.0531	1.52005	1.52823
0.434	1.55396	1.56339	2.30	1.51561	
0.508	1.54822	1.55746	2.60	1.50986	
0.5893	1.54424	1.55335	3.00	1.49953	
0.768	1.53903	1.54794	3.50	1.48451	
0.83225	1.53773	1.54661	4.00	1.46617	
0.9914	1.53514	1.54392	4.20	1.4569	
1.1592	1.53283	1.54152	5.00	1.417	
1.3070	1.53090	1.53951	6.45	1.274	
1.3958	1.52977	1.53832	7.0	1.167	
1.4792	1.52865	1.53716			

Strontium Titanate (Single Crystal, 0.4046 to 0.6907 u at 294°K, others at 299°K)

Wavelength (microns)	Refractive Index (n)	Wavelength (microns)	Refractive Index (n)	Wavelength (microns)	Refractive Index (n)
0.4046	2.6466	1.3622	2.2922	3.3033	2.2181
0.4861	2.4890	1.5295	2.2850	3.5078	2.2088
0.5893	2.4069	1.6606	2.2799	4.2566	2.1695
0.6563	2.3788	1.7092	2.2782	5.1380	2.1123
0.6907	2.3666	2.1526	2.2626	5.1472	2.1120
1.0140	2.3148	2.4374	2.2527	5.3034	2.1005
1.1286	2.3056				

TABLE 115. REFRACTIVE INDEX OF SEMICONDUCTORS.

Aluminum Antimonide (Single Crystal, 300°K). (Willardson and Beer, 1967)

Wavelength (microns)	Refractive Index (n)	Extinction Coeff. (k)	Wavelength (microns)	Refractive Index (n)	Extinction Coeff. (k)
1.1	3.445		30.2	0.246	2.643
1.5	3.382		30.5	0.330	3.616
2.0	3.300		30.7	0.453	4.482
2.5	3.291		30.9	0.719	5.722
3.0	3.245		31.0	0.990	6.612
4.0	3.182		31.1	1.509	7.838
6.0	3.173		31.2	2.710	9.627
8.0	3.127		31.3	6.278	11.73
10.0	3.100		31.4	12.23	8.261
15.0	3.080		31.5	11.29	3.291
18.0	3.037		31.6	9.541	1.633
20.0	2.995	0.001	31.7	8.373	0.984
22.0	2.931	0.002	31.8	7.581	0.666
24.0	2.828	0.004	32.0	6.583	0.372
26.0	2.637	0.009	32.5	5.406	0.148
28.0	2.158	0.031	33.5	4.547	0.053
28.5	1.893	0.051	35.0	4.080	0.022
29.0	1.435	0.102	37.5	3.784	0.010
29.4	0.623	0.349	40.0	3.652	0.006
30.0	0.223	2.097			

Gallium Antimonide (Single Crystal, 300°K). (Willardson and Beer, 1967)

Wavelength (microns)	Refractive Index (n)	Extinction Coeff. (k)	Wavelength (microns)	Refractive Index (n)	Extinction Coeff. (k)
1.80	3.820	5.51×10^{-4}	6.0	3.824	3.94×10^{-3}
2.00	3.789	9.87×10^{-5}	7.0	3.843	5.33×10^{-3}
2.5	3.749	1.65×10^{-4}	8.0	3.843	6.68×10^{-3}
3.0	3.898	3.65×10^{-4}	9.0	3.843	7.99×10^{-3}
3.5	3.861	7.46×10^{-4}	10.0	3.843	9.26×10^{-3}
4.0	3.833	1.26×10^{-3}	12.0	3.843	1.16×10^{-2}
5.0	3.824	2.53×10^{-3}	14.0	3.861	1.40×10^{-2}

Gallium Arsenide (Single Crystal, 300°K). (Willardson and Beer, 1967)

Wavelength (microns)	Refractive Index (n)	Extinction Coeff. (k)	Wavelength (microns)	Refractive Index (n)	Extinction Coeff. (k)
0.049	1.049	0.156	5.0	3.269	
0.059	1.058	0.245	6.0	3.246	
0.083	0.836	0.503	7.0	3.217	
0.138	0.901	1.136	8.0	3.183	
0.200	1.424	1.976	9.0	3.143	
0.248	1.882	3.279	10.0	3.095	
0.258	2.726	3.388	11.0	3.047	
0.310	3.513	1.857	12.0	2.991	
0.400	4.149	2.137	13.0	2.933	
0.460	4.748	0.510	14.0	2.863	
0.495	4.427	0.400	15.0	2.790	
0.540	4.127	0.313	16.0	2.707	
0.729	3.627		17.0	2.621	
0.900	3.595	0.639×10^{-4}	18.0	2.521	
1.00	3.509		19.0	2.409	
1.50	3.381		20.0	2.287	
2.0	3.314		21.0	2.151	
3.0	3.303		22.0	2.029	
4.0	3.289				

TABLE 115. (*Continued*)

Gallium Phosphide (Single Crystal, 300°K). (Willardson and Beer, 1967)

Wavelength (microns)	Refractive Index (n)	Extinction Coeff. (k)	Wavelength (microns)	Refractive Index (n)	Extinction Coeff. (k)
0.050	1.000	0.201	20.0	2.529	
0.103	0.951	1.023	22.51	2.207	
0.207	1.709	2.353	24.57	1.129	
0.310	3.770	1.649	25.00	0.100	
0.520	3.65	~9.5×10^{-4}	25.26		1.379
1.00	3.17		27.18	1.83	
1.60	3.04	4.99×10^{-4}	27.19	4.07	
2.00	3.02	1.03×10^{-3}	27.26	13.12	14.68
3.00	2.97	2.53×10^{-3}	27.32	16.58	13.89
5.00	2.94	4.22×10^{-3}	27.45	14.47	
9.00	2.91	2.69×10^{-2}	27.69	8.50	
10.0	2.90		28.21	6.19	
18.0	2.643		30.00	4.36	0.021

Indium Antimonide (Single Crystal, 300°K). (Willardson and Beer, 1967)

Wavelength (microns)	Refractive Index (n)	Extinction Coeff. (k)	Wavelength (microns)	Refractive Index (n)	Extinction Coeff. (k)
0.049	1.15	0.15	2.07	4.03	
0.103	0.75	0.51	8.00	3.995	2.3×10^{-3}
0.155	0.88		9.00	3.967	1.9×10^{-3}
0.207	1.23	1.91	15.13	3.881	
0.248	1.56	2.15	20.0	3.826	2.0×10^{-3}
0.302	2.19	3.26	25.0	3.78	5.1×10^{-3}
0.365	3.51	2.15	30.0	3.47	1.0×10^{-2}
0.477	3.42	2.06	35.0	3.25	1.4×10^{-2}
0.620	4.29	1.83	40.0	2.98	2.6×10^{-2}
0.689	5.13	1.37	45.0	2.57	5.4×10^{-2}
1.03	4.24	0.32			
1.55	4.08	0.20			

Indium Arsenide (Single Crystal, 300°K). (Willardson and Beer, 1967)

Wavelength (microns)	Refractive Index (n)	Extinction Coeff. (k)	Wavelength (microns)	Refractive Index (n)	Extinction Coeff. (k)
0.049	1.139	0.168	4.00	3.51	
0.103	0.745	0.727	10.0	3.42	
0.195	1.583	2.120	20.0	3.35	
0.282	3.800	2.735	25.0	3.26	
1.03	3.613	0.076	33.3	2.95	

Germanium (Single Crystal and Polycrystalline Material, 299°K). (Salzberg and Villa, 1958)

Wavelength (microns)	Refractive Index (n)		Wavelength (microns)	Refractive Index (n)	
	Single Xtal	Polycryst.		Single Xtal	Polycryst.
2.0581	4.1016	4.1018	4.866	4.0170	4.0167
2.1526	4.0919	4.0919	6.238	4.0094	4.0095
2.3126	4.0786	4.0785	8.66	4.0043	4.0043
2.4374	4.0708	4.0709	9.72	4.0034	4.0033
2.577	4.0609	4.0608	11.04	4.0026	4.0025
2.7144	4.0552	4.0554	12.20	4.0023	4.0020
2.998	4.0452	4.0452	13.02	4.0021	4.0018
3.3033	4.0369	4.0372			
3.4188	4.0334	4.0339			
4.258	4.0216	4.0217			

TABLE 115. (*Continued*)

Indium Phosphide (Single Crystal, 300°K). (Willardson and Beer, 1967)

Wavelength (microns)	Refractive Index (n)	Extinction Coeff. (k)	Wavelength (microns)	Refractive Index (n)	Extinction Coeff. (k)
0.062	0.793	0.494	0.459	3.754	0.599
0.071	0.840	0.469	0.620	3.430	0.298
0.103	0.771	0.899	0.975	3.346	1.13×10^{-5}
0.151	0.934	1.512	1.50	3.172	
0.175	1.174	1.882	2.00	3.134	
0.185	1.354	1.986	5.00	3.08	
0.200	1.525	1.982	6.00	3.07	
0.225	1.668	2.442	8.00	3.06	
0.248	2.451	3.166	10.00	3.05	
0.275	3.655	1.691	12.00	3.05	5.27×10^{-4}
0.302	3.162	1.389	14.00	3.04	8.86×10^{-4}
0.399	4.100	1.439	14.85	3.03	3.00×10^{-3}

Silicon (Single Crystal, 300°K). (Ballard, 1959)

Wavelength (microns)	Refractive Index (n)	Wavelength (microns)	Refractive Index (n)	Wavelength (microns)	Refractive Index (n)
1.3570	3.4975	2.4373	3.4408	5.50	3.4213
1.3673	3.4962	2.7144	3.4358	6.00	3.4202
1.3951	3.4929	3.00	3.4320	6.50	3.4195
1.5295	3.4795	3.3033	3.4297	7.00	3.4189
1.6606	3.4696	3.4188	3.4286	7.50	3.4186
1.7092	3.4664	3.50	3.4284	8.00	3.4184
1.8131	3.4608	4.00	3.4255	8.50	3.4182
1.9701	3.4537	4.258	3.4242	10.00	3.4179
2.1526	3.4476	4.50	3.4236	10.50	3.4178
2.3254	3.4430	5.00	3.4223	11.04	3.4176

Tellurium (Single Crystal, ~298°K). (Ballard, 1959)

Wavelength (microns)	Refractive Index (n)		Wavelength (microns)	Refractive Index (n)	
	E⊥c	E∥c		E⊥c	E∥c
4.0	4.929	6.372	9.0	4.802	6.253
5.0	4.864	6.316	10.0	4.796	6.246
6.0	4.838	6.286	12.0	4.789	6.237
7.0	4.821	6.270	14.0	4.785	6.230
8.0	4.809	6.257			

TABLE 116. REFRACTIVE INDEX OF MISCELLANEOUS SALTS.

Ammonium Dihydrogen Phosphate (Single Crystal, 298°K). (Zernike, 1964).

Wavelength (microns)	Refractive Index in Vacuo		Wavelength (microns)	Refractive Index in Vacuo	
	ordinary ray (n_o)	extraordinary ray (n_e)		ordinary ray (n_o)	extraordinary ray (n_e)
0.2000	1.649083	1.587632	1.2000	1.502447	1.466311
0.3000	1.563951	1.512787	1.3000	1.498888	1.464718
0.4000	1.540785	1.492571	1.4000	1.495142	1.463094
0.5000	1.530276	1.483737	1.5000	1.491187	1.461419
0.6000	1.523024	1.478828	1.6000	1.487004	1.459679
0.7000	1.519528	1.475614	1.7000	1.482580	1.457865
0.8000	1.515813	1.473227	1.8000	1.477903	1.455970
0.9000	1.512433	1.471268	1.9000	1.472965	1.453986
1.0000	1.509156	1.469530	2.0000	1.467756	1.451910
1.1000	1.505853	1.467901			

Lithium Niobate (Single Crystal, 298°K). (Boyd, 1967).

Wavelength (microns)	Refractive Index		Wavelength (microns)	Refractive Index	
	ordinary ray (n_o)	extraordinary ray (n_e)		ordinary ray (n_o)	extraordinary ray (n_e)
0.42	2.4144	2.3038	1.80	2.2074	2.1297
0.45	2.3814	2.2765	2.00	2.2015	2.1244
0.50	2.3444	2.2446	2.20	2.1948	2.1187
0.55	2.3188	2.2241	2.40	2.1882	2.1138
0.60	2.3002	2.2083	2.60	2.1814	2.1080
0.65	2.2862	2.1964	2.80	2.1741	2.1020
0.70	2.2756	2.1874	3.00	2.1663	2.0955
0.80	2.2598	2.1741	3.20	2.1580	2.0886
0.90	2.2487	2.1647	3.40	2.1493	2.0814
1.00	2.2407	2.1580	3.60	2.1298	2.0735
1.20	2.2291	2.1481	3.80	2.1299	2.0652
1.40	2.2208	2.1410	4.00	2.1193	2.0564
1.60	2.2139	2.1351			

Lithium Tantalate (Single Crystal, ~298°K). (Bond, 1965).

0.45	2.2420	2.2468	2.00	2.1066	2.1115
0.50	2.2160	2.2205	2.20	2.1009	2.1053
0.60	2.1834	2.1878	2.40	2.0951	2.0993
0.70	2.1652	2.1696	2.60	2.0891	2.0936
0.80	2.1538	2.1578	2.80	2.0825	2.0871
0.90	2.1454	2.1493	3.00	2.0755	2.0299
1.00	2.1391	2.1432	3.20	2.0680	2.0727
1.20	2.1305	2.1341	3.40	2.0601	2.0649
1.40	2.1236	2.1273	3.60	2.0513	2.0561
1.60	2.1174	2.1213	3.80	2.0424	2.0473
1.80	2.1120	2.1170	4.00	2.0335	2.0377

Potassium Dideuterium Phosphate (Single Crystal, ~298°K). (Milek and Welles, 1970).

0.4047	1.5189	1.4776	0.5779	1.5063	1.4670
0.4078	1.5185	1.4772	0.5893	1.5057	1.4677
0.4358	1.5155	1.4747	0.6234	1.5044	1.4656
0.4916	1.5111	1.4710	0.6907	1.5022	1.4639
0.5461	1.5079	1.4683	1.0000	1.47	1.44

TABLE 116. (*Continued*)

Potassium Dihydrogen Arsenate (Single Crystal, ~298°K). (Milek and Welles, 1970).

Wavelength (microns)	Refractive Index ordinary ray (n_o)	extraordinary ray (n_e)	Wavelength (microns)	Refractive Index ordinary ray (n_o)	extraordinary ray (n_e)
0.4861	1.5762	1.5252	0.5893	1.5674	1.5179
0.5460	1.5707	1.5206	0.6563	1.5632	1.5146

Potassium Dihydrogen Phosphate (Single Crystal, 298°K). (Zernike, 1964).

0.2000	1.622630	1.563913	1.2000	1.490169	1.458845
0.3000	1.545570	1.498153	1.3000	1.487064	1.457838
0.4000	1.524481	1.480244	1.4000	1.483803	1.456838
0.5000	1.514928	1.472486	1.5000	1.480363	1.455829
0.6000	1.509274	1.468267	1.6000	1.476729	1.454797
0.7000	1.505235	1.465601	1.7000	1.472890	1.453735
0.8000	1.501924	1.463708	1.8000	1.468834	1.452636
0.9000	1.498930	1.462234	1.9000	1.464555	1.451495
1.0000	1.496044	1.460993	2.0000	1.460044	1.450308
1.1000	1.493147	1.459884			

Silver Thioarsenite (Proustite) (Single Crystal, 293°K). (Hulme, 1967).

0.5876		2.7896	1.530	2.7728	2.5485
0.6328	3.0190	2.7391	1.709	2.7654	2.5423
0.6678	2.9804	2.7094	2.50	2.7478	2.5282
1.014	2.8264	2.5901	3.56	2.7379	2.5213
1.129	2.8067	2.5756	4.62	2.7318	2.5178
1.367	2.7833	2.5570			

TABLE 117. OPTICAL CONSTANTS OF METALS.
(FORSYTHE, 1969)

Metal	Wavelength (microns)	Refractive Index (n_o)	Absorption Coefficient (k)	Reflection (%)
Aluminum	0.589	1.44	5.32	83
Antimony	0.589	3.04	4.94	70
Bismuth	white	2.26		
Cadmium	0.589	1.13	5.01	85
Chromium	0.579	2.97	4.85	70
Cobalt	0.231	1.10	1.30	32
	0.275	1.41	1.52	46
	0.500	1.93	1.93	66
	0.650	2.35	1.87	69
	1.00	3.63	1.58	73
	1.50	5.22	1.29	75
	2.25	5.65	1.27	76
Copper	0.231	1.39	1.05	29
	0.347	1.19	1.23	32
	0.500	1.10	2.13	56
	0.650	0.44	7.4	86
	0.870	0.35	11.0	91

TABLE 117. (*Continued*)

Metal	Wavelength (microns)	Refractive Index (n_o)	Absorption Coefficient (k)	Reflection (%)
	1.75	0.83	11.4	96
	2.25	1.03	11.4	97
	4.00	1.87	11.4	
	5.50	3.16	9.0	
Gold	0.257	0.92	1.14	28
	0.441	1.18	1.85	42
	0.589	0.47	2.83	82
	1.00	0.24	28.0	6.7
	2.00	0.47	26.7	12.5
	3.00	0.80	24.5	19.6
	5.00	1.81	18.1	33
Iodine	0.589	3.34	0.57	30
Iridium	0.579	2.13	4.87	75
	1.00	3.6	1.60	~78
	2.00	6.0	1.48	~87
	3.00	8.0	1.37	
	5.00	12.5	1.13	
Iron	0.257	1.01	0.88	16
	0.441	1.28	1.37	28
	0.589	1.51	1.63	33
Lead	0.589	2.01	3.48	62
Magnesium	0.589	0.37	4.42	93
Manganese	0.579	2.49	3.89	64
Mercury	0.326	0.68	2.26	66
	0.441	1.01	3.42	74
	0.589	1.62	4.41	75
	0.668	1.72	4.70	77
Nickel	0.420	1.41	1.79	54
	0.589	1.79	1.86	62
	0.750	2.19	1.99	70
	1.00	2.63	2.00	74
	2.25	3.95	2.33	85
Palladium	0.579	1.62	3.41	65
Platinum	0.257	1.17	1.65	37
	0.441	1.94	3.16	58
	0.589	2.63	3.54	59
	0.668	2.91	3.66	59
	1.00	3.4	1.82	
	2.00	5.7	1.70	
	3.00	7.7	1.59	
	5.00	11.5	1.37	
Rhodium	0.579	1.54	4.67	78
Selenium	0.400	2.94	2.31	44
	0.490	3.12	1.49	35
	0.589	2.93	0.45	25
	0.760	2.60	0.06	20
Silicon	0.589	4.18	0.09	38
	1.25	3.67	0.08	33
	2.25	3.53	0.08	31
	4.0			~28
	7.0			~28
	10.0			~28

TABLE 117. (*Continued*)

Metal	Wavelength (microns)	Refractive Index (n_o)	Absorption Coefficient (k)	Reflection (%)
Silver	0.226	1.41	0.75	18
	0.293	1.57	0.62	17
	0.316	1.13	0.38	4
	0.332	0.41	1.61	32
	0.395	0.16	12.32	87
	0.500	0.17	17.1	93
	0.589	0.18	20.6	95
	0.750	0.17	30.7	97
	1.00	0.24	29.0	98
	1.50	0.45	23.7	98
	2.25	0.77	19.9	99
	3.00	1.65	12.2	
	4.50	4.49	7.42	
Sodium	0.589	0.004	2.61	99
Steel	0.226	1.30	1.26	35
	0.257	1.38	1.35	40
	0.325	1.37	1.53	45
	0.500	2.09	1.50	57
	0.650	2.70	1.33	59
	1.50	3.71	1.55	73
	2.25	4.14	1.79	80
Tantalum	0.579	2.05	2.31	44
	1.0			~78
	2.0			~90
	4.0			~93
	7.0			~94
Tellurium	0.6			~49
Tin	0.589	1.48	5.25	82
Tungsten	0.579	2.76	2.71	49
	1.00			58
	2.00			90
	4.00			95
Vanadium	0.579	3.03	3.51	58
Zinc	0.257	0.55	0.61	20
	0.441	0.93	3.19	73
	0.589	1.93	4.66	74
	0.668	2.62	5.08	73
	1.0			~80
	2.0			~92
	4.0			~97
	7.0			~98
	10.0			~98

*Liquid.

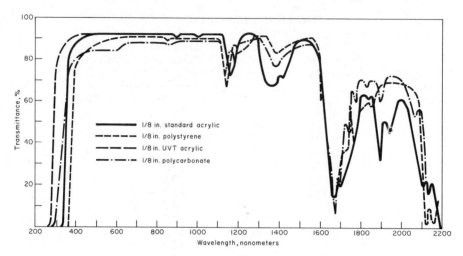

Fig. 115. Transmission of light by polymers. (Almand, 1972).

TABLE 118. REFRACTIVE INDEX OF SELECTED POLYMERS AT 0.589 MICRONS.

Polymer	Temperature (°C)	Refractive Index (n_D)	Polymer	Temperature (°C)	Refractive Index (n_D)
Acrylics	20	1.49	Polyethylene terephthalate	20	1.57–1.58
Allyl diglycol carbonate	20	1.50	Polymethyl α-chloroacrylate	20	1.517
Cellulose acetate	20	1.46–1.50	Polymethyl methacrylate	20	1.492
Cellulose acetate butyrate	20	1.46–1.49	Polypropylene	20	1.49
Cellulose acetate proprionate	20	1.46–1.49	Polypropyl methacrylate	20	1.48
Cellulose ester	20	1.47–1.50	Polystyrene	20	1.59–1.60
Cellulose nitrate	20	1.49–1.51	Polysulfone	20	1.63
Cellulose proprionate	27	1.474	Polytetrafluoroethylene	20	1.35
Diallyl isophthalate	20	1.569	Polytrifluorochloroethylene	20	1.43
Diallyl phthalate	20	1.571	Polytrifluoroethylene	20	1.35–1.37
Epoxies	20	~1.6	Polyvinyl alcohol	20	1.49–1.53
Ethyl cellulose	25	1.473	Polyvinyl acetal	20	1.48
Fluorinated ethylene propylene	20	1.34	Polyvinyl acetate	20	1.46–1.47
Phenol formaldehyde	20	1.5–1.7	Polyvinyl butyral	20	1.49
Phenoxy	20	1.598	Polyvinyl chloride	20	1.52–1.55
Polyacetal	20	1.48	Polyvinyl cyclohexene dioxide	20	1.530
Polyallyl methacrylate	20	1.519	Polyvinyl formal	20	1.60
Polyamide–Nylon 6/6	20	1.53	Polyvinyl naphthalene	20	1.680
Polyamide–Nylon 11	20	1.52	Polyvinylidene chloride	20	1.60–1.63
Polycarbonate	20	1.586	Polyvinylidene fluoride	20	1.42
Polycyclohexyl methacrylate	20	1.506	Silicone	20	1.43
Polydiallyl phthalate	20	1.566	Styrene-acrylonitrile copolymer	20	1.569
Polyester	20	1.53–1.58	Styrene-methacrylate copolymer	20	1.533
Polyester-styrene	20	1.54–1.57	Urethane, cast liquid	20	1.50–1.60
Polyethylene	20	1.51–1.54			
Polyethylene dimethacrylate	20	1.506			

TABLE 119. ELECTROOPTIC PROPERTIES OF SEMICONDUCTORS*

	Symmetry	$r_{ij}(10^{-12}\,\text{m/V})$	λ (microns)	n_i	λ (microns)	ϵ_i
ZnO	6mm	(S) $r_{33} = 2.6$ (S) $r_{13} = 1.4$ $r_{33}/r_{13} < 0$	.63 .63	$n_3 = 2.123$ $n_2 = n_1 = 2.106$ $n_3 = 2.015$ $n_3 = n_1 = 1.999$	.45 .45 .60 .60	$\epsilon = 8.15$
ZnS	$\bar{4}3m$ 6mm	(T) $r_{41} = 1.2$ 2.0 2.1 (S) $r_{33} = 1.85$ (S) $r_{13} = .92$ $r_{33}/r_{13} < 0$	.40 .546 .65 .63 .63	$n_0 = 2.471$ 2.364 2.315 $n_3 = 2.709$ $n_2 = n_1 = 2.705$ $n_3 = 2.368$ $n_2 = n_1 = 2.363$	.45 .60 .8 .36 .36 .60 .60	(T) 16 (S) 12.5 8.3
ZnSe	$\bar{4}3m$	(T) $r_{41} = 2.0$	.546	$n_0 = 2.66$	.546	9.1 8.1
ZnTe	$\bar{4}3m$	(T) $r_{41} = 4.55$ 3.95 (S) $r_{41} = 4.3$	.59 .69 .63	$n_0 = 3.1$ 2.91	.57 .70	10.1
CuCl	$\bar{4}3m$	(T) $r_{41} = 5.6$	.546	$n_0 = 1.99$	.546	(T) 10 (S) 8.3
CuBr	$\bar{4}3m$	(T) $r_{41} = .85$		$n_0 = 2.16$ 2.09	.535 .656	
GaP	$\bar{4}3m$	(S) $r_{41} = .5$ (S) $r_{41} = 1.06$	 .63	$n_0 = 3.4595$ 3.315	.54 .60	10 12
GaAs	$\bar{4}3m$	(T) $r_{41} = .27$ to 1.2 (S-T) $r_{41} = 1.3$ to 1.5 (S) $r_{41} = 1.2$ (T) $r_{41} = 1.6$	1 to 1.8 1 to 1.8 .9 to 1.09 3.39 & 10.6	$n_0 = 3.60$ 3.50 3.42 3.30	.90 1.02 1.25 5.0	(T) 12.5 (S) 10.9 (S) 11.7
CdS	6mm	(T) $r_{51} = 3.7$ (T) $r_c = 4$ (S) $r_{33} = 2.4$ (S) $r_{13} = 1.1$ $r_{33}/r_{13} < 0$	.589 .589 .63 .63	$n_3 = 2.726$ $n_2 = n_1 = 2.743$ $n_2 = n_1 = 2.493$	.515 .515 .60	(T) $\epsilon_1 = 10.6$ (T) $\epsilon_3 = 7.8$ (S) $\epsilon_1 = 8.0$ (S) $\epsilon_3 = 7.7$

(T) = constant stress, (S) = constant strain.

*Table reproduced from (Milek and Welles, 1970), based on data from: Kaminow, I. P. and E. H. Turner, "Electrooptic Light Modulators," *IEEE Proc.* 54, (10), 13474–1390 (Oct. 1966).

TABLE 1110. PROPERTIES OF LINEAR ELECTROOPTIC MODULATOR MATERIALS.
(Milek and Welles, 1970).

Material		Cuprous Chloride		Bismuth Germanium Oxide	Lithium Niobate		Lithium Tantalate	Proustite
Chemical Formula		CuCl		$Bi_{12}GeO_{20}$	$LiNbO_3$		$LiTaO_3$	Ag_3AsS_3
Crystal Symmetry		Cubic			Trigonal			
		$\bar{4}3m\,(T_d)$		23 (T)	$3m\,(C_{3v})$			
Curie Temperature (°C)					1195		660	
Transmission Range, >50% (microns)		0.4 to 20		0.5 to 7.5	0.4 to 5		0.3 to 6	0.6 to 13
$\lambda(\text{Å})$		5460		5100	6328		6328	6328
Refractive Indices		1.99		2.55	n_o	2.291	2.177	3.019
					n_e	2.200	2.182	2.739
Electrooptic Coefficients (d) (10^{-12} m/V)	Unclamped, Constant Stress (T)	r_{41}	5.6	3	r_{13}	9.6		2.8
					r_{22}	6.7		3.2
					r_{33}	31		1.6
					r_{42}	32		
					r_c	15.8	21.6	1.2
	Clamped, Constant Strain (S)	r_{41}			r_{13}	8.6	7.0	
					r_{22}	3.4		1.07
					r_{33}	30.8	30	
					r_{42}	28		
					r_c	18.6	23.5	5.1
Relative Dielectric Constants	Constant Stress (T)	10			ϵ_{33}/ϵ_o	30	45	20
					ϵ_{11}/ϵ_o	84	52	16.5
	Constant Strain (S)	8.3		38	ϵ_{33}/ϵ_o	29	43	18
					ϵ_{11}/ϵ_o	44	41	14.5
Loss Tangents	At 1 kHz						< 0.05	
	At 9.2 GHz	0.002			$\tan \delta_c$	0.01 [e]	0.002 [e]	0.005 [f]
					$\tan \delta_a$	0.08 [e]		0.02 [f]
Piezoelectric Constants (10^{-12} C/N or m/V)					d_{15}	68-74	26	
					d_{22}	21	8	
					d_{31}	-1	-3	
					d_{33}	6-16	9	
Electromechanical Coupling Coefficients				0.30	k_{15}	0.45	0.31	
					k_{22}	0.25-0.32		
					k_{31}	0.023-0.087		
					k_{33}	0.33-0.47	0.20-0.30	
Nonlinear Second Harmonic Generation Coefficients, at 1.06 microns (Relative to $d_{36}^{2\omega} = 1$ for KDP)					$d_{22}^{2\omega}$	6.3	4.3	50
					$d_{31}^{2\omega}$	11.9	2.6	30
					$d_{33}^{2\omega}$	83	40	

(a) Properties Depend upon Degree of Deuteration
(b) Below T_c
(c) Antiferroelectric Transition
(d) At Refractive Index wavelength

$$r_c = (n_e/n_o)^3\, r_{33} - r_{13}$$
$$r_{c1} = r_{22} - (n_x/n_y)^3\, r_{12}$$
$$r_{c3} = r_{22} - (n_z/n_y)^3\, r_{32}$$

(e) At 100 MHz
(f) At 20 MHz

TABLE 1110. (*Continued*)

	Ammonium Dihydrogen Arsenate (ADA)		Ammonium Dihydrogen Phosphate (ADP)	Potassium Dihydrogen Arsenate (KDA)	Potassium Dideuterium Phosphate (KDDP)[a]	Potassium Dihydrogen Phosphate (KDP)	Potassium Tantalate Niobate (KTN)			Barium Sodium Niobate		Calcium Pyroniobate	
	$NH_4H_2AsO_4$		$NH_4H_2PO_4$	KH_2AsO_4	KD_2PO_4	KH_2PO_4	$KTa_xNb_{1-x}O_3$			$Ba_2NaNb_5O_{15}$		$Ca_2Nb_2O_7$	
	Tetragonal									Orthorhombic		Monoclinic	
	$\bar{4}2m\ (D_{2d})$						$4mm\ (C_{4v})$[b]			$mm2\ (C_{2v})$		$2\ (C_2)$	
	-57[c]		-125[c]	-177	-53	-150	10 (x = 0.65)			560			
	0.3 to >0.75		0.2 to 1.2	0.3 to 1.6	0.2 to 2	0.3 to 1.4	0.2 to 1.5			0.4 to 5			
	5460		6328	5460	5460	5460	5460			6328		6328	
n_o		1.580	1.522	1.571	1.508	1.512	$T_c - T$	16°C	36°C	n_x	2.326	1.97	
n_e		1.525	1.478	1.521	1.468	1.470	n_o	2.281	2.275	n_y	2.324	2.16	
							n_e	2.318	2.318	n_z	2.221	2.17	
r_{41}			24	12.5	8.8	8.7	r_{42}	3000	14,000	r_{13}	18.3	r_{12}	
												r_{22}	
										r_{23}	13.5	r_{32}	
												r_{41}	
r_{63}		9.2	8.5	10.9	26.4	10.5	r_c	730	480	r_{33}	56	r_{c1}	12
												r_{c3}	14
r_{41}												r_{12}	6.7
												r_{22}	25
												r_{32}	6.4
r_{63}			4.5			9.5						r_{41}	2.7
												r_{c1}	20
												r_{c3}	19
ϵ_{33}/ϵ_o		14	16	21.0	51	21.6	8000 (T_c = 271°K; x = 0.66)			ϵ_{33}/ϵ_o	51		
ϵ_{11}/ϵ_o		75	56	53.7	57	44				ϵ_{11}/ϵ_o	235		
										ϵ_{22}/ϵ_o	247		
ϵ_{33}/ϵ_o		73	14	20.6	50	19.6				ϵ_{33}/ϵ_o	32	45	
ϵ_{11}/ϵ_o		74	55	53	57	42				ϵ_{11}/ϵ_o	222	45	
										ϵ_{22}/ϵ_o	227		
$\tan\delta_c$		7	0.04	0.25	0.42	0.005	0.01 (x = 0.80)						
$\tan\delta_a$		0.9	0.0046	0.14		0.0015							
$\tan\delta_c$			0.006	0.008	0.13	0.0074							
$\tan\delta_a$			0.0072	0.007	0.02	0.0045							
d_{14}		41	1.7	25	3.3	1.3				d_{15}	42		
										d_{24}	52		
d_{36}		31	49	22	58	23				d_{31}	-7		
										d_{32}	-6		
										d_{33}	37		
k_{14}		0.136	0.006	0.095		0.008				k_{15}	0.21		
										k_{24}	0.25		
k_{36}		0.24	0.33	0.13	0.22	0.121				k_{31}	0.14	k_{33}	0.30
										k_{32}	0.13		
										k_t	0.57		
$d_{14}^{2\omega}$			0.98	1.12	0.91	1.01				$d_{31}^{2\omega}$	29		
										$d_{32}^{2\omega}$	33.9		
$d_{36}^{2\omega}$			0.99	1.06	0.92	1.00				$d_{33}^{2\omega}$	34.4		

REFERENCES

Almand, P. and R. Byrd, "New Developments in Plastics for Optical Applications," *Materials Eng.*, **76**, 42–45 (Nov. 1972).

Ballard, S. S., K. A. McCarthy, and W. L. Wolfe, "Optical Materials for Infrared Instrumentation," U.S. Government Report No. AD 217367, Jan. 1959.

Bond, W. L., "Measurement of the Refractive Indices of Several Crystals," *J. Appl. Phys.* **36**, (5), 1674–1677 (May 1965).

Boyd, G. D. *et al.*, "Refractive Index as a Function of Temperature in LiNbO₃," *J. Appl. Phys.* **38**, (3), 1941–1943 (Mar. 1967).

Forsythe, W. E., "Smithsonian Physical Tables," 9th rev. ed., The Smithsonian Institution Press, Washington, 1969.

Hulme, K. F. *et al.*, "Synthetic Proustite: A New Crystal for Optical Mixing," *Appl. Phys. Letters*, **10**, (4), 133–135 (Feb. 15, 1967).

Jones, R. N., "Infrared Spectra of Organic Compounds: Summary Charts of Principal Group Frequencies," NRC Bulletin No. 6, National Research Council, Ottawa, Canada, 1959.

Milek, J. T. and S. J. Welles, "Linear Electrooptic Modulator Materials," U.S. Government Report No. AD 704556, Jan. 1970.

Salzberg, C. D. and J. J. Villa, "Index of Refraction of Germanium," *Opt. Soc. Am. J.*, **48**, (8), 579 (Aug. 1958).

Smakula, A., "A Study of the Physical Properties of High-Temperature Single Crystals," U.S. Government Report No. AD 663734, Sept. 1967.

Willardson, R. K. and A. C. Beer, Eds., "Semiconductors and Semimetals," Vol. 3, "Optical Properties of III-V Compounds," Seraphim, B. O. and H. E. Bennett, Academic Press, New York, 1967.

Wolfe, W. L., Ed., "Handbook of Military Infrared Technology," U.S. Government Printing Office, Washington, 1965.

Zernike, F. Jr., "Refractive Indices of Ammonium Dihydrogen Phosphate and Potassium Dihydrogen Phosphate between 200 Angstroms and 1.5 Microns," *Opt. Soc. Am. J.* **54**, 1215–1220 (Oct. 1964).

BIBLIOGRAPHY—OPTICAL PROPERTIES

Abeles, F., Ed., "Optical Properties of Solids," Elsevier, Amsterdam, 1972.

Albers, W. A., Jr., Ed., "The Physics of Opto-Electronic Materials," Plenum Press, New York, 1971.

Allan, W. B., "Fibre Optics-Theory and Practice," Plenum Press, New York, 1973.

Bassani, G. F. and G. P. Parravicini, "Electronic States and Optical Properties of Crystals," Pergamon Press, Elmsford, N.Y., 1973.

Greenaway, D. L. and G. Harbeke, "Optical Properties and Band Structure of Semiconductors," Pergamon Press, Elmsford, N.Y., 1968.

Jerlov, N. and E. Steeman, "Optical Aspects of Oceanography," Academic Press, London, 1973.

Kapany, N. S., "Fiber Optics." Academic Press, New York, 1967.

Kingslake, R., Ed., "Applied Optics and Optical Engineering," 5 vols. Academic Press, New York, 1965–1969.

Mayer, H., "Physik dünner Schichten," Pt. 1, Wissenschaftliche Verlagsgesellschaft m.b.H., Stuttgart, Germany, 1950.

Mitra, S. S. and S. Nudelman, Eds., "Far-Infrared Properties of Solids," Plenum Press, New York, 1970.

Nudelman, S. and S. S. Mitra, "Optical Properties of Solids," Plenum Press, New York, 1969.

Pressley, R. J., "CRC Handbook of Lasers," CRC Press, Cleveland, 1971.

Tauc, J., Ed., "The Optical Properties of Solids," Academic Press, New York, 1966.

Winchell, A. N., "The Optical Properties of Organic Compounds," Academic Press, New York, 1954.

BIBLIOGRAPHY—SPECTROSCOPY

Anan., "UV Atlas of Organic Compounds," 5 vols, Plenum Press, New York, 1966–1971.

Batterham, T. J., "NMR Spectra of Simple Heterocycles," John Wiley, New York, 1973.

Bentley, F. F., L. D. Smithson, and A. L. Rozek, "Infrared Spectra and Characteristic Frequencies 700–300 cm⁻¹," John Wiley, New York, 1968.

Berlman, I. B., "Handbook of Fluorescence Spectra of Aromatic Molecules," 2nd ed., Academic Press, New York, 1971.

Beynon, J. H. and A. E. Williams, "Mass and Abundance Tables for Use in Mass Spectrometry," American Elsevier, New York, 1963.

Bielski, B. H. and J. M. Gebicki, "Atlas of Electron Spin Resonance Spectra," Academic Press, New York, 1967.

Bovey, F. A., "NMR Data Tables for Organic Compounds," John Wiley, New York, 1967.

Fabian, D. J., Ed., "Soft X-Ray Band Spectra and the Electronic Structure of Metals and Materials," Academic Press, London, 1969.

Fang, Jen-Ho and D. F. Bloss, "X-Ray Diffraction Tables," Southern Illinois University Press, Carbondale, Ill., 1966.

Hamming, M. C. and N. G. Foster, "Interpretation of Mass Spectra of Organic Compounds," Academic Press, New York, 1972.

Harrison, G. R., "M. I. T. Wavelength Tables," John Wiley, New York, 1948.

Hirayame, K., "Handbook of Ultraviolet and Visible Absorption Spectra of Organic Compounds," Plenum Press, New York, 1967.

Holubek, J. and O. Strouf, "Spectral Data and Physical Constants of Alkaloids," Heyden & Son, Ltd., London, 1967.

Hummel, D., "Infrared Spectra of Polymers in the Medium and Long Wavelength Regions," Wiley, New York, 1966.

Hummel, D. and F. Scholl, "Infrared Analysis of Polymers, Resins and Additives: an Atlas," 2 vols, Halsted Press, New York, 1969–1972.

Johnson, L. F. and W. C. Jankowski, "Carbon-13 NMR Spectra: A Collection of Assigned, Coded, and Indexed Spectra," John Wiley, New York, 1972.

Kirschenbaum, D. M., "Atlas of Protein Spectra in the Ultraviolet and Visible Regions," 2 vols, Plenum Press, New York, 1972–1973.

Lang, D., Ed., "Absorption Spectra in the Ultraviolet and Visible Region," 17 vols, Academic Press, New York, 1963–1973.

Mecke, R. and F. Langenbucher, "Infrared Spectra of Selected Chemical Compounds," 8 vols, Heyden & Son, Ltd., London, 1967.

Meggers, W. F., C. H. Corliss and B. F. Scribner, "Tables of Spectral-line Intensities," 2 pts, National Bureau of Standards Monograph 32, U.S. Government Printing Office, Washington, 1961.

Nakamoto, K., "Infrared Spectra of Inorganic and Coordination Compounds," 2nd ed., Wiley, New York, 1970.

Nyquist, R. A. and R. O. Kagel, "Infrared Spectra of Inorganic Compounds, (3800-45 cm^{-1}), Academic Press, New York, 1971.

Parsons, M. L. and P. M. McElfresh, "Flame Spectroscopy: Atlas of Spectral Lines," Plenum Press, New York, 1971.

Porter, Q. N. and J. Baldas, "Mass Spectra of Heterocyclic Compounds," Wiley, New York, 1971.

Pouchert, C. J., "The Aldrich Library of Infrared Spectra," Aldrich Chemical Co., Milwaukee.

Sadtler Research Labs., "Catalog of Infrared Spectrograms," Sadtler Research Labs., Philadelphia.

Sadtler Research Labs., "Standard NMR Spectra," 12 vols, Sadtler Research Labs., Philadelphia, 1967.

Sadtler Research Labs., "Standard UV Spectra," 62 vols, Sadtler Research Labs., Philadelphia, 1969.

Stenhagen, E., S. Abrahamsson and F. W. McLafferty, Eds., "Atlas of Mass Spectral Data," 3 vols., John Wiley, New York, 1969.

Striganov, A. R. and N. S. Sventitskii, "Tables of Spectral Lines of Neutral and Ionized Atoms," Plenum Press, New York, 1968.

Szymanski, H. A. and R. E. Erickson, Eds., "Infrared Band Handbook," rev. ed., 2 vols., Plenum Press, New York, 1970.

Szymanski, H. A. and R. E. Yelin, "NMR Band Handbook," Plenum Press, New York, 1968.

Zaidel, A. N. et al., "Tables of Spectral Lines," 3rd rev. ed., Plenum Press, New York, 1970.

SECTION 1m
NUCLEAR RADIATION AND RADIATION EFFECTS

This Section incorporates a number of tables and graphs, describing properties of nuclear radiation and particles, radiation dosimetry and protection, and radiation effects. Additional information concerning nuclear properties of elements will be found in Chapter 2, and on nuclear particles in Section 1a. The material in Section 1m is organized as follows:

Please note that Appendix A contains a detailed table of isotopes and their properties.

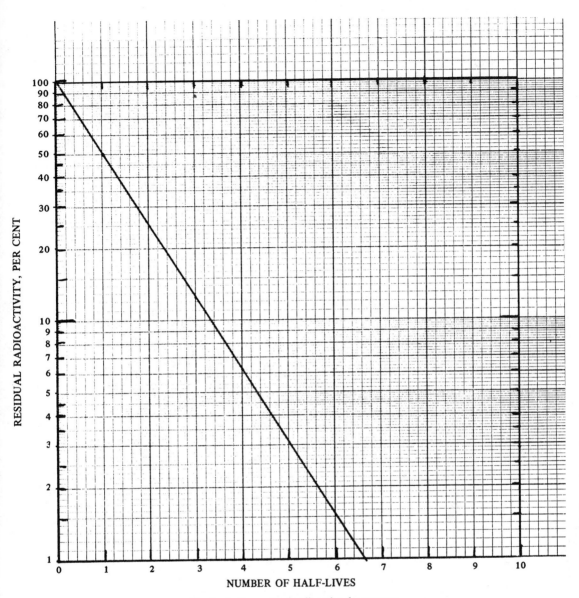

Fig. 1m1. Generalized radioactive decay curve.

TABLE 1m1. RADIOACTIVITY AND
RADIATION PROTECTION.*

Unit of activity = Curie:
   1 Ci = 3.7×10^{10} disintegrations/sec

Unit of exposure dose for x and γ radiation = Roentgen:
   1 R = 1 esu/cm^3 = 87.8 erg/g (5.49×10^7 MeV/g) of air

Unit of absorbed dose = rad:
   1 rad = 100 erg/g (6.25×10^7 MeV/g) in any material

Unit of dose equivalent (for protection) = rem:
   rems (Roentgen equivalents for man) = rads × QF,
where QF (quality factor) depends upon the type of radiation
and other factors. For γ rays and HE protons, QF ≈ 1; for
thermal neutrons, QF ≈ 3; for fast neutrons, QF ranges up
to 10; and for α particles and heavy ions, QF ranges up to 20.

Maximum permissible occupational dose for the whole body:
   5 rem/year (or ≈ 100 millirem/week)

Fluxes (per cm^2) to liberate 1R in carbon:
   3×10^7 minimum ionizing singly charged particles
   0.9×10^9 protons of 1 MeV energy
(These fluxes are correct to within a factor of 2 for all
materials.)

Natural background: 120 to 130 millirem/year
   cosmic radiation (charged particles + neutrons)   ~25 ⎫
   cosmic radiation (γ rays)   ~25 ⎬ mrem/yr
   radiation from rocks and air (γ rays)   ~73 ⎭
Cosmic ray background in counters: ~ 1/min/cm^2/ster

*Reprinted by permission of the Particle Data Group
and the Editor of *Physics Letters* from *Physics
Letters*, Vol. 50B, No. 1 (Apr. 1974).

TABLE 1m2. THE NEPTUNIUM SERIES. (GLASSTONE, 1967)

Element	Symbol	Radiation	Half-Life
Plutonium ↓	^{241}Pu	β	13.2 yr
Americium ↓	^{241}Am	α	462 yr
Neptunium ↓	^{237}Np	α	2.20×10^6 yr
Protactinium ↓	^{233}Pa	β	27.4 days
Uranium ↓	^{233}U	α	1.62×10^5 yr
Thorium ↓	^{229}Th	α	7.34×10^3 yr
Radium ↓	^{225}Ra	β	14.8 days
Actinium ↓	^{225}Ac	α	10.0 days
Francium ↓	^{221}Fr	α	4.8 min
Astatine ↓	^{217}At	α	1.8×10^{-2} sec
Bismuth 98% | 2%	^{213}Bi	β and α	47 min
Polonium	^{213}Po	α	4.2×10^{-6} sec
Thallium	^{209}Tl	β	2.2 min
Lead ↓	^{209}Pb	β	3.32 hr
Bismuth (End Product)	^{209}Bi	Stable	—

TABLE 1m3. THE THORIUM SERIES. (GLASSTONE, 1967)

Radioelement	Corresponding Element	Symbol	Radiation	Half-Life
Thorium ↓	Thorium	^{232}Th	α	1.39×10^{10} yr
Mesothorium I ↓	Radium	^{228}Ra	β	6.7 yr
Mesothorium II ↓	Actinium	^{228}Ac	β	6.13 hr
Radiothorium ↓	Thorium	^{228}Th	α	1.91 yr
Thorium X ↓	Radium	^{224}Ra	α	3.64 days
Th Emanation ↓	Radon	^{220}Rn	α	52 sec
Thorium A ↓	Polonium	^{216}Po	α	0.16 sec
Thorium B ↓	Lead	^{212}Pb	β	10.6 hr
Thorium C 66.3% | 33.7%	Bismuth	^{212}Bi	β and α	60.5 min
Thorium C′	Polonium	^{212}Po	α	3×10^{-7} sec
Thorium C″	Thallium	^{208}Tl	β	3.1 min
Thorium D (End Product)	Lead	^{208}Pb	Stable	—

TABLE 1m4. THE ACTINIUM SERIES. (GLASSTONE, 1967)

Radioelement	Corresponding Element	Symbol	Radiation	Half-Life
Actinouranium ↓	Uranium	^{235}U	α	7.13×10^8 yr
Uranium Y ↓	Thorium	^{231}Th	β	25.6 hr
Protactinium ↓	Protactinium	^{231}Pa	α	3.43×10^4 yr
Actinium 98.8% | 1.2% ↓	Actinium	^{227}Ac	β and α	21.8 yr
Radioactinium	Thorium	^{227}Th	α	18.4 days
Actinium K	Francium	^{223}Fr	β	21 min
Actinium X ↓	Radium	^{223}Ra	α	11.7 days
Ac Emanation ↓	Radon	^{219}Rn	α	3.92 sec
Actinium A ~100% | ~5 × 10⁻⁴% ↓	Polonium	^{215}Po	α and β	1.83×10^{-3} sec
Actinium B	Lead	^{211}Pb	β	36.1 min
Astatine-215	Astatine	^{215}At	α	$\sim 10^{-4}$ sec
Actinium C 99.7% | 0.3% ↓	Bismuth	^{211}Bi	α and β	2.16 min
Actinium C′	Polonium	^{211}Po	α	0.52 sec
Actinium C″	Thallium	^{207}Tl	β	4.8 min
Actinium D (End Product)	Lead	^{207}Pb	Stable	—

TABLE 1m5. THE URANIUM SERIES. (GLASSTONE, 1967)

Radioelement	Corresponding Element	Symbol	Radiation	Half-Life
Uranium I ↓	Uranium	^{238}U	α	4.51×10^9 yr
Uranium X_1 ↓	Thorium	^{234}Th	β	24.1 days
Uranium X_2* ↓	Protactinium	^{234}Pa	β	1.18 min
Uranium II ↓	Uranium	^{234}U	α	2.48×10^5 yr
Ionium ↓	Thorium	^{230}Th	α	8.0×10^4 yr
Radium ↓	Radium	^{226}Ra	α	1.62×10^3 yr
Ra Emanation ↓	Radon	^{222}Rn	α	3.82 days
Radium A 99.98% | 0.02%	Polonium	^{218}Po	α and β	3.05 min
Radium B	Lead	^{214}Pb	β	26.8 min
Astatine-218	Astatine	^{218}At	α	2 sec
Radium C 99.96% | 0.04%	Bismuth	^{214}Bi	β and α	19.7 min
Radium C'	Polonium	^{214}Po	α	1.6×10^{-4} sec
Radium C''	Thallium	^{210}Tl	β	1.32 min
Radium D ↓	Lead	^{210}Pb	β	19.4 yr
Radium E ~100% | 2×10^{-4}%	Bismuth	^{210}Bi	β and α	5.0 days
Radium F	Polonium	^{210}Po	α	138.4 days
Thallium-206	Thallium	^{206}Tl	β	4.20 min
Radium G (End Product)	Lead	^{206}Pb	Stable	—

*Uranium X_2 is an excited state of ^{234}Pa and undergoes isomeric transition to a small extent to form uranium Z (^{234}Pa in its ground state); the latter has a half-life of 6.7 hr, emitting beta radiation and forming uranium II (^{234}U).

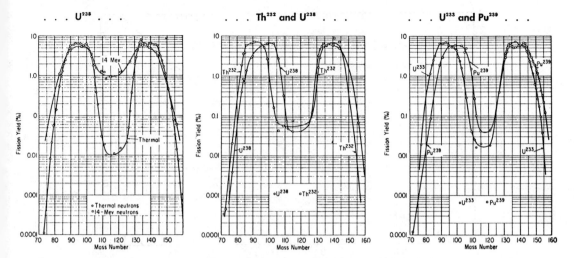

Fig. 1m2. *U, Th, and Pu graphs of fission yields.* (From: Nucleonics, *April 1958*, "*Fission-Product Yields from U, Th, and Pu,*"). Reprinted from Nucleonics magazine with permission of McGraw-Hill, Inc.

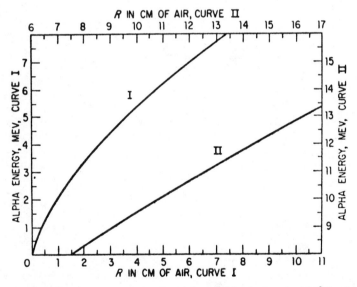

Fig. 1m3. Range-energy relation for alpha particles in dry air at 15°C and 760 torr. (Bethe, 1947).

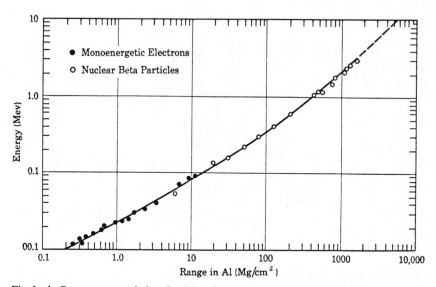

Fig. 1m4. Range-energy relation for beta particles. (Glendenin, 1948). Reprinted from Vol. 2, No. 1 Nucleonics magazine with permission from McGraw-Hill, Inc.

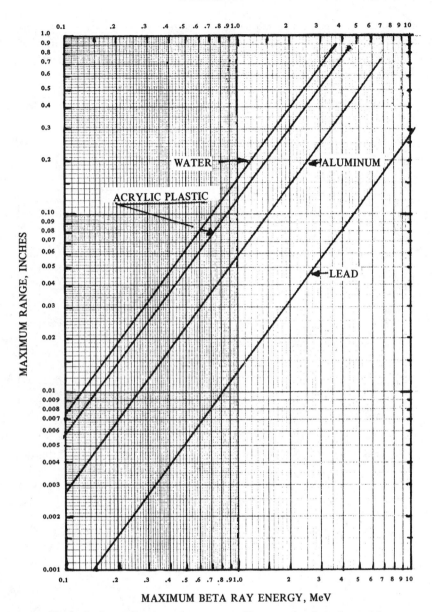

Fig. 1m5. Approximate maximum range of beta rays in selected materials.

Absorption of X- and gamma-radiation can be expressed using a variety of coefficients, with the following coefficients most often encountered:

Mass absorption coefficient, μ/s, cm^2 gm^{-1}
Linear absorption coefficient, μ, cm^{-1}

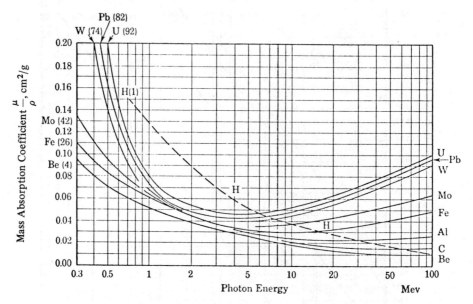

Fig. 1m6. Mass absorption coefficients for X- and gamma-radiations. (Kohl, 1961).

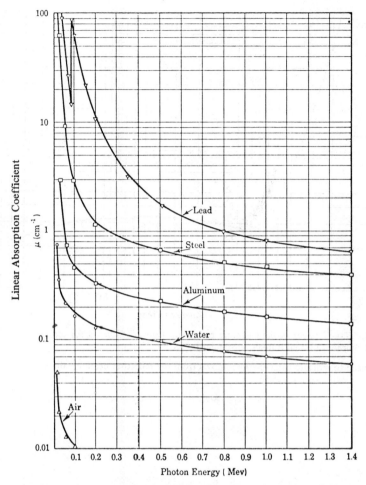

Fig. 1m7. Linear absorption coefficient for X- and gamma-radiations. Data from Gladys R. White, National Bureau of Standards. (Kohl, 1961).

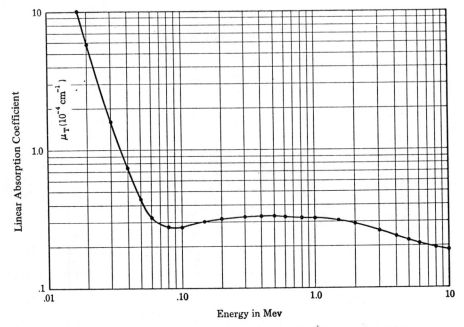

Fig. 1m8. "True" linear X- and gamma ray absorption coefficients for air at 20°C. Data from Gladys R. White, National Bureau of Standards. (Kohl, 1961).

*True is used to indicate total energy removal from the radiation beam by air.

TABLE 1m6. DOSIMETRY CONVERSION FACTORS. (KIRCHER AND BOWMAN, 1964)

To Convert	To	Multiply by
Rads	ergs g^{-1}	100
Ev g^{-1}	ergs g^{-1} (C)	1.5×10^{-12}
*Roentgen	ergs g^{-1} (C)	87.7
Rep	ergs g^{-1} (C)	84.6
Rad (tissue)	ergs g^{-1} (C)	90.9
Rad (water)	ergs g^{-1} (C)	90.0
*Mev cm^{-2a}	ergs g^{-1} (C)	4.5×10^{-8}
*Photons cm^{-2a}	ergs g^{-1} (C)	4.5×10^{-8}
Photons cm^{-2a}	rep	5×10^{-10}
*Rep hr^{-1a}	n cm^{-2} sec^{-1}	7.1×10^{4}
*Rad hr^{-1a}	n cm^{-2} sec^{-1}	8.3×10^{4}
*Rem hr^{-1a}	n cm^{-2} sec^{-1}	8.3×10^{3}
(nv_0)	rad hr^{-1}	4.2×10^{-6}

*Assumed average energy of 1 Mev.

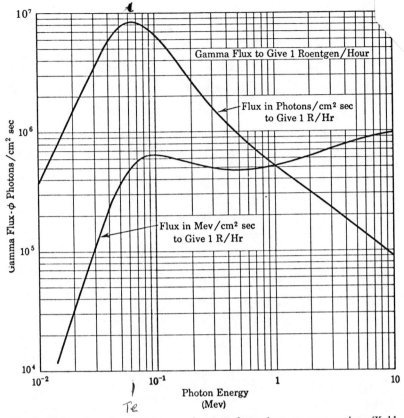

Fig. 1m9. Number of gamma rays vs. dose rate for various gamma energies. (Kohl, 1961).

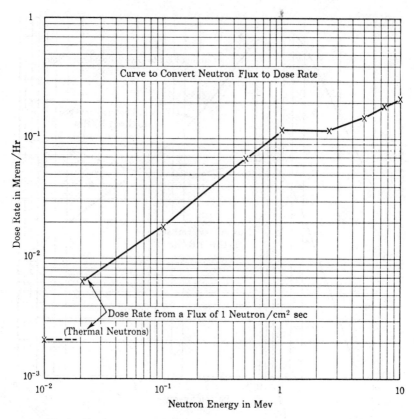

Fig. 1m10. Curve to convert Neutron Flux to Dose Rate. (Kohl, 1961).

TABLE IM7. DOSE-RATE FACTORS FOR UNSHIELDED RADIOISOTOPE SOURCES.

Radioisotope	Half-Life	Dose-Rate Factor* $(\text{rhm} \cdot \text{c}^{-1})$	Radioisotope	Half Life	Dose-Rate Factor* $(\text{rhm} \cdot \text{c}^{-1})$
Antimony-124	60 d	0.98	Iron-59	45 d	0.68
Antimony-125	2.7 y	0.72	Manganese-54	314 d	0.45
Barium-133	7.2 y	0.238	Mercury-203	47 d	0.12
Bromine-82	1.48 d	1.46	Potassium-42	12.4 h	0.15
Cerium-144	285 d	0.035	Radium-226	1620 y	0.844**
Cesium-134	2.2 y	0.82	Rubidium-86	18.7 d	0.051
Cesium-137	30 y	0.32	Scandium-46	84 d	1.1
Chromium-51	27.8 d	0.018	Selenium-75	120 d	0.30
Cobalt-58	71 d	0.54	Silver-110m	249 d	1.4
Cobalt-60	5.26 y	1.28	Sodium-22	2.58 y	1.26
Copper-64	12.9 h	0.114	Sodium-24	15.0 h	1.84
Europium-152	13 y	0.55	Tantalum-182	115 d	0.61
Gold-198	2.70 d	0.235	Thulium-170	127 d	0.004
Gold-199	3.15 d	0.042	Tin-113	118 d	0.35
Iodine-128	25 m	0.045	Zinc-65	245 d	0.27
Iodine-131	8.04 d	0.22	Zirconium-95	65 d	0.40
Iridium-192	74 d	0.50			

*Neglecting bremsstrahlung.
**With 0.5 mm Pt shielding.

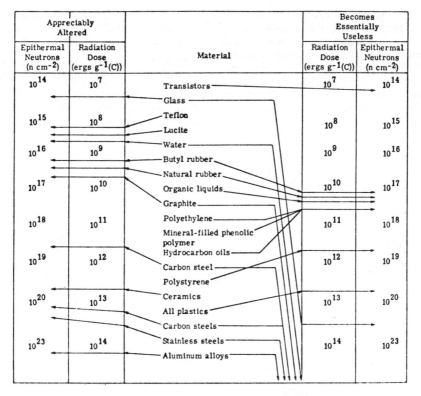

Fig. 1m11. Radiation resistance of various types of materials. (Hennig, 1957).

REFERENCES

Bethe, H. A., Report No. BNL T7, 1947.

Glasstone, S., "Sourcebook on Atomic Energy," Van Nostrand Reinhold, New York, 1967.

Glendenin, L. E., "Determination of the Energy of Beta Particles and Photons," *Nucleonics*, 2, (1), 12–32 (1948).

Hennig, G. R., Moderator, "Shields and Auxiliary Equipment," talk presented at colloqium on the effect of radiation on materials, at Johns Hopkins University, Baltimore, Maryland, 1957.

Kircher, J. F. and R. E. Bowman, Eds., "Effects of Radiation on Materials and Components," Van Nostrand Reinhold, New York, 1964.

Kohl, J., R. D. Zentner, and H. R. Lukens, "Radio-isotope Applications Engineering," Van Nostrand Reinhold, New York, 1961.

BIBLIOGRAPHY

Attix, F. H., *et al.*, Eds., "Radiation Dosimetry," 3 vols., Academic Press, New York, 1968–1969.

Baumgaertner, F., "Tables of Neutron Activation Constants," International Publications Service, New York, 1967.

Etherington, H., Ed., "Nuclear Engineering Handbook," McGraw-Hill, New York, 1958.

Fano, U., "Gamma Ray Attenuation," *Nucleonics*, 11, (8), 55–61; (9), 8–12 (1953).

Glasstone, S., "Principles of Nuclear Reactor Engineering," Van Nostrand Reinhold, New York, 1955.

Hine, G. J. and G. L. Brownell, "Radiation Dosimetry," Academic Press, New York, 1956.

Hughes, D. J. and R. B. Schwartz, "Neutron Cross Sections," Report No. BNL-325, 2nd ed., U.S. Atomic Energy Commission, Washington, 1958.

Hyde, E. K., *et al.*, "Nuclear Properties of Heavy Elements," rev. ed., 3 vols., Dover Publications, New York, 1971.

Haffner, J. W., "Radiation Shielding in Space," Academic Press, New York, 1967.

Koch, R. C., "Activation Analysis Handbook," Academic Press, New York, 1960.

Marion, J. B. and F. C. Young, "Nuclear Reaction Analysis: Graphs and Tables," North-Holland Publishing Co., Amsterdam, 1968.

Meixner, C., "Tables of Gamma Ray Energies for Activation Analysis," International Publications Service, New York, 1970.

Morgan, K. Z. and J. E. Turner, Eds., "Principles of Radiation Protection," John Wiley, New York, 1967.

Moses, A. J., "Nuclear Techniques in Analytical Chemistry," Pergamon Press, Oxford, 1964.

National Bureau of Standards, "Maximum Permissible Body Burdens and Maximum Permissible Concentrations of Radionuclides in Air and Water for Occupational Exposure," NBS Handbook 69, Superintendent of Documents, Washington, D.C., 1959.

National Bureau of Standards, "Measurement of Neutron Flux and Spectra for Physical and Biological Applications," NBS Handbook 72, Superintendent of Documents, Washington, D.C., 1960.

Rockwell, T., III, Ed., "Reactor Shielding Design Manual," Van Nostrand Reinhold, New York, 1956.

Yaffe, L., Ed., "Nuclear Chemistry," 2 vols., Academic Press, New York, 1968.

CHAPTER **II**
PROPERTIES OF
THE ELEMENTS

Chapter II provides a summary, in tabular form, of atomic, nuclear, crystallographic, thermal, thermodynamic, electrical, electronic, magnetic, optical and mechanical properties of the elements.

A list of appropriate books and reports offers the key to more detailed information.

TABLE 2.1 PROPERTIES OF THE CHEMICAL ELEMENTS

ACTINIUM

	Unit	Value(s)
ATOMIC, NUCLEAR AND CRYSTALLOGRAPHIC PROPERTIES		
Symbol: Ac		
Atomic number	–	89
Atomic weight	(based on ^{12}C)	227
Periodic classification	–	III B
Atomic volume	cm^3/g-atom	22.56
Atomic radius	Angstrom units	2.03
Ionic radius	Angstrom units (Ac^{3+})	1.11
Thermal neutron absorption -cross section	barns/atom	
Density	g./cm^3	10.1
Lattice type	–	face-centered cubic
Lattice constants	Angstrom units	
a		5.311
b		
c		
Valence electrons	–	$6d^1\ 7s^2$
THERMAL AND THERMODYNAMIC PROPERTIES		
Melting point	°C	1050 ± 50
Boiling point	°C	3200 ∓ 300
Transition point(s)	°C	
Critical temperature	°K	
Linear thermal expansion coefficient	10^{-6} cm/cm/°C	14.9
Specific heat	cal/g/°C	
Electronic specific heat	10^{-4} cal/mole/°C^2	23
Thermal conductivity	cal/cm^2/cm/sec/°C	
Latent heat of fusion	kcal/mole	3.42
Latent heat of vaporization at B.P.	kcal/mole	95.0
Heat of combustion	kcal/mole	
Heat of sublimation	kcal/mole	104
Heat capacity at constant pressure	cal/°C/mole	6.50
Heat content at M.P.	kcal/mole	8.99
Entropy	cal/°C/mole	15.00
Longitudinal sound velocity	m/sec	
ELECTRICAL, ELECTRONIC, MAGNETIC AND OPTICAL PROPERTIES		
Electrical Conductivity	% IACS	
Electrical resistivity	microohm-cm	
Photoelectric work function	eV	2.7 (estimate)
Hall coefficient	volt-cm/amp-Oe	
Curie point	°K	
First ionization potential	eV (Ac^+)	6.9
Oxidation potential	V (Ac^{3+})	2.6
Electrochemical equivalent	mg/coulomb (Ac^{3+})	0.78411
Magnetic susceptibility	c.g.s. units	
Refractive index, n_D	–	
MECHANICAL PROPERTIES		
Young's modulus of elasticity	10^{-6} kg/cm^2	0.35
Shear modulus	10^{-6} kg/cm^2	0.138
Bulk modulus	10^{-6} kg/cm^2	0.25
Poisson's ratio	–	0.269
Hardness		

TABLE 2.1 (*Continued*)

ALUMINUM

	Unit	Value(s)
ATOMIC, NUCLEAR AND CRYSTALLOGRAPHIC PROPERTIES		
Symbol: Al		
Atomic number	–	13
Atomic weight	(based on ^{12}C)	26.9815
Periodic classification	–	III A
Atomic volume	cm^3/g-atom	10.00
Atomic radius	Angstrom units	1.43
Ionic radius	Angstrom units (Al^{3+})	0.50
Thermal neutron absorption -cross section	barns/atom	0.23
Density	g./cm^3	2.6989
Lattice type	–	face-centered cubic
Lattice constants	Angstrom units	
a		4.04958
b		
c		
Valence electrons	–	$3s^2\ 3p^1$
THERMAL AND THERMODYNAMIC PROPERTIES		
Melting point	$°C$	660.24
Boiling point	$°C$	2467
Transition point(s)	$°C$	
Critical temperature	$°K$	7740
Linear thermal expansion coefficient	$10^{-6}\ cm/cm/°C$	23.9
Specific heat	cal/g/$°C$	0.215
Electronic specific heat	$10^{-4}\ cal/mole/°C^2$	3.23
Thermal conductivity	cal/cm^2/cm/sec/$°C$	0.497
Latent heat of fusion	kcal/mole	2.55
Latent heat of vaporization at B.P.	kcal/mole	78.7
Heat combustion	kcal/mole	201
Heat of sublimation	kcal/mole	78
Heat capacity at constant pressure	cal/$°C$/mole	5.820
Heat content at M.P.	kcal/mole	2.58
Entropy	cal/$°C$/mole	6.789
Longitudinal sound velocity	m/sec	5104
ELECTRICAL, ELECTRONIC, MAGNETIC AND OPTICAL PROPERTIES		
Electrical Conductivity	% IACS	65.45
Electrical resistivity	microohm-cm	2.6548
Photoelectric work function	eV	4.2
Hall coefficient	volt-cm/amp-Oe	
Curie point	$°K$	
First ionization potential	eV (Al^+)	5.986
Oxidation potential	V (Al^{3+})	1.66
Electrochemical equivalent	mg/coulomb (Al^{3+})	0.09317
Magnetic susceptibility	c.g.s. units	0.6×10^{-6}
Refractive index, n_D		1.44
MECHANICAL PROPERTIES		
Young's modulus of elasticity	$10^{-6}\ kg/cm^2$	0.724
Shear modulus	$10^{-6}\ kg/cm^2$	0.271
Bulk modulus	$10^{-6}\ kg/cm^2$	0.7358
Poisson's ratio	–	0.34
Hardness	(Brinell)	17

TABLE 2.1 (*Continued*)

AMERICIUM

	Unit	Value(s)
ATOMIC, NUCLEAR AND CRYSTALLOGRAPHIC PROPERTIES Symbol: Am		
Atomic number	–	95
Atomic weight	(based on ^{12}C)	243
Periodic classification	–	Actinide
Atomic volume	cm^3/g-atom	20.77
Atomic radius	Angstrom units	1.82
Ionic radius	Angstrom units (Am^{3+})	1.07
Thermal neutron absorption -cross section	barns/atom	
Density	g./cm^3	13.67
Lattice type	–	hexagonal
Lattice constants	Angstrom units	
a		3.47
b		
c		11.24
Valence electrons	–	$5f^7 7s^2$
THERMAL AND THERMODYNAMIC PROPERTIES		
Melting point	°C	995
Boiling point	°C	2607
Transition point(s)	°C	
Critical temperature	°K	
Linear thermal expansion coefficient	10^{-6} cm/cm/°C	
Specific heat	cal/g/°C	
Electronic specific heat	10^{-4} cal/mole/°C^2	
Thermal conductivity	cal/cm^2/cm/sec/°C	
Latent heat of fusion	kcal/mole	
Latent heat of vaporization at B.P.	kcal/mole	57.0
Heat of combustion	kcal/mole	
Heat of sublimation	kcal/mole	
Heat capacity at constant pressure	cal/°C/mole	
Heat content at M.P.	kcal/mole	
Entropy	cal/°C/mole	
Longitudinal sound velocity	m/sec	
ELECTRICAL, ELECTRONIC, MAGNETIC AND OPTICAL PROPERTIES		
Electrical Conductivity	% IACS	
Electrical resistivity	microohm-cm	
Photoelectric work function	eV	
Hall coefficient	volt-cm/amp-Oe	
Curie point	°K	
First ionization potential	eV (Am$^+$)	6.0
Oxidation potential	V (Am^{3+})	2.320
Electrochemical equivalent	mg/coulomb	
Magnetic susceptibility	c.g.s. units	
Refractive index, n_D		
MECHANICAL PROPERTIES		
Young's modulus of elasticity	10^{-6} kg/cm^2	
Shear modulus	10^{-6} kg/cm^2	
Bulk modulus	10^{-6} kg/cm^2	
Poisson's ratio	–	
Hardness		

TABLE 2.1 (*Continued*)

ANTIMONY

	Unit	Value(s)
ATOMIC, NUCLEAR AND CRYSTALLOGRAPHIC PROPERTIES		
Symbol: Sb		
Atomic number	–	51
Atomic weight	(based on ^{12}C)	121.75
Periodic classification	–	V A
Atomic volume	cm^3/g-atom	18.21
Atomic radius	Angstrom units	1.82
Ionic radius	Angstrom units ($Sb^{3+, 5+}$)	0.76, 0.62
Thermal neutron absorption -cross section	barns/atom	5
Density	g./cm^3	6.62
Lattice type	–	Rhombohedral
Lattice constants	Angstrom units	4.307
a		
b		
c		11.273
Valence electrons	–	$5s^2 \, 5p^3$
THERMAL AND THERMODYNAMIC PROPERTIES		
Melting point	°C	630.5
Boiling point	°C	1635
Transition point(s)	°C	95, 417
Critical temperature	°K	2989
Linear thermal expansion coefficient	10^{-6} cm/cm/°C	8.5
Specific heat	cal/g/°C	0.049
Electronic specific heat	10^{-4} cal/mole/°C^2	0.24
Thermal conductivity	cal/cm^2/cm/sec/°C	0.045
Latent heat of fusion	kcal/mole	4.67
Latent heat of vaporization at B.P.	kcal/mole	19.6
Heat of combustion	kcal/mole	
Heat of sublimation	kcal/mole	62.6
Heat capacity at constant pressure	cal/°C/mole	6.03
Heat content at M.P.	kcal/mole	3.97
Entropy	cal/°C/mole	10.88
Longitudinal sound velocity	m/sec (20°C)	3420.
ELECTRICAL, ELECTRONIC, MAGNETIC AND OPTICAL PROPERTIES		
Electrical Conductivity	% IACS	
Electrical resistivity	microohm-cm (0°C)	39.1
Photoelectric work function	eV	4.0
Hall coefficient	volt-cm/amp-Oe	
Curie point	°K	
First ionization potential	eV (Sb^+)	8.641
Oxidation potential	V (Sb/Sb_2O_3)	−0.152
Electrochemical equivalent	mg/coulomb (Sb^{5+})	0.25235
Magnetic susceptibility	c.g.s. units	-0.87×10^{-6}
Refractive index, n_D		3.04
MECHANICAL PROPERTIES		
Young's modulus of elasticity	10^{-6} kg/cm^2	0.560
Shear modulus	10^{-6} kg/cm^2	0.204
Bulk modulus	10^{-6} kg/cm^2	0.390
Poisson's ratio	–	0.31
Hardness	Mohs	3.0–3.5

TABLE 2.1 (Continued)

ARGON

	Unit	Value(s)
ATOMIC AND NUCLEAR PROPERTIES		
Symbol: Ar		
Atomic number	–	18
Atomic weight	(based on ^{12}C)	39.948
Periodic classification	–	rare gas
Thermal neutron absorption		
-cross section	barns/atom	0.64
SOLID STATE		
Lattice type	–	face-centered cubic
Lattice constants	Angstrom units	
a		5.43 at 40°K
b		
c		
Atomic radius	Angstrom units	1.91
Melting point	°K	83.78
Heat of fusion	kcal/mole	0.281
Entropy of fusion	cal/°K/mole	
Triple point		
Temperature	°K	83.78
Pressure	torr	516
Density of solid	g/cm^3	1.623
Density of liquid	g/cm^3	1.418
LIQUID STATE		
Boiling point	°K	87.29
Density at boiling point	g/cm^3	1.3798
Heat of vaporization at boiling point	kcal/mole	1.558
Entropy at boiling point	cal/°K/mole	
Critical Data		
Temperature	°K	150.9
Pressure	atm.	48.3
Density	g/cm^3	0.536
GASEOUS STATE*		
Density at 0°C	g/1	1.78403
Specific heat at 20°C at const. pres.	cal/g/°K	0.125
Ratio of specific heats, Cp/Cv, at 20°C	–	1.648
Thermal conductivity at 0°C	cal/cm^2/cm/sec/°K	0.000039
Heat capacity at const. press. & 25°C	cal/°K/mole	4.97
Entropy at 25°C	cal/°K/mole	36.98
Heat of combustion	kcal/mole	
Dielectric constant at 25°C	–	1.0005085
Viscosity at 25°C	micropoise	226.4
Refractive index, n_D, at 0°C	–	1.000281
Solubility in H$_2$O at 1 atm. partial		
pressure and 0°C	cm^3(STP)/1000 g.	52.4
Sound velocity at 0°C	m/sec	307.8

*Generally at one atmosphere.

TABLE 2.1 (*Continued*)

ARSENIC

	Unit	Value(s)
ATOMIC, NUCLEAR AND CRYSTALLOGRAPHIC PROPERTIES		
Symbol: As		
Atomic number	–	33
Atomic weight	(based on ^{12}C)	74.9216
Periodic classification	–	V A
Atomic volume	cm^3/g-atom	13.09
Atomic radius	Angstrom units	1.16
Ionic radius	Angstrom units (As^{3+})	0.69
Thermal neutron absorption -cross section	barns/atom	4.3
Density	g./cm^3 (metallic)	5.73
Lattice type	–	Rhombohedral
Lattice constants	Angstrom units	
a		3.760
b		
c		10.548
Valence electrons	–	4s^2 4p^3
THERMAL AND THERMODYNAMIC PROPERTIES		
Melting point	°C	817 at 28 atm.
Boiling point	°C (sublimes)	613
Transition point(s)	°C (β to α)	288
Critical temperature	°K	
Linear thermal expansion coefficient	10^{-6} cm/cm/°C	4.7
Specific heat	cal/g/°C	0.082
Electronic specific heat	10^{-4} cal/mole/°C^2	0.24
Thermal conductivity	cal/cm^2/cm/sec/°C	
Latent heat of fusion	kcal/mole	6.62
Latent heat of vaporization at B.P.	kcal/mole	
Heat of combustion	kcal/mole	
Heat of sublimation	kcal/mole	29.0
Heat capacity at constant pressure	cal/°C/mole	5.895
Heat content at M.P.	kcal/mole	3.23
Entropy	cal/°C/mole	8.40
Longitudinal sound velocity	m/sec	
ELECTRICAL, ELECTRONIC, MAGNETIC AND OPTICAL PROPERTIES		
Electrical Conductivity	% IACS	4.9
Electrical resistivity	microohm-cm (polycryst)	33.3
Photoelectric work function	eV	5.2
Hall coefficient	volt-cm/amp-Oe	
Curie point	°K	
First ionization potential	eV (As$^+$)	9.81
Oxidation potential	V (As/As$_2$O$_3$)	−0.2340
Electrochemical equivalent	mg/coulomb (As^{3+})	0.15254
Magnetic susceptibility	c.g.s. units (polycryst)	+0.015 × 10^{-6} at 240°C
Refractive index, n_D		
MECHANICAL PROPERTIES		
Young's modulus of elasticity	10^{-6} kg/cm^2	0.398
Shear modulus	10^{-6} kg/cm^2	0.149
Bulk modulus	10^{-6} kg/cm^2	0.4018
Poisson's ratio	–	0.335
Hardness	Mohs	3.5

TABLE 2.1 (*Continued*)

ASTATINE

	Unit	Value(s)
ATOMIC, NUCLEAR AND CRYSTALLOGRAPHIC PROPERTIES		
Symbol: At		
Atomic number	–	85
Atomic weight	(based on ^{12}C)	210
Periodic classification	–	VII A
Atomic volume	cm^3/g-atom	
Atomic radius	Angstrom units	
Ionic radius	Angstrom units (At$^-$)	2.3
Thermal neutron absorption -cross section	barns/atom	
Density	g./cm^3	
Lattice type	–	
Lattice constants	Angstrom units	
a		
b		
c		
Valence electrons	–	$6s^2\,6p^5$
THERMAL AND THERMODYNAMIC PROPERTIES		
Melting point	°C	302 (estimate)
Boiling point	°C	377
Transition point(s)	°C	
Critical temperature	°K	
Linear thermal expansion coefficient	10^{-6} cm/cm/°C	
Specific heat	cal/g/°C	
Electronic specific heat	10^{-4} cal/mole/°C^2	
Thermal conductivity	cal/cm^2/cm/sec/°C	
Latent heat of fusion	kcal/mole	5.7
Latent heat of vaporization at B.P.	kcal/mole	21.6
Heat of combustion	kcal/mole	
Heat of sublimation	kcal/mole	
Heat capacity at constant pressure	cal/°C/mole	14.00
Heat content at M.P.	kcal/mole	2.83
Entropy	cal/°C/mole	29.00
Longitudinal sound velocity	m/sec	
ELECTRICAL, ELECTRONIC, MAGNETIC AND OPTICAL PROPERTIES		
Electrical Conductivity	% IACS	
Electrical resistivity	microohm-cm	
Photoelectric work function	eV	
Hall coefficient	volt-cm/amp-Oe	
Curie point	°K	
First ionization potential	eV (At$^+$)	9.5
Oxidation potential	V (At$^-$/At)	−0.3
Electrochemical equivalent	mg/coulomb	
Magnetic susceptibility	c.g.s. units	
Refractive index, n_D	–	
MECHANICAL PROPERTIES		
Young's modulus of elasticity	10^{-6} kg/cm^2	
Shear modulus	10^{-6} kg/cm^2	
Bulk modulus	10^{-6} kg/cm^2	
Poisson's ratio	–	
Hardness		

TABLE 2.1 (*Continued*)

BARIUM

	Unit	Value(s)
ATOMIC, NUCLEAR AND CRYSTALLOGRAPHIC PROPERTIES		
Symbol: Ba		
Atomic number	–	56
Atomic weight	(based on ^{12}C)	137.34
Periodic classification	–	II A
Atomic volume	cm^3/g-atom	39
Atomic radius	Angstrom units	2.10
Ionic radius	Angstrom units (Ba^{2+})	1.35
Thermal neutron absorption		
-cross section	barns/atom	1.3
Density	g./cm^3	3.6
Lattice type	–	body-centered cubic
Lattice constants	Angstrom units	
a		5.015
b		
c		
Valence electrons	–	$6s^2$
THERMAL AND THERMODYNAMIC PROPERTIES		
Melting point	°C	729
Boiling point	°C	1637
Transition point(s)	°C	370
Critical temperature	°K	3920
Linear thermal expansion coefficient	10^{-6} cm/cm/°C	18.1–21.0
Specific heat	cal/g/°C	0.068
Electronic specific heat	10^{-4} cal/mole/°C^2	6.5
Thermal conductivity	cal/cm^2/cm/sec/°C	
Latent heat of fusion	kcal/mole	1.83
Latent heat of vaporization at B.P.	kcal/mole	41.74
Heat of combustion	kcal/mole	
Heat of sublimation	kcal/mole	42.5
Heat capacity at constant pressure	cal/°C/mole	6.30
Heat content at M.P.	kcal/mole	4.537
Entropy	cal/°C/mole	15.50
Longitudinal sound velocity	m/sec (20°C)	1620.
ELECTRICAL, ELECTRONIC, MAGNETIC AND OPTICAL PROPERTIES		
Electrical Conductivity	% IACS	
Electrical resistivity	microohm-cm	50
Photoelectric work function	eV	2.5
Hall coefficient	volt-cm/amp-Oe	
Curie point	°K	
First ionization potential	eV (Ba^+)	5.212
Oxidation potential	V (Ba^{2+})	2.90
Electrochemical equivalent	mg/coulomb (Ba^{2+})	0.71171
Magnetic susceptibility	c.g.s. units	0.9×10^{-6}
Refractive index, n_D		
MECHANICAL PROPERTIES		
Young's modulus of elasticity	10^{-6} kg/cm^2	0.129
Shear modulus	10^{-6} kg/cm^2	0.050
Bulk modulus	10^{-6} kg/cm^2	0.105
Poisson's ratio	–	0.28
Hardness		

TABLE 2.1 *(Continued)*

BERKELIUM

	Unit	Value(s)
ATOMIC, NUCLEAR AND CRYSTALLOGRAPHIC PROPERTIES		
Symbol: Bk		
Atomic number	–	97
Atomic weight	(based on ^{12}C)	247
Periodic classification	–	Actinide
Atomic volume	cm^3/g-atom	
Atomic radius	Angstrom units	
Ionic radius	Angstrom units (Bk^{3+})	1.00
Thermal neutron absorption -cross section	barns/atom	
Density	g./cm^3	14.78
Lattice type	–	close-packed hexagonal
Lattice constants	Angstrom units	
a		3.416
b		
c		11.069
Valence electrons	–	5f^8 6d^1 7s^2
THERMAL AND THERMODYNAMIC PROPERTIES		
Melting point	°C	
Boiling point	°C	
Transition point(s)	°C	
Critical temperature	°K	
Linear thermal expansion coefficient	10^{-6} cm/cm/°C	
Specific heat	cal/g/°C	
Electronic specific heat	10^{-4} cal/mole/°C^2	
Thermal conductivity	cal/cm^2/cm/sec/°C	
Latent heat of fusion	kcal/mole	
Latent heat of vaporization at B.P.	kcal/mole	
Heat of combustion	kcal/mole	
Heat of sublimation	kcal/mole	
Heat capacity at constant pressure	cal/°C/mole	
Heat content at M.P.	kcal/mole	
Entropy	cal/°C/mole	
Longitudinal sound velocity	m/sec	
ELECTRICAL, ELECTRONIC, MAGNETIC AND OPTICAL PROPERTIES		
Electrical Conductivity	% IACS	
Electrical resistivity	microohm-cm	
Photoelectric work function	eV	
Hall coefficient	volt-cm/amp-Oe	
Curie point	°K	
First ionization potential	eV	
Oxidation potential	V	
Electrochemical equivalent	mg/coulomb	
Magnetic susceptibility	c.g.s. units	
Refractive index, n_D	–	
MECHANICAL PROPERTIES		
Young's modulus of elasticity	10^{-6} kg/cm^2	
Shear modulus	10^{-6} kg/cm^2	
Bulk modulus	10^{-6} kg/cm^2	
Poisson's ratio	–	
Hardness		

TABLE 2.1 (*Continued*)

BERYLLIUM

	Unit	Value(s)
ATOMIC, NUCLEAR AND CRYSTALLOGRAPHIC PROPERTIES		
Symbol: Be		
Atomic number	–	4
Atomic weight	(based on ^{12}C)	9.0122
Periodic classification	–	II A
Atomic volume	cm^3/g-atom	4.898
Atomic radius	Angstrom units	1.05
Ionic radius	Angstrom units (Be^{2+})	0.31
Thermal neutron absorption -cross section	barns/atom	0.009
Density	g./cm^3	1.85
Lattice type	–	hexagonal, closed packed
Lattice constants	Angstrom units	
a		2.286
b		
c		3.584
Valence electrons	–	$2s^2$
THERMAL AND THERMODYNAMIC PROPERTIES		
Melting point	°C	1283
Boiling point	°C	2970
Transition point(s)	°C	1260
Critical temperature	°K	6153
Linear thermal expansion coefficient	10^{-6} cm/cm/°C	11.5
Specific heat	cal/g/°C	0.45
Electronic specific heat	10^{-4} cal/mole/°C^2	0.533
Thermal conductivity	cal/cm^2/cm/sec/°C	0.440 at 0°C
Latent heat of fusion	kcal/mole	3.52
Latent heat of vaporization at B.P.	kcal/mole	70.4
Heat of combustion	kcal/mole	
Heat of sublimation	kcal/mole	77.9
Heat capacity at constant pressure	cal/°C/mole	3.93
Heat content at M.P.	kcal/mole	7.92
Entropy	cal/°C/mole	2.280
Longitudinal sound velocity	m/sec	12600
ELECTRICAL, ELECTRONIC, MAGNETIC AND OPTICAL PROPERTIES		
Electrical Conductivity	% IACS	40–45
Electrical resistivity	microohm-cm	4.0
Photoelectric work function	eV	3.67
Hall coefficient	volt-cm/amp-Oe	0.0024
Curie point	°K	
First ionization potential	eV (Be^+)	9.320
Oxidation potential	V (Be^{2+})	1.85
Electrochemical equivalent	mg/coulomb (Be^{2+})	0.04674
Magnetic susceptibility	c.g.s. units	-1.00×10^{-6}
Refractive index, n_D		
MECHANICAL PROPERTIES		
Young's modulus of elasticity	10^{-6} kg/cm^2	3.04
Shear modulus	10^{-6} kg/cm^2	1.46
Bulk modulus	10^{-6} kg/cm^2	1.023
Poisson's ratio	–	0.039
Hardness	Brinell (1000 kg)	90–120

TABLE 2.1 (*Continued*)

BISMUTH

	Unit	Value(s)
ATOMIC, NUCLEAR AND CRYSTALLOGRAPHIC PROPERTIES		
Symbol: Bi		
Atomic number	–	83
Atomic weight	(based on ^{12}C)	208.980
Periodic classification	–	V A
Atomic volume	cm^3/g-atom	21.3
Atomic radius	Angstrom units	1.46
Ionic radius	Angstrom units (Bi^{5+})	0.74
Thermal neutron absorption -cross section	barns/atom	0.034
Density	g./cm^3	9.78
Lattice type	–	Rhombohedral
Lattice constants	Angstrom units	
a		4.7457
Axial angle		57° 14.2′
c		
Valence electrons	–	$6s^2\ 6p^3$
THERMAL AND THERMODYNAMIC PROPERTIES		
Melting point	°C	271.3
Boiling point	°C	1551
Transition point(s)	°C	
Critical temperature	°K	4620
Linear thermal expansion coefficient	10^{-6} cm/cm/°C	13.3
Specific heat	cal/g/°C	0.0294
Electronic specific heat	10^{-4} cal/mole/°C^2	0.049, 0.029
Thermal conductivity	cal/cm^2/cm/sec/°C	0.020
Latent heat of fusion	kcal/mole	2.60
Latent heat of vaporization at B.P.	kcal/mole	42.8
Heat of combustion	kcal/mole	
Heat of sublimation	kcal/mole	49.95
Heat capacity at constant pressure	cal/°C/mole	6.20
Heat content at M.P.	kcal/mole	1.660 (cryst.)
Entropy	cal/°C/mole	13.56
Longitudinal sound velocity	m/sec (20°C)	1790
ELECTRICAL, ELECTRONIC, MAGNETIC AND OPTICAL PROPERTIES		
Electrical Conductivity	% IACS	
Electrical resistivity	microohm-cm	106.8 at 0°C
Photoelectric work function	eV	4.4
Hall coefficient	volt-cm/amp-Oe	
Curie point	°K	
First ionization potential	eV (Bi^+)	7.289
Oxidation potential	V (Bi^{3+})	−0.22
Electrochemical equivalent	mg/coulomb (Bi^{5+})	0.43316
Magnetic susceptibility	c.g.s. units	-1.35×10^{-6}
Refractive index, n_D		
MECHANICAL PROPERTIES		
Young's modulus of elasticity	10^{-6} kg/cm^2	0.348
Shear modulus	10^{-6} kg/cm^2	0.131
Bulk modulus	10^{-6} kg/cm^2	0.321
Poisson's ratio	–	0.33
Hardness	Brinell (100 kg)	7

TABLE 2.1 (*Continued*)

BORON

	Unit	Value(s)
ATOMIC, NUCLEAR AND CRYSTALLOGRAPHIC PROPERTIES		
Symbol: B		
Atomic number	–	5
Atomic weight	(based on ^{12}C)	10.811
Periodic classification	–	III A
Atomic volume	cm^3/g-atom	4.70
Atomic radius	Angstrom units	0.97
Ionic radius	Angstrom units (B^{3+})	0.20
Thermal neutron absorption		
-cross section	barns/atom	760
Density	g./cm^3	2.34
Lattice type	–	hexagonal, closed packed
Lattice constants	Angstrom units	
a		11.98
b		
c		9.54
Valence electrons	–	$2s^2\,2p^1$
THERMAL AND THERMODYNAMIC PROPERTIES		
Melting point	°C	2027
Boiling point	°C	3802
Transition point(s)	°C	1500
Critical temperature	°K	
Linear thermal expansion coefficient	10^{-6} cm/cm/°C	8.3
Specific heat	cal/g/°C	0.309
Electronic specific heat	10^{-4} cal/mole/°C^2	3.01
Thermal conductivity	cal/cm^2/cm/sec/°C	
Latent heat of fusion	kcal/mole	5.72
Latent heat of vaporization at B.P.	kcal/mole	128.8
Heat of combustion	kcal/mole	
Heat of sublimation	kcal/mole	132.0
Heat capacity at constant pressure	cal/°C/mole	2.64
Heat content at M.P.	kcal/mole	12.08
Entropy	cal/°C/mole	1.40
Longitudinal sound velocity	m/sec	16200
ELECTRICAL, ELECTRONIC, MAGNETIC AND OPTICAL PROPERTIES		
Electrical Conductivity	% IACS	
Electrical resistivity	microohm-cm	2×10^{13}
Photoelectric work function	eV	4.5
Hall coefficient	volt-cm/amp-Oe	
Curie point	°K	
First ionization potential	eV (B^+)	8.298
Oxidation potential	V (B/H_3BO_3)	0.87
Electrochemical equivalent	mg/coulomb (B^{3+})	0.03737
Magnetic susceptibility	c.g.s. units	-0.69×10^{-6}
Refractive index, n_D		
MECHANICAL PROPERTIES		
Young's modulus of elasticity	10^{-6} kg/cm^2	4.50
Shear modulus	10^{-6} kg/cm^2	2.07
Bulk modulus	10^{-6} kg/cm^2	1.82
Poisson's ratio	–	0.089
Hardness	Mohs	9.3

TABLE 2.1 (*Continued*)

BROMINE

	Unit	Value(s)
ATOMIC AND NUCLEAR PROPERTIES		
Symbol: Br		
Atomic number	–	35
Atomic weight	(based on ^{12}C)	79.909
Periodic classification	–	VII A
Thermal neutron absorption		
-cross section	barns/atom	6.8
SOLID STATE		
Lattice type	–	Orthorhombic
Lattice constants	Angstrom units	
a		4.49 at 120°K
b		6.68 at 120°K
c		8.74 at 120°K
Atomic radius	Angstrom units	1.19
Melting point	°K	265.95
Heat of fusion	kcal/mole	2.520
Entropy of fusion	cal/°K/mole	
Triple point		
Temperature	°K	
Pressure	torr	
Density of solid	g/cm^3	
Density of liquid	g/cm^3	
LIQUID STATE		
Boiling point	°K	331.4
Density at boiling point	g/cm^3	
Heat of vaporization at boiling point	kcal/mole	7.170
Entropy at boiling point	cal/°K/mole	21.24
Critical Data		
Temperature	°K	584
Pressure	atm.	102
Density	g/cm^3	1.26
Sound volocity	m/sec	
Electrical resistivity	Microohm-cm	6.5×10^{16}
Photoelectric work function	eV	
Dielectric constant, 23°C, 10^8 Hz	–	3.2
Curie point	°K	
First ionization potential	V (Br$^+$)	11.814
Oxidation potential	V (Br$^-$/Br$_2$)	−1.07
Electrochemical equivalent	mg/coulomb (Br$^-$)	0.82815
Magnetic susceptibility	c.g.s. units	-0.38×10^{-6}
Refractive index, n_D, at 25°C	–	1.6475
Viscosity	centipoise	0.99
Solubility in H$_2$O at 25°C	g/100 g solution	3.35

TABLE 2.1 (*Continued*)

CADMIUM

	Unit	Value(s)
ATOMIC, NUCLEAR AND CRYSTALLOGRAPHIC PROPERTIES		
Symbol: Cd		
Atomic number	−	48
Atomic weight	(based on ^{12}C)	112.40
Periodic classification	−	II B
Atomic volume	cm^3/g-atom	13.0
Atomic radius	Angstrom units	1.49
Ionic radius	Angstrom units (Cd^{2+})	0.97
Thermal neutron absorption -cross section	barns/atom	2460
Density	g./cm^3	8.65
Lattice type	−	hexagonal, closed packed
Lattice constants	Angstrom units	
a		2.9787
b		
c		5.617
Valence electrons	−	5s^2
THERMAL AND THERMODYNAMIC PROPERTIES		
Melting point	°C	320.9
Boiling point	°C	767
Transition point(s)	°C	
Critical temperature	°K	1903
Linear thermal expansion coefficient	10^{-6} cm/cm/°C	29.8
Specific heat	cal/g/°C	0.055
Electronic specific heat	10^{-4} cal/mole/°C^2	1.61
Thermal conductivity	cal/cm^2/cm/sec/°C	0.222
Latent heat of fusion	kcal/mole	1.48
Latent heat of vaporization at B.P.	kcal/mole	23.87
Heat of combustion	kcal/mole	
Heat of sublimation	kcal/mole	26.8
Heat capacity at constant pressure	cal/°C/mole	6.215
Heat content at M.P.	kcal/mole	6.29
Entropy	cal/°C/mole	12.37
Longitudinal sound velocity	m/sec	2500
ELECTRICAL, ELECTRONIC, MAGNETIC AND OPTICAL PROPERTIES		
Electrical Conductivity	% IACS	25.2
Electrical resistivity	microohm-cm	6.83 at 0°C
Photoelectric work function	eV	4.07
Hall coefficient	volt-cm/amp-Oe	
Curie point	°K	
First ionization potential	eV (Cd$^+$)	8.993
Oxidation potential	V (Cd^{2+})	−0.40
Electrochemical equivalent	mg/coulomb (Cd^{2+})	0.5824
Magnetic susceptibility	c.g.s. units	−0.18 × 10^{-6}
Refractive index, n_D		1.13
MECHANICAL PROPERTIES		
Young's modulus of elasticity	10^{-6} kg/cm^2	0.635
Shear modulus	10^{-6} kg/cm^2	0.246
Bulk modulus	10^{-6} kg/cm^2	0.4766
Poisson's ratio	−	0.30
Hardness	Brinell	21

TABLE 2.1 (*Continued*)

CALCIUM

	Unit	Value(s)
ATOMIC, NUCLEAR AND CRYSTALLOGRAPHIC PROPERTIES		
Symbol: Ca		
Atomic number	–	20
Atomic weight	(based on ^{12}C)	40.08
Periodic classification	–	II A
Atomic volume	cm^3/g-atom	25.9
Atomic radius	Angstrom units	1.97
Ionic radius	Angstrom units	1.66
Thermal neutron absorption -cross section	barns/atom	0.43
Density	g./cm^3	1.54
Lattice type	–	face-centered cubic
Lattice constants	Angstrom units	
a		5.56
b		
c		
Valence electrons	–	$4s^2$
THERMAL AND THERMODYNAMIC PROPERTIES		
Melting point	°C	839
Boiling point	°C	1492
Transition point(s)	°C	464
Critical temperature	°K	3267
Linear thermal expansion coefficient	10^{-6} cm/cm/°C	22.3
Specific heat	cal/g/°C	0.149
Electronic specific heat	10^{-4} cal/mole/°C^2	6.93
Thermal conductivity	cal/cm^2/cm/sec/°C	0.3
Latent heat of fusion	kcal/mole	2.07
Latent heat of vaporization at B.P.	kcal/mole	36.7
Heat of combustion	kcal/mole	6.08
Heat of sublimation	kcal/mole	42.2
Heat capacity at constant pressure	cal/°C/mole	6.29
Heat content at M.P.	kcal/mole	6.59
Entropy	cal/°C/mole	9.95
Longitudinal sound velocity	m/sec	
ELECTRICAL, ELECTRONIC, MAGNETIC AND OPTICAL PROPERTIES		
Electrical Conductivity	% IACS	48.7
Electrical resistivity	microohm-cm	4.6
Photoelectric work function	eV	2.76
Hall coefficient	volt-cm/amp-Oe	
Curie point	°K	
First ionization potential	eV (Ca$^+$)	6.113
Oxidation potential	V (Ca^{2+})	2.87
Electrochemical equivalent	mg/coulomb (Ca^{2+})	0.20762
Magnetic susceptibility	c.g.s. units	1.39×10^{-6}
Refractive index, n_D		
MECHANICAL PROPERTIES		
Young's modulus of elasticity	10^{-6} kg/cm^2	0.200
Shear modulus	10^{-6} kg/cm^2	0.075
Bulk modulus	10^{-6} kg/cm^2	0.155
Poisson's ratio	–	0.31
Hardness	Brinell (500 kg)	17

TABLE 2.1 *(Continued)*

CALIFORNIUM

	Unit	Value(s)
ATOMIC, NUCLEAR AND CRYSTALLOGRAPHIC PROPERTIES Symbol: Cf		
Atomic number	–	98
Atomic weight	(based on ^{12}C)	251
Periodic classification	–	Actinide
Atomic volume	cm^3/g-atom	
Atomic radius	Angstrom units	
Ionic radius	Angstrom units (Cf^{3+})	0.99
Thermal neutron absorption -cross section	barns/atom	
Density	g./cm^3	
Lattice type	–	cubic
Lattice constants	Angstrom units	
a		
b		
c		
Valence electrons	–	$5f^{10}\ 7s^2$
THERMAL AND THERMODYNAMIC PROPERTIES		
Melting point	°C	
Boiling point	°C	
Transition point(s)	°C	
Critical temperature	°K	
Linear thermal expansion coefficient	10^{-6} cm/cm/°C	
Specific heat	cal/g/°C	
Electronic specific heat	10^{-4} cal/mole/°C^2	
Thermal conductivity	cal/cm^2/cm/sec/°C	
Latent heat of fusion	kcal/mole	
Latent heat of vaporization at B.P.	kcal/mole	
Heat of combustion	kcal/mole	
Heat of sublimation	kcal/mole	
Heat capacity at constant pressure	cal/°C/mole	
Heat content at M.P.	kcal/mole	
Entropy	cal/°C/mole	
Longitudinal sound velocity	m/sec	
ELECTRICAL, ELECTRONIC, MAGNETIC AND OPTICAL PROPERTIES		
Electrical Conductivity	% IACS	
Electrical resistivity	microohm-cm	
Photoelectric work function	eV	
Hall coefficient	volt-cm/amp-Oe	
Curie point	°K	
First ionization potential	eV	
Oxidation potential	V (Cf^{3+})	2.28
Electrochemical equivalent	mg/coulomb	
Magnetic susceptibility	c.g.s. units	
Refractive index, n_D	–	
MECHANICAL PROPERTIES		
Young's modulus of elasticity	10^{-6} kg/cm^2	
Shear modulus	10^{-6} kg/cm^2	
Bulk modulus	10^{-6} kg/cm^2	
Poisson's ratio	–	
Hardness		

TABLE 2.1 (Continued)

CARBON (DIAMOND)

	Unit	Value(s)
ATOMIC, NUCLEAR AND CRYSTALLOGRAPHIC PROPERTIES		
Symbol: C		
Atomic number	–	6
Atomic weight	(based on ^{12}C)	12.011
Periodic classification	–	IV A
Atomic volume	cm^3/g-atom	3.97
Atomic radius	Angstrom units	
Ionic radius	Angstrom units	
Thermal neutron absorption		
-cross section	barns/atom	0.0034
Density	g./cm^3	3.51
Lattice type	–	cubic
Lattice constants	Angstrom units	
a		3.56679
b		
c		
Valence electrons	–	$2s^2\ 2p^2$
THERMAL AND THERMODYNAMIC PROPERTIES		
Melting point	°C	3827
Boiling point	°C	4200
Transition point(s)	°C	1500 (to graphite)
Critical temperature	°K	
Linear thermal expansion coefficient	10^{-6} cm/cm/°C	1.19
Specific heat	cal/g/°C	0.125
Electronic specific heat	10^{-4} cal/mole/°C^2	~0
Thermal conductivity	cal/cm^2/cm/sec/°C	1.27
Latent heat of fusion	kcal/mole	
Latent heat of vaporization at B.P.	kcal/mole	
Heat of combustion	kcal/mole	–94.5
Heat of sublimation	kcal/mole	170.9
Heat capacity at constant pressure	cal/°C/mole	1.462
Heat content at M.P.	kcal/mole	
Entropy	cal/°C/mole	0.568
Longitudinal sound velocity	m/sec	18500
ELECTRICAL, ELECTRONIC, MAGNETIC AND OPTICAL PROPERTIES		
Electrical Conductivity	% IACS	
Electrical resistivity	microohm-cm	10^{19}
Photoelectric work function	eV	4.6
Hall coefficient	volt-cm/amp-Oe	
Curie point	°K	
First ionization potential	eV (C^+)	11.260
Oxidation potential	V (C/CO_2)	–0.394 (acid soln.)
Electrochemical equivalent	mg/coulomb (C^{4+})	0.031113
Magnetic susceptibility	c.g.s. units	-0.456×10^{-6}
Refractive index, n_D		2.4173
MECHANICAL PROPERTIES		
Young's modulus of elasticity	10^{-6} kg/cm^2	11.5
Shear modulus	10^{-6} kg/cm^2	4.6
Bulk modulus	10^{-6} kg/cm^2	5.56
Poisson's ratio	–	0.18 ± 8
Hardness	Mohs	10

TABLE 2.1 (*Continued*)

CARBON (NATURAL GRAPHITE)

	Unit	Value(s)
ATOMIC, NUCLEAR AND CRYSTALLOGRAPHIC PROPERTIES		
Symbol: C		
Atomic number	−	6
Atomic weight	(based on ^{12}C)	12.011
Periodic classification	−	IV A
Atomic volume	cm^3/g-atom	5.260
Atomic radius	Angstrom units	
Ionic radius	Angstrom units	
Thermal neutron absorption -cross section	barns/atom	0.0034
Density	g./cm^3	2.26
Lattice type	−	hexagonal
Lattice constants	Angstrom units	
a		2.4614
b		
c		6.7041
Valence electrons	−	$2s^2\ 2p^2$
THERMAL AND THERMODYNAMIC PROPERTIES		
Melting point	°C	3620 (sublimation)
Boiling point	°C	4200
Transition point(s)	°C	
Critical temperature	°K	
Linear thermal expansion coefficient	10^{-6} cm/cm/°C	1.19
Specific heat	cal/g/°C	0.172
Electronic specific heat	10^{-4} cal/mole/°C^2	0.033
Thermal conductivity	cal/cm^2/cm/sec/°C	0.02–0.6
Latent heat of fusion	kcal/mole	25 at 48 kilobars
Latent heat of vaporization at B.P.	kcal/mole	170
Heat of combustion	kcal/mole	
Heat of sublimation	kcal/mole	170.9
Heat capacity at constant pressure	cal/°C/mole	2.06
Heat content at M.P.	kcal/mole	171.299
Entropy	cal/°C/mole	1.37
Longitudinal sound velocity	m/sec (20°C)	1470
ELECTRICAL, ELECTRONIC, MAGNETIC AND OPTICAL PROPERTIES		
Electrical Conductivity	% IACS	
Electrical resistivity	microohm-cm	1375 at 0°C
Photoelectric work function	eV	4.6
Hall coefficient	volt-cm/amp-Oe	
Curie point	°K	
First ionization potential	eV (C$^+$)	11.260
Oxidation potential	V (C/CO$_2$)	−0.394 (acid soln.)
Electrochemical equivalent	mg/coulomb (C^{4+})	0.031113
Magnetic susceptibility	c.g.s. units	-21.5×10^{-6} (∥)
Refractive index, n_D		1.93–2.07
MECHANICAL PROPERTIES		
Young's modulus of elasticity	10^{-6} kg/cm^2	0.0849
Shear modulus	10^{-6} kg/cm^2	0.0332
Bulk modulus	10^{-6} kg/cm^2	0.3447
Poisson's ratio	−	0.27 ± 0.06
Hardness	Mohs	0.5–2.0

TABLE 2.1 (Continued)

CERIUM

	Unit	Value(s)
ATOMIC, NUCLEAR AND CRYSTALLOGRAPHIC PROPERTIES		
Symbol: Ce		
Atomic number	–	58
Atomic weight	(based on ^{12}C)	140.12
Periodic classification	–	Lanthanide
Atomic volume	cm^3/g-atom	20.690
Atomic radius	Angstrom units	1.824
Ionic radius	Angstrom units (Ce^{4+})	1.01
Thermal neutron absorption -cross section	barns/atom	0.6
Density	g./cm^3	6.773
Lattice type, α	–	face-centered cubic
Lattice constants	Angstrom units	
a		5.1601
b		
c		
Valence electrons	–	$4f^2\ 6s^2$
THERMAL AND THERMODYNAMIC PROPERTIES		
Melting point	°C	798
Boiling point	°C	3257
Transition point(s)	°C	-157, -94, -23, 168, 726
Critical temperature	°K	10400
Linear thermal expansion coefficient	10^{-6} cm/cm/°C	8.5
Specific heat	cal/g/°C	0.045
Electronic specific heat	10^{-4} cal/mole/°C^2	17.3
Thermal conductivity	cal/cm^2/cm/sec/°C	0.026
Latent heat of fusion	kcal/mole	1.238
Latent heat of vaporization at B.P.	kcal/mole	75.0
Heat of combustion	kcal/mole	
Heat of sublimation	kcal/mole	111.6
Heat capacity at constant pressure	cal/°C/mole	6.48
Heat content at M.P.	kcal/mole	5.822
Entropy	cal/°C/mole	16.64
Longitudinal sound velocity	m/sec	2300
ELECTRICAL, ELECTRONIC, MAGNETIC AND OPTICAL PROPERTIES		
Electrical Conductivity	% IACS	
Electrical resistivity	microohm-cm	75.3
Photoelectric work function	eV	2.7
Hall coefficient	volt-cm/amp-Oe	1.81×10^{-12}
Curie point	°K	
First ionization potential	eV (Ce^+)	5.65
Oxidation potential	V (Ce^{3+})	2.48
Electrochemical equivalent	mg/coulomb (Ce^{3+})	0.48404
Magnetic susceptibility	c.g.s. units	700×10^{-6}
Refractive index, n_D		
MECHANICAL PROPERTIES		
Young's modulus of elasticity	10^{-6} kg/cm^2	0.306
Shear modulus	10^{-6} kg/cm^2	0.122
Bulk modulus	10^{-6} kg/cm^2	0.244
Poisson's ratio	–	0.248
Hardness	Vickers, kg/mm^2	24

TABLE 2.1 (*Continued*)

CESIUM

	Unit	Value(s)
ATOMIC, NUCLEAR AND CRYSTALLOGRAPHIC PROPERTIES		
Symbol: Cs		
Atomic number	–	55
Atomic weight	(based on ^{12}C)	132.905
Periodic classification	–	I A
Atomic volume	cm^3/g-atom	69.95
Atomic radius	Angstrom units	2.62
Ionic radius	Angstrom units	1.69
Thermal neutron absorption		
-cross section	barns/atom	29.9
Density	g./cm^3	1.90 at 20°C
Lattice type	–	body-centered cubic
Lattice constants	Angstrom units	
a		6.05 at –173°C
b		
c		
Valence electrons	–	$6s^1$
THERMAL AND THERMODYNAMIC PROPERTIES		
Melting point	°C	28.5
Boiling point	°C	666
Transition point(s)	°C	
Critical temperature	°K	2073
Linear thermal expansion coefficient	10^{-6} cm/cm/°C	97
Specific heat	cal/g/°C	0.052
Electronic specific heat	10^{-4} cal/mole/°C^2	8.48
Thermal conductivity	cal/cm^2/cm/sec/°C	0.044 at 28.5°C
Latent heat of fusion	kcal/mole	0.506
Latent heat of vaporization at B.P.	kcal/mole	15.75
Heat of combustion	kcal/mole	
Heat of sublimation	kcal/mole	18.84
Heat capacity at constant pressure	cal/°C/mole	7.65
Heat content at M.P.	kcal/mole	5.084
Entropy	cal/°C/mole	20.16
Longitudinal sound velocity	m/sec	
ELECTRICAL, ELECTRONIC, MAGNETIC AND OPTICAL PROPERTIES		
Electrical Conductivity	% IACS	
Electrical resistivity	microohm-cm	20
Photoelectric work function	eV	1.81
Hall coefficient	volt-cm/amp-Oe	
Curie point	°K	
First ionization potential	eV (Cs^+)	3.894
Oxidation potential	V (Cs^+)	3.02
Electrochemical equivalent	mg/coulomb (Cs^+)	1.37731
Magnetic susceptibility	c.g.s. units	-0.10×10^{-6}
Refractive index, n_D		
MECHANICAL PROPERTIES		
Young's modulus of elasticity	10^{-6} kg/cm^2	0.0179
Shear modulus	10^{-6} kg/cm^2	0.0066
Bulk modulus	10^{-6} kg/cm^2	
Poisson's ratio	–	0.356
Hardness	Mohs	0.2

TABLE 2.1 (*Continued*)

CHLORINE

	Unit	Value(s)
ATOMIC AND NUCLEAR PROPERTIES		
Symbol: Cl		
Atomic number	–	17
Atomic weight	(based on ^{12}C)	35.453
Periodic classification	–	VII A
Thermal neutron absorption		
-cross section	barns/atom	33.3
SOLID STATE		
Lattice type	–	Tetragonal
Lattice constants (at 88°K)	Angstrom units	
a		8.56
b		
c		6.12
Atomic radius	Angstrom units	
Melting point	°K	172.16
Heat of fusion	kcal/mole	1.531
Entropy of fusion	cal/°K/mole	
Triple point		
Temperature	°K	
Pressure	torr	
Density of solid	g/cm^3	
Density of liquid	g/cm^3	
LIQUID STATE		
Boiling point	°K	239.10
Density at boiling point	g/cm^3	1.507
Heat of vaporization at boiling point	kcal/mole	4.878
Entropy at boiling point	cal/°K/mole	20.40
Critical Data		
Temperature	°K	417
Pressure	atm.	76
Density	g/cm^3	0.573
GASEOUS STATE*		
Density at 0°C and 1 atm.	g/l	3.214
Specific heat at 15°C at const. pres.	cal/g/°K	0.115
Ratio of specific heats, Cp/Cv, at 20°C	–	1.325
Thermal conductivity at 0°C	cal/cm^2/cm/sec/°K	0.0000193
Heat capacity at const. press. & –73°C	cal/°K/mole	7.576
Entropy at 25°C	cal/°K/mole	53.29
Heat of combustion	kcal/mole	
Dielectric constant at 20°C	–	1.00152
Viscosity at 20°C	micropoise	14.7
Refractive index, n_D, at 0°C	–	1.000773
Solubility in H$_2$O at 1 atm. partial		
pressure and 20°C	cm^3(STP)/1000 g.	
Sound velocity at 0°C	m/sec	206

*Generally at one atmosphere.

TABLE 2.1 (*Continued*)

CHROMIUM

	Unit	Value(s)
ATOMIC, NUCLEAR AND CRYSTALLOGRAPHIC PROPERTIES		
Symbol: Cr		
Atomic number	–	24
Atomic weight	(based on ^{12}C)	51.996
Periodic classification	–	VI B
Atomic volume	cm^3/g-atom	7.231
Atomic radius	Angstrom units	1.25
Ionic radius	Angstrom units (Cr^{6+})	0.52
Thermal neutron absorption		
-cross section	barns/atom	3.1
Density	g./cm^3	7.19
Lattice type	–	body-centered cubic
Lattice constants	Angstrom units	
a		2.8846
b		
c		
Valence electrons	–	$3d^5\ 4s^1$
THERMAL AND THERMODYNAMIC PROPERTIES		
Melting point	°C	1875
Boiling point	°C	2482
Transition point(s)	°C	
Critical temperature	°K	
Linear thermal expansion coefficient	10^{-6} cm/cm/°C	6.2
Specific heat	cal/g/°C	5.57
Electronic specific heat	10^{-4} cal/mole/°C^2	3.49
Thermal conductivity	cal/cm^2/cm/sec/°C	0.16
Latent heat of fusion	kcal/mole	3.47
Latent heat of vaporization at B.P.	kcal/mole	76.635
Heat of combustion	kcal/mole	
Heat of sublimation	kcal/mole	95.0
Heat capacity at constant pressure	cal/°C/mole	5.57
Heat content at M.P.	kcal/mole	14.85
Entropy	cal/°C/mole	5.70
Longitudinal sound velocity	m/sec (20°C)	5940
ELECTRICAL, ELECTRONIC, MAGNETIC AND OPTICAL PROPERTIES		
Electrical Conductivity	% IACS	8.8
Electrical resistivity	microohm-cm	12.9
Photoelectric work function	eV	4.58
Hall coefficient	volt-cm/amp-Oe	
Curie point	°K	
First ionization potential	eV (Cr^+)	6.766
Oxidation potential	V (Cr^{3+})	0.71
Electrochemical equivalent	mg/coulomb (Cr^{3+})	0.17965
Magnetic susceptibility	c.g.s. units	3.6×10^{-6}
Refractive index, n_D		2.97 at 0.579 μ
MECHANICAL PROPERTIES		
Young's modulus of elasticity	10^{-6} kg/cm^2	2.48
Shear modulus	10^{-6} kg/cm^2	1.19
Bulk modulus	10^{-6} kg/cm^2	1.940
Poisson's ratio	–	0.209
Hardness	Brinell (cast)	50–150

TABLE 2.1 (*Continued*)

COBALT (β)

	Unit	Value(s)
ATOMIC, NUCLEAR AND CRYSTALLOGRAPHIC PROPERTIES		
Symbol: Co		
Atomic number	—	27
Atomic weight	(based on ^{12}C)	58.9332
Periodic classification	—	VIII
Atomic volume	cm^3/g-atom	6.689
Atomic radius	Angstrom units	1.26
Ionic radius	Angstrom units (Co^{3+})	0.65
Thermal neutron absorption		
-cross section	barns/atom	37.4
Density	g./cm^3	8.90
Lattice type	—	hexagonal, close-packed
Lattice constants	Angstrom units	
a		2.5017
b		
c		4.0614
Valence electrons	—	3d^7 4s^2
THERMAL AND THERMODYNAMIC PROPERTIES		
Melting point	°C	1493
Boiling point	°C	2956
Transition point(s)	°C	417 (β to α)
Critical temperature	°K	
Linear thermal expansion coefficient	10^{-6} cm/cm/°C	13.36
Specific heat	cal/g/°C	0.1056
Electronic specific heat	10^{-4} cal/mole/°C^2	11.3
Thermal conductivity	cal/cm^2/cm/sec/°C	0.165
Latent heat of fusion	kcal/mole	3.70
Latent heat of vaporization at B.P.	kcal/mole	91.4
Heat of combustion	kcal/mole	
Heat of sublimation	kcal/mole	102.1
Heat capacity at constant pressure	cal/°C/mole	5.95
Heat content at M.P.	kcal/mole	12.4
Entropy	cal/°C/mole	7.18
Longitudinal sound velocity	m/sec	4724
ELECTRICAL, ELECTRONIC, MAGNETIC AND OPTICAL PROPERTIES		
Electrical Conductivity	% IACS	27.6
Electrical resistivity	microohm-cm	5.68 at 0°C
Photoelectric work function	eV	4.40
Hall coefficient	volt-cm/amp-Oe	
Curie point	°K	
First ionization potential	eV (Co$^+$)	7.86
Oxidation potential	V (Co^{2+})	0.277 (acid soln.)
Electrochemical equivalent	mg/coulomb (Co^{3+})	0.3054
Magnetic susceptibility	c.g.s. units	ferromagnetic
Refractive index, n_D		1.93 at 0.50 μ
MECHANICAL PROPERTIES		
Young's modulus of elasticity	10^{-6} kg/cm^2	2.10
Shear modulus	10^{-6} kg/cm^2	0.779
Bulk modulus	10^{-6} kg/cm^2	1.952
Poisson's ratio	—	0.334
Hardness, Cast	Brinell	124

TABLE 2.1 (*Continued*)

COPPER

	Unit	Value(s)
ATOMIC, NUCLEAR AND CRYSTALLOGRAPHIC PROPERTIES		
Symbol: Cu		
Atomic number	–	29
Atomic weight	(based on ^{12}C)	63.54
Periodic classification	–	I B
Atomic volume	cm^3/g-atom	7.09
Atomic radius	Angstrom units	1.276
Ionic radius	Angstrom units (Cu^{2+})	0.70
Thermal neutron absorption -cross section	barns/atom	3.8
Density	g./cm^3	8.94
Lattice type	–	face-centered cubic
Lattice constants	Angstrom units	
a		3.615
b		
c		
Valence electrons	–	$3d^{10}\, 4s^1$
THERMAL AND THERMODYNAMIC PROPERTIES		
Melting point	°C	1083
Boiling point	°C	2582
Transition point(s)	°C	
Critical temperature	°K	8280
Linear thermal expansion coefficient	10^{-6} cm/cm/°C	16.5
Specific heat	cal/g/°C	0.092
Electronic specific heat	10^{-4} cal/mole/°C^2	1.66
Thermal conductivity	cal/cm^2/cm/sec/°C	0.934
Latent heat of fusion	kcal/mole	3.110
Latent heat of vaporization at B.P.	kcal/mole	72.8
Heat of combustion	kcal/mole	
Heat of sublimation	kcal/mole	81.1
Heat capacity at constant pressure	cal/°C/mole	5.855
Heat content at M.P.	kcal/mole	20.98
Entropy	cal/°C/mole	7.97
Longitudinal sound velocity	m/sec	3500
ELECTRICAL, ELECTRONIC, MAGNETIC AND OPTICAL PROPERTIES		
Electrical Conductivity	% IACS	100
Electrical resistivity	microohm-cm	1.647
Photoelectric work function	eV	4.6
Hall coefficient	volt-cm/amp-Oe	
Curie point	°K	
First ionization potential	eV (Cu^+)	7.726
Oxidation potential	V (Cu^{2+})	–0.337
Electrochemical equivalent	mg/coulomb (Cu^{2+})	0.3294
Magnetic susceptibility	c.g.s. units	-0.086×10^{-6} at 18°C
Refractive index, n_D		1.10 at 0.50 μ
MECHANICAL PROPERTIES		
Young's modulus of elasticity	10^{-6} kg/cm^2	1.26
Shear modulus	10^{-6} kg/cm^2	0.460
Bulk modulus	10^{-6} kg/cm^2	1.335
Poisson's ratio	–	0.345
Hardness	Brinell	43 (annealed)

TABLE 2.1 (*Continued*)

CURIUM

	Unit	Value(s)
ATOMIC, NUCLEAR AND CRYSTALLOGRAPHIC PROPERTIES		
Symbol: Cm		
Atomic number	–	96
Atomic weight	(based on ^{12}C)	247
Periodic classification	–	Actinide
Atomic volume	cm^3/g-atom	18.07
Atomic radius	Angstrom units	
Ionic radius	Angstrom units (Cm^{3+})	1.01
Thermal neutron absorption		
-cross section	barns/atom	
Density	g./cm^3	13.51
Lattice type	–	close-packed hexagonal
Lattice constants	Angstrom units	
a		3.496
b		
c		11.331
Valence electrons	–	$5f^7\ 6d^1\ 7s^2$
THERMAL AND THERMODYNAMIC PROPERTIES		
Melting point	°C	1340
Boiling point	°C	
Transition point(s)	°C	
Critical temperature	°K	
Linear thermal expansion coefficient	10^{-6} cm/cm/°C	
Specific heat	cal/g/°C	
Electronic specific heat	10^{-4} cal/mole/°C^2	
Thermal conductivity	cal/cm^2/cm/sec/°C	
Latent heat of fusion	kcal/mole	
Latent heat of vaporization at B.P.	kcal/mole	
Heat of combustion	kcal/mole	
Heat of sublimation	kcal/mole	
Heat capacity at constant pressure	cal/°C/mole	
Heat content at M.P.	kcal/mole	
Entropy	cal/°C/mole	
Longitudinal sound velocity	m/sec	
ELECTRICAL, ELECTRONIC, MAGNETIC AND OPTICAL PROPERTIES		
Electrical Conductivity	% IACS	
Electrical resistivity	microohm-cm	
Photoelectric work function	eV	
Hall coefficient	volt-cm/amp-Oe	
Curie point	°K	
First ionization potential	eV	
Oxidation potential	V (Cm^{3+})	2.29
Electrochemical equivalent	mg/coulomb	
Magnetic susceptibility	c.g.s. units	
Refractive index, n_D	–	
MECHANICAL PROPERTIES		
Young's modulus of elasticity	10^{-6} kg/cm^2	
Shear modulus	10^{-6} kg/cm^2	
Bulk modulus	10^{-6} kg/cm^2	
Poisson's ratio	–	
Hardness		

TABLE 2.1 *(Continued)*

DYSPROSIUM

	Unit	Value(s)
ATOMIC, NUCLEAR AND CRYSTALLOGRAPHIC PROPERTIES		
Symbol: Dy		
Atomic number	—	66
Atomic weight	(based on ^{12}C)	162.50
Periodic classification	—	Lanthanide
Atomic volume	cm^3/g-atom	19.032
Atomic radius	Angstrom units	1.773
Ionic radius	Angstrom units (Dy^{3+})	0.908
Thermal neutron absorption -cross section	barns/atom	930
Density	g./cm^3	8.559
Lattice type	—	hexagonal, close-packed
Lattice constants	Angstrom units	
a		3.5903
b		
c		5.6475
Valence electrons	—	$4f^{10} 6s^2$
THERMAL AND THERMODYNAMIC PROPERTIES		
Melting point	°C	1407
Boiling point	°C	2335
Transition point(s)	°C	1384
Critical temperature	°K	7640
Linear thermal expansion coefficient	10^{-6} cm/cm/°C	10.0
Specific heat	cal/g/°C	0.0414
Electronic specific heat	10^{-4} cal/mole/°C^2	22.1
Thermal conductivity	cal/cm^2/cm/sec/°C	0.024
Latent heat of fusion	kcal/mole	2.49
Latent heat of vaporization at B.P.	kcal/mole	60.0
Heat of combustion	kcal/mole	222.92
Heat of sublimation	kcal/mole	71.2
Heat capacity at constant pressure	cal/°C/mole	6.72
Heat content at M.P.	kcal/mole	10.79
Entropy	cal/°C/mole	17.9
Longitudinal sound velocity	m/sec	3000
ELECTRICAL, ELECTRONIC, MAGNETIC AND OPTICAL PROPERTIES		
Electrical Conductivity	% IACS	
Electrical resistivity	microohm-cm	91
Photoelectric work function	eV	3.09
Hall coefficient	volt-cm/amp-Oe	
Curie point	°K	85
First ionization potential	eV (Dy^+)	5.93
Oxidation potential	V (Dy^{3+})	2.35
Electrochemical equivalent	mg/coulomb (Dy^{3+})	0.56069
Magnetic susceptibility	c.g.s. units	590×10^{-6}
Refractive index, n_D		
MECHANICAL PROPERTIES		
Young's modulus of elasticity	10^{-6} kg/cm^2	0.644
Shear modulus	10^{-6} kg/cm^2	0.259
Bulk modulus	10^{-6} kg/cm^2	0.3918
Poisson's ratio	—	0.243
Hardness	DPH	42

TABLE 2.1 (*Continued*)

EINSTEINIUM

	Unit	Value(s)
ATOMIC, NUCLEAR AND CRYSTALLOGRAPHIC PROPERTIES		
Symbol: Es		
Atomic number	−	99
Atomic weight	(based on ^{12}C)	254
Periodic classification	−	Actinide
Atomic volume	cm^3/g-atom	
Atomic radius	Angstrom units	
Ionic radius	Angstrom units (Es^{3+})	0.98
Thermal neutron absorption -cross section	barns/atom	
Density	g./cm^3	
Lattice type	−	
Lattice constants	Angstrom units	
a		
b		
c	−	
Valence electrons	−	$5f^{11}\ 7s^2$
THERMAL AND THERMODYNAMIC PROPERTIES		
Melting point	°C	
Boiling point	°C	
Transition point(s)	°C	
Critical temperature	°K	
Linear thermal expansion coefficient	10^{-6} cm/cm/°C	
Specific heat	cal/g/°C	
Electronic specific heat	10^{-4} cal/mole/°C^2	
Thermal conductivity	cal/cm^2/cm/sec/°C	
Latent heat of fusion	kcal/mole	
Latent heat of vaporization at B.P.	kcal/mole	
Heat of combustion	kcal/mole	
Heat of sublimation	kcal/mole	
Heat capacity at constant pressure	cal/°C/mole	
Heat content at M.P.	kcal/mole	
Entropy	cal/°C/mole	
Longitudinal sound velocity	m/sec	
ELECTRICAL, ELECTRONIC, MAGNETIC AND OPTICAL PROPERTIES		
Electrical Conductivity	% IACS	
Electrical resistivity	microohm-cm	
Photoelectric work function	eV	
Hall coefficient	volt-cm/amp-Oe	
Curie point	°K	
First ionization potential	eV	
Oxidation potential	V	
Electrochemical equivalent	mg/coulomb	
Magnetic susceptibility	c.g.s. units	
Refractive index, n_D	−	
MECHANICAL PROPERTIES		
Young's modulus of elasticity	10^{-6} kg/cm^2	
Shear modulus	10^{-6} kg/cm^2	
Bulk modulus	10^{-6} kg/cm^2	
Poisson's ratio	−	
Hardness		

TABLE 2.1 (*Continued*)

ERBIUM

	Unit	Value(s)
ATOMIC, NUCLEAR AND CRYSTALLOGRAPHIC PROPERTIES		
Symbol: Er		
Atomic number	–	68
Atomic weight	(based on ^{12}C)	167.26
Periodic classification	–	Lanthanide
Atomic volume	cm^3/g-atom	18.49
Atomic radius	Angstrom units	1.757
Ionic radius	Angstrom units (Er^{3+})	1.04
Thermal neutron absorption -cross section	barns/atom	160
Density	g./cm^3	9.045
Lattice type	–	hexagonal, close-packed
Lattice constants	Angstrom units	
a		3.559
b		
c		5.595
Valence electrons	–	$4f^{12} 6s^2$
THERMAL AND THERMODYNAMIC PROPERTIES		
Melting point	°C	1522
Boiling point	°C	2510
Transition point(s)	°C	1370
Critical temperature	°K	7250
Linear thermal expansion coefficient	10^{-6} cm/cm/°C	9.2
Specific heat	cal/g/°C	0.040 at 0°C
Electronic specific heat	10^{-4} cal/mole/°C^2	31
Thermal conductivity	cal/cm^2/cm/sec/°C	0.023
Latent heat of fusion	kcal/mole	2.62
Latent heat of vaporization at B.P.	kcal/mole	70.0
Heat of combustion	kcal/mole	
Heat of sublimation	kcal/mole	74.5
Heat capacity at constant pressure	cal/°C/mole	6.72
Heat content at M.P.	kcal/mole	10.89
Entropy	cal/°C/mole	17.48
Longitudinal sound velocity	m/sec	3100
ELECTRICAL, ELECTRONIC, MAGNETIC AND OPTICAL PROPERTIES		
Electrical Conductivity	% IACS	
Electrical resistivity	microohm-cm	87
Photoelectric work function	eV	3.12
Hall coefficient	volt-cm/amp-Oe	
Curie point	°K	19
First ionization potential	eV (Er^+)	6.10
Oxidation potential	V (Er^{3+})	2.30
Electrochemical equivalent	mg/coulomb (Er^{3+})	0.57907
Magnetic susceptibility	c.g.s. units	300×10^{-6}
Refractive index, n_D		
MECHANICAL PROPERTIES		
Young's modulus of elasticity	10^{-6} kg/cm^2	0.748
Shear modulus	10^{-6} kg/cm^2	0.302
Bulk modulus	10^{-6} kg/cm^2	0.4188
Poisson's ratio	–	0.238
Hardness	Vickers, Diamond Pyramid	44

TABLE 2.1 (Continued)

EUROPIUM

	Unit	Value(s)
ATOMIC, NUCLEAR AND CRYSTALLOGRAPHIC PROPERTIES		
Symbol: Eu		
Atomic number	−	63
Atomic weight	(based on ^{12}C)	151.96
Periodic classification	−	Lanthanide
Atomic volume	cm^3/g-atom	28.98
Atomic radius	Angstrom units	2.042
Ionic radius	Angstrom units (Eu^{3+})	1.13
Thermal neutron absorption -cross section	barns/atom	4200
Density	g./cm^3	5.259
Lattice type	−	body-centered cubic
Lattice constants	Angstrom units	
a		4.578
b		
c		
Valence electrons	−	$4f^7\,6s^2$
THERMAL AND THERMODYNAMIC PROPERTIES		
Melting point	°C	826
Boiling point	°C	1439
Transition point(s)	°C	
Critical temperature	°K	4600
Linear thermal expansion coefficient	10^{-6} cm/cm/°C	33.1
Specific heat	cal/g/°C	0.042
Electronic specific heat	10^{-4} cal/mole/°C^2	6.7
Thermal conductivity	cal/cm^2/cm/sec/°C	
Latent heat of fusion	kcal/mole	2.204
Latent heat of vaporization at B.P.	kcal/mole	42.0
Heat of combustion	kcal/mole	
Heat of sublimation	kcal/mole	42.6
Heat capacity at constant pressure	cal/°C/mole	6.48
Heat content at M.P.	kcal/mole	4.98
Entropy	cal/°C/mole	17.00
Longitudinal sound velocity	m/sec	
ELECTRICAL, ELECTRONIC, MAGNETIC AND OPTICAL PROPERTIES		
Electrical Conductivity	% IACS	
Electrical resistivity	microohm-cm	81
Photoelectric work function	eV	2.54
Hall coefficient	volt-cm/amp-Oe	
Curie point	°K	
First ionization potential	eV (Eu$^+$)	5.67
Oxidation potential	V (Eu^{3+})	2.41
Electrochemical equivalent	mg/coulomb (Eu^{3+})	0.52504
Magnetic susceptibility	c.g.s. units	224×10^{-6}
Refractive index, n_D	−	
MECHANICAL PROPERTIES		
Young's modulus of elasticity	10^{-6} kg/cm^2	0.155
Shear modulus	10^{-6} kg/cm^2	0.060
Bulk modulus	10^{-6} kg/cm^2	0.1501
Poisson's ratio	−	0.286
Hardness		

TABLE 2.1 (*Continued*)

FERMIUM

	Unit	Value(s)
ATOMIC, NUCLEAR AND CRYSTALLOGRAPHIC PROPERTIES		
Symbol: Fm		
Atomic number	–	100
Atomic weight	(based on ^{12}C)	257
Periodic classification	–	Actinide
Atomic volume	cm^3/g-atom	
Atomic radius	Angstrom units	
Ionic radius	Angstrom units (Fm^{3+})	0.97
Thermal neutron absorption -cross section	barns/atom	
Density	g./cm^3	
Lattice type	–	
Lattice constants	Angstrom units	
a		
b		
c		
Valence electrons	–	$5f^{12}\ 7s^2$
THERMAL AND THERMODYNAMIC PROPERTIES		
Melting point	°C	
Boiling point	°C	
Transition point(s)	°C	
Critical temperature	°K	
Linear thermal expansion coefficient	10^{-6} cm/cm/°C	
Specific heat	cal/g/°C	
Electronic specific heat	10^{-4} cal/mole/°C^2	
Thermal conductivity	cal/cm^2/cm/sec/°C	
Latent heat of fusion	kcal/mole	
Latent heat of vaporization at B.P.	kcal/mole	
Heat of combustion	kcal/mole	
Heat of sublimation	kcal/mole	
Heat capacity at constant pressure	cal/°C/mole	
Heat content at M.P.	kcal/mole	
Entropy	cal/°C/mole	
Longitudinal sound velocity	m/sec	
ELECTRICAL, ELECTRONIC, MAGNETIC AND OPTICAL PROPERTIES		
Electrical Conductivity	% IACS	
Electrical resistivity	microohm-cm	
Photoelectric work function	eV	
Hall coefficient	volt-cm/amp-Oe	
Curie point	°K	
First ionization potential	eV	
Oxidation potential	V	
Electrochemical equivalent	mg/coulomb	
Magnetic susceptibility	c.g.s. units	
Refractive index, n_D	–	
MECHANICAL PROPERTIES		
Young's modulus of elasticity	10^{-6} kg/cm^2	
Shear modulus	10^{-6} kg/cm^2	
Bulk modulus	10^{-6} kg/cm^2	
Poisson's ratio	–	
Hardness		

TABLE 2.1 (Continued)

FLUORINE

	Unit	Value(s)
ATOMIC AND NUCLEAR PROPERTIES		
Symbol: F		
Atomic number	–	9
Atomic weight	(based on ^{12}C)	18.9984
Periodic classification	–	VII A
Thermal neutron absorption		
-cross section	barns/atom	0.01
SOLID STATE		
Lattice type	–	
Lattice constants	Angstrom units	
a		
b		
c		
Atomic radius	Angstrom units	0.67
Melting point	°K	53.54
Heat of fusion	kcal/mole	0.122
Entropy of fusion	cal/°K/mole	6.75
Triple point		
Temperature	°K	53.4
Pressure	torr	
Density of solid	g/cm^3	
Density of liquid	g/cm^3	
LIQUID STATE		
Boiling point	°K	85.02
Density at boiling point	g/cm^3	1.108
Heat of vaporization at boiling point	kcal/mole	1.562
Entropy at boiling point	cal/°K/mole	
Critical Data		
Temperature	°K	144
Pressure	atm.	51.47
Density	g/cm^3	0.574
GASEOUS STATE*		
Density at 0°C	g/1	1.696
Specific heat at 16°C at const. pres.	cal/g/°K	0.1978
Ratio of specific heats, Cp/Cv, at 16°C	–	1.358
Thermal conductivity at 0°C	cal/cm^2/cm/sec/°K	0.0000592
Heat capacity at const. press. & 25°C	cal/°K/mole	7.49
Entropy at 25°C	cal/°K/mole	48.45
Heat of combustion	kcal/mole	
Dielectric constant at 20°C	–	
Viscosity at 0°C	micropoise	21.8
Refractive index, n_D, at 0°C	–	
Solubility in H$_2$O at 1 atm. partial		
pressure and 20°C	cm^3(STP)/1000 g.	
Sound velocity at 0°C	m/sec	

*Generally at one atmosphere.

TABLE 2.1 (*Continued*)

FRANCIUM

	Unit	Value(s)
ATOMIC, NUCLEAR AND CRYSTALLOGRAPHIC PROPERTIES		
Symbol: Fr		
Atomic number	−	87
Atomic weight	(based on ^{12}C)	223
Periodic classification	−	I A
Atomic volume	cm^3/g-atom	73.0
Atomic radius	Angstrom units	2.80
Ionic radius	Angstrom units (Fr^+)	1.76
Thermal neutron absorption		
-cross section	barns/atom	
Density	g./cm^3	
Lattice type	−	body-centered cubic
Lattice constants	Angstrom units	
a		
b		
c		
Valence electrons	−	$7s^1$
THERMAL AND THERMODYNAMIC PROPERTIES		
Melting point	°C	24
Boiling point	°C	750
Transition point(s)	°C	
Critical temperature	°K	
Linear thermal expansion coefficient	10^{-6} cm/cm/°C	102
Specific heat	cal/g/°C	
Electronic specific heat	10^{-4} cal/mole/°C^2	10
Thermal conductivity	cal/cm^2/cm/sec/°C	
Latent heat of fusion	kcal/mole	0.50
Latent heat of vaporization at B.P.	kcal/mole	15.2
Heat of combustion	kcal/mole	
Heat of sublimation	kcal/mole	18.1
Heat capacity at constant pressure	cal/°C/mole	7.60
Heat content at M.P.	kcal/mole	0.014
Entropy	cal/°C/mole	22.50
Longitudinal sound velocity	m/sec	
ELECTRICAL, ELECTRONIC, MAGNETIC AND OPTICAL PROPERTIES		
Electrical Conductivity	% IACS	
Electrical resistivity	microohm-cm	
Photoelectric work function	eV	1.8 (estimate)
Hall coefficient	volt-cm/amp-Oe	
Curie point	°K	
First ionization potential	eV	
Oxidation potential	V	
Electrochemical equivalent	mg/coulomb	
Magnetic susceptibility	c.g.s. units	
Refractive index, n_D	−	
MECHANICAL PROPERTIES		
Young's modulus of elasticity	10^{-6} kg/cm^2	0.017
Shear modulus	10^{-6} kg/cm^2	0.0063
Bulk modulus	10^{-6} kg/cm^2	0.020
Poisson's ratio	−	0.356
Hardness		

TABLE 2.1 *(Continued)*

GADOLINIUM

	Unit	Value(s)
ATOMIC, NUCLEAR AND CRYSTALLOGRAPHIC PROPERTIES		
Symbol: Gd		
Atomic number	–	64
Atomic weight	(based on ^{12}C)	157.25
Periodic classification	–	Lanthanide
Atomic volume	cm^3/g-atom	19.94
Atomic radius	Angstrom units	1.802
Ionic radius	Angstrom units (Gd^{3+})	1.11
Thermal neutron absorption -cross section	barns/atom	48000
Density	g./cm^3	7.895
Lattice type	–	hexagonal, close-packed
Lattice constants	Angstrom units	
a		3.6315
b		
c		5.777
Valence electrons	–	$4f^7\ 5d^1\ 6s^2$
THERMAL AND THERMODYNAMIC PROPERTIES		
Melting point	°C	1312
Boiling point	°C	3270
Transition point(s)	°C	1262
Critical temperature	°K	8670
Linear thermal expansion coefficient	10^{-6} cm/cm/°C	8.28
Specific heat	cal/g/°C	0.071 at 0°C
Electronic specific heat	10^{-4} cal/mole/°C^2	24
Thermal conductivity	cal/cm^2/cm/sec/°C	0.021
Latent heat of fusion	kcal/mole	2.438
Latent heat of vaporization at B.P.	kcal/mole	74.5
Heat of combustion	kcal/mole	
Heat of sublimation	kcal/mole	95.75
Heat capacity at constant pressure	cal/°C/mole	8.72
Heat content at M.P.	kcal/mole	9.48
Entropy	cal/°C/mole	15.77
Longitudinal sound velocity	m/sec	3000
ELECTRICAL, ELECTRONIC, MAGNETIC AND OPTICAL PROPERTIES		
Electrical Conductivity	% IACS	
Electrical resistivity	microohm-cm	134
Photoelectric work function	eV	3.07
Hall coefficient	volt-cm/amp-Oe	
Curie point	°K	289
First ionization potential	eV (Gd^+)	6.14
Oxidation potential	V (Gd^{3+})	2.40
Electrochemical equivalent	mg/coulomb (Gd^{3+})	0.54179
Magnetic susceptibility	c.g.s. units	ferromagnetic
Refractive index, n_D		
MECHANICAL PROPERTIES		
Young's modulus of elasticity	10^{-6} kg/cm^2	0.573
Shear modulus	10^{-6} kg/cm^2	0.227
Bulk modulus	10^{-6} kg/cm^2	0.3908
Poisson's ratio	–	0.259
Hardness		

TABLE 2.1 (*Continued*)

GALLIUM

	Unit	Value(s)
ATOMIC, NUCLEAR AND CRYSTALLOGRAPHIC PROPERTIES		
Symbol: Ga		
Atomic number	–	31
Atomic weight	(based on ^{12}C)	69.72
Periodic classification	–	III A
Atomic volume	cm^3/g-atom	11.81
Atomic radius	Angstrom units	1.33
Ionic radius	Angstrom units (Ga^{3+})	0.62
Thermal neutron absorption -cross section	barns/atom	3.1
Density	g./cm^3	5.907
Lattice type	–	Pseudotetragonal
Lattice constants	Angstrom units	
a		4.5167
b		4.5107
c		7.6448
Valence electrons	–	$4s^2\ 4p^1$
THERMAL AND THERMODYNAMIC PROPERTIES		
Melting point	°C	29.78
Boiling point	°C	2237
Transition point(s)	°C	
Critical temperature	°K	7620
Linear thermal expansion coefficient	10^{-6} cm/cm/°C	18
Specific heat	cal/g/°C	0.079
Electronic specific heat	10^{-4} cal/mole/°C^2	1.43
Thermal conductivity	cal/cm^2/cm/sec/°C	0.08 at 30°C
Latent heat of fusion	kcal/mole	1.335
Latent heat of vaporization at B.P.	kcal/mole	61.2
Heat of combustion	kcal/mole	
Heat of sublimation	kcal/mole	64.9
Heat capacity at constant pressure	cal/°C/mole	6.18
Heat content at M.P.	kcal/mole	11
Entropy	cal/°C/mole	9.82
Longitudinal sound velocity	m/sec	2740
ELECTRICAL, ELECTRONIC, MAGNETIC AND OPTICAL PROPERTIES		
Electrical Conductivity	% IACS	
Electrical resistivity	microohm-cm	56.8
Photoelectric work function	eV	3.96
Hall coefficient	volt-cm/amp-Oe	
Curie point	°K	
First ionization potential	eV (Ga^+)	5.999
Oxidation potential	V (Ga^{3+})	–0.52
Electrochemical equivalent	mg/coulomb (Ga^{3+})	0.24083
Magnetic susceptibility	c.g.s. units	-0.24×10^{-6} at 18°C
Refractive index, n_D		
MECHANICAL PROPERTIES		
Young's modulus of elasticity	10^{-6} kg/cm^2	0.944 at 0°C
Shear modulus	10^{-6} kg/cm^2	0.382 at 0°C
Bulk modulus	10^{-6} kg/cm^2	0.580 at 0°C
Poisson's ratio	–	0.235 at 0°C
Hardness	Mohs	1.5–2.5

*Data are at 20°C unless otherwise stated.

TABLE 2.1 (*Continued*)

GERMANIUM

	Unit	Value(s)
ATOMIC, NUCLEAR AND CRYSTALLOGRAPHIC PROPERTIES		
Symbol: Ge		
Atomic number	–	32
Atomic weight	(based on ^{12}C)	72.59
Periodic classification	–	IV A
Atomic volume	cm^3/g-atom	13.5
Atomic radius	Angstrom units	1.22
Ionic radius	Angstrom units (Ge^{4+})	0.53
Thermal neutron absorption -cross section	barns/atom	2.4
Density	g./cm^3	5.323
Lattice type	–	cubic, diamond
Lattice constants	Angstrom units	
a		5.658
b		
c		
Valence electrons	–	$4s^2\ 4p^2$
THERMAL AND THERMODYNAMIC PROPERTIES		
Melting point	°C	934
Boiling point	°C	2827
Transition point(s)	°C	
Critical temperature	°K	5642
Linear thermal expansion coefficient	10^{-6} cm/cm/°C	5.75
Specific heat	cal/g/°C	0.074
Electronic specific heat	10^{-4} cal/mole/°C^2	0
Thermal conductivity	cal/cm^2/cm/sec/°C	0.15
Latent heat of fusion	kcal/mole	7.6
Latent heat of vaporization at B.P.	kcal/mole	79.9
Heat of combustion	kcal/mole	
Heat of sublimation	kcal/mole	89.5
Heat capacity at constant pressure	cal/°C/mole	5.47
Heat content at M.P.	kcal/mole	5.81
Entropy	cal/°C/mole	7.43
Longitudinal sound velocity	m/sec	4914 (001)
ELECTRICAL, ELECTRONIC, MAGNETIC AND OPTICAL PROPERTIES		
Electrical Conductivity	% IACS	
Electrical resistivity	microohm-cm	47 × 10^6
Photoelectric work function	eV	4.56
Hall coefficient	volt-cm/amp-Oe	
Curie point	°K	
First ionization potential	eV (Ge$^+$)	7.899
Oxidation potential	V (Ge^{2+})	0.0
Electrochemical equivalent	mg/coulomb (Ge$^+$)	0.1881
Magnetic susceptibility	c.g.s. units	–0.10 × 10^{-6}
Refractive index, n_D		4.2 at 0.4 μ
MECHANICAL PROPERTIES		
Young's modulus of elasticity	10^{-6} kg/cm^2	1.01
Shear modulus	10^{-6} kg/cm^2	0.40
Bulk modulus	10^{-6} kg/cm^2	0.7874
Poisson's ratio	–	0.27
Hardness	Mohs	6

TABLE 2.1 (*Continued*)

GOLD

	Unit	Value(s)
ATOMIC, NUCLEAR AND CRYSTALLOGRAPHIC PROPERTIES		
Symbol: Au		
Atomic number	−	79
Atomic weight	(based on ^{12}C)	196.967
Periodic classification	−	I B
Atomic volume	cm^3/g-atom	10.22
Atomic radius	Angstrom units	1.42
Ionic radius	Angstrom units (Au^+)	1.37
Thermal neutron absorption -cross section	barns/atom	98.8
Density	g./cm^3	19.32
Lattice type	−	face-centered cubic
Lattice constants	Angstrom units	
a		4.0786
b		
c		
Valence electrons	−	$5d^{10} 6s^1$
THERMAL AND THERMODYNAMIC PROPERTIES		
Melting point	°C	1063
Boiling point	°C	2967
Transition point(s)	°C	
Critical temperature	°K	8060
Linear thermal expansion coefficient	10^{-6} cm/cm/°C	14.1
Specific heat	cal/g/°C	0.03077
Electronic specific heat	10^{-4} cal/mole/°C^2	1.79
Thermal conductivity	cal/cm^2/cm/sec/°C	0.74
Latent heat of fusion	kcal/mole	2.955
Latent heat of vaporization at B.P.	kcal/mole	77.54
Heat of combustion	kcal/mole	
Heat of sublimation	kcal/mole	87.6
Heat capacity at constant pressure	cal/°C/mole	6.065
Heat content at M.P.	kcal/mole	6.66
Entropy	cal/°C/mole	11.32
Longitudinal sound velocity	m/sec	3281
ELECTRICAL, ELECTRONIC, MAGNETIC AND OPTICAL PROPERTIES		
Electrical Conductivity	% IACS	73.4
Electrical resistivity	microohm-cm	2.032 at 0°C
Photoelectric work function	eV	5.22
Hall coefficient	volt-cm/amp-Oe	
Curie point	°K	
First ionization potential	eV (Au^+)	9.22
Oxidation potential	V (Au^{3+})	−1.50
Electrochemical equivalent	mg/coulomb (Au^{3+})	0.68117
Magnetic susceptibility	c.g.s. units	-0.14×10^{-6}
Refractive index, n_D		0.24 at 1.00 μ
MECHANICAL PROPERTIES		
Young's modulus of elasticity	10^{-6} kg/cm^2	0.794
Shear modulus	10^{-6} kg/cm^2	0.281
Bulk modulus	10^{-6} kg/cm^2	1.766
Poisson's ratio	−	0.425
Hardness	Brinell	18

TABLE 2.1 (*Continued*)

HAFNIUM

	Unit	Value(s)
ATOMIC, NUCLEAR AND CRYSTALLOGRAPHIC PROPERTIES		
Symbol: Hf		
Atomic number	−	72
Atomic weight	(based on ^{12}C)	178.49
Periodic classification	−	IV B
Atomic volume	cm^3/g-atom	13.37
Atomic radius	Angstrom units	1.539
Covalent radius	Angstrom units (Hf^{4+})	0.82
Thermal neutron absorption -cross section	barns/atom	106
Density	g./cm^3	13.29
Lattice type (α)	−	hexagonal, close-packed
Lattice constants	Angstrom units	
a		3.197
b		
c		5.057
Valence electrons	−	$5d^2\ 6s^2$
THERMAL AND THERMODYNAMIC PROPERTIES		
Melting point	°C	2222
Boiling point	°C	4606
Transition point(s)	°C	1740 (α to β)
Critical temperature	°K	
Linear thermal expansion coefficient	10^{-6} cm/cm/°C	5.9
Specific heat	cal/g/°C	0.0352
Electronic specific heat	10^{-4} cal/mole/°C^2	5.74
Thermal conductivity	cal/cm^2/cm/sec/°C	0.0533 at 50°C
Latent heat of fusion	kcal/mole	4.39
Latent heat of vaporization at B.P.	kcal/mole	158.0
Heat of combustion	kcal/mole	
Heat of sublimation	kcal/mole	145.5
Heat capacity at constant pressure	cal/°C/mole	6.10
Heat content at M.P.	kcal/mole	13.77
Entropy	cal/°C/mole	12.74
Longitudinal sound velocity	m/sec (20°C)	~3000
ELECTRICAL, ELECTRONIC, MAGNETIC AND OPTICAL PROPERTIES		
Electrical Conductivity	% IACS	
Electrical resistivity	microohm-cm	35.5
Photoelectric work function	eV	3.53
Hall coefficient	volt-cm/amp-Oe	
Curie point	°K	
First ionization potential	eV (Hf^+)	7.0
Oxidation potential	V (Hf^{4+})	1.70 (acid soln.)
Electrochemical equivalent	mg/coulomb (Hf^{4+})	0.46269
Magnetic susceptibility	c.g.s. units	0.42×10^{-6}
Refractive Index, n_D	−	
MECHANICAL PROPERTIES		
Young's modulus of elasticity	10^{-6} kg/cm^2	1.40
Shear modulus	10^{-6} kg/cm^2	0.540
Bulk modulus	10^{-6} kg/cm^2	1.11
Poisson's ratio	−	0.30
Hardness	DPH	190

TABLE 2.1 (*Continued*)

HELIUM

	Unit	Value(s)
ATOMIC AND NUCLEAR PROPERTIES		
Symbol: He		
Atomic number	–	2
Atomic weight	(based on ^{12}C)	4.0026
Periodic classification	–	rare gas
Thermal neutron absorption		
-cross section	barns/atom	0.007
SOLID STATE		
Lattice type	–	close-packed hexagonal
Lattice constants	Angstrom units	
a		3.58 at 1.7°K
b		–
c		5.84 at 1.7°K
Atomic radius	Angstrom units	
Melting point	°K	3.5 at 103 atm.
Heat of fusion	kcal/mole	0.005
Entropy of fusion	cal/°K/mole	
Triple point		
Temperature	°K	2.26
Pressure	torr	39
Density of solid	g/cm^3	
Density of liquid	g/cm^3	
LIQUID STATE		
Boiling point	°K	4.22
Density at boiling point	g/cm^3	0.1249
Heat of vaporization at boiling point	kcal/mole	0.020
Entropy at boiling point	cal/°K/mole	
Critical Data		
Temperature	°K	5.3
Pressure	atm.	2.26
Density	g/cm^3	0.0693
GASEOUS STATE*		
Density at 0°C	g/1	0.17847
Specific heat at 20°C at const. pres.	cal/g/°K	1.25
Ratio of specific heats, Cp/Cv, at 20°C	–	1.63
Thermal conductivity at 0°C	cal/cm^2/cm/sec/°K	0.000339
Heat capacity at const. press. & 25°C	cal/°K/mole	4.97
Entropy at 25°C	cal/°K/mole	30.13
Heat of combustion	kcal/mole	
Dielectric constant at 25°C	–	1.0000639
Viscosity at 20°C	micropoise	196.1
Refractive index, n_D, at 0°C	–	1.000036
Solubility in H$_2$O at 1 atm. partial		
pressure and 0°C	cm^3(STP)/1000 g.	9.78
Sound velocity at 0°C	m/sec	965

*Generally at one atmosphere.

TABLE 2.1 *(Continued)*

HOLMIUM

	Unit	Value(s)
ATOMIC, NUCLEAR AND CRYSTALLOGRAPHIC PROPERTIES		
Symbol: Ho		
Atomic number	–	67
Atomic weight	(based on ^{12}C)	164.930
Periodic classification	–	Lanthanide
Atomic volume	cm^3/g-atom	18.78
Atomic radius	Angstrom units	1.766
Ionic radius	Angstrom units (Ho^{3+})	1.05
Thermal neutron absorption		
-cross section	barns/atom	65
Density	$g./cm^3$	8.78
Lattice type	–	hexagonal close-packed
Lattice constants	Angstrom units	
a		3.578
b		
c		5.626
Valence electrons	–	$4f^{11} 6s^2$
THERMAL AND THERMODYNAMIC PROPERTIES		
Melting point	°C	1461
Boiling point	°C	2720
Transition point(s)	°C	
Critical temperature	°K	
Linear thermal expansion coefficient	10^{-6} cm/cm/°C	10.7
Specific heat	cal/g/°C	0.039
Electronic specific heat	10^{-4} cal/mole/°C^2	62
Thermal conductivity	cal/cm^2/cm/sec/°C	
Latent heat of fusion	kcal/mole	2.9
Latent heat of vaporization at B.P.	kcal/mole	60.0
Heat of combustion	kcal/mole	
Heat of sublimation	kcal/mole	70.2
Heat capacity at constant pressure	cal/°C/mole	6.49
Heat content at M.P.	kcal/mole	10.79
Entropy	cal/°C/mole	17.77
Longitudinal sound velocity	m/sec	3000
ELECTRICAL, ELECTRONIC, MAGNETIC AND OPTICAL PROPERTIES		
Electrical Conductivity	% IACS	
Electrical resistivity	microohm-cm	81
Photoelectric work function	eV	3.09
Hall coefficient	volt-cm/amp-Oe	
Curie point	°K	20
First ionization potential	eV (Ho^+)	6.02
Oxidation potential	V (Ho^{3+})	2.32
Electrochemical equivalent	mg/coulomb (Ho^{3+})	0.56974
Magnetic susceptibility	c.g.s. units	437×10^{-6}
Refractive index, n_D		
MECHANICAL PROPERTIES		
Young's modulus of elasticity	10^{-6} kg/cm^2	0.684
Shear modulus	10^{-6} kg/cm^2	0.272
Bulk modulus	10^{-6} kg/cm^2	0.4045
Poisson's ratio	–	0.255
Hardness	Vickers Diamond Pyramid	42 kg/mm^2

TABLE 2.1 (*Continued*)

HYDROGEN

	Unit	Value(s)
ATOMIC AND NUCLEAR PROPERTIES		
Symbol: H		
Atomic number	–	1
Atomic weight	(based on ^{12}C)	1.00797
Periodic classification	–	I A
Thermal neutron absorption		
-cross section	barns/atom	0.332
SOLID STATE		
Lattice type	–	close-packed hexagonal
Lattice constants	Angstrom units	
a		3.76 at 2°K
b		
c		6.13 at 2°K
Atomic radius	Angstrom units	
Melting point	°K	13.96
Heat of fusion	kcal/mole	0.028
Entropy of fusion	cal/°K/mole	
Triple point		
Temperature	°K	14.0
Pressure	torr	54
Density of solid	g/cm^3	
Density of liquid	g/cm^3	
LIQUID STATE		
Boiling point	°K	20.39
Density at boiling point	g/cm^3	0.070
Heat of vaporization at boiling point	kcal/mole	0.22
Entropy at boiling point	cal/°K/mole	
Critical Data		
Temperature	°K	33.24
Pressure	atm.	12.80
Density	g/cm^3	0.0310
GASEOUS STATE*		
Density at 0°C	g/1	0.0899
Specific heat at 20°C at const. pres.	cal/g/°K	3.44
Ratio of specific heats, Cp/Cv, at 20°C	–	1.40
Thermal conductivity at 0°C	cal/cm^2/cm/sec/°K	0.00038
Heat capacity at const. press. & 25°C	cal/°K/mole	6.89
Entropy at 25°C	cal/°K/mole	31.21
Heat of combustion	kcal/mole	0.06788
Dielectric constant at 20°C	at 2×10^6 Hz	1.000264
Viscosity at 15°C	micropoise	8.7
Refractive index, n_D, at 0°C	–	1.000132
Solubility in H_2O at 1 atm. partial		
pressure and 15°C	cm^3(STP)/1000 g.	19
Sound velocity at 0°C	m/sec	1269.5

*Generally at one atmosphere.

TABLE 2.1 (Continued)

INDIUM

	Unit	Value(s)
ATOMIC, NUCLEAR AND CRYSTALLOGRAPHIC PROPERTIES		
Symbol: In		
Atomic number	–	49
Atomic weight	(based on ^{12}C)	114.82
Periodic classification	–	III A
Atomic volume	cm^3/g-atom	15.7
Atomic radius	Angstrom units	1.51
Ionic radius	Angstrom units (In^{3+})	0.81
Thermal neutron absorption -cross section	barns/atom	194
Density	g./cm^3	7.31
Lattice type	–	face-centered tetragonal
Lattice constants	Angstrom units	
a		4.583
b		
c		4.936
Valence electrons	–	$5s^2\ 5p^1$
THERMAL AND THERMODYNAMIC PROPERTIES		
Melting point	°C	156.6
Boiling point	°C	2006
Transition point(s)	°C	
Critical temperature	°K	7050
Linear thermal expansion coefficient	10^{-6} cm/cm/°C	24.8
Specific heat	cal/g/°C	0.058
Electronic specific heat	10^{-4} cal/mole/°C^2	4.06
Thermal conductivity	cal/cm^2/cm/sec/°C	0.057
Latent heat of fusion	kcal/mole	0.78
Latent heat of vaporization at B.P.	kcal/mole	55.7
Heat of combustion	kcal/mole	
Heat of sublimation	kcal/mole	57.3
Heat capacity at constant pressure	cal/°C/mole	6.39
Heat content at M.P.	kcal/mole	0.680
Entropy	cal/°C/mole	13.82
Longitudinal sound velocity	m/sec (20°C)	~1215
ELECTRICAL, ELECTRONIC, MAGNETIC AND OPTICAL PROPERTIES		
Electrical Conductivity	% IACS	
Electrical resistivity	microhm-cm	29.0 at 22°C
Photoelectric work function	eV	3.8
Hall coefficient	volt-cm/amp-Oe	
Curie point	°K	
First ionization potential	eV (In^+)	5.786
Oxidation potential	V (In^{3+})	−0.34
Electrochemical equivalent	mg/coulomb (In^{3+})	0.39641
Magnetic susceptibility	c.g.s. units	diamagnetic
Refractive index, n_D		
MECHANICAL PROPERTIES		
Young's modulus of elasticity	10^{-6} kg/cm^2	0.107
Shear modulus	10^{-6} kg/cm^2	0.038
Bulk modulus	10^{-6} kg/cm^2	0.419
Poisson's ratio	–	0.46
Hardness	Brinell	0.9

TABLE 2.1 (*Continued*)

IODINE

	Unit	Value(s)
ATOMIC, NUCLEAR AND CRYSTALLOGRAPHIC PROPERTIES		
Symbol: I		
Atomic number	–	53
Atomic weight	(based on ^{12}C)	126.9044
Periodic classification	–	VII A
Atomic volume	cm^3/g-atom	25.7
Atomic radius	Angstrom units	1.38
Ionic radius	Angstrom units (I^-)	2.16
Thermal neutron absorption		
-cross section	barns/atom	6.2
Density	g./cm^3	4.93
Lattice type	–	Orthorhombic
Lattice constants	Angstrom units	
a		4.7761
b		7.2501
c		9.7711
Valence electrons	–	$5s^2\ 5p^2$
THERMAL AND THERMODYNAMIC PROPERTIES		
Melting point	°C	113.6
Boiling point	°C	185
Transition point(s)	°C	
Critical temperature	°K	785
Linear thermal expansion coefficient	10^{-6} cm/cm/°C	93
Specific heat	cal/g/°C	0.052
Electronic specific heat	10^{-4} cal/mole/°C^2	
Thermal conductivity	cal/cm^2/cm/sec/°C	0.0010
Latent heat of fusion	kcal/mole	1.80
Latent heat of vaporization at B.P.	kcal/mole	5.00
Heat of combustion	kcal/mole	
Heat of sublimation	kcal/mole	7.25 at 113.6°C
Heat capacity at constant pressure	cal/°C/mole	13.14
Heat content at M.P.	kcal/mole	
Entropy	cal/°C/mole	27.90
Longitudinal sound velocity	m/sec	
ELECTRICAL, ELECTRONIC, MAGNETIC AND OPTICAL PROPERTIES		
Electrical Conductivity	% IACS	
Electrical resistivity	microohm-cm	$(0.8 - 6) \times 10^{12}$ (single crystal)
Photoelectric work function	eV	2.8
Hall coefficient	volt-cm/amp-Oe	
Curie point	°K	
First ionization potential	eV (I^+)	10.451
Oxidation potential	V (I^-/I_2)	−0.5355
Electrochemical equivalent	mg/coulomb (I^-)	1.31523
Magnetic susceptibility	c.g.s. units	-0.4×10^{-6}
Refractive index, n_D	–	3.34
MECHANICAL PROPERTIES		
Young's modulus of elasticity	10^{-6} kg/cm^2	
Shear modulus	10^{-6} kg/cm^2	
Bulk modulus	10^{-6} kg/cm^2	
Poisson's ratio	–	
Hardness		

TABLE 2.1 (*Continued*)

IRIDIUM

	Unit	Value(s)
ATOMIC, NUCLEAR AND CRYSTALLOGRAPHIC PROPERTIES		
Symbol: Ir		
Atomic number	–	77
Atomic weight	(based on ^{12}C)	192.2
Periodic classification	–	VIII
Atomic volume	cm^3/g-atom	8.524
Atomic radius	Angstrom units	1.3545
Ionic radius	Angstrom units (Ir^{4+})	0.66
Thermal neutron absorption -cross section	barns/atom	426
Density	g./cm^3	22.65
Lattice type	–	face-centered cubic
Lattice constants	Angstrom units	3.8394
a		3.8394
b		
c		
Valence electrons	–	5d^7 6s^2
THERMAL AND THERMODYNAMIC PROPERTIES		
Melting point	°C	2443
Boiling point	°C	4547
Transition point(s)	°C	
Critical temperature	°K	
Linear thermal expansion coefficient	10^{-6} cm/cm/°C	6.8
Specific heat	cal/g/°C	0.031
Electronic specific heat	10^{-4} cal/mole/°C^2	7.53
Thermal conductivity	cal/cm^2/cm/sec/°C	0.354
Latent heat of fusion	kcal/mole	6.3
Latent heat of vaporization at B.P.	kcal/mole	134.7
Heat of combustion	kcal/mole	
Heat of sublimation	kcal/mole	160
Heat capacity at constant pressure	cal/°C/mole	6.10
Heat content at M.P.	kcal/mole	18.47
Entropy	cal/°C/mole	8.70
Longitudinal sound velocity	m/sec	5379
ELECTRICAL, ELECTRONIC, MAGNETIC AND OPTICAL PROPERTIES		
Electrical Conductivity	% IACS	
Electrical resistivity	microohm-cm	4.71 at 0°C
Photoelectric work function	eV	5.40
Hall coefficient	volt-cm/amp-Oe	
Curie point	°K	
First ionization potential	eV (Ir$^+$)	9.2
Oxidation potential	V (Ir^{3+})	–0.93 (acid soln.)
Electrochemical equivalent	mg/coulomb (Ir^{4+})	0.50026
Magnetic susceptibility	c.g.s. units	0.133×10^{-6}
Refractive index, n_D		3.6 at 1.00 μ
MECHANICAL PROPERTIES		
Young's modulus of elasticity	10^{-6} kg/cm^2	5.38
Shear modulus	10^{-6} kg/cm^2	2.14
Bulk modulus	10^{-6} kg/cm^2	3.62
Poisson's ratio	–	0.26
Hardness	Vickers	200

TABLE 2.1 (*Continued*)

IRON

	Unit	Value(s)
ATOMIC, NUCLEAR AND CRYSTALLOGRAPHIC PROPERTIES		
Symbol: Fe		
Atomic number	–	26
Atomic weight	(based on ^{12}C)	55.847
Periodic classification	–	VIII
Atomic volume	cm^3/g-atom	7.094
Atomic radius	Angstrom units	1.26
Ionic radius	Angstrom units (Fe^{2+})	0.75
Thermal neutron absorption -cross section	barns/atom	2.65
Density	g./cm^3	7.8733
Lattice type	–	body-centered cubic
Lattice constants	Angstrom units	
a		2.86645
b		
c		
Valence electrons	–	$3d^6\ 4s^2$
THERMAL AND THERMODYNAMIC PROPERTIES		
Melting point	°C	1535
Boiling point	°C	2887
Transition point(s)	°C	910, 1390
Critical temperature	°K	9400
Linear thermal expansion coefficient	10^{-6} cm/cm/°C	11.7
Specific heat	cal/g/°C	0.106
Electronic specific heat	10^{-4} cal/mole/°C^2	11.9
Thermal conductivity	cal/cm^2/cm/sec/°C	0.18
Latent heat of fusion	kcal/mole	3.67
Latent heat of vaporization at B.P.	kcal/mole	83.9
Heat of combustion	kcal/mole	
Heat of sublimation	kcal/mole	100
Heat capacity at constant pressure	cal/°C/mole	5.98
Heat content at M.P.	kcal/mole	13.90
Entropy	cal/°C/mole	6.491
Longitudinal sound velocity	m/sec	5130
ELECTRICAL, ELECTRONIC, MAGNETIC AND OPTICAL PROPERTIES		
Electrical Conductivity	% IACS	17.75
Electrical resistivity	microohm-cm	9.71
Photoelectric work function	eV	4.5
Hall coefficient	volt-cm/amp-Oe	
Curie point	°K	1033
First ionization potential	eV (Fe^+)	7.870
Oxidation potential	V (Fe^{2+})	−0.4402
Electrochemical equivalent	mg/coulomb (Fe^{2+})	0.2893
Magnetic susceptibility	c.g.s. units	ferromagnetic
Refractive index, n_D		1.51
MECHANICAL PROPERTIES		
Young's modulus of elasticity	10^{-6} kg/cm^2	2.14
Shear modulus	10^{-6} kg/cm^2	0.831
Bulk modulus	10^{-6} kg/cm^2	1.716
Poisson's ratio	–	0.29
Hardness	Brinell	60

TABLE 2.1 (*Continued*)

KRYPTON

	Unit	Value(s)
ATOMIC AND NUCLEAR PROPERTIES		
Symbol: Kr		
Atomic number	–	36
Atomic weight	(based on ^{12}C)	83.80
Periodic classification	–	rare gas
Thermal neutron absorption		
-cross section	barns/atom	23.9
SOLID STATE		
Lattice type	–	face-centered cubic
Lattice constants	Angstrom units	
a		5.69 at 80°K
b		
c		
Atomic radius	Angstrom units	2.01 at 88°K
Melting point	°K	115.9
Heat of fusion	kcal/mole	0.391
Entropy of fusion	cal/°K/mole	
Triple point		
Temperature	°K	116.0
Pressure	torr	548
Density of solid	g/cm^3	2.826
Density of liquid	g/cm^3	2.441
LIQUID STATE		
Boiling point	°K	119.75
Density at boiling point	g/cm^3	2.413
Heat of vaporization at boiling point	kcal/mole	2.158
Entropy at boiling point	cal/°K/mole	
Critical Data		
Temperature	°K	209.4
Pressure	atm.	54.3
Density	g/cm^3	0.908
GASEOUS STATE*		
Density at STP	g/l	3.7493
Specific heat at 20°C at const. pres.	cal/g/°K	0.06
Ratio of specific heats, Cp/Cv, at 20°C	–	1.689
Thermal conductivity at 20°C	cal/cm^2/cm/sec/°K	0.000021
Heat capacity at const. press. & 25°C	cal/°K/mole	4.97
Entropy at 25°C	cal/°K/mole	39.19
Heat of combustion	kcal/mole	
Dielectric constant at 25°C	–	1.000768
Viscosity at 20°C	micropoise	248.0
Refractive index, n_D, at 25°C	–	1.000427
Solubility in H$_2$O at 1 atm. partial		
pressure and 20°C	cm^3(STP)/1000 g.	60.5
Sound velocity at 0°C	m/sec	213

*Generally at one atmosphere.

TABLE 2.1 (*Continued*)

LANTHANUM

	Unit	Value(s)
ATOMIC, NUCLEAR AND CRYSTALLOGRAPHIC PROPERTIES		
Symbol: La		
Atomic number	—	57
Atomic weight	(based on ^{12}C)	138.91
Periodic classification	—	III B
Atomic volume	cm^3/g-atom	22.35
Atomic radius	Angstrom units	1.877
Ionic radius	Angstrom units (La^{3+})	1.061
Thermal neutron absorption		
-cross section	barns/atom	9.1
Density	$g./cm^3$	6.166
Lattice type	—	hexagonal
Lattice constants	Angstrom units	
a		3.770
b		
c		12.131
Valence electrons	—	$5d^1\ 6s^2$
THERMAL AND THERMODYNAMIC PROPERTIES		
Melting point	°C	920
Boiling point	°C	3470
Transition point(s)	°C	310, 220, 861
Critical temperature	°K	10500
Linear thermal expansion coefficient	10^{-6} cm/cm/°C	4.9
Specific heat	cal/g/°C	0.048
Electronic specific heat	10^{-4} cal/mole/°C^2	24.1
Thermal conductivity	$cal/cm^2/cm/sec/°C$	0.026
Latent heat of fusion	kcal/mole	1.482
Latent heat of vaporization at B.P.	kcal/mole	99.5
Heat of combustion	kcal/mole	257
Heat of sublimation	kcal/mole	102
Heat capacity at constant pressure	cal/°C/mole	6.25
Heat content at M.P.	kcal/mole	5.84
Entropy	cal/°C/mole	13.60
Longitudinal sound velocity	m/sec	2800
ELECTRICAL, ELECTRONIC, MAGNETIC AND OPTICAL PROPERTIES		
Electrical Conductivity	% IACS	
Electrical resistivity	microohm-cm	57
Photoelectric work function	eV	3.3
Hall coefficient	volt-cm/amp-Oe	-0.8×10^{-12}
Curie point	°K	
First ionization potential	eV (La^+)	5.577
Oxidation potential	V (La^{3+})	2.52
Electrochemical equivalent	mg/coulomb (La^{3+})	0.47986
Magnetic susceptibility	c.g.s. units	115×10^{-6}
Refractive index, n_D		
MECHANICAL PROPERTIES		
Young's modulus of elasticity	$10^{-6}\ kg/cm^2$	0.387
Shear modulus	$10^{-6}\ kg/cm^2$	0.152
Bulk modulus	$10^{-6}\ kg/cm^2$	0.2477
Poisson's ratio	—	0.288
Hardness	Brinell (as cast)	37

TABLE 2.1 (*Continued*)

LAWRENCIUM

	Unit	Value(s)
ATOMIC, NUCLEAR AND CRYSTALLOGRAPHIC PROPERTIES		
Symbol: Lw		
Atomic number	–	103
Atomic weight	(based on ^{12}C)	257
Periodic classification	–	Actinide
Atomic volume	cm^3/g-atom	
Atomic radius	Angstrom units	
Ionic radius	Angstrom units (Lw^{3+})	0.94
Thermal neutron absorption -cross section	barns/atom	
Density	g./cm^3	
Lattice type	–	
Lattice constants	Angstrom units	
a		
b		
c		
Valence electrons	–	
THERMAL AND THERMODYNAMIC PROPERTIES		
Melting point	°C	
Boiling point	°C	
Transition point(s)	°C	
Critical temperature	°K	
Linear thermal expansion coefficient	10^{-6} cm/cm/°C	
Specific heat	cal/g/°C	
Electronic specific heat	10^{-4} cal/mole/°C^2	
Thermal conductivity	cal/cm^2/cm/sec/°C	
Latent heat of fusion	kcal/mole	
Latent heat of vaporization at B.P.	kcal/mole	
Heat of combustion	kcal/mole	
Heat of sublimation	kcal/mole	
Heat capacity at constant pressure	cal/°C/mole	
Heat content at M.P.	kcal/mole	
Entropy	cal/°C/mole	
Longitudinal sound velocity	m/sec	
ELECTRICAL, ELECTRONIC, MAGNETIC AND OPTICAL PROPERTIES		
Electrical Conductivity	% IACS	
Electrical resistivity	microohm-cm	
Photoelectric work function	eV	
Hall coefficient	volt-cm/amp-Oe	
Curie point	°K	
First ionization potential	eV	
Oxidation potential	V	
Electrochemical equivalent	mg/coulomb	
Magnetic susceptibility	c.g.s. units	
Refractive index, n_D	–	
MECHANICAL PROPERTIES		
Young's modulus of elasticity	10^{-6} kg/cm^2	
Shear modulus	10^{-6} kg/cm^2	
Bulk modulus	10^{-6} kg/cm^2	
Poisson's ratio	–	
Hardness		

TABLE 2.1 (*Continued*)

LEAD

	Unit	Value(s)
ATOMIC, NUCLEAR AND CRYSTALLOGRAPHIC PROPERTIES		
Symbol: Pb		
Atomic number	—	82
Atomic weight	(based on ^{12}C)	207.19
Periodic classification	—	IV A
Atomic volume	cm^3/g-atom	18.27
Atomic radius	Angstrom units	1.75
Ionic radius	Angstrom units (Pb^{2+})	1.32
Thermal neutron absorption -cross section	barns/atom	0.18
Density	g./cm^3	11.34
Lattice type	—	face-centered cubic
Lattice constants	Angstrom units	
a		4.9489
b		
c		
Valence electrons	—	$6s^2\ 6p^2$
THERMAL AND THERMODYNAMIC PROPERTIES		
Melting point	°C	327.4
Boiling point	°C	1749
Transition point(s)	°C	
Critical temperature	°K	4760
Linear thermal expansion coefficient	10^{-6} cm/cm/°C	29.1
Specific heat	cal/g/°C	0.032
Electronic specific heat	10^{-4} cal/mole/°C^2	7.50
Thermal conductivity	cal/cm^2/cm/sec/°C	0.083
Latent heat of fusion	kcal/mole	1.14
Latent heat of vaporization at B.P.	kcal/mole	42.88
Heat of combustion	kcal/mole	
Heat of sublimation	kcal/mole	46.8
Heat capacity at constant pressure	cal/°C/mole	6.39
Heat content at M.P.	kcal/mole	2.014
Entropy	cal/°C/mole	15.49
Longitudinal sound velocity	m/sec	1200
ELECTRICAL, ELECTRONIC, MAGNETIC AND OPTICAL PROPERTIES		
Electrical Conductivity	% IACS	8.3
Electrical resistivity	microohm-cm	20.65
Photoelectric work function	eV	4.02
Hall coefficient	volt-cm/amp-Oe	
Curie point	°K	
First ionization potential	eV (Pb^+)	7.416
Oxidation potential	V (Pb^{2+})	0.126 (acid soln.)
Electrochemical equivalent	mg/coulomb (Pb^{2+})	0.53681
Magnetic susceptibility	c.g.s. units	-0.12×10^{-6}
Refractive index, n_D		2.01
MECHANICAL PROPERTIES		
Young's modulus of elasticity	10^{-6} kg/cm^2	0.160
Shear modulus	10^{-6} kg/cm^2	0.055
Bulk modulus	10^{-6} kg/cm^2	0.4382
Poisson's ratio	—	0.44
Hardness	Brinell	4.2

TABLE 2.1 *(Continued)*

LITHIUM

	Unit	Value(s)
ATOMIC, NUCLEAR AND CRYSTALLOGRAPHIC PROPERTIES Symbol: Li		
Atomic number	–	3
Atomic weight	(based on ^{12}C)	6.939
Periodic classification	–	I A
Atomic volume	cm^3/g-atom	13.02
Atomic radius	Angstrom units	1.56
Ionic radius	Angstrom units (Li^+)	0.60
Thermal neutron absorption -cross section	barns/atom	71
Density	g./cm^3	0.534
Lattice type	–	body-centered cubic
Lattice constants	Angstrom units	
a		3.502
b		
c		
Valence electrons	–	$2s^1$
THERMAL AND THERMODYNAMIC PROPERTIES		
Melting point	°C	179
Boiling point	°C	1317
Transition point(s)	°C	–196
Critical temperature	°K	3720
Linear thermal expansion coefficient	10^{-6} cm/cm/°C	56
Specific heat	cal/g/°C	0.79
Electronic specific heat	10^{-4} cal/mole/°C^2	4.04
Thermal conductivity	cal/cm^2/cm/sec/°C	0.17 at 0°C
Latent heat of fusion	kcal/mole	0.719
Latent heat of vaporization at B.P.	kcal/mole	32.8
Heat of combustion	kcal/mole	
Heat of sublimation	kcal/mole	38.6
Heat capacity at constant pressure	cal/°C/mole	5.65
Heat content at M.P.	kcal/mole	0.63
Entropy	cal/°C/mole	6.75
Longitudinal sound velocity	m/sec	6000
ELECTRICAL, ELECTRONIC, MAGNETIC AND OPTICAL PROPERTIES		
Electrical Conductivity	% IACS	
Electrical resistivity	microohm-cm	8.55 at 0°C
Photoelectric work function	eV	2.49
Hall coefficient	volt-cm/amp-Oe	
Curie point	°K	
First ionization potential	eV (Li^+)	5.363
Oxidation potential	V (Li^+)	3.045
Electrochemical equivalent	mg/coulomb (Li^+)	0.07192
Magnetic susceptibility	c.g.s. units	0.50×10^{-6}
Refractive index, n_D		
MECHANICAL PROPERTIES		
Young's modulus of elasticity	10^{-6} kg/cm^2	0.117
Shear modulus	10^{-6} kg/cm^2	0.0431
Bulk modulus	10^{-6} kg/cm^2	0.118
Poisson's ratio	–	0.362
Hardness	Mohs	0.6

TABLE 2.1 (*Continued*)

LUTETIUM

	Unit	Value(s)
ATOMIC, NUCLEAR AND CRYSTALLOGRAPHIC PROPERTIES		
Symbol: Lu		
Atomic number	–	71
Atomic weight	(based on ^{12}C)	174.99
Periodic classification	–	Lanthanide
Atomic volume	cm^3/g-atom	17.77
Atomic radius	Angstrom units	1.737
Ionic radius	Angstrom units (Lu^{3+})	0.99
Thermal neutron absorption -cross section	barns/atom	74
Density	g./cm^3	9.842
Lattice type	–	close-packed hexagonal
Lattice constants	Angstrom units	
a		3.5050
b		
c		5.5486
Valence electrons	–	$5d^1\ 6s^2$
THERMAL AND THERMODYNAMIC PROPERTIES		
Melting point	°C	1652
Boiling point	°C	3315
Transition point(s)	°C	
Critical temperature	°K	
Linear thermal expansion coefficient	10^{-6} cm/cm/°C	8.12
Specific heat	cal/g/°C	0.037 at 0°C
Electronic specific heat	10^{-4} cal/mole/°C^2	24.43
Thermal conductivity	cal/cm^2/cm/sec/°C	
Latent heat of fusion	kcal/mole	2.85
Latent heat of vaporization at B.P.	kcal/mole	59.0
Heat of combustion	kcal/mole	
Heat of sublimation	kcal/mole	102.2
Heat capacity at constant pressure	cal/°C/mole	6.46
Heat content at M.P.	kcal/mole	12.25
Entropy	cal/°C/mole	11.75
Longitudinal sound velocity	m/sec	
ELECTRICAL, ELECTRONIC, MAGNETIC AND OPTICAL PROPERTIES		
Electrical Conductivity	% IACS	
Electrical resistivity	microohm-cm	68
Photoelectric work function	eV	3.14
Hall coefficient	volt-cm/amp-Oe	
Curie point	°K	
First ionization potential	eV (Lu^+)	5.426
Oxidation potential	V (Lu^{3+})	2.25
Electrochemical equivalent	mg/coulomb (Lu^{3+})	0.60446
Magnetic susceptibility	c.g.s. units	17.9×10^{-6}
Refractive index, n_D		
MECHANICAL PROPERTIES		
Young's modulus of elasticity	10^{-6} kg/cm^2	0.86
Shear modulus	10^{-6} kg/cm^2	0.345
Bulk modulus	10^{-6} kg/cm^2	0.4193
Poisson's ratio	–	0.233
Hardness		

TABLE 2.1 (*Continued*)

MAGNESIUM

	Unit	Value(s)
ATOMIC, NUCLEAR AND CRYSTALLOGRAPHIC PROPERTIES		
Symbol: Mg		
Atomic number	—	12
Atomic weight	(based on ^{12}C)	24.312
Periodic classification	—	II A
Atomic volume	cm^3/g-atom	14.0
Atomic radius	Angstrom units	1.62
Ionic radius	Angstrom units (Mg^{2+})	0.65
Thermal neutron absorption -cross section	barns/atom	0.063
Density	g./cm^3	1.74
Lattice type	—	close-packed hexagonal
Lattice constants	Angstrom units	
a		3.203
b		
c		5.199
Valence electrons	—	3s^2
THERMAL AND THERMODYNAMIC PROPERTIES		
Melting point	°C	650
Boiling point	°C	1110
Transition point(s)	°C	
Critical temperature	°K	3530
Linear thermal expansion coefficient	10^{-6} cm/cm/°C	26.1
Specific heat	cal/g/°C	0.245
Electronic specific heat	10^{-4} cal/mole/°C^2	3.11
Thermal conductivity	cal/cm^2/cm/sec/°C	0.37
Latent heat of fusion	kcal/mole	2.14
Latent heat of vaporization at B.P.	kcal/mole	30.75
Heat of combustion	kcal/mole	
Heat of sublimation	kcal/mole	35.6
Heat capacity at constant pressure	cal/°C/mole	5.96
Heat content at M.P.	kcal/mole	4.095
Entropy	cal/°C/mole	7.81
Longitudinal sound velocity	m/sec	4800
ELECTRICAL, ELECTRONIC, MAGNETIC AND OPTICAL PROPERTIES		
Electrical Conductivity	% IACS	38.6
Electrical resistivity	microohm-cm	4.46
Photoelectric work function	eV	3.64
Hall coefficient	volt-cm/amp-Oe	
Curie point	°K	
First ionization potential	eV (Mg$^+$)	7.646
Oxidation potential	V (Mg^{2+})	2.37
Electrochemical equivalent	mg/coulomb (Mg^{2+})	126
Magnetic susceptibility	c.g.s. units	0.55 × 10^6
Refractive index, n_D		0.37
MECHANICAL PROPERTIES		
Young's modulus of elasticity	10^{-6} kg/cm^2	0.452
Shear modulus	10^{-6} kg/cm^2	0.177
Bulk modulus	10^{-6} kg/cm^2	0.3613
Poisson's ratio	—	0.28
Hardness	Brinell	30–47

TABLE 2.1 (*Continued*)

MANGANESE

	Unit	Value(s)
ATOMIC, NUCLEAR AND CRYSTALLOGRAPHIC PROPERTIES		
Symbol: Mn		
Atomic number	–	25
Atomic weight	(based on ^{12}C)	54.938
Periodic classification	–	VII B
Atomic volume	cm^3/g-atom	7.357
Atomic radius	Angstrom units	1.18
Ionic radius	Angstrom units (Mn^{2+})	0.91
Thermal neutron absorption -cross section	barns/atom	13.3
Density	g./cm^3	7.44
Lattice type (α)	–	body-centered cubic
Lattice constants	Angstrom units	
a		8.912
b		
c		
Valence electrons	–	$3d^5\ 4s^2$
THERMAL AND THERMODYNAMIC PROPERTIES		
Melting point	°C	1244
Boiling point	°C	2097
Transition point(s)	°C	700, 1079, 1140, 1244
Critical temperature	°K	6050
Linear thermal expansion coefficient	10^{-6} cm/cm/°C	22
Specific heat	cal/g/°C	0.114
Electronic specific heat	10^{-4} cal/mole/°C^2	20
Thermal conductivity	cal/cm^2/cm/sec/°C	0.0275
Latent heat of fusion	kcal/mole	3.50
Latent heat of vaporization at B.P.	kcal/mole	52.52
Heat of combustion	kcal/mole	
Heat of sublimation	kcal/mole	67.2
Heat capacity at constant pressure	cal/°C/mole	6.285
Heat content at M.P.	kcal/mole	11.78
Entropy	cal/°C/mole	7.65
Longitudinal sound velocity	m/sec (20°C)	5150
ELECTRICAL, ELECTRONIC, MAGNETIC AND OPTICAL PROPERTIES		
Electrical Conductivity	% IACS	0.9
Electrical resistivity	microohm-cm	185
Photoelectric work function	eV	3.83
Hall coefficient	volt-cm/amp-Oe	
Curie point	°K	
First ionization potential	eV (Mn^+)	7.435
Oxidation potential	V (Mn^{2+})	+1.1
Electrochemical equivalent	mg/coulomb (Mn^{2+})	0.28461
Magnetic susceptibility	c.g.s. units	9.9×10^{-6} at 18°C
Refractive index, n_D		2.49 at 0.579μ
MECHANICAL PROPERTIES		
Young's modulus of elasticity	10^{-6} kg/cm^2	2.02
Shear modulus	10^{-6} kg/cm^2	0.78
Bulk modulus	10^{-6} kg/cm^2	0.6083
Poisson's ratio	–	0.24
Hardness	Mohs	5.0

TABLE 2.1 (*Continued*)

MENDELEVIUM

	Unit	Value(s)
ATOMIC, NUCLEAR AND CRYSTALLOGRAPHIC PROPERTIES		
Symbol: Md		
Atomic number	–	101
Atomic weight	(based on ^{12}C)	256
Periodic classification	–	Actinide
Atomic volume	cm^3/g-atom	
Atomic radius	Angstrom units	
Ionic radius	Angstrom units (Md^{3+})	0.96
Thermal neutron absorption -cross section	barns/atom	
Density	g./cm^3	
Lattice type	–	
Lattice constants	Angstrom units	
a		
b		
c		
Valence electrons	–	$5f^{13}\ 7s^2$
THERMAL AND THERMODYNAMIC PROPERTIES		
Melting point	$.\,°C$	
Boiling point	$°C$	
Transition point(s)	$°C$	
Critical temperature	$°K$	
Linear thermal expansion coefficient	10^{-6} cm/cm/$°C$	
Specific heat	cal/g/$°C$	
Electronic specific heat	10^{-4} cal/mole/$°C^2$	
Thermal conductivity	cal/cm^2/cm/sec/$°C$	
Latent heat of fusion	kcal/mole	
Latent heat of vaporization at B.P.	kcal/mole	
Heat of combustion	kcal/mole	
Heat of sublimation	kcal/mole	
Heat capacity at constant pressure	cal/$°C$/mole	
Heat content at M.P.	kcal/mole	
Entropy	cal/$°C$/mole	
Longitudinal sound velocity	m/sec	
ELECTRICAL, ELECTRONIC, MAGNETIC AND OPTICAL PROPERTIES		
Electrical Conductivity	% IACS	
Electrical resistivity	microohm-cm	
Photoelectric work function	eV	
Hall coefficient	volt-cm/amp-Oe	
Curie point	$°K$	
First ionization potential	eV	
Oxidation potential	V	
Electrochemical equivalent	mg/coulomb	
Magnetic susceptibility	c.g.s. units	
Refractive index, n_D	–	
MECHANICAL PROPERTIES		
Young's modulus of elasticity	10^{-6} kg/cm^2	
Shear modulus	10^{-6} kg/cm^2	
Bulk modulus	10^{-6} kg/cm^2	
Poisson's ratio	–	
Hardness		

TABLE 2.1 (*Continued*)

MERCURY

	Unit	Value(s)
ATOMIC, NUCLEAR AND CRYSTALLOGRAPHIC PROPERTIES; Symbol: Hg		
Atomic number	–	80
Atomic weight	(based on ^{12}C)	200.59
Periodic classification	–	II B
Atomic volume	cm^3/g-atom	14.81
Atomic radius	Angstrom units	1.47
Ionic radius	Angstrom units (Hg^{2+})	1.10
Thermal neutron absorption -cross section	barns/atom	370
Density	g./cm^3	13.546 at 20°C
Lattice type	–	Rhombohedral
Lattice constants	Angstrom units	
a		3.005 at –50°C
b		
Axial angle		70° 31.7' at –50°C
Valence electrons	–	6s^2
THERMAL AND THERMODYNAMIC PROPERTIES		
Melting point	°C	–38.87
Boiling point	°C	356.58
Transition point(s)	°C	
Critical temperature	°K	1677
Expansion coefficient	10^{-6} cm/cm/°C	
Specific heat	cal/g/°C	0.033
Electronic specific heat	10^{-4} cal/mole/°C^2	5.3
Thermal conductivity	cal/cm^2/cm/sec/°C	0.022
Latent heat of fusion	kcal/mole	0.5486
Latent heat of vaporization at B.P.	kcal/mole	14.137
Heat of combustion	kcal/mole	
Heat of sublimation	kcal/mole	14.66
Heat capacity at constant pressure	cal/°C/mole	6.68
Heat content at M.P.	kcal/mole	
Entropy	cal/°C/mole	18.19
Longitudinal sound velocity	m/sec	1451
ELECTRICAL, ELECTRONIC, MAGNETIC AND OPTICAL PROPERTIES		
Electrical Conductivity	% IACS	
Electrical resistivity	microohm-cm	95.8
Photoelectric work function	eV	4.52
Hall coefficient	volt-cm/amp-Oe	
Curie point	°K	
First ionization potential	eV	10.43
Oxidation potential	V (Hg^{2+})	–0.85
Electrochemical equivalent	mg/coulomb (Hg^{2+})	2.0788
Magnetic susceptibility	c.g.s. units	–0.15 × 10^{-6}
Refractive index, n_D		1.6 to 1.9
MECHANICAL PROPERTIES		
Young's modulus of elasticity	10^{-6} kg/cm^2	0.279 at 83°K
Shear modulus	10^{-6} kg/cm^2	0.102 at 83°K
Bulk modulus	10^{-6} kg/cm^2	0.288 (solid Hg)
Poisson's ratio	–	0.364 at 83°K
Hardness	Mohs	1.5 (solid Hg)

TABLE 2.1 *(Continued)*

MOLYBDENUM

	Unit	Value(s)
ATOMIC, NUCLEAR AND CRYSTALLOGRAPHIC PROPERTIES		
Symbol: Mo		
Atomic number	–	42
Atomic weight	(based on ^{12}C)	95.94
Periodic classification	–	VI B
Atomic volume	cm^3/g-atom	9.41
Atomic radius	Angstrom units	1.36
Ionic radius	Angstrom units (Mo^{3+})	0.92
Thermal neutron absorption		
-cross section	barns/atom	2.68
Density	g./cm^3	10.22
Lattice type	–	body-centered cubic
Lattice constants	Angstrom units	
a		3.1469
b		
c		
Valence electrons	–	4d^5 5s^1
THERMAL AND THERMODYNAMIC PROPERTIES		
Melting point	°C	2610
Boiling point	°C	5560
Transition point(s)	°C	
Critical temperature	°K	16800
Linear thermal expansion coefficient	10^{-6} cm/cm/°C	5.43
Specific heat	cal/g/°C	0.066
Electronic specific heat	10^{-4} cal/mole/°C^2	5.02
Thermal conductivity	cal/cm^2/cm/sec/°C	0.34
Latent heat of fusion	kcal/mole	6.7
Latent heat of vaporization at B.P.	kcal/mole	142.00
Heat of combustion	kcal/mole	
Heat of sublimation	kcal/mole	157.5
Heat capacity at constant pressure	cal/°C/mole	5.695
Heat content at M.P.	kcal/mole	19.4
Entropy	cal/°C/mole	6.83
Longitudinal sound velocity	m/sec	6370
ELECTRICAL, ELECTRONIC, MAGNETIC AND OPTICAL PROPERTIES		
Electrical Conductivity	% IACS	34
Electrical resistivity	microohm-cm	5.78 at 27°C
Photoelectric work function	eV	4.20
Hall coefficient	volt-cm/amp-Oe	
Curie point	°K	
First ionization potential	eV (Mo$^+$)	7.099
Oxidation potential	V (Mo/MoO$_2$)	0.87 (alk. soln.)
Electrochemical equivalent	mg/coulomb (Mo^{6+})	0.1658
Magnetic susceptibility	c.g.s. units	0.93×10^{-6}
Refractive index, n_D		
MECHANICAL PROPERTIES		
Young's modulus of elasticity	10^{-6} kg/cm^2	3.34
Shear modulus	10^{-6} kg/cm^2	1.18
Bulk modulus	10^{-6} kg/cm^2	2.779
Poisson's ratio	–	0.30
Hardness	Vickers	200

TABLE 2.1 (*Continued*)

NEODYMIUM

	Unit	Value(s)
ATOMIC, NUCLEAR AND **CRYSTALLOGRAPHIC PROPERTIES**		
Symbol: Nd		
Atomic number	–	60
Atomic weight	(based on ^{12}C)	144.24
Periodic classification	–	Lanthanide
Atomic volume	cm^3/g-atom	20.60
Atomic radius	Angstrom units	1.821
Ionic radius	Angstrom units (Nd^{3+})	1.15
Thermal neutron absorption		
-cross section	barns/atom	50
Density	g./cm^3	7.004
Lattice type	–	hexagonal
Lattice constants	Angstrom units	
a		3.659
b		
c		11.799
Valence electrons	–	$4f^4 6s^2$
THERMAL AND THERMODYNAMIC **PROPERTIES**		
Melting point	°C	1024
Boiling point	°C	3027
Transition point(s)	°C	862
Critical temperature	°K	7900
Linear thermal expansion coefficient	10^{-6} cm/cm/°C	6.7
Specific heat	cal/g/°C	0.050 at 0°C
Electronic specific heat	10^{-4} cal/mole/°C^2	21.3
Thermal conductivity	cal/cm^2/cm/sec/°C	0.031
Latent heat of fusion	kcal/mole	1.71
Latent heat of vaporization at B.P.	kcal/mole	67.8
Heat of combustion	kcal/mole	
Heat of sublimation	kcal/mole	75.6
Heat capacity at constant pressure	cal/°C/mole	6.56
Heat content at M.P.	kcal/mole	8.78
Entropy	cal/°C/mole	17.54
Longitudinal sound velocity	m/sec	2700
ELECTRICAL, ELECTRONIC, MAGNETIC **AND OPTICAL PROPERTIES**		
Electrical Conductivity	% IACS	
Electrical resistivity	microohm-cm	64.3
Photoelectric work function	eV	3.3
Hall coefficient	volt-cm/amp-Oe	$+0.97 \times 10^{-12}$
Curie point	°K	3.3
First ionization potential	eV (Nd$^+$)	5.49
Oxidation potential	V (Nd^{3+})	2.44
Electrochemical equivalent	mg/coulomb (Nd^{3+})	0.49834
Magnetic susceptibility	c.g.s. units	3460×10^{-6}
Refractive index, n_D		
MECHANICAL PROPERTIES		
Young's modulus of elasticity	10^{-6} kg/cm^2	0.387
Shear modulus	10^{-6} kg/cm^2	0.148
Bulk modulus	10^{-6} kg/cm^2	0.3331
Poisson's ratio	–	0.306
Hardness	Brinell (as cast)	35

TABLE 2.1 (*Continued*)

NEON

	Unit	Value(s)
ATOMIC AND NUCLEAR PROPERTIES		
Symbol: Ne		
Atomic number	−	10
Atomic weight	(based on ^{12}C)	20.183
Periodic classification	−	rare gas
Thermal neutron absorption		
-cross section	barns/atom	0.04
SOLID STATE		
Lattice type	−	face-centered cubic
Lattice constants	Angstrom units	
a		4.53 at 5°K
b		
c		
Atomic radius	Angstrom units	
Melting point	°K	24.55
Heat of fusion	kcal/mole	0.080
Entropy of fusion	cal/°K/mole	
Triple point		
Temperature	°K	24.54
Pressure	torr	325
Density of solid	g/cm^3	1.444
Density of liquid	g/cm^3	1.247
LIQUID STATE		
Boiling point	°K	27.07
Density at boiling point	g/cm^3	1.206
Heat of vaporization at boiling point	kcal/mole	0.422
Entropy at boiling point	cal/°K/mole	
Critical Data		
Temperature	°K	44.40
Pressure	atm.	26.19
Density	g/cm^3	0.483
GASEOUS STATE*		
Density at 0°C	g/1	0.90002
Specific heat at 20°C at const. pres.	cal/g/°K	0.246
Ratio of specific heats, Cp/Cv, at 20°C	−	1.64
Thermal conductivity at 0°C	cal/cm^2/cm/sec/°K	0.00011
Heat capacity at const. press. & 25°C	cal/°K/mole	4.97
Entropy at 25°C	cal/°K/mole	34.95
Heat of combustion	kcal/mole	
Dielectric constant at 0°C	−	1.000134
Viscosity at 25°C	micropoise	317.3
Refractive index, n_D, at 25°C	−	1.000067
Solubility in H_2O at 1 atm. partial		
pressure and 20°C	cm^3(STP)/1000 g.	14.0
Sound velocity at 0°C	m/sec	433

*Generally at one atmosphere.

TABLE 2.1 (*Continued*)

NEPTUNIUM

	Unit	Value(s)
ATOMIC, NUCLEAR AND CRYSTALLOGRAPHIC PROPERTIES		
Symbol: Np		
Atomic number	–	93
Atomic weight	(based on ^{12}C)	237
Periodic classification	–	Actinide
Atomic volume	cm^3/g-atom	13.11
Atomic radius	Angstrom units	1.50
Ionic radius	Angstrom units (Np^{3+})	1.02
Thermal neutron absorption -cross section	barns/atom	
Density	g./cm^3	20.25
Lattice type	–	Orthorhombic
Lattice constants	Angstrom units	
a		4.723
b		4.887
c		6.663
Valence electrons	–	5f^4 6d^1 7s^2
THERMAL AND THERMODYNAMIC PROPERTIES		
Melting point	°C	637
Boiling point	°C	3880
Transition point(s)	°C	280, 577
Critical temperature	°K	
Linear thermal expansion coefficient	10^{-6} cm/cm/°C	27.5
Specific heat	cal/g/°C	
Electronic specific heat	10^{-4} cal/mole/°C^2	24
Thermal conductivity	cal/cm^2/cm/sec/°C	
Latent heat of fusion	kcal/mole	1.60
Latent heat of vaporization at B.P.	kcal/mole	
Heat of combustion	kcal/mole	
Heat of sublimation	kcal/mole	113
Heat capacity at constant pressure	cal/°C/mole	7.02
Heat content at M.P.	kcal/mole	
Entropy	cal/°C/mole	
Longitudinal sound velocity	m/sec	
ELECTRICAL, ELECTRONIC, MAGNETIC AND OPTICAL PROPERTIES		
Electrical Conductivity	% IACS	
Electrical resistivity	microohm-cm	116–121
Photoelectric work function	eV	
Hall coefficient	volt-cm/amp-Oe	
Curie point	°K	
First ionization potential	eV (Np$^+$)	
Oxidation potential	V (Np^{3+})	1.86 (acid soln.)
Electrochemical equivalent	mg/coulomb	
Magnetic susceptibility	c.g.s. units	
Refractive index, n_D		
MECHANICAL PROPERTIES		
Young's modulus of elasticity	10^{-6} kg/cm^2	1.02
Shear modulus	10^{-6} kg/cm^2	0.406
Bulk modulus	10^{-6} kg/cm^2	0.694
Poisson's ratio	–	0.255
Hardness		

TABLE 2.1 (*Continued*)

NICKEL

	Unit	Value(s)
ATOMIC, NUCLEAR AND CRYSTALLOGRAPHIC PROPERTIES		
Symbol: Ni		
Atomic number	−	28
Atomic weight	(based on ^{12}C)	58.71
Periodic classification	−	VIII
Atomic volume	cm^3/g-atom	6.593
Atomic radius	Angstrom units	1.24
Ionic radius	Angstrom units (Ni^{2+})	0.69
Thermal neutron absorption		
-cross section	barns/atom	4.57
Density	g./cm^3	8.908
Lattice type	−	face-centered cubic
Lattice constants	Angstrom units	
a		3.5168
b		
c		
Valence electrons	−	$3d^8\ 4s^2$
THERMAL AND THERMODYNAMIC PROPERTIES		
Melting point	°C	1453
Boiling point	°C	2782
Transition point(s)	°C	
Critical temperature	°K	11750
Linear thermal expansion coefficient	10^{-6} cm/cm/°C	13.3
Specific heat	cal/g/°C	0.109
Electronic specific heat	10^{-4} cal/mole/°C^2	17.4
Thermal conductivity	cal/cm^2/cm/sec/°C	0.198 at 100°C
Latent heat of fusion	kcal/mole	4.21
Latent heat of vaporization at B.P.	kcal/mole	88.87
Heat of combustion	kcal/mole	
Heat of sublimation	kcal/mole	102.8
Heat capacity at constant pressure	cal/°C/mole	5.95
Heat content at M.P.	kcal/mole	11.11
Entropy	cal/°C/mole	7.12
Longitudinal sound velocity	m/sec	4900
ELECTRICAL, ELECTRONIC, MAGNETIC AND OPTICAL PROPERTIES		
Electrical Conductivity	% IACS	25.2
Electrical resistivity	microohm-cm	6.844
Photoelectric work function	eV	5.0
Hall coefficient	volt-cm/amp-Oe	
Curie point	°K	626
First ionization potential	eV (Ni^+)	7.635
Oxidation potential	V (Ni^{2+})	0.250 (acid soln.)
Electrochemical equivalent	mg/coulomb (Ni^{2+})	0.30409
Magnetic susceptibility	c.g.s. units	ferromagnetic
Refractive index, n_D		1.79
MECHANICAL PROPERTIES		
Young's modulus of elasticity	10^{-6} kg/cm^2	1.97
Shear modulus	10^{-6} kg/cm^2	0.765
Bulk modulus	10^{-6} kg/cm^2	1.90
Poisson's ratio	−	0.30
Hardness	Rockwell B (annealed)	30

TABLE 2.1 (*Continued*)

NIOBIUM

	Unit	Value(s)
ATOMIC, NUCLEAR AND CRYSTALLOGRAPHIC PROPERTIES		
Symbol: Nb		
Atomic number	–	41
Atomic weight	(based on ^{12}C)	92.906
Periodic classification	–	V B
Atomic volume	cm^3/g-atom	10.83
Atomic radius	Angstrom units	1.47
Ionic radius	Angstrom units (Nb^{5+})	0.70
Thermal neutron absorption -cross section	barns/atom	1.15
Density	g./cm^3	8.57
Lattice type	–	body-centered cubic
Lattice constants	Angstrom units	
a		3.294
b		
c		
Valence electrons	–	4d^4 5s^1
THERMAL AND THERMODYNAMIC PROPERTIES		
Melting point	°C	2468
Boiling point	°C	4540
Transition point(s)	°C	
Critical temperature	°K	19000
Linear thermal expansion coefficient	10^{-6} cm/cm/°C	7.07
Specific heat	cal/g/°C	0.06430 at 0°C
Electronic specific heat	10^{-4} cal/mole/°C^2	18.3
Thermal conductivity	cal/cm^2/cm/sec/°C	0.125 at 0°C
Latent heat of fusion	kcal/mole	4.82
Latent heat of vaporization at B.P.	kcal/mole	166.5
Heat of combustion	kcal/mole	222
Heat of sublimation	kcal/mole	175
Heat capacity at constant pressure	cal/°C/mole	5.965
Heat content at M.P.	kcal/mole	17.05
Entropy	cal/°C/mole	8.73 at 15°C
Longitudinal sound velocity	m/sec	3480
ELECTRICAL, ELECTRONIC, MAGNETIC AND OPTICAL PROPERTIES		
Electrical Conductivity	% IACS	13.2
Electrical resistivity	microohm-cm	15.22 at 0°C
Photoelectric work function	eV	4.01
Hall coefficient	volt-cm/amp-Oe	
Curie point	°K	
First ionization potential	eV (Nb$^+$)	6.88
Oxidation potential	V (Nb^{5+})	0.96
Electrochemical equivalent	mg/coulomb (Nb^{5+})	0.1926
Magnetic susceptibility	c.g.s. units	$+2.28 \times 10^{-6}$
Refractive index, n_D		1.80
MECHANICAL PROPERTIES		
Young's modulus of elasticity	10^{-6} kg/cm^2	1.07
Shear modulus	10^{-6} kg/cm^2	0.382
Bulk modulus	10^{-6} kg/cm^2	1.736
Poisson's ratio	–	0.35
Hardness	Vickers (annealed)	54

TABLE 2.1 (Continued)

NITROGEN

	Unit	Value(s)
ATOMIC AND NUCLEAR PROPERTIES		
Symbol: N		
Atomic number	–	7
Atomic weight	(based on ^{12}C)	14.0067
Periodic classification	–	V A
Thermal neutron absorption -cross section	barns/atom	1.9
SOLID STATE		
Lattice type	–	close-packed hexagonal
Lattice constants	Angstrom units	
a		4.04 at 40°K
b		
c		6.60 at 40°K
Atomic radius	Angstrom units	0.71
Melting point	°K	63.18
Heat of fusion	kcal/mole	0.172
Entropy of fusion	cal/°K/mole	
Triple point		
Temperature	°K	63.166
Pressure	torr	92.5
Density of solid	g/cm^3	0.8792
Density of liquid	g/cm^3	
LIQUID STATE		
Boiling point	°K	77.36
Density at boiling point	g/cm^3	0.810
Heat of vaporization at boiling point	kcal/mole	1.335
Entropy at boiling point	cal/°K/mole	
Critical Data		
Temperature	°K	126.1
Pressure	atm.	33.50
Density	g/cm^3	0.311
GASEOUS STATE*		
Density at 0°C	g/l	1.2409
Specific heat at 20°C at const. pres.	cal/g/°K	0.247
Ratio of specific heats, Cp/Cv, at 20°C	–	1.401
Thermal conductivity at 20°C	cal/cm^2/cm/sec/°K	0.0000571
Heat capacity at const. press. & 25°C	cal/°K/mole	6.96
Entropy at 25°C	cal/°K/mole	45.77
Heat of combustion	kcal/mole	
Dielectric constant at 20°C	–	
Viscosity at 0°C	micropoise	83
Refractive index, n_D, at 0°C	–	1.000296
Solubility in H_2O at 1 atm. partial pressure and 0°C	cm^3(STP)/1000 g.	23
Sound velocity at 0°C	m/sec	334

*Generally at one atmosphere.

TABLE 2.1 (*Continued*)

NOBELIUM

	Unit	Value(s)
ATOMIC, NUCLEAR AND CRYSTALLOGRAPHIC PROPERTIES		
Symbol: No		
Atomic number	–	102
Atomic weight	(based on ^{12}C)	255
Periodic classification	–	Actinide
Atomic volume	cm^3/g-atom	
Atomic radius	Angstrom units	
Ionic radius	Angstrom units (No^{3+})	0.95
Thermal neutron absorption -cross section	barns/atom	
Density	g./cm^3	
Lattice type	–	
Lattice constants	Angstrom units	
a		
b		
c		
Valence electrons	–	$5f^{14} 7s^2$
THERMAL AND THERMODYNAMIC PROPERTIES		
Melting point	°C	
Boiling point	°C	
Transition point(s)	°C	
Critical temperature	°K	
Linear thermal expansion coefficient	10^{-6} cm/cm/°C	
Specific heat	cal/g/°C	
Electronic specific heat	10^{-4} cal/mole/°C^2	
Thermal conductivity	cal/cm^2/cm/sec/°C	
Latent heat of fusion	kcal/mole	
Latent heat of vaporization at B.P.	kcal/mole	
Heat of combustion	kcal/mole	
Heat of sublimation	kcal/mole	
Heat capacity at constant pressure	cal/°C/mole	
Heat content at M.P.	kcal/mole	
Entropy	cal/°C/mole	
Longitudinal sound velocity	m/sec	
ELECTRICAL, ELECTRONIC, MAGNETIC AND OPTICAL PROPERTIES		
Electrical Conductivity	% IACS	
Electrical resistivity	microohm-cm	
Photoelectric work function	eV	
Hall coefficient	volt-cm/amp-Oe	
Curie point	°K	
First ionization potential	eV	
Oxidation potential	V	
Electrochemical equivalent	mg/coulomb	
Magnetic susceptibility	c.g.s. units	
Refractive index, n_D	–	
MECHANICAL PROPERTIES		
Young's modulus of elasticity	10^{-6} kg/cm^2	
Shear modulus	10^{-6} kg/cm^2	
Bulk modulus	10^{-6} kg/cm^2	
Poisson's ratio	–	
Hardness		

TABLE 2.1 (Continued)

OSMIUM

	Unit	Value(s)
ATOMIC, NUCLEAR AND CRYSTALLOGRAPHIC PROPERTIES		
Symbol: Os		
Atomic number	−	76
Atomic weight	(based on ^{12}C)	190.20
Periodic classification	−	VIII
Atomic volume	cm^3/g-atom	8.441
Atomic radius	Angstrom units	1.335
Ionic radius	Angstrom units (Os^{4+})	0.67
Thermal neutron absorption -cross section	barns/atom	15.3
Density	g./cm^3	22.61
Lattice type	−	close-packed hexagonal
Lattice constants	Angstrom units	
a		2.734
b		
c		4.320
Valence electrons	−	$5d^6\ 6s^2$
THERMAL AND THERMODYNAMIC PROPERTIES		
Melting point	°C	3050
Boiling point	°C	5027
Transition point(s)	°C	
Critical temperature	°K	
Linear thermal expansion coefficient	10^{-6} cm/cm/°C	6.1
Specific heat	cal/g/°C	0.0309
Electronic specific heat	10^{-4} cal/mole/°C^2	5.62
Thermal conductivity	cal/cm^2/cm/sec/°C	0.208
Latent heat of fusion	kcal/mole	7.56
Latent heat of vaporization at B.P.	kcal/mole	150.0
Heat of combustion	kcal/mole	
Heat of sublimation	kcal/mole	187.4
Heat capacity at constant pressure	cal/°C/mole	5.95
Heat content at M.P.	kcal/mole	18.47
Entropy	cal/°C/mole	7.80
Longitudinal sound velocity	m/sec (20°C)	4940
ELECTRICAL, ELECTRONIC, MAGNETIC AND OPTICAL PROPERTIES		
Electrical Conductivity	% IACS	
Electrical resistivity	microohm-cm	8.12
Photoelectric work function	eV	4.55
Hall coefficient	volt-cm/amp-Oe	
Curie point	°K	
First ionization potential	eV (Os^+)	8.7
Oxidation potential	V (Os/OsO_4)	−0.85
Electrochemical equivalent	mg/coulomb (Os^{4+})	0.49611
Magnetic susceptibility	c.g.s. units	0.052×10^{-6}
Refractive index, n_D		
MECHANICAL PROPERTIES		
Young's modulus of elasticity	10^{-6} kg/cm^2	5.50
Shear modulus	10^{-6} kg/cm^2	2.14
Bulk modulus	10^{-6} kg/cm^2	4.26
Poisson's ratio	−	0.285
Hardness	Vickers (annealed)	300–670

TABLE 2.1 (*Continued*)

OXYGEN

	Unit	Value(s)
ATOMIC AND NUCLEAR PROPERTIES		
Symbol: O		
Atomic number	–	8
Atomic weight	(based on ^{12}C)	15.9994
Periodic classification	–	VI A
Thermal neutron absorption		
-cross section	barns/atom	0.27
SOLID STATE		
Lattice type	–	Cubic
Lattice constants	Angstrom units	
a		6.84 at 48°K
b		
c		
Atomic radius	Angstrom units	0.60
Melting point	°K	54.36
Heat of fusion	kcal/mole	0.106
Entropy of fusion	cal/°K/mole	
Triple point		
Temperature	°K	54.4
Pressure	torr	1.2
Density of solid	g/cm^3	
Density of liquid	g/cm^3	1.31
LIQUID STATE		
Boiling point	°K	90.19
Density at boiling point	g/cm^3	1.142
Heat of vaporization at boiling point	kcal/mole	1.630
Entropy at boiling point	cal/°K/mole	
Critical Data		
Temperature	°K	154.8
Pressure	atm.	50.15
Density	g/cm^3	0.430
GASEOUS STATE*		
Density at 0°C	g/1	1.429
Specific heat at 20°C at const. pres.	cal/g/°K	0.219
Ratio of specific heats, Cp/Cv, at 0°C	–	1.396
Thermal conductivity at 20°C	cal/cm^2/cm/sec/°K	0.000058
Heat capacity at const. press. & 25°C	cal/°K/mole	7.02
Entropy at 25°C	cal/°K/mole	49.01
Heat of combustion	kcal/mole	
Dielectric constant at 0°C	–	1.0005233
Viscosity at 25°C	micropoise	192
Refractive index, n_D, at 0°C	–	
Solubility in H$_2$O at 1 atm. partial		
pressure and 20°C	cm^3(STP)/1000 g.	31
Sound velocity at 25°C	m/sec	330

*Generally at one atmosphere.

TABLE 2.1 (Continued)

PALLADIUM

	Unit	Value(s)
ATOMIC, NUCLEAR AND CRYSTALLOGRAPHIC PROPERTIES		
Symbol: Pd		
Atomic number	–	46
Atomic weight	(based on ^{12}C)	106.4
Periodic classification	–	VIII
Atomic volume	cm^3/g-atom	8.879
Atomic radius	Angstrom units	1.373
Ionic radius	Angstrom units (Pd^{4+})	0.66
Thermal neutron absorption -cross section	barns/atom	8.0
Density	g./cm^3	12.02
Lattice type	–	face-centered cubic
Lattice constants	Angstrom units	
a		3.891
b		
c		
Valence electrons	–	$4d^{10}$
THERMAL AND THERMODYNAMIC PROPERTIES		
Melting point	°C	1552
Boiling point	°C	2927
Transition point(s)	°C	
Critical temperature	°K	
Linear thermal expansion coefficient	10^{-6} cm/cm/°C	11.2
Specific heat	cal/g/°C	0.058
Electronic specific heat	10^{-4} cal/mole/°C^2	23.9
Thermal conductivity	cal/cm^2/cm/sec/°C	0.182
Latent heat of fusion	kcal/mole	4.10
Latent heat of vaporization at B.P.	kcal/mole	86
Heat of combustion	kcal/mole	
Heat of sublimation	kcal/mole	90
Heat capacity at constant pressure	cal/°C/mole	6.21
Heat content at M.P.	kcal/mole	10.89
Entropy	cal/°C/mole	9.06
Longitudinal sound velocity	m/sec	5478
ELECTRICAL, ELECTRONIC, MAGNETIC AND OPTICAL PROPERTIES		
Electrical Conductivity	% IACS	16
Electrical resistivity	microohm-cm	9.93 at 0°C
Photoelectric work function	eV	4.97
Hall coefficient	volt-cm/amp-Oe	
Curie point	°K	
First ionization potential	eV (Pd^+)	8.34
Oxidation potential	V (Pd^{2+})	−0.987
Electrochemical equivalent	mg/coulomb (Pd^{4+})	0.27642
Magnetic susceptibility	c.g.s. units	5.23×10^{-6}
Refractive index, n_D		1.62 at 0.579 μ
MECHANICAL PROPERTIES		
Young's modulus of elasticity	10^{-6} kg/cm^2	1.26
Shear modulus	10^{-6} kg/cm^2	0.521
Bulk modulus	10^{-6} kg/cm^2	1.844
Poisson's ratio	–	0.375
Hardness	Vickers (annealed)	50

TABLE 2.1 (*Continued*)

PHOSPHORUS (WHITE)

	Unit	Value(s)
ATOMIC, NUCLEAR AND CRYSTALLOGRAPHIC PROPERTIES		
Symbol: P		
Atomic number	–	15
Atomic weight	(based on ^{12}C)	30.9738
Periodic classification	–	V A
Atomic volume	cm^3/g-atom	13.96
Atomic radius	Angstrom units	1.09
Ionic radius	Angstrom units (P^{5+})	0.34
Thermal neutron absorption		
-cross section	barns/atom	0.19
Density	g./cm^3	1.83
Lattice type	(α-form)	cubic
Lattice constants	Angstrom units	
a		7.18
b		
c		
Valence electrons	–	$3s^2 \ 3p^3$
THERMAL AND THERMODYNAMIC PROPERTIES		
Melting point	°C	44.1
Boiling point	°C	280
Transition point(s)	°C	−79.6
Critical temperature	°K	948
Linear thermal expansion coefficient	10^{-6} cm/cm/°C	124.5
Specific heat	cal/g/°C	0.189
Electronic specific heat	10^{-4} cal/mole/°C^2	0
Thermal conductivity	cal/cm^2/cm/sec/°C	
Latent heat of fusion	kcal/mole	0.15
Latent heat of vaporization at B.P.	kcal/mole	
Heat of combustion	kcal/mole	
Heat of sublimation	kcal/mole	7.2
Heat capacity at constant pressure	cal/°C/mole	5.63
Heat content at M.P.	kcal/mole	
Entropy	cal/°C/mole	
Longitudinal sound velocity	m/sec	
ELECTRICAL, ELECTRONIC, MAGNETIC AND OPTICAL PROPERTIES		
Electrical Conductivity	% IACS	
Electrical resistivity	microohm-cm	1×10^{17}
Photoelectric work function	eV	
Hall coefficient	volt-cm/amp-Oe	
Curie point	°K	
First ionization potential	eV (P^+)	10.486
Oxidation potential	V (P/H_3PO_4)	0.776
Electrochemical equivalent	mg/coulomb (P^{5+})	0.06421
Magnetic susceptibility	c.g.s. units	$−0.90 \times 10^{-6}$
Refractive index, n_D		
MECHANICAL PROPERTIES		
Young's modulus of elasticity	10^{-6} kg/cm^2	0.047
Shear modulus	10^{-6} kg/cm^2	0.018
Bulk modulus	10^{-6} kg/cm^2	0.0478
Poisson's ratio	–	0.335
Hardness		

TABLE 2.1 (Continued)

PLATINUM

	Unit	Value(s)
ATOMIC, NUCLEAR AND CRYSTALLOGRAPHIC PROPERTIES		
Symbol: Pt		
Atomic number	–	78
Atomic weight	(based on ^{12}C)	195.09
Periodic classification	–	VIII
Atomic volume	cm^3/g-atom	9.094
Atomic radius	Angstrom units	1.384
Ionic radius	Angstrom units (Pt^{4+})	0.64
Thermal neutron absorption -cross section	barns/atom	10
Density	g./cm^3	21.45
Lattice type	–	face-centered cubic
Lattice constants	Angstrom units	
a		3.9231
b		
c		
Valence electrons	–	$5d^9\,6s^1$
THERMAL AND THERMODYNAMIC PROPERTIES		
Melting point	°C	1769
Boiling point	°C	3827
Transition point(s)	°C	
Critical temperature	°K	8280
Linear thermal expansion coefficient	10^{-6} cm/cm/°C	9.0
Specific heat	cal/g/°C	0.03136 at 0°C
Electronic specific heat	10^{-4} cal/mole/°C^2	16.0
Thermal conductivity	cal/cm^2/cm/sec/°C	0.176
Latent heat of fusion	kcal/mole	4.7
Latent heat of vaporization at B.P.	kcal/mole	122.0
Heat of combustion	kcal/mole	
Heat of sublimation	kcal/mole	135.0
Heat capacity at constant pressure	cal/°C/mole	6.185
Heat content at M.P.	kcal/mole	
Entropy	cal/°C/mole	10.00
Longitudinal sound velocity	m/sec	4075
ELECTRICAL, ELECTRONIC, MAGNETIC AND OPTICAL PROPERTIES		
Electrical Conductivity	% IACS	
Electrical resistivity	microohm-cm	10.6
Photoelectric work function	eV	5.27
Hall coefficient	volt-cm/amp-Oe	
Curie point	°K	
First ionization potential	eV (Pt^+)	9.0
Oxidation potential	V (Pt^{2+})	–1.2
Electrochemical equivalent	mg/coulomb (Pt^{4+})	0.50578
Magnetic susceptibility	c.g.s. units	0.98×10^{-6}
Refractive index, n_D		2.63
MECHANICAL PROPERTIES		
Young's modulus of elasticity	10^{-6} kg/cm^2	1.74
Shear modulus	10^{-6} kg/cm^2	0.622
Bulk modulus	10^{-6} kg/cm^2	2.838
Poisson's ratio	–	0.38
Hardness	Vickers (annealed)	40–42

TABLE 2.1 (*Continued*)

PLUTONIUM

	Unit	Value(s)
ATOMIC, NUCLEAR AND CRYSTALLOGRAPHIC PROPERTIES		
Symbol: Pu		
Atomic number	–	94
Atomic weight	(based on ^{12}C)	244
Periodic classification	–	Actinide
Atomic volume	cm^3/g-atom	12.06
Atomic radius	Angstrom units	1.62
Ionic radius	Angstrom units (Pu^{4+})	0.86
Thermal neutron absorption -cross section	barns/atom	
Density	g./cm^3	19.84
Lattice type	–	Monoclinic
Lattice constants	Angstrom units	
a		6.182
b		4.826
c		10.956
Valence electrons	–	5f^6 7s^2
THERMAL AND THERMODYNAMIC PROPERTIES		
Melting point	°C	640
Boiling point	°C	3454
Transition point(s)	°C	110, 190, 310, 452, 480, 639.5
Critical temperature	°K	
Linear thermal expansion coefficient	10^{-6} cm/cm/°C	55
Specific heat	cal/g/°C	0.033
Electronic specific heat	10^{-4} cal/mole/°C^2	116.8
Thermal conductivity	cal/cm^2/cm/sec/°C	0.020
Latent heat of fusion	kcal/mole	0.676
Latent heat of vaporization at B.P.	kcal/mole	
Heat of combustion	kcal/mole	252.87
Heat of sublimation	kcal/mole	91.8
Heat capacity at constant pressure	cal/°C/mole	8.50
Heat content at M.P.	kcal/mole	
Entropy	cal/°C/mole	
Longitudinal sound velocity	m/sec	2250 at 30°C
ELECTRICAL, ELECTRONIC, MAGNETIC AND OPTICAL PROPERTIES		
Electrical Conductivity	% IACS	
Electrical resistivity	microohm-cm	141.4 at 107°C
Photoelectric work function	eV	
Hall coefficient	volt-cm/amp-Oe	
Curie point	°K	
First ionization potential	eV (Pu$^+$)	5.8
Oxidation potential	V (Pu^{3+})	2.07
Electrochemical equivalent	mg/coulomb	
Magnetic susceptibility	c.g.s. units	2.25 × 10^{-6}
Refractive index, n_D		
MECHANICAL PROPERTIES		
Young's modulus of elasticity	10^{-6} kg/cm^2	0.984
Shear modulus	10^{-6} kg/cm^2	0.446
Bulk modulus	10^{-6} kg/cm^2	0.546
Poisson's ratio	–	0.15
Hardness	DPH	250–283

TABLE 2.1 *(Continued)*

POLONIUM

	Unit	Value(s)
ATOMIC, NUCLEAR AND CRYSTALLOGRAPHIC PROPERTIES		
Symbol: Po		
Atomic number	–	84
Atomic weight	(based on ^{12}C)	210
Periodic classification	–	VI A
Atomic volume	cm^3/g-atom	22.53
Atomic radius	Angstrom units	1.64
Ionic radius	Angstrom units (Po^{4+})	1.02
Thermal neutron absorption -cross section	barns/atom	
Density	g./cm^3	9.4
Lattice type	–	cubic
Lattice constants	Angstrom units	
a		3.359
b		
c		
Valence electrons	–	$6s^2 6p^4$
THERMAL AND THERMODYNAMIC PROPERTIES		
Melting point	°C	254
Boiling point	°C	962
Transition point(s)	°C	36
Critical temperature	°K	2281
Linear thermal expansion coefficient	10^{-6} cm/cm/°C	23.0
Specific heat	cal/g/°C	0.030
Electronic specific heat	10^{-4} cal/mole/$°C^2$	0.24
Thermal conductivity	cal/cm^2/cm/sec/°C	0.24
Latent heat of fusion	kcal/mole	0.91
Latent heat of vaporization at B.P.	kcal/mole	14.4
Heat of combustion	kcal/mole	
Heat of sublimation	kcal/mole	34.5
Heat capacity at constant pressure	cal/°C/mole	6.30
Heat content at M.P.	kcal/mole	1.37
Entropy	cal/°C/mole	15.00
Longitudinal sound velocity	m/sec	
ELECTRICAL, ELECTRONIC, MAGNETIC AND OPTICAL PROPERTIES		
Electrical Conductivity	% IACS	
Electrical resistivity	microohm-cm	140
Photoelectric work function	eV	4.7 (estimate)
Hall coefficient	volt-cm/amp-Oe	
Curie point	°K	
First ionization potential	eV (Po^+)	8.42
Oxidation potential	V (Po^{2+})	−0.65
Electrochemical equivalent	mg/coulomb (Po^{6+})	0.36269
Magnetic susceptibility	c.g.s. units	
Refractive index, n_D		
MECHANICAL PROPERTIES		
Young's modulus of elasticity	10^{-6} kg/cm^2	0.26
Shear modulus	10^{-6} kg/cm^2	0.097
Bulk modulus	10^{-6} kg/cm^2	0.27
Poisson's ratio	–	0.338
Hardness		

TABLE 2.1 (*Continued*)

POTASSIUM

	Unit	Value(s)
ATOMIC, NUCLEAR AND CRYSTALLOGRAPHIC PROPERTIES		
Symbol: K		
Atomic number	–	19
Atomic weight	(based on ^{12}C)	39.102
Periodic classification	–	I A
Atomic volume	cm^3/g-atom	45.61
Atomic radius	Angstrom units	2.31
Ionic radius	Angstrom units (K^+)	1.33
Thermal neutron absorption -cross section	barns/atom	2.1
Density	g./cm^3	0.862
Lattice type	–	body-centered cubic
Lattice constants	Angstrom units	
a		5.334
b		
c		
Valence electrons	–	$4s^1$
THERMAL AND THERMODYNAMIC PROPERTIES		
Melting point	°C	63.4
Boiling point	°C	754
Transition point(s)	°C	
Critical temperature	°K	2140
Linear thermal expansion coefficient	10^{-6} cm/cm/°C	83
Specific heat	cal/g/°C	0.176 at 0°C
Electronic specific heat	10^{-4} cal/mole/°C^2	5.04
Thermal conductivity	cal/cm^2/cm/sec/°C	0.232 at 21°C
Latent heat of fusion	kcal/mole	0.556
Latent heat of vaporization at B.P.	kcal/mole	18.53
Heat of combustion	kcal/mole	
Heat of sublimation	kcal/mole	21.48
Heat capacity at constant pressure	cal/°C/mole	7.07
Heat content at M.P.	kcal/mole	0.013
Entropy	cal/°C/mole	15.39
Longitudinal sound velocity	m/sec	2000
ELECTRICAL, ELECTRONIC, MAGNETIC AND OPTICAL PROPERTIES		
Electrical Conductivity	% IACS	
Electrical resistivity	microohm-cm	6.15 at 0°C
Photoelectric work function	eV	2.22
Hall coefficient	volt-cm/amp-Oe	
Curie point	°K	
First ionization potential	eV (K^+)	4.341
Oxidation potential	V (K^+)	2.925
Electrochemical equivalent	mg/coulomb (K^+)	0.40514
Magnetic susceptibility	c.g.s. units	0.52×10^{-6}
Refractive index, n_D		
MECHANICAL PROPERTIES		
Young's modulus of elasticity	10^{-6} kg/cm^2	0.0361 at 90°K
Shear modulus	10^{-6} kg/cm^2	0.013 at 90°K
Bulk modulus	10^{-6} kg/cm^2	0.0324
Poisson's ratio	–	0.35
Hardness		

TABLE 2.1 (*Continued*)

PRASEODYMIUM

	Unit	Value(s)
ATOMIC, NUCLEAR AND CRYSTALLOGRAPHIC PROPERTIES		
Symbol: Pr		
Atomic number	–	59
Atomic weight	(based on ^{12}C)	140.907
Periodic classification	–	Lanthanide
Atomic volume	cm^3/g-atom	20.82
Atomic radius	Angstrom units	1.828
Ionic radius	Angstrom units (Pr^{3+})	1.16
Thermal neutron absorption		
-cross section	barns/atom	12
Density	g./cm^3	6.782
Lattice type	–	hexagonal
Lattice constants	Angstrom units	
a		3.6702
b		
c		11.828
Valence electrons	–	$4f^3\,6s^2$
THERMAL AND THERMODYNAMIC PROPERTIES		
Melting point	°C	935
Boiling point	°C	3343
Transition point(s)	°C	792
Critical temperature	°K	8900
Linear thermal expansion coefficient	10^{-6} cm/cm/°C	6.79
Specific heat	cal/g/°C	0.045
Electronic specific heat	10^{-4} cal/mole/°C^2	52.3
Thermal conductivity	cal/cm^2/cm/sec/°C	0.028
Latent heat of fusion	kcal/mole	1.652
Latent heat of vaporization at B.P.	kcal/mole	79.5
Heat of combustion	kcal/mole	
Heat of sublimation	kcal/mole	85.5
Heat capacity at constant pressure	cal/°C/mole	6.45
Heat content at M.P.	kcal/mole	7.30
Entropy	cal/°C/mole	17.45
Longitudinal sound velocity	m/sec	2700
ELECTRICAL, ELECTRONIC, MAGNETIC AND OPTICAL PROPERTIES		
Electrical Conductivity	% IACS	
Electrical resistivity	microohm-cm	68
Photoelectric work function	eV	2.7
Hall coefficient	volt-cm/amp-Oe	0.709×10^{-12}
Curie point	°K	25
First ionization potential	eV (Pr^+)	5.76
Oxidation potential	V (Pr^{3+})	2.47
Electrochemical equivalent	mg/coulomb (Pr^{3+})	0.48677
Magnetic susceptibility	c.g.s. units	5470×10^{-6}
Refractive index, n_D		
MECHANICAL PROPERTIES		
Young's modulus of elasticity	10^{-6} kg/cm^2	0.332
Shear modulus	10^{-6} kg/cm^2	0.138
Bulk modulus	10^{-6} kg/cm^2	0.3117
Poisson's ratio	–	0.305
Hardness	Vickers	43

TABLE 2.1 (*Continued*)

PROMETHIUM

	Unit	Value(s)
ATOMIC, NUCLEAR AND CRYSTALLOGRAPHIC PROPERTIES		
Symbol: Pm		
Atomic number	–	61
Atomic weight	(based on ^{12}C)	145
Periodic classification	–	Lanthanide
Atomic volume	cm^3/g-atom	20.33
Atomic radius	Angstrom units	
Ionic radius	Angstrom units (Pm^{3+})	0.98
Thermal neutron absorption -cross section	barns/atom	
Density	g./cm^3	7.3
Lattice type	–	hexagonal
Lattice constants	Angstrom units	
a		
b		
c		
Valence electrons	–	4f^5 6s^2
THERMAL AND THERMODYNAMIC PROPERTIES		
Melting point	°C	1080
Boiling point	°C	2460 (est.)
Transition point(s)	°C	912
Critical temperature	°K	
Linear thermal expansion coefficient	10^{-6} cm/cm/°C	9.0
Specific heat	cal/g/°C	
Electronic specific heat	10^{-4} cal/mole/°C^2	24
Thermal conductivity	cal/cm^2/cm/sec/°C	0.031 at 27°C
Latent heat of fusion	kcal/mole	1.94
Latent heat of vaporization at B.P.	kcal/mole	
Heat of combustion	kcal/mole	
Heat of sublimation	kcal/mole	64 (est.)
Heat capacity at constant pressure	cal/°C/mole	6.50
Heat content at M.P.	kcal/mole	
Entropy	cal/°C/mole	17.45
Longitudinal sound velocity	m/sec	
ELECTRICAL, ELECTRONIC, MAGNETIC AND OPTICAL PROPERTIES		
Electrical Conductivity	% IACS	
Electrical resistivity	microohm-cm	
Photoelectric work function	eV	
Hall coefficient	volt-cm/amp-Oe	
Curie point	°K	
First ionization potential	eV (Pm$^+$)	5.55
Oxidation potential	V (Pm^{3+})	2.42
Electrochemical equivalent	mg/coulomb (Pm^{3+})	0.50432
Magnetic susceptibility	c.g.s. units	
Refractive index, n_D		
MECHANICAL PROPERTIES		
Young's modulus of elasticity	10^{-6} kg/cm^2	0.43
Shear modulus	10^{-6} kg/cm^2	0.17
Bulk modulus	10^{-6} kg/cm^2	0.360
Poisson's ratio	–	0.278
Hardness		

TABLE 2.1 (*Continued*)

PROTACTINIUM

	Unit	Value(s)
ATOMIC, NUCLEAR AND CRYSTALLOGRAPHIC PROPERTIES		
Symbol: Pa		
Atomic number	–	91
Atomic weight	(based on ^{12}C)	231
Periodic classification	–	Actinide
Atomic volume	cm^3/g-atom	15.03
Atomic radius	Angstrom units	1.62
Ionic radius	Angstrom units (Pa^{3+})	1.06
Thermal neutron absorption -cross section	barns/atom	
Density	g./cm^3	15.37
Lattice type	–	Tetragonal
Lattice constants	Angstrom units	
a		3.925
b		
c		3.238
Valence electrons	–	$5f^2\ 6d^1\ 7s^2$
THERMAL AND THERMODYNAMIC PROPERTIES		
Melting point	°C	1425
Boiling point	°C	4410
Transition point(s)	°C	
Critical temperature	°K	
Linear thermal expansion coefficient	10^{-6} cm/cm/°C	7.3
Specific heat	cal/g/°C	
Electronic specific heat	10^{-4} cal/mole/°C^2	16.7
Thermal conductivity	cal/cm^2/cm/sec/°C	
Latent heat of fusion	kcal/mole	3.5
Latent heat of vaporization at B.P.	kcal/mole	110.0
Heat of combustion	kcal/mole	
Heat of sublimation	kcal/mole	132
Heat capacity at constant pressure	cal/°C/mole	6.79
Heat content at M.P.	kcal/mole	9.31
Entropy	cal/°C/mole	12.40
Longitudinal sound velocity	m/sec	
ELECTRICAL, ELECTRONIC, MAGNETIC AND OPTICAL PROPERTIES		
Electrical Conductivity	% IACS	
Electrical resistivity	microohm-cm	
Photoelectric work function	eV	
Hall coefficient	volt-cm/amp-Oe	
Curie point	°K	
First ionization potential	eV	
Oxidation potential	V	
Electrochemical equivalent	mg/coulomb (Pa^{5+})	0.59845
Magnetic susceptibility	c.g.s. units	2.6×10^{-6}
Refractive index, n_D	–	
MECHANICAL PROPERTIES		
Young's modulus of elasticity	10^{-6} kg/cm^2	
Shear modulus	10^{-6} kg/cm^2	
Bulk modulus	10^{-6} kg/cm^2	
Poisson's ratio	–	
Hardness		

TABLE 2.1 (*Continued*)

RADIUM

	Unit	Value(s)
ATOMIC, NUCLEAR AND CRYSTALLOGRAPHIC PROPERTIES		
Symbol: Ra		
Atomic number	–	88
Atomic weight	(based on ^{12}C)	226.05
Periodic classification	–	II A
Atomic volume	cm^3/g-atom	38.8
Atomic radius	Angstrom units	2.35
Ionic radius	Angstrom units (Ra^{2+})	1.44
Thermal neutron absorption -cross section	barns/atom (^{226}Ra)	20
Density	g./cm^3	5.0
Lattice type	–	
Lattice constants	Angstrom units	
a		
b		
c		
Valence electrons	–	$7s^2$
THERMAL AND THERMODYNAMIC PROPERTIES		
Melting point	°C	700
Boiling point	°C	1630
Transition point(s)	°C	
Critical temperature	°K	
Linear thermal expansion coefficient	10^{-6} cm/cm/°C	20.2
Specific heat	cal/g/°C	
Electronic specific heat	10^{-4} cal/mole/°C^2	7.4
Thermal conductivity	cal/cm^2/cm/sec/°C	
Latent heat of fusion	kcal/mole	1.71
Latent heat of vaporization at B.P.	kcal/mole	32.7
Heat of combustion	kcal/mole	
Heat of sublimation	kcal/mole	42
Heat capacity at constant pressure	cal/°C/mole	6.49
Heat content at M.P.	kcal/mole	4.81
Entropy	cal/°C/mole	17.00
Longitudinal sound velocity	m/sec	
ELECTRICAL, ELECTRONIC, MAGNETIC AND OPTICAL PROPERTIES		
Electrical Conductivity	% IACS	
Electrical resistivity	microohm-cm	100
Photoelectric work function	eV	3.2 (estimate)
Hall coefficient	volt-cm/amp-Oe	
Curie point	°K	
First ionization potential	eV (Ra^+)	5.279
Oxidation potential	V (Ra^{2+})	2.916
Electrochemical equivalent	mg/coulomb (Ra^{2+})	1.17124
Magnetic susceptibility	c.g.s. units	
Refractive index, n_D	–	
MECHANICAL PROPERTIES		
Young's modulus of elasticity	10^{-6} kg/cm^2	0.16
Shear modulus	10^{-6} kg/cm^2	0.061
Bulk modulus	10^{-6} kg/cm^2	0.135
Poisson's ratio	–	0.304
Hardness		

TABLE 2.1 (*Continued*)

RADON

	Unit	Value(s)
ATOMIC AND NUCLEAR PROPERTIES		
Symbol: Rn		
Atomic number	−	86
Atomic weight	(based on ^{12}C)	222
Periodic classification	−	rare gas
Thermal neutron absorption		
-cross section	barns/atom	0.7
SOLID STATE		
Lattice type	−	face-centered cubic
Lattice constants	Angstrom units	
a		
b		
c		
Atomic radius	Angstrom units	1.32
Melting point	°K	202
Heat of fusion	kcal/mole	0.693
Entropy of fusion	cal/°K/mole	
Triple point		
Temperature	°K	202
Pressure	torr	
Density of solid	g/cm^3	
Density of liquid	g/cm^3	
LIQUID STATE		
Boiling point	°K	211
Density at boiling point	g/cm^3	4.4
Heat of vaporization at boiling point	kcal/mole	3.920
Entropy at boiling point	cal/°K/mole	
Critical Data		
Temperature	°K	378
Pressure	atm.	62
Density	g/cm^3	
GASEOUS STATE*		
Density at 0°C	g/1	9.73
Specific heat at 20°C at const. pres.	cal/g/°K	
Ratio of specific heats, Cp/Cv, at 20°C	−	
Thermal conductivity at 20°C	cal/cm^2/cm/sec/°K	
Heat capacity at const. press. & 25°C	cal/°K/mole	4.97
Entropy at 25°C	cal/°K/mole	42.10
Heat of combustion	kcal/mole	
Dielectric constant at 20°C	−	
Viscosity at 20°C	micropoise	
Refractive index, n_D, at 0°C	−	
Solubility in H$_2$O at 1 atm. partial		
pressure and 0°C	cm^3(STP)/1000 g.	510

*Generally at one atmosphere.

TABLE 2.1 (*Continued*)

RHENIUM

	Unit	Value(s)
ATOMIC, NUCLEAR AND CRYSTALLOGRAPHIC PROPERTIES		
Symbol: Re		
Atomic number	–	75
Atomic weight	(based on ^{12}C)	186.2
Periodic classification	–	VII B
Atomic volume	cm^3/g-atom	8.860
Atomic radius	Angstrom units	1.3777
Ionic radius	Angstrom units (Re^{7+})	0.56
Thermal neutron absorption -cross section	barns/atom	87
Density	g./cm^3	21.02
Lattice type	–	close-packed hexagonal
Lattice constants	Angstrom units	
a		2.758
b		
c		4.454
Valence electrons	–	5d^5 6s^2
THERMAL AND THERMODYNAMIC PROPERTIES		
Melting point	°C	3160
Boiling point	°C	5762
Transition point(s)	°C	
Critical temperature	°K	20000
Linear thermal expansion coefficient	10^{-6} cm/cm/°C	6.63
Specific heat	cal/g/°C	0.03262
Electronic specific heat	10^{-4} cal/mole/°C^2	5.74
Thermal conductivity	cal/cm^2/cm/sec/°C	0.17
Latent heat of fusion	kcal/mole	7.86
Latent heat of vaporization at B.P.	kcal/mole	169.0
Heat of combustion	kcal/mole	
Heat of sublimation	kcal/mole	186.4
Heat capacity at constant pressure	cal/°C/mole	6.14
Heat content at M.P.	kcal/mole	
Entropy	cal/°C/mole	8.887
Longitudinal sound velocity	m/sec (20°C)	~4700
ELECTRICAL, ELECTRONIC, MAGNETIC AND OPTICAL PROPERTIES		
Electrical Conductivity	% IACS	9.3
Electrical resistivity	microohm-cm	19.14
Photoelectric work function	eV	5.0
Hall coefficient	volt-cm/amp-Oe	
Curie point	°K	
First ionization potential	eV (Re$^+$)	7.88
Oxidation potential	V (Re/ReO$_4^-$)	1.171 (alk. soln.)
Electrochemical equivalent	mg/coulomb (Re^{7+})	0.27581
Magnetic susceptibility	c.g.s. units	0.369×10^{-6}
Refractive index, n_D	–	
MECHANICAL PROPERTIES		
Young's modulus of elasticity	10^{-6} kg/cm^2	4.7
Shear modulus	10^{-6} kg/cm^2	1.82
Bulk modulus	10^{-6} kg/cm^2	3.79
Poisson's ratio	–	0.293
Hardness	Brinell	250

TABLE 2.1 (*Continued*)

RHODIUM

	Unit	Value(s)
ATOMIC, NUCLEAR AND CRYSTALLOGRAPHIC PROPERTIES Symbol: Rh		
Atomic number	–	45
Atomic weight	(based on ^{12}C)	102.905
Periodic classification	–	VIII
Atomic volume	cm^3/g-atom	8.292
Atomic radius	Angstrom units	1.342
Ionic radius	Angstrom units (Rh^{3+})	0.68
Thermal neutron absorption -cross section	barns/atom	148
Density	g./cm^3	12.41
Lattice type	–	face-centered cubic
Lattice constants	Angstrom units	
a		3.803
b		
c		
Valence electrons	–	$4d^8\ 5s^1$
THERMAL AND THERMODYNAMIC PROPERTIES		
Melting point	°C	1960
Boiling point	°C	3687
Transition point(s)	°C	
Critical temperature	°K	
Linear thermal expansion coefficient	10^{-6} cm/cm/°C	8.40
Specific heat	cal/g/°C	0.0589
Electronic specific heat	10^{-4} cal/mole/°C^2	11.0
Thermal conductivity	cal/cm^2/cm/sec/°C	0.358
Latent heat of fusion	kcal/mole	5.2
Latent heat of vaporization at B.P.	kcal/mole	118.4
Heat of combustion	kcal/mole	
Heat of sublimation	kcal/mole	133
Heat capacity at constant pressure	cal/°C/mole	6.11
Heat content at M.P.	kcal/mole	15.34
Entropy	cal/°C/mole	7.53
Longitudinal sound velocity	m/sec	6194
ELECTRICAL, ELECTRONIC, MAGNETIC AND OPTICAL PROPERTIES		
Electrical Conductivity	% IACS	
Electrical resistivity	microohm-cm	4.33 at 0°C
Photoelectric work function	eV	4.8
Hall coefficient	volt-cm/amp-Oe	
Curie point	°K	
First ionization potential	eV (Rh^+)	7.46
Oxidation potential	V (Rh/Rh_2O_3)	−0.04 (alk. soln.)
Electrochemical equivalent	mg/coulomb (Rh^{4+})	0.26661
Magnetic susceptibility	c.g.s. units	0.990×10^{-6}
Refractive index, n_D		1.54 at 0.579 μ
MECHANICAL PROPERTIES		
Young's modulus of elasticity	10^{-6} kg/cm^2	3.80
Shear modulus	10^{-6} kg/cm^2	1.50
Bulk modulus	10^{-6} kg/cm^2	2.758
Poisson's ratio	–	0.27
Hardness	Vickers (annealed)	100–120

TABLE 2.1 (*Continued*)

RUBIDIUM

	Unit	Value(s)
ATOMIC, NUCLEAR AND CRYSTALLOGRAPHIC PROPERTIES		
Symbol: Rb		
Atomic number	–	37
Atomic weight	(based on ^{12}C)	85.47
Periodic classification	–	I A
Atomic volume	cm^3/g-atom	55.9
Atomic radius	Angstrom units	2.43
Ionic radius	Angstrom units (Rb^+)	1.48
Thermal neutron absorption		
-cross section	barns/atom	0.4
Density	g./cm^3	1.532
Lattice type	–	body-centered cubic
Lattice constants	Angstrom units	
a		5.63 at $-173°C$
b		
c		
Valence electrons	–	5s'
THERMAL AND THERMODYNAMIC PROPERTIES		
Melting point	$°C$	38.6
Boiling point	$°C$	686
Transition point(s)	$°C$	
Critical temperature	$°K$	2102
Linear thermal expansion coefficient	10^{-6} cm/cm/$°C$	90
Specific heat	cal/g/$°C$	0.080 at $0°C$
Electronic specific heat	10^{-4} cal/mole/$°C^2$	6.02
Thermal conductivity	cal/cm^2/cm/sec/$°C$	0.07 at $39°C$
Latent heat of fusion	kcal/mole	0.56
Latent heat of vaporization at B.P.	kcal/mole	16.54
Heat of combustion	kcal/mole	
Heat of sublimation	kcal/mole	19.9
Heat capacity at constant pressure	cal/$°C$/mole	7.36
Heat content at M.P.	kcal/mole	0.014
Entropy	cal/$°C$/mole	18.22
Longitudinal sound velocity	m/sec	1300
ELECTRICAL, ELECTRONIC, MAGNETIC AND OPTICAL PROPERTIES		
Electrical Conductivity	% IACS	11.6 at $0°C$
Electrical resistivity	microohm-cm	2.09
Photoelectric work function	eV	
Hall coefficient	volt-cm/amp-Oe	
Curie point	$°K$	
First ionization potential	eV (Rb^+)	4.177
Oxidation potential	V (Rb^+)	2.9259
Electrochemical equivalent	mg/coulomb (Rb^+)	0.88580
Magnetic susceptibility	c.g.s. units	0.09×10^{-6} at $18°C$
Refractive index, n_D		
MECHANICAL PROPERTIES		
Young's modulus of elasticity	10^{-6} kg/cm^2	0.0277
Shear modulus	10^{-6} kg/cm^2	0.0102
Bulk modulus	10^{-6} kg/cm^2	0.03206
Poisson's ratio	–	0.356
Hardness	Mohs	0.3

TABLE 2.1 (*Continued*)

RUTHENIUM

	Unit	Value(s)
ATOMIC, NUCLEAR AND CRYSTALLOGRAPHIC PROPERTIES		
Symbol: Ru		
Atomic number	–	44
Atomic weight	(based on ^{12}C)	101.07
Periodic classification	–	VIII
Atomic volume	cm^3/g-atom	8.178
Atomic radius	Angstrom units	1.322
Ionic radius	Angstrom units (Ru^{4+})	0.65
Thermal neutron absorption -cross section	barns/atom	2.6
Density	g./cm^3	12.45
Lattice type	–	close-packed hexagonal
Lattice constants	Angstrom units	
a		2.706
b		
c		4.282
Valence electrons	–	$4d^7 5s^1$
THERMAL AND THERMODYNAMIC PROPERTIES		
Melting point	$°C$	2310
Boiling point	$°C$	4200
Transition point(s)	$°C$	1035, 1190, 1500
Critical temperature	$°K$	
Linear thermal expansion coefficient	10^{-6} cm/cm/$°C$	9.1
Specific heat	cal/g/$°C$	0.055 at $0°C$
Electronic specific heat	10^{-4} cal/mole/$°C^2$	11.0
Thermal conductivity	cal/cm^2/cm/sec/$°C$	0.251
Latent heat of fusion	kcal/mole	5.67
Latent heat of vaporization at B.P.	kcal/mole	135.7
Heat of combustion	kcal/mole	
Heat of sublimation	kcal/mole	155
Heat capacity at constant pressure	cal/$°C$/mole	5.80
Heat content at M.P.	kcal/mole	17.2
Entropy	cal/$°C$/mole	6.82
Longitudinal sound velocity	m/sec	6534
ELECTRICAL, ELECTRONIC, MAGNETIC AND OPTICAL PROPERTIES		
Electrical Conductivity	% IACS	
Electrical resistivity	microohm-cm	6.70
Photoelectric work function	eV	4.52
Hall coefficient	volt-cm/amp-Oe	
Curie point	$°K$	
First ionization potential	eV (Ru^+)	7.37
Oxidation potential	V (Ru/RuO_4)	–0.3 (acid soln.)
Electrochemical equivalent	mg/coulomb (Ru^{2+})	0.527
Magnetic susceptibility	c.g.s. units	0.427×10^{-6} at $18°C$
Refractive index, n_D		
MECHANICAL PROPERTIES		
Young's modulus of elasticity	10^{-6} kg/cm^2	4.20
Shear modulus	10^{-6} kg/cm^2	1.63
Bulk modulus	10^{-6} kg/cm^2	3.271
Poisson's ratio	–	0.29
Hardness	Vickers (annealed)	200–350

TABLE 2.1 (*Continued*)

SAMARIUM

	Unit	Value(s)
ATOMIC, NUCLEAR AND CRYSTALLOGRAPHIC PROPERTIES		
Symbol: Sm		
Atomic number	–	62
Atomic weight	(based on ^{12}C)	150.35
Periodic classification	–	Lanthanide
Atomic volume	cm^3/g-atom	19.95
Atomic radius	Angstrom units	1.802
Ionic radius	Angstrom units (Sm^{3+})	1.13
Thermal neutron absorption		
-cross section	barns/atom	5860
Density	g./cm^3	7.536
Lattice type	–	Rhombohedral
Lattice constants	Angstrom units	
a		3.626
b		
c		26.18
Valence electrons	–	$4f^6\ 6s^2$
THERMAL AND THERMODYNAMIC PROPERTIES		
Melting point	°C	1072
Boiling point	°C	1870
Transition point(s)	°C	917
Critical temperature	°K	5400
Linear thermal expansion coefficient	10^{-6}cm/cm/°C	10.4
Specific heat	cal/g/°C	0.042
Electronic specific heat	10^{-4}cal/mole/°C^2	25.3
Thermal conductivity	cal/cm^2/cm/sec/°C	
Latent heat of fusion	kcal/mole	2.061
Latent heat of vaporization at B.P.	kcal/mole	45.8
Heat of combustion	kcal/mole	216.94
Heat of sublimation	kcal/mole	49.9
Heat capacity at constant pressure	cal/°C/mole	6.95
Heat content at M.P.	kcal/mole	8.15
Entropy	cal/°C/mole	16.28
Longitudinal sound velocity	m/sec	2700
ELECTRICAL, ELECTRONIC, MAGNETIC AND OPTICAL PROPERTIES		
Electrical Conductivity	% IACS	
Electrical resistivity	microohm-cm	92
Photoelectric work function	eV	3.2
Hall coefficient	volt-cm/amp-Oe	-0.2×10^{-12}
Curie point	°K	14
First ionization potential	eV (Sm^+)	5.63
Oxidation potential	V (Sm^{3+})	2.41
Electrochemical equivalent	mg/coulomb (Sm^{3+})	0.51962
Magnetic susceptibility	c.g.s. units	1320×10^{-6}
Refractive index, n_D		
MECHANICAL PROPERTIES		
Young's modulus of elasticity	10^{-6} kg/cm^2	0.43
Shear modulus	10^{-6} kg/cm^2	0.129
Bulk modulus	10^{-6} kg/cm^2	0.2998
Poisson's ratio	–	0.352
Hardness	Vickers (cast)	64

TABLE 2.1 (Continued)

SCANDIUM

	Unit	Value(s)
ATOMIC, NUCLEAR AND CRYSTALLOGRAPHIC PROPERTIES		
Symbol: Sc		
Atomic number	–	21
Atomic weight	(based on ^{12}C)	44.956
Periodic classification	–	III B
Atomic volume	cm^3/g-atom	15.0
Atomic radius	Angstrom units	1.62
Ionic radius	Angstrom units (Sc^{3+})	0.81
Thermal neutron absorption		
-cross section	barns/atom	25
Density	g./cm^3	2.99
Lattice type	–	close-packed hexagonal
Lattice constants	Angstrom units	
a		3.308
b		
c		5.267
Valence electrons	–	$3d^1\ 4s^2$
THERMAL AND THERMODYNAMIC PROPERTIES		
Melting point	°C	1539
Boiling point	°C	2832
Transition point(s)	°C	1335
Critical temperature	°K	
Linear thermal expansion coefficient	10^{-6} cm/cm/°C	10.0
Specific heat	cal/g/°C	0.133
Electronic specific heat	10^{-4} cal/mole/°C^2	25.8
Thermal conductivity	cal/cm^2/cm/sec/°C	
Latent heat of fusion	kcal/mole	3.8
Latent heat of vaporization at B.P.	kcal/mole	72.85
Heat of combustion	kcal/mole	
Heat of sublimation	kcal/mole	91.0
Heat capacity at constant pressure	cal/°C/mole	6.09
Heat content at M.P.	kcal/mole	8.74
Entropy	cal/°C/mole	9.00
Longitudinal sound velocity	m/sec	
ELECTRICAL, ELECTRONIC, MAGNETIC AND OPTICAL PROPERTIES		
Electrical Conductivity	% IACS	
Electrical resistivity	microohm-cm	66
Photoelectric work function	eV	3.3
Hall coefficient	volt-cm/amp-Oe	
Curie point	°K	
First ionization potential	eV (Sc^+)	6.54
Oxidation potential	V (Sc^{3+})	2.08
Electrochemical equivalent	mg/coulomb (Sc^{2+})	0.15579
Magnetic susceptibility	c.g.s. units	8.1×10^{-6}
Refractive index, n_D		
MECHANICAL PROPERTIES		
Young's modulus of elasticity	10^{-6} kg/cm^2	0.809
Shear modulus	10^{-6} kg/cm^2	0.075
Bulk modulus	10^{-6} kg/cm^2	0.444
Poisson's ratio	–	0.269
Hardness	Brinell	50

TABLE 2.1 (*Continued*)

SELENIUM

	Unit	Value(s)
ATOMIC, NUCLEAR AND CRYSTALLOGRAPHIC PROPERTIES		
Symbol: Se		
Atomic number	–	34
Atomic weight	(based on ^{12}C)	78.96
Periodic classification	–	VI A
Atomic volume	cm^3/g-atom	16.5
Atomic radius	Angstrom units	1.15
Ionic radius	Angstrom units (Se^{2-})	1.98
Thermal neutron absorption -cross section	barns/atom	11.8
Density	g./cm^3	4.79 (gray)
Lattice type	–	hexagonal
Lattice constants	Angstrom units	
a		4.346
b		
c		4.954
Valence electrons	–	$4s^2\,4p^4$
THERMAL AND THERMODYNAMIC PROPERTIES		
Melting point	°C	217
Boiling point	°C	685
Transition point(s)	°C	31, 125, 150
Critical temperature	°K	1757
Linear thermal expansion coefficient	10^{-6} cm/cm/°C	36.9
Specific heat	cal/g/°C	0.081
Electronic specific heat	10^{-4} cal/mole/°C^2	0
Thermal conductivity	cal/cm^2/cm/sec/°C	0.0012
Latent heat of fusion	kcal/mole	1.30
Latent heat of vaporization at B.P.	kcal/mole	3.34
Heat of combustion	kcal/mole	
Heat of sublimation	kcal/mole	49.4
Heat capacity at constant pressure	cal/°C/mole	6.075
Heat content at M.P.	kcal/mole	0.65
Entropy	cal/°C/mole	10.15
Longitudinal sound velocity	m/sec (20°C)	3350
ELECTRICAL, ELECTRONIC, MAGNETIC AND OPTICAL PROPERTIES		
Electrical Conductivity	% IACS	
Electrical resistivity	microohm-cm	12 at 0°C
Photoelectric work function	eV	4.72
Hall coefficient	volt-cm/amp-Oe	
Curie point	°K	
First ionization potential	eV (Se^+)	9.752
Oxidation potential	V (Se/SeO_4^{2-})	0.316
Electrochemical equivalent	mg/coulomb (Se^{6+})	0.13637
Magnetic susceptibility	c.g.s. units	-0.32×10^{-6}
Refractive index, n_D		2.75 to 3.06
MECHANICAL PROPERTIES		
Young's modulus of elasticity	10^{-6} kg/cm^2	0.591
Shear modulus	10^{-6} kg/cm^2	0.221
Bulk modulus	10^{-6} kg/cm^2	0.09276
Poisson's ratio	–	0.338
Hardness	Mohs	2.0

TABLE 2.1 (*Continued*)

SILICON

	Unit	Value(s)
ATOMIC, NUCLEAR AND CRYSTALLOGRAPHIC PROPERTIES		
Symbol: Si		
Atomic number	—	14
Atomic weight	(based on ^{12}C)	28.086
Periodic classification	—	IV A
Atomic volume	cm^3/g-atom	12.07
Atomic radius	Angstrom units	1.18
Ionic radius	Angstrom units (Si^{4+})	0.41
Thermal neutron absorption		
-cross section	barns/atom	0.16
Density	g./cm^3	2.33
Lattice type	—	Cubic
Lattice constants	Angstrom units	
a		5.417
b		
c		
Valence electrons	—	$3s^2\ 3p^2$
THERMAL AND THERMODYNAMIC PROPERTIES		
Melting point	°C	1412
Boiling point	°C	2480
Transition point(s)	°C	—
Critical temperature	°K	4920
Linear thermal expansion coefficient	10^{-6} cm/cm/°C	4.2
Specific heat	cal/g/°C	0.162
Electronic specific heat	10^{-4} cal/mole/°C^2	0.050
Thermal conductivity	cal/cm^2/cm/sec/°C	0.20
Latent heat of fusion	kcal/mole	12.0
Latent heat of vaporization at B.P.	kcal/mole	71
Heat of combustion	kcal/mole	191 (to SiO_2)
Heat of sublimation	kcal/mole	108.4
Heat capacity at constant pressure	cal/°C/mole	4.64
Heat content at M.P.	kcal/mole	8.07
Entropy	cal/°C/mole	4.50
Longitudinal sound velocity	m/sec	9101
ELECTRICAL, ELECTRONIC, MAGNETIC AND OPTICAL PROPERTIES		
Electrical Conductivity	% IACS	
Electrical resistivity	microohm-cm	2.3×10^5
Photoelectric work function	eV	4.1
Hall coefficient	volt-cm/amp-Oe	
Curie point	°K	
First ionization potential	eV (Si^+)	8.151
Oxidation potential	V (Si/SiO_2)	1.2 (acid soln.)
Electrochemical equivalent	mg/coulomb (Si^{4+})	0.07269
Magnetic susceptibility	c.g.s. units	-0.13×10^{-6}
Refractive index, n_D		4.24
MECHANICAL PROPERTIES		
Young's modulus of elasticity	10^{-6} kg/cm^2	1.05
Shear modulus	10^{-6} kg/cm^2	0.405
Bulk modulus	10^{-6} kg/cm^2	1.008
Poisson's ratio	—	0.44
Hardness	Brinell	240

TABLE 2.1 (*Continued*)

SILVER

	Unit	Value(s)
ATOMIC, NUCLEAR AND CRYSTALLOGRAPHIC PROPERTIES		
Symbol: Ag		
Atomic number	–	47
Atomic weight	(based on ^{12}C)	107.870
Periodic classification	–	I B
Atomic volume	cm^3/g-atom	10.27
Atomic radius	Angstrom units	1.442
Ionic radius	Angstrom units (Ag^+)	1.13
Thermal neutron absorption -cross section	barns/atom	63.6
Density	g./cm^3	10.50
Lattice type	–	face-centered cubic
Lattice constants	Angstrom units	
a		4.086
b		
c		
Valence electrons	–	$4d^{10} 5s^1$
THERMAL AND THERMODYNAMIC PROPERTIES		
Melting point	°C	960.8
Boiling point	°C	2195
Transition point(s)	°C	
Critical temperature	°K	7460
Linear thermal expansion coefficient	10^{-6} cm/cm/°C	19.2
Specific heat	cal/g/°C	0.056 at 0°C
Electronic specific heat	10^{-4} cal/mole/°C^2	1.58
Thermal conductivity	cal/cm^2/cm/sec/°C	1.00
Latent heat of fusion	kcal/mole	2.78
Latent heat of vaporization at B.P.	kcal/mole	60.7
Heat of combustion	kcal/mole	
Heat of sublimation	kcal/mole	68.4
Heat capacity at constant pressure	cal/°C/mole	6.095
Heat content at M.P.	kcal/mole	6.06
Entropy	cal/°C/mole	10.20
Longitudinal sound velocity	m/sec	3636
ELECTRICAL, ELECTRONIC, MAGNETIC AND OPTICAL PROPERTIES		
Electrical Conductivity	% IACS	105
Electrical resistivity	microohm-cm	1.59
Photoelectric work function	eV	4.73
Hall coefficient	volt-cm/amp-Oe	
Curie point	°K	
First ionization potential	eV (Ag^+)	7.576
Oxidation potential	V (Ag^+)	–0.799 (acid soln.)
Electrochemical equivalent	mg/coulomb (Ag^+)	0.11793
Magnetic susceptibility	c.g.s. units	-0.20×10^{-6} at 18°C
Refractive index, n_D		0.18
MECHANICAL PROPERTIES		
Young's modulus of elasticity	10^{-6} kg/cm^2	0.822
Shear modulus	10^{-6} kg/cm^2	0.292
Bulk modulus	10^{-6} kg/cm^2	1.027
Poisson's ratio	–	0.37
Hardness	Brinell	26

TABLE 2.1 (Continued)

SODIUM

	Unit	Value(s)
ATOMIC, NUCLEAR AND CRYSTALLOGRAPHIC PROPERTIES		
Symbol: Na		
Atomic number	–	11
Atomic weight	(based on ^{12}C)	22.9898
Periodic classification	–	I A
Atomic volume	cm^3/g-atom	23.79
Atomic radius	Angstrom units	1.86
Ionic radius	Angstrom units (Na^+)	0.95
Thermal neutron absorption -cross section	barns/atom	0.534
Density	g./cm^3	0.9674
Lattice type	–	body-centered cubic
Lattice constants	Angstrom units	
a		4.2820
b		
c		
Valence electrons	–	$3s^1$
THERMAL AND THERMODYNAMIC PROPERTIES		
Melting point	°C	97.83
Boiling point	°C	881
Transition point(s)	°C	
Critical temperature	°K	2400
Linear thermal expansion coefficient	$10^{-6} cm/cm/°C$	70.6
Specific heat	cal/g/°C	0.295
Electronic specific heat	$10^{-4} cal/mole/°C^2$	3.30
Thermal conductivity	$cal/cm^2/cm/sec/°C$	0.32
Latent heat of fusion	kcal/mole	0.622
Latent heat of vaporization at B.P.	kcal/mole	20.63
Heat of combustion	kcal/mole	
Heat of sublimation	kcal/mole	25.92
Heat capacity at constant pressure	cal/°C/mole	6.745
Heat content at M.P.	kcal/mole	0.012
Entropy	cal/°C/mole	12.23
Longitudinal sound velocity	m/sec	3200
ELECTRICAL, ELECTRONIC, MAGNETIC AND OPTICAL PROPERTIES		
Electrical Conductivity	% IACS	
Electrical resistivity	microohm-cm	4.985
Photoelectric work function	eV	2.27
Hall coefficient	volt-cm/amp-Oe	
Curie point	°K	
First ionization potential	eV (Na^+)	5.139
Oxidation potential	V (Na^+)	2.714
Electrochemical equivalent	mg/coulomb	0.2384
Magnetic susceptibility	c.g.s. units	$+0.51 \times 10^{-6}$
Refractive index, n_D		0.004 (liquid)
MECHANICAL PROPERTIES		
Young's modulus of elasticity	$10^{-6} kg/cm^2$	0.0912 at 90°K
Shear modulus	$10^{-6} kg/cm^2$	0.035 at 90°K
Bulk modulus	$10^{-6} kg/cm^2$	0.06949
Poisson's ratio	–	0.315 at 90°K
Hardness	Brinell	0.07

TABLE 2.1 (*Continued*)

STRONTIUM

	Unit	Value(s)
ATOMIC, NUCLEAR AND CRYSTALLOGRAPHIC PROPERTIES		
Symbol: Sr		
Atomic number	–	38
Atomic weight	(based on ^{12}C)	87.62
Periodic classification	–	II A
Atomic volume	cm^3/g-atom	33.9
Atomic radius	Angstrom units	2.13
Ionic radius	Angstrom units (Sr^{2+})	1.13
Thermal neutron absorption -cross section	barns/atom	0.4
Density	g./cm^3	2.6
Lattice type	–	face-centered cubic
Lattice constants	Angstrom units	
a		6.087
b		
c		
Valence electrons	–	$5s^2$
THERMAL AND THERMODYNAMIC PROPERTIES		
Melting point	°C	772
Boiling point	°C	1372
Transition point(s)	°C	235, 540
Critical temperature	°K	3810
Linear thermal expansion coefficient	10^{-6} cm/cm/°C	23
Specific heat	cal/g/°C	0.176
Electronic specific heat	10^{-4} cal/mole/°C^2	8.70
Thermal conductivity	cal/cm^2/cm/sec/°C	0.024
Latent heat of fusion	kcal/mole	2.19
Latent heat of vaporization at B.P.	kcal/mole	33.5
Heat of combustion	kcal/mole	
Heat of sublimation	kcal/mole	39.2
Heat capacity at constant pressure	cal/°C/mole	6.30
Heat content at M.P.	kcal/mole	5.52
Entropy	cal/°C/mole	12.50
Longitudinal sound velocity	m/sec	
ELECTRICAL, ELECTRONIC, MAGNETIC AND OPTICAL PROPERTIES		
Electrical Conductivity	% IACS	
Electrical resistivity	microohm-cm	23
Photoelectric work function	eV	2.35
Hall coefficient	volt-cm/amp-Oe	
Curie point	°K	
First ionization potential	eV (Sr^+)	5.695
Oxidation potential	V (Sr^{2+})	2.89
Electrochemical equivalent	mg/coulomb (Sr^{2+})	0.45404
Magnetic susceptibility	c.g.s. units	-0.20×10^{-6}
Refractive index, n_D		
MECHANICAL PROPERTIES		
Young's modulus of elasticity	10^{-6} kg/cm^2	0.139
Shear modulus	10^{-6} kg/cm^2	0.0533
Bulk modulus	10^{-6} kg/cm^2	0.1184
Poisson's ratio	–	0.304
Hardness		

TABLE 2.1 *(Continued)*

SULFUR

	Unit	Value(s)
ATOMIC, NUCLEAR AND CRYSTALLOGRAPHIC PROPERTIES Symbol: S		
Atomic number	–	16
Atomic weight	(based on ^{12}C)	180.948
Periodic classification	–	VI A
Atomic volume	cm^3/g-atom	15.5
Atomic radius	Angstrom units	1.03
Ionic radius	Angstrom units (S^{6+})	0.29
Thermal neutron absorption -cross section	barns/atom	0.51
Density	g./cm^3	2.07
Lattice type	–	Orthorhombic
Lattice constants	Angstrom units	
a		10.4646
b		12.8660
c		24.4860
Valence electrons	–	$3s^2\ 3p^4$
THERMAL AND THERMODYNAMIC PROPERTIES		
Melting point	°C	115.21
Boiling point	°C	444.60
Transition point(s)	°C	95.5
Critical temperature	°K	1313
Linear thermal expansion coefficient	10^{-6} cm/cm/°C	74.33
Specific heat	cal/g/°C	0.175
Electronic specific heat	10^{-4} cal/mole/°C^2	0
Thermal conductivity	cal/cm^2/cm/sec/°C	6.31×10^{-4}
Latent heat of fusion	kcal/mole	0.336
Latent heat of vaporization at B.P.	kcal/mole	2.3
Heat of combustion	kcal/mole	
Heat of sublimation	kcal/mole	66.4
Heat capacity at constant pressure	cal/°C/mole	5.40
Heat content at M.P.	kcal/mole	0.009
Entropy	cal/°C/mole	7.62
Longitudinal sound velocity	m/sec	
ELECTRICAL, ELECTRONIC, MAGNETIC AND OPTICAL PROPERTIES		
Electrical Conductivity	% IACS	
Electrical resistivity	microohm-cm	1.9×10^{17} at 20°C
Photoelectric work function	eV	
Hall coefficient	volt-cm/amp-Oe	
Curie point	°K	
First ionization potential	eV (S^+)	10.360
Oxidation potential	V (S/SO_4^{2-})	–0.62 (acid soln.)
Electrochemical equivalent	mg/coulomb (S^{6+})	0.05537
Magnetic susceptibility	c.g.s. units	0.49×10^{-6}
Refractive index, n_D		1.957, 2.0377, 2.2454
MECHANICAL PROPERTIES		
Young's modulus of elasticity	10^{-6} kg/cm^2	0.198
Shear modulus	10^{-6} kg/cm^2	0.0737
Bulk modulus	10^{-6} kg/cm^2	0.182
Poisson's ratio	–	0.343
Hardness		

TABLE 2.1 (*Continued*)

TANTALUM

	Unit	Value(s)
ATOMIC, NUCLEAR AND CRYSTALLOGRAPHIC PROPERTIES		
Symbol: Ta		
Atomic number	–	73
Atomic weight	(based on ^{12}C)	180.948
Periodic classification	–	V B
Atomic volume	cm^3/g-atom	10.90
Atomic radius	Angstrom units	1.427
Ionic radius	Angstrom units (Ta^{5+})	0.68
Thermal neutron absorption -cross section	barns/atom	22
Density	g./cm^3	16.6
Lattice type	–	body-centered cubic
Lattice constants	Angstrom units	
a		3.296
b		
c		
Valence electrons	–	$5d^3 6s^2$
THERMAL AND THERMODYNAMIC PROPERTIES		
Melting point	°C	2996
Boiling point	°C	5487
Transition point(s)	°C	
Critical temperature	°K	22000
Linear thermal expansion coefficient	10^{-6} cm/cm/°C	6.6
Specific heat	cal/g/°C	0.034
Electronic specific heat	10^{-4} cal/mole/°C^2	14.0
Thermal conductivity	cal/cm^2/cm/sec/°C	0.130
Latent heat of fusion	kcal/mole	7.5
Latent heat of vaporization at B.P.	kcal/mole	180.0
Heat of combustion	kcal/mole	244 at 30°C
Heat of sublimation	kcal/mole	186.8
Heat capacity at constant pressure	cal/°C/mole	6.08
Heat content at M.P.	kcal/mole	
Entropy	cal/°C/mole	9.90
Longitudinal sound velocity	m/sec	3400 at 18°C
ELECTRICAL, ELECTRONIC, MAGNETIC AND OPTICAL PROPERTIES		
Electrical Conductivity	% IACS	13.0 at 18°C
Electrical resistivity	microohm-cm	12.45
Photoelectric work function	eV	4.12
Hall coefficient	volt-cm/amp-Oe	$+9.5 \times 10^{-11}$
Curie point	°K	
First ionization potential	eV (Ta^+)	7.89
Oxidation potential	V (Ta^{5+})	1.12
Electrochemical equivalent	mg/coulomb (Ta^{5+})	0.3749
Magnetic susceptibility	c.g.s. units	0.849×10^{-6}
Refractive index, n_D		2.05
MECHANICAL PROPERTIES		
Young's modulus of elasticity	10^{-6} kg/cm^2	1.85
Shear modulus	10^{-6} kg/cm^2	0.700
Bulk modulus	10^{-6} kg/cm^2	2.04
Poisson's ratio	–	0.35
Hardness		

TABLE 2.1 (*Continued*)

TECHNETIUM

	Unit	Value(s)
ATOMIC, NUCLEAR AND CRYSTALLOGRAPHIC PROPERTIES		
Symbol: Tc		
Atomic number	—	43
Atomic weight	(based on ^{12}C)	98
Periodic classification	—	VII B
Atomic volume	cm^3/g-atom	8.635
Atomic radius	Angstrom units	1.36
Ionic radius	Angstrom units (Tc^{7+})	0.58
Thermal neutron absorption -cross section	barns/atom	
Density	g./cm^3	11.49
Lattice type	—	close-packed hexagonal
Lattice constants	Angstrom units	
a		2.735
b		
c		4.391
Valence electrons	—	4d^6 5s^1
THERMAL AND THERMODYNAMIC PROPERTIES		
Melting point	°C	2170
Boiling point	°C	5030
Transition point(s)	°C	
Critical temperature	°K	7.7 (superconducting)
Linear thermal expansion coefficient	10^{-6}cm/cm/°C	8.06
Specific heat	cal/g/°C	
Electronic specific heat	10^{-4}cal/mole/°C^2	9.7
Thermal conductivity	cal/cm^2/cm/sec/°C	
Latent heat of fusion	kcal/mole	5.42
Latent heat of vaporization at B.P.	kcal/mole	138.0
Heat of combustion	kcal/mole	
Heat of sublimation	kcal/mole	152
Heat capacity at constant pressure	cal/°C/mole	5.80
Heat content at M.P.	kcal/mole	15.63
Entropy	cal/°C/mole	8.00
Longitudinal sound velocity	m/sec	
ELECTRICAL, ELECTRONIC, MAGNETIC AND OPTICAL PROPERTIES		
Electrical Conductivity	% IACS	
Electrical resistivity	microohm-cm	
Photoelectric work function	eV	4.4
Hall coefficient	volt-cm/amp-Oe	
Curie point	°K	
First ionization potential	eV (Tc$^+$)	7.28
Oxidation potential	V	
Electrochemical equivalent	mg/coulomb (Tc^{7+})	0.14478
Magnetic susceptibility	c.g.s. units	270×10^{-6}
Refractive index, n_D		
MECHANICAL PROPERTIES		
Young's modulus of elasticity	10^{-6} kg/cm^2	3.76
Shear modulus	10^{-6} kg/cm^2	1.45
Bulk modulus	10^{-6} kg/cm^2	3.03
Poisson's ratio	—	0.293
Hardness		

TABLE 2.1 (*Continued*)

TELLURIUM

	Unit	Value(s)
ATOMIC, NUCLEAR AND CRYSTALLOGRAPHIC PROPERTIES		
Symbol: Te		
Atomic number	–	52
Atomic weight	(based on ^{12}C)	127.60
Periodic classification	–	VI A
Atomic volume	cm^3/g-atom	20.45
Atomic radius	Angstrom units	1.38
Ionic radius	Angstrom units (Te^{6+})	0.56
Thermal neutron absorption		
-cross section	barns/atom	4.7
Density	g./cm^3	6.24
Lattice type	–	Hexagonal
Lattice constants	Angstrom units	
a		4.45
b		
c		5.91
Valence electrons	–	$5s^2 \, 5p^4$
THERMAL AND THERMODYNAMIC PROPERTIES		
Melting point	°C	449.6
Boiling point	°C	990
Transition point(s)	°C	
Critical temperature	°K	2329
Linear thermal expansion coefficient	10^{-6} cm/cm/°C	16.75
Specific heat	cal/g/°C	0.047
Electronic specific heat	10^{-4} cal/mole/°C^2	0
Thermal conductivity	cal/cm^2/cm/sec/°C	0.014
Latent heat of fusion	kcal/mole	4.18
Latent heat of vaporization at B.P.	kcal/mole	13.65 (Te$_2$)
Heat of combustion	kcal/mole	
Heat of sublimation	kcal/mole	46.6
Heat capacity at constant pressure	cal/°C/mole	6.14
Heat content at M.P.	kcal/mole	2.894
Entropy	cal/°C/mole	11.88
Longitudinal sound velocity	m/sec (20°C)	2610
ELECTRICAL, ELECTRONIC, MAGNETIC AND OPTICAL PROPERTIES		
Electrical Conductivity	% IACS	
Electrical resistivity	microohm-cm	436000
Photoelectric work function	eV	4.73
Hall coefficient	volt-cm/amp-Oe	
Curie point	°K	
First ionization potential	eV (Te$^+$)	9.009
Oxidation potential	V (Te/TeO$_2$)	–0.529 (acid soln.)
Electrochemical equivalent	mg/coulomb (Te^{6+})	0.22040
Magnetic susceptibility	c.g.s. units	-0.31×10^{-6} at 18°C
Refractive index, n_D		
MECHANICAL PROPERTIES		
Young's modulus of elasticity	10^{-6} kg/cm^2	0.42
Shear modulus	10^{-6} kg/cm^2	0.147
Bulk modulus	10^{-6} kg/cm^2	0.2347
Poisson's ratio	–	0.33
Hardness	Mohs	2.3

TABLE 2.1 (*Continued*)

TERBIUM

	Unit	Value(s)
ATOMIC, NUCLEAR AND CRYSTALLOGRAPHIC PROPERTIES		
Symbol: Tb		
Atomic number	−	65
Atomic weight	(based on ^{12}C)	158.93
Periodic classification	−	Lanthanide
Atomic volume	cm^3/g-atom	19.26
Atomic radius	Angstrom units	1.782
Ionic radius	Angstrom units (Tb^{3+})	1.09
Thermal neutron absorption		
-cross section	barns/atom	27
Density	g./cm^3	8.272
Lattice type	−	close-packed hexagonal
Lattice constants	Angstrom units	
a		3.599
b		
c		5.696
Valence electrons	−	4f^9 6s^2
THERMAL AND THERMODYNAMIC PROPERTIES		
Melting point	°C	1356
Boiling point	°C	3041
Transition point(s)	°C	1287
Critical temperature	°K	
Linear thermal expansion coefficient	10^{-6}cm/cm/°C	7.0
Specific heat	cal/g/°C	0.041
Electronic specific heat	10^{-4}cal/mole/°C^2	21.6
Thermal conductivity	cal/cm^2/cm/sec/°C	
Latent heat of fusion	kcal/mole	2.46
Latent heat of vaporization at B.P.	kcal/mole	70.0
Heat of combustion	kcal/mole	
Heat of sublimation	kcal/mole	93.96
Heat capacity at constant pressure	cal/°C/mole	6.92
Heat content at M.P.	kcal/mole	14.83
Entropy	cal/°C/mole	17.46
Longitudinal sound velocity	m/sec	2900
ELECTRICAL, ELECTRONIC, MAGNETIC AND OPTICAL PROPERTIES		
Electrical Conductivity	% IACS	
Electrical resistivity	microohm-cm	116
Photoelectric work function	eV	3.09
Hall coefficient	volt-cm/amp-Oe	
Curie point	°K	230
First ionization potential	eV (To$^+$)	5.85
Oxidation potential	V (Tb^{3+})	2.39
Electrochemical equivalent	mg/coulomb (Tb^{3+})	0.54991
Magnetic susceptibility	c.g.s. units	172000 × 10^{-6}
Refractive index, n_D		
MECHANICAL PROPERTIES		
Young's modulus of elasticity	10^{-6} kg/cm^2	0.586
Shear modulus	10^{-6} kg/cm^2	0.233
Bulk modulus	10^{-6} kg/cm^2	0.407
Poisson's ratio	−	0.261
Hardness	Vickers	46

TABLE 2.1 (*Continued*)

THALLIUM

	Unit	Value(s)
ATOMIC, NUCLEAR AND CRYSTALLOGRAPHIC PROPERTIES		
Symbol: Tl		
Atomic number	—	81
Atomic weight	(based on ^{12}C)	204.37
Periodic classification	—	III A
Atomic volume	cm^3/g-atom	17.22
Atomic radius	Angstrom units	1.99
Ionic radius	Angstrom units (Tl^{3+})	0.95
Thermal neutron absorption		
-cross section	barns/atom	3.7
Density	g./cm^3	11.85
Lattice type	—	close-packed hexagonal
Lattice constants	Angstrom units	
a		3.450
b		
c		5.514
Valence electrons	—	$6s^2\ 6p^1$
THERMAL AND THERMODYNAMIC PROPERTIES		
Melting point	°C	303.5
Boiling point	°C	1457
Transition point(s)	°C	230
Critical temperature	°K	3219
Linear thermal expansion coefficient	10^{-6} cm/cm/°C	28
Specific heat	cal/g/°C	0.031
Electronic specific heat	10^{-4} cal/mole/°C^2	6.76
Thermal conductivity	cal/cm^2/cm/sec/°C	0.093
Latent heat of fusion	kcal/mole	1.02
Latent heat of vaporization at B.P.	kcal/mole	38.74
Heat of combustion	kcal/mole	
Heat of sublimation	kcal/mole	43.24
Heat capacity at constant pressure	cal/°C/mole	6.29
Heat content at M.P.	kcal/mole	1.34
Entropy	cal/°C/mole	15.35
Longitudinal sound velocity	m/sec (20°C)	818
ELECTRICAL, ELECTRONIC, MAGNETIC AND OPTICAL PROPERTIES		
Electrical Conductivity	% IACS	
Electrical resistivity	microohm-cm	18 at 0°C
Photoelectric work function	eV	3.76
Hall coefficient	volt-cm/amp-Oe	
Curie point	°K	
First ionization potential	eV (Tl$^+$)	6.108
Oxidation potential	V (Tl$^+$)	−0.330
Electrochemical equivalent	mg/coulomb (Tl^{3+})	0.70601
Magnetic susceptibility	c.g.s. units	
Refractive index, n_D		
MECHANICAL PROPERTIES		
Young's modulus of elasticity	10^{-6} kg/cm^2	0.081
Shear modulus	10^{-6} kg/cm^2	0.028
Bulk modulus	10^{-6} kg/cm^2	0.3663
Poisson's ratio	—	0.46
Hardness	Brinell	2

TABLE 2.1 (*Continued*)

THORIUM

	Unit	Value(s)
ATOMIC, NUCLEAR AND CRYSTALLOGRAPHIC PROPERTIES		
Symbol: Th		
Atomic number	−	90
Atomic weight	(based on ^{12}C)	232.038
Periodic classification	−	Actinide
Atomic volume	cm^3/g-atom	19.79
Atomic radius	Angstrom units	1.81
Ionic radius	Angstrom units (Th^{4+})	1.02
Thermal neutron absorption -cross section	barns/atom	7.4
Density	g./cm^3	11.72
Lattice type	−	face-centered cubic
Lattice constants	Angstrom units	
a		5.086
b		
c		
Valence electrons	−	$6d^2\,7s^2$
THERMAL AND THERMODYNAMIC PROPERTIES		
Melting point	°C	1750
Boiling point	°C	4227
Transition point(s)	°C	1400
Critical temperature	°K	14550
Linear thermal expansion coefficient	10^{-6}cm/cm/°C	12.5
Specific heat	cal/g/°C	0.034
Electronic specific heat	10^{-4}cal/mole/°C^2	11.2
Thermal conductivity	cal/cm^2/cm/sec/°C	0.090
Latent heat of fusion	kcal/mole	3.56
Latent heat of vaporization at B.P.	kcal/mole	130.0
Heat of combustion	kcal/mole	
Heat of sublimation	kcal/mole	136.6
Heat capacity at constant pressure	cal/°C/mole	6.53
Heat content at M.P.	kcal/mole	16.44
Entropy	cal/°C/mole	12.76
Longitudinal sound velocity	m/sec	2490
ELECTRICAL, ELECTRONIC, MAGNETIC AND OPTICAL PROPERTIES		
Electrical Conductivity	% IACS	
Electrical resistivity	microohm-cm	14
Photoelectric work function	eV	3.51
Hall coefficient	volt-cm/amp-Oe	
Curie point	°K	
First ionization potential	eV (Th^+)	
Oxidation potential	V (Th^{4+})	1.90
Electrochemical equivalent	mg/coulomb (Th^{4+})	0.60135
Magnetic susceptibility	c.g.s. units	0.11×10^{-6}
Refractive index, n_D		
MECHANICAL PROPERTIES		
Young's modulus of elasticity	10^{-6} kg/cm^2	0.76
Shear modulus	10^{-6} kg/cm^2	0.284
Bulk modulus	10^{-6} kg/cm^2	0.5534
Poisson's ratio	−	0.285
Hardness	Vickers	32 to 42

TABLE 2.1 (*Continued*)

THULIUM

	Unit	Value(s)
ATOMIC, NUCLEAR AND CRYSTALLOGRAPHIC PROPERTIES		
Symbol: Tm		
Atomic number	–	69
Atomic weight	(based on ^{12}C)	168.934
Periodic classification	–	Lanthanide
Atomic volume	cm^3/g-atom	18.13
Atomic radius	Angstrom units	1.746
Ionic radius	Angstrom units (Tm^{3+})	1.04
Thermal neutron absorption		
-cross section	barns/atom	113
Density	g./cm^3	9.318
Lattice type	–	close-packed hexagonal
Lattice constants	Angstrom units	
a		3.5372
b		
c		5.5619
Valence electrons	–	4f^{13} 6s^2
THERMAL AND THERMODYNAMIC PROPERTIES		
Melting point	°C	1545
Boiling point	°C	1727
Transition point(s)	°C	
Critical temperature	°K	6430
Linear thermal expansion coefficient	10^{-6}cm/cm/°C	13.3
Specific heat	cal/g/°C	0.038
Electronic specific heat	10^{-4}cal/mole/°C^2	47.1
Thermal conductivity	cal/cm^2/cm/sec/°C	
Latent heat of fusion	kcal/mole	4.4
Latent heat of vaporization at B.P.	kcal/mole	51.0
Heat of combustion	kcal/mole	
Heat of sublimation	kcal/mole	58.3
Heat capacity at constant pressure	cal/°C/mole	6.45
Heat content at M.P.	kcal/mole	11.37
Entropy	cal/°C/mole	17.06
Longitudinal sound velocity	m/sec	
ELECTRICAL, ELECTRONIC, MAGNETIC AND OPTICAL PROPERTIES		
Electrical Conductivity	% IACS	
Electrical resistivity	microohm-cm	79
Photoelectric work function	eV	3.12
Hall coefficient	volt-cm/amp-Oe	–1.8 × 10^{-12}
Curie point	°K	51
First ionization potential	eV (Tm$^+$)	6.18
Oxidation potential	V (Tm^{3+})	2.28
Electrochemical equivalent	mg/coulomb (Tm^{3+})	0.58515
Magnetic susceptibility	c.g.s. units	152 × 10^{-6}
Refractive index, n_D		
MECHANICAL PROPERTIES		
Young's modulus of elasticity	10^{-6} kg/cm^2	0.77
Shear modulus	10^{-6} kg/cm^2	0.31
Bulk modulus	10^{-6} kg/cm^2	0.4047
Poisson's ratio	–	0.235
Hardness	Brinell (cast)	53

TABLE 2.1 (*Continued*)

TIN (GREY)

	Unit	Value(s)
ATOMIC, NUCLEAR AND CRYSTALLOGRAPHIC PROPERTIES		
Symbol: Sn, (α)		
Atomic number	–	50
Atomic weight	(based on ^{12}C)	118.69
Periodic classification	–	IV A
Atomic volume	cm^3/g-atom	20.59
Atomic radius	Angstrom units	1.40
Ionic radius	Angstrom units (Sn^{4+})	0.71
Thermal neutron absorption		
-cross section	barns/atom	0.63
Density	g./cm^3	5.77
Lattice type	–	Cubic
Lattice constants	Angstrom units	
a		6.4912
b		
c		
Valence electrons	–	$5s^2\ 5p^2$
THERMAL AND THERMODYNAMIC PROPERTIES		
Melting point	°C	231.9
Boiling point	°C	2270
Transition point(s)	°C	13.2 (cooling)
Critical temperature	°K	8000
Linear thermal expansion coefficient	10^{-6} cm/cm/°C	5.3
Specific heat	cal/g/°C	0.0493
Electronic specific heat	10^{-4} cal/mole/°C^2	0.000
Thermal conductivity	cal/cm^2/cm/sec/°C	
Latent heat of fusion	kcal/mole	1.71
Latent heat of vaporization at B.P.	kcal/mole	72.0
Heat of combustion	kcal/mole	
Heat of sublimation	kcal/mole	72.2
Heat capacity at constant pressure	cal/°C/mole	6.16
Heat content at M.P.	kcal/mole	
Entropy	cal/°C/mole	10.55
Longitudinal sound velocity	m/sec	
ELECTRICAL, ELECTRONIC, MAGNETIC AND OPTICAL PROPERTIES		
Electrical Conductivity	% IACS	
Electrical resistivity	microohm-cm	300 at 0°C
Photoelectric work function	eV	4.3
Hall coefficient	volt-cm/amp-Oe	
Curie point	°K	
First ionization potential	eV (Sn^+)	7.344
Oxidation potential	V (Sn^{2+})	0.136
Electrochemical equivalent	mg/coulomb (Sn^{4+})	0.30752
Magnetic susceptibility	c.g.s. units	-0.031×10^{-6} at 0°C
Refractive index, n_D		
MECHANICAL PROPERTIES		
Young's modulus of elasticity	10^{-6} kg/cm^2	0.534
Shear modulus	10^{-6} kg/cm^2	0.188
Bulk modulus	10^{-6} kg/cm^2	1.13
Poisson's ratio	–	0.42
Hardness		

TABLE 2.1 (*Continued*)

TIN (WHITE)

	Unit	Value(s)
ATOMIC, NUCLEAR AND CRYSTALLOGRAPHIC PROPERTIES		
Symbol: Sn, (β)		
Atomic number	–	50
Atomic weight	(based on ^{12}C)	118.69
Periodic classification	–	IV A
Atomic volume	cm^3/g-atom	16.30
Atomic radius	Angstrom units	
Ionic radius	Angstrom units	
Thermal neutron absorption -cross section	barns/atom	0.63
Density	g./cm^3	7.29
Lattice type	–	Tetragonal
Lattice constants	Angstrom units	
a		5.8314
b		
c		3.1815
Valence electrons	–	$5s^2\ 5p^2$
THERMAL AND THERMODYNAMIC PROPERTIES		
Melting point	°C	231.9
Boiling point	°C	2270
Transition point(s)	°C	13.2 (cooling)
Critical temperature	°K	9300
Linear thermal expansion coefficient	10^{-6} cm/cm/°C	21.2
Specific heat	cal/g/°C	0.054
Electronic specific heat	10^{-4} cal/mole/°C^2	4.25
Thermal conductivity	cal/cm^2/cm/sec/°C	0.150 at 0°C
Latent heat of fusion	kcal/mole	1.71
Latent heat of vaporization at B.P.	kcal/mole	69.4
Heat of combustion	kcal/mole	
Heat of sublimation	kcal/mole	72.0
Heat capacity at constant pressure	cal/°C/mole	6.30
Heat content at M.P.	kcal/mole	1.40
Entropy	cal/°C/mole	12.29
Longitudinal sound velocity	m/sec	2650
ELECTRICAL, ELECTRONIC, MAGNETIC AND OPTICAL PROPERTIES		
Electrical Conductivity	% IACS	15.6
Electrical resistivity	microohm-cm	11.0
Photoelectric work function	eV	4.3
Hall coefficient	volt-cm/amp-Oe	
Curie point	°K	
First ionization potential	eV (Sn^+)	7.344
Oxidation potential	V (Sn^{2+})	0.136
Electrochemical equivalent	mg/coulomb (Sn^{2+})	0.61505
Magnetic susceptibility	c.g.s. units	0.027×10^{-6}
Refractive index, n_D		
MECHANICAL PROPERTIES		
Young's modulus of elasticity	10^{-6} kg/cm^2	0.550
Shear modulus	10^{-6} kg/cm^2	0.208
Bulk modulus	10^{-6} kg/cm^2	0.5525
Poisson's ratio	–	0.33
Hardness	Brinell (cast)	5.3

TABLE 2.1 (*Continued*)

TITANIUM

	Unit	Value(s)
ATOMIC, NUCLEAR AND CRYSTALLOGRAPHIC PROPERTIES		
Symbol: Ti		
Atomic number	–	22
Atomic weight	(based on ^{12}C)	47.90
Periodic classification	–	IV B
Atomic volume	cm^3/g-atom	12.01
Atomic radius	Angstrom units	1.54
Ionic radius	Angstrom units (Ti^{4+})	0.68
Thermal neutron absorption		
-cross section	barns/atom	6.1
Density	g./cm^3	4.507
Lattice type, (α)	–	close-packed hexagonal
Lattice constants	Angstrom units	
a		2.9504
b		
c		4.6833
Valence electrons	–	$3d^2\ 4s^2$
THERMAL AND THERMODYNAMIC PROPERTIES		
Melting point	°C	1668
Boiling point	°C	3313
Transition point(s)	°C	882
Critical temperature	°K	
Linear thermal expansion coefficient	$10^{-6}cm/cm/°C$	8.35
Specific heat	cal/g/°C	0.125
Electronic specific heat	$10^{-4}cal/mole/°C^2$	8.15
Thermal conductivity	$cal/cm^2/cm/sec/°C$	0.041
Latent heat of fusion	kcal/mole	3.42
Latent heat of vaporization at B.P.	kcal/mole	102.5
Heat of combustion	kcal/mole	
Heat of sublimation	kcal/mole	112.7
Heat capacity at constant pressure	cal/°C/mole	5.98
Heat content at M.P.	kcal/mole	12.97
Entropy	cal/°C/mole	7.33
Longitudinal sound velocity	m/sec	5000
ELECTRICAL, ELECTRONIC, MAGNETIC AND OPTICAL PROPERTIES		
Electrical Conductivity	% IACS	3.6
Electrical resistivity	microohm-cm	42.0
Photoelectric work function	eV	4.17
Hall coefficient	volt-cm/amp-Oe	1.82×10^{-13}
Curie point	°K	
First ionization potential	eV (Ti^+)	6.82
Oxidation potential	V (Ti/TiO^{2+})	0.89 (acid soln.)
Electrochemical equivalent	mg/coulomb (Ti^{4+})	0.12409
Magnetic susceptibility	c.g.s. units	3.17×10^{-6}
Refractive index, n_D		
MECHANICAL PROPERTIES		
Young's modulus of elasticity	$10^{-6}\ kg/cm^2$	1.08
Shear modulus	$10^{-6}\ kg/cm^2$	0.401
Bulk modulus	$10^{-6}\ kg/cm^2$	1.072
Poisson's ratio	–	0.345
Hardness	Vickers	80–100

TABLE 2.1 (*Continued*)

TUNGSTEN

	Unit	Value(s)
ATOMIC, NUCLEAR AND CRYSTALLOGRAPHIC PROPERTIES		
Symbol: W		
Atomic number	–	74
Atomic weight	(based on ^{12}C)	183.85
Periodic classification	–	VI B
Atomic volume	cm^3/g-atom	9.551
Atomic radius	Angstrom units	1.37
Ionic radius	Angstrom units (W^{4+})	0.66
Thermal neutron absorption -cross section	barns/atom	18
Density	g./cm^3	19.3
Lattice type	–	body-centered cubic
Lattice constants	Angstrom units	
a		3.1585
b		
c		
Valence electrons	–	$5d^4 6s^2$
THERMAL AND THERMODYNAMIC PROPERTIES		
Melting point	°C	3380
Boiling point	°C	5727
Transition point(s)	°C	
Critical temperature	°K	23000
Linear thermal expansion coefficient	10^{-6} cm/cm/°C	4.59
Specific heat	cal/g/°C	0.0321
Electronic specific heat	10^{-4} cal/mole/°C^2	2.92
Thermal conductivity	cal/cm^2/cm/sec/°C	0.397 at 0°C
Latent heat of fusion	kcal/mole	8.42
Latent heat of vaporization at B.P.	kcal/mole	191.0
Heat of combustion	kcal/mole	
Heat of sublimation	kcal/mole	200.00
Heat capacity at constant pressure	cal/°C/mole	5.84
Heat content at M.P.	kcal/mole	
Entropy	cal/°C/mole	8.04
Longitudinal sound velocity	m/sec	5174
ELECTRICAL, ELECTRONIC, MAGNETIC AND OPTICAL PROPERTIES		
Electrical Conductivity	% IACS	
Electrical resistivity	microohm-cm	5.5
Photoelectric work function	eV	4.56
Hall coefficient	volt-cm/amp-Oe	
Curie point	°K	
First ionization potential	eV (W^+)	7.60
Oxidation potential	V (W/WO_4^{2-})	1.05 (alk. soln.)
Electrochemical equivalent	mg/coulomb (W^{6+})	0.31779
Magnetic susceptibility	c.g.s. units	0.28×10^{-6}
Refractive index, n_D		2.76 at 0.579 μ
MECHANICAL PROPERTIES		
Young's modulus of elasticity	10^{-6} kg/cm^2	4.05
Shear modulus	10^{-6} kg/cm^2	1.56
Bulk modulus	10^{-6} kg/cm^2	3.296
Poisson's ratio	–	0.284
Hardness	Rockwell C (sintered)	40

TABLE 2.1 (*Continued*)

URANIUM

	Unit	Value(s)
ATOMIC, NUCLEAR AND CRYSTALLOGRAPHIC PROPERTIES		
Symbol: U		
Atomic number	–	92
Atomic weight	(based on ^{12}C)	238.03
Periodic classification	–	Actinide
Atomic volume	cm^3/g-atom	13.16
Atomic radius	Angstrom units	1.38
Ionic radius	Angstrom units (U^{4+})	0.97
Thermal neutron absorption -cross section	barns/atom	7.6
Density	g./cm^3	19.04
Lattice type	–	Orthorhombic
Lattice constants	Angstrom units	
a		2.853
b		5.865
c		4.954
Valence electrons	–	$5f^3\ 6d^1\ 7s^2$
THERMAL AND THERMODYNAMIC PROPERTIES		
Melting point	°C	1132
Boiling point	°C	3677
Transition point(s)	°C	667, 775, 1132
Critical temperature	°K	12000
Linear thermal expansion coefficient	10^{-6} cm/cm/°C	12.6
Specific heat	cal/g/°C	0.028
Electronic specific heat	10^{-4} cal/mole/°C^2	26.1
Thermal conductivity	cal/cm^2/cm/sec/°C	0.071
Latent heat of fusion	kcal/mole	2.47
Latent heat of vaporization at B.P.	kcal/mole	101.0
Heat of combustion	kcal/mole	
Heat of sublimation	kcal/mole	125
Heat capacity at constant pressure	cal/°C/mole	6.58
Heat content at M.P.	kcal/mole	11.6
Entropy	cal/°C/mole	12.052
Longitudinal sound velocity	m/sec (20°C)	~3100
ELECTRICAL, ELECTRONIC, MAGNETIC AND OPTICAL PROPERTIES		
Electrical Conductivity	% IACS	
Electrical resistivity	microohm-cm	30
Photoelectric work function	eV	3.63
Hall coefficient	volt-cm/amp-Oe	
Curie point	°K	
First ionization potential	eV	
Oxidation potential	V (U/U^{3+})	1.80
Electrochemical equivalent	mg/coulomb (U^{6+})	0.41117
Magnetic susceptibility	c.g.s. units	
Refractive index, n_D		
MECHANICAL PROPERTIES		
Young's modulus of elasticity	10^{-6} kg/cm^2	1.90
Shear modulus	10^{-6} kg/cm^2	0.75
Bulk modulus	10^{-6} kg/cm^2	1.007
Poisson's ratio	–	0.245
Hardness	Vickers	220

TABLE 2.1 (*Continued*)

VANADIUM

	Unit	Value(s)
ATOMIC, NUCLEAR AND CRYSTALLOGRAPHIC PROPERTIES		
Symbol: V		
Atomic number	–	23
Atomic weight	(based on ^{12}C)	50.942
Periodic classification	–	V B
Atomic volume	cm^3/g-atom	8.365
Atomic radius	Angstrom units	1.32
Ionic radius	Angstrom units (V^{5+})	0.59
Thermal neutron absorption -cross section	barns/atom	5.03
Density	g./cm^3	6.11
Lattice type	–	body-centered cubic
Lattice constants	Angstrom units	
a		3.026
b		
c		
Valence electrons	–	3d^3 4s^2
THERMAL AND THERMODYNAMIC PROPERTIES		
Melting point	°C	1890
Boiling point	°C	3000
Transition point(s)	°C	
Critical temperature	°K	11200
Linear thermal expansion coefficient	10^{-6}cm/cm/°C	8.3
Specific heat	cal/g/°C	0.120
Electronic specific heat	10^{-4}cal/mole/°C^2	21.61
Thermal conductivity	cal/cm^2/cm/sec/°C	0.074 at 100°C
Latent heat of fusion	kcal/mole	3.83
Latent heat of vaporization at B.P.	kcal/mole	109.6
Heat of combustion	kcal/mole	
Heat of sublimation	kcal/mole	123
Heat capacity at constant pressure	cal/°C/mole	5.905
Heat content at M.P.	kcal/mole	14.45
Entropy	cal/°C/mole	7.05
Longitudinal sound velocity	m/sec	4560
ELECTRICAL, ELECTRONIC, MAGNETIC AND OPTICAL PROPERTIES		
Electrical Conductivity	% IACS	
Electrical resistivity	microohm-cm	24.8 at 30.6°C
Photoelectric work function	eV	4.11
Hall coefficient	volt-cm/amp-Oe	
Curie point	°K	
First ionization potential	eV (V$^+$)	6.74
Oxidation potential	V (V/V^{2+})	1.18
Electrochemical equivalent	mg/coulomb (V^{5+})	0.10560
Magnetic susceptibility	c.g.s. units	1.4 × 10^{-6}
Refractive index, n		3.03 at 0.579 μ
MECHANICAL PROPERTIES		
Young's modulus of elasticity	10^{-6} kg/cm^2	1.34
Shear modulus	10^{-6} kg/cm^2	0.474
Bulk modulus	10^{-6} kg/cm^2	1.651
Poisson's ratio	–	0.36
Hardness	Rockwell B	85

TABLE 2.1 (*Continued*)

XENON

	Unit	Value(s)
ATOMIC AND NUCLEAR PROPERTIES		
Symbol: Xe		
Atomic number	–	54
Atomic weight	(based on ^{12}C)	131.30
Periodic classification	–	rare gas
Thermal neutron absorption		
-cross section	barns/atom	25
SOLID STATE		
Lattice type	–	face-centered cubic
Lattice constants	Angstrom units	
a		6.25 at 88°K
b		
c		
Atomic radius	Angstrom units	
Melting point	°K	161.3
Heat of fusion	kcal/mole	0.549
Entropy of fusion	cal/°K/mole	
Triple point		
Temperature	°K	161.36
Pressure	torr	611
Density of solid	g/cm^3	3.540
Density of liquid	g/cm^3	3.076
LIQUID STATE		
Boiling point	°K	165.04
Density at boiling point	g/cm^3	2.987
Heat of vaporization at boiling point	kcal/mole	3.021
Entropy at boiling point	cal/°K/mole	37.66
Critical Data		
Temperature	°K	289.71
Pressure	atm.	57.64
Density	g/cm^3	1.100
GASEOUS STATE*		
Density at STP	g/l	5.8971
Specific heat at 20°C at const. pres.	cal/g/°K	0.038
Ratio of specific heats, Cp/Cv, at 20°C	–	1.666
Thermal conductivity at 20°C	cal/cm^2/cm/sec/°K	0.000012
Heat capacity at const. press. & 25°C	cal/°K/mole	4.97
Entropy at 25°C	cal/°K/mole	40.53
Heat of combustion	kcal/mole	
Dielectric constant at 25°C	–	1.001238
Viscosity at 20°C	micropoise	227.40
Refractive index, n_D, at 0°C	–	1.000702
Solubility in H$_2$O at 1 atm. partial		
pressure and 20°C	cm^3(STP)/1000 g.	108.1
Sound velocity at 0°C	m/sec	168

*Generally at one atmosphere.

TABLE 2.1 (*Continued*)

YTTERBIUM

	Unit	Value(s)
ATOMIC, NUCLEAR AND CRYSTALLOGRAPHIC PROPERTIES		
Symbol: Yb		
Atomic number	–	70
Atomic weight	(based on ^{12}C)	173.04
Periodic classification	–	Lanthanide
Atomic volume	cm^3/g-atom	24.82
Atomic radius	Angstrom units	1.940
Ionic radius	Angstrom units (Yb^{3+})	1.00
Thermal neutron absorption -cross section	barns/atom	37
Density	g./cm^3	6.977
Lattice type	–	face-centered cubic
Lattice constants	Angstrom units	
a		5.481
b		
c		
Valence electrons	–	4f^{14} 6s^2
THERMAL AND THERMODYNAMIC PROPERTIES		
Melting point	°C	824
Boiling point	°C	1193
Transition point(s)	°C	798
Critical temperature	°K	4420
Linear thermal expansion coefficient	10^{-6}cm/cm/°C	25.0
Specific heat	cal/g/°C	0.035
Electronic specific heat	10^{-4}cal/mole/°C^2	6.93
Thermal conductivity	cal/cm^2/cm/sec/°C	
Latent heat of fusion	kcal/mole	1.830
Latent heat of vaporization at B.P.	kcal/mole	37.1
Heat of combustion	kcal/mole	
Heat of sublimation	kcal/mole	38.2
Heat capacity at constant pressure	cal/°C/mole	6.16
Heat content at M.P.	kcal/mole	4.70
Entropy	cal/°C/mole	15.00
Longitudinal sound velocity	m/sec	1800
ELECTRICAL, ELECTRONIC, MAGNETIC AND OPTICAL PROPERTIES		
Electrical Conductivity	% IACS	
Electrical resistivity	microohm-cm	29.4
Photoelectric work function	eV	2.59
Hall coefficient	volt-cm/amp-Oe	3.77 × 10^{-12}
Curie point	°K	
First ionization potential	eV (Tb$^+$)	6.25
Oxidation potential	V (Yb^{3+})	2.27
Electrochemical equivalent	mg/coulomb (Yb^{3+})	0.59772
Magnetic susceptibility	c.g.s. units	81 × 10^{-6}
Refractive index, n_D		
MECHANICAL PROPERTIES		
Young's modulus of elasticity	10^{-6} kg/cm^2	0.182
Shear modulus	10^{-6} kg/cm^2	0.071
Bulk modulus	10^{-6} kg/cm^2	0.1353
Poisson's ratio	–	0.284
Hardness	Vickers	21

TABLE 2.1 (*Continued*)

YTTRIUM

	Unit	Value(s)
ATOMIC, NUCLEAR AND CRYSTALLOGRAPHIC PROPERTIES		
Symbol: Y		
Atomic number	–	39
Atomic weight	(based on ^{12}C)	88.905
Periodic classification	–	III B
Atomic volume	cm^3/g-atom	19.86
Atomic radius	Angstrom units	1.801
Ionic radius	Angstrom units (Y^{3+})	1.06
Thermal neutron absorption -cross section	barns/atom	1.28
Density	g./cm^3	4.472
Lattice type	–	close-packed hexagonal
Lattice constants	Angstrom units	
a		3.6457
b		
c		5.7305
Valence electrons	–	$4d^1 5s^2$
THERMAL AND THERMODYNAMIC PROPERTIES		
Melting point	°C	1502
Boiling point	°C	3397
Transition point(s)	°C	1459
Critical temperature	°K	8950
Linear thermal expansion coefficient	10^{-6} cm/cm/°C	10.8
Specific heat	cal/g/°C	0.071
Electronic specific heat	10^{-4} cal/mole/°C^2	24.1
Thermal conductivity	cal/cm^2/cm/sec/°C	0.035
Latent heat of fusion	kcal/mole	4.1
Latent heat of vaporization at B.P.	kcal/mole	94.0
Heat of combustion	kcal/mole	
Heat of sublimation	kcal/mole	97.8
Heat capacity at constant pressure	cal/°C/mole	6.34
Heat content at M.P.	kcal/mole	9.42
Entropy	cal/°C/mole	11.00
Longitudinal sound velocity	m/sec	4300
ELECTRICAL, ELECTRONIC, MAGNETIC AND OPTICAL PROPERTIES		
Electrical Conductivity	% IACS	
Electrical resistivity	microohm-cm	57
Photoelectric work function	eV	3.3
Hall coefficient	volt-cm/amp-Oe	-0.770×10^{-12}
Curie point	°K	
First ionization potential	eV (Y^+)	6.38
Oxidation potential	V (Y^{3+})	2.37
Electrochemical equivalent	mg/coulomb (Y^{3+})	0.3092
Magnetic susceptibility	c.g.s. units	5.3×10^{-6}
Refractive index, n_D		
MECHANICAL PROPERTIES		
Young's modulus of elasticity	10^{-6} kg/cm^2	0.661
Shear modulus	10^{-6} kg/cm^2	0.263
Bulk modulus	10^{-6} kg/cm^2	0.3733
Poisson's ratio	–	0.258
Hardness	Brinell	32

TABLE 2.1 (*Continued*)

ZINC

	Unit	Value(s)
ATOMIC, NUCLEAR AND CRYSTALLOGRAPHIC PROPERTIES		
Symbol: Zn		
Atomic number	−	30
Atomic weight	(based on ^{12}C)	65.37
Periodic classification	−	II B
Atomic volume	cm^3/g-atom	9.165
Atomic radius	Angstrom units	1.33
Ionic radius	Angstrom units (Zn^{2+})	0.74
Thermal neutron absorption -cross section	barns/atom	1.10
Density	g./cm^3	7.133
Lattice type	−	Close-packed hexagonal
Lattice constants	Angstrom units	
a		2.664
b		
c		4.9469
Valence electrons	−	$4s^2$
THERMAL AND THERMODYNAMIC PROPERTIES		
Melting point	°C	419.5
Boiling point	°C	902
Transition point(s)	°C	
Critical temperature	°K	2910
Linear thermal expansion coefficient	10^{-6} cm/cm/°C	38
Specific heat	cal/g/°C	0.0915
Electronic specific heat	10^{-4} cal/mole/°C^2	1.54
Thermal conductivity	cal/cm^2/cm/sec/°C	0.27
Latent heat of fusion	kcal/mole	1.765
Latent heat of vaporization at B.P.	kcal/mole	27.5
Heat of combustion	kcal/mole	
Heat of sublimation	kcal/mole	31.1
Heat capacity at constant pressure	cal/°C/mole	6.07
Heat content at M.P.	kcal/mole	1.94
Entropy	cal/°C/mole	9.95
Longitudinal sound velocity	m/sec (18°C)	3700
ELECTRICAL, ELECTRONIC, MAGNETIC AND OPTICAL PROPERTIES		
Electrical Conductivity	% IACS	28.27
Electrical resistivity	microohm-cm	5.916
Photoelectric work function	eV	3.74
Hall coefficient	volt-cm/amp-Oe	
Curie point	°K	
First ionization potential	eV (Zn^+)	9.394
Oxidation potential	V (Zn^{2+})	0.763
Electrochemical equivalent	mg/coulomb (Zn^{2+})	0.3388
Magnetic susceptibility	c.g.s. units	-0.139×10^{-6}
Refractive index, n_D		1.93
MECHANICAL PROPERTIES		
Young's modulus of elasticity	10^{-6} kg/cm^2	0.940
Shear modulus	10^{-6} kg/cm^2	0.379
Bulk modulus	10^{-6} kg/cm^2	0.6101
Poisson's ratio	−	0.29
Hardness	Mohs	2.5

TABLE 2.1 (Continued)

ZIRCONIUM

	Unit	Value(s)
ATOMIC, NUCLEAR AND CRYSTALLOGRAPHIC PROPERTIES Symbol: Zr		
Atomic number	—	40
Atomic weight	(based on ^{12}C)	91.22
Periodic classification	—	IV B
Atomic volume	cm^3/g-atom	14.02
Atomic radius	Angstrom units	1.61
Ionic radius	Angstrom units (Zr^{4+})	0.80
Thermal neutron absorption -cross section	barns/atom	0.182
Density	g./cm^3	6.45
Lattice type	—	close-packed hexagonal
Lattice constants	Angstrom units	
a		3.2321
b		
c		5.1474
Valence electrons	—	4d^2 5s^2
THERMAL AND THERMODYNAMIC PROPERTIES		
Melting point	°C	1852
Boiling point	°C	4377
Transition point(s)	°C	867
Critical temperature	°K	12300
Linear thermal expansion coefficient	10^{-6}cm/cm/°C	5.78
Specific heat	cal/g/°C	0.0659
Electronic specific heat	10^{-4}cal/mole/°C^2	6.95
Thermal conductivity	cal/cm^2/cm/sec/°C	0.0505
Latent heat of fusion	kcal/mole	3.74
Latent heat of vaporization at B.P.	kcal/mole	139.0
Heat of combustion	kcal/mole	
Heat of sublimation	kcal/mole	146
Heat capacity at constant pressure	cal/°C/mole	6.12
Heat content at M.P.	kcal/mole	14.22
Entropy	cal/°C/mole	9.29
Longitudinal sound velocity	m/sec (20°C)	~3800
ELECTRICAL, ELECTRONIC, MAGNETIC AND OPTICAL PROPERTIES		
Electrical Conductivity	% IACS	
Electrical resistivity	microohm-cm	44.1
Photoelectric work function	eV	4.1
Hall coefficient	volt-cm/amp-Oe	0.8×10^{-12}
Curie point	°K	
First ionization potential	eV (Zr$^+$)	6.835
Oxidation potential	V (Zr^{4+})	1.53
Electrochemical equivalent	mg/coulomb (Zr^{4+})	0.2363
Magnetic susceptibility	c.g.s. units	1.22×10^{-6}
Refractive index, n_D		
MECHANICAL PROPERTIES		
Young's modulus of elasticity	10^{-6} kg/cm^2	0.939
Shear modulus	10^{-6} kg/cm^2	0.348
Bulk modulus	10^{-6} kg/cm^2	0.8496
Poisson's ratio	—	0.34
Hardness	Rockwell B (annealed)	75–85

CHEMICAL ELEMENTS
REFERENCES AND BIBLIOGRAPHY

Adams, R. M., Ed., "Boron, Metallo-Boron Compounds and Boranes," Wiley, New York, 1964.

Bagnall, K. W., "The Chemistry of Selenium, Tellurium and Polonium," Elsevier, Amsterdam, 1966.

Bagnall, K. W.,: Chemistry of the Rare Radioelements," Butterworths, London, 1957.

Bailar, J. C. Ed. "Comprehensive Inorganic Chemistry," 5 vols., Pergamon Press, Oxford, 1973.

Beer, S. Z., "Liquid Metals: Chemistry and Physics," Marcel Dekker, New York, 1972.

Berman, R., Ed., "Physical Properties of Diamond," Clarendon Press," Oxford, 1965.

Blumenthal, W. B., "The Chemical Behavior of Zirconium," Van Nostrand, New York, 1958.

Butts, A., Ed., "Copper: the Metal, its Alloys and Compounds," Reinhold, New York, 1964.

Carr, K., "Tables of Thermionic Properties of the Elements and Compounds," U.S. Government Report No. N67-16043, 1967.

Chizhikov, D. M., "Cadmium," Pergamon Press, Oxford, 1966.

Clark, R. J. H., "The Chemistry of Titanium and Vanadium," Elsevier, Amsterdam, 1968.

Cleveland, J. M., "The Chemistry of Plutonium," Gordon & Breach, New York, 1970.

Coffinberry, A. S. and W. N. Miner, Eds., "The Metal Plutonium," Univ. of Chicago Press, Chicago, 1961.

Colton, R., "The Chemistry of Rhenium and Technetium," Interscience, London, 1965.

Connolly, T. F., Ed., "Group IV, V and VI Transition Metals and Compounds—Preparation and Properties," Plenum, New York, 1972.

Cook, G. A., Ed., "Argon, Helium and the Rare Gases," 2 vols., Wiley, New York, 1961.

Darken, L. S. and R. W. Gurry, "Physical Chemistry of Metals," McGraw-Hill, New York, 1953.

Degussa, "Edelmetall-Taschenbuch," DEGUSSA, Frankfurt am Main, Germany, 1967.

Everhart, J. L., Ed., "Engineering Properties of Nickel and Nickel Alloys," Plenum, New York, 1971.

Fairbrother, F., "The Chemistry of Niobium and Tantalum," Elsevier, London, 1967.

Filyand, M. A. and E. I. Semenova, "Handbook of the Rare Elements," Mac Donald, London, 1969.

Forsythe, W. E., "Smithsonian Physical Tables," 9th rev. ed., Smithsonian Press, Washington, 1969.

"Gmelins Handbuch der anorganischen Chemie," 8th ed., Verlag Chemie, Weinheim, Germany, 1924–1971.

Griffith, W. P., "The Chemistry of the Rarer Platinum Metals," Wiley, New York 1967.

Gschneidner, K. A. Jr., "The Application of Vacuum Metallurgy in the Purification of Rare Earth Metals," pp. 99–135, *Trans. Vacuum Metallurgy Conf., 1965*, L. M. Bianchi, Ed., American Vacuum Soc., Boston, Mass., 1966.

Gschneidner, K. A., Jr., "Physical Properties and Interrelationships of Metallic and Semielements," pp. 275–426 in *Solid State Physics*, Vol. 16, Academic Press, New York, 1964.

Gutman, V., Ed., "Halogen Chemistry," 3 vols., Academic Press, New York, 1967.

Hampel, C. A., Ed., "Rare Metals Handbook," 2nd ed., Reinhold, New York, 1961.

Hampel, C. A., "The Encyclopedia of the Chemical Elements," Reinhold, New York, 1968.

Hultgren, R., R. L. Orr, P. D. Anderson and K. K. Kelley, "Selected Values of Thermodynamic Properties of Metals and Alloys," Wiley, New York, 1963.

Hultgren, R., R. L. Orr and K. K. Kelley, "Supplement to Selected Values of Thermodynamic Properties of Metals and Alloys," University of California, Berkeley, 1971.

Jolles, Z. E., Ed., "Bromine and its Compounds," Academic Press, New York, 1966.

Jolly, W. C., "The Inorganic Chemistry of Nitrogen," Benjamin, New York, 1964.

Katz, J. J. and G. T. Seaborg, "The Chemistry of the Actinide Elements," Methuen, London, 1957.

Keller, C., "The Chemistry of the Transuranium Elements," Verlag Chemie, Weinheim, Germany, 1971.

Kelley, K. K., "Contributions to the Data on Theoretical Metallurgy," Part XIII, Bureau of Mines Bull. No. 584, U.S. Government Printing Office, Washington, 1960.

Kepert, D. L., "The Early Transition Metals," Academic Press, London, 1972.

Kleber, E. V. and B. Love, "The Technology of Scandium, Yttrium and the Rare Earth Metals, MacMillan, New York, 1963.

Latimer, W. M., "Oxidation Potentials," 2nd ed., Prentice-Hall, Englewood Cliffs, N.J., 1952.

Lee, A. G., "The Chemistry of Thallium," Elsevier, Amsterdam, 1971.

Loung, P. Y., "Graphic Handbook of Chemistry and Metallurgy," Chemical Publishing Co., New York, 1965.

Mantell, C. L., "Carbon and Graphite Handbook," Wiley, New York, 1968.

Matheson Co., "Matheson Gas Data Book," 4th ed., The Matheson Co., East Rutherford, N.J., 1966.

Mathewson, C. H., Ed., "Zinc, the Science and Technology of the Metal, its Alloys and Compounds," Reinhold, New York, 1959.

Metals Handbook Committee, "Metals Handbook," 8th ed., American Society for Metals, Novelty, Ohio, 1961.

Meyer, B., Ed., "Elemental Sulfur Chemistry and Physics," Wiley, New York, 1965.

Miller, G. L., "Tantalum and Niobium," Butterworth, London, 1959.

Miller, G. L., "Zirconium," 2nd ed., Butterworth, London, 1957.

Moeller, T., "The Chemistry of the Lanthanides," Reinhold, New York, 1963.

Moore, C. E., "Ionization Potentials and Ionization Limits Derived from the Analysis of Optical Spectra," National Bureau of Standards Report No. NSRDS-NBS 34, U.S. Government Printing Office, Washington, 1970.

Mutterties, E. L., Ed., "The Chemistry of Boron and its Compounds," Wiley, New York, 1965.

Neumark, H. R., "The Chemistry and Chemical Technology of Fluorine," Wiley, New YOrk, 1967.

Nickless, G., Ed., "Inorganic Sulfur Chemistry," Elsevier, Amsterdam, 1968.

Northcott, L., "Molybdenum," Academic Press, New York, 1956.

Peacock, R. D., "The Chemistry of Technetium and Rhenium," Elsevier, London, 1966.

Pearson, W. E., "Handbook of Lattice Spacings and Structures of Metals," 2 vols., Pergamon Press, Oxford, 1958–1967.

Perel'man, F. M., "Rubidium and Cesium," Macmillan, New York, 1965.

Promisel, N. E., Ed., "The Science and Technology of Tungsten, Tantalum, Molybdenum, Niobium and their Alloys," AGARD Conf. Oslo, 1963, Macmillan, New York, 1964.

Rieck, G., Ed., "Tungsten and its Compounds," Pergamon Press, Oxford, 1967.

Roberts, C. S., "Magnesium and its Alloys," Wiley, New York, 1960.

Samsonov, G. V., Ed., "Handbook of the Physicochemical Properties of the Elements," Plenum, New York, 1968.

Schwope, A. D., "Zirconium and Zirconium Alloys," American Society for Metals, Cleveland, 1953.

Seaborg, G. T. and J. J. Katz, Eds., "The Actinide Elements," McGraw-Hill, New York, 1954.

Seaborg, G. T., "The Transuranium Elements," Yale University Press, New Haven, 1958.

Sidgwick, N. V., "Chemical Elements and their Compounds," Oxford Univ. Press, Oxford, 1950.

Simons, J. H., Ed., "Fluorine Chemistry," Academic Press, New York, 1965.

Sisco, F. T. and E. Epremian, "Columbium and Tantalum," Wiley, New York, 1963.

Smithells, C. J., "Metals Reference Book," 4th ed., 3 vols., Plenum, New York, 1967.

Spedding, F. H. and A. H. Daane, "The Rare Earths," Wiley, New York, 1961.

Stull, D. R. and G. C. Sinke, "Thermodynamic Properties of the Elements," American Chemical Society, Washington, 1956.

Stull, D. R. and H. Prophet, "JANAF Thermochemical Tables," 2nd ed., National Bureau of Standards Report No. NSRDS-NBS37, U.S. Government Printing Office, Washington, 1971.

Sully, A. H. and E. A. Brandes, Eds., "Chromium," 2nd ed., Plenum, New York, 1968.

Sully, A. H., "Metallurgy of the Rarer Metals," Butterworth, London, 1955.

Tietz, T. E. and J. W. Wilson, "Behavior and Properties of Refractory Metals," Stanford University Press, Stanford, 1965.

Tribalat, S., "Rhénium et Technétium," Gauthier-Villars, Paris, 1957.

Van Wazer, J. R., "Industrial Chemistry and Technology of Phosphorus and Phosphorus Compounds," Wiley, New York, 1968.

Van Wazer, J. R., "Phosphorus and its Compounds," 2 vols., Wiley, New York, 1958–1961.

Wagman, D. D. et al., "Selected Values of Chemical Thermodynamic Properties," National Bureau of Standards Tech. Note No. 270-3 et seq., U.S. Government Printing Office, Washington, 1968 et seq.

Wick, O. J., "Plutonium Handbook," Gordon & Breach, New York, 1967,

Wicks, C. E. and F. E. Block, "Thermodynamic Properties of 65 Elements, their Oxides, Halides, Carbides, and Nitrides," Bureau of Mines Bull. No. 605, U.S. Government Printing Office, Washington, 1963.

Wise, E. M., Ed., "Gold: Recovery, Properties and Applications," Van Nostrand, New York, 1964.

Young, R. S., "Cobalt," Reinhold, New York, 1948.

CHAPTER III
PROPERTIES OF ORGANIC COMPOUNDS

Properties of selected organic compounds are summarized in Chapter III in the following tables:

TABLE 3-1. ACETALS AND KETALS.

Compound	Mol. Wt.	Density* (g/cm^3)	Freezing Point (°C)	Boiling Point (°C)	Refractive Index, n_D
1,1-Diethoxy ethane (Acetal)	118.2	0.8254		103.6	1.38054
1,1-Dimethoxy ethane (Dimethyl acetal)	90.1	0.8516	−113.2	64.3	1.3665
1,2-Dimethoxy ethane	90.1	0.8665	− 58	85	1.3796
Diethoxy methane (Ethylal)	104.1	0.83465^{15}	− 66.5	88.0	1.3759^{17}
Dimethoxy methane (Methylal)	76.1	0.86645$^{15}_4$	−105.0	42.3	1.35298
1,1-Diethoxy propane (Propylal)	132.2	0.834	− 97.3	137	1.393
1,3-Dioxane	88.1	1.034	− 42	105.6	1.420

*Or specific gravity.

TABLE 3-2.   ALCOHOLS

Compound	Mol. Wt.	Density* (g/cm^3)	Freezing Point ($^\circ$C)	Boiling Point ($^\circ$C)	Refractive Index, n_D
1-Butanol (*n*-Butyl alcohol)	74.1	0.8098	− 88.9	117.5	1.3993
2-Butanol (*sec*-Butyl alcohol)	74.1	0.808	− 89.3	99.5	1.397
2-Ethyl-1-butanol (Pseudohexyl alcohol)	102.2	0.8328	−114.4	146.3	1.4224
2-Methyl-1-butanol (active amyl alcohol)	88.2	0.8193	< − 70	128.0	1.4087
2-Methyl-2-butanol (*tert*-amyl alcohol)	88.2	0.8096	− 8.8	101.8	1.4052
3-Methyl-1-butanol (Isoamyl alcohol)	88.2	0.8092	−117.2	132.0	1.4053
3-Methyl-2-butanol (*sec*-Isoamyl alcohol)	88.2	0.819		111.5	1.4096
Cyclobutanol	72.1	0.9226$^{15}_{15}$		125	1.4339^{19}
1,2-Ethanediol (Ethylene glycol)	62.1	1.1135	− 13.0	197.5	1.4311
Ethanol (Ethyl alcohol)	46.1	0.7893	−114.1	78.3	1.3611
2-Aminoethanol (Ethanolamine)	61.1	1.0117	10.5	171.1	1.4539
2-Chloroethanol (Ethylene chlorohydrin)	80.5	1.2019	− 67.5	128.6	1.4422
2-Ethoxyethanol (Ethyl cellosolve)	90.1	0.9297	− 70	134.8	1.40751
2-Methoxyethanol (Methyl cellosolve)	76.1	0.96459	− 85.1	124.4	1.4017
1-Heptanol (*n*-Heptyl alcohol)	116.2	0.824	− 34.6	175.8	1.4241
1-Hexanol (*n*-Hexyl alcohol)	102.2	0.8186	− 44.6	157.2	1.41816
Cyclohexanol (Hexahydrophenol)	100.2	0.9493	25.15	161.5	1.4650
Methanol (Methyl alcohol)	32.0	0.7913	− 97.8	64.8	1.32863
2-Furyl-methanol (Furfuryl alcohol)	98.1	1.129	− 14.6	170.0	1.4873
Phenyl methanol (Benzyl alcohol)	108.1	1.0455	− 15.3	205.5	1.5403
1-Octanol (*n*-Octyl alcohol)	130.2	0.825	− 16.3	195	1.4293
1-Pentanol (*n*-Amyl alcohol)	88.2	0.8144	− 78.5	137.8	1.40994
2-Pentanol (*sec*-Amyl alcohol)	88.2	0.8103		119.3	1.4053
3-Pentanol (Diethyl carbinol)	88.2	0.8157	− 50	115.6	1.4057
Cyclopentanol	86.1	0.9488	− 19.0	139.5	1.4153
1,2-Propanediol (Propylene glycol)	76.1	1.0364	− 60	188.2	1.4331
1,3-Propanediol (Trimethylene glycol)	76.1	1.053	− 30	214.2	1.4396
1-Propanol (*n*-Propyl alcohol)	60.1	0.8044	−126.2	97.2	1.3854
2-Propanol (Isopropyl alcohol)	60.1	0.7854	− 89.5	82.3	1.3776
Cyclopropanol	58.7	0.9095		101	1.4129
2-Methyl-1-propanol (Isobutyl alcohol)	74.1	0.805	−108.9	107.9	1.3959
2-Methyl-2-propanol (*tert*-Butyl alcohol)	74.1	0.789	25.8	82.8	1.3878
1,2,3-Propanetriol (Glycerol)	92.1	1.260	18.2	290	1.4729
2-Propenol (Allyl alcohol)	58.1	0.854	−129	97.1	1.4135
2-Propyn-1-ol (Propargyl alcohol)	56.1	0.9478	− 51.8	117.6	1.4306

*Or specific gravity.

TABLE 3-3. ALDEHYDES.

Compound	Mol. Wt.	Density* (g/cm^3)	Freezing Point ($^\circ$C)	Boiling Point ($^\circ$C)	Refractive Index, n_D
Benzaldehyde	106.1	1.0447	− 56.5	179.0	1.5456
2-Hydroxy benzaldehyde (Salicylaldehyde)	122.1	1.167	− 7	197	1.5735
Butanal (Butyraldehyde)	72.1	0.8016	− 97.1	74.8	1.379
trans-2-Butenal (Crotonaldehyde)	70.1	0.8516	− 76.5	104.0	1.4373
Ethanal (Acetaldehyde)	44.0	0.7780	−123.5	20.2	1.3316
Tribromoethanal (Bromal)	280.8	2.670		174	1.5939
Trichloroethanal (Chloral)	147.4	1.510	− 57.5	97.8	1.4557
Ethandial (Glyoxal)	58.0	1.14	15	50.4	1.3626
2-Furaldehyde (Furfural)	96.1	1.1598	− 36.5	161.8	1.5243
Tetrahydrofurfural	100.1	1.107		145	1.436
Hexanal (n-Hexaldehyde)	100.2	0.816	− 56.3	128.7	1.4050
2-Ethyl hexanal (2-Ethyl Hexaldehyde)	128.2	0.8205		163.4	1.4142
Methanal (Formaldehyde)	30.0	0.8156_4^{-20}	−118	− 19.5	
Pentanal (n-Valeraldehyde)	86.1	0.8095	− 91.5	103.7	1.3944
Propanal (Propionaldehyde)	58.1	0.8071	−103.3	50.3	1.3733
2-Methyl propanal (Isobutyl aldehyde)	72.1	0.7938	− 65.9	64	1.3730
2-Propenal (Acrolein)	56.1	0.8389	− 87.7	52.5	1.4025
2,4,6-Trimethyl-1,3,5-trioxane (Paraldehyde)	132.2	0.994	12.4	128.0	1.420

*Or specific gravity.

TABLE 3-4. ALKANES.

Compound	Mol. Wt.	Density* (g/cm^3)	Freezing Point (°C)	Boiling Point (°C)	Refractive Index, n_D
Butane	58.1	0.599	−138.6	− 0.50	1.3326
2-Methyl butane (Isopentane)	72.2	0.621	−160.5	27.8	1.3537
2,2,-Dimethyl butane (Neohexane)	86.2	0.6492	− 98.2	49.7	1.3660
2,3-Dimethylbutane	86.2	0.662	−128.5	58.0	1.3750
2,2,3-Trimethyl butane (Triptan)	100.2	0.6901	− 24.9	80.9	1.3894
2,2,3,3-Tetramethyl butane (Hexamethyl ethane)	114.2	0.656_4^{101}	100.6	106.3	1.4695
Cyclobutane (Tetramethylene)	56.1	0.703_4^0	− 90.7	12.5	1.4260
1,4-Dioxane (Diethylene dioxide)	88.1	1.0336	11.8	101.4	1.4232
Decane	142.3	0.730	− 29.7	174.1	1.41189
Dodecane	170.3	0.7487	− 9.6	216.3	1.4216
Ethane	30.1	$0.546^{-88.6}$	−173.3	− 88.6	
Heptane	100.2	0.684	− 90.5	98.5	1.3877
2-Methyl heptane (Isoheptane)	114.2	0.6978	−109.6	117.6	1.3950
3-Methyl heptane	114.2	0.7058	−120.5	118.9	1.3985
4-Methyl heptane	114.2	0.7046	−121.0	117.7	1.3979
Hexane	86.2	0.6603	− 95.6	68.7	1.3751
3-Ethyl hexane	114.2	0.7136		118.5	1.40162
2-Methyl hexane (Isoheptane)	100.2	0.6789	−118.2	90.0	1.3840
3-Methyl hexane	100.2	0.68713	−119.4	91.9	1.3863
2,2-Dimethyl hexane	114.2	0.6953	−121.2	106.8	1.3908
2,3-Dimethyl hexane	114.2	0.7121		115.6	1.4011
2,4-Dimethyl hexane	114.2	0.7004		109.5	1.3955
2,5-Dimethyl hexane	114.2	0.6935	− 91.2	109.1	1.3922
3,3-Dimethyl hexane	114.2	0.7100	−126.1	112.0	1.4001
3,4-Dimethyl hexane	114.2	0.7192		117.7	1.4013
Cyclohexane	84.2	0.7785	6.5	80.8	1.4263
Butyl cyclohexane	140.3	0.7992	− 78.6	180.9	1.4408
Ethyl cyclohexane	112.2	0.7879	−111.3	131.8	1.4330
Methyl cyclohexane	98.2	0.7694	−126.4	100.8	1.4231
Methane	16.0	0.415^{-164}	−182.5	−161.5	
Nonane	138.5	0.7176	− 53.5	150.8	1.4054
Octane	114.2	0.7036	− 56.5	125.8	1.3976
Cyclo-Octatetraene	104.2	0.925	− 4.7	142	1.5394
Pentane	72.2	0.626	−129.8	36.1	1.3577
3-Ethyl pentane (Triethyl methane)	100.2	0.6982	−118.6	93.5	1.3931
2-Methyl pentane (Isohexane)	86.2	0.654	−153.7	60.2	1.3716
3-Methyl pentane	86.2	0.664	−118	63.3	1.3755
2,2-Dimethyl pentane	100.2	0.67385	−123.8	79.2	1.3824
2,3-Dimethyl pentane	100.2	0.6951	−135	89.8	1.3899
2,4-Dimethyl pentane	100.2	0.673	−123.4	80.3	1.3820
3,3-Dimethyl pentane	100.2	0.6933	−134.5	86.1	1.3921
2-Methyl-3-ethyl pentane	114.2	0.7193	−115.0	115.7	1.4040
3-Methyl-3-ethyl pentane	114.2	0.7274	− 90.9	118.3	1.40775
2,2,3-Trimethyl pentane	114.2	0.7160	−112.3	109.8	1.40295
2,2,4-Trimethyl pentane (Isooctane)	114.2	0.6918	−107.4	99.5	1.39142
2,3,4-Trimethyl pentane	114.2	0.71903	−109.3	113.4	1.40431
Cyclopentane	70.1	0.7455	− 95.0	49.3	1.4070
Butyl cyclopentane	126.2	0.7840	−108.2	156.7	1.4318
Methyl cyclopentane	84.2	0.7486	−142.4	71.8	1.4097
Propane	44.1	$0.585_4^{-44.5}$	−187.7	− 42.1	1.2898
2-Methyl propane (Isobutane)	58.1	0.557_4^2	−159.6	− 11.7	1.3169
Cyclopropane	42.1	0.720_4^{-79}	−127.4	− 32.9	

*Or specific gravity.

TABLE 3-5. ALKENES.

Compound	Mol. Wt.	Density* (g/cm^3)	Freezing Point (°C)	Boiling Point (°C)	Refractive Index, n_D
1,2-Butadiene (Methylallene)	54.1	0.676^{10}_4	−136.2	10.8	1.4205^{13}
1,3-Butadiene (Bivinyl)	54.1	0.650^{-6}_4	−108.9	− 4.5	1.4293^{-25}
2-Methyl-1,3-butadiene (Isoprene)	68.1	0.681	−146.0	34.1	1.4216
1-Butene (α-Butylene)	56.1	0.668^0_1	−185.3	− 6.5	1.3962
trans-2-Butene (trans-β-Butylene)	56.1	0.604	−105.8	0.88	1.3848^{-25}
cis-2-Butene (cis-β-Butylene)	56.1	0.6213	−139.3	3.73	1.3931^{-25}
3-Methyl butene-1 (Isopropyl ethylene)	70.1	0.6272	−168.5	20.1	1.3643
2-Methyl butene-1 (Trimethyl ethylene)	70.1	0.6623	−133.8	38.5	1.3870
Cyclobutene (Cyclobutylene)	54.1	0.733^0_4		2.4	
1-Decene (n-Decylene)	140.3	0.7408	− 66.3	172	1.4215
Ethene (Ethylene)	28.0	$0.30344^{6.5}_4$	−169.4	−103.7	1.363^{100}
1-Heptene (α-Heptylene)	98.2	0.6970	−119	94	1.3998
Cycloheptene (Suberene)	96.2	0.8228	− 56	114.5	1.4585
1-Hexene	84.2	0.6732	−139.8	63.5	1.3875
2-Hexene	84.2	0.6826	−136.5	68.5	1.3978
Cyclohexene (1,2,3,4,-Tetrahydrobenzene)	82.1	0.8098	−103.5	83.2	1.4465
4-Methyl cyclohexene	96.2	0.7991	−115.5	102.5	1.4414
1-Octene	112.2	0.7149	−101.7	121.3	1.4091
1-Pentene (Amylene)	70.1	0.6429	−165.2	30.0	1.3714
trans-2-Pentene	70.1	0.6482	−140.2	35.9	1.37921
cis-2-Pentene	70.1	0.6482	−151.4	37.0	1.3813
3-Ethyl-1-pentene	98.2	0.6962	−127.4	85.1	1.3956
3-Ethyl-2-pentene	98.2	0.7204		96.0	1.4141
Cyclopentene	68.1	0.772	−135.1	44.2	1.4220
Propene (Propylene)	42.1	0.5139	−185.2	− 47.7	1.3567
2-Methyl propene (Isobutylene)	56.1	0.5942	−140.3	− 6.9	

*Or specific gravity

TABLE 3-6. ALKYNES.

Compound	Mol. Wt.	Density* (g/cm^3)	Freezing Point (°C)	Boiling Point (°C)	Refractive Index, n_D
1-Butyne (Ethyl acetylene)	54.1	0.6682	−125.7	8.1	1.3962
2-Butyne (1,2-Dimethyl acetylene)	54.1	0.6910	− 32.3	27.0	1.3921
3-Methyl-1-butyne (Isopropyl acetylene)	68.1	0.666	− 89.7	26.4	1.3723
1-Decyne (n-Octyl acetylene)	138.3	0.7655	− 44	174	1.4265
Ethyne (Acetylene)	26.0	1.173^0	− 80.8	− 84.0	
1-Heptyne (n-Amyl acetylene)	96.2	0.7328	− 80.9	99.7	1.4087
1-Hexyne (n-Butyl acetylene)	82.2	0.7155	−131.9	71.3	1.3989
1-Nonyne (n-Heptyl acetylene)	124.2	0.7568	− 50	150.8	1.4217
1-Octyne (n-Hexyl acetylene)	110.2	0.7461	− 79.3	126.2	1.4159
1-Pentyne (n-Propyl acetylene)	68.1	0.6901	−105.7	40.2	1.3852
2-Pentyne (Isopropyl acetylene)	68.1	0.711	−109.3	56.1	1.4039
Propyne (Methyl acetylene)	40.1	0.6911^{-20}	−104.7	− 23.3	1.3725^{-20}

*Or specific gravity.

TABLE 3-7. AMIDES, IMIDES, UREAS AND THIOUREAS.

Compound	Mol. Wt.	Density* (g/cm^3)	Freezing Point ($^\circ$C)	Boiling Point ($^\circ$C)	Refractive Index, n_D
Acetamide	59.1	1.159	80.1	222.1	1.4274^{78}
Acetanilide	135.2	1.211^4	114.3	305	
Barbituric acid (Malonylurea)	128.1		245	260	
Benzamide	121.1	1.341^4	127.2	290	
Biuret (Carbamylurea)	103.1	1.467^{-5}_4	193		
N-Bromosuccinimide	178.0	2.098	173.5		
Caffeine (1,3,7-Trimethyl-xanthine)	194.2	1.23^{10}	237		
Carbanilide (1,3-Diphenylurea)	212.2	1.0415	238	263	1.462^{49}
Thiocarbanilide (sym. Diphenylthiourea)	228.3	1.32	150		
Formamide	45.0	1.3340	2.55	210.5	1.44754
Dimethylformamide	73.1	0.94873	− 60.4	149.6	1.43047
Urea (Carbamide)	60.1	1.335	132.7		1.484
1,3-Dimethylurea	88.1	1.142	107	269	
Phenylurea	136.1	1.302	147	238	
Succinimide	99.1	1.418	125	288	
Thiourea (Thiocarbamide)	76.1	1.405	182		
Uric acid	168.1	1.89	400		

*Or specific gravity

TABLE 3-8. AROMATIC AMINES.

Compound	Mol. Wt.	Density* (g/cm^3)	Freezing Point ($^\circ$C)	Boiling Point ($^\circ$C)	Refractive Index, n_D
Aniline	93.1	1.0217	− 6.3	184.1	1.5863
2-Chloroaniline	127.6	1.2127	− 1.94	208.8	1.58894
3-Chloroaniline	127.6	1.2161	− 10.3	229.9	1.59414
4-Chloroaniline	127.6	1.427	72.5	232	1.555
N-Ethylaniline	121.2	0.958	− 63.5	204.5	1.5559
2-Ethylaniline	121.2	0.983	− 45	214.3	1.5590
3-Ethylaniline	121.2	0.990^9	− 64	214.5	
4-Ethylaniline	121.2	0.9679	− 4.9	217.8	1.55535
2-Nitroaniline	138.1	1.442	71.5	284	
3-Nitroaniline	138.1	1.398	114	306	
4-Nitroaniline	138.1	1.424	147.5	331.7	
N-Methylaniline	107.1	0.9891	− 57	196.2	1.5702
Benzidine (4,4-Diamino biphenyl)	184.2	1.251	126	400	
Carbazole (Diphenyleneimine)	167.2	1.10	245	354.8	
2-Naphthylamine	143.2	1.061^{98}_2	111	306	
Piperidine (Pentanethylene imine)	85.1	0.8622	− 7	106.0	1.4534
Pyridine	79.1	0.978	− 42	116	1.5102
α-Aminotoluene (Benzylamine)	107.1	0.983		184.5	1.5401
2-Aminotoluene (o-Toluidine)	107.1	0.99843	− 14.7	200.2	1.57246
3-Aminotoluene (m-Toluidine)	107.1	0.98890	− 30.4	203.4	1.56811
4-Aminotoluene (p-Toluidine)	107.1	0.9619^{50}	43.7	200.6	1.55348^{50}

*Or specific gravity.

TABLE 3-9. PRIMARY AND SECONDARY AMINES.

Compound	Mol. Wt.	Density* (g/cm^3)	Freezing Point $(^\circ C)$	Boiling Point $(^\circ C)$	Refractive Index, n_D
Allyl amine	57.1	0.7621	− 88.2	53.3	1.42051
Benzyl amine	107.2	0.983_4^{19}	21	185	1.5401
Butyl amine (1-Aminobutane)	73.1	0.7414	− 49.1	77.8	1.4031
Dibutyl amine	129.2	0.7601	− 59.5	159.6	1.4168
Isobutyl amine	73.1	0.7346	− 84.6	68.6	1.3970
Diisobutyl amine	129.2	0.745	− 70	139.5	1.4093
sec-Butyl amine (2-Aminobutane)	73.1	0.7246	−104.5	62.5	1.3934
tert-Butyl amine (2-Amino-2-methyl propane)	73.1	0.6958	− 72.6	44.4	1.3784
N,N-Dibutylaniline	205.3	0.90368	− 32.2	274.75	1.51856
Decyl amine (1-Aminodecane)	157.3	0.7936	16.1	220.5	1.4369
Ethanolamine (2-Aminoethanol)	61.1	1.018	10.5	170.5	1.4539
Ethyl amine	45.1	0.6829	− 80.6	16.6	1.3663
Diethyl amine	73.1	0.7056	− 50.0	55.5	1.3864
Ethylene diamine	60.1	0.8977	11.0	117.3	1.45677
Heptyl amine (1-Aminoheptane)	115.2	0.7754	− 18	156.9	1.4251
Hexyl amine (1-Aminohexane)	101.2	0.763	− 19	130.2	1.4255^{17}
Cyclohexyl amine	99.2	0.8671	− 17.7	134.7	1.45926
Methyl amine	31.1	0.6624	− 92.5	− 6.8	1.3527
Dimethyl amine	45.1	0.6556	− 92.2	6.9	1.358
Nonyl amine (1-Aminononane)	143.3	0.7886	− 1	202.2	1.4336
Octyl amine (1-Aminooctane)	129.2	0.7826	0.0	179.6	1.4294
Pentyl amine (1-Aminopentane)	87.2	0.7547	− 55	104.4	1.4118
Isopentyl amine	87.2	0.749		97	1.4087
Piperidine	85.2	0.8659^{15}	− 10.5	106.4	1.4525
Propyl amine (1-Aminopropane)	59.1	0.7173	− 83.0	47.8	1.3879
Dipropyl amine	101.2	0.7400	− 63	109.0	1.4042
Isopropyl amine	59.1	0.689	−101.2	31.7	1.3742
Diisopropyl amine	101.2	0.7153	− 96.3	83.5	1.3924
Pyrrole	67.1	0.96985	− 23.4	129.8	1.51015
Pyrrolidine (Tetramethylene-imine)	71.1	0.852		89	1.4270^{15}

*Or specific gravity.

TABLE 3-10. TERTIARY AMINES.

Compound	Mol. Wt.	Density* (g/cm^3)	Freezing Point $(^\circ C)$	Boiling Point $(^\circ C)$	Refractive Index, n_D
Acridine	179.2		110	346	
Diethyl aniline	149.2	0.93507	− 21.3	217.5	1.5421
N,N'-Diethyl aniline	149.2	0.938	− 38	217	1.5409
N,N'-Dimethyl aniline	121.2	0.95198	2.4	194.3	1.5562
d,1-Nicotine	162.2	1.0097		243	1.5282
Pyridine	79.1	0.9831	− 41.7	115.2	1.51016
Quinoline	129.1	1.0929	− 15.6	237.6	1.6273
Tributyl amine	185.3	0.7782		216.5	1.4291
Triethyl amine	101.2	0.7276	−114.7	89.4	1.4003
Trimethyl amine	59.1	0.6709_4^0	−117.1	2.9	
Triphenyl amine	245.3	0.774_0^0	127	365	

*Or specific gravity.

TABLE 3-11.   ANHYDRIDES.

Compound	Mol. Wt.	Density* (g/cm^3)	Freezing Point (°C)	Boiling Point (°C)	Refractive Index, n_D
Acetic anhydride (Ethanoic anhydride)	116.2	1.0810	− 73.1	140.0	1.3904
Benzoic anhydride	226.2	1.1989[15]	42	360	
Butyric anhydride (Butanoic anhydride)	158.2	0.9668	− 56.1	186.3	1.4124
Isobutyric anhydride	158.2	0.956	− 53.5	182	
Crotonic anhydride (Butenoic anhydride)	154.2	1.040		248	1.474
Heptanoic anhydride	242.4	0.917	17	258	1.433
Hexanoic anhydride (n-Caproic anhydride)	241.3	0.920	− 40.5	245	1.430
Maleic anhydride (cis-Butenedioic anhydride)	98.1	1.509	52.8	202.0	
Octanoic anhydride	270.4	0.910	0.9	285	1.434
Pentanoic anhydride (n-Valeric anhydride)	186.2	0.925	− 56.1	218	
Phthalic anhydride (1,2-benzenedicarboxylic anhydride)	148.1	1.53	130.6	284.5	
Succinic anhydride (Butanedioic anhydride)	100.1	1.503	119.6	261	
Trichloroacetic anhydride	308.8	1.6908		223	
Trifluoroacetic anhydride	210.0	1.490	− 65	39.5	1.269[25]

*Or specific gravity.

TABLE 3-12.   CARBOHYDRATES.

Substance	Formula	Mol. Wt.	Melting Point (°C)*	Specific Rotation (in H_2O)	Density, (g/cm^3)**
Ascorbic acid	$C_6H_8O_6$	176.1	192	+24	
L-Arabinose	$C_5H_{10}O_5$	150.1	160	+190 → +105	1.585
D-Arabinose	$C_5H_{10}O_5$	150.1	159	−175 → −105	1.585
Cellobiose	$C_{12}H_{22}O_{11}$	342.3	225	+ 14 → + 34	
L-Fructose	$C_6H_{12}O_6$	180.2	102.4		1.669[18]
D-Fructose	$C_6H_{12}O_6$	180.2	102.4	−135 → − 92.4	1.669[18]
L-Galactose	$C_6H_{12}O_6$	180.2	166.5	−120 → − 78	
D-Galactose	$C_6H_{12}O_6$	180.2	166.5	+151 → + 80	
L-Glucose	$C_6H_{12}O_6$	180.2	146	− 96 → − 53	1.544[25]
D-Glucose	$C_6H_{12}O_6$	180.2	146	+111 → + 52.7	1.544[25]
Lactose	$C_{12}H_{22}O_{11}$	342.3	223	+ 34.9 → + 55.4	1.525
Maltose	$C_{12}H_{22}O_{11}$	342.3	102.5	+112 → +136	1.5470[17]
L-Mannose	$C_6H_{12}O_6$	180.2	132	+ 14 → − 14	1.539
D-Mannose	$C_6H_{12}O_6$	180.2	132	− 16 → + 14.5	1.539
L-Ribose	$C_5H_{10}O_5$	150.1	87	+ 20.3 → + 20.7	
D-Ribose	$C_5H_{10}O_5$	150.1	87	− 23.1 → − 23.7	
Sucrose	$C_{12}H_{22}O_{11}$	432.3	185	+66.5	1.588[15]
L-Xylose	$C_5H_{10}O_5$	150.1	153	− 79 → − 19	1.525
D-Xylose	$C_5H_{10}O_5$	150.1	153	+ 94 → + 18.6	1.525

*For anhydrous substances.
**d_4^{20} unless other temperature specified.

TABLE 3-13. CARBOXYLIC ACIDS.

Compound	Mol. Wt.	Density* (g/cm^3)	Freezing Point (°C)	Boiling Point (°C)	Refractive Index, n_D
1,2-Benzene dicarboxylic acid (Phthalic acid)	166.1	1.593	230		
Benzoic acid (Benzene carboxylic acid)	122.1	1.2659^{15}	122.4	250.8	1.504^{132}
2-Aminobenzoic acid (Anthranilic acid)	137.1	1.412	147		
1,2-Hydroxybenzoic acid (Salicylic acid)	138.1	1.443	158.3		1.565
3,4,5-Trihydroxy benzoic acid (Gallic acid)	170.1	1.694	254		
1,4-Butanedioic acid (Succinic acid)	118.1	1.564^{15}	182.8	235	
d1,2,3-Dihydroxybutanedioic acid (d1-Tartaric acid)	150.1	1.760	206		
Butanoic acid (*n*-Butyric acid)	88.1	0.9580	− 5.4	164.1	1.39996
2-Methylbutanoic acid (Isovaleric acid)	102.1	0.931	− 29.3	176.5	1.4043
trans-Butenedioic acid (Fumaric acid)	116.1	1.625	300		
cis-Butenedioic acid (Maleic acid)	116.1	1.59	130.5	135	
3-Butenoic acid (Vinyl acetic acid)	70.0	1.138	− 39	144	1.4306
Decanoic acid (*n*-Capric acid)	172.3	0.8858^{40}	30.7	270	1.42855^{40}
Ethanedioic acid (Oxalic acid)	126.1	$1.653^{18.5}$	189.5		
Ethanoic acid (Acetic acid)	60.0	1.04923	16.5	117.7	1.3718
Heptanoic acid	130.2	0.9181	− 7.5	223.0	1.4230
Hexadecanoic acid (Palmitic acid)	256.4	0.8414^{80}	62.3	219/17	1.4269^{80}
Hexanoic acid (*n*-Caproic acid)	116.2	0.9265	− 4.0	205.3	1.4162
Methanoic acid (Formic acid)	46.0	1.220	8.3	100.8	1.3714
cis-9-Octadecanoic acid (Oleic acid)	282.4	0.8905	13.1	360.0	1.45823
Octanoic acid (Caprylic acid)	114.2	0.90884	16.5	239.3	1.4280
Pentanoic acid (*n*-Valeric acid)	102.1	0.939	− 34.5	186.3	1.4086
Propanoic acid (Propionic acid)	74.1	0.99336	− 20.8	140.8	1.3991
2-Methly propanoic acid (Isobutyric acid)	88.1	0.950	− 46.1	154.7	1.3930
Propenoic acid (Acrylic acid)	72.1	1.0511	13.5	141.2	1.4224
2-Methyl propenoic acid (Methacrylic acid)	86.1	1.0153	15.0	160.5	1.4314

*Or specific gravity.

TABLE 3-14. ALIPHATIC BROMIDES.

Compound	Mol. Wt.	Density* (g/cm^3)	Freezing Point (°C)	Boiling Point (°C)	Refractive Index, n_D
1-Bromobutane (n-Butyl bromide)	137.0	1.2758	−112.4	101.3	1.4398
2-Bromobutane (sec-Butyl bromide)	137.0	1.2608	−112.6	91.4	1.43705
1-Bromodecane (n-Decyl bromide)	221.2	1.0702	− 29.2	240.6	1.4557
Bromoethane (Ethyl bromide)	110.0	1.430	−118.6	38.4	1.4242
1,1-Dibromoethane (Ethylidene bromide)	187.9	2.055	− 63	108.0	1.5128
1,2-Dribromoethane (Ethylene dibromide)	187.9	2.1791	9.8	131.4	1.5375
Hexabromoethane (Dicarbon hexabromide)	503.5	3.823	149	210	
Bromoethene (Vinyl bromide)	107.0	1.4933	−139.5	15.8	1.441
1-Bromoheptane (n-Heptyl bromide)	179.1	1.140	− 56.1	178.9	1.451
1-Bromohexane (n-Hexyl bromide)	165.1	1.1744	− 84.7	155.3	1.4475
Bromomethane (Methyl bromide)	95.0	1.6755	− 94.1	3.6	1.4218
Dibromomethane (Methylene bromide)	173.9	2.4956	− 52.7	97	1.5419
Tribromomethane (Bromoform)	252.8	2.890	7.5	150	1.598
Tetrabromomethane (Carbon tetrabromide)	331.7		91	189.5	2.911[99]
1-Bromononane (n-nonyl bromide)	207.2	1.0893	− 29.0	221.4	1.4542
1-Bromooctane (n-Octyl bromide)	193.1	1.1122	− 55.0	200.8	1.4524
2-Bromooctane (sec-Octyl bromide)	199.1	1.078			1.4442
1-Bromopentane (n-Amyl bromide)	151.1	1.2186	− 95.0	129.7	1.4444
1-Bromopropane (n-Propyl bromide)	123.0	1.353	−109.9	70.9	1.4341
2-Bromopropane (Isopropyl bromide)	123.0	1.3140	− 89.0	59.4	1.3254
1,1-Dibromopropane (Propylidene bromide)	201.9	1.982		133.5	1.5100
1,2-Dibromopropane (Propylene bromide)	201.9	1.933	− 55.5	142.0	1.5203
1,3-Dibromopropane (Trimethylene bromide)	201.9	1.980	− 34.2	166.7	1.5232
2-Bromo-2 methyl propane (tert-Butyl bromide)	137.0	1.2209	− 16.2	73.3	1.428
3-Bromo-propene-1	121.0	1.398	−119.4	71.3	1.4655

*Or specific gravity.

TABLE 3-15. ALIPHATIC CHLORIDES.

Compound	Mol. Wt.	Density* (g/cm^3)	Freezing Point (°C)	Boiling Point (°C)	Refractive Index, n_D
1-Chlorobutane (n-Butyl chloride)	92.6	0.8865	−123.1	78.5	1.40223
2-Chlorobutane (sec-Butyl chloride)	92.6	0.8732	−131.3	68.3	1.3971
1-Chlorodecane (n-Decyl chloride)	176.7	0.8705	− 31.3	223.4	1.4379
Chloroethane (Ethyl chloride)	64.5	0.9028[10]	−138.3	12.3	1.3738[10]
1,1-Dichloroethane (Ethylidene dichloride)	99.0	1.4164	− 96.5	57.3	1.4166
1,2-Dichloroethane (Ethylene dichloride)	99.0	1.2569	− 35.5	84.1	1.4451
1,1,1-Trichloroethane (Methyl chloroform)	133.4	1.3492	− 30.4	74.1	1.4384
1,1,2-Trichloroethane	133.4	1.4411	− 36.6	113.6	1.4708
1,1,2,2-Tetrachloroethane (Acetylene tetrachloride)	167.9	1.5953	− 43.8	146.1	1.4940
Pentachloroethane	202.3	1.6813	− 29.0	162.0	1.5030
Chloroethene (Vinyl chloride)	62.5	0.9834$_4^{-20}$	−159.7	− 13.9	1.4046[-10]
1,1-Dichloroethene (Vinylidene dichloride)	96.9	1.2129	−122.1	31.7	1.4249
cis-1,2-Dichloroethene (cis-Acetylene dichloride)	96.9	1.2818	− 80.1	60.4	1.4481
trans-1,2-Dichloroethene (trans-Acetylene dichloride)	96.9	1.2546	− 49.8	47.7	1.4456
1,1,2-Trichloroethene (1,1,2-Trichloroethylene)	131.4	1.4649	− 86.4	86.7	1.47734
Tetrachloroethene (Perchloroethylene)	165.9	1.621	− 22.4	121.1	1.5057
1-Chloroheptane (n-Heptyl chloride)	134.7	0.877	− 69.5	160.0	1.426
1-Chlorohexane (n-Hexyl chloride)	120.6	0.8785	− 94.0	134.5	1.4196
Chloromethane (Methyl chloride)	50.5	0.9159	− 97.7	− 24.2	1.3389
Dichloromethane (Methylene chloride)	84.9	1.3255	− 96.7	39.8	1.4244
Trichloromethane (Chloroform)	119.4	1.4817	− 63.5	61.1	1.4476
Tetrachloromethane (Carbon tetrachloride)	153.8	1.5947	− 23.0	76.8	1.4607
1-Chlorononane (n-Nonyl chloride)	162.7	0.8720	− 39.4	203.4	1.4345
1-Chlorooctane (n-Octyl chloride)	148.7	0.8737	− 57.8	183.8	1.4305
1-Chloropentane (n-Amyl chloride)	106.6	0.8828	− 99.0	107.8	1.4128
1-Chloropropane (n-Propyl chloride)	78.5	0.8890	−122.8	46.4	1.3886
2-Chloropropane (Isopropyl chloride)	78.5	0.8626	−117.2	34.8	1.3776
1,1-Dichloropropane (Propylidene chloride)	113.0	1.1321		88.1	1.4289
1,2-Dichloropropane (Propylene chloride)	113.0	1.1545	−100.4	96.4	1.4388
1,3-Dichloropropane (Trimethylene chloride)	113.0	1.183	− 99.5	120.8	1.4487
2,2-Dichloropropane (Isopropylidene trichloride)	113.0	1.092	− 34.6	69.7	1.412

TABLE 3-15 (*Continued*)

Compound	Mol. Wt.	Density* (g/cm^3)	Freezing Point (°C)	Boiling Point (°C)	Refractive Index, n_D
2-Chloro-2-methylpropane (*tert*-Butyl chloride)	92.6	0.846	− 25.0	51.5	1.3859
cis-1-Chloropropene (*cis*-Propenyl chloride)	76.5		− 99.0	37.4	1.4054
trans-1-Chloropropene (*trans*-Propenyl chloride)	76.5	0.935^{15}	−134.8	32.8	1.4053
3-Chloropropene-1 (Allyl chloride)	76.5	0.938	−134.5	44.6	1.4154

*Or specific gravity.

TABLE 3-16. ESTERS.

Compound	Mol. Wt.	Density* (g/cm^3)	Freezing Point (°C)	Boiling Point (°C)	Refractive Index, n_D
Allyl acetate	100.1	0.9276		104.0	1.4049
Butyl acetate	116.2	0.881	− 73.5	126.1	1.3961
Isobutyl acetate	116.2	0.8747	− 99.8	117.2	1.3901
sec-Butyl acetate	116.2	0.872		112.3	1.3865
Ethyl acetate	88.1	0.9006	− 83.6	77.1	1.3724
Methyl acetate	74.1	0.9274	− 98.7	57.3	1.3617
Pentyl acetate	130.2	0.8753	< −100	149.2	1.4028
Propyl acetate	102.1	0.8834	− 95	101.5	1.3847
Isopropyl acetate	102.1	0.872	− 73.4	88.2	1.377
Vinyl acetate	86.1	0.9312	− 92.8	72.5	1.3959
Ethyl acetoacetate	130.1	1.0282	− 39	181	1.4192
Methyl acetoacetate	116.1	1.0747	− 80	171.7	1.4186
Ethyl acrylate	100.1	0.909	− 71.2	99.3	1.406
Methyl acrylate	86.1	0.961	− 76.5	80.3	1.3984
Ethyl benzoate	150.2	1.04684	− 34.7	212.5	1.5057
Methyl benzoate	136.1	1.089	− 12.5	199.5	1.517
Phenyl benzoate	198.2	1.235	69	314	
Ethyl butyrate	116.2	0.8792	−100.8	121.6	1.400
Methyl butyrate	102.1	0.8982	− 84.8	102.7	1.3879
Butyl formate	102.1	0.8885	− 91.9	106.6	1.3894
Isobutyl formate	102.1	0.8854	− 94.5	98.4	1.3857
sec-Butyl formate	102.1	0.884		97	1.384
Ethyl formate	74.1	0.9225	− 79.4	54.2	1.3598
Methyl formate	60.0	0.97421	− 99.0	31.5	1.3440
Propyl formate	88.1	0.904	− 92.9	81.3	1.3779
Isopropyl formate	88.1	0.8728		71	1.368
Methyl methacrylate	100.1	0.9433	− 48.2	100.3	1.4146
Ethyl nitrate	91.1	1.106	−112	88.7	1.3853
Methyl nitrate	77.0	1.206	− 83.0	64.6	
Propyl nitrate	105.1	1.063		110.5	1.3979
Ethyl nitrite	75.1	0.9004$_4^{15}$		16.4	
Methyl nitrite	61.0	0.991	− 17	− 12	
Propyl nitrite	85.1	0.8864		48	1.3592
Butyl phosphate	266.3	0.9760	< − 80	289	1.42496
Allyl propionate	114.1	0.914		123	1.410
Butyl propionate	130.2	0.895	− 89.6	146.8	1.401
Isobutyl propionate	130.2	0.8884$_4^0$	− 71.4	137	1.3975
sec-Butyl propionate	130.2	0.8663		132.5	1.3952
Ethyl propionate	102.1	0.8889	− 73.9	99.1	1.3853
Methyl propionate	88.1	0.9151	− 87.5	79.4	1.3779
Propyl propionate	116.2	0.8809	− 75.9	123.2	1.3933
Isopropyl propionate	116.2	0.893^0		111	
Methyl salicylate	152.1	1.1834	− 8.6	223.0	1.535
Butyl stearate	340.6	0.8540	27.5	225	1.4441

*Or specific gravity.

TABLE 3-17. ETHERS.

Compound	Mol. Wt.	Density* (g/cm^3)	Freezing Point (°C)	Boiling Point (°C)	Refractive Index, n_D
Acetal	118.1	0.8264		103.6	1.38054
Allyl ether	98.1	0.765		94	1.4240
Anethole	148.2	0.9883	23.3	235.7	1.56145
(p-Propenylanisole)					
Anisole	108.1	0.9940	− 37.4	153.9	1.5170
(Methyl phenylether)					
Benzyl ether	198.2	1.0428	3.6	299	1.5406
Butyl ether	130.2	0.770	− 97.9	142.4	1.399
1,4-Dioxane	88.1	1.0336	11.8	101.4	1.4232
Ethyl ether	74.1	0.7135	−116.3	34.6	1.3526
Benzyl ethyl ether	136.2	0.9490		185.0	1.4955
1-Chloroethyl ethyl ether	108.6	0.966		98	1.405
2-Chloroethyl ethyl ether	108.6	0.989		107	1.411
Epichlorohydrin	92.8	1.1812	− 57.2	117.9	1.43805
Ethylene oxide	44.0	0.89713_4^0	−111.7	10.7	1.3614^4
(1,2-Epoxyethane)					
Eugenol	164.2	1.0663	− 9.1	254.8	1.5410
(4-Allyl-2-methoxyphenol)					
Furan	68.1	0.93781		31.4	1.4214
Tetrahydrofuran	72.1	0.889	−108.5	65	1.4070
Methyl ether	46.1		−141.5	− 24.8	
Ethyl methylether	60.1	0.725_0^0		7.6	
Butyl methylether	88.1	0.744	−115.5	70	1.374
Pentyl ether	158.3	0.7830	− 69.4	187.5	1.416
Phenetole	122.2	0.96514	− 30.2	170.0	1.5080
(Ethyl phenylether)					
Phenyl ether	170.2	1.073	26.9	259.0	1.5826
Phenoxy benzene)					
Propyl ether	102.2	0.7470	−122.0	90.1	1.3885
Isopropyl ether	102.2	0.7258	− 60	67.5	1.3688
Methyl propylether	74.1	0.7356_4^{13}		39.9	1.3579
Propylene oxide	58.1	0.830	−112.1	34.5	1.3664
(1,2-Epoxypropane)					
Vinyl ether	70.1	0.773		28.3	1.3989
Butyl vinylether	100.2	0.7888		93.3	1.4026
Ethyl vinylether	72.1	0.7589	−115.8	35.7	1.3768

*Or specific gravity.

TABLE 3-18.   ALIPHATIC FLUORIDES.

Compound	Mol. Wt.	Density* (g/cm^3)	Freezing Point (°C)	Boiling Point (°C)	Refractive Index, n_D
1-Fluorobutane (n-Butyl fluoride)	76.11	0.7789	−134	32.5	1.3396
2-Fluorobutane (sec-Butyl fluoride)	76.11	0.7621	−121.4	25.1	1.3326
1-Fluorodecane (n-Decyl fluoride)	160.3	0.8194	− 35	186.2	1.4085
Fluoroethane (Ethyl fluoride)	48.1	0.8182	−143.2	− 37.7	1.2656
1,1-Difluoroethane (Ethylidene fluoride)	66.1	0.95	−117.0	− 26.5	1.26
1,1,2,2-Tetrafluoroethane (Acetylene tetrafluoride)	102	–	− 89	− 19.7	–
Pentafluoroethane	120	1.53^0	−103	− 48.5	–
1,1,1-Difluorochloroethane	100.5	1.19	−131	− 9.5	–
1-Fluoroheptane (n-Heptyl fluoride)	118.2	0.8062	− 73	117.9	1.3854
1-Fluorohexane (n-Hexyl fluoride)	104.2	0.7995	−103	91.5	1.3738
Fluoromethane (Methyl fluoride)	34.0	0.5786	−142	− 78.2	1.1727
Difluoromethane (Methylene fluoride)	52.0	0.909	–	− 51.6	1.19
Trifluoromethane (Fluoroform)	70.0	–	−163	− 82.2	–
Tetrafluoromethane (Carbon tetrafluoride)	88.0	1.89^{-183}	−183.6	−127.8	–
1-Fluoropentane (n-Amyl fluoride)	90.1	0.7907	−120	62.8	1.3591
1-Fluoropropane (n-Propyl fluoride)	62.1	0.7956	−159.0	− 2.5	1.3115
1,1-Difluoropropane (Propylidene fluoride)	80.1	0.94	–	7.5	1.30

*Or specific gravity.

TABLE 3-19. AROMATIC HALIDES.

Compound	Mol. Wt.	Density* (g/cm^3)	Freezing Point (°C)	Boiling Point (°C)	Refractive Index, n_D
Bromobenzene	157.0	1.4952	− 30.8	156.2	1.5602
1,2-Dibromobenzene	235.9	1.984	7.1	225.5	1.6155
1,3-Dibromobenzene	235.9	1.952	− 7	219	1.606
1,4-Dibromobenzene	235.9	$0.964^{99.6}$	87.3	220.4	$1.5743^{99.6}$
Chlorobenzene	112.6	1.107	− 45.6	131.7	1.5248
1,2-Dichlorobenzene	147.0	1.306	− 17.5	180.5	1.5515
1,3-Dichlorobenzene	147.0	1.288	− 24.8	173.1	1.5459
1,4-Dichlorobenzene	147.0	1.2417^{60}	53.1	174.2	1.5285^{60}
1,2,3-Trichlorobenzene	181.5	1.69	52.6	221	1.5776
1,2,4-Trichlorobenzene	181.5	1.4542	16.9	213.5	1.5717
1,3,5-Trichlorobenzene	181.5		63.4	208.4	1.5662
1,2,4,5-Tetrachlorobenzene	215.9	1.734	140	240	
Pentachlorobenzene	250.3		~85	~276	
Hexachlorobenzene	284.8	2.044^{33}	228.5	322.2	
Fluorobenzene	96.1	1.024	− 41.9	84.7	1.4677
1,2-Difluorobenzene	114.1	1.496	− 34	91	1.4452
1,3-Difluorobenzene	114.1	1.1572	− 59.3	83	1.4410
1,4-Difluorobenzene	114.1	1.1716	− 13	88.8	1.4421
Pentafluorobenzene	168.1	1.531	− 48	85	1.3880
Hexafluorobenzene	186.0	1.6182	5.10	80.3	1.3769
Iodobenzene	204.0	1.8309	− 31.3	188.3	1.6200
1,2-Diiodobenzene	329.9	2.54	27	287	
1,3-Diiodobenzene	329.9	2.47	40	285	
1,4-Diiodobenzene	329.9		129.4	285	
1-Chloronaphthalene	162.6	1.1938	− 2.3	259.3	1.63321

*Or specific gravity.

TABLE 3-20. AROMATIC HYDROCARBONS.

Compound	Mol. Wt.	Density* (g/cm^3)	Freezing Point ($^\circ$C)	Boiling Point ($^\circ$C)	Refractive Index, n_D
Acenaphthene	154.2	1.024^{99}	95	277.5	
(Naphthalene ethylene)					
Anthracene	178.2	1.25^{27}	215.6	339.9	
(p-Naphthalene)					
Benzene	78.1	0.8790	5.5	80.0	1.50142
n-Butyl benzene	134.2	0.8604	− 88.5	182.1	1.4904
(1-Phenyl butane)					
Isebutyl benzene	134.2	0.853	− 51.5	172.8	1.4865
sec-Butyl benzene	134.2	0.8608	− 75.5	173.5	1.4898
(2-Phenyl butane)					
tert-Butyl benzene	134.2	0.8669	− 58.1	169.1	1.49235
(2-Methyl-2-phenyl propane)					
o-Cymene	134.2	0.876	− 71.6	175	1.500
(1-Isopropyl-2-methyl benzene)					
m-Cymene	134.2	0.8619	− 63.8	176	1.4922
(1-Isepropyl-3-methyl benzene)					
p-Cymene	134.2	0.8573	− 67.9	177.1	1.4909
(1-Isopropyl-4-methyl benzene)					
Decahydronaphthalene, mixed isomers	138.2	0.8865	−124	191.7	1.4758
Dibenzyl	182.3	0.978	52.3	284	
(1,2-Diphenyl ethane)					
1,2-Diethyl benzene	134.2	0.880	− 31.2	183.5	1.503
1,3-Diethyl benzene	134.2	0.864	− 83.9	181.1	1.495
1,4-Diethyl benzene	134.2	0.862	− 42.8	183.7	1.4948
1,2-Dimethyl benzene	106.2	0.8802	− 25.5	144.4	1.50545
(o-Xylene)					
1,3-Diethyl benzene	106.2	0.8642	− 47.9	139.3	1.4973
(m-Xylene)					
1,4-Diethyl benzene	106.2	0.8611	13.2	138.3	1.49581
(p-Xylene)					
1,2-Diphenyl benzene	230.3	1.14	57	332	
(o-Terphenyl)					
1,3-Diphenyl benzene	230.3	1.164	87	363	
(m-Terphenyl)					
1,4-Diphenyl benzene	230.3	1.236	212.5	405	
(p-Terphenyl)					
Ethyl benzene	106.2	0.8670	− 94.9	136.2	1.4959
Hexaethyl benzene	246.4	0.831^{130}	130	298	1.4736^{130}
Hexamethyl benzene	162.3		165.5	264	
Indene	116.2	0.9968	− 1.8	181.6	1.5642
(Indonaphthene)					
Naphthalene	128.2	1.400	80.2	217.9	1.0253
Phenanthrene	178.2	1.179	100.5	339	1.594
1-Phenyl propane	120.2	0.8621	− 99.5	159.2	1.4919
(n-Propyl benzene)					
2-Phenyl propane	120.2	0.8642	− 96.0	152.4	1.49146
(Cumene)					
Pyrene	202.2	1.271	150.5	404	
p-Quaterphenyl	306.4		320	428/18	
(Tetraphenyl)					
trans-Stilbene	180.2	0.9707	124	~306	1.6264^{17}
(Diphenyl ethlene)					
Styrene	104.1	0.9060	− 30.6	145.2	1.54682
1,2,3,4-Tetrahydronaphthalene	132.2	0.9702	− 35.8	207.6	1.54135
1,2,4,5-Tetramethyl benzene	134.2	0.8875	79.2	196.8	1.5116
(Durene)					
Toluene	92.1	0.8669	− 95.0	110.4	1.4961
(Methyl benzene)					
1,3,5-Trimethyl benzene	120.2	0.8637	− 44.8	164.7	1.4994
(Mesitylene)					

*Or specific gravity.

TABLE 3-21. ALIPHATIC IODIDES.

Compound	Mol. Wt.	Density* (g/cm^3)	Freezing Point (°C)	Boiling Point (°C)	Refractive Index, n_D
1-Iodobutane (n-Butyl iodide)	184.0	1.616	−103.0	130.4	1.4998
2-Iodobutane (sec-Butyl iodide)	184.0	1.592	−104	120	1.4991
1-Iododecane (n-Decyl iodide)	268.1	1.255	− 16.3	263.7	1.4858
Iodoethane (Ethyl iodide)	156.0	1.936	−111.1	72.3	1.5103
1,1-Diiodoethane (Ethylidene diiodide)	281.9	2.84_4^0	−	179	−
1,2-Diiodoethane (Ethylene diiodide)	281.9	2.132_4^{10}	82	200	1.871
1-Iodoheptane (n-Heptyl iodide)	226.1	1.373	− 48.2	204.0	1.490
1-Iodohexane (n-Hexyl iodide)	212.1	1.437	− 75	181.3	1.493
Iodomethane (Methyl iodide)	142.0	2.279	− 66.4	42.5	1.5304
Diiodomethane (Methylene iodide)	267.9	3.3345	6.1	182	1.741
Triiodomethane (Iodoform)	393.8	4.008	119	−	−
Tetraiodomethane (Carbon tetraiodide)	519.7	4.32	171	−	−
1-Iodononane (n-Nonyl iodide)	254.2	1.289	− 20	245.0	1.487
1-Iodooctane (n-Octyl iodide)	240.1	1.330	− 45.7	225.1	1.489
1-Iodopentane (n-Amyl iodide)	198.0	1.512	− 85.6	157.0	1.496
1-Iodopropane (n-Propyl iodide)	170.0	1.747	−101.3	102.4	1.5051
2-Iodopropane (Isopropyl iodide)	170.0	1.703	− 90.1	89.5	1.5024
1,3-Diiodopropane (Trimethylene iodide)	295.9	2.576	− 13	227	1.6423
3-Iodopropene	168.0	1.825	− 99.3	103.1	1.554

*Or specific gravity.

TABLE 3-22. KETONES.

Compound	Mol. Wt.	Density* (g/cm^3)	Freezing Point (°C)	Boiling Point (°C)	Refractive Index, n_D
Acetophenone (Methyl phenyl ketone)	120.1	1.026	19.7	201.7	1.5338
Benzophenone (Diphenyl ketone)	182.2	1.0976^{50}	48.1	305.4	1.5975^{25}
2-Butanone (Methyl ethyl ketone)	72.1	0.8047	−85.9	79.6	1.3785
3-Methyl-butanone-2 (Methyl isopropyl ketone)	86.1	0.802	−92	94	1.3890
Cyclobutanone	70.1	0.9383		98.9	1.4195
Cycloheptanone	112.2	0.949		180	1.4608
Cyclohexanone	98.1	0.948	−32.1	155.6	1.450
Cyclopentanone	84.1	0.951_4^{18}	−58.2	130.6	1.4366
2-Heptanone (Methyl amyl ketone)	114.2	0.8166	−35.5	151.5	1.4110
3-Heptanone (n-Butyl ethyl ketone)	114.2	0.8197	−36.7	147.8	1.4085
4-Heptanone (di-n-Propyl ketone)	114.2	0.8145	−32.6	143.7	1.4068
2,6-Dimethyl heptanone-4 (Diisobutyl ketone)	142.2	0.807	−41.5	168.1	1.4114
2-Hexanone (Methyl n-butyl ketone)	100.2	0.8209	−56.9	127.2	1.4022
3-Hexanone (Methyl n-propyl ketone)	100.2	0.8118		123	1.4002
5-Nonanone (di-n-Butyl ketone)	142.2	0.818	− 6	186	1.421
2,4-Pentanedione (Acetylacetone)	100.2	0.976	−23.1	140.5	1.4512
2-Pentanone (Methyl n-propyl ketone)	86.1	0.809	−77.8	102.4	1.3895
3-Pentanone (Diethyl ketone)	86.1	0.8144	−39.5	101.7	1.3924
3-Methyl-pentanone-2 (sec-Butyl methyl ketone)	100.2	0.8145_4^{18}		118	1.4002
4-Methyl-pentanone-2 (Methyl isobutyl ketone)	100.2	0.8006	−80.3	116.2	1.3962
2,4-Dimethyl-pentanone-3 (Diisopropyl ketone)	114.2	0.8108		123.7	1.3986
2-Propanone (Acetone)	58.1	0.7912	−94.6	56.1	1.3590

*Or specific gravity.

TABLE 3-23. NITRILE AND CYANIDES.

Compound	Mol. Wt.	Density* (g/cm^3)	Freezing Point (°C)	Boiling Point (°C)	Refractive Index, n_D
Acetonitrile (Methyl cyanide)	41.0	0.7828	− 43.9	81.5	1.3442
Acrylonitrile (Vinyl cyanide)	53.1	0.8060	− 83.6	77.3	1.3911
Adiponitrile (Tetramethylene cyanide)	108.1	0.962	1	295	1.439
Benzonitrile (Phenyl cyanide)	103.1	1.006	− 13.7	190.7	1.5289
Butyronitrile (Propyl cyanide)	69.1	0.791	−111.9	117.9	1.384
Isobutyronitrile (Isopropyl cyanide)	69.1	0.7704	− 71.5	103.8	1.3734
Capronitrile (n-Pentyl cyanide)	97.2	0.805	− 79.4	163.9	1.407
Caprylonitrile (n-Heptyl cyanide)	125.2	0.841	− 45.6	199	1.4204
Decanonitrile (n-Nonyl cyanide)	153.3	0.8199	− 14.5	244	1.4294
Ethyl cyanoacetate	113.1	1.063	− 22.5	206.0	1.4179
Formonitrile (Hydrogen cyanide)	27.0	0.6876	− 13.2	25.7	1.2614
Glutaronitrile (Trimethylene dicyanide)	94.1	0.988	− 29	286	1.429
Heptonitrile (n-Hexyl cyanide)	111.2	0.810	− 62.6	184.6	1.414
Methacrylonitrile	67.1	0.8001	− 35.8	90.3	1.4007
Methyl cyanoacetate	99.1	1.101	− 13.1	205.1	1.4174
Nononitrile (n-Octyl cyanide)	139.2	0.8178	− 34.2	224.0	1.4255
Phenyl acetonitrile (Benzyl cyanide)	117.1	1.016	− 23.8	233.5	1.523
Pimelonitrile (Pentamethylene cyanide)	122.2	0.945		169/15	1.441
Propionitrile (Ethyl cyanide)	55.1	0.783	− 92.8	97.4	1.366
Suberonitrile (Hexamethylene cyanide)	136.2	0.933		185/15	1.445
Succinonitrile (Ethylene dicyanide)	80.1	0.9867^{60}	57.1	267	1.41734^{60}
Valeronitrile (n-Butyl cyanide)	83.1	0.7993	− 96.2	141.3	1.397

*Or specific gravity.

TABLE 3-24. NITRO AND NITROSO COMPOUNDS.

Compound	Mol. Wt.	Density* (g/cm^3)	Freezing Point (°C)	Boiling Point (°C)	Refractive Index, n_D
Methyl nitrate	77.0	1.2127_{15}^{15}	− 83.0	64.6^{18}	1.37605^{18}
Nitrobenzene	123.1	1.2032	5.7	210.9	1.5529
1,2-Dinitrobenzene	168.1	1.565^{17}	118	319	
1,3-Dinitrobenzene	168.1	1.575^{18}	89.8	301	
1,4-Dinitrobenzene	168.1	1.625^{18}	173	299	
2-Nitrobenzoic acid	167.1	1.575	146		
3-Nitrobenzoic acid	167.1	1.494	142		
4-Nitrobenzoic acid	167.1	1.610	242.4		
1-Nitrobutane	103.1	0.975		153	1.4112
2-Nitrobutane	103.1	0.9854_4^{17}	−132	139	1.4057
Nitroethane	75.1	1.050	− 89.5	114.0	1.3917
2-Nitroethanol	91.1			60/0.5	1.4420
Nitroglycerin	227.1	1.601	13.1	260	
Nitromethane	61.0	1.137_4	− 28.5	101.2	1.38195
Nitronaphthalene	173.2	1.331_4	58.8	304	
2-Nitrophenol	139.1	1.3045^{30}	45.1	217.1	
3-Nitrophenol	139.1	1.485	97	194/70	
4-Nitrophenol	139.1	1.2809^{14}	114.1	279	
1-Nitropropane	89.1	1.001	−104.6	131.6	1.4018
2-Nitropropane	89.1	0.988	− 93	120.3	1.3944
2-Nitrotoluene	137.1	1.1630_4^{15}	− 4.1	222.3	1.5472
3-Nitrotoluene	137.1	1.1581	15.5	231.9	1.5459
4-Nitrotoluene	137.1	1.286	51.9	238.3	
2,4,6-Trinitrotoluene	227.1	1.654	80.1	240	
Nitrosobenzene	107.1		68	59/18	

*Or specific gravity.

TABLE 3-25. ORGANOMETALLIC COMPOUNDS.

Compound	Mol. Wt.	Density* (g/cm^3)	Freezing Point ($^\circ$C)	Boiling Point ($^\circ$C)	Refractive Index, n_D
Diethyl beryllium	67.1		-11--3	185–188	
Diethyl cadmium	170.5	1.6531	$-$ 21	64/19.5	1.5680^{15}
Diethyl tellurium	185.7	1.599_4^{15}		136–138	1.5182^{15}
Diethyl zinc	123.5	1.2065	$-$ 28	118	
Dimethyl cadmium	142.9	1.9846^{18}	$-$ 2.5	105.5	1.5849^{18}
Dimethyl zinc	95.5	1.386^{10}	$-$ 29.2	44.0	
Tetraethyl lead	323.4	1.6528	-136	200	1.5198
Tetramethyl germanium	132.7	0.9661	$-$ 88	43.4/736	1.3868
Tetramethyl lead	267.3	1.9952	$-$ 27.5	110	1.2510
Tetramethyl silane	88.2	0.648	$-$ 27	27	1.3585
Tetramethyl tin	178.9	1.2914	$-$ 54	76	
Triethyl indium	202.4	1.260	$-$ 32	184	1.538
Triisobutyl aluminum	198.3	0.787	4	45/0.8	
Trimethyl aluminum	72.1	0.752	15.4	126	1.432
Trimethyl antimony	166.9	1.528^{15}	$-$ 62	80.6	
Trimethyl arsenic	120.0	1.124	$-$ 87.3	51	1.4541
Trimethyl bismuth	254.1	2.300_4^{18}	-107.7	108	
Trimethyl gallium	114.8		$-$ 15.8	55.7	
Trimethyl indium	159.9	1.568^{10}	88.4	135.8	
Triphenyl antimony (Triphenyl stibine)	353.1	1.5^{12}	54	377	

*Or specific gravity.

TABLE 3-26. PHENOLS.

Compound	Mol. Wt.	Density* (g/cm^3)	Freezing Point ($^\circ$C)	Boiling Point ($^\circ$C)	Refractive Index, n_D
Alizarin	240.2		289	430	
2-Chlorohydroxy benzene (o-Chlorophenol)	128.6	1.235	8.7	175.6	1.5473^{40}
3-Chlorohydroxy benzene (m-Chlorophenol)	128.6	1.268	32.8	214	1.5565^{40}
4-Chlorohydroxy benzene (p-Chlorophenol)	128.6	1.306	42.9	217	1.5579^{40}
2,3-Dichlorophenol	163.0		57	206	
2,4-Dichlorophenol	163.0		43	209	
2,5-Dichlorophenol	163.0		58	212	
2,6-Dichlorophenol	163.0		67	218/20	
3,4-Dichlorophenol	163.0		68	252	
3,5-Dichlorophenol	163.0		68	233	
1,2-Dihydroxy benzene (Pyrocatechol)	110.1	1.371^{15}	105	245.6	
1,3-Dihydroxy benzene (Resorcinol)	110.1	1.285^{15}	110.0	276.5	1.578
1,4-Dihydroxy benzene (Hydroquinone)	110.1	1.332^{15}	172	285	
2,3-Dimethyl phenol (2,3-Xylenol)	122.2		75	218	
2,4-Dimethyl phenol (2,4-Xylenol)	122.2	1.036	24.5	210.9	1.5390
2,5-Dimethyl phenol (2,5-Xylenol)	122.2		74.8	212.1	
2,6-Dimethyl phenol (2,6-Xylenol)	122.2		45.6	201.0	
3,4-Dimethyl phenol (3,4-Xylenol)	122.2	1.023^{17}_{15}	65.0	226.9	
3,5-Dimethyl phenol (3,5-Xylenol)	122.2		63.2	221.7	
2,3-Dinitrophenol	184.1	1.681	144		
2,4-Dinitrophenol	184.1	1.683	114		
2,5-Dinitrophenol	184.1		105		
2,6-Dinitrophenol	184.1	1.645	64		
3,4-Dinitrophenol	184.1	1.672	134		
3,5-Dinitrophenol	184.1	1.702	126.1		
2-Ethyl hydroxy benzene (o-Ethyl phenol)	122.2	1.0374^{12}	− 45	207.5	
3-Ethyl hydroxy benzene (m-Ethyl phenol)	122.2	1.001	− 4	214	
4-Ethyl hydroxy benzene (p-Ethyl phenol)	122.2	1.054	46.7	219	1.524
Eugenol	164.2	1.0664	− 9.2	255	1.5410
4-Hydroxyaniline (p-Amino phenol)	109.1		184	284	
Hydroxy benzene (Phenol)	94.1	1.0576^{41}	40.9	181.7	1.5425^{41}
α-Hydroxynaphthalene (1-Naphthol)	144.2	$1.0954^{98.7}_{4}$	96.1	282.5	$1.6206^{98.7}$
β-Hydroxynaphthalene (2-Naphthol)	144.2	1.217	122.5	288.0	
2-Methyl hydroxy benzene (o-Cresol)	108.1	1.0273^{41}	30.9	191.0	1.5561^{45}
3-Methyl hydroxy benzene (m-Cresol)	108.1	1.034	12.2	202.2	1.5398
4-Methyl hydroxy benzene (p-Cresol)	108.1	1.0341	34.6	201.8	1.5395
2-Nitro hydroxy benzene (o-Nitrophenol)	139.1	1.3045^{30}	45.1	214.5	

TABLE 3-26 (*Continued*)

Compound	Mol. Wt.	Density* (g/cm^3)	Freezing Point ($^\circ$C)	Boiling Point ($^\circ$C)	Refractive Index, n_D
3-Nitro hydroxy benzene (*m*-Nitrophenol)	139.1	1.485	97	194/70	
4-Nitro hydroxy benzene (*p*-Nitrophenol)	139.1	1.270_4^{120}	114	279	
Phenolphthalein	318.3	1.277_4^{32}	260		
2,3,4-Trichlorophenol	197.5		83.5		
2,4,6-Trichlorophenol	197.5	1.490	68	244.5	
2,3,6-Trichlorophenol	197.5		58		
2,3,5-Trichlorophenol	197.5		62	253	
3,4,5-Trichlorophenol	197.5		101		
1,2,3-Trihydroxy benzene (Pyrogallol)	126.1	1.453_4^4	132.8	309	
3,4,5-Trihydroxy benzoic acid (Gallic acid)	170.1	1.694_4^4	254		
2,4,6-Trinitrophenol (Picric acid)	229.1	1.763	121.8		

*Or specific gravity.

TABLE 3-27. ORGANIC PHOSPHORUS COMPOUNDS.

Compound	Mol. Wt.	Density* (g/cm^3)	Freezing Point ($^\circ$C)	Boiling Point ($^\circ$C)	Refractive Indix, n_D
Diethyl phosphate	154.1	1.175		203.2	1.4148
Diethyl phosphite	138.1	1.072		87/20	1.4101
Dimethyl phosphate	126.0	1.335		172–176	1.408
Dimethyl phosphite	110.0	1.2004		171	1.4036
Tributyl phosphate	266.3	0.977	< -80	289	1.426
Tributyl phosphite	250.3	0.923		122/12	1.432
Triethyl phosphate	182.2	1.070	-56.4	216	1.405
Triethyl phosphite	166.2	0.969		157	1.414
Trimethyl phosphate	140.1	1.214		197.2	1.396
Trimethyl phosphite	124.1	1.052		112	1.410
Triphenyl phosphate	326.3	1.205_4^{50}	49.9	245/11	
Triphenyl phosphite	310.3	1.184	24	360	
Tripropyl phosphate	224.2	1.0121		252	1.4165
Tripropyl phosphite	208.2	0.952		207	1.427
Triisopropyl phosphate	224.2	0.9867		219	1.419
Triisopropyl phosphite	208.2	0.918		61/10	1.412

*Or specific gravity.

TABLE 3-28. SULFIDES.

Compound	Mol. Wt.	Density* (g/cm^3)	Freezing Point (°C)	Boiling Point (°C)	Refractive Index, n_D
Acetyl sulfide	118.1			150	1.4795
Allyl sulfide	114.2	0.88764^{27}	− 83	138.6	1.4895
Butyl sulfide	146.3	0.840	− 79.7	187	1.453
Butyl disulfide	178.4	0.938		231	1.493
Butyl ethyl sulfide (1-Ethyl thiobutane)	118.2	0.8376	− 95.1	144.2	1.4491
Carbon disulfide	76.1	1.2632	−111.5	46.2	1.62799
Dimethyl sulfoxide	78.1	1.014	18.4	190	1.4775
Ethyl sulfide (3-Thiapentane)	90.2	0.83623	−104.0	92.1	1.44298
Ethyl disulfide (3,4-Dithiahexane)	122.2	0.9927	−101.5	152.6	1.507
Ethyl methyl sulfide (2-Thiabutane)	76.2	0.84221	−105.9	66.7	1.44035
Ethylene sulfide	60.1	1.0046		56	1.4914
Methyl sulfide (2-Thiapropane)	62.1	0.84825	− 98.3	37.3	1.43547
Methyl disulfide (2,3-Dithiabutane)	94.2	1.0625	− 84.7	109.7	1.52592
Methyl-*n*-propyl sulfide (2-Thiapentane)	90.2	0.8424	−113.0	93.5	1.4442
Methyl-isopropyl sulfide (3-Methyl-2-thiabutane)	90.2	0.8291	−101.5	84.7	1.4392
2-Methyl thiophene	98.2	1.02183	− 63.5	112.4	1.52042
3-Methyl thiophene	98.2	1.0216	− 68.9	115.4	1.57042
Phenyl sulfide	186.3	1.114	− 40	296	1.6327
Phenyl disulfide	218.3	1.116	60	296	1.6312
Propyl sulfide (4-Thiaheptane)	118.2	0.8377	−102.5	142.8	1.4487
Propyl Disulfide	150.3	0.960	− 85.6	194	1.498
Isopropyl sulfide	118.2	0.8146	− 78.1	120.0	1.4388
Isopropyl disulfide	150.3	0.944		177.2	1.492
Thiophene	84.1	1.06485	− 38.3	84.2	1.52890
1,4-Thioxane	104.2	1.1177		148	1.5070

*Or specific gravity.

TABLE 3-29. SULFONIC ACIDS, SULFONES, AND SULFONAMIDES.

Compound	Mol. Wt.	Density* (g/cm^3)	Freezing Point (°C)	Boiling Point (°C)	Refractive Index, n_D
Benzene sulfonyl chloride	176.6	1.3842_{15}^{15}	14.5	119/15	1.5518
Dimethyl sulfone	94.1	1.1702_0^{10}	109	238	1.4226
Dimethyl sulfoxide	78.1	1.100^0	18.5	189.0	1.4795
Diphenyl sulfone	218.3	1.252	128	379	
Ethane sulfonic acid	110.1	1.334	− 17		1.4340^{16}
Methane sulfonic acid	96.1	1.481_4^{18}	20	167/10	1.4317^{16}
Sulfanilamide	172.2	1.485	163		
Thioacetamide	75.1		108		

*Or specific gravity.

TABLE 3-30. THIOLS.

Compound	Mol. Wt.	Density* (g/cm^3)	Freezing Point (°C)	Boiling Point (°C)	Refractive Index, n_D
Benzenethiol	110.2	1.0766	− 14.9	168.3	1.58973
1-Butanethiol (n-Butyl mercaptan)	90.2	0.8337	−115.9	98.4	1.4440
2-Butanethiol (sec-Butyl mercaptan)	90.2	0.8295	−165	85	1.4366
Cyclohexanethiol	116.2	0.9486		160	1.4933
Ethanethiol (Ethyl mercaptan)	62.1	0.83914	−147.9	35.0	1.43105
1,2-Ethanedithiol (Ethylene mercaptan)	94.2	1.1192	− 41.2	146	1.5589
1-Hexanethiol (n-Hexyl mercaptan)	118.2	0.8424	− 80	151	1.4496
Methanethiol (Methyl mercaptan)	48.1	0.8665	−123	5.95	
1-Pentanethiol (n-Amyl mercaptan)	104.2	0.84209	− 75.7	126.5	1.44692
1-Propanethiol (n-Propyl mercaptan)	76.2	0.8408	−111.5	65.3	1.4351^{25}
2-Propanethiol (Isopropyl mercaptan)	76.2	0.80895	−130.6	58	1.422
1,3-Propanedithiol	108.2	1.0772	− 79	172.9	1.5392
2-Methyl-2-propanethiol (tert. Butyl mercaptan)	90.2	0.8002	1.26	64.2	1.4232
Thiophenol	110.2	1.078	− 14.9	168.3	1.5888

*Or specific gravity.

TABLE 3-31. SOLVENT PROPERTIES COMPARISON CHART.

Copyright © 1966 by E. I. Du Pont de Nemours & Company, Wilmington, Del. 19898. Reprinted by permission of the "Freon"™ Products Division.

FLUOROCARBON SOLVENTS	DESCRIPTION		Pounds Per Gallon At 68°F	Boiling Point °F	Freezing Point °F	Evaporation Rate (Carbon Tetrachloride=100)	Coefficient Of Expansion Per °F
	Formula	Molecular Weight					
"FREON" MF	CCl$_3$F	137.38	12.42	74.87	−168	−	0.00084
"FREON" TF	CCl$_2$FCClF$_2$	187.39	13.16	117.63	−31	280[a]	0.00089
"FREON" TA	Azeotrope of "Freon" TF and Acetone		11.73$^{77°F}$	110.5	−150	−	0.00088
"FREON" TE	Azeotrope of "Freon" TF and Ethanol		12.56$^{77°F}$	112.3	−43		0.00078
"FREON" TMC	Azeotrope of "Freon" TF and Methylene Chloride		11.85$^{77°F}$	97.2	−126	−	0.00076
"FREON" T-E 35	Blend of "Freon" TF and Ethanol		9.75$^{77°F}$	119.0	−108		0.00066
"FREON" T-P 35	Blend of "Freon" TF and Isopropanol		9.60$^{77°F}$	120.0	−94	−	0.00066
"FREON" T-WD 602	Emulsion of water in "Freon" TF		12.47$^{77°F}$	112.0	32	−	−

CHLORINATED HYDROCARBONS	Formula	Molecular Weight	Pounds Per Gallon At 68°F	Boiling Range °F	Freezing Point °F	Evaporation Rate (Carbon Tetrachloride=100)	Coefficient Of Expansion Per °F	Surface Tension at 68°F Dynes /cm	Solubility % By Weight at 68°F		Flash Point (Tag Closed Cup) °F	Flammable Limits % By Volume in Air		Toxicity TLV In PPM	Specific Heat Liquid At 68°F Btu /(lb)(°F)	Latent Heat At B. P. Btu /lb	Kauri-Butanol Value cc
									In Water	Of Water		Lower	Upper				
n-PROPYL CHLORIDE	CH$_3$(CH$_2$)$_2$Cl	78.54	7.43	113–117	−189.0	187	0.00080	−	0.27	−	0	2.6	10.5	−	−	−	−
ISOPROPYL CHLORIDE	CH$_3$CHClCH$_3$	78.54	7.17	94–104	−178.6	−	0.00088	−	0.31	−	−26	2.8	10.7	−	−	−	−
METHYLENE CHLORIDE	CH$_2$Cl$_2$	84.94	11.07	102–106	−142.1	147	0.00076	28.0	1.38	0.15	None	15.5*	66.4*	500**	0.28	141	136
DICHLOROETHYLENE (1,1)	CH$_2$:CCl$_2$	96.95	10.43$^{60°F}$	99	−188.5	105	−	−	−	−	57	5.6	13.0	−	−	−	−
ETHYLENE DICHLORIDE	CH$_2$ClCH$_2$Cl	98.97	10.45	180–183	−31.5	79	0.00065	32.2	0.90	0.15	56	6.2	15.9	50**	0.31	139	−
MONOCHLOROBENZENE	C$_6$H$_5$Cl	112.56	9.23	266–273	−49.4	32	0.00051	33.2	0.05	0.047$^{20°F}$	85	1.8$^{112°F}$	9.6$^{60°F}$	75**	0.32	140	−
PROPYLENE DICHLORIDE	CH$_3$ClCHClCH$_3$	112.99	9.65	204–208	<−94.0	71	0.00064	31.4$^{77°F}$	0.27	0.06	59	3.4	14.5	75**	0.33$^{77°F}$	123	102
CHLOROFORM	CHCl$_3$	119.39	12.43	142	−82.3	118	0.00071	27.2	0.82	0.07	None	Nonflammable		50**	0.23	106	208
TRICHLOROETHYLENE	ClCH:CCl$_2$	131.40	12.22	188–190	−123.5	69[a]	0.00065	29.2	0.11	0.02	None	12	86	100**	0.23	103	130
TRICHLOROETHANE (1,1,1)	CH$_2$ClCHCl$_2$	133.42	12.04	230–239	−34.1	21	−	33.6	0.44	0.05	None	Nonflammable		10**	0.27	109	−
1,1,1, TRICHLOROETHANE (Methyl Chloroform)	CH$_3$CCl$_3$	133.42	11.03	162–190	−58.0	139	−	25.9	<0.10$^{77°F}$	−	None	−	−	350**	−	−	−
ORTHO DICHLOROBENZENE	C$_6$H$_4$Cl$_2$	147.01	10.89	351–361	1.0	7	0.00046	−	0.01$^{77°F}$	−	151	−	−	50**	0.27	117	−
CARBON TETRACHLORIDE	CCl$_4$	153.84	13.30	171–172	−9.3	100	0.00069	26.9	0.08	0.01	None	Nonflammable		10**	0.21	84	114
PERCHLOROETHYLENE	CCl$_2$:CCl$_2$	165.85	13.55	250–254	−8.2	27[a]	0.00057	32.3	0.02	0.01	None	Nonflammable		100**	0.21	90	90
TETRACHLOROETHANE (Symmetrical)	CHCl$_2$CHCl$_2$	167.86	13.35	295–297	−32.8	14	0.00067	36.0	0.29$^{77°F}$	0.13$^{77°F}$	None	Nonflammable		5**	0.27	99	−
TETRACHLOROETHANE (Unsymmetrical)	CH$_2$ClCCl$_3$	167.86	13.25	264–267	−91.7	−	−	−	0.02	−	−	−	−	−	−	−	−
TRICHLOROBENZENE (Mixed Isomers)	C$_6$H$_3$Cl$_3$	181.46	12.15$^{79°F}$	415–423	62.6	−	0.00048	38.9$^{77°F}$	<0.01$^{77°F}$	0.22$^{77°F}$	212	−	−	−	0.20	105	−

ALIPHATIC PETROLEUM	Formula	Molecular Weight	Pounds Per Gallon At 68°F	Boiling Range °F	Freezing Point °F	Evaporation Rate (Carbon Tetrachloride=100)	Coefficient Of Expansion Per °F	Surface Tension at 68°F Dynes /cm	Solubility % By Weight at 68°F		Flash Point (Tag Closed Cup) °F	Flammable Limits % By Volume in Air		Toxicity TLV In PPM	Specific Heat Liquid At 68°F Btu /(lb)(°F)	Latent Heat At B. P. Btu /lb	Kauri-Butanol Value cc
									In Water	Of Water		Lower	Upper				
n-PENTANE	CH$_3$(CH$_2$)$_3$CH$_3$	72.15	5.22	91–106	−201.5	271	0.00079	−	0.04$^{60°F}$	−	<−40	1.4	8.0	1000**	0.54	154	26.0
TEXTILE SPIRITS	−	−	5.75$^{60°F}$	145–175	−	−	0.00072	22.5	−	−	−20	1.2	6.8	−	−	−	32.8
n-HEXANE	CH$_3$(CH$_2$)$_4$CH$_3$	86.17	5.51	151–160	−139.5	139	0.00075	18.4	0.014$^{60°F}$	−	−7	1.2	6.9	500**	0.54	145	32.5
n-HEPTANE	CH$_3$(CH$_2$)$_5$CH$_3$	100.20	5.71	196–214	−131.1	97	0.00068	−	0.005$^{60°F}$	<0.02	25	1.2	6.7	500**	0.53	138	35.5
VM & P NAPHTHA	−	−	6.29$^{60°F}$	236–292	<−50	37	0.00061	25.4	−	−	48	0.99$^{120°F}$	6.0$^{135°F}$	500	−	−	38.5
HI-FLASH VM & P NAPHTHA	−	−	6.33$^{60°F}$	240–320	<−50	8	0.00061	25.8	−	−	57	1.2	6.0	500	−	−	36.4
MINERAL SPIRITS NO. 10	−	−	6.51$^{60°F}$	307–385	−92	−	0.00054	27.6	−	−	105	1.1	6.0	−	−	−	34.0
STODDARD SOLVENT	−	−	6.47$^{60°F}$	310–388	<−45	4[a]	0.00061	27.6	−	−	105	1.1	6.0	500**	−	−	34.0
MINERAL SPIRITS	−	−	6.58$^{60°F}$	314–390	<−50	−	0.00050	27.8	−	−	110	1.1	6.0	−	−	−	39.0
VARSOL-1	−	−	6.58$^{60°F}$	322–386	−	−	−	−	−	−	113	1.1	6.0	−	−	−	38.0
ODORLESS MINERAL SPIRITS	−	−	6.31$^{60°F}$	352–398	−	−	0.00061	27.0	−	−	128	1.1	6.1	−	−	−	26.2
KEROSENE	−	−	6.77$^{60°F}$	374–503	−	−	0.00050	30.0	−	0.06	145	1.2	6.0	−	−	−	29.5
INK SOLVENT	−	−	6.79$^{60°F}$	512–610	−	−	−	−	−	−	250	−	−	−	−	−	22.1

ALCOHOLS	Formula	Molecular Weight	Pounds Per Gallon At 68°F	Boiling Range °F	Freezing Point °F	Evaporation Rate (Carbon Tetrachloride=100)	Coefficient Of Expansion Per °F	Surface Tension at 68°F Dynes /cm	Solubility % By Weight At 68°F		Flash Point (Tag Closed Cup) °F	Flammable Limits % By Volume in Air		Toxicity TLV In PPM	Specific Heat Liquid At 68°F Btu /(lb)(°F)	Latent Heat At B. P. Btu /lb	Kauri-Butanol Value cc
									In Water	Of Water		Lower	Upper				
METHANOL	CH$_3$OH	32.04	6.61	147–151	−144.0	50	0.00066	22.6	∞	∞	54	6.0	36.5	200**	0.60	473	−
ETHANOL, ANHYDROUS	C$_2$H$_5$OH	46.07	6.59	171–176	−174.1	−	0.00063	22.3	∞	∞	57	3.3	19.0	1000**	0.58$^{77°F}$	361	−
ETHANOL, 95%	C$_2$H$_5$OH	46.07	6.76	166–175	−198.4	37	0.00062	22.8	∞	∞	57	3.3	19.0	1000	0.62$^{77°F}$	−	−
ISOPROPANOL, 99%	(CH$_3$)$_2$CHOH	60.09	6.55	179–181	−126.0	34	0.00062	21.7	∞	∞	56	2.5	5.2	400**	0.60	287	−
n-PROPANOL	C$_3$H$_7$OH	60.09	6.74	207–208	−196.6	34	0.00053	23.8	∞	∞	59	2.5	13.5	200**	0.57$^{77°F}$	296	−
ISOBUTYL ALCOHOL	(CH$_3$)$_2$C$_2$H$_5$OH	74.12	6.72	225–232	−162.4	16	0.00053	22.8	10	15	64	1.7	−	100**	0.58	249	−
n-BUTYL ALCOHOL	C$_4$H$_9$OH	74.12	6.75	239–245	−129.6	13	0.00052	24.6	7.7	20.1	84	1.7	18.0	100**	0.56	254	−
sec-BUTYL ALCOHOL	C$_2$H$_5$CHOHCH$_3$	74.12	6.73	208–214	−174.5	21	0.00055	23.5	20.1	36.3	76	−	−	150**	0.67	242	−
AMYL ALCOHOL (Mixed Isomers)	C$_5$H$_{11}$OH	88.15	6.76	230–293	−	7	0.00052	23.6	2.4$^{60°F}$	11.0	111	1.2	−	−	−	−	−
CYCLOHEXANOL	C$_6$H$_{11}$OH	100.16	7.91	300–324	77.2	2	0.00043	34.2$^{60°F}$	3.6	20.0	154	−	−	50**	0.42$^{60°F}$	195	−
n-HEXANOL	C$_6$H$_{13}$OH	102.17	6.83	307–320	−60.9	−	0.00050	24.5	0.58	7.2	145	−	−	100	0.50$^{77°F}$	209	−

NOTES:
*In oxygen.
**Threshold limit values for 1965, American Conference of Governmental Industrial Hygienists.
***As calculated per instructions given in American Conference of Governmental Industrial Hygienists Booklet, "Threshold Limit Value of Airborne Contaminants and Physical Agents with Intended Changes Adopted by ACGIH for 1971."

†Very slightly soluble.
–Data not available.
††Open cup.

[a] ASTM D1901-67
b none to about 30% (volume) evaporation.
c none to about 40% (volume) evaporation.

TABLE 3-31 (*Continued*)

Surface Tension At 68°F Dynes/cm	Solubility % By Weight At 68°F		Flash Point (Tag Closed Cup) °F	Flammable Limits % By Volume In Air		Toxicity TLV** In PPM	Specific Heat Liquid At 68°F Btu/(lb)(°F)	Latent Heat At B.P. Btu/lb	Kauri-Butanol Value cc
	In Water	Of Water		Lower	Upper				
18.8	0.14	0.0086	None	Nonflammable		1000	0.21	78.31	60
17.8	0.017	0.0086	None	Nonflammable		1000	0.22	63.12	31
18.4	–	$0.15^{75°F}$	None	Nonflammable		1000***	0.31	85.4***	51
18.2	–	–	None	Nonflammable		1000***	0.27	77.2***	–
21.0	–	$0.09^{75°F}$	None	Nonflammable		650***	0.26	104.0***	86
20.0	–	$6.3^{75°F}$	None^b	–	–	750***	0.47	149.0***	–
21.4	–	$9.1^{75°F}$	None^c	–	–	700***	0.36	119.0***	–
17.8	–	–	None	Nonflammable		1000***	–	–	21

ETHERS AND ETHER ALCOHOLS

ETHERS AND ETHER ALCOHOLS	Formula	Molecular Weight	Pounds Per Gallon At 68°F	Boiling Range °F	Freezing Point °F	Evaporation Rate (Carbon Tetrachloride = 100)	Coefficient Of Expansion Per °F	Surface Tension at 68°F Dynes/cm	Solubility % By Weight At 68°F In Water	Of Water	Flash Point (Tag Closed Cup) °F	Flammable Limits % By Volume In Air Lower	Upper	Toxicity TLV In PPM	Specific Heat Liquid At 68°F Btu/(lb)(°F)	Latent Heat At B.P. Btu/lb	Kauri-Butanol Value cc
ETHYL ETHER	$(C_2H_5)_2O$	74.12	5.96	93-95	-177.3	263	0.00092	17.0	6.9	1.3	-20	1.9	36.5	400**	0.55	151	–
METHYL CELLOSOLVE‡	$CH_3O(CH_2)_2OH$	76.09	8.03	250-259	-121.2	12	0.00052	$35.0^{77°F}$	∞	∞	107	$2.5^{75°F}$	$19.8^{284°F}$	25**	0.53	243	–
CELLOSOLVE‡	$C_2H_5O(CH_2)_2OH$	90.12	7.74	270-279	-94.0	9	0.00054	$32.0^{77°F}$	∞	∞	104	2.6	15.7	200**	0.56	234	–
ISOPROPYL ETHER	$(CH_3)_2CHOCH(CH_3)_2$	102.17	6.04	145-158	-121.9	187	0.00083	32.0	0.90	0.57	-18	1.4	21.0	500**	0.51	123	–
n-BUTYL CELLOSOLVE‡	$C_4H_9OC_2H_4OH$	118.17	7.51	331-343	<-40	3	0.00048	$31.5^{77°F}$	∞	∞	141	$1.1^{329°F}$	$10.6^{349°F}$	50**	0.58	171	–
DIETHYL CELLOSOLVE‡	$(C_2H_5OCH_2)_2$	118.17	7.02	243-264	-101.2	–	0.00067	–	21.0	3.4	95††	–	–	120	–	178	–
METHYL CARBITOL‡	$CH_3OC_2H_4OC_2H_4OH$	120.15	8.56	370-388	–	–	0.00048	$41.3^{77°F}$	∞	∞	200††	–	–	–	$0.51^{73°F}$	163	–
n-BUTYL ETHER	$(C_4H_9)_2O$	130.22	6.40	279-289	-139.4	–	–	22.9	0.3	0.19	77	–	–	–	$0.49^{60°F}$	122	–
CARBITOL‡	$C_2H_5O(CH_2)_2O(CH_2)_2OH$	134.17	8.55	365-401	<-104.8	–	0.00046	$35.5^{77°F}$	∞	∞	201	–	–	–	0.55	173	–
BUTYL CARBITOL‡	$C_4H_9O(CH_2)_2O(CH_2)_2OH$	162.22	7.95	428-455	-90.6	–	0.00048	$33.6^{77°F}$	∞	∞	172	–	–	–	0.55	111	–

KETONES

KETONES	Formula	Molecular Weight	Pounds Per Gallon At 68°F	Boiling Range °F	Freezing Point °F	Evaporation Rate (Carbon Tetrachloride = 100)	Coefficient Of Expansion Per °F	Surface Tension at 68°F Dynes/cm	Solubility % By Weight At 68°F In Water	Of Water	Flash Point (Tag Closed Cup) °F	Flammable Limits % By Volume In Air Lower	Upper	Toxicity TLV In PPM	Specific Heat Liquid At 68°F Btu/(lb)(°F)	Latent Heat At B.P. Btu/lb	Kauri-Butanol Value cc
ACETONE	CH_3COCH_3	58.08	6.58	132-134	-138.6	139	0.00080	23.7	∞	∞	0	2.6	12.8	1000**	0.51	224	–
METHYL ETHYL KETONE	$CH_3COC_2H_5$	72.10	6.71	174-177	-123.5	97	0.00076	24.6	26.8	11.8	28	1.8	11.5	200**	0.53	191	–
CYCLOHEXANONE	$(CH_2)_5CO$	98.14	7.88	266-343	-49.0	12	0.00051	–	2.3	8.0	145	1.1	–	50**	0.49	–	–
METHYL ISOBUTYL KETONE	$(CH_3)_2CHCH_2COCH_3$	100.16	6.68	234-244	-120.5	47	0.00063	22.7	2.0	1.8	64	1.4	7.5	100**	0.50	157	–
METHYL n-BUTYL KETONE	$CH_3COC_4H_9$	100.16	6.83	237-279	-70.4	32	0.00055	25.5	$3.47^{60°F}$	$3.77^{60°F}$	73	1.2	8.0	100**	0.55	148	–
METHYL CYCLOHEXANONE (Mixed Isomers)	$(CH_3)C_6H_9O$	112.17	7.67	237-343	–	7	0.00047	–	0.2	3.0	118	–	–	–	$0.44^{60°F}$	–	–
METHYL n-AMYL KETONE	$CH_3(CH_2)_4COCH_3$	114.18	6.81	297-309	-31.9	15	0.00057	–	0.4	1.5	120††	–	–	100**	–	149	–
DIACETONE	$(CH_3)_2C(OH)CH_2COCH_3$	116.16	7.82	266-356	-65.2	4	0.00055	29.8	∞	∞	48	–	–	50**	$0.50^{60°F}$	200	–

ESTERS

ESTERS	Formula	Molecular Weight	Pounds Per Gallon At 68°F	Boiling Range °F	Freezing Point °F	Evaporation Rate (Carbon Tetrachloride = 100)	Coefficient Of Expansion Per °F	Surface Tension at 68°F Dynes/cm	Solubility % By Weight At 68°F In Water	Of Water	Flash Point (Tag Closed Cup) °F	Flammable Limits % By Volume In Air Lower	Upper	Toxicity TLV In PPM	Specific Heat Liquid At 68°F Btu/(lb)(°F)	Latent Heat At B.P. Btu/lb	Kauri-Butanol Value cc
METHYL ACETATE	$CH_3CO_2CH_3$	74.08	7.58	127-136	-145.6	118	0.00074	24.6	24.5	8.2	15	4.1	13.9	200**	$0.47^{60°F}$	177	–
ETHYL ACETATE (99%)	$CH_3CO_2C_2H_5$	88.10	7.50	169-174	-118.5	–	0.00077	23.9	8.7	3.3	24	2.2	11.5	400**	0.46	158	–
ISOBUTYL ACETATE	$CH_3CO_2CH_2CH(CH_3)_2$	116.16	7.24	220-246	-146.0	45	0.00066	23.3	0.75	1.1	64	–	–	150**	0.46	133	–
n-BUTYL ACETATE	$CH_3CO_2C_4H_9$	116.16	7.28	244-262	-100.3	34	0.00067	$27.6^{61°F}$	0.68	1.2	72	1.7	15.0	150**	$0.51^{73°F}$	133	–
sec-BUTYL ACETATE	$CH_3CO_2CH(CH_3)C_2H_5$	116.16	7.27	219-239	-146.0	50	0.00067	–	$0.74^{77°F}$	$2.1^{77°F}$	66	1.7	–	200**	$0.46^{73°F}$	140	–
METHYL CELLOSOLVE ACETATE	$CH_3OC_2H_4OCH_3$	118.13	8.37	279-306	-85.2	12	0.00061	31.8	∞	∞	132	1.7	8.2	25**	$0.50^{60°F}$	158	–
AMYL ACETATE (Mixed Isomers)	$CH_3CO_2C_5H_{11}$	130.18	7.21	259-311	–	23	0.00061	–	1.0	1.7	77	1.1	–	–	0.49	123	–
n-BUTYL PROPIONATE	$C_2H_5CO_2C_4H_9$	130.18	7.27	255-340	-129.2	16	0.00060	–	0.15	0.8	90	–	–	–	0.46	129	–
CELLOSOLVE ACETATE	$CH_3CO_2C_2H_4OC_2H_5$	132.16	8.10	302-320	-79.1	8	0.00062	$31.8^{77°F}$	22.9	6.5	124	1.7	–	100**	0.49	145	–

AROMATICS

AROMATICS	Formula	Molecular Weight	Pounds Per Gallon At 68°F	Boiling Range °F	Freezing Point °F	Evaporation Rate (Carbon Tetrachloride = 100)	Coefficient Of Expansion Per °F	Surface Tension at 68°F Dynes/cm	Solubility % By Weight At 68°F In Water	Of Water	Flash Point (Tag Closed Cup) °F	Flammable Limits % By Volume In Air Lower	Upper	Toxicity TLV In PPM	Specific Heat Liquid At 68°F Btu/(lb)(°F)	Latent Heat At B.P. Btu/lb	Kauri-Butanol Value cc
BENZOL (Benzene)	C_6H_6	78.11	7.32	172-176	41.9	95	0.00069	28.9	0.09	0.06	12	1.4	8.0	25**	$0.42^{60°F}$	170	105-150
TOLUOL (Toluene)	$C_6H_5(CH_3)$	92.13	7.20	228-232	-139.0	58	0.00061	28.4	0.05	0.04	40	1.3	7.0	200**	0.41	156	94-105
XYLOL (Xylene) (Mixed Isomers)	$C_6H_4(CH_3)_2$	106.16	7.17	261-318	–	45	0.00055	28.9	$0.02^{60°F}$	0.02	80	1.1	7.0	100**	$0.40^{60°F}$	147	94
m-CRESOL	C_7H_7OH	108.13	8.66	396-406	51.6-53.6	–	–	37.4	2.35	–	86	$1.1^{302°F}$	–	5**	0.55	181	–

TERPENES

TERPENES	Formula	Molecular Weight	Pounds Per Gallon At 68°F	Boiling Range °F	Freezing Point °F	Evaporation Rate (Carbon Tetrachloride = 100)	Coefficient Of Expansion Per °F	Surface Tension at 68°F Dynes/cm	Solubility % By Weight At 68°F In Water	Of Water	Flash Point (Tag Closed Cup) °F	Flammable Limits % By Volume In Air Lower	Upper	Toxicity TLV In PPM	Specific Heat Liquid At 68°F Btu/(lb)(°F)	Latent Heat At B.P. Btu/lb	Kauri-Butanol Value cc
DIPENTENE (dl)	$C_{10}H_{16}$	136.23	7.10	342-374	-139.9	–	0.00048	$27.5^{60°F}$	VSS†	–	109	–	–	–	–	125	62
TURPENTINE STEAM DISTILL.	$C_{10}H_{16}$	136.23	7.13	311-343	-58	<1	0.00049	$26.1^{60°F}$	VSS†	–	95	0.8	–	100**	–	123	55

‡Cellosolve and Carbitol are trademarks of Union Carbide Corp.

REFERENCES AND BIBLIOGRAPHY

Beilstein, F. K., "Beilsteins Handbuch der organischen Chemie," 4th ed., 64 vols., J. W. Edwards, Ann Arbor, Mich., 1929–1944.

Coates, G. E., M. L. H. Green and K. Wade, "Organometallic Compounds," 3rd ed., Methuen, London, 1967–1968.

Cox, J. D. and G. Pilcher, "Thermochemistry of Organic and Organometallic Compounds," Academic Press, London, 1970.

Dawson, R. M. C. *et al.*, "Data for Biochemical Research," 2nd ed., Oxford Univ. Press, New York, 1969.

Devon, T. K. and A. I. Scott, "Handbook of Naturally Occurring Compounds," vol. 2, Academic Press, New York, 1972.

"Dictionary of Organic Compounds: The Constitution and Physical, Chemical, and other Properties of the Principal Carbon Compounds and their Derivatives," 6 vols, 4th ed., Oxford Univ. Press, Fairlawn, N.J., 1965.

Doak, G. O. and L. D. Freedman, "Organometallic Compounds of Arsenic, Antimony and Bismuth," Wiley, New York, 1970.

Dreisbach, R. R., "Physical Properties of Chemical Compounds," 3 vols., American Chemical Society, Washington, 1955–1961.

Eaborn, C., "Organosilicon Compounds," Academic Press, New York, 1960.

Ferris, S. W., "Handbook of Hydrocarbons," Academic Press, New York, 1955.

Francis, A., "Handbook of Components in Solvent Extraction," Gordon & Breach, New York, 1972.

Grasselli, J. G., Ed., "Atlas of Spectral Data and Physical Constants for Organic Compounds," CRC Press, Cleveland, 1973.

Gysel, H., "Tables of Percentage Composition of Organic Compounds," 2nd rev. ed., International Publications Service, New York, 1969.

Hagihara, M., M. Kumada and R. Okawara, Eds., "Handbook of Organometallic Compounds," Benjamin, New York, 1968.

Kaufman, H. C., "Handbook of Organo-Metallic Compounds," Van Nostrand Reinhold Co., New York, 1961.

Manske, R. H. F. and H. L. Holmes, Eds., "The Alkaloids- Chemistry and Physiology," 12 vols., Academic Press, New York, 1950–1970.

Pigman, W. and D. Horton, "The Carbohydrates," 2 vols., Academic Press, New York, 1970–1971.

Raffauf, R. F., "A Handbook of Alkaloids and Alkaloid-Containing Plants," Wiley, New York, 1970.

Rappaport, Z., Ed., "Handbook of Tables for Organic Compound Identification," 3rd ed., CRC Press, Cleveland, 1967.

Riddick, J. A. and W. B. Bunger, "Organic Solvents: Physical Properties and Methods of Purification," 3rd ed., Wiley, New York, 1971.

Rodd, E. H., "Rodd's Chemistry of Carbon Compounds," 2nd rev. ed., American Elsevier, New York, 1965–1970.

Steinberg, H., "Organoboron Chemistry," Wiley, New York, 1964.

Stout, G. H., "Composition Tables: Data for Compounds Containing C, H, N, O, S," Benjamin, New York, 1963.

Stull, D. R., E. F. Westrum and G. C. Sinke, "The Chemical Thermodynamics of Organic Compounds," Wiley, New York, 1969.

Timmermans, J., "Physico-Chemical Constants of Pure Organic Compounds," 2 vols., American Elsevier, New York, 1950, 1965.

Utermark, W. and W. Schicke, "Melting Point Tables of Organic Compounds," 2nd ed., Halsted Press, New York, 1963.

Wilhoit, R. C. and B. J. Zwolinski, "Physical and Thermodynamic Properties of Aliphatic Alcohols," *J. Phys. Chem. Ref. Data*, Vol. 2, Suppl. 1, 1973.

Winchell, A. N., "Optical Properties of Organic Compounds," Academic Press, New York, 1954.

Yukawa, Y., Ed., "Handbook of Organic Structural Analysis," Benjamin, New York, 1965.

CHAPTER **IV**
PROPERTIES
OF INORGANIC
COMPOUNDS

Chapter IV describes the properties of inorganic compounds and molten salts. For greater convenience, tables and references for these categories are presented separately.

The tables are arranged as follows:

Table 4-1   Abbreviations Used in Table 4-2.   Table 4-3   Properties of Molten Salts.
Table 4-2   Properties of Inorganic Compounds.

TABLE 4-1 ABBREVIATIONS USED IN TABLE 4-2.

Color		Crystal Symmetry	
B	brown	C	cubic
BE	blue	H	hexagonal
BK	black	M	monoclinic
CL	colorless	R	rhombic
G	gray	RH	Rhombohedral
GN	green	T	tetragonal
O	orange	TG	trigonal
P	purple	TR	triclinic
R	red		
SL	silver		
V	violet		
W	white		
Y	yellow		

Other Symbols	
d	decomposition
i	insoluble
liq	liquid
s	sublimation
sl	soluble
t	transformation
vsl	very soluble
*	explosive
AMORP	amorphous

TABLE 4-2. PROPERTIES OF INORGANIC COMPOUNDS.

Compound	Formula	Molecular Weight	Color	Crystal Symmetry	Specific Gravity	Melting Point (°C)	Boiling Point (°C)	Refractive Index n_D	Solubility (g/100 g of H_2O)
ACTINIUM									
–Bromide	$AcBr_3$	466.7	W	H	5.85	800 s			
–Chloride	$AcCl_3$	333.4	W	H	4.81	960 s			
–Fluoride	AcF_3	284.0	W	H	7.88				
–Oxide	Ac_2O_3	502.0	W	H	9.19				
ALUMINUM									
–Bromide	$AlBr_3$	266.7	CL	R	2.64^{10}	97.5	268	2.70	vsl
–Carbide	Al_4C_3	143.9	Y	H	2.36	2100	2200 d	1.56	d
–Chloride	$AlCl_3$	133.3	W	H	2.44	192.5	180 s	1.56	70^{15}
–Fluoride	AlF_3	84.0	CL	TR	2.88	1290 s		1.38	0.6^{25}
–Hydroxide	$Al(OH)_3$	78.0	W	M	2.42	300 $(-2H_2O)$			0.00015^{20}
–Iodide	AlI_3	407.7	W	R	3.95	191	386	1.54	sl
–Nitrate	$Al(NO_3)_3 \cdot 9H_2O$	375.1	CL	R	1.72	73.5	150 d		64^{25}
–Nitride	AlN	41.0	W	H	3.13	>2200			d
–Oxide	Al_2O_3	102.0	CL	H	3.97	2045	2980	1.68	i
–Phosphate	$AlPO_4$	122.0	W	R	2.57	1500		1.56	i
–Silicate	Al_2SiO_5	162.0	W	R	3.21	1810 d		1.66	i
–Sulfate	$Al_2(SO_4)_3$	342.2	W	R	2.71	760 d		1.47	31^0
–Sulfide	Al_2S_3	150.2	Y	H	2.02	1100			d
AMERICIUM									
–Oxide IV	AmO_2	275.1	B	C	11.7				
AMMONIUM									
–Bromide	NH_4Br	98.0	W	C	2.43	542 s		1.711	97^{25}
–Carbonate	$(NH_4)_2CO_3 \cdot H_2O$	114.1	W	C	1.80	58 d			100^{15}
–Chlorate	NH_4ClO_3	101.5	W	M	1.80	102 *			29^0
–Chloride	NH_4Cl	53.5	W	C	1.53	340 s		1.642	37^{20}
–Chromate	$(NH_4)_2CrO_4$	152.1	Y	M	1.89	180 d			40^{30}
–Fluoride	NH_4F	37.0	W	H	1.015	d		1.315	100^{20}
–Iodate	NH_4IO_3	192.9	W	R	3.31	150 d			2.6^{15}
–Iodide	NH_4I	144.9	W	C	2.86	551 s		1.703	173^{20}
–Nitrate	NH_4NO_3	80.0	W	C	1.725	170	210/11	1.413	188^{20}
–Nitrite	NH_4NO_2	64.0	Y	R	1.69	(60–70)*			vs

TABLE 4.2 *(Continued)*

Compound	Formula	Molecular Weight	Color	Crystal Symmetry	Specific Gravity	Melting Point (°C)	Boiling Point (°C)	Refractive Index n_D	Solubility (g/100 g of H_2O)
AMMONIUM (continued)									
−Oxalate	$(NH_4)_2C_2O_4 \cdot H_2O$	142.1	CL	R	1.50	170 (−H_2O)		1.44–1.59	2.5^0
−Perchlorate	NH_4ClO_4	117.5	W	R	1.95	130 d		1.49	25^{25}
−Hydrogen Phosphate	$(NH_4)_2HPO_4$	132.1	W	M	1.62	155 d		1.53	69^{20}
−Dihydrogen Phosphate	$NH_4H_2PO_4$	115.0	W	T	1.80	190		1.48–1.53	37^{20}
−Sulfate	$(NH_4)_2SO_4$	132.1	W	R	1.77	235 d		1.53	75^{20}
−Hydrogen sulfide	NH_4HS	51.1	W	R	1.17	120 d		1.74	128^0
−Thiocyanate	NH_4SCN	76.1	CL	M	1.30	149	170 d	1.61–1.69	120^0
ANTIMONY									
−Bromide III	$SbBr_3$	361.5	CL	R	4.15	96.6	288	1.74	d
−Chloride III	$SbCl_3$	228.1	CL	R	3.14	72.9	223	1.74	931^{20}
−Chloride V	$SbCl_5$	299.0	W	LIQ	2.34	2.8	~140	1.601^{14}	d
−Fluoride III	SbF_3	178.8	CL	R	4.38	292	319 s		443^{20}
−Fluoride V	SbF_5	216.7	CL	LIQ	2.99	7.0	149.5		sl
−Hydride III	SbH_3	124.8	CL	GAS	2.26^{-25}	−91	−18		0.41^{10}
−Iodide III	SbI_3	502.5	RD	H	4.85	170	401		d
−Iodide V	SbI_5	756.3	B			79	400.6	2.35	d
−Oxide III	Sb_2O_3	291.5	CL	R	5.67	656	1550		i
−Oxide V	Sb_2O_5	323.5	Y	C	3.78	>300 d			i
−Oxychloride III	$SbOCl$	173.2	W	M		170 d			0.0001^{18}
−Sulfate III	$Sb_2(SO_4)_3$	531.7	W		3.62	546			
−Sulfide III	Sb_2S_3	339.7	BK	R	4.64	d	~1150	4.064	
−Sulfide V	Sb_2S_5	403.8	Y		4.12	75 d			
ARSENIC									
−Acid, ortho	$H_3AsO_4 \cdot \frac{1}{2}H_2O$	151.0	CL		2.0–2.5	35.5	160 d		17^{25}
−Bromide III	$AsBr_3$	314.7	CL	R	3.54	30.8	221	1.598	d
−Chloride III	$AsCl_3$	181.3	CL	LIQ	2.16	−16.2	130.4		d
−Chloride V	$AsCl_5$	252.2	CL			~−40			
−Fluoride III	AsF_3	131.9	CL	LIQ	2.70	−5.95	62.8		d
−Fluoride V	AsF_5	169.9	CL	GAS	7.71g/l	−79.8	−52.9		d
−Hydride III	AsH_3	77.9	CL	GAS	2.70	−113.5	−62.5		
−Iodide III	AsI_3	455.6	R	H	4.39	141.8	403		8^{20}
−Iodide V	AsI_5	709.5	B	M	3.93	76			sl
−Oxide III	As_2O_3	197.2	CL	C	3.86	321 s	459		1.85^{20}

TABLE 4.2 (Continued)

Compound	Formula	Molecular Weight	Color	Crystal Symmetry	Specific Gravity	Melting Point (°C)	Boiling Point (°C)	Refractive Index n_D	Solubility (g/100 g of H_2O)
ARSENIC (continued)									
-Oxide V	As_2O_5	229.9	W		4.09	315 d			150[16]
-Sulfide II	As_2S_2	214.0	R	M	3.20	320	565	2.46–2.61	i
-Sulfide III	As_2S_3	246.0	Y	M	3.43	300	707	2.4–2.8	0.00005[18]
-Sulfide V	As_2S_5	310.2	Y	M			500 d		i
BARIUM									
-Bromate	$Ba(BrO_3)_2 \cdot H_2O$	411.2	CL	M	3.99	260 d			0.3[0]
-Bromide	$BaBr_2$	297.2	CL	R	4.78	847	d	1.75	104[20]
-Carbide	BaC_2	161.4	G	T	3.75				d
-Carbonate	$BaCO_3$	197.4	W	R	4.43	>1760 d	1450 d	1.676	0.002[18]
-Chlorate	$Ba(ClO_3)_2 \cdot H_2O$	322.3	CL	M	3.18	811 t		1.56–1.64	30[25]
-Chloride	$BaCl_2$	208.3	CL	M	3.86[24]	414	1560	1.736	27[25]
-Chromate	$BaCrO_4$	253.3	Y	R	4.50[15]	963			0.0003[20]
-Fluoride	BaF_2	175.3	CL	C	4.89	1320	2200	1.474	0.16[20]
-Hydride	BaH_2	139.4	G		4.21[0]	675 d	1400		
-Hydroxide	$Ba(OH)_2 \cdot 8H_2O$	315.5	CL	M	2.18[16]	77.9	780 d	1.502	4.7[15]
-Iodide	BaI_2	391.2	CL	M	5.15	740 d			203[20]
-Nitrate	$Ba(NO_3)_2$	261.4	CL	C	3.24	592	d		8.3[20]
-Oxalate	BaC_2O_4	225.4	W		2.66	400 d		1.572	0.008[20]
-Oxide	BaO	153.3	CL	C	5.72	1923	2000	1.98	3.5[20]
-Perchlorate	$Ba(ClO_4)_2$	336.2	CL	H	3.2	505			286[20]
-Sulfate	$BaSO_4$	233.4	W	R	4.50[15]	1580		1.636	0.0002[20]
-Sulfide	BaS	169.4	CL	C	4.25[15]	1200		2.155	d
-Titanate	$BaTiO_3$	233.3		T/H	6.0/5.8	1625		2.40	
BERYLLIUM									
-Bromide	$BeBr_2$	168.8	W	OR	3.47	490 s	520		sl
-Carbide	Be_2C	30.0	Y	H	1.90[15]	>2100 d			d
-Chloride	$BeCl_2$	79.9	W	OR	1.90	416	520		vsl
-Fluoride	BeF_2	47.0	CL	T	1.99	800	1330		vsl
-Hydroxide	$Be(OH)_2$	43.0	W	R	1.91	800	d		
-Iodide	BeI_2	262.8	CL	RH	4.33	300 d	590		d
-Nitrate	$Be(NO_3)_2 \cdot 3H_2O$	187.1	W		1.56	60	142		103[20]
-Nitride	Be_3N_2	55.1	CL	C	2.71	2200			

TABLE 4.2 (*Continued*)

Compound	Formula	Molecular Weight	Color	Crystal Symmetry	Specific Gravity	Melting Point (°C)	Boiling Point (°C)	Refractive Index n_D	Solubility (g/100 g of H_2O)
BERYLLIUM (continued)									
–Oxide	BeO	25.0	W	H	3.03	2530	4120	1.72	0.0000020^{30}
–Sulfate	BeSO$_4$	105.1	CL	T	2.43	550 d			27^0
–Sulfate	BeSO$_4 \cdot 4H_2O$	177.1	CL	T	1.71^{10}	100 (−2H$_2$O)		1.44–1.47	42.2^{25}
BISMUTH									
–Bromide III	BiBr$_3$	448.7	Y		5.60	218	461		d
–Chloride III	BiCl$_3$	315.4	W		4.75	230	447		d
–Fluoride III	BiF$_3$	266.0	G	C	8.75	727		1.74	i
–Hydroxide III	Bi(OH)$_3$	260.0	W		4.36	415 d			0.00014^{20}
–Iodide III	BiI$_3$	589.7	RD	H	5.64		439 s		i
–Nitrate III	Bi(NO$_3$)$_3 \cdot 5H_2O$	485.1	CL	TR	2.83	30 d			d
–Nitrate, Basic III	BiO(NO$_3$) $\cdot$ H$_2$O	305.0	W	H	4.93	260 d			i
–Oxide III	Bi$_2$O$_3$	466.0	Y	R	8.9	817	1890	1.91	i
–Oxide IV	Bi$_2$O$_4 \cdot 2H_2O$	518.0	B		5.6	100 d			i
–Oxide V	Bi$_2$O$_5$	498.0	B		5.10	150 d			i
–Oxychloride III	BiOCl	260.5	W	T	7.72	d		2.15	i
–Phosphate III	BiPO$_4$	304.0	W	M	6.32	418 d			i
–Sulfate III	Bi$_2$(SO$_4$)$_3$	706.1	W		5.08	685 d		1.34–1.46	d
–Sulfide III	Bi$_2$S$_3$	514.2	B	R	6.82				0.000018^{18}
BORON									
–Arsenate	BAsO$_4$	149.7	W	T	3.64	2300	2550	1.68	4.9^{20}
–Boric Acid	H$_3$BO$_3$	61.8	W	TR	1.44	169 d		1.5312^{16}	d
–Bromide	BBr$_3$	250.5	CL	LIQ	2.65	−46	90.1		i
–Carbide	B$_4$C	55.3	BK	RH	2.52	2350	>3500		d
–Chloride	BCl$_3$	117.2	CL	LIQ	1.43	−107.3	12.5		d
–Diborane	B$_2$H$_6$	27.7	CL	GAS	0.447^{-112}	−165.5	−92.5		322^0
–Fluoride	BF$_3$	67.8	CL	GAS	1.58	−128.8	−99.9		d
–Iodide	BI$_3$	391.6	W		3.35	43	210		i
–Nitride	BN	24.8	W	H	2.30	~3000 s	d		i
–Oxide	B$_2$O$_3$	69.6	W	C	1.84	450	1860		2.2^{20}
–Sulfide	B$_2$S$_3$	117.8	W		1.55	310			d
BROMINE									
–Chloride I	BrCl	115.4	R	GAS		−54	5		d

TABLE 4.2 (Continued)

Compound	Formula	Molecular Weight	Color	Crystal Symmetry	Specific Gravity	Melting Point (°C)	Boiling Point (°C)	Refractive Index n_D	Solubility (g/100 g of H_2O)
BROMINE (continued)									
− Fluoride I	BrF	98.9	B	GAS	2.49	−33	20	1.4536[25]	d
− Fluoride III	BrF_3	136.9	CL	LIQ	2.47	8.8	127.0	1.3529[25]	d
− Fluoride V	BrF_5	174.9	CL	LIQ		−61.3	40.5	1.325[10]	
− Hydride I	HBr	80.9	CL	GAS	2.16[−67]	−86.6	−66.7		70[20]
CADMIUM									
− Bromide	$CdBr_2$	272.2	W	H	5.19	566	963		sl
− Carbonate	$CdCO_3$	172.4	W	TG	4.26	357 d			i
− Chloride	$CdCl_2$	228.4	W	H	4.05	568	967		135[20]
− Fluoride	CdF_2	150.4	W	C	6.64	1100	1748	1.56	4.5[25]
− Hydroxide	$Cd(OH)_2$	146.4	W	TR	4.79	300 d			0.00032[5]
− Iodide	CdI_2	366.2	B	H	5.67	387	787		86[20]
− Nitrate	$Cd(NO_3)_2 \cdot 4H_2O$	308.5	W		2.46	59.5	132		153[20]
− Oxide	CdO	128.4	B	C	8.15	900 d			0.0005[18]
− Sulfate	$CdSO_4$	208.5	W	R	4.69	1100	1500 s		77[0]
− Sulfate	$3CdSO_4 \cdot 8H_2O$	769.6	CL	M	3.09			1.565	77[20]
− Sulfide	CdS	144.5	Y	H	4.82	1750 d	980 s	2.51	0.0001[25]
CALCIUM									
− Bromate	$CaBrO_3 \cdot H_2O$	313.9		M	3.33	180 (−H_2O)	150		vsl
− Bromide	$CaBr_2 \cdot 6H_2O$	308.0	CL	H	2.30	38.2		1.75	594[0]
− Carbide	CaC_2	64.1	CL	T	2.22[18]	2300		1.681	d
− Carbonate	$CaCO_3$	100.1	CL	R	2.93	825 d		1.52	0.0065[20]
− Chloride	$CaCl_2$	111.0	CL	C	2.15	772	>1600	1.417	74.5[20]
− Chloride	$CaCl_2 \cdot 6H_2O$	219.1	C	T	1.71[25]	29.9	200 (−6H_2O)		75[20]
− Chromate	$CaCrO_4 \cdot 2H_2O$	192.1	Y	M		200 (−2H_2O)			16[20]
− Fluoride	CaF_2	78.1	CL	C	3.18	1360	~2500	1.434	0.0016[18]
− Hydride	CaH_2	42.1	W	R	1.9	>1000	1000 d		d
− Hydroxide	$Ca(OH)_2$	74.1	CL	H	2.24	580 (−H_2O)	d	1.574	0.17[25]
− Iodide	CaI_2	293.9	W	H	3.96	740	~1100		209[20]
− Nitrate	$Ca(NO_3)_2$	164.1	CL	C	2.50	561			120[18]
− Nitrate	$Ca(NO_3)_2 \cdot 4H_2O$	236.2	CL	M	1.90	α42.7	132 d	1.498	132[20]

TABLE 4.2 (*Continued*)

Compound	Formula	Molecular Weight	Color	Crystal Symmetry	Specific Gravity	Melting Point (°C)	Boiling Point (°C)	Refractive Index n_D	Solubility (g/100 g of H_2O)
CALCIUM (continued)									
– Nitride	Ca_3N_2	148.3	B	H	2.63^{17}	1195			d
– Oxalate	CaC_2O_4	128.1	CL	C	2.2	d			0.0007^{20}
– Oxide	CaO	56.1	CL	C	3.3	2570	2850	1.838	d
– Perchlorate	$Ca(ClO_4)_2$	239.0	CL	C	2.65	270 d			189^{25}
– Peroxide	CaO_2	72.1	W	T	2.92	375 d			0.20^{25}
– Sulfate	$CaSO_4$	136.1	CL	M	2.96	1450		1.576	0.24^{25}
– Sulfate	$CaSO_4 \cdot 2H_2O$	172.2	CL	M	2.32	128 d		1.5226	d
– Sulfide	CaS	72.1	CL	C	2.59	d		2.137	
CARBON									
– Dioxide	CO_2	44.0	CL	GAS	1.107^{-37}	−56.5/5 at	−78.5 s		90 cm^3 20
– Disulfide	CS_2	76.1	CL	LIQ	1.261	−111.6	46.3	1.6290	0.18^{20}
– Monoxide	CO	28.0	CL	GAS	0.793	−205.1	−191.5		3.5 cm^3 0
– Oxybromide	$COBr_2$	187.8	CL	LIQ	2.44	−80	65		d
– Oxychloride	$COCl_2$ (Phosgene)	98.9	CL	GAS	1.38	−127.8	7.6		d
– Oxysulfide	COS	60.1	CL	GAS	1.24^{-87}	−138.2	−50.2		56 cm^3 0
CERIUM									
– Bromide III	$CeBr_3$	380.0	CL	H	5.18	733	1705		sl
– Chloride III	$CeCl_3$	246.5	W	H	3.92^{20}	817	1730		100^{20}
– Fluoride III	CeF_3	197.1	Y	H	6.16	1430	2300		i
– Iodate IV	$Ce(IO_3)_4$	839.7	Y	R	2.27	766			0.015^{20}
– Iodide III	CeI_3	520.8	Y	T	4.83	973	1397		sl
– Molybdate III	$Ce_2(MoO_4)_3$	760.0	CL			100	200 d	2.01	vsl
– Nitrate III	$Ce(NO_3)_3 \cdot 6H_2O$	434.2				(−3H$_3$O)			
– Oxide III	Ce_2O_3	328.2	GN	H	6.86	1690			i
– Oxide IV	CeO_2	172.1	W	C	7.13	2730			i
– Sulfate III	$Ce_2(SO_4)_3$	568.4	CL	M/R	3.91	920 d			10^0
– Sulfide	Ce_2S_3	376.4	Y	C	5.19	2060 d			i
CESIUM									
– Bromide	$CsBr$	212.8	CL	C	4.44	636	1300	1.642	124^{25}
– Carbonate	Cs_2CO_3	325.8	CL			610 d			261^{20}
– Chloride	$CsCl$	168.4	CL	C	3.99	645	1300	1.534;1.642	161^0
– Fluoride	CsF	151.9	CL	C	4.12	683	1250	1.48·1.58	370^{18}

TABLE 4.2 (*Continued*)

Compound	Formula	Molecular Weight	Color	Crystal Symmetry	Specific Gravity	Melting Point (°C)	Boiling Point (°C)	Refractive Index n_D	Solubility (g/100 g of H_2O)
CESIUM (continued)									
–Hydroxide	CsOH	149.9	W	C	3.68	273	1280		385[15]
–Iodide	CsI	259.8	BK	R	4.51	626		1.661; 1.788	44[0]
–Iodide III	CsI$_3$	513.7	BK	H	4.47	208			0.81[16]
–Nitrate	CsNO$_3$	194.9	W		3.69	414	d	1.55	23[20]
–Oxide	Cs$_2$O	281.8	R	R	4.25	400 d			vsl
–Perchlorate	CsClO$_4$	232.4	CL	R	3.33[4]	250 d		1.479	1.6[20]
–Periodate	CsIO$_4$	323.8	W	R	4.26				2.1[15]
–Peroxide	Cs$_2$O$_2$	297.8	Y	R	4.47	597			sl
–Sulfate	Cs$_2$SO$_4$	361.9	CL	R	4.23	1010		1.564	167[0]
–Superoxide	CsO$_2$	164.9	Y		3.77	450	d		d
–Trioxide	Cs$_2$O$_3$	313.8	B	C	4.25	400			d
CHLORINE									
–Dioxide	ClO$_2$	67.5	Y	GAS	~1.5	–59	11.0		
–Fluoride	ClF	54.5	CL	GAS	1.67[-108]	–155.6	–100.1		
–Trifluoride	ClF$_3$	92.5	CL	GAS	1.77[13]	–76.3	11.8		d
–Monoxide	Cl$_2$O	86.9	B	GAS	3.01	–11.6	3.8		
–Hydrochloric Acid	HCl	36.5	CL	GAS	1.19[-85]	–114.2	–85.1		72[20]
–Perchloric Acid	HClO$_4$	100.5	CL	LIQ	1.77	–112	39/56	1.254[10.5]	vsl
CHROMIUM									
–Bromide II	CrBr$_2$	211.8	W	M	4.36	842			
–Carbide III	Cr$_3$C$_2$	180.0	G	R	6.68	~1890			
–Chloride II	CrCl$_2$	122.9	W	R	2.88	815	1300		i
–Chloride III	CrCl$_3$	158.4	V	R	2.76	1152	1300 s		
–Fluoride II	CrF$_2$	90.0	GN	M	4.11	894	>1300		
–Fluoride III	CrF$_3$	109.0	GN	R	3.78	1100			i
–Iodide II	CrI$_2$	305.8	B	M	5.20	868			sl
–Nitrate III	Cr(NO$_3$)$_3$	238.0	GN			60 d			sl
–Nitride III	CrN	66.0		C	6.14	~1500 d			i
–Oxide II	CrO	68.0	BK	H		1550			i
–Oxide III	Cr$_2$O$_3$	152.0	GN	H	5.21	2300	d	2.551	i
–Oxide IV	CrO$_2$	84.0	B						
–Oxide VI	CrO$_3$	100.0	RD	R	2.70	198 d			168[20]
–Phosphate III	CrPO$_4 \cdot 6H_2O$	255.1	V	TR	2.12	100 (-3.5H$_2$O)			vsl

TABLE 4.2 (*Continued*)

Compound	Formula	Molecular Weight	Color	Crystal Symmetry	Specific Gravity	Melting Point (°C)	Boiling Point (°C)	Refractive Index n_D	Solubility (g/100 g of H_2O)
CHROMIUM (continued)									
–Sulfate III	$Cr_2(SO_4)_3 \cdot 18H_2O$	716.5	V	C	1.86	100 $(-12H_2O)$		1.564	122^{20}
–Sulfide II	CrS	84.1	BK	M	4.09	1560			i
–Sulfide III	Cr_2S_3	200.2	B	TG	3.97	1350 s			i
COBALT									
–Bromide II	$CoBr_2$	218.8	GN	H	4.91	678			67^{20}
–Chlorate II	$Co(ClO_3)_2 \cdot 6H_2O$	333.9	R	C	1.92	61	100 d	1.55	185^0
–Chloride II	$CoCl_2$	129.8	BE	H	3.36	735	1049		45^7
–Fluoride II	CoF_2	96.9	R	M	4.46	~1200	1400		1.4^{20}
–Fluoride III	CoF_3	115.9	B	H	3.88	d			d
–Hydroxide II	$Co(OH)_2$	92.9	R	R	3.60	d			0.00032^{18}
–Iodate II	$Co(IO_3)_2$	408.7	V		5.00	200 d			1.2^{20}
–Iodide II	CoI_2	312.7	BK	H	5.68	515	570		159^0
–Nitrate II	$Co(NO_3)_2 \cdot 6H_2O$	291.0	R	M	1.87	56 d	56 $(-3H_2O)$		100^{20}
–Oxide II	CoO	74.9	GN	C	6.43	1935			i
–Oxide III	Co_2O_3	165.9	B	R	5.18	900 d	2800 d		i
–Oxide II-III	Co_3O_4	240.8	BK	C	6.07	480 d	900 d	1.50	i
–Perchlorate II	$Co(ClO_4)_2$	257.8	R		3.33				292^{20}
–Sulfate II	$CoSO_4$	155.0	BE	C	3.71	735 d			40^{25}
–Sulfate II	$CoSO_4 \cdot 7H_2O$	281.1	R	M	1.95	96.8 d	420 d	1.48	36^{20}
–Sulfide II	CoS	91.0	R	H	5.95	1100			0.00038^{18}
–Sulfide III	Co_2S_3	214.1	BK		4.8				i
COPPER									
–Bromide I	$CuBr$	143.5	W	C	5.05	483	1345		0.0036^{25}
–Bromide II	$CuBr_2$	223.4	BK	M	4.72	498	600 d		56^{25}
–Carbonate, Basic II	$2CuCO_3 \cdot Cu(OH)_2$	344.7	BE	M	3.88	220 d		1.73, 1.84	i
–Chloride I	$CuCl$	99.0	W	C	4.14	430	1490		0.0062^{20}
–Chloride II	$CuCl_2$	134.5	Y	M	3.39	620	993 d		71^0
–Chloride II	$CuCl_2 \cdot 2H_2O$	170.5	Y	R	2.54	100 $(-2H_2O)$			110^0
–Fluoride II	$CuF_2 \cdot 2H_2O$	137.6	W	M	4.23	950 d			4.7^{20}
–Hydroxide I	$CuOH$	80.6	Y			360 $(-\frac{1}{2}H_2O)$			i
–Hydroxide II	$Cu(OH)_2$	97.6	BE		3.37	d			i

TABLE 4.2 *(Continued)*

Compound	Formula	Molecular Weight	Color	Crystal Symmetry	Specific Gravity	Melting Point (°C)	Boiling Point (°C)	Refractive Index n_D	Solubility (g/100 g of H_2O)
COPPER (continued)									
–Iodide I	CuI	190.5	W	C	5.62	605	1290	2.346	0.00008[18]
–Nitrate II	$Cu(NO_3)_2 \cdot 3H_2O$	241.6	BE		2.32	114.5	170 d		138[0]
–Oxide I	Cu_2O	143.1	R	C	6.0	1229	1800 d	2.705	i
–Oxide II	CuO	79.5	BK		6.32	1447		2.63	i
–Sulfate II	$CuSO_4$	159.6	W	R	3.60	650 d			14[0]
–Sulfate II	$CuSO_4 \cdot 5H_2O$	249.7	BE	TR	2.28	110 ($-4H_2O$)	250 ($-5H_2O$)	1.52	22[25]
–Sulfide I	Cu_2S	159.1	BK	C	5.6	1100			i
–Sulfide II	CuS	95.6	BK	H	4.64	103 t	220 d		0.000033[18]
–Thiocyanate I	CuSCN	121.6	W		2.85	1084			0.0005[18]
CURIUM									
–Bromide III	$CmBr_3$	488	W	R	6.87	400			
–Chloride III	$CmCl_3$	353	W	H	5.81	500			
–Fluoride III	CmF_3	304	B	H	9.70	1406			
–Fluoride IV	CmF_4	323	W	M	7.49				
–Iodide III	CmI_3	628		H	6.37				
DYSPROSIUM									
–Bromide	$DyBr_3$	402.3	CL	R	4.78	879	1480		
–Chloride	$DyCl_3$	268.9	Y	M	3.67	647	1630		
–Fluoride	DyF_3	219.5	CL	H	7.46	1154	2227		
–Iodide	DyI_3	543.2	GN	H	3.21	978	1320		
–Nitrate	$Dy(NO_3)_3 \cdot 5H_2O$	438.6	Y	TR		88.6 d			
–Oxide	Dy_2O_3	373.0	W	C	8.15	2340			
–Sulfate	$Dy_2(SO_4)_3 \cdot 8H_2O$	757.3	Y	M		110 d			5.1[20]
ERBIUM									
–Bromide	$ErBr_3$	407.1	V	R	4.93	923	1460		sl
–Chloride	$ErCl_3$	273.6	V	M		776	1500		
–Fluoride	ErF_3	224.3	RD	R	7.81	1140	2200		i
–Iodide	ErI_3	548.0	V	H	3.28	1015	1280		sl
–Oxide	Er_2O_3	382.6	R	C	8.64	2400			0.00049[20]
–Sulfate	$Er_2(SO_4)_3$	622.7	W		3.68	630 d			43[0]
–Sulfide	Er_2S_3	263.5	R	M	6.21	1730			sl

TABLE 4.2 (Continued)

Compound	Formula	Molecular Weight	Color	Crystal Symmetry	Specific Gravity	Melting Point (°C)	Boiling Point (°C)	Refractive Index n_D	Solubility (g/100 g of H_2O)
EUROPIUM									
–Bromide II	$EuBr_2$	311.8	G	R	5.40	702	1873		sl
–Bromide III	$EuBr_3$	391.7	W	R		705	d		sl
–Chloride II	$EuCl_2$	222.9	W	R		731	2030		sl
–Chloride III	$EuCl_3$	258.3	Y	H	4.89	830 d	d		i
–Fluoride II	EuF_2	190.0	Y	C	6.50	1371	2527		i
–Fluoride III	EuF_3	209.0	W	R	6.79	1276	2280		sl
–Iodide II	EuI_2	405.8	GN	M	5.5	580	1580		sl
–Iodide III	EuI_3	532.7	R	C		880	d		
–Oxide III	Eu_2O_3	351.9	R	M	7.42	2330			2.5^{20}
–Sulfate III	$Eu_2(SO_4)_3 \cdot 8H_2O$	736.2			375 ($-8H_2O$)				
FLUORINE									
–Dioxide	F_2O_2	70.0	B	GAS	1.45^{-57}	-163.5	-57		vsl
–Hydride	HF	20.0	CL	GAS	$0.991^{-19.9}$	-83.1	19.9		$6.8\ cm^{3\,0}$
–Oxide	F_2O	54.0	CL	GAS	1.90^{-224}	-224	-145.3		
GADOLINIUM									
–Bromide	$GdBr_3$	397.0	W	H	4.57	770	1490		
–Chloride	$GdCl_3$	263.6	W	H	4.52^0	602	1580		
–Fluoride	GdF_3	214.3	W	R	7.05	1231	2277		
–Iodide	GdI_3	538.0	Y	H	3.14	925	1340		
–Nitrate	$Gd(NO_3)_3 \cdot 6H_2O$	451.4		T	2.33	91			
–Oxide	Gd_2O_3	362.5	W	C	7.41	2395			
–Sulfate	$Gd_2(SO_4)_3$	602.7	CL		4.14				2.8^{20}
–Sulfide	Gd_2S_3	410.7	Y	C	6.15	1885			
GALLIUM									
–Arsenide III	GaAs	144.6	G	C	5.35	1237			sl
–Bromide III	$GaBr_3$	309.5	CL		3.69	122	279		d
–Chloride II	Ga_2Cl_4	281.3	W			171	200 d		vsl
–Chloride III	$GaCl_3$	176.0	CL	TR	2.47	77.9	201		0.0024^{25}
–Fluoride III	GaF_3	126.7	W	RH	4.47	950			i
–Iodide III	GaI_3	450.4	Y		4.15	212	346 s		i
–Oxide I	Ga_2O	155.4	G		4.77	>600	>500 s		
–Oxide III	Ga_2O_3	187.4	G	M (β)	5.88	1740		1.95	

TABLE 4.2 (Continued)

Compound	Formula	Molecular Weight	Color	Crystal Symmetry	Specific Gravity	Melting Point (°C)	Boiling Point (°C)	Refractive Index n_D	Solubility (g/100 g of H_2O)
GALLIUM (continued)									
–Sulfide I	Ga_2S	171.5	G		4.2	900 d	1530		d
–Sulfide III	Ga_2S_3	235.6	Y	H	3.7	1255			d
GERMANIUM									
–Bromide IV	$GeBr_4$	392.2	G	LIQ	3.13	26.1	187.1	1.627	d
–Chloride IV	$GeCl_4$	214.4	CL		1.87	–49.5	83.1	1.464	d
–Fluoride IV	GeF_4	148.6	CL	GAS	2.46^{-37}	–15	–36.5		d
–Hydride IV	GeH_4 (Germane)	76.6	CL	GAS	1.52^{-142}	–165.9	–88.4	1.00089	i
–Iodide IV	GeI_4	580.2	R	C	4.32	144.0	440 d		0.0002^{25}
–Oxide II	GeO	88.6	G		1.83	710 d			0.40^{20}
–Oxide IV	GeO_2	104.6	CL	H	4.70	1100		1.607	0.24^{20}
–Sulfide II	GeS	104.7	Y	R	4.01	530	430 s		0.45^{20}
–Sulfide IV	GeS_2	136.7	W	R	2.94^{14}	~800			
GOLD									
–Bromide I	$AuBr$	276.9	G		7.90	115 d			d
–Bromide III	$AuBr_3$	436.7	B			160 d			
–Chloride I	$AuCl$	232.4	Y	R	7.4	289 d			
–Chloride III	$AuCl_3$	303.3	R		3.9	180 s	229		68^{20}
–Hydroxide III	$Au(OH)_3$	248.0	B			100 d			i
–Iodide	AuI	323.9	Y	TR	8.25	120 d			
–Iodide III	AuI_3	577.7	G						
–Sulfate III	$Au_2(SO_4)_3 \cdot H_2O$	490.5	B						
–Sulfide I	Au_2S	426.0	B			240 d			i
–Sulfide III	Au_2S_3	490.1	B		8.75	197 d			sl
HAFNIUM									
–Bromide	$HfBr_4$	498.1	W			424.5	322 s		
–Carbide	HfC	190.5		C	12.7	3940			i
–Chloride	$HfCl_4$	320.3	W			319 s			d
–Fluoride	HfF_4	254.5	CL	M	7.13			1.56	
–Iodide	HfI_4	686.1				449			
–Nitride	HfN	192.5	Y	C	13.9	~3300			i
–Oxide	HfO_2	210.5	W	T	10.0	2812	5127		
–Sulfide	HfS_2	242.6		H	6.0				i

TABLE 4.2 (*Continued*)

Compound	Formula	Molecular Weight	Color	Crystal Symmetry	Specific Gravity	Melting Point (°C)	Boiling Point (°C)	Refractive Index n_D	Solubility (g/100 g of H₂O)
HOLMIUM									
-Bromide	HoBr₃	404.7	Y	R	4.86	919	1470		sl
-Chloride	HoCl₃	271.3	Y	M		720	1510		sl
-Fluoride	HoF₃	221.9	B	H	7.83	1143	2227		i
-Iodide	HoI₃	545.6	Y		3.24	994	1300		sl
-Oxide	Ho₂O₃	377.9		C	8.35	2396			i
HYDROGEN									
-Bromide	HBr	80.9	CL	GAS	2.16⁻⁶⁸	-86.8	-66.7		70²⁰
-Chloride	HCl	36.5	CL	GAS	1.19⁻⁸⁵	-114.2	-85.1		72²⁰
-Fluoride	HF	20.0	CL	GAS	0.991¹⁹·ᵖ	-83.1	19.9		76²⁰
-Iodide	HI	127.9	CL	GAS	2.80⁻³⁵	-50.8	-35.4		71²⁰
-Oxide	H₂O	18.0	CL	LIQ	1.00	0.0	100.0	1.466	
-Oxide-Deutero	2H₂O	20.0	CL	LIQ	1.104	3.82	101.43	1.3333	
-Peroxide	H₂O₂	34.0	CL	LIQ	1.442	-2.0	150.2	1.3284	
-Selenide	H₂Se	81.0	CL	GAS	2.12⁻⁴²	-65.7	-41.3	1.414²²	0.68²⁵
-Sulfide	H₂S	34.1	CL	GAS	0.96⁻⁶⁰	-85.5	-60.3	1.374	0.33²⁵
-Telluride	H₂Te	129.9	CL	GAS	2.57⁻²⁰	-51	-2.3		vsl
INDIUM									
-Bromide I	InBr	194.7	B		4.98	220	928		d
-Bromide III	InBr₃	354.5	CL		4.75	436	s		85²⁵
-Chloride I	InCl	150.3	R	C	4.2	225	608		d
-Chloride III	InCl₃	221.2	CL	M	3.46	586	s		198²²
-Fluoride III	InF₃	171.8	CL	H	4.39	1170	1200		0.040¹⁵
-Iodide I	InI	241.7	B		5.31	351	715		
-Iodide III	InI₃	495.5	Y	M	4.69	210			1330²²
-Oxide III	In₂O₃	277.6	Y	C	7.18	2000 d			i
-Sulfate III	In₂(SO₄)₃	517.8	W	M	3.44	1050	d		i
-Sulfide III	In₂S₃	325.8	R (β)	C	4.90				
IODINE									
-Bromide I	IBr	206.8	BK	OR	4.42	42	116		d
-Chloride I, α	ICl	162.4	R	C	3.18	27.2	101		d
-Chloride I, β	ICl	162.4	R	LIQ	3.24	13.9	101		d
-Chloride III	ICl₃	233.3	Y	R	3.19	33	77 d		d
-Fluoride V	IF₅	221.9	CL	LIQ	3.5	9.4	100.5		d

TABLE 4.2　(Continued)

Compound	Formula	Molecular Weight	Color	Crystal Symmetry	Specific Gravity	Melting Point (°C)	Boiling Point (°C)	Refractive Index n_D	Solubility (g/100 g of H_2O)
IODINE (continued)									
—Fluoride VII	IF_7	259.9	CL	GAS	2.8^6	4.5	5.5		d
—Oxide IV	I_2O_4	317.8	Y		4.2	130 d			d
—Oxide V	I_2O_5	333.8	CL		4.80	300 d			187^{13}
—Iodic Acid	HIO_3	175.9	W	R	4.63	110 d		1.466	269^{20}
—Hydrogen Iodide	HI	127.9	CL	GAS	2.80^{-35}	−50.8	−35.4		71^{20}
IRIDIUM									
—Bromide III	$IrBr_3 \cdot 4H_2O$	504.0	GN			100 ($-3H_2O$)			vsl
—Bromide IV	$IrBr_4$	511.8	BK			d			d
—Chloride III	$IrCl_3$	298.6	GN	H	5.30	763 d			i
—Chloride IV	$IrCl_4$	334.0	R	C					sl
—Fluoride VI	IrF_6	306.2	Y	T	6.0	44.4	53		d
—Iodide III	IrI_3	572.9	GN			d			sl
—Iodide IV	IrI_4	699.8	BK			100 d			i
—Oxide IV	IrO_2	224.2	BK		11.7	1100 d			0.0002^{20}
—Sulfide IV	IrS_2	256.3	BK		8.43	300 d			i
IRON									
—Arsenide	$FeAs$	130.8	W	R	7.83	1030			i
—Arsenide, di-	$FeAs_2$	205.7	G	R	7.38	919			i
—Bromide II	$FeBr_2$	215.7	GN	H	4.64	684	967		54^{25}
—Bromide III	$FeBr_3 \cdot 6H_2O$	403.7	R			27			sl
—Carbide	Fe_3C	179.6	G	C	7.4	1837			
—Carbonate II	$FeCO_3$	115.9	G		3.84	d			0.0007^{25}
—Chloride II	$FeCl_2$	126.8	G	H	2.98	677	1023		68.5^{20}
—Chloride III	$FeCl_3$	162.2	GN	H	2.90	303	317		92^{20}
—Fluoride III	FeF_3	112.9	W	R	3.18	1030 d			0.09^{125}
—Hydroxide II	$Fe(OH)_2$	89.9	GN	H	3.4	d			i
—Hydroxide III	$Fe(OH)_3$	106.9	B		3.9	500 d			0.0048^{18}
—Iodide II	FeI_2	309.7	BK	H	5.31	592	935		83.8^{20}
—Nitrate II	$Fe(NO_3)_2 \cdot 6H_2O$	288.0	GN	R		60.5 d			87.2^{25}
—Nitrate III	$Fe(NO_3)_3 \cdot 9H_2O$	404.0	CL	M	1.68	47	125 d		i
—Nitride	Fe_2N	125.7	G		6.35	200 d			i
—Oxide II	FeO	71.9	BK	C	6.04	1420	d	2.32	i
—Oxide III	Fe_2O_3	159.7	B	TG	5.25	1565 d		3.04	i

TABLE 4.2 (*Continued*)

Compound	Formula	Molecular Weight	Color	Crystal Symmetry	Specific Gravity	Melting Point (°C)	Boiling Point (°C)	Refractive Index n_D	Solubility (g/100 g of H_2O)
IRON (continued)									
– Oxide II–III	$Fe_3O_4 \cdot 2H_2O$	231.6	BK	C	5.21	1597		2.42	i
– Phosphate III	$FePO_4 \cdot 2H_2O$	186.9	W	M	2.87	d		1.35	i
– Phosphide	Fe_2P	142.7	G	H	6.56	1360			
– Sulfate II	$FeSO_4 \cdot 7H_2O$	278.0	GN	M	1.90	64 d	90 d	1.48	26.6^{20}
– Sulfate III	$Fe_2(SO_4)_3$	399.9	Y	R	3.10	480 d		1.81	26.9^{20}
– Sulfate II, Ammonium	$(NH_4)_2Fe(SO_4) \cdot 6H_2O$	392.2	GN	M	1.86	d		1.49	
– Sulfide II	FeS	87.9	BK	H	4.76	1195	d		0.00044^{18}
– Sulfide III	Fe_2S_3	207.9	BK	H	4.3	d			d
– Sulfide, di	FeS_2	120.0	Y	C	5.00	1171	d		0.0005^{18}
LANTHANUM									
– Bromate	$La(BrO_3)_3 \cdot 9H_2O$	684.8	W	H		37.5	100 $(-7H_2O)$		28.5^{18}
– Bromide	$LaBr_3$	378.6	W	H	5.07	789	1580		
– Chloride	$LaCl_3$	245.3	W	H	3.84	862	1700		
– Fluoride	LaF_3	195.9	W	H	5.94	1493	1405		0.0018^{25}
– Iodide	LaI_3	519.6	G	R	2.25	772			
– Molybdate	$La_2(MoO_4)_3$	757.6	W	T	4.77	1181			0.0004^{29}
– Oxide	La_2O_3	325.8	W	R	6.51	2315			3^0
– Sulfate	$La_2(SO_4)_3$	566.0			3.60	1150 d			
– Sulfide	La_2S_3	374.0	Y	H	4.91	~2125	4200		d
LEAD									
– Acetate II	$Pb(C_2H_3O_2)_2$	325.3	W	M	3.25	280	d		44^{20}
– Acetate IV	$Pb(C_2H_3O_2)_4$	443.4	CL		2.23	175	d		d
– Arsenate II	$Pb_3(AsO_4)_2$	899.4	W		7.80	1042			i
– Bromide II	$PbBr_2$	367.0	W	R	6.67	373	916		0.84^{20}
– Carbonate II	$PbCO_3$	267.2	CL	R	6.60	300 d		1.80–2.08	0.00015^{20}
– Chloride II	$PbCl_2$	278.1	W	R	5.85	498	956	2.22	0.97^{20}
– Chloride IV	$PbCl_4$	349.0	Y	LIQ	3.18	−15	105 d		d
– Chromate II	$PbCrO_4$	323.2	Y	M	6.12	844	d	2.33	i
– Fluoride II	PbF_2	245.2	CL	R	8.37	835	1290		0.064^{20}
– Hydroxide II	$Pb(OH)_2$	241.2	W	H		145 d			0.016^{25}
– Iodate II	$Pb(IO_3)_2$	557.0	W		6.16	300 d			0.0012^2
– Iodide II	PbI_2	461.0	Y	H	6.16	400	872	2.30	0.044^0
– Molybdate II	$PbMoO_4$	367.2	CL	T	6.92	1065			i

TABLE 4.2 (*Continued*)

Compound	Formula	Molecular Weight	Color	Crystal Symmetry	Specific Gravity	Melting Point (°C)	Boiling Point (°C)	Refractive Index n_D	Solubility (g/100 g of H_2O)
LEAD (continued)									
-Nitrate II	$Pb(NO_3)_2$	331.2	CL	C	4.53	470 d		1.782	56.5[20]
-Oxide II	PbO	223.2	R	T	9.36	890	1470		0.0052[25]
-Oxide IV	PbO_2	239.2	B	T	9.38	290 d			i
-Oxide II-IV	Pb_3O_4	685.6	R	T	9.1	500 d			i
-Phosphate, III	$Pb_3(PO_4)_2$	811.6	W	H	6.99	1014		1.95	i
-Sulfate II	$PbSO_4$	303.3	W	R	6.34	1084		1.85	0.004[15]
-Sulfide II	PbS	239.3	BK	C	7.58	1114	1290	3.911	i
-Tungstate II	$PbWO_4$	455.1	CL	M	8.46	1123			0.03[25]
LITHIUM									
-Aluminum Hydride	$LiAlH_4$	37.9	W		0.917	150 d			d
-Bromide	LiBr	86.9	W	C	3.47	552	1265	1.784	156[20]
-Carbonate	Li_2CO_3	73.9	W	M	2.11	732	1310 d	1.43; 1.57	1.33[20]
-Chloride	LiCl	42.4	W	C	2.07	610	1382	1.662	45.8[25]
-Fluoride	LiF	25.9	W	C	2.63	847	1681	1.391	0.13[25]
-Hydride	LiH	8.0	CL	C	0.82	692	1000 d		d
-Hydroxide	LiOH	24.0	W	T	2.54	462	924 d	1.46	12.4[25]
-Iodide	LiI	133.9	W	C	4.06	450	1171	1.955	163[20]
-Nitrate	$LiNO_3$	68.9	W	TG	2.38	252	600 d	1.435; 1.737	69.5[20]
-Oxide	Li_2O	29.9	W	C	2.10	1730		1.644	d
-Peroxide	Li_2O_2	45.9		H	2.36	425			
-Perchlorate	$LiClO_4$	160.4	W	H	2.43	236	430 d		60[20]
-Phosphate	Li_3PO_4	115.8	CL	R	2.54	~857			0.03[20]
-Sulfate, α	Li_2SO_4	109.9	CL	M	2.21	857		1.465 β	35[0]
-Sulfide	Li_2S	45.9	W	C	1.66	900			
LUTETIUM									
-Bromide	$LuBr_3$	414.7	W	TG	5.17	960	1410		sl
-Chloride	$LuCl_3$	281.3	W	M	3.98	892	1480		sl
-Fluoride	LuF_3	232.0	W	R	8.33	1182	2200		i
-Iodide	LuI_3	555.7	B	H	3.39	1050	1210		sl
-Oxide	Lu_2O_3	397.9		C	9.42				
MAGNESIUM									
-Aluminate	$MgO \cdot Al_2O_3$	142.3	CL	C	3.6	2115		1.723	i
-Bromide	$MgBr_2$	184.1	W	H	3.72	711	1230		138[17]

TABLE 4.2 (*Continued*)

Compound	Formula	Molecular Weight	Color	Crystal Symmetry	Specific Gravity	Melting Point (°C)	Boiling Point (°C)	Refractive Index n_D	Solubility (g/100 g of H$_2$O)
MAGNESIUM (continued)									
–Carbonate	$MgCO_3$	84.3	W	TG	3.04	350 d	900 ($-CO_2$)	1.51;1.70	0.011^{25}
–Chloride	$MgCl_2$	95.2	W	H	2.33	712	1418	1.59; 1.675	54^{20}
–Fluoride	MgF_2	62.3	CL	T	3.13	1265	2230	1.38	0.0091^{18}
–Hydroxide	$Mg(OH)_2$	58.3	CL	H	2.36	350 d		1.57	0.0009^{18}
–Iodide	MgI_2	278.2	W	H	4.2	700 d			148^{18}
–Nitrate	$Mg(NO_3)_2 \cdot 6H_2O$	256.4	CL	M	1.46	90	330 d		70.5^{20}
–Oxide	MgO	40.3	CL	C	3.65	2800	3600	1.736	0.0006^{20}
–Silicide	Mg_2Si	76.7	BE	C	1.94	1110			i
–Silicate, m	$MgSiO_3$	100.4	W	M	3.18	1524		1.66	i
–Silicate, o	Mg_2SiO_4	140.7	W	R	3.21	1910		1.65	i
–Sulfate	$MgSO_4$	120.4	CL	R	2.66	1127			27^{0}
–Sulfide	MgS	56.4	R	C	2.80	>2000 d		2.271	d
MANGANESE									
–Bromide II	$MnBr_2$	214.8	W	H	4.38	698			127^{0}
–Carbonate II	$MnCO_3$	114.9	W	R	3.12	d		1.817	0.0065^{25}
–Chloride II	$MnCl_2$	125.9	W	H	2.98	650	1190		72^{25}
–Fluoride II	MnF_2	92.9	R	T	3.98	930			0.66^{40}
–Iodide II	MnI_2	308.8	W	H	5.0	80 d			sl
–Oxide II	MnO	70.9	GN	C	5.44	80 d		2.16	i
–Oxide III	Mn_2O_3	157.9	BK	C	4.50	1080 d			i
–Oxide IV	MnO_2	86.9	BK	R	5.03	535 d			i
–Oxide II–IV	Mn_3O_4	228.8	BK	R	4.86	1705			i
–Potassium Permanganate	$KMnO_4$	158.0	P	R	2.70	<240 d		1.59	6.34^{20}
–Silicide	$MnSi$	83.0	R	C	5.90	1280			i
–Sulfate II	$MnSO_4$	151.0	R		3.25	700	850 d		52^{5}
–Sulfide II	MnS	87.0	GN	C	3.99	1530			0.0005^{18}
MERCURY									
–Bromide I	Hg_2Br_2	561.1	W	T	7.31	405	350 s		i
–Bromide II	$HgBr_2$	360.4	CL	R	5.92	237	320	1.97; 2.66	0.62^{25}
–Chloride I	Hg_2Cl_2	472.1	W	T	7.15	525 d	384 s	1.72; 1.97	0.00022^{20}
–Chloride II	$HgCl_2$	271.5	CL	R	5.53	280	303	1.645	6.6^{20}
–Cyanide II	$Hg(CN)_2$	252.7	CL	T	4.00	d			11.3^{25}
–Fluoride I	Hg_2F_2	439.2	Y	C	8.73	570	d		d

TABLE 4.2 *(Continued)*

Compound	Formula	Molecular Weight	Color	Crystal Symmetry	Specific Gravity	Melting Point (°C)	Boiling Point (°C)	Refractive Index n_D	Solubility (g/100 g or H_2O)
MERCURY (continued)									
– Fluoride II	HgF_2	238.6	CL	C	8.95	645	650		d
– Iodide I	Hg_2I_2	655.0	Y	T	7.70	~290	310 d		i
– Iodide II	HgI_2	454.4	R/Y	T/R	6.27	259	354	2.45; 2.75	i
– Nitrate I	$Hg_2(NO_3)_2 \cdot 2H_2O$	561.2	CL	M	4.79	70 d	d		d
– Nitrate II	$Hg(NO_3)_2 \cdot \frac{1}{2}H_2O$	333.6	W		4.39	79	d		
– Oxide I	Hg_2O	417.2	BK		9.8	100 d			
– Oxide II	HgO	216.6	Y/R	R	11.1	500 d		2.37; 2.65	0.005[25]
– Sulfate I	Hg_2SO_4	497.3	CL	M	7.56	d			0.06[25]
– Sulfate II	$HgSO_4$	296.7	CL	R	6.47	850 d			i
– Sulfide II	HgS	232.7	R	H	8.10	583 s		2.85; 3.20	i
MOLYBDENUM									
– Carbide II	Mo_2C	203.9	W	H	8.9	2687			i
– Carbide IV	MoC	108.0	G	H	8.40	2840	5550		i
– Chloride II	$MoCl_2$	166.9	Y		3.71	d			i
– Chloride III	$MoCl_3$	202.3	R		3.58	1027			i
– Chloride V	$MoCl_5$	273.2	BK	M	2.93	194	268		d
– Fluoride VI	MoF_6	202.9	CL		2.55	17.5	35		d
– Iodide II	MoI_2	349.8	B		5.28				d
Molybdic Acid	$H_2MoO_4 \cdot 4H_2O$	180.0	Y	M	3.12	70 $(-H_2O)$	d		0.13[18]
– Oxide IV	MoO_2	127.9	G	T	6.47				
– Oxide VI	MoO_3	143.9	CL	R	4.50	795	1280 s		0.18[23]
– Silicide IV	$MoSi_2$	152.1	G	T	6.31	2010			
– Sulfide IV	MoS_2	160.1	BK	H	4.80	1185		4.7	0.11[18]
NEODYMIUM									
– Bromide	$NdBr_3$	384.0	V	R	5.35	682	1540		97[13]
– Chloride	$NdCl_3$	250.6	V	H	4.17	758	1600		
– Fluoride	NdF_3	201.2	V	H		1374	2300		i
– Iodide	NdI_3	524.9	G	R	2.34	784	1370		
– Oxide	Nd_2O_3	336.5	BE	H	7.24	2310			0.0002[20]
– Sulfide	Nd_2S_3	384.7	GN		5.18	2010			i
NEPTUNIUM									
– Bromide II	$NpBr_3$	476.7	GN	R	6.62	800 s			

TABLE 4.2 (*Continued*)

Compound	Formula	Molecular Weight	Color	Crystal Symmetry	Specific Gravity	Melting Point (°C)	Boiling Point (°C)	Refractive Index n_D	Solubility (g/100 g of H$_2$O)
NEPTUNIUM (continued)									
-Chloride III	NpCl$_3$	343.4	GN	H	5.58	~800			
-Chloride IV	NpCl$_4$	378.8	BN	T	4.92	538			
-Fluoride III	NpF$_3$	294.0	P	H	9.12	1425			
-Fluoride VI	NpF$_6$	351.0	O	R	5.00	55.1	55.2		i
-Iodide III	NpI$_3$	617.7	B	R	6.82	770			d
-Oxide IV	NpO$_2$	269.0	GN	C	11.1				i
NICKEL									
-Arsenide	NiAs	133.6	W	H	7.57	968			i
-Bromide II	NiBr$_2$	218.5	Y		4.64	700 s			25[25]
-Carbonyl	Ni(CO)$_4$	170.7	CL	LIQ	1.32	-25	42	1.458[10]	0.018[9]
-Chloride II	NiCl$_2$	129.6	Y	H	3.55	1001			68[25]
-Fluoride II	NiF$_2$	96.7	Y	T	4.63	1450			4[25]
-Hydroxide II	Ni(OH)$_2$	92.7	GN		4.15	230 d			0.001[20]
-Iodide II	NiI$_2$	312.5	BK	H	5.83	780			144[25]
-Nitrate II	Ni(NO$_3$)$_2$ · 6H$_2$O	290.8	GN	M	2.05	57 d	137		94[20]
-Oxide II	NiO	74.7	GN	C	7.45	1960	d	2.37	i
-Phosphide	Ni$_2$P	148.4	G		6.31	1112			i
-Sulfate II	NiSO$_4$	154.8	Y	C	3.68	848 d	d		38[20]
-Sulfide II	NiS	90.8	BK	TR	5.5	790			0.0004[18]
NIOBIUM									
-Bromide	NbBr$_5$	492.5	R	R	4.44	267.5	361.6		d
-Carbide	NbC	104.9	BK	C	7.82	3497			i
-Chloride	NbCl$_5$	270.2	W	M	2.75	205	247		d
-Fluoride	NbF$_5$	187.9	CL	M	3.29	80	232		d
-Iodide	NbI$_5$	727.4	BRASS	M	5.11	280 d			
-Oxide	Nb$_2$O$_5$	265.8	W	R	4.46	1490	d		i
NITROGEN									
Ammonia	NH$_3$	17.0	CL	GAS	0.681[-33]	-77.7	-33.4	1.325	52[20]
Hydrazine	N$_2$H$_4$	32.0	CL	LIQ	1.01	1.5	113.1	1.4707	vsl
Hydrazoic Acid	HN$_3$	43.0	CL	LIQ	1.09	-80	37		vsl
Hydroxylamine	NH$_2$OH	33.0	W	R	1.20	33.1	56.5	1.44[23.5]	sl
Nitric Acid	HNO$_3$	63.0	CL	LIQ	1.50	-41.6	83	1.397[16]	vsl
-Chloride	NCl$_3$	120.4	Y	LIQ	1.65	-27	71		i

TABLE 4.2 (Continued)

Compound	Formula	Molecular Weight	Color	Crystal Symmetry	Specific Gravity	Melting Point (°C)	Boiling Point (°C)	Refractive Index n_D	Solubility (g/100 g of H_2O)
NITROGEN (continued)									
-Fluoride	NF_3	71.0	CL	GAS	1.54^{-129}	-208.5	-129		i
-Iodide	NI_3	394.7	BK			*			i
-Oxide I (nitrous-)	N_2O	44.0	CL	GAS	0.784	-90.9	-84.5		0.11^{25}
-Oxide II (nitric-)	NO	30.0	CL	GAS	1.269^{-150}	-163.6	-151.8	1.193^{16}	0.0056^{25}
-Oxide III (tri-)	N_2O_3	76.0	B	GAS	1.45^2	-111	2 d		sl
-Oxide IV (per-)	NO_2	46.0	B	GAS	1.45	-11.2	21.1		sl
-Oxide V (penta-)	N_2O_5	108.0	W	R	1.64^{18}	32.4 s	47.0		sl
-Sulfide II	N_4S_4	184.3	O	M	2.24^{18}	178 d	135 s	2.046	i
Nitrosyl Chloride	$NOCl$	65.5	O	GAS	1.42^{-12}	-64.5	-6.4		d
Nitrosyl Fluoride	NOF	49.0	CL	GAS	1.80^{-72}	-132.5	-59.9		d
Nitryl Chloride	NO_2Cl	81.5	CL	GAS	1.32^{14}	-145	-15.9		d
OSMIUM									
-Chloride IV	$OsCl_4$	332.0	R	M		450 s			
-Fluoride V	OsF_5	285.2	G			70	225.7		
-Fluoride VI	OsF_6	304.2	GN	C		32.1	45.9		
-Fluoride VIII	OsF_8	342.2	Y		3.87	34.4	47.3		d
-Iodide IV	OsI_4	697.8	BK			650 d			vsl
-Oxide IV	OsO_2	222.2	BK	T	7.91	40	130.0		5.7^{16}
-Oxide VIII	OsO_4	254.1	CL	M	4.91	d			i
-Sulfide IV	OsS_2	254.3	BK	C	9.47				
OXYGEN									
-Fluoride	OF_2	54.0	B	GAS	1.90^{-224}	-223.8	-144.8		6.8 ml^0
Ozone	O_3	48.0	CL	GAS	3.03^{-80}	-251	-112.5		
PALLADIUM									
-Bromide II	$PdBr_2$	266.6	B		5.17	d	600 d		d
-Chloride II	$PdCl_2$	177.3	R	C	4.0^{18}	940			sl
-Fluoride II	PdF_2	144.4	B	T	5.80				d
-Iodide II	PdI_2	360.2	BK		6.00	350 d			i
-Oxide II	PdO	122.4	G	T	8.31	875 d			i
-Sulfide II	PdS	138.5	BK	T	6.60	950 d			i
PHOSPHORUS									
Hypophosphorous Acid	H_3PO_2	66.0	CL		1.49	26.5	100 d		sl

TABLE 4.2 (*Continued*)

Compound	Formula	Molecular Weight	Color	Crystal Symmetry	Specific Gravity	Melting Point (°C)	Boiling Point (°C)	Refractive Index n_D	Solubility (g/100 g of H_2O)
PHOSPHORUS (continued)									
Phosphoric Acid	H_3PO_4	98.0	CL	R	1.87	42.3	261		570[25]
Phosphorous Acid	H_3PO_3	82.0	CL		1.65	73.6	180 d		309[0]
– Bromide III	PBr_3	270.7	CL	LIQ	2.85[15]	−40.5	172.8	1.6945[19.5]	d
– Bromide V	PBr_5	430.5	Y	R		<100	106 d		d
– Chloride III	PCl_3	137.3	CL	LIQ	1.57	−111.8	74.2		d
– Chloride V	PCl_5	208.3	W	T	1.6	160	167 s		d
– Fluoride III	PF_3	88.0	CL	GAS		−160	−101		d
– Fluoride V	PF_5	126.0	CL	GAS		−93.7	−84.5		d
– Hydride (Phosphine)	PH_3	34.0	CL	GAS	0.746[−90]	−133.8	−87.7		0.03[25]
– Iodide III	PI_3	411.7	R	H	4.18	61	120/15		d
– Oxide III	P_4O_6	219.9	W	M	2.13	23.8	174		d
– Oxide IV	PO_2	63.0	CL	R	2.54	>100	180		d
– Oxide V	P_2O_5	142.0	W	H	2.30	585	300 s		d
– Oxybromide V	$POBr_3$	286.7	CL		2.77[82]	56	193 d		d
– Oxychloride	$POCl_3$	153.4	CL	LIQ	1.67	1.25	105.1		d
– Oxyfluoride	POF_3	104.0	CL	GAS		−39.4	−39.8		d
– Sulfide	P_4S_7	348.4	Y		2.19	310	523		d
– Sulfide V	P_2S_5	222.3	Y		2.03	280	523		d
– Thiobromide V	$PSBr_3$	302.8	Y	C	2.85	39	212 d	1.635[25]	d
– Thiochloride V	$PSCl_3$	169.4	CL	LIQ	1.63	−36.2	125		d
PLATINUM									
– Bromide II	$PtBr_2$	354.9	B	C	6.65	250 d			i
– Bromide IV	$PtBr_4$	514.8	B		5.69	180 d			0.41[20]
– Chloride II	$PtCl_2$	260.0	GN	H	5.87	581 d			i
– Chloride IV	$PtCl_4$	336.9	B		4.30	370 d			59[25]
– Fluoride IV	PtF_4	271.2	R			600 d			d
– Fluoride VI	PtF_6	309.1	R			61.3	69.1		
– Hydroxide II	$Pt(OH)_2$	229.1	BK			d			i
– Hydroxide IV	$Pt(OH)_4$	263.1	B			100 (−$2H_2O$)			
– Iodide II	PtI_2	448.9	BK	T	6.40	300 d			i
– Oxide II	PtO	211.1	G		14.9	550 d			i
– Oxide IV	PtO_2	227.1	BK		10.2	400 d			i
– Sulfate IV	$Pt(SO_4)_2 \cdot 4H_2O$	459.4	Y						sl
– Sulfide II	PtS	227.2	BK	T	10.1	d			i
– Sulfide III	Pt_2S_3	486.6	G		5.52	d			i

TABLE 4.2 (Continued)

Compound	Formula	Molecular Weight	Color	Crystal Symmetry	Specific Gravity	Melting Point (°C)	Boiling Point (°C)	Refractive Index n_D	Solubility (g/100 g of H$_2$O)
PLATINUM (continued)									
–Sulfide IV	PtS$_2$	259.2	G		7.66	225 d			i
PLUTONIUM									
–Bromide III	PuBr$_3$	481.7	GN	R	6.83	681	1463		sl
–Carbide IV	PuC	256.0	SL	C	13.5	1650			d
–Chloride III	PuCl$_3$	346.4	GN	H	5.70	760	1727		sl
–Fluoride III	PuF$_3$	299.0	P	H	9.32	1396	1957		
–Fluoride IV	PuF$_4$	318.0	B	M	7.00	1037			
–Fluoride VI	PuF$_6$	356.0	B	R	4.86	51.6	62.2		d
–Iodide III	PuI$_3$	622.7	GN	R	6.92	777			sl
–Nitride III	PuN	256.0	BK	C	14.2	1650 d			d
–Oxide IV	PuO$_2$	274.0	GN	C	11.5	2390		2.4	i
POLONIUM									
–Bromide IV	PoBr$_4$	529.7	R	C		330	360/200		d
–Chloride II	PoCl$_2$	281.0	R	R	6.50	200 d			
–Chloride IV	PoCl$_4$	351.9	Y	M		300			
–Oxide IV	PoO$_2$	242.0	R/Y	T/C	8.96	885 s	400 s		i
POTASSIUM									
–Bromate	KBrO$_3$	167.0	CL	TR	3.24	350	370 d	1.559	6.9[20]
–Bromide	KBr	119.0	CL	C	2.76	760	1380	1.559	66[20]
–Carbonate	K$_2$CO$_3$	138.2	CL	M	2.43	897	d	1.426; 1.541	112[20]
–Chlorate	KClO$_3$	122.6	CL	M	2.32	356	400 d	1.409; 1.524	7.3[20]
–Chloride	KCl	74.6	CL	C	1.98	772	1407	1.490	35[20]
–Cyanide	KCN	65.1	CL	C	1.52	634.5		1.410	72[20]
–Dichromate	K$_2$Cr$_2$O$_7$	294.2	O	M/TR	2.69	393	500 d	1.738 TR	12[20]
–Ferrocyanide	K$_4$[Fe(CN)$_6$] · 3H$_2$O	422.4	Y	M/T	1.85	70 (–3H$_2$O)	d	1.577	28[20]
–Fluoride	KF	58.1	CL	C	2.48	857	1503	1.35	50[25]
–Hydroxide	KOH	56.1	W	C/R	2.04	410	1327		111[20]
–Iodate	KIO$_3$	214.0	CL	M	3.99	560	d		8.1[20]
–Iodide	KI	166.0	W	C	3.13	682	1324	1.677	144[20]
–Nitrate	KNO$_3$	101.1	CL	R/TR	2.11	339	400 d	1.335; 1.506	31.5[20]
–Oxide	K$_2$O	94.2	CL	C	2.32	350 d			vsl
–Perchlorate	KClO$_4$	138.6	CL	R	2.52	400 d		1.47	1.7[20]
–Periodate	KIO$_4$	230.0	CL	T	3.62	582	d	1.63	0.42[20]

TABLE 4.2 (*Continued*)

Compound	Formula	Molecular Weight	Color	Crystal Symmetry	Specific Gravity	Melting Point (°C)	Boiling Point (°C)	Refractive Index n_D	Solubility (g/100 g of H_2O)
POTASSIUM (continued)									
–Permanganate	$KMnO_4$	158.0	P	R	2.70	240 d		1.59	6.4[20]
–Peroxide	K_2O_2	110.2	Y	R	2.40	490	d		d
–Phosphate, o	K_3PO_4	212.3	CL	TR	2.26	1340			78[0]
–Sulfate	K_2SO_4	174.3	CL	R/H	2.66	1069	1698	1.495	11.1[20]
–Sulfide	K_2S	110.3	B	C	1.80	840			sl
–Superoxide	KO_2	71.1	Y	T	2.14	440	d		d
–Thiocyanate	$KSCN$	97.2	CL	R	1.89	179	500 d		177[0]
PRASEODYMIUM									
–Bromide	$PrBr_3$	380.6	GN	H	5.26	691	1550		104[14]
–Chloride	$PrCl_3$	247.3	GN	H	4.02	786	1910		
–Fluoride	PrF_3	197.9	GN	H	6.14	1395	2327		
–Iodide	PrI_3	521.6	G	R	2.31	737	1380		
–Oxide	Pr_2O_3	329.8	Y	H	7.07	2300		1.55	0.000002[20]
–Sulfate	$Pr_2(SO_4)_3 \cdot 8H_2O$	714.1	GN	M	2.83				17.4[20]
–Sulfide	Pr_2S_3	378.0	B		5.24	1795			
PROTACTINIUM									
–Bromide IV	$PaBr_4$	470.9	R	T	4.72	400 s			sl
–Chloride IV	$PaCl_4$	372.9	GN	T	6.36				i
–Fluoride IV	PaF_4	307.1	B	M					
–Iodide III	PaI_3	611.8	BK	R					
–Oxide IV	PaO_2	263.1	BK	C					
RADIUM									
–Bromide	$RaBr_2$	385.8	Y	M	5.78	728			51[20]
–Chloride	$RaCl_2$	296.1	Y	M	4.91	1000	900 s		24.5[20]
–Sulfate	$RaSO_4$	322.1	CL	R					0.0002[20]
RHENIUM									
–Bromide III	$ReBr_3$	425.9	B	H		500 s	>550		sl
–Chloride III	$ReCl_3$	292.6	R		4.9	~260	~330		sl
–Chloride V	$ReCl_5$	363.5	B	T	5.38	124.5	300 s		d
–Fluoride IV	ReF_4	262.5	GN	LIQ	3.62[19]	18.8	33.7		d
–Fluoride VI	ReF_6	300.2	Y	C		48.3	73.7		d
–Fluoride VII	ReF_7	319.2	O						

TABLE 4.2 (Continued)

Compound	Formula	Molecular Weight	Color	Crystal Symmetry	Specific Gravity	Melting Point (°C)	Boiling Point (°C)	Refractive Index n_D	Solubility (g/100 g of H_2O)
RHENIUM (continued)									
-Oxide IV	ReO_2	218.2	BK	M	11.4	1000 d			i
-Oxide VI	ReO_3	234.2	R	C	6.9–7.4	160	750		i
-Oxide VII	Re_2O_7	484.4	Y	H	8.2	300.3	360.3		sl
-Oxybromide VII	ReO_3Br	314.1	W			39.5	163		d
-Oxychloride VII	ReO_3Cl	269.7	CL	LIQ	3.87	4.5	131		i
-Sulfide IV	ReS_2	250.4	BK	H	7.51		d		i
-Sulfide VII	Re_2S_7	596.9	BK	T	4.87	d	d		i
RHODIUM									
-Chloride III	$RhCl_3$	209.3	R			430 d	800 s		i
-Fluoride II	RhF_3	159.9	R	R	5.38	>600 s			i
-Hydroxide III	$Rh(OH)_3$	155.9	Y			d			
-Oxide III	Rh_2O_3	253.8	G		8.20	1100 d			i
-Oxide IV	RhO_2	134.9	B						i
-Sulfide III	Rh_2S_3	302.0	BK		6.40	d			i
RUBIDIUM									
-Bromate	$RbBrO_3$	213.4	CL	C	3.68	430			2.9^{25}
-Bromide	$RbBr$	165.4	CL	C	3.36	682	1350	1.5530	105^{16}
-Carbonate	Rb_2CO_3	231.0	CL		3.47	837	d		450^{20}
-Chloride	$RbCl$	120.9	CL	C	2.76	717	1381	1.493	91^{20}
-Fluoride	RbF	104.5	CL	C	3.56	775	1410	1.398	131^{20}
-Hydroxide	$RbOH$	102.5	W	R	3.20	300			180^{15}
-Iodide	RbI	212.4	CL	C	3.55	641	1304	1.6474	152^{17}
-Nitrate	$RbNO_3$	147.5	CL		3.13	305	d	1.52	53^{20}
-Oxide	Rb_2O	187.0	Y	C	3.72	477			
-Perchlorate	$RbClO_4$	189.4	Y	C/R	3.01	567	d	1.4701	1.0^{20}
-Peroxide	Rb_2O_2	202.9	Y	C	3.65		d		d
-Sulfate	Rb_2SO_4	267.0	CL	R	3.61	1074	1011 d	1.513	48^{20}
-Sulfide	Rb_2S	203.0	Y		2.91	530 d	1700 d		vsl
-Superoxide	RbO_2	117.5	Y	T	3.05	412	1157 d		d
RUTHENIUM									
-Chloride III	$RuCl_3$	207.4	R	TR/H	3.11	500 d	230		d
-Fluoride V	RuF_5	196.1	GN	M	2.96	86.5			d
-Oxide IV	RuO_2	133.1	BE	T	6.97	>955 d			i

TABLE 4.2 (*Continued*)

Compound	Formula	Molecular Weight	Color	Crystal Symmetry	Specific Gravity	Melting Point (°C)	Boiling Point (°C)	Refractive Index n_D	Solubility (g/100 g of H_2O)
RUTHENIUM (continued)									
–Oxide VIII	RuO_4	165.1	Y	R	3.29	25.5	100 d		2.0[20]
–Sulfide IV	RuS_2	165.2	BK	C	6.99	1000 d			i
SAMARIUM									
–Bromate III	$Sm(BrO_3)_3 \cdot 9H_2O$	696.2	Y	H		75	150 (–9H_2O)		114[25]
–Bromide II	$SmBr_2$	310.2	B	R	5.1	669	1880		
–Bromide III	$SmBr_3$	390.1	Y	R	5.40	640	1645		
–Chloride II	$SmCl_2$	221.3	B	H	3.69	848	2030		92[10]
–Chloride III	$SmCl_3$	256.7	Y	C	4.46	682			
–Fluoride II	SmF_2	188.4	Y			1377	2427		i
–Fluoride III	SmF_3	207.4	W	R	6.64	1306	2323		i
–Iodide II	SmI_2	404.2	Y	M		520	1580		
–Iodide III	SmI_3	531.1	Y	H	3.14	850 d	d		vsl
–Nitrate III	$Sm(NO_3)_3 \cdot 6H_2O$	444.5	Y	TR	2.38	78 d		1.55	
–Oxide III	Sm_2O_3	348.7	Y	M	7.43	2350			2.7[20]
–Sulfate III	$Sm_2(SO_4)_3 \cdot 8H_2O$	733.0	Y	M	2.93	105 (–5H_2O)			
–Sulfide III	Sm_2S_3	396.9	Y	C	5.83	1780			
SCANDIUM									
–Bromide	$ScBr_3$	284.7	W	RH	3.91	948			vsl
–Chloride	$ScCl_3$	151.3	CL	RH	2.4	957	s		
–Fluoride	ScF_3	102.0	W	H		1515			vsl
–Iodide	ScI_3	425.7	CL			920			
–Nitrate	$Sc(NO_3)_3$	231.0	W			150			i
–Oxide	Sc_2O_3	137.9	CL	C	3.86				i
–Sulfate	$Sc_2(SO_4)_3$	378.1			2.58	d			39.9[25]
SELENIUM									
–Bromide I	Se_2Br_2	317.7	R	LIQ	3.60[15]		227 d		d
–Bromide IV	$SeBr_4$	398.6	B		4.03[–78]	75 d			d
–Chloride I	Se_2Cl_2	228.8	B	LIQ	2.91[17.5]	–85	100 d		d
–Chloride IV	$SeCl_4$	220.8	CL	C	3.80	305	191 s	1.807	d
–Fluoride IV	SeF_4	154.9	CL	LIQ	2.77	–9.5	106	1.895	d
–Fluoride VI	SeF_6	192.9	CL	GAS	2.26[–35]	–34.6	–46.6 s		
–Hydride II	H_2Se	81.0	CL	GAS	2.00[–42]	–65.7	–41.3		

TABLE 4.2 (Continued)

Compound	Formula	Molecular Weight	Color	Crystal Symmetry	Specific Gravity	Melting Point (°C)	Boiling Point (°C)	Refractive Index n_D	Solubility (g/100 g of H_2O)
SELENIUM (continued)									
−Oxide IV	SeO_2	111.0	CL	T	3.95	340	315 s	>1.76	38.4[14]
−Oxide VI	SeO_3	127.0	W	T		118			d
−Oxybromide	$SeOBr_2$	254.8	O	LIQ	3.38[50]	41.7	217 d		d
−Oxychloride	$SeOCl_2$	165.9	Y	LIQ	2.44[16]	10.9	177 d	1.651	d
−Oxyfluoride	$SeOF_2$	133.0	CL	LIQ	2.67	15	126		d
Selenic Acid	H_2SeO_4	145.0	W	R	3.00	58	260 d		1329[20]
Selenious Acid	H_2SeO_3	129.0	CL	H	3.00	70 d	205		167[20]
SILICON									
−Bromide	$SiBr_4$	347.7	CL	LIQ	2.77	5.2	152.8	1.5797[1]	d
−Carbide	SiC	40.1	BK	C/H	3.17	~2700	2200 d	2.67	i
−Chloride	$SiCl_4$	169.9	CL	LIQ	1.52[0]	−68	57.0		d
−Fluoride	SiF_4	104.1	CL	GAS	1.59[−78]	−90.3	−95.5 s		d
−Hydride (silane)	SiH_4	32.1	CL	GAS	0.68[−185]	−184.7	−111.4		d
−Hydride (disilane)	Si_2H_6	62.2	CL	GAS	0.69[−25]	−132.5	−14.3		d
−Hydride (trisilane)	Si_3H_8	92.3	CL	LIQ	0.73[0]	−117.4	53.0		d
−Iodide	SiI_4	535.7	CL	C	4.2	120.5	290		d
−Nitride	Si_3N_4	140.3	G	H	3.44	1900			i
−Oxide II	SiO	44.1	W	C	2.18	>1700	1880		i
−Oxide IV (amorph)	SiO_2	60.1	CL		2.63	1710	2590	1.4588	i
−Oxychloride	Si_2OCl_6	284.9	CL	LIQ		−33	135.5 s		d
−Sulfide	SiS_2	92.2	W	R	1.875	1090	1130 s		d
SILVER									
−Bromate	$AgBrO_3$	235.8	CL	T	5.21	d		1.874,1.920	0.2[25]
−Bromide	$AgBr$	187.8	Y	C	6.48	419	900 d	2.253	i
−Carbonate	Ag_2CO_3	257.8	Y		6.08	200 d			0.0032[25]
−Chlorate	$AgClO_3$	191.3	W	T	4.43	230	270 d		19[20]
−Chloride	$AgCl$	143.3	W	C	5.56	455	1550	2.071	0.00015[20]
−Cyanide	$AgCN$	133.9	W	H	3.95	320 d			i
−Fluoride	AgF	126.9	Y	C	5.85	435	1150	1.685,1.974	182[25]
−Iodate	$AgIO_3$	282.8	CL	R	5.53	>200	d		0.0044[20]
−Iodide	AgI	234.8	Y	H/C	5.67	555 d	1506	2.21	i
−Nitrate	$AgNO_3$	169.9	CL	R	4.35	209	444 d	1.74	216[20]
−Nitrite	$AgNO_2$	153.9	Y	R	4.45	140 d			0.28[15]
−Oxide	Ag_2O	231.8	B	C	7.22	300 d			0.0021[20]

TABLE 4.2 (*Continued*)

Compound	Formula	Molecular Weight	Color	Crystal Symmetry	Specific Gravity	Melting Point (°C)	Boiling Point (°C)	Refractive Index n_D	Solubility (g/100 g of H2O)
SILVER (continued)									
—Perchlorate	$AgClO_4$	207.4	W	C	2.81	486 d			525[25]
—Phosphate, o	Ag_3PO_4	418.6	Y	C	6.37	849	1085 d		0.00065[20]
—Sulfate	Ag_2SO_4	311.8	W	R	5.45	657	d		0.74[20]
—Sulfide	Ag_2S	247.8	BK	C/R	7.32	842			i
—Telluride	Ag_2Te	343.4	G	M	8.32	955			i
—Thiocyanate	$AgSCN$	166.0	CI			>120 d			i
SODIUM									
—Bicarbonate	$NaHCO_3$	84.0	W	M	2.16	270 ($-CO_2$)		1.500	9.6[20]
—Bromate	$NaBrO_3$	150.9	CL	C	3.34	381		1.594	36[20]
—Bromide	$NaBr$	102.9	CL	C	3.21	755	1393	1.6412	90[20]
—Carbonate	Na_2CO_3	106.0	W	C	2.53	851	d	1.535	7.1[0]
—Chlorate	$NaClO_3$	106.4	CL	C	2.5	248	d	1.513	96[20]
—Chloride	$NaCl$	58.4	CL	C	2.17	808	1413	1.544	36[20]
—Cyanide	$NaCN$	49.0	CL	C	1.86	563	1496	1.452	48[10]
—Fluoride	NaF	42.0	CL	C	2.78	995	1704	1.336	4.0[15]
—Hydride	NaH	24.0	SL	C	1.36	800 d		1.470	d
—Hydroxide	$NaOH$	40.0	W	R/C	2.13	318.4	1390	1.358	42[0]
—Iodate	$NaIO_3$	197.9	W	R	4.28	d			9[20]
—Iodide	NaI	149.9	CL	C	3.67	651	1304	1.775	184[20]
—Nitrate	$NaNO_3$	85.0	CL	TR	2.26	308	380 d	1.34;159	88[20]
—Nitrite	$NaNO_2$	69.0	Y	R	2.17	271	320 d		82[20]
—Oxide	Na_2O	62.0	G	C	2.27	917			d
—Perchlorate	$NaClO_4$	122.4	W	C/R	2.50	482 d		1.46	260[25]
—Periodate	$NaIO_4$	213.9	CL	T	4.17	300 d			14[25]
—Peroxide	Na_2O_2	78.0	Y	H	2.60	596	d		sl
—Phosphate, o	Na_3PO_4	163.9	W		2.54	1340			4.5[0]
—Silicate, m	Na_2SiO_3	122.1	CL	M	2.4	1027		1.52	sl
—Sulfate	Na_2SO_4	142.1	CL	R	2.68	884		1.48	5.32[25]
—Sulfide	Na_2S	78.1	W	C	1.86	1180			19[20]
—Sulfite	Na_2SO_3	126.1	W	H	2.63	d		1.5	12.5[0]
—Thiosulfate	$Na_2S_2O_3$	158.1	CL	M	1.67		d		50[0]
STRONTIUM									
—Bromide	$SrBr_2$	247.5	W	R	4.22	643	d	1.575	98[20]
—Carbonate	$SrCO_3$	147.6	CL	R	3.74	d	1350 d	1.52–1.67	0.001[20]

TABLE 4.2 (*Continued*)

Compound	Formula	Molecular Weight	Color	Crystal Symmetry	Specific Gravity	Melting Point (°C)	Boiling Point (°C)	Refractive Index n_D	Solubility (g/100 g of H$_2$O)
STRONTIUM (continued)									
– Chloride	SrCl$_2$	158.5	CL	C	3.05	870	1250	1.650	54[20]
– Fluoride	SrF$_2$	125.6	CL	C	4.24	1400	2490	1.442	0.012[20]
– Hydride	SrH$_2$	89.6	W	R	3.27	675 d	s		d
– Hydroxide	Sr(OH)$_2$	121.7	W		3.63	375	d		0.70[20]
– Iodate	Sr(IO$_3$)$_2$	437.4		TR	5.04				0.026[15]
– Iodide	SrI$_2$	341.4	CL		4.55	507	d	1.567	179[20]
– Nitrate	Sr(NO$_3$)$_2$	211.7	CL	C	2.99	645	d	1.870	71[20]
– Oxide	SrO	103.6	W	C	4.7	2460	~3000		0.85[25]
– Peroxide	SrO$_2$	119.6	CL	T	4.71	410 d			0.018[20]
– Sulfate	SrSO$_4$	183.7	CL	R	3.96	1605		1.62	0.011[20]
– Sulfide	SrS	119.7	CL	C	3.70	>2000		2.107	d
SULFUR									
– Bromide I	S$_2$Br$_2$	224.0	R	LIQ	2.64	–46	90 d	1.736	d
– Chloride I	S$_2$Cl$_2$	135.0	Y	LIQ	1.68	–80	138	1.666[14]	
– Chloride II	SCl$_2$	103.0	R	LIQ	1.62	–78	59 d	1.557	
– Chloride IV	SCl$_4$	173.9	R	LIQ			–31 d		d
– Fluoride I	S$_2$F$_2$	102.1	CL	GAS	1.5^{-100}	–105.5	–99		d
– Fluoride VI	SF$_6$	146.0	CL	GAS	1.88^{-51}	–50.8	–64 s		
– Hydride	H$_2$S	34.1	CL	GAS	0.96^{-60}	–85.5	–60.3	1.374	0.33[25]
– Oxide IV	SO$_2$	64.1	CL	GAS	1.434	–72.7	–10.2		23[0]
– Oxide VI	SO$_3$	80.1	CL	LIQ	1.97	16.8	44.5		d
– Pyrosulfuric Acid	H$_2$S$_2$O$_7$	178.1	CL	LIQ	1.89	34.8	d		d
– Sulfuric Acid	H$_2$SO$_4$	98.1	CL	LIQ	1.841	10.4	330		vsl
– Sulfuryl Chloride	SO$_2$Cl$_2$	135.0	CL	LIQ	1.67	–54.1	69.1	1.429[23]	d
– Thionyl Bromide	SOBr$_2$	207.9	Y	LIQ	2.68	–49.5	138/772	1.444[12]	d
– Thionyl Chloride	SOCl$_2$	119.0	CL	LIQ	1.64	–104.5	79/746	1.527[10]	d
TANTALUM									
– Bromide	TaBr$_5$	580.5	Y	R	4.99	280	349		d
– Carbide	TaC	193.0	BK	C	13.9	3877	5500		i
– Chloride	TaCl$_5$	358.2	Y	M	3.76	216.5	232.9		d
– Fluoride	TaF$_5$	275.9	CL	M	4.74	95.1	229.2		sl
– Iodide	TaI$_5$	815.4	BK	R	5.80	496	543		
– Nitride	TaN	194.9	BK	H	14.4	2530			i
– Oxide	Ta$_2$O$_5$	441.9	CL	R	8.0	1880			i

TABLE 4.2 *(Continued)*

Compound	Formula	Molecular Weight	Color	Crystal Symmetry	Specific Gravity	Melting Point (°C)	Boiling Point (°C)	Refractive Index n_D	Solubility (g/100 g of H_2O)
TANTALUM (continued)									
–Sulfide	Ta_2S_4	490.1	BK	H		>1300			i
TELLURIUM									
–Bromide II	$TeBr_2$	287.4	GN		5.24	210	339		d
–Bromide IV	$TeBr_4$	447.3	Y		4.31	380	~420 d		d
–Chloride II	$TeCl_2$	198.5	GN		7.05	208	324		d
–Chloride IV	$TeCl_4$	269.4	W	M	3.26	225	390		d
–Fluoride VI	TeF_6	241.6	CL	GAS	4.00^{-191}	-37.8	-38.9		d
–Hydride	H_2Te	129.6	CL	GAS	2.68^{-12}	-51	-2.3		d
–Iodide IV	TeI_4	635.2	BK	R	5.05	259	d		
–Oxide IV	TeO_2	159.6	W	T/R	5.67/5.91	733	1245	2.00–2.35	0.0007^{20}
–Oxide VI	TeO_3	175.6	Y		5.08	395 d			i
Telluric Acid, o	H_2TeO_6	229.7	W	C	3.16	136			sl
TERBIUM									
–Bromide	$TbBr_3$	398.6	W		4.67	828	1490		sl
–Chloride	$TbCl_3$	265.3	W		4.35	582	1550		
–Fluoride	TbF_3	215.9	W	R	7.24	1172	2277		i
–Iodide	TbI_3	539.6	CL	H	3.16	957	1330		
–Nitrate	$Tb(NO_3)_3 \cdot 6H_2O$	453.0	W	M		89.3			
–Oxide	Tb_2O_3	365.8	W	C	7.81	2390			
THALLIUM									
–Bromide I	$TlBr$	284.3	W	C	7.54	456	818	2.4–2.8	0.04^{20}
–Carbonate I	Tl_2CO_3	468.8	CL	M	7.11	272	d		$4.0^{15.5}$
–Chloride I	$TlCl$	239.8	W	C	7.00	430	806	2.247	0.3^{20}
–Chloride III	$TlCl_3$	310.8	W	H		25	d		vsl
–Fluoride	TlF	223.4	CL	R	8.36	327	655		79^{15}
–Hydroxide I	$TlOH$	221.4	Y	R		139 d			25^{0}
–Iodide I	TlI	331.3	Y/R	R/C	7.3/7.1	440	824	2.78	0.008^{25}
–Nitrate I	$TlNO_3$	266.4	W	C/TR	5.55	206	300 d		9.5^{20}
–Oxide I	Tl_2O	424.7	BK	RH	10.36	596	1080		vsl
–Oxide III	Tl_2O_3	456.7	CL	C	9.65	443	896		i
–Sulfate I	Tl_2SO_4	504.8	CL	R	6.77	632 d	d	1.87	4.9^{20}
–Sulfide I	Tl_2S	440.8	BK	T	8.46	433	d		0.02^{20}

TABLE 4.2 (Continued)

Compound	Formula	Molecular Weight	Color	Crystal Symmetry	Specific Gravity	Melting Point (°C)	Boiling Point (°C)	Refractive Index n_D	Solubility (g/100 g of H₂O)
THORIUM									
—Bromide	ThBr₄	551.7	W	T	5.67	679	857		sl
—Carbide	ThC₂	256.1	Y	T	8.96	2650			d
—Chloride	ThCl₄	373.9	W	T	4.60	770	922		vsl
—Fluoride	ThF₄	308.0	W	M	6.19	1110			
—Iodide	ThI₄	739.7	Y	M	6.00	566	837		sl
—Oxide	ThO₂	264.0	W	C	9.69	3050	4400		i
—Sulfate	Th(SO₄)₂	424.2	W	M	4.22	d			1.38²⁰
—Sulfide	ThS₂	296.2	BK	R	7.36	1900			i
THULIUM									
—Bromide	TmBr₃	408.7	W	H	5.02	954	1440		
—Chloride	TmCl₃	275.2	Y	M		824	1490		
—Fluoride	TmF₃	225.9	W	R	7.97	1158	2223		
—Iodide	TmI₃	549.6	Y	H	3.32	1021	1260		
—Oxide	Tm₂O₃	385.9	Y	C	8.77				
TIN									
—Bromide II	SnBr₂	278.5	Y	R	5.12	215	623		85⁰
—Bromide IV	SnBr₄	438.4	CL	R	3.35	32	203		d
—Chloride II	SnCl₂	189.6	W	R	3.95	247	623		84⁰
—Chloride IV	SnCl₄	260.5	CL	LIQ	2.23	-33	114	1.512	sl
—Fluoride II	SnF₂	156.7	W	M		213	850		30¹⁸
—Fluoride IV	SnF₄	194.7	W	M	4.78		705 s		vsl
—Hydride	SnH₄	122.7		GAS		-150	-52		
—Iodide II	SnI₂	372.5	R	R	5.28	320	720 d		0.98²⁰
—Iodide IV	SnI₄	626.3	R	C	4.70	143.5	343	2.106	d
—Oxide II	SnO	143.7	BK	T	6.45	1080 d			i
—Oxide IV	SnO₂	150.7	W	T	6.95	1130	1900 s	1.9968	sl
—Sulfide II	SnS	150.8	BK	R	5.08	882	1230		i
—Sulfide IV	SnS₂	182.8	Y	H	4.49	600 d			i
TITANIUM									
—Bromide IV	TiBr₄	367.6	O	M	3.42	39.0	230		d
—Carbide IV	TiC	59.9	G	C	4.92	3137	4820		i
—Chloride II	TiCl₂	118.8	BK	H	3.13	1025	1500		d
—Chloride III	TiCl₃	154.3	V	H	2.64	440 d	660/108		sl

TABLE 4.2 (*Continued*)

Compound	Formula	Molecular Weight	Color	Crystal Symmetry	Specific Gravity	Melting Point (°C)	Boiling Point (°C)	Refractive Index n_D	Solubility (g/100 g of H2O)
TITANIUM (continued)									
-Chloride IV	$TiCl_4$	189.7	Y	LIQ	1.73	-23.4	136.4	$1.61^{10.5}$	sl
-Fluoride IV	TiF_4	123.9	W		2.79	150	284		d
-Iodide IV	TiI_4	555.5	B	C	4.40	150	360		vsl
-Nitride	TiN	61.9	Y	C	5.21	2930			i
-Oxide II	TiO	63.9	BK	C	4.89	1750	3227		i
-Oxide IV	TiO_2	79.9	BK	T	3.84	1840	<3000	2.55	i
-Sulfide IV	TiS_2	112.0	Y	H	3.28	300 d			d
TUNGSTEN									
-Bromide V	WBr_5	583.4	B	H	16.1	276	333		d
-Carbide II	W_2C	379.7	G		15.7	2857	~6000		i
-Carbide IV	WC	195.9	G	C		2870	~6000		i
-Chloride V	WCl_5	361.1	GN		3.88	244	276		d
-Chloride VI	WCl_6	396.6	BE	C	3.52	275	347		d
-Fluoride VI	WF_6	297.8	CL	GAS	3.44^{15}	2.5	17.5		d
-Oxide IV	WO_2	215.9	B	T	12.1	1270 s	1430		i
-Oxide VI	WO_3	231.9	Y	M	7.16	1473	>1750		i
-Sulfide IV	WS_2	248.0	BK	H	7.5	1250 d			i
-Tungstic Acid	H_2WO_4	250.0	Y	R	5.5	100 d	1473	2.24	i
URANIUM									
-Bromide III	UBr_3	477.8	R	H	6.53	730			sl
-Bromide IV	UBr_4	557.7	B	M	5.55	519	792		sl
-Carbide	UC	250.0	BK	C	13.63	2380			d
-Carbide	UC_2	262.0	BK	T	11.68	2480	4370		d
-Chloride III	UCl_3	344.4	R	H	5.51	835			sl
-Chloride IV	UCl_4	379.9	GN	T	4.87	590	618 s		sl
-Fluoride IV	UF_4	314.1	GN	M	6.70	1036	1417	1.383	sl
-Fluoride VI	UF_6	352.1	Y	R	5.06	64.8	56.5 s		i
-Nitride	UN	252.0	B	C	14.31	2900			i
-Oxide IV	UO_2	270.1	BK	C	10.96	2880			i
-Oxide VI	UO_3	286.1	R	H	8.34	d			i
-Oxide IV-VI	U_3O_8	842.2	BK	R	8.39	1300 d			i
-Uranyl Acetate	$UO_2(C_2H_3O_2)_2 \cdot 6H_2O$	422.1	Y	R	2.89	110 d		1.4967	8^{15}
-Uranyl Nitrate	$UO_2(NO_3)_2 \cdot 6H_2O$	502.1	Y	R	2.81	59.5 d			119^{20}

TABLE 4.2 (*Continued*)

Compound	Formula	Molecular Weight	Color	Crystal Symmetry	Specific Gravity	Melting Point (°C)	Boiling Point (°C)	Refractive Index n_D	Solubility (g/100 g of H_2O)
VANADIUM									
–Carbide IV	VC	62.9	BK	C	5.77	2810	3900		i
–Chloride IV	VCl_4	192.7	R	LIQ	1.87	–28	154	d	d
–Fluoride III	VF_3	107.9	GN	R	3.36	800 s			i
–Fluoride V	VF_5	145.9	CL	R	2.18	19.0	47.9		
–Iodide II	VI_2	304.7	V	H	5.44	850 s	>1000 d		
–Oxide III	V_2O_3	149.9	BK	RH	4.82	1970			
–Oxide IV	VO_2	82.9	BE	T	4.65	1640			i
–Oxide V	V_2O_5	181.9	R	R	3.36	658	1750 d		0.07^{25}
–Oxychloride V	$VOCl_3$	173.3	Y	LIQ	1.83	–78.5	127		
–Sulfide II	VS	83.0	BK	H	4.20	1800 d			
XENON									
–Fluoride II	XeF_2	169.3	CL	T	4.3	129.0			d
–Fluoride IV	XeF_4	207.3	CL	M	4.1	117.1			d
–Fluoride VI	XeF_6	245.3	CL	M	3.6	49.5			d
–Oxide VI	XeO_3	179.3	CL	R	4.6	d	75.5	1.79	
YTTERBIUM									
–Bromide III	$YbBr_3$	412.8	CL	R	5.10	940 d	d		sl
–Chloride II	$YbCl_2$	244.0	GN	M	5.08	723	1927		sl
–Chloride III	$YbCl_3$	279.3	W	R		854	d		
–Fluoride III	YbF_3	230.0	W		8.17	1157	2200		i
–Iodide II	YbI_2	426.9	BK	H	5.40	527	1327		sl
–Iodide III	YbI_3	553.8	Y	H	3.33	1030	d		sl
–Oxide III	Yb_2O_3	394.1	CL	C	9.17	2350			i
–Sulfate III	$Yb_2(SO_4)_3$	634.3	CL		3.79	900 d			44^0
YTTRIUM									
–Bromide	YBr_3	328.6	W		3.95	948	1470		64^0
–Chloride	YCl_3	195.3	W	M	2.67	709	1507		78^{10}
–Fluoride	YF_3	145.9	W		5.07	1152			i
–Iodide	YI_3	469.6	W	H		965	1310		vsl
–Oxide	Y_2O_3	225.8	W		4.84	2410	4300		0.0002^{29}
–Sulfate	$Y_2(SO_4)_3$	466.0	W	C	2.61	1000 d			5.4^{25}

TABLE 4.2 (*Continued*)

Compound	Formula	Molecular Weight	Color	Crystal Symmetry	Specific Gravity	Melting Point (°C)	Boiling Point (°C)	Refractive Index n_D	Solubility (g/100 g of H$_2$O)
ZINC									
–Acetate	Zn(C$_2$H$_3$O$_2$)$_2$	183.5	CL	M	1.84	200 d	s	1.5452[18]	30[20]
–Bromide	ZnBr$_2$	225.2	CL	R	4.20	390	670		591[40]
–Carbonate	ZnCO$_3$	125.4	CL	TR	4.42	140 d			0.001[15]
–Chloride	ZnCl$_2$	136.3	W	H	2.91	270	756	1.168, 1.818	453[40]
–Fluoride	ZnF$_2$	103.4	CL	M	4.90	875	1502	1.687, 1.713	1.6[20]
–Hydroxide	Zn(OH)$_2$	99.4	CL	R	3.05	125 d			0.0002[29]
–Iodide	ZnI$_2$	319.2	CL	C	4.74	446	624 d		445[40]
–Nitrate	Zn(NO$_3$)$_2 \cdot$ 6H$_2$O	297.5	CL	T	2.07	36.4 d	105 (–6H$_2$O)		118[20]
–Oxide	ZnO	81.4	W	H	5.61	1975		2.01	i
–Sulfate	ZnSO$_4$	161.4	CL	R	3.54	600 d		1.669	42[0]
–Sulfide	ZnS	97.5	CL	C/H	4.04	1850/150 at	1185 s	2.36	i
ZIRCONIUM									
–Bromide	ZrBr$_4$	410.9	W			450	357 s		d
–Carbide	ZrC	103.2	G	C	6.73	3540	5100		i
–Chloride	ZrCl$_4$	233.1	W	C	2.80	437/25 at	331 s		sl
–Fluoride	ZrF$_4$	167.2	W	M	4.43	903		1.59	1.4[25]
–Iodide	ZrI$_4$	598.8	W			499	431 s		d
–Nitride	ZrN	105.2	B		7.09	2980			i
–Oxide	ZrO$_2$	123.2	W	M	5.73	2690	4300		i

TABLE 4-3. PROPERTIES OF MOLTEN SALTS (JANZ 1967–69, LUMSDEN 1966, CHARLOT 1969, KLEPPA 1963, BLOOM 1967)

Material	Melting Point Tm (°K)	Boiling Point (°K)	Density at Melting Point (g·cm⁻³)	Critical Temperature (°K)	Volume Change on Melting $\frac{\Delta V_f}{\Delta V_s}$ 100	Surface Tension at Melting Point (dynes·cm⁻¹)	Viscosity at Melting Point (centipoise)	Sound Velocity at Melting Point (m·sec⁻¹)	Cryoscopic Constant (°K/mole·kg)
LiF	1121	1954	1.83	4140	29.4	252		2546	2.77
NaF	1268	1977	1.96	4270	27.4	185		2080	16.6
KF	1131	1775	1.91	3460	17.2	141		1827	21.8
RbF	1048	1681	–	3280	–	167			38.4
LiCl	883	1655	1.60	3080	26.2	137	1.73	2038	13.7
NaCl	1073	1738	1.55	3400	25.0	116	1.43	1743	20.0
KCl	1043	1680	1.50	3200	17.3	99	1.38	1595	25.4
LiBr	823	1583	2.53	3020	24.3	–		1470	27.6
NaBr	1020	1665	2.36	3200	22.4	100		1325	34.0
KBr	1007	1656	2.133	3170	16.6	90		1256	55.9
NaNO₂	544	d > 593	1.81		–	120			
KNO₂	692	d623	–		–	109			
LiNO₃	527	–	1.78		21.4	116	5.46	1853	5.93
NaNO₃	583	d653	1.90		10.7	116	2.89	1808	15.4
KNO₃	610	d > 613	1.87		3.32	110	2.93	1754	30.8
RbNO₃	589	–	2.48		–0.23	109			89.0
AgNO₃	483	d > 485	3.97			148	4.25	1607	25.9
TlNO₃	480	706	4.90			94			58
Li₂SO₄	1132	–	2.00			225			142
Na₂SO₄	1157	–	2.07			192			66.3
K₂SO₄	1347	–	1.88			144			68.7
ZnCl₂	548	1005	2.39			53		1002	
HgCl₂	550	577	4.37			–			39.3
PbCl₂	771	1227	3.77			137	4.25	4952	
Na₂WO₄	969	–	3.85			202			
Na₃AlF₆	1273	–	1.84			135			
KCNS	450	–	1.60			101			12.7

Notes: (a) 5893 Å; (b) 5890 Å.

REFERENCES AND BIBLIOGRAPHY

Inorganic Compounds

Aronsson, B. I., T. Lundström and S. Rundquist, "Borides, Silicides and Phosphides," Wiley, New York, 1965.

Brown, D., "Halides of the Lanthanides and Actinides," Wiley, New York, 1968.

Canterford, J. H. and R. Colton, "Halides of the First Row Transition Metals," Wiley, New York, 1969.

Canterford, J. H. and R. Colton, "Halides of the Second and Third Row Transition Metals," Wiley, New York, 1968.

Carr, K., Tables of Thermionic Properties of the Elements and Compounds," U.S. Government Report No. N67-16043, 1967.

Connolly, T. F., Ed., "Group IV, V and VI Transition Metals and Compounds-Preparation and Properties," Plenum, New York, 1972.

Ebsworth, E. A. V., "Volatile Silicon Compounds," Macmillan, New York, 1963.

Emeleus, H. J., "The Chemistry of Fluorine," Academic Press, New York, 1969.

Eyring, L., Ed., "Progress in the Science and Technology of the Rare Earths," vol. 2, Pergamon Press, Oxford, 1966.

Forsythe, W. E., "Smithsonian Physical Tables," 9th rev. ed., Smithsonian Institution Press, Washington, 1969.

Frank, F., Ed., "Water: a Comprehensive Treatise," 5 vols., Plenum, New York, 1972-1974.

Galasso, F. S., "Structure and Properties of Inorganic Solids," Pergamon Press, Elmsford, New York, 1970.

"Gmelins Handbuch der anorganischen Chemie," 8th ed., Verlag Chemie, Weinheim, Germany, 1924-1971.

Hueckel, W., "Structural Chemistry of Inorganic Compounds," Elsevier, Amsterdam, 1951.

Hurd, D. T., "An Introduction to the Chemistry of the Hydrides," Wiley, New York, 1956.

Hyman, H. E., Ed., "Noble Gas Compounds," Univ. of Chicago Press, Chicago, 1963.

Heat Capacity, Cp (cal./°K · mole)	Heat of Fusion at Melting Point (kcal · mole⁻¹)	Entropy of Fusion at Melting Point (entropy units)	Equivalent Conductance at 1.1 Tm [(ohm)⁻¹cm² (equiv)⁻¹]	Decomposition Potential of Melt (volts)	Measurement Temperature for Decomposition Potential (°K)	Molar Refractivity at 5461 A (cm³ · mole⁻¹)	Refractive Index at 5461 A	Measurement Temperature for Refractive Index, (°K)
15.50	6.47	5.77	151	2.20	1273	2.89	1.32	1223
16.40	8.03	6.33	120	2.76	1273	3.41	1.25	1273
16.00	6.75	5.97	148	2.54	1273	5.43	1.28	1173
	6.15	5.76						
15.0	4.76	5.39	178.5	3.30	1073	8.32	1.501	883
16.0	6.69	6.23	152.3	3.25	1073	9.65	1.320	1173
16.0	6.34	6.08	122.4	3.37	1073	11.75	1.329	1173
	4.22	5.13	181	2.95	1073	11.81	1.60	843
	6.24	6.12	149	2.83	1073	13.19	1.486	1173
	6.10	6.06	108	2.97	1073	15.40	1.436	1173
			58			9.63[a]	1.416[a]	573
			~87			11.67	1.356[a]	873
26.6	5.961	11.66	44			10.74	1.467	573
37.0	3.696	6.1	58			11.54	1.431	573
29.5	2.413	4.58	46			13.57	1.426	573
	1.105	1.91	35			15.31[b]	1.431[b]	573
30.6	2.886		38			16.20[a]	1.660[a]	573
	2.264		27			21.38	1.688[b]	573
	1.975		123			14.87	1.452	1173
	5.67		90			16.53	1.395	1173
47.8	9.06		157			20.93	1.388	1173
24.1	2.45		~0.08	1.43	973	18.2	1.588	593
25.0	4.15		0.00096	0.86	973	22.9	1.661	563
	4.40		52.3	1.12	973	26.1	2.024	873
			46			24.58	1.542	1173
	27.64					17.2	1.290	1273
	3.07		17.3			19.65	1.537	573

Johnson, V. J., "Properties of Materials at Low Temperatures," Pergamon, Elmsford, New York, 1961.

Jolly, W. L., "The Chemistry of the Non-Metals," Prentice-Hall, Englewood Cliffs, N.J., 1966.

Kornilov, I. I., "The Chemistry of Metallides," Consultants Bureau, New York, 1966.

Lipscomb, W. N., "Boron Hydrides," Benjamin, New York, 1963.

Lyon, P. N., Ed., "Liquid Metals Handbook," U.S. Atomic Energy Commission and U.S. Navy Bureau of Ships, 1952; Supp. 1955.

Mackay, K. M., "Hydrogen Compounds of the Metallic Elements," E. & F. N. Spon, London, 1966.

Mandelcorn, L., Ed., "Non-Stochiometric Compounds," Academic Press, New York, 1964.

Mueller, W. M., J. P. Blackledge and G. C. Libowitz, "Metal Hydrides," Academic Press, New York, 1969.

Noll, W., "Chemistry and Technology of Silicones," Academic Press, New York, 1968.

Samsonov, G. V. and V. N. Bondarev, "Germanides," Plenum, New York, 1969.

Schumb, W. C., C. N. Satterfield and R. L. Wentworth, "Hydrogen Peroxide," Reinhold, New York, 1955.

Shaw, B. L., "Inorganic Hydrides," Pergamon Press, New York, 1967.

Sinha, S. P., "Europium," Springer-Verlag, New York, 1967.

Stephen, H. and T. Stephen, Eds., "Solubilities of Inorganic and Organic Compounds," Macmillan, New York, 1963.

Stock, A., "Hydrides of Boron and Silicon," Cornell Univ. Press, Ithaca, New York, 1933.

Stone, G. A., "Hydrogen Compounds of the Group IV Elements," Prentice-Hall, Englewood Cliffs, N.J., 1962.

Tobolsky, A. V., Ed., "The Chemistry of Sulfides," Wiley, New York, 1968.

Touloukian, Y. S. and C. Y. Ho, Eds., "Thermophysical Properties of Matter," 13 vols., Plenum, New York, 1970–1973.

Vogel, H. U., Ed., "Chemiker-Kalender," Springer, Berlin, 1966.

Vol'nov, I. I., "Peroxides, Superoxides and Ozonides of Alkali and Alkaline Earth Metals," Plenum, New York, 1966.

Westbrook, J. H., Ed., "Intermetallic Compounds," Wiley, New York, 1967.

Wiberg, E. and E. Amberger, "Hydrides of the Elements of the Main Groups I-IV," Elsevier, Amsterdam, 1971.

Molten Salts

Blander, M., Ed., "Molten Salt Chemistry," Interscience, New York, 1964.

Bloom, H., "The Chemistry of Molten Salts," Benjamin, New York, 1967.

Bockris, J. O'M., J. L. White and J. D. Mackenzie, Eds., "Physico-Chemical Measurements at High Temperatures," Butterworths, London, 1959.

Braunstein, J. and G. P. Smith, Eds., "Advances in Molten Salt Chemistry," 2 vols., Plenum, New York, 1971-1973.

Charlot, G. and B. Tremillon, "Chemical Reactions in Solvents and Melts," Pergamon Press, Oxford, 1969.

Janz, G. J., "Molten Salts Handbook," Academic Press, New York, 1967.

Janz, G. J., et al., "Molten Salts: Vol. 1, Electrical Conductance, Density, and Viscosity Data," National Bureau of Standards Report No. NSRDS-NBS 15, Superintendent of Documents, Washington, D.C., 1968.

Janz, G. J., et al., "Molten Salts: Vol. 2, Section 1, Electrochemistry, and Section 2, Surface Tension Data," National Bureau of Standards Report No. NSRDS-NBS 28, Superintendent of Documents, Washington, D.C., 1969.

Janz, G. J. and R. P. T. Tomkins, "Nonaqueous Electrolytes Handbook," 2 vols., Academic Press, New York, 1972-1973.

Kleppa, O. J. and F. G. McCarthy, "Heats of Fusion of the Monovalent Nitrates by High-Temperature Reaction Calorimetry," J. Chem. Eng. Data, 8, 331-332 (1963).

Lumsden, J., "Thermodynamics of Molten Salt Mixtures," Academic Press, New York, 1966.

Mamantov, G., Ed., "Molten Salts: Characterization and Analysis," Marcel Dekker, New York, 1969.

CHAPTER V
PROPERTIES OF ALLOYS

Chapter V contains tables describing selected physical, thermal, electrical, magnetic and mechanical properties of alloys. Owing to their unusual nature, materials for mechanical springs are tabulated separately. Chapter V contains the following tables:

*Data mainly from: Materials Selector 73 issue of *Materials Engineering*, 76 (4), Sept. 1972, published by Reinhold Publishing Co., Inc. Stamford, Conn.

TABLE 5-1.

ALUMINUM AND ALUMINUM ALLOYS–WROUGHT

Type →	EC	1060	1100	3003	3004
COMPOSITION (%)	Al 99.45 max	Al 99.60 max	Al 99.0 max	Mn 1.0–1.5	Mn 1.0–1.5, Mg 0.8–1.3
PHYSICAL, THERMAL, ELECTRICAL AND MAGNETIC PROPERTIES					
Density (lb/in.³)	0.098	0.098	0.098	0.099	0.098
Melting Range (°F)	1195–1215	1195–1215	1190–1215	1190–1210	1165–1205
Specific Heat (Btu/lb/°F)	–	–	0.22	0.22	0.22
Thermal Expansion Coeff. (μ in./in./°F) 68–212°F	13.2	13.1	13.1	12.9	13.3
1000°F	–	–	–	–	–
Thermal Conductivity (Btu/ft²/hr/°F/ft) 68–70°F	135	128	128	111	93.8
1000°F	–	–	–	–	–
Electrical Resistivity (μ ohm-cm) at 68°F	2.8	2.8	2.92	3.45	4.10
Magnetic Permeability	–	–	–	–	–
MECHANICAL PROPERTIES (annealed)					
Youngs's Modulus (10⁶ psi) 68–70°F	10	10	10	10	10
1000°F	–	–	–	–	–
Tensile Strength (1000 psi) 68–70°F	12	10	13	16	26
1000°F	–	–	–	–	–
Yield Strength (1000 psi) 68–70°F	4	4	5	6	10
1000°F	–	–	–	–	–
Elongation (in 2 in.) at 68–75°F (%)	23	43	35, 45	30, 40	20, 25
Shear Strength (1000 psi) 68–70°F	8	7	9	11	16
1000°F	–	–	–	–	–
Hardness (Brinell)	–	19	23	28	45

TABLE 5-1 (Continued)

ALUMINUM ALLOYS—WROUGHT

Type →	2011	2014	2017	2021	2024	2219		
COMPOSITION (%)	Cu 5.0–6.0, Pb 0.2–0.6, Bi 0.2–0.8	Cu 3.9–5.0, Si 0.5–1.2, Mn 0.4–1.2, Mg 0.2–0.8	Cu 3.5–4.5, Mn 0.4–1.0, Mg 0.2–0.8	Cu 5.8–6.8, Mn 0.20–0.40, Zr 0.10–0.25, Cd 0.05–0.20	Cu 3.8–4.9, Mn 0.3–0.9, Mg 1.2–1.8	Cu 5.8–6.8, Mn 0.20–0.40, Ti 0.02–0.10, V 0.05–0.15, Zr 0.10–0.25		
PHYSICAL, THERMAL, ELECTRICAL AND MAGNETIC PROPERTIES								
Density (lb/in.^{3})	0.102	0.101	0.101	0.103	0.100	0.103		
Melting Range (°F)	995–1190	950–1180	955–1185	997–1195	935–1190	1010–1190		
Specific Heat (Btu/lb/°F)	0.23	0.22	0.22	–	0.22	–		
Thermal Expansion Coeff. (μ in./in./°F)								
68–212°F	12.8	12.8	13.1	12.6	12.9	12.4		
1000°F	–	–	–	–	–	–		
Thermal Conductivity (Btu/ft^2/hr/°F/ft)								
68–70°F	82.5	111.0	99.4	–	109.2	100		
1000°F	–	–	–	–	–	–		
Electrical Resistivity (μ ohm-cm) at 68°F	4.8	3.45–4.31	3.83–5.75	–	3.45–5.75	3.90–5.7		
Magnetic Permeability	–	–	–	–	–	–		
MECHANICAL PROPERTIES (annealed)								
Youngs's Modulus (10^6 psi)								
68–70°F	10.2	10.6	10.5	10.7	10.6	10.6		
1000°F	–	–	–	–	–	–		
Tensile Strength (1000 psi)								
68–70°F	–	27	26	24	27	25		
1000°F	–	–	–	–	–	–		
Yield Strength (1000 psi)								
68–70°F	–	14 (0.2%)	10 (0.2%)	10 (0.2%)	11 (0.2%)	11 (0.2%)		
1000°F	–	–	–	–	–	–		
Elongation (in 2 in.) at 68–75°F (%)	–	18	22	23	20	18		
Shear Strength (1000 psi)								
68–70°F	–	18	18	–	18	–		
1000°F	–	–	–	–	–	–		
Hardness (Brinell)	–	45	45	44	47	–		

TABLE 5-1 (*Continued*)

ALUMINUM ALLOYS–WROUGHT

Type →	5005	5050	5052	5056	5083		
COMPOSITION (%)	Mg 0.5–1.1	Mg 1.0–1.8	Mg 2.2–2.8, Cr 0.15–0.35	Mn 0.05–0.20, Mg 4.5–5.6, Cr 0.05–0.20	Mg 4.0–4.9, Mn 0.3–1.0, Cr 0.05–0.25		
PHYSICAL, THERMAL, ELECTRICAL AND MAGNETIC PROPERTIES							
Density (lb/in.^{3})	0.097	0.097	0.097	0.095	0.096		
Melting Range (°F)	1170–1205	1160–1205	1100–1200	1055–1180	1060–1180		
Specific Heat (Btu/lb/°F)	0.23	0.22	0.22	0.22	0.23		
Thermal Expansion Coeff. (μ in./in./°F)							
68–212°F	13.3	13.2	13.2	13.4	13.2		
1000°F	–	–	–	–	–		
Thermal Conductivity (Btu/ft^2/hr/°F/ft)							
68–70°F	116	111	80.0	67.4	68		
1000°F	–	–	–	–	–		
Electrical Resistivity (μ ohm-cm) at 68°F	3.3	3.4	4.9	5.9–6.4	5.9		
Magnetic Permeability	–	–	–	–	–		
MECHANICAL PROPERTIES (annealed)							
Young's Modulus (10^6 psi)							
68–70°F	10	10	10.2	10.3	10.3		
1000°F	–	–	–	–	–		
Tensile Strength (1000 psi)							
68–70°F	18	21	28	42	42		
1000°F	–	–	–	–	–		
Yield Strength (1000 psi)							
68–70°F	6	8	13	22	21		
1000°F	–	–	–	–	–		
Elongation (in 2 in.) at 68–75°F (%)	30	24	25; 30	35	22		
Shear Strength (1000 psi)							
68–70°F	11	15	18	26	25		
1000°F	–	–	–	–	–		
Hardness (Brinell)	28	36	47	65	–		

TABLE 5-1 (*Continued*)

ALUMINUM ALLOYS—WROUGHT

Type →	5086	5090	5154	5252	5454	5456	5457
COMPOSITION (%)	Mg 3.5–4.5, Mn 0.2–0.7, Cr 0.05–0.25	Mg 6.6–7.4, Cr 0.15–0.25, B 0.005–0.015	Mg 3.1–3.9, Cr 0.15–0.35	Mg 2.5	Mn 0.50–1.0, Mg 2.4–3.0, Cr 0.05–0.20	Mg 4.7–5.5, Mn 0.5–1.0, Cr 0.05–0.20	Mn 0.15–0.45, Mg 0.8–1.2
PHYSICAL, THERMAL, ELECTRICAL AND MAGNETIC PROPERTIES							
Density (lb/in.^{3})	0.096	0.095	0.096	0.097	0.097	0.096	0.098
Melting Range (°F)	1084–1184	1040–1157	1100–1190	1100–1200	1115–1195	1060–1180	1165–1210
Specific Heat (Btu/lb/°F)	0.23	0.23	0.23	–	–	0.23 (212°F)	–
Thermal Expansion Coeff. (μ in./in./°F)							
68–212°F	13.2	13.2	13.3	13.2	13.1	13.3	13.2
1000°F	–	–	–	–	–	–	–
Thermal Conductivity (Btu/ft^2/hr/°F/ft)							
68–70°F	73	65	73	80	78	68	102
1000°F	–	–	–	–	–	–	–
Electrical Resistivity (μ ohm-cm) at 68°F	5.5	6.8	5.3	4.9	5.1	5.9	3.9
Magnetic Permeability	–	–	–	–	–	–	–
MECHANICAL PROPERTIES (annealed)							
Youngs's Modulus (10^6 psi)							
68–70°F	10.3	10.3	10.2	10	10.2	10.3	10
1000°F	–	–	–	–	–	–	–
Tensile Strength (1000 psi)							
68–70°F	38	50	35	–	36	45	13
1000°F	–	–	–	–	–	–	–
Yield Strength (1000 psi)							
68–70°F	17	22	17	–	17	23	5
1000°F	–	–	–	–	–	–	–
Elongation (in 2 in.) at 68–75°F (%)	22	25	27	–	22	24	27
Shear Strength (1000 psi)							
68–70°F	23	–	22	–	23	28	9
1000°F	–	–	–	–	–	–	–
Hardness (Brinell)	–	95	58	–	62	75	23

TABLE 5-1 (*Continued*)

ALUMINUM ALLOYS—WROUGHT

Type →	6061	6063	6066	6101	6262		
COMPOSITION (%)	Mg 0.8–1.2, Si 0.4–0.8, Cr 0.15–0.35, Cu 0.15–0.40	Mg 0.45–0.9, Si 0.2–0.6	Si 1.3, Cu 1.0, Mn 0.9, Mg 1.1	Si 0.5, Mg 0.6	Si 0.6, Cu 0.25, Mg 1.0, Cr 0.09, Pb 0.6, Bi 0.6		
PHYSICAL, THERMAL, ELECTRICAL AND MAGNETIC PROPERTIES							
Density (lb/in.^{3})	0.098	0.098	0.098	0.098	0.098		
Melting Range (°F)	1080–1200	1140–1205	1050–1200	1140–1205	1100–1205		
Specific Heat (Btu/lb/°F)	0.23	–	–	–	–		
Thermal Expansion Coeff. (μ in./in./°F)							
68–212°F	13.0	13.0	12.9	13	13		
1000°F	–	–	–	–	–		
Thermal Conductivity (Btu/ft^2/hr/°F/ft)							
68–70°F	99	111	84	125	99		
1000°F	–	–	–	–	–		
Electrical Resistivity (μ ohm-cm) at 68°F	3.8	3.3	4.7	3.0	3.9		
Magnetic Permeability	–	–	–	–	–		
MECHANICAL PROPERTIES (annealed)							
Youngs's Modulus (10^6 psi)							
68–70°F	10.0	10.0	10.0	10.0	10.0		
1000°F	–	–	–	–	–		
Tensile Strength (1000 psi)							
68–70°F	18	13	22	–	–		
1000°F	–	–	–	–	–		
Yield Strength (1000 psi) (0.2%)							
68–70°F	8	7	12	–	–		
1000°F	–	–	–	–	–		
Elongation (in 2 in.) at 68–75°F (%)	25; 30	–	18	–	–		
Shear Strength (1000 psi)							
68–70°F	12	10	14.5	–	–		
1000°F	–	–	–	–	–		
Hardness (Brinell)	30	25	43	–	–		

ALUMINUM ALLOYS–WROUGHT

TABLE 5-1 (*Continued*)

Type →	7039	7049	7075	7079	7178
COMPOSITION (%)	Mn 0.10–0.40, Mg 2.3–3.3, Cr 0.15–0.25, Zn 3.5–4.5	Zn 7.6, Mg 2.5, Cu 1.5, Cr 0.15	Zn 5.1–6.1, Mg 2.1–2.9, Cu 1.2–2.0, Cr 0.18–0.40	Zn 3.8–4.8, Mg 2.9–3.7, Cu 0.4–0.8, Mn 0.1–0.3, Cr 0.1–0.25	Zn 6.3–7.3, Mg 2.4–3.1, Cu 1.6–2.4, Mn 0.30, Cr 0.18–0.40
PHYSICAL, THERMAL, ELECTRICAL AND MAGNETIC PROPERTIES					
Density (lb/in.^{3})	0.0999	0.102	0.101	0.099	0.102
Melting Range (°F)	–	890–1160	890–1180	900–1180	890–1165
Specific Heat (Btu/lb/°F)	–	0.23	0.23	–	–
Thermal Expansion Coeff. (μ in./in./°F)					
68–212°F	–	13.0	13.1	13.1	13.0
1000°F	–	–	–	–	–
Thermal Conductivity (Btu/ft^2/hr/°F/ft)					
68–70°F	–	–	70	70	70
1000°F	–	–	–	–	–
Electrical Resistivity (μ ohm-cm) at 68°F	4.9	4.3	5.7	5.5	5.5
Magnetic Permeability	–	–	–	–	–
MECHANICAL PROPERTIES (annealed)					
Young's Modulus (10^6 psi)					
68–70°F	–	10.2	10.4	10.3	10.4
1000°F	–	–	–	–	–
Tensile Strength (1000 psi)					
68–70°F	32	–	33	32	33
1000°F	–	–	–	–	–
Yield Strength (1000 psi) (0.2%)					
68–70°F	–	–	15	14	15
1000°F	–	–	–	–	–
Elongation (in 2 in.) at 68–75°F (%)	–	–	17	16	16
Shear Strength (1000 psi)					
68–70°F	–	–	22	–	22
1000°F	–	–	–	–	–
Hardness (Brinell)	–	–	60	–	60

TABLE 5-1 (*Continued*)

ALUMINUM ALLOYS—CAST

Type →	201.0	208.0	222.0	242.0	295.0
COMPOSITION (%)	Cu 4.7, Ag 0.7	Cu 4.0, Si 3.0	Cu 10.0, Mg 0.25	Cu 4.0, Mg 1.5, Ni 2.0	Cu 4.5
PHYSICAL, THERMAL, ELECTRICAL AND MAGNETIC PROPERTIES					
Density (lb/in.^{3})	0.101	0.101	0.107	0.102	0.101
Melting Range (°F)	–	–	–	–	–
Specific Heat (Btu/lb/°F)					
Thermal Expansion Coeff. (μ in./in./°F)					
68–212°F	10.7	12.2	12.2	12.5	12.7
1000°F	–	–	–	–	–
Thermal Conductivity (Btu/ft^2/hr/°F/ft)					
68–70°F	–	70	77	87	82.5
1000°F	–	–	–	–	–
Electrical Resistivity (μ ohm-cm) at 68°F	4.3–4.5	4.3	4.2	3.9	4.1
Magnetic Permeability	–	–	–	–	–
MECHANICAL PROPERTIES (as cast)					
Young's Modulus (10^6 psi)					
68–70°F	–	–	–	–	–
1000°F	–	–	–	–	–
Tensile Strength (1000 psi)					
68–70°F	–	21	30	34	32
1000°F	–	–	–	–	–
Yield Strength (1000 psi) (0.2%)					
68–70°F	–	14	–	–	20
1000°F	–	–	–	–	–
Elongation (in 2 in.) at 68–75°F (%)	–	2.5	–	–	3.0
Shear Strength (1000 psi)					
68–70°F	–	17	–	–	–
1000°F	–	–	–	–	–
Hardness (Brinell)	–	55	–	–	–

ALUMINUM ALLOYS–CAST

TABLE 5-1 (*Continued*)

Type →	A356.0	A380.0	443.0	514.0	520.0		
COMPOSITION (%)	Si 7.0, Mg 0.50	Si 8.5, Cu 3.5, Mg 0.1, Fe 1.3	Si 5.2, Cu 0.6, Fe 0.8	Mg 4.0	Mg 10.0		
PHYSICAL, THERMAL, ELECTRICAL AND MAGNETIC PROPERTIES							
Density (lb/in.³)	0.097	0.097	0.097	0.096	0.093		
Melting Range (°F)	—	—	—	—	—		
Specific Heat (Btu/lb/°F)	—	—	—	—	—		
Thermal Expansion Coeff. (μ in./in./°F) 68–212°F	11.9	11.7	12.2	13.3	13.6		
1000°F	—	—	—	—	—		
Thermal Conductivity (Btu/ft²/hr/°F/ft) 68–70°F	92	58	84	80	51		
1000°F	—	—	—	—	—		
Electrical Resistivity (μ ohm-cm) at 68°F	3.9	4.6	4.0	4.1	4.8		
Magnetic Permeability	—	—	—	—	—		
MECHANICAL PROPERTIES (as cast)							
Young's Modulus (10⁶ psi) 68–70°F	—	—	—	—	—		
1000°F	—	—	—	—	—		
Tensile Strength (1000 psi) 68–70°F	37	47	19; 23	25	42		
1000°F	—	—	—	—	—		
Yield Strength (1000 psi) (0.2%) 68–70°F	—	23	8; 9	—	22		
1000°F	—	—	—	—	—		
Elongation (in 2 in.) at 68–75°F (%)	5.0	4	8; 10	5	12.0		
Shear Strength (1000 psi) 68–70°F	—	—	—	—	—		
1000°F	—	—	—	—	—		
Hardness (Brinell)	—	80	40; 45	50	—		

TABLE 5-2.

COBALT AND COBALT ALLOYS

Type →	Cobalt (annealed)	Ductile Cobalt (annealed)	Elgiloy™	Haynes Alloy 188™	MP35N	Nivco™	UMCo-50™
COMPOSITION (%)	Co 99.9, Ni 0.7, Fe 0.02, Cu 0.001	Co 95, Fe 5, Ni 0.01, Cu 0.002	Co 40, Cr 20, Ni 15, Mo 7, Mn 2, C 0.1, Be 0.04, Fe bal	Co 36, Cr 22, Ni 22, W 14.5, Fe 3, Mn 1.25, La 0.9, Si 0.35, C 0.1	Co 35, Ni 35, Cr 20, Mo 10	Co 73.5, Ni 22.5, Ti 1.8, Zr 1.1, Mn 0.35, Fe 0.3, Al 0.22, Si 0.15, C 0.02	Co 50.5, Cr 28, Si 0.75, Mn 0.65, P 0.02, S.002, Fe bal
PHYSICAL, THERMAL, ELECTRICAL AND MAGNETIC PROPERTIES							
Density (lb/in.^{3})	0.320	0.316	0.300	0.330	0.304	0.31	0.291
Melting Range (°F)	2723	2714	2720	–	2400–2600	2550	2540
Specific Heat (Btu/lb/°F) at 1800°F	–	–	–	0.14	–	–	–
Thermal Expansion Coeff. (μ in./in./°F) 70–1800°F	6.8	–	8.8	9.9	8.7	8.1	9.33
1000°F	–	–	–	–	–	–	–
Thermal Conductivity (Btu/ft^2/hr/°F/ft) 1300°F	–	–	7.2–10	6.3–13.3	–	16.6	5.2
1000°F	–	–	–	–	–	–	–
Electrical Resistivity (μ ohm-cm) at 68°F	6.24	–	99.5	92.2	101	23.7	81
Magnetic Permeability	–	–	–	–	–	–	–
MECHANICAL PROPERTIES (annealed)							
Young's Modulus (10^6 psi) 68–70°F	30	–	30	–	33	30	31.5
1200°F	–	–	–	–	–	–	–
Tensile Strength (1000 psi) 68–70°F	110	80	110–360	137–140	200–300	165	–
1200°F	–	–	–	103–106	–	105	–
Yield Strength (1000 psi) 68–70°F	48	26	70–290	68–70	160–290	110	88
1200°F	–	–	–	44	–	75	–
Elongation (in 2 in.) at 68–75°F (%)	22	50	34–1	56–61	9–18	25	10
Shear Strength (1000 psi) 68–70°F	–	–	–	–	–	–	–
1000°F	–	–	–	–	–	–	–
Hardness (Brinell)	–	–	–	–	–	–	–

TABLE 5-2 *(Continued)*

COBALT–BASE SUPERALLOYS–WROUGHT

Type →	S-816	V-36	Haynes Alloy 25, L-605TM	J-1570	J-1650
COMPOSITION (%)	C 0.40, Mn 1.20, Cr 20.0, Ni 20.0, Mo 4.0, W 4.0, Fe 3.0, Cb 4.0, Co bal	C 0.32, Mn 1.0, Cr 25.0, Ni 20.0, Mo 4.0, Cb 2.3, Fe 2.4, Co bal	C 0.05–0.15, Mn 1.0–2.0, Cr 19.0–21.0, Ni 9.0–11.0, W 14.0–16.0, Fe 3.0, Si 1.0, Co bal	C 0.20, Fe 2.0, Ni 28.0, Cr 20.0, W 7.0, Ti 4.0, Co bal	C 0.20, Cr 20.0, Ni 27.0, W 12.0, Ta 2.0, Ti 3.8, B 0.02, Co bal
PHYSICAL, THERMAL, ELECTRICAL AND MAGNETIC PROPERTIES					
Density (lb/in.^{3})	0.31	0.303	0.33	0.305	0.32
Melting Range (°F)	2350–2450	2350–2450	2425–2570	–	–
Specific Heat (Btu/lb/°F) at 77–1300°F	0.12	–	0.09 (80–212°F)	–	–
Thermal Expansion Coef. (μ in./in./°F)					
70–1800°F	9.3	9.1	9.4	9.0	–
1000°F	–	–	–	–	–
Thermal Conductivity (Btu/ft^2/hr/°F/ft)					
1300°F	13.0	–	13.1	11 (800°F)	–
1000°F	–	–	–	–	–
Electrical Resistivity (μ ohm-cm) at 68°F	93 (aged)	–	188.7	–	–
Magnetic Permeability	–	–	–	–	–
MECHANICAL PROPERTIES (annealed)					
Youngs's Modulus (10^6 psi)					
68–70°F	35	32	34.2	33.5	33.6
1200°F	27.0	25.9	27.4	–	–
Tensile Strength (1000 psi)					
68–70°F	140	146	146	132	174
1200°F	–	–	–	–	–
Yield Strength (1000 psi) (0.2%)					
68–70°F	70	83	67	84	116
1200°F	–	–	–	–	–
Elongation (in 2 in.) at 68–75°F (%)	35	20	64	–	8
Shear Strength (1000 psi)					
68–70°F	–	–	–	–	–
1000°F	–	–	–	–	–
Hardness (Brinell)	–	–	–	–	–

TABLE 5-2 (*Continued*)

COBALT–BASE SUPERALLOYS–CAST

Type →	Haynes Alloy 21TM	Haynes Alloy 31, X-40TM	Haynes Alloy 151TM	W1 52	Mar M 302
COMPOSITION (%)	C 0.20–0.30, Ni 1.75–3.75, Cr 25.5–29.0, Mo 5.0–6.0, Fe 2.0, Mn 1.0, Si 1.0, B 0.007, Co bal	C 0.45–0.55, Ni 9.5–11.5, Cr 24.5–26.5, Fe 2.0, Mn 1.0, Si 1.0, W 7.0–8.0, Co bal	C 0.50, Cr 20.0, W 13.0, B 0.06, Co bal	C 0.45, W 11.0, Cr 21.0, Cb + Ta 1.75, Co bal	C 0.85, Cr 21.5, W 10.0, Ta 9.0, Zr 0.20, B 0.005, Co bal
PHYSICAL, THERMAL, ELECTRICAL AND MAGNETIC PROPERTIES					
Density (lb/in.^{3})	0.30	0.31	—	0.32	0.333
Melting Range (°F)	2465	—	—	2400–2450	2400–2450
Specific Heat (Btu/lb/°F) at 77–1300°F					
Thermal Expansion Coeff. (μ in./in./°F)					
70–1800°F	8.7	9.2	—	8.8	8.7
1000°F	—	—	—	—	—
Thermal Conductivity (Btu/ft^2/hr/°F/ft)					
1300°F	11.9	12.8	—	15.8	—
1000°F	—	—	—	—	—
Electrical Resistivity (μ ohm-cm) at 68°F	87.4	97.0	—	—	—
Magnetic Permeability	—	—	—	—	—
MECHANICAL PROPERTIES (annealed)					
Youngs's Modulus (10^6 psi)					
68–70°F	36	29	—	—	—
1200°F	—	—	—	—	—
Tensile Strength (1000 psi)					
68–70°F	101	108	100	110	140
1200°F	71	76	85	108	—
Yield Strength (1000 psi) (0.2%)					
68–70°F	82	76	75	85	100
1200°F	—	—	50	53	—
Elongation (in 2 in.) at 68–75°F (%)	8.2	9.0	9	5	2
Shear Strength (1000 psi)					
68–70°F	—	—	—	—	—
1000°F	—	—	—	—	—
Hardness (Brinell)	—	—	—	—	—

TABLE 5-3

COPPER AND ALLOYS—WROUGHT

Type →	102 (OF)	104, 105, 107 (OFS)	110 (ETP)	113, 114, 116 (STP)	122 (DHP)
COMPOSITION (%)	Cu 99.5 min	Cu + Ag 99.95 min, Ag 0.027–0.085	Cu 99.90 min, O_2 0.04	Cu + Ag 99.90 min, Ag 0.027–0.085	Cu 99.90 min, P 0.015–0.040
PHYSICAL, THERMAL, ELECTRICAL AND MAGNETIC PROPERTIES					
Density (lb/in.³)	0.323	0.323	0.321–0.323	0.321–0.323	0.323
Melting Range (°F)	1981	1981	1949–1981	1980	1981
Specific Heat (Btu/lb/°F)	0.092	0.092	0.092	0.092	0.092
Thermal Expansion Coeff. (μ in./in./°F)					
68–572°F	9.8	9.8	9.8	9.8	9.8
1000°F	–	–	–	–	–
Thermal Conductivity (Btu/ft²/hr/°F/ft)					
68–70°F	226	226	226	224	196
1000°F	–	–	–	–	–
Electrical Resistivity (μ ohm-cm) at 68°F	1.71	1.71	1.71	1.72	2.03
Magnetic Permeability	–	–	–	–	–
MECHANICAL PROPERTIES (annealed)					
Youngs's Modulus (10⁶ psi)					
68–70°F	17	17	17	17	17
1000°F	–	–	–	–	–
Tensile Strength (1000 psi)					
68–70°F	32–35	32–35	32–35	32–35	32–34
1000°F	–	–	–	–	–
Yield Strength (1000 psi) (0.5%)					
68–70°F	10–11	10–11	10–11	10–11	10–11
1000°F	–	–	–	–	–
Elongation (in 2 in.) at 68–75°F (%)	45–55	45–55	45–55	45–55	45
Shear Strength (1000 psi)					
68–70°F	22–24	22–24	22–24	22–24	22–23
1000°F	–	–	–	–	–
Hardness (Rockwell)	40–45 R_F	40–45 R_F	40–45 R_F	40–45 R_F	40–45 R_F

TABLE 5-3 (*Continued*)

COPPER AND ALLOYS—WROUGHT

Type →	145 (DPTE)	147 (S-Cu)	150 (Zr-Cu)	155 (Ag-Cu)	(172 (Be-Cu)
COMPOSITION (%)	Cu + Ag + Te 99.90, P 0.004–0.012, Te 0.40–0.60	Cu 99.7, S 0.30	Cu 99.85, Zr 0.15	Cu 99.8, Mg 0.11, P 0.05, Ag 0.04	Cu 97.9, Be 1.90, Co 0.20
PHYSICAL, THERMAL, ELECTRICAL AND MAGNETIC PROPERTIES					
Density (lb/in.^{3})	0.323	0.323	0.323	0.322	0.296–0.298
Melting Range (°F)	1980	1980	1976	1972–1980	1600–1800
Specific Heat (Btu/lb/°F)	0.09	0.09	–	0.092	0.10
Thermal Expansion Coeff. (μ in./in./°F)					
68–572°F	9.9	9.8	–	9.8	9.3
1000°F	–	–	–	–	–
Thermal Conductivity (Btu/ft^2/hr/°F/ft)					
68–70°F	205	216	–	200	62–75
1000°F	–	–	–	–	–
Electrical Resistivity (μ ohm-cm) at 68°F	1.91	1.78	1.99	1.92	7.68
Magnetic Permeability	–	–	–	–	–
MECHANICAL PROPERTIES (annealed)					
Youngs's Modulus (10^6 psi)					
68–70°F	16	–	–	17	19
1000°F	–	–	–	–	–
Tensile Strength (1000 psi)					
68–70°F	–	34	–	40	60–80
1000°F	–	–	–	–	–
Yield Strength (1000 psi) (0.5%)					
68–70°F	–	10	–	18	20–30
1000°F	–	–	–	–	–
Elongation (in 2 in.) at 68–75° F (%)	–	42	–	34–40	35–50
Shear Strength (1000 psi)					
68–70°F	–	–	–	–	50–60
1000°F	–	–	–	–	–
Hardness (Rockwell)	–	–	–	F70	B50–65

PROPERTIES OF ALLOYS   765

TABLE 5-3 (*Continued*)

COPPER AND ALLOYS—WROUGHT

Type →	182 (Cr-Cu)	194 (HSM Cu)	210 (Gilding, 95%)	220 (Commercial Bronze, 90%)	226 (Jewelry Bronze, 87.5%)
COMPOSITION (%)	Cu 99.05, Cr 0.95	Cu 97.5, Fe 2.35, P 0.03, Zn 0.12	Cu 95, Zn 5	Cu 90, Zn 10	Cu 87.5, Zn 12.5
PHYSICAL, THERMAL, ELECTRICAL AND MAGNETIC PROPERTIES					
Density (lb/in.^{3})	0.321	0.317	0.320	0.318	0.317
Melting Range (°F)	2147	1992	1920–1950	1870–1910	1840–1895
Specific Heat (Btu/lb/°F)	–	–	0.09	0.09	0.09
Thermal Expansion Coeff. (μ in./in./°F) 68–572°F	9.8	9.0	10.0	10.2	10.3
1000°F	–	–	–	–	–
Thermal Conductivity (Btu/ft^2/hr/°F/ft) 68–70°F	187	150	135	109	100
1000°F	–	–	–	–	–
Electrical Resistivity (μ ohm-cm) at 68°F	–	2.54	3.08	3.92	4.31
Magnetic Permeability	–	–	–	–	–
MECHANICAL PROPERTIES (annealed)					
Youngs's Modulus (10^6 psi) 68–70°F	–	17.5	17	17	17
1000°F	–	–	–	–	–
Tensile Strength (1000 psi) 68–70°F	35;63	50	34	37;40	42
1000°F	–	–	–	–	–
Yield Strength (1000 psi) (0.5%) 68–70°F	15;45	30	10	10	15
1000°F	–	–	–	–	–
Elongation (in 2 in.) at 68–75°F (%)	40;25	27	45	45	44
Shear Strength (1000 psi) 68–70°F	–	–	–	28	31
1000°F	–	–	–	–	–
Hardness (Rockwell)	F50;B65	B45	F46	F53	F64

TABLE 5-3 (*Continued*)

COPPER AND ALLOYS—WROUGHT

Type →	230 (Red Brass, 85%)	240 (Low Brass, 80%)	260 (Cartridge Brass, 70%)	268, 170 (Yellow Brass)	280 (Muntz Metal)		
COMPOSITION (%)	Cu 85, Zn 15	Cu 80, Zn 20	Cu 70, Zn 30	Cu 65, Zn 35	Cu 60.0, Zn 40.0		
PHYSICAL, THERMAL, ELECTRICAL AND MAGNETIC PROPERTIES							
Density (lb/in.³)	0.316	0.313	0.308	0.306	0.303		
Melting Range (°F)	1810–1880	1770–1830	1680–1750	1660–1710	1650–1660		
Specific Heat (Btu/lb/°F)	0.09	0.09	0.09	0.09	0.09		
Thermal Expansion Coeff. (μ in./in./°F)							
68–572°F	10.4	10.6	11.1	11.3	11.6		
1000°F	—	—	—	—	—		
Thermal Conductivity (Btu/ft²/hr/°F/ft)							
68–70°F	92	81	70	67	71		
1000°F	—	—	—	—	—		
Electrical Resistivity (μ ohm-cm) at 68°F	4.66	5.39	6.16	6.39	6.16		
Magnetic Permeability	—	—	—	—	—		
MECHANICAL PROPERTIES (annealed)							
Youngs's Modulus (10⁶ psi)							
68–70°F	17	16	16	15	15		
1000°F	—	—	—	—	—		
Tensile Strength (1000 psi)							
68–70°F	39;41	42;44	44;48	46;50	54		
1000°F	—	—	—	—	—		
Yield Strength (1000 psi) (0.5%)							
68–70°F		11	11	14	21		
1000°F	—	—	—	—	—		
Elongation (in 2 in.) at 68–75°F (%)	48;48	52;55	66;64	65;60	45		
Shear Strength (1000 psi)							
68–70°F	31;31	32;32	—	32;34	40		
1000°F	—	—	—	—	—		
Hardness (Rockwell)	F56	F57	F54	F58	F80		

TABLE 5-3 (Continued)

COPPER AND ALLOYS—WROUGHT

Type →	314 (Leaded Commercial Bronze)	330 (Low-Leaded Brass Tube)	332 (High-Leaded Brass Tube)	335 (Low-Leaded Brass)	340 (Medium-Leaded Brass)
COMPOSITION (%)	Cu 89.0, Pb 1.75, Zn 9.25	Cu 66.0, Pb 0.5, Zn 33.5	Cu 66.0, Pb 1.6, Zn 32.4	Cu 65.0, Pb 0.5, Zn 34.5	Cu 65.0, Pb 1.0, Zn 34.0
PHYSICAL, THERMAL, ELECTRICAL AND MAGNETIC PROPERTIES					
Density (lb/in.^{3})	0.319	0.307	0.308	0.306	0.306
Melting Range (°F)	1850–1900	1660–1720	1650–1710	1650–1700	1630–1700
Specific Heat (Btu/lb/°F)	0.09	0.09	0.09	0.09	0.09
Thermal Expansion Coeff. (μ in./in./°F)					
68–572°F	10.2	11.2	11.3	11.3	11.3
1000°F	–	–	–	–	–
Thermal Conductivity (Btu/ft^2/hr/°F/ft)					
68–70°F	104	67	67	67	67
1000°F	–	–	–	–	–
Electrical Resistivity (μ ohm-cm) at 68°F	4.10	6.63	6.63	6.63	6.63
Magnetic Permeability	–	–	–	–	–
MECHANICAL PROPERTIES (annealed)					
Young's Modulus (10^6 psi)					
68–70°F	17	15	15	15	15
1000°F	–	–	–	–	–
Tensile Strength (1000 psi)					
68–70°F	–	52	52	51	51
1000°F	–	–	–	–	–
Yield Strength (1000 psi) (0.5%)					
68–70°F	–	20	20	19	19
1000°F	–	–	–	–	–
Elongation (in 2 in.) at 68–75°F (%)	–	50	50	55	53
Shear Strength (1000 psi)					
68–70°F	–	–	–	–	–
1000°F	–	–	–	–	–
Hardness (Rockwell)	–	F75	F75	F72	F72

TABLE 5-3 (*Continued*)

COPPER AND ALLOYS—WROUGHT

Type →	342, 353 (High-Leaded Brass)	356 (Extra-High-Leaded Brass)	360 (Free-Cutting Brass)	365, 366, 367, 368 (Leaded Muntz Metal)	370 (Free-Cutting Muntz Metal)
COMPOSITION (%)	Cu 65.0, Pb 2.0, Zn 33.0	Cu 63.0, Pb 2.5, Zn 34.5	Cu 61.5, Pb 3.0, Zn 35.5	Cu 60.0, Pb 0.6, Zn 39.4	Cu 60.0, Zn 39.0, Pb 1.0
PHYSICAL, THERMAL, ELECTRICAL AND MAGNETIC PROPERTIES					
Density (lb/in.^{3})	0.306	0.307	0.307	0.304	0.304
Melting Range (°F)	1630–1670	1630–1660	1630–1650	1630–1650	1630–1650
Specific Heat (Btu/lb/°F)	0.09	0.09	0.09	0.09	0.09
Thermal Expansion Coeff. (μ in./in./°F)					
68–572°F	11.3	11.4	11.4	11.6	11.6
1000°F	—	—	—	—	—
Thermal Conductivity (Btu/ft^2/hr/°F/ft)					
68–70°F	67	67	67	71	69
1000°F	—	—	—	—	—
Electrical Resistivity (μ ohm-cm) at 68°F	6.63	6.63	6.63	6.19	6.39
Magnetic Permeability	—	—	—	—	—
MECHANICAL PROPERTIES (annealed)					
Youngs's Modulus (10^6 psi)					
68–70°F	15	14	14	15	15
1000°F	—	—	—	—	—
Tensile Strength (1000 psi)					
68–70°F	49	49	49	54	54
1000°F	—	—	—	—	—
Yield Strength (1000 psi) (0.5%)					
68–70°F	17	17	18	20	20
1000°F	—	—	—	—	—
Elongation (in 2 in.) at 68–75°F (%)	52	50	53	45	40
Shear Strength (1000 psi)					
68–70°F	34	—	30	40	—
1000°F	—	—	—	—	—
Hardness (Rockwell)	F68	F68	F68	F80	F80

TABLE 5-3 (*Continued*)

COPPER AND ALLOYS–WROUGHT

Type →	377 (Forging Brass)	385 (Architectural Bronze)	405 (Tin Brass)	422 (Tin Brass)	425 (Tin Brass)
COMPOSITION (%)	Cu 59.0, Zn 39.0, Pb 2.0	Cu 57.0, Zn 40.0, Pb 3.0	Cu 95, Zn 4, Sn 1	Cu 87, Zn 12, Sn 1	Cu 88, Zn 10, Sn 2
PHYSICAL, THERMAL, ELECTRICAL AND MAGNETIC PROPERTIES					
Density (lb/in.^{3})	0.305	0.306	0.319	0.318	0.317
Melting Range (°F)	1620–1640	1610–1630	1875–1940	1868–1904	1847–1886
Specific Heat (Btu/lb/°F)	0.09	0.09	–	–	–
Thermal Expansion Coeff. (μ in./in./°F)					
68–572°F	11.5	11.6	10.2	10.2	10.2
1000°F	–	–	–	–	–
Thermal Conductivity (Btu/ft^2/hr/°F/ft)					
68–70°F	69	71	95	75	69
1000°F	–	–	–	–	–
Electrical Resistivity (μ ohm-cm) at 68°F	6.39	6.16	4.3	5.7	6.2
Magnetic Permeability	–	–	–	–	–
MECHANICAL PROPERTIES (annealed)					
Youngs's Modulus (10^6 psi)					
68–70°F	15	14	18	18	18
1000°F	–	–	–	–	–
Tensile Strength (1000 psi)					
68–70°F	52	60	42	43	44
1000°F	–	–	–	–	–
Yield Strength (1000 psi) (0.5%)					
68–70°F	20	20	13	15	16
1000°F	–	–	–	–	–
Elongation (in 2 in.) at 68–75°F (%)	45	30	47	45	49
Shear Strength (1000 psi)					
68–70°F	–	35	–	–	–
1000°F	–	–	–	–	–
Hardness (Rockwell)	F78	B65	B47	B46	B49

TABLE 5-3 *(Continued)*

COPPER AND ALLOYS—WROUGHT

Type →	442, 443, 444, 445 (Admiralty)	464, 465, 466, 467 (Naval Brass)	485 (Leaded Naval Brass)	505 (Phosphor Bronze E)	510 (Phosphor Bronze A)
COMPOSITION (%)	Cu 71, Zn 28, Sn 1	Cu 60, Zn 39.25, Sn 0.75	Cu 60, Zn 37.5, Pb 1.75, Sn 0.75	Cu 98.75, Sn 1.25, P trace	Cu 95, Sn 5
PHYSICAL, THERMAL, ELECTRICAL AND MAGNETIC PROPERTIES					
Density (lb/in.^{3})	0.308	0.304	0.305	0.321	0.320
Melting Range (°F)	1650–1720	1630–1650	1630–1650	1900–1970	1750–1920
Specific Heat (Btu/lb/°F)	0.09	0.09	0.09	0.09	0.09
Thermal Expansion Coeff. (μ in./in./°F)					
68–572°F	11.2	11.8	11.8	9.9	9.9
1000°F	–	–	–	–	–
Thermal Conductivity (Btu/ft^2/hr/°F/ft)					
68–70°F	64	67	67	120	40
1000°F	–	–	–	–	–
Electrical Resistivity (μ ohm-cm) at 68°F	6.90	6.63	6.63	3.59	11.5
Magnetic Permeability	–	–	–	–	–
MECHANICAL PROPERTIES (annealed)					
Youngs's Modulus (10^6 psi)					
68–70°F	16	15	15	17	16
1000°F	–	–	–	–	–
Tensile Strength (1000 psi)					
68–70°F	53	57	57	40	47; 50
1000°F	–	–	–	–	–
Yield Strength (1000 psi) (0.5%)					
68–70°F	22	25	25	14	19; 20
1000°F	–	–	–	–	–
Elongation (in 2 in.) at 68–75°F (%)	65	47	40	48	64; 58
Shear Strength (1000 psi)					
68–70°F	–	–	–	–	–
1000°F	–	–	–	–	–
Hardness (Rockwell)	F75	B55	B55	–	B26

TABLE 5-3 (*Continued*)

COPPER AND ALLOYS—WROUGHT

Type →	521 (Phosphor Bronze C)	524 (Phosphor Bronze D)	544 (Phosphor Bronze, Free-Cutting)	614 (Al Bronze D)	638 (Coronze)
COMPOSITION, (%)	Cu 92, Sn 8	Cu 90, Sn 10	Cu 88, Pb 4, Zn 4, Sn 4	Cu 91, Al 7, Fe 2	Cu 95.0, Al 2.8, Si 1.8, Co 0.4
PHYSICAL, THERMAL, ELECTRICAL AND MAGNETIC PROPERTIES					
Density (lb/in.^{3})	0.318	0.317	0.321	0.285	0.299
Melting Range (°F)	1620–1880	1550–1830	1700–1830	1905–1915	1854
Specific Heat (Btu/lb/°F)	0.09	0.09	0.09	0.09	–
Thermal Expansion Coeff. (μ in./in./°F)					
68–572°F	10.1	10.2	9.6	9.0	9.5
1000°F	–	–	–	–	–
Thermal Conductivity (Btu/ft^2/hr/°F/ft)					
68–70°F	36	29	50	39	–
1000°F	–	–	–	–	–
Electrical Resistivity (μ ohm-cm) at 68°F	13.3	15.7	9.07	–	17.4
Magnetic Permeability	–	–	–	–	–
MECHANICAL PROPERTIES (annealed)					
Young's Modulus (10^6 psi)					
68–70°F	16	16	15	17	16.7
1000°F	–	–	–	–	–
Tensile Strength (1000 psi)					
68–70°F	55;60	66;66	44	–	82
1000°F	–	–	–	–	–
Yield Strength (1000 psi) (0.5%)					
68–70°F	24	28	19	–	54 (0.2%)
1000°F	–	–	–	–	–
Elongation (in 2 in.) at 68–75°F (%)	70,65	68	50	–	36
Shear Strength (1000 psi)					
68–70°F	–	–	–	–	–
1000°F	–	–	–	–	–
Hardness (Rockwell)	F75	B55	F65	–	B86

TABLE 5-3 (*Continued*)

COPPER AND ALLOYS—WROUGHT

Type →	639 (Al-Si Bronze)	647 (Precip. Hard Si Bronze)	651 (Low Si Bronze B)	655 (High Si Bronze A)	669 White Alloy
COMPOSITION (%)	Cu 89.0 min, Al 6.5–8.0, Si 1.5–3.0	Cu 97.5, Ni 1.9, Si 0.6	Cu 96.0 min, Si 0.8–2.0, Mn 0.7 max, Fe 0.8 max, Zn 1.5 max, Pb 0.05 max	Cu 94.8 min, Si 2.8–3.8, Mn 1.5 max, Zn 1.5 max, Fe 1.6 max, Ni 0.6 max	Cu 63.5, Zn 24.5, Mn 12
PHYSICAL, THERMAL, ELECTRICAL AND MAGNETIC PROPERTIES					
Density (lb/in.^{3})	0.278	0.322	0.316	0.308	0.293
Melting Range (°F)	1840	1990	1890–1940	1780–1880	–
Specific Heat (Btu/lb/°F)	0.09	–	0.09	0.09	–
Thermal Expansion Coeff. (μ in./in./°F) 68–572°F	10.0	9.8	10.0	10.0	–
1000°F	–	–	–	–	–
Thermal Conductivity (Btu/ft^2/hr/°F/ft) 68–70°F	26	–	33	21	–
1000°F	–	–	–	–	–
Electrical Resistivity (μ ohm-cm) at 68°F	7	–	14.4	24.6	–
Magnetic Permeability	–	–	–	–	–
MECHANICAL PROPERTIES (annealed)					
Youngs's Modulus (10^6 psi) 68–70°F	16	18	17	15	16
1000°F	–	–	–	–	–
Tensile Strength (1000 psi) 68–70°F	85–90	65	40	55–63	61
1000°F	–	–	–	–	–
Yield Strength (1000 psi) (0.5%) 68–70°F	44–60	60	15	20–30	27
1000°F	–	–	–	–	–
Elongation (in 2 in.) at 68–75°F (%)	20–30	20	50	55–63	41
Shear Strength (1000 psi) 68–70°F	–	–	–	–	–
1000°F	–	–	–	–	–
Hardness (Rockwell)	B74–90	–	F55–60	B40–66	F94

COPPER AND ALLOYS–WROUGHT

TABLE 5-3 (*Continued*)

Type →	675 (Manganese Bronze A)	687 (Aluminum Brass)	688 (Alcoloy) TM	706 Cupro-nickel 10 TM	710 Cupro-nickel 20 TM
COMPOSITION (%)	Cu 58.5, Zn 39, Fe 1.4, Sn 1, Mn 0.1	Cu 77.5, Zn 20.5, Al 2	Cu 73.5, Zn 22.7, Al 3.4, Co 0.4	Cu 88.7, Ni 10, Fe 1.3	Cu 77.9, Ni 21, Mn 0.6, Fe 0.5
PHYSICAL, THERMAL, ELECTRICAL AND MAGNETIC PROPERTIES					
Density (lb/in.^{3})	0.302	0.301	0.296	0.323	0.323
Melting Range (°F)	1590–1630	1710–1780	1766	2010–2100	2100–2190
Specific Heat (Btu/lb/°F)	0.09	0.09	–	0.09	0.09
Thermal Expansion Coeff. (μ in./in./°F)					
68–572°F	11.8	10.3	10.1	9.5	9.1
1000°F	–	–	–	–	–
Thermal Conductivity (Btu/ft^2/hr/°F/ft)					
68–70°F	61	58	–	26	21
1000°F	–	–	–	–	–
Electrical Resistivity (μ ohm-cm) at 68°F	7.18	7.50	–	19.1	–
Magnetic Permeability	–	–	–	–	–
MECHANICAL PROPERTIES (annealed)					
Youngs's Modulus (10^6 psi)					
68–70°F	15	16	16.8	18	20
1000°F	–	–	–	–	–
Tensile Strength (1000 psi)					
68–70°F	65	60	82	44	45–51
1000°F	–	–	–	–	–
Yield Strength (1000 psi) (0.5%)					
68–70°F	30	27	55	16	–
1000°F	–	–	–	–	–
Elongation (in 2 in.) at 68–75°F (%)	33	55	36	42	27
Shear Strength (1000 psi)					
68–70°F	–	–	–	–	–
1000°F	–	–	–	–	–
Hardness (Rockwell)	B65	F77	B81	B15	–

TABLE 5-3 (*Continued*)

COPPER AND ALLOYS—WROUGHT

Type →	715 Cupro-nickel 30 TM	745 (65-10)	752 (65-18)	754 (65-15)	757 (65-12)
COMPOSITION (%)	Cu 68.9 Ni 30, Mn 0.6, Fe 0.5	Cu 65, Ni 10, Zn 25	Cu 65, Ni 18, Zn 17	Cu 65, Ni 15, Zn 20	Cu 65, Ni 12, Zn 23
PHYSICAL, THERMAL, ELECTRICAL AND MAGNETIC PROPERTIES					
Density (lb/in.^{3})	0.323	0.314	0.316	0.314	0.314
Melting Range (°F)	2140–2360	1870	1960–2030	1970	1900
Specific Heat (Btu/lb/°F)	0.09	0.09	0.09	0.09	0.09
Thermal Expansion Coeff. (μ in./in./°F)					
68–572°F	9.0	9.1	9.0	9.0	9.0
1000°F	–	–	–	–	–
Thermal Conductivity (Btu/ft^2/hr/°F/ft)					
68–70°F	17	26	19	21	23
1000°F	–	–	–	–	–
Electrical Resistivity (μ ohm-cm) at 68°F	37.5	19.2	28.7	24.6	21.6
Magnetic Permeability	–	–	–	–	–
MECHANICAL PROPERTIES (annealed)					
Youngs's Modulus (10^6 psi)					
68–70°F	22	17.5	18	18	18
1000°F	–	–	–	–	–
Tensile Strength (1000 psi)					
68–70°F	44–60	49–63	56–60	53–61	52–61
1000°F	–	–	–	–	–
Yield Strength (1000 psi) (0.5%)					
68–70°F	20–22	18–28	25–30	18–28	18–28
1000°F	–	–	–	–	–
Elongation (in 2 in.) at 68–75°F (%)	40–45	35–50	32–45	34–43	35–48
Shear Strength (1000 psi)					
68–70°F	–	41	–	41	41
1000°F	–	–	–	–	–
Hardness (Rockwell)	B37–50	B22–52	B40–55	B22–55	B22–55

TABLE 5-3 (*Continued*)

COPPER AND ALLOYS–WROUGHT

Type →	770 (55-18)					
COMPOSITION (%)	Cu 55, Ni 18, Zn 27					
PHYSICAL, THERMAL, ELECTRICAL AND MAGNETIC PROPERTIES						
Density (lb/in.^{3})	0.314					
Melting Range (°F)	1930					
Specific Heat (Btu/lb/°F)	0.09					
Thermal Expansion Coeff. (μ in./in./°F) 68–572°F	9.3					
1000°F	–					
Thermal Conductivity (Btu/ft^2/hr/°F/ft) 68–70°F	17					
1000°F	–					
Electrical Resistivity (μ ohm-cm) at 68°F	31.4					
Magnetic Permeability	–					
MECHANICAL PROPERTIES (annealed)						
Youngs's Modulus (10^6 psi) 68–70°F	18					
1000°F	–					
Tensile Strength (1000 psi) 68–70°F	60					
1000°F	–					
Yield Strength (1000 psi) (0.5%) 68–70°F	27					
1000°F	–					
Elongation (in 2 in.) at 68–75°F (%)	40					
Shear Strength (1000 psi) 68–70°F	–					
1000°F	–					
Hardness (Rockwell)	B55					

TABLE 5-3 (*Continued*)

COPPER AND ALLOYS–CAST

Type →	801	803	805, 807	809, 811
COMPOSITION (%)	Cu + Ag 99.95 min, Other 0.05 max	Cu + Ag 99.95 min, Ag 10 oz/ton min, Other 0.05 max	Cu + Ag 99.75 min, Ag 10 oz/ton min, B 0.01, Other 0.23 max	Cu + Ag 99.70 min, Ag 10 oz/ton (809 only), Other 0.30 max
PHYSICAL, THERMAL, ELECTRICAL AND MAGNETIC PROPERTIES				
Density (lb/in.^{3})	0.323	0.323	0.323	0.323
Melting Range (°F)	1948–1981	1948–1981	1948–1981	1948–1981
Specific Heat (Btu/lb/°F)	0.092	0.092	0.092	0.090
Thermal Expansion Coeff. (μ in./in./°F)				
68–572°F	9.4	9.4	9.4	9.4
1000°F	—	—	—	—
Thermal Conductivity (Btu/ft^2/hr/°F/ft)				
68–70°F	226	226	220	200
1000°F	—	—	—	—
Electrical Resistivity (μ ohm-cm) at 68°F	1.72	1.72	1.87	1.87
Magnetic Permeability	—	—	—	—
MECHANICAL PROPERTIES				
Youngs's Modulus (10^6 psi)				
68–70°F	17	17	17	17
1000°F	—	—	—	—
Tensile Strength (1000 psi)				
68–70°F	25	25	25	25
1000°F	—	—	—	—
Yield Strength (1000 psi) (0.2%)				
68–70°F	9	9	9	9
1000°F	—	—	—	—
Elongation (in 2 in.) at 68–75°F (%)	40	40	40	40
Shear Strength (1000 psi)				
68–70°F	—	—	—	—
1000°F	—	—	—	—
Hardness (Brinell) (500 kg)	44	44	44	44

TABLE 5-3 (*Continued*)

COPPER AND ALLOYS–CAST

Type →	814	815	820	822	824
COMPOSITION (%)	Cu 98.50 min Cr 0.8, Be 0.06	Cu 98.00 min Cr 1.0	Cu 96.8, Co (+ Ni) 2.6, Be 0.6	Cu 96.50 min Ni 1.5, Be 0.6	Cu 97.20 min Be 1.70, Co 0.25
PHYSICAL, THERMAL, ELECTRICAL AND MAGNETIC PROPERTIES					
Density (lb/in.^{3})	0.318	0.319	0.311	0.316	0.298
Melting Range (°F)	1950–2000	1967–1985	1780–1990	1900–2040	1650–1825
Specific Heat (Btu/lb/°F)	0.093	0.09	0.10	0.10	0.10
Thermal Expansion Coeff. (μ in./in./°F)					
68–572°F	10	9.5	9.9	9.9	9.4
1000°F	—	—	—	—	—
Thermal Conductivity (Btu/ft^2/hr/°F/ft)					
68–70°F	150	182	150	106	77
1000°F	—	—	—	—	—
Electrical Resistivity (μ ohm-cm) at 68°F	2.87	2.10	3.82	3.82	6.88
Magnetic Permeability	—	—	—	—	—
MECHANICAL PROPERTIES					
Young's Modulus (10^6 psi)					
68–70°F	16	16.5	17	16.5	18.5
1000°F	—	—	—	—	—
Tensile Strength (1000 psi)					
68–70°F	53 HT*	51 HT	50	57	72
1000°F	—	—	—	—	—
Yield Strength (1000 psi) (0.2%)					
68–70°F	36 HT	40 HT	20	30	37
1000°F	—	—	—	—	—
Elongation (in 2 in.) at 68–75°F (%)	11 HT	17 HT	20	20	20
Shear Strength (1000 psi)					
68–70°F	—	—	—	—	—
1000°F	—	—	—	—	—
Hardness (Rockwell)	B69 HT	—	B55	B60	B78

*Heat Treated

TABLE 5-3 (*Continued*)

COPPER AND ALLOYS—CAST

Type →	825	826	828	833	836
COMPOSITION (%)	Cu 97.2, Be 2.0, Co (+Ni) 0.5, Si 0.25	Cu 95.65 min, Be 2.3, Co 0.5, Si 0.25	Cu 96.6, Be 2.6, Co (+Ni) 0.5, Si 0.25	Cu 93, Zn 4, Sn 1.5, Pb 1.5	Cu 85, Zn 5, Sn 5, Pb 5
PHYSICAL, THERMAL, ELECTRICAL AND MAGNETIC PROPERTIES					
Density (lb/in.^{3})	0.292	0.292	0.292	0.318	0.318
Melting Range (°F)	1575–1800	1575–1750	1625–1710	1886–1940	1570–1850
Specific Heat (Btu/lb/°F)	0.10	0.10	0.10	0.09	0.09
Thermal Expansion Coeff. (μ in./in./°F)					
68–572°F	9.4	9.4	9.4	–	10
1000°F	–	–	–	–	–
Thermal Conductivity (Btu/ft^2/hr/°F/ft)					
68–70°F	18	20	20	32	15
1000°F	–	–	–	–	–
Electrical Resistivity (μ ohm-cm) at 68°F	9.6	8.6	8.6	5.4	11.5
Magnetic Permeability	–	–	–	–	–
MECHANICAL PROPERTIES					
Youngs's Modulus (10^6 psi)					
68–70°F	18.5	19	19.3	15	13.5
1000°F	–	–	–	–	–
Tensile Strength (1000 psi)					
68–70°F	80	82	97	32	37
1000°F	–	–	–	–	–
Yield Strength (1000 psi) (0.2%)					
68–70°F	45	47	55	10	17
1000°F	–	–	–	–	–
Elongation (in 2 in.) at 68–75°F (%)	20	20	20	35	30
Shear Strength (1000 psi)					
68–70°F	–	–	–	–	–
1000°F	–	–	–	–	–
Hardness (Rockwell)	B82	B83	B85	–	–

COPPER AND ALLOYS–CAST

TABLE 5-3 (*Continued*)

Type →	844	852	862	863	874		
COMPOSITION (%)	Cu 81, Zn 9, Pb 7, Sn 3	Cu 72, Zn 24, Pb 3, Sn 1	Cu 64, Zn 26, Al 4, Fe 3, Mn 3	Cu 63, Zn 25, Al 6, Fe 3, Mn 3	Cu 83, Zn 14, Si 3		
PHYSICAL, THERMAL, ELECTRICAL AND MAGNETIC PROPERTIES							
Density (lb/in.^{3})	0.314	0.307	0.288	0.283	0.300		
Melting Range (°F)	1549–1940	1700–1725	1650–1725	1625–1693	1510–1680		
Specific Heat (Btu/lb/°F)	0.09	0.09	0.09	0.09	0.09		
Thermal Expansion Coeff. (μ in./in./°F)							
68–572°F	10	11.5	12	12	10.9		
1000°F	–	–	–	–	–		
Thermal Conductivity (Btu/ft^2/hr/°F/ft)							
68–70°F	41.9	48.5	20.5	20.5	16.0		
1000°F	–	–	–	–	–		
Electrical Resistivity (μ ohm-cm) at 68°F	10.5	9.6	23.0	21.5	25.7		
Magnetic Permeability	–	–	–	–	–		
MECHANICAL PROPERTIES							
Youngs's Modulus (10^6 psi)							
68–70°F	13	11	15	14.2	15.4		
1000°F	–	–	–	–	–		
Tensile Strength (1000 psi)							
68–70°F	34	38	95	119	55		
1000°F	–	–	–	–	–		
Yield Strength (1000 psi) (0.5%)							
68–70°F	15	13	48 (0.2%)	83 (0.2%)	24		
1000°F	–	–	–	–	–		
Elongation (in 2 in.) at 68–75°F (%)	26	35	20	18	30		
Shear Strength (1000 psi)							
68–70°F	–	–	–	–	–		
1000°F	–	–	–	–	–		
Hardness (Brinell) (500 kg)	55	45	–	–	70		

TABLE 5-3 (*Continued*)

COPPER AND ALLOYS—CAST

Type →	903	905	916	922	923
COMPOSITION (%)	Cu 88, Sn 8, Zn 4, Ni 1 max	Cu 88, Sn 10, Zn 2 Ni 1 min	Cu 88, Sn 10, 5, Ni 1.5	Cu 88, Sn 6, Zn 4.5, Pb 1.5	Cu 87, Sn 8, Zn 4, Ni 1
PHYSICAL, THERMAL, ELECTRICAL AND MAGNETIC PROPERTIES					
Density (lb/in.^{3})	0.318	0.315	0.320	0.312	0.317
Melting Range (°F)	1570–1832	1570–1830	1575–1887	1518–1810	1570–1830
Specific Heat (Btu/lb/°F)	0.09	0.09	0.09	0.09	0.09
Thermal Expansion Coeff. (μ in./in./°F)					
68–572°F	10	11	9	10	10
1000°F	–	–	–	–	–
Thermal Conductivity (Btu/ft^2/hr/°F/ft)					
68–70°F	43.2	43.2	40.8	40.2	43.2
1000°F	–	–	–	–	–
Electrical Resistivity (μ ohm-cm) at 68°F	14.3	15.7	17.2	12.0	14.3
Magnetic Permeability	–	–	–	–	–
MECHANICAL PROPERTIES					
Youngs's Modulus (10^6 psi)					
68–70°F	14	15	16	14	14
1000°F	–	–	–	–	–
Tensile Strength (1000 psi)					
68–70°F	45	45	44	40	40
1000°F	–	–	–	–	–
Yield Strength (1000 psi) (0.5%)					
68–70°F	21	22	22	20	20
1000°F	–	–	–	–	–
Elongation (in 2 in.) at 68–75°F (%)	30	25	6	30	25
Shear Strength (1000 psi)					
68–70°F	–	–	–	–	–
1000°F	–	–	–	–	–
Hardness (Brinell) (500 kg)	70	75	85	65	70

COPPER AND ALLOYS–CAST

TABLE 5-3 (*Continued*)

Type →	932	937	947	948	952
COMPOSITION (%)	Cu 83, Pb 7, Zn 3	Cu 80, Sn 10, Pb 10	Cu 88, Sn 5, Ni 5, Zn 2 Zn 2.5 max	Cu 87, Sn 5, Ni 5,	Cu 88, Al 9, Fe 3
PHYSICAL, THERMAL, ELECTRICAL AND MAGNETIC PROPERTIES					
Density (lb/in.³)	0.322	0.320	0.320	0.320	0.276
Melting Range (°F)	1570–1790	1403–1705	1660–1880	1660–1880	1907–1913
Specific Heat (Btu/lb/°F)	0.09	0.09	0.09	0.09	0.09
Thermal Expansion Coeff. (μ in./in./°F)					
68–572°F	10	10.3	11	11	9.0
1000°F	–	–	–	–	–
Thermal Conductivity (Btu/ft²/hr/°F/ft)					
68–70°F	33.6	27.1	31.2	22.3	29.1
1000°F	–	–	–	–	–
Electrical Resistivity (μ ohm-cm) at 68°F	14.3	17.2	14.3	14.3	15.7
Magnetic Permeability	–	–	–	–	–
MECHANICAL PROPERTIES					
Young's Modulus (10⁶ psi)					
68–70°F	14.5	11	15	15	15
1000°F	–	–	–	–	–
Tensile Strength (1000 psi)					
68–70°F	35	35	50	50	80
1000°F	–	–	–	–	–
Yield Strength (1000 psi) (0.5%)					
68–70°F	18	18	23	23	27
1000°F	–	–	–	–	–
Elongation (in 2 in.) at 68–75°F (%)	20	20	35	35	35
Shear Strength (1000 psi)					
68–70°F	–	18	38	–	40
1000°F	–	–	–	–	–
Hardness (Brinell) (500 kg)	65	60	85	80	125

TABLE 5-3 (Continued)

COPPER AND ALLOYS–CAST

Type →	953	954	955	962	964		
COMPOSITION (%)	Cu 89, Al 10, Fe 1	Cu 85, Al 11, Fe 4	Cu 81, Al 11, Ni 4, Fe 4	Cu 88.6, Ni 10, Fe 1.4, Mn 1.5 max	Cu 69.1, Ni 30, Mn 1.5 max, Fe 0.9		
PHYSICAL, THERMAL, ELECTRICAL AND MAGNETIC PROPERTIES							
Density (lb/in.³)	0.272	0.269	0.272	0.323	0.323		
Melting Range (°F)	1904–1913	1880–1900	1900–1930	2010–2100	2140–2260		
Specific Heat (Btu/lb/°F)	0.09	0.10	0.10	0.09	0.09		
Thermal Expansion Coeff. (μ in./in./°F)							
68–572°F	9	9	9	9.5	9		
1000°F	–	–	–	–	–		
Thermal Conductivity (Btu/ft²/hr/°F/ft)							
68–70°F	36.3	33.9	24.2	26	17		
1000°F	–	–	–	–	–		
Electrical Resistivity (μ ohm-cm) at 68°F	13.3	13.3	20.3	15.6	34.4		
Magnetic Permeability	–	–	–	–	–		
MECHANICAL PROPERTIES							
Youngs's Modulus (10⁶ psi)							
68–70°F	16	15.5	16	18	21		
1000°F	–	–	–	–	–		
Tensile Strength (1000 psi)							
68–70°F	75	85	100	45	68		
1000°F	–	–	–	–	–		
Yield Strength (1000 psi) (0.5%)							
68–70°F	27	35	44	25	37		
1000°F	–	–	–	–	–		
Elongation (in 2 in.) at 68–75°F (%)	25	18	12	20	28		
Shear Strength (1000 psi)							
68–70°F	41	47	48	–	–		
1000°F	–	–	–	–	–		
Hardness (Brinell) (3000 kg)	140	170	195	–	140		

TABLE 5-3 (*Continued*)

COPPER AND ALLOYS–CAST

Type →	973	974	976	978	993
COMPOSITION (%)	Cu 56, Zn 20, Ni 12, Pb 10, Sn 2	Cu 59, Ni 17, Zn 16, Pb 5, Sn 3	Cu 59, Ni 17, Zn 16, Pb 5, Sn 3	Cu 66, Ni 25, Sn 5, Zn 2, Pb 2	Cu 71.8, Ni 15, Al 11, Co 1.5, Fe 0.7
PHYSICAL, THERMAL, ELECTRICAL AND MAGNETIC PROPERTIES					
Density (lb/in.^{3})	0.321	0.320	0.321	0.320	0.275
Melting Range (°F)	1850–1904	1958–2012	2027–2089	2084–2156	1955–1970
Specific Heat (Btu/lb/°F)	0.09	0.09	0.09	0.09	0.10
Thermal Expansion Coeff. (μ in./in./°F)					
68–572°F	9.0	9.2	9.3	9.7	9.2
1000°F	–	–	–	–	–
Thermal Conductivity (Btu/ft^2/hr/°F/ft)					
68–70°F	16.5	15.8	13	14.7	25
1000°F	–	–	–	–	–
Electrical Resistivity (μ ohm-cm) at 68°F	30.2	31.4	34.4	38.3	19.2
Magnetic Permeability	–	–	–	–	–
MECHANICAL PROPERTIES					
Young's Modulus (10^6 psi)					
68–70°F	16	16	19	19	18
1000°F	–	–	–	–	–
Tensile Strength (1000 psi)					
68–70°F	35	38	40	55	95
1000°F	–	–	–	–	–
Yield Strength (1000 psi) (0.5%)					
68–70°F	17	17	24	30	55
1000°F	–	–	–	–	–
Elongation (in 2 in.) at 68–75°F (%)	20	20	20	16	2
Shear Strength (1000 psi)					
68–70°F	–	–	–	–	–
1000°F	–	–	–	–	–
Hardness (Brinell) (500 kg)	55	70	80	130	200

TABLE 5-4. IRON ALLOYS AND STEELS

GRAY IRON–CAST

Type →	20	25	30	40	60
COMPOSITION (%)	C 3.50–3.80, Si 2.20–2.50, P 0.20–0.60, S 0.08–0.13, Mn 0.50–0.80	C 3.00–3.40, Si 1.90–2.40, P 0.15–0.50, S 0.08–0.12, Mn 0.50–0.80	C 2.90–3.40, Si 1.70–2.30, P 0.15–0.30, S 0–0.12, Mn 0–0.70	C 2.75–3.20, Si 1.50–2.20, P 0.07–0.25, S 0.05–0.12, Mn 0.45–0.70	C 2.50–3.00, Si 1.20–2.20, P 0.05–0.20, S 0.05–0.12, Mn 0.50–1.00
PHYSICAL, THERMAL, ELECTRICAL AND MAGNETIC PROPERTIES					
Density (lb/in.^{3})	0.25	0.25	0.26	0.26	0.27
Melting Range (°F)	–	–	–	–	–
Specific Heat (Btu/lb/°F)	–	–	–	–	–
Thermal Expansion Coeff. (μ in./in./°F)					
32–212°F	6	6	6	6	6
1000°F	–	–	–	–	–
Thermal Conductivity (Btu/ft^2/hr/°F/ft)					
212°F	28–30	28–30	28–30	28–30	28–30
1000°F	–	–	–	–	–
Electrical Resistivity (μ ohm-cm) at 68°F	ranges from	50–200 depending on composition			
Magnetic Permeability	–	–	–	–	–
MECHANICAL PROPERTIES (annealed)					
Young's Modulus (10^6 psi)					
68–70°F	12	13	15	17	20
1000°F	–	–	–	–	–
Tensile Strength (1000 psi)					
68–70°F	20–25	25–30	30–35	40–48	60–66
1000°F	–	–	–	–	–
Yield Strength (1000 psi)					
68–70°F	–	–	–	–	–
1000°F	–	–	–	–	–
Elongation (in 2 in.) at 68–75°F (%)	–	–	–	–	–
Shear Strength (1000 psi)					
68–70°F	32	37	44	57	72
1000°F	–	–	–	–	–
Hardness (Brinell)	140–180	140–190	170–220	200–240	230–290

TABLE 5-4. (*Continued*)

MALLEABLE IRON–CAST

	FERRITIC		PEARLITIC		
Type →	32510	35018	45010	45007	48004
COMPOSITION (%)	C 2.3–2.7, Si 1.5–0.8, Mn 0.55 max, P < 0.18, S < 0.15	C 2.0–2.45, Si 1.4–0.85, Mn 0.55 max, P < 0.18, S < 0.15	C 2.35–2.50, Si 1.0–1.5, Mn 0.28–0.48, P 0.06–0.10, S 0.16–0.20		
PHYSICAL, THERMAL, ELECTRICAL AND MAGNETIC PROPERTIES					
Density (lb/in.^{3})	0.259–0.263	0.259–0.263	0.265–0.268	0.265–0.268	0.265–0.268
Melting Range (°F)	–	–	–	–	–
Specific Heat (Btu/lb/°F)	–	–	–	–	–
Thermal Expansion Coeff. (μ in./in./°F)					
68–212°F	5.9	5.9	7.5	7.5	7.5
1000°F	–	–	–	–	–
Thermal Conductivity (Btu/ft^2/hr/°F/ft)					
80°F	29.5	29.5	29.5	29.5	29.5
700°F	23.0	23.0	>23	>23	>23
Electrical Resistivity (μ ohm-cm) at 68°F	14.1	14.1	15.9	15.9	15.9
Magnetic Permeability (maximum)	2300	2300	430	430	430
MECHANICAL PROPERTIES (annealed)					
Youngs's Modulus (10^6 psi)					
68–70°F	25	25	25	26	26.5
1000°F	–	–	–	–	–
Tensile Strength (1000 psi)					
68–70°F	50	53	65	68	70
1000°F	–	–	–	–	–
Yield Strength (1000 psi)					
68–70°F	32.5	35	45	45	48
1000°F	–	–	–	–	–
Elongation (in 2 in.) at 68–75°F (%)	10	18	10	7	4
Shear Strength (1000 psi)					
68–70°F (ultimate)	45–48	48–54	49	51	52.5
1000°F	–	–	–	–	–
Hardness (Brinell)	110–156	110–156	163–207	163–217	163–228

TABLE 5-4. (Continued)

PEARLITIC

MALLEABLE IRON–CAST

Type →	50007	53004	60003	80002
COMPOSITION (%)	C 2.35–2.6, Si 1.0–1.5, Mn 0.28–0.48, P 0.06–0.10, S 0.16–0.20			
PHYSICAL, THERMAL, ELECTRICAL AND MAGNETIC PROPERTIES				
Density (lb/in.^{3})	0.265–0.268	0.265–0.268	0.265–0.268	0.265–0.268
Melting Range (°F)	–	–	–	–
Specific Heat (Btu/lb/°F)	–	–	–	–
Thermal Expansion Coeff. (μ in./in./°F)				
68–212°F	7.5	7.5	7.5	7.5
1000°F	–	–	–	–
Thermal Conductivity (Btu/ft^2/hr/°F/ft)				
80°F	29.5	29.5	29.5	29.5
700°F	23	23	23	23
Electrical Resistivity (μ ohm-cm) at 68°F	15.9	15.9	16.3	16.3
Magnetic Permeability (maximum)	430	430	170	170
MECHANICAL PROPERTIES (annealed)				
Young's Modulus (10^6 psi)				
68–70°F	26.5	26.5	26–28	26–28
1000°F	–	–	–	–
Tensile Strength (1000 psi)				
68–70°F	75	80	80	100
1000°F	–	–	–	–
Yield Strength (1000 psi)				
68–70°F	50	53	60	80
1000°F	–	–	–	–
Elongation (in 2 in.) at 68–75°F (%)	7	4	3	2
Shear Strength (1000 psi)				
68–70°F	56	60	65	75
1000°F	–	–	–	–
Hardness (Brinell)	179–228	197–241	197–255	241–269

TABLE 5-4. (*Continued*)

DUCTILE OR NODULAR IRON–CAST

Type →	80–55–06	60–48–18 or 50–45–12	100–70–03	120–90–02	Austenitic
COMPOSITION (%)	C 3.3–3.8, Si 2.0–3.0, Mn 0.2–0.5, P 0.06–0.08, Ni 0–1.0, Mg 0.02–0.07.	C 3.4–4.0, Si 2.0–2.75, Mn 0.2–0.6, P 0.06–0.08, Ni 0–1.0, Mg 0.02–0.07.	C 3.4–3.8, Si 2.0–2.75, Mn 0.3–0.6, P 0.08 max, Ni 0–2.5, Mo 0–1.0, Mg 0.02–0.07.	C 3.4–3.8, Si 2.0–2.75, Mn 0.3–0.6, P 0.08 max, Ni 0–2.5, Mo 0–1.0, Mg 0.02–0.07.	C 3.0, Si 2.0–3.2, Mn 0.8–1.5, P 0.02 max, Ni 18–22, Cr 0–2.5
PHYSICAL, THERMAL, ELECTRICAL AND MAGNETIC PROPERTIES					
Density (lb/in.^{3})	0.257	0.257	0.257	0.257	0.268
Melting Range (°F)	2050–2150	2050–2150	2050–2150	2050–2150	2250
Specific Heat (Btu/lb/°F)	–	–	–	–	–
Thermal Expansion Coeff. (μ in./in./°F)					
70–400°F	6.6	6.6	6.6	6.6	10.4
1000°F	–	–	–	–	–
Thermal Conductivity (Btu/ft^2/hr/°F/ft)					
212°F	18	20	–	–	–
1000°F	–	–	–	–	–
Electrical Resistivity (μ ohm-cm) at 75°F	68	66	–	–	102
Magnetic Permeability	–	–	–	–	–
MECHANICAL PROPERTIES (annealed)					
Youngs's Modulus (10^6 psi)					
68–70°F	22–25	22–25	22–25	22–25	18.5
1000°F	–	–	–	–	–
Tensile Strength (1000 psi)					
68–70°F	90–110	60–80	100–120	120–150	58–68
1000°F	–	–	–	–	–
Yield Strength (1000 psi)					
68–70°F	60–75	45–60	75–90	90–125	32–38
1000°F	–	–	–	–	–
Elongation (in 2 in.) at 68–75°F (%)	3–10	10–25	6–10	2–7	7–40
Shear Strength (1000 psi)					
68–70°F	–	–	–	–	–
1000°F	–	–	–	–	–
Hardness (Brinell)	179–255	149–207	229–285	240–325	140–200

TABLE 5-4. (*Continued*)

WHITE AND ALLOY IRONS—CAST

Type →	White Iron	Ni-Hard	High Chromium	Molybdenum	High Silicon		
COMPOSITION (%)	C 2.8-3.6, Si 0.5-1.3, Mn 0.4-0.9.	C 2.8-3.6, Si 0.4-0.7, Mn 0.2-0.7, Ni 2.5-4.75, Cr 1.2-3.5.	C 1.8-3.5, Si 0.5-2.5, Mn 0.3-1.0, Ni 0-5.0, Cr 10-35, Cu 0-3.0, Mo 0-3.0.	C 1.7-3.7, Si 0.3-2.6, Mn 0.2-1.5, Ni 0-5.0, Cr 0-6.0, Cu 0-1.5, Mo 0.3-12.0.	C 0.4-1.0, Si 14-17, Mn 0.4-1.0, Mo 0-3.5.		
PHYSICAL, THERMAL, ELECTRICAL AND MAGNETIC PROPERTIES							
Density (lb/in.^{3})	0.274-0.281	0.275-0.280	0.264-0.280	0.275-0.285	0.252-0.254		
Melting Range (°F)	—	—	—	—	—		
Specific Heat (Btu/lb/°F)	—	—	—	—	—		
Thermal Expansion Coeff. (μ in./in./°F)							
68-70°F	5.0-5.3	4.5-5.0	5.2-5.5	—	6.70		
1000°F	—	—	—	—	—		
Thermal Conductivity (Btu/ft^2/hr/°F/ft)							
68-70°F	—	—	—	—	—		
1000°F	—	—	—	—	—		
Electrical Resistivity (μ ohm-cm) at 68°F	5.0	80	—	—	—		
Magnetic Permeability	—	—	—	—	—		
MECHANICAL PROPERTIES (annealed)							
Youngs's Modulus (10^6 psi)							
68-70°F	—	—	—	—	—		
1000°F	—	—	—	—	—		
Tensile Strength (1000 psi)							
68-70°F	20-50	40-75	32-90	25-60	13-18		
1000°F	—	—	—	—	—		
Yield Strength (1000 psi)							
68-70°F	—	—	—	—	—		
1000°F	—	—	—	—	—		
Elongation (in 2 in.) at 68-75°F (%)	—	—	—	—	—		
Shear Strength (1000 psi)							
68-70°F	—	—	—	—	—		
1000°F	—	—	—	—	—		
Hardness (Brinell)	300-575	525-600	250-700	350-700	450-500		

TABLE 5-4. (*Continued*)

ALLOY IRONS—CAST

Type →	High Silicon	High Nickel	High Aluminum			
COMPOSITION (%)	C 1.6–2.5, Si 4.0–6.0, Mn 0.4–0.8.	C 1.8–3.0, Si 1.0–2.75, Mn 0.4–1.5, Ni 14–30, Cr 0.5–5.5, Cu 0–7.0, Mo 0–1.0.	C 1.3–1.7, Si 1.3–6.0, Mn 0.4–1.0, Al 18–25.			
PHYSICAL, THERMAL, ELECTRICAL AND MAGNETIC PROPERTIES						
Density (lb/in.^{3})	0.245–0.255	0.264–0.270	0.200–0.232			
Melting Range (°F)	—	—	—			
Specific Heat (Btu/lb/°F)	—	—	—			
Thermal Expansion Coeff. (μ in./in./°F)						
68–70°F	6.0	4.5–10.7	8.5			
1000°F	—	—	—			
Thermal Conductivity (Btu/ft^2/hr/°F/ft)						
68–70°F	—	—	—			
1000°F	—	—	—			
Electrical Resistivity (μ ohm-cm) at 68°F	—	—	240			
Magnetic Permeability	—	—	—			
MECHANICAL PROPERTIES (annealed)						
Youngs's Modulus (10^6 psi)						
68–70°F	—	—	—			
1000°F	—	—	—			
Tensile Strength (1000 psi)						
68–70°F	25–45	25–45	34–90			
1000°F	—	—	—			
Yield Strength (1000 psi)						
68–70°F	—	—	—			
1000°F	—	—	—			
Elongation (in 2 in.) at 68–75°F (%)	—	—	—			
Shear Strength (1000 psi)						
68–70°F	—	—	—			
1000°F	—	—	—			
Hardness (Brinell)	170–250	130–250	180–350			

TABLE 5-4. (*Continued*)

IRON–BASE SUPERALLOYS–CAST, WROUGHT

Type →	A-286	V-57	16-25-6	Incolloy 800	Incolloy 901
COMPOSITION (%)	C 0.08M, Mn 1.35, Si 0.70, Cr 15.00, Ni 26.0, Mo 1.25, Ti 2.15, V 0.30, Al 0.20.	C 0.06, Mn 0.25, Si 0.55, Cr 15, Ni 25.5, Mo 1.25, Ti 3.0, Al 0.25, V 0.025, B 0.008.	C 0.10M, Mn 2.0M, Si 1.0 max. Cr 16.25, Ni 25.5, Mo 6.0.	C 0.05, Mn 0.75, Si 0.5, Cr 21, Ni 32.5, Ti 0.38, Al 0.38, Cu 0.38, S 0.008.	C 0.05, Mn 0.15, Si 0.40, Cr 13.0, Ni 43.0, Ti 3.00, Mo 6.0, Al 0.20, B 0.015.
PHYSICAL, THERMAL, ELECTRICAL AND MAGNETIC PROPERTIES					
Density (lb/in.^{3})	0.286	0.286	0.291	0.287	0.296
Melting Range (°F)	2500–2600	–	–	2475–2525	–
Specific Heat (Btu/lb/°F)	0.10–0.11	–	–	0.12	–
Thermal Expansion Coeff. (μ in./in./°F)					
80–1400°F	10.3	10.5	9.4	10.1	9.2
1000°F	–	–	–	–	–
Thermal Conductivity (Btu/ft^2/hr/°F/ft)					
68–70°F	–	–	–	–	–
1100°F	13.7	–	15.0	11.6	10.8
Electrical Resistivity (μ ohm-cm) at 68°F	–	–	–	–	–
Magnetic Permeability	–	–	–	–	–
MECHANICAL PROPERTIES (annealed)					
Youngs's Modulus (10^6 psi)					
68–70°F	29.1	–	28.2	28.5	29.9
1000°F	23.5	–	–	23.5	24.2
Tensile Strength (1000 psi)					
68–70°F	146	172	142	88	175
1000°F	104	129	90	55	140
Yield Strength (1000 psi), (0.2% offset)					
68–70°F	100	119	112	44	130
1000°F	88	108	75	30	115
Elongation (in 2 in.) at 68–75°F (%)	25	24	23	45	15
Shear Strength (1000 psi)					
68–70°F	–	–	–	–	–
1000°F	–	–	–	–	–
Hardness (Brinell)	–	–	–	–	–

TABLE 5-4. (*Continued*)

IRON–BASE SUPERALLOYS–WROUGHT

Type →	19-9 DL	Unitemp 212	W 545	D-979	AMS 5700
COMPOSITION (%)	C 0.32, Mn 1.15, Si 0.55, Cr 18.5, Ni 9.0, Mo 1.40, W 1.35, Cb+Ta 0.40, Ti 0.25, Cu 0.15, Fe bal	C 0.08, Mn 0.05, Si 0.15, Cr 16.0, Ni 25.0, Ti 4.0, Al 0.15, Cb+Ta 0.50, B 0.06, Zr 0.05, Fe bal	C 0.02, Mn 1.65, Si 0.40, Cr 13.5, Ni 26.0, Ti 1.75, Mo 3.00, Al 0.15, B 0.05, Fe bal	C 0.05, Mn 0.50, Si 0.50, Cr 15.0, Ni 45.0, Mo 4.0, W 4.0, Ti 3.0, Al 1.0, B 0.01, Fe bal	C 0.45, Mn 0.70, Si 0.60, P 0.030 Max, S 0.025 Max, Cr 14.00, Ni 14.00, W 2.50, Mo 0.50 Max, Fe bal
PHYSICAL, THERMAL, ELECTRICAL AND MAGNETIC PROPERTIES					
Density (lb/in.^{3})	0.287	0.286	0.285	0.295	0.29
Melting Range (°F)	2560–2615	2480	2460–2530	2225–2550	2550–2600
Specific Heat (Btu/lb/°F)	0.10	–	0.115	–	–
Thermal Expansion Coeff. (μ in./in./°F)					
70–1500°F	10	10	10.7	9.5	10
1000°F					
Thermal Conductivity (Btu/ft^2/hr/°F/ft)					
1200°F	12.2	13.7	10.7	–	10.5
1000°F					
Electrical Resistivity (μ ohm-cm) at 68°F	–	91	92.8	–	–
Magnetic Permeability	–	–	–	–	–
MECHANICAL PROPERTIES (annealed)					
Young's Modulus (10^6 psi)					
68–70°F	29.5	29.0	28.4	30.0	30.0
1000°F	23.3	23.3	23.5	26.0	–
Tensile Strength (1000 psi)					
68–70°F	114	187	181	204	110
1000°F	79	–	154	189	–
Yield Strength (1000 psi) (0.2%)					
68–70°F	71	134	133	146	48
1000°F	55	–	121	134	–
Elongation (in 2 in.) at 68–75°F (%)	41	23	19	16	40
Shear Strength (1000 psi)					
68–70°F	–	–	–	–	–
1000°F	–	–	–	–	–
Hardness (Brinell)	–	–	–	–	–

TABLE 5-4. (*Continued*)

IRON–BASE SUPERALLOYS—WROUGHT

Type →	Multimet, N-155	Refractaloy 26	S-590	19-9 DL
COMPOSITION (%)	C 0.10, Mn 1.50, Si 0.70, Cr 20.75, Ni 19.85, Co 19.50, Mo 2.95, W 2.35, Cb + Ta 1.15, Cu 0.20, Fe bal	C 0.03, Mn 0.8, Si 1.0, Cr 18.0, Ni 38.0, Co 20.0, Mo 3.2, Ti 2.6, Al 0.2, Fe bal	C 0.4, Mn 1.5, Si 0.4, Cr 20.0, Ni 20.0, Co 20.0, Mo 4.0, W 4.0, Cb 4.0, Fe bal	C 0.3, Mn 1.0, Si .50, Cr 19.0, Ni 9.0, Mo 1.4, W 1.3, Co 0.4, Ti 0.3, Fe bal
PHYSICAL, THERMAL, ELECTRICAL AND MAGNETIC PROPERTIES				
Density (lb/in.^{3})	0.296	0.296	0.301	0.287
Melting Range (°F)	2350–2470	2450	2400–2500	2600–2610
Specific Heat (Btu/lb/°F)	0.104	0.108	0.10	—
Thermal Expansion Coeff. (μ in./in./°F)				
68–70°F	9.1	8.2	8.0	9.9
1000°F	9.1	8.2	8.0	9.9
Thermal Conductivity (Btu/ft^2/hr/°F/ft)				
68–70°F	—	—	—	—
1000°F	—	—	—	—
Electrical Resistivity (μ ohm-cm) at 68°F	—	—	—	—
Magnetic Permeability	—	—	—	—
MECHANICAL PROPERTIES (annealed)				
Youngs's Modulus (10^6 psi)				
68–70°F	28.8	30.6	31.1	29.5
1000°F	24.6	26.3	—	—
Tensile Strength (1000 psi)				
68–70°F	118	154	142	118
1000°F	94	143	132	89
Yield Strength (1000 psi) (0.2%)				
68–70°F	58	91	75	69
1000°F	40	85	80	42
Elongation (in 2 in.) at 68–75°F (%)	49	19	21	56
Shear Strength (1000 psi)				
68–70°F	—	—	—	—
1000°F	—	—	—	—
Hardness (Brinell)	—	—	—	—

ALLOY STEELS–CAST

TABLE 5-4. (*Continued*)

Type →	65,000	70,000	80,000	90,000	105,000
COMPOSITION (%)					
PHYSICAL, THERMAL, ELECTRICAL AND MAGNETIC PROPERTIES					
Density (lb/in.^{3})	0.283	0.283	0.283	0.283	0.283
Melting Range (°F)	–	–	–	–	–
Specific Heat (Btu/lb/°F)	0.10–0.11	0.10–0.11	0.10–0.11	0.10–0.11	0.10–0.11
Thermal Expansion Coeff. (μ in./in./°F)					
70–1200°F	8.0–8.3	8.0–8.3	8.0–8.3	8.0–8.3	8.0–8.3
1000°F	–	–	–	–	–
Thermal Conductivity (Btu/ft^2/hr/°F/ft)					
212°F	27	27	27	27	27
1000°F	–	–	–	–	–
Electrical Resistivity (μ ohm-cm) at 68°F	15–20	15–20	15–20	15–20	15–20
Magnetic Permeability	–	–	–	–	–
MECHANICAL PROPERTIES (annealed)					
Youngs's Modulus (10^6 psi)					
68–70°F	29–30	29–30	29–30	29–30	29–30
1000°F	–	–	–	–	–
Tensile Strength (1000 psi)					
68–70°F	68	74	85	95	110
1000°F	–	–	–	–	–
Yield Strength (1000 psi)					
68–70°F	–	–	–	–	–
1000°F	–	–	–	–	–
Elongation (in 2 in.) at 68–75°F (%)	32	28	24	20	21
Shear Strength (1000 psi)					
68–70°F	–	–	–	–	–
1000°F	–	–	–	–	–
Hardness (Brinell)	137	143	170	192	217

TABLE 5-4. (*Continued*)

ALLOY STEELS–CAST

Type →	120,000	150,000	175,000	200,000
COMPOSITION (%)				
PHYSICAL, THERMAL, ELECTRICAL AND MAGNETIC PROPERTIES				
Density (lb/in.³)	0.283	0.283	0.283	0.283
Melting Range (°F)	–	–	–	–
Specific Heat (Btu/lb/°F)	0.10–0.11	0.10–0.11	0.10–0.11	0.10–0.11
Thermal Expansion Coeff. (μ in./in./°F)				
70–1200°F	8.0–8.3	8.0–8.3	8.0–8.3	8.0–8.3
1000°F	–	–	–	–
Thermal Conductivity (Btu/ft²/hr/°F/ft)				
212°F	27	27	27	27
1000°F	–	–	–	–
Electrical Resistivity (μ ohm-cm) at 68°F	15–20	15–20	15–20	15–20
Magnetic Permeability	–	–	–	–
MECHANICAL PROPERTIES (annealed)				
Youngs's Modulus (10⁶ psi)				
68–70°F	29–30	29–30	29–30	29–30
1000°F	–	–	–	–
Tensile Strength (1000 psi)				
68–70°F	128	158	179	205
1000°F	–	–	–	–
Yield Strength (1000 psi)				
68–70°F	–	–	–	–
1000°F	–	–	–	–
Elongation (in 2 in.) at 68–75°F (%)	16	13	11	8
Shear Strength (1000 psi)				
68–70°F	–	–	–	–
1000°F	–	–	–	–
Hardness (Brinell)	262	311	352	401

TABLE 5-4. (*Continued*)

CARBON STEELS–CAST

Type →	60,000	65,000	70,000	80,000	100,000
COMPOSITION (%)					
PHYSICAL, THERMAL, ELECTRICAL AND MAGNETIC PROPERTIES					
Density (lb/in.^{3})	0.283	0.283	0.283	0.283	0.283
Melting Range (°F)	–	–	–	–	–
Specific Heat (Btu/lb/°F)	0.10–0.11	0.10–0.11	0.10–0.11	0.10–0.11	0.10–0.11
Thermal Expansion Coeff. (μ in./in./°F)					
70–1200°F	8.3	8.3	8.3	8.3	8.3
1000°F	–	–	–	–	–
Thermal Conductivity (Btu/ft^2/hr/°F/ft)					
212°F	27	27	27	27	27
1000°F	–	–	–	–	–
Electrical Resistivity (μ ohm-cm) at 68°F	–	–	–	–	–
Magnetic Permeability	–	–	–	–	–
MECHANICAL PROPERTIES (annealed)					
Young's Modulus (10^6 psi)					
68–70°F	30.1	30.1	30	29.9	29.7
1000°F	–	–	–	–	–
Tensile Strength (1000 psi)					
68–70°F	63	68	75	82	105
1000°F	–	–	–	–	–
Yield Strength (1000 psi)					
68–70°F	–	–	–	–	–
1000°F	–	–	–	–	–
Elongation (in 2 in.) at 68–75°F (%)	30	28	27	23	19
Shear Strength (1000 psi)					
68–70°F	–	–	–	–	–
1000°F	–	–	–	–	–
Hardness (Brinell)	131	131	143	163	212

TABLE 5-4. (*Continued*)

CARBON STEELS–CARBURIZING GRADES

Type →	C1015	C1020	C1022	C1117	C1118
COMPOSITION (%)	C 0.13–0.18, Mn 0.30–0.60, P 0.040 max, S 0.050 max	C 0.18–0.23, Mn 0.30–0.60, P 0.040 max, S 0.050 max	C 0.18–0.23, Mn 0.70–1.00, P 0.040 max, S 0.050 max	C 0.14–0.20, Mn 1.00–1.30, P 0.040 max, S 0.08–0.13	C 0.14–0.20, Mn 1.30–1.60, P 0.040 max, S 0.08–0.13
PHYSICAL, THERMAL, ELECTRICAL AND MAGNETIC PROPERTIES					
Density (lb/in.^{3})	0.283	0.283	0.283	0.283	0.283
Melting Range (°F)	2750–2775	2750–2775	2750–2775	2750–2775	2750–2775
Specific Heat (Btu/lb/°F)	0.10–0.11	0.10–0.11	0.10–0.11	0.10–0.11	0.10–0.11
Thermal Expansion Coeff. (μ in./in./°F)					
70–1200°F	8.4	8.4	8.4	8.4	8.4
1000°F	—	—	—	—	—
Thermal Conductivity (Btu/ft^2/hr/°F/ft)					
212°F	27	27	27	27	27
1000°F	—	—	—	—	—
Electrical Resistivity (μ ohm-cm) at 68°F	—	—	—	—	—
Magnetic Permeability	—	—	—	—	—
MECHANICAL PROPERTIES (annealed)					
Young's Modulus (10^6 psi)					
68–70°F	29–30	29–30	29–30	29–30	29–30
1000°F	—	—	—	—	—
Tensile Strength (1000 psi)					
68–70°F	73	75	83	97	113
1000°F	—	—	—	—	—
Yield Strength (1000 psi)					
68–70°F	—	—	—	—	—
1000°F	—	—	—	—	—
Elongation (in 2 in.) at 68–75°F (%)	32	31	27	23	17
Shear Strength (1000 psi)					
68–70°F	—	—	—	—	—
1000°F	—	—	—	—	—
Hardness (Brinell)	149	156	163	192	229

CARBON STEELS–HARDENING GRADES

TABLE 5-4. (Continued)

Type →	C1030	C1040	C1050	C1060	C1080
COMPOSITION (%)	C 0.28–0.34, Mn 0.60–0.90, P 0.040 max, S 0.050 max	C 0.37–0.44, Mn 0.60–0.90, P 0.040 max, S 0.050 max	C 0.48–0.55, Mn 0.60–0.90, P 0.040 max, S 0.050 max	C 0.55–0.65, Mn 0.60–0.90, P 0.040 max, S 0.050 max	C 0.75–0.88, Mn 0.60–0.90, P 0.040 max, S 0.050 max
PHYSICAL, THERMAL, ELECTRICAL AND MAGNETIC PROPERTIES					
Density (lb/in.^{3})	0.283	0.283	0.283	0.283	0.283
Melting Range (°F)	2700–2750	2700–2750	–	–	–
Specific Heat (Btu/lb/°F)	0.10–0.11	0.10–0.11	0.10–0.11	0.10–0.11	0.10–0.11
Thermal Expansion Coeff. (μ in./in./°F)					
70–1200°F	8.3	8.3	8.1	8.1	8.1
1000°F	–	–	–	–	–
Thermal Conductivity (Btu/ft^2/hr/°F/ft)					
212°F	27	27	27	27	27
1000°F	–	–	–	–	–
Electrical Resistivity (μ ohm-cm) at 68°F	19	19	18	18	18
Magnetic Permeability	–	–	–	–	–
MECHANICAL PROPERTIES (annealed)					
Youngs's Modulus (10^6 psi)					
68–70°F	29–30	29–30	29–30	29–30	29–30
1000°F	–	–	–	–	–
Tensile Strength (1000 psi)					
68–70°F	122–75	113–89	143–96	160–103	190–117
1000°F	–	–	–	–	–
Yield Strength (1000 psi)					
68–70°F	–	–	–	–	–
1000°F	–	–	–	–	–
Elongation (in 2 in.) at 68–75°F (%)	18–33	19–33	10–30	12–28	12–24
Shear Strength (1000 psi)					
68–70°F	–	–	–	–	–
1000°F	–	–	–	–	–
Hardness (Brinell)	495–179	262–183	321–192	321–212	388–223

TABLE 5-4. (*Continued*)

CARBON STEELS–HARDENING GRADES

Type →	C1095	C1137	C1141	C1144			
COMPOSITION (%)	C 0.90–1.03, Mn 0.30–0.50, P 0.040 max, S 0.050 max	C 0.32–0.39, Mn 1.35–1.65, P 0.040 max, S 0.08–0.13	C 0.37–0.45, Mn 1.35–1.65, P 0.040 max, S 0.08–0.13	C 0.40–0.48, Mn 1.35–1.65, P 0.040 max, S 0.24–0.33			
PHYSICAL, THERMAL, ELECTRICAL AND MAGNETIC PROPERTIES							
Density (lb/in.^{3})	0.283	0.282	0.282	0.282			
Melting Range (°F)	–	–	–	–			
Specific Heat (Btu/lb/°F)	0.10–0.11	0.10–0.11	0.10–0.11	0.10–0.11			
Thermal Expansion Coeff. (μ in./in./°F) 70–1200°F	8.1	7.5	7.5	7.5			
1000°F	–	–	–	–			
Thermal Conductivity (Btu/ft^2/hr/°F/ft) 212°F	27	27	27	27			
1000°F	–	–	–	–			
Electrical Resistivity (μ ohm-cm) at 68°F	18	18	18	18			
Magnetic Permeability	–	–	–	–			
MECHANICAL PROPERTIES (annealed)							
Youngs's Modulus (10^6 psi) 68–70°F	29–30	29–30	29–30	29–30			
1000°F	–	–	–	–			
Tensile Strength (1000 psi) 68–70°F	188–190	158–87	237–94	128–97			
1000°F	–	–	–	–			
Yield Strength (1000 psi) 68–70°F	–	–	–	–			
1000°F	–	–	–	–			
Elongation (in 2 in.) at 68–75°F (%)	10–26	6–28	7–28	17–24			
Shear Strength (1000 psi) 68–70°F	–	–	–	–			
1000°F	–	–	–	–			
Hardness (Brinell)	401–229	352–174	461–192	277–201			

TABLE 5-4. (Continued)

FREE-CUTTING CARBON STEELS–WROUGHT

Type →	B1111, 1211	B1112, 1212	B1112, 1212
COMPOSITION (%)	C 0.13M, Mn 0.60–0.90, P 0.07–0.12, S 0.08–0.15.	C 0.13M, Mn 0.70–1.00, P 0.07–0.12, S 0.16–0.23.	C 0.13M, Mn 0.70–1.00, P 0.07–0.12, S 0.24–0.33.
PHYSICAL, THERMAL, ELECTRICAL AND MAGNETIC PROPERTIES			
Density (lb/in.^{3})	0.283	0.283	0.283
Melting Range (°F)	–	–	–
Specific Heat (Btu/lb/°F)	0.10–0.11	0.10–0.11	0.10–0.11
Thermal Expansion Coeff. (μ in./in./°F)			
70–1200°F	8.4	8.4	8.4
1000°F	–	–	–
Thermal Conductivity (Btu/ft^2/hr/°F/ft)			
212°F	27	27	27
1000°F	–	–	–
Electrical Resistivity (μ ohm-cm) at 68°F	14.3	14.3	14.3
Magnetic Permeability	–	–	–
MECHANICAL PROPERTIES (1″ diam.)			
Youngs's Modulus (10^6 psi)			
68–70°F	29	29	29
1000°F	–	–	–
Tensile Strength (1000 psi)			
68–70°F	80–105	80–105	80–105
1000°F	–	–	–
Yield Strength (1000 psi)			
68–70°F	70–90	70–90	70–90
1000°F	–	–	–
Elongation (in 2 in.) at 68–75°F (%)	12–22	12–22	12–22
Shear Strength (1000 psi)			
68–70°F	–	–	–
1000°F	–	–	–
Hardness (Brinell)	163–229	163–229	163–229

TABLE 5-4. (*Continued*)

NITRIDING STEELS—WROUGHT

Type →	135	135, Modified	N	EZ	5 Ni-2 Al
COMPOSITION (%)	C 0.30–0.40, Mn 0.40–0.70, Si 0.20–0.40, Cr 0.90–1.40, Al 0.85–1.20, Mo 0.15–0.25	C 0.38–0.45, Mn 0.40–0.70, Si 0.20–0.40, Cr 1.40–1.80, Al 0.85–1.20, Mo 0.30–0.45	C 0.20–0.27, Mn 0.40–0.70, Si 0.20–0.40, Cr 1.00–1.50, Al 0.85–1.20, Mo 0.20–0.30, Ni 3.25–3.75	C 0.30–0.40, Mn 0.50–1.10, Si 0.20–0.40, Cr 1.00–1.50, Al 0.85–1.20, Mo 0.15–0.25, Se 0.15–0.25	C 0.20–0.25, Mn 0.25–0.45, Si 0.20–0.30, Ni 4.75–5.25, Cr 0.40–0.60, Mo 0.20–0.30, Al 1.80–2.20, V 0.08–0.15
PHYSICAL, THERMAL, ELECTRICAL AND MAGNETIC PROPERTIES					
Density (lb/in.^{3})	0.283	0.283	0.283	0.283	—
Melting Range (°F)	—	—	—	—	—
Specific Heat (Btu/lb/°F)	0.11–0.12	0.11–0.12	0.11–0.12	0.11–0.12	—
Thermal Expansion Coeff. (μ in./in./°F) 32–932°F	6.5	6.5	6.5	6.5	—
1000°F	—	—	—	—	—
Thermal Conductivity (Btu/ft^2/hr/°F/ft) 212°F	30	30	30	30	—
1000°F	—	—	—	—	—
Electrical Resistivity (μ ohm-cm) at 68°F	27–29	27–29	27–29	27–29	—
Magnetic Permeability	+	+	+	+	—
MECHANICAL PROPERTIES (annealed)					
Youngs's Modulus (10^6 psi) 68–70°F	29–30	29–30	29–30	29–30	29–30
1000°F	—	—	—	—	—
Tensile Strength (1000 psi) 68–70°F	138;121	159;145	132;190	126	206
1000°F	—	—	—	—	—
Yield Strength (1000 psi) 68–70°F	120;103	141;125	114;180	90	202
1000°F	—	—	—	—	—
Elongation (in 2 in.) at 68–75°F (%)	20;23	18;20	22;15	17	15
Shear Strength (1000 psi) 68–70°F	—	—	—	—	—
1000°F	—	—	—	—	—
Hardness (Brinell)	280;230	320;285	277;415	255	44

TABLE 5-4. (Continued)

AGE HARDENABLE STAINLESS STEELS—WROUGHT, CAST

Type →	Stainless W	17-4 PH	Cast 17-4 PH	17-7 PH	PH 15-7 Mo
COMPOSITION (%)	C 0.08 max, Mn 1.0 max, Si 1.0 max, Ni 7.0, Cr 17.0, Ti 1.2, Al 0.40 Fe bal	C 0.07 max, Mn 1.0 max, Si 1.0 max, Cr 16.5, Ni 4.0, Cu 4.0, Cb + Ta 0.30	C 0.07, Mn 1.0, Si 1.0, Cr 15-17, Ni 3-5, Cu 2.3-3.0, Fe bal	C 0.09 max, Mn 1.0 max, Si 1.0 max, Cr 17.0, Ni 7.1, Al 1.0, Fe bal	C 0.09 max, Mn 1.0 max, Si 1.0 max, Cr 15.0, Ni 7.0, Mo 2.5, Al 1.0, Fe bal
PHYSICAL, THERMAL, ELECTRICAL AND MAGNETIC PROPERTIES					
Density (lb/in.^{3})	0.280	0.281	0.280	0.276	0.277
Melting Range (°F)	–	–	–	–	–
Specific Heat (Btu/lb/°F)	–	–	–	–	–
Thermal Expansion Coeff. (μ in./in./°F)					
70–200°F	5.5	6.0	6.0	5.6	5.6
1000°F	–	–	–	–	–
Thermal Conductivity (Btu/ft^2/hr/°F/ft)					
212°F	12.1	10.4	10.4	9.7	9.3
1000°F	–	–	–	–	–
Electrical Resistivity (μ ohm-cm) at 68°F	85	77	98	82	82
Magnetic Permeability	101	151	–	145	150
MECHANICAL PROPERTIES (annealed)					
Youngs's Modulus (10^6 psi)					
68–70°F	28	28.5	28.5	29	29
1000°F	–	–	–	–	–
Tensile Strength (1000 psi)					
68–70°F	195	195; 145	170 min	200; 235	210; 240
1000°F	94	100; 96	–	93	110; 130
Yield Strength (1000 psi) (0.2%)					
68–70°F	180	180; 125	140 min	185; 220	200; 225
1000°F	54	77; 92	–	76	105; 105
Elongation (in 2 in.) at 68–75°F (%)	3–15	13; 19	6 min	9; 6	7; 6
Shear Strength (1000 psi)					
68–70°F	120	130	–	136; 150	143; 160
1000°F	–	–	–	–	–
Hardness (Brinell)	–	–	–	–	–

TABLE 5-4. (*Continued*)

Type →	17-14 Cu Mo	AM-350	AM-355	Cast AM-355	AM-362
COMPOSITION (%)	C 0.12, Mn 0.75, Si 0.50, Ni 14.1, Cr 15.9, Mo 2.50, Cu 3.0, Cb 0.45, Ti 0.25, Fe bal	C 0.10, Mn 0.80, Si 0.25, Cr 16.5, Ni 4.3, Mo 2.75, N 0.10, Fe bal	C 0.13, Mn 0.95, Si 0.25, Cr 15.5, Ni 4.3, Mo 2.75, N 0.10, Fe bal	C 0.10, Mn 0.80, Si 0.60, Cr 15.0, Ni 4.2, Mo 2.3, N 0.09, Fe bal	C 0.03, Mn 0.30, Si 0.20, Cr 14.5, Ni 6.5, Ti 0.80, Fe bal
PHYSICAL, THERMAL, ELECTRICAL AND MAGNETIC PROPERTIES					
Density (lb/in.^{3})	0.287	0.282	0.282	0.282	0.281
Melting Range (°F)	–	–	–	–	–
Specific Heat (Btu/lb/°F)	–	–	–	–	–
Thermal Expansion Coeff. (μ in./in./°F) 70–200°F	8.2	6.3	6.4	6.4	5.7
1000°F	–	–	–	–	–
Thermal Conductivity (Btu/ft^2/hr/°F/ft) 212°F	8.7	8.87	9.18	9.18	–
1000°F	–	–	–	–	–
Electrical Resistivity (μ ohm-cm) at 68°F	–	78.8	75.7	90	66
Magnetic Permeability	–	–	–	–	–
MECHANICAL PROPERTIES (annealed)					
Youngs's Modulus (10^6 psi) 68–70°F	28	29.4	29.3	29.3	28.9
1000°F	–	–	–	–	–
Tensile Strength (1000 psi) 68–70°F	86	206	186	223	165
1000°F	72	106	115	129	–
Yield Strength (1000 psi) (0.2%) 68–70°F	42	173	171	183	160
1000°F	28	85	96	100	–
Elongation (in 2 in.) at 68–75°F (%)	45	13.5	19	13.7	16
Shear Strength (1000 psi) 68–70°F	–	–	–	–	–
1000°F	–	–	–	–	–
Hardness (Brinell)	–	–	–	–	–

TABLE 5-4. (*Continued*)

AGE HARDENABLE STAINLESS STEELS—WROUGHT, CAST

Type →	PH14-8 Mo	15-5 PH	16-6 PH	Custom 455	AFC-77
COMPOSITION (%)	C 0.05 max, Mn 1.00 max, Si 1.00 max, Cr 13.75–15.00, Ni 7.50–8.75, Mo 2.00–3.00, Al 0.75–1.50, Fe bal	C 0.07 max, Mn 1.00 max, Si 1.00 max, Cr 14.00–15.50, Ni 3.50–5.50, Cu 2.50–4.50, Cb + Ta 0.15–0.45, Fe bal	C 0.08, Cr 16, Ni 7, Ti 0.07.	C 0.03, Mn 0.50, Si 0.50, Ni 11.75, Cr 9.00, Ti 1.20, Cu 2.25, Cb + Ta 0.30, Fe bal	C 0.12–0.17, Cr 13.5–14.5, Mo 4.5–5.5, Co 13.0–14.0, V 0.10–0.30, Fe bal
PHYSICAL, THERMAL, ELECTRICAL AND MAGNETIC PROPERTIES					
Density (lb/in.^{3})	0.278	0.282	0.276	0.28	0.284
Melting Range (°F)	—	—	—	—	—
Specific Heat (Btu/lb/°F)	—	—	—	—	—
Thermal Expansion Coeff. (μ in./in./°F)					
68–212°F	5.3	6.2	6.2	5.9	5.87
1000°F	—	—	—	—	—
Thermal Conductivity (Btu/ft^2/hr/°F/ft)					
68–70°F	—	—	—	—	—
1000°F	—	—	—	—	—
Electrical Resistivity (μ ohm-cm) at 68°F	—	77	85	75; 133	73.5
Magnetic Permeability	—	—	—	—	—
MECHANICAL PROPERTIES (annealed)					
Youngs's Modulus (10^6 psi)					
68–70°F	—	28.5	29.5	29	—
1000°F	—	—	—	—	—
Tensile Strength (1000 psi)					
68–70°F	235; 215	190; 135	127; 190	245	252; 290
1000°F	185	—	—	—	214
Yield Strength (1000 psi) (0.2%)					
68–70°F	220	170; 105	110; 185	237	200; 214
1000°F	160	—	—	—	164
Elongation (in 2 in.) at 68–75°F (%)	5	10; 16	15; 15	11	17; 10
Shear Strength (1000 psi)					
68–70°F	—	—	—	148	—
1000°F	—	—	—	—	—
Hardness (Brinell)	—	—	—	—	—

TABLE 5-4. (Continued)

AUSTENITIC STAINLESS STEELS—WROUGHT

Type →	201,202[b]	301	302	302B	303,303 Se
COMPOSITION (%)	C 0.15 max, Mn 5.5–10.0, Cr 16.0–19.0, Ni 3.5–6.0, N 0.25 max	C 0.15, Mn 2 max, Si 1 max, P 0.045 max, S 0.03 max, Cr 16–18, Ni 6–8	C 0.15, Mn 2 max, Si 1 max, P 0.045 max, S 0.03 max, Cr 17–19, Ni 8–10	C 0.15 max, Mn 2 max, P 0.045 max, S 0.03 max, Si 2–3, Cr 17–19, Ni 8–10	C 0.15 max, Mn 2 max, P 0.20 max, S or Se 0.15 min, Si 1 max, Cr 17–19, Ni 8–10, Mo 0.60 max
PHYSICAL, THERMAL, ELECTRICAL AND MAGNETIC PROPERTIES					
Density (lb/in.^{3})	0.28	0.29	0.29	0.29	0.29
Melting Range (°F)	—	2550–2590	2550–2590	2500–2550	2550–2590
Specific Heat (Btu/lb/°F)	0.12	0.12	0.12	0.12	0.12
Thermal Expansion Coeff. (μ in./in./°F)					
32–212°F	8.7	9.4	9.6	9.0	9.6
1000°F	—	—	—	—	—
Thermal Conductivity (Btu/ft^2/hr/°F/ft)					
212°F	9.4	9.4	9.4	9.2	9.4
1000°F	—	—	—	—	—
Electrical Resistivity (μ ohm-cm) at 68°F	69	72	72	72	72
Magnetic Permeability	—	—	—	—	—
MECHANICAL PROPERTIES (annealed)					
Youngs's Modulus (10^6 psi)					
68–70°F	28.0	28.0	28.0	28.0	28.0
1000°F	—	—	—	—	—
Tensile Strength (1000 psi)					
68–70°F	115;105	110;105	90;90;85	95;90;90	90
1000°F	—	—	—	—	—
Yield Strength (1000 psi)					
68–70°F	55;55	40;40	40;35;35	40;40;40	35
1000°F	—	—	—	—	—
Elongation (in 2 in.) at 68–75°F (%)	55;55	60;55	50;60;60	55;50;50	50
Shear Strength (1000 psi)					
68–70°F	—	—	—	—	—
1000°F	—	—	—	—	—
Hardness (Brinell)	—	165	150	160;160	160

TABLE 5-4. (*Continued*)

AUSTENITIC STAINLESS STEELS–WROUGHT

Type →	304	304 L	305	308
COMPOSITION (%)	C 0.08 max, Mn 2 max, Si 1 max, P 0.045 max, S 0.030 max, Cr 18–20, Ni 8–12	C 0.030 max, Mn 2 max, P 0.045 max, S 0.030 max, Si 1 max, Cr 18–20, Ni 8–12	C 0.12 max, Mn 2 max, P 0.045 max, S 0.030 max, Si 1 max, Cr 17–19, Ni 10–13	C 0.08 max, Mn 2 max, P 0.45 max, S 0.030 max, Si 1 max, Cr 19–21, Ni 10–12
PHYSICAL, THERMAL, ELECTRICAL AND MAGNETIC PROPERTIES				
Density (lb/in.^{3})	0.29	0.29	0.29	0.29
Melting Range (°F)	2550–2650	2550–2650	2550–2650	2550–2650
Specific Heat (Btu/lb/°F)	0.12	0.12	0.12	0.12
Thermal Expansion Coeff. (μ in./in./°F)				
32–212°F	9.6	9.6	9.6	9.6
1000°F	–	–	–	–
Thermal Conductivity (Btu/ft^2/hr/°F/ft)				
212°F	9.4	9.4	9.6	9.6
1000°F	–	–	–	–
Electrical Resistivity (μ ohm-cm) at 68°F	72	72	72	72
Magnetic Permeability	–	–	–	–
MECHANICAL PROPERTIES (annealed)				
Youngs's Modulus (10^6 psi)				
68–70°F	28.0	28.8	28.0	28.0
1000°F	–	–	–	–
Tensile Strength (1000 psi)				
68–70°F	84; 82; 85	81; 81	85; 85	85; 85; 85
1000°F	–	–	–	–
Yield Strength (1000 psi)				
68–70°F	42; 35; 35	39; 33	38; 35	35; 30; 30
1000°F	–	–	–	–
Elongation (in 2 in.) at 68–75° F (%)	55; 60; 60	55; 60	50; 55	50; 55; 55
Shear Strength (1000 psi)				
68–70°F	–	–	–	–
1000°F	–	–	–	–
Hardness (Brinell)	149; 149	143	–	150; 150

TABLE 5-4. (*Continued*)

AUSTENITIC STAINLESS STEELS—WROUGHT

Type →	309,309 S	310,310 S	314	316
COMPOSITION (%)	C 0.20 max (309), C 0.08 max (309S), Mn 2 max, P 0.045 max, S 0.03 max, Si 1 max, Cr 22–24, Ni 12–15	C 0.25 max (310), C 0.08 max (310S), Mn 2 max, P 0.045 max, S 0.03 max, Si 1.5 max, Cr 24–26, Ni 19–22	C 0.25 max, Mn 2 max, P 0.45 max, S 0.030 max, Si 1.5–3.0, Cr 23–26, Ni 19–22	C 0.08 max, Mn 2 max, Si 1 max, P 0.045 max, S 0.03 max, Cr 16–18, Ni 10–14, Mo 2–3
PHYSICAL, THERMAL, ELECTRICAL AND MAGNETIC PROPERTIES				
Density (lb/in.^{3})	0.29	0.29	0.28	0.29
Melting Range (°F)	2550–2650	2550–2650	–	2500–2550
Specific Heat (Btu/lb/°F)	0.12	0.12	0.12	0.12
Thermal Expansion Coeff. (μ in./in./°F) 32–212°F	8.3	8.8	8.4	8.9
1000°F	–	–	–	–
Thermal Conductivity (Btu/ft^2/hr/°F/ft) 212°F	9	8.2	10.1	9.4
1000°F	–	–	–	–
Electrical Resistivity (μ ohm-cm) at 68°F	78	78	77	74
Magnetic Permeability	–	–	–	–
MECHANICAL PROPERTIES (annealed)				
Youngs's Modulus (10^6 psi) 68–70°F	29.0	29.0	29.0	28.0
1000°F	–	–	–	–
Tensile Strength (1000 psi) 68–70°F	90; 95; 95	95; 95; 95	100; 100; 100	84; 82; 80
1000°F	–	–	–	–
Yield Strength (1000 psi) 68–70°F	45; 40; 40	45; 45; 45	50; 50; 50	42; 36; 30
1000°F	–	–	–	–
Elongation (in 2 in.) at 68–75°F (%)	45; 45; 45	45; 50; 50	40; 45; 45	50; 55; 60
Shear Strength (1000 psi) 68–70°F	–	–	–	–
1000°F	–	–	–	–
Hardness (Brinell)	170; 160	170	180; 180	149; 149

TABLE 5-4. (*Continued*)

AUSTENITIC STAINLESS STEELS—WROUGHT

Type →	316L	317	321	347-348	384-385
COMPOSITION (%)	C 0.030 max, Mn 2 max, P 0.045 max, Si 1.00 max, Cr 16-18, Ni 10-14, Mo 2-3, Fe bal	C 0.08 max, Mn 2 max, P 0.045 max, S 0.030 max, Si 1.00 max, Cr 18-20, Ni 11-15, Mo 3-4, Fe bal	C 0.08 max, Mn 2 max, Si 1 max, P 0.04 max, S 0.030 max, Cr 17-19, Ni 9-12, Ti 5 × C, Fe bal	C 0.08 max, Mn 2, P 0.045 max, S 0.030 max, Si 1.00 max, Cr 17-19, Ni 9-13, Cb-Ta 10 × C min, Ta 0.010 max (348), Co. 0.20 max (348), Fe bal	C 0.08 max, Mn 2 max, P 0.045 max, S 0.030 max, Si 1.00 max, Cr 15-17 (384) 11.5-13.5 (385), Ni 17-19 (384) 14-16 (385), Fe bal
PHYSICAL, THERMAL, ELECTRICAL AND MAGNETIC PROPERTIES					
Density (lb/in.^{3})	0.29	0.29	0.29	0.29	0.29
Melting Range (°F)	2500-2550	2500-2550	2550-2600	2550-2600	2550-2650
Specific Heat (Btu/lb/°F)	0.12	0.12	0.12	0.12	0.12
Thermal Expansion Coeff. (μ in./in./°F)					
32-212°F	8.9	8.9	9.3	9.3	9.6; 10.4
1000°F	–	–	–	–	–
Thermal Conductivity (Btu/ft^2/hr/°F/ft)					
68-70°F	9.4	9.4	9.3	9.3	9.4; 9.6
1000°F	–	–	–	–	–
Electrical Resistivity (μ ohm-cm) at 68°F	72	74	72	73	–
Magnetic Permeability	–	–	–	–	–
MECHANICAL PROPERTIES (annealed)					
Young's Modulus (10^6 psi)					
68-70°F	–	28.0	28.0	28.0	28.0
1000°F	–	–	–	–	–
Tensile Strength (1000 psi)					
68-70°F	81; 81	90; 85; 85	90; 85; 85	95; 90; 90	75; 72
1000°F	–	–	–	–	–
Yield Strength (1000 psi)					
68-70°F	42; 34	40; 40; 40	35; 30; 35	45; 35; 35	35; 30
1000°F	–	–	–	–	–
Elongation (in 2 in.) at 68-75°F (%)	50; 55	45; 50; 50	45; 55; 55	45; 50; 50	55
Shear Strength (1000 psi)					
68-70°F	–	–	–	–	–
1000°F	–	–	–	–	–
Hardness (Brinell)	146	160; 160	160; 150	160; 160	–

TABLE 5-4. (*Continued*)

FERRITIC STAINLESS STEELS—WROUGHT

Type →	405	429	430	430F-430F Se
COMPOSITION (%)	C 0.08 max, Mn 1.00 max, P 0.040 max, S 0.030 max, Si 1.00 max, Cr 11.5-14.5, Al 0.10-0.30, Fe bal	C 0.12 max, Mn 1.00 max, P 0.040 max, S 0.030 max, Si 1.00 max, Cr 14-16. Fe bal	C 0.12 max, Mn 1.00 max, P 0.040 max, S 0.030 max, Si 1.00 max, Cr 14.0-18.0, Fe bal	C 0.12 max, Mn 1.25 max, P 0.060 max, S 0.15 min (430F) 0.060 max (430F Se) Si 1.00 max, Mo 0.60 max (430F only) opt., Se 0.15 min (430F Se only), Cr 14-18, Fe bal
PHYSICAL, THERMAL, ELECTRICAL AND MAGNETIC PROPERTIES				
Density (lb/in.^{3})	0.28	0.28	0.28	0.28
Melting Range (°F)	2700-2790	2650-2750	2600-2750	2600-2750
Specific Heat (Btu/lb/°F)	0.11	0.11	0.11	0.11
Thermal Expansion Coeff. (μ in./in./°F)				
32-212°F	6.0	5.7	5.8	5.8
1000°F	–	–	–	–
Thermal Conductivity (Btu/ft^2/hr/°F/ft)				
212°F	15.6	14.8	15.1	15.1
1000°F	–	–	–	–
Electrical Resistivity (μ ohm-cm) at 68°F	60.0	59	60.0	60.0
Magnetic Permeability	+	+	+	+
MECHANICAL PROPERTIES (annealed)				
Youngs's Modulus (10^6 psi)				
68-70°F	29	29	29	29
1000°F	–	–	–	–
Tensile Strength (1000 psi)				
68-70°F	65-70	70-74	75	80
1000°F	–	–	–	–
Yield Strength (1000 psi) (0.2%)				
68-70°F	35-40	40-45	40-45	55
1000°F	–	–	–	–
Elongation (in 2 in.) at 68-75°F (%)	25-30	30	25-30	25
Shear Strength (1000 psi)				
68-70°F	–	–	–	–
1000°F	–	–	–	–
Hardness, (Rockwell)	B75-90	–	B80	B80

TABLE 5-4. (Continued)

FERRITIC STAINLESS STEELS–WROUGHT

Type →	434	436	446
COMPOSITION (%)	C 0.12 max, Mn 1.00 max, P 0.040 max, S 0.030 max, Si 1.00 max, Mo 0.75-1.25, Cr 16-18, Fe bal	C 0.12 max, Mn 1.00 max, P 0.040 max, S 0.030 max, Si 1.00 max, Mo 0.75-1.25, Cb-Ta 5 × C min 1.25, (0.70 max), Cr 16-18, Fe bal	C 0.20 max, Mn 1.50 max, P 0.040 max, S 0.030 max, Si 1.00 max, Cr 23.0-27.0, N 0.25 max, Fe bal
PHYSICAL, THERMAL, ELECTRICAL AND MAGNETIC PROPERTIES			
Density (lb/in.³)	0.28	0.28	0.27
Melting Range (°F)	2600–2750	2600–2750	2600–2750
Specific Heat (Btu/lb/°F)	0.11	0.11	0.12
Thermal Expansion Coeff. (μ in./in./°F)			
32–212°F	6.6	5.2	5.8
1000°F	–	–	–
Thermal Conductivity (Btu/ft²/hr/°F/ft)			
212°F	15.2	13.8	12.1
1000°F	–	–	–
Electrical Resistivity (μ ohm-cm) at 68°F	60	60	67
Magnetic Permeability	+	+	+
MECHANICAL PROPERTIES (annealed)			
Youngs's Modulus (10⁶ psi)			
68–70°F	29	29	29
1000°F	–	–	–
Tensile Strength (1000 psi)			
68–70°F	77–79	77	80–85
1000°F	–	–	–
Yield Strength (1000 psi) (0.2%)			
68–70°F	53–60	53	50–55
1000°F	–	–	–
Elongation (in 2 in.) at 68–75°F (%)	23–33	23	20–25
Shear Strength (1000 psi)			
68–70°F	–	–	–
1000°F	–	–	–
Hardness, (Rockwell)	–	B83	B83–B86

TABLE 5-4. (Continued)

MARTENSITIC STAINLESS STEELS–WROUGHT

Type →	403	410	414	416, 416Sc
COMPOSITION (%)	C 0.15 max, Mn 1.00 max, P 0.040 max, S 0.030 max, Si 0.50 max, Cr 11.5–13	C 0.15 max, Cr 11.5–13.5, Mn 1.00 max, Si 1. max, P 0.040 max, S 0.030 max	C 0.15 max, Cr 11.5–13.5, Ni 1.25–2.5, Mn 1.00 max, Si 1. max, P 0.040 max, S 0.030 max	C 0.15 max, Cr 12–14, P 0.06 max, Si 1 1. max, S 0.15 min, S 0.06 max (416Sc), Se 0.15 min (416Sc)
PHYSICAL, THERMAL, ELECTRICAL AND MAGNETIC PROPERTIES				
Density (lb/in.^{3})	0.28	0.28	0.28	0.28
Melting Range (°F)	2700–2790	2700–2790	–	2700–2790
Specific Heat (Btu/lb/°F)	0.11	0.11	0.11	0.11
Thermal Expansion Coeff. (μ in./in./°F) 32–212°F	5.5	5.5	5.8	5.5
32–1200°F	–	6.5	–	6.5
Thermal Conductivity (Btu/ft^2/hr/°F/ft) 212°F	14.4	14.4	14.4	14.4
1000°F	–	–	–	–
Electrical Resistivity (μ ohm-cm) at 68°F	57	57	70	57
Magnetic Permeability	–	–	–	–
MECHANICAL PROPERTIES (annealed)				
Youngs's Modulus (10^6 psi) 68–70°F	29	29.0	29.0	29.0
1000°F	–	–	–	–
Tensile Strength (1000 psi) 68–70°F	70–75	65–75	115–120	75
1000°F	–	–	–	–
Yield Strength (1000 psi) 68–70°F	40–45	35–45	90–105	40
1000°F	–	–	–	–
Elongation (in 2 in.) at 68–75°F (%)	25–35	25–35	15–20	30
Shear Strength (1000 psi) 68–70°F	–	–	–	–
1000°F	–	–	–	–
Hardness (Brinell)	80–155	155	235	155

TABLE 5-4. *(Continued)*

MARTENSITIC STAINLESS STEELS–WROUGHT

Type →	420	420F	431	440A
COMPOSITION (%)	C > 0.15, Cr 12–14, Mn 1.00 max, P 0.040 max, S 0.030 max, Si 1.00 max, Fe bal	C > 0.15, Cr 12–14, Mn 1.25 max, Si 1.00 max, P 0.060 max, Mo 0.60 max, (opt.) S 0.15 min	C 0.20 max, Cr 15–17, Ni 1.25–2.50, Mn 1.00 max, P 0.040 max, S 0.030 max, Si 1.00 max, Fe bal	C 0.60–0.75, Cr 16–18, Mo 0.75 max, Mn 1.00 max, P 0.040 max, S 0.030 max, Si 1.00 max, Fe bal
PHYSICAL, THERMAL, ELECTRICAL AND MAGNETIC PROPERTIES				
Density (lb/in.³)	0.28	0.28	0.28	0.28
Melting Range (°F)	2650–2750	2650–2750	2650	2500–2750
Specific Heat (Btu/lb/°F)	0.11	0.11	0.11	0.11
Thermal Expansion Coeff. (μ in./in./°F)				
32–212°F	5.7	5.7	5.6	5.6
32–1200°F	6.8	–	–	–
Thermal Conductivity (Btu/ft²/hr/°F/ft)				
212°F	14.4	14.5	11.7	14.0
1000°F	–	–	–	–
Electrical Resistivity (μ ohm-cm) at 68°F	55	55	72	60
Magnetic Permeability	+	+	+	+
MECHANICAL PROPERTIES (annealed)				
Youngs's Modulus (10⁶ psi)				
68–70°F	29.0	29	29.0	29.0
1000°F	–	–	–	–
Tensile Strength (1000 psi)				
68–70°F	95	95	125	125
1000°F	–	–	–	–
Yield Strength (1000 psi)				
68–70°F	50	55	95	60
1000°F	–	–	–	–
Elongation (in 2 in.) at 68–75°F (%)	25	22	20	20
Shear Strength (1000 psi)				
68–70°F	–	–	–	–
1000°F	–	–	–	–
Hardness (Brinell)	195	220	260	215

TABLE 5-4. (Continued)

MARTENSITIC STAINLESS STEELS–WROUGHT

Type →	440B	440C	501	502
COMPOSITION (%)	C 0.75–0.95, Cr 16–18, Mo 0.75 max, Mn 1.00 max, S 0.030 max, Si 1.00 max, Fe bal	C 0.95–1.20, Cr 16–18, Mo 0.75 max, Mn 1.00 max, P 0.040 max, Si 1.00 max, Fe bal	C > 0.10, Cr 4–6, Mo 0.40–0.65, Mn 1.00 max, Si 1.00 max, P 0.040 max, S 0.030 max	C 0.10 max, Cr 4–6, Mo 0.40–0.65, Mn 1.00 max, Si 1.00 max, P 0.040 max, S 0.030 max
PHYSICAL, THERMAL, ELECTRICAL AND MAGNETIC PROPERTIES				
Density (lb/in.^{3})	0.28	0.28	0.28	0.28
Melting Range (°F)	2500–2750	2500–2700	2700–2800	2700–2800
Specific Heat (Btu/lb/°F)	0.11	0.11	0.11	0.11
Thermal Expansion Coeff. (μ in./in./°F)				
32–212°F	5.6	5.6	6.2	6.2
32–1200°F	–	–	7.3	7.3
Thermal Conductivity (Btu/ft^2/hr/°F/ft)				
212°F	14.0	14.0	21.2	21.2
1000°F	–	–	–	–
Electrical Resistivity (μ ohm-cm) at 68°F	60	69	40	40
Magnetic Permeability	+	+	+	+
MECHANICAL PROPERTIES (annealed)				
Youngs's Modulus (10^6 psi)				
68–70°F	29.0	29.0	29	29
1000°F	–	–	–	–
Tensile Strength (1000 psi)				
68–70°F	107	110	70	65–75
1000°F	–	–	–	–
Yield Strength (1000 psi)				
68–70°F	62	65	30	25–30
1000°F	–	–	–	–
Elongation (in 2 in.) at 68–75°F (%)	18	14	28	30
Shear Strength (1000 psi)				
68–70°F	–	–	–	–
1000°F	–	–	–	–
Hardness (Brinell)	220	230	160	150

TABLE 5-4. (*Continued*)

SPECIALTY STAINLESS STEELS–WROUGHT

Type →	Flo 302 HQ	JS 700	MF-1	MF-2	Pyromet 355
COMPOSITION (%)	C 0.08 max, Cr 18, Ni 9, Cu 3.5, Fe bal	C 0.03, Cr 21, Ni 25, Mo 4.5, Cb 0.30, Fe bal	C 0.045, Cr 11, Ti 0.5, Fe bal	C 0.06, Cr 12, 1 Al, 0.6 Ti, 0.20 Ni, Fe bal	C 0.12, Cr 15.5 Ni 4.5, Mo 3.0, N 0.1, Fe bal
PHYSICAL, THERMAL, ELECTRICAL AND MAGNETIC PROPERTIES					
Density (lb/in.³)	0.29	0.29	0.276	0.276	0.286
Melting Range (°F)	–	–	–	–	–
Specific Heat (Btu/lb/°F)	0.12	0.12	0.11	–	0.12
Thermal Expansion Coeff. (μ in./in./°F)					
70–212°F	9.6	9.15	6.5	5.7	8.3
1000°F	–	–	–	–	–
Thermal Conductivity (Btu/ft²/hr/°F/ft)					
212°F	6.5	8.5	–	–	–
1000°F	–	–	–	–	–
Electrical Resistivity (μ ohm-cm) at 68°F	72	–	–	–	75
Magnetic Permeability	–	–	–	–	–
MECHANICAL PROPERTIES (annealed)					
Youngs's Modulus (10⁶ psi)					
68–70°F	28	29	–	–	29.3
1000°F	–	–	–	–	–
Tensile Strength (1000 psi)					
68–70°F	73	85	65	66	186
1000°F	–	–	–	–	–
Yield Strength (1000 psi) (0.2%)					
68–70°F	27	39	34	42	55
1000°F	–	–	–	–	–
Elongation (in 2 in.) at 68–75° F (%)	65	45	32	35	29
Shear Strength (1000 psi)					
68–70°F	–	–	–	–	–
1000°F	–	–	–	–	–
Hardness, (Rockwell)	B80	–	B72	–	B100

TABLE 5-4. (*Continued*)

SPECIALTY STAINLESS STEELS–WROUGHT

Type →	Pyromet 538	Uniloy EB26-1	Uniloy 326	18 SR	18-2 Mn
COMPOSITION (%)	C 0.03 max, Mn 9.0, 1 Si, Cr 21.5 max, Ni 6.50, N 0.4 max, S 0.03 max, P 0.04 max, Fe bal	C 0.010 max, Cr 26, Mo 1, Fe bal	C 0.05 max, Cr 26, Ni 6.5, Ti 0.2, Fe bal	C 0.05, Cr 18, Al 2.0, Ti 0.4, Fe bal	C 0.10, Mn 12, Cr 18, Ni 1.6, N 0.35, Fe bal
PHYSICAL, THERMAL, ELECTRICAL AND MAGNETIC PROPERTIES					
Density (lb/in.^{3})	0.29	0.28	0.279	0.27	0.281
Melting Range (°F)	–	–	–	–	–
Specific Heat (Btu/lb/°F)	–	0.102	0.102	–	–
Thermal Expansion Coeff. (μ in./in./°F)					
70–212°F	10.1	5.9	5.9	5.9	10.3
1000°F	–	–	–	–	–
Thermal Conductivity (Btu/ft^2/hr/°F/ft)					
212°F	–	–	11.3	–	–
1000°F	–	–	–	–	–
Electrical Resistivity (μ ohm-cm) at 68°F	–	52	52	110	–
Magnetic Permeability	–	–	–	–	–
MECHANICAL PROPERTIES (annealed)					
Young's Modulus (10^6 psi)					
68–70°F	28.5	–	27	–	29
1000°F	–	–	–	–	–
Tensile Strength (1000 psi)					
68–70°F	112	68	100	85	120
1000°F	–	–	–	–	–
Yield Strength (1000 psi)					
68–70°F	65 (0.2%)	50 (0.2%)	75 (0.2%)	65	65
1000°F	–	–	–	–	–
Elongation (in 2 in.) at 68–75°F (%)	42	25	32	27	60
Shear Strength (1000 psi)					
68–70°F	–	–	–	–	–
1000°F	–	–	–	–	–
Hardness, (Rockwell)	B95	B88	B95	B90	B98

TABLE 5-4. (*Continued*)

SPECIALTY STAINLESS STEELS—WROUGHT

Type →	18-18-2	20Cb-3	203 EZ	211	216
COMPOSITION (%)	C 0.06, Si 1.9, Cr 18, Ni 18, Fe bal	C 0.07 max, Cr 20, Ni 34, Mo 2.5, Cu 3.5, Cb & Ta 1.0 max, Fe bal	C 0.08 max, Mn 6, Cr 16.5, Ni 5.5, Cu 2.0, S 0.15 min, Fe bal	C 0.05, Mn 6, Cr 17, Ni 5.5, Cu 1.5, Fe bal	C 0.08 max, Mn 8.25, Cr 19.75, Ni 6.0, Mo 2.5, N 0.37, Fe bal
PHYSICAL, THERMAL, ELECTRICAL AND MAGNETIC PROPERTIES					
Density (lb/in.³)	0.284	0.291	0.284	0.284	0.287
Melting Range (°F)	—	—	—	—	—
Specific Heat (Btu/lb/°F)	—	0.12	0.12	—	—
Thermal Expansion Coeff. (μ in./in./°F) 70–212°F	7.6	8.3	9.4	9.4	8.5
1000°F	—	—	—	—	—
Thermal Conductivity (Btu/ft²/hr/°F/ft) 212°F	—	—	9.4	—	—
1000°F	—	—	—	—	—
Electrical Resistivity (μ ohm-cm) at 68°F	86	104	—	73	—
Magnetic Permeability	—	—	—	—	—
MECHANICAL PROPERTIES (annealed)					
Young's Modulus (10⁶ psi) 68–70°F	—	28	29	28.6	—
1000°F	—	—	—	—	—
Tensile Strength (1000 psi) 68–70°F	81	90	108	87	108
1000°F	—	—	—	—	—
Yield Strength (1000 psi) 68–70°F	36	40	80	31	62
1000°F	—	—	—	—	—
Elongation (in 2 in.) at 68–75°F (%)	54	50	45	60	51
Shear Strength (1000 psi) 68–70°F	—	—	—	—	—
1000°F	—	—	—	—	—
Hardness, (Rockwell)	—	B84	B95	B74	B91

TABLE 5-4. (*Continued*)

SPECIALTY STAINLESS STEELS—WROUGHT

Type →	22-4-9	21-6-9	22-13-5	303 Pb	303 Plus-X
COMPOSITION (%)	C 0.55, Mn 8.5, Cr 21.5, Ni 4.0, N 0.4, Fe bal	C 0.08 max, Mn 9, Cr 20, Ni 6.5, N 0.30, Fe bal	C 0.06 max, Mn 5, Cr 22, Ni 12.5, Mo 2.25, N 0.30, Cb 0.20, V 0.20, Fe bal	C 0.15 max, Cr 18, Ni 9, S 0.20, Pb 0.20, Fe bal	C 0.15, Mn 3.5, Cr 18, Ni 8.5, S 0.15 min, Fe bal
PHYSICAL, THERMAL, ELECTRICAL AND MAGNETIC PROPERTIES					
Density (lb/in.^{3})	0.279	0.283	0.285	—	0.286
Melting Range (°F)	—	—	—	—	—
Specific Heat (Btu/lb/°F)	—	—	—	—	0.12
Thermal Expansion Coeff. (μ in./in./°F)					
70–212°F	7.8	8.5	9.0	—	9.6
1000°F	—	—	—	—	—
Thermal Conductivity (Btu/ft^2/hr/°F/ft)					
212°F	—	8	9	—	9.4
1000°F	—	—	—	—	—
Electrical Resistivity (μ ohm-cm) at 68°F	—	70	80	—	72
Magnetic Permeability	—	—	—	—	—
MECHANICAL PROPERTIES (annealed)					
Youngs's Modulus (10^6 psi)					
68–70°F	—	28.5	—	—	28
1000°F	—	—	—	—	—
Tensile Strength (1000 psi)					
68–70°F	—	111	121	90	90
1000°F	—	—	—	—	—
Yield Strength (1000 psi)					
68–70°F	—	64	65	35	35
1000°F	—	—	—	—	—
Elongation (in 2 in.) at 68–75°F (%)	—	42	46	50	3
Shear Strength (1000 psi)					
68–70°F	—	—	—	—	—
1000°F	—	—	—	—	—
Hardness, (Rockwell)	—	B93	B98	—	B88

TABLE 5-4. (Continued)

SPECIALTY STAINLESS STEELS—WROUGHT

Type →	304 + B	304 LN	304-N	305 H	305 MH
COMPOSITION (%)	C 0.08 max, Cr 19, Ni 13.5, B 2 max, Fe bal	C 0.03 max, Cr 19, Ni 10, N 0.12, Fe bal	C 0.06, Cr 19, Ni 8, N 0.25, Fe bal	C 0.08 max, Cr 16, Ni 18, Fe bal	C 0.08 max, Cr 12.5, Ni 15, Fe bal
PHYSICAL, THERMAL, ELECTRICAL AND MAGNETIC PROPERTIES					
Density (lb/in.³)	0.29	0.29	0.29	0.29	0.29
Melting Range (°F)	–	–	–	–	–
Specific Heat (Btu/lb/°F)	0.12	0.12	–	0.12	0.12
Thermal Expansion Coeff. (μ in./in./°F) 70–212°F	10.5	9.6	10.6	9.6	10.4
1000°F	–	–	–	–	–
Thermal Conductivity (Btu/ft²/hr/°F/ft) 212°F	–	9.4	–	9.4	–
1000°F	–	–	–	–	–
Electrical Resistivity (μ ohm-cm) at 68°F	74	72	73	79	74
Magnetic Permeability	–	–	–	–	–
MECHANICAL PROPERTIES (annealed)					
Young's Modulus (10⁶ psi) 68–70°F	30	28	29	29	29
1000°F	–	–	–	–	–
Tensile Strength (1000 psi) 68–70°F	100	85	105	80	78
1000°F	–	–	–	–	–
Yield Strength (1000 psi) 68–70°F	50	38	70	40	46
1000°F	–	–	–	–	–
Elongation (in 2 in.) at 68–75°F (%)	10	60	48	55	46
Shear Strength (1000 psi) 68–70°F	–	–	–	–	–
1000°F	–	–	–	–	–
Hardness, (Rockwell)	B95	B85	–	–	–

TABLE 5-4. (*Continued*)

SPECIALTY STAINLESS STEELS—WROUGHT

Type →	329	404	410 Cb	5 F
COMPOSITION (%)	C 0.15 max, Cr 27.5, Ni 4.5, Mo 1.5, Fe bal	C 0.05 max, Cr 11.75, Ni 1.6, N 0.03 max, Fe bal	C 0.15 max, Cr 12.5, Cb 0.25 max, Fe bal	C 0.10 max, Cr 13.75, Mo 0.5, S 0.3 max, Fe bal
PHYSICAL, THERMAL, ELECTRICAL AND MAGNETIC PROPERTIES				
Density (lb/in.^{3})	0.28	0.28	0.28	0.28
Melting Range (°F)	—	—	—	—
Specific Heat (Btu/lb/°F)	0.11	0.11	0.11	0.11
Thermal Expansion Coeff. (μ in./in./°F)				
70–212°F	8.0	4.8	6.5	6.5
1000°F	—	—	—	—
Thermal Conductivity (Btu/ft^2/hr/°F/ft)				
212°F	—	—	14.4	—
1000°F	—	—	—	—
Electrical Resistivity (μ ohm-cm) at 68°F	75	70	57	57
Magnetic Permeability	—	—	—	—
MECHANICAL PROPERTIES (annealed)				
Youngs's Modulus (10^6 psi)				
68–70°F	—	29	29	—
1000°F	—	—	—	—
Tensile Strength (1000 psi)				
68–70°F	105	—	—	—
1000°F	—	—	—	—
Yield Strength (1000 psi)				
68–70°F	80	—	—	75
1000°F	—	—	—	—
Elongation (in 2 in.) at 68–75° F (%)	25	—	—	15
Shear Strength (1000 psi)				
68–70°F	—	—	—	—
1000°F	—	—	—	—
Hardness (Brinell)	—	—	—	200

TABLE 5-4. (*Continued*)

STAINLESS STEELS—CAST

Type →	CA-15	CA-40	CB-30	CC-50
COMPOSITION (%)	C 0.15 max, Mn 1.0 max, Si 1.5 max, P 0.04 max, S 0.04 max, Cr 11.5–14.0, Ni 1.0 max, Mo 0.5a	C 0.20–0.40, Mn 1.0 max, Si 1.50 max, P 0.04 max, S 0.04 max, Cr 11.5–14.0, Ni 1.0 max, Mo 0.5a	C 0.30 max, Mn 1.0 max, Si 1.0 max, P 0.04 max, S 0.04 max, Cr 18–22, Ni 2.0 max	C 0.50 max, Mn 1.0 max, Si 1.0 max, P 0.04 max, S 0.04 max, Cr 26–30, Ni 4.0 max
PHYSICAL, THERMAL, ELECTRICAL AND MAGNETIC PROPERTIES				
Density (lb/in.^{3})	0.275	0.275	0.272	0.272
Melting Range (°F)	2750	2725	2725	2725
Specific Heat (Btu/lb/°F)	0.11	0.11	0.11	0.12
Thermal Expansion Coeff. (μ in./in./°F)				
70–1000°F	6.4	6.4	6.5	6.4
1000°F	–	–	–	–
Thermal Conductivity (Btu/ft^2/hr/°F/ft)				
212°F	14.5	14.5	12.8	12.6
1000°F	–	–	–	–
Electrical Resistivity (μ ohm-cm) at 68°F	78	76	76	77
Magnetic Permeability	ferromag.	ferromag.	ferromag.	ferromag.
MECHANICAL PROPERTIES (annealed)				
Young's Modulus (10^6 psi)				
68–70°F	29	29	29	29
1000°F	–	–	–	–
Tensile Strength (1000 psi)				
68–70°F	–	–	95	97
1000°F	–	–	–	–
Yield Strength (1000 psi) (0.2%)				
68–70°F	–	–	60	65
1000°F	–	–	–	–
Elongation (in 2 in.) at 68–75°F (%)	–	–	15	18
Shear Strength (1000 psi)				
68–70°F	–	–	–	–
1000°F	–	–	–	–
Hardness (Brinell)	–	–	195	210

TABLE 5-4. (*Continued*)

STAINLESS STEELS–CAST

Type →	CD-4MCu	CE-30	CF-3	CF-8	CF-20
COMPOSITION (%)	C 0.04 max, Mn 1.0 max, Si 1.0 max, Cr 25–27, Ni 4.75–6.0, Mo 1.75–2.25, Cu 2.75–3.25	C 0.30 max, Mn 1.50 max, Si 2.0 max, P 0.04 max, S 0.04 max, Cr 26–30, Ni 8–11	C 0.03 max, Mn 1.50 max, Si 2.0 max, Cr 17–21, Ni 8–12	C 0.08 max, Mn 1.50 max, Si 2.0 max, P 0.04 max, S 0.04 max, Cr 18–21, Ni 8–11	C 0.20 max, Mn 1.50 max, Si 2.0 max, P 0.04 max, S 0.04 max, Cr 18–21, Ni 8–11
PHYSICAL, THERMAL, ELECTRICAL AND MAGNETIC PROPERTIES					
Density (lb/in.^{3})	0.277	0.277	0.280	0.280	0.280
Melting Range (°F)	2650	2650	2625	2600	2575
Specific Heat (Btu/lb/°F)	0.12	0.14	0.12	0.12	0.12
Thermal Expansion Coeff. (μ in./in./°F)					
70–1000°F	6.5	9.6	10	10.0	10.4
1000°F	–	–	–	–	–
Thermal Conductivity (Btu/ft^2/hr/°F/ft)					
212°F	–	–	9.2	9.2	9.2
1000°F	–	–	–	–	–
Electrical Resistivity (μ ohm-cm) at 68°F	–	85	76	76	77.9
Magnetic Permeability	ferromag.	>1.5	1.0–2.0	1.0–1.3	1.01
MECHANICAL PROPERTIES (annealed)					
Youngs's Modulus (10^6 psi)					
68–70°F	29	25	27	28	28
1000°F	–	–	–	–	–
Tensile Strength (1000 psi)					
68–70°F	105;140	97	77	77	77
1000°F	–	–	–	–	–
Yield Strength (1000 psi) (0.2%)					
68–70°F	85;110	63	37	37	36
1000°F	–	–	–	–	–
Elongation (in 2 in.) at 68–75°F (%)	25;15	18	55	55	50
Shear Strength (1000 psi)					
68–70°F	–	–	–	–	–
1000°F	–	–	–	–	–
Hardness (Brinell)	260;300	170	140	140	163

PROPERTIES OF ALLOYS 821

TABLE 5-4. (*Continued*)

STAINLESS STEELS—CAST

Type →	CF-3M	CF-8M, CF-12M	CF-8C	CF-16F
COMPOSITION (%)	C 0.03 max, Mn 1.5 max, Si 1.5 max, Cr 17-21, Ni 9-13, Mo 2-3	C 0.08 max (CF-8M) or 0.12 max (CF-12M), Mn 1.5 max, Si 2.0 max, P 0.04 max, S 0.04 max, Cr 18-21, Ni 9-12, Mo 2-3	C 0.08 max, Mn 1.5 max, Si 2.0 max, P 0.04 max, S 0.04 max, Cr 18-21, Ni 9-21, Cb 1.0 max[b]	C 0.16 max, Mn 1.5 max, Si 2.0 max, P 0.17 max, S 0.04 max, Cr 18-21, Ni 9-12, Se 0.2-0.35, Mo 1.5 max
PHYSICAL, THERMAL, ELECTRICAL AND MAGNETIC PROPERTIES				
Density (lb/in.^{3})	0.280	0.280	0.280	0.280
Melting Range (°F)	2550	2550	2600	2550
Specific Heat (Btu/lb/°F)	0.12	0.12	0.12	0.12
Thermal Expansion Coeff. (μ in./in./°F) 70–1200°F	9.7	9.7	10.3	9.9
1000°F	–	–	–	–
Thermal Conductivity (Btu/ft^2/hr/°F/ft) 212°F	9.4	9.4	9.3	9.4
1000°F	–	–	–	–
Electrical Resistivity (μ ohm-cm) at 68°F	82	82	71	72
Magnetic Permeability	1.5–2.5	1.50–2.50	1.20–1.80	1.0–2.0
MECHANICAL PROPERTIES (annealed)				
Youngs's Modulus (10^6 psi) 68–70°F	27	28	28	28
1000°F	–	–	–	–
Tensile Strength (1000 psi) 68–70°F	80	80	77	77
1000°F	–	–	–	–
Yield Strength (1000 psi) (0.2%) 68–70°F	42	42	38	40
1000°F	–	–	–	–
Elongation (in 2 in.) at 68–75°F (%)	50	50	39	52
Shear Strength (1000 psi) 68–70°F	–	–	–	–
1000°F	–	–	–	–
Hardness (Brinell)	156–170	156–170	149	150

TABLE 5-4. (*Continued*)

STAINLESS STEELS–CAST

Type →	CG-8M	CH-20	CK-20	CN-7M			
COMPOSITION (%)	C 0.08 max, Mn 1.5 max, Si 1.5 max, Ni Cr 18–21, Ni 9–13, Mo 3–4	C 0.20 max, Mn 1.5 max, Si 2.0 max, P 0.04 max, S 0.04 max, Cr 22–26, Ni 12–15	C 0.20 max, Mn 1.5 max, Si 2.0 max, P 0.04 max, S 0.04 max, Cr 23–27, Ni 19–22	C 0.07 max, Mn 1.5 max, Si 1.5 max, P 0.04 max, S 0.04 max, Cr 19–22, Ni 27.5–30.5, Mo 17.5–2.50, Cu 3.0 min			
PHYSICAL, THERMAL, ELECTRICAL AND MAGNETIC PROPERTIES							
Density (lb/in.^{3})	0.281	0.279	0.280	0.289			
Melting Range (°F)	2550	2600	2600	2650			
Specific Heat (Btu/lb/°F)	0.12	0.12	0.12	0.11			
Thermal Expansion Coeff. (μ in./in./°F) 70–1000°F	9.7	9.6	9.2	9.7			
1000°F	–	–	–	–			
Thermal Conductivity (Btu/ft^2/hr/°F/ft) 212°F	9.4	8.2	8.2	12.1			
1000°F	–	–	–	–			
Electrical Resistivity (μ ohm-cm) at 68°F	82	84	90	89.6			
Magnetic Permeability	1.5–2.5	1.71	1.02	1.01–1.10			
MECHANICAL PROPERTIES (annealed)							
Youngs's Modulus (10^6 psi) 68–70°F	28	28	29	24			
1000°F	–	–	–	–			
Tensile Strength (1000 psi) 68–70°F	82	88	76	69			
1000°F	–	–	–	–			
Yield Strength (1000 psi) (0.2%) 68–70°F	43	50	38	31			
1000°F	–	–	–	–			
Elongation (in 2 in.) at 68–75°F (%)	50	38	37	48			
Shear Strength (1000 psi) 68–70°F	–	–	–	–			
1000°F	–	–	–	–			
Hardness (Brinell)	170	190	144	130			

TABLE 5-4. (*Continued*)

HEAT RESISTANT ALLOYS–CAST

Type →	HA	HC	HD	HE	HF
COMPOSITION (%)	C 0.20 max, Mn 0.35–0.65, Si 1.00 max, P0.04 max, S0.04 max, Mo 0.90–1.20, Cr 8–10	C 0.50 max, Mn 1.00 max, Si 2.00 max, P0.04 max, S0.04 max, Mo 0.5 max[a], Cr 26–30, Ni 4 max	C 0.50 max, Mn 1.50 max, Si 2.00 max, P0.04 max, S0.04 max, Mo 0.5 max[a], Cr 26–30, Ni 4–7	C 0.20–0.50, Mn 2.00 max, Si 2.00 max, P0.04 max, S0.04 max, Mo 0.5 max[a], Cr 26–30, Ni 8–11	C 0.20–0.40, Mn 2.00 max, Si 2.00 max, P0.04 max, S0.04 max, Mo 0.5 max[a], Cr 19–23, Ni 9–12
PHYSICAL, THERMAL, ELECTRICAL AND MAGNETIC PROPERTIES					
Density (lb/in.^{3})	0.279	0.272	0.274	0.277	0.280
Melting Range (°F)	2750	2725	2700	2650	2550
Specific Heat (Btu/lb/°F)	0.11	0.12	0.12	0.14	0.12
Thermal Expansion Coeff. (μ in./in./°F)					
70–212°F	7.5	6.4	8.0	9.9	10.1
1000°F	—	—	—	—	—
Thermal Conductivity (Btu/ft^2/hr/°F/ft)					
212°F	15.2	12.6	12.6	10.0	9.0
1000°F	—	—	—	—	—
Electrical Resistivity (μ ohm-cm) at 68°F	70	77	81	85	80
Magnetic Permeability	ferromag.	ferromag.	ferromag.	1.3–2.5	1.0
MECHANICAL PROPERTIES (annealed)					
Youngs's Modulus (10^6 psi)					
68–70°F	29	29	27	25	28
1000°F	—	—	—	—	—
Tensile Strength (1000 psi)					
68–70°F	95;107	70–110;115	85	95;90	85;100
1000°F	—	—	—	—	—
Yield Strength (1000 psi) (0.2%)					
68–70°F	65;81	65–75;80	48	45;55	45;50
1000°F	—	—	—	—	—
Elongation (in 2 in.) at 68–75°F (%)	34;21	2–19;18	16	20;10	35;25
Shear Strength (1000 psi)					
68–70°F	—	—	—	—	—
1000°F	—	—	—	—	—
Hardness (Brinell)	180;220	190;223	190	200;270	165;190

TABLE 5-4. (*Continued*)

HEAT RESISTANT ALLOYS–CAST

Type →	HH		HI	HK	HL
COMPOSITION (%)	C 0.20–0.50, Mn 2.00 max, Si 2.00 max, P 0.04 max, S 0.04 max, Mo 0.5 max[a], Cr 24–28, Ni 11–14, N 0.2 max		C 0.20–0.50, Mn 2.00 max, Si 2.00 max, P0.04 max, S0.04 max, Mo 0.5 max[a], Cr 26–30, Ni 14–18	C 0.20–0.60, Mn 2.00 max, Si 2.00 max, P0.04 max, S0.04 max, Mo 0.5 max[a], Cr 24–28, Ni 18–22	C 0.20–0.60, Mn 2.00 max, Si 2.00 max, P0.04 max, S0.04 max, Mo 0.5 max[a], Cr 28–32, Ni 18–22
PHYSICAL, THERMAL, ELECTRICAL AND MAGNETIC PROPERTIES					
Density (lb/in.^{3})	0.279		0.279	0.280	0.279
Melting Range (°F)	2500		2550	2550	2600
Specific Heat (Btu/lb/°F)	0.12		0.12	0.12	0.12
Thermal Expansion Coeff. (μ in./in./°F)					
70–1800°F	10.5		10.5	10.0	9.9
1000°F	–		–	–	–
Thermal Conductivity (Btu/ft^2/hr/°F/ft)					
212°F	8.2		8.2	8.2	8.2
1000°F	–		–	–	–
Electrical Resistivity (μ ohm-cm) at 68°F	75–85		–	90	94
Magnetic Permeability	1.0–1.9		1.0–1.7	1.02	1.01
MECHANICAL PROPERTIES (annealed)	Type I	Type II			
Youngs's Modulus (10^6 psi)					
68–70°F	27	27	27	29	29
1000°F	–	–	–	–	–
Tensile Strength (1000 psi)					
68–70°F	80; 86	85; 92	80; 90	75; 85	82
1400°F	33	35	38	–	50
Yield Strength (1000 psi)					
68–70°F	50; 55	40; 45	45; 65	50; 50	52
1400°F	17	18	–	–	–
Elongation (in 2 in.) at 68–75°F (%)	15; 11	15; 8	12; 6	17; 10	19
Shear Strength (1000 psi)					
68–70°F	–	–	–	–	–
1000°F	–	–	–	–	–
Hardness (Brinell)	185; 200	180; 200	180; 200	170; 190	192

PROPERTIES OF ALLOYS 825

TABLE 5-4. (*Continued*)

HEAT RESISTANT ALLOYS–CAST

Type →	HN	HT	HU	HW	HX
COMPOSITION (%)	C 0.20–0.50, Mn 2.00 max, Si 2.00 max, P 0.04 max, S 0.04 max, Mo 0.5 max[a], Cr 19–23, Ni 23–27	C 0.35–0.75, Mn 2.00 max, Si 2.50 max, P 0.04 max, S 0.04 max, Mo 0.5 max[a], Cr 13–17, Ni 33–37	C 0.35–0.75, Mn 2.00 max, Si 2.50 max, P 0.04 max, S 0.04 max, Mo 0.5 max[a], Cr 17–21, Ni 37–41	C 0.35–0.75, Mn 2.00 max, Si 2.50 max, P 0.04 max, S 0.04 max, Mo 0.5 max[a], Cr 10–14, Ni 58–62	C 0.35–0.75, Mn 2.00 max, Si 2.50 max, P 0.04 max, S 0.04 max, Mo 0.5 max[a], Cr 15–19, Ni 64–68
PHYSICAL, THERMAL, ELECTRICAL AND MAGNETIC PROPERTIES					
Density (lb/in.³)	0.283	0.286	0.290	0.294	0.294
Melting Range (°F)	2500	2450	2450	2350	2350
Specific Heat (Btu/lb/°F)	0.11	0.11	0.11	0.11	0.11
Thermal Expansion Coeff. (μ in./in./°F) 70–1800°F	–	9.8	9.6	8.8	9.2
1000°F	–	–	–	–	–
Thermal Conductivity (Btu/ft²/hr/°F/ft) 212°F	–	7.7	–	7.7	–
1000°F	–	–	–	–	–
Electrical Resistivity (μ ohm-cm) at 68°F	–	100	105	112	–
Magnetic Permeability	1.10	1.10–2.00	1.10–2.00	–	2.0
MECHANICAL PROPERTIES (annealed)					
Young's Modulus (10⁶ psi) 68–70°F	27	27	27	25	25
1000°F	–	–	–	–	–
Tensile Strength (1000 psi) 68–70°F	68	70;75	70;73	68;84	65;73
1000°F	–	–	–	–	–
Yield Strength (1000 psi) 68–70°F	38	40;45	40;43	36;52	36;44
1000°F	–	–	–	–	–
Elongation (in 2 in.) at 68–75°F (%)	17	10;5	9;5	4;4	9;9
Shear Strength (1000 psi) 68–70°F	–	–	–	–	–
1000°F	–	–	–	–	–
Hardness (Brinell)	160	180;200	170;190	185;205	176;185

TABLE 5-4. (*Continued*)

HIGH TEMPERATURE STEELS–WROUGHT

Type →	1415 NW	1430 MV	14 CVM	17-22AS			
COMPOSITION (%)	C 0.17, Mn 0.40, Si 0.30, Cr 12.75, Ni 1.95, Mo 0.15, Cu 0.13, W 3.0, Fe bal	C 0.30, Mn 1.05, Si 0.30, Cr 11.80, Ni 0.25, Mo 2.80, V 0.25, Fe bal	C 0.20, Mn 0.50, Si 0.75, Cr 1.0, Mo 1.0, V 0.10, Fe bal	C 0.30, Mn 0.55, Si 0.70, Cr 1.30, Mo 0.50, V 0.25, Fe bal			
PHYSICAL, THERMAL, ELECTRICAL AND MAGNETIC PROPERTIES							
Density (lb/in.^{3})	0.284	0.281	0.285	0.283			
Melting Range (°F)	2660–2670	2700–2750	–	2700–2750			
Specific Heat (Btu/lb/°F)	–	–	–	–			
Thermal Expansion Coeff. (μ in./in./°F)							
800°F	6.3	6.5	7.9	7.8			
1000°F	–	–	–	–			
Thermal Conductivity (Btu/ft^2/hr/°F/ft)							
800°F	–	15.8	–	17.3			
1000°F	–	–	–	–			
Electrical Resistivity (μ ohm-cm) at 68°F	–	–	–	–			
Magnetic Permeability	–	–	–	–			
MECHANICAL PROPERTIES (annealed)							
Youngs's Modulus (10^6 psi)							
68–70°F	29	30	31.6	29.5			
1000°F	21.5	22	25.4	20			
Tensile Strength (1000 psi)							
68–70°F	170	157	139	150			
1000°F	103	89(1100°F)	103	91			
Yield Strength (1000 psi) (0.2%)							
68–70°F	150	125	117	127			
1000°F	98	81(1100°F)	85	78			
Elongation (in 2 in.) at 68–75°F (%)	13.3	12	⩾8	16.5			
Shear Strength (1000 psi)							
68–70°F	–	–	–	–			
1000°F	–	–	–	–			
Hardness (Brinell)	–	–	–	–			

ULTRA HIGH STRENGTH STEELS—WROUGHT

TABLE 5-4. (Continued)

Type →	Modified H-11	MX-2	300-M	D-6A		
COMPOSITION (%)	C 0.40, Mn 0.35, Si 1.0, Cr 5.0, Mo 1.4, V 0.45, Fe bal	C 0.39, Mn 0.70, Si 1.0, Cr 1.10, Mo 0.25, V 0.15, Co 1.0, Fe bal	C 0.40, Mn 0.75, Si 1.60, Ni 1.85, Cr 0.85, Mo 0.40, V 0.08, Fe bal	C 0.46, Mn 0.75, Si 0.22, Ni 0.55, Cr 1.0, Mo 1.0, Fe bal		
PHYSICAL, THERMAL, ELECTRICAL AND MAGNETIC PROPERTIES						
Density (lb/in.^{3})	0.281	0.276	–	0.283		
Melting Range (°F)	–	–	–	–		
Specific Heat (Btu/lb/°F)	–	–	–	–		
Thermal Expansion Coeff. (μ in./in./°F)						
68–212°F	7.4	5.68	7.61	7.3		
1000°F	–	–	–	–		
Thermal Conductivity (Btu/ft^2/hr/°F/ft)						
68–70°F	17	–	21.7	–		
1000°F	–	–	–	–		
Electrical Resistivity (μ ohm-cm) at 68°F	–	–	–	–		
Magnetic Permeability	–	–	–	–		
MECHANICAL PROPERTIES (annealed)						
Youngs's Modulus (10^6 psi)						
68–70°F	30	29.4	–	30		
800°F	21.9–26.6	–	–	23.7		
Tensile Strength (1000 psi)						
68–70°F	295–311	279	289	284		
1000°F	216–220	–	–	139		
Yield Strength (1000 psi) (0.2%)						
68–70°F	241–247	239	242	250		
1000°F	172–173	–	–	121		
Elongation (in 2 in.) at 68–75°F (%)	6.6–12.0	10	10.0	7.5		
Shear Strength (1000 psi)						
68–70°F	–	–	–	–		
1000°F	–	–	–	–		
Hardness (Brinell)	–	–	–	–		

TABLE 5-4. (*Continued*)

ULTRA HIGH STRENGTH STEELS–WROUGHT

Type →	4340	18Ni[a]	9Ni-4Co-0.20C	9Ni-4Co-0.25C	9Ni-4Co-0.30C
COMPOSITION (%)	C 0.40, Mn 0.85, Si 0.20, Cr 0.75, Ni 1.80, Mo 0.25	Ni 18.5, Co 7.0, Mo 4.5, C 0.026, Mn 0.1, Si 0.11, Ti 0.22, B 0.003	Ni 9.0, Co 4.5, Mo 1.0, Cr 0.75, Mn 0.25, 0.20 C, Si 0.1 max, S, P 0.01 max	Ni 8.25, Co 4.0, Cr, Mo 0.45, C 0.27, Mn 0.25, Si 0.1, V 0.09, S, P 0.01 max	Ni 7.5, Co 4.5, Cr, Mo 1.0, C 0.31, Mn 0.25, V 0.11, Si 0.1, P, S 0.1 max
PHYSICAL, THERMAL, ELECTRICAL AND MAGNETIC PROPERTIES					
Density (lb/in.^{3})	0.283	0.290	0.283	0.283	0.283
Melting Range (°F)	—	—	—	—	—
Specific Heat (Btu/lb/°F)	—	—	—	—	—
Thermal Expansion Coeff. (μ in./in./°F)					
68–212°F	6.3	5.6	6.4	6.4	6.2
1000°F	—	—	—	—	—
Thermal Conductivity (Btu/ft^2/hr/°F/ft)					
68–70°F	—	—	—	—	—
1000°F	—	—	—	—	—
Electrical Resistivity (μ ohm-cm) at 68°F	—	—	—	—	—
Magnetic Permeability	—	—	—	—	—
MECHANICAL PROPERTIES (annealed)					
Youngs's Modulus (10^6 psi)					
68–70°F	30	26.5	28.9	28.4	28.6
1000°F	—	—	—	—	—
Tensile Strength (1000 psi)					
68–70°F	287	275	195–220	195–240	220–260
1000°F	—	154	139	120	155
Yield Strength (1000 psi) (0.2%)					
68–70°F	270	268	173–194	178–198	190–210
1000°F	—	138	107	90	140
Elongation (in 2 in.) at 68–75°F (%)	11	11	12–19	10–18	10–16
Shear Strength (1000 psi)					
68–70°F	—	—	—	—	—
1000°F	—	—	—	—	—
Hardness (Brinell)	—	—	—	—	—

Powder Metals Parts — Ferrous

Material ↓	PMPA[a] Designation	Density, g/cc	Condition	Ult Ten Str, 1000 psi	Yld Str, 1000 psi	Elong, %	Transverse Fiber Str, 1000 psi	Shear Str, 1000 psi	Impact Str, ft-lb	Hardness, Rockwell
IRONS										
99Fe Min[b]	F-0000-N	5.7-6.1	As-sintered	19	15	5	39	—	4	20R_H
99Fe Min[b]	F-0000-S	7.0	As-sintered	35	25	11	65	—	5	10R_B
99Fe Min[c]	F-0000-T	7.3	As-sintered	40	26	12	70	35	—	20R_B
99Fe Min[c]	F-0000-U	7.5	As-sintered	41	27.5	30	71	—	—	22R_B
STEELS										
99Fe-1C	F-0010-F	6.1-6.5	As-sintered	35	27	1.0	89	22	1	50R_B
99Fe-1C	F-0010-P	6.1-6.5	Heat treated	47.7	—	0.5	—	—	4.5	90R_B
99Fe-1C	F-0010-S	7.0	As-sintered	60	—	3.0	120	—	2	—
99Fe-1C	F-0010-S	7.0	Heat treated	65	—	0.5	120	—	5.0	100R_B
99Fe-1C	F-0010-T	7.3	As-sintered	68	—	3.0	140	—	3.0	—
99Fe-1C	F-0010-T	7.3	Heat treated	127	—	2.5	235	100	6.0	105R_B
Fe-1.5Ni-0.5Mo-0.6C	FN-0206-S	6.8	As-sintered	70	58	2.5	150	44	7.1	80R_B
Fe-1.5Ni-0.5Mo-0.6C	FN-0206-S	6.8	Heat treated	90	80	0.5	150	—	—	25R_C
Fe-1.5Ni-0.5Mo-0.6C	FN-0206-T	7.2	As-sintered	90	72	2.5	180	47	9.2	95R_B
Fe-1.5Ni-0.5Mo-0.6C	FN-0206-T	7.2	Heat treated	140	120	0.5	207	—	4.3	35R_C
90Fe-10Cu	FC-1000-N	5.8-6.2	As-sintered	30	25	0.5	75	17	—	—
90Fe-10Cu	FC-1000-N	5.8-6.2	Heat treated	54	—	1.0	103	—	3.5	30R_C
92Fe-7Cu-1C	FC-0710-N	5.8-6.2	As-sintered	50	40	0.5	115	35	3.0	70R_B
92Fe-7Cu-1C	FC-0710-N	5.8-6.2	Heat treated	85	—	1.5	180	—	6.0	30R_C
92Fe-7Cu-1C	FC-0710-S	6.8	As-sintered	83	63	1.0	131	57	4.0	73R_B
92Fe-7Cu-1C	FC-0710-S	6.8	Heat treated	110	—	1.5	210	—	7.0	40R_C
Fe-7Ni-2Cu-1C	FN-0710-S	6.8	As-sintered	70	50	2.5	140	50	5	70R_B
Fe-7Ni-2Cu-1C	FN-0710-S	6.8	Heat treated	135	—	1.5	262	—	6.5	42R_C
Fe-7Ni-2Cu-1C	FN-0710-T	7.2	As-sintered	92	75	3.5	180	60	11	85R_B
Fe-7Ni-2Cu-1C	FN-0710-T	7.2	Heat treated	157	—	2.0	285	—	8.6	44R_C
80Fe-20Cu	FX-2000-T	7.1 min	As-sintered	70	70	1.0	140	55	14.0	75R_B
80Fe-20Cu	FX-2000-T	7.1 min	Heat treated	128	—	0.5	210	65	11.0	35R_C
79Fe-20Cu-1C	FX-2010-T	7.1 min	As-sintered	110	90	1.0	190	66	11.0	95R_B
79Fe-20Cu-1C	FX-2010-T	7.1 min	Heat treated	152	—	1.0	—	110	—	40R_C
STAINLESS STEELS										
302	—	6.2-6.5	—	35-50	—	2.5[d]	—	—	—	40-60R_B
303L	SS-303L-P	6.0	As-sintered	35	32	2.0	—	20	—	—
303L	SS-303L-R	6.6	As-sintered	52	47	7.0	—	20	4.5	55R_B
304L	—	6.0-6.4	—	35	—	1.5[d]	—	20	—	—
316	—	6.2-6.6	—	55	—	2.0[d]	—	—	—	—
316L	SS-316L-P	6.2	As-sintered	50	—	4.0	—	31	—	42R_B
316L	SS-316L-P	6.2	As-sintered	38.5	35	2.0	95	—	2.0	55R_B
316L	SS-316L-R	6.6	As-sintered	60	—	10	—	40	—	60R_B
316L	SS-316L-R	6.65	As-sintered	58	51	8.0	135	20	4.5	65R_B
330	—	6.2-6.6	—	30-50	—	2-4	—	—	—	60-80R_B
410	SS-410-N	5.9	As-sintered	42	41	1	90	—	2.0	85R_B
410	SS-410-N	5.9	Heat treated	85	—	—	—	—	—	15R_C
410	SS-410-R	6.4	As-sintered	55	54	1	130	—	4.5	95R_B
410	SS-410-R	6.4	Heat treated	110	—	—	—	—	—	29R_C
410	—	6.6	—	65	—	1[d]	—	—	—	—
410	—	6.6	Heat treated	100	—	—	—	—	—	30-45R_C
410L	—	6.2	As-sintered	65	—	3.5	—	—	—	90R_B
410L	—	6.2	Heat treated	70	—	0.8	—	—	—	30R_C
410L	—	6.8	Heat treated	100	—	0.8	—	—	—	30R_C
430	—	6.2-6.6	—	75	—	1.5	—	—	—	—

[a] Powder Metallurgy Parts Assn. [b] Sponge iron. [c] Electrolytic iron. [d] In 1 in.

TABLE 5-5.

LEAD & ALLOYS–CAST, WROUGHT

Type →	Chemical Lead	Common Lead	Tellurium Lead	Lead 0.08 Ca	Lead 1 Sn–0.08 Ca
COMPOSITION (%)					
PHYSICAL, THERMAL, ELECTRICAL AND MAGNETIC PROPERTIES					
Density (lb/in.³)	0.41	0.41	0.41	0.41	0.41
Melting Range (°F)	618	621	617	619–621	610–640
Specific Heat (Btu/lb/°F)	0.031	0.031	0.031	0.032	0.032
Thermal Expansion Coeff. (μ in./in./°F)					
68–70°F	16.3	16.3	16	–	–
1000°F	–	–	–	–	–
Thermal Conductivity (Btu/ft²/hr/°F/ft)					
212°F	19.6	19.6	19.3	–	–
1000°F	–	–	–	–	–
Electrical Resistivity (μ ohm-cm) at 68°F	–	20.65	–	21.1	–
Magnetic Permeability	–	–	–	–	–
MECHANICAL PROPERTIES (rolled)					
Youngs's Modulus (10⁶ psi)					
68–70°F	2	2	2	–	–
1000°F	–	–	–	–	–
Tensile Strength (1000 psi)					
68–70°F	2.4	2.1	2.8	5.5	9.0
1000°F	–	–	–	–	–
Yield Strength (1000 psi) (0.5%)					
68–70°F	1.6	–	–	5.0	8.0
1000°F	–	–	–	–	–
Elongation (in 2 in.) at 68–75°F (%)	51	43	47	35	20
Shear Strength (1000 psi)					
68–70°F	–	–	–	–	–
1000°F	–	–	–	–	–
Hardness (Brinell)	4.7	–	5.5	12	17

TABLE 5-5. (*Continued*)

LEAD & ALLOYS–CAST, WROUGHT

Type →	1% Sb–Pb	4% Sb–Pb	6% Sb–Pb	8% Sb–Pb	9% Sb–Pb
COMPOSITION (%)					
PHYSICAL, THERMAL, ELECTRICAL AND MAGNETIC PROPERTIES					
Density (lb/in.³)	0.406	0.398	0.393	0.388	0.385
Melting Range (°F)	608–595	570–486	545–486	520–486	509–486
Specific Heat (Btu/lb/°F)	0.031	0.032	0.032	0.032	0.032
Thermal Expansion Coeff. (μ in./in./°F)					
68–212°F	16	15.5	15.4	14.5	14.4
1000°F	–	–	–	–	–
Thermal Conductivity (Btu/ft²/hr/°F/ft)					
212°F	19	18	17	16	16
1000°F	–	–	–	–	–
Electrical Resistivity (μ ohm-cm) at 68°F	22.0	24.0	25.3	26.5	27.1
Magnetic Permeability	–	–	–	–	–
MECHANICAL PROPERTIES (rolled)					
Youngs's Modulus (10⁶ psi)					
68–70°F	3.0	–	–	–	–
1000°F	–	–	–	–	–
Tensile Strength (1000 psi)					
68–70°F	3.0	4.0	4.2	4.6	4.7
1000°F	–	–	–	–	–
Yield Strength (1000 psi)					
68–70°F	–	–	–	–	–
1000°F	–	–	–	–	–
Elongation (in 2 in.) at 68–75°F (%)	50	50	50	30	20
Shear Strength (1000 psi)					
68–70°F	–	–	–	–	–
1000°F	–	–	–	–	–
Hardness (Brinell)	6	8	9	9	9

TABLE 5-5. (*Continued*)

LEAD–ANTIMONY–TIN ALLOYS–CAST

Type →	Alloy 7	Alloy 8	Alloy Y 10A	Alloy 13	Alloy 15
COMPOSITION (%)	Pb 75.0, Sb 15.0, Sn 10.0, Cu 0.5 max	Pb 80.0, Sb 15.0, Sn 5.0, Cu 0.5 max	Pb 90.0, Sb 10.0, Sn 0, Cu 0.5 max	Pb 84.0, Sb 10.0, Sn 6.0, Cu 0.5 max	Pb 83.0, Sb 16.0, Sn 1.0, Cu 0.6 max
PHYSICAL, THERMAL, ELECTRICAL AND MAGNETIC PROPERTIES					
Density (lb/in.^{3})	0.351	0.362	0.383	0.378	0.362
Melting Range (°F)	464–514	459–522	473–498	464–493	479–538
Specific Heat (Btu/lb/°F)	0.038	0.036	–	–	–
Thermal Expansion Coeff. (μ in./in./°F)					
68–70°F	11	13	–	–	–
1000°F	–	–	–	–	–
Thermal Conductivity (Btu/ft^2/hr/°F/ft)					
68–70°F	13.8	14.0	–	–	–
1000°F	–	–	–	–	–
Electrical Resistivity (μ ohm-cm) at 68°F	28.6	28.2	–	–	–
Magnetic Permeability	–	–	–	–	–
MECHANICAL PROPERTIES (chill-cast)					
Young's Modulus (10^6 psi)					
68–70°F	4.2	4.2	–	–	4.2
1000°F	–	–	–	–	–
Compr. Strength (1000 psi)					
68–70°F	15.65	15.6	12.9	–	–
1000°F	–	–	–	–	–
Yield Strength (1000 psi)					
68–70°F	3.55	3.40	–	–	–
1000°F	–	–	–	–	–
Elongation (in 2 in.) at 68–75°F (%)	4	5	–	–	2
Shear Strength (1000 psi)					
68–70°F	–	–	–	–	–
1000°F	–	–	–	–	–
Hardness (Brinell)	22.5	20.0	14	19.0	21.0

TABLE 5-6.

MAGNESIUM ALLOYS–WROUGHT

Type →	AZ31B-F	AZ61A-F	AZ80A-T5	ZK60A-T5	ZE10A-H24
COMPOSITION (%)	Al 2.5–3.5, Zn 0.7–1.3, Mn 0.20 min	Al 5.8–7.2, Zn 0.4–1.5, Mn 0.15 min	Al 7.8–9.2, Zn 0.2, Mn 0.15 min	Zn 4.8–6.2, Zr 0.45 min	Zn 1.0–1.5, Rare earths 0.12–0.22
PHYSICAL, THERMAL, ELECTRICAL AND MAGNETIC PROPERTIES					
Density (lb/in.^{3})	0.064	0.065	0.065	0.066	0.063
Melting Range (°F)	1050–1170	950–1140	900–1115	968–1175	1110–1195
Specific Heat (Btu/lb/°F)	0.245	0.245	0.245	0.245	0.245
Thermal Expansion Coeff. (μ in./in./°F)					
68–70°F	14	14	14	14	16
68–750°F	16	16	16	16	16
Thermal Conductivity (Btu/ft^2/hr/°F/ft)					
68–70°F	44	34	29	68–70	77
1000°F	–	–	–	–	–
Electrical Resistivity (μ ohm-cm) at 68°F	9.2	12.5	14.5	6.0–5.7	5.2
Magnetic Permeability	–	–	–	–	–
MECHANICAL PROPERTIES (extruded)					
Youngs's Modulus (10^6 psi)					
68–70°F	6.5	6.5	6.5	6.5	6.5
1000°F	–	–	–	–	–
Tensile Strength (1000 psi)					
68–70°F	36–38	41–46	50–55	50–53	34–38
1000°F	–	–	–	–	–
Yield Strength (1000 psi)					
68–70°F	24–28	24–33	38–40	40–44	19–28
1000°F	–	–	–	–	–
Elongation (in 2 in.) at 68–75°F (%)	12–16	14–17	6–8	11–14	8–12
Shear Strength (1000 psi)					
68–70°F	19	20	24	24–26	–
1000°F	–	–	–	–	–
Hardness (Brinell)	46–49	50–60	82	82	–

TABLE 5-6. (*Continued*)

MAGNESIUM ALLOYS–WROUGHT

Type →	AZ31B-H24	HK31A-H24	HM21A-T8	HM31A-T5	LA141A-T7
COMPOSITION (%)	Al 2.5–3.5, Zn 0.7–1.3, Mn 0.20	Th 2.5–4.0, Zr 0.45–1.0	Th 1.5–2.5, Mn 0.45–1.1	Th 2.5–3.5, Mn 1.2 min	Li 13.0–15.0, Al 1.0–1.5
PHYSICAL, THERMAL, ELECTRICAL AND MAGNETIC PROPERTIES					
Density (lb/in.^{3})	0.064	0.065	0.064	0.065	0.0485
Melting Range (°F)	1050–1170	1092–1195	1121–1202	1121–1202	1065–1085
Specific Heat (Btu/lb/°F)	0.245	0.245	0.245	0.245	0.346
Thermal Expansion Coeff. (μ in./in./°F)					
68–750°F	16	16	16	16	21.8
1000°F	–	–	–	–	–
Thermal Conductivity (Btu/ft^2/hr/°F/ft)					
68–70°F	44	66	79	60	25.3
1000°F	–	–	–	–	–
Electrical Resistivity (μ ohm-cm) at 68°F	9.2	6.1	5.0	6.6	15.2
Magnetic Permeability	–	–	–	–	–
MECHANICAL PROPERTIES (extruded)					
Young's Modulus (10^6 psi)					
68–70°F	6.5	8.4	6.4	6.5	6
1000°F	–	–	–	–	–
Tensile Strength (1000 psi)					
68–70°F	42; 38–39	37	34	42	18–19
1000°F	–	–	–	–	–
Yield Strength (1000 psi)					
68–70°F	32; 34–27	29	21	33	13–15
1000°F	–	–	–	–	–
Elongation (in 2 in.) at 68–75°F (%)	15; 18–19	8	10	10	10
Shear Strength (1000 psi)					
68–70°F	23	27	19	23	–
1000°F	–	–	–	–	–
Hardness (Rockwell)	E57	–	–	–	–

TABLE 5-6. *(Continued)*

MAGNESIUM ALLOYS–CAST

Type →	AZ63A	AZ81A	AZ91, AZ291B	AZ91C	AZ92A
COMPOSITION (%)	Al 5.3–6.7, Zn 2.5–3.5, Mn 0.15 min	Al 7.0–8.1, Zn 0.40–1.0, Mn 0.13 min	Al 8.3–9.7, Zn 0.4–1.0, Mn 0.13 min	Al 8.1–9.3, Zn 0.40–1.0, Mn 0.13 min	Al 8.3–9.7, Zn 1.6–2.4, Mn 0.10 min
PHYSICAL, THERMAL, ELECTRICAL AND MAGNETIC PROPERTIES					
Density (lb/in.^{3})	0.066	0.065	0.065	0.065	0.066
Melting Range (°F)	850–1130	882–1115	875–1120	875–1105	830–1110
Specific Heat (Btu/lb/°F)	0.245	0.245	0.245	0.245	0.245
Thermal Expansion Coeff. (μ in./in./°F)					
68–70°F	14	14	14	14	14
68–750°F	16	16	16	16	16
Thermal Conductivity (Btu/ft^2/hr/°F/ft)					
68–70°F	29–39	29	31	37–21	27–34
1000°F	–	–	–	–	–
Electrical Resistivity (μ ohm-cm) at 68°F	11–15	12	13	11–13	12–14
Magnetic Permeability	–	–	–	–	–
MECHANICAL PROPERTIES					
Youngs's Modulus (10^6 psi)					
68–70°F	6.5	6.5	6.5	6.5	6.5
1000°F	–	–	–	–	–
Tensile Strength (1000 psi)					
68–70°F	29	–	34	24	24
1000°F	–	–	–	–	–
Yield Strength (1000 psi)					
68–70°F	14	–	23	14	14
1000°F	–	–	–	–	–
Elongation (in 2 in.) at 68–75°F (%)	6	–	3	2	2
Shear Strength (1000 psi)					
68–70°F	18	–	20	18	18
1000°F	–	–	–	–	–
Hardness (Brinell)	50	–	60	52	65

TABLE 5-6. (*Continued*)

MAGNESIUM ALLOYS–CAST

Type →	AM100A	QE22A-T6	EZ33A-T5	HK31A-T6	HZ32A-T5
COMPOSITION (%)	Al 9.3–10.7, Zn 0.3 max, Mn 0.10 min	Ag 2.0–3.0, rare earths 1.8–2.5, Zr 0.40–1.0	Rare earths 2.5–4.0, Zn 2.0–3.1, Zr 0.50–1.0	Th 2.5–4.0, Zr 0.50–1.0	Th 2.5–4.0, Zn 1.7–2.5, Zr 0.50–1.0
PHYSICAL, THERMAL, ELECTRICAL AND MAGNETIC PROPERTIES					
Density (lb/in.^{3})	0.065	0.065	0.066	0.065	0.066
Melting Range (°F)	865–1100	1020–1190	1110–1189	1092–1204	1026–1198
Specific Heat (Btu/lb/°F)	0.245	0.245	0.245	0.245	0.245
Thermal Expansion Coeff. (μ in./in./°F)					
68–70°F	14	14	14	14	14
68–750°F	16	16	16	16	16
Thermal Conductivity (Btu/ft^2/hr/°F/ft)					
68–70°F	24–34	–	58	52	62
600°F	–	–	69	68	74
Electrical Resistivity (μ ohm-cm) at 68°F	10–15	–	7.0	7.7	6.5
Magnetic Permeability	–	–	–	–	–
MECHANICAL PROPERTIES					
Young's Modulus (10^6 psi)					
68–70°F	6.5	–	6.5	6.5	6.5
1000°F	–	–	–	–	–
Tensile Strength (1000 psi)					
68–70°F	22	40	23	31	29
600°F	–	15	12	20	12
Yield Strength (1000 psi)					
68–70°F	12	30	16	16	15
600°F	–	12	8	12	8
Elongation (in 2 in.) at 68–75°F (%)	2	4	3	6	6
Shear Strength (1000 psi)					
68–70°F	–	23	20	21	20
600°F	–	–	10	13	10
Hardness (Brinell)	–	77	50	55	57

TABLE 5-6. (*Continued*)

MAGNESIUM ALLOYS—CAST

Type →	ZE41A-T5	ZK51A-T5	ZH62A-T5	K1A-F	ZK61A-T6
COMPOSITION (%)	Zn 3.5–5.0, rare earths 0.75–1.75, Zr 0.40–1.0	Zn 3.6–5.5, Zr 0.50–1.0	Zn 5.2–6.2, Th 1.4–2.2, Zr 0.50–1.0	Zr 0.4–1.0	Zn 5.5–6.5, Zr 0.60–1.0
PHYSICAL, THERMAL, ELECTRICAL AND MAGNETIC PROPERTIES					
Density (lb/in.^{3})	0.066	0.065	0.067	0.063	0.066
Melting Range (°F)	975–1190	1020–1185	960–1169	1202	985–1175
Specific Heat (Btu/lb/°F)	0.245	0.245	0.245	—	0.245
Thermal Expansion Coeff. (μ in./in./°F)					
68–70°F	14	14	14	15	16
68–750°F	16	16	16	—	16
Thermal Conductivity (Btu/ft^2/hr/°F/ft)					
68–70°F	—	48	63	60	16
1000°F	—	—	—	—	16
Electrical Resistivity (μ ohm-cm) at 68°F	—	8.4	6.5	5.7	—
Magnetic Permeability	—	—	—	—	—
MECHANICAL PROPERTIES (annealed)					
Young's Modulus (10^6 psi)					
68–70°F	6.5	6.5	6.5	6.5	6.5
1000°F	—	—	—	—	—
Tensile Strength (1000 psi)					
68–70°F	30	40	40	25	45
1000°F	—	—	—	—	—
Yield Strength (1000 psi)					
68–70°F	20	24	25	7	28
1000°F	—	—	—	—	—
Elongation (in 2 in.) at 68–75°F (%)	3.5	8	6	19	8
Shear Strength (1000 psi)					
68–70°F	22	22	23	8	26
1000°F	—	—	—	—	—
Hardness (Brinell)	62	65	62	—	70

TABLE 5-7.

MOLYBDENUM & ALLOYS—WROUGHT

Type →	Molybdenum 99.9%	Mo–0.5Ti	TZM	AVC 70 Mo, 30W	50 Mo–50 Re
COMPOSITION (%)		Ti 0.5, W 0.02	Ti 0.5, W 0.02, Zr 0.08		
PHYSICAL, THERMAL, ELECTRICAL AND MAGNETIC PROPERTIES					
Density (lb/in.³)	0.369	0.37	0.37	0.43	0.50
Melting Range (°F)	4730	4730	4730	5150	4620
Specific Heat (Btu/lb/°F)	0.065	0.065	0.065	–	–
Thermal Expansion Coeff. (μ in./in./°F)					
68–70°F	2.7	2.7	2.7	–	–
1000°F	–	–	–	–	–
Thermal Conductivity (Btu/ft²/hr/°F/ft)					
212°F	84.5	84.5	84.5	–	–
1000°F	–	–	–	–	–
Electrical Resistivity (μ ohm-cm) at 68°F	5.2	5.2	5.2	5.3	19.4
Magnetic Permeability	–	–	–	–	–
MECHANICAL PROPERTIES (annealed)					
Youngs's Modulus (10⁶ psi)					
68–70°F	47	46	46	50	53
2400°F	21	20	20	–	–
Tensile Strength (1000 psi)					
68–70°F	95	115	125	105	240
2400°F	16	18	54	–	41
Yield Strength (1000 psi)					
68–70°F	82	104	105	95	210
2400°F	10	8	40	–	–
Elongation (in 2 in.) at 68–75°F (%)	10	10	10	–	–
Shear Strength (1000 psi)					
68–70°F	–	–	–	–	–
1000°F	–	–	–	–	–
Hardness (Vickers) (recrystall)	190	200	200	200	350

TABLE 5-8.

NICKEL AND ALLOYS—WROUGHT, CAST HIGH PURITY NICKEL

Type →	Nickel 200	Nickel 201	Nickel 270
COMPOSITION (%)	Ni 99.50, C 0.06, Mn 0.25, Fe 0.15, S 0.005, Si 0.05, Cu 0.05	Ni 99.50, C 0.01, Mn 0.20, Fe 0.15, S 0.005, Si 0.05, Cu 0.05	Ni 99.98 min
PHYSICAL, THERMAL, ELECTRICAL AND MAGNETIC PROPERTIES			
Density (lb/in.^{3})	0.321	0.321	0.321
Melting Range (°F)	2615–2635	2615–2635	2650
Specific Heat (Btu/lb/°F)	0.109	0.109	0.11
Thermal Expansion Coeff. (μ in./in./°F)			
70–200°F	7.4	7.4	7.4
70–1000°F	8.5	–	8.6
Thermal Conductivity (Btu/ft^2/hr/°F/ft)			
68–70°F	36.3	36.3	49.6
1000°F	30.8	34.2	35.5
Electrical Resistivity (μ ohm-cm) at 68°F	11.0	7.7	7.48
Magnetic Permeability	–	–	–
MECHANICAL PROPERTIES (annealed)			
Youngs's Modulus (10^6 psi)			
68–70°F	30	30	30
1000°F	–	–	–
Tensile Strength (1000 psi) (ult.)			
68–70°F	55–75	50–60	50–95
1000°F	–	–	19
Yield Strength (1000 psi)			
68–70°F	15–30	12–25	16–90
1000°F	–	–	7.5
Elongation (in 2 in.) at 68–75°F (%)	55–40	60–40	30
Shear Strength (1000 psi)			
68–70°F	–	–	–
1000°F	–	–	–
Hardness (Rockwell)	B64 max	B55 max	80–210 (Brinell)

TABLE 5-8. (Continued)

NICKEL & ALLOYS–WROUGHT MAGNETIC ALLOYS

Type →	80 Ni	80 Ni	76 Ni	50 Ni	Glass Seal Alloy, 50 Ni
COMPOSITION (%)	Ni 80, Mo 5, Mn 0.5, Si 0.15, C 0.01, Fe bal	Ni 80, Mo 4, Mn 0.35–0.50, Si 0.15–0.35, C 0.03, Fe bal	Ni 76, Cu 4–5, Cr 1.5, Fe bal	Ni 49–51, Fe bal	Ni 50–51, Fe bal
PHYSICAL, THERMAL, ELECTRICAL AND MAGNETIC PROPERTIES					
Density (lb/in.^{3})	0.316	0.316	0.307	0.301	0.301
Melting Range (°F)	2650	2650	—	—	—
Specific Heat (Btu/lb/°F)	0.118	0.118	—	0.115	0.115
Thermal Expansion Coeff. (μ in./in./°F)					
68–212°F	—	6.4	7.0	—	4.7
1000°F	—	—	—	—	—
Thermal Conductivity (Btu/ft^2/hr/°F/ft)					
68–70°F	—	20	—	—	—
1000°F	—	—	—	—	—
Electrical Resistivity (μ ohm-cm) at 68°F	65	58	60	43–45	43–45
Magnetic Permeability (10^3)	250–500	—	90	60–100	—
MECHANICAL PROPERTIES					
Young's Modulus (10^6 psi)					
68–70°F	31–34	31–34	—	—	—
1000°F	—	—	—	—	—
Tensile Strength (1000 psi)					
68–70°F	80	79	64–66	58	70–150
1000°F	—	—	—	—	—
Yield Strength (1000 psi)					
68–70°F	25	22	18.5–19.5	19	—
1000°F	—	—	—	—	—
Elongation (in 2 in.) at 68–75°F (%)	60	64	24–27	27	—
Shear Strength (1000 psi)					
68–70°F	—	—	—	—	—
1000°F	—	—	—	—	—
Hardness (Rockwell)	—	B62	—	—	—

TABLE 5-8. (*Continued*)

NICKEL & ALLOYS—WROUGHT ELECTRICAL RESISTANCE ALLOYS

Type →	80 Ni	75 Ni	70 Ni	70 Ni	60 Ni
COMPOSITION (%)	Ni 77 min, Cr 19–20, Mn 2.5 max, Si 1.5 max, C 0.25 max	Ni 75, Cr 20, Additives 5	Ni 70, Cr 30	Ni 70, Fe 30	Ni 57–61, Cr 14–18, Si 1.75 max, Mn 1 max, C 0.15 max, Fe bal
PHYSICAL, THERMAL, ELECTRICAL AND MAGNETIC PROPERTIES					
Density (lb/in.^{3})	0.3	0.293	0.293	0.305	0.298
Melting Range (°F)	2540–2600	2400	2590	2510	2550–2650
Specific Heat (Btu/lb/°F)	0.107	0.104–0.107	0.11	0.11–0.12	0.107
Thermal Expansion Coeff. (μ in./in./°F)					
68–212°F	7.3	6.9	6.8	7.2	7.2
1000°F	–	–	–	–	–
Thermal Conductivity (Btu/ft^2/hr/°F/ft)					
68–70°F	7.4	7.4	8.1	–	7.4
1000°F	–	–	–	–	–
Electrical Resistivity (μ ohm-cm) at 68°F	108	134	118	20	112
Magnetic Permeability	–	–	–	–	–
MECHANICAL PROPERTIES					
Young's Modulus (10^6 psi)					
68–70°F	31	–	24	–	30
1000°F	–	–	–	–	–
Tensile Strength (1000 psi)					
68–70°F	95–200	130–180	128–140	75–150	95–200
1000°F	–	–	–	–	–
Yield Strength (1000 psi)					
68–70°F	–	–	–	–	–
1000°F	–	–	–	–	–
Elongation (in 2 in.) at 68–75°F (%)	0–35	5–35	30–40	15–25	0–55
Shear Strength (1000 psi)					
68–70°F	–	–	–	–	–
1000°F	–	–	–	–	–
Hardness (Rockwell)	B85–B105	B75–B80	–	–	B85–B105

TABLE 5-8. (*Continued*)

NICKEL–COPPER ALLOYS–WROUGHT

Type →	Monel 400™	Monel 404™	Monel R-405™	Monel K-500™	Monel 502™
COMPOSITION (%)	Ni 66.5, Cu 31.5, Fe 1.25, Mn 1, Si 0.25, C 0.15, S 0.12	Ni 54.5, Cu 44, Fe 0.25, C 0.08, Mn 0.05, Si 0.05, Al 0.03, S 0.12	Ni 66.5, Cu 31.5, Fe 1.25, Mn 1.0, Si 0.25, C 0.15, S 0.04	Ni 66.5, Cu 29.5, Al 2.7, Fe 1.0, Mn 0.75, Ti 0.6, Si 0.25, C 0.13, S 0.005	Ni 66.5, Cu 28, Al 3.0, Fe 1.0, Mn 0.75, Ti 0.25, Si 0.25, C 0.05, S 0.005
PHYSICAL, THERMAL, ELECTRICAL AND MAGNETIC PROPERTIES					
Density (lb/in.^{3})	0.319	0.321	0.319	0.306	0.305
Melting Range (°F)	2370–2460	–	2370–2460	2400–2460	2400–2460
Specific Heat (Btu/lb/°F)	0.102	0.099	0.102	0.100	0.100
Thermal Expansion Coeff. (μ in./in./°F)					
70–200°F	7.7	7.4	7.7	7.6	7.6
70–1000°F	9.1	9.2	9.1	8.7	8.7
Thermal Conductivity (Btu/ft^2/hr/°F/ft)					
68–70°F	12.6	12.2	12.6	10.1	10.1
1000°F	22.0	–	22.0	18.3	18.3
Electrical Resistivity (μ ohm-cm) at 68°F	51	50	51	62	62
Magnetic Permeability (70°F, H-200 oersteds)	–	1.002	–	1.001	1.001
MECHANICAL PROPERTIES					
Youngs's Modulus (10^6 psi)					
68–70°F	26	25	26	26	26
1000°F	–	–	–	–	–
Tensile Strength (1000 psi) (ult.)					
68–70°F	70–100	65	70–115	90–190	87–141
1000°F	45	40	–	95	100
Yield Strength (1000 psi) (0.2%)					
68–70°F	25–100	24	25–105	40–160	37–94
1000°F	22	16	–	90	80
Elongation (in 2 in.) at 68–75°F (%)	22–60	50	15–50	13–45	25–47
Shear Strength (1000 psi)					
68–70°F	–	–	–	–	–
1000°F	–	–	–	–	–
Hardness (Brinell)	110–241	–	110–245	140–346	134–255

TABLE 5-8. (*Continued*)

NICKEL–COPPER ALLOYS–CAST

Type →	Monel 411^{TM}	Monel 505^{TM}
COMPOSITION (%)	Ni 64, Cu 31.5, Si 1.5, Fe 1.0, C 0.20	Ni 63, Cu 29.5, Si 4.0, Fe 2.0, Mn 0.8, C 0.01
PHYSICAL, THERMAL, ELECTRICAL AND MAGNETIC PROPERTIES		
Density (lb/in.³)	0.312	0.302
Melting Range (°F)	2400–2450	2300–2350
Specific Heat (Btu/lb/°F)		
Thermal Expansion Coeff. (μ in./in./°F)		
70–200°F	–	–
70–1000°F	9.1	8.9
Thermal Conductivity (Btu/ft²/hr/°F/ft)		
68–70°F	15.5	11.3
1000°F	–	–
Electrical Resistivity (μ ohm-cm) at 68°F		
Magnetic Permeability		
MECHANICAL PROPERTIES		
Youngs's Modulus (10⁶ psi)		
68–70°F	19	24.2
1000°F	–	–
Tensile Strength (1000 psi) (ult.)		
68–70°F	65–90	110–145
1000°F	–	–
Yield Strength (1000 psi) (0.2%)		
68–70°F	32–45	80–115
1000°F	–	–
Elongation (in 2 in.) at 68–75°F (%)	25–45	1.4
Shear Strength (1000 psi)		
68–70°F	–	–
1000°F	–	–
Hardness (Brinell)	125–150	275–350

TABLE 5-8. (*Continued*)

NICKEL & ALLOYS–WROUGHT LOW EXPANSION ALLOYS

Type →	36 Ni	42 Ni	50 Ni	Ni-Span-C902*TM		
COMPOSITION (%)	Ni 36, Fe bal	Ni 42, Fe bal	Ni 47–50, Fe bal	Ni 42.25, Fe 48.5, Cr 5.33, Ti 2.58, Al 0.55, Si 0.5, Mn 0.4, Cu 0.05, C 0.03, S 0.02		
PHYSICAL, THERMAL, ELECTRICAL AND MAGNETIC PROPERTIES						
Density (lb/in.^{3})	0.291	0.294	0.296	0.293		
Melting Range (°F)	2600	2600	2600	2650–2700		
Specific Heat (Btu/lb/°F)	0.123	0.121	0.120	0.12		
Thermal Expansion Coeff. (μ in./in./°F)						
0–200°F	0.70	3.18	5.55	4.2 (70–200°F)		
800–1000°F	9.48	8.55	7.26	–		
Thermal Conductivity (Btu/ft^2/hr/°F/ft)						
68–212°F	7.8	8.9	10.3	7.9 (212°F)		
1000°F	–	–	–	11.8		
Electrical Resistivity (μ ohm-cm) at 68°F	81	70	48	101.6		
Magnetic Permeability	–	–	–	–		
MECHANICAL PROPERTIES (annealed)						
Youngs's Modulus (10^6 psi)						
68–70°F	21	22	24	24–29		
1000°F	–	–	–	–		
Tensile Strength (1000 psi)						
68–70°F	71	68	77	131		
1000°F	–	–	–	–		
Yield Strength (1000 psi)						
68–70°F	40	39	33	126		
1000°F	–	–	–	–		
Elongation (in 2 in.) at 68–75°F (%)	43	49	45	6.5		
Shear Strength (1000 psi)						
68–70°F	–	–	–	–		
1000°F	–	–	–	–		
Hardness (Brinell)	132	138	144	–		

*Constant Modulus Alloy.

TABLE 5-8. (*Continued*)

NICKEL & ALLOYS–CAST, WROUGHT

Type →	Permanickel™ 300	Duranickel™ 301	Beryllium Nickels		
			Casting	Alloys	Wrought
COMPOSITION (%)	Ni (+ Co) 97.0 min, Fe 0.60, Mn 0.50, C 0.40, Si 0.35, Cu 0.25, Ti 0.20–0.60, Mg 0.2 0–0.50, S 0.01	Ni (+ Co) 93.0 min, Al 4.00–4.75, Si 1.00, Fe 0.60, Mn 0.50, C 0.30, Cu 0.25, Ti 0.25–1.00, S 0.01	Ni 97.85 nom, Be 2.15 nom	Ni 96.9 nom, Be 2.7 nom, C 0.4	Ni 97.55 nom, Be 1.95 nom, Ti 0.50 nom
PHYSICAL, THERMAL, ELECTRICAL AND MAGNETIC PROPERTIES					
Density (lb/in.³)	0.316	0.298	0.294	0.294–0.296	0.302
Melting Range (°F)	–	2550–2620	2100–2400	2100–2400	–
Specific Heat (Btu/lb/°F)	0.106	0.104	0.111	0.125	–
Thermal Expansion Coeff. (μ in./in./°F)					
70–200°F	6.8	7.2	–	–	–
70–1000°F	8.4	8.2	–	8.3	–
Thermal Conductivity (Btu/ft²/hr/°F/ft)					
68–70°F	33.3	13.7	7.3	19.6	18.3
1000°F	–	22.0	–	–	–
Electrical Resistivity (μ ohm-cm) at 68°F	15.7	42.4	40.6	22.8	23.8
Magnetic Permeability	–	–	–	–	–
MECHANICAL PROPERTIES					
Young's's Modulus (10⁶ psi)					
68–70°F	30	30	26–28	26–29	27–30
1000°F	–	–	–	–	–
Tensile Strength (1000 psi) (ult.)					
68–70°F	90–200	90–210	130–240	123–232	95–290
1000°F	125	125	180	160	150
Yield Strength (1000 psi) (0.2%)					
68–70°F	35–150	30–175	57–190	56–220	45–230
1000°F	100	100	170	130	130
Elongation (in 2 in.) at 68–75° F (%)	10–40	15–55	5–30	1–27	2–40
Shear Strength (1000 psi)					
68–70°F	–	–	–	–	–
1000°F	–	–	–	–	–
Hardness (Rockwell)	140–380 (Brinell)	B75–C40	B97–C54	B97–C55	B70–C51

TABLE 5-8. (*Continued*)

NICKEL & ALLOYS—CAST, WROUGHT CHEMICAL RESISTANT ALLOYS

Type →	Illium B^{TM}	Illium 98^{TM}
COMPOSITION (%)	C 0.05, Mn 1, Cr 28, Mo 8, Cu 5.5, Si 3.5, Fe 1.5, Ni bal	C 0.07 max, Mn 1.50 max, Cr 26-30, Mo 7.5-9, Cu 4-6.5, Si 1.25 max, Fe 1.5 max, Ni bal
PHYSICAL, THERMAL, ELECTRICAL AND MAGNETIC PROPERTIES		
Density (lb/in.³)	0.309	—
Melting Range (°F)	—	—
Specific Heat (Btu/lb/°F)	—	—
Thermal Expansion Coeff. (μ in./in./°F)		
68-212°F	—	—
1000°F	—	—
Thermal Conductivity (Btu/ft²/hr/°F/ft)		
68-70°F	—	—
1000°F	—	—
Electrical Resistivity (μ ohm-cm) at 68°F	—	—
Magnetic Permeability	—	—
MECHANICAL PROPERTIES		
Youngs's Modulus (10⁶ psi)		
68-70°F	—	—
1000°F	—	—
Tensile Strength (1000 psi) (ult.)		
68-70°F	60; 70	54; 78
1000°F	—	—
Yield Strength (1000 psi) (0.2%)		
68-70°F	50; 65	41; 43
1000°F	—	—
Elongation (in 2 in.) at 68-75°F (%)	1.0-4.5	18
Shear Strength (1000 psi)		
68-70°F	—	—
1000°F	—	—
Hardness (Brinell)	200-240	160

TABLE 5-8. (*Continued*)

NICKEL & ALLOYS–CAST, WROUGHT CHEMICAL RESISTANT ALLOYS

Type →	Hastelloy B™	Hastelloy C™ Uniloy HC™	Hastelloy™ C-276	Hastelloy D™	Hastelloy G™
COMPOSITION (%)	C 0.05–0.12, Mn 1, Si 1, Fe 4–6, Mo 26–30, Cr 1, Co 2.5, V 0.2–0.4, P 0.025, S 0.03, Ni bal	C 0.08–0.12, Mn 1, Si 1, Fe 4–7, W 3–5.25, Mo 15–18, Cr 14.5–17.5, Co 2.5, V 0.2–0.4, P 0.04, S 0.03, Ni bal	C 0.02, Mn 1, Si 0.05, Fe 4–7, W 3–4.5, Mo 15–17, Cr 14.5–16.5, Co 2.5, V 0.35, P 0.03, S 0.03, Ni bal	C 0.12, Mn 0.5–1.25, Si 8.5–10, Cu 2–4, Fe 2, Cr 1, Co 1.5, Ni bal	C 0.05–0.12, Mn 1–2, Si 1, Fe 18–21, W 1, Mo 5.5–7.5, Cr 21–23.5, Co 2.5, Cu 1.5–2.5, Cb + Ta 1.75–2.5, P 0.04, S 0.03, Ni bal
PHYSICAL, THERMAL, ELECTRICAL AND MAGNETIC PROPERTIES					
Density (lb/in.³)	0.334	0.323	0.321	0.282	0.300
Melting Range (°F)	2375–2495	2310–2450	2415–2500	2065–2220	2300–2450
Specific Heat (Btu/lb/°F)	0.091	0.092	0.102	0.109	0.093
Thermal Expansion Coeff. (μ in./in./°F)					
32–212°F	5.6	6.3	6.2 (75–200°F)	6.1	7.5 (70–200°F)
70–1200°F	6.7	7.7	7.8 (75–1200°F)	–	9.1
Thermal Conductivity (Btu/ft²/hr/°F/ft)					
392°F	7.08	6.5	7.5 (400°F)	12.08 (72°F)	7.42
1112°F	9.5	9.83	12.08 (1200°F)	–	11.08
Electrical Resistivity (μ ohm-cm) at 68°F	135	130	130	113	–
Magnetic Permeability	–	–	–	–	–
MECHANICAL PROPERTIES					
Youngs's Modulus (10⁶ psi)					
68–70°F	28.5; 31.1	29.8; 26	29.8	28.9	27.8
1000°F	26.9	24.8	25.5	–	23.4
Tensile Strength (1000 psi) (ult.)					
68–70°F	134	121; 83.4	115.6	115	102; 87.5
1000°F	113.5	99.3; 64.2	95	–	81.9; 64.9
Yield Strength (1000 psi) (0.2%)					
68–70°F	67	57.8; 48.3	58.4	–	46.2; 38.8
1000°F	48.7	43.9; 34.3	42.5	–	32.8; 25.2
Elongation (in 2 in.) at 68–75°F (%)	51	48; 12	56	1	61; 30
Shear Strength (1000 psi)					
68–70°F	–	–	–	–	–
1000°F	–	–	–	–	–
Hardness (Rockwell)	–	B91–103; B92	B92	C30–39	B84

TABLE 5-8. (*Continued*)

NICKEL & ALLOYS—CAST, WROUGHT HIGH TEMPERATURE ALLOYS

Type →	Inconel 600[TM]	Inconel 601[TM]	Inconel 625[TM]	Incoloy 804[TM]	Incoloy 825[TM]
COMPOSITION (%)	Ni 76, Cr 15.5, Fe 8, Cu 0.25, Si 0.25, Mn 0.5, C 0.08	Ni 60.5, Cr 23, Fe 14, Al 1.35, Cu 0.5, Mn 0.5, Si 0.25, S 0.007, C 0.05	Ni 61, Cr 21.5, Mo 9, Fe 2.5, Si 0.25, Mn 0.25, Al 0.2, Ti 0.2, S 0.008, C 0.05	Ni 41, Cr 29.5, Fe25.4, Mn0.75, Ti 0.6, Si 0.38, Al 0.3, Cu 0.25, S 0.008, C 0.05	Ni 42, Fe 30, Cr 21.5, Mo 3, Cu 2.25, Ti 0.9, Mn 0.5, Si 0.25, Al 0.1, S 0.015, C 0.03
PHYSICAL, THERMAL, ELECTRICAL AND MAGNETIC PROPERTIES					
Density (lb/in.^{3})	0.301	0.291	0.305	0.286	0.294
Melting Range (°F)	2470–2575	2374–2494	2350–2460	–	2500–2550
Specific Heat (Btu/lb/°F)	0.106	0.107	0.098	–	–
Thermal Expansion Coeff. (μ in./in./°F)					
70–200°F	7.70	7.6	7.1	8.0	7.8
70–1000°F	8.4	8.5	7.8	8.8	8.8
Thermal Conductivity (Btu/ft^2/hr/°F/ft)					
68–70°F	8.58	6.5	5.67	–	6.42
1000°F	13.2	11.6	10.0	–	10.9
Electrical Resistivity (μ ohm-cm) at 68°F	103	119	129	106	113
Magnetic Permeability (70°F, H = 200 oersted)	1.010	1.003	1.0006	1.003	1.005
MECHANICAL PROPERTIES					
Youngs's Modulus (10^6 psi)					
68–70°F	31.4	29.9	29.8	28	28
1000°F	27.1	–	25.6	–	–
Tensile Strength (1000 psi) (ult.)					
68–70°F	90	86.5	124	95	92
1000°F	84	78 (1200°F)	108	–	76
Yield Strength (1000 psi) (0.2%)					
68–70°F	36	28.5	53	45	36
1000°F	28	25 (1200°F)	37	–	22
Elongation (in 2 in.) at 68–75°F (%)	47	58	59	40	50
Shear Strength (1000 psi)					
68–70°F	–	–	–	–	–
1000°F	–	–	–	–	–
Hardness (Brinell)	120–170	114	145–240	–	120–180

TABLE 5-8. (*Continued*)

NICKEL & ALLOYS–CAST, WROUGHT HIGH TEMPERATURE ALLOYS

	Pyrotool M™	Pyrotool W™				
Type →						
COMPOSITION (%)	C 0.12, Mn 0.2, Si 0.2, Cr 19, Mo 10, Co 10, Al 1, Ti 2.5, Fe 2, Ni bal	C 0.05, Mn 0.2, Si 0.2, Cr 19.5, Mo 4.25, Co 13, Ti 3.1, Al 1.2, Fe 1, Ni bal				
PHYSICAL, THERMAL, ELECTRICAL AND MAGNETIC PROPERTIES						
Density (lb/in.^{3})	0.298	—				
Melting Range (°F)	—	—				
Specific Heat (Btu/lb/°F)	—	—				
Thermal Expansion Coeff. (μ in./in./°F)						
68–212°F	—	—				
1000°F	—	—				
Thermal Conductivity (Btu/ft^2/hr/°F/ft)						
68–70°F	—	—				
1000°F	—	—				
Electrical Resistivity (μ ohm-cm) at 68°F	—	—				
Magnetic Permeability	—	—				
MECHANICAL PROPERTIES						
Youngs's Modulus (10^6 psi)						
68–70°F	—	—				
1000°F	—	—				
Tensile Strength (1000 psi) (ult.)						
68–70°F	60; 70	54; 78				
1000°F	—	—				
Yield Strength (1000 psi) (0.2%)						
68–70°F	50; 65	41; 43				
1000°F	—	—				
Elongation (in 2 in.) at 68–75°F (%)	1; 0.5	18; 43				
Shear Strength (1000 psi)						
68–70°F	—	—				
1000°F	—	—				
Hardness (Brinell)	228; 221	152; 140				

TABLE 5-8. (*Continued*)

NICKEL–BASE SUPERALLOYS–CAST, WROUGHT

Type →	IN 100	IN 102	IN 162	Inconel 700™	Inconel 702™
COMPOSITION (%)	C 0.15–0.2, Mn 0.02, Cr 8–11, Co 13–17, Mo 2–4, Ti 4.5–5, Al 5–6, V 0.7–1.2, Fe 1, B0.01–0.02, Zr 0.03–0.09, Si 0.2, S 0.015, Ni bal	C 0.08, Cr 14–16, Fe 5–9, Cb + Ta 2.75–3.25, Mo 2.75–3.25, W 2.75–3.25, Al 0.3–0.6, Ti 0.4–0.7, Zr 0.01–0.05, B 0.003–0.008, Mg 0.01–0.05, Mn 0.75, P 0.01, S0.01, Si 0.04, Ni bal	C 0.12, Cr 10, Mo 4, W 2, Cb 1, Ti 1, Al 6.5, Ta 2, B 0.02, Zr 0.1, Ni bal	C 0.12, Mn 0.1, Si 0.3, Cr 15, Co 28.5, Mo 3.7, Ti 2.2, Al 3.0, Fe 0.7, Ni bal	C0.1, Mn1, Si0.7, Cr 14–17, Al 2.75–3.75, Ti 0.25–1, Fe 2, Cu 0.5, S 0.01, Ni bal
PHYSICAL, THERMAL, ELECTRICAL AND MAGNETIC PROPERTIES					
Density (lb/in.^{3})	0.280	0.309	0.292	0.295	0.304
Melting Range (°F)	2305–2435	2410–2530	2330–2380	2450–2600	—
Specific Heat (Btu/lb/°F)	—	—	—	—	—
Thermal Expansion Coeff. (μ in./in./°F)					
70–200°F	7.2	7.3	6.8	6.9	6.7
70–100°F	7.7	8.02	7.85	8.10	8.3
Thermal Conductivity (Btu/ft^2/hr/°F/ft)					
68–70°F	—	6.5	—	7.1	7.0
1000°F	—	11.3	—	9.2	17.4
Electrical Resistivity (μ ohm-cm) at 68°F	—	—	—	—	123
Magnetic Permeability (70°F, H = 200 oersteds)	—	—	—	—	1.0004
MECHANICAL PROPERTIES					
Youngs's Modulus (10^6 psi)					
68–70°F	31.2	29.7	28.5	32.5	31.5
1000°F	27.1	25.2	24.9	27.9	27.2
Tensile Strength (1000 psi) (ult.)					
68–70°F	147	139	146	170	148
1000°F	158	120	148	146	118
Yield Strength (1000 psi) (0.2%)					
68–70°F	123	73	118	103	84
1000°F	128	58	115	93	79
Elongation (in 2 in.) at 68–75°F (%)	9.0	47	710	25	35
Shear Strength (1000 psi)					
68–70°F	—	—	—	—	—
1000°F	—	—	—	—	—
Hardness (Brinell)	—	—	—	—	—

TABLE 5-8. (*Continued*)

NICKEL–BASE SUPERALLOYS–CAST, WROUGHT

Type →	Inconel 706 TM	Inconel 718 TM	Inconel 722 TM	Inconel X-750 TM	Inconel 751 TM
COMPOSITION (%)	C 0.06, Mn 0.35, Ni 39–44, Cr 14.5–17.5, Cb + Ta 2.5–3.3, Ti 1.5–2, Co 1, Al 0.4, Si 0.35, Cu 0.3, P 0.02, S 0.015, B 0.006, Fe bal	C 0.08, Mn 0.35, Ni 50–55, Cr 17–21, Cb + Ta 4.75–5.5, Mo 2.8–3.3, Co 1, Ti 0.65–1.15, Al 0.2–0.8, Si 0.35, Cu 0.3, S 0.015, P 0.015, B 0.006, Fe bal	C 0.08, Mn 1, Ni 70 min, Cr 14–17, Fe 5–9, Ti 2–2.75, Al 0.4–1, Si 0.7, Cu 0.5, S 0.01	C 0.08, Mn 1, Ni 70 min, Cr 14–17, Fe 5–9, Ti 2.25–2.75, Cb + Ta 0.7–1.2, Al 0.4–1, Cu 0.5, Si 0.5, S 0.01	C 0.1, Mn 1.0, Ni 70 min, Cr 14–17, Fe 5–9, Ti 2–2.6, Al 0.9–1.5, Cb + Ta 0.7–1.2, Cu 0.5, Si 0.5, S 0.01
PHYSICAL, THERMAL, ELECTRICAL AND MAGNETIC PROPERTIES					
Density (lb/in.³)	0.291	0.296	0.298	0.298	0.298
Melting Range (°F)	2434–2499	2300–2437	2450–2600	2540–2600	2540–2600
Specific Heat (Btu/lb/°F)	0.106	0.104	0.105	0.103	—
Thermal Expansion Coeff. (μ in./in./°F)					
70–200°F	7.8	7.1	6.5	6.96	—
70–1000°F	—	8.2	8.4	8.1	—
Thermal Conductivity (Btu/ft²/hr/°F/ft)					
68–70°F	7.25	6.5	8.5	6.92	—
1000°F	12.3	11.3	—	10.9	—
Electrical Resistivity (μ ohm-cm) at 68°F	99	125	124	122	—
Magnetic Permeability (70°F, H = 200 oersteds)	1.01	1.001	1.002	1.002	—
MECHANICAL PROPERTIES					
Young's Modulus (10⁶ psi)					
68–70°F	30.4	29.0	31.0	31.0	31.0
1000°F	—	26.7	26.4	26.7	—
Tensile Strength (1000 psi) (ult.)					
68–70°F	188	208;185	158	162;179	205
1000°F	164	185;166	125	140;146	180
Yield Strength (1000 psi) (0.2%)					
68–70°F	145	172;153	82	92;120	145
1000°F	130	154;137	76	84;109	140
Elongation (in 2 in.) at 68–75°F (%)	18	21;22	26	24;24	50
Shear Strength (1000 psi)					
68–70°F	—	—	—	—	—
1000°F	—	—	—	—	—
Hardness (Brinell)	—	—	—	—	—

TABLE 5-8. (*Continued*)

NICKEL–BASE SUPERALLOYS—CAST, WROUGHT

Type →	713 C	901	B-1900	D-979	GMR-235-D		
COMPOSITION (%)	C 0.12, Cr 12.5, Al 6.1, Mo 4.2, Cb 2, Ti 0.8, Zr 0.1, Ni bal	C 0.5, Mn 0.1, Si 0.1, Cr 12.5, Ni 42.5, Mo 5.7, Ti 2.8, Al 0.2, B 0.015, Fe bal	C 0.1, Cr 8, Co 10, Mo 6, Al 6, Ta 4, Ti 1, Zr 0.1, B 0.015, Ni bal	C 0.05, Mn 0.25, Si 0.2, Fe 27, Cr 15, Mo 4, W 4, Ti 3, Al 1, B 0.01, Ni bal	C 0.15, Cr 15.5, Mo 5, Fe 4.5, Ti 2.5, Al 3.5, B 0.05, Ni bal		
PHYSICAL, THERMAL, ELECTRICAL AND MAGNETIC PROPERTIES							
Density (lb/in.^{3})	0.286	0.297	0.297	0.296	0.291		
Melting Range (°F)	2300–2350	—	2325–2375	2225–2250	—		
Specific Heat (Btu/lb/°F)	0.14 (1000°F)	—	—	—	0.132 (1000°F)		
Thermal Expansion Coeff. (μ in./in./°F)							
70–200°F	5.92	7.75	6.5	7.60	7.1		
70–1000°F	7.52	8.50	8.32 (–1600°F)	8.25	—		
Thermal Conductivity (Btu/ft^2/hr/°F/ft)							
68–70°F	—	7.67	—	7.25	6.0		
1000°F	13.7	10.6	12.67 (1600°F)	10.7	—		
Electrical Resistivity (μ ohm-cm) at 68°F							
Magnetic Permeability							
MECHANICAL PROPERTIES							
Young's Modulus (10^6 psi)							
68–70°F	29.9	29.9	31	30.0	28.7		
1000°F	26.2	24.2	27	25.8	23.4		
Tensile Strength (1000 psi) (ult.)							
68–70°F	123	175; 166	141	204	112		
1000°F	125	149; 138	146	188	104		
Yield Strength (1000 psi) (0.2%)							
68–70°F	107	130; 107	120	146	103		
1000°F	102	113; 96	126	134	82		
Elongation (in 2 in.) at 68–75°F (%)	7.9	14; 23	8	15	3.5		
Shear Strength (1000 psi)							
68–70°F	—	—	—	—	—		
1000°F	—	—	—	—	—		
Hardness (Brinell)							

TABLE 5-8. (Continued)

NICKEL–BASE SUPERALLOYS–WROUGHT, CAST

Type →	Udimet 500™	Udimet 700™	Waspaloy™	Nicrotung™	Rene–41, R–41™
COMPOSITION (%)	C 0.15 (max), Al 2.5–3.2, Ti 2.5–3.2, Mo 3.0–5.0, Cr 15.0–20.0, Co 13.0–20.0, Fe 4.0 (max), B 0.008 (max), Ni bal	C 0.15 (max), Al 3.75–4.75, Ti 3.0–4.0, Mo 4.5–5.7, Cr 13.0–17.0, Co 17.0–20.0, Fe 1.0, B 0.10 (max), Ni bal	C 0.10 (max), Al 1.3 Ti 3.00, Mo 4.25, Cr 19.50, Co 13.5, Fe 2.0, B 0.005, Zr 0.085, Mn 0.50 (max), Si 0.75 (max), S 0.030 (max), Cu 0.10 (max), Ni bal	C 0.10, B 0.05, Zr 0.05, Cr 12.0, Co 10.0, W 8.0, Al 4.0, Ti 4.0, Ni bal	Cr 18.0–20.0, Co 10.0–12.0 (max), Mo 9.0–10.5, Fe 5.0, C 0.09–0.12, Ti Si 0.5, Mn 0.1, Ti 2.0–3.3, Al 1.4–1.6, Ni bal
PHYSICAL, THERMAL, ELECTRICAL AND MAGNETIC PROPERTIES					
Density (lb/in.³)	0.290	0.286	0.296	0.300	0.298
Melting Range (°F)	2375–2540	2200–2550	2425–2475	–	2400–2500
Specific Heat (Btu/lb/°F)	0.10–0.11	0.10–0.11	0.10–0.11	–	0.11
Thermal Expansion Coeff. (μ in./in./°F)					
70–200°F	6.75	–	6.8	–	6.63
70–1000°F	7.80	7.74	7.8	7.3	7.51
Thermal Conductivity (Btu/ft²/hr/°F/ft)					
68–70°F	6.4	11.3	6.2	–	5.2
1000°F	10.6	11.9	10.4	–	10.4
Electrical Resistivity (μ ohm-cm) at 68°F	121.5–136.5	130–148	–	–	–
Magnetic Permeability	–	–	–	–	–
MECHANICAL PROPERTIES					
Youngs's Modulus (10⁶ psi)					
68–70°F	31.2	32.1	31.9	33.5	31.8
1000°F	27.7	28.1	26.7	29.7	27.7
Tensile Strength (1000 psi) (ult.)					
68–70°F	197	205	188	130	206
1200°F	175	180	165	120	194
Yield Strength (1000 psi) (0.2%)					
68–70°F	110	140	120	120	194
1000°F	110 (1200°F)	124 (1200°F)	105	112	147
Elongation (in 2 in.) at 68–75°F (%)	18	16	29	5	14
Shear Strength (1000 psi)					
68–70°F	–	–	–	–	–
1000°F	–	–	–	–	–
Hardness (Rockwell)	C37	–	C37	C38–40	–

TABLE 5-8. (*Continued*)

NICKEL–BASE SUPERALLOYS–CAST, WROUGHT

Type →	Hastelloy™ R-235	Hastelloy X™				
COMPOSITION (%)	C 0.15, Cr 15.5, Mo 5.5, Fe 10.0, Ti 2.5, Al 2.0, bal Ni	C 0.10, Mn 0.50, Si 0.50, Cr 22.0, Co 1.5, Mo 9.0, W 0.6, Fe 18.5, bal Ni				
PHYSICAL, THERMAL, ELECTRICAL AND MAGNETIC PROPERTIES						
Density (lb/in.^{3})	0.296	0.297				
Melting Range (°F)	2465–2535	2300–2470				
Specific Heat (Btu/lb/°F)		0.116				
Thermal Expansion Coeff. (μ in./in./°F)						
70–200°F	6.70	7.70				
70–1000°F	7.98	8.39				
Thermal Conductivity (Btu/ft^2/hr/°F/ft)						
68–70°F	6.33	5.25				
1000°F	10.5	11.3				
Electrical Resistivity (μ ohm-cm) at 68°F		118.3 (72°F)				
Magnetic Permeability						
MECHANICAL PROPERTIES						
Youngs's Modulus (10^6 psi)						
68–70°F	30.5	28.6				
1000°F	26.4	23.4				
Tensile Strength (1000 psi) (ult.)						
68–70°F	169	114				
1000°F	150	94				
Yield Strength (1000 psi) (0.2%)						
68–70°F	117	52				
1000°F	113	42				
Elongation (in 2 in.) at 68–75°F (%)	21	43				
Shear Strength (1000 psi)						
68–70°F	—	—				
1000°F	—	—				
Hardness (Brinell)	—	—				

TABLE 5-9

NIOBIUM (COLUMBIUM) AND ALLOYS–WROUGHT

Type →	Columbium	Cb-1Zr	C-103	B-66	CB-752
COMPOSITION (%)	—	Zr 0.8–1.2, Nb bal	Ti 1.0, Hf 10.0, Nb bal	V 5.0, Mo 5.0, Zr 1.0, Nb bal	W 10.0, Zr 2.5, Nb bal
PHYSICAL, THERMAL, ELECTRICAL AND MAGNETIC PROPERTIES					
Density (lb/in.^{3})	0.31	0.31	0.32	0.305	0.36
Melting Range (°F)	4474	4350	4260	4300	4400
Specific Heat (Btu/lb/°F)	0.065	—	—	—	0.060
Thermal Expansion Coeff. (μ in./in./°F)					
68–70°F	3.82	—	—	—	3.8
1000°F	—	—	—	—	—
Thermal Conductivity (Btu/ft^2/hr/°F/ft)					
212°F	31.5	—	—	—	—
1000°F	—	—	—	—	—
Electrical Resistivity (μ ohm-cm) at 68°F	—	—	—	—	—
Magnetic Permeability	—	—	—	—	—
MECHANICAL PROPERTIES (annealed)					
Youngs's Modulus (10^6 psi)					
68–70°F	15	11.5	12.6	14.8	15
2400°F	—	2.7	—	—	—
Tensile Strength (1000 psi)					
68–70°F	40	48	93.5	112	85–90
2000°F	—	23	26.4	65	—
Yield Strength (1000 psi)					
68–70°F	35	135	88	90	70–75
2000°F	—	21	18.2	58	—
Elongation (in 2 in.) at 68–75°F (%)	25	15	20	15	27
Shear Strength (1000 psi)					
68–70°F	—	—	—	—	—
1000°F	—	—	—	—	—
Hardness (Brinell)	—	—	—	—	—

TABLE 5-10

PLATINUM AND ALLOYS—WROUGHT

COMPOSITION (%) Type →	Platinum 99.8%	Pt-10 Ir	Pt-10 Rh	Pt-10 Ru			
PHYSICAL, THERMAL, ELECTRICAL AND MAGNETIC PROPERTIES							
Density (lb/in.³)	0.775						
Melting Range (°F)	3217	3240	3360	3270			
Specific Heat (Btu/lb/°F)	0.031						
Thermal Expansion Coeff. (μ in./in./°F)							
68–70°F	4.9						
1000°F							
Thermal Conductivity (Btu/ft²/hr/°F/ft)							
212°F	42						
1000°F							
Electrical Resistivity (μ ohm-cm) at 68°F	10.8	25	19.2	43			
Magnetic Permeability							
MECHANICAL PROPERTIES (annealed)							
Young's Modulus (10⁶ psi)							
68–70°F	25						
1000°F							
Tensile Strength (1000 psi)							
68–70°F	18–21	55	45	83			
1000°F							
Yield Strength (1000 psi)							
68–70°F	2.0–5.5						
1000°F							
Elongation (in 2 in.) at 68–75°F (%)	30–40	26	35	31			
Shear Strength (1000 psi)							
68–70°F	—						
1000°F	—						
Hardness (Brinell)	40	130	90	190			

TABLE 5-11

TANTALUM AND ALLOYS—WROUGHT

Type →	Tantalum 99.96%	Ta-10 W	T-111	T-222
COMPOSITION (%)		W 10, Mo 0.03, Nb 0.1	W 8, Mo 0.02, Nb 0.1, Hf 2, Zr 0.1	W 10, Mo 0.02, Nb 0.1, Hf 2.5, Zr 0.1
PHYSICAL, THERMAL, ELECTRICAL AND MAGNETIC PROPERTIES				
Density (lb/in.^{3})	0.600	0.608	0.604	0.606
Melting Range (°F)	5425	5495	5400	5480
Specific Heat (Btu/lb/°F)	0.036			
Thermal Expansion Coeff. (μ in./in./°F)				
68–70°F	3.6	—	—	—
3000°F	—	3.74	4.2	—
Thermal Conductivity (Btu/ft^2/hr/°F/ft)				
68–70°F	31.5 (212°F)	—	—	—
1000°F	—	20.0	—	—
Electrical Resistivity (μ ohm-cm) at 68°F	12.5			
Magnetic Permeability	—			
MECHANICAL PROPERTIES (vacuum-recrystallized)				
Young's Modulus (10^6 psi)				
68–70°F	27	21	29	29
2400°F	—	9	—	—
Tensile Strength (1000 psi)				
68–70°F	71	80	90	110
2400°F	—	40	37	53
Yield Strength (1000 psi)				
68–70°F	48	67	82	100
2400°F	—	35	24	38
Elongation (in 2 in.) at 68–75°F (%)	28	25	29	30
Shear Strength (1000 psi)				
68–70°F	—	—	—	—
1000°F	—	—	—	—
Hardness (Vickers)	80	216	—	280

TABLE 5-12

TIN AND ALLOYS–CAST, WROUGHT

Type →	Grade A Sn	Hard Sn	White Metal	Pewter			
COMPOSITION (%)	Sn 99.8 min.	Sn 99.6, Cu 0.4.	Sn 92, Sb 8.	Sn 91, Sb 7, Cu 2.			
PHYSICAL, THERMAL, ELECTRICAL AND MAGNETIC PROPERTIES							
Density (lb/in.³)	0.264	–	0.262	0.263			
Melting Range (°F)	449.4	441–446	475	471–563			
Specific Heat (Btu/lb/°F)	0.054	–	–	–			
Thermal Expansion Coeff. (μ in./in./°F)							
32–212°F	13	–	–	–			
1000°F	–	–	–	–			
Thermal Conductivity (Btu/ft²/hr/°F/ft)							
68–70°F	37	–	–	–			
1000°F	–	–	–	–			
Electrical Resistivity (μ ohm-cm) at 68°F	11.5	–	15.5	–			
Magnetic Permeability	–	–	–	–			
MECHANICAL PROPERTIES (annealed)							
Youngs's Modulus (10⁶ psi)							
68–70°F	6–6.5	–	–	7.7			
1000°F	–	–	–	–			
Tensile Strength (1000 psi)							
68–70°F	2.2	3.3	6.7	8.6			
1000°F	–	–	–	–			
Yield Strength (1000 psi)							
68–70°F	1.3	–	–	–			
1000°F	–	–	–	–			
Elongation (in 2 in.) at 68–75°F (%)	45	–	70	40			
Shear Strength (1000 psi)							
68–70°F	–	–	–	–			
1000°F	–	–	–	–			
Hardness (Brinell)	7	–	17	13			

TABLE 5-12 (*Continued*)

TIN AND ALLOYS–CAST

Type →	Sn Babbitt #1	Sn Babbitt #2	Sn Babbitt #3	YC 135A	PY 1815A
COMPOSITION (%)	Sn 91.0, Sb 4.5, Cu 4.5, Pb 0.35 max	Sn 89.0, Sb 7.5, Cu 3.5, Pb 0.35 max	Sn 84.0, Sb 8.0, Cu 8.0, Pb 0.35 max	Sn 82.0, Sb 13.0, Cu 5.0, Pb 0.35 max	Sn 65.0, Sb 15.0, Cu 2.0, Pb 18.0
PHYSICAL, THERMAL, ELECTRICAL AND MAGNETIC PROPERTIES					
Density (lb/in.³)	0.265	0.267	0.269	0.272	0.289
Melting Range (°F)	433–700	466–669	464–792	400	358–565
Specific Heat (Btu/lb/°F)	—	—	—	—	—
Thermal Expansion Coeff. (μ in./in./°F)					
68–70°F	—	—	—	—	—
1000°F	—	—	—	—	—
Thermal Conductivity (Btu/ft²/hr/°F/ft)					
68–70°F	—	—	—	—	—
1000°F	—	—	—	—	—
Electrical Resistivity (μ ohm-cm) at 68°F	—	—	—	—	—
Magnetic Permeability	—	—	—	—	—
MECHANICAL PROPERTIES (chill-cast)					
Youngs's Modulus (10⁶ psi)					
68–70°F	7.3	7.6	—	—	—
1000°F	—	—	—	—	—
Compr. Strength (1000 psi)					
68–70°F	12.85	14.9	17.6	10.0	7.8
1000°F	—	—	—	—	—
Yield Strength (1000 psi)					
68–70°F	4.4	6.1	6.6	—	—
1000°F	—	—	—	—	—
Elongation (in 2 in.) at 68–75°F (%)	2	18	—	1	—
Shear Strength (1000 psi)					
68–70°F	—	—	—	—	—
1000°F	—	—	—	—	—
Hardness (Brinell)	17.0	24.5	27.0	29.0	27.7

TABLE 5-13

TITANIUM AND ALLOYS—WROUGHT

	Unalloyed Titanium		Alloyed Titanium		
Type →	99.9	99.0	Ti-5Al-2.5 Sn	Ti-6Al-4 V	Ti-8Mn
COMPOSITION (%)	Ti 99.9	Ti 99.0	Al 4.0–6.0, Sn 2.0–3.0, N 0.07 max, C 0.10 max, Fe 0.05 max, O 0.30 max	Al 5.5–6.75, V 3.5–4.5, N 0.07 max, C 0.10 max, Fe 0.40, O 0.30 max	Mn 6.5–9.0 max, N 0.07 max, C 0.15 max, Fe 0.50 max, O 0.30 max
PHYSICAL, THERMAL, ELECTRICAL AND MAGNETIC PROPERTIES					
Density (lb/in.^{3})	0.163	0.163	0.161	0.160	0.171
Melting Range (°F)	3035	3020	2820–3000	3000	2730–2970
Specific Heat (Btu/lb/°F)	0.125	0.125	0.125	0.135	0.171
Thermal Expansion Coeff. (μ in./in./°F)					
68–200°F	4.67	4.7	5.2	5.0	4.8
68–600°F		5.2	5.3	5.3	5.4
Thermal Conductivity (Btu/ft^2/hr/°F/ft)					
68–70°F		9.1–11.5	4.5	4.2	6.3
600°F		9.8–13.0	6.30	6.0	8.1
Electrical Resistivity (μ ohm-cm) at 68°F	42.0	48.6–54.8	157	170	90.7
Magnetic Permeability	—	—	—	—	—
MECHANICAL PROPERTIES (annealed)					
Youngs's Modulus (10^6 psi)					
68–70°F		15	16.0	16.5	16.4
600°F			13.4	13.5	14.4
Tensile Strength (1000 psi)					
68–70°F	34.0	79–95	125	135	137
600°F		33–43	82	90 (ult)	120
Yield Strength (1000 psi) (0.2%)					
68–70°F	20.0	63–80	117	131	125
600°F		19–27	65	90	85
Elongation (in 2 in.) at 68–75°F (%)	54	25–27	18	12.4	15
Shear Strength (1000 psi)					
68–70°F		50–60	100–110	100	100–105
1000°F				—	
Hardness (Rockwell)		—	C36	C36	C28–36

TABLE 5-14

TUNGSTEN AND ALLOYS—WROUGHT

COMPOSITION (%) Type →	Tungsten 99.95%	AVC TM W-25 Re				
PHYSICAL, THERMAL, ELECTRICAL AND MAGNETIC PROPERTIES						
Density (lb/in.^{3})	0.70	0.714				
Melting Range (°F)	6170	5650				
Specific Heat (Btu/lb/°F)	0.034	—				
Thermal Expansion Coeff. (μ in./in./°F)						
68–70°F	2.5	—				
1000°F		—				
Thermal Conductivity (Btu/ft^2/hr/°F/ft)						
68–70°F		—				
1000°F		—				
Electrical Resistivity (μ ohm-cm) at 68°F	5.48	20.0				
Magnetic Permeability	—	—				
MECHANICAL PROPERTIES (recrystallized)						
Youngs's Modulus (10^6 psi)						
68–70°F	59	60				
2400°F	—	—				
Tensile Strength (1000 psi)						
68–70°F	220	242				
2400°F	45	100				
Yield Strength (1000 psi)						
68–70°F	220	225				
2400°F	—	90				
Elongation (in 2 in.) at 68–75°F (%)		—				
Shear Strength (1000 psi)						
68–70°F	—	—				
1000°F	—	—				
Hardness (Vickers)	350	450				

TABLE 5-15

ZINC AND ALLOYS–ROLLED, WROUGHT

Type →	Pure Zinc	Comm. Rolled Zinc	Comm. Rolled Zinc	Comm. Rolled Zinc	Copper-Hardened Zinc
COMPOSITION (%)		Pb 0.08	Pb 0.06, Cd 0.06	Pb 0.3, Cd 0.3	Cu 1
PHYSICAL, THERMAL, ELECTRICAL AND MAGNETIC PROPERTIES					
Density (lb/in.³)	0.258	0.258	0.258	0.258	0.259
Melting Range (°F)	787	786	786	786	792
Specific Heat (Btu/lb/°F)	0.094	0.094	0.094	0.094	0.0957
Thermal Expansion Coeff. (μ in./in./°F)*					
68–212°F	22	15.5	15.5	15.9	–
1000°F	–	–	–	–	–
Thermal Conductivity (Btu/ft²/hr/°F/ft)					
68–70°F	62.2	62.2	62.2	–	–
1000°F	–	–	–	–	–
Electrical Resistivity (μ ohm-cm)* at 68°F	6.16	6.06	–	–	6.22
Magnetic Permeability	–	–	–	–	–
MECHANICAL PROPERTIES (hot-rolled)					
Youngs's Modulus (10⁶ psi)					
68–70°F		19.5; 23	21; 25		
1000°F					
Tensile Strength (1000 psi)					
68–70°F		19.5; 23	21; 25	23; 29	24; 30
1000°F		–	–	–	
Yield Strength (1000 psi)					
68–70°F					
1000°F					
Elongation (in 2 in.) at 68–75°F (%)		65; 50	52; 30	50; 32	50; 35
Shear Strength (1000 psi)					
68–70°F					
1000°F					
Hardness (Brinell)		38	43	47	52

*polycrystalline material.

TABLE 5-15 (*Continued*)

ZINC ALLOYS–CAST

Type →	Alloy AG40A (XXII)	Alloy AC41A (XXV)	Slush Casting Alloy	Slush Casting Alloy (unbreakable metal)
COMPOSITION (%)	Al 3.5–4.3 Mg 0.03–0.08, Zn bal	Al 3.5–4.3 Cu 0.75–1.25, Mg 0.03–0.08, Zn bal	Al 4.5–5.0, Cu 0.2–0.3, Zn bal	Al 5.25–5.75, Zn bal
PHYSICAL, THERMAL, ELECTRICAL AND MAGNETIC PROPERTIES				
Density (lb/in.³)	0.24	0.24	—	—
Melting Range (°F)	728	727	734	743
Specific Heat (Btu/lb/°F)	0.10	0.10	—	—
Thermal Expansion Coeff. (μ in./in./°F)				
68–212°F	15.2	15.2	—	—
1000°F	—	—	—	—
Thermal Conductivity (Btu/ft²/hr/°F/ft)				
158–284°F	65.3	62.9	—	—
1000°F	—	—	—	—
Electrical Resistivity (μ ohm-cm) at 68°F	6.37	6.54	—	—
Magnetic Permeability	—	—	—	—
MECHANICAL PROPERTIES (die-cast)				
Youngs's Modulus (10⁶ psi)				
68–70°F	—	—	—	—
1000°F	—	—	—	—
Tensile Strength (1000 psi)				
68–70°F	41	47.6	—	—
1000°F	—	—	—	—
Yield Strength (1000 psi)				
68–70°F	—	—	—	—
1000°F	—	—	—	—
Elongation (in 2 in.) at 68–75°F (%)	10	7	—	—
Shear Strength (1000 psi)				
68–70°F	31	38	—	—
1000°F	—	—	—	—
Hardness (Brinell)	82	91	—	—

TABLE 5-16

ZIRCONIUM AND ALLOYS—WROUGHT

Type →	Commercial Grade	Reactor Grade	Zircaloy-2 TM	ATR
COMPOSITION (%)	Hf 2.0, Zr bal	Hf 0.001 max, Zr bal	Sn 1.5, Fe 0.12, Cr 0.10, Ni 0.005, Zr bal	Cu 0.5, Mo 0.5, Zr bal
PHYSICAL, THERMAL, ELECTRICAL AND MAGNETIC PROPERTIES				
Density (lb/in.^{3})	0.237	0.235	0.237	0.24
Melting Range (°F)	3350	3350	3300	3300
Specific Heat (Btu/lb/°F)	–	–	–	–
Thermal Expansion Coeff. (μ in./in./°F)				
212°F	3.1	3.1	3.6	–
1000°F	–	–	–	–
Thermal Conductivity (Btu/ft^2/hr/°F/ft)				
212°F	–	9.6	8.1	–
1000°F	–	–	–	–
Electrical Resistivity (μ ohm-cm) at 68°F	40	40	74	–
Magnetic Permeability	–	–	–	–
MECHANICAL PROPERTIES (annealed)				
Young's Modulus (10^6 psi)				
68–70°F	14	14	13.8	14
1000°F	–	–	–	–
Tensile Strength (1000 psi)				
68–70°F	64	35	60	–
600°F	30	19	29	45
Yield Strength (1000 psi) (0.2%)				
68–70°F	53	15	45	–
600°F	23	10	19	42
Elongation (in 2 in.) at 68–75°F (%)	24	32	37	42 (600°F)
Shear Strength (1000 psi)				
68–70°F	–	–	–	–
1000°F	–	–	–	–
Hardness (Rockwell)	B89	B65	B89	B84

TABLE 5-17. PROPERTIES OF COMMON SPRING MATERIALS*

Common Name, Specification	Modulus of Elasticity, E 10⁶ psi	Shear Modulus, G 10⁶ psi	ρ Density lb./in.³	Electrical Conductivity % IACS	Sizes Available, in.		Fatigue Applications	Relative Strength	Max. Service Temp., °F
					Min.	Max.			
High-Carbon Steel Wires									
Music ASTM A228 . . .	30	11.5	0.284	7	0.004	0.250	Excellent	High	250
Hard-drawn ASTM A227 .	30	11.5	0.284	7	0.028	0.625	Poor	Medium	250
Oil-tempered ASTM A229 .	30	11.5	0.284	7	0.020	0.625	Poor	Medium	300
Valve-spring ASTM A230 . .	30	11.5	0.284	7	0.050	0.250	Excellent	High	300
Alloy-Steel Wires									
Chrome-vanadium AISI 6150 .	30	11.5	0.284	7	0.032	0.438	Excellent	High	425
Chrome-silicon AISI 9254 . .	30	11.5	0.284	5	0.035	0.375	Fair	High	475
Silicon-manganese AISI 9260 .	30	11.5	0.284	4.5	0.025	0.375	Fair	High	450
Stainless-Steel Wires									
Martensitic AISI 410, 420 . .	28	11	0.280	2.5	0.003	0.500	Poor	Low	500
Austenitic AISI 301, 302 . .	28	10.5	0.286	2	0.005	0.375	Good	Medium	600
Precipitation-hardening 17–7 PH	29.5	11	0.286	2	0.030	0.500	Good	High	700
Nickel-chrome A286 . . .	29	10.4	0.290	2	0.016	0.200	——	Low	950
Copper-Base Alloy Wires									
Phosphor-bronze ASTM B159 .	15	6.3	0.320	18	0.004	0.500	Good	Medium	200
Silicon-bronze ASTM B99 . .	15	6.4	0.308	7	0.004	0.500	Fair	Low	200
Beryllium-copper ASTM B197 .	18.5	7.0	0.297	21	0.003	0.500	Excellent	High	400
Nickel-Base Alloys—Wire and Strip									
Inconel 600	31	11	0.307	1.5	0.004	0.500	Fair	Low	700
Inconel X750	31.5	11.5	0.298	1	0.004	0.563	Fair	Low	1100
Ni Span C 902	27.8	9.7	0.294	1.6	0.004	0.500	Fair	Medium	200
High-Carbon Steel Strip									
AISI 1050	30	11.5	0.284	7	0.010	0.125	Poor	Low	200
AISI 1065	30	11.5	0.284	7	0.003	0.125	Fair	Medium	200
AISI 1075	30	11.5	0.284	7	0.003	0.125	Good	High	250
AISI 1095	30	11.5	0.284	7	0.003	0.125	Excellent	High	250
Stainless-Steel Strip									
Austenitic AISI 301, 302 . .	28	10.5	0.286	2	0.003	0.063	Good	Medium	600
Precipitation-hardening 17–7 PH	29.5	11	0.286	2	0.003	0.125	Good	High	700
Copper-Base Alloy Strip									
Phosphor-bronze ASTM B103 .	15	6.3	0.320	18	0.003	0.188	Good	Medium	200
Beryllium-copper ASTM B194 .	18.5	7.0	0.297	21	0.003	0.375	Excellent	High	400

*Reproduced from: "Precision Springs and Custom Metal Parts," Associated Spring Corporation, Bristol, Connecticut. Copyright by Associated Spring Corporation, Bristol, Connecticut. Used by permission.

REFERENCES AND BIBLIOGRAPHY

Butts, A., Ed., "Copper, the Metal, Its Alloys and Compounds," Van Nostrand Reinhold, New York, 1964.

Elliott, R. P., "Constitution of Binary Alloys," first suppl., McGraw-Hill, New York, 1965.

Everhart, J. L., "Engineering Properties of Nickel and Nickel Alloys," Plenum Press, New York, 1971.

Goldschmidt, H. J., "Interstitial Alloys," Butterworths, London, 1967.

Gschneider, K. A., Jr., "Rare Earth Alloys," Van Nostrand Reinhold, New York, 1961.

Hansen, M. and K. Anderko, "Constitution of Binary Alloys," 2nd ed., McGraw-Hill, New York, 1958.

Hultgren, R., R. L. Orr, P. D. Anderson and K. K. Kelly, "Selected Values of Thermodynamic Properties of Metals and Alloys," Wiley, New York, 1963.

Mathewson, C. H., Ed., "Zinc, the Science and Technology of the Metal, Its Alloys and Compounds," Van Nostrand Reinhold, New York, 1959.

Metals Handbook Committee, "Metals Handbook," 8th ed., American Society for Metals, Novelty, Ohio, 1961.

Roberts, C. S., "Magnesium and its Alloys," Wiley, New York, 1960.

Schwope, A. D., "Zirconium and Zirconium Alloys," American Society for Metals, Cleveland, 1953.

Shunk, F. A., "Constitution of Binary Alloys," 2nd suppl., McGraw-Hill, New York, 1969."

Sims, C. T. and W. C. Hagel, "The Superalloys," Wiley, New York, 1972.

Smithells, C. J., "Metals Reference Book," 4th ed., Plenum Press, New York, 1967.

Woldman, N. E., "Engineering Alloys," 4th ed., Van Nostrand Reinhold, New York, 1962.

CHAPTER VI
PROPERTIES OF GLASSES AND CERAMICS

For maximum utility, manufacturers' summaries of commercially available glasses are presented here. Data for single crystal ceramics are then given by class of ceramics.

The tables are in the following sequence:

Material for these tables has been drawn largely from the sources, given in the References; the sources under the Bibliography were also consulted, but to a much lesser degree.

TABLE 6-1. PROPERTIES OF CORNING® GLASSES*. (English Units)

1	2	3	4	5	6				7		8				9		
					Corrosion Resistance				Thermal Expansion —Multiply by 10⁻⁷ in./in./°F		UPPER WORKING TEMPERATURES (Mechanical Considerations Only)				Thermal Shock Resistance —Plates 6x6 in.		
											Annealed		Tempered		Annealed		
Glass Code	Type	Color	Principal Use	Forms Usually Available	Class	Weath- ering	Water	Acid	32 to 572°F	77°F to Setting Point	Normal Service °F	Extreme Service °F	Normal Service °F	Extreme Service °F	⅛ in. Thick °F	¼ in. Thick °F	½ in. Thick °F
0010	Potash Soda Lead	Clear	Lamp Tubing	T	I	2	2	2	52	56.2	230	716	—	—	149	122	95
0080	Soda Lime	Clear	Lamp Bulbs	BMT	I	3	2	2	52	58.3	230	860	428	482	149	122	95
0120	Potash Soda Lead	Clear	Lamp Tubing	TM	I	2	2	2	49.8	53.9	230	716	—	—	—	—	—
0330	Glass-Ceramic	Gray	Bench Tops	RS	I	—	1	3	5.4	—	1000	—	—	—	410	338	266
1720⁴	Aluminosilicate	Clear	Ignition Tube	BT	I	1	1	3	23.4	28.9	392	1202	752	842	275	239	167
1723	Aluminosilicate	Clear	Electron Tube	BT	I	1	1	3	25.6	30.	392	1202	752	842	257	212	158
1990	Potash Soda Lead	Clear	Iron Sealing	—	II	3	3	4	69	75.6	212	590	—	—	113	95	77
2405	Borosilicate	Red	General	BPU	I	—	—	—	23.9	29.5	392	896	—	—	275	239	167
2473	Soda Zinc	Red	Lamp Bulbs	B	I	2	2	2	50.4	—	230	860	—	—	149	122	95
3320	Borosilicate	Canary	Tungsten Sealing	—	I	³1	³1	³2	22.5	23.9	392	896	—	—	293	230	176
6720	Soda Zinc	Opal	General	P	I	²—	1	2	43.7	50	230	896	428	527	158	140	104
6750	Soda Barium	Opal	Lighting Ware	BPR	I	²—	2	2	49	—	230	788	428	428	149	122	95
7040	Borosilicate	Clear	Kovar Sealing	BT	II	³3	³3	³4	26.4	30	392	806	—	—	—	—	—
7050	Borosilicate	Clear	Series Sealing	T	II	³3	³3	³4	25.6	28.4	392	824	455	455	257	212	158
7052	Borosilicate	Clear	Kovar Sealing	BMPT	II	³2	³2	³4	25.6	29.5	392	788	410	410	257	212	158
7056	Borosilicate	Clear	Kovar Sealing	BTP	II	2	2	4	28.6	31.1	392	860	—	—	—	—	—
7070	Borosilicate	Clear	Low Loss Electrical	BMPT	I	³2	³2	³2	17.8	21.7	446	806	446	446	356	302	212
7251	Borosilicate	Clear	Sealed Beam Lamps	P	I	³1	³2	³2	20.4	21.2	446	860	500	500	320	266	294
7570	High Lead	Clear	Solder Sealing	—	II	1	1	4	46.7	51.1	212	572	—	—	—	—	—
7720	Borosilicate	Clear	Tungsten Sealing	BPT	I	³2	³2	³2	20	23.9	446	860	500	500	320	266	194
7740	Borosilicate	Clear	General	BPSTU	I	³1	³1	³1	18.1	19.5	446	914	500	554	320	266	194
7760⁵	Borosilicate	Clear	General	BP	I	2	2	2	18.9	20.6	446	842	482	482	320	266	194
7800	Soda Barium Borosilicate	Clear	Pharmaceutical	T	—	1	1	1	27.8	29.4	392	860	—	—	—	—	—
7913¹	96% Silica	Clear	High Temp.	BPRST	I	1	1	1	4.2	3.1*	1652	2192	—	—	—	—	—
7940	Fused Silica	Clear	Optical	U	I	1	1	1	3.1	1.9*	1652	2012	—	—	—	—	—
7971	Titanium Silicate	Clear	Optical	U	—	1	1	1	0.3	−1.1	1472	2012	—	—	—	—	—
8160	Potash Soda Lead	Clear	Electron Tubes	PT	II	2	2	3	50.6	55.5	212	716	—	—	149	122	95
8161	Potash Lead	Clear	Electron Tubes	PT	I	2	1	4	50	55	212	734	—	—	—	—	—
9606	Glass-Ceramic	White	Missile Nose Cones	C	II	—	1	4	31.7	—	1292	—	—	—	572	338	266
9608	Glass-Ceramic	White	Cooking Ware	BP	I	—	1	2	2.2-11.1	—	1292	1472	—	—	—	—	—
9741	Borosilicate	Clear	u v Transmission	BUT	II	³3	³3	³4	22	27.8	392	734	—	—	302	248	176

*Reproduced with permission by Corning Glass Works from "Properties of Glasses and Glass-Ceramics," Corning Glass Works.

COLUMN 1
¹Glasses 7905, 7910, 7911, 7912, 7913 and 7917 for special ultraviolet and infrared applications.
⁴Glass 1720 is available with improved ultraviolet transmittance (designated glass 9730).
⁵Glass 7760 also available with special transmission suitable for sun lamps.

COLUMN 5
B—Blown Ware P—Pressed Ware S—Plate Glass
M—Multiform R—Rolled Sheet T—Tubing and Rod
U—Panels C—Castings

COLUMN 6
²Since weathering is determined primarily by clouding

All data subject to normal manufacturing variations

which changes transmission, a rating for the opal glasses is omitted.
³These borosilicate glasses may rate differently if subjected to excessive heat treatment.

COLUMN 7
*Extrapolated values.
Code 9608 may be produced in a range of expansion values depending upon intended application.

COLUMN 8
Normal Service: No breakage from excessive thermal shock is assumed.
Extreme Limits: Glass will be very vulnerable to thermal shock. Recommendations in this range are based on

10	11				12	13	14		15			16			17	18
Thermal Stress Resistance °F	Viscosity Data				Knoop Hardness KHN100	Density lb/ft³	Young's Modulus Multiply by 10⁴ psi	Poisson's Ratio	Log10 of Volume Resistivity ohm-cm			Dielectric Properties at 1MHz, 68°F			Refractive Index	Glass Code
	Strain Point °F	Annealing Point °F	Softening Point °F	Working Point °F					77°F	482°F	662°F	Power Factor %	Dielectric Constant	Loss Factor %		
34	738	810	1159	1801	363	178.5	8.9	.21	17.+	8.9	7.0	.16	6.7	1.	1.539	0010
29	883	957	1285	1841	465	154	10.2	.22	12.4	6.4	5.1	.9	7.2	6.5	1.512	0080
36	743	815	1166	1805	382	190.3	8.6	.22	17.+	10.1	8.0	.12	6.7	.8	1.560	0120
320	—	—	—	—	522	158.3	12.6	.26	—	—	—	—	—	—	—	0330
50	1233	1314	1679	2196	513	157.3	12.7	.24	17.+	11.4	9.5	.38	7.2	2.7	1.530	1720
47	1229	1310	1666	2134	514	164.7	12.5	.24	17.+	13.5	11.3	.16	6.3	1.0	1.547	1723
25	644	698	932	1393	—	218.5	8.4	.25	17.+	10.1	7.7	.04	8.3	.33	—	1990
67	934	999	1409	1981	—	154.8	9.9	.21	—	—	—	—	—	—	1.507	2405
34	871	548	1287	—	—	165.3	9.5	.22	—	—	—	—	—	—	1.52	2473
77	919	1004	1436	2140	—	141.6	9.4	.19	—	8.6	7.1	.30	4.9	1.5	1.481	3320
36	941	1004	1436	1873	—	161	10.2	.21	—	—	—	—	—	—	1.507	6720
32	837	905	1249	1904	—	161.5	—	—	—	—	—	—	—	—	1.513	6750
67	840	914	1296	1976	—	139.8	8.6	.23	—	9.6	7.8	.20	4.8	1.0	1.480	7040
70	862	934	1297	1881	—	139.8	8.7	.22	16.	8.8	7.2	.33	4.9	1.6	1.479	7050
74	817	896	1314	2062	375	141.6	8.2	.22	17.	9.2	7.4	.26	4.9	1.3	1.484	7052
59	882	954	1324	1936	—	143	9.2	.21	—	10.2	8.3	.27	5.7	1.5	1.487	7056
119	853	925	—	1954	—	132.8	7.4	.22	17.+	11.2	9.1	.06	4.1	.25	1.469	7070
86	932	1011	1436	2133	—	140.4	9.3	.19	18.	8.1	6.6	.45	4.85	2.18	1.476	7251
38	648	685	824	1036	—	338	8.0	.28	17.+	10.6	8.7	.22	15.	3.3	1.86	7570
88	903	973	1391	2095	—	147.1	9.1	.20	16.	8.8	7.2	.27	4.7	1.3	1.487	7720
97	950	1040	1510	2286	418	139.3	9.1	.20	15.	8.1	6.6	.50	4.6	2.6	1.474	7740
94	892	973	1436	2188	442	139.8	9.0	.20	17.	9.4	7.7	.18	4.5	.79	1.473	7760
59	991	1069	1463	2172	—	147.3	—	—	—	7.0	5.7	—	—	—	1.491	7800
396	1634	1868	2786	—	487	136	9.8	.19	17.+	9.7	8.1	.04	3.8	.15	1.458	7913
515	1753	1983	2876	—	489	137.2	10.5	.16	17.+	11.8	10.2	.001	3.8	.0038	1.459	7940
6050	—	1832	2732	—	—	138	9.8	.17	20.3	12.2	10.1	[6]<.002	[6]4.0	[6]<.008	1.484	7971
32	747	820	1170	1783	—	185.9	—	—	17.+	10.6	8.4	.09	7.0	.63	1.553	8160
40	752	815	1112	1584	—	248	7.8	.24	17.+	12.0	9.9	.06	8.3	.50	1.659	8161
29	—	—	—	—	657	162.2	17.2	.24	16.7	10.0	8.7	.30	5.6	1.7	—	9606
—	—	—	—	—	593	156	12.5	.25	13.4	8.1	6.8	.34	6.9	2.3	—	9608
97	766	842	1301	2122	—	134.8	7.2	.23	17.+	9.4	7.6	.32	4.7	1.5	1.468	9741

mechanical stability considerations only. Tests should be made before adopting final designs. These data approximate only.

COLUMN 9
These data approximate only.
Based on plunging sample into cold water after oven heating. Resistance of 100°C (212°F) means no breakage if heated to 110°C (230°F) and plunged into water at 10°C (50°F). Tempered samples have over twice the resistance of annealed glass.

COLUMN 10
Resistance in °C (°F) is the temperature differential between the two surfaces of a tube or a constrained plate that will cause a tensile stress of 0.7 kg/mm² (1000 psi) on the cooler surface.

COLUMN 11
These data subject to normal manufacturing variations.

COLUMN 12
Determined by revised ASTM standard: number of standard not yet assigned.

COLUMN 16
⁶at 10 kHz

COLUMN 17
Refractive Index may be at either the sodium yellow line (589.3 nm) or the helium yellow line (587.6 nm). Values at these wavelengths do not vary in the first three places beyond the decimal point.

TABLE 6-2. PROPERTIES OF CORNING® GLASSES*. (Metric Units)

1	2	3	4	5	6				7		8				9		
					Corrosion Resistance				Thermal Expansion —Multiply By $10^{-7}cm/cm/°C$		Upper Working Temperatures (Mechanical Considerations Only)				Thermal Shock Resistance Plates 15 x 15 cm		
											Annealed		Tempered		Annealed		
Glass Code	Type	Color	Principal Use	Forms Usually Available	Class	Weathering	Water	Acid	0-300°C	25°C to Setting Point	Normal Service °C	Extreme Service °C	Normal Service °C	Extreme Service °C	3.2 mm Thick °C	6.4 mm Thick °C	12.7 mm Thick °C
0010	Potash Soda Lead	Clear	Lamp Tubing	T	I	2	2	2	93.5	101	110	380	—	—	65	50	35
0080	Soda Lime	Clear	Lamp Bulbs	BMT	I	3	2	2	93.5	105	110	460	220	250	65	50	35
0120	Potash Soda Lead	Clear	Lamp Tubing	TM	I	2	2	2	89.5	97	110	380	—	—	65	50	35
0330	Glass-Ceramic	Gray	Bench Tops	RS	I	—	1	3	9.7	—	538	—	—	—	—	—	—
1720[4]	Aluminosilicate	Clear	Ignition Tube	BT	I	1	1	3	42	52	200	650	400	450	135	115	75
1723	Aluminosilicate	Clear	Electron Tube	BT	I	1	1	3	46	54	200	650	400	450	125	100	70
1990	Potash Soda Lead	Clear	Iron Sealing	—	II	3	3	4	124	136	100	310	—	—	45	35	25
2405	Borosilicate	Red	General	BPU	I	—	–	–	43	53	200	480	—	—	135	115	75
2473	Soda Zinc	Red	Lamp Bulbs	B	I	2	2	2	91	—	110	460	—	—	65	50	35
3320	Borosilicate	Canary	Tungsten Sealing	—	I	3[1]	3[1]	3[2]	40	43	200	480	—	—	145	110	80
6720	Soda Zinc	Opal	General	P	I	2–	1	2	78.5	90	110	480	220	275	70	60	40
6750	Soda Barium	Opal	Lighting Ware	BPR	I	2–	2	2	88	—	110	420	220	220	65	50	35
7040	Borosilicate	Clear	Kovar Sealing	BT	II	3[3]	3[3]	3[4]	47.5	54	200	430	—	—	—	—	—
7050	Borosilicate	Clear	Series Sealing	T	II	3[3]	3[3]	3[4]	46	51	200	440	235	235	125	100	70
7052	Borosilicate	Clear	Kovar Sealing	BMPT	II	3[2]	3[2]	3[4]	46	53	200	420	210	210	125	100	70
7056	Borosilicate	Clear	Kovar Sealing	BTP	II	2	2	4	51.5	56	200	460	—	—	—	—	—
7070	Borosilicate	Clear	Low Loss Electrical	BMPT	I	3[2]	3[2]	3[2]	32	39	230	430	230	230	180	150	100
7251	Borosilicate	Clear	Sealed Beam Lamps	P	I	3[1]	3[2]	3[2]	36.7	38.1	230	460	260	260	160	130	90
7570	High Lead	Clear	Solder Sealing	—	II	1	1	4	84	92	100	300	—	—	—	—	—
7720	Borosilicate	Clear	Tungsten Sealing	BPT	I	3[2]	3[2]	3[2]	36	43	230	460	260	260	160	130	90
7740	Borosilicate	Clear	General	BPSTU	I	3[1]	3[1]	3[1]	32.5	35	230	490	260	290	160	130	90
7760[5]	Borosilicate	Clear	General	BP	I	2	2	2	34	37	230	450	250	250	160	130	90
7800	Soda Barium Borosilicate	Clear	Pharmaceutical	T		1	1	1	50	53	200	460	—	—	—	—	—
7913[1]	96% Silica	Clear	High Temp.	BPRST	I	1	1	1	7.5	5.5*	900	1200	—	—	—	—	—
7940	Fused Silica	Clear	Optical	U	I	1	1	1	5.5	3.5*	900	1100	—	—	—	—	—
7971	Titanium Silicate	Clear	Optical	U	–	1	1	1	0.5	−2	800	1100	—	—	—	—	—
8160	Potash Soda Lead	Clear	Electron Tubes	PT	II	2	2	3	91	100	100	380	—	—	65	50	35
8161	Potash Lead	Clear	Electron Tubes	PT	I	2	1	4	90	99	100	390	—	—	—	—	—
9606	Glass-Ceramic	White	Missile Nose Cones	C	II	–	1	4	57	—	700	—	—	—	200	170	130
9608	Glass-Ceramic	White	Cooking Ware	BP	I	–	1	2	4-20	—	700	800	—	—	—	—	—
9741	Borosilicate	Clear	u v Transmission	BUT	II	3[3]	3[3]	3[4]	39.5	50	200	390	—	—	150	120	80

*Reproduced with permission of Corning Glass Works from "Properties of Glasses and Glass-Ceramics," Corning Glass Works.

COLUMN 1
[1]Glasses 7905, 7910, 7911, 7912, 7913 and 7917 for special ultraviolet and infrared applications.
[4]Glass 1720 is available with improved ultraviolet transmittance (designated glass 9730).
[5]Glass 7760 also available with special transmission suitable for sun lamps.

COLUMN 5
B—Blown Ware P$_T$—Pressed Ware S—Plate Glass
M—Multiform R—Rolled Sheet T—Tubing and Rod
U—Panels C—Castings

COLUMN 6
[2]Since weathering is determined primarily by clouding

All data subject to normal manufacturing variations

which changes transmission, a rating for the opal glasses is omitted.
[3]These borosilicate glasses may rate differently if subjected to excessive heat treatment.

COLUMN 7
*Extrapolated values.
Code 9608 may be produced in a range of expansion values depending upon intended application.

COLUMN 8
Normal Service: No breakage from excessive thermal shock is assumed.
Extreme Limits: Glass will be very vulnerable to thermal shock. Recommendations in this range are based on

10	11				12	13	14		15			16			17	18
Thermal Stress Resistance °C	Viscosity Data				Knoop Hardness KHN₁₀₀	Density g/cm³	Young's Modulus Multiply By 10³ Kg/mm²	Poisson's Ratio	Log₁₀ of Volume Resistivity ohm-cm			Dielectric Properties at 1 MHz, 20°C			Refractive Index	Glass Code
	Strain Point °C	Annealing Point °C	Softening Point °C	Working Point °C					25°C	250°C	350°C	Power Factor %	Dielectric Constant	Loss Factor %		
19	392	432	626	983	363	2.86	6.3	.21	17.+	8.9	7.0	.16	6.7	1.	1.539	0010
16	473	514	696	1005	465	2.47	7.1	.22	12.4	6.4	5.1	.9	7.2	6.5	1.512	0080
20	395	435	630	985	382	3.05	6.0	.22	17.+	10.1	8.0	.12	6.7	.8	1.560	0120
178	—	—	—	—	522	2.54	8.8	.26	—	—	—	—	—	—	—	0330
28	667	712	915	1202	513	2.52	8.9	.24	17.+	11.4	9.5	.38	7.2	2.7	1.530	1720
26	665	710	908	1168	514	2.64	8.8	.24	17.+	13.5	11.3	.16	6.3	1.0	1.547	1723
14	340	370	500	756	—	3.50	5.9	.25	17.+	10.1	7.7	.04	8.3	.33	—	1990
37	501	537	765	1083	—	2.48	6.9	.21	—	—	—	—	—	—	1.507	2405
19	466	509	697	—	—	2.65	6.7	.22	—	—	—	—	—	—	1.52	2473
43	493	540	780	1171	—	2.27	6.6	.19	—	8.6	7.1	.30	4.9	1.5	1.481	3320
20	505	540	780	1023	—	2.58	7.1	.21	—	—	—	—	—	—	1.507	6720
18	447	485	676	1040	—	2.59	—	—	—	—	—	—	—	—	1.513	6750
37	449	490	702	1080	—	2.24	6.0	.23	—	9.6	7.8	.20	4.8	1.0	1.480	7040
39	461	501	703	1027	—	2.24	6.1	.22	16.	8.8	7.2	.33	4.9	1.6	1.479	7050
41	436	480	712	1128	375	2.27	5.8	.22	17.	9.2	7.4	.26	4.9	1.3	1.484	7052
33	472	512	718	1058	—	2.29	6.5	.21	—	10.2	8.3	.27	5.7	1.5	1.487	7056
66	456	496	—	1068	—	2.13	5.2	.22	17.+	11.2	9.1	.06	4.1	.25	1.469	7070
48	500	544	780	1167	—	2.25	6.5	.19	18.	8.1	6.6	.45	4.85	2.18	1.476	7251
21	342	363	440	558	—	5.42	5.6	.28	17.+	10.6	8.7	.22	15.	3.3	1.86	7570
49	484	523	755	1146	—	2.35	6.4	.20	16.	8.8	7.2	.27	4.7	1.3	1.487	7720
54	510	560	821	1252	418	2.23	6.4	.20	15.	8.1	6.6	.50	4.6	2.6	1.474	7740
52	478	523	780	1198	442	2.24	6.3·	.20	17.	9.4	7.7	.18	4.5	.79	1.473	7760
33	533	576	795	1189	—	2.36	—	—	—	7.0	5.7				1.491	7800
220	890	1020	1530	—	487	2.18	6.9	.19	17.+	9.7	8.1	.04	3.8	.15	1.458	7913
286	956	1084	1580	—	489	2.20	7.4	.16	17.+	11.8	10.2	.001	3.8	.0038	1.459	7940
3370	—	1000	1500	—	—	2.21	6.9	.17	20.3	12.2	10.1	[6]<.002	[6]4.0	[6]<.008	1.484	7971
18	397	438	632	973	—	2.98	—	—	17.+	10.6	8.4	.09	7.0	.63	1.553	8160
22	400	435	600	862	—	3.99	5.5	.24	17.+	12.0	9.9	.06	8.3	.50	1.659	8161
16	—	—	—	—	657	2.6	12	.24	16.7	10.0	8.7	.30	5.6	1.7	—	9606
—	—	—	—	—	593	2.5	8.8	.25	13.4	8.1	6.8	.34	6.9	2.3	—	9608
54	408	450	705	1161	—	2.16	5.0	.23	17.+	9.4	7.6	.32	4.7	1.5	1.468	9741

mechanical stability considerations only. Tests should be made before adopting final designs. These data approximate only.

COLUMN 9
These data approximate only.
Based on plunging sample into cold water after oven heating. Resistance of 100°C (212°F) means no breakage if heated to 110°C (230°F) and plunged into water at 10°C (50°F). Tempered samples have over twice the resistance of annealed glass.

COLUMN 10
Resistance in °C (°F) is the temperature differential between the two surfaces of a tube or a constrained plate that will cause a tensile stress of 0.7 kg/mm² (1000 psi) on the cooler surface.

COLUMN 11
These data subject to normal manufacturing variations.

COLUMN 12
Determined by revised ASTM standard: number of standard not yet assigned.

COLUMN 16
[6]at 10 kHz

COLUMN 17
Refractive index may be at either the sodium yellow line (589.3 nm) or the helium yellow line (587.6 nm). Values at these wavelengths do not vary in the first three places beyond the decimal point.

TABLE 6-3. PROPERTIES OF KIMBLE GLASSES[#]

Glass	Type	Principal Use	Working Point (°C)	Softening Point (°C)	Annealing Point (°C)	Strain Point (°C)	Expansion Coefficient (0-300°C) $\times 10^{-7}/°C$	Contraction Coefficient Annealing Point to 25°C $\times 10^{-7}/°C$	Density (g/cc)	Refractive Index Sodium (D line)	Stress Optical Coefficient (mu/cm/ kg/cm^2)
A-3004	Soda Lime	General	1045	732	548	501	84	100	2.48	1.51	2.7
Amber 203	Borosilicate	General	1130	780	570	528	50	68	2.39	1.51	3.3
CA-2	Soda Lime	Pharmaceutical Containers	1035	735	555	515	87	105	2.49	1.52	2.6
EG-6	Potash Soda Lead	Electronic Components	960	625	435	395	91	109	3.09	1.58	3.4
EG-11	Lead	Electronic Ware	980	626	434	394	94	108	2.85	1.54	2.9
EG-14	High Lead	Electronic Industry	855	600	444	411	89	104	3.93	1.67	2.1
EG-16	High Lead	Tubing and Electronics	805	580	430	390	88	101	4.30	1.69	1.7
EG-19	Lead Potash	Color Television	990	660	478	435	95	111	3.22	1.59	2.6
EM-4	Lead Free	Electronic	1000	670	468	426	89	111	2.53	1.51	2.8
EN-1	Borosilicate	†KOVAR® Sealing	1115	716	482	437	47	64	2.27	1.48	4.1
ER-7	Borosilicate	Electronic Components	1015	710	511	469	50	63	2.27	1.48	3.9
ES-1	Borosilicate	Electronic Components	1095	735	478	435	33	46	2.15	1.47	5.0
EZ-1	Alumino-silicate	Molybdenum Sealing	1200	912	707	658	43	56	2.52	1.53	2.7
IN-3	Borosilicate	General	1055	710	521	479	56	71	2.32	1.49	3.4
K-605	Borosilicate	KOVAR Sealing	990	710	511	471	49	62	2.28	1.48	3.9
K-704	Borosilicate	Tungsten and Molybdenum Sealing	1065	713	485	443	49	59	2.24	1.48	3.8
K-705	Borosilicate	Tungsten and Molybdenum Sealing	1010	715	503	460	45	57	2.25	1.48	3.9
K-772	Borosilicate	Electron Tubes	1120	755	520	478	35	51	2.35	1.48	4.0
KG-12	Potash Soda Lead	Electronic Ware	980	632	438	400	90	103	3.05	1.56	2.9
KG-22	Borosilicate	Electronic Components	1050	710	485	439	50	65	2.23	1.48	4.1
KG-33	Borosilicate	General	1260	827	565	513	32	37	2.22	1.47	3.7
KG-35	Borosilicate	General	1160	795	585	538	48	61	2.38	1.49	3.1
N-51A	Borosilicate	General	1190	798	580	538	50	66	2.36	1.49	3.4
R-6	Soda Lime	General	980	700	525	486	93	114	2.53	1.52	2.6
RN-3	Amber Boro-silicate	Pharmaceutical Containers	1070	742	556	515	59	82	2.48	1.52	3.1

Note:
Physical Properties are listed in alphabetical order. All data subject to normal manufacturing variation.
*All glasses may be supplied in ground form.
†Trademark of Westinghouse Electric Corp.
#© 1974, Owens-Illinois, Inc., Toledo, Ohio. Reproduced with permission of the copyright holder.

TABLE 6-3. PROPERTIES OF KIMBLE GLASSES[#]. (Continued)

Volume Resistivity (p) ohm–cm Expressed As Logarithm of (p)		Dielectric Properties 25°C 1 Megahertz			Estimated Δ T Thermal Shock Resistance °C			Chemical Durability USP XVIII ml N/50 H_2SO_4	Calculated Thermal Conductivity Units Expressed As Cal cm/cm^{2}s deg C × 10^{-3} Cal cm/cm^{2}s deg C × 10^{-3}			Macroscopic Thermal Neutron Cross Section (calculated)	Glass
250°C	350°C	K	Δ%	Loss Factor (%)	⅛" Thick	¼" Thick	½" Thick		−100°C	0°C	100°C		
—	—	—	—	—	70	55	40	6.5	2.1	2.6	2.8	.072	A-3004
7.0	5.7	6.0	0.90	5.4	115	95	65	0.39	2.1	2.6	2.9	2.9	Amber 203
6.8	5.4	7.2	0.75	5.4	70	55	35	5.9	2.0	2.5	2.7	0.0074	CA-2
9.5	7.7	7.2	0.11	0.8	60	50	35	4.8	1.6	2.0	2.2	0.014	EG-6
8.9	7.1	—	—	—	65	50	35	16.0	1.6	2.1	2.4	0.014	EG-11
11.7	9.6	8.9	0.09	0.8	65	50	35	0.4	1.3	1.6	1.9	0.52	EG-14
11.3	9.3	9.6	0.10	0.9	65	50	35	0.3	1.2	1.6	1.8	0.0096	EG-16
10.3	8.3	7.6	0.19	1.45	60	50	30	2.7	1.6	2.0	2.2	0.013	EG-19
9.1	7.2	6.8	0.16	1.1	65	50	35	6.3	1.9	2.4	2.6	0.71	EM-4
9.0	7.2	5.1	0.26	1.3	120	100	70	2.9	2.0	2.6	2.9	5.4	EN-1
7.7	6.2	5.1	0.55	2.8	115	95	65	25.0	1.9	2.4	2.8	6.6	ER-7
12.6	10.2	4.0	0.11	0.4	175	145	100	4.2	2.0	2.6	3.1	7.2	ES-1
10.5	8.7	6.3	0.37	2.3	135	115	75	0.65	2.8	3.3	3.1	1.4	EZ-1
7.2	5.7	5.8	0.66	3.9	105	85	55	1.3	2.0	2.5	2.8	4.9	IN-3
7.6	6.2	5.3	0.54	2.8	115	95	65	20.0	1.9	2.4	2.8	6.6	K-650
9.2	7.4	5.0	0.30	1.5	115	100	65	—	2.0	2.5	2.9	6.2	K-704
8.6	6.9	4.9	0.30	1.4	120	100	70	15.0	1.9	2.5	2.9	6.4	K-705
9.1	7.5	4.6	0.20	0.9	160	130	95	5.3	2.0	2.5	3.0	5.0	K-772
9.9	7.8	6.7	0.15	1.0	65	50	35	4.9	1.6	2.0	2.3	0.011	KG-12
9.1	7.3	—	—	—	115	95	65	17.0	2.0	2.5	2.8	6.4	KG-22
7.9	6.4	4.6	.48	2.2	180	150	100	0.26	2.1	2.7	3.1	3.7	KG-33
7.0	5.7	5.7	1.00	5.7	115	95	65	2.0	2.6	2.9	0.45	3.4	KG-35
6.8	5.4	6.0	0.95	5.7	115	95	65	0.36	2.1	2.6	2.9	3.1	N-51A
6.6	5.2	7.3	0.70	5.1	60	50	35	7.8	2.0	2.4	2.7	0.50	R-6
7.0	5.7	6.5	0.90	5.9	95	80	55	2.0	2.6	2.8	0.70	3.0	RN-3

TABLE 6-4. PROPERTIES OF SINGLE CRYSTAL CERAMICS.

ALUMINIDES

Formula	Crystal Structure	Theoretical Density (g/cm³)	Melting/Decomposition Temperature (°F)	Thermal Expansion Coefficient (10⁻⁶/°F)	Thermal Conductivity (Btu/hr ft °F)	Specific Heat (Btu/lb/°F)	Tensile Strength (1000 psi)	Compressive Strength (1000 psi)	Elastic Modulus (10⁶ psi)	Shear Modulus (10⁶ psi)	Rupture Modulus (10⁶ psi)	Microhardness (kg/mm²)	Macrohardness (Mohs)	Poisson's Ratio	Electrical Resistivity (μohm-cm)	Dielectric Constant at 1 MHz	Dielectric Strength at 1 MHz (volts/mil)	Dissipation Factor at 1 MHz
CoAl	C	6.04	2960									440						
Mo₃Al	C		3900															
NiAl	C	5.90	2980				15				0.1	330			22			
NbAl₃	T	4.46	>3190									460						
PdAl	C		2990															
TaAl₃	T	6.74	>2732									260						
Th₂Al	T	9.67										454						
TiAl	T	4.00	2660									~480						
UAl₂	C	8.3	2890															
Zr₄Al₃		5.30	2790															
ZrAl₂	H		2980									445						

TABLE 6-5. PROPERTIES OF SINGLE CRYSTAL CERAMICS.

BERYLLIDES

Formula	Crystal Structure	Theoretical Density (g/cm³)	Melting/Decomposition Temperature (°F)	Thermal Expansion Coefficient (10⁻⁶/°F)	Thermal Conductivity (Btu/hr ft °F)	Specific Heat (Btu/lb/°F)	Tensile Strength (1000 psi)	Compressive Strength (1000 psi)	Elastic Modulus (10⁶ psi)	Shear Modulus (10⁶ psi)	Rupture Modulus (10⁶ psi)	Microhardness (kg/mm²)	Macrohardness (Mohs)	Poisson's Ratio	Electrical Resistivity (μohm-cm)	Dielectric Constant at 1 MHz	Dielectric Strength at 1 MHz (volts/mil)	Dissipation Factor at 1 MHz
HfBe₁₃	C	3.93	2600															
Nb₂Be₁₇	H	3.28	3180															
Nb₂Be₁₉	H	3.15	3100								0.030							
NbBe₁₂	T	2.91	~3000	2.9				200	0.047		0.022	500			55			
TaBe₁₂	T	4.18	3360	2.6	17.7 at 1400°F	0.271 at 2000°F		150	0.045		0.031	720			43.5			
Ta₂Be₁₇	H	5.05	3610						0.045		0.030	1120						
WBe₂₂	C	3.23	~3400															
ZrBe₁₃	C	2.72	~3400	3.0	21 at 1400°F	0.417 at 2000°F		190	0.047		0.025	1000			16.1			
Zr₂Be₁₇	H	3.05	3600	2.6					0.045		0.025				—			
MoBe₁₂	–	3.03	3050		18.2 at 1600°F	0.417 at 2000°F					0.042 at 2300°F	950						

TABLE 6-6. PROPERTIES OF SINGLE CRYSTAL CERAMICS.

BORIDES

Formula	Crystal Structure	Theoretical Density (g/cm^3)	Melting/Decomposition Temperature (°F)	Thermal Expansion Coefficient ($10^{-6}/°F$)	Thermal Conductivity (Btu/hr ft °F)	Specific Heat (Btu/lb/°F)	Tensile Strength (1000 psi)	Compressive Strength (1000 psi)	Elastic Modulus (10^6 psi)	Shear Modulus (10^6 psi)	Rupture Modulus (10^6 psi)	Microhardness (kg/mm^2)	Macrohardness (Mohs)	Poisson's Ratio	Electrical Resistivity (μohm-cm)	Dielectric Constant at 1 MHz	Dielectric Strength at 1 MHz (volts/mil)	Dissipation Factor at 1 MHz
BaB_6	C	4.32	4110	3.8	27.1							3000			>200			
CaB_6	C	2.46	4050	3.6	22.7							2740						
CeB_4	T	5.74	>3812															
CeB_6	C	4.82	3975	4.0	19.6							3140			43			
Cr_4B	R	6.24	3038									1350						
Cr_2B	T	6.53	3180									1250	8		52			
CrB	R	6.20	3630			0.137						1450			64			
Cr_3B_4	R	5.76	3500															
Cr_3B_2	T	6.12	3560										>9					
CrB_2	H	5.60	3900	3.44	11.8	0.17	106		30.6			1800			21			
CoB	R	7.25																
EuB_6	C	4.94	3992	3.8	13.3										85			
GdB_6	C	5.26	3810	4.8	11.8							2340			94			
HfB	C	12.8	5250		35.9								2900 Knoop					
HfB_2	H	11.2	5880	4.15											12			
FeB	R	7.15	2800															
LaB_6	C	4.72	3900	3.5	27.6							2770			17.4			
Mo_2B	T	9.31	4135	4.3		0.09						1200	8–9		40			
α-MoB	T	8.77	3630			0.088						2500			45			
β-MoB	R		4710									1200			25			
MoB_2	H	7.78	4485			0.126						2350			45			

Mo_2B_5	H	7.48	3885	2.78	27.3						3000			25–55
NdB_6	C	4.95	4600	4.0							2530			28
NiB	R	7.13	1868											
NbB	R	7.60	4100								2200			64.5
NbB_2	H	7.21	5430	4.8	9.7	0.10						>8		32
SmB_6	C	5.07	4600	3.8	8.0						1390			200
ScB_2	H	3.67	4080	5.8							2630			
SiB_6	R	2.43	3540	3.7	26									2×10^5
SrB_6	C	3.44	4050								2900			100
TaB	R	14.28	4400			0.059					3130			
TaB_2	H	12.60	5600	4.7	6.3	0.059			37		2200	>8		70
Ta_3B_4	R	13.60	4800			0.122					3350			
ThB_4	T	8.45	4500	3.3	9.2						2043			
ThB_6	C	6.80	3900	4.3	26						2600			
TiB	C	5.26	3740	4.8							2750			40
TiB_2	H	4.52	5200	4.8	15		19	97	54–77	0.019	3400	>9		16
W_2B	T	16.72	4840	4.1		0.04					2420			
$\delta\text{-}WB$	T	16.00	~3460			0.04						8–9)		
$\beta\text{-}WB$	R	16.00	4830											86
WB_2	H	13.60	5250											
W_2B_5	H	13.10	4290	4.8		0.104					2530			39
UB_2	H	12.71	4289	3.9							1390			~21
UB_4	T	9.38	4500								2500			
UB_{12}	C	5.86	4050											
VB	R	5.44	4080											35–40
VB_2	H	5.10	4350	4.2	33	0.16			38		2080	8–9		16
YbB_6	C	5.57	>3630	3.2	14.5						3800			36.5
YB_6	C	3.70	4170	3.4	16.9						3260			40
ZrB_2	H	6.10	5500	4.6	14	0.125	29		73.5		1900	8	0.15	9.2
ZrB_{12}	C	3.63	4650		7.0									60–80

TABLE 6-7. PROPERTIES OF SINGLE CRYSTAL CERAMICS.

CARBIDES

Formula	Crystal Structure	Theoretical Density (g/cm³)	Melting/Decomposition Temperature (°F)	Thermal Expansion Coefficient (10^{-6}/°F)	Thermal Conductivity (Btu/hr. ft °F)	Specific Heat (Btu/lb/°F)	Tensile Strength (1000 psi)	Compressive Strength (1000 psi)	Elastic Modulus (10^6 psi)	Shear Modulus (10^6 psi)	Rupture Modulus (10^6 psi)	Microhardness (kg/mm²)	Macrohardness (Mohs)	Poisson's Ratio	Electrical Resistivity (µohm-cm)	Dielectric Constant at 1 MHz	Dielectric Strength at 1 MHz (volts/mil)	Dissipation Factor at 1 MHz
Al_4C	RH	2.99	5072															
BaC_2	C	3.57	3326															
Be_2C	C	2.44	4080–4350	5.83	12.1	0.334		105	45.6		0.014	2700	>9	0.10	4.5×10^3			
B_4C	RH	2.51	4400	2.63	15.7	0.443	22.5 at 1800°F	414	65	26.83	0.071	2800	9.3	0.207				
CaC_2	C	2.04	3920	5.62														
CeC_2	T	5.56	4604												59			
Cr_4C	C	6.99	2800	5.88	10.6													
Cr_4C_3	H	6.92	3150	5.67	8.7										109			
Cr_3C_2	R	6.70	3430		11.1			151	56		0.113	1450–2100		0.28	84.4			
HfC	C	12.67	7030	3.5	12.8				61.5	26		2650		0.166	45			
LaC_2	T	5.35	4420	6.7								1870	7–9		68			
Mo_2C	H	9.12	4870	3.11					~32			1480			100			
MoC	H	8.88	4880	3.2					28.6			1500	7–8		50			
NdC_2	T	6.00	4004											0.204	44			
NbC	C	7.85	6330	3.80	8.2				49.4			2470	9–10		74			
Nb_2C	H	7.85	5590									2123						

PrC_2	T	5.75	4595	6.33											25.7	
PuC	C	13.90	3000	5.95	0.45–0.60									230	
Pu_2C_3	C	12.70	3452	8.2											
ScC	C	3.06		6.3							2720			274	
α-SiC	H	3.21	5070		0.165			56			3350	9.0–9.5		4.1×10^5	10.2
β-SiC	C	3.21	3800–4350	24.2	0.288	2000	82	38–68	24.41	0.060	2500	9.2	0.192	2.1×10^6	
SmC_2	T	6.45	>3992											100	
SrC_2	C	3.04	>3500												
TaC	C	14.4	7080	12.8	0.045	2–42		52.8			1800	9–10	0.172	30	
Ta_2C	H	15.00	5880								1714			80	
ThC	C	10.67	4760	16.7							1000			25	
ThC_2	M	9.60	4810	13.8							600			30	
TiC	C	4.92	5324	10	0.201	35–40	190–560	45–60	25	0.124	2620	9–10	0.182	60–250	
UC	C	13.63	4300	13.3			51	25.0	9.7		750–935	>7	0.29	50	
U_2C_3	C	12.88	3230	18.9			66	26–32							
UC_2	T	11.68	4350		0.035					0.084	~600				
WC	H	15.77	~5000	17		50	~77	103			2400	9–10	0.26	19.2	
W_2C	H	17.34	~4970					61			2150	>9		81	
VC	C	5.48	4950	14.3		2–4	89	63			2800	9–10		98	
V_2C	H	5.75	~3930												
YC_2	T	4.53	4170								700			89	
Y_2C_3	C	3.66	3270								900			350	
ZrC	C	6.56	6440	11.9	0.049	16.0	238	~50	17.9	0.049	1830	>9	0.257	68	

TABLE 6-8. PROPERTIES OF SINGLE CRYSTAL CERAMICS.

GERMANIDES

Formula	Crystal Structure	Theoretical Density (g/cm³)	Melting/Decomposition Temperature (°F)	Thermal Expansion Coefficient (10⁻⁶/°F)	Thermal Conductivity (Btu/hr ft °F)	Specific Heat (Btu/lb/°F)	Tensile Strength (1000 psi)	Compressive Strength (1000 psi)	Elastic Modulus (10⁶ psi)	Shear Modulus (10⁶ psi)	Rupture Modulus (10⁶ psi)	Microhardness (kg/mm²)	Macrohardness (Mohs)	Poisson's Ratio	Electrical Resistivity (μohm-cm)	Dielectric Constant at 1 MHz	Dielectric Strength at 1 MHz (volts/mil)	Dissipation Factor at 1 MHz
Ce₅Ge₃	H	3.92																
Cr₃Ge	C	7.68	2768									940			30.5			
CrGe	C		>3182									~1380						
Co₂Ge	H	8.91	2192									940			250			
MoGe₂	C																	
Ni₃Ge	H	8.20	2122															
NbGe₂	H		3470									745			67			
Pu₂Ge₃	T	10.98																
PuGe₂	R	10.60	>2552															
Rh₂Ge	R	11.20	2372									700						
RhGe	T	9.70	3272									580						
Th₃Ge₂	C	10.48	3272															
ThGe	R	9.17	1112															
ThGe₂	R	8.64	2192															
TiGe₂	H	6.69	3038												1280			
U₅Ge₃	R	13.4	>3092															
V₃Ge	C														89			
ZrGe₂	R		2768									680			126			

TABLE 6-9. PROPERTIES OF SINGLE CRYSTAL CERAMICS.

NITRIDES

Formula	Crystal Structure	Theoretical Density (g/cm³)	Melting/Decomposition Temperature (°F)	Thermal Expansion Coefficient (10⁻⁶/°F)	Thermal Conductivity (Btu/hr ft °F)	Specific Heat (Btu/lb/°F)	Tensile Strength (1000 psi)	Compressive Strength (1000 psi)	Elastic Modulus (10⁶ psi)	Shear Modulus (10⁶ psi)	Rupture Modulus (10⁶ psi)	Microhardness (kg/mm²)	Macrohardness (Mohs)	Poisson's Ratio	Electrical Resistivity (μohm-cm)	Dielectric Constant at 1 MHz	Dielectric Strength at 1 MHz (volts/mil)	Dissipation Factor at 1 MHz
AlN	H	3.26	4050	3.06	17.3 at 392°F	0.196		300	50		0.04	~1200	>7		10^{17}			
α-Be₃N₂	C	2.71	3990			~0.29												
BN #	H	2.25	~5000	4.19	8.9	0.17	6-9	45	12.4		0.016	230	2		1.7×10^{19}	2.54		0.0002
CeN	C	8.09	4712	16.7											30			
CrN	C	6.14	2730	1.3	~7	0.19						1100			640			
Cr₂N	H	6.51	3022	5.2	~13	0.15						1200			76			
Co₂N	H	6.3				0.05												
Co₃N	H	7.1																
DyN	C	9.93													100			
ErN	C	10.35													79			
GdN	C	9.14																
HfN	C	13.84	5990	3.63	12.5							1650	8-9		33			
HoN	C	10.26													110			
Fe₃N	H	6.36																
Fe₄N	C	6.57		5.0														
LaN	C	6.90	3632												100			
LuN	C	11.59																

(continued)

TABLE 6-9. (continued)

NITRIDES

Formula	Crystal Structure	Theoretical Density (g/cm^3)	Melting/Decomposition Temperature (°F)	Thermal Expansion Coefficient (10^{-6}/°F)	Thermal Conductivity (Btu/hr ft °F)	Specific Heat (Btu/lb/°F)	Tensile Strength (1000 psi)	Compressive Strength (1000 psi)	Elastic Modulus (10^6 psi)	Shear Modulus (10^6 psi)	Rupture Modulus (10^6 psi)	Microhardness (kg/mm^2)	Macrohardness (Mohs)	Poisson's Ratio	Electrical Resistivity (μohm-cm)	Dielectric Constant at 1 MHz	Dielectric Strength at 1 MHz (volts/mil)	Dissipation Factor at 1 MHz
Mg_3N_2	C	2.71	2730			~0.22												
Mn_2N	H	~6.4																
MoN	H	9.18	~1380															
Mo_2N	C	>8.0	~1650	3.4	~10	0.07						650			19.8			
NbN	C	8.36	~4070	5.6	2.1							1396	>8		200			
Nb_2N	H	8.31	4350	1.8	4.8							1720			142			
NdN	C	7.70																
Ni_3N	H	7.66		7.2											75			
PrN	C	7.49													110			
PuN	C	14.25	~4800	7.67	8 at 480°F													
ScN	C	4.21	4800	1.60	16.1	0.17									$>10^{19}$	~9.4		
α-Si_3N_4	H	3.184	3452						8		0.01		9		1.5×10^6			
β-Si_3N_4	H	3.187		1.25														
SmN	C	8.50																
TaN	H	14.36	5600	2.0	4.8	~0.05						1060	>8		135			
Ta_2N	H	15.86	5400	2.9	5.8	~0.03						1220			263			
TbN	C	9.57																

ThN	C	11.56	4770	4.10									
Th_2N_3	H	10.40	3182										
TiN	C	5.43	5340	3.67	16.8	0.14	141	36		1700	8-10		22
UN	C	14.32	5260	~5.4	7.2	0.045		21.6	8.7	455		0.24	208
U_2N_3	C	11.24											
VN	C	6.08	3950	4.5	6.5	0.14				1520			86
V_3N	H	5.99								1900			
WN	H	15.94	~1100										
β-W_2N	C	>16	1470–1600										
YN	C	5.87	4840										93
YbN	C	11.30											
ZrN	C	7.32	5400	3.87	11.6	0.09	~142			1510	8-9		13.6

#Parallel to principal axis.

TABLE 6-10. PROPERTIES OF SINGLE CRYSTAL CERAMICS.

OXIDES

Formula	Crystal Structure	Theoretical Density (g/cm³)	Melting/Decomposition Temperature (°F)	Thermal Expansion Coefficient (10⁻⁶/°F)	Thermal Conductivity (Btu/hr·ft°F)	Specific Heat (Btu/lb/°F)	Tensile Strength (1000 psi)	Compressive Strength (1000 psi)	Elastic Modulus (10⁶ psi)	Shear Modulus (10⁶ psi)	Rupture Modulus (10⁶ psi)	Microhardness (kg/mm²)	Macrohardness (Mohs)	Poisson's Ratio	Electrical Resistivity (μohm-cm)	Dielectric Constant at 1 MHz	Dielectric Strength at 1 MHz (volts/mil)	Dissipation Factor at 1 MHz
α-Al_2O_3	H	3.97	3762	4.15	20.3	0.19	37	420	53	50.6	72	3,000	9	0.254	5×10^{21}	9.6	1,200	
BaO	C	5.74	3490	9.9		0.075									1×10^{13} at 741°F	34		
BaO	H	5.32	3490			0.80							3.3				34		
BeO	H	3.008	4649	4.444	126	0.238	15	170	43			1,500	9	0.34	1×10^{22}	7.66	300	
CaO	C	3.32	4737	7.00	8.7	0.18						560	4.5		1×10^{14}	11.1		
CeO_2	C	7.2	4245	5.90		0.093		85.4	26.2	10.2			6	0.311	6.5×10^{10} at 1472°F			
CoO	C	6.46	3290	7.00 at 773°F		0.23									1×10^{14}			
Cr_2O_3	H	5.21	4170	6.10 at 373°F		0.22									1.3×10^{9} at 623°F			
Dy_2O_3	C	8.30	4240	4.3														
Eu_2O_3	C	7.28	3720	3.90		0.12												
Ga_2O_3	M	5.88	3160															
Gd_2O_3	C	7.63	4230	5.8		0.066			18.0			486			5×10^{15} at 573°F			
HfO_2	M	9.68	~3150	3.25	0.66	0.067			8.2									
HfO_2	T	10.1	5140	0.73														
Fe_2O_3	C	4.59	1250															
Fe_2O_3	H	5.20	2730															
Fe_3O_4	C	4.86																
La_2O_3	C	6.51	4190	6.6		0.069									1×10^{14} at 1022°F			

La₂O₃	H	6.51			0.080					4170					9.8
MgO	C	3.57	6.4	43	0.23	14	200	44	17	5120		5.5	0.163	1.3×10^{15} at 573°F	
MnO	C	5.40								3227					
Mn₃O₄	T	4.82													
Nd₂O₃	H	7.24	6.3		0.079					4120					
NiO	C	6.80			0.140					3603		5.5		6.7×10^{9} at 1112°F	
NbO	C	6.27								3530				4 at 43°F	
Nb₂O₅	R	4.46			0.12					2710				5.5×10^{12}	
PuO₂	C	11.46	9.0							4060					
Pu₂O₃	C	10.2								4020					
Pu₂O₃	H	11.47								4020					
Sm₂O₃	C	7.62	5.7	1.2	0.079			26.5		4262	438				13.1
SrO	C	4.7	7.52		0.105					4450		3.5		1×10^{13} at 541°F	
SiO₂	H	2.65	6.7	7.2	0.188					3110				$\sim 1 \times 10^{20}$	4.5
Ta₂O₅	R	8.02			0.072					3420				1×10^{12}	
ThO₂	C	9.69	5.3	8.2	0.065	14	214	21	13.66	5828	945	6.5	0.28	4×10^{19}	
SnO₂	T	6.99	2.4							2060		6–7			
TiO	C	4.93	5.0							3180				4×10^{2}	
Ti₂O₃	H	4.60								3770				5.5×10^{5}	
TiO₂	T	4.27	5.0	7.2	0.17			36–41		3344		5.5–6		1×10^{19}	117
UO₂	C	10.96	6.2	5.8	0.056			21	10.75	5216	600	6–7	0.302	3.8×10^{10}	
VO	C	5.23								3720					
V₂O₃	H	4.87			0.105					3610				5.5×10^{3}	
VO₂	T	4.65			0.119					2815				4.5×10^{3}	
Y₂O₃	C	5.03	4.5		0.105		57	16.6	7.01	4370			0.186		
ZnO	H	5.66	3.8		0.119		300	35	14	3590		4		$\sim 10^{7}$	
ZrO₂	M	5.56	4.2		0.17	17.9		35	14	1830	1,200	6.5	0.337	2.3×10^{10} at 1292°F	10.9
ZrO₂	T	6.10	5.8							5010				7.7×10^{7} at 2192°F	

TABLE 6-11. PROPERTIES OF SINGLE CRYSTAL CERAMICS.

PHOSPHIDES

Formula	Crystal Structure	Theoretical Density (g/cm³)	Melting/Decomposition Temperature (°F)	Thermal Expansion Coefficient (10⁻⁶/°F)	Thermal Conductivity (Btu/hr ft °F)	Specific Heat (Btu/lb/°F)	Tensile Strength (1000 psi)	Compressive Strength (1000 psi)	Elastic Modulus (10⁶ psi)	Shear Modulus (10⁶ psi)	Rupture Modulus (10⁶ psi)	Microhardness (kg/mm²)	Macrohardness (Mohs)	Poisson's Ratio	Electrical Resistivity (μohm-cm)	Dielectric Constant at 1 MHz	Dielectric Strength at 1 MHz (volts/mil)	Dissipation Factor at 1 MHz
AlP	H	2.85	>2730													11.6		
BP	C	2.97	2280									3200						
Co₂P	R	7.55	2525															
CrP	R	5.49	2480									632						
Fe₂P	H	6.90	2500															
GaP	H	4.13	2732													8.4		
InP	H	4.79	1944													10.9		
LaP	C	5.18										158						
Ni₂P	H	7.33	2010									599						
NbP	T	6.54	3020									374						
TaP	T	11.15	3020									718						
TiP	H	4.27	2875									541						
UP	C	9.68	5162															
VP	H	5.00	2400															

TABLE 6-12. PROPERTIES OF SINGLE CRYSTAL CERAMICS.

SELENIDES

Formula	Crystal Structure	Theoretical Density (g/cm³)	Melting/Decomposition Temperature (°F)	Thermal Expansion Coefficient (10⁻⁶/°F)	Thermal Conductivity (Btu/hr ft °F)	Specific Heat (Btu/lb/°F)	Tensile Strength (1000 psi)	Compressive Strength (1000 psi)	Elastic Modulus (10⁶ psi)	Shear Modulus (10⁶ psi)	Rupture Modulus (10⁶ psi)	Microhardness (kg/mm²)	Macrohardness (Mohs)	Poisson's Ratio	Electrical Resistivity (μohm-cm)	Dielectric Constant at 1 MHz	Dielectric Strength at 1 MHz (volts/mil)	Dissipation Factor at 1 MHz
CeSe	C	6.55	3308												100			
Ce$_2$Se$_3$	C	6.33	2900–3720												3.3×10^3			
Ce$_3$Se$_4$	C	6.76													8.0×10^3			
Er$_2$Se$_3$	C	6.96	2770												7.9×10^6			
GdSe	C	8.2	3385												72			
La$_2$Se$_3$	C	6.15													2.4×10^4			
SmSe	C	6.42	3812												1.38×10^9			
ThSe			3420															
USe		11.3	3360–3630															
U$_2$Se$_3$			2860															
YbSe	C		3530												1.0×10^8			
ZnSe			3630															

TABLE 6-13. PROPERTIES OF SINGLE CRYSTAL CERAMICS.

SILICATES

Formula	Crystal Structure	Theoretical Density (g/cm³)	Melting/Decomposition Temperature (°F)	Thermal Expansion Coefficient (10⁻⁶/°F)	Thermal Conductivity (Btu/hr ft °F)	Specific Heat (Btu/lb/°F)	Tensile Strength (1000 psi)	Compressive Strength (1000 psi)	Elastic Modulus (10⁶ psi)	Shear Modulus (10⁶ psi)	Rupture Modulus (10⁶ psi)	Microhardness (Kg/mm²)	Macrohardness (Mohs)	Poisson's Ratio	Electrical Resistivity (μohm-cm)	Dielectric Constant at 1 MHz	Dielectric Strength at 1 MHz (volts/mil)	Dissipation Factor at 1 MHz
Sillimanite	R	3.247	2950	3.66	1.03 at 400°F													
Mullite	R	~3.19	3360	2.57	3.12		16	100-190	21									
Phenacite	H	2.99	3630	3.55		0.207												
Wollastonite	TR	2.9	2800	6.4			8											
Steatite		3.6	2830	5.9	1.45		9.3	80							10^{20}	5.9		0.0013
Forsterite	R	3.22	3470	6.1	2.13	0.201	9.5	80	15						10^{20}	6.3		0.0003
Cordierite	R	2.51	2680	1.55	1.86		7.8	50										
Zircon	T	4.68	3.05	3.30 at 390°F			11	70-100	11-15									

TABLE 6-14. PROPERTIES OF SINGLE CRYSTAL CERAMICS

SILICIDES

Formula	Crystal Structure	Theoretical Density (g/cm³)	Melting/Decomposition Temperature (°F)	Thermal Expansion Coefficient (10⁻⁶/°F)	Thermal Conductivity (Btu/hr ft °F)	Specific Heat (Btu/lb/°F)	Tensile Strength (1000 psi)	Compressive Strength (1000 psi)	Elastic Modulus (10⁶ psi)	Shear Modulus (10⁶ psi)	Rupture Modulus (10⁶ psi)	Microhardness (Kg/mm²)	Macrohardness (Mohs)	Poisson's Ratio	Electrical Resistivity (μohm-cm)	Dielectric Constant at 1 MHz	Dielectric Strength at 1 MHz (volts/mil)	Dissipation Factor at 1 MHz
B_6Si	R	2.43	3540															
B_4Si	H	2.46	2000															
B_3Si	T	2.64	3500									5350			2×10^5			
$CeSi_2$	T	5.45	3110	5.85								540			408			
Cr_3Si	C	6.45	2840									1005			45.5			
Cr_3Si_2	T	5.6	2840									1280			114			
$CrSi$	C	5.43	2810									1005			143			
$CrSi_2$	H	5.00	2800	7.2								~1000			1.4×10^3			
$CoSi_2$	C	4.94	2330									889						
$DySi_2$	T	6.68	2820															
$EuSi_2$	T	5.5	~2730															
$GdSi_2$	T	6.19	3812															
$HfSi_2$	R	8.03	3090									865						
$FeSi$	C	6.09	2750												240			
$LaSi_2$	T	5.14	2770												236			
$MnSi$	C	5.85	2325															
Mn_3Si	C	6.60	2050															
Mn_5Si_3	H	6.02	2345															

(continued)

TABLE 6-14. (*continued*)

SILICIDES

Formula	Crystal Structure	Theoretical Density (g/cm³)	Melting/Decomposition Temperature (°F)	Thermal Expansion Coefficient (10⁻⁶/°F)	Thermal Conductivity (Btu/hr, ft °F)	Specific Heat (Btu/lb/°F)	Tensile Strength (1000 psi)	Compressive Strength (1000 psi)	Elastic Modulus (10⁶ psi)	Shear Modulus (10⁶ psi)	Rupture Modulus (10⁶ psi)	Microhardness (kg/mm²)	Macrohardness (Mohs)	Poisson's Ratio	Electrical Resistivity (μohm-cm)	Dielectric Constant at 1 MHz	Dielectric Strength at 1 MHz (volts/mil)	Dissipation Factor at 1 MHz
Mo_3Si	C	8.97	3940	3.95								1300			45			
Mo_3Si_2	T	8.08	3800									1170						
$MoSi_2$	T	6.26	3650	4.51	34		40	300–350	59	23.6	0.05	1260		0.165	21.5			
$NdSi_2$	T	5.84	2780									400						
$NiSi$	R	5.92	1832									680–820						
Nb_4Si	H	8.01	3540									700						
$\alpha\text{-}Nb_5Si_3$	T	7.75	3630															
$\beta\text{-}Nb_5Si_3$	T	7.34	4420															
$PrSi_2$	T	5.64													202			
$\alpha\text{-}PuSi_2$	T	9.12																
$\beta\text{-}PuSi_2$	H	9.18																
$ReSi_2$	T	3.67	>3900															
$SmSi_2$	T	6.26																
$TaSi_2$	H	9.14	4170	5.3								1200–1600			8.5			
Ta_2Si	T	13.54	4530									1200–1500			124			

Compound											
Ta$_5$Si	H	12.86	4550						1200–1500		
Ta$_5$Si$_3$	H	13.06	~4530								
ThSi	R	9.03	>3092						696		
α-ThSi$_2$	T	7.79	2550						1120		
β-ThSi$_2$	H	8.23	3020								
TiSi	R	4.32	3200	4.9					1039		39
TiSi$_2$	R	4.15	2730	5.8					~850		123
Ti$_5$Si$_3$	H	4.32	3848	6.1					986		55
W$_3$Si$_2$	T	12.21	4210						770		
WSi$_2$	T	9.87	3840	4.6					1090		33.4
USi	R	10.40	2910						745		
α-USi$_2$	T	8.98	3092								
β-USi$_2$	H	9.25	2930						700		
USi$_3$	C	8.15	2750	9.0	8.5 at 170°F	0.073			445		~60
U$_3$Si	T	15.58	1760	8.2	8.5 at 170°F	0.033	11.3	4.8			
U$_3$Si$_2$	T	12.20	3030						796	0.17	150
V$_3$Si	C	5.74	3150	4.4					~1500		203
V$_5$Si$_3$	T	5.27	3900	5.2					~1400		
VSi$_2$	H	5.10	3090	6.2					~900		9.5
YSi$_2$	T	4.39	2770								
ZrSi	H	5.56	3850						1100		
ZrSi$_2$	R	4.88	2920	4.8					1030		161
Zr$_2$Si	T	5.99	3930						1230		49.4
Zr$_3$Si$_2$	T	5.90	4010								
Zr$_5$Si$_3$	H	5.88	4080						~1335		

TABLE 6-15. PROPERTIES OF SINGLE CRYSTAL CERAMICS.

SULFIDES

Formula	Crystal Structure	Theoretical Density (g/cm³)	Melting/Decomposition Temperature (°F)	Thermal Expansion Coefficient (10^{-6}/°F)	Thermal Conductivity (Btu/hr ft °F)	Specific Heat (Btu/lb/°F)	Tensile Strength (1000 psi)	Compressive Strength (1000 psi)	Elastic Modulus (10^6 psi)	Shear Modulus (10^6 psi)	Rupture Modulus (10^6 psi)	Microhardness (kg/mm²)	Macrohardness (Mohs)	Poisson's Ratio	Electrical Resistivity (μohm-cm)	Dielectric Constant at 1 MHz	Dielectric Strength at 1 MHz (volts/mil)	Dissipation Factor at 1 MHz
BaS	C	4.33	~4000	19		0.07									10^{15} at 800°F	19.27		
BeS	C	2.37	4975															
CdS	H	4.82	3180										3–3.5		5×10^6	10		
CaS	C	2.61											4			6.7		
CeS	C	5.98	4442	6.87	4.6	0.070						683			170			
Ce₂S₃	C	5.19	3900	5.80	2.24							403			1.2×10^{12}			
Ce₃S₄	C	5.68	3720													400		
CrS	M	4.09	2840													4×10^{-4}		
Cr₂S₃	RH	3.92														<50		
Dy₂S₃	C	6.54	2695															
Er₂S₃	M	6.21	3145															
EuS	C	5.75	3425															
Gd₂S₃	C	6.15																
Hf₂S₃	H	7.50													150			
HfS₂	H	6.03													10^9			
LaS	C	5.86	3580	6.45	13.1							677			92			
La₂S₃	C	5.01	3775	5.50								360			2×10^{12}			
LaS₂	C	4.90	3000															
MgS	C	2.68	>3630															

Formula											
MnS	C	3.99	2940							0.25–111	
MoS$_2$	H	4.80	>3310							242	
NdS	C	6.36	3880	8.54	11.1					7×10^{13}	
Nd$_2$S$_3$	C	5.34	3990	7.17	2.22		330				
PrS	C	6.03	4046	7.95	8					240	
Pr$_2$S$_3$	C	5.27	3260	6.72	1.7						
Pu$_2$S$_3$	C	8.41	~3130								
SmS	C	6.01	3520							8.5×10^4	
Sm$_2$S$_3$	C	5.73	3450							8.2×10^{13}	
SrS	C	3.67	3630			0.097				–	11.31
TaS$_2$	H		>2370								
ThS	C	9.57	4040	5.7			246				
Th$_2$S$_3$	R	7.88	3650							10^2–10^3	
ThS$_2$	R	7.36	3460			0.057				1×10^3	
TiS	H	4.46	3720							10^{-3}	
Ti$_2$S$_3$	H	3.71									
TiS$_2$	H	~3.28									
US	C	10.87	4496	6.6	5.1		200			$(1.1$–$3.6) \times 10^{-4}$	
U$_2$S$_3$	R	8.81	>3630								
β-US$_2$	R	8.03	3362								
VS	H	4.89	3450								
Yb$_2$S$_3$	RH	6.04									
YS	C	4.95	3700								
Y$_2$S$_3$	M	3.87	2910								
α-ZnS	H	4.09	3360	3.3				3.5–4	0.41		9.6
β-ZnS	C	4.10	1870	3.4					0.479	10^{16}	8.9
ZrS	C	4.56	3810							6×10^{-3}	
Zr$_2$S$_3$	C	4.29								6×10^{-2}	
ZrS$_2$	H	3.82	2820							10	

TABLE 6-16. PROPERTIES OF SINGLE CRYSTAL CERAMICS.

TELLURIDES

Formula	Crystal Structure	Theoretical Density (g/cm³)	Melting/Decomposition Temperature (°F)	Thermal Expansion Coefficient (10⁻⁶/°F)	Thermal Conductivity (Btu/hr/ft °F)	Specific Heat (Btu/lb/°F)	Tensile Strength (1000 psi)	Compressive Strength (1000 psi)	Elastic Modulus (10⁶ psi)	Shear Modulus (10⁶ psi)	Rupture Modulus (10⁶ psi)	Microhardness (Kg/mm²)	Macrohardness (Mohs)	Poisson's Ratio	Electrical Resistivity (μohm-cm)	Dielectric Constant at 1 MHz	Dielectric Strength at 1 MHz (volts/mil)	Dissipation Factor at 1 MHz
CeTe	C			2890–3430											200			
Ce₂Te	C			3030–3270											1.1×10^4			
GdTe	C			3400											700			
LaTe	C	6.68		3137											1.5×10^{11}			
NdTe	C	7.40		3710											40			
Nd₃Te₄	C	7.41		3060											350			
NbTe	C			3000														
SmTe	C			3480											1.64×10^9			
UTe		8.8		2820–3000														
YbTe	C			3160											7×10^9			
Y₂Te₃	C			2777											10^7			

REFERENCES

Bradshaw, W. G. and C. O. Matthews, "Properties of Refractory Materials: Collected Data and References," Lockheed Aircraft Corp., Sunnyvale, CA, U.S. Govt. Report No. AD 205 452 (1958).

Brixner, L. H., "Preparation and Properties of the Single Crystalline AB_2-Type Selenides and Tellurides of Niobium, Tantalum, Molybdenum and Tungsten, *J. Inorg. and Nuclear Chem.*, **24**, 257–263 (1962).

Campbell, I. E. and E. M. Sherwood, Eds., "High-Temperature Materials and Technology," John Wiley, New York, 1967.

Eyring, L., "Proceedings of the Fourth Conference on Rare Earth Research, April 22–25, 1964, Gordon and Breach, New York (1965).

Fuschillo, N. and R. A. Lindberg, "Electrical Conductors at Elevated Temperatures," Melpar, Inc., Falls Church, Va., U.S. Govt. Report No. AD 299 020, (1963).

Gschneider, K. A., Jr. and N. Kippenhan, "Thermochemistry of the Rare Earth Carbides, Nitrides and Sulfides for Steelmaking," Rare-Earth Information Center, Iowa State University, Ames, Report No. IS-RIC-5 (1971).

Hague, J. R., *et al.*, "Refractory Ceramics for Aerospace—A Materials Selection Handbook, "The American Ceramic Society, Columbus, Ohio (1964).

Kosolapova, T. A., "Carbides—Properties, Production, and Applications," Plenum Press, New York, 1971.

Samsonov, G. V. and Y. S. Umanskiy, "Hard Compounds of Refractory Metals," Translated published as U.S. Govt. Report No. NASA TT f-102, (1962).

Shaffer, P. T. B., "Plenum Press Handbooks of High-Temperature Materials, No. 1, Materials Index," Plenum Press, New York, 1964.

Sibert, M. E., "Electrical Properties of Refractory Materials," Lockheed Aircraft Corp., Sunnyvale, CA, U.S. Govt. Report No. AD 411 522 (1960).

Touloukian, Y. S. and C. Y. Ho., Eds., "Thermophysical Properties of Matter," 13 vols., Plenum, New York, 1970–1973.

Westbrook, J. H., Ed., "Intermetallic Compounds," John Wiley, New York, 1967.

BIBLIOGRAPHY

Glasses

Alper, A. M., Ed., "High Temperature Oxides, Part 4: Refractory Glasses, Glass-Ceramics, and Ceramics," Academic Press, New York, 1971.

Holland, L., "The Properties of Glass Surfaces," Halsted Press, New York, 1964.

Jones, G. O., "Glass," John Wiley, New York, 1956.

Morey, G. W., "The Properties of Glass," 2nd ed., Van Nostrand Reinhold, New York, 1954.

Porai-Koshits, E. A., Ed., "The Structure of Glass," 8 vols., Plenum Press, New York, 1958–1966.

Shand, E. B., "Glass Engineering Handbook," 2nd ed., McGraw-Hill, New York, 1958.

Stanworth, J. E., "Physical Properties of Glass," The Clarendon Press, Oxford, 1950.

Tooley, F. V., "Handbook of Glass Manufacture," Ogden Publishing Co., New York, 1953.

Ceramics

Alper, A. M., Ed., "High Temperature Oxides, Part 4: Refractory Glasses, Glass-Ceramics, and Ceramics," Academic Press, New York, 1971.

Aronsson, B., T. Lundstrom and S. Rundquist, "Borides, Silicides and Phosphides," John Wiley, New York, 1965.

Cockayne, B. and D. W. Jones, Eds., "Modern Oxide Materials," Academic Press, London, 1972.

Eitel, W., "Silicate Science," 6 vols., Academic Press, New York, 1964–1973.

Hench, L. H. and D. B. Dove, "Physics of Electronic Ceramics," 2 parts, Marcel Dekker, New York, 1971–1972.

Kornilov, I. I., "The Chemistry of Metallides," Consultants Bureau, New York, 1966.

Ryshkewitch, E., "Oxide Ceramics," Academic Press, New York, 1960.

Samsonov, G. V., Ed., "Refractory Carbides," Plenum Press, New York, 1974.

Samsonov, G. V., Ed., "The Oxide Handbook," Plenum Press, New York, 1974.

Taylor, H. F. W., Ed., "The Chemistry of Cements," Academic Press, New York, 1964.

Toth, L. E., "Transition Metal Carbides and Nitrides," Academic Press, New York, 1971.

CHAPTER VII
PROPERTIES OF COMPOSITES

Chapter VII presents a summary of data on composite or reinforced materials, omitting building materials such as plywood and concrete. The information is given in the following tables:

TABLE 7-1. PROPERTIES OF REINFORCING FILAMENTS AND WHISKERS, (SIMONDS, 1967)

Filament Reinforcement	Melting or Softening Pt. (°C)	Density (D, lb/in³)	Tensile Strength ($S_f \times 10^{-3}$ psi)	Young's Modulus ($E \times 10^{-6}$ psi)	Specific Strength ($S_f/D \times 10^{-6}$ in.)	S_f/E Ratio	Specific Modulus ($E/D \times 10^{-6}$ in.)
I. *Continuous*							
A. Glass							
E-glass	840	0.092	250	10.5	2.72	1/42	114
E/HTS	840	0.092	500	10.5	5.43	1/21	114
YM 31A	840	0.103	500	16.0	4.85	1/32	155
S–994	840	0.090	650	12.6	7.22	1/19	136
29–A	900	0.096	800	14.5	8.33	1/18	151
SiO_2	1660	0.079	850	10.5	10.75	1/12	133
B. Metal							
B	2100	0.09	300–500	55–60	4.44	1/144	639
W	3400	0.697	580	59	0.83	1/102	85
Mo	2622	0.369	320	52	0.87	1/163	141
Rene 41	1300	0.298	290	24.2	0.97	1/83	81
Steel	1400+	0.280	600	29.0	2.14	1/48	104
Be	1284	0.066	185	35	2.80	1/189	530
II. *Discontinuous* (whiskers)							
A. Ceramic							
Al_2O_3	2040	0.143	6200	62–67	43.4	1/11	524
BeO	2570	0.103	2800**	58–60	27.1	1/21	971
B_4C	2490^d	0.091	934	70	10.3	1/81	835
SiC	2690^d	0.115	1650	70+	14.3	1/75	1078
Graphite	3650^s	0.060	2845	142	47.4	1/50	2367
B. Metal							
Cu	1083	0.322	427	18	1.33	1/42	56
Ni	1455	0.324	560	31	1.73	1/55	96
Fe	1540	0.283	1900	29	6.71	1/15	102
Cr	1890	0.260	1290	35	4.96	1/27	135

**Flexure Test   d = decomposes   s = sublimes

TABLE 7-2. PROPERTIES OF SOME FIBER-REINFORCED METALS. (KOVES, 1970; BROUTMAN, 1967)

Reinforced Material		Volume Fraction of Fiber	Tensile Strength of Reinforced Material, (ksi)
Matrix	Fiber		
Aluminum	Beryllium	0.04	33.0
	Boron	0.10	43.0
	Boron Carbide	0.10	29.0
	Glass	0.50	130.0
	Quartz	0.48	440.0
	Stainless Steel	0.11	220.0
Iron	Aluminum Oxide	0.36	237.0
Nickel	Boron	0.08	384.0
	Tungsten	0.40	161.0
Silver	Aluminum Oxide	0.24	232.0
	Steel	0.44	65.0
Stainless Steel (#316)	Tungsten	0.18	58.6
Tantalum	Tantalum Carbide (Ta_2C)	0.29	155.0

Note: The strength data are greatly affected by the dimensions of the fiber, and the strengths may differ considerably from these results.

TABLE 7-3. COMPARATIVE PROPERTIES OF GLASS FIBER REINFORCED PLASTICS. (ANON. 1970)

	Polyester Thermosets				Thermoplastics									
					Acetal		Nylon		Polycarbonate		Polyethylene		Polypropylene	
	Sheet Molding Cmpd	Bulk Molding Cmpd	Matched Metal Die Preform	Spray-Up	20-40% Glass Reinf	Unrein-forced	6-50% Glass Reinf	Unrein-forced	20-40% Glass Reinf	Unrein-forced	5-40% Glass Reinf	Unrein-forced	20-40% Glass Reinf	Unrein-forced
Cost, ¢/lb	45-70	40-60	34-40	33-40	75-125	65	87.5-158	69-87.5	130-120	80-110	40-67	17-21	56-58	21-28
¢/cu in.	2.70-4.40	2 64-3.96	2.1-2.7	1.9-2.6	4.17-7.77	3.34	4.3-8.55	2.74-3.47	6.27-6.89	3.5-4.75	0.8-1.8	0.58-0.75	2.1-2.6	0.68-0.91
Specific Gravity	1.7-2.1	1.8-2.1	1.5-1.7	1.4-1.6	1.55-1.69	1.42	1.7-1 47	1.12-1.14	1.34-1.52	1.20	1.16-1.28	0.95	1.04-1.22	0.9
Heat Dist Temp @ 264 psi, F	400-500	400-450	350-400	350-400	325	230-255	300-500	122-129	285-300	265-290	200-260	93	230-300	125-140
Tensile Str, 1000 psi	8-20	4-10	25-30	9-18	11-13	8-10	13-31	9	14-21	9-11	4 4-11	4	5.5-9	3-5
Tensile Mod, 10^5 psi	16-25	16-25	9.0-20.0	8-18	8-10	4-5	2-18	2-5	10-17	3.5	9.0 (40%)	—	4.5-9	1.2
Flexural Str, 1000 psi	18-30	10-20	10-40	16-28	15-17.5	13-14	7-44	5-18	17-30	13	4.9-12	—	7-11	5-8
Flexural Mod, 10^5 psi	14-20	14-20	13-18	10-12	8-13	4	2-19	2-4	7.5-25	3	2.1-6	0.7-2.6	3.5-8.2	1.2-2.7
Compressive Str, 1000 psi	15-30	20-30	15-30	15-25	12-11	5	—	7-10	15-24	12	—	—	6.0-8	3.7-8
Izod Impact Str, lb/in. notch	8.0-22 0	2-10	10-20	4-12	0.8-2.8	1.2-2.3	0.8-4.5	1-4	1.5-4.0	16	1.2-4.0	0.6-2.0	1.0-4.0	0.5-20.0
Rockwell Hardness	H50-H112	H80-H112	H40-H105	H40-H105	M78-M94	M78-M94	—	—	M75-M100	M70	—	—	R95-R115	R50-R110
Chemical Resistance														
Acids, Weak	G-E	G-E	G-E	G-E	F	F	G	G	E	E	E	E	E	E
Acids, Strong	F	F	F	F	P	P	P	P	G[a]	G[a]	G[a]	G[a]	G[a]	G[a]
Alkalis, Weak	F	F	F	F	F	F	E	E	G	G	E	E	E	E
Alkalis, Strong	P	P	P	P	P	P	F	F	F	F	E	E	E	E
Organic Solvents	G-E	G-E	G-E	G-E	E	E	G	G	P[b]	P[b]	G[c]	P[b]	G[c]	G[e]

[a] Except to oxidizing acids.   [b] Soluble in aromatic and chlorinated hydrocarbons.   [c] Below 176°F.   [d] Styrene acrylonitrile.   [e] High impact.

TABLE 7-3. *(continued)*

	Thermoplastics										
Polystyrene		Polysulfone		ABS		PVC		Polyphenylene Oxide		SAN[d]	
20-30% Glass Reinf	Unreinforced[e]	20-40% Glass Reinf	Unreinforced	20-40% Glass Reinf	Unreinforced	15-35% Glass Reinf	Unreinforced	20-40% Glass Reinf	Unreinforced	20-40% Glass Reinf	Unreinforced
29-54	18	135-149	100	89-86	44	61-63	22-31	163-150	115	39-62	27
1.25-2.5	0.7	6.8-7.4	4.5	3.94-	1.65	3.2-3.7	1.2-1.5	7.04-8.0	4.4	1.7-3.1	1.05
				4.30							
1.20-1.29	1.05	1.38-1.55	1.24	1.23-1.38	1.05	1.45-1.62	1.4	1.20-1.38	1.06	1.22-1.40	1.03
210-220	175-203	333-350	345	233-240	215	155-165	155-165	285-365	375	210-230	190-220
10-14	3-4	15-20	10	15-20	6.6	14-18	6-7	15-22	10	13-18	9-11
8.4-12.1	3-4	15	3.6	8	—	10-18	4	—	3.7	9-18.5	5
10-17	6	21-27	1.5	23-26	9	20-25	13-16	17-31	15	15-21	9.7-17.5
8-12	4	8-15	4	9.2-18	3	9-16	4	8-15	4	8.0-18	5
—	—	21-26	14	15	—	13.4-16.8	—	18-20	15	—	—
0.4-2.5	0.7-3.6	1.3-2.5	1.3	1-2.4	2.5	0.8-1.6	2-?0	1.6-2.2	1.5-1.9	0.4-2.4	0.4
M70-M95	M12-M45	M85-M92	M69	M75-M102	R113	M80-M88	D80	M95	M75	M77-M103	M80
E	E	E	E	E	G	F	E	E	E	G	G
G[a]	G[a]	E	E	G[a]	G[h]	G	E	E	E	G[h]	G[h]
G	G	E	E	E	G	E	E	E	E	G	G
G	G	E	E	E	G	E	E	E	E	G	G
P[b]	P[b]	G	G	P[f]	P[f]	P[f]	P[f]	G[g]	G[g]	P[f]	P[f]

[f] Soluble in ketones, esters and aromatic and chlorinated hydrocarbons.   [g] Except to some aromatic and chlorinated aliphatics; resists alcohol.   [h] Except sulfuric acid.

TABLE 7-4. TYPICAL PROPERTIES OF BORON-EPOXY AND GRAPHITE EPOXY COMPOSITES[a]. (MATERIALS SELECTOR 1971)

Material	Test Direction[b]	Property	Loading		
			Tension	Compression	Shear
Boron-epoxy	0°	Modulus, 10^3 ksi	30.0	30.0	1.0
		Ultimate Strength, ksi	190	400	20.0
		Ultimate Strain, in./in.	0.0065	0.012	0.100
	90°	Modulus, 10^3 ksi	2.7	2.7	1.0
		Ultimate Strength, ksi	10.0	45.0	20.0
		Ultimate Strain, in./in.	0.0045	0.018	0.100
Graphite-epoxy	0°	Modulus, 10^3 ksi	22.5	20.0	0.6
		Ultimate Strength, ksi	212	220	14.4
		Ultimate Strain, in./in.	0.0094	0.0125	0.064
	90°	Modulus, 10^3 ksi	1.3	1.3	0.6
		Ultimate Strength, ksi	8.0	27.0	14.4
		Ultimate Strain, in./in.	0.0062	0.0028	0.064

[a] Room temperature properties of single lamina composites. From P. D. Shockey, General Dynamics, L. M. Lackman, North American Rockwell Corp., "Engineering Data and Design Allowables Criteria," Air Force Materials Symposium, 1970. [b] 0° = parallel to filaments; 90° = transverse to filaments

TABLE 7-5. TENSILE PROPERTIES OF GLASS FABRIC LAMINATES*
(RITTENHOUSE 1968)

Material	Trade Name and Manufacturer	Ultimate Tensile Strength (psi)	Ultimate Elongation (%)
Epoxy	Epon 828 TM (Shell Chemical Co.)	37,452	1.40
Phenolic	Conolon 506 TM (Narmco Div., Whittaker Corp.)	41,333	1.5
Silicone	DC-2106 TM (Dow Corning Corp.)	23,914	1.42
Phenolic	CTL-91-LD TM (U.S. Polymeric Corp.)	26,170	1.11
Phenolic	Mobiloy 81-AH7 TM (Cordo Molding Products, Inc.)	49,238	1.60
Polyester	Paraplex P-43 TM (Rohm & Haas Co.)	39,173	1.88
Silicone	DC-2104 TM (Dow Corning Corp.)	19,406	1.05
Polyester	Selectron 5003 TM (Pittsburgh Plate Glass Co.)	45,429	1.99

*Properties in air at 77°F and 760 torr.

TABLE 7-6. PROPERTIES OF LAMINATED THERMOSETTING SHEET GLASS CLOTH BASE MATERIALS. (MATERIALS SELECTOR, 1971).

ASTM/NEMA Class ➡	Glass Cloth Base						
	G-2	G-3	G-5	G-7	G-9	G-10	G-11
PHYSICAL PROPERTIES							
Density, lb/cu in.	0.054	0.060	0.069	0.060	0.069	0.065	0.065
Spec Ht, Btu/lb/°F	0.30	0.30	0.26	0.25	0.35–0.40	0.35–0.40	0.35–0.40
Ther Cond, Btu/hr/sq ft/°F/ft	–	–	0.29	0.17	0.17	0.17	0.17
Coef of Ther. Exp., 10^{-5} per °F	0.9	0.9	0.6	0.6	0.6	0.5	0.5
Insulation Res, megohms	5000	–	100	2500	10,000	200,000	200,000
MECHANICAL PROPERTIES							
Mod of Elast in Ten							
Lengthwise, 10^5 psi	18	20	23	18	23	25	25
Crosswise, 10^5 psi	12	17	20	18	20	20	20
Mod of Elast in Flex							
Lengthwise, 10^5 psi	13	15	17	14	25	27	28
Crosswise, 10^5 psi	10	12	15	12	20	22	23
Ten Str							
Lengthwise, 1000 psi	16	23	37	23	37	40	40
Crosswise, 1000 psi	11	20	30	18.5	30	35	35
Hardness (Rockwell M)	M105	M100	M120	M100	M120	M111	M112
Compr Str							
Flat, 1000 psi	38	50	70	45	70	60	60
Edge, 1000 psi	15	17.5	25	14	25	35	35
CHEMICAL RESISTANCE	All grades except G-5 and G-9 are resistant to dilute solutions of most acids; not recommended for use in alkaline solutions, except that G-5 and G-9 are resistant to dilute alkaline solutions; unaffected by most organic solvents; aromatic hydrocarbons and chlorinated aliphatics may affect G-7.						

Key to abbreviations for Tables 7-6 through 7-10:
  Spec Ht = Specific Heat
  Ther Cond = Thermal Conductivity
  Coef of Ther Exp = Coefficient of Thermal Expansion
  Insulation Res = Insulation Resistance
  Mod of Elast in Ten = Modulus of Elasticity in Tension
  Mod of Elast in Flex = Modulus of Elasticity in Flexure
  Ten Str = Tensile Strength
  Compr Str = Compressive Strength

TABLE 7-7. PROPERTIES OF LAMINATED THERMOSETTING SHEET FABRIC BASE AND ASBESTOS MATERIALS. (MATERIALS SELECTOR, 1971).

ASTM/NEMA Class ➤	Fabric Base				Asbestos Base	
	C	CE	L	LE	A	AA
PHYSICAL PROPERTIES						
Density, lb/cu in.	0.049	0.048	0.049	0.048	0.062	0.061
Spec Ht, Btu/lb/°F	0.35–0.40	0.35–0.40	0.35–0.40	0.35–0.40	0.30	0.30
Ther Cond, Btu/hr/sq ft/°F/ft	0.17	0.17	0.17	0.17	–	–
Coef of Ther Exp, 10^{-5} per °F	1.1	1.1	1.1	1.1	0.8	0.8
Insulation Res, megohms	–	–	–	30	–	–
MECHANICAL PROPERTIES						
Mod of Elast in Ten						
Lengthwise, 10^5 psi	10	9	12	10	25	17
Crosswise, 10^5 psi	9	8	9	8.5	16	15
Mod of Elast in Flex						
Lengthwise, 10^5 psi	10	9	11	10	23	16
Crosswise, 10^5 psi	9	8	8.5	8.5	14	14
Ten Str						
Lengthwise, 1000 psi	11.2	12	14	13.5	10	12
Crosswise, 1000 psi	9.5	9	10	9.5	8	10
Hardness (Rockwell M)	M103	M105	M105	M105	M111	M103
Compr Str						
Flat, 1000 psi	37	39	35	37	40	38
Edge, 1000 psi	23.5	24.5	23	25	17	21
CHEMICAL RESISTANCE	Resistant to dilute solutions of most acids; not recommended for use in alkaline solutions; unaffected by most chemical solvents.					

TABLE 7-8. PROPERTIES OF LAMINATED THERMOSETTING SHEET NYLON BASE AND FLAME RESISTANT MATERIALS (MATERIALS SELECTOR, 1971).

ASTM/NEMA Class ➤	Nylon Base	Flame Resistant			
	N-1	FR-2	FR-3	FR-4	FR-5
PHYSICAL PROPERTIES					
Density, lb/cu in.	0.42	0.048	0.051	0.067	0.067
Spec Ht, Btu/lb/°F	0.35–0.40	0.35–0.40	0.35–0.40	0.35–0.40	0.35–0.40
Ther Cond, Btu/hr/sq ft/°F/ft	–	0.17	0.17	0.17	0.17
Coef of Ther Exp, 10^{-5} per °F	–	1.1	1.1	0.6	0.5
Insulation Res, megohms	50,000	20,000	100,000	200,000	200,000
MECHANICAL PROPERTIES					
Mod of Elast in Ten					
Lengthwise, 10^5 psi	4	10	12	25	25
Crosswise, 10^5 psi	4	8	10	20	20
Mod of Elast in Flex					
Lengthwise, 10^5 psi	6	11	13	27	28
Crosswise, 10^5 psi	5	9	10	22	23
Ten Str					
Lengthwise, 1000 psi	8.5	12.5	14	40	–
Crosswise, 1000 psi	8	9.5	12	35	40
Hardness (Rockwell M)	M105	M97	M100	M111	M114
Compr Str					
Flat, 1000 psi	–	9	29	60	60
Edge, 1000 psi	–	0.25	0.1	35	35
CHEMICAL RESISTANCE	Resistant to dilute solutions of most acids; not recommended for use in alkaline solutions; unaffected by most organic solvents.				

TABLE 7-9. PROPERTIES OF LAMINATED THERMOSETTING SHEET PAPER BASE MATERIALS. (MATERIALS SELECTOR, 1971).

ASTM/NEMA Class	Paper Base									
	X	XP	XPC	XX	XXP	XXX	XXXP	ES-1	ES-2	ES-3
PHYSICAL PROPERTIES										
Density, lb/cu in.	0.049	0.048	0.048	0.048	0.048	0.047	0.048	0.052	0.051	0.050
Spec Ht, Btu/lb/°F	0.35–0.40	0.35–0.40	0.35–0.40	0.35–0.40	0.35–0.40	0.35–0.40	0.35–0.40	0.35–0.40	0.35–0.40	0.35–0.40
Ther Cond, Btu/hr/sq ft/°F/ft	0.17	0.17	0.17	0.17	0.17	0.17	0.17	0.17	0.17	0.17
Coef of Ther Exp, 10^{-5} per °F	1.1	1.1	1.1	1.1	1.1	1.1	1.1	1.1	1.1	1.1
Insulation Res, megohms	—	—	—	60	500	1000	20,000	—	—	—
MECHANICAL PROPERTIES										
Mod of Elast in Ten										
Lengthwise, 10^5 psi	19	12	10	15	10	13	10	—	—	—
Crosswise, 10^5 psi	14	9	8	12	8	10	8	—	—	—
Mod of Elast in Flex										
Lengthwise, 10^5 psi	18	12	10	14	10	13	10	—	—	—
Crosswise, 10^5 psi	13	9	8	11	7	10	7	—	—	—
Ten Str										
Lengthwise, 1000 psi	20	12	10	16	11	15	12.4	12	13	15
Crosswise, 1000 psi	16	9	8	13	8.5	12	9.5	8.5	9	12
Hardness (Rockwell M)	M110	M95	M75	M105	M100	M100	M105	M118	M118	M120
Compr Str										
Flat, 1000 psi	36	25	34	25	32	25	25	—	—	—
Edge, 1000 psi	19	—	23	—	25	—	—	—	—	—
CHEMICAL RESISTANCE	Resistant to dilute solutions of most acids; not recommended for use in alkaline solutions; unaffected by most organic solvents.									

TABLE 7-10. PROPERTIES OF VULCANIZED FIBRE SHEET.
(MATERIALS SELECTOR, 1971).

NEMA/ASTM Class ⟶	Bone	Commercial	Electrical Insulation
PHYSICAL PROPERTIES			
Density (min), lb/cu in.	0.047	0.036	0.032
Water Absorption, % wt change, 24 hr			
$1/32$ in. thick	63	68	68
$1/8$ in. thick	48	61	61
$1/2$ in. thick	25	36	–
Ther Cond (68 F), Btu/hr/sq ft/°F/ft	0.168	0.168	0.168
Coef of Ther Exp, 10^{-6} per F	1–2	1–2	1–2
Specific Heat, Btu/lb/°F	0.403	0.403	0.403
MECHANICAL PROPERTIES			
Bursting Strength (min), psi (0.030 in.)	325	375	450
Flexure Strength, 1000 psi			
Lengthwise, $1/16$ to $1/8$ in. thick	16	15	15
Lengthwise, $1/8$ to $1/2$ in. thick	15	14	–
Lengthwise, $1/2$ to 1 in. thick	–	13	–
Crosswise, $1/16$ to $1/8$ in. thick	14	13	13
Crosswise, $1/8$ to $1/2$ in. thick	13	12	–
Crosswise, $1/2$ to 1 in. thick	–	11	–
Tensile Strength, 1000 psi			
Lengthwise, up to $1/8$ in. thick	14.0	13.5	13.5
Lengthwise, $1/8$ to $1/2$ in. thick	11.0	11.0	11.0
Lengthwise, over $1/2$ in. thick	–	7.0	7.0
Crosswise, up to $1/8$ in. thick	8.0	7.5	7.5
Crosswise, $1/8$ to $1/2$ in. thick	7.0	7.0	7.0
Crosswise, over $1/2$ in. thick	–	6.0	6.0
Compr Strength, 1000 psi	30	25	–
Mod of Elast in Tens (typical), 10^5 psi			
Lengthwise, $1/4$ in. thick	12	12	12
Crosswise, $1/4$ in. thick	8	8	8
Mod of Elast in Flex (typical), 10^5 psi			
Lengthwise, $1/4$ in thick	10	10	10
Crosswise, $1/4$ in. thick	7	7	7
Hardness (min), Rockwell	R80	R50	–
Bond Strength, 1000 psi	1.0	0.8	–
Impact Strength (Izod), ft-lb/in. of notch			
Lengthwise, $1/8$ in. thick	1.4	1.6	1.6
Crosswise, $1/8$ in. thick	1.0	1.2	1.2
ELECTRICAL PROPERTIES			
Dielectric Strength, v/mil			
$1/32$ in. thick	175	175	250
$1/8$ in. thick	150	150	175
$1/2$ in. thick	50	50	–

REFERENCES

Anon., "Engineer's Guide to Glass Reinforced Plastics," *Materials Eng.* 71, (2), 34–41 (Feb. 1970).

Broutman, L. J. and R. H. Krock, "Modern Composite Materials," Addison-Wesley, Reading, Mass., 1967.

Koves, G., "Materials for Structural and Mechanical Functions," Hayden Book Co., New York, 1970.

"1971 Materials Selector," *Materials Eng.* 72, (6), 1–496, (Nov. 1970).

Rittenhouse, J. B. and J. B. Singletary, "Space Materials Handbook," third ed., U.S. Government Report No. AD 692 353, Clearinghouse for Federal Scientific & Technical Information, Springfield, Va., 1968.

Simonds, H. R. and J. M. Church, Eds., "The Encyclopedia of Basic Materials for Plastics," Van Nostrand Reinhold, New York, 1967.

BIBLIOGRAPHY

Broutman, L. J., A. C. Metcalfe and R. H. Krock, Eds., "Treatise on Composite Materials," 4 vols., Academic Press, New York, 1973.

Davis, L. W. and S. W. Bradstreet, "Metal and Ceramic Matrix Composites," Cahner's Publ. Co., Boston, 1970.

Kraus, G. Ed., "Reinforcement of Elastomers," John Wiley, New York, 1965.

Levitt, A. P., Ed., "Whisker Technology," John Wiley, New York, 1970.

Lubin, G., "Reinforced Plastics," Van Nostrand Reinhold, New York, 1969.

Mettes, D. G., T. J. Humphrey and P. Newman, "Fiberglas Reinforced Plastics," Van Nostrand Reinhold, New York, 1969.

Mohr, J. G. and S. S. Oleesky, "SPI Handbook of Technology and Engineering of Reinforced Plastics/Composites," 2nd ed., Van Nostrand Reinhold, New York, 1973.

Parratt, N. J., "Fibre Reinforced Materials Technology," Van Nostrand Reinhold, New York, 1973.

CHAPTER VIII
PROPERTIES OF POLYMERS AND ADHESIVES

Chapter VIII treats the properties of resins, plastic films, organic coatings, rubbers, plastic foams, and selected widely used inorganic cements.

The data are presented in the following tables:

TABLE 8-1 PROPERTIES OF RESINS. (MATERIALS SELECTOR)

ABS RESINS —

Molded, Extruded

PROPERTY	ASTM TEST	Medium Impact	High Impact	Very High Impact
GENERAL PROPERTIES				
Specific Gravity	D792	1.05–1.07	1.02–1.04	1.01–1.04
Thermal Conductivity, Btu/hr/sq ft/°F/ft	C177	0.08–0.18	0.12–0.16	0.01–0.14
Thermal Expansion Coefficient, 10^{-6}/°F	D696	32–48	55–60	50–60
Specific Heat, Btu/lb/°F	—	0.36–0.38	0.36–0.38	0.36–0.38
Water Absorption in 24 hr, %	D570	0.2–0.4	0.2–0.45	0.2–0.45
White Light Transmission, 0.125 in. thick, %	D1003	33.3	28.0	33.3
Haze, %	D672	100	100	100
Refractive Index, n_D	D542	—	—	—
MECHANICAL PROPERTIES				
Tensile Strength, 1000 psi	D638	6.3–8.0	5.0–6.0	4.5–6.0
Modulus of Elasticity in Tension, 10^5 psi	D638	3.3–4.0	2.6–3.1	2.0–3.1
Elongation in 2 in., %	D638	5–50	5–50	20–50
Flexural Strength, 1000 psi	D790	9.9–11.8	7.5–9.5	6.0–9.8
Flexural Modulus, 10^5 psi	D790	3.5–4.0	2.5–3.2	2.0–3.2
Compressive Strength, 1000 psi	D790	10.5–11.0	7.0–9.0	~6
Impact Strength (Izod notched), ft-lb/in.	D256	2.0–4.0	3.0–5.0	5.0–7.0
Hardness (Rockwell)–R	D785	108–115	95–105	85–105
Abrasion Resistance, (Taber CS-17 wheel), mg loss/1000 cycles	D1044	—	—	—
ELECTRICAL PROPERTIES				
Volume Resistivity, ohm-cm	D257	2.7×10^{16}	$(1-4) \times 10^{16}$	$(1-4) \times 10^{16}$
Dielectric Strength (short time), volts/mil	D149	385	350–440	300–375
Dielectric Constant, 50% RH, 73°F	D150			
60 Hz		2.8–3.2	2.8–3.2	2.8–3.5
1 MHz		2.75–3.0	2.7–3.0	2.4–3.0
Dissipation Factor, 50% RH, 73°F	D150			
60 Hz		0.003–0.006	0.005–0.007	0.005–0.010
1 MHz		0.008–0.009	0.007–0.015	0.008–0.016
Arc Resistance (Tungsten electrode), sec.	D495	70–80	70–80	70–80
HEAT RESISTANCE				
Max. Recommended Service Temperature, °F	—	160–200	160–210	140–200
Heat Deflection Temperature, °F	D648			
66 psi		215	213	208
264 psi		200–230	185–198	187–196
Flammability, inches per minute	D635	1.3	1.3	1.3
Brittleness Temperature, °F	D746	—	—	—
Softening Point, Vicat, °F	D1525	—	—	—

TABLE 8-1 (Continued)

		CAST RESIN SHEETS, RODS		MOLDINGS	
MATERIAL → ACRYLICS – (Cast, Molded, Extruded)		General Purpose Type I per (ASTM D702)	General Purpose Type II per (ASTM D702)	Grades 5, 6, 8	High Impact Grade
PROPERTY	ASTM TEST				
GENERAL PROPERTIES					
Specific Gravity	D792	1.17-1.19	1.18-1.20	1.18-1.19	1.12-1.16
Thermal Conductivity, Btu/hr/sq ft/°F/ft	C177	0.12	0.12	0.12	0.12
Thermal Expansion Coefficient, 10^{-6}/°F	D696	45	45	30-40	40-60
Specific Heat, Btu/lb/°F	—	0.35	0.35	0.35	0.34
Water Absorption in 24 hr, %	D570	0.3-0.4	0.2-0.4	0.3-0.4	0.2-0.3
White Light Transmission, 0.125 in. thick, %	D1003	92	92	92	–
Haze, %	D672	1		< 3	–
Refractive Index, n_D	D542	1.49		1.49	–
MECHANICAL PROPERTIES					
Tensile Strength, 1000 psi	D638	6-9	8-10	9.5-10.5	5.5-9.0
Modulus of Elasticity in Tension, 10^5 psi	D638	3.5-4.5	4.0-5.0	3.5-5.0	2.3-3.3
Elongation in 2 in., %	D638	2-7	2-7	3-5	> 25
Flexural Strength, 1000 psi	D790	12-14	15-17	15-16	8.7-12.0
Flexural Modulus, 10^5 psi	D790	3.5-4.5	4.0-5.0	3.5-5.0	2.8-3.6
Compressive Strength, 1000 psi (Yld, 0.1%)	D790	12-14	14-18	14.5-17	7.3-12.0
Impact Strength (Izod notched), ft-lb/in.	D256	0.4	0.4	0.2-0.4	0.8-2.3
Hardness (Rockwell)	D785	M80-90	M96-102	M80-103	L60-94
Abrasion Resistance, (Taber CS-17 wheel), mg loss/1000 cycles	D1044				
ELECTRICAL PROPERTIES					
Volume Resistivity, ohm-cm	D257	$> 10^{15}$	$> 10^{15}$	$> 10^{14}$	2.0×10^{16}
Dielectric Strength (short time), volts/mil	D149	450-530	450-500	400	400-500
Dielectric Constant, 50% RH, 73°F	D150				
60 Hz		3.5-4.5	3.5-4.5	3.5-3.9	3.5-3.9
1 MHz		2.7-3.2	2.7-3.2	2.7-2.9	2.5-3.0
Dissipation Factor, 50% RH, 73°F	D150				
60 Hz		0.05-0.06	0.05-0.06	0.04-0.06	0.03-0.04
1 MHz		0.02-0.03	0.02-0.03	0.02-0.03	0.01-0.02
Arc Resistance (Tungsten electrode), sec.	D495	no tracking	no tracking	no tracking	no tracking
HEAT RESISTANCE					
Max. Recommended Service Temperature, °F		140-160	180-200	155-190	160-185
Heat Deflection Temperature, °F	D648				
66 psi		165-235		165-225	180-225
264 psi		160-215		155-210	165-215
Flammability, inches per minute (0.125 in.)	D635	0.5-2.2	0.5-1.8	0.9-1.2	0.8-1.2
Brittleness Temperature, °F	D746				
Softening Point, Vicat, °F	D1525				

TABLE 8-1 *(Continued)*

ALKYDS and THERMOSET CARBONATE

MATERIAL →

PROPERTY	ASTM TEST	Allyl Diglycol Carbonate	ALKYDS-MOLDED		
			Putty (encapsulating)	Rope (general use)	Granular (high speed molding)
GENERAL PROPERTIES					
Specific Gravity	D792	1.32	2.05–2.15	2.20–2.22	2.21–2.24
Thermal Conductivity, Btu/hr/sq ft/°F/ft	C177	1.45	0.35–0.60	0.35–0.60	0.35–0.60
Thermal Expansion Coefficient, 10^{-6}/°F	D696	60	10–30	10–30	10–30
Specific Heat, Btu/lb/°F	–	0.3			
Water Absorption in 24 hr, %	D570	0.20	0.10–0.15	0.05–0.08	0.08–0.12
White Light Transmission, 0.125 in. thick, %	D1003	88			
Haze, %	D672	<1			
Refractive Index, n_D	D542	1.50			
MECHANICAL PROPERTIES					
Tensile Strength, 1000 psi	D638	5–6	4–5	7–8	3–4
Modulus of Elasticity in Tension, 10^5 psi	D638	3.0	20–27	19–20	24–29
Elongation in 2 in., %	D638	–	–	–	–
Flexural Strength, 1000 psi	D790	6–13	8–11	19–20	7–10
Flexural Modulus, 10^5 psi	D790	2.5–3.3		22–27	22–27
Compressive Strength, 1000 psi	D790	22.5		28	16–20
Impact Strength (Izod notched), ft-lb/in.	D256	0.2–0.4	20–25	2.2	0.30–0.35
Hardness (Barcol)	D785	M95–M100 (Rockwell)	0.25–0.35	70–80	60–70
Abrasion Resistance, (Taber CS-17 wheel), mg loss/1000 cycles	D1044		60–70		
ELECTRICAL PROPERTIES					
Volume Resistivity, ohm-cm	D257	4×10^{14}	1×10^{14}	1×10^{14}	$(1–10) \times 10^{14}$
Dielectric Strength (short time), volts/mil	D149	290	300–350	290	300–350
Dielectric Constant, 50% RH, 73° F	D150				
60 Hz		4.4	5.4–5.9	7.4	5.7–6.3
1 MHz		3.5–3.8	4.5–4.7	6.8	4.8–5.1
Dissipation Factor, 50% RH, 73° F	D150				
60 Hz		0.03–0.04	0.030–0.045	0.019	0.030–0.040
1 MHz		0.1–0.2	0.016–0.020	0.023	0.017–0.020
Arc Resistance (Tungsten electrode), sec.	D495	185	180	180	180
HEAT RESISTANCE					
Max. Recommended Service Temperature, °F	–	212	250	300	300
Heat Deflection Temperature, °F	D648				
66 psi					
264 psi		140–190	350–400	400	350–400
Flammability, inches per minute	D635	0.35	non-burning	self-ext.	self-ext.
Brittleness Temperature, °F	D746				
Softening Point, Vicat, °F	D1525				

TABLE 8-1 (*Continued*)

MATERIAL →

ALLYLIC PREPOLYMERS —
DIALLYL PHTHALATE —
DIALLYL ISOPHTHALATE —

PROPERTY	ASTM TEST	Diallyl Phthalate	Diallyl Isophthalate
GENERAL PROPERTIES			
Specific Gravity	D792	1.270	1.264
Thermal Conductivity, Btu/hr/sq ft/°F/ft	C177		
Thermal Expansion Coefficient, 10^{-6}/°F	D696		
Specific Heat, Btu/lb/°F	—		
Water Absorption in 24 hr, %	D570	0.0–0.2	0.1
White Light Transmission, 0.125 in. thick, %	D1003		
Haze, %	D672		
Refractive Index, n_D	D542	1.571	1.569
MECHANICAL PROPERTIES			
Tensile Strength, 1000 psi	D638	4.0	4.3
Modulus of Elasticity in Tension, 10^5 psi	D638		
Elongation in 2 in., %	D638		
Flexural Strength, 1000 psi	D790	7.0–9.0	7.4–8.3
Flexural Modulus, 10^5 psi	D790	5	5
Compressive Strength, 1000 psi	D790	22–24	
Impact Strength (Izod notched), ft-lb/in.	D256	0.2–0.3	0.2–0.3
Hardness (Rockwell) –M	D785	114–116	119–121
Abrasion Resistance, (Taber CS-17 wheel), mg loss/1000 cycles	D1044		
ELECTRICAL PROPERTIES			
Volume Resistivity, ohm-cm	D257	1.7×10^{16}	3.9×10^{17}
Dielectric Strength (short time), volts/mil	D149		
Dielectric Constant, 50% RH, 73°F	D150		
60 Hz		3.9	3.4
1 MHz		3.5	3.2
Dissipation Factor, 50% RH, 73°F	D150		
60 Hz		0.005	0.008
1 MHz		0.011	0.009
Arc Resistance (Tungsten electrode), sec.	D495	118	123–128
HEAT RESISTANCE			
Max. Recommended Service Temperature, °F	—		
Heat Deflection Temperature, °F	D648		
66 psi			
264 psi		155	238
Flammability, inches per minute	D635		
Brittleness Temperature, °F	D746		
Softening Point, Vicat, °F	D1525		

TABLE 8-1 (*Continued*)

CELLULOSE ACETATE —
Molded, Extruded

PROPERTY	ASTM TEST	H6-1	H4-1	H2-1	MH-1, MH-2
GENERAL PROPERTIES					
Specific Gravity	D792		1.29–1.31	1.25–1.31	1.24–1.31
Thermal Conductivity, Btu/hr/sq ft/°F/ft	C177	0.10–0.19	0.10–0.19	0.10–0.19	0.10–0.19
Thermal Expansion Coefficient, 10^{-6}/°F	D696	44–90	44–90	44–90	44–90
Specific Heat, Btu/lb/°F	—	0.3–0.42	0.3–0.42	0.3–0.42	0.3–0.42
Water Absorption in 24 hr, %	D570		1.7–2.7	1.7–2.7	1.8–4.0
White Light Transmission, 0.125 in. thick, %	D1003				
Haze, %	D672	2–15	2–15	2–10	2–10
Refractive Index, n_D	D542	1.46–1.50	1.46–1.50	1.46–1.50	1.46–1.50
MECHANICAL PROPERTIES					
Tensile Strength, 1000 psi	D638				
Modulus of Elasticity in Tension, 10^5 psi	D638				
Elongation in 2 in., %	D638				
Flexural Strength, 1000 psi	D790				
Flexural Modulus, 10^5 psi	D790				
Compressive Strength, 1000 psi	D790				
Impact Strength (Izod notched), ft-lb/in.	D256		1.1–3.1	1.5–3.9	2.5–4.9
Hardness (Rockwell) –R	D785		103–120	89–112	74–104
Abrasion Resistance, (Taber CS-17 wheel), mg loss/1000 cycles	D1044				
ELECTRICAL PROPERTIES					
Volume Resistivity, ohm-cm	D257	10^{10}–10^{13}	10^{10}–10^{13}	10^{10}–10^{13}	10^{10}–10^{13}
Dielectric Strength (short time), volts/mil	D149	250–600	250–600	250–600	250–600
Dielectric Constant, 50% RH, 73°F	D150				
60 Hz		3.5–7.5	3.5–7.5	3.5–7.5	3.5–7.5
1 MHz		3.2–7.0	3.2–7.0	3.2–7.0	3.2–7.0
Dissipation Factor, 50% RH, 73°F	D150				
60 Hz			0.01–0.06	0.01–0.06	0.01–0.06
1 MHz			0.01–0.10	0.01–0.10	0.01–0.10
Arc Resistance (Tungsten electrode), sec.	D495				
HEAT RESISTANCE					
Max. Recommended Service Temperature, °F	—		172–203	145–188	145–170
Heat Deflection Temperature, °F	D648				
66 psi			145–188	120–172	128–155
264 psi					
Flammability, inches per minute	D635	0.5–2.0	0.5–2.0	0.5–2.0	0.5–2.0
Brittleness Temperature, °F	D746				
Softening Point, Vicat, °F	D1525				

TABLE 8-1 (Continued)

CELLULOSE ACETATE PROPERTY	ASTM TEST			MS–1, MS–2	S2–1
GENERAL PROPERTIES					
Specific Gravity	D792			1.23–1.30	1.22–1.30
Thermal Conductivity, Btu/hr/sq ft/°F/ft	C177			0.10–0.19	0.10–0.19
Thermal Expansion Coefficient, 10^{-6}/°F	D696			44–90	44–90
Specific Heat, Btu/lb/°F	—			0.3–0.42	0.3–0.42
Water Absorption in 24 hr, %	D570			2.1–4.0	2.3–4.0
White Light Transmission, 0.125 in. thick, %	D1003				
Haze, %	D672			2–10	2–8
Refractive Index, n_D	D542			1.46–1.50	1.46–1.50
MECHANICAL PROPERTIES					
Tensile Strength, 1000 psi	D638				
Modulus of Elasticity in Tension, 10^5 psi	D638				
Elongation in 2 in., %	D638				
Flexural Strength, 1000 psi	D790				
Flexural Modulus, 10^5 psi	D790				
Compressive Strength, 1000 psi	D790				
Impact Strength (Izod notched), ft-lb/in.	D256			2.9–6.5	4.0–6.8
Hardness (Rockwell)	D785			54–96	49–88
Abrasion Resistance, (Taber CS-17 wheel), mg loss/1000 cycles	D1044				
ELECTRICAL PROPERTIES					
Volume Resistivity, ohm-cm	D257			10^{10}–10^{13}	10^{10}–10^{13}
Dielectric Strength (short time), volts/mil	D149			250–600	250–600
Dielectric Constant, 50% RH, 73°F	D150				
60 Hz				3.5–7.5	3.5–7.5
1 MHz				3.2–7.0	3.2–7.0
Dissipation Factor, 50% RH, 73°F	D150				
60 Hz				0.01–0.06	0.01–0.06
1 MHz				0.01–0.10	0.01–0.10
Arc Resistance (Tungsten electrode), sec.	D495				
HEAT RESISTANCE					
Max. Recommended Service Temperature, °F	—				
Heat Deflection Temperature, °F	D648				
66 psi				136–153	132–141
264 psi				123–141	117–129
Flammability, inches per minute	D635			0.5–2.0	0.5–2.0
Brittleness Temperature, °F	D746				
Softening Point, Vicat, °F	D1525				

TABLE 8-1 (Continued)

CELLULOSE ACETATE BUTYRATE —
Molded and Extruded

PROPERTY	ASTM TEST	H4	MH	S2
GENERAL PROPERTIES				
Specific Gravity	D792	1.22	1.18–1.20	1.15–1.18
Thermal Conductivity, Btu/hr/sq ft/°F/ft	C177	0.10–0.19	0.10–0.19	0.10–0.19
Thermal Expansion Coefficient, 10^{-6}/°F	D696	60–90	60–90	60–90
Specific Heat, Btu/lb/°F		0.3–0.4	0.3–0.4	0.3–0.4
Water Absorption in 24 hr, %	D570	2.0	1.3–1.6	0.9–1.3
White Light Transmission, 0.125 in. thick, %	D1003			
Haze, %	D672	2–5	2–5	2–5
Refractive Index, n_D	D542	1.46–1.49	1.46–1.49	1.46–1.49
MECHANICAL PROPERTIES				
Tensile Strength, 1000 psi (at fracture)	D638	6.9	5.0–6.0	3.0–4.0
Modulus of Elasticity in Tension, 10^5 psi	D638			
Elongation in 2 in., %	D638	9.0	5.6–6.7	2.5–3.95
Flexural Strength, 1000 psi (at yield)	D790			
Flexural Modulus, 10^5 psi	D790	8.8	5.3–7.1	2.6–4.3
Compressive Strength, 1000 psi (at yield)	D790	3.0	4.4–6.9	7.5–10.0
Impact Strength (Izod notched), ft-lb/in.	D256	114	80–100	23–42
Hardness (Rockwell)—R	D785			
Abrasion Resistance, (Taber CS-17 wheel), mg loss/1000 cycles	D1044			
ELECTRICAL PROPERTIES				
Volume Resistivity, ohm-cm	D257	10^{11}–10^{14}	10^{11}–10^{14}	10^{11}–10^{14}
Dielectric Strength (short time), volts/mil	D149	250–400	250–400	250–400
Dielectric Constant, 50% RH, 73° F	D150			
60 Hz		3.5–6.4	3.5–6.4	3.5–6.4
1 MHz		3.2–6.2	3.2–6.2	3.2–6.2
Dissipation Factor, 50% RH, 73° F	D150			
60 Hz		0.01–0.04	0.01–0.04	0.01–0.04
1 MHz		0.02–0.05	0.02–0.05	0.02–0.05
Arc Resistance (Tungsten electrode), sec.	D495			
HEAT RESISTANCE				
Max. Recommended Service Temperature, °F	—			
Heat Deflection Temperature, °F	D648			
66 psi		222	171–184	136–147
264 psi		196	146–160	118–130
Flammability, inches per minute	D635	0.5–1.5	0.5–1.5	0.5–1.5
Brittleness Temperature, °F	D746			
Softening Point, Vicat, °F	D1525			

TABLE 8-1 (Continued)

CELLULOSE ACETATE PROPRIONATE

Molded, Extruded

PROPERTY	ASTM TEST	1	2	3
GENERAL PROPERTIES				
Specific Gravity	D792	1.22	1.20-1.21	1.19
Thermal Conductivity, Btu/hr/sq ft/°F/ft	C177	0.10-0.19	0.10-0.19	0.10-0.19
Thermal Expansion Coefficient, 10^{-6}/°F	D696	60-90	60-90	60-90
Specific Heat, Btu/lb/°F	—	0.3-0.4	0.3-0.4	0.3-0.4
Water Absorption in 24 hr, %	D570	1.6-2.0	1.3-1.8	1.6
White Light Transmission, 0.125 in. thick, %	D1003			
Haze, %	D672	2-5	2-5	2-5
Refractive Index, n_D	D542	1.46-1.49	1.46-1.49	1.46-1.49
MECHANICAL PROPERTIES				
Tensile Strength, 1000 psi	D638			
Modulus of Elasticity in Tension, 10^5 psi	D638			
Elongation in 2 in., %	D638			
Flexural Strength, 1000 psi	D790			
Flexural Modulus, 10^5 psi	D790			
Compressive Strength, 1000 psi	D790			
Impact Strength (Izod notched), ft-lb/in.	D256	1.7-2.7	3.5-5.6	9.4
Hardness (Rockwell)–R	D785	100-109	92-96	57
Abrasion Resistance, (Taber CS-17 wheel), mg loss/1000 cycles	D1044			
ELECTRICAL PROPERTIES				
Volume Resistivity, ohm-cm	D257	10^{11}-10^{14}	10^{11}-10^{14}	10^{11}-10^{14}
Dielectric Strength (short time), volts/mil	D149	300-450	300-450	300-450
Dielectric Constant, 50% RH, 73°F	D150			
60 Hz		3.7-4.0	3.7-4.0	3.7-4.0
1 MHz		3.4-3.7	3.4-3.7	3.4-3.7
Dissipation Factor, 50% RH, 73°F	D150			
60 Hz		0.01-0.04	0.01-0.04	0.01-0.04
1 MHz		0.02-0.05	0.02-0.05	0.02-0.05
Arc Resistance (Tungsten electrode), sec.	D495			
HEAT RESISTANCE				
Max. Recommended Service Temperature, °F	—			
Heat Deflection Temperature, °F	D648			
66 psi		191-201	169-187	163
264 psi		163-173	141-157	129
Flammability, inches per minute	D635	0.5-1.5	0.5-1.5	0.5-1.5
Brittleness Temperature, °F	D746			
Softening Point, Vicat, °F	D1525			

TABLE 8-1 (Continued)

MATERIAL → CELLULOSE NITRATE — ETHYL CELLULOSE — Molded, Extruded				ETHYL CELLULOSE MOLDINGS		
					High Impact	
PROPERTY	ASTM TEST	Cellulose Nitrate Sheet	General Purpose	A	B
GENERAL PROPERTIES					
Specific Gravity	D792	1.35-1.40	1.10-1.16	1.10-1.16	1.10-1.16
Thermal Conductivity, Btu/hr/sq ft/°F/ft	C177	0.133	0.092-0.167	0.092-0.167	0.092-0.167
Thermal Expansion Coefficient, 10^{-6}/°F	D696	44-66	55-110	55-110	55-110
Specific Heat, Btu/lb/°F	—	0.3-0.4	0.3-0.75		
Water Absorption in 24 hr, %	D570	1.0-2.0	1.2-2.0	0.8-2.0	1.0-2.0
White Light Transmission, 0.125 in. thick, %	D1003				
Haze, %	D672	2.0-4.0	—	—	—
Refractive Index, n_D	D542	1.49-1.51	1.47	1.47	1.47
MECHANICAL PROPERTIES					
Tensile Strength, 1000 psi	D638	7-8	3-7	4-6.5	3.5
Modulus of Elasticity in Tension, 10^5 psi	D638	1.9-2.2	0.5-3.5	1.0-3.0	1.0-3.0
Elongation in 2 in., %	D638		4-10	6	4
Flexural Strength, 1000 psi	D790	9-11			
Flexural Modulus, 10^5 psi	D790	2.3-2.5			
Compressive Strength, 1000 psi	D790	22-35			
Impact Strength (Izod notched), ft-lb/in.	D256	5-7	1.7-6.0	3.5-6.0	4.0-7.0
Hardness (Rockwell)-R	D785	95-115	80-120	80-90	70-80
Abrasion Resistance, (Taber CS-17 wheel), mg loss/1000 cycles	D1044				
ELECTRICAL PROPERTIES					
Volume Resistivity, ohm-cm	D257	$(1.0-1.5) \times 10^{11}$	$10^{12}-10^{14}$	$10^{12}-10^{14}$	$10^{12}-10^{14}$
Dielectric Strength (short time), volts/mil	D149	300-600	350-500	350-500	350-500
Dielectric Constant, 50% RH, 73°F	D150				
60 Hz		7.0-7.5	3.0-4.2	2.8-3.3	3.0-3.6
1 MHz		6.4	2.8-3.5		
Dissipation Factor, 50% RH, 73°F	D150				
60 Hz		0.09-0.12	0.005-0.020	0.010-0.030	0.018-0.035
1 MHz		0.06-0.09	0.010-0.060		
Arc Resistance (Tungsten electrode), sec.	D495		60-80		
HEAT RESISTANCE					
Max. Recommended Service Temperature, °F	—	120-140	115-185	130-145	125-140
Heat Deflection Temperature, °F	D648				
66 psi		200-220	120-160		
264 psi		140-160			
Flammability, inches per minute	D635	very rapid	0.5-1.5	0.5-1.5	0.5-1.5
Brittleness Temperature, °F	D746				
Softening Point, Vicat, °F	D1525				

TABLE 8-1 (*Continued*)

CHLORINATED POLYETHER –
CHLORINATED POLYVINYL CHLORIDE –

MATERIAL →

PROPERTY	ASTM TEST	Chlorinated Polyether	Chlorinated Polyvinyl Chloride
GENERAL PROPERTIES			
Specific Gravity	D792	1.4	1.54
Thermal Conductivity, Btu/hr/sq ft/°F/ft	C177	0.076	0.079
Thermal Expansion Coefficient, 10^{-6}/°F	D696	66	44
Specific Heat, Btu/lb/°F	—		0.3
Water Absorption in 24 hr, %	D570	0.01	0.11
White Light Transmission, 0.125 in. thick, %	D1003	opaque	opaque
Haze, %	D672		
Refractive Index, n_D	D542		
MECHANICAL PROPERTIES			
Tensile Strength, 1000 psi	D638	6	7.3
Modulus of Elasticity in Tension, 10^5 psi	D638	1.5	3.7
Elongation in 2 in., %	D638	130	15–65
Flexural Strength, 1000 psi	D790	5 (0.1% offset)	14.5
Flexural Modulus, 10^5 psi	D790	1.3 (0.1% offset)	3.85
Compressive Strength, 1000 psi	D790	9.0	0.0–22.0
Impact Strength (Izod notched), ft-lb/in.	D256	0.4	6.3
Hardness (Rockwell)	D785	R100	R118
Abrasion Resistance, (Taber CS-17 wheel). mg loss/1000 cycles	D1044		
ELECTRICAL PROPERTIES			
Volume Resistivity, ohm-cm	D257	1.5×10^{16}	$(0.1-2) \times 10^{16}$
Dielectric Strength (short time), volts/mil	D149	400	1250–1550
Dielectric Constant, 50% RH, 73°F	D150		
60 Hz		3.10	3.08
1 MHz		2.92	3.2–3.6
Dissipation Factor, 50% RH, 73°F	D150		
60 Hz		0.011	0.0189–0.0208
1 MHz		0.011	0.020
Arc Resistance (Tungsten electrode), sec.	D495		
HEAT RESISTANCE			
Max. Recommended Service Temperature, °F	—	250–275	210
Heat Deflection Temperature, °F	D648		
66 psi		285	215–247
264 psi		210	202–234
Flammability, inches per minute	D635	self-ext.	non-burn.
Brittleness Temperature, °F	D746		
Softening Point, Vicat, °F	D1525		

TABLE 8-1 (*Continued*)

EPOXIES:
NOVOLACS
CYCLOALIPHATIC DIEPOXIDES

PROPERTY	ASTM TEST	Novolacs Cast, Rigid	Cycloaliphatic Diepoxides, Cast, Rigid
GENERAL PROPERTIES			
Specific Gravity	D792	1.24	1.22
Thermal Conductivity, Btu/hr/sq ft/°F/ft	C177		
Thermal Expansion Coefficient, 10^{-6}/°F	D696		16–30
Specific Heat, Btu/lb/°F	—		
Water Absorption in 24 hr, %	D570		0.1–0.7
White Light Transmission, 0.125 in. thick, %	D1003		
Haze, %	D672		
Refractive Index, n_D	D542		
MECHANICAL PROPERTIES			
Tensile Strength, 1000 psi	D638	8–12	9.6–12.0
Modulus of Elasticity in Tension, 10^5 psi	D638	4–5	4.8–5.0
Elongation in 2 in., %	D638	2–5	2.2–4.8
Flexural Strength, 1000 psi	D790	11–16	12–13
Flexural Modulus, 10^5 psi	D790	4–5	4.4–4.8
Compressive Strength, 1000 psi	D790	17–19	30–50
Impact Strength (Izod notched), ft-lb/in.	D256	0.5	
Hardness (Rockwell)	D785		
Abrasion Resistance, (Taber CS-17 wheel), mg loss/1000 cycles	D1044		
ELECTRICAL PROPERTIES			
Volume Resistivity, ohm-cm	D257	2.10×10^{14}	$> 1 \times 10^{16}$
Dielectric Strength (short time), volts/mil	D149		444
Dielectric Constant, 50% RH, 73°F	D150		
60 Hz		3.96–4.02	3.34–3.39
1 MHz		3.53–3.58	
Dissipation Factor, 50% RH, 73°F	D150		
60 Hz		0.0055–0.0074	0.001–0.007
1 MHz		0.029–0.	
Arc Resistance (Tungsten electrode), sec.	D495		120
HEAT RESISTANCE			
Max. Recommended Service Temperature, °F	—	450	450–500
Heat Deflection Temperature, °F	D648		
66 psi			
264 psi	D635	300–400	300–525
Flammability, inches per minute		self-ext.	
Brittleness Temperature, °F	D746		
Softening Point, Vicat, °F	D1525		

TABLE 8-1 (*Continued*)

PROPERTY	ASTM TEST	Cast Rigid	Cast Flexible
MATERIAL → STANDARD EPOXIES, Diglycidyl Ethers of Bisphenol A – Cast			
GENERAL PROPERTIES			
Specific Gravity	D792	1.15	1.14-1.18
Thermal Conductivity, Btu/hr/sq ft/°F/ft	C177	0.1-0.3	—
Thermal Expansion Coefficient, 10^{-6}/°F	D696	33	30-50
Specific Heat, Btu/lb/°F	—		
Water Absorption in 24 hr, %	D570	0.4-0.5	0.4-1.0
White Light Transmission, 0.125 in. thick, %	D1003	0.1-0.2	85
Haze, %	D672	90	
Refractive Index, n_D	D542	1.61	1.61
MECHANICAL PROPERTIES			
Tensile Strength, 1000 psi	D638	9.5-11.5	1.4-7.6
Modulus of Elasticity in Tension, 10^5 psi	D638	4.5	0.5-2.5
Elongation in 2 in., %	D638	4.4	1.5-60
Flexural Strength, 1000 psi	D790	14-18	1.2-12.7
Flexural Modulus, 10^5 psi	D790	4.5-5.4	0.36-3.9
Compressive Strength, 1000 psi	D790	16.5-24	
Impact Strength (Izod notched), ft-lb/in.	D256	0.2-0.5	0.3-2.0
Hardness (Rockwell)	D785	106M	50-100M
Abrasion Resistance, (Taber CS-17 wheel), mg loss/1000 cycles	D1044		
ELECTRICAL PROPERTIES			
Volume Resistivity, ohm-cm	D257	6.1×10^{15}	$(0.91\text{-}6.7) \times 10^9$
Dielectric Strength (short time), volts/mil	D149	>400	400-410
Dielectric Constant, 50% RH, 73°F	D150		
60 Hz		4.02	4.43-4.79
1 MHz		3.42	2.78-3.52
Dissipation Factor, 50% RH, 73°F	D150		
60 Hz		0.0074	0.0048-0.0380
1 MHz		0.032	0.0369-0.0622
Arc Resistance (Tungsten electrode), sec.	D495	100	75-98
HEAT RESISTANCE			
Max. Recommended Service Temperature, °F	—	175-190	100-125
Heat Deflection Temperature, °F	D648		
66 psi			
264 psi		230	
Flammability, inches per minute	D635	0.3-0.34	
Brittleness Temperature, °F	D746		90-155
Softening Point, Vicat, °F	D1525		

TABLE 8-1 (*Continued*)

FLUOROCARBONS:
POLYTRIFLUOROCHLOROETHYLENE (PTFCE) —
POLYTETRAFLUOROETHYLENE (PTFE) —
FLUORINATED ETHYLENE PROPYLENE (FEP) —
POLYVINYLIDENEFLUORIDE (PVF₂)
Molded, Extruded

PROPERTY	ASTM TEST	Polytrifluoro-chloroethylene (PTFCE)	Polytetrafluoro-ethylene (PTFE)	Fluorinated Ethylene Propylene (FEP)	Polyvinylidene-fluoride (PVF$_2$)
GENERAL PROPERTIES					
Specific Gravity	D792	2.10–2.15	2.1–2.3	2.14–2.17	1.77
Thermal Conductivity, Btu/hr/sq ft/°F/ft	C177	0.145	0.14	0.12	0.14
Thermal Expansion Coefficient, 10^{-6}/°F	D696	39	55	83–105	85
Specific Heat, Btu/lb/°F	—	0.22	0.25	0.28	0.33
Water Absorption in 24 hr, %	D570	0.00	0.01	0.01	0.03
White Light Transmission, 0.125 in. thick, %	D1003				
Haze, %	D672				
Refractive Index, n_D	D542	1.43	1.35	1.34	1.42
MECHANICAL PROPERTIES					
Tensile Strength, 1000 psi	D638	4.6–5.7	2.5–6.5	2.5–3.5	7.2–8.6
Modulus of Elasticity in Tension, 10^5 psi	D638	1.9–3.0	0.38–0.65	0.5–0.7	1.7–2
Elongation in 2 in., %	D638	125–175	250–350	250–330	200–300
Flexural Strength, 1000 psi	D790	10.7			2.0
Flexural Modulus, 10^5 psi	D790	2.54	0.807	0.95	8.7
Compressive Strength, 1000 psi	D790	4.6–7.4	1.7	2.2	3.8
Impact Strength (Izod notched), ft-lb/in.	D256	3.50–3.62	2.5–4.0	no break	
Hardness (Rockwell)–R	D785	R110–115	52D	58D	R110
Abrasion Resistance, (Taber CS-17 wheel), mg loss/1000 cycles	D1044		8.9	7.5	
ELECTRICAL PROPERTIES					
Volume Resistivity, ohm-cm	D257	1×10^{18}	$> 1 \times 10^{18}$	$> 2 \times 10^{18}$	5×10^{14}
Dielectric Strength (short time), volts/mil	D149	530–600	400–500	500–600	260
Dielectric Constant, 50% RH, 73°F	D150				
60 Hz		2.6–2.7	2.1	2.1	10.0
1 MHz		2.30–2.37	2.1	2.1	7.5
Dissipation Factor, 50% RH, 73°F	D150				
60 Hz		0.02	0.0002	0.0003	0.050
1 MHz		0.007–0.010	0.0002	0.0003	0.184
Arc Resistance (Tungsten electrode), sec.	D495	360	200	165	50–60
HEAT RESISTANCE					
Max. Recommended Service Temperature, °F	—	380	550	400	340
Heat Deflection Temperature, °F	D648				
66 psi		196–291	250	158	300
264 psi		151–178	132	124	232
Flammability, inches per minute	D635	noninflammable	noninflammable	noninflammable	self-ext.
Brittleness Temperature, °F	D746		−420	−420	<−80
Softening Point, Vicat, °F	D1525				

TABLE 8-1 (*Continued*)

PROPERTY	ASTM TEST	MELAMINE–FORMALDEHYDE – Molded Melamine, Unfilled Molded
GENERAL PROPERTIES		
Specific Gravity	D792	1.48
Thermal Conductivity, Btu/hr/sq ft/°F/ft	C177	
Thermal Expansion Coefficient, 10⁻⁶/°F	D696	
Specific Heat, Btu/lb/°F	–	
Water Absorption in 24 hr, %	D570	0.2–0.5
White Light Transmission, 0.125 in. thick, %	D1003	
Haze, %	D672	
Refractive Index, n_D	D542	
MECHANICAL PROPERTIES		
Tensile Strength, 1000 psi	D638	
Modulus of Elasticity in Tension, 10^5 psi	D638	
Elongation in 2 in., %	D638	
Flexural Strength, 1000 psi	D790	).5–14
Flexural Modulus, 10^5 psi	D790	10–13
Compressive Strength, 1000 psi	D790	40–45
Impact Strength (Izod notched), ft-lb/in.	D256	
Hardness (Rockwell)	D785	E110
Abrasion Resistance, (Taber CS-17 wheel), mg loss/1000 cycles	D1044	
ELECTRICAL PROPERTIES		
Volume Resistivity, ohm-cm	D257	
Dielectric Strength (short time), volts/mil	D149	
Dielectric Constant, 50% RH, 73°F	D150	
60 Hz		7.9–11.0
1 MHz		6.3–7.3
Dissipation Factor, 50% RH, 73°F	D150	
60 Hz		0.048–0.162
1 MHz		0.031–0.040
Arc Resistance (Tungsten electrode), sec.	D495	100–145
HEAT RESISTANCE		
Max. Recommended Service Temperature, °F	–	210
Heat Deflection Temperature, °F	D648	
66 psi		
264 psi		293–298
Flammability, inches per minute	D635	self-ext.
Brittleness Temperature, °F	D746	
Softening Point, Vicat, °F	D1525	

TABLE 8-1 (Continued)

OLEFINE–ESTER COPOLYMERS –
IONOMER:
ETHYLENE ETHYL ACETATE:
ETHYLENE ETHYL ACRYLATE:
ETHYLENE VINYL ACETATE:

PROPERTY	ASTM TEST	Ionomer	Ethylene Ethyl Acrylate	Ethylene Vinyl Acetate
GENERAL PROPERTIES				
Specific Gravity	D792	0.94	0.93	0.93
Thermal Conductivity, Btu/hr/sq ft/°F/ft	C177	0.141		
Thermal Expansion Coefficient, 10^{-6}/°F	D696	67	89–128	
Specific Heat, Btu/lb/°F	–	0.55	0.55	
Water Absorption in 24 hr, %	D570	0.1–1.4	0.04	
White Light Transmission, 0.125 in. thick, %	D1003	75–85		0–80
Haze, %	D672	3–17		2–40
Refractive Index, n_D	D542	1.51		
MECHANICAL PROPERTIES				
Tensile Strength, 1000 psi	D638	3.5–5.5	0.80–2.0	3.6
Modulus of Elasticity in Tension, 10^5 psi	D638	0.2–0.6	0.046–0.067	0.11
Elongation in 2 in., %	D638	350–450	300–700	650
Flexural Strength, 1000 psi (yield)	D790	no yield	3.0–3.6	0.01–0.20
Flexural Modulus, 10^5 psi	D790			
Compressive Strength, 1000 psi	D790			
Impact Strength (Izod notched), ft-lb/in.	D256	6.0–15.0	no break	no break
Hardness (Shore)	D785	D50–D65	D27–36	D36
Abrasion Resistance, (Taber CS-17 wheel), mg loss/1000 cycles	D1044			
ELECTRICAL PROPERTIES				
Volume Resistivity, ohm-cm	D257	$> 10^{16}$	2.4×10^9	1.5×10^8
Dielectric Strength (short time), volts/mil	D149	900–1100	450–550	525
Dielectric Constant, 50% RH, 73°F	D150			
60 Hz		2.4–2.5	2.7–2.9	2.5–3.16
1 MHz			2.7–2.8	2.6–3.2
Dissipation Factor, 50% RH, 73°F	D150			
60 Hz		0.001–0.003	0.01–0.02	0.003–0.020
1 MHz		0.0019	0.01–0.02	0.03–0.05
Arc Resistance (Tungsten electrode), sec.	D495	< 90		
HEAT RESISTANCE				
Max. Recommended Service Temperature, °F	–	160–220	190–200	140–147
Heat Deflection Temperature, °F	D648			
66 psi		100		93
264 psi		100–120		
Flammability, inches per minute	D635	1	very slow	very slow
Brittleness Temperature, °F	D746	< -160		-148
Softening Point, Vicat, °F	D1525	162		147

TABLE 8-1 *(Continued)*

MATERIAL →

PHENOLICS –
PHENOL FORMALDEHYDE

PROPERTY	ASTM TEST	Phenol-Formaldehyde (no filler)
GENERAL PROPERTIES		
Specific Gravity	D792	1.25–1.30
Thermal Conductivity, Btu/hr/sq ft/°F/ft	C177	0.087
Thermal Expansion Coefficient, 10^{-6}/°F	D696	14–33
Specific Heat, Btu/lb/°F	–	0.38–0.42
Water Absorption in 24 hr, %	D570	0.1–0.2
White Light Transmission, 0.125 in. thick, %	D1003	
Haze, %	D672	
Refractive Index, n_D	D542	1.5–1.7
MECHANICAL PROPERTIES		
Tensile Strength, 1000 psi	D638	7–8
Modulus of Elasticity in Tension, 10^5 psi	D638	7.5–10
Elongation in 2 in., %	D638	1.0–1.5
Flexural Strength, 1000 psi	D790	~ 10
Flexural Modulus, 10^5 psi	D790	~ 12
Compressive Strength, 1000 psi	D790	10–30
Impact Strength (Izod notched), ft-lb/in.	D256	0.2–0.36
Hardness (Rockwell)	D785	M124–M128
Abrasion Resistance, (Taber CS-17 wheel), mg loss/1000 cycles	D1044	
ELECTRICAL PROPERTIES		
Volume Resistivity, ohm-cm	D257	10^{11}–10^{12}
Dielectric Strength (short time), volts/mil	D149	300–400
Dielectric Constant, 50% RH, 73°F	D150	
60 Hz		5.0–6.5
1 MHz		4.5–5.0
Dissipation Factor, 50% RH, 73°F	D150	
60 Hz		0.06–0.10
1 MHz		0.015–0.03
Arc Resistance (Tungsten electrode), sec.	D495	tracks
HEAT RESISTANCE		
Max. Recommended Service Temperature, °F	–	250
Heat Deflection Temperature, °F	D648	
66 psi		
264 psi	D635	240–260
Flammability, inches per minute	D746	very slow
Brittleness Temperature, °F		
Softening Point, Vicat, °F	D1525	

TABLE 8-1 (Continued)

PROPERTY	ASTM TEST	Phenoxy		
MATERIAL →				
PHENOXY – Cast				
GENERAL PROPERTIES				
Specific Gravity	D792	1.17–1.18		
Thermal Conductivity, Btu/hr/sq ft/°F/ft	C177	0.102		
Thermal Expansion Coefficient, 10^{-6}/°F	D696	32–34		
Specific Heat, Btu/lb/°F	–	0.4		
Water Absorption in 24 hr, %	D570	0.13		
White Light Transmission, 0.125 in. thick, %	D1003			
Haze, %	D672			
Refractive Index, n_D	D542	1.5978		
MECHANICAL PROPERTIES				
Tensile Strength, 1000 psi	D638	8–9		
Modulus of Elasticity in Tension, 10^5 psi	D638	3.5–3.9		
Elongation in 2 in., %	D638	50–100		
Flexural Strength, 1000 psi (yield)	D790	12–13		
Flexural Modulus, 10^5 psi	D790	3.75–4.0		
Compressive Strength, 1000 psi	D790	10.4–12		
Impact Strength (Izod notched), ft-lb/in.	D256	1.5–2.0		
Hardness (Rockwell) –R	D785	115–123		
Abrasion Resistance, (Taber CS-17 wheel), mg loss/1000 cycles	D1044			
ELECTRICAL PROPERTIES				
Volume Resistivity, ohm-cm	D257	10^{10}–10^{13}		
Dielectric Strength (short time), volts/mil	D149	404–520		
Dielectric Constant, 50% RH, 73° F	D150			
60 Hz		4.1		
1 MHz		3.8		
Dissipation Factor, 50% RH, 73° F	D150			
60 Hz		0.0012		
1 MHz		0.03		
Arc Resistance (Tungsten electrode), sec.	D495			
HEAT RESISTANCE				
Max. Recommended Service Temperature, °F	–	170		
Heat Deflection Temperature, °F	D648			
66 psi		182–191		
264 psi		175–184		
Flammability, inches per minute	D635	slow to self-ext.		
Brittleness Temperature, °F	D746			
Softening Point, Vicat, °F	D1525			

TABLE 8-1 (*Continued*)

POLYACETALS —

PROPERTY	ASTM TEST	Homopolymer	Copolymer
GENERAL PROPERTIES			
Specific Gravity	D792	1.425	1.410
Thermal Conductivity, Btu/hr/sq ft/°F/ft	C177	0.13	0.16
Thermal Expansion Coefficient, 10^{-6}/°F	D696	45	47
Specific Heat, Btu/lb/°F	—	0.35	0.35
Water Absorption in 24 hr, %	D570	0.25	0.22
White Light Transmission, 0.125 in. thick, %	D1003		
Haze, %	D672		
Refractive Index, n_D	D542	1.48	
MECHANICAL PROPERTIES			
Tensile Strength, 1000 psi (at yield)	D638	10.0	8.8
Modulus of Elasticity in Tension, 10^5 psi	D638	5.2	4.1
Elongation in two inches, % (at yield)	D638	12	12
Flexural Strength, 1000 psi	D790	14.1	13
Flexural Modulus, 10^5 psi	D790	4.1	3.75
Compressive Strength, 1000 psi (1%)	D790	5.2	4.5
Impact Strength (Izod notched), ft-lb/in.	D256	1.4	1.2
Hardness (Rockwell)–R	D785	M94	M80
Abrasion Resistance, (Taber CS-17 wheel), mg loss/1000 cycles	D1044	14–20	14
ELECTRICAL PROPERTIES			
Volume Resistivity, ohm-cm	D257	1×10^{15}	1×10^{14}
Dielectric Strength (short time), volts/mil	D149	500	> 400
Dielectric Constant, 50% RH, 73°F	D150		
60 Hz		3.7	3.7 (100 Hz)
1 MHz		3.7	3.7
Dissipation Factor, 50% RH, 73°F	D150		
60 Hz		0.0048	0.001 (100 Hz)
1 MHz		0.0048	0.006
Arc Resistance (Tungsten electrode), sec.	D495	129	240
HEAT RESISTANCE			
Max. Recommended Service Temperature, °F	—	195	220
Heat Deflection Temperature, °F	D648		
66 psi		338	316
264 psi		255	230
Flammability, inches per minute	D635	1.1	1.1
Brittleness Temperature, °F	D746		
Softening Point, Vicat, °F	D1525		324

TABLE 8-1 (*Continued*)

POLYALLOMER –
POLYIMIDE –

PROPERTY	ASTM TEST	Polyallomer	Polyimide
GENERAL PROPERTIES			
Specific Gravity	D792	0.898–0.904	1.43
Thermal Conductivity, Btu/hr/sq ft/°F/ft	C177	0.048–0.096	0.19–0.22
Thermal Expansion Coefficient, 10^{-6}/°F	D696	50–100	28–34
Specific Heat, Btu/lb/°F	—	0.5	0.27
Water Absorption in 24 hr, %	D570	<0.01	0.32
White Light Transmission, 0.125 in. thick, %	D1003		
Haze, %	D672		
Refractive Index, n_D	D542	1.492	
MECHANICAL PROPERTIES			
Tensile Strength, 1000 psi	D638	3–4.3	11.0–14.5 (ult.)
Modulus of Elasticity in Tension, 10^5 psi	D638		3.6–3.85
Elongation in 2 in., %	D638	300–400	5.0–8.0 (ult.)
Flexural Strength, 1000 psi	D790		11.0–18.5 (ult.)
Flexural Modulus, 10^5 psi	D790	0.70–1.25	4.2–4.6
Compressive Strength, 1000 psi	D790		37–45 (50% stress)
Impact Strength (Izod notched), ft-lb/in.	D256	1.5	0.7–1.4
Hardness (Rockwell)–R	D785	R50–R85	E45–E48
Abrasion Resistance, (Taber CS-17 wheel), mg loss/1000 cycles	D1044		
ELECTRICAL PROPERTIES			
Volume Resistivity, ohm-cm	D257	>10^{15}	10^{16}–10^{17}
Dielectric Strength (short time), volts/mil	D149	500–650	560
Dielectric Constant, 50% RH, 73°F	D150		
60 Hz		2.3	3.41 (100 Hz)
1 MHz		2.3	3.4
Dissipation Factor, 50% RH, 73°F	D150		
60 Hz		<0.0005	0.0015 (100 Hz)
1 MHz		<0.0005	0.0069
Arc Resistance (Tungsten electrode), sec.	D495		230
HEAT RESISTANCE			
Max. Recommended Service Temperature, °F	—	230–280	500
Heat Deflection Temperature, °F	D648		
66 psi		165–192	
264 psi		124–133	~680
Flammability, inches per minute	D635	slow burn	non-burning
Brittleness Temperature, °F	D746	14 to –40	
Softening Point, Vicat, °F	D1525	252–277	

TABLE 8-1 (Continued)

POLYAMIDES —

NYLONS:
Type 6 and 6/6

PROPERTY	ASTM TEST	TYPE 6 General Purpose (dry, as-molded)	TYPE 6 Cast	Flexible Copolymer (dry, as molded)	TYPE 6/6 General Purpose
GENERAL PROPERTIES					
Specific Gravity	D792	1.14	1.15	1.12–1.14	1.15
Thermal Conductivity, Btu/hr/sq ft/°F/ft	C177	0.142		—	0.143
Thermal Expansion Coefficient, 10^{-6}/°F	D696	48	44	—	45
Specific Heat, Btu/lb/°F	—	0.4	0.4		0.4
Water Absorption in 24 hr, %	D570	1.6			1.5
White Light Transmission, 0.125 in. thick, %	D1003				
Haze, %	D672				
Refractive Index, n_D	D542				1.53
MECHANICAL PROPERTIES					
Tensile Strength, 1000 psi	D638	7.0–12.4	11–14	9.5	11.8
Modulus of Elasticity in Tension, 10^5 psi	D638	4.5	3.5–4.5	—	4.2
Elongation in 2 in., %	D638	100–320	10–60	200–330	60
Flexural Strength, 1000 psi	D790		16.5	3.4–16.4	
Flexural Modulus, 10^5 psi	D790	1.4–3.7	5.05	0.9–3.2	4.1
Compressive Strength, 1000 psi	D790	6.7–13.0	12	—	12.5
Impact Strength (Izod notched), ft-lb/in.	D256	0.8–1.2	1.2	1.5–19	0.9
Hardness (Rockwell)–R	D785	118–120	116	72–119	118
Abrasion Resistance, (Taber CS-17 wheel). mg loss/1000 cycles	D1044	5	2.7		
ELECTRICAL PROPERTIES					
Volume Resistivity, ohm-cm	D257	4.5×10^{13}	$(0.28\text{--}1.5) \times 10^{15}$	10^{15}	$10^{14}\text{--}10^{15}$
Dielectric Strength (short time), volts/mil	D149	385	380	440	385
Dielectric Constant, 50% RH, 73°F	D150				
60 Hz		4.0–5.3	4.0	3.2–4.0	4.0
1 MHz		3.6–3.8	3.3	3.0–3.6	3.6
Dissipation Factor, 50% RH, 73°F	D150				
60 Hz		0.06–0.014	0.015	0.007–0.010	0.014
1 MHz		0.03–0.04	0.05	0.010–0.015	0.04
Arc Resistance (Tungsten electrode), sec.	D495				
HEAT RESISTANCE					
Max. Recommended Service Temperature, °F	—	250–300	250–300	175–200	250–300
Heat Deflection Temperature, °F	D648				
66 psi		360	420	260–350	360
264 psi		155–160	410	115–130	150
Flammability, inches per minute	D635	self-ext.	self-ext.	0.6	self-ext.
Brittleness Temperature, °F	D746				–112
Softening Point, Vicat, °F	D1525				

TABLE 8-1 (*Continued*)

POLYAMIDES —

NYLONS:

Types 8, 11, 12, 6/10.

PROPERTY	ASTM TEST	Type 8 (non-cross-linked)	Type 11	Type 12	Type 6/10
GENERAL PROPERTIES					
Specific Gravity	D792	1.09	1.04	1.01	1.07–1.09
Thermal Conductivity, Btu/hr/sq ft/°F/ft	C177	—	0.125	0.142	0.125
Thermal Expansion Coefficient, 10^{-6}/°F	D696		55	72	50
Specific Heat, Btu/lb/°F	—		0.58	0.28	0.4
Water Absorption in 24 hr, %	D570	0.4	0.4	0.25	0.4
White Light Transmission, 0.125 in. thick, %	D1003	9.5			
Haze, %	D672				
Refractive Index, n_D	D542		1.52		
MECHANICAL PROPERTIES					
Tensile Strength, 1000 psi	D638		7.85	6.68–8.25	8.2
Modulus of Elasticity in Tension, 10^5 psi	D638	0.3	1.78–1.85	1.7–2.1	2.8
Elongation in 2 in., %	D638	400	300	120–350	240
Flexural Strength, 1000 psi	D790		7.8	7.0	
Flexural Modulus, 10^5 psi	D790	0.4	1.51	1.7–1.8	2.8
Compressive Strength, 1000 psi	D790		7.8	—	8.0
Impact Strength (Izod notched), ft-lb/in.	D256	—	3.3–3.6	1.2–4.2	1.2
Hardness (Rockwell)–R	D785	>16	100–108	106	111
Abrasion Resistance (Taber CS-17 wheel), mg loss/1000 cycles	D1044				
ELECTRICAL PROPERTIES					
Volume Resistivity, ohm-cm	D257	1.5×10^{11}	2×10^{13}	10^{14}–10^{15}	10^{14}–10^{15}
Dielectric Strength (short time), volts/mil	D149	340	425	407	470
Dielectric Constant, 50% RH, 73°F	D150		3.3 (1000 Hz)		
60 Hz		9.3	3.2	4.17	3.9
1 MHz		4.0		3.18	3.5
Dissipation Factor, 50% RH, 73°F	D150				
60 Hz		0.19	0.03	0.04	0.04
1 MHz		0.08	0.02	0.03	0.03
Arc Resistance (Tungsten electrode), sec.	D495			109	120
HEAT RESISTANCE					
Max. Recommended Service Temperature, °F	—		212–250	175–230	225–300
Heat Deflection Temperature, °F	D648				
66 psi			302		300
264 psi		129	131	120–131	135
Flammability, inches per minute	D635	self-ext.	self-ext.	slow burning	self-ext.
Brittleness Temperature, °F	D746				-166
Softening Point, Vicat, °F	D1525				

TABLE 8-1 *(Continued)*

POLYCARBONATE –

ABS – POLYCARBONATE –

PVC – ACRYLIC ALLOY

PROPERTY	ASTM TEST	Polycarbonate (molded)	ABS–Polycarbonate Alloy	PVC–Acrylic Alloy
GENERAL PROPERTIES				
Specific Gravity	D792	1.20	1.14	1.35
Thermal Conductivity, Btu/hr/sq ft/°F/ft	C177	0.111	0.145–0.218	0.084
Thermal Expansion Coefficient, 10^{-6}/°F	D696	39	61	35
Specific Heat, Btu/lb/°F	—	0.30	–	–
Water Absorption in 24 hr, %	D570	0.15	0.21	0.293
White Light Transmission, 0.125 in. thick, %	D1003	<89	–	0.06
Haze, %	D672	1–4	–	–
Refractive Index, n_D	D542	1.586	–	–
MECHANICAL PROPERTIES				
Tensile Strength, 1000 psi	D638	9.0–10.5	6.5–8.2	6.5 (yield)
Modulus of Elasticity in Tension, 10^5 psi	D638	3.4	2.9–3.8	3.35
Elongation in 2 in., %	D638	110	10–150	>100
Flexural Strength, 1000 psi	D790	11–13	14.3	10.7
Flexural Modulus, 10^5 psi	D790	3.4	4.0	4.0
Compressive Strength, 1000 psi	D790	11	11.1–11.8	8.4
Impact Strength (Izod notched), ft-lb/in.	D256	16	10	15
Hardness (Rockwell)–R	D785	R118	R118	R105
Abrasion Resistance, (Taber CS-17 wheel), mg loss/1000 cycles	D1044	7–11		0.073
ELECTRICAL PROPERTIES				
Volume Resistivity, ohm-cm	D257	2.1×10^{16}	2.2×10^{16}	1.5×10^{13}
Dielectric Strength (short time), volts/mil	D149	400	500	>429
Dielectric Constant, 50% RH, 73°F	D150			
60 Hz		3.17	2.74	3.86
1 MHz		2.96	2.69	3.44
Dissipation Factor, 50% RH, 73°F	D150			
60 Hz		0.0009	0.0026	0.076
1 MHz		0.010	0.0059	0.094
Arc Resistance (Tungsten electrode), sec.	D495	120	96	80
HEAT RESISTANCE				
Max. Recommended Service Temperature, °F	—	250	240	160
Heat Deflection Temperature, °F	D648			
66 psi		285	261	177
264 psi		265	246	160
Flammability, inches per minute	D635	self-ext.	0.90	nonburning
Brittleness Temperature, °F	D746			
Softening Point, Vicat, °F	D1525			

TABLE 8-1 (*Continued*)

POLYESTER — PROPERTY	ASTM TEST	CAST POLYESTER Rigid	Resilient
GENERAL PROPERTIES			
Specific Gravity	D792	1.12–1.46	1.06–1.25
Thermal Conductivity, Btu/hr/sq ft/°F/ft	C177	0.096	
Thermal Expansion Coefficient, 10^{-6}/°F	D696	39–56	
Specific Heat, Btu/lb/°F	—	0.30–0.55	0.12–2.5
Water Absorption in 24 hr, %	D570	0.20–0.60	
White Light Transmission, 0.125 in. thick, %	D1003		
Haze, %	D672		
Refractive Index, n_D	D542	1.53–1.58	1.50–1.57
MECHANICAL PROPERTIES			
Tensile Strength, 1000 psi	D638	4–10	1–8
Modulus of Elasticity in Tension, 10^5 psi	D638	1.5–6.5	0.001–0.10
Elongation in 2 in., %	D638	1.7–2.6	25–300
Flexural Strength, 1000 psi	D790	14–18	4–16
Flexural Modulus, 10^5 psi	D790	1–9	0.001–0.39
Compressive Strength, 1000 psi	D790	12–37	1–17
Impact Strength (Izod notched), ft-lb/in.	D256	0.18–0.40	4.0
Hardness (Barcol)	D785	35–50	6–40
Abrasion Resistance, (Taber CS-17 wheel), mg loss/1000 cycles	D1044		
ELECTRICAL PROPERTIES			
Volume Resistivity, ohm-cm	D257	10^{13}	10^{12}
Dielectric Strength (short time), volts/mil	D149	300–400	300–400
Dielectric Constant, 50% RH, 73° F	D150		
60 Hz		2.8–4.4	3.18–7.0
1 MHz		2.8–4.4	3.7–6.1
Dissipation Factor, 50% RH, 73° F	D150		
60 Hz		0.003–0.04	0.01–0.18
1 MHz		0.006–0.04	0.02–0.06
Arc Resistance (Tungsten electrode), sec.	D495	115–135	125–145
HEAT RESISTANCE			
Max. Recommended Service Temperature, °F	—	250–300	150–250
Heat Deflection Temperature, °F	D648		
66 psi			
264 psi		120–400	
Flammability, inches per minute	D635	0.87 to self-ext.	slow burn to self-ext.
Brittleness Temperature, °F	D746		
Softening Point, Vicat, °F	D1525		

TABLE 8-1 (*Continued*)

MATERIAL → POLYETHYLENES (molded, extruded) PROPERTY	ASTM TEST	POLYETHYLENE Low Density	Medium Density	High Density	High Molecular Weight
GENERAL PROPERTIES					
Specific Gravity	D792	0.910-0.925	0.926-0.940	0.941-0.965	0.94
Thermal Conductivity, Btu/hr/sq ft/°F/ft	C177	0.193	0.193-0.242	0.266-0.300	0.19
Thermal Expansion Coefficient, 10^{-6}/°F	D696	89-110	83-167	83-167	
Specific Heat, Btu/lb/°F	—	0.53-0.55	0.53-0.55	0.46-0.55	
Water Absorption in 24 hr, %	D570	<0.015	<0.01	<0.01	<0.01
White Light Transmission, 0.125 in. thick, %	D1003	0-75	10-80	0-40	
Haze, %	D672	4-50	2-40	10-50	
Refractive Index, n_D	D542	1.51	1.52	1.54	
MECHANICAL PROPERTIES					
Tensile Strength, 1000 psi	D638	0.9-2.5	2.0-2.4	2.9-4.4	5.4
Modulus of Elasticity in Tension, 10^5 psi	D638	0.14-0.38	0.25-0.55	0.6-1.8	1.0
Elongation in 2 in., %	D638	80-725	200-425	100-700	400
Flexural Strength, 1000 psi (yield)	D790		4.8-7.0		
Flexural Modulus, 10^5 psi	D790	0.10-0.30	0.35-0.50	0.90-1.5	0.75
Compressive Strength, 1000 psi	D790			2.7-3.6	
Impact Strength (Izod notched), ft-lb/in.	D256	no break	0.5->16	0.4-14	>20
Hardness (Shore)	D785	D45-D53	D55-D56	D60-D70	D60-D65
Abrasion Resistance, (Taber CS-17 wheel), mg loss/1000 cycles	D1044				
ELECTRICAL PROPERTIES					
Volume Resistivity, ohm-cm	D257	10^{17}-10^{19}	>10^{15}	>10^{15}	>10^{15}
Dielectric Strength (short time), volts/mil	D149	480	480	480	480
Dielectric Constant, 50% RH, 73° F	D150				
60 Hz		2.25-2.35	2.25-2.35	2.30-2.35	
1 MHz		2.25-2.35	2.25-2.35	2.30-2.35	2.30
Dissipation Factor, 50% RH, 73° F	D150				
60 Hz		<0.0005	<0.0005	<0.0005	
1 MHz		<0.0005	<0.0005	<0.0005	0.0002
Arc Resistance (Tungsten electrode), sec.	D495				
HEAT RESISTANCE					
Max. Recommended Service Temperature, °F	—	180-212	220-250	250	163
Heat Deflection Temperature, °F	D648				117
66 psi		100-121	120-165	140-190	
264 psi		90-105	105-120	110-130	
Flammability, inches per minute	D635	1.04	1.00-1.04	1.00-1.04	1.0
Brittleness Temperature, °F	D746	<-94 to +14	-148	<-76 to -180	<-100
Softening Point, Vicat, °F	D1525	176-201	215-235	240-266	277

TABLE 8-1 (*Continued*)

POLYPHENYLENE OXIDES —
POLYPHENYLENE SULFIDE —

PROPERTY	ASTM TEST	POLYPHENYLENE OXIDE		POLYPHENYLENE SULFIDE
		Polyphenylene Oxides	Modified Polyphenylene Oxides (Noryl)	Polyphenylene Sulfide
GENERAL PROPERTIES				
Specific Gravity	D792	1.06	1.06	1.34
Thermal Conductivity, Btu/hr/sq ft/°F/ft	C177	0.11	0.125	0.166
Thermal Expansion Coefficient, 10^{-6}/°F	D696	29	33	30
Specific Heat, Btu/lb/°F	—		0.32	0.26
Water Absorption in 24 hr, %	D570	0.03	0.07	
White Light Transmission, 0.125 in. thick, %	D1003			
Haze, %	D672			
Refractive Index, n_D	D542			
MECHANICAL PROPERTIES				
Tensile Strength, 1000 psi	D638	7	7.8-9.6	11 (yield)
Modulus of Elasticity in Tension, 10^5 psi	D638	3.9	3.5-3.8	4.8
Elongation in 2 in., %	D638		20-30	3 (break)
Flexural Strength, 1000 psi	D790	16.5	12.8-13.5	20 (yield)
Flexural Modulus, 10^5 psi	D790	3.75	3.6-4.0	6.0
Compressive Strength, 1000 psi	D790	16.5	16-16.4	
Impact Strength (Izod notched), ft-lb/in.	D256	1.2	1.7-1.8	
Hardness (Rockwell)—R	D785	119	118	124
Abrasion Resistance, (Taber CS-17 wheel), mg loss/1000 cycles	D1044	17	20	
ELECTRICAL PROPERTIES				
Volume Resistivity, ohm-cm	D257	10^{18}	10^{16}-10^{17}	
Dielectric Strength (short time), volts/mil	D149	500	400-550	595
Dielectric Constant, 50% RH, 73°F	D150			
60 Hz		2.58	2.65	
1 MHz		2.58	2.64	3.22
Dissipation Factor, 50% RH, 73°F	D150			
60 Hz		0.00035	0.0004	
1 MHz		0.0009	0.0009	0.0007
Arc Resistance (Tungsten electrode), sec.	D495	75	75	
HEAT RESISTANCE				
Max. Recommended Service Temperature, °F	—	225	212	500
Heat Deflection Temperature, °F	D648			
66 psi		355	279	
264 psi		345	265	278
Flammability, inches per minute	D635	self-ext.	self-ext.	non-burn
Brittleness Temperature, °F	D746			
Softening Point, Vicat, °F	D1525			

TABLE 8-1 (*Continued*)

POLYPROPYLENE –
POLY (TETRAMETHYLENE TEREPHTHALATE)

PROPERTY	ASTM TEST	POLYPROPYLENE		POLY (Tetramethylene terephthalate) Polyterephthalate® (Eastman Kodak Co.)
		General Purpose	High Impact	
GENERAL PROPERTIES				
Specific Gravity	D792	0.900–0.910	0.900–0.910	1.31
Thermal Conductivity, Btu/hr/sq ft/°F/ft	C177	0.10–0.113	0.144	
Thermal Expansion Coefficient, 10^{-6}/°F	D696	38–58	40–59	49–130
Specific Heat, Btu/lb/°F	—	0.45	0.45–0.48	0.36–0.55
Water Absorption in 24 hr, %	D570	<0.01–0.03	<0.01–0.02	0.09
White Light Transmission, 0.125 in. thick, %	D1003			
Haze, %	D672			
Refractive Index, n_D	D542	1.49		
MECHANICAL PROPERTIES				
Tensile Strength, 1000 psi	D638	4.8–5.2	2.8–4.3	8.2
Modulus of Elasticity in Tension, 10^5 psi	D638	1.6–2.2	1.3	
Elongation in two inches, % (at yield)	D638	9–15	7–13	
Flexural Strength, 1000 psi (yield)	D790	6–7	4.1	12 (not yield)
Flexural Modulus, 10^5 psi	D790	1.7–2.5	1.0–2.0	3.1
Compressive Strength, 1000 psi (yield)	D790	5.5–6.5	4.4	
Impact Strength (Izod notched), ft-lb/in.	D256	0.4–2.2	1.5–12	1.0
Hardness (Rockwell)–R	D785	80–100	28–95	117
Abrasion Resistance, (Taber CS-17 wheel), mg loss/1000 cycles	D1044	25		30 (500 g. load)
ELECTRICAL PROPERTIES				
Volume Resistivity, ohm-cm	D257	$>10^{17}$	10^{17}	
Dielectric Strength (short time), volts/mil	D149	650 (125 mil)	450–650	420
Dielectric Constant, 50% RH, 73° F	D150			
60 Hz		2.20–2.28	2.20–2.28	3.16
1 MHz		2.23–2.24	2.23–2.27	
Dissipation Factor, 50% RH, 73°F	D150			
60 Hz		0.0005–0.0007	<0.0016	0.023
1 MHz		0.0002–0.0003	0.0002–0.0003	
Arc Resistance (Tungsten electrode), sec.	D495	125–136	123–140	
HEAT RESISTANCE				
Max. Recommended Service Temperature, °F	—	230–300	190–240	302
Heat Deflection Temperature, °F	D648			
66 psi		205–230	190–235	
264 psi		135–140	120–140	122
Flammability, inches per minute	D635	0.7–1	1	1.27
Brittleness Temperature, °F	D746			
Softening Point, Vicat, °F	D1525	–4.0		

TABLE 8-1 (Continued)

POLYSTYRENES –
STYRENE ACRYLONITRILE –

PROPERTY	ASTM TEST	POLYSTYRENES General Purpose	POLYSTYRENES Medium Impact	POLYSTYRENES High Impact	STYRENE ACRYLO-NITRILE Styrene Acrylo-Nitrile (SAN)
GENERAL PROPERTIES					
Specific Gravity	D792	1.04	1.04-1.07	1.04-1.07	1.04-1.07
Thermal Conductivity, Btu/hr/sq ft/°F/ft	C177	0.058-0.090	0.024-0.090	0.024-0.090	
Thermal Expansion Coefficient, 10^{-6}/°F	D696	33-48	33-47	22-56	3.6-3.7
Specific Heat, Btu/lb/°F	–	0.30-0.35	0.30-0.35	0.30-0.35	0.33
Water Absorption in 24 hr, %	D570	0.03-0.2	0.03-0.09	0.05-0.22	0.20-0.35
White Light Transmission, 0.125 in. thick, %	D1003	88-92	35-55		78-88
Haze, %	D672	<0.1-3.0	77		
Refractive Index, n_D	D542	1.60	1.59	1.58	1.565-1.569
MECHANICAL PROPERTIES					
Tensile Strength, 1000 psi	D638	5-12	3.5-6.8	3.0-5.5	9-12
Modulus of Elasticity in Tension, 10^5 psi	D638	4.6-5.0	3.9-4.7	1.50-3.80	4.0-5.0
Elongation in 2 in., %	D638	1.0-2.5			1.5-3.7
Flexural Strength, 1000 psi (yield)	D790	10-15			14-19
Flexural Modulus, 10^5 psi	D790	4-5	3.5-5.0	2.3-4.0	<5.5
Compressive Strength, 1000 psi	D790	11.5-16.0	4-9	4-9	14-17
Impact Strength (Izod notched), ft-lb/in.	D256	0.2-0.4	0.5-0.7	0.8-1.8	0.30-0.45
Hardness (Rockwell)–M	D785	72	47-65	3-43	80-85
Abrasion Resistance, (Taber CS-17 wheel), mg loss/1000 cycles	D1044				
ELECTRICAL PROPERTIES					
Volume Resistivity, ohm-cm	D257	$>10^{16}$	$>10^{16}$	$>10^{16}$	$>10^{16}$
Dielectric Strength (short time), volts/mil	D149	>500	>425	300-650	400-500
Dielectric Constant, 50% RH, 73°F	D150				
60 Hz		2.45-2.65	2.45-4.75	2.45-4.75	2.6-3.4
1 MHz		2.45-2.65	2.4-3.8	2.5-4.0	2.6-3.0
Dissipation Factor, 50% RH, 73°F	D150				
60 Hz		0.0001-0.0003	0.0004-0.002	0.0004-0.002	>0.006
1 MHz		0.0001-0.0005	0.0004-0.002	0.0004-0.002	0.007-0.010
Arc Resistance (Tungsten electrode), sec.	D495	60-135	20-135	20-100	100-150
HEAT RESISTANCE					
Max. Recommended Service Temperature, °F	–	160-205	125-165	125-165	175-190
Heat Deflection Temperature, °F	D648				
66 psi		220 max	210 max	210 max	190-220
264 psi		220 max			210-220
Flammability, inches per minute	D635	1.0-1.5	0.5-2.0	0.5-1.5	0.8
Brittleness Temperature, °F	D746				
Softening Point, Vicat, °F	D1525	195	210	195	224

TABLE 8-1 (*Continued*)

PROPERTY	ASTM TEST	Polyvinyl Alcohol	Polyvinyl Butyral	Polyvinyl Formal
POLYVINYL ALCOHOL –				
POLYVINYL BUTYRAL –				
POLYVINYL FORMAL –				
GENERAL PROPERTIES				
Specific Gravity	D792	1.21–1.31	1.08–1.12	1.20–1.25
Thermal Conductivity, Btu/hr/sq ft/°F/ft	C177	0.46		0.089
Thermal Expansion Coefficient, 10^{-6}/°F	D696	3.88–6.55	4.4–12.7	3.55–4.27
Specific Heat, Btu/lb/°F	–	0.3	0.4	
Water Absorption in 24 hr, %	D570	> 30	0.3–0.6	1.0–1.3
White Light Transmission, 0.125 in. thick, %	D1003			
Haze, %	D672			
Refractive Index, n_D	D542	1.49–1.53	1.48–1.49	1.49–1.505
MECHANICAL PROPERTIES				
Tensile Strength, 1000 psi	D638	1–5	4–8.5	9–11
Modulus of Elasticity in Tension, 10^5 psi	D638		3.5–4.0	5–7
Elongation in 2 in., %	D638	50–250	5–60	5–60
Flexural Strength, 1000 psi	D790		10	13–18
Flexural Modulus, 10^5 psi	D790			50–60
Compressive Strength, 1000 psi	D790			
Impact Strength (Izod notched), ft-lb/in.	D256		1.2	0.4–2.0
Hardness (Rockwell)	D785	10–100 (Shore A)	L95	M80–90
Abrasion Resistance, (Taber CS-17 wheel), mg loss/1000 cycles	D1044			
ELECTRICAL PROPERTIES				
Volume Resistivity, ohm-cm	D257		$> 10^{14}$	
Dielectric Strength (short time), volts/mil	D149		400–480	310–450
Dielectric Constant, 50% RH, 73°F	D150			
60 Hz			2.7–3.3	3.2–3.5
1 MHz			2.6–2.8	2.7–3.1
Dissipation Factor, 50% RH, 73°F	D150			
60 Hz			0.0050–0.0065	0.008–0.010
1 MHz			0.013–0.027	0.018–0.023
Arc Resistance (Tungsten electrode), sec.	D495			
HEAT RESISTANCE				
Max. Recommended Service Temperature, °F	–			
Heat Deflection Temperature, °F	D648		115	130–165
66 psi				
264 psi				
Flammability, inches per minute	D635			
Brittleness Temperature, °F	D746			
Softening Point, Vicat, °F	D1525			

TABLE 8-1 (Continued)

POLYVINYL CHLORIDE and COPOLYMERS – Molded, Extruded

PROPERTY	ASTM TEST	POLYVINYL CHLORIDE, POLYVINYL CHL. ACETATE			VINYLIDENE CHLORIDE
		Nonrigid-General	Nonrigid-Electrical	Rigid-Normal Impact	Vinylidene Chloride*
GENERAL PROPERTIES					
Specific Gravity	D792	1.20–1.55	1.16–1.40	1.32–1.44	1.69–1.75
Thermal Conductivity, Btu/hr/sq ft/°F/ft	C177	0.07–0.10	0.07–0.10	0.07–0.10	0.053
Thermal Expansion Coefficient, 10^{-6}/°F	D696			28–33	87.8
Specific Heat, Btu/lb/°F	—				0.32
Water Absorption in 24 hr, %	D570	0.2–1.0	0.40–0.75	0.03–0.40	> 0.1
White Light Transmission, 0.125 in. thick, %	D1003				
Haze, %	D672				
Refractive Index, n_D	D542			1.52–1.55	1.60–1.63
MECHANICAL PROPERTIES					
Tensile Strength, 1000 psi	D638	1–3.5	2–3.2	5.5–8	4–8, 15–40
Modulus of Elasticity in Tension, 10^5 psi	D638	0.004–0.03	0.01–0.03	3.5–6.0	0.7–2.0
Elongation in 2 in., %	D638	200–450	220–360	1–10	15–25, 20–30
Flexural Strength, 1000 psi	D790			11–16	15–17, flexible
Flexural Modulus, 10^5 psi	D790			3.8–5.4	
Compressive Strength, 1000 psi	D790			11–12	
Impact Strength (Izod notched), ft-lb/in.	D256	variable	variable	0.5–10	2–8, 0.053
Hardness (Shore)	D785	A50–100	A78–100	D70–85	> A 95
Abrasion Resistance, (Taber CS-17 wheel), mg loss/1000 cycles	D1044				
ELECTRICAL PROPERTIES					
Volume Resistivity, ohm-cm	D257	$(1-700) \times 10^{12}$	$(4-300) \times 10^{11}$	$10^{14} - > 10^{16}$	$10^{14} - 10^{16}$
Dielectric Strength (short time), volts/mil	D149		24–500	725–1400	400–600
Dielectric Constant, 50% RH, 73°F	D150				
60 Hz		5.5–9.1	6.0–8.0	3.2–3.6	4.5–6.0
1 MHz				2.8–3.1	3.0–4.0
Dissipation Factor, 50% RH, 73°F	D150				
60 Hz		0.05–0.15	0.08–0.11	0.007–0.020	0.030–0.045
1 MHz				0.006–0.019	0.05–0.08
Arc Resistance (Tungsten electrode), sec.	D495			60–80	
HEAT RESISTANCE					
Max. Recommended Service Temperature, °F	—	150–220	140–220	150–165	170–212
Heat Deflection Temperature, °F	D648				
66 psi				170–185	190–210
264 psi				140–170	130–150
Flammability, inches per minute	D635	self-ext.	self-ext.	self-ext.	self-ext.
Brittleness Temperature, °F	D746				
Softening Point, Vicat, °F	D1525				240–280

*Where two values or ranges are given, they represent unoriented and oriented forms, respectively.

TABLE 8-1 *(Continued)*

MATERIAL →			
POLYSULFONE –			
POLYARYLSULFONE –			
PROPERTY	ASTM TEST	Polysulfone	Polyarylsulfone
GENERAL PROPERTIES			
Specific Gravity	D792	1.24	1.36
Thermal Conductivity, Btu/hr/sq ft/°F/ft	C177	0.15	
Thermal Expansion Coefficient, 10^{-6}/°F	D696	31	26
Specific Heat, Btu/lb/°F	–	0.24	
Water Absorption in 24 hr, %	D570	0.22	
White Light Transmission, 0.125 in. thick, %	D1003		
Haze, %	D672	5.0	
Refractive Index, n_D	D542	1.63	
MECHANICAL PROPERTIES			
Tensile Strength, 1000 psi	D638	10.2 (yield)	13
Modulus of Elasticity in Tension, 10^5 psi	D638	3.6	3.7
Elongation in two inches, % (at yield)	D638	5.6	13
Flexural Strength, 1000 psi	D790	15.5	17.2
Flexural Modulus, 10^5 psi	D790	3.9	4.0
Compressive Strength, 1000 psi	D790	13.9	17.8
Impact Strength (Izod notched), ft-lb/in.	D256	1.3	5.0
Hardness (Rockwell)	D785	R120	M110
Abrasion Resistance, (Taber CS-17 wheel), mg loss/1000 cycles	D1044	20	40
ELECTRICAL PROPERTIES			
Volume Resistivity, ohm-cm	D257	5×10^{16}	3.2×10^{16}
Dielectric Strength (short time), volts/mil	D149	425	350
Dielectric Constant, 50% RH, 73°F	D150		
60 Hz		3.06	3.94
1 MHz		3.03	3.7
Dissipation Factor, 50% RH, 73°F	D150		
60 Hz		0.0008	0.003
1 MHz		0.0034	0.012
Arc Resistance (Tungsten electrode), sec.	D495	122	67
HEAT RESISTANCE			
Max. Recommended Service Temperature, °F	–	340	500
Heat Deflection Temperature, °F	D648		
66 psi		358	
264 psi		345	525
Flammability, inches per minute	D635	self-ext.	self-ext.
Brittleness Temperature, °F	D746		
Softening Point, Vicat, °F	D1525		

TABLE 8-1 (*Continued*)

SILICONES			
	MATERIAL →		Silicone Cast Resin, Flexible
PROPERTY		ASTM TEST	
GENERAL PROPERTIES			
Specific Gravity		D792	0.99–1.50
Thermal Conductivity, Btu/hr/sq ft/°F/ft		C177	0.085–0.181
Thermal Expansion Coefficient, 10^{-6}/°F		D696	139–167
Specific Heat, Btu/lb/°F		–	
Water Absorption in 24 hr, %		D570	0.12 (7 days)
White Light Transmission, 0.125 in. thick, %		D1003	
Haze, %		D672	
Refractive Index, n_D		D542	1.43
MECHANICAL PROPERTIES			
Tensile Strength, 1000 psi		D638	0.35–1.0
Modulus of Elasticity in Tension, 10^5 psi		D638	900
Elongation in 2 in., %		D638	100–300
Flexural Strength, 1000 psi		D790	
Flexural Modulus, 10^5 psi		D790	
Compressive Strength, 1000 psi		D790	
Impact Strength (Izod notched), ft-lb/in.		D256	
Hardness (Shore)		D785	A15–65
Abrasion Resistance, (Taber CS-17 wheel), mg loss/1000 cycles		D1044	
ELECTRICAL PROPERTIES			
Volume Resistivity, ohm-cm		D257	2.0×10^{15}
Dielectric Strength (short time), volts/mil		D149	550
Dielectric Constant, 50% RH, 73°F		D150	
60 Hz			2.75–4.20
1 MHz			2.6–2.7
Dissipation Factor, 50% RH, 73°F		D150	
60 Hz			0.001–0.025
1 MHz			0.001–0.002
Arc Resistance (Tungsten electrode), sec.		D495	115–130
HEAT RESISTANCE			
Max. Recommended Service Temperature, °F		–	500
Heat Deflection Temperature, °F		D648	
66 psi			
264 psi			
Flammability, inches per minute		D635	self-ext.
Brittleness Temperature, °F		D746	
Softening Point, Vicat, °F		D1525	

TABLE 8-1 (*Continued*)

MATERIAL →

URETHANE –
Molded

PROPERTY	ASTM TEST	Cast Liquid Urethane
GENERAL PROPERTIES		
Specific Gravity	D792	1.10–1.50
Thermal Conductivity, Btu/hr/sq ft/°F/ft	C177	0.12
Thermal Expansion Coefficient, 10^{-6}/°F	D696	56–112
Specific Heat, Btu/lb/°F	—	0.42–0.44
Water Absorption in 24 hr, %	D570	0.02–1.5
White Light Transmission, 0.125 in. thick, %	D1003	
Haze, %	D672	
Refractive Index, n_D	D542	1.50–1.60
MECHANICAL PROPERTIES		
Tensile Strength, 1000 psi	D638	0.175–10
Modulus of Elasticity in Tension, 10^5 psi	D638	1.0–10.0
Elongation in 2 in., %	D638	100–1000
Flexural Strength, 1000 psi	D790	
Flexural Modulus, 10^5 psi	D790	0.1–1.0
Compressive Strength, 1000 psi	D790	20
Impact Strength (Izod notched), ft-lb/in.	D256	5.0 to flex.
Hardness (Shore)	D785	10A–80D
Abrasion Resistance, (Taber CS-17 wheel), mg loss/1000 cycles	D1044	
ELECTRICAL PROPERTIES		
Volume Resistivity, ohm-cm	D257	2.0×10^{11}–10^{15}
Dielectric Strength (short time), volts/mil	D149	400–500
Dielectric Constant, 50% RH, 73°F	D150	
60 Hz		4.0–7.5
1 MHz		6.5–7.1
Dissipation Factor, 50% RH, 73°F	D150	
60 Hz		0.015–0.017
1 MHz		0.1–0.6
Arc Resistance (Tungsten electrode), sec.	D495	
HEAT RESISTANCE		
Max. Recommended Service Temperature, °F	—	190–250
Heat Deflection Temperature, °F	D648	
66 psi		varies over
264 psi		wide range
Flammability, inches per minute	D635	slow to self-ext.
Brittleness Temperature, °F	D746	
Softening Point, Vicat, °F	D1525	

TABLE 8-2. PROPERTY OF PLASTIC FILMS. (MATERIALS SELECTOR)

PLASTIC FILMS	ASTM TEST	Cellulosics			Polypropylene
PROPERTY*		Cellulose Acetate	Cellulose Acetate Butyrate	Cellulose Triacetate	Polypropylene (unoriented)
GENERAL PROPERTIES					
Specific Gravity	D792	1.29	1.19	1.29	0.885–0.90
Water Absorption in 24 hr, %	D570	4–6	1–2	2–4	<0.005
MECHANICAL PROPERTIES					
Tensile Strength, 1000 psi	D882	13–15	7–9	12–15	4–10
Burst Strength (Mullen), psi	D774	40–60	40–70	50–70	
Tearing Strength (Elmendorf), gm/mil	D689	5–10	5–10	5–10	20–100
Folding Endurance	D643	500–600	800–1200	1000–2000	excellent
ELECTRICAL PROPERTIES					
Volume Resistivity, ohm-cm	—	7.0×10^{13}	1.4×10^{15}	1.4×10^{15}	3×10^{15}
Dielectric Strength (short time), V/mil	D149	3200	3100	3700	3000–4500
Dielectric Constant	D150				
1000 Hz		3.6	2.9	3.2	2.0–2.1
1 MHz		3.2	2.5	3.3	2.0–2.1
Dissipation Factor	D150				
1000 Hz		0.013	0.013	0.016	0.0003
1 MHz		0.038	0.044	0.033	0.0003
ENVIRONMENTAL PROPERTIES					
Max. Recommended Temperature, °F	—	150–200	120–180	300–400	275–300
Min. Recommended Temperature, °F	—	−15	—	—	0
Heat Sealing Temperature Range, °F	—	350–450	—	—	285–400
Burning Rate, In./sec.	—	0.2–2.2	0.2–1.2	0.2–0.4	slow

*0.001 in. thickness.

TABLE 8-2 (*Continued*)

PLASTIC FILMS			Fluorocarbon		
PROPERTY	ASTM TEST	Polytrifluoro-chloroethylene	Fluorinated Ethylene Propylene	Polyvinyl Fluoride	Polyvinylidene Fluoride
GENERAL PROPERTIES					
Specific Gravity	D792	2.1	2.15	1.5	1.79–1.80
Water Absorption in 24 hr, %	D570	nil	0.01	0.05	0.03
MECHANICAL PROPERTIES					
Tensile Strength, 1000 psi	D882	5–8	2.5–4.0	7–18	6–8
Burst Strength (Mullen), psi	D774	23–31	10–15	19–70	
Tearing Strength (Elmendorf), gm/mil	D689	10–26	100–150	12–100	100–500
Folding Endurance, cycles	D643		4000	5000–47000	500000
ELECTRICAL PROPERTIES					
Volume Resistivity, ohm-cm	—	10^{18}	10^{19}	3×10^{13}	2×10^{14}
Dielectric Strength at 60 Hz, V/mil	D149	1000–3700	5000	3500	260–1280
Dielectric Constant	D150				
1000 Hz		2.5–2.7	2.0	8.5	7.72
1 MHz		2.4–2.4	2.0	7.4	6.43
Dissipation Factor	D150				
1000 Hz		0.022–0.024	0.0002	1.6	0.019
1 MHz		0.009–0.017	0.0007		0.159
ENVIRONMENTAL PROPERTIES					
Max. Recommended Temperature, °F	—	300–390	400	225	300
Min. Recommended Temperature, °F	—	–320	–400	–100	–76
Heat Sealing Temperature Range, °F	—	370–500	600–700	400–425	360–500
Burning Rate, in./sec.	—	non-flammable	non-flammable	slow	non-flammable

TABLE 8-2 (Continued)

PLASTIC FILMS		Fluoro-Carbon	Polyethylene		
		Polytetra-fluoroethylene	Type I	Type II	Type III
PROPERTY	ASTM TEST				
GENERAL PROPERTIES					
Specific Gravity	D792	2.1–2.2	0.92	0.935–0.938	0.940–0.945
Water Absorption in 24 hr, %	D570	nil	0.01	<0.01	nil
MECHANICAL PROPERTIES					
Tensile Strength, 1000 psi	D882	1.5–4.0	1.6–3.0	2.5–3.5	3.5–8.0
Burst Strength (Mullen), psi	D774		10–12		
Tearing Strength (Elmendorf), gm/mil	D689	10–100	100–500	50–300	15–300
Folding Endurance	D643		good	good	good
ELECTRICAL PROPERTIES					
Volume Resistivity, ohm-cm	—	10^{18}	10^{16}	10^{16}	10^{16}
Dielectric Strength at 60 Hz, V/mil	D149	430	450	450	500
Dielectric Constant	D150				
1000 Hz		2.0–2.1	2.2	2.2	2.3
1 MHz		2.0–2.1	2.2	2.2	2.3
Dissipation Factor	D150				
1000 Hz		0.0002	0.0003	0.0003	0.0005
1 MHz		0.0002	0.0003	0.0003	0.0005
ENVIRONMENTAL PROPERTIES					
Max. Recommended Temperature, °F	—	500	180	320	250
Min. Recommended Temperature, °F	—	<-130	-70	<-100	<-100
Heat Sealing Temperature Range, °F	—		400–450	250–375	250–375
Burning Rate, in./sec.	—	self-ext.	slow	slow	slow

TABLE 8-2 (*Continued*)

PLASTIC FILMS		(Polyamide)	Nylon		
PROPERTY	ASTM TEST	Type 6	6/6	6/10	12
GENERAL PROPERTIES					
Specific Gravity	D792	1.12	1.14	1.11	1.01
Water Absorption in 24 hr, %	D570	8.0	1.5	0.4	0.25
MECHANICAL PROPERTIES					
Tensile Strength, 1000 psi	D882	9–13	12	10	7–9
Burst Strength (Mullen), psi	D774				
Tearing Strength (Elmendorf), gm/mil	D689	50	50	70	
Folding Endurance	D643	Exc	Exc	Exc	Exc
ELECTRICAL PROPERTIES					
Volume Resistivity, ohm-cm	—		4.5×10^{13}		2.5×10^{13}
Dielectric Strength at 60 Hz, V/mil	D149	480	385	470	
Dielectric Constant	D150				
1000 Hz		4.8 (60 Hz)	4.0 (60 Hz)	3.6 (60 Hz)	3.8
1 MHz		—	—	—	3.1
Dissipation Factor	D150				
1000 Hz		0.014–0.040 (60 Hz)	—	—	0.05
1 MHz		—			0.03
ENVIRONMENTAL PROPERTIES					
Max. Recommended Temperature, °F	—	380	300	300	230
Min. Recommended Temperature, °F	—	<−100	<−100	<−100	<−100
Heat Sealing Temperature Range, °F	—	400–450	490–540	420–470	350–400
Burning Rate, in./sec.	—	self-ext	self-ext	self-ext	self-ext to slow burn

TABLE 8-2 (Continued)

PLASTIC FILMS

PROPERTY	ASTM TEST	Polycarbonate	Polyethylene Terephthalate*	Polystyrene (oriented)	Cellophane (coated)	Polysulfone
GENERAL PROPERTIES						
Specific Gravity	D792	1.20	1.39	1.05–1.07	1.40–1.55	1.24–1.25
Water Absorption in 24 hr, %	D570	0.35	<0.8	0.04–0.06	45–115	0.22
MECHANICAL PROPERTIES						
Tensile Strength, 1000 psi	D882	8.4–8.8	20–40	7–12	7–16	8.4–10.6
Burst Strength (Mullen), psi	D774	25–35	45	30–60	45–70	60 (0.002 mil)
Tearing Strength (Elmendorf), gm/mil	D689	10–16	18	2–8	2–15	9–12
Folding Endurance	D643	250–400	>100,000	–	–	–
ELECTRICAL PROPERTIES						
Volume Resistivity, ohm-cm	–	10^{16}	10^{18}	10^{16}	10^{11}	5×10^{16}
Dielectric Strength at 60 Hz, V/mil	D149	3200	7500	400–600	2000 (uncoated)	7500
Dielectric Constant	D150					
1000 Hz		2.99	3.2	2.4–2.7	3.2 (uncoated)	3.07
1 MHz		2.93	3.0	2.4–2.7	–	3.03
Dissipation Factor	D150					
1000 Hz		0.0015	0.005	0.0005	0.015 (uncoated)	0.0008
1 MHz		0.010	0.016	0.0005	–	0.0034
ENVIRONMENTAL PROPERTIES						
Max. Recommended Temperature, °F	–	270–280	250	160–180	300–375	350
Min. Recommended Temperature, °F	–	<–212	–80	–70 to –94	0	–100
Heat Sealing Temperature Range, °F	–	400–430	490	220–300	200–350	500–550
Burning Rate, in./sec.	–	slow	self-ext	slow	0.7–2.3	self-ext

*polyester

TABLE 8-2 (*Continued*)

PLASTIC FILMS		Polyvinyl Chloride (incl. copolymers)		Rubber Hydrochloride	Polyimide
PROPERTY	ASTM TEST	Rigid	Nonrigid		
GENERAL PROPERTIES					
Specific Gravity	D792	1.36–1.50	1.15–1.50	1.12–1.15	1.42
Water Absorption in 24 hr, %	D570	nil	nil	5	2.9
MECHANICAL PROPERTIES					
Tensile Strength, 1000 psi	D882	6.5–8.5	1–5	5–6	24–25
Burst Strength (Mullen), psi	D774	—	9–20	—	75
Tearing Strength (Elmendorf), gm/mil	D689	20–150	30–1400	1000–1500	8
Folding Endurance	D643	—	—	250000	10000
ELECTRICAL PROPERTIES					
Volume Resistivity, ohm-cm	D149	—	—	10^{13}	10^{18}
Dielectric Strength at 60 Hz, V/mil	D150	250–1300	250–1300	—	7000
Dielectric Constant					
1000 Hz		3.0–8.0 (60 Hz)	3.0–8.0 (60 Hz)	3	3.5
1 MHz	D150	—	—	3	3.4
Dissipation Factor					
1000 Hz		0.009–0.16 (60 Hz)	0.009–0.16 (60 Hz)	—	0.003
1 MHz		—	—	0.006	0.010
ENVIRONMENTAL PROPERTIES					
Max. Recommended Temperature, °F	—	160–180	140–160	205	550
Min. Recommended Temperature, °F	—	–30	–50	–20	–450
Heat Sealing Temperature Range, °F	—	260–400	200–400	225–350	none
Burning Rate, in./sec.	—	self-ext	self-ext	self-ext	self-ext

TABLE 8-3. TRANSMISSION CONVERSION FACTORS. (BAER, 1964).

To obtain	Multiply		
	$\dfrac{\text{gm}}{24 \text{ hr} \cdot \text{m}^2}$ by	$\dfrac{\text{gm}}{24 \text{ hr} \cdot 100 \text{ in.}^2}$ by	$\dfrac{\text{grains}}{\text{hr} \cdot \text{ft}^2}$ by
$\dfrac{\text{gm}}{24 \text{ hr} \cdot \text{m}^2}$	1	15.5	16.7
$\dfrac{\text{gm}}{24 \text{ hr} \cdot 100 \text{ in.}^2}$	6.45×10^{-2}	1	1.08
$\dfrac{\text{grains}}{\text{hr} \cdot \text{ft}^2}$	5.97×10^{-2}	0.926	1

TABLE 8-4. PERMEABILITY CONVERSION FACTOR. (BAER, 1964).

To obtain	Multiply					
	$\dfrac{\text{cc} \cdot \text{mm}}{\text{cm}^2 \cdot \text{sec} \cdot \text{cm Hg}}$ by	$\dfrac{\text{cc} \cdot \text{mm}}{\text{cm}^2 \cdot \text{sec} \cdot \text{atm}}$ by	$\dfrac{\text{cc} \cdot \text{mm}}{\text{cm}^2 \cdot 24 \text{ hr} \cdot \text{atm}}$ by	$\dfrac{\text{cc} \cdot \text{mil}}{\text{cm}^2 \cdot 24 \text{ hr} \cdot \text{atm}}$ by	$\dfrac{\text{cc} \cdot \text{mil}}{100 \text{ in}^2 \cdot 24 \text{ hr} \cdot \text{atm}}$ by	$\dfrac{\text{cu in} \cdot \text{mil}}{100 \text{ in}^2 \cdot 24 \text{ hr} \cdot \text{atm}}$ by
$\dfrac{\text{cc} \cdot \text{mm}}{\text{cm}^2 \cdot \text{sec} \cdot \text{cm Hg}}$	1	1.32×10^{-2}	1.52×10^{-7}	3.87×10^{-9}	6.00×10^{-12}	9.8×10^{-11}
$\dfrac{\text{cc} \cdot \text{mm}}{\text{cm}^2 \cdot \text{sec} \cdot \text{atm}}$	76	1	1.16×10^{-5}	2.94×10^{-7}	4.56×10^{-10}	7.47×10^{-9}
$\dfrac{\text{cc} \cdot \text{mm}}{\text{cm}^2 \cdot 24 \text{ hr} \cdot \text{atm}}$	6.57×10^6	8.64×10^4	1	2.54×10^{-2}	3.90×10^{-5}	6.45×10^{-4}
$\dfrac{\text{cc} \cdot \text{mil}}{\text{cm}^2 \cdot 24 \text{ hr} \cdot \text{atm}}$	2.58×10^8	3.40×10^6	39.4	1	1.55×10^{-3}	2.54×10^{-2}
$\dfrac{\text{cc} \cdot \text{mil}}{100 \text{ in}^2 \cdot 24 \text{ hr} \cdot \text{atm}}$	1.67×10^{11}	2.19×10^9	2.54×10^4	6.45×10^2	1	16.4
$\dfrac{\text{cu in} \cdot \text{mil}}{100 \text{ in}^2 \cdot 24 \text{ hr} \cdot \text{atm}}$	1.02×10^{10}	1.34×10^8	1.6×10^3	0.394	6.10×10^{-2}	1

TABLE 8-5. GAS PERMEABILITY OF POLYMERS. (BAER, 1964)

P(At T°C) in units of cc(STP)/cm^2/mm/sec/cm Hg $\times$ 10^{10}. E_P in units of kcal/mole.

Polymer (Primary Chemical Composition)	Name	Hydrogen T	P	E_P	Helium T	P	E_P	Nitrogen T	P	E_P	Oxygen T	P	E_P	Carbon Dioxide T	P	E_P
Polybutadiene		25	420	6.6				25	64.5	8.2	25	191	7.1	25	1380	5.2
Poly[butadiene (80%)—acrylonitrile (20%)]	"Perbunan-18"	25	254	7.2	25	170	6.8	25	25	9.9	25	82	8.6	25	636	7.0
Poly[butadiene (73%)—acrylonitrile (27%)]	German "Perbunan"	25	160	7.9	25	122	7.0	25	10.6	11.4	25	39	9.7	25	310	8.1
Poly[butadiene (68%)—acrylonitrile (32%)]	"Hycar OR 25"	25	118	8.2	25	99	7.4	25	6.07	12.3	25	23.5	10.5	25	186	9.0
Poly[butadiene (61%)—acrylonitrile (39%)]	"Hycar OR 15"	25	72	8.8	25	69	7.7	25	2.36	13.8	25	9.6	12.0	25	75	10.5
Poly[butadiene-styrene]	Buna-S	25	400	6.8	25	231	6.6	25	63.5	8.7	25	172	7.3	25	1240	5.7
Polyisoprene	Natural rubber	25	500	6.9	25	308	6.3	25	84	9.0	25	230	7.1	25	1330	6.1
Poly[isoprene (74%)—acrylonitrile (26%)]		25	75	9.1	25	78	7.6	25	1.8	15.0	25	8.6	12.7	25	44	12.5
Poly[isoprene (74%)—methacrylonitrile (26%)]		25	137	8.1				25	6.0	12.8	25	23.6	11.0	25	143	10.1
Rubber hydrochloride	"Pliofilm-NO"							30	0.08		30	0.25		30	1.7	8.6
Rubber hydrochloride (plasticized)	"Pliofilm-FM"	25	16	5.6				30	1.4	10.0	30	5.4	8.4	30	12.9	8.6
Rubber hydrochloride (plasticized)	"Pliofilm-P4"	25	22	8.7				30	6.2					30	182	
Polydimethylbutadiene	Methyl rubber	25	170	8.0	25	145	6.6	25	4.8	13.3	25	21.1	11.3	25	75	11.2
Polychloroprene	Neoprene	25	136	8.1	25	45		25	11.8	10.6	25	40	9.9	25	250	8.5
Polyisobutylene		25	65	9.5	25	74	7.9	25	2.9	12.8	25	12	10.7	25	50	10.1
Poly[isobutene (98%)—isoprene (2%)]	Butyl rubber	25	73	8.7	25	66	7.6	25	3.2	12.5	25	13	10.7	25	52	9.9
Polyethylene (0.922 g/ml)		25	86	8.2	25	74	8.3	30	20	11.7	30	55	10.3	30	265	8.2
Polyethylene, irradiated (10⁸ roentgen)								30	11	11.6	30	34.8	9.5	30	152	8.6
Polyvinyl chloride		30	36	1.9				30	0.4		30	1.2		30	10.2	0.99
Polyvinyl chloride (30 parts DOP/100)		25	129					23	1.7		23	6		23	37	
Poly[vinyl chloride-vinyl acetate]		25	92	7.4	25	85	7.2	25	6	12.3	25	24	9.7	25	160	8.2
Polyvinylidene chloride	Saran	28	0.76					30	0.01	16.8	30	0.05	15.9	30	0.29	12.3
Polyvinyl fluoride	"Tedlar"	25	3.5		25	9.7		25	0.042		25	0.2		25	0.9	
Polytrifluorochloroethylene	"Kel-F"	25	9.5	7.1				30	1.3	12.5	30	5.6	10.9	30	12.5	7.4
Polytrifluorochloroethylene (30% crystalline)								40	0.2	14.3	40	0.92	11.2	40	2.11	11.8
Polytrifluorochloroethylene (80% crystalline)								40	0.09	11.9	40	0.25	10.9	40	0.48	11.1
Poly[vinylidene fluoride (3%)—trifluorochloroethylene (97%)]								25	0.11	13.6	25	0.20	13.8	25	0.80	13.1
Poly[vinylidene fluoride (70%)—trifluorochloroethylene (30%)]								25	1.62	13.5	25	5.46	13.1	25	27.2	13.9
Poly[vinylidene fluoride-hexafluoropropylene]	"Viton A"							30	4.4		30	15		30	78	
Fluorinated ethylene-propylene copolymer	"Teflon FEP"	25	140	6.3	25	400	4.9	25	21.5	7.3	25	59	5.8	25	17	
Chlorosulfonated polyethylene	"Hypalon"	23	142		23	95		23	11.6		23	28		23	208	
Polyvinyl butyral								30	2.5		27	24.9		27	260	
Polystyrene								25	3-80		25	15-250		25	75-370	
Poly[styrene-acrylonitrile]								23	0.46		23	3.5		23	10.8	
Poly[styrene-methylmethacrylate]								23	0.21		23	1.6				
Polyvinyl toluene								23	4.6		23	22		23	66	
Acetal (polyformaldehyde)	"Delrin"							30	0.22		30	3.8		30	19	
Poly[esteramide-diisocyanate]	"Vulcaprene"	25	63	9.0				25	4.9	12.2	25	15.2	10.9	25	186	9.4
Polyamide	Nylon	25	10	8.1				30	0.2	11.2	30	0.38	10.4	30	1.6	9.7
Polyethylene terephthalate	"Mylar"	25	6	5.5	25	11	4.6	25	0.05	7.5	25	0.30	6.4	25	1.0	6.2
Polycarbonate		20	136					20	3		20	20		20	85	
Polysulfide	"Thiokol" rubber	25	16	10.6							25	2.9	13.5	25	32	11.5
Polydialkylsiloxane	Silicone rubber										25	5000		25	28000	
Phenol-formaldehyde	"Bakelite"	20	0.95	5.1				20	0.095	10.5						
Cellulose acetate (unplasticized)								23	1.6		23	4.0		23	32	
Cellulose acetate	"Lumarith" P912	25	84	5.2				30	2	6.5	30	7.8	5.0	25	87	7.1
Cellulose acetate (DBP 15%)		25	123	5.3				30	5.0					25	180	4.3
Cellulose acetate-butyrate		25	210	4.1	25	140	4.6	30	16		30	60		25	310	4.2
Cellulose triacetate (43% acetyl)								30	1.7		30	10		30	57	
Methyl cellulose								23	1.74		23	5.04		23	26.3	
Ethyl cellulose	"Ethocel" 610	30	32	3.4	30	260		30	84	4.2	30	265	4.0	30	430	1.4

TABLE 8-6. SORPTION AND TRANSMISSION OF WATER IN POLYMERS. (BAER, 1964)

Polymer	% H$_2$O 24-hr Immersion $\frac{1}{8}$-Inch Thick Sample	Transmission Rate (90–95% RH) g/m^2/24 hr/mil Temp, °C	Rate
Polybutadiene	0.05–0.5	40	680
Poly(butadiene-acrylonitrile)	0.05–0.4	39	130
Poly(butadiene-styrene)	0.04–0.2	40	330
Polyisoprene (natural rubber)	0.06–0.3	39	390
Chlorinated polyisoprene		40	46
Rubber hydrochloride ("Pliofilm")*	neg–5.0	37	18–225
Polychlorobutadiene (neoprene)	0.05–0.35	39	240
Polyisobutylene	0.04–0.15	40	18
Poly(isobutene-isoprene) (butyl rubber)	0.04–0.18	39	26
Polyethylene (0.92 g/ml)	<0.01	39	28
Polyethylene (0.94 g/ml)	<0.01	37	14
Polyethylene (0.95 g/ml)	<0.01	38	6.7
Polyethylene (0.96 g/ml)	<0.01	37	4
Polypropylene	0.005–0.03	38	8.7
Chlorosulfonated polyethylene ("Hypalon")*		39	161
Polyvinyl chloride	0.03	38	32
Polyvinyl chloride, plasticized	0.4	40	88
Polyvinyl chloride-acetate	0.2–5.5	39	37–200
Polyvinylidene chloride (Saran)*	<0.1	39	1.5–7
Polyvinyl fluoride	<0.5	40	46
Polytrifluorochloroethylene ("Kel-F")*	<0.01	40	5
Fluorinated poly(ethylene-propylene)	<0.01	40	7.2
Polytetrafluoroethylene ("Teflon")*	0.00	40	5
Chlorinated polyether	0.01	36	7
Polyvinyl alcohol	>30 (80)	40	400–2000
Polyvinyl acetate		20	540
Polyvinyl formal	1.0–1.3	20	110
Polyvinyl butyral	0.3–0.6	20	90
Polystyrene	0.03–0.3	39	133
Polyacrylonitrile	1.5	40	183
Poly(styrene-acrylonitrile)		40	125
Polymethylmethacrylate	0.2–0.4	40	550
Polymethylene oxide (acetal resin)	0.12	23	47
Polyamide (nylon 6)	1.6–3.3	38	126
Polyamide (nylon 66)	1.6–3.0	39	225
Polyethylene terephthalate ("Mylar")*	0.03–2.5	39	30
Polycarbonate	0.35	20	31
Polysulfide rubber ("Thiokol")	0.24–1.2	40	668
Silicon rubber	0.1–0.15		
Epoxy resin	0.1–0.5		
Phenol-formaldehyde ("Bakelite")*	0.15–0.60		
Melamine-formaldehyde	0.3–0.5		
Cellophane, plain	45–115	38	1870
Cellophane, vinylidene chloride coated	50	39	7.5
Cellophane, nitrocellulose coated		38	33
Cellophane, polyethylene coated		38	18
Cellophane, lacquer coated	70–100	38	3–16
Cellulose acetate	1.5–3	35	160
Cellulose acetate, plasticized	5.7–7	39	1400
Cellulose acetate-butyrate	0.1–3.4	39	1500
Cellulose triacetate	3.5–4.5	39	1080
Cellulose butyrate		35	930
Ethyl cellulose	0.8–5.5	35	150–780

*Trademark.

TABLE 8-7. TRANSMISSION RATE OF ORGANIC SUBSTANCES THROUGH POLYMER FILMS. (BAER, 1964)

(J in units of grams/mil/24 hours/square meter)

Polymer	Aliphatic Hydrocarbons (X: Hexane, P: Heptane, D: Decane)			Aromatic Hydrocarbons (B: Benzene, T: Toluene)			Chlorinated Hydrocarbons (F: Chloroform, C: Carbon Tetrachloride)			Alcohols (M: Methanol, E: Ethanol)			Ketones (A: Acetone, K: Methyl-Ethyl-Ketone)			Ethyl Acetate		Other (G: Acetic acid, R: Dibutyl ether, Z: Dioxane)		
	Name	Temp °C	J	Name	Temp °C	J	Name	Temp °C	J	Name	Temp °C	J	Name	Temp °C	J	Temp °C	J	Name	Temp °C	J
Polyethylene (0.92 g/ml)	P	24	4600	B	23	7700	C	21	9500	M	23	20	A	24	300	24	650	G, R	23, 21	58, 1500
Polyethylene (0.96 g/ml)	P	23	810	T	23	1800	C	23	1500	M	23	3.3	A	23	53	23	70	G, R	23, 23	15, 330
Poly(butadiene-styrene)				B	25	940000	C	25	580000	M	25	1040	K	25	195000	25	226000			
Poly(butadiene-acrylonitrile)				B	25	685000	C	25	125000	M	25	15000	K	25	670000	25	410000			
Chlorinated polyether	D	23	2.3	T	23	390	C	23	10	M	23	8.4	A	23	11000	23	1600	R	23	1.2
Polychloroprene (neoprene)	X	23	4.2	B	25	550000	C	25	335000	M, M	25, 23	2710, 21	K	25	228000	25	125000			
Polyvinyl chloride	X	40	18	B	40	22	C	40	12	E	40	8.4	A	40	4800	40	240	G	40	3.5
Polyvinyl fluoride				B	20	23000	C	30	9000	M	20	11000				20	27000	Z	20	11000
Polyvinyl acetate				B	20	180	F	20	97000	M	20	540	A	20	7600	20	4700	Z	20	11000
Polyvinyl formal				B	20	14000	F	20	97000	M	20	7900	A	20	21000	20	15000	Z	20	6500
Polyvinyl butyral				B	35	4.3	C	35	1.7				A	35	20	35	0.96			
Polyethylene terephthalate	X	35	1.4	T	23	8.8	C	23	8.5	M	23	590	K	23	1.4	23	1.1	R	23	2.3
Polyamide (nylon 66)	X	23	5	B	40	8.6	C	40	0.43				A	25	13					
"Teflon FEP"*	X	40	1.2							E	40	0.96						G	40	0.29
Polysulfide rubber				B	25	155000	C	25	11900				K	25	79000	25	39600			
Cellophane (MSAT)	X	35	8.4	B	35	6.5	C	35	9.4	E	35	870	A	35	580	35	6.2	G	35	590

*Trademark.

TABLE 8-8. RESISTANCE OF POLYMERS TO ACIDS, ALKALIES, AND ORGANIC LIQUIDS. (BAER, 1964)

POLYMER	ACIDS			ALKALIES		HYDROCARBONS			ALCOHOLS	KETONES	ESTERS
	weak	strong	oxidizing	weak	strong	aliphatic	aromatis	halogenated			
Natural rubber	G	F–G	F	E	E	F–S	F–S	F–S	G	G	F–G
Chlorinated rubber	E	G–E		E	G		F	F		F	F
Rubber hydrochloride	G	G		G	G	G	F–G	P–S	G	F	F
Polychloroprene (neoprene)	E	G	G	E	E	G	P–S	G–S	F	P	P
Butyl rubber	E	E	G	G	F	P–S	P	P–S	G	G	G
Poly(butadiene-acrylonitrile)	G	G	F–G	G	G	G	F–G	F–G	F	F	F
Poly(butadiene-styrene)	G	G	F–G	E	E	P–S	P–S	P–S	G	F	F
Poly(acrylonitrile-butadiene-styrene)	G	F–G	F	E	G	G	F–G	P–S	G	P	F
Polyethylene	E	E	F	E	E	F	F	F	G	G	G
Polypropylene	E	E	F–P	E	E	F	F	F	G	G	G
Polystyrene	E	G–E	F–P	E	E	F	S	S	E	P–S	S
Polyacrylonitrile	E	G–E		F–G	P	G–E	G–E	G–E	G–E	G–E	G–E
Poly(styrene-acrylonitrile)	E	G	P	E	E	G–E	F–G		G–E	S	F–P
Polyvinyl chloride	G–E	G–E	F	G–E	G–E	G–E	F	F–P	E	P–S	P–S
Poly(vinyl chloride-vinyl acetate)	G–E	G–E	F	G–E	G–E	G–E	F	P–S	G	S	S
Polyvinyl alcohol	P	S	S	P	S	E	E	E	G	E	E
Polyvinyl acetate	P	P	P	P	P	G	S	P	P	S	S
Polyvinyl formal	P	P	P	G	F–P	G	S	S	F–P	F–P	G
Polyvinyl butyral	P	P	P	G	G	G	F–G	F	S	S	S
Polyvinyl fluoride	E	G–E	G	E	G	E	E	E	E	E	E
Polyvinylidene chloride	E	G	F	E	G	E	F–P	P	E	P	F–P
Polytrifluorochloroethylene	E	E	E	E	E	E	E	G–E	E	E	E
Fluorinated poly(ethylene-propylene)	E	E	E	E	E	E	E	G–E	E	E	E
Polytetrafluoroethylene	E	E	E	E	E	E	E	E	E	E	E
Polyamide (nylon)	P	S	S	E	E	E	G	G–E	G	G	G
Polyester	G	G–P	F–P	G–F	G–P	G	F	P	F	F	G
Acrylic resin	E	F	P–S	E	G–F	E	P–S	P–S	G	P–S	P–S
Polymethylmethacrylate	E	F	S	E	F	G	S	S	E	S	S
Polymethylene oxide (acetal)	G	F	P–S	G	F	E	E	E	G	E	G
Polyethylene oxide	P	P	P–S	F	F–P	G	F	F–S	S	S	S
Chlorinated polyether	E	G	P	G–E	G	E	E	E	E	E	E
Polycarbonate	E	G	F	F	F–P	F	P	P	F	S	S
Polyurethane	F	P	P	F	P	G	F		G	F–P	F–P
Diallyl phthalate	G	F		G	F	G–E	G–E	G–E	G–E	G–E	G–E
Epoxy resin	E	F–G	P	E	E	E–F	E–F		E–F	F–P	F–P
Alkyd resin	E	G		S	S	P	P		F–P		
Polysulfide rubber	E	E	E–G	E	E–G	E	G	G	E	G	E
Silicon rubber	G	F	G–F	G–F	G–F	E	P–S	P	G–P	F–P	F–P
Phenol-formaldehyde	E–F	G–P	S	F–P	S	G	F–G	F–G	G	G–P	F
Urea-formaldehyde	G–P	S	S	F–P	S	G	F–G	G	G	G	G
Melamine-formaldehyde	G–F	S	S	G–F	F	G	G	G	G	G	G
Cellophane	P	S	S	G–F	P	G–E	G–E	G–E	G–E	G–E	G–E
Cellulose acetate	G–F	S	S	F–P	S	F	F–P	P–S	P	S	P–S
Cellulose acetate-butyrate	G–F	P–S	S	G–F	P–S	G–F	F	P	F–P	S	S
Cellulose triacetate	G–F	S	S	P	S	G	G–F	G–F	F	P–S	P–S
Cellulose nitrate	G–F	F–S	F–P	G–F	P–S	G	G	F	F–P	S	S
Cellulose propionate	G	F–S	F–P	F	F–S	G	G	G	F	S	S
Methyl cellulose	F	S	S	G–E	G	soluble in cold water					
Ethyl cellulose	G–F	P–S	P–S	E	E	G	F	P	P–S	S	P–S

E: No apparent changes in appearance or physical properties.
G: Slight changes in appearance; negligible changes in physical properties.
F: Definite changes in appearance (coloration, dimensions, weight, surface layers); moderate changes in physical properties.
P: Considerable changes in appearance and physical properties; softening or mild solvent action.
S: Severe corrosive attack or solvent action.

TABLE 8-9. ELECTRICAL PROPERTIES OF ORGANIC COATINGS. (HARPER 1970, LICARI 1967, MODERN PLASTICS ENCYCLOPEDIA, 1970)

Coating	Dielectric Strength, short-time, 0.125″ thick (volts/mil)	Dielectric Constant at 125°F		Dissipation Factor at 125°F		Volume Resistivity (ohm-cm)
		60–100 Hz	10^6 Hz	60–100 Hz	10^6 Hz	
Acrylic	450–550	2.7–4	2.7–3.2	0.04–0.06	0.02–0.03	1×10^{14}
Alkyd	300–350				0.003–0.06	1×10^{14}
Chlorinated Polyether	400	3.1	2.92	0.01	0.01	1×10^{15}
Chlorosulfonated Poly-ethylene	500	6.19	5	0.03	0.07*	
Depolymerized Rubber	360–380	4.1–4.2	3.9–4.0	0.007–0.013	0.0073–0.016	
Diallyl Isophthalate	420	3.5	3.2	0.008	0.009	
Diallyl Phthalate	450	3–3.6	3.3–4.5	0.010	0.011	
Epoxy (One Component)	650–730	3.5	3.7	0.011	0.004	3.4×10^{14}
Fluorocarbon, FEP	500–600	2.1	2.1*	0.0002–0.0007**		2×10^{18}
Fluorocarbon, TFE	480		2.0–2.2*	0.003–0.004	0.003–0.004	1×10^{18}
Neoprene	150–600					
Phenolic	300–450	5–6.5	4.5–5.0	0.005–0.5	0.011	10^9–10^{12}
Polyamide	780	2.8–3.9	2.7–2.96	0.015	0.022–0.097	
Polyester	250–400	3.3–8.1	3.2–5.9	0.008–0.041		10^{12}–10^{14}
Polyethylene	500	2.3	2.3	0.00015#	0.00015#	
Polyimide	560	3.55		0.003	0.003	1.6×10^{15}
Polymethyl Methacrylate				0.06	0.02	
Polypropylene	750–800	2.22–2.28	2.22–2.28			
Polystyrene	500–700	2.45–2.65	2.4–2.65	0.0001–	0.0001–0.0004	
Polysulfide	250–600	6.9		0.0005		
Polyurethane (One Component)	450–500	4.10	3.8	0.038–0.039	0.068–0.074	1×10^{13}
Polyvinyl Butyral	400	3.6	3.33	0.007	0.0065	
Polyvinyl Chloride	300–1000	3.3–6.7	2.3–3.5	0.08–0.15	0.04–0.14	
Polyvinyl Formal	860–1000	3.7	3.0	0.007	0.02	
Polyvinylidene Fluoride	260	8.1	6.6	0.049	0.17	1×10^{14}
Polyxylylene						
Parylene N®	550	2.65	2.65	0.0002	0.0006	1.4×10^{17}
Parylene C®	550	3.15	2.95	0.019	0.013	8.8×10^{16}
Parylene D®	480	2.84	2.80	0.004	0.002	2×10^{16}
Shellac (Natural, Dewaxed)	200–600	3.6	3.3			1.8×10^9
Silicone (RTV Type)	550	3.3–4.2	3.1–4.0	0.011–0.02	0.003–0.006	1×10^{14}

*At 10^3 Hz.   **Dispersion Coating.   #Linear Polymer.   Parylene® Registered Trademark of Union Carbide Corp.

TABLE 8-10. PROPERTIES OF RUBBERS. [MAT–SELEC.]

RUBBER AND ELASTOMERS (Molded, Extruded)	ASTM TEST	Natural Rubber Polyisoprene	Synthetic Rubber Synthetic Polyisoprene	SBR Rubber Styrene Butadiene	Butyl Rubber Isobutylene Isoprene
ASTM D-2000/SAE J-200 DESIGNATION		AA	AA	AA	AA, BA
GENERAL PROPERTIES					
Specific Gravity	D792	0.92	0.91	0.94–1.10	0.90
Thermal Conductivity, Btu/hr/sq ft/°F/ft	C177	0.082	0.082	0.143	0.053
Thermal Expansion Coefficient, cubical, 10^{-6}/°F	—	370	370	370	320
Water Absorption in 24 hr, %	D570			>0.5	
Color	—				
MECHANICAL PROPERTIES					
Tensile Strength, 1000 psi (pure gum)	D412	2.5–3.5	2.5–3.5	0.2–0.3*	2.5–3.5
Elongation, % (pure gum)	D412	750–850		400–600	750–950
Flexural Strength, 1000 psi	D790			no break	
Impact Strength (Izod notched), ft-lb/in.	D256			no break	
Tearing Strength, Die "C", pli	D624				
Compressive Strength, 1000 psi	D695			no break	
Hardness, Durometer Range	D1706	A30–A90	A40–A80	A40–A90	A40–A90
ELECTRICAL PROPERTIES					
Volume Resistivity, ohm-cm	D257	1.7×10^{16}		5×10^{13}–2.5×10^{16}	
Dielectric Strength, V/mil (short-time)	D149			420–520	
Dielectric Constant	D150				
60 Hz		2.74		2.5–3.4	2.35–2.40
1 MHz				2.5–3.4	
Dissipation Factor	D150				
60 Hz				0.002–0.003	
1 MHz				0.001–0.003	
Arc Resistance, seconds	D495			95	
HEAT RESISTANCE					
Max. Recommended Service Temperatures, °F	—	180	180	180	300
Min. Recommended Service Temperature, °F	—	–60	–60	–60	–50
Stiffening Temperature, °F	—	–20 to –50	–20 to –50	–10 to –50	0 to –40
Brittle Point, °F	D746	–70	–70	–80	–75
Flame Resistance	—	poor	poor	poor	poor

*About 10 times greater for black material.

TABLE 8-10. (*Continued*)

RUBBER AND ELASTOMERS		Butadiene	EP Rubber	Neoprene	NBR Rubber
		Polybutadiene	Ethylene Propylene	Chloroprene	Butadiene-Acrylonitrile
ASTM D-2000/SAE J-200 DESIGNATION		AA	BA, DA	BC, BE	BF, BG, BK
PROPERTY	ASTM TEST				
GENERAL PROPERTIES					
Specific Gravity	D792	0.91	0.86	1.25	0.98
Thermal Conductivity, Btu/hr/sq ft/°F/ft	C177			0.112	0.143
Thermal Expansion Coefficient, cubical, 10^{-6}/°F	—	375		340	320
Water Absorption in 24 hr, %	D570				
Color	—				
MECHANICAL PROPERTIES					
Tensile Strength, 1000 psi (pure gum)	D412	0.2–1.0	<1	3–4	0.5–0.9
Elongation, %	D412	400–1000	200–600	800–900	300–700
Flexural Strength, 1000 psi	D790				
Impact Strength (Izod notched), ft-lb/in.	D256				
Tearing Strength, Die "C", pli	D624				
Compressive Strength, 1000 psi	D695				
Hardness, Durometer Range	D1706	A40–A90	A30–A90	A20–A95	A40–A95
ELECTRICAL PROPERTIES					
Volume Resistivity, ohm-cm	D257				
Dielectric Strength, V/mil	D149				
Dielectric Constant	D150				
60 Hz					
1 MHz		3.06	3.2–3.4	5.7–6.26	3.20
Dissipation Factor	D150				
60 Hz					
1 MHz					
Arc Resistance, seconds	D495				
HEAT RESISTANCE					
Max. Recommended Service Temperature, °F	—	200	350	240	300
Min. Recommended Service Temperature, °F	—	−150	−60	−40	−60
Stiffening Temperature, °F	—	−35 to −60	−20 to −50	0 to −30	15 to −35
Brittle Point, °F	D746	−90	−80	−60	−70
Flame Resistance	—	poor	poor	good	poor

TABLE 8-10. (Continued)

RUBBER AND ELASTOMERS		Polysulfide Rubber	Polyurethane	Hypalon	Acrylic Rubber
ASTM D-2000/SAE J-200 DESIGNATION		AK, BK	BG	CE	DF, DH
		Polysulfide	Diisocyanate Polyester	Chlorosulfonated Polyethylene	Polyacrylate
PROPERTY	ASTM TEST				
GENERAL PROPERTIES					
Specific Gravity	D792	1.35	1.25	1.11-1.26	1.09
Thermal Conductivity, Btu/hr/sq ft/°F/ft	C177		0.041-0.18	0.065	
Thermal Expansion Coefficient, cubical, 10^{-6}/°F	—		56-102	270	
Water Absorption in 24 hr, %	D570		0.7-0.9		
Color	—				
MECHANICAL PROPERTIES					
Tensile Strength, 1000 psi (pure gum)	D412	0.25-0.40	>5	4	0.25-0.40
Elongation, % (pure gum)	D412	450-650	540-750		450-750
Flexural Strength, 1000 psi	D790		0.7-9.0 (yield)		
Impact Strength (Izod notched), ft-lb/in.	D256		does not break		
Tearing Strength, Die "C", pli	D624		20		
Compressive Strength, 1000 psi	D695				
Hardness, Durometer Range	D1706	A40-A85	A35-A100	A45-A95	A40-A90
ELECTRICAL PROPERTIES					
Volume Resistivity, ohm-cm	D257		$(2.0-110) \times 10^{11}$		
Dielectric Strength, V/mil (short time)	D149		330-900		
Dielectric Constant	D150				
60 Hz		6.5-7.5 (1 kHz)	5.4-7.6	5.0-11.0	
1 MHz			4.21-5.10		
Dissipation Factor	D150				
60 Hz			0.015-0.048		
1 MHz			0.050-0.075		
Arc Resistance, seconds	D495		122		
HEAT RESISTANCE					
Max. Recommended Service Temperature, °F	—	250	240	<325	350
Min. Recommended Service Temperature, °F	—	-60	-65	-60	-20
Stiffening Temperature, °F	—	-10 to -40	-5 to -25	-30 to -50	+35 to +10
Brittle Point, °F	D746	-65	-60	-60	-20
Flame Resistance	—	poor	good	good	poor

TABLE 8-10. (*Continued*)

RUBBER AND ELASTOMERS		Epichlorohydrin	Silicone	Flurosilicone	Fluorocarbon
		Polyalkylene Oxide Polymer	Polysiloxane	Fluorovinyl Silane	Fluorinated Hydrocarbon
ASTM D-2000/SAE J-200 DESIGNATION		DK, DJ	FC, FE, GE	FK	HK
PROPERTY	ASTM TEST				
GENERAL PROPERTIES					
Specific Gravity	D792	1.32–1.49	1.1–1.6	1.4	1.4–1.95
Thermal Conductivity, Btu/hr/sq ft/°F/ft	C177		0.13	0.13	0.13
Thermal Expansion Coefficient, cubical, 10^{-6}/°F	—		450	450	88
Water Absorption in 24 hr, %	D570				
Color	—				
MECHANICAL PROPERTIES					
Tensile Strength, 1000 psi (pure gum)	D412		0.6–1.3#	1	<2
Elongation, %	D412		100–500#	200–400#	100–450#
Flexural Strength, 1000 psi	D790				
Impact Strength *Izod notched, ft-lb/in.	D256				
Tearing Strength, Die "C", ppi	D624		15–30	55–265 (Die B)	
Compressive Strength, 1000 psi	D695				
Hardness, Durometer Range	D1706	A30–A95	A30–A90	A40–A70	A65–A90
ELECTRICAL PROPERTIES					
Volume Resistivity, ohm-cm	D257		$(0.1–5) \times 10^{14}$	$(0.01–1.6) \times 10^{14}$	
Dielectric Strength, V/mil	D149		525–600	340–380	10–18
Dielectric Constant	D150				
60 Hz	—		2.9–3.2 (100 Hz)	6.1–7.4 (100 Hz)	
1 MHz	—		2.6–3.0	5.7–6.2	
Dissipation Factor	D150				
60 Hz	—		0.01–0.03 (100 Hz)	0.03–0.078 (100 Hz)	
1 MHz	—		0.003–0.005	0.03–0.04	
Arc Resistance, seconds	D495		90–125		
HEAT RESISTANCE					
Max. Recommended Service Temperature, °F	—	300	600	300	<500
Min. Recommended Service Temperature, °F	—	−15 to −80	−178	−15 to −80	−10
Stiffening Temperature, °F	—	−15 to −40	−65 to −180	−60 to −75	+15 to −25
Brittle Point, °F	D746	−10 to −85	−90 to −180	−90	−55
Flame Resistance	—	fair	good	poor	excellent

#Reinforced.

TABLE 8-11. PROPERTIES OF REPRESENTATIVE COMMERCIALLY AVAILABLE FOAMS (Baer, 1964)

Plastic Composition	Polystyrene						Polyurethane			Epoxy	Phenol-formaldehyde			Polyethylene			Urea-formaldehyde	Silicone		Cellulose acetate
	Extruded			Molded			Polyether Board FIP[e]	Polyether FIP[e]	Polyester FIP[e]											
Density, lb/ft³:	1.9	2.9	4.4	1.0	2.0	4.0	2.3	2.5	2.1	2.3	2.0	4.0	8.0	2.0	29[f]	30[g]	1.8	3.5	14	6-7
Mechanical Properties at 75°F																				
Compressive strength, psi	35	65	130	20	35	70	50[a]	32	37	25	25	55	140	25	670	1800	8	6.2		125
Tensile strength, psi	70	105	178	20	45	85	60	30	47	40	15	30	70				17		200	170
Flexural strength, psi	70	80	160	20	60	120	30[b]	55	60		45	90	205							147
Shear strength, psi	40	58	88								45	25	45							140
Compressive modulus, psi × 10³	1.0	3.0	5.05	.25	.75	1.75	1.0[a]													
Flexural modulus, psi × 10³	2.5	2.0	2.95	2.0	2.4	6.6	1.0													
Shear modulus, psi × 10³	.9	1.8	2.95				.5[b]			.57							.7			
Thermal Properties																				
Thermal conductivity (initial), Btu-in. °F⁻¹ft⁻²hr⁻¹	.26			.16	.16	.16	.12	.110	.110	.11	.20	.20	.27	.035			.23	.281	.3	.31
Thermal conductivity (equil.), Btu-in. °F⁻¹ft⁻²hr⁻¹	.26			.260	.240	.243	.165	.150	.157	.15										
Coefficient of thermal expansion, in. in.⁻¹ °F⁻¹ × 10⁻⁴	3.5			3.3 to 3.5			2.7				1-3	1-3	1-3							2.5
Flammability[h]					burns—can be made FR		FR				FR	FR	FR		burns		FR	FR		burns
Heat distortion temp., °F	170	170	170	175	175	175	250			300	250			160			120	650	700	350
Electrical Properties																				
Dielectric constant at 10⁶cps	<1.05	1.07	1.07	<1.017	1.03	1.06	1.04							1.05	1.50	1.55		1.09	1.25	1.12
Dissipation factor at 10⁶cps, × 10⁻⁴	<4.0	<4.0	<4.0	<1.0	7.0	7.0	13							2.0	3.3	40.0			10.2	20
Chemical Properties																				
Water absorption (10-ft head), lb/ft²	.08	.08	.08	nil	nil	nil	<.04	.06	.04	.03				.4				.284		
Water absorption, vol. %	<4.0	<4.0	<4.0	<1.0	<1.0	<1.0	<2.0	2.0	<2.0	1	100			4.0				2.3		4.5
Moisture-vapor transmission, perm-inch	1.5	1.5	1.5	2.0	2.0	2.0	2.5	1.7	1.0	1	{ .4[c], 214[d] }							41.2		
Specific heat, Btu/lb		.29								.38	.38	.38	.38				.40			

a. Load parallel to thickness dimension.
b. Load perpendicular to thickness dimension.
c. With skin.
d. Without skin.
e. FIP = foamed-in-place.
f. Prepared from low-density polyethylene.
g. Prepared from high-density polyethylene.
h. FR = flame retardant.

**TABLE 8-12. CHARACTERIZATION OF COMMERCIAL FOAMS
IN TABLE 8-11. (BAER, 1965)**

Plastic composition	Density, lb/ft^3	Cell size, inch	Gas composition	Plastic state	% Open cells
Cellulose acetate	6–7		air		
Epoxy	2.3	<.01	air-CBA		10
Phenol-formaldehyde	2.0	∼.01	air		
Phenol-formaldehyde	4.0	∼.01	air		
Phenol-formaldehyde	8.0	∼.01	air		
Polyethylene	29		air	$\rho = .920$	
Polyethylene	30		air	$\rho = .960$	
Polyethylene	2.0		air		
Polystyrene (extruded)	1.9	.02–.04	air		
Polystyrene (extruded)	2.9	<.02	air		
Polystyrene (extruded)	4.4	<.02	air		
Polystyrene (molded)	1.0	<.01	air		
Polystyrene (molded)	2.0	<.01	air		
Polystyrene (molded)	4.0	<.01	air		
Polyurethane					
Polyether board	2.3	0.1–.03	air–CBA		
Polyether FIP	2.5		air–CBA		7
Polyester FIP	2.5		air–CBA		9
Silicone	3.5		air		90
Silicone	14	.08	air		
Urea-formaldehyde	1.8	<.01	air		>50

CBA = captive blowing agent.

TABLE 8-13. ELECTRICAL PROPERTIES OF PLASTIC FOAMS. (BAER, 1964)

Plastic phase	Density, lb/ft³	Dielectric constant 10³cps	10⁴cps	10⁶cps	10⁹cps	Dissipation factor (× 10⁻⁴) 10³cps	10⁴cps	10⁵cps	10⁶cps	10⁸cps	10⁹cps
Cellulose acetate	6–7		1.10—1.12					20 to 30			
Epoxy	5.0		2.0					50			
Polyethylene	32	1.48	1.49	1.49	1.49	3.3	3.3	3.3	3.3	3.3	3.3
	2.0				1.05						2.0
	18				1.575						33
Polystyrene	1.5*	1.05				<5.0	<5.0				1.7
	1.5*		1.017		1.077**						6.0**
	9.8*				1.161			1.0			
	4.3†			<1.05	<1.05			<4.0			
	1.9†			1.07	1.07			<4.0			
	2.8†			1.07	1.07			<4.0			
	4.3†			1.09							
Silicone	3.5							28			38.0**
	9.3				1.193**						35.0**
Polyether-urethane	9.8		1.05		1.206**						
	1.5–3.0							3–13			

* Molded polystyrene foam.
† Extruded polystyrene foam.
** Measured at 24 × 10⁹ cps.

TABLE 8-14. ADHESIVES SELECTION CHART.

	Vinyl Phe-nolic	Vinyl-Butyral Phe-nolic	Neo-prene Phenolic	Nitrile Phe-nolic	Epoxy	Epoxy Phe-nolic	Ceramic	Methyl Cyano-acrylate	Neo-prene	Nitrile Rubber	Poly-sulphide Rubber (Thiokol)
Type											
Thermoplastic								NS			
Thermosetting	x	x	x	x	x	x	x	NS			
Elastomer									x	x	x
Form											
Solution	x	x	x	x	x	x	x		x	x	
Liquid					x	x		x			x
Film	x	x	x	x						x	
Powder					x	x	x				
Rod					x						
Supported Film					x						
Mix Needed					some	*	*				*
Cure Temperature											
Room					x			x	x	x	x
Elevated	x	x	x	x	x	x	x		x	x	x
Service Range (°F)											
Low	−40	−40	−70	−80	−90	−75	NS	−65	−60	−60	−60
High	300	300	180	300	300	500	850	165	200	250	180
Strength (psi)											
Tensile	to 4000	to 6500	to 5000	to 8000	to 4000	NS	NS	to 6000	NS	to 4000	NS
Shear	to 5000	to 3500	to 3000	to 6000	to 3500	to 2000	to 5000	to 2000	to 900	to 2000	to 200
Bonds											
Metal to											
Metal	x	x	x	x	x	x	x	x	x	x	
Glass (Ceramic)	x		x		x	x	x				x
Wood					x	x			x	x	x
Plastic	x	x	x	x	x	x			x	x	
Glass to											
Glass	x		x	x	x	x	x			x	x
Wood			x	x	x			x		x	
Plastic	x				x	x					
Wood to											
Wood			x	x	x	x				x	
Plastic	x		x	x	x	x				x	
Plastic to											
Plastic	x		x	x	x	x		x		x	
Resistance to											
Cold	G	G	E	P-G	G	E	NS	G	G	G	E
Heat	G	F	G	E	F-G	G	NS	F	G	G	F
Water	E	E	E	E	F-G	E	NS	F	E	E	E
Oil	E	G	E	E	E	E	NS	G	G	E	E
Gasoline	E	NS	E	E	E	E	NS	G	G	E	E
Glycol	E	NS	E	E	G	E	NS	G	NS	NS	NS

E-Exellent, G-Good, F-Fair, P-Poor, NS-Not Specified.
*Depending on specific formulation.
Reprinted from "Assembly Directory and Handbook/1967," Copyright 1967 by Hitchcock Publishing Company, Wheaton, Illinois.

TABLE 8-14. (*continued*)

Natural Rubber	Re-claimed Rubber	Butadiene Styrene Copolymer (GR-S) (SBR)	Poly-vinyl Acetate	Poly-urethene	Vinyl-Acetate Phe-nolic	Epoxy-Nylon	Epoxy-Poly-amide	Rubber-Phe-nolic	Filled Epoxy	Poly-amide	Silicone (modi-fied)	Poly-imide	Poly-benzimi-dazole
		x	x	x				x		x		x	x
x	x				x	x	x		x		x		
x	x	x	x									x	
			x	x	x	x	x		x		x	x	
				x	x	x				x			
				x						x			
										x			x
										x			
			*	*	*	*		*		*			
x	x	x		x		x			x	x	x	x	x
x	x	x	x	x	x		x		x	x	x	x	x
−60	−30	−40	NS	−425	−180	−425	−425	−100	−400	−40	NS	−425	NS
140	180	160	200	300	200	300	250	200	300	250	600	700	700
NS	NS	NS	NS	to 4000	NS	NS	to 8000		to 4000				
NS		NS	to 3000	to 1200	to 3000	to 6000	to 3500	to 4000	to 3000	to 1500	to 2000	to 3000	to 3000
				x	x	x	x		x	x	x	x	x
				x			x		x	x	x	x	
		x	x	x	x	x	x			x			
				x	x	x	x		x	x			
				x		x	x		x	x	x		
				x			x		x	x			
						x	x		x	x			
x	x	x	x	x	x		x		x	x			
			x	x	x		x		x	x			
			x	x	x	x	x	x	x	x			
G	G	G	F-G	E	G	E	E	F	G	F	G	NS	NS
P	P	F	G	G	F	G	E	F	G	P	E	E	E
E	E	E	G	G	G	G	G	NS	G	E	G	G	G
P	P	P	G	E	G	G	G	NS	E	G	G	G	G
P	P	P	F	G	G	G	G	NS	E	F	G	G	G
NS	NS	NS	NS	NS	NS	NS	NS	NS	G	NS	NS	NS	NS

TABLE 8-15. ADHESIVES CLASSIFIED BY CHEMICAL COMPOSITION. (CLAUSER, 1963)

Group →	Natural	Thermoplastic	Thermosetting	Elastomeric	Alloys[a]
Types Within Group	Casein, blood albumin, hide, bone, fish, starch (plain and modified); rosin, shellac, asphalt; inorganic (sodium silicate, litharge-glycerin)	Polyvinyl acetate, polyvinyl alcohol, acrylic, cellulose nitrate, asphalt, oleoresin	Phenolic, resorcinol, phenol-resorcinol, epoxy, epoxy-phenolic, urea, melamine, alkyd	Natural rubber, reclaim rubber, butadiene-styrene (GR-S), neoprene, acrylonitrile - butadiene (Buna-N), silicone	Penolic-polyvinyl butyral, phenolic-polyvinyl formal, phenolic-neoprene rubber, phenolic-nitrile rubber, modified epoxy
Most Used Form	Liquid powder	Liquid, some dry film	Liquid, but all forms common	Liquid, some film	Liquid, paste, film
Common Further Classifications	By vehicle (water emulsion is most common but many types are solvent dispersions)	By vehicle (most are solvent dispersions or water emulsions)	By cure requirements (heat and/or pressure most common but some are catalyst types)	By cure requirements (all are common); also by vehicle (most are solvent dispersions or water emulsions)	By cure requirements (usually heat and pressure except some epoxy types); by vehicle (most are solvent dispersions or 100% solids); and by type of adherends or end-service conditions
Bond Characteristics	Wide range, but generally low strength; good res to heat, chemicals; generally poor moisture res	Good to 150–200 F; poor creep strength; fair peel strength	Good to 200–500 F; good creep strength; fair peel strength	Good to 150–400 F; never melt completely; low strength; high flexibility	Balanced combination of properties of other chemical groups depending on formulation; generally higher strength over wider temp range

Major Type of Use[b]	Household, general purpose, quick set, long shelf life	Unstressed joints; designs with caps, overlaps, stiffeners	Stressed joints at slightly elevated temp	Unstressed joints on lightweight materials; joints in flexure	Where highest and strictest end-service conditions must be met; sometimes regardless of cost, as military uses
Materials Most Commonly Bonded	Wood (furniture) paper, cork, liners, packaging (food), textiles, some metals and plastics, Industrial uses giving way to other groups	Formulation range covers all materials, but emphasis on nonmetallics—esp wood, leather, cork, paper, etc.	Epoxy-phenolics for structural uses of most materials; others mainly for laminations; most epoxies are modified (alloys)	Few used "straight" for rubber, fabric, foil, paper, leather, plastics films; also as tapes. Most modified with synthetic resins	Metals, ceramics, glass, thermosetting plastics; nature of adherends often not as vital as design or end-service conditions (i.e., high strength, temp)

[a] "Alloy," as used here, refers to formulations containing resins from two or more *different* chemical groups. There are also formulations which benefit from compounding two resin types from the same chemical group (e.g., epoxy-phenolic).

[b] Although some uses of the "nonalloyed" adhesives absorb a large percentage of the quantity of adhesives sold, the uses are narrow in scope; from the standpoint of diversified applications, by far the most important use of any group is the forming of adhesive alloys.

TABLE 8-16. LOW TEMPERATURE MECHANICAL PROPERTIES OF SOME STRUCTURAL ADHESIVES. (RITTENHOUSE, 1968)

Adhesive*	Mfgr.	Class	Tensile Shear Strength (psi)					
			−420°F	−320°F	−100°F	−67°F	RT	+180°F
Metlbond 406 ™	N	Nylon-Epoxy	−	4360	−	5620	5980	3860
Metlbond 408 ™	N	Nylon-Epoxy	−	2800	−	6480	6490	3110
AF-40 ™	M	Nylon-Epoxy	−	−	−	5800	5000	4080
AF-41 ™	M	Nylon-Epoxy	−	−	−	6600	5700	3135
FM-1000 ™	B	Nylon-Epoxy	3370	3790	5210	−	6110	−
Resin 3135 ™	N	Epoxy-Polyamide	1640	1760	1850	−	2180	−
EC 1933B/A ™	M	Epoxy-Polyamide	1649	1775	−	3100	4000	3500
Restweld No. 4 ™	F	Epoxy-Polyamide	2010	1900	2380	−	2460	−
EC-1469 ™	M	Epoxy	2345	2535	2715	−	2715	−
Metlbond 302**™	N	Epoxy-Phenolic	3000	3135	2585	−	2595	−
Metlbond 4041 ™	N	Nitrile-Phenolic	1710	4400	5020	6200	3750	2500
Bondmaster M24B ™	R	Rubber-Epoxy Phenolic	1045	1230	1635	−	3625	−
Plastilock 601 ™	G	Rubber Resin	−	−	−	1250	3750	2000
Narmtape 111 ™	N	Nylon-Phenolic Epoxy	−	−	−	1500	3500	2750
APCO-1261 ™	A	Polyurethane	−	1940	−	−	1332	−
Swedlow 371W ™	S	Vinyl-Phenolic	1395	1795	2450	−	4855	−

TABLE 8-16. (*continued*)

Tee Peel Strength/lb-in. width						Peel Rate, (in. min.$^{-1}$)	Coefficient of Linear Expansion from −320°F to 32°F. (in./in.)
−423°F	−320°F	−67°F	−40°F	RT	+180°F		
−	−	26.6	−	60–70	49.6	2	3.06
−	−	15	−	50–60	−	2	3.24
−	−	38	54	118	90	20	3.08
−	−	11	−	111	71	20	2.79
−	−	−	−	60	−	3	2.88
0	0	−	−	−	−	2	3.21
2.25	1.7	−	−	−	−	2	2.28
−	−	−	−	−	−	−	−
−	−	−	−	−	−	−	−
−	3.66	−	−	3.33	−	2	−
−	−	5.0	5.0	45	−	2	−
−	−	−	−	−	−	−	−
−	−	1.0	1.0	18	17	2	−
−	−	3	3	22	45	2	−
−	3.33	−	−	1.92	−	2	4.10
−	−	−	−	−	−	−	−

*On aluminum alloy adherend.
RT—room temperature.
**On stainless Steel adherend.

Manufacturers:
A Applied Products Div., Hexcel Products Co.
B Bloomingdale Rubber Div., American Cyanimid Co.
F H. B. Fuller Co.
G B. F. Goodrich Co.

M 3M Co.
N Narmco Materials Div., Whittaker Corp.
R Rubber and Asbestos Corp.
S Swedlow Plastics Co.

TABLE 8-17. PROPERTIES OF SELECTED SAUEREISEN® CEMENTS.

Property	Saureisen Cement No.										
	1	6	7	8 and 9	29	31	48	54	74	78
Color	white	gray	white	white	gray	white	black	gray	gray	tan to gray
Compressive Strength (ASTM C-579), psi	3900	2700	3500	4500–5500	3900	3800	8000	1800–2550	2700	3300
Modulus of Rupture (ASTM C-580), psi	460	–	–	450	–	455	1300	450	–	–
Tensile Strength (ASTM C-608), psi	410	285	400	250	425	400	595	420–450	285	325
Shear Strength, psi	710	300	600	–	–	430	–	–	300	375
Dielectric Strength (ASTM D-149), Volts/mil										
at 70°F	12.5–51.0	12.5–51.0	12.5–51.0	76.0–101.5	25.0–51.0	12.5–38.0	–	–	12.5–51.0	12.5–51.0
at 750°F	up to 15.0	up to 15.0	up to 15.0	25.0–38.0	12.5–25.0	12.5–38.0	–	–	up to 15.0	12.5–25.0
at 1475°F	up to 1.3	up to 3.8	up to 1.3	12.5–25.0	up to 1.3	up to 2.0	–	–	up to 3.8	up to 7.5
Volume Resistivity (ASTM D-1829), ohm/cm										
at 70°F	10^8–10^9	10^7–10^8 (68°F)	10^8–10^9	10^{10}–10^{11}	10^7–10^9	10^9–10^{11}	–	–	10^7–10^8	10^8–10^9
at 750°F	10^4–10^5	10^4–10^5	10^4–10^5	10^9–10^{10}	10^4–10^6	10^7–10^8	–	–	10^4–10^5	10^5–10^6
at 1475°F	10^2–10^3	10^2–10^3	10^2–10^3	10^8–10^9	10^2–10^3	10^2–10^3	–	–	10^2–10^3	10^3–10^4
Dielectric Constant	3.5–6.0	5.0–7.0	3.5–6.0	–	5.0–7.0	–	–	–	5.0–7.0	3.5–4.5
Coefficient of Thermal Expansion, 10^{-6} in./in./°F	6.2	13.0	6.0	2.6	4.5	6.3	6.0	7.1	13.0	13.0
Maximum Service Temperature, °F	2000	2500	2500	2800	2000	2000	190	1750	2500	3000

*Data courtesy of Sauereisen Cements Company, 160 Gamma Drive, Pittsburgh, Pa. 15238.

Functions of Cements:

No.	1	Strong Inorganic Adhesive.	31	Acid-Proof Synthetic Ceramic.
	6	Air Setting Refractory.	48	Sulfur Melting Joint Compound.
	7	High Grade Silica Refractory.	54	Acid-Proof Concrete.
	8 and 9	Superior Electrical Insulation	74	High Alumina Refractory Cement.
	29	Zircon Base Cement.	78	Electrical Refractory Cement.

REFERENCES

Baer, E., "Engineering Design for Plastics," Van Nostrand Reinhold, New York, 1964.

Clauser, H., M. Riley, R. Fabian and D. Peckner, Eds., "Encyclopedia of Engineering Materials and Processes," Van Nostrand Reinhold, New York, 1963.

Harper, C. A., "Handbook of Materials and Processes in Electronics," McGraw-Hill Book Co., New York, 1970.

Licari, J. J. and E. R. Brands, "Organic Coatings for Metal and Plastic Surfaces," *Machine Design*, pp. 178–194 (May 25, 1967).

Materials Selector 1971, *Materials Eng.*, 72 (6), 1–496 (Nov. 1970).

Modern Plastics Encyclopedia 1970, *Modern Plastics*, 47, No. 14A (Oct. 1970).

Rittenhouse, J. B. and J. B. Singletary, "Space Materials Handbook," 3rd ed., U.S. Government Report AD 692 353, Clearing House for Federal Scientific & Technical Information, Springfield, Va.

BIBLIOGRAPHY

Bodnar, M. J., Ed., "Structural Adhesive Bonding," Wiley, New York, 1966.

Brandup, J. and E. H. Immergut, Eds., "Polymer Handbook," Wiley, New York, 1966.

Brydson, J. A., "Plastic Materials," 2nd ed., Van Nostrand Reinhold, New York, 1970.

Cagle, C. V., Ed., "Adhesive Bonding," McGraw-Hill, New York, 1968.

De Lollis, N. J., "Adhesives for Metals," Industrial Press, New York, 1970.

Gordon, M., "High Polymers-Structure and Physical Properties," 2nd ed., Iliffe, London, 1963.

Lee, H. L. and K. O. Neville, "Handbook of Epoxy Resins," McGraw-Hill, New York, 1966.

Mark, H. F., N. G. Gaylord and N. Bikales, Eds., "Encyclopedia of Polymer Science and Technology: Plastics, Resins, Rubber, Fibers," 16 vols., Wiley, New York, 1964–1972.

May, C. and Y. Tanaka, Eds., "Epoxy Resins: Chemistry and Technology," Marcel Dekker, New York, 1973.

Mayofis, T. M., "Plastic Insulating Materials: Chemistry, Properties and Applications," Van Nostrand Reinhold, New York, 1967.

Ogorkiewicz, R. E., Ed., "Engineering Properties of Thermoplastics," Wiley, New York, 1970.

Patrick, R. L., Ed., "Treatise on Adhesion and Adhesives: Materials," Vol. 2, Marcel Dekker, New York, 1969.

"Plastic Engineering Handbook of the SPI," 3rd ed., Van Nostrand Reinhold Co., New York, 1967.

Press, J. J., "Man-Made Textile Encyclopedia," Wiley, New York, 1959.

Ritchie, P. D., Ed., "Physics of Plastics," Van Nostrand Reinhold Co., New York, 1965.

Schramm, R. E., A. F. Clark and R. P. Reed, "A Compilation and Evaluation of Mechanical, Thermal and Electrical Properties of Selected Polymers," NBS Monograph 132, U.S. Government Printing Office, Washington, 1973.

Schwartz, M. M., "Modern Metal Joining Techniques," Wiley, New York, 1969.

Serafini, T. T. and J. L. Koenig, Eds., "Cryogenic Properties of Polymers," Marcel Dekker, New York, 1968.

Simonds, H. R. and J. M. Church, "The Encyclopedia of Basic Materials for Plastics," Van Nostrand Reinhold Co., New York, 1967.

Skeist, I., Ed., "Handbook of Adhesives," Van Nostrand Reinhold Co., New York, 1962.

CHAPTER **IX**
PROPERTIES OF SEMICONDUCTORS

The data on semiconductors, presented in this chapter, include diffusion data useful in the production and characterization of selected semiconductors, as well as resistivity, drift mobility and Hall mobility values for these semiconductors. Brief tabulation of inorganic semiconductors and information on organic solid semiconductors and liquid semiconductors complete the coverage in Chapter IX.

The data are presented in the following Figures and Tables:

TABLE 9-1. ABBREVIATIONS FOR DIFFUSION TABULATIONS.

D_0 = Frequency Factor, cm² sec⁻¹
D = Diffusion Coefficient, cm² sec⁻¹
E = Activation Energy, eV

The above quantities are related by the equation:

$$D = D_0 \exp\left(\frac{-E}{kT}\right)$$

where

$k = 8.6164 \times 10^{-5}$ eV deg⁻¹
T = temperature, °K

TABLE 9-2. VALUES FOR THE SEMICONDUCTOR DIFFUSION FREQUENCY FACTOR D_0. (YARBROUGH*, 1968).

Impurity	(See Equation 1) Diffusion Frequency Factor, D_0, cm^2 sec^{-1}						
	AlSb	GaP	GaAs	GaSb	InP	InAs	InSb
Ag			2.5, − 3(s) 4.0, − 4(f)			7.3, − 4	D
As			4.0, +21				
Au			1.0, − 3			5.8, − 3	7.0, − 4
Be			7.3, − 6				
Cd			5.0, − 2			7.4, − 4(s)	1.0, − 5
Co							2.7, − 11
Cu	3.5, − 3		3.0, − 2(i)			3.6, − 3	3.0, − 5(i)
Ga			1.0, +7	3.2, +3			
Ge						3.7, − 6	
Hg							4.0, − 6
In			D	1.2, − 7	1.0, +5		5.0, − 2
Li			4.5, +1				7.0, − 4
Mg						1.98, − 6	
Mn			6.5, − 1				
P			D		7.0, +10		
S			2.6, − 5			6.78	
Sb				8.7, − 3			5.0, − 2
Se			3.0, +3			1.3, +1	
Sn			6.0, − 4	2.4, − 5		1.5, − 6	5.5, − 8
Te				3.8, − 4		3.4, − 5	1.7, − 7
Tm			2.3, − 16				
Zn		1.0	C		C	4.2, − 3(a)	C

Impurity	ZnS	ZnSe	CdS	CdSe	CdTe	HgSe	SiC
Ag			2.5, +1				
Al							2.0, − 1
Au					6.7, +1		
B							1.6, +2
C							C
Cd			3.4				
Cr							2.28, − 1
Cu		1.7, − 5	1.5, − 3				
In					4.1, − 1		
Mn							C
Sb						6.3, − 5	
Se				2.6, − 3(b)			
Zn	1.5, +4(c)						

Explanation of Symbols and Notes

3.5, −3 means 3.5 × 10^{-3}; (s) means slow diffusion component; (f) means fast diffusion component; (i) means impurity diffusing interstitially; C means parameters not given since diffusion coefficient is concentration dependent; D means only diffusion coefficient available.

(a) Value is quoted for Zn concentration 4.5 × 10^{19} atoms/cm^3.

(b) Value is quoted for condition of minimum Cd pressure.

(c) Value is quoted for the interval 940–1030°C. D_0 = 3.0, −4 below 940°C, D_0 = 1.0, +16 above 1030°C.

*From *Solid State Technology*, Vol. 11, Nov. 1968, Binary Compound Semiconductors by D. W. Yarbrough, with permission.

TABLE 9-3. VALUES FOR THE SEMICONDUCTOR DIFFUSION ACTIVATION ENERGY. (YARBROUGH, 1968)*

Impurity	Activation Energy E (eV)						
	AlSb	GaP	GaAs	GaSb	InP	InAs	InSb
Ag			1.5(s) 0.8(f)			0.26	
As			10.2				
Au			1.0			0.65	0.32
Be			1.2				
Cd			2.6			1.15(s)	1.1
Co							0.39
Cu	0.36		0.53(i)			0.52	0.37(i)
Ga			5.6	3.14			
Ge						1.17	
Hg							1.17
In				0.53	3.85		1.81
Li			0.98				0.28
Mg						1.17	
Mn			2.49				
P					5.65		
S			1.86			2.20	
Sb				1.13			1.93
Se			4.16			2.20	
Sn			2.50	0.80		1.17	0.75
Te				1.20		1.28	0.57
Tm			−1.0				
Zn		2.1	C		C	0.96(a)	C

	ZnS	ZnSe	CdS	CdSe	CdTe	HgSe	SiC
Ag			1.2				
Al							4.9
Au					2.0		
B							5.6
Cd			2.0				
Cr							4.8
Cu		0.56	0.76				
In					1.6		
Sb						0.85	
Zn	3.25(c)				1.55(b)		

Explanation of Symbols and Notes

(s) means slow diffusion component; (f) means fast diffusion components; (i) means impurity diffusing interstitially; C means parameters not given since diffusion coefficient is concentration dependent;

(a) Value is given for Zn concentration of 4.5×10^{19} atoms/cm^3.

(b) Value is given for condition of minimum Cd pressure.

(c) Value is given for the temperature interval 940–1030°C. E = 1.52 eV below 940°C. E = 6.50 eV above 1030°C.

*From *Solid State Technology*, Vol. 11, Nov. 1968, Binary Compound Semiconductors by D. W. Yarbrough, with permission.

TABLE 9-4. VALUES OF SEMICONDUCTOR DIFFUSION COEFFICIENTS (YARBROUGH, 1968)*

Impurity	Tmin (°C)	D @ Tmin (cm²/sec)	Tmax (°C)	D @ Tmax (cm²/sec)
			AlSb	
Cu	150	1.80, −7	500	1.57, −5
			GaP	
Zn	700		1300	
			GaAs	
Ga	1120	5.50, −14	1230	1.67, −12
As	1200	5.06, −14	1220	1.49, −13
Ag(s)	500	4.16, −13	1160	1.33, −8
Ag(f)	500	2.44, −9	1160	6.15, −7
Au	740	1.06, −8	1020	1.27, −7
Be	880	4.15, −11	990	1.19, −10
Cd	850	1.08, −13	1150	3.10, −11
Cu(i)	100	2.08, −9	500	1.05, −5
In	1000	D = 7.0, −11		
Li	250	1.63, −8	500	1.84, −5
Mn	850	4.36, −12	1100	4.71, −10
P	725	5.00, −14	1125	6.00, −12
S	900	2.65, −13	1100	3.87, −12
Se	1000	1.02, −13	1200	1.76, −11
Si				
Sn	1060	2.12, −13	1200	1.68, −12
Tm	800	1.14, −11	1000	2.09, −12
Zn				
			GaSb	
Ga	650	2.30, −14	700	1.75, −13
In	320	3.76, −12	650	1.53, −10
Sb	320	2.18, −12	650	4.30, −9
Sn	320	3.82, −12	650	1.03, −9
Te	320	2.42, −16	650	1.07, −12
			InP	
In	890		920	
P	900	3.73, −14	980	1.32, −12
Zn	700		900	
			InAs	
Ag	450	1.13, −5	900	5.58, −5
Au	600	1.03, −6	840	6.61, −6
Cd(s)	650	3.90, −10	900	8.48, −9
Cu	240	2.77, −8	510	1.60, −6
Ge	600	6.59, −13	900	3.52, −11
Mg	600	3.49, −13	900	1.86, −11
S	600	1.36, −12	900	2.40, −9
Se	600	2.51, −12	900	4.43, −9
Sn	600	2.63, −13	900	1.40, −11
Te	600	1.40, −12	900	1.09, −10
Zn	600	1.21, −8	900	3.15, −7
			InSb	
In	470	2.65, −14	520	1.57, −13
Sb	470	4.07, −15	520	2.72, −14
Ag	390	D = 4, −9		
Au	140	8.74, −8	510	6.10, −6
Cd	250	2.53, −16	500	6.75, −13
Co	420	3.94, −14	500	7.74, −14
Cu(i)	230	5.90, −9	490	1.08, −7
Hg	420	1.24, −14	500	9.44, −14

TABLE 9.4. (*Continued*)

Impurity	Tmin (°C)	D @ Tmin (cm²/sec)	Tmax (°C)	D @ Tmax (cm²/sec)
Li	0	4.27, -9	210	8.40, -7
Sn	400	1.33, -13	520	9.43, -13
Te	300	1.65, -12	500	3.27, -11
			ZnS	
Zn	940	4.72, -10	1030	4.04, -9
			ZnSe	
Cu	200	1.84, -11	570	7.63, -9
			CdS	
Ag	250	6.87, -11	500	3.76, -7
Cd	750	4.78, -10	1000	4.11, -8
Cu	450	7.57, -9	750	2.71, -7
			CdSe	
Se	700	2.44, -11	1000	1.90, -9
			CdTe	
Au	600	1.91, -10	1000	8.10, -7
In	450	2.89, -12	1000	1.90, -7
O	200	D = 4, -10		
			HgSe	
Sb	540	3.39, -10	630	1.14, -9
			SiC	
Al	1800	2.45, -13	2250	3.26, -11
B	1850	8.13, -12	2250	1.04, -9
C				
Cr	1700	1.25, -13	1900	1.68, -12
N				

Explanation of Symbols

1.80, -7 means 1.80×10^{-7}; (s) —slow component; (f) —fast component; (i) —material diffusing in interstitial mode; D—a single diffusion coefficient reported.

*From *Solid State Technology*, Vol. 11, Nov. 1968, Binary Compound Semiconductors by D. W. Yarbrough, with permission.

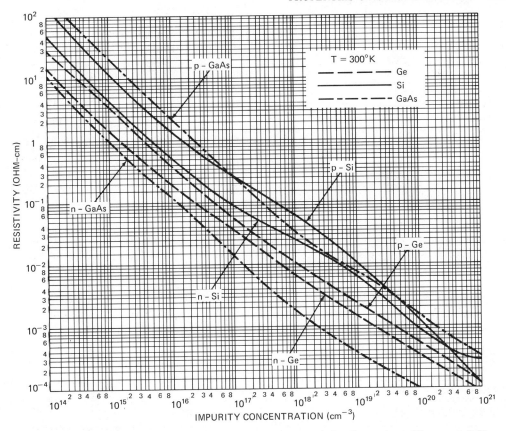

FIGURE 9-1.* Resistivity as a Function of Impurity Concentration for Germanium, Silicon, and Gallium Arsenide at 300°K (SZE 1968)

*Reprinted from *Solid State Electronics*, **11**, 599–602, 1968, with the permission of the copyright licensee, Microforms International Marketing Corporation.

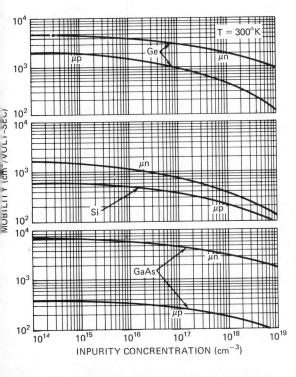

FIGURE 9-2.* Drift Mobility of Germanium and Silicon, and Hall Mobility of Gallium Arsenide, at 300°K as a Function of Impurity Concentration (SZE 1968)

*Reprinted from *Solid State Electronics*, **11**, 599–602, 1968, with the permission of the copyright licensee, Microforms International Marketing Corporation.

TABLE 9-5.   Properties of IV Solid Inorganic Semiconductors
(Neuberger, 1965, 1969, 1971), (Schultz, 1964), (Wolf, 1969), (Hampel, 1968), (Moses, 1970, 1971)

CARBON (DIAMOND)	Unit	Temp.* ($^\circ$K)	Value
PHYSICAL PROPERTIES			
Formula weight	g		12.0112
Density	g/cm^3		3.52
Melting point	$^\circ$K		4100
Melting point pressure coeff.	$^\circ$K/kbar		
Boiling point	$^\circ$K		4473
Symmetry			cubic
Lattice parameter, a_0	Å		3.56679
" " ,			
" " ,			
THERMAL PROPERTIES			
Thermal conductivity	W/cm $^\circ$K	273	6.55
" "			
" "			
Linear expansion coeff.	$10^{-6}/^\circ$K	273–351	1.19
Thermal diffusivity	cm^2/sec	293	3.62
Specific heat (C_p)	cal/g deg	298	0.122
" "			
" "			
Debye temperature	$^\circ$K		2240
Spectral emissivity	none		
Vapor pressure	torr	3719	76
" "		3403	7.6
		3137	0.6
Heat of Sublimation	kcal/g		
Heat of Fusion	cal/g		
Entropy	cal/g-atom deg	298	0.568
Zero point energy	cal/g-atom		
MAGNETIC PROPERTIES			
Magnetic susceptibility	10^{-7} emu/g	300	–4.9
Gyromagnetic properties, g factor	–		2.0031
" " , line width	Oe		
MECHANICAL PROPERTIES			
Hardness	Mohs		10
Young's modulus	10^{11} dyne/cm^2		112
Bulk modulus	10^{11} dyne/cm^2		54.5
Elastic constant (stiffness), C_{11}	10^{11} dyne/cm^2		107.6
" " C_{12}			12.50
" " C_{44}			57.58
C_{33}			
Elastic constant (compliance) S_{11}	10^{-12} cm^2/dyne		0.0953
" " S_{12}			–0.00993
" " S_{44}			0.174
S_{33}			
ACOUSTIC PROPERTIES			
Sound velocity	10^3 m/sec		18.1
" "			
" "			
ELECTRICAL/ELECTRONIC PROPERTIES			
Electrical resistivity	ohm · cm		5×10^{14}
Dielectric constant	none		5.93
Electron mobility	cm^2/V sec		2000
Hole mobility	cm^2/V sec		1500
Energy gap	eV		D7.4; I5.47
Work function	eV		6.02
Electron affinity	eV		0.9
Photoelectric threshold	eV		6.02
Photoelectric yield	eV		
Superconducting transition temp.	$^\circ$K		
Seebeck coefficient	μV/$^\circ$K	300	3.5×10^3
OPTICAL PROPERTIES			
Transmission	%		
Reflectivity	%		
Refractive index at 0.589μ	none		2.4173

*Note: Where not stated, room temperature may be assumed.

TABLE 9-5. (*Continued*)

GERMANIUM	Unit	Temp.* (°K)	Value
PHYSICAL PROPERTIES			
Formula weight	g		72.59
Density	g/cm^3	300	5.3243
Melting point	°K		1209
Melting point pressure coeff.	°K/kbar		
Boiling point	°K		3100
Symmetry			cubic
Lattice parameter, a_0	Å	300	5.65754
" " ,			
" " ,			
THERMAL PROPERTIES			
Thermal conductivity	W/cm °K	150	1.32
" "		300	0.589
" "		300	0.338
Linear expansion coeff.	10^{-6}/°K	273–373	5.75
Thermal diffusivity	cm^2/sec		0.36
Specific heat (C_p)	cal/g deg	298	0.0755
" "			
" "			
Debye temperature	°K		378
Spectral emissivity at 0.65μ	none	1000	0.56
Vapor pressure	torr	2635	76
" "		2289	7.6
" "		2024	0.76
Heat of Sublimation	kcal/g	298	1.23
Heat of Fusion	cal/g		111.5
Entropy	cal/g-atom deg	298	10.14
Zero point energy	cal/g-atom		
MAGNETIC PROPERTIES			
Magnetic susceptibility	10^{-7} emu/g	293	1.22
Gyromagnetic properties, g factor	–	1.4	g⊥1.92; g∥0.87
" " , line width	Oe		
MECHANICAL PROPERTIES			
Hardness	Mohs		6.25
Young's modulus	10^{11} dyne/cm^2		10.3
Bulk modulus	10^{11} dyne/cm^2		7.7
Elastic constant (stiffness), C_{11}	10^{11} dyne/cm^2		12.89
" " C_{12}			4.83
" " C_{44}			6.71
" " C_{33}			
Elastic constant (compliance) S_{11}	10^{-12} cm^2/dyne		0.978
" " S_{12}			–0.266
" " S_{44}			1.490
" " S_{33}			
ACOUSTIC PROPERTIES			
Sound velocity	10^3 m/sec		5.3
" "			
ELECTRICAL/ELECTRONIC PROPERTIES			
Electrical resistivity	ohm · cm	300	47.0
Dielectric constant	none	300	16.0
Electron mobility	cm^2/V sec		3900
Hole mobility	cm^2/V sec		1900
Energy gap	eV	300	D0.805; I0.664
Work function	eV		4.5
Electron affinity	eV	300	4.0
Photoelectric threshold	eV	300	4.8
Photoelectric yield	eV	300	
Superconducting transition temp.	°K		
Seebeck coefficient	μV/°K	293	–0.9
OPTICAL PROPERTIES			
Transmission at 1.8–23μ, (2 mm. thick)	%	300	⩾ 10
Reflectivity at 0.59μ	%	300	49
Refractive index at 0.59μ	none	300	5.6

*Note: Where not stated, room temperature may be assumed.

TABLE 9-5. (*Continued*)

SILICON	Unit	Temp.* ($^\circ$K)	Value
PHYSICAL PROPERTIES			
Formula weight	g		28.09
Density	g/cm^3	298	2.32831
Melting point	$^\circ$K		1685
Melting point pressure coeff.	$^\circ$K/kbar		-5.8
Boiling point	$^\circ$K		3540
Symmetry			cubic
Lattice parameter, a_0	Å		5.43089
" " ,			
THERMAL PROPERTIES			
Thermal conductivity	W/cm $^\circ$K	4.0	2.26
" "		100	8.84
" "		273	1.68
		500	0.762
Linear expansion coeff.	$10^{-6}/^\circ$K	298	2.33
Thermal diffusivity	cm^2/sec	300	0.85
Specific heat (C_p)	cal/g deg	300	0.171
" "		373	0.1840
" "		1573	0.2215
Debye temperature	$^\circ$K		689
Spectral emissivity	none		
Vapor pressure	torr	3023	76
" "		2638	7.6
		2340	0.76
Heat of Sublimation	kcal/g	298	4.91
Heat of Fusion	cal/g		397
Entropy	cal/g-atom deg	298	4.497
Zero point energy	cal/g-atom	298	1435
MAGNETIC PROPERTIES			
Magnetic susceptibility	10^{-7} emu/g	293	-1.3
Gyromagnetic properties, g factor	–	100–400	2.0058
" " , line width	Oe	100–400	~ 5.2
MECHANICAL PROPERTIES			
Hardness	Mohs		7.0
Young's modulus	10^{11} dyne/cm^2	298	10.67
Bulk modulus	10^{11} dyne/cm^2	298	9.788
Elastic constant (stiffness), C_{11}	10^{11} dyne/cm^2	300	16.58
" " C_{12}		300	6.39
" " C_{44}		300	7.96
" " C_{33}			
Elastic constant (compliance) S_{11}	10^{-12} cm^2/dyne	300	0.768
" " S_{12}		300	-0.214
" " S_{44}		300	1.26
" " S_{33}			
ACOUSTIC PROPERTIES			
Sound velocity	10^3 m/sec		8.95
" "			
ELECTRICAL/ELECTRONIC PROPERTIES			
Electrical resistivity	ohm · cm	300	2.3×10^5
Dielectric constant	none		ϵ_∞ 11.94
Electron mobility	cm^2/V sec		1880
Hole mobility	cm^2/V sec		400
Energy gap	eV	300	1.12
Work function	eV		4.83
Electron affinity	eV		3.39
Photoelectric threshold	eV	300	5.07
Photoelectric yield	eV	300	4×10^{-4} (2μ)
Superconducting transition temp.	$^\circ$K		6.7 (120 kbar)
Seebeck coefficient	μV/$^\circ$K	375	-650
OPTICAL PROPERTIES			
Transmission at 1.2–15μ, (20 mm. thick)	%	300	$\geqslant 10$
Reflectivity at 2–30μ	%	300	30
Refractive index at 10μ	none	300	3.4179

*Note: Where not stated, room temperature may be assumed.

TABLE 9-5. (*Continued*)

ALPHA TIN (grey)	Unit	Temp.* (°K)	Value
PHYSICAL PROPERTIES			
Formula weight	g		118.69
Density	g/cm^3		5.77
Melting point (transformation)	°K		286.2
Melting point pressure coeff.	°K/kbar		
Boiling point	°K		2766
Symmetry			cubic
Lattice parameter, a_0	Å		6.4892
" " ,			
THERMAL PROPERTIES			
Thermal conductivity	W/cm °K		
" "			
" "			
Linear expansion coeff.	10^{-6}/°K	215	5.3
Thermal diffusivity	cm^2/sec		
Specific heat (C_p)	cal/g deg	298	0.052
" "			
" "			
Debye temperature	°K		236
Spectral emissivity	none		
Vapor pressure	torr		
" "			
Heat of Sublimation	kcal/g		
Heat of Fusion	cal/g		
Entropy	cal/g-atom deg		
Zero point energy	cal/g-atom		
MAGNETIC PROPERTIES			
Magnetic susceptibility	10^{-7} emu/g		
Gyromagnetic properties, g factor	—		
" " , line width	Oe		
MECHANICAL PROPERTIES			
Hardness	Mohs		5.25
Young's modulus	10^{11} dyne/cm^2		11.1
Bulk modulus	10^{11} dyne/cm^2		
Elastic constant (stiffness), C_{11}	10^{11} dyne/cm^2		
" " C_{12}			
" " C_{44}			
" " C_{33}			
Elastic constant (compliance) S_{11}	10^{-12} cm^2/dyne		
" " S_{12}			
" " S_{44}			
" " S_{33}			
ACOUSTIC PROPERTIES			
Sound velocity	10^3 m/sec		
" "			
ELECTRICAL/ELECTRONIC PROPERTIES			
Electrical resistivity	ohm · cm	273	4.8 x 10^{-4}
Dielectric constant	none		
Electron mobility	cm^2/V sec	273	1400
Hole mobility	cm^2/V sec	273	1200
Energy gap	eV		0.085
Work function	eV		
Electron affinity	eV		
Photoelectric threshold	eV		
Photoelectric yield	eV		
Superconducting transition temp.	°K		
Seebeck coefficient	μV/°K		
OPTICAL PROPERTIES			
Transmission	%		
Reflectivity	%		
Refractive index	none		

*Note: Where not stated, room temperature may be assumed.

TABLE 9-6. PROPERTIES OF V SOLID INORGANIC SEMICONDUCTORS.
(Neuberger, 1970), (Hampel, 1968)

ARSENIC	Unit	Temp.* (°K)	Value
PHYSICAL PROPERTIES			
Formula weight	g		74.9216
Density	g/cm^3		5.72
Melting point	°K		1090 (28 atm)
Melting point pressure coeff.	°K/kbar		
Boiling point	°K		sublimes
Symmetry			hexagonal
Lattice parameter, a_0	Å		3.7598
" " , c_0 ,			10.547
THERMAL PROPERTIES			
Thermal conductivity	W/cm °K		
" "			
" "			
Linear expansion coeff.	10^{-6}/°K		4.7
Thermal diffusivity	cm^2/sec		
Specific heat (C_p)	cal/g deg	300–473	0.082
" "			
" "			
Debye temperature	°K		280.5
Spectral emissivity	none		
Vapor pressure	torr	783	76
" "		708	7.6
		647	0.76
Heat of Sublimation	kcal/g		0.102
Heat of Fusion	cal/g		88.5
Entropy	cal/g-atom deg		41.61
Zero point energy	cal/g-atom		
MAGNETIC PROPERTIES			
Magnetic susceptibility	10^{-7} emu/g	292	5.78‖; −2.79⊥c
Gyromagnetic properties, g factor	−	4	75
" " , line width	Oe		
MECHANICAL PROPERTIES			
Hardness	Mohs		3.5
Young's modulus	10^{11} dyne/cm^2		
Bulk modulus	10^{11} dyne/cm^2		
Elastic constant (stiffness), C_{11}	10^{11} dyne/cm^2		
" " C_{12}			
" " C_{44}			
C_{33}			
Elastic constant (compliance) S_{11}	10^{-12} cm^2/dyne		
" " S_{12}			
" " S_{44}			
S_{33}			
ACOUSTIC PROPERTIES			
Sound velocity	10^3 m/sec		
" "			
ELECTRICAL/ELECTRONIC PROPERTIES			35.6 x 10^{-6} ‖a
Electrical resistivity	ohm · cm	293	25.3 x 10^{-6} ⊥a
Dielectric constant	none	300	10.23
Electron mobility	cm^2/V sec	305	460
Hole mobility	cm^2/V sec		
Energy gap	eV	300	1.2
Work function	eV		4.79
Electron affinity	eV		
Photoelectric threshold	eV		
Photoelectric yield	eV		
Superconducting transition temp.	°K		0.5 (125 kbars)
Seebeck coefficient	μV/°K	300	18‖; −3⊥
OPTICAL PROPERTIES			
Transmission	%		
Reflectivity at 0.589μ	%		46.5 (film)
Refractive index at 0.89μ	none		4.00 (film)

*Note: Where not stated, room temperature may be assumed.

TABLE 9-7. PROPERTIES OF II–VI SOLID INORGANIC SEMICONDUCTORS.
(Neuberger, 1969A), (Abrikosov, 1969), (Hampel, 1968), (Moss, 1959), (Moses, 1970, 1971).

BARIUM OXIDE	Unit	Temp.* (°K)	Value
PHYSICAL PROPERTIES			
Formula weight	g		153.35
Density	g/cm^3		5.685
Melting point	°K		2469
Melting point pressure coeff.	°K/kbar		
Boiling point	°K		
Symmetry			cubic
Lattice parameter, a_0	Å		5.539
" " ,			
" " ,			
THERMAL PROPERTIES			
Thermal conductivity	W/cm °K	500	0.0586
" "		600	0.837
" "			
Linear expansion coeff.	$10^{-6}/°K$		17.8
Thermal diffusivity	cm^2/sec		
Specific heat (C_p)	cal/g deg	100	0.05
" "		300	0.075
" "			
Debye temperature	°K		320
Spectral emissivity	none		
Vapor pressure	torr		
" "			
Heat of Sublimation	kcal/g		
Heat of Fusion	cal/g		
Entropy	cal/g-atom deg	298	16.8
Zero point energy	cal/g-atom		
MAGNETIC PROPERTIES			
Magnetic susceptibility	10^{-7} emu/g		–1.9
Gyromagnetic properties, g factor	—		
" " , line width	Oe		
MECHANICAL PROPERTIES			
Hardness	Mohs		
Young's modulus	10^{11} dyne/cm^2		
Bulk modulus	10^{11} dyne/cm^2		5.0
Elastic constant (stiffness), C_{11}	10^{11} dyne/cm^2		
" " C_{12}			
" " C_{44}			
" " C_{33}			
Elastic constant (compliance) S_{11}	10^{-12} cm^2/dyne		
" " S_{12}			
" " S_{44}			
" " S_{33}			
ACOUSTIC PROPERTIES			
Sound velocity	10^3 m/sec		
" "			
ELECTRICAL/ELECTRONIC PROPERTIES			
Electrical resistivity	ohm · cm	667	10^7
Dielectric constant	none		ϵ_0 34; ϵ_∞ 4
Electron mobility	cm^2/V sec	400–800	10
Hole mobility	cm^2/V sec		
Energy gap	eV	123–303	3.8
Work function	eV	800	1.52
Electron affinity	eV		0.57
Photoelectric threshold	eV		5
Photoelectric yield	eV		
Superconducting transition temp.	°K		
Seebeck coefficient	$\mu V/°K$		2–3
OPTICAL PROPERTIES			
Transmission	%		
Reflectivity	%		
Refractive index at 0.4–0.7 μ	none	300	1.9

*Note: Where not stated, room temperature may be assumed.

TABLE 9-7 (*Continued*)

BARIUM SELENIDE	Unit	Temp.* (°K)	Value
PHYSICAL PROPERTIES			
Formula weight	g		216.32
Density	g/cm^3		
Melting point	°K		2103
Melting point pressure coeff.	°K/kbar		
Boiling point	°K		
Symmetry			cubic
Lattice parameter, a_0	Å		6.600
" " ,			
THERMAL PROPERTIES			
Thermal conductivity	W/cm °K		
" "			
" "			
Linear expansion coeff.	10^{-6}/°K		
Thermal diffusivity	cm^2/sec		
Specific heat (C_p)	cal/g deg		
" "			
" "			
Debye temperature	°K		
Spectral emissivity	none		
Vapor pressure	torr		
" "			
Heat of Sublimation	kcal/g		
Heat of Fusion	cal/g		
Entropy	cal/g-atom deg		
Zero point energy	cal/g-atom		
MAGNETIC PROPERTIES			
Magnetic susceptibility	10^{-7} emu/g		
Gyromagnetic properties, g factor	—		
" " , line width	Oe		
MECHANICAL PROPERTIES			
Hardness	Mohs		
Young's modulus	10^{11} dyne/cm^2		
Bulk modulus	10^{11} dyne/cm^2	296	400
Elastic constant (stiffness), C_{11}	10^{11} dyne/cm^2		
" " C_{12}			
" " C_{44}			
" " C_{33}			
Elastic constant (compliance) S_{11}	10^{-12} cm^2/dyne		
" " S_{12}			
" " S_{44}			
" " S_{33}			
ACOUSTIC PROPERTIES			
Sound velocity	10^3 m/sec		
" "			
ELECTRICAL/ELECTRONIC PROPERTIES			
Electrical resistivity	ohm · cm	300	3 x 10^{10}
Dielectric constant	none		
Electron mobility	cm^2/V sec		
Hole mobility	cm^2/V sec		
Energy gap	eV	70	D3.60
Work function	eV	1043–1373	2.07
Electron affinity	eV		0.95
Photoelectric threshold	eV		4.7
Photoelectric yield	eV		
Superconducting transition temp.	°K		
Seebeck coefficient	μV/°K		
OPTICAL PROPERTIES			
Transmission	%		
Reflectivity	%		
Refractive index at 0.589μ	none		2.268

*Note: Where not stated, room temperature may be assumed.

TABLE 9-7. (Continued)

BARIUM SULFIDE	Unit	Temp.* (°K)	Value
PHYSICAL PROPERTIES			
Formula weight	g		169.43
Density	g/cm^3		4.376
Melting point	°K		3024
Melting point pressure coeff.	°K/kbar		
Boiling point	°K		~ 3273
Symmetry			cubic
Lattice parameter, a_0	Å		6.3875
" " ,			
THERMAL PROPERTIES			
Thermal conductivity	W/cm °K		
" "			
" "			
Linear expansion coeff.	$10^{-6}/°K$	303–348	34
Thermal diffusivity	cm^2/sec		
Specific heat (C_p)	cal/g deg	100	0.053
" "		300	0.07
" "			
Debye temperature	°K		
Spectral emissivity	none		
Vapor pressure	torr		
" "			
Heat of Sublimation	kcal/g		
Heat of Fusion	cal/g		
Entropy	cal/g-atom deg		
Zero point energy	cal/g-atom		
MAGNETIC PROPERTIES			
Magnetic susceptibility	10^{-7} emu/g		
Gyromagnetic properties, g factor	—		
" " , line width	Oe		
MECHANICAL PROPERTIES			
Hardness	Mohs		
Young's modulus	10^{11} dyne/cm^2		
Bulk modulus	10^{11} dyne/cm^2		3.45
Elastic constant (stiffness), C_{11}	10^{11} dyne/cm^2		
" " C_{12}			
" " C_{44}			
C_{33}			
Elastic constant (compliance) S_{11}	10^{-12} cm^2/dyne		
" " S_{12}			
" " S_{44}			
S_{33}			
ACOUSTIC PROPERTIES			
Sound velocity	10^3 m/sec		
" "			
ELECTRICAL/ELECTRONIC PROPERTIES			
Electrical resistivity	ohm · cm	700	10^9
Dielectric constant	none		19.3
Electron mobility	cm^2/V sec		
Hole mobility	cm^2/V sec		
Energy gap	eV	70	D3.90
Work function	eV	800	2.93
Electron affinity	eV	800	0.84
Photoelectric threshold	eV		4.9
Photoelectric yield	eV		
Superconducting transition temp.	°K		
Seebeck coefficient	$\mu V/°K$	970–1270	~ 2.5
OPTICAL PROPERTIES			
Transmission	%		
Reflectivity	%		
Refractive index at 0.589μ	none		2.155

*Note: Where not stated, room temperature may be assumed.

TABLE 9-7. (*Continued*)

BARIUM TELLURIDE	Unit	Temp.* ($^\circ$K)	Value
PHYSICAL PROPERTIES			
Formula weight	g		264.97
Density	g/cm^3		
Melting point	$^\circ$K		1783
Melting point pressure coeff.	$^\circ$K/kbar		
Boiling point	$^\circ$K		
Symmetry			cubic
Lattice parameter, a_0	Å		7.004
" " ,			
" " ,			
THERMAL PROPERTIES			
Thermal conductivity	W/cm $^\circ$K		
" "			
" "			
Linear expansion coeff.	$10^{-6}/^\circ$K		
Thermal diffusivity	cm^2/sec		
Specific heat (C_p)	cal/g deg		
" "			
" "			
Debye temperature	$^\circ$K		
Spectral emissivity	none		
Vapor pressure	torr		
" "			
Heat of Sublimation	kcal/g		
Heat of Fusion	cal/g		
Entropy	cal/g-atom deg		
Zero point energy	cal/g-atom		
MAGNETIC PROPERTIES			
Magnetic susceptibility	10^{-7} emu/g		
Gyromagnetic properties, g factor	—		
" " , line width	Oe		
MECHANICAL PROPERTIES			
Hardness	Mohs		
Young's modulus	10^{11} dyne/cm^2		
Bulk modulus	10^{11} dyne/cm^2		3.4
Elastic constant (stiffness), C_{11}	10^{11} dyne/cm^2		
" " C_{12}			
" " C_{44}			
" " C_{33}			
Elastic constant (compliance) S_{11}	10^{-12} cm^2/dyne		
" " S_{12}			
" " S_{44}			
" " S_{33}			
ACOUSTIC PROPERTIES			
Sound velocity	10^3 m/sec		
" "			
ELECTRICAL/ELECTRONIC PROPERTIES			
Electrical resistivity	ohm · cm	300	7×10^7
Dielectric constant	none		
Electron mobility	cm^2/V sec		
Hole mobility	cm^2/V sec		
Energy gap	eV		D3.4
Work function	eV	800	2.72–2.88
Electron affinity	eV		1.43
Photoelectric threshold	eV		4.8
Photoelectric yield	eV		
Superconducting transition temp.	$^\circ$K		
Seebeck coefficient	μV/$^\circ$K		
OPTICAL PROPERTIES			
Transmission	%		
Reflectivity	%		
Refractive index at 0.589 μ	none		2.440

*Note: Where not stated, room temperature may be assumed.

TABLE 9-7. (*Continued*)

ALPHA BERYLLIUM OXIDE	Unit	Temp.* (°K)	Value
PHYSICAL PROPERTIES			
Formula weight	g		25.013
Density	g/cm^3	300	3.03
Melting point	°K		2843
Melting point pressure coeff.	°K/kbar		
Boiling point	°K		
Symmetry			hexagonal
Lattice parameter, a_0	Å		2.6979
" " , c_0			4.3772
"			
THERMAL PROPERTIES			
Thermal conductivity	W/cm °K	100	4.25
" "		300	2.72
" "		1000	0.47
Linear expansion coeff.	10^{-6}/°K	373	5.35
Thermal diffusivity	cm^2/sec		
Specific heat (C_p)	cal/g deg	300	0.246
" "		1000	0.471
" "			
Debye temperature	°K		
Spectral emissivity	none		
Vapor pressure	torr		
" "			
Heat of Sublimation	kcal/g		
Heat of Fusion	cal/g		
Entropy	cal/g-atom deg	298	3.37
Zero point energy	cal/g-atom		
MAGNETIC PROPERTIES			
Magnetic susceptibility	10^{-7} emu/g		
Gyromagnetic properties, g factor	—		
" " , line width	Oe		
MECHANICAL PROPERTIES			
Hardness	Mohs		9
Young's modulus	10^{11} dyne/cm^2	293	36.5
Bulk modulus	10^{11} dyne/cm^2		
Elastic constant (stiffness), C_{11}	10^{11} dyne/cm^2		
" " C_{12}			
" " C_{44}			
" " C_{33}			
Elastic constant (compliance) S_{11}	10^{-12} cm^2/dyne	300	0.23
" " S_{12}			
" " S_{44}			
" " S_{33}			
ACOUSTIC PROPERTIES			
Sound velocity at	10^3 m/sec		12
" "			
ELECTRICAL/ELECTRONIC PROPERTIES			
Electrical resistivity	ohm · cm	293	10^{16}
Dielectric constant	none	300	ϵ_0 7.66; $\epsilon_\infty \sim 3$
Electron mobility	cm^2/V sec		
Hole mobility	cm^2/V sec		
Energy gap	eV	300	D $\sim$ 11.6
Work function	eV	1400	3.8; 4.7
Electron affinity	eV		
Photoelectric threshold	eV		
Photoelectric yield	eV		
Superconducting transition temp.	°K		
Seebeck coefficient	μV/°K		
OPTICAL PROPERTIES			
Transmission	%		
Reflectivity	%		
Refractive index	none		

*Note: Where not stated, room temperature may be assumed.

TABLE 9-7. *(Continued)*

CADMIUM OXIDE	Unit	Temp.* (°K)	Value
PHYSICAL PROPERTIES			
Formula weight	g		128.41
Density	g/cm^3	300	8.15
Melting point (sublimes)	°K		1837
Melting point pressure coeff.	°K/kbar		
Boiling point	°K		
Symmetry			cubic
Lattice parameter, a_0	Å		4.6943
" " ,			
" " ,			
THERMAL PROPERTIES			
Thermal conductivity	W/cm °K	300	0.0088
" "			
" "			
Linear expansion coeff.	$10^{-6}/°K$	305–1005	14.33
Thermal diffusivity	cm^2/sec		
Specific heat (C_p)	cal/g deg	100	0.05
" "		298	0.09
" "			
Debye temperature	°K		450
Spectral emissivity	none		
Vapor pressure	torr		
" "			
Heat of Sublimation	kcal/g		
Heat of Fusion	cal/g		
Entropy	cal/g-atom deg	298	13.1
Zero point energy	cal/g-atom		
MAGNETIC PROPERTIES			
Magnetic susceptibility	10^{-7} emu/g	300	–2.32
Gyromagnetic properties, g factor	—		
" " , line width	Oe		
MECHANICAL PROPERTIES			
Hardness	Mohs		3
Young's modulus	10^{11} dyne/cm^2		
Bulk modulus	10^{11} dyne/cm^2		7.6
Elastic constant (stiffness), C_{11}	10^{11} dyne/cm^2		
" " C_{12}			
" " C_{44}			
" " C_{33}			
Elastic constant (compliance) S_{11}	10^{-12} cm^2/dyne		
" " S_{12}			
" " S_{44}			
" " S_{33}			
ACOUSTIC PROPERTIES			
Sound velocity	10^3 m/sec		
" "			
ELECTRICAL/ELECTRONIC PROPERTIES			
Electrical resistivity	ohm · cm	300	0.001
Dielectric constant	none	300	ϵ_0 21.9; ϵ_∞ 5.4
Electron mobility	cm^2/V sec		
Hole mobility	cm^2/V sec		
Energy gap	eV	295	D2.35
Work function	eV		2.43
Electron affinity	eV		
Photoelectric threshold	eV		
Photoelectric yield	eV		
Superconducting transition temp.	°K		
Seebeck coefficient	μV/°K	300	20–30
OPTICAL PROPERTIES			
Transmission	%		
Reflectivity	%		
Refractive index at 0.67μ	none	300	2.49

*Note: Where not stated, room temperature may be assumed.

TABLE 9-7. (*Continued*)

CADMIUM SELENIDE	Unit	Temp.* (°K)	Value
PHYSICAL PROPERTIES			
Formula weight	g		191.37
Density	g/cm^3		5.81
Melting point	°K		1531
Melting point pressure coeff.	°K/kbar		
Boiling point	°K		
Symmetry			hexagonal
Lattice parameter, a_0	Å		4.30
" " , c_0			7.01
THERMAL PROPERTIES			
Thermal conductivity	W/cm °K	280	0.063
" "			
" "			
Linear expansion coeff.	10^{-6}/°K		
Thermal diffusivity	cm^2/sec		
Specific heat (C_p)	cal/g deg	80	0.021
" "			
" "			
Debye temperature	°K		230
Spectral emissivity	none		
Vapor pressure	torr		
" "			
Heat of Sublimation	kcal/g		
Heat of Fusion	cal/g		
Entropy	cal/g-atom deg	298	23.1
Zero point energy	cal/g-atom		
MAGNETIC PROPERTIES			
Magnetic susceptibility	10^{-7} emu/g	300	−3.34
Gyromagnetic properties, g factor	—		$\perp$0.51; $\parallel$0.6
" " , line width	Oe		
MECHANICAL PROPERTIES			
Hardness	Mohs		3
Young's modulus	10^{11} dyne/cm^2		
Bulk modulus	10^{11} dyne/cm^2		3.8
Elastic constant (stiffness), C_{11}	10^{11} dyne/cm^2		7.41
" " C_{12}			4.52
" " C_{44}			1.37
" " C_{33}			8.36
Elastic constant (compliance) S_{11}	10^{-12} cm^2/dyne		2.338
" " S_{12}			−1.122
" " S_{44}			7.595
" " S_{33}			1.735
ACOUSTIC PROPERTIES			
Sound velocity	10^3 m/sec		
" "			
ELECTRICAL/ELECTRONIC PROPERTIES			
Electrical resistivity	ohm · cm	300	10^5
Dielectric constant	none	298	$\epsilon_0\perp$9.33; $\epsilon_0\parallel$10.02
Electron mobility	cm^2/V sec	298	580
Hole mobility	cm^2/V sec		
Energy gap	eV	0	1.88
Work function	eV		5.1
Electron affinity	eV		
Photoelectric threshold	eV		7.2
Photoelectric yield	eV		
Superconducting transition temp.	°K		
Seebeck coefficient	μV/°K	300	200
OPTICAL PROPERTIES			
Transmission	%		
Reflectivity	%		
Refractive index at 0.85μ	none	300	n_o 2.608; n_e 2.625

*Note: Where not stated, room temperature may be assumed.

TABLE 9-7. (*Continued*)

CADMIUM SULFIDE	Unit	Temp.* (°K)	Value
PHYSICAL PROPERTIES			
Formula weight	g		144.476
Density	g/cm^3		4.820
Melting point (sublimes)	°K		1173
Melting point pressure coeff.	°K/kbar		
Boiling point	°K		
Symmetry			hexagonal
Lattice parameter, a_0	Å		4.1368
" " , c_0			6.7162
THERMAL PROPERTIES			
Thermal conductivity	W/cm °K	100	0.7
" "		300	0.2
" "			
Linear expansion coeff.	$10^{-6}/°K$	300–343	4.2
Thermal diffusivity	cm^2/sec		
Specific heat (C_p)	cal/g deg	300	0.080
" "		800	0.090
" "			
Debye temperature	°K		219.3
Spectral emissivity	none		
Vapor pressure	torr		
" "			
Heat of Sublimation	kcal/g		0.35
Heat of Fusion	cal/g		
Entropy	cal/g-atom deg	298	16.5
Zero point energy	cal/g-atom		
MAGNETIC PROPERTIES			
Magnetic susceptibility	10^{-7} emu/g		–3.531
Gyromagnetic properties, g factor	—		∥–1.78; ⊥1.15
" " , line width	Oe		
MECHANICAL PROPERTIES			
Hardness	Mohs		3.0–3.5
Young's modulus	10^{11} dyne/cm^2		
Bulk modulus	10^{11} dyne/cm^2		
Elastic constant (stiffness), C_{11}	10^{11} dyne/cm^2	300	8.431
" " C_{12}		300	5.208
" " C_{44}		300	1.458
" " C_{33}		300	9.183
" " C_{13}		300	4.567
Elastic constant (compliance) S_{11}	10^{-12} cm^2/dyne	300	2.069
" " S_{12}		300	–0.999
" " S_{44}		300	6.649
" " S_{33}		300	1.697
" " S_{13}		300	–0.581
ACOUSTIC PROPERTIES			
Sound velocity	10^3 m/sec		4.17
" "			
ELECTRICAL/ELECTRONIC PROPERTIES			
Electrical resistivity	ohm · cm	300	5
Dielectric constant	none		ϵ_∞ 5.4
Electron mobility	cm^2/V sec	300	250
Hole mobility	cm^2/V sec	300	15
Energy gap	eV	300	D⊥2.41; D∥2.425
Work function	eV	300	5.01
Electron affinity	eV		4.79
Photoelectric threshold	eV		7.26
Photoelectric yield	eV		
Superconducting transition temp.	°K		
Seebeck coefficient	µV/°K	300	1
OPTICAL PROPERTIES			
Transmission at 2 − 14µ, (3 mm. thick)	%	300	> 70
Reflectivity	%		
Refractive index at 0.60µ	none	300	n_o 2.506; n_e 2.491

*Note: Where not stated, room temperature may be assumed.

TABLE 9-7. *(Continued)*

CADMIUM TELLURIDE	Unit	Temp.* (°K)	Value
PHYSICAL PROPERTIES			
Formula weight	g		240.02
Density	g/cm^3	288	6.20
Melting point	°K		1318
Melting point pressure coeff.	°K/kbar		
Boiling point	°K		1433
Symmetry			cubic
Lattice parameter, a_0	Å		6.4815
" " ,			
" " ,			
THERMAL PROPERTIES			
Thermal conductivity	W/cm °K	300	0.07
" "			
" "			
Linear expansion coeff.	10^{-6}/°K	25–250	5.5
Thermal diffusivity	cm^2/sec		
Specific heat (C_p)	cal/g deg	300	0.05
" "			
" "			
Debye temperature	°K		360
Spectral emissivity	none		
Vapor pressure	torr		
" "			
Heat of Sublimation	kcal/g		0.181
Heat of Fusion	cal/g		
Entropy	cal/g-atom deg	298	24.0
Zero point energy	cal/g-atom		
MAGNETIC PROPERTIES			
Magnetic susceptibility	10^{-7} emu/g	273–800	–1.78
Gyromagnetic properties, g factor	—		
" ", line width	Oe		
MECHANICAL PROPERTIES			
Hardness	Knoop		45
Young's modulus	10^{11} dyne/cm^2	298	3.66
Bulk modulus	10^{11} dyne/cm^2	300	4.2
Elastic constant (stiffness), C_{11}	10^{11} dyne/cm^2	77	5.351
" " C_{12}		77	3.681
" " C_{44}		77	1.994
C_{33}			
Elastic constant (compliance) S_{11}	10^{-12} cm^2/dyne	77	38.3
" " S_{12}		77	–15.8
" " S_{44}		77	51.1
S_{33}			
ACOUSTIC PROPERTIES			
Sound velocity	10^3 m/sec		
" "			
ELECTRICAL/ELECTRONIC PROPERTIES			
Electrical resistivity	ohm · cm	300	10^{10}
Dielectric constant	none		ϵ_0 10.2; ϵ_∞ 7.1
Electron mobility	cm^2/V sec	300	820–1050
Hole mobility	cm^2/V sec	300	80–100
Energy gap	eV	300	D1.58
Work function	eV		4.67
Electron affinity	eV		4.28
Photoelectric threshold	eV		5.78
Photoelectric yield	eV		
Superconducting transition temp.	°K		
Seebeck coefficient	μV/°K	300	3.5
OPTICAL PROPERTIES			
Transmission at 0.9–15μ, (2 mm. thick)	%	300	10
Reflectivity	%		
Refractive index at 1μ	none	300	2.84

*Note: Where not stated, room temperature may be assumed.

TABLE 9-7. (*Continued*)

CALCIUM OXIDE	Unit	Temp.* (°K)	Value
PHYSICAL PROPERTIES			
Formula weight	g		56.08
Density	g/cm^3		3.32
Melting point	°K	3449	
Melting point pressure coeff.	°K/kbar		
Boiling point	°K		
Symmetry			cubic
Lattice parameter, a_0	Å		4.812
$''$ $''$,			
,			
THERMAL PROPERTIES			
Thermal conductivity (powder)	W/cm °K	400	0.00356
$''$ $''$			
$''$ $''$			
Linear expansion coeff.	$10^{-6}/°K$		13.6
Thermal diffusivity	cm^2/sec		
Specific heat (C_p)	cal/g deg	87	0.058
$''$ $''$		300	0.18
$''$ $''$			
Debye temperature	°K		528
Spectral emissivity	none		
Vapor pressure	torr		
$''$ $''$			
Heat of Sublimation	kcal/g		
Heat of Fusion	cal/g		
Entropy	cal/g-atom deg	298	9.5
Zero point energy	cal/g-atom		
MAGNETIC PROPERTIES			
Magnetic susceptibility	10^{-7} emu/g	300	−2.70
Gyromagnetic properties, g factor	−		
$''$ $''$, line width	Oe		
MECHANICAL PROPERTIES			
Hardness	Mohs		3.5
Young's modulus	10^{11} dyne/cm^2		20.2
Bulk modulus	10^{11} dyne/cm^2		6.1
Elastic constant (stiffness), C_{11}	10^{11} dyne/cm^2		7.6
$''$ $''$ C_{12}			
$''$ $''$ C_{44}			
$''$ $''$ C_{33}			
Elastic constant (compliance) S_{11}	10^{-12} cm^2/dyne		
$''$ $''$ S_{12}			
$''$ $''$ S_{44}			
$''$ $''$ S_{33}			
ACOUSTIC PROPERTIES			
Sound velocity	10^3 m/sec		
$''$ $''$			
$''$ $''$			
ELECTRICAL/ELECTRONIC PROPERTIES			
Electrical resistivity	ohm · cm	300	10^8
Dielectric constant	none	300	ϵ_0 11.1; ϵ_∞ 3.33
Electron mobility	cm^2/V sec	900	5×10^4
Hole mobility	cm^2/V sec		
Energy gap	eV	300	D7.5
Work function	eV		1.9
Electron affinity	eV		0.7
Photoelectric threshold	eV		
Photoelectric yield	eV		
Superconducting transition temp.	°K		
Seebeck coefficient	$\mu V/°K$	773	9500
OPTICAL PROPERTIES			
Transmission	%		
Reflectivity	%		
Refractive index at 0.66μ	none	300	1.8238

*Note: Where not stated, room temperature may be assumed.

TABLE 9-7. (*Continued*)

MAGNESIUM OXIDE	Unit	Temp.* (°K)	Value
PHYSICAL PROPERTIES			
Formula weight	g		40.32
Density	g/cm³		3.576
Melting point	°K		3125
Melting point pressure coeff.	°K/kbar		
Boiling point	°K		3873
Symmetry			cubic
Lattice parameter, a_0	Å		4.213
" " ,			
" " ,			
THERMAL PROPERTIES			
Thermal conductivity	W/cm °K	100	2.7
" "		300	0.6
" "		500	0.32
Linear expansion coeff.	10^{-6}/°K	300	10.5
Thermal diffusivity	cm²/sec		
Specific heat (C_p)	cal/g deg	100	0.0462
" "		300	0.222
" "		1000	0.304
Debye temperature	°K		~760
Spectral emissivity	none		
Vapor pressure	torr		
" "			
" "			
Heat of Sublimation	kcal/g		
Heat of Fusion	cal/g		
Entropy	cal/g-atom deg	298	6.4
Zero point energy	cal/g-atom		
MAGNETIC PROPERTIES			
Magnetic susceptibility	10^{-7} emu/g		-2.5
Gyromagnetic properties, g factor	—		
" " , line width	Oe		
MECHANICAL PROPERTIES			
Hardness	Mohs		5.5
Young's modulus	10^{11} dyne/cm²	298	31.8 (polycr)
Bulk modulus	10^{11} dyne/cm²	303	17.2
Elastic constant (stiffness), C_{11}	10^{11} dyne/cm²		28.6
" " C_{12}			8.7
" " C_{44}			14.8
" " C_{33}			
Elastic constant (compliance) S_{11}	10^{-12} cm²/dyne		0.408
" " S_{12}			-0.095
" " S_{44}			0.676
" " S_{33}			
ACOUSTIC PROPERTIES			
Sound velocity	10^3 m/sec		9.5
" "			
ELECTRICAL/ELECTRONIC PROPERTIES			
Electrical resistivity	ohm · cm	573	1.3×10^{15}
Dielectric constant	none	300	ϵ_0 9.8; ϵ_∞ 2.95
Electron mobility	cm²/V sec		
Hole mobility	cm²/V sec		
Energy gap	eV	295	7.77
Work function	eV	1102–1182	3.55
Electron affinity	eV		
Photoelectric threshold	eV		
Photoelectric yield	eV		
Superconducting transition temp.	°K		
Seebeck coefficient	μV/°K	1120–1182	2600
OPTICAL PROPERTIES			
Transmission at 0.4–7μ, (1 mm. thick)	%	298	≥ 80
Reflectivity	%		
Refractive index at 0.546μ	none	293	1.741

*Note: Where not stated, room temperature may be assumed.

TABLE 9-7. (Continued)

MERCURY SELENIDE	Unit	Temp.* ($^\circ$K)	Value
PHYSICAL PROPERTIES			
Formula weight	g		279.51
Density	g/cm^3		7.95
Melting point	$^\circ$K		1072
Melting point pressure coeff.	$^\circ$K/kbar		
Boiling point	$^\circ$K		
Symmetry			cubic
Lattice parameter, a_0	Å		6.0854
" " ,			
" " ,			
THERMAL PROPERTIES			
Thermal conductivity	W/cm $^\circ$K	300	0.016
" "			
" "			
Linear expansion coeff.	$10^{-6}/^\circ$K	300	1.5
Thermal diffusivity	cm^2/sec		
Specific heat (C_p)	cal/g deg	300	0.047
" "			
" "			
Debye temperature	$^\circ$K		137.2
Spectral emissivity	none		
Vapor pressure	torr		
" "			
Heat of Sublimation	kcal/g		
Heat of Fusion	cal/g		
Entropy	cal/g-atom deg		
Zero point energy	cal/g-atom		
MAGNETIC PROPERTIES			
Magnetic susceptibility	10^{-7} emu/g		
Gyromagnetic properties, g factor	—		
" " , line width	Oe		
MECHANICAL PROPERTIES			
Hardness	Mohs		2.25
Young's modulus	10^{11} dyne/cm^2		5.7
Bulk modulus	10^{11} dyne/cm^2		5.95
Elastic constant (stiffness), C_{11}	10^{11} dyne/cm^2		4.307
" " C_{12}			2.2015
" " C_{44}			
" " C_{33}			
Elastic constant (compliance) S_{11}	10^{-12} cm^2/dyne		
" " S_{12}			
" " S_{44}			
" " S_{33}			
ACOUSTIC PROPERTIES			
Sound velocity	10^3 m/sec		1.6
" "			
ELECTRICAL/ELECTRONIC PROPERTIES			
Electrical resistivity	ohm · cm	300	$10^{-4}-10^{-3}$
Dielectric constant	none		14 (16.3μ)
Electron mobility	cm^2/V sec	300	12000
Hole mobility	cm^2/V sec		
Energy gap	eV	300	−0.15
Work function	eV		4.45
Electron affinity	eV		
Photoelectric threshold	eV		
Photoelectric yield	eV		
Superconducting transition temp.	$^\circ$K		
Seebeck coefficient	μV/$^\circ$K	300	−100
OPTICAL PROPERTIES			
Transmission	%		
Reflectivity	%		
Refractive index at 5μ	none		1.6

*Note: Where not stated, room temperature may be assumed.

TABLE 9-7. *(Continued)*

MERCURIC SULFIDE	Unit	Temp.* (°K)	Value
PHYSICAL PROPERTIES			
Formula weight	g		232.676
Density	g/cm^3		8.176
Melting point	°K		1723
Melting point pressure coeff.	°K/kbar		
Boiling point	°K		
Symmetry			hexagonal
Lattice parameter, a_0	Å		4.14
" " , c_0			9.49
THERMAL PROPERTIES			
Thermal conductivity	W/cm °K		
" "			
" "			
Linear expansion coeff.	$10^{-6}/°K$		‖21.5; ⊥17.9
Thermal diffusivity	cm^2/sec		
Specific heat (C_p)	cal/g deg	80	0.163
" "		273	0.051
" "		373	0.054
Debye temperature	°K		250
Spectral emissivity	none		
Vapor pressure	torr		
" "			
Heat of Sublimation	kcal/g		
Heat of Fusion	cal/g		
Entropy	cal/g-atom deg	298	19.5
Zero point energy	cal/g-atom		
MAGNETIC PROPERTIES			
Magnetic susceptibility	10^{-7} emu/g		–3.15
Gyromagnetic properties, g factor	—		
" " , line width	Oe		
MECHANICAL PROPERTIES			
Hardness	Mohs		2
Young's modulus	10^{11} dyne/cm^2		
Bulk modulus	10^{11} dyne/cm^2		
Elastic constant (stiffness), C_{11}	10^{11} dyne/cm^2		
" " C_{12}			
" " C_{44}			
" " C_{33}			
Elastic constant (compliance) S_{11}	10^{-12} cm^2/dyne	300	0.41
" " S_{12}			
" " S_{44}			
" " S_{33}			
ACOUSTIC PROPERTIES			
Sound velocity	10^3 m/sec		
" "			
ELECTRICAL/ELECTRONIC PROPERTIES			
Electrical resistivity	ohm · cm	300	10^{12}
Dielectric constant	none		18.2
Electron mobility	cm^2/V sec		39
Hole mobility	cm^2/V sec		
Energy gap	eV	300	2.0
Work function	eV		
Electron affinity	eV		
Photoelectric threshold	eV		
Photoelectric yield	eV		
Superconducting transition temp.	°K		
Seebeck coefficient	μV/°K		
OPTICAL PROPERTIES			
Transmission	%		
Reflectivity	%		
Refractive index at 0.62μ	none	298	n_o 2.90; n_e 3.26

*Note: Where not stated, room temperature may be assumed.

TABLE 9-7. (*Continued*)

MERCURY TELLURIDE	Unit	Temp.* (°K)	Value
PHYSICAL PROPERTIES			
Formula weight	g		328.22
Density	g/cm^3		8.081
Melting point	°K		943
Melting point pressure coeff.	°K/kbar		
Boiling point	°K		
Symmetry			cubic
Lattice parameter, a_0	Å		6.461
" " ,			
" " ,			
THERMAL PROPERTIES			
Thermal conductivity	W/cm °K	300	0.024
" "			
" "			
Linear expansion coeff.	$10^{-6}/°K$	300	6.8
Thermal diffusivity	cm^2/sec		
Specific heat (C_p)	cal/g deg	300	0.0395
" "		500	0.0390
" "			
Debye temperature	°K		114.3
Spectral emissivity	none		
Vapor pressure	torr		
" "			
" "			
Heat of Sublimation	kcal/g		
Heat of Fusion	cal/g		
Entropy	cal/g-atom deg	298	22.0
Zero point energy	cal/g-atom		
MAGNETIC PROPERTIES			
Magnetic susceptibility	10^{-7} emu/g	300	–3
Gyromagnetic properties, g factor	—		40
" " , line width	Oe		
MECHANICAL PROPERTIES			
Hardness	Mohs		2.5
Young's modulus	10^{11} dyne/cm^2		
Bulk modulus	10^{11} dyne/cm^2		4.55
Elastic constant (stiffness), C_{11}	10^{11} dyne/cm^2		5.48
" " C_{12}			3.81
" " C_{44}			2.04
" " C_{33}			
Elastic constant (compliance) S_{11}	10^{-12} cm^2/dyne		
" " S_{12}			
" " S_{44}			
" " S_{33}			
ACOUSTIC PROPERTIES			
Sound velocity	10^3 m/sec		1.7
" "			
ELECTRICAL/ELECTRONIC PROPERTIES			
Electrical resistivity	ohm · cm	300	0.001
Dielectric constant	none	296	ϵ_0 20; ϵ_∞ 14.1
Electron mobility	cm^2/V sec		22000
Hole mobility	cm^2/V sec		160
Energy gap	eV	300	–0.19
Work function	eV		4.13
Electron affinity	eV		
Photoelectric threshold	eV		
Photoelectric yield	eV		
Superconducting transition temp.	°K		
Seebeck coefficient	µV/°K		–130
OPTICAL PROPERTIES			
Transmission	%		
Reflectivity	%		
Refractive index	none		

*Note: Where not stated, room temperature may be assumed.

Table 9-7. (*Continued*)

STRONTIUM OXIDE	Unit	Temp.* (°K)	Value
PHYSICAL PROPERTIES			
Formula weight	g		103.63
Density	g/cm^3	298	5.023
Melting point	°K		
Melting point pressure coeff.	°K/kbar		
Boiling point	°K		
Symmetry			cubic
Lattice parameter, a_0	Å		5.1396
" " ,			
" " ,			
THERMAL PROPERTIES			
Thermal conductivity	W/cm °K	500	0.5022
" "		800	1.2585
" "			
Linear expansion coeff.	$10^{-6}/°K$	293–1148	32.5
Thermal diffusivity	cm^2/sec		
Specific heat (C_p)	cal/g deg	100	0.055
" "		300	0.105
" "		1000	0.128
Debye temperature	°K		481
Spectral emissivity	none		
Vapor pressure	torr		
" "			
" "			
Heat of Sublimation	kcal/g		
Heat of Fusion	cal/g		
Entropy	cal/g-atom deg	298	13.0
Zero point energy	cal/g-atom		
MAGNETIC PROPERTIES			
Magnetic susceptibility	10^{-7} emu/g	300	−3.4
Gyromagnetic properties, g factor	—		
" " , line width	Oe		
MECHANICAL PROPERTIES			
Hardness	Mohs		3.5
Young's modulus	10^{11} dyne/cm^2		
Bulk modulus	10^{11} dyne/cm^2		
Elastic constant (stiffness), C_{11}	10^{11} dyne/cm^2		16.55
" " C_{12}			4.64
" " C_{44}			5.26
" " C_{33}			
Elastic constant (compliance) S_{11}	10^{-12} cm^2/dyne		
" " S_{12}			
" " S_{44}			
" " S_{33}			
ACOUSTIC PROPERTIES			
Sound velocity	10^3 m/sec		
" "			
ELECTRICAL/ELECTRONIC PROPERTIES			
Electrical resistivity	ohm · cm	556	10^7
Dielectric constant	none	300	ϵ_0 13.1; ϵ_∞ 3.46
Electron mobility	cm^2/V sec		
Hole mobility	cm^2/V sec		
Energy gap	eV	113	5.71
Work function	eV		1.4
Electron affinity	eV		
Photoelectric threshold	eV		
Photoelectric yield	eV		
Superconducting transition temp.	°K		
Seebeck coefficient	μV/°K	648	3600
OPTICAL PROPERTIES			
Transmission	%		
Reflectivity	%		
Refractive index	none		

*Note: Where not stated, room temperature may be assumed.

TABLE 9-7. (*Continued*)

ZINC OXIDE	Unit	Temp.* ($^\circ$K)	Value
PHYSICAL PROPERTIES			
Formula weight	g		81.38
Density	g/cm^3		5.665
Melting point	$^\circ$K		2248
Melting point pressure coeff.	$^\circ$K/kbar		
Boiling point	$^\circ$K		
Symmetry			hexagonal
Lattice parameter, a_0	Å		3.24265
" " , c_0			5.1948
,			
THERMAL PROPERTIES			
Thermal conductivity	W/cm $^\circ$K	300	0.286
" "		500	0.168
" "			
Linear expansion coeff.	$10^{-6}/^\circ$K	291	$\parallel$4.9; $\perp$5.6
Thermal diffusivity	cm^2/sec		
Specific heat (C_p)	cal/g deg	100	0.533
" "		300	0.119
" "			
Debye temperature	$^\circ$K		920
Spectral emissivity	none		
Vapor pressure	torr		
" "			
Heat of Sublimation	kcal/g		
Heat of Fusion	cal/g		
Entropy	cal/g-atom deg	298	10.5
Zero point energy	cal/g-atom		
MAGNETIC PROPERTIES			
Magnetic susceptibility	10^{-7} emu/g	300	-3.6
Gyromagnetic properties, g factor	—		
" " , line width	Oe		
MECHANICAL PROPERTIES			
Hardness	Mohs		4.5
Young's modulus	10^{11} dyne/cm^2		
Bulk modulus	10^{11} dyne/cm^2	300	
Elastic constant (stiffness), C$_{11}$	10^{11} dyne/cm^2		20.97
" " C$_{12}$			12.11
" " C$_{44}$			4.247
" " C$_{33}$			21.09
" " C$_{13}$			10.51
Elastic constant (compliance) S$_{11}$	10^{-12} cm^2/dyne		0.7858
" " S$_{12}$			0.3432
" " S$_{44}$			0.2357
" " S$_{33}$			0.6940
" " S$_{13}$			0.2206
ACOUSTIC PROPERTIES			
Sound velocity	10^3 m/sec		
" "			
" "			
ELECTRICAL/ELECTRONIC PROPERTIES			
Electrical resistivity	ohm · cm	300	1–20
Dielectric constant	none		$\epsilon_0 \perp 7.8$; $\epsilon_0 \parallel 11.875$
Electron mobility	cm^2/V sec	300	200
Hole mobility	cm^2/V sec		180
Energy gap	eV	1.2	3.3435
Work function	eV	300	4.68
Electron affinity	eV		4.57
Photoelectric threshold	eV		7.82
Photoelectric yield	eV		
Superconducting transition temp.	$^\circ$K		
Seebeck coefficient	μV/$^\circ$K		
OPTICAL PROPERTIES			
Transmission	%		
Reflectivity at 0.41μ	%	300	90
Refractive index at 0.589μ	none	297	n_o 2.009; n_e 2.024

*Note: Where not stated, room temperature may be assumed.

TABLE 9-7. *(Continued)*

ZINC SELENIDE	Unit	Temp.* ($°K$)	Value
PHYSICAL PROPERTIES			
Formula weight	g		144.34
Density	g/cm^3		5.651
Melting point (dissociation)	$°K$		1373
Melting point pressure coeff.	$°K/kbar$		
Boiling point	$°K$		
Symmetry			cubic
Lattice parameter, a_0	Å	300	5.671
″ ″ ,			
″ ″ ,			
THERMAL PROPERTIES			
Thermal conductivity	$W/cm °K$	300	0.12
″ ″			
″ ″			
Linear expansion coeff.	$10^{-6}/°K$	300	7
Thermal diffusivity	cm^2/sec		
Specific heat (C_p)	cal/g deg	80	0.016
″ ″			
″ ″			
Debye temperature	$°K$		400
Spectral emissivity	none		
Vapor pressure	torr		
″ ″			
″ ″			
Heat of Sublimation	kcal/g		
Heat of Fusion	cal/g		
Entropy	cal/g-atom deg	298	19.8
Zero point energy	cal/g-atom		
MAGNETIC PROPERTIES			
Magnetic susceptibility	10^{-7} emu/g	20–350	−3.22
Gyromagnetic properties, g factor	—		
″ ″ , line width	Oe		
MECHANICAL PROPERTIES			
Hardness	Mohs		3–4
Young's modulus	10^{11} dyne/cm^2		7.10
Bulk modulus	10^{11} dyne/cm^2		3.90
Elastic constant (stiffness), C_{11}	10^{11} dyne/cm^2	300	8.10
″ ″ C_{12}		300	0.488
″ ″ C_{44}		300	0.441
″ ″ C_{33}			
Elastic constant (compliance) S_{11}	10^{-12} cm^2/dyne	300	2.26
″ ″ S_{12}		300	−0.85
″ ″ S_{44}		300	2.27
″ ″ S_{33}			
ACOUSTIC PROPERTIES			
Sound velocity	10^3 m/sec		
″ ″			
ELECTRICAL/ELECTRONIC PROPERTIES			
Electrical resistivity	ohm · cm	300	10^8–10^9
Dielectric constant	none		ϵ_0 9.2; ϵ_∞ 6.10
Electron mobility	cm^2/V sec	300	600
Hole mobility	cm^2/V sec		16
Energy gap	eV	297	2.67
Work function	eV	300	4.84
Electron affinity	eV		4.09
Photoelectric threshold	eV		6.82
Photoelectric yield	eV		
Superconducting transition temp.	$°K$		
Seebeck coefficient	$\mu V/°K$		
OPTICAL PROPERTIES			
Transmission at 0.5–22μ, (2 mm. thick)	%	298	> 10
Reflectivity	%		
Refractive index at 0.589μ	none	300	2.6113

*Note: Where not stated, room temperature may be assumed.

TABLE 9-7. (*Continued*)

ZINC SULFIDE	Unit	Temp.* (°K)	Value
PHYSICAL PROPERTIES			
Formula weight	g		97.446
Density	g/cm^3		4.09
Melting point	°K		1293
Melting point pressure coeff.	°K/kbar		
Boiling point	°K		
Symmetry			cubic
Lattice parameter, a_0	Å		5.4093
" " ,			
THERMAL PROPERTIES			
Thermal conductivity	W/cm °K	86.5	0.585
" "		273	0.265
" "			
Linear expansion coeff.	$10^{-6}/°K$	273	6.14
Thermal diffusivity	cm^2/sec		
Specific heat (C_p)	cal/g deg	77	0.0472
" "		300	0.116
" "			
Debye temperature	°K		300
Spectral emissivity	none		
Vapor pressure	torr		
" "			
Heat of Sublimation	kcal/g		0.71
Heat of Fusion	cal/g		
Entropy	cal/g-atom deg	298	13.8
Zero point energy	cal/g-atom		
MAGNETIC PROPERTIES			
Magnetic susceptibility	10^{-7} emu/g	300	−2.36
Gyromagnetic properties, g factor	—		2.2
" " , line width	Oe		
MECHANICAL PROPERTIES			
Hardness	Mohs		3.5–4
Young's modulus	10^{11} dyne/cm^2		1.12
Bulk modulus	10^{11} dyne/cm^2		8.4
Elastic constant (stiffness), C_{11}	10^{11} dyne/cm^2		10.79
" " C_{12}			7.22
" " C_{44}			4.12
" " C_{33}			
Elastic constant (compliance) S_{11}	10^{-12} cm^2/dyne		2.0
" " S_{12}			−0.802
" " S_{44}			2.43
" " S_{33}			
ACOUSTIC PROPERTIES			
Sound velocity	10^3 m/sec		5.22
" "			
ELECTRICAL/ELECTRONIC PROPERTIES			
Electrical resistivity	ohm · cm	300	10^{10}
Dielectric constant	none	300	ϵ_0 8.9; ϵ_∞ 5.7
Electron mobility	cm^2/V sec	300	110
Hole mobility	cm^2/V sec		
Energy gap	eV		D3.66
Work function	eV		5.4
Electron affinity	eV		3.9
Photoelectric threshold	eV		7.5
Photoelectric yield	eV		
Superconducting transition temp.	°K		
Seebeck coefficient	μV/°K		
OPTICAL PROPERTIES			
Transmission at 0.6–16μ (0.62 mm. thick)	%		$\geqslant 10$
Reflectivity	%		
Refractive index at 0.60μ	none	300	2.359

*Note: Where not stated, room temperature may be assumed.

TABLE 9-7. (*Continued*)

ZINC TELLURIDE	Unit	Temp.* ($^\circ$K)	Value
PHYSICAL PROPERTIES			
Formula weight	g		192.99
Density	g/cm^3		5.51
Melting point	$^\circ$K		1512
Melting point pressure coeff.	$^\circ$K/kbar		
Boiling point	$^\circ$K		
Symmetry			cubic
Lattice parameter, a_0	Å		6.101
" " ,			
" " ,			
THERMAL PROPERTIES			
Thermal conductivity	W/cm $^\circ$K	300	0.112
" "			
" "			
Linear expansion coeff.	$10^{-6}/^\circ$K	300	8.29
Thermal diffusivity	cm^2/sec		
Specific heat (C_p)	cal/g deg	293	0.072
" "			
" "			
Debye temperature	$^\circ$K		204.5
Spectral emissivity	none		
Vapor pressure	torr		
" "			
Heat of Sublimation	kcal/g		0.252
Heat of Fusion	cal/g		
Entropy	cal/g-atom deg	298	23.0
Zero point energy	cal/g-atom		
MAGNETIC PROPERTIES			
Magnetic susceptibility	10^{-7} emu/g	293–600	–19.8
Gyromagnetic properties, g factor	—		
" " , line width	Oe		
MECHANICAL PROPERTIES			
Hardness	Mohs		3
Young's modulus	10^{11} dyne/cm^2		4.12
Bulk modulus	10^{11} dyne/cm^2		7.13
Elastic constant (stiffness), C_{11}	10^{11} dyne/cm^2		4.07
" " C_{12}			3.12
" " C_{44}			
" " C_{33}			
Elastic constant (compliance) S_{11}	10^{-12} cm^2/dyne		2.40
" " S_{12}			–0.873
" " S_{44}			3.21
" " S_{33}			
ACOUSTIC PROPERTIES			
Sound velocity	10^3 m/sec		
" "			
ELECTRICAL/ELECTRONIC PROPERTIES			
Electrical resistivity	ohm · cm	300	100
Dielectric constant	none	300	ϵ_0 10.4; ϵ_∞ 7.3
Electron mobility	cm^2/V sec		530
Hole mobility	cm^2/V sec	300	100
Energy gap	eV	300	2.25
Work function	eV	300	5.43
Electron affinity	eV		3.53
Photoelectric threshold	eV		5.76
Photoelectric yield	eV		
Superconducting transition temp.	$^\circ$K		
Seebeck coefficient	μV/$^\circ$K	300	～200
OPTICAL PROPERTIES			
Transmission	%		
Reflectivity	%		
Refractive index at 0.60μ	none	300	3.080

*Note: Where not stated, room temperature may be assumed.

TABLE 9-8. PROPERTIES OF VI SOLID INORGANIC SEMICONDUCTORS

SELENIUM	Unit	Temp.* ($^\circ$K)	Value
PHYSICAL PROPERTIES			
Formula weight	g		78.95
Density	g/cm^3	300	4.79
Melting point	$^\circ$K		490
Melting point pressure coeff.	$^\circ$K/kbar		
Boiling point	$^\circ$K		958
Symmetry			hexagonal
Lattice parameter, a_0	Å		4.355
" " , c_0			4.9494
THERMAL PROPERTIES			
Thermal conductivity	W/cm $^\circ$K		
" "			
" "			
Linear expansion coeff.	10^{-6}/$^\circ$K	293	36.8
Thermal diffusivity	cm^2/sec		
Specific heat (C_p)	cal/g deg	323	0.080
" "			
" "			
Debye temperature	$^\circ$K		151.7
Spectral emissivity	none		
Vapor pressure	torr		
" "			
Heat of Sublimation	kcal/g	298	0.627
Heat of Fusion	cal/g		16.5
Entropy	cal/g-atom deg		
Zero point energy	cal/g-atom		
MAGNETIC PROPERTIES			
Magnetic susceptibility	10^{-7} emu/g		-3.1
Gyromagnetic properties, g factor	—		
" " , line width	Oe		
MECHANICAL PROPERTIES			
Hardness	Mohs		2.0
Young's modulus	10^{11} dyne/cm^2		5.80
Bulk modulus	10^{11} dyne/cm^2		0.91
Elastic constant (stiffness), C_{11}	10^{11} dyne/cm^2		
" " C_{12}			
" " C_{44}			
C_{33}			
Elastic constant (compliance) S_{11}	10^{-12} cm^2/dyne		
" " S_{12}			
" " S_{44}			
S_{33}			
ACOUSTIC PROPERTIES			
Sound velocity	10^3 m/sec	293	3.35
" "			
ELECTRICAL/ELECTRONIC PROPERTIES			
Electrical resistivity	ohm · cm		
Dielectric constant	none	298	6.0
Electron mobility	cm^2/V sec		2
Hole mobility	cm^2/V sec		17
Energy gap	eV	300	1.74
Work function	eV		4.42
Electron affinity	eV		
Photoelectric threshold	eV		
Photoelectric yield	eV		
Superconducting transition temp.	$^\circ$K		
Seebeck coefficient	μV/$^\circ$K		
OPTICAL PROPERTIES			
Transmission	%		
Reflectivity	%		
Refractive index	none		∥E 5.56; ⊥E 3.72

*Note: Where not stated, room temperature may be assumed.

TABLE 9-8. (*Continued*)

TELLURIUM	Unit	Temp.* (°K)	Value
PHYSICAL PROPERTIES			
Formula weight	g		127.61
Density	g/cm^3		6.25
Melting point	°K		722.8
Melting point pressure coeff.	°K/kbar		
Boiling point	°K		1263
Symmetry			hexagonal
Lattice parameter, a_0	Å		4.45
" " , c_0			5.91
THERMAL PROPERTIES			
Thermal conductivity	W/cm °K	293	0.048
" "			
" "			
Linear expansion coeff.	10^{-6}/°K	293	16.77∥c
Thermal diffusivity	cm^2/sec		
Specific heat (C_p)	cal/g deg	298	0.048
" "		200	0.046
" "		100	0.040
Debye temperature	°K		141
Spectral emissivity	none		
Vapor pressure	torr	1041	76
" "		890	7.6
" "		778	0.76
Heat of Sublimation	kcal/g	298	0.364
Heat of Fusion	cal/g		32.6
Entropy	cal/g-atom deg	298	11.88
Zero point energy	cal/g-atom		
MAGNETIC PROPERTIES			
Magnetic susceptibility	10^{-7} emu/g	291	–3.1
Gyromagnetic properties, g factor	—		
" " , line width	Oe		
MECHANICAL PROPERTIES			
Hardness	Mohs		2.3
Young's modulus	10^{11} dyne/cm^2		4.12
Bulk modulus	10^{11} dyne/cm^2		2.3
Elastic constant (stiffness), C_{11}	10^{11} dyne/cm^2		2.8
" " C_{12}			1.1
" " C_{44}			1.7
" " C_{33}			7.0
" " C_{13}			2.3
Elastic constant (compliance) S_{11}	10^{-12} cm^2/dyne		4.87
" " S_{12}			–0.69
" " S_{44}			5.81
" " S_{33}			2.34
" " S_{13}			–1.38
ACOUSTIC PROPERTIES			
Sound velocity	10^3 m/sec	293	2.61
" "			
ELECTRICAL/ELECTRONIC PROPERTIES			
Electrical resistivity	ohm · cm	298	0.436∥c
Dielectric constant	none		ϵ_0∥5.0; ϵ_∞⊥2.2
Electron mobility	cm^2/V sec		10000
Hole mobility	cm^2/V sec		1100
Energy gap	eV	300	0.38
Work function	eV		4.76
Electron affinity	eV		
Photoelectric threshold	eV		4.90
Photoelectric yield	eV		
Superconducting transition temp.	°K		
Seebeck coefficient	μV/°K		
OPTICAL PROPERTIES			
Transmission	%		
Reflectivity	%		
Refractive index at 4.0μ	none		∥E 6.31; ⊥E 4.95

*Note: Where not stated, room temperature may be assumed.

TABLE 9-9. PROPERTIES OF III–V SOLID INORGANIC SEMICONDUCTORS
(Willardson, 1962, 1966, 1966A, 1968), (Neuberger, 1971), (Hilsum, 1961), (Hampel, 1968), (Moses, 1970, 1971).

ALUMINUM ANTIMONIDE	Unit	Temp.* (°K)	Value
PHYSICAL PROPERTIES			
Formula weight	g		148.74
Density	g/cm^3	300	4.26
Melting point	°K		1353
Melting point pressure coeff.	°K/kbar		
Boiling point	°K		
Symmetry			cubic
Lattice parameter, a_0	Å		6.1355
" " ,			
" " ,			
THERMAL PROPERTIES			
Thermal conductivity	W/cm °K	300	0.56
" "			
" "			
Linear expansion coeff.	10^{-6}/°K		4.88
Thermal diffusivity	cm^2/sec		
Specific heat (C_p)	cal/g deg	298	0.0744
" "			
" "			
Debye temperature	°K		370
Spectral emissivity	none		
Vapor pressure	torr		
" "			
" "			
Heat of Sublimation	kcal/g		
Heat of Fusion	cal/g		128
Entropy	cal/g-atom deg		15.36
Zero point energy	cal/g-atom		
MAGNETIC PROPERTIES			
Magnetic susceptibility	10^{-7} emu/g	300	–1.3
Gyromagnetic properties, g factor	–		0.4
" " , line width	Oe		
MECHANICAL PROPERTIES			
Hardness	Mohs		4.8
Young's modulus	10^{11} dyne/cm^2		5.9
Bulk modulus	10^{11} dyne/cm^2	300	8.939
Elastic constant (stiffness), C_{11}	10^{11} dyne/cm^2	300	4.425
" " C_{12}		300	4.155
" " C_{44}		300	
" " C_{33}			
Elastic constant (compliance) S_{11}	10^{-12} cm^2/dyne	300	1.66
" " S_{12}		300	–0.548
" " S_{44}		300	2.41
" " S_{33}			
ACOUSTIC PROPERTIES			
Sound velocity	10^3 m/sec		4.85
" "			
ELECTRICAL/ELECTRONIC PROPERTIES			
Electrical resistivity	ohm · cm	300	5
Dielectric constant	none	300	ϵ_∞ 10.24; ϵ_0 14.4
Electron mobility	cm^2/V sec		200
Hole mobility	cm^2/V sec		300
Energy gap	eV	300	D2.218; I1.62
Work function	eV	300	4.86
Electron affinity	eV	300	3.6
Photoelectric threshold	eV	300	5.22
Photoelectric yield	eV		
Superconducting transition temp.	°K		2.8 (125 kbar)
Seebeck coefficient	μV/°K	300	200 (p-type)
OPTICAL PROPERTIES			
Transmission	%		
Reflectivity	%		
Refractive index at 0.78μ	none	300	3.4

*Note: Where not stated, room temperature may be assumed.

TABLE 9-9. (*Continued*)

ALUMINUM ARSENIDE	Unit	Temp.* ($^\circ$K)	Value
PHYSICAL PROPERTIES			
Formula weight	g		101.89
Density	g/cm^3		3.598
Melting point	$^\circ$K		2013
Melting point pressure coeff.	$^\circ$K/kbar		
Boiling point	$^\circ$K		
Symmetry			cubic
Lattice parameter, a_0	Å		5.6611
" " ,			
THERMAL PROPERTIES			
Thermal conductivity	W/cm $^\circ$K	300	0.08
" "			
" "			
Linear expansion coeff.	10^{-6}/$^\circ$K	288–1113	5.20
Thermal diffusivity	cm^2/sec		
Specific heat (C_p)	cal/g deg	298	0.108
" "			
" "			
Debye temperature	$^\circ$K		417
Spectral emissivity	none		
Vapor pressure	torr		
" "			
Heat of Sublimation	kcal/g		
Heat of Fusion	cal/g		
Entropy	cal/g-atom deg		
Zero point energy	cal/g-atom		
MAGNETIC PROPERTIES			
Magnetic susceptibility	10^{-7} emu/g		
Gyromagnetic properties, g factor	—		
" " , line width	Oe		
MECHANICAL PROPERTIES			
Hardness	Mohs		5
Young's modulus	10^{11} dyne/cm^2		
Bulk modulus	10^{11} dyne/cm^2		
Elastic constant (stiffness), C$_{11}$	10^{11} dyne/cm^2		
" " C$_{12}$			
" " C$_{44}$			
C$_{33}$			
Elastic constant (compliance) S$_{11}$	10^{-12} cm^2/dyne		
" " S$_{12}$			
" " S$_{44}$			
S$_{33}$			
ACOUSTIC PROPERTIES			
Sound velocity	10^3 m/sec		
" "			
ELECTRICAL/ELECTRONIC PROPERTIES			
Electrical resistivity	ohm · cm	300	0.1
Dielectric constant	none	300	ϵ_∞ 8.5; ϵ_0 10.9
Electron mobility	cm^2/V sec		180
Hole mobility	cm^2/V sec		
Energy gap	eV		D2.13; I2.9
Work function	eV		
Electron affinity	eV		
Photoelectric threshold	eV		
Photoelectric yield	eV		
Superconducting transition temp.	$^\circ$K		
Seebeck coefficient	μV/$^\circ$K		70
OPTICAL PROPERTIES			
Transmission	%		
Reflectivity	%		
Refractive index at 0.5μ	none		3.3

*Note: Where not stated, room temperature may be assumed.

TABLE 9-9. *(Continued)*

ALUMINUM NITRIDE	Unit	Temp.* ($^\circ$K)	Value
PHYSICAL PROPERTIES			
Formula weight	g		40.988
Density	g/cm^3		3.26
Melting point	$^\circ$K		< 2673
Melting point pressure coeff.	$^\circ$K/kbar		
Boiling point	$^\circ$K		
Symmetry			hexagonal
Lattice parameter, a_0	Å		3.111
" " , c_0			4.980
THERMAL PROPERTIES			
Thermal conductivity	W/cm $^\circ$K	473	0.301
" "			
" "			
Linear expansion coeff.	$10^{-6}/^\circ$K	300–473	4.03
Thermal diffusivity	cm^2/sec		
Specific heat (C_p)	cal/g deg	300	0.175
" "			
" "			
Debye temperature	$^\circ$K		747
Spectral emissivity	none		
Vapor pressure	torr		
" "			
Heat of Sublimation	kcal/g		
Heat of Fusion	cal/g		
Entropy	cal/g-atom deg		4.8
Zero point energy	cal/g-atom		
MAGNETIC PROPERTIES			
Magnetic susceptibility	10^{-7} emu/g		
Gyromagnetic properties, g factor	—		
" " , line width	Oe		
MECHANICAL PROPERTIES			
Hardness	Mohs		7
Young's modulus	10^{11} dyne/cm^2		
Bulk modulus	10^{11} dyne/cm^2	298	35
Elastic constant (stiffness), C_{11}	10^{11} dyne/cm^2		
" " C_{12}			
" " C_{44}			
C_{33}			
Elastic constant (compliance) S_{11}	10^{-12} cm^2/dyne		
" " S_{12}			
" " S_{44}			
S_{33}			2.8
ACOUSTIC PROPERTIES			
Sound velocity	10^3 m/sec	300	10.4
" "			
ELECTRICAL/ELECTRONIC PROPERTIES			
Electrical resistivity	ohm · cm	300	10^{12}
Dielectric constant	none	300	ϵ_0 9.14; ϵ_∞ 4.84
Electron mobility	cm^2/V sec		
Hole mobility	cm^2/V sec	290	14
Energy gap	eV	300	5.9
Work function	eV		
Electron affinity	eV		
Photoelectric threshold	eV		
Photoelectric yield	eV		
Superconducting transition temp.	$^\circ$K		
Seebeck coefficient	μV/$^\circ$K		
OPTICAL PROPERTIES			
Transmission at 3–6μ, (mm. thick)	%	300	60
Reflectivity	%		
Refractive index at 0.6μ	none	300	n_o 2.159; n_e 2.21

*Note: Where not stated, room temperature may be assumed.

TABLE 9-9. (*Continued*)

BORON ARSENIDE	Unit	Temp.* (°K)	Value
PHYSICAL PROPERTIES			
Formula weight	g		85.73
Density	g/cm^3		5.22
Melting point	°K		
Melting point pressure coeff.	°K/kbar		
Boiling point	°K		
Symmetry			cubic
Lattice parameter, a_0	Å		4.777
" " ,			
" " ,			
THERMAL PROPERTIES			
Thermal conductivity	W/cm °K		
" "			
" "			
Linear expansion coeff.	$10^{-6}/°K$		
Thermal diffusivity	cm^2/sec		
Specific heat (C_p)	cal/g deg		
" "			
" "			
Debye temperature	°K		625
Spectral emissivity	none		
Vapor pressure	torr		
" "			
Heat of Sublimation	kcal/g		
Heat of Fusion	cal/g		
Entropy	cal/g-atom deg		
Zero point energy	cal/g-atom		
MAGNETIC PROPERTIES			
Magnetic susceptibility	10^{-7} emu/g		
Gyromagnetic properties, g factor	—		
" " , line width	Oe		
MECHANICAL PROPERTIES			
Hardness	Mohs		
Young's modulus	10^{11} $dyne/cm^2$		
Bulk modulus	10^{11} $dyne/cm^2$		
Elastic constant (stiffness), C_{11}	10^{11} $dyne/cm^2$		23.35
" " C_{12}			
" " C_{44}			
" " C_{33}			
Elastic constant (compliance) S_{11}	10^{-12} $cm^2/dyne$		
" " S_{12}			
" " S_{44}			
" " S_{33}			
ACOUSTIC PROPERTIES			
Sound velocity	10^3 m/sec		
" "			
ELECTRICAL/ELECTRONIC PROPERTIES			
Electrical resistivity	ohm · cm		
Dielectric constant	none		
Electron mobility	cm^2/V sec		
Hole mobility	cm^2/V sec		
Energy gap	eV	300	1.46
Work function	eV		
Electron affinity	eV		
Photoelectric threshold	eV		
Photoelectric yield	eV		
Superconducting transition temp.	°K		
Seebeck coefficient	$\mu V/°K$		
OPTICAL PROPERTIES			
Transmission	%		
Reflectivity	%		
Refractive index	none		

*Note: Where not stated, room temperature may be assumed.

TABLE 9-9. (*Continued*)

BORON NITRIDE	Unit	Temp.* (°K)	Value
PHYSICAL PROPERTIES			
Formula weight	g		24.828
Density	g/cm^3		2.255
Melting point	°K		3273
Melting point pressure coeff.	°K/kbar		
Boiling point	°K		
Symmetry			hexagonal
Lattice parameter, a_0	Å		2.51
" " , c_0			6.69
THERMAL PROPERTIES			
Thermal conductivity	W/cm °K	473	0.8
" "			
" "			
Linear expansion coeff.	10^{-6}/°K		−2.9 (a_0)
Thermal diffusivity	cm^2/sec		
Specific heat (C_p)	cal/g deg		0.24
" "			
" "			
Debye temperature	°K		598
Spectral emissivity	none		
Vapor pressure	torr		
" "			
Heat of Sublimation	kcal/g		
Heat of Fusion	cal/g		
Entropy	cal/g-atom deg		
Zero point energy	cal/g-atom		
MAGNETIC PROPERTIES			
Magnetic susceptibility	10^{-7} emu/g	293	4.0
Gyromagnetic properties, g factor	—	77	2.0052
" " , line width	Oe		
MECHANICAL PROPERTIES			
Hardness	Mohs		2
Young's modulus	10^{11} dyne/cm^2		
Bulk modulus	10^{11} dyne/cm^2		33
Elastic constant (stiffness), C_{11}	10^{11} dyne/cm^2		
" " C_{12}			
" " C_{44}			
" " C_{33}			
Elastic constant (compliance) S_{11}	10^{-12} cm^2/dyne		
" " S_{12}			
" " S_{44}			
" " S_{33}			
ACOUSTIC PROPERTIES			
Sound velocity	10^3 m/sec		
" "			
ELECTRICAL/ELECTRONIC PROPERTIES			
Electrical resistivity	ohm · cm	298	10^{18}
Dielectric constant	none	300	ϵ_0 3.8
Electron mobility	cm^2/V sec		
Hole mobility	cm^2/V sec		
Energy gap	eV	300	3.8
Work function	eV		
Electron affinity	eV		
Photoelectric threshold	eV		
Photoelectric yield	eV		
Superconducting transition temp.	°K		
Seebeck coefficient	μV/°K		
OPTICAL PROPERTIES			
Transmission at 4-6μ, (mm. thick)	%		85–90
Reflectivity	%		
Refractive index at 0.5μ	none	300	n_O 2.20; n_e 1.66

*Note: Where not stated, room temperature may be assumed.

TABLE 9-9. *(Continued)*

BORON NITRIDE (Cubic)	Unit	Temp.* (°K)	Value
PHYSICAL PROPERTIES			
Formula weight	g		24.828
Density	g/cm³		3.45
Melting point	°K		> 2973
Melting point pressure coeff.	°K/kbar		
Boiling point	°K		
Symmetry			cubic
Lattice parameter, a_0	Å		3.615
" " ,			
" " ,			
THERMAL PROPERTIES			
Thermal conductivity	W/cm °K	273–673	3.5
" "			
" "			
Linear expansion coeff.	10^{-6}/°K	273–673	3.5
Thermal diffusivity	cm²/sec		
Specific heat (C_p)	cal/g deg	300	0.21
" "			
" "			
Debye temperature	°K		1700
Spectral emissivity	none		
Vapor pressure	torr		
" "			
Heat of Sublimation	kcal/g		
Heat of Fusion	cal/g		
Entropy	cal/g-atom deg		
Zero point energy	cal/g-atom		
MAGNETIC PROPERTIES			
Magnetic susceptibility	10^{-7} emu/g		
Gyromagnetic properties, g factor	—		
" " , line width	Oe		
MECHANICAL PROPERTIES			
Hardness	Mohs		9–10
Young's modulus	10^{11} dyne/cm²		
Bulk modulus	10^{11} dyne/cm²		
Elastic constant (stiffness), C_{11}	10^{11} dyne/cm²		71.2
" " C_{12}			
" " C_{44}			
" " C_{33}			
Elastic constant (compliance) S_{11}	10^{-12} cm²/dyne		
" " S_{12}			
" " S_{44}			
" " S_{33}			
ACOUSTIC PROPERTIES			
Sound velocity	10^3 m/sec		
" "			
" "			
ELECTRICAL/ELECTRONIC PROPERTIES			
Electrical resistivity	ohm · cm		10^{10}
Dielectric constant	none		ϵ_0 7.1; ϵ_∞ 4.5
Electron mobility	cm²/V sec		
Hole mobility	cm²/V sec		
Energy gap	eV	300	D14.5; I8.0
Work function	eV		
Electron affinity	eV		
Photoelectric threshold	eV		
Photoelectric yield	eV		
Superconducting transition temp.	°K		
Seebeck coefficient	µV/°K		
OPTICAL PROPERTIES			
Transmission at 15–50µ, (mm. thick)	%		10
Reflectivity	%		
Refractive index at 0.589µ	none	300	2.117

*Note: Where not stated, room temperature may be assumed.

TABLE 9-9. (Continued)

BORON PHOSPHIDE	Unit	Temp.* (°K)	Value
PHYSICAL PROPERTIES			
Formula weight	g		41.795
Density	g/cm^3		2.97
Melting point	°K		> 2273
Melting point pressure coeff.	°K/kbar		
Boiling point	°K		
Symmetry			cubic
Lattice parameter, a_0	Å		4.538
" " ,			
THERMAL PROPERTIES			
Thermal conductivity	W/cm °K		0.008
" "			
" "			
Linear expansion coeff.	$10^{-6}/°K$		
Thermal diffusivity	cm^2/sec		
Specific heat (C_p)	cal/g deg		
" "			
" "			
Debye temperature	°K		985
Spectral emissivity	none		
Vapor pressure	torr		
" "			
Heat of Sublimation	kcal/g		
Heat of Fusion	cal/g		
Entropy	cal/g-atom deg		
Zero point energy	cal/g-atom		
MAGNETIC PROPERTIES			
Magnetic susceptibility	10^{-7} emu/g		
Gyromagnetic properties, g factor	—		
" " , line width	Oe		
MECHANICAL PROPERTIES			
Hardness	Knoop		3200
Young's modulus	10^{11} dyne/cm^2		
Bulk modulus	10^{11} dyne/cm^2		
Elastic constant (stiffness), C_{11}	10^{11} dyne/cm^2		28.73
" " C_{12}			
" " C_{44}			
" " C_{33}			
Elastic constant (compliance) S_{11}	10^{-12} cm^2/dyne		
" " S_{12}			
" " S_{44}			
" " S_{33}			
ACOUSTIC PROPERTIES			
Sound velocity	10^3 m/sec		
" "			
" "			
ELECTRICAL/ELECTRONIC PROPERTIES			
Electrical resistivity	ohm · cm	300	0.01
Dielectric constant	none		
Electron mobility	cm^2/V sec		
Hole mobility	cm^2/V sec	300	500
Energy gap	eV		
Work function	eV	300	I2
Electron affinity	eV		
Photoelectric threshold	eV		
Photoelectric yield	eV		
Superconducting transition temp.	°K		
Seebeck coefficient	µV/°K	300	300
OPTICAL PROPERTIES			
Transmission	%		
Reflectivity	%		
Refractive index at 0.4–0.7µ	none	300	3.0–3.5

*Note: Where not stated, room temperature may be assumed.

TABLE 9-9. (*Continued*)

GALLIUM ANTIMONIDE	Unit	Temp.* (°K)	Value
PHYSICAL PROPERTIES			
Formula weight	g		191.48
Density	g/cm^3		5.613
Melting point	°K		985
Melting point pressure coeff.	°K/kbar		
Boiling point	°K		
Symmetry			cubic
Lattice parameter, a_0	Å		6.094
" " ,			
THERMAL PROPERTIES			
Thermal conductivity	W/cm °K		0.35
" "			
" "			
Linear expansion coeff.	10^{-6}/°K		6.7
Thermal diffusivity	cm^2/sec		
Specific heat (C_p)	cal/g deg	298	0.0606
" "			
" "			
Debye temperature	°K		240
Spectral emissivity	none		
Vapor pressure	torr		
" "			
Heat of Sublimation	kcal/g		
Heat of Fusion	cal/g		98
Entropy	cal/g-atom deg		18.0
Zero point energy	cal/g-atom		
MAGNETIC PROPERTIES			
Magnetic susceptibility	10^{-7} emu/g	293	−2.01
Gyromagnetic properties, g factor	—	4–300	−5.9
" " , line width	Oe		
MECHANICAL PROPERTIES			
Hardness	Mohs		4.5
Young's modulus	10^{11} dyne/cm^2		7.60
Bulk modulus	10^{11} dyne/cm^2		5.635
Elastic constant (stiffness), C_{11}	10^{11} dyne/cm^2		8.85
" " C_{12}			4.04
" " C_{44}			4.33
" " C_{33}			
Elastic constant (compliance) S_{11}	10^{-12} cm^2/dyne		1.58
" " S_{12}			−0.496
" " S_{44}			2.31
" " S_{33}			
ACOUSTIC PROPERTIES			
Sound velocity	10^3 m/sec		4.26
" "			
ELECTRICAL/ELECTRONIC PROPERTIES			
Electrical resistivity	ohm · cm	300	0.04
Dielectric constant	none		ϵ_0 15.69; ϵ_∞ 14.44
Electron mobility	cm^2/V sec	300	4000
Hole mobility	cm^2/V sec	300	1400
Energy gap	eV	300	0.67
Work function	eV	300	4.76
Electron affinity	eV	300	4.06
Photoelectric threshold	eV	300	4.76–5.24
Photoelectric yield	eV		
Superconducting transition temp.	°K		4.24
Seebeck coefficient	μV/°K	300	558
OPTICAL PROPERTIES			
Transmission at 2–17μ (0.5 mm. thick)	%	300	⩾ 10
Reflectivity	%		
Refractive index at 1.8–2.5μ	none	297	3.8

*Note: Where not stated, room temperature may be assumed.

TABLE 9-9. *(Continued)*

GALLIUM ARSENIDE	Unit	Temp.* (°K)	Value
PHYSICAL PROPERTIES			
Formula weight	g		144.63
Density	g/cm^3		5.307
Melting point	°K		1511
Melting point pressure coeff.	°K/kbar		
Boiling point	°K		
Symmetry			cubic
Lattice parameter, a_0	Å	300	5.6534
" " ,			
THERMAL PROPERTIES			
Thermal conductivity	W/cm °K	300	0.54
" "			
" "			
Linear expansion coeff.	$10^{-6}/°K$	300	6.0
Thermal diffusivity	cm^2/sec		
Specific heat (C_p)	cal/g deg	298	0.0763
" "			
" "			
Debye temperature	°K		362
Spectral emissivity	none		
Vapor pressure	torr		
" "			
" "			
Heat of Sublimation	kcal/g		
Heat of Fusion	cal/g		
Entropy	cal/g-atom deg		135
Zero point energy	cal/g-atom		15.0
MAGNETIC PROPERTIES			
Magnetic susceptibility	10^{-7} emu/g	293	−2.24
Gyromagnetic properties, g factor	−		−2.1
" " , line width	Oe		
MECHANICAL PROPERTIES			
Hardness	Mohs		4.5
Young's modulus	10^{11} dyne/cm²		
Bulk modulus	10^{11} dyne/cm²		7.54
Elastic constant (stiffness), C_{11}	10^{11} dyne/cm²		1.192
" " C_{12}			0.5986
" " C_{44}			0.538
" " C_{33}			
Elastic constant (compliance) S_{11}	10^{-12} cm²/dyne		12.64
" " S_{12}			−4.234
" " S_{44}			18.6
" " S_{33}			
ACOUSTIC PROPERTIES			
Sound velocity	10^3 m/sec		5.09
" "			
ELECTRICAL/ELECTRONIC PROPERTIES			
Electrical resistivity	ohm · cm	300	0.4
Dielectric constant	none		ϵ_0 13.18; ϵ_∞ 10.9
Electron mobility	cm²/V sec	300	8500
Hole mobility	cm²/V sec	300	400
Energy gap	eV	300	1.43
Work function	eV	300	4.35
Electron affinity	eV	300	3.58
Photoelectric threshold	eV	300	5.13
Photoelectric yield	eV	300	
Superconducting transition temp.	°K		
Seebeck coefficient	µV/°K	300	−390
OPTICAL PROPERTIES			
Transmission at 1.0–15µ, (2 mm. thick)	%	300	≥ 10
Reflectivity at 2–20µ	%	300	> 20
Refractive index at 4µ	none	300	3.31

*Note: Where not stated, room temperature may be assumed.

TABLE 9-9. (*Continued*)

GALLIUM NITRIDE	Unit	Temp.* (°K)	Value
PHYSICAL PROPERTIES			
Formula weight	g		83.728
Density	g/cm^3		6.10
Melting point	°K		1323 (vacuum)
Melting point pressure coeff.	°K/kbar		
Boiling point	°K		
Symmetry			hexagonal
Lattice parameter, a_0	Å		3.180
" " , c_0			5.166
THERMAL PROPERTIES			
Thermal conductivity	W/cm °K		
" "			
" "			
Linear expansion coeff.	10^{-6}/°K	300–900	5.59
Thermal diffusivity	cm^2/sec		
Specific heat (C_p)	cal/g deg		
" "			
" "			
Debye temperature	°K		
Spectral emissivity	none		
Vapor pressure	torr		
" "			
Heat of Sublimation	kcal/g		
Heat of Fusion	cal/g		
Entropy	cal/g-atom deg		
Zero point energy	cal/g-atom		
MAGNETIC PROPERTIES			
Magnetic susceptibility	10^{-7} emu/g	290	−139
Gyromagnetic properties, g factor	—		
" " , line width	Oe		
MECHANICAL PROPERTIES			
Hardness	Mohs		
Young's modulus	10^{11} dyne/cm^2		
Bulk modulus	10^{11} dyne/cm^2		
Elastic constant (stiffness), C_{11}	10^{11} dyne/cm^2		
" " C_{12}			
" " C_{44}			
C_{33}			
Elastic constant (compliance) S_{11}	10^{-12} cm^2/dyne		
" " S_{12}			
" " S_{44}			
S_{33}			
ACOUSTIC PROPERTIES			
Sound velocity	10^3 m/sec		
" "			
ELECTRICAL/ELECTRONIC PROPERTIES			> 10^9
Electrical resistivity	ohm · cm		ϵ_∞ 4
Dielectric constant	none		125–150
Electron mobility	cm^2/V sec	300	
Hole mobility	cm^2/V sec		
Energy gap	eV		3.39
Work function	eV		
Electron affinity	eV		
Photoelectric threshold	eV		
Photoelectric yield	eV		
Superconducting transition temp.	°K		
Seebeck coefficient	μV/°K		
OPTICAL PROPERTIES			
Transmission	%		
Reflectivity	%		n_O 2.00; n_e 2.18
Refractive index at 0.58μ	none		

*Note: Where not stated, room temperature may be assumed.

TABLE 9-9. (*Continued*)

GALLIUM PHOSPHIDE	Unit	Temp.* (°K)	Value
PHYSICAL PROPERTIES			
Formula weight	g		100.695
Density	g/cm^3	300	4.1297
Melting point	°K		1740
Melting point pressure coeff.	°K/kbar		
Boiling point	°K		
Symmetry			cubic
Lattice parameter, a_0	Å		5.4495
" " ,			
THERMAL PROPERTIES			
Thermal conductivity	W/cm °K	300	1.1 (polycr)
" "			
" "			
Linear expansion coeff.	10^{-6}/°K	211–473	5.81
Thermal diffusivity	cm^2/sec		
Specific heat (C_p)	cal/g deg	298	0.0524
" "			
" "			
Debye temperature	°K		446
Spectral emissivity	none		
Vapor pressure	torr		
" "			
" "			
Heat of Sublimation	kcal/g		
Heat of Fusion	cal/g		
Entropy	cal/g-atom deg		232
Zero point energy	cal/g-atom		
MAGNETIC PROPERTIES			
Magnetic susceptibility	10^{-7} emu/g	300	−138
Gyromagnetic properties, g factor	−		1.76
" " , line width	Oe		
MECHANICAL PROPERTIES			
Hardness	Mohs		5
Young's modulus	10^{11} dyne/cm^2		8.85
Bulk modulus	10^{11} dyne/cm^2		
Elastic constant (stiffness), C_{11}	10^{11} dyne/cm^2		14.12
" " C_{12}			6.253
" " C_{44}			7.047
" " C_{33}			
Elastic constant (compliance) S_{11}	10^{-12} cm^2/dyne	300	0.973
" " S_{12}			
" " S_{44}			
" " S_{33}			
ACOUSTIC PROPERTIES			
Sound velocity	10^3 m/sec		6.32
" "			
ELECTRICAL/ELECTRONIC PROPERTIES			
Electrical resistivity	ohm · cm	300	1
Dielectric constant	none	300	ϵ_0 11.1; ϵ_∞ 9.04
Electron mobility	cm^2/V sec		2100
Hole mobility	cm^2/V sec		1000
Energy gap	eV		D2.78; I2.261
Work function	eV		1.31
Electron affinity	eV		4.0
Photoelectric threshold	eV		3
Photoelectric yield	eV		
Superconducting transition temp.	°K		
Seebeck coefficient	μV/°K	300	1000
OPTICAL PROPERTIES			
Transmission at 0.8–3μ, (10 mm. thick)	%	300	⩾ 10
Reflectivity	%		
Refractive index at 0.545μ	none	300	3.452

*Note: Where not stated, room temperature may be assumed.

TABLE 9-9. (*Continued*)

INDIUM ANTIMONIDE	Unit	Temp.* (°K)	Value
PHYSICAL PROPERTIES			
Formula weight	g		236.58
Density	g/cm^3	300	5.7751
Melting point	°K		798
Melting point pressure coeff.	°K/kbar		
Boiling point	°K		
Symmetry			cubic
Lattice parameter, a_0	Å		6.47877
" " ,			
THERMAL PROPERTIES			
Thermal conductivity	W/cm °K	100	0.6
" "		300	0.18
" "			
Linear expansion coeff.	$10^{-6}/°K$		5.04
Thermal diffusivity	cm^2/sec		
Specific heat (C_p)	cal/g deg	298	0.050
" "			
" "			
Debye temperature	°K		235
Spectral emissivity	none		
Vapor pressure	torr		
" "			
Heat of Sublimation	kcal/g		43
Heat of Fusion	cal/g		20.6
Entropy	cal/g-atom deg		
Zero point energy	cal/g-atom		
MAGNETIC PROPERTIES			
Magnetic susceptibility	10^{-7} emu/g	300	−2.88
Gyromagnetic properties, g factor	—	1.2	−51
" " , line width	Oe		
MECHANICAL PROPERTIES			
Hardness	Mohs		220
Young's modulus	10^{11} dyne/cm^2	300	4.29; 7.42
Bulk modulus	10^{11} dyne/cm^2	300	4.33
Elastic constant (stiffness), C_{11}	10^{11} dyne/cm^2	300	6.72
" " C_{12}		300	3.67
" " C_{44}		300	3.02
" " C_{33}			
Elastic constant (compliance) S_{11}	10^{-12} cm^2/dyne	300	2.42
" " S_{12}		300	−0.855
" " S_{44}		300	3.31
" " S_{33}			
ACOUSTIC PROPERTIES			
Sound velocity	10^3 m/sec		3.66
" "			
ELECTRICAL/ELECTRONIC PROPERTIES			
Electrical resistivity	ohm · cm	300	0.06
Dielectric constant	none		ϵ_0 17.72; ϵ_∞ 15.7
Electron mobility	cm^2/V sec		10^6
Hole mobility	cm^2/V sec		1.7×10^3
Energy gap	eV		0.18
Work function	eV		4.42
Electron affinity	eV	300	4.59
Photoelectric threshold	eV		4.77
Photoelectric yield	eV		
Superconducting transition temp.	°K		
Seebeck coefficient	µV/°K	300	−400
OPTICAL PROPERTIES			
Transmission at 9–25µ, (1.0 mm. thick)	%	300	⩾ 5
Reflectivity	%	300	
Refractive index at 0.59µ	none		4.22

*Note: Where not stated, room temperature may be assumed.

TABLE 9-9. (*Continued*)

INDIUM ARSENIDE	Unit	Temp.* ($^\circ$K)	Value
PHYSICAL PROPERTIES			
Formula weight	g		189.73
Density	g/cm^3		5.667
Melting point	$^\circ$K		1216
Melting point pressure coeff.	$^\circ$K/kbar		
Boiling point	$^\circ$K		
Symmetry			cubic
Lattice parameter, a_0	Å		6.0584
" " ,			
" " ,			
THERMAL PROPERTIES			
Thermal conductivity	W/cm $^\circ$K	300	0.26
" "		700	0.11
" "			
Linear expansion coeff.	$10^{-6}/^\circ$K	293–1215	5.19
Thermal diffusivity	cm^2/sec		
Specific heat (C_p)	cal/g deg	298	0.0602
" "			
" "			
Debye temperature	$^\circ$K		280
Spectral emissivity	none		
Vapor pressure	torr		
" "			
" "			
Heat of Sublimation	kcal/g		
Heat of Fusion	cal/g		56
Entropy	cal/g-atom deg		18.1
Zero point energy	cal/g-atom		
MAGNETIC PROPERTIES			
Magnetic susceptibility	10^{-7} emu/g	300	–2.95
Gyromagnetic properties, g factor	–	300	–17
" " , line width	Oe		
MECHANICAL PROPERTIES			
Hardness	Mohs		3.8
Young's modulus	10^{11} dyne/cm^2		8.11
Bulk modulus	10^{11} dyne/cm^2		5.8
Elastic constant (stiffness), C_{11}	10^{11} dyne/cm^2		8.65
" " C_{12}			4.85
" " C_{44}			3.96
" " C_{33}			
Elastic constant (compliance) S_{11}	10^{-12} cm^2/dyne		
" " S_{12}			
" " S_{44}			
" " S_{33}			
ACOUSTIC PROPERTIES			
Sound velocity	10^3 m/sec	300	4.35
" "			
ELECTRICAL/ELECTRONIC PROPERTIES			
Electrical resistivity	ohm · cm	300	0.03
Dielectric constant	none	300	ϵ_0 14.55; ϵ_∞ 11.8
Electron mobility	cm^2/V sec	300	33000
Hole mobility	cm^2/V sec	300	460
Energy gap	eV		0.356
Work function	eV	300	4.55
Electron affinity	eV	300	4.90
Photoelectric threshold	eV	300	5.31
Photoelectric yield	eV		
Superconducting transition temp.	$^\circ$K		
Seebeck coefficient	μV/$^\circ$K	300	–150
OPTICAL PROPERTIES			
Transmission at 4–15μ, (0.17 mm. thick)	%		$\geqslant$ 20
Reflectivity	%		
Refractive index at 0.517μ	none	300	4.558

*Note: Where not stated, room temperature may be assumed.

TABLE 9-9. (*Continued*)

INDIUM PHOSPHIDE	Unit	Temp.* (°K)	Value
PHYSICAL PROPERTIES			
Formula weight	g		145.795
Density	g/cm^3		4.787
Melting point	°K		1343
Melting point pressure coeff.	°K/kbar		
Boiling point	°K		
Symmetry			cubic
Lattice parameter, a_0	Å		5.868
" " ,			
THERMAL PROPERTIES			
Thermal conductivity	W/cm °K	300	0.7
" "			
" "			
Linear expansion coeff.	10^{-6}/°K		4.5
Thermal diffusivity	cm^2/sec		
Specific heat (C_p)	cal/g deg	298	0.0745
" "			
" "			
Debye temperature	°K		420
Spectral emissivity	none		
Vapor pressure	torr		
" "			
" "			
Heat of Sublimation	kcal/g		135
Heat of Fusion	cal/g		13.4
Entropy	cal/g-atom deg		
Zero point energy	cal/g-atom		
MAGNETIC PROPERTIES			
Magnetic susceptibility	10^{-7} emu/g	290	−228
Gyromagnetic properties, g factor	—		0.6
" " , line width	Oe		
MECHANICAL PROPERTIES			
Hardness	Mohs		540
Young's modulus	10^{11} dyne/cm^2		
Bulk modulus	10^{11} dyne/cm^2	300	10.22
Elastic constant (stiffness), C_{11}	10^{11} dyne/cm^2	300	5.76
" " C_{12}		300	4.60
" " C_{44}			
" " C_{33}		300	1.645
Elastic constant (compliance) S_{11}	10^{-12} cm^2/dyne	300	−0.594
" " S_{12}		300	2.173
" " S_{44}			
" " S_{33}			
ACOUSTIC PROPERTIES			
Sound velocity	10^3 m/sec	300	3.1; 5.13
" "			
ELECTRICAL/ELECTRONIC PROPERTIES			0.008
Electrical resistivity	ohm · cm	300	ϵ_0 12.34; ϵ_∞ 9.52
Dielectric constant	none		4600
Electron mobility	cm^2/V sec	300	150
Hole mobility	cm^2/V sec	300	D1.35; 12.35
Energy gap	eV	300	4.65
Work function	eV	300	4.40
Electron affinity	eV	300	5.69
Photoelectric threshold	eV	300	
Photoelectric yield	eV		
Superconducting transition temp.	°K		
Seebeck coefficient	μV/°K	300	−600 (polycr.)
OPTICAL PROPERTIES			
Transmission at 1–12μ, (2 mm. thick)	%	100	10
Reflectivity	%		
Refractive index at 0.59μ	none	297	3.45

*Note: Where not stated, room temperature may be assumed.

TABLE 9-10. PROPERTIES OF IV-IV SOLID INORGANIC SEMICONDUCTORS
(Neuberger, 1971A), (Shmartsev, 1966), (Moss, 1959).

SILICON CARBIDE	Unit	Temp.* ($°K$)	Value
PHYSICAL PROPERTIES			
Formula weight	g		40.07
Density	g/cm^3		3.21
Melting point	$°K$		3103
Melting point pressure coeff.	$°K/kbar$		
Boiling point	$°K$		
Symmetry			hexagonal
Lattice parameter, a_0	Å		3.08065
" " , c_0			15.1174
THERMAL PROPERTIES			
Thermal conductivity	$W/cm °K$	300	4.9
" "			
" "			
Linear expansion coeff.	$10^{-6}/°K$	300	2.9
Thermal diffusivity	cm^2/sec		
Specific heat (C_p)	cal/g deg	300	0.165
" "			
" "			
Debye temperature	$°K$		1430
Spectral emissivity	none		
Vapor pressure	torr		
" "			
Heat of Sublimation	kcal/g		
Heat of Fusion	cal/g		
Entropy	cal/g-atom deg		
Zero point energy	cal/g-atom		
MAGNETIC PROPERTIES			
Magnetic susceptibility	10^{-7} emu/g		
Gyromagnetic properties, g factor	—		
" " , line width	Oe		
MECHANICAL PROPERTIES			
Hardness	Mohs		9.5–9.75
Young's modulus	10^{11} dyne/cm^2	300	38.6
Bulk modulus	10^{11} dyne/cm^2	300	9.65
Elastic constant (stiffness), C_{11}	10^{11} dyne/cm^2	300	50.0
" " C_{12}		300	9.2
" " C_{44}		300	16.8
" " C_{33}			
Elastic constant (compliance) S_{11}	10^{-12} cm^2/dyne	300	0.203
" " S_{12}		300	-0.0421
" " S_{44}		300	0.595
" " S_{33}			
ACOUSTIC PROPERTIES			
Sound velocity	10^3 m/sec	300	13.26
" "			
" "			
ELECTRICAL/ELECTRONIC PROPERTIES			
Electrical resistivity	ohm · cm		1
Dielectric constant	none		10.32; 6.7
Electron mobility	cm^2/V sec		300
Hole mobility	cm^2/V sec		50
Energy gap	eV		I2.994
Work function	eV	300	4.4
Electron affinity	eV		
Photoelectric threshold	eV	300	> 7
Photoelectric yield	eV		
Superconducting transition temp.	$°K$		
Seebeck coefficient	$\mu V/°K$		
OPTICAL PROPERTIES			
Transmission at $1-6\mu$, (0.07 mm. thick)	%	300	~ 60
Reflectivity at $1-7\mu$	%	300	⩾ 15
Refractive index at 0.589μ	none		n_O 2.65; n_e 2.69

*Note: Where not stated, room temperature may be assumed.

TABLE 9-11. PROPERTIES OF IV–VI SOLID INORGANIC SEMICONDUCTORS
(Abrikosov, 1969), (Hirayama, 1964), (Dalven, 1969), Neuberger, 1969B)

LEAD TELLURIDE	Unit	Temp.* ($^\circ$K)	Value
PHYSICAL PROPERTIES			
Formula weight	g		334.82
Density	g/cm^3		8.16
Melting point	$^\circ$K		1198
Melting point pressure coeff.	$^\circ$K/kbar		
Boiling point	$^\circ$K		
Symmetry			cubic
Lattice parameter, a_0	Å		6.452
" " ,			
THERMAL PROPERTIES			
Thermal conductivity	W/cm $^\circ$K	100	0.08
" "		300	0.0219
" "		600	0.0038
Linear expansion coeff.	$10^{-6}/^\circ$K	300	19.8
Thermal diffusivity	cm^2/sec		
Specific heat (C_p)	cal/g deg	100	0.033
" "		200	0.036
" "			
Debye temperature	$^\circ$K		125
Spectral emissivity	none		
Vapor pressure	torr		
" "			
" "			
Heat of Sublimation	kcal/g	298	0.16
Heat of Fusion	cal/g		22
Entropy	cal/g-atom deg	298	26.3
Zero point energy	cal/g-atom		
MAGNETIC PROPERTIES			
Magnetic susceptibility	10^{-7} emu/g	300	–4.1 (polycr.)
Gyromagnetic properties, g factor	–	50	$\perp$15; $\parallel$57.5
" " , line width	Oe		
MECHANICAL PROPERTIES			
Hardness	Mohs		3
Young's modulus	10^{11} dyne/cm^2		4.8–7.6
Bulk modulus	10^{11} dyne/cm^2		3.90
Elastic constant (stiffness), C_{11}	10^{11} dyne/cm^2	303	10.8
" " C_{12}		303	0.77
" " C_{44}		303	1.343
" " C_{33}			
Elastic constant (compliance) S_{11}	10^{-12} cm^2/dyne	303	0.937
" " S_{12}		303	0.061
" " S_{44}		303	7.436
" " S_{33}			
ACOUSTIC PROPERTIES			
Sound velocity	10^3 m/sec	300	2.104
" "			
ELECTRICAL/ELECTRONIC PROPERTIES			
Electrical resistivity	ohm · cm	300	0.01
Dielectric constant	none	300	ϵ_0 360; ϵ_∞ 33.4
Electron mobility	cm^2/V sec	300	1600
Hole mobility	cm^2/V sec	300	700
Energy gap	eV	300	D0.32; I0.29
Work function	eV		5
Electron affinity	eV		4.6
Photoelectric threshold	eV		
Photoelectric yield	eV		
Superconducting transition temp.	$^\circ$K		
Seebeck coefficient (n-type)	μV/$^\circ$K	900	–200
OPTICAL PROPERTIES			
Transmission	%		
Reflectivity	%		
Refractive index at 0.5μ	none		6.0

*Note: Where not stated, room temperature may be assumed.

TABLE 9-11. (*Continued*)

TIN DIOXIDE	Unit	Temp.* (°K)	Value
PHYSICAL PROPERTIES			
Formula weight	g		150.7
Density	g/cm^3		6.99
Melting point	°K		1400
Melting point pressure coeff.	°K/kbar		
Boiling point	°K		
Symmetry			tetragonal
Lattice parameter, a_0	Å		4.738
$''$ $''$, c_0			3.188
$''$ $''$,			
THERMAL PROPERTIES			
Thermal conductivity	W/cm °K	315	0.314
$''$ $''$		425	0.239
$''$ $''$			
Linear expansion coeff.	10^{-6}/°K	300	0.005
Thermal diffusivity	cm^2/sec		
Specific heat (C_p)	cal/g deg		
$''$ $''$			
$''$ $''$			
Debye temperature	°K		500
Spectral emissivity	none		
Vapor pressure	torr		
$''$ $''$			
$''$ $''$			
Heat of Sublimation	kcal/g		
Heat of Fusion	cal/g		
Entropy	cal/g-atom deg		
Zero point energy	cal/g-atom		
MAGNETIC PROPERTIES			
Magnetic susceptibility	10^{-7} emu/g		
Gyromagnetic properties, g factor	−		⊥1.876; ∥1.905
$''$ $''$, line width	Oe		
MECHANICAL PROPERTIES			
Hardness	Mohs		6–7
Young's modulus	10^{11} dyne/cm^2		
Bulk modulus	10^{11} dyne/cm^2		
Elastic constant (stiffness), C_{11}	10^{11} dyne/cm^2		
$''$ $''$ C_{12}			
$''$ $''$ C_{44}			
C_{33}			
Elastic constant (compliance) S_{11}	10^{-12} cm^2/dyne		
$''$ $''$ S_{12}			
$''$ $''$ S_{44}			
S_{33}			
ACOUSTIC PROPERTIES			
Sound velocity	10^3 m/sec		
$''$ $''$			
ELECTRICAL/ELECTRONIC PROPERTIES			
Electrical resistivity	ohm · cm	300	∼0.01
Dielectric constant	none	300	ϵ_0⊥13.5; ϵ_0∥9.0
Electron mobility	cm^2/V sec	300	20
Hole mobility	cm^2/V sec		
Energy gap	eV	300	D⊥3.69; I⊥2.55
Work function	eV		
Electron affinity	eV		
Photoelectric threshold	eV		
Photoelectric yield	eV		
Superconducting transition temp.	°K		
Seebeck coefficient	μV/°K	300–800	∼ 500
OPTICAL PROPERTIES			
Transmission	%		
Reflectivity	%		
Refractive index at 0.58μ	none	289	n_o 2.007; n_e 2.098

*Note: Where not stated, room temperature may be assumed.

TABLE 9-11. (*Continued*)

TIN DISULFIDE	Unit	Temp.* (°K)	Value
PHYSICAL PROPERTIES			
Formula weight	g		182.83
Density	g/cm^3		4.5
Melting point	°K		1143
Melting point pressure coeff.	°K/kbar		
Boiling point	°K		
Symmetry			hexagonal
Lattice parameter, a_0	Å		3.644
" " , c_0			5.884
THERMAL PROPERTIES			
Thermal conductivity	W/cm °K		
" "			
" "			
Linear expansion coeff.	10^{-6}/°K		
Thermal diffusivity	cm^2/sec		
Specific heat (C_p)	cal/g deg	569	0.0915
" "			
" "			
Debye temperature	°K		
Spectral emissivity	none		
Vapor pressure	torr		
" "			
Heat of Sublimation	kcal/g		
Heat of Fusion	cal/g		
Entropy	cal/g-atom deg		20.9
Zero point energy	cal/g-atom		
MAGNETIC PROPERTIES			
Magnetic susceptibility	10^{-7} emu/g	300	−0.58
Gyromagnetic properties, g factor	—		
" " , line width	Oe		
MECHANICAL PROPERTIES			
Hardness	Mohs		
Young's modulus	10^{11} dyne/cm^2		
Bulk modulus	10^{11} dyne/cm^2		
Elastic constant (stiffness), C_{11}	10^{11} dyne/cm^2		
" " C_{12}			
" " C_{44}			
" " C_{33}			
Elastic constant (compliance) S_{11}	10^{-12} cm^2/dyne		
" " S_{12}			
" " S_{44}			
" " S_{33}			
ACOUSTIC PROPERTIES			
Sound velocity	10^3 m/sec		
" "			
ELECTRICAL/ELECTRONIC PROPERTIES			
Electrical resistivity	ohm · cm	300	10^8–10^9
Dielectric constant	none		
Electron mobility	cm^2/V sec		
Hole mobility	cm^2/V sec		
Energy gap	eV	300	D2.88; I2.07
Work function	eV		
Electron affinity	eV		
Photoelectric threshold	eV		
Photoelectric yield	eV		
Superconducting transition temp.	°K		
Seebeck coefficient	μV/°K		
OPTICAL PROPERTIES			
Transmission	%		
Reflectivity	%		
Refractive index at 0.6μ	none		3.6

*Note: Where not stated, room temperature may be assumed.

TABLE 9-11. (*Continued*)

TIN SELENIDE	Unit	Temp.* (°K)	Value
PHYSICAL PROPERTIES			
Formula weight	g		197.66
Density	g/cm^3	273	6.179
Melting point	°K		1133
Melting point pressure coeff.	°K/kbar		
Boiling point	°K		
Symmetry			orthorhombic
Lattice parameter, a_0	Å		4.2
" " ,b_0			11.6
" " ,c_0			4.48
THERMAL PROPERTIES			
Thermal conductivity	W/cm °K		
" "			
" "			
Linear expansion coeff.	10^{-6}/°K		
Thermal diffusivity	cm^2/sec		
Specific heat (C_p)	cal/g deg	80	0.0435
" "			
" "			
Debye temperature	°K	80	210
Spectral emissivity	none		
Vapor pressure	torr		
" "			
Heat of Sublimation	kcal/g	298	0.266
Heat of Fusion	cal/g		
Entropy	cal/g-atom deg	298	20.6
Zero point energy	cal/g-atom		
MAGNETIC PROPERTIES			
Magnetic susceptibility	10^{-7} emu/g		
Gyromagnetic properties, g factor	—		
" " , line width	Oe		
MECHANICAL PROPERTIES			
Hardness	Mohs		
Young's modulus	10^{11} dyne/cm^2		
Bulk modulus	10^{11} dyne/cm^2		
Elastic constant (stiffness), C_{11}	10^{11} dyne/cm^2		
" " C_{12}			
" " C_{44}			
C_{33}			
Elastic constant (compliance) S_{11}	10^{-12} cm^2/dyne		
" " S_{12}			
" " S_{44}			
S_{33}			
ACOUSTIC PROPERTIES			
Sound velocity	10^3 m/sec		
" "			
ELECTRICAL/ELECTRONIC PROPERTIES			
Electrical resistivity	ohm · cm	300	0.1
Dielectric constant	none		
Electron mobility	cm^2/V sec		
Hole mobility	cm^2/V sec	300	200
Energy gap	eV	300	D1.2; I0.91
Work function	eV		
Electron affinity	eV		
Photoelectric threshold	eV		
Photoelectric yield	eV		
Superconducting transition temp.	°K		
Seebeck coefficient	μV/°K	300–600	400
OPTICAL PROPERTIES			
Transmission	%		
Reflectivity	%		
Refractive index at 2.5μ	none		n_o 4.6; n_e 5.0

*Note: Where not stated, room temperature may be assumed.

TABLE 9-11. (*Continued*)

TIN SULFIDE	Unit	Temp.* (°K)	Value
PHYSICAL PROPERTIES			
Formula weight	g		150.766
Density	g/cm^3		5.22
Melting point	°K		1155
Melting point pressure coeff.	°K/kbar		
Boiling point	°K		1503
Symmetry			orthorhombic
Lattice parameter, a_0	Å		4.33
" " , b_0			11.18
" " , c_0			3.98
THERMAL PROPERTIES			
Thermal conductivity	W/cm °K	100	0.028
" "		300	0.011
" "		400	0.008
Linear expansion coeff.	$10^{-6}/°K$		
Thermal diffusivity	cm^2/sec		
Specific heat (C_p)	cal/g deg	104	0.0315
" "		297	0.0780
" "			
Debye temperature	°K		270
Spectral emissivity	none		
Vapor pressure	torr		
" "			
Heat of Sublimation	kcal/g		0.35
Heat of Fusion	cal/g		50
Entropy	cal/g-atom deg		18.4
Zero point energy	cal/g-atom		
MAGNETIC PROPERTIES			
Magnetic susceptibility	10^{-7} emu/g		
Gyromagnetic properties, g factor	—		
" " , line width	Oe		
MECHANICAL PROPERTIES			
Hardness	Mohs		
Young's modulus	10^{11} dyne/cm^2		
Bulk modulus	10^{11} dyne/cm^2		
Elastic constant (stiffness), C_{11}	10^{11} dyne/cm^2		
" " C_{12}			
" " C_{44}			
C_{33}			
Elastic constant (compliance) S_{11}	10^{-12} cm^2/dyne		
" " S_{12}			
" " S_{44}			
S_{33}			
ACOUSTIC PROPERTIES			
Sound velocity	10^3 m/sec		
" "			
ELECTRICAL/ELECTRONIC PROPERTIES			
Electrical resistivity	ohm · cm	300	10^6
Dielectric constant	none		ϵ_∞ 19.5
Electron mobility	cm^2/V sec		
Hole mobility	cm^2/V sec	300	65
Energy gap	eV	300	11.08
Work function	eV		
Electron affinity	eV		
Photoelectric threshold	eV		
Photoelectric yield	eV		
Superconducting transition temp.	°K		
Seebeck coefficient	$\mu V/°K$	293–573	800
OPTICAL PROPERTIES			
Transmission	%		
Reflectivity	%		
Refractive index	none		

*Note: Where not stated, room temperature may be assumed.

TABLE 9-11. *(Continued)*

TIN TELLURIDE	Unit	Temp.* (°K)	Value
PHYSICAL PROPERTIES			
Formula weight	g		246.31
Density	g/cm^3		6.445
Melting point	°K		1079
Melting point pressure coeff.	°K/kbar		
Boiling point	°K		
Symmetry			cubic
Lattice parameter, a_0	Å		6.327
" " ,			
THERMAL PROPERTIES			
Thermal conductivity	W/cm °K	300	0.067
" "			
" "			
Linear expansion coeff.	$10^{-6}/°K$	270	21
Thermal diffusivity	cm^2/sec		
Specific heat (C_p)	cal/g deg	80	0.022
" "			
" "			
Debye temperature	°K		132
Spectral emissivity	none		
Vapor pressure	torr		
" "			
Heat of Sublimation	kcal/g		0.216
Heat of Fusion	cal/g		44
Entropy	cal/g-atom deg		24.2
Zero point energy	cal/g-atom		
MAGNETIC PROPERTIES			
Magnetic susceptibility	10^{-7} emu/g	300	−3.8
Gyromagnetic properties, g factor	—		
" " , line width	Oe		
MECHANICAL PROPERTIES			
Hardness	Mohs		
Young's modulus	10^{11} dyne/cm^2		
Bulk modulus	10^{11} dyne/cm^2	300	4.35
Elastic constant (stiffness), C_{11}	10^{11} dyne/cm^2		11.25
" " C_{12}			0.75
" " C_{44}			1.172
" " C_{33}			
Elastic constant (compliance) S_{11}	10^{-12} cm^2/dyne		
" " S_{12}			
" " S_{44}			
" " S_{33}			
ACOUSTIC PROPERTIES			
Sound velocity	10^3 m/sec		
" "			
ELECTRICAL/ELECTRONIC PROPERTIES			
Electrical resistivity	ohm · cm	300	5×10^{-5}
Dielectric constant	none		ϵ_0 1770; ϵ_∞ 46.3
Electron mobility	cm^2/V sec		
Hole mobility	cm^2/V sec		
Energy gap	eV	300	0.18–0.20
Work function	eV		5.4
Electron affinity	eV		4.5
Photoelectric threshold	eV		
Photoelectric yield	eV		
Superconducting transition temp.	°K		
Seebeck coefficient	μV/°K	300	26
OPTICAL PROPERTIES			
Transmission	%		
Reflectivity	%		
Refractive index at 2μ	none		6.5

*Note: Where not stated, room temperature may be assumed.

TABLE 9-12. PROPERTIES OF SOLID ORGANIC SEMICONDUCTORS. (Clauser, 1963)

Substance	Formula	Resistivity, ohm-cm	Band Gap	
			Conductivity, eV	Photo Conduct, eV
POLYACENES Anthracene		300	0.83	–
Tetracene		10	0.85	3.6
Pyrene		300	1.01	3.2
Perylene		10	0.98	–
Chrysene		100	1.10	3.2
Coronene		0.2	1.15	–
Pyranthrene		10^7	0.54	0.85
POLYACENES WITH QUINONOID ATTACHEMENTS Violanthrone		1000	0.39	0.84
Pyranthrone		10^6	0.54	1.14
AZO-AROMATIC COMPOUNDS Indanthrone black		3000	0.28	–

TABLE 9.12. (*Continued*)

Substance	Formula	Resis-tivity, ohm-cm	Band Gap	
			Conduc-tivity, eV	Photo Conduct, eV
1,9,4,10-Anthradipyrimidine		1000	1.61	–
PHTHALOCYANINES		10^4	1.2	1.56
FREE RADICALS α,α-Diphenyl β-pieryl hydrazyl		10^6	0.74	–

TABLE 9-13. PROPERTIES OF LIQUID SEMICONDUCTORS.

Material	Melting point (°K)	Density at °K* (g cm⁻³)		Electrical conductivity at °K*, Ω⁻¹·cm⁻¹		Atomization Energy (kcal/mole)	Heat of fusion (kcal/mole)	Entropy of fusion (e.u.)	Thermoelectric power, μV per °K		Activation energy for viscous flow (kcal per mole)	Entropy of viscous flow (e.u.)	Viscosity of liquid at °K* (centipoises)
		solid	liquid	solid	liquid				solid	liquid			
Si	1693	2.30	2.53	580	12000	204	12.1	7.1			8.63	2.1	0.348
Ge	1210	5.26	5.51	1250	14000	178	8.35	6.9	−90	0	2.74	2.85	0.135
AlSb	1353	4.18	4.72	160	9900	160	14.2	5.2	−160	−60	10	2.2	0.250
GaSb	985	5.60	6.06	280	10600	134	12.0	6.1	−60	0	2.7	5.0	0.368
InSb	809	5.76	6.48	2900	10000	121	11.6	7.2	−120	−20	2.0	9.4	0.363
GaAs	1511	5.16	5.71	300	7900	146	23.2	7.7	—	—	6.5	7.8	0.320
InAs	1215	5.5	5.89	3600	6800	130	12.6	5.2	—	—	6.2	3	0.174
ZnTe	1512					109					9.0	6.5	0.868
CdTe	1365					99					5.75	7.7	0.435
CuI	875	5.36	4.84				2.6		550	490			0.432
Ga₂Te₃	1063	5.35	5.086						−290	−85	11	7	0.546
In₂Te₃	940	5.77	5.54						−50	30	13	7.5	0.323
Mg₂Si	1375	1.84	2.27	1120	9800		20.4	5.0			13.9	2.3	0.299
Mg₂Ge	1388		3.20	1140	8400						9.5	3.8	0.311
Mg₂Sn	1051	3.45	3.52	2040	10600		11.4	3.6			9.5	5.8	0.520
Mg₂Pb	823	5.00	5.20	3530	8600		9.3	3.8			9.6	6.2	0.560
GeTe	998	5.97	5.57	2400	2600		11.3	5.7	130	21	4.70	5.8	0.375
SnTe	1063	6.15	5.85	1440	1800		8.0	3.7	140	28	4.90	6.1	0.348
PbTe	1190	7.69	7.45	420	1520		7.5	3.1	−60	−10	6.85	6.0	0.243
PbSe	1361	7.57	7.10	300	450		8.5	3.1	−120	−60	6.85	7.5	0.240
PbS	1392	7.07	6.45	250	220		8.7	3.1	−220	−200	9.80	7.5	0.319
Bi₂Se₃	979	7.27	6.97	450	900		28.35	6.6	−90	−35	9.7	5.05	0.540
Bi₂Te₃	858	7.5	7.26	1250	2580		23.65	5.3	−45	−3	2.7	7.80	0.198
Sb₂Te₃	895	6.29	6.09	900	1850				90	11	6.1	7.35	0.513
Se (hex)	493	4.69	3.975				1.5	3			3.94	6.7	6.63
Te	725	6.1	5.775				4.17	5.7			1.18	6.7	0.357

*At melting point.

TABLE 9-14. SUMMARY OF SEMICONDUCTOR ELECTRICAL PROPERTIES.

Name	Formula	Crystal	Band gap at 300°K, eV	Mobility at 300°K, cm$^2 \cdot$ sec$^{-1} \cdot$ V^{-1}	
				Electrons	Holes
III-V COMPOUNDS					
Aluminum Antimonide	AlSb	C	1.60	200	420
Aluminum Arsenide	AlAs	C	2.16	180	
Aluminum Nitride	AlN	H	3.8		
Aluminum Phosphide	AlP	C	3.0	80	
Boron Antimonide	BSb	C	2.6		
Boron Arsenide	BAs	C	3.0		
Boron Nitride	BN	C	4.6		
Boron Phosphide	BP	C	6.0		300
Gallium Antimonide	GaSb	C	0.70	4,000	1,400
Gallium Arsenide	GaAs	C	1.43	8,500	400
Gallium Nitride	GaN	H	3.25		
Gallium Phosphide	GaP	C	2.24	110	75
Indium Antimonide	InSb	C	0.18	80,000	750
Indium Arsenide	InAs	C	0.36	33,000	460
Indium Nitride	InN	H	2.4		
Indium Phosphide	InP	C	1.26	4,600	150
III-VI COMPOUNDS					
Indium Oxide	In$_2$O$_3$		3.5		
Indium Selenide	In$_2$Se$_3$		1.25	30	
Indium Sulfide	In$_2$S$_3$		2.28		
Indium Telluride	In$_2$Te$_3$		1.0	340	
II-IV COMPOUNDS					
Magnesium Germanide	Mg$_2$Ge		0.74	530	110
Magnesium Silicide	Mg$_2$Si		0.77	370	65
Magnesium Stannide	Mg$_2$Sn		0.36	210	150
II-VI COMPOUNDS					
Barium Oxide	BaO	C	3.8	<10	
Barium Telluride	BaTe	C	3.4		
Cadmium Oxide	CdO	C	2.35	100	
Cadmium Selenide	CdSe	H	1.73	800	
Cadmium Sulfide	CdS	H	2.42	300	50
Cadmium Telluride	CdTe	C	1.50	700	65
Mercuric Oxide	HgO	R	2.214		
Mercuric Selenide	HgSe	C	0.6	18,500	
Mercuric Sulfide	HgS	H	2.0	11	
Mercuric Telluride	HgTe	C	0.025	22,000	160
Zinc Oxide	ZnO	H	3.2	200	
Zinc Selenide	ZnSe	C	2.67	530	16
Zinc Sulfide	ZnS	C	3.58	165	
Zinc Telluride	ZnTe	C	2.26	530	100
V-VI COMPOUNDS					
Antimony Selenide	Sb$_2$Se$_3$		1.2	15	45
Antimony Telluride	Sb$_2$Te$_3$		0.30		270
Arsenic Selenide	As$_2$Se$_3$		1.6	15	45
Arsenic Telluride	As$_2$Te$_3$		1.0	170	80
Bismuth Selenide	Bi$_2$Se$_3$	R	0.27	600	

TABLE 9.14. (*Continued*)

Name	Formula	Crystal	Band gap at 300°K, eV	Mobility at 300°K, cm$^2 \cdot$ sec$^{-1} \cdot$ V^{-1}	
				Electrons	Holes
V-VI COMPOUNDS (*continued*)					
Bismuth Sulfide	Bi$_2$S$_3$		1.3	200	
Bismuth Telluride	Bi$_2$Te$_3$		0.15	420	400
IV-VI COMPOUNDS					
Lead Selenide	PbSe	C	0.27	1,200	1,000
Lead Sulfide	PbS	C	0.41	600	700
Lead Telluride	PbTe	C	0.31	1,800	900
Tin Selenide	SnSe	R	1.2		200
Tin Sulfide	SnS	R	1.08 I		65
Tin Telluride	SnTe	R	0.20		510
II-V COMPOUNDS					
Cadmium Antimonide	CdSb		0.48	300	1,000
Cadmium Arsenide	CdAs$_2$		1.0		100
Cadmium Arsenide	Cd$_3$As$_2$	T	0.14		15,000
Zinc Antimonide	ZnSb		0.56	10	350
Zinc Arsenide	ZnAs$_2$		0.90		10
Zinc Arsenide	Zn$_3$As$_2$	T	0.93		10
Zinc Phosphide	Zn$_3$P$_2$	T	0.49		
I-VI COMPOUNDS					
Silver Selenide	Ag$_2$Se	T	0.075	2,000	505
Silver Sulfide	Ag$_2$S	H	1.3	63.5 (100°C)	19 (100°C)
IV-IV COMPOUNDS					
Silicon Carbide	SiC	H	2.994	300	50
ELEMENTS					
Antimony	Sb		0.11		
Arsenic	As		1.2		
Boron	B		1.4	6,000	4,000
Carbon (Diamond)	C	C	5.47	1,800	1,200
Germanium	Ge	C	0.67	3,900	1,900
Iodine	I$_2$	RH	1.30		
Selenium	Se	H	1.6	2	17
Silicon	Si	C	1.106	1,500	480
α-Sulfur	S		2.6		
Tellurium	Te	H	0.37		
α-Tin (grey)	Sn	C	0.08	1,600	1,000

REFERENCES

Abrikosov, N. K., V. F. Bankina, L. V. Poret-skaya and L. E. Shelimova, "Semiconducting II-VI, IV-VI, and V-VI Compounds," Plenum Press, New York, 1969.

Clauser, H. R., E. Fabian, D. Peckner and M. Riley, Eds., "The Encyclopedia of Engineering Materials and Processes," Van Nostrand Reinhold, New York, 1963.

Dalven, R., "A Review of the Semiconducting Properties of PbTe, PbSe, PbS and PbO," *Infrared Phys.*, 9, 141–184 (1969).

Glazov, V. M., S. N. Chizhevskaya and N. N. Glagoleva, "Liquid Semiconductors," Plenum Press, New York, 1969.

Hampel, C. A., Ed., "The Encyclopedia of the Chemical Elements," Van Nostrand Reinhold, New York, 1968.

Hilsum, C. and A. C. Rose-Innes, "Semiconducting III-V Compounds," Pergamon Press, Oxford, 1961.

Hirayaya, C., "Thermodynamic Properties of Solid Monoxides, Monosulfides, Monoselenides, and Monotellurides of Ge, Sn, and Pb," *J. Chem. &. Eng. Data*, 9, 65–68 (1964).

Moses, A. J., "Refractive Index of Optical Materials in the Infrared Region," Hughes Aircraft Company, U.S. Govt. Report No. AD 704 555 (1970).

Moses, A. J., "Handbook of Electronic Materials," Vol. 1, "Optical Materials Properties," IFI/Plenum, New York, 1971.

Moss, T. S., "Optical Properties of Semi-Conductors, Butterworths Scientific Publ., London, 1959.

Neuberger, M., "Germanium," Hughes Aircraft Company, U.S. Govt. Report No. AD 610 828 (1965).

Neuberger, M. and S. J. Welles, "Silicon," Hughes Aircraft Company, U.S. Govt. Report No. AD 698 342 (1969).

Neuberger, M., "II-VI Semiconducting Compounds Data Tables," Hughes Aircraft Company, U.S. Govt. Report No. AD 698 341 (1969A).

Neuberger, M., "IV-VI Semiconducting Compounds Data Tables," Hughes Aircraft Company, U.S. Govt. Report No. AD 699 260 (1969B).

Neuberger, M., "Arsenic," Hughes Aircraft Company, Interim Report No. IR-71 (1970).

Neuberger, M., "Handbook of Electronic Materials," Vol. 2, "III-V Semiconducting Compounds," IFI/Plenum, New York, 1971.

Neuberger, M., "Handbook of Electronic Materials," Vol. 5, "Group IV Semiconducting Compounds," IFI/Plenum, New York, 1971A.

Schultz, M. L., "Silicon: Semiconductor Properties," *Infrared Phys.*, 4, 93–112 (1964).

Shmartsev, Y. V., A. Valov and A. S. Borshchevskii, "Refractory Semiconductor Materials," Goryunova, N. A. and D. N. Nasledov, Eds., Consultants Bureau, New York (1966).

Sze, S. M. and J. C. Irvin, "Resistivity, Mobility, and Impurity Levels in GaAs, Ge and Si at 300°K," *Solid State Electronics*, 11, 599–602 (1968).

Willardson, R. K. and H. L. Goering, Eds., "Compound Semi-Conductors," Vol. 1, "Preparation of III-V Compounds," Van Nostrand Reinhold, New York, 1962.

Willardson, R. K. and A. C. Beer, Eds., "Semiconductors and Semimetals," Vol. 1, "Physics of III-V Compounds," Academic Press, New York, 1966.

Willardson, R. K. and A. C. Beer, Eds., "Semiconductors and Semimetals," Vol. 2, "Physics of III-V Compounds," Academic Press, New York, 1966A.

Willardson, R. K. and A. C. Beer, Eds., "Semiconductors and Semimetals," Vol. 4, "Physics of III-V Compounds," Academic Press, New York, 1968.

Wolf, H. F., "Silicon Semiconductor Data," Pergamon Press, Oxford, 1969.

Yarbrough, D. W., "Status of Diffusion Data in Binary Compound Semiconductors," *Solid State Technol.*, 11, 23–31, 54 (Nov. 1968).

BIBLIOGRAPHY

Aigrin, P. and M. Balkanski, Eds., "Selected Constants Relative to Semiconductors," Pergamon Press, New York, 1961.

Aven, M. and J. S. Prener, Eds., "Physics and Chemistry of II-IV Compounds," North-Holland Publishing Co., Amsterdam, 1967.

Boguslavskii, L. I. and A. V. Vannikov, "Organic Semiconductors and Biopolymers," Plenum Press, New York, 1970.

Bylander, E. G., "Materials for Semiconductor Functions," Hayden Book Co., New York, 1971.

Connolly, T. F. Ed., "Semiconductors—Preparation, Crystal Growth, and Selected Properties," Plenum, New York, 1972.

Enderby, J. E. and C. J. Simmons, "The Electrical Properties of Liquid Semiconductors," *Phil. Mag.*, 20, (163), 125–134 (July 1969).

Goryunova, N. A., "The Chemistry of Diamond-like Semiconductors," Chapman & Hall, London, 1965.

Gutmann, F. and L. E. Lyons, "Organic Semiconductors," Wiley, New York, 1967.

Hannay, N. B., Ed., "Semiconductors," Van Nostrand Reinhold, New York, 1959.

Jarzebski, Z. M., "Oxide Semiconductors," Pergamon Press, Elmsford, N.Y., 1974.

Kane, P. F. and G. B. Larrabee, "Characterization of Semiconducting Materials," McGraw-Hill, New York, 1970.

Katon, J. E., Ed., "Organic Semiconducting Polymers," Marcel Dekker, New York, 1968.

Le Comber, P. G. and J. Mort, Eds., "Electronic and Structural Properties of Amorphous Semiconductors," Academic Press, London, 1973.

Long, D., "Energy Bands in Semiconductors," Wiley, New York, 1968.

Madelung, O., "Physics of III-V Compounds," Wiley, New York, 1964.

Ray, B., "II-VI Compounds," Pergamon Press, Elmsford, N.Y., 1969.

Suchet, J. P. "Crystal Chemistry and Semiconduction in Transition Metal Binary Compounds," Academic Press, New York, 1971.

Wieder, H. H., "Intermetallic Semiconducting Films," Pergamon Press, Elmsford, N.Y., 1970.

Wolf, H. F., "Semiconductors," Wiley, New York, 1971.

CHAPTER **X**
PROPERTIES OF
SUPERCONDUCTORS

The rather extensive Tables of properties of superconducting materials presented here are based on publications of the Superconductive Materials Data Center, operated for the Office of Standard Reference Data, National Bureau of Standards, by the General Electric Company, Schenectady, New York. The assistance of the Center's Director, Dr. B. W. Roberts, is hereby gratefully acknowledged.

The following Tables are presented in Chapter X:

Table 10-1 Symbols Used in Chapter X.

Table 10-2 Key to Crystal Structure Types Found in Table 10-4.

Table 10-3 Properties of Superconductive Elements.

Table 10-4 Tabulation of Superconductive Materials (Including Proven Non-Superconductors).

Table 10-5 High Magnetic Field Superconductive Materials and Selected Properties.

TABLE 10-1. SYMBOLS USED IN CHAPTER X.

T_c	Critical Temperature ($^\circ$K)
T_{obs}	Temperature of Measurement ($^\circ$K)
T_n	Lowest Temperature at Which Material Was Tested for a Superconductive Transition ($^\circ$K)
H_0	Critical Field at 0°K (oersteds)
H_{c1}, H_{c2}, H_{c3}	Critical Fields, Defining Type II Superconductors, (k · oersteds)
H_p	Paramagnetic Critical Field (oersteds)
HF	Additional Data in Table 10-5.
M	Maximum Value in a Series of Specimens or Compositions
γ	Electronic Specific Heat (mJ mole^{-1}$^\circ$K^{-2})
θ_D	Debye Temperature ($^\circ$K)

TABLE 10-2. KEY TO CRYSTAL STRUCTURE TYPES
FOUND IN TABLE 10-4.
(Courtesy of Roberts, 1969)

"Struckturbericht" Type*	Example	Class
A1	Cu	Cubic, f. c.
A2	W	Cubic, b. c.
A3	Mg	Hexagonal, close packed
A4	Diamond	Cubic, f. c.
A5	White Sn	Tetragonal, b. c.
A6	In	Tetragonal, b. c. (f. c. cell usually used)
A7	As	Rhombohedral
A8	Se	Trigonal
A10	Hg	Rhombohedral
A12	α–Mn	Cubic, b. c.
A13	β–Mn	Cubic
A15	β–W	Cubic
B1	NaCl	Cubic, f. c.
B2	CsCl	Cubic
B3	ZnS	Cubic
B4	ZnS	Hexagonal
$B8_1$	NiAs	Hexagonal
$B8_2$	Ni_2In	Hexagonal
B10	PbO	Tetragonal
B11	γ–CuTi	Tetragonal
B17	PtS	Tetragonal
B18	CuS	Hexagonal
B20	FeSi	Cubic
B27	FeB	Ortho-rhombic
B31	MnP	Ortho-rhombic
B32	NaTl	Cubic, f. c.
B34	PdS	Tetragonal
B_f	δ–CrB	Ortho-rhombic
B_g	MoB	Tetragonal, b. c.
B_h	WC	Hexagonal
B_i	γ'–MoC	Hexagonal
C1	CaF_2	Cubic, f. c.
$C1_b$	MgAgAs	Cubic, f. c.
C2	FeS_2	Cubic
C6	CdI_2	Trigonal
C11b	$MoSi_2$	Tetragonal, b. c.
C12	$CaSi_2$	Rhombohedral
C14	$MgZn_2$	Hexagonal
C15	Cu_2Mg	Cubic, f. c.
$C15_b$	$AuBe_5$	Cubic
C16	$CuAl_2$	Tetragonal, b. c.
C18	FeS_2	Ortho-rhombic
C22	Fe_2P	Trigonal
C23	$PbCl_2$	Ortho-rhombic
C32	AlB_2	Hexagonal
C36	$MgNi_2$	Hexagonal
C37	Co_2Si	Ortho-rhombic
C49	$ZrSi_2$	Ortho-rhombic
C54	$TiSi_2$	Ortho-rhombic
C_c	Si_2Th	Tetragonal, b. c.
$D0_3$	BiF_3	Cubic, f. c.
$D0_{11}$	Fe_3C	Ortho-rhombic

TABLE 10-2. (*Continued*)

"Struckturbericht" Type*	Example	Class
DO_{18}	Na_3As	Hexagonal
DO_{19}	Ni_3Sn	Hexagonal
DO_{20}	$NiAl_3$	Ortho-rhombic
DO_{22}	$TiAl_3$	Tetragonal
DO_e	Ni_3P	Tetragonal, b. c.
$D1_3$	Al_4Ba	Tetragonal, b. c.
$D1_c$	$PtSn_4$	Ortho-rhombic
$D2_1$	CaB_6	Cubic
$D2_c$	MnU_6	Tetragonal, b. c.
$D2_d$	$CaZn_5$	Hexagonal
$D5_2$	La_2O_3	Trigonal
$D5_8$	Sb_2S_3	Ortho-rhombic
$D7_3$	Th_3P_4	Cubic, b. c.
$D7_b$	Ta_3B_4	Ortho-rhombic
$D8_1$	Fe_3Zn_{10}	Cubic, b. c.
$D8_2$	Cu_5Zn_8	Cubic, b. c.
$D8_3$	Cu_9Al_4	Cubic
$D8_8$	Mn_5Si_3	Hexagonal
$D8_b$	$CrFe$	Tetragonal
$D8_i$	Mo_2B_5	Rhombohedral
$D10_2$	Fe_3Th_7	Hexagonal
$E2_1$	$CaTiO_3$	Cubic
$E9_3$	Fe_3W_3C	Cubic, f. c.
$L1_0$	$CuAu$	Tetragonal
$L1_2$	Cu_3Au	Cubic
L'_{2b}	ThH_2	Tetragonal, b. c.
L'_3	Fe_2N	Hexagonal

*See W. B. Pearson, Handbook of Lattice Spacing and Structures of Metals (Pergamon, New York, 1958), p. 79, also Vol. II (Pergamon, New York, 1967), p. 3.

TABLE 10-3.   PROPERTIES OF SUPERCONDUCTIVE ELEMENTS
(Courtesy of Roberts 1969)

Element	Critical Temperature T_C (°K) Cal.	Mag.	Critical Field H_0 (oersteds) Cal.	Mag.	Debye Temp. θ_D (°K) (See †below)	Electronic Specific Heat γ (mJ mole^{-1} deg. K^{-2}) (See ‡ below)
Al	1.183	1.196	104	99	420	1.36
Cd	0.54, 0.518	0.56	29.6	30	209	0.688
Ga	1.087, 1.078	1.091	59.4, 58.9	51	317, 324.7	0.601, 0.596
Ga (β)		6.2				
Ga (γ)		7.62		HF*		
Hg (α)	4.16	4.154	380	410.9	87, 71.9	1.81
Hg (β)		3.949		339	93	1.37
In	3.407	3.4035	282.7	293	109	1.66
Ir		0.14		19	420	3.2
La (α)	4.80	4.9			142	10.0
La (β)	5.91	6.06		1,600	132	6.7
Mo	0.915-0.918	0.92	95	98	460	1.83
Nb	9.17	9.26	1,944	1,980, HF*	277	7.79
Os		0.655		65	500	2.35
Pa		1.4				
Pb	7.23	7.193		803, HF*	96.3	3.0
Pt		<0.001				
Re	1.699	1.698	188	198	415	2.35
Rh		<0.001			500	4.7
Ru		0.49		66	550	3.0
Sb		2.6-2.7		HF*		
Sn	3.722	3.722	303	305.50	195	1.74
Ta	4.39	4.483	780	830, HF*	258	6.0
Tc		8.22, 7.92				
Th		1.368	131	162	168	4.65
Ti	0.42	0.39	56	100, HF*	425	3.32
Tl	2.38	2.39	176.5	171	78.5	1.47
U (α)		0.68, 0.23			206	12.2
U (pseudo-γ)		1.80 (extrapolated value)				
V	5.37	5.30	1,310	1,020, HF*	399	9.8
W		0.012		1.07	550	3.0
Zn	0.852	0.875	51.8	53	309	0.66
Zr		0.546		47	290	2.78
Zr (ω)		0.65				

TABLE 10-3. (*Continued*)

Element	Critical Temperature T_c (°K)		Critical Field H_0 (oersteds)[1]		Debye Temp. θ_D (°K)	Electronic Specific Heat γ (mJ mole^{-1} deg. K^{-2})
	Cal.	Mag.	Cal.	Mag.	(See † below)	(See ‡ below)
			Thin films formed at various temperatures			
Al		1.3-3.7				
Be		~6, ~8.4		$H_{c2} \gg 11,000$		
Bi		~6.0				
Ga		8.4, 7.2				
In		3.95-4.25, 3.7				
La		5.00-6.74				
Mo		~5				
Re		~7				
Sn		4.6-4.7, 4.1				
Ti		1.3M				
W		1.7-4.1				
			Under high pressure			
					Pressure[2]	
Bi II		3.916			25,000 atm	
		3.90			25,200 atm	
		3.86			26,800 atm	
Bi III		7.25		HF*	27,000-28,400 atm	
Ce		1.7			50 kbar	
Ge		4.85-5.4			~120 kbar	
Se II		6.75, 6.95			~130 kbar	
Si		7.1			120-130 kbar	
Te		~3.3		HF*	~56,000 atm	
Tl (FCC)		1.45			35 kbar	
Tl (HCP)		1.95			35 kbar	

† For another data set see Mendelssohn, K., Cryophysics, p. 178 (Interscience, New York, 1960).

‡ Parkinson, D.H., Rep. progr. Phys. <u>21</u>, 226 (1958). Also see Reference 572.

HF* See Table 3 for additional data on H_{c1}, H_{c2} and H_{c3}. M equals maximum. FCC is face-centered cubic. HCP is hexagonal close-packed.

[1] To convert "oersteds" to ampere/meters, multiply by 7.957×10^3.

[2] To convert "atm" to "newton/meter^2", multiply by 1.013×10^5.

TABLE 10-4. TABULATION OF SUPERCONDUCTIVE MATERIALS (INCLUDING PROVEN-NON-SUPERCONDUCTORS).
(Roberts, 1963, 1969, 1971)

Material	Critical Temp., T_c, (°K)	Critical Field, H_0, (oersteds)	Min. Temp.* (°K)	Structure[#]
Ag			0.30	
$Ag_{3.3}Al$			0.34	like A13, $a = 6.92$Å
Ag_2Be			1.28	
$Ag_7BF_4O_8$	0.15			cubic, $a = 9.942$
Ag_5Ba			0.34	$D2_d$, $a = 5.71$, $c = 4.64$
AgBi			1.28	
$AgBi_2$	3.0–2.78			
Ag_2Bi			1.28	
Ag_5Cd_8			1.28	
$AgCl_2$			1.90	
Ag_7FO_8	0.3			cubic, $a = 9.833$
Ag_2F	0.066	2.5		
$Ag_{0.80-0.30}Ga_{0.20-0.70}$	6.5–8			
Ag_4Ge	0.85			Hex., h.c.p.
$Ag_{0.438}Hg_{0.562}$	0.64			$D8_{1,2,3}$, complex b.c.c.
$AgIn_2$	2.30–2.46			C16
Ag_3In			1.4	
AgLa	0.92–0.96			B2
AgLa (9.5 kbar)	1.2			B2
AgLu			0.33	B2
AgMg			1.02	B2
Ag_7NO_{11}	1.04	57		cubic, $a = 9.893$
Ag_2O			1.28	
AgPb (eutectic alloy)	7.20			
Ag_xPd_y			1.00	
Ag_xPt_y			1.00	
AgS			1.90	
Ag_2S			1.28	
AgSb (eutectic alloy)			1.90	
Ag_3Sb			1.28	
Ag_2Se			1.28	
$AgSi_2$			1.4	
AgSn	3.70			
$Ag_{0.92}Sn_{0.08}$			1.26	
Ag_3Sn			1.36	
Ag_5Sr			0.34	$D2_d$, $a = 5.68$, $c = 4.62$
$AgTe_3$	2.6			cubic, primitive
Ag_2Te			1.28	
$AgTh_2$	2.26			C16, $a = 7.56$, $c = 5.84$
AgTl (eutectic alloy)	2.67			
AgY			0.33	B2
$AgZn_3$			1.28	
Ag_5Zn_8			1.28	
Ag_xZn_{1-x}	0.5–0.845			
Al	1.175	104.8		
Al 6063 alloy	1.14	94		
$AlAu_4$	0.4–0.7			like A13, $a = 6.92$
Al_2Au			1	
Al_4C_3			1.38	
$AlCl_3$			1.15	
Al_2Ca			1.02	C15, $a = 8.035$
Al_4Ca			1.02	$D1_3$

TABLE 10-4. (*Continued*)

Material	Critical Temp., T_c, (°K)	Critical Field, H_0, (oersteds)	Min. Temp.* (°K)	Structure[#]
Al_2Ce			0.34	C15
$Al_2Ce_xLa_{1-x}$	3.237–2.1			
$Al_{0.10}Cr_{0.90}$			1.4	cubic
$Al_{0.131}Cr_{0.088}V_{0.781}$	1.46			cubic
$Al_{0.999}Fe_{0.001}$	1.50			
$Al_{1-x}Ga_xNb_3$	⩾18.52			
$Al_{1-x}Ga_xV_3$	14.5–5.5			A15
$AlGd_xLa_{3-x}$	6.16–2.03			
$Al_2Gd_xLa_{1-x}$	3.237–1.0			
$Al_{3-x}Gd_xLa$	2.2–6.16			
$Al_{0.8}Ge_{0.2}Nb_3$	20.05			A15
$Al_xGe_{1-x}Nb_3$	4.2–11.4			
$Al_{1-x}Ge_xV_3$	6.5–12.3			A15
$Al_xGe_{1-x}V_3$	5.9–12			A15
$AlLa_3$	6.16			DO_{19}
Al_2La	3.23			$C15, a = 8.13$
Al_4La			1.15	
$AlLu_2$			1.02	$C15, a = 7.322$
$AlLu_3$			1.1	
Al_2Lu			1.02	C15
$AlMg_{0.0106}$	1.132			
Al_2Mg_3			0.30	$A12, a = 10.55$
Al_3Mg_2	0.84			$cubic, FC, a = 28.28$
$Al_{1-x}Mn_x$	1.17–0.12			
$AlMo_3$	0.58			$A15, a = 4.950$
Al_5Mo			1.15	
AlN	1.55			$B4, hex, a = 3.104$
Al_2NNb_3	1.3			A13
$AlNb_3$	18			A15
$Al_{0.9}Nb_{0.3}Sb_{0.1}$	18.06			A15
$Al_{0.5}Nb_3Sn_{0.5}$	16.3			A15
$AlOs$			1.02	B2
Al_2Os			1.15	
Al_3Os	5.9			
Al_3Os_2			1.15	
$AlPt$			0.34	$cubic, a = 4.85$
Al_2Pt	0.48–0.55			$Cl, a = 5.92$
$AlPu$			1.50	
Al_5Re_{24}	3.35			A12
Al_6Re	1.85			
$Al_{1-x}Sb_xV_3$	4.5–7.2			A15
Al_2Sc			1.02	$C15, a = 7.579$
$Al_{1-x}Si_xV_3$	5–14.5			A15
Al_xSn_{1-x}	3.72–3.692			
$Al_{1-x}Sn_xV_3$	4.5–6			A15
Al_2Th_3	2.6			tet.
Al_3Th	0.75			$DO_{19}, a = 6.50, c = 4.626$
Al_3Ti			1.02	$DO_{22}, a = 5.436, c = 8.596$
$Al_xTi_yV_{1-x-9}$	2.05–3.62			cubic
Al_2U			1.12	C15
Al_3U			0.07	L12
$Al_{0.108}V_{0.892}$	1.82			cubic
$Al_{0.24}V_{0.76}$	11.15			A15
AlV_3	11.65			A15

TABLE 10-4. (*Continued*)

Material	Critical Temp., T_c, (°K)	Critical Field, H_0, (oersteds)	Min. Temp.* (°K)	Structure#
AlY			1.15	
AlY_2			1.15	
AlY_3			1.1	
Al_2Y			0.34	C15, $a = 7.86$
Al_2Y_3			1.15	
Al_3Yb	0.94			L12
Al_xZn_{1-x}	0.5–0.845			
$AlZr_3$	0.73			Ll_2, $a = 4.37$
Al_2Zr	<0.3			C14, $a = 5.282$, $c = 8.748$
Al_3Zr			1.02	DO_{23}
As			1.3	
AsAu (eutectic alloy)			1.90	
AsBiPb	9.0			
AsBiPbSb	9.0			
As_2CdGe (60–70 kbar)	2.84–3.02			tet.
As_2CdSn (60 kbar)	1.79–2.29			B1
AsCo			1.1	MnP type
As_2Co			1.1	C18
AsCu (eutectic alloy)			2.20	
$AsCu_3$			1.28	
As_2Cu			1.57	
AsGe (30–65 kbar)	3–3.5			
As_2Fe_3			1.30	
As_2Mo			1.30	
AsNi			1.28	$B8_1$
$AsNi_{0.25}Pd_{0.75}$	1.34			C2
AsPb (eutectic alloy)	8.4			
AsPd			1.02	C2
$AsPd_2$ (quenched)	1.71			C22, $a = 6.65$, $c = 3.57$
$AsPd_2$ (annealed)	0.6			hex., $a = 9.79$, $c = 6.61$
$AsPd_3$			0.35	DO_e, $a = 9.986$, $c = 4.830$
As_2Pd_5	0.46			complex
As_3Pd_5	1.9			
As_2Pt			0.35	
AsRh	0.58			B31, $a = 5.65$, $b = 3.58$, $c = 6.00$
$AsRh_{1.4-1.6}$	<0.03–0.56			hex.
AsRu			0.35	
$AsRu_2$			0.35	
$As_{0.26}Sb_{0.74}$			1.32	
AsSn (eutectic alloy)	4.10			
AsV_3			1.2	A15
Au (rapid quench)			0.32	A1, f.c.c.
$AuAl_2$	0.095–0.074			
Au_5Ba	0.4–0.7			$D2_d$, $a = 5.69$, $c = 4.54$
AuBe	2.64			B20
Au_xBe_y			1.00	
Au_2Bi	1.84			C15, $a = 7.958$
Au_5Ca	0.34–0.38			$C15_b$, $a = 7.747$
Au_5Cd_8			1.28	
AuGa	1.2			B31, $a = 6.40$, $b = 6.27$, $c = 3.42$
$AuGa_2$	1.12			C1
$Au_xGa_2Pd_{1-x}$	1.25–1.79			C1

TABLE 10-4. (*Continued*)

Material	Critical Temp., T_c, (°K)	Critical Field, H_0, (oersteds)	Min. Temp.* (°K)	Structure[#]
$Au_{0.40-0.775}Ge_{0.60-0.225}$	1.63–0.99			complex
AuIn	0.4–0.6			complex
$AuIn_2$	0.096–0.093			
AuLa			0.33	
AuLu	<0.35			B2
$AuNb_3$	1.2			A2, $a = 3.29$
$AuNb_3$	11.0			A15, $a = 5.2027$
$Au_x Nb_3 Pt_{1-x}$ (annealed)	10.7–12.7			A15
$Au_x Nb_3 Pt_{1-x}$ (quenched)	8.3–9.1			A15
$Au_{0.02-0.98}Nb_3 Rh_{0.98-0.02}$	2.52–11.0			A15
AuPb (eutectic alloy)	7.0			
$AuPb_2$	3.15			
$AuPb_3$	4.40			
$Au_x Pd_y$			1.00	
$AuSb_2$	0.58			C2, $a = 6.658$
AuSn	1.25			$B8_1$, $a = 4.32$, $c = 5.52$
$AuSn_2$	2.48			
$AuSn_4$	2.38			$D1_c$
$AuTa_2$			1.2	$D8_b$
$AuTa_{4.3}$	0.58–0.51			A15
$AuTh_2$	3.08			C16, $a = 7.42$, $c = 5.95$
AuTl (eutectic alloy)	1.92			
AuV_3	0.74			A15, $a = 4.883$
$Au_x Zn_{1-x}$	0.5–0.845			
$AuZn_3$	1.21			P. cubic, $a = 7.893$
$Au_{0.05}Zr_{0.95}$	1.65			A3
$Au_{0.12}Zr_{0.88}$	2.74			A3
$AuZr_3$	0.92			A15
$B_6 Ba$			1.28	
$BCMo_2$	7.1			ortho-rhombic
BHf	3.1			cubic + extra lines
$B_6 La$	5.7			
$B_{12}Lu$	0.48			
BMo			1.74	tet.
BMo_2	5.86			
BN			1.28	
BNb	8.25			B_f
$B_2 Os$			1.02	C32, $a = 2.876$, $c = 2.871$
BPt			1.28	
BRe_2	2.80			
BRh_2			1.00	ortho-rhombic
BRu_2			1.20	
$B_3 Ru_7$	2.58			$D10_2$, $a = 7.465$, $c = 4.715$
$B_4 Sc$			1.34	
$B_{12}Sc$	0.39			
BTa	4.0			B_f, ortho-rhombic
$B_2 Ta_3$			0.1	tet.
BTh			1.20	
$B_6 Th$	0.74			
BV			1.20	B_f
$B_2 V_3$			0.1	tet.
BW_2	3.18			C16
$B_6 Y$	6.5–7.1			
$B_{12}Y$	4.7			

TABLE 10-4. (*Continued*)

Material	Critical Temp., T_c, (°K)	Critical Field, H_0, (oersteds)	Min. Temp.* (°K)	Structure#
BZr	3.4			cubic + extra lines
$B_{12}Zr$	5.82			
$BaBi_3$	5.69	740		tet., $a = 5.188, c = 5.157$
$BaRh_2$	6.0			C15, $a = 7.852$
$Be_{22}Mo$	2.51			cubic, $Be_{22}Re$ type
Be_2Nb_3	2.3			tet.
$Be_{17}Nb_2$	1.47			
BiIr			0.35	
Bi_2Ir (quenched)	3.96–3.0			
Bi_2Ir	~2.3–1.7			
BiK	3.60			
Bi_2K	3.57			C15, $a = 9.501$
BiLi (α)	2.47			$L1_0, a = 3.361, c = 4.24$
$Bi_{4-9}Mg$	~1.0–0.70			
BiMn			0.3	
Bi_3Mo	3.7–3.0			
BiNa	2.25			$L1_0, a = 3.46, c = 4.80$
$BiNb_3$ (high temp. & pressure)	3.05			A15, $a = 5.320$
$Bi_xNb_3Sn_{1-x}$	18.0–18.2			
BiNi	4.25			$B8_1, a = 4.07, c = 5.35$
Bi_3Ni	4.06			ortho-rhombic
BiOs			0.3	
BiPb (eutectic alloy)	8.8			
$Bi_{1-0}Pb_{0-1}$	7.26–9.14			
BiPbSb	8.9			
Bi_2PbSn	8.5			
BiPd	3.7			ortho-rhombic, $a = 7.203,$ $b = 8.707, c = 10.662$
$BiPd_2$	4.00			
Bi_2Pd (α)	1.70			monoclinic, $a = 12.74,$ $b = 4.25, c = 5.665,$ $\beta = 102°35'$
Bi_2Pd (β)	4.25			tet., $a = 3.362, b = 12.983$
BiPt	1.21; 2.4			$B8_1, a = 4.315, c = 5.490$
Bi_2Pt (α)			1.45	C2, $a = 6.683$
Bi_2Pt (β)	0.155	10		hex., $a = 6.44, c = 6.25$
Bi_3Pt			1.80	
Bi_2Rb	4.25			C15, $a = 9.609$
$BiRe_2$	2.20–1.9			
BiRh	2.06			$B8_1, a = 4.094, c = 5.663$
Bi_2Rh (α)			1.34	mono.
Bi_2Rh (β)			1.27	mono.
Bi_3Rh	3.2			$NiBi_3$ type
Bi_4Rh (α)			0.10	cubic, b.c.
BiRu	5.7; 4.12–3.31			
$BiRu_2$			0.35	
BiS			0.3	
BiSc			0.3	
Bi_2Se_3			1.26	
BiSn (eutectic alloy)	3.48	130		
Bi_3Sn	3.67–3.77			
Bi_3Sr	5.62	530		$L1_2, a = 5.042$
Bi_2Te_3			1.26	
Bi_3Te	~1.0–0.75			

TABLE 10-4. (*Continued*)

Material	Critical Temp., T_c, (°K)	Critical Field, H_0, (oersteds)	Min. Temp.* (°K)	Structures[#]
$BiTl_3$			1.15	
Bi_5Tl_3	6.4			
$Bi_{0.62-0.18}Tl_{0.38-0.82}$	6.6–2.3			
$Bi_{1-0.87}Tl_{0-0.13}$	6.154–6.220			
BiV_3			4.2	A15
BiW			0.3	
Bi_2Y_3	2.25			
Bi_3Zn	0.77–0.87			
$Bi_{0.3}Zr_{0.7}$	1.51			
$BiZr_3$	2.85–2.35			
$C_{1.35}Ca_{0.1}Y_{0.9}$	10.5–11.5			
C_2Ce			2.0	
CCs_x	0.020–0.135			hex.
C_8Cs (gold)	0.020–0.135			
$C_{16}Cs$ (blue)			0.011	
C_2Dy			2.0	
C_2Er			2.0	
CFe_3			1.30	
$CGaMo_2$	4.1–3.7			hex., H phase
C_2Gd			2.0	
$C_{1.5}Ge_3La_5$	3.3–3.7			cubic
$C_{1.35}Ge_{0.1}Y_{0.9}$	10.6			$D5_c$
CHf			1.23	
$CHf_{0.07}Mo_{0.93}$	8.2			B1
$CHf_{0.17}Mo_{0.83}$	8.7			B1
$CHf_{0.25}Mo_{0.75}$	6.6			B1
$CHf_{0.5}Mo_{0.5}$	3.4			B1, $a = 4.450$
$CHf_{0.8-0.2}Nb_{0.2-0.8}$	5.4–7.8			B1
$CHf_{0.9-0.1}Ta_{0.1-0.9}$	5.0–9.0			B1
C_2Ho			2.0	
CIr_2Mo_3	1.8			cubic
CIr_2W_3	2.1			cubic
CK (excess K)	0.55			hexagonal
C_8K	0.39			hexagonal
C_8K (gold)	0.55			hexagonal
$C_{16}K$ (blue)			0.011	hexagonal
C_2La	1.61			tet.
C_3La_2	5.9–11.0			$D5_c$
C_2Lu	3.33			tet.
CMo	8.0			hexagonal
CMo_2	4.0			ortho-rhombic
$CMo_{1-0}Nb_{0-1}$	10.8–14.3			B1
CMo_3Pt_2	1.1			cubic
C_2MoRe	3.8			cubic
$CMo_{0.90}Re_{0.10}$	13.8			B1
$CMo_{0.90}Ru_{0.10}$	13.6			B1
$CMo_{1-0}Ta_{0-1}$	8.3–14.3			B1
CMo_xTi_{1-x}	⩽10.2			T1
$CMo_{1-0.8}V_{0-0.2}$	14.3–12.7			B1
$C_{1.45}Mo_{0.1}Y_{0.9}$	13.8			$D5_c$
$CMo_{1-0.8}Zr_{0-0.2}$	14.3–10.9			B1
$C_xN_{1-x}Nb$	8.5–17.3			
$C_{0.83}Nb$	2.4			B1
CNb	8–10	800		cubic

TABLE 10-4. (*Continued*)

Material	Critical Temp., T_c, (°K)	Critical Field, H_0, (oersteds)	Min. Temp.* (°K)	Structure#
CNb (whiskers)	7.5–10.5			
CNb_2	9.1			
$CNb_{1-0}Ta_{0-1}$	8.9–11.1			B1
$CNb_{0.9-0.2}Ti_{0.1-0.8}$	8.8–4.4			B1
$CNb_{0.9-0.6}W_{0.1-0.4}$	11.6–12.5			B1
$C_{1.35}Nb_{0.1}Y_{0.9}$	10.8			$D5_c$
$CNb_{0.9-0.1}Zr_{0.1-0.9}$	8.4–4.2			B1
C_2Nd			2.0	
COs_2W_3	2.9			cubic
C_2Pr			2.0	
C_2PtU_2	1.47			tet.
CRb_x	0.023–0.151			hex.
C_8Rb (gold)	0.023–0.151			
$C_{16}Rb$ (blue)			0.011	
$C_{0.04}Re_{0.96}$	1.98			
$CRe_{0.08-0.1}W$	1.3–5.0			
C_2ReW	3.8			cubic
$C_{1.35}Re_{0.3}Y_{0.7}$			4.0	
$C_{1.35}Ru_{0.1}Y_{0.9}$	11.2			$D5_c$
$C_{0.96}Sc$			1.38	B1, $a = 4.54$
C_3Sc_4			1.0	cubic
CSi			1.28	
$C_{1.35}Si_{0.1}Y_{0.9}$	11.3			$D5_c$
CTa	9–11.4	810		cubic
CTa (film)	5.09			B1
CTa_2	3.2			
$Cta_{1-0.40}W_{0-0.60}$	8.5–10			B1, $a = 4.454-4.345$
$Cta_{0.9-0.3}Zr_{0.1-0.7}$	8.3–5.1			B1
C_2Tb			2.0	
CTh			1.20	
$C_{1.35}Th_{0.7-0.3}Y_{0.3-0.7}$	14.4–16.4			$D5_c$
$CTi_{0.7-0.5}W_{0.3-0.5}$	2.1–6.7			B1
CU			1.20	B1, $a = 4.952$
CW	1.0			
CW_2	2.74			L'_3
CW_2	5.2			cubic, F.C., $a = 4.25$
$C_{1.55}W_{0.1}Y_{0.9}$	14.8			$D5_c$
$C_{0.92}Y$			1.38	B1, $a = 4.68$
CY_3			1.15	
$C_{1.30}Y$	8.2			$D5_c$
$C_{1.45}Y$	11.5			$D5_c$
$C_{1.55}Y$	6.0			$D5_c$
C_2Y	3.75			tet.
C_2Y	3.88			$C11_a$
C_3Y_2	6.0–11.5			$D5_c$
$C_{1.45}Y_{0.9}Zr_{0.1}$	13.0			$D5_c$
C_2Yb			2.0	
$C_{0.992}Zr$			1.28	B1, $a = 4.68$
$CaCu_5$			0.34	$D2_d$, $a = 5.09$, $c = 4.09$
$CaGa_2$			1.02	C32, $a = 4.323$, $c = 4.323$
$CaH_{18}N_6$			1.9	
$CaIr_2$	6.15			C15, $a = 7.545$
$CaMg_2$			1.02	C14, $a = 6.23$, $c = 10.12$
CaPb	7.0			

TABLE 10-4. (*Continued*)

Material	Critical Temp., T_c, (°K)	Critical Field, H_0, (oersteds)	Min. Temp.* (°K)	Structure[#]
$CaPb_3$	0.65; 0.4			$L1_2$
$CaPd_2$			1.02	$C15, a = 7.665$
$CaPt_2$			1.02	$C15, a = 7.629$
$CaRh_2$	6.40			$C15, a = 7.525$
$CaSe$			1.70	
$CaSi_2$	1.58			
$CaTl_3$	2.04			$L1_2$
$Cd_{0.5-0.3}Hg_{0.5-0.7}$	1.92–1.70			
$Cd_{0.72-0.07}Hg_{0.28-0.93}$	1.3–3.3			tet.
$Cd_{1-0.72}Hg_{0-0.28}$	0.5–1.35			hex.
$Cd_{0.06-0}Hg_{0.94-1}$	4.09–4.15			
$CdHg$	1.77			tet., B.C.
$Cd_{11}Ia$			1.02	cubic
Cd_2Na			1.06	
Cd_6Na			1.08	
CdO			1.3	
$Cd_{0.97}Pb_{0.03}$	4.20			
$CdSn$ (eutectic alloy)	3.65	>266		
$Cd_{0.17}Tl_{0.83}$	2.30			
$CeCo_2$	0.84			$C15$
$CeCo_{1.67}Ni_{0.33}$	0.46			$C15$
$CeCo_{1.67}Rh_{0.33}$	0.47			$C15$
$Ce_{0.96}Gd_{0.08}Ru_2$	5.20			$C15$
$CeIr_2$			1.00	$C15, a = 7.571$
$CeIr_3$	3.34			
$CeIr_5$	1.82			
Ce_xLa_y	1.3–6.3			
$CeNi_2$			0.015	$C15$
$CePt_2$			0.32	$C15, a = 7.730$
$CePt_3$			0.32	$C15, a = 7.640$
$CePt_5$			0.32	$D2_d, a = 5.369, c = 4.385$
$CeRh_2$			0.30	$C15, a = 7.538$
$CeRu_2$	6.2			$C15$
$CeSn_3$			0.07	$L1_2$
$CoFeSi_2$	≤1.40			$C1$
$CoGe_3$			0.30	ortho.
$CoHf_2$			1.02	$E9_3, a = 12.067$
$CoLa_3$	4.28			$D0_{11}$
$CoLa_3$	4.01			$D0_{20}$
$CoLu_3$	~0.35			
Co_2Lu			0.32	$C15, a = 7.123$
$Co_{0.5}Mn_{0.5}U_6$	2.55			
$Co_{0.002}Mo_{0.815}Re_{0.185}$	5.8			
$Co_{0.10-0.02}Nb_3Rh_{0.90-0.98}$	1.90–2.28			A15
$CoNiSi_2$	≤1.40			$C1$
$Co_xNi_{1-x}U_6$	2.4–0.41			
$Co_{0.5}Ni_{0.5}Zr_2$	3.1			$C16$
$CoSc_2$			0.32	$C16$
$CoSc_3$			0.32	
$CoSi_2$	1.22	105		$C1$
$CoSn_2$			1.02	$C16$
$CoTe$			1.00	$B8_1$
Co_3Th_7	1.83			$D10_2, a = 9.833, c = 6.200$
Co_5Th			0.32	$D2_d$

TABLE 10-4. (*Continued*)

Material	Critical Temp., T_c, (°K)	Critical Field, H_0, (oersteds)	Min. Temp.* (°K)	Structure#
CoTi	0.71			A2
CoTi$_2$	3.44			E9$_3$
CoU	1.70			distorted B2
CoU$_6$	2.29			D2$_c$
CoV$_3$			0.015	A15
Co$_{0.28}$Y$_{0.72}$	0.34			
CoY$_2$			0.32	
CoY$_3$	<0.34			
Co$_{0.1}$Zr$_{0.9}$	3.90			A3
CoZr$_2$	5.0			C16
Cr$_{0.6}$Ir$_{0.4}$	0.4			HCP
Cr$_{0.7}$Ir$_{0.3}$	0.76			HCP
Cr$_{0.8}$Ir$_{0.2}$	0.50			A15
Cr$_{0.85}$Ir$_{0.15}$	0.77			A15
Cr$_{0.9}$Ir$_{0.1}$	0.78			A15
Cr$_3$Ir	0.45			A15
CrN			1.28	B1
Cr$_3$O			1.02	A15
Cr$_{0.80}$Os$_{0.20}$	2.5			cubic, b.c., $a = 2.925$
Cr$_{0.72}$Os$_{0.28}$	3.95			A15
Cr$_{0.67}$Os$_{0.33}$	1.03			D8$_b$
Cr$_{0.40}$Re$_{0.60}$	2.15			D8$_b$
Cr$_{0.8-0.6}$Rh$_{0.2-0.4}$	0.5–1.10			A3
Cr$_3$Rh	0.072			A15
Cr$_{0.72}$Ru$_{0.28}$	3.42			A15
Cr$_{0.85}$Ru$_{0.15}$	1.13			A15
Cr$_2$Ru	2.02			D8$_b$
Cr$_3$Ru (annealed)	3.3			A15
Cr$_3$Ru (cast)	2.15			A15
Cr$_x$Ti$_y$	⩽3.6			Cr in α – Ti
Cr$_x$Ti$_y$	⩽4.2			Cr in β – Ti
Cr$_{0.10}$Ti$_{0.30}$V$_{0.60}$	5.6	1360		
Cr$_{0.0175}$U$_{0.9825}$ (β phase)	0.75			
Cr$_x$V$_{1-x}$	1.3–5.1			A2
Cr$_{0.1}$V$_{0.9}$	3.21			
Cs$_{0.32}$O$_3$W	1.12			hex., $a = 7.4, c = 7.6$
Cu$_3$Ga			1.4	
CuLa			5.85	
Cu$_3$N			1.38	
CuO			1.28	
Cu$_2$O			1.28	
Cu$_x$Pb$_{1-x}$	5.7–7.7			
CuRh$_2$S$_4$	4.80–4.65			H1$_1$
CuS	1.62			C18
CuS$_2$	1.48–1.52			C18, $a = 5.790$
CuSSe	1.5–2.0			C18, $a = 5.923$
CuSe$_2$	2.30–2.43			C18, $a = 6.123$
CuSeTe	1.6–2.0			C18, $a = 6.302$
Cu$_x$Sn$_{1-x}$	3.2–3.7			
CuTe			1.26	
CuTe$_2$	1.25–1.3			C18, $a = 6.600$
CuTh$_2$	3.49			C16, $a = 7.28, c = 5.75$
CuTi			1.02	B1$_1$
Cu$_{0-0.027}$V	3.9–5.3			A2

TABLE 10-4. (*Continued*)

Material	Critical Temp., T_c (°K)	Critical Field, H_0, (oersteds)	Min. Temp.* (°K)	Structure[#]
$Cu_x Zn_{1-x}$	0.5–0.845			
$CuZn_3$			1.28	
$Dy_{0.01} La_{0.99}$	3.8			
$Er_{0.01} La_{0.99}$	5.3			
$FeHf_2$			1.02	$E9_3$
$Fe_{0.01} La_{0.99}$	4.85			
$Fe_{0.01} Mo_{0-0.3} Nb_{1-0.7}$	1–8			
$Fe_{0-0.04} Mo_{0.8} Re_{0.2}$	1–10			
$Fe_{0.31-0.19} Ni_{0.48-0.60} P_{0.21}$			1.02	
$Fe_{0.25} Ni_{0.75} U_6$	3.0			
$Fe_{0.5} Ni_{0.5} U_6$	2.3			
$Fe_{0.75} Ni_{0.25} U_6$	1.4			
$FeNp_6$			0.5	
FeP			0.97	B31
$Fe_2 P$			0.97	C22
FeS			1.90	
$FeSb$			1.80	
$FeSb_3$			1.45	
$FeSi$			1.28	
$Fe_3 Th_7$	1.86			D10, $a = 9.823$, $c = 6.211$
$Fe_x Ti_{1-x}$	≤3.2			Fe in α – Ti
$Fe_x Ti_{1-x}$	≤3.7			Fe in β – Ti
$Fe_{0.015} Ti_{0.985}$	2.8			
$Fe_x Ti_{0.6} V_{1-x}$	≤6.8			
FeU_6	3.86			$D2_c$
$Fe_2 U$			1.06	C15
$Fe_{0.1} Zr_{0.9}$	1			A3
$Ga_{0.5} Ge_{0.5} Nb_3$	7.3			A15, $a = 5.175$
$GaLa$			1.15	
$GaLa_3$	5.84			
$Ga_2 La$			1.4	
$GaLu_3$			1.1	
$Ga_3 Lu$	2.30			$L1_2$
$Ga_4 Mn_x Mo_{1-x}$	8.0–4.0			
$GaMo_3$	0.76			A15, $a = 4.943$
$Ga_2 Mo$	9.5			
$Ga_4 Mo$	9.8			
GaN (black)	5.85			B4, hex., $a = 3.182$, $c = 5.173$
$GaNb_3$	14.5			A15, $a = 5.171$
$GaNb_3$ (sintered)	12.5–13.2			A15
$Ga_3 Nb_5$	1.35			
$Ga_x Nb_3 Sn_{1-x}$	14.0–18.37			A15
GaP			1.68	B3, cubic ZnS
$GaPt$			0.34	B20, $a = 4.91$
$Ga_2 Pt$	1.7–1.9			
$Ga_7 Pt_3$			1.1	
GaSb (120 kbar, 77° K, annealed)	4.24			A5
GaSb (unannealed)	~5.9			
$GaSc_3$			1.1	
$Ga_{0.5} Si_{0.5} V_3$	8.6–11.9			A15
$Ga_{1-0} Sn_{0-1}$ (quenched)	4.18–3.47			
$Ga_{1-0} Sn_{0-1}$ (annealed)	3.85–2.6			
$Ga_2 Ta_3$			0.1	tet.

TABLE 10-4. (*Continued*)

Material	Critical Temp., T_c, (°K)	Critical Field, H_0, (oersteds)	Min. Temp.* (°K)	Structure#
Ga_2Th	2.56			
$GaTi_3$			0.3	DO_{19}
Ga_3Ti			1.02	DO_{22}
GaV_3	14.0			A15
$GaV_{2.1-3.5}$	6.3–14.45			A15, $a = 4.813–4.829$
GaV_4	10.1			A15
GaY			1.15	
GaY_3			1.1	
Ga_2Y	1.68			tet.
Ga_3Zr			1.02	DO_{22}
Ga_3Zr_5	3.85			
Ga_3Zr_5 (quenched)	2.5–4.0			
Gd_xLa_{1-x}	<1.0–5.5			
$Gd_xRu_2Th_{1-x}$	≤3.60			C15
$GeIr$	4.70			B31
Ge_7Ir_3	<0.87			
Ge_2La	1.49, 2.20			orthorhombic, distorted $ThSi_2$ type
$GeMo_3$	1.43			A15
Ge_2Mo	1.20			
Ge_3N_4			1.38	
$GeNb_2$	1.90			
$GeNb_3$	6.90			A15, $a = 5.166$
$GeNb_3$ (quenched)	6–17			A15
$Ge_{0.55}Nb_{3.45}$	4.9			A15
GeP (30–65 kbar, 673–1173°K)	1.8–4.2			tet.
GeP_3			1.25	rhomb.
GeP_5			1.25	rhomb.
$GePd$			0.30	B31
$GePd_2$			0.35	
$GeRh$			1.0	B31
Ge_3Rh_5	2.12			ortho., related to $InNi_2$
Ge_2Sc	1.30–1.31			
Ge_2Ta			1.20	
$Ge_{0.950}Te$	0.17–0.27			
Ge_2Ti			1.20	C54
Ge_3Ti_5			1.20	$D8_8$
Ge_3U			0.30	$L1_2$
GeV_3	6.01			A15
GeY			1.15	
$Ge_{1.62}Y$	2.4			
Ge_2Y	3.80			C_c
Ge_2Zr			0.30	C49
$H_{0.33-0.05}Nb$	7.28–7.83			cubic, b.c.
$H_{0.12-0.04}Ta$	2.81–3.62			cubic, b.c.
$HfN_{0.989}$	6.6			B1, $a = 4.50$
$Hf_{0-0.5}Nb_{1-0.5}$	8.3–9.5			A2
$Hf_{0.15}Nb_{0.85}$	9.85			
$Hf_{0.75}Nb_{0.25}$ (arc-cast & cold rolled)	>4.2			
$HfOs_2$	2.69			C14, $a = 5.184$, $c = 8.468$
$HfRe_2$	4.80			C14, $a = 5.239$, $c = 8.584$
$Hf_{0.14}Re_{0.86}$	5.86			A12
$Hf_{0.025}Re_{0.975}$	7.3			A3

TABLE 10-4. (*Continued*)

Material	Critical Temp., T_c, (°K)	Critical Field, H_0, (oersteds)	Min. Temp.* (°K)	Structure[#]
$Hf_{0.875}Re_{0.125}$	1.70			A2
HfRu			1.02	
$HfSi_2$			1.02	
$Hf_{0.55-0}Ta_{0.45-1}$	6.5–4.4			A2
HfV_2	8.9–9.5⁷			C15
HgIn	3.81			
Hg_xIn_{1-x}	3.15–4.55			
HgK			1.14	triclinic
Hg_2K	1.20			orthorhombic
Hg_3K	3.18			
Hg_4K	3.27			
Hg_8K	3.42			
$HgLi_3$			1.08	cubic, b.c.
Hg_3Li	1.7			hex.
HgNa			1.08	orthorhombic
Hg_2Na	1.62			hex.
Hg_2Na_3			1.08	
Hg_4Na	3.05			
Hg_xPb_{1-x}	4.14–7.26			
$Hg_{0.80}Pt_{0.20}$			0.32	
Hg_2Pt			1.10	
Hg_5Pt_2			1.06	
HgS			1.30	
HgSn	4.20			
Hg_xTl_{1-x}	2.30–4.109			
$HgZr_3$			0.35	A15
Hg_3Zr	3.28 ± 0.3			
Ho_xLa_{1-x}	1.3–6.3			
InHg	3.16			
$InLa_3$	10.40			$L1_2, a = 5.07$
In_3La	0.70			$L1_2$
$InNb_3$	9.2			$A15, a = 5.303$
$In_{0.3-0}Nb_3Sn_{0.7-1}$	18.0–18.9			A15
InPb	6.65			
$In_{1-0.89}Pb_{0-0.11}$	3.367–4.85			tet.
$In_{0.6}Pb_{0.4}$	6.36			
$In_{0.65-0}Pb_{0.35-1}$	6.05–7.2			
InPd	0.7			B2
InRh			0.32	B2
In_3Rh			1.02	tet.
InSb	2.1	1100		
InSb (quenched from 170 kbar into liquid nitrogen)	4.8			like A5
$In_{1-0}Sb_{1-0}Sn_{0-1}$ (25 kbar)	1.8–3.7			A5
InSbSn	2.5			A5
$InSc_2$			4.2	$B8_2$
In_xSn_{1-x}	3.4–7.3			
$In_{3(1-x)}Sn_{3x}Y$	≤1.5			$L1_2$
InTe	2.2	800 ± 50		B1
In_2Te			1.37	
In_2Te_3			1.0	
$In_{0.78-0.62}Tl_{0.22-0.38}$	3.32–2.98			
InTl	2.7			
InV_3	13.9			A15

TABLE 10-4. (*Continued*)

Material	Critical Temp., T_c, (°K)	Critical Field, H_0, (oersteds)	Min. Temp.* (°K)	Structure#
In_3Y	0.78–0.21			$L1_2$
In_3Yb			0.05	$L1_2$
$InZr_3$			1.02	$L1_2, a = 4.46$
Ir_2La	0.48			$C15, a = 7.686$
Ir_3La	2.32			$D10_2$
Ir_3La_7	2.24			$D10_2, a = 10.235,$ $c = 6.473$
Ir_5La	2.13			
$IrLu$			0.32	$B2, a = 3.330$
$IrLu_2$	<0.84			
$IrLu_3$			0.32	
Ir_2Lu	2.47			$C15, a = 7.443$
Ir_3Lu	<0.78			
Ir_3Lu_7	2.89			$C15, a = 7.434$
$IrMo$	<1.0			$A3$
$IrMo_3$	8.11			$A15$
$IrMo_3Nb_3Pt$	6.13			$A15$
$Ir_{0.1}Nb_{0.9}$	2.3			
$Ir_{0.9}Nb_{0.1}$	0.060–0.049			
$Ir_{0.99}Nb_{0.01}$	0.102–0.084			
$IrNb_3$	1.76			$A15$
Ir_2Nb_3	9.8			$D8_b, a = 9.834, c = 5.052$
$Ir_{0.02}Nb_3Rh_{0.98}$	2.43			$A15, a = 5.131$
$Ir_{0.05}Nb_3Rh_{0.95}$	2.38			$A15, a = 5.132$
$Ir_{0.287}O_{0.14}Ti_{0.573}$	5.5			$E9_3, a = 11.620$
$Ir_{0.265}O_{0.085}Ti_{0.65}$	2.30			$E9_3, a = 12.430$
$Ir_{0.6-0.2}Os_{0.4-0.8}$	0.98–0.30			
$Ir_{0.95-0.6}Os_{0.05-0.4}$	0.6–0.98			
$IrOsY$	2.60			$C15$
$Ir_{0.96-0.88}Pd_{0.04-0.12}$	<0.1			
$Ir_{0.8}Pt_{0.2}$	0.046–0.032			
$Ir_{0.9}Pt_{0.1}$	0.006–0.053			
$Ir_{0.98-0.7}Re_{0.02-0.3}$	0.1–1.7			
$Ir_{0.95-0.75}Rh_{0.05-0.25}$	<0.06			
$Ir_{0.925-0.765}Ru_{0.075-0.235}$	0.11–0.14			
$IrSc$			0.32	$B2, a = 3.205$
$IrSc_3$			0.32	
Ir_2Sc	2.07			$C15, a = 7.347$
$Ir_{2.5}Sc$	2.46			$C15, a = 7.343$
Ir_3Sc			0.32	
$IrSn_2$	0.65–0.78			$C1, a = 6.34$
Ir_2Sr	5.7			$C15, a = 7.700$
$IrTe$			0.30	$B8_1$
$Ir_{0.5}Te_{0.5}$	~3.0			
$IrTe_2$			0.32	$C6$
$IrTe_3$	1.18			$C2, a = 6.413$
$IrTh$	<0.37			B_f
Ir_2Th	6.50			$C15, a = 7.664$
Ir_3Th	4.71			
Ir_3Th_7	1.52			$D10_2$
Ir_5Th	3.93			$D2_d$
$IrTi_3$	5.40			$A15, a = 5.009$
$Ir_{0.135-0}Tl_{0.865-1}$	≤3.9			
$IrTl_3$	4.63			$A15$

TABLE 10-4. (*Continued*)

Material	Critical Temp., T_c, (°K)	Critical Field, H_0, (oersteds)	Min.* Temp.* (°K)	Structure[#]
IrTl$_3$ (as cast)	4.18			A15
Ir$_{0.99}$V$_{0.01}$	0.11–0.086			
Ir$_{0.85}$V$_{0.15}$	0.26–0.123			
Ir$_{0.33}$V$_{2.67}$	1.39			A15
Ir$_{0.25}$W$_{0.75}$	2.1–3.82			
Ir$_{0.9}$W$_{0.1}$	0.23–0.20			
Ir$_{0.987}$W$_{0.013}$	0.107			
Ir$_2$Y	2.18			C15, a = 7.500–7.520
IrY$_4$			0.32	
Ir$_2$Y	1.09			C15, a = 7.518
Ir$_2$Y$_3$	1.61			
Ir$_2$Zr	4.10			C15, a = 7.359
LaIn$_3$	0.71			L1$_2$
LaMg$_2$	1.05			C15
LaN	1.35			
La$_{0.044-0}$Nd$_{0.956-1}$	1.4–6.3			
LaOs$_2$	6.5			C15, a = 7.737
LaPb$_3$	4.07			L1$_2$
LaPd$_3$			0.32	L1$_2$
LaPt$_2$	0.46			C15, a = 7.776
La$_{0.28}$Pt$_{0.72}$	0.54			C15, a = 7.722
LaRh$_2$			0.32	C15
LaRh$_3$	2.60			
LaRh$_5$	1.62			
La$_7$Rh$_3$	2.58			D10$_2$, a = 10.145, c = 6.434
LaS			1.25	
La$_3$S$_4$	6.5			cubic, Th$_3$P$_4$ type
LaSe			1.25	
La$_2$Se$_3$			1.25	
La$_3$Se$_4$	8.6			cubic, Th$_3$P$_4$ type
LaSi$_2$	2.3			
LaSn$_3$	6.02			L1$_2$
La$_3$Te$_4$	3.75; 2.45			D7$_3$
LaTl$_3$	1.63			L1$_2$
La$_x$Y$_{1-x}$	0.1–5.4			
LaZn	1.04			B2
LiPb	7.20			
LuIn$_3$	0.14–0.24			L1$_2$
LuOs$_2$	3.49			C14
Lu$_{0.275}$Rh$_{0.725}$	1.27			C15, a = 7.355
LuRh$_5$	0.49			
Lu$_2$Rh			0.32	
Lu$_3$RH			0.32	
LuRu$_2$	0.86			C14
MnP			0.01	B31
Mn$_x$Ti$_{1-x}$	0.6–2.3			Mn in α – Ti
Mn$_x$Ti$_{1-x}$	1.1–3.0			Mn in β – Ti
Mn$_{0.14}$Tl$_{0.86}$	2.55			
MnU$_6$	2.32			D2$_c$
MoN	12.0			hex.
Mo$_2$N	5.0			cubic, FC
Mo$_x$Nb$_{1-x}$	0.016–9.22			
Mo$_{0.725}$Nb$_{0.061}$Re$_{0.187}$	5.0			

TABLE 10-4. *(Continued)*

Material	Critical Temp., T_c, (°K)	Critical Field, H_0, (oersteds)	Min. Temp.* (°K)	Structure#
MoO_2			1.30	
Mo_3Os	7.30; 11.76			A15
Mo_3P	7.00			
$Mo_{0.5}Pd_{0.5}$	3.52			A3
Mo_4Pt	4.56			A15
$Mo_{0.23}Re_{0.77}$	9.25			A12
$Mo_{0.42}Re_{0.58}$	6.35			$D8_b$
$Mo_{0.57}Re_{0.43}$	14.0			
$Mo_{0.815}Re_{0.185}$	8.27			
$Mo_{0.865}Re_{0.135}$	8.27			
$MoRe_3$	9.26, 9.89			A12
$MoRh$	1.97			A3
$Mo_{0.97-0.8}Rh_{0.03-0.2}$	1.50–8.20			cubic, B.C.
$Mo_{0.1}Ru_{0.9}$	1.0			A3
$Mo_{0.61}Ru_{0.39}$	7.18			$D8_b$
MoS_2			1.28	
Mo_3Sb_4	2.10			
$MoSe_2$			1.26	
Mo_3Si	1.3			
$MoSi_{0.7}$	1.34			
$Mo_{0.02}Si_{0.25}V_{0.73}$	10.4			A15
$Mo_{0.05}Si_{0.25}V_{0.70}$	5.59			A15, $a = 4.736$
$Mo_{0.12}Si_{0.25}V_{0.63}$	5.1			A15
$Mo_{0.15}Si_{0.25}V_{0.60}$	4.54			A15, $a = 4.758$
$Mo_{0.6-0.05}Tc_{0.4-0.95}$	15.8–10.8			
$Mo_{0.06}Ti_{0.94}$	2.04			BCC, metastable
$Mo_{0.09}Ti_{0.91}$	3.09			BCC, metastable
$Mo_{0.16}To_{0.84}$	4.18			
$Mo_{0.913}Ti_{0.087}$	2.95			
$Mo_{0.14}U_{0.86}$	2.02			$\gamma - U$
$Mo_{0.3}U_{0.7}$	1.84			$\gamma - U$
$Mo_{0.15}V_{0.85}$	2.28			
$Mo_{0.30}V_{0.70}$	0.76			
$Mo_{0.5}V_{0.5}$	0.11			
Mo_2Zr	4.75–4.27			C15
NNb (whiskers)	10–14.5			
$N_{0.0023}Nb_{0.998}$	9.20			
$N_{0.952-0.700}Nb$	15.3–11.3			
$N_{0.19}Nb$	5.72			
NNb_2	9.5			
NNb_xO_y	13.5–17.0			
$N_{0.91}Nb_{0.99}Ta_{0.01}$	15.62			B1
$N_{0.91}Nb_{0.82}Ta_{0.18}$	10.9			B1
$N_{0.47}O_{0.03}Ti_{0.50}$	2.9			cubic, $a = 4.230$
$N_{0.47}O_{0.01}Ti_{0.50}$	5.58			cubic, $a = 4.240$
$N_{0.48}O_{0.02}Ti_{0.50}$	5.58			cubic, $a = 4.240$
$N_{0.34}Re$	4–5			cubic, F.C.
NTa	6.5 ± 5			B1
NTa_2			4.2	
$N_{0.99}Ti$	4.35			
$N_{0.84}Ti$	1.2			
$N_{0.99-0.60}Ti$	1.17–4.35			B1, $a = 4.243-4.238$
Nu			1.20	B1
NV	7.50; 8.20			B1

TABLE 10-4. (*Continued*)

Material	Critical Temp., T_c, (°K)	Critical Field, H_0, (oersteds)	Min. Temp.* (°K)	Structure[#]
$N_{0.99}V$	7.9			
$N_{0.9}V$	4.8			
$N_{0.82}V$	2.9			
$N_{0.99-0.785}V$	2–8			B1, $a = 4.132–4.084$
NZr	9.8			B1, $a = 4.56$
$N_{0.984-0.932}Zr$	9.5–3.0			B1
$Na_{0.28-0.35}O_3W$	0.56			tet., $a = 12.1$, $c = 3.75$
$Na_{0.2}O_3W$	0.55			tet., $a = 12.1$, $c = 3.7$
$Na_{0.10}O_3W$			0.040	tet.
$Na_{0.28}Pb_{0.72}$	7.20			
$NaPb_3$	5.62			$L1_2$
NbO	1.25			
$Nb_{0.962}O_{0.038}$	5.840			A2
$Nb_{0.986}O_{0.014}$	8.200			A2
$Nb_{0.948}O_{0.052}$	6.650			A2
$Nb_{0.936}O_{0.064}$	9.023			A2
NbOs	1.40			$D8_b$
$NbOs_2$	2.52			A12, $a = 9.655$
Nb_3Os	0.94			A15
$Nb_{0.60}Os_{0.40}$	1.89			$D8_b$
$Nb_3Os_{0.10-0.02}Rh_{0.90-0.98}$	2.30–2.42			A15
$NbPbS_3$	2.62			tet.
$Nb_{0.67}PbS_3$	2.00			tet.
NbPd	2.0			$D8_b$
$Nb_{0.9}Pd_{0.1}$	3.5			
$Nb_{0.6}Pd_{0.4}$	1.7			A12
$Nb_3Pd_{0.10-0.02}Rh_{0.90-0.98}$	2.55–2.50			A15
NbPt	2.40			$D8_b$
$Nb_{0.9}Pt_{0.1}$	2.5			
$Nb_{0.62}Pt_{0.38}$	4.01			$D8_b$
Nb_3Pt	10.9			A15, $a = 5.1547$
$Nb_3Pt_{0.98-0.02}Rh_{0.02-0.98}$	9.6–2.52			A15
NbRe	2.0			$D8_b$
$Nb_{0.38-0.18}Re_{0.62-0.82}$	2.43–9.70			A12
NbRh	4.10			$D8_b$
$Nb_{0.9}Rh_{0.1}$	2.8			
$Nb_{0.60}Rh_{0.40}$	4.21			
Nb_3Rh	2.64			A15, $a = 5.1317$
$Nb_3Rh_{0.98-0.90}Ru_{0.02-0.10}$	2.42–2.44			A15
$Nb_{0.9}Ru_{0.1}$	2.8			
$Nb_{0.6}Ru_{0.4}$	1.2			
$Nb_{0.4}Ru_{0.6}$	2.5			
NbS			1.28	
NbS_2	6.1–6.3			hex., 2 layer $NbSe_2$ type
NbS_2	5.0–5.5			hex., 3 layer type
$NbSb_2$			1.15	
$Nb_{0.83}Sb_{0.17}$	1.95; 2.0			A15
$Nb_{0.9-0.7}Sb_{0.1-0.3}$	5.8–<0.5			A15
Nb_3Sb			1.02	A15, $a = 5.262$
$Nb_3Sb_{0.3-0}Sn_{0.7-1}$	18.0–14.7			A15, $a = 5.270–5.263$
$Nb_3Sb_{0.7-0}Sn_{0.3-1}$	18.0–6.8			A15, $a = 5.292–5.270$
$NbSe_2$	7.0			hex.
$Nb_{1.05-1}Se_2$	7.0–2.2			hex., NbS_2 type

TABLE 10-4. (*Continued*)

Material	Critical Temp., T_c, (°K)	Critical Field, H_0, (oersteds)	Min. Temp.* (°K)	Structure[#]
$Nb_{0.339}Se_{0.661}$	6.1			
$NbSe_{2(1-x)}Te_{2x}$	0.74–2.7			
$NbSi_{0.6}$			1.20	
$NbSi_2$			1.20	
Nb_2Si			1.20	
Nb_3Si_5			1.02	tet.
$Nb_3Si_{0.5}Sn_{0.5}$	8.3			A15
$Nb_3Si_{0.6}Sn_{0.4}$	6.5			A15
Nb_3SiSnV_3	4.0			
$Nb_{0.8}Sn_{0.2}$ (1823°K-4 hours)	5.5			A15, a = 5.283
$NbSn_2$	2.68			
Nb_3Sn	18.05			A15, a = 5.289
Nb_3Sn_2	2.60	620		orthorhombic
Nb_xSn_{1-x}	8.2–17.9			
$NbSnTa_2$	10.8			A15, a = 5.280
Nb_2SnTa	16.4			A15, a = 5.289
$Nb_{3x}SnTa_{3(1-x)}$	6.0–18.0			
$NbSnTaV$	6.2			A15, a = 5.175
$Nb_2SnTa_{0.5}V_{0.5}$	12.2			A15
$NbSnV_2$	5.5			A15, a = 5.115
Nb_2SnV	9.8			A15, a = 5.171
$Nb_{2.5}SnV_{0.5}$	14.2			A15
$Nb_3Sn_2V_3$	7.4			A15
$Nb_{2.70}SnZr_{0.30}$	18.01			A15
Nb_xTa_{1-x}	4.4–9.2			A2
$Nb_{1-0}Ta_{0-1}$	9.18–4.33			
$Nb_{1-0.6}Ta_{0-0.4}$	9.23–6.56			
$NbTc_3$	10.5			A12, a = 9.625
$NbTe_2$	0.6			
Nb_3Te_4	1.49			
$Nb_{1-0.25}Ti_{0-0.75}$	9.6–6.7			A2
$Nb_{0.25-0}Ti_{0.75-1}$	6.3–1.6			hex.
$Nb_{0.5}Ti_{0.5}$	9.5			
$Nb_{0.75}Ti_{0.15}Zr_{0.10}$	9.7			
$Nb_{0.57}Ti_{0.33}Zr_{0.10}$	9.6			
$Nb_{0.35}Ti_{0.15}Zr_{0.50}$	8.6			
$Nb_{0.65}Ti_{0.15}Zr_{0.20}$	9.8			
$Nb_{0.21}Ti_{0.61}Zr_{0.18}$	7.21			
$Nb_{0.222}U_{0.778}$ (γ)	1.98			A3, a = 3.45
Nb_xU_{1-x}	0.9–1.0			Nb in α – U
Nb_xU_{1-x}	1.8–2.0			Nb in γ – U
$Nb_{0.88}V_{0.12}$	5.7			A2
Nb_xV_{1-x}	4.1–9.2			A2
$NbZn_3$			1.02	L1$_2$, a = 3.932
$Nb_{0.88-0.06}Zr_{0.12-0.94}$	10.5–10			
$Nb_{0.06-0.0125}Zr_{0.94-0.9875}$	10.0–3.2			
$Nb_{0.0125-0}Zr_{0.9875-1}$	3.2–1.2			A3
$Nb_{0.75}Zr_{0.25}$	10.8			
$Nb_{0.20}Zr_{0.80}$	⩽8.5			
Nd_2S_3			1.68	cubic, b.c.
NiO			1.28	
Ni_2P			1.01	C22
Ni_3P			1.01	D0$_e$
$Ni_{0.05}Pd_{0.95}Te_2$	1.40			C6

TABLE 10-4. (*Continued*)

Material	Critical Temp., T_c, (°K)	Critical Field, H_0, (oersteds)	Min. Temp.* (°K)	Structure[#]
$Ni_{0.1}Pd_{0.9}Te_2$	1.30			C6
NiS			1.28	
NiSb			0.30	$B8_1$
NiSi			1.90	
$NiSi_2$			1.00	C1
NiTe			1.00	$B8_1$
$NiTe_2$			1.2	C6
Ni_3Th_7	1.98			$D10_2$
NiTi			1.02	A2
NiU			1.12	complex
NiU_6	0.41			
$Ni_{0.20}V_{0.80}$	0.57			A15
$Ni_{0.225}V_{0.775}$	0.30			A15
$Ni_{0.175}V_{0.825}$	0.78			A15
$Ni_{0.1}Zr_{0.9}$	1.50			A3
$NiZr_2$	1.52			
O_2Mo			1.28	
O_5Mo_2			1.28	
O_2Pb			1.02	
$O_{0.105}Pd_{0.285}Zr_{0.61}$	2.09			$E9_3, a = 12.470$
$O_3Rb_{0.27-0.29}W$	1.98			hex., $a = 7.4, c = 7.6$
OReTi	5.74			
O_3Rh_2			1.28	
$O_{0.14}Rh_{0.287}Ti_{0.573}$	3.37			$E9_3, a = 11.588$
$O_{0.105}Rh_{0.285}Zr_{0.61}$	11.8			$E9_3, a = 12.408$
$O_3Sr_{0.08}W$	2.0–4.0			hex., $a = 7.414, c = 7.569$
$O_{0.006}Ta_{0.994}$	4.185			A2
$O_{0.017}Ta_{0.983}$	3.78			A2
$O_{0.028}Ta_{0.972}$	3.48			A2
OTi	0.58			
O_3Ti_2			1.30	
$O_3Tl_{0.30}W$	2.00–2.14			hex., $a = 7.344, c = 7.482$
O_2U			1.28	
OV			1.20	B1
$O_{0.03}V_{0.97}$	1.8–2.4			
OV_3Zr_3	7.5			$E9_3, a = 12.160$
OW_3 (film)	1.1; 3.35			A15
O_2W			0.3	
O_3W			0.3	
$Os_{0.055}Re_{0.945}$	1.93			
OsReY	2.00			C14
OsTa	1.95			$A12, a = 9.773$
Os_2Th			1.02	$C15, a = 7.704$
Os_3Th_7	1.51			$D10_2, a = 10.02, c = 6.285$
OsTi	0.46			$B2, a = 3.077$
Os_2U			1.02	$C15, a = 7.509$
$Os_{0.55}V_{0.45}$	5.04			A15
OsW	4.40			$D8_b$
$Os_{0.075}W_{0.925}$	0.0			A2
$Os_{0.15}W_{0.85}$	2.2			
$Os_{0.34}W_{0.66}$	3.81			$D8_b$
$Os_{0.45-0.37}W_{0.55-0.63}$	4.1–3.7			
$Os_{0.7-0.52}W_{0.3-0.48}$	3.7–0.9			A3
OsW_3	2.21–3.02			

TABLE 10-4. (*Continued*)

Material	Critical Temp., T_c, (°K)	Critical Field, H_0, (oersteds)	Min. Temp.* (°K)	Structure[#]
Os_2Y	4.7			C14
OsZr	1.5–5.6			
Os_2Zr	3.0			C14, $a = 5.219$, $c = 8.538$
PPb	7.8			
$PPd_{3.2-3.0}$	0.7–<0.35			DO_{11}
P_3Pd_7 (high temperature)	1.00			rhombohedral
P_3Pd_7 (low temperature)	0.70			complex
PRh_2	1.3			C1, $a = 5.516$
PRu			0.35	
PRu_2			0.35	
PV			1.01	B31
PV_3			1.00	
PW	1.01			B31
PW_3	2.26			DO_e
Pb_2Pd	2.95			C16
Pb_4Pt	2.80			
Pb_2Rh	2.66			C16
PbS			1.28	
PbSb (eutectic alloy)	6.6			
PbSe			1.26	
$Pb_{0.01-0}Sn_{0.99-1}$	3.734–3.731			tet.
PbTe			1.28	
$PbTl_{0.27-0.04}$	6.43–7.06			
$Pb_{1-0.26}Tl_{0-0.74}$	7.20–3.68			
$Pb_{1-0}Tl_{0-1}$	7.26–2.38			
$PbTl_2$	3.75; 4.09			
Pb_3Y	4.72			$L1_2$
Pb_3Yb	0.23 ± 0.10			$L1_2$
$Pd_{0.4}Pt_{0.1}Rh_{0.5}$			0.015	
$Pd_{0.25}Pt_{0.25}Rh_{0.5}$			0.015	
$Pd_{0.9}Pt_{0.1}Te_2$	1.65			C6
$Pd_{0.95}Pt_{0.05}Te_2$	1.71			C6
$Pd_{0.75}Rh_{0.25}$			0.015	
$Pd_{0.5}Rh_{0.5}$			0.015	
$Pd_{0.95}Rh_{0.05}Te_2$	1.65			C6
PdS			0.35	
$Pd_{2.2}S$ (quenched)	1.63			cubic, $a = 8.93$
$Pd_{2.8}S$			0.35	
Pd_4S			0.32	tet.
PdSb	1.50			$B8_1$
$PdSb_2$	1.25			C2, $a = 6.459$
$Pd_{0.52-0.49}Sb_{0.48-0.51}$	1.44–1.67			
$PdSc_2$			0.32	$E9_3$, $a = 12.442$
PdSe			0.32	B34
Pd_2Se	2.20			
Pd_4Se	0.42			tet., $a = 5.234$, $c = 5.6470$
$Pd_{6-7}Se$	0.66			
$Pd_{17}Se_{15}$			0.32	cubic, $a = 10.606$
PdSeTe			1.2	C6
PdSi	0.93			B31
PdSn	0.41			B31
$PdSn_4$			1.35	$D1_c$
Pd_2Sn	0.41			C37
Pd_3Sn_2	0.47–0.64			$B8_2$

TABLE 10-4. (*Continued*)

Material	Critical Temp., T_c, (°K)	Critical Field, H_0, (oersteds)	Min.* Temp.* (°K)	Structure[#]
PdTe	3.85			$B8_1$
$PdTe_{1.08-1.02}$	1.88–2.56			$B8_1$
$Pd_{1.75-1.05}Te_2$ (annealed)	2.25–1.77			C6
$Pd_{1.75-1.05}Te_2$ (unannealed)	1.93–1.74			C6
$PdTe_2$	1.69			C6
$PdTe_{2.1}$	1.89			C6
$PdTe_{2.3}$	1.85			C6
Pd_3Te	0.76			
Pd_4Te			0.32	cubic, $a = 12.674$
$PdTh_2$	0.85			C16
Pd_5Th			0.32	
$Pd_{0.25}V_{0.75}$	0.08			A15
PdV_3	0.082			A15
$Pd_{1-0.75}W_{0-0.25}$			0.2	A1
$Pd_{0.74-0.56}W_{0.26-0.44}$	0.1–1.6			A1
$Pd_{0.1}Zr_{0.9}$	7.5			A3
Pr_2S_3			1.68	cubic, b.c.
$Pt_{0.2}Rh_{0.8}$			0.015	
PtSb	2.10			$B8_1$
PtSc			0.32	B2, $a = 3.268$
$PtSc_4$			0.32	
Pt_3Sc			0.32	$L1_2$, $a = 3.958$
PtSi	0.88			B31
PtSn	0.37			$B8_1$
$PtSn_4$			1.3	$D1_c$
Pt_2Sr			1.02	C15, $a = 7.777$
PtTa	1.0			$D8_b$
$Pt_{0.15}Ta_{0.85}$	0.40			A15
Pt_3Ta_7	1.2–1.5			$D8_b$
PtTe	0.59			orthorhombic
$PtTe_2$			1.2	C6
PtTh	0.44			B_f
Pt_3Th_7	0.98			$D10_2$
$Pt_{2-4}Th$			0.32	
Pt_5Th	3.13			
$Pt_{0.02}U_{0.98}$	0.87			β-phase
$Pt_{0.46}U_{0.54}$			0.3	
$Pt_{0.222}V_{0.778}$	0.98			A15
$Pt_{0.25}V_{0.75}$	3.20			A15
$Pt_{0.28}V_{0.72}$	1.50			A15
$PtV_{2.5}$	1.36			A15
PtV_3	2.87–3.20			
$PtV_{3.5}$	1.26			A15
$Pt_{0.5}W_{0.5}$	1.45			A1
$Pt_{0.9-0.63}W_{0.1-0.37}$	2.55–2.57			A1 and A2
$Pt_{0.98-0.95}W_{0.02-0.05}$	1.1–2.2			A2
$Pt_{0.60-0.30}W_{0.40-0.70}$	0.4–2.15			A1
PtY			0.32	
Pt_2Y	1.57			C15, $a = 7.590$
$Pt_{2.2}Y$	1.70			C15, $a = 7.576$
Pt_3Y			0.32	$L1_2$, $a = 4.075$
Pt_3Y_7	0.82			$D10_2$
Pt_5Y			0.32	
PtZr	3.0			A3

TABLE 10-4. (*Continued*)

Material	Critical Temp., T_c, (°K)	Critical Field, H_0, (oersteds)	Min. Temp.* (°K)	Structure#
ReSe$_2$			1.15	
ReSi$_2$			1.15	
ReTa	1.3			D8$_b$
Re$_{0.2}$Ta$_{0.8}$	0.21			
Re$_{0.6}$Ta$_{0.4}$	1.4			D8$_b$
Re$_{0.64}$Ta$_{0.36}$	1.46			A12, a = 9.765
Re$_2$Th	5.05			
Re$_x$Ti$_{1-x}$	⩽6.6			
Re$_{24}$Ti$_5$	6.6			A12, a = 9.587
Re$_2$U			1.02	orthorhombic
Re$_{0.9}$V$_{0.1}$	9.4			
Re$_{0.92}$V$_{0.08}$	6.8			A3
Re$_{0.76}$V$_{0.24}$	4.52			D8$_b$
Re$_{0.50}$V$_{0.50}$	5.12			D8$_b$
Re$_{0.50}$W$_{0.50}$	5.12			D8$_b$
Re$_{1-0.84}$W$_{0-0.16}$	8.0–1.6			A3
Re$_{0.7-0.5}$W$_{0.3-0.5}$	5.2–4.8			σ-phase
Re$_{0.4-0.15}$W$_{0.6-0.85}$	4.0–2.3			A2
ReW	5.20			D8$_b$
Re$_3$W	9.0			A12
Re$_3$W$_2$	6.0			
Re$_2$Y	1.83			C14
Re$_2$Zr	6.8			C14
Re$_6$Zr	7.40			A12, a = 9.698
Rh$_{1-x}$Ru$_x$Se$_2$	4.5–<0.05			
Rh$_{17}$S$_{15}$	5.8			cubic, a = 9.911
RhSb			0.30	B31
RhSc			1.02	B2, a = 3.206
Rh$_{0.25}$Sc$_{0.75}$	~0.88			
Rh$_3$Sc			0.32	L1$_2$, a = 3.898
Rh$_x$Se$_{1-x}$	⩽6.0			
Rh$_{0.53}$Se$_{0.47}$	6.0			C2
Rh$_2$Se$_5$			1.04	
RhSi			0.30	B20, a = 4.673
Rh$_2$Sr	6.2			C15, a = 7.706
RhTa	2.0			D8$_b$
Rh$_{0.4}$Ta$_{0.6}$	2.35			D8$_b$
RhTe			1.06	B8$_1$
Rh$_{0.67}$Te$_{0.33}$	0.49			
RhTe$_2$	1.51			C2, a = 6.442
RhTh	0.36			B$_f$
Rh$_2$Th			0.32	
Rh$_3$Th			0.32	L1$_2$
Rh$_3$Th$_7$	2.15			D10$_2$
Rh$_5$Th	1.07			
Rh$_x$Ti$_{1-x}$	2.25–3.95			
Rh$_{0.88}$Ti$_{0.12}$	4.0			cubic
Rh$_{0.12}$Ti$_{0.88}$	4.0			cubic
Rh$_{0.04}$Ti$_{0.96}$	2.0			cubic
Rh$_{0.02}$U$_{0.98}$	0.96			
Rh$_{0.25}$V$_{0.75}$			0.015	A15
RhV$_3$			0.30	
RhW	2.64–3.37			A3
RhY			0.32	B2

TABLE 10-4. (*Continued*)

Material	Critical Temp., T_c, (°K)	Critical Field, H_0, (oersteds)	Min. Temp.* (°K)	Structure[#]
RhY_2			0.32	
RhY_3	0.65			
Rh_2Y			0.32	C15
Rh_2Y_3	1.48			
Rh_3Y	1.07			C15, $a = 7.424$
Rh_3Y_7			0.32	$D10_2$
Rh_5Y	0.56			
RhZr	2.7			
$Rh_{0.005-0.027}Zr_{0.995-0.973}$	3.5–4.8			A3
$Rh_{0.035-0.09}Zr_{0.965-0.91}$	5.0–11.0			cubic
$Rh_{0.12}Zr_{0.88}$	11.0			cubic
$Rh_{0.4}Zr_{0.6}$	6.4			
RuSb	1.27			
Ru_2Sb			0.35	
Ru_2Sc	2.24			C15
$Ru_2Sc_{1.2}$	1.67			C14
$RuSe_2$			0.32	C2, $a = 5.934$
Ru_2Th	3.56			C15, $a = 7.651$
RuTi	1.07			B2, $a = 3.067$
$Ru_{0.1}Ti_{0.9}$	3.5			
$Ru_{0.05}Ti_{0.95}$	2.5			
$Ru_xTi_{0.6}V_{0.4}$	≤6.6			
$Ru_{0.45}V_{0.55}$	4.0			B2
RuW	7.5			A3
$Ru_{0.58}W_{0.42}$	5.20			$D8_b$
$Ru_{0.4}W_{0.6}$	4.67			$D8_b$
Ru_2Y	2.42			C15
$Ru_{0.1}Zr_{0.9}$	5.7			A3
Ru_2Zr	1.84			C14
S_3Sb			1.28	
SSn			1.28	
S_2Ta	0.7–1			hex.
SbSn	1.30–2.37			
$Sb_{0.08-0.02}Sn_{0.92-0.98}$	3.89–2.64	304–345		
$Sb_{0.04-0}Sn_{0.96-1}$	3.739–3.730			tet.
Sb_2Sn_3	3.80			
$SbTa_3$	0.59–0.72			A15
$SbTi_3$	5.80			A15, $a = 5.217$
Sb_2Tl	5.20			
Sb_2Tl_7	5.20			
$Sb_{0.03-0.01}V_{0.97-0.99}$	2.63–3.76			A2
SbV_3	0.80			A15, $a = 4.941$
SbY			1.02	B1
Sb_2Zr_3	1.74			
$ScSi_2$			1.00	
Se_4Nb_3	1.61			
Se_2Ta	0.2			
Si_3Sr_2	~0.55			
SiTa	4.25–4.38			hex.
$SiTa_5$			1.20	
Si_2Ta			1.20	
Si_2Ta_3			1.20	
Si_3Ta_5			1.20	
Si_2Th (α)	3.16			C_c

TABLE 10-4. (*Continued*)

Material	Critical Temp., T_c, (°K)	Critical Field, H_0, (oersteds)	Min. Temp.* (°K)	Structure#
Si_2Th (β)	2.41			C32
Si_2Th_3			1.20	
Si_2Ti			1.20	
SiU_3			1.10	tet.
Si_2U_3			0.1	tet.
$Si_{0.4}V_{0.6}$			1.20	
$Si_{0.38-0.1}V_{0.62-0.99}$	<17			
$Si_{0.263}V_{0.737}$	15.8			A15, a = 4.726
$Si_{0.206}V_{0.794}$	14.5			A15, a = 4.729
SiV_3	17.0			A15, a = 4.728
Si_2V			1.20	
Si_3V_5			0.30	
$Si_{0.9}V_3Al_{0.1}$	14.05			A15, a = 4.727
$SiV_{2.7}Cr_{0.3}$	11.3			A15, a = 4.697
$Si_{0.9}V_3Ge_{0.1}$	14.0			A15, a = 4.731
$SiV_{2.7}Mo_{0.3}$	11.7			A15, a = 4.732
$SiV_{2.7}Nb_{0.3}$	12.8			A15, a = 4.756
$SiV_{2.7}Ru_{0.3}$	2.9			A15, a = 4.707
$SiV_{2.7}Ti_{0.3}$	10.9			A15, a = 4.736
$SiV_{2.7}Zr_{0.3}$	13.2			A15, a = 4.724
Si_2W_3	2.8			
SiY			1.15	
Si_2Y			1.20, 0.30	
$SiZr$			1.20	
$SiZr_2$			1.20	
$SiZr_4$			1.20	
Si_2Zr			1.02	C49
Si_2Zr_3			1.20	
Si_3Zr_4			1.20	
Si_3Zr_5			1.1	D8$_8$
Si_5Zr_6			1.20	
$Sn_{0.174-0.104}Ta_{0.826-0.896}$	6.5-<4.2			A15
$SnTa_3$ (low state of order)	6.2			A15, a = 5.277
$SnTa_3$ (high state of order)	8.35			A15, a = 5.280
$SnTa_2V$	3.7			A15, a = 5.174
$SnTaV_2$	2.8			A15, a = 5.041
Sn_3Th	3.33			L1$_2$
Sn_xTl_{1-x}	2.37-5.2			
$Sn_{0.65}Tl_{0.35}$	6-7.1			
$Sn_{0.02}V_{0.98}$	2.87			A2
$Sn_{0.04}V_{0.96}$	1.86			A2
$Sn_{0.057}V_{0.943}$	~1.6			A2
SnV_3	3.8			A15, a = 4.96
Sn_3V_2			1.15	
SnY_2			1.15	
Sn_3Y_5			1.4	
$SnZr_4$	0.79-0.92			A15
$TaTe_2$			0.05	
$Ta_{0.65-0}Ti_{0.35-1}$	7.8-4.4			
$Ta_{1-0.7}Ti_{0-0.3}$	4.3-6.5			
$Ta_{0.52}Ti_{0.48}$	7.86			
$Ta_{0.025}Ti_{0.975}$	1.3			hex.
$Ta_{0.05}Ti_{0.95}$	2.9			hex.
$Ta_{0.75}V_{0.25}$	2.65			A2, a = 3.254

TABLE 10-4. (*Continued*)

Material	Critical Temp., T_c, (°K)	Critical Field, H_0, (oersteds)	Min. Temp.* (°K)	Structure#
$Ta_{0.50}V_{0.50}$	2.35			A2, $a = 3.182$
$Ta_{0.25}V_{0.75}$	2.80			A2, $a = 3.111$
$Ta_{0.05}V_{0.95}$	4.30			A2
$Ta_{1-0.8}W_{0-0.2}$	4.4–1.2			A2
$Tc_{0.60-0.40}W_{0.40-0.60}$	7.88–7.18			
$Tc_{0.30-0.10}W_{0.70-0.90}$	5.75–1.25			cubic
Tc_6Zr	9.7			A12, $a = 9.636$
TeY			1.02	B1
$Th_{0.55-0}Y_{0.45-1}$	1.8–1.2			
$Th_{0.5}Y_{0.5}$	1.25			
$Th_{0.25}Y_{0.75}$	1.8			
TiU_2			1.06	C32
$Ti_{0.80-0.15}V_{0.20-0.85}$	7.30–3.5			A2
$Ti_{0.5-0}V_{0.5-1}$	6.7–5.3	1050–1250		A2
$Ti_{0.80}V_{0.20}$	3.37–3.65			
$Ti_{0.66}Zr_{0.34}$	1.36			A3
$Ti_{0.5}Zr_{0.5}$ (annealed)	1.23			
$Ti_{0.5}Zr_{0.5}$ (quenched)	2.0			
$Ti_{0.33}Zr_{0.67}$	1.35			A3
$Ti_{0.18}Zr_{0.82}$	1.03			A3
Tl_3Y	1.52			$L1_2$
$TlZr_4$			0.35	A15
$V_{0.6}Zr_{0.4}$	8.3			
$V_{0.4}Zr_{0.6}$	~7.8			
$V_{0.26}Zr_{0.74}$	≈5.9			
V_2Zr	8.8			C15, $a = 7.439$
W_2Zr	2.16			C15, $a = 7.621$
YZn			0.33	B2

*T_n lowest temperature at which material was checked for a superconductive transition.
#Cell edges in Angstrom Units.

TABLE 10-5. HIGH MAGNETIC FIELD SUPERCONDUCTIVE MATERIALS AND SELECTED PROPERTIES. (ROBERTS, 1963, 1969, 1971)

Material	T_c (°K)	H_{c1} (kOe)	H_{c2} (kOe)	H_{c3} (kOe)	T_{obs} (°K)#
Al_2CMo_3	9.2		101		4.2
$Al_{3-x}Gd_xLa$	2.2–6.16		1.3–13.6		0
$Al_{2.968}Gd_{0.032}La$	3.00		2.09		0
$Al_{2.966}Gd_{0.034}La$	2.05		1.30		0
$Al_{2.988}Gd_{0.012}La$	5.00		13.55		0
$Al_{0.66}Ge_{0.33}Nb_{2.5}$	19.6–20.1		~380		0
$Al_{0.75}Ge_{0.25}Nb_3$	18.5		420		4.2
$Al_{0.8}Ge_{0.2}Nb_3$	10.7		130		4.2
$AlLa_3$	6.16		7.92		0
$AlNb_3$	17.14		246		0
$AlNb_3$	≈18.7		295		4.2
$Al_{0.0015}Sn_{0.9985}$			0.0175		3.595
AuV_3	2.55		~9		2.25
AuV_3	0.86–2.980		22–37		0
$BCMo_2$	7.1		28		4.2
$Ba_xO_3Sr_{1-x}Ti$	0.50	0.0039			
$Bi_{0.5}Cd_{0.1}Pb_{0.27}Sn_{0.13}$			>24	3.06	
Bi_2K	3.57				
$Bi_{0.40-0.05}Pb_{0.60-0.95}$	8.4–7.35	0.122	~30		4.2
$Bi_{0.56-0}Pb_{0.44-1}$			13.8–0.53		4.2
BiPb (eutectic alloy)	8.7		>22		4.24
$Bi_{0.53}Pb_{0.32}Sn_{0.16}$			>22		4.24
Bi_5Tl_3	6.4		>4.08		4.23
CK (excess K)	0.55		0.160(H ⊥ C)		
	0.55		0.730(H ∥ C)		
C_8K	0.39		0.025(H ⊥ C)		0.32
	0.39		0.250(H ∥ C)		0.32
C_8K (excess K)	0.55		0.160(H⊥C)		0.32
	0.55		0.730(H∥C)		0.32
$C_{0.44}Mo_{0.56}$	12.5–13.5	0.087	98.5(H_p = 238)		1.2
CNb (whiskers)	7.5–10.5				
CNb	8–10	0.12	16.9(H_p = 130)		4.2
$CNb_{0.4}Ta_{0.6}$	10–13.6	0.19	14.1(H_p = 214)		1.2
CTa	9–11.4	0.22	4.6(H_p = 185)		1.2
$C_{0.52}Ti$	3.42		48		1.6
$C_{0.46}Ti$	3.32		45		1.6
$Ca_xO_3Sr_{1-x}Ti$	<0.1–0.55	0.00215, 0.0038			
$CaSi_2$	1.58		1.0		0.35
			0.32		1.0
$Cd_{0.1}Hg_{0.9}$ (by weight)		0.23	0.34		2.04
$Cd_{0.05}Hg_{0.95}$ (by weight)		0.28	0.31		2.16
$Co_{0.002}Mo_{0.815}Re_{0.185}$	5.8		6.1		0
Cr_3Ir	0.168		10.5		0
Cr_3Rh	0.072		9.1		0
$Cr_{0.10}Ti_{0.30}V_{0.60}$	5.6	0.071	84.4		0
Cr_xV_{1-x}	1.3–5.1				
$Fe_xMo_{0.865}Re_{0.135}$	2.1–6.1		1.7–3.6		0
$Fe_{0.0006}Mo_{0.865}Re_{0.135}$			1.44		1.53
$Fe_xMo_{0.87}Re_{0.13}$			1.7–3.1		5.55
$Fe_{0.05}Nb_{0.38}Ti_{0.57}$			≤83		4.2
$Ga_4Mn_xMo_{1-x}$	4.0–8.0		25–74		0
Ga_xNb_{1-x}			>28		4.2
GaSb (annealed)	4.24		2.64		3.5

TABLE 10-5. (*Continued*)

Material	T_c (°K)	H_{c1} (k Oe)	H_{c2} (k Oe)	H_{c3} (k Oe)	T_{obs} (°K)#
$GaV_{1.95}$	5.3		~73		0
GaV_3	14.1		208		0
			200		4.2
$GaV_{4.5}$	8.6		95		4.2
$Hf_{0.75}Nb_{0.25}$ (arc cast)	>4.2		⩾26		4.2
$Hf_{0.75}Nb_{0.25}$ (cold rolled)	>4.2		>28		4.2
Hf_xTa_y			>28–86		1.2
$Hg_{0.15}Pb_{0.85}$	~6.75		13		2.93
$Hg_{0.05}Pb_{0.95}$		0.235	2.3		
$In_{0.98}Pb_{0.02}$	3.45	0.1		0.12	2.76
$In_{0.96}Pb_{0.04}$	3.68	0.1	0.12	0.25	2.94
$In_{0.94}Pb_{0.06}$	3.90	0.095	0.18	0.35	3.12
$In_{0.913}Pb_{0.087}$	4.2	~0.17	0.55	2.65	
$In_{0.6}Pb_{0.4}$	6.36	0.630	3.250		0
$In_{0.35}Pb_{0.965}$		0.6	1.75		0
$In_{0.139}Pb_{0.661}$		0.22	2.45		4.2
$In_{0.037}Pb_{0.963}$		0.39	0.97		4.2
$In_{1.000}Te_{1.002}$	3.5–3.7		1.2		
$In_{0.95}Tl_{0.05}$		0.263	0.263		3.3
$In_{0.9}Tl_{0.1}$		0.257	0.257		3.25
$In_{0.83}Tl_{0.17}$		0.242	0.39		3.21
$In_{0.75}Tl_{0.25}$		0.216	0.50		3.16
LaN	1.35	0.45			0.76
La_3S_4	6.5	0.15	>25		1.3
La_3Se_4	8.6	0.2	>25		1.25
La_3Te_4	3.75	0.060	12.5		1.4
	2.45	0.020	8		1.4
$Mo_{0.725}Nb_{0.061}Re_{0.187}$	5.0		2.65		0
$Mo_{0.865}Re_{0.135}$	6.1		1.57		4.2
$Mo_{0.815}Re_{0.185}$	8.27		7.0		0
$Mo_{0.5}Re_{0.5}$	12.6		27		
$Mo_{0.913}Ti_{0.087}$	2.95	0.060	~15		4.2
$Mo_{0.16}Ti_{0.84}$	4.18	0.028	98.7		0
			38		3.0
$Mo_{0.305-0.116}U_{0.695-0.874}$	1.85–2.06		>25		
$N_{0.93}Nb$	15.85		158		0
NNb (whiskers)	10–14.5				
NNb (wires)	16.10		53		12
			95		8
			132		4.2
			153 ± 3		0
$N_{0.92}Nb_{0.946}Ta_{0.054}$	14.41		135		0
$N_{0.91}Nb_{0.99}Ta_{0.01}$	15.62		135		0
$N_{0.91}Nb_{0.82}Ta_{0.18}$	10.9		100		0
$N_{0.90}Nb_{0.114}Ti_{0.886}$	10.1		100		0
$N_{0.88}Nb_{0.256}Ti_{0.744}$	14.72		104		0
$N_{0.85}Nb_{0.66}Ti_{0.34}$	17.61		119		0
$N_{0.93}Nb_{0.85}Zr_{0.15}$	13.8		>130		
$N_{0.85}Nb_{0.75}Zr_{0.25}$	12.96		116		0
$N_{0.76}Nb_{0.85}Zr_{0.15}$	14.16		132		0
$N_{0.74}Nb_{0.9}Zr_{0.1}$	14.42		136		0
$N_{0.73}Nb_{0.95}Zr_{0.05}$	15.42		146		0
NNb_xZr_{1-x}	9.8–13.8		4–>130		
$Na_{0.086}Pb_{0.914}$		0.19	6.0		
$Na_{0.016}Pb_{0.984}$		0.28	2.05		

TABLE 10-5. (Continued)

Material	T_c (°K)	H_{c1} (kOe)	H_{c2} (kOe)	H_{c3} (kOe)	T_{obs} (°K)#
$Nb_{0.9926}O_{0.0084}$			7.74	~13	4.2
$Nb_{0.993}O_{0.007}$	8.7_8		7	11.1	4.2
$Nb_{0.985}O_{0.0152}$	8.0_4		9.6	11.5	4.2
Nb_3Os	0.943		1.26		0
Nb_3Sn	18.0		235		4.2
Nb_3Sn (clad)	18.00		260		0
$Nb_{2.85}SnZr_{0.15}$ (clad)	18.07		260		0
$Nb_{0.96}Ta_{0.04}$	8.87		6.14		0
$Nb_{0.9844}Ta_{0.0156}$	8.76	1.70	4.50		0
$Nb_{0.9378}Ta_{0.0622}$	8.42	1.12	5.56		0
$Nb_{0.87}Ta_{0.13}$	8.15	0.91	7.08		0
$Nb_{0.79}Ta_{0.21}$	7.51	0.83	7.93		0
$Nb_{0.67}Ta_{0.33}$	6.81	0.55	8.73		0
$Nb_{0.54}Ta_{0.46}$	6.25	0.48	8.60		0
$Nb_{0.37}Ta_{0.63}$	5.31	0.37	6.75		0
$Nb_{0.17}Ta_{0.83}$	4.65	0.33	4.26		0
$Nb_{0.10}Ta_{0.90}$		0.084	0.154		4.195
$Nb_{0.73}Ta_{0.02}Zr_{0.25}$			>70		4.2
$Nb_{0.9}Ti_{0.1}$	>4.2		>30–32		
$Nb_{0.55}Ti_{0.45}$	9.4		108		4.2
$Nb_{0.20}Ti_{0.80}$	7.5	1.12		80	4.2
$Nb_{0.75}Ti_{0.15}Zr_{0.10}$	9.7		57		4.2
$Nb_{0.62}Ti_{0.14}Zr_{0.24}$	9.6		69; 76		4.2
$Nb_{0.57}Ti_{0.33}Zr_{0.10}$	9.6		78		4.2
$Nb_{0.53}Ti_{0.18}Zr_{0.29}$	9.1		81		4.2
(after anneal)	9.0		80		4.2
$Nb_{0.35}Ti_{0.15}Zr_{0.50}$	8.6		79		4.2
(after anneal)	9.3		77		4.2
$Nb_{0.222}U_{0.778}$	1.98		≥25		
NbZr	10.8		92		0
$Nb_{0.75}Zr_{0.25}$	10.6		81.9		0
$Nb_{0.20}Zr_{0.80}$		1.12	80		4.2
O_3SrTi	0.33	0.00195	0.420		0
$PbSb_{1\ w/o}$ (annealed)			>0.7		4.2
$PbSb_{2.8\ w/o}$ (quenched)			>2.3		4.2
$PbSb_{2.8\ w/o}$ (annealed)			>0.7		4.2
$Pb_{0.965}Sn_{0.035}$		0.53	0.56		
$Pb_{0.871}Sn_{0.129}$		0.45	1.1		
$Pb_{0.965}Tl_{0.035}$		0.8	1.5		0
$PbTl_{29.9\ w/o}$			2.927	4.751	4.2
$PbTl_{19.9\ w/o}$			2.580	4.404	4.2
$PbTl_{10.1\ w/o}$			1.691	2.974	4.2
$PbTl_{4.87\ w/o}$			1.048	1.844	4.2
$PbTl_{2.9\ w/o}$				1.415	4.2
$PbTl_{1.06\ w/o}$				0.906	4.2
$Pb_{1-0.26}Tl_{0-0.74}$	7.20–3.68		2–6.96		0
$Re_{0.26}W_{0.74}$			>30		
SiV_3	16.9	0.55	156		
$SnTa_3$ (high order)	8.35		72.5		4.2
$SnTa_3$ (low order)	6.2		15.5		
SnTe	0.034–0.214	0.0005–0.0019	~0.005–0.09		0
$Ta_{0.5}Nb_{0.5}$			3.55		4.2
$Ta_{0.9}Ti_{0.1}$			>16		1.2
$Ta_{1-0.7}Ti_{0-0.3}$	4.3–6.5		>42–108		1.2
$Ta_{0.65-0}Ti_{0.35-1}$	4.4–7.8		>14–138		1.2

TABLE 10.5. (*Continued*)

Material	T_c (°K)	H_{c1} (k Oe)	H_{c2} (k Oe)	H_{c3} (k Oe)	T_{obs} (°K)#
$Ta_{0.52}Ti_{0.48}$	7.86				
$Tc_{0.60}W_{0.40}$	7.88		43.5		4.2
$Tc_{0.50}W_{0.50}$	7.52		29.0		4.2
$Tc_{0.40}W_{0.60}$	7.18		19.0		4.2
$Tc_{0.30}W_{0.70}$	5.75		7.5		4.2
$Ti_{0.775}V_{0.225}$	4.7	0.024	172(H_p = 86.5)		0
$Ti_{0.75}V_{0.25}$	5.3	0.029	199(H_p = 97.5)		0
$Ti_{0.615}V_{0.385}$	7.07	0.050	~34		4.2
$Ti_{0.516}V_{0.484}$	7.20	0.062	~28		4.2
$Ti_{0.415}V_{0.585}$	7.49	0.078	~25		4.2
$Ti_{0.12}V_{0.88}$			17.3	28.1	4.2
$Ti_{0.09}V_{0.91}$			14.3	16.4	4.2
$Ti_{0.06}V_{0.94}$			8.2	12.7	4.2
$Ti_{0.03}V_{0.97}$			3.8	6.8	4.2
$V_{0.26}Zr_{0.74}$	5.9	0.238			1.05
	5.9	0.227			1.78
	5.9	0.185			3.04
	5.9	0.165			3.5

#Temperature at which critical field data were measured.

REFERENCES

Roberts, B. W., "Superconductive Materials and Some of Their Properties," General Electric Co., U.S. Govt. Report No. AD 428 672 (1963).

Roberts, B. W., "Superconductive Materials and Some of Their Properties," NBS Technical Note 482, U.S. Govt. Printing Office, Washington, D.C. (1969).

Roberts, B. W., "Superconductive Materials and Some of Their Properties," General Electric Co., Final Report, Contract No. NBS 32-70-5 (1971).

BIBLIOGRAPHY

Anderson, D. E., "Superconductivity," in *Magnetic Materials Digest 1964*, M. W. Lads, Philadelphia, Pa., pp. 196–217 (1964).

Bardenn, J., "Critical Fields and Currents in Superconductors," *Rev. Mod. Phys.*, **34**, 667 (1962).

Chester, P. F., "Superconducting Magnets," *Repts. Progr. Phys.*, **30**, Part II, p. 561 (1967).

Fishlock, D., Ed., "A Guide to Superconductivity," American Elsevier, New York, 1969.

Gaballe, T. H. and B. T. Matthias, "Superconductivity," in *Ann. Rev. Phys. Chem.*, **14**, 141–160 (1964).

Glover, R. E., III, "Superconductivity Above the Transition Temperature," *Progr. in Low Temp. Phys.*, **6**, 291–332 (1970).

Hulm, J. K., M. Ashkin, D. W. Deis and C. K. Jones, "Superconductivity in Semiconductors and Semi-Metals," *Prog. in Low Temp. Phys.*, **6**, 205–242 (1970).

Luiten, A. L., "Superconducting Magnets," *Phillips Tech. Rev.*, **29**, 309–322 (1968).

Matthias, B. T., "Superconductivity and the Periodic System," *Am. Scientist*, **58**, 80 (1970).

McClintock, M., "Cryogenics," Van Nostrand Reinhold, New York, 1964.

Parks, R. D., Ed., "Superconductivity," Vols. I and II, Marcel Dekker, New York, 1969.

Volger, J., "Superconductivity," *Philips Tech. Rev.*, **29**, 1–16 (1968).

Weis, O., "The Physical Properties of Superconductive Metals," *Chemiker Zeitung*, **95**, 168 (1971).

CHAPTER **XI**
PROPERTIES OF
THE ENVIRONMENT

Chapter XI describes the properties of the environment, beginning with the interior of the earth, leading to the surface of the earth—both in solid form and as the Sea, then to the various regions of the atmosphere, and finally to land in outer space among the other planets, stars and asteroids.

The information is presented in the following Figures and Tables:

TABLE 11-1. DENSITY, GRAVITY, PRESSURE AND SEISMIC VELOCITY IN THE EARTH*

Depth in Earth (km)	Density (g/cm^3)	Gravity (cm/sec^2)	Pressure (10^{12} dynes/cm^2)	Seismic Velocity (km/sec) Longitudinal	Seismic Velocity (km/sec) Transverse
33	3.32	985	0.009	7.75	4.35
100	3.38	989	0.031	7.95	4.45
200	3.47	992	0.065	8.26	4.60
300	3.55	995	0.100	8.58	4.76
413	3.64	998	0.141	8.97	4.96
600	4.13	1001	0.213	10.25	5.66
800	4.49	999	0.300	11.00	6.13
1000	4.68	995	0.392	11.42	6.36
1200	4.80	991	0.49	11.71	6.50
1400	4.91	988	0.58	11.99	6.62
1600	5.03	986	0.68	12.26	6.73
1800	5.13	985	0.78	12.53	6.83
2000	5.24	986	0.88	12.79	6.93
2200	5.34	990	0.99	13.03	7.02
2400	5.44	998	1.09	13.27	7.12
2600	5.54	1009	1.20	13.50	7.21
2800	5.63	1026	1.32	13.64	7.30
2898	5.68	1037	1.37	13.64	7.30
2898	9.43	1037	1.37	8.10	—
3000	9.57	1019	1.47	8.22	—
3200	9.85	979	1.67	8.47	—
3400	10.11	936	1.85	8.76	—
3600	10.35	892	2.04	9.04	—
3800	10.56	848	2.22	9.28	—
4000	10.76	803	2.40	9.51	—
4200	10.94	758	2.57	9.70	—
4400	11.11	716	2.73	9.88	—
4600	11.27	677	2.88	10.06	—
4800	11.41	646	3.03	10.25	—
4982	11.54	626	3.17	10.44	—
5121	14.2	585	3.27	9.7	—
5121	16.8	—	—	11.16	—
5400	—	460	3.41	—	—
5700	—	320	3.53	11.26	—
6371	17.2	0	3.64	11.31	—

*Based on data by K. E. Bullen and H. Jeffreys, reported in: Jacobs, J. A., "The Earth's Core and Geomagnetism," Pergamon Press, Oxford, 1963.

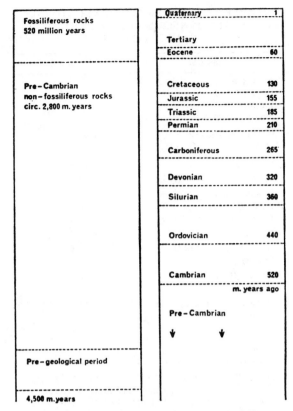

Fossiliferous rocks 520 million years	Quaternary	1
	Tertiary	
	Eocene	60
Pre-Cambrian non-fossiliferous rocks circ. 2,800 m. years	Cretaceous	130
	Jurassic	155
	Triassic	185
	Permian	210
	Carboniferous	265
	Devonian	320
	Silurian	360
	Ordovician	440
	Cambrian	520
		m. years ago
	Pre-Cambrian	
Pre-geological period		
4,500 m.years		

Fig. 11-1. The geological systems. (DAY, 1963)

TABLE 11-2. ESTIMATED COMPOSITION OF THE EARTH'S CRUST AND OF SOME OF ITS MAJOR MINERALS*

Con-stituent	Peridotite (ultra-basic rock), %	Basaltic rock (basic rock), %	Inter-mediate rock, %	Granitic rock, %	Crust, %	Shale, %
SiO_2	43.5	48.5	54.5	69.1	58.7	58.1
TiO_2	0.8	1.8	1.5	0.5	1.2	0.7
Al_2O_3	2.	15.5	16.4	14.5	15.	15.4
Fe_2O_3	2.5	2.8	3.3	1.7	2.3	4.
FeO	9.9	8.1	5.2	2.2	5.2	2.5
MnO	0.2	0.17	0.15	0.07	0.12	
MgO	37.	8.6	3.8	1.1	4.9	2.4
CaO	3.	10.7	6.5	2.6	6.7	3.1
Na_2O	0.4	2.3	4.2	3.9	3.1	1.3
K_2O	0.1	0.7	3.2	3.8	2.3	3.2

*From "Distribution of the Elements in Our Planet," by L. H. Ahrens. Copyright 1965 McGraw-Hill Book Company. Used with permission of McGraw-Hill Book Company.

TABLE 11-3. MINERALS CONSTITUTING THE IGNEOUS ROCK IN THE EARTH'S CRUST*

Mineral	Percent proportion	
Alkali feldspar (1 part orthoclase, $KAlSi_3O_8$, and 1 part Albite, $NaAlSi_3O_8$)	31.	Total feldspar (60.2)
Plagioclase feldspar ($NaAlSi_3O_8 \rightarrow CaAl_2Si_2O_8$)	29.2	
Quartz (SiO_2)	12.4	
Pyroxene [$Ca(Mg,Fe)Si_2O_6$]	12.	
Oxides of iron and titanium (magnetite, Fe_3O_4; hematite, Fe_2O_3; ilmenite, $Fe^{II}Ti^{IV}O_3$)	4.1	
Biotite mica (complex K,Mg,Fe,Al,Ti, hydroxy-fluo silicate)	3.8	
Olivine [$(Fe,Mg)_2SiO_4$]	2.6	
Muscovite (complex K,Al hydroxy-fluo silicate)	1.4	
Other minerals	3.5	
Total	100.	

*From "Distribution of the Elements in Our Planet," by L. H. Ahrens. Copyright 1965 McGraw-Hill Book Company. Used with permission of McGraw-Hill Book Company.

TABLE 11-4. AVERAGE COMPOSITION OF CHONDRITES*

Compound or Element	Percent
SiO_2	38.04
MgO	23.84
FeO	12.45
Al_2O_3	2.50
CaO	1.95
Na_2O	0.98
K_2O	0.17
Cr_2O_3	0.36
MnO	0.25
TiO_2	0.11
P_2O_5	0.21
Fe	11.76
Ni	1.34
Co	0.08
FeS	5.73
Total Fe	25.07

*From "Distribution of the Elements in Our Planet," by L. H. Ahrens. Copyright 1965 McGraw-Hill Book Company. Used with permission of McGraw-Hill Book Company.

TABLE 11-5. ESTIMATED ABUNDANCES OF SOME ELEMENTS IN THE EARTH'S CRUST AND IN CHONDRITES*†

Element	Chondrites	Crust	Element	Chondrites	Crust
Li	2.5	20	Ag	0.03–0.1	0.07
Be	0.04	2.8	Cd	0.06	0.2
B	0.43	10	In	0.001	0.1
F	30–130	625	Sn	0.43	2
Na	0.68%	2.4%	Sb	0.10	0.2
Mg	14.4%	1.95%	Te	0.55	
Al	1.3%	8.2%	I	0.04	0.5
Si	17.8%	28.2%	Cs	0.1	3
P	0.11%	1,050	Ba	4.5	425
S	2.3%	260	La	0.34	30
Cl	160	130	Ce	1.1	60
K	0.09%	2.1%	Pr	0.13	8.2
Ca	1.4%	4.2%	Nd	0.61	28
Sc	8.5	22	Sm	0.23	6
Ti	850	0.57%	Eu	0.08	1.2
V	65	135	Gd	0.34	5.4
Cr	3,000	100	Tb	0.052	0.9
Mn	2,600	950	Dy	0.34	3
Fe	25.1%	5.6%	Ho	0.08	1.2
Co	520	25	Er	0.24	2.8
Ni	1.35%	75	Tm	0.034	0.5
Cu	90	55	Yb	0.20	3
Zn	54	70	Lu	0.035	0.5
Ga	5.3	15	Hf	0.19	3
Ge	9.5	1.5	Ta	0.023	2
As	2.2	1.8	W	0.14	1.5
Se	8.5	0.05	Re	0.057	
Br	?	2.5	Os	0.80	
Rb	3	90	Ir	0.60	
Sr	11	375	Pt	1.2	
Y	2	33	Au	0.16	0.004
Zr	12	165	Hg		0.08
Nb	?	20	Tl	0.001	0.45
Mo	1.5	1.5	Pb	0.18	12.5
Ru	1		Bi	0.0025	0.17
Rh	0.17		Th	0.40	9.6
Pd	0.6		U	0.014	2.7

*Unless otherwise indicated, the values are in parts per million.
†From "Distribution of the Elements in Our Planet," by L. H. Ahrens. Copyright 1965 McGraw-Hill Book Company. Used with permission of McGraw-Hill Book Company.

TABLE 11-6. ABBREVIATIONS FOR MINERALS PROPERTIES TABLES

Colors:

b	brown		p	purple
be	blue		r	red
bk	black		w	white
c	colorless		y	yellow
g	gray		M-	metallic
gn	green		N-	non-metallic
o	orange			

NOTE: most colors of minerals are from Dietrich, 1969; where neither M- nor N- is stated, the mineral has either both appearances or is not listed by said author.

Temperature:
r = room temperature, assumed to be 25 5°C

Other Quantities:

$S^{\circ}_{298.15}$ entropy at reference conditions (298.15°K)

H°_f heat of formation from reference state, cal gfw^{-1}

G°_f Gibbs free energy of formation from reference state, cal gfw^{-1}

gfw   Gram formula weight

Z   Gram formula weights per unit cell

NOTE: (1) Space Group and its number taken from: "Symmetry Groups," Vol. 1, "International Tables for X-Ray Crystallography," Henry, N. F. M. and K. Lonsdale, Eds., Kynoch Press, Birmingham, England, 1952.

(2) Cell dimensions in Å (10^{-8} cm).

TABLE 11-7. X-RAY CRYSTALLOGRAPHIC DATA OF MINERALS (ROBIE 1967)

Name and formula	Crystal system	Space group	Structure type	Z	a_0	b_0	c_0	α_0	β_0	γ_0	Temp. °C
Sulfides, arsenides, tellurides, selenides, and sulfosalts											
Shandite β-$Ni_3Pb_2S_2$*	hex-R	$R\bar{3}m$(166)		3	5.576 ± .010		13.658 ± .010				r
High-Argentite Ag_2S I	cubic			4	6.269 ± .020						600
Argentite Ag_2S II	cubic			2	4.870 ± .008						189
Acanthite Ag_2S III	mon.	$P2_1/c$(14)		4	4.228 ± .002	6.928 ± .005	7.862 ± .003		99.58 ± .30		25
High-Naumanite Ag_2Se	cubic			2	4.993 ± .016						170
Ag_2Te I	cubic			2	5.29 ± .01						825
Ag_2Te II	cubic			4	6.585 ± .010						250
Hessite Ag_2Te III	mon.	$P2_1/c$(14)		4	8.09 ± .02	4.48 ± .01	8.96 ± .02		123.33 ± .30		r
$Ag_{1.55}Cu_{45}S$ I	cubic			4	6.110 ± .010						300
$Ag_{1.55}Cu_{45}S$ II	cubic			2	4.825 ± .005						116
Jalpaite $Ag_{1.55}Cu_{45}S$ III	tet.			16	8.673 ± .004		11.756 ± .006				r
$Ag_{.93}Cu_{1.07}S$ I	cubic			4	5.961 ± .009						196
$Ag_{.93}Cu_{1.07}$	hex.			2	4.138 ± .004		7.105 ± .007				100
Stromeyerite $Ag_{.93}Cu_{1.07}S$ III	orth.	$Cmcm$(63)		4	4.066 ± .002	6.628 ± .003	7.972 ± .004				r
Eucairite AgCuSe	orth.	pseudo $P4/nmm$(129)		10	4.105 ± .010	20.35 ± .02	6.31 ± .01				r
Petzite Ag_3AuTe_2*	cubic	$I4_132$(214)		8	10.38 ± .02						r

	system	space group	type	Z	a	b	c	β	T (°C)
Maldonite Au_2Bi	cubic	$Fd3m$(227)	Cu_2Mg	8	7.958 ±.002				r
High-Digenite Cu_2S I	cubic			4	5.725 ±.010				465
High-Chalcocite Cu_2S II	hex.			2	3.961 ±.004		6.722 ±.007		152
Chalcocite Cu_2S III	orth.	$Ab2m$(39)		96	11.881 ±.004	27.323 ±.010	13.491 ±.004		r
Digenite $Cu_{1.79}S$ (Cu rich side)	cubic		deformed fluorite	4	5.5695 ±.0010				25
Digenite $Cu_{1.77}S$ (S rich side)	cubic		deformed fluorite	4	5.5542 ±.0010				25
Berzelianite Cu_2Se	cubic			4	5.85 ±.01				170
Cu_2Se	cubic			1	5.50 ±.01				240
High-Bornite Cu_5FeS_4*	cubic			8	10.94 ±.02				r
Metastable Bornite Cu_5FeS_4	tet.	$P\bar{4}2_1c$(144)		16	10.94 ±.02		21.88 ±.04		r
Low-Bornite Cu_5FeS_4*	tet.	$P4/mmm$(123)		2	6.402 ±.010		4.276 ±.010		
Umangite Cu_3Se_2	tet.			2					r
Heazelwoodite Ni_3S_2	hex-R	$R32$(155)		3	5.746 ±.001		7.134 ±.002		r
Maucherite $Ni_{11}As_8$	tet.	$P4_12_12$(92)		4	6.870 ±.001		21.81 ±.01		r
Pentlandite $Fe_{5.25}Ni_{3.75}S_8$	cubic	$Fm3m$(225)		4	10.196 ±.010				r
Pentlandite $Fe_{3.75}Ni_{5.25}S_8$	cubic	$Fm3m$(225)		4	10.095 ±.010				r
Sternbergite $AgFe_2S_3$*	orth.	$Ccmm$(63)		8	11.60 ±.02	12.675 ±.020	6.63 ±.01		r
Argentopyrite $AgFe_2S_3$*	orth.	$Pmmm$(47)		4	6.64 ±.01	11.47 ±.02	6.45 ±.02		r
Realgar AsS*	mon.	$P2_1/m$(11)		16	9.29 ±.05	13.53 ±.05	6.57 ±.03	106.55 ±.30	r
Oldhamite CaS	cubic	$Fm3m$(225)	rock salt	4	5.689 ±.006				r

TABLE 11-7 (Continued)

Sulfides, arsenides, tellurides, selenides, and sulfosalts—Continued

Name and formula	Crystal system	Space group	Structure type	Z	a_0	b_0	c_0	α_0	β_0	γ_0	Temp. °C
Greenockite CdS	hex.	P6$_3$mc(186)	zincite	2	4.1354 ±.0010		6.7120 ±.0010				r
Hawleyite CdS	cubic	F$\overline{4}$3m(216)	sphalerite	4	5.833 ±.002						r
(hypothetical) CdS	cubic	Fm3m(225)	rock salt	4	5.516 ±.002						r
Cadmoselite CdSe	hex.	P6$_3$mc(186)	zincite	2	4.2977 ±.0010		7.0021 ±.0010				r
CdTe	cubic	F$\overline{4}$3m(216)	sphalerite	4	6.4805 ±.0006						25
(hypothetical) CoS	cubic	F$\overline{4}$3m(216)	sphalerite	4	5.339 ±.001						r
Chalcopyrite (CuFeS$_2$) CuFeS$_{1.90}$	tet.	I$\overline{4}$2d(122)		4	5.2988 ±.0010		10.434 ±.005				r
Cubanite CuFe$_2$S$_3$*	orth.	Pcmn(62)		4	6.46 ±.01	11.12 ±.01	6.23 ±.01				r
Covellite CuS	hex.	P6$_3$/mmc(194)		6	3.792 ±.001		16.34 ±.01				r
Klockmannite CuSe	hex.		deformed covellite	78	14.206 ±.010		17.25 ±.05				r
Troilite FeS	hex.	P6$_3$/mmc(194)	niccolite	2	3.446 ±.003		5.877 ±.001				28
Pyrrhotite Fe$_{.980}$S	hex.	P6$_3$mmc(194)	defect niccolite	2	3.446 ±.001		5.848 ±.002				28
Pyrrhotite Fe$_{.885}$S	hex.	P6$_3$/mmc(194)	defect niccolite	2	3.440 ±.001		5.709 ±.003				28
(hypothetical) FeS	cubic	F$\overline{4}$3m(216)	sphalerite	4	5.455 ±.001						r
(hypothetical) FeS	hex.	P6$_3$mc(186)	zincite	2	3.872 ±.001		6.345 ±.002				r
Cinnabar HgS	hex.	P3$_1$21(152) P3$_2$1(154)	cinnabar	3	4.149 ±.001		9.495 ±.002				r

Mineral	Formula	System	Space group	Structure type	Z	a_0	b_0	c_0	Ref.
Metacinnabar	HgS	cubic	F$\bar{4}$3m(216)	sphalerite	4	5.8517 ± .0010			r
Tiemannite	HgSe	cubic	F$\bar{4}$3m(216)	sphalerite	4	6.0853 ± .0050			r
Coloradoite	HgTe	cubic	F$\bar{4}$3m(216)	sphalerite	4	6.4600 ± .0006			r
Alabandite	MnS	cubic	Fm3m(225)	rock salt	4	5.2234 ± .0005			r
(hypothetical)	MnS	cubic	F$\bar{4}$3m(216)	sphalerite	4	5.611 ± .002			r
(hypothetical)	MnS	hex.	P6$_3$mc(186)	zincite	2	3.986 ± .001		6.465 ± .002	r
Niccolite	NiAs	hex.	P6$_3$/mmc(194)	niccolite	2	3.618 ± .001		5.034 ± .001	r
Millerite	NiS	hex-R	R3m(160)	niccolite	9	9.616 ± .001		3.152 ± .001	r
Breithauptite	NiSb	hex.	P6$_3$/mmc(194)	niccolite	2	3.942 ± .001		5.155 ± .001	26
Galena	PbS	cubic	Fm3m(225)	rock salt	4	5.9360 ± .0005			r
Clausthalite	PbSe	cubic	Fm3m(225)	rock salt	4	6.1255 ± .0005			r
Teallite	PbSnS$_2$	orth.	Pbnm(62)	GeS	2	4.266 ± .003	11.419 ± .007	4.090 ± .002	r
Altaite	PbTe	cubic	Fm3m(225)	rock salt	4	6.4606 ± .0005			r
Cooperite	PtS	tet.	P4$_2$/mmc(131)		2	3.4699 ± .0006		6.1098 ± .0010	r
Herzenbergite	SnS	orth.	Pbnm(62)	GeS	4	4.328 ± .002	11.190 ± .004	3.978 ± .001	r
Sphalerite	ZnS	cubic	F$\bar{4}$3m(216)	sphalerite	4	5.4093 ± .0005			r
Wurtzite	ZnS	hex.	P6$_3$mc(186)	zincite	2	3.8230 ± .0010		6.2565 ± .0010	r
Stilleite	ZnSe	cubic	F$\bar{4}$3m(216)	sphalerite	4	5.6685 ± .0005			r
	ZnTe	cubic	F$\bar{4}$3m(216)	sphalerite	4	6.1020 ± .0006			r

Table 11-7 (*Continued*)

Sulfides, arsenides, tellurides, selenides, and sulfosalts—Continued

Name and formula	Crystal system	Space group	Structure type	Z	a_0	b_0	c_0	α_0	β_0	γ_0	Temp. °C
Orpiment As_2S_3*	mon.	$P2_1/n$(14)		4	11.49 ± .02	9.59 ± .02	4.25 ± .01		90.45 ± .30		r
Bismuthinite Bi_2S_3	orth.	Pbnm(62)	stibnite	4	11.150 ± .004	11.300 ± .004	3.981 ± .001				26
Tellurobismuthite Bi_2Te_3	hex-R	$R\bar{3}m$(166)	Bi_2Te_2S	3	4.3835 ± .0020		30.487 ± .003				25
Stibnite Sb_2S_3	orth.	Pbnm(62)	stibnite	4	11.229 ± .004	11.310 ± .004	3.8389 ± .0010				25
Linnaeite Co_3S_4	cubic	Fd3m(227)	spinel	8	9.401 ± .001						r
Greigite Fe_3S_4	cubic	Fd3m(227)	spinel	8	9.876 ± .002						r
Daubreeite $FeCr_2S_4$	cubic	Fd3m(227)	spinel	8	9.966 ± .005						r
Violarite $FeNi_2S_4$	cubic	Fd3m(227)	spinel	8	9.464 ± .005						r
Polymidite Ni_3S_4	cubic	Fd3m(227)	spinel	8	9.480 ± .001						r
Co-Safflorite $CoAs_2$	mon.		deformed marcasite	2	5.049 ± .002	5.872 ± .002	3.127 ± .001		90.45 ± .20		26
Safflorite $(Co_{.5}Fe_{.5})As_2$	orth.	Pnmm(58)	marcasite	2	5.231 ± .002	5.953 ± .002	2.962 ± .002				26
Cobaltite $CoAsS$*	cubic	$P2_13$(198)	NiSbS	4	5.60 ± .05						r
Glaucodot $(Co,Fe)AsS$*	orth.	Cmmm(65)		24	6.64 ± .05	28.39 ± .10	5.64 ± .05				r
Cattierite CoS_2	cubic	Pa3(205)	pyrite	4	5.5345 ± .0005						r
Trogtalite $CoSe_2$	cubic	Pa3(205)	pyrite	4	5.8588 ± .0010						r
Loellingite $FeAs_2$	orth.	Pnmm(58)	marcasite	2	5.300 ± .002	5.981 ± .002	2.882 ± .001				26

Mineral	Formula	system	space group	structure type	Z	a	b	c	α	β	γ	ref
Arsenopyrite	$FeAsS$*	tri.	$P\bar{1}(2)$		4	5.760 ±.010	5.690 ±.005	5.785 ±.005	90.00 ±.20	112.23 ±.20	90.00 ±.20	r
Gudmundite	$FeSbS$*	mon.	$B2_1/d(14)$		8	10.00 ±.05	5.93 ±.03	6.73 ±.03	90.00 ±.20	90.00 ±.50		r
Pyrite	FeS_2	cubic	$Pa3(205)$	pyrite	4	5.4175 ±.0005						r
Marcasite	FeS_2*	orth.	$Pnnm(58)$	marcasite	2	4.443 ±.002	5.423 ±.002	3.3876 ±.0015				25
Ferroselite	$FeSe_2$	orth.	$Pnnm(58)$	marcasite	2	4.801 ±.005	5.778 ±.005	3.587 ±.004				r
Frohbergite	$FeTe_2$	orth.	$Pnnm(58)$	marcasite	2	5.265 ±.005	6.265 ±.005	3.869 ±.002				r
Hauerite	MnS_2	cubic	$Pa3(205)$	pyrite	4	6.1014 ±.0006						28
Molybdenite	MoS_2	hex.	$P6_3/mmc(194)$	molybdenite	2	3.1604 ±.0010		12.295 ±.002				26
Rammelsbergite	$NiAs_2$	orth.	$Pnnm(58)$	marcasite	2	4.757 ±.002	5.797 ±.004	3.542 ±.002				26
Pararammelsbergite	$NiAs_2$	orth.	$Pbca(61)$		8	5.75 ±.01	5.82 ±.01	11.428 ±.02				r
Gersdorffite	$NiAsS$	cubic	$P2_13(198)$		4	5.693 ±.001						26
Vaesite	NiS_2	cubic	$Pa3(205)$	pyrite	4	5.6873 ±.0005						r
	$NiSe_2$	cubic	$Pa3(205)$	pyrite	4	5.9604 ±.0010						20
Melonite	$NiTe_2$	hex.	$P\bar{3}m1(164)$	cadmium iodide	1	3.869 ±.010		5.308 ±.010				84
Sperrylite	$PtAs_2$	cubic	$Pa3(205)$	pyrite	4	5.968 ±.005						r
Laurite	RuS_2	cubic	$Pa3(205)$	pyrite	4	5.60 ±.02						r
Tungstenite	WS_2	hex.	$P6_3/mmc(194)$	molybdenite	2	3.154 ±.001		12.362 ±.004				26
Co-Skutterudite $CoAs_{3-x}$	$CoAs_{2.95}$	cubic	$Im3(204)$		8	8.2060 ±.0010						r
Fe-Skutterudite $FeAs_{3-x}$	$FeAs_{2.95}$	cubic	$Im3(204)$		8	8.1814 ±.0010						r

TABLE 11-7 (*Continued*)

Name and formula	Crystal system	Space group	Structure type	Z	a_0	b_0	c_0	α_0	β_0	γ_2	Temp. °C
Sulfides, arsenides, tellurides, selenides, and sulfosalts—Continued											
Ni-Skutterudite $NiAs_{3-x}$ $NiAs_{2.95}$	cubic	Im3(204)		8	8.3300 ±.0010						r
Tennantite $Cu_{12}As_4S_{13}$	cubic	I$\bar{4}$3m(217)	tetrahedrite	2	10.190 ±.004						r
Tetrahedrite $Cu_{12}Sb_4S_{13}$	cubic	I$\bar{4}$3m(217)	tetrahedrite	2	10.327 ±.004						r
Enargite Cu_3AsS_4	orth.	Pnn2(34)		2	6.426 ±.005	7.422 ±.005	6.144 ±.005				26
Luzonite Cu_3AsS_4*	tet.	I$\bar{4}$2m(121)		2	5.289 ±.005		10.440 ±.008				26
Famatimite Cu_3SbS_4*	tet.	I$\bar{4}$m(121)		2	5.384 ±.005		10.770 ±.008				26
Proustite Ag_3AsS_3	hex-R	R3c(161)		6	10.816 ±.001		8.6948 ±.0013				26
Pyrargyrite Ag_3SbS_3	hex-R	R3c(161)		6	11.052 ±.002		8.7177 ±.0020				26
Miargyrite $AgSbS_2$*	mon.	Cc(9)		8	12.862 ±.013	4.111 ±.004	13.220 ±.010		98.63 ±.15		r
Oxides and hydroxides											
Corundum Al_2O_3	hex-R	R$\bar{3}$c(167)	corundum	6	4.7591 ±.0004		12.9894 ±.0030				25
Boehmite AlO(OH)*	orth.	Cmcm(63)	lepidocrocite	4	2.868 ±.003	12.227 ±.003	3.700 ±.003				26
Diaspore AlO(OH)*	orth.	Pbnm(62)		4	4.401 ±.005	9.421 ±.005	2.845 ±.002				r
Gibbsite $Al(OH)_3$	mon.	P2$_1$/n(14)		8	9.719 ±.002	5.0705 ±.0010	8.6412 ±.0010		94.57 ±.25		r
Arsenolite As_2O_3	cubic	Fd3m(227)	diamond	16	11.074 ±.005						r
Claudetite As_2O_3	mon.	P2$_1$/n(14)		4	5.339 ±.002	12.984 ±.005	4.5405 ±.0010		94.27 ±.10		25

Name	System	Space group	Structure	Z	a	b	c	angle	Ref
Bromellite BeO	hex.	P6$_3$mc(186)	zincite	2	2.6979 ±.0005		4.3772 ±.0005		26
Bismite α-Bi$_2$O$_3$	mon.	P2$_1$/c(14)	pseudo orthorhombic	8	8.166 ±.005	13.827 ±.010	5.850 ±.004	90.00 ±.20	25
Lime CaO	cubic	Fm3m(225)	rock salt	4	4.8108 ±.0005				26
Portlandite Ca(OH)$_2$	hex.	P$\bar{3}$m1(164)	CdI$_2$	1	3.5933 ±.0005		4.9086 ±.0020		27
Monteponite CdO	cubic	Fm3m(225)	rock salt	4	4.6953 ±.0010				26
Cerianite CeO$_2$	cubic	Fm3m(225)	fluorite	4	5.4110 ±.0020				26
CoO	cubic	Fm3m(225)	rock salt	4	4.260 ±.002				26
Eskolaite Cr$_2$O$_3$	hex-R	R$\bar{3}$c(167)	corundum	6	4.9607 ±.0020		13.599 ±.010		r
Tenorite CuO	mon.	C2/c(15)		4	4.684 ±.005	3.425 ±.005	5.129 ±.005	99.47 ±.17	26
Cuprite Cu$_2$O	cubic	Pn3m(224)		2	4.2696 ±.0010				26
Wustite Fe$_{.953}$O	cubic	Fm3m(225)	defect rock salt	4	4.3088 ±.0003				17
Hematite Fe$_2$O$_3$	hex-R	R$\bar{3}$c(167)	corundum	6	5.0329 ±.0010		13.7492 ±.0010		25
Magnetite Fe$_3$O$_4$	cubic	Fd3m(227)	spinel	8	8.3940 ±.0005				22
Goethite α-FeO(OH)*	orth.	Pbnm(62)		4	4.596 ±.005	9.957 ±.010	3.021 ±.003		r
Lepidocrocite γ-FeO(OH)*	orth.	Amam(63)		4	3.868 ±.010	12.525 ±.010	3.066 ±.003		r
α-Ga$_2$O$_3$	hex-R	R$\bar{3}$c(167)	corundum	6	4.9793 ±.0010		13.429 ±.003		24
Low-germania GeO$_2$	tet.	P4/mnm(136)	rutile	2	4.3963 ±.0010		2.8626 ±.0010		25
High-germania GeO$_2$	hex.	P3$_1$21(152) P3$_2$21(154)	α-quartz	3	4.987 ±.002		5.652 ±.002		26
Ice H$_2$O	hex.	P6$_3$/mmc(194)		4	4.5212 ±.0010		7.3666 ±.0010		0

TABLE 11-7 (*Continued*)

Name and formula	Crystal system	Space group	Structure type	Z	a_0	b_0	c_0	α_0	β_0	γ_0	Temp. °C
							Oxides and hydroxides—Continued				
Hafnia HfO$_2$	mon.	P2$_1$/c(14)	baddeleyite	4	5.1156 ±.0010	5.1722 ±.0010	5.2948 ±.0010		99.18 ±.08		r
Montroydite HgO	orth.	Pnma(62)		4	6.608 ±.003	5.518 ±.003	3.519 ±.003				25
Periclase MgO	cubic	Fm3m(225)	rock salt	4	4.2117 ±.0005						25
Brucite Mg(OH)$_2$	hex.	P$\bar{3}$m1(164)	CdI$_2$	1	3.147 ±.004		4.769 ±.004				26
Manganosite MnO	cubic	Fm3m(225)	rock salt	4	4.4448 ±.0005						26
Pyrolusite MnO$_2$	tet.	P4/mnm(136)	rutile	2	4.388 ±.003		2.865 ±.002				r
Bixbyite Mn$_2$O$_3$	cubic	Ia3(206)	Tl$_2$O$_3$	16	9.411 ±.005						25
Hausmanite Mn$_3$O$_4$	tet.	I4$_1$/amd(141)		8	8.136 ±.005		9.422 ±.005				20
Molybdite MoO$_3$	orth.	Pbnm(62)		4	3.962 ±.002	13.858 ±.005	3.697 ±.004				26
Bunsenite NiO	cubic	Fm3m(225)	rock salt	4	4.177 ±.002						26
Litharge PbO red	tet.	P4/nmm(129)		2	3.9759 ±.0040		5.023 ±.004				27
Massicot PbO yellow	orth.	Pb2a(32)		4	5.489 ±.003	4.755 ±.004	5.891 ±.004				27
Minium Pb$_3$O$_4$	tet.	P4$_2$/mbc(135)		4	8.815 ±.005		6.565 ±.003				25
Senarmontite Sb$_2$O$_3$	cubic	Fm3m(225)	arsenic trioxide	16	11.152 ±.003						26
Valentinite Sb$_2$O$_3$	orth.	Pccn(56)	antimony trioxide	4	4.914 ±.002	12.468 ±.005	5.421 ±.004				25
Cervantite Sb$_2$O$_4$	cubic	Fd3m(227)		8	10.305 ±.005						26

Name	System	Space group	Structure type	Z	a	b	c	angle	Temp
Selenolite SeO₂	tet.	P4₂/mbc(135) P4₂bc(106)		8	8.35 ± .01		5.08 ± .01		26
α-Quartz SiO₂*	hex.	P3₁21(152) P3₂21(154)		3	4.9136 ± .0001		5.4051 ± .0001		25
β-Quartz SiO₂*	hex.	P6₄22(181) P6₂22(180)		3	4.999 ± .001		5.4592 ± .0020		575
α-Cristobalite SiO₂	tet.	P4₁2₁2(92) P4₃2₁2(96)		4	4.971 ± .003		6.918 ± .003		25
β-Cristobalite SiO₂	cubic	Fd3m(227)		8	7.1382 ± .0010				405
Keatite SiO₂	tet.	P4₁2₁2(92) P4₃2₁2(96)		12	7.456 ± .003		8.604 ± .005		r
β-Tridymite SiO₂	hex.	P6̄2c(172) P6₃/mmc(194)		4	5.0463 ± .0020		8.2563 ± .0030		405
Coesite SiO₂*	mon.	B2/b(15)		16	7.152 ± .001	12.379 ± .002	7.152 ± .001	120.00 ± .17	25
Stishovite SiO₂*	tet.	P4/mnm(136)	rutile	2	4.1790 ± .0010		2.6649 ± .0010		r
Melanophlogite SiO₂*	cubic	Pm3n(223)	clathrate type	46	13.402 ± .004				r
Cassiterite SnO₂	tet.	P4/mnm(136)	rutile	2	4.738 ± .003		3.188 ± .003		26
Tellurite TeO₂*	orth.	Pbca(61)	tellurite	8	5.607 ± .003	12.034 ± .005	5.463 ± .003		25
Paratellurite TeO₂	tet.	P4₁2₁2(92) P4₃2₁2(96)		4	4.810 ± .002		7.613 ± .002		25
Thorianite ThO₂	cubic	Fm3m(225)	fluorite	4	5.5952 ± .0005				25
Rutile TiO₂	tet.	P4/mnm(136)		2	4.5937 ± .0005		2.9618 ± .0010		25
Anatase TiO₂	tet.	I4₁/amd(141)		4	3.785 ± .002		9.514 ± .006		r
Brookite TiO₂*	orth.	Pcab(61)		8	5.456 ± .002	9.182 ± .005	5.143 ± .003		r
Titanium sesquioxide Ti₂O₃	hex–R	R3̄c(167)	corundum	6	5.149 ± .002		13.642 ± .010		r
Uraninite UO₂	cubic	Fm3m(225)	fluorite	4	5.4682 ± .0010				26

TABLE 11-7 (*Continued*)

Name and formula	Crystal system	Space group	Structure type	Z	a_0	b_0	c_0	α_0	β_0	γ_0	Temp. °C
Oxides and hydroxides—Continued											
Karelianite V_2O_3	hex-R	R̄3c(167)	corundum	6	4.952 ±.002		14.002 ±.010				r
Zincite ZnO	hex.	P6₃mc(186)	zincite	2	3.2495 ±.0005		5.2069 ±.0005				25
Baddeleyite ZrO_2	mon.	P2₁/c(14)	baddeleyite	4	5.1454 ±.0010	5.2075 ±.0010	5.3107 ±.0010		99.23 ±.08		r
Multiple oxides											
Spinel $MgAl_2O_4$	cubic	Fd3m(227)	spinel	8	8.080 ±.002						26
Hercynite $FeAl_2O_4$	cubic	Fd3m(227)	spinel	8	8.150 ±.004						25
Galaxite $MnAl_2O_4$	cubic	Fd3m(227)	spinel	8	8.258 ±.002						25
Gahnite $ZnAl_2O_4$	cubic	Fd3m(227)	spinel	8	8.0848 ±.0020						26
Magnetite $FeFe_2O_4$	cubic	Fd3m(227)	spinel	8	8.3940 ±.0005						22
Jacobsite $MnFe_2O_4$	cubic	Fd3m(227)	spinel	8	8.499 ±.002						25
Trevorite $NiFe_2O_4$	cubic	Fd3m(227)	spinel	8	8.339 ±.003						25
Picrochromite $MgCr_2O_4$	cubic	Fd3m(227)	spinel	8	8.333 ±.003						26
Ilmenite $FeTiO_3$	hex-R	R̄3(148)	ilmenite	6	5.093 ±.005		14.055 ±.020				r
Geikielite $MgTiO_3$	hex-R	R̄3(148)	ilmenite	6	5.054 ±.005		13.898 ±.010				26
Pyrophanite $MnTiO_3$	hex-R	R̄3(148)	ilmenite	6	5.155 ±.001		14.18				r
Cobalt Titanate $CoTiO_3$	hex-R	R̄3(148)	ilmenite	6	5.066 ±.001		13.918 ±.005				r

Mineral / Formula	System	Space group	Structure	Z	a	b	c	Ref
Perovskite $CaTiO_3$	orth.	Pcmn(62)	perovskite	4	5.3670 ±.0010	7.6438 ±.0010	5.4439 ±.0010	r
Chrysoberyl $BeAl_2O_4$	orth.	Pmnb(62)	olivine	4	5.4756 ±.0020	9.4041 ±.0030	4.4267 ±.0020	25
Halides								
Halite $NaCl$	cubic	Fm3m(225)	rock salt	4	5.6402 ±.0002			26
Sylvite KCl	cubic	Fm3m(225)	rock salt	4	6.2931 ±.0002			25
Villiaumite NaF	cubic	Fm3m(225)	rock salt	4	4.6342 ±.0005			25
Chlorargyrite $AgCl$	cubic	Fm3m(225)	rock salt	4	5.5491 ±.0005			26
Bromargyrite $AgBr$	cubic	Fm3m(225)	rock salt	4	5.7745 ±.0005			26
Nantockite $CuCl$	cubic	F$\bar{4}$3m(216)	sphalerite	4	5.416 ±.003			25
Marshite CuI	cubic	F$\bar{4}$3m(216)	sphalerite	4	6.0507 ±.0010			26
Miersite AgI	cubic	F$\bar{4}$3m(216)	sphalerite	4	6.4963 ±.0010			r
Iodargyrite AgI	hex.	P6$_3$mc(186)	zincite	2	4.5955 ±.0010		7.5005 ±.0033	25
Calomel $HgCl$	tet.	I4/mm(139)		4	4.478 ±.005		10.910 ±.005	26
Fluorite CaF_2	cubic	Fm3m(225)	fluorite	4	5.4638 ±.0004			25
Sellaite MgF_2	tet.	P4$_2$/mnm(136)	rutile	2	4.621 ±.001		3.050 ±.001	18
Chloromagnesite $MgCl_2$	hex-R	R$\bar{3}$m(166)		3	3.632 ±.004		17.795 ±.016	r
Lawrencite $FeCl_2$	hex-R	R$\bar{3}$m(166)		3	3.593 ±.003		17.58 ±.09	r
Scacchite $MnCl_2$	hex-R	R$\bar{3}$m(166)		3	3.711 ±.002		17.59 ±.07	r
Cotunnite $PbCl_2$	orth.	Pnmb(62)		4	4.535 ±.005	7.62 ±.01	9.05 ±.01	26

TABLE 11-7 (*Continued*)

Name and formula	Crystal system	Space Group	Structure type	Z	a_0	b_0	c_0	α_0	β_0	γ_0	Temp. °C
	Halides—Continued						Halides—Continued				
Matlockite $PbFCl$	tet.	P4/nmm(129)		2	4.106 ± .005		7.23 ± .01				26
Cryolite Na_3AlF_6*	mon.	P2$_1$/n(14)		2	5.40 ± .01	5.60 ± .01	7.776 ± .010		90.18 ± .25		r
Neighborite $NaMgF_3$	orth.	Pcmn(62)	perovskite	4	5.363 ± .001	7.676 ± .001	5.503 ± .001				18
	Carbonates and nitrates						Carbonates and nitrates				
Calcite $CaCO_3$	hex-R	R$\bar{3}$c(167)	calcite	6	4.9899 ± .0010		17.064 ± .002				26
Otavite $CdCO_3$	hex-R	R$\bar{3}$c(167)	calcite	6	4.9204 ± .0010		16.298 ± .003				26
Cobalticalcite $CoCO_3$	hex-R	R$\bar{3}$c(167)	calcite	6	4.6581 ± .0010		14.958 ± .003				26
Siderite $FeCO_3$	hex-R	R$\bar{3}$c(167)	calcite	6	4.6887 ± .0010		15.373 ± .003				26
Magnesite $MgCO_3$	hex-R	R$\bar{3}$c(167)	calcite	6	4.6330 ± .0010		15.016 ± .003				26
Rhodochrosite $MnCO_3$	hex-R	R$\bar{3}$c(167)	calcite	6	4.7771 ± .0010		15.664 ± .003				26
Nickelous Carbonate $NiCO_3$	hex-R	R$\bar{3}$c(167)	calcite	6	4.5975 ± .0010		14.723 ± .002				26
Smithsonite $ZnCO_3$	hex-R	R$\bar{3}$c(167)	calcite	6	4.6528 ± .0010		15.025 ± .003				26
Dolomite $CaMg(CO_3)_2$*	hex-R	R$\bar{3}$(148)	calcite	3	4.8079 ± .0010		16.010 ± .003				26
Huntite $Mg_3Ca(CO_3)_4$*	hex-R	R32(155)		3	9.498 ± .003		7.816 ± .004				26
Norsethite $BaMg(CO_3)_2$*	hex-R	R32(155)	calcite	3	5.020 ± .005		16.75 ± .02				r
Vaterite $CaCO_3$	hex.			6	7.135 ± .005		8.524 ± .007				r

Mineral	System	Space group	Z	a	b	c	β	Ref
Witherite BaCO$_3$	orth.	Pnam(62)	4	6.430 ± .005	8.904 ± .005	5.314 ± .005		26
Aragonite CaCO$_3$	orth.	Pnam(62)	4	5.741 ± .005	7.968 ± .005	4.959 ± .005		26
Cerussite PbCO$_3$	orth.	Pnam(62)	4	6.152 ± .005	8.436 ± .005	5.195 ± .005		26
Strontianite SrCO$_3$	orth.	Pnam(62)	4	6.029 ± .005	8.414 ± .005	5.107 ± .005		26
Shortite Na$_2$Ca$_2$(CO$_3$)$_3$	orth.	Amm2(38)	2	4.961 ± .005	11.03 ± .02	7.12 ± .01		r
Malachite Cu$_2$(OH)$_2$CO$_3$	mon.	P2$_1$/a(14)	4	9.502 ± .007	11.974 ± .007	3.240 ± .003	98.75 ± .25	25
Azurite Cu$_3$(OH)$_2$(CO$_3$)$_2$	mon.	P2$_1$/a(14)	2	5.008 ± .005	5.844 ± .005	10.336 ± .005	92.45 ± .25	25
Niter KNO$_3$	orth.	Pnam(62)	4	6.431 ± .005	9.164 ± .005	5.414 ± .005		26
Soda Niter NaNO$_3$	hex-R	R$\bar{3}$c(167)	6	5.0696 ± .0010	16.829 ± .005			25
Gerhardite Cu$_2$(NO$_3$)(OH)$_3$	orth.	P2$_1$2$_1$2$_1$(19)	4	6.075 ± .004	13.812 ± .008	5.592 ± .004		r

Sulfates and borates

Mineral	System	Space group	Z	a	b	c	β	Ref
Barite BaSO$_4$	orth.	Pnma(62)	4	8.878 ± .005	5.450 ± .005	7.152 ± .003		26
Anhydrite CaSO$_4$	orth.	Amma(63) Ccmm(63)	4	6.991 ± .005	6.996 ± .005	6.238 ± .005		26
Anglesite PbSO$_4$	orth.	Pnma(62)	4	8.480 ± .005	5.398 ± .005	6.958 ± .003		25
Celestite SrSO$_4$	orth.	Pnma(62)	4	8.359 ± .005	5.352 ± .005	6.866 ± .005		26
Zinkosite ZnSO$_4$	orth.	Pnma(62)	4	8.588 ± .008	6.740 ± .006	4.770 ± .005		25
Arcanite K$_2$SO$_4$	orth.	Pnma(62)	4	5.772 ± .005	10.072 ± .005	7.483 ± .004		25
Mascagnite (NH$_4$)$_2$SO$_4$	orth.	Pnma(62)	4	7.782 ± .005	5.993 ± .005	10.636 ± .005		25
Thenardite Na$_2$SO$_4$	orth.	Fddd(70)	8	5.863 ± .005	12.304 ± .005	9.821 ± .005		25

TABLE 11-7 (*Continued*)

Name and formula	Crystal system	Space group	Structure type	Z	a_0	b_0	c_0	α_0	β_0	γ_0	Temp. °C
		Sulfates and borates—Continued									
Gypsum $CaSO_4 \cdot 2H_2O$*	mon.	C2/c(15)		4	5.68 ± .01	15.18 ± .01	6.29 ± .01		113.83 ± .22		r
Epsomite $MgSO_4 \cdot 7H_2O$	orth.	$P2_1 2_1 2_1(19)$		4	11.86 ± .01	11.99 ± .01	6.858 ± .007				25
Goslarite $ZnSO_4 \cdot 7H_2O$	orth.	$P2_1 2_1 2_1(19)$	epsomite	4	11.779 ± .005	12.050 ± .005	6.822 ± .003				25
Mirabilite $Na_2SO_4 \cdot 10H_2O$	mon.	$P2_1/c(14)$		4	11.51 ± .01	10.38 ± .01	12.83 ± .01		107.75 ± .17		24
Chalcanthite $CuSO_4 \cdot 5H_2O$	tri.	$P\bar{1}(2)$		2	6.1045 ± .0050	10.72 ± .01	5.949 ± .007	97.57 ± .17	107.28 ± .17	77.43 ± .17	r
Brochantite $Cu_4SO_4(OH)_6$*	mon.	$P2_1/c(14)$		4	13.066 ± .010	9.85 ± .01	6.022 ± .010		103.27 ± .25		r
Syngenite $K_2Ca(SO_4)_2 \cdot H_2O$	mon.	$P2_1/m(11)$		2	9.775 ± .005	7.156 ± .005	6.251 ± .005		104.00		r
Alunite $KAl_3(SO_4)_2(OH)_6$	hex-R	R3m(160)		3	6.982 ± .005		17.32 ± .01				r
Natroalunite $NaAl_3(SO_4)_2(OH)_6$	hex-R	R3m(160)		3	6.974 ± .005		16.69 ± .01				r
Hexahydrite $MgSO_4 \cdot 6H_2O$	mon.	C2/c(15)		8	10.110 ± .005	7.212 ± .004	24.41 ± .01		98.30 ± .10		r
Leonhardite $MgSO_4 \cdot 4H_2O$	mon.	$P2_1/n(14)$		4	5.922 ± .006	13.604 ± .004	7.905 ± .005		90.85 ± .20		r
Melanterite $FeSO_4 \cdot 7H_2O$	mon.	$P2_1/c(14)$		4	14.072 ± .010	6.503 ± .007	11.041 ± .010		105.57 ± .15		r
Vanthoffite $MgSO_4 \cdot 3Na_2SO_4$	mon.	$P2_1/c(14)$		2	9.797 ± .003	9.217 ± .003	8.199 ± .003		113.50 ± .10		r
Dolerophanite $Cu_2O(SO_4)$	mon.	C2/m(15)		4	9.355 ± .010	6.312 ± .005	7.628 ± .005		122.29 ± .10		r
Retgersite $NiSO_4 \cdot 4H_2O$	tet.	$P4_1 2_1 2(92)$ $P4_3 2_1(96)$		4	6.782 ± .004		18.28 ± .01				25
Colemanite $CaB_3O_4(OH)_3 \cdot H_2O$*	mon.	$P2_1/a(14)$		4	8.743 ± .004	11.264 ± .002	6.102 ± .003		110.12 ± .08		r

Mineral / Formula	System	Space group	Structure	Z	a	b	c	α	β	γ	T
Borax Na$_2$B$_4$O$_7\cdot$10H$_2$O	mon.	C2/c(15)		4	11.858 ±.005	10.674 ±.005	12.197 ±.005		106.68 ±.03		r
Kernite Na$_2$B$_4$O$_7\cdot$4H$_2$O	mon.	P2$_1$/c(14)		4	7.022 ±.003	9.151 ±.004	15.676 ±.008		108.83 ±.25		r
Hambergite Be$_2$BO$_3\cdot$(OH$_1$F)*	orth.	Pbca(61)		8	9.755 ±.001	12.201 ±.001	4.426 ±.001				r
Phosphates, molybdates, and tungstates											
Berlinite AlPO$_4$	hex.	P3$_1$21(152) P3$_2$21(154)	α-quartz	3	4.942 ±.005		10.97 ±.007				25
Xenotime YPO$_4$	tet.	I4$_1$/amd(141)	zircon	4	6.885 ±.005		5.982 ±.005				26
Hydroxylapatite Ca$_5$(PO$_4$)$_3$OH	hex.	P6$_3$/m(176)	apatite	2	9.418 ±.003		6.883 ±.003				r
Fluorapatite Ca$_5$(PO$_4$)$_3$F	hex.	P6$_3$/m(176)	apatite	2	9.3684 ±.0030		6.8841 ±.0030				25
Chlorapatite Ca$_5$(PO$_4$)$_3$Cl	hex.	P6$_3$/m(176)	apatite	2	9.629 ±.005		6.777 ±.003				r
Carbonate-apatite Ca$_{10}$(PO$_4$)$_6$CO$_3$H$_2$O	hex.	P6$_3$/m(176)	apatite	1	9.436 ±.010		6.883 ±.010				r
Turquois CuAl$_6$(PO$_4$)$_4$(OH)$_8\cdot$4H$_2$O*	tri.	P$\bar{1}$(2)		1	7.424 ±.008	7.629 ±.008	9.910 ±.010	68.61 ±.20	69.71 ±.20	65.08 ±.20	r
Powellite CaMoO$_4$	tet.	I4$_1$/a(100)	scheelite	4	5.226 ±.005		11.43 ±.007				25
Wulfenite PbMoO$_4$	tet.	I4$_1$/a(100)	scheelite	4	5.435 ±.005		12.110 ±.007				25
Scheelite CaWO$_4$	tet.	I4$_1$/a(100)	scheelite	4	5.242 ±.005		11.372 ±.005				25
Stolzite PbWO$_4$	tet.	I4$_1$/a(100)	scheelite	4	5.4616 ±.0030		12.046 ±.005				25
Ferberite FeWO$_4$	mon.	P2/c(13)	wolframite	2	4.732 ±.004	5.708 ±.003	4.965 ±.004		90.00 ±.05		r
Huebnerite MnWO$_4$	mon.	P2/c(13)	wolframite	2	4.834 ±.004	5.758 ±.005	4.999 ±.004		91.18 ±.10		r
Wolframite Fe$_{.5}$Mn$_{.5}$WO$_4$	mon.	P2/c(13)	wolframite	2	4.782 ±.004	5.731 ±.004	4.982 ±.004		90.57 ±.10		r
Sanmartinite ZnWO$_4$	mon.	P2/c(13)	wolframite	2	4.691 ±.003	5.720 ±.003	4.925 ±.003		89.36 ±.20		25

TABLE 11-7 (*Continued*)

Name and formula	Crystal system	Space group	Structure type	Z	a_0	b_0	c_0	α_0	β_0	γ_0	Temp. °C
Ortho and ring structure silicates											
Forsterite Mg_2SiO_4	orth.	Pbnm(62)	olivine	4	4.758 ± .002	10.214 ± .003	5.984 ± .002				25
Fayalite Fe_2SiO_4	orth.	Pbnm(62)	olivine	4	4.817 ± .005	10.477 ± .005	6.105 ± .010				r
Tephroite Mn_2SiO_4*	orth.	Pbnm(62)	olivine	4	4.871 ± .005	10.636 ± .005	6.232 ± .005				r
Lime Olivine γCa_2SiO_4	orth.	Pbnm(62)	olivine	4	5.091 ± .010	11.371 ± .020	6.782 ± .010				r
Nickel Olivine Ni_2SiO_4	orth.	Pbnm(62)	olivine	4	4.727 ± .002	10.121 ± .005	5.915 ± .002				r
Cobalt Olivine Co_2SiO_4	orth.	Pbnm(62)	olivine	4	4.782 ± .002	10.301 ± .005	6.003 ± .002				r
Monticellite $CaMgSiO_4$	orth.	Pbnm(62)	olivine	4	4.827 ± .005	11.084 ± .005	6.376 ± .005				r
Kerschsteinite $CaFeSiO_4$	orth.	Pbnm(62)	olivine	4	4.886 ± .005	11.146 ± .005	6.434 ± .010				r
Knebelite $MnFeSiO_4$*	orth.	Pbnm(62)	olivine	4	4.854 ± .010	10.602 ± .010	6.162 ± .010				r
Glauchroite $CaMnSiO_4$	orth.	Pbnm(62)	olivine	4	4.944 ± .004	11.19 ± .01	6.529 ± .005				r
Fluor-Norbergite $Mg_2SiO_4 \cdot MgF_2$	orth.	Pnmb(62)		4	8.727 ± .005	10.271 ± .010	4.709 ± .002				25
Chondrodite $2Mg_2SiO_4 \cdot MgF_2$*	mon.	P2$_1$/c(14)		2	7.89 ± .03	4.743 ± .020	10.29 ± .03		109.03 ± .30		r
Fluor-Humite $3Mg_2SiO_4 \cdot MgF_2$	orth.	Pnma(62)		4	10.243 ± .005	20.72 ± .02	4.735 ± .002				25
Clinohumite $4Mg_2SiO_4 \cdot MgF_2$*	mon.	P2$_1$/c(14)		2	13.68 ± .04	4.75 ± .02	10.27 ± .02		100.83 ± .50		r
Grossularite $Ca_3Al_2Si_3O_{12}$	cubic	Ia3d(230)	garnet	8	11.851 ± .001						25
Uvarovite $Ca_3Cr_2Si_3O_{12}$	cubic	Ia3d(230)	garnet	8	11.999 ± .002						26

Mineral / Formula	System	Space group	Structure	Z	a	b	c	α	β	γ	T
Andradite $Ca_3Fe_2Si_3O_{12}$	cubic	Ia3d(230)	garnet	8	12.048 ±.001						25
Goldmanite $Ca_3V_2Si_3O_{12}$	cubic	Ia3d(230)	garnet	8	12.070 ±.005						r
Almandite $Fe_3Al_2Si_3O_{12}$	cubic	Ia3d(230)	garnet	8	11.526 ±.001						25
Pyrope $Mg_3Al_2Si_3O_{12}$	cubic	Ia3d(230)	garnet	8	11.459 ±.001						25
Spessartite $Mn_3Al_2Si_3O_{12}$	cubic	Ia3d(230)	garnet	8	11.621 ±.001						25
Zircon $ZrSiO_4$*	tet.	I4/amd(141)	zircon	4	6.604 ±.005		5.979 ±.005				25
Thorite $ThSiO_4$	tet.	I4/amd(141)	zircon	4	7.143 ±.004		6.327 ±.003				r
Coffinite $USiO_4$	tet.	I4/amd(141)	zircon	4	6.995 ±.004		6.263 ±.005				r
Kyanite Al_2SiO_5*	tri.	P$\bar{1}$(2)		4	7.123 ±.001	7.848 ±.002	5.564 ±.008	89.92 ±.15	101.25 ±.08	105.97 ±.08	25
Andalusite Al_2SiO_5*	orth.	Pnnm(58)		4	7.7959 ±.0050	7.8983 ±.0020	5.5583 ±.0020				25
Sillimanite Al_2SiO_5*	orth.	Pbnm(62) Pnma(62)		4	7.4843 ±.0030	7.6730 ±.0030	5.7711 ±.0040				25
3.2 Mullite $3Al_2O_3 \cdot 2SiO_2$	orth.			3/4	7.557 ±.002	7.6876 ±.0020	2.8842 ±.0010				r
2.1 Mullite $2Al_2O_3 \cdot SiO_2$	orth.	Pbam(55)		6/5	7.5788 ±.0020	7.6909 ±.0020	2.8883 ±.0010				r
Staurolite $Fe_2Al_9Si_4O_{22}(OH)_2$*	mon.	C1/m(15)		2	7.90 ±.10	16.65 ±.15	5.63 ±.10		90.00 ±.25		r
Topaz $Al_2(SiO_4)(OH)$*	orth.	Pmnb(62)		4	8.394 ±.005	8.792 ±.007	4.649 ±.003				26
Phenacite Be_2SiO_4*	hex-R	R$\bar{3}$(148)	phenacite	18	12.472 ±.005		8.252 ±.005				25
Willemite Zn_2SiO_4	hex-R	R$\bar{3}$(148)	phenacite	18	13.94 ±.01		9.309 ±.003				25
Dioptase CuH_2SiO_4*	hex-R	R$\bar{3}$(148)	phenacite	18	14.61 ±.02		7.80 ±.01				r
Larnite β-Ca_2SiO_4*	mon.	P2_1/n(14)		4	5.48 ±.02	6.76 ±.02	9.28 ±.02		94.55 ±.33		r

TABLE 11-7 (*Continued*)

Name and formula	Crystal system	Space group	Structure type	Z	a_0	b_0	c_0	α_0	β_0	γ_0	Temp. °C
Ortho and ring structure silicates—Continued					Ortho and ring structure silicates—Continued						
Akermanite Ca$_2$MgSi$_2$O$_7$	tet.	P$\bar{4}$2$_1$m(113)	melilite	2	7.8435 ±.0030		5.010 ±.003				r
Gehlenite Ca$_2$Al$_2$SiO$_7$	tet.	P$\bar{4}$2$_1$m(113)	melilite	2	7.690 ±.003		5.0675 ±.0030				r
Fe-Gehlenite Ca$_2$Fe$_2$SiO$_7$	tet.	P$\bar{4}$2$_1$m(113)	melilite	2	7.54 ±.01		4.855 ±.005				r
Hardystonite Ca$_2$ZnSi$_2$O$_7$*	tet.	P$\bar{4}$2$_1$m(113)	melilite	2	7.87 ±.03		5.01 ±.02				r
Sodium Melilite NaCaAlSi$_2$O$_7$	tet.	P$\bar{4}$2$_1$m(113)	melilite	2	8.511 ±.005		4.809 ±.003				r
Beryl Be$_3$Al$_2$(Si$_6$O$_{18}$)*	hex.	P6/mmc(192)	beryl	2	9.215 ±.005		9.192 ±.005				25
Indialite high Cordierite Mg$_2$Al$_3$(AlSi$_5$O$_{18}$)	hex.	P6/mmc(192)	beryl	2	9.7698 ±.0030		9.3517 ±.0030				25
Low Cordierite Mg$_2$Al$_3$(AlSi$_5$O$_{18}$)	orth.	Cccm(66)	cordierite	4	9.721 ±.003	17.062 ±.006	9.339 ±.003				25
Fe-Indialite Fe$_2$Al$_3$(AlSi$_5$O$_{18}$)	hex.	P6/mmc(192)	beryl	2	9.860 ±.010		9.285 ±.010				r
Fe-Cordierite Fe$_2$Al$_3$(AlSi$_5$O$_{18}$)	orth.	Cccm(66)	cordierite	4	9.726 ±.010	17.065 ±.010	9.287 ±.010				r
Mn-Indialite Mn$_2$Al$_3$(AlSi$_5$O$_{18}$)	hex.	P6/mmc(192)	beryl	2	9.925 ±.010		9.297 ±.010				r
Sapphirine Mg$_2$Al$_4$O$_6$SiO$_4$*	mon.	P2$_1$/c(14)		8	11.26 ±.03	14.46 ±.03	9.95 ±.02		125.33 ±.50		r
Elbaite NaLiAl$_{7.67}$B$_3$Si$_6$O$_{27}$(OH)$_4$*	hex-R	R3m(160)	tourmaline	3	15.842 ±.010		7.009 ±.010				r
Schorl NaFe$_3$Al$_6$B$_3$Si$_6$O$_{27}$(OH)$_4$*	hex-R	R3m(160)	tourmaline	3	16.032 ±.010		7.149 ±.010				r
Dravite NaMg$_3$Al$_6$B$_3$Si$_6$O$_{27}$(OH)$_4$	hex-R	R3m(160)	tourmaline	3	15.942 ±.010		7.224 ±.010				r
Uvite CaMg$_4$Al$_5$B$_3$Si$_6$O$_{27}$(OH)$_4$	hex-R	R3m(160)	tourmaline	3	15.86 ±.01		7.19 ±.01				r

Mineral / Formula	System	Space group	Z	a	b	c	α	β	γ	Type	Ref.
Sphene $CaTiSiO_5$*	mon.	A2/a(15)	4	7.07 ±.01	8.72 ±.01	6.56 ±.01		113.95 ±.25			r
Datolite $CaBSiO_4(OH)$*	mon.	P2₁/c(14)	4	9.62 ±.03	7.60 ±.03	4.84 ±.02		90.15 ±.25			r
Euclase $AlBeSiO_4(OH)$*	mon.	P2₁/a(14)	4	4.763 ±.005	14.29 ±.02	4.618 ±.005		100.25 ±.10			r
Chloritoid $H_2FeAl_2SiO_7$*	mon.	C2/c(15)	8	9.48 ±.01	5.48 ±.01	18.18 ±.01		101.77 ±.25			r
Hemimorphite $Zn_4(OH)_2Si_2O_7 \cdot H_2O$*	orth.	Imm2(35)	2	8.370 ±.005	10.719 ±.005	5.120 ±.005					25
Zoisite $Ca_2Al_3(SiO_4)_3OH$	orth.	Pnma(62)	4	16.15 ±.01	5.581 ±.005	10.06 ±.01					r
Clinozoisite $Ca_2Al_3(SiO_4)_3OH$	mon.	P2₁/m(11)	2	8.887 ±.007	5.581 ±.005	10.14 ±.01		115.93 ±.33			r
Epidote $Ca_2Al_{1.5}Fe_{1.5}(SiO_4)_3OH$*	mon.	P2₁/m(11)	2	8.89 ±.02	5.63 ±.01	10.19 ±.02		115.40 ±.30			r
Piemontite $Ca_2Al_{1.5}Mn_{1.5}(SiO_4)_3OH$*	mon.	P2₁m(11)	2	8.95 ±.02	5.70 ±.01	9.41 ±.02		115.70 ±.50			r
Lawsonite $CaAl_2Si_2O_7(OH)_2 \cdot H_2O$	orth.	Cccm(63)	4	8.787 ±.005	5.836 ±.005	13.123 ±.008					r

Chain and band structure silicates

Mineral / Formula	System	Space group	Z	a	b	c	α	β	γ	Type	Ref.
Enstatite $MgSiO_3$*	orth.	Pcab(61)	16	8.829 ±.010	18.22 ±.01	5.192 ±.005					r
Clinoenstatite $MgSiO_3$	mon.	P2₁/c(15)	8	9.620 ±.005	8.825 ±.005	5.188 ±.005		108.33 ±.17			r
Protoenstatite $MgSiO_3$	orth.	Pbcn(60)	8	9.25 ±.01	8.74 ±.01	5.32 ±.01					r
High Clinoenstatite $MgSiO_3$	tri.		8	10.000 ±.005	8.934 ±.004	5.170 ±.003	88.27 ±.05	70.03 ±.04	91.01 ±.04		r
Clinoferrosilite $FeSiO_3$	mon.	P2₁/c(15)	8	9.7085 ±.0010	9.0872 ±.001	5.2284 ±.004		108.43 ±.05			r
Orthoferrosilite $FeSiO_3$	orth.	Pcab(61)	16	9.080 ±.002	18.431 ±.004	5.238 ±.001				enstatite	r
Diopside $CaMg(SiO_3)_2$	mon.	C2/c(15)	4	9.743 ±.005	8.923 ±.005	5.251 ±.003		105.93 ±.25		diopside	r
Hedenbergite $CaFe(SiO_3)_2$*	mon.	C2/c(15)	4	9.854 ±.010	9.024 ±.010	5.263 ±.010		104.23 ±.33		diopside	r

TABLE 11-7 (Continued)

Name and formula	Crystal system	Space group	Structure type	Z	a_0	b_0	c_0	α_0	β_0	γ_0	Temp. °C
Chain and band structure silicates—Continued											
Johannsenite CaMn(SiO₃)₂ *	mon.	C2/c(15)	diopside	4	9.83 ± .03	9.04 ± .03	5.27 ± .02		105.00 ± .50		r
Ureyite NaCr(SiO₃)₂	mon.	C2/c(15)	diopside	4	9.550 ± .016	8.712 ± .007	5.273 ± .008		107.44 ± .16		r
Jadeite NaAl(SiO₃)₂ *	mon.	C2/c(15)	diopside	4	9.409 ± .005	8.564 ± .005	5.220 ± .005		107.50 ± .20		r
Acmite (Aegirine) NaFe(SiO₃)₂	mon.	C2/c(15)	diopside	4	9.658 ± .005	8.795 ± .005	5.294 ± .005		107.42 ± .20		r
CaTschermak Molecule CaAl₂SiO₆	mon.	C2/c(15)	diopside	4	9.615 ± .005	8.661 ± .005	5.272 ± .003		106.12 ± .20		r
Spodumene LiAl(SiO₃)₂	mon.	C2/c(15)	diopside	4	9.451 ± .002	8.387 ± .002	5.208 ± .001		110.07 ± .03		r
β-Spodumene LiAl(SiO₃)₂	tet.	P4₃2₁2(96) P4₁2₁2(92)		4	7.5332 ± .0008		9.1540 ± .0008				r
Pectolite Ca₂NaH(SiO₃)₃ *	tri.	P1̄(2)		2	7.99 ± .01	7.04 ± .01	7.02 ± .01	90.05 ± .25	95.27 ± .25	102.47 ± .25	r
Wollastonite CaSiO₃ *	tri.	P1̄(2)		6	7.94 ± .01	7.32 ± .01	7.07 ± .01	90.03 ± .25	95.37 ± .25	103.43 ± .25	r
Parawollastonite CaSiO₃ *	mon.	P2₁(4)		12	15.417 ± .004	7.321 ± .002	7.066 ± .003		95.40 ± .10		r
Pseudowollastonite CaSiO₃ *	tri.			24	6.90 ± .02	11.78 ± .02	19.65 ± .02	90.00 ± .30	90.80 ± .30	90.00 ± .30	r
Rhodonite MnSiO₃ *	tri.	P1̄(2)		10	7.682 ± .002	11.818 ± .003	6.707 ± .002	92.36 ± .05	93.95 ± .05	105.66 ± .05	r
Bustamite CaMn(SiO₃)₂ *	tri.	A1̄(2)		6	7.736 ± .003	7.157 ± .003	13.824 ± .010	90.52 ± .25	94.58 ± .25	103.87 ± .25	r
Pyroxmangite MnFe(SiO₃)₂ *	tri.	P1̄(2)		7	7.56 ± .02	17.45 ± .05	6.67 ± .02	84.00 ± .30	94.30 ± .30	113.70 ± .30	r
Tremolite Ca₂Mg₅[Si₈O₂₂](OH)₂ *	mon.	C2/m(12)	tremolite	2	9.840 ± .010	18.052 ± .020	5.275 ± .010		104.70 ± .25		r
Fluor-tremolite Ca₂Mg₅[Si₈O₂₂]F₂	mon.	C2/m(12)	tremolite	2	9.781 ± .007	18.01 ± .01	5.267 ± .005		104.52 ± .25		20

Mineral / Formula	System	Space group	Structure	Z	a	b	c	α	β	γ	Ref.
Ferrotremolite $Ca_2Fe_5[Si_8O_{22}](OH)_2$	mon.	C2/m(12)	tremolite	2	9.97 ± .01	18.34 ± .02	5.30 ± .01		104.50 ± .10		r
Grunerite $Fe_7[Si_8O_{22}](OH)_2$	mon.	C2/m(12)	tremolite	2	9.572 ± .005	18.44 ± .01	5.342 ± .007		101.77 ± .25		r
Cummingtonite (hypo.) $Mg_7[Si_8O_{22}](OH)_2$	mon.	C2/m(12)	tremolite	2	9.476 ± .010	17.935 ± .010	5.292 ± .005		102.23 ± .25		r
Riebeckite $Na_2Fe_3Fe_2[Si_8O_{22}](OH)_2$	mon.	C2/m(12)	tremolite	2	9.729 ± .020	18.065 ± .020	5.334 ± .010		103.31 ± .25		r
Magnesioriebeckite $Na_2Mg_3Fe_2[Si_8O_{22}](OH)_2$	mon.	C2/m(12)	tremolite	2	9.733 ± .010	17.946 ± .020	5.299 ± .010		103.30 ± .25		r
Gaucophane I $Na_2Mg_3Al_2[Si_8O_{22}](OH)_2$	mon.	C2/m(12)	tremolite	2	9.748 ± .010	17.915 ± .020	5.273 ± .010		102.78 ± .25		r
Glaucophane II $Na_2Mg_3Al_2[Si_8O_{22}](OH)_2$	mon.	C2/m(12)	tremolite	2	9.663 ± .010	17.696 ± .020	5.277 ± .010		103.67 ± .10		r
Fluor-edenite $NaCa_2Mg_5[AlSi_7O_{22}]F_2$	mon.	C2/m(12)	tremolite	2	9.847 ± .005	18.00 ± .01	5.282 ± .005		104.83 ± .25		r
Fluor-richterite $Na_2CaMg_5[Si_8O_{22}]F_2$	mon.	C2/m(12)	tremolite	2	9.823 ± .005	17.96 ± .01	5.268 ± .005		104.33 ± .25		r
Anthophyllite $Mg_7[Si_8O_{22}](OH)_2$	orth.	Pnma(62)		4	18.61 ± .02	18.01 ± .06	5.24 ± .01				r

Framework structure silicates

Mineral / Formula	System	Space group	Structure	Z	a	b	c	α	β	γ	Ref.
Microcline $KAlSi_3O_8$	tri.	C$\bar{1}$(2)		4	8.582 ± .002	12.964 ± .005	7.222 ± .002	90.62 ± .10	115.92 ± .10	87.68 ± .10	r
High Sanidine $KAlSi_3O_8$	mon.	C2/m(12)		4	8.615 ± .002	13.031 ± .003	7.177 ± .002		115.98 ± .10		r
Orthoclase $KAlSi_3O_8$*	mon.	C2/m(12)		4	8.562 ± .003	12.996 ± .004	7.193 ± .003		116.02 ± .15		r
Fe-Sanidine $KFeSi_3O_8$	mon.	C2/m(12)		4	8.689 ± .008	13.12 ± .01	7.319 ± .007		116.10 ± .30		r
Fe-Microcline $KFeSi_3O_8$	tri.	C$\bar{1}$(2)		4	8.68 ± .01	13.10 ± .01	7.340 ± .007	90.75 ± .25	116.05 ± .25	86.23 ± .25	r
Low Albite $NaAlSi_3O_8$	tri.	C$\bar{1}$(2)		4	8.139 ± .002	12.788 ± .003	7.160 ± .002	94.27 ± .10	116.57 ± .10	87.68 ± .10	26
High Albite (Analbite) $NaAlSi_3O_8$	tri.	C$\bar{1}$(2)		4	8.160 ± .002	12.870 ± .003	7.106 ± .002	93.54 ± .10	116.36 ± .10	90.19 ± .10	r
Anorthite $CaAl_2Si_2O_8$	tri.	P$\bar{1}$(2)	primitive cell	8	8.177 ± .002	12.877 ± .003	14.169 ± .003	93.17 ± .02	115.85 ± .02	91.22 ± .02	r

TABLE 11-7 (*Continued*)

Name and formula	Crystal system	Space group	Structure type	Z	a_0	b_0	c_0	α_0	β_0	γ_0	Temp. °C
Framework structure silicates—Continued											
Synthetic $CaAl_2Si_2O_8$	hex.	P6₃/mcm(193)		2	5.10 ± .02		14.72 ± .02				r
Synthetic $CaAl_2Si_2O_8$	orth.	P2₁2₁2(18)		2	8.22 ± .02	8.60 ± .02	4.83 ± .01				r
Celsian $BaAl_2Si_2O_8$ *	mon.	I2₁/c(15)		8	8.627 ± .010	13.045 ± .010	14.408 ± .020		115.20 ± .25		r
Paracelsian $BaAl_2Si_2O_8$ *	mon.	P2₁/a(14)		4	8.58 ± .02	9.583 ± .020	9.08 ± .02		90.00 ± .50		r
Banalsite $BaNa_2Al_4Si_4O_{16}$ *	orth.			4	8.50 ± .02	9.97 ± .02	16.72 ± .03				r
Danburite $CaB_2Si_2O_8$ *	orth.	Pnam(62)		4	8.04 ± .02	8.77 ± .02	7.74 ± .02				r
Low Nepheline $NaAlSiO_4$	hex.	C6₃(178)		8	9.986 ± .005		8.330 ± .004				r
High Carnegieite $NaAlSiO_4$	cubic			4	7.325 ± .004						750
Kaliophilite natural $KAlSiO_4$ *	hex.	P6₃22(182)		54	26.930 ± .010		8.522 ± .004				r
Kaliophilite synthetic $KAlSiO_4$	hex.	P6₃(173) P6₃22(182)		2	5.180 ± .002		8.559 ± .004				r
Kalsilite $KAlSiO_4$	hex.	P6₃(173)		2	5.1597 ± .0020		8.7032 ± .0030				r
Leucite $KAlSi_2O_6$	tet.	I4₁/a(100)		16	13.074 ± .003		13.738 ± .003				25
High Leucite $KAlSi_2O_6$ *	cubic	Ia3d(230)		16	13.43 ± .05						625
Fe-Leucite $KFeSi_2O_6$	tet.	I4₁/a(100)		16	13.205 ± .002		13.970 ± .003				25
Petalite $LiAlSi_4O_{10}$ *	mon.	P2₁/n(14)		2	11.32 ± .03	5.14 ± .01	7.62 ± .01		105.90 ± .20		r
Marialite $Na_4Al_3Si_9O_{24}Cl$	tet.	I4/m(87) P4/m(83)		2	12.064 ± .008		7.514 ± .004				r

Mineral / Formula	System	Space group	Polytype	Z	a	b	c	α	β	γ	ref
Meionite $Ca_4Al_6Si_6O_{24}CO_3$	tet.	I4/m(87) P4/m(83)		2	12.174 ±.008		7.652 ±.015				r

Sheet structure silicates

Mineral / Formula	System	Space group	Polytype	Z	a	b	c	α	β	γ	ref
Muscovite $KAl_2[AlSi_3O_{10}](OH)_2$*	mon.	C2/c(15)	2M₂ mica	4	5.203 ±.005	8.995 ±.005	20.030 ±.010		94.47 ±.33		r
Paragonite $NaAl_2[AlSi_3O_{10}](OH)_2$*	mon.	C2/c(15)	2M₁ mica	4	5.13 ±.03	8.89 ±.05	19.32 ±.10		95.17 ±.50		r
Lepidolite $K_2Al_3Li_2[AlSi_7O_{20}](OH)_4$*	mon.	C2/c(15)	2M₂ mica	2	9.2 ±.1	5.3 ±.1	20.0 ±.2		98.00 ±.50		r
Phlogopite $KMg_3[AlSi_3O_{10}](OH)_2$	mon.	Cm(8)	1M mica	2	5.326 ±.010	9.210 ±.010	10.311 ±.010		100.17 ±.10		r
Fluor-phlogopite $KMg_3[AlSi_3O_{10}]F_2$	mon.	Cm(8)	1M mica	2	5.299 ±.005	9.188 ±.005	10.135 ±.005		99.92 ±.10		r
Annite $KFe_3[AlSi_3O_{10}](OH)_2$	mon.	Cm(8)	1M mica	2	5.391 ±.010	9.350 ±.005	10.313 ±.020		99.70 ±.25		r
Ferriannite $KFe_3[FeSi_3O_{10}](OH)_2$	mon.	C2/m(12)		2	5.430 ±.002	9.404 ±.003	10.341 ±.006		100.07 ±.20		r
Margarite $CaAl_2[Al_2Si_2O_{10}](OH)_2$*	mon.	C2/c(15)	2M mica	4	5.13 ±.02	8.92 ±.03	19.50 ±.05		95.00 ±.50		r
Talc $Mg_3Si_4O_{10}(OH)_2$*	mon.	C2/c(15)	2M₁	4	5.287 ±.007	9.158 ±.010	18.95 ±.01		99.50 ±.20		r
Pyrophyllite $Al_2Si_4O_{10}(OH)_2$*	mon.	C2/c(15)	2M₁	4	5.14 ±.02	8.90 ±.02	18.55 ±.03		99.92 ±.20		r
Minnesotaite $Fe_3Si_4O_{10}(OH)_2$*	mon.	C2/c(15)		4	5.4 ±.1	9.42 ±.04	19.4 ±.1		100.00 ±.50		r
Dickite $Al_2Si_2O_5(OH)_4$*	mon.	Cc(9)		4	5.150 ±.002	8.940 ±.003	14.736 ±.005		103.58 ±.10		r
Kaolinite $Al_2Si_2O_5(OH)_4$*	tri.	P1(1)		2	5.155 ±.007	8.959 ±.010	7.407 ±.008	91.68 ±.35	104.87 ±.35	89.93 ±.35	r
Nacrite $Al_2Si_2O_5(OH)_4$*	mon.	Cc(9)		4	8.909 ±.010	5.146 ±.010	15.697 ±.020		113.70 ±.25		r

Zeolites

Mineral / Formula	System	Space group	Polytype	Z	a	b	c	α	β	γ	ref
Analcite $NaAlSi_2O_6 \cdot H_2O$	cubic	Ia3d(230)		16	13.733 ±.005						r
Natrolite $Na_2Al_2Si_3O_{10} \cdot 2H_2O$*	orth.	Fdd2(43)		8	18.30 ±.02	18.63 ±.02	6.60 ±.01				r

TABLE 11-8. PROPERTIES OF ARSENIDES, SELENIDES, SULFIDES AND TELLURIDES (ROBIE, 1967, 1968), (DIETRICH, 1969)[#]

Name and Formula	Gram Formula Weight	Entropy S^0 (cal deg. gfw^{-1})	Molar Volume (cm^3)	Heat of Formation ΔH^0 (cal gfw^{-1})	Free Energy of Formation ΔG^0 (cal gfw^{-1})	Crystal System	Cell Volume (10^{-24} cm^3)	X-ray Density (g cm^{-3})	Refractive Index, n_D	Color
Acanthite Ag_2S III	247.804	34.14	34.19	−7731	−9562	mon.	115.5	7.125		M-bk
Realgar AsS	106.986	15.18	29.80	−17050	−16806	mon.	791.6	3.591	2.590β (Li)	N-r, y, o
Orpiment As_2S_3	246.035	39.1	70.51	−40400	−40250	mon.	468.4	3.490	2.72β (Li)	M-y, b
Bismuthinite Bi_2S_3	514.152	47.9	75.52	−33900	−33298	orth.	501.59	6.8081		M-y, g, w
Oldhamite CaS	72.144	13.54	27.722	−114265	−113070	cubic	184.12	2.602	2.137	M-b
Greenockite CdS	144.464	16.80	29.934	−35755	−34807	hex.	99.407	4.8261	2.506	N-o, y
Covellite CuS	95.604	15.93	20.42	−11610	−11720	hex.	203.48	4.682	1.45	r, y, be
Chalcocite Cu_2S III	159.144	28.86	27.475	−19148	−20734	ort.	4379.5	5.7924		M-g, bk
Troilite FeS	87.911	14.42	18.20	−24130	−24219	hex.	60.439	4.830		b
Pyrrhotite $Fe_{0.877}S$	81.042	14.53				hex.		4.58–4.79		M-y, b
Pyrite FeS_2	119.975	12.65	23.940	−41000	−38296	cubic	159.00	5.0116		M-y
Ferroselite $FeSe_2$	213.767	20.76	29.96			orth.	99.50	7.134		
Frohbergite $FeTe_2$	311.047	23.94	38.43			orth.	127.62	8.094		
Cinnabar HgS (red)	232.654	19.72	28.416	−13900	−12096	hex.	141.55	8.187	2.854	r
Metacinnabar HgS (black)	232.654	23.0	30.169	−11170	−10344	cubic	200.38	7.712		M-gbk

Alabandite MnS	87.002	18.69	21.46	−51115	−52140	cubic	142.51	4.0546	2.700 (Li)	bk, b
Molybdenite MoS$_2$	160.068	14.96	32.02	−73200	−71086	hex.	106.35	4.9982	4.7	M-g
Millerite NiS	90.774	15.80	16.89	−20284	−20600	hex-R	252.41	5.3743		M-y
Galena PbS	239.254	21.84	31.49	−23353	−22962	cubic	209.16	7.5973	3.912	M-g
Clausthalite PbSe	286.15	24.48	34.61	−24662	−24300	cubic	229.84	8.2690		M-g, be
Altaite PbTe	334.79	26.26	40.60	−16949	−16600	cubic	269.66	8.2459		M-y, w
Cooperite PtS	227.154	13.16	22.15	−19700	−18391	tet.	73.56	10.254		M-g
Stibnite Sb$_2$S$_3$	339.692	43.5	73.41	−41800	−41460	orth.	487.54	4.6276	4.303β	M-g
Herzenbergite SnS	150.754	18.36	29.01	−25464	−24999	orth.	192.66	5.197		b
Tungstenite WS$_2$	247.978	22.7	32.07	−71300	−71210	hex.	106.50	7.7325		M-g
Sphalerite ZnS	97.434	13.77	23.83	−49750	−48623	cubic	158.28	4.0885	2.37-2.47	M-b, bk, y, gn, w
Wurtzite ZnS	97.434	16.56	23.846	−46095	−45760	hex.	79.19	4.0859	2.356	N-bk, b

#Some information also drawn from (Forsythe, 1969), (Kerr, 1959), (Smith, 1953), (Wahlstrom, 1955), (Winchell, 1951, 1939).

TABLE 11-9. PROPERTIES OF CARBONATES (ROBIE 1967, 1968), (DIETRICH 1969)[#]

Name and Formula	Gram Formula Weight	Entropy S^0 (cal deg^{-1} gfw^{-1})	Molar Volume (cm^3)	Heat of Formn. ΔH^0 (cal gfw^{-1})	Free Energy of Formn. ΔG^0 (cal gfw^{-1})	Crystal System	Cell Volume (10^{-24} cm^3)	X-ray Density (g cm^{-3})	Refractive Index, n_D	Color
Witherite $BaCO_3$	197.349	26.8	45.81	-297460	-278359	orth.	304.24	4.308	1.676	N-y, gn, c, w, g, b
Aragonite $CaCO_3$	100.089	21.18	34.15	-288651	-269678	orth.	226.85	2.930	1.682	N-r, y, gn, be, p, w, g
Calcite $CaCO_3$	100.089	22.15	36.934	-288592	-269908	hex-R	367.96	2.7100	1.658	N-r, o, y, gn, be, p, c, w, g, bk, b
Dolomite $CaMg(CO_3)_2$	184.411	37.09	64.34	-557613	-518734	hex-R	320.50	2.8661	1.681	N-r, y, gn, c, w, g, b
Otavite $CdCO_3$	172.409	23.3	34.300	-179030	-159964	hex-R	341.72	5.0265		w, r
Malachite $Cu_2(OH)_2CO_3$	221.104	—	54.86	—	-216440	mon.	364.35	4.030	1.875β	N-g, bk
Azurite $Cu_3(OH)_2(CO_3)_2$	344.653	—	91.01	—	-343730	mon.	302.22	3.787	1.758	N, be
Siderite $FeCO_3$	115.856	25.1	29.378	-177812	-161030	hex-R	292.68	3.9436	1.875	N-r, y, gn, w, g, b
Magnesite $MgCO_3$	84.321	15.7	28.018	-266081	-246112	hex-R	279.13	3.0095	1.700	N, y, c, w, g, b
Huntite $Mg_3Ca(CO_3)_4$	353.053	67.0	122.58	-1086960	-1007700	hex-R	610.63	2.880		
Rhodochrosite $MnCO_3$	114.947	23.90	31.073	-212521	-195045	hex-R	309.57	3.6992	1.818	N-r, g, b
Cerussite $PbCO_3$	267.199	31.3	40.59	-167951	-150325	orth.	269.61	6.582	2.076	N-gn, be, c, w, g, bk
Strontianite $SrCO_3$	147.629	23.2	39.01	-294581	-275450	orth.	259.07	3.785	1.667	N-r, y, gn, c, w, g, b
Smithsonite $ZnCO_3$	125.379	19.70	28.275	-194200	-174786	hex-R	281.69	4.4343	1.818	N-y, gn, be, c, w, g, b

[#]Some information also drawn from (Forsythe, 1969), (Kerr, 1959), (Smith, 1953), (Wahlstrom, 1955), (Winchell, 1951, 1939).

TABLE 11-10. PROPERTIES OF HALIDES (ROBIE 1967, 1968), (DIETRICH 1969)[#]

Name and Formula	Gram Formula Weight	Entropy S^0 (cal deg gfw^{-1})	Molar Volume (cm^3)	Heat of Formn. ΔH^0 (cal gfw^{-1})	Free Energy of Formn. ΔG^0 (cal gfw^{-1})	Crystal System	Cell Volume (10^{-24} cm^3)	X-ray Density (g cm^{-3})	Refractive Index, n_D	Color
Bromargyrite $AgBr$	187.779	25.60	28.991	−23990	−23158	cubic	192.55	6.4772	2.253	N-y, gn, c, g
Chlorargyrite $AgCl$	143.323	23.00	25.727	−30370	−25242	cubic	170.87	5.5710	2.061	N-y, p, c, g, b
Hydrophilite $CaCl_2$	110.986	27.2	50.75	−190000	−179255	orth.		2.2	1.605	w, p
Lawrencite $FeCl_2$	126.753	28.19	39.46	−81700	−72273	hex-R	196.55	3.212	1.57	N-g, w, b
Molysite $FeCl_2$	162.206	34.02	57.86	−95460	−79827	hex.		~2.9		N-r, y, g, p, b
Calomel $HgCl$	236.043	23.08	32.939	−31695	−25215	tet.	218.77	7.166	1.973	N-y, c, w, b, g
Sylvite KCl	74.555	19.73	37.524	−104370	−97693	cubic	249.23	1.9868	1.490	N-r, y, be, c, w, g
Chloromagnesite $MgCl_2$	95.218	21.42	40.81	−153350	−141521	hex-R	203.29	2.333	1.675	c
Scacchite $MnCl_2$	125.844	28.26	42.11	−115038	−105295	hex-R	209.79	2.988		N-r, b, c
Salammoniac NH_4Cl	53.492	22.7	35.06	−75180	−48572	cubic		~1.528	1.639	N-y, c, w, g, b
Halite $NaCl$	58.443	17.24	27.015	−98260	−91807	cubic	179.43	2.1634	1.5444	N-r, y, be, p, c, w, g
Cotunnite $PbCl_2$	278.096	32.5	47.09	−86200	−75366	orth.	312.74	5.906	2.217β	N-y, gn, c, w
Fluorite CaF_2	78.077	16.46	24.542	−290300	−277799	cubic	163.11	3.1792	1.434	N-r, y, gn, be, p, c, w, g, bk, b
Sellaite MgF_2	62.309	13.68	19.61	−268700	−256008	tet.	65.13	3.177	1.378	N-c, 2
Villiaumite NaF	41.988	12.26	14.984	−137027	−129812	cubic	99.523	2.8021	1.328	N-r, c, w

TABLE 11-10 (Continued)

Name and Formula	Gram Formula Weight	Entropy S^0 (cal deg-gfw^{-1})	Molar Volume (cm^3)	Heat of Formn. ΔH^0 (cal gfw^{-1})	Free Energy of Formn. ΔG^0 (cal gfw^{-1})	Crystal System	Cell Volume (10^{-24} cm^3)	X-ray Density (g cm^{-3})	Refractive Index, n_D	Color
Cryolite Na$_3$AlF$_6$	209.941	56.98	70.81	-790000	-750695	mon.	235.1	2.965	1.339	N-r, c, w, b, bk
Iodargyrite AgI	234.774	27.60	41.301	-14780	-15830	hex.	137.18	5.683	2.210	N-y, gn, c, b, g
Coccinite HgI$_2$	454.399	42.4	71.84	-25200	-24148					

#Some information also drawn from (Forsythe, 1969), (Kerr, 1959), (Smith, 1953), (Wahlstrom, 1955), (Winchell, 1951, 1939).

TABLE 11-11. PROPERTIES OF NITRATES (ROBIE 1967, 1968), (DIETRICH 1969)#

Name and Formula	Gram Formula Weight	Entropy S^0 (cal deg-gfw^{-1})	Molar Volume (cm^3)	Heat of Formation ΔH^0 (cal gfw^{-1})	Free Energy of Formation ΔG^0 (cal gfw^{-1})	Crystal System	Cell Volume (10^{-24} cm^3)	X-ray Density (g cm^{-3})	Refractive Index, n_D	Color
Nitrobarite BaNO$_3$	261.350	51.14	80.58	-237060	-190066	cubic		3.25	1.572	c, w
Niter KNO$_3$	101.107	31.81	48.04	-117760	-93893	orth.	319.07	2.105	1.505β	N-c, w, g
Ammonia-Niter NH$_4$NO$_3$	80.043	36.11	46.49	-87373	-43971	rhomb.		~1.73		c, w
Soda Niter NaNO$_3$	84.995	27.85	37.60	-111540	-87459	hex-R	374.57	2.2606	1.586	N-r, y, c,

#Some information also drawn from (Forsythe, 1969), (Kerr, 1959), (Smith, 1953), (Wahlstrom, 1955), (Winchell, 1951, 1939).

TABLE 11-12. PROPERTIES OF OXIDES AND HYDROXIDES (ROBIE 1967, 1968), (DIETRICH 1969)#

Name and Formula	Gram Formula Weight	Entropy S^0 (cal deg⁻¹ gfw⁻¹)	Molar Volume (cm³)	Heat of Formation ΔH^0 (cal gfw⁻¹)	Free Energy of Formation ΔG^0 (cal gfw⁻¹)	Crystal System	Cell Volume (10⁻²⁴ cm³)	X-ray Density (g cm⁻³)	Refractive Index, n_D	Color
Corundum Al₂O₃	101.961	12.18	25.575	−400400	−378082	hex-R	254.78	3.9869	1.769	N-r, y, gn, be, p, c
Boehmite AlO(OH)	59.988	11.58	19.535	−235500	−217674	orth.	129.75	3.071	1.645β	N-w, b
Diaspore AlO(OH)	59.988	8.43	17.760			orth.	117.96	3.378	1.722β	N-r, y, gn, be, p, c, w, g, b
Gibbsite Al(OH)₃	78.004	16.75	31.956	−306380	−273486	mon.	424.49	2.441	1.554–1.567β	N-r, y, gn, w, g,
Arsenolite As₂O₃	197.841	25.6	51.118	−157020	−137731	cubic	1358.0	3.870	1.754	N-r, y, be, w
Claudetite As₂O₃	197.841	28.0	47.26	−156483	−137910	mon.	313.88	4.1863	1.920β	N-c, w
Boric Oxide B₂O₃	69.620	12.90	27.22	−303640	−284729					
Bromellite BeO	25.012	3.37	8.309	−143100	−136121	hex.	27.59	3.0104	1.719	N-w
Bismite α-Bi₂O₃	465.958	36.2	49.73	−137160	−117955	mon.	660.53	9.371	2.42	N-y, gn, g
Lime CaO	56.079	9.5	16.764	−151790	−144352	cubic	111.34	3.3453	1.838	N-w, g
Portlandite Ca(OH)₂	74.095	19.93	33.056	−235610	−214673	hex.	54.89	2.2315	1.575	N-c
Monteponite CdO	128.399	13.1	15.585	−61200	−54111	cubic	103.51	8.2386		N-bk
Cerianite CeO₂	172.119	14.89	23.853	−260180	−245450	cubic	158.43	7.216		
Cobalt oxide CoO	74.933	12.66	11.64	−57100	−51430	cubic	77.31	6.438		
Eskolaite Cr₂O₃	151.990	19.4	29.090	−272700	−253203	hex-R	289.82	5.225		
Tenorite CuO	79.539	10.19	12.22	−37140	−30498	mon.	81.16	6.509	2.84	M-g, bk

TABLE 11-12 (*Continued*)

Name and Formula	Gram Formula Weight	Entropy S^0 (cal deg⁻¹ gfw⁻¹)	Molar Volume (cm³)	Heat of Formation ΔH^0 (cal gfw⁻¹)	Free Energy of Formation ΔG^0 (cal gfw⁻¹)	Crystal System	Cell Volume (10^{-24} cm³)	X-ray Density (g cm⁻³)	Refractive Index, n_D	Color
Cuprite Cu_2O	143.079	22.4	23.437	−40400	−35022	cubic	77.83	6.1047	2.849	r, bk
Wustite $Fe_{0.947}O$	68.887	13.76	12.04	−63640	−58599	cubic			2.32	
Hematite Fe_2O_3	159.692	20.89	30.274	−197300	−177728	hex-R	301.61	5.2749	3.220	r, g
Magnetite Fe_3O_4	231.539	36.03	44.524	−267400	−243094	cubic	591.43	5.2003	2.42	bk, b
Goethite α-FeO(OH)	88.854		20.82	−133750		orth.	138.2	4.269	2.350β	r, y, bk, b
Germanium Dioxide GeO_2 (Quartz form)	104.589	13.21	24.44	−129080	−116195	hex.	121.73	4.2797		
Hafnia HfO_2	210.489	14.18	20.823	−266050	−252566	mon.	138.30	10.108		
Montroydite HgO	216.589	16.80	19.32	−21711	−13998	orth.	128.3	11.21	2.500β	N-r, b
Periclase MgO	40.311	6.44	11.248	−143800	−136087	cubic	74.709	3.5837	1.736	N-c, g, w
Brucite $Mg(OH)_2$	58.327	15.09	24.63	−221200	−199460	hex.	40.90	2.368	1.566	N-gn, be, g
Manganosite MnO	70.937	14.27	13.221	−92050	−86720	cubic	87.813	5.3653	2.160	N-gn, bk
Pyrolusite MnO_2	86.937	12.68	16.61	−124450	−111342	tet.	55.16	5.234		M-be, bk, g
Bixbyite Mn_2O_3	157.874	26.40	31.37	−228700	−210097	cubic	833.5	5.032		bk
Hausmanite Mn_3O_4	228.812	36.8	46.95	−331400	−306313	tet.	623.68	4.873	2.46	bk, b
Molybdite MoO_3	143.938	18.58	30.56	−178160	−159745	orth.	202.98	4.710		
Bunsenite NiO	74.709	9.08	10.97	−57300	−50574	cubic	72.88	6.809	2.18 (Li)	N-gn

Litharge PbO (red)	223.189	15.6	23.91	-52410	-45121	tet.	79.40	9.334	2.65 (Li)	N-r
Massicot PbO (yellow)	223.189	16.1	23.15	-52070	-44930	orth.	153.8	9.641	2.610	N-r, y
Minium Pb$_3$O$_4$	685.568	50.5	76.81	-171700	-143632	tet.	510.13	8.926	2.40 (Li)	N-r, y, b
Valentinite Sb$_2$O$_3$	291.498	29.4	50.01	-169350	-149692	orth.	332.13	5.8292	2.350β	N-r, y, c, w, g, b
α-quartz SiO$_2$	60.085	9.88	22.688	-217650	-204646	hex.	113.01	2.6483	1.544	N-r, o, y, gn, be, p, c, w, g, bk
α-Cristobalite SiO$_2$	60.085	10.38	25.739	-216930	-204075	tet.	170.95	2.3344	1.484	N-w
α-Tridymite SiO$_2$	60.085	10.50	26.53	-216895	-204076	orth.		~2.26	1.469β	N-c, w
Coesite SiO$_2$	60.085	9.65	20.641	-216440	-203367	mon.	548.37	2.9110	1.599	N-c, w
Stishovite SiO$_2$	60.085	6.64	14.014	-205860	-191890	tet.	46.54	4.2874		
Cassiterite SnO$_2$	150.689	12.5	21.55	-138820	-124266	tet.	71.57	6.992	1.996	r, y, c, w, g, bk, b
Tellurite TeO$_2$	159.599	16.8	27.75	-77740	-64600	orth.	368.61	5.7514	2.18 (Li)	N-y, w
Thorianite ThO$_2$	264.037	15.59	26.373	-293200	-279436	cubic	175.16	10.012	2.20	g, bk
Rutile TiO$_2$	79.899	12.04	18.820	-225760	-212559	tet.	62.50	4.2453	2.616	r, y, gn, be, p, bk, b
Anatase TiO$_2$	79.899	11.93	20.52	-225860	-212626	tet.	136.30	3.893	2.554	r, y, gn, be, p, bk, b
Uraninite UO$_2$	270.029	18.63	24.618	-259200	-246569	cubic	163.51	10.969		bk, b
Karelianite V$_2$O$_3$	149.882	23.53	29.85	-291290	-272273	hex-R	297.36	5.0216		
Zincite ZnO	81.369	10.43	14.338	-83250	-76089	hex.	47.615	5.6750	2.013	N-r, o, y
Baddeleyite ZrO$_2$	123.219	12.04	21.15	-262300	-248505	mon.	140.46	5.8267	2.190	N-r, y, gn, c, bk, b

#Some information also drawn from (Forsythe, 1969), (Kerr, 1959), (Smith, 1953), (Wahlstrom, 1955), (Winchell, 1951, 1939).

TABLE 11-13. PROPERTIES OF MULTIPLE OXIDES (ROBIE 1967, 1968), (DIETRICH 1969)[#]

Name and Formula	Gram Formula Weight	Entropy S^0 (cal deg⁻¹ gfw⁻¹)	Molar Volume (cm³)	Heat of Formation ΔH^0 (cal gfw⁻¹)	Free Energy of Formation ΔG^0 (cal gfw⁻¹)	Crystal System	Cell Volume (10^{-24} cm³)	X-ray Density (g cm⁻³)	Refractive Index, n_D	Color
Chrysoberyl $BeAl_2O_4$	126.973	15.58	34.320	–	–	orth.	227.94	3.6997	1.748β	N-r, y, gn
Perovskite $CaTiO_3$	135.978	22.4	33.626	-396900	-376517	orth.	223.33	4.0439	2.380	y, g, b, bk
Hercynite $FeAl_2O_4$	173.808	25.4	40.75	–	–	cubic	541.3	4.265	1.800	N-bk
Chromite $FeCr_2O_4$	223.837	34.90	44.01	–	–	cubic		~4.5	2.070	M-bk
Ilmenite $FeTiO_3$	151.745	25.3	31.69	-295560	-277065	hex-R	315.73	4.788		g, bk
Spinel $MgAl_2O_4$	142.273	19.26	39.71	-552800	-522961	cubic	527.5	3.583	1.720	N-r, y, gn, be, c, bk, b
Picrochromite $MgCr_2O_4$	192.302	25.3	43.56	–	–	cubic	578.6	4.415	2.00	bk, b
Magnesioferrite $MgFe_2O_4$	200.004	29.60	44.57	-341720	-315113	cubic		~4.5	2.35 (Li)	bk, b
Geikielite $MgTiO_3$	120.210	17.82	30.86	-375900	-354790	hex-R	307.44	3.896	2.310	bk, b
Trevorite $NiFe_2O_4$	234.402	31.5	43.65	–	–	cubic	579.9	5.370	2.3	bk, b

[#]Some information also drawn from (Forsythe, 1969). (Kerr, 1959), (Smith, 1953), (Wahlstrom, 1955), (Winchell, 1951, 1939).

TABLE 11-14. PROPERTIES OF PHOSPHATES, MOLYBDATES, AND TUNGSTATES (ROBIE 1967, 1968), (DIETRICH 1969)#

Name and Formula	Gram Formula Weight	Entropy S^0 (cal deg⁻¹ gfw⁻¹)	Molar Volume (cm³)	Heat of Formation ΔH^0 (cal gfw⁻¹)	Free Energy of Formation ΔG^0 (cal gfw⁻¹)	Crystal System	Cell Volume (10^{-24} cm³)	X-ray Density (g cm⁻³)	Refractive Index, n_D	Color
Berlinite $AlPO_4$	121.953	21.70	46.58	−414400	−388010	hex.	232.03	2.618	1.529	N-c, r, g
Whitlockite $Ca_3(PO_4)_2$	310.183	57.58	97.62	−986200	−932785	hex.		~3.12	1.629	N-c, y, g, w
Hydroxylapatite $Ca_5(PO_4)_3OH$	502.322	93.30	159.6	−3215000	−3123504	hex.	528.7	3.155	1.651	N-c, w, g, b, be, r, gn, p
Fluoroapatite $Ca_5(PO_4)_3F$	504.313	92.70	157.56	—	—	hex.	523.25	3.2007	1.634	N-r, gn, be, p, c, w, g, b
Strengite $Fe(PO_4) \cdot 2H_2O$	186.849	40.93	64.54	−451500	−397700	orth.		~2.87	1.708β	N-r, p, c
Powellite $CaMoO_4$	200.018	29.3	47.00	−369500	−344011	tet.	312.17	4.256	1.974	N-y, gn, be, g, w, b, bk
Wulfenite $PbMoO_4$	367.128	39.7	53.86	—	—	tet.	357.72	6.816	2.402 (Li)	N-o, r, y, gn, g, b
Scheelite $CaWO_4$	287.928	30.2	47.05	−402410	−376906	tet.	312.49	6.120	1.918	N-r, o, y, gn, c, w, g, b
Ferberite $FeWO_4$	303.695	31.5	40.38	—	—	mon.	134.11	7.520	2.40β (Li)	M-bk
Stolzite $PbWO_4$	455.038	40.2	54.10	—	—	tet.	359.32	8.4110	2.269	N-r, y, gn, b, g

#Some information also drawn from (Forsythe, 1969), (Kerr, 1959), (Smith, 1953), (Wahlstrom, 1955), (Winchell, 1951, 1939).

TABLE 11-15. PROPERTIES OF CHAIN AND BAND STRUCTURE SILICATES (ROBIE 1967, 1968), (DIETRICH 1969)[#]

Name and Formula	Gram Formula Weight	Entropy S^0 (cal deg⁻¹ gfw⁻¹)	Molar Volume (cm³)	Heat of Formation ΔH^0 (cal gfw⁻¹)	Free Energy of Formation ΔG^0 (cal gfw⁻¹)	Crystal System	Cell Volume (10^{-24} cm³)	X-ray Density (g cm⁻³)	Refractive Index, n_D	Color
Wollastonite $CaSiO_3$	116.164	19.60	39.93	−390640	−370313	tri.	397.82	2.909	1.632β	N-gn, c, w, g
Pseudowollastonite $CaSiO_3$	116.164	20.90	40.08	−389070	−369031	tri.	1597.0	2.899	1.611β	N-c
Diopside $CaMg(SiO_3)_2$	216.560	34.20	66.09	−767390	−725784	mon.	438.97	3.277	1.671β	N-gn, c, w
Spodumene $LiAl(SiO_3)_2$	186.090	—	58.37	−727735	—	mon.	387.7	3.188	1.666	N-y, gn, c, p, g, w
Clinoenstatite $MgSiO_3$	100.396	16.22	31.47	−370140	−349394	mon.	418.10	3.190	1.654	N-y, gn, g, c, bk, b
Rhodonite $MnSiO_3$	131.022	24.50	35.16	−315620	−297390	tri.	583.77	3.727	1.726β	N-r, b
Jadeite $NaAl(SiO_3)_2$	202.140	31.90	60.40	−719871	−677206	mon.	401.15	3.347	1.659–1.674β	N-gn, be, c, w
Tremolite $Ca_2Mg_5[Si_8O_{22}](OH)_2$	812.410	131.19	272.92	−2952935	−2779137	mon.	906.34	2.977	1.613–1.644β	N-c, g

[#]Some information also drawn from (Forsythe, 1969), (Kerr, 1959), (Smith, 1953), (Wahlstrom, 1955), (Winchell, 1951, 1939).

TABLE 11-16. PROPERTIES OF FRAMEWORK STRUCTURE SILICATES (ROBIE 1967, 1968), (DIETRICH 1969)[#]

Name and Formula	Gram Formula Weight	Entropy S^0 (cal deg⁻¹ gfw⁻¹)	Molar Volume (cm³)	Heat of Formation ΔH^0 (cal gfw⁻¹)	Free Energy of Formation ΔG^0 (cal gfw⁻¹)	Crystal System	Cell Volume (10^{-24} cm³)	X-ray Density (g cm⁻³)	Refractive Index, nD	Color
Anorthite $CaAl_2Si_2O_8$	278.210	48.45	100.79	-1009300	-955626	tri.	1338.9	2.760	1.584β	N-c, w, g
Hexagonal Anorthite $CaAl_2Si_2O_8$	278.210	45.84	99.85	-1004410	-949958	hex.	331.57	2.786		
Leonhardite $Ca_2Al_4Si_8O_{24} \cdot 7H_2O$	922.867	220.4	404.4	-3397535	-3146948	mon.			1.524β	N-r, y, c, b, w
Microcline $KAlSi_3O_8$	278.337	52.47	108.72	-946265	-892817	tri.	722.06	2.560	1.526β	N-r, o, y, gn, c, w, b, g
High Sanidine $KAlSi_3O_8$	278.337	56.94	109.04	-944378	-892263	mon.	724.28	2.552	1.525β	N-w, c, g, b
Adularia $KAlSi_3O_8$	278.337	55.99	108.29	-94500	-892602	mon.	719.25	2.570	1.523	N-c, w
Kaliophilite $KAlSiO_4$	158.167	31.85	59.89	-503926	-476230	hex.	5352.4	2.650	1.532	N-c
Leucite $KAlSi_2O_6$	218.252	44.05	88.39	-721650	-681642	tet.	2348.23	2.469	1.509	N-w, g
Low Albite $NaAlSi_3O_8$	262.224	50.20	100.07	-937146	-883988	tri.	664.65	2.620	1.529β	N-be, c, g, w
High Albite (Analbite) $NaAlSi_3O_8$	262.224	54.67	100.43	-934513	-882687	tri.	667.00	2.611		
Nepheline $NaAlSiO_4$	145.055	29.72	54.16	-497029	-469664	hex.	719.38	2.623	1.542	N-c, w, g
Analcime $NaAlSi_2O_6 \cdot H_2O$	220.155	56.03	97.49	-786341	-734262	cubic		~2.25	1.487	N-r, w, g

[#]Some information also drawn from (Forsythe, 1969), (Kerr, 1959), (Smith, 1953), (Wahlstrom, 1955), (Winchell, 1951, 1939).

TABLE 11-17. PROPERTIES OF ORTHO- AND RING STRUCTURE SILICATES (ROBIE 1967, 1968), (DIETRICH 1969)#

Name and Formula	Gram Formula Weight	Entropy S^0 (cal deg^{-1} gfw^{-1})	Molar Volume (cm^3)	Heat of Formation ΔH^0 (cal gfw^{-1})	Free Energy of Formation ΔG^0 (cal gfw^{-1})	Crystal System	Cell Volume (10^{-24} cm^3)	X-ray Density (g cm^{-3})	Refractive Index, n_D	Color
Kyanite Al_2SiO_5	162.046	20.02	44.09	-619930	-584000	tri.	292.83	3.675	1.720β	N-g, be, g, w
Andalusite Al_2SiO_5	162.046	22.28	51.53	-619390	-584134	orth.	342.25	3.145	1.638β	N-r, p, w, g
Sillimanite Al_2SiO_5	162.046	22.97	49.90	-618650	-583600	orth.	331.42	3.248	1.642β	N-y, c, b, w
3.2 Mullite $3Al_2O_3 \cdot 2SiO_2$	426.053	64.43	134.55	-1629543	-1539002	orth.	167.56	3.166	1.644β	N-r, c, w, g
Larnite β-Ca_2SiO_4	172.244	30.50	51.60	-551420	-524022	mon.	342.7	3.338	1.715	w, g
Calcium Olivine γ-Ca_2SiO_4	172.244	28.80	59.11	-553973	-526069	orth.	392.61	2.914		
Gehlenite $Ca_2Al_2SiO_7$	274.205	47.4	90.24	-952740	-904432	tet.	299.67	3.0387	1.669	N-g, c, g, b
Grossularite $Ca_3Al_2Si_3O_{12}$	450.454	57.7	125.30	-1588393	-1500986	cubic	1664.43	3.595	1.736	N-y, g, w, b
Lawsonite $CaAl_2Si_2O_7(OH)_2 \cdot H_2O$	314.241	56.79	101.32	-1161315	-1076910	orth.	672.96	3.101	1.674β	N-be, c, w
Monticellite $CaMgSiO_4$	156.476	24.5	51.36	-540800	-512252	orth.	341.13	3.046	1.662β	N-c, g
Merwinite $Ca_3Mg(SiO_4)_2$	328.719	60.5	104.4	-1091490	-1037184	mon.		~3.15	1.711	N-gn, c
Akermanite $Ca_2MgSi_2O_7$	272.640	50.03	92.81	-926510	-879353	tet.	308.22	2.9375	1.633	N-gn, c, g, b
Sphene $CaTiSiO_5$	196.063	30.88	55.65	-622050	-588246	mon.	369.61	3.523	1.894-1.921β	N-y, gn, b, c
Fayalite Fe_2SiO_4	203.778	35.45	46.39	-353544	-329668	orth.	308.11	4.3928	1.838-1.877β	N-y, gn
Forsterite Mg_2SiO_4	140.708	22.75	43.79	-520370	-491938	orth.	290.81	3.2136	1.651-1.660β	N-y, gn

Name and Formula	Gram Formula Weight	Entropy S^0 (cal deg⁻¹ gfw⁻¹)	Molar Volume (cm³)	Heat of Formation ΔH^0 (cal gfw⁻¹)	Free Energy of Formation ΔG^0 (cal gfw⁻¹)	Crystal System	Cell Volume (10^{-24} cm³)	X-ray Density (g cm⁻³)	Refractive Index, n_D	Color
Cordierite $Mg_2Al_3(AlSi_5O_{18})$	584.969	97.33	233.22	—	—	orth.	1548.96	2.508	1.538β	N-be, g
Tephroite Mn_2SiO_4	201.960	39.00	48.61	-413520	-390028	orth.	322.87	4.1545	1.786β	N-be, g
Willemite Zn_2SiO_4	222.824	31.40	52.42	-391140	-364011	hex-R	1566.6	4.251	1.691	N-r, y, gn, w
Zircon $ZrSiO_4$	183.304	20.08	39.26	—	—	tet.	260.76	4.669	1.923	N-r, o, y, gn, g, b, c

#Some information also drawn from (Forsythe, 1969), (Kerr, 1959), (Smith, 1953), (Wahlstrom, 1955), (Winchell, 1951, 1939).

TABLE 11-18. PROPERTIES OF SHEET STRUCTURE SILICATES (ROBIE 1967, 1968), (DIETRICH 1969)#

Name and Formula	Gram Formula Weight	Entropy S^0 (cal deg⁻¹ gfw⁻¹)	Molar Volume (cm³)	Heat of Formation ΔH^0 (cal gfw⁻¹)	Free Energy of Formation ΔG^0 (cal gfw⁻¹)	Crystal System	Cell Volume (10^{-24} cm³)	X-ray Density (g cm⁻³)	Refractive Index, n_D	Color
Dickite $Al_2Si_2O_5(OH)_4$	258.161	47.10	99.30	-979165	-902142	mon.	659.49	2.600	1.562β	N-y, w
Kaolinite $Al_2Si_2O_5(OH)_4$	258.161	48.53	99.52	-979465	-902868	tri.	330.48	2.594	1.565β	N-r, be, w, b
Halloysite $Al_2Si_2O_5(OH)_4$	258.161	48.6	99.3	-974995	-898419	am.		~2.1	1.549–1.561β	N-w
Muscovite $KAl_2[AlSi_3O_{10}](OH)_2$	398.313	69.0	140.71	-1421180	-1330103	mon.	934.57	2.831	1.587–1.607β	N-r, gn, c, b
Phlogopite $KMg_3[AlSi_3O_{10}](OH)_2$	417.29	76.4	149.91	—	—	mon.	497.83	2.784	1.598–1.606β	N-r, y, gn, c, b
Fluor-Phlogopite $KMg_3[AlSi_3O_{10}]F_2$	421.268	75.90	146.37	-1522020	-1439522	mon.	486.07	2.878		
Talc $Mg_3Si_4O_{10}(OH)_2$	379.289	62.34	136.25	-1416205	-1324486	mon.	904.94	2.784	1.590	N-gn, c, b, w

#Some information also drawn from (Forsythe, 1969), (Kerr, 1959), (Smith, 1953), (Wahlstrom, 1955), (Winchell, 1951, 1939).

TABLE 11-19. PROPERTIES OF SULFATES AND BORATES (ROBIE 1967, 1968), (DIETRICH 1969)#

Name and Formula	Gram Formula Weight	Entropy S^0 (cal deg^{-1} gfw^{-1})	Molar Volume (cm^3)	Heat of Formation ΔH^0 (cal gfw^{-1})	Free Energy of Formation ΔG^0 (cal gfw^{-1})	Crystal System	Cell Volume (10^{-24} cm^3)	X-ray Density (g cm^{-3})	Refractive Index n_D	Color
Barite $BaSO_4$	233.402	31.6	52.10	−352131	−325300	orth.	346.05	4.480	1.637	N-r, y, gn, be, c, w, g
Anhydrite $CaSO_4$	136.142	25.5	45.94	−343321	−316475	orth.	305.09	2.964	1.576	N-r, be, p, c, g, b
Gypsum $CaSO_4 \cdot 2H_2O$	172.172	46.36	74.69	−483981	−430137	mon.	496.1	2.305	1.523	N-y, c, w, g, b
Chalcanthite $CuSO_4 \cdot 5H_2O$	249.678	73.0	108.97	−544340	−449203	tri.	361.88	2.2912	1.539β	N-gn, be
Brochantite $Cu_4SO_4(OH)_6$	452.266	–	113.6	–	−435310	mon.	754.3	3.982	1.778	N-gn, bk
Szomolnokite $FeSo_4 \cdot H_2O$	169.924	–	55.9	−297305	–	mon.		3.05	1.623	y, b
Melanterite $FeSO_4 \cdot 7H_2O$	278.016	97.8	146.54	−720470	−599942	mon.	973.29	1.8972	1.478β	N-y, gn, be, w
Arcanite K_2SO_4	174.266	42.0	65.50	−343481	−315290	orth.	435.03	2.661	1.494	N-c, w
Alunite $K_2Al_6(OH)_{12}(SO_4)_4$	828.440	156.8	293.6	–	–	hex–R	146.8	2.822	1.572	N-r, y, be, w, g
Epsomite $MgSO_4 \cdot 7H_2O$	246.481	–	146.83	−808700	–	orth.	975.22	1.679	1.455β	N-r, gn, c, w
Thenardite Na_2SO_4	142.041	35.73	53.33	−331528	−303400	orth.	708.47	2.663	1.474β	N-r, y, c, w, g, b
Mirabilite $Na_2SO_4 \cdot 10H_2O$	322.195	141.46	219.8	−1034237	−871545	mon.	1459.9	1.466	1.396β	N-c, w
Mascagnite $(NH_4)_2SO_4$	132.139	52.6	74.68	−282230	−215565	orth.	496.04	1.7693	1.523β	N-y, c, g
Retgersite α-$NiSO_4 \cdot 6H_2O$	262.864	79.94	126.59	–	–	tet.	840.80	2.076	1.510	N-gn, be
Morenosite $NiSO_4 \cdot 7H_2O$	280.879	90.57	143.82	–	–	orth.		~2.0	1.489	N-gn, w

Anglesite $PbSO_4$	303.252	35.51	−219891	−194360	orth.	318.50	6.324	1.882β	N-y, gn, be, c, w, g
Celestite $SrSO_4$	183.682	28.2	−346646	−319830	orth.	307.17	3.972	1.624β	N-r, gn, be, c, w, b
Zinkosite $ZnSO_4$	161.432	26.4	−233600	−207022	orth.	276.10	3.883	1.669	w
Goslarite $ZnSO_4 \cdot 7H_2O$	287.539	92.9	—	—	orth.	968.29	1.9723	1.480β	N-gn, be, c, w, b
Borax $Na_2B_4O_7 \cdot 10H_2O$	381.373	—	−1497200	—	mon.	1478.8	1.7128	1.470	N-gn, be, c, w, g

#Some information also drawn from (Forsythe, 1969), (Kerr, 1959), (Smith, 1953), (Wahlstrom, 1955), (Winchell, 1951, 1939).

TABLE 11-20. SOME PHYSICAL PROPERTIES OF SEA WATER AND SEA ICE
(FAIRBRIDGE, 1966)

SEA WATER

Freezing point: $T_f(°C) = -0.055\ S_w\ (S_w$ in g salt/1000 g water)
Thermal conductivity: 0.00139 cal/cm sec °C (at 0°C and 35‰ salinity)
Diffusivity for NaCl: 0.0000068 cm²/sec (at 0°C and 35‰ salinity)
Initial solid precipitation of $Na_2SO_4 \cdot 10H_2O$ at $-\ 8.2°C$
$\qquad\qquad\qquad\qquad$ $NaCl \cdot 2H_2O \qquad -22.9°C$
$\qquad\qquad\qquad\qquad$ $MgCl_2 \cdot 12H_2O \qquad -36.0°C$

SEA ICE

Density: in nature mostly between 0.89 and 0.93 g/cm³, depending on temperature (T, °C), salinity (S_i, g salt/1000 g ice), and air content (v_a, cm³ air/1000 cm³ ice)

$$\rho_i(g/cm^3) = (1 - \frac{v_a}{1000})\ (1 - \frac{0.00456S_i}{T})0.917$$

Brine volume: $v_b \approx -55\ S_i/T\ (v_b$ and S_i in ‰, T in °C)

Thermal conductivity: k (cal/cm sec °C) $\approx 0.00486 + \dfrac{0.00025\ S_i}{T}$ (S_i in ‰, T in °C)

Specific heat: c_i (cal/g °C) $\approx 0.5 + \dfrac{4.1S_i}{T^2}$ (S_i in ‰, T in °C)

Tensile strength: $1-1.7 \times 10^7$ dynes/cm² between -8 and $-25°C$ and $S_i < 10$‰. The strength approaches zero close to 0°C because of the rapid increase of brine volume (see above)
Shear strength: somewhat smaller than tensile strength
Crushing strength: order of magnitude 10^8 dynes/cm²
Young's modulus: E (dynes/cm²) $= (9.75 - 0.242\ S_i) \times 10^{10}$ at $-20°C$ (S_i in ‰)
Poisson ratio: ~ 0.35
Longitudinal bulk wave velocity: 2800 m/sec at $-2°C$, 3500 m/sec at $-15°C$
Shear wave velocity: 1550 m/sec at $-2°C$, 1850 m/sec at $-15°C$
Electrical resistivity: greatly variable, no exact relationships established, changes from 25 to 25,000 ohm-meters between -1 and $-30°C$
Infrared emissivity: 0.99
Reflectivity (albedo) for visible light:

sludge	0.2
bare ice 5–10 cm thick	0.3–0.4
bare ice thicker than 50 cm	0.5–0.6
meltwater puddle	0.2–0.3
overall summer (bare and puddles)	0·5
overall winter (snow-covered)	0.85

coefficient for visible light: 0.05–0.015 cm⁻¹
Hydrodynamic roughness parameter:

upper surface	0.02 cm
lower surface	2. cm

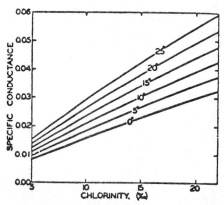

Fig. 11-2. Specific conductance, reciprocal ohms/cm³, of seawater as a function of temperature and chlorinity (Sverdrup, 1942). (By permission of Prentice-Hall, Englewood Cliffs, N.J.)

TABLE 11-21. LIGHT ATTENUANCE IN PURE SEAWATER (FAIRBRIDGE 1966)

Wavelength (microns)	Attenuance (observed) (%/m)	Scatterance (theoretical) (%/m)	Absorptance (computed) (%/m)
0.375	4.4	0.7	3.7
0.400	4.2	0.5	3.7
0.425	3.2	0.4	2.8
0.450	1.9	0.3	1.6
0.475	1.8	0.2	1.6
0.500	3.5	0.2	3.3
0.525	4.0	0.2	3.8
0.550	6.7	0.1	6.6
0.575	8.7	0.1	8.6
0.600	16.7	0.1	16.6
0.625	20.4	0.1	20.3
0.650	25.0	0.1	24.9
0.675	30.7	0.1	30.6
0.700	39.3	0.0	39.3

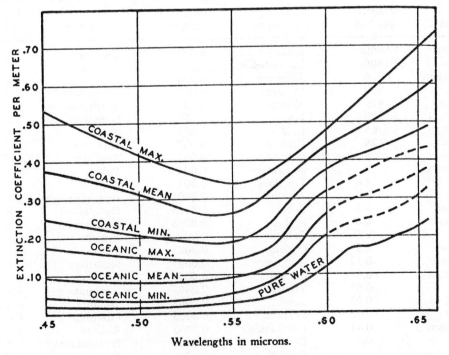

Fig. 11-3. Extinction Coefficients of Visible Light in Water. (Reproduced from: Forsythe, W. E., "Smithsonian Physical Tables," 9th rev. ed., Smithsonian Institution Press, Washington, 1969.

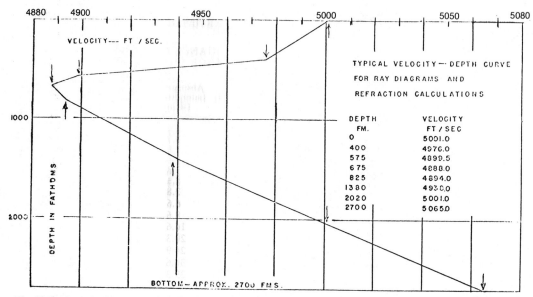

Fig. 11-4. Sound velocity profile [after Ewing, Worzel, and Pekeris, "Propagation of Sound in the Ocean," *Geol. Soc. Am., Mem.* 27 (1948)]. (Note: 1 fathom = 1.8288 meters; 1 ft/sec = 0.30480 m/sec; thus at 5000 meters, the velocity is approximately 1520 m/sec.) (Fairbridge 1966.)

TABLE 11-22. ABUNDANCE OF KNOWN ELEMENTS IN SEA WATER (JOSEPHS 1964)

Element	Milligrams per liter	Element	Milligrams per liter	Element	Milligrams per liter
Chlorine	19,000	Zinc	0.01	Tungsten	1×10^{-4}
Sodium	10,600	Molybdenum	0.01	Germanium	1×10^{-4}
Magnesium	1,300	Selenium	0.004	Xenon	1×10^{-4}
Sulfur	900	Copper	0.003	Chromium	5×10^{-5}
Calcium	400	Arsenic	0.003	Beryllium	5×10^{-5}
Potassium	380	Tin	0.003	Scandium	4×10^{-5}
Bromine	65	Lead	0.003	Mercury	3×10^{-5}
Carbon	28	Uranium	0.003	Niobium	1×10^{-5}
Oxygen	8	Vanadium	0.002	Thallium	1×10^{-5}
Strontium	8	Manganese	0.002	Helium	5×10^{-6}
Boron	4.8	Titanium	0.001	Gold	4×10^{-6}
Silicon	3.0	Thorium	0.0007	Praseodymium	2×10^{-7}
Fluorine	1.3	Cobalt	0.0005	Gadolinium	2×10^{-7}
Nitrogen	0.8	Nickel	0.0005	Dysprosium	2×10^{-7}
Argon	0.6	Gallium	0.0005	Erbium	2×10^{-7}
Lithium	0.2	Cesium	0.0005	Ytterbium	2×10^{-7}
Rubidium	0.12	Antimony	0.0005	Samarium	2×10^{-7}
Phosphorus	0.07	Cerium	0.0004	Holmium	8×10^{-8}
Iodine	0.05	Yttrium	0.0003	Europium	4×10^{-8}
Barium	0.03	Neon	0.0003	Thulium	4×10^{-8}
Indium	0.02	Krypton	0.0003	Lutetium	4×10^{-8}
Aluminum	0.01	Lanthanum	0.0003	Radium	3×10^{-11}
Iron	0.01	Silver	0.0003	Protactinium	2×10^{-12}
		Bismuth	0.0002	Radon	9×10^{-15}
		Cadmium	0.0001		

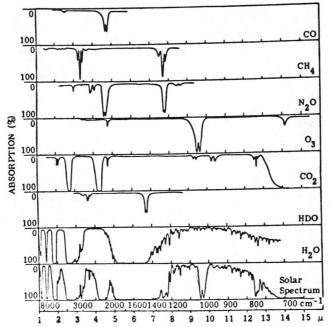

Fig. 11-5. Low-resolution solar spectrum from 1 to 24 μ. (Wolfe 1965.)

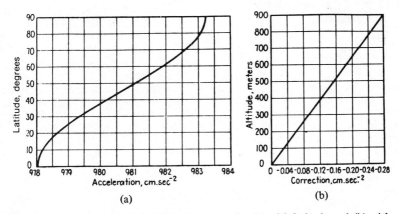

(a)

(b)

Fig. 11-6. Variation of acceleration due to gravity (a) with latitude and (b) with altitude (Byers, 1944). (Fairbridge 1966.)

TABLE 11-23. NORMAL COMPOSITION OF CLEAN, DRY ATMOSPHERIC AIR NEAR SEA LEVEL. (FROM *U. S. STANDARD ATMOSPHERE*, *1962*.)

Constituent Gas	Gas Symbol	Content (% by volume)
Nitrogen	N_2	78.084
Oxygen	O_2	20.9476
Argon	Ar	0.934
†Carbon dioxide	CO_2	0.0314
Neon	Ne	0.001818
Helium	He	0.000524
Krypton	Kr	0.000114
Xenon	Xe	0.0000087
Hydrogen	H_2	0.00005
†Methane	CH_4	0.0002
Nitrous oxide	N_2O	0.00005
†Ozone	O_3	Summer: 0 to 0.000007
		Winter: 0 to 0.000002
†Sulfur dioxide	SO_2	0 to 0.0001
†Nitrogen dioxide	NO_2	0 to 0.000002
†Ammonia	NH_3	0 to trace
†Carbon monoxide	CO	0 to trace
†Iodine	I_2	0 to 0.000001

†Variable.

TABLE 11-24. REGIONS OF THE ATMOSPHERE

Region	Altitude, km
Troposphere	0–11
Stratosphere	11–50
Mesosphere	50–85
Thermosphere	85–500
Exosphere	500

TABLE 11-25. REGIONS OF THE IONOSPHERE

Region	Altitude, km
D-Region	50–85
E-Region	85–140
F_1-Region	140–200
F_2-Region	200–1300

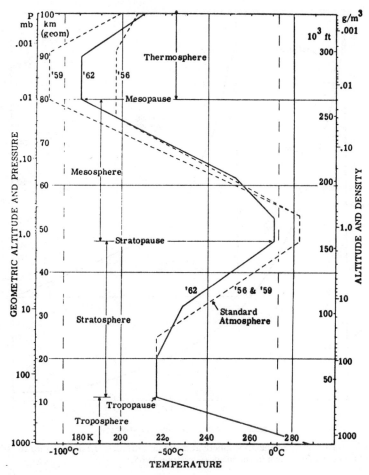

Fig. 11-7. Atmospheric temperature profiles from 0 to 100 km. Based on U. S. Standard Atmosphere, 1962 [1]. (Wolfe 1965.)

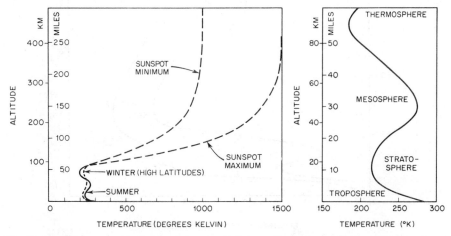

Fig. 11-8. Temperature of the atmosphere at different altitudes (after F. S. Johnson); the temperatures in the lower atmosphere are shown at the right. (Glasstone 1965.)

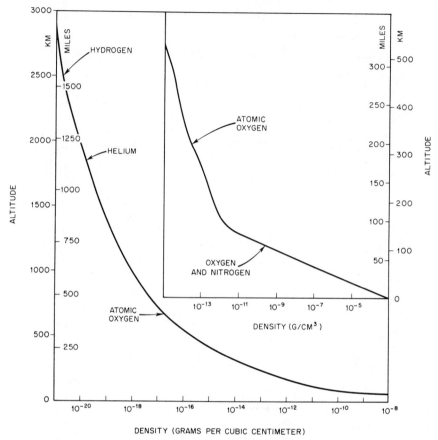

Fig. 11-9. Mass density of the atmosphere at different altitudes; the densities up to altitudes of 500 kilometers are shown in the inset. (Glasstone 1965.)

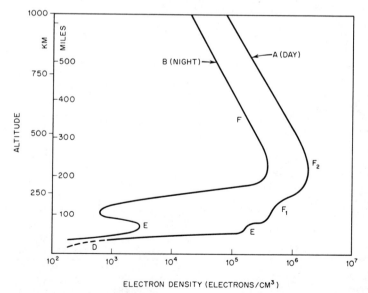

Fig. 11-10. Electron density profiles during summer at middle latitudes around sunspot maximum (after W. B. Hanson). (Glasstone 1965.)

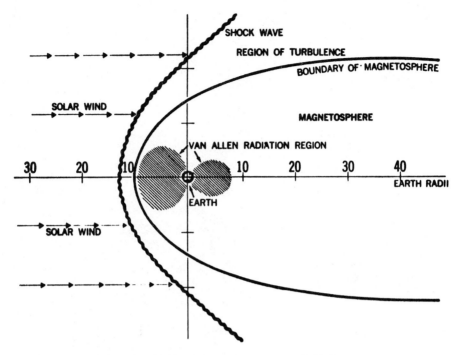

Fig. 11-11. Magnetosphere (Sanders 1966).*

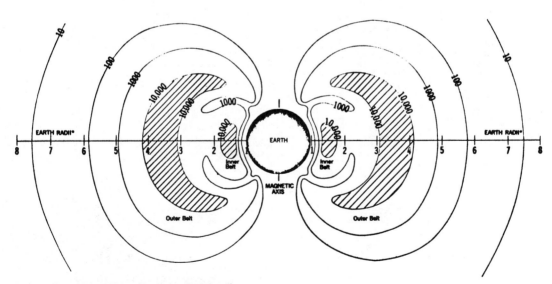

*Radius of the earth is 6370 km. Numbers indicated on the shells are the counts per second along contours of constant radiation intensity.

Fig. 11-12. Van Allen region (Sanders 1966).**

*Reprinted with permission from Sanders, H. J., "Chemistry and the Atmosphere," *Chem. & Eng. News*, **44**, 1A-54A, March 28, 1966. Copyright by the American Chemical Society.

Reprinted with permission from Sanders, H. J., "Chemistry and the Atmosphere," *Chem. & Eng. News*, **44, 1A-54A, March 28, 1966. Copyright by the American Chemical Society.

TABLE 11-26. SOLAR-SYSTEM DATA (JOHNSON 1961)*

Member of Solar System	Solar Distance Mean (A.U.)	Solar Distance Minimum (km)	Solar Distance Maximum (km)	Orbit Inclination	Orbit Eccentricity	Mean Radius (km)	Mass (g)
Mercury	0.387	4.59×10^7	6.97×10^7	7°00'	0.206	2.42×10^3	3.58×10^{26}
Venus	0.723	1.07×10^8	1.088×10^8	3°23'	0.007	6.16×10^3	4.90×10^{27}
Earth	1.000	1.458×10^8	1.520×10^8	0°00'	0.017	6.37×10^3	5.975×10^{27}
Mars	1.524	2.06×10^8	2.49×10^8	1°51'	0.093	3.33×10^3	6.58×10^{26}
Jupiter	5.203	6.39×10^8	8.14×10^8	1°18'	0.048	6.99×10^4	5.69×10^{28}
Saturn	9.539	1.344×10^9	1.502×10^9	2°29'	0.056	5.75×10^4	5.69×10^{28}
Moon	0.00258[a]	3.63×10^{5a}	4.04×10^{5a}	...	0.055	1.738×10^3	7.343×10^{25}
Sun	...	...	...	...	...	1.3914×10^6	1.987×10^{33}

Member of Solar System	Visual Albedo	Surface Temperature (°K)	Density (g/cm³)	Surface Gravity[d]	Rotation Period	Sidereal Revolution Period (yr)	Mean Orbit Velocity (km/sec)	Escape Velocity (km/sec)
Mercury	0.058	690	6.2	0.30	88 d (?)	0.241	47.85	4.2
Venus	0.76	293 (?)	5.0	0.90	30 hr (?)	0.615	35.01	10.3
Earth	0.39	288	5.52	1.00	23.93 hr	1.000	29.76	11.2
Mars	0.148	235	3.81	0.38	24.6 hr	1.88	24.11	5.0
Jupiter	0.51	135	1.36	2.65	9.9 hr	11.86	13.05	61
Saturn	0.50	120	0.72	1.14	10.2 hr	29.46	9.64	37
Moon	0.072	120–400	3.3	0.17	27.3 d	0.075	1.03[c]	2.4
Sun	...	5800	1.41	27.9	27 d[b]	...	...	620

[a] Distance from Earth.
[b] 24.65 near equator, increasing with latitude to 30.9 d at 60° and about 34 d near the poles.
[c] Velocity relative to Earth.
[d] Gravity relative to Earth.

*Reprinted with permission of Lockheed Aircraft Corporation, copyright holder.

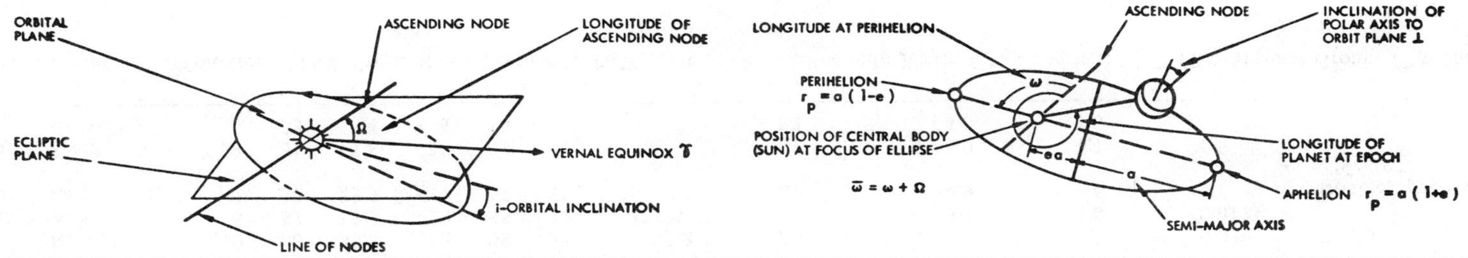

Symbol Planet	☿ Mercury	♀ Venus	⊕ Earth	♂ Mars	♃ Jupiter	♄ Saturn	♅ Uranus	♆ Neptune	♇ Pluto
Semi-Major Axis–a–AU	0.387099	0.723332	1.000000	1.523691	5.202694	9.538836	19.25290	30.04427	39.64092
Perihelion Distance–AU = a (1 – e)	0.307501	0.718422	0.983277	1.381416	4.951857	9.012979	18.285114	29.765219	29.719670
Aphelion Distance–AU = a (1 + e)	0.466697	0.728242	1.016723	1.665966	5.453531	10.064693	20.220686	30.323321	49.562170
Orbital Eccentricity e	0.205628	0.006788	0.016723	0.093375	0.048213	0.055128	0.050267	0.009288	0.250278
Mean Orbital Velocity–(⊕ = 1)	1.607271	1.175794	1.00	0.806855	0.438411	0.323782	0.2283249	0.1823988	0.1590757
km/sec	47.90	35.05	29.77	24.02	13.05	9.64	6.797	5.43	4.73
NM/sec	25.87	18.92	16.08	12.97	7.047	5.20	3.6705	2.93	2.55
ft/sec	157,186.0	114,958.0	97,702.1	78,805.7	42,817.6	31,595.2	22,302.0	17,802.7	15,560
Sidereal Mean Daily Motion–sec	14,732.4	5767.670	3548.193	1886.519	299.128	120.455	42.235	21.532	14.283
Period of Revolution–(⊕ = 1)	0.2411	0.6156	1.00	1.8822	11.86	29.46	84.0	164.8	247.7
Orbital Inclination i to Ecliptic–deg	7.00412	3.39431	0	1.84989	1.30601	2.48875	0.77250	1.77310	17.12631
Inclination of Equatorial Plane to Orbit–deg		<10	23.443597	23.99	3.067	26.733	97.884	28.8	
Mean Longitude of Ascending Node–Ω–deg	47.94364	76.38541		49.30530	100.1176	113.4342	73.9208	131.3902	109.7729
True Longitude of Perihelion– $\bar{\omega}$–deg (Epoch 1967 April 20.0)	76.94664	131.11097	102.3781	335.45705	13.5550	92.3756	169.1552	56.2011	222.7818
Mean Anomaly M–deg	253.906	353.764	105.02313	240.222	108.0054	277.3342	3.3525	176.5732	328.0351
Axial Rotational Period	$59^d \pm 2^d$	$42.6^d \pm 6^d$ (retrograde)	$23^h 56.07^m$	$24^h 37.38^m$	$9^h 50^m$	$10^h 14^m$	$10^h 49^m$	$15^h 48^m$	6^d
Escape Velocity $\left(\frac{2\mu}{R}\right)^{1/2}$ – ft/sec	13,600	33,500	36,675	16,500	197,500	119,500	72,500	82,400	31,300

Fig. 11-13. Orbital characteristics of the planets (Kendrick 1967).*

*Reprinted with permission from "TRW Space Data," third ed., TRW Systems Group, Redondo Beach, CA.
Copyright 1967 TRW Systems Group, TRW Inc.

TABLE 11-27. PROPERTIES OF 25 BRIGHTEST STARS (KENDRICK 1967)

Star	Name	Mean Places of Stars, 1966.0 (for January 0^d.799)						Distance Light-Years	Other Properties				Remarks
		Right Ascension			Declination				Classification of Spectrum	Absolute Visual Magnitude M	Apparent Visual Magnitude m	Heliocentric Parallax p arc sec	
		h	m	s	°	′	″						
α CMa	Sirius	6	43	39.0	−16	40	05	8.6	A0	+1.5	−1.6	0.379	
α Car	Canopus	6	23	11.8	−52	40	36	181.0	F0	−4.4	−0.9	0.08	
α Cen	α Centauri	14	37	11.6	−60	41	46	4.3	G0	+4.1	0.1	0.760	Double
α Boo	Arcturus	14	14	06.6	+19	21	31	36.2	K0	−0.3	0.2	0.090	
α Lyr	Vega	18	35	47.2	+38	45	03	26.6	A0	+0.5	0.1	0.123	
α Aur	Capella	5	14	10.4	+45	57	56	44.7	G0	−0.6	0.2	0.073	Spectroscopic binary, double
β Ori	Rigel	5	12	54.2	− 8	14	23	650.0	B8 I$_p$	−6.4	0.3	0.005	Double
α CMi	Procyon	7	37	31.4	+ 5	18	48	11.3	F5	+2.7	0.5	0.288	
α Eri	Achernar	1	36	27.0	−57	24	33	142.0	B5	−2.7	0.6	0.023	
β Cen	Agena (Nadar)	14	01	24.4	−60	12	36	204.0	B1	−3.3	0.9	0.016	Double
α Ori	Betelgeuse	5	53	19.8	+ 7	24	08	650.0	M0	−5.8	0.1	0.005	Variable
α Aql	Altair	19	49	07.4	+ 8	46	38	16.5	A7	+2.3	0.9	0.198	Variable, double
α Tau	Aldebaran	4	33	58.0	+16	26	33	68.0	K5 III	−0.7	0.9	0.048	Double
α Cru	Acrux	12	24	41.5	−62	54	39	218.0	B0	−3.2	1.0	0.015	Double, variable
α Sco	Antares	16	27	19.1	−26	21	30	172.0	Mi, A3	−2.6	1.2	0.019	
α Vir	Spica	13	23	23.9	−10	59	04	155.0	B2	−2.4	1.2	0.021	Spectroscopic binary
α PsA	Fomalhaut	22	55	46.5	−29	48	11	22.7	A3	+2.0	1.3	0.144	
β Gem	Pollux	7	43	14.3	+28	06	35	35.1	K0	+1.0	1.2	0.093	
α Cyg	Deneb	20	40	16.3	+45	09	29	544.0	A2$_p$	−4.8	1.3	0.006	
β Cru	β Crux	12	45	43.3	−59	30	12	(297)	B1	−3.5	1.5	(0.011)	
α Leo	Regulus	10	06	33.8	+12	08	03	83.8	B8	−0.7	1.3	0.039	Double
ε CMa	Adhara	6	57	17.3	−28	55	29	(272)	B1	−3.1	1.6	(0.012)	
α Gem	Castor	7	32	25.9	+31	57	51		A0	+0.9	1.6		Double, spectroscopic binary
λ Sco	Shaula	17	31	17.8	−37	04	51		B2	−3.0	1.7		
γ Ori	Bellatrix	5	23	18.4	+ 6	19	13		B2	−4.0	1.6		

TABLE 11-28. ELEMENTS OF TYPICAL COMETS (KENDRICK 1967)*

Period-Years		Name of Comet	Perihelion q-AU	Orbital Eccentricity e	Remarks
Short Period	3.3	Encke	0.34	0.847	Has been observed on 48 passages. Year of last passage 1964.
	6.12	Pons-Winnechke	1.23	0.641	Currently suffering strong Jupiter perturbations. Last passage 1964.
	76.0	Halley	0.59	0.967	First comet to have had its return predicted. Last passage 1910.
Long Period	$>10^3$	1843I Brilliant	0.005	Almost indistinguishable from parabolas with e = 1	Had a tail 2 AU long, and grazed the sun
	$>10^3$	1916I Taylor	1.55		Split in two during passage
	$>10^3$	1957III Arend-Roland	0.316		Developed several tails and an antitail
	$>10^3$	1965 f Ikeya-Seki	0.00778		Split into two pieces after passage

*Reprinted with permission from "TRW Space Data," third ed., TRW Systems Group, Redondo Beach, CA. Copyright 1967 TRW Systems Group, TRW Inc.

TABLE 11-29. ELEMENTS OF TEN NOTABLE ASTEROIDS (KENDRICK 1967)*

Catalog No. and Name of Asteroid	Period About Sun (years)	Mean Distance to Sun, AU		Orbital Eccentricity	Orbital Inclination to Ecliptic (deg)	Photo-magnitude at Mean Opposition
		Perihelion	Aphelion			
1 Ceres	4.60	2.56	2.98	0.076	10.0	7.4
2 Pallas	4.61	2.12	3.42	0.234	34.8	8.5
3 Juno	4.36	1.98	3.36	0.258	13.0	9.6
4 Vesta	3.63	2.15	2.57	0.089	7.1	6.8
1566 Icarus	1.12	0.19	1.97	0.827	23.0	12.4
Apollo	1.81	0.64	2.33	0.566	6.4	19.0
588 Achilles	11.90	4.44	5.98	0.148	10.3	16.0
617 Patroclus	11.88	4.47	5.94	0.141	22.1	15.8
433 Eros	1.76	1.13	1.78	0.223	10.8	11.4
944 Hidalgo	13.95	1.99	9.59	0.656	42.5	19.2

*Reprinted with permission from "TRW Space Data," third ed., TRW Systems Group, Redondo Beach, CA. Copyright 1967 TRW Systems Group, TRW Inc.

TABLE 11-30. SPECTRAL CLASS AND SURFACE TEMPERATURE OF MAIN-SEQUENCE STARS (GLASSTONE 1965)

Spectral Class	Temper- ($°K$)	Color Index	Spectral Class	Temper- ($°K$)	Color Index
O5	50,000	–	F5	6600	+0.44
B0	25,000	−0.32	G0	6000	+0.60
B5	15,600	−0.16	G5	5520	+0.68
A0	11,000	0.00	K0	5120	+0.82
A5	8700	+0.15	K5	4400	+1.18
F0	7600	+0.30	M0	3600	+1.45

TABLE 11-31. MAXIMUM AGES OF MAIN-SEQUENCE STARS (GLASSTONE 1965)

Mass (solar masses)	Approximate Maximum Age (years)	Spectral Class
25	2×10^7	B0
4	4×10^8	A0
2	2×10^9	F0
1	10^{10}	G2
0.6	(10^{11})*	K5
0.2	(10^{12})*	M5

*These maxima are probably greatly in excess of the actual ages; there is no evidence of any star with an age exceeding about 1 or 2×10^{10} years.

REFERENCES

Ahrens, L. H., "Distribution of the Elements in our Planet," McGraw-Hill, New York, 1965.

Centeno, M. 1941.

Gutenberg, B., "Physics of the Earth's Interior," Academic Press, New York, 1959.

Day, F. H., "The Chemical Elements in Nature," Van Nostrand Reinhold, New York, 1964.

Defant, A., "Physical Oceanography," Vol. 1, Pergamon Press, Oxford, 1961.

Dietrich, R. V., "Mineral Tables," McGraw-Hill, New York, 1969.

Fairbridge, R. W., Ed., "The Encyclopedia of Oceanography," Van Nostrand Reinhold, New York, 1966.

Forsythe, W. E., "Smithsonian Physical Tables," 9th Rev. Ed., Smithsonian Institution Press, Washington, D.C., 1969.

Glasstone, S., "Sourcebook on the Space Sciences," Van Nostrand Reinhold, New York, 1965.

Josephs, M., "Chemistry and the Oceans," *Chem. & Eng. News*, **42**, 1A-48A (June 1, 1964).

Johnson, F. S., Ed., "Satellite Environment Handbook," Stanford University Press, Stanford, Calif., 1965.

Kendrick, J. B., Ed., "TRW Space Data," Third ed., TRW Systems Group, Redondo Beach, Calif., 1967.

Kerr, P. F., "Optical Mineralogy," Third ed., McGraw-Hill, New York, 1959.

Robie, R. A., P. M. Bethke and K. M. Beardsley," Selected X-Ray Crystallographic Data Molar Volumes, and Densities of Minerals and Related Substances," *Geol. Survey Bull.* 1248, U.S. Govt. Printing Office, Washington, D.C. (1967).

Robie, R. A. and D. R. Waldbaum, "Thermodynamic Properties of Minerals and Related Substances at 298.15°K (25.0°C) and One Atmosphere (1.013 Bars) Pressure and at Higher Temperatures," Geol. Survey Bull. 1259, U.S. Govt. Printing Office, Washington, D.C. (1968).

Sanders, H. J., "Chemistry and the Atmosphere," *Chem. & Eng. News*, **44**, 1A-54A (March 28, 1966).

Sanders, H. J., "Chemistry and the Solid Earth," *Chem. & Eng. News*, **45**, 1A-48A (October 2, 1967).

Smith, O. C., "Identification and Qualitative Chemical Analysis of Minerals," 2nd. Ed., Van Nostrand Reinhold, New York, 1953.

Sverdrup, H. U., M. W. Johnson and R. H. Fleming, "The Oceans," Prentice Hall, New York, 1942.

Wahlstrom, E. E., "Petrographic Mineralogy," John Wiley, New York, 1955.

Winchell, A. N. and H. Winchell, "Elements of Optical Mineralogy," 4th Ed., Part II, "Description of Minerals," John Wiley, New York, 1951.

Winchell, A. N., "Elements of Optical Mineralogy," 2nd. Ed., Part III, "Determinative Tables," John Wiley, New York, 1939.

Wolfe, W. L., Ed., "Handbook of Military Infrared Technology," Office of Naval Research, Washington, D.C. (1965).

BIBLIOGRAPHY

Berry, L. G. and R. M. Thompson, "X-Ray Powder Data for Ore Minerals," Memoir 85, Geological Society of America, New York, 1962.

Bialek, E. L., "Handbook of Oceanographic Tables 1966," Publ. No. SP-68, U.S. Navy Oceanographic Office, Washington, D.C., 1966.

Clark, S. P., Ed., "Handbook of Physical Constants," Memoir 97, Geological Society of America, New York, 1966.

Fairbridge, R. W., Ed., "The Encyclopedia of Geochemistry and Environmental Sciences," Van Nostrand Reinhold, New York, 1972.

Fisher, W., H. Burzlaff, E. Hellner and J. D. H. Donnay, "Space Groups and Lattice Complexes," NBS Monograph No. 134, U.S. Government Printing Office, Washington, D.C., 1973.

Ford, W. E., "Cana's Textbook of Mineralogy," Wiley, New York, 1953.

Geophysics Research Directorate, Air Research and Development Command, USAF, "Handbook of Geophysics," rev. ed., Macmillan, New York, 1960.

Kuiper, G. P., Ed., "The Atmospheres of the Earth and Planets," University of Chicago Press, Chicago, 1952.

Morris, D. F. C. and E. L. Short, "Handbook of Geochemistry," Springer-Verlag, New York, 1969.

Nairn, A. E. M. and F. G. Stehli, Eds., "The Ocean Basins and Margins," 7 vols., Plenum, New York, 1973-1974.

Parkomenko, E. I., "Electrical Properties of Rocks," Plenum, New York, 1967.

Rasool, S. I., Ed., "Chemistry of the Lower Atmosphere," Plenum, New York, 1973.

Roederer, J. G., Ed., "Physics and Chemistry in Space," 7 vols., Springer-Verlag, New York, 1970-1973.

Roll, H. E., "Physics of the Marine Atmosphere," Academic Press, New York, 1965.

Roth, G. D., Ed., "Astronomical Handbook," Springer-Verlag, New York, 1973.

Smith, J. V., "Feldspar Minerals," 2 vols., Springer-Verlag, New York, 1974.

Wedepohl, K. H., Ed., "Handbook of Geochemistry," 2 vols., Springer-Verlag, New York, 1969-1972.

CHAPTER XII
PROPERTIES OF MISCELLANEOUS MATERIALS

This chapter is concerned with materials that are for reason of convenience not integrated in earlier chapters. The materials are felts, micas and woods, whose properties are offered in the following Tables:

TABLE 12-1. PROPERTIES OF WOOL FELTS—ROLL[a] (LOFARO , 1971)

Commercial Standard 185-52 Grade →	18R1	16R1	16R2	16R3	12R1	12R2
SAE Spec. No. →	—	F-1	F-2	F-3	F-5	F-6
GENERAL PROPERTIES						
Wool Content (fiber basis), min %	100	100	100	95	100	100
Standard Thickness Range, in.	⅛ to 1	⅛ to 1	⅛ to 1	⅛ to 1	⅛ to 1	⅛ to 1
Standard Width, in.	60	60	60	60	60	60 or 72
Texture	Fine	Fine	Med fine	Medium	Fine	Med fine
Color[b]	White	White	Any, except gray or blk	Gray	White	Gray
PHYSICAL PROPERTIES						
Specific Gravity	0.384	0.342	0.342	0.330	0.262	0.262
Density (1 in.), lb/sq yd	18.00	16.00	16.00	15.60	12.24	12.24
Operating Temp Range, F[c]	−80 to +200	−80 to +200	−80 to +200	−80 to +200	−80 to +200	−80 to +200
Ther Cond (70 F), Btu/hr/sq ft/°F/in.[d]	0.39	0.36	0.36	0.35	0.30	0.30
Coef of Ther Exp, per °F	0	0	0	0	0	0
Air Perm (1/16 in.), cfm/sq ft/0.5 in. H_2O	5-15	10-30	10-30	15-35	20-50	20-50
Liquid Absorption, %						
By Weight (1.0 sp gr liquid)	>125	>175	>175	>190	>250	>225
By Volume	71	74	74	76	80	80
Capillarity (wicking height, 575 SSU, 70)F, in.	4.5	4.0	4.0	4.0	3.0	3.0
Coef of Friction[e]	0.37	0.37	0.37	0.37	0.37	0.37
Vibration Absorption[f]						
Static Load Bearing Cap, per Unit Area	High	High	High	High	Medium	Medium
Dynamic Stress Endurance	High	High	High	High-med	High	High
Coef of Noise Reduction (1 in. thick)[d]	0.45	0.50	0.50	0.52	0.58	0.58
MECHANICAL PROPERTIES						
Ten Str (min), psi	600	500	500	400	400	275
Elong (at 100 psi), %	9	13	14	16	16	18
Mullen Burst Str (1/8 in. thick), psi	300	250	225	200	175	150
Split Res (min), lb/2-in. width	35	33	28	22	18	16
Hardness Range, Shore A	40-50	30-40	30-40	30-40	20-30	20-30
Compressibility (at 10% defl), psi	37	21	21	13	6	6
Recovery (within 1 min after 10% defl), %	99	99	99	99	99	99
Vibration Disintegration	None	None	None	None	None	None
Collapse When Wet	None	None	None	None	None	None
Abrasion Resistance[g]	Excellent	Excellent	Excellent	Excellent	Good	Good
Flexibility (fold endurance)	¼-in. thick felt exceeds 3 million 180 deg. flexes					
CHEMICAL AND ENVIRONMENTAL PROPERTIES[h]						
Effect of Sunlight and Oxidation	None	None	None	None	None	None
Solvent Res, Stability in Oil	Excellent	Excellent	Excellent	Excellent	Excellent	Excellent
Acid Resistance						
Dilute	Excellent	Excellent	Excellent	Excellent	Excellent	Excellent
Concentrated	Good-fair	Good-fair	Good-fair	Good-fair	Good-fair	Good-fair
Alkali Resistance						
Dilute	Fair	Fair	Fair	Fair	Fair	Fair
Concentrated	Poor	Poor	Poor	Poor	Poor	Poor

[a] Industrial, mechanical and filter felts; three dimensional fibrous structure.
[b] Colors available on special order.
[c] Felts are flameproofed to meet government and industrial specifications.
[d] For 1-in. felts. Felts blended with kapok fiber have a k factor of 0.21 and a coefficient of sound absorption of 0.80 at 512 cps.
[e] Depends upon condition of contact surface, but can be moderately controlled by altering surface finish of the felt.
[f] Up to 85% under appropriate design conditions.
[g] Increases with density.
[h] Treated to resist moths, fungus, mildew and vermin. Data obtained by ASTM D-461 test methods.

TABLE 12-1. (*Continued*)

12R3	9R1	9R2	9R3	9R4	9R5	8R5	16R1X	16R3X	12R3X
F-7	F-10	F-11	F-12	F-13	F-15	F-26	F-50	F-51	F-55

GENERAL PROPERTIES

12R3	9R1	9R2	9R3	9R4	9R5	8R5	16R1X	16R3X	12R3X
85	100	100	90	80	60	50	100	100	80
⅛ to 1	⅛ to 1	⅛ to 1	⅛ to 1	⅛ to 1	⅛ to 1	⅛ to 1	3/64 to 3/32	3/64 to 3/32	1/16 to 3/32
72	72	72	72	72	72	72	60 or 72	60 or 72	60 or 72
Medium	Fine	Med fine	Med fine	Medium	Medium	Medium	Fine	Med fine	Medium
Gray	White	Gray	Gray	Gray	Gray	Gray	White	Gray	Gray or blk

PHYSICAL PROPERTIES

12R3	9R1	9R2	9R3	9R4	9R5	8R5	16R1X	16R3X	12R3X
0.262	0.181	0.181	0.181	0.181	0.181	0.154	0.330	0.330	0.256
12.24	8.48	8.48	8.48	8.48	8.48	7.20	15.60	15.60	12.00
−80 to +200	−80 to +200	−80 to +200	−80 to +200	−80 to +200	−80 to +200	−80 to +200	−80 to +200	−80 to +200	−80 to +200
0.30	0.30	0.24	0.24	0.24	0.24	0.25	0.32	0.32	0.30
0	0	0	0	0	0	0	0	0	0
20-50	75-150	75-150	75-150	75-150	75-150	100-200	15-25	15-40	20-50
>225	>400	>375	>350	>350	>350	>400	>180	>170	>225
80	88	88	88	88	88	92	75	75	81
3.0	2.5	2.5	2.5	2.5	2.5	—	4.0	4.0	3.0
0.37	0.37	0.37	0.37	0.37	0.37	0.37	0.37	0.37	0.37
Medium	Low	Low	Low	Low	Low	Very low	—	—	—
Medium	High	High-med	Medium	Low	Low	Very low	—	—	—
0.58	0.58	0.64	0.64	0.64	0.64	0.65	0.55	0.55	0.58

MECHANICAL PROPERTIES

12R3	9R1	9R2	9R3	9R4	9R5	8R5	16R1X	16R3X	12R3X
250	225	200	100	75	75	—	500	200	200
21	33	35	35	37	39	—	8	9	25
125	75	60	55	50	40	25	225	225	200
12	8	6	3	2	2	—	—	—	—
20-30	15-25	15-25	15-25	15-25	15-25	5-15	—	—	—
6	4	4	3	3	3	1	—	—	—
99	99	99	99	99	99	99	99	99	99
None	None	None	None	None	None	None	None	None	None
None	None	None	None	None	None	Slight	None	None	None
Good	Fair	Fair	Fair	Fair	Fair	Poor	Excellent	Excellent	Good

¼-in. thick felt exceeds 3 million 180 deg flexes

CHEMICAL AND ENVIRONMENTAL PROPERTIES

12R3	9R1	9R2	9R3	9R4	9R5	8R5	16R1X	16R3X	12R3X
None	None	None	None	None	None	None	None	None	None
Excellent	Excellent	Excellent	Excellent	Excellent	Excellent	Excellent	Excellent	Excellent	Excellent
Good	Excellent	Excellent	Good	Fair	Fair	Fair	Excellent	Excellent	Good
Fair	Fair-good	Fair-good	Fair-good	Fair	Fair	Poor	Fair-good	Fair-good	Fair
Fair	Fair	Fair	Fair	Fair	Fair	Fair	Fair	Fair	Fair
Poor	Poor	Poor	Poor	Poor	Poor	Poor	Poor	Poor	Poor

TABLE 12-2.* PROPERTIES OF MAN-MADE ORGANIC FIBERS

Fiber ↓ Property →	Form	Length, in. (mm)	Width, microns	Cross section	Specific gravity	Breaking tenacity, (70 F, 65% RH), gm/den	Breaking tenacity (wet) gm/den
NYLON							
6/6, regular tenacity	Filament	Cont.	11-43	round	1.14	4.6-5.9	4.0-5.2
6/6, high tenacity	Filament	Cont.	16-43	round	1.14	5.9-8.8	5.1-7.6
6/6, staple	Staple	—	14-43	round	1.14	4.0-4.7	3.5-4.2
6, regular tenacity	Filament	Cont.	—	round	1.14	4.5-5.8	4.3-5.3
6, high tenacity	Filament	Cont.	—	round	1.14	6.8-8.6	5.4-7.5
6, staple	Staple	—	—	round	1.14	3.8-5.5	—
Qiana	Filament	Cont.	19-35	trilobal	1.03	2.7-3.8	—
CELLULOSE ESTERS							
Cellulose acetate	Fil, staple	—	11-46	clover leaf	1.32	1.3-1.5	0.8-1.2
Cellulose triacetate	Fil, staple	—	—	clover leaf	1.3	1.2-1.4	0.8-1.0
REGENERATED CELLULOSE							
Viscose, reg to med ten	Fil, staple	—	4.3-8.4	irreg	1.46-1.52	1.5-3.2	0.7-1.9
Viscose, high ten	Filament	Cont.	10-15	irreg	—	3.0-5.0	1.9-3.6
Saponified acetate (Fortisan)	Filament	Cont.	3-9	irreg	1.5	6-7	5.1-6.0
(Fortisan 36)	Filament	Cont.	3-9	irreg	1.5	8	6.0-6.4
ARAMIDS							
Nomex'	Fil, staple	1-2.5 (25.4-63.5)	—	round	1.38	5.3	4.1
Kevlar 29	Filament	Cont.	11.9	round	1.44	21.5	21.5
Kevlar 49	Filament	Cont.	11.9	round	1.45	21.5	21.5
POLYESTERS (Dacron)							
Regular	Fil, staple	1.5-4.0 (38-102)	11-28	round	1.38	4.4-5.0	4.4-5.0
High tenacity	Fil, staple	1.5-4.0 (38-102)	11-28	round	1.38	6-7	6-7
Staple	Fil, staple	1.5-4.0 (38-102)	18-25	—	1.38	3.8-4.3	3.8-4.3
ACRYLICS							
Polyacrylonitrile	Staple	—	14-27	dogbone	1.14	2.2-2.6	1.8-2.1
Acrylonitrile vinyl chloride derivatives	Staple	—	15-30	—	1.17	2.5	2.0
Acrylonitrile vinyl chloride	Staple	—	—	—	1.3	2.5-3.3	2.5-3.3
Modacrylic	Staple	—	—	⌐	1.37	2.5-2.8	2.4-2.7
Acrylonitrile base	—	—	—	—	1.17	3.3	3.3
Dinitrile"	—	—	—	—	1.18	1.75	1.5
Nitrile alloy'	—	—	—	—	1.19	3.5	3.1
POLYOLEFINS							
Polyethylene, Type I	Monofil	—	250-1300	round	0.92	1.0-3.0	1.0-3.0
Polyethylene, Type II	Monofil	—	250-1300	round	0.95-0.96	5.0-7.3	5.0-7.3
Polypropylene	Monofil	—	—	round	0.90-0.91	5.5-7.0	5.5-7.0

TABLE 12-2.* (Continued)

Ten str, 1000 psi (MPa)	Breaking Elong, (70 F, 65% RH), %	Breaking Elong, (wet), %	Stiffness[a] (avg) gm/den	Strain recovery (at 2%), %[b]	Toughness, (avg) gm-cm/den-cm	Moisture regain, %
67-86 (462-593)	26-32	30-37	—	100, 100	0.76	4.5
86-128 (593-882)	18-28	21-32	—	100 (4%)	0.85	4.5
58-69 (400-476)	38-42	42-46	—	—	0.87	4.5
73-84 (503-579)	24-34	28-38	—	—	0.67	4.0
10 -125 (751-862)	16-17.5	19-24	—	100, 100	0.75	4.0
70-80 (483-552)	37-40	42-46	—	100 (4%)	0.64-0.78	4.0
35-50 (241-345)	26-36	—	7.5-15.0	100	0.85-1.0	2.5
22-28 (152-193)	23-34	30-45	5.5	48-65, (4%)	0.17	6
20-26 (138-179)	22-28d	30-40	5.2	88 (3%) 43 (10%)	0.16	3.2
29-65 (200-448)	15-30	17-40	11.1-16.6	30-97	0.19-0.21	13
65-105 (448-724)	9-22	14-30	25.5-29	70-100	0.22-0.30	13
136 (938)	6	6	117	100 (20%) 60 (40%)	0.21	10.7
155 (1069)	6.2	6	135	85, 70 (5%)	0.26	9.6
95.5 (658)	12.2	16	—	—	—	5.0
400 (2758)	3-4	4	460	—	—	5.0
400 (2758)	2.25	2.25	1000	—	—	3.5
77-88 (531-607)	19-25	19-25	—	97, 80 (8%)	0.78	0.4
106-123 (731-848)	9-11	9-11	—	100, 90 (8%)	0.50	0.4
67-76 (462-524)	30-36	30-36	—	—	1.03	0.4
32-39 (221-269)	20-28	26-34	10	—	0.40	1.5
37-40 (255-276)	36	44	7	99, 89 (5%)	0.46	1.2
40-57 (276-393)	30-42	30-42	8.2	94	0.53	0.3-0.4
42-57 (290-393)	33-35	32-34	8.0	88 (4%) 55 (10%)	0.46	3.5-4
41 (283)	32	32	10.3	90 (1%)	0.53	1.3
26 (179)	30	30	6	100 (3%)	0.3	2-3
53 (365)	33	33	11	99, 72 (10%)	0.58	2.5
11-35 (76-241)	20-80	20-80	2-12	90-95 (5%)	0.3	none
50-90 (345-620)	10-40	10-40	20-50	slow	—	none
—	12-25	—	—	—	—	—

TABLE 12-2.* *(Continued)*

Fiber ↓ Property →	Form	Length, in. (mm)	Width, microns	Cross section	Specific gravity	Breaking tenacity, (70 F, 65% RH), gm/den	Breaking tenacity (wet) gm/den
VINYL DERIVATIVES							
Vinyl chloride acetate	Staple	—	16-18	barbell	1.33-1.35	0.7-1.0	0.7-1.0
Vinylidene chloride, saran							
Monofilament	—	—	1300	round	1.7	1.2-2.3	1.2-2.3
Filament	—	—	50	round	1.7	up to 2	up to 2
Staple	—	—	—	round	1.7	up to 1.5	up to 1.5
Polyvinyl alcohol	Staple	—	—	peanut	1.26-1.3	4.4-6.0	3.7-5.0
NOVOLOID							
Kynol	Fil, staple	—	14	—	1.25	1.7	1.56
FLUOROCARBON							
TFE	Fil, staple	4.5 (114)	—	round	2.3	1.6	1.6
SPANDEX	Filament	—	—	barbell	1.0	0.7	—

*Reproduced from: Anon, "Primer on Organic Man-Made Fibers," *Materials Eng. 80*, No. 8, 34–39, August 1974.

[a] Ratio of breaking stress to breaking strain (i.e., gm/den to rupture divided by strain in cm/gage cm at breaking stress.

[b] Recovery after 2% strain except where specific percentage strain is given in parenthesis.

[c] 20 denier yarn with 3 yarns/in. of twist.

[d] For filament; 35–40% for staple fiber.

[e] Though not an acrylic (actually a polymer of vinylidene cyanide) it is listed here because of similar properties.

[f] Nitrile alloy based on acrylonitrile.

TABLE 12-2.* (*Continued*)

Ten str, 1000 psi (MPa)	Breaking Elong, (70 F, 65% RH), %	Breaking Elong, (wet), %	Stiffness" (avg) gm/den	Strain recovery (at 2%), %ᵇ	Toughness, (avg) gm-cm/den-cm	Moisture regain, %
12-17 (83-117)	100-120	100-120	1.5	—	1.3	up to 1.5
15-45 (103-310)	20-30	20-30	7-10	95 (10%)	0.17-0.27	none
44 (303)	15-25	15-25	8-12	95 (10%)	0.125	none
33 (227)	15-25	15-25	8-12	95 (10%)	0.125	none
—	15-17	—	24-40	—	0.79-0.92	4.5-5.0
27 (186)	35	37	45	95	0.5	6
47 (324)	13	13	12	88 (3%)	0.12	None
—	520-610	—	—	—	—	0.3

TABLE 12-3. PROPERTIES OF TYPICAL WOOL FELTS[i]—SHEET (LOFARO, 1971)

Commercial Standard Grade →	32S1	26S1	20S1	16S1	12S1
GENERAL PROPERTIES					
Wool Content (fiber basis), min %	100	100	100	100	100
Standard Thickness Range, in.	¼ to 3	¼ to 3	¼ to 3	¼ to 3	¼ to 3
Standard Width, in.	36 x 36	36 x 36	36 x 36	36 x 36	36 x 36
Texture	Extra fine	Extra fine	Extra fine	Extra fine	Extra fine
Color[h]	White	White	White	White	White
PHYSICAL PROPERTIES					
Specific Gravity	0.682	0.555	0.426	0.342	0.256
Density (1 in.) lb/sq yd	32.00	26.00	20.00	16.00	12.00
Operating Temp Range, F[c]	−80 to +200	−80 to +200	−80 to +200	−80 to +200	−80 to +200
Ther Cond (70 F), Btu/hr/sq ft/°F/in.[d]	0.91	0.63	0.45	0.36	0.30
Coef of Ther Exp, per °F	0	0	0	0	0
Air Perm (1/16 in.), cfm/sq ft/0.5 in. H_2O	<1	<3	<10	<20	<30
Liquid Absorption, %					
By weight (1.0 sp gr liquid)	>50	>75	>100	>175	>250
By Volume	48	58	66	74	81
Capillarity (wicking height, 575 SSU, 70 F), in.	5.5	5.0	4.5	4.0	3.0
Coef of Friction[a]	0.42	0.42	0.42	0.37	0.37
Vibration Absorption[f]					
Static Load Bearing Cap, per Unit Area	Ultra high	Very high	High	High	Medium
Dynamic Stress Endurance	High	High	High	High	High
Coef of Noise Reduction (1 in. thick)[d]	0.05	0.28	0.41	0.50	0.58
MECHANICAL PROPERTIES					
Ten Str (min), psi	800	700	600	500	400
Elong (at 100 psi), %	2	4	9	12	28
Mullen Burst Str (⅛ in. thick), psi	>500	>400	325	250	150
Split Res (min), lb/2-in. width	50	48	44	32	18
Hardness Range, Shore A	75-85	55-65	45-55	30-40	20-30
Compressibility (at 10% defl), psi	121	86	58	32	18
Recovery (within 1 min after 10% defl), %	99	99	99	99	99
Vibration Disintegration	None	None	None	None	None
Collapse When Wet	None	None	None	None	None
Abrasion Resistance[g]	Excellent	Excellent	Excellent	Excellent	Good
Flexibility (fold endurance)	¼-in. thk felt exceeds 3 million 180 deg flexes				
CHEMICAL AND ENVIRONMENTAL PROPERTIES					
Effect of Sunlight and Oxidation	None	None	None	None	None
Solvent Res and Stability in Oil	Excellent	Excellent	Excellent	Excellent	Excellent
Acid Resistance					
Dilute	Excellent	Excellent	Excellent	Excellent	Excellent
Concentrated	Fair-good	Fair-good	Fair-good	Fair-good	Fair-good
Alkali Resistance					
Dilute	Fair	Fair	Fair	Fair	Fair
Concentrated	Poor	Poor	Poor	Poor	Poor

[i]See Table 12-1 for footnotes.

TABLE 12-4. PROPERTIES OF TYPICAL SYNTHETIC FIBER PAD FELTS[a]
(LOFARO 1971)

Fiber Type (all 100%) →	Dacron[b] Polyester	Nylon	Rayon Viscose	Polypropylene	TFE[b]-Fluoro-carbon	Nomex[b] polyamide
Texture →	Medium	Fine	Fine	Fine	Fine	Fine
GENERAL PROPERTIES						
Fiber Denier Per Filament	6	3	3	3	—	2
Density (1 in. nominal thk), lb/sq yd	3.5	8.0	12.0	6.0	18.0	4.0
Standard Thickness Range, in.	1/16 to 1	1/16 to 1	1/16 to 1	1/16 to 1	1/16 to 1	1/16 to 1
Standard Width, in.	72	72	72	72	72	72
Color	White[g]	White[g]	White[g]	White	Brown or bleached white	Cream
PHYSICAL PROPERTIES						
Specific Gravity	0.08	0.17	0.26	0.13	0.38	0.09
Normal Service Temp Range, F	To 300	To 225	To 225	To 200	To 450	To 350
Ther Cond (70 F), Btu/hr/sq ft/°F/in.	0.21	0.26	0.28	0.26	0.42	0.22
Thermal Expansion	None	None	None	None	None	None
Air Perm (1/16 in.), cfm/sq ft/0.5 in. H_2O	450	350	50	400	75	100
Liquid Absorption, %						
By Weight[c]	1300	500	330	670	210	1100
By Volume[d], in. in 2 hr[e]	95	85	83	86	83	94
Capillarity (wicking height)	1	1½	2¾	3¼	3¼	2
Coef of Friction	0.20	0.15	0.25	0.15	0.12	0.20
Vibration Absorption						
Static Load-Bearing Capac/Unit Area	Low	Medium	Medium	Medium	Medium	Low
Dynamic Stress Endurance	Low	Medium	Medium	Medium	Medium	Low
Noise Reduction Coef (1 in. thk)	0.69	0.62	0.59	0.61	0.43	0.61
Flame Resistance	Shrinks, melts	Shrinks, melts	Can be treated for flame retardance	Burns	Decomposes	Self extinguishing
MECHANICAL PROPERTIES						
Tensile Strength, psi	125	600	200	1000	900	570
Elongation at Break, %	80	100	45	60	60	90
Shore A Durometer Hardness (avg range, ±10)	5	20	20	25	20	1
Compressibility (at 8 psi), %	60	25	21	18	12	47
Recovery After 8 Psi Compress, (within 1 min), %	70	91	95	94	90	81
CHEMICAL AND ENVIRONMENTAL PROPERTIES						
Sunlight Resistance	Very good	Very good	Good	Good	Excellent	Poor
Solvent Resistance and Stability in Oil	Excellent	Very good	Excellent	Good	Excellent	Excellent
Acid Resistance						
Dilute	Excellent	Good	Fair	Good[f]	Excellent	Good
Concentrated	Excellent	Poor	Poor	Good[f]	Excellent	Good
Alkali Resistance						
Dilute	Good	Excellent	Very good	Excellent	Excellent	Good
Concentrated	Poor	Good	Good	Excellent	Excellent	Poor
Biological Stability	Inert to insects, vermin, fungi, bacteria	Inert to insects, vermin, fungi	Inert to moths, can be treated for vermin and fungi resistance	Inert	Inert	Inert

[a] Available as rolls, sheets, strips, washers, channels, die cut parts.
[b] Du Pont fiber.
[c] % by wt. of liquid of 1.0 specific gravity.
[d] Porosity.
[e] For 575 SSU (at 70 F) oil.
[f] Except nitric acid.
[g] Colors available on special order.

TABLE 12-5. PROPERTIES OF NATURAL MICA (KOHL, 1967)

Property	Unit	Muscovite	Phlogopite
Color	—	light red, green, brown	green-yellow, brown
Density	g/cc	2.6–3.2	2.6–3.2
Mohs hardness		2.8–3.2	2.5–3.0
Shore hardness		80–150	70–100
Tensile strength	kg/mm^2 (psi $\times$ 10^3)	25–30 (35.5–42.6)	—
Compressive strength	kg/mm^2 (psi $\times$ 10^3)	17.7 (25.1)	10.8 (15.3)
Shear strength (d = 50–500μ)	kg/mm^2 (psi $\times$ 10^3)	23.5–26.5 (33.4–37.6)	10–13 (14.2–18.5)
Yield point (d = 250μ)	kg/mm^2 (psi $\times$ 10^3)	35–39 (49.7–55.4)	20–28 (28.4–39.8)
Young's modulus (d = 250μ)	kg/mm^2 (psi $\times$ 10^6)	16,000–21,000 (22.7–29.8)	15,700–19,400 (22.3–27.6)
Onset of calcination	°C	550–650	750–900
Calcination temperature	°C	600–800	900–1100
Melting point	°C	1200–1300	1200–1300
Safe operating temp. *in vacuo*	°C	350–450	—
Specific heat	cal/g°C	0.208	0.206
Thermal conductivity	cal/cm sec °C $\times$ 10^{-3}	0.6–1.4	—
Coeff. of thermal expansion	cm/cm °C $\times$ 10^{-7}	‖83–98 (20–200°C) ⊥170–250 (20–300°C)	‖135 (20–200°C) ⊥162–183 (20–400°C) (sometimes very much more)
Volume resistivity	Ohm. cm	10^{16}–10^{17} (20°C) 10^8–10^{10} (500°C)	lower than muscovite at low temps., higher at high temps.
Surface resistivity	Ohms per square	10^{13}	2 $\times$ 10^{12}
Te-value	°C	>750	—
Dielectric constant (1Gc)	—	7	6
Loss tangent $\times$ 10^{-4}	—	1 kc:6 100 Mc:2 1 Mc:3 3 Gc:3	10–100 (1 Mc)
Dielectric strength (d = 1–3 mil)	kV/mil	3–6	3–4.2
Refractive index	—	1.6	1.6
Optical axis	degrees	55–75	5–25
Transmittance for heat radiation	—	~60% of window glass	same
Transmittance to UV radiation	—	above 2500–3000 A	same
Water absorption	—	practically zero	
Gas evolution *in vacuo*	—	excessive above 750°C	practically none to 800°C
Gas permeation	—	practically zero	

TABLE 12-6. PROPERTIES OF SYNTHETIC MICA (KOHL, 1967)

Property	Unit	Fluor-Phlogopite Mica
Color	—	clear
Density	g/cc	2.9
Mohs hardness	—	3–4
Tensile strength	psi $\times$ 10^3	40–50
Young's modulus	psi $\times$ 10^6	26
Onset of calcination	°C	900–1000
Calcination temperature	°C	1100–1200
Melting point	°C	1365
Safe operating temperature	°C	1000
Specific heat	cal/g	0.2
Thermal conductivity	cal/cm^2/sec/°C/cm	0.0012
Coefficient of thermal expansion	cm/cm/°C $\times$ 10^{-7}	80–130
Volume resistivity	Ohm-cm	10^{16}–10^{17}
Surface resistivity	Ohm per square	10^{13}
Dielectric constant (1 Mc)	—	5–6
Loss tangent $\times$ 10^{-4} (1 Mc)	—	4–7
Dielectric strength (d = 1–3 mils)	kV/mil	2–3
Refractive index	—	1.534–1.566
Optical axis	—	
Transmittance for heat rad.	—	~20% at 6,500 A
Transmittance for UV rad.[26]	—	80% 2500–4500 A
		85% 3000–4000 A
Water absorption	—	practically zero
Gas evolution *in vacuo*	—	practically zero to 800°C
Gas permeation	—	practically zero to 750°C

TABLE 12-7. PHYSICAL PROPERTIES OF WOOD (PISANI, 1964)

Wood	Tensile Strength, psi.	Compressive Strength, psi.	Shearing Strength (Along grain), psi.	Modulus of Rupture, psi.	Modulus of Elasticity, psi.
Ash (black)	15,000	6,000	700	13,000	1,600,000
Basswood (linden)	. . .	4,300		7,000	1,000,000
Beech	11,500	7,500	5,000	10,500	1,500,000
Birch	15,000	8,000	650	11,700	1,600,000
Cedar (white)	11,400	4,800		7,400	600,000
Chestnut	11,500	6,500		10,600	1,200,000
Cypress		6,000		9,800	1,200,000
Elm	13,500	7,200		9,000	1,100,000
Hemlock	8,500	5,800		8,800	950,000
Mahogany	10,000	8,200		11,000	1,300,000
Maple (rock)	10,600	8,000	450	15,000	1,900,000
Oak (white)	10,200	5,200	800	10,600	1,800,000
Pine (yellow)	13,000	8,500	450	14,000	2,000,000
Pine (white)	9,500	4,500	310	8,500	1,100,000

TABLE 12-8. WORKING STRESSES FOR STRUCTURAL WOOD (PISANI, 1964)

| Species | Bending psi. | Hori-zontal Shear psi. | Compression psi. | | Modulus of Elasticity psi. | Modulus of Rupture, psi. | Wgt. (Dry), lb. per cu. ft. |
			With Grain	Across Grain			
Ash, Commercial White	1400	125	1100	500	1,500,000	15,400	42
Cedar, Western Red	900	80	700	200	1,000,000	7,700	23
Cypress	1300	100	1100	350	1,200,000	10,000	32
Douglas Fir, Rocky Mt. Type	1100	90	800	345	1,600,000	9,600	30
Douglas Fir, Coast Type	1600	105	1200	380	1,600,000	11,700	34
White Fir	1100	70	700	300	1,000,000	9,300	26
Hemlock, Western	1300	75	900	300	1,400,000	10,100	29
Hemlock, Eastern	1100	70	700	300	1,000,000	8,900	28
Maple Sugar, Black	1500	125	1200	500	1,600,000	15,800	44
Oak, Commercial	1400	125	1000	500	1,500,000	15,200	48
Pine, Southern Yellow	1600	110	1175	345	1,600,000	12,800	36
Pine, Southern Yellow Dense	1750	125	1300	380	1,600,000	14,700	41
Redwood	1200	70	1000	250	1,200,000	10,000	28
Spruce	1100	80	800	250	1,200,000	10,300	28

REFERENCES

Kohl, W. H., "Handbook of Materials and Techniques for Vacuum Devices," Van Nostrand Reinhold, New York, 1967.

Lofaro, H. W., "Engineer's Guide to Felt," *Materials Eng.*, **74**, (6), 34–42 (Nov. 1971).

Pisani, T. J., "Essentials of Strength of Materials," 3rd ed., Van Nostrand Reinhold, New York, 1964.

BIBLIOGRAPHY

Cook, J. G., "Handbook of Textile Fibers," 2 vols., 4th ed., Textile Book Service, Plainfield, N.J.

Hall, A. J., "The Standard Handbook of Textiles," Chemical Publishing Co., New York, 1970.

Kollmann, F. F. P. and W. A. Cote, Jr., "Principles of Wood Science and Technology," 2 vols., Springer-Verlag, New York, 1968–1974.

Wenzl, H. F., "The Chemical Technology of Wood," Academic Press, New York, 1970.

APPENDIX A
LIST OF REGISTERED TRADEMARKS

This Appendix lists trademarks, used in this book, and not identified when used.

AF-40	3M Co.	INCOLOY	International Nickel Co.
ALUMEL	Hoskins Manufacturing Co.	INCONEL	International Nickel Co.
AM-350	Aluminum Company of America	KEL-F	3M Co.
		KEVLAR	Du Pont
APCO-1261	Hexcel Products Co.	KYNAR	Pennwalt Corp.
AVC	Hudson Wire Co.	KYNOL	Carborundum Co.
BAKELITE	Union Carbide Corp.	LUCITE	Du Pont
BONDMASTER	Rubber and Asbestos Corp.	LUMARITH	Celanese Corporation of America
CHROMEL	Hoskins Manufacturing Co.		
CONSTANTAN	Driver-Harris Co.	METLBOND	Whittaker Corp.
CUPRONICKEL	Chemetron Welding Products	MONEL	International Nickel Co.
		MULTIMET	Cabot Corp.
DACRON	Du Pont#	MYLAR	Du Pont
DELRIN	Alto Corp.	NARMTAPE	Whittaker Corp.
DURANICKEL	International Nickel Co.	NICROTUNG	Westinghouse Electric Corp.
EC 1933	3M Co.	NI-SPAN	International Nickel Co.
ELGILOY	Elgin National Watch Co.	NIVCO	Westinghouse Electric Corp.
EPON	Shell Chemical Co.	NOMEX	Du Pont
ETHOCEL	Dow Chemical Co.	NORYL	General Electric Co.
FORTISAN	Calenese Fibers Co.	PARYLENE	Union Carbide Corp.
HASTELLOY	Union Carbide Corp.	PERBUNAN	Farbenwerke Bayer A.G.
HAYNES	Cabot Corp.	PERMANICKEL	International Nickel Co.
HYCAR	B. F. Goodrich Chemical Co.	PLASTILOCK	B. F. Goodrich Co.
		PLIOFILM	Goodyear Tire & Rubber Co.
HYPALON	Du Pont		
ILLIUM	Stainless Foundry & Engineering, Inc.	PYREX	Corning Glass Works
		PYROMET	Carpenter Technology Corp.

#E. I. du Pont de Nemours & Co.

PYROTOOL	Carpenter Technology Corp.	UDIMET	Special Metals Corp.
REFRACTALOY	Westinghouse Electric Corp.	UMCO-50	Union Minière du Haut-
RENÉ	General Electric Corp.		Katanga
RESTWELD	H. B. Fuller	UNILOY	Cyclops Corp.
SARAN	Dow Chemical Co.	UNITEMP	Cyclops Corp.
SAUEREISEN	Sauereisen Cement Co.	VITON	Du Pont
SWEDLOW	Swedlow Plastics Co.	VULCAPRENE	Essex International, Inc.
TEDLAR	Du Pont	ZIRCALOY	Westinghouse Electric Corp.
TEFLON	Du Pont		

APPENDIX B
TABLE OF ISOTOPES

The material in this section is taken from: Lederer, C. M., J. M. Hollander, and I. Perlman, "Table of Isotopes," John Wiley and Sons, New York (1967), and is reproduced with the permission of the Office of Technical Services, U.S. Atomic Energy Commission, which holds all rights to this material.

Introduction

TABLE I. RADIOISOTOPE DATA

This table displays all radioactive and stable nuclei arranged according to atomic number with increasing mass number for each element. The criterion for the selection of data on each radioactive isotope has been that of identifying it in terms of its rate and mode of decay, principal radiations, and how it is prepared. The data are arranged in six columns, each of which receives comment below.

Note on references. References to the original publications are coded according to the first author and the year of publication. Example: the symbol AagP57 permits the appropriate journal reference to be found readily in the alphabetical listing in the bibliography. If the reader is already familiar with the work, he will recognize this symbol as referring to a 1957 paper of P. Aagard and co-workers.

Column 1—Isotope. The symbols here give the isotopic assignments in usual form. Stable or long-lived naturally occurring isotopes are indicated by underlining. The superscript m following the mass number refers to a metastable, or isomeric, state which has a sufficiently long half-life to be investigated independently from its ground state. Likewise, the designations m_1 and m_2 refer to several metastable states of a nucleus. When it is not established which of several isomers is the ground state, each isomer is referred to by the same symbol without the m; for example, Eu^{150} (12.6 h) and Eu^{150} (≈ 5 y).

Generally, isomeric states are included in Table I if their half-lives exceed ≈ 1 s; exceptions are made for a few chemically or genetically identified isomers of somewhat shorter half-life. The half-lives of many short-lived excited states have been measured because of their importance to nuclear structure. They are not listed in Table I as isomeric states but can be found in Table II, under the listing of the ground state of the appropriate isotope.

The historical names for the naturally occurring activities Th^{232}, U^{235}, U^{238}, and their descendents are given in Column 1 beneath the isotopic assignment.

Column 2—Half-life. An attempt has been made to list the most accurate value first, usually inferred from the stated precision. Unless otherwise stated, the value listed is the total half-life, which is the entity measured when the decay is followed. When a nucleus has more than one mode of decay, the percentage of each mode is given in Column 3.

An exception is made for those heavy nuclei that have measurable spontaneous-fission rates. The appropriate spontaneous-fission half-life is listed in Column 2 and designated by the symbol $t_{1/2}(SF)$. In a number of cases no radioactivity has been observed, although sought, and the lower limit of the half-life is listed for the mode of decay looked for ($\beta \equiv \beta$ decay, $\beta\beta$ $\equiv$ simultaneous emission of two β particles, EC $\equiv$ electron capture, $\alpha \equiv \alpha$ decay).

If there is no special designation after the listed half-life, it may be assumed that the determination was made by direct decay measurement. (For the very short lifetimes the timing is done electronically rather than mechanically.) For indirect half-life determinations, the methods are described by the following symbols:

sp act (+ mass spect)   Determination of disintegration rate of a sample containing a known weight of the active substance (mass spectographic analysis of the sample to correct for other isotopes present).
genet   Decay of parent substance, followed by the periodic removal of a decay product which can be measured. (genet $\equiv$ genetic relation).
yield   Measurement of radioactivity from a sample containing a number of atoms calculated according to the expected yield of the reaction by which it was produced.
est   In a few instances (α emitters) the half-lives are estimated from the energies of the measured radiations.
delay coinc   Several isotopes are short-lived products of longer lived parents. Those whose half-lives are in the millisecond range or shorter were measured by recording the time-interval distribution between the emissions from the parent substance and the daughter product.

Column 3—Type of decay. Because many classes of data are included in this column, the entry denoting *type of decay* is preceded by the special symbol for radiation, ☢. When the mode of decay is enclosed in square brackets, that mode is inferred or assumed, not directly measured. When independent modes of decay have been measured, the branching ratios are entered as percentages. Symbols used are

β^-	Negative β-particle (negatron) emission
β^+	Positive β-particle (positron) emission
EC	Orbital electron capture
α	Alpha-particle emission
IT	Isomeric transition (decay from an excited metastable state to a lower state)
SF	Spontaneous fission. Listings are made here only if the branching is about 1% or more. For others the

partial half-lives for spontaneous fission are entered in Column 2.

n   Neutron emission from excited states promptly following β decay to those levels. Entry is made in conjunction with the β emitter.

p   Proton emission from excited states promptly following β decay to those levels. Entry is made in conjunction with the β emitter.

Wherever experimenters have searched for and failed to find a particular mode of decay, the indication is, for example, "no β^+." Experimental limits are given but no limits predicted from theory. Limits of detection in cases in which *no* radioactivity has been observed are listed in Column 2 in terms of a lower limit on the half-life.

Among the α emitters in the heavy element region closed decay cycles may almost always be employed to determine whether a nucleus is β stable without resort to specific experimental evidence. Those that are known to be β stable are designated by the entry β *stable (cons energy)* to indicate that the principle of conservation of energy underlies the calculations.

Percent abundance. The isotopic abundances listed are on an "atom percent" basis and refer to the elements as they exist in the earth's crust. Some of the light elements have variations in composition outside the accuracy of determination. For these elements ranges are given with references to the publications in which the variations are discussed. Particular values are also given for some specific sources of the specimens analyzed.

Isotopic mass. The atomic masses of all species measured by mass spectrometry or calculated from reaction energies are entered in the form of the mass excess, $\triangle (\equiv M\text{-}A)$; the unified mass scale ($\triangle(C^{12}) = 0$) is employed. It will be noted that these mass excess values are in units of million electron volts. Most of the data were taken from the compilation of Mattauch, Theile, and Wapstra (MTW), which should be consulted for the accuracy attached to them. The experimental decay energies of radioactive species on which many of their masses are based may be found as Q values on the decay schemes in Table II.

Cross sections. It is not possible to list all known reaction cross sections in a table such as this, but values are given for the neutron-capture reaction (σ_c) and for neutron-induced fission (σ_t) in units of 10^{-24} cm^2 (barns). Most of the cross sections shown are taken from a compilation by D. T. Goldman and M. D. Goldberg (GoldmDT64) and refer to neutrons with velocity 2200 meters/sec. The reader is cautioned to note that many nuclei have strong resonances in the epithermal region, and because "thermal" reactors contain epithermal neutrons in the irradiation positions the effective cross sections for certain nuclei can be larger than those indicated here.

Our symbol σ_c refers to that part of the capture reaction in which fission does not result. Unless otherwise stated, σ_c applies to the (n, γ) reaction. For some light nuclei the principal reaction with thermal neutrons may be (n, p) or some other reaction. Wherever such a reaction is referred to, it is so indicated.

Column 4—class; identification; genetic relationships.
Class. The degree of certainty of each isotopic assignment is indicated by a letter according to the following code:

A   Element and mass number certain
B   Element certain and mass number probable
C   Element probable and mass number certain or probable
D   Element certain but mass number not well established
E   Element probable and mass number not well established
F   Insufficient evidence
G   Probably in error.

These "ratings" should not be read as levels of confidence in the experiments but rather as an indication of the limitations of the experiments as they relate isotopic assignments to the radioactive properties discerned. In some instances a simple cross bombardment (production of an isotope in two or more ways) results in an unambiguous assignment. In others much more elaborate experiments are insufficient. Among the factors that can limit the certainty of an assignment based on its means of production are targets of mixed isotopic composition, low cross sections, the possibility of isomerism, similarity of properties to other isotopes, and absence of knowledge of neighboring isotopes.

Identification. The means by which the isotopic assignments were established are tabulated next. In general, several references are combined, and among them the first refers to the discovery of the isotope (except for classical natural radioactivities). Indication of the experimental methods used in making the various assignments may be had from the following symbols:

chem   Chemical separations establishing the chemical identity (atomic number) of the isotope.
genet   Established decay relationship (by chemical or other means) with another isotope whose mass assignment is known.
excit   Refers broadly to energy considerations in the production of the isotope, some of which are
    (1)   excitation-function or yield experiments to establish the nuclear reaction which produced the isotope;
    (2)   limitation of products formed by limiting the energy of bombarding particles;
    (3)   making use of a calculated Q value;
    (4)   in a few instances use of fission-yield data to limit mass assignments.
cross bomb   Arrival at an assignment by producing the isotope in different ways.

n-capt   Key evidence supplied by production with slow neutrons from which it is usually inferred that the (n, γ) reaction was observed.

sep isotopes   The use of target elements enriched or depleted in a particular isotope.

mass spect   Mass number determined by mass spectrometry.

decay charac   Identification of predicted decay properties such as decay energy or energy-level pattern.

genet energy levels   Energy levels of daughter nucleus agree with those from decay of another isotope whose isotopic assignment and mode of decay are known or with levels observed in nuclear reactions.

atomic level spacing   Atomic number of decay product established by measuring the characteristic energy differences between internal-conversion electron lines from a particular γ transition converted in different shells.

critical abs   Identification of the atomic number of the decay product by critical absorption of X-rays accompanying the decay process.

Genetic relationships. Below the designation of how the isotope was identified are listed specifically those genetic (or parent-daughter) relations established by chemical or physical separation and radiochemical characterization of the daughter atoms. Among other things, this list also gives the reader some warning that radiations from decay products may be present with those from the parent.

Column 5—Major radiations. The purpose of this list is to acquaint the reader at a glance with the principal radiations associated with each isotope. The radiations shown will often be sufficient to identify the isotope. Because it is the purpose here to delineate what is actually seen when a particular isotope is encountered, the X-rays and annihilation radiation (0.511-MeV γ rays from the annihilation of positrons, designated by the symbol $\gamma\pm$), are indicated if they are prominent in the electromagnetic spectrum. If essentially all the decays proceed by positron emission, the notation 0.511 (200%, $\gamma\pm$) will appear. (Several per cent of the positrons annihilate in flight, which means that a corresponding number of photons will not have 0.511 MeV energy.) The notation "L X-rays" is used only when K X-rays are absent or very weak. Similarly, conversion electrons are listed if they are prominent in the electron spectrum. Auger electrons (electrons emitted in the de-excitation of atomic levels) are not listed explicitly; they will always accompany the emission of X-rays. Continuous β^- or β^+ spectra are usually represented by the endpoint of the highest energy beta group followed by the notation "max." When the highest energy group is of low intensity, so that a spectrometer of low resolving power (such as a scintillator) would also detect the presence of a continuous spectrum with a lower endpoint energy, this

is also indicated. Thus the notation "β^- 1.176 max (7%), 0.514 max" means that there is a continuous spectrum with endpoint 1.176 MeV and 7% intensity, but the major portion of the β^- spectrum (which may be composed of one *or more* beta groups) has an endpoint energy of 0.514 MeV. Decay products can often give rise to radiations that soon become prominent, and this is indicated by the notation "daughter radiations from . . ." so that the reader will look up the radiations that arise from these sources. The data in this column are derived from the references listed in Table II. *Quantities enclosed in square brackets are calculated or inferred, not measured.*

The term "major radiations," as used here, requires some explanation. In each of the three general categories of radiation, α particles, β particles and electrons, and γ rays and X-rays, we have listed the most prominent radiations, even though they may be of relatively low intensity. For example, with an α emitter may be listed a γ ray of only $10^{-5}\%$ intensity relative to the α intensity if that γ ray is the most intense in its energy range. Conversion electrons are listed according to the actual energies of the electron lines and not in terms of the transitions that give rise to the lines.

The intensities of radiations when expressed as *percentages* without other qualifications refer to percentages of the total decay events. Another way of expressing *relative* intensities is also sometimes employed. A number following the dagger(†) symbol is the relative intensity for the particular mode of decay beside which the † appears.

The terms "doublet" and "complex" are used to indicate γ rays which would be unresolved or incompletely resolved by instruments of moderately low resolving power such as scintillators. It is *not* indicated when an electron line is complex. Because of conversion in different atomic shells and subshells, many of the electron lines listed in Column 5 are complex.

The reader is referred to Table II for a more detailed account of radiations accompanying the decay of each isotope and for references to the original literature.

Column 6—principal means of production. The methods for producing each isotope selected for inclusion here are those that have given the highest yield and those that permit greatest isotopic purity. These listings will serve principally as references to the original literature in which important aspects of the preparations such as experimental conditions, yields, and purity of product are discussed.

The methods fall into three main categories. For ordinary nuclear reactions in which a target isotope is bombarded with charged particles or neutrons the usual system of abbreviations is employed. For example, to make Pu^{237}, the reaction Np^{237} $(d, 2n)$ appears;

this means that Np^{237} is the target, deuterons (d) are the projectiles, and two neutrons $(2n)$ are emitted. When the target material is not isotopically pure, the experimenter must be concerned with radioactive substances produced from other components of the target. A second category of production consists of the separation of the isotope in question from a radioactive parent. Such an isotope is indicated as the daughter of another. Finally, with the advent of very high fluxes of neutrons it has become possible to prepare isotopes by the successive capture of neutrons (with intervening β^- decay in some cases). Such preparations have been designated by "multiple n-capt from —," where the dash refers to the starting material.

TABLE 1

Radioisotope data

Half-life — type of decay — isotopic abundance — atomic mass — neutron cross-section (capture and fission) — class (assignment rating) — means of identification — genetic relationships — major radiations — means of production

Isotope Z A	Half-life	Type of decay (♠); % abundance; Mass excess ($\triangle$=M−A), MeV (C¹²=0); Thermal neutron cross section (σ), barns	Class; Identification; Genetic relationships	Major radiations: approximate energies (MeV) and intensities	Principal means of production
$_0$n^1	11.7 m (SosA59, SosA58, SosA59a, ProkY62) 12.8 m (RobsJ51) 12 m (HameM56a) others (SneA50)	♠ β⁻ (ChadJ35, SneA50) △ 8.0714 (MTW)	A recoil nuclei, conservation of momentum (ChadJ32) observation of (n, α) reaction (FeaN32, HarkW33) parent H¹ (SneA50, RobsJ50)	β⁻ 0.78 max	fission, H³(d, α), Be⁹(α, n), H²(d, He³) Be⁹(γ, n) (photons from electron generator)
$_1$H^1		% 99.9852 (Lake Michigan water); 99.9842 to 99.9877 (other sources) (BegF59a) 99.9849 to 99.9861 (KirI51) △ 7.2890 (MTW) σ_c 0.332 (GoldmDT64)			
H^2		% 0.0148 (Lake Michigan water); 0.0123 to 0.0158 (other sources) (BegF59a) 0.0139 to 0.0151 (KirI51) △ 13.1359 (MTW) σ_c 0.0005 (GoldmDT64)			
H^3	12.262 y genet (JonWM55) 12.46 y genet (JenkG50) 12.6 y (PopM58) others (JonWM51, NoviA47, AlvL39, AlvL40, HugD48a, ONeaR40, CornR41)	♠ β⁻ (AlvL39, AlvL40) △ 14.9500 (MTW) σ_c <6.7 x 10⁻⁶ (GoldmDT64) (absorption not possible)	A chem, sep isotopes, excit (AlvL39, AlvL40)	β⁻ 0.0186 max average β⁻ energy: 0.0057 calorimetric (PilW61) 0.0055 calorimetric (PopM58) others (GregD58) γ no γ	Li⁶(n, α) (ONeaR40)
$_2$He^3		% 1.3 x 10⁻⁴ (atmosphere) 1.7 x 10⁻⁵ (wells) (AldL46, CoonJ49) △ 14.9313 (MTW) σ (n, p) 5330 (GoldmDT64)			H³(β⁻)
He^4		% ≈100 △ 2.4248 (MTW) σ_c (total absorption) 0 (GoldmDT64)			
He^6	0.797 s (BieJ62) 0.799 s (KliR54) 0.85 s (BornG62, VeeN56) 0.83 s (HerrmW58, AlleJS59) 0.86 s (MalmS62) 0.82 s (HolmJ49) others (SomH46, RusB55, BattM53, VenG52, ShelR52a, PolA37, DewJ52)	♠ β⁻ (BjeT36b) △ 17.598 (MTW)	A chem (BjeT36, BjeT36a) cross bomb, excit, chem (SomH46)	β⁻ 3.508 max γ no γ	Be⁹(n, α) (RusB55, BjeT36, PolA37, SomH46, KnoW48, PerezV50) Li⁷(γ, p) (ShelR52a)
He^8	0.122 s (PosA65a) 0.03 s (NefB63a)	♠ β⁻ 100%, n 12% (PosA65a) △ 31.7 (CerJ66a)	B chem, excit, cross bomb (PosA65)	β⁻ [9.7 max] γ 0.98 (88%) daughter radiations from Li⁸	protons on C, O (PosA65a)
$_3$Li^6		% 7.42 (OmuI58, HigM55, OrdK55) 7.29–7.42 (CamAE55) △ 14.088 (MTW) σ (n, α) 953 (GoldmDT64)			
Li^7		% 92.58 (OmuI58, HigM55, OrdK55) 92.58–92.71 (CamAE55) △ 14.907 (MTW) σ_c 0.037 (GoldmDT64)			

TABLE 1 *(Continued)*

Isotope Z A	Half-life	Type of decay (☢); % abundance; Mass excess (△≡M–A), MeV (C¹²=0); Thermal neutron cross section (σ), barns	Class; Identification; Genetic relationships	Major radiations: approximate energies (MeV) and intensities	Principal means of production
₃Li⁸	0.841 s (KliR54) 0.83 s (RalW51) 0.88 s (BayD37, OglW47, ConnD59) 0.87 s (BretP53) 0.85 s (ShelR52a) others (HugD47a, WinnM54, BunbD53, NefB53a)	☢ β⁻, 2α (LewisW37) △ 20.946 (MTW)	A excit (CranH35a) n-capt, sep isotopes, genet (HugD47a)	β⁻ 13 max α 1.6 (broad peak, with 2.90 level of Be⁸)	Li⁷(n, γ) (ImhW59) Li⁷(d, p) (CranH35a, DellL35, FowW37, BayD37, LewisW37, HornW50, YafL50)
Li⁹	0.176 s (DosI65) 0.168 s (GardW51, ReaD53) 0.170 s (HoltR52) others (AlbuD63a, NefB63a, SchoR65, ShelR52a, BendP55)	☢ β⁻, n, [2α] (GardW51, HoltR52) △ 24.97 (MTW)	A excit, cross bomb (GardW51) genet energy levels (AlbuD63a)	β⁻ 13.61 max n 0.76 α [0.05 (with ground state of Be⁸)]	Be⁹(n, p) (AlbuD63a) Be⁹(d, 2p) (GardW51, SchoR65)
₄Be⁶	≈0.4 s (TyrH54)	☢ (TyrH54) △ 18.37 (MTW)	C excit (TyrH54) nucleus is particle-unstable (AjzF59)		protons on Li, Be (TyrH54)
Be⁷	53.6 d (KraJJ53a) 52.9 d (SegE49a) 53.1 d (EnglJ65) 53.0 d (RobeJ59, BourR56, BouR47) 53.5 d (WriH57)	☢ EC (RumL38) △ 15.769 (MTW) σ (n, p) 54,000 (GoldmDT64)	A chem, cross bomb, excit (RumL38)	γ 0.477 (10.3%)	Li⁶(d, n) (RumL38, RobeR38, ZloI42) B¹⁰(p, α) (RobeR38, MaiH39) C¹²(He³, 2α) (EnglJ65)
Be⁹		% 100 (NierA37a) △ 11.351 (MTW) σ_c 0.009 (GoldmDT64)			
Be¹⁰	2.5 x 10⁶ y sp act + mass spect (MMilE47) 2.9 x 10⁶ y yield (HugD47)	☢ β⁻ (MMilE46) △ 12.607 (MTW)	A chem (MMilE46) chem, mass spect (PierAK46)	β⁻ 0.555 max γ no γ	Be⁹(n, γ) (HugD47, AlbuD50, BellP50c) Be⁹(d, p) (MMilE46, LeviJ47)
Be¹¹	13.6 s (WilkD59, NefB63a, AlbuD58c) 14.1 s (NurM58a)	☢ β⁻ (AlbuD58c, WilkD59) △ 20.18 (MTW)	A excit, genet energy levels (AlbuD58c, WilkD59)	β⁻ 11.5 max γ 2.14 (32%), 4.67 (2.1%), 5.85 (2.4%), 6.79 (4.4%), 7.99 (1.7%)	B¹¹(n, p) (WilkD59, AlbuD58c)
Be¹²	0.0114 s (PosA65)	☢ [β⁻], n (PosA65) △ 25 (PosA65, MTW)	C cross bomb (PosA65)		protons on O¹⁸, N¹⁵, F¹⁹, Na²³, Al²⁷, O¹⁶ (PosA65)
₅B⁸	0.77 s (MattE64) 0.78 s (DunnK58) others (ShelR52a)	☢ β⁺, 2α (AlvL50) △ 22.923 (MTW)	A excit, cross bomb (AlvL50)	β⁺ [14.0 max] α 1.6 (broad peak, with 2.90 level of Be⁸) γ [0.511 (200%, γ±)]	Li⁶(He³, n) (DunnK58, MattE64)
B¹⁰		% 19.6–19.8 (NewD59) 19.58 (ShiuV55) 18.45–18.98 (ThodH48) 19.3 (BentP58) 20.0 (LehW59) △ 12.052 (MTW) σ (n, α) 3837 (GoldmDT64)			
B¹¹		% 80.2–80.4 (NewD59) 80.42 (ShiuV55) 80.0 (LehW59) 81.7 (BentP58) 81.02–81.55 (ThodH48) △ 8.6677 (MTW) σ_c 0.005 (GoldmDT64)			
B¹²	0.0203 s (FishT63, SchaA61) 0.0202 s (PeteRW63) 0.0189 s (KreW59) others (NorE56, BretP53, JelJ48a, BrolJ51, CookB56, CookB57)	☢ β⁻ (CranH35) β⁻ 100%, 3α 1.5% AlbuD63, CookCW57, CookCW58) △ 13.370 (MTW)	A excit (CranH35, FowW36)	β⁻ 13.37 max γ 4.43 (1.3%) α 0.195 (1.5%), broad distribution to ≈3 MeV	B¹¹(d, p) (CranH35, FowW36, BrolJ51)

TABLE 1 (*Continued*)

Isotope Z A	Half-life	Type of decay (☢); % abundance; Mass excess (Δ≡M–A), MeV (C^{12}=0); Thermal neutron cross section (σ), barns	Class; Identification; Genetic relationships	Major radiations: approximate energies (MeV) and intensities	Principal means of production
$_5B^{13}$	0.0186 s (MarqA62)	☢ β^- (MarqA62) no n, lim 0.3% (PosA65) Δ 16.562 (MTW)	B excit (HubbE53, NorE56) excit, genet energy levels (MarqA62)	β^- 13.44 max γ 3.68 (7%)	B^{11}(t, p) (MarqA62)
$_6C^9$	0.127 s (HardJ65a)	☢ [β^+], p, [2α] (HardJ65a) Δ 29.0 (CerJ66)	B excit, cross bomb (HardJ65a)	p 8.2 (60%), 1.1 (40%), both peaks broad α [0.05, 1.6 (broad peak, with 2.90 level of Be^8)]	B^{10}(p, 2n) (HardJ65a) B^{11}(p, 3n) (HardJ65a)
C^{10}	19.48 s (EarL62) 19.3 s (BartiF63) 19.1 s (SherrR49)	☢ β^+ (SherrR49) Δ 15.66 (MTW)	A chem, sep isotopes (SherrR48, SherrR49)	β^+ 1.87 max γ 0.511 (200%, $\gamma^\pm$), 0.717 (100%), 1.023 (1.7%)	B^{10}(p, n) (SherrR48, SherrR49)
C^{11}	20.34 m (KavT64) 20.4 m (FolK62, SmiJ41) 20.5 m (SolA41, PerlmM48, ChrisD50) 20.1 m (ArnS58) 20.3 m (MartiW52) others (KunD53, PoolM52, SiegK44a, DicksJ51, PatJ65)	☢ β^+ 99+%, EC(K) 0.19% (ScoJ57a) Δ 10.648 (MTW)	A excit (CranH34) chem, excit (BarkW39)	β^+ 0.97 max γ 0.511 (200%, $\gamma^\pm$)	B^{11}(p, n) (BarkW39) B^{10}(p, γ) (CranH34a, BarkW39) B^{10}(d, n) (CocJ35, YosD35, FowW36) N^{14}(p, α) (BarkW39)
<u>C^{12}</u>		% 98.892 (limestone CO_2) (NierA50) Δ ≡0 σ_c 0.0034 (GoldmDT64)			
<u>C^{13}</u>		% 1.108 (limestone CO_2) (NierA50) Δ 3.125 (MTW) σ_c 0.0009 (GoldmDT64)			
C^{14}	5730 y (GodH62) 5745 y (HugE64, MannWB61) 5680 y (OlsI62) 5568 y (LibW55) (all values by sp act) others (WatD61, EngeA50, JonWM49, MillWW50, ManoG51, HawR49, ReidA46, HawR48, NorL48, YafL48a, CaswR54)	☢ β^- (KameM40) Δ 3.0198 (MTW)	A chem, cross bomb, excit (RubeS41)	β^- 0.156 max average β^- energy: 0.045 calorimetric (JenkG52) γ no γ	N^{14}(n, p) (RubeS41, LibW55)
C^{15}	2.5 s (NelJB64) 2.25 s (DouR56) 2.4 s (HudE50a)	☢ β^- (HudE50) Δ 9.873 (MTW)	A excit, sep isotopes (HudE50) genet energy levels (WarbE65)	β^- 9.82 max (32%), 4.51 max (68%) γ 5.299 (68%)	C^{14}(d, p) (HudE50, HudE50a, AlFuD59a)
C^{16}	0.74 s (HinS61a)	☢ [β^-], n (HinS61a) Δ 13.69 (MTW)	C excit, decay charac (HinS61a)		C^{14}(t, p) (HinS61a)
$_7N^{12}$	0.01095 s (FishT63) 0.0110 s (PeteRW63) 0.0125 s (AlvL49a)	☢ β^+, 3α (AlvL50) β^+ 100%, 3α 3.0% (MayT62, GlasN63) Δ 17.36 (MTW)	A excit, sep isotopes (AlvL49a) genet energy levels (MayT62, WilkD63a, GlasN63, PeteRW63)	β^+ 16.4 max γ 0.511 (200%, $\gamma^\pm$), 4.43 (2.4%) α 0.195 (3%), broad distribution to ≈3 MeV	C^{12}(p, n) (AlvL49a, AlvL50) B^{10}(He^3, n) (PeteRW63)
N^{13}	9.96 m (EbrT65, ArnS58, DaniH58, DaniH57b) 10.05 m (FolK62, BormM65, ChurJ53) 10.08 m (WilkD55) 9.93 m (WardAG39a)	☢ β^+ (CranH34) Δ 5.345 (MTW)	A excit (CuriI34, CranH34)	β^+ 1.20 max γ 0.511 (200%, $\gamma^\pm$)	B^{10}(α, n) (CuriI34, ElliC35, RideL37a) C^{12}(d, n) (CranH34, HafL35, YosD35, FowW36, CocJ35) C^{13}(p, n) (AdaRE50) C^{12}(p, γ) (HafL35, CocJ35)
<u>N^{14}</u>		% 99.635 (NierA50) Δ 2.8637 (MTW) σ (n, p) 1.81 (GoldmDT64)			

TABLE 1 (*Continued*)

Isotope Z A	Half-life	Type of decay (☙); % abundance; Mass excess ($\triangle \equiv$M–A), MeV (C^{12}=0); Thermal neutron cross section (σ), barns	Class; Identification; Genetic relationships	Major radiations: approximate energies (MeV) and intensities	Principal means of production
$_7N^{15}$		% 0.365 (NierA50) $\triangle$ 0.100 (MTW) σ_c 2.4 x 10^{-5} (GoldmDT64)			
N^{16}	7.14 s (BieJ64) 7.35 s (ElliJ59, BleE47) 7.16 s (GrayP65a) 7.31 s (MalmS62) 7.22 s (PinI62) others (MartiHC54, NelJB64, SomH46, CrePA65)	☙ β^- (LivM34a, FermE34) α 0.0006% (SegR61, SegR61b) α 0.0012% (KauW61) 0.0003% (AlbuD61) $\triangle$ 5.685 (MTW)	A excit (LivM34a, FermE34)	β^- 10.40 max (26%), 4.27 max γ 2.75 (1%), 6.13 (69%), 7.11 (5%) α 1.7	N^{15}(d, p) (AlbuD59a, FowW36) O^{16}(n, p) (ChanW37, BleE47) F^{19}(n, α) (LivM34a, FermE34, NahM36, PolA37) N^{15}(n, γ) (PinI62)
N^{17}	4.16 s (DosI65) 4.14 s (KnaK48) 4.15 s (StepW51)	☙ β^-, n (KnaK48) $\triangle$ 7.87 (MTW)	A chem, cross bomb (AlvL49, KnaK48, ChupW48)	β^- 8.68 max (1.6%), 7.81 max (2.6%), 4.1 max (95%) γ 0.87 (3%), 2.19 (0.5%) n 0.40 (45%), 1.21 (45%), 1.81 (5%)	N^{15}(t, p) (SilM64) C^{14}(a, p) (StepW51) O^{17}(n, p) (CharR49)
N^{18}	0.63 s (ChasL64)	☙ β^- (ChasL64) $\triangle$ 13.1 (ChasL64, MTW)	A sep isotopes, genet energy levels (ChasL64)	β^- 9.4 max γ 0.82 (59%), 1.65 (59%), 1.98 (100%), 2.47 (41%)	O^{18}(n, p) (ChasL64)
$_8O^{13}$	0.0087 s (MPheR65a)	☙ [β^+], p (MPheR65a) $\triangle$ 23.1 (CerJ66)	C excit, genet energy levels (MPheR65a, BartoR63)	p 6.40 (↑ 100), 6.97 (↑ 24)	N^{14}(p, 2n) (MPheR65a)
O^{14}	70.91 s (HendD61) 71.0 s (BardR62) 71.3 s (FrickG63) others (BardR60, GerhJ54, SherrR49, BromD57a, KuaH64a)	☙ β^+ (SherrR49) $\triangle$ 8.0080 (MTW)	A chem, excit (SherrR49) genet energy levels (SherrR53)	β^+ 4.12 max (0.6%), 1.811 max (99%) γ 0.511 (200%, $\gamma^\pm$), 2.312 (99%)	N^{14}(p, n) (SherrR49)
O^{15}	123 s (NelJW63) 124 s (PenJ57, KliR54, FolK62) 125 s (CsiJ63a) others (PerezV49, BashS55, KisO57, MMilE35a, BotW39, DuncD51, VasiSS63a)	☙ β^+ (LivM34) $\triangle$ 2.860 (MTW)	A chem, excit (LivM34, MMilE35a) excit (FowW36, KinL39a)	β^+ 1.74 max γ 0.511 (200%, $\gamma^\pm$)	N^{14}(d, n) (LivM34, MMilE35a, FowW36, BrowH50) N^{14}(p, γ) (DubL38, DuncD51) O^{16}(He^3, α) (WarbE65) C^{12}(α, n) (KinL39a, VasiSS63a)
O^{16}		% 99.759 (air O_2) (NierA50) O^{16}/O^{18} variation $\leq$4% (ThodH49, KameM46) $\triangle$ –4.7366 (MTW) σ_c 0.00018 (GoldmDT64)			
O^{17}		% 0.037 (air O_2) (NierA50) $\triangle$ –0.808 (MTW) σ (n, α) 0.24 (GoldmDT64)			
O^{18}		% 0.204 (air O_2) (NierA50) $\triangle$ –0.7824 (MTW) σ_c 0.00021 (GoldmDT64)			
O^{19}	29.1 s (MalmS62) 27.2 s (BormM65) 29.4 s (FulH44) 27.0 s (BleE47a)	☙ β^- (MarsJ43) $\triangle$ 3.333 (MTW)	A excit (NahM36) n-capt (MarsJ43)	β^- 4.60 max γ 0.197 (97%), 1.37 (59%)	O^{18}(n, γ) (MarsJ43, SerL47b, SerL46) O^{18}(d, p) (AlbuD59a)
O^{20}	14 s (SchaG60)	☙ [β^-] (SchaG60) $\triangle$ 3.80 (MTW)	B sep isotopes, excit, genet (SchaG60) parent F^{20} (SchaG60)	β^- [2.75 max] γ 1.06 (100%) daughter radiations from F^{20}	O^{18}(t, p) (SchaG60)

TABLE 1 (*Continued*)

Isotope Z A	Half-life	Type of decay (☙); % abundance; Mass excess ($\triangle \equiv M-A$), MeV ($C^{12} \equiv 0$); Thermal neutron cross section (σ), barns	Class; Identification; Genetic relationships	Major radiations: approximate energies (MeV) and intensities	Principal means of production
$_9F^{17}$	66.6 s (ArnS58) 66 s (KoesL54, WonC54a) others (WarrJ54, NewH35, PerezV50b, HorsR52, PerlmM48, DubL38, HestR58, VasiSS62c)	☙ β^+ (NewH35) $\triangle$ 1.952 (MTW)	A cross bomb (Werl34, ElliC34a) chem, excit (NewH35, HaxO35, Dubl38)	β^+ 1.74 max γ 0.511 (200%, $\gamma^\pm$)	O^{16}(d, n) (NewH35, FowW36, PerezV50b) N^{14}(a, n) (WerL34, ElliC34a, RideL37a)
F^{18}	109.7 m (MahJ64) 109.9 m (EbrT65) others (BendW58, CarlC59, BegK63, HofmI64, BlasJ49, PerlmM48, KriR41, JarN55, BormM65, HubeO43, DubL38, SneA37a)	☙ β^+ 97%, EC 3% (DreR56) $\triangle$ 0.872 (MTW)	A chem (SneA37a) chem, sep isotopes, excit (DubL38)	β^+ 0.635 max γ O X-rays, 0.511 (194%, $\gamma^\pm$)	O^{18}(p, n) (DubL38) O^{16}(t, n) (MahJ64) O^{16}(He^3, p) (MahJ64) F^{19}(n, 2n) (BormM65) F^{19}(d, t) (KriR41) Ne^{20}(d, a) (SneA37a)
$\underline{F^{19}}$		% 100 (AstF20) $\triangle$ −1.486 (MTW) σ_c 0.010 (GoldmDT64)			
F^{20}	11.56 s (MalmS62) 11.4 s (GliS63) 11.2 s (SchaG60) 10.7 s (SnoS50) others (CranH35a, VasiSS59)	☙ β^- (CranH35a) $\triangle$ −0.012 (MTW)	A excit (CranH35, FowW36, NahM36) daughter O^{20} (SchaG60)	β^- 5.41 max γ 1.63 (100%)	F^{19}(n, γ) (SerL47b, GliS63, NahM36) F^{19}(d, p) (CranH35a, FowW36, SnoS50, JelJ50, NemY50)
F^{21}	4.35 s (ForJ65) 4.6 s (KieP63) 5 s (CamE52)	☙ β^- (KieP63) $\triangle$ −0.05 (MTW)	A cross bomb (CamE52) genet energy levels (KieP63)	β^- 5.4 max γ 0.350 (↑ 100), 1.38 (↑ 13)	O^{18}(a, p) (ForJ65) F^{19}(t, p) (KieP63, HorvP64, HinS62, SilM61a)
F^{22}	4.0 s (VauF65a)	☙ β^- (VauF65a) $\triangle$ 4 (VauF65a, MTW)	B sep isotopes, genet energy levels (VauF65a)	β^- 11 max γ 1.28 (100%), 2.06 (67%)	Ne^{22}(n, p) (VauF65a)
$_{10}Ne^{17}$	0.10 s (MPheR64)	☙ [β^+], p (MPheR64, BartoR63) $\triangle$ 33.9 (MPheR64, MTW)	B excit, genet energy levels (MPheR64, BartoR63)	p 4.59	F^{19}(p, 3n) (MPheR64)
Ne^{17}	0.69 s (DAurJ64)	☙ [β^+], p (DAurJ64)	G cross bomb (DAurJ64) activity not observed (EstR66)		
Ne^{18}	1.5 s (ButlJW61a, FrickG63) 1.6 s (GowJ54) others (EccD61)	☙ β^+ (GowJ54) $\triangle$ 5.319 (MTW)	B excit, cross bomb (GowJ54)	β^+ 3.42 max γ 0.511 (200%, $\gamma^\pm$), 1.04 (7%)	F^{19}(p, 2n) (GowJ54) O^{16}(He^3, n) (FrickG63)
Ne^{19}	17.4 s (EarL62, AlleJS59) 17.7 s (PenJ57) 18.5 s (SchrG52) 18.6 s (BlasJ51b) 18.3 s (AlfWP57) 18.2 s (SherrR49) others (WhiM39, NahM54c, WallR60, VasiSS64)	☙ β^+ (WhiM39) $\triangle$ 1.752 (MTW)	A cross bomb, excit (WhiM39)	β^+ 2.22 max γ 0.511 (200%, $\gamma^\pm$)	F^{19}(p, n) (WhiM39, BlasJ51b, SchrG52)
$\underline{Ne^{20}}$		% 90.92 (NierA50a) variations in Ne^{20}/Ne^{21} and Ne^{20}/Ne^{22} (WetG54) $\triangle$ −7.042 (MTW)			
$\underline{Ne^{21}}$		% 0.257 (NierA50a) $\triangle$ −5.730 (MTW)			
$\underline{Ne^{22}}$		% 8.82 (NierA50a) $\triangle$ −8.025 (MTW) σ_c 0.04 (GoldmDT64)			

TABLE 1 (*Continued*)

Isotope Z A	Half-life	Type of decay (☘); % abundance; Mass excess (△≡M−A), MeV (C¹²=0); Thermal neutron cross section (σ), barns	Class; Identification; Genetic relationships	Major radiations: approximate energies (MeV) and intensities	Principal means of production
$_{10}Ne^{23}$	37.6 s (PenJ57) 37.5 s (AlleJS59, BurmRL59) 38.0 s (NurM58) 40.2 s (BrowH50a) others (HubeO44, RidlB58, AmaE35, BjeT37)	☘ β⁻ (PolE40) △ −5.148 (MTW)	A excit (AmaE35) chem (BjeT37, PolA37)	β⁻ 4.38 max γ 0.439 (33%), 1.64 (0.9%)	Ne²²(n, γ) (LancH65) Ne²²(d, p) (PolE40, BrowH50a, PerezV50a) Na²³(n, p) (AmaE35, NahM36, PolA37, BjeT37, CarlT63)
Ne²⁴	3.38 m (DroB56)	☘ β⁻ (DroB56) △ −5.95 (MTW)	B chem, genet (DroB56) ancestor Na²⁴, parent Na²⁴ᵐ (DroB56)	β⁻ 1.99 max γ 0.472 (100%, with Na²⁴ᵐ), 0.88 (8%) daughter radiations from Na²⁴	Ne²²(t, p) (DroB56)
$_{11}Na^{20}$	0.39 s (MacfR64a, BirgA52a) 0.23 s (ShelR51) 0.25 s (AlvL50)	☘ β⁺, α (AlvL50, ShelR51) △ 7.0 (PehR65b)	A excit (AlvL50) excit, cross bomb (MacfR64a) daughter Mg²⁰ (MacfR64a)	β⁺ [11.4 max] γ [0.511 (200%, γ±), 1.63] α 2.14 (↑ 100), 2.49 (↑ 5), 4.44 (↑ 21)	Ne²⁰(p, n) (AlvL50) C¹²(B¹⁰, 2n), C¹²(B¹¹, 3n) (MacfR64a)
Na²¹	23.0 s (ArnS58) 21.6 s (WallR60) 22.8 s (SchrG52) 23 s (CreEC40c)	☘ β⁺ (PolE40) △ −2.19 (MTW)	A excit (CreEC40c)	β⁺ 2.52 max γ 0.350 (2.3%), 0.511 (200%, γ±)	Mg²⁴(p, α) (BradHu48) Ne²⁰(p, γ) (BrosK47) Ne²⁰(d, n) (PolE40)
Na²²	2.62 y (WyaE61) 2.58 y (MerW57) 2.60 y (LasL49) others (SahN39)	☘ β⁺ 90.6%, EC 9.4% (WilliA64) β⁺ 90%, EC 10% (KoniJ58c, SherrR54) β⁺ 89%, EC 11% (AlleR55, KreW54a, HageH57) △ −5.182 (MTW)	A chem, excit (FrisO35)	β⁺ 1.820 max (0.05%), 0.545 max γ Ne X-rays, 0.511 (180%, γ±), 1.275 (100%)	F¹⁹(α, n) (FrisO35, LasL37, MagC37) Mg²⁴(d, α) (LasL37, AlbuD49)
Na²³		% 100 (SamM36a, WhiF56) △ −9.528 (MTW) σc 0.40 (to Na²⁴ᵐ) 0.53 (to Na²⁴ by direct production + production via Na²⁴ᵐ) (GoldmDT64)			
Na²⁴	14.96 h (CamP58) 14.95 h (WolfG60) 15.05 h (WyaE61, JozE61, MonaJ62) 14.97 h (LocE53) 15.06 h (SreJ51) 15.10 h (CobJ50) 15.04 h (SolA50) 14.90 h (TobJ55) others (PouA59, LovG60, SinW51, WilsR49, ForS52, WriH57)	☘ β⁻ (LawE35) △ −8.418 (MTW)	A chem, excit (FermE34, LawE35) descendant Ne²⁴ (DroB56)	β⁻ 4.17 max (0.003%), 1.389 max (100%) γ 1.369 (100%), 2.754 (100%)	Na²³(n, γ) (AmaE35, SerL47b)
Na²⁴ᵐ	0.0203 s (AlexKF63) 0.0199 s (SchaA61) others (GlagV61, AlexKF60, GlagV59, DroB56)	☘ IT, β⁻ (DroB56) △ −7.945 (LHP, MTW)	A genet (DroB56) n-capt (FetP62a) daughter Ne²⁴ (DroB56)	β⁻ 6 max γ 0.472	daughter Ne²⁴ (DroB56) Na²³(n, γ) (CamE59, AlexKF60) Na²³(d, p) (SchaA61)
Na²⁵	60 s (RieW44, IweJ55, NahM56) 61 s (HubeO44) 62 s (PerlmM48, BaldG46) 58 s (BleE47a)	☘ β⁻ (HubeO43b) △ −9.36 (MTW)	A excit (HubeO43b) genet energy levels (MaeD55)	β⁻ 3.83 max γ 0.39 (14%), 0.58 (14%), 0.98 (15%), 1.61 (6%)	Mg²⁵(n, p) (HubeO43b, BleE47a)
Na²⁶	1.04 s (NurM58) 1.03 s (RobiE61)	☘ β⁻ (NurM58) △ −7.7 (MTW)	B excit (NurM58) genet energy levels (RobiE61)	β⁻ 6.7 max γ 1.82 (100%)	Mg²⁶(n, p) (NurM58, RobiE61)
$_{12}Mg^{20}$	0.6 s (MacfR64a)	☘ [β⁺] (MacfR64a) △ 16 (MacfR64a, PehR65b)	C genet (MacfR64a) parent Na²⁰ (MacfR64a)		Ne²⁰ on Al²⁷ (MacfR64a)
Mg²¹	0.121 s (MPheR65)	☘ [β⁺], p (MPheR65, BartoR63) △ 10.9 (MPheR65, MTW)	C excit, cross bomb (MPheR65, BartoR63)	p 3.3, 3.8, 4.58, 6.14	Na²³(p, 3n) (MPheR65)

TABLE 1 (*Continued*)

Isotope Z A	Half-life	Type of decay (☢); % abundance; Mass excess (△≡M-A), MeV (C¹²=0); Thermal neutron cross section (σ), barns	Class; Identification; Genetic relationships	Major radiations: approximate energies (MeV) and intensities	Principal means of production
$_{12}Mg^{22}$ (or Al^{23})	0.13 s (TyrH54)	☢ △ -0.38 (CerJ66a)	F excit (TyrH54)		protons on Mg (TyrH54)
Mg^{23}	12.1 s (MihM58) 11.9 s (WallR60, HubeO43) 12.3 s (BolF51) 11.6 s (WhiM39) 11 s (HunS54)	☢ β⁺ (WhiM39) △ -5.472 (MTW)	A excit, cross bomb (WhiM39)	β⁺ 3.03 max γ 0.44 (9%), 0.511 (200%), γ±)	$Na^{23}(p, n)$ (WhiM39, DubL40a)
Mg^{24}		% 78.60 (WhiJ48) 78.8 (WhiF56) △ -13.933 (MTW) σ (total absorption) 0.03 (GoldmDT64)			
$\underline{Mg^{25}}$		% 10.11 (WhiJ48) 10.2 (WhiF56) △ -13.191 (MTW) σ (total absorption) 0.3 (GoldmDT64)			
$\underline{Mg^{26}}$		% 11.29 (WhiJ48) 11.1 (WhiF56) △ -16.214 (MTW) σc 0.027 (GoldmDT64)			
Mg^{27}	9.46 m (PouA59) 9.51 m (DaniH53) 9.45 m (SargB53) 9.39 m (LocE53) 9.5 m (ElliJ59, BonaG64) 9.6 m (EklS43, ForS52, SalS65) others (CriE39, HendM35)	☢ β⁻ (HendM35) △ -14.583 (MTW) σc <0.030 (GoldmDT64)	A chem, excit (AmaE35, HendM35)	β⁻ 1.75 max γ 0.18 (0.7%), 0.84 (70%), 1.013 (30%)	$Mg^{26}(n, \gamma)$ (AmaE35, SerL47b)
Mg^{28}	21.2 h (LindnM53) 21.3 h (ShelR53) 21.8 h (IweJ53) 22.1 h (JonJW53) 20.8 h (MarqL53) 21.4 h (WapA53c)	☢ β⁻ (LindnM53, ShelR53) △ -15.02 (MTW)	A chem, genet (LindnM53, ShelR53) parent Al^{28} (LindnM53, ShelR53)	β⁻ 0.46 max e⁻ 0.030 γ 0.031 (96%), 0.40 (30%), 0.95 (30%), 1.35 (70%) daughter radiations from Al^{28}	$Mg^{26}(t, p)$ (IweJ53, MidR64b) $Mg^{26}(\alpha, 2p)$ (WapA53c, ShelR53, ShelR54)
$_{13}Al^{23}$ (or Mg^{22})	0.13 s (TyrH54)	☢ △ -0.38 (CerJ66a)	F excit (TyrH54)		protons on Mg (TyrH54)
Al^{24}	2.10 s (GlasN53) 2.0 s (BrecS54) 2.3 s (BirgA52)	☢ β⁺ 100%, α ≈10⁻²% (GlasN55) △ -0.1 (MTW)	A excit, decay charac (BirgA52)	β⁺ 8.5 max γ 0.511 (200%, γ±), 1.368, 2.754, 4.2, 5.3, 7.1 α 2	$Mg^{24}(p, n)$ (BirgA52, BrecS54, GlasN53)
Al^{25}	7.24 s (MullT58a) 7.1 s (ArnS58) 7.3 s (WallR60, BradHu48) 7.6 s (HunS54a, ChurJ53)	☢ β⁺ (BradHu48) △ -8.93 (MTW)	B excit, sep isotopes (BradHu48)	β⁺ 3.24 max γ 0.511 (200%, γ±)	$Mg^{24}(p, \gamma)$ (HunS54a, ArnS58, MullT58a) $Mg^{25}(p, n)$ (BradHu48)
Al^{26}	7.4 x 10⁵ y sp act + mass spect (RigR58) 8 x 10⁵ y sp act + mass spect (FishP58) others (RigR57)	☢ β⁺ 85%, EC 15% (RigR59) △ -12.211 (MTW)	A chem, decay charac (SimaJ54) chem, cross bomb, mass spect (RigR58)	β⁺ 1.17 max γ Mg X-rays, 0.511 (170%, γ±), 1.12 (4%), 1.81 (100%)	$Mg^{26}(p, n)$ (HandT55a) $Mg^{25}(d, n)$ (RigR59, FergJ58) $Si^{28}(d, \alpha)$ (LauM55)
Al^{26m}	6.37 s (FreeJ65, FreeJ62a) 6.28 s (MullT58a) 6.74 s (MihM58) 6.5 s (KatzL51a, HasR54, ArnS58) 6.7 s (HunS54a, ChurJ53) others (FrickG63, WhiM39, AllaH48, PerlmM48, WafH48)	☢ β⁺ (FrisO34) △ -11.982 (LHP, MTW)	A excit (FrisO34) cross bomb (HubeO43, BradHu48)	β⁺ 3.21 max γ 0.511 (200%, γ±)	$Na^{23}(\alpha, n)$ (FrisO34, MagC37)

TABLE 1 (*Continued*)

Isotope Z A	Half-life	Type of decay (☢); % abundance; Mass excess (△≡M−A), MeV (C¹²=0); Thermal neutron cross section (σ), barns	Class; Identification; Genetic relationships	Major radiations: approximate energies (MeV) and intensities	Principal means of production
$_{13}Al^{27}$		% 100 (BaiK50, WhiF56) △ −17.196 (MTW) σ_c 0.235 (GoldmDT64)			
Al^{28}	2.31 m (ElliJ59, MalmS62) 2.27 m (BarthR53b) 2.30 m (EklS43) others (CohAV56, SzaA48, IweJ53, FlorJ62)	☢ β⁻ (MMilE35) △ −16.855 (MTW)	A chem, excit (CuriI34b, CuriI34a, FermE34) chem, cross bomb (AmaE35) daughter Mg²⁸ (LindnM53, ShelR53)	β⁻ 2.85 max γ 1.780 (100%)	Al²⁷(n, γ) (AmaE35, SerL47b, OrsA49, HumV51, MotH52a) daughter Mg²⁸ (LindnM53, ShelR53)
Al^{29}	6.6 m (SeiL49) 6.7 m (BetH39) 6.4 m (HendW39) others (MeyA37, IweJ53)	☢ β⁻ (BetH39) △ −18.22 (MTW)	A excit, cross bomb (BetH39)	β⁻ 2.40 max γ 1.28 (94%), 2.43 (6%)	Mg²⁶(α, p) (ElliC36, BetH39, HendW39, SeiL49)
Al^{30}	3.3 s (RobiE61b) 3 s (PeeE63)	☢ β⁻ (RobiE61b) △ −17.2 (MTW)	C excit, genet energy levels (RobiE61b)	β⁻ 5.0 max γ [1.27 (46%)], 2.23 (61%), 3.51 (39%)	Si³⁰(n, p) (RobiE61b, PeeE63)
Al^{30}	72 s (PeeE63)	☢ IT (?) (PeeE63)	C chem, sep isotopes (PeeE63)	γ 2.23, 3.51	Si³⁰(n, p) (PeeE63)
$_{14}Si^{25}$	0.23 s (MPheR65)	☢ [β⁺], p (BartoR63, MPheR56) △ 4.0 (MPheR65, MTW)	C excit, cross bomb (BartoR63, MPheR65)	p 3.34, 4.08, 4.68, 5.39	Al²⁷(p, 3n) (MPheR65)
Si^{26}	2.1 s (FrickG63, RobiE60) 1.7 s (TyrH54)	☢ β⁺ (RobiE60, FrickG63) △ −7.13 (MTW)	C excit (RobiE60)	β⁺ 3.83 max γ 0.511 (200%, γ±), 0.82 (34%) daughter radiations from Al²⁶ᵐ	Mg²⁴(He³, n) (RobiE60, FrickG63)
Si^{27}	4.14 s (MihM58, KusI57) 4.22 s (BubI65) 4.45 s (SumR53) 4.1 s (WallR60, HunS54, VasiSJ60a) others (ElliD41a, WafH48, BolF51)	☢ β⁺ (MCreR40) △ −12.386 (MTW)	A excit (KueG39)	β⁺ 3.85 max γ 0.511 (200%, γ±)	Al²⁷(p, n) (KueG39, MCreR40, BarkW40a, CassJ51)
Si^{28}		% 92.18 (ReynJH53) 92.27 (BaiK50) △ −21.490 (MTW) σ (total absorption) 0.08 (GoldmDT64)			
Si^{29}		% 4.71 (ReynJH53) 4.68 (BaiK50) △ −21.894 (MTW) σ (total absorption) 0.3 (GoldmDT64)			
Si^{30}		% 3.12 (ReynJH53) 3.05 (BaiK50) △ −24.439 (MTW) σ_c 0.11 (GoldmDT64)			
Si^{31}	2.62 h (CicJ38, WenA51, DVriL52) 2.65 h (MotH52) 2.59 h (LusE50) others (NewH37, AlleW40, ForS52)	☢ β⁻ (NewH35a) △ −22.96 (MTW)	A n-capt (AmaE35) chem, excit (NewH35a)	β⁻ 1.48 max γ 1.26 (0.07%)	Si³⁰(n, γ) (Ama E35, SerL47b)
Si^{32}	≈650 y yield (GeiD62) ≈710 y yield (LindnM53) others (TurA53, RoyL57)	☢ β⁻ (LindnM53) △ −24.08 (BrodR64, MTW)	A chem, genet (LindnM53, TurA54, BrodR64) parent P³² (LindnM53, TurA54, BrodR64)	β⁻ 0.21 max no γ daughter radiations from P³²	Si³⁰(t, p) (GeiD62) protons on Cl (LindnM53, BrodR64)
$_{15}P^{28}$	0.28 s (GlasN55) 0.29 s (BrecS54) 0.27 s (TyrH54)	☢ β⁺, no α (GlasN55, GlasN53, BrecS54) △ −7.7 (MTW)	B excit, decay charac (GlasN53, BrecS54)	β⁺ 11.0 max γ 0.511 (200%, γ±), 1.780 (75%), 2.6, 4.44 (10%), 4.9, 6.1, 6.7, 7.0, 7.6 (5%)	Si²⁸(p, n) (GlasN55, BrecS54, TyrH54)

TABLE 1 (*Continued*)

Isotope Z A	Half-life	Type of decay (☢); % abundance; Mass excess (Δ≡M–A), MeV (C¹²=0); Thermal neutron cross section (σ), barns	Class; Identification; Genetic relationships	Major radiations: approximate energies (MeV) and intensities	Principal means of production
$_{15}P^{29}$	4.45 s (RoderH55, RoderH53) 4.2 s (WallR60) 4.6s (WhiM41)	☢ β^+ (WhiM41) Δ –16.95 (MTW)	A excit (WhiM41) genet energy levels (RoderH55)	β^+ 3.95 max γ 0.511 (200%, γ$^\pm$), 1.28 (0.8%), 2.43 (0.2%)	$Si^{28}(d,n)$ (RoderH55)
P^{30}	2.50 m (MDonW63) 2.49 m (EbrT65) 2.51 m (ArnS58) 2.55 m (KoesL54) others (RideL37a, VasiSS62c, FrickG63, BaskK52, CicJ38)	☢ β^+ (CuriI34) Δ –20.20 (MTW)	A excit (CuriI34, FrisO34)	β^+ 3.24 max γ 0.511 (200%, γ$^\pm$), 2.23 (0.5%)	$Al^{27}(\alpha,n)$ (FrisO34, CuriI34, RideL37a) $S^{32}(d,\alpha)$ (VasiSS62c, SagR36) $Si^{29}(p,\gamma)$ (BotW39, BaldG46, PerlmM48)
$\underline{P^{31}}$		% 100 (AstF20, KerL54) Δ –24.438 (MTW) σ_c 0.19 (GoldmDT64)			
P^{32}	14.28 d (MaraP61) 14.22 d (AndeO57) 14.30 d (CacB38, BayJ50) 14.58 d (RobeJ59) 14.60 d (SinW51) 14.50 d (LocE53) 14.35 d (KlemE48) others (MuldD40)	☢ β^- (LymE37) Δ –24.303 (MTW)	A chem, n-capt (AmaE35) daughter Si^{32} (LindnM53, TurA54, BrodR64)	β^- 1.710 max average β^- energy: 0.69 calorimetric (ShimN56a, HovV62) 0.70 ion ch (CaswR52, BrabJ53)	$P^{31}(n,\gamma)$ (SerL47b) $S^{34}(d,\alpha)$ (SagR36) $S^{32}(n,p)$ (AmaE35)
P^{33}	24.4 d (NicR54) 25.2 d (FogI60) 24.8 d (JensE52) 25 d (WestT52, ShelR51a)	☢ β^- (JensE52, ShelR51a) Δ –26.335 (MTW)	A chem, cross bomb (ShelR51a)	β^- 0.248 max γ no γ	$S^{33}(n,p)$ (ShelR51a, JensE52, WestT52, NicR54, FogI60) $Cl^{37}(\gamma,\alpha)$ (ShelR51a)
P^{34}	12.4 s (BleE46) 12.7 s (CorkJ40a) 12.5 s (ScaR58)	☢ β^- (ZunW45) Δ –24.8 (MTW)	B excit (CorkJ40a) chem, excit, cross bomb (BleE46)	β^- 5.1 max γ 2.13 (25%), 4.0 (0.2%)	$Cl^{37}(n,\alpha)$ (ZunW45, HubeO45, BleE46, ScaR58) $S^{34}(n,p)$ (CorkJ40a, ZunW45, BleE46)
$_{16}S^{29}$	0.19 s (HardJ64)	☢ $[\beta^+]$, p (HardJ64) Δ –2.9 (HardJ64, MTW)	C excit, cross bomb (HardJ64)	p 3.73, 5.40	$P^{31}(p,3n)$ (HardJ64)
S^{30}	1.4 s (FrickG63, RobiE61a)	☢ β^+ (RobiE61a) Δ –14.09 (MTW)	C excit, genet energy levels (RobiE61a)	β^+ 5.09 max (20%), 4.42 max (80%) γ 0.511 (200%, γ$^\pm$), 0.687 (80%) daughter radiations from P^{30}	$Si^{28}(He^3,n)$ (RobiE61a, FrickG63)
S^{31}	2.72 s (MihM58) 2.66 s (HasR52) 2.61 s (LindeKH60) 2.6 s (WallR60, NelJW63, MElhJ49) 2.4 s (HunS54) others (ElliD41a, WhiM41, BolF51, VasiSS63)	☢ β^+ (WhiM41) Δ –18.99 (MTW)	A excit, cross bomb (WhiM41, ElliD41a)	β^+ 4.42 max γ 0.511 (200%, γ$^\pm$), 1.27 (1.1%)	$P^{31}(p,n)$ (WhiM41) $Si^{28}(\alpha,n)$ (ElliD41, ElliD41a, KinL40)
$\underline{S^{32}}$		% 95.0 (BradP56) 95.018 (meteoritic sulfur) (MacnJ50a) terrestrial S^{32}/S^{34} variation ≤5% (TudA50) S^{32}/S^{34} variation (KulJ56) Δ –26.013 (MTW)			
$\underline{S^{33}}$		% 0.760 (BradP56) 0.750 (meteoritic sulfur) (MacnJ50a) Δ –26.583 (MTW)			
$\underline{S^{34}}$		% 4.22 (BradP56) 4.215 (meteoritic sulfur) (MacnJ50a) Δ –29.934 (MTW) σ_c 0.27 (GoldmDT64)			

TABLE 1 (*Continued*)

Isotope Z A	Half-life	Type of decay (☀); % abundance; Mass excess (△ =M−A), MeV (C¹²=0); Thermal neutron cross section (σ), barns	Class; Identification; Genetic relationships	Major radiations: approximate energies (MeV) and intensities	Principal means of production
$_{16}S^{35}$	87.9 d (FlyK65a) 86.4 d (CoopR59) 87.2 d (SelH58) 89 d (WyaE61, CaliJ59) 87 d (HendR43) 88 d (LeviH40, KameM41) others (SerL47b, CoolR39, MauW49, RudG52)	☀ β⁻ (LibW39) △ −28.847 (MTW)	A chem, excit (AndeEB36a) chem, cross bomb, excit (KameM41) sep isotopes (KameM42)	β⁻ 0.167 max average β⁻ energy: 0.0488 calorimetric (ConnR57, HovV64) γ no γ	$S^{34}(n, γ)$ (SerL47b) $Cl^{37}(d, α)$ (KameM41)
$\underline{S^{36}}$		% 0.014 (BradP56) 0.017 (meteoritic sulfur) (MacnJ50a) △ −30.66 (MTW) $σ_c$ 0.14 (GoldmDT64)			
S^{37}	5.07 m (ElliJ59) 5.04 m (BleE46) others (ScaR58)	☀ β⁻ (ZunW45) △ −27.0 (MTW)	B chem, excit, cross bomb (ZunW45, BleE46)	β⁻ 4.7 max (10%), 1.6 max (90%) γ 3.09 (90%)	$S^{36}(n, γ)$ $Cl^{37}(n, p)$ (BleE46, ZunW45, ScaR58)
S^{38}	2.87 h (NetD58)	☀ β⁻ (NetD58) △ −26.8 (MTW)	B chem, genet (NetD58) parent Cl^{38}, not parent Cl^{38m} (NetD58)	β⁻ 3.0 max (5%), 1.1 max γ 1.88 (95%) daughter radiations from Cl^{38}	$Cl^{37}(α, 3p)$ (NetD58)
$_{17}Cl^{32}$	0.306 s (GlasN53) 0.32 s (BrecS54) 0.28 s (TyrH54) others (LeiO56)	☀ β⁺, α ≈0.01% (GlasN53) △ −12.8 (MTW)	B excit, genet energy levels (GlasN53, GlasN55, TyrH54)	β⁺ 9.9 max γ 0.511 (200%, γ±), 2.24 (70%), 4.29 (7%), 4.77 (14%)	$S^{32}(p, n)$ (GlasN53)
Cl^{33}	2.53 s (MullT58a) 2.9 s (WallR60) 2.4 s (WhiM41) 2.8 s (HoaJ40, ScheIA48) others (VasisSS62c, BolF51, TyrH54)	☀ β⁺ (WhiM41) △ −21.01 (MTW)	A excit (HoaJ40, WhiM41)	β⁺ 4.55 max γ 0.511 (200%, γ±), 2.9 (0.3%)	$S^{32}(d, n)$ (HoaJ40, ScheIA48) $S^{33}(p, n)$ (WhiM41)
Cl^{34}	1.56 s (FreeJ65, JaneJ61) 1.61 s (MihM58) 1.53 s (KliR54) others (StahP53, ArbW53a, ScaR58)	☀ β⁺ (StahP53a, ArbW53) △ −24.45 (MTW)	A genet (ArbW53, StahP53a) excit (FreeJ65) daughter Cl^{34m} (ArbW53a)	β⁺ 4.46 max γ 0.511 (200%, γ±)	daughter Cl^{34m} (ArbW53a) $P^{31}(α, n)$ (JaneJ61)
Cl^{34m}	31.99 m (EbrT65) 32.40 m (GreeD56) 32.5 m (HinN52a) 33.2 m (WafH48) 33.0 m (PerlmM48) others (ScaR58, TohT60, SagR36, BranH38)	☀ β⁺ ≈50%, IT ≈50% (ArbW53, StahP53a) △ −24.31 (LHP, MTW)	A chem, excit (FrisO34, SagR36) parent Cl^{34} (ArbW53a)	β⁺ 2.48 max e⁻ 0.142 γ Cl X-rays, 0.145 (45%), 0.511 (100%, γ±), 1.17 (12%), 2.12 (38%), 3.30 (12%) daughter radiations from Cl^{34}	$P^{31}(α, n)$ (FrisO34, RideL37a, BranH38)
$\underline{Cl^{35}}$		% 75.53 (BoydA55) 75.79 (ShieW62) 75.4 (NierA36) Cl^{35}/Cl^{37} variation <0.2% (OweH55) △ −29.015 (MTW) $σ_c$ 44 (GoldmDT64)			
Cl^{36}	3.08 x 10⁵ y sp act + mass spect (BarthR55) 2.6 x 10⁵ y sp act, yield (WriH57) 4.4 x 10⁵ y sp act (WuC49) others (SerL47b)	☀ β⁻ 98.1%, EC 1.9%, β⁺ 0.0012% (DreR55, DouP62a) β⁺ 0.002% (BereD62a) △ −29.520 (MTW) $σ_c$ 100 (GoldmDT64)	A chem, n-capt (GrahD41)	β⁻ 0.714 max γ S X-rays, 0.511 (0.003%, γ±)	$Cl^{35}(n, γ)$ (GrahD41, SerL47b)
$\underline{Cl^{37}}$		% 24.47 (BoydA55) 24.6 (NierA36) 24.20 (ShieW62) Cl^{35}/Cl^{37} variation <0.2% (OweH55) △ −31.765 (MTW) $σ_c$ 0.4 (to Cl^{38}) 0.005 (to Cl^{38m}) (GoldmDT64)			

TABLE 1 *(Continued)*

Isotope Z A	Half-life	Type of decay (☢); % abundance; Mass excess (△≡M–A), MeV (C¹²=0); Thermal neutron cross section (σ), barns	Class; Identification; Genetic relationships	Major radiations: approximate energies (MeV) and intensities	Principal means of production
$_{17}Cl^{38}$	37.29 m (CobJ50) 37.1 m (MonaJ62) others (VVooS36, HoleN46, HurD37, MacqP55, CurrS40a, SlaH45, MacqP54a)	☢ β^- (KuriF36) △ −29.80 (MTW)	A chem, n-capt (AmaE35) chem, sep isotopes (KenJ40) daughter S^{38} (NetD58)	β^- 4.91 max γ 1.60 (38%), 2.170 (47%)	$Cl^{37}(n,\gamma)$ (AmaE35, KenJ40, SerL47b, AkaH41)
Cl^{38m}	0.74 s (KieP62b) 1.0 s (SchaG54)	☢ IT (KieP62b) △ −29.13 (LHP, MTW)	C n-capt, sep isotopes (SchaG54)	γ 0.66 (100%) e⁻ 0.66	$Cl^{37}(n,\gamma)$ (KieP62b, SchaG54)
Cl^{39}	55.5 m (HasR49) others (RudG52, MillDR48a)	☢ β^- (HasR49) △ −29.80 (MTW)	A chem (MillDR48a) chem, excit (HasR49)	β^- 3.45 max (7%), 2.18 max (8%), 1.91 max γ 0.246 (44%), 1.27 (50%), 1.52 (42%)	$Ar^{40}(\alpha,\alpha p)$ (PenJ56) $Ar^{40}(\gamma,p)$ (HasR49, HasR50)
Cl^{40}	1.4 m (MoriH56)	☢ β^- (MoriH56) △ −27.5 (MTW)	B chem, genet energy levels (MoriH56)	β^- 7.5 max γ 1.46 (↑ 100), 2.83 (↑ 100), 3.10, 5.8	$Ar^{40}(n,p)$ (GrayP65, MoriH56)
$_{18}Ar^{33}$	0.18 s (ReeP64, HardJ65)	☢ [β^+], p (ReeP64, HardJ65) △ −9.5 (ReeP64, MTW)	C excit, decay charac (ReeP64)	p 3.16	$Cl^{35}(p,3n)$ (HardJ65) $S^{32}(He^3,2n)$ (ReeP64)
Ar^{35}	1.83 s (KisO56, AlleJS59) 1.76 s (NelJW63) 1.88 s (ElliD41) 1.84 s (SchelA48) 1.8 s (WallR60)	☢ β^+ (ElliD41, WhiM41) △ −23.05 (MTW)	A excit (WhiM41, KinL40)	β^+ 4.94 max γ 0.511 (200%, γ±), 1.22 (5%), 1.76 (2%)	$S^{32}(\alpha,n)$ (KinL40, SchelA48) $Cl^{35}(p,n)$ (WhiM41)
$\underline{Ar^{36}}$		% 0.337 (NierA50) Ar^{36}/Ar^{38} variations (WetG54, FleW53) △ −30.232 (MTW) σ_c 6 (GoldmDT64)			
Ar^{37}	35.1 d (StoeR65) 34.3 d (KisR59) 35.0 d (MiskJ52, PerlmM53) 34.1 d (WeimP44) 32 d (AndeC53)	☢ EC (WeimP44, RodebG52) △ −30.951 (MTW)	A chem, cross bomb (WeimP41)	γ Cl X-rays, continuous bremsstrahlung to 0.81 (weak)	$Cl^{37}(p,n)$, $Cl^{37}(d,2n)$, $S^{34}(\alpha,n)$, $K^{39}(d,\alpha)$, $Cl^{37}(d,2n)$, $Ca^{40}(n,\alpha)$ (WeimP44, WeimP41) $Ar^{36}(n,\gamma)$
$\underline{Ar^{38}}$		% 0.063 (NierA50) Ar^{36}/Ar^{38} variations (WetG54, FleW53) △ −34.718 (MTW) σ_c 0.8 (GoldmDT64)			
Ar^{39}	269 y sp act (StoeR65) ≈265 y sp act (ZelH52)	☢ β^- (BrosA50) △ −33.24 (MTW)	B chem, excit (ZelH52)	β^- 0.565 max γ no γ	neutrons on KCl (ZelH52) $Ar^{38}(n,\gamma)$ (KatcS52)
$\underline{Ar^{40}}$		% 99.600 (NierA50) △ −35.038 (MTW) σ_c 0.61 (GoldmDT64)			
Ar^{41}	1.83 h (HalgW51, PauH64, KatcS52, SneA36) 1.82 h (BleE46b) 1.85 h (SchwaA56)	☢ β^- (SneA36) △ −33.061 (PauH64, MarlK65, MTW) σ_c 0.5 (StoeR65)	A chem, excit (SneA36) mass spect (AndeG54)	β^- 2.49 max (0.8%), 1.198 max γ 1.293 (99%)	$Ar^{40}(n,\gamma)$ (SneA36)
Ar^{42}	33 y sp act (StoeR65) others (HonM64, KatcS52)	☢ [β^-] (KatcS52) △ −34.42 (MTW)	B chem, genet (KatcS52) parent K^{42} (KatcS52)	[daughter radiations from K^{42}]	$Ar^{40}(n,\gamma)Ar^{41}(n,\gamma)Ar^{42}$ (KatcS52) $Ar^{40}(t,p)$ (JarN61)
$_{19}K^{37}$	1.23 s (SchweF58) 1.25 s (KavR64a) others (SunC58, WallR60, BolF51, LangmR48, TyrH54)	☢ β^+ (BolF51) △ −24.79 (KavR64a, MTW)	C excit (LangmR48)	β^+ 5.14 max γ 0.511 (200%, γ±), 2.79 (2.0%)	$Ca^{40}(p,\alpha)$ (WallR60, SunC58, SchweF58, KavR64a)

TABLE 1 (*Continued*)

Isotope Z A	Half-life	Type of decay (⚛); % abundance; Mass excess (△≡M–A), MeV (C¹²=0); Thermal neutron cross section (σ), barns	Class; Identification; Genetic relationships	Major radiations: approximate energies (MeV) and intensities	Principal means of production
$_{19}K^{38}$	7.71 m (EbrT65) 7.67 m (BormM65) 7.7 m (HurD37, RideL37a, GreeD56) others (RamsM47, PerlmM48, SalG63, PhiE65a)	⚛ β⁺ (HurD37) △ −28.79 (MTW)	A chem, cross bomb (HurD37, HendW37)	β⁺ 2.68 max γ 0.511 (200%, γ±), 2.170 (100%)	Cl³⁵(α,n) (HurD37, RideL37a, HendW37, RamsM47) Ca⁴⁰(d,α) (HurD37)
K^{38m}	0.95 s (JaneJ61, StahP53b) 0.94 s (LindKH60, KliR54) 0.97 s (MihM58)	⚛ β⁺ (StahP53, StahP53b) no IT (GoldmD62) △ −28.66 (LHP, MTW)	C excit (StahP53, StahP53b, KliR54)	β⁺ 5.0 max γ 0.511 (200, γ±)	Cl³⁵(α,n) (LindKH60, JaneJ61) K³⁹(γ,n) (StahP53b, KliR54, GoldmD62) Ca⁴⁰(d,α) (JaneJ63, MicS65, HasY59)
K^{39}		% 93.22 (KenB60) 93.08 (NierA50) others (WhiF56, ReuC56, ReuC52, CookK43) △ −33.803 (MTW) σ_c 2.0 (GoldmDT64)			
K^{40}	t₁/₂ 1.26 x 10⁹ y assuming t₁/₂ (β⁻) = 1.42 x 10⁹ y and β⁻/(β⁻ + EC) = 0.89 t₁/₂ (β⁻) sp act: 1.415 x 10⁹ y (LeuH65a) 1.42 x 10⁹ y (GleL61) 1.37 x 10⁹ y (BrinGA65, KonoS55) 1.45 x 10⁹ y (MNaiA56) 1.47 x 10⁹ y (KellWH59) 1.48 x 10⁹ y (FleD62) 1.35 x 10⁹ y (SutA55) others (WetG56, SawG50, HouF50, SmaB50, GooML51a, GrafT48, FloyJ49, StouR49, SpierF50, FauWR50, DelC51, MNaiA55) sp act of 1.460 γ: (WetG57, BackeG55a, BurcP53, AhrL48, SutA55, FauWR50, HouF50, SawG49, SpierF50) sp act of EC(K): (HeiJ54)	⚛ β⁻ 89%, EC 11%, β⁺ 0.0010% (MNaiA56, EngeD62) β⁻ 89.5%, EC 10.3%, β⁺ 0.00013% (LeuH65a) others (MNaiA55, IngM50b, GrafT51, SutA55, SpierF50, SawG50, CecM50, FauWR50, HouF50, MousuA52, ShilH54, WasG55, AldL56, WasG54, RusR53, ShilH54a, WetG56) % 0.118 (KenB60, ReuC52, ReuC56, WhiF56) 0.119 (NierA50) △ −33.533 (MTW) σ_c 70 (GoldmDT64)	A chem (ThomJ05, CamN06) chem, mass spect (SmyW37)	β⁻ 1.314 max β⁺ 0.483 max γ Ar X-rays, 1.460 (11%)	
K^{41}		% 6.77 (KenB60) 6.91 (NierA50) △ −35.552 (MTW) σ_c 1.2 (GoldmDT64)			
K^{42}	12.36 h (MerJ62) 12.52 h (BurcP53) 12.4 h (SiegK47c, KahB53, MackJ59, HurD37) 12.5 h (WriH57, MonaJ62, SinW51)	⚛ β− (KuriF36) △ −35.02 (MTW)	A chem, n-capt (AmaE35) chem, cross bomb (HevG35, HevG36) mass spect (AndeG54) daughter Ar⁴² (KatcS52)	β− 3.52 max γ 0.31 (0.2%), 1.524 (18%)	K⁴¹(n,γ) (AmaE35, HurD37, SerL47b)
K^{43}	22.4 h (OveR49, AndeG54) 22.0 h (LindqT54)	⚛ β⁻ (OveR49) △ −36.58 (MTW)	A chem, excit (OveR49) mass spect (AndeG54)	β⁻ 1.82 max (1%), 1.2 max (3%) 0.83 max γ 0.220 (3%), 0.373 (85%), 0.39 (18%, doublet), 0.59 (13%), 0.619 (81%), 1.01 (2%)	Ar⁴⁰(α,p) (LasN64, OveR49, BencN59)
K^{44}	22.0 m (CohB54, HilleP61) 22.3 m (SugiyK60) others (WalkH37a, WalkH40b)	⚛ β⁻ (WalkH37a) △ −36.3 (HilleP61, MTW)	A chem, excit (WalkH37a) chem, sep isotopes, cross bomb (CohB54) mass spect (AndeG54)	β⁻ 5.2 max γ 1.156 (61%), 1.74 (8%), 2.1 (37%, complex), 2.6 (7%), 3.7 (4%)	Ca⁴⁴(n,p) (CohB54, WalkH37a, WalkH40b, HilleP61, SugiyK60)
K^{45}	16.3 m (ChacK65) 20 m (MoriH64) 34 m (AndeG54)	⚛ β⁻ (MoriH64) △ −36.6 (MTW)	B chem, genet energy levels (MoriH64) mass spect (AndeG54)	β⁻ 4.0 max, 2.1 max γ 0.175 (strong), 0.50, 0.95 (complex?), 1.23, 1.71 (strong), 1.90, 2.10, 2.35, 2.60, 3.1	Ca⁴⁸(d,αn) (MoriH64)

TABLE 1 (*Continued*)

Isotope Z A	Half-life	Type of decay (☢); % abundance; Mass excess ($\triangle \equiv$ M–A), MeV (C^{12}=0); Thermal neutron cross section (σ), barns	Class; Identification; Genetic relationships	Major radiations: approximate energies (MeV) and intensities	Principal means of production
$_{19}$K^{47}	17.5 s (KuroT64)	☢ β^- (KuroT64) $\triangle$ −36.3 (MTW)	B chem, sep isotopes, excit (KuroT64)	β^- 6.1 max (1%), 4.1 max γ 2.0 (84%), 2.6 (15%)	Ca^{48}(γ, p) (KuroT64)
$_{20}$Ca^{37}	0.173 s (HardJ64a) 0.170 s (ReeP64)	☢ [β^+], p (HardJ64a, ReeP64) $\triangle$ −13.3 (ReeP64, MTW)	C excit, decay charac (ReeP64, HardJ64a)	p 3.10	K^{39}(p, 3n) (HardJ64a) Ca^{40}(p, d2n) (HardJ64a) Ar^{36}(He^3, 2n) (ReeP64)
Ca^{38}	0.66 s (CliJ57)	☢ β^+ (CliJ57) $\triangle$ −22 (MTW)	C excit, decay charac (CliJ57)	γ 0.511 [200%, $\gamma^\pm$], 3.5 [daughter radiations from K^{38m}]	Ca^{40}(γ, 2n) (CliJ57)
Ca^{39}	0.87 s (LindKH60) 0.86 s (MihM58) 0.88 s (KisO58) 0.90 s (KliR54) others (WallR60, SumR53, BraaR53, HubeO43, BagJ64)	☢ β^+ (HubeO43) $\triangle$ −27.30 (MTW)	B excit (HubeO43, MElhJ49)	β^+ 5.49 max γ 0.511 (200%, $\gamma^\pm$)	K^{39}(p, n) (KisO58, WallR60) Ca^{40}(γ, n) (MihM58, WafH48, HubeO43, MElhJ49, KliR54)
<u>Ca^{40}</u>		% 96.97 (NierA38a) $\triangle$ −34.848 (MTW) σ (total absorption) 0.23 (GoldmDT64)			
Ca^{41}	8 x 10^4 y yield (DroJ62) others (BrowF53b)	☢ EC (BrowF51) $\triangle$ −35.125 (JohnCH64, MTW)	B chem, n-capt, sep isotopes (BrowF51) others (SaiV51)	γ potassium X-rays	Ca^{40}(n, γ) (BrowF51, SaiV51, BrowF53, DroJ62)
<u>Ca^{42}</u>		% 0.64 (NierA38a) $\triangle$ −38.540 (MTW) σ (total absorption) 42 (GoldmDT64)			
<u>Ca^{43}</u>		% 0.145 (NierA38a) $\triangle$ −38.396 (MTW)			
<u>Ca^{44}</u>		% 2.06 (NierA38a) $\triangle$ −41.460 (MTW) σ_c 0.7 (GoldmDT64)			
Ca^{45}	165 d (WyaE61) 167 d (CaliJ59) 153 d (ThirH57) 164 d (DelC53) others (MatthD47, WalkH40b)	☢ β^- (WalkH40b) $\triangle$ −40.809 (MTW)	A chem, excit, cross bomb (WalkH40b)	β^- 0.252 max average β^- energy: 0.075 ion ch (CaswR52)	Ca^{44}(n, γ) (WalkH40b, SerL47b)
<u>Ca^{46}</u>		% 0.0033 (NierA38a) $\triangle$ −43.14 (MTW) σ_c 0.3 (GoldmDT64)			
Ca^{47}	4.535 d (GilmC64) 4.53 d (WyaE61) 4.56 d (GleG64) 4.7 d (LangeL63a, LidL56) others (BatzR51a, MarqL53a, CorkJ53a, LyoW55c)	☢ β^- (MatthD47) $\triangle$ −42.35 (MTW)	A chem, genet (BatzR51a) parent Sc^{47} (BatzR51a, CookL53)	β^- 1.98 max (18%), 0.67 max γ 0.49 (5%), 0.815 (5%), 1.308 (74%) daughter radiations from Sc^{47}	Ca^{46}(n, γ) (CorkJ53e, CookL53)
<u>Ca^{48}</u>	$t_{1/2}$ (β^-) >1.1 x 10^{18} y sp act (AwsM56) $t_{1/2}$ ($\beta\beta$) >7 x 10^{18} y sp act (DobE59) others (BeliV58, JonJW52, MCarJ55, FremJ52, DobE57, AwsM56)	% 0.185 (NierA38a) $\triangle$ −44.22 (MTW) σ_c 1.1 (GoldmDT64)			
Ca^{49}	8.8 m (OKelG56) 8.9 m (MartiDW56a) 8.5 m (DMatE50)	☢ β^- (DMatE50) $\triangle$ −41.29 (MTW)	A chem, n-capt, sep isotopes (DMatE50)	β^- 1.95 max γ 3.10 (89%), 4.1 (10%) daughter radiations from Sc^{49}	Ca^{48}(n, γ) (DMatE50)

TABLE 1 (Continued)

Isotope Z A	Half-life	Type of decay (☢); % abundance; Mass excess (△≡M–A), MeV (C¹²=0); Thermal neutron cross section (σ), barns	Class; Identification; Genetic relationships	Major radiations: approximate energies (MeV) and intensities	Principal means of production
$_{20}Ca^{50}$	9 s (ShidY64a)	☢ [β⁻] (ShidY64a) △ 41 (ShidY64a, MTW)	C excit, decay charac (ShidY64a)	γ 0.072, 0.258 (with Sc⁵⁰ᵐ) daughter radiations from Sc⁵⁰	Ca⁴⁸(t, p) (ShidY64a)
$_{21}Sc^{40}$	0.179 s (SchweF62) 0.22 s (GlasN55) others (TyrH54)	☢ β⁺ (GlasN55) △ –20.3 (RickM65, MTW)	C excit (GlasN55)	β⁺ 9.1 max γ 0.511 (200%, γ±), 3.75 [100%]	Ca⁴⁰(p, n) (GlasN55, SchweF62)
Sc^{41}	0.60 s (YouD65) 0.55 s (CramJ62) 0.87 s (MartiW52, ElliD41a, WallR60)	☢ β⁺ (ElliD41) △ –28.63 (MTW, JohnCH64)	B excit (ElliD41a, YouD65)	β⁺ 5.47 max γ [0.511 (200%, γ±)]	Ca⁴⁰(p, γ) (YouD65) Ca⁴⁰(d, n) (ElliD41a, ElliD41, CramJ62)
Sc^{42}	0.683 s (FreeJ65a) 0.65 s (NelJW65) 0.68 s (CloJ57) 0.69 s (JaneJ61) 0.62 s (MoriH55)	☢ β⁺ (MoriH55) △ –32.109 (FreeJ65a, MTW)	C decay charac (MoriH55) excit (CloJ57, NelJW65)	β⁺ [5.41 max] γ 0.511 (200%, γ±)	K³⁹(α, n) (MoriH55, JaneJ61, NelJW65)
Sc^{42m}	60.6 s (NelJW65) 62.0 s (RogeP63)	☢ β⁺ (RogeP63) △ –31.58 (LHP, MTW, FreeJ65a)	A chem, cross bomb, excit, genet energy levels (RogeP63)	β⁺ 2.82 max γ 0.438 (100%), 0.511 (200%, γ±), 1.22 (100%), 1.52 (100%)	K³⁹(α, n) (RogeP63, NelJW65)
Sc^{43}	3.92 h (HibC45) 3.95 h (DuvJ53) 3.84 h (AndeG54) others (WalkH40)	☢ β⁺ (FrisO35), [EC] △ –36.17 (MTW)	A chem, excit (FrisO35) mass spect (AndeG54)	β⁺ 1.20 max γ [Ca X-rays], 0.375 (22%), 0.511 [176%, γ±]	Ca⁴⁰(α, p) + Ca⁴⁰(α, n)Ti⁴³(β⁻) (FrisO35, WalkH40)
Sc^{44}	3.92 h (HibC45) 3.90 h (AndeG54) others (BruneJ50, WalkH40, SmiG42)	☢ β⁺, EC (HibC45) EC ≈5% (DilL63) ≈3% (KoniJ58c) ≈7% (BluJ55) △ –37.81 (MTW)	A chem, excit (CorkJ38) mass spect (AndeG54) daughter Ti⁴⁴ (SharpRA54)	β⁺ 1.47 max γ 0.511 (188%, γ±), 1.159 (100%)	daughter Ti⁴⁴ (SharpRA54, DilL63) daughter Sc⁴⁴ᵐ (KliiJ63) K⁴¹(α, n) (BruneJ50, WalkH40, HibC43)
Sc^{44m}	2.44 d (HibC45) 2.46 d (AndeG54) others (BruneJ50, WalkH40, RudG52)	☢ IT 98.6%, EC 1.4% (DilL63) △ –37.54 (LHP, MTW)	A chem, excit, cross bomb (WalkH37) mass spect (AndeG54)	γ Sc X-rays, 0.271 (86%), 1.02 (1.3%), 1.14 (2.7%, doublet) e⁻ 0.267 daughter radiations from Sc⁴⁴	K⁴¹(α, n) (BruneJ50, WalkH40, HibC43)
$\underline{Sc^{45}}$		% 100 (LelW50, HollaR64) △ –41.061 (MTW) σ_c 13 (to Sc⁴⁶) 10 (to Sc⁴⁶ᵐ) (GoldmDT64)			
Sc^{46}	83.9 d (GeiKW57) 84.1 d (SchumR56) 84.2 d (WriH57) others (MurH54, AzuR55, WalkH39)	☢ β⁻, no EC (MillA47) no β⁺, lim 0.0016% (MimW51) △ –41.756 (MTW)	A n-capt, chem (HevG36) chem, excit, cross bomb (WalkH37b)	β⁻ 1.48 max (0.004%), 0.357 max γ Ti X-rays, 0.889 (100%), 1.120 (100%)	Sc⁴⁵(n, γ) (HevG36, WalkH37b, SerL47b)
Sc^{46m}	19.5 s (DMatE51) 20 s (HammB52a, GoldhM48)	☢ IT (GoldhM48) △ –41.614 (LHP, MTW)	A n-capt, neutron resonance activation (GoldhM48)	γ [Sc X-rays], 0.142 e⁻ [0.138]	Sc⁴⁵(n, γ) (GoldhM48)
Sc^{47}	3.43 d (KriN49) 3.44 d (MarqL53a, DuvJ53) 3.40 d (CorkJ53e, MisrS64)	☢ β⁻ (HibC45a) △ –44.326 (MTW)	A chem, cross bomb (HibC45a) sep isotopes (KriN49) mass spect (AndeG54) daughter Ca⁴⁷ (BatzR51a, CookL53)	β⁻ 0.600 max γ 0.160 (73%)	daughter Ca⁴⁷ (BatzR51a, CookL53)
Sc^{48}	1.83 d (WalkH40, KriN49, PouA59, AndeG54, RudG52) 1.84 d (HillmM63) 1.81 d (HibC45a) others (MandeC42)	☢ β⁻ (WalkH37c) △ –44.51 (MTW)	A chem, excit (WalkH37c) sep isotopes (KriN49) mass spect (AndeG54)	β⁻ 0.65 max γ 0.175 (6%), 0.983 (100%), 1.040 (100%), 1.314 (100%)	V⁵¹(n, α) (WalkH37c, PoolM37, WalkH40) Ti⁵⁰(d, α) (KriN49) Ca⁴⁸(p, n) (HibC45a) Ca⁴⁸(d, 2n) (SmiG42, MandeC42, MandeC43a)
Sc^{49}	57.5 m (RezI61a) 57 m (WalkH40, OKelG56, KoesL54)	☢ β⁻ (WalkH40) △ –46.55 (MTW)	A chem, excit, cross bomb (WalkH40) sep isotopes (KoesL54, OKelG56)	β⁻ 2.01 max γ 1.76 (0.03%)	Ca⁴⁸(d, n) (WalkH40)

TABLE 1 (*Continued*)

Isotope Z A	Half-life	Type of decay (☢); % abundance; Mass excess (△≡M−A), MeV (C''=0); Thermal neutron cross section (σ), barns	Class; Identification; Genetic relationships	Major radiations: approximate energies (MeV) and intensities	Principal means of production
$_{21}Sc^{50}$	1.72 m (KantJ63b) 1.7 m (ChilG63) others (KoehD63, MoriH55a) no 23 m activity (KantJ63b, KoehD63)	☢ β⁻ (MoriH55a) △ −45.0 (MTW)	C excit (MoriH55a) excit, sep isotopes (KoehD63)	β⁻ 3.6 max γ 0.520 (100%), 1.12 (100%), 1.55 (100%)	Ti^{50}(n, p) (KoehD63, ChilG63, MoriH55a) Ca^{48}(t, n) (ShidY64a)
Sc^{50m}	0.35 s (KarrM63a, KantJ63b)	☢ IT, no β⁻, lim 10% (KarrM63a) △ −44.7 (LHP, MTW)	C excit, sep isotopes (KarrM63a'	γ 0.258 daughter radiations from Sc^{50}	Ti^{50}(n, p) (KarrM63a) Ca^{48}(t, n) (ShidY64a)
$_{22}Ti^{41}$	0.090 s (ReeP64)	☢ [β⁺], p (ReeP64) △ −15.9 (ReeP64, MTW)	C excit, decay charac (ReeP64)	p 2.3 (↑ 8), 3.05 (↑ 17), 3.68 (↑ 16), 4.12 (↑ 4), 4.64 (↑ 50), 5.30 (↑ 5)	Ca^{40}(He³, 2n) (ReeP64)
Ti^{43}	0.56 s (JaneJ61) 0.58 s (TyrH54) others (SchelA48, VasiSS61)	☢ β⁺ (JaneJ61) △ −29.3 (MTW)	C excit (SchelA48) excit, decay charac (JaneJ61)	β⁺ 5.8 max γ [0.511 (200, γ±)]	Ca^{40}(α, n) (SchelA48, JaneJ61, VasiSS63)
Ti^{44}	48 y (MorelP65) 46 y (WingJ65) others (HuiJ57)	☢ EC (SharpRA54) △ −37.66 (MTW)	A chem, genet (SharpRA54, HuiJ57, DilL63) parent Sc^{44}, not parent Sc^{44m} (SharpRA54, DilL63, HuiJ57)	γ [Sc X-rays], 0.068 (90%), 0.078 (98%) e⁻ 0.065, 0.073 daughter radiations from Sc^{44}	Sc^{45}(p, 2n) (SharpRA54, MorelP65) Sc^{45}(d, 3n) (HuiJ57, WingJ65)
Ti^{45}	3.09 h (KunD50a) 3.10 h (RudG52) 3.05 h (TPogM50) others (AlleJS41, PouA59)	☢ β⁺, EC (KunD50a) △ −39.002 (MTW)	A chem, cross bomb, excit (AlleJS41) mass spect (AndeG54)	β⁺ 1.04 max γ Sc X-rays, γ± [170%], 0.718 (0.4%), 1.408 (0.3%)	Sc^{45}(p, n) (AlleJS41, TPogM50, KunD50a) Sc^{45}(d, 2n) (AlleJS41, TPogM50)
$\underline{Ti}^{46}$		% 7.99 (HogJ54) 7.95 (NierA38a) △ −44.123 (MTW) σ (total absorption) 0.6 (GoldmDT64)			
$\underline{Ti}^{47}$		% 7.32 (HogJ54) 7.75 (NierA38a) △ −44.927 (MTW) σ (total absorption) 1.7 (GoldmDT64)			
$\underline{Ti}^{48}$		% 73.99 (HogJ54) 73.45 (NierA38a) △ −48.483 (MTW) σ (total absorption) 8.0 (GoldmDT64)			
$\underline{Ti}^{49}$		% 5.46 (HogJ54) 5.51 (NierA38a) △ −48.558 (MTW) σ (total absorption) 1.9 (GoldmDT64)			
$\underline{Ti}^{50}$		% 5.25 (HogJ54) 5.34 (NierA38a) △ −51.431 (MTW) σ_c 0.14 (GoldmDT64)			
Ti^{51}	5.79 m (SargB53) 5.80 m (BunkM55) others (HammWR53, AteA53b, SegeE49, DMatE50, SerL47b)	☢ β⁻ (SerL47b) △ −49.74 (MTW)	A n-capt (SerL47b) cross bomb (HammWR53)	β⁻ 2.14 max γ 0.320 (95%), 0.605 (1.5%), 0.928 (5%)	Ti^{50}(n, γ) (SerL47b, DMatE50)
$_{23}V^{46}$	0.426 s (FreeJ65a) 0.44 s (MillJH58) 0.40 s (MartiW52) 0.37 s (LeiO56) others (TyrH54)	☢ β⁺ (MartiW52) △ −37.069 (FreeJ65a, MTW)	B excit (MartiW52) sep isotopes, excit (JaneJ63a)	β⁺ 6.03 max γ 0.511 (200%, γ±)	Ti^{46}(p, n) (JaneJ63a, MartiW52, TyrH54, MillJH58)

TABLE 1 (*Continued*)

Isotope Z A	Half-life	Type of decay (☢); % abundance; Mass excess (Δ≡M–A), MeV (C¹²=0); Thermal neutron cross section (σ), barns	Class; Identification; Genetic relationships	Major radiations: approximate energies (MeV) and intensities	Principal means of production
$_{23}V^{47}$	33 m (BaskK62a, KriN49, OConJ42, WalkH37c) 31.1 m (KoesL54) 31 m (DaniH54a)	☢ β⁺ (WalkH37c), [EC] Δ −42.01 (MTW)	A chem, excit, cross bomb (OConJ42) chem, sep isotopes (KriN49) mass spect (AndeG54)	β⁺ 1.89 max γ 0.511 [192%, γ±], 1.5 ? (0.7%), 1.80 (0.5%), 2.16 ? (0.2%)	Sc⁴⁵(α, 2n) Ti⁴⁷(p, n) (KriN49, OConJ42) Ti⁴⁶(d, n) (WalkH37c, OConJ42) Ti⁴⁷(d, 2n) (RuaJ62)
V^{48}	16.0 d (KafP56, WalkH37c) 16.3 d (BurgW54) 16.4 d (MeyPe53) 16.2 d (VNooB57)	☢ β⁺ 49%, EC 51% (CassH53) β⁺ 56%, EC 44% (VNooB57, HageL57) β⁺ 61%, EC 39% (RisR63) others (GooW46, SterM53) Δ −44.470 (MTW)	A chem, excit, cross bomb (WalkH37b, WalkH37c) daughter Cr⁴⁸ (RudG52)	β⁺ 0.696 max γ Ti X-rays, 0.511 (100%, γ±), 0.945 (10%), 0.983 (100%), 1.312 (97%), 2.241 (3%)	Sc⁴⁵(α, n) (WalkH37b) Ti⁴⁸(p, n) (DubL40a, TicH52) Ti⁴⁸(d, 2n) (WalkH37c) Cr⁵⁰(d, α) (WalkH37c, PeaW46a)
V^{49}	330 d (HaywR56a, LyoW55)	☢ EC (WalkH39) Δ −47.950 (MTW)	B chem (WalkH39, TurL40) chem, excit (HaywR56a, LyoW55)	γ Ti X-rays, continuous bremsstrahlung to 0.60	Cr⁵²(p, α) (LyoW55) Ti⁴⁸(d, n) (WalkH39)
$\underline{V^{50}}$	6 x 10¹⁵ y sp act (WatD62) 5 x 10¹⁴ y sp act (BaumE58) t½ (EC) >8 x 10¹⁵ y sp act (MNaiA61) t½ (β⁻) >1.2 x 10¹⁶ y sp act (MNaiA61) others (GloR57a, HeiJ55, CohS52, BaumR56)	☢ EC ≈70%, β⁻ ≈30% (WatD62) % 0.25 (WhiF56) 0.24 (HessD49, LelW49a) Δ −49.216 (MTW) σc 130 (GoldmDT64)	B chem (WatD62)	γ [Ti X-rays], 0.783 (30%), 1.55 (70%)	
$\underline{V^{51}}$		% 99.75 (WhiF56) 99.76 (HessD49, LelW49a) Δ −52.199 (MTW) σc 4.9 (GoldmDT64)			
V^{52}	3.75 m (BormM65, LBlaJ54, AmaE35) 3.77 m (KoesL54) 3.76 m (SargB53, MalmS63) 3.74 m (MarteJ47) others (KohW65)	☢ β⁻ (AmaE35) Δ −51.44 (MTW)	A chem, n-capt (AmaE35) cross bomb, excit (WalkH37c)	β⁻ 2.47 max γ 1.434 (100%)	V⁵¹(n, γ) (AmaE35, WalkH37c, PoolM37, SerL47b)
V^{53}	2.0 m (KumI60, SchaA56)	☢ β⁻ (SchaA56) Δ −51.8 (SchaA56, LHP, MTW)	C decay charac (SchaA56)	β⁻ 2.50 max γ 1.00 [100%]	Cr⁵³(n, p) (SchaA56)
V^{54}	55 s (SchaA56)	☢ β⁻ (SchaA56) Δ −50 (MTW)	C decay charac (SchaA56)	β⁻ 3.3 max γ 0.84 (100%), 0.99 (100%), 2.21 [100%]	Cr⁵⁴(n, p) (SchaA56)
$_{24}Cr^{46}$	1.1 s (TyrH54)	☢	F excit (TyrH54)		protons on Cr, V (TyrH54)
Cr^{47} (or V^{46})	0.4 s (TyrH54)	☢	F excit (TyrH54)		protons on Cr, V (TyrH54)
Cr^{48}	23 h (VLieR55) 24 h (ShelR55)	☢ EC, no β⁺, lim 2% (VLieR55, ShelR55) Δ −43.1 (MTW)	A chem, genet (RudG52) parent V⁴⁸ (RudG52)	γ V X-rays, 0.116 (98%), 0.31 (99%) e⁻ 0.111, 0.31 daughter radiations from V⁴⁸	Ti⁴⁶(α, 2n) (ShelR55)
Cr^{49}	41.9 m (OConJ42) 41.7 m (CrasB53a)	☢ β⁺ (OConJ42), [EC] Δ −45.39 (MTW)	A chem, excit, cross bomb (OConJ42)	β⁺ 1.54 max e⁻ 0.058, 0.084, 0.148 γ V X-rays, 0.063 (14%), 0.091 (28%), 0.153 (13%), 0.511 ([186%], γ±)	Ti⁴⁸(α, 3n), Ti⁴⁷(α, 2n) (CrasB53a, NusR54) Ti⁴⁶(α, n) (OConJ42)
$\underline{Cr^{50}}$		% 4.31 (WhiJ48) Δ −50.249 (MTW) σc 17 (GoldmDT64)			

TABLE 1 (*Continued*)

Isotope Z A	Half-life	Type of decay (☢); % abundance; Mass excess (Δ≡M–A), MeV (C¹²=0); Thermal neutron cross section (σ), barns	Class; Identification; Genetic relationships	Major radiations: approximate energies (MeV) and intensities	Principal means of production
$_{24}Cr^{51}$	27.8 d (SchumR56, GleG64, LyoW52, WriH57) 27.9 d (KafP56) 27.5 d (SalS65)	☢ EC (BradH45b, WalkH40a) no β⁺ (BradH45b, KerB49, LyoW52) Δ −51.447 (MTW)	A chem, excit, cross bomb (WalkH40a) daughter Mn⁵¹ (BurgW50)	γ V X-rays, 0.320 (9%) e⁻ 0.315	Cr⁵⁰(n, γ) (SerL47b, WalkH40a)
Cr^{52}		% 83.76 (WhiJ48) Δ −55.411 (MTW) σ_c 0.8 (GoldmDT64)			
Cr^{53}		% 9.55 (WhiJ48) Δ −55.281 (MTW) σ_c 18 (GoldmDT64)			
Cr^{54}		% 2.38 (WhiJ48) Δ −56.931 (MTW) σ_c 0.38 (GoldmDT64)			
Cr^{55}	3.52 m (FlaA52b) 3.6 m (BazG54) 3.59 m (KohW65)	☢ β⁻ (FlaA52b) Δ −55.11 (MTW)	B chem, cross bomb (FlaA52b)	β⁻ 2.59 max γ no γ, lim 10%	Cr⁵⁴(n, γ) (FlaA52b)
Cr^{56}	5.9 m (DroB60)	☢ β⁻ (DroB60) Δ −55.3 (MTW)	A chem, genet (DroB60) parent Mn⁵⁶ (DroB60)	β⁻ 1.5 max e⁻ [0.020, 0.077] γ [Mn X-rays], 0.026, 0.083 daughter radiations from Mn⁵⁶	Cr⁵⁴(t, p) (DroB60)
$_{25}Mn^{49}$ (or Cr⁴⁷, V⁴⁶)	0.4 s (TyrH54)	☢	F excit (TyrH54)		protons on Cr (TyrH54)
Mn^{50}	0.286 s (FreeJ65a) 0.28 s (MartiW52, MillJH58) 0.27 s (TyrH54)	☢ β⁺ (MartiW52) Δ −42.618 (FreeJ65a, MTW)	B excit (MartiW52, MillJH58, FreeJ65a)	β⁺ 6.61 max γ [0.511 (200%, γ±)]	Cr⁵⁰(p, n) (MartiW52, MillJH58, TyrH54)
Mn^{50}	2 m (SutD59)	☢ β⁺ (SutD59), [EC]	E excit (SutD59)	γ 0.511 (198%, γ±), 0.66 (25%), 0.783 (100%), 1.11 (100%), 1.28 (25%), 1.45 (75%)	Cr⁵⁰(p, n) (SutD59)
Mn^{51}	45.2 m (KoesL54) 44.3 m (BurgW50) 44 m (NozM60) others (MillDR48, LivJ38d)	☢ β⁺ (LivJ37a), [EC] Δ −48.26 (MTW)	A chem, cross bomb (LivJ37a, LivJ38d) chem, genet (BurgW50) parent Cr⁵¹ (BurgW50)	β⁺ 2.17 max γ 0.511 [194%, γ±], 1.56 (?), 2.03 (?)	Cr⁵⁰(d, n) (LivJ38d, BurgW50) Cr⁵⁰(p, γ) (DubL38, DelL39)
Mn^{52}	5.60 d (BurgW54) 5.69 d (KafP56) 5.72 d (BackoE55)	☢ EC 66%, β⁺ 34% (KoniJ58c, KoniJ58a) EC 71%, β⁺ 29% (RemL63, WilsRR62) others (GooW46, SehR54) Δ −50.70 (MTW)	A chem, excit, cross bomb (LivJ37a, LivJ38d)	β⁺ 0.575 max γ Cr X-rays, 0.511 (67%, γ±), 0.744 (82%), 0.935 (84%), 1.434 (100%)	Cr⁵²(p, n) (HemA40) Cr⁵²(d, 2n) (PeaW46a, KoniJ58a)
Mn^{52m}	21.1 m (JuliaJ59a) 21.3 m (HemA40) 22.1 m (KayG65)	☢ β⁺, IT 2% (KatoT60), [EC] Δ −50.32 (LHP, MTW)	A chem (DarB37) chem, excit, cross bomb (LivJ37a, LivJ38d) daughter Fe⁵² (MillDR48)	β⁺ 1.63 max γ 0.383 (2%), 0.511 (193%, γ±), 1.434 (100%)	daughter Fe⁵² (MillDR48, JuliaJ59a)
Mn^{53}	1.9 x 10⁶ y geochemical method (KayJ65) ≈2 x 10⁶ y yield (ShelR57, calc from WilkJR55, DobW56a)	☢ EC (WilkJR55) Δ −54.683 (JohnCH64, MTW) σ_c ≈170 (GoldmDT64)	B chem, decay charac (WilkJR55)	γ Cr X-rays	Cr⁵³(p, n) (WilkJR55) Cr⁵²(d, n) (DobW56a)
Mn^{54}	303 d (MartiWH64) 291 d (BackoE55) 313 d (WyaE61) 278 d (SchumR56) 290 d (KafP56) 300 d (WriH57) others (LivJ38d, SuwS53, SalS65)	☢ EC (AlvL38) no β⁺, no β⁻ (LivJ38d, DeuM44) Δ −55.55 (MTW)	A chem, excit, cross bomb (LivJ37a, LivJ38d)	γ Cr X-rays, 0.835 (100%) e⁻ 0.829	Fe⁵⁶(d, α) (LivJ38d, DeuM44) V⁵¹(α, n) (LivJ38d) Cr⁵³(d, n) (LivJ38d) Cr⁵⁴(p, n) (DubL40a)

TABLE 1 (*Continued*)

Isotope Z A	Half-life	Type of decay ($\maltese$); % abundance; Mass excess $(\triangle \equiv M-A)$, MeV $(C^{12}=0)$; Thermal neutron cross section (σ), barns	Class; Identification; Genetic relationships	Major radiations: approximate energies (MeV) and intensities	Principal means of production
$_{25}Mn^{55}$		% 100 (SamM36a, WhiF56) $\triangle$ -57.705 (MTW) σ_c 13.3 (GoldmDT64)			
Mn^{56}	2.576 h (BarthR53a, BarthR53b) 2.574 h (LocE53) 2.586 h (BisG50) others (LivJ38d, BonaG64, BieJ64a, SalS65)	$\maltese$ β^- (AmaE35) $\triangle$ -56.904 (MTW)	A chem, n-capt (AmaE35) daughter Cr^{56} (DroB60)	β^- 2.85 max γ 0.847 (99%), 1.811 (29%), 2.110 (15%)	Mn^{55}(n, γ) (AmaE35, SerL47b, OrsA49, HumV51)
Mn^{57}	1.7 m (CohB54a, KumI60) 1.9 m (VasiSS63)	$\maltese$ β^- (CohB54a) $\triangle$ -57.5 (MTW)	B chem, excit (CohB54a)	β^- 2.55 max γ [Fe X-rays, 0.014], 0.122 (strong), 0.136 (strong), 0.22, 0.353, 0.692	Cr^{54}(α, p) (VasiSS63) Fe^{57}(n, p) (CohB54a)
Mn^{57}	7 d (SharmH51)	$\maltese$ β^- (SharmH51)	G chem, cross bomb (SharmH51) activity not observed (CohB54a, NelM50)		alphas on Cr, Mn (SharmH51)
Mn^{58}	1.1 m (ChitD61)	$\maltese$ β^- (ChitD61) $\triangle$ -56 (MTW)	B chem, sep isotopes (ChitD61)	γ 0.36, 0.41, 0.52, 0.57, 0.82, 1.0, 1.25, 1.4, 1.6, 2.2, 2.8	Fe^{58}(n, p) (ChitD61)
$_{26}Fe^{52}$	8.2 h (JuliaJ59a) 7.8 h (MillDR48)	$\maltese$ β^+ 56%, EC 44% (JuliaJ59a) others (ArbE56, FrieG51a) $\triangle$ -48.33 (MTW)	A chem, genet (MillDR48) parent Mn^{52m} (MillDR48) not parent Mn^{52}, lim 5% (FrieG51a)	β^+ 0.80 max γ Mn X-rays, 0.165 (100%), 0.511 (112%, $\gamma^\pm$) daughter radiations from Mn^{52m} Mn^{52}	Cr^{50}(α, 2n) (FrieG51a)
Fe^{53}	8.51 m (EbrT65) 8.9 m (RideL37a, LivJ38b, JuliaJ59a) 8.6 m (SalS65)	$\maltese$ β^+ (RideL37a), [EC] $\triangle$ -50.70 (MTW)	A chem (RideL37a) chem, excit, cross bomb (LivJ38b)	β^+ 3.0 max γ 0.38 (32%), 0.511 (196%, $\gamma^\pm$)	Cr^{50}(α, n) (NelM50, RideL37a, LivJ38b) Cr^{52}(α, 3n) (JuliaJ59a)
Fe^{54}		% 5.84 (ValleG41a) $\triangle$ -56.246 (MTW) σ_c 2.9 (GoldmDT64)			
Fe^{55}	2.60 y (SchumR56) 2.94 y (BrowG50) others (SchumR51a)	$\maltese$ EC, no β^+ (BradH46b, MaeD51a, PortF53) $\triangle$ -57.474 (MTW)	A chem, excit (LivJ39c) daughter Co^{55} (LivJ41)	γ Mn X-rays, continuous bremsstrahlung to 0.23 (0.004%)	Fe^{54}(n, γ) (EmmW54a)
Fe^{56}		% 91.68 (ValleG41a) $\triangle$ -60.605 (MTW) σ_c 2.7 (GoldmDT64)			
Fe^{57}		% 2.17 (ValleG41a) $\triangle$ -60.176 (MTW) σ_c 2.5 (GoldmDT64)			
Fe^{58}		% 0.31 (ValleG41a) $\triangle$ -62.147 (MTW) σ_c 1.1 (GoldmDT64)			
Fe^{59}	45.6 d (PierA59) 44.5 d (GleG64) 45.1 d (SchumR51a) 45.0 d (TobJ53, TobJ51) 45.5 d (GovJ43) 44.3 d (WriH57) others (WorD63, HeaR60, FusE60, WahA53)	$\maltese$ β^- (LivJ38b) $\triangle$ -60.660 (MTW)	A chem, excit, cross bomb (LivJ38b)	β^- 1.57 max (0.3%), 0.475 max γ 0.143 (0.8%), 0.192 (2.8%), 1.095 (56%), 1.292 (44%)	Fe^{58}(n, γ) (SerL47b)
Fe^{60}	3 x 10^5 y yield (RoyJ57)	$\maltese$ [β^-] (RoyJ57) $\triangle$ -61.51 (MTW)	B chem, genet (RoyJ57) parent Co^{60m} (RoyJ57)	daughter radiations from Co^{60m} Co^{60}	protons on Cu (RoyJ57)

TABLE 1 (*Continued*)

Isotope Z A	Half-life	Type of decay (☢); % abundance; Mass excess (△≡M−A), MeV (C¹²=0); Thermal neutron cross section (σ), barns	Class; Identification; Genetic relationships	Major radiations: approximate energies (MeV) and intensities	Principal means of production
₂₆Fe⁶¹	6.0 m (StraJ66, RiccE57) others (RiccE55)	☢ β⁻ (RiccE55, RiccE57) △ −59 (MTW)	A chem, genet (RiccE55, RiccE57, StraJ66) parent Co⁶¹ (RiccE55, RiccE57, StraJ66)	β⁻ 2.8 max γ 0.13 (↑ 11), 0.30 (↑ 48), 1.03 (↑ 98), 1.20 (↑ 100) daughter radiations from Co⁶¹	Ni⁶⁴(n, α) (RiccE57) Ni⁶⁴(d, αp) (RiccE57)
₂₇Co⁵⁴	0.194 s (FreeJ65) others (MartiW52, LeiO56, TyrH54)	☢ β⁺ (MartiW52) △ −47.99 (MTW)	C excit (MartiW52, FreeJ65)	β⁺ [7.23 max] γ [0.511 (200%, γ±)]	Fe⁵⁴(p, n) (FreeJ65, MartiW52)
Co⁵⁴	1.5 m (SutD59)	☢ β⁺ (SutD59)	E excit (SutD59, FreeJ65)	β⁺ 4.3 max γ 0.41 (100%), 0.511 (200%, γ±), 1.14 (100%), 1.41 (100%)	Fe⁵⁴(p, n) (FreeJ65)
Co⁵⁵	18.2 h (DarB37) 17.9 h (RudG52) 18.0 h (LivJ41)	☢ β⁺ 81%, EC 19% (MukA58) β⁺ ≈60%, EC ≈40% (calc from DeuM49) △ −54.01 (MTW)	A chem (DarB37) chem, cross bomb, genet (LivJ41) parent Fe⁵⁵ (LivJ41)	β⁺ 1.50 max γ Fe X-rays, 0.480 (12%), 0.511 (160%, γ±), 0.930 (80%), 1.41 (13%)	Fe⁵⁴(d, n) (DarB37, LivJ41, DeuM49) Fe⁵⁴(p, γ) (LivJ41) Fe⁵⁶(p, 2n) (MukA58)
Co⁵⁶	77.3 d (WriH57) 77 d (BurgW54) others (CookCS42, LivJ41)	☢ EC 80%, β⁺ 20% (CookCS56) △ −56.03 (MTW)	A chem, excit, cross bomb (LivJ41) daughter Ni⁵⁶ (ShelR52, WorW52)	β⁺ 1.49 max γ Fe X-rays, 0.511 (40%, γ±), 0.847 (100%), 1.04 (15%), 1.24 (66%), 1.76 (15%), 2.02 (11%), 2.60 (17%), 3.26 (13%)	Fe⁵⁶(p, n) (KieP59, GrabZ60a, SakM54) Mn⁵⁵(α, 3n) (ChenL52a) daughter Ni⁵⁶ (ShelR52, WorW52) Fe⁵⁶(d, 2n) (LivJ41, JensA41, PleE42, ElliL43a) Ni⁵⁸(d, α) (LivJ41, CookCS42, ElliL43a)
Co⁵⁷	270 d (LivJ41) 267 d (CorkJ55)	☢ EC, no β⁺, lim 0.002% (CrasB55) △ −59.339 (MTW)	A chem, excit, cross bomb (LivJ41) daughter Ni⁵⁷ (FrieG52)	γ Fe X-rays, 0.014 (9%), 0.122 (87%), 0.136 (11%), 0.692 (0.14%) e⁻ 0.007, 0.013, 0.115, 0.129	Ni⁵⁸(γ, p); Fe⁵⁶(d, n) (LivJ38a, PerrC38, BarrG39, LivJ41) Fe⁵⁶(p, γ) (LivJ41) Mn⁵⁵(α, 2n) (ChenL52a)
Co⁵⁸	71.3 d (SchumR56) 71.0 d (CorkJ55) 72 d (LivJ41, HoffD52, PreiI60)	☢ EC 85%, β⁺ 15% (GooW46, CookCS56) △ −59.84 (MTW) σ_c 2500 (GoldmDT64)	A chem, excit, cross bomb (LivJ41)	β⁺ 0.474 max γ Fe X-rays, 0.511 (30%, γ±), 0.810 (99%), 0.865 (1.4%), 1.67 (0.6%)	Mn⁵⁵(α, n) (LivJ38a, LivJ41)
Co⁵⁸ᵐ	9.2 h (ChrisD50) 9.0 h (PreiI60) 8.8 h (StraK50)	☢ IT, no β⁺ (StraK50) △ −59.81 (LHP, MTW) σ_c 1.4 x 10⁵ (GoldmDT64)	A chem, excit (StraK50)	γ Co X-rays e⁻ 0.017, 0.024	Mn⁵⁵(α, n) (StraK50)
Co⁵⁹		% 100 (MitJ41) △ −62.233 (MTW) σ_c 19 (to Co⁶⁰) 18 (to Co⁶⁰ᵐ) (GoldmDT64)			
Co⁶⁰	5.263 y (GorbS63) 5.24 y (GeiKW57) 5.20 y (LocE56) 5.21 y (KasJ53a) 5.27 y (TobJ55, TobJ51) others (LocE53, LivJ41, BrowG50, SinW51)	☢ β⁻ (RisJ37) △ −61.651 (MTW) σ_c 6 (GoldmDT64)	A n-capt (SamM36) chem, excit, cross bomb (LivJ41)	β⁻ 1.48 max (0.12%), 0.314 max (99+%) γ 1.173 (100%), 1.332 (100%)	Co⁵⁹(n, γ) (RisJ37, LivJ38a, LivJ41, SerL47b, YafL51)
Co⁶⁰ᵐ	10.47 m (BarthR53b) 10.3 m (SchmW63) 10.5 m (PreiI60) 10.7 m (LivJ41)	☢ IT 99+%, β⁻ 0.25% (SchmW63) IT 99+%, β⁻ 0.28% (DeuM51) △ −61.593 (LHP, MTW) σ_c 100 (GoldmDT64)	A n-capt (HeyF37a) chem, excit, cross bomb (LivJ41) daughter Fe⁶⁰ (RoyJ57)	β⁻ 1.55 max e⁻ 0.051, 0.058 γ Co X-rays, 0.059 (2.1%), 1.33 (0.25%)	Co⁵⁹(n, γ) (HeyF37a, LivJ37a, LivJ41, SerL47b)

TABLE 1 (*Continued*)

Isotope Z A	Half-life	Type of decay (☢); % abundance; Mass excess (△≡M–A), MeV (C¹²=0); Thermal neutron cross section (σ), barns	Class; Identification; Genetic relationships	Major radiations: approximate energies (MeV) and intensities	Principal means of production
$_{27}Co^{61}$	99.0 m (SmiL51, NerW55) 95 m (StraJ66) 100 m (NusR56) 104 m (ValtA62) others (ParmT49, BrowF53a, HopH50, PreiI60)	☢ β⁻ (ParmT47) △ –62.93 (MTW)	A chem, excit, cross bomb, sep isotopes, mass spect (ParmT47) daughter Fe⁶¹ (RiccE55, RiccE57, StraJ66)	β⁻ 1.22 max e⁻ [0.059] γ [Ni X-rays], 0.067 (89%)	Ni⁶⁴(p, α), Ni⁶⁴(d, αn), Ni⁶¹(n, p) (ParmT47, ParmT49) Co⁵⁹(t, p) (KunD48)
Co⁶²	13.9 m (ParmT49, GardD57, ValtA62) 13.8 m (PreiI60)	☢ β⁻ (ParmT49) △ –61.53 (MTW)	A chem, sep isotopes (ParmT49, GardD57)	β⁻ 2.88 max γ 1.17 (180%, complex), 1.47 (20%), 1.74 (19%), 2.03 (7%)	Ni⁶⁴(d, α) (ParmT49, GardD57) Ni⁶²(n, p) (ParmT49, ValtA62)
Co⁶²	1.5 m (ValtA62) 1.6 m (ParmT49) 1.9 m (PreiI60)	☢ β⁻ (ParmT49)	C cross bomb, sep isotopes (ParmT49)	γ γ rays observed	Ni⁶⁴(d, α) (ParmT49) Ni⁶²(n, p) (ValtA62, PreiI60, ParmT49)
Co⁶³	52 s (MoriH60)	☢ β⁻ (MoriH60) △ –61.9 (MTW)	E chem, excit (MoriH60)	β⁻ 3.6 max γ no γ, lim 10%	Ni⁶⁴(γ, p) (MoriH60)
Co⁶³	1.40 h (PreiI60) 2.0 h (ValtA62)	☢	G sep isotopes (PreiI60) activity assigned to Co⁶¹ (StraJ66)		Ni(n, np) (PreiI60, ValtA62)
Co⁶⁴	28 s (StraJ66)	☢	F excit (StraJ66)	γ 0.095	Ni⁶⁴(n, p) (StraJ66)
Co⁶⁴	7.8 m (PreiI60) others (ValtA62, ParmT49)	☢	G sep isotopes (PreiI60) activity not observed (StraJ66) others (ParmT49)		neutrons on Ni⁽⁶⁴⁾ (PreiI60)
Co⁶⁴	2.0 m (PreiI60) others (ValtA62, ParmT49)	☢	G sep isotopes (PreiI60) activity not observed (StraJ66) others (ParmT49)		neutrons on Ni⁽⁶⁴⁾ (PreiI60)
$_{28}Ni^{56}$	6.10 d (WelD63) 6.4 d (ShelR52) 6.0 d (WorW52)	☢ EC, no β⁺, lim 1% (ShelR52) △ –53.92 (MTW)	A chem (WorW52) chem, sep isotopes, genet (ShelR52) parent Co⁵⁶ (ShelR52, WorW52)	γ Co X-rays, 0.163 (99%), 0.276 (31%), 0.472 (35%), 0.748 (48%), 0.812 (85%), 1.56 (14%) e⁻ 0.155 daughter radiations from Co⁵⁶	Fe⁵⁴(α, 2n) (ShelR52, WorW52, OhnH65, JenkR64)
Ni⁵⁷	36.0 h (EbrT65) 35.7 h (RudG64) others (MaiF49, LivJ38, FrieG50, ChilG62, RoaJ59, PauA65)	☢ EC 54%, β⁺ 46% (KoniJ58c, KoniJ58) EC 50%, β⁺ 50% (FrieG50) EC 63%, β⁺ 37% (ChilG62) △ –56.10 (MTW)	A chem, excit, cross bomb (LivJ38) parent Co⁵⁷ (FrieG52)	β⁺ 0.85 max γ Co X-rays, 0.127 (14%), 0.511 (92%, γ±), 1.37 (86%), 1.89 (14%) daughter radiations from Co⁵⁷	Co⁵⁹(p, 3n) (WagG52) Fe⁵⁴(α, n) (LivJ38, DorR41, NelM42, MaiF49, FrieG50, CanR51c)
<u>Ni⁵⁸</u>		% 67.76 (WhiJ48) △ –60.23 (MTW) σ_c 4.4 (GoldmDT64)			
Ni⁵⁹	8 x 10⁴ y yield (BrosA51) 1 x 10⁵ y yield (SaraB56) 8 x 10⁵ y yield (WilsH51a)	☢ EC (WilsH51a) no β⁺, lim 2 x 10⁻³% (EmmW54a) △ –61.159 (MTW)	A chem, cross bomb, n-capt (CamM45) chem, sep isotopes, n-capt (BrosA51)	γ Co X-rays, continuous bremsstrahlung to 1.06	Ni⁵⁸(n, γ) (BrosA51, CamM45, WilsH50) Co⁵⁹(d, 2n) (BrosA51)
<u>Ni⁶⁰</u>		% 26.16 (WhiJ48) △ –64.471 (MTW) σ_c 2.6 (GoldmDT64)			
<u>Ni⁶¹</u>		% 1.25 (WhiJ48) △ –64.22 (MTW) σ_c 2 (GoldmDT64)			
<u>Ni⁶²</u>		% 3.66 (WhiJ48) △ –66.75 (MTW) σ_c 15 (GoldmDT64)			

TABLE 1 (*Continued*)

Isotope Z A	Half-life	Type of decay (♥); % abundance; Mass excess ($\triangle \equiv M-A$), MeV ($C^{12}=0$); Thermal neutron cross section (σ), barns	Class; Identification; Genetic relationships	Major radiations: approximate energies (MeV) and intensities	Principal means of production
$_{28}Ni^{63}$	92 y sp act (HorrD62) 125 y sp act (MMulC56) 85 y yield (BrosA51) 61 y yield (WilsH51a)	♥ β^- (BrosA51) $\triangle$ −65.52 (MTW)	A chem, n-capt, sep isotopes (BrosA51)	β^- 0.067 max γ no γ	$Ni^{62}(n,\gamma)$ (BrosA51, WilsH49, WilsH50)
$\underline{Ni^{64}}$		% 1.16 (WhiJ48) $\triangle$ −67.11 (MTW) σ_c 1.5 (GoldmDT64)			
Ni^{65}	2.564 h (SilL51) 2.55 h (CliJ63a) 2.56 h (ScaR58) 2.50 h (RiccR60b) others (BonaG64, LivJ38, MaiF49, ForS52, NelM42, GrenH65a)	♥ β^- (HeyF37b) $\triangle$ −65.14 (MTW)	A n-capt (RotJ36) chem, sep isotopes, excit (SwarJ46, ConnE46)	β^- 2.13 max γ 0.368 (4.5%), 1.115 (16%), 1.481 (25%)	$Ni^{64}(n,\gamma)$ (HeyF37b, ConnE46, DorR41, NelM42, SerL47b, MaiF49)
Ni^{66}	54.8 h (KjeA56) 55 h (JohnN56) 56 h (HopH50, GoeR49)	♥ β^- (GoeR49) $\triangle$ −66.06 (MTW)	A chem, genet (GoeR49) parent Cu^{66} (GoeR49)	β^- 0.20 max γ no γ, lim 1% daughter radiations from Cu^{66}	fission (KjeA56, GoeR49, JohnN56)
Ni^{67}	50 s (MeaJL65)	♥ β^- (MeaJL65) $\triangle$ −63.2 (MeaJL65, MTW)	C excit (MeaJL65)	β^- 4.1 max γ 0.90 (51%, doublet), 1.26 (15%)	$Zn^{70}(n,\alpha)$ (MeaJL65)
$_{29}Cu^{57}$ (or Co^{54})	0.18 s (TyrH54) 0.14 s (MartiW52)	♥	F excit (TyrH54, MartiW52)		protons on Ni (TyrH54, MartiW52)
Cu^{58}	3.20 s (FreeJ65) 3.04 s (MartiW52) 3.3 s (GerhJ58) others (TyrH54)	♥ β^+ (MartiW52) $\triangle$ −51.66 (MTW)	C excit (TyrH54, MartiW52, FreeJ65)	β^+ 8.2 max γ [0.511 (200%, $\gamma^\pm$)]	$Ni^{58}(p,n)$ (FreeJ65, MartiW52, TyrH54)
Cu^{58}	9.5 m (YuaT55a) 7.9 m (DelL39) 10.0 m (LeiC47)	♥ β^+ (DelL39, YuaT55a)	G chem (DelL39) chem, excit, sep isotopes (LeiC47) activity cannot be assigned to Cu^{58} from threshold considerations (NDS)		protons on Ni^{58} (LeiC47, DelL39) deuterons on Ni^{58} (YuaT55a)
Cu^{59}	81.5 s (ButlJW58) 81 s (LindnL55, DelL39, LeiC47) others (BudA62, YuaT55a)	♥ β^+, no EC, lim 5% (YuaT55b) $\triangle$ −56.36 (MTW)	B chem (DelL39) excit, sep isotopes (LeiC47) genet energy levels (CohB62a, ButlJW58)	β^+ 3.7 max γ 0.343 (5%), 0.463 (5%), 0.511 (197%, $\gamma^\pm$), 0.872 (9%), 1.305 (11%), 1.70 (1%)	$Ni^{58}(p,\gamma)$ (LeiC47, DelL39, ButlJW58) $Ni^{58}(d,n)$ (LindnL55)
Cu^{60}	23.4 m (NusR54b) 24.6 m (LeiC47) 24 m (BudA62)	♥ β^+ 93%, EC 7% (NusR54b) $\triangle$ −58.35 (MTW)	A chem, excit, sep isotopes, mass spect (LeiC47) daughter Zn^{60} (LindnL55a)	β^+ 3.92 max (6%), 3.00 max (18%), 2.00 max γ Ni X-rays, 0.511 (186%, $\gamma^\pm$), 0.85 (15%), 1.332 (80%), 1.76 (52%), 2.13 (6%, doublet), 2.64 (5%), 3.13 (4%), 2.52 (2%), 4.0 (1.0%)	$Ni^{60}(p,n)$ (LeiC47) $Ni^{60}(d,2n)$ (BudA62, LeiC47, LeviN58)
Cu^{61}	3.32 h (BermA54) 3.33 h (CookCS48b) 3.35 h (BudA62, BoeF50) 3.4 h (ThorRL37, RideL37a, KunD50a) 3.3 h (HopH50)	♥ β^+ 60%, EC 40% (NusR56) others (CookCS51, BouR50, HubeO49, KuzM57) $\triangle$ −61.98 (MTW)	A chem, excit (RideL37) chem, excit, sep isotopes (LeiC47, KunD50a) daughter Zn^{61} (LindnL55a, CumJ55, CumJ59)	β^+ 1.22 max e^- 0.059 γ Ni X-rays, 0.067 (4%), 0.284 (12%), 0.38 (3%), 0.511 (120%, $\gamma^\pm$), 1.19 (5%)	$Ni^{60}(d,n)$ (ThorRL37) $Co^{59}(\alpha,2n)$
Cu^{62}	9.76 m (EbrrT65) 9.73 m (BermA54) 9.9 m (CriE39, ButlJW58a, PerlmM48, ForS52) 10.1 m (LeiC47, NusR54c) 10.0 m (RideL37a) others (HeyF37a)	♥ β^+ (HeyF37a), [EC] $\triangle$ −62.81 (MTW)	A excit (HeyF37a) excit, cross bomb (RideL37a, StraC38, BotW39) chem, sep isotopes (LeiC47) daughter Zn^{62} (MillDR48)	β^+ 2.91 max γ 0.511 (195%, $\gamma^\pm$), 0.88 (0.3%), 1.17 (0.5%, complex)	daughter Zn^{62} (MillDR48) $Co^{59}(\alpha,n)$ (RideL37a) $Ni^{62}(p,n)$ (StraC38)
$\underline{Cu^{63}}$		% 69.1 (BrowHS47) $\triangle$ −65.583 (MTW) σ_c 4.5 (GoldmDT64)			

TABLE 1 (*Continued*)

Isotope Z A	Half-life	Type of decay (♥); % abundance; Mass excess (Δ≡M–A), MeV (C¹²=0); Thermal neutron cross section (σ), barns	Class; Identification; Genetic relationships	Major radiations: approximate energies (MeV) and intensities	Principal means of production
$_{29}Cu^{64}$	12.80 h (TobJ55, RabE50) 12.88 h (SilL51) 12.87 h (WriH57) 12.7 h (SchumR51a) others (BonaG64, BatzR51a, KunD51, HubeO43a, HubeO44a, JohnH50, PerlmM49, StraK51, EdwL52, MillDR48, VVooS36a, HopH50, BeydJ57a, PauA65, ZinH65)	♥ EC 43%, β⁻ 38%, β⁺ 19% (NDS) (β⁺ + EC)/β⁻ 1.6 (ReynJH50) Δ −65.428 (MTW)	A chem, n-capt (AmaE35) excit (VVooS36a) chem, excit (DelL39)	β⁻ 0.573 max β⁺ 0.656 max e⁻ 1.33 γ Ni X-rays, 0.511 (38%, γ±), 1.34 (0.5%)	Cu^{63}(n, γ) (HeyF37b, SerL47b)
Cu^{65}		% 30.9 (BrowHS47) Δ −67.27 (MTW) σ_c 2.3 (GoldmDT64)			
Cu^{66}	5.10 m (SargB53) 5.07 m (BarthR53b) 5.12 m (SchumR51a) 5.20 m (KoesL54) 5.2 m (RoderH51, CamAG50) 5.3 m (BormM65) others (FrieG51)	♥ β⁻ (AmaE35) Δ −66.26 (MTW) σ_c 130 (GoldmDT64)	A n-capt (AmaE35) excit (ChanW37) daughter Ni⁶⁶ (GoeR49)	β⁻ 2.63 max γ 1.039 (9%)	Cu^{65}(n, γ) (AmaE35, SerL47b, OrsA49, HumV51) daughter Ni⁶⁶ (GoeR49)
Cu^{67}	58.5 h (KunD50a) 61 h (HopH50, EwaG53) 56 h (GoeR49)	♥ β⁻ (GoeR49) Δ −67.29 (MTW)	A chem (GoeR49) chem, cross bomb, sep isotopes (KunD50a)	β⁻ 0.57 max e⁻ 0.082, 0.091 γ Zn X-rays, 0.092 (23%, doublet), 0.184 (40%)	Ni^{64}(α, p) (KunD50a) Zn^{67}(n, p) (KunD50a) Cu^{65}(t, p) (KunD51)
Cu^{68}	30 s (BakH64) 32 s (FlaA53a)	♥ β⁻ (FlaA53a) Δ −65.4 (MTW)	B chem, excit (FlaA53a) genet energy levels (BakH64)	β 3.5 max γ 0.80 (17%), 1.078 (95%), 1.24 (3%), 1.88 (5%)	Ga^{71}(n, α) (FlaA53a, BakH64) Zn^{68}(n, p) (FlaA53a, YthC60c, BakH64)
$_{30}Zn^{60}$	2.1 m (LindnL55a)	♥ [EC, β⁺] (LindnL55a)	B chem, genet (LindnL55a) parent Cu⁶⁰ (LindnL55a)		Ni^{58}(α, 2n) (LindnL55a)
Zn^{61}	1.48 m (LindnL55a, CumJ59)	♥ β⁺, [EC] (CumJ55, LindnL55a, CumJ59), Δ −56.6 (MTW)	A chem, genet (CumJ55, LindnL55a, CumJ59) parent Cu⁶¹ (CumJ55, LindnL55a, CumJ59)	β⁺ 4.4 max γ 0.48 (11%), 0.511 (198%, γ±), 0.98 (3%), 1.64 (6%)	Ni^{58}(α, n) (CumJ55, LindnL55a, CumJ59)
Zn^{62}	9.13 h (RudG64) 9.33 h (HaywR50a) 9.3 h (NusR54c) 9.5 h (MillDR48) others (KunD53, PoolM52)	♥ EC ≈82%, β⁺ ≈18% (NDS) EC ≈90%, β⁺ ≈10% (HaywR50a) Δ −61.12 (MTW)	A chem, genet (MillDR48) excit (GhoS50) parent Cu⁶² (MillDR48)	β⁺ 0.66 max e⁻ 0.033 γ Cu X-rays, 0.042 (20%), 0.51 (47%, doublet, includes γ±), 0.59 (22%) daughter radiations from Cu⁶²	Cu^{63}(p, 2n) (GhoS50) Cu^{63}(d, 3n) (NusR54c)
Zn^{63}	38.4 m (CumJ61) 38.1 m (RiccR59a) 38.3 m (HubeO47, StraC38, WafH48) 38.5 m (DelL39) 37.6 m (VasiSS61b) others (BotW39, PauA65)	♥ β⁺ 93%, EC 11% (HubeO47) Δ −62.22 (MTW)	A chem, excit (BotW37a, HeyF37b, RideL37a) daughter Ga⁶³ (NurM65)	β⁺ 2.34 max γ Cu X-rays, 0.511 (186%, γ±), 0.669 (8%), 0.962 (6%), 1.42 (0.9%)	Ni^{60}(α, n) (GhoS50, RideL37a) Cu^{63}(p, n) (StraC38, DelL39, BlasJ51, GhoS50, CumJ61) Cu^{63}(d, 2n) (LivRS40, TownA41)
Zn^{64}	$t_{1/2}$ (EC EC) >8 × 10¹⁵ y sp act (BertA53)	% 48.89 (BaiK50) Δ −66.000 (MTW) σ_c 0.46 (GoldmDT64)			
Zn^{65}	245 d (TobJ53, PerrC38) 244 d (GeiKW57) 246 d (WriH57, EasH60) 250 d (TatV61, AgarI61)	♥ EC 98.3%, β⁺ 1.7% (GleG59, RiccR60b) β⁺ 1.2% (BereD62b) Δ −65.92 (MTW)	A chem (PerrC38) chem, excit, cross bomb (LivJ39a) daughter Ga⁶⁵ (LivJ39d)	β⁺ 0.327 max e⁻ 1.106 γ Cu X-rays, 0.511 (3.4%, γ±), 1.115 (49%)	Zn^{64}(n, γ) (SagR39, SerL47b)
Zn^{66}		% 27.81 (BaiK50) Δ −68.88 (MTW)			
Zn^{67}		% 4.11 (BaiK50) Δ −67.86 (MTW)			

TABLE 1 (*Continued*)

Isotope Z A	Half-life	Type of decay (☢); % abundance; Mass excess (△≡M–A), MeV (C¹²=0); Thermal neutron cross section (σ), barns	Class; Identification; Genetic relationships	Major radiations: approximate energies (MeV) and intensities	Principal means of production
$_{30}Zn^{68}$		% 18.56 (BaiK50) △ –69.99 (MTW) σ_c 1.0 (to Zn^{69}) 0.1 (to Zn^{69m}) (GoldmDT64)			
Zn^{69}	57 m (LivJ39a) 51 m (HopH48) 52 m (HansA49)	☢ β^- (HeyF37b) △ –68.43 (MTW)	A chem, n-capt (HeyF37b) chem, excit, cross bomb (LivJ39a, KenJ39) daughter Zn^{69m} (KenJ39)	β^- 0.90 max γ no γ	daughter Zn^{69m} (KenJ39) Zn^{68}(n, γ) (HeyF37b, HeyF36, SerL47b, HumV51, SagR39) Ga^{71}(d, α) (LivJ39a)
Zn^{69m}	13.8 h (LivJ39a) others (HopH50, HopH48)	☢ IT (KenJ39) △ –67.99 (LHP, MTW)	A chem, excit (ThorRL38) chem, excit, cross bomb (LivJ39a, KenJ39) parent Zn^{69} (KenJ39)	γ Zn X-rays, 0.439 (95%) e⁻ 0.429 daughter radiations from Zn^{69}	Zn^{68}(n, γ) (ThorRL38, LivJ39a, SerL47b) Ga^{71}(d, α) (LivJ39a)
Zn^{70}	$t_{1/2}$ (ββ) >10¹⁵ y sp act (FremJ52)	% 0.62 (BaiK50) △ –69.55 (MTW) σ_c 0.10 (to Zn^{71}) 0.01 (to Zn^{71m}) (GoldmDT64)			
Zn^{71}	2.4 m (ThwT61) 2.2 m (LBlaJ55, HugD46)	☢ β^- (HugD46) △ –67.5 (MTW)	C n-capt, cross bomb (HugD46) n-capt, sep isotopes (LBlaJ55)	β^- 2.61 max γ 0.120 (0.9%), 0.39 (1.3%), 0.510 (13%), 0.92 (3%), 1.12 (1.3%)	Zn^{70}(n, γ) (HugD46, LBlaJ55, ThwT61)
Zn^{71m}	3.92 h (LevkV58) 4.1 h (SonT64) 4.0 h (ThwT61)	☢ β^- (LBlaJ55) △ –67.2 (LHP, MTW)	A sep isotopes, n-capt (LBlaJ55) chem (SonT64)	β^- 1.46 max γ 0.13 (9%), 0.385 (94%), 0.495 (75%), 0.609 (65%), 0.76 (5%), 0.99 (8%), 1.11 (4%)	Zn^{70}(n, γ) (LBlaJ55, ThwT61, TanP64, SonT64)
Zn^{72}	46.5 h (ThwT63) 49 h (SiegJ51) 37 h (IshM63)	☢ β^- (SiegJ51) △ –68.14 (MTW)	A chem, genet (SiegJ46, SiegJ51) parent Ga^{72} (SiegJ51)	β^- 0.30 max e⁻ 0.005, 0.014 γ Ga X-rays, 0.015 (8%), 0.046 (weak), 0.145 (90%), 0.192 (10%) daughter radiations from Ga^{72}	fission .(SiegJ51, SteinE51c, GoeR49, FolR51, TurA51a, ThwT63, KjeA63)
$_{31}Ga^{63}$	33 s (NurM65)	☢ [β^+, EC] (NurM65) △ –57 (MTW)	B chem, excit, cross bomb, genet (NurM65) parent Zn^{63} (NurM65)		Cu^{63}(α, 4n) (NurM65) Ni^{60}(Li⁶, 3n) + Ni^{58}(Li⁶, n) (NurM65)
Ga^{64}	2.6 m (CrasB53) 2.5 m (CohB53)	☢ β^+, (CrasB53), [EC] △ –58.93 (MTW)	B chem, cross bomb (CrasB53) chem, excit, sep isotopes (CohB53)	β^+ 6.05 max (33%), 2.8 max γ 0.511 (196%, γ±), 0.80 (15%), 0.992 (43%), 1.25 (7%), 1.38 (14%), 1.56 (7%), 1.78 (5%), 2.18 (11%), 2.34 (9%), 3.32 (18%)	Cu^{63}(α, 3n) (CrasB53) Zn^{64}(p, n) (CohB53, JacoT60) Zn^{64}(d, 2n) (CrasB53)
Ga^{65}	15.2 m (DaniH57a) 15 m (AlvL38, LivJ39d, CrasB54, KoesL54, PoolM52)	☢ EC (AlvL38) β^+ >50% (AteA52) △ –62.66 (MTW)	A chem, genet (LivJ39d) parent Zn^{65} (LivJ39d) daughter Ge^{65} (PoriN58)	β^+ 2.24 max (12%), 2.11 max e⁻ 0.044, 0.053, 0.105 γ Zn X-rays, 0.054 (8%), 0.061 (12%), 0.115 (55%), 0.152 (10%), 0.206 (4%), 0.511 (180%, γ±), 0.75 (10%), 0.93 (3%)	Cu^{63}(α, 2n), Zn^{64}(d, n), Zn^{64}(p, γ) (MorrD59)
Ga^{65}	8.0 m (CrasB54)	☢	G chem, excit, cross bomb (CrasB54) activity not observed (MorrD59)		alphas on Cu, protons on Zn (CrasB54)
Ga^{66}	9.45 h (LangeL50d) 9.3 h (RudG64) 9.5 h (CarvJ59) 9.4 h (RideL37a, BucJ38) 9.2 h (MukA50, MannW37) others (FrauH57b)	☢ β^+ 57%, EC 43% (CamD63) △ –63.71 (MTW)	A chem, excit (MannW37, RideL37) daughter Ge^{66} (HopH49)	β^+ 4.153 max γ Zn X-rays, 0.511 (114%, γ±), 0.828 (5%), 1.039 (37%), 1.91 (3%), 2.183 (5%), 2.748 (25%), 4.30 (5%)	Cu^{63}(α, n) (MannW37, RideL37a, LangeL50d)

TABLE 1 (Continued)

Isotope Z A	Half-life	Type of decay (☢); % abundance; Mass excess (△≡M-A), MeV (C¹²=0); Thermal neutron cross section (σ), barns	Class; Identification; Genetic relationships	Major radiations: approximate energies (MeV) and intensities	Principal means of production
$_{31}$Ga^{67}	77.9 h (TobJ55, TobJ51) 79.2 h (RudG64) 78.2 h (MCowD48) others (HopH50, HopH48, MannW38)	☢ EC (AlvL38) no β⁺, lim 0.01% (MeyW53) △ −66.87 (MTW)	A chem, excit (MannW38, MannW38a) chem, excit, cross bomb (AlvL38) daughter Ge^{67} (HopH49)	γ Zn X-rays, 0.093 (40%), 0.184 (24%), 0.296 (22%), 0.388 (7%) e⁻ 0.084, 0.092	Zn^{67}(d, n) (AlvL38, ValleG39) Cu^{65}(α, 2n) (HubbJ57)
Ga^{68}	68.3 m (EbrT65) 68.2 m (BormM65) 68 m (RideL37a, PerlmM48, KoesL54)	☢ β⁺ 88%, EC 12% (RamasM59a, TayH63a) △ −67.07 (MTW)	A chem, excit (BotW37a, RideL37a) daughter Ge^{68} (HopH48, HopH50)	β⁺ 1.90 max γ Zn X-rays, 0.511 (176%, γ±), 0.80 (0.4%), 1.078 (3.5%), 1.24 (0.14%), 1.87 (0.15%)	daughter Ge^{68} (HopH48) Cu^{65}(α, n) (RideL37a, MannW37) Zn^{68}(p, n) (DubL38, BucJ38, MukA50) Zn^{67}(d, n) (ValleG39)
Ga^{69}		% 60.2 (IngM48b) 60.5 (AntkS53) △ −69.326 (MTW) σ$_c$ 1.9 (GoldmDT64)			
Ga^{70}	21.1 m (BunkM57) 20 m (AmaE35, MannW38)	☢ β⁻ (DubL38) △ −68.90 (MTW)	A chem, n-capt (AmaE35) chem, excit (DubL38)	β⁻ 1.65 max γ 0.173 (0.16%), 1.040 (0.5%)	Ga^{69}(n, γ) (AmaE35, SerL47b)
Ga^{71}		% 39.8 (IngM48b) 39.5 (AntkS53) △ −70.135 (MTW) σ$_c$ 5.0 (GoldmDT64)			
Ga^{72}	14.12 h (WyaE61) 14.08 h (BisG50) 14.3 h (SiegJ51, MandeC43a) 14.1 h (SagR39) others (LangeL60)	☢ β⁻ (SagR39) △ −68.58 (MTW)	A chem, n-capt, excit (LivJ38b, SagR39) daughter Zn^{72} (SiegJ51)	β⁻ 3.15 max γ 0.601 (8%), 0.630 (27%), 0.835 (96%), 0.894 (10%), 1.050 (7%), 1.465 (3.5%), 1.60 (5%, complex), 1.860 (5%), 2.201 (26%), 2.50 (20%, doublet)	Ga^{71}(n, γ) (SagR39, SerL47b, SiegJ51)
Ga^{73}	4.9 h (YthC58) 5.1 h (MarqL59) 5.0 h (SiegJ51)	☢ β⁻ (SiegJ51) △ −69.74 (MTW)	A chem, excit (SiegJ46, SiegJ51) chem, sep isotopes, cross bomb (YthC58)	β⁻ 1.19 max e⁻ 0.012, 0.043, 0.053 γ Ge X-rays, 0.054 (9%), 0.295 (94%), 0.74 (6%) daughter radiations from Ge^{73m} included in above listing	Ge^{73}(n, p) (SiegJ51, YthC58) Ge^{76}(d, αn) (YthC58)
Ga^{74}	8.0 m (YthC59b) 7.8 m (EicE58) others (MarinJ60, MoriH56)	☢ β⁻ (EicE58) △ −67.8 (MTW)	A decay charac, excit (MoriH56) chem, sep isotopes, excit, genet energy levels (EicE58, EicE62)	β⁻ 2.5 max γ 0.50 (11%, complex?), 0.60 (100%, doublet), 0.87 (9%, doublet), 1.11 (5%), 1.20 (8%, doublet), 1.33 (5%), 1.46 (8%, doublet), 1.76 (7%, doublet), 2.35 (45%)	Ge^{76}(d, α) (YthC59b) Ge^{74}(n, p) (MarinJ60, EicE62, EicE58, YthC59b)
Ga^{75}	2.0 m (MoriH60) 1.5 m (YthC60a)	☢ β⁻ (MoriH60, YthC60a) △ −68.5 (MoriH60, MTW)	D chem (YthC60a)	β⁻ 3.3 max γ 0.36 ? (1%), 0.58 (3%) [daughter radiations from Ge^{75}]	Ge^{76}(n, pn) (YthC60a) Ge^{76}(γ, p) (MoriH60)
Ga^{76}	32 s (TakaK61)	☢ β⁻ (TakaK61)	C genet energy levels (TakaK61)	β⁻ 6 max γ 0.563, 0.96, 1.12	Ge^{76}(n, p) (TakaK61)
$_{32}$Ge^{65}	1.5 m (PoriN58)	☢ β⁺ (PoriN58), [EC] △ −56 (MTW)	A chem, excit, sep isotopes, genet (PoriN58) parent Ga^{65} (PoriN58)	β⁺ 3.7 max γ 0.511 (197%, γ±), 0.67 (3%), 1.72 (2%) daughter radiations from Ga^{65}	Zn^{64}(α, 3n) (PoriN58)
Ge^{66}	2.4 h (RiccR60a) 2.5 h (HopH50) others (RiccR56, ZinH65)	☢ β⁺ ≈62%, EC ≈38% (RiccR60a) EC(K) ≈48% (ZinH65) △ −60.7 (MTW)	A chem, genet (HopH49) parent Ga^{66} (HopH49)	β⁺ 2.0 max (<10%), 1.3 max γ Ga X-rays, 0.046 (37%), 0.068 (11%), 0.114 (22%), 0.185 (23%), 0.245 (7%), 0.27 (19%), 0.30 (6%), 0.34 (19%), 0.38 (48%, doublet?), 0.40 (6%), 0.47 (19%), 0.511 (124%, γ±) daughter radiations from Ga^{66}	Zn^{64}(α, 2n) (RiccR60a)

TABLE 1 (*Continued*)

Isotope Z A	Half-life	Type of decay (☢); % abundance; Mass excess (△≡M-A), MeV (C¹²=0); Thermal neutron cross section (σ), barns	Class; Identification; Genetic relationships	Major radiations: approximate energies (MeV) and intensities	Principal means of production
$_{32}Ge^{67}$	18.7 m (RiccR59) 18.6 m (CogM65) 21 m (HopH50, VasiSS64) 19 m (AteA53) others (RiccR56)	☢ β⁺, EC (HopH50, RiccR59) △ -62.5 (MTW)	A chem, genet (HopH49) parent Ga⁶⁷ (HopH49)	β⁺ 3.1 max γ 0.170 (105%, doublet), 0.511 (170%, γ±), 0.84 (4%), 0.92 (7%), 1.48 (5%) daughter radiations from Ga⁶⁷	Zn⁶⁴(α, n) (RiccR59)
Ge^{68}	275 d (CrasB56) 250 d (HopH50)	☢ EC (HopH48) no β⁺, lim 0.4% (RamasM59a) △ -67 (MTW)	A chem (MannW38) chem, genet (HopH48) parent Ga⁶⁸ (HopH48, HopH50)	γ Ga X-rays daughter radiations from Ga⁶⁸	Zn⁶⁶(α, 2n) (MannW38, RamasM59a, HoreD59)
Ge^{69}	36 h (TemJ65) 40.4 h (NusR57) 38.5 h (SchweC63) 40 h (MCowD48, HopH50) others (MannW38, HubeO44a)	☢ EC ≈67%, β⁺ ≈33% (MCowD48) EC(K) ≈55% (ZinH65) △ -67.101 (MTW)	A chem (MannW38) chem, excit, cross bomb (MCowD48) daughter As⁶⁹ (ButeF55)	β⁺ 1.22 max γ Ga X-rays, 0.511 (68%, γ±), 0.573 (13%), 0.872 (10%), 1.107 (28%), 1.335 (3%)	Ga⁶⁹(d, 2n) (SeaG41, MCowD48, HudC51, TemJ65)
$\underline{Ge^{70}}$		% 20.55 (IngM48e) △ -70.558 (MTW) σ_c 3.2 (GoldmDT64)			
Ge^{71}	11.4 d (MCowD48) 11 d (MandeC49, SeaG41)	☢ EC, no β⁺ (MCowD48, MandeC49) △ -69.90 (MTW)	A chem, excit, cross bomb (SeaG41) sep isotopes, n-capt (ReynS50) daughter As⁷¹ (HopH49)	γ Ga X-rays	Ge⁷⁰(n, γ) (SerL47b, MCowD48, MandeC49, ReynS50) Ga⁷¹(d, 2n) (SeaG41, MCowD48)
$Ge^{?}$	14 d (LangeM56, LangeM54b)	☢ EC (LangeM56, (LangeM54b)	E chem, critical abs (LangeM56, LangeM54b)	γ Ga X-rays, continuous internal bremsstrahlung to 0.15	neutrons on Ge (LangeM54b, LangeM56)
$\underline{Ge^{72}}$		% 27.37 (IngM48e) △ -72.579 (MTW) σ_c 1.0 (GoldmDT64)			
$\underline{Ge^{73}}$		% 7.67 (IngM48e) △ -71.293 (MTW) σ_c 14 (GoldmDT64)			
Ge^{73m}	0.53 s (CamE57)	☢ IT (CamE57) △ -71.226 (LHP, MTW)	A n-capt, chem, genet (CamE57) daughter As⁷³ (CamE57)	γ Ge X-rays, 0.054 (9%) e⁻ 0.012, 0.043, 0.053	daughter As⁷³ (CamE57)
$\underline{Ge^{74}}$		% 36.74 (IngM48e) △ -73.419 (MTW) σ_c 0.3 (to Ge⁷⁵) 0.2 (to Ge⁷⁵ᵐ) (GoldmDT64)			
Ge^{75}	82 m (MCowD48) 89 m (SeaG41) 79 m (ReynS50)	☢ β⁻ (SeaG41) △ -71.83 (MTW)	A chem, excit, cross bomb (SeaG41) n-capt, sep isotopes (ReynS50)	β⁻ 1.19 max γ 0.066 (0.3%), 0.199 (1.4%), 0.265 (11%), 0.427 (0.3%), 0.477 (0.3%), 0.628 (0.1%)	Ge⁷⁴(n, γ) (ReynS50, SmiA52c, SagR39, SagR41, SerL47b) As⁷⁵(n, p) (SagR41, SeaG41, MCowD48)
Ge^{75m}	48 s (SmiA52c) 49 s (BursS54a) 42 s (FlaA52)	☢ IT (FlaA52) △ -71.69 (LHP, MTW)	A excit (FlaA52) cross bomb, n-capt, sep isotopes (SmiA52c)	γ Ge X-rays, 0.139 (34%) e⁻ 0.128, 0.138 daughter radiations from Ge⁷⁵	Ge⁷⁴(n, γ) (SmiA52c, FlaA52) As⁷⁵(n, p) (FlaA52, SmiA52c)
$\underline{Ge^{76}}$	$t_{1/2}$ (ββ) >2 x 10¹⁶ y sp act (FremJ52)	% 7.67 (IngM48a) △ -73.209 (MTW) σ_c 0.1 (to Ge⁷⁷) 0.1 (to Ge⁷⁷ᵐ) (GoldmDT64)			

TABLE 1 (*Continued*)

Isotope Z A	Half-life	Type of decay (☢); % abundance; Mass excess ($\triangle \equiv$ M–A), MeV (C^{12}=0); Thermal neutron cross section (σ), barns	Class; Identification; Genetic relationships	Major radiations: approximate energies (MeV) and intensities	Principal means of production
$_{32}$Ge^{77}	11.3 h (LyoW57) 12 h (SeaG41, SteinE51)	☢ β^- (SagR41) $\triangle$ –71.2 (MTW)	A chem, excit, cross bomb (SeaG41) parent As^{77} (SteinE46, SteinE51)	β^- 2.2 max e$^-$ 0.198, 0.253 γ As X-rays, 0.21 (61%, doublet), 0.263 (45%), 0.368 (15%), 0.417 (25%), 0.563 (18%), 0.632 (11%), 0.73 (14%, complex), 0.80 (6%, complex), 0.93 (5%, complex), 1.09 (6%), others to 2.4 daughter radiations from As^{77}	Ge^{76}(n, γ) (LyoW57, ReynS50, SagR39, SagR41, SerL47b)
Ge^{77m}	54 s (LyoW57) 52 s (BursS54a) 59 s (ArnJ47, VKooJ65) others (ReynS50)	☢ β^- 76%, IT 24% (VKooJ65) β^- 73%, IT 27% (LyoW57) β^- ≈85%, IT ≈15% calc from (BursS54a) $\triangle$ –71.0 (LHP, MTW)	A cross bomb, genet, n-capt (ArnJ47) sep isotopes (ReynS50) parent As^{77} (ArnJ47, ReynS50)	β^- 2.9 max e$^-$ 0.148, 0.158 γ Ge X-rays, 0.159 (12%), 0.215 (21%)	Ge^{76}(n, γ) (LyoW57, ReynS50, ArnJ47)
Ge^{78}	1.47 h (FritK65, KvaE65) 1.43 h (SuganN53) 2.1 h (YthC59a, SteinE51)	☢ β^- (SteinE51) $\triangle$ –71.8 (KvaE65, MTW)	B chem, genet (SteinE46, SteinE51, YthC59a) parent As^{78} (SteinE46, SteinE51, SuganN53, YthC59a)	β^- 0.71 max γ 0.277 (94%) daughter radiations from As^{78}	fission (SteinE51, FritK65, KvaE65) Se^{82}(n, αn) (YthC59a)
$_{33}$As^{68}	≈7 m (ButeF55)	☢	E chem, excit (ButeF55)		Ge^{70}(p, 3n) (ButeF55)
As^{69}	15 m (ButeF55)	☢ β^+ (ButeF55), [EC] $\triangle$ –63.2 (MTW)	B chem, genet (ButeF55) parent Ge^{69} (ButeF55)	β^+ 2.9 max γ 0.23, 0.511 ($\gamma^\pm$) daughter radiations from Ge^{69}	Ge^{70}(p, 2n) (ButeF55)
As^{70}	52 m (HopH50, VerkB52) 47 m (SouA55)	☢ β^+ (HopH50) no EC, lim 20% (VerkB52) $\triangle$ –64.32 (MTW)	A chem (HopH49, HopH50) chem, decay charac (SouA55) chem (VerkB52) chem, excit (ButeF55) daughter Se^{70} (HopH50)	β^+ 2.89 max (6%), 2.14 max γ [Ge X-rays], 0.511 (183%, $\gamma^\pm$), 0.60 (23%), 0.67 (25%), 0.75 (23%), 0.91 (17%), 1.040 (78%), 1.12 (23%), 1.36 (12%), 1.42 (10%), 1.54 (7%), 1.71 (22%), 1.80 (6%), 2.03 (19%), others to 4.7	Ge^{70}(p, n) (ButeF55) Ge^{70}(d, 2n) (VerkB52, BornP63)
As^{71}	62 h (GravW55) 60 h (HopH50, StokP53, BeydJ57a) 65 h (AttH53, ThuS54b)	☢ EC ≈70%, β^+ ≈30% (ThuS54b) EC(K) ≈54% (ZinH65) $\triangle$ –67.89 (MTW)	A chem (SagR41) chem, genet (HopH49) mass spect (BracD52) parent Ge^{71} (HopH49)	β^+ 0.81 max e$^-$ 0.012, 0.022, 0.164 γ Ge X-rays, 0.175 (90%), 0.511 (60%, $\gamma^\pm$) daughter radiations from Ge^{71}	Ga^{69}(α, 2n) (MeiJ50) Ge^{70}(d, n) (GravW55, ThuS54b, BracD52, MCowD48a)
As^{72}	26 h (MCowD48a) 27 h (HopH50)	☢ EC, β^+ (MCowD48a) EC(K) <30% (ZinH65) $\triangle$ –68.22 (MTW)	A chem, excit (MitA47) chem, excit, sep isotopes (MCowD48a) daughter Se^{72} (HopH48, HopH50)	β^+ 3.34 max (17%), 2.50 max e$^-$ 0.679 γ Ge X-rays, 0.511 (150%, $\gamma^\pm$), 0.630 (8%), 0.835 (78%), other weak γ's to 3.7 (each <3%)	daughter Se^{72} (HopH48) Ga^{69}(α, n) (MitA47, MCowD48a, MeiJ50, BrunE56)
As^{73}	80.3 d (GleG64) 76 d (MCowD48a) others (SagR39a, MeiJ50)	☢ EC, no β^+, lim 2% (MCowC48a, ElliL43b) $\triangle$ –70.92 (MTW)	A chem (SagR39a) chem, excit, cross bomb, sep isotopes (MCowC48a) mass spect (JohaS51a) parent Ge^{73m} (CamE57)	γ Ge X-rays, 0.054 (9%) e$^-$ 0.012, 0.043, 0.053 daughter radiations from Ge^{73m} included in above listing	Ge^{72}(d, n) (SagR39a, JohaS52)
As^{74}	17.9 d (GleG64) 17.5 d (MCowD48a) others (HopH50, SagR39a, MocD48)	☢ β^+ 29%, EC 39%, β^- 32% (GrigE58d) others (GriR59j, HoreD59a, JohaS51, ScoJ57, MeiJ50) $\triangle$ –70.855 (MTW)	A excit (CurtB38) chem, excit (SagR39a)	β^- 1.36 max β^+ 1.54 max (3%), 0.95 max (26%) γ Ge X-rays, 0.511 (59%, $\gamma^\pm$), 0.596 (61%), 0.635 (14%)	Ga^{71}(α, n) (MCowD48a, HoreD59a)
As^{74m}	8.0 s (SchaA61a)	☢ IT (SchaA61a) $\triangle$ –70.572 (LHP, MTW)	B sep isotopes, cross bomb, excit (SchaA61a)	γ 0.283	Ge^{74}(p, n) (SchaA61a) Ge^{73}(p, γ) (SchaA61a)
As^{75}		% 100 (NierA37a) $\triangle$ –73.031 (MTW) σ_c 4.5 (GoldmDT64)			

TABLE 1 (*Continued*)

Isotope Z A	Half-life	Type of decay (☢); % abundance; Mass excess ($\triangle \equiv M-A$), MeV ($C^{12}=0$); Thermal neutron cross section (σ), barns	Class; Identification; Genetic relationships	Major radiations: approximate energies (MeV) and intensities	Principal means of production
$_{33}As^{76}$	26.4 h (HubeP53, HubeP52), 26.5 h (DzhB55), 26.3 h (MitA40), 26.8 h (WriH57, WeilG42), 26.1 h (PhiK48)	☢ β^-, no β^+, lim 0.03% (BarbW47), no EC(K), lim 0.02% (ScoJ57), $\triangle$ −72.29 (MTW)	A chem, n-capt (AmaE35)	β^- 2.97 max, γ 0.559 (43%), 0.657 (6%), 1.22 (5%, doublet), 1.44 (0.7%, doublet), 1.789 (0.3%), 2.10 (0.9%, doublet)	$As^{75}(n,\gamma)$ (AmaE35, CurtB38, OrsA49, HumV51)
As^{77}	38.7 h (BunkM53, SchmJ55), 38 h (SuganN53, TurA51a), 39 h (EndP54, ReynS53), others (SteinE51)	☢ β^-, $\triangle$ −73.92 (MTW)	A chem, genet (SteinE46, SteinE51), daughter Ge^{77} (SteinE51, SteinE46), daughter Ge^{77m} (ArnJ47, ReynS50)	β^- 0.68 max, γ 0.086 (0.1%), 0.239 (2.5%), 0.522 (0.8%), daughter radiations from Se^{77m}	$Ge^{76}(n,\gamma)Ge^{77}+Ge^{77m}$ (β^-) (LyoW57, ArnJ47, ReynS57)
As^{78}	91 m (SuganN53, KjeA59), 90 m (SteinE51, BrigR51), 88 m (CunJ53), others (SneA37, SagR39a, CurtB38)	☢ β^- (SneA37), $\triangle$ −72.8 (MTW)	B chem (SneA37), excit (CurtB38), daughter Ge^{78} (SteinE46, SteinE51, SuganN53, YthC59a)	β^- 4.1 max, γ 0.614 (↑ 42), 0.70 (↑ 15), 0.83 (↑ 8), 1.31 (↑ 11)	$Br^{81}(n,\alpha)$ (SneA37, SagR39a, BrigR51), fission (SteinE46, SteinE51), $Se^{78}(n,p)$ (NemY58a)
As^{78m}	6 m (NemY58a)	☢ IT (?) (NemY58a)	G excit (NemY58a), activity not observed (FritK65a)		neutrons on Se^{78} (NemY58a)
As^{79}	9.0 m (CunJ53), 9.1 m (YthC54)	☢ β^- (VHaaP52), $\triangle$ −73.7 (MTW)	A chem (ButeF50), chem, genet (YthC54, CunJ53), parent Se^{79m} (YthC54, CunJ53)	β^- 2.15 max, γ 0.36 (2%), 0.43 (2%), 0.54 (0.5%), 0.73 (0.5%), 0.89 (1%), daughter radiations from Se^{79m}	$Se^{82}(n,\alpha)[Ge^{79}](\beta^-)$ (YthC61, YthC54), $Se^{80}(n,pn)$ (VHaaP52, YthC61), $Se^{80}(\gamma,p)$ (KuroT61a)
As^{80}	15.3 s (MeaRE59), others (YthC54)	☢ β^- (MeaRE59), $\triangle$ −71.8 (MTW)	C chem, excit (YthC54), excit, sep isotopes (MeaRE59)	β^- 6.0 max, γ 0.666 (42%), 0.8 (1.4%, complex), 1.22 (4%), 1.64 (4%), 1.77 (1.7%)	$Se^{80}(n,p)$ (MeaRE59, YthC54)
As^{81}	33 s (YthC60), 31 s (MoriH60)	☢ β^- (YthC60, MoriH60), $\triangle$ −72.6 (MoriH60, MTW)	B chem, excit (MoriH60, YthC60)	β^- 3.8 max, γ no γ	$Se^{82}(n,pn)$ (YthC60), $Se^{82}(\gamma,p)$ (MoriH60)
As^{85}	0.43 s (WanR55)	☢ $[\beta^-]$, n (WanR55)	F excit (WanR55)		fission (WanR55)
$_{34}Se^{70}$	≈44 m (HopH50)	☢ β^+ (HopH50), [EC]	D chem (HopH49, HopH50), parent As^{70} (HopH50)	γ [As X-rays, 0.511 ($\gamma^{\pm}$)], [daughter radiations from As^{70}]	$As^{75}(d,7n)$ (HopH50)
Se^{71}	4.5 m (AteA57), 5 m (BeydJ57)	☢ β^+ (BeydJ57), [EC], $\triangle$ −63.5 (MTW)	B chem, excit (BeydJ57, AteA57)	β^+ 3.4 max, γ 0.16, 0.511 ($\gamma^{\pm}$, [195%])	$Ge^{70}(\alpha,3n)$ (AteA57), N^{14} on Cu (BeydJ57)
Se^{72}	8.4 d (CumJ58), 9.7 d (HopH50)	☢ EC (HopH50), no β^+, lim 0.1% (CumJ58), $\triangle$ −68 (MTW)	A chem, genet (HopH48), parent As^{72} (HopH48, HopH50)	γ As X-rays, 0.046 (59%), e⁻ 0.034, 0.044, daughter radiations from As^{72}	$As^{75}(d,5n)$ (HopH48, HopH50), $Ge^{70}(\alpha,2n)$ (CumJ58)
Se^{73}	7.1 h (CowW48, ScoF51, HaywR56, RiccR60c), others (HopH50)	☢ β^+ 65%, EC 35% (HaywR56, LHP), others (KuzM57, RiccR60c), no IT (RiccR60c), $\triangle$ −68.17 (MTW)	A chem (HopH48), chem, excit, sep isotopes (CowW48)	β^+ 1.66 max? (≤0.7%), 1.30 max, e⁻ 0.054, 0.064, 0.347, γ As X-rays, 0.066 (65%), 0.359 (99%), 0.511 (130%, $\gamma^{\pm}$), daughter radiations from As^{73}	$Ge^{70}(\alpha,n)$ (CowC48, ScoF51, RiccR60c), $As^{75}(d,4n)$ (HopH50)
Se^{73}	42 m (RiccR60c), 44 m (HooF53)	☢ β^+, EC (HooF53, RiccR60c), $\triangle$ −68.2 (RiccR60c, MTW)	B chem, excit (ScoF53)	β^+ 1.7 max, γ As X-rays, 0.088 ? (6%), 0.251 ? (14%), 0.58 ? (6%)	$Ge^{70}(\alpha,n)$ (RiccR60c), $Ge^{72}(\alpha,3n)$ (HooF53)
$\underline{Se^{74}}$		% 0.87 (WhiJ48), $\triangle$ −72.212 (MTW), σ_c 30 (GoldmDT64)			

TABLE 1 *(Continued)*

Isotope Z A	Half-life	Type of decay (☢); % abundance; Mass excess (△≡M–A), MeV (C¹²=0); Thermal neutron cross section (σ), barns	Class; Identification; Genetic relationships	Major radiations: approximate energies (MeV) and intensities	Principal means of production
₃₄Se⁷⁵	120.4 d (EasH60) 120 d (WriH57, HopH50) 127 d (CowW48) others (CorkJ50f, FrieH47)	☢ EC, no β⁺ (FrieH47, CowW48, CorkJ50f) △ −72.166 (MTW)	A chem, excit (DubL40a, KenC42) sep isotopes, n-capt (CorkJ50f)	γ As X-rays, 0.066 (1.0%), 0.097 (3.3%), 0.121 (17%), 0.136 (57%), 0.265 (60%), 0.280 (25%), 0.401 (12%) e⁻ 0.085, 0.095, 0.109, 0.124, 0.253	Se⁷⁴(n, γ) (CorkJ50f, SerL47b, FrieH47) As⁷⁵(d, 2n) (KenC42, FrieH47, CowW48, HopH50) As⁷⁵(p, n) (DubL40a)
Se⁷⁶		% 9.02 (WhiJ48) △ −75.26 (MTW) σ_c 63 (to Se⁷⁷) 22 (to Se⁷⁷ᵐ) (GoldmDT64)			
Se⁷⁷		% 7.58 (WhiJ48) △ −74.60 (MTW) σ_c 42 (GoldmDT64)			
Se⁷⁷ᵐ	17.5 s (ArnJ47, CanR51a, RutW52) 17.4 s (FlaA50) 17.7 s (AlexKF63) 18.8 s (MalmS62)	☢ IT (ArnJ47) △ −74.44 (LHP, MTW)	A n-capt (ArnJ47) sep isotopes, n-capt (GoldhM48a) genet (CanR51a) daughter Br⁷⁷ (CanR51a, CanR51c)	γ Se X-rays, 0.161 (50%) e⁻ 0.148, 0.160	Se⁷⁶(n, γ) (GoldhM48a, ArnJ47) daughter Br⁷⁷ (CanR51a, CanR51c)
Se⁷⁸		% 23.52 (WhiJ48) △ −77.021 (MTW) σ_c 0.05 (to Se⁷⁹) 0.36 (to Se⁷⁹ᵐ) (GoldmDT64)			
Se⁷⁹	≤6.5 x 10⁴ y sp act (est fission yield) (ParkG49a)	☢ β⁻ (ParkG49a) △ −75.921 (MTW)	B chem, decay charac (ParkG49a)	β⁻ 0.16 max γ no γ	fission (ParkG49a)
Se⁷⁹ᵐ	3.91 m (YthC54) 3.88 m (CunJ53)	☢ IT (FlaA50a) △ −75.825 (LHP, MTW)	A excit, n-capt (FlaA50, FlaA50a) n-capt, sep isotopes (RutW52) daughter As⁷⁹ (YthC54, CunJ53)	γ Se X-rays, 0.096 (9%) e⁻ 0.083, 0.095	Se⁷⁸(n, γ) (RutW52, FlaA50, FlaA50a)
Se⁸⁰		% 49.82 (WhiJ48) △ −77.753 (MTW) σ_c 0.5 (to Se⁸¹) 0.1 (to Se⁸¹ᵐ) (GoldmDT64)			
Se⁸¹	18.6 m (ApeD57) 18.2 m (YthC54) others (GleL51b, FlaA50, LangsA40, RutW52, WafH48)	☢ β⁻ (LangsA40) △ −76.40 (MTW)	A chem, genet (LangsA40) daughter Se⁸¹ᵐ (LangsA40)	β⁻ 1.58 max γ 0.030 (0.06%), 0.28 (0.9%, complex), 0.56 (0.3%, complex), 0.83 (0.2%)	Se⁸⁰(n, γ), daughter Se⁸¹ᵐ (SneA37, LangsA40, SerL47b, SunR62)
Se⁸¹ᵐ	56.8 m (YthC54) 56.5 m (WafH48) 62 m (ApeD57) 57 m (SneA37, LangsA40) 61 m (YthC54) others (GleL51b, RutW52, BergI49b)	☢ IT, no β⁻ (SunR62) IT, [β⁻] (YthC59) △ −76.29 (LHP, MTW)	A chem, excit, cross bomb (SneA37) sep isotopes, n-capt (LeviHA47) mass spect (BergI49b) parent Se⁸¹ (LangsA40)	γ Se X-rays, 0.103 (8%) e⁻ 0.090, 0.102 daughter radiations from Se⁸¹	Se⁸⁰(n, γ) (SneA37, HeyF37, SerL47b, LevyHA47)
Se⁸²	>10¹⁷ y genet (SharmH53)	% 9.19 (WhiJ48) △ −77.59 (MTW) σ_c 0.004 (to Se⁸³) 0.05 (to Se⁸³ᵐ) (GoldmDT64)			
Se⁸³	25 m (GleL51a) 26 m (RutW52) others (LangsA40, YthC54)	☢ β⁻ (SneA37) △ −75.4 (CocR59, MTW)	A chem, excit, cross bomb (SneA37) chem, genet (LangsA40) parent Br⁸³ (LangsA40, GleL51a)	β⁻ 1.8 max γ 0.22 (44%), 0.36 (69%), 0.52 ? (59%), 0.71 ? (25%), 0.83 ? (41%, complex), 1.06 ? (16%), 1.31 ? (25%), 1.88 (16%), 2.29 (9%) daughter radiations from Br⁸³, Kr⁸³ᵐ	Se⁸²(n, γ) (SneA37, LangsA40, SerL47b, CocR59)

TABLE 1 (*Continued*)

Isotope Z A	Half-life	Type of decay (☙); % abundance; Mass excess (Δ≡M–A), MeV (C¹²=0); Thermal neutron cross section (σ), barns	Class; Identification; Genetic relationships	Major radiations: approximate energies (MeV) and intensities	Principal means of production
$_{34}Se^{83m}$	70 s (CocR58) 69 s (RutW52) 67 s (ArnJ47)	☙ β⁻ (ArnJ47) Δ –75.2 (CocR59, MTW)	A chem, genet (ArnJ47) parent Br⁸³ (ArnJ47)	β⁻ 3.8 max γ 0.35 (↑ 16), 0.65 (↑ 20), 1.01 (↑ 100, complex), 2.02 (↑ 40) daughter radiations from Br⁸³, Kr⁸³ᵐ	Se⁸²(n, γ) (ArnJ47, CocR58)
Se⁸⁴	3.3 m (SatJ60)	☙ [β⁻] (SatJ60)	A chem, genet (GleL46) parent 31.8 m Br⁸⁴ (GleL51, EdwR51, SatJ60) not parent 6.0 m Br⁸⁴ (SatJ60)		fission (SatJ60)
Se⁸⁵	39 s genet (SatJ60)	☙ [β⁻] (SatJ60)	B chem, genet (SatJ60) parent Br⁸⁵ (SatJ60)		fission (SatJ60)
Se⁸⁷	16 s (SatJ60)	☙ [β⁻] (SatJ60)	D chem, genet (SatJ60) parent Br⁸⁷ (or Br⁸⁶) (SatJ60)	daughter radiations from Br⁸⁷	fission (SatJ60)
$_{35}Br^{<74}$	4 m (HollaJ53)	☙ (HollaJ53)	E chem, excit (HollaJ53)		C¹² on Cu (HollaJ53)
Br⁷⁴	36 m (HollaJ53, GrayJH60) 26 m (ButeF60a) 42 m (BeydJ57a)	☙ β⁺, [EC] (HollaJ53) Δ –65 (MTW)	B chem, excit (HollaJ53) chem, genet energy levels (BeydJ57a) daughter Kr⁷⁴ (20 m) (GrayJH60) daughter Kr⁷⁴ (12 m) (ButeF60a)	β⁺ 4.7 max γ 0.511 (γ±), 0.64	Cu⁶⁵(C¹², 3n) (HollaJ53)
Br⁷⁵	1.7 h (BaskK61, WoodwL48a) 1.6 h (HollaJ53, BeydJ57a)	☙ β⁺ ≈90%, EC ≈10% (BaskK61) Δ –69.44 (MTW)	B chem, cross bomb, sep isotopes (WoodwL48a) daughter Kr⁷⁵ (ButeF60a)	β⁺ 1.70 max γ [Se X-rays], 0.285, 0.511 (180%, γ±), 0.62	Se⁷⁴(d, n) (WoodwL48a, FulS52, BaskK61) Se⁷⁴(p, γ) (WoodwL48a) Cu⁶⁵(C¹², 2n) (HollaJ51)
Br⁷⁶	16.1 h (GirR59c) 16.2 h (DosI63) 16.3 h (ButeF60a) 17.2 h (FulS52) 17.5 h (ThuS55)	☙ β⁺ ≈62%, EC ≈38% (DosI63) [β⁺ 67%, EC 33%] (GirR59c) EC(K) 20% (KuzM57) Δ –70.6 (MTW)	A chem (HopH48a) chem, sep isotopes (FulS52) chem, mass spect (ThuS55) daughter Kr⁷⁶ (CareA54, ThuS55, DosI63)	β⁺ 3.6 max γ Se X-rays, 0.511 (133%, γ±), 0.559 (63%), 0.65 (19%), 0.75 (6%), 0.85 (7%), 1.21 (13%), 1.37 (5%), 1.47 (7%), 1.86 (11%), 2.10 (7%), 2.39 (4%), 2.78 (5%), 2.97 (8%), 3.57 (2%)	As⁷⁵(α, 3n) (GirR59c)
Br⁷⁷	57 h (HollaJ51) 58 h (WoodwL48a)	☙ EC 99%, β⁺ 1% (SehR54) others (WoodwL48a) Δ –73.24 (MTW)	A chem, sep isotopes (WoodwL48a) parent Se⁷⁷ᵐ (CanR51c, CanR51a)	β⁺ 0.34 max e⁻ 0.229, 0.287, 0.508 γ Se X-rays, 0.24 (30%, complex), 0.300 (6%), 0.52 (24%), 0.58 (7%), 0.75 (2%), 0.82 (3%), 1.00 (1.3%) daughter radiations from Se⁷⁷ᵐ	As⁷⁵(α, 2n) (HollaJ51, CanR51a, MonaS63)
Br⁷⁷ᵐ	4.2 m (GooA59)	☙ IT (GooA59) Δ –73.13 (LHP, MTW)	B excit, sep isotopes (GooA59)	γ [Br X-rays], 0.108 e⁻ 0.094, 0.106 (these radiations were formerly assigned to Br⁷⁸)	Se⁷⁶(p, γ) (GooA59)
Br⁷⁸	6.5 m (SchaA61a, RikR61) 6.4 m (SneA37) 6.2 m (PierW60)	☙ β⁺ [92%], EC [8%] (RikR61, PierW60) Δ –73.45 (MTW)	A chem, excit (SneA37) cross bomb (PierW60)	β⁺ 2.55 max γ Se X-rays, 0.511 (184%, γ±), 0.614 (14%)	As⁷⁵(α, n) (SneA37) Se⁷⁸(p, n) (SchaA61a, RikR61, PierW60, BucJ38, ValleG39) Se⁷⁷(p, γ) (SchaA61a) Se⁷⁷(d, n) (SneA37, VasiSS62c)
Br⁷⁸	<6 m (SneA37)	☙ β⁺ (SneA37)	G [genet] (StahP53a) activity not observed (SchaA61a, PierW60)		[daughter Br⁷⁸] (StahP63a)
<u>Br⁷⁹</u>		% 50.52 (WilliD46) 50.56 (CamAE55a) Δ –76.075 (MTW) σ_c 8.5 (to Br⁸⁰) 2.9 (to Br⁸⁰ᵐ) (GoldmDT64)			

TABLE 1 (*Continued*)

Isotope Z A	Half-life	Type of decay (☢); % abundance; Mass excess (Δ≡M–A), MeV (C¹²=0); Thermal neutron cross section (σ), barns	Class; Identification; Genetic relationships	Major radiations: approximate energies (MeV) and intensities	Principal means of production
$_{35}Br^{79m}$	4.8 s (GooA59) 5.0 s (SchaG54)	☢ IT (SchaG54) Δ −75.87 (LHP, MTW)	B excit (SchaG54) excit, sep isotopes (GooA59)	γ [Br X-rays], 0.21	$Se^{78}(p, γ)$ (GooA59) $Br^{79}(n, n')$ (SchaG54)
Br^{80}	17.6 m (KinA57) 18 m (SneA37, SegE39, AmaE35)	☢ β⁻ 92%, β⁺ 2.6%, EC 5.7% (TrehP62) others (MimW51, ReynJH50, LabJ51, BarbW47) Δ −75.882 (MTW)	A chem, n-capt (AmaE35) chem, excit, cross bomb (SneA37) chem, genet (SegE39) daughter Br^{80m} (SegE39, DVauD40, SidR41)	β⁻ 2.00 max β⁺ 0.87 max γ Se X-rays, 0.511 (5%, γ±), 0.618 (7%), 0.666 (1.0%)	$Br^{79}(n, γ)$, daughter Br^{80m} (SneA37, SerL47b, OrsA49, AliA36, SegE39)
Br^{80m}	4.38 h (KinA57) 4.40 h (SchmW60) 4.6 h (MimW51) others (SneA37, BucJ38, BotW39)	☢ IT (SegE39) Δ −75.796 (LHP, MTW)	A chem, n-capt (AmaE35) chem, excit, cross bomb (SneA37) parent Br^{80} (SegE39, DVauD40, SidR41)	γ Br X-rays, 0.037 (36%) e⁻ 0.024, 0.036, 0.047 daughter radiations from Br^{80}	$Br^{79}(n, γ)$ (AliA36, SneA37, SegE39, SerL47b)
$\underline{Br^{81}}$		% 49.48 (WilliD46) 49.44 (CamAE55a) Δ −77.97 (MTW) $σ_c$ 3 (GoldmDT64)			
Br^{82}	35.34 h (MerJ62) 35.9 h (CobJ50) 35.1 h (WintF51) 36.0 h (BerneE50) 35.5 h (WyaE61) 35.7 h (SinW51)	☢ β⁻ (KurtB35) no EC or β⁺, lim 0.03% (ReynJH50) no β⁺, lim 0.02% (MimW51) Δ −77.50 (MTW)	A chem, n-capt (KurtB35) chem, excit, cross bomb (SneA37) daughter Br^{82m} (EmeJ65, AndeO65)	β⁻ 0.444 max γ 0.554 (66%), 0.619 (41%), 0.698 (27%), 0.777 (83%), 0.828 (25%), 1.044 (29%), 1.317 (26%), 1.475 (17%)	$Br^{81}(n, γ)$ (SneA37, KurtB35, SerL47b, EmeJ65)
Br^{82m}	6.05 m (AndeO65) 6.20 m (EmeJ65) 6.2 m (IyeR65)	☢ IT 97.6%, β⁻ 2.4% (EmeJ65) IT, β⁻ ≥0.18% (AndeO65) Δ −77.45 (LHP, MTW)	A chem, genet, sep isotopes (AndeO65) genet (EmeJ65) parent Br^{82} (EmeJ65, AndeO65)	γ Br X-rays, 0.046 (0.3%), 0.777 (0.15%), 1.475 (0.009%) β⁻ [3.138 max] e⁻ [0.033, 0.044] daughter radiations from Br^{82}	$Br^{81}(n, γ)$ (EmeJ65, AndeO65)
Br^{83}	2.41 h (BowleB61) 2.39 h (PastM63) 2.30 h (SwiP53) 2.4 h (GleL51a, SneA37, VasiI58) 2.3 h (LangsA40, HasR51)	☢ β⁻ (SneA37) Δ −79.02 (MTW)	A chem, excit (SneA37) daughter Kr^{83m}, parent Kr^{83m} (LangsA40, StraF40, MoussA41, GleL51a) daughter Se^{83m} (ArnJ47)	β⁻ 0.93 max γ 0.530 (1.4%) daughter radiations from Kr^{83m}	$Se^{82}(n, γ)Se^{83}(β⁻)$ (SneA37, LangsA40, GleL51a, BowleB61)
Br^{84}	31.8 m (JohnN57) 31.7 m (SatJ60) others (StraF40, DufR51, KatcS51)	☢ β⁻ (DodR39) Δ −77.7 (MTW)	A chem (DodR39) chem, excit (BornH43) daughter Se^{84} (GleL51, EdwR51, SatJ60) not parent 6.0 m Br^{84} (SatJ60)	β⁻ 4.68 max γ 0.81 (9%), 0.88 (51%), 1.01 (10%), 1.21 (4%), 1.90 (18%), 2.47 (8%), 3.93 (13%)	$Rb^{87}(n, α)$ (BornH43, SatJ60) fission (DodR39, HahO39c, HahO39e, StraF40, MoussA41, BornH43, KatcS51)
Br^{84}	6.0 m (SatJ60)	☢ β⁻ (SatJ60)	A chem, excit, sep isotopes (SatJ60) not daughter Se^{84} (SatJ60) not daughter 31.8 m Br^{84} (SatJ60)	β⁻ 1.9 max γ 0.44 (68%), 0.88 (75%), 1.46 (75%), 1.89 (16%)	$Rb^{87}(n, α)$ (SatJ60) fission (SatJ60)
Br^{85}	3.00 m (SugaN49) 3.0 m (StraF40, BornH43)	☢ β⁻ (StraF40) Δ −78.7 (MTW)	A chem (StraF40) chem, genet (SeeW43) parent Kr^{85m} (SeeW43, SugaN49) daughter Se^{85} (SatJ60)	β⁻ 2.5 max γ no γ daughter radiations from Kr^{85m}	fission (StraF40, BornH43, SeeW43, SugaN49)
Br^{86}	54 s (StehA62, WilliE63)	☢ β⁻ (StehA62) no n, lim 0.25% (SteinE63) Δ −76 (MTW)	B chem, excit, sep isotopes (StehA62)	β⁻ 7.1 max γ 1.29 (↑ 12), 1.36 (↑ 39), 1.56 (↑ 100), 1.97 (↑ 20), 2.34 (↑ 20), 2.75 (↑ 36)	$Kr^{86}(n, p)$ (StehA62)
Br^{87}	55.6 s (n) (HugD48) 54.5 s (n) (KeeG57, PerloG59) 55.0 s (n) (RedW47) 56.1 s (β⁻) (SugaN49) 55.4 s (n) (WilliE63)	☢ β⁻, β⁻n (≈2%) (LeviJ51, StehA53) Δ −74.6 (WilliE63, MTW)	A chem (StraF40) chem, genet (BornH43, SugaN49) parent Kr^{87} (BornH43, SeeW43, SugaN49) parent Kr^{86} (2%) (SneA47a, SugaN49) daughter Se^{87} (?) (SatJ60)	β⁻ 8.0 max(?), 2.6 max n 0.3 (mean energy) γ 1.44 (↑ 100), 1.85 (↑ 18), 2.48 (↑ 18), 2.64 (↑ 16), 2.98 (↑ 25), 3.18 (↑ 16), 3.80 (↑ 11), 4.19 (↑ 21), 4.8 (↑ 17), 5.0 (↑ 17), 5.2 (↑ 12) daughter radiations from Kr^{87}	fission (StraF40, SneA47a, SugaN47, SugaN49, RedW47, HugD48)

TABLE 1 (*Continued*)

Isotope Z A	Half-life	Type of decay (☢); % abundance; Mass excess (△≡M–A), MeV (C¹²=0); Thermal neutron cross section (σ), barns	Class; Identification; Genetic relationships	Major radiations: approximate energies (MeV) and intensities	Principal means of production
$_{35}Br^{88}$	15.5 s (SugaN49) 16.3 s (PerloG59) others (PerloG57, KeeG57)	☢ β^- (SugaN49) n (weak) (PerloG59, PerloG57)	A chem, genet (SugaN49) parent Kr^{88} (SugaN49)	γ 0.76	fission (SugaN49, KeeG57, PerloG59, PerloG57)
Br^{89}	4.5 s (n) (HugD48, RedW47) 4.4 s (n) (PerloG59)	☢ β^-, β^-n (SneA47, HugD48)	D chem (SneA47) parent Kr^{89} (?), parent Kr^{88} (?) (CoryC51)	n 0.5 (mean energy)	fission (SugaN47, SneA47, SugaN49, RedW47, HugD48)
Br^{90}	1.6 s (PerloG59)	☢ [β^-], n (PerloG59)	D chem, decay charac (PerloG59)		fission (PerloG59)
$_{36}Kr^{74}$	20 m (GrayJH60) 12 m (ButeF60a)	☢ β^+, [EC] (GrayJH60) △ –62 (MTW)	B chem, genet (GrayJH60, ButeF60a) parent Br^{74} (36 m) (GrayJH60) parent Br^{74} (26m) (ButeF60a)	β^+ 3.1 max γ 0.511 ($\gamma^{\pm}$) daughter radiations from Br^{74}	protons on Br (GrayJH60) protons on Sr (ButeF60a)
Kr^{75}	5.5 m (ButeF60a) <1 m (GrayJH60)	☢ [B^+, EC] (ButeF60a) △ –64 (MTW)	E chem, genet (ButeF60a) activity not observed (GrayJH60) parent Br^{75} (ButeF60a)		protons on Br (ButeF60a)
Kr^{76}	14.8 h genet (DosI63) 9.7 s (CareA54) 11 h (ThuS55)	☢ EC, no β^+, lim 1% (DosI63) no EC(K) (CareA54) △ –69 (MTW)	A chem, genet (CareA54) chem, mass spect (ThuS55) parent Br^{76} (CareA54, ThuS55, DosI63)	γ [Kr X-rays], 0.039, 0.104, 0.135, 0.267, 0.316, 0.407, 0.452 daughter radiations from Br^{76}	Br^{79}(p, 4n) (ThuS55) Se^{74}(α, 2n) (DosI63)
Kr^{77}	1.19 h (ButeF60a) others (ThuS55, WoodwL48a, BeydJ57a)	☢ EC ≈20%, β^+ ≈80% (ThuS55) others (WoodwL48a) △ –70.4 (MTW)	A chem, sep isotopes (WoodwL48) chem, mass spect (ThuS55)	β^+ 1.86 max e^- 0.011, 0.023, 0.094 (with Br^{77m}), 0.106 (with Br^{77m}), 0.118, 0.136 γ Br X-rays, 0.024, 0.108 (with Br^{77m}), 0.131, 0.149, 0.665 daughter radiations from Br^{77}	Br^{79}(p, 3n) (ThuS55)
$\underline{Kr^{78}}$		% 0.354 (NierA50a) △ –74.14 (MTW) σ_c 2 (to Kr^{79}) (GoldmDT64)			
Kr^{79}	34.92 h (BonaE64) 34.5 h (RadP52) others (WoodwL48, CreEC40a, ChacK61)	☢ EC 92%, β^+ 8% (NDS, BonaE64) others (RadP52a, RadP52b, RadP55, LangeM54, BergI51d, ThuS54c) △ –74.46 (MTW)	A chem (CreEC40a) chem, sep isotopes (WoodwL48) mass spect (BracD52) daughter Rb^{79} (ChacK61)	β^+ 0.60 max e^- 0.031, 0.043, 0.123, 0.204, 0.248, 0.384 γ Br X-rays, 0.136 (0.7%), 0.261 (9%), 0.398 (10%), 0.511 (15%, $\gamma^{\pm}$), 0.606 (10%), 0.836 (2.0%), 1.119 (0.5%), 1.336 (0.5%)	Br^{79}(p, n) (CreEC40a) Br^{79}(d, 2n) (ClarE44, BonaE64) Kr^{78}(n, γ) (HoaE51a, BergI51d)
Kr^{79m}	55 s (CreEC40a)	☢ IT (?), no β^+ (CreEC40a) △ –74.33 (LHP, MTW)	D chem (CreEC40a)	γ Kr X-rays, 0.127 e^- 0.113, 0.125	Br^{79}(p, n) (CreEC40a)
$\underline{Kr^{80}}$		% 2.27 (NierA50a) △ –77.89 (MTW) σ_c 15 (GoldmDT64)			
Kr^{81}	2.1 x 10⁵ y sp act, mass spect (EasT64a, ReynJH50a)	☢ EC (ReynJH50a) △ –77.7 (MTW)	A chem, mass spect (ReynJH50a)	γ Br X-rays	Kr^{80}(n, γ) (ReynJH50a, EasT64a)
Kr^{81m}	13 s (ChacK61, CreEC40a) others (KarrD50)	☢ IT, no β^+ (CreEC40a) △ –77.5 (LHP, MTW)	A chem (CreEC40a) genet (KarrD50) daughter Rb^{81} (KarrD50)	γ Kr X-rays, 0.190 (65%) e^- 0.176, 0.188	daughter Rb^{81} (KarrD50)
$\underline{Kr^{82}}$		% 11.56 (NierA50a) △ –80.589 (MTW) σ_c 42 (to Kr^{83}) 3 (to Kr^{83m}) (GoldmDT64)			

TABLE 1 (*Continued*)

Isotope Z A	Half-life	Type of decay (☢); % abundance; Mass excess ($\Delta \equiv M-A$), MeV ($C^{12}=0$); Thermal neutron cross section (σ), barns	Class; Identification; Genetic relationships	Major radiations: approximate energies (MeV) and intensities	Principal means of production
$_{36}Kr^{83}$		% 11.55 (NierA50a) Δ −79.985 (MTW) σ_c 180 (GoldmDT64)			
Kr^{83m}	1.86 h (DVriL52) 1.90 h (BergI51b) 1.88 h (LangsA40) others (RieW46)	☢ IT (LangsA40) Δ −79.943 (LHP, MTW)	A: chem, genet (LangsA40) mass spect (BergI50) daughter Br^{83} (LangsA40) daughter Rb^{83} (CastS50)	γ Kr X-rays, 0.009 (9%) e^- 0.007, 0.018, 0.031	daughter Rb^{83} (CastS50)
Kr^{84}		% 56.90 (NierA50a) Δ −82.433 (MTW) σ_c 0.04 (to Kr^{85}) 0.10 (to Kr^{85m}) (GoldmDT64)			
Kr^{85}	10.76 y (LernJ63) 10.3 y (WanR53) 9.4 y (ThodH48a) others (HoaE51)	☢ β^- (HoaE51) Δ −81.48 (MTW) σ_c <15 (GoldmDT64)	A: chem (HoaE51) chem, mass spect (ThodH47)	β^- 0.67 max γ 0.514 (0.41%)	Kr^{84}(n, γ) (HoaE51a) fission (ThodH47, HoaE51)
Kr^{85m}	4.4 h (KocJ49, WoodwL48) 4.5 h (HoaE51a, SneA37) 4.6 h (RieW46, SeeW43)	☢ β^- 77%, IT 23% (BergI51) β^- 78%, IT 22% (BladA55) Δ −81.18 (LHP, MTW)	A: chem (SneA37) chem, mass spect (KocJ49) daughter Br^{85} (SeeW43, SugaN49)	β^- 0.82 max e^- 0.134, 0.291 γ Kr X-rays, 0.150 (74%), 0.305 (13%)	Kr^{84}(n, γ) (RieW46, HoaE51a) fission (SeeW43, SugaN49) Se^{82}(α, n) (WoodwL48)
Kr^{86}		% 17.37 (NierA50a) Δ −83.259 (MTW) σ_c 0.06 (GoldmDT64)	daughter Br^{87} (2%) (SneA47a, SugaN49)		
Kr^{87}	76 m (ClarW64) 78 m (KocJ49) 74 m (SneA37) 75 m (SeeW43, SugaN49)	☢ β^- (SneA37) Δ −80.70 (MTW) σ_c <600 (GoldmDT64)	A: chem (SneA37) chem, mass spect (KocJ49) daughter Br^{87} (SeeW43, BornH43, SugaN49)	β^- 3.8 max γ 0.403 (84%), 0.85 (16%), 2.57 (35%)	Kr^{86}(n, γ) (RieW46, HoaE51a) fission, daughter Br^{87} (BornH43, SeeW43, SugaN49)
Kr^{88}	2.80 h (ClarW64) 2.77 h (KocJ49) others (GlasG40, SugaN49)	☢ β^- (LangsA39) Δ −79.9 (MTW)	A: chem (HeyF39) chem, genet (LangsA39) chem, mass spect (KocJ49) parent Rb^{88} (LangsA39, AteA39, HeyF39, GlasG40, HahO40, HahO40b) daughter Br^{88} (SugaN49)	β^- 2.8 max e^- 0.013 γ 0.028, 0.166 (7%), 0.191 (35%), 0.36 (5%), 0.85 (23%), 1.55 (14%), 2.19 (≤18%), 2.40 (35%) daughter radiations from Rb^{88}	fission (HeyF39, HahO40, GlasG40, HahO40b)
Kr^{89}	3.18 m (KofO51b) 3.2 m (OckD62) 2.6 m (DilC51a) others (HahO43b)	☢ β^- (GlasG40) Δ −78 (MTW)	A: chem, genet (GlasG40, SeeW40) mass spect (KofO51b) parent Rb^{89} (GlasG40, SeeW40, HahO40b, HahO43, BradE51, KofO51b)	β^- 4.0 max γ 0.23 († 85), 0.36 († 28), 0.43 († 29), 0.51 († 42), 0.60 († 100), 0.74 († 32), 0.88 († 65), 1.12 († 45), 1.29 († 31), 1.51 († 88, complex?), 1.71 († 34), 1.93 († 10), 2.04 († 16), 2.23 († 10), 2.42 († 22), 2.57 († 10), 2.84 († 25), (some of these may be sum peaks) daughter radiations from Rb^{89}	fission (GlasG40, SeeW40, HahO40b, HahO43, BradE51, KofO51b, AdaRM51)
Kr^{90}	33 s (KofO51b) 35 s (OckD62)	☢ β^- (DilC51) Δ −74.8 (MTW)	A: chem, genet (DilC51) mass spect (KofO51b) parent Rb^{90} (KofO51b) ancestor Sr^{90} (DilC51, DilC51a)	β^- 2.80 max γ 0.105 (15%), 0.120 (65%), 0.236 (16%), 0.495 (12%), 0.536 (48%), 1.11 (48%), 1.54 (17%), 1.79 (11%), 2.48 (4%) daughter radiations from Rb^{90}	fission (DilC51, DilC51a, KofO51b, OckD62, GooR64)
Kr^{91}	9.8 s (DilC51a) 10 s (KofO51b) 6 s (OveR51)	☢ β^- (HahO40c)	A: chem, genet (HahO40c) mass spect (KofO51b) parent 1.2 m Rb^{91}, parent 14 m Rb^{91} (KofO51b) ancestor Y^{91} (HahO40c, BradE51, DilC51, DilC51a)	β^- 3.6 max γ no γ daughter radiations from 1.2 m Rb^{91}	fission (HahO40c, DilC51a, BradE51, DilC51, AdaRM51)

TABLE 1 (*Continued*)

Isotope Z A	Half-life	Type of decay (☙); % abundance; Mass excess (Δ≡M–A), MeV (C¹²=0); Thermal neutron cross section (σ), barns	Class; Identification; Genetic relationships	Major radiations: approximate energies (MeV) and intensities	Principal means of production
₃₆Kr⁹²	3.0 s (DilC51a)	☙ β⁻ (HahO40)	B chem, genet (HahO40, DilC51a) ancestor Y⁹², parent Rb⁹² (DilC51a)		fission (HahO40, DilC51a)
Kr⁹³	2.0 s (DilC51a)	☙ β⁻ (HahO42)	B chem, genet (HahO42, SelB51) parent Rb⁹³ (BradE51, DilC51a, DilC51) ancestor Y⁹³ (SelB51)		fission (HahO42, DilC51a, SelB51, BradE51)
Kr⁹⁴	1.4 s (DilC51a)	☙ β⁻ (HahO43b)	B chem, genet (HahO43b, DilC51a) parent Rb⁹⁴ (HahO43, HahO43b, DilC51) ancestor Y⁹⁴ (HahO43b, DilC51a)		fission (HahO43b, DilC51a, HahO43)
Kr⁹⁵	short (DilC51)	☙ [β⁻] (DilC51)	F chem, genet (DilC51) parent Rb⁹⁵, ancestor Zr⁹⁵ (DilC51)		fission (DilC51)
Kr⁹⁷	≈1 s (DilC51)	☙ β⁻ (AdaRM51)	G chem, genet (AdaRM51, DilC51) activity not observed (WahA62)		fission (DilC51, AdaRM51)
₃₇Rb⁷⁹	24 m (BeydJ57a, ChamiR57) 21 m genet (ChacK61)	☙ β⁺ (BeydJ57), [EC]	A chem (BeydJ57, ChamiR57) chem, genet (ChacK61) parent Kr⁷⁹ (ChacK61)	γ [Kr X-rays], 0.15 (73%), 0.19 (29%), 0.511 (γ⁺, [180%]), daughter radiations from Kr⁷⁹	Cu⁶⁵(O¹⁶, 2n) (BeydJ57, ChamiR57) Br⁷⁹(He³, 3n) (ChacK61)
Rb⁸⁰	34 s (HoffR61)	☙ β⁺, [EC] (HoffR61) Δ –73 (MTW)	A chem, mass spect (HoffR61) daughter Sr⁸⁰ (HoffR61)	β⁺ 4.1 max γ 0.511 (γ⁺, [195%]), 0.618 (39%)	daughter Sr⁸⁰ (HoffR61)
Rb⁸¹	4.7 h (KarrD50, DogW56, CastS52)	☙ EC 87%, β⁺ 13% (KarrD50) Δ –75.4 (MTW)	A chem, mass spect (ReynF49) parent Kr⁸¹ᵐ (KarrD50) daughter Sr⁸¹ (CastS52, CastS52) daughter Rb⁸¹ᵐ (DogW56) descendant Zr⁸¹ (ZaitN65)	β⁺ 1.03 max γ Kr X-rays, 0.253, 0.450, 0.511 (26%, γ⁺), 1.10 daughter radiations from Kr⁸¹ᵐ	Br⁷⁹(α, 2n) (ReynF49, KarrD50)
Rb⁸¹ᵐ	31 m (DogW56)	☙ β⁺, [EC], IT (DogW56) Δ –75.3 (LHP, MTW)	B chem, genet (DogW56) parent Rb⁸¹ (DogW56)	β⁺ 1.4 mag spect e⁻ 0.071, 0.083 γ [Rb X-rays, Kr X-rays, 0.085, 0.511 (γ⁺)] daughter radiations from Rb⁸¹ Kr⁸¹ᵐ	Br⁷⁹(α, 2n) (DogW56)
Rb⁸²	1.25 m (LitL53) 1.3 m (KruP53) 1.1 m (KurcB55)	☙ β⁺ 96%, EC 4% (SakM62) Δ –76.42 (MTW)	A chem, genet (LitL53, KruP53) daughter Sr⁸² (LitL53, KruP53, KurcB55)	β⁺ 3.15 max γ Kr X-rays, 0.511 (192%, γ⁺), 0.777 (9%)	daughter Sr⁸² (LitL53, KruP53, KurcB55)
Rb⁸²ᵐ	6.3 h (KarrD50) 6.5 h (HancJ40)	☙ EC 94%, β⁺ 6% (KarrD50) [EC 79%, β⁺ 21%] (NDS) Δ –76.14 (LHP, MTW)	A chem (HancJ40) chem, mass spect (ReynF49) not daughter Sr⁸², lim 0.1% (LitL53, CastS52)	β⁺ 0.78 max γ Sr X-rays, 0.511 (γ⁺), 0.554 (66%), 0.619 (41%), 0.698 (27%), 0.777 (83%), 0.828 (25%), 1.044 (29%), 1.317 (26%), 1.475 (17%)	Br⁷⁹(α, n) (HancJ40, ReynF49, KarrD50) Kr⁸²(d, 2n) (HancJ40)
Rb⁸³	83 d (CastS50) 100 d (KurcB55) 107 d (KarrD50)	☙ EC (KarrD50) no β⁺ (PerlmM55) Δ –79 (MTW)	A chem, mass spect (KarrD50) daughter Sr⁸³, parent Kr⁸³ᵐ (CastS50)	γ Kr X-rays, 0.53 (93%, 3 γ rays), 0.79 (0.9%) e⁻ 0.007, 0.52 daughter radiations from Kr⁸³ᵐ	Br⁸³(α, 2n) (KarrD50) daughter Sr⁸³ (CastS50, DosI64a)
Rb⁸⁴	33.0 d (WelJ55) 34 d (KarrD50)	☙ EC 76%, β⁺ 21%, β⁻ 3% (NDS) Δ –79.753 (MTW)	A chem, cross bomb (BarbW47) chem, mass spect (KarrD50)	β⁺ 1.66 max β⁻ 0.91 max γ Kr X-rays, 0.511 (42%, γ⁺), 0.88 (74%), 1.01 (0.5%), 1.90 (0.8%)	Br⁸¹(α, n) (KarrD50)

TABLE 1 (*Continued*)

Isotope Z A	Half-life	Type of decay (☢); % abundance; Mass excess (△≡M-A), MeV (C¹²=0); Thermal neutron cross section (σ), barns	Class; Identification; Genetic relationships	Major radiations: approximate energies (MeV) and intensities	Principal means of production
₃₇Rb⁸⁴ᵐ	20 m (CohL58, HancJ40) 21 m (CaiR53) 23 m (FlaA50b)	☢ IT, EC (weak) (CaiR53) △ -79.289 (LHP, MTW)	B chem (HancJ40) chem, excit (FlaA50b)	γ Rb X-rays, 0.216 (37%), 0.250 (65%), 0.464 (32%) e⁻ 0.201, 0.214, 0.449	Br⁸¹(α,n) (HancJ40) Rb⁸⁵(n,2n) (FlaA50b)
Rb⁸⁵		% 72.15 (NierA50a) △ -82.16 (MTW) σc 0.9 (to Rb⁸⁶) 0.1 (to Rb⁸⁶ᵐ) (GoldmDT64)			
Rb⁸⁶	18.66 d (EmeE55a, EmeE55) 18.64 d (NidJ55) 18.7 d (WriH57) 18.8 d (GleG64) others (HelmhA41, RobiR58a)	☢ β⁻ (HelmhA41) △ -82.72 (MTW)	A chem, n-capt (SneA37) chem, excit (HelmhA41)	β⁻ 1.78 max γ 1.078 (8.8%)	Rb⁸⁵(n,γ) (SneA37, ScheiH38, SerL47b)
Rb⁸⁶ᵐ	1.02 m (SchwaR53) 1.06 m (FlaA51)	☢ IT (SchwaR53) △ -82.16 (LHP, MTW)	B chem, excit, n-capt (FlaA51)	γ [Rb X-rays], 0.56	Rb⁸⁵(n,γ) (FlaA51, SchwaR53)
Rb⁸⁷	4.8 x 10¹⁰ y sp act (KovA65) 4.7 x 10¹⁰ y sp act (FlyK59, GleL61) 5.2 x 10¹⁰ y sp act (MNaiA61a, BrinGA65) 5.8 x 10¹⁰ y sp act (EgeK61, LeuH62a) 5.0 x 10¹⁰ y Sr⁸⁷/Rb⁸⁷ ratio (AldL56, OvcG60) 6.2 x 10¹⁰ y sp act (MGreM54, CurrS51, FliJ54*) 5.1 x 10¹⁰ y sp act (LibW57) 5.9 x 10¹⁰ y sp act (LewisG52) 4.3 x 10¹⁰ y sp act (GeeI54) others (FritK56, StraF38, HaxO48a, HaxO48, KemM49, CharG51, EklS46, BahI52) *corrected for 27.85% abundance (NDS)	☢ β⁻ (ThomJ05, CamN06) % 27.85 (NierA50a) △ -84.591 (MTW) σc 0.12 (GoldmDT64)	A chem (ThomJ05, CamN06) chem, genet (HahO37, MattaJ37) chem, mass spect (HemA37) parent Sr⁸⁷ (mass spect) (HahO37, MattaJ37)	β⁻ 0.274 max γ no γ	
Rb⁸⁸	17.8 m (GlasG40, BunkM51) 17.7 m (ThuS52b) 17.5 m (WeilG42) 18 m (HahO40b, SneA37)	☢ β⁻ (HahO39c) △ -82.7 (MTW) σc 1.0 (GoldmDT64)	A chem (SneA37) chem, genet (LangsA39, GlasG40, HahO39c) daughter Kr⁸⁸ (HeyF39, LangsA39, GlasG40, HahO40, HahO40b, AteA39)	β⁻ 5.3 max γ 0.898 (13%), 1.863 (21%), 2.68 (2.3%)	Rb⁸⁷(n,γ) (SneA37, PoolM37, ScheiH38, SerL47b) fission, daughter Kr⁸⁸ (HeyF39, LangsA39, GlasG40, HahO40, HahO40b)
Rb⁸⁹	15.4 m (GlasG40) 14.9 m (OKelG56a) 15.5 m (HahO40b)	☢ β⁻ (GlasG40) △ -82.3 (MTW)	A chem, genet (GlasG40, SeeW40) daughter Kr⁸⁹ (GlasG40, SeeW40, HahO40b, HahO43, BradE51, KofO51b) parent Sr⁸⁹ (GlasG40, HahO40, HahO43, HahO40b, GrumW46)	β⁻ 3.92 max (7%), 2.9 max (5%), 1.6 max γ 0.66 (17%), 1.05 (75%), 1.26 (54%), 2.20 (14%), 2.59 (13%)	fission (GlasG40, SeeW40, HahO40b, HahO43, BradE51)
Rb⁹⁰	2.91 m (JohnN64) 2.74 m (KofO51b) 2.8 m (OckD62)	☢ β⁻ (KofO51b) △ -79.3 (MTW)	A chem, genet (KofO51b) daughter Kr⁹⁰, parent Sr⁹⁰ (DilC51, DilC51a, KofO51b)	β⁻ 6.6 max γ 0.53 (4%), 0.83 (61%, doublet), 1.03 (5%), 1.11 (7%), 1.40 (5%), 1.70 (3%), 3.07 (5%), 3.34 (15%, doublet), 3.54 (5%), 4.13 (11%), 4.34 (18%, doublet), 4.60 (5%), 5.2 (4%)	fission (KofO51b, DilC51, DilC51a, JohnN64, OckD62)
Rb⁹¹	1.2 m (JohnN64, WahA62) 1.7 m (KofO51b)	☢ β⁻ (KofO51b) △ -78 (MTW)	A chem, genet (KofO51b) daughter Kr⁹¹, parent Sr⁹¹ (KofO51b) ancestor Y⁹¹ (DilC51, HahO40c)	β⁻ 4.6 max	fission (KofO51b, DilC51, HahO40c, WahA62, JohnN64)

TABLE 1 (*Continued*)

Isotope Z A	Half-life	Type of decay (☢); % abundance; Mass excess (△≡M−A), MeV (C¹²=0); Thermal neutron cross section (σ), barns	Class; Identification; Genetic relationships	Major radiations: approximate energies (MeV) and intensities	Principal means of production
$_{37}Rb^{91}$	14 m (KofO51b) activity not observed (WahA62)	☢ β⁻ (KofO51b)	E chem, genet (KofO51b) daughter Kr⁹¹, parent Sr⁹¹ (KofO51b) no 14 m Rb parent of Sr⁹¹ (WahA62)		fission (KofO51b)
Rb^{92}	5.3 s genet (FritK60) others (BradE51, DilC51a, HahO40)	☢ [β⁻] (DilC51a) △ −75 (MTW)	B genet (DilC51a) chem, genet (FritK60) daughter Kr⁹², ancestor Y⁹² (DilC51a) parent Sr⁹² (FritK60)		fission (FritK60) DilC51a)
Rb^{93}	5.6 s (FritK60)	☢ [β⁻] (HahO42)	B chem, genet (FritK60) parent Sr⁹³, ancestor Y⁹³ (FritK60) daughter Kr⁹³ (BradE51, DilC51a, DilC51)		fission (FritK60)
Rb^{94}	2.9 s (FritK61) others (DilC51a, HahO43b, HahO43)	☢ [β⁻] (HahO43b, HahO43, FritK61)	B chem, genet (FritK61) ancestor Y⁹⁴ (FritK61) daughter Kr⁹⁴, ancestor Y⁹⁴ (HahO43, HahO43b, DilC51)		fission (FritK61)
Rb^{95}	<2.5 s (FritK61)	☢ [β⁻] (DilC51)	F genet (DilC51) daughter Kr⁹⁵, ancestor Zr⁹⁵ (DilC51)		fission (DilC51)
$_{38}Sr^{80}$	1.7 h (HoffR61)	☢ EC (HoffR61)	A chem, genet (HoffR61) parent Rb⁸⁰ (HoffR61)	γ [Rb X-rays], 0.58 daughter radiations from Rb⁸⁰	N¹⁴ on Ga (HoffR61)
Sr^{81}	29 m (CastS50, CastS52)	☢ EC, β⁺ (CastS50)	B chem, genet (CastS50, CastS52) parent Rb⁸¹ (CastS50, CastS52) descendant Zr⁸¹ (ZaitN65)	γ [Rb X-rays, 0.511 (Y±)] daughter radiations from Rb⁸¹ Kr⁸¹ᵐ	Rb⁸⁵(p, 5n) (CastS50, CastS52)
Sr^{82}	25.0 d (SanV58) 25.5 d (KruP53) others (MacIK52, LitL53, CastS50)	☢ EC, no β⁺, lim 5% (KurcB55) △ −76 (MTW)	A chem, excit (CastS50) mass spect (MLurK52) parent Rb⁸², not parent Rb⁸²ᵐ, lim 0.1% (CastS52, LitL53, KruP53, KurcB55) daughter Y⁸² (MaxV62, ButeF63) descendant Zr⁸² (ZaitN65)	γ Rb X-rays daughter radiations from Rb⁸²	Rb⁸⁵(p, 4n) (CastS50, CastS52) As⁷⁵(C¹², 5n)Y⁸²(EC) (MaxV62)
Sr^{83}	32.4 h (DosI64) 32.9 h (KuroT61) others (KurcB55, CastS50, MacIK52, ButeF63, MaxV62)	☢ EC 84%, β⁺ 16% (KuroT61) △ −77 (MTW)	A chem, genet (CastS50) mass spect (MLurK52) parent Rb⁸³ (CastS50) daughter Y⁸³ (MaxV62, DosI64a, NiecW65) descendant Zr⁸³ (ZaitN65)	β⁺ 1.15 max e⁻ 0.025, 0.040 γ Rb X-rays, 0.040 (24%), 0.38 (35%), 0.511 (32%, Y±), 0.76 (40%), 1.16, 1.52 daughter radiations from Rb⁸³	Rb⁸⁵(p, 3n) (CastS52)
$\underline{Sr^{84}}$		% 0.56 (NierA38b) 0.55 (AldL53) △ −80.638 (MTW) σc 0.8 (to Sr⁸⁵) 0.65 (to Sr⁸⁵ᵐ) (GoldmDT64)			
Sr^{85}	64.0 d (WriH57) 64.9 d (GleG64) 63.9 d (SatA62a) 65 d (HerrmG56, TPogM51) 66 d (DubL40)	☢ EC (TPogM51, BisA56f) no β⁺ (TPogM51) △ −81.05 (MTW)	A chem, excit (DubL40) daughter Y⁸⁵ (DosI63a, CareA52, NiecW65)	γ Rb X-rays, 0.514 (100%) e⁻ 0.499	Sr⁸⁴(n, γ) (SatA62a) Rb⁸⁵(p, n) (DubL40) Rb⁸⁵(d, 2n) (TPogM51, EmmW52)
Sr^{85m}	70 m (DubL40)	☢ IT 86%, EC 14% (SunA52) △ −80.81 (LHP, MTW)	A chem, excit (DubL40) daughter Y⁸⁵ᵐ (MaxV62, DosI63a, NiecW65) descendant 15 m Zr⁸⁵ (ButeF63)	γ Rb X-rays, Sr L X-rays, 0.150 (14%), 0.231 (85%) e⁻ 0.005, 0.134, 0.215	Sr⁸⁴(n, γ) (SunA52) Rb⁸⁵(p, n) (DubL40) Rb⁸⁵(d, 2n) (TPogM51)

TABLE 1 (*Continued*)

Isotope Z A	Half-life	Type of decay (☢); % abundance; Mass excess (△≡M−A), MeV (C¹²=0); Thermal neutron cross section (σ), barns	Class; Identification; Genetic relationships	Major radiations: approximate energies (MeV) and intensities	Principal means of production
₃₈Sr⁸⁶		% 9.86 (NierA38b) 9.87 (AldL53) △ −84.499 (MTW) σ_c 1.3 (to Sr⁸⁷ᵐ) (GoldmDT64)			
Sr⁸⁷		% 7.02 (NierA38b, AldL53) △ −84.865 (MTW)	daughter Rb⁸⁷ (mass spect) (HahO37, MattaJ37)		
Sr⁸⁷ᵐ	2.83 h (BormM65) 2.80 h (MannL51, HydE51) 2.88 h (GravG52) others (HerrmG56, DubL40)	☢ IT 99+%, EC(K) 0.6% (SunA60) △ −84.477 (LHP, MTW)	A chem, excit (StewD37) chem, excit, cross bomb, genet (DubL40) daughter Y⁸⁷ (DubL39, DubL40, MannL50, MannL51, LindnM50a, HydE51)	γ Sr X-rays, 0.388 (80%) e⁻ 0.372, 0.386	daughter Y⁸⁷ (DubL39, MannL50, MannL51) Sr⁸⁶(n, γ) (StewD37, DubL39, RedH40, RedH40a) Rb⁸⁷(p, n) (DubL39)
Sr⁸⁸		% 82.56 (NierA38b, AldL53) △ −87.89 (MTW) σ_c 0.006 (GoldmDT64)			
Sr⁸⁹	52.7 d (FlyK65a) 50.4 d (OsmR59) 53.6 d (SatA62) 50.5 d (HerrmG54, HerrmG55) others (KjeA56, NoveT51, LieC39, StewD39, GoeR49, GrumW46)	☢ β⁻ (StewD37) △ −86.22 (MTW) σ_c 0.4 (GoldmDT64)	A chem, excit (StewD37) chem, mass spect (HaydR48) daughter Rb⁸⁹ (GlasG40, HahO40, HahO40b, HahO43, GrumW46) parent Y⁸⁹ᵐ 0.009% (SatA62); 0.02% (LyoW55b); 0.01% (HerrmG56); <0.0005% (BisA55d)	β⁻ 1.463 max γ 0.91 (0.009%, with Y⁸⁹ᵐ)	Sr⁸⁸(d, p) (StewD37, StewD39) Sr⁸⁸(n, γ) (SerL47b, StewD37, StewD39)
Sr⁸⁹ᵐ	10 d (HerrmG54, HerrmG55)		G activity not observed (HerrmG56, SatA62, FleJ62)		
Sr⁹⁰	27.7 y sp act, mass spect (WileDM55) 28.0 y (FlyK65) 28.4 y (ReeG55) others (AniM58, PowR50)	☢ β⁻ (NotR51) △ −85.95 (MTW, LHP) σ_c 1 (GoldmDT64)	A chem, genet (HahO42) chem, mass spect (HaydR48) daughter Rb⁹⁰ (DilC51, DilC51a, KofO51b) parent Y⁹⁰ (HahO42, HahO43, GrumW46, NotR51) descendant Kr⁹⁰ (DilC51, DilC51a)	β⁻ 0.546 max γ no γ daughter radiations from Y⁹⁰	fission (DilC51, DilC51a, KofO51b, GrumW46, GrumW48)
Sr⁹¹	9.67 h (AmeD53) 9.7 h (HerrmG54, HerrmG55, FinB51, Vasil58, BakH65) others (HahO43)	☢ β⁻ (GotH41) △ −83.68 (MTW)	A chem, genet (GotH41) chem, excit (SeeW43b) parent Y⁹¹ᵐ, parent Y⁹¹ (GotH41, HahO43, FinB51) daughter 1.2 m Rb⁹¹, daughter 14 m Rb⁹¹ (KofO51b) no 14 m Rb parent of Sr⁹¹ (WahA62)	β⁻ 2.67 max γ 0.645 (15%), 0.748 (27%), 0.93 (3%), 1.025 (30%), 1.413 (5%) daughter radiations from Y⁹¹ᵐ, Y⁹¹	fission (GotH41, HahO43, FinB51, KatcS48, FinB51c) Zr⁹⁴(n, α) (SeeW43b)
Sr⁹²	2.71 h (FritK60) 2.60 h (HerrmG56) 2.7 h (GotH41)	☢ β⁻ (GotH41) △ −82.9 (MTW)	A chem, genet (GotH41) parent Y⁹² (GotH41, HoaE51b) daughter Rb⁹² (FritK60)	β⁻ 1.5 max (10%), 0.55 max γ 0.23 (3%), 0.44 (4%), 1.37 (90%) daughter radiations from Y⁹²	fission (HahO40, HahO43b, KatcS51a, BradE51, KatcS48)
Sr⁹³	8.3 m (ValliD61) 7.5 m (FritK60) 8.5 m (BakH65) 8 m (KniJD59) 7 m (LieC39)	☢ β⁻ (LieC39) △ −79.4 (MTW, SteinE65)	A chem (LieC39, HahO43) chem, sep isotopes (BakH65) parent Y⁹³ (HahO43, HahO43b, KniJD59) daughter Rb⁹³ (FritK60)	β⁻ 4.8 max ? (weak), 2.9 max γ 0.60, 0.8, 1.2, others between 0.2 and 3.0 daughter radiations from Y⁹³	Zr⁹⁶(n, α) (ValliD61, BakH65) fission (LieC39, HahO42, HahO43, KniJD59)
Sr⁹⁴	1.35 m (FritK61) 1.2 m (HovD64) 1.3 m (KniJD59)	☢ β⁻ (HahO43b, HahO43) △ −78.8 (MTW)	A chem, genet (HahO43b, HahO43) parent Y⁹⁴ (HahO43, HahO43b, KniJD59)	β⁻ 2.1 max γ 1.42 (100%) daughter radiations from Y⁹⁴	fission (HahO43, HahO43b, DilC51, KniJD59, FritK61, HovD64)

TABLE 1 (*Continued*)

Isotope Z A	Half-life	Type of decay (☢); % abundance; Mass excess (△≡M−A), MeV (C¹²=0); Thermal neutron cross section (σ), barns	Class; Identification; Genetic relationships	Major radiations: approximate energies (MeV) and intensities	Principal means of production
$_{38}Sr^{95}$	0.8 m genet (FritK61)	☢ β⁻ (DilC51)	B genet (DilC51) chem (FritK61) ancestor Zr⁹⁵, descendant Kr⁹⁵ (DilC51) parent Y⁹⁵ (FritK61)		fission (DilC51, FritK61)
$_{39}Y^{82}$	12.3 m genet (ButeF63) 9 m genet (MaxV62) <1.5 m genet (not observed) (NiecW65)	☢ [EC, β⁺] (MaxV62, ButeF63)	B chem, genet (MaxV62, ButeF63) parent Sr⁸² (MaxV62, ButeF63) daughter Zr⁸² (ZaitN65)		As⁷⁵(C¹², 5n) (MaxV62) protons on Y⁸⁹ (ButeF63)
Y⁸²	70 m (CareA52)		G chem, genet (CareA52) activity not observed (MaxV62, ButeF63)		protons on Y (CareA52)
Y⁸³	7.4 m genet (DosI64a) 7.5 m genet (NiecW65) 8 m genet (MaxV62)	☢ [EC, β⁺] (MaxV62)	A chem, genet (MaxV62, DosI64a, NiecW65) parent Sr⁸³ (MaxV62, DosI64a, NiecW65)		As⁷⁵(C¹², 4n) (MaxV62) Sr⁸⁴(p, 2n) (DosI64a)
Y⁸³	3.5 h (CareA52)		G chem, genet (CareA52) activity not observed (DosI64a, NiecW65)		protons on Y (CareA52)
Y⁸³	35 m (ButeF63)		G chem, genet (ButeF63) activity not observed (DosI64a, NiecW65)		protons on Y (ButeF63)
Y⁸⁴	43 m (YamaT62) 39 m (MaxV62)	☢ β⁺, [EC] (YamaT62) △ −74.3 (MTW)	A chem, excit, cross bomb (MaxV62) chem, excit (YamaT62) daughter Zr⁸⁴ (ZaitN65)	β⁺ 3.5 max γ [Sr X-rays], 0.511 (strong, γ±), 0.590 (15%), 0.795 (100%), 0.982 (100%), 1.041 (50%), 1.27 (9%), 1.47 (6%)	As⁷⁵(C¹², 3n) (MaxV62) Sr⁸⁴(d, 2n) (MaxV62) Sr⁸⁸(p, 5n) (YamaT62)
Y⁸⁴	3.7 h (RobeB49) 2.6 h (ButeF63)		G chem, excit, sep isotopes (RobeB49) assigned to Y⁸⁵ᵐ (MaxV62, YamaT62)		deuterons, protons on Sr⁸⁴ (RobeB49)
Y⁸⁵	5.0 h (DosI63a) 4.9 h (NiecW65) 5 h (CareA52)	☢ β⁺ 70%, EC 30%, no IT, lim 1% (DosI63a) △ −77.79 (MTW)	A chem, genet (DosI63a, CareA52) parent Sr⁸⁵ (DosI63a, CareA52, NiecW65) daughter 1.4 h Zr⁸⁵ (ZaitN65)	β⁺ 2.24 max e⁻ 0.215 γ Sr X-rays, 0.231 (13%), 0.511 (140%, γ±), 0.77 (8%), 2.16 (9%) daughter radiations from Sr⁸⁵	Sr⁸⁴(d, n) (DosI63a)
Y⁸⁵ᵐ	2.68 h (DosI63a) 2.5 h (NiecW65) others (MaxV62, PatA62a, ButeF63)	☢ β⁺ 55%, EC 45%, no IT, lim 1% (DosI63a) △ −77.75 (LHP, MTW)	A chem, genet (MaxV62, DosI63a) parent Sr⁸⁵ᵐ (MaxV62, DosI63a, NiecW65) daughter Zr⁸⁵ (ButeF63, DosI63a, ZaitN65)	β⁺ 1.54 max γ Sr X-rays, 0.51 (200%, complex, includes Y±), 0.92 (9%) daughter radiations from Sr⁸⁵ᵐ Sr⁸⁵	Sr⁸⁴(d, n) (DosI63a) Sr⁸⁴(p, γ) (PatA62a)
Y⁸⁶	14.6 h (HydE51, CastS51, ButeF63)	☢ [EC 74%, β⁺ 26%] (VNooB65) [EC 72%, β⁺ 28%] (YamaT62a) △ −79.23 (MTW)	A chem, excit, sep isotopes (CastS51) genet energy levels (VNooB65, HarpJ63) daughter Zr⁸⁶ (HydE51) daughter Y⁸⁶ᵐ (HasL61, KimY62)	β⁺ 3.15 max (0.5%), 2.34 max γ Sr X-rays, 0.443 (14%), 0.511 (35%, doublet, includes Y±), 0.63 (37%, doublet), 0.704 (14%), 0.778 (21%), 0.836 (7%), 1.026 (10%), 1.077 (82%), 1.16 (35%, doublet), 1.857 (18%), 1.925 (24%)	Rb⁸⁵(α, 3n) (YamaT62a) Sr⁸⁶(p, n) (VNooB65, YamaT62a) Sr⁸⁸(p, 3n) (CastS51)
Y⁸⁶ᵐ	48 m (KimY62) 49 m (HasL61)	☢ IT (HasL61) △ −79.01 (LHP, MTW)	A chem, cross bomb, genet (HasL61) chem, cross bomb, sep isotopes, genet (KimY62) parent Y⁸⁶ (HasL61, KimY62)	γ L X-rays, 0.208 (94%) e⁻ 0.008 daughter radiations from Y⁸⁶	Rb⁸⁵(α, 3n) (HasL61, KimY62)

TABLE 1 (*Continued*)

Isotope Z A	Half-life	Type of decay (☢); % abundance; Mass excess (△≡M–A), MeV (C¹²=0); Thermal neutron cross section (σ), barns	Class; Identification; Genetic relationships	Major radiations: approximate energies (MeV) and intensities	Principal means of production
$_{39}Y^{87}$	80 h (MannL51, HydE51, DubL40)	☢ EC 99+%, β⁺ ≈0.3% (MannL51) △ –83.2 (MTW)	A chem (StewD37) chem, excit, cross bomb (DubL40) daughter Y⁸⁷ᵐ (MannL50, MannL51, HydE51) parent Sr⁸⁷ᵐ (DubL40, DubL39, LindnM50a, MannL50, HydE51, MannL51)	β⁺ 0.7 max (?) γ Sr X-rays, 0.483 daughter radiations from Sr⁸⁷ᵐ	Sr⁸⁶(d, n) (StewD37, DubL40, MannL51, MannL50) Sr⁸⁷(p, n) (DubL40, MannL51)
Y^{87m}	14 h (DubL40, HydE51, MannL51) 13 h (VanR65)	☢ IT (DubL40) β⁺ ≈5% (YamaT62a) no β⁺ (HydE51) △ –82.8 (LHP, MTW)	A chem (StewD37) chem, excit, cross bomb (DubL40) daughter Zr⁸⁷ (HydE51) parent Y⁸⁷ (MannL50, HydE51, MannL51)	β⁺ 1.60 max (?) e⁻ 0.364, 0.379 γ X-rays, 0.381 (74%) daughter radiations from Y⁸⁷	Sr⁸⁶(d, n) (StewD37, DubL40, MannL50, MannL51) Y⁸⁹(p, 3n) Zr⁸⁷(β⁻) (ButeF63, AwaY64)
Y^{88}	108.1 d (WyaE61) 105 d (DubL40)	☢ EC 99+%, β⁺ 0.20% (RhoJ63) △ –84.27 (MTW)	A chem (DubL40) chem, excit (HelmhA42) mass spect (HaydR48) daughter Zr⁸⁸ (HydE51)	β⁺ 0.76 max γ Sr X-rays, 0.898 (91%), 1.836 (100%)	Sr⁸⁸(p, n) (DubL40) Sr⁸⁸(d, 2n) (PecC40, HelmhA42, GamG44, BradE50)
Y^{89}		% 100 (DempA39, CollT57) △ –87.678 (MTW) σ_c 1.3 (to Y⁹⁰) 0.001 (to Y⁹⁰ᵐ) (GoldmDT64)			
Y^{89m}	16.1 s (SwanC55) 16.5 s (SatA62) 16.8 s (BroaK65) 16 s (BramE62, BramE63) others (GoldhM51)	☢ IT (GoldhM51) △ –86.77 (LHP, MTW)	A chem, genet (GoldhM51) daughter Zr⁸⁹ (GoldhM51) daughter Sr⁸⁹ 0.009% (SatA62); 0.02% (LyoW55b); 0.01% (HerrmG56); <0.0005% (BisA55d)	γ 0.91 (99%) e⁻ 0.89	daughter Zr⁸⁹ (GoldhM51)
Y^{90}	64.0 h (PepD57, HeaR61) 63.7 h (VGunH63) 64.8 h (HerrmG56, MaraE55) 64.2 h (VolH55, SchmP55) 64.3 h (RobeJ59a) 64.6 h (ChetA54) 64.4 h (WriH57) 64.9 h (BiryE61a)	☢ β⁻ (StewD37) △ –86.50 (LHP, MTW)	A chem, excit, cross bomb (StewD37) chem, mass spect (HaydR48) daughter Sr⁹⁰ (HahO42, HahO43, GrumW46, NotR51) daughter Y⁹⁰ᵐ (HasL61, AlfWL61)	β⁻ 2.27 max average β⁻ energy: 0.93 calorimeter (BiryE61a) 0.90 ion ch (CaswR52) γ no γ	Y⁸⁹(n, γ) (StewD37, SagR38, SerL47b) daughter Sr⁹⁰ (HahO42, HahO43, CrumW46, NotR51)
Y^{90m}	3.1 h (AlfWL61, LyoW61a, HeaR61) 3.2 h (HasL61, BacM60, CartC61, DavP64) 3.0 h (BramE62) others (FergJ61a)	☢ IT 99.6%, β⁻ 0.4% (DavP64) △ –85.81 (LHP, MTW)	A chem, cross bomb, sep isotopes, genet, excit, n-capt (LyoW61a, HasL61, HeaR61, AlfWL61, FergJ61a, CartC61) parent Y⁹⁰ (HasL61, AlfWL61)	γ Y X-rays, 0.202 (97%), 0.482 (91%), 2.315 (0.4% with Zr⁹⁰ᵐ) e⁻ 0.185, 0.465 daughter radiations from Y⁹⁰	Rb⁸⁷(α, n) (CartC61, HasL61) Y⁸⁹(n, γ) (HeaR61, FergJ61a, LyoW61a) Y⁸⁹(d, p) (CartC61) Nb⁹³(n, α) (BramE62, AlfWL61, LyoW61a)
Y^{91}	58.8 d (HoffD63) 59.1 d (WyaE61) 57.5 d (KahB55) 58.3 d (HerrmG56) others (GrumW46, LangeL49, BolF53, GotH41, HahO40c, JoliF44)	☢ β⁻ (HahO40c) △ –86.35 (MTW) σ_c 1.4 (GoldmDT64)	A chem, genet (HahO40c, HahO43) chem, mass spect (BradE51a, HaydR48) daughter Sr⁹¹ (GotH41, HahO43, FinB51) descendant Kr⁹¹ (HahO40c, BradE51, DilC51, DilC51a)	β⁻ 1.545 max γ 1.21 (0.3%)	fission (GotH41, HahO43, FinB51, FinB51c, EngeD51c)
Y^{91m}	50.3 m (AmeD53) 51.0 m (FinB51) 50 m (GotH41)	☢ IT, no β⁻, lim 1.5% (AmeD53) △ –85.80 (LHP, MTW)	A chem, genet (GotH41) daughter Sr⁹¹ (GotH41, HahO43, FinB51)	γ Y X-rays, 0.551 (95%) e⁻ 0.534	fission, daughter Sr⁹¹ (GotH41, HahO43, FinB51)
Y^{92}	3.53 h (FritK60) 3.50 h (BunkM62) 3.5 h (AgeM43, HahO43b, LieC39)	☢ β⁻ (LieC39) △ –84.83 (MTW)	A chem (LieC39) fission fragment range (KatcS48) chem, sep isotopes (SchoG53) daughter Sr⁹² (GotH41, HoaE51b) descendant Kr⁹², descendant Rb⁹² (DilC51a)	β⁻ 3.63 max γ 0.448 (2.3%), 0.560 (2.6%), 0.934 (14%), 1.40 (4.7%), 1.83 (0.4%)	Zr⁹⁴(d, α) (SchoG53, CassW55) fission, daughter Sr⁹² (GotH41, HoaE51b, BunkM62, KatcS48) Zr⁹²(n, p) (SagR40a, SeeW43b, AgeM43)

TABLE 1 (*Continued*)

Isotope Z A	Half-life	Type of decay (☢); % abundance; Mass excess (△≡M−A), MeV (C¹²=0); Thermal neutron cross section (σ), barns	Class; Identification; Genetic relationships	Major radiations: approximate energies (MeV) and intensities	Principal means of production
₃₉Y⁹³	10.3 h (KniJD59) 10.1 h (FritK60) others (BallN51a, HahO43)	☢ β⁻ (BallN51a) △ −84.22 (MTW, SteinE65)	A chem (HahO43, BallN46, BallN51a, SelB51) fission fragment range (KatcS48) genet (HahO43b, KniJD59) daughter Sr⁹³ (HahO43, HahO43b, KniJD59) descendant Kr⁹³ (SelB51) descendant Rb⁹³ (FritK60)	β⁻ 2.89 max γ 0.267 (6%), 0.67 (0.7%), 0.94 (2.3%), 1.42 (0.7%), 1.90 (1.8%), 2.18 (0.3%, doublet)	fission (HahO43, HahO43b, BallN51a, FritK61, KniJD59)
Y⁹⁴	20.3 m (FritK61) 20 m (KniJD59, DilC51b, HahO43) 16 m (BrowLJ49)	☢ β⁻ (HahO43, HahO43b) △ −82.3 (MTW)	A chem (HahO43, HahO43a) fission fragment range (KatcS48) chem, sep isotopes (SchoG53) daughter Sr⁹⁴ (HahO43, HahO43b, KniJD59) descendant Kr⁹⁴ (HahO43b, DilC51a) descendant Rb⁹⁴ (FritK61, HahO43, HahO43b, DilC51)	β⁻ 5.0 max γ 0.56 (6%), 0.92 (43%), 1.13 (5%), 1.65 (2.4%), 1.90 (1.6%), 2.13 (2.4%), 2.57 (1.5%, complex), 3.06 (1.3%), 3.53 (1.1%)	fission (HahO43, HahO43b, KatcS48, KniJD59, FritK61, DilC51a) Zr⁹⁶(d, α) (SchoG53)
Y⁹⁵	10.9 m (FritK61) 10.5 m (KniJD49)	☢ β⁻ (KniJD49) △ −81 (MTW)	B chem, sep isotopes, excit (KniJD49) daughter Sr⁹⁵ (FritK61)	γ 1.30 (?), 1.80 (?)	fission (FritK61, KniJD59) Zr⁹⁶(γ, p) (KniJD49)
Y⁹⁶	2.3 m (ValliD61)	☢ β⁻ (ValliD61) △ −79 (MTW)	B chem, excit (ValliD61)	β⁻ 3.5 max γ 0.7, 1.0, 1.5 (complex)	Zr⁹⁶(n, p) (ValliD61)
₄₀Zr⁸¹	7–15 m genet (ZaitN65)	☢ [β⁺, EC] (ZaitN65)	E chem, genet (ZaitN65) ancestor Sr⁸¹, Rb⁸¹ (ZaitN65)		protons on Y⁸⁹ (ZaitN65)
Zr⁸²	10 m genet (ZaitN65)	☢ [β⁺, EC] (ZaitN65)	D chem, genet (ZaitN65) parent Y⁸², ancestor Sr⁸² (ZaitN65)		protons on Y⁸⁹ (ZaitN65)
Zr⁸³	5–10 m genet (ZaitN65)	☢ [EC, β⁺] (ZaitN65)	E chem, genet (ZaitN65) ancestor Sr⁸³ (ZaitN65)		protons on Y⁸⁹ (ZaitN65)
Zr⁸⁴	16 m genet (ZaitN65)	☢ [EC, β⁺] (ZaitN65)	B chem, genet (ZaitN65) parent Y⁸⁴ (ZaitN65)		protons on Y⁸⁹ (ZaitN65)
Zr⁸⁵	15 m (ZaitN65) 6 m (ButeF63)	☢ [EC, β⁺] (ButeF63)	B chem, genet (ButeF63, ZaitN65) parent Y⁸⁵ᵐ, ancestor Sr⁸⁵ᵐ (ButeF63, DosI63a, ZaitN65)		Y⁸⁹(p, 5n) (ButeF63)
Zr⁸⁵	1.4 h genet (ZaitN65)	☢ [EC, β⁺] (ZaitN65)	B chem, genet (ZaitN65) parent Y⁸⁵ (ZaitN65)		protons on Y⁸⁹ (ZaitN65)
Zr⁸⁶	16.5 h (AwaY64) 17 h genet (HydE51) 15 h genet (ZaitN65)	☢ EC, no β⁺, lim 0.1% (HydE66, HydE54a) △ −78 (MTW)	A chem, genet (HydE51) parent Y⁸⁶ (HydE51)	γ Y X-rays, 0.028 (20%), 0.243 (96%), 0.612 (5%) e⁻ [0.015] daughter radiations from Y⁸⁶	Y⁸⁹(p, 4n) (AwaY64)
Zr⁸⁷	1.6 h (HydE51) 1.5 h (ButeF63, HoltzR52, ZaitN65) 2.0 h (RobeB49)	☢ β⁺, EC (RobeB49) [β⁺ 83%, EC 17%] (NDS) △ −79.7 (MTW)	A chem, excit, sep isotopes (RobeB49) chem, genet (HydE51) parent Y⁸⁷ᵐ (HydE51)	β⁺ 2.10 max γ Y X-rays, 0.511 (γ±, [166%]), 1.2, 2.2 daughter radiations from Y⁸⁷ᵐ, Y⁸⁷	Y⁸⁹(p, 3n) (ButeF63, AwaY64)
Zr⁸⁸	85 d (HydE53a)	☢ EC (HydE51) no β⁺ (HydE55) △ −84 (MTW)	B chem, genet (HydE51) parent Y⁸⁸ (HydE51) descendant Mo⁸⁸ (ButeF64c)	γ Y X-rays, 0.394 (97%) e⁻ 0.377 daughter radiations from Y⁸⁸	protons on Nb (HydE51, HydE55)
Zr⁸⁹	78.4 h (VPatD64) 79.0 h (HamiJ60) 79.3 h (ShuK51) others (HydE51, KatzL53, DubL40, ShoF53, HowD62)	☢ EC 78%, β⁺ 22% (VPatD64, MonaS61) △ −84.85 (MTW)	A chem excit (SagR38, DubL40) parent Y⁸⁹ᵐ (GoldhM51) daughter Nb⁸⁹ (DiarR54, MathH55) descendant Mo⁸⁹ (ButeF64c)	β⁺ 0.90 max e⁻ 0.89 (with Y⁸⁹ᵐ) γ Y X-rays, 0.511 (44%, γ±), 0.91 (99%, with Y⁸⁹ᵐ), 1.71 (1%)	Y⁸⁹(p, n) (DubL40, VPatD64) Y⁸⁹(d, 2n) (GoldhM51, HamiJ60, MonaS61)

TABLE 1 (*Continued*)

Isotope Z A	Half-life	Type of decay (☢); % abundance; Mass excess (Δ≡M−A), MeV (C¹²=0); Thermal neutron cross section (σ), barns	Class; Identification; Genetic relationships	Major radiations: approximate energies (MeV) and intensities	Principal means of production
$_{40}Zr^{89m}$	4.18 m (VPatD64) 4.4 m (ShoF53, ShoF51, MangS63) 4.3 m (KatzL53) 4.5 m (DubL40)	☢ IT 94%, EC 4.7%, β⁺ 1.4% (VPatD64) IT 93%, EC 5.6%, β⁺ 1.8% (ShoF53) Δ −84.26 (LHP, MTW)	A chem, excit (DubL40) daughter Nb⁸⁹ᵐ (DiaR54, MathH55)	β⁺ 2.40 max (0.2%), 0.89 max (1.2%) e⁻ 0.570 γ Zr, Y X-rays, 0.588 (87%), 1.51 (6%)	Y⁸⁹(p, n) (VPatD64, DubL40)
$\underline{Zr^{90}}$		% 51.46 (WhiJ48) Δ −88.770 (MTW) σ_c 0.1 (GoldmDT64)			
Zr^{90m}	0.80 s (WagR63) 0.83 s (SchmW63, CamE55) 0.86 s (WhiW62)	☢ IT (CamE55) Δ −86.45 (LHP, MTW)	A excit (CamE55) genet energy levels (SchmW63, BjoS59)	γ Zr X-rays, 0.133 (4%), 2.18 (14%), 2.32 (86%) e⁻ 0.115, 0.130	Nb⁹³(p, α) (WhiJ62) Zr⁹⁰(n, n') (CamE55, WagR63, SchmW63)
$\underline{Zr^{91}}$		% 11.23 (WhiJ48) Δ −87.893 (MTW) σ_c 1 (GoldmDT64)			
$\underline{Zr^{92}}$		% 17.11 (WhiJ48) Δ −88.462 (MTW) σ_c 0.2 (GoldmDT64)			
Zr^{93}	1.5 x 10⁶ y sp act (SteinE65)	☢ β⁻ (SteinE50) Δ −87.11 (SteinE65, MTW) σ_c <4 (GoldmDT64)	A chem (SteinE50) mass spect (GleL53) parent Nb⁹³ᵐ (GleL53)	β⁻ 0.060 max γ no γ daughter radiations from Nb⁹³ᵐ	fission (SteinE50)
$\underline{Zr^{94}}$		% 17.40 (WhiJ48) Δ −87.267 (MTW) σ_c 0.08 (GoldmDT64)			
$\underline{Zr^{95}}$	65.5 d (FlyK65a) 65 d (BradE51a, GrumW46, CorkJ53b) 66 d (GrossA48) 63 d (SagR40a)	☢ β⁻ (SagR40a) Δ −85.663 (MTW)	A chem (GrossA40, SagR40a) chem, genet (GoldsB51) parent Nb⁹⁵ᵐ, parent Nb⁹⁵ (HudJ49, BradE51a, JacoL51, SteinE51a) descendant Kr⁹⁵, descendant Rb⁹⁵ (DilC51)	β⁻ 0.89 max (2%), 0.396 max γ 0.724 (49%), 0.756 (49%) daughter radiations from Nb⁹⁵, Nb⁹⁵ᵐ	Zr⁹⁴(n, γ) (SagR40a, SerL47b) fission (HudJ49, BradE51a, JacoL51, SteinE51a, FinB51c)
$\underline{Zr^{96}}$	t₁/₂(β⁻) >3.6 x 10¹⁷ y sp act (AwsM56) t₁/₂ (ββ) >5 x 10¹⁷ y sp act (AwsM56) t₁/₂ (ββ) 6 x 10¹⁶ y sp act (MCarJ53)	% 2.80 (WhiJ48) Δ −85.430 (MTW) σ_c 0.05 (GoldmDT64)			
Zr^{97}	17.0 h (BurgW50a, MandeC52, GrossA40, KatcS51b, VasiI58)	☢ β⁻ (GrossA40) Δ −82.93 (MTW)	A chem (GrossA40) chem, n-capt, sep isotopes (BurgW50a, MandeC52) parent Nb⁹⁷ᵐ (BurgW50a)	β⁻ 1.91 max γ 0.747 (92%, with Nb⁹⁷ᵐ) daughter radiations from Nb⁹⁷	Zr⁹⁶(n, γ) (BurgW50a, MandeC52, SagR40a, SerL47b) fission (GrossA40, HahnO41, KatcS48)
Zr^{98}	1 m (OrtC60)	☢ [β⁻] (OrtC60) Δ −82 (MTW)	E chem, genet (OrtC60) [parent <2 m Nb⁹⁸], not parent 51 m Nb⁹⁸ (OrtC60)		fission (OrtC60)
Zr^{99}	35 s genet (OrtC60)		G chem, genet (OrtC60) activity not observed, t₁/₂ ≤1.6 s genet (TroD63)		fission (OrtC60)
$_{41}Nb^{88}$	14 m (KorR64, HydE65) 21 m (ButeF64b)	☢ β⁺ (HydE65), [EC] Δ −77 (MTW)	B chem, genet (KorR64, HydE65, ButeF64b) daughter Mo⁸⁸ (ButeF64c)	β⁺ 3.2 max γ 0.076, 0.141, 0.272, 0.399, 0.511 (γ±), 0.671, 1.058, 1.083	Br⁷⁹(C¹², 3n) (KorR64, HydE65)
Nb^{89}	1.9 h (HydE65, DiaR54, MathH55) 2.0 h (ButeF64b)	☢ β⁺ (DiaR54), [EC] Δ −81.0 (MTW)	A chem, genet (DiaR54, MathH55) parent Zr⁸⁹ (DiaR54, MathH55)	β⁺ 2.9 max γ 0.511 (γ±), 1.626, 3.577, 3.838 daughter radiations from Zr⁸⁹	C¹² on Br (MathH55, HydE65) Y⁸⁹(α, 4n) (MathH55)

TABLE 1 (*Continued*)

Isotope Z A	Half-life	Type of decay ($\bigtriangledown$); % abundance; Mass excess (Δ ≡M–A), MeV (C¹²=0); Thermal neutron cross section (σ), barns	Class; Identification; Genetic relationships	Major radiations: approximate energies (MeV) and intensities	Principal means of production
$_{41}$Nb^{89m}	42 m (ButeF64b) ≈48 m (DiaR54)	$\bigtriangledown$ β^+ (DiaR54), [EC] Δ −80.2 (LHP, MTW)	A chem, genet (DiaR54, MathH55) parent Zr^{89m} (DiaR54, MathH55) daughter Mo^{89} (ButeF64c)	β^+ 3.1 max e^- 0.570 (with Zr^{89m}) γ 0.511 ($\gamma^\pm$), 0.588 (93%, with Zr^{89m}) daughter radiations from Zr^{89}	C^{12} on Br (MathH55) protons on Zr (DiaR54)
Nb^{90}	14.6 h (OngP54a, ShelR57a) 14.7 h (DiaR53, ButeF64b) others (KunD49, JacoL51)	$\bigtriangledown$ β^+, EC (BjoS59, LazN58, ShelR57a) EC(K) ≈50% (KuzM57) Δ −82.66 (MTW)	A chem, excit, cross bomb (BjoS59) chem, sep isotopes, cross bomb (KunD49) descendant Mo^{90} (DiaR53, MathH55b)	β^+ 1.50 max e^- 0.115, 0.123 γ Zr X-rays, 0.142 (75%), 0.511 ($\gamma^\pm$), 1.14 (97%), 2.18 (14%), 2.32 (82%) daughter radiations from Zr^{90m} included in above listing	Zr^{90}(p, n) (BjoS59, LazN58) Zr^{90}(d, 2n) (KunD49, JacoL51) descendant Mo^{90} (ButeF64b, DiaR53)
Nb^{90m}	24 s (MathH55b)	$\bigtriangledown$ IT (MathH55b) Δ −82.54 (LHP, MTW)	A chem, genet (MathH55b) daughter Mo^{90} (MathH55b)	γ Nb X-rays, 0.122 (71%) e^- 0.104, 0.120	daughter Mo^{90} (MathH55b)
Nb^{91}	long (OvaJ51)	$\bigtriangledown$ [EC] (OvaJ51) Δ −86.8 (MTW)	B genet (OvaJ51) [daughter Nb^{91m}] (OvaJ51)	γ [Zr X-rays]	Zr^{90}(d, n) (OvaJ51)
Nb^{91m}	64 d (BoydG49) 60 d (JacoL51)	$\bigtriangledown$ IT 97%, EC 3% (NDS) Δ −86.6 (LHP, MTW)	A chem, excit (JacoL51) chem, sep isotopes (OvaJ51)	γ Nb X-rays, 0.104 (0.5%), 1.21 (3%) e^- 0.086, 0.102	Y^{89}(α, 2n) (HaywR55a) Zr^{90}(d, n) (OvaJ51, HaywR55a, JacoL51)
Nb^{92}	>350 y or <1 h (BunkM62)	Δ −86.45 (ShelR64, MTW)	F levels observed in Nb^{93}(d, t) reaction (ShelR64) and in Nb^{93}(p, d) reaction (SweR64)		
Nb^{92m}	10.16 d (BunkM62) 10.15 d (WestH59) others (GlagV61, MacD48, SagR40b, SagR38a)	$\bigtriangledown$ EC 99+%, β^+ 0.06% (WestH59, BunkM62) no β^-, lim 0.05% (PreiP51) Δ −86.32 (ShelR64, MTW)	A chem, excit (SagR38a)	γ Zr X-rays, 0.934 (99%)	Y^{89}(α, n) (BunkM62)
Nb^{92}	13 h (JameR54)		G chem, excit (JameR54) activity not observed (SilE58, BramE62, BunkM62, BosH64b)		protons on Nb^{93} (JameR54)
<u>Nb^{93}</u>		% 100 (SamM36a, WhiF56) Δ −87.204 (MTW) σ_c 0.1 (to Nb^{94}) 1 (to Nb^{94m}) (GoldmDT64)			
Nb^{93m}	13.6 y (FlyK65a) ≈4 y (SchumR54)	$\bigtriangledown$ IT (SchumR54) Δ −87.173 (LHP, MTW)	A chem, genet (GleL53) daughter Zr^{93} (85%) (GleL53) daughter Mo^{93} (HohK64)	γ Nb X-rays e^- 0.011, 0.028	daughter Zr^{93} (GleL53) Nb^{93}(n, n') (SchumR54, HohK64)
Nb^{94}	2.0 x 10^4 y sp act, mass spect (SchumR59a) 1.8 x 10^4 y sp act (RolM55) 2.2 x 10^4 y sp act (DouDL53)	$\bigtriangledown$ β^-, no EC (DouDL53) no EC(K), lim 6% (SchumR59a) Δ −86.35 (MTW) σ_c ≈15 (GoldmDT64)	A n-capt (GoldhM46a) chem, n-capt (HeiR52)	β^- 0.49 max γ 0.702 (100%), 0.871 (100%)	Nb^{93}(n, γ) (GoldhM46a, HeiR52)
Nb^{94m}	6.29 m (KilP62) 6.6 m (SagR40b)	$\bigtriangledown$ IT 99+%, β^- 0.2% (ReicC63, YinL62) IT 99+%, β^- 0.5% (KilP62) Δ −86.31 (LHP, MTW)	A n-capt, excit (PoolM37, SagR38a, GoldhM48a, KunD46)	γ Nb X-rays, 0.871 (0.2%) e^- 0.023, 0.039	Nb^{93}(n, γ) (PoolM37, SagR38a, SagR40b, SerL47b)
Nb^{95}	35.0 d (WyaE61) 35.6 d (PierA59) 35 d (CorkJ53a, EngeD51) others (JacoL51, LangeL63, FlyK65a)	$\bigtriangledown$ β^- (GoldsB51) Δ −86.784 (MTW) σ_c ≈7 (GoldmDT64)	A chem (GoldsB46, GoldsB51) chem, excit, cross bomb (JacoL51) daughter Zr^{95} (HudJ49, BradE51a, SteinE51a, JacoL51) daughter Nb^{95m} (SteinE51a, LeviJ51a)	β^- 0.160 max γ 0.765 (100%)	daughter Zr^{95} (HudJ49, BradE51a, JacoL51, SteinE51b)

TABLE 1 (*Continued*)

Isotope Z A	Half-life	Type of decay (☢); % abundance; Mass excess (△≡M–A), MeV (C¹²=0); Thermal neutron cross section (σ), barns	Class; Identification; Genetic relationships	Major radiations: approximate energies (MeV) and intensities	Principal means of production
$_{41}$Nb^{95m}	90 h (SteinE51a, HudJ49, DrabG55) 84 h (SlaH52a, SlaH53)	☢ IT (SteinE51a) △ −86.549 (LHP, MTW)	A chem (EngeD46, EngeD51a) chem, genet (SteinE51a) daughter Zr^{95} (HudJ49, BradE51a, JacoL51, SteinE51a) parent Nb^{95} (SteinE51a, LeviJ51a)	γ Nb X-rays, [0.235] e⁻ 0.216 daughter radiations from Nb^{95}	daughter Zr^{95} (HudJ49, BradE51a, JacoL51, SteinE51a) Mo^{97}(d, α) (JacoL51, BoydG49) Zr^{94}(d, n) (JacoL51)
Nb^{96}	23.35 h (KunD49) 23.5 h (MonaS62)	☢ β⁻ (KunD49) △ −85.64 (MTW)	A chem, excit, sep isotopes (KunD49)	β⁻ 0.7 max γ 0.459 (28%), 0.569 (59%), 0.778 (97%), 0.811 (14%), 0.851 (22%), 1.092 (49%), 1.200 (21%)	Zr^{96}(p, n) (KunD49) Mo^{98}(d, α) (BornP63c)
Nb^{97}	72 m (MandeC52) 74 m (BurgW50a) 75 m (GrossA40)	☢ β⁻ (GrossA40) △ −85.61 (MTW)	A chem, genet (GrossA40) daughter Nb^{97m} (SaraB55a)	β⁻ 1.27 max γ 0.665 (98%)	descendant Zr^{97} (GrossA40, BurgW50a)
Nb^{97m}	1.0 m (BurgW50a)	☢ IT (BurgW50a) △ −84.86 (LHP, MTW)	A chem, excit, sep isotopes, genet (BurgW50a) daughter Zr^{97} (BurgW50a) parent Nb^{97} (SaraB55a)	γ 0.747 (98%) e⁻ 0.728 daughter radiations from Nb^{97}	daughter Zr^{97} (BurgW50a)
Nb^{98}	51 m (OrtC60, WahA62, TakaK61) others (BoydG49)	☢ β⁻ (BoydG49) △ −83.5 (OrtC60, MTW)	B chem, sep isotopes (BoydG49) chem, genet energy levels (OrtC60) not daughter Zr^{98} (OrtC60)	β⁻ 3.1 max γ 0.330 (9%), 0.720 (75%), 0.787 (100%), 1.16 (30%), 1.44 (10%), 1.52 (4%), 1.68 (10%), 1.88 (4%), 1.93 (8%)	Mo^{98}(n, p) (OrtC60, TakaK61, WahA62)
Nb^{98}	<2 m (OrtC60)	☢ β⁻ (OrtC60)	F genet, excit (OrtC60) [daughter Zr^{98}] (OrtC60)	β⁻ high-energy β	fission, daughter Zr^{98} (OrtC60)
Nb^{99}	2.4 m (OrtC60) 2.3 m (TroD63) 2.5 m (DufR50)	☢ β⁻ (DufR50) △ −83 (MTW)	A chem, excit, sep isotopes (DufR50) chem, genet (OrtC60) parent Mo^{99} (OrtC60)	β⁻ 3.2 max γ 0.100 (↑ 1), 0.260 (↑ 1)	fission (OrtC60, TroD63) Mo^{100}(γ, p) (DufR50)
Nb^{99}	10 s genet (TroD63)	☢ β⁻ >52% (TroD63)	C chem, genet (TroD63) parent Mo^{99} (TroD63)		fission (TroD63)
Nb^{100}	3.0 m (OrtC60)	☢ [β⁻] (OrtC60) △ −80 (MTW)	B chem, genet energy levels (OrtC60)	γ 0.140 (↑ 10), 0.36 (↑ 55), 0.45 (↑ 40), 0.53 (↑ 100, complex), 0.65, 2.2, 2.3, 2.65, 2.85	fission (OrtC60)
Nb^{100}	11 m (TakaK61)	☢ β⁻ (TakaK61) △ −80 (MTW)	C chem, genet energy levels (TakaK61)	β⁻ 4.2 max (≤10%), 3.5 max γ 0.535 (↑ 100), 0.62 (↑ 60), 1.04 (↑ 10), 1.15 (↑ 10), 1.47 (↑ 5)	Mo^{100}(n, p) (TakaK61)
Nb^{101}	1.0 m genet (OrtC60)	☢ [β⁻] (OrtC60)	B chem, genet (OrtC60) parent Mo^{101} (OrtC60)		fission (OrtC60)
$_{42}$Mo^{88}	27 m (ButeF64c)	☢ β⁺ (ButeF64c), [EC]	B chem, genet (ButeF64c) parent Nb^{88}, ancestor Zr^{88} (ButeF64c)	β⁺ 2.5 max γ 0.511 (γ±), 2.69 daughter radiations from Nb^{88}	protons on Nb, Mo (ButeF64c)
Mo^{89}	7 m (ButeF64c)	☢ β⁺ (ButeF64c), [EC]	B chem, genet (ButeF64c) parent Nb^{89m}, ancestor Zr^{89} (ButeF64c)	β⁺ 4.9 max γ 0.511 (γ±) daughter radiations from Nb^{89m}	protons on Mo (ButeF64c)
Mo^{90}	5.67 h (PettH66) 5.7 h (DiaR53) 6.3 h (KuzM57) others (KurcB55)	☢ EC 75%, β⁺ 25% (CoopJ65) △ −80.17 (PettH66, MTW)	A chem, genet (DiaR53, MathH55b) ancestor Nb^{90} (DiaR53, MathH55b) parent Nb^{90m} (MathH55b)	β⁺ 1.2 max e⁻ 0.104, 0.120, 0.239, 0.255 γ Nb X-rays, 0.122 (71%), 0.257 (85%), 0.445 (9%), 0.511 (50%, γ±), 0.945 (10%), 1.273 (8%), 1.389 (4%), 1.46 (4%, doublet) daughter radiations from Nb^{90} (daughter radiations from Nb^{90m} included in above listing)	Nb^{93}(p, 4n) (DiaR53, MathH55b, CoopJ65) Zr^{90}(α, 4n) (CoopJ65)

TABLE 1 (*Continued*)

Isotope Z A	Half-life	Type of decay (☼); % abundance; Mass excess ($\triangle \equiv M-A$), MeV ($C^{12}=0$); Thermal neutron cross section (σ), barns	Class; Identification; Genetic relationships	Major radiations: approximate energies (MeV) and intensities	Principal means of production
$_{42}\text{Mo}^{91}$	15.49 m (EbrT65) 15.5 m (DufR49b, WafH48, KatzL53) others (AxeP55, BotW39, SagR38)	☼ β^+ (SagR38), [EC] $\triangle$ −82.3 (MTW)	A excit (BotW37) chem, excit (SagR38) chem, sep isotopes, excit (KunD49a, DufR49b)	β^+ 3.44 γ Nb X-rays, 0.511 ($\gamma^\pm$)	Mo^{92}(n, 2n) (KunD49a, HeyF37, SagR38, SagR40a, BrolJ52, EbrT65)
Mo^{91m}	64 s (PrenJ57) 66 s (KatzL53, AxeP55) 73 s (WafH48) 75 s (DufR49b)	☼ IT ≈57%, β^+ + EC ≈43% (SmiF56) IT ≈70%, β^+ + EC ≈30% (AxeP55) $\triangle$ −81.6 (LHP, MTW)	B chem, sep isotopes (DufR49b)	β^+ 3.99 max (↑ 15), 2.78 max (↑ 100) e^- 0.638 γ Nb X-rays, Mo X-rays, 0.511 ($\gamma^\pm$ [76%]), 0.658 (54%), 1.21 (22%), 1.53 (15%) daughter radiations from Mo^{91}	Mo^{92}(γ, n) (DufR49b)
$\underline{\text{Mo}^{92}}$	$t_{1/2}$ (ECEC) >4 x 10^{18} y (WintR55)	% 15.86 (WilliD46) $\triangle$ −86.804 (MTW) σ_c <0.3 (to Mo^{93}) <0.006 (to Mo^{93m}) (GoldmDT64)			
Mo^{93}	>100 y genet (HohK64)	☼ EC (BoydG49a) $\triangle$ −86.79 (MTW)	A chem, n-capt (BoydG49a) genet (HohK64) parent Nb^{93m} (85%) (HohK64)	γ Nb X-rays daughter radiations from Nb^{93m}	Mo^{92}(n, γ) (BoydG49a) Nb^{93}(p, n) (HohK64)
Mo^{93m}	6.95 h (BoydG52b) 6.75 h (KunD50)	☼ IT (KunD50) $\triangle$ −84.36 (LHP, MTW)	A chem, excit (KunD46) chem, excit, cross bomb, sep isotopes (KunD50) chem, excit (BoydG52b) chem, mass spect (AlbuD53, BernaR53) not daughter Tc^{93} (BoydG50)	γ Mo X-rays, 0.264 (58%), 0.685 (100%), 1.479 (100%) e^- 0.244, 0.261	Nb^{93}(d, 2n) (AlbuD53, KunD46, WieM46, KunD50a) Zr^{90}(α, n) (KunD50) Nb^{93}(p, n) (BoydG52b, ForC53)
$\underline{\text{Mo}^{94}}$		% 9.12 (WilliD46) $\triangle$ −88.407 (MTW)			
$\underline{\text{Mo}^{95}}$		% 15.70 (WilliD46) $\triangle$ −87.709 (MTW) σ_c 14 (GoldmDT64)			
$\underline{\text{Mo}^{96}}$		% 16.50 (WilliD46) $\triangle$ −88.794 (MTW) σ_c 1 (GoldmDT64)			
$\underline{\text{Mo}^{97}}$		% 9.45 (WilliD46) $\triangle$ −87.539 (MTW) σ_c 2 (GoldmDT64)			
$\underline{\text{Mo}^{98}}$		% 23.75 (WilliD46) $\triangle$ −88.110 (MTW) σ_c 0.51 (GoldmDT64)			
Mo^{99}	66.7 h (CrowP65) 66.0 h (GunS57) 67.0 h (WriH57) others (SeaG39, CorkJ49a, VasiI58, WafH48, SagR40a)	☼ β^- (SagR38) $\triangle$ −85.96 (MTW)	A chem, n-capt, excit (SagR38, SagR40a) parent Tc^{99m} (SeaG39, SagR40a, MedH49, GleL51d, MihJ51) daughter 2.4 m Nb^{99} (OrtC60) daughter 10 s Nb^{99} (TroD63) ancestor Tc^{99} (MotE47a)	β^- 1.23 max γ Tc X-rays, 0.041 (2%), 0.181 (7%), 0.372 (1%), 0.740 (12%), 0.780 (4%) daughter radiations from Tc^{99m}	Mo^{98}(n, γ) (SagR40, SagR40a, MauW41, SerL47b, HumV51) fission (HahO39b, SagR40a, KatcS51c, KatcS48, FinB51c)
$\underline{\text{Mo}^{100}}$	$t_{1/2}$ (ββ) ≥3 x 10^{17} y sp act (WintR55) others (FremJ52)	% 9.62 (WilliD46) $\triangle$ −86.185 (MTW) σ_c 0.2 (GoldmDT64)			
Mo^{101}	14.6 m (MauW41, WileDR54, OKelG57)	☼ β^- (SagR40a) $\triangle$ −83.50 (MTW)	A chem, n-capt (SagR40a) parent Tc^{101} (SagR40, BotW41, HahO41a, HahO41b, MauW41) daughter Nb^{101} (OrtC60)	β^- 2.23 max e^- 0.170 γ 0.191 (25%), 0.51 (15%), 0.59 (21%), 0.70 (11%), 0.89 (15%), 1.02 (25%), 1.18 (11%), 1.38 (9%), 1.56 (11%), 2.08 (16%) daughter radiations from Tc^{101}	Mo^{100}(n, γ) (SagR40, SagR40a, MauW41, SerL47b, HumV51)

TABLE 1 (*Continued*)

Isotope Z A	Half-life	Type of decay (☢); % abundance; Mass excess (△≡M–A), MeV (C¹²=0); Thermal neutron cross section (σ), barns	Class; Identification; Genetic relationships	Major radiations: approximate energies (MeV) and intensities	Principal means of production
$_{42}$Mo^{102}	11.5 m (FleJ54) 11.0 m (WileDR54a) 12 m (HahO41a)	☢ β⁻ (HahO41a) △ –84 (MTW)	D chem (HahO41a) parent 5 s Tc^{102} (HahO41a, HahO41b, FleJ54)	β⁻ 1.2 max daughter radiations from 5 s Tc^{102}	fission (HahO41a, FleJ54, WileDR54a)
Mo^{103}	62 s genet (VBaeA65) 70 s (KieP63a)	☢ [β⁻] (KieP63a)	B chem, genet (KieP63a) parent Tc^{103} (KieP63a)		fission (KieP63a, VBaeA65)
Mo^{104}	1.1 m (KieP62) 1.6 m (TerG64)	☢ β⁻ (TerG64)	B chem, genet (KieP62) chem, excit (TerG64) parent Tc^{104} (KieP62)	β⁻ 4.8 max γ 0.070 daughter radiations from Tc^{104}	fission (TerG64, KieP62)
Mo^{105}	40 s (KieP62a) 42 s genet (VBaeA65) others (FleJ55a, FleJ56a, SeeW47)	☢ β⁻ (BornH43b)	B chem, genet (BornH43b, KieP62a) ancestor Ru^{105} (BornH43b, KieP62a) parent Tc^{105}, ancestor Rh^{105} (KieP62a, BornH43b, FleJ55a)		fission (BornH43b, FleJ55a, FleJ56a, KieP62a, VBaeA65)
$_{43}$Tc^{92}	4.4 m (VLieR64)	☢ β⁺ ≈92%, EC ≈8% (VLieR64) △ –78.8 (MTW)	B chem, sep isotopes (MotE48, VLieR64)	β⁺ 4.1 max γ Mo X-rays, 0.090 (20%), 0.14 (67%), 0.24 (30%), 0.33 (90%), 0.511 (184%, γ±), 0.79 (95%), 1.54 (100%)	Mo^{92}(d,2n) (MotE48, VLieR64)
Tc^{93}	2.75 h (KunD48a) 2.7 h (VinG62, MotE48, DelL39)	☢ EC 87%, β⁺ 13% (VinG62, LeviC54a) △ –83.60 (MTW)	A chem (SeaG39) chem, excit, sep isotopes (KunD48a) not parent Mo^{93m} (BoydG50)	β⁺ 0.80 max γ Mo X-rays, 0.511 (26%, γ±), 1.35 (65%), 1.49 (33%)	Mo^{92}(d,n) (KunD48a, MotE48, SeaG39, VinG62) Mo^{92}(p,γ) (KunD48a, DelL39)
Tc^{93m}	43 m (MedH50, VinG62) 47 m (KunD48a)	☢ IT 82%, EC 18% (VinG62) △ –83.21 (LHP, MTW)	A chem, excit, sep isotopes (KunD48a) mass spect (BernaR54) chem, mass spect (LeviC54a)	γ Tc X-rays, Mo X-rays, 0.390 (63%), 2.66 (18%) e⁻ 0.369 daughter radiations from Tc^{93}	Mo^{92}(d,n) (EasH53, BernaR54, VinG62) Mo^{92}(p,γ) (EasH53) Nb^{93}(α,4n) (EasH53)
Tc^{94}	293 m (MatuJ63) 270 m (MonaS62a)	☢ EC 89%, β⁺ 11% (HamiJ64) EC 93%, β⁺ 7% (MatuJ63) EC 86%, β⁺ 14% (MonaS62a) △ –84.15 (MTW)	A excit (MonaS62) chem, excit, cross bomb (MatuJ63)	β⁺ 0.816 max γ Mo X-rays, 0.511 (22%, γ±), 0.702 (100%), 0.849 (100%), 0.871 (100%)	Nb^{93}(α,3n) (MatuJ63) Mo^{94}(d,2n) (MatuJ63, MonaS62a, HamiJ64)
Tc^{94m}	53 m (MedH50, MonaS62) 50 m (MotE48a)	☢ β⁺ 66%, EC 34% (HamiJ64) β⁺ 72%, EC 28% (MonaS62a) β⁺ 61%, EC 39% (MatuJ63) △ –84.04 (LHP, MTW)	A chem, excit (GugP47) chem, excit, sep isotopes (MotE48a) genet energy levels (HamiJ64) daughter Ru^{94} (VWieA52)	β⁺ 2.47 max γ Mo X-rays, 0.511 (132%, γ±), 0.871 (91%), 1.53 (10%), 1.87 (9%), 2.73 (5%), 3.20 (2%)	Nb^{93}(α,3n) (MatuJ63) Mo^{94}(d,2n) (MotE48a, MonaS62, MatuJ63, HamiJ64) Mo^{94}(p,n) (GugP47, HubeO48a, MedH50)
Tc^{95}	20.0 h (VinG62, EggD48) 20 h (MotE48a)	☢ EC (EggD48) no β⁺ (MedH50) △ –86.05 (MTW)	A chem, sep isotopes (EggD48, MotE48a)	γ Mo X-rays, 0.768 (82%), 0.84 (11%), 1.06 (4%)	Mo^{95}(p,n) (EggD48, MedH50) Mo^{94}(d,n) (VinG62) Mo^{95}(d,2n) (MotE48a)
Tc^{95m}	61 d (UniJ59) 60 d (MedH50) 62 d (CacB39) 52 d (EdwJ47)	☢ EC 95%, β⁺ 0.42%, IT 4% (UniJ59, MedH50, MedH50a, CreT65a) △ –86.01 (LHP, MTW)	A chem (CacB37, CacB39) chem, sep isotopes (MotE48b)	β⁺ 0.68 max e⁻ 0.019, 0.036, 0.184 γ Mo X-rays, 0.204 (70%), 0.584 (36%), 0.78 (12%, complex), 0.823 (9%), 0.838 (27%), 1.042 (4%) daughter radiations from Tc^{95}	Mo^{95}(p,n) (EdwJ47) Mo^{94}(d,n) (CacB37, CacB39, UniJ59) Mo^{95}(d,2n) (MotE48b)
Tc^{96}	4.35 d (MedH50) 4.20 d (CobJ50) 4.3 d (MonaS62, EdwJ47) 4.2 d (MotE48b)	☢ EC (MotE48b) no β⁺ (MedH50) △ –85.9 (MTW)	A chem (EwiD39) chem, excit, cross bomb (EdwJ47) chem, excit, sep isotopes (MedH52)	γ Mo X-rays, 0.32 (5%), 0.778 (100%), 0.81 (84%), 0.851 (100%), 1.12 (16%) e⁻ 0.30, 0.75, 0.79, 0.82	Nb^{93}(α,n) (EdwJ47)

TABLE 1 *(Continued)*

Isotope Z A	Half-life	Type of decay (☢); % abundance; Mass excess (△≡M–A), MeV (C¹²=0); Thermal neutron cross section (σ), barns	Class; Identification; Genetic relationships	Major radiations: approximate energies (MeV) and intensities	Principal means of production
$_{43}Tc^{96m}$	52m (MedH50, EasH53)	☢ IT (MedH50) β⁺ ≈0.01% (EasH53) △ –85.8 (LHP, MTW)	B chem, excit (MedH50) chem, excit, sep isotopes (MedH52)	γ Tc X-rays e⁻ 0.013, 0.032 daughter radiations from Tc⁹⁶	Nb⁹³(α,n) (EasH53)
Tc^{97}	2.6 x 10⁶ y yield (KatcS58a) others (BoydG54)	☢ EC (BoydG54) △ –87 (MTW)	A genet (BoydG51a) chem (KatcS58a) [daughter Tc⁹⁷ᵐ] (BoydG51a) daughter Ru⁹⁷ (99+%) (KatcS58a)	γ Mo X-rays	Ru⁹⁶(n,γ)Ru⁹⁷(β⁻) (KatcS58a) Mo⁹⁷(d,2n) (BoydG54)
Tc^{97m}	91 d (BoydG54, HelmhA41a, EdwJ47) 90 d (MotE48b, GugP47, CacB37) 87 d (UniJ59) 95 d (EdwJ47)	☢ IT (HelmhA41a, EdwJ47) △ –87 (LHP, MTW)	A chem (PerrC37, CacB37) chem, genet (MotE47) excit, sep isotopes (MotE48b) daughter Ru⁹⁷ (0.04%) (KatcS58a)	γ Tc X-rays e⁻ 0.075, 0.094	Mo⁹⁶(d,n) (CacB37, PerrC37, CacB39) Mo⁹⁷(p,n) (EdwJ47) Mo⁹⁷(d,2n) (MotE48b) Ru⁹⁶(n,γ)Ru⁹⁷(β⁻) (KatcS58a)
Tc^{98}	1.5 x 10⁶ y sp act (OKelG56b) others (KatcS55)	☢ β⁻ (KatcS55) △ –86.5 (MTW) σ_c 3 (to Tc⁹⁹ᵐ) (GoldmDT64)	A chem, mass spect (BoydG55)	β⁻ 0.30 max γ 0.66 (100%), 0.76 (100%)	Mo⁹⁸(p,n) (BoydG55) Ru⁹⁶(n,γ)Ru⁹⁷(β⁻) Tc⁹⁷(n,γ) (KatcS55, KatcS58a)
Tc^{99}	2.12 x 10⁵ y sp act (FrieS51) 2.15 x 10⁵ y sp act (BoydG60)	☢ β⁻ (LincD51, SchumR51) △ –87.33 (MTW) σ_c 22 (GoldmDT64)	A chem (LincD46, SchumR46) chem, mass spect (IngM47g) daughter Tc⁹⁹ᵐ (SeaG39, HahO41a) descendant Mo⁹⁹ (MotE47a)	β⁻ 0.292 max γ no γ	fission (IngM47g, LincD51, SchumR51) Mo⁹⁸(n,γ)Mo⁹⁹(β⁻) (MotE47a)
Tc^{99m}	6.049 h (GleG64) 6.00 h (ByeD58) others (GleL51d, BaiK53, PortR60, CreT65)	☢ IT (SeaG39) △ –87.18 (LHP, MTW)	A chem, genet (SeaG39) daughter Mo⁹⁹ (SeaG39, SagR40a, MedH49, GleL51d, MihJ51) parent Tc⁹⁹ (SeaG39, HahO41a)	γ Tc X-rays, 0.140 (90%) e⁻ 0.001, 0.119	daughter Mo⁹⁹ (SeaG39, SagR40a, MedH49, GleL51d, MihJ51)
Tc^{100}	15.8 s (BoydG52a) 17.5 s (HouR52) 17 s (CsiG63)	☢ β⁻ (HouR52) △ –85.9 (MTW)	A sep isotopes (HouR52) sep isotopes, n-capt (BoydG52a)	β⁻ 3.38 max γ 0.540 (strong), 0.60 (strong), 0.71, 0.81, 0.89, 1.01, 1.31, 1.49, 1.8	Tc⁹⁹(n,γ) (BoydG52a, OKelG58) Mo¹⁰⁰(p,n) (HouR52) Rh¹⁰³(n,α) (CsiG63)
Tc^{101}	14.0 m (OKelG57, MauW41, HahO41b) 14.3 m (WileDR54) 14.5 m (PerlmM48) 16.5 m (MacD48)	☢ β⁻ (SagR40) △ –86.32 (MTW)	A chem, genet (SagR40) daughter Mo¹⁰¹ (BotW41, HahO41a, HahO41b, MauW41, SagR40)	β⁻ 1.32 max γ 0.13 (3%, complex), 0.307 (γ 91%), 0.545 (γ 8%)	Mo¹⁰⁰(n,γ)Mo¹⁰¹(β⁻) (SagR40, SagR40b, MauW41)
Tc^{102}	4.5 m (FleJ54, FleJ57)	☢ β⁻ (FleJ56a) △ –85 (MTW)	B chem, genet energy levels (FleJ56a, FleJ57)	β⁻ 2 max γ 0.47	Ru¹⁰²(n,p) (FleJ57) fission (FleJ56a)
Tc^{102}	5 s (FleJ54) others (HahO41a)	☢ β⁻ (HahO41a) △ –85 (MTW)	C chem, genet (HahO41a, FleJ54) daughter Mo¹⁰² (HahO41a, HahO41b, FleJ54)	β⁻ 4.4 max	daughter Mo¹⁰² (HahO41a, HahO41b, FleJ54)
Tc^{103}	50 s (KieP63a, VBaeA65) 72 s (FleJ57)	☢ β⁻ (KieP63b) △ –84.9 (MTW)	B excit (FleJ57) chem, genet (KieP63a) [parent Ru¹⁰³] (KieP63a) daughter Mo¹⁰³ (KieP63a)	β⁻ 2.2 max γ 0.135 (↑ 17), 0.21 (↑ 10), 0.35	fission (KieP63a, KieP63b, VBaeA65) Ru¹⁰⁴(n,np) (FleJ57)
Tc^{104}	18 m (FleJ56a, KieP62)	☢ β⁻ (FleJ56a, KieP62) △ –82.2 (MTW)	B chem (FleJ56a) chem, genet energy levels (KieP62) daughter Mo¹⁰⁴ (KieP62)	β⁻ [5.8 max] (weak), 4.6 max γ 0.36, 0.53, 0.89, 1.15, 1.25, 1.37, 1.6 (complex), 1.9, 2.2, 2.7, 3.2, 3.4, 3.7, 4.0, 4.4, 4.7	fission (FleJ56a, KieP62) Ru¹⁰⁴(n,p) (FleJ57)
Tc^{105}	7.7 m (KieP62a) 7.8 m (VBaeA65) 10 m genet (FleJ55a, FleJ56a)	☢ β⁻ (BornH43b) △ –82.6 (MTW)	B chem, genet (BornH43b) parent Ru¹⁰⁵, daughter Mo¹⁰⁵ (BornH43b, FleJ55a, KieP62a) ancestor Rh¹⁰⁵ (KieP62a)	β⁻ 3.4 max γ 0.110 daughter radiations from Ru¹⁰⁵	fission (BornH43b, FleJ55a, FleJ56a, KieP62a, VBaeA65)

TABLE 1 (*Continued*)

Isotope Z A	Half-life	Type of decay (☢); % abundance; Mass excess (△≡M–A), MeV (C¹²=0); Thermal neutron cross section (σ), barns	Class; Identification; Genetic relationships	Major radiations: approximate energies (MeV) and intensities	Principal means of production
$_{43}$Tc^{106}	37 s (VBaeA65)	☢ [β⁻] (VBaeA65)	B chem, genet (VBaeA65) parent Ru^{106} (VBaeA65)		fission (VBaeA65)
Tc^{107}	29 s (VBaeA65) others (BornH43b)	☢ [β⁻] (VBaeA65)	B chem, genet (VBaeA65) ancestor Rh^{107} (VBaeA65)		fission (VBaeA65)
$_{44}$Ru^{93}	50 s (AteA55a)	☢ β⁺ (?) (AteA55a)	E chem, excit (AteA55a)		Mo^{92}(α, 3n) (AteA55a)
Ru^{94}	57 m genet (VWieA52)	☢ EC (VWieA52)	D chem, genet (VWieA52) parent Tc^{94m} (VWieA52)	γ [Tc X-rays] daughter radiations from Tc^{94m}	Mo^{92}(α, 2n) (VWieA52)
Ru^{95}	1.65 h (SchaE56, EggD48) 1.7 h (KurcB55) 1.6 h (MocD48)	☢ EC 85%, β⁺ 15% (RieP63) △ –84.02 (MTW)	A chem, cross bomb, sep isotopes (EggD48)	β⁺ 1.33 max γ Tc X-rays, 0.340 (70%), 0.511 (30%, γ±), 0.625 (13%), 1.09 (21%), 1.43 (5%) daughter radiations from Tc^{95}	Mo^{92}(α, n) (EggD48) Ru^{96}(n, 2n) (EggD48, SchaE56, RieP63)
Ru^{96}		% 5.46 (OrdK60) 5.57 (WhiF56) 5.50 (FrieL53) 5.7 (EwaH44) △ –86.07 (MTW) σ$_c$ 0.2 (GoldmDT64)			
Ru^{97}	2.88 d (KatcS58a) 2.8 d (MocD48, SulW46, AteA55b, ShpV56) 2.44 d (CorkJ55a)	☢ EC (SulW46) △ –86 (MTW)	A chem, excit (SulW46) chem, cross bomb, sep isotopes (EggD48) parent Tc^{97m} (0.04%), parent Tc^{97} (99+%) (KatcS58a) daughter 32 m Rh^{97} (AteA55b)	γ Tc X-rays, 0.215 (91%), 0.324 (8%) e⁻ 0.194	Ru^{96}(n, γ) (SulW46, KatcS58a, CorkJ55a) Mo^{94}(α, n) (EggD48)
Ru^{98}		% 1.868 (OrdK60) 1.86 (WhiF56) 1.91 (FrieL53) 2.2 (EwaH44) △ –88.222 (MTW) σ$_c$ <8 (GoldmDT64)			
Ru^{99}		% 12.63 (OrdK60) 12.7 (WhiF56, FrieL53) 12.8 (EwaH44) △ –87.619 (MTW) σ$_c$ 11 (GoldmDT64)			
Ru^{100}		% 12.53 (OrdK60) 12.7 (FrieL53) 12.6 (WhiF56) △ –89.219 (MTW) σ$_c$ 10 (GoldmDT64)			
Ru^{101}		% 17.02 (OrdK60) 17.0 (EwaH44, FrieL53) 17.1 (WhiF56) △ –87.953 (MTW) σ$_c$ 3 (GoldmDT64)			
Ru^{102}		% 31.6 (OrdK60, WhiF56) 31.5 (FrieL53) 31.3 (EwaH44) △ –89.098 (MTW) σ$_c$ 1.4 (GoldmDT64)			
Ru^{103}	39.5 d (FlyK65a) 39.8 d (KondE50a) 39.4 d (CaliJ59) others (WriH57, SulW51d, BohE45, HoleN48a, GleL51e, MocD48, NisY42)	☢ β⁻ (NisY42) △ –87.27 (MTW)	A excit (LivJ36) chem (NisY42, GoldsB46) chem, excit (SulW51d, SulW51f) parent Rh^{103m} (SulW51f) [daughter Tc^{103}] (KieP63a)	β⁻ 0.70 max (3%), 0.21 max γ 0.497 (88%), 0.610 (6%) daughter radiations from Rh^{103m}	Ru^{102}(n, γ) (SulW51d, DVriH38) fission (NisY41, NisY42, GoldsB51a, SulW51e, FinB51c)

TABLE 1 (*Continued*)

Isotope Z A	Half-life	Type of decay (☢); % abundance; Mass excess (△≡M–A), MeV (C¹²=0); Thermal neutron cross section (σ), barns	Class; Identification; Genetic relationships	Major radiations: approximate energies (MeV) and intensities	Principal means of production
$_{44}$Ru^{104}		% 18.87 (OrdK60) 18.5 (WhiF56) 18.7 (FrieL53) 18.3 (EwaH44) △ –88.090 (MTW) σ$_c$ 0.48 (GoldmDT64)			
Ru^{105}	4.44 h (RiccR60) 4.43 h (BranHW62) others (SleN51, SulW51, SulW51b, BohE45, ShpV56)	☢ β⁻ (NisY41) △ –86.00 (MTW) σ$_c$ 0.2 (GoldmDT64)	A chem (SegE41) chem, excit (SulW51a) daughter Tc^{105} (BornH43b, FleJ55a, KieP62a) parent Rh^{105m} (DufR51) parent Rh^{105m} (25%) (BranHW62); (27%) (NeesJ65) descendant Mo^{105} (BornH43b, KieP62a) ancestor Rh^{105} (NisY41, BohE45, SleN51, SulW51a)	β⁻ 1.87 max (11%), 1.15 max γ 0.263 (6%), 0.317 (11%, doublet), 0.40 (6%, doublet), 0.475 (20%, doublet), 0.67 (16%, doublet), 0.726 (48%) daughter radiations from Rh^{105m} Rh^{105}	Ru^{104}(n, γ) (DVriH38, SulW51a)
Ru^{106}	368 d (FlyK65a) 367 d (SchumR56) 366 d (EasH60) 371 d (WyaE61) others (MerW57, GleL51e, SeeW46)	☢ β⁻ (GoldsB51a, GleL51e) △ –86.33 (MTW) σ$_c$ 0.15 (GoldmDT64)	A chem (GoldsB46, GleL46a) chem, mass spect (HaydR48) parent 30 s Rh^{106} (SeeW46, GrumW46, GleL51e) not parent 130 m Rh^{106} (BaraG55) daughter Tc^{106} (VBaeA65)	β⁻ 0.039 max γ no γ daughter radiations from 30 s Rh^{106}	fission (GleL51e, HaydR48, GrumW48, FinB51c)
Ru^{107}	4.2 m (PierW62) 4.8 m (BaumF58) 4 m (GleL51f, BornH43b)	☢ β⁻ (BornH43b) △ –83.7 (MTW)	B chem (BornH43b, GleL51f, BaumF58) chem, genet (PierW62) parent Rh^{107} (PierW62, GleL51f, BornH43b, BaroG55a) [daughter Tc^{107}] (BornH43b)	β⁻ 3.2 max γ 0.195 (14%), 0.37 (weak), 0.48 (weak), 0.86 (7%), 0.93 (4%), 1.03 (4%), 1.29 (4%) daughter radiations from Rh^{107}	Pd^{110}(n, α) (BaumF58, BaroG55a) fission (BornH43b, GleL51f, BaroG55a, BaumF58, PierW62)
Ru^{108}	4.5 m (PierW62) 4.4 m (BaumF58) others (BaroG55a)	☢ β⁻ (BaroG55a) △ –84 (MTW)	B chem, excit (BaroG55a) chem, genet (BaumF58, PierW62) parent Rh^{108} (BaumF58, PierW62, BaroG55a)	β⁻ 1.3 max γ 0.165 (28%) daughter radiations from Rh^{108}	fission (BaroG55a, BaumF58, PierW62)
$_{45}$Rh^{97}	32 m (BasuB62a, EggD49) 37 m (ChikV62) 35 m (AteA55b)	☢ β⁺ (AteA52a, [EC]) △ –83 (MTW)	A chem, genet (AteA55b) chem, excit (ChikV62) excit, sep isotopes (BasuB62a) parent Ru^{97} (AteA55b)	β⁺ 2.47 max γ Ru X-rays, 0.08, 0.187, 0.255, 0.420, 0.511 (γ±), 0.86, 1.18, 1.57, 1.70, 1.96, 2.16 daughter radiations from Ru^{97}	Ru^{96}(d, n) (AteA55b, AteA52a, ChikV62) Ru^{96}(p, γ) (BasuB62a)
Rh^{97}	1.0 m (BasuB62a)	☢ β⁺ ? (BasuB62a)	F sep isotopes (BasuB62a)	γ 0.75	Ru^{96}(p, γ) (BasuB62a)
Rh^{98}	8.7 m (KatcS56a) 9 m (AteA55)	☢ β⁺ (AteA52a), [EC] △ –84.0 (MTW)	B chem, excit (AteA52a, AteA53d, AteA55b) daughter Pd^{98} (AteA55b, KatcS56a)	β⁺ 2.5 max γ [Ru X-rays, 0.511 (γ±)], 0.65 (100%)	daughter Pd^{98} (AteA55b, KatcS56a)
Rh^{99}	16.1 d (TownCW59) 15.0 d (FarmD55)	☢ β⁺, EC (FarmD55, HisK56) △ –85.57 (NDS, MTW)	B chem (FarmD55, HisK56) genet energy levels (TemG56a, MatthE65)	β⁺ 1.03 max γ Ru X-rays, 0.090, 0.175, 0.31 (complex), 0.354, 0.444, 0.48 (complex), 0.511 (γ±), 0.529, others to 2.7	Ru^{99}(p, n) (FarmD55, MatthE65)
Rh^{99}	4.7 h (KatcS56a) 4.5 h (ScoC52)	☢ EC 90%, β⁺ 10% (KatcS56a) △ –85.52 (LHP, NDS, MTW)	B chem, excit (EggD49) daughter Pd^{99} (KatcS56a, AteA55b)	β⁺ 0.74 max γ Ru X-rays, 0.34 (70%), 0.511 (20%, γ±), 0.62 (20%), 0.89, 1.26, 1.41	Ru^{99}(p, n) (EggD49, ScoC52) Ru^{98}(d, n) (ScoC52, EggD49)
Rh^{100}	20.8 h (MarqL53a) 19.4 h (LindnM48a) 18 h (AntoN64b) 21 h (SulW51k)	☢ EC 93%, β⁺ 7% (KoiM64) △ –85.58 (MTW)	A chem (SulW51k, LindnM48a) excit, sep isotopes (BasuB62) daughter Pd^{100} (LindnM48a)	β⁺ 2.62 max e⁻ 0.516 γ Ru X-rays, 0.444 (8%), 0.511 (13%, γ±), 0.540 (88%), 0.820 (25%), 1.11 (13%), 1.35 (20%), 1.55 (23%), 1.93 (10%), 2.37 (39%), all γ rays complex	daughter Pd^{100} (LindnM48a, KoiM64) Ru^{100}(p, n) (KoiM64) Ru^{99}(d, n) (SulW51e) Ru^{99}(p, γ) (BasuB62)

TABLE 1 (*Continued*)

Isotope Z A	Half-life	Type of decay (☢); % abundance; Mass excess ($\Delta \equiv M{-}A$), MeV ($C^{12}{=}0$); Thermal neutron cross section (σ), barns	Class; Identification; Genetic relationships	Major radiations: approximate energies (MeV) and intensities	Principal means of production
$_{45}\mathrm{Rh}^{101}$	3.0 y (HisK65) 5 y (FarmD55) 10 y (PerrN60)	☢ [EC] (FarmD55) Δ −87.39 (MTW)	B chem (FarmD55) genet energy levels, excit (SharmB60)	γ [Ru X-rays], 0.127 (88%), 0.198 (75%), 0.325 (11%) e⁻ 0.105, 0.124, 0.176	$\mathrm{Ru}^{101}(p,n)$ (SharmB60, FarmD55, PerrN56)
Rh^{101m}	4.4 d (EvaJS65) 4.7 d (KatcS56a) 4.3 d (FarmD55, LindnM48a) 4.5 d (ScoC52) 5.9 d (SulW51j)	☢ EC 90%, IT 10% (EvaJS65) no β⁺ (KatcS56a, LindnM48a) Δ −87.24 (LHP, MTW)	A chem, excit (SulW51j) genet energy levels, excit (SharmB60) daughter Pd¹⁰¹ (LindnM48a, EvaJS65)	γ Ru X-rays, Rh X-rays, 0.307 (83%), 0.545 (6%) e⁻ 0.134, 0.154	$\mathrm{Ru}^{101}(p,n)$ (ScoC52, FarmD55, SharmB60) $\mathrm{Ru}^{100}(d,n)$ (SulW51j, ScoC52)
Rh^{102}	206 d (HisK61) 210 d (MinaO41) 205 d (MGowF61a) others (HoleN47)	☢ EC, β⁺, β⁻; β⁺/β⁻ 0.75 (HisK61) 0.84 (MarqL54) Δ −86.77 (MTW)	A chem, excit (MinaO41)	β⁻ 1.15 max β⁺ 1.29 max γ Ru X-rays, 0.475 (57%), 0.511 (25%, γ±), 0.628 (4%), 1.103 (3%), 1.37 (0.5%), 1.57 (0.2%)	$\mathrm{Ru}^{102}(p,n)$ (FarmD55, HisK61, MGowF61a) $\mathrm{Ru}^{101}(d,n)$, $\mathrm{Ru}^{102}(d,2n)$ (BesD55, BornP61, SulW51i) $\mathrm{Rh}^{103}(n,2n)$ (MinaO41, HoleN45a)
Rh^{102}	2.9 y (BornP63a) others (MGowF61a, HisK65)	☢ EC (MGowF61a, BornP63a)	B chem, excit (MGowF61a)	γ Ru X-rays, 0.418 (13%), 0.475 (95%), 0.632 (54%, doublet), 0.698 (41%), 0.768 (30%), 1.05 (41%), 1.11 (22%, doublet)	$\mathrm{Ru}^{102}(p,n)$ (MGowF61a) deuterons on Ru (BornP63a)
$\underline{\mathrm{Rh}^{103}}$		% 100 (CohAA43) Δ −88.014 (MTW) σ_c 144 (to Rh¹⁰⁴) 11 (to Rh¹⁰⁴ᵐ) (GoldmDT64)			
Rh^{103m}	57.5 m (JonG56) 57 m (GleL51e) 56 m (MeiJ50a) 45 m (WieM45b) others (FlaA47a, FlaA44)	☢ IT (FlaA44, WieM45b) Δ −87.974 (LHP, MTW)	A chem, excit (FlaA44) chem (GleL46a, GleL51e) chem, genet (SulW51f) daughter Ru¹⁰³ (SulW51f) daughter Pd¹⁰³ (MeiJ50a, BrosA46)	γ Rh X-rays, 0.040 (0.4%) e⁻ 0.017, 0.037	daughter Ru¹⁰³ (SulW51f) daughter Pd¹⁰³ (MeiJ50a)
Rh^{104}	43 s (CsiJ63) 44 s (AmaE35, PonB38a) 42 s (CriE39)	☢ β⁻ (PonB38a) EC 0.5% (FrevL65a) no β⁺, lim 5 x 10⁻⁴% (LanghH61b) Δ −86.95 (MTW) σ_c 40 (GoldmDT64)	A n-capt (AmaE35) genet (PonB38a) daughter Rh¹⁰⁴ᵐ (PonB38a, FlaA47a)	β⁻ 2.44 max γ Ru X-rays, 0.56 (2.0%), 1.24 (0.13%)	daughter Rh¹⁰⁴ᵐ Rh¹⁰³(n, γ) (AmaE35, PoolM37, PoolM38, GrumW46, SerL47b, PonB38a, FlaA47a, HumV51)
Rh^{104m}	4.41 m (ElliJ59) 4.3 m (CsiJ63) 4.4 m (CriE39) others (DMatE51, FlaA47a)	☢ IT 99+%, β⁻ 0.18% (WieK63) Δ −86.82 (LHP, MTW) σ_c 800 (GoldmDT64)	A n-capt (AmaE35) parent Rh¹⁰⁴ (PonB38a, FlaA47a)	γ Rh X-rays, 0.051 (47%), 0.078 (2.5%), 0.097 (2.6%), 0.56 (0.18%), 0.77 (0.24%, doublet) e⁻ 0.028, 0.054, 0.074 β⁻ [0.5 max] daughter radiations from Rh¹⁰⁴	Rh¹⁰³(n, γ) (AmaE35, PoolM37, PonB38a, GrumW46, SerL47b, HumV51)
Rh^{105}	35.88 h (BranHW62) 36.2 h (DufR51) 36.5 h (SulW51a) others (BohE45, NisY41, KunD48, MandeC51)	☢ β⁻ (NisY41) Δ −87.87 (MTW) σ_c 6,000 (to 30 s Rh¹⁰⁶) 15,000 (to 130 m Rh¹⁰⁶) (GoldmDT64)	A chem, genet (NisY41, SulW51a) daughter Rh¹⁰⁵ᵐ (DufR51) descendant Ru¹⁰⁵ (NisY41, BohE45, SleN51, SulW51a) descendant Tc¹⁰⁵, descendant Mo¹⁰⁵ (KieP62a)	β⁻ 0.568 max γ 0.306 (5%), 0.319 (19%)	Ru¹⁰⁴(n, γ) Ru¹⁰⁵(β⁻) (SulW51a)
Rh^{105m}	45 s (DufR51)	☢ IT (DufR51) Δ −87.74 (LHP, MTW)	A chem, genet (DufR51) daughter Ru¹⁰⁵, parent Rh¹⁰⁵ (DufR51) daughter Ru¹⁰⁵ (25%) (BranHW62): (27%) (NeesJ65)	γ Rh X-rays, 0.129 e⁻ 0.106, 0.126	daughter Ru¹⁰⁵ (DufR51)
Rh^{106}	30 s (GleL51e) 40 s (SeeW46)	☢ β⁻ (GleL51e) Δ −86.37 (MTW)	A chem, genet (GleL46a, GleL51e) daughter Ru¹⁰⁶ (SeeW46, GrumW46, GleL51e)	β⁻ 3.54 max γ 0.512 (21%), 0.622 (11%, doublet), 1.05 (1.5%, doublet), 1.13 (0.5%, doublet), 1.55 (0.2%)	daughter Ru¹⁰⁶ (SeeW46, GrumW46, GleL51e)

TABLE 1 (*Continued*)

Isotope Z A	Half-life	Type of decay (☢); % abundance; Mass excess (Δ≡M−A), MeV (C^{12}=0); Thermal neutron cross section (σ), barns	Class; Identification; Genetic relationships	Major radiations: approximate energies (MeV) and intensities	Principal means of production
$_{45}Rh^{106}$	130 m (MayS58) 133 m (SegO60a) others (BaroG55, NerW55)	☢ β^- (BaroG55) Δ −86.3 (SegO60a, MTW)	A chem, excit (BaroG55, NerW55) genet energy levels (MayS58, SegO60a) not daughter Ru^{106} (BaroG55)	β^- 1.62 max (10%), 1.1 max γ 0.220 (18%, complex), 0.406 (18%), 0.451 (35%), 0.512 (88%), 0.616 (29%), 0.735 (41%), 0.82 (35%), 1.046 (25%), 1.128 (12%), 1.223 (17%), 1.56 (18%)	Pd^{108}(d, α) (BaroG55, MayS58, SegO60a) Ag^{109}(n, α) (MayS58)
Rh^{107}	21.7 m (PierW62) 24 m (BornH43b) 25 m (NerW55) 23.0 m (MallC56, BaroG55a) others (GleL51f)	☢ β^- (BornH43b) Δ −86.86 (MTW)	A chem (BornH43b) chem, sep isotopes, excit (PierW62) daughter Ru^{107} (PierW62, BornH43b, GleL51f, BaroG55a) descendant Tc^{107} (VBaeA65)	β^- 1.20 max γ 0.305 (73%), 0.390 (11%), 0.68 (3%)	Ru^{104}(α, p) (PierW46) fission (BornH43a, GleL51f, PierW62)
Rh^{108}	16.8 s (PierW62) 17.5 s (BaumF58) 18 s (BaroG55a)	☢ β^- (BaroG55a) Δ −85 (MTW)	B chem (BaroG55a) chem, genet energy levels (PierW62) daughter Ru^{108} (BaumF58, PierW62, BaroG55a)	β^- 4.5 max γ 0.434 (43%), 0.51 (10%, complex), 0.62 (22%)	fission, daughter Ru^{108} (BaroG55a, BaumF58, PierW62)
Rh^{109}	<1 h (SeiJ51)	☢ [β^-] (SeiJ51) Δ −85 (MTW)	F genet (SeiJ51) [parent Pd^{109}] (SeiJ51)		fission (SeiJ51)
Rh^{110}	5 s (KarrM63a)	☢ β^- (KarrM63a) Δ −83 (MTW)	C sep isotopes, genet energy levels (KarrM63a)	β^- 5.5 max γ 0.374	Pd^{110}(n, p) (KarrM63a)
$_{46}Pd^{98}$	17.5 m genet (KatcS56a) 17 m genet (AteA53b)	☢ [EC] (AteA53d)	B chem, genet (AteA53d, AteA55b) parent Rh^{98} (KatcS56a, AteA53d)	γ [Rh X-rays], 0.132 (?) daughter radiations from Rh^{98}	Ru^{96}(α, 2n) (AteA55b, KatcS56a)
Pd^{99}	22 m (KatcS56a) 24 m (AteA55b)	☢ β^+ (KatcS56a), [EC] Δ −81.7 (MTW)	B chem, excit (AteA55b, KatcS56a) parent 4.7 h Rh^{99} (KatcS56a, AteA55b)	β^+ 2.0 max γ Rh X-rays, 0.140, 0.275, 0.420, 0.511 ($\gamma^{\pm}$), 0.67 daughter radiations from 4.7 h Rh^{99}	Ru^{96}(α, n) (KatcS56a)
Pd^{100}	4.0 d (LindnM48a) 4.1 d (KurcB55) 3.7 d (AntoN64a)	☢ EC, no β^+ (LindnM48a) Δ −85 (MTW)	A chem, excit, genet (LindnM48a) parent Rh^{100} (LindnM48a)	γ Rh X-rays, 0.074 (34%), 0.084 (49%), 0.126 (16%), 0.159 (4%) e^- 0.010, 0.019, 0.052, 0.061, 0.071, 0.081 daughter radiations from Rh^{100}	Rh^{103}(p, 4n) (KoiM64, EvaJS65a) Rh^{103}(d, 5n) (LindnM48a)
Pd^{101}	8.4 h (EvaJS65) 8.5 h (KatcS56a) others (LindnM50a)	☢ EC 97.5%, β^+ 2.5% (EvaJS65) others (KatcS56a) Δ −85.40 (EvaJS65)	A chem, genet (LindnM48a, EvaJS65) parent Rh^{101m} (LindnM48a, EvaJS65)	γ Rh X-rays, 0.270 (8%), 0.296 (30%), 0.511 (5%, $\gamma^{\pm}$), 0.566 (7%), 0.590 (24%), 0.723 (5%), 0.993 (1.7%), 1.20 (3.3%, complex), 1.30 (3.3%, doublet) β^+ 0.78 max e^- 0.021 daughter radiations from Rh^{101m}	Rh^{103}(p, 3n) (EvaJ65) Ru^{99}(α, 2n) (KatcS56a)
<u>Pd^{102}</u>		% 0.96 (SitJ53) 0.8 (SamM36a) Δ −87.92 (MTW) σ_c 4.8 (GoldmDT64)			
Pd^{103}	17.0 d (MatthD47, BrosA46, MeiW53) 17.5 d (RieL54)	☢ EC (BrosA46) Δ −87.46 (MTW)	A chem, genet (BrosA46) chem, excit (MatthD47) parent Rb^{103m} (BrosA46, MeiJ50a) daughter Ag^{103} (HaldB54)	γ Rh X-rays, 0.297 (0.011%), 0.362 (0.06%), 0.498 (0.011%) daughter radiations from Rh^{103m}	Pd^{102}(n, γ) (BrosA46) Rh^{103}(d, 2n) (MatthD47, LindnM48a) Rh^{103}(p, n) (MatthD47)
<u>Pd^{104}</u>		% 10.97 (SitJ53) 9.3 (SamM36a) Δ −89.41 (MTW)			
<u>Pd^{105}</u>		% 22.2 (SitJ53) 22.6 (SamM36a) Δ −88.43 (MTW)			

TABLE 1 (*Continued*)

Isotope Z A	Half-life	Type of decay (☢); % abundance; Mass excess ($\triangle \equiv M-A$), MeV ($C^{12}=0$); Thermal neutron cross section (σ), barns	Class; Identification; Genetic relationships	Major radiations: approximate energies (MeV) and intensities	Principal means of production
$_{46}Pd^{106}$		% 27.3 (SitJ53) 27.2 (SamM36a) $\triangle$ -89.91 (MTW) σ_c 0.29 (GoldmDT64)			
Pd^{107}	$\approx 7 \times 10^6$ y sp act (ParkG49)	☢ β^- (ParkG49) $\triangle$ -88.368 (MTW)	B chem (ParkG49)	β^- 0.04 max γ no γ	fission (ParkG49)
Pd^{107m}	21.3 s (StriT57a) 23 s (SchinU58, FlaA52a)	☢ IT (FlaA52a) $\triangle$ -88.16 (LHP, MTW)	A excit (FlaA52a) n-capt, sep isotopes (SchinU58, WeirW64) genet energy levels (CujB63)	γ Pd X-rays, 0.21 e^- 0.19, 0.21	$Pd^{106}(n, \gamma)$, $Pd^{108}(n, 2n)$ (SchinU58, WeirW64)
Pd^{108}		% 26.7 (SitJ53) 26.8 (SamM36a) $\triangle$ -89.52 (MTW) σ_c 12 (to Pd^{109}) 0.2 (to Pd^{109m}) (GoldmDT64)			
Pd^{109}	13.47 h (BranHW62) 13.6 h (MeiW53, BonaG64) 13.1 h (WafH48) 14.1 h (MacD48) others (KraJD37, SeiJ51, KondE52, DzaB57)	☢ β^- (KraJD37) $\triangle$ -87.60 (MTW)	A n-capt (AmaE35) chem, excit (KraJD37) chem, mass spect (RalW46, BergI49) parent Ag^{109m} (SegE41, SiegK49a, SeiJ51) [daughter Rh^{109}] (SeiJ51)	β^- 1.028 max e^- 0.062 (with Ag^{109m}), 0.084 (with Ag^{109m}) γ Ag X-rays, 0.088 (5%, with Ag^{109m}), 0.129 (0.013%), 0.31 (0.010%, doublet), 0.41 (0.010%, doublet), 0.60 (0.03%), 0.64 (0.010%)	$Pd^{108}(n, \gamma)$ (AmaE35, KraJD37, SerL47b, OrsA49, HumV51)
Pd^{109m}	4.69 m (StarJ59) 4.75 m (StriT57a) others (FlaA52a, MangS62, OkaM63)	☢ IT (KahJ51, FlaA52a) $\triangle$ -87.41 (LHP, MTW)	A n-capt (KahJ51) excit, cross bomb, n-capt (FlaA52a) n-capt, sep isotopes, excit (SchinU58) genet energy levels (CujB63)	γ Pd X-rays, 0.188 (58%) e^- 0.164, 0.185	$Pd^{108}(n, \gamma)$ (FlaA52a, SchinU58)
Pd^{110}		% 11.8 (SitJ53) 13.5 (SamM36a) $\triangle$ -88.34 (MTW) σ_c 0.2 (to Pd^{111}) 0.04 (to Pd^{111m}) (GoldmDT64)			
Pd^{111}	22 m (DzaB57, MGinC52) others (SegE41)	☢ β^- (KraJD37) $\triangle$ -86.0 (MTW)	A n-capt (AmaE35) chem, genet (SegE41) parent Ag^{111} (KraJD37, SegE41, JohaS50) parent Ag^{111m} (SchinU57)	β^- 2.2 max γ 0.38 ($\uparrow$ 5), 0.60 ($\uparrow$ 13, doublet), 0.81 ($\uparrow$ 1), 1.4 ($\uparrow$ 8, doublet) daughter radiations from Ag^{111m}	$Pd^{110}(n, \gamma)$, daughter Pd^{111m} (AmaE35, KraJD37, SerL47b)
Pd^{111m}	5.5 h (MGinC52, DzaB57)	☢ IT 75%, β^- 25% (MGinC52) $\triangle$ -85.8 (LHP, MTW)	A chem, genet (MGinC52, DzaB57) parent Ag^{111} (MGinC52, DzaB57)	β^- 2.0 max e^- 0.148, 0.169 γ Pd X-rays, 0.17 daughter radiations from Pd^{111}, Ag^{111m}, Ag^{111}	$Pd^{110}(n, \gamma)$ (DzaB57, PraW60) $Pd^{110}(d, p)$ (MGinC52, EccS62)
Pd^{112}	21.0 h (GirR59k) 21 h (SeiJ51)	☢ β^- (NisY40b) $\triangle$ -86.27 (MTW)	A chem, genet (NisY40b, SegE41) parent Ag^{112} (NisY40b, NisY40, SegE41, SeiJ51)	β^- 0.28 max e^- [0.016] γ [Pd L X-rays], 0.019 (20%) daughter radiations from Ag^{112}	fission (SegE41, TurA51a, KatcS48, NisY40b, NisY40, SeiJ51, GoeR49, NewA49)
Pd^{113}	1.4 m (AlexJ58) 1.5 m (HicH54, PouA60)	☢ [β^-] (HicH54)	A chem, genet (HicH54, AlexJ58) parent 5.3 h Ag^{113} (HicH54, AlexJ58) parent 1.2 m Ag^{113} (AlexJ58)	γ no γ daughter radiations from 5.3 h Ag^{113} and 1.2 m Ag^{113}	fission (AlexJ58, HicH54) $Cd^{116}(n, \alpha)$ (PouA60)
Pd^{114}	2.4 m (AlexJ58)	☢ [β^-] (AlexJ58)	D chem, genet (AlexJ58) parent 5 s Ag^{114} (AlexJ58) not parent 2 m Ag^{114} (AlexJ58)	γ no γ	fission (AlexJ58)

TABLE 1 (*Continued*)

Isotope Z A	Half-life	Type of decay (☢); % abundance; Mass excess (Δ≡M−A), MeV (C^{12}=0); Thermal neutron cross section (σ), barns	Class; Identification; Genetic relationships	Major radiations: approximate energies (MeV) and intensities	Principal means of production
$_{46}Pd^{115}$	45 s genet (AlexJ58)	☢ [β^-] (AlexJ58)	B chem, genet (AlexJ58) parent 20 m Ag^{115}, parent 20 s Ag^{115} (AlexJ58)		fission (AlexJ58)
$_{47}Ag^{102}$	15 m (AmeO60) 16 m (EnnT39)	☢ [EC, β^+] (EnnT39, AmeO60) Δ −83 (MTW)	C excit (EnnT39) excit, sep isotopes (AmeO60)		Pd^{102}(p, n) (AmeO60, EnnT39)
Ag^{103}	66 m (PatA62b, HaldB54, BendW53) 69 m (PreiI60a) 59 m (JohnFA55)	☢ β^+, EC (HaldB54) EC(K) ≈70% (KuzM57) Δ −84.9 (MTW)	A chem (BendW53) chem, genet (HaldB54) chem, excit (GirR59e) excit, sep isotopes (AmeO60, PatA62b) parent Pd^{103} (HaldB54) daughter Cd^{103} (PreiI60a)	β^+ 1.6 max γ Pd X-rays, 0.12 († 26, doublet), 0.15 († 23), 0.24 († 10), 0.27 († 34), 0.511 († 100, $\gamma^\pm$), 1.01 († 10, complex), 1.16 († 9), 1.28 († 13) daughter radiations from Pd^{103}	Rh^{103}(α, 4n) (GirR59e) Pd^{104}(p, 2n) (AmeO60) Pd^{102}(d, n) (BendW53) Pd^{102}(p, γ) (PatA62b)
Ag^{103m}	5.7 s (WhiW62)	☢ IT (WhiW62) Δ −84.7 (LHP, MTW)	C excit (WhiW62)	γ Ag X-rays, 0.138 e^- [0.113, 0.135]	Pd^{104}(p, 2n) (WhiW62)
Ag^{104}	66 m (NutH60) 70 m (GirR59e) 69 m (AmeO60) others (EnnT39)	☢ β^+, EC (LindnM50a) Δ −85.14 (MTW)	A excit (EnnT39) chem, excit (GirR59e) sep isotopes, excit (AmeO60)	β^+ 0.99 max e^- 0.532, 0.743 γ Pd X-rays, 0.511 ($\gamma^\pm$), 0.556 (84%), 0.764 (48%), 0.854 (30%), 1.34 (8%), 1.53 (7%), 1.62 (8%), 1.81 (7%)	Rh^{103}(α, 3n) (GirR59e, NutH60, EwbW59)
Ag^{104m}	29.8 m (NutH60) 27 m (GirR59e, AmeO60, JohnFA55)	☢ β^+, EC (JohnFA55, GirR59e) IT 20-40% (AmeO60) Δ −85.12 (LHP, MTW)	A chem (JohnFA55) excit (GirR59) excit, sep isotopes (AmeO61) daughter Cd^{104} (JohnFA55, PreiI60a)	β^+ 2.70 max e^- 0.532 γ Pd X-rays, 0.511 (120%, $\gamma^\pm$), 0.556 (100%) daughter radiations from Ag^{104}	Rh^{103}(α, 3n) (GirR59e, NutH60, EwbW59) daughter Cd^{104} (JohnFA55, PreiI60a)
Ag^{105}	40 d (GumJ50) others (EnnT39)	☢ EC, no β^+ (GumJ50) Δ −87 (MTW)	A excit (EnnT39) chem, excit (BradH47a)	γ Pd X-rays, 0.064 (10%), 0.280 (32%), 0.344 (42%, complex), 0.443 (10%), 0.62-0.68 (12%, complex), 1.088 (2%) e^- 0.040, 0.060, 0.256, 0.320	Rh^{103}(α, 2n) (BradH47a, GumJ50, MeiJ50b) protons, deuterons on Pd (EnnT39, GumJ50, MeiJ50b, SutT61a, BoeR58, EwbW63)
Ag^{106}	23.96 m (EbrT65) 24.3 m (MocD48) 24.0 m (BendW51, BendW53) others (PoolM38, ForS52, DubL38, EnnT39)	☢ β^+ (KraJD37) β^+, EC, β^- (?) ≈1% (BendW53) Δ −86.94 (MTW)	A chem, excit (BotW37, HeyF37) chem, excit, cross bomb (KraJD37, PoolM38)	β^+ 1.96 max γ Pd X-rays, 0.511 (140%, 0.512 γ + $\gamma^\pm$)	Rh^{103}(α, n) (PoolM38, BradH47a)
Ag^{106m}	8.5 d (SmiW61b) 8.2 d (PoolM38) 8.4 d (RobiR60)	☢ EC (HurL44) no β^+, lim 0.1% (BendW53) Δ −86.6 (LHP, MTW)	A chem, excit, cross bomb (KraJD37, PoolM38)	γ Pd X-rays, 0.221 (9%), 0.451 (9%), 0.512 (86%), 0.616 (23%), 0.717 (31%, complex), 0.748 (13%), 0.80 (41%, complex), 1.046 (29%), 1.128 (9%), 1.199 (9%), 1.528 (15%), 1.58 (8%), 1.83 (3%) e^- 0.197, 0.382, 0.405, 0.426, 0.487, 0.508, 0.592, 0.693	Rh^{103}(α, n) (PoolM38, BradH47a, MeiJ50b, SmiW61b)
<u>Ag^{107}</u>		% 51.35 (WhiJ48) Δ −88.403 (MTW) σ_c 35 (to Ag^{108}) (GoldmDT64)			
Ag^{107m}	44.3 s (BradH47a, BradH45b) others (WoliEJ51, AlvL40a)	☢ IT (AlvL40a) Δ −88.310 (LHP, MTW)	A chem, genet (AlvL40a, HelmhA41b) daughter Cd^{107} (AlvL40a, HelmhA41b, BradH45a, HelmhA46, BradH47a)	γ Ag X-rays, 0.094 (5%) e^- 0.068, 0.090	daughter Cd^{107} (AlvL40a, HelmhA41b, BradH45a, HelmhA46, BradH47a)
Ag^{108}	2.42 m (WahM60) 2.41 m (EbrT65) others (SehM57, AmaE35, PerlmM48, MocD48, BotW39, FlaA44)	☢ β^- 97.5%, EC 2.2%, β^+ 0.28% (FrevL65, FrevL62) β^- 95.7%, EC 3.9%, β^+ 0.36% (WahM60) Δ −87.61 (MTW)	A chem, n-capt (AmaE35) excit, cross bomb (PoolM38) daughter Ag^{108m} (WahM60)	β^- 1.64 max β^+ 0.90 max γ Pd X-rays, 0.434 (0.45%), 0.511 (0.56%, $\gamma^\pm$), 0.615 (0.18%), 0.632 (1.7%)	daughter Ag^{108m} (WahM60) Ag^{107}(n, γ) (FlaA44b, AmaE35, FlaA44, SerL47b)

TABLE 1 (*Continued*)

Isotope Z A	Half-life	Type of decay (☢); % abundance; Mass excess ($\triangle \equiv M-A$), MeV ($C^{12}=0$); Thermal neutron cross section (σ), barns	Class; Identification; Genetic relationships	Major radiations: approximate energies (MeV) and intensities	Principal means of production
$_{47}\text{Ag}^{108m}$	>5 y (WahM60)	☢ EC 90%, IT 10% (WahM60) $\triangle$ -87.50 (LHP, MTW)	A chem, n-capt, genet (WahM60) parent Ag^{108} (WahM60)	γ Pd X-rays, Ag X-rays, 0.080 (5%), 0.434 (89%), 0.614 (90%), 0.722 (90%) e^- 0.027 daughter radiations from Ag^{108}	$\text{Ag}^{107}(n,\gamma)$ (WahM60)
Ag^{109}		% 48.65 (WhiJ48) $\triangle$ -88.717 (MTW) σ_c 89 (to Ag^{110}), 3 (to Ag^{110m}) (GoldmDT64)			
Ag^{109m}	39.2 s (BradH46, BradH47a) 40 s (WoliEJ51, WieM45, SchinU57)	☢ IT (HelmhA41b) $\triangle$ -88.630 (LHP, MTW)	A chem, genet (HelmhA41b) daughter Pd^{109} (SegE41, SiegK49a, SeiJ51) daughter Cd^{109} (HelmhA41b), BradH46, HelmhA46, BradH45a)	γ Ag X-rays, 0.088 (5%) e^- 0.062, 0.084	daughter Cd^{109} (HelmhA41b, BradH46, HelmhA46) daughter Pd^{109} (SegE41, SiegK49a, SeiJ51)
Ag^{110}	24.4 s (MalmS62) 24.5 s (HirzO46) others (SehM57, BolF54, ThieP62, AmaE35, PoolM38, FlaA44, GaeE36, SerL47b, HirzO47a)	☢ β^- (PoolM38) EC 0.3% (FrevL65) no β^+, lim 10^{-3}% (BereD62b) $\beta^+ \approx 6 \times 10^{-4}$% (BadN62) $\triangle$ -87.47 (MTW)	A n-capt (AmaE35) sep isotopes, n-capt (FlaA44b) chem, genet (MiskJ50) daughter Ag^{110m} (MiskJ50)	β^- 2.87 max γ 0.658 (4.5%)	daughter Ag^{110m} (MiskJ50) $\text{Ag}^{109}(n,\gamma)$ (AmaE35, GaeE36, FlaA44, SerL47b, FrevL63)
Ag^{110m}	255 d (EasH60) 253 d (GeiKW57, ThirH57) 249 d (NilR62) others (CaliJ59, SchinJ64, GumJ50, ColoJ64, CorkJ50h, LivJ38c, CorkJ48b)	☢ β^- 98.7%, IT 1.3% (calc from SutT63, NewW64, GeiJ65 by LHP) $\triangle$ -87.35 (LHP, MTW) σ_c 80 (GoldmDT64)	A chem, n-capt (RedH38) resonance neutron activation (GoldhM46) chem, mass spect (BergI49) parent Ag^{110} (MiskJ50)	β^- 1.5 max (0.6%), 0.53 max (31%), 0.087 max e^- 0.090, 0.113 γ 0.658 (96%), 0.68 (16%, doublet), 0.706 (19%), 0.764 (23%), 0.818 (8%), 0.885 (71%), 0.937 (32%), 1.384 (21%), 1.505 (11%) daughter radiations from Ag^{110}	$\text{Ag}^{109}(n,\gamma)$ (RedH38, LivJ38c, AlexK38, MitA38, SerL47b)
Ag^{111}	7.5 d (JohaS50, KraJD37, PoolM38, StorA50) 7.6 d (SteinE51b) 7.3 d (DzaB57) others (KunD47, HirzO47a, DufR49, LindnM50a, GoeR49, DConP48, NisY40b, TurA51a, FinB51c)	☢ β^- (KraJD37) $\triangle$ -88.20 (MTW)	A chem, excit (KraJD37) chem, excit, cross bomb (PoolM38) daughter Pd^{111} (KraJD37, SegE41, JohaS50) daughter Pd^{111m} (MGinC52, DzaB57)	β^- 1.05 max average β^- energy: 0.38 ion ch (BrabJ53) γ 0.247 (1%), 0.342 (6%)	$\text{Pd}^{110}(n,\gamma)\text{Pd}^{111}$ + $\text{Pd}^{111m}(\beta^-)$ (KraJD37) $\text{Pd}^{110}(d,n)$ (KraJD37, PoolM38, ZimK49)
Ag^{111m}	74 s (SchinU57)	☢ IT, no β^-, lim 1% (SchinU57) $\triangle$ -88.13 (LHP, MTW)	B chem, genet (SchinU57) daughter Pd^{111} (SchinU57)	γ [Ag X-rays], 0.065 e^- [0.040, 0.062]	daughter Pd^{111} (SchinU57)
Ag^{112}	3.14 h (InoH62) 3.2 h (PoolM38, HirzO47a)	☢ β^- (PoolM38a) $\triangle$ -86.57 (MTW)	A chem, excit, cross bomb (PoolM38) daughter Pd^{112} (NisY40b, NisY40, SegE41, SeiJ51)	β^- 3.94 max γ 0.617 (41%), 1.40 (5%), 1.63 (3%), 2.11 (3%), 2.55 (2%), many others between 0.3 and 3.3	daughter Pd^{112} (NisY40b, NisY40, SegE41, SeiJ51) $\text{In}^{115}(n,\alpha)$ (PoolM38) $\text{Cd}^{114}(d,\alpha)$ (InoH62)
Ag^{113}	5.3 h (AlexJ58, TurA47, DufR49, VasiI58)	☢ β^- (TurA47) $\triangle$ -87.04 (MTW)	A chem (TurA47) chem, sep isotopes, excit (DufR49) daughter Pd^{113} (HicH54, AlexJ58)	β^- 2.0 max γ 0.12 (↑ 10), 0.30 (↑ 100), 0.58 (↑ 5), 0.67 (↑ 17), 0.88 (↑ 4), 0.98 (↑ 5), 1.18 (↑ 4)	fission (TurA47, FolR51) $\text{Cd}^{114}(\gamma,p)$ (DufR49)
Ag^{113}	1.2 m (AlexJ58)	☢ β^- (AlexJ58)	B chem, genet (AlexJ58) daughter Pd^{113} (AlexJ58)	β^- <2.0 max γ 0.14, 0.30, 0.39, 0.56, 0.70	fission (AlexJ58)
Ag^{114}	4.5 s (PouA60) 5s (AlexJ58)	☢ β^- (AlexJ58) $\triangle$ -85.4 (MTW)	C chem, genet (AlexJ58) daughter Pd^{114} (AlexJ58)	β^- 4.6 max γ 0.57	fission, daughter Pd^{114} (AlexJ58) $\text{Cd}^{114}(n,p)$ (PouA60)
Ag^{114}	2 m (DufR49) 3 m (SeeW47)	☢ β^- (DufR49)	E chem (TurA47, SeeW47) chem, excit, sep isotopes (DufR49) not daughter Pd^{114} (AlexJ58)	β^- hard β^-	$\text{Cd}^{114}(n,p)$ (DufR49) fission (TurA47, SeeW47) not observed in $\text{Cd}^{114}(n,p)$ (AlexJ58)

TABLE 1 (*Continued*)

Isotope Z A	Half-life	Type of decay (☢); % abundance; Mass excess (△≡M–A), MeV (C¹²=0); Thermal neutron cross section (σ), barns	Class; Identification; Genetic relationships	Major radiations: approximate energies (MeV) and intensities	Principal means of production
$_{47}$Ag^{115}	20.0 m (BahE64) 21.1 m (AlexJ58) others (DufR49, SeeW47, WahA52)	☢ β⁻ (TurA47) △ -84.8 (MTW)	A chem (TurA47, SeeW47) chem, excit, sep isotopes (DufR49) parent Cd^{115} (91%), parent Cd^{115m} (9%) (WahA52) parent Cd^{115} (92%), parent Cd^{115m} (8%) (HicH55) daughter Pd^{115} (AlexJ58)	β⁻ 3.2 max γ 0.14 (12%, complex), 0.22 (49%, complex), 0.28 (13%), 0.36 (11%), 0.42 (7%), 0.47 (10%), 0.64 (4%, complex), 1.48 (11%), 1.66 (8%), 1.89 (10%, complex), 2.12 (13%)	fission (TurA47, SeeW47, BahE64, AlexJ58) Cd^{116}(γ, p) (DufR49)
Ag^{115}	≈20 s (AlexJ58)	☢ [β⁻] (AlexJ58)	B chem, genet (AlexJ58) daughter Pd^{115}, parent Cd^{115} (AlexJ58)		fission (AlexJ58)
Ag^{116}	2.5 m (AlexJ58)	☢ β⁻ (AlexJ58) △ -83 (MTW)	D chem (AlexJ58)	β⁻ 5.0 max γ 0.52, 0.70	fission (AlexJ58)
Ag^{117}	1.1 m (AlexJ58)	☢ [β⁻] (AlexJ58)	B chem, genet (AlexJ58) parent Cd^{117} and/or Cd^{117m} (AlexJ58)		fission (AlexJ58)
$_{48}$Cd^{103}	10 m (PreiI60a)	☢ β⁺, [EC] (PreiI60a)	A chem, genet (PreiI60a) parent Ag^{103} (PreiI60a)	γ Ag X-rays, 0.22, 0.511 (γ±), 0.63, 0.85 daughter radiations from Ag^{103}	O^{16} on Mo (PreiI60a)
Cd^{104}	57 m (PreiI60a) 54 m (KurcB55) 59 m (JohnFA55)	☢ EC, no β⁺ (JohnFA55) △ -84 (MTW)	A chem, genet, excit (JohnFA55) parent Ag^{104m} (JohnFA55, PreiI60a)	γ Ag X-rays, 0.084 e⁻ 0.041, 0.058, 0.080 daughter radiations from Ag^{104m} Ag^{104}	Ag^{107}(p, 4n) (JohnFA55) O^{16} on Mo (PreiI60a)
Cd^{105}	55 m (JohnFA53) 57 m (GumJ50)	☢ EC, β⁺ (GumJ50) △ -84 (MTW)	B cross bomb (GumJ50) chem, excit (JohnFA53)	β⁺ 1.69 max e⁻ 0.282, 0.295, 0.321, 0.408, others γ [Ag X-rays, 0.308, 0.320, 0.347, 0.433, 0.511 (γ±), others to 2.3] daughter radiations from Ag^{105}	Pd^{102}(α, n) (GumJ50) Ag^{107}(p, 3n) (JohnFA53)
Cd^{106}		% 1.22 (LelW48) △ -87.128 (MTW) σc 1 (GoldmDT64)			
Cd^{107}	6.49 h (LarN62) 6.7 h (DelL39, HelmhA41b) 6.4 h (ValleG39)	☢ EC 99+%, β⁺ 0.28% (LarN62) △ -86.99 (MTW)	A chem (DelL39) chem, n-capt, sep isotopes (HelmhA46) parent Ag^{107m} (AlvL40a, HelmhA41b, BradH45a, HelmhA46, BradH47a)	β⁺ 0.302 max γ Ag X-rays, 0.511 (0.56%, γ±), 0.796 (0.08%), 0.829 (0.21%) daughter radiations from Ag^{107m}	Cd^{106}(n, γ) (HelmhA46) Ag^{107}(d, 2n) (AlvL40a, KriR39, KriR40a, HelmhA41b) Ag^{107}(p, n) (DelL39, ValleG39)
Cd^{108}		% 0.88 (LelW48) △ -89.248 (MTW) σc 3 (GoldmDT64)			
Cd^{109}	453 d (LeuH65) 470 d (GumJ50) others (MangS62, BradH46)	☢ EC (HelmhA41b) no β⁺ (DreB51) △ -88.55 (MolR65, MTW)	A chem (KriR40a) chem, n-capt, sep isotopes (HelmhA46) parent Ag^{109m} (HelmhA41b, BradH45a, HelmhA46, BradH46)	γ Ag X-rays, 0.088 (with Ag^{109m}) e⁻ 0.062 (with Ag^{109m}), 0.084 (with Ag^{109m})	Cd^{108}(n, γ) (HelmhA46, CorkJ50g) Ag^{109}(d, 2n) (KriR40a, HelmhA41b, GumJ50)
Cd^{110}		% 12.39 (LelW48) △ -90.342 (MTW) σc 0.1 (to Cd^{111m}) (GoldmDT64)			
Cd^{111}		% 12.75 (LelW48) △ -89.246 (MTW)			
Cd^{111m}	48.6 m (MGinC51) 48.7 m (WieM45)	☢ IT (FelJ41, WieM45) △ -88.850 (LHP, MTW)	A chem (DodM38) chem, sep isotopes, n-capt (GoldhM48a) daughter In^{111} (0.01%) (MGinC51a)	γ Cd X-rays, 0.150 (30%), 0.247 (94%) e⁻ 0.123, 0.146	Cd^{110}(n, γ) (GoldhM48a, DodM38, HoleN48b) daughter In^{111} (MGinC51a)

TABLE 1 *(Continued)*

Isotope Z A	Half-life	Type of decay (☘); % abundance; Mass excess ($\triangle \equiv M-A$), MeV ($C^{12}=0$); Thermal neutron cross section (σ), barns	Class; Identification; Genetic relationships	Major radiations: approximate energies (MeV) and intensities	Principal means of production
$_{48}Cd^{112}$		☘ 24.07 (LelW48) $\triangle$ -90.575 (MTW) σ_c 0.03 (to Cd^{113m}) (GoldmDT64)			
Cd^{113}	$t_{1/2}$ >1.3 x 10^{15} y sp act (WatD62a)	% 12.26 (LelW48) $\triangle$ -89.041 (MTW) σ_c 20,000 (GoldmDT64)			
Cd^{113m}	13.6 y (FlyK65a) 14 y (WahA59) 5 y (CarsW50)	☘ β^- (CarsW50) IT weak (DMatE56) $\triangle$ -88.77 (LHP, MTW)	A chem, excit, sep isotopes (CarsW50)	β^- 0.58 max γ [Cd X-rays], 0.265 ($\approx$0.1%)	$Cd^{112}(n,\gamma)+Cd^{113}(n,n')$ (CarsW50) fission (WahA52, WahA59)
Cd^{114}		% 28.86 (LelW48) $\triangle$ -90.018 (MTW) σ_c 1.1 (to Cd^{115}) 0.14 (to Cd^{115m}) (GoldmDT64)			
Cd^{115}	53.5 h (WyaE61) 53 h (WahA52, VasiI58) 54 h (CorkJ50g, BedaA64) others (LawJL40, MetR51a)	☘ β^- (CorkJ37) $\triangle$ -88.09 (MTW)	A chem (CorkJ37) chem, genet (GoldhM38) chem, sep isotopes, n-capt (CorkJ50g) parent In^{115m} (GoldhM38, CorkJ39, NisY40, MetR51a, WahA52, LangeL52a) daughter 20 m Ag^{115} (91%) (WahA52) daughter 20 m Ag^{115} (92%) (HicH55) daughter $\approx$20 s Ag^{115} (AlexJ58)	β^- 1.11 max γ In X-rays, 0.230 (0.6%), 0.262 (2%), 0.49 (10%), 0.53 (26%) daughter radiations from In^{115m}	$Cd^{114}(n,\gamma)$ (GoldhM38, MitA37, SerL47b)
Cd^{115m}	43 d (SerL47, CorkJ50g) 44 d (GleL51g, WahA59)	☘ β^- (CorkJ39) $\triangle$ -87.91 (LHP, MTW)	A chem, excit (SerL47) chem, sep isotopes, n-capt (CorkJ50g) daughter 20 m Ag^{115} (9%) (WahA52) daughter 20 m Ag^{115} (8%) (HicH55)	β^- 1.62 max γ 0.485 (0.31%), 0.935 (1.9%), 1.29 (0.9%)	$Cd^{114}(n,\gamma)$ (SerL47b, SerL47, CorkJ50g)
Cd^{116}	$t_{1/2}$ ($\beta\beta$) >10^{17} y sp act (WintR55)	% 7.58 (LelW48) $\triangle$ -88.712 (MTW) σ_c 1.4 (to Cd^{117}) (GoldmDT64) 0.7 (to Cd^{117m}) (TanC66a, GoldmDT64)			
Cd^{117}	2.4 h (TanC66) $\approx$3 h (SharmR64, MancR65) others (CoryC53, AteA52, LawJL40, MetR51b)	☘ β^- (SharmR64) $\triangle$ -86.41 (MTW)	A chem, genet, n-capt (SharmR64, TanC66) parent In^{117m} (93%), parent In^{117} (7%) (TanC66) not daughter Cd^{117m} (SharmR64) others (CorkJ39, GoldhM38, LawJL40, MetR51b, MGinC55)	β^- 2.23 max e^- 0.286 (with In^{117m}) γ In X-rays (with In^{117m}), 0.089 (7%), 0.273 (31%), 0.314 (16%, with In^{117m}), 0.345 (18%), 0.434 (13%), 0.832 (4%), 0.880 3%), 0.95 (4%, doublet), 1.052 (5%), 1.303 (19%), 1.577 (17%) daughter radiations from In^{117m}, In^{117}	$Cd^{116}(n,\gamma)$ (TanC66a) $Cd^{116}(d,p)$ (TanC66a)
Cd^{117m}	3.4 h (TanC66) $\approx$3 h (SharmR64, MancR65) others (CoryC53, AteA52, LawJL40, MetR51b)	☘ β^- (SharmR64) $\triangle$ -86.27 (LHP, MTW)	A chem, genet, n-capt (SharmR64, TanC66) parent In^{117} (56%), parent In^{117m} (44%) (TanC66) not parent Cd^{117} (SharmR64) others (CorkJ39, GoldhM38, LawJL40, MetR51b, MGinC55)	β^- [1.91 max (weak)], 0.67 max e^- 0.286 (with In^{117m}) γ In X-rays (with In^{117m}), 0.273 (18%), 0.314 (8%, with In^{117m}), 0.345 (4%), 0.434 (4%), 0.565 (6%), 0.715 (4%), 0.880 (10%), 1.065 (9%), 1.117 (4%), 1.24 (11%, complex), 1.338 (8%), 1.408 (8%), 1.433 (10%), 1.562 (6%), 1.998 (15%), 2.319 (3%)	$Cd^{116}(n,\gamma)$ (TanC66a) $Cd^{116}(d,p)$ (TanC66a)
Cd^{117}	$\approx$50 m (CoryC53)		G chem, genet (CoryC53) activity not observed (SharmR64, TanC66)		

TABLE 1 *(Continued)*

Isotope Z A	Half-life	Type of decay (☢); % abundance; Mass excess (△≡M–A), MeV (C¹²=0); Thermal neutron cross section (σ), barns	Class; Identification; Genetic relationships	Major radiations: approximate energies (MeV) and intensities	Principal means of production
$_{48}Cd^{118}$	49 m (GleC61)	☢ β⁻ (CoryC53) △ −87 (MTW)	B chem, excit (CoryC53) chem, genet (GleC61) parent 5.0 s In¹¹⁸ (CoryC53, GleC61) not parent 4.4 m In¹¹⁸ (CoryC53, GleC61)	daughter radiations from 5.0 s In¹¹⁸	fission (CoryC53, GleC61)
Cd¹¹⁹	2.7 m (GleC61a)	☢ β⁻ (GleC61a) △ −84.1 (MTW)	B chem, genet (GleC61a) parent In¹¹⁹, parent In¹¹⁹m (GleC61a)	β⁻ 3.5 max daughter radiations from In¹¹⁹m, In¹¹⁹	fission (GleC61a)
Cd¹¹⁹	10 m (NusN57, GleC61a)	☢ β⁻ (NusN57, GleC61a) △ −84.1 (MTW)	B chem, genet (NusN57, GleC61a) parent In¹¹⁹m (NusN57, GleC61a)	β⁻ 3.5 max daughter radiations from In¹¹⁹m, In¹¹⁹	Sn¹²²(d, αp) (NusN57) fission (GleC61a)
Cd¹²¹	12.8 s (WeisH65)	☢ [β⁻] (WeisH65)	B chem, genet (WeisH65) ancestor Sn¹²¹ (WeisH65)		fission (WeisH65)
Cd⁽¹²¹?⁾	3.5 m (NusN57)	☢ [β⁻] (NusN57)	G chem, excit (NusN57) parent 11.5 m In⁽¹²¹?⁾ and 32 m In⁽¹²¹?⁾ (NusN57) Daughter In isotopes are probably incorrectly assigned (NDS, YutH60)		deuterons on Sn (NusN57)
$_{49}In^{106}$	5.3 m (CatR62) others (CatR65)	☢ β⁺ (CatR62), [EC] △ −80.6 (MTW)	A chem, excit, sep isotopes (CatR62)	β⁺ 4.9 max γ [Cd X-rays], 0.511 (γ±), 0.63, 1.65, 1.85, many others	Cd¹⁰⁶(p, n) (CatR62)
In¹⁰⁷	33 m (MallE49) 31 m (BasuB63) 30 m (MaclK52)	☢ β⁺, EC (BasuB63) △ −83.5 (MTW)	A chem, sep isotopes (MallE49) mass spect (MaclK52)	β⁺ 2.2 max γ Cd X-rays, 0.22 (46%), 0.32, 0.511 (γ±), 0.73, 0.84, 0.94, 1.05, 1.25 daughter radiations from Cd¹⁰⁷ Ag¹⁰⁷m	Cd¹⁰⁶(d, n) (MallE49, CassW55a) Cd¹⁰⁶(p, γ) (MallE49, BasuB63)
In¹⁰⁸	57 m (KatoT63) 55 m (MeaS55, MallE49) others (KatoT62b, MGinC51)	☢ EC, β⁺ (KatoT62b) △ −84.14 (KatoT62b, MTW)	A chem, sep isotopes (MallE49) mass spect (MaclK52)	β⁺ 1.29 max e⁻ 0.123, 0.147, 0.216, 0.238, 0.260, 0.606, 0.845 γ Cd X-rays, 0.150, 0.175, 0.243, 0.511 (γ±), 0.633, 0.872	Ag¹⁰⁷(α, 3n) (KatoT62a, KatoT62b)
In¹⁰⁸	39 m (KatoT63) 40 m (MeaS55, KatoT62b)	☢ EC, β⁺ (KatoT62b) △ −84.10 (KatoT62b, MTW)	B chem, excit (MeaS55) genet energy levels (KatoT62b) daughter Sn¹⁰⁸ (MeaS55)	β⁺ 3.50 max e⁻ 0.606 γ Cd X-rays, 0.383, 0.511 (γ±), 0.633, 0.842	Ag¹⁰⁷(α, 3n) (KatoT62a, KatoT62b)
In¹⁰⁹	4.3 h (MallE49, NozM62) 4.2 h (MGinC51) 5.2 h (GhoS48) others (TenD47a)	☢ EC 94%, β⁺ 6% (PetrM56a) △ −86.53 (MTW, MolR65)	A chem, excit (TenD47a) chem, mass spect (GhoS48) chem, excit, sep isotopes (MallE49) descendant Sn¹⁰⁹ (PetrM56a)	β⁺ 0.79 max e⁻ 0.033, 0.056, 0.178, 0.201 γ Cd X-rays, 0.205, 0.28 (complex), 0.35 (complex), 0.65 (complex), 0.91 (complex)	Ag¹⁰⁷(α, 2n) (NozM62, KatoT62a, TenD47a)
In¹⁰⁹m₁	1.3 m (AlexKF65) <2m (PetrM56a)	☢ IT (PetrM56a) △ −85.87 (LHP, MTW)	C genet (PetrM56a) daughter Sn¹⁰⁹ (PetrM56a)	γ 0.658 e⁻ 0.630	daughter Sn¹⁰⁹ (PetrM56a)
In¹⁰⁹m₂	0.20 s (AlexKF65) 0.21 s (DemiA65) 0.22 s (PoeG63)	☢ IT (AlexKF65, DemiA65) △ −84.42 (LHP, MTW)	C excit, cross bomb (AlexKF65, DemiA65, PoeG63)	γ 0.17 (12%), 0.21 (12%), 0.40 (20%), 0.68 (100%), 1.04 (20%), 1.43 (77%)	Ag¹⁰⁷(α, 2n) (AlexKF65, DemiA65, PoeG63) Rh¹⁰³(C¹², α2n) (AlexKF65)
In¹¹⁰	66 m (KatoT62a, BarnS39a) 69 m (HamiJ63) 65 m (GhoS48)	☢ β⁺ 71%, EC 29% (NaiT64) △ −86.41 (MTW)	A chem (BarnS39a) chem, excit, mass spect (GhoS48) daughter Sn¹¹⁰ (MeaS55)	β⁺ 2.25 max e⁻ 0.631 γ Cd X-rays, 0.511 (142%, γ±), 0.658 (95%)	daughter Sn¹¹⁰ (NaiT64) Ag¹⁰⁷(α, n) (KatoT62a) Ag¹⁰⁹(α, 3n) (FukS65)

TABLE 1 (*Continued*)

Isotope Z A	Half-life	Type of decay (☢); % abundance; Mass excess ($\Delta \equiv M-A$), MeV ($C^{12}=0$); Thermal neutron cross section (σ), barns	Class; Identification; Genetic relationships	Major radiations: approximate energies (MeV) and intensities	Principal means of production
$_{49}$In^{110}	4.9 h (BleE51, KatoT62a) 5.0 h (MGinC51) others (GhoS48)	☢ EC, β^+ ? (weak) (KatoT62a) no IT, lim 0.008% (HamiJ63)	A chem (GhoS48) chem, genet energy levels (MGinC51a, BleE51) not daughter Sn^{110} (MeaS55)	γ Cd X-rays, 0.66 († 160, complex), 0.91 († 110, complex) e^- 0.094, 0.558, 0.615, 0.631, 0.653, 0.680, 0.858, 0.910	Ag^{109}(α, 3n) (FukS65, KatoT62a)
In^{111}	2.81 d (MaiA57) 2.84 d (MGinC51) others (BarnS39a, CorkJ39)	☢ EC (LawJL40) no β^+, lim 0.06% (MGinC51) Δ −88.2 (MTW)	A chem (CorkJ39) chem, excit (TenD47, GhoS48) mass spect (GhoS48) parent Cd^{111m} (0.01%) (MGinC51a)	γ Cd X-rays, 0.173 (89%), 0.247 (94%) e^- 0.146, 0.220, 0.243	Ag^{109}(α, 2n) (FukS65, LawJL40, TenD47, GhoS48, MGinC51)
In^{112}	14.4 m (FukS65) 12 m (RuaJ62a) 11 m (GirR59i) 15 m (BleE53)	☢ β^- 44%, β^+ 22%, EC 34% (calc) (RuaJ62a) others (BleE53) Δ −87.98 (MTW)	A chem, cross bomb, excit (SmiRN42) chem, excit (TenD47) daughter In^{112m} (SmiRN42, TenD47, GoldsG50)	β^- 0.66 max β^+ 1.56 max γ Cd X-rays, 0.511 (44%, $\gamma^{\pm}$), 0.617 (6%)	Ag^{109}(α, n) (FukS65, SmiRN42, TenD47, RuaJ62a, KatoT62a)
In^{112m}	20.7 m (BleE53) others (RuaJ62a, GirR59i, BarnS39a, TenD47)	☢ IT (SmiRN42, TenD47) Δ −87.83 (LHP, MTW)	A chem (BarnS39a) chem, cross bomb, excit (SmiRN42) chem, excit (TenD47) parent In^{112} (SmiRN42, TenD47, GoldsG50)	γ In X-rays, 0.156 (9%) e^- 0.128, 0.152 daughter radiations from In^{112}	Ag^{109}(α, n) (SmiRN42, TenD47, RuaJ62a, KatoT62a)
<u>In^{113}</u>		% 4.23 (WhiJ48) 4.33 (WhiF56) Δ −89.34 (MTW) σ_c 4 (to In^{114}) 8 (to In^{114m}) (GoldmDT64)			
In^{113m}	99.8 m (GleG64) 104 m (LawJL40) 103 m (GirR58) others (BarnS39a, CatR65)	☢ IT (BarnS39a) Δ −88.95 (LHP, MTW)	A chem, excit, genet (BarnS39a) daughter Sn^{113} (BarnS39a)	γ In X-rays, 0.393 (64%) e^- 0.365, 0.389	daughter Sn^{113} (GirR58, BarnS39a)
In^{114}	72 s (LawJL37, BarnS39a)	☢ β^- 98%, EC 1.9%, β^+ 0.004% (GrodL56) β^+ 0.0039% (DzhB57c) Δ −88.58 (MTW)	A excit (ChanW37, BotW37, LawJL37) n-capt, sep isotopes (GoldhM48a) daughter In^{114m} (GoldsG50)	β^- 1.988 max β^+ 0.42 max γ Cd X-rays, 1.299 (0.17%)	daughter In^{114m} (GoldsG50) In^{113}(n, γ) (GoldhM48a)
In^{114m}	50.0 d (WriH57) 50.1 d (CaliJ59) others (BendW58, BoeF49a, HoffK57, BarnS39a, MaiF49, LawJL40)	☢ IT 96.5%, EC 3.5% (GrodL56) Δ −88.39 (LHP, MTW)	A chem, n-capt, excit (LawJL37, MitA38) parent In^{114} (GoldsG50)	γ In X-rays, 0.192 (17%), 0.558 (3.5%), 0.724 (3.5%) e^- 0.164, 0.188 daughter radiations from In^{114}	In^{113}(n, γ) (LawJL37, MitA38, MaiF49)
<u>In^{115}</u>	6×10^{14} y sp act (MarteE50) 5.1×10^{14} y sp act (WatD62a) 7×10^{14} y sp act (BearG61a) others (CohS51)	☢ β^- (MarteE50, CohS51) % 95.77 (WhiJ48) 95.67 (WhiF56) Δ −89.54 (MTW) σ_c 45 (to In^{116}) 154 (to In^{116m_1}) 4 (to In^{116m_2}) (GoldmDT64)	A chem, sep isotopes (MarteE50)	β^- 0.48 max γ no γ	
In^{115m}	4.50 h (DunwJ47) 4.53 h (LawJL40) 4.48 h (SalS65)	☢ IT 95%, β^- 5% (LangeL52a) Δ −89.21 (LHP, MTW)	A chem, excit (GoldhM38) daughter Cd^{115} (GoldhM38, CorkJ39, NisY40, MetR51a, WahA52, LangeL52a)	β^- 0.83 max e^- 0.308, 0.331 γ In X-rays, 0.335 (50%)	Cd^{114}(n, γ) Cd^{115}(β^-) (GoldhM38, SehM62) In^{115}(n, n') (GoldhM38, CohS48) In^{115}(p, p') (BarnS39a, BarnS39) In^{115}(α, α') (LarK39)
In^{116}	13.4 s (DomF60) 14.0 s (DucA60) 14.5 s (CapP57) 15.6 s (BrzJ65) 13 s (AmaE35, CorkJ39, WilhZ53, LawJL37)	☢ β^- (LawJL37) Δ −88.20 (MTW)	A n-capt (AmaE35) excit, n-capt (LawJL37)	β^- 3.3 max γ 0.434 (0.12%), 0.95 (0.1%), 1.293 (1.2%)	In^{115}(n, γ) (AmaE35, LawJL37, SerL47b)

TABLE 1 (*Continued*)

Isotope Z A	Half-life	Type of decay (☢); % abundance; Mass excess (Δ=M−A), MeV (C¹²=0); Thermal neutron cross section (σ), barns	Class; Identification; Genetic relationships	Major radiations: approximate energies (MeV) and intensities	Principal means of production
$_{49}$In^{116m1}	54.0 m (LocE53, GravA47) 53.9 m (SilL51, DomF60) 55.1 m (CapP57) 57 m (BrzJ65)	☢ β⁻ (LawJL37) no IT, lim 0.5% (ColaJ60) Δ −88.14 (LHP, MTW)	A chem, n-capt (AmaE35) chem, excit, n-capt (LawJL37)	β⁻ 1.00 max γ 0.138 (3%), 0.417 (36%), 0.819 (17%), 1.09 (53%), 1.293 (80%), 1.508 (11%), 2.111 (20%)	In^{115}(n, γ) (AmaE35, MitA38a, SerL47b, HumV51, BolH64)
In^{116m2}	2.16 s (AlexKF63) 2.2 s (HecP61) 2.5 s (AlexKF60, FetP62a) 2.3 s (WhiW62)	☢ IT (AlexKF60, FetP62a) Δ −87.98 (LHP, MTW)	A n-capt, sep isotopes (AlexKF60, HecP61, FetP62a) excit, sep isotopes, cross bomb (WhiW62)	γ In X-rays, 0.164 e⁻ 0.138, 0.160	In^{115}(n, γ) (AlexKF60, HecP61, FetP62a, WhiW62, AlexKF63)
In^{117}	45 m (NeedJ63, BrzJ65) 38 m (DudN61) 43 m (WolfeJ61) others (MGinC55, CoryC53)	☢ β⁻ (MGinC55) Δ −88.93 (MTW)	A chem, genet (CoryC53) daughter Cd^{117m}, daughter Cd^{117} (TanC66, CoryC53) not parent Sn^{117m}, lim 1% (MGinC55) daughter In^{117m} (MGinC55)	β⁻ 0.74 max e⁻ 0.132 γ Sn X-rays, 0.158 (87%), 0.565 (100%)	Cd^{116}(n, γ) Cd$^{117, 117m}$ (β⁻); daughter Cd^{117m} (TanC66a)
In^{117m}	1.93 h (DudN61, BrzJ65) 1.96 h (NeedJ63) 1.90 h (MGinC55, MetR51b) 1.95 h (LawJL40) others (WolfeJ61, CoryC53)	☢ IT 47%, β⁻ 53% (TanC66b) IT 28%, β⁻ 72% (WolfeJ61) IT 22%, β⁻ 78% (MGinC55) Δ −88.61 (LHP, MTW)	A chem, excit (CorkJ39) daughter Cd^{117}, daughter Cd^{117m} (TanC66, MGinC55) parent In^{117} (MGinC55)	β⁻ 1.78 max e⁻ 0.286 γ In X-rays, 0.158 (14%), 0.314 (31%) daughter radiations from In^{117}	Cd^{116}(n, γ) Cd$^{117, 117m}$ (β⁻) (TanC66a)
In^{118}	5.7 s (BrzJ65) 5.0 s (KantJ64a) 5.1 s (GleC61)	☢ β⁻ (CoryC53) Δ −87.5 (MTW)	B genet (CoryC53) chem, genet energy levels (GleC61) excit, sep isotopes (KantJ64a) daughter Cd^{118} (CoryC53, GleC61)	β⁻ 4.2 max γ 1.230 (15%)	daughter Cd^{118} (CoryC53, GleC61) Sn^{118}(n, p) (KantJ64a)
In^{118}	4.35 m (KantJ64a) 4.5 m (WilhZ53, DufR49a) 4.7 m (MeyP65) 4.9 m (BrzJ65)	☢ β⁻ (DufR49a) Δ −87.4 (KantJ64a, MTW)	B excit, sep isotopes (DufR49a) excit, sep isotopes, genet energy levels (KantJ64a) not daughter Cd^{118} (CoryC53, GleC61)	β⁻ 2.0 max γ 0.69 (41%), 1.05 (80%), 1.230 (97%), 2.04 (3%)	Sn^{118}(n, p) (KantJ64a)
In^{119}	2.1 m (KuoC60) 2.0 m (GleC61a) 2.3 m (YutH60) 2.8 m (BrzJ65)	☢ β⁻ (KuoC60, YutH60, GleC61a) Δ −87.6 (MTW)	B sep isotopes, excit (KuoC60, YutH60) chem, genet (GleC61a) daughter In^{119m} (GleC61a) daughter 2.7 m Cd^{119} (GleC61a)	β⁻ 1.6 max γ 0.82 (95%)	Sn^{120}(γ, p) (KuoC60, YutH60) daughter In^{119m}, fission (GleC61a)
In^{119m}	17.5 m (KuoC60) 18 m (DufR49a, GleC61a) 22.6 m (BrzJ65)	☢ β⁻ 95%, IT 5% (GleC61a) Δ −87.3 (LHP, MTW)	B chem, excit, sep isotopes (DufR49a) parent In^{119} (GleC61a) daughter 10 m Cd^{119} (NusN57, GleC61a) daughter 2.7 m Cd^{119} (GleC61a)	β⁻ 2.7 max γ [In X-rays, Sn L X-rays], 0.024, 0.30, 0.91 (doublet) daughter radiations from In^{119}	Sn^{120}(γ, p) (DufR49b, KuoC60) fission (GleC61a)
In^{120}	3.2 s (KantJ64a) 3 s (PouA60)	☢ β⁻ (KantJ64a) Δ −86 (KantJ64a, MTW)	B sep isotopes, cross bomb (PouA60)	β⁻ 5.6 max γ 1.171 (15%)	Sn^{120}(n, p) (PouA60, KantJ64a) Sb^{123}(n, α) (PouA60)
In^{120}	44 s (KantJ64a) 48 s (MeyP65) 50 s (PouA60) ≈55 s (MGinC58)	☢ β⁻ (PouA60) Δ −85.8 (KantJ64a, MTW)	B excit (MGinC58) sep isotopes, genet energy levels (PouA60)	β⁻ 3.1 max γ 0.090 (12%), 0.198 (9%), 0.71 (12%), 0.86 (34%), 0.94 (12%), 1.02 (61%), 1.171 (100%), 1.28 (14%), 1.47 (6%), 1.87 (7%), 2.01 (6%)	Sn^{120}(n, p) (MGinC58, PouA60, KantJ64a)
In^{121}	30 s (YutH60)	☢ [β⁻] (YutH60) Δ −86 (MTW)	C excit, sep isotopes (YutH60)	γ 0.94	Sn^{122}(γ, p) (YutH60)
In^{121}	3.1 m (YutH60, WeisH65a)	☢ β⁻ (YutH60) Δ −86 (MTW)	C excit, sep isotopes (YutH60)	β⁻ 3.7 max	Sn^{122}(γ, p) (YutH60)

TABLE 1 (*Continued*)

Isotope Z A	Half-life	Type of decay (☢); % abundance; Mass excess (Δ≡M-A), MeV (C¹²=0); Thermal neutron cross section (σ), barns	Class; Identification; Genetic relationships	Major radiations: approximate energies (MeV) and intensities	Principal means of production
$_{49}$In$^{(121?)}$	11.5 m (NusN57)	☢ β⁻ (NusN57)	G chem, genet (NusN57) daughter 3.5 m Cd$^{(121?)}$ (NusN57) Assignment probably incorrect (NDS, YutH60)	γ 0.85	deuterons on Sn (NusN57)
In$^{(121?)}$	32 m (NusN57)	☢ β⁻ (NusN57)	G chem, genet (NusN57) daughter 3.5 m Cd$^{(121?)}$ (NusN57) Assignment probably incorrect (NDS, YutH60)	γ 0.52	deuterons on Sn (NusN57)
In^{122}	8 s (KantJ63a)	☢ β⁻ (KantJ63a) Δ -83 (MTW)	B sep isotopes, genet energy levels (KantJ63a)	β⁻ 5 max γ 0.99, 1.14	Sn^{122}(n, p) (KantJ63a)
In^{123}	36 s (YutH60)	☢ β⁻ (YutH60) Δ -83 (MTW)	E excit, sep isotopes (YutH60)	β⁻ 4.6 max	Sn^{124}(γ, p) (YutH60)
In^{123}	10 s (YutH60)	☢ [β⁻] (YutH60) Δ -83 (MTW)	F excit, sep isotopes (YutH60) May be identical to 8 s In^{122} (LHP)	γ 1.1	Sn^{124}(γ, p) (YutH60)
In^{124}	≈3.6 s (KarrM64)	☢ β⁻ (KarrM64) Δ -81 (MTW)	B sep isotopes, genet energy levels (KarrM64)	β⁻ 5 max γ 0.99 (↑ 3), 1.13 (↑ 10), 3.21 (↑ 3)	Sn^{124}(n, p) (KarrM64)
$_{50}$Sn^{108}	9.2 m (HahR65) 9 m genet (MeaS55)	☢ [EC] (MeaS55)	A genet (MeaS55) chem, excit (HahR65) parent 39 m In^{108} (MeaS55)	γ In X-rays, 0.28, 0.42 daughter radiations from 39 m In^{108}	Cd^{106}(α, 2n) (HahR65)
Sn^{109}	18.1 m (PetrM56a)	☢ EC, β⁺ (PetrM56a)	B chem, genet (PetrM56a) ancestor In^{109}, parent In^{109m} (PetrM56a)	β⁺ 1.6 max e⁻ 0.305, 0.491, 0.86, 1.09 γ In X-rays, 0.335, 0.521, 0.89, 1.12 daughter radiations from In^{109m}, In^{109}	Cd^{106}(α, n) (PetrM56a)
Sn^{110}	4.0 h (MeaS55, MGinC51) 4.5 h (MallE49)	☢ EC (MallE49)	A chem, sep isotopes (MallE49) chem, genet (MeaS55, NaiT64) parent 67 m In^{110}, not parent 4.9 h In^{110} (MeaS55, NaiT64)	γ In X-rays, 0.283 (95%) e⁻ 0.255 daughter radiations from 67 m In^{110}	In^{115}(p, 6n) (NaiT64) Cd^{108}(α, 2n) (MeaS55, MallE49)
Sn^{111}	35.0 m (HinR49) 35 m (MGinC51, SnyJ65)	☢ EC 73%, β⁺ 27% (SnyJ65) EC 71%, β⁺ 29% (MGinC51) Δ -85.6 (MTW)	A chem, sep isotopes (HinR49) excit, cross bomb (SnyJ65)	β⁺ 1.51 max γ In X-rays, 0.511 (54%, γ±), 0.75 (1.1%), 0.97 (0.7%), 1.14 (1.8%), 1.54 (0.5%), 1.59 (0.6%) (0.9%), 1.89 (1.0%), 2.11 (0.3%), 2.32 (0.2%) daughter radiations from In^{111}	Cd^{110}(α, 3n) (MGinC51)
Sn^{112}		% 0.95 (BaiK50) Δ -88.64 (MTW) σ$_c$ 0.9 (to Sn^{113}) 0.4 (to Sn^{113m}) (GoldmDT64)			
Sn^{113}	115 d (GleG64) 118 d (CorkJ51f) 119 d (AviP56) 130 d (GardG56) others (DesY53, BarnS39a)	☢ EC, no β⁺ (BarnS39a) Δ -88.32 (MTW)	A chem, excit (BarnS39a, LivJ39b) parent In^{113m} (BarnS39a)	γ In X-rays, 0.255 (1.8%) daughter radiations from In^{113m}	Sn^{112}(n, γ) (NelC50, CorkJ51f, SerL47b, BoweJ51) In^{113}(p, n) (BarnS39a) In^{113}(d, 2n) (ColeK47, GirR58)
Sn^{113m}	20 m (SchmM61) 27 m (SelI60)	☢ IT 91%, EC 9%, no β⁺, lim 10^{-3}% (SchmM61) Δ -88.24 (LHP, MTW)	A chem, genet (SelI60) crit abs (SchmM61) daughter Sb^{113} (SelI60)	γ Sn X-rays, In X-rays, 0.079 (0.6%) e⁻ 0.050, 0.075	Sn^{112}(n, γ) (SchmM61) Sn^{112}(d, n)Sb^{113}(EC), Sn^{114}(p, 2n)Sb^{113}(EC) (SelI60, SelI59)
Sn^{114}		% 0.65 (BaiK50) Δ -90.57 (MTW)			

TABLE 1 (*Continued*)

Isotope Z A	Half-life	Type of decay (☢); % abundance; Mass excess (△≡M–A), MeV (C¹²=0); Thermal neutron cross section (σ), barns	Class; Identification; Genetic relationships	Major radiations: approximate energies (MeV) and intensities	Principal means of production
$_{50}Sn^{115}$		% 0.34 (BaiK50) △ −90.03 (MTW)			
Sn^{116}		% 14.24 (BaiK50) △ −91.523 (MTW) σ_c 0.006 (to Sn¹¹⁷m) (GoldmDT64)			
Sn^{117}		% 7.57 (BaiK50) △ −90.392 (MTW)			
Sn^{117m}	14.0 d (CorkJ51f, MihJ50)	☢ IT (MallE50) △ −90.075 (LHP, MTW)	A chem (LivJ39b) chem, sep isotopes, cross bomb (MallE50) not daughter In¹¹⁷ (MGinC55)	γ Sn X-rays, 0.158 (87%) e⁻ 0.130, 0.155	Sn¹¹⁶(n, γ) (MihJ50) Cd¹¹⁴(α, n) (LivJ39b)
Sn^{118}		% 24.01 (BaiK50) △ −91.652 (MTW) σ_c 0.01 (to Sn¹¹⁹m) (GoldmDT64)			
Sn^{119}		% 8.58 (BaiK50) △ −90.062 (MTW)			
Sn^{119m}	≈250 d (MihJ50)	☢ IT (MihJ50) △ −89.973 (LHP, MTW)	A chem, n-capt, sep isotopes (MihJ50)	γ Sn X-rays, 0.024 (16%) e⁻ 0.020, 0.026, 0.061	Sn¹¹⁸(n, γ) (MihJ50, NelC50, SchaG51a, BoweJ51)
Sn^{120}		% 32.97 (BaiK50) △ −91.100 (MTW) σ_c 0.14 (to Sn¹²¹) ≈0.001 (to Sn¹²¹m) (GoldmDT64)			
Sn^{121}	27.5 h (NelC50) 27 h (MajN63) others (LeeJ49, LivJ39b)	☢ β⁻ (LivJ39b) △ −89.21 (MTW)	A chem, excit (LivJ39b) chem, sep isotopes (LindnM48) descendant 13 s Cd¹²¹ (WeisH65)	β⁻ 0.383 max	Sn¹²⁰(n, γ) (LeeJ49, DufR49c, NelC50, LivJ39b, SerL47b) Sb¹²³(d, α) (LindnM50a)
Sn^{121m}	76 y (FlyK65a) ≈25 y (DroB62)	☢ β⁻ (NelC50) △ −89.14 (LHP, MTW)	D sep isotopes, n-capt (NelC50) chem (DroB62)	β⁻ 0.42 max e⁻ [0.007, 0.033] γ Sb X-rays, 0.037	Sn¹²⁰(n, γ) (NelC50, SnyR65) fission (DroB62)
Sn^{122}		% 4.71 (BaiK50) △ −89.943 (MTW) σ_c 0.001 (to Sn¹²³) 0.2 (to Sn¹²³m) (GoldmDT64)			
Sn^{123}	125 d (CorkJ51f) 130 d (LeeJ49, LeadG51) 126 d (NelC50) 136 d (GrumW46)	☢ β⁻ (LeadG51) △ −87.80 (MTW)	A chem (LeadG46, LeadG51) chem, sep isotopes, cross bomb (LeeJ49)	β⁻ 1.42 max γ 1.08 ? (weak)	Sn¹²²(n, γ) (LeeJ49, NelC50)
Sn^{123m}	39.5 m (DufR49c) 40 m (LivJ39b, LeeJ49, NelC50, MajN63) 41.5 m (MocD48)	☢ β⁻ (LivJ39b) △ −87.78 (LHP, MTW)	A chem (LivJ39b) chem, sep isotopes, excit (LeeJ49, NelC50)	β⁻ 1.26 max e⁻ [0.130] γ Sb X-rays, 0.160 [84%]	Sn¹²²(n, γ) (SerL47b, DufR49c, LeeJ49, NelC50) Sn¹²⁴(n, 2n) (PoolM37, LeeJ49)
Sn^{124}	t_{1/2} (ββ) >2 × 10¹⁷ y sp act (KalkM52, FireE52, HogB52)	% 5.98 (BaiK50) △ −88.237 (MTW) σ_c 0.004 (to Sn¹²⁵) 0.1 (to Sn¹²⁵m) (GoldmDT64)			

TABLE 1 (*Continued*)

Isotope Z A	Half-life	Type of decay (☢); % abundance; Mass excess ($\triangle \triangleq M-A$), MeV ($C^{12}=0$); Thermal neutron cross section (σ), barns	Class; Identification; Genetic relationships	Major radiations: approximate energies (MeV) and intensities	Principal means of production
$_{50}Sn^{125}$	9.4 d (NelC50) 10.0 d (LeeJ49)	☢ β^- (LivJ39b) $\triangle$ -85.93 (MTW)	A chem (LivJ39b) chem, excit, sep isotopes (LeeJ49) chem, sep isotopes, n-capt, genet (NelC50) parent Sb^{125} (NelC50)	β^- 2.34 max γ 0.342 (0.3%), 0.468 (0.4%), 0.811 (1.5%), 0.904 (1.4%), 1.068 (4%), 1.17 (0.14%), 1.41 (0.14%), 1.97 (0.6%), 2.23 (0.05%) daughter radiations from Sb^{125}	$Sn^{124}(n, \gamma)$ (LeeJ49, NelC50, LivJ39b, SerL47b)
Sn^{125m}	9.5 m (NelC50) 9.8 m (LeeJ49) 9.7 m (MajN63)	☢ β^- (LivJ39b) $\triangle$ -85.91 (LHP, MTW)	A chem, excit, n-capt (LivJ39b) chem, sep isotopes (DufR50a, LeeJ49)	β^- 2.04 max γ 0.325 (97%)	$Sn^{124}(n, \gamma)$ (LeeJ49, NelC50, DufR50a, LivJ39b, SerL47b)
Sn^{126}	$\approx 10^5$ y yield (DroB62)	☢ $[\beta^-]$ (DroB62) $\triangle$ -86 (MTW)	B chem, genet (DroB62) parent 19 m Sb^{126}, ancestor 12.5 d Sb^{126} (DroB62)	γ 0.060, 0.067, 0.092	fission (DroB62)
Sn^{126}	~50 m yield (BarnJ51)	☢ β^- (BarnJ51)	G chem, genet (BarnJ51) reassigned to Sn^{128} (DroB62)		fission (BarnJ51)
Sn^{127}	2.05 h (CarmH56) 2.10 h (UhlJ62) 2.2 h (DroB62, HageE62) others (DMarP62, MajN63)	☢ β^- (BarnJ51) $\triangle$ -84 (MTW)	A chem, genet (BarnJ51, CarmH56, DroB62, HageE62) chem, mass spect (UhlJ62) parent Sb^{127} (BarnJ51, CarmH56, DroB62, HageE62)	β^- 1.45 max ? γ 0.44, 0.49, 0.82, 1.10, 2.00, 2.32, 2.58, 2.68, 2.82 daughter radiations from Sb^{127}	fission (BarnJ51, DroB62, HageE62, UhlJ62) $Te^{130}(n, \alpha)$ (CarmH56, MajN63)
Sn^{127}	4.1 m (KauP65) 4.6 m genet (HageE62) ≈ 2.5 m genet (DroB62)	☢ β^- (KauP65) $\triangle$ -83.5 (KauP65, MTW)	A chem, genet (HageE62, DroB62) chem, sep isotopes (KauP65) parent Sb^{127} (HageE62, DroB62)	β^- 2.7 max γ 0.49 (100%)	fission (HageE62, DroB62) $Te^{130}(n, \alpha)$ (KauP65)
Sn^{128}	59 m (UhlJ62) 57 m (FranI55, HageE62) 62 m (DMarP62) 58 m (DroB62)	☢ β^- (DMarP62) $\triangle$ -83.4 (MTW)	A chem, genet (FranI55, HageE62, DroB62) chem, mass spect (UhlJ62) parent 11 m Sb^{128} (FranI55, DroB62, HageE62, UhlJ62, DMarP62) ancestor 9 h Sb^{128} ($\approx 3\%$) (FranI56, DroB62) not ancestor 9 h Sb^{128}, lim 5% (HageE62)	β^- 0.80 max γ Sb X-rays, 0.044 (7%), 0.072 (19%), 0.50 (61%), 0.57 (22%) daughter radiations from 11 m Sb^{128}	fission (FranI55, DroB62, HageE62, DMarP62, UhlJ62)
Sn^{129}	9 m genet (HageE62, DroB62) 6 m (DroB62)	☢ $[\beta^-]$ (HageE62, DroB62)	B chem (DroB62) chem, genet (HageE62) parent Sb^{129} (HageE62)	γ 1.15, others daughter radiations from Sb^{129}	fission (HageE62, DroB62)
Sn^{129}	1.0 h genet (HageE62)	☢ $[\beta^-]$ (HageE62)	B chem, genet (HageE62) parent Sb^{129} (HageE62)	daughter radiations from Sb^{129}	fission (HageE62)
Sn^{130}	2.6 m (PapA56)	☢ $[\beta^-]$ (PapA56)	D chem, genet (PapA56) parent 7 m Sb^{130} (PapA56, DroB62) not parent 35 m Sb^{130}, lim 10% (DroB62)	daughter radiations from 7.1 m Sb^{130}	fission (PapA56, FranI55, DroB62)
Sn^{131}	3.4 m (PapA56) <2 m (DroB62)	☢ $[\beta^-]$ (PapA56)	E chem, genet (PapA56) activity not observed (DroB62) parent Sb^{131} (PapA56)		fission (PapA56)
Sn^{132}	2.2 m genet (PapA56)	☢ $[\beta^-]$ (PapA56)	B chem, genet (PapA56) parent Sb^{132} (PapA56)		fission (PapA56)
$_{51}Sb^{112}$	0.9 m (SelI59)	☢ β^+, EC (SelI59)	B chem, excit (SelI59)	γ Sn X-rays, 0.511 ($\gamma^{\pm}$), 1.27	$Sn^{112}(p, n)$ (SelI59)

TABLE 1 (*Continued*)

Isotope Z A	Half-life	Type of decay (☢); % abundance; Mass excess (△=M-A), MeV (C¹²=0); Thermal neutron cross section (σ), barns	Class; Identification; Genetic relationships	Major radiations: approximate energies (MeV) and intensities	Principal means of production
$_{51}Sb^{113}$	6.4 m (PatA62) 7 m (SeII58, SeII59)	☢ EC, β⁺ (SeII58, SeII59, SeII60) △ -83.85 (MTW)	A chem (RhoA57) chem, excit, sep isotopes cross bomb (SeII60, SeII59, SeII58) excit, sep isotopes (PatA62) parent Sn¹¹³ᵐ (SeII60)	β⁺ 2.42 max γ Sn X-rays, 0.32, 0.511 (γ±), 0.6-0.9 (complex), 1.03, 1.2 (complex), 1.52 ? daughter radiations from Sn¹¹³ᵐ	Sn¹¹²(d,n) (SeII58, SeII60, RhoA57) Sn¹¹⁴(p,2n) (SeII59)
Sb^{114}	3.3 m (SeII59)	☢ β⁺, EC (SeII59) △ -84.3 (MTW)	B chem, excit, sep isotopes (SeII59)	β⁺ 2.7 max γ Sn X-rays, 0.9, 1.30	Sn¹¹⁴(p,n), Sn¹¹⁵(p,2n) (SeII59)
Sb^{115}	31 m (SeII58, SeII59) 36 m (FinR61) 32 m (SehM62)	☢ EC 67%, β⁺ 33% (VarN63) EC 65%, β⁺ 35% (SeII60) EC 88%, β⁺ 12% (SehM62) △ -87.00 (MTW)	A chem (RhoA57) chem, sep isotopes, excit, cross bomb (SeII58, SeII59, SeII61) chem, mass spect (FinR61) daughter Te¹¹⁵ (SeII60a, ReisR65)	β⁺ 1.51 max γ Sn X-rays, 0.499 (100%), 0.511 (67%, γ±), 0.98 (5%), 1.24 (5%), 2.22 (1%)	Sn¹¹⁴(d,n) (SeII58, SeII61) Sn¹¹⁶(p,2n) (SeII59) In¹¹³(α,2n) (SehM62)
Sb^{116}	16 m (StahP53a) 14 m (AteA54) 15 m (KuzM58)	☢ EC 72%, β⁺ 28% (FinR61) △ -87.0 (MTW)	A chem, excit (StahP53a) genet (FinR61) daughter Te¹¹⁶ (FinR61)	β⁺ 2.3 max γ Sn X-rays, 0.511 (γ±, 56%), 0.93 (26%), 1.293 (85%), 2.23 (14%)	daughter Te¹¹⁶ (FinR61) In¹¹⁵(α,3n) (AteA54)
Sb^{116m}	60 m (TemG49, AteA54)	☢ EC 81%, β⁺ 19% (BolH64a) △ -86.5 (LHP, MTW)	A chem, excit, mass spect (TemG49) not daughter Te¹¹⁶ (FinR61)	β⁺ 1.16 max e⁻ 0.070, 0.095, 0.111 γ Sn X-rays, 0.099 (30%), 0.140 (30%), 0.406 (36%), 0.511 (38%, γ±), 0.545 (68%), 0.96 (75%), 1.06 (27%), 1.293 (100%)	In¹¹⁵(α,3n) (TemG49) In¹¹³(α,n) (JensB60)
Sb^{117}	2.8 h (FinR61, ColeK47, TemG49, KuzM58)	☢ EC 97.4%, β⁺ 2.6% (MGinC55) EC 97.7%, β⁺ 2.3% (BaskK64) △ -88.57 (MTW)	A chem (LivJ39) chem, excit, mass spect (TemG49) daughter Te¹¹⁷ (FinR61)	β⁺ 0.57 max γ Sn X-rays, 0.158 (87%), 0.511 (5%, γ±)	In¹¹⁵(α,2n) (TemG49)
Sb^{117m}	1.6 x 10⁻⁴ s delay coinc (GhoA63)	☢	F crit abs (GhoA63) same as 0.726 level of Sn¹¹⁵?	γ 0.080 (↑ 10), 0.17 (↑ 8), 0.24 (↑ 9), 0.46 (↑ 24) scint spect (GhoA63)	protons on Sb (GhoA63) not produced by protons on Sn (GritV65a)
Sb^{118}	3.5 m (LindnM48, FinR61) 3.6 m (RisJ40)	☢ EC, β⁺ (FinR61) △ -87.96 (MTW)	A excit (RisJ40) chem (LarK39) genet (FinR61, LindnM48) daughter Te¹¹⁸ (LindnM48, LindnM50a, FinR61)	β⁺ 2.67 max γ Sn X-rays, 0.511 (150%, γ±), 0.83 (0.4%), 1.230 (3%, doublet)	daughter Te¹¹⁸ (LindnM48a, FinR61) In¹¹⁵(α,n) (LarK39, RisJ40)
Sb^{118m_1}	5.1 h (ColeK47, TemG49)	☢ EC 99+%, β⁺ 0.16% (BolH61) no β⁺, lim 0.1% (JensB60) △ -87.77 (LHP, MTW)	A chem, cross bomb (ColeK47) chem, excit, mass spect (TemG49) not daughter Te¹¹⁸ (FinR61)	γ Sn X-rays, 0.041 (29%), 0.254 (93%), 1.049 (100%), 1.230 (100%) e⁻ 0.012, 0.036, 0.223	In¹¹⁵(α,n) (ColeK47, TemG49, BolH61, RamasM61a, BodE62a)
Sb^{118m_2}	0.87 s (WhiW62)	☢ [IT] (WhiW62)	E excit (WhiW62)	γ 0.14 (↑ 4), 0.30 (↑ 10), 0.38 (↑ 10)	protons on Sb (WhiW62)
Sb^{119}	38.0 h (OlsJ57) others (ZaitN60a, ColeK47, LindnM48)	☢ EC (ColeK47) △ -89.48 (MTW)	A chem, cross bomb (ColeK47) chem, genet energy levels (OlsJ57) daughter Te¹¹⁹ᵐ (LindnM48, LindnM50a, FinR61) daughter Te¹¹⁹ (FinR61)	γ Sn X-rays, 0.024 (16%) e⁻ 0.020	Sb¹²¹(p,3n)Te¹¹⁹(EC) (FinR61) Sn¹¹⁹(p,n), Sn¹¹⁸(d,n) (ColeK47)
Sb^{120}	15.89 m (EbrT65) 16.4 m (JohnH50) 16.6 m (PerlmM48, StahP53a) 17 m (HeyF37, LivJ38c)	☢ β⁺, EC (BlasJ50) △ -88.42 (MTW)	A chem, excit (BotW39, HeyF37, ChanW37) chem, excit, cross bomb (LivJ37)	β⁺ 1.70 max γ Sn X-rays, 0.511 (87%, γ±), 1.171 (1.3%)	Sn¹²⁰(p,n) (BlasJ50) Sn¹²⁰(d,2n) (LindnM48) Sn¹¹⁹(d,n) (LivJ39)
Sb^{120}	5.8 d (MGinC55a) 6.0 d (LindnM48)	☢ EC (LindnM48) no β⁺ or IT, lim 0.3% (MGinC55a) △ -88.42 (MTW)	A chem, sep isotopes (LindnM48) chem, cross bomb (MGinC55a) chem, mass spect (JensB60)	γ Sn X-rays, 0.090 (81%), 0.200 (88%), 1.03 (99%), 1.171 (100%) e⁻ 0.061, 0.096, 0.171, 0.196	Sn¹¹⁹(d,n) (JensB60) Sn¹²⁰(d,2n) (LindnM48)

TABLE 1 (*Continued*)

Isotope Z A	Half-life	Type of decay (☣); % abundance; Mass excess (Δ≡M-A), MeV (C¹²=0); Thermal neutron cross section (σ), barns	Class; Identification; Genetic relationships	Major radiations: approximate energies (MeV) and intensities	Principal means of production
$_{51}$Sb^{121}		% 57.25 (WhiJ48) Δ -89.593 (MTW) σ$_c$ 6 (to Sb^{122}) 0.06 (to Sb^{122m}) (GoldmDT64)			
Sb^{122}	2.80 d (BlasJ51a) 2.75 d (CorkJ54) 2.73 d (PerlmM58)	☣ β⁻ 97%, EC 3.0%, β⁺ 0.006% (GlauM55, PerlmM58) β⁻ 97%, EC 3.1% (FarrB55) Δ -88.32 (MTW)	A chem (AmaE35) chem, cross bomb (LivJ39)	β⁻ 1.97 max β⁺ 0.56 max γ Sn X-rays, 0.564 (66%), 0.686 (3.4%), 1.140 (0.7%), 1.26 (0.7%)	Sb^{121}(n, γ) (AmaE35, LivJ39, SerL47b, HumV51)
Sb^{122m}	4.2 m (DMatE63, EngeR62) others (DMatE47, VanJ62)	☣ IT (DMatE47) no β⁺, no β⁻, lim 0.5% (DMatE62) Δ -88.16 (LHP, MTW)	A chem, n-capt, sep isotopes (DMatE47)	γ Sb X-rays, 0.061 (50%), 0.075 (17%) e⁻ 0.021, 0.030, 0.045, 0.056, 0.071	Sb^{121}(n, γ) (DMatE51)
$\underline{Sb^{123}}$	t$_{1/2}$ >1.3 x 10^{16} y sp act (WatD62a)	% 42.75 (WhiJ48) Δ -89.224 (MTW) σ$_c$ 3.3 (to Sb^{124}) 0.03 (to Sb^{124m}1) 0.015 (to Sb^{124m}2) (GoldmDT64)			
Sb^{124}	60.4 d (MackR57a) 60.9 d (WriH57) 60.1 d (CaliJ59) 59.9 d (JohnCH58) others (BrzJ65)	☣ β⁻ (LivJ39) no EC, no β⁺ (LangeL50c) Δ -87.58 (MTW) σ$_c$ 2000 (GoldmDT64)	A chem (LivJ37) chem, excit, cross bomb (LivJ39)	β⁻ 2.31 max γ 0.603 (97%), 0.644 (7%), 0.72 (14%, doublet), 0.967 (2.4%), 1.048 (2.4%), 1.31 (3%, doublet), 1.37 (5%, doublet), 1.45 (2%), 1.692 (50%), 2.088 (7%)	Sb^{123}(n, γ) (LivJ39, SerL47b)
Sb^{124m}1	93 s (VanJ62a) 96 s (BrzJ63, BrzJ65) ≈78 s (DMatE47)	☣ IT 80%, β⁻ 20% (VanJ62a) Δ -87.57 (LHP, MTW)	A chem, n-capt, sep isotopes (DMatE47) genet energy levels (VanJ62a) daughter Sb^{124m}2 (VanJ62a)	β⁻ 1.19 max e⁻ 0.006, 0.009 γ Sb L X-rays, 0.505 (20%), 0.603 (20%), 0.644 (20%)	Sb^{123}(n, γ) (VanJ62a, DMatE47)
Sb^{124m}2	21 m (VanJ62a, DMatE47, BrzJ65)	☣ IT (VanJ62a) Δ -87.55 (LHP, MTW)	A chem, n-capt, sep isotopes (DMatE47) genet (VanJ62a) parent Sb^{124m}1 (VanJ62a)	e⁻ 0.021, 0.024 γ Sb L X-rays daughter radiations from Sb^{124m}1	Sb^{123}(n, p) (VanJ62a, DMatE57)
Sb^{125}	2.71 y (FlyK65a) 2.78 y (WyaE61) 2.6 y (KlehE60) 2.0 y (LazN56a) others (LeadG51a)	☣ β⁻ (CamG51) Δ -88.28 (MTW) σ$_c$ <20 (GoldmDT64)	A chem (LivJ39) chem, n-capt (StanlC51) daughter Sn^{125} (NelC50) parent Te^{125m} (FrieG48, KerB49)	β⁻ 0.61 max e⁻ 0.004, 0.030, 0.144, 0.395 γ Te X-rays, 0.176 (6%), 0.427 (31%), 0.463 (10%), 0.599 (24%, doublet), 0.634 (11%), 0.66 (3%, doublet) daughter radiations from Te^{125m}	Sn^{124}(n, γ)Sn^{125}(β⁻) (SiegK49, FrieG48, StanlC51)
Sb^{126}	12.5 d (DroB62) others (GrumW46, BarnJ51)	☣ β⁻ (DroB62) Δ -86.3 (MTW)	B chem, genet (DroB62) descendant Sn^{126} (DroB62)	β⁻ 1.9 max γ 0.41, 0.69 (complex, 3 γ rays)	fission, descendant Sn^{126} (DroB62)
Sb^{126}	19.0 m (DroB62) 19 m (FranI56a, FranI58)	☣ β⁻ (FranI56a) β⁻, [IT] (DroB62)	B chem (FranI56a) chem, sep isotopes (FranI58) chem, genet (DroB62) daughter Sn^{126} (DroB62)	β⁻ 1.9 max γ 0.41, 0.67 (complex, 2 γ rays)	Te^{126}(n, p) (FranI56a, FranI58) fission, daughter Sn^{126} (DroB62)
Sb^{126}	9 h (BarnJ51)	☣ β⁻ (BarnJ51)	G chem, excit (BarnJ51) reassigned to Sb^{128} (DroB62)		fission (BarnJ51)
Sb^{127}	93 h (DroB62, SeiJ51b) 94 h (UhlJ62) 88 h (BosH57) 95 h (GrumW46) others (AbeP39)	☣ β⁻ (AbeP39) Δ -86.70 (MTW)	A chem, genet (AbeP39) chem, mass spect (UhlJ62) parent Te^{127} (AbeP39, GleL51h) parent Te^{127} (84%), parent Te^{127m} (16%) (BeydJ48) daughter 2.1 h Sn^{127} (BarnJ51, CarmH56, DroB62, HageE62) daughter 4 m Sn^{127} (HageE62, DroB62)	β⁻ 1.5 max γ 0.060, 0.25, 0.41, 0.46, 0.68, 0.77, 0.92, 1.10, 1.34 daughter radiations from Te^{127}, Te^{127m}	fission, daughter Sn^{127} (AbeP39, SieN51b, GrumW46, BarnJ51, DroB62, UhlJ62, KatcS48)

TABLE 1 (*Continued*)

Isotope Z A	Half-life	Type of decay (☢); % abundance; Mass excess (△≡M–A), MeV (C¹²=0); Thermal neutron cross section (σ), barns	Class; Identification; Genetic relationships	Major radiations: approximate energies (MeV) and intensities	Principal means of production
$_{51}Sb^?$	6.2 d (BosH57)	☢ (BosH57)	G chem (BosH57) activity is probably a mixture of Sb^{126} (12.5 d) and Sb^{127} (LHP)		fission (BosH57)
Sb^{128}	10.8 m (DMarP62) 10.1 m (DroB62) 10.3 m (FranI56, BrzJ63) 10.7 m (HageE62) others (UhlJ62, BarnJ51, BrzJ65)	☢ β⁻ (FranI55) △ -84.7 (MTW)	A chem (FranI56) chem, sep isotopes, genet energy levels (HageE62) chem, mass spect (UhlJ62) daughter Sn^{128} (FranI55, DroB62, HageE62, UhlJ62, DMarP62)	β⁻ 2.6 max γ 0.320 (83%), 0.75 (200%, doublet), 1.07 (4%)	fission, daughter Sn^{128} (FranI55, DroB62, HageE62, UhlJ62, DMarP62) Te^{128}(n, p) (HageE62, BrzJ63)
Sb^{128}	8.6 h (UhlJ62) 8.9 h (DroB62) 9.6 h (FranI56, BrzJ65) 9.9 h (HageE62)	☢ β⁻ (FranI56)	A chem (FranI56) chem, sep isotopes, genet energy levels (HageE62) chem, mass spect (UhlJ62) descendant Sn^{128} (FranI56, DroB62) not descendant Sn^{128} (HageE62)	β⁻ 1 max γ 0.314, 0.53, 0.64, 0.75 (complex)	fission (FranI55, UhlJ62) Te^{128}(n, p) (HageE62)
Sb^{129}	4.3 h (UhlJ62) 4.2 h (DroB62, AbeP39, Vasil58)	☢ β⁻ (AbeP39) △ -85 (MTW)	A chem, genet (AbeP39) chem, mass spect (UhlJ62) parent Te^{129} (AbeP39) daughter 9 m Sn^{129}, daughter 1.0 h Sn^{129} (HageE62)	β⁻ 1.87 max γ 0.073, 0.34, 0.460, 0.540, 0.81, 0.91, 1.04, 1.24 daughter radiations from Te^{129}	fission (AbeP39, KatcS48, DroB62, HageE62, UhlJ62)
Sb^{130}	33 m (HageE62) 36 m (BrzJ63, BrzJ65) 37 m (DroB62) others (BarnJ62)	☢ β⁻ (BarnJ52) △ -82 (MTW)	A chem, excit (fission yield) (BarnJ52) chem, sep isotopes (HageE62, BrzJ63) chem, genet energy levels (DroB62) not daughter Sn^{130}, lim 10% (DroB62)	γ 0.19, 0.33, 0.82 (complex), 0.94	Te^{130}(n, p) (HageE62, BrzJ63) fission (BarnJ52, DroB62)
Sb^{130}	7.1 m (HageE62) 6 m (DroB62) 10 m (BarnJ52) 12 m (BrzJ65)	☢ β⁻ (BarnJ52) △ -82 (MTW)	A chem (PapA56, BarnJ52) chem, sep isotopes (HageE62, BrzJ63) chem, genet energy levels (DroB62) daughter Sn^{130} (PapA56, DroB62)	γ 0.20, 0.82 (complex), 1.03, 1.16	Te^{130}(n, p) (HageE62, BrzJ63) fission (BarnJ52, DroB62) daughter Sn^{130} (PapA56, DroB62)
Sb^{131}	26 m (CoopJ64, UhlJ62, DMarP62) 23 m (PapA51) others (CookG51)	☢ β⁻ (PapA51)	A chem, genet (PapA51, CookG51) parent Te^{131}, parent Te^{131m} (PapA51, CookG51) parent Te^{131} (93%), parent Te^{131m} (7%) (SaraD65) daughter Sn^{131} (PapA56)	γ 0.64 (37%), 0.95 (48%) daughter radiations from Te^{131}, Te^{131m}	fission (PapA51, CookG51, CoopJ64)
Sb^{132}	2.1 m (PapA56) others (AbeP39, CookG51)	☢ β⁻ (AbeP39)	B chem, genet (AbeP39) parent Te^{132} (AbeP39) daughter Sn^{132} (PapA56)		fission (AbeP39, PapA56, CookG51)
Sb^{133}	4.2 m (CookG51) 4.4 m (PapA51)	☢ β⁻ (PapA51)	B chem, genet (PapA51) parent Te^{133m} (PapA51)		fission (PapA51, CookG51)
Sb^{134}	<1.5 s (BemC64)		F genet (activity not observed) (BemC64)		fission (BemC64)
$Sb^{134?}$	≈50 s (PapA51) 45 s (CookG51)	☢ β⁻ (PapA51)	G chem (PapA51) not ancestor I^{134}; may be an isomer of Sb^{132} (BemC64)		fission (PapA51, CookG51)
Sb^{135}	2 s genet (BemC64)	☢ [β⁻] (BemC64)	B chem, genet (BemC64) ancestor I^{135} (BemC64)		fission (BemC64)
$_{52}Te^{107}$	2.2 s (MacfR65)	☢ α (MacfR65)	B excit, cross bomb, sep isotopes (MacfR65)	α 3.28	$Ru^{96}(O^{16}, 5n)$ (MacfR65)
Te^{108}	5.3 s (MacfR65)	☢ α (MacfR65) [β⁺, EC], p (SiiA65)	B excit, cross bomb, sep isotopes (MacfR65, SiiA65)	α 3.08 p 2.6 (broad peak), 3.4, 3.7	$Ru^{196}(O^{16}, 4n)$ (MacfR65, SiiA65)

TABLE 1 (*Continued*)

Isotope Z A	Half-life	Type of decay (☢); % abundance; Mass excess ($\triangle \equiv$M–A), MeV (C^{12}=0); Thermal neutron cross section (σ), barns	Class; Identification; Genetic relationships	Major radiations: approximate energies (MeV) and intensities	Principal means of production
$_{52}$Te$^{(<113?)}$	16 m (RhoA57)	☢ β^+, [EC] (RhoA57)	F chem (RhoA57)		alphas on Sn (RhoA57)
Te^{115}	6.0 m (ReisR65) 6 m (SelI60a) 5–6 m (FinR61)	☢ β^+ ≈80%, EC ≈20% (ReisR65) $\triangle$ –82.5 (ReisR65, MTW)	B chem, excit, sep isotopes, genet (SelI60a, ReisR65) parent Sb^{115} (SelI60a, ReisR65)	β^+ 2.8 max γ Sb X-rays, 0.511 (160%, $\gamma^\pm$), 0.72 (34%), 0.96 (6%), 1.08 (24%), 1.28 (32%), 1.38 (32%), 1.58 (6%) daughter radiations from Sb^{115}	Sn^{112}(α, n) (SelI60a, ReisR65)
Te$^{114,115?}$	1.4 h (RhoA57)	☢ β^+, [EC] (RhoA57)	F chem (RhoA57) may be Te^{116} + Sb^{116} (+ Te^{117}?) (LHP)	γ 0.10 (?), 0.12 (?), 0.511 ($\gamma^\pm$), 0.75, 1.0 (?), 1.3 (?)	alphas on Sn (RhoA57)
Te$^?$	8 h (RhoA57)	☢	F chem (RhoA57)	γ 0.67	alphas on Sn (RhoA57)
Te^{116}	2.50 h (FinR61) others (LindnM48, KuzM58)	☢ EC, β^+ (?) (FinR61) β^+ (LindnM48) no α, lim 1 x 10^{-7}% (KarrM63) $\triangle$ –85.4 (MTW)	A chem (LindnM48) chem, mass spect (FinR61) parent Sb^{116} (FinR61) not parent Sb^{116m} (FinR61)	γ Sb X-rays, 0.094 e$^-$ 0.063, 0.089 β^+ 0.44 max (?) daughter radiations from Sb^{116}	protons on Sb (FinR61, LindnM48, KuzM58)
Te^{117}	61 m (FinR61) 65 m (KhuD62) 66 m (VarN61, ButeF65a)	☢ EC 70%, β^+ 30% (FinR61, KhuD62) no α, lim 0.005% (KarrM63) $\triangle$ –85.1 (MTW)	A chem, mass spect (FinR61) parent Sb^{117} (FinR61) daughter 14.5 m I^{117} (ButeF65a)	β^+ 1.81 max γ Sb X-rays, 0.511 (60%, $\gamma^\pm$), 0.72 (65%), 0.93 (6%), 1.78 (9%) daughter radiations from Sb^{117}	Sn^{114}(α, n) (VarN61, KhuD62) protons on Sb (FinR61)
Te^{117}	1.9 h (ButeF65a)	☢ β^+ (ButeF65a)	E chem, decay charac (ButeF65a) daughter 14.5 m I^{117} (ButeF65a)	β^+ 1.7 max	daughter 14.5 m I^{117} (ButeF65a)
Te^{118}	6.00 d (FinR61) others (LindnM48, AndeG65)	☢ EC (LindnM48) no α, lim 2 x 10^{-6}% (KarrM63) $\triangle$ –88 (MTW)	A chem (LindnM48) chem, mass spect (FinR61) parent Sb^{118} (LindnM48, LindnM50a, FinR61) not parent Sb^{118m1} (FinR61) daughter I^{118} (ZaitN60a)	γ Sb X-rays daughter radiations from Sb^{118}	protons on Sb (FinR61) Sb^{121}(d, 5n) (LindnM48, LindnM50a)
Te^{119}	15.9 h (FinR61) others (ZaitN60, ZaitN60a, KocC60)	☢ EC (FinR61) β^+ 5% (KocC60) $\triangle$ –87.19 (MTW)	A chem, excit, sep isotopes (KocC60) chem, mass spect, genet (FinR61) parent Sb^{119} (FinR61) daughter I^{119} (ZaitN60, ZaitN60a)	β^+ 0.627 max γ Sb X-rays, 0.645 (85%), 0.70 (11%), 1.76 (3.6%) daughter radiations from Sb^{119}	Sb^{121}(p, 3n) (FinR61) Sn^{116}(α, n) (KocC60)
Te^{119m}	4.68 d (KantJ63) others (SorA60, FinR61, KocC60, ZaitN60, ZaitN60a, LindnM48)	☢ EC (LindnM48) β^+ ≲0.5% (KantJ63) no α, lim 4 x 10^{-5}% (KarrM63) $\triangle$ –86.9 (LHP, MTW)	A chem, genet (LindnM48) chem, genet, mass spect (FinR61) parent Sb^{119} (LindnM48, LindnM50a, FinR61)	γ Sb X-rays, 0.153 (62%), 0.270 (25%), 0.92–1.14 (36%, complex), 1.221 (67%), 2.09 (4%) e$^-$ 0.122, 0.133, 0.148, 0.240, 0.266 daughter radiations from Sb^{119}	Sb^{121}(p, 3n) (FinR61) Sb^{121}(d, 4n) (LindnM48, LindnM50) Sn^{116}(α, n) (KocC60)
Te^{120}		% 0.089 (BaiK50) $\triangle$ –89.40 (MTW) σ_c 0.3 (to Te^{121}) 2.0 (to Te^{121m}) (GoldmDT64)			
Te^{121}	17 d (EdwJ46, ZaitN60a, BhaR63) others (BursS46)	☢ EC (EdwJ46) no β^+, lim 0.1% (ChuY64) $\triangle$ –88.31 (MTW)	A chem, genet (EdwJ46, BursS46) daughter Te^{121m} (BursS46) daughter I^{121} (MarqL50)	γ Sb X-rays, 0.508 (18%), 0.573 (80%) e$^-$ 0.007, 0.033, 0.543	Sb^{121}(α, 4n)I^{121}(β^+) (MarqL50) Sb^{121}(d, 2n) (EdwJ46, AubR64) Sb^{121}(p, n) (EdwJ46, AubR64) Te^{120}(n, γ) (HillR49a, AubR64) daughter Te^{121m} (BursS46)

TABLE 1 (*Continued*)

Isotope Z A	Half-life	Type of decay (☢); % abundance; Mass excess (△≡M-A), MeV (C¹²=0); Thermal neutron cross section (σ), barns	Class; Identification; Genetic relationships	Major radiations: approximate energies (MeV) and intensities	Principal means of production
$_{52}Te^{121m}$	154 d (HillR51, BhaR63) 143 d (EdwJ46) 125 d (SeaG40) 140 d (CorkJ51f)	☢ IT 90%, EC 10% (ChuY64) β⁺ ≈0.003% (AubR64) △ -88.01 (LHP, MTW)	A chem, excit, cross bomb (SeaG40) chem, n-capt, sep isotopes (CorkJ51f) parent Te¹²¹ (BursS46)	γ Te X-rays, Sb X-rays, 0.212 (82%), 1.10 (3%) e⁻ 0.007, 0.050, 0.077, 0.180 daughter radiations from Te¹²¹	Sb¹²¹(d, 2n) (SeaG40, EdwJ46, AubR64) Sb¹²¹(p, n) (SeaG40, EdwJ46, AubR64) Te¹²⁰(n, γ) (CorkJ51f, AubR64)
Te^{122}		% 2.46 (BaiK50) △ -90.29 (MTW) σ_c 2 (to Te¹²³) 1 (to Te¹²³ᵐ) (GoldmDT64)			
Te^{123}	t₁/₂ (EC_K) 1.2 x 10¹³ y sp act (WatD62a) t₁/₂ (EC) >10¹³ y sp act (HeiJ55)	☢ EC (WatD62a) % 0.87 (BaiK50) △ -89.16 (MTW) σ_c 400 (GoldmDT64)	B chem (WatD62a)	γ Sb X-rays	
Te^{123m}	117 d (AndeG65) 104 d (HillR51) 121 d (CorkJ51f)	☢ IT (HillR49a) △ -88.92 (LHP, MTW)	A chem, n-capt, sep isotopes (HillR49a)	γ Te X-rays, 0.159 (84%) e⁻ 0.057, 0.084, 0.127	Te¹²²(n, γ) (HillR49a, KatzR50, HammB51, CorkJ51f) Sb¹²³(d, 2n) (KatzR50)
Te^{124}		% 4.61 (BaiK50) △ -90.50 (MTW) σ_c 2 (to Te¹²⁵) 5 (to Te¹²⁵ᵐ) (GoldmDT64)			
Te^{125}		% 6.99 (BaiK50) △ -89.03 (MTW) σ_c 1.5 (GoldmDT64)			
Te^{125m}	58 d (HillR51, AndeG65)	☢ IT (FrieG48) △ -88.89 (LHP, MTW)	A chem, genet (FrieG48) daughter Sb¹²⁵ (FrieG48, KerB49) not daughter I¹²⁵, lim 0.05% (FrieG51a)	e⁻ 0.004, 0.030, 0.078, 0.105 γ Te X-rays, 0.035 (7%), 0.110 (0.3%)	daughter Sb¹²⁵ (FrieG48, KerB49) Te¹²⁴(n, γ) (HillR49a)
Te^{126}		% 18.71 (BaiK50) △ -90.05 (MTW) σ_c 0.9 (to Te¹²⁷) 0.1 (to Te¹²⁷ᵐ) (GoldmDT64)			
Te^{127}	9.4 h (KniJD56, MajN63) 9.3 h (SeaG40, MangS62) 9.5 h (BonaG64)	☢ β⁻ (AbeP39) △ -88.30 (MTW)	A chem (TapG38, AbeP39) chem, excit, cross bomb (SeaG40) daughter Te¹²⁷ᵐ (SeaG40, GleL51h, WilliRR51) daughter Sb¹²⁷ (84%) (AbeP39, GleL51h, BeydJ48)	β⁻ 0.70 max γ I X-rays, 0.058 (0.010%), 0.21 (0.03%, doublet), 0.360 (0.05%), 0.417 (0.3%)	Te¹²⁶(n, γ), daughter Te¹²⁷ᵐ (SeaG40, SerL47b) fission (AbeP39, SeaG40, WilliRR48, GleL51h)
Te^{127m}	109 d (AndeG65) 105 d (KniJD56) 115 d (CorkJ51f) 90 d (SeaG40)	☢ IT 99.2%, β⁻ 0.8% (AubR65) IT 98%, β⁻ 2% (KniJD56) △ -88.21 (LHP, MTW)	A chem, excit, genet (SeaG40) parent Te¹²⁷ (SeaG40, GleL51h, WilliRR51) daughter Sb¹²⁷ (16%) (BeydJ48)	γ Te X-rays, 0.059 (0.19%), 0.089 (0.08%), 0.67 (0.004%) e⁻ 0.057, 0.084 β⁻ [0.73 max] daughter radiations from Te¹²⁷	Te¹²⁶(n, γ) (HillR49a, SeaG40, SerL47b) fission (GrumW46, GleL51h, WilliRR48, GrumW48)
Te^{128}		% 31.79 (BaiK50) △ -88.98 (MTW) σ_c 0.14 (to Te¹²⁹) 0.017 (to Te¹²⁹ᵐ) (GoldmDT64)			

TABLE 1 (Continued)

Isotope Z A	Half-life	Type of decay (☢); % abundance; Mass excess (Δ≡M–A), MeV (C¹²=0); Thermal neutron cross section (σ), barns	Class; Identification; Genetic relationships	Major radiations: approximate energies (MeV) and intensities	Principal means of production
$_{52}Te^{129}$	68.7 m (BrzJ63, BrzJ65) 67 m (WafH48, MajN63) 72 m (SeaG40, BonaG64) 70 m (AbeP39, GleL51h, MangS62) 74 m (GravW56)	☢ β⁻ (SeaG40) Δ −87.02 (MTW)	A chem, excit (BotW39, SeaG40) daughter Te^{129m} (SeaG40, GrumW46, WilliRR51) daughter Sb^{129} (AbeP39)	β⁻ 1.45 max e⁻ 0.022, 0.026 γ I X-rays, 0.027 (19%), 0.275 (1.7%, doublet), 0.455 (15%), 0.81 (0.5%, complex), 1.08 (1.5%)	daughter Te^{129m} (SeaG40, GrumW46, WilliRR51) $Te^{128}(n,γ)$ (MangS62, SeaG40, SerL47b) fission (AbeP39, HahO43a, GrumW48, WilliRR48, NoveT51a)
Te^{129m}	34.1 d (AndeG65) 33.5 d (CorkJ51f) 33 d (MajN63) 32 d (BrzJ65) others (SeaG40, NoveT51b, GravW56, WafH48)	☢ IT 64%, β⁻ 36% (DevaS64a) IT 68%, β⁻ 32% (AndeG62) Δ −86.92 (LHP, MTW)	A chem, genet (SeaG40) parent Te^{129} (SeaG40, GrumW46, WilliRR51)	β⁻ 1.60 max e⁻ 0.074, 0.102 γ Te X-rays, 0.69 (6%) daughter radiations from Te^{129}	$Te^{128}(n,γ)$ (HillR49a, SeaG40, SerL47b) fission (HahO43a, AbeP39, WilliRR48, NoveT51b, PapA51a, GrumW48)
$\underline{Te^{130}}$	$t_{1/2}$ (ββ) 8 x 10²⁰ y, Xe ratios, mass spect (TakaN65) 1 x 10²¹ y Xe ratios, mass spect (IngM50) others (FremJ52, SharmH53, FulH52)	% 34.49 (BaiK50) Δ −87.34 (MTW) $σ_c$ 0.2 (to Te^{131}) 0.04 (to Te^{131m})			
Te^{131}	24.8 m (GeiK52) others (MangS62, SeaG40, AbeP39)	☢ β⁻ (SeaG40) Δ −85.16 (MTW)	A chem, excit (SeaG40) daughter Te^{131m} (AbeP39, SeaG40, WilliRR51) parent I^{131} (AbeP39, SeaG40, PapA51, CookG51, LivJ38e, HahO39c) daughter Sb^{131} (PapA51, CookG51, SaraD65)	β⁻ 2.14 max e⁻ 0.116, 0.144 γ I X-rays, 0.150 (68%), 0.453 (16%), 0.493 (5%), 0.603 (4%), 0.95 (3%, complex), 1.00 (4%, doublet), 1.147 (6%)	$Te^{130}(n,γ)$ (SeaG40, SerL47b, GeiK52) daughter Te^{131m} (AbeP39, SeaG40, WilliRR51)
Te^{131m}	30 h (AbeP39, SeaG40)	☢ β⁻ 82%, IT 18% (BedeA61, DevaS65) β⁻ 78%, IT 22% (HebeE55) Δ −84.98 (LHP, MTW)	A chem, genet (SeaG40) parent Te^{131} (AbeP39, SeaG40, WilliRR51) daughter Sb^{131} (CookG51, PapA51, SaraD65)	β⁻ 2.46 max (5%), 0.9 max e⁻ 0.048, 0.069, 0.149, 0.177 γ Te X-rays, I X-rays, 0.081 (2%), 0.102 (5%), 0.200 (8%), 0.241 (8%), 0.336 (9%), 0.78 (60%, complex), 0.85 (31%, doublet), 1.127 (13%), 1.206 (11%), 1.629 (3%), 1.860 (1%), 1.965 (2%) daughter radiations from Te^{131} I^{131}	$Te^{130}(n,γ)$ (SeaG40, SerL47b) fission (SaraD65, AbeP39, HahO39c, KatcS51d, WilliRR51, PapA51a)
Te^{132}	77.7 h (PapA51a) 78 d (AndeG65) others (AbeP39, CheeG58, FleW56, HahO39b)	☢ β⁻ (AbeP39) Δ −85.21 (MTW)	A chem, genet (AbeP39) fission fragment range (KatcS48) parent I^{132} (AbeP39, HahO39c, HahO39b, NoveT51a, WinsW51) daughter Sb^{132} (AbeP39)	β⁻ 0.22 max e⁻ 0.020, 0.048, 0.197 γ I X-rays, 0.053 (17%), 0.230 (90%) daughter radiations from I^{132}	fission (AbeP39, HahO39a, HahO39b, PapA51a, KatcS48)
Te^{133}	12.5 m (PruS65)	☢ [β⁻] (PruS65)	B chem, genet (PruS65) daughter Te^{133m}, parent I^{133} (PruS65)	γ 0.15, 0.31, 0.41, 0.73, 1.02, 1.33, 1.71, 1.85	fission, daughter Te^{133m} (PruS65, SaraD65)
Te^{133m}	50 m (FergJ62) 63 m (PapA52) 53 m (AlvT57) 60 m (AbeP39, WuC40)	☢ β⁻ 87%, IT 13% (AlvT57)	A chem, genet (AbeP39) parent 12.5 m Te^{133} (PruS65) daughter Sb^{133} (PapA51) ancestor I^{133} (AbeP39, HahO39c, SegE40, WuC40, WuC45, PapA51)	β⁻ 2.4 max e⁻ 0.303 γ Te X-rays, 0.31 (21%), 0.432 (50%), 0.47 (22%), 0.557 (35%), 0.63 (18%), 0.70 (24%), 0.754 (85%), 0.91 (57%), 1.01 (10%), 1.33, 1.71, 1.85 daughter radiations from I^{133} daughter radiations from Te^{133} included in above listing	fission (AbeP39, HahO39c, SegE40, WuC40, PapA51, KatcS48, SaraD65)
Te^{133}	2 m (PapA52)	☢ β⁻ (PapA52)	G chem, genet (PapA52) activity not observed (PruS65)		daughter Te^{133m} (PapA52)
Te^{134}	42 m (FergJ62) 44 m (PapA51a) 43 m (AbeP39)	☢ β⁻ (AbeP39)	A chem, genet (AbeP39) parent I^{134} (AbeP39, HahO39c, PapA51a) others (KatcS48, PolA40a)	γ I X-rays, 0.08 (13%), 0.17 (16%), 0.204 (21%), 0.262 (19%) daughter radiations from I^{134}	fission (KatcS48, HahO39c, AbeP39, PolA40a, PapA51a, FergJ62)

TABLE 1 (*Continued*)

Isotope Z A	Half-life	Type of decay (☢); % abundance; Mass excess (△≡M–A), MeV (C¹²=0); Thermal neutron cross section (σ), barns	Class; Identification; Genetic relationships	Major radiations: approximate energies (MeV) and intensities	Principal means of production
$_{52}\text{Te}^{135}$	<2 m (GleL51i, DodR40, KatcS51f)	☢ β⁻ (DodR40)	E genet (DodR40) parent I¹³⁵ (GleL51i, KatcS51f)		fission (GleL51i, DodR40, KatcS51f)
Te⁷	≈1 m (HahO43a)	☢ β⁻ (HahO43a)	E chem (HahO43a)		fission (HahO43a)
$_{53}\text{I}^{117}$	7 m (AndeG65)	☢ β⁺ (AndeG65), [EC]	C mass spect, [chem] (AndeG65)	Y 0.16, 0.34, 0.522 (Y⁺)	protons on La (AndeG65)
I¹¹⁷	14.5 m genet (ButeF65a)	☢ [β⁺] (ButeF65a)	F chem, genet (ButeF65a) parent 61 m Te¹¹⁷, parent 1.9 h Te¹¹⁷ (ButeF65a)		protons on I (ButeF65a)
I¹¹⁸	13.9 m (AndeG65) 17 m (ZaitN60a, ButeF65a) others (AagP57)	☢ β⁺ ≈54%, EC ≈46% (AndeG65) △ –81 (ButeF65a, AndeG65, MTW)	B mass spect (AagP57) chem, genet (ZaitN60a) parent Te¹¹⁸ (ZaitN60a) daughter Xe¹¹⁸ (AndeG65)	Y Te X-rays, 0.511 (108%, Y⁺), 0.55, 0.60, 1.15	protons on I (ZaitN60a, ButeF65a)
I¹¹⁹	19.5 m (AndeG65) 18 m (RosG54) 21 m genet (ZaitN60, ZaitN60a) 19 m (AagP57) 26 m (ButeF65a)	☢ β⁺ 51%, EC 49% (AndeG65)	A chem (MarqL50) mass spect (AagP57) chem, genet (ZaitN60a) parent Te¹¹⁹ (ZaitN60, ZaitN60a) daughter Xe¹¹⁹ (AndeG65)	Y Te X-rays, 0.26, 0.511 (102%, Y⁺), 0.78 daughter radiations from Te¹¹⁹, Sb¹¹⁹	N¹⁴ on Pd (RosG54) protons on I (ZaitN60, ZaitN60a)
I¹²⁰	1.35 h (AndeG65) 1.30 h (ButeF65) 1.4 h (AagP57)	☢ EC 54%, β⁺ 46% (AndeG65) △ –83.8 (ButeF65, AndeG65, MTW)	A mass spect, chem (AagP57, AndeG65) chem, genet (ButeF65) daughter Xe¹²⁰ (ButeF65)	β⁺ 4.0 max Y Te X-rays, 0.511 (92%, Y⁺), 0.56, 0.62, 1.52	protons on I, daughter Xe¹²⁰ (ButeF65)
I¹²⁰	30 m (MarqL50, KuzM58a)	☢ β⁺ (MarqL50)	G chem (MarqL50, KuzM58a) activity not observed (AndeG65)		alphas on Sb (MarqL50) protons on I (KuzM58a)
I¹²¹	2.12 h (AndeG65) 2.0 h (AagP57, ButeF65) 1.5 h (MathH54a, DroB52) 2.1 h (ZaitN60) 1.4 h (RosG54) 1.8 h (MarqL50, KuzM58a)	☢ EC 91%, β⁺ 9% (AndeG65) △ –86.0 (MTW)	A chem, genet (MarqL50) mass spect (AagP57) parent Te¹²¹ (MarqL50) daughter Xe¹²¹ (MathH54a, DroB52)	β⁺ 1.2 max Y Te X-rays, 0.212 (90%), 0.27 (3%), 0.32 (6%), 0.511 (18%, Y⁺)	Sb¹²¹(α, 4n) (MarqL50)
I¹²²	3.5 m (MathH54a) 3.4 m (DroB52) 3.6 m (YouJ51) 4 m (MarqL50)	☢ β⁺ (MarqL50), [EC] △ –86.15 (MTW)	A chem, excit (MarqL50) sep isotopes (YouJ51) daughter Xe¹²² (TilDE52, DroB52)	β⁺ 3.1 max Y Te X-rays, 0.511 [130%, Y⁺], 0.564, 0.69, 0.78	Sb¹²¹(α, 3n) (MarqL50) Te¹²²(p, n) (YouJ51)
I¹²³	13.3 h (AndeG65) 13.0 h (MitA49a) 13 h (MarqL50, MathH54a, KuzM58a)	☢ EC (MarqL50) no β⁺ (MitA59) △ –88 (MTW)	A chem, excit (MarqL50) chem, sep isotopes (MitA49a) daughter Xe¹²³ (DroB52, MathH54a, TilDE52)	Y Te X-rays, 0.159 (83%) e⁻ 0.127	Sb¹²¹(α, 2n) (MarqL50, MitA49a, MitA59, GupR60b)
I¹²⁴	4.15 d (AndeG65) 4.2 d (DysN58, MitA59) 4.1 d (GirR59g) 4.0 d (LivJ38e) 4.5 d (MarqL50) 3.4 d (AagP57)	☢ EC 74%, β⁺ 26% (DysN58) EC 75%, β⁺ 25% (GirR59g) EC 71%, β⁺ 29% (MitA59) no β⁻, lim 0.1% (MerC61) EC(K)/EC(L) 9 (MitA59) △ –87.33 (MTW)	A chem, excit, cross bomb (LivJ38e)	β⁺ 2.14 max Y Te X-rays, 0.511 (50%, Y⁺), 0.605 (67%), 0.644 (12%), 0.73 (14%), 1.37 (3%), 1.51 (4%), 1.69 (14%), 2.09 (2.0%), 2.26 (1.5%)	Sb¹²¹(α, n) (MarqL50, LivJ38e) Sb¹²³(α, 3n) (MarqL50)
I¹²⁵	60.2 d (LeuH64, GleG64) 60.0 d (FrieG51a) 57.4 d (MatthC60) others (KuzM58a, ReidA46a)	☢ EC, no β⁺ (ReidA46a, GleL47) EC(L+M+...)/EC(K) 0.254 (LeuH64) △ –88.88 (MTW) σc 900 (GoldmDT64)	A chem (ReidA46a) chem, excit (GleL47) genet (BergI51c) daughter Xe¹²⁵ (BergI51c) not parent Te¹²⁵ᵐ, lim 0.05% (FrieG51a)	Y Te X-rays, 0.035 (7%) e⁻ 0.004, 0.030	Sb¹²³(α, 2n) (MarqL50) daughter Xe¹²⁵ (BergI51c) deuterons on Te (ReidA46a, GleL47, FleP58)

TABLE 1 (*Continued*)

Isotope Z A	Half-life	Type of decay (☢); % abundance; Mass excess ($\triangle \equiv$M–A), MeV (C^{12}=0); Thermal neutron cross section (σ), barns	Class; Identification; Genetic relationships	Major radiations: approximate energies (MeV) and intensities	Principal means of production
$_{53}$I^{126}	12.8 d (AndeG65) 13.3 d (PerlmM54) 13.0 d (LivJ38e) 13.1 d (MocD48, AagP57)	☢ EC 55%, β⁻ 44%, β⁺ 1.3% (PerlmM54) EC 55%, β⁻ 44%, β⁺ 1.2% (KoerL55) EC(K) 51% (EroJ57a) $\triangle$ –87.90 (MTW)	A chem (TapG38) chem, excit, cross bomb (LivJ38e)	β⁻ 1.25 max β⁺ 1.13 max γ Te X-rays, 0.386 (34%), 0.667 (33%)	Sb^{123}(α, n) (LivJ38e, MarqL50) Te^{125}(d, n) (LivJ38e) Te^{126}(p, n) (DubL40a)
I^{126}	2.6 h (AagP57)		G mass spect (AagP57) activity not observed (NaraV65)		fission (AagP57)
I^{127}		% 100 (NierA37a) $\triangle$ –88.984 (MTW) σ_c 6.4 (GoldmDT64)			
I^{128}	24.99 m (HulO41) others (AagP57, LivJ38e)	☢ β⁻ 93.6%, EC 6.4% (BencN56) β⁺ 3 × 10⁻³% (LanghH61b) $\triangle$ –87.71 (MTW)	A chem, n-capt (AmaE35)	β⁻ 2.12 max γ Te X-rays, 0.441 (14%), 0.528 (1.4%), 0.743 (0.2%), 0.969 (0.3%)	I^{127}(n, γ) (AmaE35, TapG38, SerL47b, SiegK46c, OrsA49, HumV51)
I^{129}	1.7 × 10⁷ y sp act (KatcS51k) others (PurB56)	☢ β⁻ (KatcS47) $\triangle$ –88.50 (MTW) σ_c 28 (GoldmDT64)	A chem, n-capt (KatcS47) chem, mass spect (KatcS51k)	β⁻ 0.150 max e⁻ 0.005, 0.034 γ Xe X-rays, 0.040 (9%)	fission (KatcS47, KatcS51k)
I^{130}	12.3 h (AndeG65) 12.5 h (AagP57) 12.6 h (LivJ38e)	☢ β⁻ (LivJ38e) $\triangle$ –86.89 (DaniH65, MTW) σ_c 18 (GoldmDT64)	A chem, cross bomb (LivJ38e) chem, mass spect (AagP57)	β⁻ 1.7 max (0.4%), 1.04 max γ 0.419 (35%), 0.538 (99%), 0.669 (100%), 0.743 (87%), 1.15 (12%)	I^{129}(n, γ) (SmiW59) Te^{130}(d, 2n) (LivJ38e) Te^{130}(p, n) (GarvH58b) Cs^{133}(n, α) (WuC40)
I^{131}	8.05 d (BurkL58, BarthR53, GleG64) 8.07 d (KeeJ58, SelH53) 8.06 d (LocE53) 8.14 d (SreJ51a) 8.04 d (SinW51)	☢ β⁻ (LivJ38e) $\triangle$ –87.441 (MTW) σ_c ~0.7 (GoldmDT64)	A chem (LivJ38e) chem, genet (SeaG40) daughter Te^{131} (LivJ38e, AbeP39, HahO39c, SeaG40, PapA51, CookG51) parent Xe^{131m} (≈1%) (BrosA49, BergI50c)	β⁻ 0.806 max (0.6%), 0.606 max average β⁻ energy: 0.19 ion ch (CaswR52) e⁻ 0.046, 0.330 γ Xe X-rays, 0.080 (2.6%), 0.284 (5.4%), 0.364 (82%), 0.637 (6.8%), 0.723 (1.6%) daughter radiations from Xe^{131m}	fission (AbeP39, HahO39c, GrumW46, SulW51g, YafL47, GrumW48, FinB51c)
I^{132}	2.26 h (EmeE54) 2.34 h (AndeG65) 2.30 h (WahA55) 2.5 h (WillaD62) others (AagP57, AbeP39, HahO39b)	☢ β⁻ (AbeP39) $\triangle$ –85.71 (MTW)	A chem, genet (AbeP39) chem, mass spect (AagP57) daughter Te^{132} (AbeP39, HahO39c, HahO39b, NoveT51a, WinsW51)	β⁻ 2.12 max γ 0.24 (1%), 0.52 (20%, complex), 0.67 (144%, complex), 0.773 (89%), 0.955 (22%), 1.14 (6%, doublet), 1.28 (7%), 1.40 (14%, complex), 1.45 (1%), 1.91 (1.3%), 1.99 (1.3%)	daughter Te^{132}, from fission (AbeP39, HahO39c, HahO39b, NoveT51a, WinsW51)
I^{133}	20.3 h (AndeG65) 20.8 h (KatcS53) 20.9 h (WahA55) 20.5 h (VasiI58) 22.4 h (PapA51a)	☢ β⁻ (AbeP39, HahO39c) $\triangle$ –85.9 (MTW)	A chem (AbeP39) chem, genet (WuC40) descendant Te^{133m} (AbeP39, HahO39c, SegE40, WuC40, WuC45, PapA51) daughter 12.5 m Te^{133} (PruS65) parent Xe^{133} (SegE40, WuC40, WuC45) parent Xe^{133m} (2.4%) (ZelH51, KetB51a)	β⁻ 1.27 max γ 0.53 (90%) daughter radiations from Xe^{133}, Xe^{133m}	fission (AbeP39, HahO39c, SegE40, WuC40, PapA51, SulW51h, FinB51c, HolmG59)
I^{134}	52.0 m (AndeG65) 52.8 m (JohnN61) 52.5 m (PapA51a) 52.4 m (WahA55) others (AbeP39, AagP57)	☢ β⁻ (AbeP39) $\triangle$ –84.0 (MTW)	A chem (AbeP39) fission fragment range (KatcS48) chem, mass spect (AagP57) daughter Te^{134} (HahO39c, AbeP39, PapA51a)	β⁻ 2.43 max γ 0.135 (3%), 0.41 (8%, complex), 0.55 (8%), 0.61 (18%), 0.85 (95%), 0.89 (65%), 1.07 (1.4%), 1.15 (10%), 1.46 (4%), 1.62 (5%), 1.79 (5%)	fission (YafL47, HahO39c, AbeP39, PolA40a, PolA40, LidL49, KatcS51e, PapA51a, KatcS48, FinB51c)
I^{135}	6.68 h (PeaW47a) 6.7 h (GleL51i, KatcS51f) 6.8 h (WahA55) others (DodR40)	☢ β⁻ (WahA55) $\triangle$ –84 (MTW)	A chem, genet (DodR40, SegE40) parent Xe^{135m} (30%), parent Xe^{135} (70%) (PeaW47a) daughter Te^{135} (GleL51i, KatcS51f) descendant Sb^{135} (BemC64) others (SegE40, DodR40, GotH40, WuC45, BallN51h, WuC40, FinB51c, AagP57)	β⁻ 1.4 max γ 0.42 (7%), 0.86 (11%), 1.04 (9%), 1.14 (37%), 1.28 (34%), 1.46 (12%), 1.72 (19%), 1.80 (11%) daughter radiations from Xe^{135m}, Xe^{135}	fission (SegE40, WuC40, DodR40, WuC45, PeaW47a, GleL51i, KatcS51f, FinB51c)

TABLE 1 (Continued)

Isotope Z A	Half-life	Type of decay (☢); % abundance; Mass excess (△≡M−A), MeV (C¹²=0); Thermal neutron cross section (σ), barns	Class; Identification; Genetic relationships	Major radiations: approximate energies (MeV) and intensities	Principal means of production
$_{53}I^{136}$	83 s (JohnN59) 86 s (StanlC49) others (StraF40)	☢ β⁻ (StraF40) △ −79.4 (MTW)	B chem (StraF40) chem, decay charac (JohnN59)	β⁻ 7.0 max (≤6%), 5.6 max γ 0.20 (12%), 0.27 (18%), 0.39 (19%), 1.32 (95%, complex), 2.3 (19%, complex), 2.63 (10%), 2.8 (8%), 3.2 (5%)	fission (StraF40, SeeW43, StanlC49, JohnN59)
I^{137}	22.0 s (n) (HugD48) 24.4 s (n) (PerloG59) 22.5 s (RedW47) 19.3 s genet (SugaN49) others (CoxS58, KeeG57, SneA47a)	☢ β⁻, β⁻n (≈6%) (LeviJ51)	A chem (StraF40) chem, genet (SeeW43, SugaN49) parent Xe¹³⁷ (SeeW43, SugaN49)	n average energy 0.6 daughter radiations from Xe¹³⁷	fission (StraF40, SeeW43, RedW47, SugaN49, HugD48, PerloG59, SneA47a, SugaN47)
I^{138}	5.9 s (SugaN49) others (PerloG59, KeeG57)	☢ β⁻ (SugaN49) n (KeeG57, PerloG59)	B chem, genet (SugaN49) ancestor Cs¹³⁸ (SugaN49)		fission (SugaN49, PerloG59, KeeG57)
I^{139}	2.7 s (SugaN49) 2.0 s (PerloG59, CoxS58)	☢ β⁻ (SugaN49) n (PerloG59, CoxS58)	B chem, genet (SugaN49) parent Xe¹³⁹, ancestor Ba¹³⁹ (SugaN49)	daughter radiations from Xe¹³⁹	fission (SugaN49, CoxS58, PerloG59)
$_{54}Xe^{118}$	6 m (AndeG65)	☢ β⁺, [EC] (AndeG65)	B chem, mass spect (AndeG65) parent I¹¹⁸ (AndeG65)	γ 0.05, 0.511 (γ±) daughter radiations from I¹¹⁸	protons on La (AndeG65)
Xe^{119}	6 m (AndeG65)	☢ β⁺, [EC] (AndeG65)	B chem, mass spect (AndeG65) parent I¹¹⁹ (AndeG65)	γ [I X-rays], 0.10, 0.511 (γ±) daughter radiations from I¹¹⁹	protons on La (AndeG65)
Xe^{120}	40 m (AndeG65) 43 m (ButeF65)	☢ [EC] (AndeG65)	A chem, mass spect (AndeG65) chem, genet (ButeF65) parent I¹²⁰ (ButeF65)	γ I X-rays, 0.055, 0.073, 0.176, 0.76 daughter radiations from I¹²⁰	protons on I (ButeF65)
Xe^{121}	39 m (AndeG65) 40 m (DroB52, MooR60, MathH54a) others (TilDE52, ButeF65)	☢ β⁺ (MathH54a), [EC] △ −82.2 (MTW)	A chem, genet (DroB52, TilDE52) chem, mass spect (AndeG65) parent I¹²¹ (MathH54a, DroB52)	β⁺ 2.8 max γ [I X-rays], 0.080, 0.096, 0.132, 0.437, 0.511 (γ±) daughter radiations from I¹²¹	I¹²⁷(p, 7n) (TilDE52, DroB52, MooR60)
Xe^{122}	20.1 h (AndeG65) 19.5 h (TilDE52) 18.5 h (MooR60) 20.0 h (DroB52) 19 h (MathH54a)	☢ EC (MathH54a)	A chem, genet (TilDE52, DroB52, MathH54a) chem, mass spect (AndeG65) parent I¹²² (TilDE52, DroB52)	γ I X-rays, 0.060, 0.090, 0.110, 0.148, 0.180, 0.345 e⁻ 0.058, 0.116 daughter radiations from I¹²²	I¹²⁷(p, 6n) (TilDE52, DroB52)
Xe^{123}	2.08 h (AndeG65) 1.85 h (MooR60, ButeF65) 1.8 h (MathH54a, PreiI62) 1.7 h (DroB52) 2.1 h (TilDE52)	☢ EC, β⁺ (MathH54a) △ 85 (MTW)	A chem, genet (TilDE52, DroB52, MathH54a) chem, mass spect (AndeG65) parent I¹²³ (TilDE52, DroB52, MathH54a) daughter Cs¹²³ (MathH54a, MathH54, PreiI62)	β⁺ 1.51 max e⁻ 0.115, 0.144, 0.295 γ I X-rays, 0.090, 0.110, 0.149, 0.178, 0.329, 0.511 (γ±), 0.68, 0.90, 1.10 daughter radiations from I¹²³	I¹²⁷(p, 5n) (TilDE52, DroB52, MathH54a)
$\underline{Xe^{124}}$		% 0.096 (NierA50a) △ −87.5 (MTW) σc 110 (GoldmDT64)			
Xe^{125}	16.8 h (AndeG65) 18.0 h (BergI52) 17 h (MooR60) 20 h (AndeDL50)	☢ EC, no β⁺ (BergI51c, AndeDL50) △ −87 (MTW)	A chem, sep isotopes (AndeDL50) chem, mass spect (BergI51c) parent I¹²⁵ (BergI51c) daughter Cs¹²⁵ (MathH54)	γ I X-rays, 0.055, 0.188, 0.242 e⁻ 0.022, 0.050, 0.154, 0.182, 0.209 daughter radiations from I¹²⁵	I¹²⁷(p, 3n) (MooR60) Xe¹²⁴(n, γ) (BergI51c)
Xe^{125m}	55 s (MathH54) 60 s (MooR60)	☢ IT (?) (MathH54)	B genet (MathH54) daughter Cs¹²⁵ (≈0.1%) (MathH54)	γ [Xe X-rays], 0.075, 0.111	daughter Cs¹²⁵ (MathH54) I¹²⁷(p, 3n) (MooR60)
$\underline{Xe^{126}}$		% 0.090 (NierA50a) △ −89.15 (MTW) σc ≈2 (GoldmDT64)			

TABLE 1 (Continued)

Isotope Z A	Half-life	Type of decay (☢); % abundance; Mass excess ($\Delta \equiv M-A$), MeV ($C^{12}=0$); Thermal neutron cross section (σ), barns	Class; Identification; Genetic relationships	Major radiations: approximate energies (MeV) and intensities	Principal means of production
$_{54}\text{Xe}^{127}$	36.41 d (BaleS54) others (BresM64, ForR58, CreEC40a, ArteK50, BergI51)	☢ EC, no β^+ (MathH55a, ForR58) Δ −88.54 (WintG65a, MTW)	A chem (CreEC40a) chem, sep isotopes (ArteK50) mass spect (BergI51a) daughter Cs^{127} (FinR50a)	γ I X-rays, 0.058 (1.4%), 0.145 (4.2%), 0.172 (22%), 0.203 (65%), 0.375 (20%) e^- 0.024, 0.112, 0.139, 0.170, 0.198	$\text{I}^{127}(p,n)$ (CreEC40a, MathH55a, ForR58) $\text{I}^{127}(d,2n)$ (BaleS54, ForR58) $\text{Xe}^{126}(n,\gamma)$ (CamM44, BergI51a)
Xe^{127m}	75 s (CreEC40a)	☢ IT (CreEC40a, MathH54) 	B chem (CreEC40a) genet (MathH54) daughter Cs^{127} (0.01%) (MathH54)	γ Xe X-rays, 0.125, 0.175	$\text{I}^{127}(p,n)$ (CreEC40a) daughter Cs^{127} (MathH54)
$\underline{\text{Xe}^{128}}$		% 1.919 (NierA50a) Δ −89.85 (MTW) σ_c <5 (GoldmDT64)			
$\underline{\text{Xe}^{129}}$		% 26.44 (NierA50a) Δ −88.692 (MTW) σ_c 25 (GoldmDT64)			
Xe^{129m}	8.0 d (BergI51a)	☢ IT (BergI51a) Δ −88.456 (LHP, MTW)	A chem, mass spect (BergI51a)	γ Xe X-rays, 0.040 (9%), 0.197 (6%) e^- 0.005, 0.034, 0.162, 0.191	$\text{Xe}^{128}(n,\gamma)$ (BergI51a)
$\underline{\text{Xe}^{130}}$		% 4.08 (NierA50a) Δ −89.88 (MTW) σ_c <5 (GoldmDT64)			
$\underline{\text{Xe}^{131}}$		% 21.18 (NierA50a) Δ −88.411 (MTW) σ_c 85 (GoldmDT64)			
Xe^{131m}	11.8 d (AndeG65) 12.0 d (BergI50c, PerlmM53) others (BrosA49, CamM44)	☢ IT (BrosA49, CamM44) Δ −88.247 (LHP, MTW)	A chem (CamM44) chem, genet (BrosA49) mass spect (BergI50c) daughter I^{131} (≈1%) (BrosA49, BergI50c) not daughter Cs^{131} (CanR51b, SaraB54)	γ Xe X-rays, 0.164 (2%) e^- 0.129, 0.159	$\text{Xe}^{130}(n,\gamma)$ (CamM44, BergI50c)
$\underline{\text{Xe}^{132}}$		% 26.89 (NierA50a) Δ −89.272 (MTW) σ_c 0.2 (to Xe^{133}) <5 (to Xe^{133m}) (GoldmDT64)			
Xe^{133}	5.270 d (MacnJ50) 5.4 d (BergI52)	☢ β^- (DodR40) Δ −87.73 (MTW) σ_c 190 (GoldmDT64)	A chem (LangsA39, DodR40, SegE40) chem, excit (WuC40) mass spect (ThodH47, ThuS49) daughter I^{133} (SegE40, WuC40, WuC45)	β^- 0.346 max e^- 0.045, 0.075 γ Cs X-rays, 0.081 (37%)	fission (SegE40, DodR40, WuC40, BornH43a, WuC45, ThodH47, BehH51, EngeD51b) $\text{Xe}^{132}(n,\gamma)$ (RieW43, AlvT58, ThieP62, BrowF61)
Xe^{133m}	2.26 d (ErmP61) 2.35 d (BergI52) 2.1 d (KetB51a) others (BergI51b)	☢ IT (KetB50a) Δ −87.50 (LHP, MTW)	A chem (KetB50a) mass spect (BergI51b) daughter I^{133} (2.4%) (ZelH51, KetB51a)	γ Xe X-rays, 0.233 (14%) e^- 0.198, 0.227 daughter radiations from Xe^{133}	fission (KetB51a, BergI50b) $\text{Xe}^{132}(n,\gamma)$ (BergI51b, ErmP61)
$\underline{\text{Xe}^{134}}$		% 10.4 (NierA50a) Δ −88.121 (MTW) σ_c 0.2 (to Xe^{135}) <5 (to Xe^{135m}) (GoldmDT64)			

TABLE 1 (*Continued*)

Isotope Z A	Half-life	Type of decay (☢); % abundance; Mass excess (△≡M−A), MeV (C¹²=0); Thermal neutron cross section (σ), barns	Class; Identification; Genetic relationships	Major radiations: approximate energies (MeV) and intensities	Principal means of production
₅₄Xe¹³⁵	9.14 h (AndeG65) 9.13 h (BrowF53) 9.19 h mass spect (ClarW64) 9.2 h (NewA51, HoaE51c, BergI52, GleL51i) others (RieW43, DodR40, WuC40, ClanE41)	☢ β⁻ (SegE40) △ −86.6 (MTW) σ_c 2.7 x 10⁶ (GoldmDT64)	A chem (SegE40, DodR40) chem, excit (WuC40) mass spect (ThuS49) daughter I¹³⁵ (70%) (PeaW47a) daughter Xe¹³⁵ᵐ (WuC45) parent Cs¹³⁵ (SugaN49a) others (SegE40, DodR40, GotH40, RieW43, ClanE41, SeeW43a, BehH51)	β⁻ 0.92 max e⁻ 0.214 γ 0.250 (91%), 0.61 (3%)	Xe¹³⁴(n, γ) (RieW43) fission (SegE40, DodR40, BehH51) Ba¹³⁸(n, α) (WuC40, SeeW43a, WuC45)
Xe¹³⁵ᵐ	15.6 m (KotK60, RieW43) 15.8 m (AlvT60) 15.3 m (PeaW47a) others (NoveT51c)	☢ IT (WuC45) no β⁻, lim 10% (AlvT60) △ −86.1 (MTW, LHP)	A chem, genet (GotH40, WuC45) daughter I¹³⁵ (30%) (PeaW47a) parent Xe¹³⁵ (WuC45) others (GotH40, WuC45, RieW46, SeeW43a, ThodH47)	γ Xe X-rays, 0.527 (80%) e⁻ 0.493, 0.522 daughter radiations from Xe¹³⁵	daughter I¹³⁵ (PeaW47a, GotH40, WuC45, AlvT60) Xe¹³⁴(n, γ) (RieW43) fission (GotH40, WuC45, ThodH47, KotK60, AlvT60) Ba¹³⁸(n, α) (SeeW43a)
Xe¹³⁶		% 8.87 (NierA50a) △ −86.42 (MTW) σ_c 0.15 (GoldmDT64)			
Xe¹³⁷	3.9 m (SugaN49, OnegR64, HolmGB63) 3.8 m (SeeW43) 3.4 m (RieW43)	☢ β⁻ (SeeW43) △ −82.8 (MTW)	A chem (SeeW43) mass spect (ThuS49) daughter I¹³⁷ (SeeW43, SugaN49) parent Cs¹³⁷ (TurA51, GleL51k)	β⁻ 4.1 max γ 0.455 (33%)	Xe¹³⁶(n, γ) (RieW43, SeeW43a, SugaN49) fission (SeeW43, SugaN49, GleL51k)
Xe¹³⁸	17.5 m (OckD62) 14.0 m (ClarW64) 17 m (GlasG40) others (HahO40, AndeG65)	☢ β⁻ (HahO39c) △ −80.9 (NDS, MTW)	A chem (HahO39c) mass spect (ThuS49) parent Cs¹³⁸ (HahO39c, HahO40, GlasG40, SeeW43a)	β⁻ 2.4 max γ 0.16 (↑ 33), 0.26 (↑ 100), 0.42 (↑ 40), 0.51 (↑ 8), 1.78 (↑ 66), 2.02 (↑ 58) daughter radiations from Cs¹³⁸	fission (HahO39c, HahO40, GlasG40, SeeW43a, ThuS49, ThuS55, NasS55)
Xe¹³⁹	43 s (OckD62) 41 s (DilC51a)	☢ β⁻ (HahO39c, HeyF39) △ −76.5 (MTW)	A chem, genet (HahO39c, HeyF39) daughter I¹³⁹ (SugaN49) parent Cs¹³⁹ (HahO39c, HeyF39, HahO40a, HahO40) ancestor Ba¹³⁹ (HahO39c, HeyF39, DilC51a)	γ 0.18 (↑ 41), 0.22 (↑ 100), 0.30 (↑ 57), 1.15 (↑ 23) daughter radiation from Cs¹³⁹	fission (HahO39c, HeyF39, HahO40a, HahO40, SugaN49, DilC51a, OckD62)
Xe¹⁴⁰	16.0 s (DilC51a) 10 s (OveR51) ≈15 s (OckD62) others (HahO40a)	☢ β⁻ (HahO40)	A chem, genet (HahO40) ancestor Ba¹⁴⁰ (HahO40, DilC51, DilC51a, OveR51, BradE51)	γ 0.13 daughter radiations from Cs¹⁴⁰	fission (HahO40a, HahO40, DilC51, DilC51a, OveR51, BradE51, OckD62)
Xe¹⁴¹	1.7 s (KatcS46, OveR51) 3 s (DilC51a)	☢ β⁻ (BradE51)	B chem, genet (BradE51) ancestor La¹⁴¹ (BradE51) ancestor Ce¹⁴¹ (DilC51, DilC51a, OveR51) ancestor Ba¹⁴¹ (BradE51, OveR51, DilC51a)		fission (BradE51, DilC51, DilC51a, OveR51)
Xe¹⁴²	≈1.5 s (WolfsK60)	☢ [β⁻] (WolfsK60)	B chem, genet (WolfsK60) ancestor La¹⁴² (WolfsK60)		fission (WolfsK60)
Xe¹⁴³	1.0 s (DilC51a)	☢ β⁻ (BradE51)	B chem, genet (BradE51) ancestor Ce¹⁴³ (BradE51, DilC51a)		fission (DilC51a, BradE51)
Xe¹⁴⁴	≈1 s (DilC51a)	☢ β⁻ (DilC51)	B chem, genet (DilC51) ancestor Ce¹⁴⁴ (DilC51, DilC51a)		fission (DilC51, DilC51a)
₅₅Cs¹²³	8.0 m (PreiI62) 6 m (MathH54)	☢ β⁺ (MathH54), [EC]	B chem, genet (MathH54, PreiI62) parent Xe¹²³ (MathH54, MathH54a, PreiI62) daughter Ba¹²³ (PreiI62)		In¹¹⁵(C¹², 4n) (PreiI62) I¹²⁷(α, 8n) (MathH54)

TABLE 1 (*Continued*)

Isotope Z A	Half-life	Type of decay (♣); % abundance; Mass excess (△≡M–A), MeV (C¹²=0); Thermal neutron cross section (σ), barns	Class; Identification; Genetic relationships	Major radiations: approximate energies (MeV) and intensities	Principal means of production
₅₅Cs¹²⁵	45 m (MathH54) 49 m (PreiI62)	♣ EC 51%, β⁺ 49% (FrieG62) △ -84 (MTW)	A chem, mass spect (MathH54, MicM54) parent Xe¹²⁵ᵐ (≈0.1%), parent Xe¹²⁵ (MathH54) daughter Ba¹²⁵ (PreiI62) [descendant La¹²⁵] (PreiI63)	β⁺ 2.05 max e⁻ 0.077, 0.107 γ Xe X-rays, 0.112, 0.511 (98%, γ±) daughter radiations from Xe¹²⁵ Xe¹²⁵ᵐ	I¹²⁷(α, 6n) (MathH54) In¹¹⁵(N¹⁴, 4n)Ba¹²⁵(β⁺) (PreiI62)
Cs¹²⁶	1.6 m (KalkM54)	♣ β⁺ 82%, EC 18% (KalkM54) △ -84.4 (MTW)	A chem, mass spect (KalkM54) daughter Ba¹²⁶ (KalkM54)	β⁺ 3.8 max γ Xe X-rays, 0.386 (38%), 0.511 (164%, γ±)	daughter Ba¹²⁶ (KalkM54)
Cs¹²⁷	6.2 h (MathH54, PreiI63) 6.1 h (MicM54, NijG55) 5.5 h (FinR50a)	♣ EC 96.5%, β⁺ 3.5% (FrieG62) △ -86.4 (MTW, WintG65a)	A chem, mass spect (FinR50a, MicM54) parent Xe¹²⁷ (FinR50a) parent Xe¹²⁷ᵐ (0.01%) (MathH54) daughter Ba¹²⁷ (LindnM52, PreiI62) descendant La¹²⁷ (YafL63)	γ Xe X-rays, 0.125 (10%), 0.406 (72%), 0.511 (7%, γ±) e⁻ 0.090, 0.119, 0.371 β⁺ 1.08 max daughter radiations from Xe¹²⁷	I¹²⁷(α, 4n) (FinR50a, MicM54, MathH54)
Cs¹²⁸	3.8 m (LindnM52) 3.9 m (WapA53a) 3.5 m (FinR53) 2.5 m (MurA55)	♣ β⁺ ≈51%, EC ≈49% (JhaS61) β⁺ 75%, EC 25% (HollaJ55) △ -85.92 (MTW)	B chem, genet (FinR51) daughter Ba¹²⁸ (FinR51, LindnM52, HollaJ55) descendant La¹²⁸ (YafL63)	β⁺ 2.9 max e⁻ 0.407 γ Xe X-rays, 0.441 (27%), 0.511 110%, γ±), 0.528, 0.576, 0.97 (1%), 1.12 (1%) See also γ's of Ba¹²⁸	daughter Ba¹²⁸ (FinR51, LindnM52, HollaJ55)
Cs¹²⁹	32.1 h (SheraE65) 30.7 h (NijG55) 31 h (FinR50a)	♣ EC, no β⁺ (FinR50a) △ -88 (MTW)	A chem, mass spect (FinR50a, MicM54) daughter Ba¹²⁹ (ThomC50, FinR50)	γ Xe X-rays, 0.040 (2%), 0.280 (3%), 0.320 (4%), 0.375 (48%), 0.416 (25%), 0.550 (5%) e⁻ 0.005, 0.034, 0.057, 0.336, 0.376	I¹²⁷(α, 2n) (FinR50a, JhaS60a, NierW58) daughter Ba¹²⁹ (ThomC50, FinR50)
Cs¹³⁰	30 m (SmiA52a, MicM54) others (FinR50a)	♣ β⁺, EC, β⁻ (β⁺/β⁻ 27.5) (SmiA52a) △ -86.89 (MTW)	A chem, excit (SmiA52a) chem, mass spect (MicM54)	β⁺ 1.97 max β⁻ 0.442 max γ Xe X-rays, 0.511 (γ±)	I¹²⁷(α, n) (FinR50a, SmiA52a, NierW58)
Cs¹³¹	9.70 d (GleG64) 9.69 d (LarN60) others (LyoW63, YafL49, KatcS47a, YuF49, KondE50, JosB60)	♣ EC, no β⁺ (FinB47, CanR51b, KondE50) △ -88.06 (MTW)	A chem, genet (KatcS47a) chem, mass spect (KarrD49) daughter Ba¹³¹ (KatcS47a, YuF47, YafL49, CanR51b) not parent Xe¹³¹ᵐ (CanR51b, SaraB54)	γ Xe X-rays	Ba¹³⁰(n, γ)Ba¹³¹(EC) (KatcS47a, YuF47, YafL49, CanR51b)
Cs¹³²	6.59 d (DeaP64) 6.54 d (RobiR62a) 6.48 d (WhyG60) others (CamM44)	♣ EC 97%, β⁺ 0.6%, β⁻ 2% (RobiR62a, TayH63) β⁺ 1.2% (JhaS61b) △ -87.19 (MTW)	A chem, excit (CamM44) genet energy levels (BhaK56, RobiR62a)	β⁺ 0.40 max β⁻ [0.7 max] γ Xe X-rays, 0.48 (4%, complex), 0.668 (99%), 1.138 (0.5%), 1.320 (0.6%)	Cs¹³³(p, pn) (JhaS61b, RobiR62a, TayH63) Xe¹³²(p, n) (NierW58) Cs¹³³(n, 2n) (CamM44, LangeL51a)
<u>Cs¹³³</u>		% 100 (NierA37a, WhiF56) △ -88.16 (MTW) σc 28 (to Cs¹³⁴) 2.6 (to Cs¹³⁴ᵐ) (GoldmDT64)			
Cs¹³⁴	2.046 y (DieL63) 2.05 y (EasH60) 1.99 y (FlyK65a) 2.07 y (WyaE61, GeiKW57) 2.19 y (MerW57) 2.26 y (EdwJ58) others (BayJ58, GleL51m, KalbD40, ScheiH38, SerL47b)	♣ β⁻ (KalbD40) no EC, lim 1% (KeiG55) no β⁺, lim 0.009% (MimW51) △ -86.79 (MTW) σc 136 (GoldmDT64)	A n-capt (AlexK38) chem, n-capt, excit (KalbD40)	β⁻ 0.662 max γ 0.57 (23%, complex), 0.605 (98%), 0.796 (99%, complex), 1.038 (1.0%), 1.168 (1.9%), 1.365 (3.4%)	Cs¹³³(n, γ) (AlexK38, ScheiH38, KalbD40, SerL47b)
Cs¹³⁴ᵐ	2.895 h (KeiB61) 2.91 h (BaeA60, WarhH64) others (SlaH45, KalbD40, SerL47b)	♣ IT (GoldhM48a, CaldR50) β⁻ ≈1% (KeiG55) △ -86.65 (MTW, LHP)	A chem, n-capt (AmaE35, MLenJ35a) chem, excit, n-capt (KalbD40)	γ Cs X-rays, 0.128 (14%) e⁻ 0.005, 0.009, 0.092, 0.122 β⁻ 0.55 max	Cs¹³³(n, γ) (AmaE35, MLenJ35a, KalbD40, SerL47b)

TABLE 1 (*Continued*)

Isotope Z A	Half-life	Type of decay (☢); % abundance; Mass excess (△≡M–A), MeV (C¹²=0); Thermal neutron cross section (σ), barns	Class; Identification; Genetic relationships	Major radiations: approximate energies (MeV) and intensities	Principal means of production
$_{55}$Cs^{135}	3.0 x 10^6 y sp act (ZelH49) 2.1 x 10^6 y yield (SugaN49a)	☢ β⁻ (SugaN49a) △ -87.8 (MTW) σ$_c$ 8.7 (GoldmDT64)	A chem, genet (SugaN49a) chem, mass spect (IngM49) daughter Xe^{135} (SugaN49a)	β⁻ 0.21 max γ no γ	daughter Xe^{135} (SugaN49a) fission (ZelH49)
Cs^{135m}	53 m (WarhH62, HalleI64)	☢ IT (WarhH62) △ -86.2 (MTW, LHP)	A chem, sep isotopes, cross bomb, crit abs (WarhH62) chem, mass spect (HalleI64)	γ Cs X-rays, 0.781 (100%), 0.840 (96%) e⁻ 0.745, 0.775, 0.804	Xe^{134}(d, n) (WarhH62) Xe^{132}(α, p) (WarhH62) Ba^{135}(n, p) (WarhH62) protons on Ba (HalleI64)
Cs^{136}	13.7 d (GleL49) 12.9 d (OlsJ54a) 13.5 d (WilleR60)	☢ β⁻ (GleL51ℓ) △ -86.6 (LHP, MTW)	A chem (GleL46, GleL51ℓ) chem, excit (GleL49) chem, mass spect (OlsJ54a)	β⁻ 0.657 max (7%), 0.341 max γ 0.116, 0.126, 0.158, 0.302 γ Ba X-rays, 0.067 (11%), 0.086 (6%), 0.16 (36%, complex), 0.273 (18%), 0.340 (53%), 0.818 (100%), 1.05 (82%), 1.25 (20%) daughter radiations from Ba^{136m} included in above listing	La^{139}(n, α) (CamM44, GleL49, BernsH61) Ba^{138}(d, α) (GirR59, GrabZ60b)
Cs^{137}	30.0 y (weighted average by FlyK65) 29.7 y (GorbS63) 30.4 y mass spect (FarrH61, DieL63) 29.2 y mass spect (RideB63) 30.0 y sp act, mass spect (BrowF55) others (FlyK65, FleD62a, WileDM55a, GlazM61, WileDR53, GleL51j)	☢ β⁻ (MelhM41) △ -86.9 (MTW) σ$_c$ 0.11 (GoldmDT64)	A chem, genet (MelhM41) chem, mass spect (HaydR46a, IngM49) daughter Ba^{137} (TurA51, GleL51k) parent Ba^{137m} (TownJ48)	β⁻ 1.176 max (7%), 0.514 max e⁻ 0.624, 0.656 γ Ba X-rays, 0.662 (85%) daughter radiations from Ba^{137m} included in above listing	fission (HaydR48, IngM49, GleL51j, GrumW48, FinB51c)
Cs^{138}	32.2 m (BarthR56) 32.1 m (BunkM56) others (GlasG40, WilleR60, EvaHB51, AteA39, HahO39a, GleL51k, OckD62, LangeL53a)	☢ β⁻ (HahO39c) △ -83.7 (NDS, MTW)	A chem (HahO39c, HeyF39) chem, mass spect (ThuS49) descendant I^{138} (SugaN49) daughter Xe^{138} (HahO39c, HahO40, GlasG40, SeeW43a)	β⁻ 3.40 max γ 0.463 (23%), 0.55 (8%), 1.01 (25%), 1.426 (73%), 2.21 (18%), 2.63 (9%)	fission (HahO39c, HahO40a, HeyF39, HahO40, BunkM56) Ba^{138}(n, p) (WilleR60, SeeW43a)
Cs^{139}	9.5 m (SugaN50, ZheE63) others (AteA39, HeyF39, OckD62, HahO40)	☢ β⁻ (HahO39c) △ -81.1 (MTW)	A chem, genet (HahO39c, HeyF39) daughter Xe^{139} (HahO39c, HeyF39, HahO40a, HahO40) parent Ba^{139} (HahO39c, HeyF39, HahO40a, HahO40, SugaN50)	γ 0.50, 0.63, 0.80, 1.28 (strong), 1.65 (complex), 1.90, 2.08 daughter radiations from Ba^{139}	fission (HahO39c, HahO40a, AteA39, SugaN50, HahO40a, HahO40, AksV62, ZheE63, OckD62)
Cs^{140}	66 s (SugaN50) 63 s (ZheE63)	☢ β⁻ (HahO40) △ -77 (MTW)	A chem (HahO40) chem, genet (SugaN50) parent Ba^{140} (SugaN50)	γ 0.59, 0.88, 1.14, 1.62, 1.85, 2.06, 2.32, 2.72, 3.15	fission (HahO40, SugaN50, ZheE63)
Cs^{141}	24 s (FritK62a) 25 s (WahA62)	☢ [β⁻] (BradE51)	A chem, genet (WahA62, FritK62a) parent Ba^{141} (WahA62, HahO42a) ancestor Ce^{141} (FritK62a)		fission (BradE51, DilC51a, OveR51, WahA62, FritK62a)
Cs^{142}	2.3 s (FritK62a) others (WahA62, HahO42a)	☢ [β⁻] (FritK62a)	B chem, genet (FritK62a) ancestor La^{142} (FritK62a)		fission (FritK62a)
Cs^{143}	2.0 s (FritK62a)	☢ [β⁻] (BradE51)	B genet (BradE51) chem, genet (FritK62a) ancestor La^{143} (FritK62a)		fission (BradE51, DilC51a)
Cs^{144}	short (DilC51, DilC51a)	☢ [β⁻] (DilC51)	F genet (DilC51) [descendant Xe^{144}, ancestor Ce^{144}] (DilC51)		descendant Xe^{144} from fission (DilC51, DilC51a)
$_{56}$Ba^{123}	2.0 m (PreiI62)	☢ [β⁺, EC] (PreiI62)	B chem, cross bomb, genet (PreiI62) parent Cs^{123} (PreiI62)		O^{16} on In, Sn (PreiI62) N^{14} on In (PreiI62) C^{12} on Sn (PreiI62)

TABLE 1 (*Continued*)

Isotope Z A	Half-life	Type of decay (☢); % abundance; Mass excess (△≡M-A), MeV (C¹²=0); Thermal neutron cross section (σ), barns	Class; Identification; Genetic relationships	Major radiations: approximate energies (MeV) and intensities	Principal means of production
₅₆Ba¹²⁵	6.5 m (PreiI62)	☢ [EC, β⁺] (PreiI62)	B chem, cross bomb, genet (PreiI62) parent Cs¹²⁵ (PreiI62)		In¹¹⁵(N¹⁴, 4n) C¹² on Sn, O¹⁶ on In (PreiI62)
Ba¹²⁶	97 m (KalkM54) 103 m (PreiI62)	☢ EC (KalkM54)	A chem, genet (KalkM54) chem, cross bomb (PreiI62) parent Cs¹²⁶ (KalkM54) daughter La¹²⁶ (PreiI63, ShelR61)	Y 0.23 (↑ 100), 0.70 (↑ 33), 0.9 (weak) daughter radiations from Cs¹²⁶	In¹¹⁵(N¹⁴, 3n) (KalkM54, PreiI62) C¹², O¹⁶ on Sn (PreiI62)
Ba¹²⁷	10.0 m genet (PreiI62) 12 m (KalkM54, LindnM52)	☢ β⁺ (LindnM52), [EC] △ -83 (MTW)	A chem, genet (LindnM52) chem, genet, cross bomb (PreiI62) parent Cs¹²⁷ (LindnM52, PreiI62) daughter La¹²⁷ (PreiI63)		O¹⁶, N¹⁴ on In; C¹², O¹⁶ on Sn (PreiI62) Cs¹³³(d, 8n) (LindnM52)
Ba¹²⁸	2.43 d (YafL63) 2.4 d (PreiI63, FinR50, ThomC50)	☢ EC (FinR53, LindnM52) △ -85 (MTW)	B chem (FinR50, ThomC50) parent Cs¹²⁸ (FinR51, LindnM52, HollaJ55) daughter La¹²⁸ (YafL63, PreiI63)	Y Cs X-rays, 0.134, 0.278 e⁻ 0.128, 0.242 (above radiations with Ba¹²⁸ or Cs¹²⁸) daughter radiations from Cs¹²⁸	Cs¹³³(p, 6n) (FinR50, ThomC50, LindnM52) Cs¹³³(d, 7n) (LindnM52)
Ba¹²⁹, Ba¹²⁹ᵐ?	2.61 h (β⁺) (ArbE61) 2.0 to 2.4 h (conv, e⁻) (ArbE61) 2.20 h (YafL63) 2.45 h (HenkW59)	☢ EC 94%, β⁺ 6% (ArbE61) △ -85 (MTW)	A chem, genet (ThomC50, FinR50) parent Cs¹²⁹ (ThomC50, FinR50) probable isomerism shown by different half-lives of electron lines (ArbE61) daughter La¹²⁹ (PreiI63, LavA63, YafL63)	β⁺ 1.42 max e⁻ 0.017, 0.048, 0.093, 0.142, 0.171 others to 1.5 Y Cs X-rays, 0.129 (↑ 26), 0.182 (↑ 100), 0.21 (↑ 65, complex), 0.511 (Y±), 1.45 (↑ 42) daughter radiations from Cs¹²⁹	Cs¹³³(p, 5n) (ThomC50, FinR50, ArbE61)
Ba¹³⁰		% 0.101 (NierA38b) 0.13 (AkiP56) △ -87.33 (MTW) σ_c 8.8 (GoldmDT64)			
Ba¹³¹	12.0 d (KatcS47a, WriH57, LyoW63, SmiKM63) 11.5 d (BegW56) 11.8 d (CorkJ53c) 11.7 d (YuF47)	☢ EC (KatcS47a) no β⁺ (YuF47, FinB47) △ -86.89 (MTW)	A chem, n-capt, excit (KatcS47a) parent Cs¹³¹ (KatcS47a, YuF47, YafL49, CanR51b) daughter La¹³¹ (YafL63) daughter Ba¹³¹ᵐ (TilR63)	Y Xe X-rays, 0.124 (28%, complex), 0.216 (19%), 0.25 (5%, complex), 0.373 (13%), 0.496 (48%, complex), 0.60 (3%, doublet), 0.924 (0.8%), 1.048 (1.3%) e⁻ 0.019, 0.042, 0.049, 0.088, 0.097, 0.118, 0.180, 0.460 daughter radiations from Cs¹³¹	Ba¹³⁰(n, Y) (KatcS47a, YuF47, YafL49, DalE50, ZimE50, CanR51b) Cs¹³³(p, 3n) (HiroT64)
Ba¹³¹ᵐ	14.6 m (HoreD63a) 14.5 m (TilR63)	☢ IT, no EC, lim 0.1% (TilR63) △ -86.71 (LHP, MTW)	A chem, excit, cross bomb, genet (TilR63) parent Ba¹³¹ (TilR63) not daughter La¹³¹, lim 1% (HoreD63a)	Y Ba X-rays, 0.107 (40%) e⁻ [0.041, 0.071, 0.101]	Cs¹³³(p, 3n) (TilR63)
Ba¹³²		% 0.097 (NierA38b) 0.19 (AkiP56) △ -88.4 (MTW) σ_c 7 (to Ba¹³³) <0.2 (to Ba¹³³ᵐ) (GoldmDT64)			
Ba¹³³	7.2 y (KatcS56a) 10.7 y (WyaE61)	☢ EC (KatcS47a) no β⁺, lim 0.1% (LangeM56) △ -87.67 (MTW)	A chem, n-capt, excit (KatcS47a) chem, genet (YuF48) daughter Ba¹³³ᵐ (YuF48)	Y Cs X-rays, 0.080 (36%, complex), 0.276 (7%), 0.302 (14%), 0.356 (69%), 0.382 (8%) e⁻ 0.045, 0.075, 0.266, 0.319	Ba¹³³(n, Y) (KatcS47a, CrasB57) Cs¹³³(p, n) (GupR58)
Ba¹³³ᵐ	38.9 h (WilleR60, YuF48) others (MocD48)	☢ IT (CorkJ41) △ -87.39 (LHP, MTW)	A chem, excit (CorkJ41, DubL40a) parent Ba¹³³ (YuF48)	Y Ba X-rays, 0.276 (17%) e⁻ 0.006, 0.011, 0.238, 0.270	Cs¹³³(p, n) (DubL40a) Cs¹³³(d, 2n) (CorkJ41, HillR51b, HillR51d) Ba¹³²(n, Y) (YuF48)

TABLE 1 (*Continued*)

Isotope Z A	Half-life	Type of decay ($\maltese$); % abundance; Mass excess ($\Delta \equiv$ M–A), MeV (C^{12}=0); Thermal neutron cross section (σ), barns	Class; Identification; Genetic relationships	Major radiations: approximate energies (MeV) and intensities	Principal means of production
$_{56}$Ba^{134}		% 2.42 (NierA38b) 2.60 (AkiP56) Δ –88.85 (MTW) σ_c <4 (to Ba^{135}) 0.16 (to Ba^{135m}) (GoldmDT64)			
Ba^{135}		% 6.59 (NierA38b) 6.7 (AkiP56) Δ –88.0 (MTW) σ_c 5 (GoldmDT64)			
Ba^{135m}	28.7 h (WilleR60, YuF48)	$\maltese$ IT (WeimK43a, YuF48) Δ –87.7 (MTW, LHP)	A chem (KalbD40) chem, n-capt, sep isotopes (HillR51b) not daughter La^{135} (MoriS65)	γ Ba X-rays, 0.268 (16%) e$^-$ 0.231, 0.262	Ba^{134}(n, γ) (HillR51b, KalbD40)
Ba^{135}	0.32 s (FetP62a) others (CamE59)		G sep isotopes (FetP62a) assigned to Ba^{136m} (RudF65)		neutrons on Ba^{135} (FetP62a, CamE59)
Ba^{136}		% 7.81 (NierA38b) 8.1 (AkiP56) Δ –89.1 (MTW) σ_c <1 (to Ba^{137}) 0.010 (to Ba^{137m}) (GoldmDT64)			
Ba^{136m}	0.32 s (FetP62a) 0.37 s (RudF65) others (CamE59)	$\maltese$ IT (RudF65) Δ –87.1 (LHP, MTW)	B chem, genet, genet energy levels (RudF65)	γ Ba X-rays, 0.164 (40%), 0.818 (100%), 1.05 (100%) e$^-$ [0.126, 0.158]	daughter Cs^{136} (RudF65)
Ba^{137}		% 11.32 (NierA38b) 11.9 (AkiP56) Δ –88.0 (MTW) σ_c 4 (GoldmDT64)			
Ba^{137m}	2.554 m (MerJ65) 2.60 m (MitA49) 2.6 m (TownJ48, WilleR60)	$\maltese$ IT (TownJ48) Δ –87.4 (LHP, MTW)	A n-capt (AmaE35) chem, genet (TownJ48) daughter Cs^{137} (TownJ48)	γ Ba X-rays, 0.662 (89%) e$^-$ 0.624, 0.656	daughter Cs^{137} (TownJ48)
Ba^{138}		% 71.66 (NierA38b) 70.4 (AkiP56) Δ –88.5 (MTW) σ_c 0.4 (GoldmDT64)			
Ba^{139}	82.9 m (ButlJP58, FritK62) 84.0 m (BaeA57) 85.0 m (DilC51c) others (WilleR60, ShepL48, HahO40, KellWH60, PoolM37a)	$\maltese$ β^- (PoolM37a) Δ –85.1 (MTW) σ_c 4 (GoldmDT64)	A chem, n-capt (AmaE35) chem, excit (PoolM38a) daughter Cs^{139} (HahO39c, HeyF39, HahO40a, HahO40, SugaN50) descendant Xe^{139} (HahO39c, HeyF39, DilC51a) descendant I^{139} (SugaN49)	β^- 2.3 max e$^-$ 0.126, 0.159 γ La X-rays, 0.166 (23%), 1.43 (0.4%)	Ba^{138}(n, γ) (AmaE35, PoolM37, SerL47b, YafLA9a) fission (HeyF39, HahO39c, DilC51a, KatcS48, FinB51c)
Ba^{140}	12.80 d (EngeD51c) 12.8 d (Vasil58)	$\maltese$ β^- (HahO39c) Δ –83.31 (MTW) σ_c <20 (GoldmDT64)	A chem, genet (HahO39, HahO39c) parent La^{140} (HahO39, HahO39c, HahO40, GlasG40, HahO42a, GrumW46, FinB51b) daughter Cs^{140} (SugaN50) descendant Xe^{140} (HahO40, BradE51, DilC51a, DilC51, OveR51)	β^- 1.02 max e$^-$ 0.024, 0.029 γ La X-rays, 0.030 (11%), 0.163 (6%), 0.305 (6%), 0.438 (5%), 0.537 (34%) daughter radiations from La^{140}	fission (HahO39, HeyF39, HahO40, GlasG40, GrumW46, SugaN50, DilC51a, DilC51, BradE51, OveR51, WilkR51, EngeD51c, EngeD51d, KatcS48, FinB51c)
Ba^{141}	18 m (SchumR59, FritK62, HahO42a, GoldsA51)	$\maltese$ β^- (HahO42a) Δ –80.1 (MTW)	A chem, genet (HahO42a) daughter Cs^{141} (HahO42a, WahA62) parent La^{141} (HahO62a) descendant Xe^{141} (BradE51, OveR51, DilC51a) others (HahO39a, HahO39, GoldsA51a, LangeA40)	β^- 3.0 max γ La X-rays, 0.118 ($\uparrow$ 10), 0.193 ($\uparrow$ 100), 0.28 ($\uparrow$ 50), 0.31 ? ($\uparrow$ 60), 0.35 ($\uparrow$ 20), 0.46 ($\uparrow$ 30, complex), 0.64 ($\uparrow$ 20, complex?), 0.73 ($\uparrow$ 7), 0.86 ($\uparrow$ 6), 0.93 ($\uparrow$ 3), 1.19 ($\uparrow$ 8), 1.29 ($\uparrow$ 3), 1.42 ($\uparrow$ 4), 1.65 ($\uparrow$ 3) daughter radiations from La^{141}	fission (HahO42a, GoldsA51, GoldsA51a, BradE51, OveR51, DilC51a, SchumR59, FritK62, NagaK60)

TABLE 1 (*Continued*)

Isotope Z A	Half-life	Type of decay (☢); % abundance; Mass excess (Δ≡M–A), MeV (C^{12}=0); Thermal neutron cross section (σ), barns	Class; Identification; Genetic relationships	Major radiations: approximate energies (MeV) and intensities	Principal means of production
$_{56}Ba^{142}$	11 m (SchumR59, FritK62a) others (HahO42a)	☢ β^- (HahO42a) Δ –77.9 (MTW)	B chem, genet (HahO42a) parent La^{142} (HahO42a) others (HahO39a, HahO39, LangeA40)	β^- 1.7 max γ La X-rays, 0.080 (↑ 30), 0.26 (↑ 100), 0.89 (↑ 40), 0.97 (↑ 15), 1.08 (↑ 10), 1.20 (↑ 35) daughter radiations from La^{142}	fission (SchumR59, FritK62, HahO42a)
Ba^{143}	12 s (WahA62)	☢ β^- (HahO42a)	B chem, genet (HahO42a) chem (WahA62) parent La^{143} (HahO42a)		fission (HahO42a, WahA62, FritK62a)
Ba^{144}	short (DilC51a, DilC51)	☢ [β^-] (DilC51a)	F genet (DilC51a) [descendant Xe^{144}, ancestor Ce^{144}] (DilC51a, DilC51)		descendant Xe^{144} from fission (DilC51, DilC51a)
$_{57}La^{125}$	<1 m (PreiI63)	☢	F chem, genet [ancestor Cs^{125}] (PreiI63)		O^{16} on In (PreiI63)
La^{126}	1.0 m (ShelR61, PreiI63)	☢ [β^+, EC] (ShelR61)	B chem, cross bomb, genet (ShelR61) chem (PreiI63) parent Ba^{126} (PreiI63, ShelR61)	γ Ba X-rays, 0.256, 0.511 ($\gamma^\pm$)	$In^{115}(O^{16}, 5n)$ (ShelR61, PreiI63) $Sb^{121}(C^{12}, 7n)$ (ShelR61)
La^{127}	3.5 m genet (YafL63) 3.8 m genet (PreiI63)	☢ [β^+, EC] (PreiI63, YafL63)	B chem, genet (PreiI63, YafL63) parent Ba^{127} (PreiI63) ancestor Cs^{127} (YafL63)		C^{12} on Sb (YafL63) O^{16} on In (PreiI63)
La^{128}	4.2 m (PreiI63) 4.6 m (YafL63) 6 m (ShelR61)	☢ [β^+, EC] (ShelR61)	B chem, cross bomb (ShelR61) chem, genet (YafL63, PreiI63) parent Ba^{128} (PreiI63, YafL63) ancestor Cs^{128} (YafL63)	γ Ba X-rays, 0.279, 0.511 ($\gamma^\pm$)	$Sb^{121}(C^{12}, 5n)$ (ShelR61, YafL63) $Sb^{123}(C^{12}, 7n)$ (ShelR61, YafL63) $In^{115}(O^{16}, 3n)$ (ShelR61, PreiI63)
La^{129}	10.0 m (YafL63) 7.2 m genet (PreiI63) ≈24 m (LavA63)	☢ [β^+, EC] (PreiI63, LavA63, YafL63) Δ –81 (MTW)	A chem, genet (PreiI63, LavA63) chem, sep isotopes, cross bomb, genet (YafL63) parent Ba^{129} with $t_{1/2}$ 2.20 h (YafL63), 2.1 to 2.4 h (LavA63) daughter Ce^{129} (LavA63)		C^{12} on Sb (YafL63) O^{16} on In (PreiI63)
La^{130}	8.7 m (YafL63) 9 m (ShelR61)	☢ β^+, EC (ShelR61, YafL63) Δ –82 (MTW)	A chem, cross bomb, genet energy levels (ShelR61) chem, sep isotopes (YafL63)	γ Ba X-rays, 0.356, 0.45, 0.511 ($\gamma^\pm$), 0.55, 0.72, 0.81, 0.91, 1.01, 1.19, 1.45, 1.55	$Ba^{130}(p,n)$ (YafL63) $Sb^{121}(C^{12}, 3n)$ (ShelR61) $Sb^{123}(C^{12}, 5n)$ (ShelR61)
La^{131}	56 m genet (YafL63) 61 m (CreC60) 58 m (GranM51)	☢ EC 72%, β^+ 28% (CreC60) Δ –83.9 (MTW)	A chem, mass spect (GranM51) chem, genet (YafL63) parent Ba^{131} (YafL63) not parent Ba^{131m}, lim 1% (HoreD63a)	β^+ 1.94 max e^- 0.078, others γ Ba X-rays, 0.115 (23%), 0.169 (5%), 0.214 (8%), 0.285 (17%), 0.364 (20%), 0.417 (20%), 0.455 (8%), 0.511 (56%, $\gamma^\pm$), 0.597 (7%), 0.878 (4%)	$Ba^{130}(d,n)$ (CreC60) $Sb^{123}(C^{12}, 4n)$ (YafL63, HoreD63a)
La^{132}	4.5 h (GranM51) 4.8 h (WareW60) 4.2 h (GrigE60)	☢ β^+ (GranM51), [EC] Δ –83.1 (LHP, MTW)	A chem, mass spect (GranM51) daughter Ce^{132} (WareW60)	β^+ 3.8 max γ Ba X-rays, 0.47, 0.511 ($\gamma^\pm$), 0.56, 0.66, 0.90 (doublet), 1.03, 1.22, 1.58, 1.92	protons on Ba (GranM51)
La^{133}	4.0 h (NauR50)	☢ EC, β^+ (weak) (NauR50) Δ –85.5 (MTW)	A chem, mass spect (NauR50) daughter Ce^{133} (StovB51)	γ Ba X-rays, 0.511 ($\gamma^\pm$), 0.8 β^+ 1.2 max e^- 0.26	$Cs^{133}(\alpha, 4n)$ (NauR50)
La^{134}	6.8 m (GirR59a) 6.5 m (StovB51)	☢ β^+ 62%, EC 38% (GirR59a) β^+ ≈44%, EC ≈56% (StovB51) Δ –85.1 (MTW)	B chem, genet (StovB51) daughter Ce^{134} (StovB51)	β^+ 2.7 max γ Ba X-rays, 0.511 (124%, $\gamma^\pm$), 0.605 (6%)	daughter Ce^{134} (StovB51) $Cs^{133}(\alpha, 3n)$ (GirR59a)

TABLE 1 (*Continued*)

Isotope Z A	Half-life	Type of decay (☢); % abundance; Mass excess (△≡M-A), MeV (C¹²=0); Thermal neutron cross section (σ), barns	Class; Identification; Genetic relationships	Major radiations: approximate energies (MeV) and intensities	Principal means of production
₅₇La¹³⁵	19.4 h (MoriS65) 19.8 h (MitA58) 19.5 h (ChubJ48) others (NauR50, WeimK43)	☢ EC (MounK42, ChubJ48) no β⁺, lim 0.002% (MoriS65) others (GrenH65, MitA58) △ -87.0 (MoriS65, MTW)	A chem (MounK42) chem, excit (ChubJ48) chem, mass spect (NauR50) daughter Ce¹³⁵ (ChubJ48) not parent Ba¹³⁵ᵐ (MoriS65)	γ Ba X-rays, 0.481 (1.9%), 0.588 (0.13%), 0.87 (0.24%, complex) e⁻ 0.181, 0.444, 0.475	Cs¹³³(α, 2n) (ChubJ48, NauR50, MitA58) Ba¹³⁴(d, n) (MounK42, WeimK43) Ba¹³⁸(p, 4n) (MoriS65) Ba¹³⁵(p, n) (WeimK43)
La¹³⁶	9.5 m (NauR50) 9.0 m (RobeB50) 10.0 m (GirR59) others (MauW47)	☢ EC ≈67%, β⁺ ≈33% (NauR50) △ -86.3 (MTW)	A chem (MauW47) chem, excit, sep isotopes (RobeB50)	β⁺ 1.9 max γ Ba X-rays, 0.511 (66%, γ±), 0.818 (2.5%)	Cs¹³³(α, 2n) (RobeB50, NauR50, GirR59) Ba¹³⁵(d, n), Ba¹³⁶(d, 2n) (RobeB50)
La¹³⁷	6 x 10⁴ y sp act (BrosA56) others (ChubJ48, IngM48c, BrosA55)	☢ EC (BrosA56) △ -88 (MTW)	A mass spect (IngM48c) chem (BrosA56)	γ Ba X-rays	Ce¹³⁶(n, γ) Ce¹³⁷(β⁻) (IngM48c, BrosA56, BrosA55, ChubJ48)
La¹³⁸	1.12 x 10¹¹ y sp act (GloR57) 1.1 x 10¹¹ y sp act (TurW56) others (PriR51, MulhG52a)	☢ EC ≈70%, β⁻ ≈30% (GloR57) EC 53%, β⁻ 47% (TurW56) EC ≈94%, β⁻ ≈6% (MulhG52a) % 0.089 (IngM47e, WhiF56) △ -86.7 (MTW)	A chem, mass spect (IngM47e)	β⁻ 0.21 max γ Ba X-rays, 0.81 (30%), 1.426 (70%)	
La¹³⁹		% 99.911 (WhiF56, IngM47e) △ -87.43 (MTW) σ_c 8.9 (GoldmDT64)			
La¹⁴⁰	40.22 h (KirH54) 40.27 h (PepD57) 40.3 h (YafL54a) 40.0 h (BallN51b, BisG50, WeimK43)	☢ β⁻ (PoolM38a) △ -84.36 (MTW)	A n-capt (MarsJK35) chem, excit, n-capt (PoolM38a) chem, mass spect (HaydR48) daughter Ba¹⁴⁰ (HahO39, HahO39c, HahO40, GlasG40, HahO42a, GrumW46, FinB51b)	β⁻ 2.175 max (6%), 1.69 max (15%), 1.36 max γ 0.329 (20%), 0.487 (40%), 0.815 (19%), 0.923 (10%), 1.596 (96%), 2.53 (3%)	La¹³⁹(n, γ) (MarsJK35, PoolM38a, GotH42, WeimK43, SerL47b) fission, daughter Ba¹⁴⁰ (HahO39, HahO39c, HahO40, GlasG40, HahO42a, GrumW46, FinB51b, GrumW48, GrumW47, FinB51c)
La¹⁴¹	3.87 h (AlsJ60) 3.90 h (FritK62) others (SchumR59, RydH58, KatcS51i, HahO42a)	☢ β⁻ (HahO42a) △ -83.06 (MTW)	A chem (HahO42a) chem, genet (BurgW51, DufR51a) daughter Ba¹⁴¹ (HahO42a) parent Ce¹⁴¹ (BurgW51, DufR51a) descendant Xe¹⁴¹ (BradE51) others (KatcS49, CuriI39, BallN51h)	β⁻ 2.43 max γ 1.37 (2%) daughter radiations from Ce¹⁴¹	fission (HahO42a, KatcS51i, Schum R59, AlsJ60, FritK62)
La¹⁴²	92.5 m (FritK62) 81 m (RydH58) 77 m (KatcS51i, BosA53, WilleR60) others (HahO42a)	☢ β⁻ (KatcS51i) △ -80.1 (MTW)	A chem (HahO42a, PresW64) sep isotopes, excit (WolfsK60) genet energy levels (PresW64, HansO63) daughter Ba¹⁴² (HahO42a) descendant Cs¹⁴² (FritK62a) descendant Xe¹⁴² (WolfsK60)	β⁻ 4.51 max γ 0.65 (48%), 0.90 (9%), 1.01 (5%), 1.06 (4%), 1.55 (5%, complex), 1.74 (5%), 1.91 (9%), 2.06 (6%), 2.41 (15%), 2.55 (11%), 2.99 (5%), 3.31 (1.9%), 3.65 (2.3%)	fission (PresW64, HahO42a, KatcS51i, HahO43a, GesH51, RydH58, BosA53, FritK62, SchumR59) Ce¹⁴²(n, p) (WilleR60, WolfsK60)
La¹⁴³	14.0 m (FritK61a) others (HahO43a, GesH51)	☢ β⁻ (GesH51) △ -78.4 (MTW)	A chem, genet (GesH51) parent Ce¹⁴³ (GesH51) daughter Ba¹⁴³ (HahO42a) descendant Cs¹⁴³ (FritK62a)	β⁻ 3.3 max γ 0.62 (↑ 100), 0.80 (↑ 44), 1.07 (↑ 26), 1.17 (↑ 57), 1.58 (↑ 28), 1.98 (↑ 35), 2.56 (↑ 27)	fission (HahO42a, HahO43a, GesH51)
La¹⁴⁴	short (DilC51)	☢ [β⁻] (DilC51) △ -75 (MTW)	F genet (DilC51) [descendant Xe¹⁴⁴, ancestor Ce¹⁴⁴] (DilC51)		descendant Xe¹⁴⁴ from fission (DilC51)
₅₈Ce¹²⁹	≈13 m (LavA63)	☢ [β⁺, EC] (LavA63)	E chem, genet (LavA63) parent La¹²⁹ (LavA63)	γ La X-rays, 0.080, 0.32, 0.75 daughter radiations from La¹²⁹, Ba¹²⁹	protons on Pr (LavA63)
Ce¹³⁰	30 m (AlboG65, WareW60)	☢ [EC, β⁺] (AlboG65, GersG65)	B chem, mass spect (AlboG65)	γ [La X-rays], 0.13 daughter radiations from La¹³⁰	La¹³⁹(p, 10n) (GersG65)

TABLE 1 (*Continued*)

Isotope Z A	Half-life	Type of decay (☢); % abundance; Mass excess ($\Delta \equiv M-A$), MeV ($C^{12}=0$); Thermal neutron cross section (σ), barns	Class; Identification; Genetic relationships	Major radiations: approximate energies (MeV) and intensities	Principal means of production
$_{58}Ce^{132}$	4.2 h genet (WareW60)	☢ [EC] (WareW60) Δ -82 (MTW)	B chem, genet (WareW60) parent La^{132} (WareW60)	γ [La X-rays], 0.18 daughter radiations from La^{132}	protons on Ce (WareW60)
Ce^{133}	6.3 h (StovB51)	☢ EC, β$^+$ (StovB51) Δ -83 (MTW)	B chem, genet (StovB51) parent La^{133} (StovB51)	β$^+$ 1.3 max γ La X-rays, 0.511 (Y$^\pm$), 1.8 daughter radiations from La^{133}	La^{139}(p, 7n) (StovB51)
Ce^{134}	72.0 h (StovB51) 72 h (LavA60)	☢ EC (StovB51) Δ -84.9 (MTW)	B chem, excit (StovB51) parent La^{134} (StovB51) daughter Pr^{134} (LavA60, LavA63)	γ La X-rays, 0.44? daughter radiations from La^{134}	La^{139}(p, 6n) (StovB51)
Ce^{135}	17.0 h (DzhB63a) 17.6 h (TakaKa64) others (StovB51, ChubJ48)	☢ EC, β$^+$ <1% (StovB51) Δ -85 (MTW)	A chem, genet (ChubJ48) parent La^{135} (ChubJ48) daughter Pr^{135} (HandT54c)	γ La X-rays, 0.205 (↑ 17), 0.265 (↑ 100), 0.300 (↑ 56), 0.39 (↑ 10, complex), 0.52 (↑ 46, complex), 0.59 (↑ 98, complex), 0.777 (↑ 22), 0.821 (↑ 22), 0.865 (↑ 14), 0.901 (↑ 10) e$^-$ 0.048, 0.078, 0.166, 0.225, 0.25 β$^+$ 0.81 max daughter radiations from La^{135}	La^{139}(p, 5n) (StovB51, TakaKa64) La^{139}(d, 6n) (ChubJ48)
$\underline{Ce^{136}}$	$t_{1/2}$ (EC$_K$) >2.9 x 10^{11} y sp act (HohK65)	% 0.193 (IngM47e) Δ -86.6 (MTW) σ_c 6.0 (to Ce^{137}) 0.6 (to Ce^{137m}) (GoldmDT64)			
Ce^{137}	9.0 h (DanbG58) 8.7 h (BrosA55)	☢ EC 99+%, β$^+$ ≤0.009% (StonN65a, LHP) Δ -86 (MTW)	A chem, n-capt (BrosA55) chem, genet (DanbG58) daughter Ce^{137m} (DanbG58) daughter Pr^{137} (DanbG58, DahC58)	γ La X-rays, 0.446 (2.3%, complex), 0.481 (0.06%, complex), 0.698 (0.04%), 0.92 (0.10%, complex) e$^-$ [0.004, 0.009], 0.408	daughter Pr^{137} (DanbG58, DahC58) La^{139}(p, 3n) (DanbG58) Ce^{136}(n, γ) (FranR64) alphas on Ba (BrosA55)
Ce^{137m}	34.4 h (DanbG58) others (BrosA55, DanbG56, ChubJ48)	☢ IT 99.4%, EC 0.6% (StonN65a, LHP) Δ -87 (LHP, MTW)	A chem, excit (ChubJ48) n-capt, sep isotopes (HillR51a) parent Ce^{137} (DanbG58) not daughter Pr^{137} (DanbG58)	γ Ce X-rays, 0.168 (0.4%), 0.255 (11%), 0.762 (0.16%), 0.825 (0.5%, complex) e$^-$ 0.214, 0.248 daughter radiations from Ce^{137}	La^{139}(p, 3n) (DanbG58) Ce^{136}(n, γ) (HillR51a, KellH51, FranR64) alphas on Ba (BrosA55)
$\underline{Ce^{138}}$		% 0.250 (IngM47e) Δ -87.7 (MTW) σ_c 1.0 (to Ce^{139}) 0.04 (to Ce^{139m}) (GoldmDT64)			
Ce^{139}	140 d (PoolM48, PoolM43) others (WilleR60)	☢ EC (EC(L)/EC(K) 0.37) (KetB56) EC(L)/EC(K) 0.21 (PruC54) Δ -87.16 (MTW)	A chem (PoolM43) chem, excit, cross bomb (PoolM48) n-capt, sep isotopes (HillR51a) daughter Pr^{139} (StovB51, HandT54c, DanbG58) descendant Nd^{139m} (StovB51)	γ La X-rays, 0.165 (80%) e$^-$ 0.126, 0.159	Ce^{138}(n, γ) (HillR51a, KellH51, MosA50) La^{139}(d, 2n) (PoolM43, PoolM48)
Ce^{139m}	54 s (JameR60) 60 s (KotK60) 55 s (KetB56)	☢ IT (KetB56) Δ -86.41 (LHP, MTW)	B n-capt (KetB56) not daughter Pr^{139} (DanbG58)	γ Ce X-rays, 0.746 (93%) e$^-$ 0.706, 0.740	Ce^{138}(n, γ) (KetB56) La^{139}(p, n) (JameR60)
$\underline{Ce^{140}}$		% 88.48 (IngM47e) Δ -88.13 (MTW) σ_c 0.6 (GoldmDT64)			
Ce^{141}	32.5 d (FreeM50a) 33.1 d (WalkD49a) others (PoolM48, WilleR60)	☢ β$^-$ (HahO40c) Δ -85.49 (MTW) σ_c 30 (GoldmDT64)	A chem (HahO40c) chem, excit, n-capt, cross bomb (PoolM43, BallN51d) chem, mass spect (HaydR48) daughter La^{141} (BurgW51, DufR51a) descendant Cs^{141} (FritK62a) descendant Xe^{141} (OveR51, DilC51, DilC51a)	β$^-$ 0.581 max e$^-$ 0.104, 0.139 γ Pr X-rays, 0.145 (48%)	Ce^{140}(n, γ) (PoolM43, BallN51d, IngM48c) daughter La^{141} (BurgW51, DufR51a) Pr^{141}(n, p) (PoolM43)

TABLE 1 (*Continued*)

Isotope Z A	Half-life	Type of decay (☢); % abundance; Mass excess (△≡M–A), MeV (C¹²=0); Thermal neutron cross section (σ), barns	Class; Identification; Genetic relationships	Major radiations: approximate energies (MeV) and intensities	Principal means of production
$_{58}$Ce^{142}	$t_{1/2}$ (α) >5 x 10^{16} y sp act (MacfR61a) others (SenF59, RieW57)	% 11.07 (IngM47e) ☢ no α (MacfR61a, SenF59) α (RieW57) △ –84.63 (MTW) σ 1 (GoldmDT64)			
Ce^{143}	33 h (VasiI58, MartiDW56, BallN51d, StovB50, BotW46a) 34 h (KondE51c, WilleR60) others (BunyD49, PoolM43)	☢ β⁻ (SugaN46) △ –81.67 (MTW) σ$_c$ 6 (GoldmDT64)	A chem (SugaN46, PoolM43) chem, cross bomb (PoolM48) chem, genet (BallN51d) mass spect (IngM48c) daughter La^{143} (GesH51) parent Pr^{143} (PoolM43, BotW46a, BallN51d) descendant Xe^{143} (BradE51, DilC51a)	β⁻ 1.39 max e 0.015, 0.051, 0.252 Pr X-rays, 0.057 (11%), 0.293 (46%), 0.493 (2.4%), 0.668 (7%), 0.725 (8%), 0.88 (1.4%), 1.10 (0.6%) daughter radiations from Pr^{143}	Ce^{142}(n, γ) (KellH51, PoolM43, BotW46a, PoolM48, BallN51d)
Ce^{144}	284 d (FlyK65a) 285 d (SchumR56, MerW57) 277 d (EasH60) others (BurgW51a, JoliF44)	☢ β⁻ (HahO40c) △ –80.49 (MTW) σ$_c$ 1.0 (GoldmDT64)	A chem (HahO40c) chem, mass spect (HaydR48) parent Pr^{144} (HahO43a, NewA51a) descendant Xe^{144} (DilC51)	β⁻ 0.31 max e 0.038, 0.092 γ Pr X-rays, 0.080 (2%), 0.134 (11%) daughter radiations from Pr^{144}	fission (HahO40c, BornH43a, DilC51a, NewA51a, BurgW51a, GrumW48, FinB51c)
Ce^{145}	3.0 m (MarkS54) 3.1 m (WilleR60)	☢ β⁻ (MarkS54) △ –77 (MTW)	B chem, excit, genet (MarkS54) parent Pr^{145} (MarkS54)	β⁻ 2.0 max γ γ rays reported	fission (MarkS54) Nd^{148}(n, α) (WilleR60)
Ce^{146}	14 m (CareA53) 15 m (SchumR45) others (GotH46)	☢ β⁻ (GotH43) △ –75.8 (MTW)	B chem, genet (GotH43) parent Pr^{146} (GotH43, HahO43a, GotH46, CareA53)	β⁻ 0.7 max γ Pr X-rays, 0.110 (↑ 20), 0.142 (↑ 42), 0.22 (↑ 50), 0.27 (↑ 12), 0.32 (↑ 100) daughter radiations from Pr^{146}	fission (GotH43, HahO43a, SchumR45, GotH46, BernsW54)
Ce^{147}	65 s genet (HoffD64)	☢ β⁻ (HoffD64)	B chem, genet (HoffD64) parent Pr^{147} (HoffD64)		fission (HoffD64)
Ce^{148}	≈43 s genet (HoffD64)	☢ β⁻ (HoffD64)	B chem, genet (HoffD64) parent Pr^{148} (HoffD64)		fission (HoffD64)
$_{59}$Pr^{134}	17 m (ClarJ65) 40 m genet (LavA63) others (LavA60)	☢ β⁺ (ClarJ65), [EC]	B chem, genet (LavA60, LavA63) chem, excit, genet energy levels (ClarJ65) parent Ce^{134} (LavA60, LavA63)	γ Ce X-rays, 0.22, 0.30, 0.409, 0.511 (γ$^{\pm}$), 0.639, 0.96 daughter radiations from Ce^{134}, La^{134}	I^{127}(C^{12}, 5n) (ClarJ65) protons on Pr (LavA63)
Pr^{135}	22 m (HandT54c)	☢ β⁺, EC (HandT54c)	B chem, excit, genet (HandT54c) parent Ce^{135} (HandT54c)	β⁺ 2.5 max γ Ce X-rays, 0.080, 0.22, 0.30, 0.511 (γ$^{\pm}$) daughter radiations from Ce^{135}	Ce^{136}(p, 2n) (HandT54c)
Pr^{136}	1.2 h (HandT54c) 1.0 h (DanbG58)	☢ EC ≈67%, β⁺ ≈33% (DanbG58)	A chem, excit (HandT54c) chem, mass spect (DanbG58)	β⁺ 2.0 max γ Ce X-rays, 0.17?, 0.511 (66%, γ$^{\pm}$)	Ce^{136}(p, n) (HandT54c) protons on Ce, Pr (DanbG58)
Pr^{137}	1.5 h (DanbG58, DahC58)	☢ EC 73%, β⁺ 27% (DanbG58) △ –84 (MTW)	B chem, mass spect (DanbG58, DahC58) parent Ce^{137}, not parent Ce^{137m} (DanbG58) daughter Nd^{137} (GromK65)	β⁺ 1.7 max γ Ce X-rays, 0.511 (54%, γ$^{\pm}$), no other γ's (lim 6%) daughter radiations from Ce^{137}	protons on Ce (DanbG58, DahC58)
Pr^{138}	2.10 h (DanbG58) 2.2 h (FujiM64) 2.0 h (StovB51, HandT54c)	☢ EC 77%, β⁺ 23% (FujiM64) EC 84%, β⁺ 16% (DanbG58) others (StovB51) △ –82.9 (FujiM64, MTW)	A chem, excit (StovB51) chem, mass spect (DanbG58)	β⁺ 1.65 max e 0.258, 0.292 γ Ce X-rays, 0.298 (77%), 0.40 (9%), 0.511 (46%, γ$^{\pm}$), 0.79 (100%), 1.04 (100%)	Ce^{140}(p, 3n) (StovB51, DanbG58, FujiM64) Ce^{138}(p, n) (HandT54c)
Pr^{138}	short (GromK64)	☢ (GromK64)	F genet (GromK64) [daughter ≈5 h Nd^{138}] (GromK64)		daughter ≈5 h Nd^{138} (GromK64)

TABLE 1 (*Continued*)

Isotope Z A	Half-life	Type of decay ($\frac{a}{b}$); % abundance; Mass excess ($\triangle \equiv$ M–A), MeV (C^{12}=0); Thermal neutron cross section (σ), barns	Class; Identification; Genetic relationships	Major radiations: approximate energies (MeV) and intensities	Principal means of production
$_{59}$Pr^{139}	4.5 h (DanbG58, StovB51, HandT54c) 4.9 h (BiryE63a)	EC 89%, β^+ 11% (BiryE63a) EC 93%, β^+ 7% (DanbG58) EC ≈94%, β^+ ≈6% (StovB51) others (BoreO61) $\triangle$ -85.0 (BiryE63a, MTW)	A chem, genet (StovB51) chem, mass spect, genet (DanbG58) parent Ce^{139} (StovB51, HandT54c, DanbG58) not parent Ce^{139m} (DanbG58)	β^+ 1.09 max γ Ce X-rays, 0.511 (18%, $\gamma^\pm$), 1.35 (0.5%), 1.61 (0.3%)	Pr^{141}(p, 3n) Nd^{139}(β^-) (DanbG58) Ce^{140}(p, 2n) (StovB51, DanbG58)
Pr^{139m}?	≈6 m (KolG63)		F genet (KolG63) daughter Nd^{139} or Nd^{139m}? (KolG63)		daughter Nd^{139m} (KolG63)
Pr^{140}	3.39 m (EbrT65) others (DWirJ42, HandT54c, PoolM38a, BiryE62, WilleR60, StovB51, HubeO45, PerlmM49)	EC 50%, β^+ 50% (BrabV60) EC(K)/EC(L) 8 (BiryE60) others (BiryE60, BiryE62, BrowCI52) $\triangle$ -84.78 (HisK64, MTW)	A excit (AmaE35) excit (PoolM38a) daughter Nd^{140} (WilkG49c, BrowCI52)	β^+ 2.32 max e$^-$ 1.862 (0.07%) γ Ce X-rays, 0.511 (100%, $\gamma^\pm$), 1.596 (0.3%)	daughter Nd^{140} (WilkG49c, BrowCI52, HisK64)
Pr^{141}	$t_{1/2}$ (α) >2 x 10^{16} y sp act (PorsW54)	% 100 (IngM48a, CollT57) $\triangle$ -86.07 (MTW) σ_c 12 (GoldmDT64)			
Pr^{142}	19.2 h (WyaE61, BotW46a) 19.3 h (DWirJ42) 19.1 h (JensE50) others (WilleR60)	β^- (DWirJ42) no EC or β^+, lim 0.5% (ReynJH50b) $\triangle$ -83.85 (MTW) σ_c 20 (GoldmDT64)	A n-capt (AmaE35, MarsJK35)	β^- 2.16 max γ 1.57 (3.7%)	Pr^{141}(n, γ) (AmaE35, MarsJK35, PoolM37, PoolM38a, DWirJ42, SerL47b)
Pr^{143}	13.59 d (PepD57) 13.76 d (WriH57) 13.6 d (HoffD63) others (FelL49, BallN51f, RoyL56, PoolM48, MartiDW56)	β^- (BallN51e, JoliF44) mass spect (HaydR46a) $\triangle$ -83.11 (MTW) σ_c 89 (GoldmDT64)	A chem (BallN51e, JoliF44) daughter Ce^{143} (PoolM43, BotW46a, BallN51d) others (HahO43a, FinB51c)	β^- 0.933 max average β^- energy: 0.31 calorimetric (HovV64) γ no γ	Ce^{142}(n, γ) Ce^{143}(β^-) (PoolM43, BotW46a, BallN51d) fission (HahO43a, JoliF44, BallN51e, FinB51c)
Pr^{144}	17.27 m (PepD57) 17.30 m (HoffD63) others (NewA51a, SeiJ51b, HahO43a, GrumW46)	β^- (NewA51a) $\triangle$ -80.81 (MTW)	A chem, genet (NewA51a, HahO43a) daughter Ce^{144} (HahO43a, NewA51a)	β^- 2.99 max γ 0.695 (1.5%), 1.487 (0.29%), 2.186 (0.7%)	daughter Ce^{144} (HahO43a, NewA51a)
Pr^{145}	5.98 h (DroB59) 5.9 h (MarkS54, AlsJ60)	β^- (MarkS54) $\triangle$ -79.66 (MTW)	B chem, excit (MarkS54) chem, sep isotopes (HoffD64) daughter Ce^{145} (MarkS54)	β^- 1.80 max γ 0.072, 0.68, 0.75, 0.92, 0.98, 1.05, 1.16	fission (MarkS54, DroB59, AlsJ60, HoffD64) Nd^{146}(γ, p) (HoffD64)
Pr^{146}	24.0 m (HoffD64) others (SchumR45a, CareA53, GotH46)	β^- (GotH43) $\triangle$ -76.8 (MTW)	B chem, genet (GotH43) daughter Ce^{146} (GotH43, HahO43a, GotH46, CareA53)	β^- 3.7 max γ 0.455 (77%), 0.74 (16%), 0.78 (15%), 0.92 (6%), 1.37 (6%), 1.51 (27%), 1.72 (4%), 2.23 (4%), 2.39 (3%), 2.73 (1.7%)	fission (GotH43, HahO43a, SchumR45, GotH46, BernsW54, HoffD64) Nd^{146}(n, p) (RamayA65)
Pr^{147}	12.0 m (HoffD64) 12 m (WilleR60)	β^- (HoffD64) $\triangle$ -75.5 (HoffD64, MTW)	B chem, genet (HoffD64) parent Nd^{147}, daughter Ce^{147} (HoffD64)	β^- 2.1 max γ 0.078 (17%, complex?), 0.127 (9%, complex?), 0.32 (47%, complex), 0.56 (39%), 0.61 (10%), 0.65 (24%), 1.26 (11%)	Nd^{148}(γ, p), fission (HoffD64)
Pr^{148}	2.0 m (HoffD64)	β^- (HoffD64) $\triangle$ -72.9 (HoffD64, MTW)	B chem, genet energy levels (HoffD64) daughter Ce^{148} (HoffD64)	β^- 4.2 max γ 0.30	fission (HoffD64)
Pr^{149}	2.3 m (HoffD64)	β^- (HoffD64)	E excit, sep isotopes (HoffD64)	β^- 2.8 max γ 0.08, 0.155, 0.325, 0.36, 0.745	Nd^{150}(γ, p) (HoffD64)
$_{60}$Nd^{137}	55 m (GromK65)	β^+, [EC] (GromK65)	B chem, atomic level spacing, genet (GromK65) parent Pr^{137} (GromK65)	β^+ 3 max e$^-$ 0.067 γ [Pr X-rays, 0.109, 0.511 ($\gamma^\pm$), 0.55 (complex)] daughter radiations from Pr^{137}, Ce^{137}	protons on Ta, Er (GromK65)

TABLE 1 (*Continued*)

Isotope Z A	Half-life	Type of decay (☣); % abundance; Mass excess (△≡M-A), MeV (C¹²=0); Thermal neutron cross section (σ), barns	Class; Identification; Genetic relationships	Major radiations: approximate energies (MeV) and intensities	Principal means of production
$_{60}Nd^{138}$	22 m (StovB51)	☣ β^+ (StovB51), [EC]	D chem, excit (StovB51)	β^+ 2.4 max γ [Pr X-rays, 0.511 (γ*)]	Pr^{141}(p, 4n) (StovB51)
Nd^{138}	≈5 h (GromK64)	☣ (GromK64)	F chem (GromK64)		protons on Ta, Er (GromK64)
Nd^{139}	[<<5 h] (GromK63b)	☣ [EC, β^+] △ -82 (MTW)	F [genet] (GromK63b) [daughter Nd^{139m}] (GromK63b)	β^+, γ see Nd^{139}m	[daughter Nd^{139m}] (GromK63b)
Nd^{139m}	5.5 h (StovB51) 5.2 h (BoncN61)	☣ IT (+EC+β^+?) (GromK63b) EC ≈90%, β^+ ≈10% (with Nd^{139}) (StovB51) △ -82 (LHP, MTW)	B chem, genet (StovB51) atomic level spacing (GromK63b) ancestor Ce^{139} (StovB51)	β^+ 3.1 max e^- 0.072, 0.107, 0.189, 0.226 γ Nd X-rays, Pr X-rays, 0.114 (↑ 80), 0.327 (↑ 50), 0.511 (↑ 1400), 0.73 (↑ 210, complex), 0.82 (↑ 70, complex), 0.90 (↑ 25), 0.983 (↑ 70), 1.03 (↑ 30), 1.10 (↑ 30), 1.24 (↑ 20), 1.34 (↑ 20), 1.48 (↑ 10), 1.58 (↑ 8), 2.05 (↑ 10) daughter radiations from Pr^{139} daughter radiations from Nd^{139} included in above listing	Pr^{141}(p, 3n) (StovB51)
Nd^{140}	3.3 d (WilkG49c)	☣ EC (BrowCI52) EC(K)/EC(L) 6 (BiryE60) △ -84 (MTW)	A chem, excit, genet (WilkG49c) parent Pr^{140} (WilkG49c, BrowCI52)	γ Pr X-rays daughter radiations from Pr^{140}	Pr^{141}(p, 2n) (StovB51) Pr^{141}(d, 3n) (WilkG49c, BrowCI52)
Nd^{141}	2.42 h (WilkG49c) 2.5 h (KurbJ42) 2.6 h (BiryE63) others (WilleR60)	☣ EC 96%, β^+ 4% (BiryE63) EC 98%, β^+ 2% (PolH58) others (AlfWL63) △ -84.27 (MTW)	A excit (KurbJ42) chem, excit (WilkG49c) others (PoolM38a)	β^+ 0.79 max γ Pr X-rays, 0.145 (0.2%), 0.511 (6%, γ*), 1.14 (2%, complex?), 1.30 (1%)	Pr^{141}(p, n) (KurbJ42, WilkG49c) Pr^{141}(d, 2n) (WilkG49c, PolH58)
Nd^{141m}	64 s (JameR60) 61 s (KotK60)	☣ [IT] (KotK60) △ -83.52 (LHP, MTW)	C excit (JameR60) chem (KotK60)	γ 0.755	Pr^{141}(p, n) (JameR60)
$\underline{Nd^{142}}$		% 27.13 (IngM48a) 27.09 (WalkW53) 27.3 (WhiF56) △ -86.01 (MTW) σ_c 17 (GoldmDT64)			
$\underline{Nd^{143}}$		% 12.20 (IngM48a) 12.14 (WalkW53) 12.32 (WhiF56) △ -84.04 (MTW) σ_c 330 (GoldmDT64)			
$\underline{Nd^{144}}$	2.4 x 10¹⁵ y sp act (MacfR61a) 2.1 x 10¹⁵ y sp act (IsolA65) 5 x 10¹⁵ y sp act (PorsW56, PorsW54) 2 x 10¹⁵ y sp act (WaldE54)	☣ α (WaldE54, PorsW54, PorsW56) % 23.87 (IngM48a) 23.83 (WalkW53) 23.8 (WhiF56) others (IngM50a) △ -83.80 (MTW) σ_c 5 (GoldmDT64)	A sep isotopes, decay charac, chem (PorsW56, MacfR61a)	α 1.83	
$\underline{Nd^{145}}$	$t_{1/2}$ (α) >6 x 10¹⁶ y (IsolA65)	% 8.29 (WhiF56, WalkW53) 8.30 (IngM48a) △ -81.47 (MTW) σ_c 50 (GoldmDT64)			
$\underline{Nd^{146}}$		% 17.18 (IngM48a) 17.26 (WalkW53) 17.1 (WhiF56) others (IngM50a) △ -80.96 (MTW) σ_c 2 (GoldmDT64)			

TABLE 1 (*Continued*)

Isotope Z A	Half-life	Type of decay (☢); % abundance; Mass excess (△≡M-A), MeV (C¹²=0); Thermal neutron cross section (σ), barns	Class; Identification; Genetic relationships	Major radiations: approximate energies (MeV) and intensities	Principal means of production
$_{60}$Nd^{147}	11.06 d (WriH57) 11.02 d (HoffD63) 11.1 d (AlsJ60) others (KondE51a, RutW52, MarinJ51, EmmW51, BotW46a)	☢ β⁻ (MarinJ47, MarinJ51) △ −78.18 (MTW)	A chem, genet (MarinJ47, MarinJ51a) parent Pm^{147} (MarinJ47, MarinJ51a) daughter Pr^{147} (HoffD64)	β⁻ 0.81 max e⁻ 0.046, 0.084 γ 0.091 (28%), 0.319 (3%), 0.43 (4%, complex), 0.533 (13%) daughter radiations from Pm^{147}	Nd^{146}(n, γ) (BotW46a, MarinJ47, CorkJ48a, MarinJ51c) fission (MarinJ51)
Nd^{148}		% 5.72 (IngM48a) 5.74 (WalkW53) 5.67 (WhiF56) others (IngM50a) △ −77.44 (MTW) σ$_c$ 4 (GoldmDT64)			
Nd^{149}	1.8 h (RutW52, WilleR60, HoffD64) 2.0 h (BotW46a, PoolM38a) others (MarinJ51c)	☢ β⁻ (PoolM38a) △ −74.41 (MTW)	A excit (PoolM38a) chem, genet (MarinJ51c) parent Pm^{149} (KruP52, MarinJ51c)	β⁻ 1.5 max e⁻ 0.051, 0.068, 0.079, 0.090, 0.165, 0.195 γ Pm X-rays, 0.114 (18%), 0.156 (4%), 0.210 (27%), 0.27 (26%, complex), 0.327 (5%), 0.424 (9%), 0.541 (10%), 0.654 (9%) daughter radiations from Pm^{149}	Nd^{148}(n, γ) (PoolM38a, BotW46a, MarinJ51c, GopK64)
Nd^{150}	t$_{1/2}$ (β) >10^{16} y sp act (DixD54a) t$_{1/2}$ (ββ) >2 x 10^{18} y sp act (CowC56) others (MulhG52)	% 5.60 (IngM48a) 5.63 (WalkW53) 5.56 (WhiF56) others (IngM50a) △ −73.67 (MTW) σ$_c$ 1.5 (GoldmDT64)			
Nd^{151}	12 m (RutW52, MarinJ51c) others (WilleR60)	☢ β⁻ (RutW52) △ −71.0 (MTW)	B n-capt (MarinJ51c) sep isotopes, n-capt, atomic level spacing (RutW52) parent Pm^{151} (RutW52)	β⁻ 2.0 max e⁻ 0.072 γ Pm X-rays, 0.086 (5%), 0.118 (40%), 0.138 (6%), 0.174 (10%, complex), 0.256 (11%), 0.425 (5%), 0.737 (5%), 0.797 (3%), 1.122 (2%), 1.180 (9%)	Nd^{150}(n, γ) (RutW52, MarinJ51c, SchmL59a, FosD65)
$_{61}$Pm^{141}	22 m (GratI59) 20 m (FiscV52)	☢ β⁺ 57%, EC 43% (GratI59) △ −80.7 (MTW)	A chem, excit (FiscV52) mass spect (GratI59)	β⁺ 2.6 max γ Nd X-rays, 0.195 (13%), 0.511 (114%, γ⁺) daughter radiations from Nd^{141}	Pr^{141}(α, 4n) (GratI59) Nd^{142}(p, 2n) (FiscV52)
Pm^{142}	40 s (GratI59) others (MarsT58)	☢ β⁺ ≈95%, EC ≈5% (GratI59) △ −81.2 (MTW)	B chem, genet (MarsT58) excit (Gra'I59) daughter Sm^{142} (MarsT58)	β⁺ 3.78 max (MarsT58) γ Nd X-rays, 0.511 (190%, γ⁺)	Nd^{142}(α, 4n)Sm^{142}(EC) (GratI59, MarsT58) Nd^{142}(p, n) (GratI59)
Pm^{143}	0.73 y (PagI63, BunnL64, FunE60) 0.78 y (WilkG50e)	☢ EC (WilkG50e) △ −82.9 (MTW)	A chem, excit (WilkG50e) chem, mass spect (BallN58)	γ Nd X-rays, 0.742 (47%) e⁻ 0.698	Sm^{144}(p, 2n)[Eu^{143}](EC) Sm^{143}(EC) (FunE60) Pr^{141}(α, 2n) (WilkG50e, FiscV52, OfeS59, BunnL64) Nd^{143}(p, n) (PagI63)
Pm^{144}	0.96 y (BunnL64) 1.03 y (PagI63) 1.1 y (FunE60) 1.2 y (TotK59c) others (FiscV52)	☢ EC (FiscV52) no β⁺, lim 0.2% (OfeS59) △ −82 (MTW)	A chem (FiscV52) chem, mass spect (BallN58) excit (OfeS59)	γ Nd X-rays, 0.474 (45%), 0.615 (99%), 0.695 (99%) e⁻ 0.430, 0.571, 0.651	Pr^{141}(α, n) (OfeS59, TotK59c, FiscV52) Nd^{144}(p, n) (PagI63, SugiyK61, FiscV52)
Pm^{144}?	60 d (PagI63)	☢ (PagI63)	F sep isotopes (PagI63)	γ γ spectrum may be identical to 1.1 y Pm^{144} (PagI63)	Nd^{144}(p, n) (PagI63)
Pm^{145}	17.7 y (BrosA59) others (ButeF51)	☢ EC (ButeF51) α 3 x 10⁻⁷% (NurM62) △ −81.33 (MTW)	A chem, genet (ButeF51) chem, mass spect (BallN58) daughter Sm^{145} (ButeF51)	γ Nd X-rays, 0.067 (1.0%), 0.072 (2.3%) e⁻ 0.023, 0.028, 0.061	Sm^{144}(n, γ)Sm^{145}(EC) (ButeF51, BrosA59)
Pm^{145}	16 d (LongJ52a)	☢ β⁺ (LongJ52a)	F sep isotopes (LongJ52a)	β⁺ 0.45 max	protons on Nd (LongJ52a)

TABLE 1 (*Continued*)

Isotope Z A	Half-life	Type of decay (☢); % abundance; Mass excess ($\Delta \equiv M-A$), MeV ($C^{12}=0$); Thermal neutron cross section (σ), barns	Class; Identification; Genetic relationships	Major radiations: approximate energies (MeV) and intensities	Principal means of production
$_{61}Pm^{146}$	4.4 y (PagI63) 1.9 y (FunE60) 1 y (FiscV52) 1-2 y (LongJ52a)	☢ EC 65%, β^- 35% (FunE60) EC 69%, β^- 31% (PagI63) Δ −79.52 (MTW)	A chem, excit (FiscV52) chem, sep isotopes, genet energy levels (FunE60, FunE62)	β^- 0.78 max γ Nd X-rays, 0.453 (65%), 0.75 (65%, doublet)	Nd^{146}(p, n) (PagI63, FiscV52, LongJ52a) Nd^{148}(p, 3n) (FunE60)
Pm^{147}	2.62 y (WheeE65) 2.60 y (FlyK65a) 2.64 y (MerW57) 2.66 y (SchumR56) others (MelaE55, IngM50a, SchumR51a)	☢ β^- (BallN51g) Δ −79.08 (MTW) σ_c 120 (to Pm^{148}) 110 (to Pm^{148m}) (GoldmDT64)	A chem (MarinJ47, MarinJ51a) mass spect (HaydR48) daughter Nd^{147} (MarinJ47, MarinJ51a) parent Sm^{147} (RasJ50)	β^- 0.224 max average β^- energy: 0.070 calorimetric (HovV62) γ no γ	Nd^{146}(n, γ)Nd^{147}(β^-) (MarinJ47, MarinJ51a) fission (BallN51g, SeiJ51c, MarinJ51a, GrumW48, IngM50a)
Pm^{148}	5.4 d (ReicC62, EldJ61) others (SchweC62a, ParkG47, KurbJ43, BhaS59)	☢ β^- (KurbJ43) Δ −76.89 (BabC63a, MTW) σ_c ≈2000 (GoldmDT64)	A chem, n-capt, mass spect (ParkG47) daughter Pm^{148m} (BabC63a)	β^- 2.48 max γ 0.551 (27%), 0.914 (15%), 1.465 (23%)	Nd^{148}(p, n) (LongJ52, FiscV52, KurbJ43, SchweG62a) Nd^{148}(d, 2n) (KurbJ42, KurbJ43, BabC63a) Pm^{147}(n, γ) (ParkG47, ReicC62)
Pm^{148m}	41.8 d (EldJ61) 40.6 d (ReiC62) 45.5 d (SchweC62a) others (FiscV52, FolR51, LongJ52)	☢ β^- 93%, IT 7% (BabC63a) others (ReiC62, SchweC62a) Δ −76.75 (LHP, MTW) σ_c 30, 000 (GoldmDT64)	A excit, sep isotopes (LongJ52) chem (FolR51) chem, mass spect, genet (BabC63a) parent Pm^{148} (BabC63a)	β^- 0.69 max e^- 0.031, 0.053, 0.091, 0.242, 0.503, 0.583 γ Pm X-rays, Sm X-rays, 0.289 (13%), 0.413 (17%), 0.551 (95%), 0.630 (87%), 0.727 (36%), 0.916 (21%), 1.015 (20%) daughter radiations from Pm^{148}	Nd^{148}(p, n) (LongJ52, FiscV52, SchweC62a) Nd^{148}(d, 2n) (BabC63a) Pm^{147}(n, γ) (ReiC62)
Pm^{149}	53.1 h (HoffD63, BunnL60) others (ArtnA60, FiscV52, IngM47d, RutW52, KondE51c, BotW46a, MarinJ51b)	☢ β^- (MarinJ47) Δ −76.07 (MTW)	A chem (MarinJ47, MarinJ51b) chem, mass spect (IngM47d) daughter Nd^{149} (KruP52, MarinJ51c)	β^- 1.07 max γ 0.286 (2%), 0.58 (0.1%), 0.85 (0.2%)	Nd^{148}(n, γ)Nd^{149}(β^-) (KruP52, MarinJ47, SchmL60a, BunnL60)
Pm^{150}	2.68 h (FiscV52) 2.7 h (LongJ52)	☢ β^- (LongJ52) Δ −73.6 (MTW)	A excit, sep isotopes (LongJ52) chem, excit, sep isotopes (FiscV52)	β^- 3.05 max γ 0.334 (71%), 0.406 (7%), 0.71 (8%), 0.831 (18%), 0.88 (12%), 1.165 (23%), 1.33 (22%), 1.75 (10%), 1.96 (2.5%), 2.06 (1.2%), 2.53 (0.9%)	Nd^{150}(p, n) (LongJ52, FiscV52)
Pm^{151}	27.8 h (HoffD63) 28.4 h (BunnL60) 27.5 h (RutW52)	☢ β^- (RutW52) Δ −73.40 (MTW)	A genet, atomic level spacing (RutW52) chem (BunnL60) daughter Nd^{151} (RutW52)	β^- 1.19 max e^- 0.003, 0.018, 0.053, 0.058 γ Sm X-rays, 0.07 (5%, complex), 0.10 (7%, doublet), 0.17 (18%, complex), 0.24 (5%, complex), 0.275 (6%), 0.340 (21%), 0.45 (5%, complex), 0.66 (3%, complex), 0.72 (6%, complex), others to 0.96	Nd^{150}(n, γ)Nd^{151}(β^-) (RutW52, BunnL60)
$Pm^{?}$	12.5 h (FolR51, (PoolM38a)	☢ β^- (PoolM38a)	E (PoolM38a) chem (FolR51)		deuterons on Nd (PoolM38a) fission (FolR51)
Pm^{152}	6.5 m (WilleR58, WilleR60)	☢ β^- (WilleR58) Δ −71 (MTW)	B sep isotopes, excit (WilleR58) genet energy levels (AteA59)	β^- 2.2 max γ [Sm X-rays], 0.122, 0.245	Sm^{152}(n, p) (WilleR58, WilleR60, AteA59)
Pm^{153}	5.5 m (KotK62)	☢ β^- (KotK62) Δ −70.8 (MTW)	E excit, sep isotopes (KotK62)	β^- 1.65 max γ 0.090 (?), 0.12, 0.18	Sm^{154}(γ, p) (KotK62)
Pm^{154}	2.5 m (WilleR58, WilleR60)	☢ β^- (WilleR60)	C excit, sep isotopes (WilleR58)	β^- 2.5 max	Sm^{154}(n, p) (WilleR58, WilleR60)
$_{62}Sm^{142}$	73 m (GratI59) 72 m (MarsT58)	☢ EC ≈50%, β^+ ≈50% (DCapG59)	B chem (MarsT58) excit (GratI59) parent Pm^{142} (MarsT58)	γ Pm X-rays, 0.15-0.35 (complex), 0.511 (100%, $\gamma^{\pm}$) daughter radiations from Pm^{142}	Nd^{142}(α, 4n) (GratI59, MarsT58)
Sm^{143}	9.0 m (SilE56) 8.9 m (AlfWL63a) 8.6 m (GratI59) 8.5 m (WilleR60) 8.3 m (MirM56) 8.8 m (KotK60) others (ButeF50)	☢ EC 52%, β^+ 48% (DCapG59) EC ≈63%, β^+ ≈37% (GratI59) others (SilE56, MirM56) Δ −79.6 (MTW)	B chem (ButeF50) excit (SilE56) chem, sep isotopes (MirM56)	γ Pm X-rays, 0.511 (100%, $\gamma^{\pm}$)	Nd^{142}(α, 3n) (GratI59) Sm^{144}(n, 2n) (WilleR60, MirM56, AlfWL63a) Sm^{144}(γ, n) (SilE56, ButeF50, KotK60, DCapG59)

TABLE 1 (*Continued*)

Isotope Z A	Half-life	Type of decay (☢); % abundance; Mass excess (Δ≡M−A), MeV (C¹²=0); Thermal neutron cross section (σ), barns	Class; Identification; Genetic relationships	Major radiations: approximate energies (MeV) and intensities	Principal means of production
$_{62}$Sm^{143m}	64 s (KotK60) 65 s (AlfWL63a) 61 s (BroaK65) others (JameR60)	☢ [IT] (KotK60) Δ -78.8 (LHP, MTW)	C chem (KotK60) excit (AlfWL63a)	γ 0.748	Sm^{144}(n, 2n) (AlfWL63a) Sm^{144}(γ, n) (KotK60) Sm^{144}(p, pn) (JameR60)
Sm^{144}		% 3.16 (IngM48) 3.15 (CollT57) 3.02 (AitK57) Δ -81.98 (MTW) σ$_c$ ≈0.7 (GoldmDT64)			
Sm^{145}	340 d (BrosA59) others (ButeF51, CorkJ48a, IngM47c)	☢ EC (ButeF51, RutW52) Δ -80.67 (MTW) σ$_c$ ≈100 (GoldmDT64)	A mass spect (IngM47c) chem (ButeF51) parent Pm^{145} (ButeF51)	γ Pm X-rays, 0.061 (13%), 0.485 (3 x 10^{-3}%) e⁻ 0.016, 0.054 daughter radiations from Pm^{145}	Sm^{144}(n, γ) (ButeF51, RutW52, IngM47c, BrosA59)
Sm^{146}	7 x 10^7 y sp act (NurM64) 5 x 10^7 y yield (DunlD53)	☢ α (DunlD53) % <2 x 10^{-7} (MacfR60) Δ -81.05 (MTW)	B chem, decay charac (DunlD53)	α 2.46	Sm^{147}(n, 2n) (NurM64) alphas on Nd (DunlD53)
Sm^{147}	1.05 x 10^{11} y sp act (WriP61) others (DonhD64, MacfR61a, GraeG61, BearG54, BearG58, KarrM60, KarrM60a, LatC47, HosR35, PicE49)	☢ α (HevG32, LibW33) % 15.07 (IngM48) 15.1 (CollT57) 14.9 (AitK57) Δ -79.30 (MTW) σ$_c$ ≈90 (GoldmDT64)	A chem (HevG32) sep isotopes, mass spect WeaB50) chem, genet, mass spect (RasJ50) daughter Pm^{147} (RasJ50)	α 2.23	
Sm^{148}	t$_{1/2}$ (α) >2 x 10^{14} y sp act (MacfR61a) t$_{1/2}$ (α) 1.2 x 10^{13} y sp act (KarrM60)	% 11.27 (IngM48) 11.35 (CollT57) 11.22 (AitK57) ☢ no α (MacfR61a) α (KarrM60) Δ -79.37 (MTW)			
Sm^{149}	>1 x 10^{15} y sp act (MacfR61a) 4 x 10^{14} y sp act (KarrM60)	% 13.82 (AitK57) 13.84 (IngM48) 14.0 (CollT57) ☢ no α (MacfR61a) α (KarrM60) Δ -77.15 (MTW) σ$_c$ 41, 500 (GoldmDT64)			
Sm^{150}		% 7.47 (IngM48, CollT57) 7.40 (AitK57) Δ -77.06 (MTW) σ$_c$ 100 (GoldmDT64)			
Sm^{151}	≈87 y (FlyK65a) ≈93 y yield + mass spect (MelaE55) ≈73 y (KarrD52) ≈120 y yield (IngM50a)	☢ β⁻ (IngM47c) Δ -74.59 (MTW) σ$_c$ 15, 000 (GoldmDT64)	A mass spect (IngM47c, IngM50a) chem (MarinJ49a)	β⁻ 0.076 max e⁻ 0.014, 0.020 γ Eu L X-rays, 0.022 (4%)	fission (IngM50a, MarinJ49a, AchW59) Sm^{150}(n, γ) (IngM47c)
Sm^{152}		% 26.63 (IngM48) 26.6 (CollT57) 26.8 (AitK57) Δ -74.75 (MTW) σ$_c$ 210 (GoldmDT64)			
Sm^{153}	46.8 h (WyaE61) 47.1 h (CorkJ58, CabM62) 46.2 h (GreeRE61) 46.5 h (HoffD63) 47.0 h (LeeM54) others (KurbJ42, BotW46a, WinsL51, RutW52)	☢ β⁻ (KurbJ42) Δ -72.56 (MTW)	A n-capt, excit (PoolM38a) mass spect (HaydR46, IngM47d) chem (WinsL51)	β⁻ 0.80 max e⁻ 0.022, 0.055, 0.062, 0.095, 0.101 γ Eu X-rays, 0.070 (5.4%), 0.103 (28%), 0.41 to 0.64 (0.6%, 16 γ rays)	Sm^{152}(n, γ) (HevG36, PoolM38a, HaydR46, SerL47b, WinsL51) Nd^{150}(α, n) (KurbJ42)
Sm^{154}		% 22.53 (IngM48) 22.4 (CollT57) 22.9 (AitK57) others (IngM50a) Δ -72.39 (MTW) σ$_c$ 5 (GoldmDT64)			

TABLE 1 *(Continued)*

Isotope Z A	Half-life	Type of decay (⚛); % abundance; Mass excess (△≡M–A), MeV (C¹²=0); Thermal neutron cross section (σ), barns	Class; Identification; Genetic relationships	Major radiations: approximate energies (MeV) and intensities	Principal means of production
$_{62}$Sm^{155}	23.5 m (RutW52) 21.9 m (SunR60) others (WinsL51a, PoolM38a)	⚛ β⁻ (KurbJ42) △ –70.14 (MTW)	A n-capt (AmaE35, MarsJK35) chem (WinsL51a) sep isotopes (SunR60, SchmL59b) parent Eu^{155} (IngM47c)	β⁻ 1.53 max e⁻ 0.056, 0.097, 0.103 γ Eu X-rays, 0.104 (73%), 0.246 (4%)	Sm^{154}(n,γ) (AmaE35, MarsJK35, HevG36, PoolM38a, SerL47b, IngM47c, WinsL51a, SunR60, SchmL59b)
Sm^{156}	9.4 h (GunR63) 9 h (AlsJ60)	⚛ β⁻ (WinsL51c) △ –69.33 (MTW)	B chem, genet (WinsL51c) parent Eu^{156} (WinsL51c)	β⁻ 0.72 max e⁻ 0.014, 0.021, 0.030, 0.039 γ Eu X-rays, 0.088 (30%), 0.166 (10%), 0.204 (20%), 0.25 (5%, complex), 0.291 (3%) daughter radiations from Eu^{156}	fission (WinsL51c, AlsJ60, GunR63)
Sm^{157}	0.5 m (WilleR60)	⚛ [β⁻] (WilleR60)	C sep isotopes, cross bomb (WilleR60)	γ 0.57	Gd^{160}(n, α) (WilleR60)
$_{63}$Eu^{143}	2.3 m (KotK65)	⚛ β⁺ (KotK65), [EC]	E excit, decay charac (KotK65)	β⁺ 4.0 max γ 0.511 (γ±)	Sm^{144}(d, 3n) (KotK65)
Eu^{144}	10.5 s (MesR65)	⚛ β⁺ (MesR65), [EC] △ –75.66 (MesR65, MTW)	C excit, decay charac (MesR65)	β⁺ 5.2 max γ 0.511 (γ±)	Sm^{144}(p, n) (MesR65)
Eu^{144}	18 m (HoffR52)	⚛ β⁺ (HoffR52)	G excit, sep isotopes (HoffR52) activity not observed (OlkJ59b, MesR65)		protons on Sm^{144} (HoffR52)
Eu^{145}	5.9 d (FrieA63) 5.6 d (GrovJ59) others (HoffR51)	⚛ EC 99%, β⁺ 1% (FrieA63) △ –77.9 (MTW)	A chem, excit, sep isotopes (GrovJ59) chem, mass spect (FrieA63) daughter Gd^{145} (GrovJ59) daughter Tb^{149} (HoffR51)	γ Sm X-rays, 0.23?, 0.33?, 0.53 (complex), 0.656 (↑ 30), 0.766 (↑ 10), 0.894 (↑ 100), 1.66 (↑ 16), 2.00 (↑ 8) e⁻ 0.063, 0.103, 0.847 daughter radiations from Sm^{145}	Sm^{144}(α, 3n)Gd^{145}(EC) (GrovJ59, OlkJ59b, FrieA63) Sm^{144}(d, n) (GrovJ59)
Eu^{146}	4.59 d (TakekE64) others (FrieA63, GrovJ59, FunE62, GoroG58, AntoN59a, GoroG57a)	⚛ EC 96.5%, β⁺ 3.5% (FunE62) EC 95.5%, β⁺ 4.5% (TakekE64) others (FrieA63) △ –77.18 (MTW)	A chem, genet (GoroG57a, GoroG58, GrovJ59) chem, mass spect (FrieA63) daughter Gd^{146} (GoroG58, GrovJ59)	γ Sm X-rays, 0.511 (7%, γ±), 0.634 (77%, doublet), 0.666 (12%), 0.71 (13%, complex), 0.749 (100%), 0.90 (8%, complex), 1.058 (7%), 1.16 (6%, complex), 1.298 (6%), 1.408 (5%), 1.535 (8%), others to 2.93 β⁺ 2.11 max (0.14%), 1.47 max (3.3%) e⁻ 0.586, 0.702	Sm^{144}(α, 2n)Gd^{146}(EC) (GrovJ59, FrieA63)
Eu^{146}?	38 h (HoffR51) others (FunE62)	⚛ (HoffR51)	E excit, sep isotopes (HoffR51) chem (FunE62) not daughter 50 d Gd^{146} (FrieA63, AntoN61) daughter 7 h Gd^{146}? (GuseI57)	γ γ-ray spectrum may be identical to that of 4.59 d Eu^{146}	Sm^{147}(d, 3n), alphas on Sm^{144} (HoffR51) Sm^{147}(p, 2n) (FunE62)
Eu^{147}	21.5 d (FrieA63) 24 d (SchweC62, HoffR51, RasJ53, MackRC53) 25 d (AntoN58c)	⚛ EC 99.5%, β⁺ 0.5% (MNulJ64) α 0.002% (SiiA62, TotK64) others (HoffR51, FrieA63) △ –77.5 (MTW)	A chem, excit, sep isotopes (HoffR51) chem, mass spect (FrieA63) daughter Gd^{147} (GoroG57a)	γ Sm X-rays, 0.122 (20%), 0.198 (24%), 0.600 (7%), 0.680 (11%), 0.800 (6%), 0.957 (9%), 1.079 (9%), 1.25 (1.2%) e⁻ 0.030, 0.075, 0.114, 0.151 α 2.91	Sm^{147}(p, n) (HoffR51, RasJ53, SchweC62) Sm^{148}(p, 2n) (MNulJ64) deuterons on Sm (RasJ53)
Eu^{148}	54 d (WilkG50c) 50 d (HoffR51) 58 d (SchweC62a) 53 d (MarinJ51d)	⚛ EC 99+%, β⁺ 0.13% (BabC63b) α 9 x 10⁻⁷% (TotK64) △ –76.26 (BabC63b, MTW)	A chem (MarinJ51d) excit, sep isotopes (HoffR51, MackRC52) mass spect (BabC63b)	γ Sm X-rays, 0.413 (18%, complex), 0.551 (120%, complex), 0.62 (90%, complex), 0.72 (18%, complex), 0.872 (7%), 0.917 (5%), 0.967 (5%), 1.033 (7%), 1.16 (5%, complex), 1.345 (8%), 1.62 (11%, complex) e⁻ 0.02–0.04, 0.51, 0.193, 0.366, 0.505, 0.544, 0.584 β⁺ 0.92 max α 2.63	Sm^{148}(p, n) (HoffR51, MackRC52, WilkG50c, SchweC62a) Sm^{147}(d, n) (KurbJ43, MarinJ51d) Sm^{148}(d, 2n) (BabC63b)
Eu^{149}	106 d (HarlO61) others (AntoN59, DzhB62d, WanF62)	⚛ EC (HarlO61, HarmB61, AntoN59) no α, lim 4 x 10⁻⁷% (SiiA62) △ –76 (MTW)	A sep isotopes, excit (HoffR52) chem, excit (MackRC53, HarlO61, HarlO63) genet energy levels (JhaS62b, AlfV64)	γ Sm X-rays, 0.277 (↑ 10), 0.328 (↑ 10) e⁻ 0.015, 0.021, 0.230, 0.281	Sm^{149}(p, n) (HoffR52, HarlO61, HarlO63) Sm^{150}(p, 2n) (HarmB61, HarlO61)

TABLE 1 (Continued)

Isotope Z A	Half-life	Type of decay (☢); % abundance; Mass excess (△≡M−A), MeV (C¹²=0); Thermal neutron cross section (σ), barns	Class; Identification; Genetic relationships	Major radiations: approximate energies (MeV) and intensities	Principal means of production
Eu¹⁵⁰ 63	12.55 h (SiiA62) 12.8 h (YosY63) 13.7 h (MackRC53) 14.0 h (RiccR62) 15.0 h (WilkG50c) others (WilleR60, ButeF50)	☢ β⁻ 90%, EC 9%, β⁺ 0.4% (GutM65) β⁻ 95%, EC 4%, β⁺ 1% (YosY63) β⁻ 95%, EC 5% (SiiA62) △ −74.81 (MTW)	A chem, excit (ButeF50) chem, excit, sep isotopes (HoffR52) excit, sep isotopes (MackRC52) parent Gd¹⁵⁰ (KarrM61, SiiA62)	β⁻ 1.01 max β⁺ 1.24 max γ Sm X-rays, 0.334 (4%), 0.406 (3%), 0.511 (0.8%, γ⁺), 0.619 (0.2%), 0.713 (0.2%), 0.831 (0.5%), 0.921 (0.4%, doublet), 1.165 (0.4%), 1.224 (0.4%), 1.224 (0.3%), 1.630 (0.09%), 1.964 (0.2%)	Sm¹⁵⁰(p,n) (HoffR52, MackRC52, WilkG50c, HarmB61, YosY63) Sm¹⁵⁰(d,2n) (YosY63)
Eu¹⁵⁰	≈5 y (GutM61) >5 y (HarmB61)	☢ EC (HarmB61, GutM61)	A chem, genet energy levels (HarmB61, GutM61)	γ Sm X-rays, 0.334 (96%), 0.439 (86%), 0.584 (60%), 0.74 (21%, doublet), 1.049 (9%), 1.248 (5%), 1.347 (4%) e⁻ 0.287, 0.327, 0.392	Sm¹⁵⁰(p,n) (HarmB61, GutM61)
Eu¹⁵¹		% 47.77 (HessD48) 47.86 (CollT57) △ −74.67 (MTW) σc 5900 (to Eu¹⁵²) 2800 (to Eu¹⁵²m₁) (GoldmDT64)			
Eu¹⁵²	12.7 y (LocE56, LocE53) 12.2 y (GeiKW57) others (KarrD52, KasJ53)	☢ EC 72%, β⁻ 28%, β⁺ 0.021% (LHP) △ −72.89 (MTW) σc 5000 (GoldmDT64)	A n-capt, mass spect (IngM47) chem (MarinJ49)	β⁻ 1.48 max e⁻ 0.075, 0.115, 0.120 β⁺ 0.71 max γ Gd X-rays, Sm X-rays, 0.122 (37%), 0.245 (8%), 0.344 (27%), 0.779 (14%), 0.965 (15%), 1.087 (12%), 1.113 (14%), 1.408 (22%)	Eu¹⁵¹(n,γ) (IngM47, SerL47b)
Eu¹⁵²m₁	9.3 h (BotW46a, ChilG61a) 9.2 h (PoolM38a, HaydR49, AntoS59)	☢ β⁻ 77%, EC 23%, β⁺ 0.011% (NDS) no IT, lim 0.003% (TakaK65) △ −72.84 (LHP, MTW)	A n-capt (MarsJK35) n-capt, excit (PoolM38a) mass spect (HaydR46, HaydR49)	β⁻ 1.88 max e⁻ 0.075, 0.115, 0.120 β⁺ 0.89 max γ 0.122 (8%), 0.344 (2.5%), 0.842 (13%), 0.963 (12%), 1.315 (1.2%), 1.389 (1.1%)	Eu¹⁵¹(n,γ) (MarsJK35, PoolM38a, HevG36, FajK41, SerL47b, HaydR49)
Eu¹⁵²m₂	96 m (KirP63)	☢ IT (KirP63) no β⁻, no EC, lim 5% (KirP63) △ −72.74 (LHP, MTW)	A chem, excit, sep isotopes, cross bomb (KirP63)	γ Eu X-rays, 0.090 (74%) e⁻ 0.010, 0.016, 0.032, 0.039	Sm¹⁵⁴(p,3n) (KirP63) Sm¹⁵²(p,n) (KirP63) Eu¹⁵¹(n,γ) (TakaK65)
Eu¹⁵³		% 52.23 (HessD48) 52.14 (CollT57) △ −73.36 (MTW) σc 320 (GoldmDT64)			
Eu¹⁵⁴	16 y (KarrD52) others (HaydR49, GeiKW57, KasJ53)	☢ β⁻ (HaydR49) no β⁺, lim 0.003% (AlbuD58b) △ −71.68 (MTW) σc 1400 (GoldmDT64)	A n-capt (ScheiH38) mass spect (IngM47, HaydR49) chem (KarrD52)	β⁻ 1.85 max (10%), 0.87 max e⁻ 0.073, 0.115, 0.122 γ Gd X-rays, 0.123 (38%), 0.248 (7%), 0.593 (6%), 0.724 (21%), 0.759 (5%), 0.876 (12%), 1.00 (31%, doublet), 1.278 (37%)	Eu¹⁵³(n,γ) (ScheiH38, FajK39, FajK41a, SerL47b)
Eu¹⁵⁵	1.811 y (PierrA59) others (RutW52, WinsL51d, HaydR49)	☢ β⁻ (WinsL51d) △ −71.79 (MTW) σc 13,000 (GoldmDT64)	A chem (WinsL51d) mass spect (HaydR49) daughter Sm¹⁵⁵ (IngM47c)	β⁻ 0.25 max e⁻ 0.011, 0.017, 0.036, 0.054, 0.078, 0.082 γ Gd X-rays, 0.087 (32%), 0.105 (20%)	Sm¹⁵⁴(n,γ)Sm¹⁵⁵(β⁻) (IngM47c)
Eu¹⁵⁶	15.4 d (WinsL51c, IngM47c)	☢ β⁻ (WinsL51c) △ −70.05 (MTW)	A chem (WinsL51c) mass spect (IngM47d, IngM47c) daughter Sm¹⁵⁶ (WinsL51c)	β⁻ 2.45 max e⁻ 0.039, 0.081, 0.087 γ Gd X-rays, 0.089 (8%), 0.646 (7%), 0.723 (6%), 0.812 (9%), 1.07 (11%, complex), 1.15 (14%, complex), 1.24 (16%, complex), 1.97 (7%, complex), 2.098 (3%), 2.19 (5%, complex)	Sm¹⁵⁴(n,γ)Sm¹⁵⁵(β⁻) Eu¹⁵⁵(n,γ) (EwaG62, CliJ61) daughter Sm¹⁵⁶ (WinsL51c)
Eu¹⁵⁷	15.1 h (DaniW63) 15.4 h (WinsL51b)	☢ β⁻ (WinsL51b) △ −69.43 (LHP, MTW)	A chem (WinsL51b) genet energy levels (HarmB62) cross bomb (DaniW63) sep isotopes (ShidY64)	β⁻ 1.3 max e⁻ 0.004, 0.014, 0.046, 0.056 γ Gd X-rays, 0.055 (5%), 0.064 (27%), 0.32 (5%, doublet), 0.37 (14%, doublet), 0.413 (27%), 0.477 (5%), 0.623 (6%)	Gd¹⁶⁰(p,α) (HarmB62) neutrons on Gd (KantJ64)

TABLE 1 (*Continued*)

Isotope Z A	Half-life	Type of decay (☢); % abundance; Mass excess (△≡M−A), MeV (C¹²=0); Thermal neutron cross section (σ), barns	Class; Identification; Genetic relationships	Major radiations: approximate energies (MeV) and intensities	Principal means of production
$_{63}$Eu^{158}	46 m (MunH65, SchimF65a, DaniW63) 60 m (WinsL51b)	☢ β⁻ (WinsL51b) △ −67.1 (MTW)	B chem (WinsL51b) chem, genet energy levels (DaniW63)	β⁻ 2.5 max e⁻ [0.049, 0.072] γ 0.080 (↑ 100), 0.182, 0.52 (↑ 25, complex), 0.61 (↑ 8), 0.95 (↑ 95, complex), 1.11 (↑ 11), 1.19 (↑ 16)	Gd^{160}(d, α) (DaniW63) fission (WinsL51b)
Eu^{159}	18.1 m (MunH65) 19.0 (IwaT65) others (IwaT64, ButeF50, KuroT61b)	☢ β⁻ (KuroT61b) △ −66.02 (IwaT65, MTW)	C excit (ButeF50) sep isotopes, genet (IwaT64) parent Gd^{159} (IwaT64)	β⁻ 2.6 max γ 0.07 (42%), 0.09 (18%), 0.15 (14%), 0.22 (5%), 0.67 (21%), 0.73 (10%), 0.8 (11%, complex?), 1.1 (11%, complex), 1.5 (5%, complex?)	Gd^{160}(γ, p) (IwaT64, KuroT61b, ButeF50)
Eu^{160}	≈2.5 m (TakaK61)	☢ β⁻ (TakaK61) △ −64 (MTW)	F decay charac (TakaK61)	β⁻ 3.6 max γ no γ	Gd^{160}(n, p) (TakaK61)
$_{64}$Gd^{145}	25 m (GrovJ59) others (OlkJ59b)	☢ EC, β⁺ (GrovJ59, OlkJ59b)	A chem, excit, sep isotopes, genet (GrovJ59) parent Eu^{145} (GrovJ59)	β⁺ 2.4 max γ Eu X-rays, 0.511 (γ±), 0.80 (↑ 9), 1.03 (↑ 10), 1.75 (↑ 100, complex?)	Sm^{144}(α, 3n) (GrovJ59, OlkJ59b)
Gd^{146}	50 d (FrieA63) 46 d (GrovJ59) others (AntoN59a, GoroG58, GoroG57a, OlkJ59)	☢ EC (GoroG58) EC ≈99%, β⁺ ≈1% (FrieA63) △ −76 (MTW)	A chem, genet (GoroG57a, GoroG58) chem, excit, sep isotopes (GrovJ59) chem, mass spect (FrieA63) parent Eu^{146} (GoroG58, GrovJ59)	γ Eu X-rays, 0.078 (↑ 30), 0.115 (↑ 100, complex), 0.155 (↑ 45) e⁻ 0.066, 0.106 daughter radiations from 4.59 d Eu^{146}	Sm^{144}(α, 2n) (GrovJ59, FrieA63)
Gd$^{146?}$	7 h (OlkJ59, SunK51a) 12 h genet (GuseI57)	☢ α (SunK51a) α, [EC] (OlkJ59)	F chem (GuseI57, OlkJ57) parent 38 h Eu$^{146?}$ (GuseI57)	γ 0.22, 0.34, 0.55, 0.72	alphas on Sm (SunK51a) protons on Tb (OlkJ59) protons on Ta (GuseI57)
Gd^{147}	35 h (AntoN58c) 22 h (FrieA63) 29 h (ShirV57)	☢ EC, no β⁺, lim 1.2% (ShirV57) β⁺ (weak) (FrieA63) △ −75 (MTW)	A chem, genet (GoroG57a) chem, excit (ShirV57) chem, mass spect (FrieA63) parent Eu^{147} (GoroG57a) daughter Tb^{147} (TotK60)	γ Eu X-rays, 0.229 (↑ 150), 0.39 (↑ 85, complex), 0.64 (↑ 70, complex), 0.77 (↑ 60, complex), 0.932 (↑ 60), 1.10 (↑ 19, complex) e⁻ 0.181, 0.221, 0.321, 0.348, 0.388 daughter radiations from Eu^{147}	Sm^{144}(α, n) (FrieA63) Sm^{147}(α, 4n) (ShirV57)
Gd^{148}	84 y (SiiA62) others (RasJ53, SurY57)	☢ α (RasJ53) △ −76.29 (MTW)	B chem, excit, sep isotopes (RasJ53)	α 3.18	Sm^{147}(α, 3n), Eu^{151}(p, 4n) (RasJ53)
Gd^{149}	9.5 d (PraH62a) 9.3 d (ShirV57) others (HoffR51, AntoN58b)	☢ EC 99+%, α ≈0.0007%, no β⁺, lim 0.4% (ShirV57, RasJ53) α 0.0005% (SiiA65a) △ −75.2 (MTW)	A chem, excit, sep isotopes, cross bomb (HoffR51) chem, excit (ShirV57) chem, sep isotopes (PraH62a)	γ Eu X-rays, 0.150 (48%), 0.299 (26%), 0.347 (25%), 0.750 (11%), 0.790 (10%), 0.94 (5%, complex) e⁻ 0.101, 0.142, 0.250, 0.298 α 3.01 daughter radiations from Eu^{149}	Eu^{151}(p, 3n) (HoffR51, PraH62a) Sm^{147}(α, 2n) (RasJ53, ShirV57)
Gd^{150}	2.1 x 10⁶ y sp act (SiiA62) 1.4 x 10⁶ y sp act (OgaI65) 1.2 x 10⁵ y sp act (FrieA63b) ≈1 x 10⁵ y (KarrM61)	☢ α (RasJ53) △ −75.82 (MTW)	A chem (RasJ53) mass spect (FrieA63b) daughter 12.6 h Eu^{150} (KarrM61, SiiA62)	α 2.73	daughter 12.6 h Eu^{150} (KarrM61, SiiA62) Eu^{151}(d, 3n) (RasJ53) alphas on Sm (FrieA63b)
Gd^{151}	120 d (AntoN58a) 150 d (HeiR50)	☢ EC, no β⁺ (HeiR50) α ≈8 x 10⁻⁷% (SiiA65a) △ −74 (MTW)	A chem, excit (HeiR50) chem, genet energy levels (BisA57, ShirV58) daughter Tb^{151} (BaranV58)	γ Eu X-rays, 0.0216 (3%), 0.154 (7%), 0.175 (3%), 0.244 (7%), 0.308 (1%) e⁻ 0.014, 0.020, 0.105, 0.127, 0.167 α 2.60	Eu^{151}(p, n) (ShirV58, SiiA65a) Eu^{151}(d, 2n) (FajK41, ShirV58, KriN48, HeiR50, SteicE63)
Gd^{152}	1.1 x 10¹⁴ y sp act (MacfR61a) ≈10¹⁵ y (RieW59)	% 0.20 (BaiK50) 0.21 (CollT57) ☢ α (RieW59, MacfR61a) △ −74.71 (MTW) σc <180 (GoldmDT64)	A chem, sep isotopes (RieW59, MacfR61a)	α 2.1	

TABLE 1 (*Continued*)

Isotope Z A	Half-life	Type of decay (☢); % abundance; Mass excess (△≡M−A), MeV (C¹²=0); Thermal neutron cross section (σ), barns	Class; Identification; Genetic relationships	Major radiations: approximate energies (MeV) and intensities	Principal means of production
$_{64}Gd^{153}$	242 d (HoffD63) 236 d (HeiR50)	☢ EC, no β⁺ (HeiR50) △ −73.12 (MTW)	A mass spect (IngM47c) chem, n-capt (HeiR50) daughter Tb¹⁵³ (MihJ57a, BaraV58)	γ Eu X-rays, 0.070 (2.4%), 0.099 (55%, complex) e⁻ 0.021, 0.049, 0.065, 0.101	Gd¹⁵²(n, γ) (IngM47c, CorkJ48a, HeiR50) Eu¹⁵³(d, 2n) (HeiR50)
$\underline{Gd^{154}}$		% 2.15 (BaiK50) 2.23 (CollT57) △ −73.65 (MTW)			
$\underline{Gd^{155}}$		% 14.7 (BaiK50) 15.1 (CollT57) 15.0 (LowW59) △ −72.04 (MTW) σ_c 58,000 (GoldmDT64)			
$\underline{Gd^{156}}$		% 20.47 (BaiK50) 20.6 (CollT57) △ −72.49 (MTW)			
$\underline{Gd^{157}}$		% 15.68 (BaiK50) 15.7 (CollT57) others (LowW59) △ −70.77 (MTW) σ_c 2.4 x 10⁵ (GoldmDT64)			
$\underline{Gd^{158}}$		% 24.9 (BaiK50) 24.5 (CollT57) △ −70.63 (MTW) σ_c 3.4 (GoldmDT64)			
Gd^{159}	18.0 h (KriN48, ButeF50, ButeF49, BarlR55a, WilleR60) others (TotK60a, TakaK62, SerL47b)	☢ β⁻ (KriN48) △ −68.59 (MTW)	A n-capt (SerL47b) chem (ButeF49, HeiR50) genet energy levels (JorW53a) mass spect (NielK58a) daughter Eu¹⁵⁹ (IwaT64)	β⁻ 0.95 max e⁻ 0.006, 0.049, 0.056 γ Tb X-rays, 0.058 (3%), 0.363 (9%)	Gd¹⁵⁸(n, γ) (SerL47b, ButeF49, HeiR50)
$\underline{Gd^{160}}$		% 21.9 (BaiK50) 21.6 (CollT57) △ −67.89 (MTW) σ_c 0.8 (GoldmDT64)			
Gd^{161}	3.6 m (ButeF49) 3.7 m (JorW53a) others (KriN48, WilleR60)	☢ β⁻ (KetB49c) △ −65.5 (MTW)	A n-capt (IngM46) n-capt, excit (ButeF49) n-capt, sep isotopes (SchmL59) parent Tb¹⁶¹ (KetB49c)	β⁻ 1.6 max e⁻ 0.005, 0.026, 0.049, 0.055, 0.263, 0.309 γ Tb X-rays, 0.102 (11%), 0.284 (8%), 0.315 (25%), 0.361 (66%)	Gd¹⁶⁰(n, γ) (IngM46, ButeF49, KetB49b, SchmL59)
Gd^{162}	several years (?) (FalK57)	☢ [β⁻] (FalK57) △ −64 (MTW)	F chem (FalK57) not parent Tb¹⁶² (FalK57)		Gd¹⁶⁰(n, γ)Gd¹⁶¹(n, γ) (FalK57)
$_{65}Tb^{147}$	24 m (TotK60)	☢ EC, β⁺ (TotK60)	C excit, genet (TotK60) parent Gd¹⁴⁷ (TotK60)	γ Gd X-rays, 0.305, 0.511 (γ±) daughter radiations from Gd¹⁴⁷	Pr¹⁴¹(C¹², 6n) (TotK60)
Tb^{148}	70 m (TotK60) 66 m (BoncN61)	☢ EC, β⁺ (TotK60) △ −70.7 (MTW)	B chem, excit (TotK60)	β⁺ 4.6 max γ Gd X-rays, 0.511 (γ±), 0.78, 1.12	Pr¹⁴¹(C¹², 5n) (TotK60)
$Tb^{[<157]}$	17 h (RolM53)	☢ β⁻ (RolM53)	G chem (RolM53) existence of a Tb isotope with A <162, t₁/₂ ≈17 h, and Q_β⁻ >2 is highly improbable (LHP)	β⁻ 2.34 max	alphas on Eu (RolM53)
$Tb^{[<157]}$	>17 h (RolM53)	☢ β⁺ (RolM53)	G chem (RolM53) probably a mixture of Tb¹⁵², Tb¹⁵⁵, and Tb¹⁵⁶ (LHP)	β⁺ 3.1 max e⁻ 0.076, 0.088, 0.126, 0.153, 0.20	alphas on Eu (RolM53)

TABLE 1 (*Continued*)

Isotope Z A	Half-life	Type of decay (☢); % abundance; Mass excess (△≡M–A), MeV (C¹²≡0); Thermal neutron cross section (σ), barns	Class; Identification; Genetic relationships	Major radiations: approximate energies (MeV) and intensities	Principal means of production
$_{65}$Tb^{149}	4.10 h (TotK60a) 4.2 h (BruniE65) others (RasJ53, SurY57)	☢ EC 84%, α ≈16%, no β⁺ (TotK60a, RasJ53, RolM53) △ –71.4 (MTW)	A chem, mass spect (RasJ50, TotK60a) parent Eu¹⁴⁵ (HoffR51) daughter Dy¹⁴⁹ (TotK59) daughter Tb¹⁴⁹ᵐ (MacfR62) descendant Er¹⁵³ (MacfR63a)	γ Gd X-rays, 0.16, 0.35 e⁻ 0.115, 0.127, 0.157, 0.301, 0.338, 0.587 α 3.95 daughter radiations from Gd¹⁴⁹	Pr¹⁴¹(C¹², 4n) (TotK59) Eu¹⁵¹(α, 6n) (RasJ53)
Tb¹⁴⁹ᵐ	4.3 m (MacfR62, MacfR64)	☢ [IT+EC+β⁺] 99+%, α 0.025% (MacfR64)	B excit, cross bomb, genet (MacfR62) parent Tb¹⁴⁹ (MacfR62)	γ [Tb X-rays] α 3.99 daughter radiations from Tb¹⁴⁹	La¹³⁹(O¹⁶, 6n) (MacfR62, MacfR64)
Tb¹⁵⁰	3.1 h (TotK59d, TotK60a, BoncN61)	☢ EC, β⁺ (TotK59d, TotK60, BoncN61) no α, lim 0.05% (TotK60a) △ –71.03 (MTW)	A chem, mass spect (TotK59d, TotK60a)	β⁺ 3.6 max γ Gd X-rays, 0.511 (↑ 100, γ±), 0.637 (↑ 100), 0.93 (↑ 35)	protons on Gd (TotK59d, TotK60a)
Tb¹⁵¹	18 h (TotK60a, BaranV58) 19 h (RasJ53) 20 h (MihJ57a) others (TotK58a, AntoN58)	☢ EC 99+%, α 0.0005% (MacfR64) △ –71.6 (MTW)	A chem, excit (RasJ53, MihJ57a, TotK58a) chem, genet (BaranV58) chem, mass spect (TotK60a) parent Gd¹⁵¹ (BaranV58)	γ Gd X-rays, 0.108 (35%), 0.18 (18%, doublet), 0.252 (35%), 0.288 (32%), 0.40 (complex), 0.44 (complex), 0.48 (complex), 0.60 (complex), 0.72 (complex), 0.87 e⁻ 0.058, 0.100, 0.130, 0.202, 0.237 α 3.42	Eu¹⁵¹(α, 4n) (TotK58a, MacfR64) protons on Gd (TotK60a, HarmB62)
Tb¹⁵²	17.4 h (TotK60a) 18.5 h (TotK59b) 19.6 h (StriA62) others (BoncN60, BoncN61, AbdurA60a)	☢ EC ≈80%, β⁺ ≈20% (GromK65a) no α, lim 10⁻⁵% (TotK59b) △ –70.5 (MTW)	A chem, genet energy levels (TotK59b) chem, mass spect (TotK60a, StriA62) daughter Dy¹⁵² (BasiA60a)	β⁺ 2.82 max e⁻ 0.221, 0.263, 0.294, 0.336, 0.382, 0.536, 0.565, 0.607 γ Gd X-rays, 0.271 (↑ 13), 0.344 (↑ 100), 0.411 (↑ 6), 0.586 (↑ 14), 0.779 (↑ 14), 0.974 (↑ 10), 1.12 (↑ 10, complex), 1.31 (↑ 11, complex), 1.60 (↑ 7, complex), 1.95 (↑ 8, complex), 2.40 (↑ 9, complex), 2.70 (↑ 6, complex)	Eu¹⁵¹(α, 3n) (TotK59b) protons on Gd (TotK60a, StriA62)
Tb¹⁵²	4.0 m (OlkJ59a)	☢ EC, β⁺, α 0.002% (OlkJ59a)	C excit, cross bomb, sep isotopes (OlkJ59a)	γ Tb X-rays, 0.14, 0.23, 0.511 (γ±)	Eu¹⁵¹(α, 3n), Gd¹⁵²(p, n) (OlkJ59a)
Tb¹⁵³	55 h (TotK60a) 63 h (StriA61) 62 h (MihJ57a) others (TotK59a, BaraV58, AntoN58)	☢ EC (MihJ57a) △ –71 (MTW)	A chem, excit, genet (MihJ57a) chem, genet (BaraV58) chem, mass spect (TotK60a) parent Gd¹⁵³ (MihJ57a, BaraV58) daughter Dy¹⁵³ (DobA58)	γ Gd X-rays, 0.083 (11%, complex), 0.11 (12%, complex), 0.17 (9%, complex), 0.212 (30%), 0.250, 0.33, 0.88 e⁻ 0.012, 0.034, 0.037, 0.040, 0.044, 0.052, 0.057, 0.162 daughter radiations from Gd¹⁵³	protons on Gd (MihJ57a, HarmB62, TotK60a)
Tb¹⁵⁴	21.0 h (TotK60a) 17 h (WilkG50c, RolM53, HandT55b) others (MihJ57a, AntoN58, HenrR59, TotK59a)	☢ EC, β⁺ ≈0.5% (?) (WilkG50c) △ –70 (MTW)	A chem, excit (WilkG50c) chem, genet energy levels (MihJ57a) chem, excit, sep isotopes (HandT55b) chem, mass spect (TotK60a) not daughter Dy¹⁵⁴ (MacfR61)	γ Gd X-rays, 0.123, 0.18?, 0.248, 0.30 (complex), 0.347, 0.53 (complex), 0.65 (complex), others to 2.5 e⁻ 0.073, 0.115, 0.122, 0.198	Eu¹⁵¹(α, n) (WilkG50c) Eu¹⁵³(α, 3n) (TotK59a) protons on Gd (HandT55b, MihJ57a, TotK60a)
Tb¹⁵⁴	8.5 h (TotK60a) ≈7.5 h (HandT55b) 8 h (MihJ57a)	☢ EC, β⁺ (?) (HandT55b) △ –70 (MTW)	A chem, excit (HandT55a) chem, genet energy levels (MihJ57a) chem, mass spect (TotK60a) not daughter Dy¹⁵⁴ (MacfR61)	γ Gd X-rays, 0.123, 0.18?, 0.248, 0.53 (complex), 0.65 (complex) e⁻ 0.073, 0.115, 0.122, 0.198	protons on Gd (HandT55b, MihJ57a, TotK60a)
Tb¹⁵⁵	5.6 d (MihJ57a) 5.4 d (TotK60a) 4.5 d (DzhB58) others (AntoN58)	☢ EC (MihJ57a, HarmB62) △ –71 (MTW)	A chem, excit (WilkG50a) chem, sep isotopes, genet energy levels (MihJ57a) chem, mass spect (TotK60a) others (HandT55b) daughter Dy¹⁵⁵ (GoroG57a, DobA58, MayM64)	γ Gd X-rays, 0.087 (37%), 0.105 (25%), 0.163 (8%, complex), 0.180 (8%), 0.262 (7%), 0.368 (4%) e⁻ 0.011, 0.034, 0.053, 0.078, 0.110, 0.129, 0.210	protons on Gd (MihJ57a, HandT55b, TotK60a)

TABLE 1 (*Continued*)

Isotope Z A	Half-life	Type of decay (☢); % abundance; Mass excess (Δ≡M–A), MeV (C¹²=0); Thermal neutron cross section (σ), barns	Class; Identification; Genetic relationships	Major radiations: approximate energies (MeV) and intensities	Principal means of production
$_{65}Tb^{156}$	5.1 d (TotK60a) 5.3 d (HenrR59) 5.6 d (MihJ57a) others (HandT55b, WilkG50a, ButeF49, AntoN58, HolloJ59)	☢ EC, β⁻ (weak), no β⁺ (HandT55b) no β⁺ (HolloJ59, OfeS59a) Δ −70 (MTW)	A chem, excit (HandT55b) chem, genet energy levels (MihJ57a)	γ Gd X-rays, 0.089 (17%), 0.199 (40%), 0.356 (13%), 0.535 (70%), 1.065 (12%), 1.16 (17%, complex), 1.22 (29%), 1.42 (15%), 1.65 (5%), 1.85 (4%) e⁻ 0.039, 0.081, 0.087, 0.149	Eu¹⁵³(α, n) (HansP59, OfeS59a, WilkG50a) Gd¹⁵⁶(p, n) (WilkG50c)
Tb^{156m}	5.5 h (MihJ57, HandT55b) 5.0 h (WilkG50a)	☢ IT (MihJ57, MihJ57a) EC, β⁺ <25% (WilkG50a) β⁻ (weak), no β⁺ (HandT55b) Δ −70 (LHP, MTW)	B chem, excit (WilkG50a, HandT55b) chem, sep isotopes (MihJ57) chem, mass spect (TotK60)	γ [Tb L X-rays, Tb K X-rays (weak), 0.088 (weak)] e⁻ 0.036, 0.081 daughter radiations from Tb¹⁵⁶	Gd¹⁵⁶(p, n) (HandT55b, MihJ57)
Tb^{157}	1.5 x 10² y sp act (FujI64) 3 x 10² y sp act (GrigE64) others (IwaS63)	☢ EC (BhaM62, FujI64, IwaS63) Δ −70.71 (MTW)	A chem, mass spect (NauR60a, TotK60a) chem, sep isotopes, cross bomb (BhaM62) daughter Dy¹⁵⁷ (IwaS63, FujI64)	γ Gd X-rays	Dy¹⁵⁶(n, γ) Dy¹⁵⁷(EC) (NauR60a, BhaM62) Gd¹⁵⁷(p, n) (BhaM62) Gd¹⁵⁶(α, 3n) Dy¹⁵⁷(β⁻) (IwaS63, FujI64)
Tb^{158}	1.2 x 10³ y (LewisH61) others (TotK60a, HandT55b, GovN58)	☢ EC 86%, β⁻ 14%, no β⁺, lim 2% (BhaM62) Δ −69.43 (MTW)	A chem (ButeF60) chem, mass spect (NauR60a) chem, cross bomb, sep isotopes (BhaM62)	β⁻ 0.85 max e⁻ 0.029, 0.044, 0.072, 0.078, 0.092, 0.132 γ Gd X-rays, 0.080 (12%), 0.182 (10%), 0.782 (10%), 0.95 (69%, doublet), 1.110 (2.2%), 1.190 (1.8%)	Dy¹⁵⁶(n, γ) Dy¹⁵⁷(EC) Tb¹⁵⁷(n, γ) (NauR60a, BhaM62, LewisH61, NauR62)
Tb^{158m}	10.5 s (SchmW65, GovN58) 11.0 s (HammC57) 10.2 s (BroaK65) others (HandT55b, PoolM38)	☢ IT (HandT55b) no β⁻ (lim 0.6%), no β⁺ (lim 0.04%), no EC (lim 1.5%) (SchmW65) Δ −69.32 (LHP, MTW)	C excit (GovN58, HammC57)	e⁻ 0.060, 0.102 γ Tb X-rays, 0.110 (0.5%)	Tb¹⁵⁹(n, 2n) (SchmW65) Tb¹⁵⁹(γ, n) (GovN58, HammC57)
$\underline{Tb^{159}}$	$t_{1/2}$ (α) >5 x 10¹⁶ y sp act (PorsW54)	% 100 (HessD48, CollT57) Δ −69.53 (MTW) σ_c 46 (GoldmDT64)			
Tb^{160}	72.1 d (HoffD63) 72.3 d (KreeK54) 73.0 d (ThirH57) others (BotW46a, BursS50, SmiRR56, IngM47c, KriN48, CorkJ50e, CorkJ48a)	☢ β⁻ (BotW43) no EC(K), lim 0.5% (ClarM57) Δ −67.85 (LHP, MTW) σ_c 525 (GoldmDT64)	A n-capt (BotW43) mass spect (IngM47c) chem (FolR51)	β⁻ 1.74 max (0.4%), 0.86 max e⁻ 0.033, 0.079, 0.085 γ Dy X-rays, 0.087 (12%), 0.197 (6%), 0.299 (30%), 0.879 (31%), 0.966 (31%, complex), 1.178 (15%), 1.272 (7%)	Tb¹⁵⁹(n, γ) (BotW43, BotW46a, SerL47b)
Tb^{161}	6.9 d (HoffD63, BisA56) 6.8 d (ButeF49, SmiRR56) 7.2 d (BaranS58, FunL64, HeirR50, CorkJ56a) others (CorkJ52c, BarlR55a)	☢ β⁻ (KriN48) Δ −67.47 (MTW)	A excit (KriN48) chem, excit (KetB49c) genet energy levels (CorkJ56a, SmiW56b) daughter Gd¹⁶¹ (KetB49c)	β⁻ 0.59 max (10%), 0.52 max e⁻ 0.017, 0.040, 0.048 γ Dy X-rays, 0.026 (21%), 0.049 (19%), 0.057 (5%), 0.075 (10%)	Gd¹⁶⁰(n, γ) Gd¹⁶¹(β⁻) (KetB49c, KetB49c)
Tb^{162}	7.48 m (SchnT65)	☢ [β⁻] (SchnT65) Δ −65 (MTW)	B genet energy levels, excit (SchnT65)	γ Dy X-rays, 0.040 (↑ 17), 0.081 (↑ 8), 0.140 (↑ 6), 0.180 (↑ 26), 0.258 (↑ 100), 0.81 (↑ 44), 0.89 (↑ 54) e⁻ [0.027, 0.072]	Dy¹⁶²(n, p) (SchnT65)
Tb^{162}	2.24 h (SchnT65) 2 h (FalK57)	☢ [β⁻] (FalK57) Δ −65 (MTW)	C chem, excit, sep isotopes (FalK57)		Gd¹⁶⁰(α, pn) (FalK57)
Tb^{163}	6.5 h (AlsJ60, TakaK62) others (FalK57)	☢ [β⁻] (TakaK62) Δ −64.7 (MTW)	B chem, excit (fission yield) (AlsJ60) sep isotopes (TakaK62)	β⁻ 1.65 max γ Dy X-rays, 0.025, 0.235, 0.330, 0.510	Gd¹⁶⁰(α, p) (FalK57) Dy¹⁶⁴(γ, p) (TakaK62) high energy fission (AlsJ60)
Tb^{163}	7 m (WilleR60)	☢ [β⁻] (WilleR60)	E sep isotopes, excit (WilleR60) possibly identical to 7.5 m Tb¹⁶²	γ 0.18	Dy¹⁶³(n, p) (WilleR60)
$Tb^{162, 163}$	14 m (ButeF50)		F excit (ButeF50)		gammas on Dy (ButeF50)

TABLE 1 (*Continued*)

Isotope Z A	Half-life	Type of decay (☢); % abundance; Mass excess (△ ≡M−A), MeV (C¹²=0); Thermal neutron cross section (σ), barns	Class; Identification; Genetic relationships	Major radiations: approximate energies (MeV) and intensities	Principal means of production
$_{65}Tb^{164}$	23 h (AlsJ60)	☢ [β⁻] (AlsJ60) △ −62 (MTW)	D chem, excit (fission yield) (AlsJ60)		high energy fission (AlsJ60)
$_{66}Dy^{149}$	10–20 m (TotK59, TotK58a)	☢ EC (TotK58a, TotK59)	C excit, genet (TotK59, TotK58a) parent Tb¹⁴⁹ (TotK59)		Pr¹⁴¹(N¹⁴, 6n) (TotK59, TotK58a)
Dy^{150}	7.2 m (MacfR64) 8 m (TotK59) 7 m (RasJ53)	☢ EC, β⁺, α (TotK59) EC+β⁺ 82%, α 18% (MacfR64) △ −69 (MTW)	C cross bomb (RasJ53) excit (TotK59) daughter Ho¹⁵⁰ (MacfR63) daughter Er¹⁵⁴ (MacfR63)	γ Tb X-rays, 0.39, 0.511 (γ±) α 4.23 daughter radiations from Tb¹⁵⁰	Pr¹⁴¹(N¹⁴, 5n) (TotK59) Ce¹⁴⁰(O¹⁶, 6n) (MacfR64) Tb¹⁵⁹(p, 10n) (RasJ53)
Dy^{151}	18.0 m (MacfR64) 19 m (TotK59, RasJ53)	☢ β⁺ + EC 94%, α 6% (MacfR64) △ −69 (MTW)	B cross bomb (RasJ53) excit (TotK59) daughter 35.6 s Ho¹⁵¹ (MacfR63)	α 4.06 γ Tb X-rays, 0.145, 0.511 (γ±) daughter radiations from Tb¹⁵¹	Pr¹⁴¹(N¹⁴, 4n) (TotK59) Ce¹⁴⁰(O¹⁶, 5n) (MacfR64) Tb¹⁵⁹(p, 9n) (RasJ53)
Dy^{152}	2.41 h (SiiA62) 2.3 h (MacfR64, RasJ53, SurY57, BasiA60a) 2.5 h (TotK58a)	☢ EC, β⁺ (?), α (RasJ53, TotK59) α 0.05% (MacfR64) △ −70.11 (MTW)	A chem, excit (RasJ53, TotK59) chem, genet (BasiA60a) parent 18 h Tb¹⁵² (BasiA60a) daughter 52.35 Ho¹⁵² (MacfR63)	γ Tb X-rays, 0.257, 0.511 ? (γ±) α 3.65 daughter radiations from 18 h Tb¹⁵²	Pr¹⁴¹(N¹⁴, 3n) (TotK59) Gd¹⁵²(α, 4n) (TotK58a, MacfR64)
Dy^{153}	6.4 h (MacfR64) 5.5 h (RydH62) 5.0 h (TotK58a) 6.4 h (DzhB61a) others (DobA58, GoroG57a)	☢ EC, α 0.0030% (MacfR64) △ −69.2 (MTW)	A chem, excit, sep isotopes (TotK58a) chem, mass spect, genet (DobA58) parent Tb¹⁵³ (DobA58)	γ Tb X-rays, 0.08 (complex), [0.25 (complex)], others e⁻ 0.029, 0.047, 0.072, 0.091, 0.192, 0.202 α 3.48 daughter radiations from Tb¹⁵³	Gd¹⁵²(α, 3n) (TotK58a, MacfR64)
Dy^{154}	$t_{1/2}$ >10 y (MacfR61) $t_{1/2}$ (α) ≈1 x 10⁶ y sp act (MacfR61)	☢ α (MacfR61) △ −70.5 (MTW)	B chem, excit (MacfR61) not parent 21 h or 8.5 h Tb¹⁵⁴ (MacfR61)	α 2.85	Gd¹⁵⁴(α, 4n) (MacfR61)
Dy^{154m}	13 h (TotK58a)	☢ α (TotK58a)	B chem, excit, sep isotopes (TotK58a)	α 3.37	Gd¹⁵⁴(α, 4n) (TotK58a)
Dy^{155}	10.2 h (PersL63c, PersL64a) others (MayM64, TotK58a, GoroG57a, BoncN60, DzhB58a, DobA58, MihJ57a)	☢ EC (TotK58a) β⁺ 2% (PersL63c) △ −69 (MTW)	A chem, excit (MihJ57a) chem, mass spect (DobA58) parent Tb¹⁵⁵ (GoroG57a, DobA58, MayM64) daughter Ho¹⁵⁵ (DalB60a, KalyA59, BasiA61)	γ Tb X-rays, 0.227 (68%), 0.52 (8%, complex), 0.65 (5%, complex), 0.74 (4%, complex), 0.91 (5%, complex), 1.000 (6%), 1.091 (5%), 1.16 (6%, complex), 1.250 (4%), 1.39 (3%), 1.45 (4%), 1.66 (2%) β⁺ 1.08 max (0.14%), 0.85 max (2%) e⁻ 0.013, 0.038, 0.057, 0.175 daughter radiations from Tb¹⁵⁵	Tb¹⁵⁹(p, 5n) (MihJ57a, PersL64a) Gd¹⁵³(α, 2n), Gd¹⁵⁴(α, 3n) (TotK58a) Gd¹⁵²(α, n) (TotK61)
$\underline{Dy^{156}}$	$t_{1/2}$ (α) >1 x 10¹⁸ y sp act (RieW58)	% 0.0524 (IngM48d) 0.057 (CollT57) △ −70.9 (MTW) σ_c ≈3 (GoldmDT64)			
Dy^{157}	8.1 h (PersL63b) 8.2 h (MayM64, HandT53, RayG63) others (DobA58, GoroG57a)	☢ EC, no β⁺ (HandT53) △ −70 (MTW)	A chem, excit (HandT53) chem, sep isotopes (TotK61) chem, mass spect (DobA58) parent Tb¹⁵⁷ (IwaS63, FujI64)	γ Tb X-rays, 0.326 (91%) e⁻ 0.009, 0.031, 0.052, 0.074, 0.274	Tb¹⁵⁹(p, 3n) (HandT53, PersL63b) Gd¹⁵⁴(α, n) (TotK61)
$\underline{Dy^{158}}$		% 0.0902 (IngM48d) 0.100 (CollT57) △ −70.37 (MTW) σ_c 100 (GoldmDT64)			
Dy^{159}	144 d (KetB59) 151 d (HoffD63) 138 d (RayG63, MayM64) others (ButeF51a, KetB49, BjoS61, GrigE60a)	☢ EC (KetB49) △ −69.15 (MTW)	A chem, n-capt (KetB49) chem, cross bomb (ButeF51a) genet energy levels (MihJ57a)	γ Tb X-rays, 0.058 (4%), 0.348 (9 x 10⁻⁴%) e⁻ 0.006, 0.049, 0.056	Dy¹⁵⁸(n, γ) (SerL47b, ButeF49, HeiR50) Tb¹⁵⁹(d, 2n) (ButeF51a) Tb¹⁵⁹(p, n) (KetB59)

TABLE 1 *(Continued)*

Isotope Z A	Half-life	Type of decay (☢); % abundance; Mass excess ($\Delta \equiv$M–A), MeV (C^{12}=0); Thermal neutron cross section (σ), barns	Class; Identification; Genetic relationships	Major radiations: approximate energies (MeV) and intensities	Principal means of production
$_{66}Dy^{160}$		% 2.294 (IngM48d) 2.35 (CollT57) Δ –69.67 (MTW)			
Dy^{161}		% 18.88 (IngM48d) Δ –68.05 (MTW) σ_c 600 (GoldmDT64)			
Dy^{162}		% 25.53 (IngM48d) 25.5 (CollT57) Δ –68.18 (MTW) σ_c 140 (GoldmDT64)			
Dy^{163}		% 24.97 (IngM48d) 24.9 (CollT57) Δ –66.36 (MTW) σ_c 130 (GoldmDT64)			
Dy^{164}		% 28.18 (IngM48d) 28.1 (CollT57) Δ –65.95 (MTW) σ_c 800 (to Dy^{165}) 2000 (to Dy^{165m}) (GoldmDT64)			
Dy^{165}	139.2 m (SherR52) 139.0 m (PersL63) others (BotW46a, KetB49, SerL47b, MangS62, SlaH46, MayE54)	☢ β^- (PoolM38a) Δ –63.51 (MTW) σ_c 4700 (GoldmDT64)	A n–capt (HevG36, MarsJK35) n–capt, sep isotopes (IngM47f) mass spect (IngM47a)	β^- 1.29 max e^- 0.039, 0.085 γ Ho X-rays, 0.095 (4%), 0.280 (0.6%), 0.361 (1.1%), 0.633 (0.7%), 0.716 (0.7%) others to 1.08	$Dy^{164}(n,\gamma)$ (MarsJK35, HevG36, PoolM38a, MeiL40, SerL47b, KetB49)
Dy^{165m1}	1.26 m (HardR64) others (FlaA46, FlaA44a, HoleN48a)	☢ IT (FlaA44a) β^- 2.5% (HardR64) β^- 2.4% (TorR60) others (JorW53b) Δ –63.40 (LHP, MTW)	A n–capt (FlaA44a) n–capt, sep isotopes (IngM47f)	β^- 1.04 max (0.4%), 0.89 max e^- 0.054, 0.100, 0.106 γ Dy X-rays, 0.108 (3%), 0.152 (0.3%), 0.362 (0.6%), 0.514 (1.8%) daughter radiations from Dy^{165}	$Dy^{164}(n,\gamma)$ (FlaA44a, FlaA46, SerL47b, CaldR50, HardR64)
Dy^{165m2}	32 s (HardR64)	☢ [IT] (HardR64)	C n–capt, sep isotopes (HardR64)	γ complex spectrum to 1.1	$Dy^{164}(n,\gamma)$ (HardR64)
Dy^{166}	81.5 h (HoffD63) 81.8 h (GunR62) others (HelmeR60, ButeF50a, KetB49)	☢ β^- (KetB49) Δ –62.59 (MTW)	A chem, genet (KetB49) parent Ho^{166} (KetB49, ButeF50a)	β^- 0.48 max (5%), 0.40 max e^- 0.019, 0.027, 0.046 γ Ho X-rays, 0.082 (12%), 0.372 (0.5%), 0.426 (0.5%) daughter radiations from Ho^{166}	$Dy^{164}(n,\gamma)Dy^{165}(n,\gamma)$ (KetB49, ButeF50a, RusL60, HelmeR60, GunR62, BrabV64, HoffD63)
Dy^{167}	4.4 m (WilleR60)	☢ [β^-] (WilleR60)	C sep isotopes, excit (WilleR60)		$Er^{170}(n,\alpha)$ (WilleR60)
$_{67}Ho^{150}$	≈20 s (MacfR63)	☢ [EC, β^+] (MacfR63)	F genet (MacfR63) parent Dy^{150} (MacfR63)		$Pr^{141}(O^{16},7n)$ (MacfR63)
Ho^{151}	35.6 s (MacfR63)	☢ β^+ + EC 80%, α 20% (MacfR63)	B excit, cross bomb, genet (MacfR63) parent Dy^{151} (MacfR63)	α 4.51 γ [Dy X-rays, 0.511 ($\gamma^\pm$)] daughter radiations from Dy^{151}, Tb^{147}	$Pr^{141}(O^{16},6n)$ (MacfR63)
Ho^{151}	42 s (MacfR63)	☢ α ≈30%, β^+ + EC ≈70% (MacfR64)	C excit, cross bomb (MacfR63)	α 4.60 γ [Dy X-rays, 0.511 ($\gamma^\pm$)] daughter radiations from Dy^{151}, Tb^{147}	O^{16} on Nd^{142} (MacfR63)
Ho^{152}	52.3 s (MacfR63)	☢ [EC+β^+] 81%, α 19% (MacfR63)	B excit, genet (MacfR63) parent Dy^{152} (MacfR63)	α 4.45	$Pr^{141}(O^{16},5n)$ (MacfR63)

TABLE 1 (*Continued*)

Isotope Z A	Half-life	Type of decay (☢); % abundance; Mass excess (△=M-A), MeV (C¹²=0); Thermal neutron cross section (σ), barns	Class; Identification; Genetic relationships	Major radiations: approximate energies (MeV) and intensities	Principal means of production
$_{67}$Ho^{152}	2.4 m (MacfR63) ≈4 m (RasJ53)	☢ [EC+β⁺] ≈70%, α ≈30% (MacfR63) △ -63.8 (MTW)	C excit (RasJ53) excit, cross bomb (MacfR63, MacfR64b) daughter Er^{152} (MacfR63a)	α 4.38	Pr^{141}(O^{16}, 5n) (MacfR63)
Ho^{153}	9 m (MacfR63)	☢ [EC+β⁺], α 0.3% (MacfR63) △ -65.0 (MTW)	C excit (MacfR63)	α 3.92	Pr^{141}(O^{16}, 4n) (MacfR63)
Ho^{153}	27 m (MayM64)	☢ [α] (MayM64)	F genet (MayM64) ancestor Eu^{145} (MayM64)		protons on Dy (MayM64)
Ho^{154}	7 m (LagP66)	☢ β⁺, [EC] (LagP66) △ -65 (MTW)	B chem, mass spect (LagP66)	γ [Dy X-rays], 0.335, 0.511 (γ⁺)	protons on Dy (LagP66)
Ho^{155}	50 m (LagP66, KalyA59) 46 m (DalB60a)	☢ [EC], β⁺ (KalyA59)	A chem, genet (KalyA59, DalB60a, BasiA61) mass spect (LagP66) parent Dy^{155} (DalB60a, KalyA59, BasiA61)	β⁺ 2.1 max γ Dy X-rays, 0.092, 0.138, 0.511 (γ⁺) daughter radiations from Dy^{155}	protons on Dy, Ho (LagP66)
Ho^{156}	55 m (LagP66, BasiA61) 57 m (GrigE60d) others (MihJ57a)	☢ [EC] (MihJ57a) β⁺ (GrigE60d)	A chem, sep isotopes (MihJ57a) chem, mass spect (LagP66)	γ [Tb X-rays], 0.138 (↑ 100), 0.266 (↑ 99), 0.367 (↑ 23), 0.511 (γ⁺), 0.685, 0.89, 1.20, 1.41 e⁻ 0.084, 0.130, 0.213 β⁺ 2.9 max (↑ 1), 1.8 max (↑ 18)	Dy^{156}(p, n) (MihJ57a)
Ho^{157}	14 m (LagP66)	☢ β⁺, [EC] (LagP66)	B [chem], mass spect (LagP66)	γ Dy X-rays, 0.087, 0.152, 0.190, 0.227, 0.511 (γ⁺), 0.71, 0.86, 0.90, 1.20 daughter radiations from Dy^{157}	protons on Dy, Ho (LagP66)
Ho^{158}	11.5 m (SchepH62) 11 m (StenT65a)	☢ EC, no β⁺, lim 10% (SchepH62) △ -66.33 (MTW)	A chem (DneI60) chem, excit (SchepH62) chem, genet (StenT65a) daughter Ho^{158m} (StenT65a)	γ Dy X-rays, 0.099, 0.218, 0.329, 0.412, 0.52, 0.647, 0.73, 0.86, 0.940, 1.21, 1.47, 1.6, 1.8, 2.05, 2.21, 2.87, 3.1 e⁻ 0.045, 0.062, 0.091, 0.097, 0.164	Tb^{159}(α, 5n) (SchepH62)
Ho^{158m}	29 m (SchepH62) 27 m (DneI60, GromK61a) 22 m (LagP66) others (BasiA61, BoncN61a)	☢ IT (AbdurA61, GromK61a) [EC], β⁺ (BoncN61a) △ -66.26 (LHP, MTW)	A chem (DneI60) chem, excit (SchepH62) mass spect (LagP66) daughter Er^{158} (GromK61a, BoncN61a, AbdurA61) parent Ho^{158} (StenT65a)	γ Dy X-rays, Ho L X-rays, 0.099, 0.218, 0.32 (complex), 0.356, 0.412, 0.46 (complex), 0.52, 0.63 (complex), 0.73 (complex), 0.85 (complex), 0.95 (complex), 1.21, 1.47, 1.60, 1.80, 2.06, 2.20, 2.62 e⁻ 0.029, 0.044, 0.072, 0.078, 0.092, 0.132 β⁺ 1.32 max daughter radiations from Ho^{158} included in above listing	Tb^{159}(α, 5n) (SchepH62)
Ho^{159}	33 m (LagP66, TotK58) 35 m (MayM64)	☢ EC (TotK58) △ -67 (MTW)	A chem, excit (TotK58) chem, sep isotopes (MayM64) daughter Er^{159} (AbdurA61a)	γ Dy X-rays, 0.057, 0.080, 0.13, 0.18 (complex?), 0.253, 0.309 e⁻ [0.026], 0.048, 0.071, 0.121, 0.198, 0.243, 0.256, 0.300	Tb^{159}(α, 4n) (TotK58) Dy^{160}(p, 2n) (MayM64)
Ho^{159m}	6.9 s (BorgJ66)	☢ IT (BorgJ66) △ -67 (LHP, MTW)	A excit, sep isotopes, genet energy levels (BorgJ66)	γ Ho X-rays, 0.206 e⁻ 0.150, 0.197	daughter Er^{159} (AbdurA61a, LagP66) Dy^{160}(p, 2n) (BorgJ66)
Ho^{160}	25.6 m (StenT65, StenT65a) 28 m (TotK58, MayM64) 22.5 m (WilkG50a) ≈33 m (GoroG57a) ≈22 m (HandT54a)	☢ EC 99+%, β⁺ ≈0.4% (GrigE59d) others (WilkG50a) △ -66.4 (MTW)	A excit (WilkG50c) chem (HandT54a) chem, sep isotopes, excit (MayM64) daughter Ho^{160m} (GrigE62b) not daughter Er^{160}, lim 5% (DzhB63e)	see radiations of Ho^{160m}	daughter Ho^{160m} (GrigE62b) Tb^{159}(α, 3n) (WilkG50a, TotK58) protons on Dy (MayM64)

TABLE 1 (Continued)

Isotope Z A	Half-life	Type of decay (⚛); % abundance; Mass excess (Δ=M-A), MeV (C¹²=0); Thermal neutron cross section (σ), barns	Class; Identification; Genetic relationships	Major radiations: approximate energies (MeV) and intensities	Principal means of production
$_{67}$Ho^{160m}	5.0 h (StenT65, NerW55, MihJ57, HandT54a, RayG63) 4.8 h (GrigE60a) 4.6 h (WilkG50a) 5.3 h (DzhB57) others (DzhB57g)	⚛ IT 66%, EC+β⁺ 34% (NDS) β⁺ ≈0.1% (GrigE59d) Δ -66.3 (LHP, MTW)	A chem, genet (NerW55) chem, sep isotopes (MihJ57) chem, excit, sep isotopes (MayM64) daughter Er^{160} (NerW55) parent Ho^{160} (GrigE62b)	γ Dy X-ray, 0.087 (14%), 0.197 (20%), 0.539 (5%), 0.646 (20%), 0.729 (50%), 0.880 (26%), 0.965 (37%, complex), others to 2.8 e⁻ 0.033, 0.051, 0.058, 0.079, 0.085, 0.144, 0.188 β⁺ 1.9 max daughter radiations from Ho^{160} included in above listing	Tb^{159}(α,3n) (TotK58, TotK59a, WilkG50a) daughter Er^{160} (BjoS61, RayG63, NerW55, GrigE62b) protons on Dy (MayM64)
Ho^{161}	2.4 h (DneI58) 2.5 h (RayG63, HandT54a, HandT54) others (BjoS61, BasiA61, WilkG50c)	⚛ EC (HandT54a, HandT54) Δ -67 (MTW)	A chem, genet, excit (HandT54a, HandT54) daughter Er^{161} (HandT54, HandT54a)	γ Dy X-rays, 0.026 (23%), 0.075 (15%), 0.157 (1%), 0.176 (2%) e⁻ 0.017, 0.024, 0.049, 0.069, 0.076	Tb^{159}(α,2n) (WilkG50a) protons on Dy (MayM64)
Ho^{161m}	6.1 s (BorgJ66) 6.8 s (StenT65a)	⚛ IT (StenT65a, BorgJ66) Δ -67 (LHP, MTW)	A chem, genet (StenT65a, StenT65) excit, sep isotopes (BorgJ66) daughter Er^{161} (StenT65a, StenT65)	γ Ho X-rays, 0.211 (53%) e⁻ 0.155, 0.202	daughter Er^{161} (StenT65, StenT65) Dy^{162}(p,2n) (BorgJ66)
Ho^{162}	15 m (StenT65, StenT65a) 12 m (JorM61)	⚛ EC 95%, β⁺ 5% (JorM61) Δ -66.02 (MTW)	A genet (JorM61) chem, genet (StenT65, StenT65a) daughter Ho^{162m} (JorM61, StenT65, StenT65a)	γ Dy X-rays, 0.081 (8%), 0.511 (9%, γ±) β⁺ 1.10 max e⁻ 0.027, 0.072, 0.079	daughter Ho^{162m} (JorM61, HarmB61)
Ho^{162m}	68 m (JorM61, MayM64) 67 m (MihJ57a)	⚛ IT 63%, EC 37% (JorM61) Δ -65.92 (LHP, MTW)	A chem, sep isotopes (MihJ57a) chem, mass spect (JorM61) others (HandT54a, WilkG50a) parent Ho^{162} (JorM61, StenT65, StenT65a)	γ Ho X-rays, Dy X-rays, 0.081 (10%), 0.185 (26%), 0.283 (12%), 0.940 (13%), 1.224 (24%) e⁻ 0.027, 0.036, 0.048, 0.072, 0.079, 0.131, 0.177 daughter radiations from Ho^{162}	Tb^{159}(α,n) (JorM61) protons on Dy (MayM64)
Ho^{163}	t$_{1/2}$ >10³ y sp act (NauR60) others (BjoS61)	Δ -66.35 (MTW)	A chem, mass spect (NauR60)		Er^{162}(n,γ)Er^{163}(EC) (NauR60)
Ho^{163m}	1.1 s (BorgJ66) 0.8 s (HammC57)	⚛ IT (GovN58) Δ -66.05 (LHP, MTW)	B excit (GovN58) excit, sep isotopes (BorgJ66)	γ Ho X-rays, 0.305 e⁻ 0.249, 0.296	Ho^{165}(γ,2n) (HammC57, GovN58)
Ho^{164}	36.7 m (BrowHN54) 34.0 m (WilkG50a) 41.5 m (WafH50) 47 m (PoolM38a) others (HandT54a)	⚛ β⁻ 53%, EC 47%, no β⁺, lim 0.05% (BrowHN54) Δ -64.84 (MTW)	A excit (PoolM38a)	β⁻ 0.99 max e⁻ 0.019, 0.034, 0.065, 0.071, 0.083, 0.089 γ Dy, Er X-rays, 0.073, 0.091	protons on Dy (WilkG50a, MihJ57a) Ho^{165}(γ,n) (WafH48, BrowHN54) Ho^{165}(n,2n) (PoolM38a, WafH50)
Ho^{165}	t$_{1/2}$ (α) >6 x 10¹⁶ y sp act (PorsW54)	% 100 (LelW50, CollT57) Δ -64.81 (MTW) σ$_c$ 64 (to Ho^{166}) ≈1 (to Ho^{166m}) (GoldmDT64)			
Ho^{166}	26.9 h (GranP49, CorkJ58) 27.0 h (HoffD63) others (FunL63, IngM47, BotW46a, AntoN50, AntoN50a, KetB49b, CorkJ49b)	⚛ β⁻ (HevG36) Δ -63.07 (MTW)	A n-capt (HevG36) mass spect (IngM47) chem (KetB49b) daughter Dy^{166} (KetB49, ButeF50a)	β⁻ 1.84 max e⁻ 0.023, 0.072, 0.078 γ Er X-rays, 0.081 (5.4%), 1.380 (0.9%), 1.582 (0.20%), 1.663 (0.10%)	Ho^{165}(n,γ) (HevG36, PoolM38a, MeiL40, SerL47b) daughter Dy^{166} (KetB49, ButeF50a, HoffD63)
Ho^{166m}	1.2 x 10³ y sp act, mass spect (FalK65) others (ButeF52)	⚛ β⁻ (ButeF52) Δ -63.06 (LHP, MTW)	A chem, excit (ButeF52) chem, genet energy levels (MiltJ55)	β⁻ [0.07 max] e⁻ 0.023, 0.072, 0.078, 0.127, 0.175 γ Er X-rays, 0.081 (12%), 0.184 (90%), 0.280 (30%), 0.412 (12%), 0.532 (12%), 0.711 (58%), 0.810 (60%), 0.830 (11%), others to 1.43	Ho^{165}(n,γ) (ButeF52)
Ho^{167}	3.1 h (WilleR60) 3.0 h (HandT55)	⚛ β⁻ (HandT55) Δ -62.3 (MTW)	A chem, excit (HandT55) genet energy levels (HarmB62)	β⁻ 0.96 max e⁻ 0.024, 0.048, 0.073, 0.150, 0.180, 0.199, 0.263 γ Er X-rays, [0.079, 0.083, 0.208, 0.238, 0.321, 0.348, 0.387]	Er^{170}(p,α) (HandT55) Er^{167}(n,p) (WilleR60, HandT55)

TABLE 1 (*Continued*)

Isotope Z A	Half-life	Type of decay (☢); % abundance; Mass excess ($\triangle \equiv$M–A), MeV (C^{12}=0); Thermal neutron cross section (σ), barns	Class; Identification; Genetic relationships	Major radiations: approximate energies (MeV) and intensities	Principal means of production
$_{67}Ho^{168}$	3.3 m (WilleR60) 3.5 m (TakaK61)	☢ β^- TakaK61) $\triangle$ –59.7 (MTW)	C sep isotopes, cross bomb (WilleR60)	β^- 2.2 max γ 0.85	Er^{168}(n, p) (WilleR60, TakaK61)
Ho^{169}	4.8 m (MiyK63)	☢ β^- (MiyK63) $\triangle$ –58.8 (MTW)	C excit, sep isotopes, decay charac (MiyK63)	β^- 1.95 max γ 0.15, 0.68, 0.76, 0.84, 0.92	Er^{170}(γ, p) (MiyK63)
Ho^{169}	96 m (ButeF50)	☢ (ButeF50)	G excit (ButeF50) possibly Er^{163} (LHP)		gammas on Er (ButeF50)
Ho^{170}	45 s (TakaK61) 40 s (WilleR60)	☢ β^- (TakaK61) $\triangle$ –55.8 (MTW)	C excit, sep isotopes (WilleR60)	β^- 3.1 max γ 0.43	Er^{170}(n, p) (WilleR60, TakaK61)
$_{68}Er^{152}$	10.7 s (MacfR63a)	☢ $\alpha \approx$90%, [EC+β^+] $\approx$10% (MacfR63a)	C excit, cross bomb (MacfR63a, MacfR64b) parent 2.4 m Ho^{152} (MacfR63a)	α 4.80	Pr^{141}(F^{19}, 8n), Nd^{142}(O^{16}, 6n), Ce^{140}(Ne^{20}, 8n) (MacfR64b, MacfR63a)
Er^{153}	36 s (MacfR63a)	☢ α >75%, EC+β^+ <25% (MacfR63a)	B excit, cross bomb, genet (MacfR63a, MacfR64b) ancestor Tb^{149} (MacfR63a)	α 4.67	Nd^{142}(O^{16}, 5n) (MacfR63a) Pr^{141}(F^{19}, 7n), Ce^{140}(Ne^{20}, 7n) (MacfR64b)
Er^{154}	5 m (MacfR63a)	☢ α (MacfR63a) $\triangle$ –63 (MTW)	C excit, genet (MacfR63a) parent Dy^{150} (MacfR63a)	α 4.15 daughter radiations from Dy^{150}	Nd^{142}(O^{16}, 4n) (MacfR63a)
Er^{157}	$\approx$25 m (LagP66)	☢ β^+, [EC] (LagP66)	B [chem], mass spect (LagP66)	γ Ho X-rays, 0.117, 0.386, 0.511 ($\gamma^\pm$), 1.32, 1.66, 1.82, 2.0 daughter radiations from Ho^{157}	Ho^{165}(p, 9n) (LagP66)
Er^{158}	2.3 h (StenT65, GromK61a) 2.4 h (DneI60) 2.5 h (BoncN61a)	☢ EC, β^+ (BoncN61a)	B chem, genet (GromK61a, BoncN61a) parent Ho^{158m} (GromK61a, BoncN61a, AbdurA61)	γ Ho X-rays, 0.072, 0.250, 0.315, 0.387, 0.511 ($\gamma^\pm$), 0.875, 0.906 0.978 e^- 0.058, 0.065 β^+ 0.8 max daughter radiations from Ho^{158m}, Ho^{158}	protons on Ta (GromK61a, AbdurA61, BoncN61a, DneI60) Ho^{165}(p, 8n) (LagP66)
Er^{159}	36 m (LagP66) 1 h (AbdurA61a)	☢ [EC, β^+] (AbdurA61a)	A chem, atomic level spacing, genet (AbdurA61a) mass spect (LagP66) parent Ho^{159} (AbdurA61a)	γ Ho X-rays, 0.206, 0.37, 0.511 ($\gamma^\pm$), 0.62 (complex), 0.84, 1.20, 1.40, 1.80, 2.60 e^- 0.150, 0.197 daughter radiations from Ho^{159} daughter radiations from Ho^{159m} included in above listing	Ho^{165}(p, 7n) (LagP66) protons on Ta (AbdurA61a)
Er^{160}	29.4 h (NerW55) 28.7 h (BjoS61) 29.5 h (RayG63) others (MicM54, DzhB57, GoroG57a, LagP66)	☢ [EC], no β^+ (NerW55)	A chem, mass spect (NerW55, MicM54) parent Ho^{160m} (NerW55) not parent Ho^{160}, lim 5% (DzhB63e)	γ Ho X-rays daughter radiations from Ho^{160m} and Ho^{160}	protons on Er (RayG63, BjoS61)
Er^{161}	3.1 h (NerW55, RayG63, GrenH61) 3.2 h (BjoS61, GromK61a, DneI60a) others (HandT54, MicM54)	☢ [EC], β^+ (NerW55) EC, no β^+, lim 3% (HandT54, GrenH61) $\triangle$ –65 (MTW)	A chem, cross bomb, excit (HandT54) chem, mass spect (MicM54, NerW55) parent Ho^{161} (HandT54, HandT54a) daughter Tm^{161} (ButeF60, RayG63) parent Ho^{161m} (StenT65a, StenT65)	γ Ho X-rays, 0.211 (9%), 0.305 (3%), 0.592 (8%), 0.826 (63%), 1.17 (8%, complex), 1.37 (5%, complex), 1.66 (2%, complex) e^- 0.059, 0.065, 0.155, 0.202 β^+ 1.2 max daughter radiations from Ho^{161} daughter radiations from Ho^{161m} included in above listing	protons on Er (RayG63, HarmB59, BjoS61, ButeF60)
$\underline{Er^{162}}$		% 0.136 (HaydR50) $\triangle$ –66.4 (MTW) σ_c 2 (GoldmDT64)			
Er^{163}	75.1 m (PersL63d) others (HandT53a, BjoS61, StenT65)	☢ EC 99+%, β^+ 0.004% (PersL63d) $\triangle$ –65.14 (MTW)	A chem, excit (HandT53a, PersL63d) chem, genet (ButeF60, BjoS61) daughter Tm^{163} (ButeF60, BjoS61)	γ Ho X-rays, 0.43 (0.06%), 1.10 (0.04%) β^+ 0.19 max	Ho^{165}(p, 3n) (HandT53a, PersL63d)

TABLE 1 (*Continued*)

Isotope Z A	Half-life	Type of decay (☢); % abundance; Mass excess (Δ≡M–A), MeV (C¹²=0); Thermal neutron cross section (σ), barns	Class; Identification; Genetic relationships	Major radiations: approximate energies (MeV) and intensities	Principal means of production
₆₈Er¹⁶⁴		% 1.56 (HaydR50) Δ –65.87 (MTW) σ_c 1.7 (GoldmDT64)			
Er¹⁶⁵	10.34 h (RydH63) 10.3 h (StenT65) 10.4 h (ZylJ63) others (RayG63, BjoS61, SchoR63, ButeF50b, GrigO58, GoroG57)	☢ EC (ButeF50b) Δ –64.44 (MTW)	A chem, excit (ButeF50b) chem, mass spect (NierW56, BjoS61) daughter Tm¹⁶⁵ (HandT53a, NerW54)	γ Ho X-rays, continuous bremsstrahlung to 0.37	Ho¹⁶⁵(d, 2n) (RydH63) Ho¹⁶⁵(p, n) (RayG63) Er¹⁶⁴(n, γ) (SchoR63)
Er¹⁶⁶		% 33.41 (HaydR50) Δ –64.92 (MTW) σ_c 12 (GoldmDT64)			
Er¹⁶⁷		% 22.94 (HaydR50) Δ –63.29 (MTW) σ_c 700 (GoldmDT64)			
Er¹⁶⁷ᵐ	2.3 s (AlexKF63) 2.5 s (DMatE49, HammC57)	☢ IT (DMatE49) Δ –63.08 (LHP, MTW)	B n-capt (DMatE49) excit (HammC57) genet (MihJ57a) daughter Tm¹⁶⁷ (MihJ57a)	γ Er X-rays, 0.208 (43%) e⁻ 0.150, 0.199	daughter Tm¹⁶⁷ (MihJ57, MihJ57a) daughter Ho¹⁶⁷ (HarmB62) Er¹⁶⁶(n, γ) (DMatE49, AlexKF63)
Er¹⁶⁸		% 27.07 (HaydR50) Δ –62.98 (MTW) σ_c 2 (GoldmDT64)			
Er¹⁶⁹	9.6 d (BjoS61) 9.0 d (RayG63) 9.4 d (KetB48) 9.0 d (BisA56e, ButeF50) others (WilleR60)	☢ β⁻ (KetB48) Δ –60.91 (MTW)	A chem, n-capt (KetB48) genet energy levels (HatE56a) chem, mass spect (BjoS61)	β⁻ 0.34 max e⁻ 0.006 γ [Tm M X-rays], 0.008 (0.3%)	Er¹⁶⁸(n, γ) (KetB48)
Er¹⁷⁰		% 14.88 (HaydR50) Δ –60.0 (MTW) σ_c 9 (GoldmDT64)			
Er¹⁷¹	7.52 h (CranF58) others (KellH51, KetB48)	☢ β⁻ (KetB48) Δ –57.6 (MTW)	A n-capt (HevG36, NeunE35) chem, genet (KetB48) chem, mass spect (NetD56) parent Tm¹⁷¹ (KetB48)	β⁻ 1.49 max (2.3%), 1.06 max e⁻ 0.004, 0.052, 0.065, 0.102, 0.115 γ Tm X-rays, 0.112 (25%), 0.124 (9%), 0.296 (28%), 0.308 (63%), others to 0.96	Er¹⁷⁰(n, γ) (HevG36, PoolM38a, KetB48, BotW46a, NeunE35)
Er¹⁷²	49.5 h (HansP61a) 48.7 h (GunR62) others (NetD56, OrtC61)	☢ β⁻ (OrtC61) Δ –56.5 (MTW)	A chem, genet (NetD56) parent Tm¹⁷² (NetD56)	β⁻ 0.89 max (<10%), 0.37 max e⁻ 0.010, 0.020, 0.049, 0.058, 0.348 γ Tm X-rays, 0.407 (40%), 0.610 (40%) daughter radiations from Tm¹⁷²	Er¹⁷⁰(n, γ) Er¹⁷¹(n, γ) (NetD56, OrtC61, HelmeR61b, HansP61a, GunR62)
Er¹⁷³ (or Tm¹⁷⁶, Yb¹⁷²)	2.0 m (WilleR60)	☢ β⁻ or IT (WilleR60)	F sep isotopes (WilleR60)	γ 0.18, 0.25, 0.36	neutrons on Yb¹⁷⁶ (WilleR60)
₆₉Tm¹⁵³	1.6 s (MacfR64b)	☢ α (MacfR64b)	C excit, cross bomb (MacfR64b)	α 5.10	Pr¹⁴¹(Ne²⁰, 8n), Nd¹⁴²(F¹⁹, 8n) (MacfR64b)
Tm¹⁵⁴	3.0 s (MacfR64b)	☢ α (MacfR64b)	C excit, cross bomb (MacfR64b)	α 5.04	Pr¹⁴¹(Ne²⁰, 7n), Nd¹⁴²(F¹⁹, 7n) (MacfR64b)
Tm¹⁵⁴	5 s (MacfR64b)	☢ α (MacfR64b)	E excit, cross bomb (MacfR64b)	α 4.96	Pr¹⁴¹(Ne²⁰, 7n), Nd¹⁴²(F¹⁹, 7n) (MacfR64b)

TABLE 1 (*Continued*)

Isotope Z A	Half-life	Type of decay (☢); % abundance; Mass excess (Δ≡M–A), MeV (C¹²=0); Thermal neutron cross section (σ), barns	Class; Identification; Genetic relationships	Major radiations: approximate energies (MeV) and intensities	Principal means of production
$_{69}$Tm^{161}	32 m (ButeF60) 30 m (HarmB59) 20 to 30 m (RayG63) 44 m (GromK63)	☢ EC (HarmB59) Δ –62 (MTW)	A chem, sep isotopes (HarmB59) chem, genet (ButeF60) chem, excit, sep isotopes, genet (RayG63) parent Er^{161} (ButeF60, RayG63)	γ Er X-rays, 0.084, 0.106, 0.112, 0.145 (complex), 0.172, others e⁻ 0.027, 0.036, 0.050, 0.055, 0.065, 0.075, 0.089, 0.115, others daughter radiations from Er^{161}	Er^{162}(p, 2n) (RayG63, HarmB59)
Tm^{162}	77 m (WilsRG60g) 90 m (RayG63) activity not observed, $t_{1/2}$ <45 m (BjoS61)	☢ EC (WilsRG60g) Δ –61.5 (MTW)	B excit, sep isotopes (WilsRG60g) chem, excit, sep isotopes (RayG63)	γ Er X-rays, 0.102 (↑ 20), 0.236 (↑ 10)	protons on Er (RayG63, WilsRG60g)
Tm^{162}	22 m (AbdumA63)	☢ β⁺, EC (AbdumA63) Δ –61.5 (MTW)	D chem (AbdumA63) daughter Yb^{162} (AbdumA63)	β⁺ 3.82 max e⁻ 0.045, 0.093, 0.100 γ [Er X-rays, 0.102, 0.511 (γ±)]	daughter Yb^{162} (AbdumA63)
Tm^{163}	1.8 h (BjoS61, GromK63, RayG63) others (HarmB59, BoncN60, ButeF60)	☢ EC (HarmB59) β⁺ (BoncN60) Δ –62.87 (MTW)	A chem, sep isotopes (HarmB59) chem, mass spect (BjoS61) chem, sep isotopes, excit (RayG63) parent Er^{163} (ButeF60, BjoS61)	γ Er X-rays, 0.104 (↑ 8), 0.17 (↑ 1, complex), 0.240 (↑ 5, complex), 0.29 (↑ 3, complex), 0.34 (↑ 3, complex) e⁻ 0.047, 0.095, 0.184 β⁺ 1.1 max daughter radiations from Er^{163}	Er^{164}(p, 2n) (RayG63)
Tm^{164}	2.0 m (WilsRG60g) 1.8 m (RayG63)	☢ EC 50%, β⁺ 50% (WilsRG60g) Δ –61.91 (MTW)	A chem, genet energy levels (DalB60, AbdurA60, AbdurA60b) excit, sep isotopes (RayG63, WilsRG60g) daughter Yb^{164} (DalB60, AbdurA60, AbdurA60b)	γ Er X-rays, 0.091 (4%), 0.356, 0.361, 0.391, 0.511 (100%, γ±), 0.773, 0.862, 0.907, 0.930 β⁺ 2.94 max e⁻ 0.034, 0.083, 0.089	Er^{164}(p, n) (RayG63, WilsRG60g)
Tm^{165}	30.1 h (BjoS61) others (MicM54, RayG63, GoroG57, HandT53a)	☢ EC, no β⁺ (HandT53a) β⁺ 0.007% (PreiZ65) Δ –62.87 (PreiZ65, MTW)	A chem, excit (HandT53a) chem, mass spect (MicM54) parent Er^{165} (HandT53a, NerW54)	γ Er X-rays, 0.054, 0.113, 0.243 (↑ 50), 0.297 (↑ 35, complex), 0.34 (↑ 10, complex), 0.44 (↑ 5, complex), 0.70 (↑ 2), 0.807 (↑ 15), 1.13 (↑ 5), 1.30 (↑ 1) e⁻ 0.038, 0.045, 0.052, 0.056, 0.068, 0.161, 0.185, 0.233, 0.240 β⁺ 0.30 max daughter radiations from Er^{165}	protons on Er (RayG63)
Tm^{166}	7.7 h (WilsRG60d, GrigE60a, WilkG49b, RayG63, MicM54) others (BjoS61, BoncN60, PariP63)	☢ EC 98.2%, β⁺ 2% (GrigE61) others (WilsRG60d, WilkG49b) Δ –61.88 (LHP, MTW)	A chem, excit (WilkG49a) chem, mass spect (MicM54) daughter Yb^{166} (FolR51, NerW55, GoroG57)	β⁺ 1.94 max e⁻ 0.023, 0.072, 0.079, 0.127 γ Er X-rays, 0.081, 0.19 (doublet), 0.215, 0.46, 0.60 (complex), 0.69 (complex), 0.78 (complex), 1.180, 1.277, 1.378, 1.873, 2.06 (doublet)	Ho^{165}(α, 3n) (WilkG49b) protons on Yb (WilkG49b, RayG63, WilsRG60d)
Tm^{167}	9.6 d (NaraH60, WilkG49b, NerW55, RayG63) 9.3 d (BjoS61, BonnN62)	☢ EC, no β⁺ (WilkG49b) no β⁺, lim 0.3% (GromK62) Δ –62.13 (GromK62, MTW)	A chem, excit (WilkG49a, RayG63) chem, mass spect (MicM54, NerW55, BjoS61) parent Er^{167m} (MihJ57a) daughter Yb^{167} (WilsRG60f)	γ Er X-rays, 0.057 (4%), 0.208 (43%), 0.532 (2%) e⁻ 0.048, 0.150, 0.199 daughter radiations from Er^{167m} included in above listing	Ho^{165}(α, 2n) (RayG63) protons on Er (RayG63)
Tm^{168}	85 d (WilkG49b) 86 d (RayG63) 87 d (HandT54b) 93 d (BonnN62) others (BjoS61, GoroG57)	☢ EC, β⁻ (?) ≈2% (WilkG49b) Δ –61.27 (MTW)	A chem, excit (WilkG49b, RayG63) chem, mass spect (BjoS61)	γ Er X-rays, 0.080 (11%), 0.19 (77%, complex), 0.448 (27%), 0.63 (14%, complex), 0.73 (40%, complex), 0.82 (88%, complex), 0.917 (4%), 1.280 (3%) e⁻ 0.022, 0.071, 0.077, 0.127, 0.141	Er^{170}(p, 3n) (RayG63) Ho^{165}(α, n) (WilkG49b) Er^{168}(p, n) (RayG63)
Tm^{169}	$t_{1/2}$ (α) >5 x 10^{16} y sp act (PorsW54)	% 100 (LagC50, CollT57) Δ –61.25 (MTW) σ_c 125 (GoldmDT64)			
Tm^{170}	134 d (FlyK65a) 125 d (BonnN62) others (BotW46, CaldR50, KetB49b)	☢ β⁻ (BotW46) EC(K) 0.15% (DayP56) no EC(K), lim 0.3%, no β⁺, lim 0.01% (GrahR52) Δ –59.6 (MTW) σ_c 150 (GoldmDT64)	A n-capt (NeunE36) chem (KetB48a)	β⁻ 0.97 max e⁻ 0.023, 0.075, 0.082 γ Yb X-rays, 0.084 (3.3%)	Tm^{169}(n, γ) (HevG36, NeunE36, SerL47b) Er^{170}(p, n) (RayG63)

TABLE 1 (*Continued*)

Isotope Z A	Half-life	Type of decay (☢); % abundance; Mass excess (△≡M–A), MeV (C¹²=0); Thermal neutron cross section (σ), barns	Class; Identification; Genetic relationships	Major radiations: approximate energies (MeV) and intensities	Principal means of production
$_{69}Tm^{171}$	1.92 y (FlyK65a) 1.9 y (KetB49b)	☢ β⁻ (KetB48) △ –59.1 (MTW)	A chem, genet (KetB48) chem, mass spect (NetD56) daughter Er¹⁷¹ (KetB48)	β⁻ 0.097 max e⁻ 0.057, 0.065 γ Yb X-rays, 0.067	Er¹⁷⁰(n, γ) Er¹⁷¹(β⁻) (KetB48)
Tm^{172}	63.6 h (NetD56) 63.5 h (HansP61a) others (KuroT61b, FolR51)	☢ β⁻ (FolR51) △ –57.4 (MTW)	A chem (FolR51) chem, n-capt, mass spect (NetD56) daughter Er¹⁷² (NetD56)	β⁻ 1.88 max γ Yb X-rays, 0.079 (5%), 0.181 (2.2%), 0.91 (1.4%), 1.09 (7%), 1.39 (7%), 1.46 (7%), 1.53 (6%), 1.61 (5%)	daughter Er¹⁷² (NetD56, HelmeR61a, HansP61a, OrtC61)
Tm^{173}	8.2 h (OrtC63, KuroT63) others (KuroT61b)	☢ β⁻ (KuroT61b) △ –56.4 (MTW)	B chem, sep isotopes, cross bomb (OrtC63)	β⁻ 1.3 max (2%), 0.89 max e⁻ 0.008, 0.056, 0.064 γ Yb X-rays, 0.066 (1.1%), 0.399 (89%), 0.465 (8%)	Er¹⁷⁰(α, p) (OrtC63) Yb¹⁷³(n, p) (OrtC63) Yb¹⁷⁴(γ, p) (KuroT63, OrtC63, KuroT61b)
Tm^{174}	5.5 m (WilleR60) 5 m (TakaK61)	☢ β⁻ (TakaK61) △ –54.6 (TakaK61, MTW)	E sep isotopes (WilleR60) decay charac (TakaK61)	β⁻ 2.5 max γ no γ	Yb¹⁷⁴(n, p) (WilleR60, TakaK61)
Tm^{174}	5.2 m (KantJ64c)	☢ β⁻ (KantJ64c) △ –54.1 (MTW)	B genet energy levels (KantJ64c, OrtC64)	β⁻ 1.2 max e⁻ 0.015, 0.067, 0.074 γ Yb X-rays, 0.176 (67%), 0.273 (85%), 0.366 (93%), 0.50 (15%), 0.99 (89%)	Yb¹⁷⁴(n, p) (KantJ64b)
Tm^{175}	20 m (KuroT61b) 19 m (ButeF50)	☢ β⁻ (KuroT61b) △ –52.3 (LHP, MTW)	E excit (ButeF50) excit, decay charac (KuroT61b)	β⁻ 2.0 max γ 0.51	Yb¹⁷⁶(γ, p) (KuroT61b)
Tm^{176}	1.5 m (TakaK61)	☢ β⁻ (TakaK61) △ –49.2 (MTW)	F decay charac (TakaK61)	β⁻ 4.2 max γ no γ	Yb¹⁷⁶(n, p) (TakaK61)
Tm^{176} (or Er¹⁷³, Yb¹⁷⁷)	2.0 m (WilleR60)	☢ IT or β⁻ (WilleR60)	F sep isotopes (WilleR60)	γ 0.18, 0.25, 0.36	neutrons on Yb¹⁷⁶ (WilleR60)
$_{70}Yb^{154}$	0.39 s (MacfR64b)	☢ α (MacfR64b)	C excit, cross bomb (MacfR64b)	α 5.33	Sm¹⁴⁴(O¹⁶, 6n), Nd¹⁴²(Ne²⁰, 8n) (MacfR64b)
Yb^{155}	1.6 s (MacfR64b)	☢ α (MacfR64b)	C excit, cross bomb (MacfR64b)	α 5.21	Sm¹⁴⁴(O¹⁶, 5n), Nd¹⁴²(Ne²⁰, 7n) (MacfR64b)
Yb^{162}	≈24 m (AbdumA63)	☢ [EC] (AbdumA63)	D chem (AbdumA63) parent 22 m Tm¹⁶² (AbdumA63)	γ [Tm X-rays] e⁻ 0.032, 0.039 daughter radiations from 22 m Tm¹⁶²	protons on Ta (AbdumA63)
Yb^{164}	75 m (DalB60, AbdurA60b, AbdurA60) 78 m (PariP64) 74 m (ButeF60) others (NerW55, KalyA59)	☢ EC (DalB60, AbdurA60, AbdurA60b)	A chem (NerW55) chem, genet (AbdurA60, DalB60, AbdurA60b) chem, mass spect (PariP64) parent Tm¹⁶⁴ (AbdurA60b, DalB60, AbdurA60)	γ Tm X-rays daughter radiations from Tm¹⁶⁴	Tm¹⁶⁹(p, 6n) (ButeF60, PariP64)
Yb^{165}	10.5 m (PariP64)	☢ [EC, β⁺] (PariP64) △ –60 (MTW)	C mass spect (PariP64)		Tm¹⁶⁹(p, 5n) (PariP64)
Yb^{166}	57.5 h (PariP63) 54 h (NerW55) 62 h (FolR51) 60 h (GoroG57)	☢ EC (FolR51) △ –61.6 (MTW)	A chem, genet (FolR51) chem, mass spect (MicM54, NerW55) parent Tm¹⁶⁶ (FolR51, NerW55, GoroG57)	γ Tm X-rays, 0.082 (17%) e⁻ 0.023, 0.072 daughter radiations from Tm¹⁶⁶	Tm¹⁶⁹(p, 4n) (PariP63)
Yb^{167}	17.7 m (WilsRG60f) 17.3 m (WanC64) others (HandT54b, BasiA60b)	☢ EC, no β⁺ (HandT54b) β⁺ 0.4% (WanC64) β⁺ 0.2% (TamT65) △ –60.17 (MTW, GromK62)	B chem, excit (HandT54b) genet (WilsRG60f) parent Tm¹⁶⁷ (WilsRG60f) daughter Lu¹⁶⁷ (AroP58, ButeF60)	γ Tm X-rays, 0.113 (90%, complex), 0.176 (15%) e⁻ 0.047, 0.055, 0.096	daughter Lu¹⁶⁷ (HarmB59) Tm¹⁶⁹(p, 3n) (HandT54b) Er¹⁶⁴(α, n) (WilsRG60f)

TABLE 1 (*Continued*)

Isotope Z A	Half-life	Type of decay (☢); % abundance; Mass excess (Δ≡M−A), MeV (C^{12}=0); Thermal neutron cross section (σ), barns	Class; Identification; Genetic relationships	Major radiations: approximate energies (MeV) and intensities	Principal means of production
$_{70}Yb^{168}$		% 0.140 (BaiK50) 0.135 (CollT57) Δ −61.3 (MTW) σ_c 11, 000 (GoldmDT64)			
Yb^{169}	31.8 d (WalkD49a) 30.6 d (CorkJ56) 33 d (BotW46, MartiDS51, HandT54b)	☢ EC (BotW46) Δ −60 (MTW)	A n−capt (BotW46) chem, excit (KetB48a) mass spect (MicM54) daughter Lu^{169} (GoroG57b, MerE61)	γ Tm L X-rays (56%), Tm K X-rays (185%), 0.063 (45%), 0.110 (18%), 0.131 (11%), 0.177 (22%), 0.198 (35%), 0.308 (10%) e^- 0.004−0.011, 0.034, 0.050, 0.053, 0.071, 0.100, 0.118, 0.121, 0.139	Yb^{168}(n, γ) (AttH45, BotW46) Tm^{169}(d, 2n) (KetB48a)
Yb^{169m}	46 s (HoffK60a) 50 s (DMatE49)	☢ IT (DMatE49) Δ −60 (LHP, MTW)	B n−capt (DMatE49) n−capt, sep isotopes (HoffK60a) daughter Lu^{169} (HarmB60)	γ Yb L X-rays e^- 0.014, 0.022	Yb^{168}(n, γ) (DMatE49, HoffK60a)
Yb^{170}		% 3.03 (BaiK50) 3.14 (CollT57) Δ −60.5 (MTW)			
Yb^{171}		% 14.31 (BaiK50) 14.4 (CollT57) Δ −59.2 (MTW)			
Yb^{171m}	<<8 d (MihJ57)	☢ IT (MihJ57a, MihJ57, Δ −59.1 (LHP, MTW)	D chem (MihJ57) daughter Lu^{171} (MihJ57a, MihJ57, HarmB60)	γ Yb L X-rays, 0.019, 0.076 e^- 0.010, 0.017, 0.067, 0.074	daughter Lu^{171} (MihJ57, MihJ57a)
Yb^{172}		% 21.82 (BaiK50) 21.9 (CollT57) Δ −59.3 (MTW)			
Yb^{173}		% 16.13 (BaiK50) 16.2 (CollT57) Δ −57.7 (MTW)			
Yb^{174}		% 31.84 (BaiK50) 31.6 (CollT57) Δ −57.1 (MTW) σ_c 9 (to Yb^{175}) 46 (to 0.513 level of Yb^{175}) (GoldmDT64)			
Yb^{175}	101 h (AttH45, CorkJ56) 102 h (IngM47a) 99 h (BotW46)	☢ β^- (AttH45) Δ −54.8 (MTW)	A n−capt (BotW46, AttH45) mass spect (IngM47a) chem (KetB49b)	β^- 0.466 max γ Lu X-rays, 0.114 (1.9%), 0.283 (3.7%), 0.396 (6.0%) e^- 0.051, 0.102, 0.112, 0.333	Yb^{174}(n, γ) (AttH45, BotW46, IngM47a)
Yb^{176}		% 12.73 (BaiK50) 12.6 (CollT57) Δ −53.4 (MTW) σ_c 7 (GoldmDT64)			
Yb^{176m}	11.7 s (KantJ62) 11 s (VergM65)	☢ [IT] (KantJ62) Δ −52.4 (LHP, MTW)	B sep isotopes, excit (KantJ62) genet energy levels (DBoeJ64, KantJ62)	γ Yb X-rays, 0.19, 0.29, 0.39	Yb^{176}(n, n') (KantJ62)
Yb^{177}	1.9 h (CorkJ56, AttH45) 2.4 h (BotW46)	☢ β^- (BotW46) Δ −50.8 (JohaH64, MTW)	A n−capt (MarsJK35, HevG36) chem, genet (BetR58) parent Lu^{177} (BetR58)	β^- 1.40 max γ Lu X-rays, 0.122 (3%), 0.151 (16%), 1.080 (5%), 1.241 (3%) e^- 0.059, 0.075, 0.088, 0.110, 0.140	Yb^{176}(n, γ) (MarsJK35, HevG36, PoolM38a, BotW46, IngM47a)
Yb^{177m}	6.5 s (FetP62a, CamE59) 6.4 s (HoffK60a) others (DMatE49, KahJ51)	☢ IT (HoffK60a, FetP62a, DMatE49) Δ −50.5 (LHP, MTW)	A n−capt (DMatE49) n−capt, sep isotopes (HoffK60a, FetP62a)	γ Yb X-rays, 0.104 (65%), 0.228 (13%) e^- 0.043, 0.094, 0.167, 0.219	Yb^{176}(n, γ) (HoffK60a, FetP62a, CamE59)
Yb^m	0.15 s (KahJ52)	☢ [IT] (KahJ51)	F n−capt (KahJ51)	γ 0.455 (KahJ52)	neutrons on Yb (KahJ51)

TABLE 1 (*Continued*)

Isotope Z A	Half-life	Type of decay (☢); % abundance; Mass excess (△≡M–A), MeV (C¹²=0); Thermal neutron cross section (σ), barns	Class; Identification; Genetic relationships	Major radiations: approximate energies (MeV) and intensities	Principal means of production
$_{70}$Yb^{177} (or Er^{173}, Tm^{176})	2.0 m (WilleR60)	☢ IT or β⁻ (WilleR60)	F sep isotopes (WilleR60)	γ 0.18, 0.25, 0.36	neutrons on Yb^{176} (WilleR60)
$_{71}$Lu^{155}	0.07 s (MacfR65a)	☢ α (MacfR65a)	C cross bomb, excit (MacfR65a)	α 5.63	Sm^{144}(F^{19}, 8n) (MacfR65a)
Lu^{156}	0.23 s (MacfR65a)	☢ α (MacfR65a)	C cross bomb, excit (MacfR65a)	α 5.54	Sm^{144}(F^{19}, 7n) (MacfR65a)
Lu^{156}	0.5 s (MacfR65a)	☢ α (MacfR65a)	C cross bomb, excit (MacfR65a)	α 5.43	Sm^{144}(F^{19}, 7n) (MacfR65a)
Lu^{167}	54 m (HarmB59, ButeF60) 55 m (AroP58) others (BasiA60, BoncN60, KalyA59)	☢ EC (AroP58, HarmB59) β⁺ ≈1% (BoncN60) △ –57.1 (MTW, GromK62)	B chem, genet (AroP58, ButeF60) parent Yb^{167} (AroP58, ButeF60)	γ Yb X-rays, 0.030, 0.18–0.24 (complex), 0.278, 0.372, 0.402, 0.511 (γ⁺) e⁻ 0.020, 0.028, 0.039, 0.069, 0.076, 0.152, 0.178 β⁺ 1.5 max daughter radiations from Yb^{167}	Yb^{168}(p, 2n) (HarmB59)
Lu^{168}	7.1 m (WilsRG60b) 7.0 m (MerE61)	☢ EC, no β⁺ lim 1% (WilsRG60b) EC, β⁺ ≈12% (MerE61) △ –57 (MTW)	A sep isotopes, excit (WilsRG60b) chem (MerE61) daughter Hf^{168} (MerE61)	γ Yb X-rays, 0.087 (7%), 0.223, 0.71, 0.90 (10%), 0.99 (13%), 1.41, 1.81, 2.1 e⁻ [0.026, 0.078, 0.085] β⁺ 1.2 max	Yb^{168}(p, n) (WilsRG60b) Lu^{175}(p, 8n)Hf^{168}(EC) (MerE61)
Lu^{169}	34 h (DzhB64a) others (MerE61, DzhB59g, GoroG57b, NerW55)	☢ EC (GoroG57) β⁺ (DzhB59g) △ –58 (MTW)	A chem, excit (NerW55) chem, genet (GoroG57b, MerE61) parent Yb^{169} (GoroG57b, MerE61) parent Yb^{169m} (HarmB60) daughter Hf^{169} (MerE61)	γ Yb X-rays, 0.063, 0.111, 0.191, 0.577, many others to 2.2 e⁻ 0.010, 0.014, 0.022, 0.026, 0.050, 0.053, 0.060, 0.066, 0.077, others to 2.2 β⁺ 1.2 max daughter radiations from Yb^{169} daughter radiations from Yb^{169m} included in above listing	protons on Yb (HarmB59) daughter Hf^{169} (MerE61)
Lu^{169m}	2.7 m (BjoS65)	☢ IT (BjoS65) △ –58 (LHP, MTW)	B excit, sep isotopes (BjoS65)	γ [Lu L X-rays] e⁻ 0.019, 0.027	Yb^{170}(p, 2n) (BjoS65)
Lu^{170}	2.05 d (WilsRG60e) 2.0 d (DzhB64a) others (MerE61, MihJ57a, DzhB59g, WilkG51)	☢ EC (WilkG51) β⁺ (DzhB59g, MerE61) △ –57.1 (HansP65a, MTW)	A chem, excit (WilkG51) chem, mass spect (MicM56) daughter Hf^{170} (ValenJ62, MerE61)	γ Yb X-rays, 0.084 (13%), 0.193, 0.24, 1.01, 1.03, 1.17, 1.27, 1.41, 2.03, 2.32, 2.67, 2.89, 3.09, many others to 3.2 e⁻ 0.023, 0.075, 0.082, others to 3.2 β⁺ 2.4 max	Tm^{169}(α, 3n) (WilkG51) daughter Hf^{170} (ValenJ62, DzhB64a) protons on Yb (WilsRG60e, HarmB60)
Lu^{170m}	0.7 s (BjoS65)	☢ IT (BjoS65, ValenJ65) △ –57.0 (LHP, MTW)	B excit, sep isotopes, genet energy levels (BjoS65)	γ Lu L X-rays e⁻ 0.036, 0.044	daughter Hf^{170} (ValenJ65) Yb^{170}(p, n) (BjoS65)
Lu^{171}	8.3 d (WilsRG60h) 8.2 d (BonnN62) others (RaoC63, WilkG51, MihJ57a, ValenJ62)	☢ EC (WilkG51) β⁺ ≈0.007% (VitV65a, LHP) △ –58 (MTW)	A chem, excit (WilkG51) excit, sep isotopes (WilsRG60h) genet energy levels (IodM60a, ChupE58a) parent Yb^{171m} (MihJ57a, MihJ57, HarmB60) daughter Hf^{171} (WilkG51)	γ Yb X-rays, 0.019 (20%), 0.075 (8%, complex), 0.668 (14%), 0.741 (68%), 0.842 (7%) e⁻ 0.010, 0.017, 0.057, 0.066, 0.074, others to 0.85	Tm^{169}(α, 2n) (WilkG51) Yb^{171}(p, n) (WilkG51, WilsRG60h)
Lu^{171m}	76 s (BjoS65)	☢ IT (BjoS65) △ –58 (LHP, MTW)	B excit, sep isotopes (BjoS65) genet energy levels (BjoS65, BarnD65)	γ Lu X-rays, 0.071 (0.2%) e⁻ 0.061, 0.069	daughter Hf^{171} (BarnD65) Yb^{171}(p, n) (BjoS65)
Lu^{172}	6.70 d (WilkG51, WilsRG60a) others (BonnN62, RaoC63)	☢ EC (WilkG51) △ –57 (MTW)	A chem, excit (WilkG51) sep isotopes, excit (WilsRG60a) daughter Hf^{172} (WilkG51, ValenJ62b, RaoC63)	γ Yb X-rays, 0.079 (13%, complex), 0.182 (26%), 0.81 (21%), 0.90 (45%, complex), 1.09 (60%) e⁻ 0.017, 0.029, 0.069, 0.077, 0.081, 0.120, others to 2.1	Yb^{172}(p, n) (WilkG51, WilsRG60a) Tm^{169}(α, n) (WilkG51)

TABLE 1 *(Continued)*

Isotope Z A	Half-life	Type of decay (☢); % abundance; Mass excess (△≡M−A), MeV (C¹²=0); Thermal neutron cross section (σ), barns	Class; Identification; Genetic relationships	Major radiations: approximate energies (MeV) and intensities	Principal means of production
$_{71}$Lu^{172m}	3.7 m (ValenJ62b)	☢ IT (ValenJ62b) △ −57 (LHP, MTW)	B chem, genet (ValenJ62b) daughter Hf^{172} (ValenJ62b)	γ Lu L X-rays e⁻ 0.032, 0.040	daughter Hf^{172} (ValenJ62b)
Lu^{172}	4.0 h (WilkG51)	☢ β⁺, EC (WilkG51)	G chem, excit (WilkG51) activity not observed (WilsRG60a)		alphas on Tm, protons on Lu (WilkG51)
Lu^{173}	1.37 y (BonnN62) 1.4 y (WilkG51, MihJ57a) 1.3 y (BicJ59, GrigE60a) 1.7 y (WilsRG60a) others (GoroG58a)	☢ EC (WilkG51) △ −57.0 (MTW)	A chem, excit (WilkG51) sep isotopes (WilsRG60a) daughter Hf^{173} (WilkG51)	γ Yb L X-rays, Yb K X-rays (150%), 0.079 (14%), 0.101 (7%), 0.17 (5%, complex), 0.272 (18%), 0.637 (1.5%) e⁻ 0.017, 0.039, 0.068, 0.077, 0.090	Yb^{173}(p, n) (WilkG51, BicJ59, WilsRG60a) Lu^{175}(p, 3n)Hf^{173}(EC) (BicJ59, WilkG51)
Lu^{174}	3.6 y (BonnN62) >800 d (BalaV64) <<160 d (HarmB60) others (WilkG51, WilleR60)	☢ EC, no β⁻, β⁺ (WilkG51) others (WilkG51) △ −55.6 (MTW)	A chem, excit (WilkG51) excit, sep isotopes (WilsRG60) daughter Lu^{174m} (HarmB60)	γ Yb X-rays, 0.076 (6%), 1.24 (9%) e⁻ 0.015, 0.067, 0.074	Yb^{174}(p, n) (WilsRG60, HarmB60, PraH62)
Lu^{174m}	140 d (BonnN62) 150 d (BalaV64) others (WilkG51, WilleR60)	☢ IT (HarmB60, RomV60) EC (FunL65, RiccR65a) △ −55.4 (LHP, MTW)	B chem, genet (HarmB60, RomV60) chem (BonnN62) parent Lu^{174} (HarmB60)	γ Lu L X-rays, 0.067, 0.176, 0.273, 0.994 e⁻ 0.004, 0.034, 0.050, 0.057 daughter radiations from Lu^{174}	Yb^{174}(p, n) (WilsRG60, HarmB60, PraH62)
Lu^{175}	$t_{1/2}$ (α) >1 × 10^{17} y sp act (PorsW54)	% 97.40 (HaydR50) 97.41 (CollT57) △ −55.3 (MTW) σ$_c$ 5 (to Lu^{176}) 18 (to Lu^{176m}) (GoldmDT64)			
Lu^{176}	2.2 × 10^{10} y sp act (DonhD64) 3.6 × 10^{10} y sp act (MNaiA61b, BrinGA65) 2.4 × 10^{10} y sp act (ArnJ54) 2.1 × 10^{10} y sp act (GloR57b) 4.6 × 10^{10} y sp act (DixD54) others (HerrW58a, LibW39a)	☢ β⁻, no EC, lim 10% (ArnJ54) no EC (GloR57b) EC(K) 3% (DixD54) % 2.60 (HaydR50) 2.59 (CollT57) △ −53.4 (MTW) σ$_c$ 2100 (to Lu^{177}) ≈1 (to Lu^{177m}) (GoldmDT64)	A chem (HeyM38) mass spect (MattaJ39)	β⁻ 0.43 max e⁻ 0.023, 0.078, 0.086, 0.137 γ Hf X-rays, 0.088 (15%), 0.202 (85%), 0.306 (95%)	natural source
Lu^{176m}	3.69 h (SchmL60) others (BetR58, AttH45, BotW46)	☢ β⁻, no IT (SchaG52) no β⁺, lim 0.0005% (LanghH61b) △ −53.1 (LHP, MTW)	A n-capt (MLenJ35b, MarsJK35) chem, excit (WilkG48a)	β⁻ 1.31 max e⁻ 0.023, 0.078, 0.086 γ Hf X-rays, 0.088 (10%)	Lu^{175}(n, γ) (MLenJ35b, MarsJK35, HevG36, FlaA43, BotW46, AttH45, SerL47b, AntoN50a)
Lu^{177}	6.74 d (SchmL60) others (BetR58, BotW46, WilkG48a, DouDG49, CorkJ49b, FlaA43, AttH45)	☢ β⁻ (BotW46) △ −52.2 (MTW)	A n-capt (HevG36) mass spect (IngM47a) chem, excit (WilkG48a) daughter Yb^{177} (BetR58)	β⁻ 0.497 max γ Hf X-rays, 0.113 (2.8%), 0.208 (6.1%) e⁻ 0.048, 0.103, 0.111, 0.143	Lu^{176}(n, γ) (HevG36, FlaA43, AttH45, BotW46, SerL47b, AntoN50a, AlexP64)
Lu^{177m}	155 d (JorM62)	☢ β⁻ 78%, IT 22% (KriL64) △ −51.3 (LHP, MTW)	A chem, n-capt, mass spect (JorM62) parent Hf^{177m} (BodE66)	γ Lu X-rays, Hf X-rays, 0.105 (13%), 0.113 (23%), 0.128 (17%), 0.153 (17%), 0.174 (13%), 0.208 (62%), 0.228 (37%), 0.281 (14%), 0.319 (10%), 0.327 (18%), 0.378 (29%), 0.414 (17%), 0.418 (21%), many others between 0.05 and 0.47 β⁻ [0.165 max] e⁻ very complex spectrum between 0 and 0.47 daughter radiations from Lu^{177} daughter radiations from Hf^{177m} included in above listing	Lu^{176}(n, γ) (JorM62, AlexP64)
Lu^{178}	30 m (KuroT61b)	☢ β⁻ (KuroT61b) △ −50.0 (MTW)	F decay charac (KuroT61b)	β⁻ 2.25 max γ no γ	Hf^{179}(γ, p) (KuroT61b)

TABLE 1 (*Continued*)

Isotope Z A	Half-life	Type of decay (☢); % abundance; Mass excess (△=M–A), MeV (C¹²=0); Thermal neutron cross section (σ), barns	Class; Identification; Genetic relationships	Major radiations: approximate energies (MeV) and intensities	Principal means of production
$_{71}Lu^{178}$	22.0 m (PouA60) 18.7 m (StriT57) 19 m (GleP61) 22 m (ButeF50) 16 m (KuroT61b) 30 m (BakH64a)	☢ β⁻ (KuroT61b) △ –49.6 (LHP, MTW)	B chem (ButeF50) chem, genet energy levels (KuroT61b)	β⁻ 1.50 max e⁻ 0.023, 0.028, 0.077, 0.083, 0.091, 0.148, 0.204 γ Hf X-rays, 0.089, 0.214, 0.326, 0.427 daughter radiations from Hf¹⁷⁸ᵐ included in above listing	$Ta^{181}(n,\alpha)$ (GleP61, PouA60, BakH64a, StriT57)
Lu^{178}	5 m (BakH64a)	☢ β⁻ (BakH64a)	F chem (BakH64a)	β⁻ 2.25 γ 0.090, 0.22, 0.33, 0.43	$Ta^{181}(n,\alpha)$ (BakH64a)
Lu^{179}	4.6 h (StenW63) 7.5 h (KuroT61b)	☢ β⁻ (KuroT61b) △ –48.9 (MTW)	B decay charac (KuroT61b) chem, sep isotopes, decay charac (StenW63)	β⁻ 1.35 max γ 0.213	$Hf^{180}(\gamma,p)$ (StenW63, KuroT61b)
Lu^{180}	2.5 m (TakaK61)	☢ β⁻ (TakaK61) △ –46.2 (MTW)	F decay charac (TakaK61)	β⁻ 3.3 max γ no γ	$Hf^{180}(n,p)$ (TakaK61)
$_{72}Hf^{157}$	0.12 s (MacfR65a)	☢ α (MacfR65a)	C cross bomb, excit (MacfR65a)	α 5.68	$Sm^{144}(Ne^{20},7n)$ (MacfR65a)
Hf^{158}	3 s (MacfR65a)	☢ α (MacfR65a)	C cross bomb, sep isotopes (MacfR65a)	α 5.27	$Sm^{144}(Ne^{20},6n)$ (MacfR65a)
Hf^{168}	22 m (MerE61)	☢ [EC], β⁺ ? ≈2% (MerE61)	B chem, genet (MerE61) parent Lu¹⁶⁸ (MerE61)	γ Lu X-rays, 0.129, 0.17 β⁺ ? 1.7 max daughter radiations from Lu¹⁶⁸	$Lu^{175}(p,8n)$ (MerE61)
Hf^{169}	1.5 h (MerE61) others (WilkG51)	☢ EC, β⁺ (MerE61)	B chem, genet (MerE61) parent Lu¹⁶⁹ (MerE61)	γ Lu X-rays, 0.115 β⁺ 1.3 max daughter radiations from Lu¹⁶⁹	$Lu^{175}(p,7n)$ (MerE61)
Hf^{170}	12.2 h (ValenJ62) 9 h (MerE61)	☢ EC (ValenJ62)	A chem, genet (MerE61) chem, genet, mass spect (ValenJ62) parent Lu¹⁷⁰ (MerE61, ValenJ62)	γ Lu X-rays, 0.120, 0.165, 0.99, 1.28, 0.65, 2.03, 2.36, 2.52, 2.94 e⁻ 0.035, 0.057, 0.102, 0.145, others between 0 and 3 daughter radiations from Lu¹⁷⁰	$Lu^{175}(p,6n)$ (MerE61, ValenJ62)
Hf^{171}	10.7 h (ValenJ62) 16.0 h (WilkG51) 12 h (NerW55) 13 h (BaranV59a) others (BrabV61a, RaoC63)	☢ EC (WilkG51)	B chem, genet, excit (WilkG51) chem, mass spect (ValenJ62) parent Lu¹⁷¹ (WilkG51)	γ Lu X-rays, 0.122, 0.188, 0.29, 0.34, 0.47, 0.66, 0.86, 1.07 daughter radiations from Lu¹⁷¹	$Lu^{175}(p,5n)$ (WilkG51, ValenJ62) alphas on Yb (WilkG51)
Hf^{172}	5 y (RaoC63, WilkG51)	☢ EC (WilkG51)	A chem, genet (WilkG51) chem, sep isotopes (ValenJ62b) parent Lu¹⁷² (WilkG51, ValenJ62b, RaoC63) parent Lu¹⁷²ᵐ (ValenJ62b)	γ Lu X-rays, 0.024 (22%), 0.082 (10%), 0.125 (21%, complex) e⁻ 0.014, 0.018, 0.032, 0.040, 0.063 daughter radiations from Lu¹⁷² daughter radiations from Lu¹⁷²ᵐ included in above listing	$Lu^{175}(p,4n)$ (WilkG51) alphas on Yb (WilkG51, ValenJ62b)
Hf^{173}	23.6 h (WilkG51) 24 h (RaoC63, ValenJ62a, MalyT62, BaranV59a) others (NerW55, WapA54c)	☢ EC (WilkG51)	A chem, excit, genet (WilkG51) parent Lu¹⁷³ (WilkG51) daughter Ta¹⁷³ (FalK60, RaoC63, MalyT62)	γ Lu X-rays, 0.13 (96%, complex), 0.162 (5%), 0.30 (52%, complex), 0.55 (1.1% complex), 0.898 (1.9%), 1.04 (1.0%, complex), 1.20 (0.4%, complex) e⁻ 0.060, 0.072, 0.076, 0.113, 0.127, others between 0 and 1.1	$Lu^{175}(p,3n)$ (WilkG51, BicJ59) alphas on Yb (WilkG51, ValenJ62a)
$\underline{Hf^{174}}$	2.0 x 10¹⁵ y sp act (MacfR61a) 4 x 10¹⁵ y sp act (RieW59)	☢ α (RieW59, MacfR61a) % 0.163 (WhiF56) 0.20 (ReynJH53) △ ~55.6 (MTW) σ_c 400 (GoldmDT64)	A sep isotopes, decay charac (MacfR61a)	α 2.50	
Hf^{175}	70 d (WilkG49)	☢ EC (WilkG49) △ –54.7 (FunL65f, MTW)	A chem, excit (WilkG49) n-capt, sep isotopes (BursS51) mass spect (HedA51) daughter Ta¹⁷⁵ (RaoC63, FalK60)	γ Lu X-rays, 0.089 (3.4%), 0.343 (85%), 0.433 (1.4%) e⁻ 0.026, 0.079, 0.280, 0.333	$Hf^{174}(n,\gamma)$ (HedA51, HatE56, MizJ55) $Lu^{175}(d,2n)$, $Lu^{175}(p,n)$ (WilkG49)

TABLE 1 (*Continued*)

Isotope Z A	Half-life	Type of decay (☢); % abundance; Mass excess (△≡M-A), MeV (C¹²=0); Thermal neutron cross section (σ), barns	Class; Identification; Genetic relationships	Major radiations: approximate energies (MeV) and intensities	Principal means of production
$_{72}$Hf^{176}		% 5.21 (WhiF56) 5.23 (ReynJH53) △ -54.4 (MTW) σ$_c$ <30 (GoldmDT64)			
Hf^{177}		% 18.56 (WhiF56) 18.6 (ReynJH53) △ -52.7 (MTW) σ$_c$ 370 (to Hf^{178}) 1.4 (to Hf^{178m}) (GoldmDT64)			
Hf^{177m}	1.1 s (BodE66)	☢ IT (BodE66) △ -51.4 (LHP, MTW)	A chem, genet (BodE66) daughter Lu^{177m} (BodE66)	γ Hf X-rays, 0.105 (17%), 0.113 (30%), 0.128 (21%), 0.153 (22%), 0.174 (16%), 0.208 (81%), 0.228 (48%), 0.281 (18%), 0.327 (23%), 0.378 (37%), 0.418 (27%), many others between 0 and 0.47 e⁻ very complex spectrum between 0 and 0.47	daughter Lu^{177m} (BodE66)
Hf^{178}		% 27.1 (WhiF56) 27.2 (ReynJH53) △ -52.3 (MTW) σ$_c$ 30 (to Hf^{179}) 50 (to Hf^{179m}) (GoldmDT64)			
Hf^{178m}	4.3 s (AlexKF62) 4.8 s (FelF58) 3.5 s (CamE59, FetP62a)	☢ IT (FelF58) △ -51.1 (MTW)	A chem, genet (FelF58) n-capt, sep isotopes (FetP62a) daughter 2.1 h Ta^{178} (FelF58)	γ Hf X-rays, 0.089 (54%), 0.093 (14%), 0.214 (75%), 0.326 (94%), 0.427 (97%) e⁻ 0.023, 0.028, 0.077, 0.083, 0.091, 0.148, 0.204	daughter Ta^{178} (FelF58) Hf^{177}(n, γ) (FetP62a)
Hf^{179}		% 13.75 (WhiF56) 13.7 (ReynJH53) △ -50.3 (MTW) σ$_c$ 65 (to Hf^{180}) 0.2 (to Hf^{180m}) (GoldmDT64)			
Hf^{179m}	18.6 s (HoffK59) others (FlaA44a, DMatE51a, AlexKF62)	☢ IT (FlaA46) △ -49.9 (LHP, MTW)	A n-capt (FlaA44a) n-capt, sep isotopes (BursS51, DMatE51a)	γ Hf X-rays, 0.217 (94%) e⁻ 0.096, 0.150	Hf^{178}(n, γ) (FlaA44a, FlaA46, DMatE51a, BursS51)
Hf^{180}		% 35.22 (WhiF56) △ -49.5 (MTW) σ$_c$ 10 (GoldmDT64)			
Hf^{180m}	5.5 h (BursS51) others (RaoC63)	☢ IT (BursS51) no β⁻, lim 5% (Gal1C62) △ -48.4 (LHP, MTW)	A chem, n-capt, sep isotopes (BursS51) genet energy levels (MihJ54b)	γ Hf X-rays, 0.058 (48%), 0.093 (16%), 0.215 (93%), 0.333 (93%), 0.444 (80%), 0.501 (17%) e⁻ 0.028, 0.047, 0.055, 0.083, 0.091, 0.150, 0.206, 0.267	Hf^{179}(n, γ) (BursS51)
Hf^{181}	42.5 d (LindnM60) 44.6 d (WriH57) 45.5 d (CaliJ59) others (MurH53, CorkJ50d, BeneJ48a, SerL47b)	☢ β⁻ (HevG38) △ -47.41 (MTW) σ$_c$ ≈40 (GoldmDT64)	A chem, n-capt (HevG38) mass spect (HedA51) sep isotopes, n-capt (BursS51)	β⁻ 0.41 max e⁻ 0.066, 0.069, 0.122, 0.415 γ Ta X-rays, 0.133 (48%, complex), 0.346 (13%), 0.482 (81%)	Hf^{180}(n, γ) (HevG38, SerL47b, BursS51, LindnM60)
Hf^{182}	9 x 10^6 y sp act (HutWH61, WingJ61) ≈8 x 10^6 y sp act (NauR61)	☢ β⁻ (HutWH61, WingJ61, NauR61) △ -45.8 (LHP, MTW)	A chem, mass spect, genet (HutW61, WingJ61, NauR61) parent Ta^{182} (HutW61, WingJ61, NauR61)	β⁻ [0.5 max] γ 0.271 (84%) daughter radiations from Ta^{182}	Hf^{180} + 2n (HutW61, WingJ61, NauR61)
Hf^{183}	65 m (BlacJe65) 64 m (GatO56, GatO58)	☢ β⁻ (GatO56, GatO58) △ -43.0 (MTW)	D chem (GatO56, GatO58)	β⁻ 1.6 max γ 0.46 (↑ 58), 0.82 (↑ 100)	W^{186}(n, α) (GatO56, GatO58, BlacJe65)

TABLE 1 (*Continued*)

Isotope Z A	Half-life	Type of decay (☢); % abundance; Mass excess ($\Delta \equiv M-A$), MeV ($C^{12}=0$); Thermal neutron cross section (σ), barns	Class; Identification; Genetic relationships	Major radiations: approximate energies (MeV) and intensities	Principal means of production
$_{73}Ta^{172}$	44 m (AboH64a) 24 m (ButeF61)	☢ β^+, EC (AboH64a)	B chem (ButeF61) chem, mass spect (AboH64a)	γ Hf X-rays, 0.092, 0.208, 0.511 ($\gamma^{\pm}$), others to 3.3	protons on Hf (AboH64a, ButeF61)
Ta^{173}	3.7 h (FalK60, SanA63, RaoC63) 3.5 h (MalyT62) 2.5 h (HarmB60)	☢ EC, β^+ (FalK60) EC, no β^+ (SanA63)	A chem, excit, genet (FalK60, RaoC63) chem, genet (MalyT62) parent Hf^{173} (FalK60, RaoC63, MalyT62) daughter W^{173} (SanA63)	γ Hf X-rays, 0.090 (complex), 0.170 (complex), 0.64, 1.00 e^- 0.059, 0.069, 0.095, 0.107, 0.161 daughter radiations from Hf^{173}	$Ho^{165}(N^{14}, 6n)W^{173}$(EC) (FalK60) protons on Ta^{181} (RaoC63, SanA63)
Ta^{174}	1.2 h (DemeI65) 1.3 h (FalK60, RaoC63) 1.1 h (ButeF61)	☢ EC, β^+ (FalK60)	A chem, excit (FalK60, RaoC63) chem, mass spect (AboH65) daughter W^{174} (DemeI65)	γ Hf X-rays, 0.091, 0.125, 0.160, 0.205, 0.280, 0.350, 0.511 ($\gamma^{\pm}$) e^- 0.026, 0.081, 0.089	$Ho^{165}(N^{14}, 5n)W^{174}$(EC) (FalK60) protons on Hf (HarmB60, ButeF61) protons on Ta^{181} (RaoC63)
Ta^{175}	10.5 h (SanA63) 11 h (FalK60, RaoC63)	☢ EC (FalK60)	A chem, cross bomb, excit, genet (FalK60) chem, excit, genet (RaoC63) parent Hf^{175} (RaoC63, FalK60) daughter W^{175} (SanA63)	γ Hf X-rays, 0.08, 0.13, 0.21, 0.27, 0.35, 0.45, 0.60, 0.83, 1.2, 1.4, 1.7, all complex e^- 0.016, 0.039, 0.061, 0.070, 0.116, 0.202, others between 0 and 1.6	$Lu^{175}(\alpha, 4n)$ (FalK60) $Hf^{176}(p, 2n)$ (HarmB60) $Ho^{165}(N^{14}, 4n)W^{175}$(EC) (FalK60) protons on Ta^{181} (RaoC63, SanA63)
Ta^{176}	8.0 h (WilkG50d)	☢ EC (WilkG50d) no β^+, lim 0.2% (FelF56) Δ −51 (NDS, MTW)	A chem, excit (WilkG48a, WilkG50d) genet energy levels (FelF56) daughter W^{176} (WilkG50d)	γ Hf X-rays, 0.088, 0.202, many others to 3.0 e^- 0.023, 0.078, 0.086, 0.137, others to 3.0	$Lu^{175}(\alpha, 3n)$ (WilkG50d, VerhH63, HasA63) $Hf^{176}(p, n)$ (HarmB60)
Ta^{177}	56.6 h (WestH61) 56 h (RaoC63) 53 h (WilkG50d)	☢ EC (WilkG50d) Δ −51.6 (MTW)	A chem, excit (WilkG48a, WilkG50d) genet energy levels (WestH61, HarmB60)	γ Hf X-rays, 0.113 (6%), 0.208 (1.0%), 0.425 (0.13%), 0.509 (0.10%), 0.746 (0.22%), 1.058 (0.30%), others between 0.07 and 0.95 e^- 0.048, 0.102, 0.111, others between 0 and 1.06	$Lu^{175}(\alpha, 2n)$ (WilkG50d, WestH61) protons on Hf (WilkG50d, HarmB60) $Ta^{181}(p, 5n)W^{177}$(EC) (WilkG50d)
Ta^{178}	9.35 m (WilkG50d) 9.5 m (CarvJ58)	☢ EC 99%, β^+ 1% (GallC61a) others (FelF58, BisA56b, WilkG50d) Δ −50.4 (MTW)	A chem, genet (WilkG50d) daughter W^{178} (WilkG50d)	γ Hf X-rays, 0.093 († 100), 0.511 ($\gamma^{\pm}$, † 10), 1.10 († 11), 1.18 († 4, complex), 1.35 († 46, complex), 1.45 († 9, complex) β^+ 0.89 max e^- 0.028, 0.082	daughter W^{178} (WilkG50d, GallC61a, BodE62, KarlE62a)
Ta^{178}	2.1 h (WilkG50d, RaoC63) 2.5 h (CarvJ58)	☢ EC, no β^+, lim 2% (CarvJ58) EC ≈97%, β^+ ≈3% (WilkG50d)	A chem, excit (WilkG50d, RaoC63) chem, cross bomb, genet (FelF58) parent Hf^{178m} (FelF58)	γ Hf X-rays, 0.089 (54%), 0.093 (14%), 0.214 (75%), 0.328 (120%, complex), 0.427 (97%) e^- 0.023, 0.028, 0.077, 0.083, 0.091, 0.148, 0.204, 0.263 daughter radiations from Hf^{178m} included in above listing	$Lu^{175}(\alpha, n)$ (WilkG50d, GallC62a, FelF58) deuterons on Hf (FelF58) protons on Hf (WilkG50d)
Ta^{179}	≈600 d (WilkG50d)	☢ EC (WilkG50d) Δ −50.2 (MTW)	B chem, excit (WilkG50d, RaoC63) excit (CarvJ58)	γ Hf X-rays	protons on Ta^{181} (RaoC63) $Lu^{176}(\alpha, n)$ (WilkG50d)
Ta^{180}	$t_{1/2}$ (β^-): >1 x 10^{12} y sp act (CarvJ58) >1 x 10^{13} y sp act (BaumE58) $t_{1/2}$ (EC): >2 x 10^{13} y sp act (BaumE58) >4 x 10^{9} y sp act (CarvJ58) others (EberP55, EberP58)	% 0.0123 (WhiF56) Δ −48.86 (MTW)			
Ta^{180m}	8.15 h (BrowHN51) 8.00 h (WilkG50d) 8.1 h (RaoC63) others (OldO38)	☢ EC 87%, β^- 13% (GallC62) EC ≈79%, β^- ≈21%, no β^+, lim 0.005% (BrowHN51) Δ −48.65 (LHP, MTW)	A chem, excit (OldO38)	β^- 0.71 max e^- 0.028, 0.083, 0.091 γ Hf X-rays, 0.093 (4%), 0.103 (0.6%)	$Hf^{180}(d, 2n)$ (GallC62) $Ta^{181}(n, 2n)$ (PoolM37, OldO38, WilkG50d) $Ta^{181}(\gamma, n)$ (GelK60, GusaM58)

TABLE 1 (*Continued*)

Isotope Z A	Half-life	Type of decay (☢); % abundance; Mass excess (Δ≡M–A), MeV (C¹²=0); Thermal neutron cross section (σ), barns	Class; Identification; Genetic relationships	Major radiations: approximate energies (MeV) and intensities	Principal means of production
₇₃Ta¹⁸¹		% 99.9877 (WhiF56, WhiF55) 100 (WhiJ48) Δ –48.43 (MTW) σ_c 21 (to Ta¹⁸²) 0.07 (to Ta¹⁸²ᵐ) (GoldmDT64)			
Taᵐ	0.33 s (CamE49, GooM50, KahJ51)	☢ IT (GooM50)	E excit (CamE49) critical abs (GooM50)	γ Ta L X-rays	neutrons on Ta (CamE49, GooM50, KahJ51)
Ta¹⁸²	115.1 d (WriH57) others (EicG52, SinW51, SerL47b)	☢ β⁻ (HouF40) Δ –46.35 (HansP64, MTW) σ_c 8000 (GoldmDT64)	A chem, n-capt (FomV36, OldO38) daughter Hf¹⁸² (HutW61, WingJ61, NauR61)	β⁻ 1.71 max (0.3%), 0.522 max e⁻ 0.030, 0.044, 0.054, 0.073, 0.089, 0.110, many others between 0 and 1.6 γ W X-rays, 0.068 (42%), 0.100 (14%), 0.152 (7%), 0.222 (8%), 1.122 (34%), 1.189 (16%), 1.222 (27%), 1.231 (13%), many others between 0 and 1.6	Ta¹⁸¹(n, γ) (FomV36, OldO38, HouF40, SerL47b, MeiL48)
Ta¹⁸²ᵐ	16.5 m (HoleN48b) 16.2 m (SerL47b) others (WilkG50d)	☢ IT (HoleN48b) no β⁻ (SunA61) Δ –45.84 (LHP, MTW)	A chem, n-capt (SerL47b, HoleN48b)	γ Ta X-rays, 0.147 (40%), 0.172 (40%), 0.184 (20%), 0.319 (5%), 0.356 (0.3%) e⁻ 0.080, 0.105, 0.117, 0.173	Ta¹⁸¹(n, γ) (SerL47b, HoleN48b, SunA61)
Ta¹⁸³	5.0 d (PoeA55) 5.2 d (MurJ55, DMonJ53) others (SumO57a, MosA51)	☢ β⁻ (ButeF50, PoeA55) Δ –45.20 (MTW)	A chem, excit (ButeF50) n-capt, chem, genet energy levels (MurJ55) parent W¹⁸³ᵐ (GallC61)	β⁻ 0.62 max γ W X-rays, 0.046 (5%), 0.053 (5%), 0.099 (7%), 0.108 (11%), 0.161 (17%, complex), 0.246 (33%, complex), 0.30 (11%, complex), 0.354 (11%) e⁻ 0.034–0.043, 0.050, 0.073, 0.088, 0.093, 0.177, many others between 0 and 0.40 daughter radiations from W¹⁸³ᵐ included in above listing	Ta¹⁸¹(n, γ)Ta¹⁸²(n, γ) (MurJ55)
Ta¹⁸⁴	8.7 h (ButeF55a)	☢ β⁻ (ButeF55a) Δ –42.9 (MTW)	B chem, sep isotopes (ButeF55a)	β⁻ 2.64 max (0.2%), 1.76 max (0.9%), 1.19 max e⁻ [0.042, 0.100] γ W X-rays, 0.111 (21%), 0.16 (7%), 0.21 (7%), 0.25 (42%), 0.30 (24%), 0.41 (71%), 0.53 (19%), 0.79 (16%, complex), 0.90 (49%, complex), 0.95 (15%), 1.16 (12%)	W¹⁸⁶(d, α) (VerhH64) W¹⁸⁴(n, p) (ButeF55a)
Ta¹⁸⁵	50 m (PoeA55) 48 m (MosA51, ButeF50) others (DufR50)	☢ β⁻ (DufR50) Δ –41.3 (NDS, MTW)	B chem, excit (ButeF50) excit, sep isotopes (DufR50) not parent W¹⁸⁵ᵐ (PoeA55)	β⁻ 1.7 max γ W X-rays, 0.075 (5%), 0.100 (6%), 0.175 (60%), 0.245 (5%)	W¹⁸⁶(γ, p) (DufR50, ButeF50, MoriH60a) W¹⁸⁶(n, pn) (PoeA55)
Ta¹⁸⁶	10.5 m (PoeA55)	☢ β⁻ (PoeA55) Δ –38.7 (MTW)	C sep isotopes, cross bomb (PoeA55)	β⁻ 2.2 max γ W X-rays, 0.123 (18%), 0.20 (74%), 0.30 (18%), 0.41 (15%), 0.51 (33%), 0.61 (33%), 0.73 (48%), 0.94 (11%)	W¹⁸⁶(n, p) (PoeA55)
₇₄W¹⁶⁰?		☢ α (MacfR65a)	F excit (MacfR65a)	α 5.75	S³² on Sm¹⁴⁴ (MacfR65a)
W¹⁷³	16.5 m (SanA63)	☢ EC (SanA63)	B chem, excit, genet (SanA63) parent Ta¹⁷³ (SanA63)		Ta¹⁸¹(p, 9n) (SanA63)
W¹⁷⁴	31 m genet (DemeI65)	☢ [EC] (DemeI65)	B chem, genet (DemeI65) parent Ta¹⁷⁴ (DemeI65)		C¹² on Er (DemeI65)
W¹⁷⁵	34 m (SanA63)	☢ EC (SanA63)	A chem, mass spect, genet (SanA63) parent Ta¹⁷⁵ (SanA63)	γ Ta X-rays, 0.26, 0.80, 1.3, 1.6 daughter radiations from Ta¹⁷⁵	Ta¹⁸¹(p, 7n) (SanA63)
W¹⁷⁶	2.3 h (ValenJ63) 2.7 h (RaoC63) others (GrigE62)	☢ EC 99+%, β⁺ ≈0.5% (WilkG50d) Δ –50 (NDS, MTW)	A chem, genet (WilkG50d, GrigE62) chem, mass spect (ValenJ63) parent Ta¹⁷⁶ (WilkG50d)	γ Ta X-rays, 0.034, 0.100 e⁻ 0.017, 0.023, 0.027, 0.033, 0.050, 0.083 daughter radiations from Ta¹⁷⁶	Ta¹⁸¹(p, 6n) (RaoC63, WilkG50d)

TABLE 1 (*Continued*)

Isotope Z A	Half-life	Type of decay (☢); % abundance; Mass excess ($\triangle \equiv M{-}A$), MeV ($C^{12}{=}0$); Thermal neutron cross section (σ), barns	Class; Identification; Genetic relationships	Major radiations: approximate energies (MeV) and intensities	Principal means of production
$_{74}W^{177}$	135 m (SanA63) 130 m (WilkG50d) 132 m (RaoC63) others (MalyT63a)	☢ EC (WilkG50d) $\triangle$ -50 (NDS, MTW)	A chem, genet (WilkG50d) chem, mass spect (SanA63) chem, excit (RaoC63) parent Ta^{177} (WilkG50d) daughter Re^{177} (HaldB57)	γ Ta X-rays, 0.20, 0.42, 0.62, 0.83, 1.00 e⁻ 0.020, 0.028, 0.048, 0.059, 0.068, 0.075, 0.088, 0.119, 0.360 daughter radiations from Ta^{177}	Ta^{181}(p, 5n) (RaoC63, SanA63, WilkG50d)
W^{178}	21.5 d (WilkG50d) 22.0 d (BisA56b)	☢ EC (WilkG50d) $\triangle$ -50 (NDS, MTW)	A chem (WilkG50d) chem, excit (RaoC63) parent 9.35 m Ta^{178} (WilkG50d)	γ Ta X-rays daughter radiations from 9.35 m Ta^{178}	Ta^{181}(p, 4n) (RaoC63, WilkG50d)
W^{179}	37.5 m (ValenJ63a) 38 m (SanA63) others (RaoC63, WilkG50d, RocT56)	☢ EC (WilkG50d) $\triangle$ -49 (NDS, MTW)	A chem, excit (RaoC63, WilkG50d) chem, sep isotopes (HarmB60) chem, mass spect (SanA63, ValenJ63a)	γ Ta X-rays, 0.031 (22%) e⁻ 0.020, 0.029	Ta^{181}(p, 3n) (RaoC63, WilkG50d) W^{180}(p, 2n)Re^{179}(EC) (HarmB60)
W^{179m}	5.2 m (WilkG50d) ≈7 m (RocT56) activity not observed (SofS55)	☢ IT (HarmB60) $\triangle$ -49 (NDS, MTW)	B chem, excit (WilkG50d) genet energy levels (HarmB60)	γ W X-rays, 0.222 e⁻ 0.152, 0.211 daughter radiations from W^{179}	daughter Re^{179} (HarmB60) Ta^{181}(p, 3n) (WilkG50d)
$\underline{W^{180}}$	$t_{1/2}$ (α): >1.1 x 10^{15} y sp act (BearG60) >9 x 10^{14} y sp act (MacfR61a)	% 0.135 (WilliD46) $\triangle$ -49.37 (MTW) σ_c <20 (GoldmDT64)			
$W^{[180]}$	$t_{1/2}$(α) <2 x 10^{17} y sp act (PorsW56)	☢ α (PorsW56)	G (PorsW56) activity not observed (BearG60, MacfR61a)	α 3.0	natural source (PorsW56)
W^{181}	140 d (RaoC63, WilkG47, SinB59) 120 d (GodK61) 126 d (KreW60) 145 d (BisA56b)	☢ EC (WilkG47) no β⁺ (BisA56b, BisA55) $\triangle$ -48.24 (MTW)	A chem, excit (WilkG47) chem, n-capt (LindnM51a) daughter Re^{181} (GallC57)	γ Ta X-rays, 0.006 (1%), 0.136 (0.1%), 0.152 (0.1%) e⁻ 0.004, 0.006	Ta^{181}(d, 2n) (WilkG47) Ta^{181}(p, n) (MuiA61) W^{180}(n, γ) (MuiA61, LindnM51a, CorkJ53d)
$\underline{W^{182}}$	$t_{1/2}$ (α) >2 x 10^{17} y sp act (BearG60)	% 26.4 (WilliD46) $\triangle$ -48.16 (MTW) σ_c 20 (to W^{183}) 0.5 (to W^{183m}) (GoldmDT64)			
$\underline{W^{183}}$	$t_{1/2}$ (α) >1.1 x 10^{17} y sp act (BearG60)	% 14.4 (WilliD46) $\triangle$ -46.27 (MTW) σ_c 11 (GoldmDT64)			
W^{183m}	5.3 s (GallC61) 5.1 s (SchmW61) 5.5 s (DMatE49)	☢ IT (DMatE49) $\triangle$ -45.96 (LHP, MTW)	A sep isotopes, n-capt (DMatE49) chem, genet, genet energy levels (GallC61) daughter Ta^{183} (GallC61)	γ W X-rays, 0.046 (8%), 0.053 (11%), 0.099 (9%), 0.102 (4%), 0.108 (19%), 0.160 (6%) e⁻ 0.034, 0.040	daughter Ta^{183} (GallC61) W^{182}(n, γ) (SchmW61, DMatE49)
$\underline{W^{184}}$		% 30.6 (WilliD46) $\triangle$ -45.62 (MTW) σ_c 2.1 (to W^{185}) 0.01 (to W^{185m}) (GoldmDT64)			
W^{185}	75 d (AndeR64, FajK40a, KreW55) others (ThirH57, GodK61, DoyW63a)	☢ β⁻ (MinaO40) $\triangle$ -43.30 (MTW)	A chem, excit, n-capt (MinaO40) mass spect (BisA58a)	β⁻ 0.429 max average β⁻ energy: 0.14 calorimetric (ShimN56a) γ no γ	W^{184}(n, γ) (MinaO40, FajK40a, SerL47b, CorkJ49a) Re^{187}(d, α) (FajK40a)
W^{185m}	1.62 m (PoeA55) 1.55 m (MangS62) 1.85 m (DufR50)	☢ IT (DufR50) $\triangle$ -42.93 (LHP, MTW)	B excit, sep isotopes (DufR50, PoeA55) not daughter Ta^{185} (PoeA55)	γ W X-rays, 0.075 (↑ 8), 0.100 (↑ 16), 0.13 (↑ 70), 0.17 (↑ 100)	W^{184}(n, γ) (PoeA55) W^{186}(γ, n) (DufR50, MoriH60a)

TABLE 1 (*Continued*)

Isotope Z A	Half-life	Type of decay (☢); % abundance; Mass excess (△≡M–A), MeV (C¹²=0); Thermal neutron cross section (σ), barns	Class; Identification; Genetic relationships	Major radiations: approximate energies (MeV) and intensities	Principal means of production
$_{74}W^{186}$	$t_{1/2}$ ($\beta\beta$) >6 × 10¹⁵ y sp act (FremJ52)	% 28.4 (WilliD46) △ –42.44 (MTW) σ_c 40 (GoldmDT64)			
W^{187}	23.9 h (EicG53) 23.7 h (AndeR64) 24.0 h (WriH57) others (MinaO40, CorkJ53, FajK40a)	☢ β^- (MinaO40) △ –39.83 (MTW) σ_c ≈90 (GoldmDT64)	A chem, n-capt (AmaE35) chem, n-capt, excit (MinaO40)	β^- 1.31 max (15%), 0.63 max e^- 0.063, 0.122, others between 0 and 0.8 γ Re X-rays, 0.072 (11%), 0.134 (9%), 0.479 (23%), 0.552 (5%), 0.618 (6%), 0.686 (27%), 0.773 (4%)	W^{186}(n, γ) (MinaO40, AmaE35, MLenJ35, FajK40a, SerL47b, CorkJ49a)
W^{188}	69.4 d (RoyJ62) others (LindnM51a)	☢ β^- (LindnM51a) △ –38.44 (BursS64, MTW)	A chem, genet (LindnM51a, RoyJ62) parent Re¹⁸⁸ (RoyJ62, LindnM51a, LindnM51)	β^- 0.349 max γ Re X-rays, 0.227 (0.22%), 0.290 (0.40%) daughter radiations from Re¹⁸⁸	W^{186}(n, γ)W^{187}(n, γ) (LindnM51a, LindnM51, RoyJ62)
W^{189}	11.5 m (KauP65a) 11 m (FleJ63)	☢ β^- (FleJ63) △ –35.3 (KauP65a, MTW)	A chem, sep isotopes, genet (FleJ63) chem, genet (KauP65a) parent Re¹⁸⁹ (FleJ63, KauP65a)	β^- 2.5 max (weak), 2.0 max γ Re X-rays, 0.032 (?), 0.130 (↑ 12), 0.178 (↑ 13), 0.258 (↑ 100), 0.417 (↑ 96), 0.55 (↑ 28), 0.86 (↑ 20) 0.96 (↑ 17)	Os¹⁹²(n, α) (FleJ63)
$_{75}Re^{177}$	17 m (HaldB57)	☢ β^+ (HaldB47), [EC] △ –47 (NDS, MTW)	B chem, genet (HaldB57) parent W¹⁷⁷ (HaldB57)	γ [W X-rays, 0.511 (γ⁺)] daughter radiations from W¹⁷⁷	protons on W (HaldB57)
Re^{178}	15 m (HaldB57)	☢ β^+ (HaldB57), [EC] △ –47 (NDS, MTW)	D chem, sep isotopes (HaldB57)	β^+ 3.1 max γ [W X-rays, 0.511 (γ⁺)]	protons on W, Re (HaldB57)
Re^{179}	20 m (HarmB60) 18 m (FosJ58)	☢ EC (HarmB60) △ –46 (NDS, MTW)	B chem, sep isotopes (HarmB60) others (FosJ58)	γ W X-rays daughter radiations from W¹⁷⁹ᵐ W¹⁷⁹	W^{180}(p, 2n) (HarmB60)
Re^{180}	2.4 m (FiscV55)	☢ β^+, EC (FiscV55)	C excit (FiscV55)	β^+ 1.1 max γ [W X-rays], 0.11, 0.511 (γ⁺), 0.88	W^{182}(p, 3n) (FiscV55)
Re^{180}	20 h·(HaldB57)	☢ β^+ (HaldB57), [EC]	D chem, decay charac, cross bomb (HaldB57)	β^+ 1.9 max γ [W X-rays, 0.511 (γ⁺)]	protons on W, Re (HaldB57)
Re^{180}	18 m (FosJ58)	☢ [EC] (FosJ58)	G chem, excit, sep isotopes (FosJ58) activity assigned to Re¹⁷⁹ (HarmB60)		protons on Re (FosJ58)
Re^{181}	18 h (GranG63) 19 h (FosJ58) 20 h (GallC57)	☢ EC (GallC57) △ –47 (NDS, MTW)	B chem, excit, genet (GallC57) parent W¹⁸¹ (GallC57) daughter 23 m Os¹⁸¹ (FosJ58) daughter 2.7 h Os¹⁸¹ (SurY60)	γ W X-rays, 0.365, many others between 0 and 1.5 e^- 0.008, 0.040, 0.053, 0.296, many others between 0 and 1.5	Ta¹⁸¹(α, 4n) (GallC57) W¹⁸²(p, 2n) (HarmB60)
Re^{182}	12.7 h (WilkG50) 13 h (GallC59)	☢ EC (WilkG50) β^+ 0.3% (BadN63) △ –45.30 (MTW)	A chem, excit (WilkG50) chem, genet energy levels (GallC59) daughter Os¹⁸² (StovB50, FosJ58)	γ W X-rays, 0.068, 0.100, 1.122, 1.189, 1.23 (complex), 2.01, 2.05, many others between 0 and 2.05 β^+ 1.74 max e^- 0.015, 0.031, 0.056, 0.089, 0.098, many others between 0 and 2.05	Ta¹⁸¹(α, 3n) (WilkG50, GallC59) W¹⁸²(p, n) (WilkG50, HarmB61) daughter Os¹⁸² (FosJ58, StovB50)
Re^{182}	64.0 h (WilkG50) 60 h (GallC58a)	☢ EC (WilkG50) no β^+, lim 5 × 10⁻⁴% (BadN63)	A chem, excit (WilkG50) chem, genet energy levels (GallC58a)	γ W X-rays (very strong), 0.068, 0.100, 0.15–0.36 (complex), 1.08, 1.112 (complex), 1.19, 1.22 (complex), 1.43, many others between 0 and 1.4 e^- 0.015, 0.031, 0.044, 0.061, 0.089, 0.098, 0.100, 0.122, 0.160, 0.187, many others between 0 and 1.4	Ta¹⁸¹(α, 3n) (WilkG50, GallC58a) W¹⁸²(p, n) (WilkG50, HarmB60)

TABLE 1 (*Continued*)

Isotope Z A	Half-life	Type of decay (☢); % abundance; Mass excess (△≡M–A), MeV (C¹²=0); Thermal neutron cross section (σ), barns	Class; Identification; Genetic relationships	Major radiations: approximate energies (MeV) and intensities	Principal means of production
₇₅Re¹⁸³	71 d (BliP65, GallC58) 68 d (FosJ58) others (ThuS56, TurS51, StovB50)	☢ EC (WilkG50) △ –45 (MTW)	A chem, excit (WilkG50) chem, genet energy levels (ThuS56) daughter Os¹⁸³ (StovB50)	γ W X-rays, 0.046, 0.053, 0.109 (complex), 0.209 (strong), 0.246, 0.292 e⁻ 0.030, 0.034, 0.040, 0.088, 0.093, many others between 0 and 0.40	Ta¹⁸¹(α, 2n) (WilkG50, ThuS56)
Re¹⁸⁴	38 d (BodE60, DzhB62b) 34 d (BliP65) 33 d (JohnN63) others (WilkG50, TurS51)	☢ EC (WilkG50) △ –44 (MTW)	A chem, excit (FajK40a) chem, excit (WilkG50) chem, genet energy levels (GallC58)	γ W X-rays, 0.111, 0.78 (complex), 0.90 (complex) e⁻ 0.042, 0.100	Ta¹⁸¹(α, n) (WilkG50) deuterons on W (BisK63a, BodE60, DzhB62b, GallC58) protons on W (WilkG50, HarmB64) Re¹⁸⁵(n, 2n) (GallC58, JohnN63)
Re¹⁸⁴ᵐ	169 d (JohnN63) 160 d (HarmB64) 166 d (BliP65) others (DzhB62b)	☢ IT 70%, EC 30% (HarmB64) △ –44 (LHP, MTW)	A chem, genet energy levels (JohnN63, HarmB64)	γ Re X-rays, W X-rays, 0.111, 0.78 (complex), 0.90 (complex) e⁻ 0.035, 0.042, 0.073, 0.081, 0.100 daughter radiations from Re¹⁸⁴	See Re¹⁸⁴
Re¹⁸⁴?	2.2 d (WilkG50)	☢ EC or IT (WilkG50)	D chem, excit (WilkG50)	γ 0.159	Ta¹⁸¹(α, n) (WilkG50) W¹⁸⁴(p, n) (WilkG50)
Re¹⁸⁵		% 37.07 (WhiJ48) △ –43.73 (MTW) σc 110 (GoldmDT64)			
Re¹⁸⁶	88.9 h (PortF56) 92.8 h (GooLJ47) 91 h (CorkJ48b) 90 h (SinK39)	☢ β⁻ 95%, EC 5% (MalyL64) others (PortF56, JohnM56, MetF51) no β⁺, lim 10⁻⁵% (MetF51) △ –41.9 (MTW)	A n-capt (KurtI35) n-capt, excit (SinK39) chem, n-capt, excit (FajK40a) mass spect (HessD47)	β⁻ 1.07 max e⁻ 0.063, 0.125 γ W X-rays, Os X-rays, 0.137 (9%), 0.632 (0.032%), 0.768 (0.035%)	Re¹⁸⁵(n, γ) (KurtI35, SinK39, FajK40a, SerL47b)
Re¹⁸⁶?	1 h (HaldB57)	☢ (HaldB57)	D chem (HaldB57)		protons on Re, W (HaldB57)
Re¹⁸⁷	4.3 x 10¹⁰ y genet (HirtB63) 1.2 x 10¹¹ y sp act (WolfC62) others (HerrW58, WatD62a, HinH54, SutA54, DixD54a, NalS48, SugaN48)	☢ β⁻ (NalS48) % 62.93 (WhiJ48) △ –41.14 (MTW) σc 70 (to Re¹⁸⁸) 1.3 (to Re¹⁸⁸ᵐ) (GoldmDT64)	A chem (NalS48)	β⁻ 0.003 max (in about 1/3 of the decays the electron goes into a stable atomic orbit)	
Re¹⁸⁸	16.7 h (FlaA53, AjzF56, DzhB54) 16.9 h (LindnM51a) 18.9 h (GooLJ47) others (PoolM37, DoyW63a)	☢ β⁻ (SinK39) △ –38.79 (MTW) σc <2 (GoldmDT64)	A chem, n-capt (AmaE35) n-capt, excit (SinK39) chem, n-capt, excit (FajK40) mass spect (HessD47) daughter W¹⁸⁸ (LindnM51a, LindnM51, RoyJ62) daughter Re¹⁸⁸ᵐ (HerrW52)	β⁻ 2.12 max e⁻ 0.081, 0.143 γ Os X-rays, 0.155 (10%), 0.478 (0.6%), 0.633 (0.9%), 0.829 (0.3%), 0.932 (0.4%), other weak γ's to 2.0	Re¹⁸⁷(n, γ) (KurtI35, AmaE35, SinK39, FajK40a, SerL47b)
Re¹⁸⁸ᵐ	18.7 m (TakaK64, FlaA53) others (ButeF50, MihJ53b)	☢ IT (MihJ53b) △ –38.62 (LHP, MTW)	A n-capt, sep isotopes (MihJ53b) chem, genet (HerrW52) parent Re¹⁸⁸ (HerrW52)	γ Re X-rays, 0.092 (5%), 0.106 (10%) e⁻ 0.004, 0.013, 0.021, 0.034, 0.051, 0.061, 0.080, 0.093 daughter radiations from Re¹⁸⁸	Re¹⁸⁷(n, γ) (MihJ53b)
Re¹⁸⁹	24.3 h (BliP65) 23 h (CrasB63)	☢ β⁻ (CrasB63) △ –37.8 (MTW)	A chem, excit, cross bomb (CrasB63) genet energy levels (CrasB63, ResD61) daughter W¹⁸⁹ (FleJ63, KauP65a)	β⁻ 1.00 max e⁻ 0.023, 0.028, 0.057, 0.074, 0.112, 0.143, others between 0 and 0.25 γ Os X-rays, 0.150 (4%, doublet), 0.187 (3%, doublet), 0.218 (10%, doublet), 0.245 (4%)	W¹⁸⁶(α, p) (CrasB63) Os¹⁸⁹(n, p) + Os¹⁹⁰(n, pn) (CrasB63) Os¹⁹²(d, αn) (FleJ63)
Re¹⁸⁹	140 d (BliP65) 150 d (LindnM51a)	☢ β⁻ (LindnM51a, TurS51) β⁻, IT (?) (BliP65)	F chem (LindnM51a, TurS51) chem, genet energy levels (BliP65) activity assigned to Re¹⁸⁴ᵐ (CrasB63, JohnN63)	γ 0.211, 0.57, 0.67	W¹⁸⁶(α, p) (BliP65, TurS51)

TABLE 1 (*Continued*)

Isotope Z A	Half-life	Type of decay (☢); % abundance; Mass excess (△≡M-A), MeV (C¹²=0); Thermal neutron cross section (σ), barns	Class; Identification; Genetic relationships	Major radiations: approximate energies (MeV) and intensities	Principal means of production
$_{75}Re^{[189?]}$	≥5 y (LindnM51a)	☢ β⁻ (LindnM51a)	F chem (LindnM51a) activity not observed (SmiRR56a)	β⁻ 0.75 max activity not observed	neutrons on Re (LindnM51a)
Re^{190}	2.8 m (AteA55) others (BaroG62)	☢ β⁻ (AteA55) △ -35.4 (MTW)	B chem, genet energy levels, cross bomb (AteA55)	β⁻ 1.6 max γ Os X-rays, 0.191 (↑ 10), 0.392 (↑ 10), 0.57 (↑ 10), 0.83 (↑ 3)	$Os^{192}(d,α)$, $Os^{190}(n,p)$ (AteA55)
Re^{190m}	2.8 h (FleJ64, BaroG62)	☢ [IT] (FleJ64, BaroG62)	B chem, cross bomb, sep isotopes (FleJ64, BaroG62)	β⁻ 1.6 max γ [Os] X-rays, 0.12, 0.19, 0.23, 0.38 (complex), 0.56 (complex), 0.82 (these are probably daughter radiations of 2.8 m Re^{190} according to FleJ64)	$Os^{192}(d,α)$, $Os^{190}(n,p)$, $Ir^{193}(n,α)$ (FleJ64, BaroG62)
$Re^{[191]}$	9.8 m (AteA53c)	☢ β⁻ (AteA53c) △ -34.6 (NDS, MTW)	D chem (AteA53c) excit (AteA55) decay charac (CrasB63)	β⁻ 1.8 max	[$Os^{192}(n,np)$] (AteA53c)
Re^{192}	6 s (BlacJe65a)	☢ β⁻ (BlacJe65a)	C sep isotopes, genet energy levels (BlacJe65a)	β⁻ 2.5 max γ 0.20, 0.29, 0.37, 0.48, 0.57	$Os^{192}(n,p)$ (BlacJe65a)
$_{76}Os^{181}$	23 m (FosJ58)	☢ [EC] (FosJ58) △ -44 (NDS, MTW)	B chem, excit, sep isotopes, genet (FosJ58) activity not observed (SurY60) parent Re^{181} (FosJ58)	γ [Re X-rays], others e⁻ 0.093, 0.101 daughter radiations from Re^{181}	$Re^{185}(p,5n)$ (FosJ58)
Os^{181}	2.7 h (SurY60)	☢ [EC] (SurY60)	E chem, genet (SurY60) parent Re^{181} (SurY60)	γ Re X-rays, 0.23 daughter radiations from Re^{181}	protons on Au (SurY60)
Os^{182}	21.9 h (FosJ58) 21.1 h (NewJ60a) 20 h (GranG63) others (StovB50)	☢ EC, no β⁺ (StovB50) △ -44 (NDS, MTW)	A chem, genet (StovB50) chem, excit, sep isotopes (NewJ60a) parent 12.7 h Re^{182} (StovB50, FosJ58) daughter Ir^{182} (DiaR61)	γ Re X-rays, 0.180 (↑ 7), 0.263 (↑ 1.4), 0.510 (↑ 10) e⁻ 0.015, 0.025, 0.043, 0.052, 0.108, 0.438 daughter radiations from 12.7 h Re^{182}	$Re^{185}(p,4n)$ (StovB50) $W^{182}(α,4n)$ (NewJ60a)
Os^{183}	12.0 h (NewJ60a, StovB50) 15.4 h (FosJ58) others (GranG63, SurY60)	☢ EC (StovB50) △ -43 (NDS, MTW)	A chem, genet (StovB50) parent Re^{183} (StovB50) daughter Ir^{183} (DiaR61, LavA61)	γ Re L X-rays, Re K X-rays (170%), 0.114 (27%), 0.168 (10%), 0.236 (5%), 0.382 (90%), 0.48 (9%, complex), 0.86 (5%, complex), 1.44 (1%) e⁻ 0.043, 0.102, many others between 0 and 1.4, all weak	$Re^{185}(p,3n)$ (FosJ58, StovB50) alphas on W (NewJ60a) daughter Ir^{183} from $Lu^{175}(C^{12},4n)$ (DiaR61)
Os^{183m}	9.9 h (NewJ60a) 10 h (FosJ58)	☢ EC ≈54%, IT ≈46% (NewJ60a, NewJ60b) △ -43 (NDS, MTW)	A chem, excit, sep isotopes (FosJ58, NewJ60a) genet (DiaR61) daughter Ir^{183} (DiaR61)	γ Os X-rays, 1.035 (6%), 1.105 (48%, complex) e⁻ 0.055, 0.096, 0.158, 0.168 daughter radiations from Os^{183}	$Re^{185}(p,3n)$ (FosJ58) alphas on W (NewJ60a) daughter Ir^{183} from $Lu^{175}(C^{12},4n)$ (DiaR61)
Os^{184}		% 0.018 (NierA37) △ -44.0 (MTW) $σ_c$ <200 (GoldmDT64)			
Os^{185}	93.6 d (JohnM57) others (FosJ58, GooLJ47, KatziL48, TurS51, SurY60, GranG63)	☢ EC (MillM51a) no β⁺, lim 4 x 10⁻⁴% (MaliS58) △ -42.74 (MTW)	A chem, cross bomb (GooLJ47, KatziL48) chem, genet energy levels (MartyN57)	γ Re X-rays, 0.646 (80%), 0.875 (14%, complex) e⁻ 0.059, 0.091, 0.574, 0.634	$Re^{185}(d,2n)$ (GooLJ47, ChuT50) $Os^{184}(n,γ)$ (KatziL48) $Re^{185}(p,n)$ (FosJ58, StovB50)
Os^{186}		% 1.59 (NierA37) △ -43.0 (MTW)			
Os^{187}		% 1.64 (NierA37) △ -41.14 (MTW)			
Os^{187m}	39 h (GreeG56) 35 h (ChuT50)	☢ (ChuT50)	G chem (ChuT50) activity not observed (NewJ60a, MerE63)		

TABLE 1 (*Continued*)

Isotope Z A	Half-life	Type of decay (♥); % abundance; Mass excess (△≡M–A), MeV (C¹²=0); Thermal neutron cross section (σ), barns	Class; Identification; Genetic relationships	Major radiations: approximate energies (MeV) and intensities	Principal means of production
$_{76}Os^{188}$		% 13.3 (NierA37) △ –40.91 (MTW)			
Os?	26 d (GreeG56)	♥ (GreeG56)	F chem (GreeG56)	γ X-rays	N^{14} on Os (GreeG56)
Os^{189}		% 16.1 (NierA37) △ –38.8 (MTW) σ_c 0.008 (to Os^{190m}) (GoldmDT64)			
Os^{189m}	5.7 h (SchaG58) others (ChuT50, GreeG56)	♥ IT (SchaG58) △ –38.8 (LHP, MTW)	A chem (ChuT50, GreeG56) chem, genet (SchaG58) genet energy levels (NewJ60c, CrasB63) daughter Ir^{189}(SchaG58)	γ Os L X-rays e⁻ 0.019, 0.028	daughter Ir^{189} (SchaG58)
Os^{190}		% 26.4 (NierA37) △ –38.5 (MTW) σ_c 3.9 (to Os^{191}) 8.6 (to Os^{191m}) (GoldmDT64)			
Os^{190m}	9.9 m (SchaG58) others (ChuT50, AteA55c, MalyT61, MangS62)	♥ IT (SchaG58, AteA55c) △ –36.8 (LHP, MTW)	A chem, genet (ChuT50, AteA55c) genet energy levels (SchaG58, ResD61) daughter Ir^{190m_2} (ChuT50, AteA55c)	γ Os X-rays, 0.187 (70%), 0.361 (94%), 0.502 (98%), 0.616 (99%) e⁻ 0.026, 0.036, 0.113, 0.175	daughter Ir^{190m_2} (ChuT50, AteA55c, SchaG58)
Os^{191}	15.0 d (KatziL48) 16.0 d (ChuT50) 14.6 d (NabS58)	♥ β⁻ (SeaG41b) △ –36.4 (MTW)	A n-capt (ZinE40) chem, n-capt (SeaG41b) chem, excit (SwanJ52) daughter Os^{191m} (SwanJ52) parent Ir^{191m} (NauR54a, CamE56)	β⁻ 0.143 max e⁻ 0.030, 0.042, 0.053, 0.116, 0.127 γ Ir X-rays, 0.129 (25%) daughter radiations from Ir^{191m} included in above listing	Os^{190}(n, γ) (SeaG41b, ZinE40, SerL47b, SwanJ52)
Os^{191m}	13.0 h (PlaZ63) 14 h (SwanJ52)	♥ IT, no β⁻ (lim 5%) (SwanJ52) △ –36.3 (LHP, MTW)	A chem, genet (SwanJ52) parent Os^{191} (SwanJ52)	γ Os L X-rays e⁻ 0.062, 0.072 daughter radiations from Os^{191}	Os(n, γ) (SwanJ52)
Os^{192}	$t_{1/2}$ (ββ) >10^{14} y sp act (FremJ52)	% 41.0 (NierA37a) △ –35.9 (MTW) σ_c 1.6 (GoldmDT64)			
Os^{193}	31.5 h (NabS58) 30.6 h (ChuT50) others (GooLJ47, SeaG41b, ZinE40)	♥ β⁻ (SeaG41b) △ –33.32 (MTW) σ_c 200 (GoldmDT64)	A n-capt (KürtI35, ZinE40) chem, n-capt (SeaG41b) chem, excit (SwanJ52)	β⁻ 1.13 max e⁻ 0.060, 0.070 γ Ir X-rays, 0.139 (3%), 0.28 (2.1%, complex), 0.322 (1.4%), 0.38 (2.0%, complex), 0.460 (3.9%), 0.558 (2.1%)	Os^{192}(n, γ) (KürtI35, ZinE40, SeaG41b, SerL47b)
Os^{194}	6.0 y (JohnN65b) 5.8 y (WilliDC64) others (LindnM51a)	♥ β⁻ (WilliDC64) △ –32.39 (MTW)	A chem, genet (LindnM50) chem, genet, n-capt (WilliDC64) parent Ir^{194} (LindnM50, LindnM51a, WilliDC64)	β⁻ 0.053 max e⁻ [0.029, 0.040] γ Ir X-rays, 0.043 (10%), 0.078 (0.03%) daughter radiations from Ir^{194}	Os^{192}(n, γ)Os^{193}(n, γ) WilliDC64, LindnM50, LindnM51a)
Os^{195}	6.5 m (BaroG57, ReyP57)	♥ β⁻ (BaroG57, ReyP57) △ –30 (MTW)	B chem, genet (BaroG57, ReyP57) parent Ir^{195} (BaroG57, ReyP57)	β⁻ 2 max	Pt^{198}(n, α) (BaroG57, ReyP57)
$_{77}Ir^{182}$	15 m (DiaR61)	♥ EC, [β⁺] (DiaR61) △ –39 (NDS, MTW)	A chem, cross bomb, genet (DiaR61) parent Os^{182} (DiaR61)	γ Os X-rays, 0.133, 0.278, 0.510, others to ≈4	$Lu^{175}(C^{12}, 5n)$, $Tm^{169}(O^{16}, 3n)$ (DiaR61)
Ir^{183}	0.9 h (DiaR61) 1.0 h (LavA61) others (SurY60)	♥ EC (DiaR61, LavA61)	A chem, genet (DiaR61, LavA61) parent Os^{183} (DiaR61, LavA61) parent Os^{183m} (DiaR61)	γ Os X-rays, 0.24 daughter radiations from Os^{183m} Os^{183}	$Lu^{175}(C^{12}, 4n)$ (DiaR61)

TABLE 1 (Continued)

Isotope Z A	Half-life	Type of decay (☢); % abundance; Mass excess (Δ≡M-A), MeV (C¹²=0); Thermal neutron cross section (σ), barns	Class; Identification; Genetic relationships	Major radiations: approximate energies (MeV) and intensities	Principal means of production
$_{77}$Ir^{184}	3.2 h (DiaR61) 3.1 h (BaranV60)	☢ EC, β⁺ (DiaR61) Δ −40 (NDS, MTW)	B chem, decay charac (BaranV60) chem, excit, decay charac (DiaR61) daughter 42 m Pt^{184} (QaiS65)	γ Os X-rays, 0.125 (↑ 100), 0.267 (↑ 200), 0.392 (↑ 90), 0.51 (γ±?), 0.83, 0.96, 1.09, others to 4.3	Lu^{175}(C^{12}, 3n) (DiaR61)
Ir^{185}	14 h (EmeG63) 15 h (DiaR58)	☢ EC (DiaR58) Δ −40 (NDS, MTW)	B chem, excit (DiaR58) sep isotopes (HarmB62) daughter Pt^{185} (QaiS65)	γ Os X-rays, 0.101, 0.254, others e⁻ 0.024, 0.034, 0.047, 0.085, 0.180	Re^{185}(α, 4n) (DiaR58, EmeG63) Os^{186}(p, 2n) (HarmB62)
Ir^{186}	15.8 h (EmeG63) 14 h (SmiW55) 16 h (DiaR58) others (MalyT60, KryL61)	☢ EC 97%, β⁺ 3% (EmeG63) Δ −39.1 (MTW)	A chem, excit (DiaR58) genet energy levels (EmeG63) daughter Pt^{186} (SmiW55, QaiS65)	γ Os X-rays, 0.137 (45%), 0.297 (74%), 0.434 (35%), 0.511 (6%, γ±), 0.64 (9%, complex), 0.77 (8%, complex), 1.60-1.75 (4%, complex), many others between 0 and 3.0 β⁺ 1.94 max e⁻ 0.063, 0.125, 0.135, 0.226	Re^{185}(α, 3n) (DiaR58, EmeG63)
Ir^{186}	1.7 h (MalyT63) 2.0 h (BoncN62, GranG63)	☢ β⁺, EC (BoncN62, GranG63)	B chem (BoncN62, MalyT63) chem, excit (GranG63) not daughter Pt^{186} (QaiS65)	γ Os X-rays, 0.137, 0.295, 0.511 (γ±), 0.630, 0.77, 0.99, others β⁺ 2.6 max e⁻ 0.063, 0.125	Ir^{191}(p, p5n) (GranG63)
Ir^{187}	10.5 h (EmeG63) others (DiaR58, MalyT60, KryL61)	☢ EC (DiaR58) Δ −40 (MTW)	B chem, excit (DiaR58) daughter Pt^{187} (BaranV60)	γ Os X-rays, 0.18 (↑ 45), 0.31 (↑ 14), 0.41 (↑ 100), 0.50 (↑ 35), 0.61 (↑ 45), 0.90 (↑ 40), 0.98 (↑ 50), all γ rays complex, many others e⁻ 0.007, 0.013, 0.053, 0.063, 0.073, 0.104, many others between 0 and 1.1	Re^{185}(α, 2n) (DiaR58, EmeG63)
Ir^{188}	41.5 h (ChuT50) others (SmiW55, NauR54, GranG63, KryL61, MalyT60)	☢ EC 99+%, β⁺ ≈0.3% (ChuT50) Δ −38.08 (MTW)	A chem, excit, sep isotopes (ChuT50) genet energy levels (GrahR62, MarkI63) daughter Pt^{188} (NauR54, SmiW55)	γ Os X-rays, 0.155 (34%), 0.478 (16%), 0.633 (29%, doublet), 0.829 (7%), 1.210 (7%), 1.717 (4%), 2.08 (16%, complex), 2.217 (13%), many others between 0 and 2.7 β⁺ 1.66 max e⁻ 0.081, 0.143, many others between 0 and 2.7	alphas on Re (ChuT50, WarnL62, YamaT63) Os^{189}(p, 2n) (HarmB64) deuterons on Os (ChuT50)
Ir^{189}	13.3 d (GranG63, LewisH64) others (ChuT50, SmiW55, MalyT60, KryL61)	☢ EC (SmiW55) Δ −38 (MTW)	A chem, genet (SmiW55) daughter Pt^{189} (SmiW55) parent Os^{189m} (SchaG58)	γ Os X-rays, 0.245 (18%) e⁻ 0.023, 0.046, 0.058, 0.067, 0.171, many others between 0 and 0.27	Ir^{191}(p, 3n) Pt^{189}(EC) (GranG63, LewisH64) Re^{187}(α, 2n) (DiaR58) Os^{190}(p, 2n) (HarmB62)
Ir^{190}	11 d (GranG63, AteA55c) 10.7 d (GooLJ47) 12.3 d (KaneW60) 12.6 d (ChuT50)	☢ EC (AteA55c) no β⁺, lim 0.002% (KaneW60) Δ −36.5 (MTW)	A chem, excit, cross bomb (GooLJ47, AteA55c) genet energy levels (KaneW60, ResD61)	γ Os X-rays, 0.187 (51%), 0.37 (39%, complex), 0.40 (39%, complex), 0.518 (39%), 0.56 (72%, complex), 0.604 (47%), others to 1.7 e⁻ 0.113, 0.175, others to 1.7	Re^{187}(α, n) (ChuT50) Os^{189}(d, n) (GooLJ47) Os^{190}(p, n) (HarmB64)
Ir^{190m_1}	1.2 h (HarmB64)	☢ IT (HarmB64) Δ −36.5 (LHP, MTW)	B chem, sep isotopes, excit (HarmB64)	γ Ir L X-rays e⁻ 0.015, 0.024 daughter radiations from Ir^{190}	Os^{190}(p, n) (HarmB64)
Ir^{190m_2}	3.2 h (ChuT50) 3.0 h (GranG63)	☢ EC 94%, IT 6% (HarmB64) EC 90%, β⁺ 10% (AteA55c) Δ −36.3 (LHP, MTW)	A chem, excit, sep isotopes (ChuT50) chem, cross bomb (AteA55c) genet energy levels (HarmB64) parent Os^{190m} (ChuT50, AteA55c)	γ Os X-rays, Ir X-rays, 0.187 (66%), 0.361 (88%), 0.502 (92%), 0.616 (93%) e⁻ 0.026, 0.036, 0.113, 0.175 daughter radiations from Ir^{190m_1}, Ir^{190} daughter radiations from Os^{190m} included in above listing	Re^{187}(α, n) (ChuT50) deuterons on Os (ChuT50) Os^{190}(p, n) (HarmB64)
Ir^{191}		% 38.5 (SamM36a) Δ −36.7 (MTW) σc 750 (to Ir^{192}) 250 (to Ir^{192m_1}) 0.3 (to Ir^{192m_2}) (GoldmDT64)			

TABLE 1 (*Continued*)

Isotope Z A	Half-life	Type of decay (☢); % abundance; Mass excess (Δ=M–A), MeV (C^{12}=0); Thermal neutron cross section (σ), barns	Class; Identification; Genetic relationships	Major radiations: approximate energies (MeV) and intensities	Principal means of production
$_{77}Ir^{191m}$	4.9 s (FiscV55, CamE56) 4.5 s (CloJ58) others (NauR54a, MihJ54a)	☢ IT (NauR54a) Δ –36.5 (LHP, MTW)	A chem, genet (NauR54a, CamE56) daughter Os^{191} (NauR54a, CamE56)	γ Ir X-rays, 0.129 (25%) e⁻ 0.030, 0.042, 0.053, 0.116, 0.127	daughter Os^{191} (NauR54a, CamE56) Os^{192}(p, 2n) (CloJ58)
Ir^{192}	74.2 d (AlliJ60) 74.4 d (KasJ51) others (WyaE61, HarbG63, SinW51, ChuT50)	☢ β⁻ 95.5%, EC 4.5% (BashA56) β⁻ 96.5%, EC 3.5% (BagL55) β⁺ 1.5 x 10⁻⁵% (AntoS60) Δ –34.7 (MTW) σ_c 700 (to Ir^{193m}) (GoldmDT64)	A n-capt (AmaE36) mass spect (RalW46) chem (WilkG48) daughter Ir^{192m_1}, daughter Ir^{192m_2} (SchaG59)	β⁻ 0.67 max e⁻ 0.217, 0.230, 0.239, 0.390 γ Os X-rays, Pt X-rays, 0.296 (29%), 0.308 (30%), 0.317 (81%), 0.468 (49%), 0.589 (4%), 0.604 (9%), 0.612 (6%)	Ir^{191}(n, γ) (AmaE36, MMilE37, JaeR38, SerL47b) Os^{192}(d, 2n) (GooLJ47, ChuT50)
Ir^{192m_1}	1.42 m (HoleN48b, MizJ54) 1.45 m (WebG53) others (SchaG61, MMilE37)	☢ IT 99+% β⁻ 0.017% (SchaG61, SchaG59) Δ –34.7 (LHP, MTW, NDS)	A n-capt (MMilE37) resonance neutron activation (GoldhM47) parent Ir^{192} (SchaG59) not daughter Ir^{192m_2} (SchaG59)	γ Ir L X-rays, 0.058 (0.005%), 0.317 (0.008%), 0.612 (0.003%) e⁻ 0.046, 0.056 β⁻ 1.5 max	Ir^{191}(n, γ) (MMilE37, GoldhM47, SerL47b) Os^{192}(d, 2n) (GooLJ47) ChuT50)
Ir^{192m_2}	>5 y (SchaG59)	☢ IT (SchaG59) Δ –34.6 (LHP, MTW, NDS)	B genet, n-capt (SchaG59) parent Ir^{192} (SchaG59) not parent Ir^{192m_1} (SchaG59)	γ Ir K X-rays (weak), Ir L X-rays e⁻ 0.149, 0.158 daughter radiations from Ir^{192}	Ir^{191}(n, γ) (SchaG59)
$\underline{Ir^{193}}$		% 61.5 (SamM36a) Δ –34.45 (MTW) σ_c 110 (GoldmDT64)			
Ir^{193m}	11.9 d (BoeF57)	☢ IT (BoeF57) Δ –34.37 (LHP, MTW)	B chem, n-capt (BoeF57)	γ Ir L X-rays e⁻ 0.069, 0.078	Ir^{191}(n, γ)Ir^{192}(n, γ) (BoeF57)
Ir^{194}	17.4 h (PeiM64) 19.0 h (GooLJ47) others (WitC41, AmaE35, MMilE37, SerL47b)	☢ β⁻ (MMilE37) Δ –32.49 (MTW)	A n-capt (AmaE35) mass spect (RalW46) chem (WilkG48) daughter Os^{194} (LindnM50, LindnM51a, WilliDC64)	β⁻ 2.24 max γ 0.328 (10%), 0.64 (1.0%, doublet), 0.939 (0.4%), 1.16 (0.8%, complex), 1.48 (0.6%, complex), 1.7 (0.2%, complex), many others	Ir^{193}(n, γ) (AmaE35, PoolM37, SerL47b, MMilE37, JaeR38) daughter Os^{194} (PeiM64)
Ir^{194m}	47 s (HennH60, HennH60a)	☢ β⁻, IT (HennH60, HennH60a)	G n-capt, decay charac (HennH60, HennH60a, HennH61) activity not observed (SchaG61) activity produced by thermal neutrons on Ir, but not with enriched Ir^{193} (FetP62a)	β⁻ 2.3 max (HennH60a) γ 0.13, 0.32, 0.63 (HennH60a)	neutrons on Ir (HennH60, HennH60a)
Ir^{195}	4.2 h (ClafA62) 2.3 h (ButeF54) 2.7 h (ChrisD52)	☢ β⁻ (ChrisD52) Δ –31.8 (MTW)	B chem, excit (ChrisD52, ButeF54, HomS61) sep isotopes (ClafA62) daughter Os^{195} (BaroG57, ReyP57)	β⁻ 1.0 max γ Pt X-rays, 0.10, 0.13, 0.33, 0.37, 0.43, 0.66	Pt^{195}(n, p) (ButeF54) Pt^{196}(γ, p) (ChrisD52, HomS61) Os^{192}(α, p) (ClafA62)
Ir^{196}	120 m (BisW65)	☢ β⁻ (BisW65) Δ –29.23 (BisW65, MTW)	B chem, genet energy levels, sep isotopes (BisW65)	β⁻ 0.95 max γ 0.100 (33%), 0.356 (94%), 0.39 (95%), 0.44 (95%), 0.522 (99%), 0.65 (100%)	Pt^{198}(d, α) (BisW65)
Ir^{196}	9.7 d (ButeF54)	☢ β⁻ (ButeF54)	G chem, cross bomb (ButeF54) activity assigned to Ir^{189} + Ir^{190} (GardD57); not produced by Pt^{194}(d, α) (GardD57)		
Ir^{197}	7 m (ChrisD52, ButeF54, HomS61)	☢ β⁻ (ButeF54) Δ –28.4 (MTW)	D chem, excit (ChrisD52) chem, cross bomb (ButeF54)	β⁻ 2.0 max γ 0.50	Pt^{198}(n, pn) (ButeF54) Pt^{198}(γ, p) (ChrisD52, HomS61)
Ir^{198}	50 s (ButeF54)	☢ β⁻ (ButeF54) Δ –25.5 (MTW)	C excit, cross bomb (ButeF54)	β⁻ 3.6 max γ 0.78	Pt^{198}(n, p) (ButeF54)

TABLE 1 (*Continued*)

Isotope Z A	Half-life	Type of decay (☢); % abundance; Mass excess (△≡M–A), MeV (C¹²=0); Thermal neutron cross section (σ), barns	Class; Identification; Genetic relationships	Major radiations: approximate energies (MeV) and intensities	Principal means of production
$_{78}Pt^{173}$	short (SiiA66)	α (SiiA66)	F cross bomb, excit (SiiA66)	α 6.19	O^{16} on Yb, Ne^{20} on Er (SiiA66)
Pt^{174}	0.7 s (SiiA66)	α 80%, [EC+β⁺] 20% (SiiA66)	B cross bomb, excit (SiiA66)	α 6.03	O^{16} on Yb, Ne^{20} on Er (SiiA66)
Pt^{175}	2.1 s (SiiA66)	α (SiiA66)	B cross bomb, excit (SiiA66)	α 5.95	O^{16} on Yb, Ne^{20} on Er (SiiA66)
Pt^{176}	6.0 s (SiiA66)	α 1.4%, [EC+β⁺] 98.6% (SiiA66)	B cross bomb, excit (SiiA66)	α 5.74	O^{16} on Yb, Ne^{20} on Er (SiiA66)
Pt^{177}	6.6 s (SiiA66)	α 0.3%, [EC+β⁺] 99+% (SiiA66)	B cross bomb, excit (SiiA66)	α 5.51	O^{16} on Yb, Ne^{20} on Er (SiiA66)
Pt^{178}	21 s (SiiA66)	α 1.3%, [EC+β⁺] 98.7% (SiiA66)	B cross bomb, excit (SiiA66)	α 5.44	O^{16} on Yb, Ne^{20} on Er (SiiA66)
Pt^{179}	33 s (SiiA66)	α 0.1%, [EC+β⁺] 99+% (SiiA66)	B cross bomb, excit (SiiA66)	α 5.15	O^{16} on Yb, Ne^{20} on Er (SiiA66)
Pt^{180}	50 s (SiiA66)	α 0.3%, [EC+β⁺] 99+% (SiiA66)	B cross bomb, excit (SiiA66)	α 5.14	O^{16} on Yb, Ne^{20} on Er (SiiA66)
Pt^{181}	51 s (SiiA66)	α 0.0006%, [EC+β⁺] 99+% (SiiA66)	B cross bomb, excit (SiiA66)	α 5.02	O^{16} on Yb, Ne^{20} on Er (SiiA66)
Pt^{182}	3.0 m (SiiA66) 2.5 m (GraeG63)	α 0.02%, [EC+β⁺] 99+% (GraeG63, SiiA66) △ −36 (NDS, MTW)	B chem, decay charac (GraeG63) cross bomb, excit (SiiA66)	α 4.84 daughter radiations from Ir^{182}	O^{16} on Yb, Ne^{20} on Er (SiiA66) protons on Ir (GraeG63)
Pt^{183}	6.5 m (GraeG63) 7 m (SiiA66)	α 0.001%, [EC+β⁺] 99+% (GraeG63, SiiA66)	B chem, decay charac (GraeG63) cross bomb, excit (SiiA66)	α 4.73	O^{16} on Yb, Ne^{20} on Er (SiiA66) protons on Ir (GraeG63)
Pt^{184}	20 m (GraeG63) 16 m (SiiA66)	α 0.0015%, [EC+β⁺] 99+% (GraeG63, SiiA66)	B chem, decay charac (GraeG63) cross bomb, excit (SiiA66)	α 4.50	O^{16} on Yb, Ne^{20} on Er (SiiA66) Ir^{193}(p, 10n) (GraeG63)
Pr^{184}	42 m (QaiS65)	EC (QaiS65)	D chem, genet (QaiS65) parent Ir^{184} (QaiS65)	Y [Ir X-rays], 0.68, 1.72, 1.85 daughter radiations from Ir^{184}	N^{14} on Ta (QaiS65)
Pt^{185}	1.2 h (AlboG60) 1.0 h (QaiS65)	[EC] (AlboG60)	C genet (AlboG60) chem, genet (QaiS65) daughter 7 m Au^{185} (AlboG60) parent Ir^{185} (QaiS65)	Y [Ir X-rays], 0.035, 0.63, 1.56 daughter radiations from Ir^{185}	descendant Hg^{185} (AlboG60) N^{14} on Ta (QaiS65)
Pt^{186}	3.0 h (GranG63) 2.9 h (AlboG60) 2.8 h (QaiS65) 2.5 h (SmiW55) 2.0 h (α) (GraeG63)	EC (SmiW55, AlboG60) α 1.4 x 10⁻⁴% (GraeG63)	B chem, genet (SmiW55, AlboG60) chem, excit (GranG63) parent 16 h Ir^{186} (SmiW55, QaiS65) not parent 1.7 h Ir^{186} (QaiS65) daughter Au^{186} (SmiW55)	Y Ir X-rays, 0.67 α 4.23 daughter radiations from 16 h Ir^{186}	protons on Ir (GranG63)
Pt^{187}	2.0 h (BaranV60) 2.1 h (QaiS65) 3.1 h (GranG63) 2.2 h (AlboG60) others (KryL61, MalyT60)	EC (BaranV60)	B chem, genet (BaranV60) chem, excit (GranG63) parent Ir^{187} (BaranV60) daughter Au^{187} (AlboG60)	Y Ir X-rays, 0.11 (?), 0.18 (?), 2.0 daughter radiations from Ir^{187}	protons on Ir (GranG63)
Pt^{188}	10.2 d (GraeG63) 10.0 d (SmiW55) others (NauR54, KarrM63, GranG63)	EC (NauR54) α 3 x 10⁻⁵% (GraeG63) α 5 x 10⁻⁵% (KarrM63) △ −37.6 −MTW)	A chem, genet (NauR54, SmiW55) parent Ir^{188} (NauR54, SmiW55) daughter Au^{188} (SmiW55)	Y Ir X-rays, 0.140 (↑ 22), 0.19 (↑ 100, complex), 0.38 (↑ 15), 0.42 (↑ 7) e⁻ 0.042, 0.111, 0.119, others between 0 and 0.4 daughter radiations from Ir^{188} α 3.93	Ir^{191}(p, 4n) (GranG63)

TABLE 1 (*Continued*)

Isotope Z A	Half-life	Type of decay; % abundance; Mass excess ($\Delta \equiv M-A$), MeV ($C^{12}=0$); Thermal neutron cross section (σ), barns	Class; Identification; Genetic relationships	Major radiations: approximate energies (MeV) and intensities	Principal means of production
$_{78}Pt^{189}$	10.9 h (LewisH64) 10.5 h (GrigE62) 11.1 h (AndeG61) others (KryL61, GranG63, PofN60, AlboG60, SmiW55, QaiS65)	EC (SmiW55, AlboG60) Δ -37 (MTW)	A chem, excit, genet (SmiW55) chem, excit (GranG63) parent Ir^{189} (SmiW55) daughter Au^{189} (SmiW55, ChacK57) descendant Hg^{189} (AndeG61, PofN60, AlboG60)	Y Ir X-rays, 0.094 (↑ 120), 0.114 (↑ 61), 0.141 (↑ 124), 0.187 (↑ 137), 0.243 (↑ 100), 0.31 (↑ 96, complex), 0.404 (↑ 32), 0.56 (↑ 230, complex), 0.61 (↑ 180, complex), 0.722 (↑ 156), 0.80 (↑ 27, complex) e⁻ 0.037, 0.058, 0.068, 0.082, 0.092, 0.168, 0.231, 0.241, many others between 0 and 0.8 daughter radiations from Ir^{189}	Ir^{191} (p, 3n) (GranG63)
Pt^{190}	6.9×10^{11} y sp act (MacfR61a) 5.4×10^{11} y sp act (GraeG63) others (PetrK61, GraeG61, PorsW56, PorsW54)	α (PorsW54) % 0.0127 (WhiF56) Δ -37.3 (MTW) σ_c ≈150 (GoldmDT64)	A decay charac (PorsW56) chem, sep isotopes (MacfR61a)	α 3.18	
Pt^{191}	3.00 d (WilkG49a) others (CorkJ54a, SwanJ53a, SmiW55, LindsJ62, KryL61, GranG63)	EC (WilkG48) Δ -36 (MTW)	A chem, excit (WilkG48) genet energy levels (GillL54) daughter Au^{191} (SmiW55)	Y Ir X-rays, 0.096 (1%), 0.129 (2%), 0.269 (1%), 0.36 (5%, complex), 0.410 (3%), 0.457 (1%), 0.539 (9%), 0.624 (1%) e⁻ 0.020, 0.053, 0.069, 0.080, others between 0 and 0.6	protons on Ir (GranG63, HarmB62) Ir^{191} (d, 2n) (WilkG49a)
Pt^{192}	$\approx 10^{15}$ y sp act (PorsW56) $>10^{14}$ y sp act (GraeG63)	α (PorsW56) % 0.78 (WhiF56) Δ -36.2 (MTW) σ_c <14 (to Pt^{193}), 2 (to Pt^{193m}) (GoldmDT64)	E decay charac (PorsW56)	α 2.6 ?	
Pt^{193}	<500 y yield (NauR56) >74 d, or <1 h (no activity observed (SwanJ53a)	EC (L/K>1000), no β⁻, no β⁺ (NauR56) Δ -34.41 (MTW)	B n-capt, chem (NauR56)	Y Ir L X-rays	Pt^{192}(n, Y) (NauR56)
Pt^{193m}	4.3 d (WilkG49a) 3.4 d (CorkJ54a) 4.4 d (EwaG57) 4.5 d (SwanJ53a) 3.5 d (BrunnJ55)	IT (SwanJ53a) Δ -34.26 (LHP, MTW)	B chem, excit (WilkG48) daughter Au^{193m} (0.03%) (BrunnJ55) daughter Au^{193} (WilkG49a)	Y Pt X-rays e⁻ 0.01, 0.057, 0.124, 0.133	Ir^{193}(d, 2n), Pt^{192}(n, Y) (WilkG49a)
Pt^{194}		% 32.9 (WhiF56) Δ -34.72 (MTW) σ_c 1.1 (to Pt^{195}) 0.09 (to Pt^{195m}) (GoldmDT64)			
Pt^{195}		% 33.8 (WhiF56) Δ -32.78 (MTW) σ_c 27 (GoldmDT64)			
Pt^{195m}	4.1 d (BresM60) others (HoleN48b, DShaA52, HaldB52, MMilE37, MalyT60)	IT (DShaA52) Δ -32.52 (LHP, MTW)	A chem (MMilE37) chem, excit (?) (DShaA52) genet energy levels (CorkJ54a, BernsE55)	Y Pt X-rays, 0.099 (11%), 0.129 (1%) e⁻ 0.018, 0.028, 0.051, 0.085, 0.116, 0.126	Pt^{194}(n, Y) (MandeC48d, HaldB52, DShaA52, MMilE37, PoolM37, SerL47b, HubeO51) Pt^{194}(d, p) (KriR41c)
Pt^{196}		% 25.2 (WhiF56) Δ -32.63 (MTW) σ_c 0.9 (to Pt^{197}) 0.05 (to Pt^{197m}) (GoldmDT64)			
Pt^{197}	18 h (MMilE37) 20.0 h (BresM60) 17.4 h (CorkJ52a)	β⁻ (MMilE37) Δ -30.42 (MTW)	A chem (CorkJ36) chem, excit (MMilE37)	β⁻ 0.670 max e⁻ 0.063, 0.074, 0.110 Y Au X-rays, 0.077 (20%), 0.191 (6%)	Pt^{196}(n, Y) (MMilE37, SherrR41, SerL47b, HaldB52)

TABLE 1 (*Continued*)

Isotope Z A	Half-life	Type of decay (♥); % abundance; Mass excess ($\triangle \equiv M-A$), MeV ($C^{12}=0$); Thermal neutron cross section (σ), barns	Class; Identification; Genetic relationships	Major radiations: approximate energies (MeV) and intensities	Principal means of production
$_{78}Pt^{197m}$	78 m (HoleN48b) 80 m (SherrR41, MangS62) 88 m (ChrisD52)	♥ IT (HoleN48b) β^- 3% (HavA65) $\triangle$ -30.02 (LHP, MTW)	A chem (SherrR41) chem, excit, cross bomb (ChrisD52) genet, genet energy levels (HavA65) parent Au^{197m} (PraK64, HavA65)	γ Pt X-rays, 0.279 (2.6%), 0.346 (13%) e^- 0.040, 0.050, 0.268, 0.332 β^- 0.737 max (3%) daughter radiations from Pt^{197} daughter radiations from Au^{197m} included in above listing	$Pt^{196}(n,\gamma)$ (HavA65) $Pt^{196}(d,p)$ (SherrR41)
Pt^{198}	$t_{1/2}$ ($\beta\beta$) >10^{15} y sp act (FremJ52)	% 7.19 (WhiF56) $\triangle$ -29.91 (MTW) σ_c 4 (to Pt^{199}) 0.03 (to Pt^{199m}) (GoldmDT64)			
Pt^{199}	31 m (MMilE37) 30 m (LBiaJ56) 29 m (SherrR41)	♥ β^- (MMilE37) $\triangle$ -27.40 (MTW) σ_c ≈15 (GoldmDT64)	A n-capt (MLenJ35, AmaE35) chem, n-capt, excit (SherrR41) parent Au^{199} (MMilE37, BeacL49, MeeJ49, HillR50a)	β^- 1.69 max γ 0.075 + Au K X-ray (9%), 0.197 (9%), 0.245 (4%), 0.32 (8%, doublet), 0.475 (12%, doublet), 0.540 (24%), 0.715 (3%), 0.790 (2%), 0.960 (2%)	$Pt^{198}(n,\gamma)$ (AmaE35, MLenJ35, MMilE37, SherrR41, SerL47b, HumV51)
Pt^{199m}	14.1 s (WahM59)	♥ IT (WahM59) $\triangle$ -26.98 (LHP, MTW)	B n-capt, sep isotopes (WahM59)	γ Pt X-rays, 0.393 (90%) e^- 0.018, 0.029, 0.315, 0.381	$Pt^{198}(n,\gamma)$ (WahM59)
Pt^{200}	11.5 h (RoyL57a)	♥ β^- (RoyL57a) $\triangle$ -27 (MTW)	B n-capt, chem, genet (RoyL57a) parent Au^{200} (RoyL57a)	daughter radiations from Au^{200}	$Pt^{198}(n,\gamma)Pt^{199}(n,\gamma)$ (RoyL57a)
Pt^{201}	2.3 m (FacJ62) 2.5 m (GopK63)	♥ β^- (FacJ62, GopK63) $\triangle$ -23.5 (MTW)	B chem, genet (FacJ62) parent Au^{201} (FacJ62)	β^- 2.66 max γ 0.15, 0.23, 1.76 daughter radiation from Au^{201}	$Hg^{204}(n,\alpha)$ (FacJ62, GopK63)
$_{79}Au^{177}$	1.4 s (SiiA65b)	♥ α (SiiA65b)	C excit, sep isotopes (SiiA65b)	α 6.11	F^{19} on Yb (SiiA65b)
Au^{178}	2.7 s (SiiA65b)	♥ α (SiiA65b)	C excit, sep isotopes (SiiA65b)	α 5.91	F^{19} on Yb (SiiA65b)
Au^{179}	7.1 s (SiiA65b)	♥ α (SiiA65b)	C excit, sep isotopes (SiiA65b)	α 5.84	F^{19} on Yb (SiiA65b)
Au^{181}	10 s (SiiA65b)	♥ α (SiiA65b)	C excit, sep isotopes (SiiA65b)	α 5.60, 5.47	F^{19} on Yb (SiiA65b)
Au^{183}	44 s (SiiA65b)	♥ α (SiiA65b)	C excit, sep isotopes (SiiA65b)	α 5.34	F^{19} on Yb (SiiA65b)
Au^{185}	7 m (AlboG60)	♥ [EC] (AlboG60)	C genet (AlboG60) daughter Hg^{185}, parent Pt^{185} (AlboG60) possibly identical to 4.3 m Au^{185} (LHP)		daughter Hg^{185} (AlboG60)
Au^{185}	4.33 m (SiiA65b) 4.3 m (RasJ53)	♥ EC, β^+, α ≈0.01% (ThomS49, RasJ53)	B chem, excit (ThomS49) excit, sep isotopes (SiiA65b)	α 5.07	F^{19} on Yb (SiiA65b) protons on Pt, Au (ThomS49, RasJ53)
Au^{186}	12 m (AlboG60) ≈15 m (SmiW55)	♥ EC (SmiW55, AlboG60)	B chem, genet (SmiW55, AlboG60) parent Pt^{186} (SmiW55) daughter Hg^{186} (AlboG60)	γ Pt X-rays, 0.16, 0.22, 0.30, 0.40 daughter radiations from Pt^{186}	daughter Hg^{186} (AlboG60)
Au^{187}	8 m (AlboG60)	♥ EC (AlboG60)	C genet (AlboG60) parent Pt^{187}, daughter Hg^{187} (AlboG60)	γ Pt X-rays daughter radiations from Pt^{187}	daughter Hg^{187} (AlboG60)
Au^{188}	8 m (PofN60, AlboG60) ≈10 m (SmiW55) 4.5 m (ChacK57)	♥ EC (SmiW55, PofN60, AlboG60) β^+ (ChacK57)	B chem, genet (SmiW55, PofN60, AlboG60) chem, excit (ChacK57) parent Pt^{188} (SmiW55) daughter Hg^{188} (PofN60, AlboG60)	γ Pt X-rays, 0.25, 0.33, 0.63	$Ta^{181}(C^{12},5n)$ (ChacK57) protons on Pt (SmiW55) daughter Hg^{188} (PofN60, AlboG60)

TABLE 1 *(Continued)*

Isotope Z A	Half-life	Type of decay (☢); % abundance; Mass excess ($\triangle \equiv$ M–A), MeV (C¹²=0); Thermal neutron cross section (σ), barns	Class; Identification; Genetic relationships	Major radiations: approximate energies (MeV) and intensities	Principal means of production
$_{79}$Au^{189}	30 m (PofN60, AlboG60) <<40 m, activity not observed (LilG64) 42 m (SmiW55)	☢ [EC] no α, lim 3 × 10^{-5}% (KarrM63)	B chem, genet, cross bomb (SmiW55) chem, mass spect (KilPi65) parent Pt^{189}, daughter Hg^{189} (SmiW55, ChacK57)	γ Pt X-rays e⁻ 0.027, 0.036, 0.088, 0.137, 0.154, 0.166, 0.269 daughter radiations from Pt^{189}	Au^{197}(p, 9n)Hg^{189}(EC) (PofN60, AlboG60) Ta^{181}(C^{12},4n) (SmiW55)
Au^{190}	39 m (AndeG61, JasJ61a) 45 m (PofN60)	☢ EC (AlboG59, AlboG60, PofN60) EC 98%, β⁺ 2% (JasJ61a) β⁺ <1% (AlboG59) no α, lim 1 × 10^{-6}% (KarrM63) Δ -33 (MTW)	B genet (AndeG61, JasJ61a) daughter Hg^{190} (AndeG61)	γ Pt X-rays, 0.29 († 100, complex), 0.60 († 5, complex), other weak γ's to 3.5 e⁻ 0.22, 0.29	daughter Hg^{190} (AndeG61, JasJ61a)
Au$^{191-193}$	2.0 s (HenrA53)	☢ (HenrA53)	F excit (HenrA53)		protons on Tl, Hg (HenrA53)
Au^{191}	3.2 h (AndeG61a) others (SmiW55, GillL54)	☢ EC (SmiW55) no α, lim 5 × 10^{-6}% (KarrM63) Δ -34 (MTW)	A chem, genet (SmiW55, GillL54) parent Pt^{191} (SmiW55) daughter Hg^{191} (SmiW55, GillL54)	γ Pt X-rays, 0.14 († 10), 0.30 († 60), 0.39 († 5), 0.48 († 4), 0.60 († 10), all γ's complex e⁻ 0.035, 0.046, 0.054, 0.080, 0.089 many others between 0 and 2.0 daughter radiations from Pt^{191}	protons on Pt (MarkI62) Ir^{191}(α, 4n) (WilkG49a, EwbW60) Pt^{192}(d, 3n) (WilkG49a)
Au^{192}	4.1 h (FinR52) others (WilkG49a, EngeT53)	☢ EC, β⁺ ≈1% (WilkG49a) Δ -33.0 (MTW)	A chem, excit (WilkG49a) chem, genet (FinR52, GillL54) genet energy levels (GillL54) daughter Hg^{192} (FinR52, GillL54)	γ Pt X-rays, 0.137, 0.158, 0.296, 0.308, 0.317, others between 0.1 and 1.2 e⁻ 0.032, 0.143, 0.23, 0.30 β⁺ 2.2 max	daughter Hg^{192} (HuqM57, GillL54) Ir^{191}(α, n)` (WilkG49a)
Au^{193}	15.8 h (WilkG49a) 17.5 h (EwaG57) 15.3 h (FinR52)	☢ EC, no β⁺ (lim 0.08%) (EwaG57) no α, lim 1 × 10^{-5}% (KarrM63) Δ -33 (MTW)	B chem, genet (WilkG49a) daughter Hg^{193} (GillL54, FinR52) parent Pt^{193m} (WilkG49a)	γ Pt X-rays, 0.114 (5%, complex), 0.18 (11%, complex), 0.26 (9%, doublet), 0.378 (1.4%), 0.440 (3%) e⁻ 0.034, 0.095, 0.108, 0.177	Ir^{191}(α, 2n) (WilkG49a) deuterons on Pt (WilkG49a) daughter Hg^{193} (EwaG57) protons on Pt (MarkI62)
Au^{193m}	3.9 s (FiscV55) 3.8 s (BrunnJ55)	☢ IT (FiscV55, BrunnJ55, GillL54) EC 0.03% (BrunnJ55) Δ -33 (LHP, MTW)	B genet (BrunnJ55) daughter Hg^{193m} (GillL54, BrunnJ55) parent Pt^{193m} (0.03%) (BrunnJ55)	γ Au X-rays, 0.258 (65%) e⁻ 0.019, 0.030	daughter Hg^{193m} (BrunnJ55) protons on Pt (FiscV55)
Au^{194}	39.5 h (WilkG49a) others (StefR49)	☢ EC ≈97%, β⁺ ≈3% (WilkG49a) Δ -32.21 (MTW)	A chem, excit (WilkG49a) genet energy levels (ThieM56a) daughter Hg^{194} (BrunnJ55a, MerE61a, BellL64)	β⁺ 1.49 max e⁻ 0.250, 0.315, many others between 0.02 and 2.4 γ Pt X-rays, 0.294 (12%), 0.328 (68%), 1.469 (8%), 1.596 (3%), 1.887 (4%), 2.044 (4%), many others between 0.1 and 2.4	deuterons on Pt (WilkG49a) Ir^{193}(α, 3n) (WilkG49a) protons on Pt (StefR49)
Au^{195}	183 d (HarbG63) 185 d (BonnN62) 192 d (BisA59) 199 d (BresM60) others (StefR49, WilkG49a)	☢ EC (WilkG49a) Δ -32.55 (LHP, MTW)	A chem, genet (WilkG49a) descendant Hg^{195m} (BradC54) daughter Hg^{195} (GillL54)	γ Pt X-rays, 0.099 (10%), 0.129 (1%) e⁻ 0.018, 0.028, 0.085	deuterons on Pt (WilkG49a) Ir^{193}(α, 2n) (WilkG49a) Pt^{195}(p, n) (StefR49)
Au^{195m}	30.6 s (FiscV55) others (HubeO52)	☢ IT (HubeO52a) Δ -32.23 (LHP, MTW)	B chem, genet (HubeO52a) excit (FiscV55) daughter Hg^{195m} (HubeO52a, JolyR55) not daughter Hg^{195} (HubeO53, GillL54)	γ Au X-rays, 0.261 (77%) e⁻ 0.044, 0.056, 0.180	daughter Hg^{195m} (HubeO52a, JolyR55) protons on Pt (FiscV55)
Au^{196}	6.18 d (IkeH63) others (BonnN62, WapA62, TilR63a, LingE62, BakM60, WilkG49a, StefR49, WafH48, KriR41c)	☢ EC 93.8%, β⁻ 6.2% (BergO61) β⁺ 5 × 10^{-5}% (IkeH63) others (StefR49, WilkG49a, ThieM56) Δ -31.15 (MTW)	A chem, excit (MMilE37)	β⁻ 0.259 max (6%) e⁻ 0.255, 0.277, 0.343 γ Pt X-rays, 0.333 (25%), 0.356 (94%), 0.426 (6%), 1.091 (0.2%)	Pt^{196}(d, 2n) (WapA62) Pt^{196}(p, n) (StefR49, IkeH63, MarkI62) Pt^{195}(d, n) (KriR41c, WilkG49a, StahP52) Ir^{193}(α, n) (EwbW60) Au^{197}(n, 2n) (MMilE37, WilkG49a, WapA62)

TABLE 1 (*Continued*)

Isotope Z A	Half-life	Type of decay (☢); % abundance; Mass excess ($\triangle \equiv$ M-A), MeV (C^{12}=0); Thermal neutron cross section (σ), barns	Class; Identification; Genetic relationships	Major radiations: approximate energies (MeV) and intensities	Principal means of production
$_{79}$Au^{196m}	9.7 h (BonnN62) others (KavT60, BakM60, AdemM60, VLieR59, TilR63a, WilkG49a, MMilE37)	☢ IT (WapA62a) $\triangle$ -30.56 (LHP, MTW)	A chem, excit (MMilE37, TilR63a)	γ Au X-rays, 0.148 (42%), 0.188 (32%), 0.285 (5%), 0.316 (5%) e$^-$ 0.069, 0.081, 0.094, 0.108, 0.135, 0.160 daughter radiations from Au^{196}	Pt^{196}(d, 2n) (WapA62a, VLieR59) Au^{197}(n, 2n) (MMilE37, WilkG49a, VLieR59) Au^{197}(p, pn) (TilR63a)
Au^{197}		% 100 $\triangle$ -31.17 (MTW) σ_c 98.8 (GoldmDT64)			
Au^{197m}	7.2 s (FiscV55) 7.4 s (FrauH47) 7.5 s (WieM45a)	☢ IT (WieM45a) $\triangle$ -30.76 (LHP, MTW)	A excit (WieM45a) daughter Hg^{197m} (FrauH50a, DShaA52, HavA65) daughter Pt^{197m} (PraK64, HavA65)	γ Au X-rays, 0.130 (8%), 0.279 (75%) e$^-$ 0.050, 0.117, 0.127, 0.198, 0.265	daughter Hg^{197m}, Pt^{197m} (HavA65)
Au^{198}	2.697 d (LocE53, JohaK56) 2.699 d (BellRE54, RobeJ60) 2.687 d (StarS63) 2.686 d (TobJ55) 2.704 d (KeeJ58) others (SasC56, SinW51, SilL51, DieG46, HumV51, SerL47b, SherrR41, PoolM37, WriH57)	☢ β$^-$ (MMilE37) no EC(K) lim 0.01% (BashA56) no β$^+$, lim 0.003% (MimW51) $\triangle$ -29.59 (MTW) σ_c 26,000 (GoldmDT64)	A chem, n-capt (AmaE35, MMilE37)	β$^-$ 0.962 max average β$^-$ energy: 0.32 calorimetric (ShimN56a) 0.29 calorimetric (LecM64) e$^-$ 0.329, 0.398 γ 0.412 (95%), 0.676 (1%), 1.088 (0.2%)	Au^{197}(n, γ) (AmaE35, MMilE37, PoolM37, DzhB41, SerL47b, HumV51) Pt^{198}(p, n) (StefR49, StefR48)
Au^{199}	3.15 d (BellRE55) others (WriH57, DShaA52, MMilE37, GleG64)	☢ β$^-$ (KriR41c) $\triangle$ -29.09 (MTW) σ_c ≈30 (GoldmDT64)	A chem, genet (MMilE37) daughter Pt^{199} (MMilE37, BeacL49a, MeeJ49, HillR50a)	β$^-$ 0.46 max (6%), 0.30 max γ Hg X-rays, 0.158 (37%), 0.208 (8%) e$^-$ 0.075, 0.125, 0.145	Pt^{198}(n, γ)Pt^{199}(β$^-$) (MMilE37, HahR63, LindsJ63a) Au^{197}(n, γ)Au^{198}(n, γ) (HillR50) Pt^{198}(d, n) (KriR41c)
Au^{200}	48.4 m (RoyJ59) others (ButeF52a, MauW42, GirR60)	☢ β$^-$ (SherrR41) $\triangle$ -27.3 (MTW)	B chem (SherrR41) chem, sep isotopes, excit (ButeF52a) daughter Pt^{200} (RoyL57a)	β$^-$ 2.2 max γ 0.368 (24%), 1.227 (23%), 1.593 (1%)	Hg^{202}(d, α) (GirR60) Tl^{203}(n, α) (ButeF52a) Hg^{201}(γ, p) (ButeF52a)
Au^{201}	26 m (ErdP57, ButeF52a) others (FacJ62, EutP62)	☢ β$^-$ (ButeF52a) $\triangle$ -26.2 (MTW)	B chem, excit, sep isotopes (ButeF50, ButeF52a) daughter Pt^{201} (FacJ62)	β$^-$ 1.5 max γ 0.53	Hg^{202}(γ, p) (ButeF50, ButeF52a, EutP62)
Au$^{202, 204}$	≈25 s (ButeF52a)	☢ β$^-$ or IT (ButeF52a)	E excit (ButeF52a)		Hg$^{202, 204}$(n, p) (ButeF52a)
Au^{203}	55 s (ButeF52a)	☢ β$^-$ (ButeF52a) $\triangle$ -23 (MTW)	B chem, excit, sep isotopes (ButeF52a)	β$^-$ 1.9 max γ 0.69	Hg^{204}(γ, p) (ButeF52a)
$_{80}$Hg$^{<195}$	0.7 m (RasJ53)	☢ α (RasJ53)	E chem (ThomS49, RasJ53) probably Hg^{185} or Hg^{186} (LHP)	α 5.6	deuterons on Au^{197} (RasJ53)
Hg^{185}	50 s (AlboG60)	☢ [EC] (AlboG60)	C chem, mass spect (AlboG60) parent 7 m Au^{185} (AlboG60)		Au^{197}(p, 13n) (AlboG60)
Hg^{186}	1.5 m (AlboG60)	☢ EC (AlboG60)	B chem, mass spect (AlboG60) parent Au^{186} (AlboG60)	γ Au X-rays, 0.125, 0.27, 0.35, 0.44 daughter radiations from Au^{186}	Au^{197}(p, 12n) (AlboG60)
Hg^{187}	3 m (AlboG60)	☢ EC (AlboG60) α? (KarrR63)	B chem, mass spect (AlboG60) parent Au^{187} (AlboG60)	γ Au X-rays, 0.175, 0.255, 0.40 daughter radiations from Au^{187}	Au^{197}(p, 11n) (AlboG60)
Hg^{188}	3.7 m (PofN60, AlboG60) 3.0 m (α) (KarrM63)	☢ EC (PofN60, AlboG60) α (?) (KarrM63)	B chem, mass spect (PofN60, AlboG60) parent Au^{188} (PofN60, AlboG60)	γ Au X-rays, 0.14 α 5.14 (? may be Hg^{187}) daughter radiations from Au^{188}	Au^{197}(p, 10n) (PofN60, AlboG60, KarrM63a)

TABLE 1 (*Continued*)

Isotope Z A	Half-life	Type of decay (☢); % abundance; Mass excess ($\Delta \equiv M-A$), MeV ($C^{12}=0$); Thermal neutron cross section (σ), barns	Class; Identification; Genetic relationships	Major radiations: approximate energies (MeV) and intensities	Principal means of production
$_{80}Hg^{189}$	9.6 m (AndeG61) 9 m (PofN60, AlboG60)	☢ EC, β^+ ? (PofN60, AlboG60, AndeG61) no α, lim 3 x 10^{-5}% (KarrM63)	A chem, mass spect (PofN60, AlboG60, AndeG61) parent Au^{189} (SmiW55, ChacK57) ancestor Pt^{189} (PofN60, AlboG60, AndeG61)	γ Au X-rays, 0.165, 0.24, 0.32, 0.50 daughter radiations from Au^{189}	Au^{197}(p, 9n) (PofN60, AlboG60, AndeG61)
Hg^{190}	20 m (AndeG61, JasJ64) 21 m (AlboG59, AlboG60, PofN60) others (GillL54, ChacK57, SmiW55)	☢ EC (AlboG59, AlboG60, PofN60) no β^+, lim 1% (AlboG59) no α, lim 5 x 10^{-5}% (KarrM63) Δ -31 (NDS, MTW)	A chem, mass spect (AlboG59, AndeG61, JasJ61b) parent Au^{190} (AndeG61)	γ Au X-rays, 0.14 (complex) e^- 0.015, 0.026, 0.049, 0.062, 0.076 daughter radiations from Au^{190}	Au^{197}(p, 8n) (AlboG59, AndeG61, JasJ61b, AlboG60, PofN60)
$Hg^{<191}$	90 m (GillL54)	☢ (GillL54)	F excit (GillL54)		protons on Au^{197} (GillL54)
$Hg^{<191}$	≈3 h (GillL54)	☢ (GillL54)	F excit (GillL54)	e^- 0.088	protons on Au^{197} (GillL54)
Hg^{191}	55 m (PofN60, SmiW55) 57 m (GillL54) no 12 h Hg^{191} observed (SmiW55)	☢ EC (SmiW55)	A excit (GillL54) chem, genet (SmiW55) mass spect (AndeG61a, PofN60) parent Au^{191} (SmiW55, GillL54)	γ Au X-rays, 0.26 (complex) e^- 0.170, 0.191, 0.239 daughter radiations from Au^{191}	Au^{197}(p, 7n) (GillL54, AndeG61a, PofN60)
Hg^{192}	4.8 h (JasJ61) 5.7 h (FinR52) 6.3 h (VinA55a)	☢ EC, β^+ (FinR52) β^+ <1% (JasJ61) no α, lim 4 x 10^{-6}% (KarrM63) Δ -32 (MTW)	B chem, excit (FinR52, GillL54) parent Au^{192} (FinR52, GillL54)	γ Au X-rays, 0.114 († 10), 0.157 († 20), 0.274 († 100) e^- 0.017, 0.028, 0.034, 0.039, 0.077 daughter radiations from Au^{192}	Au^{197}(p, 6n) (GillL54, HuqM57)
Hg^{193}	≈6 h (GillL54) 4 h (MalyT58)	☢ EC (GillL54) Δ -31 (MTW)	B genet (GillL54) daughter Hg^{193m} (GillL54, BrunnJ55) parent Au^{193} (GillL54, FinR52)	γ Au X-rays, 0.187, 0.574, 0.762, 0.855, 1.04, 1.08 e^- 0.025, 0.035, 0.108, 0.174 daughter radiations from Au^{193}	Au^{197}(p, 5n) (FireE52, GillL54, EwaG57)
Hg^{193m}	10.0 h (FireE52) 11 h (BrunnJ58) others (VinA55a, GillL54)	☢ EC 84%, IT 16% (GillL54) β^+ 1.5% (BrunnJ58) EC(K)/EC(L) 7.3 (BrunnJ58) no α, lim 1 x 10^{-5}% (KarrM63) Δ -31 (LHP, MTW)	B chem, excit (FireE52, GillL54) parent Hg^{193} (GillL54) parent Au^{193m} (GillL54, BrunnJ55)	γ Hg X-rays, Au X-rays, 0.218, 0.258, 0.574, many others between 0.1 and 1.6 e^- 0.020, 0.025, 0.029, 0.036, 0.087, 0.178, 0.243, many others between 0 and 1.6 daughter radiations from Hg^{193} daughter radiations from Au^{193m} included in above listing	Au^{197}(p, 5n) (FireE52, GillL54, EwaG57)
Hg^{194}	1.9 y (BellL64) 0.40 y (same activity?) (MerE61a) ≈1.6 y (BrunnJ58) 0.4 y (BrunnJ55a, MalyT58)	☢ EC(L), no EC(K) (BellL64) EC(K) (MerE61a) no β^+, lim 1% (MerE61a) Δ -32.2 (BellL64, MTW)	B chem, genet (BrunnJ55a, MerE61a, BellL64) parent Au^{194} (MerE61a, BrunnJ55a, BellL64)	γ Au X-rays daughter radiations from Au^{194}	Au^{197}(p, 4n) (BrunnJ55a, BellL64)
Hg^{194m}	0.4 s (HenrA53)	☢ [IT or EC]	E excit (HenrA53)	γ 0.048, 0.134	protons on Au and Hg (HenrA53)
Hg^{195}	9.5 h (JolyR55, BrunnJ54, HubeO53)	☢ EC (JolyR55) Δ -31 (MTW)	A chem, genet, excit (GillL54) mass spect (JunB61a) daughter Hg^{195m} (GillL54) daughter Tl^{195} (KniJD55) parent Au^{195} (GillL54) not parent Au^{195m} (HubeO53, GillL54)	γ Au X-rays, 0.20 (complex), 0.261, 0.59 (doublet), 0.780, 0.930, 1.110, 1.172 e^- 0.048, 0.058, 0.099	daughter Tl^{195} (KniJD55, JunB61a) Au^{197}(p, 3n) (TilR63a, GillL54)
Hg^{195m}	40.0 h (HubeD53, JolyR55, BrunnJ54) others (TilR63a)	☢ EC 50%, IT 50% (JolyR55, BrunnJ54) EC 52%, IT 48% (GillL54) Δ -31 (LHP, MTW)	A chem, excit (FinR52) chem, excit, genet (GillL54) mass spect (JunB61a) parent Au^{195m} (HubeO52, JolyR55) parent Hg^{195} (GillL54) not daughter Tl^{195} (KniJD55) ancestor Au^{195} (BradC54)	γ Hg X-rays, Au X-rays, 0.200 (35%), 0.261 (20%), 0.560 (20%) e^- 0.0014, 0.013, 0.022, 0.034, 0.043, 0.048, 0.053, 0.058, 0.109, 0.120, 0.180 daughter radiations from Hg^{195} daughter radiations from Au^{195m} included in above listing	Au^{197}(p, 3n) (TilR63a, GillL54)

TABLE 1 *(Continued)*

Isotope Z A	Half-life	Type of decay (☿); % abundance; Mass excess (△≡M–A), MeV (C¹²=0); Thermal neutron cross section (σ), barns	Class; Identification; Genetic relationships	Major radiations: approximate energies (MeV) and intensities	Principal means of production
$_{80}$Hg^{196}	$t_{1/2}$ (α) >1 x 10^{14} y sp act (MacfR61a)	% 0.146 (NierA50a) △ −31.84 (MTW) σ$_c$ 880 (to Hg^{197}) 25 (to Hg^{197m}) (GoldmDT64)			
Hg^{197}	65 h (HubeO51, TilR63a) others (CorkJ52, FrieG43, SherrR41, KriR40b, KriR41a)	☿ EC (FrieG43) △ −30.75 (DWitS65, MTW)	A chem, excit, cross bomb (WuC41, FrieG43) daughter Hg^{197m} (HubeO53) daughter Tl^{197} (KniJD55)	γ Au X-rays, 0.077 (18%), 0.191 (2%), 0.268 (0.15%) e$^-$ 0.064, 0.074	Au^{197}(p, n) (TilR63a) Au^{197}(d, 2n) (FrieG43, WuC41)
Hg^{197m}	24 h (BradC54, TilR63a) others (FrieG43, HubeO51, MMilE37)	☿ IT 94%, EC 6% (HavA65) others (DShaA52, JolyR55) △ −30.45 (LHP, MTW)	A n-capt (AndeEB36) chem (MMilE37) chem, excit, cross bomb (WuC41, FrieG43) parent Au^{197m} (FrauH50a, DShaA52, HavA65) parent Hg^{197} (HubeO53) not daughter Tl^{197} (KniJD55)	γ Hg X-rays, 0.134 (42%), 0.279 (7%) e$^-$ 0.051, 0.082, 0.120, 0.131, 0.152, 0.162 daughter radiations from Hg^{197} daughter radiations from Au^{197m} included in above listing	Au^{197}(p, n) (TilR63a) Au^{197}(d, 2n) (WuC41, FrieG43)
Hg^{198}		% 10.02 (NierA50a) △ 30.97 (MTW) σ$_c$ 0.02 (to Hg^{199m}) (GoldmDT64)			
Hg^{199}		% 16.84 (NierA50a) △ −29.55 (MTW) σ$_c$ 2000 (GoldmDT64)			
Hg^{199m}	43 m (SmeF65, MMilE37, HeyF37) 44 m (HoleN47a, MacD48) others (PoolM37, WuC41, SherrR41, WieM45a)	☿ IT (FrieG43) △ −29.01 (LHP, MTW)	A chem, excit (HeyF37, MMilE37) mass spect (BergI49a) not daughter Tl^{199} (BergI53)	γ Hg X-rays, 0.158 (53%), 0.375 (15%) e$^-$ 0.075, 0.144, 0.285, 0.354	Hg^{198}(d, p) (KriR40b) Pt^{196}(α, n) (SherrR41) Hg^{200}(n, 2n) (MMilE37, HeyF37) Hg^{199}(n, n) (FrieG43, WuC41, BergI49a)
Hg^{200}		% 23.13 (NierA50a) △ −29.50 (MTW) σ$_c$ <50 (GoldmDT64)			
Hg^{201}		% 13.22 (NierA50a) △ −27.66 (MTW) σ$_c$ <50 (GoldmDT64)			
Hg^{202}		% 29.80 (NierA50a) △ −27.35 (MTW) σ$_c$ 4 (GoldmDT64)			
Hg^{203}	46.9 d (EicG56) 46.6 d (GleG64) 47.9 d (CorkJ52) others (LyoW51, WilsH51, WriH57, CaliJ59, SherrR41, IngmM47b, SerL47b, MauW42)	☿ β$^-$ (FrieG43) △ −25.26 (MTW)	A excit (KriR40b) chem, excit, n-capt (WuC41, FrieG43) mass spect (SlaH49a, BergI49)	β$^-$ 0.214 max e$^-$ 0.194, 0.264, 0.275 γ 0.279 (77%)	Hg^{202}(n, γ) (FrieG43, WuC41, IngmM47b, SerL47b)
Hg^{204}		% 6.85 (NierA50a) △ −24.69 (MTW) σ$_c$ 0.4 (GoldmDT64)			
Hg^{205}	5.5 m (MauW42, KriR40b) 5.6 m (LyoW51) others (WuC41, FrieG43)	☿ β$^-$ (KriR40b) △ −22.2 (MTW)	A n-capt, excit (KriR40b, KriR42) sep isotopes, n-capt (LyoW51)	β$^-$ 1.7 max γ 0.205	Hg^{204}(n, γ) (LyoW51) Hg^{204}(d, p) (KriR40b, KriR42)
Hg^{206}	8.1 m (WolfGK64) 8.5 m (KauP62) others (NurM61)	☿ β$^-$ (NurM61) △ −20.95 (MTW)	A chem, genet (NurM61, KauP62) daughter Pb^{210} (RaD), parent Tl^{206} (NurM61, KauP62, WolfGK64)	β$^-$ [1.3 max] γ 0.31 daughter radiations from Tl^{206}	daughter Pb^{210} (NurM61, KauP62, WolfGK64) Pb^{208}(p, 3p) (KauP62)

TABLE 1 (*Continued*)

Isotope Z A	Half-life	Type of decay (☢); % abundance; Mass excess (Δ≡M−A), MeV (C¹²=0); Thermal neutron cross section (σ), barns	Class; Identification; Genetic relationships	Major radiations: approximate energies (MeV) and intensities	Principal means of production
$_{81}Tl^{191}$	10 m (ChacK60) <10 m (AndeG61a)	EC, β⁺ (ChacK60)	B chem, sep isotopes (ChacK60) chem, mass spect (AndeG61a)	γ Hg X-rays, 0.511 (γ±)	W¹⁸²(N¹⁴, 5n) (ChacK60) protons on Hg (AndeG61a)
Tl^{192}	11 m (AndeG61a) 10 m (DiaR63a)	[EC, β⁺] (AndeG61a)	B chem, mass spect (AndeG61a) excit, cross bomb (DiaR63a)	γ [Hg X-rays], 0.424, [0.511 (γ±)] e⁻ 0.341	Ta¹⁸¹(O¹⁶, 5n) (DiaR63a) C¹² on Re (DiaR63a) protons on Hg (AndeG61a)
Tl^{193}	23 m (AndeG61a) 30 m (ChacK60)	EC, β⁺ (ChacK60, AndeG61a) no α, lim 2 x 10⁻⁴% (KarrM63)	B chem, sep isotopes (ChacK60) chem, mass spect (AndeG61a)	γ Hg X-rays, 0.158, 0.169, 0.178, 0.187, 0.208, 0.216, 0.238, 0.247, 0.511 (γ±), if electrons observed by (AndeG61a) are all K-lines converted in Hg e⁻ 0.24	W¹⁸⁴(N¹⁴, 5n) (ChacK60) protons on Hg (AndeG61a)
Tl^{193m}	2.1 m (DiaR63a)	[IT] (DiaR63a)	C excit, cross bomb (DiaR63a)	γ Tl X-rays, 0.365 e⁻ 0.280	Ta¹⁸¹(O¹⁶, 4n), Re¹⁸⁵(C¹², 4n) (DiaR63a)
Tl^{194}	33.0 m (JunB60)	EC (JunB60) no α, lim 1 x 10⁻⁷% (KarrM63) Δ -26 (MTW)	A chem, mass spect, genet (JunB60) daughter Pb¹⁹⁴ (JunB60)	γ Hg X-rays, 0.427 e⁻ 0.344	protons on Hg (JunB60) daughter Pb¹⁹⁴ (JunB60)
Tl^{194m}	32.8 m (JunB60)	EC, no IT observed (JunB60)	B chem, mass spect (JunB60) not daughter Pb¹⁹⁴ (JunB60)	γ Hg X-rays, 0.097 e⁻ 0.083	protons on Hg (JunB60)
Tl^{195}	1.16 h (JunB61a) others (KniJD55, AndeG57)	EC (AndeG57) β⁺ (weak) (JunB61a) no α, lim 3 x 10⁻⁷% (KarrM63) Δ -28 (MTW)	B chem, genet (KniJD55) mass spect, genet energy levels (AndeG57) parent Hg¹⁹⁵ (KniJD55) not parent Hg¹⁹⁵ᵐ (KniJD55)	γ Hg L X-rays, others e⁻ 0.022, 0.034 β⁺ 1.8 max daughter radiations from Hg¹⁹⁵	Hg¹⁹⁶(d, 3n) (KniJD55) protons on Hg (JunB61a)
Tl^{195m}	3.5 s (AndeG57a) 3.6 s (DiaR63a)	IT (AndeG57a) Δ -28 (LHP, MTW)	B chem (AndeG57a) excit (DiaR63a) daughter Pb¹⁹⁵ (AndeG57a)	γ Tl L X-rays, 0.383 (95%) e⁻ 0.084, 0.096	daughter Pb¹⁹⁵ (AndeG57a) Re¹⁸⁷(C¹², 4n) (DiaR63a)
Tl^{196}	1.84 h (JunB60) others (AndeG58, VVijR63)	EC (AndeG55) Δ -27.2 (MTW)	A chem, genet energy levels, mass spect (AndeG58, AndeG55, AndeG57, JunB60) daughter Pb¹⁹⁶ (AndeG57)	γ Hg X-rays, 0.426 e⁻ 0.343	daughter Pb¹⁹⁶ (AndeG57, AndeG58, JunB60) protons on Hg (JunB60) Au¹⁹⁷(α, 5n) (VVijR63)
Tl^{196m}	1.41 h (JunB60)	EC 96%, IT 4% (JunB60) Δ -26.8 (LHP, MTW)	A chem, mass spect, genet energy levels (JunB60) excit (VVijR63) not daughter Pb¹⁹⁶ (JunB60)	γ Hg X-rays, 0.426, others e⁻ 0.071, 0.081, 0.107, others daughter radiations from Tl¹⁹⁶	protons on Hg (JunB60) Au¹⁹⁷(α, 5n) (VVijR63)
Tl^{197}	2.84 h (JunB61) others (KniJD55, AndeG57, AndeG55)	EC (AndeG55) Δ -28.5 (MTW, DWitS65)	A chem, excit, genet (KniJD55) mass spect, genet energy levels (AndeG55) parent Hg¹⁹⁷ (KniJD55) not parent Hg¹⁹⁷ᵐ (KniJD55)	γ Hg X-rays, 0.152, 0.426 e⁻ 0.067, 0.137 daughter radiations from Hg¹⁹⁷	Au¹⁹⁷(α, 4n) (VVijR63, KniJD55) Hg¹⁹⁸(d, 3n) (KniJD55)
Tl^{197m}	0.54 s (HenrA53) 0.55 s (SchmW65a) others (DiaR63a, AndeG57a)	IT (AndeG57a) Δ -27.9 (LHP, MTW)	A excit (HenrA53) chem (AndeG57a) excit, genet energy levels (DiaR63a)	γ Tl X-rays, 0.222 (40%), 0.385 (90%) e⁻ 0.136, 0.207, 0.219, 0.300	daughter Pb¹⁹⁷ᵐ (AndeG55, AndeG57) Au¹⁹⁷(α, 4n) (DiaR63a)
Tl^{198}	5.3 h (MicM54) others (BergI53)	EC (AndeG55) β⁺ ≈0.7% (GupR61) no α, lim 3 x 10⁻⁷% (KarrM63) Δ -27.5 (MTW)	A chem, genet energy levels (BergI53) excit (VVijR63) mass spect (MicM54) genet (JunB59, GupR61, LindgI58) daughter Pb¹⁹⁸ (JunB59, GupR61, LindgI58) descendant Po¹⁹⁸ (BrunC65a)	γ Hg X-rays, 0.412 (90%), 0.65 (40%, complex), 1.20 (21%), 1.42 (24%), 2.01 (15%), 2.45 (5%), 2.78 (2%) β⁺ 2.4 max e⁻ 0.111, 0.201, 0.317, 0.329, others	daughter Pb¹⁹⁸ (JunB59, GupR61, LindgI58) Au¹⁹⁷(α, 3n) (VVijR63) deuterons on Hg (BergI53)

TABLE 1 (*Continued*)

Isotope Z A	Half-life	Type of decay (☢); % abundance; Mass excess ($\Delta \equiv$M–A), MeV (C^{12}=0); Thermal neutron cross section (σ), barns	Class; Identification; Genetic relationships	Major radiations: approximate energies (MeV) and intensities	Principal means of production
$_{81}Tl^{198m}$	1.87 h (JunB60) 1.90 h (FiscP56) others (OrtD49, BergI53)	☢ IT 55%, EC 45% (JunB60) others (FiscP56, BergI53) Δ -27.0 (LHP, MTW)	A chem, excit (OrtD49, BergI53) mass spect (MicM54, JunB60) genet energy levels (FiscP56) daughter Pb^{198m} (NeumH50a, KarrD51)	γ Hg X-rays, Tl X-rays, 0.283 (30%), 0.412 (45%), 0.586 (35%), 0.635 (35%) e^- 0.033, 0.046, 0.175, 0.197, 0.246 daughter radiations from Tl^{198}	Au^{197}(α, 3n) (FiscP56, MicM54, BrinGO57)
Tl^{199}	7.4 h (JunB60a, MicM54) others (OrtD49)	☢ EC (OrtD49) no β^+ (IsrH51) Δ -28.5 (MTW)	A chem (KriR40b) chem, excit (OrtD49) mass spect (MicM54, JunB60a) daughter Pb^{199} (NeumH50a) not parent Hg^{199m} (BergI53) descendant Po^{199}, Po^{199m} (BrunC65a)	γ Hg X-rays, 0.158 (5%), 0.208 (12%), 0.247 (9%), 0.455 (14%) e^- 0.035, 0.125, 0.161, 0.193	Au^{197}(α, 2n) (VVijR63) Hg^{199}(d, 2n) (KriR40b)
Tl^{200}	26.1 h (JansJ62) others (HerrlC57, OrtD49, MicM54)	☢ EC (OrtD49) β^+ 0.37% (VNooB62, LHP) Δ -27.05 (MTW)	A chem, excit (OrtD49) mass spect (MicM54) daughter Pb^{200} (NeumH50a) descendant Po^{200} (BrunC65a)	γ Hg X-rays, 0.368 (88%), 0.579 (10%), 0.829 (8%), 1.21 (35%, complex), 1.364 (4%), 1.410 (1.6%), 1.517 (4%), others β^+ 1.44 max (0.06%), 1.07 max (0.3%) e^- 0.285, 0.354	deuterons on Hg (KriR40b, VNooB62, GupR60a) Au^{197}(α, n) (OrtD49) Tl^{203}(p, 4n)Pb^{200}(β^-) (SakM65)
Tl^{201}	74 h (HerrlC60) 72 h (NeumH50a) others (KriR40b)	☢ EC (NeumH50a) Δ -27.3 (MTW)	A chem, mass spect, genet (JohaB59, HerrlC60) chem, excit, cross bomb (NeumH50a) daughter Pb^{201} (NeumH50a, JohaB59, HerrlC60) descendant Po^{201}, Po^{201m} (BrunC65a)	γ Hg X-rays, 0.135 (2%), 0.167 (8%) e^- 0.016, 0.052, 0.084	daughter Pb^{201} (NeumH50a) deuterons on Hg (KriR40b, LingdI58)
Tl^{202}	12.0 d (HameH57) others (MartiHC52, WilkG50b, FajK41a)	☢ EC (KriR40b, MauW42) no β^+, β^- (WilkG50b) Δ -26.13 (MTW)	A chem, excit (KriR40b, FajK41a) daughter Pb^{202} (HuiJ54)	γ Hg X-rays, 0.439 (95%), 0.522 (0.1%), 0.961 (0.07%) e^- 0.356	Hg^{202}(d, 2n) (KriR40b) Hg^{201}(d, n), Tl^{203}(d, t) (BornP59)
<u>Tl^{203}</u>		% 29.50 (BaiK50) Δ -25.75 (MTW) σ_c 11 (GoldmDT64)			
Tl^{204}	3.81 y (LeuH62) 3.80 y (HarbG63) 3.78 y (FinR59) 3.91 y (WahA59, NilR62) 3.68 y (FlyK65a) others (EdwJ58, MerW57, TobJ55c, WyaE61, HorrD54) SpenH64)	☢ β^- 97.9%, EC 2.1% (LeuH62) β^- 97.5%, EC 2.5% (ChrisP64) others (LidL52, DMatE52) Δ -24.34 (MTW)	A chem, n-capt (FajK40) mass spect (MicM54)	β^- 0.766 max γ Hg X-rays	Tl^{203}(n, γ) (FajK40, SerL47b)
<u>Tl^{205}</u>		% 70.50 (BaiK50) Δ -23.81 (MTW) σ_c 0.11 (GoldmDT64)			
Tl^{206}	4.19 m (SargB53) 4.23 m (FajK40) others (PouA59, AlbuD51a, PoolM37, HeyF37)	☢ β^- (FajK40, KriR42) Δ -22.26 (MTW)	A n-capt (PreiP35) chem, genet (BrodE47) excit, sep isotopes (NeumH50) daughter Bi^{210} (RaE) (BrodE47) daughter Bi^{210m} (NeumH50, LevyHB54) daughter Hg^{206} (NurM61, KauP62, WolfGK64)	β^- 1.52 max γ no γ	Tl^{205}(n, γ) (PreiP35, PoolM37, HeyF37, NeumH50) daughter Bi^{210m} from Bi^{209}(n, γ) (NeumH50)
Tl^{207} (AcC″)	4.79 m (SargB53) 4.76 m (CuriM31, SargB39a) others (FajK40, BretE40, BaldG46)	☢ β^- Δ -21.01 (MTW)	A chem, genet (CuriM31) daughter Bi^{211} (AcC)	β^- 1.44 γ 0.897 (0.16%)	descendant Ac^{227} (HydE64)
Tl^{207m}	1.3 s (EccD65)	☢ IT (EccD65) Δ -19.67 (LHP, MTW)	E excit (EccD65)	γ 0.35, 1.00	Pb^{208}(t, α) (EccD65)

TABLE 1 (*Continued*)

Isotope Z A	Half-life	Type of decay (☢); % abundance; Mass excess (△≡M-A), MeV (C¹²=0); Thermal neutron cross section (σ), barns	Class; Identification; Genetic relationships	Major radiations: approximate energies (MeV) and intensities	Principal means of production
$_{81}Tl^{208}$ (ThC")	3.10 m (BaulD57) others (CuriM31)	☢ β⁻ △ -16.76 (MTW)	A chem, genet (CuriM31) daughter Bi²¹² (ThC)	β⁻ 1.80 max e⁻ 0.187, 0.423, 0.495 γ 0.511 (23%), 0.583 (86%), 0.860 (12%), 2.614 (100%)	natural source, descendant Th²²⁸ (HydE64)
Tl^{209}	2.2 m (HageF50a)	☢ β⁻ (HageF50a) △ -13.65 (MTW)	A chem, genet (HageF50a) daughter Bi²¹³ (HageF47, EnglA47, HageF50a) parent Pb²⁰⁹ (HageF47, EnglA47)	β⁻ 1.99 max e⁻ 0.03, 0.10 γ Pb X-rays, 0.12 (50%), 0.45 (100%), 1.56 (100%) daughter radiations from Pb²⁰⁹	descendant U²³³ Th²²⁹ Ac²²⁵ (HydE64)
Tl^{210} (RaC")	1.32 m (CuriM31) others (BisG50, DevoS37)	☢ β⁻; n ≈0.02% (KogA56, KogA57) △ -9.23 (MTW)	A chem, genet (CuriM31) daughter Bi²¹⁴ (RaC), parent Pb²¹⁰ (RaD)	β⁻ 2.3 max e⁻ 0.208, 0.28 γ 0.296 (80%), 0.795 (100%), 1.08 (19%, complex), 1.21 (17%), 1.31 (21%), 2.01 (7%), 2.09 (5%), 2.36 (8%), 2.43 (9%)	descendant Ra²²⁶ (HydE64)
$_{82}Pb^{194}$	11 m (JunB60)	☢ EC (JunB60)	A chem, mass spect, genet (JunB60) parent Tl¹⁹⁴, not parent Tl¹⁹⁴ᵐ (JunB60)	γ 0.204 daughter radiations from Tl¹⁹⁴	protons on Tl (JunB60)
Pb^{195}	17 m (AndeG57)	☢ EC (AndeG57)	B chem, mass spect (AndeG57) parent Tl¹⁹⁵ᵐ (AndeG57)	γ Tl X-rays, 0.39 (doublet) e⁻ 0.084, 0.096, 0.30 daughter radiations from Tl¹⁹⁵ daughter radiations from Tl¹⁹⁵ᵐ included in above listing	Tl²⁰³(p, 9n) (AndeG57)
Pb^{196}	37 m (AndeG57, SveJ61)	☢ EC (AndeG57) no α, lim 3 x 10⁻⁵% (KarrM63) △ -24 (MTW)	A chem, genet (AndeG57) chem, mass spect (SveJ61) parent Tl¹⁹⁶ (AndeG57) not parent Tl¹⁹⁶ᵐ (JunB60)	γ Tl X-rays, 0.192, 0.240, 0.253, 0.367, 0.503, others e⁻ 0.155, 0.168, others daughter radiations from Tl¹⁹⁶	Tl²⁰³(p, 8n) (AndeG57, SveJ61)
Pb^{197}		☢ [EC] △ -24 (MTW)	F [AndeG57]	γ Tl X-rays, 0.386 (doublet)	[daughter Pb¹⁹⁷ᵐ]
Pb^{197m}	42 m (AndeG55)	☢ EC 80%, IT 20% (AndeG57) no α, lim 3 x 10⁻⁴% (KarrM63) △ -24 (LHP, MTW)	A chem, mass spect (AndeG55, JunB62)	γ Tl and Pb X-rays, 0.085, 0.222, 0.234, 0.386 (doublet) e⁻ 0.069, 0.136, 0.146, 0.207, 0.219, 0.300 (doublet) daughter radiations from Pb¹⁹⁷, Tl¹⁹⁷ᵐ included in above listing	Tl²⁰³(p, 7n) (AndeG55, AndeG57)
Pb^{198}	2.4 h (JunB59, AndeG57)	☢ EC (AndeG55) no α, lim 1 x 10⁻⁷% (KarrM63) △ -26 (MTW)	A chem, mass spect (AndeG55, JohaB59, JunB59) parent Tl¹⁹⁸ (JunB59, GupR61, LindgI58)	γ Tl X-rays, 0.117 (3%), 0.173 (28%), 0.259 (8%), 0.290 (16%), 0.38 (40%, complex), 0.575 (4%), 0.649 (2%), 0.865 (6%) e⁻ 0.031, 0.088, 0.159, 0.172, 0.205, 0.270, others daughter radiations from Tl¹⁹⁸	Tl²⁰³(p, 6n) (AndeG55, JohaB59, JunB59)
Pb^{198m}	25 m (KarrD51)	☢ EC (KarrD51)	G chem, genet (KarrD51) activity not observed (AndeG57)		protons on Tl (KarrD51)
Pb^{199}	90 m (AndeG55) ≈80 m (NeumH50a)	☢ EC (NeumH50a) β⁺ (weak) (AndeG57) △ -25 (MTW)	A chem, genet (NeumH50a) chem, mass spect (AndeG55) parent Tl¹⁹⁹, daughter Bi¹⁹⁹ (NeumH50a) descendant Bi¹⁹⁹ (NeumH50a)	γ Tl X-rays, 0.353 (17%), 0.367 (80%), 0.720 (10%) e⁻ 0.267 β⁺ 2.8 max (?) daughter radiations from Tl¹⁹⁹	Tl²⁰³(p, 5n) (JohaB59, AndeG55, AndeG57)
Pb^{199m}	12.2 m (AndeG55) others (StocR56)	☢ IT (AndeG55) △ -25 (LHP, MTW)	B chem, mass spect (AndeG55) daughter Bi¹⁹⁹ (SiiA64)	γ Pb X-rays, 0.424 (20%) e⁻ 0.336, 0.409 daughter radiations from Pb¹⁹⁹	Tl²⁰³(p, 5n) (AndeG55)

TABLE 1 (*Continued*)

Isotope Z A	Half-life	Type of decay (☡); % abundance; Mass excess (Δ≡M–A), MeV (C¹²=0); Thermal neutron cross section (σ), barns	Class; Identification; Genetic relationships	Major radiations: approximate energies (MeV) and intensities	Principal means of production
₈₂Pb²⁰⁰	21.5 h (BergK55) others (JohaB59, GerhT56a, NeumH50a, BelyB61)	☡ EC (NeumH50a) Δ –26 (MTW)	A chem, genet (NeumH50a) chem, mass spect (WirB63) parent Tl²⁰⁰, daughter Bi²⁰⁰ (NeumH50a) daughter Po²⁰⁴ (KarrD51)	γ Tl X-rays, 0.109, 0.146 (doublet), 0.236, 0.26 (complex), 0.290 (doublet), 0.450, 0.605 e⁻ 0.024, 0.06, 0.133, 0.150, 0.172, 0.183, many others daughter radiations from Tl²⁰⁰	Tl²⁰³(p, 4n) (JohaB59, BashE60, WirB63)
Pb²⁰¹	9.4 h (BergK55) others (WapA54d, NeumH50a)	☡ EC (NeumH50a) β⁺ (weak) (AndeG57, BergK57) Δ –25 (MTW)	A chem, mass spect (JohaB59) chem, genet (NeumH50a) parent Tl²⁰¹, daughter Bi²⁰¹, daughter Bi²⁰¹ᵐ (NeumH50a) parent Tl²⁰¹ (JohaB59, HerrlC60) daughter Po²⁰⁵ (KarrD51)	γ Tl X-rays, 0.330, 0.361, 0.406, 0.585, 0.766, 0.907, 0.946, 1.30, 1.40, others e⁻ 0.244, 0.275, 0.316 β⁺ 0.55 max daughter radiations from Tl²⁰¹	Tl²⁰³(p, 3n) (JohaB59, LindsJ60) Tl²⁰³(d, 4n) (WapA54d)
Pb²⁰¹ᵐ	61 s (StocR56) others (FiscV55, HopN52)	☡ IT (HopN52) Δ –25 (LHP, MTW)	B chem, excit (HopN52) chem, genet (StocR56) daughter Bi²⁰¹ (StocR56)	γ Pb X-rays, 0.629 (51%) e⁻ 0.541, 0.614	daughter Bi²⁰¹ (StocR56) Tl²⁰³(p, 3n) (HopN52)
Pb²⁰²	≈3 x 10⁵ y yield (HuiJ54) others (TemD47a, NeumH50a)	☡ EC(L), no EC(K), lim 0.5% (HuiJ54) Δ –26.08 (MTW)	A chem, genet, mass spect (HuiJ54) parent Tl²⁰² (HuiJ54)	γ Tl L X-rays daughter radiations from Tl²⁰²	Tl²⁰³(d, 3n) (HuiJ54)
Pb²⁰²ᵐ	3.62 h (AstB57a) others (MaeD54a, MaeD54b)	☡ IT 90%, EC 10% (MDonJ57) Δ –23.91 (LHP, MTW)	A chem, excit (MaeD54a, MaeD54b) chem, mass spect (MDonJ57)	γ Tl, Pb X-rays, 0.390 (7%), 0.422 (90%), 0.460 (8%), 0.490 (10%), 0.658 (35%), 0.787 (45%), 0.961 (90%) e⁻ 0.115, 0.126, 0.302, 0.334, 0.699, 0.772 daughter radiations from Tl²⁰²	Tl²⁰³(d, 3n) (MaeD54a, MaeD54b)
Pb²⁰³	52.1 h (BartlA58, PersL61a) others (FajK40, TemD47a, KriR40b, FajK41a, BaldG46)	☡ EC (MauW42) Δ –24.94 (MTW)	A chem, excit (MauW42) chem, excit, cross bomb (TemD47a) genet energy levels (WapA54d) mass spect (PersL61a) daughter Bi²⁰³ (NeumH50a)	γ Tl X-rays, 0.279 (81%), 0.401 (5%), 0.680 (0.9%) e⁻ 0.193, 0.264	Tl²⁰³(d, 2n) (TemD47a)
Pb²⁰³ᵐ	6.1 s (AstB57a) others (StocR56, FiscV55, BergI55, FritA58, HopN52)	☡ IT (HopN52) Δ –24.11 (LHP, MTW)	A excit (HopN52) chem, genet (StocR56, FritA58) daughter Bi²⁰³ (StocR56, FritA58)	γ Pb X-rays, 0.825 (70%) e⁻ 0.737, 0.810	daughter Bi²⁰³ (StocR56, FritA58)
Pb²⁰⁴		% 1.40 (WhiF56) 1.36 (CollC52) 1.48 (NierA38) Δ –25.11 (MTW) σ_c 0.7 (GoldmDT64)			
Pb²⁰⁴ᵐ	66.9 m (BartlA58) 67.5 m (HerrlC56) others (MauW42, FajK41a, DVriH39, BaldG46)	☡ IT (MauW42) Δ –22.92 (LHP, MTW)	A chem (FajK41a) chem, excit, genet (TemD47a, KarrD51) mass spect (MaeD54a) daughter Bi²⁰⁴ (TemD47a, SunA50, KarrD51)	γ Pb X-rays, 0.375 (93%), 0.90 (189%, doublet) e⁻ 0.287, 0.360, 0.824, 0.897	daughter Bi²⁰⁴ (20%) (StocR58, TemD47a, SunA50, KarrD51) Tl²⁰³(d, n) (FajK41a)
Pb²⁰⁵	3.0 x 10⁷ y sp act (WingJ58)	☡ EC(L) (HuiJ56) no EC(K), lim 0.06% (WingJ58) Δ –23.77 (MTW)	A chem, genet (HuiJ56) chem, mass spect (WingJ58) daughter Bi²⁰⁵ (HuiJ56)	γ Tl L X-rays	Pb²⁰⁴(n, γ) (WingJ58)
Pb²⁰⁶		% 25.1 (CollC52) 25.2 (WhiF56) 23.6 (NierA38) Δ –23.79 (MTW) σ_c 0.03 (GoldmDT64)			
Pb²⁰⁷		% 21.7 (WhiF56) 21.3 (CollC52) 22.6 (NierA38) Δ –22.45 (MTW) σ_c 0.72 (GoldmDT64)			

TABLE 1 (*Continued*)

Isotope Z A	Half-life	Type of decay (☢); % abundance; Mass excess (△≡M–A), MeV (C¹²=0); Thermal neutron cross section (σ), barns	Class; Identification; Genetic relationships	Major radiations: approximate energies (MeV) and intensities	Principal means of production
$_{82}Pb^{207m}$	0.80 s (BendW55, HopN52, FariU58) 0.81 s (GlagV61) 0.82 s (LasJ51) others (CamE56, ReidJ54, IamP55, StelP55, VeeN56)	☢ IT (CamE50) △ −20.81 (LHP, MTW)	A excit, sep isotopes (CamE50) chem, genet (FrieG53) daughter Bi²⁰⁷ (FrieG53, CamE56, MGowF53, WapA54b) daughter Po²¹¹ᵐ (JentW54) not daughter Po²¹¹, lim 0.005% (FrieG53)	γ 0.570 (98%), 1.064 (83%) e⁻ 0.482, 0.975, 1.048	daughter Bi²⁰⁷ (FrieG53) Pb²⁰⁷(n,n'), Pb²⁰⁸(n,2n) (GlagV59) Pb²⁰⁸(γ,n) (FariU58)
$\underline{Pb^{208}}$		% 52.3 (CollC52, NierA38) 51.7 (WhiF56) △ −21.75 (MTW) σ_c 0.0005 (GoldmDT64)			
Pb^{209}	3.30 h (WapA53) others (FajK41a, KriR42, MauW42, KriR40b)	☢ β⁻ (KriR40b, FajK41a) △ −17.63 (MTW)	A chem (ThorRL37a, KriR40b) chem, sep isotopes (FajK41b) daughter Po²¹³ (HageF47, HageF50, EnglA47, MeiW49, MeiW51) daughter Tl²⁰⁹ (EnglA47, HageF47)	β⁻ 0.635 max γ no γ	Pb²⁰⁸(d,p) (Raml W59) descendant U²³³, Th²²⁹, Ac²²⁵ (HydE64) Pb²⁰⁸(n,γ) (MauW42)
Pb^{210} (RaD)	20.4 y (HarbG59) 22.0 y (RamtH64) 22.8 y (ImrL63) 21.4 y (EckW60) 19.4 y (TobJ55b) 23.3 y (PatB59) others (CuriM31)	☢ β⁻; α 1.7 × 10⁻⁶% (KauP62) α 2 × 10⁻⁶% (WolfGK64) others (NurM61) △ −14.73 (MTW)	A chem, genet (CuriM31) daughter Tl²¹⁰ (RaC'') daughter Po²¹⁴ (RaC'), parent Bi²¹⁰ (RaE); not parent Bi²¹⁰ᵐ, lim 10⁻⁴% (LevyHB54) parent Hg²⁰⁶ (NurM61, KauP62, WolfGK64)	β⁻ 0.061 max e⁻ 0.030, 0.043 Bi L X-rays, 0.047 (4%) α 3.72 daughter radiations from Bi²¹⁰, Po²¹⁰	descendant Ra²²⁶ (HydE64)
Pb^{211} (AcB)	36.1 m (SargB39a, NurM65a) 36.0 m (CuriM31)	☢ β⁻ △ −10.46 (MTW)	A chem, genet (CuriM31) daughter Po²¹⁵ (AcA); parent Bi²¹¹ (AcC)	β⁻ 1.36 max γ 0.405 (3.4%), 0.427 (1.8%), 0.702 (0.4%), 0.766 (0.6%), 0.832 (3.4%) daughter radiations from Bi²¹¹, Tl²⁰⁷, Po²¹¹	descendant Ac²²⁷ (HydE64)
Pb^{212} (ThB)	10.64 h (TobJ55a, MarinP53) others (ButtH52, CuriM31, DzhB55)	☢ β⁻ △ −7.55 (MTW)	A chem, genet (CuriM31) daughter Po²¹⁶ (ThA), parent Bi²¹² (ThC)	β⁻ 0.58 max e⁻ 0.148, 0.222 γ Bi X-rays, 0.239 (47%), 0.300 (3.2%) daughter radiations from Bi²¹², Po²¹², Tl²⁰⁸	descendant Th²²⁸ (HydE64)
Pb^{213}	10.2 m (ButeF64a)	☢ β⁻ (ButeF64a) △ −3 (MTW)	B chem, genet (ButeF64a) parent Bi²¹³ (ButeF64a)	daughter radiations from Bi²¹³, Po²¹³, Pb²⁰⁹, Tl²⁰⁹	descendant Rn²²¹ (ButeF64a)
Pb^{214} (RaB)	26.8 m (CuriM31)	☢ β⁻ (SargB33, RasF36) △ −0.15 (MTW)	A chem, genet (CuriM31) daughter Po²¹⁸ (RaA), parent Bi²¹⁴ (RaC)	β⁻ 1.03 max (6%), 0.67 max e⁻ 0.037, 0.049 γ 0.053 (≈1%), 0.242 (4%), 0.295 (19%), 0.352 (36%) daughter radiations from Bi²¹⁴, Po²¹⁴	descendant Ra²²⁶ (HydE64)
$_{83}Bi^{≤198}$	1.7 m (NeumH50a)	☢ α (TemD48)	E (TemD48) chem (NeumH50a)	α 6.2	deuterons on Pb (TemD48, NeumH50a)
$Bi^{197?}$	8.0 m (SiiA64) 7 m genet (NeumH50a)	☢ EC 99+%, α 0.05% (NeumH50a)	D chem (TemD48, NeumH50a) parent "25 m Pb" (NeumH50a) decay charac (SiiA64) formerly assigned to Bi¹⁹⁸ (NeumH50a)	α 5.81	protons on Pb (TemD48, NeumH50a)
Bi^{199}	24.4 m (SiiA64) others (NeumH50a)	☢ EC 99+%, α ≈0.01% (NeumH50a, SiiA64) △ −20 (MTW)	A chem (TemD48) chem, genet (NeumH50a, SiiA64) mass spect (SiiA64) ancestor Pb¹⁹⁹ (NeumH50a) parent Pb¹⁹⁹ᵐ (SiiA64) possible existence of 2 isomers noted by SiiA64	γ Pb X-rays α 5.53 daughter radiations from Pb¹⁹⁹, Pb¹⁹⁹ᵐ	protons on Pb (NeumH50a, TemD48, SiiA64)

TABLE 1 (*Continued*)

Isotope Z A	Half-life	Type of decay (☠); % abundance; Mass excess (Δ=M–A), MeV (C^{12}=0); Thermal neutron cross section (σ), barns	Class; Identification; Genetic relationships	Major radiations: approximate energies (MeV) and intensities	Principal means of production
$_{83}Bi^{200}$	35 m genet (NeumH50a) others (VinA55)	☠ EC (NeumH50a) Δ –20 (MTW)	B chem, genet (NeumH50a) parent Pb^{200} (NeumH50a) daughter Po^{200} (KarrD51a)		protons on Pb (NeumH50a)
Bi^{201}	1.85 h (StocR56) others (NeumH50a, VinA55)	☠ EC (NeumH50) Δ –21 (MTW)	A chem, genet (NeumH50a) chem, mass spect (SiiA64) parent Pb^{201} (NeumH50a) parent Pb^{201m} (StocR56) daughter Po^{201} (?) (KarrD51a)	γ Pb X-rays daughter radiations from Pb^{201m}, Pb^{201}, Tl^{201}	protons on Pb (NeumH50a)
Bi^{201m}	52 m (SiiA64) others (NeumH50a, VinA55)	☠ α/KX-rays 0.02% (SiiA64) EC 99+%, α 0.003% (NeumH50a)	A chem, mass spect (SiiA64) chem, genet (NeumH50a), parent Pb^{201} (NeumH50a) daughter Po^{201} (SiiA64, KarrD51a)	γ Pb X-rays α 5.28 daughter radiations from Pb^{201}, Tl^{201}	protons on Pb, Bi (SiiA64, NeumH50a)
Bi^{202}	95 m (KarrD51) others (VinA55)	☠ EC (KarrD51) Δ –21 (MTW)	A chem, genet (KarrD51) daughter Po^{202} (KarrD51)	γ Pb X-rays, 0.422, 0.961	daughter Po^{202} (KarrD51)
Bi^{203}	11.8 h (StocR60a) others (StocR56, FritA58, NeumH50a)	☠ EC (NeumH50a) β⁺ weak (NovaT58) no α, lim 6 × 10⁻⁷% (NDS) α ≈10⁻⁵% (DunlD52a) Δ –21.8 (MTW)	A chem, genet (NeumH50) parent Pb^{203} (NeumH50) parent Pb^{203m} (StocR56, FritA58) daughter Po^{203} (KarrD51) daughter At^{207} (BartoG51)	β⁺ 1.35 max e⁻ 0.045, 0.098, 0.112, 0.176, 0.737 γ Pb X-rays, 0.186 (6%), 0.264 (6%), 0.381 (9%), 0.82 (78%, complex), 1.034 (16%), 1.52 (31%, complex), 1.87 (35%, doublet) daughter radiations from Pb^{203} daughter radiations from Pb^{203m} included in above listing	Pb^{206}(p, 4n) (NovaT58a, StocR60a)
Bi^{204}	11.2 h (StocR60a) 11.6 h (WerG56) 11.0 h (FritA58) others (StocR56, TemD47a)	☠ EC, no β⁺ (TemD47a) no β⁺, lim 0.07% (StocR58) Δ –21 (MTW)	A chem, sep isotopes, cross bomb, genet (TemD47a) parent Pb^{204m} (21%) (TemD47a, SunA50, KarrD51, StocR58) daughter Po^{204} (KarrD51)	γ Pb X-rays, 0.21 (complex), 0.375, 0.671, 0.91 (complex), 0.98, 1.21 (complex), many others e⁻ 0.063, 0.075, 0.087, 0.128, 0.133, 0.161, 0.201, 0.287, 0.360, 0.583, 0.811, 0.824, 0.897, many others daughter radiations from Pb^{204m} included in above listing	Pb^{206}(p, 3n) (StocR60a) Tl^{203}(α, 3n) (StocR58) Pb^{204}(d, 2n) (TemD47a, SunA50)
Bi^{205}	15.31 d (BrunnJ61) others (FritA58, KarrD51, VinA55)	☠ EC (KarrD51) β⁺ 0.06% (PerdC62) Δ –21.07 (MTW)	A chem, genet, sep isotopes (KarrD51) daughter Pb^{205} (KarrD51) daughter At^{209} (BartoG51) parent Pb^{205} (HuiJ56)	β⁺ 0.98 max e⁻ 0.011, 0.023, others γ Pb X-rays, 0.26 (3%, complex), 0.51 (4%, complex), 0.57 (14%, complex), 0.703 (28%), 0.911 (4%), 0.988 (17%), 1.044 (8%), 1.615 (4%), 1.766 (27%), 1.864 (6%), 1.906 (2%)	Pb^{206}(d, 3n) (HerrlC61, StocR60, Bergl62, BonaE62) Bi^{209}(p, 5n)Po^{205}(EC) (BonaE62)
Bi^{206}	6.243 d (BrunnJ61) others (ArbE57, AlbuD51, KriR40b)	☠ EC (LutA44, AlbuD51) β⁺ 8 × 10⁻⁴% (PerdC62) Δ –20.18 (MTW)	A chem, sep isotopes (FajK41b, TemD47a) genet energy levels (AlbuD54a, StelP55b) daughter Po^{206} (TemD47) daughter At^{210} (NeumH50b)	γ Pb X-rays, 0.184 (21%), 0.343 (26%), 0.398 (10%), 0.497 (18%), 0.516 (46%), 0.538 (34%), 0.803 (99%), 0.880 (72%), 0.895 (19%), 1.019 (8%), 1.099 (13%), 1.596 (8%), 1.720 (36%) e⁻ 0.096, 0.168, 0.255	Pb^{206}(d, 2n) (FajK41b, WieR63)
Bi^{207}	30.2 y (HarbG59) 28 y (SosJ59) 38 y (AppE61) others (AlbuD55, NeumH51)	☠ EC (GermL50, NeumH51) Δ –20.04 (MTW)	A chem, genet (MGowF53a) daughter At^{211} (NeumH51) parent Pb^{207m} (MGowF53, FrieG53, WapA54b, CamE56)	γ Pb X-rays, 0.570 (98%), 1.063 (77%), 1.771 (9%) e⁻ 0.482, 0.975, 1.048 daughter radiations from Pb^{207m} included in above listing	Pb(d, xn), daughter At^{211} (HydE64)
Bi^{208}	3.68 × 10⁵ y sp act, mass spect (HalpJ64) others (RoyJ58, MillC59)	☠ EC, no β⁺ lim 0.3% (MillC59) Δ –18.88 (MTW)	B chem (NeumH51) excit, genet energy levels (RoyJ58, MillC59)	γ Pb X-rays, 2.614 (100%)	Bi^{209}(n, 2n) (RoyJ58, HalpJ64)
$\underline{Bi^{209}}$	>2 × 10¹⁸ y sp act (HinE58) 2 × 10¹⁷ y sp act (RieW52, PorsW56) others (FaraH51a)	☠ no α (HinE58) α (FaraH51, PorsW56) % 100 (NierA38) Δ –18.26 (MTW) σ_c 0.015 (to Bi^{210}) 0.019 (to Bi^{210m}) (GoldmDT64)		α ? 3.0	

TABLE 1 (*Continued*)

Isotope Z A	Half-life	Type of decay (☢); % abundance; Mass excess (Δ≡M–A), MeV (C¹²=0); Thermal neutron cross section (σ), barns	Class; Identification; Genetic relationships	Major radiations: approximate energies (MeV) and intensities	Principal means of production
$_{83}Bi^{210}$ (RaE)	5.013 d (RobeJ56) others (LocE53, BegF52, SiegK47, CuriM31, HoleN45, TemD47a, SerL47b, LivJ36, CorkJ40, HurD40)	☢ β^- 99+%, α 1.3 x 10^{-4}% (KauP62) others (NurM61, BrodE47) Δ -14.79 (MTW)	A chem, genet (CuriM31) daughter Pb^{210} (RaD), parent Po^{210} (RaF); parent Tl^{206} (BrodE47)	β^- 1.160 max α 4.69 (5 x 10^{-5}%), 4.65 (7 x 10^{-5}%) γ Po X-rays (weak)	Bi^{209}(n, γ) (SiegK47b) descendant Ra^{226} (HydE64)
Bi^{210m}	≈2.6 x 10^6 y yield (HugD53)	☢ α 99.6%, β^- 0.4% (LevyHB54) Δ -14.52 (LHP, MTW)	A chem, genet (NeumH50) chem, mass spect (LevyHB54) parent Po^{210} (RaF) (0.4%), parent Tl^{206} (99.6%) (LevyHB54) not daughter Pb^{210}, lim 10^{-4}% (LevyHB54) others (NeumH50)	α 4.96 (58%), 4.92 (36%), 4.57 (6%) γ 0.262 (45%), 0.30 (23%), 0.34, 0.61 daughter radiations from Tl^{206}	Bi^{209}(n, γ) (NeumH50)
Bi^{211} (AcC)	2.16 m (CuriM31) 2.15 m (SpiesF54) 2.13 m (NurM65a)	☢ α 99+%, β^- 0.27% (NurM65a) α 99+%, β^- 0.29% (GiaM62a) Δ -11.84 (MTW)	A chem, genet (CuriM31) daughter Pb^{211} (AcB), parent Po^{211} (AcC'), parent Tl^{207} (AcC"); daughter At^{215} (KarlB44)	α 6.62 (84%), 6.28 (16%) γ 0.351 (14%) e^- 0.265 daughter radiations from Tl^{207}, Po^{211}	descendant Ac^{227} (HydE64)
Bi^{212} (ThC)	60.60 m (AppK61) 60.5 m (CuriM31)	☢ β^- 64.0%, α 36.0% (WalkJ65) β^- 64.2%, α 35.8% (BertG62, BertG60) others (SchupG60, BarkS61, RiceP58a, SenF56, MarinP53, FlaF62, ProsD58, FerrJ61, KovAF38) Δ -8.13 (MTW)	A chem, genet (CuriM31) daughter Pb^{212} (ThB), parent Po^{212} (ThC') and Tl^{208} (ThC"); daughter At^{216} (KarlB43a, GhiA48, MeiW51)	β^- 2.25 max e^- 0.025, 0.036 α 6.09 (10%), 6.05 (25%) γ Tl X-rays, 0.040 (2%), 0.288 (0.5%), 0.46 (0.8%, complex), 0.727 (7%), 0.785 (1.1%), 1.620 (1.8%) daughter radiations from Tl^{208}, Po^{212}	descendant Th^{228} (HydE64)
Bi^{213}	47 m (HageF47) 46 m (EnglA47)	☢ β^- 97.8%, α 2.2% (GraeG64, ValliK64) Δ -5.24 (MTW)	A chem, genet (EnglA47, HageF47) daughter At^{217}, parent Po^{213} (HageF47, EnglA47, HageF50) parent Tl^{209} (HageF50a, HageF47, EnglA47) daughter Pb^{213} (ButeF64a)	β^- 1.39 max γ 0.437 α 5.87 daughter radiations from Po^{213}, Pb^{209}, Tl^{209}	descendant U^{233}, Th^{229}, Ac^{225} (HydE64)
Bi^{214} (RaC)	19.7 m (CuriM31) 19.9 m (DaniH56)	☢ β^- 99+% (CuriM31) α 0.021% (WaleR60) Δ -1.19 (MTW)	A chem, genet (CuriM31) daughter Pb^{214} (RaB), daughter At^{218}, parent Po^{214} (RaC'), parent Tl^{210} (RaC"); descendant Fr^{222} (HydE50a, HydE51a)	β^- 3.26 max γ 0.609 (47%), 0.769 (5%), 0.935 (3%), 1.120 (17%), 1.238 (6%), 1.378 (5%), 1.40 (4%, complex), 1.509 (2%), 1.728 (3%), 1.764 (17%), 1.848 (2%), 2.117 (1%), 2.204 (5%), 2.445 (2%) α 5.51 (0.008%), 5.45 (0.012%) daughter radiations from Po^{214}	descendant Ra^{226} (HydE64)
Bi^{215}	7 m (NurM65a) 8 m (HydE53)	☢ β^- (HydE53) Δ 1.7 (MTW)	A chem, genet (HydE53) daughter At^{219}, parent Po^{215} (AcA) (HydE53)	daughter radiations from Po^{215}, Po^{211}	descendant Ac^{227}, natural source (HydE53, HydE64)
$_{84}Po^{193}$	short (SiiA65b)	☢ α (SiiA65b)	E excit, decay charac (SiiA65b)	α 7.0	F^{19} on Re (SiiA65b)
Po^{194}	0.5 s (SiiA65b) others (TovP58)	☢ α (SiiA65b)	B excit, decay charac (SiiA65b)	α 6.85	F^{19} on Re (SiiA65b)
Po^{195}	3 s (SiiA65b) others (TovP58)	☢ α (SiiA65b)	B excit, decay charac (SiiA65b)	α 6.63	F^{19} on Re (SiiA65b)
Po^{195m}	1.4 s (SiiA65b)	☢ α (SiiA65b)	B excit, decay charac (SiiA65b)	α 6.72	F^{19} on Re (SiiA65b)
Po^{196}	6 s (SiiA65b) 4 s (TovP58)	☢ α (SiiA65b, TovP58)	B excit, decay charac (TovP58, SiiA65b) formerly assigned to Po^{193} (TovP58)	α 6.53	Bi^{209}(p, 14n) (TovP58) F^{19} on Re (SiiA65b)

TABLE 1 (*Continued*)

Isotope Z A	Half-life	Type of decay (☢); % abundance; Mass excess ($\triangle$=M–A), MeV (C^{12}=0); Thermal neutron cross section (σ), barns	Class; Identification; Genetic relationships	Major radiations: approximate energies (MeV) and intensities	Principal means of production
$_{84}Po^{197}$	54 s (SiiA65b) 58 s (BrunC65a)	☢ α (SiiA65b, BrunC65a)	B excit (SiiA65b) chem (BrunC65a)	α 6.30	F^{19} on Re (SiiA65b) Bi^{209}(p, 13n) (BrunC65a)
Po^{197m}	25 s (SiiA65b) 29 s (BrunC65a) others (TovP58, AttH59a)	☢ α (SiiA65b, BrunC65a)	B decay charac (TovP58) chem (AttH59a, BrunC65a) excit (SiiA65b)	α 6.39	Bi^{209}(p, 13n) (BrunC65a, TovP58) F^{19} on Re (SiiA65b) $Ne^{20,22}$ on W (AttH59a)
Po^{198}	1.7 m (SiiA65b, BrunC65a, BrunC64) 1.8 m (AttH59a, AttH59) others (AttH56)	☢ α >34% (BrunC65a)	B chem (AttH56, AttH59a) excit (SiiA65b) chem, genet (BrunC65a) ancestor Tl^{198} (BrunC65a) formerly assigned to Po^{196} (AttH56, AttH59a, AttH59, BrunC64)	α 6.16	Bi^{209}(p, 12n) (BrunC65a) C^{12} on Pt (AttH59) F^{19} on Re (SiiA65b) $Ne^{20,22}$ on W (AttH59a)
Po^{199}	5.0 m (TieE65) 5.2 m (BrunC65a) others (RosS54b, AttH59, BrunC64)	☢ EC 97.3%, α 2.7% (BrunC65a)	A chem (RosS54b) chem, mass spect (TieE65) chem, genet (BrunC65a) ancestor Tl^{199} (BrunC65a) formerly assigned to Po^{198} (RosS54b, AttH59, BrunC64)	α 5.94	Bi^{209}(p, 11n) (BrunC65a, TieE65)
Po^{199m}	4.2 m (SiiA65b) 4.1 m (TieE65, BrunC65a) others (RosS54b, AttH59, AttH59a, BrunC64)	☢ EC 74%, α 26% (BrunC65a)	A chem (RosS54b) excit (SiiA65b) chem, mass spect (TieE65) chem, genet (BrunC65a) ancestor Tl^{199} (BrunC65a) formerly assigned to Po^{197} (RosS54, AttH59, AttH59a, BrunC64)	α 6.05	Bi^{209}(p, 11n) (TieE65, BrunC65a) F^{19} on Re (SiiA65b)
Po^{200}	10.5 m (HoffR63) 11.4 m (SiiA65b, TieE65, BrunC65, BrunC65a) others (KarrD51a, AttH59, ForW61a, RosS54b, BrunC64)	☢ EC 88%, α 12% (BrunC65a) $\triangle$ –16 (MTW)	A chem (KarrD51a) chem, mass spect (ForW61a, TieE65) parent Bi^{200} (KarrD51a) daughter At^{200} (HoffR63) ancestor Tl^{200} (BrunC65a) formerly assigned to Po^{199} (RosS54b, AttH59, ForW61a, BelyB61, BelyB62, BrunC64)	α 5.86	C^{12} on Pt (ForW61a, AttH59, BrunC65) Au^{197}(C^{12}, 9n)At^{200}(EC) (HoffR63) Bi^{209}(p, 10n) (BrunC64, BrunC65, TieE65)
Po^{201}	15.1 m (TieE65) 15 m (HoffR63) others (ForW61a, BelyB61, AttH59, BrunC65a, BrunC65, KarrD51a, BrunC64)	☢ EC 98.9%, α 1.1% (BrunC65a) EC 99.2%, α 0.8% (BelyB61, BelyB62) $\triangle$ –16 (MTW)	A chem, genet (KarrD51a, SiiA64) chem, mass spect (ForW61a, TieE65) parent Bi^{201m} (SiiA64, KarrD51a) parent Bi^{201} (?) (KarrD51a) daughter At^{201} (HoffR63) ancestor Tl^{201} (BrunC65a)	α 5.68 daughter radiations from Bi^{201m}	Bi^{209}(p, 9n) (BrunC64, BrunC65a, TieE65) C^{12} on Pt (AttH59, ForW61a) daughter At^{201} (HoffR63)
Po^{201m}	8.9 m (TieE65) 9 m (HoffR63, BrunC65a, BrunC65) others (BrunC65, RosS54b)	☢ α 3%, EC 97% (BrunC65a) α (RosS54b, HoffR63)	A chem (RosS54b) excit, decay charac (HoffR63) chem, mass spect (TieE65) ancestor Tl^{201} (BrunC65a) formerly assigned to Po^{200} (RosS54b, AttH59, ForW61a, BelyB61, BrunC64)	α 5.78	B^{209}(p, 9n) (TieE65, BrunC65a) C^{12} on Pt (BrunC65)
Po^{202}	45 m (BelyB61, HoffR63, TieE65) others (StonA57, RosS54b, BurcW54, AttH59, ForW61a, BrunC64, BrunC65, BrunC65a)	☢ EC 98%, α 2% (StonA57) $\triangle$ –18 (MTW)	A chem, genet, excit (KarrD51) chem, genet, mass spect (ForW61a, ForW61) parent Bi^{202} (KarrD51) daughter At^{202} (ForW61, HoffR63) daughter Rn^{206} (StonA57, MomF55a)	α 5.58 daughter radiations from Bi^{202}, Pb^{198}	Bi^{209}(p, 8n) (BrunC64, BrunC65a) C^{12} on Pt (ForW61a, BrunC65) Au^{197}(C^{12}, 7n)At^{202}(EC) (ForW61, HoffR63)
Po^{203}	42 m (BellRE56) 47 m (KarrD51)	☢ EC 99+%, α 0.02% (BelyB62, BelyB61) $\triangle$ –17 (MTW)	A chem, genet (ForW61, KarrD51) parent Bi^{203} (KarrD51) daughter At^{203} (ForW61)	α 5.49 daughter radiations from Bi^{203}	Bi^{209}(p, 7n) (BellRE56, KarrD51) Au^{197}(C^{12}, 6n)At^{203}(EC) (ForW61)

TABLE 1 (*Continued*)

Isotope Z A	Half-life	Type of decay (☢); % abundance; Mass excess ($\triangle$≡M–A), MeV (C¹²=0); Thermal neutron cross section (σ), barns	Class; Identification; Genetic relationships	Major radiations: approximate energies (MeV) and intensities	Principal means of production
$_{84}$Po^{204}	3.6 h (ForW61a) others (BelyB61, KarrD51, RossS54b, BurcW56)	☢ EC 99+%, α 0.6% (BelyB63) $\triangle$ –18 (MTW)	A chem, genet (KarrD51) daughter Rn^{208} (MomF55a) parent Bi^{204}, parent Pb^{200} (KarrD51) daughter At^{204} (ThorP64)	α 5.38 daughter radiations from Bi^{204}	Bi^{209}(p, 6n) (AxeS61) Au^{197}(C^{12}, 5n) At^{204}(EC) (HoffR63, ForW61, LatR61, ThorP64) Pt^{196}(C^{12}, 4n) (AttH59, ForW61a) alphas on Pb (KarrD51)
Po^{205}	1.8 h (BellRE56) others (KarrD51)	☢ EC 99+%, α 0.07% (HallK51) $\triangle$ –18 (MTW)	A chem, genet, sep isotopes, excit (KarrD51) chem, mass spect (ForW61a) parent Bi^{205}, parent Pb^{201} (KarrD51) daughter At^{205} (BartoG51)	α 5.25	Bi^{209}(p, 5n) (BellRE56, AxeS61) Pb^{204}(α, 3n) (KarrD51)
Po^{206}	8.8 d (ArbE57, JohnW56) others (TemD47, BarabS57, BurcW54)	☢ EC 95%, α 5% (MomF55a) no β$^+$, lim 0.1% (ArbE57) others (TemD47) $\triangle$ –18.33 (MTW)	A chem, genet, sep isotopes (TemD47) chem, mass spect (ForW61a) parent Bi^{206} (TemD47) daughter Rn^{210} (MomF55a, MomF52) daughter At^{206} (ThorP64)	γ Bi X-rays, 0.286 (↑ 35), 0.338 (↑ 40), 0.51 (↑ 100, complex), 0.807 (↑ 60), 1.02 (↑ 85, complex) e$^-$ 0.045, 0.196, 0.248 α 5.22 (5%) daughter radiations from Bi^{206}	Bi^{209}(p, 4n) (AxeS61) Pb^{204}(α, 2n) (TemD47) Pb^{206}(α, 4n) (JohnW56)
Po^{207}	5.7 h (BellRE56, TemD47) 6.2 h (JohnW56)	☢ EC 99+%, α ≈0.01% (TemD47) β$^+$ 0.5% (ArbE58a) $\triangle$ –17.14 (MTW)	A chem, excit, sep isotopes (TemD47) chem, genet (StonA56) daughter Rn^{211} (StonA56) daughter At^{207} (BartoG51)	γ Pb X-rays, 0.25 (↑ 5), 0.35 (↑ 4), 0.41 (↑ 13), 0.74 (↑ 36), 0.95 (↑ 84), 1.15 (↑ 6), 1.37 (↑ 4), 2.06 (↑ 1.6), others, all γ rays complex e$^-$ 0.159, 0.255, 0.315, 0.652, 0.902, many others β$^+$ 1.14 max α 5.11	Bi^{209}(p, 3n) (BellRE56) Pb^{206}(α, 3n) (JohnW56)
Po^{207m}	2.8 s (HargC62)	☢ IT (HargC62) $\triangle$ –15.75 (LHP, MTW)	B excit, critical abs (HargC62)	γ Po X-rays, 0.26 (42%), 0.31 (40%), 0.82 (100%) e$^-$ 0.22, 0.24	Bi^{209}(p, 3n) (HargC62)
Po^{208}	2.93 y (TemD50)	☢ α (TemD47) EC ≈0.006% (AsaF57a) $\triangle$ –17.47 (MTW)	A chem, excit, sep isotopes (TemD47) chem, mass spect (ForW61) daughter Rn^{212}, daughter At^{208} (HydE50, MomF52)	α 5.11 γ Bi X-rays, 0.285 (0.003%), 0.60 (0.006%, complex)	Bi^{209}(d, 3n) (RamlW59) Bi^{209}(p, 2n) (AndrC56)
Po^{209}	103 y sp act (AndrC56)	☢ α 99+%, EC ≈0.5% (PerlmI50, AsaF57a) $\triangle$ –16.37 (MTW)	A chem, excit (KellE49) daughter At^{209} (BartoG51)	α 4.88 (99%) γ Bi X-rays, 0.261 (0.4%, complex), 0.91 (0.5%) e$^-$ 0.173	Bi^{209}(d, 2n) (RamlW59) Bi^{209}(p, n) (AndrC56)
Po^{210} (RaF)	138.40 d (EicJ54) others (CurtM53, GinD53, BeamW49, TemD47, HurD40, CorkJ40, CuriM31)	☢ α; β stable (cons energy) $\triangle$ –15.95 (MTW) σ_c <0.03 (to Po^{211}), <0.0005 (to Po^{211m}) (GoldmDT64)	A chem, genet (CuriM31) daughter Bi^{210} (RaE); daughter Bi^{210m} (0.4%) (LevyH54) daughter At^{210} (KellE49, BartoG51)	α 5.305 (100%) γ 0.803 (0.0011%)	daughter Bi^{210} from natural source or Bi^{209}(n, γ) Bi^{210}(β$^-$) (HydE64)
Po^{211} (AcC')	0.52 s (SpiesF54, LeiR51) others (TovP58, WinnM54a)	☢ α; β stable (cons energy) (ForB58) $\triangle$ –12.43 (MTW)	A genet (CuriM31) daughter Bi^{211} (AcC); daughter At^{211} (CorsD40, CorsD40a) daughter Rn^{215} (MeiW52) not parent Pb^{207m}, lim 0.005% (FrieG53) not daughter Po^{211m}, lim 1% (JentW54)	α 7.45 (99%) γ 0.570 (0.5%), 0.90 (0.5%)	descendant Ac^{227} (HydE64)
Po^{211m}	25 s (JentW54, SpiesF54, KarnV62) others (WinnM54a)	☢ α (SpiesF54) $\triangle$ –11.00 (LHP, MTW)	A chem, excit (SpiesF54) genet energy levels (JentW54) parent Pb^{207m} (JentW54) not parent Po^{211}, lim 1% (JentW54) not daughter At^{211}, lim 0.01% (SpiesF54)	α 8.88 (7%), 7.28 (91%) γ 0.570 (92%), 1.063 (77%) e$^-$ [0.482, 0.975, 1.048]	Pb^{208}(α, n) (SpiesF54) Bi^{209}(α, pn) (PerlmI62)

TABLE 1 (*Continued*)

Isotope Z A	Half-life	Type of decay (☢); % abundance; Mass excess (Δ=M–A), MeV (C¹²=0); Thermal neutron cross section (σ), barns	Class; Identification; Genetic relationships	Major radiations: approximate energies (MeV) and intensities	Principal means of production
$_{84}Po^{212}$ (ThC')	3.04×10^{-7} s delay coinc (BunyD49) others (FlaF62, HillJ48, JelJ48, VNamF49, DunwJ39, BradH43, HayaT53)	☢ α; β stable (cons energy) (ForB58) Δ –10.37 (MTW)	A genet (CuriM31) daughter Bi²¹² (ThC); daughter Rn²¹⁶ (MeiW49, MeiW51)	α 8.78 (100%); also long range α's following decay of Bi²¹² parent	descendant Th²²⁸ (HydE64)
Po^{212m}	45 s (PerlmI62) others (KarnV62)	☢ α, no IT, lim 1.5% (PerlmI62) Δ –7.44 (LHP, MTW)	A chem, cross bomb, genet energy levels (PerlmI62)	α 11.65 (97%) γ 0.57 (2%), 2.61 (2.6%)	Bi²⁰⁹(α, p), Pb²⁰⁸(B¹¹, Li⁷) (PerlmI62)
Po^{213}	4.2×10^{-6} s delay coinc (JelJ48)	☢ α (HageF47, EnglA47) Δ –6.66 (MTW)	A genet (HageF47, EnglA47) daughter Bi²¹³, parent Pb²⁰⁹ (HageF47, EnglA47, HageF50) daughter Rn²¹⁷, parent Pb²⁰⁹ (MeiW49, MeiW51)	α 8.38	daughter Bi²¹³ (HydE64)
Po^{214} (RaC')	1.64×10^{-4} s delay coinc (DobT61, DarG50) others (OgiK60, BallR53, DunwJ39, RotJ41a, WardAG42, JacoJ43, LunA47, BunyD48, RowS47)	☢ α; β stable (cons energy) (ForB58) Δ –4.47 (MTW)	A genet (CuriM31) daughter Bi²¹⁴ (RaC), parent Pb²¹⁰ (RaD) daughter Rn²¹⁸ (StuM48)	α 7.69 (100%); also long range α's, principally 9.06 (0.0022%), following decay of Bi²¹⁴ parent γ 0.799 (0.014%)	descendant Ra²²⁶, from natural source descendant U²³⁰ (HydE64)
Po^{215} (AcA)	1.778×10^{-3} s delay coinc (VolY61) others (WardAG42)	☢ α 99+%, β⁻ 0.00023% (AviP50) α 99+%, β⁻ 0.0005% (KarlB44) Δ –0.52 (MTW)	A genet (CuriM31) daughter Rn²¹⁹ (An), parent Pb²¹¹ (AcB); parent At²¹⁵ (KarlB44) daughter Bi²¹⁵ (HydE53)	α 7.38 (100%) daughter radiations from Pb²¹¹, etc.	descendant Ac²²⁷, from Ra²²⁶(n, γ) Ra²²⁷)β⁻), or natural source (HydE64)
Po^{216} (ThA)	0.145 s (DiaH63) others (WardAG42)	☢ α; β stable (cons energy) (ForB58) others (KarlB43a) Δ 1.78 (MTW)	A genet (CuriM31) daughter Rn²²⁰ (Tn), parent Pb²¹² (ThB)	α 6.78 (100%) daughter radiations from Pb²¹², etc.	descendant Th²²⁸ (HydE64)
Po^{217}	<10 s (MomF56)	☢ α (MomF56) no β⁻, lim 0.1% (ValliK64) Δ 6 (MTW)	B genet (MomF56) daughter Rn²²¹ (MomF56, MomF52)	α 6.55	daughter Rn²²¹ (MomF52)
Po^{218} (RaA)	3.05 m (CuriM31)	☢ α 99+%, β⁻ 0.0185% (WaleR59a) others (HieF52) Δ 8.38 (MTW)	A chem, genet (CuriM31) daughter Rn²²² (Rn), parent Pb²¹⁴ (RaB); parent At²¹⁸ (KarlB43)	α 6.00 (100%) daughter radiations from Pb²¹⁴, Bi²¹⁴, Po²¹⁴	descendant Ra²²⁶, from natural source (HydE64)
$_{85}At^{200}$	0.9 m (HoffR63) others (BartoG51)	☢ α (HoffR63) α, EC (BartoG51)	B chem, excit (BartoG51) chem, excit, genet (HoffR63) parent Po²⁰⁰ (HoffR63)	α 6.47, 6.42	Au¹⁹⁷(C¹², 9n) (HoffR63)
At^{201}	1.5 m (HoffR63) others (BartoG51)	☢ α, EC (HoffR63)	A chem, excit, genet (HoffR63) parent Po²⁰¹ (HoffR63) daughter Fr²⁰⁵ (GrifR64)	α 6.35 daughter radiations from Bi¹⁹⁷, Po²⁰¹	Au¹⁹⁷(C¹², 8n) (ThomT62)
At^{202}	3.0 m (LatR61, HoffR63) others (ForW61)	☢ EC 88%, α 12% (LatR61) Δ –10 (MTW)	A chem, mass spect (ForW61) chem, excit, genet (HoffR63) parent Po²⁰² (ForW61, HoffR63)	α 6.23 (4.3%), 6.12 (7.7%) daughter radiations from [Bi¹⁹⁸], Po²⁰²	Au¹⁹⁷(C¹², 7n) (ThomT62)
At^{203}	7.4 m (LatR61, HoffR63) others (ForW61, BartoG51, BurcW56)	☢ EC 86%, α 14% (LatR61) Δ –11 (MTW)	A chem, excit (BartoG51, MillJF50) chem, mass spect (ForW61) parent Po²⁰³ (ForW61)	α 6.09 daughter radiations from Po²⁰³, Bi¹⁹⁹, etc.	Au¹⁹⁷(C¹², 6n) (ThomT62)
At^{204}	9.3 m (LatR61, HoffR63) 8.9 m genet (ThorP64) others (ForW61)	☢ EC 95.5%, α 4.52% (LatR61) Δ –11 (MTW)	A chem, mass spect (ForW61) chem, genet (ThorP64) chem, excit (HoffR63) parent Po²⁰⁴ (ThorP64)	α 5.95 [daughter radiations from Po²⁰⁴, Bi²⁰⁰]	Au¹⁹⁷(C¹², 5n) (HoffR63, ForW61, LatR61) Bi²⁰⁹(α, 9n) (ThorP64)

TABLE 1 (*Continued*)

Isotope Z A	Half-life	Type of decay (❤); % abundance; Mass excess (△≡M−A), MeV (C¹²=0); Thermal neutron cross section (σ), barns	Class; Identification; Genetic relationships	Major radiations: approximate energies (MeV) and intensities	Principal means of production
₈₅At²⁰⁴	≈25 m genet (BartoG51)	❤ EC (BartoG51)	G chem, excit, genet (BartoG51) activity not observed (ThorP64, · LatR61)		alphas on Bi²⁰⁹ (BartoG51)
At²⁰⁵	26.2 m (HoffR63, LatR61) others (BartoG51, BurcW54, ForW61, BurcW56)	❤ EC 82%, α 18% (LatR61) △ −13 (MTW)	A chem, mass spect (ForW61) chem, excit, genet (BartoG51, MillJF50) parent Po²⁰⁵ (BartoG51) daughter Fr²⁰⁹ (GrifR64)	α 5.90 daughter radiations from Po²⁰⁵ Bi²⁰¹, Pb²⁰¹, Pb²⁰¹ᵐ, Tl²⁰¹	Au¹⁹⁷(C¹², 4n) (ThomT62) Bi²⁰⁹(α, 8n) (BartoG51) Au¹⁹⁷(N¹⁴, 6n)[Rn²⁰⁵] (EC) (HoffR63, ForW61)
At²⁰⁶	32.8 m (ThorP64) 29.5 m (LatR61) 31 m (HoffR63)	❤ α ≈88%, EC ≈12% (LatR61) △ −12 (MTW)	A chem, mass spect (ForW61) chem, genet (ThorP64) parent Po²⁰⁶ (ThorP64)	α 5.70 (88%) γ Bi X-rays, Po X-rays, 0.068 (10%) e⁻ 0.052, 0.064 daughter radiations from Bi²⁰² Po²⁰⁶	Au¹⁹⁷(C¹², 3n) (ForW61, LatR61, HoffR63) Au¹⁹⁷(N¹⁴, 5n)Rn²⁰⁶(EC) (HoffR63) Bi²⁰⁹(α, 7n) (ThorP64)
At²⁰⁶	2.9 h (StonA56) 2.6 h (BartoG51)	❤ EC (BartoG51)	G chem, excit, genet (BartoG51) activity not observed (ThorP64)		alphas on Bi (BartoG51)
At²⁰⁷	1.8 h (BurcW54, StonA57, ForW61) 2.0 h (BartoG51)	❤ EC ≈90%, α ≈10% (BartoG51, TemD48a) △ −13.41 (LHP, MTW)	A chem, excit, genet (TemD48a, BartoG51) parent Po²⁰⁷, parent Bi²⁰³ (BartoG51) daughter Rn²⁰⁷ (BurcW54, StonA57) daughter Fr²¹¹ (GrifR64)	α 5.76 daughter radiations from Po²⁰⁷ Bi²⁰³, Pb²⁰³	Bi²⁰⁹(α, 6n) (TemD48a, BartoG51) Au¹⁹⁷(N¹⁴, 4n)Rn²⁰⁷(EC) (HoffR63)
At²⁰⁸	1.6 h (StonA56, ForW61) 1.7 h (BartoG51)	❤ EC 99+%, α 0.5% (HydE50) △ −12 (MTW)	A chem, genet (HydE50, ThorP64) chem, mass spect (ForW61) daughter Fr²¹², parent Po²⁰⁸ (HydE50, MomF52)	γ Po X-rays, 0.18 (25%), 0.25, 0.66 (100%) α 5.65 daughter radiations from Bi²⁰⁴	Bi²⁰⁹(α, 5n) (ThorP64)
At²⁰⁸	6.3 h genet (BartoG51)	❤ EC (BartoG51)	G chem, excit, genet (BartoG51) activity not observed (ThorP64)		alphas on Bi²⁰⁹ (BartoG51)
At²⁰⁹	5.5 h (ForW61, BartoG51)	❤ EC ≈95%, α ≈5% (BartoG51) △ −12.89 (MTW)	A chem, genet, excit (BartoG51) chem, mass spect (ForW61) parent Po²⁰⁹, parent Bi²⁰⁵ (BartoG51) daughter Rn²⁰⁹ (MomF52, MomF55a) daughter Fr²¹³ (GrifR64)	γ Po K X-rays, 0.195 (23%), 0.545 (62%), 0.780 (94%) e⁻ 0.076, 0.102, 0.178, 0.451, 0.686 α 5.65 (5%)	Bi²⁰⁹(α, 4n) (RamlW59)
At²¹⁰	8.3 h (KellE49)	❤ EC 99+%. α 0.17% (HoffR53) △ −12.12 (MTW)	A chem, genet, excit (KellE49) parent Po²¹⁰ (RaF) (KellE49, BartoG51) parent Bi²⁰⁶ (NeumH50b)	γ Po X-rays, 0.245 (79%), 1.180 (100%), 1.436 (29%), 1.483 (48%), 1.599 (14%) e⁻ 0.023, 0.031, 0.043, 0.152, 0.229 α 5.52 (0.05%), 5.44 (0.05%), 5.36 (0.06%)	Bi²⁰⁹(α, 3n) (RamlW59)
At²¹¹	7.21 h (AppE61) others (GrayP56, CorsD40, KellE49, CrofP64)	❤ α 40.9%, EC 59.1% (NeumH51) △ −11.64 (MTW)	A chem, excit, genet (CorsD40, KellE49) parent Bi²⁰⁷ (NeumH51) daughter Rn²¹¹ (MomF55a, MomF52) parent Po²¹¹ (AcC′) (CorsD40, CorsD40a) not parent Po²¹¹ᵐ, lim 0.01% (SpiesF54)	α 5.868 γ Po X-rays, 0.67 (weak) daughter radiations from Po²¹¹	Bi²⁰⁹(α, 2n) (RamlW59)
At²¹²	0.30 s (JonWB63) others (RitzJ62, WinnM54a)	❤ α (JonWB63) EC unstable (cons energy) (MTW) △ −8.64 (MTW)	B excit, decay charac (JonWB63)	α 7.66 (80%), 7.60 (20%) e⁻ 0.047, 0.059	Bi²⁰⁹(α, n) (JonWB63)

TABLE 1 *(Continued)*

Isotope Z A	Half-life	Type of decay (☢); % abundance; Mass excess (Δ≡M–A), MeV (C¹²=0); Thermal neutron cross section (σ), barns	Class; Identification; Genetic relationships	Major radiations: approximate energies (MeV) and intensities	Principal means of production
$_{85}At^{212m}$	0.12 s (JonWB63) others (RitJ62)	☢ α, no IT, lim 1% (JonWB63) β⁻, EC unstable (cons energy) (MTW) Δ -8.42 (LHP, MTW)	B excit, decay charac (JonWB63)	α 7.88 (20%), 7.82 (80%) e⁻ 0.047, 0.059	Bi²⁰⁹(α, n) (JonWB63)
At^{213}	[short] (KeyJ51)	☢ α (KeyJ51) Δ -6.5 (MTW)	E genet, decay charac (KeyJ51) descendant Pa²²⁵ (KeyJ51)	α 9.2	descendant Pa²²⁵ (KeyJ51)
At^{214}	≈2 x 10⁻⁶ s est (MeiW51)	☢ α (MeiW49) EC unstable (cons energy) (MTW) Δ -3.42 (MTW)	B genet (MeiW49) daughter Fr²¹⁸ (MeiW49, MeiW51)	α 8.78 (99%)	descendant Pa²²⁶ (MeiW49, MeiW51)
At^{215}	≈10⁻⁴ s delay coinc (GhiA48, MeiW51)	☢ α (KarlB44, GhiA48) Δ -1.25 (MTW)	A genet (KarlB44, GhiA48) daughter Fr²¹⁹, parent Bi²¹¹ (AcC) (GhiA48, MeiW51, MeiW49) daughter Po²¹⁵ (AcA), parent Bi²¹¹ (AcC) (KarlB44)	α 8.01 daughter radiations from Bi²¹¹, etc.	descendant Pa²²⁷ (HydE64)
At^{216}	≈3 x 10⁻⁴ s delay coinc (MeiW49, MeiW51)	☢ α (KarlB43a, GhiA48) β⁻, EC unstable (cons energy) (MTW) Δ 2.25 (MTW)	A genet (GhiA48) daughter Fr²²⁰, parent Bi²¹² (ThC) (GhiA48, MeiW51) parent Bi²¹² (ThC) (KarlB43a)	α 7.80 (97%)	descendant Pa²²⁸ (HydE64)
At^{217}	0.0323 s delay coinc (DiaH63) others (HageF47, HageF50, EnglA47)	☢ α (EnglA47, HageF47) β⁻ unstable (cons energy) (MTW) Δ 4.38 (MTW)	A genet (EnglA47, HageF47) daughter Fr²²¹, parent Bi²¹³ (EnglA47, HageF47, HageF50, CranT48)	α 7.07 (99+%) daughter radiations from Bi²¹³, etc.	descendant Ac²²⁵ (EnglA47, HageF47)
At^{218}	1.5-2.0 s (WaleR48) others (KarlB43)	☢ α (KarlB43) α 99+%, β⁻ 0.1% (WaleR48) Δ 8.11 (MTW)	B genet (KarlB43, WaleR59a) daughter Po²¹⁸ (RaA), parent Bi²¹⁴ (RaC) (KarlB43, WaleR48, WaleR59a)	α 6.70 (94%), 6.65 (6%) daughter radiations from Rn²¹⁸, Bi²¹⁴, etc.	daughter Po²¹⁸ (KarlB43, WaleR48)
At^{219}	0.9 m (HydE53)	☢ α ≈97%, β⁻ ≈3% (HydE53) Δ 10.5 (MTW)	B chem, genet (HydE53) daughter Fr²²³ (AcK), parent Rn²¹⁹ (An), parent Bi²¹⁵ (HydE53)	α 6.28 daughter radiations from Bi²¹⁵, Rn²¹⁹, etc.	descendant Ac²²⁷, natural source (HydE53)
$_{86}Rn^{<202}$	short (NurM66)	☢ α (NurM66)	F excit (NurM66)	α 6.90	O¹⁶ on Pt, N¹⁴ on Au, C¹² on Hg (NurM66)
$Rn^{<202}$	1 s (NurM66)	☢ α (NurM66)	F excit (NurM66)	α 6.85	O¹⁶ on Pt, N¹⁴ on Au, C¹² on Hg (NurM66)
$Rn^{201?}$	3 s (NurM66)	☢ α (NurM66)	E cross bomb, excit (NurM66)	α 6.77	Au¹⁹⁷(N¹⁴, 10n), O¹⁶ on Pt (NurM66)
$Rn^{<202}$	<1 s (NurM66)	☢ α (NurM66)	F excit (NurM66)	α 6.69	O¹⁶ on Pt, N¹⁴ on Au, C¹² on Hg (NurM66)
Rn^{202}	13 s (NurM66)	☢ α (NurM66)	D cross bomb, excit (NurM66)	α 6.64	Au¹⁹⁷(N¹⁴, 9n), O¹⁶ on Pt, C¹² on Hg (NurM66)
Rn^{203}	45 s (NurM66)	☢ α (NurM66)	D cross bomb, excit (NurM66)	α 6.50	Au¹⁹⁷(N¹⁴, 8n), O¹⁶ on Pt, C¹² on Hg (NurM66)
Rn^{203m}	28 s (NurM66)	☢ α (NurM66)	D cross bomb, excit (NurM66)	α 6.55	Au¹⁹⁷(N¹⁴, 8n), O¹⁶ on Pt, C¹² on Hg (NurM66)

TABLE 1 (*Continued*)

Isotope Z A	Half-life	Type of decay (☢); % abundance; Mass excess ($\triangle \equiv$M–A), MeV (C^{12}=0); Thermal neutron cross section (σ), barns	Class; Identification; Genetic relationships	Major radiations: approximate energies (MeV) and intensities	Principal means of production
$_{86}$Rn^{204}	75 s (NurM66)	☢ α (NurM66) Δ –7 (MTW)	B cross bomb, excit (NurM66)	α 6.42 [daughter radiations from Po^{200}]	Au^{197}(N^{14}, 7n), O^{16} on Pt, C^{12} on Hg (NurM66)
Rn^{205}	1.8 m (NurM66)	☢ α (NurM66) Δ –7 (MTW)	B cross bomb, excit (NurM66) 6.29 α (t$_{1/2}$ 3 m) formerly assigned to Rn^{204} (StonA57, MomF55)	α 6.26	Au^{197}(N^{14}, 6n), O^{16} on Pt, C^{12} on Hg (NurM66)
Rn^{206}	6.5 m (BurcW54, NurM66) others (StonA57, BarabS57, WinnM54a)	☢ α 65%, EC 35% (StonA57, MomF66) α 22%, EC 78% (BarabS57) Δ –9 (MTW)	B chem, genet (BurcW54, NurM66) parent Po^{202} (StonA57, MomF55a) 6.29 α (t$_{1/2}$ 3 m) formerly assigned to Rn^{204}	α 6.26 daughter radiations from At^{206}, Po^{202}, etc.	Au^{197}(N^{14}, 5n) (BurcW54, StonA57, NurM66) C^{16} on Hg, O^{16} on Pt (NurM66)
Rn^{207}	11 m (BurcW54) 10 m (StonA57)	☢ EC 96%, α 4% (StonA57, MomF55) Δ –9 (MTW)	A chem, genet (BurcW54) parent At^{207} (BurcW54, StonA57)	α 6.15 daughter radiations from At^{207}, Po^{203}, etc.	Au^{197}(N^{14}, 4n) (StonA57, BurcW54)
Rn^{208}	23 m (MomF55a) 21 m (StonA57)	☢ EC ≈80%, α ≈20% (MomF55a, StonA57) Δ –10 (MTW)	B chem, genet (MomF55a) parent Po^{204} (MomF55a)	α 6.15 daughter radiations from At^{208}, Po^{204}, Bi^{204}	Au^{197}(N^{14}, 3n) (StonA57) protons on Th^{232} (MomF55a)
Rn^{209}	30 m (MomF55a)	☢ EC 83%, α 17% (MomF55a) Δ –9 (MTW)	B chem, genet (MomF55a) daughter Ra^{213}, parent At^{209} (MomF55a, MomF52)	α 6.04 daughter radiations from Po^{205}, At^{209}	daughter Ra^{213}, from protons on Th^{232} (MomF55a)
Rn^{210}	2.42 h (CrofP64) 2.7 h (MomF52, MomF55a) 2.1 h (GhiA49)	☢ α ≈96%, EC ≈4% (MomF55a) Δ –9.74 (MTW)	A chem, genet (MomF55a, MomF52) parent Po^{206} (MomF52, MomF55a)	α 6.04 daughter radiations from At^{210}, Po^{206}	protons on Th^{232} (MomF52, MomF55a)
Rn^{211}	15 h (CrofP64) 16 h (MomF52, MomF55a)	☢ EC 74%, α 26% (MomF52) Δ –8.75 (MTW)	A chem, genet (MomF52) mass spect (AstG63) parent At^{211} (MomF52, MomF55a) parent Po^{207} (StonA56)	α 5.85 (9%), 5.78 (17%) γ At X-rays, 0.445 (29%), 0.680 (74%), 0.865 (18%), 0.946 (21%), 1.13 (23%), 1.37 (38%) e$^-$ 0.053, 0.065, 0.073, 0.153, 0.168, 0.200, 0.237, 0.349, 0.584, 0.665 daughter radiations from At^{211}, Po^{211}	protons on Th^{232} (MomF55, MomF55a)
Rn^{212}	25 m (CrofP64) 23 m (GhiA49, HydE50, MomF52)	☢ α (HydE50) Δ –8.66 (MTW)	A chem, genet (HydE50, GhiA49) daughter Fr^{212}, parent Po^{208} (HydE50, MomF52)	α 6.27	daughter Fr^{212} (HydE50)
Rn^{215}	≈10^{-6} s est (MeiW52)	☢ α (MeiW52) Δ –1.2 (MTW)	B genet (MeiW52) daughter Ra^{219}, parent Po^{211} (AcC') (MeiW52)	α 8.6 daughter radiations from Po^{211}	descendant U^{227} (HydE64)
Rn^{216}	4.5 × 10^{-5} s delay coinc (RuiC61)	☢ α (MeiW49, MeiW51) β stable (cons energy) (ForB58) Δ 0.25 (MTW)	A genet (MeiW49, MeiW51) daughter Ra^{220}, parent Po^{212} (ThC') (MeiW49, MeiW51)	α 8.05 daughter radiations from Po^{212}	descendant U^{228} (HydE64)
Rn^{217}	5.4 × 10^{-4} s delay coinc (RuiC61) others (MeiW51)	☢ α (MeiW51) β stable (cons energy) (ForB58) Δ 3.65 (MTW)	A genet (MeiW49, MeiW51) daughter Ra^{221}, parent Po^{213} (MeiW49, MeiW51)	α 7.74 daughter radiations from Po^{213}	descendant U^{229} (MeiW49, MeiW51, HydE64)
Rn^{218}	0.035 s delay coinc (DiaH63) others (RuiC61, StuM48)	☢ α (StuM48) β stable (cons energy) (ForB58) Δ 5.22 (MTW)	A genet (StuM48) daughter Ra^{222}, parent Po^{214} (RaC') (StuM48)	α 7.14 (99.8%) γ 0.609 (0.2%) daughter radiations from Po^{214}	descendant U^{230} (HydE64)

TABLE 1 (*Continued*)

Isotope Z A	Half-life	Type of decay (☢); % abundance; Mass excess (△=M–A), MeV (C¹²=0); Thermal neutron cross section (σ), barns	Class; Identification; Genetic relationships	Major radiations: approximate energies (MeV) and intensities	Principal means of production
$_{86}Rn^{219}$ (An)	4.00 s (RodenH61) 3.92 s (CuriM31)	☢ α; β⁻ unstable (cons energy) (MTW) △ 8.85 (MTW)	A chem, genet (CuriM31) daughter Ra²²³ (AcX), parent Po²¹⁵ (AcA) daughter At²¹⁹ (HydE53)	α 6.82 (81%), 6.55 (11%), 6.42 (8%) γ Po X-rays, 0.272 (9%), 0.401 (5%) e⁻ 0.179, 0.255, 0.308 daughter radiations from Po²¹⁵, etc.	descendant Th²²⁷ (HydE64)
Rn^{220} (Tn)	55.3 s (GinJ63) 56.3 s (RodenH61) 54.5 s (CuriM31) 51.5 s (SchmH55)	☢ α; β stable (cons energy) (ForB58) △ 10.61 (MTW) σ_c <0.2 (GoldmDT64)	A chem, genet (CuriM31) daughter Ra²²⁴ (ThX), parent Po²¹⁶ (ThA)	α 6.29 (100%) γ 0.55 (0.07%) daughter radiations from Po²¹⁶	natural source, descendant Th²²⁸ (HydE64)
Rn^{221}	25 m (MomF56)	☢ β⁻ ≈80%, α ≈20% (MomF56) △ 14 (MTW)	B chem, genet (MomF56, MomF52) parent Fr²²¹, parent Po²¹⁷ (MomF56, MomF52)	α 6.0 daughter radiations from Fr²²¹, Po²¹⁷, etc.	protons on Th²³² (MomF52)
Rn^{222} (Rn)	3.8229 d (MarinP56) 3.825 d (TobJ55, TobJ51, RobeJ56a, CuriM31) no β⁻, lim 1 x 10⁻⁴% (KarlB46)	☢ α; β stable (cons energy) (ForB58) △ 16.39 (MTW) σ_c 0.7 (GoldmDT64)	A chem, genet (CuriM31) daughter Ra²²⁶, parent Po²¹⁸ (RaA)	α 5.49 (100%) γ 0.510 (0.07%) daughter radiations from Po²¹⁸, etc.	natural source (HydE64)
Rn^{223}	43 m (ButeF64)	☢ [β⁻] (BellA61)	B genet, chem (BellA61, ButeF64) ancestor Ra²²³ (AcX) (BellA61, ButeF64)	daughter radiations from Fr²²³	protons on Th²³² (BellA61, ButeF64)
Rn^{224}	1.9 h (ButeF64)	☢ [β⁻] (BellA61)	B genet, chem (BellA61, ButeF64) ancestor Ra²²⁴ (ThX) (BellA61, ButeF64)		protons on Th²³² (BellA61)
$_{87}Fr^{204}$	2.0 s (GrifR64)	☢ α (GrifR64)	C excit, decay charac (GrifR64)	α 7.03	Au¹⁹⁷(O¹⁶, 9n) (GrifR64)
Fr^{205}	3.7 s (GrifR64)	☢ α (GrifR64)	B excit, genet (GrifR64) parent At²⁰¹ (GrifR64)	α 6.92 daughter radiations from At²⁰¹	Au¹⁹⁷(O¹⁶, 8n) (GrifR64)
Fr^{206}	15.8 s (GrifR64)	☢ α (GrifR64) △ -0 (MTW)	B excit, cross bomb (GrifR64)	α 6.80	Au¹⁹⁷(O¹⁶, 7n), Tl²⁰³(C¹², 9n) (GrifR64)
Fr^{207}	19 s (GrifR64)	☢ α (GrifR64) △ -2 (MTW)	B excit, cross bomb (GrifR64)	α 6.78	Au¹⁹⁷(O¹⁶, 6n), Tl²⁰³(C¹², 8n) (GrifR64)
Fr^{208}	37 s (GrifR64)	☢ α (GrifR64) △ -2 (MTW)	B excit, cross bomb (GrifR64)	α 6.66	Au¹⁹⁷(O¹⁶, 5n), Tl²⁰³(C¹², 7n) (GrifR64)
Fr^{209}	55 s (GrifR64)	☢ α (GrifR64) △ -3 (MTW)	B genet, excit, cross bomb (GrifR64) parent At²⁰⁵ (GrifR64)	α 6.66	Au¹⁹⁷(O¹⁶, 4n), Tl²⁰³(C¹², 6n) (GrifR64)
Fr^{210}	2.6 m (GrifR64)	☢ α (GrifR64) △ -3 (MTW)	B excit, cross bomb (GrifR64)	α 6.56	Tl²⁰³(C¹², 5n), Tl²⁰⁵(C¹², 7n), Au¹⁹⁷(O¹⁶, 3n) (GrifR64)
Fr^{211}	3.1 m (GrifR64)	☢ α (GrifR64) EC unstable (cons energy) (MTW) △ -4.3 (MTW)	B chem, genet, excit (GrifR64) parent At²⁰⁷ (GrifR64)	α 6.56 daughter radiations from At²⁰⁷	Tl²⁰³(C¹², 4n), Tl²⁰⁵(C¹², 6n) (GrifR64)
Fr^{212}	19.3 m (HydE50)	☢ EC 56%, α 44% (HydE50) △ -4 (MTW)	A chem, genet (HydE50) chem, mass spect (MomF52) parent Rn²¹², parent At²⁰⁸ (HydE50, MomF52)	α 6.42 (16%), 6.39 (17%), 6.35 (11%) daughter radiations from Rn²¹², At²⁰⁸	protons on Th²³² (HydE50)

TABLE 1 (*Continued*)

Isotope Z A	Half-life	Type of decay (☢); % abundance; Mass excess (△=M-A), MeV (C¹²=0); Thermal neutron cross section (σ), barns	Class; Identification; Genetic relationships	Major radiations: approximate energies (MeV) and intensities	Principal means of production
$_{87}Fr^{213}$	34 s (GrifR64)	☢ α 99+%, EC 0.5% (GrifR64) △ -3.55 (MTW)	A chem, genet, excit, cross bomb (GrifR64) parent At²⁰⁹ (GrifR64)	α 6.78	Tl²⁰⁵(C¹², 4n), Pb²⁰⁸(B¹¹, 6n) (GrifR64)
Fr^{217}	[short] (KeyJ51)	☢ α (KeyJ51) EC unstable (cons energy) (MTW) △ 4.4 (MTW)	E genet, decay charac (KeyJ51) descendant Pa²²⁵ (KeyJ51)	α 8.3	descendant Pa²²⁵ (KeyJ51)
Fr^{218}	5 x 10⁻³ s est (MeiW51)	☢ α (MeiW51) EC unstable (cons energy) (MTW) △ 7.00 (MTW)	B genet (MeiW49, MeiW51) daughter Ac²²², parent At²¹⁴ (MeiW49, MeiW51)	α 7.85 (93%) daughter radiations from At²¹⁴	descendant Pa²²⁶ (MeiW49, MeiW51)
Fr^{219}	0.02 s delay coinc (MeiW51)	☢ α (GhiA48) β stable (cons energy) (ForB58) △ 8.61 (MTW)	A genet (GhiA48) daughter Ac²²³, parent At²¹⁵ (GhiA48, MeiW49, MeiW51)	α 7.31 daughter radiations from At²¹⁵	descendant Pa²²⁷ (HydE64)
Fr^{220}	27.5 s (MeiW51)	☢ α (GhiA48) β⁻, EC unstable (cons energy) (MTW) △ 11.47 (MTW)	A genet (GhiA48) daughter Ac²²⁴, parent At²¹⁶ (GhiA48, MeiW49, MeiW51)	α 6.68 (85%), 6.64 (13%) daughter radiations from At²¹⁶, etc.	descendant Pa²²⁸ (HydE64)
Fr^{221}	4.8 m (HageF50) others (EnglA47)	☢ α (EnglA47, HageF47) no β⁻, lim 0.1% (ValliK64) β⁻ unstable (cons energy) (MTW) △ 13.27 (MTW)	A chem, genet (HageF47, EnglA47) daughter Ac²²⁵, parent At²¹⁷ (EnglA47, HageF47, CranT48, HageF50) daughter Rn²²¹ (MomF56, MomF52)	α 6.34 (82%), 6.12 (15%) γ At X-rays, 0.218 (14%) e⁻ 0.122, 0.202 daughter radiations from At²¹⁷, etc.	ancestor Th²²⁹ (EnglA47, HageF47, HageF50)
Fr^{222}	14.8 m (HydE50a)	☢ β⁻ 99+%, α 0.01-0.1% (HydE51a) △ 16.34 (MTW)	B chem, genet (HydE50a) parent Ra²²², ancestor Bi²¹⁴ (RaC) (HydE50a, HydE51a)	daughter radiations from Ra²²², etc.	protons on Th²³² (HydE50a)
Fr^{223} (AcK)	22 m genet (PereyM56, AdlJ55, PereyM39)	☢ β⁻ (PereyM39a, GuiM47) α =4 x 10⁻³% (HydE53) α =6 x 10⁻³% (AdlJ55, PereyM56) △ 18.40 (MTW)	A chem, genet (PereyM39, PereyM39b) daughter Ac²²⁷, parent Ra²²³ (AcX) (PereyM39, PereyM39a, PereyM39b, PereyM41, PereyM46, GuiM47, LecM50) parent At²¹⁹ (HydE53)	β⁻ 1.15 max e⁻ 0.031, 0.045, 0.062, 0.075 γ Ra L X-rays, 0.050 (40%), 0.080 (13%), 0.234 (4%)	natural source (HydE64)
Fr^{224}	<2m (ButeF64)	☢ [β⁻] (BellA61) △ 22 (MTW)	F genet (BellA61) daughter Rn²²⁴, parent Ra²²⁴ (ThX) (BellA61)		daughter Rn²²⁴ (BellA61)
$_{88}Ra^{213}$	2.7 m (MomF55a)	☢ α (MomF52) △ -0 (MTW)	B chem, genet (MomF52) parent Rn²⁰⁹ (MomF52, MomF55a)	α 6.91	Pb²⁰⁶(C¹², 5n), protons on Th²³² (MomF52, MomF55)
Ra^{219}	≈10⁻³ s est (MeiW52)	☢ α (MeiW52) △ 9.4 (MTW)	B genet (MeiW52) daughter Th²²³, parent Rn²¹⁵ (MeiW52)	α 8.0 daughter radiations from Rn²¹⁵, Po²¹¹	descendant U²²⁷ (HydE64)
Ra^{220}	0.023 s (RuiC61)	☢ α (MeiW51) △ 10.27 (MTW)	A genet (MeiW49, MeiW51) daughter Th²²⁴, parent Rn²¹⁶ (MeiW49, MeiW51)	α 7.46 (99%) γ 0.465 (1%) daughter radiations from Rn²¹⁶, Po²¹²	descendant U²²⁸ (HydE64)
Ra^{221}	30 s (MeiW51) 28 s (TovP58)	☢ α (MeiW51) β stable (cons energy) (ForB58) △ 12.96 (MTW)	A chem, genet (MeiW49, MeiW51) daughter Th²²⁵, parent Rn²¹⁷ (MeiW49, MeiW51)	α 6.76 (30%), 6.67 (20%), 6.61 (34%), 6.59 (8%) γ Rn X-rays, 0.091 (3.5%), 0.151 (13%), 0.175 (2%) daughter radiations from Rn²¹⁷, etc.	descendant U²²⁹ (MeiW49, MeiW51, RuiC61)

TABLE 1 (*Continued*)

Isotope Z A	Half-life	Type of decay (☢); % abundance; Mass excess ($\triangle$≡M–A), MeV (C¹²=0); Thermal neutron cross section (σ), barns	Class; Identification; Genetic relationships	Major radiations: approximate energies (MeV) and intensities	Principal means of production
$_{88}$Ra^{222}	38 s (StuM48) 37 s (AsaF56)	☢ α (StuM48) β stable (cons energy) (ForB58) $\triangle$ 14.32 (MTW)	A chem, genet (StuM48) daughter Th^{226}, parent Rn^{218} (StuM48) daughter Fr^{222} (HydE50a, HydE51a)	α 6.56 (96%) γ 0.325 (4%), 0.473 (0.007%), 0.52 (0.004%),	0.85 (0.003%) daughter radiations from Rn^{218}, etc.	descendant U^{230} (StuM48)
Ra^{223} (AcX)	11.435 d (KirH65) 11.2 d (CuriM31) 11.7 d (HageG54) others (BaeA53, SeaG47a)	☢ α; β stable (cons energy) (ForB58) $\triangle$ 17.26 (MTW) σ_c 130 (GoldmDT64)	A chem, genet (CuriM31) daughter Th^{227} (RdAc), parent Rn^{219} (An); daughter Ac^{223} (MeiW51) daughter Fr^{223} (AcK) (PereyM39, PereyM39a, PereyM39b, PereyM41, PereyM46, GuiM47, LecM50) descendant Rn^{223} (BellA61, ButeF64)	α 5.75 (9%), 5.71 (54%), 5.61 (26%), 5.54 (9%) γ Rn X-rays, 0.149 (10%, complex), 0.270 (10%), 0.33 (6%, complex) e⁻ 0.024, 0.046, 0.056, 0.126, 0.136, 0.171 daughter radiations from Rn^{219}, Po^{215}, Pb^{211}, etc.	daughter Th^{227} (HydE64)
Ra^{224} (ThX)	3.64 d (CuriM31) others (SeaG47a)	☢ α; β stable (cons energy) (ForB58) $\triangle$ 18.82 (MTW) σ_c 12 (GoldmDT64)	A chem, genet (CuriM31) daughter Th^{228} (RdTh), parent Rn^{220} (Tn); daughter Ac^{224} (GhiA48, MeiW49, MeiW51) descendant Rn^{224} (BellA61, ButeF64)	α 5.68 (94%), 5.45 (6%) γ Rn X-rays, 0.241 (3.7%), 0.29 (0.008%), 0.41 (0.004%), 0.65 (0.009%) e⁻ 0.144, 0.225 daughter radiations from Rn^{220}, Po^{216}, Pb^{212}, etc.	daughter Th^{228}, from natural source (HydE64)
Ra^{225}	14.8 d (HageF50) others (EnglA47)	☢ β⁻ (EnglA47, HageF47) no α, lim 10⁻⁴% (MalkL60) others (MomF56) $\triangle$ 22.01 (MTW)	A chem, genet (EnglA47, HageF47) daughter Th^{229}, parent Ac^{225} (EnglA47, HageF47, HageF50)	β 0.36 max e⁻ 0.021, 0.035 γ Ac L X-rays, 0.040 (33%) daughter radiations from Ac^{225}, etc.	descendant U^{233}, Th^{229} (HydE64)
Ra^{226}	1602 y sp act (MartiG59) 1622 y sp act (KohT49) 1617 y sp act (SebW56) 1590 y sp act (CuriM31) others (GorsG58, GorsG59)	☢ α; β stable (cons energy) (ForB58) $\triangle$ 23.69 (MTW) σ_c 20 (GoldmDT64)	A chem, genet (CuriM31) daughter Th^{230} (Io), parent Rn^{222} (Rn)	α 4.78 (95%), 4.60 (6%) γ Rn X-rays, 0.186 (4%), 0.26 (0.007%), 0.42 (2 x 10⁻⁴%), 0.61 (2 x 10⁻⁴%) e⁻ 0.087, 0.170 daughter radiations from Rn^{222}, Po^{218}, Pb^{214}, Bi^{214}, Po^{214}	natural source (HydE64)
Ra^{227}	41.2 m (ButlJP53)	☢ β⁻ (PeteS49) $\triangle$ 27.18 (MTW)	A n-capt, genet (PeteS49) parent Ac^{227} (PeteS49)	β 1.31 max e⁻ 0.008, 0.023 γ [Ac X-rays], 0.291 (4%), 0.498 (0.6%)	Ra^{226} (n, γ) (PeteS49)
Ra^{228} (MsTh$_1$)	6.7 y (CuriM31)	☢ β⁻; no α, lim 2 x 10⁻⁶% (FeaN57) $\triangle$ 28.96 (MTW) σ_c ≈36 (GoldmDT64)	A chem, genet (CuriM31) daughter Th^{232}, parent Ac^{228} (MsTh$_2$)	β 0.05 max e⁻ 0.005 daughter radiations from Ac^{228}, Th^{228}, Ra^{224}, etc.	natural source (HydE64)
Ra^{229}	[short] (DepF52)	☢ [β⁻] (DepF52)	F n-capt, genet (DepF52) [parent Ac^{229}] (DepF52)		Ra^{228}(n, γ) (DepF52)
Ra^{230}	1 h (JenkW52)	☢ β⁻ (JenkW52) $\triangle$ 35 (LHP, MTW)	D chem (JenkW52) parent Ac^{230} (JenkW52)	β 1.2 max	[Th^{232}(d, 3pn)] (JenkW52)
$_{89}$Ac^{221}	[short] (KeyJ51)	☢ α (KeyJ51) EC unstable (cons energy) (MTW) $\triangle$ 14.6 (MTW)	E genet, decay charac (KeyJ51) descendant Pa^{225} (KeyJ51)	α 7.6	ancestor Pa^{225} (KeyJ51)
Ac^{222}	5.5 s (MeiW52) 4.2 s (TovP58)	☢ α (MeiW51) EC unstable (cons energy) (MTW) $\triangle$ 16.55 (MTW)	B genet (MeiW49, MeiW51) daughter Pa^{226}, parent Fr^{218} (MeiW49, MeiW51, MeiW52)	α 7.00 (93%) daughter radiations from Fr^{218}, etc.	daughter Pa^{226} (MeiW49, MeiW51, MeiW52) Ra^{226}(p, 5n) (TovP58)

TABLE 1 (Continued)

Isotope Z A	Half-life	Type of decay (☢); % abundance; Mass excess (△=M–A), MeV (C¹²=0); Thermal neutron cross section (σ), barns	Class; Identification; Genetic relationships	Major radiations: approximate energies (MeV) and intensities	Principal means of production
$_{89}Ac^{223}$	2.2 m (MeiW51)	☢ α 99%, EC 1% (MeiW51) △ 17.82 (MTW)	A genet (GhiA48) daughter Pa²²⁷, parent Fr²¹⁹, parent Ra²²³ (AcX) (GhiA48, MeiW49, MeiW51)	α 6.66 (38%), 6.65 (42%), 6.57 (13%) γ Fr L X-rays, 0.082 (0.2%), 0.096 (0.2%) daughter radiations from Fr²¹⁹, etc.	daughter Pa²²⁷ (MeiW51)
Ac^{224}	2.9 h (MeiW51)	☢ EC ≈90%, α ≈10% (MeiW51) β⁻ unstable (cons energy) (MTW) △ 20.21 (MTW)	A chem, genet (GhiA48) daughter Pa²²⁸, parent Fr²²⁰, parent Ra²²⁴ (GhiA48, MeiW49, MeiW51)	γ Ra X-rays, 0.132 (28%), 0.217 (62%) e⁻ 0.067, 0.080 α 6.20 (3%), 6.14 (3%), 6.04 (3%) daughter radiations from Fr²²⁰, etc.	daughter Pa²²⁸ (MeiW51)
Ac^{225}	10.0 d (HageF50, EnglA47)	☢ α (EnglA47, HageF47) β stable (cons energy) (ForB58) △ 21.62 (MTW)	A chem, genet (HageF47, EnglA47) daughter Ra²²⁵, parent Fr²²¹ (HageF47, EnglA47, HageF50, CranT48) daughter Pa²²⁹ (HydE49a) daughter Th²²⁵ (MeiW49, MeiW51)	α 5.83 (54%), 5.79 (28%), 5.73 (10%, doublet) γ Fr X-rays, 0.099, 0.150, 0.187 e⁻ 0.020, 0.032, 0.044, 0.081 daughter radiations from Fr²²¹, At²¹⁷, etc.	descendant U²³³, Th²²⁹ Ra²²⁶(d, 3n) (HydE64)
Ac^{226}	29 h (StreK50)	☢ β⁻ ≈80%, EC ≈20% (StepF57d) α ? (weak) (MCoyJ64) △ 24.31 (MTW)	A chem, genet (StreK48) daughter Th²²⁶, parent Th²²⁶ (StreK48, StreK50, MeiW50)	β⁻ 1.2 max e⁻ 0.053, 0.067 γ Th L X-rays, Ra X-rays, 0.158 (32%), 0.185 (9%), 0.230 (47%), 0.253 (11%) α 5.44 ? daughter radiations from Th²²⁶, etc.	Ra²²⁶(d, 2n) (HydE64)
Ac^{227}	21.6 y (TobJ55) 22.0 y (HollaJ50) 21.7 y (CuriI44) 21.2 y (ShimN56b) others (CuriM31)	☢ β⁻ 99% (PereyM39, PeteS49a) α 1.4% (NurM65a) α 1.2% (MeyS14, PereyM39, PereyM46, PeteS49a) △ 25.87 (MTW) σ_c 830 (GoldmDT64)	A chem, genet (CuriM31) daughter Pa²³¹, parent Th²²⁷ (RdAc); parent Fr²²³ (PereyM39, PereyM46, GuiM47, LecM50) daughter Ra²²⁷ (PeteS49)	β⁻ 0.046 max e⁻ 0.005, 0.010 γ Th L X-rays, 0.070 [0.08%], 0.166, 0.190 α 4.95 (1.2%, doublet), 4.86 (0.18%, doublet) daughter radiations from Th²²⁷, Ra²²³, Fr²²³, etc.	Ra²²⁶(n, γ)Ra²²⁷(β⁻) (PeteS49) natural source (HydE64)
Ac^{228} (MsTh₂)	6.13 h (CuriM31)	☢ β⁻; △ 28.91 (MTW)	A chem, genet (CuriM31) daughter Ra²²⁸ (MsTh₁), parent Th²²⁸ (RdTh)	β⁻ 2.11 max e⁻ 0.040, 0.054, 0.110 γ Th X-rays, 0.34 (15%, complex), 0.908 (25%), 0.96 (20%, complex)	natural source (HydE64)
Ac^{229}	66 m (DepF52)	☢ β⁻ (DepF52) △ 31 (MTW)	B chem, n-capt (DepF52) daughter Ra²²⁹ (DepF52)		Ra²²⁸(n, γ)[Ra²²⁹]β⁻ (DepF52)
Ac^{230}	<1 m genet (JenkW52)	☢ β⁻ (JenkW52) △ 34 (MTW)	F genet (JenkW52) daughter Ra²³⁰ (JenkW52)	β⁻ 2.2 max	daughter Ra²³⁰ (JenkW52)
Ac^{231}	15 m (TakaK60a)	☢ β⁻ (TakaK60a) △ 35.9 (MTW)	C excit (TakaK60a)	β⁻ 2.1 max γ 0.185, 0.28, 0.39, 0.71	Th²³²(γ, p) (TakaK60a)
$_{90}Th^{223}$	0.9 s (TovP58) ≈0.1 s est (MeiW51)	☢ α (MeiW52) EC unstable (cons energy) (MTW) △ 19.5 (MTW)	B genet (MeiW52) daughter U²²⁷, parent Ra²¹⁹ (MeiW52)	α 7.56 [daughter radiations from Ra²¹⁹, etc.]	daughter U²²⁷ (MeiW52)
Th^{224}	1.05 s (TovP58)	☢ α (MeiW51) β stable (cons energy) (ForB58) △ 20.00 (MTW)	A genet (MeiW49, MeiW51) daughter U²²⁸, parent Ra²²⁰ (MeiW49, MeiW51)	α 7.18 (79%), 6.91 (19%) γ Ra X-rays, 0.177 (9%), 0.235 (0.4%), 0.297 (0.3%), 0.410 (0.8%) daughter radiations from Ra²²⁰, etc.	daughter U²²⁸ (MeiW51, RuiC61)

TABLE 1 (*Continued*)

Isotope Z A	Half-life	Type of decay (☢); % abundance; Mass excess (Δ≡M–A), MeV (C^{12}=0); Thermal neutron cross section (σ), barns	Class; Identification; Genetic relationships	Major radiations: approximate energies (MeV) and intensities	Principal means of production
$_{90}$Th^{225}	8.0 m (MeiW51)	☢ α ≈90%, EC ≈10% (MeiW51) Δ 22.30 (MTW)	A chem, genet (MeiW49, MeiW51) daughter U^{229}, parent Ra^{221}, parent Ac^{225} (MeiW49, MeiW51)	α 6.80 (8%), 6.75 (6%), 6.50 (12%), 6.48 (39%), 6.44 (13%) γ [Ac X-rays], Ra X-rays, 0.246 (5%), 0.322 (27%), 0.362 (5%), 0.45 (1%), 0.49 (1%) daughter radiations from Ra^{221}, etc.	daughter U^{229} (MeiW49, MeiW51)
Th^{226}	30.9 m (StuM48)	☢ α (StuM48) β stable (cons energy) (ForB58) Δ 23.19 (MTW)	A chem, genet (StuM48) daughter U^{230}, parent Ra^{222} (StuM48) daughter Ac^{226} (StreK48, StreK50)	α 6.34 (79%), 6.22 (19%) γ Ra X-rays, 0.111 (3.4%), 0.131 (0.34%), 0.20 (0.4%, complex), 0.242 (1.2%) e$^-$ 0.094, 0.107 daughter radiations from Ra^{222}, Rn^{218}, etc.	daughter U^{230} (HydE64)
Th^{227} (RdAc)	18.2 d (HageG54) others (PeteS49b, CuriM31)	☢ α; β stable (cons energy) (ForB58) Δ 25.82 (MTW) σ_f ≈1500 (GoldmDT64)	A chem, genet (CuriM31) daughter Ac^{227}, parent Ra^{223} (AcX) daughter Pa^{227} (MeiW51, GhiA48) daughter U^{231} (CranW50)	α 6.04 (23%), 5.98 (24%), 5.76 (21%), 5.72 (14%, doublet) γ Ra X-rays, 0.050 (8%), 0.237 (15%, complex), 0.31 (8%, complex) e$^-$ 0.013, 0.026, 0.044, others daughter radiations from Ra^{223}, Rn^{219}, Po^{215}, etc.	daughter Ac^{227}, from natural source or from Ra^{226}(n, γ)Ra^{227}(β$^-$) (HydE64)
Th^{228} (RdTh)	1.910 y (KirH56) others (CuriM31)	☢ α; β stable (cons energy) (ForB58) Δ 26.77 (MTW) σ_c 123 (GoldmDT64) σ_f <0.3 (GoldmDT64)	A chem, genet (CuriM31) daughter Ac^{228} (MsTh$_2$); parent Ra^{224} (ThX); daughter U^{232} (GofJ49) daughter Pa^{228} (MeiW51)	α 5.43 (71%), 5.34 (28%) γ Ra L X-rays, 0.084 (1.6%), 0.132 (0.2%), 0.167 (0.1%), 0.214 (0.3%) e$^-$ 0.067, 0.080 daughter radiations from Ra^{224}, Rn^{220}, Po^{216}, etc,	natural source daughter U^{232} Ra^{226}(n, γ)Ra^{227}(β$^-$) Ac^{227}(n, γ)Ac^{228}(β$^-$) (HydE64)
Th^{229}	7340 y genet (HageF50) others (EnglA47)	☢ α; β stable (cons energy) (ForB58) Δ 29.61 (MTW) σ_f 32 (GoldmDT64)	A chem, genet (EnglA47, HageF47, HageF50) daughter U^{233}, parent Ra^{225} (EnglA47, HageF47, HageF50)	α 5.05 (7%), 4.97 (complex, 10%), 4.90 (11%), 4.84 (58%), 4.81 (11%) γ Ra X-rays, 0.137 (≈3%, complex), 0.20 (≈10%, doublet) e$^-$ 0.006–0.090 daughter radiations from Ra^{225}, Ac^{225}, etc.	daughter U^{233} (HydE64)
Th^{230} (Io)	8.0 x 10^4 y sp act (HydE49) 7.5 x 10^4 y sp act (AttR62) 8.2 x 10^4 y genet (CuriM30) $t_{1/2}$ (SF) ≥1.5 x 10^{17} y (SegE52)	☢ α; β stable (cons energy) (ForB58) Δ 30.87 (MTW) σ_c 23 (GoldmDT64) σ_f ≤0.001 (GoldmDT64)	A chem, genet (CuriM31) daughter U^{234} (U$_{II}$), parent Ra^{226}; daughter Pa^{230} (StuM48a)	α 4.68 (76%), 4.62 (24%) γ Ra L X-rays, 0.068 (0.6%), 0.142 (0.07%), 0.184 (0.014%), 0.253 (0.017%) e$^-$ 0.051, 0.064 daughter radiations from Ra^{226}, Rn^{222}, etc.	natural source (HydE64)
Th^{231} (UY)	25.52 h (CabM58) 25.6 h (JafAH51) 25.5 h (KniG49) others (CuriM31, GratO32, NisY38)	☢ β$^-$; Δ 33.83 (MTW)	A chem, genet (CuriM31) daughter U^{235} (AcU), parent Pa^{231}	β$^-$ 0.30 max e$^-$ 0.040, 0.054, 0.061 γ Pa L X-rays, 0.026 (2%), 0.084 (10%, complex)	Th^{230}(n, γ) (BaranS60, HoltzM66) daughter U^{235}
Th^{232}	1.41 x 10^{10} y sp act, (FarlT60) others (KovAF38, PicE56, MackR56, SenF56) $t_{1/2}$ (SF): >10^{21} y (FleG58) others (PocA55, SegE52)	☢ α; β stable (cons energy) (ForB58) % 100 (AstF35, DempA36) Δ 35.47 (MTW) σ_c 7.4 (GoldmDT64) σ_f <0.0002 (GoldmDT64)	A chem, genet (CuriM31) parent Ra^{228} (MsTh$_1$)	α 4.01 (76%), 3.95 (24%) γ [Ra L X-rays] e$^-$ 0.042, 0.055 daughter radiations from Ra^{228}, Ac^{228}, Th^{228}, Ra^{224}, etc.	natural source (HydE64)
Th^{233}	22.12 m (JenkE55) 22.4 m (DroB57) 22.3 m (BunkM50a) 22.5 m (SeaG47) others (RutW52, GrossA41)	☢ β$^-$ (SeaG47) Δ 38.76 (MTW) σ_c 1500 (GoldmDT64) σ_f 15 (GoldmDT64)	A chem, n-capt (MeiL38) parent Pa^{233} (MeiL38, GrossA41, SeaG41a, HahO41, SeaG47)	β$^-$ 1.23 max e$^-$ 0.009, 0.024, 0.036, 0.051, 0.067, 0.082 γ Pa X-rays, 0.029 (2.1%), 0.087 (2.7%), 0.171 (0.7%), 0.195 (0.3%), 0.453 (1%), 0.67 (0.25%), 0.895 (0.14%)	Th^{232} (n, γ) (MeiL38, SeaG47, SeaG41a, GrossA41)

TABLE 1 (*Continued*)

Isotope Z A	Half-life	Type of decay (☢); % abundance; Mass excess (△≡M-A), MeV (C¹²=0); Thermal neutron cross section (σ), barns	Class; Identification; Genetic relationships	Major radiations: approximate energies (MeV) and intensities	Principal means of production
$_{90}Th^{234}$ (UX₁)	24.10 d (KniG48) others (SargB39a, CuriM31)	☢ β⁻; no α, lim 10⁻⁴% (DeuS55) △ 40.64 (MTW) σ_c 1.8 (GoldmDT64) σ_f <0.01 (GoldmDT64)	A chem, genet (CuriM31) daughter U²³⁸, parent Pa²³⁴ᵐ (UX₂); ancestor Pa²³⁴ (UZ) (ZijW54)	β⁻ 0.191 max e⁻ 0.012, 0.025, 0.072, 0.088 γ Pa L X-rays, 0.063 (3.5%, doublet), 0.093 (4%, doublet) daughter radiations from Pa²³⁴ᵐ	natural source (HydE64)
Th²³⁵	<<10 m genet (HarvB50)	☢ [β⁻] (HarvB50)	F n-capt, genet (HarvB50) [parent Pa²³⁵] (HarvB50)		Th²³⁴(n, γ) (HarvB50)
$_{91}Pa^{224}$	0.6 s (TovP58)	☢ α (TovP58)	F decay charac (TovP58)		Th²³²(p, 9n) (TovP58)
Pa²²⁵	0.8 s (TovP58) 2.0 s (KeyJ51)	☢ α (KeyJ51) △ 25 (MTW)	E excit, decay charac (KeyJ51, TovP58) ancestor Ac²²¹, Fr²¹⁷, At²¹³ (KeyJ51)		Th²³²(p, 8n) (TovP58, KeyJ51)
Pa²²⁶	1.8 m (MeiW51)	☢ α 74%, EC 26% (MCoyJ64) △ 25.96 (MTW)	B chem, genet (MeiW49, MeiW51) parent Ac²²² (MeiW49, MeiW51, MeiW52)	α 6.86 (38%), 6.82 (34%) daughter radiations from Ac²²², Th²²⁶, etc.	Th²³²(p, 7n) (MeiW49, MeiW51, MeiW52)
Pa²²⁷	38.3 m (MeiW51) others (OConP48)	☢ α ≈85%, EC ≈15% (MeiW51) △ 26.83 (MTW)	A chem, genet (GhiA48) parent Ac²²³, parent Th²²⁷ (RdAc) (GhiA48, MeiW51) daughter Np²³¹ (MagL50)	γ [Th X-rays], Ac L X-rays, 0.065 (6%, complex), 0.110 (2%) α 6.47 (43%), 6.42 (23%, complex), 6.40 (8%), 6.36 (7%) daughter radiations from Ac²²³, etc.	Th²³²(d, 7n) (MeiW56, SubV63) Th²³²(p, 6n) (MeiW56, HillM58)
Pa²²⁸	22 h (MeiW51)	☢ EC ≈98%, α ≈2% (MeiW51) △ 28.86 (MTW)	A chem, genet (GhiA48) daughter U²²⁸, parent Ac²²⁴, parent Th²²⁸ (RdTh) (GhiA48, MeiW49, MeiW51)	γ Th X-rays, 0.14 (3%), 0.20 (9%), 0.28 (5%), 0.33 (18%), 0.41 (13%), 0.46 (32%), 0.95 (93%), 1.57 (7%), 1.85 (4%), all γ's complex e⁻ 0.040, 0.054, 0.110 α 6.11 (1%, complex), 6.08 (0.4%), 6.03 (0.2%), 5.80 (0.2%), others daughter radiations from Ac²²⁴, etc.	Th²³²(p, 5n) (ArbE60) Th²³²(d, 6n) (HydE64) Th²³⁰(d, 4n) (HillM58)
Pa²²⁹	1.5 d (HydE49b)	☢ EC 99+%, α 0.25% (SlaLM51) others (MeiW51) △ 29.88 (MTW)	A chem, genet (HydE49a) parent Ac²²⁵ (HydE49a) daughter U²²⁹ (MeiW51, MeiW49)	γ Th X-rays e⁻ 0.023, 0.038 α 5.67 (0.05%), 5.62 (0.07%, complex), 5.58 (0.10%), 5.54 (0.03%)	daughter U²²⁹ (MeiW51, SubV63) Th²³⁰(d, 3n) (HydE49a) Th²³²(p, 4n) (SubV63)
Pa²³⁰	17.7 d (OsbD49) 17.0 d (StuM48) others (HydE49a, HydE49b)	☢ EC 89.6%, β⁻ 10.4%, α 0.0032% (BastG65a) β⁺ ? (≈0.03%) (OngP55a) others (BriaJ65a, MCoyJ64, MeiW51) △ 32.17 (MTW) σ_f 1500 (GoldmDT64)	A chem, excit, genet (StuM48) parent U²³⁰ (StuM48, OsbD49) parent Th²³⁰ (Io) (StuM48a) parent Ac²²⁶ (MeiW50)	β⁻ 0.41 max e⁻ 0.034, 0.048 γ Th X-rays, 0.45 (18%, complex), 0.51 (8%, complex), 0.91 (24%, complex), 0.954 (50%) α 5.26–5.34 (complex) daughter radiations from U²³⁰, Th²²⁶, etc.	Th²³²(p, 3n) (TewH55, MeiW56) Th²³²(d, 4n) (MeiW56) Th²³⁰(d, 2n) (HydE64)
Pa²³¹	3.25 x 10⁴ y sp act (KirH61) 3.43 x 10⁴ y sp act (VWinQ49) 3.2 x 10⁴ y sp act (GrossA30)	☢ α; β stable (cons energy) (ForB58) △ 33.44 (MTW) σ_c 200 (GoldmDT64) σ_f 0.010 (GoldmDT64)	A chem, genet (CuriM31) daughter Th²³¹ (UY), parent Ac²²⁷ daughter U²³¹ (CranW50)	α 5.06 (10%), 5.02 (23%), 5.01 (24%), 4.95 (22%), 4.73 (11%) γ Ac X-rays, 0.027 (6%), 0.29 (6%, complex) e⁻ 0–0.10, 0.195, 0.323, 0.350 daughter radiations from Ac²²⁷, Th²²⁷, Fr²²³, Ra²²³, etc.	natural source (HydE64)
Pa²³²	1.31 d (BrowCI54) others (JafAH50, OsbD49, GofJ49, StuM48)	☢ β⁻ (GofJ49) no EC, lim 2% (BrowCI52a) △ 35.95 (MTW) EC unstable (cons energy) (MTW) σ_c ≈760 (GoldmDT64) σ_f ≈700 (GoldmDT64)	A chem, genet (GofJ49) parent U²³² (GofJ49, OsbD49)	β⁻ 1.3 max (0.7%), 0.32 max e⁻ 0.028, 0.043, 0.091 γ U X-rays, 0.107 (5%, doublet), 0.150 (12%), 0.39 (9%, doublet), 0.46 (9%, doublet), 0.57 (8%, doublet), 0.87 (51%, complex), 0.971 (40%)	Pa²³¹(n, γ), Th²³²(d, 2n) (HydE64) Th²³²(p, n) (TewH55)

TABLE 1 (*Continued*)

Isotope Z A	Half-life	Type of decay (☢); % abundance; Mass excess (△≡M–A), MeV (C¹²=0); Thermal neutron cross section (σ), barns	Class; Identification; Genetic relationships	Major radiations: approximate energies (MeV) and intensities	Principal means of production
$_{91}Pa^{233}$	27.0 d (MIsaL56, WriH57) 27.4 d (GrossA41) others (StuM48, GofJ49)	☢ β⁻ (MeiL38, GrossA41, SeaG41a) △ 37.51 (MTW) σ_c 21 (to Pa²³⁴) 22 (to Pa²³⁴ᵐ) (GoldmDT64) σ_f <0.1 (GoldmDT64)	A chem, genet (MeiL38, GrossA41, SeaG41a) daughter Th²³³ (MeiL38, GrossA41, SeaG41a, HahO41, SeaG47) parent U²³³ (SeaG47) daughter Np²³⁷ (HageF47, MagL47)	β⁻ 0.568 max (5%), 0.257 max e⁻ 0.013, 0.023, 0.036, 0.054, 0.065, 0.185, 0.197, 0.291 γ U X-rays, 0.31 (44%, complex)	Th²³²(n, γ) Th²³³(β⁻) (MeiL38, GrossA41, SeaG41a, HahO41, SeaG47) Th²³²(d, n) (StuM48, GofJ49)
Pa²³⁴ (UZ)	6.75 h (BjoS62) 6.66 h (ZijW54) 6.7 h (CuriM31)	☢ β⁻; △ 40.38 (MTW) σ_f <5000 (GoldmDT64)	A chem, genet (CuriM31) parent U²³⁴ (U_II); daughter Pa²³⁴ᵐ (UX₂) (ZijW54)	β⁻ 1.3 max (≤2%), 1.13 max (13%), 0.53 max e⁻ 0.024, 0.039, 0.080, 0.095, 0.112 γ U X-rays, 0.100 (50%), 0.126 (26%), 0.22 (14%), 0.36 (13%), 0.56 (15%), 0.70 (24%), 0.90 (70%), 1.08 (12%), (many of the γ rays are complex)	natural source (HydE64)
Pa²³⁴ᵐ (UX₂)	1.175 m (BareF51) 1.14 m (CuriM31)	☢ β⁻ 99+%, IT 0.13% (BjoS63a) others (FeaN38a, BradH45d, ZijW54) △ 40.45 (LHP, MTW) σ_f <500 (GoldmDT64)	A chem, genet (CuriM31) daughter Th²³⁴ (UX₁), parent U²³⁴ (U_II); parent Pa²³⁴ (UZ) (ZijW54)	β⁻ 2.29 max γ U L X-rays, 0.765 (0.30%), 1.001 (0.60%)	natural source (HydE64)
Pa²³⁵	23.7 m (MeiW50) others (HarvB50)	☢ β⁻ (MeiW50, HarvB50) △ 42.3 (MTW)	B chem, excit, sep isotopes (MeiW50) genet (HarvB50) [daughter Th²³⁵] (HarvB50)	β⁻ 1.4 max γ no γ	Th²³⁴(n, γ) [Th²³⁵]β⁻ (HarvB50)
Pa²³⁶	12 m (WolzG63) others (CranW54)	☢ β⁻ (WolzG63) △ 45 (MTW)	D chem, decay charac (WolzG63)	β⁻ 3.3 max γ U L X-rays	U²³⁸(d, α) (WolzG63)
Pa²³⁷	39 m (TakaK60)	☢ β⁻ (TakaK60) △ 47.7 (MTW)	B chem, excit (TakaK60)	β⁻ 2.3 max γ U X-rays, 0.090 (↑ 50), 0.145 (↑ 45), 0.205 (↑ 55), 0.275 (↑ 20), 0.330 (↑ 40), 0.405 (↑ 30), 0.46 (↑ 100), 0.55 (↑ 30), 0.59 (↑ 25), 0.75 (↑ 50), 0.80 (↑ 45), 0.87 (↑ 100), 0.92 (↑ 100), 1.04 (↑ 35), 1.32 (↑ 10), 1.42 (↑ 15)	U²³⁸(γ, p) (TakaK60)
$_{92}U^{227}$	1.3 m (MeiW52)	☢ α (MeiW52) △ 29 (MTW)	B chem, genet (MeiW52) parent Th²²³ (MeiW52)	α 6.8 daughter radiations from Th²²³, etc.	Th²³²(α, 9n) (MeiW52)
U²²⁸	9.1 m (RuiC61) others (MeiW51)	☢ α ≥95%, EC ≤5% (RuiC61) others (MeiW51) EC unstable (cons energy) (MTW) △ 29.23 (MTW)	A chem, genet (MeiW49, MeiW51) parent Th²²⁴, parent Pa²²⁸ (MeiW49, MeiW51) daughter Pu²³² (JameR48, OrtD51a)	α 6.69 (↑ 70), 6.60 (↑ 29) γ Th X-rays, 0.152 (0.2%), 0.187 (0.3%), 0.246 (0.4%) daughter radiations from Th²²⁴, etc.	Th²³²(α, 8n) (RuiC61)
U²²⁹	58 m (MeiW51)	☢ EC ≈80%, α ≈20% (MeiW51) △ 31.20 (MTW)	A chem, genet (MeiW49, MeiW51) parent Th²²⁵, parent Pa²²⁹ (MeiW49, MeiW51) daughter Pu²³³ (ThomT57)	γ Pa X-rays α 6.36 (13%), 6.33 (4%), 6.30 (3%) daughter radiations from Th²²⁵, Pa²²⁹, etc.	Th²³²(α, 7n) (MeiW49, MeiW51)
U²³⁰	20.8 d (StuM48)	☢ α (StuM48) β stable (cons energy) (ForB58) △ 31.60 (MTW) σ_f 25 (GoldmDT64)	A chem, genet (StuM48) daughter Pa²³⁰ (StuM48, OsbD49) daughter Pu²³⁴ (PerlmI49, OrtD51a) parent Th²²⁶ (StuM48)	α 5.89 (67%), 5.82 (32%) γ Th L X-rays, 0.072 (0.54%), 0.156 (doublet, 0.034%), 0.231 (0.18%) e⁻ 0.054, 0.068 daughter radiations from Th²²⁶, Ra²²², etc.	daughter Pa²³⁰ (HydE64)
U²³¹	4.3 d (CranW50) 4.2 d (OsbD49)	☢ EC 99+%, α 0.0055% (CranW50) △ 33.8 (MTW) σ_f ≈400 (GoldmDT64)	A chem, sep isotopes, genet (OsbD49) genet (CranW50) parent Th²²⁷ (RdAc), parent Pa²³¹ (CranW50)	γ Pa X-rays, 0.026, 0.084 (7%), 0.218 (1%) e⁻ 0.040, 0.054, 0.063 α 5.46	Th²³⁰(α, 3n) (HollaJ56c) Pa²³¹(d, 2n) (OsbD49) Th²³²(α, 5n) (CranW50)

TABLE 1 (*Continued*)

Isotope Z A	Half-life	Type of decay (☢); % abundance; Mass excess (Δ≡M−A), MeV (C^{12}=0); Thermal neutron cross section (σ), barns	Class; Identification; Genetic relationships	Major radiations: approximate energies (MeV) and intensities	Principal means of production
$_{92}U^{232}$	72 y sp act, calorim (ChilJ64) others (SelP54, JameR49, GofJ49) $t_{1/2}$ (SF) ≈8 x 10^{13} y (HydE57)	☢ α (GofJ49) β stable (cons energy) (ForB58) Δ 34.60 (MTW) σ_c 78 (GoldmDT64) σ_f 77 (GoldmDT64)	A chem, genet (GofJ49) daughter Pa^{232} (GofJ49, OsbD49) daughter Pu^{236} (JameR49) parent Th^{228} (RdTh) (GofJ49)	α 5.32 (68%), 5.27 (32%) γ Th L X-rays, 0.058 (0.21%), 0.129 (0.082%), 0.270 (0.0038%), 0.328 (0.0034%) e^- 0.040, 0.054 daughter radiations from Th^{228}, Ra^{224}, Rn^{220}, etc.	daughter Pa^{232} (GofJ49) Th^{232}(α, 4n) (HydE64)
U^{233}	1.62 x 10^5 y sp act + mass spect (HydE52) 1.63 x 10^5 y sp act + mass spect (DokY59a, LineG45) 1.61 x 10^5 y sp act (PopD61) others (SeaG52)	☢ α (SeaG52) β stable (cons energy) (ForB58) Δ 36.94 (MTW) σ_c 49 (GoldmDT64) σ_f 524 (GoldmDT64)	A chem, genet (SeaG47, SeaG52) daughter Pa^{233} (SeaG47) parent Th^{229} (EnglA47, HageF47, HageF50)	α 4.82 (83%), 4.78 (15%) γ Th X-rays, 0.029 († 60), 0.042 († 310), 0.055 († 68), 0.097 († 100), 0.119 († 40, complex), 0.146 († 35, doublet), 0.164 († 27), 0.22 († 45, complex), 0.291 († 23), 0.32 († 43, doublet) e^- 0.023, 0.038 daughter radiations from Th^{229}, Ra^{225}, Ac^{225}, etc.	Th^{232}(n, γ) Th^{233}(β^-) Pa^{233}(β^-) (SeaG47)
U^{234} (U_{II})	2.47 x 10^5 y sp act (FleE52, WhiP65) others (KieC52, KieC49, GoldiA49, ChambO46) $t_{1/2}$ (SF) 2 x 10^{16} y (GhiA52)	☢ α; β stable (cons energy) (ForB58) % 0.0057 (LouM56) others (WhiF56) Δ 38.16 (MTW) σ_c 95 (GoldmDT64)	A chem, genet, mass spect (CuriM31) daughter Pa^{234m} (UX_2), daughter Pa^{234} (UZ), parent Th^{230} (Io)	α 4.77 (72%), 4.72 (28%) γ Th L X-rays, 0.053 (0.2%), 0.117, 0.48 (4 x 10^{-5}%, complex), 0.58 (1.2 x 10^{-5}%) daughter radiations from Th^{230}, Ra^{226}, Rn^{222}, etc.	daughter Pu^{238} descendant Th^{234} (HydE64)
<u>U^{235}</u> (AcU)	7.1 x 10^8 y sp act (FleE52, WhiP65) 7.1 x 10^8 y radiogenic Pb ratios (NierA39) 6.9 x 10^8 y sp act (DerA65) 6.8 x 10^8 y sp act (WurE57) $t_{1/2}$ (SF) 1.9 x 10^{17} y (SegE52) others (BaldE54)	☢ α; β stable (cons energy) (ForB58) % 0.7196 (GrunB61) others (LouM56, WhiF56) Δ 40.93 (MTW) σ_c 101 (GoldmDT64) σ_f 577 (GoldmDT64)	A chem, mass spect (CuriM31) parent Th^{231} (UY)	α 4.58 (8%, doublet), 4.40 (57%), 4.37 (18%) γ Th X-rays, 0.143 (11%), 0.185 (54%), 0.204 (5%) daughter radiations from Th^{231}, etc.	natural source
U^{235m}	26.1 m (ShimS65) 26.5 m (AsaF57) 26.6 m* (HuiJ57a)	☢ IT (AsaF57, HuiJ57a) Δ 40.93 (LHP, MTW)	A genet (AsaF57) chem, genet (HuiJ57a) daughter Pu^{239} (AsaF57, HuiJ57a) not daughter Np^{235}, lim 2% (GinJ58)	e^- ≤0.0001 (100 eV)	daughter Pu^{239} (AsaF57, HuiJ57a)
U^{236}	2.39 x 10^7 y sp act (FleE52) 2.46 x 10^7 y sp act (JafAH51a) $t_{1/2}$ (SF) 2 x 10^{16} y (HydE57)	☢ α (GhiA51a) β stable (cons energy) (ForB58) Δ 42.46 (MTW) σ_c 6 (GoldmDT64)	A chem, n-capt, mass spect (GhiA51a)	α 4.49 (76%), 4.44 (24%) γ [Th L X-rays] e^- 0.032, 0.045	U^{235}(n, γ) (HydE64)
U^{237}	6.75 d (WagF53) 6.63 d (MelaL48) others (WahA48, JameR49, ShermL58)	☢ β^- (NisY40a, MMilE40a) Δ 45.41 (MTW)	A chem, excit (NisY40a, MMilE40a) parent Np^{237} (WahA48) daughter Pu^{241} (SeaG49a)	β^- 0.248 max e^- 0.008, 0.011, 0.038, 0.089, 0.186 γ 0.026 (2%), 0.060 (36%), 0.165 (2.0%), 0.208 (23%), 0.267 (0.76%), 0.332 (1.4%, doublet), 0.370 (0.17%, doublet)	U^{236}(n, γ) (RasJ57, YamaT66) U^{238}(n, 2n) (MMilE40a, NisY40a, WahA48)
<u>U^{238}</u>	4.51 x 10^9 y sp act (KovAF55, NierA39) others (KieC49, LeacR57) $t_{1/2}$ (SF): 6.5 x 10^{15} y sp act (KuzB59) 1.0 x 10^{16} y sp act (FleR64, KuroP56) 8.0 x 10^{15} y sp act (SegE52, SchaG46, ParkPL58) 5.8 x 10^{15} y sp act (GerlE59)	☢ α; β stable (cons energy) (ForB58) % 99.276 (WhiF56) others (LouM56) Δ 47.33 (MTW) σ_c 2.73 (GoldmDT64) σ_f <0.0005 (GoldmDT64)	A chem, genet, mass spect (CuriM31) parent Th^{234} (UX_1) (BecH1896)	α 4.20 (75%), 4.15 (25%) γ [Th L X-rays] e^- 0.030, 0.043 daughter radiations from Th^{234}, Pa^{234m}	natural source (HydE64)

TABLE 1 (*Continued*)

Isotope Z A	Half-life	Type of decay (⚛); % abundance; Mass excess (△≡M–A), MeV (C¹²=0); Thermal neutron cross section (σ), barns	Class; Identification; Genetic relationships	Major radiations: approximate energies (MeV) and intensities	Principal means of production
$_{92}U^{239}$	23.54 m (MitA43) 23.5 m (FeaN47a, MelaL47) others (IrvJ39, SeaG49)	⚛ β⁻ (MMilE39) △ 50.60 (MTW) σ_c 22 (GoldmDT64) σ_f 14 (GoldmDT64)	A n-capt (MeiL37) parent Np²³⁹ (MMilE40, StarK42)	β⁻ 1.29 max e⁻ 0.011, 0.023, 0.052, 0.069 γ Np L X-rays, 0.044 (4%), 0.075 (51%) daughter radiations from Np²³⁹	U²³⁸(n, γ) (MeiL37, IrvJ39, MMilE39, StarK42)
U²⁴⁰	14.1 h (KniJD53)	⚛ β⁻ (KniJD53) △ 52.74 (MTW)	A chem, n-capt (StuM49) parent Np²⁴⁰ᵐ (KniJD53, HydE48a) daughter Pu²⁴⁴ (ButlJP56a, DiaH56)	β⁻ 0.36 max e⁻ 0.022, 0.038 γ Np L X-rays daughter radiations from Np²⁴⁰ᵐ	U²³⁸(n, γ)U²³⁹(n, γ) (HydE64)
$_{93}Np^{231}$	≈50 m (MagL50)	⚛ α (MagL50) △ 35.7 (MTW) EC unstable (cons energy) (MTW)	B chem, genet, excit, sep isotopes (MagL50) parent Pa²²⁷ (MagL50)	α 6.29 daughter radiations from Pa²²⁷ etc.	U²³³(d, 4n) (MagL50)
Np²³²	≈13 m (MagL50)	⚛ EC (MagL50) △ 37 (MTW)	D chem (MagL50)	γ U X-rays, hard γ rays (MagL50)	U²³⁵(d, 5n), U²³⁸(d, 8n), U²³³(d, 3n) (MagL50)
Np²³³	35 m (MagL50)	⚛ EC 99+%, α ≈10⁻³% (MagL50) △ 38 (MTW)	B chem, excit, sep isotopes (MagL50)	α 5.54 γ U X-rays, γ rays observed	U²³³(d, 2n), U²³⁵(d, 4n) (MagL50)
Np²³⁴	4.40 d (HydE49b) others (OsbD49)	⚛ EC (OrtD51a) no α, lim 0.01% (HydE49b) β⁺ ≈0.05% (PresRJ55) △ 40.0 (MTW) σ_f ≈900 (GoldmDT64)	A chem, excit, genet, sep isotopes (JameR49) daughter Pu²³⁴ (PerlmI49, OrtD51a)	γ U X-rays, 0.109, 0.23, 0.25, 0.45, 0.50, 0.75, 0.95, 1.21, 1.56 (all radiations complex) e⁻ 0.024, 0.039, 0.696 β⁺ 0.8 max	U²³³(d, n) (HydE64) U²³⁵(d, 3n) (HydE64) U²³⁵(p, 2n) (HydE64) U²³³(α, p2n) (VanR58a, HydE64)
Np²³⁵	410 d (JameR52) others (HydE49b)	⚛ EC 99+%, α 1.6 x 10⁻³% (GinJ58) others (HoffR56) △ 41.05 (MTW)	A chem, excit, sep isotopes (JameR49) not parent U²³⁵ᵐ, lim 2% (GinJ58)	γ U L X-rays, U K X-rays (weak) α 5.02	U²³⁵(d, 2n) (HydE64) daughter Pu²³⁵ (HydE64) U²³³(α, pn) (VanR58a, HydE64) U²³⁵(α, p3n) (HydE64)
Np²³⁶	22 h (JameR49)	⚛ EC 51%, β⁻ 49% (GinJ59a) EC(K)/β⁻ 0.75 (GrayP56) others (OrtD51) △ 43.41 (MTW)	A chem, genet, sep isotopes, excit (JameR49) parent Pu²³⁶ (JameR49, JameR49a, HydE49b, GhiA52)	β⁻ 0.52 max e⁻ 0.025, 0.040 γ U X-rays, 0.642, 0.688	U²³⁵(d, n) (HydE64) U²³⁵(α, p2n) (HydE64)
Np²³⁶	t₁/₂ (β⁻) >5 x 10³ y sp act (StuM55)	⚛ β⁻ (?), no α observed (StuM55) σ_f 2500 (GoldmDT64)	A chem, mass spect (GinJ58, StuM55)		U²³⁵(d, n) (GinJ58, StuM55)
Np²³⁷	2.14 x 10⁶ y sp act (BrauF60) 2.2 x 10⁶ y sp act (MagL48) t₁/₂ (SF) >10¹⁸ y (DruV61a)	⚛ α (WahA48); β stable (cons energy) (ForB58) △ 44.89 (MTW) σ_c 170 (GoldmDT64) σ_f 0.019 (GoldmDT64)	A chem, genet, excit (WahA48) daughter U²³⁷ (WahA48) parent Pa²³³ (MagL47, HageF47)	α 4.78 (75%, complex), 4.65 (12%, doublet) γ Pa L X-rays, 0.030 (14%), 0.086 (14%), 0.145 (1%) e⁻ 0.009, 0.024, 0.036, 0.051, 0.067, 0.082 daughter radiations from Pa²³³, U²³³, etc.	U²³⁸(n, 2n)U²³⁷(β⁻) (WahA48)
Np²³⁸	2.10 d (FreeM50) others (SeaG49, JameR49a)	⚛ β⁻ (SeaG46, SeaG49) no EC(K), lim 1% (RasJ55a) EC unstable (cons energy) (MTW) △ 47.47 (MTW) σ_f 1600 (GoldmDT64)	A chem, genet, n-capt, sep isotopes (SeaG46) parent Pu²³⁸ (SeaG46, KenJ49a, JafAH49, JameR49, SeaG46a) daughter Am²⁴²ᵐ (SeaG49a, StreK50a, AsaF60)	β⁻ 1.25 max e⁻ 0.022, 0.039 γ 1.01 (42%, complex)	Np²³⁷(n, γ) (HydE64) U²³⁸(d, 2n) (SeaG46) U²³⁸(p, n) (MCorG54)
Np²³⁹	2.346 d (WisL56) 2.37 d (CohD59) 2.34 d (ConnR59) others (PhiK46, DavD65, SeaG46, JameR49)	⚛ β⁻ (MMilE40) △ 49.32 (MTW) σ_c 25 (to Np²⁴⁰) 35 (to Np²⁴⁰ᵐ) (GoldmDT64) σ_f <1 (GoldmDt64)	A chem, n-capt, genet, excit (MMilE39, MMilE40) daughter U²³⁹ (MMilE40, StarK42) parent Pu²³⁹ (KenJ49, SeaG49) daughter Am²⁴³ (StreK50a)	β⁻ 0.713 max (11%), 0.437 max e⁻ 0.02-0.04, 0.048, 0.088, 0.106, 0.156 γ Pu X-rays, 0.106 (23%), 0.209 (4%), 0.228 (12%), 0.278 (14%)	U²³⁸(n, γ)U²³⁹(β⁻) (MMilE40, StarK42)

TABLE 1 (*Continued*)

Isotope Z A	Half-life	Type of decay (☢); % abundance; Mass excess (Δ≡M−A), MeV (C¹²=0); Thermal neutron cross section (σ), barns	Class; Identification; Genetic relationships	Major radiations: approximate energies (MeV) and intensities	Principal means of production
$_{93}Np^{240}$	63 m (LesR60) others (OrtD51a)	☢ β^- (OrtD51a) Δ 52.2 (MTW)	A chem, cross bomb (OrtD51a) chem, mass spect (LesR60) not daughter Np^{240m}, lim 5% (LesR60)	β^- 0.89 max γ 0.16, 0.25, 0.44, 0.56, 0.60, 0.92, 1.00, 1.16	$U^{238}(\alpha, pn)$ (VanR58a, HydE64)
Np^{240m}	7.3 m (KniJD53, HydE48a)	☢ β^- (HydE48a) Δ 52.3 (LHP, MTW)	A chem, genet (HydE48a, KniJD53) daughter U^{240} (HydE48a, KniJD53) not parent Np^{240}, lim 5% (LesR60) descendant Pu^{244} (ButlJP56a, DiaH56)	β^- 2.16 max e^- 0.022, 0.038 γ 0.56 (21%), 0.60 (13%), 0.92 (3%, complex), 1.5 (3%, complex)	daughter U^{240} (HydE64)
Np^{241}	16 m (VanR59, LesR60)	☢ β^- (VanR59) Δ 54.3 (MTW)	A chem, mass spect (LesR60)	β^- 1.4 max	$U^{238}(\alpha, p)$ (VanR59, LesR60)
Np^{241}	3.4 h (LesR60)	☢ $[\beta^-]$ (LesR60)	B chem, mass spect (LesR60)		$U^{238}(\alpha, p)$ (LesR60)
$_{94}Pu^{232}$	36 m (OrtD51a)	☢ $\alpha \geq 2\%$, EC ≤98% (OrtD51a) Δ 38.4 (MTW)	B chem, sep isotopes, excit, genet (OrtD51a) parent U^{228} (OrtD51a, JameR48)	α 6.59 daughter radiations from Np^{232}, U^{228}, etc	$U^{233}(\alpha, 5n)$ (ThomT57) $U^{235}(\alpha, 7n)$ (HydE64)
Pu^{233}	20 m (ThomT57)	☢ EC 99+%, α 0.1% (ThomT57) Δ 40.04 (MTW)	B chem, excit, genet (ThomT57) parent U^{229} (ThomT57)	α 6.31 daughter radiations from Np^{233}, U^{229}, Th^{225}, etc.	$U^{233}(\alpha, 4n)$ (ThomT57)
Pu^{234}	9.0 h (OrtD51a) 8.5 h (PerlmI49) others (HigG52a)	☢ EC 94%, α 6% (AsaF57a) Δ 40.34 (MTW)	A chem, genet, sep isotopes, excit (HydE49b, PerlmI49) parent U^{230}, parent Np^{234} (PerlmI49, OrtD51a) daughter Cm^{238} (HigG52a)	α 6.20 (4%), 6.15 (1.9%) γ Np X-rays daughter radiations from Np^{234}, U^{230}, etc.	$U^{233}(\alpha, 3n)$ (VanR58a) $U^{235}(\alpha, 5n)$ (HydE64)
Pu^{235}	26 m (OrtD51a, ThomT57)	☢ EC 99+%, α 0.003% (ThomT57) Δ 42.2 (MTW)	B chem, excit, sep isotopes (OrtD51a, ThomT57)	γ Np X-rays α 5.86	$U^{235}(\alpha, 4n)$, $U^{233}(\alpha, 2n)$ (ThomT57, OrtD51a)
Pu^{236}	2.85 y (HoffD57) others (JameR49) $t_{1/2}$ (SF) 3.5×10^9 y (GhiA52)	☢ α (JameR49) β stable (cons energy) (ForB58) Δ 42.90 (MTW) σ_f 170 (GoldmDT64)	A chem, excit, sep isotopes, cross bomb, genet (JameR49) parent U^{232} (JameR49) daughter Cm^{240} (SeaG49b) daughter 22 h Np^{236} (JameR49, JameR49a, HydE49b, GhiA52)	α 5.77 (69%), 5.72 (31%) γ U L X-rays, 0.048 (0.31%), 0.109 (0.012%) e^- 0.028, 0.043 daughter radiations from U^{232}, etc.	daughter Np^{236} (HydE64) $U^{235}(\alpha, 3n)$ (VanR58a)
Pu^{237}	45.6 d (HoffD57a) 44 d (ThomT57) 40 d (HoffR53) others (JameR49a)	☢ EC 99+%, α 0.0033% (ThomT57) EC 99+%, α 0.002% (HoffD57a) Δ 45.12 (MTW) σ_f 2500 (GoldmDT64)	A chem, sep isotopes, cross bomb, genet energy levels (HoffD58) chem, mass spect (ThomT57)	γ Np X-rays, 0.060 (5%) e^- 0.026, 0.032, 0.038, 0.042, 0.056 α 5.66 (↑ 21), 5.37 (↑ 79)	$U^{235}(\alpha, 2n)$ (VanR58a) $Np^{237}(d, 2n)$ (JameR49a)
Pu^{237m}	0.18 s (StepF57a)	☢ IT (StepF57a) Δ 45.26 (MTW)	A genet (StepF57a) daughter Cm^{241} (StepF57a)	γ Pu L X-rays, 0.145 (2%) e^- 0.125 (75%), 0.140 (23%)	daughter Cm^{241} (StepF57a)
Pu^{238}	86.4 y genet (HoffD57b) others (SeaG49b, JafAH49) $t_{1/2}$ (SF) 4.9×10^{10} y (HydE57) others (DruV61a, SegE52)	☢ α (SeaG46) β stable (cons energy) (ForB58) Δ 46.18 (MTW) σ_c 500 (GoldmDT64) σ_f 16.8 (GoldmDT64)	A chem, sep isotopes, excit (SeaG46, SeaG46a, SeaG49) daughter Np^{238} (JameR49, JafAH49, SeaG46a, KenJ49a, SeaG46) daughter Cm^{242} (SeaG49b)	α 5.50 (72%), 5.46 (28%) γ U L X-rays, 0.099 (8 × 10⁻³%), 0.150 (1 × 10⁻³%), 0.77 (5 × 10⁻⁵%, complex) e^- 0.024, 0.039	daughter Np^{238} from $Np^{237}(n, \gamma)$ (HydE64) daughter Cm^{242} (HydE64)
Pu^{239}	24,390 y sp act (DokY59) 24,413 y sp act (MarkT59) 24,181 y calorimeter (DetF65, StouJ47) others (FarwG54, CunB49) $t_{1/2}$ (SF) 5.5×10^{15} y (SegE52)	☢ α (KenJ49) β stable (cons energy) (ForB58) Δ 48.60 (MTW) σ_c 274 (GoldmDT64) σ_f 741 (GoldmDT64)	A chem, genet, mass spect (KenJ49) daughter Np^{239} (KenJ49, SeaG49) parent U^{235m} (AsaF57, HuiJ57a)	α 5.16 (88%, doublet), 5.11 (11%) γ U X-rays, 0.039 (0.007%), 0.052 (0.020%), 0.129 (0.005%), 0.375 (0.0012%), 0.414 (0.0012%), 0.65 (8 × 10⁻⁵%, complex), 0.77 (2 × 10⁻⁵%, doublet) e^- 0.008, 0.019, 0.033, 0.047	$U^{238}(n, \gamma) U^{239}(\beta^-)$ $Np^{239}(\beta^-)$ (KenJ49, SeaG49)

TABLE 1 (*Continued*)

Isotope Z A	Half-life	Type of decay (☢); % abundance; Mass excess (△≡M–A), MeV (C¹²=0); Thermal neutron cross section (σ), barns	Class; Identification; Genetic relationships	Major radiations: approximate energies (MeV) and intensities	Principal means of production
$_{94}Pu^{240}$	6580 y genet (IngM51) others (DokY59, ButlJP56a, WestE51, FarwG54) $t_{1/2}$ (SF): 1.34 x 10^{11} y (WatD62b) 1.45 x 10^{11} y (MalkL63) others (BarcF54, ChambO54)	☢ α (JameR49) β stable (cons energy) (ForB58) △ 50.14 (MTW) σ_c 286 (GoldmDT64) σ_f <0.08 (GoldmDT64)	A chem, n-capt, mass spect (ChambO44, FarwG46, BartlA44) daughter Cm^{244} (FrieA54)	α 5.17 (76%), 5.12 (24%) γ U L X-rays, 0.65 (complex, 2 x 10^{-5}%) e⁻ 0.026, 0.040	multiple n-capt from U^{238}, Pu^{239} (HydE64)
Pu^{241}	13.2 y (BrowF60) others (HallG56, MKenD53, RosB56, SmiH61, ThomS50d)	☢ β⁻ 99+%, α 2.3 x 10^{-3}% (BrowF60, SmiH61) others (AsaF57a, SeaG49a, GhiA50, IvaR63) △ 52.98 (MTW) σ_c 425 (GoldmDT64) σ_f 950 (GoldmDT64)	A chem, n-capt, mass spect, excit, genet (SeaG49a, SeaG49, GhiA50) parent Am^{241} (SeaG49a, CunB49a) parent U^{237} (SeaG49a) daughter Cm^{245} (FrieA54)	β⁻ 0.021 max α 4.90 (0.0019%), 4.85 (0.0003%) γ U X-rays, 0.145 (1.6 x 10^{-4}%) daughter radiations from Am^{241}	multiple n-capt from U^{238}, Pu^{239}, etc. (HydE64)
Pu^{242}	3.79 x 10^5 y sp act (ButlJP56a) 3.73 x 10^5 y sp act (ButlJP56) others (MecJ56, ThomS50d) $t_{1/2}$ (SF): 7.1 x 10^{10} y (MecJ56) 7.4 x 10^{10} y (MalkL63) 6.6 x 10^{10} y (ButlJP56) others (DruV61a)	☢ α (ThomS50d) β stable (cons energy) (ForB58) △ 54.74 (MTW) σ_c 19 (GoldmDT64) σ_f <0.2 (GoldmDT64)	A chem, mass spect, n-capt, genet (ThomS50d) daughter Am^{242} (AsaF60, OKelG50) daughter Cm^{246} (FrieA54)	α 4.90 (76%), 4.86 (24%) γ [U L X-rays]	multiple n-capt from U^{238}, Pu^{239}, etc. (HydE64) daughter Am^{242} (ButlJP56, HydE64)
Pu^{243}	4.98 h (EngeD53) others (SulJ51, ThomS51)	☢ β⁻ (SulJ51) △ 57.77 (MTW) σ_c 170 (GoldmDT64)	A chem, n-capt, cross bomb (SulJ51) genet (ThomS51) parent Am^{243} (ThomS51)	β⁻ 0.58 max e⁻ 0.019, 0.036 γ Am L X-rays, 0.084 (21%), 0.381 (0.7%)	Pu^{242}(n, γ) (HydE64, SulJ51, ThomS51)
Pu^{244}	≈7.6 x 10^7 y genet (DiaH56) ≈7.5 x 10^7 y genet (ButlJP56a) $t_{1/2}$ (SF) 2.5 x 10^{10} y (FieP55a)	☢ [α] (StuM54a) β stable (cons energy) (ForB58) △ 59.83 (MTW) σ_c 1.8 (GoldmDT64)	A chem, n-capt, mass spect, genet (StuM54a, ButlJP56a, DiaH56) ancestor Np^{240m}, parent U^{240} (ButlJP56a, DiaH56) daughter Am^{244m} (FieP55a)	α [4.58] daughter radiations from U^{240}, Np^{240m}	multiple n-capture from U^{238}, Pu^{239}, etc. (HydE64, EngeD55, StuM54a)
Pu^{245}	10.1 h (FieP55) 10.6 h genet (ButlJP56a) others (BrowCI55)	☢ β⁻ (FieP55) △ 63 (MTW) σ_c ≈260 (GoldmDT64)	B chem, n-capt (FieP55, BrowCI55) parent Am^{245} (ButlJP56a, FieP55)	daughter radiations from Am^{245}	Pu^{244}(n, γ); multiple n-capt from Pu^{239}, etc. (HydE64, ButlJP56a)
Pu^{246}	10.85 d (HoffD56) others (EngeD55)	☢ β⁻ (EngeD55) △ 65.3 (MTW)	A chem, n-capt, mass spect (EngeD55) parent Am^{246} (EngeD55)	β⁻ 0.33 max (10%), 0.15 max e⁻ 0.020, 0.038, 0.055, 0.156 γ Am X-rays, 0.044 (30%), 0.180 (10%), 0.224 (25%) daughter radiations from Am^{246}	multiple n-capt from U^{238} (EngeD55, HydE64)
$_{95}Am^{237}$	≈1.3 h (HigG52a)	☢ EC 99+%, α 0.005% (HigG52a) △ 47 (MTW)	B chem, excit (HigG52a)	α 6.02	Pu^{239}(p, 3n), Pu^{239}(d, 4n) (HigG52a)
Am^{238}	1.9 h (GlasR60) others (HigG52a)	☢ EC (StreK50a) no α, lim 3 x 10^{-4}% (HigG52a) △ 48 (MTW)	B chem, excit (StreK50a)	γ Pu X-rays, 0.36 (12%), 0.58 (29%), 0.98 (80%, doublet), 1.35 (76%)	Pu^{239}(p, 2n) (GlasR60) Pu^{239}(d, 3n) (StreK50a, HydE64) Np^{237}(α, 3n) (HydE64)
Am^{239}	12.1 h (GlasR60) 12 h (SeaG49a)	☢ EC 99+%, α 0.005% (GlasR60) EC 99+%, α 0.003% (HigG52a) △ 49.41 (MTW)	A chem, excit (SeaG49a) genet energy levels (SmiW57) daughter Bk^{243} (ThomS50b)	γ Pu X-rays, 0.209 (5%), 0.228 (18%, doublet), 0.278 (17%) e⁻ 0.02-0.04, 0.048, 0.088, 0.106, 0.156 α 5.78	Pu^{239}(p, n) (StreK50a) Pu^{239}(d, 2n) (GlasR60, HigG52a, SeaG49a) Np^{237}(α, 2n) (SeaG49a)

TABLE 1 (*Continued*)

Isotope Z A	Half-life	Type of decay (☸); % abundance; Mass excess ($\triangle$=M-A), MeV (C^{12}=0); Thermal neutron cross section (σ), barns	Class; Identification; Genetic relationships	Major radiations: approximate energies (MeV) and intensities	Principal means of production
$_{95}$Am^{240}	51.0 h (GlasR60) others (SeaG49a)	☸ EC (SeaG49a) no α, lim 0.2% (HigG52a) $\triangle$ 51 (MTW)	A chem, excit (SeaG49) chem, excit, cross bomb (StreK50a) genet energy levels (SmiW57)	γ Pu X-rays, 0.90 (23%), 1.00 (77%) e⁻ 0.022, 0.038, 0.079, 0.094	Pu^{239}(d,n) (StreK50a) Pu^{239}(α,p2n) (GlasR56, VanR58) Np^{237}(α,n) (StreK50a, HydE64)
Am^{241}	458 y sp act (HallG57, WallJ58, HallG56) others (HarvB52) $t_{1/2}$ (SF) 2 x 10^{14} y (DruV61) others (MikV59)	☸ α (SeaG49a) β stable (cons energy) (ForB58) $\triangle$ 52.96 (MTW) σ_c 700 (to Am^{242}) 100 (to Am^{242m}) (GoldmDT64) σ_f 3.0 (GoldmDT64)	A chem, n-capt, excit, mass spect (SeaG49a) daughter Pu^{241} (SeaG49a, CunB49a)	α 5.49 (85%), 5.44 (13%) γ Np L X-rays, 0.060 (36%), 0.101 (0.04%, complex), 0.208 (6 x 10^{-4}%), 0.335 (8 x 10^{-4}%, complex), 0.37 (4 x 10^{-4}%, complex), 0.663 (5 x 10^{-4}%), 0.722 (3 x 10^{-4}%) e⁻ 0.022, 0.038, 0.054	daughter Pu^{241} (HydE64)
Am^{242}	16.01 h (KeeT53) others (BaranS55, SeaG49b)	☸ β⁻ 84%, EC 16% (HoffR59) others (BarnR59, HoffR55, BaranS55) $\triangle$ 55.48 (MTW) σ_f 2900 (GoldmDT64)	A chem, n-capt, genet (MannWM49, SeaG49b) parent Cm^{242} (MannWM49, SeaG49b, AsaF60) parent Pu^{242} (OKelG50, AsaF60) daughter Am^{242m} (AsaF60)	β⁻ 0.67 max e⁻ 0.021, 0.037 γ Pu X-rays, Cm L X-rays	Am^{241}(n,γ), or multiple n-capt from U^{238}, Pu^{239}, etc (HydE64)
Am^{242m}	152 y (BarnR59) others (StreK50a)	☸ IT 99+%, α 0.48% (BarnR59, AsaF60) $\triangle$ 55.52 (LHP, MTW) σ_c 2000 (GoldmDT64) σ_f 6000 (GoldmDT64)	A chem, mass spect, n-capt (SeaG49a, StreK50a) parent Am^{242} (AsaF60) parent Np^{238} (SeaG49a, StreK50a, AsaF60)	α 5.21 (0.41%) e⁻ 0.028, 0.044 γ Am L X-rays, Np X-rays, 0.049 (0.20%), 0.087 (0.036%), 0.110 (0.025%), 0.163 (0.025%) daughter radiations from Am^{242}, Np^{238}	Am^{241}(n,γ) (SeaG49a, MannWM49, AsaF60)
Am^{243}	7.95 x 10^3 y sp act (WallJ58) 7.65 x 10^3 y sp act (BeadA60) others (BarnR59, ButlJP57, HulE57, AsaF54, DiaH53)	☸ α (StreK50a) β stable (cons energy) (ForB58) $\triangle$ 57.18 (MTW) σ_c 74 (GoldmDT64) σ_f <0.07 (GoldmDT64)	A chem, mass spect (StreK50a) parent Np^{239} (StreK50a) daughter Pu^{243} (ThomS51)	α 5.28 (87%), 5.23 (11.5%) γ Np L X-rays, 0.044 (4%), 0.075 (50%) e⁻ [0.011, 0.023, 0.052, 0.069] daughter radiations from Np^{239}	multiple n-capt from U^{238}, Pu^{239}, etc. (HydE64, StreK50a)
Am^{244}	10.1 h (VanS62)	☸ β⁻ (VanS62) $\triangle$ 59.90 (MTW) σ_f 2300 (GoldmDT64)	A chem, n-capt, sep isotopes, genet (VanS62) parent Cm^{244} (VanS62)	β⁻ 0.387 max e⁻ 0.020, 0.037, 0.077, 0.094 γ Cm X-rays, 0.099 (5%), 0.154 (19%), 0.746 (66%), 0.900 (25%)	Am^{243}(n,γ) (VanS62)
Am^{244m}	26 m (GhiA54a)	☸ β⁻ 99%, EC 0.039% (FieP55a) $\triangle$ 60.02 (LHP, MTW)	A chem, n-capt (StreK50a) chem, genet (FieP55a) parent Cm^{244} (ReynF50, FieP55a) parent Pu^{244} (FieP55a)	β⁻ 1.50 max e⁻ 0.020, 0.037 γ Cm L X-rays	Am^{243}(n,γ) (StreK50a)
Am^{245}	2.07 h (ButlJP56a) others (BrowCl55, FieP55)	☸ β⁻ (BrowCl55, FieP55) $\triangle$ 61.93 (MTW)	B chem, genet (BrowCl55, FieP55) daughter Pu^{245} (FieP55, ButlJP56a)	β⁻ 0.91 max e⁻ 0.125 γ Cm X-rays, 0.253	daughter Pu^{245} (ButlJP56a, FieP55, BrowCl55, HydE64)
Am^{246}	25.0 m (EngeD55) others (BrowCl55)	☸ β⁻ (EngeD55, BrowCl55) $\triangle$ 64.9 (MTW)	A chem, genet (BrowCl55, EngeD55) parent Cm^{246} (BrowCl55) daughter Pu^{246} (EngeD55)	β⁻ 2.10 max (7%), 1.60 max γ Cm X-rays, 0.799 (29%), 1.07 (65%, complex)	daughter Pu^{246} (EngeD55, HydE64)
$_{96}$Cm^{238}	2.5 h (StreK48)	☸ EC <90%, α >10% (CarrR52) $\triangle$ 49.39 (MTW)	B chem (StreK48) chem, genet (HigG52a) parent Pu^{234} (HigG52a)	α 6.51 daughter radiations from Pu^{234}	Pu^{239}(α,5n) (GlasR56, StreK48) Pu^{238}(α,4n) (GlasR56)
Cm^{239}	2.9 h (VanR58) 3 h (CarrR52)	☸ EC, no α (lim 0.1%) (CarrR52) $\triangle$ 51 (MTW)	B chem, excit (CarrR52) chem, genet energy levels (VanR58)	γ Am X-rays, 0.188 daughter radiations from Am^{239}	Pu^{239}(α,4n) (CarrR52)

TABLE 1 (*Continued*)

Isotope Z A	Half-life	Type of decay (⚹); % abundance; Mass excess (△ ≡M–A), MeV (C¹²=0); Thermal neutron cross section (σ), barns	Class; Identification; Genetic relationships	Major radiations: approximate energies (MeV) and intensities	Principal means of production
$_{96}$Cm^{240}	26.8 d (SeaG49b) $t_{1/2}$ (SF) 7.9 x 10^5 y (GhiA52)	⚹ α (SeaG49b) no EC, lim 0.5% (HigG52) △ 51.72 (MTW)	A chem, genet (SeaG49b) parent Pu^{236} (SeaG49b) daughter Cf^{244} (ChetA56)	α 6.29 (72%), 6.25 (28%) daughter radiations from Pu^{236}	Pu^{239}(α, 3n) (GlasR56)
Cm^{241}	35 d (HigG52)	⚹ EC 99%, α 1.0% (GlasR56) △ 53.73 (MTW)	A chem, excit, cross bomb (SeaG49b, HigG52, GlasR59) parent Pu^{237m} (StepF57a)	γ Am X-rays, 0.475 (95%), 0.60 e⁻ 0.123, 0.350 α 5.94 daughter radiations from Pu^{237} daughter radiations from Pu^{237m} included in above listing	Pu^{239}(α, 2n) (GlasR56)
Cm^{242}	162.5 d (GloK54, HannG50) 164.4 d (FlyK65a) others (HutWP54) $t_{1/2}$ (SF) 7.2 x 10^6 y (HannG51)	⚹ α (SeaG49b) β stable (cons energy) (ForB58) △ 54.82 (MTW) σ$_c$ 20 (GoldmDT64) σ$_f$ <5 (GoldmDT64)	A chem, genet (SeaG49b) mass spect (ReynF50) daughter Am^{242} (AsaF60, MannWM49, SeaG49b) daughter Cf^{246} (HulE51) parent Pu^{238} (SeaG49b)	α 6.12 (74%), 6.07 (26%) γ Pu L X-rays, 0.044 (0.041%), 0.102 (4 x 10⁻³%), 0.158 (2.5 x 10⁻³%), 0.58 (3.2 x 10⁻⁴%, complex), 0.89 (3 x 10⁻⁵%) e⁻ 0.022, 0.039 daughter radiations from Pu^{238}	daughter Am^{242m}, from Am^{241}(n, γ), or multiple n-capt from U^{238}, Pu^{239}, etc. (HydE64)
Cm^{243}	32 y sp act + mass spect (AsaF57a, HydE64) others (ThomS50b)	⚹ α (ReynF50) EC 0.3% (ChoG58) △ 57.19 (MTW) σ$_c$ 250 (GoldmDT64) σ$_f$ 660 (GoldmDT64)	A chem, mass spect, genet (ReynF50) daughter Bk^{243} (ThomS50b)	α 6.06 (6%, doublet), 5.99 (6%, doublet), 5.79 (73%), 5.74 (11.5%) γ Pu X-rays, 0.209 (4%), 0.228 (12%), 0.278 (14%) e⁻ 0.02–0.04, 0.048, 0.088, 0.106, 0.156	multiple n-capt from U^{238}, Pu^{239}, etc. (HydE64, ReynF50)
Cm^{244}	17.6 y sp act + mass spect (CarnW61) others (FrieA54, StevC54) $t_{1/2}$ (SF) 1.31 x 10^7 y (MetD65) 1.46 x 10^7 y (MalkL63a) others (GhiA52)	⚹ α (ReynF50) β stable (cons energy) (ForB58) △ 58.47 (MTW) σ$_c$ 15 (GoldmDT64)	A chem, mass spect (ReynF50) daughter Am^{244m} (ReynF50, FieP55a) daughter Am^{244} (VanS62) daughter Bk^{244} (GuseL56, ChetA56b) daughter Cf^{248} (HulE54) parent Pu^{240} (FrieA54)	α 5.81 (77%), 5.77 (23%) γ Cm L X-rays, 0.043 (0.02%), 0.100 (0.0015%), 0.150 (0.0013%), 0.262 (1.4 x 10⁻⁴%), 0.59 (2.5 x 10⁻⁴%, doublet), 0.82 (7 x 10⁻⁵%) e⁻ 0.022, 0.038	multiple n-capt from U^{238}, Pu^{239}, Am^{243} etc. (HydE64)
Cm^{245}	9.3 x 10^3 y genet, mass spect (CarnW61) others (HuiJ57b, BrowCI55, FrieA54)	⚹ α (HulE51) β stable (cons energy) (ForB58) △ 61.02 (MTW) σ$_c$ 200 (GoldmDT64) σ$_f$ 1900 (GoldmDT64)	A chem, decay charac, genet (HulE51) chem, mass spect (StevC54, HulE54) daughter Bk^{245} (HulE54, HulE51) parent Pu^{241} (FrieA54)	α 5.36 (80%), 5.31 (7%) γ Pu X-rays, 0.13 (5%), 0.173 (14%) daughter radiations from Pu^{241}, Am^{241}	multiple n-capt from U^{238}, Pu^{239}, Am^{243}, Cm^{244}, etc. (StevC54, FieP56) daughter Bk^{245} (HulE51, HulE54) (HydE64)
Cm^{246}	5.5 x 10^3 y genet (CarnW61) others (ButlJP56b, BrowCI55, FrieA54) $t_{1/2}$ (SF) 1.7 x 10^7 y (MetD65) others (FrieS56)	⚹ α (FrieA54, StevC54) β stable (cons energy) (ForB58) △ 62.64 (MTW) σ$_c$ 15 (GoldmDT64)	A chem, mass spect (StevC54, FieP56) parent Pu^{242} (FrieA54) daughter Am^{246} (BrowCI55) daughter Cf^{250} (ButlJP56b)	α 5.39 (81%), 5.34 (19%) γ [Pu L X-rays]	multiple n-capt from U^{238}, Pu^{239}, Cm^{244} etc. (HydE64, StevC54, FieP56) daughter Cf^{250} (ButlJP56b)
Cm^{247}	$t_{1/2}$ (α) 1.6 x 10^7 y genet + mass spect (FieP63) $t_{1/2}$ (α) >4 x 10^7 y genet + mass spect (DiaH57, StevC54)	⚹ [α] (DiaH57, StevC54) △ 65.56 (MTW) σ$_c$ 180 (GoldmDT64)	A chem, mass spect (StevC54, DiaH57) daughter Cf^{251} (EasT57)		multiple n-capt from U^{238}, Pu^{239}, Cm^{244} etc. (HydE64, DiaH57, StevC54)
Cm^{248}	4.7 x 10^5 y sp act (ButlJP56b) $t_{1/2}$ (SF) 4.6 x 10^6 y (ButlJP56b)	⚹ α 89%, SF 11% (ButlJP56b) β stable (cons energy) (ForB58) △ 67.43 (MTW) σ$_c$ 6 (GoldmDT64)	B chem, genet (ButlJP56b) daughter Cf^{252} (ButlJP56b)	α 5.08 (82%), 5.04 (18%) γ [Pu L X-rays] SF fission fragments, neutrons, γ rays, electrons, daughter radiations	daughter Cf^{252} (ButlJP56b) multiple n-capt from U^{238}, Pu^{239}, Cm^{244}, etc. (HydE64)
Cm^{249}	64 m (EasT58) 65 m (FieP56)	⚹ β⁻ (FieP56) △ 70.8 (MTW)	B n-capt, chem (FieP56)	β⁻ 0.9 max	Cm^{248}(n, γ) (EasT58) multiple n-capt from U^{238}, Pu^{239}, Cm^{244}, etc. (ThomS54, FieP56, HydE64)

TABLE 1 (*Continued*)

Isotope Z A	Half-life	Type of decay (☢); % abundance; Mass excess (Δ≡M−A), MeV (C^{12}=0); Thermal neutron cross section (σ), barns	Class; Identification; Genetic relationships	Major radiations: approximate energies (MeV) and intensities	Principal means of production
$_{96}Cm^{250}$	$t_{1/2}$ (SF): 1.7 x 10^4 y (GrouCR66) 2 x 10^4 y (HuiJ57b) others (FieP56)	☢ SF (HuiJ57b) Δ 73 (MTW)	A chem, decay charac (HuiJ57b) chem, mass spect (GrouCR66)	SF fission fragments, neutrons, γ rays, electrons, daughter radiations	multiple n-capt from U^{238} (HuiJ57b, HydE64)
$_{97}Bk^{243}$	4.6 h (ThomS50b, GhiA54) 4.5 h (ChetA56b) others (HulE51)	☢ EC 99+%, α 0.15% (ChetA56b) Δ 58.70 (MTW)	A chem, genet (ThomS50, ThomS50b) parent Cm^{243} (ThomS50b) parent Am^{239} (ThomS50b)	α 6.76 (0.023%), 6.72 (0.019%), 6.57 (0.038%), 6.54 (0.029%), 6.21 (0.020%) γ Cm X-rays, 0.755, 0.84, 0.946	Am^{241}(α, 2n) (ThomS50b) Cm^{242}(d, n) (HulE51) Am^{243}(α, 4n) (ChetA56b) (HydE64)
Bk^{244}	4.4 h (ChetA56b)	☢ EC 99+%, α 0.006% (ChetA56b) Δ 61 (MTW)	B chem, excit, genet (ChetA56b) parent Cm^{244} (ChetA56b, GuseL56)	α 6.67 (0.003%), 6.62 (0.003%) γ Cm X-rays, 0.145 († 7), 0.188 († 16), 0.218 († 100), 0.334 († 10), 0.490 († 14), 0.892 († 88), 0.922 († 17), 1.16 († 11, doublet)	Am^{243}(α, 3n) (ChetA56b) [Cm^{244}(d, 2n)], [Cm^{244}(p, n)], Am^{241}(α, n) (HydE64)
Bk^{245}	4.98 d (MagL56) others (HulE51)	☢ EC 99+%, α 0.11% (MagL56) Δ 61.84 (MTW)	A chem, excit, decay charac (HulE51) daughter Cf^{245} (ChetA56) parent Cm^{245} (HulE51, HulE54)	α 6.36 (0.018%), 6.32 (0.017%), 6.15 (0.021%), 6.12 (0.016%), 5.89 (0.024%) γ Cm X-rays, 0.253 (31%), 0.39 (3%, doublet) e^- 0.125	Am^{243}(α, 2n) (ChetA56b) Cm^{244}(d, n) (HulE51) Cm^{242}(α, p) (HulE51) (HydE64)
Bk^{246}	1.8 d (HulE54)	☢ EC (HulE54) Δ 64 (MTW)	B chem, decay charac, excit (HulE54, ChetA56b)	γ Cm X-rays, 0.800 (40%), 1.07 (12%, complex)	Cm^{244}(α, pn), Am^{243}(α, n) (HulE54, ChetA56b, HydE64)
Bk^{247}	1.4 x 10^3 y (MilsJ65) others (ChetA56b)	☢ α, no EC (ChetA56b) Δ 65.47 (MTW)	B chem, decay charac (ChetA56b)	α 5.68 (37%), 5.52 (58%) γ Am X-rays, 0.084 (40%), 0.27 (30%) daughter radiations from Am^{243} etc.	daughter Cf^{247}, Cm^{244}(α, p), Cm^{245-6}(α, pxn) (HydE64, ChetA56b)
Bk^{248}	16 h (ChetA56b) 23 h genet (HulE56)	☢ $β^-$ 70%, EC 30% (ChetA56b) Δ 67.9 (MTW)	B n-capt, chem, genet (ChetA56b) parent Cf^{248} (HulE56, ChetA56b)	$β^-$ 0.65 max γ Cm X-rays daughter radiations from Cf^{248}	Bk^{247}(n, γ) (ChetA56b) Cm^{245}(α, p) (HulE56) (HydE64)
Bk^{248}	>9 y sp act + mass spect (MilsJ65) $t_{1/2}$ ($β^-$) >10^4 y genet (MilsJ65)	☢ ?	B chem, mass spect (MilsJ65)		Cm^{246}(α, pn) (MilsJ65)
Bk^{249}	314 d (EasT57) others (MagL54, DiaH54) $t_{1/2}$ (SF): 6 x 10^8 y (HydE57) >1.5 x 10^9 y (EasT57)	☢ $β^-$ 99+%, α 0.0022% (EasT57) others (MagL54, DiaH54) Δ 69.86 (MTW) $σ_c$ 500 (GoldmDT64)	A chem, genet (ThomS54, GhiA54a, DiaH54) chem, mass spect (FieP56) parent Cf^{249} (GhiA54a, MagL54)	$β^-$ 0.125 max α 5.42 (0.0015%) γ 0.32 (3 x 10^{-5}%, doublet) daughter radiations from Cf^{249}, Am^{245}	multiple n-capt from U^{238}, Pu^{239}, Cm^{244}, etc. (ThomS54, DiaH54, MagL54, FieP56, HydE64)
Bk^{250}	193.3 m (VanS59) others (GhiA54a, MagL54)	☢ $β^-$ (GhiA54a) Δ 72.95 (MTW)	A n-capt, chem, genet (GhiA54a) parent Cf^{250} (GhiA54a) daughter Es^{254} (HarvB55, JonM56)	$β^-$ 1.76 max (11%), 0.73 max e^- 0.019, 0.036 γ Cf L X-rays, 0.990 (47%), 1.032 (39%)	Bk^{249}(n, γ) (GhiA54a) daughter Es^{254} (HarvB55, JonM56) (HydE64)
$_{98}Cf^{244}$	25 m (ChetA56) others (ThomS50c, ThomS50a, GhiA51, GhiA54, GuseL56)	☢ α (ChetA56) Δ 61.43 (MTW)	A chem, excit, genet (ThomS50a, ChetA56) parent Cm^{240} (ChetA56) daughter Fm^{248} (GhiA58)	α 7.18	Cm^{244}(α, 4n) (ChetA56) Cm^{242}(α, 2n) (ChetA56) U^{238}(C^{12}, 6n) (HydE64)
Cf^{245}	44 m (ThomS50c) others (ThomS50a, GhiA51, GhiA54)	☢ EC 70%, α 30% (ChetA56) Δ 63.38 (MTW)	B chem, excit, genet (ChetA56) parent Bk^{245} (ChetA56) not parent Cm^{240} (ChetA56) daughter Fm^{249} (PerelV59)	α 7.12 daughter radiations from Bk^{245} Cm^{241}	Cm^{244}(α, 3n) (ChetA56) Cm^{242}(α, n) (ChetA56) U^{238}(C^{12}, 5n) (GhiA51, GhiA54) (HydE64)
Cf^{246}	35.7 h (HulE51) $t_{1/2}$ (SF) 2.1 x 10^3 y (HulE53)	☢ α (GhiA51) Δ 64.11 (MTW)	A chem, genet (GhiA51) parent Cm^{242} (HulE51) daughter Es^{246} (GhiA54)	α 6.76 (78%), 6.72 (22%) γ Cm L X-rays daughter radiations from Cm^{242}	Cm^{244}(α, 2n) (ChetA56, HulE51) U^{238}(C^{12}, 4n) (GhiA51) (HydE64)

TABLE 1 (*Continued*)

Isotope Z A	Half-life	Type of decay (☢); % abundance; Mass excess (△≡M–A), MeV (C¹²=0); Thermal neutron cross section (σ), barns	Class; Identification; Genetic relationships	Major radiations: approximate energies (MeV) and intensities	Principal means of production
$_{98}Cf^{247}$	2.5 h (HulE54, ChetA56b) others (GhiA54)	☢ EC (HulE54) △ 66 (MTW)	B chem (HulE54) chem, excit (ChetA56b)	γ Bk X-rays, 0.295 (1%), 0.417, 0.460 e⁻ 0.164	Cm²⁴⁴(α,n) (HulE54) Cm²⁴⁵⁻⁶(α,xn) (HydE64) U²³⁸(N¹⁴,p4n) (GhiA54)
Cf^{248}	350 d genet (HulE57a) others (GhiA54) $t_{1/2}$ (SF) ≥1.5 x 10⁴ y (HulE57a)	☢ α (GhiA54, HulE54) β stable (cons energy) (ForB58) △ 67.26 (MTW)	A chem, genet (GhiA54, HulE54) parent Cm²⁴⁴ (HulE54) daughter 16 h Bk²⁴⁸ (HulE56, ChetA56b) daughter Fm²⁵² (FrieA56) daughter Es²⁴⁸ (ChetA56a)	α 6.27 (82%), 6.22 (18%) γ [Cm L X-rays]	Cm²⁴⁵⁻²⁴⁸(α,xn) (HulE54) U²³⁸(N¹⁴,p3n) (GhiA54) daughter Bk²⁴⁸, Es²⁴⁸, Fm²⁵² (HydE64)
Cf^{249}	360 y genet (EasT57) others (MagL54, GhiA54a) $t_{1/2}$ (SF): 1.5 x 10⁹ y (HydE57) others (DiaH54, MagL54)	☢ α (ThomS54) β stable (cons energy) (ForB58) △ 69.74 (MTW) σ_c 270 (GoldmDT64) σ_f 1735 (GoldmDT64)	A chem, genet (ThomS54, GhiA54a) chem, genet, mass spect (DiaH54, MagL54, FieP56) daughter Bk²⁴⁹ (GhiA54a, MagL54)	α 5.81 (84%) γ Cm X-rays, 0.333 (16%), 0.388 (72%)	daughter Bk²⁴⁹ (GhiA54a, DiaH54, MagL54, HydE64) multiple n-capt from U²³⁸, Pu²³⁹, Cm²⁴⁴, etc. (HydE64)
Cf^{250}	13.2 y genet (MetD65) 13 y (PhiL63) others (EasT57, MagL54, GhiA54a) $t_{1/2}$ (SF) 1.7 x 10⁴ y (MetD65, PhiL63) others (MagL54, DiaH54, GhiA54a)	☢ α (GhiA54a) β stable (cons energy) (ForB58) △ 71.19 (MTW) σ_c 1500 (GoldmDT64) σ_f <350 (GoldmDT64)	A chem, genet (ThomS54, GhiA54a) chem, mass spect (DiaH54, MagL54) daughter Bk²⁵⁰ (GhiA54a) daughter Fm²⁵⁴ (PhiL63) parent Cm²⁴⁶ (ButlJP56b)	α 6.03 (83%), 5.99 (17%) e⁻ 0.023, 0.038 γ [Cm L X-rays]	multiple n-capt from U²³⁸, Pu²³⁹, Cm²⁴⁴, etc. (MagL54) daughter Bk²⁵⁰ (GhiA54a, PhiL63) daughter Fm²⁵⁴ (LedC63) (HydE64)
Cf^{251}	≈800 y genet (EasT57) others (MagL54)	☢ α (EasT57) β stable (cons energy) (ForB58) △ 74.15 (MTW) σ_c 3000 (GoldmDT64) σ_f 3000 (GoldmDT64)	A chem, mass spect (DiaH54, MagL54) parent Cm²⁴⁷ (EasT57)	α 5.85 (45%), 5.67 (55%) γ Cm X-rays, 0.18	multiple n-capt from U²³⁸, Pu²³⁹, Cm²⁴⁴, etc. (EasT57, MagL54, DiaH54, HydE64)
Cf^{252}	2.646 y (MetD65) others (MagL54, EasT57, FieP56, GhiA54a) $t_{1/2}$ (SF): 85 y (MetD65) others (GhiA54a, EasT57, MagL54, SevK61)	☢ α 96.9%, SF 3.1% (MetD65) α 97.0%, SF 3.0% (AsaF66a) β stable (cons energy) (ForB58) △ 76.05 (MTW) σ_c 30 (GoldmDT64)	A chem (ThomS54, GhiA54a) chem, mass spect (StuM54, MagL54, DiaH54) parent Cm²⁴⁸ (ButlJP56b)	α 6.12 (82%), 6.08 (15%) e⁻ 0.022, 0.038 γ Cm L X-rays SF fission fragments, neutrons, γ rays, electrons, daughter radiations	multiple n-capt from U²³⁸, Pu²³⁹, Cm²⁴⁴, etc. (GhiA54, DiaH54, MagL54, FieP56, HydE64)
Cf^{253}	17.6 d genet (MetD65) 17 d genet (EasT57) 18 d (DiaH54, MagL54) others (ChoG54)	☢ β⁻ 99+%, α 0.31% (GrouCR66) △ 79.3 (MTW)	A chem, genet (ChoG54, DiaH54, MagL54) chem, mass spect (FieP56) parent Es²⁵³ (ChoG54, MagL54) [daughter Fm²⁵⁷] (HulE64)	β⁻ 0.27 max α 5.98 daughter radiations from Es²⁵³	multiple n-capt from U²³⁸, Pu²³⁹, Cm²⁴⁴, Cf²⁵², etc. (MagL54, ThomS54, ChoG54, HydE64)
Cf^{254}	60.5 d (PhiL63, MetD65) others (HuiJ57b, FieP56, HarvB55)	☢ SF 99+%, α ≈0.2% (AsaF66a) β stable (cons energy) (ForB58) △ 81 (MTW) σ_c <2 (GoldmDT64)	A chem, genet (HarvB55) chem, mass spect (FieP56) daughter Es²⁵⁴m (HarvB55, FieP56) not daughter Fm²⁵⁷ (HulE64)	SF fission fragments, neutrons, γ rays, electrons, daughter radiations α 5.84	multiple n-capt from U²³⁸, Pu²³⁹, Cm²⁴⁴, Cf²⁵², etc. (FieP56, DiaH60) daughter Es²⁵⁴m (0.08%) (HarvB55, FieP56) (HydE64)
$_{99}Es^{245}$	1.3 m (GhiA61a, MikV66)	☢ α 17%, EC 83% (MikV66) △ 66 (MTW)	B cross bomb (GhiA61a) cross bomb, excit, genet (MikV66) parent Cf²⁴⁵ (MikV66)	α 7.70 daughter radiations from Cf²⁴⁵	U²³⁵(N¹⁴,4n), U²³⁸(N¹⁴,7n) (MikV66) Np²³⁷(C¹²,4n), Pu²⁴⁰(B¹⁰,5n) (GhiA61a)
Es^{246}	7.3 m (GhiA54) 7.7 m (MikV66) others (GuseL56)	☢ α 10%, EC 90% (MikV66) △ 68 (MTW)	D chem, decay charac, genet (GhiA54) excit, genet (MikV66) parent Cf²⁴⁶ (GhiA54, MikV66)	α 7.33	U²³⁸(N¹⁴,6n) (GhiA54, MikV66, HydE64)

TABLE 1 (*Continued*)

Isotope Z A	Half-life	Type of decay ($\frac{r}{r}$); % abundance; Mass excess ($\triangle \equiv$ M–A), MeV (C¹²=0); Thermal neutron cross section (σ), barns	Class; Identification; Genetic relationships	Major radiations: approximate energies (MeV) and intensities	Principal means of production
$_{99}Es^{247}$	5.0 m (MikV66)	α ≈7%, EC ≈93% (MikV66) Δ 68 (MTW)	C excit (MikV66)	α 7.33	$U^{238}(N^{14}, 5n)$ (MikV66)
Es^{248}	25 m (ChetA56a)	EC 99+%, α ≈0.3% (ChetA56a) Δ 70 (MTW)	B chem, excit, genet (ChetA56a) parent Cf^{248} (ChetA56a)	α 6.88	$Cf^{249}(d, 3n)$ (ChetA56a) (HydE64)
Es^{249}	2 h (HarvB56)	EC 99+%, α 0.13% (HarvB56) Δ 71.15 (MTW)	B chem, excit (HarvB56)	α 6.77	$Bk^{249}(α, 4n)$ (HarvB56) $Cf^{249}(d, 2n)$ (ChetA56a) $Cf^{249}(α, p3n)$ (HarvB56) (HydE64)
Es^{250}	8 h (HarvB56)	EC (HarvB56) Δ 73 (MTW)	B chem, excit (HarvB56)	γ [Cf X-rays]	$Bk^{249}(α, 3n)$, $Cf^{249}(d, n)$, $Cf^{249}(α, t)$ (HydE64)
Es^{251}	1.5 d (HarvB56)	EC 99+%, α 0.53% (HarvB56) Δ 74.5 (MTW)	B chem, excit (HarvB56)	α 6.49	$Bk^{249}(α, 2n)$ (HarvB56)
Es^{252}	≈140 d (HarvB56)	α, no β⁻, lim 3%, no EC (HarvB56) EC and β⁻ unstable (cons energy) (MTW) Δ 77.1 (MTW)	B chem, excit (HarvB56)	α 6.64 (82%), 6.58 (13%) γ Bk X-rays, 0.074 (0.07%), 0.154 (0.07%), 0.198 (0.08%), 0.228 (0.23%), 0.278 (0.21%), 0.40 (1.1%, complex)	$Bk^{249}(α, n)$ (HarvB56) $Cf^{252}(d, 2n)$ (MHarW65)
Es^{253}	20.47 d (HalvS66) 20.7 d (GrouCR66) 20.03 d (JonM56) others (FieP54, ChoG54) $t_{1/2}$ (SF): 6.4 x 10⁵ y (MetD65) 7 x 10⁵ y (JonM56) others (FieP54, StuM54)	α (ThomS54) β stable (cons energy) (ForB58) Δ 79.03 (MTW) σ_c 300 (to Es^{254m})	A chem, genet (ThomS54, ChoG54, StuM54) daughter Cf^{253} (ChoG54, MagL54) daughter Fm^{253} (AmiS57) descendant Fm^{257} (SikT65)	α 6.64 (90%) e⁻ 0.017, 0.027, 0.035, 0.040 γ Bk X-rays, 0.387 (0.05%, complex), 0.429 (0.008%, doublet)	daughter Cf^{253} (from multiple n-capt) (JonM56, StuM54, ThomS54, HydE64)
Es^{254}	276 d (UniJ66) 480 d (SchumR58, JonM56) others (HarvB55) $t_{1/2}$ (SF) 7 x 10⁵ y (MHarW65)	α, no β⁻, lim 3 x 10⁻⁴% (MHarW66) Δ 82.00 (MTW) σ_c <40 (GoldmDT64)	A chem, genet (HarvB55, JonM56) parent Bk^{250} (HarvB55, JonM56) not parent Fm^{254}, lim 3 x 10⁻⁴% (MHarW66)	α 6.44 (93%) γ Bk X-rays, 0.063 (2.0%), 0.27 (0.12%, complex), 0.31 (0.22%, doublet), 0.39 (0.07%, complex) e⁻ 0.011, 0.018, 0.030, 0.037 daughter radiations from Bk^{250}, Cf^{250}	multiple n-capt from U^{238}, Pu^{239}, Cm^{244}, Cf^{252}, Es^{253}, etc. (JonM56, HarvB55, HydE64)
Es^{254m}	39.3 h (UniJ62) others (FieP54, ChoG54) $t_{1/2}$ (SF) >10 y (FieP54)	β⁻ 99+%, EC 0.08% (PhiL63) others (HarvB55) Δ 82.10 (MTW)	A n-capt, chem, decay charac (FieP54, ChoG54, HarvB55) parent Fm^{254} (FieP54, ChoG54) parent Cf^{254} (HarvB55, FieP56)	β⁻ 1.13 max (25%), 0.43 max e⁻ 0.020, 0.038 γ Fm X-rays, 0.65 (31%), 0.69 (38%, complex) daughter radiations from Fm^{254}	multiple n-capt from U^{238}, Pu^{239}, Cm^{244}, Cf^{252}, Es^{253}, etc. (FieP54, ChoG54, HydE64)
Es^{255}	38.3 d (HalvS66) others (GrouCR66, MHarW66, JonM56, ChoG54) $t_{1/2}$ (SF) >170 y (GrouCR66)	β⁻ 91.5%, α 8.5% (GrouCR66) Δ 84 (MTW)	B chem, genet (ChoG54, JonM56) parent Fm^{255} (ChoG54, JonM56)	α 6.31 daughter radiations from Fm^{255} [Bk^{251}]	multiple n-capt from U^{238}, Pu^{239}, Cm^{244}, Cf^{252}, Es^{253}, etc. (JonM56, ChoG54, DiaH60, FieP56, GhiA55a, HydE64)
Es^{256}	short (ChoG55)	[β⁻] (ChoG55)	F (ChoG55)		$E^{255}(n, γ)$ (ChoG55, HydE64)
$_{100}Fm^{248}$	0.6 m genet (GhiA58) others (GuseL56)	[α] (GhiA58) Δ 72 (MTW)	B genet, chem (GhiA58) parent Cf^{244} (GhiA58) daughter 102^{252} (MikV66a, GhiA67)		$Pu^{240}(C^{12}, 4n)$ (GhiA58) $U^{238}(O^{16}, 6n)$ (GuseL56) (HydE64)
Fm^{249}	≈2.5 m (PerelV59)	α (PerelV59) β⁻ unstable (cons energy) (MTW) Δ 73.8 (MTW)	B genet, excit, decay charac (PerelV59) parent Cf^{245} (PerelV59)	α 7.9	$U^{238}(O^{16}, 5n)$ (PerelV59)

TABLE 1 (*Continued*)

Isotope Z A	Half-life	Type of decay (☢); % abundance; Mass excess (△≡M–A), MeV (C¹²=0); Thermal neutron cross section (σ), barns	Class; Identification; Genetic relationships	Major radiations: approximate energies (MeV) and intensities	Principal means of production
$_{100}Fm^{250}$	30 m (AmiS57a, AttH54) others (DoneE62)	☢ α, EC ? (AmiS57a) △ 74.10 (MTW)	B chem, excit (AttH54, AmiS57a) daughter 102^{254} (GhiA58, DoneE65, MikV66a, GhiA67)	α 7.44	$Cf^{249}(α, 3n)$ (AmiS57a) $U^{238}(O^{16}, 4n)$ (AttH54) (HydE64)
Fm^{251}	7 h (AmiS57a)	☢ EC ≈99%, α ≈1% (AmiS57a) △ 76 (MTW)	B chem, excit (AmiS57a)	α 6.89 γ [Es X-rays] daughter radiations from Es^{251}	$Cf^{249}(α, 2n)$ (AmiS57a)
Fm^{252}	22.7 h (FrieA56) others (AmiS57a) $t_{1/2}$ (SF) >8 y (FrieA56)	☢ α (FrieA56) β stable (cons energy) (ForB58) △ 76.84 (MTW)	B chem, genet (FrieA56) chem, excit (AmiS57a) parent Cf^{248} (FrieA56) daughter 102^{256} (DoneE64)	α 7.05	$Cf^{250-252}(α, xn)$ (FrieA56) $Cf^{249}(α, n)$ (AmiS57a)
Fm^{253}	3 d (AmiS57) >10 d (FrieA56)	☢ EC 89%, α 11% (AmiS57) △ 80 (MTW)	B chem (FieP56) chem, genet (AmiS57) parent Es^{253} (AmiS57)	α 6.96 (9%), 6.91 (2%) daughter radiations from Es^{253}	$Cf^{252}(α, 3n)$ (FrieA56, AmiS57)
Fm^{254}	3.24 h (JonM56) others (FieP54, StuM54, ChoG54, HarvB54) $t_{1/2}$ (SF): 246 d (JonM56) 220 d (FieP54) 200 d (ChoG54)	☢ α 99+%, SF 0.055% (JonM56) β stable (cons energy) (ForB58) △ 80.93 (MTW)	A chem, genet (HarvB54, ChoG54, FieP54, StuM54) daughter Es^{254m} (ChoG54, FieP54) not daughter Es^{254}, lim 3 x 10^{-4}% (MHarW66) parent Cf^{250} (PhiL63)	α 7.20 (82%), 7.16 (17%) γ Cf L X-rays e⁻ 0.019, 0.036	daughter Es^{254m} (StuM54a, ChoG54, HydE64)
Fm^{255}	20.1 h (AsaF64) others (JonM56, ChoG54) $t_{1/2}$ (SF): 1 x 10^4 y (PhiL63) others (HydE57)	☢ α (ChoG54) β stable (cons energy) (ForB58) △ 83.82 (MTW)	B chem, genet (ChoG54) daughter Es^{255} (ChoG54, JonM56) daughter Md^{255} (PhiL58)	α 7.03 (93%) γ Cf L X-rays, 0.059 (0.9%, doublet), 0.081 (1.1%, doublet) e⁻ 0.032, 0.05-0.07	daughter Es^{255} (ChoG54, JonM56, HydE64)
Fm^{256}	2.7 h (PhiL58, SikT65) others (ChoG55)	☢ SF 97%, α 3% (SikT65) β stable (cons energy) (ForB58) △ 85.44 (MTW)	B chem, decay charac (ChoG55) SF daughter Md^{256} (PhiL58)	fission fragments, neutrons, γ rays, electrons, daughter radiations α 6.86	$Es^{255}(n,γ)[Es^{256}](β^-)$ (ChoG55, HydE64) daughter Md^{256} (PhiL58, SikT65)
Fm^{257}	80 d (SikT65) 79 d (HulE64) 94 d (GrouCR66) others (AsaF66b) $t_{1/2}$ (SF) 100 y (HulE64) 94 y (AsaF66b) others (GrouCR66)	☢ α (HulE64) △ 88.6 (MTW)	B chem, [genet], excit [parent Cf^{253}], not parent Cf^{254} (HulE64) ancestor Es^{253}, daughter Md^{257} (SikT65)	α 6.53 (94%) γ Cf X-rays, 0.180 (8%), 0.242 (10%) e⁻ 0.037, 0.045, 0.055, 0.106 daughter radiations from Cf^{253}, Es^{253}	multiple n-capt from Pu^{242}, Am^{243}, Cm^{244} etc. (HulE64, AsaF66b)
$Fm^{258?}$	≈11 d (GatR63) ≲2 h (GrouCR66)	☢ SF (GatR63)	G chem, decay charac (GatR63) activity not observed (GrouCR66)		multiple n-capt from Cm^{244} (GatR63)
$_{101}Md^{255}$	0.6 h (SikT65) ≈0.5 h (PhiL58)	☢ EC 90%, α 10% (SikT65) △ 84.4 (MTW)	B chem, genet (PhiL58) parent Fm^{255} (PhiL58)	α 7.34 daughter radiations from Fm^{255}	$Es^{253}(α, 2n)$ (PhiL58) B^{11}, C^{12}, C^{13} on Cf^{252} (SikT65)
Md^{256}	1.5 h (PhiL58, SikT65) others (GhiA55)	☢ EC 97%, α 3% (SikT65) △ 86.9 (MTW)	B chem (GhiA55) chem, genet (PhiL58) parent Fm^{256} (PhiL58)	α 7.18 daughter radiations from Fm^{256}	$Es^{253}(α, n)$ (GhiA55) B^{11}, C^{12}, C^{13} on Cf^{252} (SikT65)
Md^{257}	3 h (SikT65)	☢ EC ≈92%, α ≈8%, no SF, lim 10% (SikT65) △ 89 (MTW)	D chem, excit, decay charac (SikT65) parent Fm^{257} (SikT65)	α 7.25 ?, 7.08	B^{11}, C^{12}, C^{13} on Cf^{252} (SikT65)
102^{251}	0.8 s (GhiA67)	☢ α (GhiA67)	E excit, decay charac, cross bomb (GhiA67)	α 8.68 ? (20%), 8.58 (80%)	$Cm^{244}(C^{12}, 5n)$ (GhiA67)
102^{252}	2.1 s (GhiA67) 5 s (MikV66a) 3 s (GhiA58, GhiA59)	☢ α ≈70%, SF ≈ 30% (GhiA59) α (MikV66a) △ 83 (LHP, MTW)	C excit, decay charac (GhiA59) excit, genet, cross bomb, decay charac (MikV66a, GhiA67) parent Fm^{248} (MikV66a, GhiA67) formerly assigned to 102^{254} (GhiA58, GhiA59)	α 8.41	$Cm^{244}(C^{12}, 4n)$ (GhiA67, GhiA59) $Cm^{244}(C^{13}, 5n)$ (GhiA67) $Pu^{239}(O^{18}, 5n)$ (MikV66a)

TABLE 1 (*Continued*)

Isotope Z A	Half-life	Type of decay (☢); % abundance; Mass excess (Δ=M−A), MeV (C^{12}=0); Thermal neutron cross section (σ), barns	Class; Identification; Genetic relationships	Major radiations: approximate energies (MeV) and intensities	Principal means of production
102^{253}	95 s (MikV66a) 100 s (GhiA67)	☢ α (MikV66a, GhiA67) Δ 84 (LHP, MTW)	C excit, cross bomb, genet (MikV66a) excit, cross bomb, genet (GhiA67) parent Fm^{249} (MikV66a, GhiA67)	α 8.02	Cm^{244}(C^{13}, 4n), Cm^{246}(C^{12}, 5n) (GhiA67) Pu^{242}(O^{16}, 5n), Pu^{239}(O^{18}, 4n) (MikV66a)
102^{254}	55 s (GhiA67) 50 s (DubG66) 75 s (MikV66a) others (DoneE65, ZagB65)	☢ α (ZagB65, GhiA67, MikV66a) no SF, lim 0.06% (FleG66) Δ 84.8 (LHP, MTW)	C genet (GhiA58, GhiA59) genet, excit (DoneE65) excit, decay charac, cross bomb (MikV66a, GhiA67) parent Fm^{250} (GhiA58, GhiA59, DoneE65, MikV66a, GhiA67)	α 8.10	Cm^{246}(C^{12}, 4n) (GhiA67, GhiA58, GhiA59) Cm^{246}(C^{13}, 5n), Cm^{244}(C^{13}, 3n) (GhiA67) Pu^{242}(O^{16}, 4n) (MikV66a) Am^{243}(N^{15}, 4n) (DoneE65, ZagB65, MikV66) U^{238}(Ne^{22}, 6n) (DoneE65)
102^{255}	180 s (DubG66, GhiA67) 2 m (AkaGN66)	☢ α (AkaGN66, DubG66, GhiA67) Δ 87 (LHP, MTW)	C excit, cross bomb, decay charac (AkaGN66) excit, cross bomb, decay charac (DubG66, GhiA67)	α 8.11	Cm^{246}(C^{13}, 4n), Cm^{248}(C^{12}, 5n) (GhiA67) Pu^{242}(O^{18}, 5n) (DubG66) U^{238}(Ne^{22}, 5n) (AkaGN66, DubG66)
102^{256}	2.7 s (GhiA67) 6 s (AkaGN66) 9 s (DubG66) 8 s (KuzV65, DoneE64)	☢ α (DoneE64, AkaGN66, GhiA67) SF 0.5% (KuzV65) Δ 87.83 (LHP, MTW)	C genet, excit (DoneE64) excit, cross bomb, decay charac (DubG66, GhiA67) chem (?) (ChubY66) parent Fm^{252} (DoneE64)	α 8.43	Cm^{248}(C^{12}, 4n), Cm^{248}(C^{13}, 5n), Cm^{246}(C^{13}, 3n) (GhiA67) Pu^{242}(O^{18}, 4n) (KuzV65) U^{238}(Ne^{22}, 4n) (DoneE64, AkaGN66)
102^{257}	20 s (GhiA67)	☢ α (GhiA67) Δ 90 (LHP, MTW)	E excit, cross bomb, decay charac (GhiA61, GhiA67)	α 8.27 (50%), 8.23 ? (50%)	Cm^{248}(C^{13}, 4n), Cm^{248}(C^{12}, 3n) (GhiA67) B^{10}, B^{11} on Cf$^{250-252}$ (GhiA61)
103Lw^{256}	$\approx$45 s (DubG66)	☢ α, EC (?) (DubG66)	F excit (DubG66)		Am^{243}(O^{18}, 5n) (DubG66)
103Lw^{258,} ^{259}	8 s (GhiA61)	☢ α (GhiA61)	E cross bomb, excit, decay charac (GhiA61, GhiA67a) formerly assigned to Lw^{257} (GhiA61)	α 8.6	B^{10}, B^{11} on Cf$^{250-252}$ (GhiA61)
104^{260}	0.3 s (FleG64)	☢ SF (FleG64) no α, lim 50% (DruV66)	E excit, cross bomb (FleG64) chem (?) (ZvaI66)		Pu^{242}(Ne^{22}, 4n) (FleG64)

INDEX

INDEX

U. S. DEPARTMENT OF COMMERCE
National Bureau of Standards
Washington, D. C. 20234

NBS Special Publication 365
Issued July 1972

METRIC CONVERSION CARD

Approximate Conversions
to Metric Measures

Symbol	When You Know	Multiply by	To Find	Symbol
		LENGTH		
in	inches	*2.5	centimeters	cm
ft	feet	30	centimeters	cm
yd	yards	0.9	meters	m
mi	miles	1.6	kilometers	km
		AREA		
in^2	square inches	6.5	square centimeters	cm^2
ft^2	square feet	0.09	square meters	m^2
yd^2	square yards	0.8	square meters	m^2
mi^2	square miles	2.6	square kilometers	km^2
	acres	0.4	hectares	ha
		MASS (weight)		
oz	ounces	28	grams	g
lb	pounds	0.45	kilograms	kg
	short tons (2000 lb)	0.9	tonnes	t
		VOLUME		
tsp	teaspoons	5	milliliters	ml
Tbsp	tablespoons	15	milliliters	ml
fl oz	fluid ounces	30	milliliters	ml
c	cups	0.24	liters	l
pt	pints	0.47	liters	l
qt	quarts	0.95	liters	l
gal	gallons	3.8	liters	l
ft^3	cubic feet	0.03	cubic meters	m^3
yd^3	cubic yards	0.76	cubic meters	m^3
		TEMPERATURE (exact)		
°F	Fahrenheit temperature	5/9 (after subtracting 32)	Celsius temperature	°C